Some Measurements Used in Biology

QUANTITY	NAME OF UNIT	SYMBOL	DEFINITION
Length	meter (*also* metre)	m	A base unit. 1 m = 100 cm = 39.37 inches
	kilometer	km	1 km = 1000 m = 10^3 m
	centimeter	cm	1 cm = $\frac{1}{100}$ m = 10^{-2} m
	millimeter	mm	1 mm = $\frac{1}{1000}$ m = 10^{-3} m
	micrometer	μm	1 μm = $\frac{1}{1000}$ mm = 10^{-6} m
	nanometer	nm	1 nm = $\frac{1}{1000}$ μm = 10^{-9} m
Area	square meter	m²	Area encompassed by a square, each side of which is 1 m in length
	hectare	ha	1 ha = 10,000 m² = 10^4 m² (2.47 acres)
	square centimeter	cm²	1 cm² = $\frac{1}{10,000}$ m² = 10^{-4} m²
Volume	liter (*also* litre)	l	1 l = $\frac{1}{1000}$ m³ = 10^{-3} m³ (1.057 qts)
	milliliter	ml	1 ml = $\frac{1}{1000}$ l = 10^{-3} l = 1 cm³ = 1 cc
	microliter	μl	1 μl = $\frac{1}{1000}$ ml = 10^{-3} ml = 10^{-6} l
Mass	kilogram	kg	A basic unit. 1 kg = 1000 g = 2.20 lbs
	gram	g	1 g = $\frac{1}{1000}$ kg = 10^{-3} kg
	milligram	mg	1 mg = $\frac{1}{1000}$ g = 10^{-3} g = 10^{-6} kg
Time	second	s	A basic unit. 1 s = $\frac{1}{60}$ min
	minute	min	1 min = 60 s
	hour	h	1 h = 60 min = 3,600 s
	day	d	1 d = 24 h = 86,400 s
Temperature	kelvin	K	A basic unit. 0 K = −273.15°C = absolute zero
	degree Celsius	°C	0°C = 273.15 K = melting point of ice
Heat, Work	calorie	cal	1 cal = heat necessary to raise 1 gram of pure water from 14.5°C to 15.5°C = 4.184 J
	kilocalorie	kcal	1 kcal = 1000 cal = 10^3 cal = (in nutrition) 1 Calorie
	joule	J	1 J = 0.2389 cal (The joule is now the accepted unit of heat in most sciences.)
Electric potential	volt	V	A unit of potential difference or electromotive force
	millivolt	mV	1 mV = $\frac{1}{1000}$ V = 10^{-3} V

Life

The Science of Biology

FIFTH EDITION

Life

The Science of Biology

William K. Purves
Harvey Mudd College
Claremont, California

Gordon H. Orians
The University of Washington
Seattle, Washington

H. Craig Heller
Stanford University
Stanford, California

David Sadava
The Claremont Colleges
Claremont, California

SINAUER ASSOCIATES, INC.

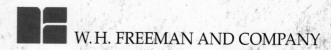

W. H. FREEMAN AND COMPANY

The Cover

Grizzly bears (*Ursus arctos*) take many years to mature reproductively, and cubs remain with their mothers for several years. If their populations are to persist, adult bears must have access to rich food resources. Grizzly bears that live in coastal Alaska depend on salmon that swim up the rivers to spawn. Given that abundant, high-quality food, they grow to become the world's largest carnivorous mammal. Photograph by Michio Hoshino/Minden Pictures.

The Frontispiece

A sunset scene with nesting painted storks (*Mycteria leucocephalus*) taken in Bhakatpur, India. Photograph by Mike Powles/Woodfall Wild Images.

LIFE: The Science of Biology, Fifth Edition

Copyright © 1998 by Sinauer Associates, Inc. All rights reserved.
This book may not be reproduced in whole or in part without permission.

Address editorial correspondence to Sinauer Associates, Inc., 23 Plumtree Road, Sunderland, Massachusetts 01375 U.S.A.

www.sinauer.com

Address orders to W. H. Freeman and Co. Distribution Center, 4419 West 1980 South, Salt Lake City, Utah 84104 U.S.A.

Examination copy information: 1-800-446-8923
Orders: 1-800-877-5351

www.whfreeman.com

Library of Congress Cataloging-in-Publication Data

Life, the science of biology / William K. Purves ... [et al.] —
    5th ed.
                    p.   cm.
        Rev. ed. of: Life, the science of biology / William K. Purves,
    Gordon H. Orians, H. Craig Heller. 4th ed. c1995.
        Includes bibliographical references and index.
        ISBN 0-7167-2869-9 (hardcover)
        1. Biology   I. Purves, William K. (William Kirkwood), 1934–   .
    II. Purves, William K. (William Kirkwood), 1934–   Life, the science
    of biology

QH308.2.L565   1997   79-34772
579—dc21   CIP

Printed in U.S.A.
First printing 1998 RRD

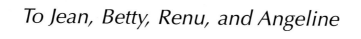

To Jean, Betty, Renu, and Angeline

About the Authors

Bill Purves is Professor Emeritus of Biology as well as founder and former chair of the Department of Biology at Harvey Mudd College in Claremont, California. He received his Ph.D. from Yale University in 1959 under Arthur Galston. A fellow of the American Association for the Advancement of Science, Professor Purves has served as head of the Life Sciences Group at the University of Connecticut, Storrs, and as chair of the Department of Biological Sciences, University of California, Santa Barbara, where he won the Harold J. Plous Award for teaching excellence. His research interests focus on the chemical and physical regulation of plant growth and flowering. Professor Purves elected early retirement in 1995, after teaching introductory biology for 34 consecutive years, in order to turn his skills to writing and producing multimedia for introductory biology students.

Craig Heller is the Lorry Lokey / Business Wire Professor of Biological Sciences and Human Biology at Stanford University. He has served as Director of the popular interdisciplinary undergraduate program in Human Biology and is now Chairman of Biological Sciences. He is a fellow of the American Association for the Advancement of Science and received the Walter J. Gores Award for Excellence in Teaching. Dr. Heller received his Ph.D. from Yale University in 1970 and did postdoctoral work at Scripps Institution of Oceanography on how the brain regulates body temperature of mammals. His current research is on the neurobiology of sleep and circadian rhythms. Over the years Dr. Heller has done research on systems ranging from sleeping college students to diving seals to hibernating bears to meditating yogis. He teaches courses on animal and human physiology and neurobiology in Stanford's introductory core curriculum.

Gordon Orians is Professor Emeritus of Zoology at the University of Washington. He received his Ph.D. from the University of California, Berkeley, in 1960 under Frank Pitelka. Professor Orians has been elected to the National Academy of Sciences, the American Academy of Arts and Sciences, and is a Foreign Fellow of the Royal Netherlands Academy of Arts and Sciences. He was President of the Organization for Tropical Studies, 1988–1994, and President of the Ecological Society of America, 1995–1996. He is a recipient of the Distinguished Service Award of the American Institute of Biological Sciences. Professor Orians is a leading authority in ecology and evolution, with research experience in behavioral ecology, plant–herbivore interactions, community structure, the biology of rare species, and environmental policy. He elected early retirement to be able to devote more time to writing and environmental policy activities.

David Sadava is the Pritzker Family Foundation Professor of Biology at Claremont McKenna, Pitzer, and Scripps, three of the Claremont Colleges. He received his Ph.D. from the University of California, San Diego in 1972 and has been at Claremont ever since. The author of textbooks on cell biology and on plants, genes and agriculture, Professor Sadava has done research in many areas of cell biology and biochemistry, ranging from developmental biology, to human diseases, to pharmacology. His current research concerns human lung cancer and its resistance to chemotherapy. Virtually all of the research articles he has published have undergraduates as coauthors. Professor Sadava teaches introductory biology and has recently developed a new course on the biology of cancer. For the last 15 years, Dr. Sadava has been a visiting professor in the Department of Molecular, Cellular, and Developmental Biology at the University of Colorado, Boulder.

Preface

This is an exciting time to be a biologist: our knowledge of living systems is expanding rapidly and our technologies for research improve daily. This fifth edition of *Life: The Science of Biology* has been an opportunity for us to communicate to students the excitement of modern biology by expanding and refining our coverage, by finding new ways to make important concepts more understandable and memorable, and by conveying the sense of adventure in biological research.

Our overriding goal continues to be to stimulate students' interests in biology. We have tried to do this by making underlying concepts clear and easy to grasp and showing their relevance to medical, agricultural, and environmental issues. Also, we want students to appreciate *how* we know rather than just *what* we know. To that end, we discuss scientific methods and show how experiments, field observations, and comparative methods help biologists formulate and test hypotheses. In the preparation of this edition, we have tried to introduce opportunities for students to think about concepts rather than just learning facts.

Themes and approaches that characterize the new edition

Throughout the book, we use several themes to link chapters and provide continuity. These themes, which are introduced in Chapter 1, include evolution, the experimental foundations of our knowledge, the flow of energy in the living world, the application and influence of molecular techniques, and human health considerations

One of our approaches is to show how basic principles presented in earlier chapters apply in later chapters. For example, programmed cell death, also called apoptosis, has been a major focus of biological research in the past few years. This process is first presented in the context of cell reproduction (Chapter 9). Then we show its applications in development (Chapter 15), cancer (Chapter 17), and the immune system (Chapter 18). Another example is cladistics, introduced first in Chapter 22, and applied in subsequent chapters to show how evolutionary relationships help us understand a wide variety of biological problems.

A new organization enhances accessibility

In chapter after chapter, we have concentrated on making the descriptions, explanations, and applications more accessible to student readers. We have rewritten obscure or difficult passages, deleted some details, simplified the writing and illustrations, and shortened both paragraphs and sections. We have tried to tighten connections, improve transitions, and sharpen the focus. Many changes have been made in how information is distributed among the text, captions, figure labels, and a new feature of the illustrations—"balloon captions."

We have also taken a new approach to headings. We have tried to offer the reader more guidance in identifying, understanding, and interrelating key topics. We use two levels of heads (although occasionally a third level is introduced). Major heads divide the chapters into discrete topics, and second-level heads, now full sentences, identify the explicit focus of each subsection. In addition to providing a clear outline and introduction to covered topics, these "sentence heads" are useful to students for study and review.

To further guide the reader, we have provided explicit forecasts of concepts about to be discussed, both as part of the introduction to each chapter and as part of the introductions to most of the major sections within each chapter. This forecasting allows students to read with expectation and direction, better equipped to appreciate the implications of early topics and to see relationships among topics across the entire chapter.

Different students have different learning styles: some are more image-focused, others more text-focused. Line drawings and photographs have the advantages of directness, emphasis, and drama; on the other hand, text explanations provide explicit information and better describe events that occur through time. We have combined the strengths of both text and graphics through the abundant use of what we're calling "balloon captions." These brief statements are incorporated directly into the graphics and go beyond mere labeling to describe, define, or explain graphic elements. Thus, text becomes more intimately related to graphic representations and the graphics take on more significance. Balloon captions, sometimes numbered to clarify a sequence, guide the reader through the inevitable complexities of some figures; in other figures, balloons emphasize the most important features. This new feature has drawn extensive praise during the development of this edition, and we believe that students will find them highly effective aids to their learning.

A new format for the chapter summaries emphasizes the chapter outline, using major heads to distinguish and identify summary statements. The summary emphasizes major points but also includes specific references to key figures and tables where supporting details are found.

The seven parts: Content, changes, and themes

Each section of the book has undergone important changes. In Part One, The Cell, we eliminated some details and advanced topics, notably in Chapter 6 (Energy, Enzymes, and Metabolism), allowing us to develop certain key concepts such as allostery and cooperativity more clearly. New developments in such areas as protein folding are now introduced in a broad context so the student can relate them to other topics. When appropriate, we have tried to link biochemical and cellular phenomena to specific conditions and diseases that affect human health and well-being.

In Part Two, Information and Heredity, the first six chapters (Chapters 9–14) describe what we know and how we have gained some of this knowledge, and the final four (Chapters 15–18) describe its biological applications. The expression of DNA is dealt with separately in prokaryotes (Chapter 13) and eukaryotes (Chapter 14), and these principles are then used to describe the molecular analysis of development (Chapter 15), the manufacture of useful products via biotechnology (Chapter 16), the diagnosis and treatment of human genetic diseases (Chapter 17), and the production of antibodies (18). Because of its centrality to genetics and molecular biology, we now devote separate chapters to the structure and the role of DNA (Chapters 11 and 12, respectively).

The chapter on development (Chapter 15) in Part Two now concentrates entirely on molecular and genetic aspects of development; the cellular and tissue aspects of embryology are presented in Chapter 40. In addition to applying the principles of molecular biology to recombinant DNA technologies, Chapter 16 emphasizes how these technologies are being used in agriculture and medicine. The "molecular revolution" that is just beginning in medicine, including the Human Genome Project, is the subject of an extensively updated chapter (Chapter 17).

In Part Three, Evolutionary Processes, we have expanded the treatment of cladistic methods to assess evolutionary relationships and show how cladograms are constructed and why knowing evolutionary relationships helps us better understand a wide array of biological problems, including human health problems. With this background, we are able to use phylogenetic trees in subsequent chapters to illustrate evolutionary patterns that range from individual molecules to phyla.

Part Three also includes an entirely new chapter (Chapter 23) on molecular evolution, one of the most exciting and vigorous fields in contemporary biology. Contributed by Peg Riley of Yale University, this chapter emphasizes both detailed molecular comparisons among species and their implications as to why and how molecules change over evolutionary time as organisms encounter and survive environmental challenges.

The results of molecular evolutionary studies have led us to a new emphasis on lineages in Part Four, The Evolution of Diversity, especially in our treatment of bacteria, archaea, and protists. Systematics is in ferment, and we try to impart some sense of current controversies in the field in Chapters 25 and 26. We explicitly treat today's diversity of organisms as the product of evolution.

In Part Five of the fourth edition, we introduced a new chapter, The Biology of Flowering Plants, on plant responses to environmental challenges. It was so well received that we have enriched it with an up-to-date treatment of plant–pathogen interactions. This topic and others continue to emphasize the theme of evolution. Part Five also includes new findings on multiple phytochromes and on developmental mutants in *Arabidopsis*.

In response to requests from instructors, Part Six, The Biology of Animals, now features a chapter (Chapter 40) on animal embryology, which follows the chapter on animal reproduction. The coverage of neurobiology (Chapters 41–44) has been redesigned and expanded to include a new chapter (Chapter 43) on the organization and higher functions of the mammalian brain.

Our theme of human health concerns is manifest throughout Part Six. Chapter 47, on animal nutrition, includes new material on environmental toxicology, an emerging discipline we feel will be of increasing importance to the well-being of our planet.

In Part Seven, Ecology and Biogeography, we have further expanded our coverage of the role of experiments in helping biologists understand the complex interactions among organisms that structure ecological systems. New materials illustrate the role of phylogenetic analyses in behavioral ecology and biogeography. In Chapter 54 we have designed an original graphic method of displaying material on Earth's biomes. This new and striking presentation enables students to visualize and quantify the differences and similarities in the dominant features of Earth's major biomes.

We wish to thank a lot of people

We were all students and teachers long before we were textbook authors, and we want to help students in every way possible. In the next section, "To the Student," we offer some advice that many of our own students have told us they found helpful.

Again, we have been fortunate to receive cogent and significant advice from the more than 60 colleagues who reviewed chapters or whole sections of the book. Their names are listed after this Preface. Their reviews helped to shape many of the changes described above, ranging from the addition of new chapters to the many ways in which we worked to sharpen our story. We thank them all, and hope this new edition measures up to their expectations.

We were already indebted to J/B Woolsey Associates for the elegance and effectiveness of the art programs they developed for the third and fourth editions of this textbook. They have, of course, produced many new illustrations for this edition. However, rather than limiting ourselves to incremental changes in the existing art program, we have taken a major step forward this time with the introduction of the balloon captions. The success of this approach is the result of many factors. James Funston worked with authors and illustrators, offering input to virtually every pixel in the entire art program. John Woolsey and a dedicated team of artists led by Michael Demaray turned our ideas and suggestions into exciting new art.

James Funston, the developmental editor we chose to work with us on this edition, paid close attention to clarity and pedagogical focus. Stephanie Hiebert provided rigorous copy editing from beginning to end. Her sharp eye extended to the illustrations, and her polishing of and additions to the balloon copy often enhanced the clarity of the presentation. Carol Wigg once again coordinated and checked every change made by editors, artists and authors—indeed, she coordinated the entire preproduction process, and she applied her knowledge and talent to writing captions that tightly link the ilustations to key points in the text. We owe her more than we can say for her patience, persistence, and skill. Jane Potter, as photo researcher, found many new and exciting photographs to enhance the learning experience and enliven the appearance of the book as a whole.

We wish to thank the dedicated professionals in W. H. Freeman's marketing and sales group. Their efficiency and enthusiasm has helped bring *Life* to a wider audience. We appreciate their constant support and valuable marketing feedback. A large share of *Life*'s success is due to their efforts in this publishing partnership.

Sinauer Associates provided the best publishing environment we can imagine. Their years of success in publishing biology books at the introductory, intermediate, and advanced levels result from their ability to envision a product and to guide, assist, and motivate authors through the long, demanding process. Remarkably, Andy Sinauer never ceases to extend helpful, and, above all, warm support to his authors.

Bill Purves   Gordon Orians   Craig Heller   David Sadava

November, 1997

Reviewers for the Fifth Edition

Henry W. Art, Williams College

Carla Barnwell, University of Illinois

Judith L. Bronstein, University of Arizona

Robert J. Brooker, University of Minnesota

Steven B. Carroll, Northeast Missouri State University

James J. Champoux, University of Washington

William A. Clemens, University of California/Berkeley

Frederick M. Cohan, Wesleyan University

Newton Copp, The Claremont Colleges

D. Andrew Crain, University of Florida

Joe W. Crim, University of Georgia

Rowland H. Davis, University of California/Irvine

Patrick E. Elvander, University of California/Santa Cruz

Wayne R. Fagerberg, University of New Hampshire

Michael Feldgarden, Yale University

Rachel D. Fink, Mt. Holyoke College

Barbara Fishel, University of Arizona

William Fixsen, Harvard University

Cecil H. Fox, Molecular Histology, Inc.

Stephen A. George, Amherst College

Wayne Goodey, University of British Columbia

Deborah Gordon, Stanford University

David M. Green, McGill University

Adrian Hayday, Yale University

Joseph Heilig, University of Colorado

Walter S. Judd, University of Florida

Mark V. Lomolino, University of Oklahoma

Michael A. Lydan, University of Toronto/Erindale

Denis H. Lynn, University of Guelph

Laura MacIntosh, Stanford University

James Manser, Harvey Mudd College

John M. Matter, Juniata College

Larry R. McEdward, University of Florida

Michael Meighan, University of California/Berkeley

Melissa Michaël, University of Illinois

Charles W. Mims, University of Georgia

Anthony G. Moss, Auburn University

Shahid Naeem, University of Minnesota

Peter Nonacs, University of California/Los Angeles

Barry M. O'Connor, University of Michigan

Ron O'Dor, Dalhousie University

Richard Olmstead, University of Washington

Laura J. Olsen, University of Michigan

Judith A. Owen, Haverford College

Randall W. Phillis, University of Massachusetts

Lorraine Pillus, University of Colorado

Ellen Porzig, Stanford University

Thomas L. Poulson, University of Illinois at Chicago

Loren Reiseberg, Indiana University

Wayne C. Rosing, Middle Tennessee State University

Albert Ruesink, Indiana University

C. Thomas Settlemire, Bowdoin College

Joan Sharp, Simon Fraser University

Esther Siegfried, Pennsylvania State University

Anne Simon, University of Massachusetts

Mitchell L. Sogin, Marine Biological Laboratory, Woods Hole

Collette St. Mary, University of Florida

Millard Susman, University of Wisconsin

Elizabeth Vallen, Swarthmore College

Elizabeth Van Volkenburgh, University of Washington

Gary Wagenbach, Carleton College, Minnesota

Bruce Walsh, University of Arizona

Mark Wheelis, University of California/Davis

Brian White, Massachusetts Institute of Technology

Fred Wilt, University of California/Berkeley

Gregory A. Wray, State University of New York at Stony Brook

To the Student

Welcome to the study of life! In our student days—and ever since—we have enjoyed studying the fascinating and fast-changing field of biology, and we hope that you will, too.

There are a few things you can do to help you get the most from this book and from your course. For openers, read the book actively—don't just read passively, but do things that force you to think as you read. If we pose questions, stop and think about them. If a passage reminds you of something that has gone before, think about that, or even check back to refresh your memory. Ask questions of the text as you go. Do you understand what is being said? Does it relate to something you already know? Is it supported by experimental or other evidence? Does that evidence convince you? How does this passage fit into the chapter as a whole? Annotate the book—write down comments in the margins about things you don't understand, or about how one part relates to another, or even when you find an idea particularly interesting. The point of doing these things is that they will help you learn. People remember things they think about much better than they remember things they have read passively. Highlighting is passive; copying is drudge work; questioning and commenting are active and well worthwhile.

"Read" the illustrations actively too. You will find the balloon captions in the illustrations especially useful—they are there to guide you through the complexities of some topics and to highlight the major points.

The chapter summaries will help you quickly review the high points of what you have read. A summary identifies particular illustrations that you should study to help organize the material in your mind. It is essential that you study the cited illustrations and their captions as you review because important information that is covered in illustrations has been left out of the summary statements. Add concepts and details to the framework by reviewing the text. A way to review the material in slightly more detail after reading the chapter is to go back and look at the boldfaced terms. You can use the boldfaced terms to pose questions—and see if you can answer those questions. The boldfacing will probably be more useful on a second reading than on the first.

Use the self-quizzes and "Applying Concepts" questions at the end of each chapter. The self-quizzes are meant to help you understand some of the more detailed material and to help you sort out the information we have laid before you. Answers to all self-quizzes are in the back of the book. The concept questions, on the other hand, are often fairly open-ended and are intended to cause you to reflect on the material.

Two parts of a textbook that are, unfortunately, often underused or even ignored are the glossary and the index. Both can help you a great deal. When you are uncertain of the meaning of a term, check the glossary first—there are more than 1,500 definitions in it. If you don't find a term in the glossary, or if you want a more thorough discussion of the term, use the index to find where it's discussed.

What if you'd like to pursue some of the topics in greater detail? At the end of each chapter there is a short, annotated list of supplemental readings. We have tried to choose readings from books and magazines, especially *Scientific American*, that should be available in your college library.

To provide another kind of help for students, we commissioned a CD-ROM (*Life 5.0*) covering the subject matter of Parts One and Two of this text-book. *Life 5.0* introduces and illustrates (often with unique animations) over 1700 key terms and concepts. You can access this information in several ways: via *Life* chapter reviews; via minicourses such as "Molecular Structure," "The Cell Cycle," and "DNA Replication"; or via a hyperlinked index. There are also several hundred self-quiz items and dozens of thought problems. You may have a copy of the disk inside the front cover of this book; if not, and if you would like to purchase one, contact **www.mona-group.com**. If you use the disk, explore its contents to see which of its tools best correspond to your needs.

Most students occasionally have difficulty in courses, including biology courses. If you find that you are slipping behind in the course, or if a partic-ular topic is giving you an unreasonable amount of trouble, here are some useful steps you might take. First, the basics: attend class, take careful lec-ture notes, and read the textbook assignments. Second, note that one of the most important roles of studying is to discover what you don't know, so that you can do something about it. Use the index, the glossary, the chapter sum-maries, and the text itself to try to answer any questions you have and to help you organize the material. Make a habit of looking over your lecture notes within 24 hours of when you take them—find out right away what points are unclear, and get them straightened out in your mind. The CD-ROM can help by providing a different perspective.

If none of these self-help remedies does the trick, get help! Other students are often a good source of help, because they are dealing with the material at the same level as you are. Study groups can be very useful, as long as the participants are all committed to learning the material. Tutors are almost al-ways helpful and useful, as are faculty members. The main thing is to get help when you need it. It is not a good idea to be strong and silent and drift into a low grade.

But don't make the grade the point of this or any other course. You are in college to learn, to pursue interesting subjects, and to enjoy the subjects you are pursuing. We hope you'll enjoy the pursuit of biology.

Bill Purves   Gordon Orians   Craig Heller   David Sadava

Contents in Brief

Contents

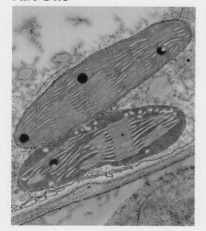

Part One

The Cell

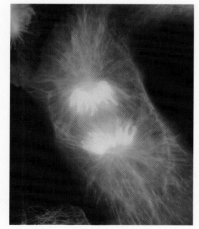

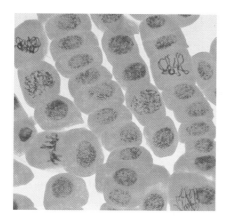

Part Three

Evolutionary Processes

Part Six

The Biology of Animals

Part Seven

Ecology and Biogeography

Chapter 1

An Evolutionary Framework for Biology

We live on an ancient planet. People in some cultures believe that Earth has always existed—that it is eternal. In the Western world, however, people have long believed that Earth had a beginning, and a relatively recent one. In 1650 Irish Archbishop James Ussher, estimating from his close study of the Bible, calculated that Earth was created in 4004 B.C. Although not everyone agreed with his calculations, until the nineteenth century most people in the Western world shared Bishop Ussher's view that Earth was relatively young and that its entire history was chronicled in ancient texts.

During the nineteenth century, geologists and biologists accumulated evidence that Earth was much older, although they could not say exactly how old. Their evidence for an ancient Earth came primarily from the remains of organisms found in sedimentary rocks. The geologists' guiding concepts were simple: Rocks form slowly by the piling up of sediments, and younger rocks are deposited on top of older ones. A great canyon carved into sedimentary rocks may have a visible record of more than a billion years.

Preserved within some rocks were **fossils**—the remains of organisms that lived while the sediments were accumulating. When they compared older rocks with younger ones, geologists could detect slight but significant differences among similar fossil organisms. Furthermore, they found fossils of similar organisms at widely separated locations. By assuming that rocks at different locations containing the same type of fossil were of approximately the same age, early geologists determined the general order of events in the history of life on Earth. Although they could establish a sequence, these geologists had no method for determining the absolute ages of fossils.

One of the triumphs of twentieth-century science has been the development of methods to date materials formed in the past. The discovery that unstable forms (radioactive isotopes) of familiar atoms such as carbon and phosphorus decay at constant rates made it possible to date materials. Radioactive isotopes are incorporated into rocks and fossils in proportion to their presence in the environment when the rock solidified. Each type of radioactive isotope then begins to decay at its own constant rate, eventually becoming stable. Scientists can calculate the absolute ages of rocks from the proportions of radioactive and stable isotopes they contain.

Meanwhile, scientists in the fields of astronomy and physics, using data from the powerful telescopes and space probes that became available in the latter half of the twentieth century, have come to believe that our planet formed approximately 4 billion years ago. The earliest known fossils have been dated, using radioisotopes, as being 3.8 billion years old, so we know that life arose early in the history of Earth.

Evolutionary Milestones

The fullness of time is difficult for people to grasp. We all understand time spans measured in seconds, minutes, hours, days, years, and decades, but we find it difficult to comprehend millions, much less billions, of years. The following overview of the major evolutionary milestones is intended to provide a framework that presents life's characteristics as they will be covered in this book, and an overview of how these characteristics evolved during the history of life on Earth.

Life arose from nonlife

The first life must have come from nonlife. All matter, living and nonliving, is made up of chemicals. The smallest chemical units are atoms, which bond together into molecules (the properties of these molecules are the subject of Chapter 2). We think that the processes leading to life began nearly 4 billion years ago with interactions among small molecules that stored information in easy-to-copy sequences.

Chemical information became more complex when the information stored in these simple sequences resulted in the synthesis of larger molecules with complex but relatively stable shapes. Because they were both complex and stable, these molecules could participate in increasing numbers and kinds of chemical reactions. Certain types of large molecules—carbohydrates, lipids, proteins, and nucleic acids—are formed only by living systems, and they are found in all living systems. The properties and functioning of these complex *organic molecules* are the subject of Chapter 3.

About 3.8 billion years ago interacting systems of molecules came to be enclosed in compartments surrounded by *membranes*. Within these membrane-enclosed units, or *cells*, control was exerted over the entrance and retention of molecules, the chemical reactions taking place within the cell, and the exit of molecules. Cells and membranes are the subjects of Chapters 4 and 5.

Cells are so effective at capturing energy and replicating themselves—two fundamental characteristics of life—that since they evolved, cells have apparently outcompeted any noncellular life. The cell is the unit on which all life has been built.

An Englishman, Robert Hooke, built a simple microscope in 1665 and was the first person to observe cells. The cells he saw were those of cork, wood, and other dead plant materials, and they were empty. Living organisms that were fully contained in a single cell were first observed a few years later by the Dutch naturalist Antoni van Leeuwenhoek.

By 1839, microscopes had improved and enough living material had been observed that the German physiologist Theodor Schwann could assert that *all organisms consist of cells*. In 1858, the German physician Rudolf Virchow suggested that *all cells come from preexisting cells*. Experiments by the French chemist and microbiologist Louis Pasteur between 1859 and 1861 convinced most scientists that cells do not arise from noncellular material, but must come from other cells. In the modern world, life no longer arises from nonlife.

The first organisms were single cells

For 2 billion years all cells were small. They lived mostly autonomous lives, each separate from the other. Their lives were confined to the oceans, where they were shielded from lethal ultraviolet light. The relatively small amounts of genetic information that allowed these **prokaryotic cells** to replicate themselves and the biochemical machinery by which they obtained energy floated loose within an outer membrane. Some prokaryotes living today are similar to those that existed early in the evolution of living cells, several billion years ago (Figure 1.1).

To maintain themselves, to grow, and to reproduce, all organisms—whether they consist of one cell or tril-

1.1 Early Life May Have Resembled These Cells   These "rock-eating" bacteria, appearing red in the artificially colored micrograph, were discovered in pools of water trapped between layers of rock more than 1,000 meters below Earth's surface. Deriving chemical nutrients from the rocks and living in an environment devoid of oxygen, they may resemble some of the earliest prokaryotic cells.

lions of cells—must obtain raw materials and energy from the environment. These raw materials are chemicals that are digested; the products are used to synthesize large carbon-based molecules. The energy obtained from chemical digestion is used to power the synthetic reactions. These conversions of matter and energy are called **metabolism**.

All organisms can be viewed as devices to capture, process, and convert matter and energy from one form to another; these conversions are the subjects of Chapters 6 and 7. *A major theme in the evolution of life is the development of increasingly diverse ways of capturing external energy and using it to drive biologically useful reactions.*

The earliest cells derived their energy from simple chemical compounds because complex molecules were scarce in their environment. On early Earth, volcanoes poured large quantities of methane and hydrogen sulfide into the atmosphere. Early prokaryotes evolved the ability to ingest these molecules and use them as sources of energy.

Photosynthesis and sex changed the course of evolution

Two powerful evolutionary events took place in the first billion years. One, the evolution of photosynthesis, created new metabolic pathways. The other, the evolution of sex, stimulated the evolution of the almost unimaginable diversity of organisms on Earth.

PHOTOSYNTHESIS CHANGED EARTH'S ENVIRONMENT.   About 2.5 billion years ago, some prokaryotes evolved the ability to use the energy of sunlight to power their metabolism. Although raw chemicals were still taken up from the environment, the energy used to metabolize these chemicals came directly from the sun.

The early photosynthetic prokaryotes were probably similar to present-day cyanobacteria (Figure 1.2). The energy-capturing process they used—**photosynthesis**—is the basis of nearly all life on Earth today. As you will learn in Chapter 8, photosynthesis is a complex process made up of many chemical reactions. The ability to perform the photosynthetic reactions probably accumulated gradually during the first billion years or so of evolution, but once the ability had evolved, the effects of photosynthesis were dramatic.

Photosynthetic prokaryotes were so successful that they released vast quantities of oxygen gas (O_2) into the atmosphere. The presence of oxygen opened up new avenues of evolution. Metabolic pathways based on O_2—*aerobic metabolism*—came to be used by most organisms on Earth. The air we breathe today would not exist without photosynthesis.

Over a much longer time frame, the vast quantities of oxygen liberated by photosynthesis had another effect. A form of oxygen we call ozone (O_3) began to accumulate along with the O_2 in the atmosphere. The ozone slowly formed a dense layer that acted as a shield, intercepting much of the sun's deadly ultraviolet radiation. Eventually (although only within the last 800 million years of evolution) the presence of this shield allowed organisms to leave the protection of the ocean and find new lifestyles on Earth's land surfaces.

SEX CHANGED EVOLUTIONARY RATES.   The earliest unicellular organisms reproduced by dividing. Progeny cells were identical to parent cells. But **sexual recombination**—the combining of genes from two cells in one cell—appeared early during the evolution of life. Sex is advantageous because an organism that receives genetic information from another individual produces

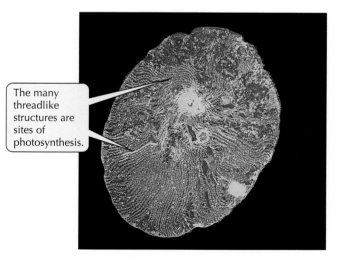

The many threadlike structures are sites of photosynthesis.

1.2 Oxygen Produced by Prokaryotes Changed Earth's Atmosphere   This modern cyanobacterium is probably very similar to early photosynthetic prokaryotes.

offspring that are more variable. *Reproduction with variation is a major characteristic of life.* Because environments continuously vary, individuals that produce variable offspring rather than genetically identical "clones" are more likely to produce at least some offspring that *adapted* to changes in the environment.

Adaptation to environmental change is one of life's most distinctive features. An organism is adapted to its environment when it possesses features that enhance its survival and ability to reproduce in a given environment. Sex is so adaptive that today nearly all organisms on Earth engage in sex at least occasionally. By creating increased variation, sexual recombination increased the rate of evolutionary change.

Early prokaryotes engaged in sex (exchanging genetic material) and reproduction (cell division) at different times. Even today in many unicellular organisms, sex and reproduction occur at different times (Figure 1.3). But a different kind of organism evolved that would require a more complicated sex life.

Eukaryotes are "cells within cells"

As the ages passed, some prokaryotic cells became large enough to attack and consume smaller cells, becoming the first *predators.* Usually the smaller cells were destroyed within the predators' cells, but some of these smaller cells survived and became permanently integrated into the operation of their hosts' cells. In this manner, cells with complex internal compartments arose. We call these cells **eukaryotic cells.** Their appearance slightly more than 1.5 billion years ago opened more new evolutionary pathways.

Prokaryotic cells—including all the early bacteria and archaea—have only one membrane, the one that surrounds them. Eukaryotic cells, on the other hand, are filled with membrane-enclosed compartments. In eukaryotic cells, genetic material—*genes* and *chromo-*

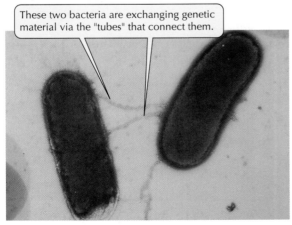

1.3 Sex Between Prokaryotes Genetic exchange produces variation that leads to adaptive evolution.

These two bacteria are exchanging genetic material via the "tubes" that connect them.

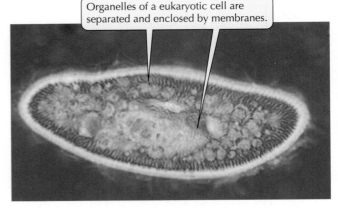

Organelles of a eukaryotic cell are separated and enclosed by membranes.

1.4 Multiple Compartments Characterize Eukaryotic Cells The nucleus and other specialized compartments (known as organelles) probably evolved from small prokaryotes that were ingested by a larger prokaryotic cell.

somes—became contained within a discrete **nucleus** and became increasingly complex. Other compartments became specialized for other purposes, such as photosynthesis (Figure 1.4). We refer to these specialized compartments as **organelles.**

Cells evolved the ability to change their structures and specialize

Until slightly more than 1 billion years ago, only unicellular organisms existed. Two key developments made the evolution of **multicellular organisms**—organisms consisting of more than one cell—possible. One was the ability of a cell to change its structure and functioning to meet the challenges of a changing environment. Prokaryotes accomplished this when they evolved the ability to change from rapidly growing cells into resting *spores* that could survive harsh environmental conditions. The second development allowed cells to stick together after they divided, forming a multicellular organism.

Once organisms began to be composed of many cells, it became possible for the cells to specialize. Certain cells, for example, could be specialized to perform photosynthesis. Other cells might become specialized to transport chemical raw materials, such as oxygen, from one part of an organism to another. Very early in the evolution of multicellular life, certain cells began to be specialized for sex. As multicellular life evolved, sex and reproduction became linked. In almost all present-day multicellular organisms, sex and reproduction occur together.

With more complicated and specialized sex cells, sex itself became more complicated. Simple cell division, which we know as **mitosis,** was and is sufficient for the needs of most cells. But a whole new method of cell division—**meiosis**—evolved that opened up new

1.5 Organisms May Change Dramatically during Their Lives The caterpillar, pupa, and adult are all stages in the life cycle of a monarch butterfly. The transition from one stage to another is triggered by internal signals.

realms of recombination possibilities for the specialized sex cells, or *gametes*. Mitosis and meiosis are explained and compared in Chapter 9.

The cells of an organism are constantly adjusting

Both the emergence of multicellular life and the changes in Earth's atmosphere that allowed life to move out of the oceans and exploit the environments of the land masses quickened the pace of evolution. Photosynthetic green plants colonized the land and provided a rich source of energy for a vast array of organisms that consumed them. But whether it is made up one cell or many, an organism must respond appropriately to many signals emanating from its external and internal environments.

The external environment can change rapidly and unpredictably in ways that are outside of the organism's control. An organism can remain healthy only if its internal environment remains within a given range of physical and chemical conditions. Organisms maintain a relatively constant internal environment by adjusting their metabolism in response to external and internal signals indicating such things as a change in temperature, the presence or absence of sunlight, the presence or absence of specific chemicals, the need for nutrients (food) and water, or the presence of a foreign agent inside the organism's body. Maintenance of a relatively stable internal condition is called **homeostasis**.

The adjustments that organisms make to maintain constant internal conditions are usually minor; they are not obvious, because nothing appears to change. However, at some time during their lives many organisms respond to signals not by maintaining their status, but by undergoing major physical reorganization. We mentioned in the previous section the ability of prokaryotes to change from rapidly growing cells into dormant spores in response to environmental stresses. A striking example that evolved much later is *metamorphosis*, seen in many modern insects, such as butterflies. In response to internal chemical signals, a caterpillar changes into a pupa and then into an adult butterfly (Figure 1.5).

A major theme in the evolution of life is the development of increasingly complicated systems for responding to sig-

nals and maintaining homeostasis. Indeed, some animals exhibit a widespread and important biological process called *learning*, in which important changes in the internal environment result from responses to external signals (such as this textbook).

Multicellular organisms develop and grow

Multicellular organisms cannot achieve their adult shapes or function effectively unless their growth is carefully regulated. Uncontrolled growth, one example of which is cancer, ultimately destroys life. The functioning of a multicellular organism requires a sequence of events leading from a single cell to a multicellular adult. *A vital characteristic of living organisms is regulated growth.*

The activation of information within cells, and the exchange of information among many cells, produce the well-timed events that are required by the transition from single cell to adult form. Genes control the metabolic processes necessary for life. The astounding nature of the genetic material that controls these life-long events has been understood only within the twentieth century; it is the story to which much of Part Two of this book is devoted.

Altering the timing of developmental processes can produce striking changes. Just a few genes can control processes that result in dramatically different adult organisms. Chimpanzees and humans share more than 97 percent of their genes, but the differences between

1.6 Genetically Similar but Quite Distinct By looking at the two, you would never guess that chimpanzees and humans share more than 97 percent of their genes.

the two in form and in behavioral abilities, most notably speech, are dramatic (Figure 1.6). When we realize how little information it sometimes takes to create major transformations, the still-mysterious process of *speciation* becomes a little less of a mystery.

Speciation has resulted in the diversity of life

All organisms on Earth today are the descendants of a unicellular organism that lived almost 4 billion years ago. The preceding sections of this chapter described the changes that led to the diversity present in life today. The course of this evolution has been accompanied by the storage of larger and larger quantities of information and increasingly complex mechanisms for using it. But if that were the entire story, only one kind of organism might exist on Earth today. Instead, Earth is populated by many millions of kinds of organisms that are genetically different from one another. We call these genetically independent groups **species**.

As long as individuals within a population mate at random and reproduce, structural and functional changes may occur, but only one species will exist. However, if a population becomes divided into two or more groups, individuals can mate only with individuals in their own group. When this happens, differences may accumulate with time, and the groups may evolve into different species.

The splitting of groups of organisms into separate species has resulted in the great richness and variety of life found on Earth today, as described in Chapter 19. How species form is explained in Chapters 20 and 21. From a single ancestor, many species may arise as a result of the repeated splitting of populations. How bi-

ologists determine which species have descended from a particular ancestor is discussed in Chapter 22.

Sometimes humans refer to species as "primitive" or "advanced." These and similar terms, such as "lower" and "higher," are best avoided because they imply that some organisms function better than others. The abundance of prokaryotes—all of which are relatively simple—readily demonstrates that they are highly functional, despite their relative simplicity. Therefore, in this book, we usually use the terms "ancestral" and "derived" to describe characteristics that appeared earlier and later in the evolution of lineages of life, respectively, recognizing that all organisms that have survived are successfully adapted to their environments. The wings that allow a bird to fly or the structures that allow green plants to survive in environments where water is either scarce or overabundant are examples of the rich array of adaptations found among organisms (Figure 1.7).

Biological Diversity: Domains and Kingdoms

As many as 30 million species of organisms inhabit Earth today. Many times that number lived in the past but are now extinct. To help us understand the past and current diversity of organisms, biologists use classification systems that group organisms according to their evolutionary relationships.

In the classification system used by most modern biologists, organisms are grouped into three **domains** and six **kingdoms** (Figure 1.8). Organisms belonging to a particular domain have been evolving separately from organisms in other domains for more than a billion years. Organisms in the domains **Archaea** and **Bacteria** are prokaryotes, single cells that lack a nucleus and other internal compartments found in the cells from other kingdoms.

Archaea and Bacteria differ so fundamentally from one another in the chemical pathways by which they function and in the products they produce that they are believed to have separated into distinct evolutionary lineages very early during the evolution of life. Each of these domains consists of a single kingdom, Archaebacteria and Eubacteria, respectively. These kingdoms are covered in Chapter 25.

Members of the other domain—**Eukarya**—have eukaryotic cells with nuclei and complex cellular compartments called *organelles*. The Eukarya are divided into four kingdoms—Protista, Plantae, Fungi, and Animalia. The kingdom **Protista** (protists), the subject of Chapter 26, contains mostly single-celled organisms. The remaining three kingdoms, nearly all of whose members are multicellular, are believed to have arisen from ancestral protists.

Most members of the kingdom **Plantae** (plants) convert light energy to chemical energy by photosynthesis.

1.7 Adaptations to the Environment   (a) The long, point-ed wings of the peregrine falcon allow it to accelerate rapidly as it dives on its prey. (b) The wings of an Arctic tern allow it to hover above the water while searching for fish. (c) In a water-limited environment, this saguaro cactus stores water in its fleshy trunk. Its roots spread broadly to quickly extract water immediately after it rains. (d) The above-ground root system of red mangroves is a modification that allows them to thrive while inundated by water—an environment that would kill most terrestrial plants.

The biological molecules that they and some protists synthesize are the primary food for nearly all other living organisms. This kingdom is covered in Chapter 27.

The kingdom **Fungi**, the subject of Chapter 28, includes molds, mushrooms, yeasts, and other similar organisms, all of which are *heterotrophs*—that is, they require a food source of energy-rich molecules synthesized by other organisms. Fungi absorb food substances from their surroundings and digest them within their cells. Many are important as decomposers of the dead bodies of other organisms.

Members of the kingdom **Animalia** (animals) are heterotrophs that ingest their food source, break down (digest) the food outside their cells, and then absorb the products. Animals eat other forms of life for their raw materials and energy. Perhaps because we are animals ourselves, we are often drawn to study members of this kingdom, which is covered in this book in Chapters 29 and 30.

Biologists recognized that organisms were adapted for life in differing environments long before they understood how adaptation came about. Nearly 150 years ago, Charles Darwin and Alfred Russel Wallace proposed the first scientifically testable theory about adaptation. Their suggestion—that *adaptation is the result of evolution by natural selection*—has guided biological investigations ever since.

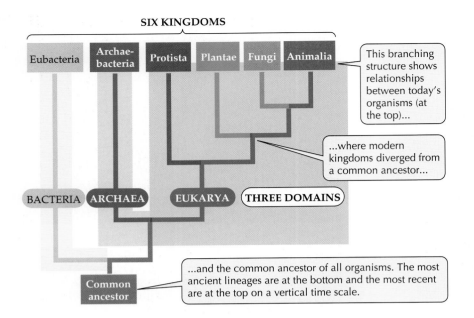

SIX KINGDOMS

Eubacteria Archae-bacteria Protista Plantae Fungi Animalia

This branching structure shows relationships between today's organisms (at the top)...

...where modern kingdoms diverged from a common ancestor...

BACTERIA ARCHAEA EUKARYA THREE DOMAINS

...and the common ancestor of all organisms. The most ancient lineages are at the bottom and the most recent are at the top on a vertical time scale.

Common ancestor

1.8 Domains and Kingdoms In the classification system used in this book, Earth's organisms are divided into three domains and six kingdoms.

The World Into Which Darwin Led Us

Long before scientists understood how biological evolution happened, they suspected that living organisms had evolved from organisms no longer alive on Earth. In the 1760s, the French naturalist Count George-Louis Leclerc de Buffon (1707–1788) wrote his *Natural History of Animals*, which contained a clear statement of the possibility of evolution.

Buffon originally believed that each species had been divinely created for different ways of life, but as he studied animal anatomy, doubts arose. He observed that the limb bones of all mammals, no matter what their way of life, were remarkably similar in many details (Figure 1.9). Buffon also noticed that the legs of certain animals, such as pigs, have toes that never touch the ground and appear to be of no use. He found it difficult to explain the presence of these seemingly useless small toes by special creation.

However, both these troubling facts could be explained if mammals had not been specially created in their present forms but had been modified from a common ancestor. Buffon suggested that the limb bones of mammals might all have been inherited, and that pigs might have functionless toes because they inherited them from ancestors with fully formed and functional toes. This was an early statement of evolution (descent with modification), although Buffon did not attempt to explain how such changes took place.

Buffon's student Jean Baptiste de Lamarck (1744–1829) wrote extensively about evolution and was the first person to propose a mechanism of evolutionary change. Lamarck suggested that lineages of organisms may change gradually over many generations as

offspring inherit structures that have become larger and more highly developed as a result of continued use or, conversely, have become smaller and less developed as a result of disuse.

For example, Lamarck suggested that aquatic birds extend their toes while swimming, stretching the skin between them. This stretched condition, he thought, could be inherited by the offspring, who would further stretch their skin during their lifetimes and would also pass this condition along to their offspring. According to Lamarck, birds with webbed feet would thereby evolve over a number of generations. He explained many other examples of adaptations in a similar way.

Today scientists do not believe that evolutionary changes are produced by inheritance resulting from use and disuse, as Lamarck suggested. But Lamarck did realize that species evolve with time, and his ideas

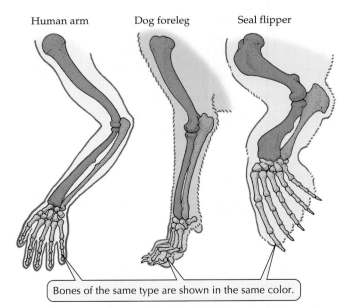

Human arm Dog foreleg Seal flipper

Bones of the same type are shown in the same color.

1.9 All Mammals Have Similar Limb Bones Mammalian forelimbs have different purposes: Humans use theirs for manipulating objects, dogs use theirs for walking, and seals use theirs for swimming. But the number and type of their bones are similar, indicating that they have been modified over time from a common ancestor.

deserved more attention than they received from his contemporaries, most of whom believed in a relatively young and unchanging universe. After Lamarck, other naturalists, scientists, and thinkers speculated that with time species change.

By 1858, the climate of opinion (among biologists, at least) was receptive to a theory of evolutionary processes proposed independently by Charles Darwin and Alfred Russel Wallace. By then geologists had shown that Earth had changed over millions of years, not merely a few thousand years. Thus, the presentation in the latter half of the nineteenth century of a well-documented and thoroughly scientific argument for evolution triggered a transformation of biology.

Darwin initiated the scientific study of evolution

Charles Darwin (1809–1882) based his approach to evolution on the following hypotheses:

1. Earth is very old, and organisms have been changing steadily throughout the history of life.
2. All organisms are descendants of a single common ancestor.
3. Species multiply by splitting into daughter species; such speciation has resulted in the great diversity of life found on Earth.
4. Evolution proceeds via gradual changes in populations, not by the sudden production of individuals of dramatically different types.
5. The major agent of evolutionary change is natural selection.

These five hypotheses have all been supported by the mass of research that has been conducted since Darwin published his book *The Origin of Species* in 1859.

Darwin's major insight was to perceive the significance of facts that were familiar to most of his fellow biologists. He understood that populations of all species have the potential for exponential increases in numbers. To illustrate this point, Darwin used the following example:

Suppose … there are eight pairs of birds, and that only four pairs of them annually … rear only four young, and that these go on rearing their young at the same rate, then at the end of seven years (a short life, excluding violent deaths for any bird) there will be 2048 birds instead of the original sixteen.

Charles Darwin

1.10 Many Organisms Have High Birth and Death Rates
This pair of sergeant majors is guarding the thousands of reddish eggs the female laid in a compact mat. If all the offspring of sergeant majors grew to adulthood and reproduced, the world's population would be overwhelming within a few years. However, most of these eggs and the fish that hatch from them will not survive.

Yet such rates of increase are rarely achieved in nature; the numbers of individuals of most species are relatively stable through time. Therefore, death rates in nature must be high (Figure 1.10). Without high death rates, even the most slowly reproducing species would quickly reach enormous population sizes.

Darwin also observed that, although offspring tend to resemble their parents, the offspring of most organisms are not identical to each other or to their parents (Figure 1.11). He suggested that slight variations among individuals significantly affect the chance that a given individual will survive and reproduce. He called this differential reproductive success of individuals **natural selection**.

Darwin probably used the words "natural selection" because he was familiar with the artificial selection practices of animal and plant breeders. Many of Darwin's observations on the nature of variation came from domesticated plants and animals. Darwin himself was a pigeon breeder, and he knew firsthand the astonishing diversity in color, size, form, and behavior that could be achieved by artifical human selection of which pigeons to mate (Figure 1.12). He recognized close parallels between artificial selection by breeders and selection in nature.

1.11 Offspring Differ from Their Parents These ducklings are members of a brood that hatched from a single clutch of eggs laid by this patterned female and fathered by a single male. Genetic variability among offspring of two parents is the norm.

Darwin states his case for natural selection

In *The Origin of Species* Darwin argued his case for natural selection:

> How can it be doubted, from the struggle each individual has to obtain subsistence, that any minute variation in structure, habits or instincts, adapting that individual better to the new conditions, would tell upon its vigour and health? In the struggle it would have a better chance of surviving; and those of its offspring which inherited the variation, be it ever so slight, would have a better chance.

That statement, written more than 100 years ago, still stands as a good expression of the idea of evolution by natural selection. Since Darwin wrote these words, biologists have developed a much deeper understanding of the genetic basis of evolutionary change and

have assembled a rich array of examples of how natural selection acts.

Biology began a major conceptual shift a little more than a century ago with the general acceptance of long-term evolutionary change and the recognition that natural selection is the primary agent that adapts organisms to their environments. The shift took a long time because it required abandoning many components of an earlier worldview. The pre-Darwinian view held that the world was young, and that organisms had been created in their current forms. In the Darwinian view, the world is ancient, and both Earth and its inhabitants have been changing from forms very different from the ones they now have.

Accepting this new world view means accepting not only the processes of evolution, but also the view that the living world is constantly evolving, but without any "goals." The idea that evolutionary change is not directed toward a final goal or state has been more difficult for some people to accept than the process of evolution itself.

Asking the Questions "How?" and "Why?"

Biological processes and products can be viewed from two different but complementary perspectives. We ask functional questions: How does it work? We also ask adaptive questions: Why has it evolved to work in that way?

Suppose, for example, some marine biologists out on their research vessel are suddenly surrounded by dolphins leaping completely out of the ocean (Figure 1.13). Two obvious questions to ask are, *How* do these marine mammals achieve such a jump?, and *Why* do they do it? An answer to the how question would deal with the molecular mechanisms underlying muscular contraction, nerve and muscle interactions, and the receipt of stimuli by the dolphins' brains.

An investigation to answer the why question would attempt to determine why leaping out of the water is

1.12 Many Types of Pigeons Have Been Produced by Artificial Selection
Charles Darwin, who raised pigeons as a hobby, saw similar forces at work in artificial and natural selection.

1.13 How and Why do Dolphins Leap? Scientists from different disciplines focus on only one of these questions, and their answers are certain to be very different.

adaptive—that is, why it improves the survival and reproductive success of dolphins. For this particular instance, the why question is more obscure, and scientists still have no clear-cut answer. Some animals, such as frogs, evolved jumping behavior because it increased their chances of escape from predators; for others, such as mountain goats, jumping allowed them to cross obstacles like ravines and ridges. Neither of these explanations seems to apply to dolphins, who do not encounter barriers in the ocean and who jump when no predators are chasing them. Jumping may allow them to see better, or it may help dislodge parasites from their bodies; but neither of these possible benefits is established. Perhaps *you* can come up with an explanation of the behavior; in the meantime, scientists (along with most other humans) will enjoy watching it.

Is either of the two types of question more basic or important than the other? Is any one of the answers more fundamental or more important than the others? Not really. The richness of possible answers to apparently simple questions makes biology a complex science, but also an exciting field. Whether we're talking about molecules bonding, cells dividing, blood flowing, dolphins leaping, or forests growing, we are constantly posing both how and why questions. To answer these questions, scientists generate hypotheses that can be tested.

Hypothesis testing guides scientific research

Underlying all scientific research is the **hypothetico-deductive approach** with which scientists ask questions and test answers. This method allows scientists to modify and correct their beliefs as new observations and information become available. The method has five stages:

1. Making observations.
2. Asking questions.
3. Forming **hypotheses**, which are tentative answers to the questions.
4. Making predictions based on the hypotheses.
5. Testing the predictions by making additional observations or conducting experiments. The additional data gained may support or contradict the predictions being tested.

If the data support the hypothesis, it is subjected to additional predictions and tests. If they continue to support it, confidence in its correctness increases and the hypothesis comes to be considered a theory. If the data do not support the hypothesis, it is abandoned or modified in accordance with the new information. Then new predictions are made, and more tests are conducted. An example will illustrate this process.

APPLYING THE HYPOTHETICO-DEDUCTIVE METHOD. Biologists have long known that some caterpillars are conspicuously colored but that other caterpillars blend in with their backgrounds (Figure 1.14). Conspicuously colored caterpillars often live in groups but are seldom attacked by birds. These initial observations suggested questions that were used to develop hypotheses, make predictions, and devise an experiment to test the predictions. Let's examine how this was done.

GENERATING A HYPOTHESIS. The bright colors of some caterpillars, together with the observation that potential predators usually avoid brightly colored caterpillars, suggested a question that became a hypothesis: The bright color patterns of these caterpillars signal to potential predators that the caterpillars are distasteful or toxic. A companion hypothesis is that inconspicuous caterpillars are good to eat (palatable), and their coloration thus reduces the chance that predators will discover and eat them.

For each hypothesis of an effect there is a corresponding **null hypothesis** asserting that the proposed effect is absent. The null hypothesis for the hypotheses we have just stated is that there is no difference in palatability between colorful and camouflaged caterpillars.

Notice that these hypotheses depend on certain assumptions or on previous knowledge. For example, we assume that birds have color vision and can learn about the qualities of their prey by encountering and tasting them. If such assumptions are uncertain, they should be tested before other experiments are performed.

(a)

(b)

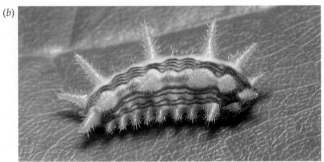

1.14 Caterpillars Can Be Easy or Hard to See (a) Many caterpillars blend into their surroundings, like these catocala moth larvae camouflaged by the bark of an oak tree. (b) This colorful butterfly larva, a stinging rose, contrasts with its leafy environment. Its name implies that its effect on a predator is unpleasant.

MAKING AND TESTING PREDICTIONS. The hypotheses about colorful and inconspicuous caterpillars led to predictions that were tested by an experiment. Captive birds, blue jays, were presented with both brightly colored monarch butterfly caterpillars and caterpillars that blended into their backgrounds. The blue jays were first deprived of food long enough to make them hungry, so they readily attacked the caterpillars. Ingesting even part of one monarch caterpillar caused a blue jay to vomit.

Because the birds were housed individually, the experimenters knew which ones had previously tasted monarchs and which ones had not. They found that a single experience with a monarch caterpillar was enough to cause a blue jay to reject all other monarch caterpillars presented to it. The camouflaged caterpillars, on the other hand, were readily attacked and eaten, and the jays continued to eat these caterpillars without showing signs of sickness or discontent.

These results supported the palatability hypothesis. The null hypothesis, that there is no difference in palatability between colorful and camouflaged caterpillars, was thus rejected.

Experiments are powerful tools

Scientists use a variety of methods to test predictions from their hypotheses. Among these are laboratory and field experiments and carefully focused observations. Each method has its strengths and weaknesses. The key feature of **experimentation** is the control of most factors so that the influence of a single factor can be seen clearly. In the experiments with blue jays, all birds were equally hungry and they were presented with caterpillars in the same way. By controlling these conditions, the experimenters could reject alternative explanations. Their results, for example, could not have been due to lack of hunger on the part of the birds or to the birds' failure to see the caterpillars.

The advantage of working in a laboratory is that control of the environment is easier. Field experiments are more difficult because it is usually impossible to control more than a small part of the total environment. But field experiments have one important advantage: Their results are more readily applicable to what happens where the organisms actually live and evolve. Just because an organism does something in the laboratory does not mean that it behaves the same way in nature. Because biologists usually wish to explain nature, not processes in the laboratory, combinations of laboratory and field experiments are needed to test most hypotheses about what organisms do (Figure 1.15).

A single piece of evidence supporting a hypothesis rarely leads to widespread acceptance of the hypothesis. Similarly, a single contrary result rarely leads to abandonment of a hypothesis. Negative results can be obtained for many reasons, only one of which is that the hypothesis is wrong. For example, the error may be that incorrect predictions were made from a correct hypothesis. A negative result can result from poor experimental design or because an inappropriate organism was chosen for the test. For example, a predator lacking color vision, or one that uses primarily its sense of smell, would not be appropriate for testing hypotheses about the colors of caterpillars.

A general textbook like this one presents hypotheses and theories that have been extensively tested and that are generally accepted. When possible in this text, we illustrate hypotheses and theories with observations and experiments that support them, but we cannot, because of space constraints, detail all the evidence. Remember as you read that statements of biological "fact" are mixtures of observations, predictions, and interpretations.

SOMETIMES ORGANISMS MUST BE SACRIFICED. Obtaining answers to many of the questions posed by biologists requires manipulating and sometimes sacrificing living organisms. To study the antipredator adaptations of caterpillars, the investigators had to keep blue jays in cages, make them hungry by depriving them of food, and then feed them caterpillars. This procedure resulted in the deaths of some caterpillars and temporary stress for some of the birds. To study their

1.15 Experimentation Is Essential in Biology Research and experimentation in biology are carried out in the field and in laboratories. (*a*) Biologists who study the canopies of rainforest trees use special climbing equipment that allows them to collect data and carry out vital studies in the field. (*b*) Many scientific experiments take place within the laboratory. Work with cells and many tiny organisms requires the use of microscopes.

detailed structure and functioning, scientists must often kill organisms.

No amount of observation without intervention could possibly substitute for experimental manipulation. However, this does not mean that scientists are insensitive to the welfare of the organisms with which they work. Most scientists who work with animals are continually alert to finding ways of getting answers that use the smallest number of experimental subjects and that cause the subjects the least pain and suffering.

Not all forms of inquiry are scientific

If you understand the methods of science, you can distinguish science from nonscience. Recently some people have claimed that "creation science," sometimes called "scientific creationism," is a legitimate science that deserves to be taught in schools together with the evolutionary view of the world presented in this book. In spite of these claims, creation science is not science.

Science begins with observations and the formulation of testable hypotheses that can be rejected by contrary evidence. Creation "science" begins with the unsubstantiated assertion that Earth is only about 4,000 years old and that all species of organisms were created in approximately their present forms. This assertion is not presented as a hypothesis from which testable predictions are derived. Advocates of creation science do not believe that tests are needed, because they assume the assertion to be true, nor do they suggest what evidence would refute it.

In this chapter we have outlined the hypothesis that Earth is about 4 billion years old, that today's living organisms evolved from single-celled ancestors, and that many organisms dramatically different from those we see today lived on Earth in the remote past. The rest of this book will provide evidence supporting this scenario. To reject this view of Earth's history, a person must reject not only evolutionary biology, but also modern geology, astronomy, chemistry, and physics. All of this extensive scientific evidence is rejected or misinterpreted by proponents of creation "science" in favor of a religious belief held by a very small proportion of the world's people.

Evidence gathered by scientific procedures does not diminish the value of religious accounts of creation. Religious beliefs are based on faith—not on falsifiable hypotheses, as science is. They serve different purposes, giving meaning and spiritual guidance to human lives. They form the basis for establishing values—something science cannot do. The legitimacy and value of both religion and science is undermined when a religious belief is called science.

Life's Emergent Properties

Biologists study structures and processes ranging from the simple to the complex and from the small to the large. They study the structure and functioning of small parts of cells, as well as the interactions among the hundreds or thousands of different types of organisms that live together in a particular region. Biology can be visualized as ordered into a hierarchy in which the units, from the smallest to the largest, are molecules, cells, tissues, organs, organisms, populations, communities, and biomes (Figure 1.16).

We have already identified the cell as the fundamental unit of life. A **tissue** is a group of cells with similar and coordinated functions. Several different tissues are usually joined to form larger structures called **organs**, such as hearts, brains, kidneys, and lungs. Organs often are joined to form **organ systems**, such as the nervous system of animals and the vascular system of plants. Organs and organ systems perform distinct functions for the **organism** in which they are found.

The organization of living systems extends beyond the individual organism. Organisms living in the same area that are capable of interbreeding with one another form a **population**. All of the populations of a particular kind of organism, whether or not they live in the same area, constitute a species. Individuals of many different species typically live together and interact to form an ecological **community**. Ecological communities are grouped by their distinctive vegetation into **biomes**. All the biomes on Earth constitute the **biosphere**.

The organism is the central unit of study in biology, and Parts Five and Six of this book discuss organismal biology in detail. But to understand organisms, biologists must study life at all its levels of organization. Biologists must study molecules, chemical reactions, and cells to understand the operations of tissues and organs. They study organs and organ systems to determine how whole organisms function and maintain internal homeostasis. At higher levels in the hierarchy, biologists study how organisms interact with one another to form social systems, populations, ecological communities, and biomes, which are the subjects of Part Seven of this text.

Each level of biological organization has properties, called **emergent properties**, that are not found at lower levels. For example, cells and multicellular organisms have processes and characteristics that are not shown by the molecules of which they are composed. Emergent properties arise in two ways.

First, many emergent properties of systems result from interactions among their parts. For example, at the organismal level, developmental interactions of cells result in a multicellular organism whose adult features are vastly richer than those of the single cell from which it developed. Other examples of properties that emerge through complex interactions are memory, learning, consciousness, and emotions such as hate, fear, envy, anger, and love. These properties result from interactions in the human brain among the 10^{12} (trillion) cells with 10^{15} (a quadrillion) connections. No single cell, or even small group of cells, possesses them.

Second, emergent properties arise because aggregations have collective properties that their individual units lack. For example, individuals are born and they die—they have a life span. An individual does not have a birth rate or a death rate, but a population does. Death rate is an emergent property of a population. Other emergent properties of populations include age distribution and density. Evolution is an emergent property of populations that depends on differences in birth and death rates of individuals. Ecological communities possess emergent properties such as species richness.

Emergent properties do not violate principles that operate at lower levels of organization. However, emergent properties usually cannot be detected or even suspected by studying lower levels. Biologists could never discover the existence of human emotions by studying single nerve cells, even though they may eventually be able to explain those emotions in terms of interactions among many nerve cells.

Curiosity Motivates Scientists

The most important motivator of most biologists is curiosity. People are fascinated by the richness and diversity of life and want to learn more about organisms and how they function. Curiosity is probably an adaptive trait. Humans who were motivated to learn about their surroundings are likely to have survived and reproduced better, on average, than their less curious relatives. We hope this book will help you share the excitement biologists feel as they develop and test hypotheses. There are vast numbers of how and why questions for which we do not have answers, and new discoveries usually engender questions no one thought to ask before. Perhaps *your* curiosity will lead to an important new idea.

Summary of "An Evolutionary Framework for Biology"

• Earth is an ancient planet.
• Geologists in the nineteenth century provided the first evidence of Earth's antiquity and determined the sequence of events in life's evolution.
• The discovery of radioisotopes in the twentieth century enabled evolutionary events to be dated accurately.

Evolutionary Milestones

• Life arose from nonlife about 3.8 billion years ago when interacting systems of molecules became enclosed in membranes to form cells.
• All living organisms contain the same types of large molecules—carbohydrates, lipids, proteins, and nucleic acids. These organic molecules are formed only by living systems.
• All organisms consist of cells, and all cells come from pre-existing cells. Life no longer arises from nonlife.

1.16 From Molecules to the Biosphere: The Hierarchy of Life

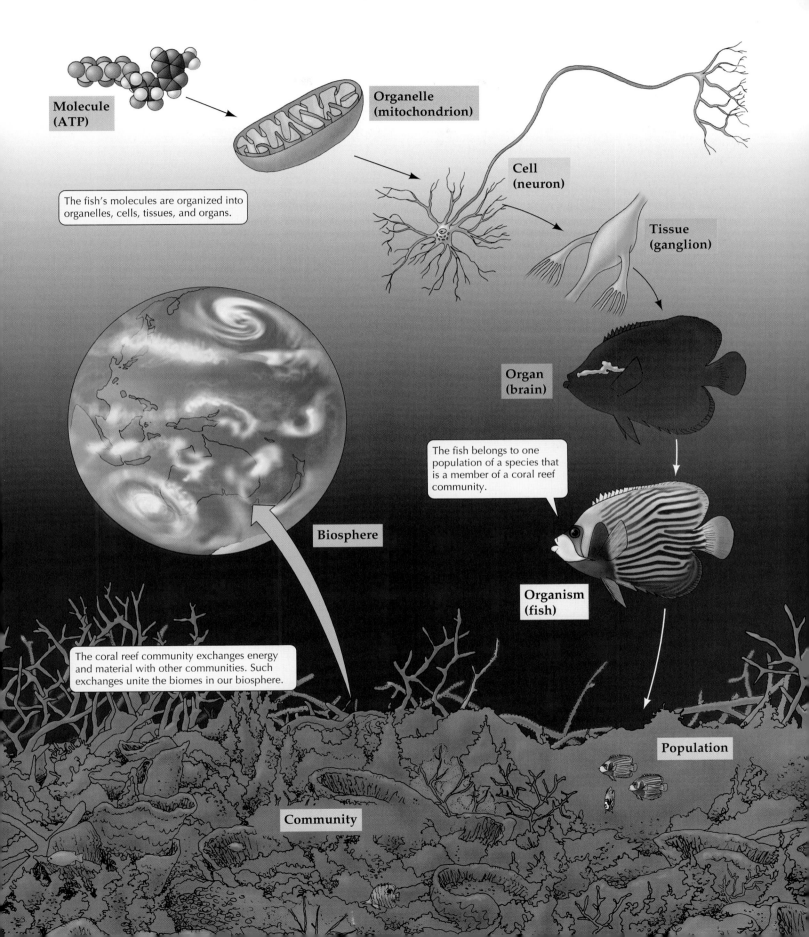

Molecule (ATP)

Organelle (mitochondrion)

Cell (neuron)

The fish's molecules are organized into organelles, cells, tissues, and organs.

Tissue (ganglion)

Organ (brain)

The fish belongs to one population of a species that is a member of a coral reef community.

Biosphere

Organism (fish)

The coral reef community exchanges energy and material with other communities. Such exchanges unite the biomes in our biosphere.

Population

Community

- A major theme in the evolution of life is the development of increasingly diverse ways of capturing external energy and using it to drive biologically useful reactions.
- Photosynthetic single-celled organisms released large amounts of O_2 into Earth's atmosphere, making possible the oxygen-based metabolism of large cells and, eventually, multicellular organisms.
- Reproduction with variation is a major characteristic of life. The evolution of sex greatly increased the rate of life's evolution.
- Complex eukaryotic cells evolved and were able to "stick together" after they divided, forming multicellular organisms. The individual cells of multicellular organisms became modified for specific functions within the organism.
- A major theme in the evolution of life is the development of increasingly complicated systems for responding to signals from the internal and external environments and for maintaining homeostasis.
- Regulated growth is a vital characteristic of life.
- Speciation resulted in the millions of species living on Earth today.
- Adaptation to environmental change is one of life's most distinctive features and is the result of evolution by natural selection. This principle guides virtually all biological investigation today.

Biological Diversity: Domains and Kingdoms

- Species are classified into three domains: Archaea, Bacteria, and Eukarya. The domains Archaea and Bacteria consist of prokaryotic cells and each contains one kingdom, Archaebacteria and Eubacteria, respectively. The domain Eukarya contains the kingdoms Protista, Plantae, Fungi, and Animalia, all of which have eukaryotic cells. **Review Figure 1.8**

The World Into Which Darwin Led Us

- Evolution is the theme that unites all of biology. The idea and evidence for evolution existed before Darwin.
- Darwin based his approach to evolution on five major hypotheses: (1) Earth is very old, (2) all organisms are descendants of a single common ancestor, (3) species multiply by splitting into daughter species, (4) evolution proceeds via gradual changes in populations, and (5) the major agent of evolutionary change is natural selection.

Asking the Questions "How?" and "Why?"

- Biologists ask two kinds of questions. Functional questions concern *how* organisms work. Adaptive questions concern *why* they evolved to work in that way.
- Both how and why questions are usually answered using a hypothetico-deductive approach. Hypotheses are tentative answers to the questions. Predictions are made on the basis of a hypothesis; then the predictions are tested by observations and experiments, which may support or refute the hypothesis.
- Science is based on the formulation of testable hypotheses that can be rejected by contrary evidence. The acceptance on faith of already refuted, untested, or untestable assumptions is not science.

Life's Emergent Properties

- Biology is organized into a hierarchy from molecules to the biosphere. Each level has emergent properties that are not found at the lower levels. **Review Figure 1.16**

Curiosity Motivates Scientists

- The most important motivator of most biologists is curiosity. There are vast numbers of unanswered—and in many cases, unasked—questions still to be explored.

Applying Concepts

1. According to the theory of evolution by natural selection, an organism evolves certain features because they improve the chances that it will survive and reproduce. There is no evidence, however, that evolutionary mechanisms have foresight or that organisms can anticipate future conditions. What do biologists mean when they say, for example, that wings are "for flying"?

2. Why is it so important in science that we design and perform tests capable of rejecting a hypothesis?

3. One hypothesis about the conspicuous coloration of caterpillars was described in this chapter, and some tests were mentioned. Suggest some other plausible hypotheses for conspicuous coloration in these animals. Develop some critical tests for one of these alternatives. What are the appropriate associated null hypotheses?

4. Some philosophers and scientists believe that it is impossible to prove any scientific hypothesis—that we can only fail to find a cause to reject it. Evaluate this view. Can you think of reasons why we can be more certain about rejecting a hypothesis than about accepting it?

Readings

Cziko, G. 1995. *Without Miracles: Universal Selection Theory and the Second Darwinian Revolution.* MIT Press, Cambridge, MA. An in-depth analysis of how an evolutionary view of the world can account for all of the features of life, including human self-awareness.

Darwin, C. 1859. *The Origin of Species by Means of Natural Selection.* John Murray, London. The book that set the world to thinking about evolution; still well worth reading. Many reprinted versions are available.

Futuyma, D. J. 1995. *Science on Trial: The Case for Evolution*, Revised Edition. Sinauer Associates, Sunderland, MA. A thorough presentation, for a general audience, of the scientific arguments that support evolution and its status as the single most fundamental principle in biology.

Margulis, L. and K. V. Schwartz. 1998. *Five Kingdoms: An Illustrated Guide to the Phyla of Life on Earth*, 3rd Edition. W. H. Freeman, New York. A good introduction to the kingdoms of organisms, in which the two kingdoms of prokaryotes are united into one. Excellent examples and illustrations.

Mayr, E. 1991. *One Long Argument: Charles Darwin and the Genesis of Modern Evolutionary Thought.* Harvard University Press, Cambridge, MA. An excellent account of the history of evolutionary thinking during the past century, written by a prominent exponent of the modern, or neo-Darwinian, synthesis of evolution.

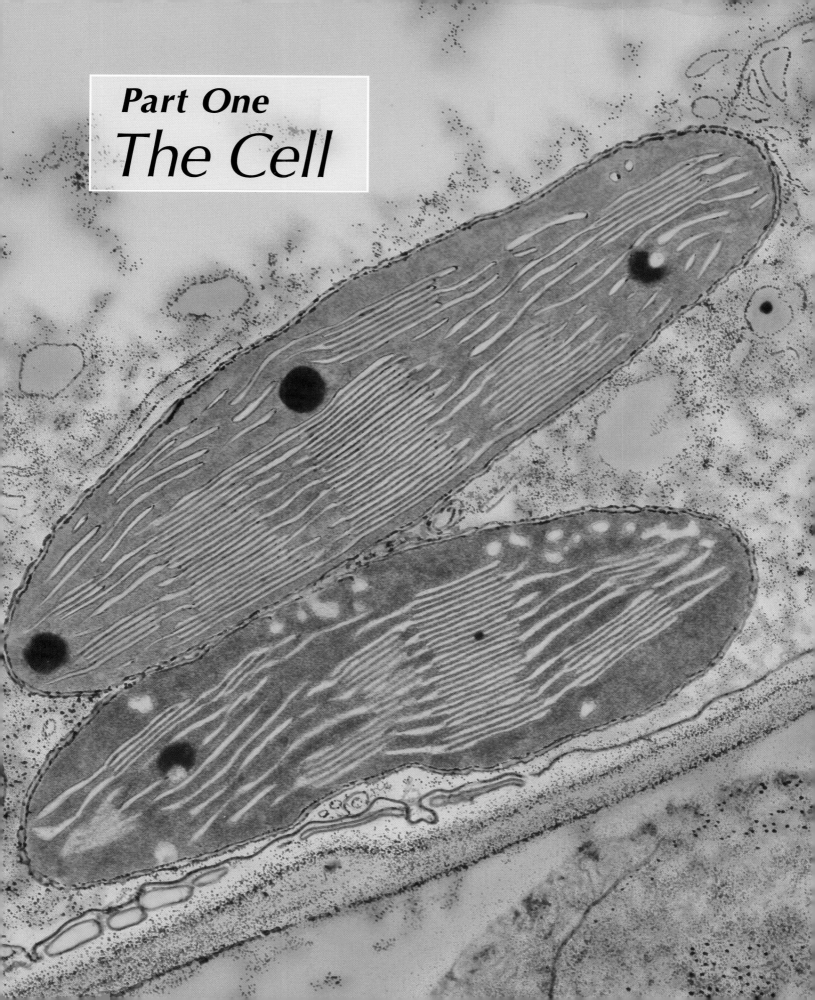

Part One
The Cell

Chapter 2

Small Molecules: Structure and Behavior

The Source of Life
Water, most of it in the oceans, covers three-fourths of Earth's surface. The oceans teem with life.

Water spews from the Earth as geysers. It circles the globe as clouds and falls from the sky as the gentle rain or in thundering torrents. Frozen, it covers parts of Antarctica to a depth of 3,000 meters (10,000 feet) or more. Vast oceans of it submerge much of the planet. Water is one of the key ingredients that makes life on Earth possible. It makes up as much as 95 percent of the weight of some living things.

Water is a simple substance containing only three atoms—two of hydrogen and one of oxygen. What does the composition of water tell us about its characteristics, and why is water so important to living systems? We can't answer these questions without more information about atoms in general, and about hydrogen and oxygen atoms in particular. But it is not only the constituents of matter that are important. To understand the behavior of something as apparently simple as water, we need to know how its constituent atoms are linked together. The same is true of the other small molecules that are essential to living systems.

The first part of this chapter will address the constituents of matter: atoms—their variety, properties, and capacity to combine with other atoms. Then we'll consider how matter changes. In addition to changes in state (solid to liquid to gas), substances undergo changes that transform both their composition and their characteristic properties. When cells use oxygen to "burn" glucose, the products are water, carbon dioxide, and energy to power life activities. This transformation is similar to what happens when the fuel propane is burned in a stove, a combustion reaction.

By studying simple systems such as the combustion of propane, we can understand better what happens in systems as complicated as living cells. The discussion of general chemical principles and their application to small molecules aids our understanding of the large molecules that form the basis for the life of an organism.

Later in this chapter, we return to a consideration of the structure and properties of water and its relationship to acids and bases. We close with a bridge to the next chapter—a consideration of characteristic groups of atoms that contribute specific properties to larger molecules of which they are part.

Atoms: The Constituents of Matter

More than a million million (1×10^{12}) atoms could fit in a single layer over the period at the end of this sentence. Each atom consists of a dense, positively charged nucleus, around which one or more negatively charged electrons move. The nucleus contains one or more protons and may contain one or more neutrons. Atoms and their component particles have mass. Mass is a property of all matter. Measuring mass measures the quantity of matter present.* The greater the mass, the greater the quantity of matter.

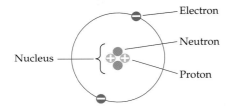

The mass of a proton serves as a standard unit: the atomic mass unit (amu), or dalton (named after the English chemist John Dalton). A single proton or neutron has the mass of 1 dalton, which is 1.7×10^{-24} grams (0.0000000000000000000000017 g). The mass of an electron is 9×10^{-28} g (0.0005 dalton). Because the mass of an electron is so much less than the mass of a proton or a neutron, the contribution of electrons to the mass of an atom can usually be ignored.

The positive electric charge on a proton is defined as a unit of charge. An electron has a charge equal and opposite to that of a proton. Thus the charge of a proton is +1 unit, that of an electron is –1 unit. Unlike charges attract each other; like charges repel. The neu-

tron, as its name suggests, is electrically neutral, so its charge is 0 unit. Because the number of protons in an atom equals the number of electrons, the atom itself is electrically neutral.

An element is made up of only one kind of atom

The element hydrogen consists only of hydrogen atoms; the element iron consists only of iron atoms. An element is a pure substance that contains only one type of atom. The atoms of each element have certain characteristics or properties that distinguish them from the atoms of other elements. The more than 100 elements found in the universe are arranged in the periodic table (Figure 2.1). The periodic table arranges elements with similar properties in vertical columns in order of their increasing size. Although there are more than 100 elements in the world, about 98 percent of the mass of a living organism (bacterium, turnip, or human) is composed of just six elements—carbon, hydrogen, nitrogen, oxygen, phosphorus, and sulfur. The chemistry of these elements will be our primary concern.

A substance (such as oxygen gas) that contains only one kind of atom is an **elemental substance**. A substance that contains more than one kind of atom is a **compound**. Most substances of biological interest are compounds.

The number of protons identifies the element

An atom is distinguished from other atoms by the number of its protons, which does not change. This number is called the **atomic number**. An atom of hydrogen contains 1 proton, a helium atom has 2 protons, carbon has 6 protons, and plutonium has 94. The atomic numbers of these elements are thus 1, 2, 6, and 94, respectively.

Every atom except hydrogen has one or more neutrons in its nucleus. The **mass number** of an atom equals the total number of protons and neutrons in its nucleus. Because the mass of an electron is infinitesimal compared with that of a neutron or proton, electrons are ignored in calculating the mass number. The nucleus of a helium atom contains 2 protons and 2 neutrons; oxygen has 8 protons and 8 neutrons. Helium, therefore, has a mass number of 4 and oxygen a mass number of 16. The mass number may be thought of as the weight of the atom, in daltons.

Each element has its own one- or two-letter symbol. For example, H stands for hydrogen, He for helium, and O for oxygen. Some symbols come from other languages: Fe (from Latin *ferrum*) stands for iron, Na (Latin *natrium*) for sodium, and W (German *Wolfram*) for tungsten. The periodic table (Figure 2.1) gives the symbols for all of the 92 natural elements, as well as those for 14 elements that do not occur naturally. In text, the atomic number and mass number of

---

* The term "weight" is sometimes substituted for the term "mass," but the two concepts are not identical. Weight is the measure of the Earth's gravitational attraction for mass. On another planet, the same quantity of mass will have a different weight. On Earth, however, the term "weight" is often used as a measure of mass.

The six elements highlighted in yellow make up 98% of the mass of any living organism.

Chemical symbol
Atomic number
Atomic weight

Elements shown in orange are present in tiny amounts in many organisms.

Vertical columns have elements with similar properties.

1 H 1.0079																	2 He 4.003
3 Li 6.941	4 Be 9.012											5 B 10.81	6 C 12.011	7 N 14.007	8 O 15.999	9 F 18.998	10 Ne 20.179
11 Na 22.990	12 Mg 24.305											13 Al 26.982	14 Si 28.086	15 P 30.974	16 S 32.06	17 Cl 35.453	18 Ar 39.948
19 K 39.098	20 Ca 40.08	21 Sc 44.956	22 Ti 47.88	23 V 50.942	24 Cr 51.996	25 Mn 54.938	26 Fe 55.847	27 Co 58.933	28 Ni 58.69	29 Cu 63.546	30 Zn 65.38	31 Ga 69.72	32 Ge 72.59	33 As 74.922	34 Se 78.96	35 Br 79.909	36 Kr 83.80
37 Rb 85.4778	38 Sr 87.62	39 Y 88.906	40 Zr 91.22	41 Nb 92.906	42 Mo 95.94	43 Tc (99)	44 Ru 101.07	45 Rh 102.906	46 Pd 106.4	47 Ag 107.870	48 Cd 112.41	49 In 114.82	50 Sn 118.69	51 Sb 121.75	52 Te 127.60	53 I 126.904	54 Xe 131.30
55 Cs 132.905	56 Ba 137.34	57–71 La–Lu	72 Hf 178.49	73 Ta 180.948	74 W 183.85	75 Re 186.207	76 Os 190.2	77 Ir 192.2	78 Pt 195.08	79 Au 196.967	80 Hg 200.59	81 Tl 204.37	82 Pb 207.19	83 Bi 208.980	84 Po (209)	85 At (210)	86 Rn (222)
87 Fr (223)	88 Ra 226.025	89–103 Ac–Lr	104	105	106	107	108	109									

Lanthanide series

57 La 138.906	58 Ce 140.12	59 Pr 140.9077	60 Nd 144.24	61 Pm (145)	62 Sm 150.36	63 Eu 151.96	64 Gd 157.25	65 Tb 158.924	66 Dy 162.50	67 Ho 164.930	68 Er 167.26	69 Tm 168.934	70 Yb 173.04	71 Lu 174.97

Actinide series

89 Ac 227.028	90 Th 232.038	91 Pa 231.0359	92 U 238.02	93 Np 237.0482	94 Pu (244)	95 Am (243)	96 Cm (247)	97 Bk (247)	98 Cf (251)	99 Es (252)	100 Fm (257)	101 Md (258)	102 No (259)	103 Lr (260)

2.1 The Periodic Table The periodic table of the elements reflects a periodicity of physical and chemical properties among the elements. Elements with similar properties are found in vertical columns.

an element are written to the left of the element's symbol:

Mass number 12
Atomic number 6
$^{12}_{6}C$ Symbol of element

Thus hydrogen, carbon, and oxygen are written as $^{1}_{1}H$, $^{12}_{6}C$, and $^{16}_{8}O$.

Isotopes differ in number of neutrons

We have been speaking of hydrogen and oxygen as if each had only one atomic form. But this is not true. Not all atoms of an element have the same mass number. The different atomic forms of a single element are called **isotopes** of the element (Figure 2.2). Isotopes of an element differ in the number of neutrons in the atomic nucleus. The common form of hydrogen is ^{1}H, but about one out of every 6,500 hydrogen atoms on Earth has a neutron as well as a proton in its nucleus and is thus ^{2}H, called deuterium. Furthermore, it is possible to create ^{3}H, tritium, which has *two* neutrons and a proton in its nucleus. Because all three types of

hydrogen atoms have only one proton, they all have the atomic number 1. Deuterium, tritium, and common hydrogen have virtually identical chemical properties, although ^{2}H is twice and ^{3}H three times as heavy as ^{1}H.

In nature, many elements exist as several isotopes. For example, the natural isotopes of carbon are ^{12}C, ^{13}C, and ^{14}C. Unlike the hydrogen isotopes, the isotopes of most other elements do not have distinct names. Rather they are written in the form shown here and are referred to as carbon-12, carbon-13, and carbon-14, respectively. Most carbon atoms are ^{12}C, about 1.1 percent are ^{13}C, and a tiny fraction are ^{14}C. An element's **atomic mass** (**atomic weight**) is the average of the mass numbers of a representative sample of atoms of the element, with all isotopes in their normal proportions. For example, the atomic mass of carbon is 12.011. In biology, one encounters the terms "weight" and "atomic weight" more frequently than "mass" and "atomic mass"; therefore, we will use "weight" for the remainder of this book. Thus we say that the atomic weight of carbon is 12.011.

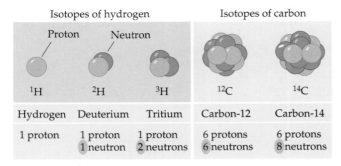

Isotopes of hydrogen			Isotopes of carbon	
Proton	Neutron			
^1H	^2H	^3H	^{12}C	^{14}C
Hydrogen	Deuterium	Tritium	Carbon-12	Carbon-14
1 proton	1 proton 1 neutron	1 proton 2 neutrons	6 protons 6 neutrons	6 protons 8 neutrons

2.2 Isotopes Have Different Numbers of Neutrons

Some isotopes, called **radioisotopes**, are unstable and spontaneously give off energy as α (alpha), β (beta), or γ (gamma) radiation from the atomic nucleus. Such radioactive decay transforms the original atom into another type, usually of another element. For example, uranium-238 loses an alpha particle to form thorium-234, and carbon-14 loses a beta particle to form nitrogen-14. Biologists can incorporate radioisotopes into molecules and use the emitted radiation as a tag to identify changes that the molecules undergo in the body or to identify the locations of molecules within the cell (Figure 2.3). Some radioisotopes commonly used in biological experiments are ^3H (tritium), ^{14}C (carbon-14), and ^{32}P (phosphorus-32).

Black dots identify the location of radioisotope.

2.3 Probing an Embryo with a Radioisotope Developmental biologists treated a fruit fly embryo with a substance labeled with a radioisotope. Later, they pressed the embryo against X-ray film and left the embryo and film in the dark for several days. Wherever a radioactive atom decayed, the emitted particle exposed the film. When the film was developed, silver grains appeared over the parts of the embryo that contained the radioactive substance.

Although radioisotopes are useful for experiments and medicine, even low doses of radiation from radioisotopes have the potential to damage molecules and cells. Gamma radiation from cobalt-60 (^{60}Co) is used medically to damage or kill rapidly dividing cancer cells. In addition to these applications, radioisotopes can be used to date fossils (see Chapter 19).

Electron behavior determines chemical bonding

In atoms, biologists are concerned primarily with electrons. To understand living organisms, biologists must study chemical changes that occur in living cells. These changes, called **chemical reactions** or just **reactions**, are changes in the atomic composition of substances. They occur because of the way in which electrons behave. The characteristic number of electrons in each atom of an element determines how the atom reacts with other atoms. All chemical reactions involve changes in the relationships of electrons with each other.

The location of a given electron in an atom at any given time is impossible to determine. We can only describe a volume of space within the atom where the electron is likely to be. The region of space within which the electron is found at least 90 percent of the time is the electron's **orbital** (Figure 2.4). In an atom, a given orbital can be occupied by at most two electrons. Thus any atom larger than helium (atomic number 2) must have electrons in two or more orbitals. As Figure 2.4 shows, the different orbitals have characteristic forms and orientations in space.

The orbitals constitute a series of **electron shells**, or energy levels, around the nucleus (Figure 2.5). The innermost electron shell, called the K shell, consists of only one orbital, called an *s* orbital. The *s* orbital fills first, and its electrons have the lowest energy. Hydrogen ($_1H$) has one K shell electron; helium ($_2He$) has two. All other atoms have two K shell electrons, as well as electrons in other shells. The L shell is made up of four orbitals (an *s* orbital and three *p* orbitals) and hence can hold up to eight electrons. The M, N, O, P, and Q shells have different numbers of orbitals, but the outermost orbitals can usually hold only eight electrons.

In any atom, the outermost shell determines how the atom combines with other atoms; that is, these outermost electrons determine how an atom behaves chemically. When an outermost shell consisting of four orbitals contains eight electrons, the atom is stable and will not react with other atoms. Examples of some chemically inert elements are helium, neon, and argon, in each of which the outermost shell contains eight electrons (see Figure 2.5). The atoms of other elements seek to attain the stable condition of having eight electrons in their outer orbitals. They attain this stability by sharing electrons with other atoms or by gaining or

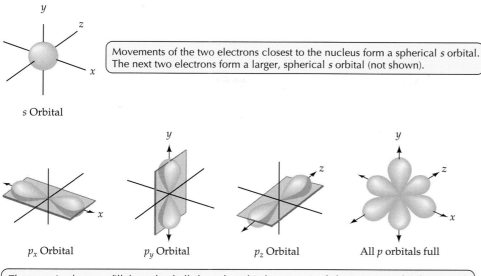

s Orbital

Movements of the two electrons closest to the nucleus form a spherical s orbital. The next two electrons form a larger, spherical s orbital (not shown).

p_x Orbital p_y Orbital p_z Orbital   All p orbitals full

The next six electrons fill three dumbell-shaped p orbitals, one pair of electrons per orbital, oriented on the x, y, and z axes through a point in the center of the atom.

2.4 Electron Orbitals

Orbitals are the regions around an atom's nucleus where electrons are most likely to be found.

Many atoms in biologically important molecules follow the octet rule—for example, carbon (C) and nitrogen (N). However, some biologically important atoms such as hydrogen and phosphorus are exceptions to the rule. Hydrogen (H) attains stability when two electrons occupy its outermost orbital; phosphorus (P) is stable when its outermost orbitals contain ten electrons.

losing one or more electrons from their outermost orbitals.

When they share electrons, atoms are bonded together. Such bonds create stable associations of atoms called molecules. A **molecule** can be defined as two or more atoms linked by chemical bonds. The tendency of atoms in stable molecules to have eight electrons in their outermost orbitals is known as the **octet rule**.

Chemical Bonds: Linking Atoms Together

A **chemical bond** is an attractive force that links two atoms to form a molecule. There are different kinds of chemical bonds (Table 2.1), but all strong chemical

2.5 Electron Shells Determine the Reactivity of Atoms

Each shell can hold a specific maximum number of electrons. The K shell holds two; the L and M shells each hold eight. An atom with room for more electrons in its outermost shell may react with other atoms.

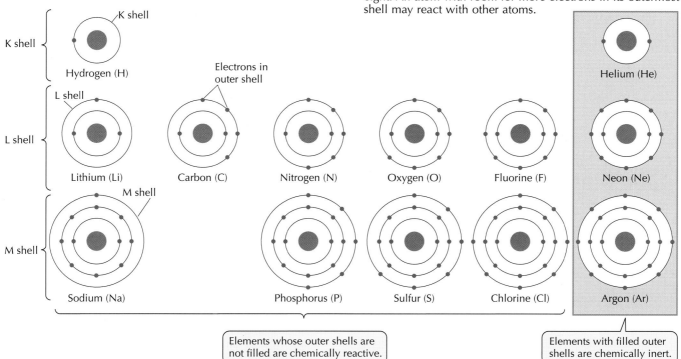

K shell

Hydrogen (H)

L shell

Lithium (Li)   Carbon (C)   Nitrogen (N)   Oxygen (O)   Fluorine (F)

Electrons in outer shell

M shell

Sodium (Na)   Phosphorus (P)   Sulfur (S)   Chlorine (Cl)

Helium (He)

Neon (Ne)

Argon (Ar)

Elements whose outer shells are not filled are chemically reactive.

Elements with filled outer shells are chemically inert.

TABLE 2.1 Chemical Bonds

TYPE OF BOND	BASIS OF BONDING	ENERGY	BOND LENGTH
Covalent bond	Sharing of electron pairs	50–110 kcal/mol[a]	≈0.1 nm[b]
Ionic bond	Attraction of opposite charges	3–7 kcal/mol	0.28 nm (optimal)
Hydrogen bond	Sharing of H atom	3–7 kcal/mol	0.26–0.31 nm (between atoms that share H)
van der Waals interaction	Interaction of electron clouds	≈1 kcal/mol	0.24–0.4 nm

[a]kcal/mol = kilocalories per mole; for other abbreviations of units of measurement see the inside front cover.
[b]For comparison, the radii of the H and C atoms are 0.2 and 0.08 nm, respectively.

bonds result from an atom's tendency to attain stability by filling its outermost electron orbitals. Atoms can gain stability in the outermost orbitals by sharing electrons or by losing or gaining one or more electrons. In this section, we will first discuss covalent bonds, the strong bonds that result from sharing of electrons. Then we'll examine hydrogen bonds, which are weaker than covalent bonds but enormously important to biology. Finally, we'll consider ionic bonding, which results when ions form as a consequence of the complete loss or gain of electrons by atoms.

Covalent bonds consist of shared pairs of electrons

When two atoms attain stable electron numbers in their outer shells by sharing one or more pairs of electrons, a **covalent bond** forms (Figure 2.6). A hydrogen atom has one electron in its only shell, but two electrons would be a more stable condition. Imagine two hydrogen atoms, initially far apart but coming closer and closer, until they begin to interact. The negatively charged electron in each hydrogen atom is attracted by the positively charged proton in the nucleus of the other hydrogen atom. When the two atoms are close enough, the two electrons spend time between both nuclei, and the two atoms are covalently bonded together, forming a molecule of hydrogen gas (H_2). The two hydrogen nuclei share the two electrons equally and completely.

The two atoms do not come *too* close together, because their positively charged nuclei strongly repel each other. A certain distance between the coupled atoms gives the most stable arrangement. Pulling the atoms slightly farther apart would require an input of energy because of the "gluing" effect of the shared electrons. Pushing the atoms closer together would require energy because of the mutual repulsion of the protons. So the most stable arrangement of the covalently bonded hydrogens can also be described as an arrangement that has a minimum amount of energy and that is less reactive than are the individual atoms alone, each of which has an incompletely filled orbital in the K shell.

A carbon atom has a total of six electrons; two electrons fill its inner shell and four are in its outer L shell. Because the L shell can hold up to eight electrons, this atom can share electrons with up to four other atoms. Thus it can form four covalent bonds. When an atom of carbon reacts with four hydrogen atoms, a substance called methane (CH_4) forms, resulting from the overlapping of electron orbitals (Figure 2.7a and b). Thanks to electron sharing, the outer shell of methane's carbon atom is filled with eight electrons, and the outer shell of each hydrogen atom is also filled. Thus four covalent bonds—each bond consisting of a shared pair of electrons—hold methane together. Table 2.2 shows the covalent bonding capacities for some biologically significant elements.

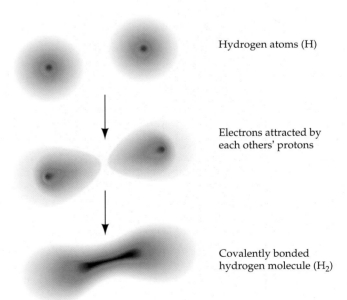

Hydrogen atoms (H)

Electrons attracted by each others' protons

Covalently bonded hydrogen molecule (H_2)

2.6 Electrons Are Shared in Covalent Bonds Two hydrogen atoms combine to form a hydrogen molecule. Each electron is attracted to both protons, but the two protons cannot come too close together, because they strongly repel each other. A covalent bond forms when the electron orbitals in the K shells of the two atoms overlap.

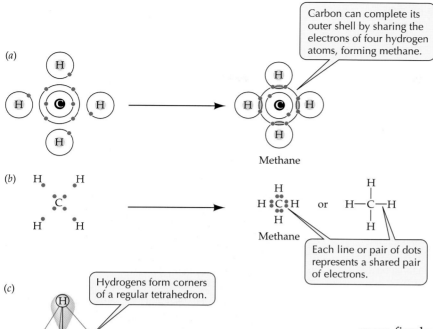

(a)

Carbon can complete its outer shell by sharing the electrons of four hydrogen atoms, forming methane.

Methane

(b)

Each line or pair of dots represents a shared pair of electrons.

Methane

(c)

Hydrogens form corners of a regular tetrahedron.

Methane

2.7 Covalent Bonding with Carbon Different representations of covalent bond formation in methane (CH_4). (a) A representation illustrating filling of a carbon molecule's outer electron shells. (b) Two common ways of representing bonds. (The two electrons in the K shell of carbon are not shown; these renderings indicate only the initially unfilled shells.) (c) Representation of the spatial orientation of methane's bonds.

ORIENTATION OF BONDS IN SPACE. Not only the number, but also the spatial orientation, of bonds is important. The four filled orbitals around the carbon nucleus of methane distribute themselves in space so that the bonded hydrogens are directed to the corners of a regular tetrahedron with carbon in the center (Figure 2.7c). Although the orientation of orbitals and shapes of molecules differ depending on the kinds of atoms and how they are linked together, it is essential to remember that all molecules occupy space and have three-dimensional shapes. The shapes of molecules contribute to their biological functions, as we will see in Chapter 3.

MULTIPLE COVALENT BONDS. A covalent bond is represented by a line between the chemical symbols for the atoms. Bonds in which a single pair of electrons is shared are called **single bonds** (for example, H—H, C—H). When four electrons (two pairs) are shared, the link is a **double bond** (C=C). In the gas ethylene ($H_2C=CH_2$), two carbon atoms share two pairs of electrons. **Triple bonds** (six shared electrons) are rare,

but there is one in nitrogen gas (N≡N), the chief component of the air we breathe. In the covalent bonds in these five examples, the electrons are shared more or less equally between the nuclei; consequently all regions of the bonds are identical. However, when electrons are shared unequally in a covalent bond, regions of partial electric charge exist.

UNEQUAL SHARING OF ELECTRONS. So far we have discussed the covalent bonds that result from the equal sharing of electrons between two nuclei. Now we want to consider the kind of covalent bond that results from unequal sharing of the electrons.

Some atoms hold electrons to themselves more firmly than other atoms do. This characteristic is called **electronegativity**. Highly electronegative atoms that form covalent bonds include oxygen and nitrogen. When these atoms are covalently bonded to atoms with weaker electronegativity, such as carbon and hydrogen, the bonding pair of electrons is unequally shared between the two atoms, and the result is a **polar covalent bond** (Figure 2.8). For example, when oxygen is bonded to hydrogen, the bonding electrons spend much more time near the oxygen nucleus than near the hydrogen nucleus. Consequently, the oxygen end of the bond is slightly negative (symbolized δ– and spoken as "delta negative," meaning a partial unit charge), and the hydrogen end is slightly positive (δ+). The bond is polar because these opposite charges are separated at the two ends of the bond. The partial charges that result from polar covalent bonds produce **polar molecules** or polar regions of large molecules. Polar bonds greatly influence the interactions between molecules that contain them, as we see in the interaction of water molecules in the liquid state.

TABLE 2.2 Typical Bonding Capabilities of Some Biologically Important Elements

ELEMENT	NUMBER OF COVALENT BONDS
Hydrogen (H)	1
Oxygen (O)	2
Sulfur (S)	2
Nitrogen (N)	3
Carbon (C)	4
Phosphorus (P)	5

2.8 The Polar Covalent Bond in the Water Molecule

A covalent bond between atoms with different electronegativities is a polar covalent bond with partial (δ) electric charges at the ends. Two ways of representing the water molecule are shown. (*a*) The lines between the symbols represent covalent bonds. (*b*) Four pairs of electrons are identified by dots, and the dots for the covalent bonds are displaced toward oxygen and away from the hydrogens.

(*a*)

$$\delta^+ H - O^{\delta^-}$$
$$\overset{|}{\underset{\delta^+}{H^+}}$$

Water has polar covalent bonds.

(*b*)

Water's bonding electrons are shared unequally; electron density is greatest around the oxygen atom.

Hydrogen bonds may form between molecules

In liquid water, the negatively charged oxygen ($\delta-$) atom of one water molecule is attracted to the positively charged hydrogen ($\delta+$) of another water molecule. (Remember, negative charges attract positive charges.) The bond resulting from this attraction is called a **hydrogen bond** and is usually symbolized by a series of dots (Figure 2.9). Hydrogen bonds are not restricted to water molecules. They may form between any covalently bonded hydrogen and an electronegative atom, usually oxygen or nitrogen: —H···O— or —H···N—. Hydrogen bonds form between small molecules or between different parts of large molecules. Covalent bonds and polar covalent bonds, on the other hand, are always found *within* molecules.

A hydrogen bond is a weak bond; it has about one-twentieth (5 percent) of the strength of a covalent bond between a hydrogen atom and an oxygen atom. However, where many hydrogen bonds form, they have considerable strength and greatly influence the structure and properties of substances. Later in this chapter we'll discuss further how hydrogen bonding in water contributes to many of the properties of water that are significant for living systems. Hydrogen bonds also play important roles in determining and maintaining the three-dimensional shapes of giant molecules such as DNA and protein (see Chapter 3).

Ions form bonds by electrical attraction

When one interacting atom is much more electronegative than the other, a complete transfer of one or more electrons may take place. For example, a sodium atom has only one electron in its outermost shell; this condition is unstable. A chlorine atom has seven electrons in its outer shell, another unstable condition. The reaction between sodium and chlorine makes both atoms more stable. When the two atoms meet, the highly electronegative chlorine atom takes the single unstable electron from the sodium atom (Figure 2.10). The result is two electrically charged particles, called ions. **Ions** are electrically charged particles that form when atoms gain or lose one or more electrons.

The sodium ion (Na^+) has a +1 unit charge because it has one less electron than it has protons. The outermost electron shell of the sodium ion is full, with eight electrons, so the ion is stable. The chloride ion (Cl^-) has a –1 unit charge because it has one more electron than it has protons. This additional electron gives Cl^- an outer shell with a stable load of eight electrons. Negatively charged ions are called **anions**; positively charged ions are called **cations**.

Some elements form ions with multiple charges by losing or gaining more than one electron to achieve a stable electron configuration in their outer shell. Examples are Ca^{2+} (the calcium ion, created from a cal-

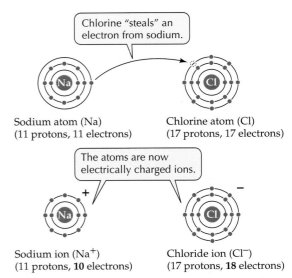

Chlorine "steals" an electron from sodium.

Sodium atom (Na)
(11 protons, 11 electrons)

Chlorine atom (Cl)
(17 protons, 17 electrons)

The atoms are now electrically charged ions.

Sodium ion (Na^+)
(11 protons, **10** electrons)

Chloride ion (Cl^-)
(17 protons, **18** electrons)

2.10 Formation of Sodium and Chloride Ions When it reacts with a sodium atom, a chlorine atom, because it is much more electronegative, "steals" an electron and becomes a chloride ion (Cl^-). This ion is negatively charged because it contains one more electron than it does protons. The sodium atom, upon losing the electron, becomes a sodium ion (Na^+).

2.9 Hydrogen Bonding

Hydrogen bonds form between a slightly positive hydrogen and a slightly negative oxygen in separate molecules. The same type of bond could form between hydrogen and nitrogen.

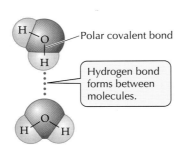

Polar covalent bond

Hydrogen bond forms between molecules.

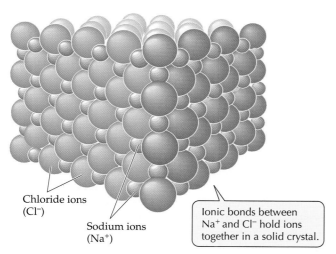

Chloride ions
(Cl⁻)

Sodium ions
(Na⁺)

Ionic bonds between
Na⁺ and Cl⁻ hold ions
together in a solid crystal.

2.11 Ionic Bonding in a Solid Ionic bonds are electrical attractions between ions with opposite electric charges.

IONIC BONDS: ELECTRICAL ATTRACTIONS. **Ionic bonds** are the bonds formed by electrical attractions between ions bearing opposite charges. In solids such as table salt ($NaCl$), the cations and anions are held together by ionic bonds (Figure 2.11). In solids, the ionic bonds are strong because the ions are close together. However, when ions are dispersed in water, the distance between them can be large; the strength of their attraction is thus greatly reduced. Under the conditions that exist in the cell, an ionic bond is less than one-tenth as strong as a covalent bond that shares electrons equally, so an ionic bond can be broken much more readily than a covalent bond.

Not surprisingly, ions with one or more unit charges can interact with polar substances as well as with other ions. Such interaction results when table salt or any other ionic solid dissolves in water (Figure 2.12). The hydrogen bond that we described earlier is a weak type of ionic bond, because it is formed by electrical attractions. However, it is weaker than most ionic bonds because the hydrogen bond is formed by partial charges ($\delta-$ and $\delta+$) rather than by whole unit charges (+1 unit, −1 unit).

Nonpolar substances have no attraction for polar substances

We have been discussing the bonds that result from electrical attractions between positive and negative charges (ionic bonds and hydrogen bonds). Now let's return to a brief consideration of substances that have "pure" covalent bonds (Figure 2.13). These bonds

cium atom that has lost two electrons), Mg^{2+} (magnesium ion), and Al^{3+} (aluminum ion). Two biologically important elements each yield more than one stable ion: Iron yields Fe^{2+} (ferrous ion) and Fe^{3+} (ferric ion), and copper yields Cu^+ (cuprous ion) and Cu^{2+} (cupric ion). Groups of covalently bonded atoms that carry an electric charge are called **complex ions**; examples include NH_4^+ (ammonium ion), SO_4^{2-} (sulfate ion), and PO_4^{2-} (phosphate ion).

Once they form, ions are usually stable, and no more electrons are lost or gained. As stable entities, ions can enter into stable associations through ionic bonding. Thus stable solids such as sodium chloride ($NaCl$) and potassium phosphate (K_3PO_4) are formed. Although a very complex solid, bone has as one of its major components the simple ionic compound $Ca_3(PO_4)_2$.

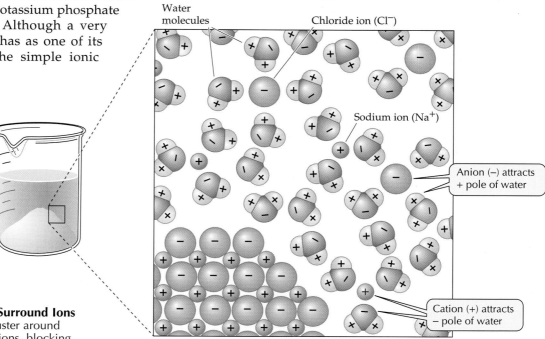

Water molecules

Chloride ion (Cl⁻)

Sodium ion (Na⁺)

Anion (−) attracts + pole of water

Cation (+) attracts − pole of water

Undissolved sodium chloride

2.12 Water Molecules Surround Ions
Polar water molecules cluster around cations or anions in solutions, blocking their reassociation into a solid.

Water, a polar molecule   Ethane, a nonpolar molecule

2.13 Nonpolar Molecules: Hydrocarbons The electrons are uniformly distributed around the entire hydrocarbon molecule (ethane, CH_3—CH_3). The molecule is nonpolar, with no electrical attraction to substances with electric charges or polar bonds.

form between atoms that have equal or nearly equal electronegativities—such as carbon and hydrogen—which share the bonding electrons equally. Such bonds are abundant in the compounds of hydrogen and carbon—the **hydrocarbons**. Molecules such as ethane (CH_3—CH_3) and butane (CH_3—CH_2—CH_2—CH_3) are small hydrocarbons, but in living systems, molecules exist with hydrocarbon chains consisting of 16 or more carbon atoms.

ATTRACTIONS BETWEEN NONPOLAR MOLECULES. Nonpolar substances such as oils and fats show **van der Waals attractions** between molecules. These attractive forces operate only when nonpolar substances come very close to each other. The random variations in the electron distribution in one molecule create an opposite charge distribution in the adjacent molecule, and the result is a brief, weak attraction. Although each such interaction is brief and weak at any one site, the summation of many such interactions over the entire span of a nonpolar molecule can produce substantial attraction. Thus van der Waals interactions are important in

holding together the long hydrocarbon chains that make up the inner portion of biological membranes. They also stabilize portions of the DNA double helix and the intricate folded structure of proteins (see Chapters 3 and 11).

POLAR AND NONPOLAR INTERACTIONS. When electrons are shared equally, the resultant covalent bonds are nonpolar and they do not interact with the charges of polar covalent bonds. Substances with only nonpolar bonds, such as butane and oils, will not interact with substances that have polar bonds or ionic bonds. This explains why oils will not dissolve in water. Oils are nonpolar hydrocarbons, while water is a highly polar substance. Substances with nonpolar covalent bonds are said to be **hydrophobic** (literally "water-fearing"), which refers to the fact that there is no attraction between nonpolar substances and water or other electrically charged substances. Hydrophobic substances are also called **nonpolar substances**.

When hydrocarbons are dispersed in water, they slowly come together—dispersed molecules form droplets that form larger droplets. The forces that bring about this combining of molecules are sometimes called *hydrophobic interactions*, but this term is somewhat misleading. The interactions that combine nonpolar substances have less to do with forces between the nonpolar molecules than with the hydrogen bonding of the water that surrounds the molecules (Figure 2.14).

When nonpolar substances such as hydrocarbons are introduced into water, they cause a disruption in the usual hydrogen bonding between water molecules. In the vicinity of the hydrocarbon, the water molecules form a hydrogen-bonded "cage" that surrounds the nonpolar hydrocarbons and pushes them together.

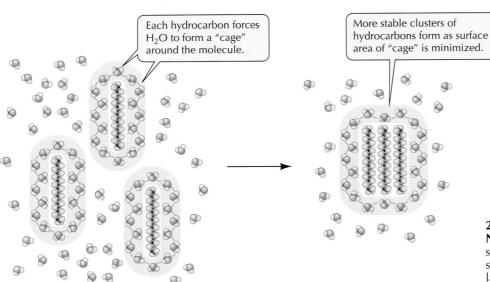

Each hydrocarbon forces H_2O to form a "cage" around the molecule.

More stable clusters of hydrocarbons form as surface area of "cage" is minimized.

2.14 Interactions of Water and Nonpolar Substances Nonpolar substances dispersed in water will slowly cluster together, forming larger and larger droplets.

These water cages can bring together dispersed nonpolar molecules into larger groups (see Figure 2.14).

Eggs by the Dozen, Molecules by the Mole

In modern biology, as in chemistry, the question "How much?" is as important as "What kind?" In this section, we will examine briefly how chemists and biologists deal quantitatively with atoms and molecules.

The **molecular formula** uses chemical symbols to identify different atoms, and subscript numbers to show how many atoms are present. For example, the molecular formula for methane is CH_4 (each molecule contains one carbon atom and four hydrogen atoms), that for oxygen gas is O_2, and that for sucrose (table sugar) is $C_{12}H_{22}O_{11}$. The hormone insulin is represented by the molecular formula $C_{254}H_{377}N_{65}O_{76}S_6$! Although molecular formulas tell us what kinds of atoms and how many of each kind are present in the molecule, they tell us nothing about which atoms are linked to which. **Structural formulas** give us this information (see Figure 2.7).

Each compound has a **molecular weight** (molecular mass): the sum of the atomic weights of the atoms in the molecule. The atomic weights of hydrogen, carbon, and oxygen are 1.008, 12.011, and 16.000, respectively. Thus the molecular weight of water (H_2O) is $(2 \times 1.008) + 16.000 = 18.016$, or about 18. What is the molecular weight of sucrose ($C_{12}H_{22}O_{11}$)? Your calculations should tell you that the answer is approximately 342. If you remember the molecular weights of a few representative biological compounds, you will be able to picture the relative sizes of molecules that interact with one another (Figure 2.15). Experiments require quantitative information. Suppose we want to compare how sodium chloride (NaCl), potassium chloride (KCl), and lithium chloride (LiCl) affect a biological process. At first you might think we could simply give, say, 2 grams (g) of NaCl to one set of subjects, 2 g of KCl to another, and 2 g of LiCl to the third. But because the molecular weights of NaCl, KCl, and LiCl are different, 2-gram samples of each of these substances contain different numbers of molecules. The comparison would thus not be legitimate. Instead, we want to give *equal numbers of molecules* of each substance so that we can compare the activity of one molecule of one substance with that of one molecule of another. How can we measure out equal numbers of molecules?

We can calculate numbers of molecules by weighing

Measuring the number of molecules is essential to studying how many molecules take part in chemical reactions. Consider two large barrels, one filled with bolts and the other with pennies. How can we determine the number of units in each barrel? We could count them, but that would be tedious (and it is impossible to count individual molecules directly). However, if we know the weight of one bolt and one penny, we can calculate the number of units by dividing the total weight by the weight of each unit. The same principles apply to determining the number of molecules in a weighed sample.

To determine the number of molecules in a sample of a pure substance, we first determine the weight of the substance in grams, then we divide the grams by the relative weight of one molecule (the molecular weight defined earlier). Therefore, to measure out quantities of substances containing equal numbers of molecules, we weigh out in grams a quantity of each substance equivalent to its molecular weight. The

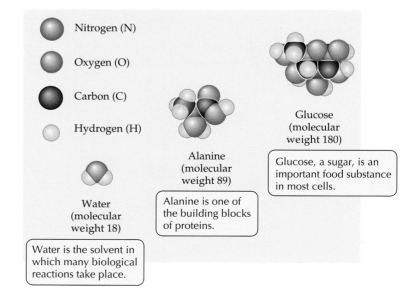

2.15 Molecular Weight and Size The relative sizes of three common molecules and their molecular weights. These space-filling models are the most realistic representations of molecules we can create; you will continue to see them in this and subsequent chapters. The color conventions are standard for the atoms (yellow is used for sulfur and phosphorus atoms, which are not shown here).

Nitrogen (N)

Oxygen (O)

Carbon (C)

Hydrogen (H)

Glucose (molecular weight 180)

Glucose, a sugar, is an important food substance in most cells.

Alanine (molecular weight 89)

Alanine is one of the building blocks of proteins.

Water (molecular weight 18)

Water is the solvent in which many biological reactions take place.

molecular weight of NaCl is 58.45. Therefore 58.45 g of NaCl will have the same number of molecules as 74.55 g of KCl, whose molecular weight is 74.55.

If we divide the 74.55 g by the molecular weight of 74.55, the answer is 1. But one what? Surely not one molecule! The calculated numbers here are not the exact number, because we did not divide by the actual weight in grams of a molecule of potassium chloride (which would be about 1×10^{-22} g). Instead we divided by the relative weight. Therefore, instead of knowing the exact number of molecules, we know the relative number, which is called a **mole**. *One mole of a substance is an amount whose weight in grams is numerically equal to the molecular weight of the substance.* Potassium chloride (KCl) has a molecular weight of 74.55, so 1 mole of KCl weighs 74.55 g. Likewise, 1 mole of NaCl weighs 58.45 g, and 1 mole of LiCl, 42.40 g.

A mole of one substance contains the same number of molecules as does a mole of any other substance. This number, known as **Avogadro's number**, is 6.023×10^{23} molecules per mole. The concept of the mole is important for biology because it enables us to work easily with known numbers of molecules. Since we can neither weigh nor count individual molecules, we work with moles. The mole concept is analogous to the concept of a dozen or a gross. We buy some things, such as eggs, by the dozen (or another similar unit, such as the gross) because the individual items are inconvenient to count. Just as a dozen of anything contains twelve of that thing, a mole of any substance contains Avogadro's number of molecules of that substance.

Reactions take place in solutions

In cells and in the laboratory, reactions take place in solutions formed when substances (solutes) are dissolved in water (the solvent). The amount of substance dissolved in a given amount of water is the concentration of the solution. Knowing the concentrations of solutions is essential in performing experiments and interpreting their results.

A solution that contains 1 mole of solute per liter of solution has a one-molar concentration, abbreviated 1

M. A solution containing a half mole per liter is referred to as 0.5 *M*, or half-molar. How would you make 100 milliliters (ml) of a 0.5 *M* sucrose solution? The molecular weight of sucrose is 342, so 1 liter (1,000 ml) of a 1 *M* sucrose solution contains 1 mole, or 342 g, of sucrose. You were asked to make just 100 ml of 0.5 *M* solution. Since 34.2 g of sucrose would make 100 ml of 1 *M* sucrose, to make 100 ml of a 0.5 *M* sucrose solution, you would use 0.5×34.2 g = 17.1 g of sucrose.

Chemical Reactions: Atoms Change Partners

When atoms combine or change bonding partners, a chemical reaction is occurring. Consider the combustion reaction that takes place in the flame of a propane stove. When propane (C_3H_8) reacts with oxygen gas (O_2), the carbon atoms become bonded to oxygen atoms instead of to hydrogen atoms, and the hydrogen atoms become bonded to oxygen instead of to carbon (Figure 2.16). As the covalently bonded atoms change bonding partners, the composition of the matter changes, and propane and oxygen gas become carbon dioxide and water. This chemical reaction can be represented by the balanced equation

$$C_3H_8 + 5 O_2 \rightarrow 3 CO_2 + 4 H_2O$$

In this equation, the propane and oxygen are the **reactants**, and the carbon dioxide and water are the **products**. The arrow symbolizes the chemical reaction. The numbers preceding the molecular formulas balance the equation and indicate how many molecules react or are produced. In this and all other chemical reactions, matter is neither created or destroyed. The total number of carbons on the left equals the total number on the right. However, there is another product of this reaction: energy.

The heat of the stove's flame and its blue light reveal that the reaction of propane and oxygen releases a great deal of energy. **Energy** is defined as the capacity

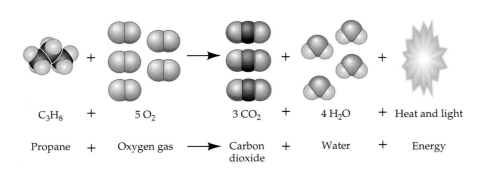

C_3H_8 + 5 O_2 → 3 CO_2 + 4 H_2O + Heat and light

Propane + Oxygen gas → Carbon dioxide + Water + Energy

2.16 Bonding Partners and Energy May Change in a Chemical Reaction One molecule of propane reacts with five molecules of oxygen gas to give three molecules of carbon dioxide and four molecules of water. This reaction releases energy, in the form of heat and light.

to do work, but on a more intuitive level, it can be thought of as the capacity to change. Chemical reactions do not create or destroy energy, but *changes* in energy usually accompany chemical reactions. The energy released as heat and light was present in the reactants in another form, called *potential chemical energy*. In some chemical reactions, energy must be supplied from the environment (for example, some substances will react only after being heated), and some of this supplied energy becomes stored as potential chemical energy in the bonds formed in the reactants.

We can measure the energy associated with chemical reactions using the unit called a **calorie** (**cal**). A calorie is the amount of heat energy needed to raise the temperature of 1 g of pure water from 14.5°C to 15.5°C. (The nutritionist's Calorie, with a capital C, is what biologists call a kilocalorie (kcal) and is equal to 1,000 heat-energy calories.) Another unit of energy that is increasingly used is the **joule** (**J**). When you compare data on energy, always compare joules to joules and calories to calories. The two units can be interconverted: 1 J = 0.239 cal, and 1 cal = 4.184 J. Thus, for example, 486 cal = 2033 J, or 2.033 kJ. Although defined in terms of heat, the calorie and the joule are measures of any form of energy—mechanical, electric, or chemical. We'll discuss energy changes further in Chapter 6.

Within living cells, chemical reactions called oxidation–reduction reactions take place that have much in common with this combustion of propane. The fuel for these biological reactions is different (the sugar glucose, rather than propane), and the reactions proceed by many intermediate steps that permit the energy released from the glucose to be harvested and put to use by the cell. But the products are the same: carbon dioxide and water. We will present and discuss oxidation–reduction reactions and several other types of chemical reactions that are prevalent in living systems in the chapters that follow.

Water: Structure and Properties

Water, like all other matter, can exist in three states: solid (ice), liquid, and gas (vapor) (Figure 2.17). Liquid water is the medium in which life originated on Earth more than 3.5 billion years ago, and it is in water that life evolved for about a billion years. Today water covers three-fourths of Earth's surface, and all active organisms contain between 45 and 95 percent water. No organism can remain biologically active without water, which is important in both the outer and inner environments. Within cells, water participates directly in many chemical reactions, and it is the medium (or solvent) in which most reactions take place. In this section we will consider the structure and interactions of water molecules, exploring how these generate properties essential to life.

Water has a unique structure and special properties

The shape of a water molecule is determined by the distribution in space of the four pairs of electrons in the outer shell of the oxygen atom. Each pair of electrons is confined to an orbital. Two of these pairs are bonded to hydrogens, but the other two pairs (nonbonding pairs) also influence shape. Because of their negatively charged electrons, the four orbitals, each containing a pair of electrons, repel one another. In seeking to be as far apart as possible, they give the water molecule a nearly tetrahedral shape (Figure 2.18).

The shape of the water molecule, its polar nature, and the formation of hydrogen bonds between water molecules or with other substances give water its unusual properties. For example, ice floats, and compared to other liquids, water is an excellent solvent, making it an ideal medium for biochemical reactions. Water is both cohesive (sticking to itself) and adhesive (sticking to other things). And the energy changes that

2.17 Water: Solid and Liquid Solid water from a glacier floats in its liquid form. The clouds are also water, but not in its gaseous phase: They are composed of fine drops of liquid water.

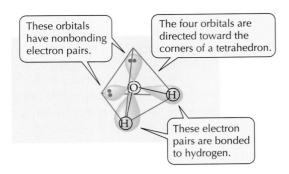

These orbitals have nonbonding electron pairs.

The four orbitals are directed toward the corners of a tetrahedron.

These electron pairs are bonded to hydrogen.

2.18 The Shape of the Water Molecule   The four pairs of electrons in the outer shell of oxygen repel each other, producing a tetrahedral shape.

accompany its transitions from solid to liquid to gas are significant for living systems.

ICE FLOATS.   In its solid state (ice), water is held by its hydrogen bonds in a rigid, crystalline structure in which each water molecule is hydrogen-bonded to four other molecules (Figure 2.19*a*). Although these molecules are held firmly in place, they are not as tightly packed as they are in liquid water (Figure 2.19*b*). In other words, *solid water is less dense than liquid water*, which is why ice floats in water. If ice sank in water, as almost all other solids do in their corresponding liquids, ponds and lakes would freeze from the bottom up, becoming solid blocks of ice in winter and killing most of the organisms living in them. Once the whole pond had frozen, its temperature could drop well below the freezing point of water. However, because ice floats, it forms a protective insulating layer on the top of the pond, reducing heat flow to the cold air above. Thus fish, plants, and other organisms in

the pond can survive the winter at temperatures no lower than 0°C, the freezing point of pure water.

MELTING AND FREEZING.   Compared to other nonmetallic substances of the same size, ice requires a great deal of heat energy to melt. Melting 1 mole of water molecules requires the addition of 5.9 kJ of energy. This value is high because more than a mole of hydrogen bonds must be broken for 1 mole of water to change from solid to liquid. In the opposite process, freezing, a great deal of energy must be lost for water to transform from liquid to solid. These properties help make water a moderator of temperature changes.

Another property of water that moderates temperature is the high heat capacity of liquid water. **Heat capacity** is the amount of heat energy that is required to raise the temperature of a substance 1°C. Raising the temperature of liquid water takes a relatively large amount of heat. The temperature of a given quantity of water is raised only 1°C by an amount of heat that would increase the temperature of the same quantity of ethyl alcohol by 2°C, or of chloroform by 4°C. This phenomenon contributes to the surprising constancy of the temperature of the oceans and other large bodies of water through the seasons of the year. The temperature changes in coastal land masses are also moderated by large bodies of water. Indeed, water helps minimize variations in atmospheric temperature throughout the planet.

COHESION AND SURFACE TENSION.   In liquid water, the molecules are free to move about. The hydrogen bonds between the water molecules continually form and break. In other words, liquid water has a dynamic structure. On the average, every water molecule forms 3.4 hydrogen bonds with other water molecules. This number represents fewer bonds than exist in ice, but it is still a high number.

(a) Ice (solid water)

(b) Liquid water

Hydrogen bonds continually break and form as water molecules move.

In ice, water molecules are held in a rigid state by hydrogen bonds.

2.19 Hydrogen Bonding in Liquid Water and Ice
Hydrogen bonding exists between the molecules of water in both its liquid and solid states. (*a*) Solid water. (*b*) Liquid water. Although more structured, ice is less dense than liquid water, so it floats.

These hydrogen bonds explain the cohesive strength of water. The **cohesive strength** of water is what permits narrow columns of water to stretch from the roots to the leaves of trees more than 100 meters high. When water evaporates from leaves, the entire column moves upward in response to the pull of the molecules at the top.

Water also has a high **surface tension**, which means that the surface of water exposed to the air is difficult to puncture. The water molecules in this surface layer are hydrogen-bonded to other water molecules below. The surface tension of water permits a container to be filled slightly above its rim without overflowing, and it permits small animals to walk on the surface of water (Figure 2.20).

EVAPORATION AND COOLING. Water has a high **heat of vaporization**, which means a lot of heat is required to change water from its liquid state to its gaseous state (the process of evaporation). This heat is absorbed from the environment in contact with the water. Evaporation thus has a cooling effect on the environment—whether a leaf, a forest, or an entire land mass. This effect explains why sweating cools the human body: As the sweat evaporates off the skin, it takes with it some of the adjacent body heat.

Water molecules sometimes form ions

The water molecule has a slight but significant tendency to come apart into a hydroxide ion (OH⁻) and a hydrogen ion (a proton, H⁺). Actually, *two* water molecules participate in ionization. One of the two molecules "captures" a hydrogen ion from the other, forming a hydroxide ion and a hydronium ion:

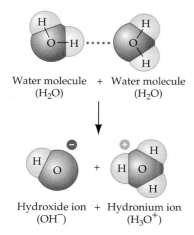

Water molecule + Water molecule
(H_2O)   (H_2O)

Hydroxide ion + Hydronium ion
(OH⁻)   (H_3O^+)

The hydronium ion is in effect a hydrogen ion bound to a water molecule. For simplicity, biochemists tend to use a modified representation of the ionization of water:

$$H_2O \rightarrow H^+ + OH^-$$

2.20 Surface Tension Water striders "skate" along, supported by the surface tension of the water that is their home.

Even though only about one water molecule in 500 million is ionized at any given time, the transformation is significant because H⁺ and OH⁻ ions participate in many important biochemical reactions.

Acids, Bases, and the pH Scale

The ionization of water is very important for all living creatures. This fact may seem surprising, since the ionization is so slight—only one ionization out of 5×10^8 water molecules. But we are less surprised if we focus on the abundance of water in living systems and the reactive nature of H⁺ produced by that ionization. Remember that H⁺ is a proton, a tiny bit of charged matter—smaller than any atom or molecule in the cell. Usually it attaches to another water molecule, but in the cell it can attach to other molecules and substantially change their properties. Because acids and bases donate and accept H⁺, they can profoundly alter the water environment in which the chemical reactions and processes of life take place.

In this section we'll examine acids, bases, and their measurement using the pH scale. We'll close by looking at how buffers limit the changes in pH.

Acids donate H⁺, bases accept H⁺

In pure water, the concentration of hydrogen ions exactly equals that of hydroxide ions (OH⁻), and this "solution" is said to be **neutral**. Now suppose we add some HCl (hydrochloric acid). As it dissolves, the HCl ionizes, releasing H⁺ and Cl⁻ ions:

$$HCl \rightarrow H^+ + Cl^-$$

Now there are more H^+ than OH^- ions. Such a solution is *acidic*. A *basic*, or alkaline, solution is one in which there are more OH^- than H^+ ions. A basic solution can be made from water by adding, for example, sodium hydroxide (NaOH), which ionizes to yield OH^- and Na^+ ions, thus making the concentration of OH^- ions greater than that of H^+ ions.

$$NaOH \rightarrow Na^+ + OH^-$$

An **acid** is any compound that can *release* H^+ ions in solution. HCl is an acid, as is H_2SO_4 (sulfuric acid). One molecule of sulfuric acid may ionize to yield two H^+ ions and one SO_4^{2-} ion. Biological compounds such as acetic acid and pyruvic acid, which contain —COOH (the carboxyl group; see Figure 2.23) are also acids, because —COOH $\rightarrow$ —COO$^-$ + H^+.

Bases are compounds that can *accept* H^+ ions. These include the bicarbonate ion (HCO_3^-), which can accept a H^+ ion and become carbonic acid (H_2CO_3); ammonia (NH_3), which can accept a H^+ ion and become an ammonium ion (NH_4^+); and many others.

Note that although —COOH is an acid, —COO$^-$ is a base, because —COO$^-$ + H^+ $\rightarrow$ —COOH. Acids and bases exist as pairs, such as —COOH and —COO$^-$, because any acid becomes a base when it releases a proton, and any base becomes an acid when it gains a proton.

You may have noticed that the two reactions just discussed are the opposites of each other. The reaction that yields —COO$^-$ and H^+ is reversible and may be expressed as

$$—COOH \rightleftharpoons —COO^- + H^+$$

A **reversible reaction** is one that can proceed in either direction—left to right or right to left—depending on the relative starting concentrations of reacting substances and products. In principle, *all* chemical reactions are reversible. We will look at some consequences of this reversibility in Chapter 6.

pH is the measure of hydrogen ion concentration

The terms "acid*ic*" and "bas*ic*" refer only to *solutions*. How acidic or basic a solution is depends on the relative concentrations of H^+ and OH^- ions in it. "Acid" and "base" refer to *compounds* and *ions*. A compound or ion that is an acid can donate H^+ ions; one that is a base can accept H^+ ions.

How do we specify how acidic or basic a solution is? First, let's look at the H^+ ion concentrations of a few contrasting solutions. In pure water, the H^+ concentration is 10^{-7} M. In 1 M hydrochloric acid, the H^+ concentration is 1 M; and in 1 M sodium hydroxide, the H^+ concentration is 10^{-14} M. Because its values range so widely—from more than 1.0 M to less than 10^{-14} M—

the H^+ concentration itself is an inconvenient quantity. It is easier to work with the logarithm of the concentration, because logarithms compress this range. We indicate how acidic or basic a solution is by its **pH** (a term derived from "*potential of Hydrogen*"). The pH value is defined as the negative logarithm of the hydrogen ion concentration in moles per liter (molar concentration). In chemical notation, molar concentration is often indicated by putting brackets around the symbol for a substance; thus [H^+] stands for the molar concentration of H^+. The equation for pH is

$$pH = -\log_{10}[H^+]$$

Since the H^+ concentration of pure water is 10^{-7} M, its pH is $-\log(10^{-7}) = -(-7)$, or 7. A smaller negative logarithm means a larger number. In practical terms, a lower pH means a higher H^+ concentration, or greater acidity. In 1 M HCl, the H^+ concentration is 1 M, so the pH is the negative logarithm of 1 ($-\log 10^0$), or 0. The pH of 1 M NaOH is the negative logarithm of 10^{-14}, or 14. A solution with a pH of less than 7 is acidic: It contains more H^+ ions than OH^- ions. A solution with a pH of 7 is neutral. And a solution with a pH value greater than 7 is basic. Because the pH scale is logarithmic, the values are exponential: A solution with a pH of 5 is 10 times more acidic than one with a pH of 6 (it has ten times as great a concentration of H^+); a solution with a pH of 4 is 100 times more acidic than one with a pH of 6. Figure 2.21 shows the pH values of some common substances.

Buffers minimize pH changes

An organism must control the chemistry of its cells—in particular, the pH of the separate compartments within cells. Animals must also control the pH of their blood. The normal pH of human blood is 7.4, and deviations of even a few tenths of a pH unit can be fatal. The control of pH is made possible in part by **buffers**, systems that maintain a relatively constant pH even when substantial amounts of acid or base are added. A buffer is a mixture of an acid that does not ionize completely in water and its corresponding base—for example, carbonic acid (H_2CO_3) and bicarbonate ions (HCO_3^-). If acid is added to this buffer, not all the H^+ ions from the acid stay in solution. Instead, many of the added H^+ ions combine with bicarbonate ions to produce more carbonic acid, thus using up some of the H^+ ions in the solution and decreasing the acidifying effect of the added acid:

$$HCO_3^- + H^+ \rightleftharpoons H_2CO_3$$

If base is added, the reaction reverses. Some of the carbonic acid ionizes to produce bicarbonate ions and more H^+, which counteracts some of the added base.

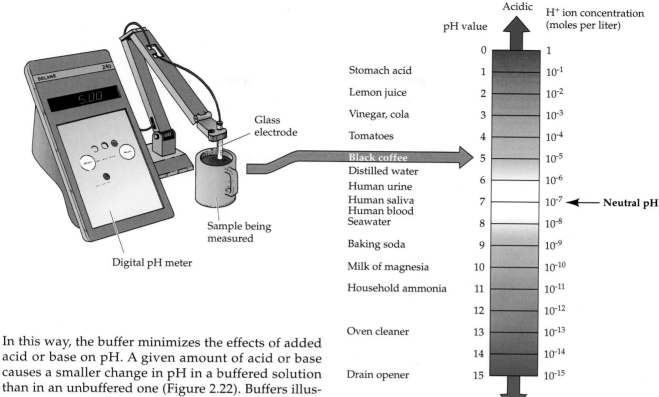

In this way, the buffer minimizes the effects of added acid or base on pH. A given amount of acid or base causes a smaller change in pH in a buffered solution than in an unbuffered one (Figure 2.22). Buffers illustrate the reversibility of chemical reactions: Addition of acid drives the reaction in one direction; addition of base drives it in the other.

2.21 pH Values of Some Familiar Substances A pH meter such as the one on the left tells us the pH of a solution. This scale reads from low (acidic) pH values at the top to high (basic) pH values at the bottom.

The Properties of Molecules

Molecules vary in size. Some are small, such as H_2 and CH_4. Others are larger, such as a molecule of table sugar (sucrose), which has 45 atoms. Still other molecules, such as proteins, are gigantic, sometimes containing tens of thousands of atoms bonded together in specific ways. Whether large, medium, or small, most of the molecules in living systems contain carbon atoms and are thus referred to as **organic molecules**. Most organic molecules include hydrogen and oxygen atoms, and many also include nitrogen and phosphorus.

All molecules have a specific three-dimensional shape. For example, the orientation of the bonding orbitals around the carbon atom gives the molecule of methane (CH_4) the shape of a regular tetrahedron (see Figure 2.7c), while in carbon dioxide (CO_2) the three atoms are in line with one another. Larger molecules have specific complex shapes that result from the number and kinds of atoms present and the ways in which these atoms are linked together. Some large molecules have compact ball-like shapes. Others are long, thin, ropelike structures. The shapes relate to the roles these molecules play in living cells.

In addition to size and shape, molecules have certain properties that characterize them and determine

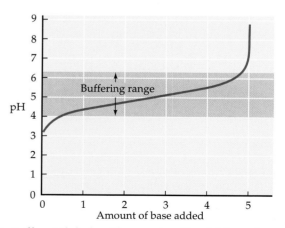

2.22 Buffers Minimize Changes in pH Adding a base increases the pH of a solution. With increasing amounts of added base, the overall slope of a graph of pH is upward. In the buffering range, however, the slope is shallow. At high and low values of pH, where the buffer is ineffective, the slopes are much steeper.

Functional group	Class of compounds	Structural formula	Example	Ball-and-stick model
Hydroxyl —OH	Alcohols	R—OH	$$\begin{matrix} & H & H \\ & | & | \\ H-&C-&C-OH \\ & | & | \\ & H & H \end{matrix}$$ Ethanol	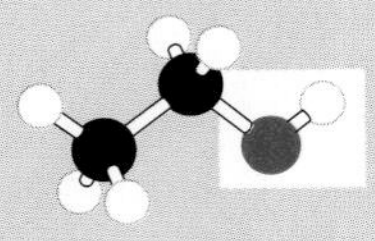
Carbonyl —CHO	Aldehydes	$$\begin{matrix} & O \\ & \| \\ R-&C \\ & \backslash \\ & H \end{matrix}$$	$$\begin{matrix} & H & & O \\ & | & & \| \\ H-&C-&C \\ & | & & \backslash \\ & H & & H \end{matrix}$$ Acetaldehyde	
Carbonyl \\CO/	Ketones	$$\begin{matrix} & O \\ & \| \\ R-&C-R \end{matrix}$$	$$\begin{matrix} H & & O & & H \\ | & & \| & & | \\ H-C-&&C-&&C-H \\ | & & & & | \\ H & & & & H \end{matrix}$$ Acetone	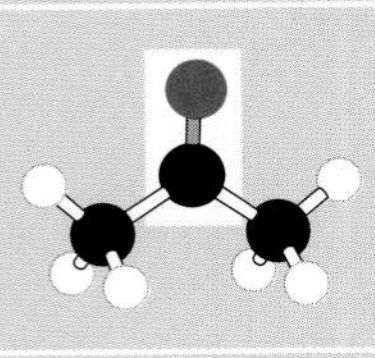
Carboxyl —COOH	Carboxylic acids	$$\begin{matrix} & O \\ & \| \\ R-&C \\ & \backslash \\ & OH \end{matrix}$$	$$\begin{matrix} & H & & O \\ & | & & \| \\ H-&C-&C \\ & | & & \backslash \\ & H & & OH \end{matrix}$$ Acetic acid	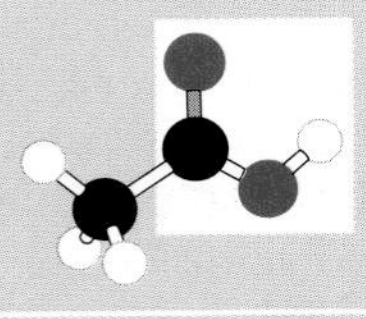
Amino —NH₂	Amines	$$\begin{matrix} & H \\ & | \\ R-&N \\ & \backslash \\ & H \end{matrix}$$	$$\begin{matrix} H & & H \\ | & & | \\ H-C-&&N \\ | & & \backslash \\ H & & H \end{matrix}$$ Methylamine	
Phosphate —OPO₃²⁻	Organic phosphates	$$\begin{matrix} & O \\ & \| \\ R-O-&P-O^- \\ & | \\ & O^- \end{matrix}$$	$$\begin{matrix} HO & \; O \\ \backslash & \| \\ & C \\ & | \\ H-&C-OH \quad\;\; O \\ & | \qquad\quad\; \| \\ H-&C-O-P-O^- \\ & | \qquad\quad | \\ & H \qquad\quad O^- \end{matrix}$$ 3-Phosphoglyceric acid	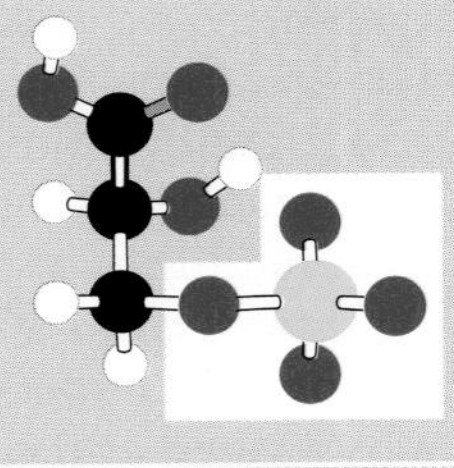
Sulfhydryl —SH	Thiols	R—SH	$$\begin{matrix} & H & H \\ & | & | \\ HO-&C-&C-SH \\ & | & | \\ & H & H \end{matrix}$$ Mercaptoethanol	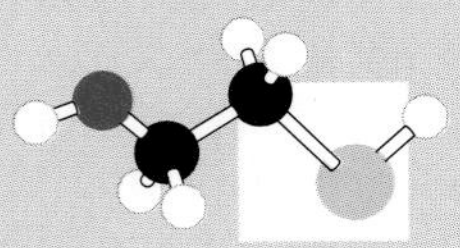

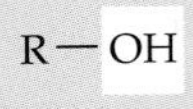

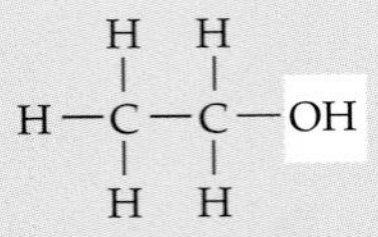

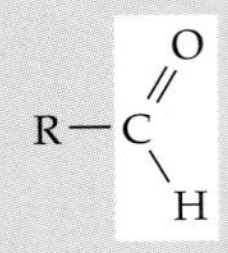

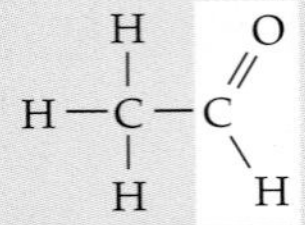

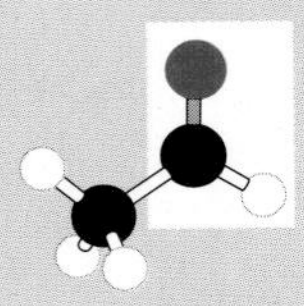

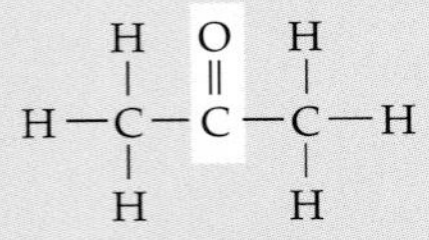

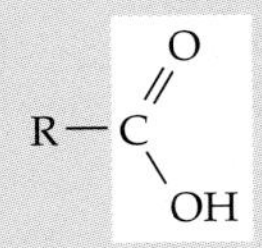

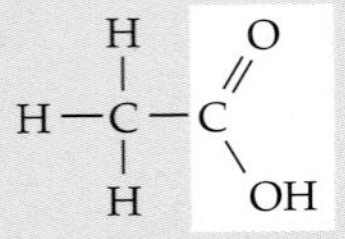

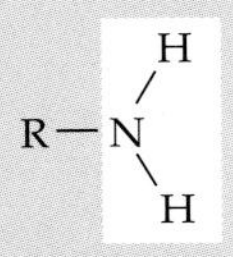

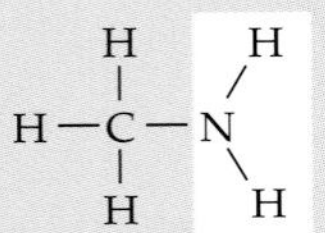

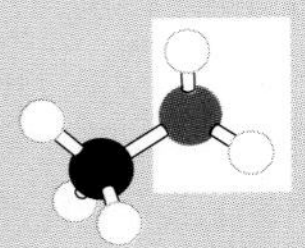

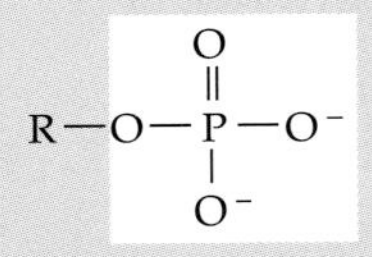

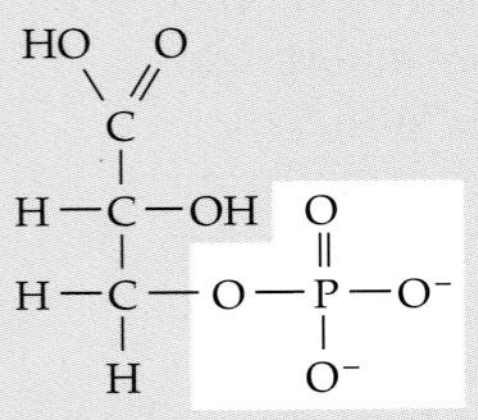

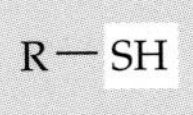

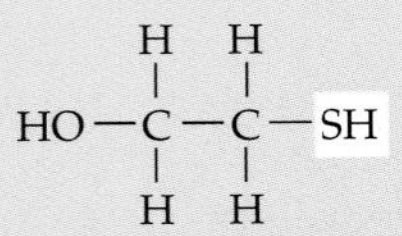

2.23 Some Functional Groups Important to Living Systems Compounds of the types shown here will appear throughout this book. The functional groups (highlighted) are the most common ones found in biologically important molecules. The letter *R* represents the remainder of the molecule, which may consist of any of a large number of carbon skeletons or other chemical groupings.

their biological roles. Chemists can use the characteristics of composition, structure (three-dimensional shape), reactivity, and solubility to distinguish a sample of one pure compound from another. For example, a compound such as sugar is soluble in water, which means that the solid disperses to form a uniform homogeneous mixture, but the same compound is insoluble in oil (the solid sugar remains a solid), and the resulting mixture is heterogeneous. Another substance, such as butane, is insoluble in water but highly soluble in oil. These solubility differences are due to the different atoms present and to their arrangement in these two kinds of molecules.

The presence of polar or charged sites on a molecule plays important roles in determining the molecule's solubility in water. Such sites can also determine the kinds of chemical reactions in which the molecule participates. The sizes, shapes, solubilities, and reactivities of molecules are significant to understanding the structures and operations of cells and organisms. Certain groups of atoms found together in a variety of molecules simplify our understanding of the reactions that molecules undergo.

Functional groups give specific properties to molecules

On the basis of atomic composition, structure, and reactivity, we can distinguish families of molecules from other families. Each member within a family of compounds has some characteristics in common with the other members of that family but differs from them in other characteristics. For example, organic acids are a family of carbon compounds that are all acidic but differ in other ways. They all contain a characteristic group of atoms called the **carboxyl group**,

$$\overset{\displaystyle O}{\overset{\displaystyle \|}{-\,C\,-\,OH}},\text{ or COOH}$$

that is the source of the H^+ that defines an acid:

$$\overset{\displaystyle O}{\overset{\displaystyle \|}{-\,C\,-\,OH}} \;\rightarrow\; \overset{\displaystyle O}{\overset{\displaystyle \|}{-\,C\,-\,O^-}} + H^+$$

The carboxyl group is a functional group. **Functional groups** are groups of atoms that are part of a larger molecule and have particular reactive characteristics. The same functional group may be part of very different molecules. In addition to the carboxyl group, you will encounter several other functional groups in your study of biology (Figure 2.23).

Several classes of biologically important compounds are defined by the functional groups they contain. When the functional group is a **hydroxyl group** (—OH), the product is an **alcohol**. Perhaps the most familiar alcohol is **ethanol** (also called ethyl alcohol, CH_3CH_2OH). Small alcohols like ethanol are soluble in water because of the hydrogen bonding possible between the polar hydroxyl group and water molecules, but larger alcohols are not soluble in water because of their long hydrocarbon chains.

Sugars contain both hydroxyl and carbonyl groups. The **carbonyl group** has a central carbon atom with a double bond to an oxygen atom ($C{=}O$). If one of the other two bonds of the carbon atom in a carbonyl group is attached to a hydrogen atom, the compound is an **aldehyde**.

Carboxyl groups are found in organic acids. Some organic bases, called amines, possess an **amino group** (—NH_2), which has a tendency to react with H^+ to produce the positively charged ammonium group (—NH_3^+). This H^+-accepting characteristic accounts for the classification of amines as bases.

Amino acids are important compounds that possess both a carboxyl group *and* an amino group attached to the same carbon atom, the α (alpha) carbon. Also attached to the α carbon atom are a hydrogen atom and a side chain designated by the letter R (Figure 2.24). Different side chains have different chemical compositions, structures, and properties. Each of the 20 amino acids found in proteins has a different side chain that gives it its distinctive chemical properties, as we'll see in Chapter 3. Because they possess both carboxyl and amino groups, amino acids are simultaneously acids and bases. At the pH values commonly found in cells, both the carboxyl and the amino groups are ionized: The carboxyl group has lost a proton, and the amino group has gained one (see Figure 2.24).

2.24 Amino Acid Structure Two depictions of the general structure of an amino acid. The side chain attached to the α carbon differs from one amino acid to another. At pH values found in living cells, both the carboxyl group and the amino group of an amino acid are ionized, as shown in the right-hand model.

Conventional depiction

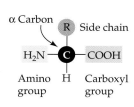

Three-dimensional depiction

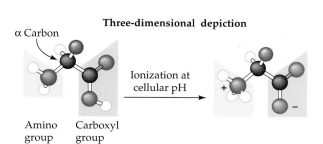

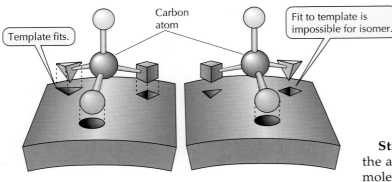

Template fits.

Carbon atom

Fit to template is impossible for isomer.

2.25 Optical Isomers Optical isomers are mirror images of each other. They result when four different groups are attached to a single carbon atom. If a template is laid out to match the groups on one carbon atom, the groups on the mirror-image isomer cannot be rotated to fit the same template.

Two other functional groups deserve our attention: the sulfhydryl group and the phosphate group (see Figure 2.23). The **sulfhydryl group** (—SH) is an important constituent of the side chains of two amino acids. When two of these sulfhydryl groups react, the two hydrogen atoms are lost and a covalent bond, called a disulfide bond, is formed between the sulfur atoms (—S—S—). As we'll see in the next chapter, disulfide bonds help maintain the three-dimensional structure of proteins that is required for normal protein functioning.

The **phosphate group** (—OPO_3^{2-}) is found on many different kinds of molecules. It plays important structural roles in DNA and RNA and participates in reactions that transfer energy. The removal (hydrolysis) of phosphate groups from some molecules releases energy that can be used to fuel other energy-requiring reactions.

Isomers have different arrangements of the same atoms

Isomers are compounds that have the same chemical formula but different arrangements of the atoms. (The prefix "iso-" means "same" and is encountered in many technical terms.) Of the different kinds of isomers, we will consider two: structural isomers and optical isomers.

Structural isomers are isomers that differ in how the atoms are joined together. Consider two simple molecules, each composed of four carbon and ten hydrogen atoms bonded covalently, with the formula C_4H_{10}. These atoms can be linked in two alternative ways in which carbon can form four bonds and hydrogen one bond. These two forms are called butane and isobutane. Their different bonding relationships are distinguished in structural formulas, and they have different chemical properties.

$$H_3C - \underset{\underset{H}{|}}{\overset{\overset{H}{|}}{C}} - \underset{\underset{H}{|}}{\overset{\overset{H}{|}}{C}} - CH_3 \qquad\qquad H_3C - \underset{\underset{H}{|}}{\overset{\overset{CH_3}{|}}{C}} - CH_3$$

Butane   Isobutane

Many molecules of biological importance, particularly the sugars and amino acids, have **optical isomers**. Optical isomers (also called enantiomers) are related to each other in the way an object is related to its mirror image (Figure 2.25). Optical isomers occur whenever a carbon atom has four *different* atoms or groups attached to it. These instances allow two different ways of making the attachments, each the mirror image of the other. Such a carbon atom is an asymmetric carbon, and the pair of compounds are optical isomers of each other. Your right and left hands are optical isomers. Just as a glove is specific for a particular hand, so some biochemical molecules can interact with a specific optical isomer of a compound but are unable to "fit" the other.

The α carbon in an amino acid is an asymmetric carbon because it is bonded to four different groups. Therefore, amino acids exist in two isomeric forms, called D-amino acids and L-amino acids (Figure 2.26). "D" and "L" are abbreviations for right and left, respectively. Only L-amino acids are commonly found in most proteins of living things.

The compounds discussed in this chapter include some of the more common ones found in organisms.

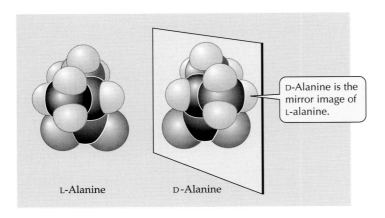

D-Alanine is the mirror image of L-alanine.

L-Alanine   D-Alanine

2.26 Isomers of an Amino Acid Space-filling models of the D and L forms of the amino acid alanine. Only L-alanine is commonly found in living things.

Between these small molecules and the world of the living stands another level, that of the giant macromolecules. These huge molecules—the proteins, lipids, carbohydrates, and nucleic acids—are the subject of the next chapter.

Summary of "Small Molecules: Structure and Behavior"

Atoms: The Constituents of Matter

• Matter is composed of atoms. Each atom is a positively charged nucleus of protons and neutrons surrounded by electrons bearing negative charges. Electrons are distributed in shells consisting of orbitals. Each orbital contains a maximum of two electrons. **Review Figures 2.4, 2.5**

• Isotopes of an element differ in their numbers of neutrons. Some isotopes are radioactive, emitting radiation as they decay. **Review Figure 2.2**

• In losing, gaining, or sharing electrons to become more stable, an atom can combine with other atoms to form molecules. **Review Figures 2.6, 2.7, 2.8**

Chemical Bonds: Linking Atoms Together

• Covalent bonds are strong bonds formed when two atomic nuclei share one or more pairs of electrons. Covalent bonds have spatial orientations that give molecules three-dimensional shapes. **Review Figure 2.7**

• Nonpolar covalent bonds occur when the electronegativity of both atoms is equal. When atoms with high electronegativity (e.g., oxygen) bond to atoms with weaker electronegativity (e.g., hydrogen), a polar covalent bond is formed in which one end is δ+ and the other is δ–. **Review Figure 2.8**

• Hydrogen bonds are weak electrical attractions that form between a δ+ hydrogen and a δ– nitrogen or oxygen atom in another molecule or in another part of a large molecule. Hydrogen bonds are abundant in water. **Review Figure 2.9**

• Ions are electrically charged bodies that form when an atom gains or loses one or more electrons. Ionic bonds are electrical attractions between oppositely charged ions. Ionic bonds are strong in solids, but weaker when the ions are separated from each other in a water solution. **Review Figures 2.10, 2.11, 2.12**

• Molecules with no electric charge are nonpolar and do not interact directly with polar or charged substances, including water. Nonpolar molecules are attracted to each other by very weak bonds called van der Waals attractions. In water, nonpolar molecules are pushed together because of cages of water molecules that form around them. **Review Figures 2.13, 2.14**

Eggs by the Dozen, Molecules by the Mole

• Molecular formulas identify the kind and number of atoms in a molecule; structural formulas tell which atoms are linked together.

• A mole of any substance equals its molecular weight in grams and contains the same number of molecules (6.023×10^{23}) as a mole of any other substance.

• Solutions are produced when substances dissolve in water. The concentration of a solution is the amount of substance in a given amount of solution.

Chemical Reactions: Atoms Change Partners

• In chemical reactions, substances change their atomic compositions and properties. Energy is released in some reactions, whereas in others energy must be provided. Neither matter nor energy is created or destroyed in a chemical reaction, but both change form.

• Combustion reactions are oxidation–reduction reactions in which a fuel is converted to carbon dioxide and water, and energy is released as heat and light. In living cells, the same kind of reaction takes place in multiple steps so that the energy released can be harvested for cellular activities. **Review Figure 2.16**

Water: Structure and Properties

• Water's molecular structure and its capacity to form hydrogen bonds give it unusual properties that are significant for life. Water is an excellent solvent; solid water floats in liquid water; and water loses a great deal of heat when it condenses from vapor or freezes, a property that moderates environmental temperature changes.

• The cohesion of water molecules permits liquid water to rise to great heights in narrow columns and produces a high surface tension. Water's high heat of vaporization assures effective cooling when water evaporates.

• The ionization of water molecules produces negatively charged hydroxide ions and positive hydrogen ions. These ions participate in many important chemical reactions. **Review Figures 2.18, 2.19**

Acids, Bases, and the pH Scale

• Acids are substances that donate hydrogen ions. Bases are substances that accept hydrogen ions.

• Hydrogen ion concentration is measured by the pH scale, which has values from less than 0 to more than 14.

• The pH of a solution is the negative logarithm of the hydrogen ion concentration. Values lower than pH 7 indicate an acidic solution; values above pH 7 indicate a basic solution. **Review Figure 2.21**

• Buffers are systems of weak acids and bases that limit the change in pH when H^+ ions are added or removed. **Review Figure 2.22**

The Properties of Molecules

• Molecules vary in size, shape, complexity, reactivity, solubility, and other chemical properties.

• Functional groups are part of a larger molecule and have particular chemical properties. The consistent chemical behavior of functional groups helps us understand the properties of molecules that contain them. **Review Figure 2.23**

• Structural and optical isomers have the same kinds and numbers of atoms, but differ in their structures and properties. **Review Figures 2.25, 2.26**

Self-Quiz

1. The atomic number of an element
   a. equals the number of neutrons in an atom.
   b. equals the number of protons in an atom.
   c. equals the number of protons minus the number of neutrons.
   d. equals the number of neutrons plus the number of protons.
   e. depends on the isotope.

2. The atomic weight (atomic mass) of an element
   a. equals the number of neutrons in an atom.
   b. equals the number of protons in an atom.
   c. equals the number of electrons in an atom.
   d. equals the number of neutrons plus the number of protons.
   e. depends on the relative abundances of its isotopes.

3. Which of the following statements about all the isotopes of an element is *not* true?
   a. They have the same atomic number.
   b. They have the same number of protons.
   c. They have the same number of neutrons.
   d. They have the same number of electrons.
   e. They have identical chemical properties.

4. Which of the following statements about a covalent bond is *not* true?
   a. It is stronger than a hydrogen bond.
   b. One can form between atoms of the same element.
   c. Only a single covalent bond can form between two atoms.
   d. It results from the sharing of electrons by two atoms.
   e. One can form between atoms of different elements.

5. Hydrophobic interactions
   a. are stronger than hydrogen bonds.
   b. are stronger than covalent bonds.
   c. can hold two ions together.
   d. can hold two nonpolar molecules together.
   e. are responsible for the surface tension of water.

6. Which of the following statements about water is *not* true?
   a. It releases a large amount of heat when changing from liquid into vapor.
   b. Its solid form is less dense than its liquid form.
   c. It is the most effective solvent known.
   d. It is typically the most abundant substance in an active organism.
   e. It takes part in some important chemical reactions.

7. A solution with a pH of 9
   a. is acidic.
   b. is more basic than a solution with a pH of 10.
   c. has 10 times the hydrogen ion concentration of a solution with a pH of 10.
   d. has a hydrogen ion concentration of 9 M.
   e. has a hydroxide ion concentration of 9 M.

8. Which of the following compounds is an alcohol?
   a. O_2
   b. $CH_3CH_2CH_2OH$
   c. CH_3COOH
   d. C_3H_8
   e. CH_3COCH_3

9. Which of the following statements about the carboxyl group is *not* true?
   a. It has the chemical formula —COOH.
   b. It is an acidic group.
   c. It can ionize.
   d. It is found in amino acids.
   e. It has an atomic weight of 45.

10. Which of the following statements about amino acids is *not* true?
    a. They are the building blocks of proteins.
    b. They contain carboxyl groups.
    c. They contain amino groups.
    d. They do not ionize.
    e. They have both L and D isomers.

Applying Concepts

1. Would you expect the elemental composition of Earth's crust to be the same as that of the human body? How could you find out? Try to find and discuss the answer.

2. Lithium (Li) is the element with atomic number 3. Draw the structures of the Li atom and of the Li^+ ion.

3. Draw the structure of a pair of water molecules held together by a hydrogen bond. Your drawing should indicate the covalent bonds.

4. The molecular weight of sodium chloride (NaCl) is 58.45. How many grams of NaCl are there in 1 liter of a 0.1 M NaCl solution? How many in 0.5 l of a 0.25 M NaCl solution?

5. The two optical isomers of alanine are shown in Figure 2.27. The side chain of the amino acid glycine is simply a hydrogen atom (—H). Are there two optical isomers of glycine? Explain.

Readings

Atkins, P. W. and L. L. Jones. 1997. *Chemistry: Molecules, Matter, and Change*, 3rd Edition. W. H. Freeman, Inc., New York. This and the next two listings are popular and accessible textbooks.

Brown, T. L., H. E. LeMay, Jr. and B. E. Bursten. 1997. *Chemistry: The Central Science*, 7th Edition. Prentice Hall, Englewood Cliffs, NJ.

Chang, R. 1994. *Chemistry*, 5th Edition. McGraw-Hill, New York.

Henderson, L. J. 1958. *The Fitness of the Environment*. Beacon Press, Boston. An essay written in 1912 about physical properties of water and carbon dioxide in relation to life. Includes a thought-provoking introduction.

Lancaster, J. R. 1992. "Nitric Oxide in Cells." *American Scientist*, vol. 80, pages 248–259. A readable account of the multiple biological effects of this very small molecule.

Mertz, W. 1981. "The Essential Trace Elements." *Science*, vol. 213, pages 1332–1338. A review of the roles of more than a dozen elements that are necessary in small amounts for animals to function normally.

Zumdahl, S. 1995. *Chemical Principles*, 2nd Edition. D. C. Heath, Lexington, MA. A higher-level text for students with strong math backgrounds.

Chapter 3

Macromolecules: Their Chemistry and Biology

A Gigantic Molecule
The protein actin, shown here in a computer-rendered graphic, is an example of a macromolecule. The thousands of atoms that make up the protein are shown in different colors.

The lives of cells are dances and dramas with tens of thousands of different kinds of molecules. Their dramatic choreography is what scientists reveal as they investigate the molecules of living systems, their physical properties, and their chemical reactions. What molecules are present? What are their chemical structures and properties? And what biological functions do these molecules perform? In different ways, we will be concerned with these questions throughout much of this book.

In this chapter, we'll look at gigantic molecules called macromolecules (*macro-*, "large") and the subunits from which they are constructed. Macromolecules may contain hundreds or thousands of atoms and have very large molecular weights. For example, human hemoglobin has a molecular weight of 64,500 and contains more than 6,000 atoms. The molecular complexity of such a structure can be understood in terms of the structures and properties of its subunits, which are fewer in number and easier to understand. Macromolecules are chains of small, individual units called monomers (*mono-*, "one"; *-mer*, "unit") covalently bonded together to form a polymer (*poly-*, "many").

In living systems there are three types of macromolecules: polysaccharides, proteins, and nucleic acids. *Polysaccharides* are constructed from sugar monomers such as glucose; *proteins* are constructed from amino acids; and *nucleic acids* are composed of nucleotides. Another important group of molecules that can form large structures is the *lipids*. But because the individual components

41

3.1 Building Blocks Diverse structures can be assembled from a few simple units. In a child's play, these units may be Lego® pieces. In the cell, these building blocks are monomers such as glucose, amino acids, and nucleotides.

(the lipids) do not covalently bond together, the resulting large structures are not true macromolecules.

The monomers that form the basis of polysaccharides, proteins, and nucleic acids are identical in different species. For example, a molecule of glucose in human blood is identical to a molecule of glucose from a reptile or a cabbage plant. The same group of 20 different amino acids is used to construct proteins for all living things—from bacteria to birds, bats, and humans. However, this cannot always be said of the larger molecular forms built from these monomers.

Proteins and nucleic acids in particular are informational macromolecules that are different in different species. For example, human hemoglobin is different from the hemoglobin found in fish or birds. No animal acquires its macromolecules directly from its food. Instead, it uses the subcomponents of its food to construct new macromolecules suited to its unique needs. For example, we eat proteins constructed by other animals and plants, but we break these proteins down into their amino acids and then reassemble the amino acids into the chemically different proteins of our own bodies. This process is like picking up Lego toys that somebody else has made, taking them apart, and putting the parts together again to make the toys *we* want (Figure 3.1).

Before we turn to the main focus of this chapter—the macromolecules: carbohydrates, proteins, and nucleic acids—let's first take a look at lipids and the large macromolecule-like structures that they form.

Lipids: Water-Insoluble Molecules

Lipids are a chemically diverse group of hydrocarbons. The property they all share is an insolubility in water that is due to the presence of many nonpolar covalent bonds. Nonpolar molecules can associate together and form massive structures, but these structures are not considered macromolecules, because the bonds between the separate molecules are not covalent bonds. As we saw in Chapter 2, these nonpolar hydrocarbon molecules are literally pushed together by surrounding water molecules, which are not attracted to the nonpolar substance. When the nonpolar molecules are sufficiently close together, weak but additive van der Waals forces hold them together. Although insoluble in water, lipids are soluble in other lipids or in nonpolar solvents such as ether or benzene.

In addition to their diverse chemical structures, lipids have many different biological roles. Some of them store energy (the fats and oils). Others play important structural roles in cell membranes (phospholipids) (Figure 3.2). The carotenoids help plants capture light energy, and the steroids and some lipids play regulatory roles. We'll begin our discussion by considering the structures and properties of the fats and oils.

Fats and oils store energy

Chemically, fats and oils are **triglycerides**, also known as *simple lipids*. Triglycerides that are solid at room temperature (20°C) are called **fats**; those that are liquid at room temperature are called **oils**. Triglycerides

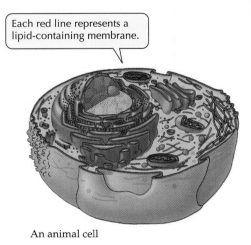

Each red line represents a lipid-containing membrane.

An animal cell

3.2 Lipids Form Cellular Membranes Membranes made of lipids separate the cell from its environment; they also separate the contents of internal compartments from the rest of the cell. Materials that do not dissolve in lipids usually cannot pass through membranes, but lipid-soluble materials move through with relative ease.

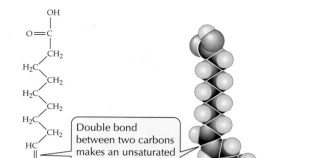

(a) Palmitic acid   (b) Oleic acid

All bonds between carbon atoms are single, making a saturated fatty acid.

Double bond between two carbons makes an unsaturated fatty acid.

3.3 Fatty Acids   (a) The straight-chain configuration seen in the model of palmitic acid is characteristic of saturated fatty acids. (b) The double bond between two carbons in the chain of an unsaturated fatty acid such as oleic acid causes the molecule to bend, obstructing the packing of the hydrocarbon chains.

are composed of two types of building blocks: fatty acids and glycerol. **Glycerol** is a small molecule with three hydroxyl (—OH) groups. **Fatty acids** have long nonpolar hydrocarbon tails and a polar carboxyl functional group (—COOH). Fatty acids can be of different lengths (usually 16 to 20 carbon atoms) and may be either saturated or unsaturated.

In **saturated fatty acids**, all the bonds between the carbon atoms in the hydrocarbon chain are single bonds—there are no double bonds. That is, all the bonds are *saturated* with hydrogen atoms (Figure 3.3a). Saturated fatty acids include palmitic acid, which has 16 carbon atoms, and stearic acid, which has 18. These molecules are relatively rigid and straight, and they pack together tightly, like pencils in a box.

In **unsaturated fatty acids**, the hydrocarbon chain contains one or more double bonds. Oleic acid is a monounsaturated fatty acid that

has 18 carbon atoms, and one double bond near the middle of the hydrocarbon chain causes a kink in the molecule (Figure 3.3b). Fatty acids, such as linoleic acid, that have more than one double bond are **polyunsaturated** and have multiple kinks. These kinks prevent the molecules from packing together tightly. The two fatty acids that humans cannot synthesize and must obtain from their diet are both unsaturated fatty acids. Diets favoring unsaturated fatty acids over saturated ones tend to reduce the incidence of arteriosclerosis, which is a narrowing of the arterial wall that restricts blood flow to the heart and may lead to heart attacks (see Chapter 46). Diets rich in saturated fatty acids have the opposite effect.

Three fatty acid molecules combine with a molecule of glycerol to form a molecule of a triglyceride (Figure 3.4). The carboxyl group of each fatty acid reacts with a hydroxyl group of the glycerol to form an

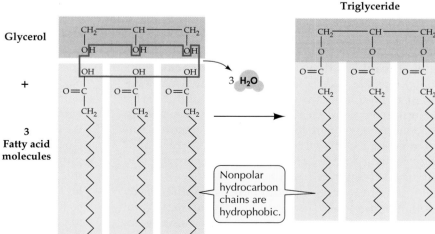

3.4 A Triglyceride and Its Components   Three fatty acid molecules and a molecule of glycerol react to form a triglyceride molecule and three molecules of water (H_2O). The long hydrocarbon chains from the fatty acid make triglyceride a very hydrophobic substance. In living things the reaction is more complex, but the end result is as shown here.

Glycerol

+

3 Fatty acid molecules

3 H_2O

Triglyceride

Nonpolar hydrocarbon chains are hydrophobic.

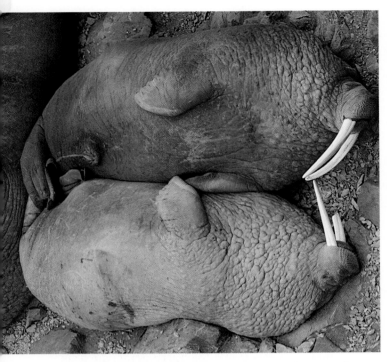

3.5 Energy to Fight Cold Temperatures   These Alaskan walrus spend much of their time in frigid water. Layers of body fat insulate their bodies against the cold and store energy efficiently.

ester (the reaction product of an acid and an alcohol) and water. The ester linkage looks like this:

$$\overset{\overset{\textstyle O}{\|}}{-C} - O - C -$$

The three fatty acids in one triglyceride molecule need not all have the same length, nor do they all have to be either saturated or unsaturated. The kinks associated with double bonds are important in determining the fluidity and melting point of a lipid. Animal fats such as lard and tallow are usually solids, and their triglycerides tend to have many long-chain saturated fatty acids. These fats are solids at room temperature because their fatty acids pack well together. The triglycerides of plants tend to have short or unsaturated fatty acids. Because of their kinks, these fatty acids pack poorly together and these triglycerides are usually liquid at room temperature. Natural peanut butter, for example, contains a great deal of oil. However, to market a more convenient and solid product, manufacturers often subject the oil to hydrogenation, which adds hydrogens to double bonds and converts unsaturated fatty acids to saturated fatty acids.

Lipids are marvelous storehouses for energy. By taking in excess food, many animal species deposit fat (lipid) droplets in their cells as a means for storing en-

ergy (Figure 3.5). Some plant species, such as olives, avocados, sesame, castor beans, and all nuts, have substantial amounts of lipids in their seeds or fruits that serve as energy reserves for the next generation.

Phospholipids form the core of biological membranes

Because lipids and water do not interact, a mixture of water and lipids forms two distinct phases. Many biologically important substances—such as ions, sugars, and free amino acids—that are soluble in water are insoluble in lipids. These two properties—water solubility and lipid insolubility—have proved essential to distinguishing the interior of cells from their external environment.

Suppose that you must design water-filled compartments, separated from each other and from their environment by barriers that limit the passage of materials. Given the properties of lipids, a seemingly effective way to accomplish this task is to use membranes that contain a special class of lipid called the phospholipid (Figure 3.6). This is the system that has evolved in nature. Molecular traffic within an organism or into and out of its compartments is constrained by the properties of the lipid portion of the surrounding membrane (see Figure 3.2). Compounds that dissolve readily in lipids can move rapidly through biological membranes, but compounds that are insoluble in lipids are prevented from passing through the membrane or must be transported across the membrane by specific proteins.

Like triglycerides, **phospholipids** have fatty acids bound to glycerol by ester linkages. In phospholipids, however, any one of several phosphate-containing compounds may replace one of the fatty acids (see Figure 3.6). Many phospholipids are important constituents of biological membranes. If you think carefully about the structure and properties of phospholipids, you will find it easy to understand how they are oriented in membranes. The phosphate functional group has negative electric charges, so this portion is **hydrophilic,** attracting polar water molecules. But the two fatty acids are **hydrophobic**, so they are pushed together by water (see Chapter 2).

In a biological membrane, phospholipids line up in such a way that the nonpolar, hydrophobic "tails" pack tightly together to form the interior of the membrane, and the phosphate-containing "heads" face outward (some to one side of the membrane and some to the other), where they interact with water, which is excluded from the interior of the membrane. The phospholipids thus form a bilayer, a sheet two molecules thick (Figure 3.7). Because membranes are so important, we will devote all of Chapter 5 to them.

Because the word "lipid" defines compounds in terms of their solubility rather than their structural

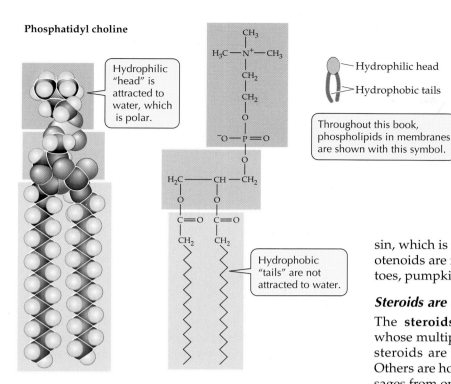

Phosphatidyl choline

Hydrophilic "head" is attracted to water, which is polar.

CH_3
H_3C — N^+ — CH_3
CH_2
CH_2
O
^-O — P = O
O
H_2C — CH — CH_2
O O
C = O C = O
CH_2 CH_2

Hydrophilic head

Hydrophobic tails

Throughout this book, phospholipids in membranes are shown with this symbol.

Hydrophobic "tails" are not attracted to water.

3.6 Phospholipid Structure Every phospholipid molecule has two hydrophobic "tails" and a hydrophilic "head." The two fatty acids bonded to a glycerol provide the hydrophobic tails, while the phosphorus-containing compound linked to the third —OH of the glycerol provides the hydrophilic head. In phosphatidyl choline, the hydrophilic head consists of phosphate and choline. In other phospholipids, the amino acid serine, the sugar alcohol inositol, or another compound can replace choline.

similarity, a great variety of different chemical structures are included as lipids. The next two lipid classes we'll discuss—the carotenoids and the steroids—have chemical structures very different from the structures of triglycerides and phospholipids and from the structures of each other.

Carotenoids trap light energy

The **carotenoids** are a family of light-absorbing pigments found in plants and animals (Figure 3.8). Beta-carotene (β-carotene) is one of the pigments that traps light energy in leaves during photosynthesis (see Chapter 8). It is the β-carotene in plants that senses light and causes their parts to grow toward or away from the light (a behavior called phototropism, discussed in Chapter 35). In humans, a molecule of β-carotene can be broken down into two vitamin A molecules, from which we make the pigment rhodop-

sin, which is required for vision (see Chapter 42). Carotenoids are responsible for the color of carrots, tomatoes, pumpkins, egg yolks, and butter.

Steroids are signal molecules

The **steroids** are a family of organic compounds whose multiple rings share carbons (Figure 3.9).Some steroids are important constituents of membranes. Others are hormones, chemical signals that carry messages from one part of the body to another. *Testosterone* and the *estrogens* are steroid hormones that regulate sexual development in vertebrates. *Cortisol* and related hormones play many regulatory roles in the digestion of carbohydrates and proteins, in the maintenance of salt balance and water balance, and in sexual development.

Cholesterol is synthesized in the liver and contributes to the structure of some cellular membranes. It is the starting material for making testosterone and other steroid hormones, as well as the bile salts that help break down dietary fats so that they can be digested. We absorb cholesterol from foods such as milk, butter, and animal fats. When we have too much cholesterol in our blood, it is deposited in our arteries (along with other substances), a condition that may lead to arteriosclerosis and heart attack.

3.7 Phospholipids in Biological Membranes In a water environment, hydrophobic interactions bring the "tails" of phospholipids together in the interior of a phospholipid bilayer. The hydrophilic "heads" face outward on both sides of the membrane, where they interact with the surrounding water molecules.

Water

Water

Hydrophilic "head"

Hydrophobic fatty acid "tails"

Hydrophilic "head"

Phospholipid bilayer of biological membrane

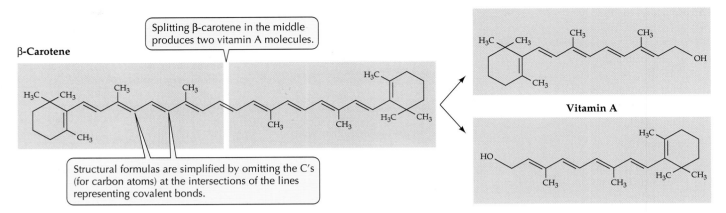

β-Carotene

Splitting β-carotene in the middle produces two vitamin A molecules.

Structural formulas are simplified by omitting the C's (for carbon atoms) at the intersections of the lines representing covalent bonds.

Vitamin A

3.8 β-Carotene Is the Source of Vitamin A The carotenoid β-carotene is symmetrical around the central double bond; when split, β-carotene becomes two vitamin A molecules. The simplified structural formula is standard chemical shorthand for large structures with many carbon atoms.

Some lipids are vitamins

A large group of fat-soluble substances, including the carotenoids and steroids, are synthesized by covalent linking and chemical modification of isoprene to form a series of isoprene units:

$$CH_2=C-CH=CH_2$$

with CH_3 on the central carbon

The fat-soluble vitamins (A, D, E, and K) are formed in this manner by plants and bacteria. These substances are not synthesized by humans and must be acquired from dietary sources. **Vitamin A** is formed from β-carotene found in green and yellow vegetables (see Figure 3.8). Among other roles, vitamin A is directly involved in the reception of light by our eyes. Deficiency in vitamin A leads to dry skin, eyes, and internal body surfaces; retarded growth and development; and night blindness, which is a diagnostic symptom for the deficiency. **Vitamin D** regulates the absorption of calcium from the intestines. It is necessary for the proper deposition of calcium in bones; a deficiency of vitamin D can lead to rickets, a bone-softening disease.

Vitamin E is not a single vitamin, but a group of related lipids that seem to protect cells from damaging effects of oxidation–reduction reactions. These lipids appear to have an important role in preventing unhealthy changes in the double bonds in the unsaturated fatty acids of membrane phospholipids. Commercially, vitamin E is added to some foods to slow spoilage. **Vitamin K** is found in green leafy plants and is also synthesized by bacteria normally present in the human intestine. This vitamin is essential to the formation of blood clots. Predictably, a deficiency of vitamin K leads to slower clot formation and potentially fatal bleeding from a wound.

Because of their insolubility in water, in the body some lipids may require water-soluble carrier proteins in order to be transported in the blood, which is mostly water. Among the lipids, only triglycerides and phospholipids assemble into large structures in cells:

3.9 All Steroids Have the Same Ring Structure These steroids, all important in vertebrates, are composed of carbon and hydrogen and are highly hydrophobic. However, small chemical variations, such as the presence or absence of a methyl or hydroxyl group, can produce enormous functional differences.

Cholesterol is a constituent of membranes and the source of steroid hormones.

Vitamin D₂ can be produced in the skin by the action of light on a cholesterol derivative.

Cortisol is a hormone secreted by the adrenal glands.

Testosterone is a male sex hormone.

triglycerides form droplets of stored fat, and phospholipids form the bilayers of membranes. But in spite of their size, these droplets and bilayers are not usually considered macromolecules.

Now that we have seen the macromolecule-like structures that lipids form, let's turn to the large structures that constitute true macromolecules.

Macromolecules: Giant Polymers

As noted earlier, **macromolecules** are giant *polymers* constructed by the covalent linking of smaller molecules called *monomers*. These monomers may or may not be identical, but they always have similar chemical structures. Molecules with molecular weights exceeding 1,000 are usually considered macromolecules, and the polysaccharides, proteins, and nucleic acids of living systems certainly fall into this category.

In the cell, each type of macromolecule performs some combination of a diversity of functions: energy storage, structural support, protection, catalysis, transport, defense, regulation, movement, and heredity. These roles are not necessarily exclusive. For example, both carbohydrates and proteins can play structural roles—supporting and protecting cells and organisms. However, only nucleic acids specialize in information and function as hereditary material, carrying both species and individual traits from generation to generation.

The functions of macromolecules are directly related to their shapes and the chemical properties of their monomers. Some macromolecules, such as catalytic and defensive proteins, fold into compact spherical forms with surface features that make them water-soluble and capable of intimate interaction with other molecules. Other proteins and carbohydrates form long, fibrous systems that provide strength and rigidity to cells and organisms. Still other long, thin assemblies of proteins can contract and cause movement.

Because macromolecules are so large, they contain many different functional groups. For example, a large protein may contain hydrophobic, polar, and charged groups. These groups give specific properties to local sites on a macromolecule. As we will see, this diversity of properties determines the shapes of macromolecules and their interactions with both other macromolecules and smaller molecules.

Macromolecules form by condensation reactions

The polymers of living things are constructed by a series of reactions called **condensation reactions,** or dehydration reactions (both words refer to the loss of water). Condensation reactions covalently bond monomers. Water is lost in the reaction:

$$A\text{—}H + B\text{—}OH \rightarrow A\text{—}B + H_2O$$

A—H is a molecule with a reactive hydrogen; B—OH is a molecule with a hydroxyl group. The products of the reaction are A—B and H_2O. The atoms that make up the water molecule are derived from the reactants: one hydrogen atom from one reactant, and an oxygen atom and the other hydrogen atom from the other reactant. Condensation reactions are not limited to polymer formation; for example, we encountered them in the linkage (esterification) of fatty acids to glycerol (see Figure 3.4).

The condensation reactions that produce the different kinds of macromolecules differ in detail, but in all cases polymers will form only if energy is added to the system. In living systems, specific energy-rich molecules supply the energy, and there are additional steps to the reaction.

The reverse of a condensation reaction is a **hydrolysis reaction** (*hydro-*, "water"; *-lysis,* "breakage"). These reactions digest polymers and produce monomers. Water reacts with the bonds that link the monomers together, and the products are free monomers. The elements of the reactant H_2O become part of the products. Hydrolysis reactions of all sorts are very important in cellular functioning and are not limited to the digestion of polymers.

Carbohydrates: Sugars and Sugar Polymers

Carbohydrates are a diverse group of compounds based on the general formula CH_2O. Some are relatively small, with molecular weights less than 100. Others are true macromolecules, with molecular weights of hundreds of thousands. There are four categories of biologically important carbohydrates. *Monosaccharides* (*mono-*, "one"; *saccharide*, "sugar")—such as glucose or fructose—are the monomers out of which the larger forms are constructed. *Disaccharides* (*di-*, "two") consist of two monosaccharides. *Oligosaccharides* (*oligo-*, "several") have several monosaccharides (3 to 20). *Polysaccharides* (*poly-*, "many")—such as starch, glycogen, and cellulose—are composed of hundreds of thousands of glucose units.

The relative proportions of carbon, hydrogen, and oxygen indicated by the general formula for carbohydrates, CH_2O, is true for monosaccharides. However, for disaccharides, oligosaccharides, and polysaccharides, these proportions differ slightly from this general formula because two hydrogens and an oxygen are lost during the condensation reactions.

Monosaccharides are simple sugars

All living cells contain the monosaccharide **glucose**, whose formula is $C_6H_{12}O_6$. Green plants produce glucose by photosynthesis, and other organisms acquire it directly or indirectly from plants. Cells use glucose as an energy source, changing it through a series of reac-

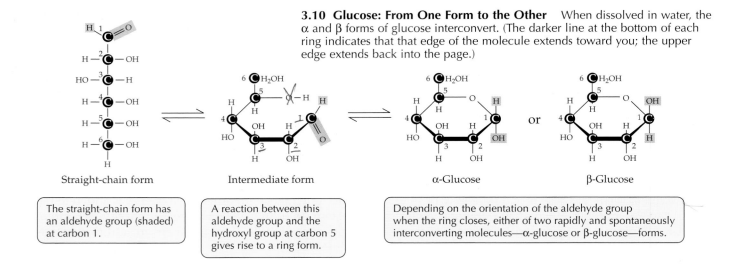

3.10 Glucose: From One Form to the Other   When dissolved in water, the α and β forms of glucose interconvert. (The darker line at the bottom of each ring indicates that that edge of the molecule extends toward you; the upper edge extends back into the page.)

Straight-chain form

The straight-chain form has an aldehyde group (shaded) at carbon 1.

Intermediate form

A reaction between this aldehyde group and the hydroxyl group at carbon 5 gives rise to a ring form.

α-Glucose   or   β-Glucose

Depending on the orientation of the aldehyde group when the ring closes, either of two rapidly and spontaneously interconverting molecules—α-glucose or β-glucose—forms.

tions that release stored energy and produce water and carbon dioxide.

Two forms of glucose, the straight chain and the ring, exist in equilibrium with each other when dissolved in water, but the ring form predominates (>99%) (Figure 3.10). The two distinct ring forms (α- and β-glucose) differ only in the placement of the —H and —OH attached to carbon 1. (The convention for numbering carbons shown in Figure 3.10 is used throughout this book.)

Most of the monosaccharides found in living systems belong to the D series of optical isomers (see Chapter 2). But there are structural isomers—composed of the same kinds and numbers of atoms, but with the atoms combined differently in each. All **hex-**

oses (*hex-*, "six"), a group of structural isomers, have the formula $C_6H_{12}O_6$. Included among the hexoses are fructose (so named because it was first found in fruits), mannose, and galactose (Figure 3.11).

Pentoses (*pent-*, "five") are five-carbon sugars. Some pentoses are found primarily in the cell walls of plants, as are several of the hexoses. Two pentoses are of particular importance: **Ribose** and **deoxyribose** form part of the backbones of RNA and of DNA, respectively. These two pentoses are not isomers; rather, one oxygen atom is missing from carbon 2 in deoxyribose (*de-*, "absent") (see Figure 3.11). As we will see in Chapter 12, the absence of this oxygen atom has enormous consequences for the functional distinction of RNA and DNA.

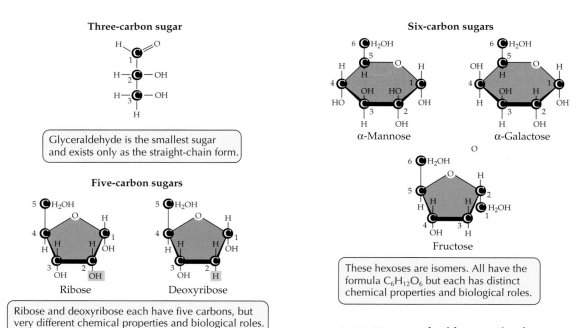

Three-carbon sugar

Glyceraldehyde is the smallest sugar and exists only as the straight-chain form.

Five-carbon sugars

Ribose

Deoxyribose

Ribose and deoxyribose each have five carbons, but very different chemical properties and biological roles.

Six-carbon sugars

α-Mannose

α-Galactose

Fructose

These hexoses are isomers. All have the formula $C_6H_{12}O_6$ but each has distinct chemical properties and biological roles.

3.11 Monosaccharides Are Simple Sugars

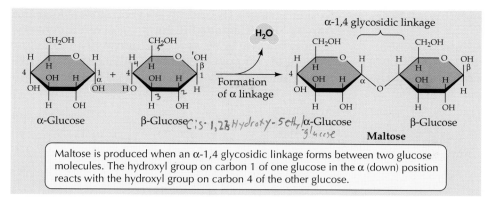

α-1,4 glycosidic linkage

α-Glucose   β-Glucose   Formation of α linkage   α-Glucose   β-Glucose
Maltose

Maltose is produced when an α-1,4 glycosidic linkage forms between two glucose molecules. The hydroxyl group on carbon 1 of one glucose in the α (down) position reacts with the hydroxyl group on carbon 4 of the other glucose.

3.12 Disaccharides Are Formed by Glycosidic Linkages

Glycosidic linkages between two monosaccharides create many different disaccharides. Which disaccharide is formed depends on the nature of the linked monosaccharides (glucose, galactose, fructose) and the site (carbon atoms) and form (α or β) of the linkage. These are condensation reactions, creating a water molecule as each linkage forms.

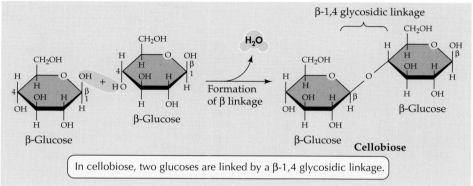

β-1,4 glycosidic linkage

β-Glucose   β-Glucose   Formation of β linkage   β-Glucose   β-Glucose
Cellobiose

In cellobiose, two glucoses are linked by a β-1,4 glycosidic linkage.

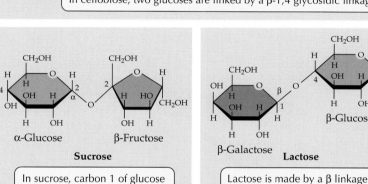

α-Glucose   β-Fructose
Sucrose

In sucrose, carbon 1 of glucose is joined by an α-1,2 linkage to carbon 2 of fructose.

β-Galactose   β-Glucose
Lactose

Lactose is made by a β linkage between carbon 1 of galactose and carbon 4 of glucose.

Glycosidic linkages bond monosaccharides together

Monosaccharides are covalently bonded together by condensation reactions that form **glycosidic linkages.** Such a linkage between two simple sugars forms a **disaccharide** (Figure 3.12). For example, a molecule of *sucrose* (table sugar)—the sugar that is transported to different parts of plants—is formed from a glucose and a fructose molecule, while *lactose* (milk sugar) contains glucose and galactose. The disaccharide *maltose* contains two glucose molecules.

But maltose is not the only disaccharide that can be made from two glucoses. When glucose molecules form glycosidic linkages, as shown in Figure 3.12, the disaccharide product must be one of two types: α-linked or β-linked, depending on whether the molecule that bonds by its 1 carbon is α-glucose or β-glucose. An α

linkage with carbon 4 of a second glucose molecule gives *maltose*, whereas a β linkage gives *cellobiose*. Maltose and cellobiose are disaccharide isomers, and both have the formula $C_{12}H_{22}O_{11}$. However, they are different compounds with different properties. They undergo different chemical reactions and are recognized by different catalytic proteins (enzymes).

Oligosaccharides contain several monosaccharides linked by glycosidic linkages at various sites. Many oligosaccharides gain additional functional groups, which give them special properties. Oligosaccharides are often covalently bonded to proteins and lipids on the outer cell surface, where they serve as cell recognition signals.

Polysaccharides are energy stores or structural materials

Polysaccharides are giant chains of monosaccharides connected by glycosidic linkages. **Starch** is a polysaccharide of glucose with linkages in the α orientation. **Cellulose** is a similar giant polysaccharide made up solely of glucose, but its individual units are connected by β linkages instead of α linkages (Figure 3.13*a*). Cellulose is the predominant component of plant cell walls and is by far the most abundant organic compound on this planet. Both starch and cellulose are composed of nothing but glucose, but their biological functions and chemical and physical properties are entirely different.

Starch can be more or less easily degraded by the actions of chemicals or catalytic proteins (enzymes). Cellulose, however, is very stable because of its β glycosidic linkages. Thus starch is a good storage form that can be easily degraded to supply glucose for energy-producing reactions, while cellulose is an excellent structural component that can withstand harsh environmental conditions without changing.

(a) **Molecular structure**

Cellulose

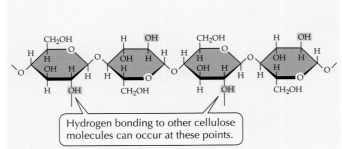

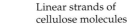

Hydrogen bonding to other cellulose molecules can occur at these points.

Cellulose is an unbranched polymer of glucose with β-1,4 glycosidic linkages that are chemically very stable.

Starch and glycogen

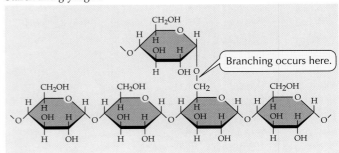

Branching occurs here.

Glycogen and starch are polymers of glucose with α-1,4 glycosidic linkages. α-1,6 glycosidic linkages produce branching at carbon 6.

(b) **Macromolecular structure**

Linear strands of cellulose molecules

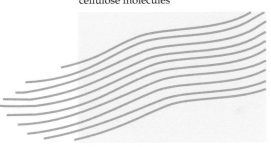

Parallel cellulose molecules hydrogen-bond together to form long thin fibrils.

Branched starch molecule

Branching limits the number of hydrogen bonds that can form in starch molecules, making starch less compact than cellulose.

Highly branched glycogen molecule

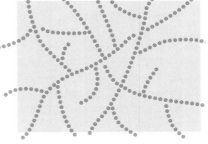

The high amount of branching in glycogen makes its solid deposits less compact than starch.

(c) **Polysaccharides in cells**

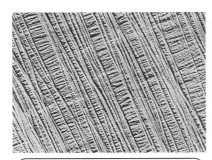

Layers of cellulose fibrils, as seen in this scanning electron micrograph, give plant cell walls great strength.

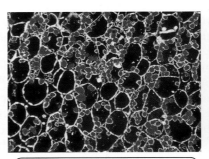

Dyed red in this micrograph, starch deposits have a large granular shape within cells.

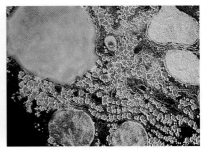

Colored pink in this electron micrograph of human liver cells, glycogen deposits have a small granular shape.

3.13 Representative Polysaccharides Cellulose, starch, and glycogen demonstrate different levels of branching and compaction in polysaccharides.

Starch is not just one chemical substance, but a large family of giant molecules of broadly similar structure. All starches are large polymers of glucose with α linkages, but different starches can be distinguished by the amount of branching between carbons 1 and 6 (Figure 3.13*b*). Some starches are highly branched; others are not. Plant starches, called **amylose**, are not highly branched. The polysaccharide **glycogen**, which stores glucose in animal livers and muscles, is highly branched.

What do we mean when we say that starch and glycogen are storage compounds for energy? Very simply, these compounds can readily be hydrolyzed to yield glucose monomers. Glucose, in turn, can enter into a series of reactions that liberate its stored energy in forms that can be used for cellular activities. However, glucose can also serve the cell's need for raw materials—that is, carbon atoms. Glucose can undergo other chemical reactions that disassemble parts of the molecule and rearrange its carbon atoms to form the skeletons of other compounds needed by cells. Glycogen and starch are thus storage depots for carbon atoms as well as for energy.

Derivative carbohydrates contain other elements

Sometimes carbohydrates are modified by chemical changes in their structure or by the addition of functional groups such as phosphate and amino groups, thus becoming **derivative carbohydrates** (Figure 3.14). For example, carbon 6 in glucose may be oxidized from —CH$_2$OH to a carboxyl group (—COOH), producing glucuronic acid. Or a phosphate group may be added to one or more of the —OH sites. Some of these *sugar phosphates*, such as fructose 1,6-bisphosphate, are important intermediates in cellular energy reactions (see Chapters 7 and 8). When an amino group is substituted for an —OH group, *amino sugars* such as glucosamine and galactosamine are produced. Galactosamine is a major component of cartilage, the material that forms caps on the ends of bones and stiffens the protruding parts of the ears and nose. A derivative of glucosamine produces the polymer **chitin**, which is the principal structural polysaccharide in the skeletons of insects, crabs, and lobsters, as well as in the cell walls of fungi. Fungi and insects (and their relatives) constitute more than 80 percent of the species ever described, and chitin is one of the most abundant substances on Earth.

Proteins: Amazing Polymers of Amino Acids

Proteins are an extraordinary group of macromolecules. Constructed of amino acids, proteins are fascinating because they have such intricate and diverse structures and because they perform so many different functions for cells. Among the cellular functions of macromolecules listed earlier, only energy storage and

(*a*) **Sugar phosphate**

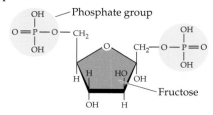

Fructose 1,6-bisphosphate is involved in the reactions that liberate energy from glucose. (The numbers in its name refer to the carbon sites of phosphate bonding; *bis-* indicates that two phosphates are present.)

(*b*) **Amino sugars**

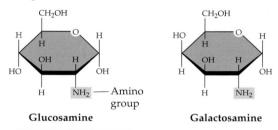

The monosaccharides glucosamine and galactosamine are amino sugars with an amino group in place of a hydroxyl group.

(*c*) **Chitin**

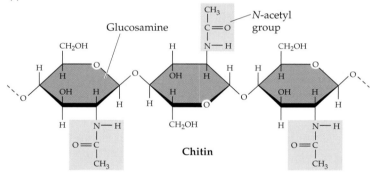

Chitin is a polymer of *N*-acetylglucosamine; *N*-acetyl groups provide additional sites for hydrogen bonding between the polymers.

3.14 Derivative Carbohydrates Added functional groups modify the form and properties of a carbohydrate.

heredity are not functions of proteins. Proteins are involved in structural support, protection, catalysis, transport, defense, regulation, and movement. Of particular importance are the catalytic proteins, called **enzymes**, that increase the rates of chemical reactions in cells. In general, each chemical reaction requires a different enzyme, because proteins show great specificity for the smaller molecules with which they will interact.

Proteins range in size from the small RNA-digesting enzyme *ribonuclease A*, which has a molecular weight of 5,733 and 51 amino acid residues, to gigantic molecules such as the cholesterol transport protein *apolipoprotein B*, which has a molecular weight of 513,000 and 4,636 amino acid residues. (The word "residue" is used for a monomer when it is part of a polymer.) Each of these proteins consists of one chain of amino acids folded into a specific three-dimensional shape that is required for protein function. Some proteins have more than one polymer chain. For example, the oxygen-carrying protein *hemoglobin* has four chains that are folded separately and associate together to make the functional protein. The largest protein complex known has more than 40 separate chains of amino acids.

All of these different proteins have a characteristic amino acid composition, but every protein contains neither all 20 amino acids nor an equal number of different amino acids. Nor is there a simple regular sequence in which the amino acids are linked. The diversity in amino acid content and sequence is the source of the diversity in protein structures and functions.

In some proteins, different kinds of chemical substances, called **prosthetic groups**, may be attached to the protein. These prosthetic groups include carbohydrates, lipids, phosphate groups, the iron-containing heme group, and metal ions such as copper and zinc. Whether they are small or large, and whether or not they have prosthetic groups, there is nothing "casual" about the structures of proteins. Proteins have specific three-dimensional shapes that are necessary for their specific functions.

To understand this stunning variety of functions, we must explore protein structure. First, we will examine the properties of the 20 amino acids and the characteristics of how they link to form proteins. Then we will systematically examine the four levels of protein structure and look at how a linear chain of amino acids is consistently folded into a compact three-dimensional shape.

Proteins are composed of amino acids

The 20 different amino acids commonly found in proteins show a wide variety of properties. In Chapter 2, we considered the structure of amino acids and identified four different groups attached to a central carbon atom: a hydrogen atom, an amino group, a carboxyl group, and a side chain, or R group (see Figure 2.24). Since amino acids are linked together by reactions between their amino and carboxyl groups, these groups are not exposed and do not give the protein its distinguishing properties and functional specificity—the side chains fill this role.

The **side chains** of amino acids are a protein's reactive groups and show a wide variety of chemical properties. Side chains control the function of a protein and contribute to its structure. The sequence of side chains

determines how a protein folds into a three-dimensional shape (which we will discuss shortly). Although side chains are very important, they are often omitted from diagrams of proteins when the focus is on other aspects of structure. In these cases, the side chains are identified by the letter *R* (for "residue") and are sometimes called R groups. Side chains are highlighted in white in Table 3.1.

As Table 3.1 shows, one useful classification of amino acids is based on whether their side chains are electrically charged (+1, –1), polar ($\delta+$, $\delta-$), or nonpolar and hydrophobic. The five amino acids that have electrically charged side chains attract water and oppositely charged ions of all sorts. The four amino acids that have polar side chains tend to form weak hydrogen bonds with water and with other polar or charged substances. Eight amino acids have side chains that are nonpolar hydrocarbons or very slightly modified hydrocarbons. In the watery environment of the cell, the hydrophobic side chains may cluster together.

Three amino acids—cysteine, glycine, and proline—are special cases, although their side chains are generally hydrophobic. Two *cysteine* side chains, which have terminal —SH groups, can react to form a covalent bond in a **disulfide bridge** (—S—S—) (Figure 3.15). Hydrogen bonds and disulfide bridges help determine how a protein chain folds. When cysteine is not part of a disulfide bridge, its side chain is very hydrophobic. The *glycine* side chain consists of a single hydrogen atom; thus glycines may fit into tight corners in the interior of a protein molecule, where a larger side chain could not fit. *Proline* differs from other amino acids because it possesses a modified amino group (see Table 3.1).

Peptide linkages covalently bond amino acids together

When amino acids polymerize, the carboxyl group of one amino acid reacts with the amino group of another, undergoing a condensation reaction that forms a **peptide linkage**. Figure 3.16 gives a simplified description of the reaction. (In living cells, other molecules must activate the reactants in order for this reaction to proceed, and there are intermediate steps.) A linear polymer of amino acids connected by peptide linkages is a **polypeptide**. A **protein** is made up of one or more polypeptides.

At one end of the polypeptide is a free amino group. This end is the N terminus, named for the nitrogen atom in the amino group. At the other end of the polypeptide—the C terminus—is a free carboxyl group. The other amino and carboxyl groups are bound in peptide linkages. Thus a protein has direction. For example, the dipeptide glycine–alanine, in which glycine has the free amino group, differs from alanine–glycine, in which alanine has the free amino

TABLE 3.1 Twenty amino acids found in proteins

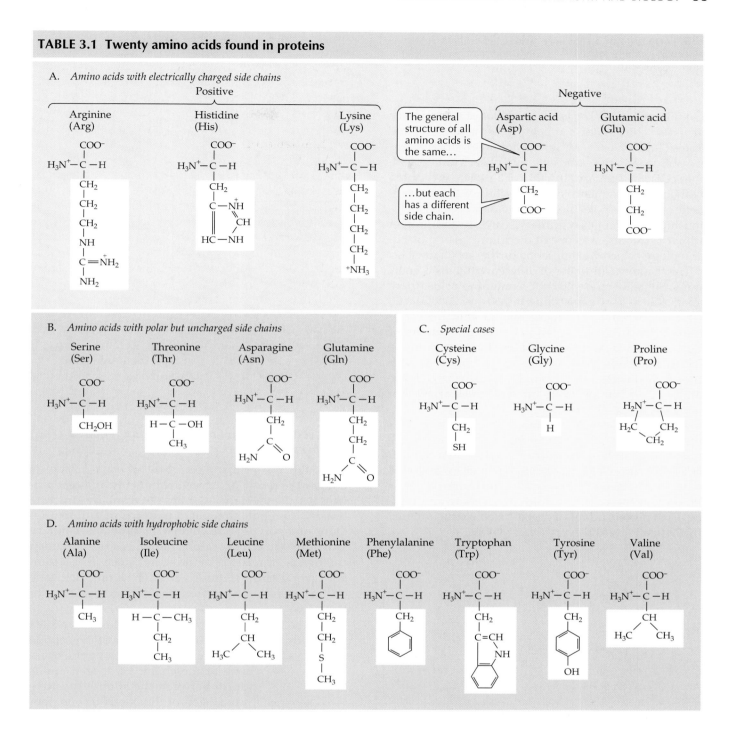

A. *Amino acids with electrically charged side chains*

B. *Amino acids with polar but uncharged side chains*

C. *Special cases*

D. *Amino acids with hydrophobic side chains*

group. In cells, the synthesis of polypeptide chains begins with the N terminus.

In the peptide linkage, the C=O oxygen carries a slight negative charge ($\delta-$), whereas the N—H hydrogen is slightly positive ($\delta+$). This asymmetry of charge favors hydrogen bonding (see Chapter 2) within the protein molecule itself and with other molecules, contributing to both the structure and the function of many proteins.

The primary structure of a protein is its amino acid sequence

Protein structure is elegant and complex—so complex that it is described as consisting of four different levels: *primary*, *secondary*, *tertiary*, and *quaternary* (Figure 3.17). However, proteins are always linear chains; there are no branches. The precise sequence of amino acids in a polypeptide constitutes the **primary structure** of a pro-

3.15 Formation of a Disulfide Bridge Disulfide bridges (—S—S—) are important in maintaining the proper three-dimensional shapes of some protein molecules.

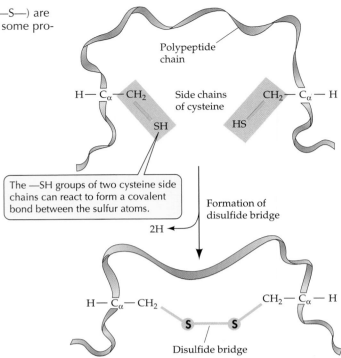

The —SH groups of two cysteine side chains can react to form a covalent bond between the sulfur atoms.

Formation of disulfide bridge

Disulfide bridge

tein (Figure 3.17*a*). The *peptide backbone* of this primary structure consists of a repeating sequence of three atoms (—N—C—C—) from the amino group, the central carbon, and the carboxyl group of each amino acid.

In cells, the primary structure of a protein is dictated by the precise sequence of nucleotides in a linear segment of a DNA molecule. The elucidation of this relationship between DNA primary structure and protein primary structure was one of the triumphs of molecular biology, which we'll describe further in Chapter 12.

The theoretical number of different proteins is enormous. Since there are 20 different amino acids, there are $20 \times 20 = 400$ distinct dipeptides, and $20 \times 20 \times 20 = 8,000$ different tripeptides. Imagine this process of multiplying by 20 extended to a protein made up of 100 amino acids (which is considered a small protein): There could be 20^{100} of these small proteins, each with its own distinctive primary structure. How large is the number 20^{100}? In the entire universe there aren't that many electrons!

At the higher levels of protein structure, local coiling and folding give the final functional shape of the molecule, but all of these levels derive from the primary structure. The different properties associated with a precise sequence of amino acids determine how the protein can twist and fold. By twisting and folding, each protein adopts a specific stable structure that distinguishes it from every other protein.

The secondary structure of a protein requires hydrogen bonding

Although the primary structure of each protein is unique, the secondary structure of many different proteins may be the same. A protein's **secondary structure** consists of regular, repeated patterns in different regions of a polypeptide chain. One type of secondary structure, the **α helix** (alpha helix), is a right-handed coil that is "threaded" in the same direction as a standard wood screw (Figure 3.17*b*). The amino acid side chains extend outward from the peptide backbone of the helix.

This helical structure of a polypeptide chain results from hydrogen bonds between elements of the peptide bonds that are distributed along the backbone of the chain. Hydrogen bonds form between the slightly positive hydrogen of the N—H of one peptide bond and the slightly negative oxygen of the C=O of another peptide bond (Figure 3.17*b*). When this pattern of hydrogen bonding is established repeatedly over a segment of the protein, it stabilizes the twisted form, resulting in an α helix. However, the ability of a protein to

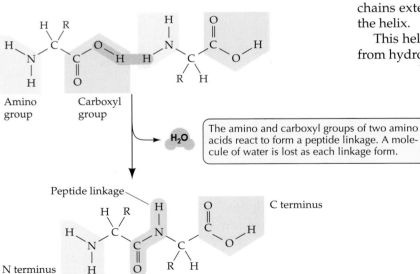

Amino group

Carboxyl group

The amino and carboxyl groups of two amino acids react to form a peptide linkage. A molecule of water is lost as each linkage form.

Peptide linkage

C terminus

N terminus

Repetition of this reaction links many amino acids together into a polypeptide.

3.16 Formation of Peptide Linkages In living things the reaction has many intermediate steps, but the reactants and products are the same as shown here.

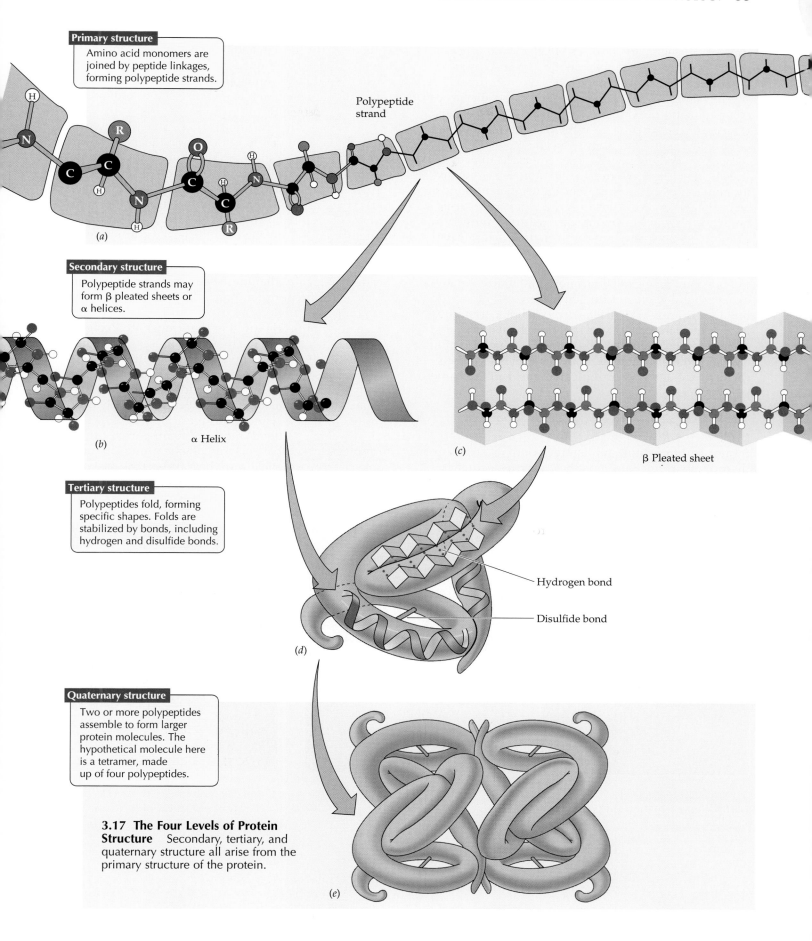

Primary structure

Amino acid monomers are joined by peptide linkages, forming polypeptide strands.

Polypeptide strand

(a)

Secondary structure

Polypeptide strands may form β pleated sheets or α helices.

(b)   α Helix

(c)   β Pleated sheet

Tertiary structure

Polypeptides fold, forming specific shapes. Folds are stabilized by bonds, including hydrogen and disulfide bonds.

Hydrogen bond

Disulfide bond

(d)

Quaternary structure

Two or more polypeptides assemble to form larger protein molecules. The hypothetical molecule here is a tetramer, made up of four polypeptides.

3.17 The Four Levels of Protein Structure Secondary, tertiary, and quaternary structure all arise from the primary structure of the protein.

(e)

form an α helix depends on its primary structure. Amino acids with larger side chains that distort the coil or otherwise prevent the formation of the necessary hydrogen bonds will keep the α helix from forming.

Alpha helical secondary structure is particularly evident in the fibrous structural proteins called keratins. Keratins constitute many of the protective materials found in mammals and birds, such as fingernails and claws, skin, hair, wool, and feathers. Hair can be stretched because this stretching requires that only hydrogen bonds in an α helix, and not covalent bonds, be broken; when the tension on the hair is released, both the helix and the hydrogen bonds re-form.

Another type of secondary structure, the β **pleated sheet**, is found in the protein silk. In silk, rather than being coiled, the protein chains are almost completely extended and lie next to one another, stabilized by hydrogen bonds between the elements of the peptide linkages (Figure 3.17c). The β pleated sheet may be found between separate polypeptide chains, as in silk, or between different regions of the same polypeptide that is bent back on itself. Many enzymes contain regions of α helix and of β pleated sheet in the same polypeptide chain.

The tertiary structure of a protein is formed by bending and folding

In most proteins, to establish the compact structure the polypeptide chain must be bent at specific sites and folded back and forth. The resulting overall shape of a protein is its **tertiary structure** (Figure 3.17d). Although the α helices and β pleated sheets contribute to the tertiary structure, more frequently only limited portions of the molecule have these secondary structures, and large regions consist of structures unique to a particular protein.

A complete description of the tertiary structure specifies the location of every atom in the molecule in three-dimensional space, in relation to all the other atoms. The tertiary structure of the protein *lysozyme* is represented in Figure 3.18. Bear in mind that this tertiary structure and the secondary structure derive from the protein's primary structure. If lysozyme is heated carefully, causing only the tertiary structure to break down, the protein will return to its normal tertiary structure when it cools. The only information needed to specify the unique shape of the lysozyme molecule is the information contained in its primary structure.

Hemoglobin is the protein that transports oxygen (O_2) from the lungs to the tissues. In muscle tissue, hemoglobin delivers O_2 to myoglobin, a smaller but similar protein, for storage. The hemoglobin molecule consists of four similar polypeptide chains that have tertiary structures made up almost entirely of α helices but that lack any disulfide bridges (Figure 3.19). The helical segments bend and fold against each other, forming a pocket that encloses a **heme group**, an iron-containing prosthetic group that binds O_2. The stabilizing interaction of hydrophobic side chains on the

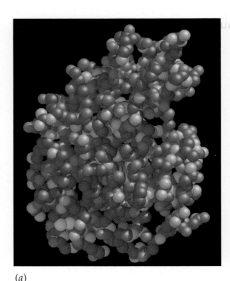

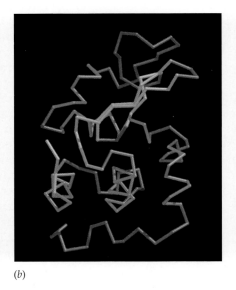

(a)   (b)   (c)

3.18 Representations of Lysozyme Different molecular representations emphasize different aspects of tertiary structure. These three representations of lysozyme are similarly oriented. (a) This computer drawing gives the most realistic impression of lysozyme's tertiary structure, which is densely packed. (b) Another computer drawing emphasizes the backbone of the folded polypeptide. (c) The green coils here represent the α helices, and orange arrows represent the β pleated sheet.

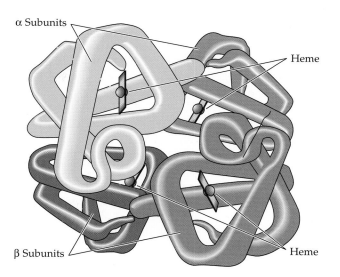

3.19 Quaternary Structure of a Protein   Hemoglobin consists of four folded polypeptide subunits that assemble themselves into the quaternary structure shown here. In these two graphic representations, each subunit is a different color. The heme groups contain iron and are the oxygen-carrying sites.

inner sides of the α helices helps ensure that the helices fold against one another correctly as each polypeptide assumes its tertiary structure.

The quaternary structure of a protein consists of subunits

As we mentioned earlier, some proteins have two or more polypeptide chains folded into their own unique tertiary structures. Each of these folded polypeptides is considered a protein subunit. **Quaternary structure** results from the ways in which these protein subunits fit together and interact; it can be illustrated by hemoglobin (see Figure 3.19). Hydrophobic interactions, hydrogen bonds, and ionic bonds all help hold the four subunits together to form the functional hemoglobin molecule. As the hemoglobin molecule takes up one O_2 molecule, the four subunits shift their relative positions slightly, changing the quaternary structure. Ionic bonds are broken, exposing buried side chains that enhance the binding of additional O_2 molecules.

Each subunit of hemoglobin is folded like a myoglobin molecule, suggesting that both hemoglobin and myoglobin are evolutionary descendants of the same oxygen-binding ancestral protein. But on the surfaces where its subunits come in contact with each other—regions that on myoglobin are exposed to aqueous surroundings and are hydrophilic—hemoglobin has hydrophobic side chains. Again, the chemical nature of side chains on individual amino acids determines how the molecule folds and packs in three dimensions.

Molecular chaperones help shape proteins

The primary structure of a protein constrains the secondary, tertiary, and quaternary structures (if subunits exist). By determining the primary structure, DNA also determines the higher levels of structure. However, other factors also affect the tertiary structure that is required for proper protein function.

Elevated temperatures, pH changes, or altered salt concentrations can cause a protein to adopt a different, biologically inactive tertiary structure. Biological function depends on a specific three-dimensional structure. Increased temperature causes more rapid molecular movement and thus can break weak hydrogen bonds and hydrophobic interactions. Altered pH can change the pattern of ionization of carboxyl and amino groups in the side chains of amino acids, thus disrupting the pattern of ionic attractions and repulsions that contribute to normal tertiary structure.

The loss of appropriate tertiary structure is called **denaturation**, and it is always accompanied by a loss of the normal biological function of the protein (Figure 3.20). Denaturation can be caused by heat or high concentrations of polar substances such as urea that disrupt the hydrogen bonding that is crucial to protein structure. Nonpolar solvents may also disrupt normal structure. Usually denaturation is irreversible, particularly if many denatured proteins interact in random nonspecific ways. However, in some cases denaturation is reversible, and upon return to normal environmental conditions, the protein may return to its active form, as does lysozyme (see Figure 3.18). This return is called **renaturation**.

How can such a complicated molecule spontaneously fold into its normal tertiary structure? Biologists are just beginning to be able to predict how a protein will fold, given its primary structure. The study of protein folding has a long way to go. The sit-

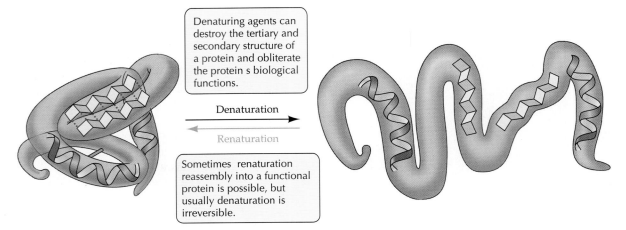

Denaturing agents can destroy the tertiary and secondary structure of a protein and obliterate the protein s biological functions.

Denaturation →

← Renaturation

Sometimes renaturation reassembly into a functional protein is possible, but usually denaturation is irreversible.

3.20 Denaturation Is the Loss of Protein Structure and Function Agents that can cause denaturation include high temperatures and certain chemicals.

uation in the living cell is complex—even more complex than we first imagined. The protein doesn't form all at once. In a human cell it may take minutes from the beginning of synthesis until the entire protein squeezes into the compartment where it folds—and things are very crowded in that compartment. There, other proteins present inappropriate potential partners for interaction (in the form of amino acid side chains).

A special group of proteins, called the **chaperone proteins**, at least in part help prevent newly formed proteins from reacting inappropriately. The chaperones attach to a new protein while it is forming and protect it from interactions with other proteins until the new molecule has its correct shape. Most chaperone proteins can act on many different forming proteins. Some act on only a few, and others may act on only a single, specific protein.

What if the chaperone proteins fail to do their job, or if for other reasons a protein folds abnormally? Evidence suggests that abnormal protein folding underlies certain infectious diseases, including mad cow disease and, in humans, Creutzfeldt–Jakob disease. Alzheimer's disease may also result from protein misfolding.

Nucleic Acids: Informational Macromolecules

The nucleic acids are linear polymers specialized for the storage, transmission, and use of information. There are two types of nucleic acids: DNA (deoxyribonucleic acid) and RNA (ribonucleic acid). **DNA** molecules are giant polymers that encode hereditary information and pass it from generation to generation (through reproduction). The information encoded in

DNA is also used to make specific proteins through the intermediate RNA. **RNA** molecules of various types copy the information in segments of DNA to specify the sequence of amino acids in proteins. Information flows from DNA to DNA in reproduction. But for nonreproductive activities of the cell, information flows from DNA to RNA to proteins, which ultimately carry out these functions. What compositions, structures, and properties of nucleic acids permit them to play these fundamental roles in living systems?

The nucleic acids have characteristic structures and properties

Nucleic acids are composed of monomers called **nucleotides**, each of which consists of a pentose sugar, a phosphate group, and a nitrogen-containing base—either a pyrimidine or a purine (Figure 3.21). Molecules consisting of a pentose sugar and a nitrogenous base, but no phosphate group, are called nucleo*sides*. In DNA, the pentose sugar is deoxyribose, which differs from the ribose found in RNA by only one oxygen atom (see Figure 3.11).

In both RNA and DNA, the backbone of the molecule consists of alternating sugars and phosphates (—sugar—phosphate—sugar—phosphate—). Bases are attached to the sugars and project from the chain (Figure 3.22). The nucleotides are joined by covalent bonds in what are called **phosphodiester linkages** between the sugar of one nucleotide and the phosphate of the next ("-diester" refers to the two bonds formed by reacting —OH groups with acidic phosphate groups). The phosphate groups link carbon 3' in one ribose to carbon 5' in the adjacent ribose.

Most RNA molecules are *single-stranded*, consisting of only one polynucleotide chain. DNA, however, is usually *double-stranded*; it has two polynucleotide chains held together by hydrogen bonding between their complementary nitrogenous bases. The two strands of DNA run in opposite directions. You can see what this means by drawing an arrow through the

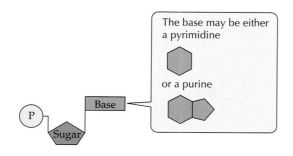

A nucleotide consists of a phosphate, a pentose sugar, and a nitrogen-containing base.

The base may be either a pyrimidine or a purine

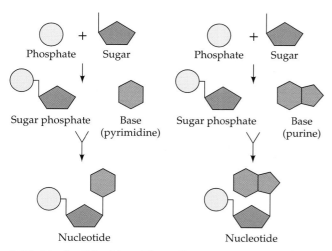

3.21 Nucleotides Have Three Components A nucleotide consists of a phosphate group, a pentose sugar, and a nitrogen-containing base—all linked together by covalent bonds. The nitrogenous bases fall into two categories: Purines have two fused rings, and the smaller pyrimidines have a single ring.

phosphate group from carbon 3′ to carbon 5′ in the next ribose. If you do this for both strands, the arrows point in opposite directions. This antiparallel orientation is necessary for the strands to fit together.

The uniqueness of a nucleic acid resides in its base sequence

Only four nitrogenous bases—and thus only four nucleotides—are found in DNA. The DNA bases and their abbreviations are: adenine (A), cytosine (C), guanine (G), and thymine (T). A key to understanding the structures and functions of nucleic acids is the principle of **complementary base pairing** through hydrogen bond formation. In double-stranded DNA, adenine and thymine always pair (AT), and cytosine and guanine always pair (CG).

Base pairing is complementary because of two factors: the corresponding sites for hydrogen bonding and the molecular sizes of the paired bases. Adenine

and guanine are both purines consisting of two fused rings. Thymine and cytosine are both pyrimidines consisting of only one ring. The pairing of a large purine with a small pyrimidine ensures a stable and consistent dimension to the double-stranded molecule of DNA. In Chapter 11, we'll discuss in more detail how complementary strands of DNA separate and are faithfully copied.

Ribonucleic acids also have four different monomers, but the nucleotides differ from those of DNA. In RNA the nucleotides are termed **ribonucleotides.** They contain ribose rather than deoxyribose, and instead of the base thymine, RNA uses the base uracil (Table 3.2). The other three bases are the same as in DNA.

Although RNA is generally single-stranded, complementary hydrogen bonding between ribonucleotides can take place. These bonds play important roles in determining the shapes of some RNA molecules and in the associations between RNA molecules during protein synthesis. During the DNA-directed synthesis of RNA, complementary base pairing also takes place between ribonucleotides and the bases of DNA. In RNA, guanine and cytosine pair as in DNA, but adenine pairs with uracil. Adenine in an RNA strand can pair either with uracil (in another RNA strand) or with thymine (in a DNA strand).

The three-dimensional appearance of DNA is strikingly regular. The segment shown in Figure 3.23 could be from any DNA molecule. Through hydrogen bonding, the two complementary polynucleotide strands pair and twist to form a **double helix.** Compared to the complex and varied tertiary structures of different proteins, this formation seems amazingly regular. But this structural contrast makes sense in terms of the functions of these two classes of macromolecules.

DNA is a purely informational molecule. *The information in DNA is encoded in the sequence of bases carried in its chains.* This sequence is, in a sense, like the tape of a tape recorder. The message must be read easily and reliably, in a specific order. A uniform molecule like DNA can be interpreted by standard molecular

TABLE 3.2	Distinguishing RNA from DNA	
NUCLEIC ACID	**SUGAR**	**BASES**
RNA	Ribose	Adenine Cytosine Guanine **Uracil**
DNA	Deoxyribose	Adenine Cytosine Guanine **Thymine**

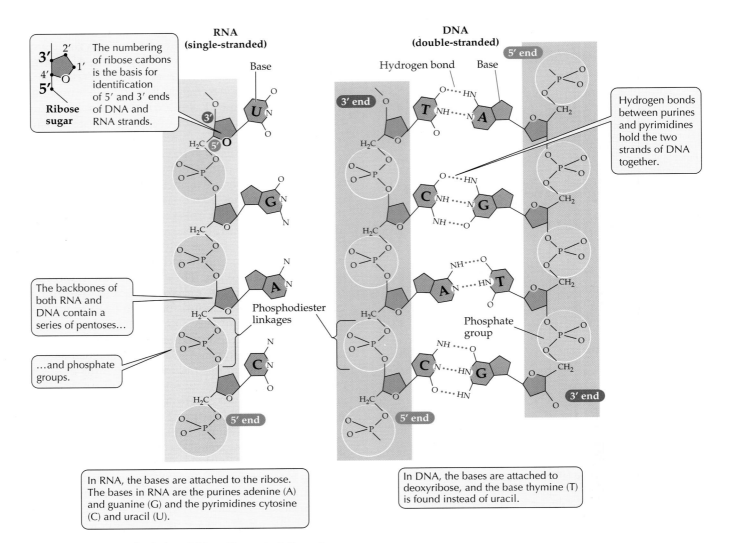

3.22 Common and Distinguishing Characteristics of DNA and RNA RNA is usually a single strand. DNA usually consists of two strands running in opposite directions.

machinery, and any cell's machinery can read any molecule of DNA—just as a tape player can play any tape of the right size.

Proteins, on the other hand, have good reason to be so varied. In particular, different enzymes must recognize their own specific "target" molecules. They do this by having a unique three-dimensional form that can match at least a portion of the surface of their target molecules. In other words, structural diversity in the molecules with which enzymes react requires corresponding diversity in the structure of the enzymes themselves. *In DNA the information is in the sequence of bases; in proteins the information is in the shape.*

DNA is a guide to evolutionary relationships

Because DNA carries hereditary information between generations, a series of DNA molecules with changes in base sequences stretches back through time. Of

course, we cannot study all of these DNA molecules, because many of their organisms have become extinct. However, we can study the DNA of living organisms, some of which are judged to belong to lineages that are more ancient than those of other living forms. Comparisons and contrasts of these DNAs add to the evolutionary record.

Closely related living species should have more similar base sequences than species judged by other criteria to be more distantly related. Indeed this is the case. The examination of base sequences confirms many of the evolutionary relationships that have been inferred from the study of microscopic or macroscopic structures or studies of biochemistry and physiology. For example, the closest living relative of humans (*Homo sapiens*) is the chimpanzee (genus *Pan*), which shares more than 98 percent of its DNA base sequence with human DNA.

Confirmation of well-established evolutionary relationships gives credibility to the use of DNA to elucidate relationships when studies of structure are not

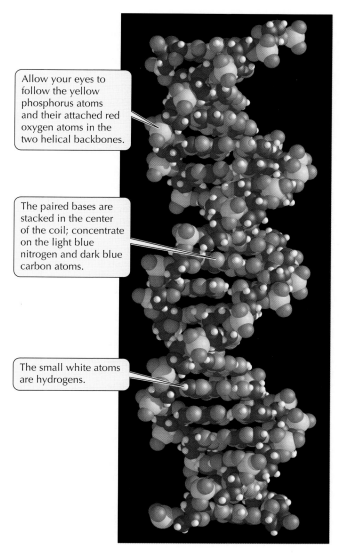

Allow your eyes to follow the yellow phosphorus atoms and their attached red oxygen atoms in the two helical backbones.

The paired bases are stacked in the center of the coil; concentrate on the light blue nitrogen and dark blue carbon atoms.

The small white atoms are hydrogens.

3.23 The Double Helix of DNA The backbones of the two strands in a DNA molecule are coiled in a double helix.

possible or are not conclusive. For example, DNA studies revealed a close evolutionary relationship between starlings and mockingbirds that was not expected on the basis of studies of anatomy or behavior. DNA studies support the division of the prokaryotes into two kingdoms (Eubacteria and Archaebacteria). Each of these two groups of prokaryotes is as distinct from the other as either is from the other four kingdoms into which living things are classified. In addition, DNA comparisons support the hypothesis that certain subcellular compartments of eukaryotes (the organelles called mitochondria and chloroplasts) evolved from early bacteria that established a stable and mutually beneficial way of life inside larger cells.

The Interactions of Macromolecules

We have been treating the classes of macromolecules as if each were completely separate from the others. In cells, however, certain macromolecules of different classes may be covalently bonded to one another. Proteins with attached oligosaccharides are called *glycoproteins* (*glyco-*, "sugar"). The specific oligosaccharide chain determines the placement of a glycoprotein within the cell.

Other carbohydrate chains bind to lipids, resulting in *glycolipids*, which reside in the cell surface membrane, with the carbohydrate chain extending out into the cell's environment. In humans, the substances that determine the ABO blood types are carbohydrates attached to either proteins or lipids. When a human body cell becomes cancerous, the glycolipids and glycoproteins on the cell surface are modified, and these changes serve as a recognition signal for the body defenses to destroy the abnormal cell.

Not all the associations among macromolecules or between a macromolecule and a smaller molecule are covalent. Weaker linkages such as ionic bonds, hydrogen bonds, and even van der Waals forces may also be involved. Enzymes bind to their reactants using a variety of weak bonds. Ionic bonds link proteins to DNA, forming *nucleoproteins* that regulate the activities of DNA. Still other proteins, in combination with cholesterol and other lipids, form *lipoproteins*. The lipoproteins make it possible to move very hydrophobic lipids through water-rich environments such as the blood and tissue fluid and deliver cholesterol to all the body's cells.

Summary of "Macromolecules: Their Chemistry and Biology"

Lipids: Water-Insoluble Molecules
• Although lipids can form gigantic structures such as lipid droplets and membranes, these aggregations are not considered macromolecules because they are not linked by covalent bonds.
• Saturated fatty acids have a reactive carboxyl group and a nonpolar hydrocarbon chain with no double bonds. The hydrocarbon chains of unsaturated fatty acids have one or more double bonds that bend the chain, making close packing less possible. **Review Figure 3.3**
• In certain cells, deposits of fat or oil store energy. Fats and oils are composed of three fatty acids covalently bonded to a glycerol molecule by ester linkages. **Review Figure 3.4**
• Phospholipids form the core of biological membranes. Phospholipids have a hydrophobic "tail" and a hydrophilic "head." **Review Figures 3.2, 3.6**

- In water, the interactions of the hydrophobic tails and hydrophilic heads in membranes generate a phospholipid bilayer that is two molecules thick. The head groups are directed outward, where they interact with the surrounding water. The tails are packed together in the interior of the bilayer. **Review Figure 3.7**
- Carotenoids trap light energy in green plants. β-Carotene can be split to form vitamin A, a lipid vitamin. The other lipid vitamins are D, E, and K. **Review Figure 3.8**
- Vitamins are substances that are required for normal functioning but that must be acquired from the diet.
- Some steroids, such as testosterone, function as hormones. Cholesterol is synthesized by the liver and has a role in some cell membranes, as well as in the digestion of other fats. Too much cholesterol in the diet can lead to arteriosclerosis. **Review Figure 3.9**

Macromolecules: Giant Polymers

- Macromolecules are constructed by the formation of covalent bonds between smaller molecules called monomers. Macromolecules include polysaccharides, proteins, and nucleic acids.
- Macromolecules have specific, characteristic three-dimensional shapes that depend on the structures, properties, and sequence of their monomers. Different functional groups give local sites on macromolecules specific properties that are important for their biological functioning and interactions with other macromolecules.
- Monomers join by condensation reactions, which lose a molecule of water for each bond formed. Hydrolysis reactions use water to digest polymers into monomers.

Carbohydrates: Sugars and Sugar Polymers

- All carbohydrates approximate multiples of the general formula CH_2O.
- Hexoses contain six carbon atoms. Examples of hexoses include glucose, galactose, and fructose, which can exist as chains or rings. **Review Figures 3.10, 3.11**
- The pentoses are five-carbon monosaccharides; two pentoses, ribose and deoxyribose, are components of the nucleic acids RNA and DNA, respectively. **Review Figure 3.11**
- Glycosidic linkages may have either α or β orientation in space. They covalently link monosaccharides into larger units such as disaccharides (for example, sucrose, lactose, maltose, and cellobiose), oligosaccharides, and polysaccharides. **Review Figures 3.12, 3.13**
- Cellulose, a very stable glucose polymer, is the principal component of the cell walls of plants. It is formed by glucose units linked together by β-glycosidic linkages between carbons 1 and 4. **Review Figure 3.13**
- Starches, less dense and less stable than cellulose, store energy in plants. Starches are formed by α-glycosidic linkages between carbons 1 and 4 and are distinguished by the amount of branching that occurs through glycosidic bond formation at carbon 6. **Review Figure 3.13**
- Glycogen contains α-1,4 glycosidic linkages and is highly branched. Glycogen stores energy in animal livers and muscles. **Review Figure 3.13**
- Derivative monosaccharides include the sugar phosphates and amino sugars. One such derivative, N-acetylglucosamine, polymerizes to form the polysaccharide chitin, which is found in the cell walls of fungi and the exoskeletons of insects. **Review Figure 3.14**

Proteins: Amazing Polymers of Amino Acids

- The functions of proteins include support, protection, catalysis, transport, defense, regulation, and movement. Protein function sometimes requires an attached prosthetic group, such as the heme group.
- There are 20 amino acids found in proteins. Each amino acid consists of an amino group, a carboxyl group, a hydrogen, and a side chain bonded to a central carbon atom. **Review Table 3.1 and Figure 2.24**
- The side chains of amino acids may be charged, polar, or hydrophobic; there are also "special cases," such as the —SH groups that can form disulfide bridges. The side chains give different properties to each of the amino acids. **Review Table 3.1 and Figure 3.15**
- Amino acids are covalently bonded together by peptide linkages that form by condensation reactions between the carboxyl and amino groups. **Review Figure 3.16**
- The polypeptide chains of proteins are folded into specific three-dimensional shapes. Four levels of structure are possible: primary, secondary, tertiary, and quaternary.
- The primary structure of a protein is the sequence of amino acids bonded by peptide linkages. This primary structure determines both the higher levels of structure and protein function. **Review Figure 3.17a**
- Secondary structures of proteins, such as α helices and β pleated sheets, are maintained by hydrogen bonding between atoms of the peptide linkages. **Review Figure 3.17b**
- The tertiary structure of a protein is generated by bending and folding of the polypeptide. **Review Figures 3.17d, 3.18**
- The quaternary structure of a protein is the arrangement of polypeptides in a single functional unit consisting of more than one polypeptide subunit. **Review Figures 3.17e, 3.19**
- Molecular chaperones assist protein folding by preventing random, nonspecific interactions that have no biological function.
- Proteins denatured by heat, acid, or reagents lose tertiary and secondary structure as well as biological function. Renaturation is not always possible. **Review Figure 3.20**

Nucleic Acids: Informational Macromolecules

- In cells, DNA is the hereditary material. Both DNA and RNA play roles in the formation of proteins. Information flows from DNA to RNA to protein.
- Nucleic acids are polymers of nucleotides that consist of a phosphate group, a sugar (ribose in RNA and deoxyribose in DNA), and a nitrogen-containing base. In DNA the bases are adenine, guanine, cytosine, and thymine, but in RNA uracil substitutes for thymine. **Review Figure 3.21 and Table 3.2**
- In the nucleic acids, the bases extend from a sugar–phosphate backbone. The information content of DNA and RNA resides in their base sequences.
- RNA is single-stranded. DNA is a double-stranded helix in which there is complementary, hydrogen-bonded base pairing between adenine and thymine (AT) and guanine and cytosine (GC). The two strands of the DNA double helix run in opposite directions. **Review Figures 3.22, 3.23**
- Comparing of DNA base sequences of different living species provides information on their evolutionary relatedness.

The Interactions of Macromolecules

- Both covalent and noncovalent linkages are found between the various classes of macromolecules.
- Glycoproteins contain an oligosaccharide "label" that directs the protein to the proper cell destination. The carbohydrate groups of glycolipids are displayed on the cell's outer surface, where they serve as recognition signals.
- Weaker forces bind a catalytic protein (enzyme) to a reactant. Hydrophobic interactions bind cholesterol to the protein that transports it in the blood.

Self-Quiz

1. All lipids
    a. are triglycerides.
    b. are polar.
    c. are hydrophilic.
    d. are polymers.
    e. are more soluble in nonpolar solvents than in water.

2. Which of the following is *not* a lipid?
    a. A steroid
    b. A fat
    c. A triglyceride
    d. A biological membrane
    e. A carotenoid

3. All carbohydrates
    a. are polymers.
    b. are simple sugars.
    c. consist of one or more simple sugars.
    d. are found in biological membranes.
    e. are more soluble in nonpolar solvents than in water.

4. Which of the following is *not* a carbohydrate?
    a. Glucose
    b. Starch
    c. Cellulose
    d. Hemoglobin
    e. Deoxyribose

5. All proteins
    a. are enzymes.
    b. consist of one or more polypeptides.
    c. are amino acids.
    d. have quaternary structures.
    e. are more soluble in nonpolar solvents than in water.

6. Which of the following statements about the primary structure of a protein is *not* true?
    a. It may be branched.
    b. It is determined by the structure of the corresponding DNA.
    c. It is unique to that protein.
    d. It determines the tertiary structure of the protein.
    e. It is the sequence of amino acids in the protein.

7. The amino acid leucine (see Table 3.1)
    a. is found in all proteins.
    b. cannot form peptide linkages.
    c. is likely to appear in the part of a membrane protein that lies within the phospholipid bilayer.
    d. is likely to appear in the part of a membrane protein that lies outside the phospholipid bilayer.
    e. is identical to the amino acid lysine.

8. The quaternary structure of a protein
    a. consists of four subunits—hence the name *quaternary*.
    b. is unrelated to the function of the protein.
    c. may be either α or β.
    d. depends on covalent bonding among the subunits.
    e. depends on the primary structures of the subunits.

9. All nucleic acids
    a. are polymers of nucleotides.
    b. are polymers of amino acids.
    c. are double-stranded.
    d. are double-helical.
    e. contain deoxyribose.

10. Which of the following statements about condensation reactions is *not* true?
    a. Protein synthesis results from them.
    b. Polysaccharide synthesis results from them.
    c. Nucleic acid synthesis results from them.
    d. They consume water as a reactant.
    e. Different condensation reactions produce different kinds of macromolecules.

Applying Concepts

1. Phospholipids make up a major part of every biological membrane; cellulose is the major constituent of the cell walls of plants. How do the chemical structures and physical properties of phospholipids and cellulose relate to their functions in cells?

2. Suppose that, in a given protein, one lysine is replaced by aspartic acid (see Table 3.1). Does this change occur in the primary structure or in the secondary structure? How might it result in a change in tertiary structure? In quaternary structure?

3. If there are 20 different amino acids commonly found in proteins, how many different dipeptides are there? How many different tripeptides? How many different trinucleotides? How many different single-stranded RNAs composed of 200 nucleotides?

4. Contrast the following three structures: hemoglobin, a DNA molecule, and a protein that spans a biological membrane.

Readings

Doolittle, R. F. 1985. "Proteins." *Scientific American*, October. A strikingly illustrated treatment of protein structure and evolution.

Horgan, J. 1993. "Stubbornly Ahead of His Time." *Scientific American*, March. About Linus Pauling who developed the theory of covalent bonding, discovered α-helical protein structure, and won Nobel prizes for chemistry and peace.

Lehninger, A. L., D. L. Nelson and M. M. Cox. 1993. *Principles of Biochemistry,* 2nd Edition. Worth, New York. A balanced textbook that uses energetic, evolutionary, regulatory, and structure-function themes to place biochemistry in physical, chemical, and biological contexts.

Stryer, L. 1995. *Biochemistry*, 4th Edition. W. H. Freeman, New York. A relatively advanced but beautiful reference on the subjects of this chapter; outstanding illustrations, concise descriptions, clear prose.

Taubes, G. 1996. "Misfolding the Way to Disease." *Science,* vol. 271, pages 1493–1495. A brief overview of protein misfolding, with good references.

Chapter 4

The Organization of Cells

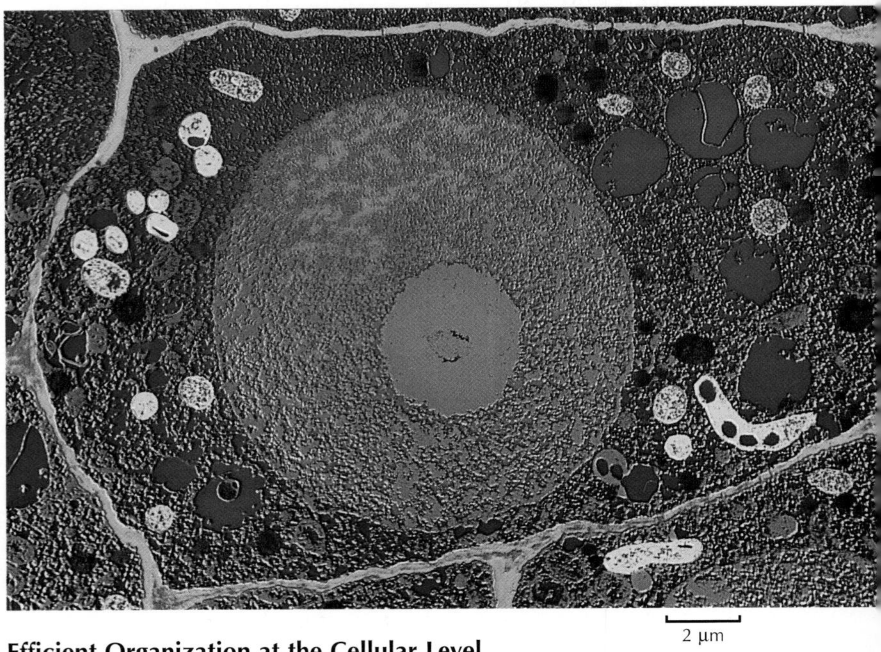

Efficient Organization at the Cellular Level
This cell in the tip of a corn root is surrounded by a firm wall. The circular nucleus is separated from the rest of the cell by membranes, and other membranes enclose other functional parts of the cell.

2 μm

You enter a room on campus and are greeted by a tuba player practicing full blast, workers from a dining hall washing dishes, a professor lecturing earnestly, a karate class in full kick, and six students holding their ears and trying to study while another group plans a ski trip. Is this any way to run a campus?

Campuses are organized so that their many functions can proceed efficiently. They have special rooms for special purposes and different employees to provide different services. Without these divisions, there would be chaos. Like the campus, cells are also divided into special structures that carry out special functions. A cell is not simply a bag of enzymes and other molecules; it is a highly ordered, efficient workplace.

In this chapter, we will examine the general structure of cells. We'll start with the distinctions between *prokaryotic cells* and *eukaryotic cells* and describe the microscopic methods that reveal their structure. Then we will turn to a detailed examination of the components of eukaryotic cells that are involved in information storage and processing, energy transformations, and support and protection. After contrasting cells to *viruses*, we'll close the chapter with a description of the experimental techniques by which cells are taken apart and their components isolated for study. Such chemical studies complement the structural knowledge from microscopy and have revealed much about how living cells function.

64

The Cell: The Basic Unit of Life

All organisms are composed of cells, and all cells come from preexisting cells. These two statements constitute the **cell theory.** Cells show the characteristics by which living systems are recognized: They use DNA as hereditary material and proteins as catalysts. In addition, most cells reproduce, transform matter and energy, and respond to their environment.

Viruses, on the other hand, show only a few of these characteristics and thus are not usually considered alive. Viruses depend entirely on living cells to reproduce, and they neither transform matter and energy nor respond to the environment. Some cells can develop into entire multicellular organisms, but no *part* of a cell can do that—nor can viruses.

Most cells are tiny. The volume of cells ranges from 1 to 1,000 μm³ (Figure 4.1). The eggs of some birds are enormous exceptions, to be sure, and individual cells of several types of algae are large enough to be viewed with the unaided eye. And although neurons (nerve cells) have a volume that is within the "normal cell" range, they often have fine projections that may extend for meters, carrying signals from one part of a large animal to another. But by and large, cells are minuscule. Why?

Why are cells so small?

Whether they are individual unicellular organisms or parts of multicellular organisms consisting of millions and millions of cells, cells are small. Why can't a single cell be the size of a human body? The answer relates to the change in the surface area-to-volume ratio (SA/V) of any object as it increases in size.

As a cell increases in volume, its surface area increases also, but not to the same extent. Consider the example illustrated in Figure 4.2. The surface area of a cube with sides measuring 4 mm is 96 mm² ($SA = 6 \times 4^2$ mm²), and its volume is 64 mm³ ($V = 4^3$ mm³). However, if the same volume—64 mm³—is contained in 64 smaller cubes, each of which has sides measuring 1 mm, the surface area is 384 mm² ($SA = 64 \times 6 \times 1^2$ mm²). In other words, although the total volume remains unchanged, the surface area has increased fourfold (384 = 4 × 96). And the surface area-to-volume ratio has increased from 1.5/1 to 6/1. This phenomenon has great biological significance.

The volume of a cell is related to the amount of chemical activity the cell carries out per unit time. However, the surface area of the cell limits the exchange of nutrients and waste products with its environment. As the living cell grows larger, its rate of pro-

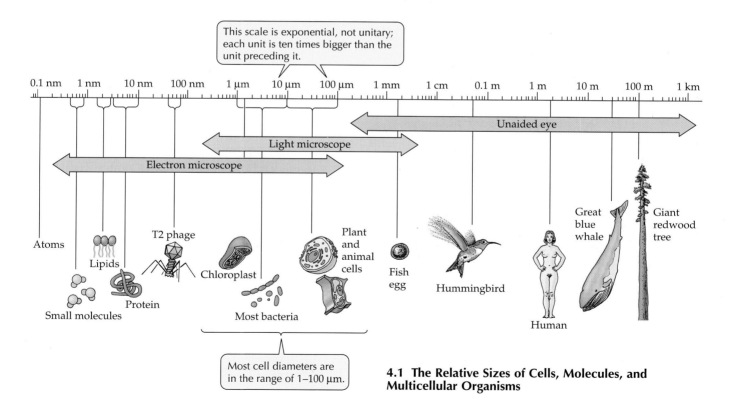

4.1 The Relative Sizes of Cells, Molecules, and Multicellular Organisms

4.2 The Ratio of Surface Area to Volume Explains Why Cells Are So Small Although they contain equal volumes, a cube measuring 4 mm on an edge has a smaller surface area than the total surface area of 64 cubes each measuring 1 mm on an edge. In cells, a large surface-area-to-volume ratio is needed for adequate exchange of nutrients and wastes with the environment.

	One 4-mm cube	Eight 2-mm cubes	Sixty-four 1-mm cubes
Surface area (mm^2)	96	192	384
Volume (mm^3)	64	64	64
Surface area-to-volume ratio	1.5/1	3/1	6/1

duction of wastes and its need for resources increase faster than the surface area through which the cell must obtain resources and excrete wastes. The more limited increase in surface area restricts the increase in volume as the cell grows—this explains why large organisms must consist of many small cells. *Cells are small in volume and thus maintain a large surface area-to-volume ratio.*

In a multicellular organism, the large surface area represented by the multitude of small cells that make up the whole organism enables the myriad functions required for survival of the organism. Special structures transport food, oxygen, and waste materials to and from the small cells that are distant from the external surface of the organism.

Cells show two organizational patterns: Prokaryotic and eukaryotic

Cells must carry out many functions in order to survive. They must obtain and process energy; they must convert the genetic information of DNA into protein; and they must keep certain biochemical reactions separate from other reactions—incompatible reactions that must occur simultaneously. Specific structures within the cell—organized into two general patterns: *prokaryotic* and *eukaryotic*—perform these functions.

Prokaryotic cell organization is characteristic of the kingdoms Eubacteria (Domain bacteria) and Archaebacteria (Domain Archaea). Organisms in these kingdoms are called *prokaryotes*. Their cells do not have membrane-enclosed internal compartments.

Eukaryotic cell organization is found in the other four kingdoms of living things (Protista, Plantae, Fungi, and Animalia, which comprise the domain Eukarya). The DNA of eukaryotic cells is contained in a special membrane-enclosed compartment called the nucleus. Eukaryotic cells also contain other membrane-enclosed compartments in which specific chemical reactions take place. These compartments isolate certain molecules and chemical reactions from other molecules and reactions. Organisms with this type of cell are known as *eukaryotes*.

Both prokaryotes and eukaryotes have prospered through many hundreds of millions of years of evolution, and both are great success stories. Let's look first at prokaryotic cells.

Prokaryotic Cells

Prokaryotes are very diverse in their metabolic capabilities. They can live off a greater diversity of energy sources than any other living creatures, and they inhabit greater environmental extremes. The vast diversity within the prokaryotic kingdoms is the subject of Chapter 25. Some archaea are found in sulfurous hot springs at temperatures that would denature the proteins of most other life-forms. Some bacteria are photosynthetic and use light energy to synthesize needed materials from CO_2. Other prokaryotes are able to oxidize inorganic ions to obtain energy for the synthesis of cell-specific materials from raw materials.

Prokaryotic cells are generally smaller than eukaryotic cells, ranging in dimensions from 0.25×1.2 μm to 1.5×4 μm (see Figure 4.1). Although each prokaryote is a single cell, many types of prokaryotes are usually seen in chains, small clusters, or even colonies containing hundreds of individuals.

In this section, we will first consider the features that cells in the kingdoms Eubacteria and Archaebacteria have in common. Then we will examine structural features that are found in some, but not all, prokaryotes.

All prokaryotic cells share certain features

All prokaryotic cells have a plasma membrane, a nucleoid, and cytoplasm filled with ribosomes. The plasma membrane encloses the cell, separating it from its environment and regulating the traffic of materials into and out of the cell. The **nucleoid** is a relatively clear area (as seen under the electron microscope) that contains the hereditary material (DNA) of the cell. The rest of the material within the plasma membrane—called the **cytoplasm**—is made up of two parts: the cytosol and insoluble suspended particles.

The **cytosol** consists mostly of water that contains dissolved ions, small molecules, and soluble macromolecules such as enzymes. The insoluble suspended particles are revealed at high magnification. For example, high magnification shows that the cytoplasm contains many minute, roughly spherical structures called **ribosomes**. Although the ribosomes seem small in comparison to the cell in which they are contained, on a macromolecular scale they are gigantic structures assembled from both RNA and proteins. The ribosomes coordinate the synthesis of proteins.

Although structurally less complicated than eukaryotic cells, prokaryotic cells are functionally complex. The enzymes in prokaryotic cells catalyze thousands of chemical reactions. In addition to making these thousands of enzymes, prokaryotic cells are capable of shutting off the synthesis of enzymes that are not needed in a particularly rich nutrient environment.

Some prokaryotic cells have specialized features

Many prokaryotic cells have at least a few more structural complexities. For example, many prokaryotes have a **cell wall** located outside the plasma membrane (Figure 4.3). The rigidity of the cell wall supports the cell and determines its shape. The cell walls of most bacteria, but not archaea, contain **peptidoglycan**, a polymer of amino sugars, cross-linked to form a single molecule around the entire cell. In some bacteria, another layer—the **outer membrane** (a polysaccharide-rich phospholipid membrane)—encloses the cell wall. Unlike the plasma membrane, this outer membrane is not a major permeability barrier, and some of its polysaccharides are disease-causing toxins.

Enclosing the cell wall and outer membrane in many bacteria is a layer of slime composed mostly of polysaccharide and referred to as a **capsule**. The capsules of some bacteria may protect them from attack by white blood cells in the animals they infect. The capsule helps keep the cell from drying out, and sometimes it traps other cells for the bacterium to attack. Many prokaryotes produce no capsule at all, and those that do have capsules can survive even if they lose them, so the capsule is not a structure essential to cell life.

Some groups of bacteria—the cyanobacteria and some others—carry on photosynthesis. In *photosynthesis*, the energy of sunlight is converted to chemical energy that can be used for a variety of energy-requiring reactions, such as the synthesis of cellular proteins and DNA. In these photosynthetic bacteria, the plasma membrane folds into the cytoplasm—often

4.3 A Prokaryotic Cell The bacterium *Pseudomonas aeruginosa* illustrates typical prokaryotic cell structures. The electron micrograph on the left is magnified about 80,000 times. Note the existence of several protective structures external to the plasma membrane.

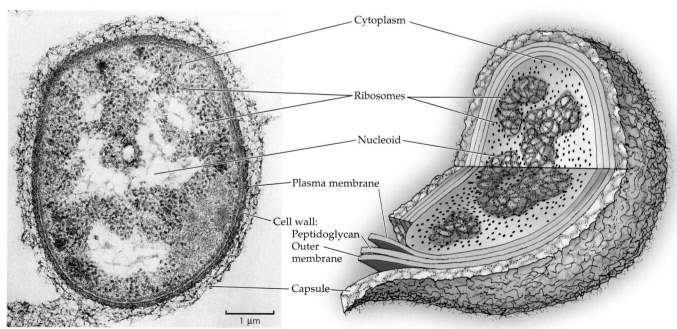

Cytoplasm

Ribosomes

Nucleoid

Plasma membrane

Cell wall:
Peptidoglycan
Outer membrane

Capsule

1 µm

(a)

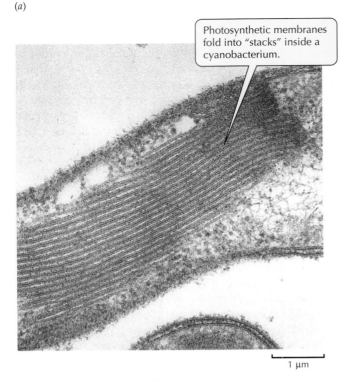

Photosynthetic membranes fold into "stacks" inside a cyanobacterium.

1 μm

(b)

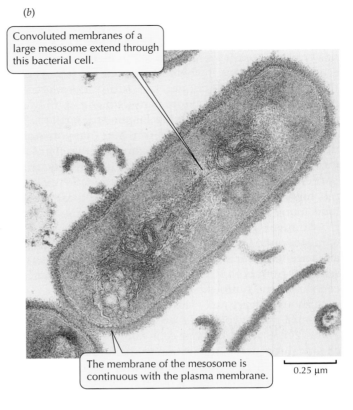

Convoluted membranes of a large mesosome extend through this bacterial cell.

The membrane of the mesosome is continuous with the plasma membrane.

0.25 μm

4.4 Some Prokaryotes Have Internal Systems of Membranes The presence of internal membranes in prokaryotic cells contradicts the mistaken notion that prokaryotes are nothing more than tiny bags of molecules. (a) Photosynthetic membranes contain compounds needed for photosynthesis. (b) The convoluted membranes of a mesosome contain enzymes involved in cellular respiration.

very extensively—to form an internal membrane system that contains bacterial chlorophyll and other compounds needed for photosynthesis (Figure 4.4a).

Other groups of prokaryotes possess different sorts of membranous structures called **mesosomes**, which may function in cell division or in various energy-releasing reactions (Figure 4.4b). Like the photosyn-

4.5 Prokaryotic Projections Surface projections such as these bacterial flagella (a,b) and pili (c) contribute to movement, to adhesion, and to the complexity of prokaryotic cells.

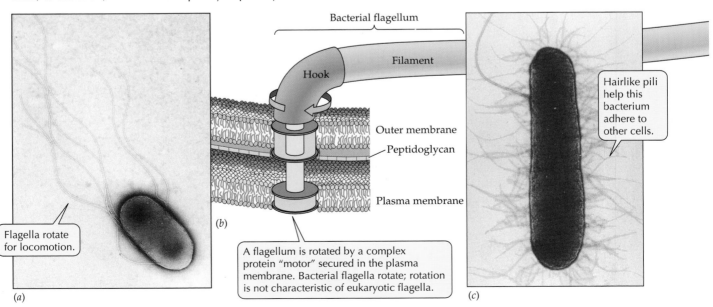

Bacterial flagellum

Filament

Hook

Outer membrane

Peptidoglycan

Plasma membrane

Flagella rotate for locomotion.

(a)

(b)

A flagellum is rotated by a complex protein "motor" secured in the plasma membrane. Bacterial flagella rotate; rotation is not characteristic of eukaryotic flagella.

Hairlike pili help this bacterium adhere to other cells.

(c)

thetic membrane systems, mesosomes are formed by infolding of the plasma membrane. They remain attached to the plasma membrane and never form the free-floating, isolated, membranous organelles that are characteristic of eukaryotic cells.

Some prokaryotes swim by using appendages called **flagella** (Figure 4.5*a*). A single flagellum, made of a protein called flagellin, looks at times like a tiny corkscrew. It spins on its axis like a propeller, driving the cell along. Ring structures anchor the flagellum to the plasma membrane and, in some bacteria, to the outer membrane of the cell wall (Figure 4.5*b*). Flagella are known to cause the motion of the cell because if they are removed, the cell cannot move.

Pili project from the surface of some groups of bacteria (Figure 4.5*c*). Shorter than flagella, these thread-like structures seem to help bacteria adhere to one another during mating, as well as to animal cells for protection and food.

Microscopes: Revealing the Subcellular World

The preceding discussion included images of prokaryotes produced by microscopes. How are these images produced? In this section, we examine the principles and technologies of microscopy by means of visible light and by beams of electrons: light and electron microscopy, respectively (Figure 4.6).

Many significant advances in our knowledge of cells have depended on the **resolution**, or resolving power, of the instruments we use to magnify tiny objects. The resolving power of a lens or microscope is the smallest distance separating two objects that allows them to be seen as two distinct things rather than as a single entity. For example, most humans see two fine parallel lines as distinct markings if they are separated by at least 0.1 mm; if they are any closer together, we see them as a single line. Thus, the resolving power of the human eye is about 0.1 mm, which is the approximate diameter of the human egg. To see anything smaller, we must use some form of microscope.

Development of the light microscope made the study of cells possible

The **light microscope** uses glass lenses and visible light to form a magnified image of an object (see Figure 4.6*a*). In its contemporary form, the light microscope has a resolving power of about 200 nm (that is, 0.2 µm, or 0.0002 mm), so it gives a useful view of cells and can reveal features of some of the subcellular organelles of eukaryotes. Today, half a century after the invention of the electron microscope, which has more resolving power than the light microscope, the light microscope still remains an important tool for the biologist. Many of the illustrations in this book are pho-

4.6 Light and Electron Microscopes Both light and electron microscopes magnify a specimen, permitting more detail to be observed. Light microscopes use glass lenses to focus visible light, creating an enlarged image. Electron microscopes use magnets to focus a beam of electrons to create an image. Because of their greater resolution, electron microscopes reveal more internal detail than light microscopes can.

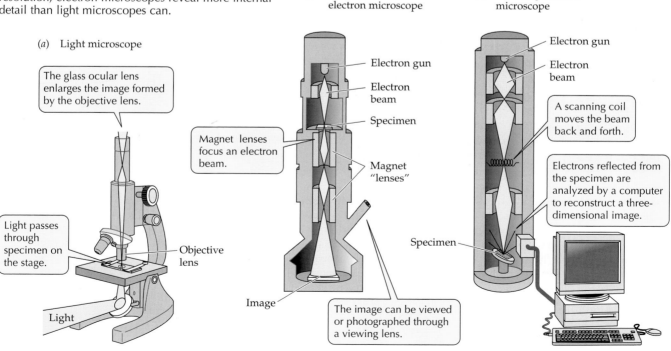

(*a*) Light microscope

The glass ocular lens enlarges the image formed by the objective lens.

Light passes through specimen on the stage.

Objective lens

Light

(*b*) Transmission electron microscope

Electron gun

Electron beam

Specimen

Magnet lenses focus an electron beam.

Magnet "lenses"

Image

The image can be viewed or photographed through a viewing lens.

(*c*) Scanning electron microscope

Electron gun

Electron beam

A scanning coil moves the beam back and forth.

Electrons reflected from the specimen are analyzed by a computer to reconstruct a three-dimensional image.

Specimen

tographs taken through the light microscope; they are called photomicrographs or just *micrographs*.

The light microscope has its limitations—principally its 200-nm resolving power. This resolving power cannot be improved by the addition of lenses or by the enlargement of micrographs. Although micrographs can be enlarged, such enlargements do not increase the resolution; as the images become larger, they simply become fuzzier.

One way to make images clearer is to kill the cells and stain them before examining them in the light microscope. But killing is not always necessary. One of

TABLE 4.1 Systems of Light Microscopy

THE FOLLOWING FORMS OF LIGHT MICROSCOPY CAN BE USED TO VIEW LIVING CELLS.[a]

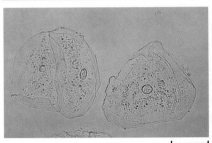

25 μm

In **bright-field microscopy**, light passes directly through the cells. Unless natural pigments are present, there is little contrast and details are not distinguished.

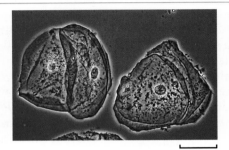

25 μm

In **phase-contrast microscopy**, contrast in the image is increased by emphasizing differences in refractive index (the capacity to bend light), thereby enhancing light and dark regions in the cell.

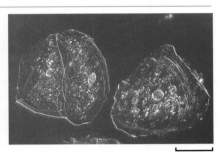

25 μm

Differential interference-contrast microscopy (Nomarski optics) uses two beams of plane-polarized light. Changes in the phase of these two beams as they pass through adjacent parts of a cell result in a bright image (if the beams are in phase) or a dark image (if the beams are out of phase). The observed image looks as if the cell were casting a shadow on one side.

[a]Certain dyes, called vital dyes, can sometimes be used to stain cells to enhance contrast without killing the cells.

THE FOLLOWING STAINING PROCEDURES USUALLY REQUIRE NONLIVING, PRESERVED CELLS.

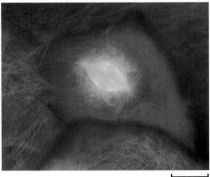

40 μm

In **fluorescent microscopy**, a natural substance in the cell or a fluorescent dye that binds to a specific cell material is stimulated by a beam of light, and the longer wavelength fluorescent light is observed coming directly from the dye.

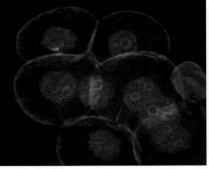

40 μm

Confocal microscopy uses fluorescent materials but adds a system of focusing both the stimulating and emitted light so that a single plane through the cell is viewed. The result is a sharper two-dimensional image than is obtained with standard fluorescent microscopy. Electronic analysis of light from multiple planes through a living cell can provide a three-dimensional image.

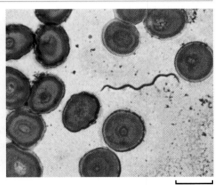

75 μm

In **stained bright-field microscopy**, a stain added to preserved cells enhances contrasts and reveals detail not otherwise visible. Stains differ greatly in their chemistry and capacity to bind to cell materials. so many choices are available.

the great advantages of light microscopy over electron microscopy is that *living* cells can be examined. Many different kinds of light microscopy have been invented and used in the past 50 years in order to examine living cells more effectively (Table 4.1).

Electron microscopy has expanded our knowledge of cellular structures

Some cellular structures are far too small to be resolved with the light microscope. Ribosomes, for example, are 20 nm or less in diameter and cannot be resolved as individual objects under the light microscope. The electron microscope, however, can readily resolve ribosomes.

An **electron microscope** uses powerful magnets as lenses to focus an electron beam (see Figure 4.6b), much as the light microscope employs glass lenses to focus a beam of light. Since we cannot see electrons, the electron microscope directs them at a fluorescent screen or a photographic film to create an image we can see. These images are called *electron micrographs.* Figures 4.3, 4.4, and 4.5 are examples of electron micrographs.

The resolving power of modern electron microscopes is about 0.2 nm, but no biological specimen has yet been seen in such detail. One reason is that the energy of the electron beam at that power is so great that it destroys biological molecules before they can be seen. Because of this and other technical limitations, most electron micrographs resolve detail no finer than 2 nm, and even the best micrographs rarely resolve detail as fine as 1 nm. The corresponding resolving power is about 100,000 times finer than that of the human eye.

There are two types of electron microscopy: transmission and scanning. In **transmission electron microscopy**, which produced the electron micrographs in Figures 4.4 and 4.5, electrons pass *through* a sample. Transmission electron microscopy is used to examine thin slices of objects. Extremely thin sections are shaved off the material to be examined and placed on a grid, which supports the material in the beam of electrons. The grid is comparable to the glass slide that supports the material observed in the light microscope.

In **scanning electron microscopy**, electrons are directed at the surface of the sample, where they cause other electrons to be emitted; the scanning electron microscope focuses these secondary electrons on a viewing screen. Scanning electron microscopy reveals the *surface* structures of three-dimensional objects, such as the egg and the amoeba shown in Figure 4.22. A scanning electron microscope has a resolving power no better than 10 nm, so scanning electron micrographs are usually at a somewhat lower magnification than transmission electron micrographs.

You might think that with such resolving power the electron microscope would be used for all microscopic studies, but this is not so. For some applications, electron microscopy would be sheer overkill—like using a magnifying glass to look at an elephant. A significant limitation of electron microscopy is that biological samples must be killed and dehydrated before they can be examined. In addition, for transmission electron microscopy, cells must be stained or shadowed with heavy metals that will deflect electrons. And the cells must be sliced in thin sections before being placed in the electron beam.

To obtain a reasonable three-dimensional view of large cells or tissues with a microscope, one must look at many successive slices, rather like examining successive slices of Swiss cheese in order to "see" one of the holes. When you look at a transmission electron micrograph, keep in mind that the image is a two-dimensional slice through a three-dimensional reality. The imaginative reconstruction of the third dimension is particularly important with eukaryotic cells because, as we will see in the next section, they have many internal compartments.

Eukaryotic Cells

Animals, plants, fungi, and protists have cells that are larger and structurally more complex than those of the prokaryotes. To get a sense of the most prominent differences, compare the plant and animal cells in Figure 4.7 (eukaryotic cells) with Figure 4.3 (a prokaryotic cell).

Eukaryotic cells generally have dimensions ten times greater than those of prokaryotes; for example, the spherical yeast cell has a diameter of 8 μm. Like prokaryotic cells, eukaryotic cells have a plasma membrane, cytosol, and ribosomes. Unlike prokaryotes, however, eukaryotic cells have an internal cytoskeleton that maintains cell shape and moves materials. And eukaryotic cells contain many different kinds of membranous compartments that carry on particular biochemical functions.

The compartmentalization of eukaryotic cells is the key to their function

The compartments of eukaryotic cells are defined by one or two membranes that regulate what enters and leaves the interior. The membranes ensure that conditions inside the compartment are different from those in the surrounding cytoplasm. Some of the compartments are like little factories that make specific products. Others are like power plants that take energy in one form and convert it to a more useful form. These membranous compartments, as well as other structures (such as ribosomes) that lack membranes but possess distinctive shapes and functions, are called **organelles** (see Figure 4.7).

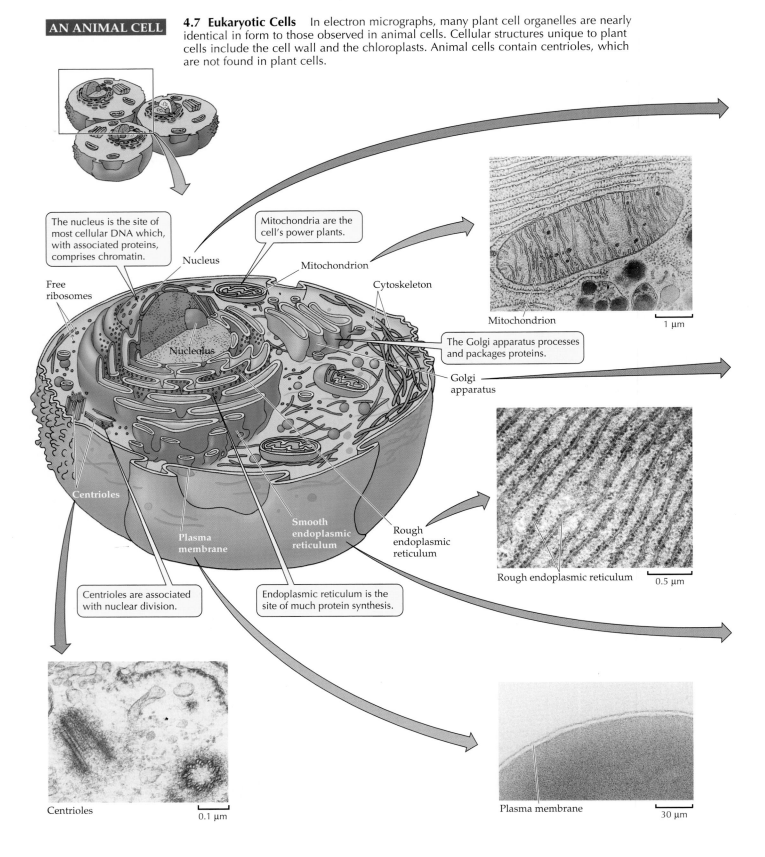

AN ANIMAL CELL

4.7 Eukaryotic Cells In electron micrographs, many plant cell organelles are nearly identical in form to those observed in animal cells. Cellular structures unique to plant cells include the cell wall and the chloroplasts. Animal cells contain centrioles, which are not found in plant cells.

The nucleus is the site of most cellular DNA which, with associated proteins, comprises chromatin.

Mitochondria are the cell's power plants.

Free ribosomes

Nucleus

Mitochondrion

Cytoskeleton

Nucleolus

The Golgi apparatus processes and packages proteins.

Golgi apparatus

Centrioles

Centrioles are associated with nuclear division.

Plasma membrane

Smooth endoplasmic reticulum

Endoplasmic reticulum is the site of much protein synthesis.

Rough endoplasmic reticulum

Mitochondrion   1 μm

Rough endoplasmic reticulum   0.5 μm

Centrioles   0.1 μm

Plasma membrane   30 μm

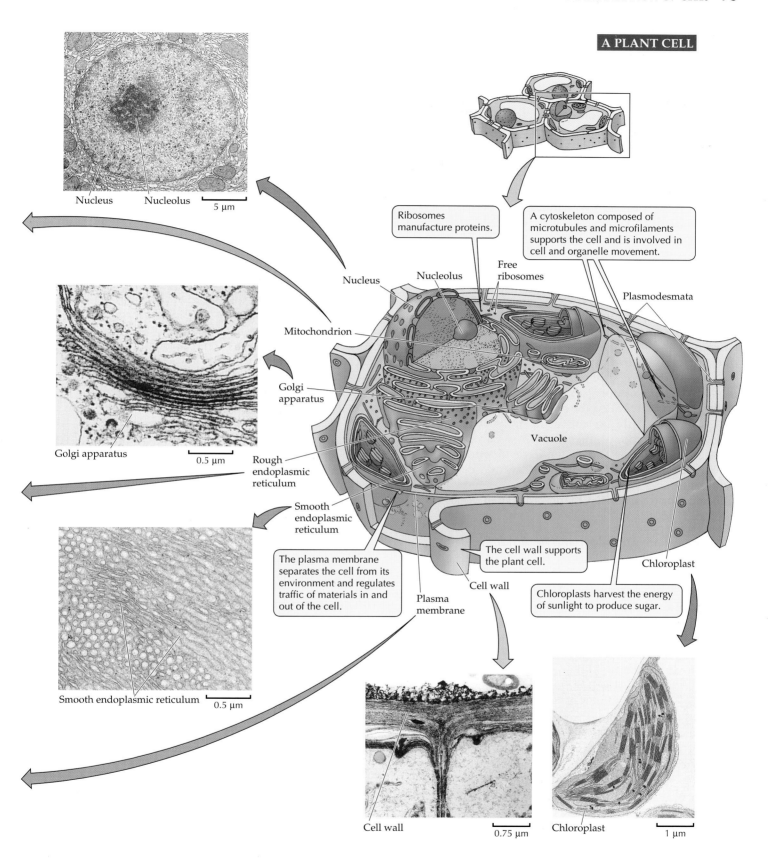

A PLANT CELL

Nucleus   Nucleolus   5 µm

Golgi apparatus   0.5 µm

Smooth endoplasmic reticulum   0.5 µm

Ribosomes manufacture proteins.

A cytoskeleton composed of microtubules and microfilaments supports the cell and is involved in cell and organelle movement.

Nucleus

Nucleolus

Free ribosomes

Plasmodesmata

Mitochondrion

Golgi apparatus

Vacuole

Rough endoplasmic reticulum

Smooth endoplasmic reticulum

The plasma membrane separates the cell from its environment and regulates traffic of materials in and out of the cell.

Plasma membrane

The cell wall supports the plant cell.

Cell wall

Chloroplast

Chloroplasts harvest the energy of sunlight to produce sugar.

Cell wall   0.75 µm

Chloroplast   1 µm

The organelles of cells from different eukaryotes are similar in molecular composition and structure, and they look similar in the electron microscope. Thus, if you know the characteristic appearance of the *nucleus, mitochondrion, endoplasmic reticulum,* and *Golgi apparatus,* you can recognize them in electron micrographs of different eukaryotic cells.

The membranes of all these organelles have similar structure (a phospholipid bilayer), dimensions, and appearance. But there are important differences between the organelles of a plant cell and those of an animal cell (see Figure 4.7). The differences between nuclei from plant and animal cells, for example, reside in the informational content of their DNA. The Golgi apparatuses in both plant and animal cells look very similar and always function in the storage, internal transport, and processing of proteins and other substances. However, the nature of these proteins and other substances cannot be determined by electron microscopy; instead, chemical methods of extraction, purification, and identification are necessary to answer questions about function.

Not all organelles are present or equally abundant in all cells. For example, some plant cells have a specialized organelle, called the *chloroplast* (a type of plastid), that is the site for capturing light energy and using it to bring about the synthesis of carbohydrate from carbon dioxide and water. Cells in animals lack chloroplasts. In another example, animal cells that synthesize and secrete digestive enzymes have more ribosomes and associated membranes than do less active cell types.

Membranes are abundant and important in eukaryotic cells

In 1952, the first people to look at clear electron micrographs of eukaryotic cells were stunned by the complexity of what they saw. No one expected that cells would contain so many internal membranes or that these membranes would be arranged in such diverse forms. What do all these membranes do? How do they function? What is their structure? How do they differ? We will deal with these questions in detail in Chapter 5; here we discuss a few of the most basic ideas.

Biological membranes form compartments whose internal contents and conditions differ from those of the surrounding environment. Although they are a barrier, membranes also permit the passage of certain materials. Thus transport is a major function of all membranes: They regulate molecular traffic between an inside and an outside.

The hydrophobic interior of the membrane serves as a barrier to the passage of charged or polar materials that are readily soluble in water (see Figure 3.2).

Many polar molecules and ions are transported through the membrane with the help of highly specific protein molecules inserted in the phospholipid bilayer. In addition, larger quantities of materials enter the cell by a process in which part of the plasma membrane folds inward and detaches from the rest of the membrane to form a small spherical compartment, a *vesicle,* in the cytoplasm.

Membranes participate in many activities besides transport. They are staging areas for interactions between cells. For example, in the human body, defensive white blood cells recognize and interact with their targets by means of recognition signals built into their plasma membranes (see Chapter 18). And the proper development and organization of multicellular animals depends on recognition of such surface recognition signals (see Chapter 15).

The mitochondrial membranes carry enzymes responsible for energy transformations in cells. In the chloroplasts of green plant cells, tightly stacked membranes contain chlorophyll and carotenoid molecules that capture light energy for photosynthesis. In many respects, a discussion of eukaryotic cells is a discussion of membranes and compartments that are specialized for various cellular functions.

Organelles That Process Information

Living things depend on accurate, appropriate information: internal signals, environmental cues, and stored instructions. Information is *stored* as the sequence of bases in DNA molecules. Most DNA in eukaryotic cells resides in the nucleus. Information is *translated* from the language of DNA into the language of proteins on the surface of the ribosomes. (This process is described in detail in Chapter 12.)

The nucleus stores most of the cell's information

The **nucleus** is usually the largest organelle in a cell (see Figure 4.7). The nucleus of most animal cells is approximately 5 μm in diameter—substantially larger than most entire prokaryotic cells.

The possession of a membrane-enclosed nucleus is the defining property of the eukaryotic cell—prokaryotes have no membranes separating their DNA from the surrounding cytoplasm. As viewed under the electron microscope, a nucleus is surrounded by *two* membranes, which together are called the **nuclear envelope** (Figure 4.8). During most of the life cycle of the cell, the nuclear envelope is a stable structure. However, before duplicated DNA is distributed to separate nuclei during division, the nuclear envelope fragments into vesicles. It re-forms when nuclear division is completed.

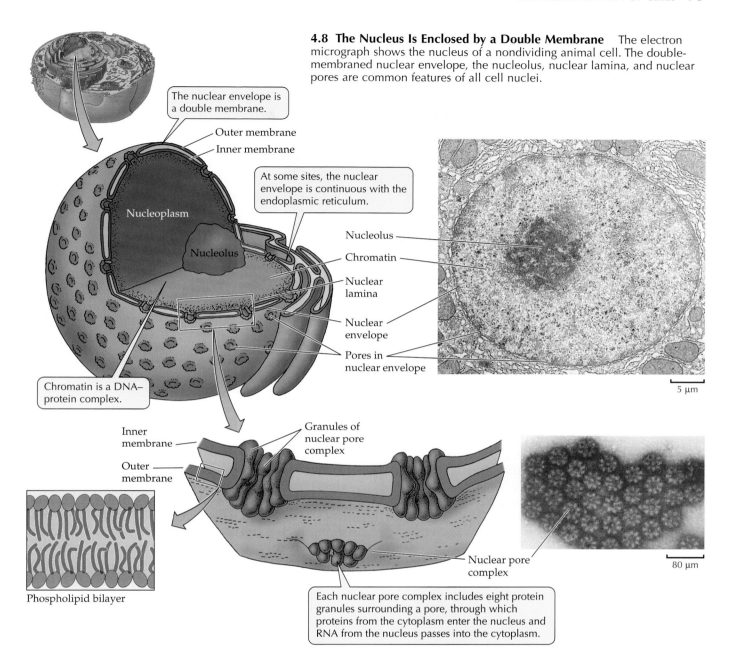

4.8 The Nucleus Is Enclosed by a Double Membrane The electron micrograph shows the nucleus of a nondividing animal cell. The double-membraned nuclear envelope, the nucleolus, nuclear lamina, and nuclear pores are common features of all cell nuclei.

The nuclear envelope is a double membrane.

Outer membrane

Inner membrane

Nucleoplasm

Nucleolus

At some sites, the nuclear envelope is continuous with the endoplasmic reticulum.

Nucleolus

Chromatin

Nuclear lamina

Nuclear envelope

Pores in nuclear envelope

Chromatin is a DNA–protein complex.

5 μm

Inner membrane

Outer membrane

Granules of nuclear pore complex

Nuclear pore complex

80 μm

Phospholipid bilayer

Each nuclear pore complex includes eight protein granules surrounding a pore, through which proteins from the cytoplasm enter the nucleus and RNA from the nucleus passes into the cytoplasm.

The membranes of the nuclear envelope are separated by only a few tens of nanometers and are perforated by **nuclear pores** approximately 9 nm in diameter that connect the interior of the nucleus with the cytoplasm (see Figure 4.8). At these pores the outer membrane of the nuclear envelope is continuous with the inner membrane. Each pore is surrounded by eight large protein granules arranged in an octagon where the inner and outer membranes merge. RNA and water-soluble molecules pass through these pores to enter or leave the nucleus.

At certain sites, the outer membrane of the nuclear envelope folds outward into the cytoplasm and is continuous with the membrane of another organelle, the endoplasmic reticulum (discussed later in the chapter). The endoplasmic reticulum and, to a lesser extent, the cytoplasmic surface of the nuclear envelope often carry great numbers of ribosomes. But there are no ribosomes on the other membrane surfaces of the nuclear envelope.

Inside the nucleus, DNA combines with proteins to form a fibrous complex called **chromatin**. Surround-

4.9 The Nuclear Lamina The shape of the nucleus is maintained by a meshwork of proteins, the nuclear lamina. Both the lamina and the nuclear envelope break down before the nucleus divides, then re-form after nuclear division.

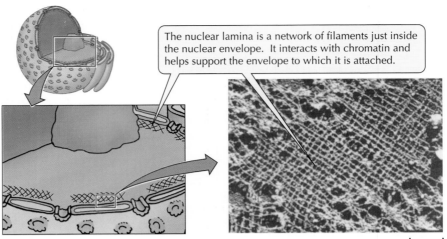

The nuclear lamina is a network of filaments just inside the nuclear envelope. It interacts with chromatin and helps support the envelope to which it is attached.

0.5 μm

ing the chromatin are water and dissolved substances collectively referred to as the **nucleoplasm**. At the periphery of the nucleus, chromatin attaches to a protein meshwork, called the **nuclear lamina**, which is formed by the polymerization of proteins called *lamins* into filaments (Figure 4.9). The nuclear lamina maintains the shape of the nucleus by its attachment to both chromatin and the nuclear envelope. When the nuclear envelope breaks down in preparation for nuclear division, the nuclear lamina also depolymerizes, but after nuclear division it re-forms just as the nuclear envelope re-forms.

Throughout most of the life cycle of the cell, chromatin exists as exceedingly long, thin, entangled threads that cannot be clearly distinguished by any

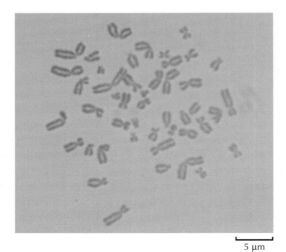

5 μm

4.10 Humans Have 46 Chromosomes The chromosome complement of a eukaryotic cell can be observed early in the process of nuclear division. If a nucleus about to divide is ruptured and treated with certain stains, its chromosomes are readily visible under the light microscope, as this human cell shows.

microscope. However, when the nucleus is about to divide (that is, to undergo mitosis or meiosis; see Chapter 9), the chromatin condenses and coils tightly to form a precise number of readily visible objects called **chromosomes** (Figure 4.10). Each chromosome contains one long molecule of DNA. The chromosomes are the bearers of hereditary instructions; their DNA carries the information required to perform the synthetic functions of the cell and to endow the cell's descendants with the same instructions.

Usually, dense, roughly spherical bodies called **nucleoli** (singular nucleolus) are visible in the nucleus (see Figure 4.8). Taken together, the nucleoli contain from 10 to 20 percent of a cell's RNA. Ribosomal subunits are assembled in the nucleolus from ribosomal RNA and specific proteins. Then they move out of the nucleus into the cytoplasm, where ribosome assembly is completed. Each nucleus must have at least one nucleolus, and the nuclei of some species have several. The exact number of nucleoli in its cells is characteristic of a species. As nuclear and cell division approach, the nucleoli become smaller and disappear, but after division they reappear.

Ribosomes are the sites of protein synthesis

In both eukaryotic and prokaryotic cells, proteins are synthesized on ribosomes. Ribosomes reside in three places in almost all eukaryotic cells: free in the cytoplasm, attached to the surface of endoplasmic reticulum (as will be described later in this chapter), and contained in the mitochondria, where energy is processed. Ribosomes are also found in chloroplasts, the photosynthetic organelles of plant cells. In each of these locations, the ribosomes provide the site where proteins are synthesized under the direction of nucleic acids (see Chapter 12).

The ribosomes of prokaryotes and of eukaryotes are similar in that both consist of two different-sized subunits. Eukaryotic ribosomes are somewhat larger, but the structure of prokaryotic ribosomes is better understood. Chemically, ribosomes consist of a special type of RNA, called ribosomal RNA, to which more than 50 different protein molecules are bound. The ribosome temporarily binds two other types of RNA molecules (messenger RNA and transfer RNA) as hereditary information from the DNA is translated into the primary structure of protein.

Organelles That Process Energy

In addition to information, cells require energy and raw materials. A cell uses energy to transform raw materials into cell-specific materials that it can use for ac-

tivities such as growth, reproduction, and movement. Energy is transformed from one form to another in the mitochondria found in all eukaryotic cells and in the chloroplasts of cells that harvest energy from sunlight (see Figure 4.7). In contrast, energy transformations in prokaryotic cells are associated with enzymes attached to the inner surface of the plasma membrane or extensions of the plasma membrane that protrude into the cytoplasm.

Mitochondria are energy transformers

In eukaryotic cells, utilization of food molecules such as glucose begins in the cytosol (the liquid part of the cytoplasm). The fuel molecules that result from partial degradation of this food enter **mitochondria** (singular mitochondrion), whose primary function is to convert the potential chemical energy of fuel molecules into a form that the cell can use: the energy-rich molecule called **ATP**, or *adenosine triphosphate* (see Chapter 6).

ATP is not a long-term energy storage form but a kind of energy currency. Its role in the cell is analogous to the role of paper money and coins in an economy. Chemically, ATP can participate in a great number of different cellular reactions and processes that require energy. In the mitochondria, the production of ATP using fuel molecules and O_2 is called *cellular respiration*.

Typical mitochondria are small—somewhat less than 1.5 μm in diameter and 2 to 8 μm in length—about the size of many bacteria. Mitochondria are visible with a light microscope, but almost nothing was known of their precise structure until they were examined with the electron microscope. Electron micrographs show that mitochondria have two membranes: an outer membrane and an inner membrane. The **outer membrane** is smooth and protective, and it offers little resistance to the movement of substances into and out of the mitochondrion.

Immediately inside the outer mitochondrial membrane is an **inner membrane** that folds inward in many places, giving it a much greater surface area than that of the outer membrane (Figure 4.11). In animal cells, these folds tend to be quite regular, giving rise to shelflike structures called **cristae**. The mitochondria of plants also have cristae, but plant cristae tend to be much less regular in size and structure. Special techniques and electron microscopy show that the inner

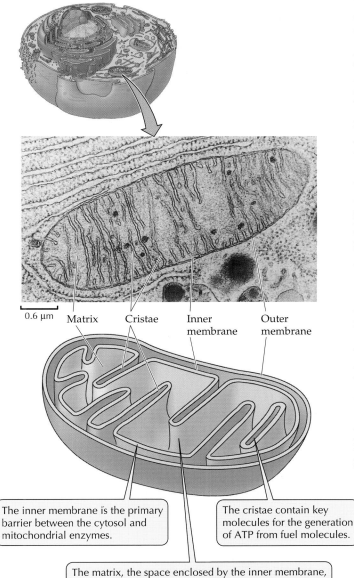

0.6 μm   Matrix   Cristae   Inner membrane   Outer membrane

The inner membrane is the primary barrier between the cytosol and mitochondrial enzymes.

The cristae contain key molecules for the generation of ATP from fuel molecules.

The matrix, the space enclosed by the inner membrane, contains several of the enzymes for cellular respiration. It also contains ribosomes and DNA.

4.11 A Mitochondrion Converts Energy from Fuel Molecules into ATP The electron micrograph is a two-dimensional slice through a three-dimensional reality. As this drawing emphasizes, the cristae are extensions of the inner mitochondrial membrane.

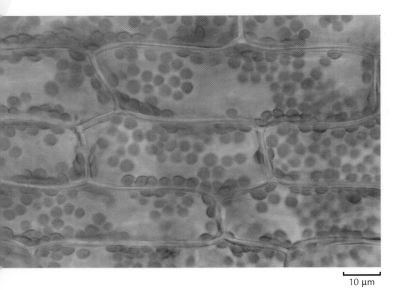

some enzymes, the matrix contains some ribosomes and DNA that make some of the proteins needed for cellular respiration. In some ways, the mitochondria function as semiautonomous "organisms" within the cytoplasm of eukaryotic cells. With their own DNA and the capacity to use its coded information for the synthesis of some of their own proteins, these organelles show striking similarities to bacteria. We will explore the evolutionary origin of mitochondria and chloroplasts a little later in this chapter.

Almost all eukaryotes have mitochondria. The number of mitochondria per cell ranges from one contorted giant in some unicellular protists to a few hundred thousand in large egg cells. An average human liver cell contains more than a thousand mitochondria. Cells that require the most chemical energy tend to have the most mitochondria per unit volume. In Chapter 7 we will see how different parts of the mitochondrion work together in cellular respiration.

10 μm

4.12 Chloroplasts in Cells   These plant cells contain many chloroplasts, the green organelles that carry on photosynthesis.

Plastids photosynthesize or store materials

CHLOROPLASTS.   One class of organelles—the **plastids**—is produced only in plants and certain protists. The most familiar of the plastids is the **chloroplast**, which contains the green pigment *chlorophyll* and is the site of photosynthesis (Figure 4.12). In photosynthesis, light energy is converted into the energy of chemical bonds. The molecules formed in photosynthesis provide food

mitochondrial membrane contains many large protein molecules that participate in cellular respiration and the production of ATP. The inner membrane exerts much more control over what enters and leaves the mitochondrion than does the outer membrane.

The region enclosed by the inner membrane is referred to as the **mitochondrial matrix**. In addition to

4.13 The Chloroplast: The Organelle That Feeds the World   An electron micrograph of a chloroplast from a leaf of corn. Chloroplasts are large compared to mitochondria and contain an extensive network of photosynthetic thylakoids surrounded by two membranes.

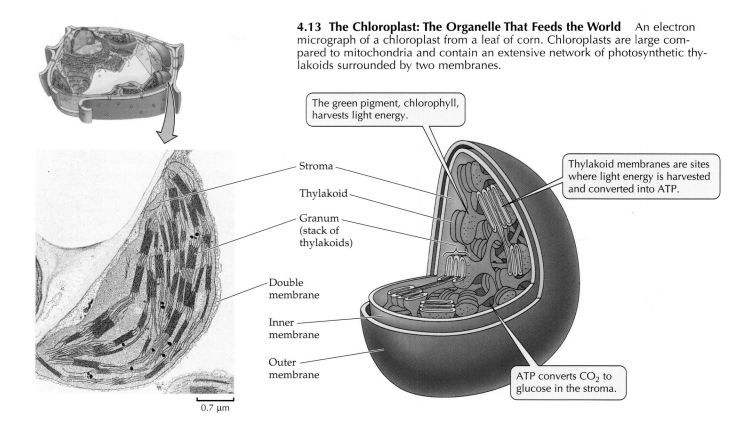

The green pigment, chlorophyll, harvests light energy.

Thylakoid membranes are sites where light energy is harvested and converted into ATP.

Stroma

Thylakoid

Granum (stack of thylakoids)

Double membrane

Inner membrane

Outer membrane

ATP converts CO_2 to glucose in the stroma.

0.7 μm

for the plant itself and for other organisms that eat plants. Directly or indirectly, photosynthesis is the energy source for most of the living world.

Like the mitochondrion, the chloroplast is surrounded by two membranes. Arising from the inner membrane is a series of discrete internal membranes whose structure and arrangement vary from one group of photosynthetic organisms to another. As an introduction, we concentrate on the chloroplasts of the flowering plants. Even these show some variation, but the pattern shown in Figure 4.13 is typical.

As seen in electron micrographs, chloroplasts contain membrane structures that look like stacks of pancakes. These stacks, called **grana** (singular granum), consist of a series of flat, closely packed, circular sacs called **thylakoids**. Thylakoids are surrounded by a single membrane composed of the usual membrane components (phospholipids and proteins), to which have been added chlorophyll and carotenoids, which are needed to harvest light energy and use it to produce glucose from CO_2 and water. All the cell's chlorophyll is contained in the thylakoid membranes. Thylakoids of one granum may be connected to those of other grana (see Figure 4.13), making the interior of the chloroplast a highly developed network of membranes.

The fluid in which the grana are suspended is referred to as **stroma**. Like the mitochondrial matrix, the chloroplast stroma contains ribosomes and DNA, and these are used to synthesize some, but not all, of the proteins that make up the chloroplast.

Not all plant cells contain chloroplasts. Most root cells, for example, lack chloroplasts, although the DNA in their nuclei has the information for chloroplast construction. By not forming chloroplasts in cells that do not receive light, the plant conserves energy.

Animal cells do not *produce* chloroplasts, but some do *contain* functional chloroplasts. These are taken up either as free chloroplasts derived from the partial digestion of green plants, or as bound chloroplasts contained within unicellular algae that live within the animal's tissues. The green color of some corals and sea anemones results from chloroplasts in algae that live within the animals (Figure 4.14). The animals derive some of their nutrition from the photosynthesis that these chloroplast-containing "guests" carry out.

OTHER TYPES OF PLASTIDS. Chloroplasts are not the only plastids found in plants. The red color of a ripe tomato results from the presence of legions of plastids called **chromoplasts**. Just as chloroplasts derive their color from chlorophyll, chromoplasts are red, orange, or yellow depending on the kinds of carotenoid pigments present (see Chapter 3). The chromoplasts have no known chemical function in the cell, but the colors they give to some petals and fruits probably help attract ani-

4.14 Living Together: Anemone–Alga Symbiosis
Symbiosis is the coexistence of organisms, sometimes for mutual benefit, as in this case. This giant sea anemone, an animal, owes its green color to the chloroplasts in a unicellular alga that lives and carries on photosynthesis within the anemone's tissues.

mals that assist in pollination or seed dispersal. (On the other hand, carrot roots gain no apparent advantage from being orange.) Other plastids, called **leucoplasts**, are storage depots for starch and fats.

All plastids develop from small proplastids, which are very simple in structure. And all plastid types are related to one another. For example, chromoplasts are formed from chloroplasts that lose their chlorophyll and undergo changes in internal structure.

Some organelles have an endosymbiotic origin

Chloroplasts and mitochondria are about the size of whole prokaryotes; they contain DNA and have ribosomes that are similar to prokaryotic ribosomes. And these organelles divide within the cell to produce additional mitochondria and chloroplasts. Given these facts, might they not be treated like little cells in their own right?

At one time, biologists believed that it might be possible to grow chloroplasts or mitochondria in culture, outside the cells they normally inhabit. These efforts failed because organelles depend on the cell's nucleus and cytoplasm for some essential components. But the experiments did help nurture thoughts about another important question: How did the eukaryotic cell with its organelles arise in the first place?

As we have seen, prokaryotic cells are generally simpler in structure than eukaryotic cells, precisely because prokaryotes *lack* membrane-enclosed organelles. Prokaryotic fossils have been found in sediments well over 3 billion years old, whereas the earliest known eukaryotic fossils date back to only 1.4 billion years ago. Biologists thus generally agree that eukaryotes evolved from prokaryotes. But how?

An explanation that has gained acceptance in recent decades is the **endosymbiosis theory** of the origin of mitochondria and chloroplasts. An important current champion of and contributor to this theory is Lynn Margulis of the University of Massachusetts, Amherst, who proposed the following idea.

Picture a time, around 2 billion years ago, when only prokaryotes inhabited Earth. Some of them absorbed their food directly from the environment. Others were photosynthetic. Still others fed on smaller prokaryotes by engulfing them (Figure 4.15). Under these conditions, suppose that a small, photosynthetic prokaryote was *in*gested by a larger one but was not *di*gested. Instead it survived trapped within a vesicle in the cytoplasm of the larger cell.

Suppose further that the smaller prokaryote divided at about the same rate as the larger one, so successive generations of the larger cell also contained the offspring of the smaller one. We would call this phenomenon *endosymbiosis* (*endo-*, "within"; *symbiosis*, "living together"), comparable to the algae that live within sea anemones seen in Figure 4.14. The endosymbiosis described here provided benefits for both organisms: The larger cell obtained the photosynthetic products from the smaller cell, and the smaller cell was protected by the larger one.

Could the little green prokaryote that took up residence in the larger prokaryote have been the first chloroplast? Present-day chloroplasts are surrounded by a double membrane, a structure that might have arisen when, in the process of engulfing the photosynthetic cell, the plasma membrane of the larger cell extended around the plasma membrane of the smaller cell (see Figure 4.15). But there is even stronger evidence for the endosymbiosis theory.

The fact that chloroplasts contain ribosomes and DNA is consistent with the endosymbiosis theory. Additional evidence comes from studies of ribosome structures. In composition, structure, and size, the ribosomes of chloroplasts resemble the smaller ribosomes of prokaryotes more than they resemble the ribosomes in the eukaryotic cytoplasm.

Similar evidence and arguments also support the proposition that mitochondria are the descendants of respiring prokaryotes engulfed by, and ultimately endosymbiotic with, larger prokaryotes. In addition to having prokaryote-like ribosomes, the mitochondrial inner membrane shows striking similarities to some of the energy-transforming bacterial membranes, and certain mitochondrial enzymes have primary structures similar to those of prokaryotic enzymes. In the case of mitochondria, the benefits of the initial endosymbiotic relationship might have been due to the

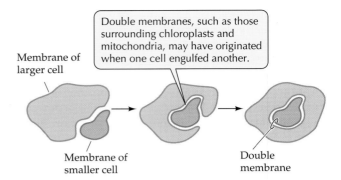

4.15 Origin of an Organelle's Double Membrane
The double membrane that encloses mitochondria and chloroplasts may have arisen from two different sources: the outer membrane from the engulfing cell's plasma membrane and the inner membrane from the engulfed cell's plasma membrane.

capacity of the engulfed prokaryote to detoxify molecular oxygen (O_2), which was increasing in the atmosphere because of photosynthesis.

The fact that a few modern cells contain other, smaller cells as endosymbionts supports the endosymbiosis theory of the origin of the eukaryotic cell by showing that engulfment can lead to a stable condition. But mitochondria and chloroplasts are not enough to make a prokaryote into a eukaryote. The theory is still incomplete. For example, the origins of the nuclear envelope and other important structures—including those responsible for nuclear division—still need to be understood better.

We discuss further aspects of the origin of the eukaryotic cell in Chapter 26. Is the endosymbiosis theory true? Almost certainly. The endosymbiosis theory is a good example of creative biological thinking that makes logical sense out of a variety of facts about organelles and bacteria.

The Endomembrane System

Much of the volume of a eukaryotic cell is taken up by its extensive membrane systems. All of these membranes look similar in electron micrographs. In addition, when organelles are extracted from whole cells and separated from other cells, their lipid compositions are found to be similar, if not identical. These observations suggest that some of the organelles are parts of a single system, called the **endomembrane system**. Evidence for the interrelationship of the cellular membranes exists in electron micrographs that show connections between different parts of the endomembrane system.

In this section, we'll examine the functional significance of these interrelationships and discover that materials synthesized in the endoplasmic reticulum can be transferred to another organelle, the Golgi apparatus, for further processing, storage, or transport. We will also describe the role of the lysosome in cell digestion.

The endoplasmic reticulum is a complex factory

Electron micrographs reveal a network of membranes branching throughout the cytoplasm. These membranes form tubes and flattened sacs called the **endoplasmic reticulum**, or **ER**, in which the interior compartment, referred to as the *lumen*, is separate and distinct from the surrounding cytoplasm (Figure 4.16). At certain sites, the ER is continuous with the outer membrane of the nuclear envelope.

Parts of the ER are liberally sprinkled with ribosomes, which are attached to the outer faces of the flattened sacs. Because of their appearance in the electron microscope, these regions are called **rough ER** (see the top halves of Figure 4.16). The attached ribosomes are sites for the synthesis of proteins that function outside the cytosol—that is, proteins that are to be exported from the cell, incorporated into membranes, or moved into organelles of the endomembrane system. These proteins enter the lumen of the ER as they are synthesized, directed there by a special sequence of amino acids known as the *signal sequence*.

Once in the lumen of the ER, these proteins undergo several changes, including the formation of disulfide bridges and folding into their tertiary structures (see Figure 3.18). Some proteins gain carbohydrate groups in the rough ER, thus becoming glycoproteins. The carbohydrate groups are part of an "addressing" system that ensures that the right proteins are directed to the right parts of the cell.

Proteins that remain within the cytosol or move into mitochondria and chloroplasts are synthesized on "free" ribosomes in the cytosol—ribosomes that are *not* attached to the ER. These proteins lack the signal sequence that would otherwise direct them into the ER (see Chapter 12).

Some parts of the endoplasmic reticulum, called the **smooth ER**, are more tubular (less like flattened sacs) and lack ribosomes (see the bottom halves of Figure 4.16). Within the lumen of the smooth ER, proteins that have been synthesized on the rough ER are chemically modified. The smooth ER is also the site at which phospholipids, steroids, and fatty acids are synthesized, some carbohydrates are metabolized, and toxic substances such as drugs are rendered inert. Lots of things happen in the smooth ER.

Cells that synthesize a lot of protein for export are usually packed with ER. Examples include glandular cells that secrete digestive enzymes (see Chapter 47) and plasma cells that secrete antibodies (see Chapter

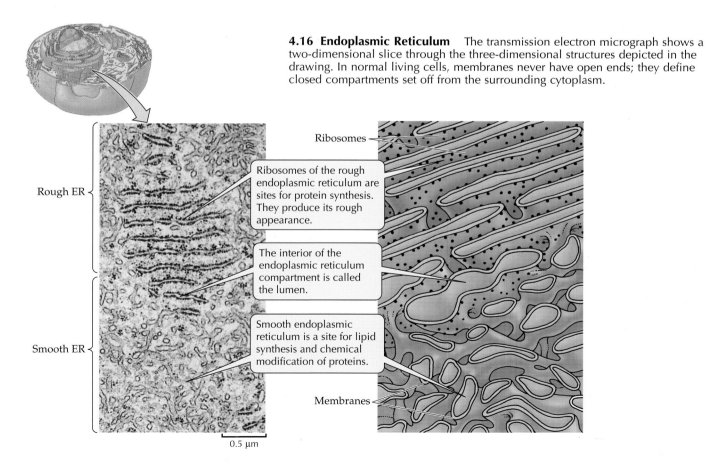

4.16 Endoplasmic Reticulum The transmission electron micrograph shows a two-dimensional slice through the three-dimensional structures depicted in the drawing. In normal living cells, membranes never have open ends; they define closed compartments set off from the surrounding cytoplasm.

Ribosomes

Rough ER

Ribosomes of the rough endoplasmic reticulum are sites for protein synthesis. They produce its rough appearance.

The interior of the endoplasmic reticulum compartment is called the lumen.

Smooth ER

Smooth endoplasmic reticulum is a site for lipid synthesis and chemical modification of proteins.

Membranes

0.5 μm

18). In contrast, cells with less work to do (such as storage cells) contain very little ER.

The Golgi apparatus stores, modifies, and packages proteins

In 1898 the Italian microscopist Camillo Golgi discovered a delicate structure in nerve cells, which came to be known as the **Golgi apparatus**. Because of the resolution limits of light microscopy, and because the staining technique often failed to reveal the structure, many biologists regarded the structure as a figment of Golgi's imagination. In the late 1950s, however, the electron microscope showed clearly that the Golgi apparatus does exist—and not just in nerve cells, but in most eukaryotic cells.

The exact appearance of the Golgi apparatus varies from species to species, but it always consists of flattened membranous sacs called *cisternae* and small membrane-enclosed *vesicles*. The flattened sacs are always seen lying together like a stack of saucers (Figure 4.17*a*). In the cells of plants, protists, fungi, and many invertebrate animals, these stacks are individual units scattered throughout the cytoplasm. In vertebrate cells, a few such stacks usually form a larger, more complex Golgi apparatus. The bottom saucers, constituting the *cis* region of the Golgi apparatus, lie nearest the nucleus or a patch of rough ER (Figure 4.17*b*). The top saucers, constituting the *trans* region, lie closest to the surface of the cell. The saucers in the middle make up the *medial* region of the complex. These three parts of the Golgi apparatus contain different enzymes and perform different functions.

What are the functions of the organelle that Golgi discovered? The first clue comes from observing the relationships between the Golgi apparatus and other parts of the cell. Vesicles form from the rough ER, move through the cytoplasm, and fuse with the *cis* region of the Golgi apparatus, where their contents are released into the lumen of the Golgi. Other small vesicles move between the flattened sacs, always in the direction *cis* to *trans*, transporting proteins. Associated with the sacs, particularly those toward the *trans* region, are tiny vesicles that pinch off from the sacs and move to other sacs or away from the Golgi (see Figure 4.17*b*).

The membranes of two vesicles can sometimes make contact with each other and fuse, resulting in a larger vesicle and a mixing of the contents. Vesicles may also fuse with other organelles or with the plasma membrane, where they release their contents

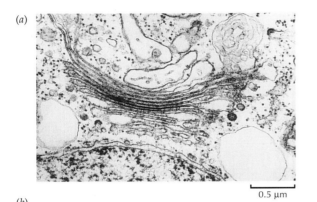

(a)

0.5 μm

(b)

4.17 The Golgi Apparatus The Golgi apparatus modifies incoming proteins and "targets" them to the correct addresses.

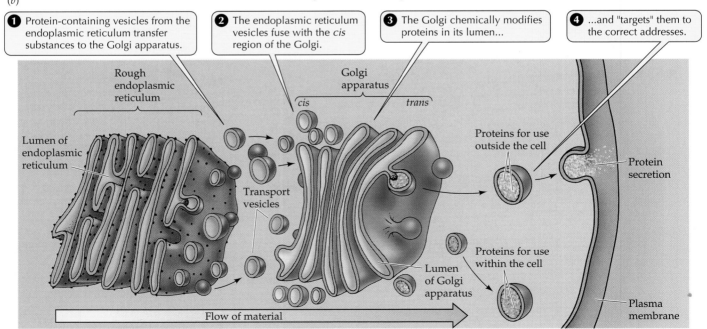

1 Protein-containing vesicles from the endoplasmic reticulum transfer substances to the Golgi apparatus.

2 The endoplasmic reticulum vesicles fuse with the *cis* region of the Golgi.

3 The Golgi chemically modifies proteins in its lumen...

4 ...and "targets" them to the correct addresses.

Rough endoplasmic reticulum

Golgi apparatus

cis *trans*

Lumen of endoplasmic reticulum

Transport vesicles

Proteins for use outside the cell

Protein secretion

Proteins for use within the cell

Lumen of Golgi apparatus

Plasma membrane

Flow of material

to the outside of the cell. The formation, transport, and fusing behavior of vesicles is essential to the function of the Golgi apparatus: The Golgi apparatus serves as a sort of postal depot in which some of the proteins synthesized by ribosomes on the rough ER are stored, chemically modified, and packaged for delivery to the outside of the cell or to other organelles within the cell.

How does the Golgi apparatus send the right proteins to the right destinations? The delivery system consists of a series of chemical reactions in which proteins gain specific chemical "address tags." For example, a protein destined for use in an organelle called a lysosome (discussed in the next section) is given a certain signal sequence that directs it into the lumen of the rough ER as the protein is synthesized. In the lumen of the ER, the signal sequence is removed, and an oligosaccharide "address label" is added in the form of a glycoprotein. This tag identifies the proteins that are to be secreted from the cell. But before the Golgi secretes these proteins, its enzymes modify the oligosaccharide by adding a phosphate group (that is, by phosphorylating it).

When a phosphorylated glycoprotein reaches the *trans* region of the Golgi apparatus, it binds to a specific receptor protein in the Golgi membrane. A vesicle containing the bound and phosphorylated glycoprotein separates from the Golgi apparatus and delivers its contents to the developing lysosome. Comparable mechanisms deliver other proteins to other parts of the cell. The Golgi apparatus exports some proteins constantly but retains others, releasing them only at the appropriate time. In these ways the Golgi apparatus directs the molecular "mail" of the cell.

Lysosomes contain digestive enzymes

Originating in part from the Golgi apparatus, organelles called **lysosomes** contain and transport digestive enzymes that accelerate the breakdown of proteins, polysaccharides, nucleic acids, and lipids. Lysosomes are surrounded by a single membrane and have a densely staining, featureless interior (Figure 4.18a).

Lysosomes are sites for the breakdown of food and foreign objects taken up by phagocytosis. In **phagocytosis** (*phago-*, "eating"; *cytosis*, "cellular"), a pocket

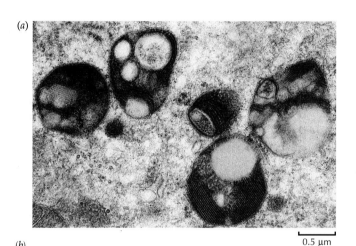

(a)

0.5 μm

(b)

4.18 Lysosomes Isolate Digestive Enzymes from the Cytoplasm (a) In this electron micrograph of a rat cell, the darkly stained organelles are secondary lysosomes in which digestion is taking place. (b) The origin and action of lysosomes and lysosomal digestion.

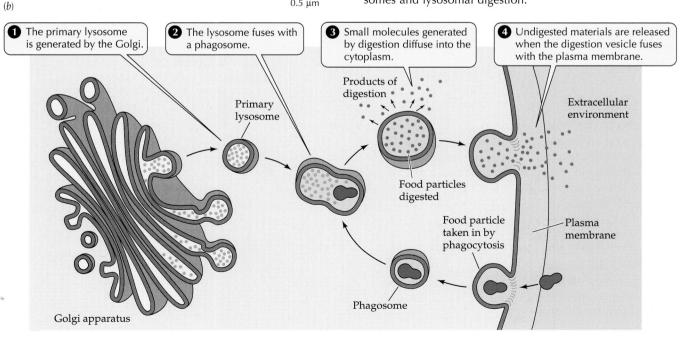

1 The primary lysosome is generated by the Golgi.

2 The lysosome fuses with a phagosome.

3 Small molecules generated by digestion diffuse into the cytoplasm.

4 Undigested materials are released when the digestion vesicle fuses with the plasma membrane.

Products of digestion

Food particles digested

Food particle taken in by phagocytosis

Phagosome

Primary lysosome

Golgi apparatus

Extracellular environment

Plasma membrane

forms in the plasma membrane and eventually deepens and encloses material from outside the cell. The pocket becomes a small vesicle and breaks free of the plasma membrane to move into the cytoplasm as a **phagosome** that contains food or other material for chemical digestion (Figure 4.18b). The phagosome fuses with a primary lysosome pinched off from the Golgi apparatus, to form a **secondary lysosome**, in which digestion takes place.

The effect of this fusion is rather like releasing hungry foxes into a chicken coop. The enzymes in the secondary lysosome quickly hydrolyze the food particles. The activity of the enzymes is enhanced by the mild acidity of the lysosome's interior, where the pH is lower than in the surrounding cytoplasm. The products of digestion exit through the membrane of the lysosome, providing fuel molecules and raw materials for other cell processes. The "used" secondary lysosome containing undigested particles then moves to the plasma membrane, fuses with it, and releases the undigested contents to the environment.

Phagocytosis and lysosomal digestion are important activities of some of the white blood cells in humans and other vertebrates. These cells identify and attack foreign cells, abnormal cells, and cell debris resulting from trauma, disease, or normal wear and tear.

Lysosomal compartmentalization is an effective arrangement to prevent the digestive enzymes from attacking the contents of the cytosol and the other organelles. These enzymes are isolated in the lysosome and cannot escape. The consequences of digestive enzymes escaping from the lysosomes can be severe. But such digestive activity is sometimes appropriate, as during the development of a frog from a tadpole. The fleshy tail of the tadpole, which no longer exists on the mature frog, disappears in part because lysosomes within the tail cells of the tadpole break down, releasing enzymes into the cytoplasm that digest the cells themselves.

Other Organelles

In addition to the information-processing organelles (nucleus and ribosomes), the energy-processing organelles (mitochondria and chloroplasts), and the organelles that form the endomembrane system of the cell (endoplasmic reticulum, Golgi apparatus, and lysosomes), there are two other kinds of membrane-enclosed organelles: peroxisomes and vacuoles.

Peroxisomes house specialized chemical reactions

As seen with the electron microscope, **peroxisomes** are small organelles—0.2 to 1.7 μm in diameter—and they have a single membrane and a granular interior (Figure 4.19). Peroxisomes form on the rough ER and

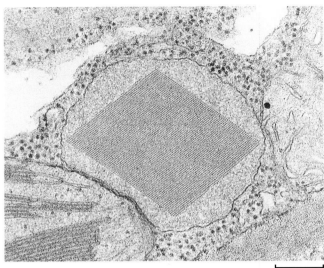

0.25 μm

4.19 A Peroxisome A diamond-shaped crystal, composed of an enzyme, almost entirely fills this rounded peroxisome in a leaf cell. The enzyme catalyzes one of the reactions fulfilling the special function of the peroxisome.

are found at one time or another in at least some of the cells of almost every eukaryotic species.

Peroxisomes are organelles within which toxic peroxides (such as hydrogen peroxide, H_2O_2) are formed as unavoidable side products of chemical reactions. Subsequently, the peroxides are safely broken down within the peroxisomes without mixing with other parts of the cell.

A structurally similar organelle, the **glyoxysome**, is found only in plants. Glyoxysomes, which are most prominent in young plants, are the sites where stored lipids are converted into carbohydrates for transport to growing cells.

Vacuoles are filled with water and soluble substances

Many eukaryotic cells, but particularly those of plants and protists, contain membrane-enclosed organelles that look empty under the electron microscope. These organelles are called **vacuoles** (Figure 4.20). They are not actually empty; rather they are filled with aqueous solutions that contain many dissolved substances.

Despite their structural simplicity, vacuoles have a variety of functions in the lives of cells. For example, like animals and other organisms, plant cells produce a number of toxic by-products and waste materials. Animals have specialized excretory mechanisms for getting rid of such wastes, but plants are not equipped in the same way. Although plants can secrete some wastes to their environment, many toxic and waste

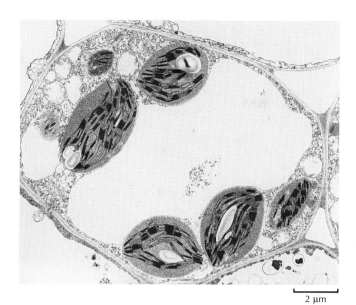

4.20 Vacuoles in Plant Cells Are Usually Large The large central vacuole in this cell is typical of mature plant cells. Smaller vacuoles are visible toward each end of the cell.

materials they simply store within vacuoles. And since they are poisonous or distasteful, these stored materials deter some animals from eating the plants. Thus stored wastes may contribute to plant survival for reproduction.

In many plant cells, enormous vacuoles take up more than 90 percent of the cellular volume and grow as the cell grows. But vacuoles are by no means a waste of space, for the dissolved substances in the vacuole, working together with the vacuolar membrane, provide the turgor, or stiffness, of the cell, which in turn provides support for the structure of nonwoody plants. Vacuoles even play a role in the sex life of plants. Some pigments (especially blue and pink ones) in petals and fruits are contained in vacuoles. These pigments—the anthocyanins—are visual cues that encourage animals to visit flowers and thus aid in pollination, or to eat fruits and thus aid in seed dispersal.

Some unicellular protists, simple multicellular organisms such as sponges, and some of the other ancient invertebrate animals obtain nutrients directly by phagocytosis. Particles from the environment are trapped and engulfed by phagosomes, which in these cells are called *food vacuoles.*

Many freshwater protists also have a highly specialized **contractile vacuole** (see Chapter 26). Its function is to rid the cell of excess water that rushes in because of the imbalance in salt concentration between the relatively salty interior of the cell and its freshwater environment. The contractile vacuole visibly enlarges as water enters, then abruptly contracts, forcing the water out of the cell through a special pore structure.

The Cytoskeleton

As you have discovered, membranes divide the cytoplasm of eukaryotic cells into numerous compartments. But membrane-enclosed organelles and ribosomes are not the only large constituents of the cytoplasm. Also present are a dynamic set of long, thin fibers called the **cytoskeleton**, which fills at least two important roles: maintaining cell shape and support, and providing for various types of cell movement (Figure 4.21). Some fibers act as tracks or supports for "motor proteins" that help a cell move or that move things within the cell. In the discussion that follows, we'll look at three components of the cytoskeleton that are visible in electron micrographs: microfilaments, intermediate filaments, and microtubules. These structures are illustrated in detail in Figure 4.21.

Microfilaments function in support and movement

Microfilaments (actin filaments) exist as single filaments, in bundles, or in networks. They stabilize cell shape and help the entire cell or parts of the cell contract. In muscle cells, actin fibers are associated with myosin fibers, and their interactions account for the contraction of muscles. In other cells, actin fibers are associated with localized changes of shape in cells. For example, microfilaments are involved in a flowing movement of the cytoplasm called *cytoplasmic streaming*, in movements of specific organelles and particles within cells, and in "pinching" contractions that divide an animal cell into two daughter cells (Figure 4.22a). Microfilaments are also involved in the formation of cellular extensions, called *pseudopodia* (*pseudo-*, "false;" *podia*, "feet"), that enable amoebas to move (Figure 4.22b).

Microfilaments are assembled from a protein called *G actin.* G actin is a globular protein that has distinct "head" and "tail" sites for interacting with other molecules of G actin to assemble into a long chain (see Figure 4.21). Two of these chains interact to form the double helical structure that is a microfilament, which is about 7 nm in diameter and several micrometers long. The polymerization of G actin into microfilaments is reversible, and microfilaments can disappear from cells by breaking down into free units of G actin that are too small to see with the electron microscope.

In some cells (animal muscle cells, for example) microfilaments are very stable. For instance, microfilaments provide an internal structure to maintain the form of tiny extensions called *microvilli* (singular microvillus) that increase the surface area of cells special-

The cytoskeleton

Components of the cytoskeleton maintain cell shape, reinforce the cell, and contribute to cell movements.

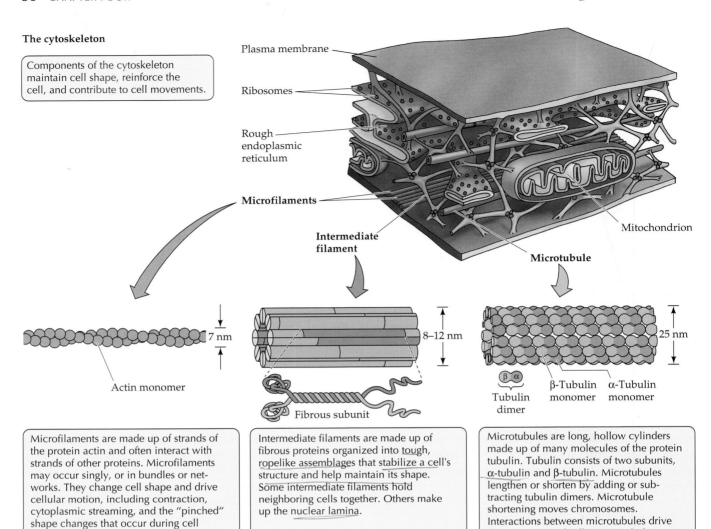

Plasma membrane

Ribosomes

Rough endoplasmic reticulum

Microfilaments

Intermediate filament

Mitochondrion

Microtubule

7 nm

8–12 nm

25 nm

Actin monomer

Fibrous subunit

β α

Tubulin dimer

β-Tubulin monomer

α-Tubulin monomer

Microfilaments are made up of strands of the protein actin and often interact with strands of other proteins. Microfilaments may occur singly, or in bundles or networks. They change cell shape and drive cellular motion, including contraction, cytoplasmic streaming, and the "pinched" shape changes that occur during cell division. Microfilaments and myosin strands together drive muscle action.

Intermediate filaments are made up of fibrous proteins organized into tough, ropelike assemblages that stabilize a cell's structure and help maintain its shape. Some intermediate filaments hold neighboring cells together. Others make up the nuclear lamina.

Microtubules are long, hollow cylinders made up of many molecules of the protein tubulin. Tubulin consists of two subunits, α-tubulin and β-tubulin. Microtubules lengthen or shorten by adding or subtracting tubulin dimers. Microtubule shortening moves chromosomes. Interactions between microtubules drive the movement of cells. Microtubules serve as "tracks" for the movement of vesicles.

4.21 The Cytoskeleton Three highly visible and important structural components of the cytoskeleton are shown in detail.

ized for absorption (Figure 4.23). Microfilaments can form networks when other specific proteins, called *actin-binding proteins*, interact with G actin units in specific ways. Such a network forms directly beneath the plasma membrane and helps determine cell shape, as described in more detail in the next section.

MEMBRANE INTEGRITY UNDER STRESS. Red blood cells appear fragile, yet they survive repeated compression and deformation as they squeeze through the finest of capillaries. The red blood cell gets this surprising resilience from certain proteins associated with its plasma membrane (Figure 4.24). The protein spectrin

4.22 Microfilaments for Motion
The contraction of microfilaments contributes (*a*) to the division of animal cells and (*b*) to amoeboid motion.

(*a*)

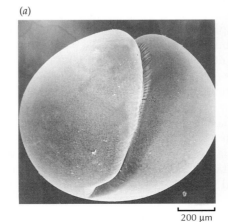

200 μm

(*b*)

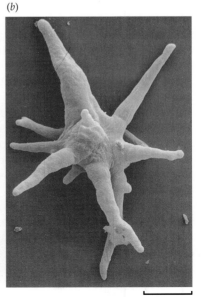

25 μm

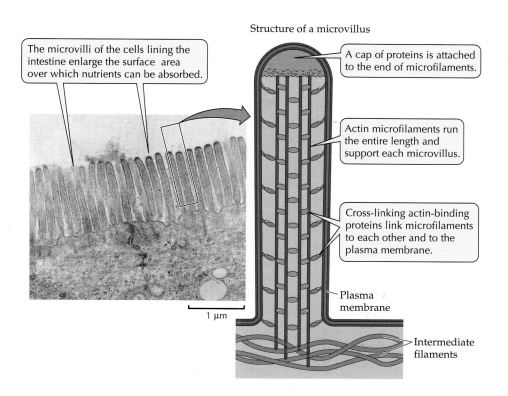

Structure of a microvillus

The microvilli of the cells lining the intestine enlarge the surface area over which nutrients can be absorbed.

A cap of proteins is attached to the end of microfilaments.

Actin microfilaments run the entire length and support each microvillus.

Cross-linking actin-binding proteins link microfilaments to each other and to the plasma membrane.

Plasma membrane

Intermediate filaments

1 μm

4.23 Microfilaments for Support Microfilaments form the backbone of the microvilli that increase the surface area of some cells.

forms a meshwork of microfibrils on the cytoplasmic surface of the plasma membrane. This spectrin meshwork provides structural support. It is anchored to the actin filaments of the cytoskeleton. Another peripheral (surface) protein, ankyrin, anchors the spectrin to the membrane at many points by binding both the spectrin and an anion channel that is a protein embedded in the membrane.

Genetic defects in spectrin and others of these proteins result in abnormal red blood cells and thus in various diseases. Mice with hemolytic anemia have spherical, fragile red blood cells. Their red blood cells have very little spectrin, but the cells take on a normal shape if provided with spectrin.

Other linkages between the cytoskeleton and the plasma membrane can be very important. For exam-

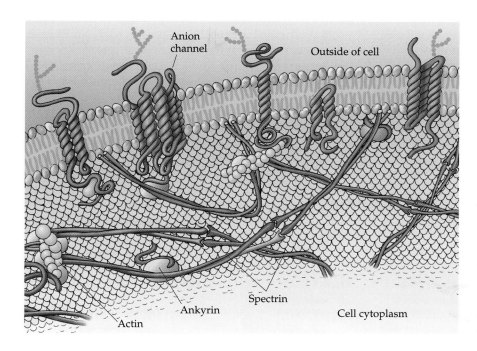

Anion channel

Outside of cell

Ankyrin

Spectrin

Actin

Cell cytoplasm

4.24 Some Proteins of the Red Blood Cell Membrane Many proteins contribute to this resilient structure. Spectrin does not bind the membrane directly but connects by way of linker proteins; it also binds actin filaments of the cytoskeleton. The structures of this cell membrane depiction will be discussed in detail in Chapter 5.

ple, a genetic deficiency in one protein, called *dystrophin*, that normally links actin filaments to the plasma membrane in muscle cells, is the cause of the most common form of muscular dystrophy, an inherited human disease that involves progressive loss of muscle cells and affects one out of every 3,500 male children.

Intermediate filaments are tough supporting elements

Filaments of another type, the **intermediate filaments**, are found only in multicellular organisms and play more static roles: They stabilize cell structure and resist tension (see Figure 4.21). Although there are at least five distinct types of intermediate filaments, all share the same general structure and are composed of fibrous proteins of the keratin family similar to the protein that makes up hair and fingernails. In cells, these proteins are organized into tough, ropelike assemblages 8 to 12 nm in diameter.

In some cells, intermediate filaments end at the nuclear envelope and may maintain the positions of the nucleus and other organelles in the cell. The lamins of the nuclear lamina are intermediate filaments. Other kinds of intermediate filaments help hold a complex apparatus of microfilaments in place in muscle cells (see Chapter 44). Still other kinds stabilize and help maintain rigidity in surface tissues by connecting "spot welds" called *desmosomes* between adjacent cells (see Figure 5.6b). Rapidly growing or newly formed cells do not contain intermediate filaments.

Microtubules are long and hollow

Microtubules are long, hollow, unbranched cylinders about 25 nm in diameter and up to several micrometers long. Many of the microtubules in a cell radiate from a region called the *microtubule organizing center*. They are assembled from molecules of the protein **tubulin**.

Tubulin itself is a dimer made up of two polypeptide subunits, called α-tubulin and β-tubulin. Thirteen rows, or protofilaments, of tubulin dimers surround the central cavity of the microtubule (see Figure 4.21). Microtubules have polarity: One end is called the + end, the other the – end. Tubulin dimers can be added or subtracted from the + end, lengthening or shortening the microtubule to affect the cell in various ways. The capacity to change length rapidly makes microtubules dynamic structures.

In plants, microtubules help control the arrangement of the fibrous components of the cell wall. Electron micrographs of plants frequently show microtubules lying just inside the plasma membrane of cells that are forming or extending their cell walls. Disruption of the cell's microtubules leads to a disordered arrangement of newly synthesized fibers in the cell wall.

In animal cells, microtubules are often found in the parts of the cell that are changing shape. In some cells, microtubules serve as tracks along which motor proteins carry protein-laden vesicles from one part of the cell to another. Microtubules are essential in distributing chromosomes to daughter cells during cell division (see Chapter 9). And they are intimately associated with movable cell appendages: the flagella and cilia.

FLAGELLA AND CILIA. Many eukaryotic cells possess whiplike appendages, the flagella and cilia. These organelles push or pull the cell through its aqueous environment, or they may move surrounding liquid over the surface of the cell (Figure 4.25a). Cilia and eukaryotic flagella are both assembled from specialized microtubules and have identical internal structures, but they differ in their relative lengths and their patterns of beating.

Flagella are longer than cilia and are usually found singly or in pairs; waves of bending propagate from one end of a flagellum to the other in snakelike undulation. **Cilia** are shorter appendages and are usually present in great numbers. They beat stiffly in one direction and recover flexibly in the other direction (like a swimmer's arm), so the recovery stroke does not undo the work of the power stroke (see Chapter 44).

In cross section, a typical cilium or eukaryotic flagellum is seen to be covered by the plasma membrane and to contain what is usually called a "9 + 2" array of microtubules. As Figure 4.25b shows, there are actually nine fused pairs of microtubules—called **doublets**—forming an outer cylinder, and one pair of unfused microtubules running up the center. The motion of cilia and flagella results from the sliding of the microtubules past one another (as described in Chapter 44).

But what is the "motor" that drives this sliding? It is a protein called **dynein**, which can undergo changes in tertiary structure driven by energy from ATP. Dynein molecules on one microtubule bind a neighboring microtubule. Then, as the dynein molecules change shape, they "row" one microtubule past its neighbor (Figure 4.26a).

Dynein and another motor protein, **kinesin**, are responsible for moving cell organelles in opposite directions along microtubules. Recall that microtubules have a + end and a – end. The dynein moves attached vesicles and other organelles toward the – end of the tubule, while the kinesin moves them toward the + end. These motor proteins attach both to the microtubules, which guide their motion, and to the vesicles that transport materials from the ER to the Golgi apparatus and to the plasma membrane for secretion (Figure 4.26b).

Some prokaryotes have flagella, but prokaryotic flagella lack microtubules and dynein. The flagella of

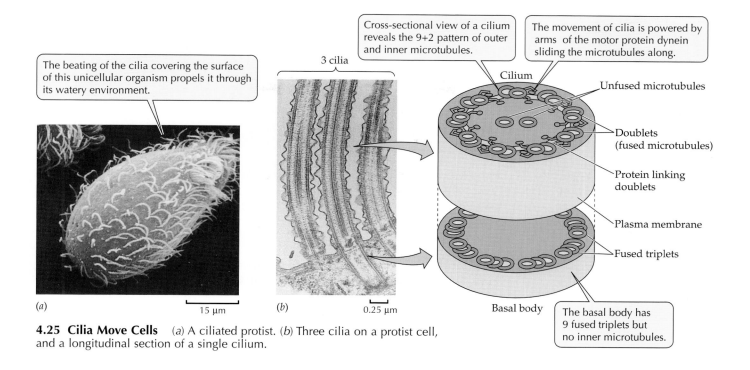

The beating of the cilia covering the surface of this unicellular organism propels it through its watery environment.

Cross-sectional view of a cilium reveals the 9+2 pattern of outer and inner microtubules.

The movement of cilia is powered by arms of the motor protein dynein sliding the microtubules along.

3 cilia

Cilium

Unfused microtubules

Doublets (fused microtubules)

Protein linking doublets

Plasma membrane

Fused triplets

Basal body

The basal body has 9 fused triplets but no inner microtubules.

4.25 Cilia Move Cells (*a*) A ciliated protist. (*b*) Three cilia on a protist cell, and a longitudinal section of a single cilium.

(*a*) 15 µm

(*b*) 0.25 µm

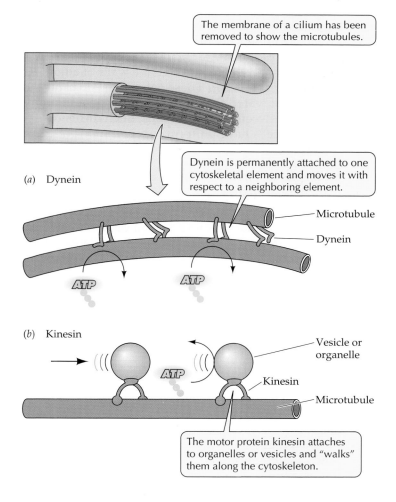

The membrane of a cilium has been removed to show the microtubules.

(*a*) Dynein

Dynein is permanently attached to one cytoskeletal element and moves it with respect to a neighboring element.

Microtubule

Dynein

ATP   ATP

(*b*) Kinesin

Vesicle or organelle

ATP

Kinesin

Microtubule

The motor protein kinesin attaches to organelles or vesicles and "walks" them along the cytoskeleton.

4.26 Motor Proteins Use Energy from ATP to Move Things (*a*) Dynein operates in muscle contraction and flagellar movement. (*b*) Kinesin delivers vesicles to various parts of the cell. All motor proteins work by undergoing reversible shape changes powered by energy from ATP.

prokaryotes are neither structurally nor evolutionarily related to those of eukaryotes. The prokaryotic flagellum is assembled from a protein called flagellin, and it has a much simpler structure and a smaller diameter than those of a single microtubule. And whereas eukaryotic flagella beat in a wavelike motion, prokaryotic flagella rotate (see Figure 4.5).

At the base of every eukaryotic flagellum or cilium is an organelle called a **basal body** (see Figure 4.25*b*). The nine microtubule doublets extend into the basal body. In the basal body, each doublet is accompanied by another microtubule, making nine sets of *three* microtubules. The central, unfused microtubules of the cilium or flagellum do not extend into the basal body.

Centrioles are organelles that are almost identical to basal bodies. Centrioles are found in all eukaryotes except for cells of the flowering plants, the pines and their relatives, and some protists. Under the light microscope, a centriole looks like a small, featureless particle, but the electron microscope reveals that it is made up of a precise bundle of microtubules, arranged as nine sets of three

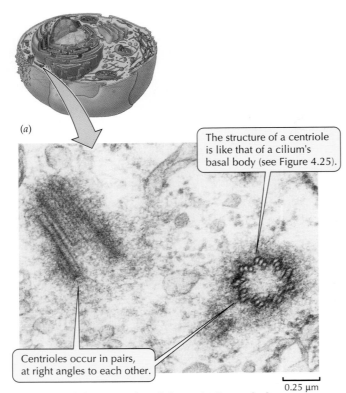

(a)

The structure of a centriole is like that of a cilium's basal body (see Figure 4.25).

Centrioles occur in pairs, at right angles to each other.

0.25 μm

4.27 Centrioles Contain Triplets of Microtubules
Centrioles are found in the microtubule organizing center, a region near the nucleus. The electron micrograph (a) shows a pair of centrioles at right angles to each other. Nine sets of three fused microtubules are evident in the centriole on the right, which is seen in cross section. The diagram (b) emphasizes the three-dimensional structure of a centriole.

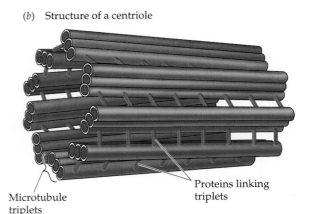

(b)   Structure of a centriole

Microtubule triplets

Proteins linking triplets

fused microtubules each (Figure 4.27). Centrioles lie in the microtubule organizing center in cells that are about to undergo division.

Extracellular Structures

Although the plasma membrane is the functional barrier between the inside and outside of a cell, many structures outside the plasma membrane are produced by the cell and play essential roles in protecting, supporting, or attaching cells. These structures are said to be **extracellular**. In bacteria, the peptidoglycan cell wall is the extracellular structure that plays these roles. In eukaryotes, a variety of extracellular structures play these roles: in plants, the cellulose cell wall; in animals, the extracellular matrix between cells in multicellular organisms and tissues.

The plant cell wall consists largely of cellulose

The plant **cell wall** is a semirigid structure outside the plasma membrane (Figure 4.28). It consists primarily of polysaccharides, the most prominent of which is cellulose (see Chapter 3). The cell wall provides support for the cell and limits its volume by remaining rigid. In some instances in which hydrophobic substances are added to the cellulose, the cell wall can prevent water from reaching the plasma membrane. Modifications of the cell wall such as the addition of substances are important in determining the functional roles of some plant cells (see Chapter 31).

Because of their thick cell walls, under the light microscope plant cells appear entirely isolated from each other, but electron microscopy reveals that this is not the case. The cytoplasm of adjacent plant cells is connected by numerous plasma membrane-lined channels, called **plasmodesmata**, that are about 20 to 40 nm in diameter and extend through the walls of adjoining cells (see Figure 4.28). These connections permit the diffusion of water, ions, small molecules, and many proteins between connected cells. Such diffusion ensures that cells have uniform concentrations of these substances.

Multicellular animals have an elaborate extracellular matrix

In multicellular animals, cells lack the semirigid cell wall that is characteristic of plant cells, but many animal cells are surrounded by or are in contact with an **extracellular matrix**. This matrix is composed of proteins such as collagen (the most abundant protein in mammals) and glycoproteins. These proteins and other substances are secreted by cells that are present in or near the matrix. In the human body, some tissues, such as those in the brain, have very little extracellular matrix. Other tissues, such as bone and cartilage, have large amounts of extracellular matrix.

The cells embedded in bone and cartilage secrete and maintain the characteristic composition and structure of the extracellular material. Bone cells are embedded in an extracellular matrix that consists primarily of collagen and substantial amounts of the ionic solid calcium phosphate. This matrix gives bone its familiar rigidity. Epithelial cells, which line body cavi-

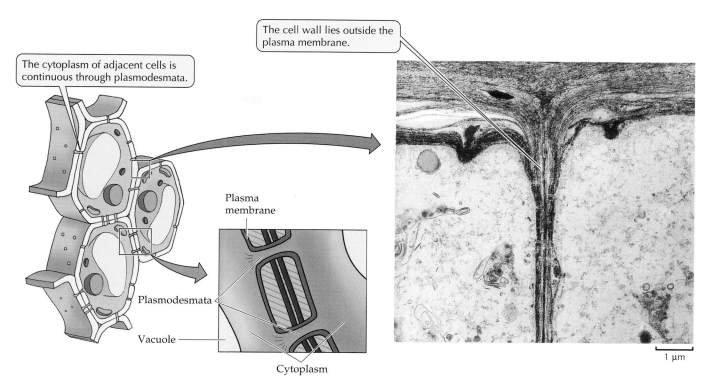

The cytoplasm of adjacent cells is continuous through plasmodesmata.

The cell wall lies outside the plasma membrane.

Plasma membrane

Plasmodesmata

Vacuole

Cytoplasm

1 μm

4.28 The Plant Cell Wall   This semirigid structure provides support for plant cells. Plasmodesmata allow water and other molecules to cross the wall and move from cell to cell.

ties, lie together as a sheet spread over the **basal lamina**, or basement membrane, a form of extracellular matrix (Figure 4.29).

In some cases the extracellular matrix is made up, in part, of one of the most spectacular molecules ever seen, an enormous **proteoglycan** (see Figure 4.29). A single molecule of this proteoglycan consists of many hundreds of polysaccharides attached to about a hundred proteins, all of which are attached to one enormous polysaccharide. The molecular weight of the proteoglycan can exceed 100 million; the molecule takes up as much space as an entire prokaryotic cell. What do such fantastic molecules do for the organism?

The proteoglycans and other components of the extracellular matrix contribute to the physical properties of cartilage, skin, and other tissues. They help filter materials passing between the blood and urine. They help orient cell migration during embryonic development and during tissue regeneration following injury. They play roles in chemi-

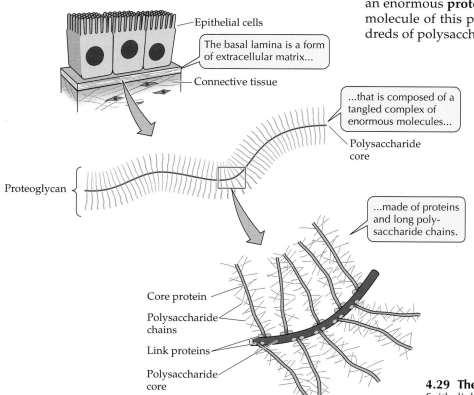

Epithelial cells

The basal lamina is a form of extracellular matrix...

Connective tissue

...that is composed of a tangled complex of enormous molecules...

Polysaccharide core

Proteoglycan

...made of proteins and long polysaccharide chains.

Core protein

Polysaccharide chains

Link proteins

Polysaccharide core

4.29 The Extracellular Matrix of Animal Cells
Epithelial cells secrete a basal lamina, which anchors them to the connective tissue.

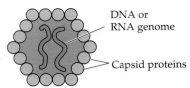

Nonenveloped virus

DNA or
RNA genome

Capsid proteins

4.30 Viruses Are Incapable of Reproducing on Their Own Viruses lack the ribosomes, enzymes, and other cellular constituents necessary for independent reproduction and processing of raw materials and energy. Some viruses are enveloped by a membrane, whereas others lack such membranes.

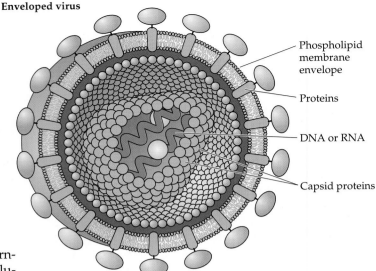

Enveloped virus

Phospholipid
membrane
envelope

Proteins

DNA or RNA

Capsid proteins

cal signaling from one cell to another. We are still learning about the structure and functions of the extracellular matrix.

Cells versus Viruses

As you have discovered, in contrast to prokaryotes, eukaryotes have a true nucleus and other membrane-enclosed organelles that allow various cellular activities to be concentrated in specialized compartments. In addition, eukaryotes have many specialized molecules, such as tubulin and actin, that are not found in prokaryotes. The cell walls of prokaryotes differ structurally and chemically from those of eukaryotes. And, as we'll see in Chapter 14, prokaryotic DNA and eukaryotic DNA have very different organization.

Do such differences mean that eukaryotes are more advanced, or "higher," or more successful than prokaryotes? Not at all. Every surviving species is the product of eons of natural selection and is superbly adapted to its environment. Each species has characteristics that enable it to live where and how it does, and to compete successfully against other species. But where do viruses fit into this picture?

Viruses are "gigantic" macromolecular assemblies of proteins and nucleic acid, some of which are surrounded by a phospholipid membrane (Figure 4.30). But they are not cells. Viruses lack many of the important characteristics of living systems. They cannot reproduce outside of the living cells that supply the raw materials and energy from which new viruses are constructed. Specific kinds of viruses infect animal, plant, fungal, protist, and prokaryotic cells. When a virus infects a host cell, the viral nucleic acid enters the cell and subverts the host's metabolic machinery to make new viruses (see Chapter 13).

Unlike cells, viruses do not transform energy, nor are they responsive to environmental stimuli as are organisms. But viruses do reproduce, and they evolve. Many different viruses infect humans, causing both mild diseases such as the common cold and life-threat-

ening conditions such as AIDS. Some viruses have been implicated in certain forms of cancer in humans.

Isolating Organelles for Study

During the early days of cell biology in the nineteenth century, all that anyone could do with organelles—and only the largest ones at that—was to observe them with a light microscope. The refinement of the electron microscope in the middle of the twentieth century made it possible to view cells and their organelles at higher magnification and greater resolution.

However, to answer questions about the composition and function of such organelles, it was necessary to perform chemical analyses, and such analyses required that relatively large quantities of the different organelles be obtained in pure form. To accomplish this task, the processes of cell fractionation were developed. In **cell fractionation**, cells are first ruptured; then a system of centrifugation separates the organelles from one another.

Depending on the type of cell or tissue, various methods are available to break cells open without excessively damaging organelles or proteins. The simplest methods use an old-fashioned mortar and pestle or a hand-operated glass homogenizer (which squeezes and shears cells between two tightly fitting, counter-rotating ground-glass surfaces). Motor-driven homogenizers or blenders are also commonly used.

All these methods break open plasma membranes and (if present) cell walls, liberating the cytoplasm into a solution that contains a substantial concentration of solutes. This solution prevents the organelles from bursting as they would in a more dilute solution or in pure water (see Figure 5.12). Cooling with ice further minimizes damage to the organelles and proteins.

The techniques described here reduce biological tissue to a crude suspension of mixed organelles, unbroken cells, and debris. To separate the components of such a suspension requires one of two types of *centrifugation*. The **centrifuge** is a laboratory instrument that can spin materials extremely rapidly about a fixed axis. This rapid rotation generates large centrifugal forces that cause components of the suspension to sediment according to mass, size, or density. The two common types of centrifugation are differential centrifugation and equilibrium centrifugation.

Differential centrifugation is a process of repeated centrifugations at increasing speeds, at higher forces (designated as multiples of the force of gravity, *g*), and for longer periods of time (Figure 4.31). After cells are ruptured in an appropriate solution, the mixture is centrifuged briefly at low speed (and hence low relative centrifugal force). The largest and densest particles sediment out, forming a *pellet* in the bottom of the tube and a liquid *supernatant fluid* above. This pellet is left behind for further study when the supernatant fluid and its suspended contents are poured off. The supernatant fluid is collected and spun at a higher speed and for a longer time, causing other organelles to sediment to the bottom. By repeating this procedure with ever-increasing centrifugal force, one can separate out many organelles.

After cell fractionation is completed, the nuclei, mitochondria, components of the ER, and free ribosomes are isolated from each other in separate tubes for further study. The contents of each tube can be purified further by repeating the centrifugation routine. The contents and purity of each pellet are confirmed by the electron microscope or by enzyme assays. Variations on this procedure have revealed extensive information about the activities and interrelationships of the different organelles and about the life of the cell.

Equilibrium centrifugation differs from differential centrifugation not in its goals but in its use of a solution that varies in density from the top to the bottom

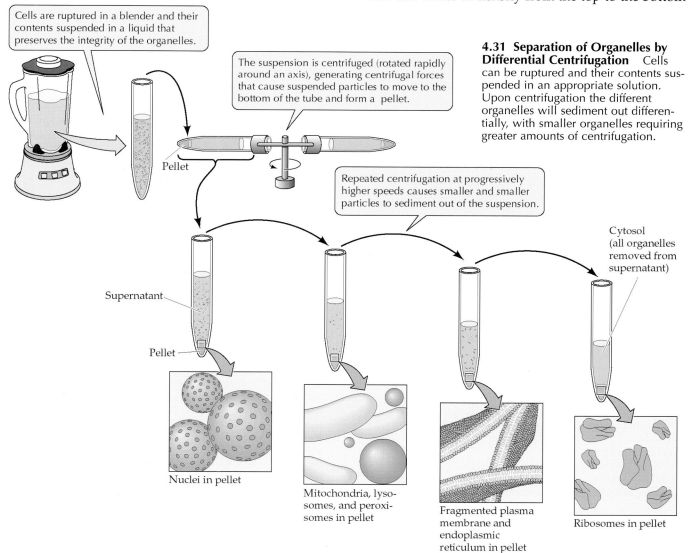

Cells are ruptured in a blender and their contents suspended in a liquid that preserves the integrity of the organelles.

The suspension is centrifuged (rotated rapidly around an axis), generating centrifugal forces that cause suspended particles to move to the bottom of the tube and form a pellet.

Pellet

Repeated centrifugation at progressively higher speeds causes smaller and smaller particles to sediment out of the suspension.

Cytosol (all organelles removed from supernatant)

Supernatant

Pellet

Nuclei in pellet

Mitochondria, lysosomes, and peroxisomes in pellet

Fragmented plasma membrane and endoplasmic reticulum in pellet

Ribosomes in pellet

4.31 Separation of Organelles by Differential Centrifugation   Cells can be ruptured and their contents suspended in an appropriate solution. Upon centrifugation the different organelles will sediment out differentially, with smaller organelles requiring greater amounts of centrifugation.

4.32 Equilibrium Centrifugation A density gradient of sucrose solution is prepared as described in the text and a sample of the suspension to be fractionated is added. During centrifugation, particles of different density move to discrete bands in the gradient and can be recovered in separate tubes.

Organelles in suspension

A density gradient of sucrose solution is prepared. A sample of the suspension to be fractionated is gently added to the top of the tube.

During centrifugation, organelles of different density will move to the location in the tube where the sucrose density is equal to that of the organelle, forming discrete bands.

Sucrose gradient

Centrifuge

Lower density

Higher density

A small hole is punched in the bottom of the tube and successive layers or fractions are collected in separate tubes.

of a centrifuge tube (Figure 4.32). This system permits the separation of substances that have different densities. A density gradient is prepared in a plastic centrifuge tube by first adding a small amount of a highly concentrated (and hence very dense) sugar solution— say, 60 percent sucrose. Next, a small amount of a 50 percent sucrose solution is layered on top, then 40 percent, and so on. In practice, such a **density gradient** is usually constructed by an automatic device, so changes in density are smooth rather than abrupt.

After the gradient is established, the mixture of organelles to be separated is carefully layered on the top of the sugar solution, and the tube is centrifuged. Under the forces generated by centrifugation, organelles sediment into the gradient as long as they are more dense than the surrounding liquid. Once an organelle reaches that part of the gradient where its density matches the density of the solution, it stops sedimenting—it has reached buoyant equilibrium. If two kinds of organelles have different average densities, they form two separate bands, which can be collected by using a pipette or by punching a hole in the bottom of the tube and collecting drops from it in separate tubes.

Summary of "The Organization of Cells"

The Cell: The Basic Unit of Life

• All cells come from preexisting cells and have certain processes, types of molecules, and structures in common.
• To maintain adequate exchanges with its environment, a cell's surface area must be large compared to its volume. **Review Figure 4.2**
• Prokaryotic cell organization is characteristic of the kingdoms Eubacteria and Archaebacteria. Prokaryotes lack internal compartments such as a nucleus. **Review Figure 4.3**

• Eukaryotic cell organization is characteristic of cells in the other four kingdoms. Eukaryotic cells have many membrane-enclosed compartments, including a nucleus that contains DNA. **Review Figure 4.7**

Prokaryotic Cells

• All prokaryotic cells have a plasma membrane, a nucleoid region with DNA, and a cytoplasm that contains ribosomes, dissolved enzymes, water, and small molecules. Some prokaryotes have additional protective structures: cell wall, outer membrane, and capsule. Some prokaryotes contain photosynthetic membranes, and some show mesosomes. **Review Figures 4.3, 4.4**
• Projecting from the surface of some prokaryotes, rotating flagella move prokaryotic cells from place to place. Pili are sites at which prokaryotic cells attach to one another or to environmental surfaces. **Review Figure 4.5**

Microscopes: Revealing the Subcellular World

• Because of their greater resolving power, electron microscopes enable observation of greater detail than can be seen with light microscopes. Whereas light microscopy can be used for viewing either dead or living cells, electron microscopy can be used only with preserved dead material. **Review Table 4.1**
• Scanning electron microscopy gives a three-dimensional view of surfaces. Transmission electron microscopy gives a two-dimensional view of a three-dimensional reality. **Review Figures 4.3, 4.4**

Eukaryotic Cells

• Like prokaryotic cells, eukaryotic cells have a plasma membrane, cytoplasm, and ribosomes. However, eukaryotic cells are larger and contain many membrane-enclosed organelles, such as the nucleus (containing DNA), mitochondria, endoplasmic reticulum, and Golgi apparatus. **Review Figure 4.7**

• Plant cells have chloroplasts and cellulose cell walls not found in animal cells.

• The membranes that envelop organelles in the eukaryotic cell are partial barriers, ensuring that the chemical composition of the interior of the organelle differs from the chemical composition of the surrounding cytoplasm.

Organelles That Process Information

• The nucleus is usually the largest organelle in a cell and is bounded by two membranes—together called the nuclear envelope, which disassembles during nuclear division. Within the nucleus, the nucleoli are the source of the ribosomes found in the cytoplasm. Ribosomes participate in protein synthesis and are sometimes attached to the endoplasmic reticulum. **Review Figure 4.8**

• The nucleus contains most of the cell's DNA, which is associated with protein to form chromatin. Chromatin has a diffuse appearance, but before nuclear division it condenses to form chromosomes. **Review Figure 4.10**

• Nuclear pores have complex structures that govern what enters and leaves the nucleus. **Review Figure 4.8**

Organelles That Process Energy

• Mitochondria are enclosed by an outer membrane and an inner membrane that folds inward to form cristae. Mitochondria contain the enzymes for cellular respiration and the generation of ATP. **Review Figure 4.11**

• Almost all eukaryotic cells contain mitochondria. However, green plant cells also contain chloroplasts, which are enclosed by two membranes and contain an internal system of thylakoids organized as grana. **Review Figures 4.7, 4.13**

• Thylakoids within chloroplasts contain the chlorophyll and proteins that harvest light energy for the synthesis of glucose from carbon dioxide. **Review Figure 4.13**

• Both mitochondria and chloroplasts contain their own DNA and ribosomes and are capable of making some of their own proteins.

• The endosymbiosis theory of the evolutionary origin of mitochondria and chloroplasts states that mitochondria and chloroplasts originated when larger prokaryotes engulfed but did not digest smaller prokaryotes. Mutual benefits permitted this symbiotic relationship to be maintained and to evolve into the eukaryotic organelles observed today. **Review Figure 4.15**

The Endomembrane System

• The endomembrane system is a series of interrelated membranes and compartments.

• The rough ER has attached ribosomes that synthesize proteins. These proteins enter the lumen of the ER, where they are chemically processed and sorted into vesicles. The smooth ER lacks ribosomes and is associated with the synthesis of lipids. **Review Figures 4.7, 4.17**

• The Golgi apparatus receives materials from the rough ER vesicles that fuse with the *cis* region of the Golgi. Within the lumen of the Golgi, signal molecules are added to proteins, directing them to their proper destinations. Vesicles originating from the *trans* region of the Golgi contain proteins for different cellular functions. **Review Figures 4.7, 4.17**

• Some vesicles fuse with the plasma membrane and release their contents outside the cell. Other vesicles, such as lysosomes, are retained within the cell. **Review Figure 4.17**

• Lysosomes contain many digestive enzymes. Lysosomes fuse with the phagosomes produced by phagocytosis to form secondary lysosomes in which engulfed materials are digested. Undigested materials are excreted from the cell when the secondary lysosome fuses with the plasma membrane. **Review Figure 4.18**

Other Organelles

• Membrane-enclosed organelles include peroxisomes and glyoxysomes. These organelles contain special enzymes and carry out specialized chemical reactions for the cell.

• Vacuoles are prominent in many plant cells and consist of a membrane-enclosed compartment that contains water and dissolved substances. By receiving water, vacuoles enlarge and provide the pressure needed to stretch the cell wall during plant cell growth.

The Cytoskeleton

• The cytoskeleton within the cytoplasm of eukaryotic cells provides shape, strength, and movement. It consists of three interacting types of protein fibers with different diameters. **Review Figure 4.21**

• Microfilaments consist of two helical chains of G actin units. Microfilaments strengthen cellular structures such as microvilli and provide movement involved in animal cell division, cytoplasmic streaming, and pseudopod extension. Microfilaments are found as independent fibers, bundles of fibers, or networks of fibers joined by linking proteins. **Review Figures 4.21, 4.23**

• Intermediate filaments are formed of proteins such as keratin and are organized into tough ropelike structures that add strength to cell attachments in multicellular organisms. **Review Figure 4.21**

• Microtubules are composed of dimers of the protein tubulin. They can lengthen and shorten by adding and losing tubulin dimer units and are involved in the distribution of chromosomes during nuclear division. They are involved in the structure and function of cilia and flagella, both of which have a characteristic 9 + 2 pattern of microtubules. The movements of cilia and flagella are due to the motor protein dynein. Basal bodies and centrioles are also constructed of microtubules. Microtubules also move cellular organelles through the action of motor proteins, including kinesin and dynein, that use energy from ATP. **Review Figures 4. 25, 4.26, 4.27**

Extracellular Structures

• Materials external to the plasma membrane provide protection, support, and attachment for cells in multicellular systems.

• The cell wall of plants consists principally of cellulose and is pierced by plasmodesmata that join the cytoplasm of adjacent cells. **Review Figure 4.28**

• In multicellular animals, the extracellular matrix consists of different kinds of proteins, including proteoglycan. In bone and cartilage the protein collagen predominates. **Review Figure 4.29**

Cells versus Viruses

• Although prokaryotes and eukaryotes differ in specific ways, they share characteristics that identify them as living cells. Viruses are not living. They are smaller than the smallest cells and consist of nucleic acid, protein, and sometimes a membrane. Like cells, viruses use nucleic acid as hereditary material, but they require a living cell in order to reproduce. And viruses do not transform matter and energy or respond to the environment. **Review Figure 4.30**

Isolating Organelles for Study

• To study an organelle chemically, one must isolate it from other cell contents by rupturing the cells and suspending their contents in a liquid medium, then subjecting this suspension to centrifugation.
• In differential centrifugation, a series of rotations at increasing speeds followed by separation of supernatant fluid and pellet results in the isolation of organelles in separate tubes. **Review Figure 4.31**
• In equilibrium centrifugation, cellular materials are added to a sucrose solution that has a density gradient and centrifuged so that cellular materials with different densities come to reside at different sites in the gradient. **Review Figure 4.32**

Self-Quiz

1. Which statement is true of both prokaryotic and eukaryotic cells?
   a. They contain ribosomes.
   b. They have peptidoglycan cell walls.
   c. They contain membrane-enclosed organelles.
   d. They contain true nuclei.
   e. Their flagella have the 9 + 2 structure.

2. Which statement about the nuclear envelope is *not* true?
   a. It is continuous with the endoplasmic reticulum.
   b. It has pores.
   c. It consists of two membranes.
   d. RNA and some proteins pass through it to move in and out of the nucleus.
   e. Its inner membrane bears ribosomes.

3. Which statement about mitochondria is *not* true?
   a. Their inner membrane folds to form cristae.
   b. They are usually 1 μm or less in diameter.
   c. They are green because of the chlorophyll they contain.
   d. Energy-rich substances from the cytosol are oxidized in them.
   e. Much ATP is synthesized in them.

4. Which statement about plastids is true?
   a. They are found in prokaryotes.
   b. They are surrounded by a single membrane.
   c. They are the sites of cellular respiration.
   d. They are found in fungi.
   e. They are of several types, with different functions.

5. Which statement about the endoplasmic reticulum is *not* true?
   a. It is of two types: rough and smooth.
   b. It is a network of tubes and flattened sacs.
   c. It is found in all living cells.
   d. Some of it is sprinkled with ribosomes.
   e. Parts of it modify proteins.

6. The Golgi apparatus
   a. is found only in animals.
   b. is found in prokaryotes.
   c. is the appendage that moves a cell around in its environment.
   d. is a site of rapid ATP production.
   e. packages and modifies proteins.

7. Which organelle is *not* surrounded by one or more membranes?
   a. Ribosome
   b. Chloroplast
   c. Mitochondrion
   d. Peroxisome
   e. Vacuole

8. Eukaryotic flagella
   a. are composed of a protein called flagellin.
   b. rotate like propellers.
   c. cause the cell to contract.
   d. have the same internal structure as cilia.
   e. cause the movement of chromosomes.

9. Microfilaments
   a. are composed of polysaccharides.
   b. are composed of actin.
   c. provide the motive force for cilia and flagella.
   d. make up the spindle that aids the movement of chromosomes.
   e. maintain the position of the nucleus in the cell.

10. Which statement about the plant cell wall is *not* true?
    a. Its principal chemical components are polysaccharides.
    b. It lies outside the plasma membrane.
    c. It provides support for the cell.
    d. It completely isolates adjacent cells from one another.
    e. It is semirigid.

Applying Concepts

1. Which organelles and other structures are found in both plant and animal cells? Which are found in plant but not animal cells? Which in animal but not plant cells? Discuss, in relation to the activities of plants and animals.

2. Through how many membranes would a molecule have to pass in going from the interior of a chloroplast to the interior of a mitochondrion? From the interior of a lysosome to the outside of a cell? From one ribosome to another?

3. How does the possession of double membranes by chloroplasts and mitochondria relate to the endosymbiosis theory of the origins of these organelles? What other evidence supports the theory?

4. What sorts of cells and subcellular structures would you choose to examine by transmission electron microscopy? By scanning electron microscopy? By light microscopy? What are the advantages and disadvantages of each of these modes of microscopy?

5. Some organelles that cannot be separated from one another by equilibrium centrifugation can be separated by differential centrifugation. Other organelles cannot be separated from one another by differential centrifugation but can be separated by equilibrium centrifugation. Explain these observations.

Readings

Alberts, B., D. Bray, J. Lewis, M. Raff, K. Roberts and J. D. Watson. 1994. *Molecular Biology of the Cell*, 3rd Edition. Garland, New York. An outstanding book in which to pursue the topics of this chapter in greater detail; authoritative treatment of modern cell biology and its experimental basis.

Allen, R. D. 1987. "The Microtubule as an Intracellular Engine." *Scientific American*, February. A description of how microtubules enable two-way transport of materials in cells.

Brandt, W. H. 1975. *The Student's Guide to Optical Microscopes.* William Kaufmann, Los Altos, CA. A short, programmed guide for those interested in learning how to use a light microscope.

Cooper, G. M. 1997. *The Cell: A Molecular Approach*. ASM Press/Sinauer Associates, Sunderland, MA. An excellent textbook—up-to-date, well illustrated, and concise.

De Duve, C. 1975. "Exploring Cells with a Centrifuge." *Science*, vol. 189, pages 186–194. A discussion by a Nobel laureate of the uses of centrifugation in studies of cells.

Fawcett, D. W. 1981. *The Cell*, 2nd Edition. Saunders, Philadelphia. Beautiful electron micrographs of subcellular structures in animal cells.

Glover, D. M., C. Gonzalez and J. W. Raff. 1993. "The Centrosome." *Scientific American*, June. A report of new findings about the structure and function of the organelle that directs the assembly of the cytoskeleton and controls cell division—when it is present.

Howells, M. R., J. Kirz and D. Sayre. 1991. "X-Ray Microscopes." *Scientific American*, February. A description of novel methods of microscopy that afford striking improvements in resolution.

Lodish, H., D. Baltimore, A. Berk, S. L. Zipursky, P. Matsudaira and J. Darnell. 1995. *Molecular Cell Biology*, 3rd Edition. Scientific American Books, New York. Another excellent middle-level book; fine illustrations.

Margulis, L. 1993. *Symbiosis in Cell Evolution*, 2nd Edition. W. H. Freeman, New York. An authoritative and thought-provoking reference on the origin and evolution of eukaryotic cells by a leading student of the problem.

Rothman, J. E. 1985. "The Compartmental Organization of the Golgi Apparatus." *Scientific American*, September. A discussion of the structure and function of the Golgi apparatus.

Weber, K. and M. Osborn. 1985. "The Molecules of the Cell Matrix." *Scientific American*, October. A clear treatment of microfilaments, intermediate filaments, tubulin, and the ways in which they are studied.

Chapter 5

Cellular Membranes

Complex Membranes That Process Solar Energy
The distinct proteins that appear as dots embedded in these thylakoids from a spinach chloroplast, magnified about 80,000 times, are necessary for photosynthesis.

*B*iological membranes are both abundant and essential in cells. Membranes form the compartments that isolate the interior of cells from their outside environments, and they isolate internal cell processes in different organelles. Biological membranes function in this capacity because their lipid composition and hydrophobic bilayer structure make them effective barriers to the passage of many hydrophilic substances. However, some substances move across membranes in spite of these restrictions, and this chapter will describe both passive and active transport processes.

Biological membranes are more than just barriers that regulate the passage of substances into and out of the cell. In addition to defining compartments and restricting movement between them, membranes process materials, energy, and information. Many plasma membranes are veritable antennae for information in their environment. They respond to some signals by puckering up and nibbling at bits of the environment. Other signals—chemical messengers from other parts of the body—cause dramatic changes in the "receiving" cell. A flash of light falling on a rod cell in our eye causes the cell's plasma membrane to become less permeable to sodium ions; this is one of the first steps leading to our seeing the flash. Poke an electrode through the plasma membrane of any living cell and you discover that the membrane is electrically charged, with the interior of the cell electrically more negative than the exterior. This electric charge permits nerve cell membranes to carry messages.

In this chapter, we'll examine the molecular composition and structure of membranes and consider specialized structures in plasma membranes. Then we'll explore the passive and active processes by which substances move across membranes to enter and leave cells and their compartments. Finally we'll con-

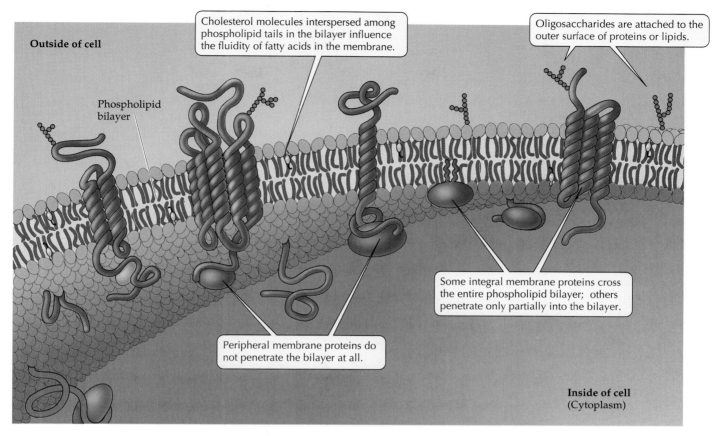

5.1 The Fluid Mosaic Model The general molecular structure of biological membranes is a continuous phospholipid bilayer in which proteins are embedded. On the outer surface, carbohydrates may be attached to proteins or phospholipids.

sider how membranes contribute to processing information and transforming energy, as well as the dynamic interrelationships between membranes.

Membrane Composition and Structure

The plasma membrane and other cellular **membranes** are thin, pliable bilayers of phospholipids with embedded proteins (Figure 5.1). (The phospholipid composition and bilayer architecture of biological membranes were introduced in Chapter 3, and membrane-enclosed organelles were described in Chapter 4.) As we consider membranes, the relationship between what a membrane does (its functions) and its chemical composition and physical structure is particularly obvious. Because of their composition and structure, biological membranes are **selectively permeable**; that is, they permit some substances to pass through but block others.

The chemical makeup, physical organization, and functioning of a biological membrane depend on three classes of biochemical compounds: lipids, proteins, and carbohydrates (see Figure 5.1). The phospholipids establish the physical integrity of the membrane and are an effective barrier to the passage of many hydrophilic materials. In addition to serving as a barrier, the phospholipid bilayer is a lipid "lake" in which a variety of proteins "float." This general design is known as the **fluid mosaic model** of the membrane. It applies to all biological membranes, although the membranes of archaea differ from the model in certain details.

Membrane proteins stretch across the phospholipid bilayer and protrude on both sides. They are responsible for many of the specific tasks membranes perform. Certain membrane proteins allow materials to pass through the membrane that cannot pass through the pure lipid bilayer. Other proteins receive chemical signals from the cell's external environment and respond by regulating certain processes inside the cell. Still other proteins function as enzymes and accelerate chemical reactions on the membrane surface.

Like some proteins, carbohydrates—the third class of compounds important in membranes—are crucial in recognizing specific molecules. The carbohydrates attach either to lipid or to protein molecules on the outside of the plasma membrane, where they protrude into the environment, away from the cell.

Let's look at each of the three major components of membranes—lipids, proteins, and carbohydrates—in more detail.

Lipids constitute the bulk of a membrane

Nearly all lipids in biological membranes are **phospholipids**. Recall from our discussion in Chapter 3 that some compounds are hydrophilic ("water loving") and others are hydrophobic ("water fearing"). Phospholipids are both: They have both hydrophilic regions and hydrophobic regions.

The long, nonpolar fatty acid parts of phospholipids are hydrophobic and associate easily with other nonpolar materials, but they do not dissolve in water or associate with hydrophilic substances. The phosphorus-containing region of the phospholipid is electrically charged and hence very hydrophilic. As a consequence, one way for phospholipids and water to coexist is for the phospholipids to form a double layer, with the fatty acids of the two layers interacting with each other and the polar regions facing the outside water environment (Figure 5.2). It is easy to make artificial membranes with the same two-layered arrangement in the laboratory. Both artificial and natural membranes form continuous sheets. Because of the tendency of the fatty acids to associate with one another and exclude water, small holes or rips in a membrane seal themselves spontaneously. This property helps membranes fuse during vesicle fusion, phagocytosis, and related processes.

The phospholipid bilayer stabilizes the entire membrane structure. At the same time, the fatty acids of the phospholipids make the membrane somewhat fluid—about as fluid as lightweight machine oil—so some material can move laterally within the plane of the membrane. As we will see, some membrane proteins migrate about relatively freely, and individual phospholipid molecules may also "travel."

A given phospholipid molecule in the plasma membrane of a bacterium may travel from one end of the bacterium to the other in a little more than a second. On the other hand, seldom does a phospholipid molecule in one half of the bilayer flop over to the other side and trade places with another phospholipid molecule. For such a swap to happen, the polar part of each molecule would have to move through the hydrophobic interior of the membrane. Since phospholipid flip-flops are rare, the two halves of the bilayer may be quite different in the kinds of phospholipids present.

All biological membranes are similar, but membranes from different cells or organelles may differ greatly in their lipid composition. For example, 25 percent of the lipid in many membranes is *cholesterol* (see Chapter 3), but some membranes have no cholesterol at all. When present, cholesterol is commonly situated

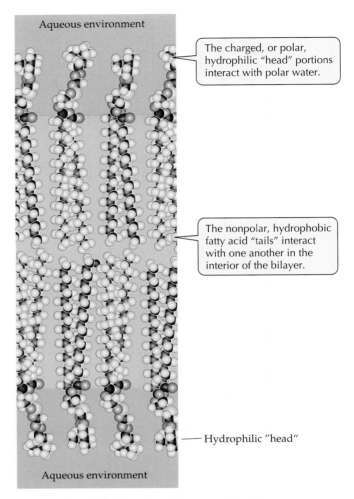

The charged, or polar, hydrophilic "head" portions interact with polar water.

The nonpolar, hydrophobic fatty acid "tails" interact with one another in the interior of the bilayer.

Aqueous environment

Aqueous environment

Hydrophilic "head"

5.2 A Phospholipid Bilayer Separates Two Aqueous Regions The eight phospholipid molecules shown here represent a small section of a membrane bilayer.

next to an unsaturated fatty acid, and its polar region extends into the surrounding aqueous layer (see Figure 5.1). Cholesterol plays more than one role in determining the fluidity of the membrane: It may either increase or decrease membrane fluidity, depending on the circumstances. Shorter fatty acid chains make for a more fluid membrane, as do unsaturated fatty acids.

Organisms can modify their membrane lipid composition, thus changing membrane fluidity, to compensate for changes in temperature. For example, some houseplants can survive both indoor and outdoor temperatures. But when accustomed to indoor temperatures, they can die if suddenly placed outdoors at cooler, but nonfreezing, temperatures. The sudden change in temperature does not give the plants sufficient time to adjust their membrane lipid composition. Lipids constitute a major fraction of all membranes, and they always form the continuous matrix into which the other chemical components become inserted.

Membrane components are revealed by freeze-fracturing

To determine how proteins and other components are inserted in the membrane, we use techniques that begin by freezing the tissue sample. The frozen tissue is then fractured (broken), splitting the lipid bilayer of the membrane and exposing the integral membrane proteins as bumps. The figure at the beginning of this chapter was produced by this *freeze-fracturing* technique. As an analogy, consider that slicing a chocolate-almond candy bar with a sharp razor gives one view of the interior; but if you break the bar instead, you reveal the almonds as protruding lumps.

We can reveal further detail in tissue by putting a freeze-fractured sample under a high vacuum and allowing water to evaporate. This *freeze-etching* technique reveals more texture. Freeze-etched samples can also be sprayed with metals such as platinum (shadowcasting) to reveal shadow patterns that further enhance contrast (Figure 5.3).

Membrane proteins are asymmetrically distributed

Biological membranes possess two types of proteins: integral membrane proteins and peripheral membrane proteins (see Figure 5.1). **Integral membrane proteins** penetrate the phospholipid bilayer, and many are *transmembrane* proteins, extending from one side of the membrane to the other. **Peripheral membrane**

5.3 Membrane Proteins Revealed by Freeze-Etching
The outer membrane of this mitochondrion has been fractured away, exposing the inner membrane. The image has been magnified about 65,000 times. The particles giving the inner membrane a grainy appearance are proteins necessary for cellular respiration.

proteins are not embedded in the bilayer. Instead, they are attached to exposed parts of the integral membrane proteins or phospholipid molecules by weak (noncovalent) bonds. These two types of proteins play a variety of different roles in membrane function and cellular processes.

The membranes of the various organelles differ sharply in protein composition, and different organelles house quite different chemical reactions (many of them requiring membrane-bound enzymes). In cellular respiration and photosynthesis, membrane-bound enzymes carry electrons from a donor to an acceptor molecule. Accordingly, both mitochondria and chloroplasts have highly specialized internal membranes, and these differ markedly (see Figure 5.3 and the figure at the beginning of the chapter).

Many membrane proteins move relatively freely within the phospholipid bilayer. Experiments using the technique of cell fusion illustrate this migration dramatically. In the laboratory, specially treated cells from humans and mice can be fused so that one continuous plasma membrane surrounds the combined cytoplasm and both nuclei. Initially, the membrane proteins from the two different cells reside in distinguishable halves of the joint plasma membrane. However, the membrane proteins of the two cells migrate, and after about 40 minutes they are uniformly dispersed (Figure 5.4). Although many membrane proteins are mobile in the membrane, there is also good evidence that other membrane proteins are not free to migrate. These proteins are "anchored" by components of the cytoskeleton as described in Chapter 4 (see Figure 4.25).

Proteins are asymmetrically distributed in membranes. Many transmembrane proteins show different "faces" on the two membrane surfaces. Such proteins have certain specific domains (or regions) of their primary structure on one side of the membrane, other domains within the membrane, and still other domains on the other side of the membrane. Peripheral membrane proteins are localized on one side of the membrane or the other, but not both. This arrangement gives the two sides or surfaces of the membrane different properties.

In addition to surface differences, there are regional differences. Some membrane proteins are confined to one part of the cell surface rather than being scattered evenly. This segregation of different proteins contributes to the functional specialization of various regions on the cell surface. For example, in certain muscle cells, the plasma membrane protein receptor for the chemical signal from nerve cells is normally found only at the site where a nerve cell meets the muscle cell. None of this protein is found elsewhere on the surface of the muscle cell. However, if the nerve is experimentally separated from the muscle, the protein receptor molecules become evenly distributed through-

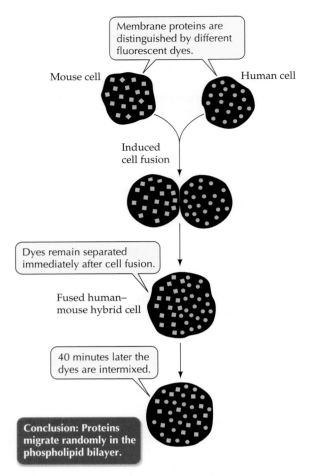

Membrane proteins are distinguished by different fluorescent dyes.

Mouse cell   Human cell

Induced cell fusion

Dyes remain separated immediately after cell fusion.

Fused human–mouse hybrid cell

40 minutes later the dyes are intermixed.

Conclusion: Proteins migrate randomly in the phospholipid bilayer.

5.4 Proteins Move Around in Membranes When treated mouse and human cells join, one continuous plasma membrane surrounds the fused hybrid cell and membrane proteins become randomly distributed.

out the plasma membrane of the muscle cell. If the nerve regenerates its attachment to the muscle, then the protein is once again limited to the junction area. This experiment not only attests to the mobility of membrane proteins and their regional grouping; it also shows how cell interactions can determine membrane characteristics of a cell.

What determines whether a particular membrane protein is integral or peripheral? If it is integral, what controls whether it reaches all the way through the membrane or is limited to one side? What keeps it in the bilayer, and what determines how far in it reaches? All these questions are answered in terms of the location of particular amino acids in the tertiary structure of the protein. Recall that the side chains of the various amino acids in a protein differ chemically—some of the side chains are hydrophilic and others are hydrophobic.

After a polypeptide chain folds into its tertiary structure, the protein may have both hydrophilic an-

hydrophobic surfaces. If one end of a folded protein is hydrophilic and the other hydrophobic, it will be an integral membrane protein, sticking out of one side of the membrane. Many integral membrane proteins that stretch across the entire membrane have long hydrophobic α-helical regions that penetrate the entire depth of the phospholipid bilayer. They also have hydrophilic ends that protrude into the aqueous environments on either side of the membrane (Figure 5.5). Integral membrane proteins like the one illustrated in Figure 5.5 resist being removed from the membrane. If the hydrophobic surface of such a protein is withdrawn partway out of the phospholipid bilayer, it is repelled by the aqueous environment. If an integral membrane protein is pushed farther into the membrane, its hydrophilic end is pushed back by the hydrophobic fatty acid region of the lipids. Thus such a protein may migrate laterally in the membrane sheet, but it may not push through the membrane or pop out of it.

How does an integral membrane protein get into the membrane in the first place? Clearly, it cannot fold into its final shape and then be pushed into the membrane—for the same reasons that it cannot push through or pop out of the membrane. Thus an integral protein is usually inserted into the membrane while the protein is being synthesized.

Specific sequences of hydrophobic amino acids in the growing protein interact with receptors on the endoplasmic reticulum membrane, causing the protein to penetrate the hydrophobic bilayer. However, the presence of additional sequences of hydrophilic amino acids may prevent the growing protein from extending farther into the bilayer and entering the lumen of the ER. Instead, it remains lodged in the membrane. Final folding is achieved after the protein is completely synthesized and all its regions have taken up residence either in the hydrophobic bilayer or in the hydrophilic regions on either surface.

Membrane carbohydrates are recognition sites

All plasma membranes and some other membranes contain significant amounts of carbohydrate along with the lipids and proteins. For example, the plasma membrane of the human red blood cell consists by weight of approximately 40 percent lipid, 52 percent protein, and 8 percent carbohydrate. The carbohydrates are located on the outer surface and serve as recognition sites (see Figure 5.1).

Some of the membrane carbohydrate is bound to lipids, forming *glycolipids*. These carbohydrate units often serve as recognition signals for interactions between cells. For example, the carbohydrate component of some glycolipids changes when a cell becomes cancerous. This change may identify the cancer cell for destruction by white blood cells.

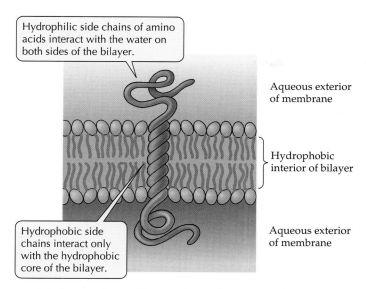

Hydrophilic side chains of amino acids interact with the water on both sides of the bilayer.

Aqueous exterior of membrane

Hydrophobic interior of bilayer

Hydrophobic side chains interact only with the hydrophobic core of the bilayer.

Aqueous exterior of membrane

5.5 Interactions of Integral Membrane Proteins This integral membrane protein is held in the membrane by the distribution of the hydrophilic and hydrophobic side chains of its amino acids.

Most of the carbohydrate in membranes is bound to proteins, forming *glycoproteins.* These bound carbohydrates are *oligosaccharide chains,* usually not exceeding 15 monosaccharide units in length. Glycoproteins enable a cell to recognize foreign substances. The oligosaccharide chains are added to the membrane proteins inside the endoplasmic reticulum and are modified in the Golgi apparatus.

A few different monosaccharides can provide an alphabet to generate a diversity of messages. Different messages are formed when different kinds of monosaccharides bond together at different sites and in different numbers. Recall that monosaccharides can link together at any of several different carbons to form branched oligomers. The possibility of different branching patterns in itself greatly increases the specificity and diversity of signals that oligosaccharides can provide.

Carbohydrates associated with the plasma membrane are always on the outer surface. In this location, their structural diversity is important for binding reactions at the cell surface whereby cells recognize and react with specific substances (see Figure 5.1).

Animal Cell Junctions

Animal cells form special junctions to help them adhere to other cells. In a multicellular system such as a tissue, direct links between the cells and sometimes between the cells and components of the extracellular matrix establish stable cohesion. The integral membrane proteins that accomplish this cohesion are called *cell adhesion proteins.* In a kind of "hook and eye" inter-

action, the external domain of a cell adhesion protein from one cell forms a strong link with that of an adjacent cell.

Generally, such adhesions are too small to be directly observed in electron micrographs, but the *intercellular space* between adjacent cells can be observed as an open region between the cells. Materials moving out of one cell and into the adjacent cell must cross this space. This intercellular space is also the region where all sorts of molecules can move around cells. In some cases, adjacent cells show specialized structures that link them or that restrict movement through the intercellular space.

Specialized structures that join cells are observed in electron micrographs of epithelial tissues, which are layers of cells that line body cavities or cover body surfaces. We will examine three general types of **cell surface junctions** that enable cells to make direct physical contact and link with one another: tight junctions, desmosomes, and gap junctions (see Figure 5.6).

Tight junctions seal tissues and prevent leaks

Tight junctions are specialized structures that link adjacent epithelial cells lining a hollow structure such as the intestine. Tight junctions result from the mutual binding of strands of specific membrane proteins that form belts encircling the epithelial cells in the region surrounding the lumen or cavity (Figure 5.6*a*). Tight junctions have two functions: to limit the passage of materials from the lumen through the intercellular space, and to restrict the migration of proteins in the cells' plasma membranes.

Because they are restricted from moving through the intercellular space, substances from the lumen must pass through the epithelial cell. This restriction gives epithelial cells control over what enters the body. The tight junctions also restrict the migration of membrane proteins and phospholipids from one region of the cell to other regions.

Thus, the membrane proteins and phospholipids in the *apical region* facing the lumen are different from the proteins and phospholipids in the *basolateral regions* facing the sides and bottom of the cell. By forcing materials to enter some cells, and by allowing different ends of cells to have different membrane proteins, tight junctions help ensure the directional movement of materials into the body.

Desmosomes rivet cells together

Desmosomes are specialized structures associated with the plasma membrane at certain sites in epithelial tissues. They hold adjacent cells firmly together, acting like spot welds or rivets at individual points (Figure 5.6*b*). Each desmosome consists of a dense *plaque* on the cytoplasmic surface of the plasma membrane that is attached to *keratin fibers* in the cytoplasm and *cell ad-*

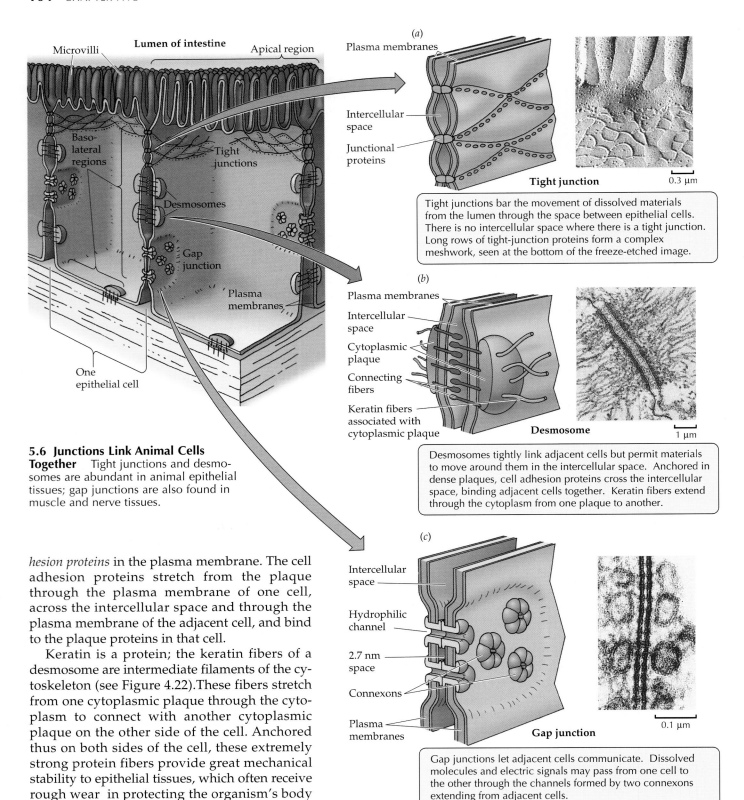

Microvilli **Lumen of intestine** Apical region

Baso-lateral regions

Tight junctions

Desmosomes

Gap junction

Plasma membranes

One epithelial cell

(a) Plasma membranes

Intercellular space

Junctional proteins

Tight junction

0.3 µm

Tight junctions bar the movement of dissolved materials from the lumen through the space between epithelial cells. There is no intercellular space where there is a tight junction. Long rows of tight-junction proteins form a complex meshwork, seen at the bottom of the freeze-etched image.

(b) Plasma membranes

Intercellular space

Cytoplasmic plaque

Connecting fibers

Keratin fibers associated with cytoplasmic plaque

Desmosome

1 µm

Desmosomes tightly link adjacent cells but permit materials to move around them in the intercellular space. Anchored in dense plaques, cell adhesion proteins cross the intercellular space, binding adjacent cells together. Keratin fibers extend through the cytoplasm from one plaque to another.

(c) Intercellular space

Hydrophilic channel

2.7 nm space

Connexons

Plasma membranes

Gap junction

0.1 µm

Gap junctions let adjacent cells communicate. Dissolved molecules and electric signals may pass from one cell to the other through the channels formed by two connexons extending from adjacent cells.

5.6 Junctions Link Animal Cells Together Tight junctions and desmosomes are abundant in animal epithelial tissues; gap junctions are also found in muscle and nerve tissues.

hesion proteins in the plasma membrane. The cell adhesion proteins stretch from the plaque through the plasma membrane of one cell, across the intercellular space and through the plasma membrane of the adjacent cell, and bind to the plaque proteins in that cell.

Keratin is a protein; the keratin fibers of a desmosome are intermediate filaments of the cytoskeleton (see Figure 4.22).These fibers stretch from one cytoplasmic plaque through the cytoplasm to connect with another cytoplasmic plaque on the other side of the cell. Anchored thus on both sides of the cell, these extremely strong protein fibers provide great mechanical stability to epithelial tissues, which often receive rough wear in protecting the organism's body integrity.

Gap junctions are a means of communication

Whereas tight junctions and desmosomes have mechanical roles, **gap junctions** facilitate communication between cells. Gap junctions are made up of specialized protein channels, called *connexons*, that span the plasma membranes of adjacent cells and the intercellular space between them (Figure 5.6c). Through these channels pass a variety of small molecules and ions but not proteins, nucleic acids, or cellular organelles.

In some nerve cells and in vertebrate heart and smooth muscle, gap junctions allow the direct passage of an electric signal from one cell to the next. Gap junctions are also important in normal embryonic development, for they appear at a specific developmental stage.

If animal tissues are experimentally disrupted to dissociate the cells, the cells quickly form new gap junctions as they reassociate. In fact, cells isolated from one species of vertebrate readily form gap junctions with cells from other vertebrate species. Cancer cells, however, never develop gap junctions, and this failure of cancer cells to communicate with other cells contributes to their abnormal uncontrolled growth.

We have examined the general structure of biological membranes and the specialized junctions found between some animal cells: tight junctions, desmosomes, and gap junctions. Now we want to consider the more general functions of biological membranes. One of the most important of these functions is *selective permeability*—the ability to allow some substances, but not others, to pass through the membrane to enter or leave the cell.

Passive Processes of Membrane Transport

There are two fundamentally different classes of processes by which substances cross biological membranes to enter and leave cells or organelles: passive processes and active processes. We'll discuss active processes later in the chapter; here we focus on the passive processes, all of which are different types of diffusion: simple diffusion through the phospholipid bilayer, diffusion through channel proteins, and facilitated diffusion by means of carriers. However, before considering diffusion as it works across a membrane, we must understand the basic principles of diffusion.

The physical nature of diffusion

Nothing in this world is ever absolutely at rest. Everything is in motion, though the motions may be very small. The constant jiggling of molecules and ions in solution increases as the temperature rises. An important consequence of the random jiggling is that all the components of a solution tend eventually to become evenly distributed throughout the system. For example, if a drop of ink is allowed to fall into a container of water, the pigment molecules of the ink are initially very concentrated. Without human intervention such as stirring, the pigment molecules of the ink move about at random, spreading slowly through the water until eventually the concentration of pigment—and thus the intensity of color—is exactly the same in every drop of liquid in the container. A solution in which the particles are uniformly distributed is said to be at equilibrium because there will be no future net change in concentration. **Diffusion** is the process of random movement toward a state of equilibrium.

Although the motion of each individual particle is absolutely random, in diffusion the *net movement* of particles is directional until equilibrium is reached. Diffusion is this net movement from regions of *greater concentration* to regions of *lesser concentration* (Figure 5.7). In a complex solution (one with many different solutes), the diffusion of each substance is independent of that of the other substances. How fast substances diffuse depends on four factors: (1) the diameter of the molecules or ions; (2) the temperature of the solution; (3) the electric charge, if any, of the diffusing material; and (4) the **concentration gradient** in the system. The concentration gradient is the change in concentration with distance in a given direction. The greater the concentration gradient, the more rapidly a substance diffuses.

DIFFUSION WITHIN CELLS. Within cells, where distances are very short, solutes distribute rapidly by diffusion. Small molecules and ions may move from one end of an organelle to another in a millisecond (10^{-3} s). The diffusion of a chemical signal from one nerve cell to another takes less than one millionth of a second ($<10^{-6}$ s). On the other hand, the usefulness of diffusion as a transport mechanism declines drastically as distances become greater. In the absence of mechanical stirring, diffusion across more than a centimeter may take an hour or more, and diffusion across meters may take years! Diffusion would not be adequate to distribute materials over the length of the human body, but within our cells or across layers of one or two cells, diffusion is rapid enough to distribute small molecules and ions almost instantaneously.

DIFFUSION ACROSS MEMBRANES. In a solution without barriers, all the solutes diffuse at rates determined by their physical properties and the concentration gradient of each solute. If a biological membrane is introduced as a barrier, the movement of the different solutes can be affected. The membrane is said to be *permeable* to solutes that can cross it more or less easily, but *impermeable* to substances that cannot move across it. Molecules to which the membrane is permeable diffuse from one compartment to the other until their concentrations are equal on both sides of the membrane. Molecules to which the membrane is impermeable remain in distinct compartments, so their concentrations remain different on the two sides of the membrane.

Equilibrium is reached when the concentrations of the diffusing substance are identical on both sides of the membrane. Individual molecules are still passing through the membrane when equilibrium is established, but equal numbers of molecules are moving in each direction, so there is no net change in concentra-

(a)

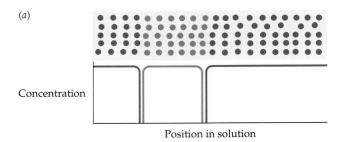

Initially, three different solutes are concentrated in different parts of the solution.

(b)

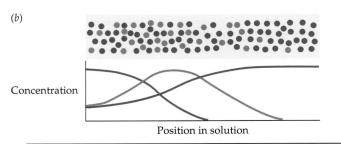

As the solutes diffuse, each is at a higher concentration—its peak on the graph—in one part of the solution, but it is also present elsewhere.

(c)

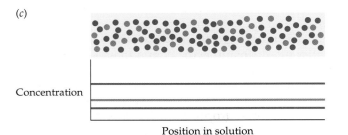

At equilibrium, all three solutes are uniformly distributed throughout the solution, as shown by the lack of peaks on the graph.

5.7 Diffusion Leads to Uniform Distribution of Solutes
Diffusion is the net movement of a solute from regions of greater concentration to regions of lesser concentration. The speed of diffusion varies with the substances involved, but it continues until the solution reaches an equilibrium.

tion. The two primary ways in which substances diffuse through biological membranes are simple diffusion and facilitated diffusion.

Simple diffusion is passive and unaided

In **simple diffusion**, small, nonpolar molecules pass through the lipid bilayer of the membrane. The more lipid-soluble the molecule, the more rapidly it diffuses through a biological membrane (Figure 5.8). This statement holds true over a wide range of molecular weights. Only certain ions and the smallest of molecules seem to deviate from this rule; materials such as

water, potassium ions (K^+), and chloride ions (Cl^-) pass through membranes much more rapidly than their solubilities in lipid would predict. How does a membrane's composition and molecular structure affect diffusion?

The key feature of membrane architecture, as we have seen, is the phospholipid bilayer that forms the framework of the membrane. The inner portion of the bilayer consists of the fatty acid chains of the phospholipids, along with cholesterol and other highly hydrophobic, nonpolar materials. When a hydrophilic molecule or ion moves into such a hydrophobic region, it is "rejected" by the lipid layer and forced back out again. Such a molecule seldom enters the hydrophobic region; it penetrates the membrane only when energy is available to push it in. On the other hand, a molecule that is itself hydrophobic, and hence soluble in lipids, enters the membrane readily and is thus able to pass through it.

This explanation accounts for most of the information in Figure 5.8, but the problem of how polar water molecules and charged ions can move across biological membranes remains. The passive diffusion of water into and out of cells occurs through the process of *osmosis* and will be discussed later in the chapter. The rapidity of osmosis is still something of a mystery, and several alternative mechanisms have been suggested to explain it. However, we *do* understand the rapid movement of certain ions through biological membranes. These ions pass through specific, water-filled protein pores in the membrane. Such **channel proteins** are known to allow the diffusion of ions such as potassium, sodium, calcium, and chloride.

Membrane transport proteins are of several types

The channel proteins through which certain ions diffuse and the carriers for facilitated diffusion and active transport are both **membrane transport proteins**—integral membrane proteins that stretch from one side of the membrane to the other. Different membrane transport proteins allow specific substances to pass through in various ways (Figure 5.9). The pore size and other properties of different channel proteins are different, permitting certain ions to pass through and excluding others. Some channel proteins have an aqueous region through which ions can diffuse.

There are two classes of membrane transport facilitated by **carrier proteins**: uniport and coupled transport. In **uniport**, a single type of solute is transported in one direction. In **coupled transport**, two or more different solutes are transported, but neither solute can move through the membrane unless the other solute is also present. Coupled transport is further divided into two types: symport and antiport.

In **symport**, the two coupled solutes are transported in the same direction. For example, one system for the

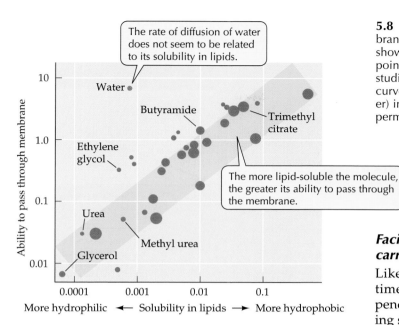

The rate of diffusion of water does not seem to be related to its solubility in lipids.

Water

Butyramide

Trimethyl citrate

Ethylene glycol

The more lipid-soluble the molecule, the greater its ability to pass through the membrane.

Urea

Methyl urea

Glycerol

More hydrophilic ← Solubility in lipids → More hydrophobic

5.8 Membrane Permeability Most substances cross membranes at rates determined by their solubility in lipids, as shown by the data points on this graph. The sizes of the points correspond roughly to the sizes of the molecules studied. The assortment of molecules of all sizes along the curve (that is, the fact that like sizes are not clumped together) indicates that size alone is not an important factor for permeability.

Facilitated diffusion is passive but uses carrier proteins

Like simple diffusion, **facilitated diffusion** (sometimes called *carrier-mediated diffusion* because it depends on the carrier proteins described in the preceding section) involves movement down a concentration gradient, from the side of higher concentration to the side of lower concentration. Eventually both processes produce equal concentrations of solute on the two sides of a membrane. The difference is that in facilitated diffusion the solute molecules do not diffuse through the phospholipid bilayer or through pores formed by channel proteins. Rather, they combine with carrier proteins in the membrane.

Carrier proteins have an abundance of hydrophobic amino acid side chains on the surface of the molecule in contact with the hydrophobic core of the membrane bilayer. However, within each carrier protein is a highly specific hydrophilic region to which a solute molecule binds. This binding appears to cause the protein to undergo a slight but significant change in shape (tertiary or quaternary structure) that moves the solute

uptake of glucose by the cell is a symport system in which the entrance of glucose is coupled to the entrance of a sodium ion. In **antiport**, the two substances are transported in opposite directions. For example, the plasma membrane of red blood cells contains about a million antiport transport proteins for the exchange of bicarbonate and chloride ions. This particular exchange is important in transporting carbon dioxide, which is present in the bloodstream as bicarbonate ions, from working tissues where carbon dioxide is produced to the lungs where it is released.

The specific tertiary structures of membrane transport proteins contribute to their transport specificities. A given membrane transport protein will carry only one particular solute or only closely related solutes. The great diversity of protein structures allows this high specificity for different transported solutes, just as it allows enzymes to recognize only specific reactants in the chemical reactions they accelerate.

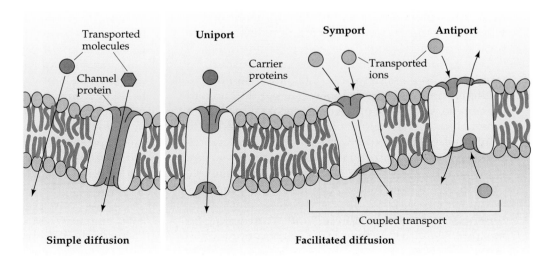

5.9 Membrane Transport Proteins Substances cross biological membranes by many mechanisms, most of which involve membrane transport proteins.

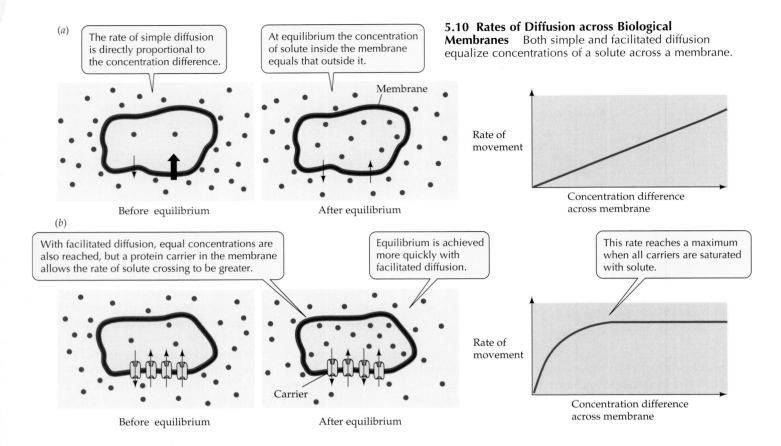

5.10 Rates of Diffusion across Biological Membranes Both simple and facilitated diffusion equalize concentrations of a solute across a membrane.

(a)

The rate of simple diffusion is directly proportional to the concentration difference.

At equilibrium the concentration of solute inside the membrane equals that outside it.

Membrane

Rate of movement

Concentration difference across membrane

Before equilibrium

After equilibrium

(b)

With facilitated diffusion, equal concentrations are also reached, but a protein carrier in the membrane allows the rate of solute crossing to be greater.

Equilibrium is achieved more quickly with facilitated diffusion.

This rate reaches a maximum when all carriers are saturated with solute.

Rate of movement

Concentration difference across membrane

Carrier

Before equilibrium

After equilibrium

across the bilayer and releases it on the other side. By this mechanism, still not completely understood, these carrier molecules speed (or facilitate) the passage of the solute molecules across the membrane.

In both simple diffusion and facilitated diffusion, the *rate* of movement depends on the concentration difference across the membrane. In simple diffusion, the net rate of movement is directly proportional to the concentration difference across the membrane (Figure 5.10*a*). In facilitated diffusion, the rate of movement also increases with the difference in solute concentration across the membrane, but a point is reached at which further increases in concentration difference are not accompanied by an increased rate (Figure 5.10*b*).

At this concentration, the facilitated diffusion system is said to be *saturated*. Because there are only a limited number of carrier molecules per unit area of membrane, the rate of movement reaches a maximum when all the carrier molecules are fully loaded with solute molecules. In other words, when the differences in the solute concentration across the membrane are sufficiently high, not enough carrier molecules are free at a given moment to handle all the solute molecules.

In facilitated diffusion, the carrier proteins enable the solutes to pass in both directions. The net movement is toward the side where the solute concentration is lower.

Osmosis is passive water movement through a membrane

Water moves through membranes by a diffusion process called **osmosis**. This completely passive process uses no metabolic energy and can be understood in terms of the solute concentrations of solutions. Osmosis depends on the *number* of solute particles present—not the kind of particles. We will describe osmosis using red blood cells and plant cells.

Red blood cells are normally suspended in a fluid called *plasma*, which contains salts, proteins, and other solutes. If a drop of blood is examined under the light microscope, the red cells are seen to have their characteristic shape. If pure water is added to the drop of blood, the cells quickly swell and burst (Figure 5.11*a*). Similarly, if slightly wilted lettuce is put in pure water, it soon becomes crisp; by weighing it before and after, we can show that it has taken up water (Figure 5.11*b*). If, on the other hand, the red blood cells or crisp lettuce leaves are placed in a relatively concentrated solution of salt or sugar, the leaves become limp and the red blood cells pucker and shrink.

From analyses of such observations, we learn that solute concentration is the principal factor in what is called the **osmotic potential** of a solution. *The greater the solute concentration, the more negative the osmotic potential of the solution.* Pure water has nothing dissolved

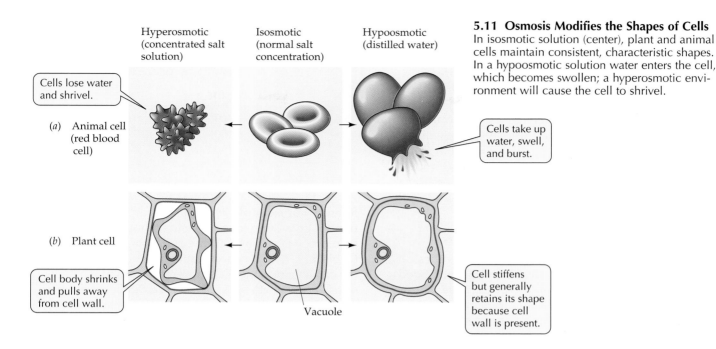

5.11 Osmosis Modifies the Shapes of Cells
In isosmotic solution (center), plant and animal cells maintain consistent, characteristic shapes. In a hypoosmotic solution water enters the cell, which becomes swollen; a hyperosmotic environment will cause the cell to shrivel.

Hyperosmotic (concentrated salt solution)

Isosmotic (normal salt concentration)

Hypoosmotic (distilled water)

Cells lose water and shrivel.

(a) Animal cell (red blood cell)

Cells take up water, swell, and burst.

(b) Plant cell

Cell body shrinks and pulls away from cell wall.

Vacuole

Cell stiffens but generally retains its shape because cell wall is present.

---

in it, so its osmotic potential equals zero. Other things being equal, if two unlike solutions are separated by a membrane that allows water to pass through but not solutes, water molecules will move across the membrane toward the solution with the more negative osmotic potential. This movement is called osmosis. In other words, water will move from a region of its higher concentration (lower concentration of solutes) to a region of its lower concentration (higher concentration of solutes).

If two solutions have identical osmotic potentials, they are **isosmotic** to one another (*iso-*, "same"). (This is true even if their chemical compositions are very different.) If the two solutions are not isosmotic, then solution A with a more negative osmotic potential (with a higher concentration of solutes) is said to be **hyperosmotic** to solution B, and solution B is **hypoosmotic** to solution A (*hyper-*, "more, high"; *hypo-*, "less, low").

All three of these terms are strictly relative. They can be used only in comparing the osmotic potentials of two specific solutions. For instance, no solution can be called hyperosmotic except in comparison with another solution that has a less negative osmotic potential. A related set of terms—iso-, hypo-, and hypertonic—refers to *solute concentrations* of solutions. Two solutions with equal total solute concentrations are *isotonic*. If solution A has a higher solute concentration than solution B, solution A is *hypertonic* to B, which is *hypotonic* to A.

Osmotic potentials determine the direction of osmosis in all animal cells. A red blood cell takes up water from a solution that is hypoosmotic to the cell's contents. The cell bursts because its plasma membrane cannot withstand the swelling of the cell (see Figure 5.11*a*). The integrity of red blood cells and other blood cells is absolutely dependent on the maintenance of a constant osmotic potential in the plasma in which they are suspended: The plasma must be isosmotic with the cells if the cells are not to burst or shrink.

In contrast to animal cells, the cells of plants, archaea, bacteria, fungi, and some protists have cell walls that limit the volume of the cells and keep them from bursting. Unlike the red blood cell, the wilted lettuce leaf becomes firm and crisp when placed in pure water (see Figure 5.11*b*). Cells with sturdy cell walls take up a limited amount of water and, in so doing, build up an internal pressure against the cell wall that prevents further water from entering. This pressure within the cell, called the **turgor pressure**, is the driving force for growth of plant cells—it is a normal and essential component of plant development.

In other words, in cells with walls, osmosis is regulated not only by osmotic potentials but also by the opposed turgor pressure. A salt solution sitting in a glass has a turgor pressure of zero. Enclose some of it with a rigid membrane that is impermeable to salt but permeable to water, drop this package into pure water, and the enclosed salt solution will have a turgor pressure that increases as water moves by osmosis through the membrane. Osmotic phenomena in plants are discussed in Chapter 32.

Active Processes of Membrane Transport

Earlier in this chapter we distinguished passive transport processes from active transport processes. As we have seen, the passive processes are driven by a concentration difference for the material being transported. Passive transport processes always operate from regions of greater to regions of lesser concentration, and no external source of energy is required.

Active transport systems, on the other hand, require a source of energy to move materials. In this section we will examine two different classes of transport phenomena that require the input of energy. First, we'll consider highly specific systems located in membranes that operate on one or a few molecules (or ions) at a time. Second, we'll consider systems of membrane fusion that enable a cell to engulf or secrete bulk quantities of material.

Active transport requires energy and carriers

Like facilitated diffusion, **active transport** relies on carrier molecules that are proteins. Unlike facilitated diffusion, this process requires energy to move ions or molecules across a biological membrane from regions of lower concentration to regions of greater concentration—that is, to move solutes *up* a concentration gradient. Sometimes, but not always, the immediate source of energy for this process is energy provided by the hydrolysis of ATP. There are two basic types of active transport: primary and secondary.

Primary active transport requires the direct participation of adenosine triphosphate (ATP). In primary active transport, energy released from ATP drives the movement of specific ions against a concentration dif-

TABLE 5.1 Concentration of Major Ions Inside and Outside the Nerve Cell of a Squid

ION	CONCENTRATION (MOLAR)	
	INSIDE	OUTSIDE
K^+	0.400	0.020
Na^+	0.050	0.440
Cl^-	0.120	0.560

ference. For example, we can compare the concentrations of potassium ions (K^+) and sodium ions (Na^+) inside a nerve cell and in the fluid bathing the nerve (Table 5.1). The K^+ concentration is much higher inside the cell, whereas the Na^+ concentration is much higher outside. Nevertheless, the nerve cells continue to pump Na^+ out and K^+ in, against these concentration differences, ensuring that the differences are maintained.

This **sodium–potassium pump** is found in all animal cells and is an integral membrane *glycoprotein*. Breaking down a molecule of ATP to ADP and phosphate (P_i), the sodium–potassium pump brings two K^+ ions into the cell and exports three Na^+ ions (Figure 5.12). The sodium–potassium pump is thus an antiport transport system. Different pumps are responsible for the transport of several other ions, but only cations are

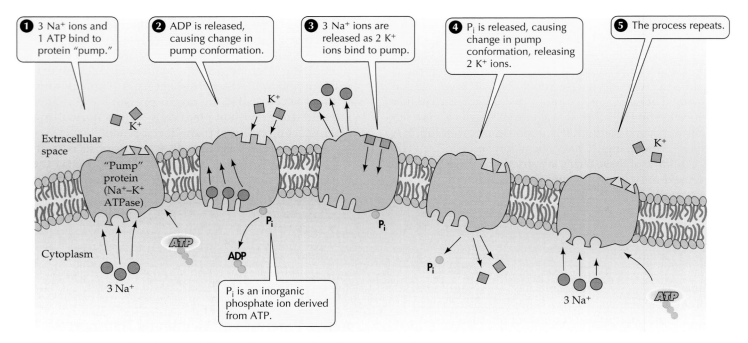

❶ 3 Na⁺ ions and 1 ATP bind to protein "pump."

❷ ADP is released, causing change in pump conformation.

❸ 3 Na⁺ ions are released as 2 K⁺ ions bind to pump.

❹ P_i is released, causing change in pump conformation, releasing 2 K⁺ ions.

❺ The process repeats.

P_i is an inorganic phosphate ion derived from ATP.

5.12 Primary Active Transport: The Sodium–Potassium Pump   In active transport, energy is used to move a solute against its concentration gradient. Even though the Na^+ concentration is higher outside the cell and the K^+ concentration is higher inside the cell, for each molecule of ATP used, two K^+ ions are pumped into the cell and three Na^+ ions are pumped out of the cell.

transported directly by pumps in primary active transport. Other solutes are transported by secondary active transport.

Unlike primary active transport, **secondary active transport** does not use ATP directly; rather, the transport of the solute is tightly coupled to the difference in ion concentration established by primary active transport. The movement of particular solutes, such as sugars and amino acids, is regulated by coupled transport systems that move these specific solutes against their concentration difference, using energy "regained" by letting Na^+ or other ions diffuse across the membrane from regions of higher to regions of lower concentration—that is, to move *down* their concentration gradient (Figure 5.13).

Both types of coupled transport—symport and antiport—are used for secondary active transport. Putting primary and secondary active transport together, we see that energy from ATP is used in one example of primary active transport to establish concentration differences of potassium and sodium ions; then the passive transport of some sodium ions in the opposite direction provides energy for the secondary active transport of the sugar glucose (see Figure 5.13). Other secondary active transporters aid in the uptake of amino acids and sugars, which are essential raw materials for cell maintenance and growth.

Macromolecules and particles enter the cell by endocytosis

A process called **endocytosis** brings macromolecules, large particles, and small cells into the eukaryotic cell (Figure 5.14*a*). There are three types of endocytosis: phagocytosis, pinocytosis, and receptor-mediated endocytosis. In all three, the plasma membrane folds inward, making a small pocket. The pocket deepens, forming a vesicle whose contents are materials from the environment. This vesicle separates from the surface of the cell and migrates to the cell's interior.

We encountered phagocytosis briefly in Chapter 4. In this process, part of the plasma membrane engulfs fairly large particles or even entire cells. Phagocytosis is a cellular feeding process found in unicellular protists and in some white blood cells that defend the body against foreign cells and substances (see Chapter 18). The food vacuole or phagosome formed usually fuses with a lysosome, and its contents are digested (see Figure 4.19). Vesicles also form in pinocytosis ("cellular drinking"). However, these vesicles are smaller, and the process operates to bring in small dis-

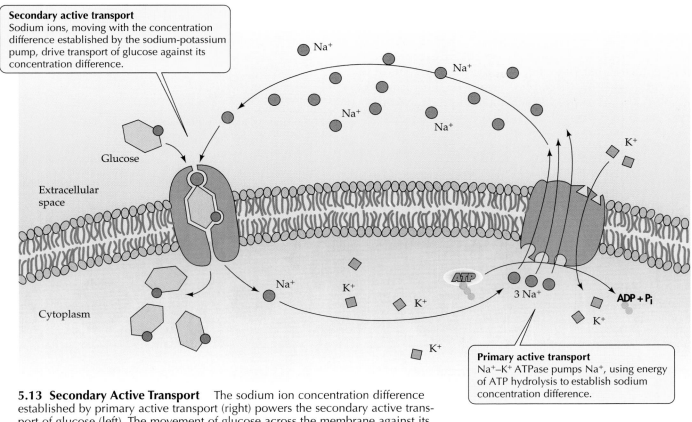

Secondary active transport
Sodium ions, moving with the concentration difference established by the sodium-potassium pump, drive transport of glucose against its concentration difference.

Primary active transport
Na^+–K^+ ATPase pumps Na^+, using energy of ATP hydrolysis to establish sodium concentration difference.

5.13 Secondary Active Transport The sodium ion concentration difference established by primary active transport (right) powers the secondary active transport of glucose (left). The movement of glucose across the membrane against its concentration difference is coupled by a symport protein to the diffusion of Na^+ into the cell.

(a) **Endocytosis**

In endocytosis, the plasma membrane surrounds a part of the exterior environment and the whole unit buds off to the interior as a vesicle.

(b) **Exocytosis**

In exocytosis, a membrane-enclosed vesicle containing substances for export fuses with the plasma membrane. The contents of the vesicle scatter, and the vesicle membrane becomes part of the plasma membrane.

Secretory vesicle

Plasma membrane

Vesicle

5.14 Endocytosis and Exocytosis Endo- and exocytosis are used by all eukaryotic cells to take up and eliminate substances from and to the outside environment.

solved substances or fluids. The third type of endocytosis, receptor-mediated endocytosis, involves specific reactions at the cell surface that trigger uptake. Let's take a closer look at this process.

Receptor-mediated endocytosis is highly specific

Most animal cells have a mechanism called **receptor-mediated endocytosis** that captures specific macromolecules from the cell's environment. The uptake is similar to endocytosis as already described, except that in receptor-mediated endocytosis receptor proteins at particular sites on the outer surface of the plasma membrane bind to specific substances in the environ-

ment outside the cell. These sites are called **coated pits** because they have a slight depression of the plasma membrane whose inner surface is coated by fibrous proteins, such as *clathrin*.

When a receptor protein binds the appropriate macromolecule outside the cell, the associated coated pit invaginates (folds inward) and forms a **coated vesicle** around the bound macromolecules. Strengthened and stabilized by clathrin molecules, this vesicle carries the macromolecules into the cell (Figure 5.15). Once inside, the vesicle loses its coat and may fuse with another vesicle for the processing and release into the cytoplasm of the engulfed material. Because of its specificity for particular macromolecules, receptor-mediated endocytosis is a more rapid and efficient method of taking up what may be minor constituents of the cell's environment.

5.15 Formation of a Coated Vesicle in Receptor-Mediated Endocytosis The receptor proteins in a coated pit bind specific macromolecules, which are then carried into the cell by the coated vesicle.

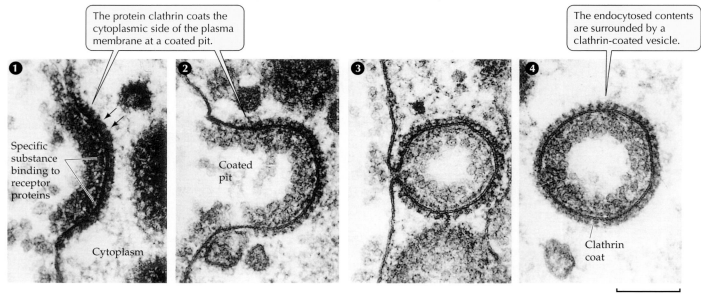

The protein clathrin coats the cytoplasmic side of the plasma membrane at a coated pit.

The endocytosed contents are surrounded by a clathrin-coated vesicle.

❶ Specific substance binding to receptor proteins

Cytoplasm

❷ Coated pit

❸

❹ Clathrin coat

0.1 μm

Receptor-mediated endocytosis is the method by which cholesterol is taken up by most mammalian cells. Water-insoluble cholesterol is synthesized in the liver and transported in the blood attached to a protein, forming a lipoprotein such as **low-density lipoprotein**, or **LDL** (Figure 5.16). The uptake mechanism for cholesterol includes the binding of LDL to specific receptor proteins in coated pits. After being engulfed by endocytosis, LDL particles are freed from the receptors, which then segregate to a region that buds off to form a new vesicle that is recycled to the plasma membrane. The freed LDL particles remain in a vesicle that fuses with a lysosome in which the LDL particles are digested and the cholesterol made available for cell use. In the inherited disease *hypercholesterolemia* (*-emia*, "blood"), patients have dangerously high levels of cholesterol in their blood because of a deficient receptor for LDL.

Exocytosis moves materials out of the cell

In eukaryotic cells (but not in prokaryotic cells) the plasma membrane regulates ongoing traffic: Membrane-enclosed "packages" enter the cell by endocytosis and leave by exocytosis. **Exocytosis** is the process by which materials packaged in vesicles are secreted from the cell when the vesicle membrane fuses with the plasma membrane (see Figure 5.14*b*). The phospholipid regions of the two membranes merge, and an

opening to the outside of the cell develops. The contents of the vesicle are released to the environment, and the vesicle membrane is smoothly incorporated into the plasma membrane.

In Chapter 4 we encountered exocytosis as the last step in processing the material engulfed by phagocytosis: the excretion of indigestible materials to the environment. Exocytosis is also important in the secretion of many different substances, including digestive enzymes from the pancreas, neurotransmitters from nerve cells, and materials for plant cell growth.

Membranes Are Not Simply Barriers

We have discussed some functions of membranes—the compartmentalization of cells, the regulation of traffic between compartments, and the active pumping of solutes—but there are more. As discussed in Chapter 4, the membrane of rough endoplasmic reticulum serves as a site for ribosome attachment. Newly formed proteins are passed from the ribosomes through the membrane and into the interior of the ER for delivery to other parts of the cell. The membranes of nerve cells, muscle cells, some eggs, and other cells are also electrically excitable. In nerve cells, the plasma membrane is the conductor of the nerve impulse from one end of the cell to the other (see Chapter 41). Numerous other biological activities and properties

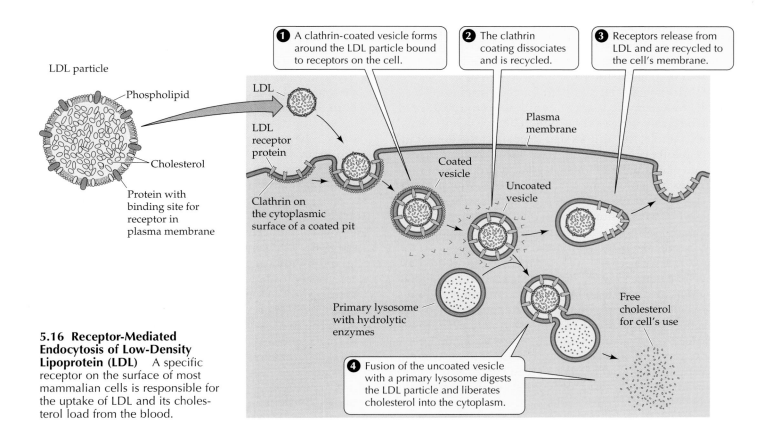

5.16 Receptor-Mediated Endocytosis of Low-Density Lipoprotein (LDL) A specific receptor on the surface of most mammalian cells is responsible for the uptake of LDL and its cholesterol load from the blood.

discussed in the chapters to follow are integrally associated with membranes.

To elaborate further on the versatility of biological membranes, we will describe three additional cellular functions of membranes: their roles in processing information from the environment, their roles in energy-trapping and energy-releasing reactions, and their roles in helping cells associate as tissues.

Some membranes process information

The plasma membranes at cell surfaces and the membranes within cells have protruding portions of integral membrane proteins or attached carbohydrates that recognize and bind to specific substances in their environment. The general term for such a substance is **ligand**.

The binding of a receptor to a ligand initiates specific change in the function of the cell or organelle to which the receptor is attached. In this type of information processing, specificity in binding is essential. We have already seen the role of a specific receptor in the endocytosis of LDL and its cargo of cholesterol (review Figure 5.16), but there are many other examples.

RECOGNITION AND BINDING BY PROTEINS AND CARBOHYDRATES. Antibodies are specific molecules synthesized by the body to defend against invading foreign substances and cells (see Chapter 18). Antibodies begin defensive operations by reacting with the specific proteins or carbohydrates protruding from the plasma membranes of many foreign or cancerous cells. Here, as in other membrane signaling, specificity is important; antibodies must not destroy normal body cells. Another example of the specificity of signals on cell surfaces is found in the initial steps of virus infection.

Viruses may begin attacking their intended host cells by attaching to carbohydrates on the surface of the host cell. Viruses show a high degree of specificity. For example, the viruses that infect bacteria do not infect plants or animals. A plant virus that infects tobacco plants will not infect begonias, and some animal viruses infect just one or a few species. The measles virus, for example, infects only humans, while the rabies virus infects humans and other mammals.

Many hormones, including insulin, are recognized by membrane proteins that serve as receptors on the hormone's target cells (see Chapter 38). And the passage of a nerve impulse from a nerve cell to a neighboring nerve or muscle cell by a neurotransmitter requires a receptor molecule on the plasma membrane of the target cell that is specific for the particular neurotransmitter that is released (see Chapter 41). Again specificity is essential; receptors for one transmitter will not accept another kind of transmitter. One of the most important classes of receptors consists of those that bind substances called growth factors, which regulate cell reproduction and differentiation (see Chapter 15).

Receptor proteins are integral membrane proteins that span the entire thickness of the plasma membrane. Receptor carbohydrates are bound to such proteins or to phospholipid molecules. The specificity of receptors (both proteins and carbohydrates) resides in their particular tertiary structure, that is, in their three-dimensional shape. Some portion of the receptor protein or carbohydrate fits the ligand like a hand fits a glove. We will deal with the molecular nature of such binding specificities in the next chapter, and this topic will arise again in Chapter 18 when we discuss immune responses.

When a ligand binds to its specific receptor on the cell surface, changes occur within the cell, on the cytoplasmic side of the membrane. How can such a visitor produce an effect inside the cell if it does not actually enter it? The answer is signal transduction.

SIGNAL TRANSDUCTION BY PROTEINS. *Signal transduction* is the conversion of one type of signal to another type of signal. When a ligand binds to a receptor protein on the outside surface of a cell, the receptor's tertiary structure changes—there is a slight *conformational* change. In an integral membrane protein that spans the entire bilayer, this change is transmitted to the part of the protein exposed to the cytoplasm (Figure 5.17). In many cases the change in protein activates some dormant function of the protein. For example, this change may allow the ligand–receptor complex to activate another membrane protein, called a G protein.

The **G protein**, once activated, diffuses in the membrane to activate or inactivate another membrane protein. This last protein may be an enzyme or a transport protein. Thus, the signal represented by the ligand is transduced by the membrane proteins into an altered flow of ions or a changed rate of chemical reaction. Both of these results have still broader consequences for cell functioning. Several different G proteins play roles in hormone action, vision, and other important processes to be discussed later. G proteins are so named because their activation requires GTP, a close chemical analog of ATP.

Some membranes transform energy

In a variety of cells, the membranes of organelles are specialized to process energy. For example, the thylakoid membranes of chloroplasts participate in the conversion of light energy to the energy of chemical bonds, and the inner mitochondrial membrane helps convert the energy of fuel molecules to the energy in ATP. The two characteristics of membranes that enable them to participate in these processesare their structural organization and the separation of their electric charges.

ORGANIZING CHEMICAL REACTIONS. Many processes in cells depend on a series of enzyme-catalyzed reac-

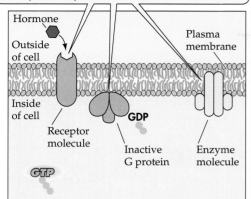
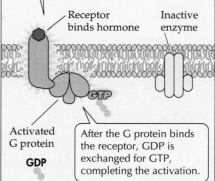
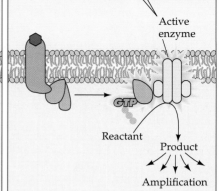

The actions of several membrane-associated proteins are required to convert the signal from a hormone to an amplified response in the cell.

Hormone binding provides a signal that is converted to another type of signal: the activated G protein.

Part of the activated G protein activates an enzyme that converts thousands of reactants to products, thus amplifying the action of a single hormone molecule.

After the G protein binds the receptor, GDP is exchanged for GTP, completing the activation.

5.17 Signal Transduction by a G Protein In signal transduction, one type of signal (the binding of a hormone) is converted to another type of signal (the activation of a G protein). The G protein then activates an enzyme that repeatedly catalyzes a reaction.

tions in which the products of one reaction serve as the reactants for the next. For such a reaction to occur, all the necessary molecules must collide. In a solution, the reactants and enzymes are all randomly distributed and collisions are random. For this reason, a complete series of chemical reactions in solution may occur very slowly.

However, if the different enzymes are bound to a membrane in sequential order, the product of one reaction can be released close to the enzyme for the next reaction. With such an "assembly line," reactions proceed more rapidly and efficiently. In this sense, the membrane is a pegboard for the orderly attachment of specific proteins and other molecules.

SEPARATING ELECTRIC CHARGES. A biological membrane operates like an electric battery. Work can be obtained from a battery by letting electrons flow from one of its terminals to the other by way of a device, such as a motor or a light bulb, that uses the electric current. Something similar takes place in both photosynthesis and cellular respiration (see Chapters 7 and 8).

Because of the impermeability of the mitochondrial and chloroplast membranes to hydrogen ions (H^+) and because of the activities of certain electron carriers in those membranes, it is possible to separate charges and to concentrate H^+. In other words, a substantial gradient of both electric charge and pH is established across membranes. When electric charge flows back across the membrane, its diffusion is coupled to a protein that catalyzes the synthesis of ATP from ADP and phosphate, thus capturing the energy of a concentration gradient as chemical bond energy in ATP. Without

a membrane to allow the separation of charge, these reactions would not proceed.

Both the structure of membranes and their ability to separate electric charges relate to the properties of the two bulk components of membranes: lipids and proteins. The pegboard effect of the membrane's structure comes from the ability of the phospholipid bilayer to hold certain proteins in a defined plane so that they do not diffuse freely throughout the cell. Certain membrane proteins create the separation of charges, which is then maintained by the selective permeability of the phospholipid bilayer.

Cell adhesion molecules organize cells into tissues

During the growth of an animal embryo, the **cell adhesion proteins** mentioned earlier are crucial to the formation of the adult organism. As the embryo develops, its cells move about and associate with other specific cell types. This behavior is mediated by cell adhesion molecules in the plasma membranes, which play roles in organizing the cells into tissues. One type of cell adhesion molecule, for example, organizes individual nerve cells into nerve cell bundles. Groups of cells that are about to migrate within the embryo lose their specific cell adhesion molecules; when they reach their new location, they regain their cell adhesion molecules and reorganize into tissue.

Membranes Are Dynamic

As we have seen in this chapter, membranes participate in numerous physiological and biochemical processes. Membranes are dynamic in another sense as well: They are constantly forming, transforming from one type to another, fusing with one another, and breaking down.

In eukaryotes, phospholipids are synthesized on the surface of the endoplasmic reticulum and rapidly

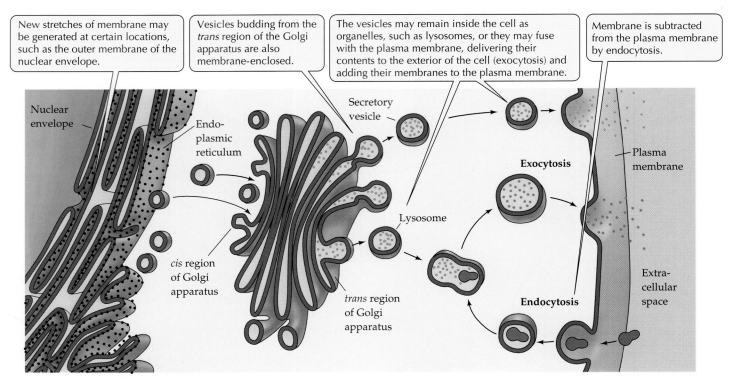

New stretches of membrane may be generated at certain locations, such as the outer membrane of the nuclear envelope.

Vesicles budding from the *trans* region of the Golgi apparatus are also membrane-enclosed.

The vesicles may remain inside the cell as organelles, such as lysosomes, or they may fuse with the plasma membrane, delivering their contents to the exterior of the cell (exocytosis) and adding their membranes to the plasma membrane.

Membrane is subtracted from the plasma membrane by endocytosis.

Nuclear envelope

Endo-plasmic reticulum

Secretory vesicle

Exocytosis

Plasma membrane

cis region of Golgi apparatus

Lysosome

trans region of Golgi apparatus

Endocytosis

Extra-cellular space

5.18 Dynamic Continuity of Cellular Membranes The arrows trace how membranes form, move, and fuse in cells.

distributed to membranes throughout the cell as vesicles form from the ER, move away, and fuse with other organelles. Membrane proteins are inserted into the lumen of the rough endoplasmic reticulum as they form on ribosomes. Sugars may be added to the proteins while they are in the endoplasmic reticulum. Next the proteins are found in the Golgi apparatus, where other carbohydrates are added to some of them. The proteins then travel in Golgi-derived vesicles to the plasma membrane and are incorporated into it (see Figure 4.19).

Functioning membranes move about within eukaryotic cells. For example, portions of the rough ER bud away from the ER and join the *cis* faces of the Golgi apparatus (see Chapter 4). Rapidly—often in less than an hour—these segments of membrane find themselves in the *trans* regions of the Golgi, from which they bud away to join the plasma membrane (Figure 5.18). Bits of membrane are constantly merging with the plasma membrane by exocytosis, but this process is largely balanced by the removal of membrane in endocytosis. This removal by endocytosis affords a recovery path by which internal membranes are replenished. In sum, there is a steady flux of membranes as well as membrane components in cells.

Because we know about the constant interconversion of membranes, we might expect all subcellular membranes to be chemically identical. As you already know, this is not the case, for there are major chemical differences among the membranes of even a single cell. Apparently membranes are changed chemically when they form parts of certain organelles. In the Golgi apparatus, for example, the membranes of the *cis* face closely resemble those of the endoplasmic reticulum, but the *trans*-face membranes are more similar in composition to the plasma membrane.

Ceaselessly moving, constantly carrying out functions vital to the life of the cell, biological membranes certainly are not the static, stodgy structures they once were thought to be.

Summary of "Cellular Membranes"

Membrane Composition and Structure

• Biological membranes consist of lipids, proteins, and carbohydrates. The fluid-mosaic model of membrane structure describes a phospholipid bilayer in which integral membrane proteins can move about in the plane of the membrane. **Review Figures 5.1, 5.4, 5.5**
• The two surfaces of a membrane have different properties because of their different phospholipid composition, exposed domains of integral membrane proteins, and peripheral membrane proteins. Defined regions of a plasma membrane may have different membrane proteins. **Review Figures 5.1, 5.2**

• Carbohydrates attached to proteins or phospholipids project from the external surface of the plasma membrane and function as recognition signals for interactions between cells. **Review Figure 5.1**

Animal Cell Junctions

• Tight junctions prevent the passage of molecules through the space around cells, and they define functional regions of the plasma membrane by restricting the migration of membrane proteins uniformly over the cell surface. Desmosomes allow cells to adhere strongly to one another. Gap junctions provide channels for chemical and electrical communication between adjacent cells. **Review Figure 5.6**

Passive Processes of Membrane Transport

• Solutes are passively transported across a membrane by diffusion from a region with a greater solute concentration to a region of lesser solute concentration. Equilibrium is reached when the concentrations of a given solute are identical on both sides of the membrane. **Review Figures 5.7, 5.8**
• Substances can diffuse across a membrane by three processes: unaided diffusion through the phospholipid bilayer, diffusion through protein channels that form specific pores, or diffusion by means of a carrier protein (facilitated diffusion). **Review Figure 5.9**
• Channel proteins and carrier proteins may function in uniport or in coupled transport systems (symport or antiport). **Review Figure 5.9**
• The rate of simple diffusion is directly proportional to the concentration difference. The rate of facilitated diffusion reaches a maximum when a solute concentration is reached that saturates the carrier proteins so that no further increase in rate is observed with increases in solute concentration. **Review Figure 5.10**
• In osmosis, water diffuses from regions of less negative osmotic potential to regions of more negative osmotic potential. Isosmotic environmental conditions must be maintained to prevent animal cells from destructive loss or gain of water. In hypoosmotic solutions cells tend to take up water, while in hyperosmotic solutions cells tend to lose water. **Review Figure 5.11a**
• The cell walls of plants and some other organisms prevent the cells from bursting under hypoosmotic conditions. The turgor pressure (internal pressure) that develops under these conditions keeps plants upright and stretches the cell wall during plant cell growth. **Review Figure 5.11b**

Active Processes of Membrane Transport

• Active transport always requires energy to move substances across a membrane. Active transport systems moving single solute molecules or ions include primary and secondary active transport.
• In primary active transport, energy from the hydrolysis of ATP is used to move Na^+ out of cells and K^+ into cells against their concentration gradients. **Review Figure 5.12**
• Secondary active transport couples the passive movement of Na^+ down its concentration gradient with the movement of a sugar molecule up its concentration gradient. Energy from ATP is used indirectly to establish the Na^+ concentration gradient. **Review Figure 5.13**
• Energy is required for the many cellular processes that contribute to endocytosis, which transports macromolecules, large particles, and small cells into eukaryotic cells by means of engulfment and vesicle formation from the plasma membrane. Phagocytosis and pinocytosis are both nonspecific types of endocytosis. In receptor-mediated endocytosis, a specific membrane receptor binds to a particular macromolecule. **Review Figures 5.14, 5.15, 5.16**
• In exocytosis, materials in vesicles are secreted from the cell when the vesicles fuse with the plasma membrane. **Review Figure 5.14**

Membranes Are Not Simply Barriers

• Membranes also function as sites for protein or carbohydrate recognition signals for hormones, neurotransmitters, and growth factors. Conformational changes in membrane proteins transmit signals from outside the cell to enzymes within the cell by means of G proteins. In mitochondria and chloroplasts, membranes binding enzymes ensure faster and more efficient operation of a series of chemical reactions. **Review Figure 5.17**
• Membranes impermeable to H^+ separate electric charges and provide the basis for synthesis of ATP from ADP and phosphate in mitochondria and chloroplasts.
• Cell adhesion proteins in plasma membranes play crucial roles in organizing cells into tissues during embryonic development.

Membranes Are Dynamic

• Although not all cellular membranes are identical, ordered modifications in membrane composition accompany the conversions of one type of membrane into another type. **Review Figure 5.18**

Self-Quiz

1. Which statement about membrane phospholipids is *not* true?
   a. They associate to form bilayers.
   b. They have hydrophobic "tails."
   c. They have hydrophilic "heads."
   d. They give the membrane fluidity.
   e. They flop readily from one side of the membrane to the other.

2. The phospholipid bilayer
   a. is readily permeable to large, polar molecules.
   b. is entirely hydrophobic.
   c. is entirely hydrophilic.
   d. has different lipids in the two layers.
   e. is made up of polymerized amino acids.

3. Which statement about membrane proteins is *not* true?
   a. They all extend from one side of the membrane to the other.
   b. Some serve as channels for ions to cross the membrane.
   c. Many are free to migrate laterally within the membrane.
   d. Their position in the membrane is determined by their tertiary structure.
   e. Some play roles in photosynthesis.

4. Which statement about membrane carbohydrates is *not* true?
   a. Most are bound to proteins.
   b. Some are bound to lipids.
   c. They are added to proteins in the Golgi apparatus.
   d. They show little diversity.
   e. They are important in recognition reactions at the cell surface.

5. Which statement about animal cell junctions is *not* true?
   a. Tight junctions are barriers to the passage of molecules between cells.
   b. Desmosomes allow cells to adhere strongly to one another.
   c. Gap junctions block communication between adjacent cells.
   d. Connexons are made of protein.
   e. The fibers associated with desmosomes are made of protein.

6. Which statement about diffusion is *not* true?
   a. It is the movement of molecules or ions to a state of even distribution.
   b. At the subcellular level it is a slow process.
   c. The motion of each molecule or ion is random.
   d. The diffusion of each substance is independent of that of other substances.
   e. Diffusion across meters takes years.

7. Which statement about membrane channels is *not* true?
   a. They are pores in the membrane.
   b. They are proteins.
   c. All ions pass through the same type.
   d. Movement through them is from high concentration to low.
   e. Movement through them is by simple diffusion.

8. Facilitated diffusion and active transport
   a. both require ATP.
   b. both require the use of proteins as carriers.
   c. both carry solutes in only one direction.
   d. both increase without limit as the solute concentration increases.
   e. both depend on the solubility of the solute in lipid.

9. Primary and secondary active transport
   a. both generate ATP.
   b. both are based on passive movement of sodium ions.
   c. both include the passive movement of glucose molecules.
   d. both use ATP directly.
   e. both can move solutes against their concentration gradients.

10. Which statement about osmosis is *not* true?
    a. It obeys the laws of diffusion.
    b. In animal tissues, water moves to the cell with the most negative osmotic potential.
    c. Red blood cells must be kept in a plasma that is hypoosmotic to the cells.
    d. Two cells with identical osmotic potentials are isosmotic to each other.
    e. Solute concentration is the principal factor in osmotic potential.

Applying Concepts

1. In Chapter 44 we will see that the functioning of muscles requires calcium ions to be pumped into a subcellular compartment against a calcium concentration gradient. What types of chemical substances are required for this to happen?

2. Some algae have complex glassy structures in their cell walls. The structures form within the Golgi apparatus. How do these structures reach the cell wall without having to pass through a membrane?

3. Organisms that live in fresh water are almost always hyperosmotic to their environment. In what way is this a serious problem? How do some organisms cope with this problem?

4. Contrast simple endocytosis and receptor-mediated endocytosis with respect to mechanism and to performance.

Readings

Alberts, B., D. Bray, J. Lewis, M. Raff, K. Roberts and J. D. Watson. 1994. *Molecular Biology of the Cell*, 3rd Edition. Garland Publishing, New York. An outstanding general text in modern cell and molecular biology. Chapters 10 and 11 are particularly suitable for further study of biological membranes.

Bretscher, M. S. 1985. "The Molecules of the Cell Membrane." *Scientific American*, October. A fine treatment of membrane chemistry, cell junctions, endocytosis, and other topics.

Cooper, G. M. 1997. *The Cell: A Molecular Approach*. ASM Press, Washington, DC, and Sinauer Associates, Sunderland, MA. An excellent short cell biology text. See Chapter 12 in particular.

Horwitz, A. E. 1997. "Integrins and Health." *Scientific American*, May. These cell adhesion molecules play many roles in human cells; malfunctioning integrins result in disease.

Lodish, H., D. Baltimore, A. Berk, S. L. Zipursky, P. Matsudaira and J. Darnell. 1995. *Molecular Cell Biology*, 3rd Edition. Scientific American Books, New York. A fine general text. See Chapters 14 and 15; Chapter 16 is also relevant.

Stryer, L. 1995. *Biochemistry*, 4th Edition. W. H. Freeman, New York. A reference for the major types of molecules found in membranes, with outstanding illustrations of phospholipids.

Unwin, N. and R. Henderson. 1984. "The Structure of Proteins in Biological Membranes." *Scientific American*, February. A clear presentation of how the proteins in membranes transport molecules.

Chapter 6

Energy, Enzymes, and Metabolism

Change is everywhere. Wherever we look, we observe changes occurring in both the physical and the biological worlds. Some changes are slow, others are fast. Some are gigantic, others are small. Some are complex, others are simple. The concept of energy is profoundly helpful in understanding how changes are different and how they are all similar. In the preceding chapters, we often referred to energy, and in Chapter 2 we defined it as the capacity to do work. Now we want to focus on the different forms of energy and their role in accomplishing the activities characteristic of living systems.

The rock climber shown opposite has expended a lot of energy—done a lot of work—to get where she is. Looking at her, you can practically *feel* that energy expenditure. A climber has more reason than most people to be concerned about her potential energy—energy of position or state. As she climbs higher, her potential energy becomes greater. If she slips and falls, her potential energy will convert to kinetic energy—the energy of motion. Her climb has increased her potential energy; a fall would decrease it. To accomplish her climb, the cells in her body have converted the potential energy of chemical bonds (chemical energy) into the kinetic energy of muscle contractions.

In this chapter, we will examine the nature of energy and its relationship to work. We will apply physical principles to the chemical substances and chemical changes that constitute living systems. Then we will examine the role of en-

States of Energy
This rock climber's precarious position was achieved only with a great deal of expended energy. The potential energy of the position, should she fall, would be equally great.

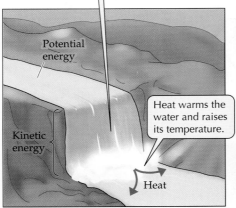

The water's potential energy is converted to kinetic energy, which is converted to heat at the base of the waterfall.

The dam converts the kinetic energy of a flowing river to potential energy by backing up the water.

Electric energy can be transmitted, stored, and used in a variety of ways to do work.

Potential energy

Kinetic energy

Heat warms the water and raises its temperature.

Heat

Dam

Light

Stored in batteries

Motor

A generator converts the movement of water released from the dam (kinetic energy) into electric energy.

6.1 Energy Conversions and Work
The kinetic energy of a flowing river can be converted to potential energy by a dam. Release of water from the dam converts the potential energy back into kinetic energy, which a generator can convert into electric energy.

zymes in speeding up chemical reactions in living systems. We'll close the chapter by focusing on metabolism, showing the roles of energy and enzymes in establishing and regulating the sequences of chemical reactions, called *pathways*, that transform matter and energy to produce the activities characteristic of life.

Energy Conversions: The Laws of Thermodynamics

All living things must obtain energy from the environment—no living cell manufactures energy. However, energy can be transformed from one kind into another. These energy transformations are linked to transformations of matter that occur as chemical reactions in cells, so changes in energy are associated with changes in matter. These changes in matter constitute metabolism. **Metabolism** is the total chemical activity of a living organism; at any instant, metabolism consists of thousands of individual chemical reactions.

Changes in energy are related to changes in matter

Energy comes in many forms (Figure 6.1). Heat, light, electric, and mechanical energy, among others, are forms of **kinetic energy**. Energy associated with position and gravity (think of the rock climber) is one type of **potential energy**. Potential energy is also associated with chemical bonds. If the rock climber jumps down to a ledge below, she loses some of her potential energy in a burst of movement—kinetic energy. Plants use light energy from the sun to drive the chemical re-

actions that convert the simple molecules CO_2 and H_2O into sugars, which are more complex. Some of the kinetic energy of light is stored as potential energy in the chemical bonds of the sugars. As you have seen in earlier chapters, these sugars can be stored as the polymer starch, or they can be degraded and the energy released in the process of cellular respiration.

In all cells (cells of plants, animals, fungi, protists, archaea, and bacteria), two types of metabolic reactions are found: anabolic and catabolic. **Anabolic reactions** (anabolism) link together simple molecules to form more complex molecules; the photosynthetic production of sugar is an anabolic reaction. Anabolic reactions store energy in the chemical bonds formed. **Catabolic reactions** (catabolism) break down complex molecules into simpler ones and release stored energy.

Catabolic and anabolic reactions are often linked: The energy released in the former is used to do biological work and drive the latter (Figure 6.2). In these energy conversions, there is a direction to the flow of energy. Energy flows through the biosphere in an irreversible conversion of light energy, by way of chemical energy, to heat. In general, the heat released by chemical reactions cannot be used for work in a biological system. Heat cannot do work unless it flows from one part of a system to another, which can happen only if one part of the system is hotter than another. Since all the parts of a cell are at the same temperature, heat cannot do work in cells.

The cellular activities of growth, motion, and active transport of ions across a membrane all require energy. None of these cellular activities or other forms of work would proceed without a source of energy. However,

6.2 Biological Energy Transformations Cavorting lionesses convert chemical energy, obtained from the prey they have eaten, into a burst of kinetic energy of motion. Their prey obtained chemical energy by consuming plants. The plants trapped light energy and produced the prey's food by photosynthesis.

some kinds of energy cannot be used for these tasks. In the discussion that follows, you will discover the physical laws that govern all energy transformations, identify the energy available to do work, and consider the direction of energy flow.

The first law: Energy is neither created nor destroyed

When all forms of energy are accounted for, the total amount of energy in the universe is constant. In any conversion of energy from one form to another (light to chemical, mechanical to electric), *energy is neither created nor destroyed*. This is the **first law of thermodynamics** (Figure 6.3a). The first law applies to the universe as a whole or to any *closed system* within the universe. By "system" we mean any part of the universe with specified matter and energy. A **closed system** is one that is not exchanging energy with its surroundings. In considering the origin, evolution, and maintenance of living organisms on Earth, the appropriate closed system consists of the solar system, in particular the sun, Earth, and surrounding space (Figure 6.4).*

A closed system may contain parts that are not closed. **Open systems**, such as living cells, exchange

*Some energy does enter and leave the solar system—after all, we *can* see the light of stars—but the system is very nearly closed. In the same way, a thermos bottle is *almost* a closed system, although it does slowly gain or lose heat. The universe itself is a perfectly closed system.

matter and energy with their surroundings. Does this mean that cells disobey the first law or that the first law does not apply to living organisms? Not at all. It means that an open system is merely one part of a larger closed system and receives energy from other parts of that larger system.

The second law: Not all energy can be used, and disorder tends to increase

The second law concerns what happens when energy changes form. The law applies to all changes, but we will be concerned primarily with the changes that involve chemical reactions in living systems.

NOT ALL ENERGY CAN BE USED. The **second law of thermodynamics** states that, although energy cannot be created or destroyed, *when energy is converted from one form to another, some of the energy becomes unavailable to do work* (see Figure 6.3b). In other words, no physical process or chemical reaction is 100 percent efficient, and not all the energy released can be converted to work. Some energy is lost as a form associated with disorder.

In any system, the total energy includes the *usable energy* that can be used to do work and the *unusable energy* that is lost to disorder.

$$\text{total energy} = \text{usable energy} + \text{unusable energy}$$

In biological systems, the total energy is called **enthalpy** (**H**). The usable energy that can do work is called the **free energy** (**G**). Free energy is what cells require for all the chemical reactions of cell growth, cell division, and the maintenance of cell health. The unusable energy is represented by **entropy** (**S**), which is the disorder of the system, multiplied by the absolute temperature (**T**). Thus we can rewrite the word equation above more precisely as

$$H = G + TS$$

Because we are interested in the usable energy, we rearrange this expression:

$$G = H - TS$$

Although we cannot measure G, H, or S absolutely, we can determine the *change* of each at a constant temperature. These energy changes are measured as calories (cal) or joules (J) (see Chapter 2). A change in a value is represented by the Greek letter delta (Δ), and it can be negative or positive. Therefore, the change in free energy (ΔG) of any reaction is defined in terms of the change in total energy (ΔH) and the change in entropy (ΔS):

$$\Delta G = \Delta H - T\Delta S$$

(a)

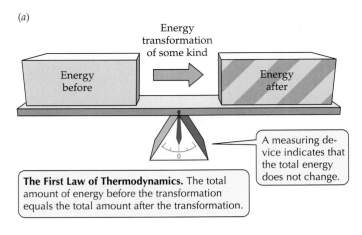

The First Law of Thermodynamics. The total amount of energy before the transformation equals the total amount after the transformation.

A measuring device indicates that the total energy does not change.

(b)

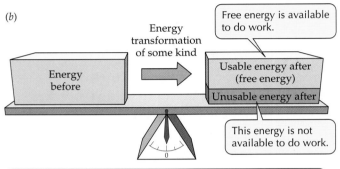

Free energy is available to do work.

This energy is not available to do work.

The Second Law of Thermodynamics. Although a transformation does not change the total amount of energy within a closed system, after any transformation the amount of free energy available to do work is always less than the original amount of energy.

(c)

Closed system

Unusable energy

Free energy

Another statement of the **Second Law** is that in any closed system, with repeated energy transformations free energy decreases and unusable energy–a function of entropy–increases.

This is an important relationship. It tells us whether free energy is *released* by a reaction, in which case ΔG is negative ($-\Delta G$), or free energy is *required* for a reaction, in which case ΔG is positive ($+\Delta G$). The sign and magnitude of ΔG depend on the two factors on the right of the equation: ΔH and $T\Delta S$.

For a chemical reaction, ΔH is the amount of energy added to the system ($+\Delta H$) or released ($-\Delta H$). In deter-

6.3 The Laws of Thermodynamics Energy cannot be created or destroyed, but during energy transformations energy available to do work is lost.

mining the free energy released, a negative value for ΔH is obviously important. But the term $-T\Delta S$ must also be considered—it involves a change in entropy. If entropy increases for a chemical reaction, the products are more disordered or random. When are products more disordered?

If there are more products than reactants, as in the hydrolysis of a protein to its amino acids, the disorder and the entropy will be greater, and the change in entropy (ΔS) will be positive. If there are fewer products, and they are more constrained in their movements than the reactants, ΔS will be negative. Looking at the equation, we see that the negative sign in front of the entropy term means that a positive entropy (increasing disorder) will tend to make ΔG negative and large.

DISORDER TENDS TO INCREASE. The second law of thermodynamics also states that *disorder tends to increase in the universe or a closed system*. Chemical changes, physical changes, biological processes—and anything else you can think of—all tend toward disorder, or randomness (Figure 6.3c). This tendency for disorder to increase gives a *directionality* to physical processes and chemical reactions. It explains why some reactions proceed in one direction rather than another. For example, consider the dissolving of table salt in a glass of water.

A solid crystal of sodium chloride is a highly ordered structure with each ion held rigidly in place; in other words, a crystal has low entropy. However, when a crystal dissolves in water, the ions have a more random relationship to one another (see Chapter 2): Disorder of the ions increases, randomness increases, and entropy increases ($+\Delta S$). A sodium chloride solution, on the other hand, will not spontaneously reorder itself into a crystal of salt and pure water. But if energy is added to the sodium chloride solution, the water can be evaporated and the salt crystals will reform. Without such intervention, crystals will not form spontaneously. This directionality is true of all reactions. For example, you know you must put energy into cleaning your room. Without that input of energy, disorder will increase.

Be sure you understand the second law correctly. It is *not* saying that ordered systems cannot be formed inside a large, complex closed system (Figure 6.4). In a closed system such as our solar system, free energy can be used to create order, just as an input of energy results in the formation of sodium chloride crystals from a solution, or an input of free energy maintains the order in your cells and in your room.

(*a*)   A closed system may be large, for example the solar system, or small like a thermos bottle, sealed and insulated from its surroundings.

In a closed system, no energy or matter enters or leaves.

(*b*)   A living cell is an open system that must obtain energy and raw material from its surroundings.

In an open system, energy and matter can enter and leave.

Raw materials enter

Waste materials leave

Energy enters

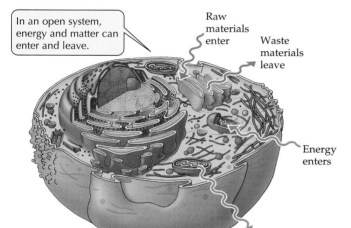

Energy leaves

(*c*)   Within a closed system, an open system can increase in order and complexity using free energy from elsewhere in the closed system.

Open systems, such as a cell, can be part of larger closed systems.

Chemical Reactions Release or Take Up Energy

The laws of thermodynamics are directly relevant to our understanding of the chemical reactions that occur in living things. Chemical reactions are accompanied by exchanges of energy with the environment, some reactions releasing free energy and others taking it up. The amount of energy released ($-\Delta G$) or taken up ($+\Delta G$) is related directly to the tendency of a reaction to run to completion.

In principle, all chemical reactions can run both forward and backward. For example, if compound A can be converted into compound B (A $\rightarrow$ B), then B in principle can be converted into A (B $\rightarrow$ A), although *at given concentrations* of A and B only one of these directions will be favored. Think of the overall reaction as resulting from competition between forward and reverse reactions. Increasing the concentration of the reactants (A) speeds up the forward reaction, and increasing the concentration of the products (B) favors the reverse. At some point the forward and reverse reactions take place at the same rate. At this point, no

6.4 Closed Systems, Open Systems, and Living Systems
(*a*) In considering the origin, evolution, and maintenance of life as we know it, the sun and Earth can be considered a closed system. A sealed thermos is a small closed system. (*b*) Living systems are open systems. (*c*) Open systems may be part of larger closed systems.

further change in the system is observable, although individual molecules are still forming and breaking apart. This balance between forward and reverse reactions is known as **chemical equilibrium**.

Exergonic reactions release free energy

When a reaction goes more than halfway to completion without an input of energy, we say that it is a *spontaneous reaction* (Figure 6.5a). Spontaneous reactions release free energy. A reaction that releases free energy is said to be **exergonic** and has a negative ΔG. A reaction that proceeds only with the *addition* of free energy from the environment is said to be **endergonic** and has a positive ΔG (Figure 6.5b). It is a *nonspontaneous reaction*. If a reaction runs spontaneously in one direction (from reactant A to product B, for example), then the reverse reaction (from B to A) requires a steady supply of energy to drive it: if

A → B is spontaneous, exergonic (−ΔG), then
B → A is nonspontaneous, endergonic (+ΔG).

For example, starch breaks down spontaneously in water, producing glucose. In contrast, glucose monomers do not form starch spontaneously.

Entire chemical pathways also have a spontaneous direction. For example, the complete catabolism (breakdown) of glucose to carbon dioxide and water uses a sequence of many chemical reactions, and the overall process is spontaneous (−ΔG), releasing energy for cellular work. On the other hand, the anabolism (synthesis) of glucose from carbon dioxide and water by photosynthesis is nonspontaneous (+ΔG), and it must be driven by energy from the absorption of light.

In this context, the designation "spontaneous" has nothing to do with speed. A reaction may be extremely slow, yet still be spontaneous, as is the breakdown of starch to glucose. The burning of a newspaper and the slow browning of pages in old library books are both spontaneous processes, but they take place at different speeds.

Chemical equilibrium and free energy are related

Every chemical reaction proceeds to a certain extent, but not necessarily to completion. In other words, all the reactants present are not necessarily converted to products. Each reaction has a specific equilibrium point that is also related to the free energy released by the reaction under specified conditions. To understand the principle of equilibrium, consider the following example.

Every living cell contains glucose 1-phosphate, which is converted to glucose 6-phosphate. In our example, a solution of glucose 1-phosphate has an initial concentration of 0.02 *M*. (*M* stands for molar concentration; see Chapter 2.) As the reaction comes to equilibrium, the final concentration of the product, glucose

(a) **Exergonic reaction**
(spontaneous; energy-releasing)

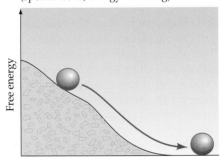

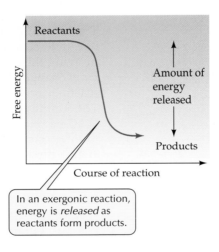

In an exergonic reaction, energy is *released* as reactants form products.

(b) **Endergonic reaction**
(not spontaneous; energy-requiring)

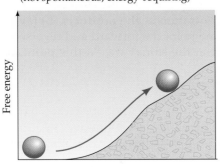

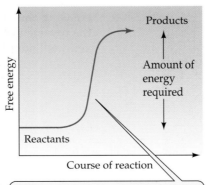

Energy is *required* for an endergonic reaction, in which reactants with a low energy content are converted to products with a higher energy level.

6.5 Exergonic and Endergonic Reactions (a) In a spontaneous reaction, the reactants behave like a ball rolling down a hill and energy is released. (b) The ball will not roll *up* the hill spontaneously. Driving an endergonic reaction, like moving a ball uphill, requires the input of free energy.

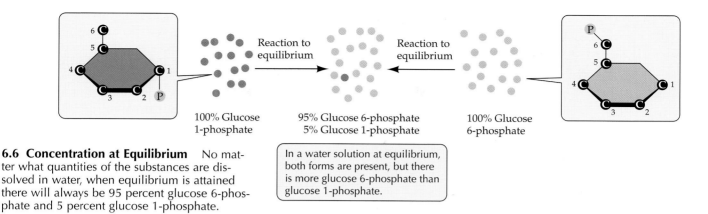

100% Glucose
1-phosphate

95% Glucose 6-phosphate
5% Glucose 1-phosphate

100% Glucose
6-phosphate

6.6 Concentration at Equilibrium   No matter what quantities of the substances are dissolved in water, when equilibrium is attained there will always be 95 percent glucose 6-phosphate and 5 percent glucose 1-phosphate.

In a water solution at equilibrium, both forms are present, but there is more glucose 6-phosphate than glucose 1-phosphate.

6-phosphate, rises from 0 to 0.019 M, while the glucose 1-phosphate concentration falls to 0.001 M. The reaction proceeds until equilibrium is reached at these concentrations (Figure 6.6). From then on, the reverse reaction, from glucose 6-phosphate to glucose 1-phosphate, progresses at the same rate as the forward reaction to glucose 6-phosphate. At equilibrium, then, the forward reaction has gone 95 percent of the way to completion; thus the forward reaction is a spontaneous reaction. This result is obtained every time the experiment is run under the same conditions—namely, at 25°C and pH 7. The reaction is described by the equation

Glucose 1-phosphate $\rightleftharpoons$ glucose 6-phosphate

The change in free energy (ΔG) is related directly to the point of equilibrium for any reaction. The farther toward completion the point of equilibrium lies, the more free energy is given off. In an exergonic reaction such as the conversion of glucose 1-phosphate to glucose 6-phosphate, ΔG is a negative number (in this example, $\Delta G = -1.7$ kcal/mol, or -7.1 kJ/mol). Recall that equilibrium for this reaction has a product-to-reactant ratio of 19:1 (0.019/0.001)—that is, the reaction goes nearly to completion ("to the right" as written).

A large, positive ΔG for a reaction means that it proceeds hardly at all to the right (A → B). But if the product is present, such a reaction runs backward, or "to the left" (A ← B), to near completion (nearly all B is converted to A). A ΔG value near zero is characteristic of a readily reversible reaction: Reactants and products have almost the same free energies.

The principles of thermodynamics we have been discussing apply to all energy exchanges in the universe—no exceptions have ever been found. Thus these principles are very powerful and useful. Next, we'll apply these principles and the concept of free energy to reactions in cells that involve the energy currency ATP (see Chapter 4).

ATP: Transferring Energy in Cells

Previous chapters have mentioned ATP and its role in cells. All living cells rely on **ATP**, adenosine triphosphate, for the capture, transfer, and application of free energy to do chemical work and maintain cellular activities. ATP operates as a kind of energy currency. That is, some of the free energy released by certain exergonic reactions is captured in ATP, which can then release free energy to drive endergonic reactions. How does the chemistry of ATP permit it to serve these functions in cells?

ATP is rich in energy

An ATP molecule consists of the base adenine bonded to ribose (a sugar), which is attached to a sequence of three phosphate groups (Figure 6.7). The hydrolysis of ATP yields **ADP** (adenosine diphosphate) and an inorganic phosphate ion (abbreviated P_i, short for HPO_4^{2-}) and free energy:

$$ATP + H_2O \rightarrow ADP + P_i + \text{free energy}$$

For the hydrolysis of ATP to ADP and phosphate, the change in free energy (ΔG) is about -12 kcal/mol (-50 kJ/mol) at the temperature, pH, and substrate concentrations typical of living cells. Thus this reaction is highly exergonic. (The "standard" ΔG for ATP hydrolysis is -7.3 kcal/mol or -30 kJ/mol, but that value is valid only at pH 7 and with ATP, ADP, and phosphate present at concentrations of 1 M—concentrations that differ greatly from those found in cells.) The hydrolysis of most phosphate esters—the closest analogs to the linkages in ATP—produces considerably less than half as much free energy as the hydrolysis of ATP. For a dramatic example of how free energy from ATP hydrolysis can be used, see Figure 6.8.

What characteristics of ATP account for the large amount of free energy released by the hydrolysis of its phosphates? Under conditions in the cell, the phosphate groups are ionized and bear negative charges that repel one another (see Figure 6.7). When ATP is

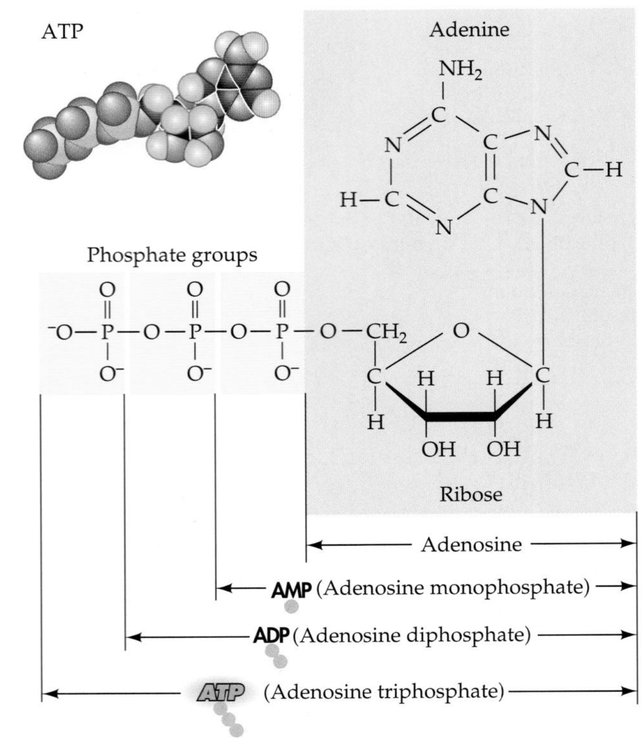

6.7 Structures of ATP, ADP, and AMP ATP is richer in energy than its relatives ADP and AMP. The hydrolysis of ATP releases free energy.

hydrolyzed, the charges are spread over two molecules—ADP and inorganic phosphate—which can then move farther apart, producing a more stable situation. The hydrolysis of ADP to AMP (adenosine monophosphate) and inorganic phosphate liberates an even slightly greater amount of free energy than does that of ATP to ADP, although this reaction does not occur often in cells. Hydrolyzing the last phosphate group (that is, converting AMP to adenosine) does not spread the negative charges any farther apart, so the change in free energy is low, similar to that for any other ester hydrolysis.

Because of their larger free energies of hydrolysis, the first and second bonds broken in ATP are sometimes called high energy bonds, although "high energy" refers to the energy of hydrolysis and not to any intrinsic energy of the bond itself.

6.8 Using ATP to Make Light Fireflies convert the energy of chemical bonds in ATP into light energy, emitting rhythmic flashes that signal the insect's readiness to mate. Very little of the energy in this conversion is lost as heat.

ATP couples exergonic and endergonic reactions

The hydrolysis of ATP is exergonic and yields ADP, P_i, and free energy. The formation of ATP from ADP and P_i is endergonic and consumes as much free energy as is released by the breakdown of ATP.

$$ADP + P_i + \text{free energy} \rightarrow ATP + H_2O$$

Many different enzyme-catalyzed reactions in the cell can provide the energy to convert ADP to ATP. In eukaryotes, the main ones are the reactions of cellular respiration, in which energy is released from fuel molecules and trapped in ATP when it is formed from ADP. The formation and hydrolysis of ATP constitute what might be called an "energy-coupling cycle" in which ATP shuttles energy form exergonic reactions to endergonic reactions.

How does this ATP cycle trap and release energy? An exergonic reaction is coupled to the reaction that forms ATP from ADP and P_i (Figure 6.9). While forming, ATP captures free energy. ATP then diffuses to another site in the cell, where its hydrolysis releases free energy to drive an endergonic reaction. We'll take a closer look at coupling later in this chapter, but now we want to focus on how ATP can drive an endergonic reaction.

A common way in which ATP drives endergonic reactions proceeds in two steps. Consider the endergonic reaction in Figure 6.10*a*, which will not proceed without the input of free energy from ATP. The overall reaction proceeds in two steps and involves two specific enzymes and a *phosphorylated intermediate*, common to both reactions (Figure 6.10*b*). In the first reaction, ATP transfers a phosphate group to glutamate (glutamic acid), producing the higher-energy product glutamyl phosphate (Figure 6.10*b*). In the second reaction, the hydrolysis of the glutamyl phosphate (the phosphorylated intermediate) provides sufficient free energy to drive the reaction with the ammonium ion (NH_4^+). The total change in free energy for the coupled reactions is negative. Hence the reactions proceed spontaneously when coupled, and the amino acid glutamine is synthesized.

A cell uses much of its ATP to meet its ongoing needs to do work, but it diverts some of its ATP into

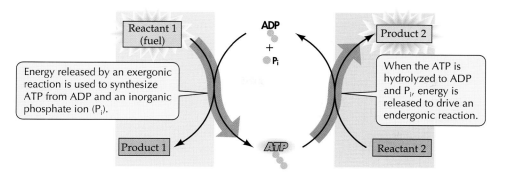

6.9 Formation and Use of ATP
Synthesis of ATP from ADP and P_i requires the input of energy; hydrolysis of ATP to ADP and P_i releases energy.

synthesizing long-term energy storage compounds. For this purpose, plants synthesize starch, a long-chain polymer of glucose, and sometimes fats. Animals store energy in glycogen (another glucose polymer) and fats (see Chapter 3). Of course, any large molecule synthesized by the cell (a protein, for example) is a storehouse of energy, but energy storage may not be its primary function.

ATP can be considered the circulating currency of energy exchange in living organisms, and starches and fats the savings accounts. When organisms need en-

ergy, they draw on their deposits of fat and carbohydrates. They break down these deposits into carbon dioxide and water, using the energy released in the process to make ATP. ATP can then "plug into" many different energy-requiring reactions to meet the cell's need to move, to actively transport substances across its membrane, and to synthesize new molecules such as proteins and nucleic acids.

An active cell requires millions of molecules of ATP per second to drive its biochemical machinery. An ATP molecule is consumed within a minute following its

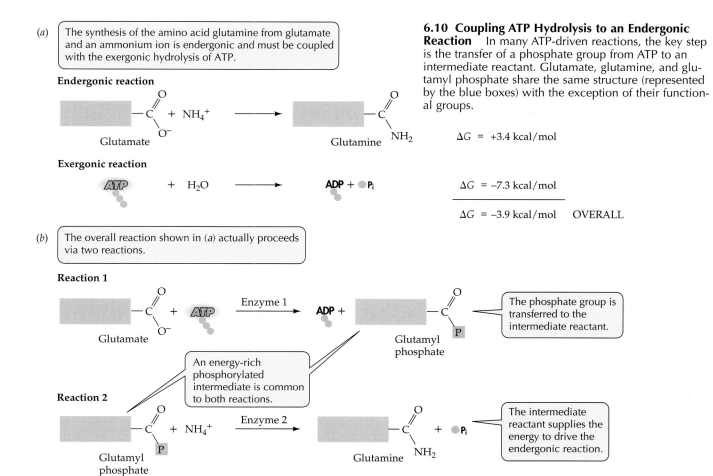

(a) The synthesis of the amino acid glutamine from glutamate and an ammonium ion is endergonic and must be coupled with the exergonic hydrolysis of ATP.

Endergonic reaction

Glutamate $+ NH_4^+ \longrightarrow$ Glutamine $\Delta G = +3.4$ kcal/mol

Exergonic reaction

ATP $+ H_2O \longrightarrow$ ADP $+ P_i$ $\Delta G = -7.3$ kcal/mol

$\Delta G = -3.9$ kcal/mol OVERALL

(b) The overall reaction shown in (a) actually proceeds via two reactions.

Reaction 1

Glutamate $+$ ATP $\xrightarrow{\text{Enzyme 1}}$ ADP $+$ Glutamyl phosphate

The phosphate group is transferred to the intermediate reactant.

An energy-rich phosphorylated intermediate is common to both reactions.

Reaction 2

Glutamyl phosphate $+ NH_4^+ \xrightarrow{\text{Enzyme 2}}$ Glutamine $+ P_i$

The intermediate reactant supplies the energy to drive the endergonic reaction.

6.10 Coupling ATP Hydrolysis to an Endergonic Reaction In many ATP-driven reactions, the key step is the transfer of a phosphate group from ATP to an intermediate reactant. Glutamate, glutamine, and glutamyl phosphate share the same structure (represented by the blue boxes) with the exception of their functional groups.

formation, on average. At rest, an average person hydrolyzes about 40 kg of ATP per day—as much as some people weigh!

Enzymes: Biological Catalysts

When we know the change in free energy (ΔG) of a reaction, we know where the equilibrium of the reaction lies. The more negative ΔG is, the farther the reaction proceeds toward completion. However, ΔG tells us nothing about the **rate of a reaction**—the *speed* at which the reaction moves toward equilibrium. Some exergonic reactions are very rapid; others are slower. What determines the rate of a chemical reaction? In living cells, biological catalysts increase the rate of chemical reactions, and thereby exert enormous control over cellular activities.

A **catalyst** is any substance that speeds up a chemical reaction without itself being used up. *A catalyst does not cause a reaction to take place that would not take place eventually without it.* It merely speeds up the rates of forward and backward reactions, allowing equilibrium to be approached faster. Most biological catalysts, as you have read in preceding chapters, are proteins called **enzymes**. Although we will focus on enzymes, you should know that certain RNA molecules, called *ribozymes*, are also catalytic. Indeed, in the evolution of life, catalytic RNA may have preceded catalytic proteins (see Chapter 24).

In the discussion that follows, we will identify the energy barrier that controls the rate of reactions. Then we'll focus on the role of enzymes: how enzymes interact with reactants, how they lower the barrier, and how they permit reactions to proceed faster. After exploring the nature and significance of enzyme specificity, we'll look at how enzymes contribute to the coupling of reactions.

For a reaction to proceed, an energy barrier must be overcome

An exergonic reaction may release a great deal of free energy, but the reaction may take place very slowly. The key to understanding reaction rates is to recognize that in every reaction there is an energy barrier between reactants and products. Think about a butane lighter. The burning of the butane is obviously exergonic—heat and light are released. Once started, the reaction goes to completion; that is, all the butane reacts with oxygen to form carbon dioxide and water vapor. Because burning butane liberates so much energy, you might expect this reaction to proceed rapidly when butane is exposed to oxygen. But this does not happen. Simply mixing butane with air produces no reaction. Butane will start burning only if a spark—a tiny amount of energy—is provided. The need for this spark to start the reaction shows that there is an energy barrier between reactants and products.

In general, exergonic reactions proceed only after they are pushed over the energy barrier by bits of energy called **activation energy**, which is symbolized E_a (Figure 6.11*a*). Recall the ball rolling down the hill in Figure 6.5. The ball has more potential energy at the top of the hill. However, if the ball is in a small depression, it won't roll down the hill, even though that action is exergonic (Figure 6.11*b*). To start the ball rolling, a small amount of energy (activation energy) is needed to get the ball out of the depression (Figure 6.11*c*).

In a chemical reaction, the energy barrier is the activation energy. It is the energy needed to change reac-

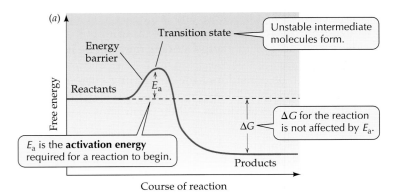

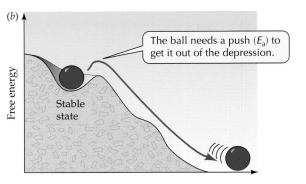

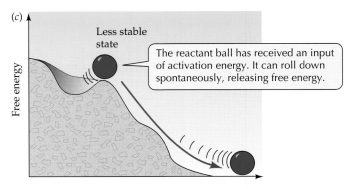

6.11 Activation Energy Initiates Reactions In any chemical reaction, an initial stable state must become less stable before change is possible.

tants into unstable molecular forms called *transition-state species*. Transition-state species have higher free energies than either the reactants or the products. Although the activation energy needed for different reactions varies in size, it is often small compared to the change in free energy of the reaction. The activation energy that starts a reaction is recovered during the ensuing "downhill" phase of the reaction, so it is not a part of the free energy released, $-\Delta G$ (see Figure 6.11*a*).

With this energy barrier between reactants and products, how do reactions ever take place? In any collection of reactants, some molecules have more kinetic energy than others. Picture a mixture of reactant molecules with various kinetic energies. Some of the molecules in the mixture have enough energy to surmount the energy barrier and enter the transition state, and some of these molecules react, yielding products (Figure 6.12).

A reaction with a low activation energy proceeds more rapidly because more of the reactant molecules have enough energy to get over the initial hump. When activation energy is high, the reaction is slow unless more energy is provided, usually as heat. If the system is heated, all the reactant molecules move faster and have more kinetic energy. Since more of them have energy exceeding the required activation energy, the reaction speeds up.

Adding heat to increase the average kinetic energy of the molecules is not an effective option for living systems. Such a general, nonspecific approach would accelerate all the reactions, including destructive reactions, such as the denaturation of proteins that contribute to the structural integrity of the cell (see Chapter 3). Another, biologically more effective way to speed up a reaction is to lower the activation energy. In living cells, catalysts, most of which are enzymes, accomplish this task.

Enzymes bind specific reactant molecules

All types of catalysts speed chemical reactions. Most nonbiological catalysts are nonspecific; that is, they work on a diversity of reactants. For example, a powdered form of the metal platinum, called platinum black, catalyzes virtually any reaction in which molecular hydrogen (H_2) is a reactant. In contrast, most biological catalysts are members of the class of proteins called enzymes, and they are *highly specific*. An enzyme usually recognizes and binds to one or only a few closely related reactants, and it catalyzes only a single chemical reaction. The specificity of enzymes to the molecules they bind has been described as resembling the specificity of a lock and key.

The names of enzymes reflect the specificity of their functions. For example, the enzyme *RNA polymerase* will not work on DNA, and the enzyme *hexokinase* ac-

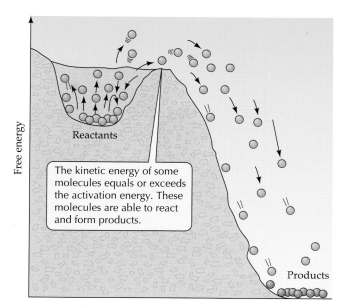

6.12 Over the Energy Barrier Some molecules surmount the energy barrier and react, forming products.

celerates the phosphorylation of hexose sugars but not pentose sugars. Most, but not all, names of enzymes end in the suffix "-ase."

In an enzyme-catalyzed reaction, the reactants are called **substrates**. Substrate molecules *bind* to a particular site on the protein surface, called the **active site**, where catalysis takes place (Figure 6.13). The binding of the substrate to the active site of the enzyme depends on the same kinds of forces that maintain the tertiary structure of the enzyme: hydrogen bonds, the attraction and repulsion of electrically charged groups, and hydrophobic interactions (see Chapter 2). The specificity of an enzyme results from the exact three-dimensional structure of its active site. Only one substrate fits precisely into the active site. Other molecules—with different shapes, different functional groups, and different properties—cannot properly fit and bind to the active site.

The binding of a substrate to the active site produces an **enzyme–substrate complex** held together by one or more means, such as hydrogen bonding, ionic attraction, or covalent bonding. The enzyme–substrate complex may form product and free enzyme:

$$E + S \rightarrow ES \rightarrow E + P$$

where E is the enzyme, S is the substrate, P is the product, and ES is the enzyme–substrate complex. Note that E, the free enzyme, is in the same chemical form at the end of the reaction as at the beginning. While bound to the substrate, it may change chemically, but by the end of the reaction it has been restored to its initial form.

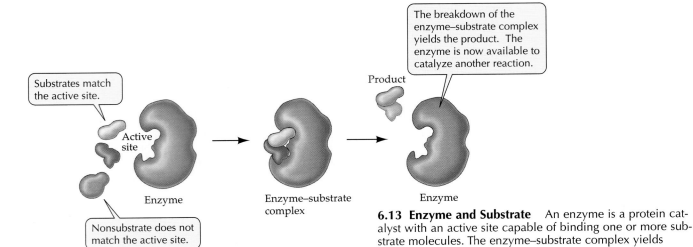

Substrates match the active site.

Active site

Enzyme

Nonsubstrate does not match the active site.

Enzyme–substrate complex

The breakdown of the enzyme–substrate complex yields the product. The enzyme is now available to catalyze another reaction.

Product

Enzyme

6.13 Enzyme and Substrate An enzyme is a protein catalyst with an active site capable of binding one or more substrate molecules. The enzyme–substrate complex yields product and free enzyme.

Enzymes lower the activation energy

The enzyme–substrate complex has a lower free energy than does the transition-state species of the corresponding uncatalyzed reaction (Figure 6.14). Thus the enzyme provides the reaction with a lower energy barrier—it offers an easier path. When an enzyme lowers the activation energy, both the forward and the reverse reactions speed up, so the enzyme-catalyzed overall reaction proceeds toward equilibrium more rapidly than the uncatalyzed reaction. Recall that the final equilibrium is the same with or without the enzyme. Adding an enzyme to a reaction does not change the difference in free energy (ΔG) between the reactants and the products; it changes only the activation energy and, consequently, the rate of reaction.

What are the chemical events at active sites of enzymes?

When the enzyme–substrate complex has formed, several types of chemical events in the active site contribute directly to the breaking of old bonds and the formation of new ones. We'll consider three types of catalysis: acid–base catalysis, covalent catalysis, and metal ion catalysis. In catalyzing a particular reaction, an enzyme may use more than one of these.

In *acid–base catalysis*, the acidic or basic side chains of amino acids forming the active site transfer H^+ to or from the substrate, destabilizing a covalent bond and

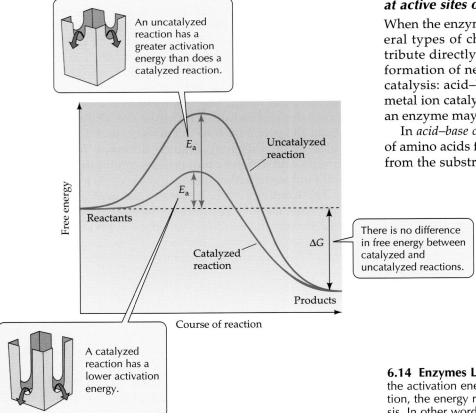

An uncatalyzed reaction has a greater activation energy than does a catalyzed reaction.

E_a

Uncatalyzed reaction

E_a

Reactants

Catalyzed reaction

ΔG

There is no difference in free energy between catalyzed and uncatalyzed reactions.

Products

Course of reaction

A catalyzed reaction has a lower activation energy.

6.14 Enzymes Lower the Activation Energy Although the activation energy is lower in an enzyme-catalyzed reaction, the energy released is the same with or without catalysis. In other words, E_a is lower but ΔG is unchanged.

permitting it to break more readily. In *covalent catalysis*, a functional group in a side chain forms a temporary covalent bond with a portion of the substrate. For example, consider the general case of hydrolysis:

$$A—B + H_2O \rightarrow A—OH + BH$$

With an enzyme, this reaction may proceed in two steps:

$$A—B + enzyme—OH \rightarrow A—OH + enzyme—B$$
$$enzyme—B + H_2O \rightarrow enzyme—OH + BH$$

where the enzyme has a hydroxyl group that enters into the catalytic process. Such a path to reaction is effective only if the combined activation energies of its steps are lower than the total activation energy for the overall uncatalyzed reaction.

In *metal ion catalysis*, metal ions such as copper, zinc, iron, and manganese that are firmly bound to side chains of the protein contribute in several ways. All of these ions can lose or gain electrons without altering the bonds that hold them to the protein. This ability makes them important participants in oxidation–reduction reactions, which involve loss or gain of electrons. Metal ions are present in about a third of the enzymes that have been studied.

Substrate concentration affects reaction rate

For a reaction of the type A → B, the rate of the uncatalyzed reaction is directly proportional to the concentration of A (Figure 6.15). Addition of the appropriate enzyme speeds up the reaction, of course, but it also changes the shape of the plot of rate versus substrate concentration. At first, the rate of the enzyme-catalyzed reaction increases as the substrate concentration increases, but then the reaction rate levels off. Further increases in the substrate concentration do not increase the reaction rate. Since the concentration of the enzyme is usually much lower than that of the substrate, what we are seeing is a saturation phenomenon like the ones that occur in facilitated diffusion (see Figure 5.10*b*). When all the enzyme molecules are bound to substrate molecules, nothing is gained by adding more substrate, because no enzyme molecules are left to act as catalysts.

The study of the rates of enzyme-catalyzed reactions is called *enzyme kinetics*. As we will see later in this chapter, some graphs of rate versus substrate concentration are quite different from the one pictured in Figure 6.15. Such graphs tell us a great deal about the nature of the enzyme-catalyzed reaction.

Some enzymes couple reactions

Some of the most important reactions in living organisms are endergonic but proceed because specific enzymes couple them with other reactions that are exer-

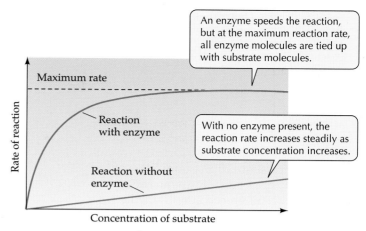

6.15 Enzymes Speed Up Reaction Rates Because there is usually less enzyme than substrate present, the reaction rate levels off when the enzyme becomes saturated.

gonic (recall Figure 6.10). Consider, for example, a pair of coupled reactions that occur in mitochondria. The first reaction, which converts succinate to fumarate, is highly exergonic. The second reaction, the hydrogenation of FAD (flavin adenine dinucleotide) to $FADH_2$, is endergonic, requiring a large input of free energy. The catalyst that couples these two reactions is the enzyme succinate dehydrogenase (Figure 6.16).

In a mitochondrion, the two hydrogen atoms that are removed from succinate are transferred to a molecule of a carrier substance, FAD. Succinate dehydrogenase couples the exergonic reaction to the endergonic one by ensuring that hydrogen atoms liberated by succinate are used to make $FADH_2$. One site on the enzyme surface binds succinate; a nearby second site binds FAD. Every time a succinate ion reacts with succinate dehydrogenase, much of the free energy that is released by this highly exergonic process is immediately trapped and used to synthesize $FADH_2$ from FAD. $FADH_2$ acts as a carrier of the hydrogen and the chemical free energy for use in an endergonic reaction (see Chapter 7).

In Chapter 5, you encountered other instances of coupled reactions. In animals the sodium–potassium pump (for primary active transport) is an enzyme that couples the exergonic hydrolysis of ATP to the endergonic pumping of Na^+ and K^+ against their concentration differences (see Figure 5.13):

$$ATP + H_2O \rightarrow ADP + P_i$$
$$3\ Na^+_{in} \rightarrow 3\ Na^+_{out}$$
$$2\ K^+_{out} \rightarrow 2\ K^+_{in}$$

The contractile proteins of muscle couple the exergonic breakdown of ATP to the performance of mechanical work against a load (see Chapter 44).

6.16 Succinate Dehydrogenase Couples Two Reactions

In mitochondria, the enzyme succinate dehydrogenase binds to both succinate and FAD, coupling an energy-producing reaction (succinate → fumarate) with an energy-requiring reaction (FAD → FADH₂).

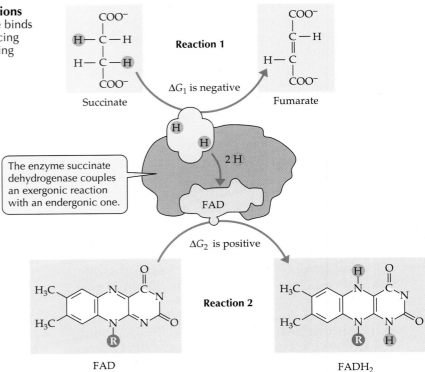

The enzyme succinate dehydrogenase couples an exergonic reaction with an endergonic one.

Succinate

Reaction 1

ΔG_1 is negative

Fumarate

2 H

FAD

ΔG_2 is positive

FAD

Reaction 2

FADH₂

Overall reaction: $\Delta G = \Delta G_1 + \Delta G_2$
$\Delta G \approx 0$

ANOTHER WAY TO COUPLE REACTIONS. In metabolic pathways, there is another type of coupling, in which successive enzyme-catalyzed steps share compounds. A reaction A + B → C may be endergonic (ΔG = +10 kJ/mol) but still proceed rapidly if the next step C + D → E is so exergonic (ΔG = –13 kJ/mol) that the overall reaction (A + B + D → E) is exergonic (+10 kJ/mol – 13 kJ/mol = –3 kJ/mol).

Another way to look at this example is in terms of the effect of the second reaction on the equilibrium of the first reaction. As the highly exergonic reaction C + D → E proceeds toward its equilibrium (far to the right), it reduces the concentration of the substance C, so an equilibrium for the first reaction is never established. Thus more C is produced, since the reaction A + B → C progresses toward equilibrium.

These examples illustrate an important generalization: *Coupled reactions are the major means by which energy-requiring reactions are carried out in cells.* We will encounter many coupled reactions in the next two chapters.

Molecular Structure Determines Enzyme Function

Until the 1960s, biochemists knew little about the behavior of enzymes at the molecular level. It was generally agreed that the substrates of enzymes bind to an active site on the surface of the enzyme molecule, but the structure of an active site was not understood. The remarkable ability of an enzyme to select exactly the right substrate was explained by the assumption that the binding of the substrate to the site depends on a precise interlocking of molecular shapes. In 1894 the German chemist Emil Fischer compared the fit between an enzyme and substrate to that of a lock and key. Fischer's model persisted for more than half a century with only indirect evidence to support it.

The first direct evidence came in 1965, when David Phillips and his colleagues at the Royal Institution in London succeeded in crystallizing the enzyme lysozyme and determined its structure using the techniques of X ray crystallography (to be described in Chapter 11). The tertiary structure of lysozyme is shown in Figure 6.17. Lysozyme is an enzyme that

protects the animals that produce it by destroying invading bacteria. To destroy the bacteria, it cleaves certain polysaccharide chains in their cell walls. Lysozyme is found in tears and other bodily secretions, and it is particularly abundant in the whites of bird eggs. In Figure 6.17, the active site of lysozyme appears as a large indentation filled with the substrate (shown in yellow).

In the discussion that follows, we'll consider how the binding of substrate to an enzyme may change the enzyme and contribute to catalysis. Then we will examine cofactors that are sometimes bound to proteins required for reactions.

Binding at the active site may cause enzymes to change shape

The entire structure of an enzyme molecule is the source of the impeccable specificity of enzymes, on which all metabolism depends. As Fischer suggested in his "lock and key" model, the structure of the active site fits the substrate molecule. However, we now know that after a substrate binds to an enzyme's active site, the entire enzyme may undergo a change in shape to accomplish the "fit" at the active site. This change in enzyme shape caused by substrate binding is called **induced fit**. Induced fit can be observed in the enzyme hexokinase when it is studied with and

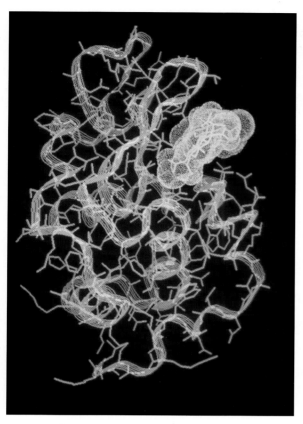

6.17 Tertiary Structure of Lysozyme The polysaccharide substrate, shown in yellow, is bound to a lysozyme molecule, shown in blue. The magenta ribbon highlights the backbone of the protein.

without its substrate glucose (Figure 6.18). Induced fit brings reactive side chains from the enzyme's active site into alignment with the substrate, facilitating the catalytic events described earlier (see Figure 6.13).

To operate, some enzymes require cofactors

Whether they consist of a single folded polypeptide chain or several chains, many enzymes require non-protein molecules, called *cofactors*, in order to function. In addition to the bound metal ions mentioned earlier, some enzymes require other cofactors to catalyze the reaction; prosthetic groups and coenzymes are examples of such cofactors. *Prosthetic groups* are bound to the enzyme and include the heme groups that are attached to the protein in the oxygen-carrying protein hemoglobin (see Figure 3.19). *Coenzymes* are generally relatively small (compared to the enzyme) carbon-containing molecules that are required for the action of one or more enzymes (Figure 6.19).

Since coenzymes are not permanently bound to the enzyme, they must react with it as a substrate does. For the catalyzed reaction to proceed, coenzymes must collide with the enzyme and bind to its active site just as the substrate must. A coenzyme can be considered a "cosubstrate," because it changes chemically during the reaction and separates from the enzyme to participate in other reactions. Coenzymes move from enzyme molecule to enzyme molecule.

ATP and ADP can be considered coenzymes: They are necessary for reaction, are changed by the reaction, and bind and detach from the enzyme. In the next chapter, we will encounter coenzymes that function by accepting or donating electrons or hydrogen atoms. In animals, some coenzymes are produced from vitamins that must be obtained from food—they cannot be synthesized by the body.

6.18 Some Enzymes Change Shape When Substrate Binds to Them Shape changes result in an induced fit between enzyme and substrate, improving the catalytic ability of the enzyme–substrate complex.

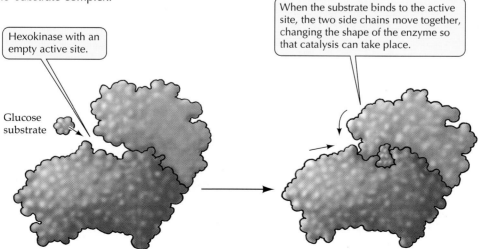

Hexokinase with an empty active site.

Glucose substrate

When the substrate binds to the active site, the two side chains move together, changing the shape of the enzyme so that catalysis can take place.

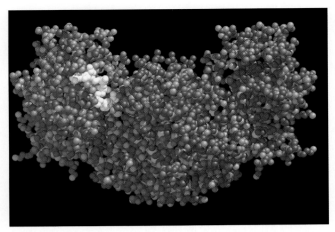

6.19 A Coenzyme Some enzymes require coenzymes in order to function. This illustration shows the relative sizes of the enzyme (red and blue) and coenzyme (white).

Metabolism and the Regulation of Enzymes

All organisms need to maintain stable internal conditions, or *homeostasis* (which will be covered in detail in Chapter 37). If homeostasis breaks down because the concentrations of some compounds rise or fall too much, illness results. Thus we and all other organisms must regulate our metabolism, and the regulation of the rates at which thousands of different enzymes operate contributes to metabolic homeostasis.

In the remainder of this chapter we'll investigate the role of enzymes in organizing and regulating metabolism. The activity of enzymes can be inhibited in various ways, so the presence of an enzyme does not necessarily ensure that it is functioning in a cell. We'll discover how the rate at which some enzymes catalyze reactions can be altered, making enzymes the target points at which entire pathways can be regulated. We'll close with an examination of how temperature and pH affect enzyme action.

Metabolism is organized into metabolic pathways

An organism's metabolism is the totality of the biochemical reactions that take place within it. Metabolism transforms raw materials and stored potential energy into forms that can be used by living cells. These reactions proceed down **metabolic pathways**, which are series of enzyme-catalyzed reactions. In these sequences, the product of one reaction is the substrate for the next:

$$A \rightarrow B \rightarrow C \rightarrow D$$

Some pathways synthesize the important chemical building blocks from which macromolecules are built. For example, one group of metabolic pathways forms the various amino acids. Another group produces the nucleotides for the nucleic acids. Some pathways harvest energy. The pathway called *glycolysis*, for example, partly degrades glucose and generates some ATP. The pathway called *cellular respiration* completes the catabolism of glucose and produces a great deal more ATP. Chapter 7 explores these pathways in more detail. To have adequate amounts of all the necessary building blocks without wasting raw materials or energy, a cell must regulate all its metabolic pathways constantly.

Enzyme activity is subject to regulation

Various substances, called **inhibitors**, bind to enzymes, decreasing the rates of enzyme-catalyzed reactions. Some inhibitors occur naturally in cells; others are artificial. Naturally occurring inhibitors regulate metabolism; the artificial ones are used either to treat disease or to study how enzymes work. Some inhibitors irreversibly inhibit the enzyme by permanently binding to it. Others have reversible effects; that is, these inhibitors can become unbound. The removal of a natural reversible inhibitor increases an enzyme's rate of catalysis.

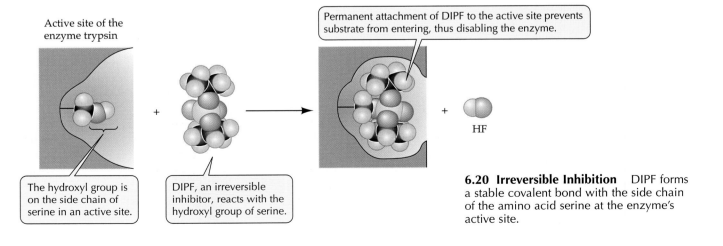

Active site of the enzyme trypsin

The hydroxyl group is on the side chain of serine in an active site.

DIPF, an irreversible inhibitor, reacts with the hydroxyl group of serine.

Permanent attachment of DIPF to the active site prevents substrate from entering, thus disabling the enzyme.

HF

6.20 Irreversible Inhibition DIPF forms a stable covalent bond with the side chain of the amino acid serine at the enzyme's active site.

6.21 Reversible Inhibition Enzyme inhibition is sometimes reversible. *(a,b)* In competitive inhibition, an inhibitor binds temporarily to the active site. For example, succinate dehydrogenase is subject to competitive inhibition by oxaloacetate. *(c)* A noncompetitive inhibitor binds away from the active site.

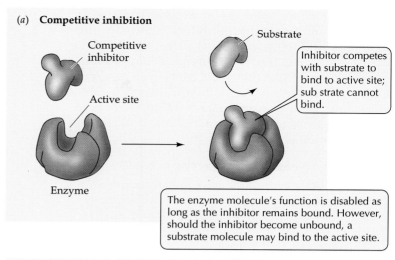

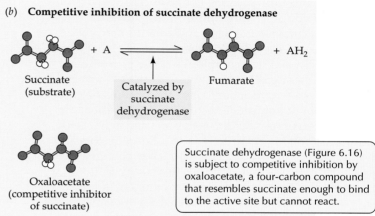

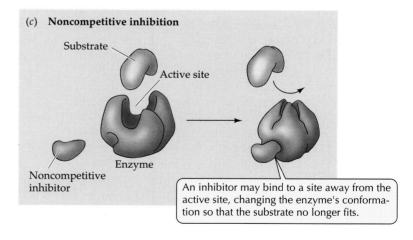

IRREVERSIBLE INHIBITION. Some inhibitors bond irreversibly to certain side chains at active sites of enzymes, thereby inactivating the enzymes by destroying their capacity to interact with the normal substrate. An example of such an **irreversible inhibitor** is DIPF (diisopropylphophorofluoridate). DIPF reacts with a hydroxyl group belonging to the amino acid serine (see Table 3.1) at an enzyme's active site, preventing the use of this side chain in the catalytic mechanism (Figure 6.20).

DIPF is an irreversible inhibitor for the protein-digesting enzyme trypsin and for many other enzymes whose active sites contain serine. Among these is acetylcholinesterase, an enzyme that is essential for the orderly propagation of impulses from one nerve cell to another (see Chapter 41). Because of their effect on acetylcholinesterase, DIPF and other similar compounds are classified as nerve gases.

REVERSIBLE INHIBITION. Not all inhibitory action is irreversible. Some inhibitor molecules are similar enough to a particular enzyme's natural substrate to bind to the active site, yet different enough that the enzyme catalyzes no chemical reaction. While such a molecule is bound to the enzyme, the natural substrate cannot enter the active site; thus the intruder effectively wastes the enzyme's time, inhibiting its catalytic action. These molecules are called **competitive inhibitors** because they compete with the natural substrate for the active site and block the action of the enzyme (Figure 6.21*a*); the blockage is reversible. When the concentration of the competitive inhibitor is reduced, it detaches from the active site and the enzyme is again active.

The enzyme *succinate dehydrogenase* is subject to competitive inhibition. Recall that this enzyme, found in all mitochondria, catalyzes the conversion of the compound succinate to another compound, fumarate (see Figure 6.16). The compound oxaloacetate is similar to succinate and can act as a competitive inhibitor of succinate dehydrogenase by binding to the active site (Figure 6.21*b*). However,

having bound to oxaloacetate, the enzyme can do nothing more with it—no reaction occurs. An enzyme molecule cannot bind a succinate molecule until the inhibitor molecule has moved out of the active site. The inhibitor *can* move out of the site, because binding of a competitive inhibitor is reversible, *as is binding of the substrate.*

Some inhibitors that do not react with the active site are called **noncompetitive inhibitors**. Noncompetitive inhibitors bind to the enzyme at a site away from the active site. Their binding causes a conformational change in the protein that alters the active site (Figure 6.21c). The active site may still bind substrate molecules, but the rate of product formation is reduced. Noncompetitive inhibitors can become unbound, so their effects are reversible.

Allosteric enzymes have interacting subunits

Many important enzymes have a quaternary structure consisting of two or more polypeptide subunits, each with a molecular weight in the tens of thousands (see Chapter 3). These subunits are bound together by various weak bonds that permit changes in the shape of one subunit to influence the shape and properties of the other subunits. Multisubunit enzymes that undergo such changes in shape and function are called **allosteric enzymes** (allo-, "different"; -steric, "shape").

The activity of these complex enzymes is controlled by molecules called **effectors**, which may have no similarity either to the reactants or to the products of the reaction being catalyzed. Effectors bind to an **allosteric site** that is separate from the active site and enhance or diminish reactions at the active site. Thus, effectors are activators or inhibitors. Their binding changes the structure of the enzyme and thus its activity.

Allosteric enzymes and single-subunit enzymes differ greatly in their reaction rates when the substrate concentration is low. Graphs of rate plotted against substrate concentration show this relationship. For an enzyme with a single subunit, the plot looks like that in Figure 6.22a. The reaction rate first increases very sharply with increasing substrate concentration, then tapers off to a constant maximum rate as the supply of enzyme becomes saturated with substrate. The plot for many allosteric enzymes is radically different, with a sigmoidal (S-shaped) appearance (Figure 6.22b).

With sigmoid kinetics, the increase in rate with increasing substrate concentration is slight at low substrate concentrations, but within a certain range the reaction rate is extremely sensitive to relatively small changes in the substrate concentration. Because of this sensitivity, allosteric enzymes are important in fine-tuning the activities of a cell. We can understand this behavior in terms of interactions between the different kinds of subunits that make up an allosteric enzyme, which we'll examine next.

Catalytic and regulatory subunits interact and cooperate

An allosteric enzyme not only has more than one subunit, it has more than one *type* of subunit: A **catalytic subunit** has an active site that binds the enzyme's substrate; a **regulatory subunit** has one or more allosteric

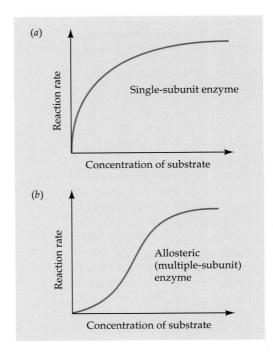

6.22 Allostery and Reaction Rate   How the rate of an enzyme-catalyzed reaction changes with increasing substrate concentration depends on whether the enzyme consists of one or more than one polypeptide subunit.

sites that bind specific effector molecules (Figure 6.23). Binding of either a substrate or an effector affects the structure of the enzyme as a whole. An allosteric enzyme can exist in two forms. The *active form* has catalytic activity, whereas the *inactive form* lacks activity. An allosteric enzyme usually consists of two or more catalytic subunits and one or more regulatory subunits. The existence of two or more linked catalytic subunits allows for *cooperativity*, which works as follows.

When a molecule of substrate binds to the active site of one catalytic subunit of an allosteric enzyme, it causes a favorable, cooperative change in the other catalytic subunits, making it easier for substrate to bind them (Figure 6.23c). Conversely, when an allosteric inhibitor binds to the allosteric site of a regulatory subunit, it causes an unfavorable change in the catalytic subunits, making it harder for substrate to bind them.

Cooperativity makes an enzyme exquisitely sensitive to its environment. The binding of one substrate molecule makes it easier for further substrate molecules to react.

Allosteric effects control metabolism

Some metabolic pathways are branched. At a branch point an intermediate substance is acted on by more than one enzyme and thus is sent through more than one metabolic branch (Figure 6.24). Two different pathways emerge from a branch.

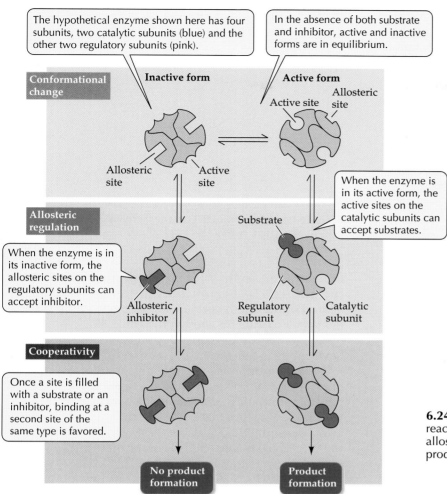

The hypothetical enzyme shown here has four subunits, two catalytic subunits (blue) and the other two regulatory subunits (pink).

In the absence of both substrate and inhibitor, active and inactive forms are in equilibrium.

Conformational change

Inactive form

Active form

Active site

Allosteric site

Allosteric site

Active site

Allosteric regulation

When the enzyme is in its inactive form, the allosteric sites on the regulatory subunits can accept inhibitor.

When the enzyme is in its active form, the active sites on the catalytic subunits can accept substrates.

Substrate

Allosteric inhibitor

Regulatory subunit

Catalytic subunit

Cooperativity

Once a site is filled with a substrate or an inhibitor, binding at a second site of the same type is favored.

No product formation

Product formation

6.23 Allosteric Regulation of Enzymes
The hypothetical enzyme shown here has four subunits: two catalytic (blue), the other two regulatory. When the enzyme is in its active form, the active sites on the catalytic subunits can accept substrate. When the enzyme is in its inactive form, the allosteric sites on the regulatory subunits can accept inhibitor.

6.24 Feedback in Metabolic Pathways The first reaction following a branch point is catalyzed by an allosteric enzyme that can be inhibited by the end product of the pathway.

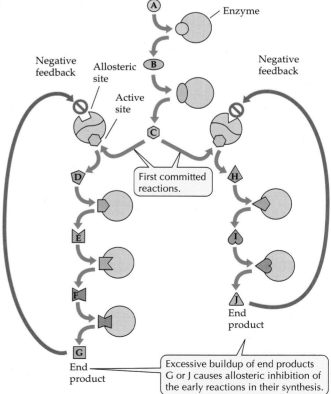

Enzyme

Negative feedback

Allosteric site

Active site

Negative feedback

First committed reactions.

End product

End product

Excessive buildup of end products G or J causes allosteric inhibition of the early reactions in their synthesis.

At the branching points where metabolic pathways diverge, **regulatory enzymes** catalyze reactions. These regulatory enzymes are like track switches in a railroad: Increasing or decreasing the rates of the reactions that they catalyze determines what fraction of the flow of material goes through which branch. What turns such a switch? The end product of a branch pathway may damp the initial step in that branch pathway, reducing the formation of the end product (see Figure 6.24). This is the principle of **feedback inhibition**, also called end product inhibition.

The end product of a particular pathway typically is an allosteric inhibitor of the regulatory enzyme that catalyzes the first **committed step** in its own synthesis—that is, the earliest step in the branched pathway that leads to the synthesis of only that end product and no other. The committed steps in metabolic pathways are particularly effective points for feedback control. For instance, inhibition of the C-to-D step in Figure 6.24 shunts all the reactants from the G pathway into the J pathway.

Allosteric regulation is very effective. It allows rapid adjustment to short-term changes in metabolism or in the environment. The activities of enzyme molecules are adjusted by their interactions with small molecules, the end products. However, if a particular enzyme were not needed, wouldn't it be a good idea simply to stop making it until it was needed? Wouldn't it be advantageous to regulate enzyme *production* as well as enzyme *activity*? The answer is yes, and the regulation of enzyme synthesis plays an important role in controlling metabolism and development (see Chapters 13, 14, and 15).

Enzymes are sensitive to their environment

Enzymes enable cells to perform chemical reactions and carry out complex processes without using the extremes of temperature and pH employed by chemists in the laboratory. Enzymes themselves are extremely sensitive to changes in the medium around them, particularly temperature and pH.

pH AFFECTS ENZYME ACTIVITY. The rates of most enzyme-catalyzed reactions depend on the pH of the medium in which they occur. Each enzyme is most active at a particular pH; its activity decreases as the solution is made more acidic or more basic than its ideal pH (Figure 6.25).

Several factors contribute to this effect. One is the ionization of carboxyl, amino, and other groups on either the substrate or the enzyme. Carboxyl groups (—COOH) ionize to become negatively charged carboxylate groups (—COO⁻) in neutral or basic solutions. Similarly, amino groups (—NH$_2$) accept H⁺ ions

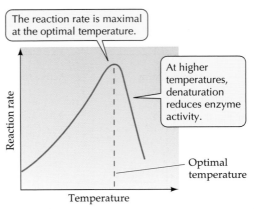

6.26 Temperature Affects Enzyme Activity   An enzyme is most active at a particular temperature.

in neutral or acidic solutions, becoming positively charged —NH$_3^+$ groups (see Chapter 2). Thus, in a neutral solution a molecule with an amino group is attracted electrically to another molecule that has a carboxyl group because both groups are ionized and they have opposite charges.

Evolution has matched enzymes to their environment. For example, the protein-digesting enzyme pepsin, found only in the stomach, works best at the very low pH values that prevail in the stomach after a meal. In contrast, salivary amylase works best at neutral pH, which is characteristic of the mouth.

TEMPERATURE AFFECTS ENZYME ACTIVITY. At low temperatures, warming increases the rate of an enzyme-catalyzed reaction because at higher temperatures a greater fraction of the reactant molecules have enough energy to provide the activation energy of the reaction (Figure 6.26). Temperatures that are too high, however, inactivate enzymes, because at high temperatures enzyme molecules vibrate and twist so rapidly that some of their noncovalent bonds break. When heat destroys their tertiary structure, enzyme molecules become inactivated, or **denatured** (see Chapter 3). Some enzymes denature at temperatures only slightly above that of the human body, but a few are stable even at the boiling point of water.

Individual organisms adapt to changes in the environment in many ways, one of which is based on groups of enzymes, called **isozymes**, that catalyze the same reaction but have different chemical compositions and physical properties. Within a given group, different isozymes may have different optimal temperatures. In rainbow trout, an example is the isozymes of the enzyme acetylcholinesterase, whose operation is essential to normal transmission of nerve impulses. If a rainbow trout is transferred from warm water to near-freezing water (2°C), the fish produces an isozyme of acetylcholinesterase that is different from

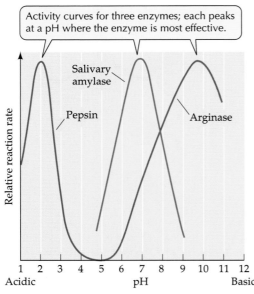

6.25 pH Affects Enzyme Activity   Each enzyme catalyzes at a maximum rate at a particular pH.

the acetylcholinesterase produced at the higher temperature. The new isozyme has a lower optimal temperature, which helps the fish to perform normally in the colder water.

Summary of "Energy, Enzymes, and Metabolism"

Energy Conversions: The Laws of Thermodynamics
• Energy is the capacity to do work. Potential energy is the energy of state or position; it includes the energy stored in chemical bonds. Kinetic energy is energy of motion (and related forms such as electric energy, light, and heat).
• Potential energy can be converted to kinetic energy, which can do work. Biological work includes synthesis, growth, movement, transport, and maintenance. **Review Figure 6.1**
• The first law of thermodynamics tells us that energy cannot be created or destroyed. The second law of thermodynamics tells us that, in any closed system, the quantity of energy available to do work (free energy) decreases and unusable energy associated with entropy increases.
• Living things, like everything else, obey the laws of thermodynamics. Organisms are open systems that are part of a larger closed system. **Review Figures 6.3, 6.4**
• Changes in free energy, total energy, temperature, and entropy are related by the equation $\Delta G = \Delta H - T\Delta S$.

Chemical Reactions Release or Take Up Energy
• Spontaneous, exergonic reactions release free energy and have a negative ΔG. Nonspontaneous, endergonic reactions take up free energy and have a positive ΔG. Endergonic reactions proceed only if free energy is provided. **Review Figure 6.5**
• The change in free energy of a reaction determines its point of chemical equilibrium, at which the forward and reverse reactions proceed at the same rate. For spontaneous, exergonic reactions, equilibrium lies toward completion (products). **Review Figure 6.6**

ATP: Transferring Energy in Cells
• ATP (adenosine triphosphate) is an energy transfer molecule in cells. Hydrolysis of the terminal phosphate releases a relatively large amount of free energy ($\Delta G = -12$ kcal/mol). **Review Figure 6.7**
• The ATP cycle couples exergonic and endergonic reactions, transferring free energy from the exergonic to the endergonic reaction. **Review Figures 6.9, 6.10**

Enzymes: Biological Catalysts
• The rate of a chemical reaction is independent of ΔG but is determined by the size of the activation energy barrier. Catalysts speed reactions by lowering the activation energy. **Review Figures 6.11, 6.12**
• Enzymes are biological catalysts, proteins that are highly specific for their substrates. Substrates bind to the active site, where catalysis takes place, forming an enzyme–substrate complex. **Review Figure 6.13**

• At the active site, amino acid side chains and in some cases attached metal ions participate in chemical events such as acid–base catalysis, covalent catalysis, or metal ion catalysis that lower the activation energy for reactions. **Review Figure 6.14**
• Substrate concentration affects the rate of an enzyme-catalyzed reaction. **Review Figure 6.15**
• Some enzymes couple exergonic and endergonic reactions by catalyzing both reactions. Other reactions are coupled if a product of the endergonic reaction is a reactant for an exergonic reaction catalyzed by a second enzyme; this coupling works if the overall ΔG for the two reactions is negative. **Review Figures 6.10, 6.16**

Molecular Structure Determines Enzyme Function
• The active site where substrate binds determines the specificity of an enzyme. Upon binding to substrate, some enzymes change shape, facilitating catalysis. **Review Figures 6.13, 6.18**
• Some enzymes require cofactors to carry out catalysis. Prosthetic groups, such as heme, are permanently bound to the enzyme. Coenzymes are not usually bound to the enzyme, they enter into the reaction as a "cosubstrate," and they appear in modified form as a product. **Review Figure 6.19**

Metabolism and the Regulation of Enzymes
• Metabolism is organized into pathways, in which the product of one reaction is a reactant for the next reaction. Each reaction is catalyzed by an enzyme.
• Enzyme activity is subject to regulation. Some compounds react irreversibly with enzymes and reduce their catalytic activity. Others react reversibly, inhibiting enzyme action only temporarily. A compound closely similar in structure to an enzyme's normal substrate may competitively inhibit the action of the enzyme. **Review Figures 6.20, 6.21**
• For allosteric enzymes, the plots of reaction rate versus substrate concentration are sigmoidal, in contrast to plots of the same variables for single-subunit enzymes. **Review Figure 6.22**
• Allosteric inhibitors bind to a site different from the active site and stabilize the inactive form of the enzyme. The multiple catalytic subunits of many allosteric enzymes interact cooperatively. **Review Figure 6.23**
• Allosteric effects control metabolism. The end product of a metabolic pathway feeds back on the enzyme that catalyzes the first committed step in that branch, inhibiting that allosteric enzyme and preventing excessive buildup of the end product. **Review Figure 6.24**
• Enzymes are sensitive to their environment. Both pH and temperature affect enzyme activity. **Review Figures 6.25, 6.26**

Self-Quiz

1. Which statement about energy is *not* true?
   a. It can neither be created nor destroyed.
   b. It is the capacity to do work.
   c. All of its conversions are fully reversible.
   d. In the universe as a whole, the amount of free energy decreases.
   e. In the universe as a whole, the amount of entropy increases.

2. Which statement about thermodynamics is *not* true?
   a. Free energy is given off in an exergonic reaction.
   b. Free energy can be used to do work.
   c. A spontaneous reaction is exergonic.
   d. Free energy tends always to a minimum.
   e. Entropy tends always to a minimum.

3. In a chemical reaction,
   a. the rate depends on the value of ΔG.
   b. the rate depends on the activation energy.
   c. the entropy change depends on the activation energy.
   d. the activation energy depends on the value of ΔG.
   e. the change in free energy depends on the activation energy.

4. Which statement about enzymes is *not* true?
   a. They consist of proteins, with or without a nonprotein part.
   b. They change the rate of the catalyzed reaction.
   c. They change the value of ΔG of the reaction.
   d. They are sensitive to heat.
   e. They are sensitive to pH.

5. The active site of an enzyme
   a. never changes shape.
   b. forms no chemical bonds with substrates.
   c. determines, by its structure, the specificity of the enzyme.
   d. looks like a lump projecting from the surface of the enzyme.
   e. changes ΔG of the reaction.

6. A prosthetic group
   a. is a tightly bound, nonprotein part of an enzyme.
   b. is composed of protein.
   c. does not participate in chemical reactions.
   d. is present in all enzymes.
   e. is an artificial enzyme.

7. The rate of an enzyme-catalyzed reaction
   a. is constant under all conditions.
   b. decreases as substrate concentration increases.
   c. cannot be measured.
   d. depends on the value of ΔG.
   e. can be reduced by inhibitors.

8. Which statement about enzyme inhibitors is *not* true?
   a. A competitive inhibitor binds the active site of the enzyme.
   b. An allosteric inhibitor binds a site on the active form of the enzyme.
   c. A noncompetitive inhibitor binds a site other than the active site.
   d. Noncompetitive inhibition cannot be completely overcome by the addition of more substrate.
   e. Competitive inhibition can be completely overcome by the addition of more substrate.

9. Which statement about feedback inhibition of enzymes is *not* true?
   a. It is exerted through allosteric effects.
   b. It is directed at the enzyme that catalyzes the first committed step in a branch of a pathway.
   c. It affects the rate of reaction, not the concentration of enzyme.
   d. It acts very slowly.
   e. It is an example of negative feedback.

10. Which statement about temperature effects is *not* true?
    a. Raising the temperature may reduce the activity of an enzyme.
    b. Raising the temperature may increase the activity of an enzyme.
    c. Raising the temperature may denature an enzyme.
    d. Some enzymes are stable at the boiling point of water.
    e. The isozymes of an enzyme have the same optimal temperature.

Applying Concepts

1. How can endergonic reactions proceed in organisms?

2. Consider two proteins: One is an enzyme dissolved in the cytosol; the other is an ion channel in a membrane. Contrast the structures of the two proteins, indicating at least two important differences.

3. Plot free energy versus the course of an endergonic reaction and that of an exergonic reaction. Include the activation energy in both plots. Label E_a and ΔG on both graphs.

4. Consider an enzyme that is subject to allosteric regulation. If a competitive inhibitor (not an allosteric inhibitor) is added to a solution of such an enzyme, the ratio of enzyme molecules in the active form to those in the inactive form increases. Explain this observation.

Readings

Dickerson, R. E. and I. Geis. 1969. *The Structure and Action of Proteins*. W. A. Benjamin, Menlo Park, CA. This classic volume presents the structure of enzymes as high art.

Karplus, M. and J. A. MacCammon. 1986. "The Dynamics of Proteins." *Scientific American*, April. This article will correct any misconception of proteins as rigid molecules; it describes the constant, rapid changes in local shape that underlie the functioning of proteins.

Kauffman, S. A. 1993. *The Origins of Order*. Oxford University Press, New York. A discussion of thermodynamics, chaos, and life.

Koshland, D. E., Jr. 1973. "Protein Shape and Biological Control." *Scientific American*, October. This paper shows that the ability of proteins to change shape in specific circumstances underlies the control and coordination of biological processes.

Morowitz, H. J. 1978. *Foundations of Bioenergetics*. Academic Press, New York. An excellent advanced text on thermodynamics in biology.

Stryer, L. 1995. *Biochemistry*, 4th Edition. W. H. Freeman, New York. Good discussions of enzymes, the basic concepts of metabolism, and protein structure.

Chapter 7

Cellular Pathways That Harvest Chemical Energy

A Cold New World
Newborn humans experience a sudden drop in temperature when they emerge from the womb into the world. They adapt by metabolizing reserves of a special type of body fat.

A human fetus develops at the mother's body temperature. At birth, most newborns experience a drop in temperature, and their bodies must quickly do something about it. In fact what they do is the same thing a hibernating mammal does as it rouses itself from its winter "snooze." During hibernation, body temperature is low. In order to move about and take care of itself once awake again, an animal that has been hibernating must raise its body temperature.

Both the newly awakened mammal and the newborn baby respond to their temperature challenges—the low temperature of the hibernating body and the rapidly dropping temperature of the baby—by starting to metabolize stored "brown fat" reserves. Usually reserves are metabolized to generate ATP for work. In this case, however, a temperature-triggered signal tells the brown fat cells *not* to form ATP. Instead, these fat cells "waste" their stored energy as heat: They generate *lots* of heat and thus warm the body.

Cells convert energy from one form to another in order to carry out biological work (mechanical work, transport, and synthesis) and, sometimes, to generate heat. Most of the energy in the biosphere derives from the sun. Energy flows from the sun through photosynthetic *autotrophs* (plants and some protists and bacteria) to *heterotrophs* (organisms that must obtain their energy in the form of food—organic compounds) and back to the environment as heat. Photosynthetic autotrophs, as we will see in the next chapter, convert light energy into chemical energy—food for themselves and for the organisms that eat them. In this chapter we are concerned with how organisms process the chemical energy of food.

Cells of all living things can process this energy. Prokaryotes and eukaryotes share some energy-processing metabolic pathways. In this chapter we'll examine the pathways that operate to extract the energy from glucose by a series of oxidation–reduction reactions. The evolutionarily most ancient pathways probably operated without oxygen; the controlled chemical breakdown of glucose to pyruvate is part of this ancient process. However, as oxygen gas was produced by photosynthesis, chemical reactions using it proved to have a great advantage because they liberated more free energy, and thus aerobic metabolism evolved.

In this chapter, we'll emphasize energy metabolism that uses oxygen, starting with the chemical breakdown of glucose to produce pyruvate and the three sequential pathways by which pyruvate is completely oxidized, step-by-step, to water and carbon dioxide, and the released energy is captured as ATP. In the absence of oxygen, oxidation is less complete, and glucose molecules are converted to waste products, releasing only some of the potential energy. Toward the end of the chapter, we'll examine how the pathways of energy metabolism connect with other pathways for the catabolism and anabolism of other molecular raw materials of the cell. Then we'll see how these processes are regulated.

Obtaining Energy and Electrons from Glucose

What is the food that fuels all living cells? The most common food is the sugar glucose. Many other compounds serve as foods, but almost all of them yield their energy after being converted to glucose or to compounds intermediate in the metabolism of glucose. How do cells obtain energy from glucose?

Cells trap free energy while metabolizing glucose

Glucose burns readily, yielding carbon dioxide, water, and a lot of energy—but only if oxygen gas is present:

$$C_6H_{12}O_6 + 6\,O_2 \rightarrow 6\,CO_2 + 6\,H_2O + \text{energy}$$
$$\text{(heat and light)}$$

It wouldn't do cells a lot of good just to burn their glucose. They need to trap as much as possible of the chemical energy of glucose in usable form rather than as heat or light. Cells trap some of this energy in the energy storage compound ATP (adenosine triphosphate) (see Chapter 6).

Most kinds of organisms can metabolize their glucose completely in the processes of glycolysis and cellular respiration:

$$C_6H_{12}O_6 + 6\,O_2 \rightarrow 6\,CO_2 + 6\,H_2O + \text{energy}$$
$$\text{(ATP and heat)}$$

The change in free energy (ΔG) for the complete conversion of glucose and oxygen to carbon dioxide and water, whether by combustion or by complete metabolism, is -686 kcal/mol ($-2{,}870$ kJ/mol). Thus the overall reaction is highly exergonic and can drive the endergonic formation of a great deal of ATP. Some other kinds of organisms, unable to obtain or use oxygen gas, metabolize the glucose incompletely, thus obtaining less ATP per glucose molecule. Not all of the carbon atoms of glucose are converted to carbon dioxide in this incomplete metabolism, which is called *fermentation.*

Three metabolic processes play roles in the utilization of glucose for energy: glycolysis, cellular respiration, and fermentation (Figure 7.1). These processes consist of metabolic pathways made up of many distinct, but coupled, chemical reactions.

Glycolysis is a series of reactions that begins the metabolism of glucose in all cells and produces the product *pyruvate.* What happens to pyruvate depends on the type of organism that is extracting the energy and on whether the environment is **aerobic** (containing oxygen gas, O_2) or **anaerobic** (lacking oxygen gas). Glycolysis takes place under both conditions. However, fermentation occurs only under anaerobic conditions, and cellular respiration takes place only under aerobic conditions.

Less energy is captured as ATP during fermentation than during cellular respiration, and energy-rich carbon compounds such as lactic acid or ethanol (ethyl alcohol) are produced as waste products of fermentation. Cellular respiration releases much more energy from each glucose molecule than does fermentation. Importantly, glycolysis and cellular respiration operate together, but fermentation is an entirely separate process, operating only in the absence of O_2.

The combined operation of glycolysis and cellular respiration is the biological equivalent of burning glucose. When glucose is burned with a match, it releases energy rapidly as heat and light. If the burning is coupled to a mechanical device, some of the released energy can be used to do work. In cellular respiration, glucose is broken down to the same products (CO_2 and H_2O) as it is when it is simply burned, but much of the released energy is trapped in ATP. Both burning and cellular respiration are chemical reactions known as oxidation–reduction (redox) reactions.

Redox reactions transfer electrons and energy

Chemical reactions in which one substance transfers one or more electrons to another substance are called oxidation–reduction reactions, or **redox reactions**. The *gain* of one or more electrons by an atom, ion, or molecule is called **reduction**. The *loss* of one or more electrons is called **oxidation**. Although oxidation and reduction are always defined in terms of traffic in

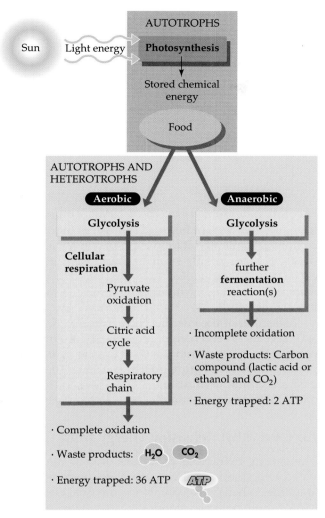

7.1 Energy for Life In photosynthesis, autotrophic organisms use light energy to synthesize food compounds. Heterotrophic and autotrophic organisms both process these food compounds by glycolysis, fermentation, and cellular respiration. Glycolysis precedes both fermentation and cellular respiration.

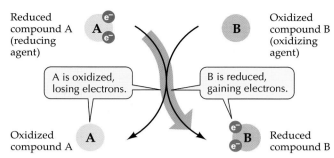

7.2 Oxidation and Reduction Are Coupled Compound A is oxidized and compound B reduced in a redox reaction. In the process, A loses electrons and B gains electrons.

duced an *oxidizing agent* and the one that becomes oxidized a *reducing agent* (Figure 7.2). An oxidizing agent accepts electrons; a reducing agent gives up electrons. In the process of oxidizing the reducing agent, the oxidizing agent itself becomes reduced. Conversely, the reducing agent becomes oxidized as it reduces the oxidizing agent. In the burning or metabolism of glucose, glucose is the reducing agent and oxygen gas the oxidizing agent.

Energy is transferred in a redox reaction: Some of the energy originally present in the reducing agent becomes associated with the reduced product. The overall ΔG is negative. As we will see, some of the key reactions of glycolysis and cellular respiration are highly exergonic redox reactions.

At some early stage in evolution, cells began to use reducing agents and oxidizing agents. Natural selection favored the use of certain of these agents (the ones whose redox reactions have suitable values of ΔG) as a system for the orderly exchange of electrons, analogous to the use of the ATP–ADP system for the orderly transfer of energy (see Figure 6.9). We have already encountered an example of such agents at work in cells: In Chapter 6 we saw that FAD accepts hydrogens during the respiratory conversion of succinate to fumarate (see Figure 6.16). In that reaction, FAD is an oxidizing agent and $FADH_2$ is a reducing agent. NAD is another important oxidizing agent that functions as a coenzyme in many reactions of glycolysis, cellular respiration, and fermentation.

NAD is a key electron carrier in redox reactions

The main pair of oxidizing and reducing agents in cells is based on the compound **NAD** (nicotinamide adenine dinucleotide). NAD exists in two chemically distinct forms, one oxidized (NAD^+) and the other reduced ($NADH + H^+$; Figure 7.3). NAD^+ and $NADH + H^+$ participate in biological redox reactions. The reduction

$$NAD^+ + 2\,H \rightarrow NADH + H^+$$

electrons, we must also think in these terms when hydrogen atoms (not hydrogen ions) are gained or lost, because transfers of hydrogen atoms involve transfers of electrons. Thus when a molecule loses hydrogen atoms, it becomes oxidized:

$$\overbrace{AH_2 + B \rightarrow \underbrace{BH_2 + A}_{\text{reduction}}}^{\text{oxidation}}$$

Oxidation and reduction *always* occur together: As one material is oxidized, the electrons it loses are transferred to another material, reducing that material. In a redox reaction, we call the reactant that becomes re-

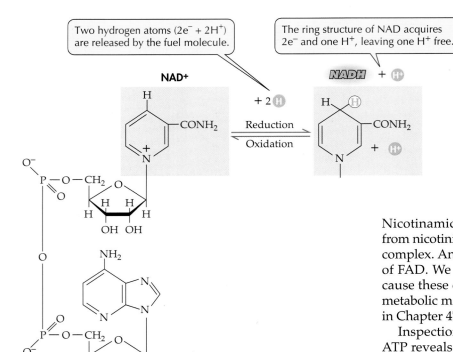

Two hydrogen atoms ($2e^- + 2H^+$) are released by the fuel molecule.

The ring structure of NAD acquires $2e^-$ and one H^+, leaving one H^+ free.

7.3 Oxidized and Reduced Forms of NAD NAD^+ is the oxidized form and NADH the reduced. As the shaded part of the NAD^+ molecule is reduced by acquiring two electrons (e^-) and a hydrogen ion (H^+), to yield a neutral NADH molecule, the second H^+ is released to the surroundings. The unshaded portion of the molecule remains unchanged by the reaction.

Nicotinamide, which is part of NAD, forms directly from nicotinic acid, or niacin, a member of the vitamin B complex. Another B vitamin is riboflavin, which is part of FAD. We need only small amounts of vitamins because these carrier molecules are recycled through the metabolic machinery. Vitamins are discussed more fully in Chapter 47.

Inspection of the chemical structures of NAD^+ and ATP reveals several common features. Both molecules contain the base adenine, the sugar ribose, and phosphate groups (see Figure 6.7). These components suggest a common evolutionary origin for both NAD^+ and ATP in an ancient and less efficient molecule.

is formally equivalent to the transfer of two hydrogen atoms ($2 H^+ + 2 e^-$). However, what is actually transferred is a hydride ion (H^-, a proton and two electrons), leaving a free proton (H^+). The oxidation of NADH + H^+ by oxygen gas is highly exergonic:

$$NADH + H^+ + \tfrac{1}{2} O_2 \rightarrow NAD^+ + H_2O$$

$$\Delta G = -52.4 \text{ kcal/mol} \ (-219 \text{ kJ/mol})$$

(Note that the oxidizing agent appears here as "$\tfrac{1}{2} O_2$" instead of "O." This notation emphasizes that it is oxygen gas, O_2, that acts as the oxidizing agent.) In the same way that ATP can be thought of as a means of packaging free energy in bundles of about 12 kcal/mol (50 kJ/mol), NAD can be thought of as a means of packaging free energy in bundles of approximately 50 kcal/mol (200 kJ/mol) (Figure 7.4).

Besides NAD and FAD, there are several other biologically important electron carriers. The molecular structures of NAD and some of the others include components that we humans need but cannot synthesize for ourselves; these are classified as vitamins.

An Overview: Releasing Energy from Glucose

The energy-extracting processes of cells may be divided into distinct pathways that we can consider one at a time. When O_2 is available as the final electron acceptor, four pathways operate: glycolysis, pyruvate oxidation, the citric acid cycle, and the respiratory chain. When O_2 is unavailable, pyruvate oxidation, the citric acid cycle, and the respiratory chain do not function, and additional *fermentation reactions* are added to the glycolytic pathway. Figure 7.5 summarizes the

7.4 NAD Is an Energy Carrier Thanks to its ability to carry free energy and electrons, NAD is a major and universal energy intermediary in cells.

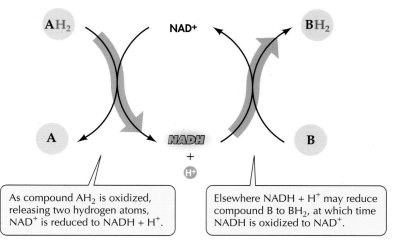

As compound AH_2 is oxidized, releasing two hydrogen atoms, NAD^+ is reduced to NADH + H^+.

Elsewhere NADH + H^+ may reduce compound B to BH_2, at which time NADH is oxidized to NAD^+.

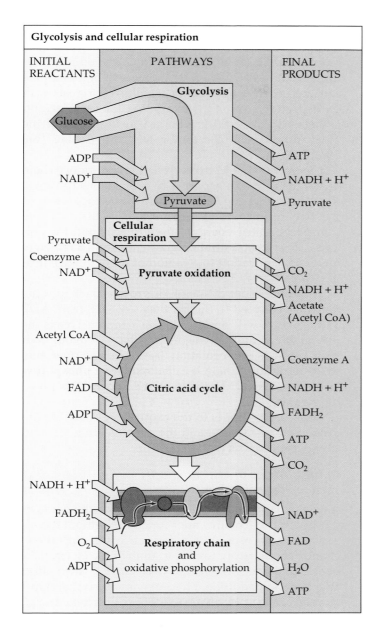

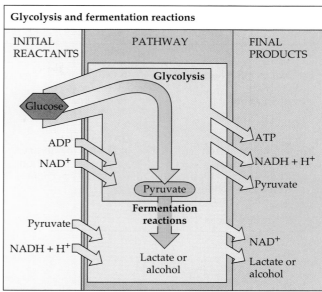

7.5 An Overview of the Cellular Energy Pathways The energy-producing reactions can be grouped into five pathways: glycolysis, pyruvate oxidation, the citric acid cycle, the respiratory chain, and fermentation. The three middle pathways occur only in the presence of oxygen and are collectively referred to as cellular respiration. Each pathway will be depicted in more detailed figures later in this chapter.

starting reactants and products of these five pathways.

These distinct chemical pathways are separated in the cell (Table 7.1). In eukaryotes, glycolysis and fermentation take place in the cytoplasm outside of mitochondria. The enzymes for these pathways were once believed to be soluble in the cytosol, but more recent discoveries suggest that at least some of them may be bound to components of the cytoskeleton. The other reactions are associated with the mitochondria. Pyruvate oxidation and the respiratory chain are both associated with the inner membrane of mitochondria, where their enzymes are bound. The enzymes and reactions of the citric acid cycle are found in the matrix of mitochondria. These different locations reflect separate evolutionary origins of these processes.

In prokaryotes, the enzymes used in glycolysis, fermentation, and the citric acid cycle are soluble in the cytosol. Enzymes involved in pyruvate oxidation and the respiratory chain are associated with the inner surface of the plasma membrane or inward elaborations of that membrane (see Chapter 2).

In the presence of O_2, glycolysis begins the breakdown of glucose

Glycolysis (the glycolytic pathway) is a sequence of 10 separate chemical reactions. This near-universal process was probably the first energy-releasing pathway to evolve; if any earlier pathway existed, it has disappeared from Earth. Today virtually all living cells, even the most evolutionarily ancient, use glycolysis.

Glycolysis is the pathway in which glucose is incompletely oxidized, to **pyruvate** (pyruvic acid).* Glycolysis may operate in the presence or the absence of O_2. In either case, it contains an oxidative step in which the electron carrier NAD$^+$ becomes reduced,

*We tend to use words like "pyruvate" and "pyruvic acid" interchangeably. However, at pH values commonly found in cells, the ionized form—pyruvate—is present rather than the acid—pyruvic acid. Similarly, all carboxylic acids are present as ions (the "-ate" forms) at these pH values.

TABLE 7.1 Cellular Locations for Energy Pathways in Eukaryotes and Prokaryotes

EUKARYOTES	PROKARYOTES
External to mitochondrion	**In cytoplasm**
Glycolysis	Glycolysis
Fermentation	Fermentation
	Citric acid cycle
Inside mitochondrion	**On inner face of plasma membrane**
Inner membrane	Pyruvate oxidation
Pyruvate oxidation	Respiratory chain
Respiratory chain	
Matrix	
Citric acid cycle	

acquiring electrons. Each molecule of glucose processed through glycolysis produces a net yield of two molecules of ATP. The major products of glycolysis are ATP (which the cell will use to drive endergonic reactions), pyruvate, and the electrons acquired by NAD. Both the pyruvate and the electrons must be processed further.

Cellular respiration operates when O_2 is available, yielding CO_2 and H_2O as products. It is made up of three pathways: pyruvate oxidation, the citric acid cycle, and the respiratory chain. In **pyruvate oxidation**, the end product of glycolysis (pyruvate) loses two hydrogen atoms and a carboxyl group ($—COO^-$) as CO_2, forming the two-carbon molecule acetate (acetic acid), which is activated by the addition of a coenzyme (coenzyme A).

The **citric acid cycle** (also called the Krebs cycle or the tricarboxylic acid cycle) is a cyclical series of reactions in which the product of pyruvate oxidation, acetate, becomes *completely* oxidized, forming CO_2 and transferring electrons (along with their hydrogen nuclei) to carrier molecules (FAD and NAD^+). The citric acid cycle produces many more electrons than are produced in glycolysis. And as we are about to see, harvesting more electrons means a greater ultimate harvest of ATP.

In glycolysis, pyruvate oxidation, and the citric acid cycle, the molecules that become reduced acquire hydrogen atoms. Through these pathways, energy originally present in the covalent bonds of glucose becomes associated with reduced forms of NAD and FAD ($NADH + H^+$ and $FADH_2$). Hydrogen is an outstanding fuel. When it reacts with O_2, a great deal of free energy is released; better still, the "waste" product of this reaction—water—is not toxic to the environment or to any organism that produces it.

The fourth energy-extracting pathway for aerobic cells is the **respiratory chain**, whose principal role is to release energy from reduced NAD in such a way that it may be used to form ATP. This pathway is a series of redox reactions in which electrons derived from hydrogen atoms are passed from one type of membrane carrier to another and finally are allowed to react with O_2 to produce water. In eukaryotes, the carriers (and associated enzymes) are bound to the folds of the inner mitochondrial membranes, the **cristae** (see Figure 4.13).

The transfer of electrons along the respiratory chain drives the active transport of hydrogen ions (protons) from the mitochondrial matrix into the space between the two mitochondrial membranes. This is active transport because energy is used to transport the protons *against* a concentration gradient—from a region of lower concentration to a region of higher concentration (see Chapter 5).

The subsequent diffusion of protons back into the matrix is coupled to the synthesis of ATP from ADP and P_i, as we will see later in this chapter. This is the way in which the vast majority of the ATP in our bodies is formed. The formation of ATP during operation of the respiratory chain is called **oxidative phosphorylation** because NADH is oxidized and ADP is phosphorylated.

Overall, the inputs to the respiratory chain are hydrogen atoms and O_2, and the outputs are water and energy captured as ATP.

In the absence of O_2, some cells carry on fermentation

If we are deprived of O_2 for too long, we die because the respiratory chain cannot function. Without oxygen molecules as receptors, the carriers in our mitochondrial cristae are unable to unload the electrons bound to them. Soon, no oxidized carriers are available to accept additional hydrogens. When that happens, glycolysis, pyruvate oxidation, and the citric acid cycle stop. Without these processes, and with no respiratory chain activity, we have insufficient ATP for our cells to maintain their structure and metabolism, and we die.

Not all cells require O_2. **Fermentation** utilizes glycolysis and an additional reaction. It produces only a fraction of the energy produced when cells utilize O_2. Our muscle cells have such an alternative way to rid themselves of the hydrogen atoms produced during glycolysis: The hydrogens produced by glycolysis are passed to the end product of glycolysis (pyruvate), and lactic acid (lactate) is formed.

$$\text{pyruvate} + \text{NADH} + H^+ \rightarrow \text{lactate} + NAD^+$$

This reaction recycles the NAD needed in glycolysis. Thus even in the absence of oxygen, glycolysis continues (often at an increased rate), without the activity of the citric acid cycle or the respiratory chain, and ATP continues to be produced. The cells that have the enzyme necessary for this reaction can function for a time in the absence of oxygen. However, eventually the concentration of lactic acid in muscles reaches a toxic level. Such anaerobic production of ATP is called fermentation. For some organisms that live entirely without oxygen, fermentation is the sole pathway to trap energy in ATP.

In the discussion that follows, we will examine in more detail the four pathways of aerobic energy metabolism (glycolysis, pyruvate oxidation, the citric acid cycle, and the respiratory chain) and fermentation.

Glycolysis: From Glucose to Pyruvate

In glycolysis, glucose is only partly oxidized. A molecule of glucose taken in by a cell enters the glycolytic pathway, which consists of 10 reactions that convert the six-carbon glucose molecule, step-by-step, into two molecules of the three-carbon compound pyruvic acid (Figure 7.6). These reactions are accompanied by the *net* formation of two molecules of ATP and by the reduction of two molecules of NAD^+ to two molecules of $NADH + H^+$. At the end of the pathway, then, ready energy is located in ATP, and four hydrogen atoms are located in $NADH + H^+$.

The fate of the pyruvic acid depends on the type of cell carrying out glycolysis and on whether the environment is aerobic or anaerobic. The fate of the NADH + H$^+$, too, varies. In most cases, NADH + H$^+$ is oxidized through the respiratory chain to yield H_2O and NAD^+—a chain of reactions that results in the formation of much more ATP (three molecules of ATP per molecule of NADH + H$^+$). In fermentation, however, NADH + H$^+$ is reoxidized to NAD^+ either by pyruvic acid itself or by one of its metabolites, with no further ATP production. In either case, glycolysis may be regarded as a series of *preparatory reactions*, to be followed either by the fermentation reactions or by cellular respiration (pyruvate oxidation, citric acid cycle, and respiratory chain).

Glycolysis can be divided into two groups of reactions: reactions that invest energy from ATP hydrolysis and energy-harvesting reactions that produce ATP.

The energy-investing reactions of glycolysis require ATP

Using Figure 7.6, we can trace our way through the glycolytic pathway. The first five reactions are endergonic, taking up free energy; that is, the cell is *investing* free energy rather than gaining it during the early reactions of glycolysis. In separate reactions, two molecules of ATP are invested in attaching two phosphate groups to the sugar (reactions 1 and 3), thereby raising its free energy by about 15 kcal/mol (62.7 kJ/mol)

(Figure 7.7). Later, these phosphate groups will be transferred to ADP to make new molecules of ATP.

Although both of these first steps of glycolysis use ATP as one of the substrates, each is catalyzed by a different, specific enzyme. The enzyme hexokinase catalyzes reaction 1, in which glucose receives a phosphate group from ATP. (A *kinase* is any enzyme that catalyzes the transfer of a phosphate group from ATP to another substrate.) In reaction 2, the six-membered glucose ring is rearranged to a five-membered fructose ring. Then, in reaction 3, the enzyme phosphofructokinase adds a second phosphate (taken from another ATP) to the sugar ring.

The fourth reaction opens the sugar ring with its two phosphates, and the six-carbon sugar bisphosphate* is cleaved to give two different three-carbon sugar phosphates. In reaction 5, one of these sugar phosphates (dihydroxyacetone phosphate) is converted into a second molecule of the other (glyceraldehyde 3-phosphate).

By this time, the halfway point in glycolysis, the following things have happened: Two molecules of ATP have been invested, and the six-carbon glucose molecule has been converted into two molecules of a three-carbon sugar phosphate. No ATP has been gained, and nothing has been oxidized.

The energy-harvesting reactions of glycolysis yield ATP and NADH + H$^+$

Now things begin to happen, including a key redox reaction. In what follows, remember that each step occurs twice for each glucose molecule going through glycolysis, because in the first five reactions of glycolysis each glucose molecule has been split into two molecules of three-carbon sugar phosphate, both of which go through the remaining steps of glycolysis.

Reaction 6 is a two-step reaction catalyzed by the enzyme triose phosphate dehydrogenase. The end product of reaction 6 is 1,3-bisphosphoglycerate (or 1,3-bisphosphoglyceric acid). A phosphate ion has been snatched from the surroundings (but not, this time, from ATP) and tacked onto the three-carbon compound. Figure 7.7 shows that this reaction is accompanied by an enormous drop in free energy—more than 100 kcal of energy per mole of glucose is released in this extremely exergonic reaction. What has happened here? Why the big energy change?

Reaction 6 is the conversion of a sugar to an acid:

$$R-\overset{\overset{\textstyle O}{\|}}{C}-H + (O) \rightarrow R-\overset{\overset{\textstyle O}{\|}}{C}-OH$$

*The root *bis-* means "two." A sugar bisphosphate has two phosphate groups.

The pathways of glycolysis and cellular respiration (pyruvate oxidation, citric acid cycle and respiratory chain) are represented by a "road map" of symbols that guide you to better understand the relationship of these pathways in the illustrations that follow.

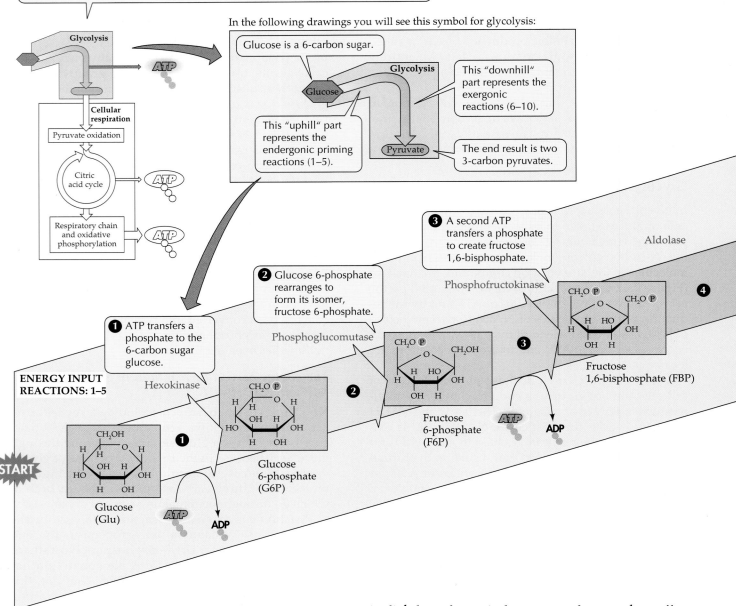

In the following drawings you will see this symbol for glycolysis:

Glucose is a 6-carbon sugar.

Glycolysis

This "downhill" part represents the exergonic reactions (6–10).

This "uphill" part represents the endergonic priming reactions (1–5).

The end result is two 3-carbon pyruvates.

Glycolysis

Cellular respiration

Pyruvate oxidation

Citric acid cycle

Respiratory chain and oxidative phosphorylation

❸ A second ATP transfers a phosphate to create fructose 1,6-bisphosphate.

Aldolase

❷ Glucose 6-phosphate rearranges to form its isomer, fructose 6-phosphate.

Phosphofructokinase

❹

Phosphoglucomutase

❶ ATP transfers a phosphate to the 6-carbon sugar glucose.

ENERGY INPUT REACTIONS: 1–5

Hexokinase

❸

Fructose 1,6-bisphosphate (FBP)

❷

Fructose 6-phosphate (F6P)

❶

Glucose 6-phosphate (G6P)

START

Glucose (Glu)

ATP → ADP

ATP → ADP

Since this is an oxidation reaction, it is very exergonic. (Note that here we do *not* write $1/2$ O_2, because in this case oxygen gas does not participate in the reaction.) The formation of the phosphate ester from the acid:

$$R - \overset{\overset{O}{\|}}{C} - OH + HPO_4^{2-} \rightarrow R - \overset{\overset{O}{\|}}{C} - O - \overset{\overset{O}{\|}}{\underset{\underset{O^-}{|}}{P}} - O^- + H_2O$$

is slightly endergonic, but not nearly enough to offset the drop in free energy from the oxidation.

If this big energy drop were simply the loss of heat, glycolysis would not provide useful energy to the cell. However, rather than being lost, this energy is used to make two molecules of NADH + H$^+$ from two molecules of NAD$^+$. This stored energy is regained later—either in the respiratory chain, by the formation of ATP, or in the last step of fermentation, when pyruvate or its product is reduced and the two molecules of NADH + H$^+$ are restored once again to NAD$^+$. This cycling of NAD is necessary to keep glycolysis going; if all the NAD$^+$ is converted to NADH + H$^+$, glycolysis comes to a halt.

7.6 Glycolysis Converts Glucose to Pyruvate   Starting with hexokinase, ten enzymes catalyze ten reactions in turn. Along the way, ATP is produced (reactions 7 and 10), and two NAD$^+$ are reduced to two NADH + 2 H$^+$ (reaction 6).

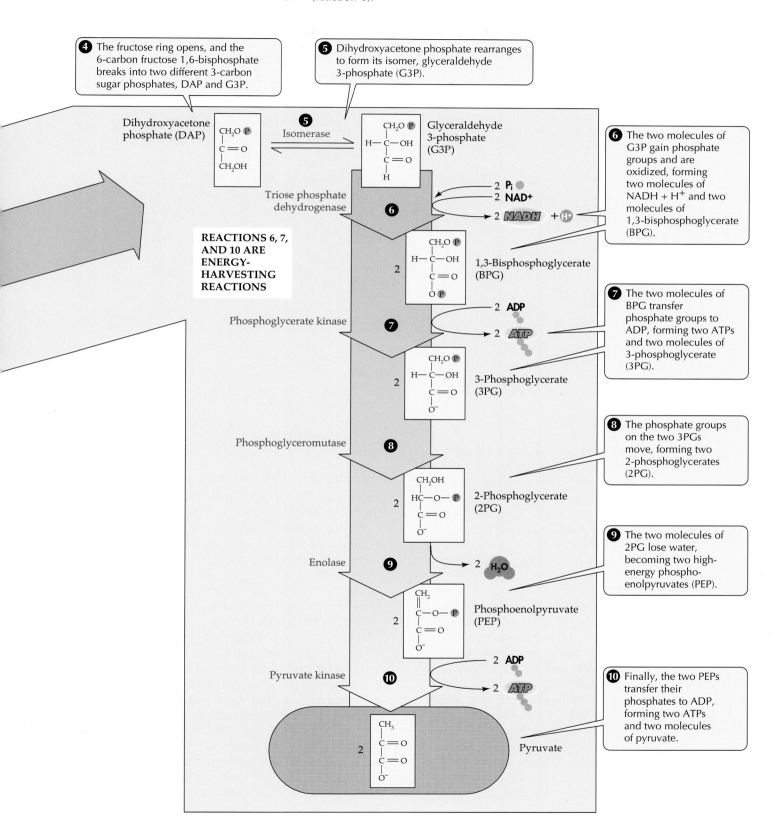

4 The fructose ring opens, and the 6-carbon fructose 1,6-bisphosphate breaks into two different 3-carbon sugar phosphates, DAP and G3P.

5 Dihydroxyacetone phosphate rearranges to form its isomer, glyceraldehyde 3-phosphate (G3P).

6 The two molecules of G3P gain phosphate groups and are oxidized, forming two molecules of NADH + H$^+$ and two molecules of 1,3-bisphosphoglycerate (BPG).

7 The two molecules of BPG transfer phosphate groups to ADP, forming two ATPs and two molecules of 3-phosphoglycerate (3PG).

8 The phosphate groups on the two 3PGs move, forming two 2-phosphoglycerates (2PG).

9 The two molecules of 2PG lose water, becoming two high-energy phosphoenolpyruvates (PEP).

10 Finally, the two PEPs transfer their phosphates to ADP, forming two ATPs and two molecules of pyruvate.

REACTIONS 6, 7, AND 10 ARE ENERGY-HARVESTING REACTIONS

Dihydroxyacetone phosphate (DAP)

Isomerase

Glyceraldehyde 3-phosphate (G3P)

Triose phosphate dehydrogenase

2 P$_i$
2 NAD$^+$
2 NADH + H$^+$

1,3-Bisphosphoglycerate (BPG)

Phosphoglycerate kinase

2 ADP
2 ATP

3-Phosphoglycerate (3PG)

Phosphoglyceromutase

2-Phosphoglycerate (2PG)

Enolase

2 H$_2$O

Phosphoenolpyruvate (PEP)

Pyruvate kinase

2 ADP
2 ATP

Pyruvate

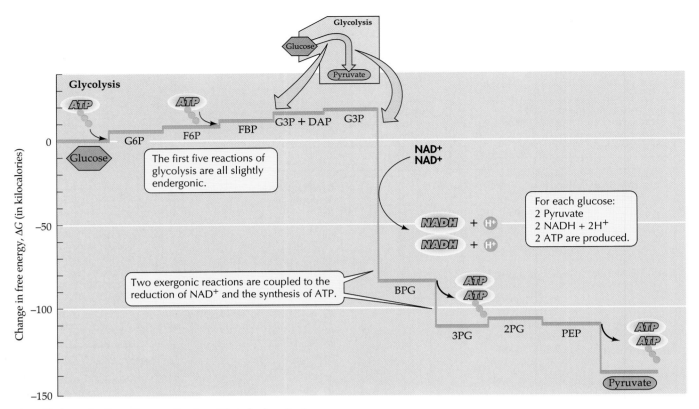

7.7 Changes in Free Energy During Glycolysis Each reaction of glycolysis changes the free energy available, as shown by the different energy levels of the series of reactants and products from glucose to pyruvate.

The remaining steps of glycolysis (see Figure 7.6) are simpler. The two phosphate groups of 1,3-bisphosphoglycerate are transferred, one at a time, to molecules of ADP, with a rearrangement in between. More than 20 kcal (83.6 kJ/mol) of free energy is stored in ATP for every mole of 1,3-bisphosphoglycerate broken down. Finally, we are left with two molecules of pyruvate for every molecule of glucose that entered glycolysis.

SUBSTRATE-LEVEL PHOSPHORYLATION. The enzyme-catalyzed transfer of phosphate groups from donor molecules to ADP molecules is called **substrate-level phosphorylation**. This process is driven by energy obtained from oxidation. For example, when glyceraldehyde 3-phosphate reacts with P_i and NAD^+, becoming 1,3-bisphosphoglycerate, an aldehyde is oxidized to a carboxylic acid, with NAD^+ acting as the oxidizing agent. The oxidation provides so much energy that the newly added phosphate group is linked to the rest of the molecule by a bond that has even higher energy than the high-energy bond of ATP (Figure 7.8a). A second enzyme catalyzes the transfer of this phosphate group

from 1,3-bisphosphoglycerate to ADP, forming ATP (Figure 7.8b). Both reactions are exergonic, even though a substantial amount of energy is consumed in the formation of ATP.

Reviewing glycolysis and fermentation

A review of the glycolytic reactions shows that at the beginning of glycolysis, two molecules of ATP are used per molecule of glucose, but that ultimately four are produced (two for each of the two 1,3-bisphosphoglycerates)—a net gain of two ATP molecules and two $NADH + H^+$.

In fermentation (anaerobic conditions), the total usable energy yield is just two ATP molecules per glucose molecule. Under these anaerobic conditions, the $NADH + H^+$ is rapidly recycled to NAD^+ by the reduction of pyruvate. The NAD^+ is then available for the glycolytic reaction catalyzed by the enzyme triose phosphate dehydrogenase (reaction 6 in Figure 7.6). On the other hand, in the presence of oxygen (aerobic conditions), eukaryotes and some bacteria are able to reap far more energy by the complete oxidation of pyruvate and by oxidizing the $NADH + H^+$ of glycolysis through the respiratory chain, as we will see in the sections that follow. In eukaryotes, these reactions take place in the mitochondria.

(a) **Oxidation of substrate**

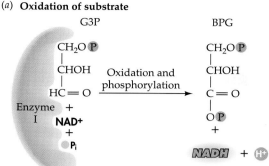

Glyceraldehyde 3-phosphate is both oxidized and phosphorylated to become 1,3-bisphosphoglycerate.

(b) **Transfer of phosphate to ADP**

Hydrolysis of 1,3-bisphosphoglycerate releases enough energy to transfer a phosphate group to ADP, forming ATP.

7.8 Substrate-Level Phosphorylation Two enzymes collaborate to catalyze substrate-level phosphorylation.

Pyruvate Oxidation

The oxidation of pyruvate to acetate is a complex multistep reaction catalyzed by an enormous enzyme complex that is attached to the mitochondrial inner membrane. The three-carbon compound, pyruvate, is oxidized to the two-carbon compound, acetate (CH_3COO^-), yielding free energy and CO_2. In this process, the acetate is linked to a coenzyme, called **coenzyme A (CoA)**, producing the energy-rich compound **acetyl coenzyme A**, or acetyl CoA (Figure 7.9). Acetyl CoA has 7.5 kcal/mol (31.4 kJ/mol) more energy than simple acetate. (Acetyl CoA can donate acetate to acceptors such as oxaloacetate, much as ATP can donate phosphate to various acceptors.)

There are three steps in this short pathway: (1) Pyruvate is oxidized to acetate, and CO_2 is released.

(2) Part of the energy from the oxidation in step 1 is saved by the reduction of NAD^+ to $NADH + H^+$. (3) Some of the remaining energy is stored temporarily by the combining of the acetate with CoA. An analogous three-step reaction occurs in glycolysis when glyceraldehyde 3-phosphate is converted to 1,3-bisphosphoglycerate (reaction 6 in Figure 7.6). In that reaction, an aldehyde group is oxidized to an acid, some of the energy released by oxidation is stored in $NADH + H^+$, and some of the remaining energy is preserved in a second phosphate bond in the molecule. As the similarity between these two three-step reactions shows, a good metabolic idea is likely to appear more than once; we will see this one yet again, in the citric acid cycle.

As you might suspect, a complex set of steps such as those in the reaction from pyruvate to acetyl CoA requires more than one type of catalytic protein. This reaction is catalyzed by the *pyruvate dehydrogenase com-*

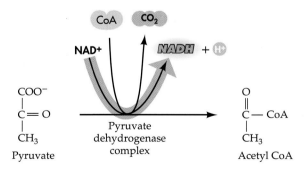

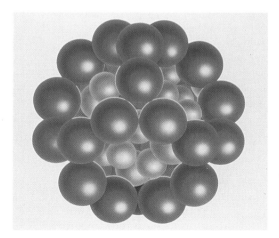

7.9 The Pyruvate Dehydrogenase Complex Catalyzes Pyruvate Oxidation The massive complex transfers electrons, removes a carboxyl group, and adds a coenzyme.

plex, which consists of 72 subunits—24 each of three different types of protein, for a total molecular weight of 4.6 million (Figure 7.9b). The three component enzymes use a total of five different coenzymes—four of which are vitamins or contain vitamins (thiamin, riboflavin, pantothenic acid, and niacin). This enzyme complex is an impressive example of biological organization.

The Citric Acid Cycle

Acetyl CoA is the starting point for the citric acid cycle (Figure 7.10). In this section we examine the citric acid cycle, in which the two-carbon molecule of acetate is oxidized to two molecules of carbon dioxide.

Figure 7.7 shows that the metabolism of glucose to pyruvate is accompanied by a drop in free energy of about 140 kcal/mol (585 kJ/mol). About a third of this energy is captured in the formation of ATP and reduced NAD (NADH + H$^+$). Oxidizing the pyruvate to acetate yields additional free energy for biological work. The citric acid cycle takes acetate and breaks it down to CO_2, using the hydrogen atoms to reduce carrier molecules and to pass chemical free energy to those carriers. The reduced carriers are later oxidized in the respiratory chain, which transfers an enormous amount of free energy to ATP.

The principal inputs to the citric acid cycle are acetate in the form of acetyl CoA, water, and oxidized electron carriers. The principal outputs are carbon dioxide and reduced electron carriers. Overall, for each molecule of acetate, during the citric acid cycle two carbons are removed as CO_2 and four pairs of hydrogen atoms are used to reduce carrier molecules that trap energy for use later in the synthesis of ATP. The energy-trapping reactions of the cycle are a major reason for its existence.

The citric acid cycle produces two CO₂ molecules and reduced carriers

At the beginning of the citric acid cycle, acetyl CoA, which has two carbon atoms in its acetate group, reacts with a four-carbon acid (oxaloacetate) to form the six-carbon compound citric acid (citrate). The remainder of the cycle consists of a series of enzyme-catalyzed reactions in which citric acid is degraded to a new four-carbon molecule of oxaloacetate. This new oxaloacetate can react with a second acetyl CoA, producing a second molecule of citrate and thus enabling the cycle to continue. Acetyl CoA enters the cycle from pyruvate, and CO_2 exits.

As we describe the citric acid cycle in detail, concentrate on how it is maintained in a steady state—that is, with material entering and leaving and with intermediate compounds like succinate and malate turning over constantly, but without changing concentration. Pay close attention to the numbered reactions

7.10 Pyruvate Oxidation and the Citric Acid Cycle ▶
Pyruvate diffuses into the mitochondrion and is oxidized to acetyl CoA, which enters the citric acid cycle. Notice that the two carbons from acetyl CoA are traced with color through reaction 4, after which they may be at either end of the molecule (note the symmetry of succinate and fumarate). For each glucose molecule, the cycle operates twice, producing 4 CO_2, 6 NADH + 6 H$^+$, 2 FADH$_2$, and 2 ATP in all.

in Figure 7.10 as you read the next several paragraphs.

The energy temporarily stored in acetyl CoA helps drive the formation of *citrate* from *oxaloacetate* (reaction 1). During this reaction, the coenzyme A molecule falls away, to be recycled. In reaction 2, citrate is rearranged to *isocitrate*. In reaction 3, a CO_2 molecule and two hydrogen atoms are removed in the conversion of isocitrate to *α-ketoglutarate*. As Figure 7.11 indicates, this reaction produces a large drop in free energy. The released energy is stored in NADH + H$^+$ and can be recovered later in the respiratory chain, when the NADH + H$^+$ is reoxidized.

Like the oxidation of pyruvate to acetyl CoA, reaction 4 of the citric acid cycle is complex. The five-carbon α-ketoglutarate molecule is oxidized to the four-carbon molecule *succinate*, CO_2 is given off, some of the oxidation energy is stored in NADH + H$^+$, and some of the energy is preserved temporarily by combining succinate with CoA to form *succinyl CoA*. In reaction 5, the energy in succinyl CoA is harvested to make GTP (guanosine triphosphate) from GDP and P$_i$, which is another example of substrate-level phosphorylation. Then GTP is used to make ATP from ADP.

Free energy is released in reaction 6, when two hydrogens are transferred to an enzyme that contains FAD. After a molecular rearrangement (reaction 7), one more NAD$^+$ reduction occurs, producing oxaloacetate from *malate* (reaction 8). The oxaloacetate produced by all these reactions is ready to combine with another acetate from acetyl CoA and go around the cycle again. Bear in mind that the citric acid cycle operates twice for each glucose molecule that enters glycolysis.

Although most of the enzymes of the citric acid cycle are dissolved in the mitochondrial matrix, there are two exceptions: succinate dehydrogenase, which catalyzes reaction 6, and the enormous complex that catalyzes reaction 4. These enzymes are integral membrane proteins of the inner mitochondrial membrane.

The Respiratory Chain: Electrons, Proton Pumping, and ATP

Without the oxidizing agents NAD$^+$ and FAD, the oxidative steps of glycolysis, pyruvate oxidation, and the citric acid cycle could not occur. Once reduced forms of these carriers have been produced, they must have

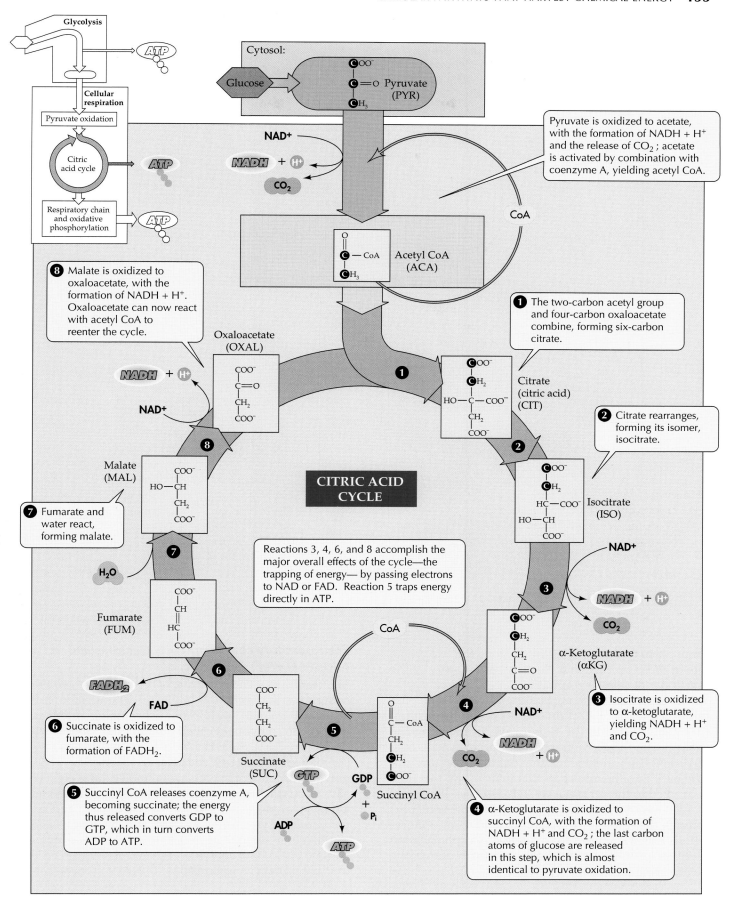

Glycolysis

ATP

Cellular respiration

Pyruvate oxidation

Citric acid cycle

ATP

Respiratory chain and oxidative phosphorylation

ATP

Cytosol:

Glucose → COO⁻ — C=O — CH₃ Pyruvate (PYR)

NAD⁺

NADH + H⁺

CO_2

Pyruvate is oxidized to acetate, with the formation of NADH + H⁺ and the release of CO_2; acetate is activated by combination with coenzyme A, yielding acetyl CoA.

CoA

C — CoA / CH₃ Acetyl CoA (ACA)

8 Malate is oxidized to oxaloacetate, with the formation of NADH + H⁺. Oxaloacetate can now react with acetyl CoA to reenter the cycle.

1 The two-carbon acetyl group and four-carbon oxaloacetate combine, forming six-carbon citrate.

Oxaloacetate (OXAL)

COO⁻ / C=O / CH₂ / COO⁻

NADH + H⁺

NAD⁺

8

Citrate (citric acid) (CIT)

COO⁻ / CH₂ / HO—C—COO⁻ / CH₂ / COO⁻

2 Citrate rearranges, forming its isomer, isocitrate.

2

Malate (MAL)

COO⁻ / HO—CH / CH₂ / COO⁻

CITRIC ACID CYCLE

Isocitrate (ISO)

COO⁻ / CH₂ / HC—COO⁻ / HO—CH / COO⁻

NAD⁺

7 Fumarate and water react, forming malate.

7

Reactions 3, 4, 6, and 8 accomplish the major overall effects of the cycle—the trapping of energy— by passing electrons to NAD or FAD. Reaction 5 traps energy directly in ATP.

3

NADH + H⁺

CO_2

H_2O

Fumarate (FUM)

COO⁻ / CH / HC / COO⁻

CoA

COO⁻ / CH₂ / CH₂ / C=O / COO⁻

α-Ketoglutarate (αKG)

6

FADH₂

FAD

Succinate (SUC)

COO⁻ / CH₂ / CH₂ / COO⁻

5

O / C — CoA / CH₂ / CH₂ / COO⁻

Succinyl CoA

4

NAD⁺

NADH

CO_2 + H⁺

3 Isocitrate is oxidized to α-ketoglutarate, yielding NADH + H⁺ and CO_2.

6 Succinate is oxidized to fumarate, with the formation of FADH₂.

5 Succinyl CoA releases coenzyme A, becoming succinate; the energy thus released converts GDP to GTP, which in turn converts ADP to ATP.

GTP

GDP + P$_i$

ADP

ATP

4 α-Ketoglutarate is oxidized to succinyl CoA, with the formation of NADH + H⁺ and CO_2; the last carbon atoms of glucose are released in this step, which is almost identical to pyruvate oxidation.

7.11 The Citric Acid Cycle Releases Much More Free Energy Than Glycolysis Does Electron carriers (NAD in glycolysis; NAD and FAD in the citric acid cycle) are reduced and ATP is generated in reactions coupled to other reactions producing major drops in free energy.

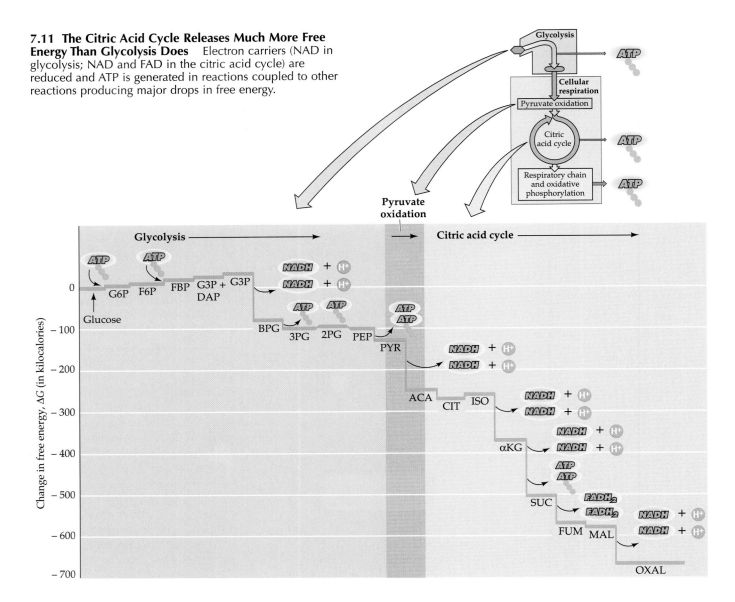

some place to donate their hydrogens (H$^+$ + e$^-$). The fate of these protons and electrons is the story of the remainder of cellular respiration.

The story has three parts: First, the electrons pass through a series of electron carriers called the respiratory chain. Second, the flow of electrons along the chain causes the active transport of protons across the inner mitochondrial membrane. Third, the protons diffuse back into the mitochondrial matrix, and this flow is coupled to the synthesis of ATP. The overall process of ATP synthesis resulting from electron transport through the respiratory chain is oxidative phosphorylation. The mechanism for oxidative phosphorylation involves proton transfer back and forth across the membrane and is called *chemiosmosis*.

The respiratory chain transports electrons and releases energy

The respiratory chain contains (1) three large protein complexes that span the inner mitochondrial membrane, (2) a small protein (cytochrome *c*), and (3) a nonprotein component called ubiquinone (abbreviated Q) (Figure 7.12). NADH + H$^+$ passes hydrogens to Q by way of **NADH-Q reductase**, a complex of 26 polypeptide subunits, with a total molecular weight of 850,000. **Cytochrome reductase**, with 10 subunits and a molecular weight of 280,000, lies between Q and cytochrome *c*. **Cytochrome oxidase**, with 8 subunits and a molecular weight of 160,000, lies between cytochrome *c* and oxygen. Different subunits within each of these protein complexes bear different electron carriers, so electrons are transported *within* each complex, as well as from complex to complex.

Cytochrome *c* is a peripheral membrane protein that lies in the space between the inner and outer mitochondrial membranes, loosely attached to the inner membrane. Ubiquinone is a small, nonpolar molecule that moves within the hydrophobic interior of the phospholipid bilayer of the inner membrane (see Figure 7.12).

Why should the respiratory chain have so many links? Why, for example, don't cells just use the following single step?

$$NADH + H^+ + \tfrac{1}{2} O_2 \rightarrow NAD^+ + H_2O$$

Wouldn't this accomplish the same thing, and more efficiently? To begin with, no enzyme will catalyze the direct oxidation of NADH + H⁺ by oxygen. More fundamentally, this would be an untamable reaction. It would be terrifically exergonic—rather like setting off a stick of dynamite in the cell. There is no biochemical way to harvest that burst of energy efficiently and put it to physiological use (that is, no metabolic reaction is so endergonic as to consume a significant fraction of that energy in a single step).

To control the release of energy during oxidation of glucose in a cell, evolution has produced the lengthy respiratory chain we observe today: a *series* of reactions, each releasing a small, manageable amount of energy. Electron transport within each of the three protein complexes results, as we'll see, in the pumping of protons across the inner mitochondrial membrane, and the return of the protons across the membrane leads to the formation of ATP. Thus the energy originally contained in glucose and other foods is finally tucked into the cellular energy currency, ATP. For each pair of electrons passed along the respiratory chain from NADH + H⁺ to oxygen, three molecules of ATP are formed.

The electron carriers of the respiratory chain (including those contained in the three protein complexes) differ as to how they change when they become reduced. NAD⁺, for example, accepts H⁻ (a hydride ion—one proton and two electrons), leaving the proton from the other hydrogen atom to float free: NADH + H⁺. Other carriers, including Q, bind both protons and both electrons, becoming, for example, QH₂. The remainder of the chain, however, is only an electron transport process. Electrons, but not protons, are passed

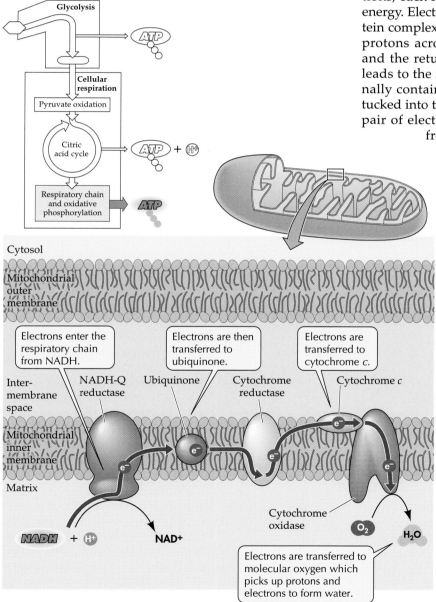

7.12 The Oxidation of NADH + H⁺ Electrons from NADH + H⁺ are passed through the respiratory chain, a series of carrier molecules in the inner mitochondrial membrane (or the plasma membrane of an aerobic prokaryote). The carriers gain free energy when they become reduced and release free energy when they are oxidized.

from Q to cytochrome *c*. An electron from QH$_2$ reduces a cytochrome's Fe^{3+} to Fe^{2+}.

Electrons pour into the pool of Q molecules from the NADH + H$^+$ pathway, or they can come from another source: the succinate-to-fumarate reaction of the citric acid cycle (reaction 6 in Figure 7.10). Another protein complex, **succinate-Q reductase**, links the oxidation of succinate to the reduction of Q (Figure 7.13). The enzyme that constitutes the first part of succinate-Q reductase has attached to it an FAD carrier molecule, which is reduced by succinate to FADH$_2$. Later, hydrogen atoms are transferred to the Q molecules. No protons are pumped and hence no ATP is generated in the succinate-to-Q branch of the respiratory chain.

If only electrons are carried through the final reactions of the respiratory chain, what happens to H$^+$? And how is electron transport coupled to ATP production?

Active proton transport is followed by diffusion coupled to ATP synthesis

All the carriers and enzymes of the respiratory chain (except cytochrome *c*) are embedded in the inner mito-

chondrial membrane (see Figure 7.12). The operation of the respiratory chain results in the transport of protons (H$^+$), against their concentration gradient, across the inner membrane of the mitochondrion from inside to outside ("outside" being the space between the two mitochondrial membranes).

Three integral membrane protein complexes of the inner mitochondrial membrane (NADH-Q reductase, cytochrome reductase, and cytochrome oxidase) accomplish this active transport of protons across the membrane (Figure 7.14; see also Figure 7.12). Because of the charge on the proton (H$^+$), this transport also causes a difference in electric charge across the membrane. Together, the proton concentration gradient and the charge difference constitute a **proton-motive force** that tends to drive the protons back across the membrane, just as the charge on a battery drives the flow of electrons, discharging the battery.

The discharge of the proton-motive force is prevented by the proton impermeability of the phospholipid bilayer of the inner mitochondrial membrane. Protons can diffuse across the membrane only by passing through specific channel proteins, called **ATP synthases**, that couple proton movement to the synthesis of ATP from ADP and P$_i$. This coupling by an ATP synthase of the exergonic discharge of the proton-motive force to the endergonic formation of ATP is called the **chemiosmotic mechanism**. The ATP synthase complex is an integral membrane protein that spans the phospholipid bilayer of the inner mitochondrial membrane. The catalytic part of the complex is visible in electron micrographs as large knobs protruding from the inner mitochondrial membrane into the matrix (see Figure 7.14).

THE CHEMIOSMOTIC MECHANISM SUMMARIZED. The enzymes and other electron carriers of the respiratory chain are part of the inner mitochondrial membrane or are closely associated with it. The flow of electrons down the respiratory chain is an exergonic process. At the beginning of the chain, the electrons are at a higher energy than at its end. As the electrons lose energy, some of the energy is captured by proton pumps that actively transport H$^+$ across the inner mitochondrial membrane, whose phospholipid bilayer is impermeable to protons.

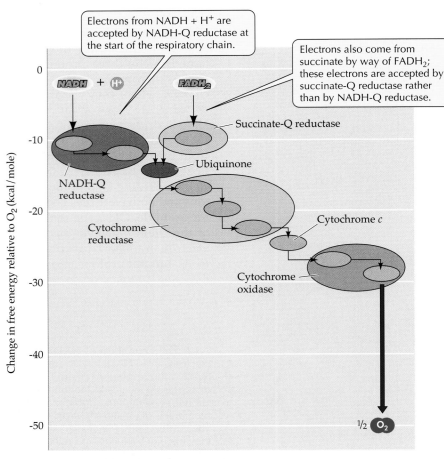

7.13 The Complete Respiratory Chain Electrons enter the chain from two sources, but they follow the same path from ubiquinone on.

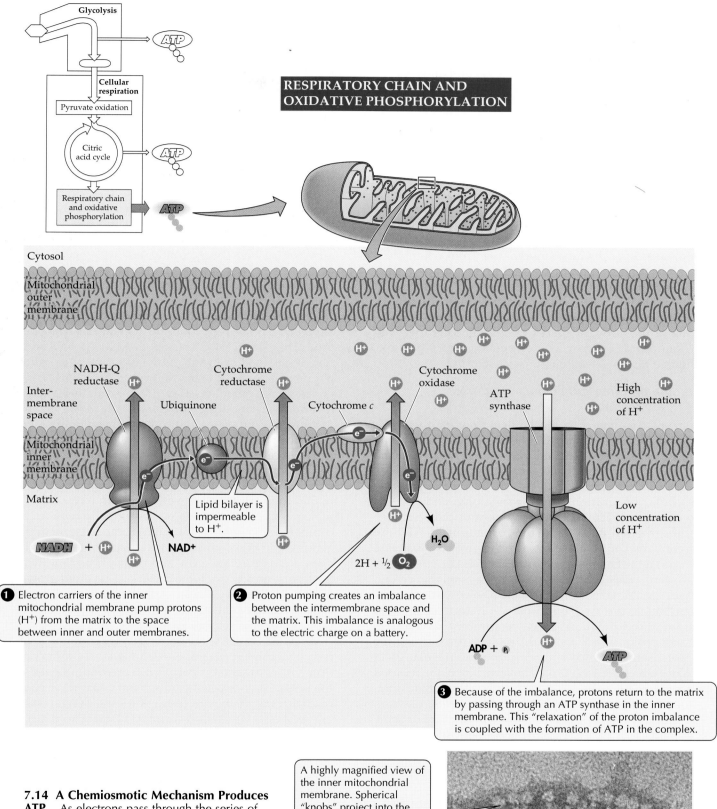

RESPIRATORY CHAIN AND OXIDATIVE PHOSPHORYLATION

Glycolysis

ATP

Cellular respiration

Pyruvate oxidation

Citric acid cycle

ATP

Respiratory chain and oxidative phosphorylation

ATP

Cytosol

Mitochondrial outer membrane

Inter-membrane space

Mitochondrial inner membrane

Matrix

NADH-Q reductase

Ubiquinone

Cytochrome reductase

Cytochrome c

Cytochrome oxidase

ATP synthase

H+

High concentration of H+

Low concentration of H+

NADH + H+

NAD+

Lipid bilayer is impermeable to H+.

$2H + \frac{1}{2}$ O$_2$

H$_2$O

ADP + P$_i$

ATP

1 Electron carriers of the inner mitochondrial membrane pump protons (H+) from the matrix to the space between inner and outer membranes.

2 Proton pumping creates an imbalance between the intermembrane space and the matrix. This imbalance is analogous to the electric charge on a battery.

3 Because of the imbalance, protons return to the matrix by passing through an ATP synthase in the inner membrane. This "relaxation" of the proton imbalance is coupled with the formation of ATP in the complex.

7.14 A Chemiosmotic Mechanism Produces ATP As electrons pass through the series of carriers in the respiratory chain, protons are pumped from the mitochondrial matrix to the intermembrane space. As the protons return to the matrix through an ATP synthase, ATP forms.

A highly magnified view of the inner mitochondrial membrane. Spherical "knobs" project into the mitochondrial matrix; these knobs catalyze the synthesis of ATP.

This active transport establishes a proton concentration gradient. The potential energy of this proton gradient is used to synthesize ATP from ADP and P_i when protons diffuse back into the mitochondrial interior through the membrane protein ATP synthase.

An experiment demonstrates the chemiosmotic mechanism

According to the chemiosmotic model, one would expect that the mitochondrion could be "fooled" into making more ATP by the following clever trick. A sample of isolated mitochondria is maintained in a solution at pH 8 (slightly basic) until it is fully adapted; then suddenly the mitochondria are transferred into a second solution at pH 4 (fairly acidic) containing ADP and P_i. This transfer should lead to an excess of protons on the outside of the inner membrane, from where they should be able to proceed through the proposed ATP synthase channels, causing a burst of ATP production.

And that's exactly what happens. This acid-induced ATP production by isolated mitochondria stands as one of the stronger pieces of evidence favoring the chemiosmotic theory as the explanation for how oxidative phosphorylation proceeds in cells.

Proton diffusion can be uncoupled from ATP production

For the chemiosmotic mechanism to work, the phospholipid bilayer of the mitochondrial inner membrane must be quite impermeable to protons; otherwise, the protons would leak back across the bilayer as fast as they were pumped out by the respiratory chain. Cellular respiration would race along, and no ATP would form. The mechanism operates only because the diffusion of H^+ and the formation of ATP are tightly *coupled*: To move inward, the protons must pass through the channel protein ATP synthase.

Many years ago, biochemists discovered that certain hydrophobic compounds that are also very weak acids, such as *dinitrophenol*, uncouple respiratory metabolism from ATP formation. These **respiratory uncouplers** carry protons back across the bilayer, discharging the proton-motive force without capturing the energy in ATP. As a result, food is metabolized, but all the released energy is lost as heat. One naturally occurring respiratory uncoupler, a protein called *thermogenin*, plays an important role in regulating the temperature of some mammals: In "brown fat," the uncoupling of respiration raises the body temperature, as mentioned at the beginning of this chapter (see Chapter 37 for more detail).

Fermentation: ATP from Glucose, without O_2

Suppose that the supply of oxygen to a respiring cell is cut off, perhaps by drowning or by extreme exertion. As we can see in Figure 7.14, the first consequence of an insufficient supply of O_2 is that the cell cannot reoxidize cytochrome *c*, so all of that compound is soon in the reduced form. When this happens, there is no oxidizing agent to reoxidize QH_2, and soon all the Q is in the reduced form. So it goes, until the entire respiratory chain is reduced. Under these circumstances, no NAD^+ and no FAD are generated from their reduced forms; therefore, the oxidative steps in glycolysis, pyruvate oxidation, and the citric acid cycle also stop. If the cell has no other way to obtain energy from its food, it will die.

Some cells, such as muscle cells, have the enzymes necessary for fermentation and can thus switch from aerobic metabolism to fermentation. Fermentation has two defining characteristics. First, a fermentative reaction uses $NADH + H^+$ to reduce pyruvate or one of its metabolites, and consequently NAD^+ is regenerated. Once the cell has some NAD^+, it can carry more glucose through glycolysis (that is, through the early steps of fermentation).

The second characteristic of fermentation is that, by allowing glycolysis to continue, it enables a sustained production of ATP. Only as much ATP is produced as can be obtained from substrate-level phosphorylation in glycolysis—not the much greater yield of ATP obtained by the operation of pyruvate oxidation, the citric acid cycle, and the respiratory chain. However, this amount of ATP is enough to keep the cell alive, because the rate of glycolysis increases.

When cells capable of fermentation become anaerobic, the rate of glycolysis speeds up 10-fold or even more. Thus a substantial rate of ATP production is maintained, although the efficiency in terms of ATP molecules per glucose molecule is greatly reduced as compared to aerobic respiration. Some bacteria of the genus *Clostridium*, while growing anaerobically in the presence of glucose, grow and multiply as rapidly as the fastest-growing aerobic bacteria. This rapid growth is made possible by the fact that the *Clostridium* bacteria are running the glycolytic reactions much more rapidly than the aerobes do.

As noted earlier, some organisms are confined to totally anaerobic environments and use only fermentation. Other organisms carry on fermentation even in the presence of oxygen. And several bacteria carry on *cellular respiration*—not fermentation—*without using oxygen gas as an electron acceptor*. Instead, to oxidize their cytochromes these bacteria reduce nitrate ions (NO_3^-) to nitrite ions (NO_2^-). We'll put that observation into a broader context in Chapter 25.

Some fermenting cells produce lactic acid

Many different types of fermentation are found in different bacteria and some body cells. These different fermentations are identified by the final product produced. For example, in *lactic acid fermentation*, pyru-

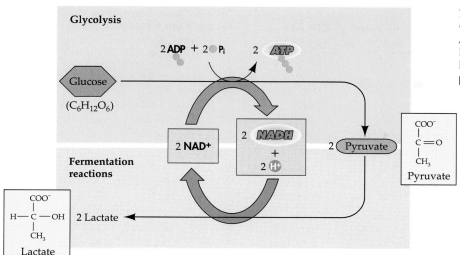

7.15 Lactic Acid Fermentation
Glycolysis produces pyruvate, as well as ATP and NADH + H⁺, from glucose. Lactic acid fermentation, using NADH + H⁺ as the reducing agent, then reduces pyruvate to lactic acid (lactate).

vate is reduced to lactic acid (Figure 7.15). Lactic acid fermentation takes place in many microorganisms and in our muscle cells. Unlike muscle cells, however, nerve cells (neurons) are incapable of fermentation because they lack the enzyme that reduces pyruvate to lactate. For this reason, without adequate oxygen our nervous system (including the brain) is rapidly destroyed; it is the first part of the body to die.

Other fermenting cells produce alcohol

Certain yeasts and some plant cells in anaerobic conditions carry on a process called **alcoholic fermentation** (Figure 7.16). This process requires two enzymes to metabolize pyruvate. First, carbon dioxide is removed from pyruvate, leaving the compound *acetaldehyde.* Second, the acetaldehyde is reduced by NADH + H⁺, producing NAD⁺ and *ethyl alcohol* (ethanol). Remember that recycling NAD in glycolysis (reaction 6 in Figure

7.6) allows the fermenting cell to produce ATP by substrate-level phosphorylation (see Figure 7.8). The brewing industry relies on alcoholic fermentation to produce wine and beer.

Contrasting Energy Yields

The total yield of stored energy from fermentation is two molecules of ATP per molecule of glucose oxidized. In contrast, the maximum yield that can be obtained from glycolysis followed by complete aerobic respiration of a molecule of glucose is much greater— about 36 molecules of ATP (Figure 7.17). (Study Figures 7.6, 7.10, and 7.14 to review where the ATP molecules come from.)

Why is so much more ATP produced by aerobic respiration? Because carriers (mostly NAD⁺) are reduced in pyruvate oxidation and the citric acid cycle and then oxidized by the respiratory chain, with the accompanying production of ATP (three for each NADH + H⁺ and two for each FADH₂) by the chemiosmotic mechanism. In an aerobic environment, a species capable of this type of metabolism will be at an advantage (in terms of energy availability per glucose molecule) over one limited to fermentation.

The total gross yield of ATP from one molecule of glucose taken through glycolysis and respiration is 38. However, we must subtract two from that gross—for a net yield of 36

7.16 Alcoholic Fermentation In alcoholic fermentation (the basis for the brewing industry), pyruvate from glycolysis is converted to acetaldehyde, and CO_2 is released. The NADH + H⁺ from glycolysis acts as a reducing agent, reducing acetaldehyde to ethanol.

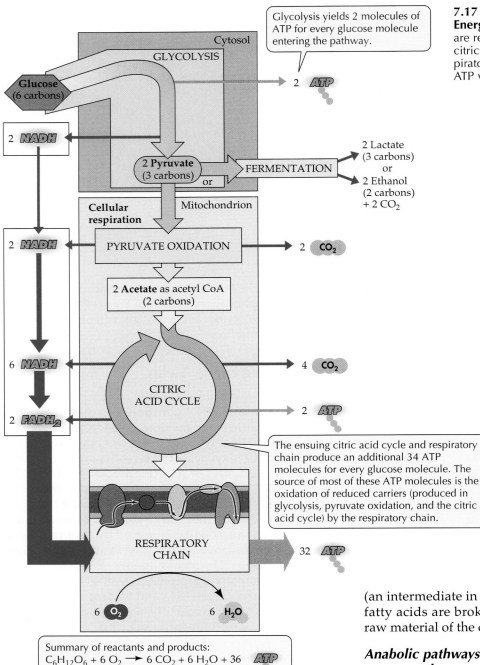

Glycolysis yields 2 molecules of ATP for every glucose molecule entering the pathway.

7.17 Cellular Respiration Yields More Energy Than Gylcolysis Does Carriers are reduced in pyruvate oxidation and the citric acid cycle, then oxidized by the respiratory chain. These reactions produce ATP via the chemiosmotic mechanism.

The ensuing citric acid cycle and respiratory chain produce an additional 34 ATP molecules for every glucose molecule. The source of most of these ATP molecules is the oxidation of reduced carriers (produced in glycolysis, pyruvate oxidation, and the citric acid cycle) by the respiratory chain.

Summary of reactants and products:
$$C_6H_{12}O_6 + 6\ O_2 \rightarrow 6\ CO_2 + 6\ H_2O + 36\ ATP$$

an interchange, with traffic flowing in both directions. These energy pathways are connected to both catabolic (degradative) and anabolic (synthetic) pathways.

Catabolic pathways feed into respiratory metabolism

In addition to glucose, other sugars, fats, and even proteins can serve as the starting materials for respiratory ATP production. Other monosaccharides may be used, after being converted to glucose. Polymers such as starch and glycogen are **digested** (hydrolyzed) to glucose and subsequently metabolized to yield ATP. Fats are first digested to yield glycerol and fatty acids (see Chapter 3). The glycerol is then readily converted to glyceraldehyde 3-phosphate (an intermediate in glycolysis; see Figure 7.6), and the fatty acids are broken down to form acetyl CoA (the raw material of the citric acid cycle; see Figure 7.10).

Anabolic pathways use intermediates from energy pathways

Each of the reactions just described also operates in reverse. Thus, in the synthesis of fats, fatty acids form from acetyl CoA and glycerol forms from glyceraldehyde 3-phosphate. These reverse reactions occur only when the cell has an adequate energy supply; otherwise the acetyl CoA would be needed strictly for the citric acid cycle and ATP formation. However, with an abundant supply of glucose, the cell can divert some acetyl CoA to fatty acid production and some glyceraldehyde 3-phosphate to glycerol formation. The fat that forms on our bodies, adding baggage, is a result of this diversion.

ATP—because the inner mitochondrial membrane is impermeable to NADH, and a "toll" of one ATP must be paid for each NADH (produced in glycolysis) that is shuttled into the mitochondrial matrix.

Connections with Other Pathways

Glycolysis and the respiratory pathways do not operate in isolation from the rest of metabolism. Rather, there is

Some intermediates of the citric acid cycle are used in the synthesis of various important cellular constituents. Succinyl CoA (succinyl coenzyme A) is a starting point for chlorophyll synthesis, and α-ketoglutarate and oxaloacetate are starting materials for the synthesis of certain amino acids required for protein production. Still other amino acids derive from pyruvate, from glycolysis. Acetyl CoA has many different fates: In addition to its role in fatty acid production, it is a building block for various pigments, plant growth substances, rubber, and the steroid hormones of animals—among other functions.

Regulating the Energy Pathways

Glycolysis, cellular respiration, and fermentation are energy-harvesting pathways. Cells use the ATP produced by these pathways to do work: biosynthesis of macromolecules and other important compounds, active transport of materials across membranes, and movement of cellular components and entire cells. This work constitutes the life of the cell. To accomplish these activities, the cell must regulate its energy metabolism.

When a yeast cell switches from aerobic respiration to anaerobic fermentation at low concentrations of oxygen, it must metabolize glucose 18 times faster to obtain the same amount of energy. But as soon as aerobic respiration begins again, glycolysis in the yeast cell slows down. The amount of glucose used is only as much as is needed for energy production under the existing conditions, anaerobic or aerobic. The slowing down of glycolysis in response to the resumption of aerobic respiration is called the **Pasteur effect**, after its discoverer, Louis Pasteur. What is the mechanism that slows down glycolysis when the respiratory chain begins to operate?

Allostery regulates respiratory metabolism

Glycolysis, the citric acid cycle, and the respiratory chain are regulated by allosteric control of the enzymes involved (see Chapter 6). Some products of later reactions, if they are in oversupply, can suppress the action of enzymes that catalyze earlier reactions. On the other hand, an excess of the products of one branch of a synthetic chain can speed up reactions in another branch, diverting raw materials away from synthesis of the former (Figure 7.18).These negative and positive feedback control mechanisms are used at many points in the energy-extracting processes, which are summarized in Figure 7.19.

The main control point in glycolysis is the enzyme *phosphofructokinase* (reaction 3 in Figure 7.6). This enzyme is allosterically inhibited by ATP and activated by ADP or AMP. As long as fermentation proceeds, yielding a relatively small amount of ATP, phospho-

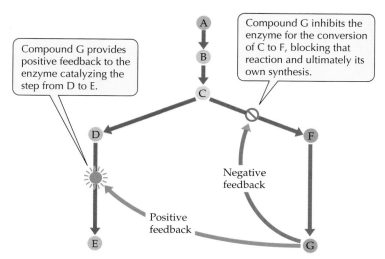

Compound G provides positive feedback to the enzyme catalyzing the step from D to E.

Compound G inhibits the enzyme for the conversion of C to F, blocking that reaction and ultimately its own synthesis.

Negative feedback

Positive feedback

7.18 Regulation by Negative and Positive Feedback Allosteric regulation plays an important role in respiratory metabolism. Excess accumulation of some products can shut down their continued synthesis or stimulate the synthesis of other products.

fructokinase operates at full efficiency. But when aerobic respiration begins producing ATP 18 times faster than before, the excess ATP allosterically inhibits the conversion of fructose 6-phosphate to fructose 1,6-bisphosphate, and the rate of glucose utilization drops.

The main control point in the citric acid cycle is the enzyme *isocitrate dehydrogenase*, which converts isocitrate to α-ketoglutarate (reaction 3 in Figure 7.10). ATP and NADH + H$^+$ are feedback inhibitors of this reaction; ADP and NAD$^+$ are activators. If too much ATP is accumulating, or if NADH + H$^+$ is being produced faster than it can be used by the respiratory chain, the isocitrate reaction is almost completely blocked and the citric acid cycle is essentially shut down. A shutdown of the citric acid cycle would cause large amounts of isocitrate and citrate to accumulate, except that the conversion of acetyl CoA to citrate is also slowed by ATP and NADH + H$^+$.

The negative effects of halting the isocitrate reaction are thus spread backward through the chain of reactions. However, a certain excess of citrate does accumulate, and this excess acts as a negative-feedback inhibitor to slow the fructose 6-phosphate reaction early in glycolysis. Consequently, if the citric acid cycle has been slowed down because of an excess of ATP (and not because of a lack of oxygen), glycolysis is shut down as well. Both processes resume when the ATP level falls and they are needed. Allosteric control keeps the process in balance.

Another control point in Figure 7.19 involves a method for storing excess acetyl CoA. If too much ATP is being made and the citric acid cycle shuts down, the accumulation of citrate switches acetyl CoA to the syn-

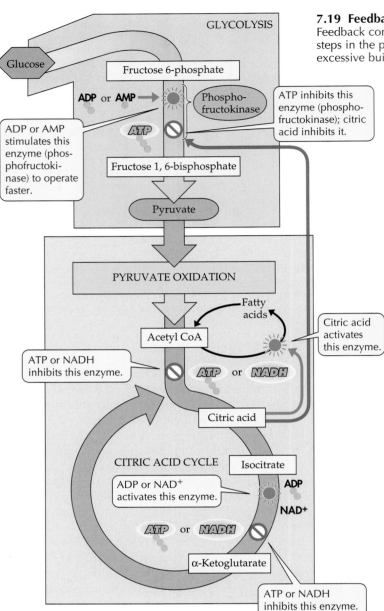

GLYCOLYSIS

Glucose

Fructose 6-phosphate

ADP or **AMP** →

Phospho-fructokinase

ATP

ADP or AMP stimulates this enzyme (phos-phofructoki-nase) to operate faster.

ATP inhibits this enzyme (phospho-fructokinase); citric acid inhibits it.

Fructose 1, 6-bisphosphate

Pyruvate

PYRUVATE OXIDATION

Fatty acids

Acetyl CoA

Citric acid activates this enzyme.

ATP or NADH inhibits this enzyme.

ATP or NADH

Citric acid

CITRIC ACID CYCLE

Isocitrate

ADP or NAD+ activates this enzyme.

ADP

NAD+

ATP or NADH

α-Ketoglutarate

ATP or NADH inhibits this enzyme.

7.19 Feedback Regulation of Glycolysis and the Citric Acid Cycle

Feedback controls glycolysis and the citric acid cycle at crucial early steps in the pathways, increasing their efficiency and preventing the excessive buildup of intermediates.

the energy-processing pathways present in cells today, glycolysis and fermentation are the most ancient. These pathways appeared when the planetary environment was strictly anaerobic and all life was prokaryotic—before there were cells with membrane-enclosed organelles. To this day, the enzymes of glycolysis and fermentation are located in the cytoplasm.

Eventually some cells gained the capacity to perform photosynthesis, a form of which added O_2 to the atmosphere, rendering most environments aerobic. Evolution in the aerobic environment led to the appearance of pyruvate oxidation and the citric acid cycle. Elaboration of membranes, especially in eukaryotic cells, allowed the evolution of chemiosmotic mechanisms for coupling electron transport to ATP production, as in oxidative phosphorylation.

There have also been evolutionary refinements within the pathways themselves. Eukaryotic cells, but not prokaryotic ones, have a cytoskeleton based on microtubules and actin microfilaments (see Figure 4.23). With the appearance of a cytoskeleton, some glycolytic enzymes became attached to cytoskeletal components, which thus organized the enzymes into efficient associations that allow molecules to move from one enzyme in the pathway to the next. In eukaryotes hexokinase, the first glycolytic enzyme, binds to the outer mitochondrial membrane, giving the enzyme immediate access to ATP produced within the mitochondria. In metabolism as in all the rest of biology, evolution leads to adaptation.

thesis of fatty acids for storage. These fatty acids may be metabolized later to produce more acetyl CoA.

Evolution has led to metabolic efficiency

Allosteric control of the sort illustrated in Figure 7.19 is one of the most impressive examples of the tight organization that can evolve through natural selection when selection favors efficient operation in the competition among organisms for limited resources. Each of the feedback controls regulates a part or various parts of the energy-releasing pathways and keeps them operating in harmony and balance.

Not just the regulatory systems have evolved; the pathways themselves are the products of evolution. Of

Summary of "Cellular Pathways That Harvest Chemical Energy"

Obtaining Energy and Electrons from Glucose

• When glucose burns, energy is released as heat and light: $C_6H_{12}O_6 + 6 O_2 \rightarrow 6 CO_2 + 6 H_2O + energy$. The same equation applies to the complete aerobic metabolism of glucose by cells, but the reaction is accomplished in many separate steps so that the energy can be captured as ATP for cellular work. **Review Figure 7.1**

• As a material is oxidized, the electrons it loses are transferred to another material, which is thereby reduced. Redox reactions transfer large amounts of energy. Much of the energy liberated in an oxidation is captured in the reduction of the oxidizing agent, such as NAD^+ or FAD. **Review Figures 7.2, 7.3, 7.4**

An Overview: Releasing Energy from Glucose

• Glycolysis is the most ancient energy pathway and operates in the presence or absence of O_2. Glycolysis breaks glucose down into two pyruvate molecules and produces two ATPs. Under aerobic conditions, cellular respiration continues the breakdown process. **Review Figure 7.5**
• Pyruvate oxidation and the citric acid cycle produce CO_2 and considerable numbers of hydrogen atoms carried by NADH and $FADH_2$. The respiratory chain combines these hydrogens with O_2, releasing enough energy for the synthesis of substantial amounts of ATP. **Review Figure 7.5**
• In some cells under anaerobic conditions, pyruvate can be reduced by NADH to form lactate and regenerate the NAD needed to sustain glycolysis. **Review Figure 7.5**
• In eukaryotes, glycolysis and fermentation take place in the cytoplasm outside of the mitochondria; pyruvate oxidation, the citric acid cycle, and the respiratory chain operate in association with mitochondria. In prokaryotes, glycolysis, fermentation, and the citric acid cycle take place in the cytoplasm; and pyruvate oxidation and the respiratory chain operate in association with the plasma membrane. **Review Table 7.1**

Glycolysis: From Glucose to Pyruvate

• Glycolysis is a pathway of 10 enzyme-catalyzed reactions located in the cytoplasm. Glycolysis provides starting materials for both cellular respiration and fermentation. **Review Figure 7.6**
• Energy-investing reactions use two ATPs per glucose molecule and eventually yield two glyceraldehyde 3-phosphate molecules. In the energy-harvesting reactions, two NADH molecules are produced, and four ATP molecules are generated by substrate-level phosphorylation. Two pyruvates are produced for each glucose molecule. **Review Figures 7.6, 7.7, 7.8**

Pyruvate Oxidation

• The pyruvate dehydrogenase complex, an assemblage of many polypeptide subunits, catalyzes three reactions: (1) Pyruvate is oxidized to acetate, releasing one CO_2 molecule and considerable energy. (2) Some of this energy is captured when NAD^+ is reduced to NADH + H^+. (3) The remaining energy is captured when acetate is combined with coenzyme A, yielding energy-rich acetyl CoA. **Review Figure 7.9**

The Citric Acid Cycle

• The energy in acetyl CoA drives the reaction of acetate with oxaloacetate to produce citric acid. The citric acid cycle is a series of reactions in which citrate is oxidized and oxaloacetate regenerated (hence a "cycle"). It produces two CO_2, one $FADH_2$, three NADH, and one ATP for each acetyl CoA. **Review Figures 7.10, 7.11**

The Respiratory Chain: Electrons, Proton Pumping, and ATP

• NADH + H^+ and $FADH_2$ from glycolysis, pyruvate oxidation, and the citric acid cycle are oxidized by the respiratory chain, regenerating NAD^+ and FAD. With the exception of cytochrome *c*, the enzymes and other electron carriers of the respiratory chain are part of the inner mitochondrial membrane. Oxygen (O_2) is the final acceptor of electrons and protons, forming water (H_2O). **Review Figures 7.12, 7.13**
• The chemiosmotic mechanism couples proton transport to oxidative phosphorylation. As the electrons move along the respiratory chain, they lose energy, which is captured by proton pumps that actively transport H^+ out of the mitochondrion, establishing a gradient of both proton concentration and electric charge—the proton-motive force. This force is used to produce ATP when protons diffuse back into the mitochondrial interior through the membrane protein ATP synthase. The phospholipid bilayer of the inner membrane is impermeable to protons. **Review Figure 7.14**

Fermentation: ATP from Glucose, without O_2

• Many organisms and some cells live without O_2, deriving all their energy from glycolysis and fermentation reactions. Together, these pathways partly oxidize glucose and generate energy-containing waste molecules such as lactic acid or ethanol. Fermentation reactions anaerobically oxidize the NADH + H^+ produced in glycolysis. Some fermenting cells produce lactic acid; others produce ethanol and CO_2. **Review Figures 7.15, 7.16**

Contrasting Energy Yields

• For each molecule of glucose used, fermentation yields 2 molecules of ATP. In contrast, glycolysis operating with pyruvate oxidation, the citric acid cycle, and the respiratory chain yields up to 36 molecules of ATP per molecule of glucose. **Review Figure 7.17**

Connections with Other Pathways

• Catabolic pathways feed into respiratory metabolism. Starch and glycogen produce glucose. Other monosaccharides can be converted to glucose for respiratory metabolism. Glycerol from fats enters glycolysis; acetate from fatty acid degradation forms acetyl CoA to enter the citric acid cycle.
• Anabolic pathways use intermediate components of respiratory metabolism to synthesize fats, amino acids, and other essential building blocks for cellular structure and function.

Regulating the Energy Pathways

• When aerobic respiration is operating, glycolysis slows down (the Pasteur effect). The rates of glycolysis and the citric acid cycle are increased or decreased by the actions of ATP, ADP, NAD^+, or NADH + H^+ on allosteric enzymes.
• Inhibition of the glycolytic enzyme phosphofructokinase by abundant ATP from oxidative phosphorylation slows down glycolysis. ADP activates this enzyme, speeding up glycolysis. The citric acid cycle enzyme isocitrate dehydrogenase is inhibited by ATP and NADH and activated by ADP and NAD^+. **Review Figures 7.18, 7.19**

Self-Quiz

1. Which statement about ATP is *not* true?
   a. It is formed only under aerobic conditions.
   b. It is used as an energy currency by all cells.
   c. Its formation from ADP and P_i is an endergonic reaction.
   d. It provides the energy for many different biochemical reactions.
   e. Some ATP is used to drive the synthesis of storage compounds.

2. Oxidation and reduction
   a. entail the gain or loss of proteins.
   b. are defined as the loss of electrons.
   c. are both endergonic reactions.
   d. always occur together.
   e. proceed only under aerobic conditions.

3. NAD^+
   a. is a type of organelle.
   b. is a protein.
   c. is an oxidizing agent.
   d. is a reducing agent.
   e. is formed only under aerobic conditions.

4. Glycolysis
   a. takes place in the mitochondrion.
   b. produces no ATP.
   c. has no connection with the respiratory chain.
   d. is the same thing as fermentation.
   e. reduces two molecules of NAD^+ for every glucose molecule processed.

5. Fermentation
   a. takes place in the mitochondrion.
   b. takes place in all animal cells.
   c. does not require O_2.
   d. requires lactic acid.
   e. prevents glycolysis.

6. Which statement about pyruvate is *not* true?
   a. It is the end product of glycolysis.
   b. It becomes reduced during fermentation.
   c. It is a precursor of acetyl CoA.
   d. It is a protein.
   e. It contains three carbon atoms.

7. The citric acid cycle
   a. takes place in the mitochondrion.
   b. produces no ATP.
   c. has no connection with the respiratory chain.
   d. is the same thing as fermentation.
   e. reduces two molecules of NAD^+ for every glucose molecule processed.

8. Which statement about the respiratory chain is *not* true?
   a. It operates in the mitochondrion.
   b. It uses O_2 as an oxidizing agent.
   c. It leads to the production of ATP.
   d. It regenerates oxidizing agents for glycolysis and the citric acid cycle.
   e. It operates simultaneously with fermentation.

9. Which statement about the chemiosmotic mechanism is *not* true?
   a. Protons are pumped across a membrane.
   b. Protons return through the membrane by way of a channel protein.
   c. ATP is required for the protons to return.
   d. Proton pumping is associated with the respiratory chain.
   e. The membrane in question is the inner mitochondrial membrane.

10. Which statement about oxidative phosphorylation is *not* true?
   a. It is the formation of ATP during operation of the respiratory chain.
   b. It is brought about by the chemiosmotic mechanism.
   c. It requires aerobic conditions.
   d. In eukaryotes, it takes place in mitochondria.
   e. Its functions can be served equally well by fermentation.

Applying Concepts

1. Trace the sequence of chemical changes that occurs in mammalian brain tissue when the oxygen supply is cut off. (The first change is that the cytochrome oxidase system becomes totally reduced, because electrons can still flow from cytochrome *c* but there is no oxygen to accept electrons from cytochrome oxidase. What are the remaining steps?)

2. Trace the sequence of chemical changes that occurs in mammalian *muscle* tissue when the oxygen supply is cut off. (The first change is exactly the same as that in Question 1.)

3. Some cells that use the citric acid cycle and the respiratory chain can also thrive by using fermentation under anaerobic conditions. Given the lower yield of ATP (per molecule of glucose) in fermentation, why can these cells function so efficiently under anaerobic conditions?

4. Describe the mechanisms by which the rates of glycolysis and of aerobic respiration are kept in balance with one another.

Readings

Alberts, B., D. Bray, J. Lewis, M. Raff, K. Roberts and J. D. Watson. 1994. *Molecular Biology of the Cell*, 3rd Edition. Garland Publishing, New York. Chapter 14 develops the themes introduced in this chapter; Chapter 2 is also useful as an introduction.

Hinkle, P. C. and R. E. McCarty. 1978. "How Cells Make ATP." *Scientific American*, March. Discussion of the chemiosmotic mechanism, in which ATP is formed by protons passing back through a membrane after being pumped out by the respiratory chain.

Lehninger, A. L., D. L. Nelson, and M. M. Cox. 1993. *Principles of Biochemistry*, 2nd Edition. Worth, New York. A balanced textbook stressing the themes of energy, evolution, regulation, and structure–function.

Lodish, H., D. Baltimore, A. Berk, S. L. Zipursky, P. Matsudaira and J. Darnell. 1995. *Molecular Cell Biology*, 3rd Edition. Scientific American Books, New York. Chapter 17 gives an excellent, more detailed treatment of the topics in this chapter.

Stryer, L. 1995. *Biochemistry*, 4th Edition. W. H. Freeman, New York. Although more advanced than this chapter, the section on glycolysis and respiration is straightforward and does not demand an advanced knowledge of chemistry.

Voet, D. and J. G. Voet. 1995. *Biochemistry*, 2nd Edition. John Wiley & Sons, New York. A general textbook with a full discussion of energy, enzymes, and catalysis. Outstanding illustrations.

Chapter 8

Photosynthesis: Energy from the Sun

Star of Life
Light energy from the sun sustains life on Earth.

We are creatures of the sun. Its light is the source—direct or indirect—of the free energy that powers life on Earth. This is the important message—perhaps already familiar to you—about photosynthesis.

As biologists, however, we might express things differently. We might say, "We are creatures of the chloroplasts." This abundant organelle, which gives green plants their color, captures the sun's energy for us. Inside plants, within stacks of connected, inner membranes that are wrapped in two outer membranes, chemical reactions convert and store the energy that we draw upon. **Photosynthesis** is the conversion of light energy to chemical energy. This process has profound significance for living systems on Earth.

Our own dependence on the sun, although absolute, is indirect. Like the other animals, the fungi, many protists, and most prokaryotes, humans depend on a ready supply of partly reduced, carbon-containing compounds as a food source. From such compounds we obtain all the free energy that keeps us alive and functioning. These compounds are also the source of the carbon atoms used in every organic molecule in our bodies. In a word, we are *heterotrophs*: We need to feed on something else. In a world populated exclusively by heterotrophs, all life would grind to an end as the food gradually disappeared. For all life to be self-sustained, some organisms must make food for others.

Our world owes the continued existence of life to the presence of *autotrophs*—organisms that do not need previously formed organic substances from their environment. For autotrophs, an energy source (such as light) and an inorganic carbon source (such as carbon dioxide gas) suffice as a diet. From these simple ingredients, autotrophs make the reduced carbon compounds from which their

165

bodies are built and their food needs met. By feeding on autotrophs, the heterotrophs of the world meet their needs for energy and matter.

The principal autotrophs are photosynthetic organisms that use visible light as their energy source. From light, carbon dioxide, and water, they begin the chemistry that sustains almost the entire biosphere. The worldwide extent of photosynthetic activity is stunning: Each year, tens of billions of tons of carbon atoms are taken from carbon dioxide and incorporated into molecules of sugars, amino acids, and the other vital compounds of the biosphere.

In this chapter, we will examine the details of photosynthesis. What are the overall reactants and products of this process? What is the nature of light, how does it interact with chlorophyll, and what chemical changes does it cause? We'll examine the light-dependent production of ATP and reducing agents. Then we'll turn to the light-*independent* use of these chemical resources for the fixation of carbon dioxide and its reduction to sugars such as glucose. How are the pigments and enzymes for these processes arranged on the thylakoid membranes of the chloroplast? We close the chapter by looking at a key enzyme with a strange property that limits the effectiveness of photosynthesis, examining how different groups of plants have adapted to this limitation.

Identifying Photosynthetic Reactants and Products

By the beginning of the nineteenth century, scientists understood the broad outlines of photosynthesis. It was known to use three principal ingredients—water, carbon dioxide (CO_2), and light—and to produce not only food but also oxygen gas (O_2). Scientists had learned that the water for photosynthesis comes primarily from the soil (for plants living on land) and must travel from the roots to the leaves; that carbon dioxide is taken in and water and O_2 are released through tiny openings in leaves, called **stomata** (singular stoma), which can open and close (Figure 8.1); and that light is absolutely necessary for the production of oxygen and food.

The last of the important early discoveries, made during the first decade of the nineteenth century, was that carbon dioxide uptake and oxygen release are closely related and that both depend on light action.

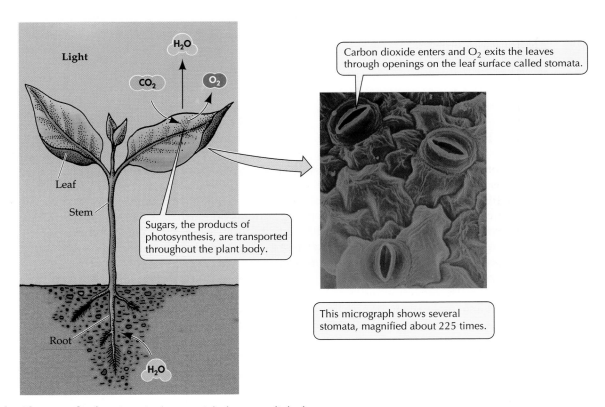

8.1 Ingredients for Photosynthesis A typical terrestrial plant uses light from the sun, water from the soil, and carbon dioxide from the atmosphere to form organic compounds by photosynthesis.

By 1804, scientists could summarize photosynthesis in plants as follows:

Carbon dioxide + water + light energy → sugar + oxygen

which turned into the balanced equation that is the reverse of the overall equation for cellular respiration given in Chapter 7:

$$6\ CO_2 + 6\ H_2O \rightarrow C_6H_{12}O_6 + 6\ O_2$$

Although true, these statements say nothing about the details of the process. What roles does light play? How do the carbons become linked? And where does the oxygen come from?

Almost a century and a half passed before the source of the O_2 released during photosynthesis—CO_2 or water—was determined. The direct demonstration depended on one of the first uses of an isotopic tracer in biological research. In the experiments, two groups of green plants were allowed to carry on photosynthesis. Plants in the first group were supplied with water containing the heavy-oxygen isotope ^{18}O and with CO_2 containing only the common oxygen isotope ^{16}O; plants in the second group were supplied with CO_2 labeled with ^{18}O and water containing only ^{16}O.

When oxygen gas was collected from each group of plants and analyzed, it was found that O_2 containing ^{18}O was produced in abundance by the plants that had been given ^{18}O-labeled water but not by the plants given labeled CO_2. From these results, scientists concluded that all the oxygen gas produced during photosynthesis comes from water (Figure 8.2). This discovery is reflected in a revised balanced equation:

$$6\ CO_2 + 12\ H_2O \rightarrow C_6H_{12}O_6 + 6\ O_2 + 6\ H_2O$$

Water appears on both sides of the equation because water is used as a reactant (the 12 molecules on the left) and released as a product (the 6 new ones on the right). In this equation there are now sufficient water molecules to account for all the oxygen gas produced.

The photosynthetic production of oxygen by green plants is an important source of atmospheric oxygen, which most organisms—including plants themselves—require in order to complete their respiratory chains and thus obtain the energy to live.

The Two Pathways of Photosynthesis: An Overview

The overall photosynthetic reaction just described cannot proceed in a single step. In fact, there is no precedent in all of chemistry for the completion of such a complex reaction in a single step. Rather, a whole series

Water and carbon dioxide provided	Photosynthesis	Oxygen released

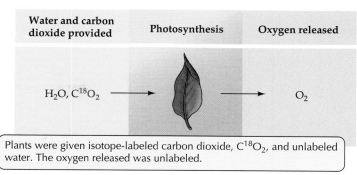

$H_2O, C^{18}O_2$ O_2

Plants were given isotope-labeled carbon dioxide, $C^{18}O_2$, and unlabeled water. The oxygen released was unlabeled.

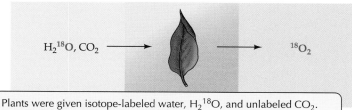

$H_2{}^{18}O, CO_2$ $^{18}O_2$

Plants were given isotope-labeled water, $H_2{}^{18}O$, and unlabeled CO_2. The oxygen released was labeled.

8.2 Water Is the Source of the Oxygen Produced by Photosynthesis Because only plants given isotope-labeled water released labeled O_2, this experiment showed that water is the source of the oxygen released during photosynthesis.

of simpler steps is required. By the middle of the twentieth century, it was clear that photosynthesis includes two pathways: One pathway, driven by light, produces ATP and a reduced electron carrier (NADPH + H+). The other pathway, which does not use light directly, uses the ATP, NADPH + H+, and CO_2 to produce sugar (Figure 8.3).

Just as NAD (nicotinamide adenine dinucleotide) bridges the pathways of cellular respiration (see Chapter 7), a very similar compound, **NADP** (nicotinamide adenine dinucleotide phosphate) bridges the two pathways of photosynthesis. NADP is identical to NAD, except that it has another phosphate group attached to the ribose. Whereas NAD participates in metabolic breakdown reactions and energy transfers (catabolism), NADP participates in synthetic reactions (anabolism) that require energy and reducing power.

Like NAD, NADP exists in two forms. One (NADP+) is an oxidizing agent; the other (NADPH + H+) is a reducing agent (see Figure 7.4). NADPH + H+ is an intermediary for energy and reducing power. ATP and NADPH + H+ are carriers of reducing power because reduction is always an endergonic process that requires both energy and electrons.

Light energy is used to produce ATP from ADP and P_i. The reactions of this pathway, termed the *light reactions*, are mediated by molecular assemblies called **photosystems** (which we'll discuss in detail later), and

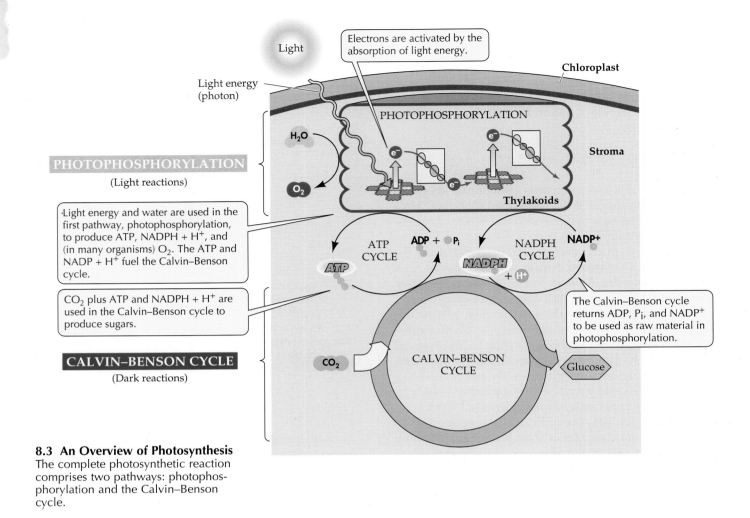

8.3 An Overview of Photosynthesis
The complete photosynthetic reaction comprises two pathways: photophosphorylation and the Calvin–Benson cycle.

Labels within figure:

Light

Light energy (photon)

Electrons are activated by the absorption of light energy.

Chloroplast

PHOTOPHOSPHORYLATION

Stroma

H_2O

O_2

Thylakoids

PHOTOPHOSPHORYLATION
(Light reactions)

·Light energy and water are used in the first pathway, photophosphorylation, to produce ATP, NADPH + H$^+$, and (in many organisms) O_2. The ATP and NADP + H$^+$ fuel the Calvin–Benson cycle.

ATP CYCLE

ADP + P$_i$

NADPH CYCLE

NADP$^+$

ATP

NADPH + H$^+$

CO_2 plus ATP and NADPH + H$^+$ are used in the Calvin–Benson cycle to produce sugars.

The Calvin–Benson cycle returns ADP, P$_i$, and NADP$^+$ to be used as raw material in photophosphorylation.

CO_2

CALVIN–BENSON CYCLE

Glucose

CALVIN–BENSON CYCLE
(Dark reactions)

---

the pathway itself is called **photophosphorylation** (see Figure 8.3). There are two different systems of photophosphorylation: *noncyclic photophosphorylation*, carried out by plants, which produces NADPH and ATP, and *cyclic photophosphorylation*, which produces only ATP.

The NADPH + H$^+$ and ATP produced in the first pathway of photosynthesis are used in the second pathway, where reactions trap CO_2 and reduce the resulting acid to sugar. These sugar-producing reactions constitute the Calvin–Benson cycle (see Figure 8.3), also known as the photosynthetic carbon reduction cycle, or simply the *dark reactions* (because none of them uses light directly). The reactions of both pathways proceed within the chloroplast, but, as we will see, they reside in different parts of the organelle. *Both pathways stop in the dark* because ATP synthesis and NADP$^+$ reduction require light. The rate of each set of reactions depends on the rate of the other. They are tied together by the exchange of ATP and ADP, and of NADP$^+$ and NADPH.

Properties of Light and Pigments

The living world makes marvelous use of light. Light is a source of both energy and information. In later chapters, we'll examine the many roles of light and pigments in the transmission of *information*. In this chapter, our focus is on light as a source of *energy*. However, before we can consider the reactions in the chloroplasts that use light energy, we need to examine the physical nature of light: its properties and energy content, and how it interacts with molecules.

Light comes in packets called photons

Light is a form of radiant energy. It comes in discrete packets called **photons**. Light also behaves as if it were propagated in waves. The **wavelength** of light is the distance from the peak of one wave to the peak of the next (Figure 8.4). Different colors result from different wavelengths. Light and other forms of radiant energy—cosmic rays, gamma rays, X rays, ultraviolet radiation, infrared radiation, microwaves, and radio

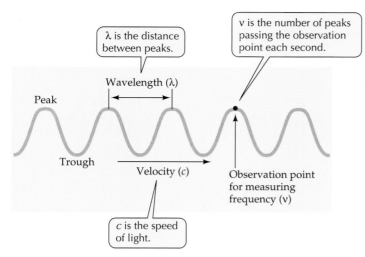

8.4 Light Has Wavelike Properties Light can be envisioned as a series of waves whose peaks pass a fixed observation point with uniform frequency.

waves—are **electromagnetic radiation**. We have listed these forms of radiation in order of increasing wavelength and of decreasing energy per photon. Visible light fits into this **electromagnetic spectrum** of wavelengths between ultraviolet and infrared radiation (Figure 8.5).

The speed of light in a vacuum is one of the universal constants of nature. In a vacuum, light travels at 3×10^{10} cm/s (186,000 miles/s), a value symbolized as c. In air, glass, water, and other media, light travels slightly more slowly.

Let's consider light as a long train of waves moving in a straight line and see what the train would look like to a stationary observer. Successive peaks of the waves pass the observer with a uniform **frequency** determined by the wavelength and the speed of light. The exact relationship is $\nu = c/\lambda$, where ν (the Greek letter nu) is the frequency; c is the speed of light; and λ (Greek lambda) is the wavelength. Often ν is expressed in hertz (Hz), c in centimeters per second (cm/s), and λ in nanometers (nm). (1 nm = 10^{-9} m or 10^{-7} cm; see the conversion table inside the back cover.)

Humans perceive light as having distinct colors (the reason for this will be explained in Chapter 42). The colors relate to the wavelengths of the light, as shown in Figure 8.5. Most of us can see electromagnetic radiation in the range of wavelengths from 400 to 700 nm. The wavelength at 400 nm marks the blue end of the visible spectrum, at 700 nm the red end. Wavelengths in the range from about 100 to 400 nm are ultraviolet radiation; those immediately above 700 nm are referred to as infrared.

The amount of energy, E, contained in a single photon is directly proportional to its frequency. The constant of proportionality that describes this relation-

ship, h, is named Planck's constant after Max Planck, who first introduced the concept of the photon. With this information we can write the equation $E = h\nu$, where ν is the frequency in Hz. Substituting c/λ for ν (from the earlier equation relating λ, ν, and c), we see that $E = hc/\lambda$. Thus shorter wavelengths mean greater energies; that is, energy is inversely proportional to wavelength. A photon of red light of wavelength 660 nm has less energy than a photon of blue light at 430 nm; an ultraviolet photon of wavelength 284 nm is much more energetic than either of these. For any light-driven biological process—such as photosynthe-

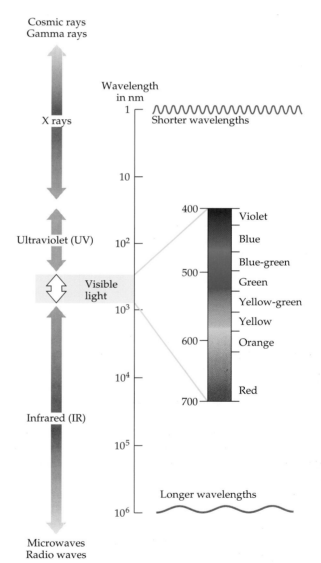

8.5 The Electromagnetic Spectrum A portion of the electromagnetic spectrum in the vicinity of light that is visible to humans is represented here. Visible light comprises wavelengths between about 400 and 700 nm. Ultraviolet radiation extends from the short-wavelength end of the visible spectrum, infrared radiation from the long-wavelength end.

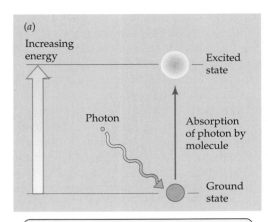

(a)

Increasing energy

Excited state

Photon

Absorption of photon by molecule

Ground state

When a molecule, initially in the ground state, absorbs a photon, the molecule is raised to an excited state and possesses more energy.

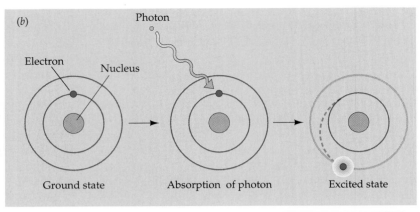

(b)

Photon

Electron

Nucleus

Ground state

Absorption of photon

Excited state

The absorption of the photon boosts one of the molecule's electrons to an orbital farther from the nucleus.

8.6 Exciting a Molecule After absorbing the energy of a photon (*a*), one of the molecule's electrons is boosted to a higher orbital (*b*) and, in this excited state, is held less firmly by the molecule.

sis—a photon can be active only if it consists of enough energy to perform the work required.

The brightness, or **intensity**, of light at a given point is the amount of energy falling on a defined area—such as 1 cm²—per second. Light intensity is usually expressed in energy units (such as calories) per square centimeter per second, but pure light of a single wavelength may also be expressed in terms of photons per square centimeter per second.

Absorption of a photon puts a pigment in an excited state

When a photon meets a molecule, one of three things happens. The photon may bounce off the molecule or it may pass through it; in other words, it may be reflected or transmitted. Neither of these causes any change in the molecule, and neither has any chemical consequences. The third possibility is that the photon may be *absorbed* by the molecule. In this case, the photon disappears. Its energy, however, cannot disappear, because energy is neither created nor destroyed.

The molecule acquires the energy of the absorbed photon and is thereby raised from a **ground state** (lower energy) to an **excited state** (higher energy) (Figure 8.6*a*). The difference in energy between this excited state and the ground state is precisely equal to the energy of the absorbed photon. The increase in energy boosts one of the electrons in the molecule into an orbital farther from the nucleus; this electron is now held less firmly by the molecule (Figure 8.6*b*). We will see the chemical consequence of this looser hold on the electron later in this chapter.

All molecules absorb electromagnetic radiation. The specific wavelengths absorbed by a particular molecule are characteristic of that type of molecule. Some molecules cannot absorb wavelengths in the visible region; those that can are called **pigments**.

When a beam of white light (light containing visible light of all wavelengths) falls on a pigment, certain wavelengths of the light are absorbed. The remaining wavelengths, which are reflected or transmitted, make the pigment appear to us to be colored. For example, if a pigment absorbs both blue and red light, as does the pigment chlorophyll, what we see is the remaining light—primarily green. The fact that chlorophyll absorbs light in both the blue and the red region of the spectrum indicates that it has two excited states of differing energy levels, both close enough to the ground state to be reached with the energy of photons of visible light.

Light absorption and biological activity vary with wavelength

A given type of molecule can absorb radiant energy of only certain wavelengths. If we plot a compound's absorption of light as a function of the wavelength of the light, the result is an **absorption spectrum** (Figure 8.7). Absorption spectra are good "fingerprints" of compounds; sometimes an absorption spectrum contains enough information to enable us to identify an unknown compound.

Light may be analyzed for its *biological effectiveness*, the magnitude of its effect on a particular activity such as photosynthesis. We may plot the effectiveness of light as a function of wavelength. The resulting graph is an **action spectrum**. Figure 8.8 shows the action spectrum for photosynthesis by *Anacharis*, a freshwater plant. All wavelengths of visible light are at least somewhat effective in causing photosynthesis, although some are more effective than others.

8.7 Photosynthetic Pigments Have Distinct Absorption Spectra
Photosynthesis uses most of the visible spectrum, because the participating pigments absorb photons most strongly at different wavelengths.

Because light must be absorbed in order to produce a chemical change or biological effect, action spectra are helpful in determining what pigment or pigments are used in a particular photobiological process, such as photosynthesis. We should be able to find which pigment or pigments have absorption spectra that match the action spectrum of the process.

Photosynthesis uses chlorophylls and accessory pigments

Certain pigments are important in biological reactions, and we will discuss them as they appear in the book. Here we discuss the pigments, found in leaves and in other parts of photosynthetic organisms, that play roles in photosynthesis. Of these, the most important ones are the **chlorophylls**. Chlorophylls occur universally in the plant kingdom, in photosynthetic protists, and in photosynthetic bacteria. A mutant individual that lacks chlorophyll is unable to perform photosynthesis and will starve to death.

In green plants, two chlorophylls predominate, **chlorophyll *a*** and **chlorophyll *b*;** they differ only slightly in structure. Both have a complex ring struc-

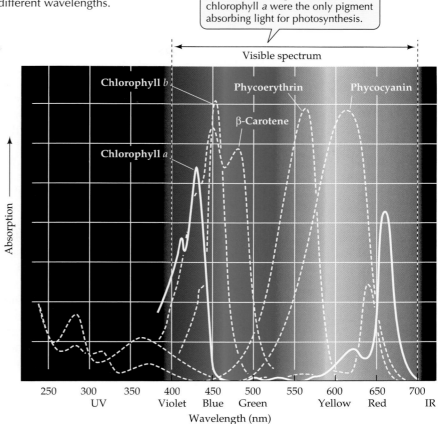

Notice how much of the visible spectrum would go to waste if chlorophyll *a* were the only pigment absorbing light for photosynthesis.

ture of a type referred to as a chlorin,* a lengthy hydrocarbon "tail," and a central magnesium atom in the chlorin ring (Figure 8.9).

We saw in Figure 8.7 that the chlorophylls absorb blue and red wavelengths, which are near the two ends of the visible spectrum. Thus if *only* chlorophyll pigments were active in photosynthesis, much of the visible spectrum would go unused. However, all photosynthetic organisms possess **accessory pigments** that absorb photons intermediate in energy between the red and the blue wavelengths and then transfer a portion of the energy to chlorophyll to use in photosynthesis.

*In Chapters 6 and 7 we learned about hemoglobin and cytochrome *c*. Both contain ring structures called porphyrins, which are very similar in structure to chlorins.

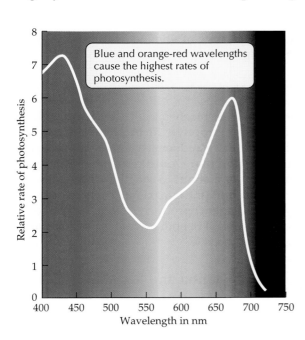

Blue and orange-red wavelengths cause the highest rates of photosynthesis.

8.8 Action Spectrum of Photosynthesis
An action spectrum plots the biological effectiveness of wavelengths of radiation against the wavelength. Here the rate of photosynthesis in the freshwater plant *Anacharis* is plotted against wavelengths of visible light. If we compare this action spectrum with the absorption spectra of specific pigments, such as those in Figure 8.7, we can identify which pigments are responsible for the process in *Anacharis*.

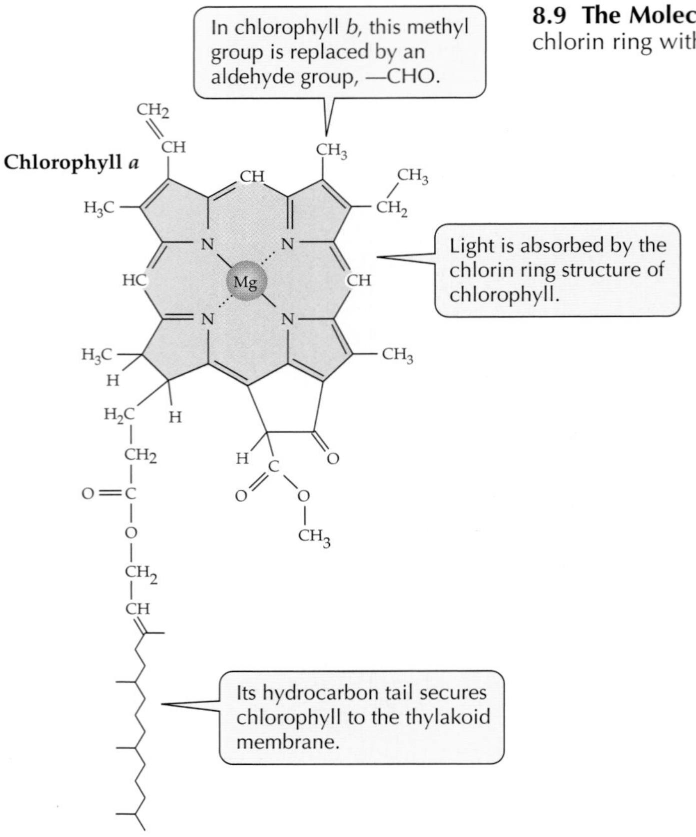

In chlorophyll *b*, this methyl group is replaced by an aldehyde group, —CHO.

Chlorophyll *a*

Light is absorbed by the chlorin ring structure of chlorophyll.

Its hydrocarbon tail secures chlorophyll to the thylakoid membrane.

8.9 The Molecular Structure of Chlorophyll Chlorophyll consists of a chlorin ring with magnesium (shaded area), plus a hydrocarbon "tail."

Among these accessory pigments are **carotenoids** such as β-carotene (see Figure 3.8); the carotenoids absorb photons in the blue and blue-green wavelengths and appear deep yellow. The **phycobilins** (phycocyanin and phycoerythrin), which are found in red algae and in cyanobacteria (contributing to their respective colors), absorb various yellow-green, yellow, and orange wavelengths (see Figure 8.7). Such accessory pigments, in collaboration with the chlorophylls, constitute an energy-absorbing *antenna system* covering much of the visible spectrum.

In the energy-absorbing antenna, any pigment molecule with a suitable absorption spectrum may absorb an incoming photon and become excited. The excitation passes from one pigment molecule to another in the antenna, moving to pigments that absorb longer wavelengths (lower energies) of light. Thus the excitation must end up in the one pigment molecule in the antenna that absorbs the longest wavelength—the molecule that occupies the **reaction center** of the antenna.

The reaction center is the part of the antenna that converts light absorption into chemical energy. In plants, the pigment molecule in the reaction center is always a molecule of chlorophyll *a*. There are many other chlorophyll *a* molecules in the antenna, but all of them absorb light at shorter wavelengths than does the molecule in the reaction center.

Excited chlorophyll acts as a reducing agent

A pigment molecule enters an excited state when it absorbs a photon (see Figure 8.6). The molecule usually does not stay in the excited state very long. It may return to the ground state, emitting light as **fluorescence**, or it may reduce another substance. In fluorescence, the boosted electron falls back from its higher orbital to the original, lower one. This process is accompanied by a loss of energy, which is given off as another photon (Figure 8.10*a*). The molecule absorbs one photon and within approximately 10^{-9} seconds emits another photon of longer wavelength than the one absorbed. The emitted light is fluorescence. If energy is simply absorbed and then rapidly returned as a photon of light, there can be no chemical changes or biological consequences—no chemical work is done.

For biological work to be done, something must happen to transfer energy in some other way during the billionth of a second before a photon is emitted as fluorescence. The excited pigment molecule may pass the energy to another pigment molecule (Figure 8.10*b*), as explained in the preceding section. Ultimately, however, photosynthesis conserves energy by using the excited chlorophyll molecule in the reaction center as a reducing agent (Figure 8.10*c*).

Ground state chlorophyll (symbolized as Chl) is not much of a reducing agent, but excited chlorophyll (Chl*) is a good one. To understand the reducing capability of Chl*, recall that in an excited molecule, one of the electrons is zipping about in an orbital farther from its nucleus than it was before. Less tightly held, this electron can be passed on in a redox reaction to an oxidizing agent. Thus Chl* (but not Chl) can react with an oxidizing agent A in a reaction like this:

$$Chl^* + A \rightarrow Chl^+ + A^-$$

This, then, is the first biochemical consequence of light absorption by chlorophyll in the chloroplast: The chlorophyll becomes a reducing agent and participates in a redox reaction that would not have occurred in the dark. As we are about to see, the further adventures of that electron (the one passed from chlorophyll to the oxidizing agent) produce ATP and a stable reducing agent (NADPH), both of which are required in the Calvin–Benson cycle.

Photophosphorylation and Reductions

In our earlier overview of photosynthesis, we distinguished the light-independent reactions (dark reac-

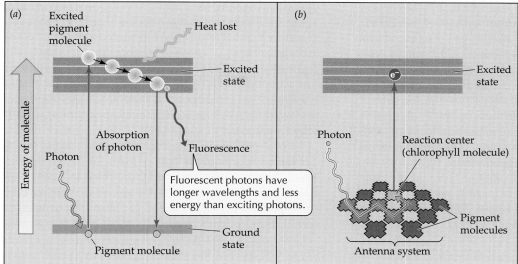

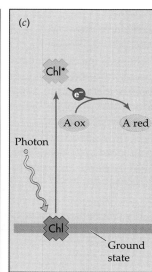

(a) When a pigment molecule absorbs a photon, boosting an electron to a higher orbital, the pigment moves to an excited state from its ground state. The molecule spends very little time in the excited state. Some of the absorbed energy is dissipated as heat, so that when the molecule returns to the ground state, a lesser amount of light energy is emitted as fluorescence.

(b) Rather than waste the absorbed energy in fluorescence, the excited pigment molecule may pass the excitation energy to another molecule, such as chlorophyll, in the antenna system of pigments. The excitation ends up in the reaction center—the molecule that absorbs the longest wavelength

(c) When a ground state chlorophyll molecule (Chl) becomes excited (Chl*), it becomes a reducing agent; the electron boosted to a higher orbital may pass to an oxidized electron carrier (A), reducing the carrier. Thus much of the energy of the excited state is chemically preserved rather than being lost, as it is in fluorescence.

8.10 Excitation, Energy Transfer, and Electron Transfer
A molecule spends very little time in the excited state. It may emit some of its acquired energy as fluorescence (a), but in order for work to be done, the energy must be transferred to the reaction center (b). In plants, the molecule in the reaction center is always chlorophyll a. This excited chlorophyll molecule is a reducing agent and participates in a redox reaction that is the initial step toward the production of the ATP and NADPH + H$^+$ needed for the Calvin–Benson cycle.

tions) from the light-dependent reactions (light reactions). Now we want to return to focus on the light reactions and examine how they function to provide the ATP and NADPH + H$^+$ needed to drive the dark reactions. We'll consider first the noncyclic reactions and cyclic reactions before considering the role of chemiosmosis in phosphorylation—a process that is very similar to that discovered for oxidative phosphorylation in mitochondria (see Chapter 7).

Noncyclic photophosphorylation produces ATP and NADPH

With the appearance of **noncyclic photophosphorylation,** the evolution of life on Earth made a crucial advance, because noncyclic photophosphorylation uses light energy not only to form ATP and NADPH + H$^+$, but also to release O$_2$. The primitive atmosphere of Earth contained little O$_2$ before it began being released as a by-product of photosynthesis. The accumulation of O$_2$ in the atmosphere forever changed the Earth and

the course of evolution, making possible the use of aerobic metabolism and a remarkable diversification of life. The origin of O$_2$ from water requires us to look at the role of water in the production of NADPH and ATP.

In noncyclic photophosphorylation, water is oxidized, and the electrons from water replenish the electrons that chlorophyll molecules lose when they are excited by light. These electrons are transferred to oxidizing agents and ultimately to NADP$^+$, reducing it to NADPH + H$^+$. As the electrons are passed from water to chlorophyll, and ultimately to NADP, they go through a series of electron carriers. These spontaneous redox reactions are exergonic, and some of the free energy released is used ultimately to form ATP by a chemiosmotic mechanism.

Noncyclic photophosphorylation requires the participation of two distinct molecules of chlorophyll. These are associated with two different photosystems that consist of many chlorophyll molecules and accessory pigments in separate energy-absorbing antennas (Figure 8.11). **Photosystem I** makes a reducing agent strong enough to reduce NADP$^+$ to NADPH + H$^+$. The reaction center of the antenna for photosystem I contains a chlorophyll a molecule in a form called P$_{700}$ because it can absorb light of wavelength 700 nm.

Photosystem II uses light to oxidize water molecules, producing electrons, protons (H$^+$), and O$_2$.

Electrons from water are passed to a series of redox carriers located in the thylakoid membranes of the chloroplast. Some of the energy lost in this redox process is used in the conversion of ADP + P_i to ATP. The reaction center of photosystem II contains a chlorophyll *a* molecule in a form called P_{680} because it absorbs light maximally at 680 nm. Thus photosystem II requires photons that are somewhat more energetic than those required by photosystem I. To keep noncyclic photophosphorylation going, both photosystems I and II must constantly be absorbing light, thereby boosting electrons to higher orbitals from which they may be captured by specific oxidizing agents.

We can follow the noncyclic pathway from water to NADP in Figure 8.11. Photosystem II (P_{680}) absorbs photons, sending electrons from P_{680} to an oxidizing agent (pheophytin-I) and causing P_{680} to become oxidized to P_{680}^+. Electrons from the oxidation of water are passed to P_{680}^+ of photosystem II, reducing it once again to P_{680}, which can absorb photons, and so on. The electron donated by photosystem II to its oxidizing agent passes through a series of exergonic redox reactions coupled to proton pumping that stores energy that is used to form ATP.

In photosystem I, P_{700} absorbs photons, becoming excited to P_{700}^*, which then reduces its own oxidizing agent (ferredoxin) while being oxidized to P_{700}^+. Then P_{700}^+ returns to the ground state by accepting electrons passed through the redox chain from photosystem II. Now photosystem II is accounted for, and we

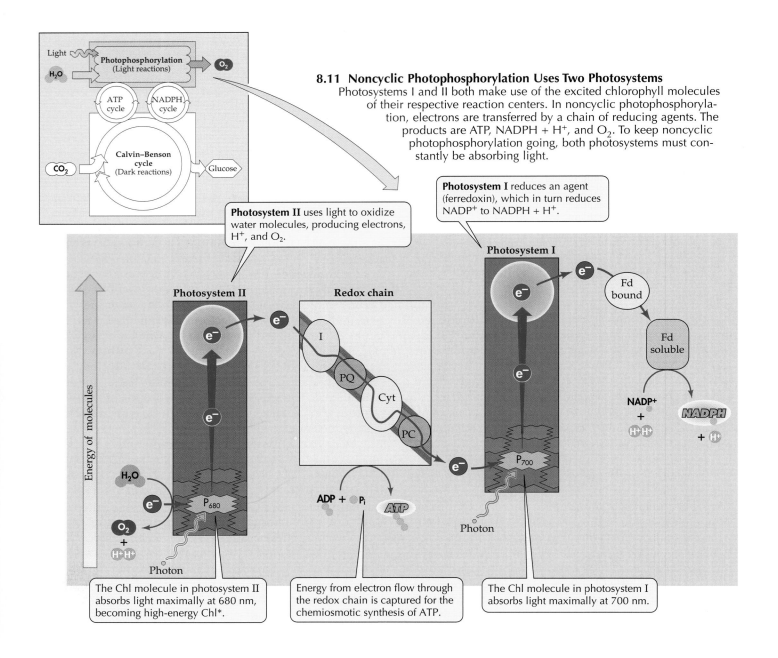

8.11 Noncyclic Photophosphorylation Uses Two Photosystems
Photosystems I and II both make use of the excited chlorophyll molecules of their respective reaction centers. In noncyclic photophosphorylation, electrons are transferred by a chain of reducing agents. The products are ATP, NADPH + H⁺, and O₂. To keep noncyclic photophosphorylation going, both photosystems must constantly be absorbing light.

Photosystem II uses light to oxidize water molecules, producing electrons, H⁺, and O₂.

Photosystem I reduces an agent (ferredoxin), which in turn reduces NADP⁺ to NADPH + H⁺.

The Chl molecule in photosystem II absorbs light maximally at 680 nm, becoming high-energy Chl*.

Energy from electron flow through the redox chain is captured for the chemiosmotic synthesis of ATP.

The Chl molecule in photosystem I absorbs light maximally at 700 nm.

must consider only the electrons from photosystem I. These are used in the last step of noncyclic photophosphorylation, in which two electrons and two protons (from two operations of the noncyclic scheme) are used to reduce a molecule of $NADP^+$ to $NADPH + H^+$.

In sum, noncyclic photophosphorylation uses a molecule of water, four photons (two each absorbed by photosystems I and II), one molecule each of $NADP^+$ and ADP, and one P_i. From these ingredients it produces one molecule each of $NADPH + H^+$ and ATP, and half a molecule of oxygen (review Figure 8.11). A substantial fraction of the light energy absorbed in noncyclic photophosphorylation is lost as heat, but another significant fraction is trapped in ATP and $NADPH + H^+$.

Cyclic photophosphorylation produces ATP but no NADPH

Noncyclic photophosphorylation produces equal quantities of ATP and $NADPH + H^+$. However, as we will see, the Calvin–Benson cycle uses more ATP than $NADPH + H^+$. In order to keep things in balance, plants sometimes make use of a supplementary form of photophosphorylation that does not generate $NADPH + H^+$.

Photophosphorylation that produces only ATP is called **cyclic photophosphorylation** because an elec-

tron passed from an excited chlorophyll molecule at the outset cycles back to the same chlorophyll molecule at the end of the chain of reactions (Figure 8.12). Water, which supplies electrons to restore chlorophyll molecules to the ground state in noncyclic photophosphorylation, does not enter these reactions; thus they produce no O_2.

Before cyclic photophosphorylation begins, P_{700}, the reaction center chlorophyll of photosystem I, is in the ground state. It absorbs a photon and becomes the reducing agent P_{700}^*. The P_{700}^* then reacts with oxidized ferredoxin (Fd_{ox}) to produce reduced ferredoxin (Fd_{red}). The reaction is spontaneous; that is, it is exergonic, releasing free energy.

In noncyclic photophosphorylation, Fd_{red} reduces $NADP^+$ to form $NADPH + H^+$. However, Fd_{red} can also pass its added electron to a *different* oxidizing agent, plastoquinone (PQ, a small organic molecule). This is what happens in cyclic photophosphorylation (see Figure 8.12), which occurs in some organisms when the ratio of $NADPH + H^+$ to $NADP^+$ in the chloroplast is high.

Fd_{red} reduces PQ (which is part of the redox chain that connects photosystems I and II; see Figure 8.11) in the reaction $Fd_{red} + PQ_{ox} \rightarrow Fd_{ox} + PQ_{red}$. Acting as if the electron passed to it came from pheophytin-I (as happens in noncyclic photophosphorylation), PQ_{red} passes the electron to a cytochrome complex (Cyt). The electron continues down the chain until it completes its cycle by returning to P_{700}. This cycle is a series of redox reactions, each exergonic, and the released energy is stored in a form that ultimately can be used to produce ATP.

Remember that when P_{700}^* passed its electron on to Fd, we were left with a molecule of positively charged P_{700}^+ (having

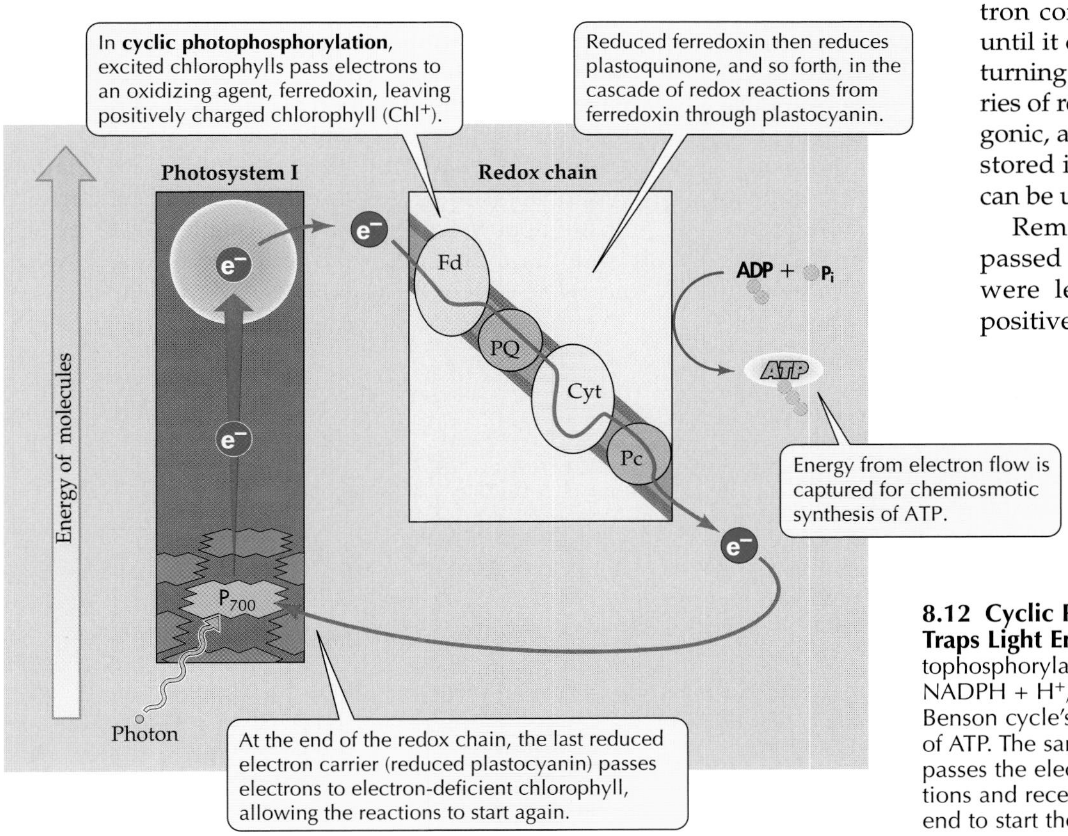

In **cyclic photophosphorylation**, excited chlorophylls pass electrons to an oxidizing agent, ferredoxin, leaving positively charged chlorophyll (Chl^+).

Reduced ferredoxin then reduces plastoquinone, and so forth, in the cascade of redox reactions from ferredoxin through plastocyanin.

Energy from electron flow is captured for chemiosmotic synthesis of ATP.

At the end of the redox chain, the last reduced electron carrier (reduced plastocyanin) passes electrons to electron-deficient chlorophyll, allowing the reactions to start again.

8.12 Cyclic Photophosphorylation Traps Light Energy as ATP Cyclic photophosphorylation produces ATP but no $NADPH + H^+$, thus balancing the Calvin–Benson cycle's need for greater amounts of ATP. The same chlorophyll molecule passes the electrons that start the reactions and receives the electrons at the end to start the process over again.

lost an electron, the chlorophyll has one unbalanced positive charge). In due course, P_{700}^+ interacts with a reducing agent that donates an electron, converting it back to uncharged P_{700}. This reducing agent, plastocyanin (PC), is the last member of the redox chain in Figure 8.12. By the time the electron (passed from P_{700}^* through the redox chain) comes back to P_{700}^+ and reduces it, all the energy from the original photon has been released. In each of the redox reactions, some free energy is lost, until all of the original energy has been converted to heat *except* for that used to form ATP.

Comparing the cartoon in Figure 8.13 with Figure 8.12 may help make the concept of cyclic photophosphorylation clearer to you.

Chemiosmosis is the source of ATP in photophosphorylation

How does electron transport in photosystems I and II form ATP? In Chapter 7 we considered the **chemiosmotic mechanism** for ATP formation in the mitochondrion. The chemiosmotic mechanism also operates in photophosphorylation. In chloroplasts, as in mitochondria, electrons move through a series of redox reactions releasing energy that is used to transport protons (H^+) across a membrane. This active proton transport results in a *proton-motive force*—a difference in pH and in electric charge across the membrane.

In the mitochondrion, protons are pumped from the matrix, across the internal membrane, and into the space between the inner and outer mitochondrial membranes (see Figure 7.14). In the chloroplast, the electron carriers are located in the thylakoid membranes (see Figure 4.13). The electron carriers are oriented so that protons move into the interior of the thylakoid, and the inside becomes acidic with respect to the outside. This difference in pH leads to the passive movement of protons back out of the thylakoid, through specific protein channels in the membrane. These proteins are the enzymes, ATP synthases, that couple the formation of ATP to the diffusion of protons back across the membrane, just as in mitochondria (Figure 8.14).

The hypothesis that this chemiosmotic mechanism is responsible for the formation of ATP in chloroplasts was tested by Andre Jagendorf (of Cornell University) and Ernest Uribe (now at Washington State University) in the following way. Chloroplast thylakoids were isolated from spinach leaves and then kept in the dark, so

that there would be no light energy to drive the production of ATP (Figure 8.15). The thylakoids were moved from a neutral solution to one with a low pH so that by diffusion the interior of the thylakoids would become acidic. Then they were transferred to a solution that had a higher pH, so that the interiors of the thylakoids would be more acidic than the outsides—mimicking the situation created by light-driven pumping of protons into the interiors.

This final step immediately resulted in the formation of ATP, even though no light was available to serve as the energy source—precisely the result predicted by the chemiosmotic model. (Recall that a very similar experiment, using mitochondria, pH changes, and ATP formation, was described in Chapter 7.)

Photosynthetic pathways are the products of evolution

Photosystem I evolved before photosystem II; thus, cyclic photophosphorylation evolved before noncyclic photophosphorylation. Early in evolutionary history, photosynthetic bacteria used photosystem I and cyclic photophosphorylation to make ATP. This form of photosynthesis evolved long before Earth's atmosphere contained significant quantities of oxygen gas.

Nearly 3 billion years ago the cyanobacteria evolved photosystem II, thus gaining the ability to perform noncyclic photophosphorylation—and to extract electrons from water and use them to reduce $NADP^+$ while producing O_2. Over hundreds of millions of years, noncyclic photophosphorylation by cyanobacteria poured enough oxygen gas into the atmosphere to make possible the evolution of cellular respiration.

Today the evolutionarily ancient photosynthetic bacteria still have only cyclic photophosphorylation. Cyanobacteria, algae, and plants, which perform mostly noncyclic photophosphorylation, still perform cyclic photophosphorylation to produce ATP when their ratio of $NADPH + H^+$ to $NADP^+$ is high.

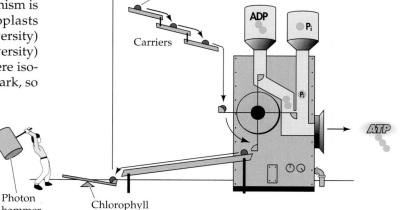

8.13 Cyclic Photophosphorylation Cycles Electrons A mythical "ATP machine" illustrates the concept of cyclic photophosphorylation.

Making Sugar from CO_2: The Calvin–Benson Cycle

The second main pathway of photosynthesis is the **Calvin–Benson cycle** of reactions, which incorporates CO_2 into sugars. In the chloroplast, most of the enzymes that catalyze the reactions of this pathway are dissolved in the stroma (see Figure 4.14), and this is where the reactions take place. These reactions are sometimes called the *dark reactions* because they do not directly require light energy. However, they require it indirectly, and *they take place only in the light*.

In this section, we will identify the methods and examine the experimental re-

8.14 Chloroplasts Form ATP Chemiosmotically   Protons (H^+) pumped across the thylakoid membrane from the stroma during photophosphorylation make the interior of the thylakoid more acidic than the stroma. Driven by this pH difference, the protons then diffuse to the stroma through ATP synthase channels, which couple the energy of proton flow to the formation of ATP from ADP + P_i.

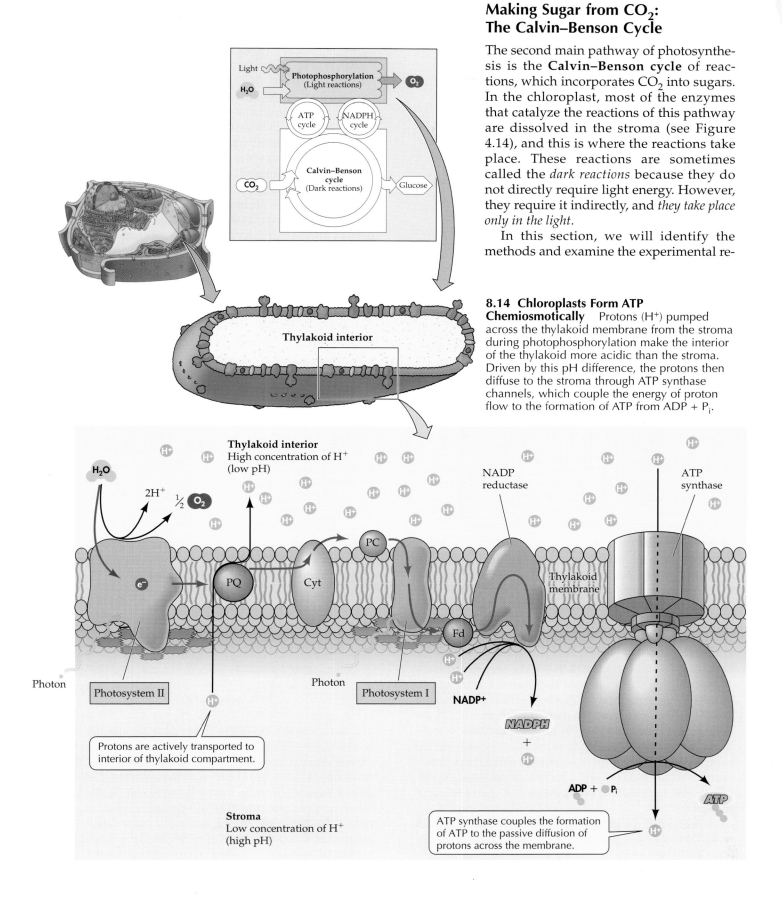

Light

Photophosphorylation (Light reactions) O_2

H_2O

ATP cycle

NADPH cycle

CO_2

Calvin–Benson cycle (Dark reactions)   Glucose

Thylakoid interior

Thylakoid interior
High concentration of H^+
(low pH)

H_2O

$2H^+$ ½ O_2

NADP reductase

ATP synthase

e^-

PQ   Cyt   PC

Thylakoid membrane

Fd

Photon   Photosystem II

Photon   Photosystem I

NADP+

NADPH
+

Protons are actively transported to interior of thylakoid compartment.

Stroma
Low concentration of H^+
(high pH)

ADP + P_i

ATP

ATP synthase couples the formation of ATP to the passive diffusion of protons across the membrane.

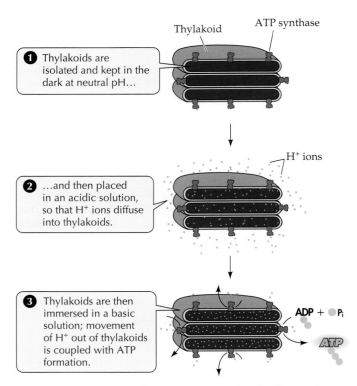

8.15 An Experiment Shows ATP Formation in the Dark
This experiment (described in the text) artificially induced ATP formation even in the absence of light energy, confirming the prediction of the chemiosmotic mechanism.

sults and critical thinking that were used to discover the sequence of reactions by which the gas CO_2 is incorporated into sugars.

Elucidation of the Calvin–Benson cycle required radioactive carbon

Real progress in understanding the dark reactions came only after World War II, when radioactive isotopes became available. A group of scientists at the University of California, Berkeley, led by Melvin Calvin and including Andrew Benson and James Bassham, wanted to identify the biochemical steps between the uptake of CO_2 and the appearance of the first complex carbohydrates in the chloroplast.

Accomplishing this goal depended on three advances in experimental methods: (1) the discovery and availability of a radioactive carbon isotope, ^{14}C; (2) the development of *paper partition chromatography*, a technique permitting the rapid separation of individual compounds from complex solutions; and (3) the development of *autoradiography*, a technique for locating colorless but radioactive compounds on a paper chromatogram.

Armed with these tools, the Berkeley group set out to investigate how photosynthetic organisms metabolize CO_2. They worked mostly with unicellular aquatic algae, such as the green alga *Chlorella*. Algae were grown in dense suspensions in a flattened flask (called a "lollipop" because of its shape) between two bright lights, ensuring a rapid rate of photosynthesis (Figure 8.16*a*). At the start of an experiment, a solution containing dissolved $^{14}CO_2$ was suddenly squirted into the lollipop. At a carefully measured time after this squirt, a sample of the culture was rapidly drained into a container of boiling ethanol.

The ethanol rapidly penetrated the algal cell walls, performing two functions: It killed the algae, stopping photosynthesis immediately, and it extracted the ^{14}C-containing intermediates of $^{14}CO_2$ metabolism from the algae (along with many other compounds). A drop of this ethanol extract was placed on filter paper for paper chromatography followed by autoradiography (Figure 8.16*b*). Many spots corresponding to radioactive substances appeared, indicating that many biochemical reactions had taken place during that short interval.

The first stable product of CO_2 fixation is the compound 3PG

The first reaction incorporating carbon dioxide into a larger compound is called **carbon fixation**. During 30 seconds of continuous exposure to $^{14}CO_2$, many different compounds were formed in the cells in Calvin's lollipop (see Figure 8.16*b*). But which of these compounds was the first to form?

To determine the carbon fixation reaction, the experiment was repeated several times, using ever-shorter exposures. Even after exposures of less than 2 seconds, half a dozen or more labeled compounds appeared in the autoradiographs. One, however, was produced most rapidly and in greatest abundance: **3PG** (3-phosphoglycerate, also called 3-phosphoglyceric acid), which we have already encountered as an intermediate in glycolysis (see Figure 7.6).

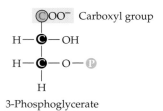

3-Phosphoglycerate

The compound 3PG is the first stable product of CO_2 fixation. The Berkeley group isolated the individual carbon atoms from the 3PG produced in the lollipop experiments and found that the carbon of the carboxyl group was much more intensely radioactive than the other two carbon atoms. (A single atom either is or is not radioactive. By "more intensely radioactive" we mean that in a population of molecules, the fraction of carboxyl carbons labeled with radioactivity was

(a)

The thin flask was filled with a suspension of algae and illuminated from both sides. After injection of $^{14}CO_2$, a sample was drained from the flask into boiling ethanol to kill the cells rapidly, stop enzymatic action, and extract molecules.

(b)

MALIC ACID

CITRIC ACID

ALANINE

GLUTAMIC ACID

GLYCINE

ASPARTIC ACID

SERINE

PEPA

SUCROSE

TRIOSE PHOSPHATE

PGA

30 SEC PHOTOSYNTHESIS WITH C^{14}O$_2$
CHLORELLA

SUGAR
PHOSPHATES

A chromatogram showing the products of algal photosynthesis. The dark spots are compounds containing ^{14}C—all formed in the 30 seconds following injection of $^{14}CO_2$.

8.16 A Lollipop and Its Products (a) The "lollipop" apparatus used in early experiments on photosynthesis. (b) In this reproduction of an original chromatogram, the spot labeled PGA (an older notation) corresponds to the position of 3PG.

greater than the fractions of radioactive carbons in the other two carbon positions.)

The Berkeley group drew two important conclusions from this discovery. First, the heavy labeling showed that the carboxyl carbon is obtained directly from CO_2. Second, because radioactive carbon appears in the other carbon atoms of 3PG, they concluded that some kind of *cyclical* process is involved, a process by which 3PG is made by adding CO_2 to another compound that is itself produced from photosynthetic 3PG.

The CO_2 acceptor is the compound RuBP

What is the compound, obtained from the metabolism of 3PG, that forms a covalent bond with CO_2 to make more 3PG? Given the structure of 3PG, it seemed reasonable to expect that the mysterious CO_2 acceptor would be a *two*-carbon compound that could react with CO_2 to become a *three*-carbon compound—3PG—with the CO_2 becoming a carboxyl group.

If this idea were correct, it should have been possible for the Berkeley group to find, on their chromatograms, a compound with only two carbon atoms, both of which were radioactive after a lollipop experiment. However, they did *not* find such a compound; thus the problem became more difficult. The "obvious" answer—that the CO_2 acceptor is a two-carbon compound—was wrong. Where would they go from here?

BASING A MODEL ON A CYCLE OF SUGARS. At this point in the investigation, concentrating on a photosynthetic *cycle* became useful. Consider a tentative cycle of the sort shown in Figure 8.17a: CO_2 reacts with X, the CO_2 acceptor, to produce 3PG. From the 3PG, photosynthetic organisms make products (things like glucose) and more X; this new X can react with another molecule of CO_2 and keep the cycle going. But what is X, and what are the other intermediates in the cycle?

It was observed that 3PG was the only acid phosphate produced in significant amount, whereas many kinds of *sugar* phosphates appeared on the chromatograms. On this basis, the Berkeley group guessed that the first thing to happen to 3PG is its conversion to a three-carbon sugar phosphate (glyceraldehyde 3-phosphate, which we will call G3P). Such a reaction (acid to aldehyde) is a reduction, and since reductions are highly endergonic, the Berkeley group proposed that this reaction would require ATP (Figure 8.17b).

They supported this proposal with two lollipop experiments, one in the light and one in the dark. When

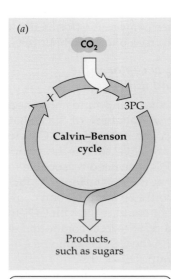

(a)

CO$_2$

Calvin–Benson cycle

X

3PG

Products, such as sugars

After it became apparent that CO$_2$ combines with some molecule to form 3PG and that a cyclical process is involved, a pathway could be devised in which a molecule of compound X combines with CO$_2$ to form 3PG and in which further reactions regenerate X.

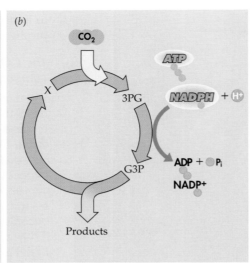

(b)

CO$_2$

ATP

X

3PG

NADPH + H$^+$

ADP + P$_i$

NADP$^+$

G3P

Products

Ensuing speculation and experimentation led to the proposition that 3PG was reduced to a sugar phosphate, glyceraldehyde 3-phosphate (G3P); this endergonic reaction, a reduction, would require ATP, as shown here.

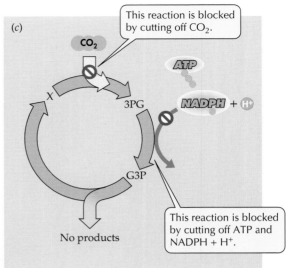

(c)

This reaction is blocked by cutting off CO$_2$.

CO$_2$

ATP

X

3PG

NADPH + H$^+$

G3P

No products

This reaction is blocked by cutting off ATP and NADPH + H$^+$.

In the proposed model, a cutoff of CO$_2$ blocks the formation of 3PG from compound X and CO$_2$; a cutoff of ATP (by turning off the light) blocks the formation of G3P from 3PG.

8.17 Manipulating the Reactions of Photosynthesis
The Berkeley group manipulated the photosynthetic cycle in their laboratory by two simple means. First, turning off the light blocked the reaction 3PG → G3P. Second, the reaction X → 3PG was blocked by cutting off CO$_2$.

they supplied $^{14}CO_2$ to the algae in the lollipop, 3PG rapidly became radioactively labeled in both experiments, but G3P became radioactively labeled only in the light, showing that the reduction does require energy—presumably in the form of ATP generated in the light.

TWEAKING THE LIGHT AND CO$_2$ SUPPLIES. At this point an extremely clever suggestion was made: If we assume that Figure 8.17b accurately models what occurs in the chloroplast, then it should be possible to regulate this cycle in the laboratory by two simple means. First, turning off the light would specifically block the step from the acid, 3PG, to the aldehyde, G3P, because the necessary ATP and NADPH + H$^+$ can be produced only with an input of energy, and in photosynthesis the energy source is light. Therefore ATP should be made in the chloroplast only when the light is on. Second, the reaction from X to 3PG could easily be blocked by cutting off the supply of CO$_2$ to the lollipop (Figure 8.17c).

Assume that photosynthesis is proceeding at a steady pace, so the concentrations of 3PG and the CO$_2$

acceptor X in the cells are constant. The lights are on, of course, and there is plenty of CO$_2$. Suddenly the investigator turns off the lights, thus blocking the cycle as proposed in Figure 8.17c: How does the concentration of 3PG change immediately? How does the concentration of X change immediately? *Stop.* Think about this before you read on.

When the light is turned off, no more ATP is made. Without ATP, the reaction from 3PG to the three-carbon sugar G3P cannot take place. Therefore, 3PG is no longer being used up. However, there is nothing to stop the formation of 3PG from incoming CO$_2$ and X, as long as X is around. Therefore, the immediate consequence of turning off the light is an *increase* in the level of 3PG.

On the other hand, X continues to be used up, because its reaction with CO$_2$ does not depend on light or ATP, but the formation of X slows down because 3PG can no longer be reduced and ultimately form new X. Therefore, the concentration of X *decreases* (Figure 8.18). These are the changes we would expect to see *if* our model (see Figure 8.17c) were correct. Similar reasoning should convince you that when the CO$_2$ supply is cut off (with the lights on), the concentration of X rises and that of 3PG falls. With no CO$_2$ available, X is no longer used up and 3PG is no longer formed, but 3PG can still be reduced to G3P.

Having devised this model, the Berkeley group proceeded to study the effects of changes in light intensity and CO$_2$ supply on the concentrations of all the major radioactive compounds found in the lollipop experi-

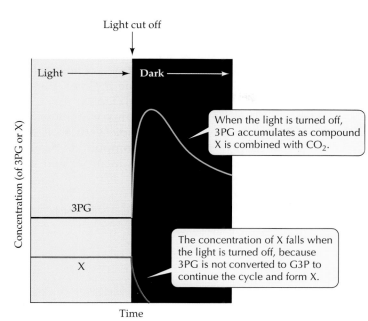

8.18 Dark-Induced Changes Predicted by the Model
Light is necessary to provide the ATP for the endergonic reaction 3PG → G3P.

When the light is turned off, 3PG accumulates as compound X is combined with CO_2.

The concentration of X falls when the light is turned off, because 3PG is not converted to G3P to continue the cycle and form X.

FINDING THE ENZYME THAT CATALYZES THE CAPTURE OF CO_2.
The best way to prove that a proposed reaction takes place is to find an *enzyme* that catalyzes it. In this case, such an enzyme was soon discovered, and from more than one source. The enzyme, RuBP carboxylase, now commonly called **rubisco**, was found in the algae studied by the Berkeley group and also in the leaves of spinach. Of most importance, studies of spinach revealed that rubisco is found in only one part of the cell: the chloroplast—exactly where one should find an enzyme concerned with photosynthesis. The investigators concluded, then, that RuBP is the CO_2 acceptor, the previously unknown compound X.

Identifying intermediate reactions of the Calvin–Benson cycle

Having discovered the first product of CO_2 fixation (3PG) and the CO_2 acceptor (RuBP), the Berkeley group proceeded to work out the remaining reactions of the cycle. They found some relatively complicated steps between G3P and RuBP; among the intermediates are sugar phosphates with four, five, six, and seven carbon atoms. All the proposed intermediates have been found in chloroplasts, as have all the necessary enzymes.

The group also discovered that ATP is needed for an additional reaction in the Calvin–Benson cycle: the reaction producing RuBP (ribulose *bis*phosphate) from RuMP (ribulose *mono*phosphate). This additional ATP requirement makes it even easier to understand why turning off the light drastically reduces the concentration of RuBP (X in Figure 8.18).

Figure 8.20 summarizes the key features of the Calvin–Benson cycle. RuBP reacts with CO_2 to form 3PG. Then 3PG is reduced—in a reaction requiring ATP, as well as hydrogens provided by NADPH + H^+—to a three-carbon sugar phosphate (G3P). What follows this step is a complex sequence of reactions with two principal outcomes: the formation of more RuBP, and the release of products such as glucose. The

ments. The first thing they noticed was that only one compound exhibited the concentration changes proposed for 3PG—and that was 3PG itself. This result showed that their model was likely to be accurate.

But would any compound behave in the way predicted for the mysterious compound X, the CO_2 acceptor? Yes—and only one: a five-carbon sugar phosphate called **RuBP** (ribulose bisphosphate). It seemed, then, that instead of the originally proposed reaction (in which CO_2 was thought to react with a two-carbon sugar to form 3PG), there must be a reaction in which CO_2, with its single carbon, combines with the five-carbon RuBP to give two molecules of the three-carbon 3PG (Figure 8.19).

The fate of the carbon atom in CO_2 is followed in red.

Carbon dioxide

Ribulose bisphosphate (RuBP)

The enzyme rubisco catalyzes the reaction of CO_2 with RuBP.

Six-carbon skeleton of reaction intermediate

The reaction intermediate splits into two molecules of 3-phosphoglyceric acid (3PG).

8.19 RuBP Is the Carbon Dioxide Acceptor Ribulose bisphosphate (RuBP) is the CO_2-accepting compound X in Figures 8.17 and 8.18.

production of one molecule of glucose ($C_6H_{12}O_6$) requires the Calvin–Benson cycle to operate six times on six CO_2 molecules. Just as the respiration of one mole of glucose *yields* 686 kcal (2,867 kJ/mol) of energy (see Chapter 7), the production of one mole of glucose from CO_2 *requires* 686 kcal (2,867 kJ/mol).

Figure 8.21 gives a general summary of photosynthesis. The glucose produced in photosynthesis is subsequently used to make other compounds besides sugars. The carbon of glucose is incorporated into

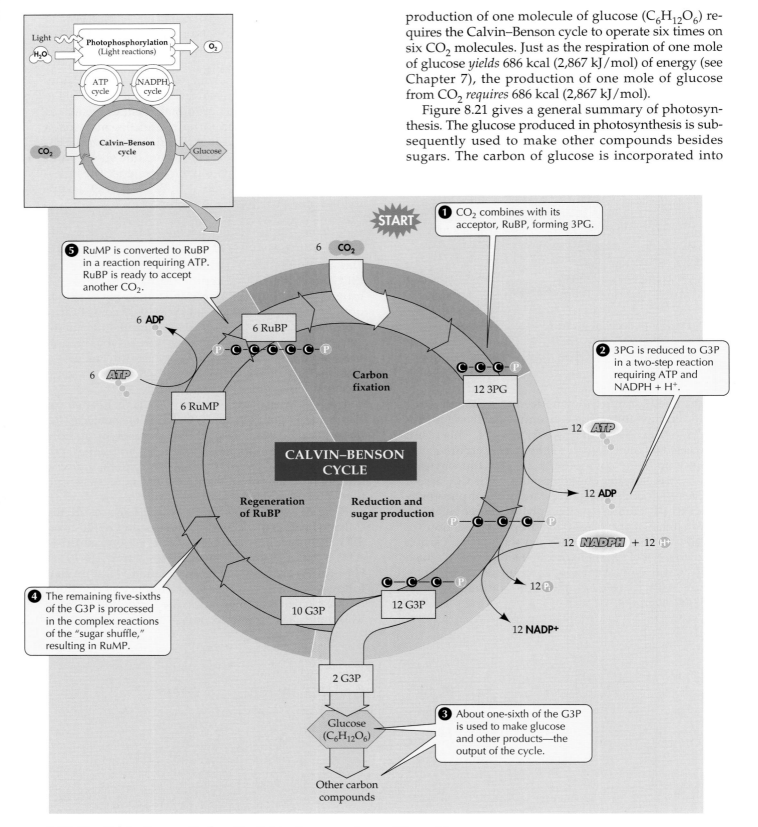

8.20 The Calvin–Benson Cycle The Calvin–Benson cycle uses CO_2 and the ATP and NADPH + H⁺ generated in the light reactions to produce glucose. This diagram shows only the key steps; the values given are those necessary to make one molecule of glucose, which requires six "turns" of the cycle.

amino acids, lipids, and the building blocks of the nucleic acids.

The products of the Calvin–Benson cycle are of crucial importance to the entire biosphere, for they serve as the food for all of life. Their covalent bonds represent the total energy yield from the harvesting of light by plants. Most of this stored energy is released by the plants themselves in their own glycolysis and cellular respiration. However, much plant matter ends up being consumed by animals. Glycolysis and cellular

respiration in the animals then releases free energy from the plant matter for use in the animal cells.

Photorespiration and Its Evolutionary Consequences

The enzyme rubisco is the most abundant protein in the biosphere. Its properties are remarkably identical in all photosynthetic organisms, from bacteria to flowering plants. Its operation is the basis for the life of nearly

8.21 An Overview of the Photosynthetic Reactions

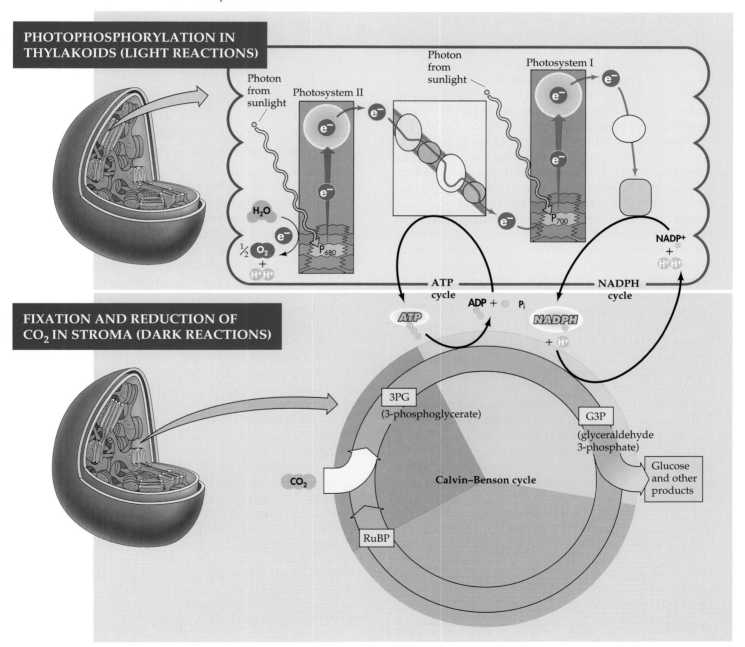

all heterotrophic organisms. However, some properties of this enzyme severely limit its effectiveness.

In the discussion that follows, we will identify and explore some of these limitations and see how evolution has constructed bypasses around them. First we'll look at photorespiration, and then we'll examine alternative pathways and plant anatomy that compensate for the limitations of rubisco and photorespiration.

In photorespiration, RuBP reacts with O_2

The substrate specificity of the enzyme rubisco is not limited to CO_2. Rubisco also functions as an oxidase, catalyzing the reaction of RuBP with O_2. (Indeed, the name *rubisco* stands for *ribulose bis*phosphate *carboxy*lase/*o*xygenase.) Because this reaction requires light and takes up oxygen, it has been termed **photorespiration**.

The oxygenase function of rubisco is favored at higher temperatures (above 28°C) when CO_2 levels are low or O_2 levels are high. This temperature sensitivity of CO_2 fixation by rubisco will be important later in understanding how photosynthesis is accomplished under hot, dry conditions.

When RuBP and O_2 react, one of the products is glycolate, a two-carbon compound that leaves the chloroplasts and diffuses into membrane-enclosed organelles called *peroxisomes* (Figure 8.22). In peroxisomes, glycolate is oxidized by O_2 in a pathway whose product diffuses into mitochondria, where it is acted on and releases CO_2—undoing the carbon-fixing work of the Calvin–Benson cycle. Chloroplasts, peroxisomes, and mitochondria all play parts in photorespiration.

Photorespiration interferes with photosynthesis. In fact, it apparently reverses it—but without resulting in ATP formation as does cellular respiration. One estimate suggests that photorespiration reduces the rate of photosynthesis by 25 percent. This is a very significant detriment to plant growth and, put in human terms, crop productivity.

The role of photorespiration in the life of the plant is unknown. Perhaps it has no positive role and is merely wasteful. Perhaps it is an evolutionary "leftover" from the days when O_2 was still building up in Earth's atmosphere and hence was less concentrated. With this in mind, many scientists are attempting to develop a gene that codes for a form of rubisco that recognizes only CO_2 as its substrate, and to insert that gene into crop plants. (See Chapter 16 for a discussion of this recombinant-DNA technology.)

It seems odd that rubisco, so abundant in the living world, apparently functions less than optimally. Most types of plants photorespire away a substantial fraction of the CO_2 initially fixed in photosynthesis. But, as we are about to see, some plants have minimized photorespiration, thus maximizing the efficiency of their photosynthesis.

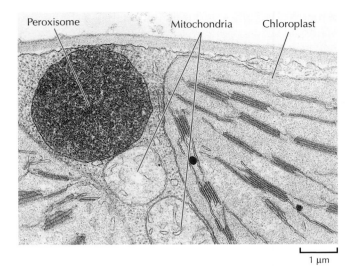

Peroxisome   Mitochondria   Chloroplast

1 µm

8.22 Photorespiration: Chloroplasts, Peroxisomes, and Mitochondria Within the peroxisome, glycolate from chloroplasts is oxidized by O_2. The product diffuses into mitochondria, where it is acted on and releases CO_2.

Some plants have evolved systems to bypass photorespiration

The discoveries of the Berkeley group who elucidated the Calvin–Benson cycle led to the expectation that the exposure of a plant to both light and $^{14}CO_2$ would always lead to the appearance of the three-carbon compound 3-phospho[^{14}C]glycerate (3PG) as the first labeled product of CO_2 fixation. Thus scientists were surprised when they learned that in many plants, the first products of CO_2 fixation are four-carbon compounds. They named these plants **C_4 plants** in contrast to the C_3 plants, which make the three-carbon 3PG as their first products.

C_4 plants perform the normal Calvin–Benson cycle, but they have an additional early reaction that fixes CO_2 without losing carbon to photorespiration, greatly increasing the overall photosynthetic yield. Because this initial CO_2 fixation step functions even at low levels of CO_2 and high temperatures, C_4 plants very effectively optimize photosynthesis under conditions that inhibit the photosynthesis of C_3 plants.

C_4 plants live in dry environments exposed to high temperatures during daylight hours. Under these conditions, C_4 plants such as corn, sugar cane, and crabgrass maintain high rates of photosynthesis and growth, even though their stomata must close at times to limit water loss. In C_3 plants, which use only the Calvin–Benson cycle, any daytime closure of the stomata to limit water loss also limits the access of cells to CO_2 from the air, thus limiting their ability to photosynthesize. Why can C_4 plants tolerate a reduced CO_2 supply better than C_3 plants?

The leaves of C_4 plants contain the enzyme **PEP carboxylase** (phosphoenolpyruvate carboxylase), which catalyzes the reaction of PEP (phosphoenolpyruvate, a

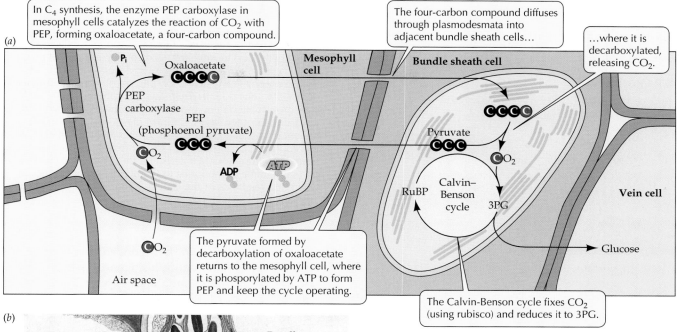

(a)

In C$_4$ synthesis, the enzyme PEP carboxylase in mesophyll cells catalyzes the reaction of CO$_2$ with PEP, forming oxaloacetate, a four-carbon compound.

The four-carbon compound diffuses through plasmodesmata into adjacent bundle sheath cells...

...where it is decarboxylated, releasing CO$_2$.

Mesophyll cell

Bundle sheath cell

P$_i$

Oxaloacetate
CCCC

PEP carboxylase

PEP (phosphoenol pyruvate)
CCC

CO$_2$

ADP

ATP

CCCC

Pyruvate
CCC

CO$_2$

RuBP

Calvin–Benson cycle

3PG

Vein cell

Glucose

The pyruvate formed by decarboxylation of oxaloacetate returns to the mesophyll cell, where it is phosporylated by ATP to form PEP and keep the cycle operating.

CO$_2$

Air space

The Calvin-Benson cycle fixes CO$_2$ (using rubisco) and reduces it to 3PG.

(b)

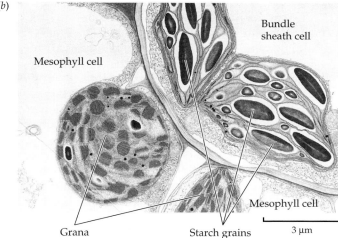

Bundle sheath cell

Mesophyll cell

Mesophyll cell

Grana

Starch grains

3 µm

8.23 C$_4$ Photosynthesis (a) The biochemistry of C$_4$ photosynthesis allows these plants to fix carbon and optimizes their fitness in hot, dry conditions. (b) In this electron micrograph, the chloroplasts in the mesophyll cells (left) have numerous grana and few starch grains. In contrast, the chloroplasts in the bundle sheath cell (right) have abundant starch grains, indicating that the enzymes of the Calvin–Benson cycle are actively forming 3PG, from which glucose and then startch are formed.

three-carbon acid) with CO$_2$ to yield the four-carbon compound *oxaloacetate* as the first product of CO$_2$ fixation (Figure 8.23). PEP carboxylase has a greater affinity for CO$_2$ than does rubisco, allowing C$_4$ plants to trap CO$_2$ more effectively than C$_3$ plants when stomatal closure depletes the CO$_2$ supply. And, because it lacks an oxygenase function, PEP carboxylase does not support photorespiration.

In C$_4$ plants, the two processes—initial CO$_2$ fixation and the Calvin–Benson cycle—take place in different cells. For the Calvin–Benson cycle to operate, oxaloacetate or another four-carbon compound derived from it diffuses from mesophyll cells to bundle sheath cells (see Figure 8.23).

The leaf anatomy of C$_4$ plants differs from that of C$_3$ plants

C$_3$ plants have only one type of cell capable of photosynthesis (Figure 8.24*a*). In contrast, the leaves of C$_4$ plants have two classes of photosynthetic cells, each with a distinctive type of chloroplast. The cells are arranged as shown in Figure 8.24*b*: a photosynthetic **mesophyll** layer surrounds an inner layer of photosynthetic **bundle sheath cells**. In C$_4$ plants, the four-carbon compound that is the product of initial CO$_2$ fixation diffuses out of the mesophyll cells and into the bundle sheath cells (see Figure 8.23).

In the bundle sheath cells, the four-carbon compound loses a carboxyl group to release CO$_2$, which is recaptured by rubisco and used in the Calvin–Benson cycle—the cycle that in C$_3$ plants takes place entirely within the mesophyll cells (see Figure 8.23). The chloroplasts in the bundle sheath cells of C$_4$ plants lack well-developed grana but typically have numerous, large starch grains, indicating that they (rather than the mesophyll cells) are the sites where sugars and starches form.

To summarize, in C$_4$ plants, the mesophyll cells pump CO$_2$ from a region where its concentration is low (the intercellular spaces within the leaf) to the

(a) **Arrangement of cells in a C₃ leaf**

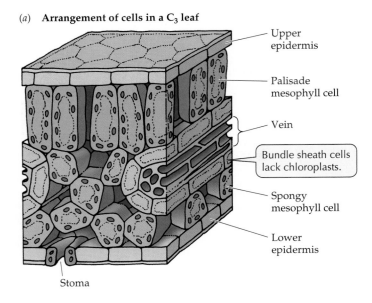

- Upper epidermis
- Palisade mesophyll cell
- Vein
- Bundle sheath cells lack chloroplasts.
- Spongy mesophyll cell
- Lower epidermis
- Stoma

(b) **Arrangement of cells in a C₄ leaf**

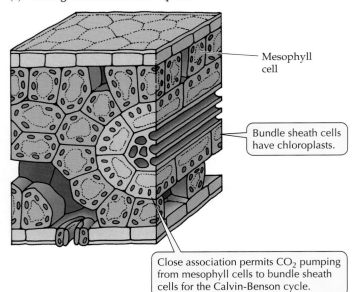

- Mesophyll cell
- Bundle sheath cells have chloroplasts.
- Close association permits CO_2 pumping from mesophyll cells to bundle sheath cells for the Calvin-Benson cycle.

8.24 The Different Leaf Anatomy of C₃ and C₄ Plants
(a) In the leaf of a C₃ plant, the bundle sheath cells surrounding the vascular elements of a vein are relatively small; the upper part of the leaf is filled with upright palisade mesophyll cells; and loosely arranged spongy mesophyll cells allow gases to circulate in the lower layers within the leaf. Both mesophyll layers carry on photosynthesis, but the bundle sheath cells do not. (b) In a C₄ leaf, the bundle sheath cells are usually larger and contain prominent chloroplasts toward their outer edges; uniform mesophyll cells surround the entire vascular bundle. This arrangement facilitates the incorporation of carbon from CO_2 into four-carbon compounds by the mesophyll cells and the passage of these carbon-containing compounds to the bundle sheath cells, where the reactions of the Calvin–Benson cycle take place.

bundle sheath cells, so their concentration of CO_2 is high enough to maintain photosynthesis even on the hot, dry days when stomata are often closed and the temperature normally would favor the oxygenase function of rubisco. This system is effective because PEP carboxylase in the mesophyll cells can fix CO_2 at temperatures too high for rubisco to fix it effectively, and because PEP carboxylase, unlike rubisco, does not function as an oxygenase and hence its action does not lead to wasteful photorespiration. Table 8.1 compares C₃ and C₄ photosynthesis.

A related but distinguishable system called crassulacean acid metabolism (CAM) functions in certain plants that face frequent water shortages.

CAM plants also use PEP carboxylase

Some plants use PEP carboxylase to fix and accumulate CO_2 while conserving water during hot, dry daylight hours—including some succulents (water-storing plants of some families, such as the Crassulaceae), many cacti, pineapples, and several other kinds of flowering plants. These plants conserve water by keeping their stomata closed during the daylight hours, thus minimizing water loss by evaporation. How, then, can they perform photosynthesis? Their

TABLE 8.1 Comparison of Photosynthesis in C₃ and C₄ Plants

VARIABLE	C₃ PLANTS	C₄ PLANTS
Photorespiration	Extensive	Minimal
Perform Calvin–Benson cycle?	Yes	Yes
Primary CO_2 acceptor	RuBP	PEP
CO_2-fixing enzyme	Rubisco (RuBP carboxylase/oxygenase)	PEP carboxylase
First product of CO_2 fixation	3PG (3-carbon compound)	Oxaloacetate (4-carbon compound)
Affinity of carboxylase for CO_2	Moderate	High
Leaf anatomy: photosynthetic cells	Mesophyll	Mesophyll + bundle sheath
Classes of chloroplasts	One	Two

trick is to open their stomata at night and store CO_2 by a different mechanism.

The CO_2 metabolism of these plants is called **crassulacean acid metabolism**, or **CAM**, after the succulents in which it was first discovered. CAM is much like the metabolism of C_4 plants: CO_2 is initially fixed into four-carbon compounds that accumulate. In CAM plants, however, the two processes (initial CO_2 fixation and the Cal-vin–Benson cycle) are separated temporally (in *time*) rather than spatially (in *space*—in separate cells; Figure 8.25). And CAM plants lack the specialized cell relationships of C_4 plants.

In CAM plants, CO_2 is fixed initially in mesophyll cells during the *night*, when stomata are open and water loss is less of a problem. The products of CO_2 fixation accumulate in the vacuoles of mesophyll cells. When daylight arrives, the accumulated four-carbon compounds are shipped to the chloroplasts, where decarboxylation supplies the CO_2 for operation of the Calvin–Benson cycle, and the light reactions supply the necessary ATP and NADPH + H$^+$. We will discuss the stomatal behavior of CAM plants more thoroughly in Chapter 32.

Plants Perform Cellular Respiration

In plants, cellular respiration takes place both in the light and in the dark, whereas photosynthesis takes place strictly in the light. The site of glycolysis is the cytosol, that of respiration is the mitochondria, and that of photosynthesis is the chloroplasts. Thus photosynthesis and respiration can proceed simultaneously but in different organelles.

For a plant to live, it must photosynthesize more than it respires, giving it a net gain of carbon dioxide and energy from the environment. The excess of photosynthesis over respiration is great enough that plants—as well as photosynthetic bacteria and photosynthetic protists—can export food—and oxygen—to the animal kingdom and to all other nonphotosynthetic organisms.* Animals require both food and oxygen; they return carbon dioxide that plants may use in photosynthesis. Thus both carbon dioxide and oxygen have natural cycles.

Photosynthesis and respiration have important similarities. In eukaryotes, both processes reside in specialized organelles that have complex systems of inter-

*The exceptions are a few forms of archaea and bacteria that oxidize minerals deep in caves or on the ocean floor where light does not penetrate, and they support animal life in these isolated habitats.

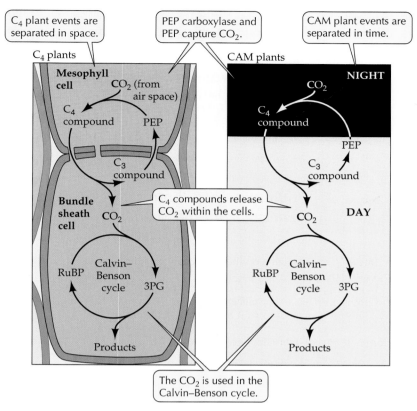

Sorghum stalks (C_4)

Sempervivum tectorum (CAM)

8.25 CAM and C_4 Plants Separate Two Sets of Reactions Differently Both plant types utilize four-carbon compounds whose production is separate from the Calvin–Benson cycle. The separation is spatial in C_4 plants, temporal in CAM plants.

nal membranes. ATP synthesis in both processes relies on the chemiosmotic mechanism, involving the pumping of protons across a membrane. Another key feature of both respiration and photosynthesis is electron transport, that is, the passing of electrons from carrier to carrier in a series of exergonic redox reactions.

In respiration, the carriers receive electrons from high-energy food molecules and pass them ultimately to oxygen, forming water. Photosynthesis, on the other hand, requires an input of light energy to make chlorophyll a reducing agent that is strong enough to initiate the transfer of electrons. In photosynthesis, water is the source of the electrons, and oxygen is released from water in a very early step. The electrons from water end up in NADPH and, finally, in food molecules.

Summary of "Photosynthesis: Energy from the Sun"

• Life on Earth depends on the absorption of light energy from the sun.
• In plants, photosynthesis takes place in chloroplasts.
• Photosynthetic autotrophs such as plants trap solar energy to nourish themselves and heterotrophs. Heterotrophs such as animals must obtain energy from autotrophs, or from heterotrophs that have eaten autotrophs.

Identifying Photosynthetic Reactants and Products

• Photosynthesizing plants take in CO_2, water, and light energy, producing O_2 and energy-rich, reduced carbon compounds. The overall reaction is $6 CO_2 + 12 H_2O + \text{light} \rightarrow C_6H_{12}O_6 + 6 O_2 + 6 H_2O$. The oxygen atoms in O_2 come from water, not from CO_2. **Review Figures 8.1, 8.2**

The Two Pathways of Photosynthesis: An Overview

• In the light-dependent reactions of photosynthesis, photophosphorylation produces ATP and reduces $NADP^+$ to $NADPH + H^+$. **Review Figure 8.3**
• ATP and $NADPH + H^+$ are needed for the reactions that fix and reduce CO_2 in the Calvin–Benson cycle, forming sugars. **Review Figure 8.3**

Properties of Light and Pigments

• Light energy comes in packets called photons, but it also has wavelike properties. **Review Figure 8.4**
• Pigments absorb light in the visible spectrum. **Review Figure 8.5**
• Absorption of a photon puts a pigment molecule in an excited state that has more energy than the ground state. **Review Figure 8.6**
• Each compound has a characteristic absorption spectrum. An action spectrum reveals the biological effectiveness of different wavelengths of light. **Review Figures 8.7, 8.8**
• Chlorophyll is essential for the photosynthesis of plants and algae. Chlorophyll and accessory pigments form antenna systems for absorption of light energy. If antenna systems funnel sufficient energy to a reactive chlorophyll molecule, it gives up an electron, reducing a carrier. **Review Figures 8.9, 8.10**
• An excited pigment molecule may lose its energy by fluorescence, by transferring it to another pigment molecule, or by transferring an electron to an oxidizing agent. **Review Figure 8.10**
• Energized electrons from excited chlorophyll are used to make ATP and $NADPH + H^+$.

Photophosphorylation and Reductions

• Noncyclic photophosphorylation uses two photosystems (I and II), producing ATP, $NADPH + H^+$, and O_2. Photosystem II uses P_{680} chlorophyll, from which light-excited electrons are passed to a redox chain that drives chemiosmotic ATP production. Light-driven oxidation of water releases O_2 and passes electrons from water to the P_{680} chlorophyll. Photosystem I passes electrons from P_{700} chlorophyll to another redox chain and then to $NADP^+$, forming $NADPH + H^+$. **Review Figure 8.11**
• Cyclic photophosphorylation uses P_{700} chlorophyll and produces only ATP. Its operation maintains the proper balance of ATP and $NADPH + H^+$ in the chloroplast. **Review Figures 8.12, 8.13**
• Chemiosmosis is the source of ATP in photophosphorylation. Electron transport pumps protons from the stroma into the thylakoids, establishing a proton-motive force. Diffusion of the protons back to the stroma via ATP synthase channels drives ATP formation from ADP and P_i. **Review Figures 8.14, 8.15**
• Cyclic photophosphorylation evolved before noncyclic photophosphorylation. Both are evolutionarily more ancient than cellular respiration, which requires the O_2 that noncyclic photophosphorylation added to the atmosphere.

Making Sugar from CO_2: The Calvin–Benson Cycle

• The Calvin–Benson cycle makes sugar from CO_2. The reactions of the cycle were elucidated by a combination of theorizing and experimentation using a radioactive carbon isotope, paper partition chromatography, and autoradiography. **Review Figures 8.16, 8.17, 8.18**
• The Calvin–Benson cycle consists of three phases: fixation of CO_2, reduction and sugar production, and regeneration of RuBP. RuBP is the initial CO_2 acceptor, and 3PG is the first stable product of CO_2 fixation. The enzyme rubisco catalyzes the reaction of CO_2 and RuBP to form 3PG. The remaining intermediates of the Calvin–Benson cycle include a variety of sugar phosphates with four, five, six, and seven carbon atoms, and some of the reactions use ATP and $NADPH + H^+$. **Review Figures 8. 19, 8.20, 8.21**

Photorespiration and Its Evolutionary Consequences

• The enzyme rubisco can catalyze a reaction between O_2 and RuBP in addition to the reaction between CO_2 and RuBP. This consumption of O_2 is called photorespiration and significantly reduces the efficiency of photosynthesis. The reactions that constitute photorespiration are distributed over three organelles: chloroplasts, peroxisomes, and mitochondria.
• At high temperatures and low CO_2 concentrations, the oxygenase function of rubisco is favored, and the dominance of photorespiration would prevent plants from living in arid climates were it not for bypass reactions.
• C_4 plants bypass photorespiration with special chemical reactions and specialized leaf anatomy. In C_4 plants, PEP carboxylase in mesophyll chloroplasts initially fixes CO_2 in four-carbon acids that then diffuse into bundle sheath cells, where their decarboxylation produces locally high concentrations of CO_2 for rubisco operation and normal functioning of the Calvin–Benson cycle. **Review Figures 8.23, 8.24**
• CAM plants operate much like C_4 plants, but their initial CO_2 fixation by PEP carboxylase is temporally separated from the Calvin–Benson cycle, rather than spatially separated as in C_4 plants. **Review Figure 8.25**

Plants Perform Cellular Respiration

- Plants respire both in the light and in the dark but photosynthesize only in the light. To survive, a plant must photosynthesize more than it respires, giving it a net gain of reduced energy-rich compounds.
- Photosynthesis and respiration have several similarities. Both use systems of electron transport. Both rely on the chemiosmotic mechanism for the synthesis of ATP. However, photosynthesis traps light energy and converts it to chemical forms (sugars). Cellular respiration converts one form of chemical energy (sugar) to another (ATP).

Self-Quiz

1. Which statement about light is *not* true?
   a. Its velocity in a vacuum is constant.
   b. It is a form of energy.
   c. The energy of a photon is directly proportional to its wavelength.
   d. A photon of blue light has more energy than one of red light.
   e. Different colors correspond to different frequencies.

2. Which statement about light is true?
   a. An absorption spectrum is a plot of biological effectiveness versus wavelength.
   b. An absorption spectrum may be a good means of identifying a pigment.
   c. Light need not be absorbed to produce a biological effect.
   d. A given kind of molecule can occupy any energy level.
   e. A pigment loses energy as it absorbs a photon.

3. Which statement about chlorophylls is *not* true?
   a. They absorb light near both ends of the visible spectrum.
   b. They can accept energy from other pigments, such as carotenoids.
   c. Excited chlorophyll can either reduce another substance or fluoresce.
   d. Excited chlorophyll is an oxidizing agent.
   e. They contain magnesium.

4. In cyclic photophosphorylation
   a. oxygen gas is released.
   b. ATP is formed.
   c. water donates electrons and protons.
   d. NADPH + H$^+$ forms.
   e. CO$_2$ reacts with RuBP.

5. Which of the following does *not* happen in noncyclic photophosphorylation?
   a. Oxygen gas is released.
   b. ATP forms.
   c. Water donates electrons and protons.
   d. NADPH + H$^+$ forms.
   e. CO$_2$ reacts with RuBP.

6. In the chloroplast
   a. light leads to the pumping of protons out of the thylakoids.
   b. ATP forms when protons are pumped into the thylakoids.
   c. light causes the stroma to become more acidic than the thylakoids.
   d. protons return passively to the stroma through protein channels.
   e. proton pumping requires ATP.

7. Which statement about the Calvin–Benson cycle is *not* true?
   a. CO$_2$ reacts with RuBP to form 3PG.
   b. RuBP forms by the metabolism of 3PG.
   c. ATP and NADPH + H$^+$ form when 3PG is reduced.
   d. The concentration of 3PG rises if the light is switched off.
   e. Rubisco catalyzes the reaction of CO$_2$ and RuBP.

8. In C$_4$ photosynthesis
   a. 3PG is the first product of CO$_2$ fixation.
   b. rubisco catalyzes the first step in the pathway.
   c. four-carbon acids are formed by PEP carboxylase in bundle sheath cells.
   d. photosynthesis continues at lower CO$_2$ levels than in C$_3$ plants.
   e. CO$_2$ released from RuBP is transferred to PEP.

9. C$_4$ photosynthesis and crassulacean acid metabolism differ in that
   a. only C$_4$ photosynthesis uses PEP carboxylase.
   b. CO$_2$ is trapped by night in CAM plants and by day in C$_4$ plants.
   c. four-carbon acids are formed only in C$_4$ photosynthesis.
   d. only Crassulaceae commonly grow in dry or salty environments.
   e. only C$_4$ photosynthesis helps conserve water.

10. Photorespiration
    a. takes place only in C$_4$ plants.
    b. includes reactions carried out in peroxisomes.
    c. increases the yield of photosynthesis.
    d. is catalyzed by PEP carboxylase.
    e. is independent of light intensity.

Applying Concepts

1. Both photophosphorylation and the Calvin–Benson cycle stop when the light is turned off. Which specific reaction stops first? Which stops next? Continue answering the question "Which stops next?" until you have explained why both pathways have stopped.

2. In what principal ways are the reactions of photophosphorylation similar to the respiratory chain and oxidative phosphorylation discussed in Chapter 7? Differentiate between cyclic and noncyclic photophosphorylation in terms of (1) the products and (2) the source of electrons for the reduction of oxidized chlorophyll.

3. Draw a cartoon representation of noncyclic photophosphorylation using elements similar to those in Figure 8.13. (*Hint:* You have to add a second "photon hammer.")

4. The development of what three experimental techniques made it possible to elucidate the Calvin–Benson cycle? How were these techniques used in the investigation?

5. If water labeled with ^{18}O is added to a suspension of photosynthesizing chloroplasts, which of the following compounds will first become labeled with ^{18}O: ATP, NADPH, O$_2$, or 3PG? If water labeled with ^{3}H is added to a suspension of photosynthesizing chloroplasts, which of the same compounds will first become radioactive? If CO$_2$ labeled with ^{14}C is added to a suspension of photosynthesizing chloroplasts, which of those compounds will first become radioactive?

Readings

Alberts, B., D. Bray, J. Lewis, M. Raff, K. Roberts and J. D. Watson. 1994. *Molecular Biology of the Cell*, 3rd Edition. Garland Publishing, New York. Chapter 14, on energy conversion, contains a good discussion of photosynthesis.

Bazzazz, F. A. and E. D. Fajer. 1992. "Plant Life in a CO_2-Rich World." *Scientific American*, January. A comparison of C_3 and C_4 plants and their prospects.

Govindjee, and W. J. Coleman. 1990. "How Plants Make Oxygen." *Scientific American*, February. A description of the "clock" in photosystem II that splits water into oxygen gas, protons, and electrons.

Hall, D. O. and K. K. Rao. 1987. *Photosynthesis*, 4th Edition. Edward Arnold, New York. An intermediate-level treatment of all the major topics in photosynthesis and an excellent bibliography, all in 100 pages.

Stryer, L. 1995. *Biochemistry*, 4th Edition. W. H. Freeman, New York. Chapter 26 gives an advanced but clear treatment of topics in photosynthesis.

Voet, D. and J. G. Voet. 1995. *Biochemistry*, 2nd Edition. John Wiley & Sons, New York. Chapter 22 discusses photosynthesis.

Weinberg, C. J. and R. H. Williams. 1990. "Energy from the Sun." *Scientific American*, September. Photosynthesis and biomass technology, along with other solar-derived technologies such as wind and solar-thermal, are considered as sources of energy for industrial and other uses.

Youvan, D. C. and B. L. Marrs. 1987. "Molecular Mechanisms of Photosynthesis." *Scientific American*, June. A difficult but interesting article on events in the first fraction of a millisecond of photosynthesis in a bacterium. Part of the article is better read after reading Part Two of this book.

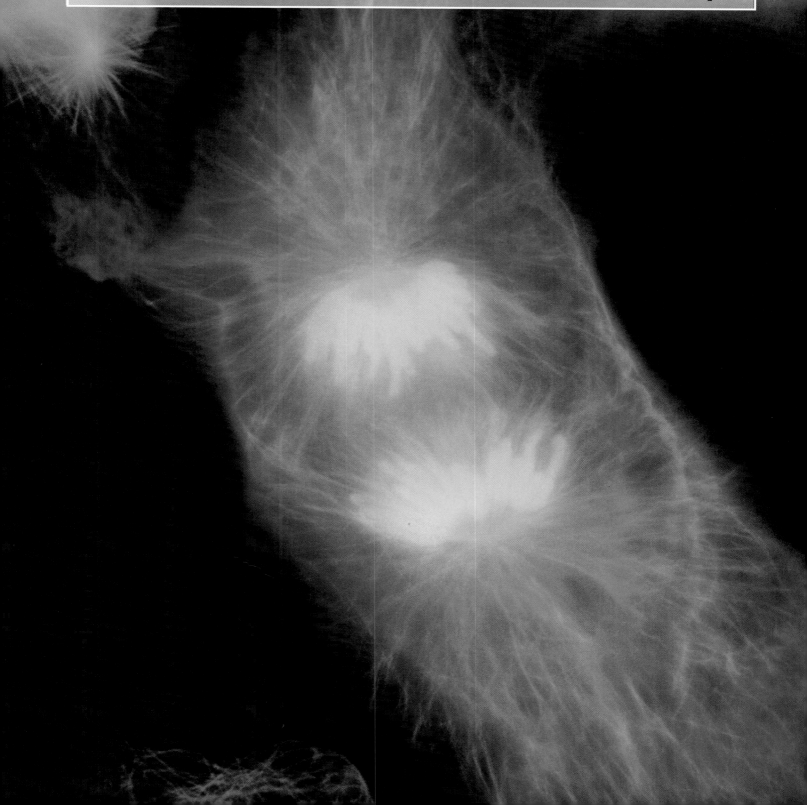

Part Two
Information and Heredity

Chapter 9

Chromosomes, the Cell Cycle, and Cell Division

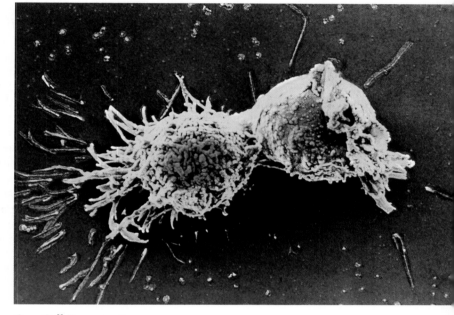

One Cell Becomes Two
The two human cells in this scanning electron micrograph are in the final stage of mitosis—the division of a single cell into two daughter cells—and will soon separate from each other.

O f the more than 100 trillion (10^{14}) cells that make up an adult human, just the right cells must divide, at just the right time, to produce and maintain a healthy individual. Cell division is initiated by both internal and external signals, and these signals are highly regulated. When signals go wrong and cells divide inappropriately, a disorganized mass of cells—a tumor—may result. But in both tumor cells and normal cells, cell division consists of the same precise and elegant sequence of events. The nuclear DNA that controls cell function makes two copies of itself, and these copies are exactly distributed to the two new cells. The cytoplasm and its associated organelles are also parceled out to the two new cells. Finally, the two cells are separated by plasma membranes and, in some cases, cell walls.

Whereas unicellular organisms use cell division only as a means to reproduce, in multicellular organisms cell division also plays important roles in the growth and repair of tissues (Figure 9.1). The fact that a single cell, the fertilized egg, gives rise to a newborn baby with trillions of cells implies that development includes a lot of cell division. Less obvious but no less significant is the need to replace billions of blood cells and epithelial cells that die every day. In addition, a special type of reductive cell division is involved in forming the specialized reproductive cells—eggs and sperm—in complex eukaryotes.

In this chapter, after briefly discussing cell division in prokaryotes, we will describe the two types of eukaryotic cell division—mitosis and meiosis—and relate these to asexual and sexual reproduction.

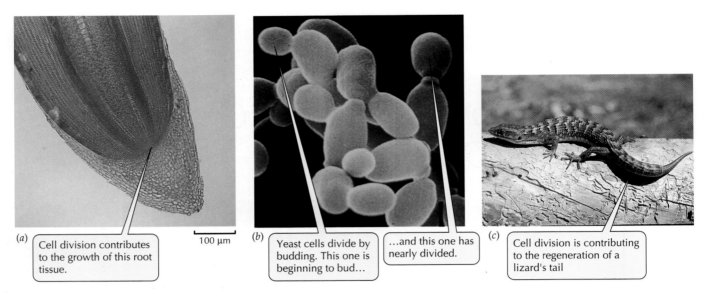

(a) Cell division contributes to the growth of this root tissue.

100 μm

(b) Yeast cells divide by budding. This one is beginning to bud...

...and this one has nearly divided.

(c) Cell division is contributing to the regeneration of a lizard's tail

9.1 Important Consequences of Cell Division Cell division is the basis for growth, reproduction, and regeneration.

Systems of Cell Division

Prokaryotes typically have a single chromosome

Prokaryotic cells, by definition, lack nuclei. Hence, they do not use the same mechanisms for cell division that eukaryotic cells use. Even so, when a prokaryote divides, by a process called **fission**, genetic information is distributed in an orderly fashion to its daughter cells.

A **chromosome** is a DNA molecule that contains the genetic information for an organism. Most prokaryotes have only one chromosome, a single long DNA molecule with proteins bound to it. In the bacterium *Escherichia coli*, the chromosome is a circular molecule of DNA about 1.6 million nm (1.6 mm) long. The bacterium itself is about 1 μm (1,000 nm) in diameter and about 4 μm long. Thus the space into which the long thread of DNA is packed in the bacterial nucleoid is very small relative to the length of the DNA molecule. It is thus not surprising that the molecule usually appears in electron micrographs as a hopeless tangle of fibers (Figure 9.2). The DNA molecule accomplishes some packing by folding on itself, and proteins bound to DNA contribute to this packing.

When prokaryotic cells are gently lysed (broken open) to release their contents, the chromosome sometimes becomes less tangled and spreads out. Several techniques have shown that the prokaryotic chromosome is a closed circle rather than the linear structure found in eukaryotes. Circular chromosomes are probably to be found in all prokaryotes, as well as in some viruses and in the chloroplasts and mitochondria of eukaryotic cells.

The prokaryotic chromosome is attached to the plasma membrane. When a new DNA molecule forms from the old one, it, too, attaches to the membrane. As new membrane material is added between the two sites of attachment, the two DNA molecules gradually separate. As the cell divides, new wall and membrane are inserted between the two chromosomes to form

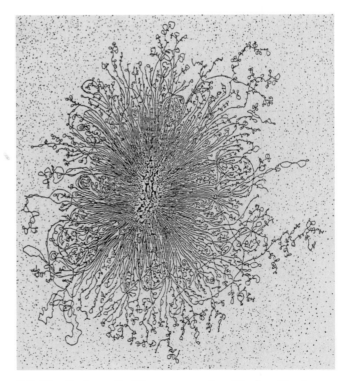

9.2 The Prokaryotic Chromosome Is a Circle The long, looping fibers of DNA from this cell of the bacterium *Escherichia coli* are all part of one continuous circular chromosome.

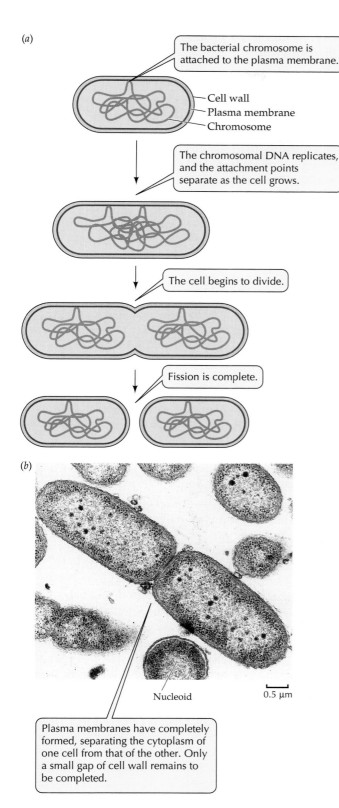

(a)

The bacterial chromosome is attached to the plasma membrane.

Cell wall
Plasma membrane
Chromosome

The chromosomal DNA replicates, and the attachment points separate as the cell grows.

The cell begins to divide.

Fission is complete.

Nucleoid

0.5 μm

Plasma membranes have completely formed, separating the cytoplasm of one cell from that of the other. Only a small gap of cell wall remains to be completed.

9.3 Prokaryotic Cell Division (a) The steps of cell division in prokaryotes. (b) These two cells of the bacterium *Pseudomonas aeruginosa* have almost completed fission. Each cell contains a complete chromosome, visible as the nucleoid in the center of the cell.

two daughter cells, each with an identical chromosome (Figure 9.3).

These three processes—DNA replication, chromosome separation, and cell division—occur in about 30 minutes total in an adequately nourished prokaryotic cell. This is a sharp contrast to the hours and even days it takes for eukaryotes to complete cell division.

Eukaryotic cells divide by mitosis or meiosis

The intricate mechanism mentioned at the beginning of this chapter that parcels out the duplicated copies of genetic material for cell division in eukaryotes is called **mitosis**. This replicating process produces identical nuclei for all the cells of an organism's adult body. A second mechanism for nuclear division, called *meiosis*, generates diversity among nuclei and plays a key role in sexual life cycles. (We will discuss meiosis in detail later in the chapter.)

The duplication of a eukaryotic cell typically consists of three steps: (1) the replication of the genetic material within the nucleus, (2) the packaging and separation of the genetic material into two new nuclei, and (3) the division of the cytoplasm. Between divisions—that is, for most of its life—a eukaryotic cell is in a condition called **interphase**.

What determines whether a cell in interphase will divide? How does mitosis lead to exact copies, and to diversity of products? Why do we need both exact copies and diverse products? Why do most organisms have sex in their life cycles? In the pages that follow we will describe the details of mitosis, meiosis, and interphase, as well as their consequences for heredity, development, and evolution.

Interphase and the Control of Cell Division

A cell lives and functions until it divides or dies—or, if it is a sex cell, until it fuses with another sex cell. Some, such as red blood cells, muscle cells, and nerve cells, lose the capacity to divide as they mature. Other cell types, such as cortical cells in plant stems, divide only rarely. Most cells, however, have some probability of dividing, and some are specialized for rapid division. For many kinds of cells we may speak of a **cell cycle** that has two phases: mitosis and interphase.

A given cell lives for one turn of the cell cycle and then becomes two cells. The cycle repeats again and again, a constant source of renewal. Even in tissues engaged in rapid growth, however, the cell cycle consists mainly of the time spent in interphase. Examination of any collection of cells, such as the tip of a root or a slice of liver, reveals that most of the cells are in interphase most of the time. Only a small percentage of the cells are in mitosis at any given moment. We can confirm this fact, in certain cultures of cells, by watching a single cell through its entire cycle.

In this section, we will describe the cell cycle events that occur during interphase, especially the "decision" to enter mitosis.

The subphases of interphase have different functions

Interphase consists of three subphases, identified as S, G1, and G2. The cell's DNA replicates during the **S phase** (the *S* stands for *synthesis*). The gap between the end of mitosis and the onset of S phase is called **G1**, or Gap 1. Another gap—**G2**—separates the end of the S phase and the beginning of mitosis. Mitosis itself is referred to as the **M phase** of the cell cycle (Figure 9.4). If a cell is not going to divide, it may remain in G1 for weeks or even years until it dies; the cell seemingly will not waste matter and energy replicating its genetic material in the S phase if it is not going to divide. (There are some exceptions in which cells that will not divide do synthesize DNA and are thus stuck in G2, but the continuation of G1 is the rule in the vast majority of nondividing cells.)

Although one key event—DNA replication—dominates and defines the S phase, important cell cycle events take place in the gap phases as well. G1 is quite variable in length; some rapidly dividing embryonic cells have dispensed with it entirely. The biochemical hallmark of a G1 cell is that it is preparing for S phase. It is at this time that the "decision" to enter another cell cycle is made. During G2, the cell makes preparations for mitosis, synthesizing components of the microtubules that form the spindle, for example.

Cyclins and other proteins signal events in the cell cycle

How are *appropriate* decisions to enter the S or M phases made? These transitions—from G1 to S and from G2 to M—require the activation of a protein complex called **cyclin-dependent kinase**, or **Cdk** (Figure 9.5). A *kinase* is an enzyme that catalyzes the transfer of a phosphate group from ATP to another molecule; such phosphate transfer is called *phosphorylation*. Cdk is a kinase that catalyzes the phosphorylation of certain amino acids in proteins. Phosphorylation by Cdk changes the three-dimensional structure of a targeted protein, sometimes simultaneously changing that protein's function. Active Cdk's catalyze the action of proteins that are important in initiating progress through the cell cycle.

The discovery that Cdk's induce mitosis is a beautiful example of research on different organisms and cells converging on a single mechanism. One group of scientists was studying immature sea urchin eggs, trying to find out how they are stimulated to divide and form mature eggs. A protein called maturation promoting factor was purified from the maturing eggs, which by itself prodded the immature eggs into division. At the same time, other scientists studying the cell cycle in yeasts, which are single-celled eukaryotes, found a strain that was stalled at the G1–S boundary. What this strain lacked was a Cdk, and it turned out that this Cdk was very similar to the sea urchin's maturation promoting factor. Similar Cdk's were soon found to control the G1–S transition in many other organisms, including humans.

But Cdk's are not active by themselves. They must be bound with a second type of protein, called **cyclins**. It is the cyclin–Cdk *complex* (in humans, cyclin D and Cdk4) that triggers protein kinase activity and the transition from G1 to S phase with the resulting DNA replication. Then the cyclin breaks down and the Cdk becomes inactive. Later in the cell cycle, another specific cyclin–Cdk partnership (cyclin B and Cdk2) takes over, activating a kinase activity that promotes the transition from G2 to chromosome condensation and mitosis.

What do cyclin–Cdk complexes target for phosphorylation? Not all such targets are known, but some are important for progression in the cell cycle. For ex-

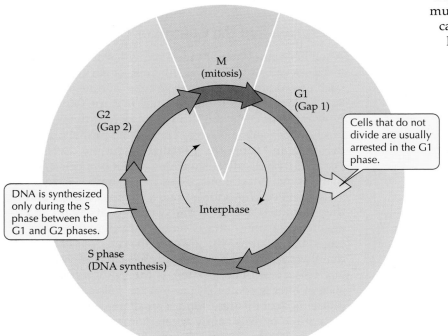

DNA is synthesized only during the S phase between the G1 and G2 phases.

Cells that do not divide are usually arrested in the G1 phase.

M (mitosis)

G1 (Gap 1)

G2 (Gap 2)

Interphase

S phase (DNA synthesis)

9.4 The Eukaryotic Cell Cycle A cell's life history is made up of a short mitosis (purple) and a longer interphase (green). Interphase has three subphases (G1, S, and G2) in cells that divide.

9.5 Cyclin-Dependent Kinase and Cyclin Trigger Decisions in the Cell Cycle

A human cell makes the "decision" to enter the cell cycle during G1 when cyclin D binds to a cyclin-dependent kinase (Cdk4). There are four such cyclin–Cdk controls during the typical cell cycle in humans.

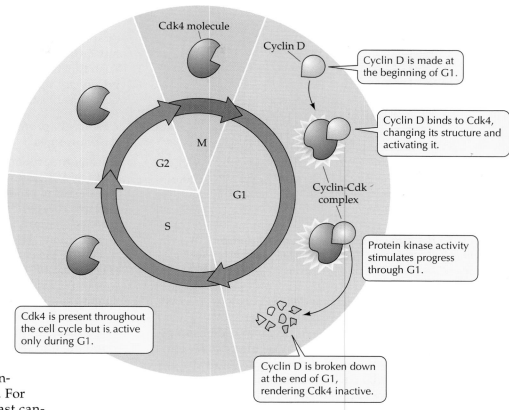

Cdk4 molecule

Cyclin D

Cyclin D is made at the beginning of G1.

Cyclin D binds to Cdk4, changing its structure and activating it.

Cyclin-Cdk complex

Protein kinase activity stimulates progress through G1.

Cdk4 is present throughout the cell cycle but is active only during G1.

Cyclin D is broken down at the end of G1, rendering Cdk4 inactive.

ample, the cyclin B–Cdk2 complex catalyzes the phosphorylation of certain target proteins, causing them to bind to DNA and initiate chromosome condensation. Phosphorylation of other target proteins results in the disaggregation of the nuclear envelope early in mitosis.

Since cancer results from inappropriate cell division, it is not surprising that the cyclin–Cdk controls are disrupted in cancer cells. For example, some fast-growing breast cancers have too much cyclin D, which overstimulates Cdk4 and cell division. Tumors of the parathyroid gland often make an entirely new cyclin that stimulates kinase activity in Cdk4. Finally, as we will describe in Chapter 17, a major protein in normal cells that prevents them from dividing is p53, which leads to inhibition of Cdk's. More than half of all human cancers have defective p53, resulting in the absence of cell cycle controls.

Mitotic inducers can stimulate cells to divide

Cyclin–Cdk complexes provide an *internal* control for progress through the cell cycle. But there are situations in the body where cells that are slowly cycling, or not cycling at all, must be stimulated to divide through *external* controls. When a person is cut and bleeds, specialized cell fragments called platelets gather at the wound and help initiate blood clotting. The platelets also produce and release a protein, *platelet-derived growth factor*, that diffuses to the adjacent cells in the skin and stimulates them to divide and heal the wound.

Other **growth factors** include *interleukins*, which are made by one type of white blood cell and promote cell division in other cells that are essential for the body's immune system defenses, and *erythropoietin*, made by the kidney to stimulate the division of bone marrow cells and the production of red blood cells. In addition, many hormones promote division in specific cells.

We will describe the different physiological roles of these external mitotic inducers in later chapters, but all growth factors act in similar ways. Growth factors bind to their target cells via specialized receptor proteins on the target cell surface. The specific binding event triggers events within the cell so that it begins a cell cycle. Cancer cells often cycle inappropriately because either they make their own growth factors or they no longer require growth factors to start cycling.

Eukaryotic Chromosomes

Most human cells other than eggs and sperm contain two full sets of genetic information, one from the mother and the other from the father. Eggs and sperm, however, contain only a single set; any particular egg or sperm in your body contains some information from your mother and some from your father. As in prokaryotes, the genetic information consists of molecules of DNA packaged as chromosomes.

However, eukaryotes have more than one chromosome, and during interphase these chromosomes reside within the cell's nucleus. Seen through a light microscope, the nucleus appears relatively featureless (except for the dark nucleolus) during most of the life of a cell (see Figure 4.8); the chromosomes cannot be seen. Mitosis has been defined historically as the stage

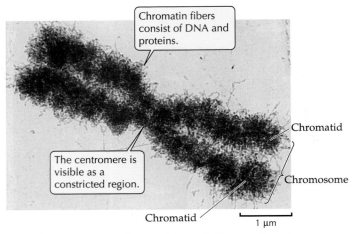

Chromatin fibers consist of DNA and proteins.

The centromere is visible as a constricted region.

Chromatid

Chromosome

Chromatid

1 μm

9.6 Chromosomes, Chromatids, and Chromatin   A human chromosome, shown as the cell prepares to divide.

of the cell cycle at which condensed chromosomes become visible in the microscope.

The basic unit of the eukaryotic chromosome is a gigantic, linear, double-stranded molecule of DNA complexed with many proteins. During most of the eukaryotic cell cycle, each chromosome contains only one such double-stranded DNA molecule. However, when the DNA molecule duplicates, the chromosome consists of two joined **chromatids**, each made up of one double-stranded DNA molecule complexed with proteins (together known as *chromatin*; Figure 9.6). At the time chromosomes become visible in a microscope, the two chromatids are joined at a specific small region of the chromosome called the **centromere**. This single-centromere structure, whether it contains one or two DNA molecules, is properly called a *mitotic* or *meiotic chromosome.*

Chromatin consists of DNA and proteins

The complex of DNA and proteins in a eukaryotic chromosome is referred to as **chromatin**. The DNA carries the genetic information; the proteins organize the chromosome physically and regulate the activities of the DNA. By mass, the amount of chromosomal proteins is equivalent to that of DNA.

Chromatin changes form dramatically during mitosis and meiosis. During interphase, the chromatin is strung out so thinly that the chromosome cannot be seen as a defined body under the light microscope. But during most of mitosis and meiosis, the chromatin is so highly coiled and compacted that the chromosome appears as a dense, bulky object (see Figure 9.6).

This alternation of forms relates to the function of chromatin during different phases of the cell cycle. Before each mitosis the genetic material is duplicated. Remember that mitosis separates the duplicated genetic material into two new nuclei. This separation is easier to accomplish if the DNA is neatly arranged in compact units rather than being tangled up like a plate of spaghetti. During interphase, however, the DNA must direct the growth and other activities of the cell. Such functions require that DNA be unwound and exposed so that it can interact with enzymes (see Chapters 12 and 14).

Chromatin proteins organize the DNA in chromosomes

Chromatin proteins associate closely with chromosomal DNA. Chromosomes contain large quantities of five classes of proteins called **histones**. Histones have a positive charge at the pH levels found in the cell. The positive charge is a result of their particular amino acid compositions. Histone molecules join together to produce complexes around which the DNA is wound (Figure 9.7). Eight histone molecules, two of each of four of the histone classes, unite to form a core or spool shaped so that the DNA molecule fits snugly in a coil around it. The other class of histone proteins (histone H1) appears to fit on the outside of the DNA, perhaps clamping it to the histone core. The resulting beadlike units, called **nucleosomes**, provide the major structure for packing the DNA in chromatin (Figure 9.8).

A chromatid has a single DNA molecule running through vast numbers of nucleosomes. The many nucleosomes of a mitotic chromatid may pack together and coil. During both mitosis and meiosis, the chromatin becomes ever more coiled and condensed, with further folding of the chromatin continuing up to the time at which chromosomes begin to move apart. Some diverse acidic proteins are also present in small quantities in chromosomes. Some of these proteins are involved in chromosome packaging; others are involved in the expression of DNA sequences.

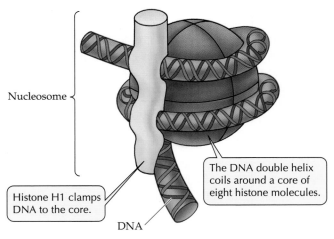

Nucleosome

Histone H1 clamps DNA to the core.

The DNA double helix coils around a core of eight histone molecules.

DNA

9.7 DNA Interacts with Histones to Form Nucleosomes   In some cases, histone H1 may be on the inside of the core.

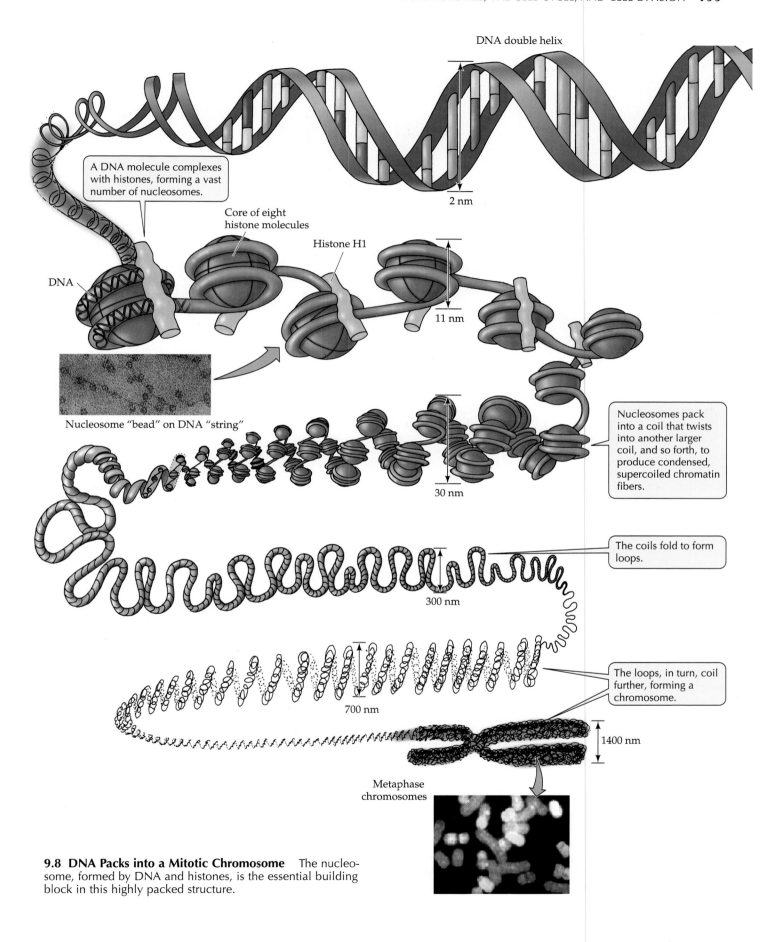

DNA double helix

2 nm

A DNA molecule complexes with histones, forming a vast number of nucleosomes.

Core of eight histone molecules

Histone H1

DNA

11 nm

Nucleosome "bead" on DNA "string"

Nucleosomes pack into a coil that twists into another larger coil, and so forth, to produce condensed, supercoiled chromatin fibers.

30 nm

The coils fold to form loops.

300 nm

The loops, in turn, coil further, forming a chromosome.

700 nm

1400 nm

Metaphase chromosomes

9.8 DNA Packs into a Mitotic Chromosome The nucleosome, formed by DNA and histones, is the essential building block in this highly packed structure.

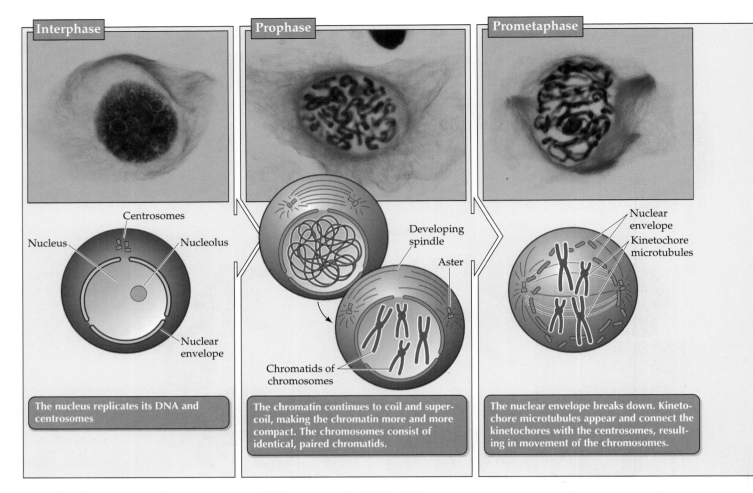

9.9 Mitosis Mitosis results in two new nuclei that are genetically identical to one another and to the nucleus from which they formed. These photomicrographs are of plant nuclei, which lack centrioles and asters. The diagrams are of corresponding phases in animal cells and introduce the structures not found in plants. In the micrographs, the red dye stains microtubules (and thus the spindle); the blue dye stains the chromosomes. In the diagrams, the chromosomes are stylized to emphasize the fates of the individual chromatids.

Although we know less about the organization of interphase chromatin than about that of mitotic chromatin, we know that interphase chromatin has nucleosomes that are spaced at the same intervals as in supercoiled chromatin (as shown in Figure 9.8). During interphase, DNA thus remains associated with histone molecules while it replicates and directs the synthesis of RNA. As we will show in Chapter 14, these protein–DNA interactions are important in regulating interphase activities.

Mitosis: Distributing Exact Copies of Genetic Information

Eukaryotic cells contain multiple chromosomes. Mitosis ensures the accurate distribution of these chromosomes to the daughter nuclei. During mitosis, a sin-

gle nucleus gives rise to two nuclei that are genetically identical to each other and to the parent nucleus. In reality, mitosis is a continuous process in which each event flows smoothly into the next. For discussion, however, it is convenient to look at mitosis—the M phase of the cell cycle—as a series of separate events, or subphases: prophase, prometaphase, metaphase, anaphase, and telophase (Figure 9.9).

During interphase, the nucleus is between divisions. At this stage we see the nuclear envelope, the nucleoli, and a barely discernible tangle of chromatin. Once the decision to enter mitosis has been made, DNA replicates, and during G2 a pair of **centrosomes** forms from a single centrosome that lies near the nucleus. The centrosomes are regions of the cell that initiate the formation of microtubules and thus help orchestrate chromosomal movement. The regions are not

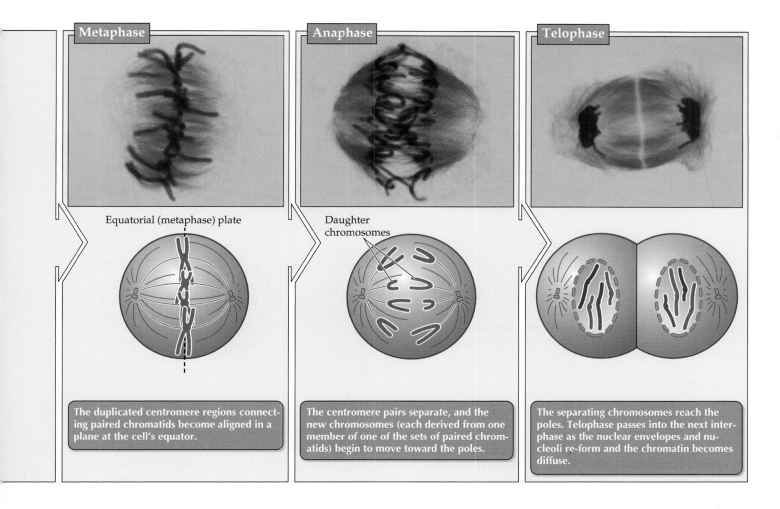

Metaphase

Equatorial (metaphase) plate

The duplicated centromere regions connecting paired chromatids become aligned in a plane at the cell's equator.

Anaphase

Daughter chromosomes

The centromere pairs separate, and the new chromosomes (each derived from one member of one of the sets of paired chromatids) begin to move toward the poles.

Telophase

The separating chromosomes reach the poles. Telophase passes into the next interphase as the nuclear envelopes and nucleoli re-form and the chromatin becomes diffuse.

enclosed by membranes and are not visible as discrete objects, but their positions are made evident by the arrangement of nearby microtubules.

In many organisms, each centrosome contains a pair of **centrioles**. However, the centrosomes of seed plants and some other organisms do not have centrioles. Where present, each centriole pair consists of one "parent" and one smaller "daughter" centriole at right angles to the parent centriole (see Figure 4.27).

The spindle forms during prophase

The appearance of the nucleus changes as the cell enters **prophase**—the beginning of mitosis (see Figure 9.9). The centrosomes move away from each other toward opposite ends, or *poles*, of the cell. Each centrosome then serves as a **mitotic center** that organizes microtubules. Some of the microtubules, called **polar microtubules**, run between the mitotic centers and make up the developing **spindle** (Figure 9.10).

The spindle is actually two *half spindles*: Each polar microtubule runs from one mitotic center to the middle of the spindle, where it overlaps with polar microtubules of the other half spindle. Polar microtubules are initially unstable, constantly forming and falling

apart until they contact polar microtubules from the other half spindle and become more stable. The spindle is responsible for the movement and distribution of chromosomes during mitosis and meiosis.

A prophase chromosome consists of two chromatids

The chromatin also changes during prophase. The extremely long, thin fibers take on a more orderly form as a result of coiling and compacting (see Figures 9.8 and 9.9). Under the light microscope, each prophase chromosome is seen to consist of two chromatids held tightly together over much of their length. The two chromatids of a single mitotic chromosome are identical in structure, chemistry, and the hereditary information they carry because of the way in which DNA replicates during the S phase that precedes mitosis.

Within the region of tight binding of the chromatids lies the centromere, which is where chromatids become associated with the microtubules of the spindle. Very late in prophase, specialized three-layered structures called **kinetochores** develop in the centromere region, one on each chromatid (see Figure 9.10). This is where the microtubules actually attach.

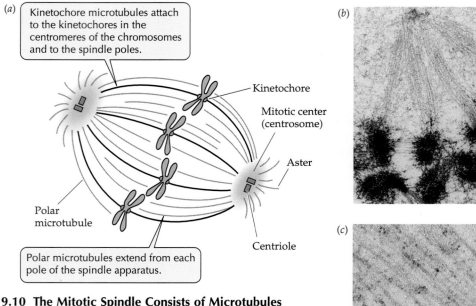

(a) Kinetochore microtubules attach to the kinetochores in the centromeres of the chromosomes and to the spindle poles.

Kinetochore

Mitotic center (centrosome)

Aster

Centriole

Polar microtubule

Polar microtubules extend from each pole of the spindle apparatus.

(b)

1.2 μm

(c)

Kinetochore microtubules

Kinetochore

0.6 μm

9.10 The Mitotic Spindle Consists of Microtubules
(a) Diagram of the spindle apparatus in a cell at metaphase.
(b) An electron micrograph of the stage shown in (a).
(c) The kinetochore at high magnification.

Chromosome movements are highly organized

The somewhat condensed chromosomes start to move at the end of prophase, which is the beginning of **prometaphase** (see Figure 9.8). The nuclear lamina disintegrates and the nuclear envelope breaks into small vesicles, allowing the developing spindle to "invade" the nuclear region.

Some polar microtubules now attach to chromosomes at their kinetochores and so are called **kinetochore microtubules** (see Figure 9.10). The kinetochore of one chromatid is attached to microtubules coming from one pole, while the kinetochore of its sister chromatid is attached to microtubules emanating from the other pole. At the attachment points are proteins that act as "molecular motors"; these *motor proteins* have the ability to hydrolyze ATP to ADP and phosphate, thus releasing energy to move the chromosomes along microtubule "railroad tracks" toward the poles.

The movement of chromosomes toward the poles is counteracted by two factors. First, there seems to be a repulsive force that keeps the chromosomes in the middle of the spindle region. Second, the two chromatids are held together at their centromere. So chromosomes during prometaphase appear to move aimlessly back and forth between the centrosomes at the poles and the middle of the spindle. Gradually, the kinetochores approach this middle region, called the **equatorial plate** or **metaphase plate**, halfway between the poles (see Figure 9.9).

The cell is said to be in **metaphase** when all the kinetochores arrive at the equatorial plate. Metaphase lasts up to an hour and is the best time to see the sizes and shapes of chromosomes. At the end of metaphase the centromeres separate, possibly because a protein holding the chromatids together breaks down. Separation of the centromeres marks the beginning of **anaphase**, the phase of mitosis during which the two sister chromatids of each chromosome—now called *daughter chromosomes*, each containing one double-stranded DNA molecule—move to opposite ends of the spindle.

What propels this highly organized mass migration, which takes about 10 minutes, is not clear. Two things seem to move the chromosomes along. First, motor proteins propel them, and second, the kinetochore microtubules shorten from the poles, drawing the chromosomes toward them.

During anaphase the poles of the spindle are pushed farther apart, doubling the distance between them. The distance between poles increases because polar microtubules from opposite ends of the spindle have motors that cause them to slide, pushing the poles apart in much the same way that microtubules slide in cilia and flagella (see Chapter 4). This polar separation contributes to the separation of one set of daughter chromosomes from the other.

Nuclei re-form at the end of mitosis

When the chromosomes stop moving at the end of anaphase, the cell enters **telophase** (the final frame in Figure 9.9). Two nuclei with identical DNA, carrying identical sets of hereditary instructions, are at the opposite ends of the spindle, which begins to break down. The chromosomes begin to uncoil, continuing until they become the diffuse tangle of chromatin that is characteristic of interphase. The materials of nuclear envelopes and nucleoli, which were disaggregated during prophase, coalesce and form their respective structures. When these and other changes are complete, telophase—and mitosis—is at an end, and each of the daughter nuclei enters another interphase.

During interphase, the DNA duplicates and new chromatids form, so that each chromosome consists of two chromatids. The duplication of DNA is a major topic and will be discussed in Chapter 11. Centrioles, if present, replicate during interphase: The two paired centrioles first separate, and then each acts as a "parent" for the formation of a new "daughter" centriole at right angles to it.

Mitosis is beautifully precise. Its result is two nuclei that are *identical to each other* and to the parent nucleus in chromosomal makeup and hence in genetic constitution.

Cytokinesis: The Division of the Cytoplasm

Mitosis refers only to the division of the nucleus. The division of the cell's cytoplasm is accomplished by **cytokinesis**. Cytokinesis generally, but not always, follows mitosis. Animal cells usually divide by a furrowing of the membrane, as if an invisible thread were tightening between the two parts (Figure 9.11a). The "invisible" threads are microfilaments of actin and myosin located in a ring just beneath the plasma membrane. These two proteins interact to produce a contraction, just as they do in muscles, thus pinching the cell in two (see Figure 4.22a).

Plant cell cytoplasm divides differently, because plants have cell walls. As the spindle breaks down after mitosis, membranous vesicles derived from the Golgi apparatus appear in the equatorial region roughly midway between the two daughter nuclei of a dividing plant cell. With the help of microtubules, the vesicles fuse to form new plasma membrane and contribute their contents to a cell plate, which is the beginning of a new cell wall (Figure 9.11b).

Following cytokinesis, both daughter cells contain all the components of a complete cell. The precise distribution of chromosomes is ensured by mitosis. Organelles such as ribosomes, mitochondria, and chloroplasts need not be distributed equally between daughter cells, as long as some of each are present in both cells; accordingly, there is no mechanism with precision comparable to that of mitosis to provide for their equal allocation to daughter cells.

Mitosis: Asexual Reproduction and Genetic Constancy

The cell cycle repeats itself. By this process, a single cell can give rise to a vast number of others. The cell could be an entire organism reproducing with each cycle, or a cell that divides to produce a multicellular organism. The multicellular organism, in turn, may be able to reproduce itself by releasing one or more of its cells, derived from mitosis and cytokinesis, as a spore or by having a multicellular piece break away and grow on its own (Figure 9.12). The unicellular organ-

9.11 Cytokinesis Differs in Animal and Plant Cells
Plant cells form cell walls and thus must divide differently from animal cells. (a) A sea urchin egg has just completed cytokinesis at the end of the first division in its development into an embryo. (b) A dividing plant cell in late telophase.

(a)

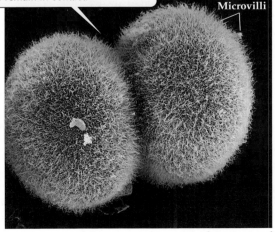

The division furrow has completely separated the cytoplasm of one daughter cell from another, although their surfaces remain in contact.

Microvilli

170 μm

(b)

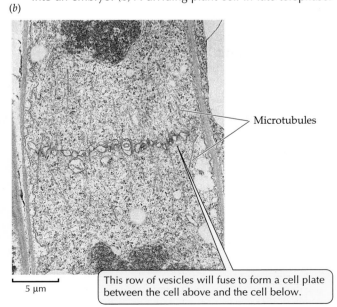

Microtubules

This row of vesicles will fuse to form a cell plate between the cell above and the cell below.

5 μm

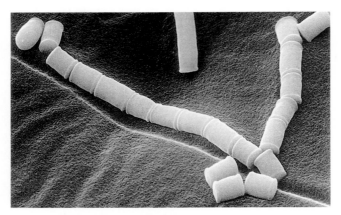

9.12 Asexual Reproduction These spool-shaped cells are asexual spores formed by a fungus. Each spore contains a nucleus produced by a mitotic division. A spore is the same genetically as the parent that fragmented to produce it.

ism and the multicellular organism reproducing by releasing cells provide examples of **asexual reproduction**, sometimes called vegetative reproduction. This mode of reproduction is based on mitotic division of the nucleus and, accordingly, produces offspring that

are genetically identical with the parent. Asexual reproduction is a rapid and effective means of making new individuals, and it is common in nature.

The uniformity of asexual reproduction, which leads to the production of a **clone** of genetically identical progeny, is very different from the situation in sexual reproduction. In sexual reproduction two parents, each contributing one cell, produce offspring that differ from each parent and from each other. This variety of genetic combinations results in a variety of offspring, some of whom may be better adapted to reproduce in a particular environment. Asexually reproducing organisms can also produce varied offspring. The cause of variation in asexual reproduction is primarily mutations, or changes, in the genetic material.

Meiosis: Sexual Reproduction and Diversity

Diversity is fostered by **sexual reproduction**, which combines genetic information from two different cells. Sexual life cycles are summarized in Figure 9.13. In the reproduction of most animal species, the two cells are contributed by two separate parents. Each parent provides a sex cell, or **gamete**. Each gamete is **haploid**,

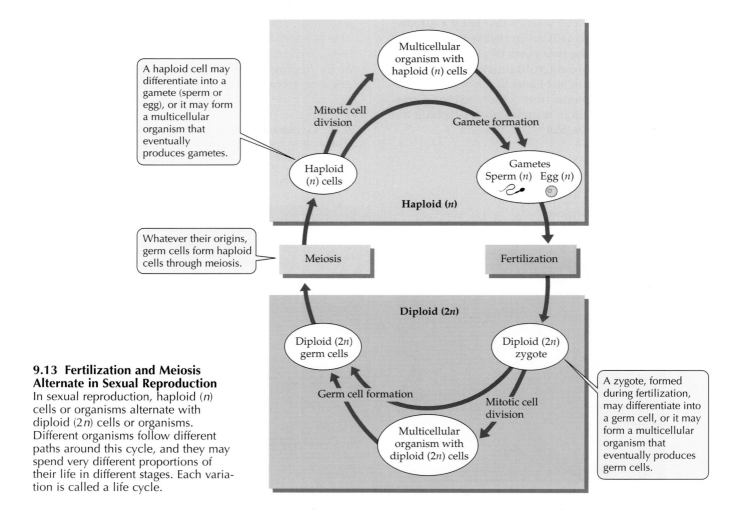

9.13 Fertilization and Meiosis Alternate in Sexual Reproduction
In sexual reproduction, haploid (n) cells or organisms alternate with diploid (2n) cells or organisms. Different organisms follow different paths around this cycle, and they may spend very different proportions of their life in different stages. Each variation is called a life cycle.

A haploid cell may differentiate into a gamete (sperm or egg), or it may form a multicellular organism that eventually produces gametes.

Whatever their origins, germ cells form haploid cells through meiosis.

A zygote, formed during fertilization, may differentiate into a germ cell, or it may form a multicellular organism that eventually produces germ cells.

Multicellular organism with haploid (n) cells

Mitotic cell division

Gamete formation

Haploid (n) cells

Gametes
Sperm (n)   Egg (n)

Haploid (n)

Meiosis

Fertilization

Diploid (2n)

Diploid (2n) germ cells

Diploid (2n) zygote

Germ cell formation

Mitotic cell division

Multicellular organism with diploid (2n) cells

meaning that it contains a single set of chromosomes; the number of chromosomes in such a single set is denoted by n. The two gametes—often identifiable as a female egg and a male sperm—fuse to produce a single cell, the **zygote**, or fertilized egg. This fusion is called **fertilization**.

As a consequence of fertilization, the zygote contains genetic information from both gametes and, hence, from both parents. In addition, the zygote has two sets of chromosomes; it is thus said to be **diploid**, denoted $2n$. In many species, including all animals, the zygote develops by mitotic divisions into a multicellular adult. Because the zygotic nucleus is diploid, all the body cells produced by mitosis also have two sets of chromosomes.

What happens when the adult from this zygote reproduces? If the gametes were produced by mitosis in this diploid parent, they too would be diploid. Fusion of two diploid ($2n$) gametes would result in a $4n$ zygote. Such a situation would not be reasonable, because subsequent generations would contain more and more chromosomes. To avoid increasing chromosome numbers in the sexual life cycle, a special type of nuclear division reduces the chromosome number in the gametes from diploid to haploid. This *reduction division* in sexually reproducing organisms is called **meiosis**.

Meiosis in animal cells produces haploid gametes directly. In plants and some fungi, however, meiosis gives rise to haploid **spores**, which undergo mitosis, producing multicellular haploid bodies (find the correct path in Figure 9.13). Certain cells in these haploid bodies ultimately give rise by mitosis to haploid gametes, and the life cycle continues. Details of some eukaryotic life cycles are given in Part Four of this book.

The simplest sexual life cycle is one in which two haploid (n) gametes fuse to give one diploid ($2n$) zygote and this zygote immediately undergoes meiosis, yielding a new set of haploid (n) gametes. Embellishments on this scheme consist mainly of the addition of mitotic divisions leading to multicellularity in the haploid phase, the diploid phase, or both. In animals only the diploid phase is multicellular.

Sex and reproduction are not the same thing. Sex is the combining of genetic material from two cells. Reproduction is the formation of new individuals, whether of unicellular or multicellular organisms. Sex and reproduction can be widely separated in time, as illustrated by the life cycle of the protist *Paramecium*, described in Chapter 26.

The essence of sexual reproduction is the selection of half of a parent's diploid chromosome set to make a haploid gamete, followed by the fusion of two such haploid gametes to produce a diploid cell that contains genetic information from both gametes. Both of these steps contribute to a shuffling of genetic information in the population, so no two individuals have exactly the same genetic constitution. The resulting genetic diversity is the opposite of the situation with asexual reproduction, which perpetuates genetic uniformity. The diversity provided by sexual reproduction opened up enormous opportunities for evolution.

The number, shapes, and sizes of the metaphase chromosomes constitute the karyotype

When nuclei are in metaphase of mitosis, it is often possible to count and characterize the individual chromosomes. This is a relatively simple process in some organisms, thanks to techniques that can capture cells in metaphase and spread out the chromosomes. A photograph of the entire set of chromosomes can then be made, and the images of the individual chromosomes can be cut out and pasted together in an orderly arrangement. Such a rearranged photograph reveals the number, shapes, and sizes of chromosomes in a cell, all of which constitute its **karyotype** (Figure 9.14).

Individual chromosomes in mitosis can be recognized by their lengths, the positions of their centromeres, and characteristic banding when they are stained and observed at high magnification. When the cell is diploid, the karyotype consists of pairs of chromosomes—23 pairs for a total of 46 chromosomes in humans, and greater or smaller numbers of pairs in other diploid species. There is no simple relationship between the size of an organism and its chromosome number (Table 9.1).

In each recognizable pair of chromosomes, one chromosome comes from one parent and one from the other. The members of such a **homologous pair** are identical in size and appearance (with the exception of the so-called sex chromosomes in some species; see Chapter 10), and the two chromosomes (the homologs) of a homologous pair bear corresponding,

Table 9.1 Numbers of Pairs of Chromosomes in Some Plant and Animal Species

COMMON NAME	SPECIES	NUMBER OF CHROMOSOME PAIRS
Mosquito	*Culex pipiens*	3
Housefly	*Musca domestica*	6
Toad	*Bufo americanus*	11
Rice	*Oryza sativa*	12
Frog	*Rana pipiens*	13
Alligator	*Alligator mississippiensis*	16
Rhesus monkey	*Macaca mulatta*	21
Wheat	*Triticum aestivum*	21
Human	*Homo sapiens*	23
Potato	*Solanum tuberosum*	24
Donkey	*Equus asinus*	31
Horse	*Equus caballus*	32
Dog	*Canis familiaris*	39
Carp	*Cyprinus carpio*	52

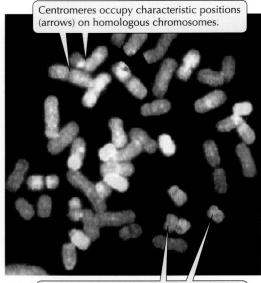

Centromeres occupy characteristic positions (arrows) on homologous chromosomes.

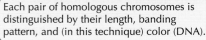

Each pair of homologous chromosomes is distinguished by their length, banding pattern, and (in this technique) color (DNA).

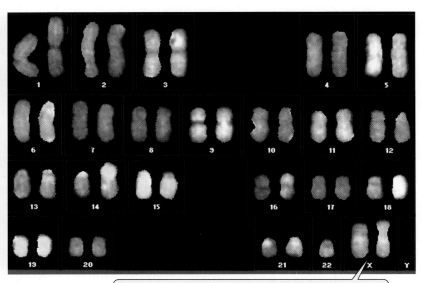

The karyotype shows 23 pairs of chromosomes, including the sex chromosomes. This female's sex chromosomes are X and X; a male would have X and Y chromosomes.

9.14 Human Cells Have 46 Chromosomes
Chromosomes from a human cell in metaphase of mitosis. In this "chromosome painting" technique, each homologous pair shares a distinctive color. The karyotype on the right is produced by computerized analysis of the image on the left.

though generally not identical, types of genetic information. Haploid cells contain only one homolog from each pair of chromosomes. Thus when haploid gametes fuse in fertilization, the resulting diploid zygote has two homologs of each type.

Meiosis: A Pair of Nuclear Divisions

Meiosis consists of two nuclear divisions that reduce the number of chromosomes to the haploid number in preparation for sexual reproduction. To understand the process and its specific details, it is useful to keep in mind the overall functions of meiosis:

- To reduce the chromosome number from diploid to haploid
- To ensure that each of the haploid products of meiosis has a complete set of chromosomes
- To promote genetic diversity among the products

Pay particular attention to the fact that, although the nucleus divides twice during meiosis, the DNA is replicated only once.

Two unique features characterize the first meiotic division, **meiosis I**. The first feature is that homologous chromosomes pair along their entire lengths, a process called **synapsis** that lasts from prophase to the

end of metaphase of meiosis I. The second key feature is that after this metaphase, homologous chromosomes separate. The individual chromosomes, each consisting of two joined chromatids, remain intact until the end of the metaphase of **meiosis II**, the second meiotic division. In the discussion that follows, refer to Figure 9.15 (on pages 208–209) to help you visualize each step.

The first meiotic division reduces the chromosome number

Meiosis I is preceded by an interphase during which each chromosome is replicated, with the result that each chromosome consists of two sister chromatids. Meiosis I begins with a long **prophase I** (the first three frames of Figure 9.15) during which the chromosomes change markedly. A key change is that *homologous chromosomes synapse*; they are already tightly joined as soon as they can be clearly seen under the light microscope. Throughout prophase I and metaphase I, the chromatin continues to coil and compact progressively, so the chromosomes appear ever thicker.

Partway through prophase I, the homologous chromosomes seem to *repel* each other, especially near the centromeres, but they are held together by physical attachments. Regions having these attachments take on an X-shaped appearance and are called **chiasmata** (singular chiasma, meaning "cross" in Greek; Figure 9.16). A chiasma reflects an exchange of material between chromatids on homologous chromosomes—what geneticists call crossing over (Figure 9.17).

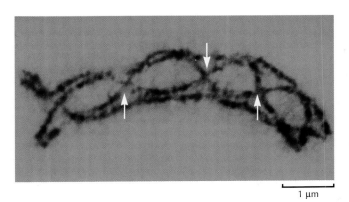

9.16 Chiasmata: Evidence of Exchange between Chromatids Chiasmata are visible near the middle of some of the chromatids in this micrograph, and near the ends of others. Three chiasmata are indicated with arrows.

Crossing over increases the genetic differences among the products of meiosis. We will have a great deal to say about crossing over and its genetic consequences in coming chapters. The chromosomes exchange material shortly after synapsis begins, but the chiasmata do not become visible until later, when homologs are repelling each other.

There seems to be plenty of time for the complicated events of prophase I to occur. Whereas mitotic prophases are usually measured in minutes and all of mitosis seldom takes more than an hour or two, meiosis can take a long time. In human males, the cells in the testis that undergo meiosis take about a week for prophase I and about a month for the entire meiotic cycle. In the cells that will become eggs, prophase I begins long before a woman's birth, during early fetal development, and ends as much as decades later, during the monthly ovarian cycle.

Prophase I is followed by prometaphase I (not pictured in Figure 9.15), during which the nuclear envelope and the nucleoli disappear. A spindle forms, and microtubules become attached to the kinetochores of the chromosomes. In meiosis I, there is only one kinetochore per chromosome, not one per chromatid as in mitosis. Thus, the entire chromosome, consisting of two chromatids, will migrate to one pole in the meiotic cell.

By metaphase I, the chromosomal kinetochores have become connected to the microtubules from the poles. All the chromosomes have moved to the equatorial plate, and the homologous chromosomes are about to be pulled apart. Until this point, they have been held together by chiasmata.

Homologous chromosomes separate in anaphase I, when individual *chromosomes*, each still consisting of two chromatids, are pulled to the poles, one homolog of a pair going to one pole and the other homolog

going to the opposite pole. (Note that this differs from the separation of *chromatids* during mitotic anaphase.) Each of the two daughter nuclei from this division is haploid; that is, it contains only one set of chromosomes, compared to the two sets of chromosomes that were present in the original diploid nucleus. However, because they consist of two chromatids rather than just one, each of these chromosomes has twice the mass that a chromosome at the end of a mitotic division has.

In some species, but not in others, there is a telophase I, with the reappearance of nuclear envelopes and so forth. When there is a telophase I, it is followed by an interkinesis phase similar to mitotic interphase. During interkinesis the chromatin is somewhat but not completely uncoiled; however, there is no replication of the genetic material, because each chromosome already consists of two chromatids. In contrast to mitotic interphase, the sister chromatids in meiotic interkinesis are generally not genetically identical, because crossing over in prophase I has reshuffled genetic material between maternal and paternal chromosomes.

The second meiotic division is similar to, and different from, mitosis

Meiosis II is similar to mitosis. In each nucleus produced by meiosis I, the chromosomes line up at equatorial plates in metaphase II; the chromatids—each having a centromere—separate, and new daughter chromosomes (consisting now of single chromatids) move to the poles in anaphase II.

9.17 Crossing Over Forms Genetically Diverse Chromosomes The exchange of genetic material by crossing over may result in new combinations of genetic determinants on the recombinant chromosomes.

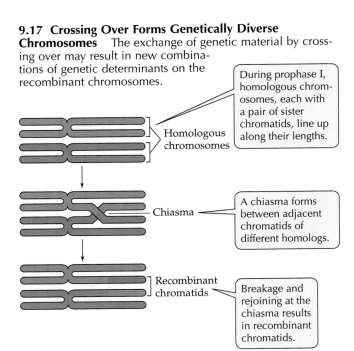

During prophase I, homologous chromosomes, each with a pair of sister chromatids, line up along their lengths.

Homologous chromosomes

A chiasma forms between adjacent chromatids of different homologs.

Chiasma

Recombinant chromatids

Breakage and rejoining at the chiasma results in recombinant chromatids.

MEIOSIS I

Middle Prophase I *most important*

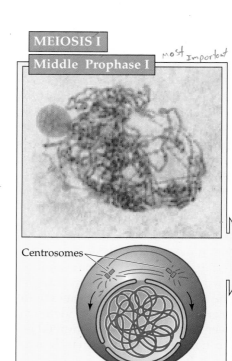

Centrosomes

The chromatin begins to condense following interphase.

Later Prophase I

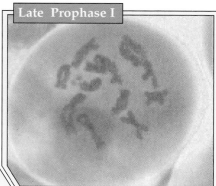

Pairs of homologs

Chiasmata

Synapsis aligns homologs, and chromosomes shorten. Homologs are shown in different colors indicating those coming from each parent. In reality, their differences are very small, usually comprising different alleles of some genes.

Late Prophase I

The chromosomes continue to coil and shorten. Crossing-over at chiasmata results in an exchange of genetic material.

MEIOSIS II

Prophase II

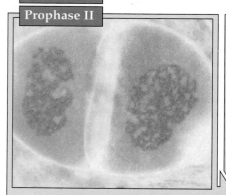

The chromosomes condense again, following a brief interphase in which DNA does not replicate.

Metaphase II

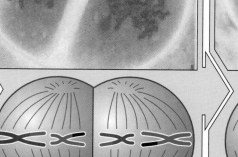

Kinetochores of the paired chromatids line up across the equator of each cell.

Anaphase II

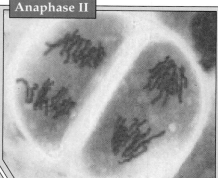
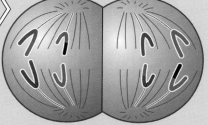

The chromatids of the chromosomes finally separate, becoming chromosomes in their own right, and are pulled to opposite poles. Because of crossing over in Prophase I, each cell-to-be will have a different genetic makeup.

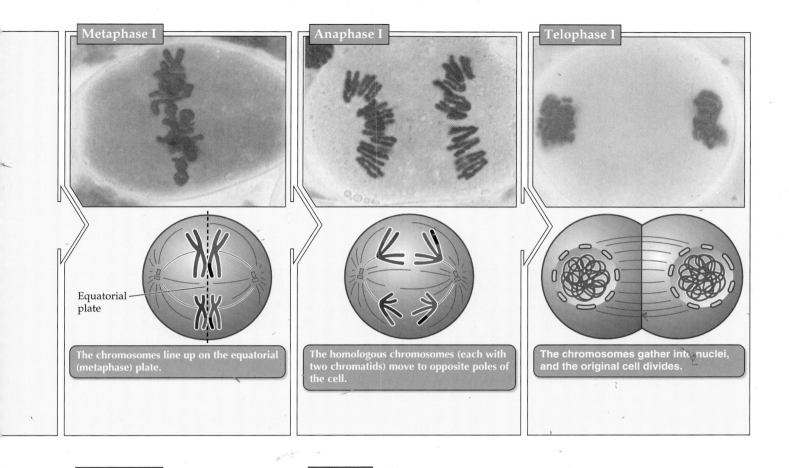

Metaphase I

Equatorial plate

The chromosomes line up on the equatorial (metaphase) plate.

Anaphase I

The homologous chromosomes (each with two chromatids) move to opposite poles of the cell.

Telophase I

The chromosomes gather into nuclei, and the original cell divides.

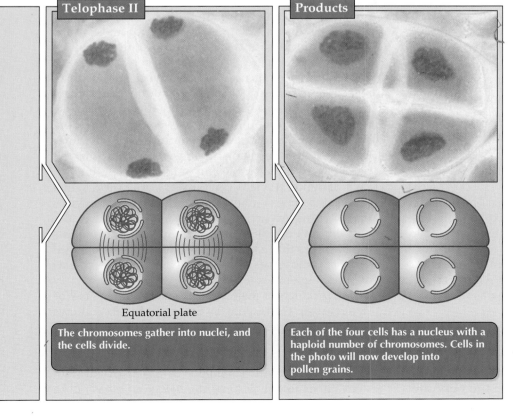

Telophase II

Equatorial plate

The chromosomes gather into nuclei, and the cells divide.

Products

Each of the four cells has a nucleus with a haploid number of chromosomes. Cells in the photo will now develop into pollen grains.

9.15 Meiosis   In meiosis, two sets of chromosomes are divided among four cells, each of which then has half as many chromosomes as the original cell. These four haploid cells are the result of two successive nuclear divisions. The photomicrographs shown here are of meiosis in the male reproductive organ of a lily. As in Figure 9.9, the diagrams show corresponding phases in an animal.

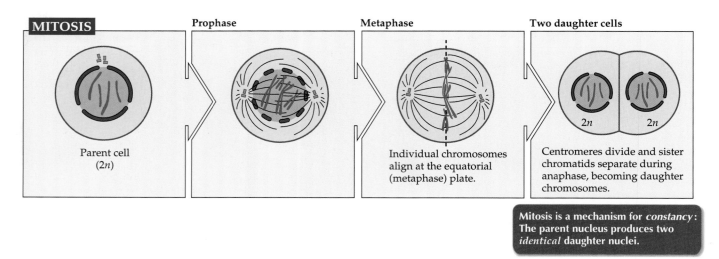

MITOSIS	Prophase	Metaphase	Two daughter cells
Parent cell (2*n*)		Individual chromosomes align at the equatorial (metaphase) plate.	Centromeres divide and sister chromatids separate during anaphase, becoming daughter chromosomes.

Mitosis is a mechanism for *constancy*: The parent nucleus produces two *identical* daughter nuclei.

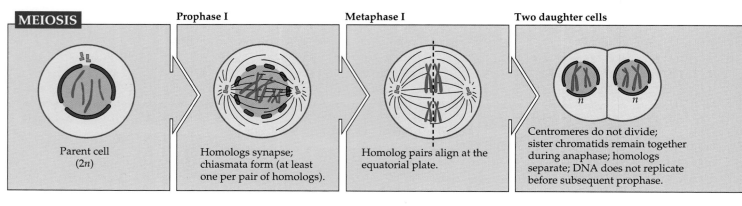

MEIOSIS	Prophase I	Metaphase I	Two daughter cells
Parent cell (2*n*)	Homologs synapse; chiasmata form (at least one per pair of homologs).	Homolog pairs align at the equatorial plate.	Centromeres do not divide; sister chromatids remain together during anaphase; homologs separate; DNA does not replicate before subsequent prophase.

The two processes are also different. The three major differences between meiosis II and mitosis are:

- DNA replicates before mitosis but not before meiosis II.
- In mitosis, the sister chromatids that make up a given chromosome are identical; in meiosis II they differ over part of their length if they participated in crossing over during prophase of meiosis I.
- The number of chromosomes on the equatorial plate of each of the two nuclei in meiosis II is half the number in the single mitotic nucleus.

Figure 9.18 compares mitosis and meiosis. The result of meiosis is four nuclei; each nucleus is haploid and has a single full set of chromosomes that differs from other such sets in its exact genetic composition. The differences, to repeat a very important point, result from crossing over during prophase I and from the segregation of homologous chromosomes during anaphase I.

Meiosis leads to genetic diversity

What are the consequences of the synapsis and separation of homologous chromosomes during meiosis? In mitosis, each chromosome behaves independently of its homolog; its two chromatids are sent to opposite

poles at anaphase. If we start a mitotic division with *x* chromosomes, we end up with *x* chromosomes in each daughter nucleus, and each chromosome consists of one chromatid. In meiosis, things are very different.

In meiosis, synapsis organizes things so that chromosomes of maternal origin pair with the paternal homologs. Then the separation during meiotic anaphase I ensures that each pole receives one member from each pair of homologous chromosomes. (Remember that each chromosome still consists of two chromatids.) For example, at the end of meiosis I in humans, each daughter nucleus contains 23 out of the original 46 chromosomes—one member of each homologous pair. In this way, the chromosome number is decreased from diploid to haploid. Furthermore, meiosis I guarantees that each daughter nucleus gets one full set of chromosomes, for it must have one of each pair of homologous chromosomes.

The products of meiosis I are genetically diverse for two reasons. First, synapsis during prophase I allows the maternal chromosome to interact with the paternal one; if there is crossing over, the recombinant chromatids contain some genetic material from each chromosome. Second, which member of a pair of chromosomes goes to which daughter cell at anaphase I is a matter of pure chance. If there are two pairs of chromosomes in the diploid parent nucleus, a particular daughter nucleus could get paternal chromosome 1 and maternal chromosome 2, or paternal 2 and maternal 1, or both maternals, or both paternals. It all depends on the random way in which the homologous pairs line up at metaphase I.

Note that of the four possible chromosome combinations just described, two produce daughter nuclei that

9.18 Mitosis and Meiosis Compared Meiosis differs from mitosis by synapsis and by the failure of the centromeres to separate at the end of metaphase I.

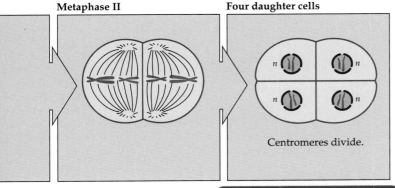

Metaphase II   Four daughter cells

Centromeres divide.

Meiosis is a mechanism for *diversity*: The parent nucleus produces four daughter nuclei, each *different* from the parent and from its sisters.

are the same as one of the parental types (except for any material exchanged by crossing over). The greater the number of chromosomes, the less probable that the original parental combinations will be reestablished. Most species of diploid organisms do, indeed, have more than two pairs. In humans, with 23 chromosome pairs, 2^{23} different combinations can be produced.

Meiotic Errors:
The Source of Chromosomal Disorders

A pair of homologous chromosomes may fail to separate during meiosis I, or sister chromatids may fail to separate during meiosis II or during mitosis. This phenomenon is called **nondisjunction**, and it results in the production of aneuploid cells (Figure 9.19). **Aneuploidy** is a condition in which one or more chromosomes or pieces of chromosomes are either lacking or are present in excess.

If, for example, the chromosome 21 pair fails to separate during the formation of a human egg (and thus both members of the pair go to one pole during anaphase I), the resulting egg contains either two copies of chromosome 21 or none at all. If an egg with two of these chromosomes is fertilized by a normal sperm, the resulting zygote and infant has three copies of the chromosome: He or she is **trisomic** for chromosome 21. As a result of carrying an extra chromosome 21, such a child demonstrates the symptoms of Down syndrome: impaired intelligence; characteristic abnormalities of the hands, tongue, and eyelids; and an increased susceptibility to cardiac abnormalities and diseases such as leukemia.

Other abnormal events can also lead to aneuploidy. In a process called *translocation*, a piece of a chromosome may break away and become attached to another chromosome. For example, a particular large part of one chromosome 21 may be translocated to another chromosome. Individuals who inherit this translocated piece along with two normal chromosomes 21 have Down syndrome.

Other human disorders result from particular chromosomal abnormalities. Sex chromosome aneuploidy causes disorders such as Turner syndrome and Klinefelter syndrome, discussed in Chapter 10 in connection with sex determination. Deletion of a portion of chromosome 5 results in cri du chat (French for "cat's cry") syndrome, so named because an afflicted infant's cry sounds like that of a cat. Symptoms of this syndrome include severe mental retardation.

Trisomies (and the corresponding monosomies) are surprisingly common in human zygotes, but most of the embryos that develop from such zygotes do not survive to birth. Trisomies for chromosomes 13, 15, and 18 greatly reduce the probability that an embryo will survive to birth, and virtually all infants who are born with such trisomies die before the age of 1 year.

Trisomies and monosomies for other chromosomes are lethal to the embryo. At least one-fifth of all recognized pregnancies spontaneously terminate during the first two months, largely because of such trisomies and monosomies. (The actual proportion of spontaneously terminated pregnancies is certainly higher, because the earliest ones often go unrecognized.)

Polyploids can have difficulty in cell division

Both diploid and haploid nuclei divide by mitosis. Multicellular diploid and multicellular haploid individuals develop from single-celled beginnings by mitotic divisions. Mitosis may proceed in diploid organisms even when a chromosome from one of the haploid sets is missing or when there is an extra copy of one of the chromosomes (as in Down syndrome).

Under some circumstances triploid ($3n$), tetraploid ($4n$), and higher-order polyploid nuclei form. Each of these *ploidy levels* represents an increase in the number

9.19 Nondisjunction during Meiosis I Leads to Aneuploidy
Nondisjunction can also occur during meiosis II.

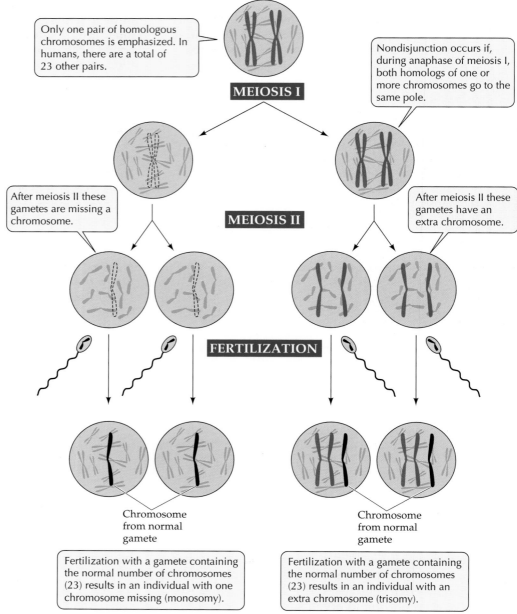

Only one pair of homologous chromosomes is emphasized. In humans, there are a total of 23 other pairs.

MEIOSIS I

Nondisjunction occurs if, during anaphase of meiosis I, both homologs of one or more chromosomes go to the same pole.

After meiosis II these gametes are missing a chromosome.

MEIOSIS II

After meiosis II these gametes have an extra chromosome.

FERTILIZATION

Chromosome from normal gamete

Chromosome from normal gamete

Fertilization with a gamete containing the normal number of chromosomes (23) results in an individual with one chromosome missing (monosomy).

Fertilization with a gamete containing the normal number of chromosomes (23) results in an individual with an extra chromosome (trisomy).

of complete sets of chromosomes present. If, through accident, the nucleus has one or more extra full sets of chromosomes—that is, if it is triploid, tetraploid, or of still higher ploidy—this abnormally high ploidy in itself does not prevent mitosis. In mitosis, each chromosome behaves independently of the others.

In meiosis, by contrast, chromosomes *synapse* to begin division. If even one chromosome has no homolog, anaphase I cannot send representatives of that chromosome to both poles. A diploid nucleus can undergo normal meiosis; a haploid one cannot. A tetraploid nucleus has an even number of each kind of chromosome, so each chromosome can pair with its homolog. But a triploid nucleus cannot undergo nor-

mal meiosis, because one-third of the chromosomes would lack partners.

This limitation has important consequences for the fertility of triploid, tetraploid, and other chromosomally unusual organisms that may be produced by plant breeding or by natural accidents. Modern bread wheat plants are hexaploids, the result of the accidental crossing of three different grasses, each having its own diploid set of 14 chromosomes.

Cell Death

As we mentioned at the start of the chapter, an essential role of cell division in complex eukaryotes is to re-

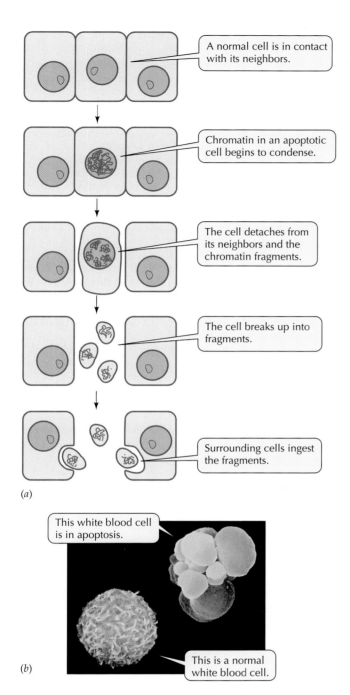

A normal cell is in contact with its neighbors.

Chromatin in an apoptotic cell begins to condense.

The cell detaches from its neighbors and the chromatin fragments.

The cell breaks up into fragments.

Surrounding cells ingest the fragments.

(a)

This white blood cell is in apoptosis.

(b)

This is a normal white blood cell.

9.20 Apoptosis: Programmed Cell Death Many cells are genetically programmed to "self destruct" when they are no longer needed, or when they have lived long enough to accumulate a burden of damage that might harm the organism.

place cells that die. In humans, billions of cells die each day, mainly in blood and epithelia lining organs such as the intestine. Cells die in one of two ways. The first, **necrosis**, occurs when cells either are damaged by poisons or are starved of essential nutrients. The scab that forms around a wound is a familiar example

of necrotic tissue. More typical in an organism is **apoptosis**, a series of events that constitute genetically programmed cell death (Figure 9.20).

Why would a cell initiate apoptosis, which is essentially "cell suicide"? One reason is that the cell in question is no longer needed by the organism. For example, before birth a human fetus has weblike hands, with connective tissue between the fingers. As development proceeds, this unneeded tissue disappears as the cells undergo apoptosis (see Figure 15.12).

A second reason for apoptosis is that the longer cells live, the more prone they are to damage that could lead to cancer. This is especially true of cells in the blood and intestine, which are exposed to high levels of toxic substances. In these cases, cells sacrifice their lives to the good of the organism.

Like the cell cycle, the events of apoptosis are very similar in most organisms. The cell becomes isolated from its neighbors, chops up its chromatin into nucleosome-sized pieces, and then fragments itself. In a remarkable example of the economy of nature, the surrounding living cells ingest the remains of the dead cell. The genetic signals that lead to apoptosis are common to many organisms. We will describe some of them in Chapter 15.

Summary of "Chromosomes, the Cell Cycle, and Cell Division"

• Cell division consists of three steps in the following order: replication of the genetic material (DNA), partitioning of the two DNA molecules to separate places in the cell, and division of the cytoplasm.
• Cell division is an important mechanism for reproduction, growth, and repair. **Review Figure 9.1**

Systems of Cell Division

• In prokaryotes, cellular DNA is a single molecule, or chromosome. Prokaryotes divide by cell fission. **Review Figure 9.3**
• In eukaryotes, nuclei divide by either mitosis or meiosis.

Interphase and the Control of Cell Division

• The mitotic cell cycle has two main phases: interphase (during which cells are not dividing) and mitosis (when cells divide). During most of the cell cycle the cell is in interphase, which is divided into three subphases: S, G1, and G2. **Review Figure 9.4**
•Cyclin–Cdk complexes regulate the passage of cells from G1 into S phase and from G2 into M phase. **Review Figure 9.5**
• In addition to the internal cyclin–Cdk complexes, controls external to the cell, such as growth factors and hormones, can also stimulate the cell to begin a division cycle.

Eukaryotic Chromosomes

• Chromosomes contain DNA and proteins. At mitosis, chromosomes initially appear double because two sister chromatids are held together at the centromere. Each sister chromatid consists of one double-stranded DNA molecule complexed with proteins and referred to as chromatin. **Review Figure 9.6**

• During interphase, the DNA is wound around cores of histones to form nucleosomes. DNA folds over and over again, packing itself within the nucleus. When mitotic chromosomes form, it folds even more. **Review Figures 9.7, 9.8**

Mitosis: Distributing Exact Copies of Genetic Information

• After DNA is replicated during S phase, the first visible sign of mitosis is the separation of centrosomes, which initiate microtubule formation for the spindle. **Review Figures 9.9, 9.10**

• Mitosis consists of prophase, prometaphase, metaphase, anaphase, and telophase. During prophase, the chromosomes condense and appear as paired chromatids. During prometaphase, the chromosomes move toward the middle of the spindle. In metaphase, the doublet chromosomes gather at the middle of the cell with their centromeres on the metaphase plate. At the end of metaphase, the centromeres holding the chromatid pairs together separate, and during anaphase each member of the pair, now called a daughter chromosome, migrates to its pole along the microtubule track. During telophase, the chromosomes become less condensed; the nuclear envelopes and nucleoli re-form, thus producing two nuclei whose chromosomes are identical to each other and to those of the cell that began the cycle. **Review Figure 9. 9**

Cytokinesis: The Division of the Cytoplasm

• Nuclear division is usually followed by cytokinesis. Animal cell cytoplasm usually divides by a furrowing of the plasma membrane, caused by the contraction of cytoplasmic microfilaments. In plant cells cytokinesis is accomplished by vesicle fusion and the synthesis of new cell wall material. **Review Figure 9.11**

Mitosis: Asexual Reproduction and Genetic Constancy

• The cell cycle can repeat itself many times, forming a clone of genetically identical cells.

• Asexual reproduction produces a new organism that is genetically identical to the parent. Any genetic variety is the result of genetic changes, or mutations, in the parent cells.

Meiosis: Sexual Reproduction and Diversity

• In sexual reproduction, two haploid gametes—one from each parent—unite in fertilization to form a genetically unique, diploid zygote. **Review Figure 9.13**

• The number, shapes, and sizes of the chromosomes constitute the karyotype of an organism. **Review Figure 9.14**

• To form gametes, certain cells in the adult undergo meiosis, in which a diploid cell produces haploid gametes. Each gamete contains a random mix of one of each pair of homologous chromosomes from the parent. The more chromosome pairs there are in a diploid cell, the greater the diversity of chromosome combinations generated by meiosis.

Meiosis: A Pair of Nuclear Divisions

• Meiosis reduces the chromosome number from diploid to haploid and ensures that each haploid cell contains one copy of each chromosome. It consists of two nuclear divisions. **Review Figure 9.15**

• DNA is replicated during an interphase before meiosis I. During prophase I, homologous chromosomes pair up with each other, and material may be exchanged by crossing over between nonsister chromatids of two adjacent homologs. In metaphase I, the paired homologs gather at the metaphase plate—each chromosome having only one kinetochore and associating with spindle fibers for one pole. In anaphase I, entire chromosomes, each with two chromatids, migrate to the poles. By the end of meiosis I, there are two nuclei, each with the haploid number of chromosomes with two sister chromatids.

• In meiosis II, the sister chromatids separate. No DNA replication precedes this division, which is similar to mitosis. The result is a cell with a haploid chromosome and DNA content. **Review Figures 9.15, 9.17**

• Meiosis II is similar to mitosis in chromosome movements, including the separation of the sister chromatids. **Review Figures 9.15, 9.17, 9.18**

• Both crossing over during prophase I and the random selection of which homolog of a pair migrates to which pole during anaphase I ensure that the genetic composition of each haploid gamete is very different from that of the parent, and from that of the other gametes. **Review Figure 9.17**

Meiotic Errors: The Source of Chromosomal Disorders

• In nondisjunction, one member of a homologous pair of chromosomes fails to separate from the other and both go to the same pole. This event leads to one gamete with an extra chromosome and the other lacking that chromosome. Fertilization with a fully haploid gamete then results in aneuploidy and genetic abnormalities that are invariably harmful or lethal to the organism. **Review Figure 9.19**

Cell Death

• Cells may die by necrosis or may kill themselves by apoptosis, a genetically programmed series of events that includes the detachment of the cell from its neighbors and the fragmentation of its nuclear DNA. **Review Figure 9.20**

Self-Quiz

1. Which statement about eukaryotic chromosomes is *not* true?
   a. They sometimes consist of two chromatids.
   b. They sometimes consist of a single chromatid.
   c. They normally possess a single centromere.
   d. They consist of chromatin.
   e. They are always clearly visible as defined bodies under the light microscope.

2. Nucleosomes
   a. are made of chromosomes.
   b. consist entirely of DNA.
   c. consist of DNA wound around a histone core.
   d. are present only during mitosis.
   e. are present only during prophase.

3. Which statement about the cell cycle is *not* true?
   a. It consists of mitosis and interphase.
   b. The cell's DNA replicates during G1.
   c. A cell can remain in G1 for weeks or much longer.
   d. Most proteins are formed throughout all subphases of interphase.
   e. Histones are synthesized primarily during S phase.

4. Which statement about mitosis is *not* true?
   a. A single nucleus gives rise to two identical daughter nuclei.
   b. The daughter nuclei are genetically identical to the parent nucleus.
   c. The centromeres separate at the onset of anaphase.
   d. Homologous chromosomes synapse in prophase.
   e. Mitotic centers organize the microtubules of the spindle fibers.

5. Which statement about cytokinesis is true?
   a. In animals, a cell plate forms.
   b. In plants, it is initiated by furrowing of the membrane.
   c. It generally immediately follows mitosis.
   d. In plant cells, actin and myosin play an important part.
   e. It is the division of the nucleus.

6. In sexual reproduction
   a. gametes are usually haploid.
   b. gametes are usually diploid.
   c. the zygote is usually haploid.
   d. the chromosome number is reduced during mitosis.
   e. spores are formed during fertilization.

7. In meiosis
   a. meiosis II reduces the chromosome number from diploid to haploid.
   b. DNA replicates between meiosis I and II.
   c. the chromatids that make up a chromosome in meiosis II are identical.
   d. each chromosome in prophase I consists of four chromatids.
   e. homologous chromosomes separate from one another in anaphase I.

8. In meiosis
   a. a single nucleus gives rise to two identical daughter nuclei.
   b. the daughter nuclei are genetically identical to the parent nucleus.
   c. the centromeres separate at the onset of anaphase I.
   d. homologous chromosomes synapse in prophase I.
   e. no spindle forms.

9. Which statement about aneuploidy is *not* true?
   a. It results from chromosomal nondisjunction.
   b. It does not happen in humans.
   c. An individual with an extra chromosome is trisomic.
   d. Trisomies are common in human zygotes.
   e. A piece of one chromosome may translocate to another chromosome.

10. In prokaryotes
    a. there are no meiotic divisions.
    b. mitosis proceeds as in eukaryotes.
    c. the genetic information is not carried in chromosomes.
    d. the chromosomes are identical to those of eukaryotes.
    e. cell division follows division of the nucleus.

Applying Concepts

1. How does a nucleus in the G2 phase of the cell cycle differ from one in the G1 phase?

2. What is a chromatid? When does a chromatid become a chromosome?

3. Compare and contrast mitosis (and subsequent cytokinesis) in animals and plants.

4. Suggest two ways in which, with the help of a microscope, one might determine the relative durations of the various phases of mitosis.

5. Contrast mitotic prophase and prophase I of meiosis. Contrast mitotic anaphase and anaphase I of meiosis.

6. Compare the sequence of events in the mitotic cell cycle with the sequence in programmed cell death.

Readings

Alberts, B., D. Bray, J. Lewis, M. Raff, K. Roberts and J. D. Watson. 1994. *Molecular Biology of the Cell*, 3rd Edition. Garland Publishing, New York. An outstanding book in which to pursue the topics of this chapter in greater detail.

Baserga, R. 1986 *The Biology of Cell Reproduction.* Harvard University Press, Cambridge, MA. A fine review of cell cycles in animal cells, with special emphasis on tumor cells.

The Cell Cycle. 1991. Cold Spring Harbor Symposia on Quantitative Biology, vol. 56. Cold Spring Harbor Laboratory, Cold Spring Harbor, NY. Summaries of research on cell cycle control by the scientists who did the work.

Cooper, G. M. 1997. *The Cell: A Molecular Approach.* ASM Press, Washington, DC, and Sinauer Associates, Sunderland, MA. A compact treatment of cell biology. Chapter 14 has a summary of the cell cycle.

Duke, R. C., D. M. Ojcius and J. D.-E. Young. 1996. "Cell Suicide in Health and Disease." *Scientific American*, December. A discussion of apoptosis.

Glover, D. M., C. Gonzalez and J. W. Raff. 1993. "The Centrosome." *Scientific American*, June. A description of the structure and function of the organelle that directs the assembly of the cytoskeleton and controls cell division.

Murray, A. and T. Hunt. 1993. *The Cell Cycle: An Introduction.* W. H. Freeman, New York. An excellent treatment of cell cycles, with historical background.

Nigg, E. 1993. "Targets of Cyclin-Dependent Protein Kinases." *Current Opinions in Cell Biology*, vol. 5, pages 187–193. This journal does what its title indicates: it gives up-to-date views of the field. The cell cycle is summarized each year.

Chapter 10

Transmission Genetics: Mendel and Beyond

A Royal Family Stalked by Hemophilia
Although they did not suffer from the disease themselves, five of the women in this 1895 photograph of the British royal family carried the gene for the disease hemophilia, which afflicted several of their sons and grandsons. Because Victoria's son Edward VII was not affected and did not carry the gene, his descendants—including the present royal family of England—are free of the disease.

*I*n 1895, Queen Victoria and her numerous progeny sat for a family portrait. She and four other women in this photo shared a characteristic: All five carried a gene for the disease hemophilia, even though none of them had the disease. However, more than a dozen of Victoria's male descendants suffered from hemophilia, a failure of the blood to clot normally following an injury. All the family members with hemophilia were males; all the carriers—healthy individuals capable of transmitting the disease to their children—were women.

How can we explain this apparent correlation between sex and an inherited disease? How can we account for, and predict, patterns of inheritance in general? Experimental work in Mendelian genetics, begun in the nineteenth century and running through the first half of the twentieth century, enabled us to understand these patterns. The molecular details underlying the processing of inherited information have occupied us ever since the work of Mendelian genetics was completed.

The term "Mendelian genetics" refers to certain basic inheritance patterns and the origin of our concept of a particulate unit (the gene) responsible for those patterns. The term honors the Austrian monk Gregor Johann Mendel (1822–1884), the person who first made rigorous, quantitative observations of the patterns of inheritance and proposed plausible mechanisms to explain them. In organisms that reproduce sexually and have more than one chromosome (and orderly meiosis), many traits pass from parent to offspring in accord with these patterns.

In this chapter we will discuss how genes are transmitted from generation to generation of plants and animals, and show how many of the rules that govern genetics can be explained by the behavior of chromosomes during meiosis. We will also describe the interactions of genes with one another and with the environment, and the consequences of the fact that genes occupy specific positions in chromosomes.

Mendel's Work

Early genetic research did not use humans as subjects; there are obvious limitations to any such program. Even without considering ethical issues, we can readily see that the nine-month period of embryonic development, the comparatively small number of offspring each human produces, and the long time it takes to reach sexual maturity all argue against the use of our species for breeding experiments.

Much of the early genetic work was done with plants because of their economic importance; records show that people were deliberately cross-breeding date palm trees as early as 5,000 years ago. By the early 1800s, plant breeding was widespread, especially with ornamental flowers such as tulips. Half a century later, Gregor Mendel used the knowledge of plant reproduction to design and conduct experiments on inheritance between generations. Although his published results were neglected by scientists for 40 years, they ultimately became the foundation for the science of genetics.

Plant breeders control which plants mate

Plants are easily grown in large quantities, they often produce large numbers of offspring (in the form of seeds), and many have relatively short generation times. In many plant species, the same individuals have both male and female reproductive organs, in which case each plant may reproduce as a male, as a female, or as both. Best of all, it is often easy to control which individuals mate (Figure 10.1).

Some observations that Mendel found useful in his studies had been made in the late eighteenth century by a German botanist, Josef Gottlieb Kölreuter. Kölreuter attempted many crosses between plants, producing many **hybrids** (the offspring of genetically different parents) and learning a great deal about the biology of plant reproduction. He was the first to demonstrate that the two parents play equal roles in heredity, and he studied the offspring from **reciprocal**

10.1 A Controlled Cross between Two Plants Mendel used the pea plant, *Pisum sativum*, in many of his experiments.

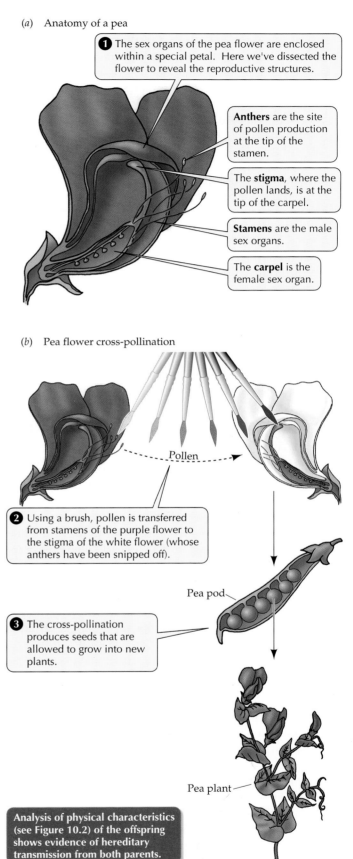

(a) Anatomy of a pea

❶ The sex organs of the pea flower are enclosed within a special petal. Here we've dissected the flower to reveal the reproductive structures.

Anthers are the site of pollen production at the tip of the stamen.

The **stigma**, where the pollen lands, is at the tip of the carpel.

Stamens are the male sex organs.

The **carpel** is the female sex organ.

(b) Pea flower cross-pollination

Pollen

❷ Using a brush, pollen is transferred from stamens of the purple flower to the stigma of the white flower (whose anthers have been snipped off).

Pea pod

❸ The cross-pollination produces seeds that are allowed to grow into new plants.

Pea plant

Analysis of physical characteristics (see Figure 10.2) of the offspring shows evidence of hereditary transmission from both parents.

crosses—matings made in two directions. For example, in one set of crosses, males that have white flowers are crossed with females that have red flowers, while in a complementary set of crosses red-flowered males and white-flowered females are the parents. In Kölreuter's experience, such reciprocal crosses always gave identical results.

Although the concept of equal parental contributions of genetic determinants was an important discovery, the nature of these determinants remained unknown. Laws of inheritance proposed at the time favored the concept of *blending*. If a plant that had one form of a characteristic (say, red flowers) were crossed with one that had a different form of that characteristic (blue flowers), the offspring would be a blended combination of the two parents (purple flowers).

And following the blending motif, it was thought that once heritable elements were combined, they could not be separated again (like combined inks). The red and blue genetic determinants were thought to be forever blended into the new purple one. Then, about a century after Kölreuter completed his work, Mendel began his.

G. Mendel

Mendel's discoveries lay dormant for decades

When Mendel began his research, it was known that one female gamete combines with one male gamete to bring about fertilization. However, the role of the chromosomes as bearers of genetic information was unknown, and mitosis and meiosis were yet to be discovered.

Mendel himself was well qualified to make the big step forward. Although in 1850 he had failed an examination for a teaching certificate in natural science, he later undertook intensive studies in physics, chemistry, mathematics, and various aspects of biology at the University of Vienna. His work in physics and mathematics probably led to his applying experimental and quantitative methods to the study of heredity—and these were the key ingredients in his success.

Mendel worked out the basic principles of the heredity of plants and animals over a period of about nine years, the work culminating in a public lecture in 1865 and a detailed written account published in 1866. Mendel's paper on plant hybridization appeared in a journal that was received by 120 libraries, and he sent reprinted copies (of which he had obtained 40) to several distinguished scholars. However, his theory was not accepted. In fact, it was ignored.

Perhaps the chief difficulty was that the physical basis of his theory was not understood until the discovery of meiosis, some years later. Furthermore, the most prominent biologists of Mendel's time were not in the habit of thinking in mathematical terms, even

the simple terms used by Mendel. Whatever the reasons, Mendel's pioneering paper had no discernible influence on the scientific world for more than 30 years.

Then, in 1900, Mendel's discoveries burst into prominence as a result of independent experiments by the Dutch Hugo de Vries, the German Karl Correns, and the Austrian Erich von Tschermak. Each of these scientists carried out crossing experiments and obtained quantitative data about the progeny; each published his principal findings in 1900; each cited Mendel's 1866 paper. At last the time was ripe for biologists to appreciate the significance of what these four geneticists had discovered—largely because meiosis had by then been described. That Mendel was able to make his discoveries *before* the discovery of meiosis was due in part to the methods of experimentation he used.

Mendel's Experiments and Laws of Inheritance

Mendel's work is a fine example of preparation, execution, and interpretation in the experimental method. Let's see how he approached each of these steps.

Mendel prepares to experiment

Mendel chose the garden pea for his studies because of its ease of cultivation, the feasibility of controlled pollination (see Figure 10.1), and the availability of varieties with differing traits. He controlled pollination, and thus fertilization, by manually moving pollen from one plant to another; thus he knew the parentage of the offspring in his experiments. If untouched, the peas Mendel studied naturally self-pollinate—that is, the female organs of flowers receive pollen from the male organs of the same flowers—and he made use of this natural phenomenon in some of his experiments.

Mendel began by examining varieties of peas in a search for heritable characters and traits suitable for study. A **character** is a feature such as flower color; a **trait** is a particular form of a character, such as white flowers; a **heritable** character trait is one that is passed from parent to offspring. Suitable characters for Mendel's experiments were those that had well-defined, contrasting alternatives, such as purple flowers versus white flowers, and that were *true-breeding*.

To be considered true-breeding, the observed trait must be the only form present in many generations. In other words, peas with white flowers, when crossed with one another, would have to give rise *only* to progeny with white flowers for many generations; tall plants bred to tall plants would have to produce *only* tall progeny.

Mendel isolated each of his true-breeding strains by repeated inbreeding (crossing of seemingly identical individuals, or of individuals with themselves) and se-

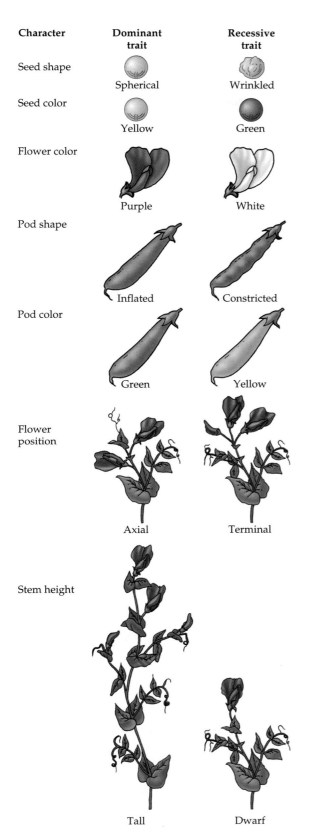

Character	Dominant trait	Recessive trait
Seed shape	Spherical	Wrinkled
Seed color	Yellow	Green
Flower color	Purple	White
Pod shape	Inflated	Constricted
Pod color	Green	Yellow
Flower position	Axial	Terminal
Stem height	Tall	Dwarf

10.2 True-Breeding Traits Traits that are the only form present over many generations are considered to be true-breeding. Mendel's studies focused on seven contrasting pairs of such traits in *Pisum sativum*.

lection. For most of his work, Mendel concentrated on the seven pairs of contrasting traits shown in Figure 10.2. Before performing a given cross, he made sure that each potential parent was from a true-breeding strain—an essential point in his analysis of his experimental results.

Mendel then placed pollen he collected from one parental strain onto the stigma (female organ) of flowers of the other strain. The plants providing and receiving the pollen were the **parental generation**, designated **P**. In due course, seeds formed and were planted. The resulting new plants constituted the **first filial generation, F$_1$**. Mendel and his assistants examined each F$_1$ plant to see which traits it bore and then recorded the number of F$_1$ plants expressing each trait. In some experiments the F$_1$ plants were allowed to self-pollinate and produce a **second filial generation, or F$_2$**. Again, each F$_2$ plant was characterized and counted.

In sum, Mendel devised a well-organized plan of research, pursued it faithfully and carefully, recorded great amounts of quantitative data, and analyzed the numbers he recorded to explain the relative proportions of the different kinds of progeny. His 1866 paper stands to this day as a model of clarity. His results and the conclusions to which they led are the subject of the next few sections.

Mendel's Experiment 1 examined a monohybrid cross

"Experiment 1" in Mendel's paper included a monohybrid cross—one in which the parents were true-breeding for a given character but each displayed a different form of that character. He took pollen from plants of a true-breeding strain with wrinkled seeds and placed it on the stigmas of flowers of a true-breeding, spherical-seeded strain (Figure 10.3). He also performed the reciprocal cross, placing pollen from the spherical-seeded strain on the stigmas of flowers of the wrinkled-seeded strain.

In both cases, all the F$_1$ seeds that were produced were spherical—it was as if the wrinkled trait had disappeared completely. The following spring Mendel grew 253 F$_1$ plants from these spherical seeds, each of which was allowed to self-pollinate—this was the monohybrid cross—to produce F$_2$ seeds. In all, there were 7,324 F$_2$ seeds, of which 5,474 were spherical and 1,850 wrinkled (Figure 10.4).

Mendel observed that the spherical seed trait was **dominant**: In the F$_1$ generation, it was always expressed rather than the wrinkled seed trait, which he called **recessive**. In each of the other six pairs of traits studied by Mendel, one proved to be dominant over the other. When he crossed plants differing in one of these traits, only one of each pair of traits was evident in the F$_1$ generation. However, the trait that was not seen in the F$_1$ *reappeared* in the F$_2$.

10.3 Contrasting Traits In Experiment 1, Mendel studied the inheritance of seed shape. We know today that the wrinkled seeds possess an abnormal form of starch. Contrast their appearance with that of the spherical seeds below.

Of most importance, the ratio of the two traits in the F_2 generation was always the same—approximately 3:1; that is, three-fourths of the F_2 showed the dominant trait and one-fourth showed the recessive trait (Table 10.1). In Mendel's Experiment 1, the ratio was 5,474:1,850 = 2.96:1. Reciprocal crosses in the parental generation gave similar outcomes in the F_2.

How Mendel interpreted his results

By themselves, the results from Experiment 1 disproved the widely held belief that inheritance is a blending phenomenon. According to the blending theory, Mendel's F_1 seeds should have had an appearance intermediate between those of the two parents—in other words, they should have been *slightly* wrinkled. Furthermore, the blending theory offered no explanation for the *reappearance* of the wrinkled trait in the F_2 seeds after its apparent absence in the F_1 seeds.

From his results Mendel proposed that the units responsible for the inheritance of specific traits are present as discrete units (particles) that occur in pairs and segregate (separate) from one another during the formation of gametes. In this particulate theory, the hereditary carriers retain their integrity in the presence of other units. This is a sharp contrast to blending, in which the carriers were believed to lose their identities when mixed together.

As he wrestled mathematically with his data, Mendel reached the conclusion that each pea has two units for each character, one from each parent. During the production of gametes, only one of the paired units for a given character is given to a gamete. Hence each gamete contains one unit, and the resulting zygote contains two, because it was produced by the fusion of two gametes. This conclusion is the core of Mendel's model of inheritance. Mendel's unit is now called a **gene**.

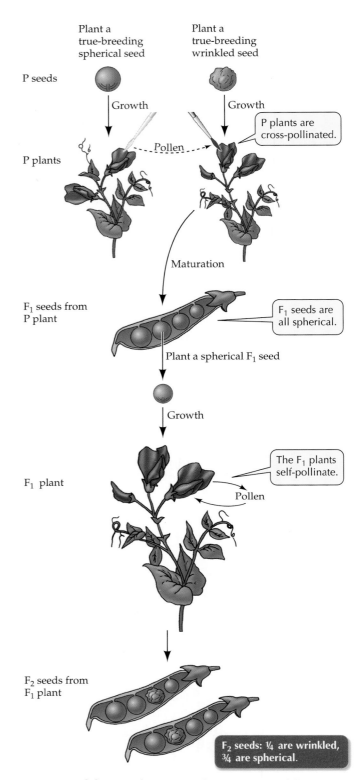

10.4 Mendel's Experiment 1 The pattern Mendel observed in the F_2 generation—¼ of the seeds wrinkled, ¾ spherical—was the same no matter which variety contributed the pollen in the parental generation.

TABLE 10.1   Mendel's Results from Monohybrid Crosses

P			F₂			
DOMINANT	×	RECESSIVE	DOMINANT	RECESSIVE	TOTAL	RATIO
Spherical	×	Wrinkled seeds	5,474	1,850	7,324	2.96:1
Yellow	×	Green seeds	6,022	2,001	8,023	3.01:1
Purple	×	White flowers	705	224	929	3.15:1
Inflated	×	Constricted pods	882	299	1,181	2.95:1
Green	×	Yellow pods	428	152	580	2.82:1
Axial	×	Terminal flowers	651	207	858	3.14:1
Tall	×	Dwarf stems	787	277	1,064	2.84:1

Mendel reasoned that in Experiment 1, the spherical-seeded parent had a pair of genes of the same type, which we will call *S*, and the parent with wrinkled seeds had two *s* genes. The *SS* parent produced gametes each containing a single *S*, and the *ss* parent produced gametes each with a single *s*. Each member of the F₁ generation had an *S* from one parent and an *s* from the other; an F₁ could thus be described as *Ss*. We say that *S* is dominant over *s* because *s* is not evident when both forms of the gene are present.

The different forms of a gene (*S* and *s* in this case) are called **alleles**. Individuals that are true-breeding for a trait contain two copies of the same allele. For example, all the individuals in a population of a strain of true-breeding peas with wrinkled seeds must have the allele pair *ss*; if *S* were present, the plants would produce spherical seeds.

We say that individuals that produce wrinkled seeds are **homozygous** for the allele *s*, meaning that they have two copies of the same allele (*ss*). Some peas with spherical seeds—the ones with the genotype *SS*—are also homozygous. However, not all plants with spherical seeds have the *SS* genotype. Some spherical-seeded plants are **heterozygous**: They have two different alleles of the gene in question (in this case *Ss*).

To illustrate these terms with a more complex example, one in which there are three gene pairs, an individual with the genotype *AABbcc* is homozygous for the *A* and *C* genes—because it has two *A* alleles and two *c* alleles—but heterozygous for the *B* gene because it contains the *B* and *b* alleles. An individual that is homozygous for a character is sometimes called a *homozygote*; a *heterozygote* is heterozygous for the character in question.

The physical appearance of an organism is its **phenotype**. Mendel correctly supposed the phenotype to be the result of the **genotype**, or genetic constitution, of the organism showing the phenotype. In Experiment 1 we are dealing with two phenotypes (spherical seeds and wrinkled seeds). As we will see in the next section, the F₂ generation contains these two phenotypes and three genotypes: The wrinkled-seed phenotype is produced only by the genotype *ss*, whereas the spherical-seed phenotype may be produced by the genotypes *SS* or *Ss*.

Mendel's first law says that alleles segregate

How does Mendel's model of inheritance explain the composition of the F₂ generation in Experiment 1? Consider first the F₁, which has the spherical-seeded phenotype and the *Ss* genotype. According to the model, when any individual produces gametes, the alleles **segregate**, or separate, so each gamete receives only *one* member of the pair of alleles. This is Mendel's first law, the **law of segregation**. In Experiment 1, half the gametes produced by the F₁ contain the *S* allele and half the *s* allele. Years after Mendel's death, geneticists determined that the basis of segregation was the separation of homologous chromosomes during meiosis I (see Chapter 9).

During plant pollination, the random combination of these gametes produces the F₂ generation (Figure 10.5). Three different F₂ genotypes are possible: *SS*, *Ss* (which is the same thing as *sS*), and *ss*. Our quantitative way of looking at things may lead us to wonder what proportions of these genotypes we might expect to observe in the F₂ progeny. The expected frequencies of these three genotypes in our example may be determined by using the Punnett square, devised in 1905 by the British geneticist Reginald Crundall Punnett.

The **Punnett square** is a device to remind us to consider all possible combinations of gametes. The square looks like this:

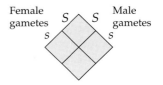

It is a simple grid with all possible sperm genotypes shown across one side and all possible egg genotypes along another side. To complete the grid, fill each

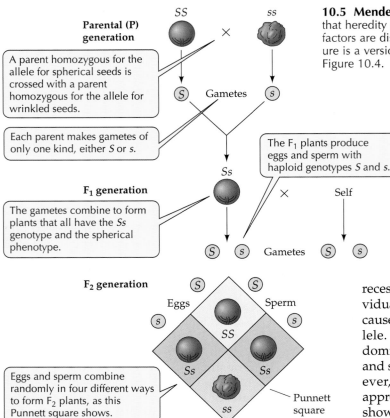

Parental (P) generation

A parent homozygous for the allele for spherical seeds is crossed with a parent homozygous for the allele for wrinkled seeds.

Each parent makes gametes of only one kind, either *S* or *s*.

Gametes

The F₁ plants produce eggs and sperm with haploid genotypes *S* and *s*.

F₁ generation

The gametes combine to form plants that all have the *Ss* genotype and the spherical phenotype.

Gametes

F₂ generation

Eggs   Sperm

Eggs and sperm combine randomly in four different ways to form F₂ plants, as this Punnett square shows.

Punnett square

10.5 Mendel's Explanation of Experiment 1 Mendel concluded that heredity depends on factors from each parent, and that these factors are discrete units that do not blend in the offspring. This figure is a version of Mendel's explanation of the experiment shown in Figure 10.4.

Mendel verified his hypothesis by performing a test cross

The **test cross** is a way to test whether a given individual showing a dominant trait is homozygous or heterozygous. In a test cross, the individual in question is crossed with an individual known to be homozygous for the recessive trait—an easy individual to identify because in order to have the recessive phenotype it must be homozygous; that is, it must contain alleles only for the recessive trait.

For the gene that we have been considering, the recessive homozygote for the test cross is *ss*. The individual being tested may be described initially as *S*– because we do not yet know the identity of the second allele. If the individual being tested is homozygous dominant (*SS*), all offspring of the test cross will be *Ss* and show the dominant trait (spherical seeds). If, however, the tested individual is heterozygous (*Ss*), then approximately half of the offspring of the test cross will show the dominant trait, but the other half will be homozygous for the recessive trait (Figure 10.7). These are exactly the results that are obtained; thus Mendel's model accurately predicts the results of such test crosses.

Mendel's second law says that alleles of different genes assort independently

What happens if two parents that differ at two or more loci are crossed? When a double heterozygote (for example, *AaBb*) makes gametes, do the alleles of maternal origin go together to one gamete and those of paternal origin to another gamete? Or does a single gamete receive some maternal and some paternal alleles? To answer these questions Mendel performed a series of **dihybrid crosses**, crosses made between parents that are identical double heterozygotes.

In these experiments Mendel began with peas that differed for two characters of the seeds: seed shape and seed color. One true-breeding strain produced only spherical, yellow seeds (*SSYY*) and the other strain produced only wrinkled, green ones (*ssyy*). A cross between these two strains produces an F₁ generation in which all the plants are *SsYy*. Because the *S* and *Y* alleles are dominant, these F₁ seeds would all be yellow and spherical.

Mendel continued this experiment to the next generation—the dihybrid cross. There are two ways in which these doubly heterozygous plants might produce gametes, as Mendel saw it. (Remember that he had never heard of chromosomes, let alone of meio-

square with the corresponding sperm genotype and egg genotype, giving the diploid genotype of one member of the F₂ generation. For example, to fill the rightmost square, put in the *S* from the egg (female gamete) and the *s* from the sperm (male gamete), yielding *Ss*.

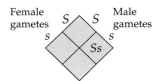

Female gametes   Male gametes

Examination of the Punnett square in Figure 10.5 reveals that self-pollination of the F₁ genotype *Ss* will give the three F₂ genotypes in the expected ratio 1 *SS*: 2 *Ss*:1 *ss*. Because *S* is dominant and *s* recessive, only two phenotypes result, in the ratio of 3 spherical (*SS* and *Ss*) to 1 wrinkled (*ss*), just as Mendel observed.

Mendel did not live to see his theory placed on a sound physical footing based on chromosomes and DNA. Genes are now known to be portions of the DNA molecules in chromosomes. More specifically, a gene is a portion of the DNA that resides at a particular site, called a **locus**, within the chromosome and that encodes a particular function. Mendel arrived at his law of segregation with no knowledge of chromosomes or meiosis, but today we can picture the alleles segregating as chromosomes separate into gametes in meiosis (Figure 10.6).

Diploid parent

Homologous chromosomes

Ss

Meiotic interphase

This site on the chromosome is the locus of the gene with the alleles *S* and *s*.

Ss

Meiosis I

The alleles have segregated.

Because of meiosis, each gamete contains one member of each pair of homologous chromosomes, and thus one allele for each pair of genes.

Meiosis II

Gametes

10.6 Meiosis Accounts for the Segregation of Alleles Although Mendel had no knowledge of chromosomes or meiosis, we now know that a pair of alleles resides on homologous chromosomes and that meiosis segregates those alleles.

---

sis.) First, if the alleles maintain the associations they had in the original parents (that is, if they are linked), then only two types of gametes would be produced (*SY* and *sy*); and the F$_2$ progeny resulting from self-pollination of the F$_1$ plants would consist of three times as many plants bearing spherical, yellow seeds as ones with wrinkled, green seeds. Were such results to be obtained, there would be no reason to suppose that seed shape and seed color were really regulated by two different genes, because spherical seeds would always be yellow, and wrinkled ones always green.

The second possibility is that the segregation of *S* from *s* is *independent* of the segregation of *Y* from *y* during the production of gametes (that is, they are un-linked). In this case, four kinds of gametes would be produced, and in equal numbers: *SY*, *Sy*, *sY*, and *sy*. When these gametes combined at random, they would produce an F$_2$ of nine different genotypes. The progeny can have any of three possible genotypes for shape (*SS*, *Ss*, or *ss*) and any of three genotypes for color (*YY*, *Yy*, or *yy*). The combined nine genotypes

10.7 Homozygous or Heterozygous? A plant with a dominant phenotype may be homozygous or heterozygous. Its genotype can be determined by making a test cross, which means observing the phenotypes of progeny produced by crossing it with a homozygous recessive plant.

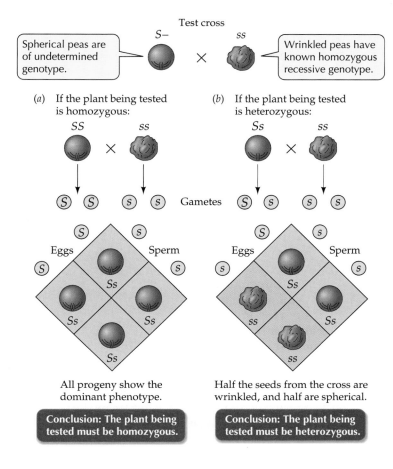

Test cross

Spherical peas are of undetermined genotype.

S–   *ss*

Wrinkled peas have known homozygous recessive genotype.

(*a*) If the plant being tested is homozygous:

SS   *ss*

Gametes

(*b*) If the plant being tested is heterozygous:

Ss   *ss*

Eggs   Sperm

Eggs   Sperm

All progeny show the dominant phenotype.

Conclusion: The plant being tested must be homozygous.

Half the seeds from the cross are wrinkled, and half are spherical.

Conclusion: The plant being tested must be heterozygous.

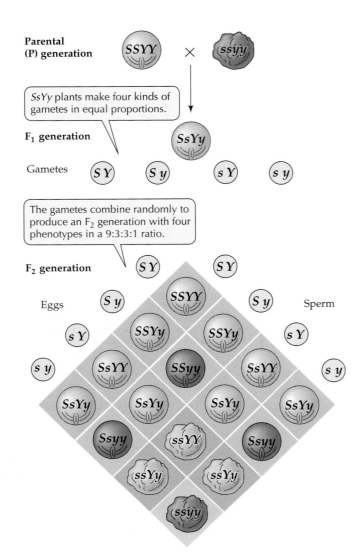

Parental (P) generation

SsYy plants make four kinds of gametes in equal proportions.

F₁ generation

Gametes

The gametes combine randomly to produce an F₂ generation with four phenotypes in a 9:3:3:1 ratio.

F₂ generation

Eggs

Sperm

10.8 Independent Assortment The 16 possible combinations of gametes result in 9 different genotypes. Because *S* and *Y* are dominant over *s* and *y*, respectively, the 9 genotypes determine 4 phenotypes in the ratio of 9:3:3:1.

would produce just four phenotypes (spherical yellow, spherical green, wrinkled yellow, wrinkled green). By using a Punnett square, we can show that these four phenotypes would be expected to occur in a ratio of 9:3:3:1. (Figure 10.8)

Mendel's dihybrid crosses produced the results predicted by the second possibility. Four different phenotypes appeared in the F₂ in a ratio of about 9:3:3:1, rather than only the two parental types as predicted from the first possibility. The parental traits appeared in new combinations in two of the phenotypic classes (spherical green and wrinkled yellow). These new combinations are called **recombinant phenotypes**.

These results led Mendel to the formulation of what is now known as Mendel's second law: *Alleles of differ-*

ent genes assort independently of one another during gamete formation. This **law of independent assortment** is not as universal as the law of segregation, because it applies to genes that lie on separate chromosomes, but not necessarily to those that lie on the same chromosome. However, it is correct to say that *chromosomes* assort independently during the formation of gametes (Figure 10.9).

Punnett squares or probability calculations? You choose!

Many people find it easiest to solve genetics problems using probability calculations, perhaps because the basic underlying considerations are familiar. When we flip a coin, for example, we expect it has an equal probability of landing "heads" or "tails."

For a given toss, the probability of heads is independent of what happened in all the previous tosses. For a fair coin, a run of ten straight heads implies nothing about the next toss. No "law of averages" increases the likelihood that the next toss will come up tails, and no "momentum" makes an eleventh occurrence of heads any more likely. On the eleventh toss, the odds are still 50:50.

The basic conventions of probability are simple: If an event is absolutely certain to happen, its probability is 1. If it cannot happen, its probability is 0. Otherwise, its probability lies between 0 and 1. A coin toss results in heads approximately half the time, and the probability of heads is ½—as is the probability of tails.

MULTIPLYING PROBABILITIES. If *two* coins (a penny and a dime, say) are tossed, each acts independently of the other. What, then, is the probability of both coins coming up heads? Half the time, the penny comes up heads; of that fraction, half the time the dime also comes up heads. Therefore, the joint probability of two heads is half of one-half, or ½ × ½ = ¼. To find the joint probability of *independent* events, then, the general rule is to *multiply* the probabilities of the individual events (Figure 10.10).

THE MONOHYBRID CROSS. To apply a probabilistic approach to genetics problems, we need only deal with gamete formation and random fertilization. A homozygote can produce only one type of gamete, so, for example, an *SS* individual has a probability equal to 1 of producing gametes with the genotype *S*. The heterozygote *Ss* produces *S* gametes with a probability of ½, and *s* gametes with a probability of ½.

Consider the F₂ progeny of the cross in Figure 10.5. They are obtained by self-pollination of F₁ hybrids of genotype *Ss*. The probability that an F₂ plant has the genotype *SS* must be ½ × ½ = ¼ because there is a 50:50 chance that the sperm will have the genotype *S*, and this chance is independent of the 50:50 chance that

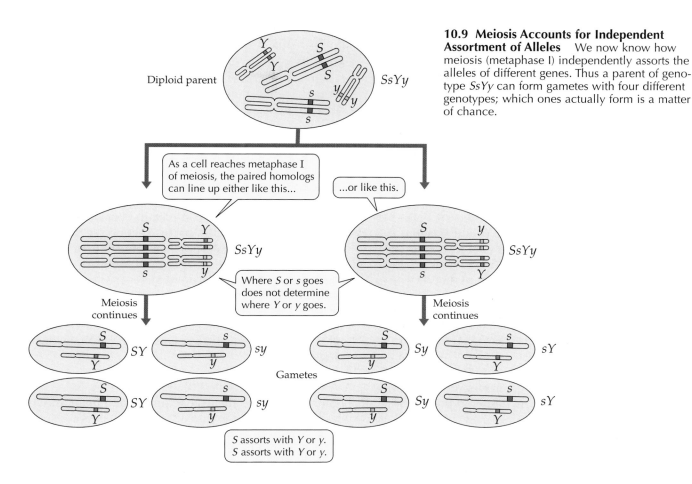

10.9 Meiosis Accounts for Independent Assortment of Alleles We now know how meiosis (metaphase I) independently assorts the alleles of different genes. Thus a parent of genotype *SsYy* can form gametes with four different genotypes; which ones actually form is a matter of chance.

the egg will have the genotype *S*. Similarly, the probability of *ss* offspring is ½ × ½ = ¼.

ADDING PROBABILITIES. The probability of getting *S* from the sperm and *s* from the egg is also ¼, but remember that the same genotype can also result from *s* in the sperm and *S* in the egg, with a probability of ¼. The probability of an event that can occur in two or more different ways is the sum of the individual probabilities of those ways. Thus the probability that an F₂ plant is a heterozygote is ¼ + ¼ = ½ (see Figure 10.10). The three genotypes are expected in the ratio ¼ *SS*: ½ *Ss*:¼ *ss*—hence the 1:2:1 ratio of genotypes and the 3:1 ratio of phenotypes seen in Figure 10.5.

10.10 Joint Probabilities of Independent Events Like two tosses of a coin, the segregation of alleles during sperm and egg formation are independent events. The probability (*P*) of any given combination of alleles from a sperm and an egg is obtained by multiplying the probabilities of each event; this is the probability of producing a homozygote. Since a heterozygote can be formed in two ways, the two probabilities are added together.

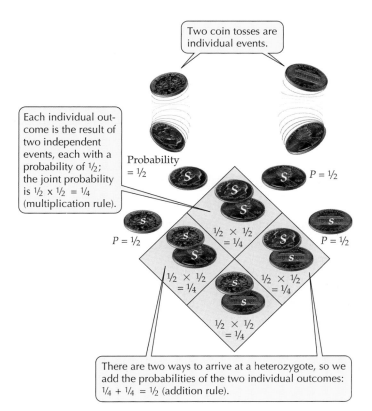

THE DIHYBRID CROSS. If F_1 plants heterozygous for two independent characters self-pollinate, the resulting F_2 plants express four phenotypes. The proportions of these phenotypes are easily determined by probabilities. Let's see how this works for the experiment of Figure 10.8.

The probability that a seed will be spherical is $\frac{3}{4}$, as we have just seen. By the same reasoning, the probability that a seed will be yellow is also $\frac{3}{4}$. The two characters are determined by separate genes and are independent of one another, so the joint probability that a seed will be both spherical and yellow is $\frac{3}{4} \times \frac{3}{4} = \frac{9}{16}$. For the wrinkled, yellow members of the F_2 generation, the probability of being yellow is again $\frac{3}{4}$; the probability of being wrinkled is $\frac{1}{2} \times \frac{1}{2} = \frac{1}{4}$. The joint probability that a seed will be both wrinkled and yellow, then, is $\frac{3}{4} \times \frac{1}{4} = \frac{3}{16}$. The same probability applies, for similar reasons, to the spherical, green F_2 seeds. Finally, the probability that F_2 seeds will be both wrinkled and green must be $\frac{1}{4} \times \frac{1}{4} = \frac{1}{16}$. Looking at all four phenotypes, we see they are expected in the ratio of 9:3:3:1.

Probability calculations and Punnett squares give the same results. Learn to do genetics problems both ways, and then decide which method you prefer.

Alleles and Their Interactions

Mendel successfully interpreted the monohybrid crosses, test crosses, and dihybrid crosses he performed, and he gave us his laws of segregation and independent assortment. Decades later, genes became defined as chemical entities—DNA sequences—that are expressed as proteins.

Let's move on to the extensions to Mendelian genetics that have been developed by other workers, mostly in the early part of the twentieth century. Recall that alleles are alternative forms of the same gene. In the next chapter we'll see the molecular basis for the distinction between alleles. In this section we deal with how alleles relate to one another, some of their general properties, and how they arise.

In many cases, alleles do not show simple relationships between dominance and recessiveness. In others, a single allele may have multiple phenotypic effects when it is expressed. Existing alleles can form new alleles by mutation, so there can be many alleles for a single character.

Dominance is usually not complete

Some genes have alleles that are not dominant and recessive to each other. Instead, the heterozygotes show an intermediate phenotype—at first glance like that predicted by the old blending theory of inheritance.

For example, if a true-breeding red snapdragon is crossed with a true-breeding white one, all the F_1 flowers are pink. That this phenomenon can still be explained in terms of Mendelian genetics rather than of a blending theory is readily demonstrated by a further cross.

If one of these pink F_1 snapdragons is crossed with a true-breeding white one, the blending theory predicts that all the offspring would be a still-lighter pink. In fact, approximately $\frac{1}{2}$ of the offspring are white and $\frac{1}{2}$ the same pink as the original F_1. Suppose now that the F_1 pink snapdragons are self-pollinated. The resulting F_2 plants are distributed in a ratio of 1 red:2 pink:1 white (Figure 10.11). Clearly the hereditary particles—the genes—have *not* blended; in the F_2 they are readily sorted out in their original forms.

We can understand these results in terms of the Mendelian model. When a

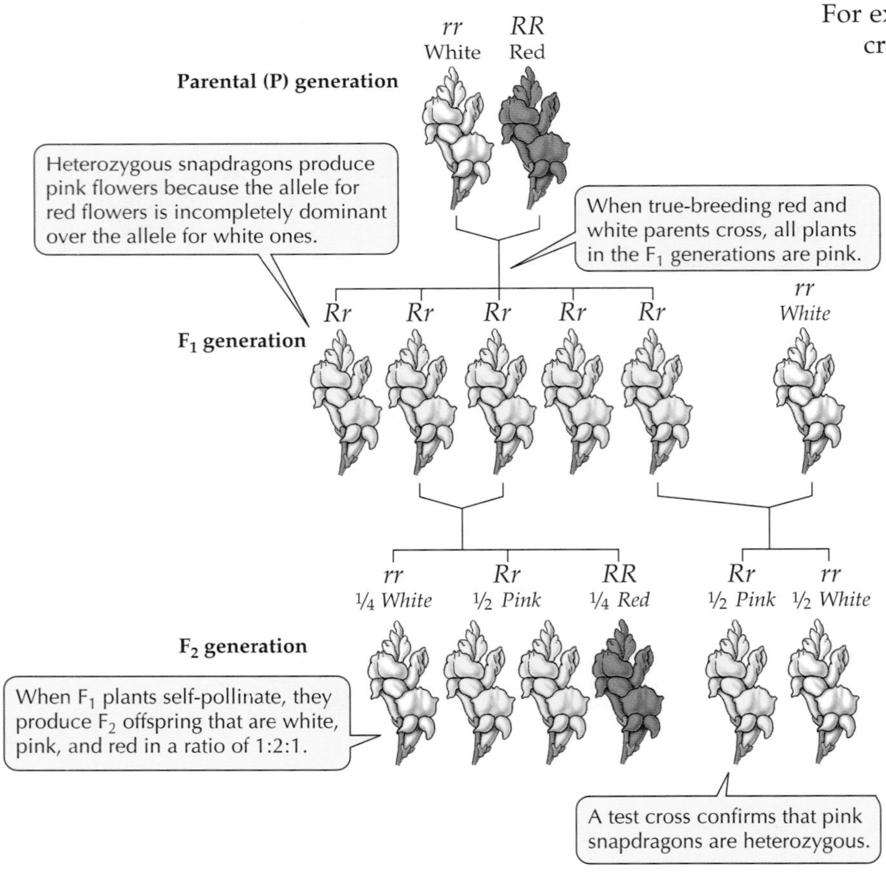

Parental (P) generation

rr White *RR* Red

Heterozygous snapdragons produce pink flowers because the allele for red flowers is incompletely dominant over the allele for white ones.

When true-breeding red and white parents cross, all plants in the F_1 generations are pink.

F₁ generation

Rr *Rr* *Rr* *Rr* *Rr*   *rr* White

F₂ generation

rr ¼ White   *Rr* ½ Pink   *RR* ¼ Red   *Rr* ½ Pink   *rr* ½ White

When F_1 plants self-pollinate, they produce F_2 offspring that are white, pink, and red in a ratio of 1:2:1.

A test cross confirms that pink snapdragons are heterozygous.

10.11 Incomplete Dominance Follows Mendel's Laws The study of incomplete dominance clearly disproves the idea of "blended" inheritance. In the F_2 generation, the genes are readily sorted out in their original forms.

heterozygous phenotype is intermediate, as in the snapdragon example, the gene is said to be governed by **incomplete dominance**. All we need to do in cases like this is recognize that the heterozygotes show a phenotype intermediate between those of the two homozygotes. Genes code for the production of specific proteins, many of which are enzymes. Different alleles at a locus code for alternative forms of a protein. When the protein is an enzyme, the different forms often have different degrees of catalytic activity.

In the snapdragon example, one allele codes for an enzyme that catalyzes a reaction leading to the formation of a red pigment in the flowers. The alternative allele codes for an altered protein that lacks catalytic activity for pigment production. Plants homozygous for this alternative allele cannot synthesize red pigment, and their flowers are white. Heterozygous plants, with only one allele for the functional enzyme, produce just enough red pigment that their flowers are pink. Homozygous plants that have two alleles for the functional enzyme produce more red pigment, resulting in red flowers.

There are more examples of incomplete dominance than of complete dominance. Thus an unusual feature of Mendel's report is that all seven of the examples he described (see Table 10.1) are characterized by complete dominance. For dominance to be complete, a single copy of the dominant allele must produce enough of its protein product to give the maximum phenotypic response.

For example, just one copy of the dominant allele T at one of the loci studied by Mendel leads to the production of enough of a growth-promoting chemical that the Tt heterozygotes are as tall as the homozygous dominant plants (TT)—the second copy of T causes no further growth of the stem. The homozygous recessive plants (tt) are much shorter because the allele t does not lead to the production of the growth promoter.

Sometimes two alleles at a locus produce two different phenotypes that *both* appear in heterozygotes. An example of this phenomenon, called **codominance**, is seen in the ABO blood group system in humans. One allele of the gene determines the A blood type, and another determines the B blood type. When both alleles are present, both are expressed, resulting in the AB blood type, which includes proteins produced by both alleles. We will return to other features of the ABO blood system shortly.

Some alleles have multiple phenotypic effects

When a single allele has more than one distinguishable phenotypic effect, we say that the allele is **pleiotropic**. A familiar example of pleiotropy is the allele responsible for the coloration pattern (light body, darker extremities) of Siamese cats, discussed later in this chapter. The same allele is also responsible for the characteristic crossed eyes of Siamese cats. Although these effects appear to be unrelated, both result from the same protein produced under the influence of that allele.

New alleles arise by mutation

Why does a gene have different alleles? Different alleles exist because any gene is subject to **mutation**, which means that it can be changed to a *stable, heritable* new form. In other words, an allele can mutate to become a different allele.

One particular allele of a gene may be defined as the **wild type**, or standard, because it is present in most individuals in nature and gives rise to an expected trait or phenotype. Other forms of that same gene, often called **mutant** alleles, may produce a different phenotype. The wild-type and mutant alleles reside at the same locus and are inherited according to the rules set forth by Mendel. A genetic locus with a wild-type allele present less than 99 percent of the time (the rest of the alleles being mutant) is said to be **polymorphic** (from the Greek *poly*, "many," and *morph*, "form").

Mutation, to be discussed in detail in Chapter 12, is a random process; different copies of the same gene may be changed in different ways, depending on how and where the DNA changes. Genetic mutation is the raw material for the process of evolution—the subject of Part Three—and is the ultimate source of the vast biological diversity described in Part Four.

Many genes have multiple alleles

A group of individuals may have more than two alleles of a given gene. (Any one individual has only two alleles, of course—one from the mother and one from the father.) In fact, there are many examples of such **multiple alleles**. In the fruit fly *Drosophila melanogaster*, many alleles at one locus affect eye color by determining the amount of pigment produced (Figure 10.12). The exact color of the fly's eyes depends on which two alleles are inherited.

The ABO blood group system in humans is determined by a set of three alleles (I^A, I^B, and I^O) at one locus, which determines certain proteins (antigens) on the surface of red blood cells. Different combinations of these alleles in different people produce four different blood types, or phenotypes: A, B, AB, and O (Figure 10.13). Early attempts at blood transfusion—made before these blood types were understood—frequently killed the patient. Around 1900, however, the Austrian scientist Karl Landsteiner mixed blood cells and serum (blood from which cells have been removed) from different individuals. He found that only certain combinations of blood types are compatible. In other combinations, the red blood cells form clumps

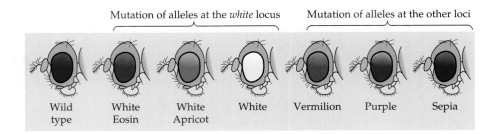

Mutation of alleles at the *white* locus — Mutation of alleles at the other loci

Wild type — White Eosin — White Apricot — White — Vermilion — Purple — Sepia

10.12 Multiple Alleles Govern Eye Color in *Drosophila* Wild-type flies have dark red eyes, but many alleles affect eye color. Mutations can produce a wide variety of eye colors by changing the amount of pigment produced.

because of the presence in the serum of specific proteins, called antibodies (see Chapter 18), that react with foreign, or "nonself," cells. This discovery led to our ability to administer compatible blood transfusions that do not kill the recipient.

The Rh factor, so named because it was first found in rhesus monkeys, is another substance on the surface of red blood cells. In most human populations, almost 100 percent of the individuals have the Rh factor, and their blood is said to be Rh⁺ (Rh-positive). Among Caucasians, however, only 83 percent are Rh⁺; the others lack the Rh factor, and their blood is called Rh⁻ (Rh-negative).

Like the ABO blood types, the Rh factor is genetically determined. A single locus with at least eight multiple alleles is responsible. Certain dominant alleles cause the production of the Rh factor; Rh⁻ individuals are homozygous recessives who lack these alleles. People who are Rh⁻ make antibodies that react with blood cells from Rh⁺ individuals. Because the blood systems of mother and newborn mix, if a woman who is Rh⁻ has an Rh⁺ child, the mother may make antibodies that react adversely with the blood of her newborn infant and can be fatal to the child (or, more commonly, to any subsequent Rh⁺ children she may have). Thus, pregnancies with Rh incompatibility are closely monitored.

Another system of multiple alleles is illustrated by the scallops in "Applying Concepts" Question 2 at the end of this chapter. The question of how differing alleles may be maintained in a population through time will be examined in Chapter 20.

Gene Interactions

Thus far we have treated the phenotype of an organism, with respect to a given character, as a simple result of its genotype, and we have implied that a single trait results from the alleles of a single gene. In fact, several genes may interact to determine a trait's phenotype. For example, height in people is determined by the actions of many genes, such as those that determine bone growth, hormones, and other aspects of development. Sometimes several genes act additively, so the phenotype can be predicted by how many of these genes are active. To complicate things further, the physical environment may interact with the genetic constitution of an individual in determining the phenotype. People's height, for example, is not determined only by their genes. Nutrition, one environmental factor, undoubtedly has a strong influence.

Some genes alter the effects of other genes

In **epistasis**, one gene alters the effect of another gene. For example, several genes determine coat color in mice. The wild-type color is agouti, a grayish pattern resulting from bands on the individual hairs. The dominant allele *B* determines that the

Blood type of cells	Genotype	Antibodies made by cells	Reaction to added antibodies Anti-A	Anti-B
A	$I^A I^A$ or $I^A I^O$	Anti-B		
B	$I^B I^B$ or $I^B I^O$	Anti-A		
AB	$I^A I^B$	Neither anti-A nor anti-B		
O	$I^O I^O$	Both anti-A and anti-B		

Red blood cells that do not react with antibody remain evenly dispersed.

Red blood cells that react with antibodies clump together (speckled appearance in the photograph).

10.13 ABO Blood Reactions Are Important in Transfusions Cells of blood types A, B, AB, and O were mixed with anti-A or anti-B antibodies. Note that anti-A reacts with A and AB cells but not with B or O. Which blood types do anti-B antibodies react with? As you look down the columns, note that each of the types, when mixed separately with anti-A and with anti-B, gives a unique pair of results; this is the basic method by which blood is typed.

Mice with genotype *aa* are albino regardless of their genotype for the other locus, because the *aa* genotype blocks all pigment production.

10.14 Genes May Interact Epistatically Epistasis occurs when one gene alters the effect of another gene. In these mice, the presence of the recessive genotype (*aa*) at one locus blocks pigment production, producing an albino mouse no matter what the genotype is at the second locus.

Mice with *bb* genotypes are black unless they are also *aa* (which makes them albino).

Mice that have at least one dominant allele at each locus are agouti.

hairs will have bands and thus that the color will be agouti, whereas the homozygous recessive genotype *bb* results in unbanded hairs and the mouse appears black. On another chromosome, a second locus affects an early step in the formation of hair pigments. The dominant allele *A* at this locus allows normal color development, but *aa* blocks all pigment production. Thus, *aa* mice are all-white albinos, irrespective of their genotype at the *B* locus (Figure 10.14).

If a mouse with genotype *AABB* (and thus the agouti phenotype) is crossed with an albino of genotype *aabb*, the F$_1$ is *AaBb* and of the agouti phenotype. If the F$_1$ mice are crossed with each other to produce an F$_2$, the epistasis of *aa* will result in an expected phenotypic ratio of 9 agouti:3 black:4 albino. Can you show why? The underlying ratio is the usual 9:3:3:1 for a dihybrid cross with unlinked genes, but be sure to look closely at each genotype and watch out for epistasis.

In another form of epistasis, two genes are mutually dependent: The expression of each depends on the alleles of the other. The epistatic action of such complementary genes may be explained as follows: Suppose the dominant gene *A* codes for enzyme A on the pathway for purple pigment in flowers and gene *B* codes for enzyme B:

colorless precursor —enzyme A→ colorless intermediate —enzyme B→ purple pigment

In order for the pigment to be produced, *both* reactions must take place. The alleles *a* and *b* code for nonfunc-

tional enzymes. If a plant is homozygous for either *a* or *b*, the corresponding reaction will not occur, no purple pigment will form, and the flowers will be white.

Polygenes mediate quantitative inheritance

Individual heritable characters are often found to be controlled by many genes, **polygenes**, of which each allele intensifies or diminishes the observed phenotype. As a result, variation in such characters is **continuous** rather than, as in the examples we have been considering, **discontinuous** (or discrete). Many characters that vary continuously—such as height and other aspects of size, or skin color—are under genetic control. Polygenes affecting a particular quantitative character are common on many different chromosomes.

Humans differ with respect to the amount of a dark pigment, melanin, in their skin (Figure 10.15). There is great variation in the amount of melanin among different people, but much of this variation is determined by alleles at just four (possibly three) loci. No alleles at these loci demonstrate dominance. Of course, skin color is not entirely determined by the genotype, since exposure to sunlight in fair-skinned people can cause the production of more melanin—that is, a suntan.

The environment affects gene action

Environmental variables such as light, temperature, and nutrition can sharply affect the translation of a genotype into a phenotype. A familiar example is the Siamese cat (Figure 10.16). This handsome animal normally has darker fur on its ears, nose, paws, and tail than on the rest of its body. The darkened parts have a lower temperature than the light parts.

A few simple experiments show that the Siamese cat has a genotype that results in dark fur, but only at temperatures below the general body temperature. If some dark fur is removed from the tail and the cat is kept at higher-than-usual temperatures, the new fur that grows in is light. Conversely, removal of light fur from the back, followed by local chilling of the area, causes the spot to fill with dark fur.

Genotype and environment interact to determine the phenotype of an organism. It is sometimes possible to determine the proportion of individuals in a group with a given genotype that actually show the expected phenotype. This proportion is called the **penetrance** of the genotype. The environment may also affect the **expressivity** of the genotype—that is, the degree to which it is expressed. For an example of environmental

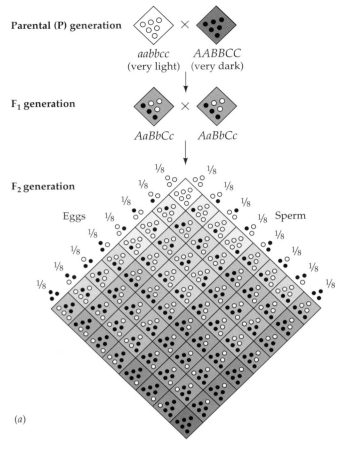

(a)

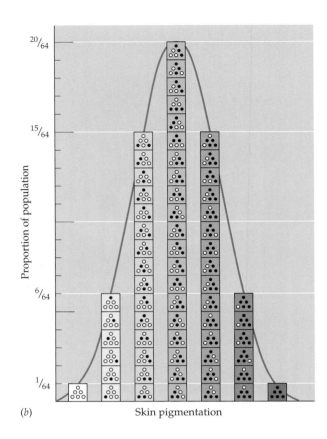

(b) Skin pigmentation

10.15 Polygenes Determine Human Skin Pigmentation
A model of polygenic inheritance based on three genes. The alleles A, B, and C contribute dark pigment to the skin, but the alleles a, b, and c do not. The more A, B, and C alleles an individual possesses, the darker that person's skin will be. If both members of a couple have intermediate pigmentation (AaBbCc, for example), they are unlikely to have children with either very light or very dark skin.

effects on expressivity, consider how Siamese cats kept indoors or outdoors in different climates might look.

Uncertainty over how much of the observed variation is due to the environment and how much to the effects of the several polygenes complicates the analysis of quantitative inheritance. A useful approach that avoids this difficulty is to study identical twins. Since such twins are genetically identical, any differences between them are attributed to environmental effects.

The phenotype of an organism depends on its total genetic makeup and on its environment. Some of the interactive effects will become more obvious when we focus, in Chapter 14, on the regulation of gene expression.

10.16 The Environment Affects the Phenotype This Siamese cat has dark fur on its extremities, where the temperature is below the general body temperature.

Genes and Chromosomes

The recognition that genes occupy characteristic positions on chromosomes and thus are segregated by meiosis enabled Mendel's successors to provide a physical explanation for his model of inheritance. It soon became apparent that the association of genes with chromosomes has other genetic consequences as well.

In this section we will address the following questions: What is the pattern of inheritance of genes that occupy nearby loci on the same chromosome? How do

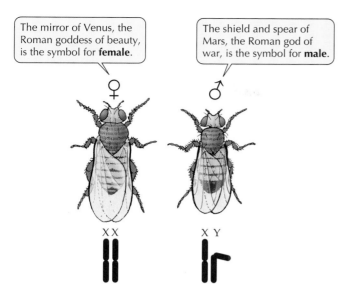

The mirror of Venus, the Roman goddess of beauty, is the symbol for **female**.

The shield and spear of Mars, the Roman god of war, is the symbol for **male**.

X X

X Y

10.17 *Drosophila melanogaster,* the Star of Morgan's Fly Room The fruit fly (whose Latin name means "vinegar-loving, dark-bodied fly") has a short generation time—a major reason for its widespread use as a laboratory organism in genetics experiments.

we determine the order of genes on a chromosome—and the distances between them? Why were all the carriers of hemophilia in Queen Victoria's family women, and why were all of her descendants who had hemophilia men?

The answers to these and many other genetic questions were worked out in studies of the fruit fly *Drosophila melanogaster* (Figure 10.17). Its small size, its ease of cultivation, and its short generation time made it an attractive experimental subject. Beginning in 1909, Thomas Hunt Morgan and his students established *Drosophila* as a highly useful laboratory organism in Columbia University's famous "fly room," where

10.18 Alleles That Don't Assort Independently
Morgan's studies showed that the genes for body color and wing size in *Drosophila* are linked, so their alleles do not assort independently. Linkage accounts for the departure of the actual phenotypes observed from the results predicted by Mendel's laws.

the group discovered the phenomena described in this section. *Drosophila* remains extremely important in studies of chromosome structure, population genetics, the genetics of development, and the genetics of behavior.

Linked genes are near each other on a chromosome

In the immediate aftermath of the rediscovery of Mendel's laws, the second law—independent assortment—was considered to be generally applicable. However, some investigators, including R. C. Punnett (the inventor of the square), began to observe strange deviations from the expected 9:3:3:1 ratio in some dihybrid crosses. T. H. Morgan, too, obtained data not in accord with Mendelian ratios and specifically not in accord with the law of independent assortment.

T. H. Morgan

Morgan crossed *Drosophila* of two known genotypes, $BbVgvg \times bbvgvg$, where B, the wild-type (gray) body, is dominant over b (black body), and Vg (wild-type wing) is dominant over vg (*vestigial*, a very small wing). Do you recognize this type of cross? It is a test cross (see Figure 10.7) for the two gene pairs. Morgan expected to see four phenotypes in a ratio of 1:1:1:1, but this was not what he observed. The body color gene and the wing size gene were not assorting independently; rather, they were for the most part inherited together (Figure 10.18).

These results became understandable to Morgan when he assumed that the two loci are on the *same*

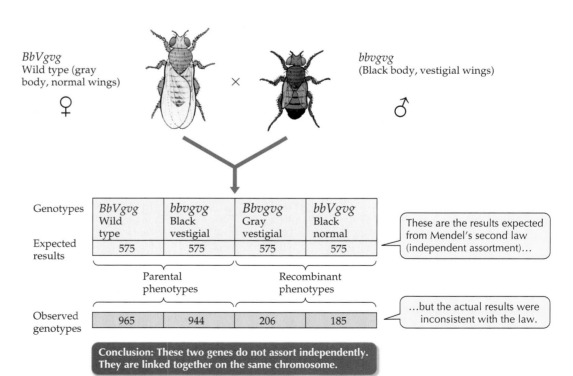

BbVgvg
Wild type (gray body, normal wings)
♀

×

bbvgvg
(Black body, vestigial wings)
♂

Genotypes	*BbVgvg* Wild type	*bbvgvg* Black vestigial	*Bbvgvg* Gray vestigial	*bbVgvg* Black normal
Expected results	575	575	575	575

These are the results expected from Mendel's second law (independent assortment)…

Parental phenotypes — Recombinant phenotypes

Observed genotypes	965	944	206	185

…but the actual results were inconsistent with the law.

Conclusion: These two genes do not assort independently. They are linked together on the same chromosome.

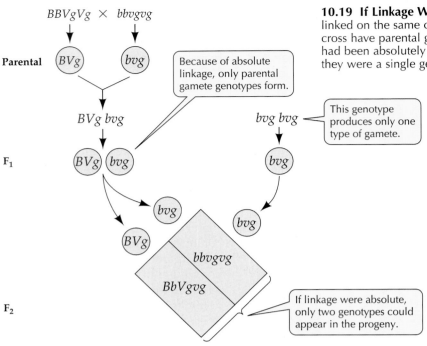

$BBVgVg \times bbvgvg$

Parental

Because of absolute linkage, only parental gamete genotypes form.

This genotype produces only one type of gamete.

$BVg \; bvg$

$bvg \; bvg$

F_1

F_2

If linkage were absolute, only two genotypes could appear in the progeny.

bbvgvg

BbVgvg

10.19 If Linkage Were Absolute If two genes are absolutely linked on the same chromosome, all the F_2 offspring from a dihybrid cross have parental genotypes. If the genes in Morgan's experiment had been absolutely linked, they would have been inherited as if they were a single gene; such linkage is extremely rare.

Genes can be exchanged between chromatids

Complete linkage is extremely rare. If linkage were absolute, Mendel's second law (independent assortment of alleles of different loci) would apply only to loci on different chromosomes. What actually happens is more complex and therefore more interesting. The chromosome is not unbreakable, so **recombination** can occur. Genes at different loci on the same chromosome do sometimes separate from one another during meiosis.

chromosome—that is, they are **linked**. After all, since the number of genes in a cell far exceeds the number of chromosomes, each chromosome must contain many genes. (Today we know that the human genome consists of perhaps 50,000 genes, distributed over 23 pairs of chromosomes.) The full set of loci on a given chromosome constitutes a *linkage group*. The number of linkage groups in a species equals the number of homologous chromosome pairs.

Suppose, now, that the *Bb* and *Vgvg* loci are on the same chromosome. If we assume that the linkage is absolute, we expect to see just *two* types of progeny from Morgan's cross (Figure 10.19). However, this is not always the case.

Without crossing over

Paired homologous chromosomes

Meiosis I continues

Meiosis II

All gametes parental

With crossing over

Crossover (chiasma)

Meiosis I continues

Recombinant chromosomes

Meiosis II

Half parental, half recombinant gametes

10.20 Crossing Over Results in Genetic Recombination Chromosomes are breakable. Genes at different loci on the same chromosome can separate from one another and recombine by crossing over. Recombination occurs at a chiasma during prophase I of meiosis.

Genes on the same chromosome pair recombine by **crossing over**; that is, two homologous chromosomes physically exchange corresponding genetic segments during prophase I of meiosis (Figure 10.20). In other words, recombination occurs at a chiasma (see Figure 9.16) when homologous chromosomes are paired.

Recall that the DNA has duplicated by this stage, and each chromosome consists of two chromatids. The exchange event at any point along the length of the chromosome involves only two of the four chromatids, one from each member of the chromosome pair. The lengths of chromosome are exchanged reciprocally, so both chromatids involved in crossing over become recombinant (that is, each chromatid contains genes from both parents).

When crossing over takes place between linked genes, not all progeny of a cross are of parental types. Instead, recombinant offspring appear as well, and they appear in repeatable proportions called **recombinant frequencies**, which equal the number of recombinant progeny divided by the total number of progeny (Figure 10.21). Recombinant frequencies will be greater for loci that are far apart than for loci that are closer together. Recombination is a random event, much as if you were to close your eyes, pick up scissors, and try to cut a string held by a friend. You probably would not cut it if the string were very short, but if the string were long, the probability would be greater.

Geneticists make maps of eukaryotic chromosomes

If two loci are very close together on a chromosome, the odds for crossing over between them are small. In contrast, if two loci are far apart, crossing over could occur at many points. In 1911 Alfred Sturtevant, then an undergraduate student in T. H. Morgan's fly room, realized how that simple insight could be used to show where different genes lie on the chromosome in relation to one another. He suggested that the farther apart two genes are on a chromosome, the greater the likelihood that they will separate and recombine in meiosis.

The Morgan group had determined recombinant frequencies for many pairs of linked genes. Sturtevant used these recombinant frequencies to create genetic maps that indicated the arrangement of genes along the chromosome (Figure 10.22). Ever since Sturtevant demonstrated this important point, geneticists have mapped the chromosomes of eukaryotes, prokaryotes, and viruses, assigning distances in **map units**. A map unit corresponds to a recombinant frequency of 0.01; it is also referred to as a centimorgan (cM), in honor of the founder of the fly room. You, too, can work out a genetic map (Figure 10.23).

Sex is determined in different ways in different species

In Kölreuter's experience, and later in Mendel's, reciprocal crosses apparently always gave identical results.

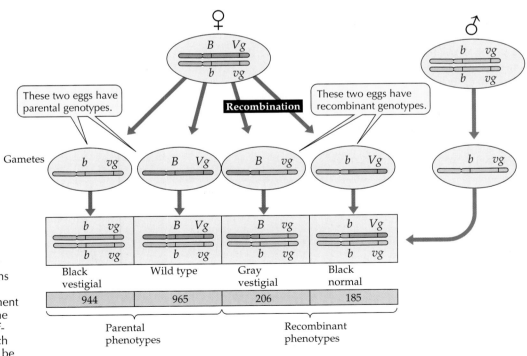

10.21 Recombinant Frequencies Crossing over between linked genes explains the nonparental genotypes Morgan found in the experiment described in Figure 10.18. The frequency of recombinant offspring can be calculated; such recombinant frequences will be larger for loci that are far apart than for those that are close together on the chromosome.

$$\text{RECOMBINANT FREQUENCY} = \frac{391 \text{ recombinants}}{2,300 \text{ total offspring}} = 0.17$$

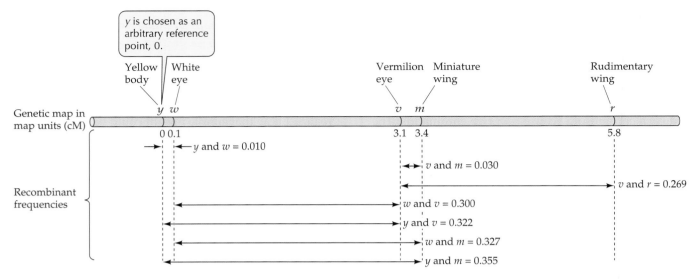

10.22 Steps toward a Genetic Map Because the chance of a recombinant genotype occurring increases the farther apart two loci fall, Sturtevant was able to derive this partial map of a *Drosophila* chromosome from the Morgan lab's data on the recombinant frequencies of five recessive traits. He assigned an arbitrary unit of distance—the map unit or centimorgan (cM)—equivalent to a recombinant frequency of 0.01.

At the outset, there appear to be several possible sequences (*a-b-c, a-c-b, b-a-c*), and we have no idea of the individual distances.

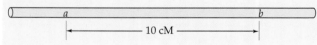

We make a cross *AABB* × *aabb*, and obtain an F$_1$ generation with a genotype *AaBb*. We test cross these *AaBb* individuals. How? By crossing them with *aabb*. Here are the genotypes of the first 1,000 progeny:

450 *AaBb*, 450 *aabb*, 50 *Aabb*, and 50 *aaBb*.

How far apart are the *a* and *b* genes? Well, what is the recombinant frequency? Which are the recombinant types, and which are the parental types?

Recombinant frequency (*a* to *b*) = (50 + 50)/1,000 = 0.1
So the map distance is
Map distance = 100 × recombinant frequency = 100 × 0.1 = 10 cM

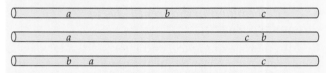

Now we make a cross *AACC* × *aacc*, obtain an F$_1$ generation, and test cross it, obtaining:

460 *AaCc*, 460 *aacc*, 40 *Aacc*, and 40 *aaCc*.

How far apart are the *a* and *c* genes?

Recombinant frequency (*a* to *c*) = (40 + 40)/1,000 = 0.08
Map distance = 100 × recombinant frequency = 100 × 0.08 = 8 cM

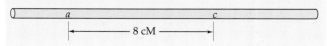

Now can you tell the order of the three genes on the chromosome? Why not? What else do you need to know?

We make a cross *BBCC* × *bbcc*, obtain an F$_1$ generation, and test cross it, obtaining:

490 *BbCc*, 490 *bbcc*, 10 *Bbcc*, and 10 *bbCc*.

Determine the map distance between *b* and *c*.

Recombinant frequency (*b* to *c*) = (10 + 10)/1,000 = 0.02
Map distance = 100 × recombinant frequency = 100 × 0.02 = 2 cM

Which of the three genes is between the other two, then? Because *a* and *b* are the farthest apart, *c* must be between them.

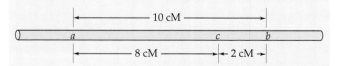

These numbers add up perfectly, but in most real cases they don't add up perfectly because of double crossovers.

10.23 Help Map These Genes We want to determine the order of three loci (*a, b,* and *c*) on a chromosome, as well as the map distances (in cM) between them. How do we determine a map distance?

The reason is that in diploid organisms, chromosomes come in pairs. One member of each chromosome pair derives from each parent; it does not matter, in general, whether a dominant allele was contributed by the mother or by the father. But this is not always the case. Sometimes the parental origin of a chromosome does matter. To understand the types of inheritance in which parental origin is important, we must consider the ways in which sex is determined in different species.

In corn (maize), a plant much studied by geneticists, every diploid adult has both male and female reproductive structures. These two types of tissue are genetically identical, just as roots and leaves are genetically identical. Plants such as maize and Mendel's pea plants, and animals such as earthworms, which produce both male and female gametes in the same organism, are said to be **monoecious** (from the Greek for "single house"). Other plants, such as date palms and oak trees, and most animals are **dioecious** ("two houses"), meaning that some of the individuals can produce only male gametes and the others can produce only female gametes.

In most dioecious organisms, sex is determined by differences in the chromosomes; but such determination operates in various different ways. For example, the sex of a honeybee (Figure 10.24*a*) depends on whether it develops from a fertilized or an unfertilized egg. A fertilized egg is diploid and gives rise to a female bee—either a worker or a queen, depending on the diet during larval life (again, note how the environment affects the phenotype). An unfertilized egg is haploid and gives rise to a male drone.

In many other animals, including ourselves, sex is determined by a single **sex chromosome** (or by a pair of them). Both males and females have two copies of each of the rest of the chromosomes, which are called **autosomes**.

For example, female grasshoppers have two **X chromosomes**, whereas males have only one. Females form eggs that contain one copy of each autosome and one X chromosome. The males form approximately equal amounts of two types of sperm: One type contains an X chromosome and one copy of each autosome; the other type contains only autosomes. Female grasshoppers are described as being XX (ignoring the autosomes) and males as XO (pronounced "ex-oh"). When an X-bearing sperm fertilizes an egg, the zygote is XX and develops into a female. When a sperm without an X fertilizes an egg, the zygote is XO and develops into a male. This chromosomal mechanism ensures that the two sexes are produced in approximately equal numbers.

As in grasshoppers, female mammals have two X chromosomes and males have one (Figure 10.24*b*). However, male mammals also have a sex chromosome that is not found in females: the **Y chromosome**. Females may be represented as XX and males as XY.

(a) Honeybees

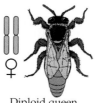

Diploid worker   Diploid queen   Haploid drone

In honeybees, fertilized eggs develop into diploid females and unfertilized eggs develop into haploid males.

(b) Humans

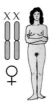
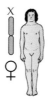
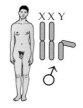

Normal   Abnormal (Turner syndrome), sterile   Abnormal (Klinefelter syndrome), sterile   Normal

Normal human females (far left) carry two X chromosomes; normal males (far right) carry one X and one Y chromosome. Persons who have some other number of sex chromosomes may develop abnormally.

(c) *Drosophila*

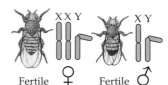

Fertile   Sterile   Fertile   Fertile

Drosophila (fruit fly) females have two X chromosomes and may also have a Y chromosome; males have an X chromosome and, if they are fertile, a Y chromosome.

(d) Chickens

In birds, the males carry two identical sex chromosomes (ZZ) and females have two differing ones (ZW).

10.24 Sex Is Determined in Different Ways For traits transmitted on the sex chromosomes, the parental origin of the gene is crucial, and varies from species to species.

The males produce two kinds of gametes: Each has a complete set of autosomes, but half the gametes carry an X chromosome and the rest carry a Y. When an X-bearing sperm fertilizes an egg, the resulting XX zygote is female; when a Y-bearing sperm fertilizes an egg, the XY zygote is male.

X and Y chromosomes have different functions

Some subtle but important differences show up clearly in mammals with abnormal sex chromosome constitutions. These conditions, resulting from nondisjunctions as described in Chapter 9, tell us something about the functions of the X and Y chromosomes. In humans, XO individuals sometimes appear. Human XO individuals are females who are physically moderately abnormal but mentally normal; usually they are also sterile. The XO condition in humans is called Turner syndrome. It is the only known case in which a human can live with only one member of a chromosome pair (here, the XY pair), although most XO conceptions terminate spontaneously early in development.

XXY individuals are also possible, a condition known as Klinefelter syndrome. They are sometimes taller than average, always sterile, and always male. Research with these individuals and related studies show that in humans the Y chromosome carries the genes that determine maleness.

The specific gene that determines maleness has been identified from observations of people with chromosome abnormalities. For example, some XY individuals who are women have been studied; a small portion of their Y chromosome was missing. In other cases, men who were genetically XX had a small part of the Y chromosome present, attached to another chromosome. The missing and present Y fragment in these two examples, respectively, contained the maleness-determining gene, which was identified as *SRY* (*s*ex-determining *r*egion on the *Y* chromosome).

The *SRY* gene codes for a protein for *primary sex determination*. In the presence of functional SRY protein, the embryo develops sperm-producing testes. If SRY protein is absent, the primary sex determination is female: the presence of ovaries and eggs. *Secondary sex determination*, the outward manifestations of maleness and femaleness (such as body type, breast development, body hair, and voice) are not determined directly by the presence or absence of the Y chromosome. Rather, they are determined by the actions of hormones, such as testosterone and estrogen.

The Y chromosome functions differently in *Drosophila melanogaster* (Figure 10.24c). Superficially, *Drosophila* follows the same pattern of sex determination as mammals: Females are XX and males are XY. However, XO individuals are males (rather than females as in mammals) and almost always are indistinguishable from normal XY males except that they are sterile. XXY *Drosophila* are normal, fertile females. Thus, in *Drosophila*, sex is determined strictly by the ratio of X chromosomes to autosome sets. If there is one X chromosome for each set of autosomes, the individual is a female; if there is only one X chromosome for the two sets of autosomes, the individual is a male. The Y chromosome plays no sex-determining role in *Drosophila*, but it is needed for male fertility.

In birds, moths, and butterflies, males are XX and females are XY. To avoid confusion, these forms are usually expressed as ZZ (male) and ZW (female) (Figure 10.24d). In these organisms, the female produces two types of gametes. Thus the egg determines the sex of the offspring, rather than the sperm as in humans and fruit flies.

Genes on sex chromosomes are inherited in special ways

How does the existence of sex chromosomes affect patterns of inheritance? In *Drosophila* and in humans, the Y chromosome carries few known genes, but a substantial number of genes affecting a great variety of characters are carried on the X chromosome. The result of this arrangement is an important deviation from the usual Mendelian ratios for the inheritance of genes located on the X chromosome. Any such gene is present in two copies in females, but in only one copy in males. Therefore, females may be heterozygous for genes that are on the X chromosome, but males will always be **hemizygous** for genes on the X chromosome—they will have only one of each.

Kölreuter's historic reciprocal crosses, mentioned at the beginning of this chapter, always gave the same outcome regardless of which parent displayed which trait. However, reciprocal crosses of parents that have different alleles on their sex chromosomes do not give identical results; this is a sharp deviation from the rules governing the inheritance of alleles on autosomes.

The first and still one of the best examples of **sex-linked inheritance**—inheritance of characters governed by loci on the sex chromosomes—is that of eye color in *Drosophila*. The wild-type eye color of these flies is red. In 1910, Morgan discovered a mutation that causes white eyes. He experimented by crossing flies of the wild-type and mutant phenotypes. His results demonstrated that the eye color locus is on the X chromosome.

When homozygous red-eyed females were crossed with (hemizygous) white-eyed males, all the sons and daughters had red eyes, because red is dominant over white and all the progeny had inherited a wild-type X chromosome from their mothers (Figure 10.25a). However, in the reciprocal cross, in which a white-eyed female was mated with a red-eyed male, all the sons were white-eyed and all the daughters red-eyed (Figure 10.25b).

The sons from the reciprocal cross inherited their only X chromosome from their white-eyed mother; the Y chromosome they inherited from their father does not carry the eye color locus. The daughters, on the

10.25 Eye Color Is a Sex-Linked Trait in *Drosophila*
Thomas Hunt Morgan demonstrated that the mutant allele that causes white eyes in *Drosophila* is carried on the X chromosome.

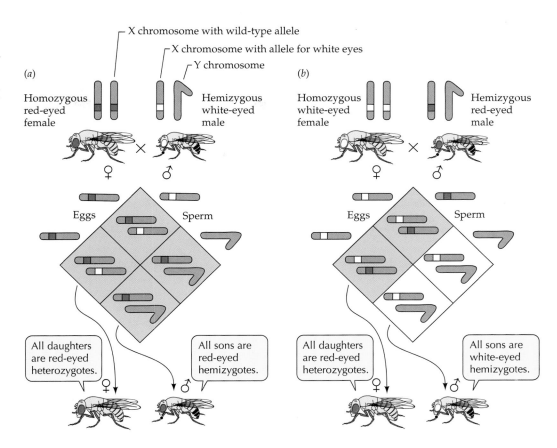

other hand, got an X chromosome with the white allele from their mother and an X chromosome bearing the red allele from their father; they were therefore red-eyed heterozygotes.

When Morgan mated heterozygous females with red-eyed males, he observed that half their sons had white eyes, but all their daughters had red eyes. Thus, in this case, eye color was carried on the X chromosome and not on the Y.

HUMANS BEINGS DISPLAY MANY SEX-LINKED CHARACTERS. The human X chromosome carries many genes. The alleles at these loci follow the same pattern of inheritance as those for white eyes in *Drosophila*. One human X chromosome gene, for example, has an allele that leads to red-green color blindness, a hereditary disorder. Red-green color blindness appears in individuals who are homozygous or hemizygous for a mutant recessive allele.

A color-blind man married to a homozygous normal woman will produce no color-blind children (Figure 10.26a). The sons inherit a single, normal X from their mother and will neither have the disorder nor transmit it to their children. The daughters get an X chromosome bearing a normal allele from their mother and one bearing the allele for color blindness from their father. Because color blindness is recessive, the daughters will not be color-blind (Figure 10.26b). However, they will be heterozygous *carriers*.

Female carriers of an X-linked trait will transmit the disorder to half their sons and the carrier role to half their daughters if the male parent has normal color vision. What parental genotypes would produce a color-blind female? Her father would have to be color-blind, and her mother either a carrier or herself color-blind. Because color blindness is rare, two such people are unlikely to meet and produce children. As a result, very few women are color-blind. This pattern of inher-

itance is the same as for hemophilia, the serious blood disorder discussed at the start of this chapter.

The small human Y chromosome carries very few genes. Among them are the maleness determinants, whose existence was suggested by the phenotypes of the XO and XXY individuals described earlier (see Figure 10.24b). The pattern of inheritance of Y-linked alleles should be easy for you to work out. Give it a try now. (Hint: Use a Punnett square to determine how the Y chromosome is inherited.)

Mendelian ratios are averages, not absolutes

You have been introduced to the basic phenotypic ratios observed for traits inherited following Mendelian patterns: 3:1, 1:1, 9:3:3:1. You will discover others as you do homework or test problems. It is essential to remember, however, that these ratios represent highest probabilities, not invariant rules.

The X–Y system of sex determination in our species results in roughly equal numbers of males and females in a substantial population, but you know that a given family of four children may not consist of two girls and two boys. It is not unusual for four children to be of the same sex; in fact, among families with four children, one in eight will have all boys or all girls. How do we know if this deviation is simply the result of chance?

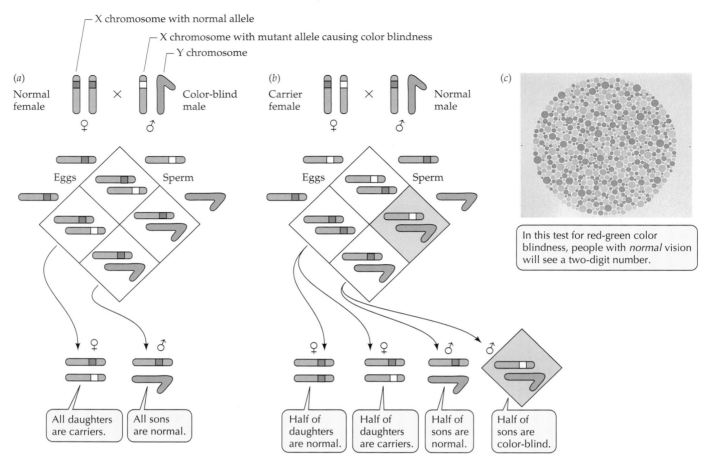

10.26 Red-Green Color Blindness Is a Sex-Linked Trait in Humans The pattern of transmission of red-green color blindness is the same as that for hemophilia, discussed at the beginning of this chapter.

When we are trying to understand the genetic basis for the results of a cross, it is important to know how much deviation from a predicted ratio is reasonably expected as a result of normal chance. Statistical methods have been devised for determining whether the observed deviation from an expected ratio can be attributed to chance variation or whether the deviation is large enough to suggest that the observed ratio has a specific cause.

Statistical methods take several factors into consideration, but one of the most important is **sample size**. If we expect to find a 3:1 ratio between two phenotypes and we look at a sample of only 8 progeny, we should not be surprised to find 7 individuals of one phenotype and 1 of the other (rather than the expected 6 and 2). It would be surprising, however, in a sample of 80 to find 70 of one phenotype and 10 of the other—you would question whether that really represented a 3:1 ratio. In genetics experiments—and in quantitative biology in general—sample sizes should be large so that the data are easy to evaluate with confidence.

Cytoplasmic Inheritance

You have studied the basic patterns of Mendelian inheritance in terms of chromosomal behavior. Does all inheritance in eukaryotes conform to the Mendelian pattern?

Consider the four-o'clock plant shown in Figure 10.27. This particular four-o'clock shows three different patterns of chlorophyll distribution in three different parts of the shoot. One branch is all white; it has no chlorophyll. Another branch is entirely green. The rest of the shoot is variegated; that is, some patches of cells have chlorophyll and others have none. Each part of the plant has flowers, so we can do the following experiment.

Take pollen (which produces male gametes) from some flowers and transfer it to the female parts of flowers on another part of the plant, thus performing a cross. Table 10.2 shows the six possible crosses and their outcomes. Study the table for a moment before reading further and try to discern the basic pattern of inheritance. Do you see the surprising feature? Look again. What should become obvious is that the phenotype of the "father" (the parent producing the pollen) is irrelevant to the outcome. Only the "mother" (the

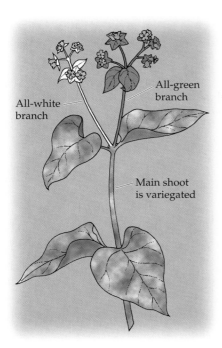

10.27 A Variegated Four-o'Clock with Green and White Branches This is the plant used for the experiment summarized in Table 10.2. Pollen is transferred from a flower on one part of the plant to a flower on another part.

parent producing the egg) seems to play a role in determining the phenotype of the offspring.

This pattern of inheritance, in which the progeny's phenotype is unaffected by the male parent, is found for particular characteristics in all sorts of eukaryotes and is referred to as **maternal inheritance**. How does it work?

The essence of Mendelian inheritance is that information carried on chromosomes is partitioned with great precision during meiosis, but eukaryotic cells have other self-reproducing entities besides the nuclear chromosomes. Chloroplasts and mitochondria carry some genetic information in small circular chromosomes (see Chapter 4). The DNA of these organelles is subject to mutation just as the DNA in the chromosomes of the nucleus is, so we may speak of alleles of nonnuclear genes.

Nonnuclear genes are not inherited in the same way as nuclear chromosomal genes, because meiosis does not segregate organelles precisely; cytokinesis does—randomly. Because the eggs of most species contain large amounts of cytoplasm and sperm contain almost no cytoplasm, the mitochondria in a zygote come from the cytoplasm of the female parent's egg, even though half the zygote's nuclear chromosomes come from the male parent.

In plant zygotes, all the chloroplasts come from the maternal cytoplasm. Hence any particle that is inherited through the cytoplasm is said to be maternally inherited. In such cases, reciprocal crosses give results that differ considerably from the predictions made by Mendel, as we saw in Table 10.2.

Chloroplasts and mitochondria are complex organelles, and only a minority of their functions are maternally inherited. For example, most components of the electron transport system are encoded by nuclear genes. Since mitochondria are essential to providing cellular energy, it is not surprising that inherited human diseases whose phenotypes show up as abnormal mitochondria often include serious muscle or nervous system abnormalities among their symptoms. Some of these genetic diseases are maternally inherited; others show typical Mendelian inheritance patterns.

TABLE 10.2 Results of the Four-o'Clock Experiments

PHENOTYPE OF BRANCH WITH FEMALE PARENT	PHENOTYPE OF BRANCH WITH MALE PARENT	PHENOTYPE OF PROGENY
White	White	White
White	Green	White
White	Variegated	White
Green	White	Green
Green	Green	Green
Green	Variegated	Green

Summary of "Transmission Genetics: Mendel and Beyond"

Mendel's Work
• Plant breeders can control which plants mate. Although it has long been known that both parent plants contribute equivalent determinants to the character traits of their offspring, before Mendel's time it was believed that, once they were brought together, these determinants blended and could never be separated. **Review Figure 10.1**
• Although Mendel's work was meticulous and well documented, his discoveries, reported in the 1860s, lay dormant until decades later, when (after meiosis was known) others rediscovered them.

Mendel's Experiments and Laws of Inheritance
• Mendel used garden pea plants for his studies because they were easily cultivated and crossed, and because they showed numerous characters (such as seed shape) with clearly different traits (spherical or wrinkled). **Review Figures 10.1, 10.2**
• In a monohybrid cross, the offspring showed only one of the two traits. Mendel proposed that the trait observed in the first generation (F_1) was dominant and the other was recessive. **Review Table 10.1**

• When the F_1 offspring were interbred, the resulting F_2 generation showed a 3:1 phenotypic ratio, with the recessive phenotype present in one-fourth of the offspring. This reappearance of the recessive phenotype refuted the blending hypothesis. **Review Figure 10.4 and Table 10.1**

• Because some alleles are dominant and some recessive, the same phenotype can result from different genotypes. Homozygous genotypes have two copies of the same allele; heterozygous genotypes have two different alleles. Heterozygous genotypes yield phenotypes that show the dominant trait.

• On the basis of many crosses with different characters, Mendel proposed his first law: that genetic determinants (now known as genes) are particulate, that there are two copies (alleles) of a gene in every parent, and that during gamete formation the two alleles for a character segregate from one another. **Review Figure 10.5**

• Geneticists who followed Mendel showed that genes are carried on chromosomes and that alleles are segregated during meiosis I. **Review Figure 10.6**

• Using a test cross, Mendel was able to determine whether a plant showing the dominant character was homozygous or heterozygous. The appearance of the recessive phenotype in half of the offspring of such a cross indicates that the dominant-appearing parent was heterozygous. **Review Figure 10.7**

• From studies of the simultaneous inheritance of two characters (seed shape and seed color, for example) on the same plants, Mendel concluded that alleles of different genes assort independently in these dihybrid crosses. **Review Figures 10.8, 10.9**

• We can predict the results of hybrid crosses either by using a Punnett square or by calculating probabilities. To determine the joint probability of independent events, multiply the individual probabilities. To determine the probability of an event that can occur in two or more different ways, add the individual probabilities. **Review Figure 10.10**

Alleles and Their Interactions

• Dominance is usually not complete, since both alleles in a heterozygous organism may be expressed in the phenotype. **Review Figure 10.11**

• Some alleles have multiple phenotypic effects. Alleles arise by mutation, and many genes (such as those that determine blood type) have multiple alleles. **Review Figures 10.12, 10.13**

Gene Interactions

• In epistasis, the products of different genes interact to produce a phenotype. **Review Figure 10.14**

• In some cases, the phenotype is the result of the additive effects of many genes (polygenes), and inheritance is quantitative. **Review Figure 10.15**

• Environmental variables such as temperature, nutrition, and light affect gene action.

Genes and Chromosomes

• Each chromosome carries many genes. Genes located near each other are said to be linked, and they are usually inherited together. **Review Figures 10.18, 10.19**

• Linked genes recombine by crossing over in prophase I of meiosis. The result is recombinant gametes, which have new combinations of linked genes because of the exchange. **Review Figures 10.20, 10.21**

• The distance between genes is proportional to the frequency of crossing over between them. Genetic maps are based on recombinant frequencies. **Review Figures 10.22, 10.23**

• Sex chromosomes carry genes that determine whether male or female gametes are produced. The specific functions of X and Y chromosomes differ depending on the species. **Review Figure 10.24**

• Genes on the sex chromosomes can be followed in crosses in fruit flies and mammals because the X chromosome carries many genes, but its homolog, the Y chromosome, has few. Males have only one allele for most X-linked genes, so rare alleles will show up phenotypically more often in males than in females. **Review Figures 10.25, 10.26**

Cytoplasmic Inheritance

• Cytoplasmic organelles such as chloroplasts and mitochondria have some genes.

• Cytoplasmic inheritance is generally by way of the egg (maternal), because male gametes contribute only their nucleus and no cytoplasm to the zygote at fertilization.

Self-Quiz

1. Which statement about Mendel's cross of *TT* peas with *tt* peas is *not* true?
   a. Each parent can produce only one type of gamete.
   b. F_1 individuals produce gametes of two types, each gamete being *T* or *t*.
   c. Three genotypes are observed in the F_2 generation.
   d. Three phenotypes are observed in the F_2 generation.
   e. This is an example of a monohybrid cross.

2. The phenotype of an individual
   a. depends at least in part on the genotype.
   b. is either homozygous or heterozygous.
   c. determines the genotype.
   d. is the genetic constitution of the organism.
   e. is either monohybrid or dihybrid.

3. Which statement about alleles is *not* true?
   a. They are different forms of the same gene.
   b. There may be several at one locus.
   c. One may be dominant over another.
   d. They may show incomplete dominance.
   e. They occupy different loci on the same chromosome.

4. Which statement about an individual that is homozygous for an allele is *not* true?
   a. Each of its cells possesses two copies of that allele.
   b. Each of its gametes contains one copy of that allele.
   c. It is true-breeding with respect to that allele.
   d. Its parents were necessarily homozygous for that allele.
   e. It can pass that allele to its offspring.

5. Which statement about a test cross is *not* true?
   a. It tests whether an unknown individual is homozygous or heterozygous.
   b. The test individual is crossed with a homozygous recessive individual.
   c. If the test individual is heterozygous, the progeny will have a 1:1 ratio.
   d. If the test individual is homozygous, the progeny will have a 3:1 ratio.
   e. Test cross results are consistent with Mendel's model of inheritance.

6. Linked genes
   a. must be immediately adjacent to one another on a chromosome.
   b. have alleles that assort independently of one another.
   c. never show crossing over.
   d. are on the same chromosome.
   e. always have multiple alleles.

7. In the F_2 generation of a dihybrid cross
   a. four phenotypes appear in the ratio 9:3:3:1 if the loci are linked.
   b. four phenotypes appear in the ratio 9:3:3:1 if the loci are unlinked.
   c. two phenotypes appear in the ratio 3:1 if the loci are unlinked.
   d. three phenotypes appear in the ratio 1:2:1 if the loci are unlinked.
   e. two phenotypes appear in the ratio 1:1 whether or not the loci are linked.

8. The sex of a honeybee is determined by
   a. ploidy, the male being haploid.
   b. X and Y chromosomes, the male being XY.
   c. X and Y chromosomes, the male being XX.
   d. the number of X chromosomes, the male being XO.
   e. Z and W chromosomes, the male being ZZ.

9. In epistasis
   a. nothing changes from generation to generation.
   b. one gene alters the effect of another.
   c. a portion of a chromosome is deleted.
   d. a portion of a chromosome is inverted.
   e. the behavior of two genes is entirely independent.

10. Individual heritable traits
   a. are always determined by dominant and recessive alleles.
   b. always vary discontinuously.
   c. can sometimes be controlled by many genes.
   d. were first studied in the twentieth century.
   e. do not exist outside the laboratory.

Applying Concepts

1. Using the Punnett squares below, show that for typical dominant and recessive autosomal traits, it does not matter which parent contributes the dominant allele and which the recessive allele. Cross true-breeding tall plants (*TT*) with true-breeding dwarf plants (*tt*).

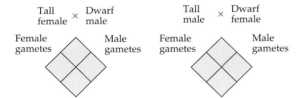

2. The photograph shows the shells of 15 bay scallops, *Argopecten irradians*. These scallops are hermaphroditic; that is, a single individual can reproduce sexually, as did the pea plants of the F_1 generation in Mendel's experiments. Three color schemes are evident: yellow, orange, and black and white. The color-determining gene has three alleles. The top row shows a yellow scallop and a representative sample of its offspring, the middle row shows a black-and-white scallop and its offspring, and the bottom row shows an orange scallop and its offspring. Assign a suitable symbol to each of the three alleles participating in color control; then determine the genotype of each of the three parent individuals and tell what you can about the genotypes of the different offspring. Explain your results carefully.

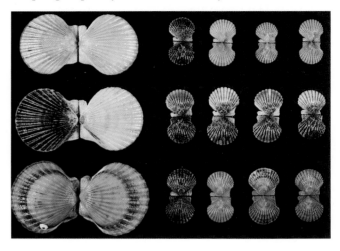

3. Show diagrammatically what occurs when the F_1 offspring of the cross in Question 1 self-pollinate.

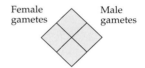

4. A new student of genetics suspects that a particular recessive trait in fruit flies (dumpy wings, which are somewhat smaller and more bell-shaped than the wild-type) is sex-linked. A single mating between a fly having dumpy wings (*dp*; female) and a fly with wild-type wings (*Dp*; male) produces 3 dumpy-winged females and 2 wild-type males. On the basis of these data, is the trait sex-linked or autosomal? What were the genotypes of the parents? Explain how these conclusions can be reached on the basis of so few data.

5. The sex of fishes is determined by the same X–Y system as in humans. An allele of one locus on the Y chromosome of the fish *Lebistes* causes a pigmented spot to appear on the dorsal fin. A male fish that has a spotted dorsal fin is mated with a female fish that has an unspotted fin. Describe the phenotypes of the F_1 and the F_2 generations from this cross.

6. In *Drosophila melanogaster*, the recessive allele *p*, when homozygous, determines pink eyes. *Pp* or *PP* results in wild-type eye color. Another gene, on another chromosome, has a recessive allele, *sw*, that produces short wings when homozygous. Consider a cross between females of genotype *PPSwSw* and males of genotype *ppswsw*. Describe the phenotypes and genotypes of the F_1 generation and of the F_2 generation produced by allowing the F_1 progeny to mate with one another.

7. On the same chromosome of *Drosophila melanogaster* that carries the *p* (pink eyes) locus, there is another locus that affects the wings. Homozygous recessives, *byby*, have blistery wings, while the dominant allele *By* produces wild-type wings. The *P* and *By* loci are very close together on the chromosome; that is, the two loci are tightly linked. In answering these questions, assume that no crossing over occurs.
   a. For the cross *PPByBy* × *ppbyby*, give the phenotypes and genotypes of the F_1 and of the F_2 generations produced by interbreeding of the F_1 progeny.
   b. For the cross *PPbyby* × *ppByBy*, give the phenotypes and genotypes of the F_1 and of the F_2 generations.
   c. For the cross of Question 7b, what further phenotype(s) would appear in the F_2 generation if crossing over occurred?
   d. Draw a nucleus undergoing meiosis, at the stage in which the crossing over (Question 7c) occurred. In which generation (P, F_1, or F_2) did this crossing over take place?

8. Consider the following cross of *Drosophila melanogaster* with alleles as described in Question 6. Males with genotype *Ppswsw* are crossed with females of genotype *ppSwsw*. Describe the phenotypes and genotypes of the F_1 generation.

9. In the Andalusian fowl, a single pair of alleles controls the color of the feathers. Three colors are observed: blue, black, and splashed white. Crosses among these three types yield the following results:

PARENTS	PROGENY
Black × blue	Blue and black (1:1)
Black × splashed white	Blue
Blue × splashed white	Blue and splashed white (1:1)
Black × black	Black
Splashed white × splashed white	Splashed white

   a. What progeny would result from the cross blue × blue?
   b. If you wanted to sell eggs, all of which would yield blue fowl, how should you proceed?

10. In *Drosophila melanogaster*, white (*w*), eosin (*w^e*), and wild-type red (*w^+*) are multiple alleles of a single locus for eye color. This locus is on the X chromosome. A female that has eosin (pale orange) eyes is crossed with a male that has wild-type eyes. All the female progeny are red-eyed; half the male offspring have eosin eyes, and half have white eyes.
   a. What is the order of dominance of these alleles?
   b. What are the genotypes of the parents and progeny?

11. Color blindness is a recessive trait. Two people with normal vision have two sons, one color-blind and one with normal vision. If the couple also has daughters, what proportion of them will have normal vision? Explain.

12. A mouse with an agouti coat is mated with an albino mouse of genotype *aabb*. Half of the offspring are albino, one-fourth are black, and one-fourth are agouti. What are the genotypes of the agouti parents and of the various kinds of offspring? (*Hint*: See the section on epistasis.)

13. The disease Leber's optic neuropathy is caused by a mutation in a gene carried on mitochondrial DNA. What would be the result in their first child if a man with this disease married a woman who did not have the disease? What would be the result if the wife had the disease and the husband did not?

Readings

Griffiths, A. J. F., J. H. Miller, D. T. Suzuki, R. C. Lewontin and W. M. Gelbart. 1996. *An Introduction to Genetic Analysis*, 6th Edition. W. H. Freeman, New York. An excellent textbook of modern genetics. Chapters 2 and 3 are particularly relevant to this chapter; Chapters 4 and 5 are also useful.

Mange, E. J. and A. P. Mange. 1994. *Basic Human Genetics*. Sinauer Associates, Sunderland, MA. An introductory-level study of genetics, especially chromosomal inheritance, using humans as examples.

McKusick, V. A. 1981. "The Anatomy of the Human Genome." *Hospital Practice*, April. An article on the increasingly detailed maps of human chromosomes. Explains how the maps are created and what they mean to an understanding of human genetic diseases.

Russell, P. J. 1996. *Genetics*, 4th Edition. Harper/Collins, New York. A well-balanced treatment of a broad range of topics in genetics. Highly recommended.

Sapienza, C. 1990. "Parental Imprinting of Genes." *Scientific American*, October. A discussion of what happens when reciprocal crosses are not equivalent.

Stern, C. and E. R. Sherwood (Eds.). 1966. *The Origin of Genetics: A Mendel Source Book*. W. H. Freeman, New York. A collection of the writings of researchers at the dawn of the science of genetics, including translations of Mendel's papers and letters. The last two articles discuss the likelihood that Mendel fudged his data.

Sturtevant, A. H. and G. W. Beadle. 1962. *An Introduction to Genetics*. Dover, New York. First published by W. B. Saunders in 1939. Though old, this text holds up as a fine introduction to formal chromosome genetics. Sturtevant was a key member of Morgan's Fly Room.

Chapter 11

DNA and Its Role in Heredity

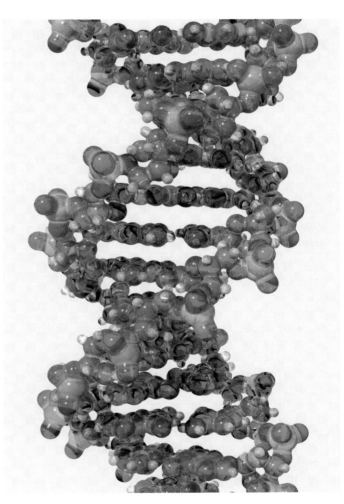

The Double Helix of DNA
A computer-generated model of DNA. The discovery of this "master molecule" transformed biology and medicine during the latter half of the twentieth century, and many aspects of everyday life are now affected by our knowledge and manipulation of its properties.

*F*or half a century, "DNA" has been a magic word—first to the early molecular biologists, then to biology as a whole, and eventually to medicine, the courtroom, and the stock market. In this chapter, DNA makes its grand entrance, as we see how Mendel's genes were recognized as DNA, how the mystery of the chemical structure of DNA was solved, how DNA replicates to carry the hereditary information from generation to generation, and how the cell monitors and repairs its DNA to avert disaster. DNA occupies center stage for the rest of Part Two. In the next few chapters we examine many aspects of DNA transmission and function.

As you learn about DNA, try to focus on the key role of DNA *structure* and the "codes" buried in it. DNA structure determines how the molecule replicates and how it controls cellular events—and, as we'll see in Chapter 14, it tells the tiny cell how to make its gigantic DNA molecules fold up and fit inside. The coding capacity of DNA is linear and one-dimensional.

DNA: The Genetic Material

During the first half of the twentieth century, the hereditary material was generally assumed to be protein. The impressive chemical diversity of proteins made this assumption seem reasonable. In addition, some proteins—notably enzymes

243

and antibodies—show great specificity. By contrast, nucleic acids were known to have only a few components and seemed too simple to carry the complex information expected in the genetic material.

Circumstantial evidence, however, pointed to DNA. It was in the right place, as an important component of the nucleus and chromosomes, which were found to carry genes. And it was present in the right amounts. During the 1920s a dye was developed that bound specifically to DNA and turned red in direct proportion to the amount of DNA present. When different cells were stained with this Feulgen dye, as it is known, and their color intensity (and DNA content) measured, each species appeared to have its own specific nuclear DNA content, and the quantity in somatic cells was twice that in eggs or sperm—as might be expected for diploid and haploid cells, respectively. These two observations were consistent with DNA as the genetic material.

But circumstantial evidence is not a scientific demonstration of cause and effect. After all, proteins also are present in nuclei. The convincing demonstration that DNA is the gene came from two lines of experiments, one on bacteria and the other on bacterial viruses.

DNA from one type of bacterium genetically transforms another type

The history of biology is filled with incidents in which research on one specific topic has—with or without answering the question originally asked—contributed richly to another, apparently unrelated area. Such a case is the work of Frederick Griffith, an English physician.

In the 1920s Griffith was studying the disease-causing behavior of the bacterium *Streptococcus pneumoniae*, or pneumococcus, one of the agents that cause pneumonia in humans. He was trying to develop a vaccine against this devastating illness, because antibiotics had not yet been invented. He worked with two strains of pneumococcus. A bacterial strain is a population of cells, descended from a single parent cell; strains differ in one or more inherited characteristics. Griffith's strains were designated S and R because one produces shiny, smooth (S) colonies when grown in the laboratory, and the other produces colonies that look rough (R).

When the S strain was injected into mice, the mice died within a day, and the hearts of the dead mice were found to be teeming with the deadly bacteria. When the R strain was injected instead, the mice did not become diseased. In other words, the S strain is virulent (disease-causing) and the R strain is nonvirulent. The virulence of the S strain is caused by a polysaccharide capsule that protects the bacterium from the defense mechanisms of the host (see Chapter 18).

The R strain lacks this capsule, so the R strain cells can be inactivated by the defenses of a mouse.

In hopes of developing a vaccine against pneumonia, Griffith inoculated other mice with heat-killed S pneumococci. Recall that heat denatures proteins, thus killing cells. Neither heat-killed S nor living R pneumococci produced infection. However, when Griffith inoculated mice with a mixture of living R bacteria and heat-killed S bacteria, to his astonishment all these mice died of pneumonia. When he examined blood from the hearts of these mice, he found it full of living bacteria, many of them belonging to the virulent S strain! Griffith concluded that, in the presence of the dead S pneumococci, some of the inoculated R pneumococci had been transformed into virulent organisms (Figure 11.1).

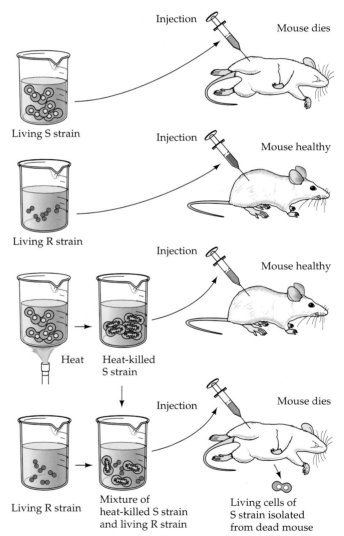

11.1 Genetic Transformation of Nonvirulent R Pneumococci Frederick Griffith's experiments demonstrated that a particular factor in the virulent S strain could transform nonvirulent R strain bacteria into a lethal form, even when the S strain bacteria had been killed by high temperatures.

We now call the phenomenon of genetically altering another organism **transformation**. In terms of Griffith's observations, one could say that transformation is the uptake of information from the environment. As we'll soon see, today's definition of transformation is more precise. For now, note that living R pneumococci had gained the trait of virulence from their environment.

Did transformation of the bacteria depend on something the mouse did? No. It was shown that simply incubating living R and heat-killed S bacteria together in a test tube yielded the same transformation. Next it was discovered that a cell-free extract of heat-killed S cells also transforms R cells. (A cell-free extract contains all the contents of ruptured cells, but no intact cells.) This result demonstrated that a particular substance—called at the time a chemical **transforming principle**—from the dead S pneumococci could cause an inherited change in the affected R cells. From these observations some scientists concluded that the transforming principle carried heritable information and thus was genetic material.

The transforming principle is DNA

A crucial step in the history of biology was the identification of the transforming principle, accomplished over a period of several years by Oswald T. Avery and his colleagues at what is now Rockefeller University. They treated samples of the transforming principle in a variety of ways to destroy different types of substances—proteins, nucleic acids, carbohydrates, and lipids—and then tested the treated samples to see if they had retained transforming activity.

The answer was always the same: If the DNA in the sample was destroyed, transforming activity was lost; everything else was dispensable. As a final step, Avery, with Colin MacLeod and MacLyn McCarty, isolated virtually pure DNA from a sample of pneumococcal transforming principle and showed that it was highly active in causing bacterial transformation.

The work of Avery, MacLeod, and McCarty, published in 1944, was a milestone in establishing that DNA is the genetic material in cells. However, it had little impact at the time, for two reasons. First, most scientists did not believe that DNA was chemically complex enough to be the hereditary material, especially given the great chemical complexity of proteins. Second, and perhaps of more importance, it was not yet obvious that bacteria even had genes—bacterial genetics was still to be discovered (see Chapter 13).

Viral replication confirms that DNA is the genetic material

A report published in 1952 by Alfred D. Hershey and Martha Chase of the Carnegie Laboratory of Genetics

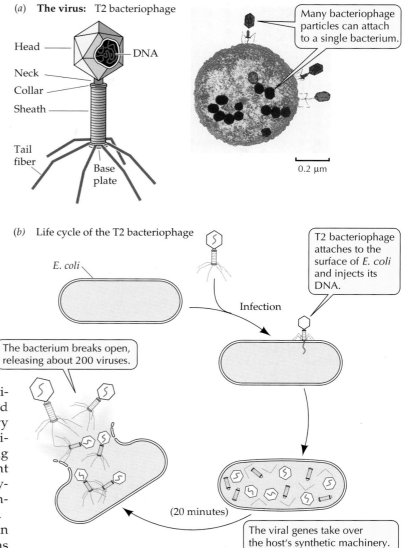

(a) **The virus:** T2 bacteriophage

Head
DNA
Neck
Collar
Sheath
Tail fiber
Base plate

Many bacteriophage particles can attach to a single bacterium.

0.2 μm

(b) Life cycle of the T2 bacteriophage

E. coli

Infection

T2 bacteriophage attaches to the surface of *E. coli* and injects its DNA.

The bacterium breaks open, releasing about 200 viruses.

(20 minutes)

The viral genes take over the host's synthetic machinery.

11.2 T2 and the Bacteriophage Reproduction Cycle
(a) The external structures of the bacteriophage T2 consist entirely of protein. This cutaway view shows a strand of DNA within the head. (b) T2 is parasitic on *E. coli*, depending on the bacterium to produce new viruses.

had a much greater immediate impact than did Avery's 1944 paper. The Hershey–Chase experiment was carried out with a virus that infects bacteria. This virus, called T2 bacteriophage, consists of a DNA core packed within a protein coat (Figure 11.2a). The virus is thus made of the two materials that were, at the time, the leading candidates for the genetic material.

When a T2 bacteriophage attacks a bacterial cell, part—but not all—of the virus enters the cell. Hershey and Chase set out to determine which part of the virus—protein or DNA—is the hereditary material that enters the bacterial cell. The idea was to trace

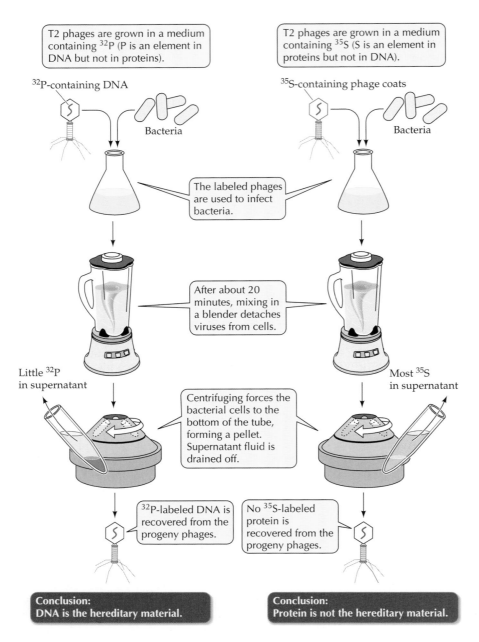

T2 phages are grown in a medium containing ^{32}P (P is an element in DNA but not in proteins).

^{32}P-containing DNA

Bacteria

T2 phages are grown in a medium containing ^{35}S (S is an element in proteins but not in DNA).

^{35}S-containing phage coats

Bacteria

The labeled phages are used to infect bacteria.

After about 20 minutes, mixing in a blender detaches viruses from cells.

Little ^{32}P in supernatant

Most ^{35}S in supernatant

Centrifuging forces the bacterial cells to the bottom of the tube, forming a pellet. Supernatant fluid is drained off.

^{32}P-labeled DNA is recovered from the progeny phages.

No ^{35}S-labeled protein is recovered from the progeny phages.

Conclusion:
DNA is the hereditary material.

Conclusion:
Protein is not the hereditary material.

11.3 The Hershey–Chase Experiment
Because progeny generations of viruses incorporated the radioactively tagged DNA but not the radioactively tagged proteins, the experiment demonstrated that DNA, not protein, is the hereditary material.

In separate experiments, Hershey and Chase combined radioactive viruses containing either ^{32}P or ^{35}S with bacteria (Figure 11.3). After a few minutes, they agitated the mixtures vigorously, using a kitchen blender, which (without bursting the bacteria) stripped away the parts of the virus coats that had not penetrated the bacteria. Then, using a centrifuge, Hershey and Chase separated the bacteria from the rest of the material. They found that most of the ^{35}S (and thus the protein) had separated from the bacteria, and that most of the ^{32}P (and thus the DNA) had stayed with the bacteria. These results suggested that the DNA was transferred to the bacteria while the protein remained outside.

Hershey and Chase then performed similar but longer experiments, allowing a progeny generation of viruses to be collected. The number of viruses produced by these "blended" bacteria was the same as if the blender had not been used. The resulting T2 progeny contained almost none of the original ^{35}S but about one-third of the original ^{32}P—and thus, presumably, one-third of the DNA. This result suggested that T2 injects the DNA from its head into a bacterium while the external protein structures remain outside of the bacterial cell. Because DNA was carried over in the virus from generation to generation, whereas protein was not, a logical conclusion was that the hereditary information of the viruses is contained in the DNA.

Although these results were less clear-cut than Avery's, the Hershey–Chase experiment convinced most scientists that DNA is the carrier of hereditary information. By this time, other workers had identified mutations, and therefore genes, in viruses and bacteria. The time was ripe to talk about viruses and hereditary information in the same breath.

these two components during the life cycle of the virus (Figure 11.2b), so Hershey and Chase labeled each with a specific radioactive tracer.

All proteins contain some sulfur (in the amino acids cysteine and methionine), an element not present in DNA, and sulfur has a radioactive isotope, ^{35}S. The deoxyribose–phosphate "backbone" of DNA, on the other hand, is rich in phosphorus (see Chapter 3), an element that is not present in most proteins—and phosphorus also has a radioactive isotope, ^{32}P. Hershey and Chase grew one batch of T2 in a bacterial culture in the presence of ^{32}P, so that all the viral DNA was labeled with ^{32}P. Similarly, the proteins of another batch of T2 were labeled with ^{35}S.

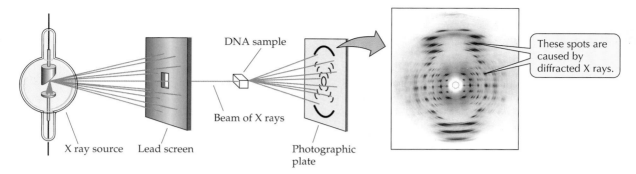

11.4 X Ray Crystallography Revealed the Basic Helical Nature of DNA Structure The positions of atoms in DNA can be inferred by the pattern of diffraction of X rays passed through it, although the task requires tremendous skill.

The Structure of DNA

Once scientists agreed that the genetic material is DNA, they wanted to learn its precise chemical structure. From the structure of DNA, they hoped to find clues to two questions: how the molecule is replicated between nuclear divisions, and how it causes the synthesis of specific proteins. Both expectations were fulfilled. X ray studies provided clues about the dimensions of DNA and hinted that it had a helical form. Dimensionally accurate model building by James Watson and Francis Crick completed the picture.

X ray crystallography provided clues to DNA structure

The structure of DNA was deciphered only after many types of experimental evidence and theoretical considerations were combined. The most crucial evidence was obtained by X ray crystallography (Figure 11.4). The positions of atoms in a crystalline substance can be inferred by the pattern of diffraction of X rays passed through it, but even today this is not an easy task when the substance is of enormous molecular weight.

In the early 1950s, even a highly talented X ray crystallographer could (and did) look at the best available images from DNA preparations and fail to see what they meant. Nonetheless, the attempt to characterize DNA would have been impossible without the crystallographs prepared by the English chemist Rosalind Franklin. Franklin's work, in turn, depended on the success of the English biophysicist Maurice Wilkins in preparing very uniformly oriented DNA fibers, which made samples for diffraction that were far better than previous samples.

The chemical composition of DNA was known

The chemical composition of DNA also provided important clues about its structure. Biochemists knew that DNA was a polymer of nucleotides. Each nucleotide of DNA consists of a molecule of the sugar deoxyribose, a phosphate group, and a nitrogen-containing base (see Figures 3.21 and 3.22). The only differences among the four nucleotides of DNA are their nitrogenous bases: the purines **adenine** and **guanine** and the pyrimidines **cytosine** and **thymine**.

In 1950 Erwin Chargaff at Columbia University reported observations of major importance. He and his colleagues had found that DNA from many different species—and from different sources within a single organism—exhibits certain regularities. In almost all DNA the following rule holds: The amount of adenine equals the amount of thymine, and the amount of guanine equals the amount of cytosine (Figure 11.5). As a result, the total abundance of purines equals the total abundance of pyrimidines. The structure of DNA could not have been worked out without this information, yet its significance was overlooked for at least 3 years.

Watson and Crick described the double helix

A solution to the puzzle of the structure of DNA was accelerated by the technique of model building—the assembling of three-dimensional representations of possible molecular structures. Such models use correct relative dimensions and correct bond angles. This technique, originally exploited in structural studies by the American chemist Linus Pauling, was used by the English physicist Francis Crick and the American geneticist James D. Watson, then both at the Cavendish Laboratory of Cambridge University.

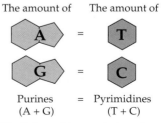

11.5 Chargaff's Rules The total abundances of purines and pyrimidines are equal in DNA.

Watson and Crick attempted to combine all that had been learned so far about DNA structure into a single coherent model. The crystallographers' results (see Figure 11.4) convinced Watson and Crick that the DNA molecule is **helical** (cylindrically spiral) and provided the values of certain distances within the helix. The results of density measurements and model building suggested that there are two polynucleotide chains in the molecule. The modeling studies also led to the conclusion that the two chains in DNA run in opposite directions—that is, that they are **antiparallel**. (We'll clarify this point in the next section.)

Crick and Watson attempted several models. Late in February of 1953, they built the one that established the general structure of DNA. There have been minor amendments to their first published structure, but the principal features remain unchanged.

Four key features define DNA structure

Four features summarize the molecular architecture of DNA. The DNA molecule is (1) a double-stranded helix, (2) of uniform diameter, (3) that twists to the right (that is, twists in the same direction as the threads on most screws), (4) with the two strands running in opposite directions. The sugar–phosphate backbones of the polynucleotide chains coil around

the outside of the helix, and the nitrogenous bases point toward the center (Figure 11.6). The two chains are held together by hydrogen bonding between specifically paired bases.

Consistent with Chargaff's studies, adenine (A) pairs with thymine (T) by forming two hydrogen bonds, and guanine (G) pairs with cytosine (C) by forming three hydrogen bonds (Figure 11.7). Every base pair consists of one purine (A or G) and one pyrimidine (C or T). Because the AT and GC pairs, like rungs of a ladder, are of equal length and fit identically into the double helix, the diameter of the helix is uniform. The base pairs are flat, and their stacking in the center of the molecule is stabilized by hydrophobic interactions (see Chapter 2), contributing to the overall stability of the double helix.

What does it mean to say that the two DNA strands run in opposite directions? The direction of a polynucleotide can be defined by looking at the linkages (called phosphodiester bonds) between adjacent nucleotides. In the sugar–phosphate backbone of DNA, the phosphate groups connect to the 3′ carbon of one deoxyribose molecule and the 5′ carbon of the next, linking successive sugars together (see Figure 11.7). The prime (′) designates the position of a carbon atom in the 5-carbon sugar, deoxyribose, in each nucleotide in DNA.

Thus the two ends of a polynucleotide differ. Polynucleotides have a free (not connecting to another nucleotide) 5′ phosphate ($-OPO_3^{3-}$) group at one end—the **5′ end**—and a free 3′ hydroxyl ($-OH$) group

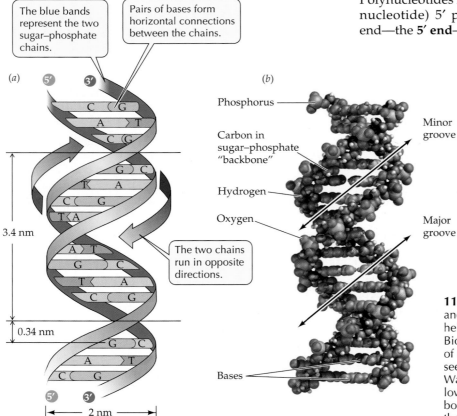

Francis Crick   James Watson

11.6 DNA Is a Double Helix (*a*) Watson and Crick proposed that DNA is a double helical molecule. (Also see Figure 3.23.) (*b*) Biochemists can now pinpoint the position of every atom in a DNA macromolecule. To see that the essential features of the original Watson–Crick model have been verified, follow with your eyes the double helical ribbons of sugar–phosphate groups and note the horizontal rungs of the bases.

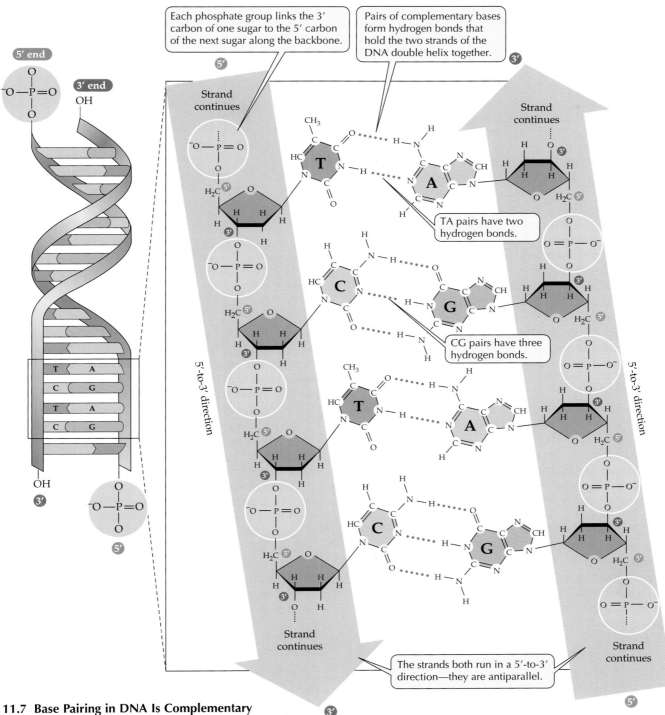

Each phosphate group links the 3′ carbon of one sugar to the 5′ carbon of the next sugar along the backbone.

Pairs of complementary bases form hydrogen bonds that hold the two strands of the DNA double helix together.

TA pairs have two hydrogen bonds.

CG pairs have three hydrogen bonds.

The strands both run in a 5′-to-3′ direction—they are antiparallel.

11.7 Base Pairing in DNA Is Complementary
The purines (A and G) pair with the pyrimidines (T and C, respectively) to form equal-size "rungs" on a ladder attached to a sugar–phosphate backbone. The ladder "twists" in a double helix structure.

at the other—the **3′ end**. The 5′ end of one strand in a DNA double helix is paired with the 3′ end of the other strand, and vice versa; that is, the strands run in opposite directions (see Figure 11.6).

The double helical structure of DNA is essential to its function

Watson and Crick's double-helix model hints at how DNA carries out its biological role. First, the molecule is linear. It runs on and on, nucleotide pair after nucleotide pair, with no kinks and no bulges. Such a molecule can carry and convey information in only one way: The information must lie in the linear sequence of the nitrogenous bases.

An implication of **complementary base pairing**, A with T and G with C, was pointed out by Crick and Watson in the original publication of their findings in the journal *Nature* (1953): "It has not escaped our notice that the specific pairing we have postulated immediately suggests a possible copying mechanism for the genetic material."

Each strand of the double helix is complementary to the other; that is, at each point, the base on one strand is complementary with the base on the other strand. If the spirals unwound, each strand could serve as a guide for the synthesis of a new one; thus each single strand of the parent double helix could produce a new double-stranded molecule identical to the original. All the information needed in new DNA molecules is present in a single strand of DNA, because that information is the sequence of bases.

The double helical structure of DNA also suggested an explanation for some mutations: They might simply be changes in the linear sequence of nucleotide pairs. In sum, genetic information, gene function, gene replication, and gene mutation could all be accounted for by the double helical structure of DNA.

DNA Replication

Watson and Crick made DNA replication sound simple. In reality, the mechanics of replication are complex, and they ensure that life goes on. First, DNA was shown to replicate from template strands in the test tube with simple substrates and an enzyme from bacteria. Then, an elegant experiment showed that replication is semiconservative, as predicted.

Three modes of DNA replication appeared possible

Just three years after Watson and Crick published their paper in *Nature*, their prediction that the DNA molecule contains the information for its own replication was demonstrated by the work of Arthur Kornberg, then at Washington University in St. Louis. Kornberg showed that DNA can replicate in a test tube with no cells present. The only requirements are a mixture containing DNA, a specific enzyme (which he called *DNA polymerase*), and a mixture of the four precursors: the nucleoside triphosphates deoxy-ATP, deoxy-CTP, deoxy-GTP, and deoxy-TTP (Figure 11.8). If any one of the four nucleoside triphosphates is omitted from the reaction mixture, DNA does not replicate. Somehow the DNA itself serves as a template for the reaction—a guide to the exact placement of nucleotides in the new strand. Where there is a T in the template, there must be an A in the new strand, and so forth. How does DNA perform the template function; that is, how exactly does the molecule replicate?

In their paper, Watson and Crick had suggested that DNA replication is *semiconservative* (Figure 11.9a): that

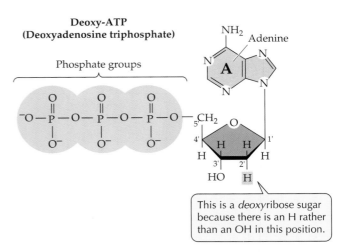

11.8 A Nucleoside Triphosphate Nucleoside triphosphates such as deoxy-ATP, shown here, form the building blocks of DNA. Deoxy-ATP differs from ATP in lacking an —OH group at carbon 2' (compare with Figure 6.7). The other DNA building blocks contain guanine, thymine, or cytosine in place of adenine.

each parent strand serves as a template for a new strand, and that the two new DNAs each have one old and one new strand. But there were two other possibilities. In the model of *conservative* replication (Figure 11.9b), the original double helix would serve as a template but either would be reconstituted or perhaps would never unwind at all. Thus the new molecule would contain none of the atoms of the original and the old one would be fully retained. According to the model of *dispersive* replication (Figure 11.9c), fragments of the original molecule would serve as templates, assembling two molecules, each containing old and new parts, perhaps at random.

Kornberg's experiment did not provide a basis for choosing among the three mechanisms. But the experimental work of Meselson and Stahl, described next, confirmed the model of semiconservative replication.

Meselson and Stahl demonstrated that DNA replication is semiconservative

A clever experiment conducted by Matthew Meselson and Franklin Stahl convinced the scientific community that semiconservative replication is the correct model. Working at the California Institute of Technology in 1957, they devised a simple way to distinguish old strands of DNA from new ones: density labeling.

The key to the experiment was to use a "heavy" isotope of nitrogen. Heavy nitrogen (^{15}N) is a rare, nonradioactive isotope that makes molecules containing it more dense than chemically identical molecules containing the common isotope ^{14}N. To distinguish DNA of different densities (that is, DNA containing ^{15}N versus DNA containing ^{14}N), Meselson, Stahl, and Jerome

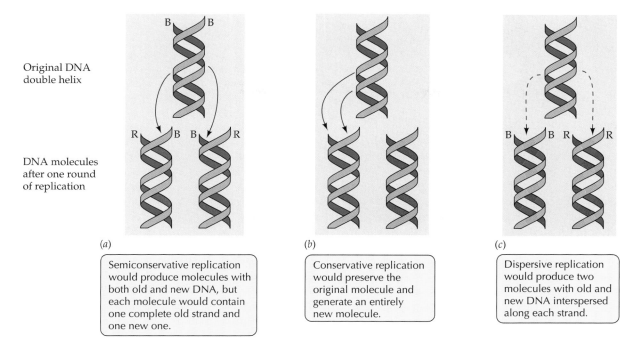

Original DNA double helix

DNA molecules after one round of replication

(a)

Semiconservative replication would produce molecules with both old and new DNA, but each molecule would contain one complete old strand and one new one.

(b)

Conservative replication would preserve the original molecule and generate an entirely new molecule.

(c)

Dispersive replication would produce two molecules with old and new DNA interspersed along each strand.

11.9 Three Models for DNA Replication All Obeyed Base-Pairing Rules In each model, the stretches of original DNA are shown in blue and newly synthesized DNA in red. Watson and Crick's original paper suggested that DNA replication is semiconservative.

Vinograd invented a new centrifugation procedure using a cesium chloride (CsCl) solution.

A concentrated solution of CsCl has a density very close to that of DNA. At the high gravitational forces produced in an ultracentrifuge, cesium ions sediment to some extent, establishing a gradient from low density at the top of the tube to high density at the bottom. When a DNA sample is dissolved in CsCl and centrifuged at about 100,000 times the force of gravity, the DNA gathers in a layer in the centrifuge tube at a position where the density of the CsCl solution equals that of the DNA (Figure 11.10).

After developing this method of distinguishing DNA densities, Meselson and Stahl grew a culture of the bacterium *Escherichia coli* for 17 generations on a medium in which the nitrogen source (ammonium chloride, NH_4Cl) was made with ^{15}N instead of ^{14}N. As a result, all the DNA in the bacteria was "heavy." They grew another culture on medium with ^{14}N, and they extracted DNA from both cultures. When these DNA extracts were combined and centrifuged with CsCl, two separate DNA bands formed, showing that this method would work for separating DNA samples of slightly different densities.

11.10 Density Gradient Centrifugation The two strands of each DNA molecule stay together during extraction and centrifugation. Labeled ("heavy") DNA will separate from lighter DNA in the gradient formed by a cesium chloride solution.

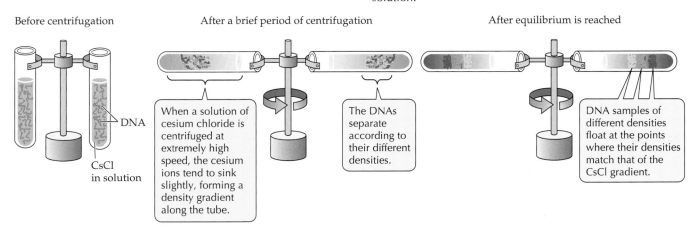

Before centrifugation

DNA

CsCl in solution

After a brief period of centrifugation

When a solution of cesium chloride is centrifuged at extremely high speed, the cesium ions tend to sink slightly, forming a density gradient along the tube.

The DNAs separate according to their different densities.

After equilibrium is reached

DNA samples of different densities float at the points where their densities match that of the CsCl gradient.

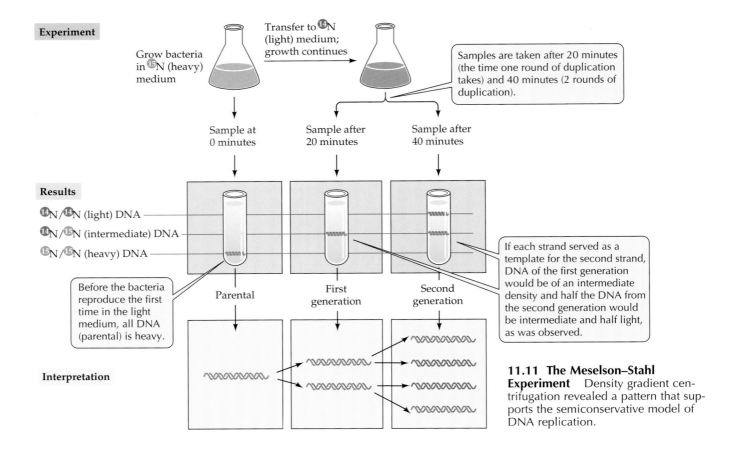

11.11 The Meselson–Stahl Experiment Density gradient centrifugation revealed a pattern that supports the semiconservative model of DNA replication.

Next, Meselson and Stahl grew another *E. coli* culture on ^{15}N medium, then *transferred* it to normal ^{14}N medium and allowed the bacteria to continue growing (Figure 11.11). *E. coli* reproduces every 20 minutes, and Meselson and Stahl collected some of the bacteria from each division after the transfer. They extracted DNA from the samples. After DNA was duplicated and the cells divided to produce each new generation, the DNA banding in the density gradient was different from the original banding.

Initially the DNA was uniformly labeled with ^{15}N and hence was relatively dense. After one generation, when the DNA had been duplicated once, all the DNA was of an intermediate density. After two generations, there were two equally large DNA bands: one of low density and one of intermediate density. In samples from subsequent generations, the proportion of low-density DNA increased steadily.

The data gathered in this experiment can be explained by the semiconservative model of DNA replication. The high-density DNA had two ^{15}N strands; the intermediate-density DNA had one ^{15}N and one ^{14}N strand; and the low-density DNA had two ^{14}N strands. In the first round of DNA replication, the strands of the double helix—both heavy with ^{15}N— separated. While separating, each strand acted as the template for a second strand, which contained only

^{14}N and hence was less dense. Each double helix then consisted of one ^{15}N and one ^{14}N strand and was of intermediate density. In the second replication, the ^{14}N-containing strands directed the synthesis of partners with ^{14}N, creating low-density DNA, and the ^{15}N strands got new ^{14}N partners (see Figure 11.11).

If the DNA had replicated in accord with either of the other models, the results would have been quite different (see "Applying Concepts" Question 3 at the end of this chapter). The crucial observation proving the semiconservative model was that intermediate-density DNA (^{15}N–^{14}N) appeared in the first generation and continued to appear in subsequent generations.

The Mechanism of DNA Replication

How does the enzymatic machinery replicate DNA semiconservatively? At a minimum, each DNA strand must function as a template, to which new bases are added by the AT and GC base-pairing rules; the substrates (deoxyribonucleoside triphosphates dATP, dGTP, dCTP, and dTTP) must be available; there should be a suitable DNA polymerase to bring the substrates to the template; and there must be a source of chemical energy to drive this highly endergonic synthesis reaction.

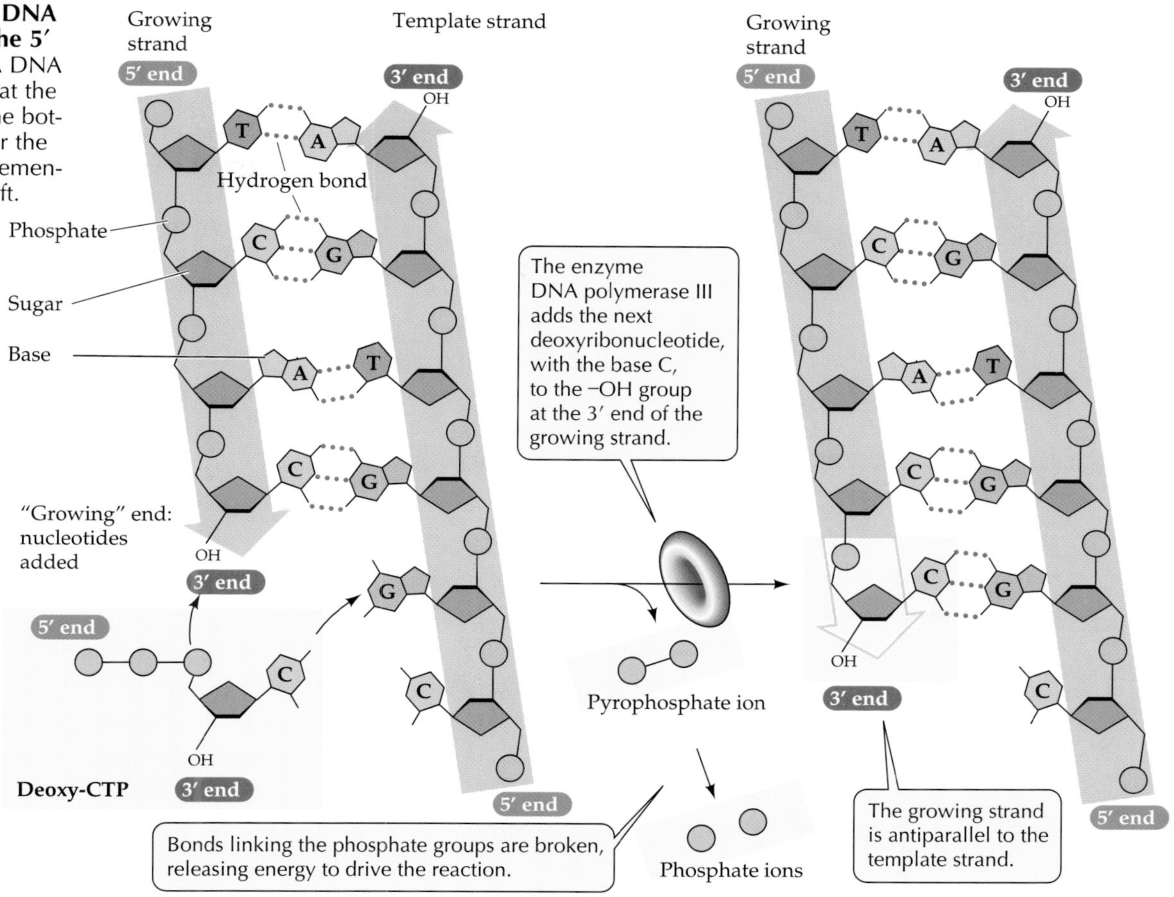

11.12 A Replicating DNA Strand Grows from the 5′ End to the 3′ End A DNA strand, with its 3′ end at the top and its 5′ end at the bottom, is the template for the synthesis of the complementary strand at the far left.

A key observation of virtually all DNA replication is that nucleotides are always added to the growing strand at the same end: the 3′ end—the end at which the DNA strand has a free hydroxyl (—OH) group on the 3′ carbon of its terminal deoxyribose (Figure 11.12). This hydroxyl group reacts with a phosphate group on the 5′ carbon of the deoxyribose of a deoxyribonucleoside triphosphate (see Figure 11.8), and thus the chain grows. A bond linking the terminal two phosphate groups to the rest of the deoxyribonucleoside triphosphate breaks and thereby releases energy for this reaction. The pyrophosphate ion, consisting of the two terminal phosphate groups, also breaks, forming two phosphate ions and in the process releasing additional free energy. The phosphate group still on the nucleotide becomes part of the sugar–phosphate backbone of the growing DNA molecule.

Helicases initiate DNA replication

In order to function as templates, the two individual strands of the double helix must be made available. The strands must be unwound, and then they must be kept from rejoining. Enzymes called **helicases** unwind the double helix, using energy from ATP, so that the bases on the two strands can take new partners. Once the single strands are exposed, special proteins bind to them to stabilize them and prevent them from reassociating to re-form the parent double helix.

Where does the unwinding begin? At one end of the DNA molecule? The DNA molecules—chromosomes—of bacteria don't *have* ends; they are circular! Helicases act at a specific point on the prokaryotic DNA circle called an **origin of replication** (Figure 11.13a). Replication then proceeds in both directions around the circle, meeting about halfway around and resulting in two circular daughter molecules.

Although eukaryotic DNA molecules are linear rather than circular, their replication also does not begin at the ends of the molecule. These molecules are much larger than prokaryotic DNA molecules, and replication would require far too much time if it proceeded only from the ends. Instead, each eukaryotic chromosome has many origins of replication (Figure 11.13b).

In both prokaryotes and eukaryotes, replication proceeding from an origin results in a bubble with **replication forks** at each growing point (Figure 11.13c). The replication forks are moving, Y-shaped structures that are the regions where new DNA strands are being synthesized. We will describe the

(a)

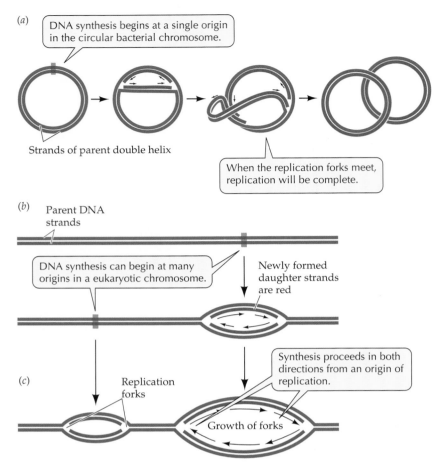

DNA synthesis begins at a single origin in the circular bacterial chromosome.

Strands of parent double helix

When the replication forks meet, replication will be complete.

(b)

Parent DNA strands

DNA synthesis can begin at many origins in a eukaryotic chromosome.

Newly formed daughter strands are red

(c)

Replication forks

Synthesis proceeds in both directions from an origin of replication.

Growth of forks

11.13 Origins of Replication (a) In bacteria, replication proceeds in both directions around the circular chromosome from a single point of origin. The result is two circular daughter molecules. (b) Because eukaryotic DNA molecules are so large, there are many origins of replication. (c) New strands of DNA are synthesized at replication forks.

further details of DNA replication as it takes place in prokaryotes; some of the details differ in eukaryotes.

Most DNA polymerases need a primer

DNA polymerase *elongates* a polynucleotide strand, but it cannot *start* a strand from scratch. DNA polymerases require the assistance of a previously existing strand of DNA or RNA to which they can add new nucleotides. Such a helper strand is called a **primer**. DNA polymerases add nucleotides to the 3' end of the primer.

In DNA replication, the primer is a short single strand of RNA (Figure 11.14). This RNA, complementary to the DNA template strand, is formed by an en-

zyme called a **primase**, which is one of several polypeptides bound together in an aggregate called a **primosome**. A single primer suffices for synthesis of the leading strand, but each Okazaki fragment in the lagging strand requires its own primer. (We will discuss leading and lagging strands shortly.) When DNA replication is complete, each daughter molecule consists only of DNA. The question that remains then is, What happens to all those RNA primers?

DNA polymerase III extends the new DNA strands

Bacterial cells contain more than one DNA polymerase. The one that Kornberg discovered, appropriately called DNA polymerase I, does not catalyze the replication of the *E. coli* chromosome. Instead, it is involved in the DNA repair mechanisms that will be discussed later in the chapter. **DNA polymerase III** catalyzes the elongation of the new DNA strands at the replication forks in prokaryotes. Two molecules of DNA polymerase III clamp together at the replication fork, each acting on one of the strands (see Figure 11.12). Both of the separated strands of the parent molecule act as templates, and the formation

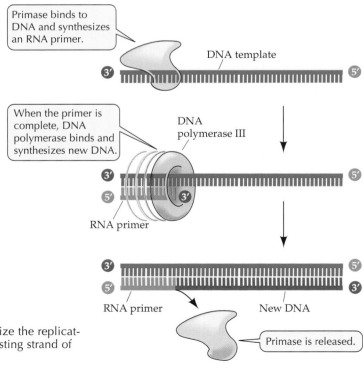

Primase binds to DNA and synthesizes an RNA primer.

DNA template

When the primer is complete, DNA polymerase binds and synthesizes new DNA.

DNA polymerase III

RNA primer

RNA primer

New DNA

Primase is released.

11.14 No DNA Forms without a Primer In order to synthesize the replicating strand, DNA polymerases require a primer—a previously existing strand of DNA or RNA to which they can add new nucleotides.

11.15 Many Proteins Collaborate at the Replication Fork Shown here are some of the many proteins involved in DNA replication.

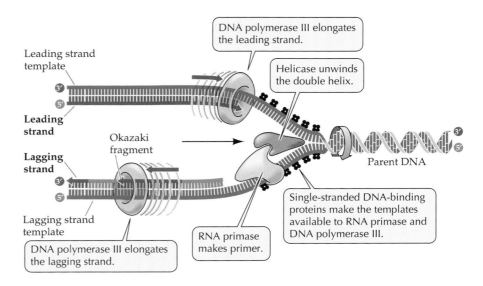

of the new strands is guided by complementary base pairing.

One parent strand is being exposed in the 3′ end, which presents no problem—its complementary strand is synthesized continuously as the replication fork proceeds. This daughter strand, called the **leading strand**, elongates from its own 5′ to its own 3′ end, as do all other nucleic acids.

Besides DNA polymerase III, various other proteins play roles in replacement of the RNA primer and in other replication tasks; some of these are shown in Figure 11.15. Recall that the two original strands are antiparallel; that is, the 3′ end of one strand is paired with the 5′ end of the other. Thus, the second daughter strand—called the **lagging strand**—is antiparallel to the leading strand. The template for the lagging strand is exposed from its 5′ end toward its 3′ end, yet that template must direct the 5′-to-3′ synthesis of the lagging strand. How can it do this, given that new nucleotides are added only at the 3′ end of a polynucleotide chain?

The lagging strand looks like a problem

Synthesis of the lagging strand requires working in relatively small, backward-directed bits. This strand is produced in discontinuous spurts (100 to 200 nucleotides at a time in eukaryotes; 1,000 to 2,000 at a time in prokaryotes). The discontinuous stretches are synthesized just as the leading strand is, by adding new nucleotides one at a time to the 3′ end of the daughter strand, but the synthesis of this new daughter DNA moves in the direction opposite to that in which the replication fork is moving (Figure 11.16).

These stretches of new DNA for the lagging strand are called **Okazaki fragments** after their discoverer, the Japanese biochemist Reiji Okazaki. While the leading strand grows continuously "forward," the lagging strand grows in shorter, "backward" stretches with gaps between them.

11.16 The Two Daughter Strands Form in Different Ways
As the original DNA unwinds, both daughter strands are synthesized in the 5′-to-3′ direction, although their template strands are antiparallel. The leading strand grows continuously "forward," but the lagging strand grows in shorter, "backward" stretches called Okazaki fragments. Eukaryotic Okazaki fragments are hundreds of nucleotides long, with gaps between them.

When the next Okazaki fragment forms, DNA polymerase I removes the old RNA primer and replaces it with DNA adjacent to the new Okazaki fragment. Left behind is a tiny gap—the final phosphodiester bond between the adjacent Okazaki fragments is missing (Figure 11.17). Another enzyme, DNA ligase, catalyzes

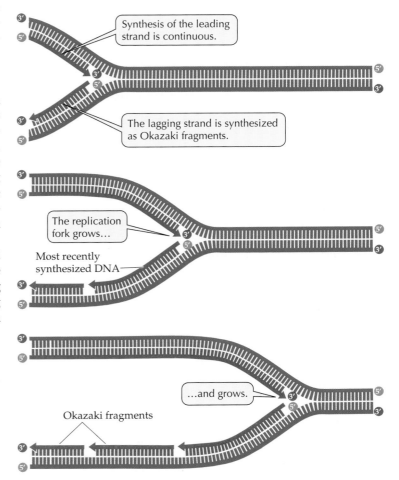

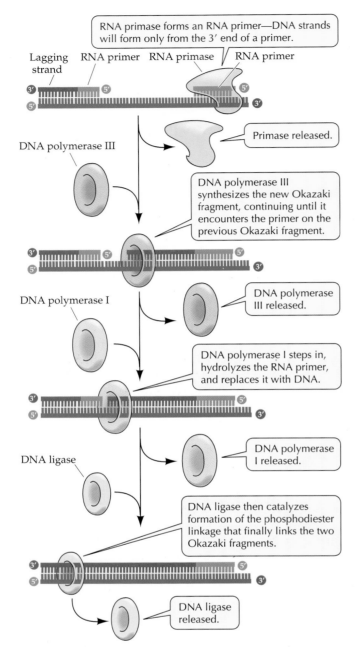

RNA primase forms an RNA primer—DNA strands will form only from the 3' end of a primer.

Lagging strand · RNA primer · RNA primase · RNA primer

Primase released.

DNA polymerase III

DNA polymerase III synthesizes the new Okazaki fragment, continuing until it encounters the primer on the previous Okazaki fragment.

DNA polymerase I

DNA polymerase III released.

DNA polymerase I steps in, hydrolyzes the RNA primer, and replaces it with DNA.

DNA ligase

DNA polymerase I released.

DNA ligase then catalyzes formation of the phosphodiester linkage that finally links the two Okazaki fragments.

DNA ligase released.

11.17 The Lagging Strand Story The lagging strand forms from left to right, while each Okazaki fragment forms from right to left—from its own 5' end to its own 3' end. Other proteins join those shown in Figure 11.15 to complete the complex task of replication.

the formation of that bond, linking the fragments and making the lagging strand whole.

Working together, helicase, the two DNA polymerases, primase, DNA ligase, and the other proteins do the complex job of DNA synthesis with a speed and accuracy that are almost unimaginable. In *E. coli*, the complex makes new DNA at a rate in excess of 1,000 base pairs per second, committing errors in fewer than one base in 10^8 to 10^{12}.

DNA Proofreading and Repair

DNA must be faithfully replicated and maintained. The price of failure can be great—perhaps even death. The transmission of genetic information is at stake, as is the functioning of a cell or multicellular organism. Yet the replication of DNA is *not* perfectly accurate, and the DNA of nondividing cells is subject to damage by environmental agents. In the face of these threats, how has life gone on so long?

The preservers of life are DNA repair mechanisms. Although replication yields mistakes in fewer than one base in 10^8 to 10^{12}, DNA polymerases initially make a significant number of mistakes in assembling polynucleotide strands. In *E. coli*, for example, the error rate would result in flaws in approximately one out of every ten genes each time the cell divided. In humans, about 1,000 genes in every cell would be affected each time the cell divided.

Fortunately, our cells normally have some DNA repair mechanisms at their disposal. These include a "proofreading" function that corrects errors as DNA polymerase makes them; a mismatch repair function that scans DNA after it has been made and corrects any base-pairing mismatches; and excision repair, in which abnormal bases that have formed because of chemical damage are removed and replaced with functional bases.

Proofreading, mismatch repair, and excision repair

After introducing a new nucleotide into a growing polynucleotide strand, the DNA polymerases perform a **proofreading** function (Figure 11.18*a*). When a DNA polymerase recognizes a mispairing of bases, it removes the improperly introduced nucleotide and tries again. (Helicase, DNA ligase, and other proteins also play roles in this key mechanism.) The polymerase likely will be successful in inserting the correct monomer the second time, because the error rate for this process is only about 1 in 10,000 base pairs. This proofreading mechanism greatly lowers the overall error rate for replication.

After new DNA has been replicated and during genetic recombination, a second mechanism surveys the molecule and looks for mismatches (Figure 11.18*b*). For example, this **mismatch repair** system might detect an AC base pair instead of an AT pair. Since both AT and GC obey the base-pairing rules, how does the repair mechanism "know" whether the AC pair should be repaired to remove the C and replace it with T, for instance, instead of removing the A and replacing it with G?

The repair mechanism can detect the "wrong" base because a newly synthesized DNA strand has not yet been covalently modified with methyl groups on its cytosines (see Chapter 14). Thus, the relatively unmethylated strand must be the one with the errors.

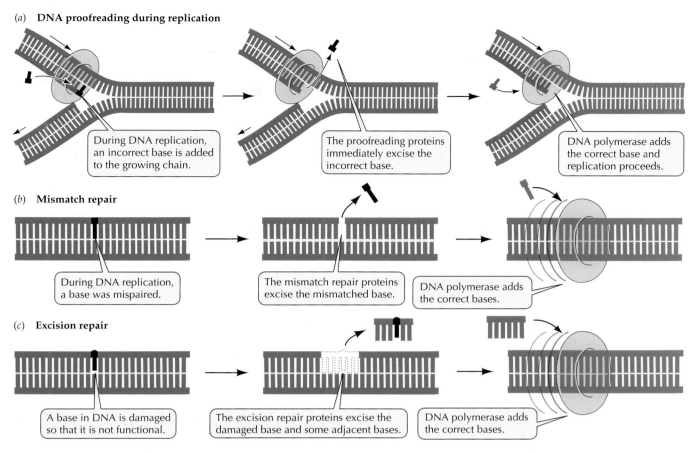

(a) **DNA proofreading during replication**

During DNA replication, an incorrect base is added to the growing chain.

The proofreading proteins immediately excise the incorrect base.

DNA polymerase adds the correct base and replication proceeds.

(b) **Mismatch repair**

During DNA replication, a base was mispaired.

The mismatch repair proteins excise the mismatched base.

DNA polymerase adds the correct bases.

(c) **Excision repair**

A base in DNA is damaged so that it is not functional.

The excision repair proteins excise the damaged base and some adjacent bases.

DNA polymerase adds the correct bases.

11.18 DNA Repair Mechanisms The proteins of DNA replication also play roles in the life-preserving repair mechanisms, helping to ensure the exact replication of template DNA.

When mismatch repair fails, DNA sequences are altered. One form of colon cancer arises in part from a failure of mismatch repair.

What if a DNA molecule becomes damaged during the life of a cell? Some cells live and play important roles for many years, even though their DNA is constantly at risk from hazards such as high-energy radiation, chemicals that induce mutations, and random spontaneous chemical reactions. Cells owe their lives to DNA repair mechanisms. For example, in **excision repair**, certain enzymes "inspect" the cell's DNA (Figure 11.18c). When they find mispaired bases, chemically modified bases, or points at which one strand has more bases than the other (with the result that one or more bases of one strand form an unpaired loop), these enzymes cut the defective strand. Another enzyme cuts away the bases adjacent to and including the offending base, and DNA polymerase and DNA ligase synthesize and seal up a new (usually correct) piece to replace the excised one.

Our dependence on this repair mechanism is underscored by our susceptibility to various diseases that arise from excision repair defects. One example is the skin disease xeroderma pigmentosum. People with this disease lack a mechanism that normally repairs damage caused by the ultraviolet radiation in sunlight. Without this mechanism, a person exposed to sunlight develops skin cancer.

DNA repair requires energy

What does it cost the cell to keep its DNA accurate and ensure that it replicates properly? At first glance, you might expect DNA polymerization to be fairly "neutral" energetically, because adding a new monomer to the chain requires the formation of a new phosphodiester bond but is supported by the hydrolysis of one of the high-energy bonds in the nucleoside triphosphate (see Figure 11.12). Overall, however, this reaction is slightly endergonic. But help is available in the form of the pyrophosphate ion released in the polymerization reaction. The enzyme pyrophosphatase cleaves the high-energy bond in the pyrophosphate. Coupling this reaction to the polymerization gives it a big boost.

Noncovalent bonds also play a major role in favoring DNA polymerization. Hydrogen bonds form between the complementarily paired bases, and other weak interactions form as the bases stack in the middle of the double helix. These bonds and interactions stabilize the DNA molecule and help drive the polymerization reaction. Thus DNA synthesis itself does not take a tremendous toll in energy.

DNA repair processes, however, are far from cheap energetically. Some are very inefficient. Nonetheless the cell deploys many DNA repair mechanisms, some overlapping in function with others. Why? Perhaps because the cell simply can't afford to leave its genetic information unprotected, regardless of the cost.

Normal, undamaged DNA is vital to life and its continuation. But what exactly is the function of DNA? What do genes do? What do they control? These are the topics of Chapter 12.

Summary of "DNA and Its Role in Heredity"

DNA: The Genetic Material

• In addition to circumstantial evidence (the location and quantity of DNA in the cell) that genes are made of DNA, two experiments provided convincing demonstration.
• In one experiment, DNA from a virulent strain of pneumococcus bacteria genetically transformed nonvirulent bacteria into virulent bacteria. **Review Figure 11.1**
• In the second set of experiments, labeled viruses were allowed to attach to host bacteria. The labeled protein coats of the viruses were shaken off the bacterial surface, but the labeled viral DNA entered the host cells, where it took control and caused the bacterial cells to produce hundreds of new viruses. **Review Figures 11.2, 11.3**

The Structure of DNA

• X ray crystallography showed that the DNA molecule is a helix. **Review Figure 11.4**
• Biochemical analysis revealed that DNA is composed of nucleotides, each containing one of four bases—adenine, cytosine, thymine, or guanine—and that the amount of adenine equals the amount of thymine and the amount of guanine equals the amount of cytosine. **Review Figure 11.5**
• Putting the accumulated data together, Watson and Crick proposed that DNA is a double-stranded helix in which the strands are antiparallel and the bases form opposite strands held together by hydrogen bonding. This model accounts for the genetic information, function, mutation, and replication of DNA. **Review Figures 11.6, 11.7**

DNA Replication

• Three possible models for DNA replication were hypothesized: semiconservative, conservative, and dispersive. **Review Figure 11.9**
• An experiment by Meselson and Stahl proved the replication of DNA to be semiconservative. Each parent strand acts as a template for the synthesis of a new strand; thus, the two replicated DNA helices contain one parent strand and one newly synthesized strand each. **Review Figures 11.10, 11.11**

The Mechanism of DNA Replication

• In DNA replication, the enzyme DNA polymerase catalyzes the addition of nucleotides to the 3' end of each strand. The nucleotides are added according to base-pairing rules of the template strand of DNA. The substrates are deoxyribonucleoside triphosphates, which are hydrolyzed when they are added to the growing chain, releasing energy that fuels the synthesis of DNA. **Review Figure 11.12**
• Many proteins assist in DNA replication. DNA helicases unwind the double helix, and the template strands are stabilized by other proteins.
• Prokaryotes have a single origin of replication; eukaryotes have many. Replication in both cases proceeds in both directions from an origin of replication. **Review Figure 11.13**
• An RNA primase catalyzes the synthesis of short RNA primers, to which the nucleotides are added as the chain grows. **Review Figure 11.14**
• Using DNA polymerase, the leading strand grows continuously in the 5'-to-3' direction until the replication of that section of DNA has been completed. Then the RNA primer is degraded and DNA added in its place. On the lagging strand, which grows in the other direction, DNA is still made in the 5'-to-3' direction (toward the fork). But synthesis of the lagging strand is discontinuous: The DNA is added as short fragments to primers; then the polymerase skips toward the 5' end to make the next fragment. **Review Figures 11.15, 11.16, 11.17**

DNA Proofreading and Repair

• The machinery of DNA replication makes about one error in 100,000 nucleotides added. These errors are repaired by three different mechanisms: proofreading, mismatch repair, and excision repair. DNA repair mechanisms lower the overall error rate of replication to about one base in 10 billion. **Review Figure 11.18**
• Although energetically costly and somewhat redundant, DNA repair is crucial to the survival of the cell.

Self-Quiz

1. Griffith's studies of *Streptococcus pneumoniae*
    a. showed that DNA is the genetic material of bacteria.
    b. showed that DNA is the genetic material of bacteriophages.
    c. demonstrated the phenomenon of bacterial transformation.
    d. proved that prokaryotes reproduce sexually.
    e. proved that protein is not the genetic material.

2. In the Hershey–Chase experiment
    a. DNA from parent bacteriophages appeared in progeny bacteriophages.
    b. most of the phage DNA never entered the bacteria.
    c. more than three-fourths of the phage protein appeared in progeny phages.
    d. DNA was labeled with radioactive sulfur.
    e. DNA formed the coat of the bacteriophages.

3. Which statement about complementary base pairing is *not* true?
    a. It plays a role in DNA replication.
    b. In DNA, T pairs with A.
    c. Purines pair with purines, and pyrimidines pair with pyrimidines.

d. In DNA, C pairs with G.

e. The base pairs are of equal length.

4. In semiconservative replication of DNA
   a. the original double helix remains intact and a new double helix forms.
   b. the strands of the double helix separate and act as templates for new strands.
   c. polymerization is catalyzed by RNA polymerase.
   d. polymerization is catalyzed by a double helical enzyme.
   e. DNA is synthesized from amino acids.

5. Which of the following does not occur during DNA replication?
   a. Unwinding of the parent double helix
   b. Formation of short pieces that are united by DNA ligase
   c. Complementary base pairing
   d. Use of a primer
   e. Polymerization in the 3′-to-5′ direction

6. The primer used for DNA replication
   a. is a short strand of RNA added to the 3′ end.
   b. is present only once on the leading strand.
   c. remains on the DNA after replication.
   d. ensures that there will be a free 5′ end to which nucleotides can be added.
   e. is added to only one of the two template strands.

7. The 3′ end of a DNA strand is defined as the place where
   a. the phosphate group is not bound to another nucleotide.
   b. both DNA strands end opposite each other.
   c. DNA polymerase binds to begin replication.
   d. there is a free —OH group at the 3′ carbon of deoxyribose.
   e. three A residues are present.

8. The role of DNA ligase in DNA replication is to
   a. add more nucleotides to the growing chain one at a time.
   b. open up the two DNA strands to expose template strands.
   c. ligate base to sugar to phosphate in a nucleotide.
   d. bond Okazaki fragments to one another.
   e. remove incorrectly paired bases.

9. Incorrect bases that are added to DNA
   a. can be repaired by proofreading.
   b. cannot have been added by DNA polymerases, since these enzymes make no errors.
   c. do not result in mispairing.
   d. are replaced along with adjacent nucleotides.
   e. are methylated.

10. The following events occur in excision repair of DNA. What is their proper order?
    1 Base-paired DNA is made complementary to the template.
    2 Damaged bases are recognized.
    3 DNA ligase seals the new strand to existing DNA.
    4 Part of a single strand is excised.
    a. 1234 *b.* 2134 *c.* 2413
    d. 3421 *e.* 4231

Applying Concepts

1. Outline a series of experiments using radioactive isotopes to show that bacterial DNA and not protein enters the host cell and is responsible for bacterial transformation.

2. Suppose that Meselson and Stahl had continued their experiment on DNA replication for another ten bacterial generations. Would there still have been any ^{14}N–^{15}N hybrid DNA present? Would it still have appeared in the centrifuge tube? Explain.

3. If DNA replication were conservative rather than semiconservative, what results would Meselson and Stahl have observed? Diagram the results using the conventions of Figure 11.11.

4. Using the following information, calculate the number of origins of DNA replication on a human chromosome: DNA polymerase adds nucleotides at 3,000 base pairs per minute in one direction; replication is bidirectional; S phase lasts 300 minutes; there are 120 million base pairs per chromosome. With a typical chromosome 3 μm long, how many origins are there per micrometer?

5. The drug dideoxycytidine (used to treat certain viral infections) is a nucleotide made with 2′,3′-dideoxyribose. This sugar lacks —OH groups at both the 2′ and the 3′ positions. Explain why this drug would stop the growth of a DNA chain if added to the DNA.

Readings

Felsenfeld, G. 1985. "DNA." *Scientific American*, October. A well-illustrated description of DNA structure and function.

Griffiths, A. J. F., J. H. Miller, D. T. Suzuki, R. C. Lewontin and W. M. Gelbart. 1996. *An Introduction to Genetic Analysis*, 6th Edition. W. H. Freeman, New York. An excellent textbook of modern genetics. Chapters 11, 12, and 13 are particularly relevant.

Judson, H. F. 1996. *The Eighth Day of Creation: Makers of the Revolution in Biology*, Expanded Edition. CSHL Press, Plainview, NY. A sparkling history of molecular biology, with the best available description of the events surrounding the discovery of the structure of DNA.

Modrich, P. 1994. "Mismatch Repair, Genetic Stability, and Cancer." *Science*, vol. 266, pages 1959–1960. This brief article is from the *Science* issue recognizing DNA repair as the "Molecule of the Year."

Radman, M. and R. Wagner. 1988. "The High Fidelity of DNA Duplication." *Scientific American*, August. A description of how error avoidance and error correction work that poses the question, Why don't they work even better?

Sancar, A. 1994. "Mechanisms of DNA Excision Repair." *Science*, vol. 266, pages 1954–1956. This brief article is from the *Science* issue recognizing DNA repair as the "Molecule of the Year."

Stent, G. S. and R. Calendar. 1978. *Molecular Genetics*, 2nd Edition. W. H. Freeman, New York. A brilliant technical and historical introduction to molecular genetics and the role of DNA.

Upton, A. C. 1982. "The Biological Effects of Low-Level Ionizing Radiation." *Scientific American*, February. An explanation of how radiation leads to mutations.

Watson, J. D. 1968. *The Double Helix*. Atheneum, New York. A captivating and, to some, infuriating book in which Watson describes the events leading to the discovery of DNA structure.

Chapter 12

From DNA to Protein:
Genotype to Phenotype

F lying alone in your corporate jet, you make a crash landing on an uncharted island in the North Sea. The batteries in your radio are dying, but you manage to send off a frantic message. A ham radio operator on the northern German coast, who does not know English, hears your message and writes it down phonetically. He gives it to an English-speaking friend, who deciphers the words and converts them into German. A message is sent to a German ship, and you are rescued.

This story and its happy ending illustrate the main themes of this chapter, which is concerned with what DNA *does* (Chapter 11 covered what DNA *is*). The information content of a gene consists of its base sequence, just as your frantic message consisted of the words you spoke. In the cell, the information in DNA is *transcribed* into information in RNA, as your spoken words were transcribed into phonetic English. A remarkable team of proteins and RNAs, some of which are constituents of ribosomes, *translates* the information from a sequence of bases in the RNA into a sequence of amino acids in an

One Member of the Translation Team
Transfer RNA molecules recognize the genetic message encoded in the nucleotide sequence of DNA and simultaneously carry specific amino acids, enabling them to translate the language of DNA into the language of proteins.

enzyme or other protein, just as the ham radio operator's friend converted your English into German. The enzyme produces the final effect—catalysis of a chemical reaction—just as the German translation of your message led to effective action.

Of course, both our adventure story and the cellular processing of DNA depend on accurate functioning of all components of the system. The origi-

260

nal message itself may be erroneous—DNA is subject to mutation, and you might use a wrong word in your haste. Errors can also arise in transcription or translation.

In this chapter we will focus first on how it became established that the gene in DNA expresses itself as the phenotype in a polypeptide. Next we will describe how this expression occurs: First, the information content of DNA is converted to mRNA (transcription); then mRNA travels to the ribosome, where its information determines the order of specific amino acids in a polypeptide (translation). Finally, we will describe the genetic concept of mutation in more precise terms than we have up to now, defining alleles in terms of alterations in the nucleotide sequences of DNA.

Genes and the Synthesis of Polypeptides

There are many steps between genotype and phenotype. Genes cannot, all by themselves, directly produce a phenotypic result such as a particular eye color, a specific seed shape, or a cleft chin, any more than a compact disk can play a symphony without the help of a CD player.

With the gene defined as DNA, the first step in relating it to its phenotype is to define phenotypes in chemical terms. This chemical definition of phenotypes was developed actually before the discovery of DNA as the genetic material. Using organisms as diverse as humans and bread molds, scientists studied the chemical differences between organisms carrying wild-type and mutant alleles and found that the major phenotypic differences were in specific proteins.

Some hereditary diseases feature defective enzymes

The first hints as to how genes are expressed came early in the twentieth century from the work of the English physician Archibald Garrod. Alkaptonuria is a hereditary disease in which the patient's urine turns black when exposed to air. Garrod recognized that this symptom showed that the biochemistry of the affected individual was different from that of other people. He suggested in 1908 that alkaptonuria and some other hereditary diseases are consequences of "inborn errors of metabolism."

Garrod proposed that what makes the urine dark is a defect in an enzyme that metabolizes the amino acid tyrosine. He studied the pattern of inheritance of alkaptonuria and reasoned that it affects individuals who are homozygous for the recessive allele of a particular gene (see Chapter 10), which in normal individuals codes for active enzyme. His studies and proposal explicitly linked genotype and phenotype by means of enzymes.

However, like Mendel's explanation of inheritance in the garden pea, Garrod's ideas were too advanced for their time and sat almost unappreciated for more than 30 years. We will return to Garrod's work and human genetic diseases in Chapter 17.

The one-gene, one-polypeptide hypothesis

A series of experiments performed by George W. Beadle and Edward L. Tatum at the California Institute of Technology in the 1940s confirmed and extended Garrod's ideas. Beadle and Tatum experimented with the bread mold *Neurospora crassa.* The nuclei in the mass of the mold are haploid (*n*), as are the reproductive spores. This fact is important because it means that even recessive mutant alleles are easy to detect in experiments.

Neurospora can be grown on a simple, completely defined medium (that is, one in which all the ingredients are known) that contains inorganic ions, a simple source of nitrogen (such as ammonium chloride), an organic source of energy and carbon (such as glucose), and a single vitamin (biotin). From this minimal medium, the enzymes of wild-type *Neurospora* can catalyze the metabolic reactions needed to make all the chemical constituents of its cells.

Beadle and Tatum hypothesized that mutations might lead to altered enzymes that could no longer do their jobs. In that case, mutants of *Neurospora* might be found that could not make certain compounds they needed; such mutants would grow only on media to which those compounds were added. We call mutants of this type **auxotrophs** ("increased eaters"), in contrast to the wild-type **prototrophs** ("original eaters") that constituted the original *Neurospora* population. Whereas prototrophs can grow on minimal medium, auxotrophs require specific additional nutrients, such as a particular amino acid, a vitamin, or a purine or pyrimidine.

Beadle and Tatum isolated a number of auxotrophic mutant strains of *Neurospora*. These auxotrophs did not grow on the minimal medium that supported the growth of the wild-type strain. But they did grow on a complete medium, to which all supplements (amino acids, nucleotides, vitamins, and so on) had been added. For each strain, Beadle and Tatum were able to find a single compound that, when added to the minimal medium, supported the growth of that mutant. This result supported the idea that mutations have simple effects—and, perhaps, the idea that each mutation causes a defect in only one enzyme in the metabolic pathway leading to the synthesis of the required nutrient.

One group of auxotrophs could grow on minimal medium supplemented with the amino acid arginine. They were classified as *arg* mutants. Mapping studies established that some of the *arg* mutations were at different loci on a chromosome or were on different chromosomes. Beadle and Tatum concluded from this observation that different genes can participate in governing a single biosynthetic pathway—in this case, the pathway leading to arginine.

They grew 15 different *arg* mutants in the presence of various suspected intermediates in the synthetic metabolic pathway for arginine:

$$X \rightarrow \text{ornithine} \rightarrow \text{citrulline} \rightarrow \text{arginine}$$

Some of the mutants were able to grow on different intermediates, as well as on arginine-supplemented medium (Figure 12.1). For example, some mutants could grow on either arginine or citrulline, and some could grow on arginine, citrulline, or ornithine.

Beadle and Tatum concluded that the mutants that grew on arginine- or citrulline-supplemented medium but not on ornithine-supplemented medium were deficient in an enzyme that catalyzes the conversion of ornithine to citrulline. Similarly, these investigators were able to relate each of the other *arg* mutants to a particular enzyme. Subsequent analysis of cell extracts of the various mutant strains for the relevant enzymes confirmed their hypotheses.

This work led Beadle and Tatum to formulate the one-gene, one-enzyme hypothesis. According to this hypothesis, the function of a gene is to control the production of a single, specific enzyme. This proposal strongly influenced the subsequent development of the sciences of genetics and molecular biology. Garrod had pointed in the same direction more than three decades earlier, but only after Beadle and Tatum's research were other scientists prepared to act on the suggestion.

Many enzymes are composed of more than one polypeptide chain (that is, they have quaternary structure). Each chain is specified by its own separate gene. Thus, it is more correct to speak of a **one-gene, one-polypeptide hypothesis**: The function of a gene is to control the production of a single, specific polypeptide.

Much later, it was discovered that some genes code for forms of RNA that do not become translated into polypeptides, and still other genes are sequences of DNA involved in controlling which DNA sequences are expressed. But these discoveries did not invalidate the relations between other genes and polypeptides.

DNA, RNA, and the Flow of Information

Now we turn our attention to the mechanisms by which a gene expresses itself as a polypeptide. The first mechanism, transcription, transcribes the information of DNA (the gene) into corresponding information in an RNA sequence. The second mechanism, translation, translates this RNA information into an appropriate amino acid sequence in a polypeptide.

RNA differs from DNA

To understand the transcription and translation of genetic information, you need to know about RNA. **RNA** (ribonucleic acid) is a polynucleotide similar to DNA (see Figure 3.23) but different in three ways.

1. RNA generally consists of only one polynucleotide strand (thus Chargaff's equalities, G = C and A = T [see Figure 11.5], are true only for DNA and not for RNA).
2. The sugar molecule found in ribonucleotides is ribose rather than the deoxyribose found in DNA.
3. Although three of the nitrogenous bases (adenine, guanine, and cytosine) in ribonucleotides are identical to the bases in deoxyribonucleotides, the fourth base in RNA is uracil (U), which is similar to thymine but lacks the methyl ($-CH_3$) group.

RNA can base-pair with single-stranded DNA, and this pairing obeys the AT, UA, and GC hydrogen-bonding rules. RNA can also fold over and base-pair within its own sequence, as we will see with tRNA later in this chapter.

Information flows in one direction when genes are expressed

Francis Crick proposed what he called the **central dogma** of molecular biology. The central dogma is, simply, that DNA codes for the production of RNA (transcription), RNA codes for the production of protein (translation), and protein does *not* code for the production of protein, RNA, or DNA (Figure 12.2). In Crick's words, "once 'information' has passed into protein *it cannot get out again*."

Crick contributed two key ideas to the development of the central dogma. The first solved a difficult problem: How could one explain the relationship between a specific nucleotide sequence (in DNA) and a specific amino acid sequence (in protein), since the nucleotides of DNA do not attach to amino acids? Crick made a clever suggestion: He proposed that an adapter molecule carries a specific amino acid at one end and recognizes a sequence of nucleotides with another region.

In due course, other molecular biologists found and characterized these adapter molecules, called transfer RNAs, or **tRNAs**. Because they recognize the genetic message (a series of nucleotides) and simultaneously carry specific amino acids, tRNAs can translate the language of DNA into the language of proteins.

Crick's second major contribution to the central dogma addressed another problem: How does the genetic information get from the nucleus to the cytoplasm? (Most of the DNA of a eukaryotic cell is confined to the nucleus, but proteins are synthesized in the cytoplasm.) Crick, together with the South African geneticist Sydney Brenner and the French molecular

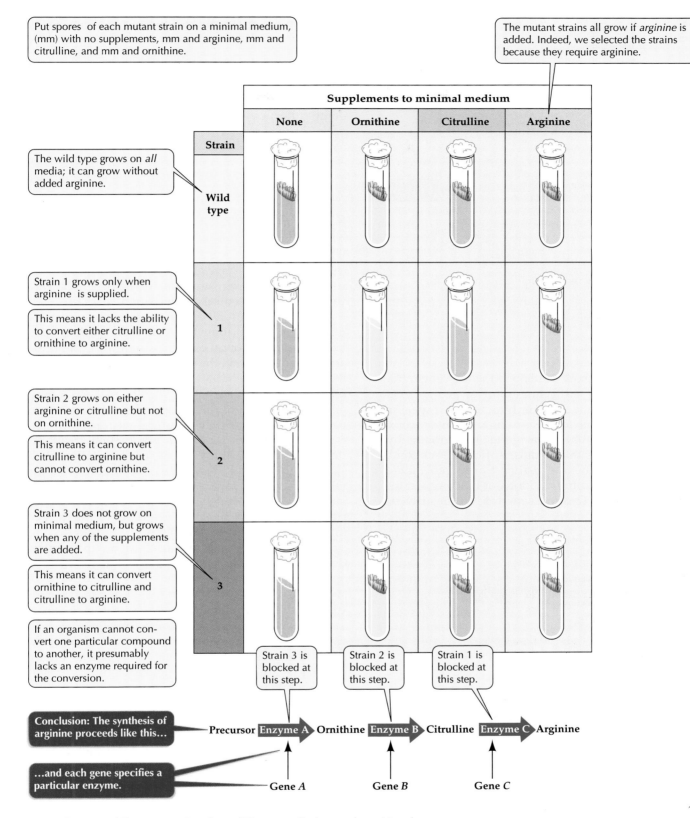

Put spores of each mutant strain on a minimal medium, (mm) with no supplements, mm and arginine, mm and citrulline, and mm and ornithine.

The mutant strains all grow if *arginine* is added. Indeed, we selected the strains because they require arginine.

The wild type grows on *all* media; it can grow without added arginine.

Strain 1 grows only when arginine is supplied.

This means it lacks the ability to convert either citrulline or ornithine to arginine.

Strain 2 grows on either arginine or citrulline but not on ornithine.

This means it can convert citrulline to arginine but cannot convert ornithine.

Strain 3 does not grow on minimal medium, but grows when any of the supplements are added.

This means it can convert ornithine to citrulline and citrulline to arginine.

If an organism cannot convert one particular compound to another, it presumably lacks an enzyme required for the conversion.

Supplements to minimal medium

Strain	None	Ornithine	Citrulline	Arginine
Wild type				
1				
2				
3				

Strain 3 is blocked at this step.

Strain 2 is blocked at this step.

Strain 1 is blocked at this step.

Conclusion: The synthesis of arginine proceeds like this…

Precursor → Enzyme A → Ornithine → Enzyme B → Citrulline → Enzyme C → Arginine

…and each gene specifies a particular enzyme.

Gene A   Gene B   Gene C

12.1 Genes and Enzymes   Beadle and Tatum studied several nutritional mutants of *Neurospora,* as shown here. Wild-type, prototrophic strains grow on the minimal medium, but different auxotrophic mutants required the addition of different nutrients in order to grow; step through the figure to follow the reasoning that upheld the "one-gene, one-enzyme" hypothesis.

(a)

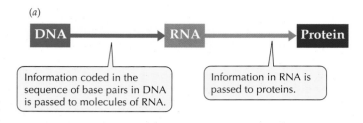

Information coded in the sequence of base pairs in DNA is passed to molecules of RNA.

Information in RNA is passed to proteins.

(b)

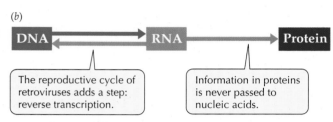

The reproductive cycle of retroviruses adds a step: reverse transcription.

Information in proteins is never passed to nucleic acids.

12.2 The Central Dogma (a) Information flows from DNA to proteins, as indicated by the arrows. (b) The reproductive cycle of retroviruses adds a step, reverse transcription, to the central dogma.

biologist François Jacob, developed the messenger hypothesis in response to this question.

According to the messenger hypothesis, a specific type of RNA molecule forms as a complementary copy of one strand of a particular gene. The process by which RNA forms is called **transcription**. If each RNA molecule contains the information from a gene, there should be as many different kinds of RNA molecules as there are genes. This messenger RNA, or **mRNA**,

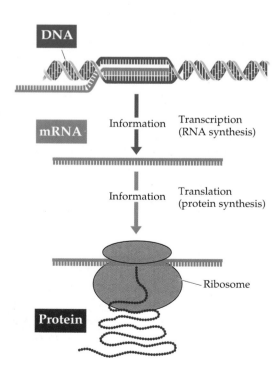

then travels from the nucleus to the cytoplasm. In the cytoplasm, mRNA serves as a template on which the tRNA adapters line up so that the amino acids are in the proper sequence for a growing polypeptide chain—a process called **translation** (Figure 12.3).

Summarizing the main features of the central dogma, the messenger hypothesis, and the adapter hypothesis, we may say that *a given gene is transcribed to produce a messenger RNA (mRNA) complementary to one of the DNA strands,* and that *transfer RNA (tRNA) molecules translate the sequence of bases in the mRNA into the appropriate sequence of amino acids.*

RNA viruses modify the central dogma

According to the central dogma of molecular biology, DNA codes for RNA and RNA codes for protein. All cellular organisms have DNA as their hereditary material. Only among viruses (which are not cellular) are variations on the central dogma found.

Many viruses, such as the tobacco mosaic virus, have RNA rather than DNA as their nucleic acid. Heinz Fraenkel-Conrat of the University of California at Berkeley separated the protein and RNA fractions of the tobacco mosaic virus and then recombined them to obtain active virus particles. When he took RNA from one mutant strain of this virus and combined it with protein from another, the resulting viruses replicated to produce more virus particles like the first (the RNA-donating) strain. With this experiment Fraenkel-Conrat showed that RNA is the genetic material of the tobacco mosaic virus. RNA itself is the template for the synthesis of the next generation of viral RNA and viral proteins. In this virus, DNA is left out of the flow of information (which is normally from DNA to RNA to protein). A more radical variation on the central dogma is seen in the retroviruses.

MAKING DNA FROM RNA: RETROVIRUSES. Rous sarcoma virus is an RNA virus that causes a cancer in chickens. The virus enters a chicken cell and subsequently causes the cell to make a DNA "transcript" of the viral RNA, the reverse of the usual process (see Figure 12.2*b*). The afflicted cell does not burst, but it changes permanently in shape, metabolism, and growth habit. The new DNA becomes part of the hereditary apparatus of the infected chicken cell.

In 1964, Howard Temin of the University of Wisconsin hypothesized that the virus carries an enzyme

12.3 From Gene to Phenotype in Prokaryotes This figure summarizes the messenger hypothesis as it appears in prokaryotes. In eukaryotes the process is more complex in that transcription forms a pre-mRNA in the nucleus. Pre-mRNA must then be processed to form the mature mRNA that is translated in the cytoplasm (see Chapter 14).

for the manufacture of DNA, using viral RNA as the information template. Five years later he and David Baltimore simultaneously but independently discovered the enzyme, which was named **reverse transcriptase** because it transcribes DNA from RNA rather than RNA from DNA. The DNA copy of the viral RNA can then use cellular machinery to make more viral RNA. Viruses that employ reverse transcriptase are known as *retroviruses*.

The central dogma requires slight modification to account for the flow of information in retroviruses and their hosts. However, the stipulation that information does not flow from protein back to the nucleic acids still holds.

Transcription: DNA-Directed RNA Synthesis

Transcription, the formation of a specific RNA under the control of a specific DNA, requires the enzyme **RNA polymerase**. It also requires the appropriate ribonucleoside triphosphates (ATP, GTP, CTP, and UTP) and the DNA template. In a given region of DNA, such as a gene, only *one* of the strands—the **template strand**—is transcribed. The other, complementary DNA strand remains untranscribed. For different genes in the same DNA molecule, different strands may be transcribed. That is, the strand that is the complementary strand in one gene may be the template strand in another.

Not only mRNA is produced by transcription. The same process is responsible for the synthesis of tRNA and ribosomal RNA (**rRNA**), which constitutes a major fraction of the ribosome. Like mRNA, these other forms of ribonucleic acid are encoded by specific genes. In prokaryotes, most of the DNA acts as a template for the production of mRNA, tRNA, or rRNA. The situation in eukaryotes is more complicated, as will be explained later in this chapter and in more detail in Chapter 14.

In DNA replication, the two strands of the parent molecule unwind, and each strand becomes paired with a new strand. In transcription, DNA partly unwinds so that it may serve as a template for RNA synthesis. As the RNA transcript forms, it peels away, allowing the DNA that has already been transcribed to rewind into the double helix (Figure 12.4).

Transcription can be divided into three distinct processes: initiation, elongation, and termination. Let's consider each of these in turn.

Initiation of transcription requires a promoter and an RNA polymerase

RNA polymerase has a relatively weak attraction for any DNA sequence, but it binds very tightly and is effective at beginning transcription only at special sequences on DNA called **promoters**. These sequences can be identified in several ways. For example, DNA can be chopped up into short stretches and presented to RNA polymerase in the test tube; sequences to which RNA polymerase binds tightly are promoters. Or, a promoter region can be detected by a change in its DNA sequence; in some cases a single base-pair change results in a promoter that can no longer bind tightly to the RNA polymerase.

There is one promoter for each gene (or, in prokaryotes, each set of genes) to be transcribed into mRNA. Promoters serve as punctuation marks, telling the RNA polymerase where to start and which strand of DNA to read. A promoter, being a specific sequence in the DNA and reading in a particular direction, orients the RNA polymerase and thus "aims" it at the correct strand to use as a template. Part of each promoter is the initiation site, where transcription begins. Farther toward the 3' end of the promoter lie groups of nucleotides that help the RNA polymerase bind. RNA polymerase moves in a 3'-to-5' direction along the template strand (see Figure 12.4).

Not all promoters are identical. One promoter may bind RNA polymerase very effectively and therefore trigger frequent transcription of its gene; in other words, it competes effectively for the available RNA polymerase. Another promoter may bind the polymerase poorly, and its genes will rarely be transcribed. The efficiency of the promoter sets a limit on how often each gene can be transcribed. An enzyme that is needed in large amounts is encoded by a gene whose promoter is efficient, but the synthesis of an enzyme that is needed only in tiny amounts is controlled by an inefficient promoter. We will consider prokaryotic promoters in more detail in Chapter 13 and eukaryotic promoters in Chapter 14.

Not all RNA polymerases are identical. Prokaryotes have a single RNA polymerase that produces mRNA, tRNA, and rRNA. Eukaryotes have three different RNA polymerases with distinct roles. Of these, RNA polymerase II is responsible for mRNA production.

Initiation of transcription in eukaryotes differs from that in prokaryotes in another respect. As we will see in slightly more detail in Chapter 14, eukaryotic RNA polymerases cannot bind the promoter and start the process of transcription until other proteins bind to sites in the promoter and thus prepare a docking site for the RNA polymerase. For now, recognize that the requirement for these proteins affords a way to regulate the transcription of particular genes, augmenting the differences already imposed by the varying efficiency of promoters.

Transcription can proceed as soon as RNA polymerase binds to the promoter and starts unwinding the DNA strands at the initiation site. The next stage is elongation of the RNA transcript, in which additional ribonucleotides are added to the growing chain.

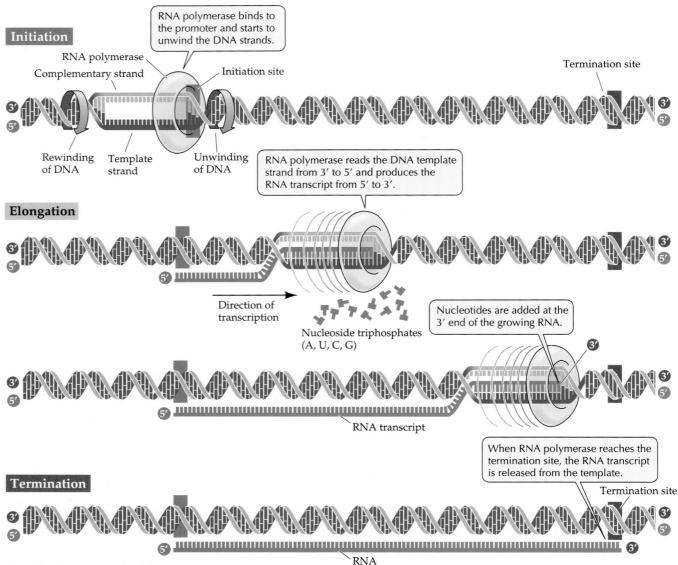

Initiation

RNA polymerase binds to the promoter and starts to unwind the DNA strands.

RNA polymerase

Complementary strand

Initiation site

Termination site

Rewinding of DNA

Template strand

Unwinding of DNA

Elongation

RNA polymerase reads the DNA template strand from 3′ to 5′ and produces the RNA transcript from 5′ to 3′.

Direction of transcription

Nucleoside triphosphates (A, U, C, G)

Nucleotides are added at the 3′ end of the growing RNA.

RNA transcript

When RNA polymerase reaches the termination site, the RNA transcript is released from the template.

Termination site

Termination

Termination site

RNA

12.4 DNA Is Transcribed into RNA DNA partially unwinds to serve as a template for RNA synthesis. The RNA transcript forms and then peels away, allowing the DNA that has already been transcribed to rewind into a double helix. Three distinct processes—initiation, elongation, and termination—comprise DNA transcription. The RNA polymerase is much larger in reality than indicated here, covering about 50 base pairs.

RNA polymerases elongate the transcript

RNA polymerase unwinds the DNA about 20 base pairs at a time and reads the template strand of DNA in the 3′-to-5′ direction (see Figure 12.4). Like DNA polymerases, RNA polymerases add new nucleotides to the 3′ end of the growing strand. That is, the new RNA grows from its own 5′ end to its 3′ end. The RNA transcript is thus antiparallel to the DNA template strand.

As the RNA polymerase moves along the template strand, unwinding the next stretch of DNA, it rewinds the stretch of DNA that it just processed. Transcription, like DNA replication, is a lot of work, requiring a lot of energy. As with replication, transcription draws on energy released by both the removal and the breakdown of the pyrophosphate group from each added nucleotide.

Unlike DNA polymerases, RNA polymerases do not inspect and correct their work. Transcription errors occur at a rate of one mistake for every 10^4 to 10^5 bases.

Transcription terminates at particular base sequences

What tells RNA polymerase to stop adding nucleotides to a growing transcript? Just as initiation sites specify the start of transcription, particular base sequences in the DNA specify its termination. The mechanisms of termination are complex and of more than one kind. For some genes the newly formed transcript simply falls away from the DNA template and the RNA polymerase. For others, a helper protein pulls the transcript away.

In prokaryotes the translation of mRNA often begins (at the 5′ end of the mRNA) before transcription of the mRNA molecule is complete. In eukaryotes the situation is more complicated. First, there is a spatial separation of transcription (in the nucleus) and translation (in the cytoplasm). Second, the first product of transcription is a pre-mRNA that is longer than the final mRNA and must undergo considerable processing before it becomes the mRNA and can be translated. The reasons for this processing, and its mechanisms, will be discussed in Chapter 14.

The Genetic Code

You can think of the genetic information transcribed in an mRNA molecule as a series of three-letter "words." Each sequence of three nucleotides (the "letters") along the chain specifies a particular amino acid. The three-letter "word" is called a **codon**. That codon is complementary to the corresponding codon in the DNA molecule from which it was transcribed.

The complete genetic code is shown in Figure 12.5. Notice that there are many more RNA codons than there are different amino acids in proteins. Combinations of the four "letters" (the bases) give 64 (4^3) different three-letter codons, yet these determine only 20 amino acids. AUG, which codes for methionine, is also the **start codon**, the initiation signal for translation. Three of the codons (UAA, UAG, UGA) are **stop codons**, or chain terminators; when the translation machinery reaches one of these codons, translation stops and the polypeptide is released from the translation complex.

After describing the properties of the genetic code, we will examine

some of the scientific thinking and experimentation that went into deciphering it.

The genetic code is degenerate but not ambiguous

After the start and stop codons, the remaining 60 codons are far more than enough to code for the other 19 amino acids—and indeed there are repeats. Thus we say that the code is **degenerate**; that is, an amino acid may be represented by more than one codon. The degeneracy is not evenly divided among the amino acids. For example, methionine and tryptophan are represented by only one codon each, whereas leucine is represented by six different codons (see Figure 12.5).

The term "degeneracy" should not be confused with "ambiguity." To say that the code was ambiguous would mean that a single codon could specify either of two (or more) different amino acids; there would be doubt whether to put in, say, leucine or something else. The genetic code is not ambiguous. Degeneracy in the code means that there is more than one clear way to say, "Put leucine here." In other words, a given amino acid may be encoded by more than one codon, but a codon can code for only one amino acid. But just as people in different places prefer different ways of saying the same thing—"Good-bye!" "See you!" "Ciao!" and "So long!" have the same meaning—different organisms prefer one or others of the degenerate codons. These preferences are important in genetic engineering (see Chapter 16).

The code appears to be relatively universal, applying to all the species on our planet. Thus the code must be an ancient one that has been maintained intact throughout the evolution of living things. Exceptions are known: Within mitochondria and chloroplasts the code differs slightly from that in

12.5 The Universal Genetic Code
Genetic information is encoded in mRNA in three-letter units—codons—made up of the bases uracil (U), cytosine (C), adenine (A), and guanine (G). To decode a codon, find its first letter in the left column, then read across the top to its second letter, then read down the right column to its third letter. The amino acid the codon specifies is given in the corresponding row. For example, AUG codes for methionine, and GUA codes for valine.

Second letter

First letter	U	C	A	G	
U	UUU UUC Phenyl-alanine / UUA UUG Leucine	UCU UCC UCA UCG Serine	UAU UAC Tyrosine / UAA Stop codon UAG Stop codon	UGU UGC Cysteine / UGA Stop codon UGG Tryptophan	U C A G
C	CUU CUC CUA CUG Leucine	CCU CCC CCA CCG Proline	CAU CAC Histidine / CAA CAG Glutamine	CGU CGC CGA CGG Arginine	U C A G
A	AUU AUC AUA Isoleucine / AUG Methionine; start codon	ACU ACC ACA ACG Threonine	AAU AAC Asparagine / AAA AAG Lysine	AGU AGC Serine / AGA AGG Arginine	U C A G
G	GUU GUC GUA GUG Valine	GCU GCC GCA GCG Alanine	GAU GAC Aspartic acid / GAA GAG Glutamic acid	GGU GGC GGA GGG Glycine	U C A G

prokaryotes and elsewhere in eukaryotic cells; in one group of protists, UAA and UAG code for glutamine rather than functioning as stop codons. The significance of these differences is not yet clear. What is clear is that the exceptions are few and slight.

You should remember that the codons in Figure 12.5 are mRNA codons. The master codons on the DNA strand that was transcribed to produce the mRNA are complementary and antiparallel to these codons. Thus, for example, AAA in the template DNA strand corresponds to phenylalanine (which is coded for by the mRNA codon UUU), and CCA in the DNA template corresponds to tryptophan (which is coded for by the mRNA codon UGG). If the last example surprised you, note that we normally list the base sequences of nucleic acids in a 5′-to-3′ order.

Does this code really work? Convincing evidence comes from experiments in which artificial DNA of a known base sequence is introduced into prokaryotes and the prokaryotes are induced to produce the specific protein encoded by that DNA (see Chapter 16). We can now program bacteria to synthesize proteins that no organism has ever made before.

Biologists broke the genetic code by translating artificial messengers

Molecular biologists broke the code in which genetic information is stored in the early 1960s. The problem seemed difficult: How could more than 20 "code words" be written with an "alphabet" consisting of only four "letters"? How, in other words, could four bases code for 20 or so different amino acids?

The idea that the code was a triplet code, based on three-letter codons, was considered likely. With only four letters (A, G, C, U), a one-letter code clearly could not unambiguously encode 20 amino acids; it could encode only four of them. A two-letter code could contain only $4 \times 4 = 16$ codons—still not enough. But a triplet code could contain up to $4 \times 4 \times 4 = 64$ codons.

Marshall W. Nirenberg and J. H. Matthaei, at the National Institutes of Health, made the first "decoding" breakthrough in 1961 when they realized that they could use a very simple artificial polynucleotide instead of a complex, natural mRNA as a messenger. They could then identify the polypeptide that the artificial messenger encoded.

Nirenberg had prepared an artificial mRNA in which all the bases were uracil: poly U. When poly U was added to a reaction mixture containing all the ingredients necessary for cell-free protein synthesis (ribosomes, amino acids, activating enzymes, tRNAs, and other factors), a polypeptide formed. This polypeptide contained only one kind of amino acid: phenylalanine (Phe). Poly U coded for poly Phe! Accordingly, UUU appeared to be the mRNA code word—the codon—for phenylalanine. Following up

on this success, Nirenberg and Matthaei soon showed that CCC codes for proline and AAA for lysine. (Poly G presented some chemical problems and was not tested initially.) UUU, CCC, and AAA were three of the easiest codons; different approaches were required to work out the rest.

Other scientists later found that simple "mRNAs," only three nucleotides long and each amounting to a codon, can bind to ribosomes and that the resulting complex can then cause the binding of the corresponding charged tRNA. Thus, for example, simple UUU causes the tRNA charged with phenylalanine to bind to the ribosome. After this discovery, complete deciphering of the code book was relatively simple. To find the "translation" of a codon, Nirenberg could use a sample of that codon as an artificial mRNA and see which amino acid became bound.

The Key Players in Translation

Prokaryotic mRNAs are ready to be translated as they peel away from the DNA template strand, and they degrade within minutes. Eukaryotic pre-mRNAs are processed extensively before they become translatable mRNAs, and these mRNAs continue to be functional for many minutes to several hours. In both cases, however, translation is rapid, taking only a few minutes to make a polypeptide of hundreds of amino acids. This complex process occurs at the ribosome, which binds to mRNA, carrying the genetic code from DNA. Before translation begins, however, each amino acid becomes attached to its specific tRNA.

Transfer RNAs carry specific amino acids

Before we examine translation, let's see how a codon is related to the amino acid for which it codes. As predicted by Crick, the codon and the amino acid are related by way of an adapter—a specific type of tRNA. For each of the 20 amino acids, there is at least one specific tRNA molecule.

A tRNA molecule is small, consisting of only about 75 to 80 nucleotides (Figure 12.6). At the 3′ end of every tRNA molecule is a site to which the amino acid attaches. At about the midpoint is a group of three bases, called the **anticodon**, that constitutes the point of contact with mRNA. At contact, the tRNA and mRNA are antiparallel to each other. Each tRNA species has a unique anticodon, allowing it to unite by complementary base pairing with a particular codon. Complementary base pairing is what enables translation to be so specific.

Recall that 61 different codons encode the 20 amino acids in proteins. Does this mean that the cell must produce 61 different tRNA species, each with a different anticodon? No. The cell gets by with about two-thirds that number of tRNA species, because the speci-

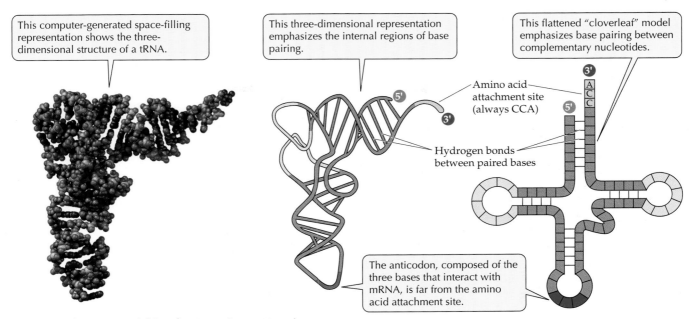

This computer-generated space-filling representation shows the three-dimensional structure of a tRNA.

This three-dimensional representation emphasizes the internal regions of base pairing.

This flattened "cloverleaf" model emphasizes base pairing between complementary nucleotides.

Amino acid attachment site (always CCA)

Hydrogen bonds between paired bases

The anticodon, composed of the three bases that interact with mRNA, is far from the amino acid attachment site.

12.6 Transfer RNA: Crick's Adapter   The tRNA molecules carry amino acids, associate with mRNA molecules, and interact with ribosomes. There is at least one specific tRNA molecule for each of the amino acids.

ficity for the base at the 3' end of the codon (and 5' end of the anticodon) is relaxed. This phenomenon, called *wobble*, allows the alanine codons GCA, GCC, and GCU all to be recognized by the same tRNA. Wobble is allowed in some matches but not in others; of most importance, it does not allow the genetic code to be ambiguous!

The three-dimensional shape of tRNAs (see Figure 12.6) allows them to combine specifically with binding sites on ribosomes. The structure of tRNA molecules relates clearly to their functions: They carry amino acids, associate with mRNA molecules, and interact with ribosomes.

Activating enzymes link the right tRNAs and amino acids

How does a tRNA molecule combine with the correct amino acid? A family of **activating enzymes**, known more formally as aminoacyl-tRNA synthetases, accomplishes this task (Figure 12.7). Each activating enzyme is specific for one amino acid and for one tRNA. The enzyme has a three-part active site that recognizes three smaller molecules: a specific amino acid, ATP, and a specific tRNA.

The enzyme reacts first with a molecule of amino acid and a molecule of ATP, producing a high-energy amino acid–AMP (adenosine monophosphate, which remains bound to the enzyme). The high energy results from the breaking of the bonds in the ATP—the high-energy bond between AMP and the terminal py-

rophosphate group (see Figure 6.7), and then the high-energy bond in the pyrophosphate (PP_i). The energy is conserved in the bond between the amino acid (AA) and AMP:

$$\text{enzyme} + \text{ATP} + \text{AA} \rightarrow \text{enzyme—AMP—AA} + PP_i$$

The enzyme then catalyzes a shifting of the amino acid from the AMP to the 3'-terminal nucleotide of the tRNA:

$$\text{enzyme—AMP—AA} + \text{tRNA} \rightarrow$$
$$\text{enzyme} + \text{AMP} + \text{tRNA—AA}$$

The activating enzyme finally releases this charged tRNA (tRNA with its attached amino acid) and can then charge another tRNA molecule. The bond between the amino acid and tRNA is a high-energy bond; it provides the energy for the synthesis of a peptide bond joining adjacent amino acids.

A clever experiment showed the importance of the specificity of the attachment of tRNA to its amino acid—a specificity that has been called the "second genetic code." The amino acid cysteine, already properly attached to its tRNA, was chemically modified to become a different amino acid, alanine. Which component—the amino acid or the tRNA—would be recognized when the hybrid charged tRNA was put into a protein-synthesizing system? The answer was, the latter: Everywhere in the synthesized protein where cysteine was supposed to be, alanine appeared instead. The cysteine-specific tRNA delivered its cargo (alanine) to every address where cysteine was called for.

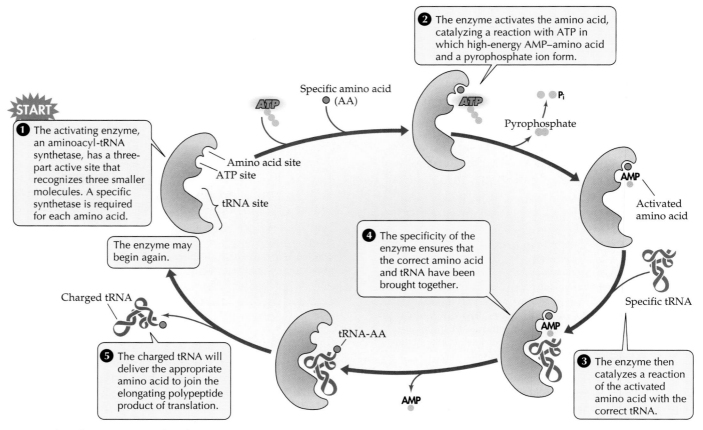

START

❶ The activating enzyme, an aminoacyl-tRNA synthetase, has a three-part active site that recognizes three smaller molecules. A specific synthetase is required for each amino acid.

Amino acid site
ATP site

tRNA site

❷ The enzyme activates the amino acid, catalyzing a reaction with ATP in which high-energy AMP–amino acid and a pyrophosphate ion form.

Specific amino acid (AA)

ATP

ATP

Pyrophosphate

Pᵢ

AMP

Activated amino acid

The enzyme may begin again.

❹ The specificity of the enzyme ensures that the correct amino acid and tRNA have been brought together.

Specific tRNA

Charged tRNA

AMP

tRNA-AA

❸ The enzyme then catalyzes a reaction of the activated amino acid with the correct tRNA.

❺ The charged tRNA will deliver the appropriate amino acid to join the elongating polypeptide product of translation.

AMP

12.7 Charging a tRNA Molecule Each activating enzyme must make the correct association of an amino acid and its tRNA; the enzyme is the essential link between nucleic acid "language" and protein "language."

The ribosome is the staging area for translation

Ribosomes are required for translation of the genetic information into a polypeptide chain. Although ribosomes are the smallest cellular organelles, their mass of several million daltons makes them large in comparison with the charged tRNAs.

Each ribosome consists of two subunits, a large one and a small one (Figure 12.8). In eukaryotes, the large subunit consists of three different molecules of rRNA (ribosomal RNA) and about 45 different protein molecules, arranged in a precise pattern. The small subunit in eukaryotes consists of one rRNA molecule and 33 different protein molecules. These different proteins and RNAs are held together by ionic and hydrophobic forces, not covalent bonds. If these forces are dis-

rupted by detergents, for example, the proteins and rRNAs separate from each other. When the detergent is removed, the entire complex structure *self-assembles*. This is like separating the pieces of a jigsaw puzzle and having them fit together again without human hands to guide them.

The ribosomes of prokaryotes are somewhat smaller than those of eukaryotes. Mitochondria and chloroplasts also contain ribosomes, some of which are even smaller than those of prokaryotes. When not active in the translation of mRNA, the ribosomes exist as sepa-

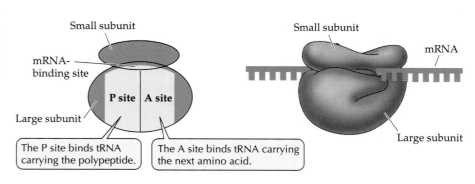

Small subunit

mRNA-binding site

Large subunit

P site | **A site**

The P site binds tRNA carrying the polypeptide.

The A site binds tRNA carrying the next amino acid.

Small subunit

mRNA

Large subunit

12.8 Ribosome Structure Each ribosome consists of a large and a small subunit, which separate when they are not in use.

rated subunits. Each ribosome has two tRNA-binding sites (P and A) that participate in translation. The ribosome also binds to the mRNA that it is translating.

A given ribosome is not specifically adapted to produce just one kind of protein. A ribosome can combine with any mRNA and all tRNAs and thus can be used to make different polypeptide products. The mRNA contains the information that specifies the polypeptide sequence. The ribosome is simply the molecular machine that accomplishes the task. Its structure enables it to hold the mRNA and tRNAs in the right positions, thus allowing the growing polypeptide to be assembled efficiently.

Translation: RNA-Directed Polypeptide Synthesis

We have been working our way through the steps by which the sequence of bases in the template strand of a DNA molecule specifies the sequence of amino acids in a protein (see Figure 12.3). We are now at the last step: translation, the RNA-directed assembly of a protein. Like transcription, translation occurs in three steps: initiation, elongation, and termination. Most of these reactions are catalyzed by ribosomal proteins. A polypeptide contains information for its three-dimensional shape, as well as for its ultimate cellular destination.

Translation begins with an initiation complex

The translation of mRNA begins with the formation of an initiation complex, which consists of a charged tRNA bearing the first amino acid and a small ribosomal subunit, both bound to the starting point on the mRNA chain (Figure 12.9). The small ribosomal unit binds to a recognition sequence on the mRNA. Recall that the start codon in the genetic code is AUG. Thus

the first amino acid is methionine (see Figure 12.5). The anticodon of a methionine-charged tRNA binds to the appropriate point on the mRNA by complementary base pairing with AUG, the initiation codon. (Not all proteins have methionine as their N-terminal amino acid. In many cases the initiator methionine is removed by an enzyme.)

After the methionine-charged tRNA has bound to the mRNA, the large subunit of the ribosome joins the complex. The ribosome has two tRNA-binding sites: the A site (which accepts a tRNA molecule bearing one *a*mino acid) and the P site (which carries a tRNA molecule bearing a growing *p*olypeptide chain). The first charged tRNA, bearing methionine, now lies in the P site of the ribosome, and the A site is aligned with the second codon.

How are all these ingredients—mRNA, two ribosomal subunits, and methionine-charged tRNA—put together properly? A group of proteins called **initiation factors** help direct the process, using GTP as an energy supply.

The polypeptide elongates from the N terminus

During translation the ribosome moves along the mRNA in the 5'-to-3' direction (Figure 12.10). A charged tRNA whose anticodon is complementary to the second codon enters the open A site. The large subunit then catalyzes the formation of a peptide bond between the amino acid on the P site and the amino acid on the A site in such a way that the first amino acid is the N terminus of the new protein, while the second amino acid remains attached to its tRNA by its carboxyl group (—COOH).

In 1992, Harry Noller and his colleagues at the University of California at Santa Cruz found that if they removed almost all the proteins in the large ribosomal subunit, it still catalyzed peptide bond formation. But if the rRNA was destroyed, so was the catalytic activity. Part of rRNA in the large subunits interacts with the end of the charged tRNA where the amino acid is attached. Thus rRNA appears to be the catalyst for peptide bond formation.

The idea that RNA—instead of the usual protein—can act as a catalyst, or **ribozyme**, is not unprecedented. For example, we

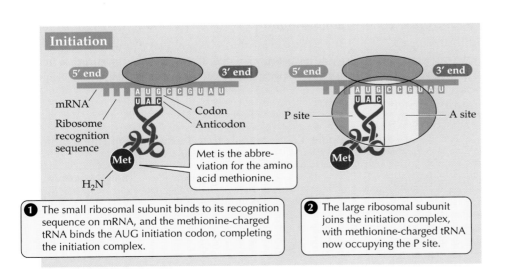

❶ The small ribosomal subunit binds to its recognition sequence on mRNA, and the methionine-charged tRNA binds the AUG initiation codon, completing the initiation complex.

❷ The large ribosomal subunit joins the initiation complex, with methionine-charged tRNA now occupying the P site.

12.9 The Initiation of Translation Translation begins with the formation of an initiation complex.

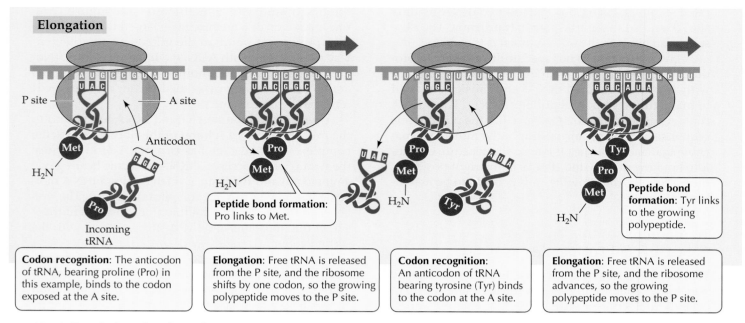

Elongation

Codon recognition: The anticodon of tRNA, bearing proline (Pro) in this example, binds to the codon exposed at the A site.

Elongation: Free tRNA is released from the P site, and the ribosome shifts by one codon, so the growing polypeptide moves to the P site.

Codon recognition: An anticodon of tRNA bearing tyrosine (Tyr) binds to the codon at the A site.

Elongation: Free tRNA is released from the P site, and the ribosome advances, so the growing polypeptide moves to the P site.

Peptide bond formation: Pro links to Met.

Peptide bond formation: Tyr links to the growing polypeptide.

12.10 Translation: The Elongation Stage The polypeptide chain elongates as the mRNA is translated.

will describe catalytic RNAs that act in mRNA splicing in Chapter 14. Because of its base pairing, RNA can fold into three-dimensional shapes (see tRNA, Figure 12.6) and bind substrates, just as protein-based enzymes do.

Elongation continues and the polypeptide grows

After the first tRNA releases its amino acid, it dissociates from the complex, returning to the cytosol to become charged with another amino acid of the same kind. The second tRNA, now bearing a *dipeptide*, shifts to the P site of the ribosome, which moves along the mRNA by another triplet codon. Energy for this movement comes from the hydrolysis of another molecule of GTP.

The process continues: (1) The next charged tRNA enters the open A site; (2) its amino acid forms a peptide bond, picking up the growing polypeptide chain from the tRNA in the P site; and (3) the entire tRNA–polypeptide complex, along with its codon, moves to the newly vacated P site. All these steps are assisted by proteins called elongation factors. How does the cycle end?

A release factor terminates translation

When a stop codon—UAA, UAG, or UGA—enters the A site, translation terminates (Figure 12.11). These codons encode no amino acids, nor do they bind any tRNA. Rather, they bind a protein **release factor**, which causes a water molecule instead of an amino acid to attach to the forming protein.

The newly completed protein thereupon separates from the ribosome. Its C terminus is the last amino acid to join the chain. Its N terminus, at least initially, is methionine, as a consequence of the AUG start codon. Table 12.1 summarizes the initiation and termination of transcription and translation.

Regulation of Translation

As in any factory, the machinery of translation can work at varying rates. For example, externally applied chemicals such as some antibiotics can stop translation. The presence of more than one ribosome on an mRNA can speed up protein synthesis. And the endoplasmic reticulum (ER) can be used to segregate a protein as it is being made.

Some antibiotics work by inhibiting translation

Antibiotics are defense molecules produced by microorganisms such as certain bacteria and fungi. These substances often destroy other microbes, which might

Table 12.1	Signals That Start and Stop Transcription and Translation	
	TRANSCRIPTION	**TRANSLATION**
Initiation	Promoter sequence in DNA	AUG start codon in mRNA
Termination	Terminator sequence in DNA	UAA, UAG, or UGA stop codon in mRNA

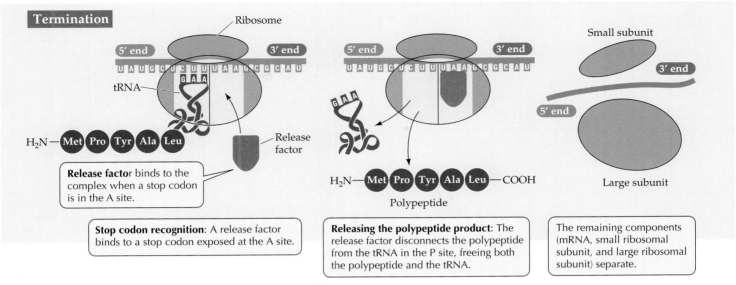

Termination

Release factor binds to the complex when a stop codon is in the A site.

Stop codon recognition: A release factor binds to a stop codon exposed at the A site.

Releasing the polypeptide product: The release factor disconnects the polypeptide from the tRNA in the P site, freeing both the polypeptide and the tRNA.

The remaining components (mRNA, small ribosomal subunit, and large ribosomal subunit) separate.

12.11 The Termination of Translation Translation terminates when the ribosome encounters a stop signal on the mRNA.

compete with the defender for nutrients. Since the 1940s, scientists have isolated increasing numbers of antibiotics, and physicians use them to treat a great variety of infectious diseases, ranging from bacterial meningitis to pneumonia to gonorrhea.

The key to antibiotic action is specificity: An antibiotic must work to destroy the microbial invader but not harm the human host. One way in which antibacterials accomplish this task is to block the synthesis of the bacterial cell wall, something that is essential to the microbe but that is not part of human biochemistry. Penicillin works in this fashion.

Another way is to inhibit bacterial protein synthesis. Recall that the bacterial ribosome is smaller and has a different collection of proteins than the eukaryotic ribosome has. Some antibiotics bind only to bacterial ribosomal proteins that are important in protein synthesis (Table 12.2). So, without the ability to make proteins, the bacterial invaders die and the infection is stemmed.

Polysome formation increases the rate of protein synthesis

Several ribosomes can work simultaneously at translating a single mRNA molecule to produce multiple molecules of the protein at the same time. As soon as the first ribosome has moved far enough from the initiation point, a second initiation complex can form, then a third, and so on. The assemblage of a thread of mRNA with its beadlike ribosomes and their growing polypeptide chains is called a polyribosome, or **polysome** (Figure 12.12).

A polysome is like a cafeteria line, where patrons follow each other, adding items to their trays. The person at the start has a little food (initiation); the person at the end has a complete meal (completed protein). Cells that are actively synthesizing proteins contain large numbers of polysomes and fewer free ribosomes or ribosomal subunits.

A signal sequence leads a protein through the ER

As a polypeptide chain forms on the ribosome, it spontaneously folds into its three-dimensional shape. As described in Chapter 3, this shape is determined by the particular order of amino acids that make up the protein, as well as factors such as their polarity and charge. Ultimately, this shape allows the polypeptide to interact with other molecules in the cell, such as a substrate if it acts as an enzyme. In addition to this structural information, the amino acid sequence contains information indicating where in the cell the polypeptide belongs.

As you learned in Chapter 4, an important difference between prokaryotes and eukaryotes is that eu-

Table 12.2	Antibiotics That Inhibit Bacterial Protein Synthesis
ANTIBIOTIC	**STEP INHIBITED**
Chloromycetin	Formation of peptide bonds
Erythromycin	Translocation of mRNA along ribosome
Neomycin	Interactions between tRNA and mRNA
Streptomycin	Initiation of translation
Tetracycline	Binding of tRNA to ribosome

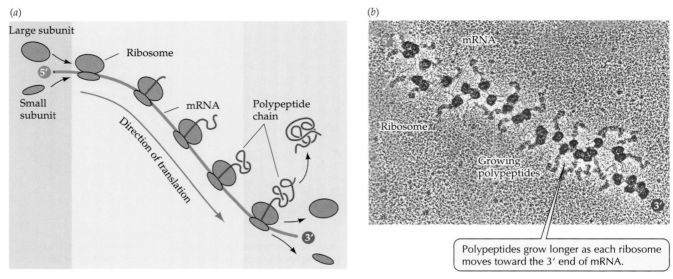

12.12 A Polysome (*a*) A polysome consists of ribosomes and their growing polypeptide chains moving in single file along an mRNA molecule. (*b*) An electron microscope gave us this detailed view of a polysome.

karyotic cells have many individual compartments. Different compartments need different proteins. Are proteins synthesized where they are needed, or are they transported from a synthesis site? How are particular proteins targeted to the correct site—electron transport chain components to the mitochondria, histones to the nucleus, and so forth?

Proteins that are to remain soluble within the cell are synthesized on "free" ribosomes—that is, ribosomes that are not attached to the endoplasmic reticulum (ER). Proteins that are to become parts of membranes, or are to be exported from the cell, or are to end up in lysosomes or peroxisomes are synthesized on ribosomes of the rough ER. All protein synthesis, however, *begins* on free ribosomes. The first few amino acids of a polypeptide chain determine whether production of the protein will be completed on the rough ER or on free ribosomes.

If a specific sequence of amino acids, the **signal sequence**, is present at the beginning of the chain, the finished product will be a membrane protein or a protein destined for export. The signal sequence attaches to a *signal recognition particle* composed of protein and RNA (Figure 12.13). This attachment blocks further protein synthesis until the ribosome can become attached to a specific receptor protein in the membrane of the ER. The receptor protein becomes a channel through which the growing polypeptide is extruded, either into the membrane itself or into the interior of the ER, as synthesis continues.

An enzyme within the ER interior then removes the signal sequence from the new protein, which ends up either built into the ER membrane or retained

within the ER rather than in the cytosol. From the ER the newly formed protein can be transported to its appropriate location—to other cellular compartments or to the outside of the cell—without mixing with other molecules in the cytoplasm (see Figure 12.13). Signals, consisting of amino acid sequences or sugars added in the ER and the Golgi apparatus, determine the cellular destination of a protein, much as postal zip codes direct mail.

Mutations: Heritable Changes in Genes

Accurate DNA replication, transcription, and translation all depend on the reliable pairing of complementary bases. Errors occur, though infrequently, in all three processes. In particular, errors in the DNA replication during the production of the gametes produce mutations. **Mutations** are heritable changes in genetic information.

Minute changes in the genetic material often lead to easily observable changes in the phenotype. Some effects of mutation in humans are readily detectable—dwarfism, for instance, or the presence of more than five fingers on each hand. A mutant genotype in a microorganism may be obvious if, for example, it results in a change in nutritional requirements, as we described for *Neurospora* (see Figure 12.1).

However, other mutations may be unobservable. In humans, for example, a particular mutation drastically lowers the level of an enzyme called glucose 6-phosphate dehydrogenase that is present in many tissues, including red blood cells. The red blood cells of a person carrying the mutant gene are abnormally sensitive

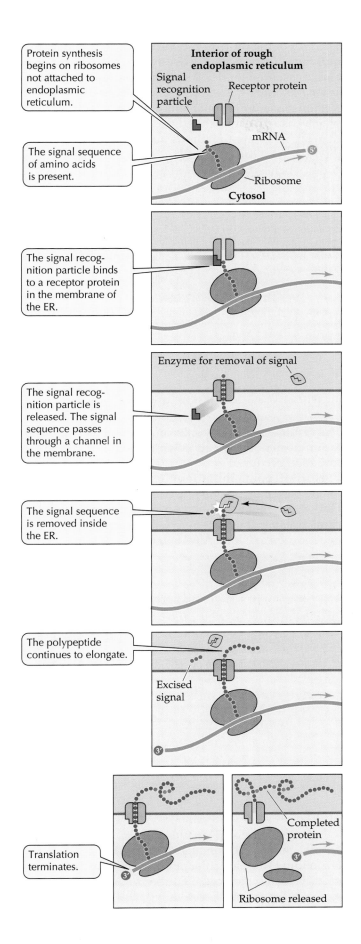

Protein synthesis begins on ribosomes not attached to endoplasmic reticulum.

The signal sequence of amino acids is present.

Interior of rough endoplasmic reticulum

Signal recognition particle

Receptor protein

mRNA

Ribosome

Cytosol

The signal recognition particle binds to a receptor protein in the membrane of the ER.

The signal recognition particle is released. The signal sequence passes through a channel in the membrane.

Enzyme for removal of signal

The signal sequence is removed inside the ER.

The polypeptide continues to elongate.

Excised signal

Translation terminates.

Completed protein

Ribosome released

12.13 A Signal Sequence Moves a Polypeptide into the ER When a signal sequence of amino acids is present at the beginning of the chain, the polypeptide will be taken into the endoplasmic reticulum and elongation completed there. The finished protein is thus segregated from the cytosol.

to an antimalarial drug called primaquine; when such people are treated with this drug, their red blood cells rupture, causing serious medical problems. People with the normal allele have no such problem. Before the drug came into use, no one was aware that such a mutation existed. Similarly, distinguishing a mutant bacterium from a normal bacterium may require sophisticated chemical methods, not just visual inspection.

Some mutations cause their phenotypes only under certain restrictive conditions and are not detectable under other, permissive conditions. We call organisms carrying such mutations *conditional mutants*. Many conditional mutants are temperature-sensitive, unable to grow at a particular restrictive temperature, such as 37°C, but able to grow normally at a lower, permissive temperature, such as 30°C. The mutant allele in such an organism may code for an enzyme with an unstable tertiary structure that is altered at the restrictive temperature.

All mutations are alterations in the nucleotide sequence in DNA. We divide mutations into two categories: point mutations and chromosomal mutations. **Point mutations** are mutations of single genes: One allele becomes another because of small alterations in the sequence or number of nucleotides—even as small as the substitution of one nucleotide for another. **Chromosomal mutations** are more extensive alterations. They may change the position or direction of a DNA segment without actually removing any genetic information, or they may cause a segment of DNA to be irretrievably lost. Both point mutations and chromosomal mutations are heritable.

Point mutations may be silent, missense, nonsense, or frame-shift

Many point mutations consist of the substitution of one base for another in the DNA and hence in the mRNA. Because of the degeneracy of the genetic code, some of these mutations result in no change in amino acids after the altered mRNA is translated; for this reason they are called **silent mutations**.

For example, four mRNA codons code for the amino acid proline: CCA, CCC, CCU, and CCA (see Figure 12.5). If the template strand of DNA for this particular region has the sequence GGC, it will be transcribed to CCG in mRNA and proline-charged *tRNA will bind at the ribosome*. But if there is a mutation in the DNA such that the triplet in the template DNA reads GGA, the mRNA codon will be CCU—and the tRNA that binds will still bring proline:

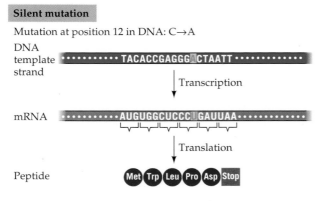

Silent mutation

Mutation at position 12 in DNA: C→A

Result: No change in amino acid sequence

Silent mutations are quite common and account for genetic diversity that is not expressed as phenotypic differences.

In contrast to silent mutations, some base substitution mutations may change the genetic message such that one amino acid substitutes for another in the protein. This is a **missense mutation**:

Missense mutation

Mutation at position 14 in DNA: T→A

Result: Amino acid change at position 5: Asp → Val

A specific example of such a mutation is the sickle allele for human β-globin. Sickle-cell anemia is the consequence of a recessive allele that, when homozygous, results in defective red blood cells. Where oxygen is abundant, as in the lungs, the cells are normal in structure and function. But at the low oxygen levels characteristic of working muscles, the red blood cells collapse into the shape of a sickle (Figure 12.14).

The disease results from a defect in hemoglobin, a protein that carries oxygen. One of the polypeptides in hemoglobin differs by one amino acid between normal and sickle-cell hemoglobin. A missense mutation such as this may sometimes cause the protein not to function, but often the effect is only to reduce the functional

efficiency of the protein. Individuals carrying missense mutations may survive, even though the affected protein is essential to life. Through evolution, some missense mutations even improve functional efficiency.

Nonsense mutations, another type of mutation in which bases are substituted, are more often disruptive than are missense mutations. In a nonsense mutation, the base substitution causes a chain terminator (stop) codon, such as UAG, to form in the mRNA product.

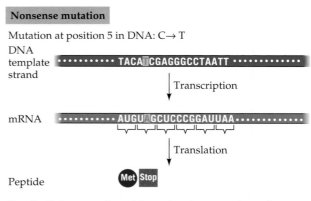

Nonsense mutation

Mutation at position 5 in DNA: C→ T

Result: Only one amino acid translated; no protein made

The result is a shortened protein product, since translation does not proceed beyond the point where the mutation occurred.

Not all point mutations are base substitutions. Single base pairs may be inserted into or deleted from DNA. Such mutations are known as **frame-shift mutations** because they interfere with the decoding of the genetic message by throwing it out of register:

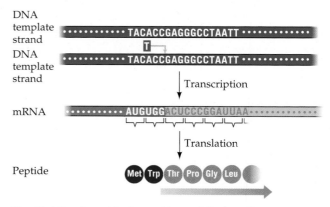

Frame-shift mutation

Mutation by insertion of T between bases 6 and 7 in DNA

Result: All amino acids changed beyond the insertion

Think again of codons as three-letter words, each corresponding to a particular amino acid. Translation proceeds codon by codon; if a base is added to the mes-

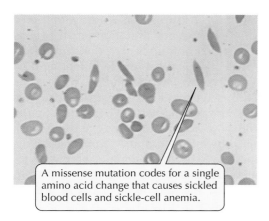

A missense mutation codes for a single amino acid change that causes sickled blood cells and sickle-cell anemia.

12.14 Sickled Blood Cells   This misshapen red blood cell is caused by a mutation that substitutes an incorrect amino acid in one of the two polypeptides of hemoglobin.

sage or subtracted from it, translation proceeds perfectly until it comes to the one-base insertion or deletion. From that point on, the three-letter words in the message are one letter out of register. In other words, such mutations shift the "reading frame" of the genetic message. Frame-shift mutations almost always lead to the production of completely nonfunctional proteins.

Chromosomal mutations are extensive changes

Genetic strands can break and rejoin, grossly disrupting the sequence of genetic information. There are four types of such chromosomal mutations: deletions, duplications, inversions, and translocations (Figure 12.15).

Deletions remove part of the genetic material (Figure 12.15a). Like frame-shift point mutations, they cause death unless they affect unnecessary genes or are masked by the presence, in the same cell, of normal copies of the deleted genes. It is easy to imagine one mechanism that could produce deletions: A DNA molecule might break at two points and the two end pieces might rejoin, leaving out the DNA between the breaks.

Another mechanism by which deletion mutations might arise would lead simultaneously to the production of a second kind of chromosomal mutation: a **duplication** (Figure 12.15b). Duplication would arise if homologous chromosomes broke at

12.15 Chromosomal Mutations   Chromosomes may break during replication, and parts of chromosomes may then rejoin incorrectly. Letters on the colored chromosomes distinguish segments and identify consequences of duplications, deletions, inversions, and reciprocal translocations.

different positions and then reconnected to the wrong partners. One of the two molecules produced by this mechanism would lack a segment of DNA (it would have a deletion), and the other would have two tandem copies (a duplication) of the information that was deleted from the first.

Breaking and rejoining can also lead to **inversion**—the removal of a segment of DNA and its reinsertion into the same location, but "flipped" end for end so that it runs in the opposite direction (Figure 12.15c). If an inversion includes part of a segment of DNA that codes for a protein, the resulting protein will be drastically altered and almost certainly nonfunctional.

The fourth type of chromosomal mutation, called **translocation**, results when a segment of DNA breaks, moves from a chromosome, and is inserted into a different chromosome. Translocations may be reciprocal, as in Figure 12.15d, or nonreciprocal, as the mutation involving duplication and deletion in Figure 12.15b illustrates. Translocations can make synapsis in meiosis difficult and thus sometimes lead to aneuploidy (the lack or excess of chromosomes; see Chapter 9).

Some chemicals induce mutations and cancer

Mutagens are agents that cause mutations. Among the chemical mutagens are *base analogs*—purines or pyrimidines that are not found in natural DNA but are enough like the natural bases that they can be incorporated into DNA. Base analogs are mutagenic presumably because they are more likely than natural DNA bases to mispair. For example, 5-bromouracil is very

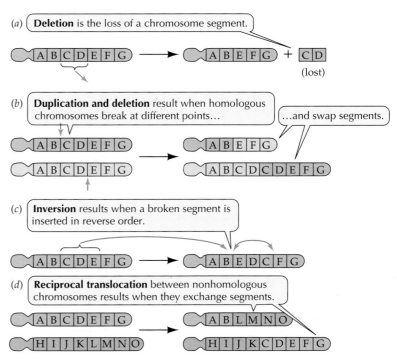

(a) **Deletion** is the loss of a chromosome segment.

(b) **Duplication and deletion** result when homologous chromosomes break at different points… …and swap segments.

(c) **Inversion** results when a broken segment is inserted in reverse order.

(d) **Reciprocal translocation** between nonhomologous chromosomes results when they exchange segments.

similar to thymine, and it is easily incorporated into DNA in place of thymine:

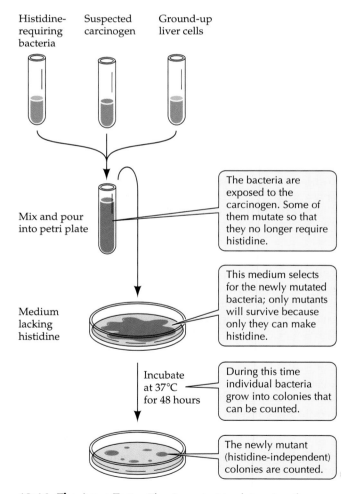

Thymine
(Base)

5-Bromouracil
(Base analog)

But 5-bromouracil is much more likely than thymine to engage in an abnormal pairing with guanine, and it therefore induces mutations of AT to GC and GC to AT. Thus 5-bromouracil is a potent chemical mutagen.

Cancerous cells have lost control of their cell cycle (see Chapter 9). Biomedical scientists have traced many kinds of cancers to mutations in particular genes, as we will see in Chapter 17. In addition, many substances present in polluted air, tobacco smoke, and foods we eat are known to be chemical **carcinogens**—cancer producers. Many known carcinogens either are mutagens or are converted to mutagens by enzymes in the endoplasmic reticulum.

The most widely used method to determine whether a substance is carcinogenic is called the **Ames test**, named for its inventor Bruce Ames and based on the idea that liver cells convert carcinogens to mutagens. To perform this test, we combine the suspected carcinogenic compound with a suspension of ground-up liver cells (the source of modifying enzymes) and mutant bacterial cells that cannot grow in the absence of, say, a particular amino acid. We then look for the appearance of bacteria that can grow in the absence of that amino acid (Figure 12.16). Such bacteria must result from a mutation that reverses the original mutation; thus their presence indicates that the compound added to the suspension is mutagenic. There is a correlation between activity in the Ames test and tumor-forming activity: If the Ames test indicates that a substance is mutagenic, that substance may also be carcinogenic. Thus we can test many compounds for carcinogenic activity without unnecessarily sacrificing the lives of large numbers of rodents or other test animals.

Mutations are the raw material of evolution

Without mutation, there would be no evolution. As we will see in Part Three, mutation does not drive evolution, but it provides the genetic diversity on which natural selection and other agents of evolution act.

All mutations are rare events, but mutation frequencies vary from organism to organism and from gene to gene within a given organism. The frequency of mutation is usually much lower than one mutation per 10^4 genes per DNA duplication, and sometimes the frequency is as low as one mutation per 10^9 genes per duplication. Most mutations are point mutations in which one nucleotide is substituted for another during the synthesis of a new DNA strand.

Most mutations harm the organism that carries them, and some are neutral (they have no effect on the organism's ability to survive or produce offspring). Once in a while, however, a mutation improves an organism's adaptation to its environment or becomes favorable when environmental variables change. Duplication mutations may be the source of "extra" genes. Most of the complex creatures living on Earth have more DNA and therefore more genes than the simpler creatures do. Humans, for example, have 1,000 times more genetic material than prokaryotes have.

How do new genes arise? If whole genes were sometimes duplicated by the mechanism described in the previous section, the bearer of the duplication

Histidine-
requiring
bacteria

Suspected
carcinogen

Ground-up
liver cells

Mix and pour
into petri plate

The bacteria are exposed to the carcinogen. Some of them mutate so that they no longer require histidine.

Medium
lacking
histidine

This medium selects for the newly mutated bacteria; only mutants will survive because only they can make histidine.

Incubate
at 37°C
for 48 hours

During this time individual bacteria grow into colonies that can be counted.

The newly mutant (histidine-independent) colonies are counted.

12.16 The Ames Test   The Ames test to determine the ability of a suspected carcinogen to cause mutations works because liver cells convert carcinogens to mutagens.

would have a surplus of genetic information that might be turned to good use. Subsequent mutations in one of the two copies of the gene might not have an adverse effect on survival, because the other copy of the gene would continue to produce functional protein. The extra gene might mutate over and over again without ill effect because its function would be fulfilled by the original copy.

If the random accumulation of mutations in the extra gene led to the production of a useful protein (for example, an enzyme with an altered specificity for the substrates it binds, allowing it to catalyze different—but related—reactions), natural selection would tend to perpetuate the existence of this new gene. New copies of genes also arise through the activity of transposable elements, which are discussed in Chapters 13 and 14.

Summary of "From DNA to Protein: Genotype to Phenotype"

Genes and the Synthesis of Polypeptides

• Genes are made up of DNA and are expressed in the phenotype as polypeptides (proteins).
• Certain genetic diseases in humans were initially found to be caused by abnormal alleles expressed as defective enzymes. This discovery led to the one-gene, one-polypeptide hypothesis.
• Experiments with the bread mold *Neurospora* confirmed the one-gene, one-polypeptide hypothesis. Mutant strains of this haploid mold were found to be blocked, each at a specific enzymatic step, along a biochemical pathway. **Review Figure 12.1**

DNA, RNA, and the Flow of Information

• A gene is expressed in two steps: First, DNA is transcribed to RNA; then RNA is translated into protein. RNA differs from DNA in three ways: It is single-stranded, its sugar molecule is ribose rather than deoxyribose, and its fourth base is uracil rather than thymine.
• The central dogma of molecular biology is DNA → RNA → protein. In retroviruses, the rule for transcription is reversed: RNA → DNA. Other RNA viruses exclude DNA altogether, going directly from RNA to protein. **Review Figures 12.2, 12.3**

Transcription: DNA-Directed RNA Synthesis

• RNA is transcribed from a DNA template strand after the bases of DNA are exposed by unwinding of the double helix.
• In a given region of DNA, only one of the two strands (the template strand) can act as a template for transcription.
• RNA polymerase catalyzes transcription from the template strand of DNA.

• The initiation of transcription requires that RNA polymerase recognize and bind tightly to a special promoter sequence on DNA. RNA is synthesized (elongates) in a 5'-to-3' direction, reading in an antiparallel (3'-to-5') way from the template DNA. Special sequences and protein helpers terminate transcription. **Review Figure 12.4**

The Genetic Code

• The genetic code is present in the RNA transcripts (mRNA) that contain information for protein synthesis.
• The code is read sequentially, as nonoverlapping triplets of nucleotides (codons). Since there are four bases, there are 64 possible codons.
• One codon indicates the start of translation (protein synthesis) and codes for methionine. Three codons indicate the end of translation. The other 60 codons code only for particular amino acids.
• Since there are only 20 different amino acids, the genetic code is degenerate; that is, there is more than one codon for many amino acids. But the code is not ambiguous: A single codon does not specify more than one amino acid. **Review Figure 12.5**

The Key Players in Translation

• In prokaryotes, translation begins before the mRNA is completed. In eukaryotes, transcription occurs in the nucleus and translation occurs in the cytoplasm.
• Translation requires three special components to function properly: tRNAs, activating enzymes, and ribosomes.
• In translation, amino acids are linked in an order specified by the triplet codons in mRNA. This task is achieved by an adapter, transfer RNA (tRNA), which binds the correct amino acid and has an anticodon complementary to the mRNA codon. **Review Figure 12.6**
• The aminoacyl-tRNA synthetases, a family of activating enzymes, attach specific amino acids to their appropriate tRNAs. Each amino acid–identifying codon has a specific enzyme. **Review Figure 12.7**
• The mRNA with its DNA-directed codons meets the tRNAs with their amino acids at the ribosome, which contains most of the molecules that catalyze events in translation. The ribosome has two tRNA-binding sites: The P site binds tRNA containing the growing polypeptide; the A site binds the tRNA carrying the next amino acid. **Review Figure 12.8**

Translation: RNA-Directed Polypeptide Synthesis

• An initiation complex consisting of an amino acid–charged tRNA and a small ribosomal subunit bound to mRNA triggers the beginning of translation. **Review Figure 12.9**
• Polypeptides grow from the N terminus toward the C terminus. The ribosome moves along the mRNA one triplet codon at a time. **Review Figure 12.10**
• The presence of a stop codon in the A site of the ribosome causes translation to terminate. **Review Figure 12.11**

Regulation of Translation

• Some antibiotics work by blocking events in translation. **Review Table 12.2**
• In a polysome, more than one ribosome moves along the mRNA at one time. **Review Figure 12.12**

• As polypeptides are made, their amino acid sequences contain the information to fold them into three-dimensional shapes. Amino acid sequences also contain information on the cellular destination of proteins. All protein synthesis begins on free ribosomes in the cytoplasm. However, if translation produces a signal sequence of amino acids, the ribosomal complex binds to the outside of the ER. The signal sequence routes proteins into the lumen of the endoplasmic reticulum, which is the first destination of proteins headed to membrane-enclosed organelles. **Review Figure 12.13**

Mutations: Heritable Changes in Genes

• Mutations in DNA are often expressed as abnormal proteins. However, the result may not be easily observable phenotypic changes. Some mutations appear only under certain conditions, such as exposure to a certain environmental agent (such as a drug) or condition (such as temperature).
• Point mutations (silent, missense, nonsense, or frame-shift) result from alterations in single base pairs of DNA. **Review Page 276**
• Chromosomal mutations (deletions, duplications, inversions, or translocations) involve large regions of a chromosome. **Review Figure 12.15**
• Some chemicals induce mutations and may also cause cancer. These substances may chemically alter bases or substitute for them. The Ames test uses bacteria and liver enzymes to rapidly screen for mutagens (and hence carcinogens). **Review Figure 12.16**
• Mutation is essential to evolution: It is the source of the genetic variation on which natural selection and other evolutionary agents act.

Self-Quiz

1. Which of the following is *not* a difference between RNA and DNA?
   a. RNA has uracil and DNA has thymine.
   b. RNA has ribose and DNA has deoxyribose.
   c. RNA has 5 bases and DNA has 4.
   d. RNA is a single polynucleotide strand and DNA is a double strand.
   e. RNA is relatively smaller than human chromosomal DNA.

2. A *Neurospora* mutant deficient in the enzyme that catalyzes the conversion of citrulline to arginine (see the pathway of arginine synthesis in Figure 12.1) would grow on
   a. minimal medium.
   b. minimal medium + ornithine.
   c. minimal medium + citrulline.
   d. minimal medium + arginine.
   e. both *b* and *c*.

3. A region of DNA template strand has the sequence 3'-ATTCGC-5'. What is the sequence of RNA transcribed from this DNA?
   a. 3'-AUUCGC-5'
   b. 3'-TAAGCG-5'
   c. 5'-UAAGCG-3'
   d. 5'-AUUCGC-3'
   e. 5'-ATTCGC-3'

4. Which of the following is *not* a type of chromosome mutation?
   a. duplication
   b. inversion
   c. deletion
   d. frame-shift
   e. translocation

5. At a certain location in a gene, the template strand of DNA has the sequence GAA. A mutation alters the triplet to GAG. This type of mutation is called
   a. silent.
   b. missense.
   c. nonsense.
   d. frame-shift.
   e. translocation.

6. Transcription
   a. produces only mRNA.
   b. requires ribosomes.
   c. requires tRNAs.
   d. produces RNA growing from the 5' end to the 3' end.
   e. takes place only in eukaryotes.

7. Which statement about translation is *not* true?
   a. It is RNA-directed polypeptide synthesis.
   b. An mRNA molecule can be translated by only one ribosome at a time.
   c. The same genetic code operates in all organisms.
   d. Any ribosome can be used in the translation of any mRNA.
   e. There are both start and stop codons.

8. Which statement is *not* true?
   a. Transfer RNA functions in translation.
   b. Ribosomal RNA functions in translation.
   c. RNAs are produced in transcription.
   d. Messenger RNAs are produced on ribosomes.
   e. DNA codes for mRNA, tRNA, and rRNA.

9. The genetic code
   a. is different for prokaryotes and eukaryotes.
   b. has changed during the course of recent evolution.
   c. has 64 codons that code for amino acids.
   d. is degenerate.
   e. is ambiguous.

10. A mutation that results in the codon UAG where there had been UGG
    a. is a nonsense mutation.
    b. is a missense mutation.
    c. is a frame-shift mutation.
    d. is a large-scale mutation.
    e. is unlikely to have a significant effect.

Applying Concepts

1. The genetic code is described as degenerate. What does this mean? How is it possible that a point mutation, consisting of the replacement of a single nitrogenous base in DNA by a different base, might not result in an error in protein production?

2. Har Gobind Khorana, at the University of Wisconsin, synthesized artificial mRNAs such as poly CA (CACACA…) and poly CAA (CAACAACAA…). He found that poly CA codes for a polypeptide consisting of threonine (Thr) and histidine (His), in alternation (His–Thr–His–Thr…). There are two possible codons in poly CA, CAC and ACA. One of these must code for histidine and the other for threonine—but which is which? The answer comes from results with poly CAA, which produces three different polypeptides: poly Thr, poly Gln (glutamine), and poly Asn (asparagine). (An artificial messenger can be read, inefficiently, beginning at any point in the chain; there is no specific initiator region.) Thus poly CAA can be read as a polymer of CAA, of ACA, or of AAC. Compare the results of the poly CA and poly CAA experiments, and determine which codon codes for threonine and which for histidine.

3. Look back at Question 2. Using the genetic code (Figure 12.5) as a guide, deduce what results Khorana would have obtained had he used poly UG and poly UGG as artificial messengers. In fact, very few such artificial messengers would have given useful results. For an example of what could happen, consider poly CG and poly CGG. If poly C were the messenger, a mixed polypeptide of arginine and alanine (Arg–Ala–Ala–Arg . . .) would be obtained; poly CGG would give three polypeptides: poly Arg, poly Ala, and poly Gly (glycine). Can any codons be determined from only these data? Explain.

4. Errors in transcription occur about 100,000 times as often as do errors in DNA replication. Why can this high rate be tolerated in RNA synthesis but not in DNA synthesis?

Readings

Cooper, G. M. 1997. *The Cell: A Molecular Approach*. ASM Press, Washington, DC, and Sinauer Associates, Sunderland, MA. Chapter 7 is a concise, up-to-date treatment of protein synthesis.

Griffiths, A. J. F., J. H. Miller, D. T. Suzuki, R. C. Lewontin and W. M. Gelbart. 1996. *An Introduction to Genetic Analysis*, 6th Edition. W. H. Freeman, New York. Genetics texts, and this is one of the best, have descriptions of how the one-gene, one-polypeptide hypothesis was developed, as well as the use of genetics in deciphering the genetic code.

Hill, W. E., E. A. Dahlberg, R. Garrett, P. B. Moore, D. Schlesinger and J. R. Warner. 1990. *The Ribosome: Structure, Function and Evolution*. American Society for Microbiology, Washington, DC. A detailed summary of knowledge of this organelle and its role in protein synthesis.

Judson, H. F. 1996. *The Eighth Day of Creation: Makers of the Revolution in Biology*, Expanded Edition. CSHL Press, Plainview, NY. A sparkling history of molecular biology, including the discovery of messenger RNA.

Lodish, H., D. Baltimore, A. Berk, S. L. Zipursky, P. Matsudaira and J. Darnell. 1995. *Molecular Cell Biology*, 3rd Edition. Scientific American Books, New York. Superb, detailed chapters on protein synthesis.

Saks, M. E., J. R. Sampson and J. N. Abelson. 1995. "The Transfer RNA Identity Problem." *Science*, vol. 263, pages 191–197. A look at how the correct amino acid becomes attached to the correct tRNA.

Stent, G. S. and R. Calendar. 1978. *Molecular Genetics*, 2nd Edition. W. H. Freeman, New York. A brilliant technical and historical introduction to molecular genetics and the role of DNA.

Upton, A. C. 1982. "The Biological Effects of Low-Level Ionizing Radiation." *Scientific American*, February. A discussion of how radiation leads to mutations.

Watson, J. D., M. Gilman, J. Witkowski and M. Zoller. 1992. *Recombinant DNA*, 2nd Edition. W. H. Freeman, New York. A superbly written outline of molecular biology, including protein synthesis.

Chapter 13

The Genetics of Viruses and Prokaryotes

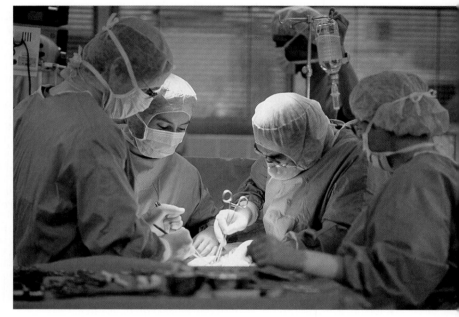

Are There Uninvited Guests Here?
A masked team performs surgery on a patient. But have harmful, drug-resistant bacteria invaded the surgical suite?

*A*member of her university's cross-country team, Janet entered the hospital just after final exams for some long-delayed surgery on a tendon in her knee. The tendon repair went well, but she left the hospital with something new: Bacteria called *Pseudomonas aeruginosa* had infected the surgical wounds. The antibiotics typically used to kill these bacteria did not work. She ended up back in the hospital two weeks later, where she received intensive antibiotic therapy and ultimately recovered.

Janet developed what is called a *nosocosomial infection*—an infection acquired as a result of a hospital stay. Why would a hospital, which we think of as a place to get better, sometimes—in fact for about 10% of all patients—be a place where we get sick? Of course, the stresses of Janet's surgery could have reduced her immunity to the bacteria that are common everywhere in our environment; but, increasingly, the heavy use of antibiotics in hospitals makes them breeding grounds for bacteria that have genes for resistance to these antibiotics.

What's going on here? How have bacteria acquired antibiotic resistance so rapidly, and how do they pass that acquired resistance along to other bacteria? To answer these questions, we will discuss remarkable pieces of DNA known as R factors. But first we need to introduce the ways in which viruses and prokaryotes undergo genetic recombination and how their genetic information is organized.

Viruses are not cells but intracellular parasites that can reproduce only within living cells. As we saw in Chapter 11, the molecular biology of bacterial viruses has been important in determining the principles of molecular biology. Viruses have also been useful in genetic research. In this chapter we will examine the structures, classification, reproduction, and genetics of viruses.

Bacteria and other prokaryotes are a lot more complex than viruses, but still are relatively simple when compared to multicellular eukaryotes. Their genetics, too, has been extensively studied. The elegance of transcriptional controls in prokaryotes has provided valuable lessons for the more complicated situation in eukaryotes. In this chapter we will explore reproduction, mating, and genetic recombination in bacteria. We will see how bacteria acquire genes from other prokaryotes, from viruses, and from naked nucleic acids.

Using Prokaryotes and Viruses for Genetic Experiments

Prokaryotes such as *Escherichia coli* and viruses such as bacteriophages have often been easier to study than eukaryotes. What are the advantages of working with prokaryotes and viruses?

First, it is easier to work with small amounts of DNA than with large amounts. A typical bacterium contains about 1/1,000 as much DNA as a single human cell, and a typical bacterial virus contains about 1/100 as much DNA as a bacterium. Second, data on large numbers of organisms can be obtained easily from prokaryotes, but not from most eukaryotes. A single milliliter of medium can contain more than 10^9 *E. coli* cells or 10^{11} bacteriophages. In addition, most prokaryotes grow rapidly. A culture of *E. coli* can be grown under conditions that allow cells to double every 20 minutes. By contrast, 10^9 mice would cost more than 10^9 dollars and would require a cage that would cover about 3 square miles, and growth of a generation of mice takes about 3 months instead of 20 minutes. Third, prokaryotes and viruses are usually haploid, facilitating genetic analyses.

The ease of growing and handling bacteria and their viruses permitted the explosion of genetics and molecular biology that began shortly after the mid-twentieth century (you read about some of these discoveries in Chapters 11 and 12). The relative biological simplicity of bacteria and bacteriophages contributed immeasurably to the discoveries about the genetic material, the replication of DNA, and the mechanisms of gene expression. Later these bacteria and bacteriophages were the first subjects of recombinant-DNA technology (see Chapter 16).

Questions of interest to all biologists continue to be studied in prokaryotes, and prokaryotes continue to be important tools for biotechnology and for research on eukaryotes.

Viruses: Reproduction and Recombination

When Oswald Avery and his colleagues showed that DNA is the hereditary material of bacteria, the finding did not create a sensation. The results of the Hershey–Chase experiment, on the other hand, which supported the hypothesis that DNA is the hereditary material of certain viruses, captured the imagination of scientists, including James Watson and Francis Crick. What made the difference? In the decade between those two discoveries, scientists had shown that bacteria and viruses have genes, and that they undergo genetic recombination. These discoveries all increased the value of using bacteria and viruses for genetic study.

Although there are many kinds of viruses, they are usually composed of just nucleic acid and a few proteins, and they have relatively simple means of infecting their targeted host cells. Some viruses—the best studied are certain bacteriophages—can infect a cell but postpone reproduction until conditions are favorable. In these cases, we will discover how viral genetic information is inserted into the host chromosome. Finally, we will describe the simplest infective agents: *viroids*, which are made up only of genetic material, and *prions*, which apparently have no nucleic acid but are made entirely of protein.

Scientists studied viruses before they could see them

Most viruses are much smaller than most bacteria (Table 13.1). Viruses have become well understood only within the last half century, but the first step on this path of discovery was taken by the Russian botanist Dmitri Ivanovsky in 1892. He was trying to find the cause of *tobacco mosaic disease*, which results in the destruction of photosynthetic tissues and can devastate a tobacco crop. Ivanovsky passed an infectious extract of diseased tobacco leaves through a fine porcelain filter, a technique that had been used previously by physicians and veterinarians to isolate disease-causing bacteria, which would stay on the filter.

To Ivanovsky's surprise, the disease agent in this case did not stick to the filter: It passed through, and the liquid filtrate still caused tobacco mosaic disease. Clearly, the agent was smaller than a bacterium. Other

TABLE 13.1 Common Sizes of Microorganisms

MICROORGANISM	TYPE	TYPICAL SIZE RANGE (μm^3)
Protists	Eukaryote	5,000–50,000
Photosynthetic bacteria	Prokaryote	5–50
Spirochetes	Prokaryote	0.1–2
Mycoplasmas	Prokaryote	0.01–0.1
Poxviruses	Virus	0.01
Influenza virus	Virus	0.0005
Poliovirus	Virus	0.00001

workers soon showed that similar filterable agents, so tiny that they cannot be seen under the light microscope, cause several plant and animal diseases. And alcohol, which kills bacteria, does not destroy the ability of these tiny agents to cause disease.

In 1935, Wendell Stanley, working at what is now Rockefeller University, was the first to succeed in crystallizing viruses. The crystalline viral preparation became infectious again when it was dissolved. It was soon shown that crystallized viral preparations consist of protein and nucleic acid. Finally, direct observation of viruses with electron microscopes showed clearly how much they differ from bacteria and other organisms.

Viruses reproduce only with the help of living cells

Unlike the organisms that make up the six taxonomic kingdoms of the living world, viruses are **acellular**; that is, they are not cells and do not consist of cells. Unlike cellular creatures, viruses do not metabolize energy—they neither produce ATP nor conduct fermentation, cellular respiration, or photosynthesis.

Whole viruses never arise directly from preexisting viruses. They are *obligate intracellular parasites*; that is, they develop and reproduce only within the cells of specific hosts. The cells of animals, plants, fungi, protists, and prokaryotes (both bacteria and archaea) serve as hosts to viruses. When they reproduce, viruses usually destroy the host cells, releasing progeny viruses that then seek new hosts. Many diseases of humans, animals, and plants are caused by viruses.

Because they lack the distinctive cell wall and ribosomal biochemistry of bacteria, viruses are not affected by antibiotics. Few specific drugs affect viruses.

The best way a person can stem a viral infection is to rely on the immune system (we will describe how this works in Chapter 18). The best way to prevent the spread of a viral infection in a population is to contain it. If there are no nearby susceptible hosts, the viral infection will run its course and the particles will die because they have no place to reproduce. This was the strategy used to eliminate smallpox, which used to be a worldwide scourge and is now extinct. Currently an effort is under way to do the same for polio.

Viruses outside of host cells exist as individual particles called **virions** (Figure 13.1). The virion, the basic unit of a virus, consists of a central core of either DNA or RNA (but not both) surrounded by a **capsid**, or coat, which is composed of one or more proteins. The way in which these proteins are assembled gives the virion a characteristic shape. In addition, many animal viruses have a lipid and protein membrane acquired from host cell membranes in the course of viral reproduction and release. Many bacterial viruses (bacteriophages) have specialized "tails" made of protein. The complex architecture of HIV, the AIDS virus, is shown in Figure 13.2.

There are many kinds of viruses

A common way to classify viruses separates them by whether they have DNA or RNA and then by whether their nucleic acid is single- or double-stranded. Some of the RNA viruses have more than one molecule of RNA, and the DNA of one virus family is circular. Further levels of classification depend on factors such as the overall shape of the virus and the symmetry of the capsid.

Most capsids may be categorized as **helical** (coiled like a spring; Figure 13.1*a*), **icosahedral** (a regular solid

(*a*)

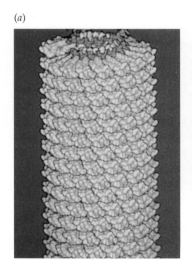

(*b*)

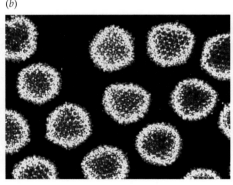

(*c*)

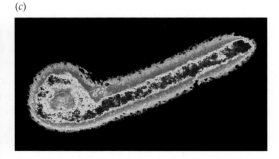

13.1 Virions Come in Various Shapes   (*a*) A computer model of the tobacco mosaic virus, a plant virus. The model shows only about one-seventh of this long virus, which consists of an inner helix of RNA covered with a helical array of protein molecules. (*b*) Many animal viruses, such as the adenovirus modeled here, have an icosahedral (20-sided) capsid as an outer shell. Inside the shell is a spherical mass of proteins and DNA. (*c*) Not all virions are regularly shaped. Wormlike virions of the Ebola virus infect humans, causing internal hemorrhaging that is usually fatal.

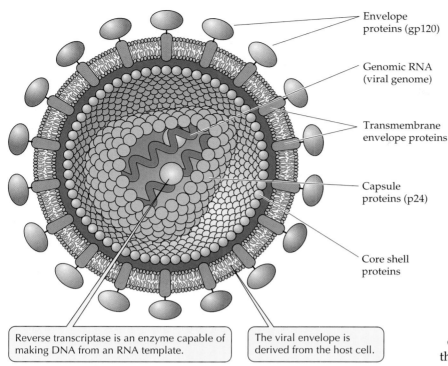

Envelope proteins (gp120)

Genomic RNA (viral genome)

Transmembrane envelope proteins

Capsule proteins (p24)

Core shell proteins

Reverse transcriptase is an enzyme capable of making DNA from an RNA template.

The viral envelope is derived from the host cell.

13.2 The Structure of HIV, the AIDS Virus HIV has a core containing two molecules of genomic RNA (it is diploid) and various proteins, including the enzyme reverse transcriptase. Another protein (shown in blue) surrounds the core. The surface of the virus is complex.

with 20 faces; Figure 13.1*b*), or **binal** (having a polyhedral, or many-faced, head and a helical tail; see the structure of the T2 bacteriophage in Figure 11.2). Another level of classification is based on the presence or absence of a membranous envelope around the virion; still further subdivision is based on capsid size.

The first level of virus classification is based on the type of host. Let's see how reproductive cycles and other properties vary among the major groups of viruses: bacterial, animal, and plant viruses.

Bacteriophages reproduce by either a lytic cycle or a lysogenic cycle

Viruses that parasitize bacteria are known as **bacteriophages**. Bacteriophages recognize their hosts by means of a specific interaction between the proteins of the bacteriophage and the molecules of the host bacteria's cell wall. The virions, which also must penetrate cell walls, are often equipped with tail assemblies that inject their nucleic acid through the cell wall into the host bacterium or archaeon. After the phage has injected its nucleic acid into the host, one of two things happens, depending on the kind of phage.

We saw one type of viral reproductive cycle when we studied the Hershey–Chase experiment (see Figure 11.3). That was the **lytic cycle**, in which the infected bacterium lyses (bursts), releasing progeny phage. In the **lysogenic cycle** the infected bacterium does not lyse, but instead harbors the viral nucleic acid for many rounds of bacterial cell division before one or more of the progeny lyse.

THE LYTIC CYCLE. Phages that reproduce only by the lytic cycle are called **virulent viruses**. The phage attacks a prokaryotic cell by specific combination of part of the capsid with a receptor protein on the cell surface (Figure 13.3). The phage injects its nucleic acid into the host cell, where the phage nucleic acid takes over the host's synthetic machinery, making it copy the phage nucleic acid and produce the phage proteins. In one clever mechanism for this process, the phage genetic material has a promoter sequence that attracts the host's RNA polymerase. One of the proteins made by the translation of phage mRNAs destroys the host cell's own DNA, thus removing a potential competitor for the viral nucleic acid.

As synthesis proceeds, viral parts accumulate within the infected cell. The new nucleic acid molecules are loaded into the "head" proteins, and the other proteins (tails, tail fibers) join the complex. At this point the host cell is teeming with new phages. One of the phage genes encodes an enzyme that attacks the host cell wall, causing lysis and the release of many new infectious phages. The whole process—from attachment and infection to lysis of the host—takes about half an hour.

THE LYSOGENIC CYCLE. Phage infection does not always result in lysis of the bacteria. Some phages seem to disappear from the culture, leaving the bacteria immune to further attack by the same strain of phage. In such cultures, however, a few free phages are always present. Bacteria harboring phages that are not lytic are called **lysogenic**, and the phages are called **temperate viruses**. When lysogenic bacteria are combined with other bacteria that are sensitive to the phages, they cause the sensitive cells to lyse.

Experiments revealed that the lysogenic bacteria contain a noninfective entity called a prophage. A **prophage** is a molecule of phage DNA that has been integrated into the bacterial chromosome (Figure 13.4). As part of the bacterial genome, the prophage can remain quiet within bacteria through many cell divisions. However, an occasional lysogenic cell can be in-

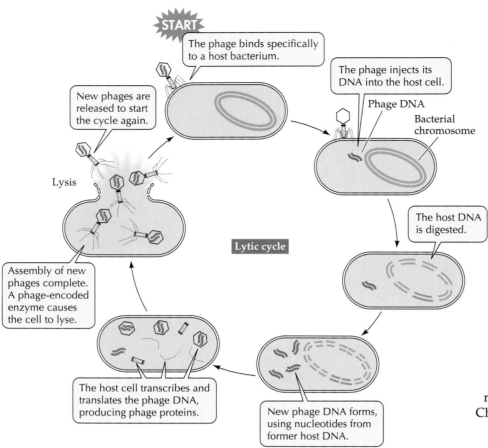

START

The phage binds specifically to a host bacterium.

The phage injects its DNA into the host cell.

Phage DNA

Bacterial chromosome

New phages are released to start the cycle again.

Lysis

The host DNA is digested.

Lytic cycle

Assembly of new phages complete. A phage-encoded enzyme causes the cell to lyse.

The host cell transcribes and translates the phage DNA, producing phage proteins.

New phage DNA forms, using nucleotides from former host DNA.

13.3 The Lytic Cycle of a Virulent Bacteriophage In the lytic cycle, infection by viral DNA leads directly to the multiplication of the virus and lysis of the host bacterial cell.

phages to a culture of *E. coli* to produce substantial numbers of phages of both parental types.

However, when *E. coli* were infected with both phage strains, not only the parental types appeared, but also many phages of genotypes h^+r^+ and hr—that is, recombinant phages (Figure 13.5). Thus, the *map distance* between the two genetic markers could be calculated in a way very similar to the method of mapping in eukaryotes (see Chapter 10):

$$\text{Map distance} = \frac{\text{number of recombinant phages}}{\text{total number of phages}}$$

$$= \frac{(h^+r^+)+(hr)}{(h^+r^+)+(hr)+(h^+r)+(hr^+)}$$

As more genes in such phage crosses were studied, a map began to take form. In due course, it was established that the phage has a single chromosome.

Animal viruses have diverse reproductive cycles

Among the more common viral diseases of humans are the common cold and influenza. Almost all vertebrates are susceptible to viral infections, but among invertebrates, such infections are common only in arthropods (a group that includes insects, crustaceans, and some other animals). A group of viruses called **arboviruses** (short for "*arthropod-borne viruses*") causes serious diseases, such as encephalitis, in humans and other mammals. Arboviruses are transmitted to the mammalian host through a bite (certain arboviruses are carried by mosquitoes, for example). Although carried within the arthropod host's cells, arboviruses apparently do not affect that host severely; they affect only the bitten and infected mammal.

duced to lyse, releasing a large number of free phages, which can then infect other uninfected bacteria and renew the reproductive cycle.

This "*genetic switch*" from the lysogenic to the lytic cycle (and vice versa) is very useful to the phage, whose purpose is to reproduce as many offspring as possible. When the host cell of an infecting phage is growing slowly and is low on energy, the phage becomes lysogenic. Then, when the host's health is restored to a level that allows maximal phage reproduction, the prophage is released from its dormant state and the lytic cycle proceeds. In the laboratory, ultraviolet radiation is a potent inducer of the switch to the lytic cycle.

Bacteriophage genes can recombine

When investigators simultaneously infected *E. coli* with *two* different mutant strains of the bacteriophage T2, they observed genetic recombination. In one such experiment, one of the phage strains was genotypically h^+r and the other was hr^+. (We need not worry here about what the phenotypes were; just note that h^+ and h are alleles at one locus and r^+ and r are alleles at another locus.) We would expect the addition of these

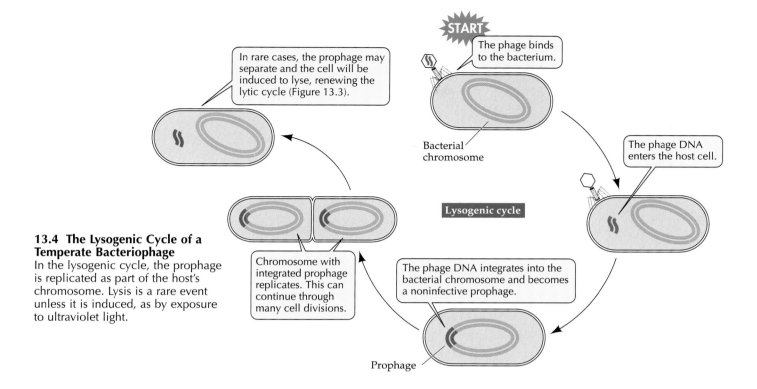

START

The phage binds to the bacterium.

Bacterial chromosome

The phage DNA enters the host cell.

Lysogenic cycle

In rare cases, the prophage may separate and the cell will be induced to lyse, renewing the lytic cycle (Figure 13.3).

The phage DNA integrates into the bacterial chromosome and becomes a noninfective prophage.

Chromosome with integrated prophage replicates. This can continue through many cell divisions.

Prophage

13.4 The Lysogenic Cycle of a Temperate Bacteriophage
In the lysogenic cycle, the prophage is replicated as part of the host's chromosome. Lysis is a rare event unless it is induced, as by exposure to ultraviolet light.

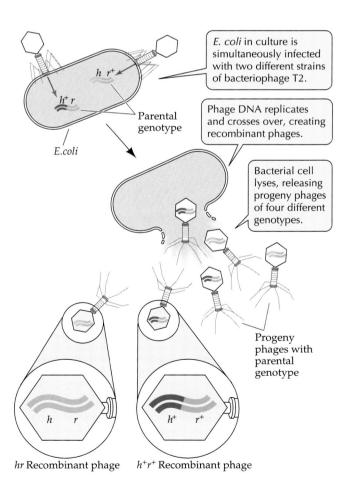

E. coli in culture is simultaneously infected with two different strains of bacteriophage T2.

Parental genotype

E.coli

Phage DNA replicates and crosses over, creating recombinant phages.

Bacterial cell lyses, releasing progeny phages of four different genotypes.

Progeny phages with parental genotype

hr Recombinant phage

h⁺r⁺ Recombinant phage

Animal viruses are very diverse. Table 13.2 shows only the major levels of animal virus classification and some examples.

Animal viruses begin the process of infection by attaching to the plasma membrane of the host cell. The viruses themselves may have membranes, which derive from the host cell that produced them. Some, such as herpesviruses, obtain their membranes from the host's nuclear envelope. Others, such as HIV (the AIDS virus), obtain their membranes from the host plasma membrane.

Animal viruses without membranes are taken up by endocytosis, which traps them within a membranous vesicle inside the host cell. Then the membrane of the vesicle breaks down, releasing the virion, and the host cell digests the protein capsid, liberating the viral nucleic acid, which takes charge. This general pattern of events is common, but the details vary sharply among the different types of animal viruses.

In the case of DNA viruses with membranes, viral membranes are studded with glycoproteins that bind to receptors on the host's plasma membrane. The host

13.5 Genetic Recombination in Bacteriophages If a bacterial culture is exposed simultaneously to a sufficient concentration of bacteriophages of two different strains, viruses of both strains may infect the same bacterium. This—a phage cross—results in recombinant progeny as well as parental types.

TABLE 13.2 A Classification Scheme for Some Animal Viruses

| VIRUS GROUP | NUCLEIC ACID | | | VIRION | | |
	MOL. WT. (millions)	TYPE	NUMBER OF STRANDS	SHAPE	SIZE (nm)	NOTES
Families of viruses that affect both vertebrates and other hosts						
Poxviridae	160–200	DNA	2	Brick-shaped	300 × 240 × 100	Poxviruses
Parvoviridae	1.2–1.8	DNA	1	Icosahedral	20	Hosts include rats and insects
Reoviridae	15	RNA	2	Icosahedral	50–80	Vertebrate, insect, and plant hosts
Rhabdoviridae	4	RNA	1	Bullet-shaped	175 × 70	Rabies, vesicular stomatitis
Families of viruses that affect only vertebrates						
Herpetoviridae	100–200	DNA	2	Icosahedral	150	Herpes
Adenoviridae	20–29	DNA	2	Icosahedral	70–80	Adenovirus
Papovaviridae	3–5	DNA	2	Icosahedral	45–55	Papillomas
Retroviridae	10–12	RNA	1	Spherical	100–200	Tumor viruses
Paramyxoviridae	7	RNA	1	Spherical	100–300	Measles, Newcastle disease
Orthomyxoviridae	5	RNA	1	Spherical	80–120	Influenza
Togaviridae	4	RNA	1	Spherical	40–60	Rubella, hog cholera, arboviruses
Coronaviridae	?	RNA	1	Spherical	80–120	Common cold
Arenaviridae	3.5	RNA	1	Spherical	85–120	Lassa fever
Picornaviridae	2.6–2.8	RNA	1	Icosahedral	20–30	Digestive and respiratory diseases
Bunyaviridae	6	RNA	1	Spherical	90–100	Encephalitis

and viral membranes fuse, releasing the rest of the virion into the cell. The host cell then replicates the viral nucleic acid and synthesizes new capsid protein as directed by the viral nucleic acid. New capsids and new viral nucleic acid combine spontaneously, and in due course, the host cell releases the new virions (Figure 13.6). Such viruses usually escape from the host cell by budding through virus-modified areas of the plasma membrane. During this process the completed virions acquire a membrane similar to that of the host cell.

Retroviruses such as HIV have a much more complex reproductive cycle (Figure 13.7). A major feature of that cycle, the reverse transcription of retroviral RNA to produce a DNA provirus, was the notable exception to Francis Crick's "central dogma" of molecular biology (see Figure 12.2). The resulting DNA transcript is integrated into a host chromosome, where it may reside permanently, occasionally being expressed to produce new virions. We discuss other aspects of HIV and AIDS in Chapter 18.

As noted in Chapter 12, there are RNA viruses other than the retroviruses. These other RNA viruses have a simpler reproductive cycle, lacking the reverse transcription step (step 3 in Figure 13.7). In cases such as poliovirus, the viral RNA serves as the mRNA for virally coded proteins, which the host cell synthesizes following infection. Some of these proteins are the coat and packaging proteins for the virus; others are involved in viral RNA replication. In other RNA viruses, such as the ones that cause influenza, the replication enzymes are part of the virus particle.

Many plant viruses spread with the help of vectors

Viral diseases of flowering plants are very common. After reproduction, plant viruses must pass through a cell wall and through the host plasma membrane. Most plant viruses accomplish this penetration through their association with **vectors**—intermediate carriers of disease from one organism to another. Infection of a plant usually results from attack by a virion-laden insect vector.

The insect vector uses its proboscis (snout) to penetrate the cell wall, allowing the virions to move via the proboscis from the insect into the plant. Plant viruses, such as tobacco mosaic virus, can be introduced artificially without insect vectors: First a leaf or other plant part is bruised; then it is exposed to a suspension of virions. Viral infections may also be inherited vegetatively or sexually, in which cases a vector is not needed.

13.6 The Reproductive Cycle of an Animal Virus with a Membrane The infected host cell replicates the viral genome and synthesizes capsid proteins, which combine spontaneously. The new viral membrane is aquired as the virus leaves the cell.

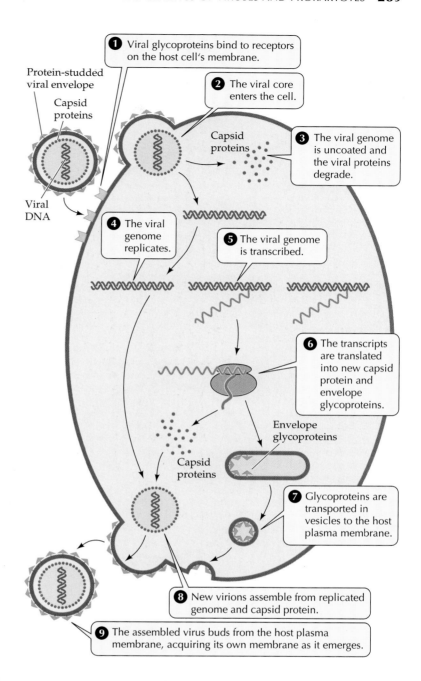

Once inside a plant cell, the virus reproduces and now must spread to other cells in the plant. Within an organ such as a leaf, the virus spreads through the plasmodesmata, the cytoplasmic connections between cells (see Chapter 4). However, some modification is needed. Because the viruses are too large to go through these channels, special proteins bind to them and help change their shape to squeeze the viruses through the pores.

Viroids are infectious agents consisting entirely of RNA

Pure viral nucleic acids can produce viral infections under laboratory conditions, although inefficiently. Might there be infectious agents in nature that consist of nucleic acid without a protein capsid? In 1971, Theodore Diener of the U.S. Department of Agriculture reported the isolation of agents of this type, called viroids. **Viroids** are circular, single-stranded RNA molecules consisting of a few hundred nucleotides. They are one-thousandth the size of the smallest viruses. These RNAs are most abundant in the nuclei of infected cells.

Viroids have been found only in plants, where they produce a variety of diseases. Two mechanisms are known by which viroids may be transmitted from plant to plant. Like tobacco mosaic virus, viroids can be transmitted *horizontally*, from one plant to another, if a bruised infected plant contacts an injured uninfected one. They can also be transmitted *vertically*, from parent to offspring. When a pollen grain or an ovule produced by an infected plant contains viroids, these infect the daughter plant produced by fertilization.

There is no evidence that viroids are translated to synthesize proteins, and it is not known how they cause disease. Viroids are replicated by the enzymes of their plant hosts. Similarities in base sequences between viroids and certain nontranslated sequences (introns) of plant genes suggest that viroids evolved from introns. This conclusion is supported by the fact that viroids, although made of RNA, are catalytically active in the way that some introns are (see Chapter 14).

Prions may be infectious proteins

If an RNA molecule by itself can be infectious, what about a protein? A class of protein fibrils, called **prions**, appear to cause certain degenerative diseases of mammalian central nervous systems. These fibrils consist entirely of protein, with no nucleic acid component. Prions appear to cause scrapie, a disease of sheep and goats. Prions have also been identified in connection with "mad cow disease," which is thought to have been spread to humans when people ate contaminated beef from cows that had eaten the remains of infected sheep. Prions are involved in other, similar

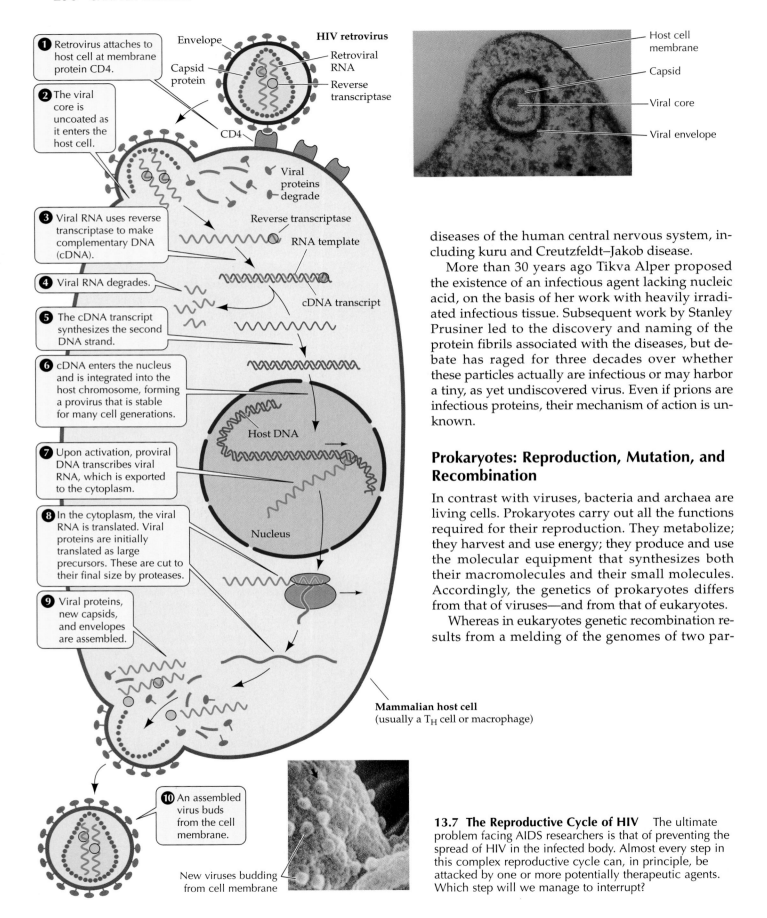

❶ Retrovirus attaches to host cell at membrane protein CD4.

❷ The viral core is uncoated as it enters the host cell.

❸ Viral RNA uses reverse transcriptase to make complementary DNA (cDNA).

❹ Viral RNA degrades.

❺ The cDNA transcript synthesizes the second DNA strand.

❻ cDNA enters the nucleus and is integrated into the host chromosome, forming a provirus that is stable for many cell generations.

❼ Upon activation, proviral DNA transcribes viral RNA, which is exported to the cytoplasm.

❽ In the cytoplasm, the viral RNA is translated. Viral proteins are initially translated as large precursors. These are cut to their final size by proteases.

❾ Viral proteins, new capsids, and envelopes are assembled.

❿ An assembled virus buds from the cell membrane.

Envelope

Capsid protein

HIV retrovirus

Retroviral RNA

Reverse transcriptase

CD4

Viral proteins degrade

Reverse transcriptase

RNA template

cDNA transcript

Host DNA

Nucleus

Host cell membrane

Capsid

Viral core

Viral envelope

Mammalian host cell
(usually a T_H cell or macrophage)

New viruses budding from cell membrane

diseases of the human central nervous system, including kuru and Creutzfeldt–Jakob disease.

More than 30 years ago Tikva Alper proposed the existence of an infectious agent lacking nucleic acid, on the basis of her work with heavily irradiated infectious tissue. Subsequent work by Stanley Prusiner led to the discovery and naming of the protein fibrils associated with the diseases, but debate has raged for three decades over whether these particles actually are infectious or may harbor a tiny, as yet undiscovered virus. Even if prions are infectious proteins, their mechanism of action is unknown.

Prokaryotes: Reproduction, Mutation, and Recombination

In contrast with viruses, bacteria and archaea are living cells. Prokaryotes carry out all the functions required for their reproduction. They metabolize; they harvest and use energy; they produce and use the molecular equipment that synthesizes both their macromolecules and their small molecules. Accordingly, the genetics of prokaryotes differs from that of viruses—and from that of eukaryotes.

Whereas in eukaryotes genetic recombination results from a melding of the genomes of two par-

13.7 The Reproductive Cycle of HIV The ultimate problem facing AIDS researchers is that of preventing the spread of HIV in the infected body. Almost every step in this complex reproductive cycle can, in principle, be attacked by one or more potentially therapeutic agents. Which step will we manage to interrupt?

ents, in prokaryotes it results from the interaction of the genome of a parent cell with a much smaller sample of genes from another cell. There are three ways in which the smaller sample is introduced into the parent cell: conjugation, transformation, and transduction.

In *conjugation*, or "bacterial sex," two bacteria come close to each other and one produces a hollow tube through which some DNA can pass between the cells. In *transformation*, which was described in Chapter 11, a fragment of DNA from one cell is taken up from the environment by another cell. In *transduction*, a bacteriophage actually carries bacterial DNA from one bacterial cell to another. Small extra pieces of DNA, called *plasmids*, are present in some bacteria. These plasmids are involved in recombination, as well as in the antibiotic resistance we described at the beginning of this chapter. We will discuss each mode of genetic recombination in more detail in this section.

The reproduction of prokaryotes gives rise to clones

All prokaryotes reproduce by the division of single cells into two identical offspring (see Figure 9.3). A single cell gives rise to a **clone**—a population of genetically identical individuals. As long as conditions remain favorable, a population of *E. coli*, for example, can double every 20 minutes.

Pure cultures of *E. coli* can be grown on the surface of a solid medium that contain a sugar, minerals, a nitrogen source such as ammonium chloride (NH_4Cl), and a solidifying agent such as agar (Figure 13.8). If the number of cells spread on the surface is small, each cell will give rise to a small, rapidly growing colony. If an extremely large number of cells is poured onto the solid medium, their growth will produce one continuous colony—a bacterial "lawn." Bacteria can also be grown on a liquid nutrient medium. We'll see examples of all these techniques in this chapter.

Prokaryotic genes mutate

We can do an experiment to demonstrate that prokaryotic mutants arise. First we mix a large sample of the bacterium *E. coli* with a suspension of a suitable bacteriophage and pour the mixture over solid growth medium in a wide, shallow dish called a petri plate. Wherever the virus finds a bacterial cell, it attaches to it, infects it, and eventually causes it to burst, killing the bacterium and releasing many new viruses. These viruses, in turn, attack neighboring cells. Soon circular clearings, called **plaques**, begin to appear in the lawn wherever the viruses have killed bacteria (Figure 13.9). The plaques, which are caused by the virus-induced bursting, or lysis, of the bacteria, grow larger and larger.

However, if you scan several such petri plates, here and there you will find a bacterial colony growing in the midst of a plaque, in spite of the surrounding hordes of viruses. Each of these colonies has arisen from a mutant bacterium that is resistant to the virus. We know that resistance is a heritable trait because the mutant bacteria give rise to colonies of cells that are similarly resistant to the virus.

The preceding description of the growth of bacteria and bacteriophages in the laboratory and the consequence of a certain mutation illustrates evolution on a small scale. In nature, *E. coli* constantly mutate to a virus-resistant form at a low rate. The mutants normally do not take over the entire population but exist in low frequency as members of the bacterial population. However, when the environment favors one genotype in a population over others, the proportions of the different genotypes in the population change. In the case of *E. coli*, the virus kills the wild type but not

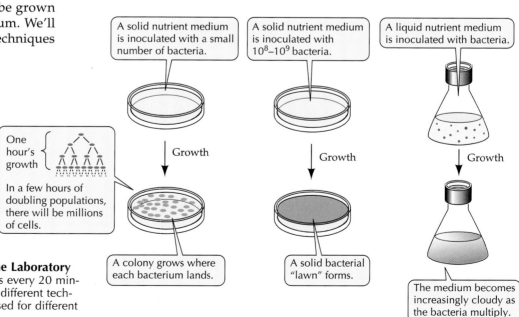

13.8 Growing Bacteria in the Laboratory
A population of *E. coli* doubles every 20 minutes in laboratory culture. The different techniques of culture shown are used for different applications.

One hour's growth

In a few hours of doubling populations, there will be millions of cells.

A solid nutrient medium is inoculated with a small number of bacteria.

A solid nutrient medium is inoculated with 10^8–10^9 bacteria.

A liquid nutrient medium is inoculated with bacteria.

Growth

Growth

Growth

A colony grows where each bacterium lands.

A solid bacterial "lawn" forms.

The medium becomes increasingly cloudy as the bacteria multiply.

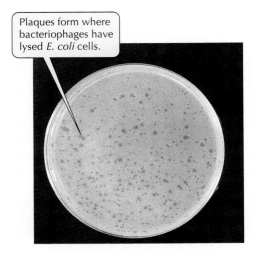

Plaques form where bacteriophages have lysed *E. coli* cells.

13.9 Bacteriophages Clear Bacterial Lawns The dark circles are clear plaques in an opaque lawn of *E. coli*.

13.10 New Prototrophic Colonies Appear After growing together, a mixture of complementary auxotrophic strains contains a few cells that can give rise to new prototrophic colonies. This experiment proved that genetic recombination takes place in prokaryotes.

the resistant strain, so the resistant strain soon predominates.

Some bacteria conjugate, recombining their genes

The existence and heritability of mutations in bacteria attracted the attention of geneticists to these microbes. But if there were no form of exchange of genetic information between individuals, bacteria would not be useful for genetic analysis. Luckily, in 1946 Joshua Lederberg and Edward Tatum demonstrated that such exchanges do occur; in *E. coli*, however, genetic recombination is a rare event.

Lederberg and Tatum used two nutrient-requiring, or *auxotrophic*, strains of *E. coli* as parents. Like the *Neurospora* in Figure 12.1, these strains will not grow on normal media, but require supplementation with a substance they cannot synthesize because of an enzyme defect. *E. coli* strain 1 requires the amino acid methionine and the vitamin biotin for growth, and its genotype is symbolized as *met⁻bio⁻*. Strain 2 requires neither of these substances but could not grow without the amino acids threonine and leucine. Considering all four factors, we note that strain 1 is *met⁻bio⁻thr⁺leu⁺* and strain 2 is *met⁺bio⁺thr⁻leu⁻*.

These two mutant strains were mixed and cultured together for several hours on a medium supplemented with methionine, biotin, threonine, and leucine, so that both strains could grow. The bacteria were then removed from the medium by centrifugation, washed,

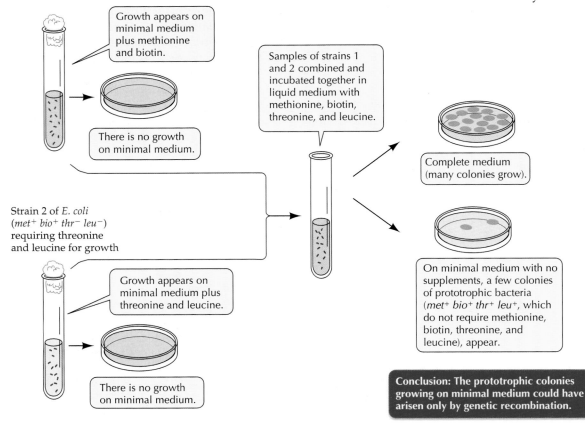

Strain 1 of *E. coli* (*met⁻ bio⁻ thr⁺ leu⁺*) requires methionine and biotin for growth

Growth appears on minimal medium plus methionine and biotin.

There is no growth on minimal medium.

Strain 2 of *E. coli* (*met⁺ bio⁺ thr⁻ leu⁻*) requiring threonine and leucine for growth

Growth appears on minimal medium plus threonine and leucine.

There is no growth on minimal medium.

Samples of strains 1 and 2 combined and incubated together in liquid medium with methionine, biotin, threonine, and leucine.

Complete medium (many colonies grow).

On minimal medium with no supplements, a few colonies of prototrophic bacteria (*met⁺ bio⁺ thr⁺ leu⁺*, which do not require methionine, biotin, threonine, and leucine), appear.

Conclusion: The prototrophic colonies growing on minimal medium could have arisen only by genetic recombination.

and transferred to minimal medium, which lacked all four supplements. Neither parent strain could grow on this medium, because of their nutritional requirements. However, a few colonies *did* appear on the plates. Because they grew in the minimal medium, these colonies must have consisted of bacteria that were *met⁺bio⁺thr⁺leu⁺*; that is, they were prototrophic (Figure 13.10). These colonies appeared at a rate of approximately 1 for every 10 million cells put on the plates ($1/10^7$).

How might these prototrophic colonies have arisen? Lederberg and Tatum were able to rule out mutation, and other investigators ruled out bacterial transformation (see Chapter 11). A third possibility is that the bacteria had **conjugated** in pairs, allowing their genetic material to mix and recombine to produce prototrophic colonies from cells containing *met⁺* and *bio⁺* alleles from strain 2 and *thr⁺* and *leu⁺* alleles from strain 1 (see Figure 13.10).

Later experiments showed that conjugation had indeed occurred, and that the two cells of differing genotype mated. One bacterial cell—the *recipient*—received DNA from the other cell—the *donor*—that included the two wild-type alleles for the loci in the recipient. Recombination then created a genotype with four wild-type alleles. The physical contact required for conjugation could be observed under the electron microscope (Figure 13.11).

What sort of a process brings about the recombination of genes after bacteria conjugate? We will learn about this shortly, but first we want to examine what distinguishes a recipient cell from a donor cell in conjugation.

Male bacteria have a fertility plasmid

The transfer of genetic material during conjugation in *E. coli* is a one-way process from a donor to a recipient.

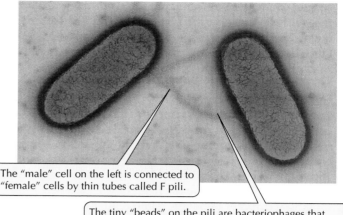

The "male" cell on the left is connected to "female" cells by thin tubes called F pili.

The tiny "beads" on the pili are bacteriophages that attach specifically to F pili, making the pili more visible.

13.11 Bacterial Conjugation   After cells are joined by F pili, they are drawn into closer contact, and DNA is transferred from one cell to the other.

English microbiologist William Hayes characterized many strains and found that each strain is either a recipient or a donor, which can also be called female and male, respectively.

In many cases, the female bacterium becomes male after conjugation—in bacteria, maleness is an infectious venereal disease! Maleness in bacteria is defined by the presence of a fertility factor, called the **F plasmid**. A **plasmid** is a small, circular piece of DNA that is separate from the main chromosome. The F plasmid contains about 25 genes. Males, which possess the plasmid, are F⁺; females, lacking the plasmid, are F⁻. In a cross of F⁺ × F⁻, the F plasmid replicates and a copy is transferred to the female, thus rendering the female F⁺, while the original male remains F⁺ (Figure 13.12).

The F plasmid can replicate itself and persist in the cell population as if it were a second chromosome independent of the normal bacterial chromosome. Genes on the F plasmid direct the formation of long, thin structures called **F pili** (singular *pilus*, "hair") projecting from the surface of the male bac-

Conjugating cells: A single strand of F plasmid is transferred to the F⁻ cell.

The F⁺ plasmid replicates the remaining strand.

F⁺ E. coli (male)

F (fertility) plasmid

Bacterial chromosomes

Conjugation tube

Cells, now both F⁺males, separate.

F⁻ E. coli (female)

The F⁻ cell becomes F⁺ as it obtains a copy of the F⁺ plasmid.

13.12 Transfer of the F Plasmid in *E. coli*   During conjugation, the F⁻ recipient cell receives a copy of the F plasmid—an extra piece of DNA—from the donor cell by way of a connecting tube (the conjugation tube) and becomes F⁺.

terium. The ends of these F pili attach to the surface of female cells (see Figure 13.11). Although an F pilus makes the initial contact, a subsequent contact allows DNA to be transferred.

Genes from the male integrate into the female's chromosome

The discovery that genetic recombination could follow conjugation opened the possibility of mapping the genetic material of bacteria by determining recombinant frequencies (see Chapter 10). However, early attempts at mapping were complicated by the fact that very few recombinant offspring arose from F+ × F− crosses, thus making it difficult to obtain reliable quantitative data.

The situation changed with the discovery of certain mutant male strains that gave more abundant recombinant offspring. These strains were called **Hfr mutants** (for "high frequency of recombination"). Hfr males, unlike ordinary F+ males, do not generally transfer their F plasmid to the female. Furthermore, only certain genes are transferred with high frequency—they transfer other genes no more frequently than ordinary F+ males do. We know now that in Hfr strains the F plasmid is actually integrated into the bacterial chromosome.

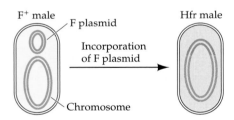

Work in 1955 by the French biologists Elie Wollman and François Jacob explained these observations. Wollman and Jacob showed that the genes from an Hfr male enter the female in sequence. In their most dramatic experiments, they used the technique of **interrupted mating**.

Wollman and Jacob allowed a brief time for the male and female bacteria to attach and transfer their DNA; then they separated the two cells and observed what happened. They mixed Hfr and F− E. coli bacteria at high concentration to initiate conjugation; at various times thereafter

they diluted samples of the mixture and agitated them in a kitchen blender for 2 minutes. Such agitation separates conjugating bacteria but does not damage them.

The number of Hfr genes passed to the female depended on the length of time allowed before conjugation was interrupted—the longer the conjugation, the more genes were transferred (Figure 13.13b). The genes always entered in the same order from any particular Hfr strain. The Hfr mutant almost never transferred the F plasmid.

Jacob and Wollman recognized immediately that this interrupted mating technique provided a simple way to map the chromosome. They prepared different mutant strains and crossed pairs of strains; then they

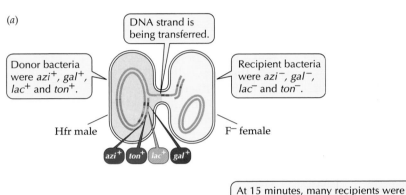

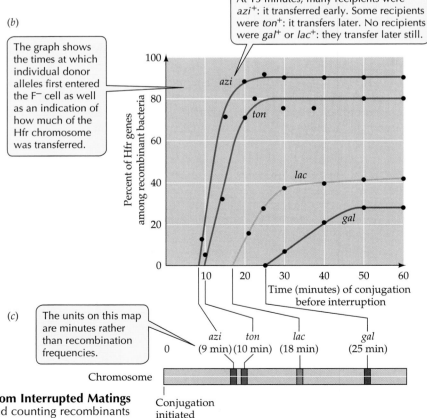

13.13 Creating a Chromosome Map of *E. coli* from Interrupted Matings
By interrupting conjugating cells at various times and counting recombinants from the matings, we can prepare chromosome maps.

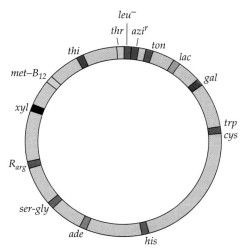

13.14 The *E. coli* Chromosome This map of the *E. coli* chromosome summarizes data from interrupted-conjugation experiments with many Hfr strains (see Figure 13.13). This early version of the map shows only a few genes, but well over half of the estimated 3,000 *E. coli* genes have been mapped. The three-letter abbreviation for each gene is derived from its mutant phenotype: *leucine*-requiring, *azide*-resistant, and so on.

interrupted successive samples from the crosses. Because the genes are transferred in a particular sequence, the length of mating time required before a particular gene is transferred and thus available to appear in recombinant progeny is a measure of its location on the chromosome (Figure 13.13*c*).

Different Hfr strains had different orders for gene transfer. Strain 1 might transfer in the order *azi-ton-lac-gal*, while strain 2 transferred in the order *lac-gal-azi-ton* and strain 3 in the order *ton-lac-gal-azi*. These and other observations led Jacob and Wollman to the following conclusions: (1) The *E. coli* chromosome is circular (Figure 13.14), (2) Hfr males have the F plasmid incorporated into their chromosome, (3) the location where the F plasmid is integrated in the chromosome varies, giving rise to different Hfr strains (thus the different orders of gene transfer), (4) the inserted F plasmid marks the point at which the chromosome "opens" as conjugation begins (see Figure 13.15), and (5) one end—always the same one—of the opened chromosome leads the way into the female. The piece of chromosome continues to move through the conjugation tube until mating is interrupted naturally or otherwise. At the extreme end of the opened chromosome lies the portion of the F plasmid that determines maleness; this will be the last portion of the donor chromosome to be transferred.

What moves from the Hfr to the F⁻ is just one strand of the double-stranded Hfr DNA. Transfer is initiated when one strand within the sequence of the F plasmid is cut once and opens (is "nicked"; Figure

13.15). The 5′ end of the nicked strand begins to unravel from the chromosome and moves to the F⁻ cell. Meanwhile, the transferred strand is replaced in the Hfr cell by DNA synthesis at the 3′ end of the nick, using the intact circular strand as the template. Thus the male still contains a complete double-stranded set of DNA sequences, even after donating some DNA to the F⁻ cell.

As it enters the F⁻ cell, the nicked DNA strand functions as a template, becoming double-stranded. Genes on this piece of DNA will give rise to recombinant bacteria only if the genes become integrated into the F⁻ chromosome by crossing over (Figure 13.16). About half of the transferred Hfr genes become integrated in this way. The DNA containing the other, unincorporated genes does not replicate when the cell divides, and it is eventually destroyed.

Prokaryotic conjugation can be summarized as follows: A female cell (F⁻) is contacted by an F pilus growing on a male cell (either F⁺ or Hfr), and the male cell transfers some DNA to the female. An F⁺ cell transfers the F plasmid, turning the F⁻ cell into F⁺. An Hfr cell transfers DNA other than the F plasmid. The transferred DNA that integrates into the F⁻ chromosome becomes part of the genome of the F⁻ cell.

As we suggested earlier, there are other, quite different ways to transfer genes from one bacterium to

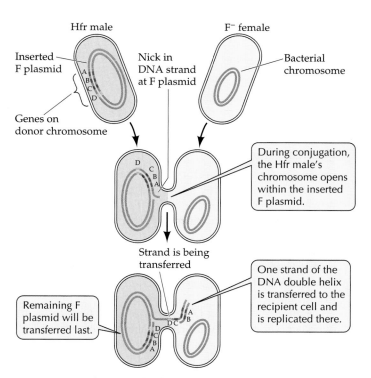

13.15 The Behavior of Hfr Males Because most of the F plasmid is the last DNA to be transferred, the recipient cell usually does not become a male, since the complete chromosome is rarely transferred.

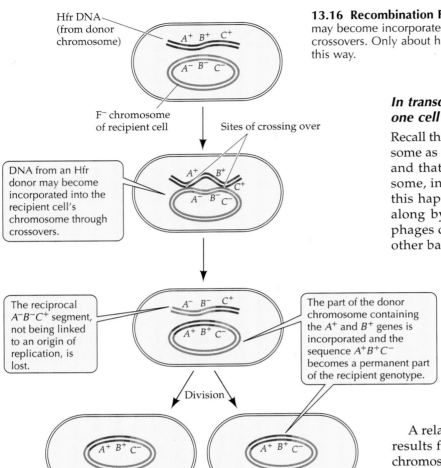

Hfr DNA (from donor chromosome)

A^+ B^+ C^+

F⁻ chromosome of recipient cell

A^- B^- C^-

Sites of crossing over

DNA from an Hfr donor may become incorporated into the recipient cell's chromosome through crossovers.

A^+ B^+ C^+

A^- B^- C^-

The reciprocal $A^-B^-C^+$ segment, not being linked to an origin of replication, is lost.

A^- B^- C^+

A^+ B^+ C^-

The part of the donor chromosome containing the A^+ and B^+ genes is incorporated and the sequence $A^+B^+C^-$ becomes a permanent part of the recipient genotype.

Division

A^+ B^+ C^-

A^+ B^+ C^-

13.16 Recombination Following Conjugation DNA from an Hfr donor may become incorporated into the recipient cell's chromosome through crossovers. Only about half the transferred Hfr genes become integrated in this way.

another, that also result in genetic recombination. Next, we will examine transformation.

In transformation, cells pick up genes from their environment

Frederick Griffith obtained the first evidence for the transfer of prokaryotic genes more than 70 years ago when he discovered **transformation** (see Figure 11.1). Recall that living bacteria of a nonvirulent strain of *Pneumococcus* picked up heritable information from dead virulent pneumococci. We now know that DNA had leaked from the dead cells and was taken up as free DNA by the living cells, within the body of a living mouse.

Once transforming DNA is inside the host cells, an event very similar to the recombination of Hfr and F⁻ occurs, and new genes can be spliced into the host chromosome. Bacteria can also take up and express DNA from a solution in a test tube. As we'll see in Chapter 16, on genetic engineering, we can transform bacteria with recombinant DNA, inducing cells to make proteins they have never seen before.

In transduction, viruses carry genes from one cell to another

Recall that a temperate phage can integrate its chromosome as a prophage into a host bacterial chromosome, and that the prophage can escape from the chromosome, initiating a lytic cycle (see Figure 13.4). When this happens, bacterial genes are occasionally taken along by the departing phage DNA. The resulting phages can then introduce these bacterial genes into other bacteria that they infect, and the genes may be randomly integrated into the chromosomes of the new hosts. This phenomenon is called **specialized transduction** (here "transduce" means "transfer") (Figure 13.17a). Transducing phages, which carry bacterial genes, cause newly infected bacteria to become lysogenic. In specialized transduction, only the chromosomal genes that are adjacent to the site of integration of the prophage may be taken along with the phage DNA.

A related phenomenon, **generalized transduction**, results from the incorporation of part of the *bacterial* chromosome, *without* the prophage DNA, into a phage coat (Figure 13.17b). The resulting particle, even though it lacks any phage genes, can infect another bacterium, injecting the piece of DNA from its former host. A bacterium thus infected with a piece of foreign bacterial DNA does *not* become lysogenic, nor does it form new phages and burst as in the lytic cycle (see Figure 13.3). It simply contains extrachromosomal bacterial DNA, as if it had conjugated with an Hfr cell.

If crossing over takes place between the host chromosome and the transduced DNA, the transfer of genes is completed. In contrast to specialized transduction, generalized transduction can move any small part of the bacterial chromosome. There is no limitation on which bacterial genes might become enclosed in a phage coat.

The phage particle is big enough to house several adjacent bacterial genes. Generalized transduction is thus another powerful tool for mapping the bacterial chromosome—with viral assistance. If two genes are transduced together, they must be close together on the chromosome. See Question 2 in "Applying Concepts" for an example of mapping by generalized transduction.

Transduction is also an important method for genetic engineering. As you will see in Chapter 16, viruses are used as vectors to transfer DNA into a cell, which becomes genetically transformed.

13.17 Transduction Phages may carry bacterial DNA from cell to cell.

Resistance factors are plasmids carrying harmful genes

The F plasmid and viral prophages are examples of episomes. **Episomes** are nonessential genetic elements that can exist in either of two states: independently replicating within a cell, or integrated into the main chromosome. Episomes cannot arise by mutation; they must be obtained by infection from outside the bacterium. The infection can come from a virus or from another bacterium. As we have seen, episomes may be used as vehicles for transferring genes from one bacterium to another.

Plasmids, free circles of DNA such as the F plasmid we encountered earlier (see Figure 13.12), are also nonessential genetic elements; they usually are not incorporated into the bacterial chromosome. (An episome is simply a plasmid that has the possibility of becoming part of the chromosome.) Every plasmid is a replicon, which means that it contains a sequence where DNA replication starts, and so is capable of independent replication.

Resistance factors, or **R factors**, are plasmids that carry genes for resistance to antibiotics. R factors first came to the attention of biologists in 1957 during an epidemic in Japan, when it was discovered that some strains of the *Shigella* bacterium, which causes dysentery, were resistant to several antibiotics. Researchers found that resistance to the entire spectrum of antibiotics could be transferred by conjugation even when no genes on the main chromosome were transferred. Also, F⁻ cells could serve as donors, indicating that the genes for antibiotic resistance were not carried by the F plasmid.

Eventually it was shown that the genes for antibiotic resistance are carried on plasmids, but not the F plasmid. Each of these plasmids (the R factor) carries one or more genes conferring resistance to particular antibiotics, as well as genes that code for proteins involved in the transfer of DNA to a recipient bacterium. As far as biologists can determine, R factors appeared long before antibiotics were discovered, but they seem to have become more abundant in modern times. Can you propose a hypothesis to explain why R factors might be more widespread now than they were in the past?

R factors pose a serious threat to hospital patients, as we noted at the beginning of this chapter. The heavy use of antibiotics in the hospital environment selects for bacterial strains bearing the R factors, and the unfortunate infected patient becomes home to bacterial strains that can't be knocked out by antibiotics.

(a) **Specialized transduction**

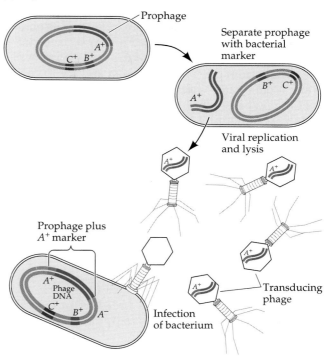

An A^- bacterial cell becomes A^+ when a transducing phage introduces the A^+ gene to the recipient cell when the phage DNA inserts into the host chromosome.

(b) **Generalized transduction**

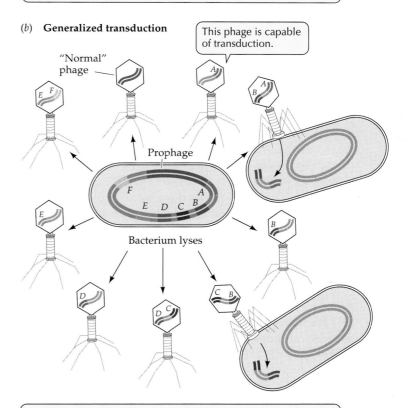

Rarely during phage production, parts of the bacterial chromosome are incorporated into phage particles. This DNA may be injected into a recipient bacterial cell and may become part of the recipient's chromosome.

R factors also pose a threat to people in the general clinical setting if antibiotics are used inappropriately. You probably have gone to see a physician with a sore throat, which can have either a viral or a bacterial cause. The best way to know is for the doctor to take a small sample from the inflamed throat and try to culture and identify any bacteria that are present. But you cannot wait another day. Impatient, you ask the doctor to give you something to cure the disease. She prescribes an antibiotic, which you take. The sore throat gradually gets better, and you think that the antibiotic did the job.

But suppose the infection was viral? In this case, the antibiotic did nothing to combat the disease, which just ran its normal course. However, the antibiotic may have done something harmful: By killing many normal bacteria in your body, it may have exerted a selection for bacteria harboring R factors. These bacteria reproduce in the presence of the antibiotic, and soon could be quite numerous. Then the next time you got a *bacterial* infection, there would be a ready supply of R factors to be transferred to the invading bacteria and antibiotics would be ineffective.

Transposable elements move genes among plasmids and chromosomes

As we have seen, plasmids, viruses, and even phage coats (in the case of transduction) can transport genes from one bacterial cell to another. Another type of "gene transport" within the individual cell relies on segments of chromosomal or plasmid DNA called **transposable elements**. Copies of transposable elements can be inserted at new locations in the same chromosome, or into another chromosome. Insertion often produces multiple physiological effects by disrupting the genes into which the transposable elements are inserted (Figure 13.18*a*).

The first transposable elements to be discovered in prokaryotes were large pieces of DNA, typically 1,000 to 2,000 base pairs long, found in many places in the *E. coli* chromosome. In one mechanism of transposition, the sequence of a transposable element can replicate independently of the rest of the chromosome. The copy then inserts itself at other, seemingly random places in the chromosome. The genes encoding the enzymes necessary for this insertion are found within the transposable element itself. Some other transposable elements are cut from their original sites and inserted elsewhere without replication. Many transposable elements discovered later were longer (about 5,000 base pairs) and carried one or more additional genes. These longer elements with additional genes are called **transposons** (Figure 13.18*b*).

What do transposons and other transposable elements have to do with the genetics of prokaryotes—or with hospitals? Transposable elements have contributed

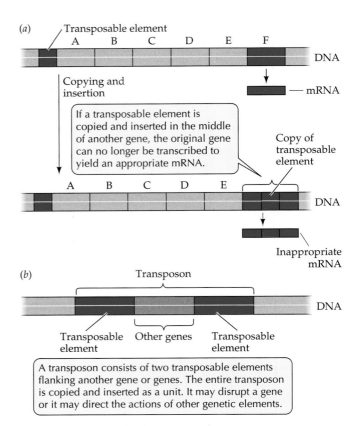

13.18 Transposable Elements and Transposons
(*a*) Transposable elements are segments of DNA that can be inserted at new locations, either on the same chromosome or on a different chromosome. (*b*) Transposons consist of transposable elements and other genes.

to the evolution of plasmids. R plasmids originally gained their genes for antibiotic resistance through the activity of transposable elements; one piece of evidence for this conclusion is that each resistance gene in an R plasmid is part of a transposon. Transposons on the F plasmid and on the bacterial chromosome interact to direct the insertion of the F plasmid into the chromosome in the formation of an Hfr male.

We will have more to say about transposable elements in the next chapter, which is concerned with the genetics of eukaryotes.

Regulation of Gene Expression in Prokaryotes

We have now seen that prokaryotes have genes, that prokaryotic genes mutate, that they are exchanged among cells by three basic mechanisms, that they may be parts of chromosomes or plasmids, and that they may move among these DNA molecules. In preceding chapters we learned how genes replicate and how they are expressed. But are all prokaryotic genes expressed all the time? If not, why not? And if not, how then does the prokaryotic cell regulate gene expression?

13.19 An Inducer Stimulates the Synthesis of an Enzyme

It is most efficient for a cell to produce an enzyme only when it is needed; some enzymes are induced by the presence of the substance they act upon (for example, the induction of β-galactosidase by the presence of lactose).

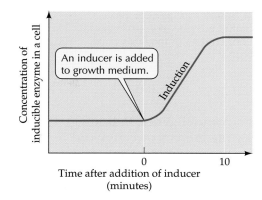

Regulation of transcription conserves energy

To express all the genes of a cell all the time would be inefficient in terms of energy. Why should a cell use valuable ATP and other resources to produce an enzyme when that enzyme is not needed? In fact, it does not. Prokaryotes produce some enzymes only when their substrates are present in the environment, and they turn off the synthesis of others when the products of those enzymes build to excessive concentrations.

As a normal inhabitant of the human intestine, *E. coli* has to adjust to sudden changes in its chemical environment. Its host may present it with one foodstuff one hour and another the next. For example, these bacteria may suddenly be deluged with milk, the main carbohydrate of which is the sugar lactose. Lactose is a β-galactoside—a disaccharide containing galactose β-linked to glucose (see Chapter 3). Before lactose can be of any use to the bacteria, it must first be taken into their cells by a membrane transport carrier called β-galactoside permease. Then it must be hydrolyzed to glucose and galactose by the enzyme β-galactosidase. A third protein, the enzyme thiogalactoside transacetylase, is also required for lactose metabolism.

When *E. coli* is grown on a medium that does not contain lactose or other β-galactosides, the levels of all three of these proteins within the bacterial cell are extremely low—the cell does not waste energy and material making the unneeded proteins. If, however, the environment changes such that lactose is the predominant sugar and very little glucose is present, the synthesis of all three of these proteins begins promptly

and they increase in abundance. There are only two molecules of β-galactosidase in an *E. coli* cell when glucose is in the medium. But when it is absent, lactose can induce the synthesis of 3,000 molecules of β-galactosidase per cell! Regulation of enzyme synthesis by the genes that code for them thus promotes efficiency in the cell.

Compounds that stimulate the synthesis of an enzyme (as does lactose in this example) are called **inducers** (Figure 13.19). The enzymes that are produced are called **inducible enzymes**, whereas enzymes that are made all the time at a constant rate are called **constitutive enzymes**. If lactose is removed from *E. coli*'s environment, synthesis of the three enzymes stops almost immediately. The enzyme molecules that have already formed do not disappear; they are merely diluted during subsequent growth and reproduction until their concentration falls to the original low level within each bacterium.

We have now seen two basic ways to regulate the rates of metabolic pathways. In Chapter 6, we described allosteric regulation of enzyme *activity* (the rate of enzyme-catalyzed reactions); this mechanism allows rapid, fine-tuning of metabolism. Regulation of gene expression—that is, regulation of the *concentration* of enzyme—is slower but produces a greater savings of energy. Figure 13.20 compares these two modes of regulation.

Transcriptional regulation uses some familiar tools and some new tools

The blueprints for the synthesis of the three proteins that process lac-

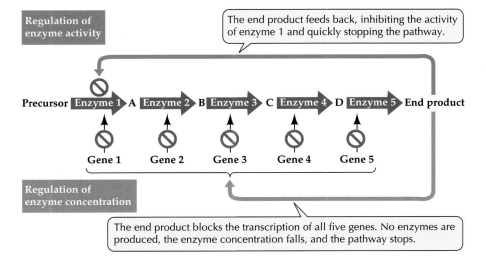

13.20 Two Ways to Regulate a Metabolic Pathway
Feedback from the end product can block enzyme activity, or it can stop the transcription of genes that code for the enzymes.

13.21 Repressor Bound to Operator Blocks Transcription Portions of the repressor bind to the major and minor grooves in the DNA helix, preventing transcription.

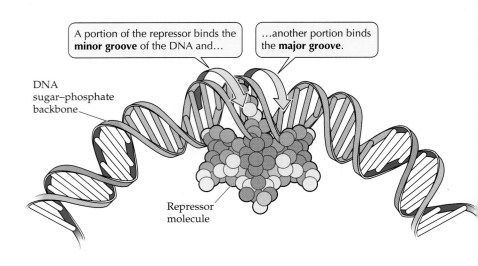

A portion of the repressor binds the **minor groove** of the DNA and...

...another portion binds the **major groove**.

DNA sugar–phosphate backbone

Repressor molecule

tose are called **structural genes**, indicating that they specify the primary structure (the amino acid sequence) of a protein molecule. In other words, structural genes can be transcribed into mRNA. When Jacob, Wollman, and Monod mapped the particular structural genes coding for proteins involved in the metabolism of lactose, they discovered that all three lie close together in a region that covers about 1 percent of the *E. coli* chromosome.

It is no coincidence that these three genes lie next to one another. The information from them is transcribed into a single, continuous molecule of mRNA. Because this particular messenger governs the synthesis of all three lactose-metabolizing enzymes, either all or none of the enzymes are made, depending on whether their common message—their mRNA—is present in the cell.

The three genes share a single promoter. Recall from Chapter 12 that a promoter is a site on DNA where RNA polymerase binds to initiate transcription. The promoter for these three structural genes is very efficient so that the maximum rate of mRNA synthesis can be high, but there must also be a way to shut down mRNA synthesis when the enzymes are not needed. Such flexibility requires some new tools.

Operons are units of transcription in prokaryotes

Prokaryotes shut down transcription by placing an obstacle between the promoter and its structural genes. A short stretch of DNA called the **operator** can bind very tightly a special type of protein molecule, a **repressor**, thereby creating the obstacle. When the repressor protein is bound to the operator region DNA, it blocks the transcription of mRNA (Figure 13.21). When the repressor is not attached to the operator, messenger synthesis proceeds rapidly.

The whole unit, consisting of closely linked structural genes and the stretches of DNA that control their transcription, is called an **operon** (Figure 13.22). An operon always consists of a promoter, an operator, and one or more structural genes. The promoter and operator are binding sites on DNA and are not transcribed.

13.22 The *lac* Operon of *E. coli* and Its Regulator The *lac* operon is a segment of DNA that includes a promoter, an operator, and the three structural genes that code for the *lac*tose-metabolizing enzymes.

E. coli has three different ways of controlling the transcription of operons. Two depend on interactions of repressor with the operator, and the third depends on interactions of other proteins with the promoter. Let's consider each of the three control systems in turn.

Operator–repressor control that induces transcription: The *lac* operon

The operon that controls and contains the genes for the three *lac*tose-metabolizing enzymes is called the *lac* **operon** (see Figure 13.22). As we just learned, RNA polymerase can bind to the promoter, and a repressor can bind to the operator. How is the operon controlled?

The key lies in the repressor and its binding to the operator. The repressor is a protein that has two binding sites: one for the operator and the other for small molecules called *inducers*. Inducers of the *lac* operon are molecules of lactose and certain other β-galactosides. Binding of the inducer changes the shape of the repressor (by allosteric modification; see Chapter 6). This change in shape prevents the repressor from binding to the operator (Figure 13.23). At this point,

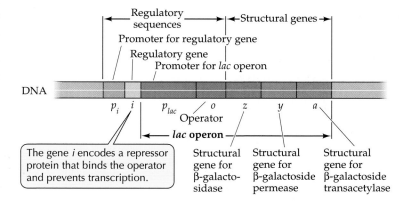

Regulatory sequences

Structural genes

Promoter for regulatory gene

Regulatory gene

Promoter for *lac* operon

DNA

p_i i p_{lac} o z y a

Operator

lac operon

The gene *i* encodes a repressor protein that binds the operator and prevents transcription.

Structural gene for β-galactosidase

Structural gene for β-galactoside permease

Structural gene for β-galactoside transacetylase

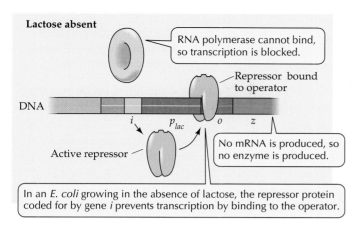

Lactose absent

RNA polymerase cannot bind, so transcription is blocked.

Repressor bound to operator

DNA

i | p_{lac} | *o* | *z*

Active repressor

No mRNA is produced, so no enzyme is produced.

In an *E. coli* growing in the absence of lactose, the repressor protein coded for by gene *i* prevents transcription by binding to the operator.

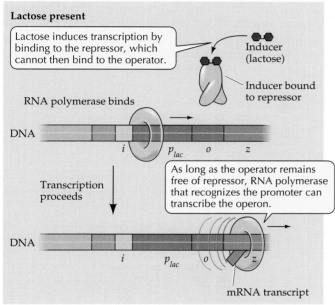

Lactose present

Lactose induces transcription by binding to the repressor, which cannot then bind to the operator.

Inducer (lactose)

Inducer bound to repressor

RNA polymerase binds

DNA

i | p_{lac} | *o* | *z*

Transcription proceeds

As long as the operator remains free of repressor, RNA polymerase that recognizes the promoter can transcribe the operon.

DNA

i | p_{lac} | *o* | *z*

mRNA transcript

13.23 The *lac* Operon: Transcription Induced by the Removal of a Repressor Lactose (the inducer) leads to enzyme production by preventing the repressor protein (which would have stopped transcription) from binding to the operator.

RNA polymerase can bind to the promoter and start transcribing the structural genes of the *lac* operon. The mRNA transcribed from these genes is translated on ribosomes, synthesizing the three proteins required for metabolizing lactose.

What happens if the concentration of lactose drops? As the lactose concentration decreases, the inducer (lactose) molecules separate from the repressor. Free of lactose molecules, the repressor returns to its original shape and binds to the operator, and transcription of the *lac* operon stops. Translation stops soon thereafter, because the mRNA that is already present breaks down quickly. Thus, the inducer regulates the binding of the repressor to the operator.

Repressor proteins are coded by **regulatory genes**. The regulatory gene that codes for the repressor of the *lac* operon is called the *i* (inducibility) gene. The *i* gene happens to lie close to the operon that it controls, but some other regulatory genes are distant from their operons. Like all other genes, the *i* gene itself has a promoter, which can be designated p_i. Because this promoter does not bind RNA polymerase very effectively, only enough mRNA to synthesize about ten molecules of repressor protein per cell per generation is produced. This quantity of the repressor is enough to regulate the operon effectively—to produce more would be a waste of energy. There is no operator between p_i and the *i* gene. Therefore, the repressor of the *lac* operon is constitutive; that is, it is made at a constant rate that is not subject to environmental control.

Let's review the important features of inducible systems such as the *lac* operon. In the absence of inducer the *lac* operon is turned *off*. Adding inducer turns the operon *on*. Control is exerted by a regulatory protein— the repressor—that turns the operon *off*. Some genes, such as *i*, produce proteins whose sole function is to regulate the expression of other genes; certain other DNA sequences (namely, operators and promoters) do not code for proteins but are binding sites for regulatory proteins.

Operator–repressor control that represses transcription: The trp operon

We have seen that *E. coli* benefits from having an inducible system for lactose metabolism. Only when lactose is present does the system switch on. Equally valuable to a bacterium is the ability to switch *off* the synthesis of certain enzymes in response to the excessive accumulation of their products. For example, if the amino acid tryptophan, an essential constituent of proteins, is present in ample concentration, it is advantageous to stop making the enzymes for tryptophan synthesis. When the formation of an enzyme can be turned off in response to such a biochemical cue, synthesis of the enzyme is said to be **repressible**.

Monod realized that repressible systems, such as the *trp* operon for *tryp*tophan synthesis, could work by mechanisms similar to those of inducible systems, such as the *lac* operon. In repressible systems, the repressor cannot shut off its operon unless it first binds to a **corepressor**, which may be either the metabolite itself (tryptophan in this case) or an analog of it (Figure 13.24). If the metabolite is absent, the operon is transcribed at a maximum rate. If the metabolite is present, the operon is turned off.

The difference between inducible and repressible systems is small but significant. In inducible systems, a substance in the environment (the inducer) interacts with the regulatory-gene product (the repressor), rendering it *incapable* of binding to the operator and thus

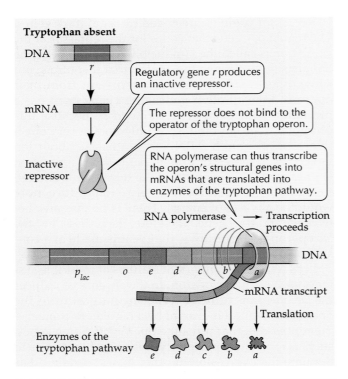

Tryptophan absent

Regulatory gene *r* produces an inactive repressor.

The repressor does not bind to the operator of the tryptophan operon.

RNA polymerase can thus transcribe the operon's structural genes into mRNAs that are translated into enzymes of the tryptophan pathway.

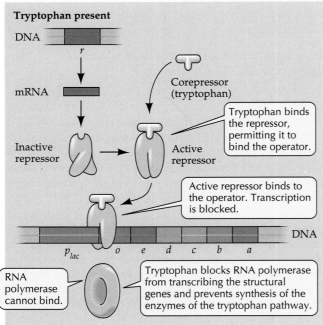

Tryptophan present

Tryptophan binds the repressor, permitting it to bind the operator.

Active repressor binds to the operator. Transcription is blocked.

RNA polymerase cannot bind.

Tryptophan blocks RNA polymerase from transcribing the structural genes and prevents synthesis of the enzymes of the tryptophan pathway.

13.24 The *trp* Operon: Transcription Repressed by the Binding of a Repressor Because tryptophan activates an otherwise inactive repressor, it is called a corepressor.

incapable of blocking transcription. In repressible systems, a substance in the cell (the corepressor) interacts with the regulatory-gene product to make it *capable* of binding to the operator and blocking transcription. Although the effects of the substances are exactly opposite, the systems as a whole are strikingly similar.

In both the inducible lactose system and the repressible tryptophan system, the regulatory molecule functions by binding the operator. Next we'll consider an example of control by binding the *promoter*.

Protein synthesis can be controlled by increasing promoter efficiency

A prokaryotic cell has the means to increase the transcription of certain relevant genes when it needs a new energy source. *E. coli* prefers to get its energy from glucose in the environment. When glucose is unavailable, *E. coli* must get energy from another source, such as lactose or certain other sugars or even amino acids. The alternative energy source must be catabolized (broken down) by reactions requiring catabolic enzymes. Specific regulatory molecules enhance the transcription of operons that contain the genes for these enzymes. The mechanism that produces this effect is

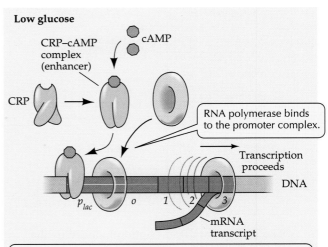

Low glucose

RNA polymerase binds to the promoter complex.

When supplies of glucose are low, a receptor protein (CRP) and cAMP form a complex that binds to the promoter and activates it, allowing transcription of structural genes that encode enzymes for catabolizing the alternative energy source.

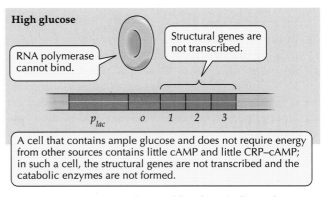

High glucose

RNA polymerase cannot bind.

Structural genes are not transcribed.

A cell that contains ample glucose and does not require energy from other sources contains little cAMP and little CRP–cAMP; in such a cell, the structural genes are not transcribed and the catabolic enzymes are not formed.

13.25 Transcription Enhanced by the Binding of CRP at the Promoter Site The structural genes of this operon encode enzymes that break down a food source other than glucose.

entirely different from the two operator–repressor mechanisms just discussed, which turn the operon on or off. This type of mechanism makes the promoter function more efficiently.

Suppose that a bacterial cell lacks a glucose supply but has access to another food that can be catabolized to yield energy. In operons containing genes for catabolic enzymes, the promoters bind RNA polymerase in a series of steps (Figure 13.25). First, a special protein (abbreviated **CRP**, for *cAMP receptor protein*) binds the low-molecular-weight compound adenosine 3′, 5′-cyclic monophosphate, better known as cyclic AMP or cAMP. Next, the CRP–cAMP complex binds close to the binding site of the RNA polymerase and enhances the binding of the polymerase to the promoter 50-fold. The promoters of the *lac* operon and many other genes responsible for sugar metabolism are activated in this way.

When glucose becomes abundant in the medium, alternative food molecules do not need to be broken down, so the cell diminishes or ceases synthesizing the enzymes that catabolize these alternative sources. Glucose decreases synthesis of these enzymes—a phenomenon called *catabolite repression*—by lowering the cellular concentration of cAMP.

The *lac* and *trp* systems—the two operator–repressor systems—are examples of *negative* control of transcription because the regulatory molecule (the repressor) in each case *prevents* transcription. The promoter system is an example of *positive* control of transcription because the regulatory molecule (the CRP–cAMP complex) *enhances* transcription.

As we will see in Chapter 13, the regulation of gene expression in eukaryotes is somewhat different than in prokaryotes.

Summary of "The Genetics of Viruses and Prokaryotes"

Using Prokaryotes and Viruses for Genetic Experiments

• Bacteria and viruses are useful to the study of genetics and molecular biology because they contain much less DNA than complex eukaryotes have, they grow and reproduce rapidly, often in less than an hour, and they are haploid, which means their genetics is simpler.

Viruses: Reproduction and Recombination

• Viruses were discovered as disease-causing agents small enough to pass through a filter that retains bacteria. They consist of a nucleic acid genome that codes for a few proteins, and a protein coat. Some viruses also have a lipid membrane that is derived from host membranes.

• Viruses are obligate intracellular parasites, needing the biochemical machinery of living cells to reproduce.
• There are many types of viruses, classified by their size, shape, and genetic material (RNA or DNA), or host (plant, animal, protist, fungus, or prokaryote). **Review Figure 13.1**
• Bacteriophages infect bacteria. In the lytic cycle, about 30 minutes after phage infection the host cell breaks open, releasing many new phage particles. Some phages can undergo a lysogenic cycle, in which their DNA is inserted into the host chromosome and replicates for generations. When conditions are appropriate, the lysogenic DNA exits the host chromosome and enters a lytic cycle. **Review Figures 13.3, 13.4**
• If two phages infect the same host cell simultaneously, their DNAs can come into close proximity and genetic recombination can occur. The resulting data make it possible to generate a map of the phage chromosome. **Review Figure 13.5**
• Most of the many types of RNA and DNA viruses that infect animals cause diseases. Some animal viruses are surrounded by membranes. Retroviruses, such as HIV, have RNA and reproduce their genomes through a DNA intermediate. Other RNA viruses use their RNA as mRNA to code for enzymes and replicate their genomes without using DNA. **Review Figures 13.6, 13.7, and Table 13.2**
• Many plant viruses are spread by other organisms, such as insects.
• Viroids are made only of RNA and infect plants, where they are replicated by the plant's enzymes.
• Prions appear to be infective proteins and cause diseases in animals, including humans.

Prokaryotes: Reproduction, Mutation, and Recombination

• When bacteria divide, they form clones of identical cells that can be observed as colonies when grown on solid media. **Review Figure 13.8**
• Prokaryotic genes can mutate, as shown by experiments in which prokaryotes demonstrate resistance to bacteriophages. **Review Figure 13.9**
• A bacterium can transfer its genes to another bacterium by conjugation, transformation, or transduction.
• In conjugation, a bacterium attaches to another bacterium and passes a partial copy of its DNA to the adjacent cell. **Review Figures 13.10, 13.11**
• The F plasmid codes for the conjugation tube. A bacterium harboring F is male (F⁺). Without F, the cell is female (F⁻). During conjugation, an F⁺ cell can transfer a copy of its F plasmid to an F⁻ cell, converting it to an F⁺ cell. **Review Figure 13.12**
• If the F plasmid integrates into the main chromosome (as in Hfr males), it can prompt the transfer of part of that chromosome into a recipient cell, where the extra DNA can undergo genetic recombination with the host chromosome, transferring new genes to the host chromosome permanently. The order of transfer reflects the positions of different markers on the circular bacterial chromosome. **Review Figures 13.13, 13.14, 13.15, 13.16**
• In transformation, genes are transferred between prokaryotes when fragments of bacterial DNA are taken up by a cell from the medium. These genetic fragments may recombine with the host chromosome, thereby permanently adding new genes.

• In transduction, phage particles carry bacterial DNA from one bacterium to another. This transfer can include a few specific bacterial genes attached to a lysogenic phage genome when it leaves the bacterial chromosome. Or it can involve fragments of bacterial DNA that are packaged by themselves into phage particles. In both cases, the new bacterial DNA can recombine with the chromosome of the recipient bacteria. **Review Figure 13.17**

• Plasmids are independently replicating, extrachromosomal DNA elements that usually are not incorporated into the main chromosome. Bacterial R factors are plasmids that carry genes for antibiotic resistance, as well as genes coding for interbacterial transfer by conjugation. R factors are a serious public health threat.

• Transposable elements are movable stretches of DNA that can jump from one place to another on the bacterial chromosome—either by actually moving or by making a new copy, which inserts at a new location. **Review Figure 13.18**

Regulation of Gene Expression in Prokaryotes

• In bacteria, constitutive genes are constantly expressed, and their products are essential to the cells at all times. The expression of other genes is regulated; their products are made only when they are needed. Genes are regulated by inducers (to turn on expression) or corepressors (to turn off expression). **Review Figures 13.19, 13.20, 13.24**

• The functionally related bacterial genes contained in an operon are transcribed into a single mRNA, so they are under the same regulatory control. **Review Figure 13.22**

• The expression of prokaryotic genes is regulated by three different mechanisms: inducible operator–repressor systems, repressible operator–repressor systems, and enhancement systems that increase the efficiency of a promoter.

• The *lac* operon is an example of an inducible system whose proteins allow bacteria to use lactose. When glucose is the energy source, very few of the *lac* operon proteins are present in the cell. But if glucose is absent, lactose can induce the syntheses of these proteins. **Review Figure 13.19**

• A promoter controls RNA polymerase–catalyzed transcription of the entire *lac* operon. Between the promoter and the structural genes lies a stretch of DNA called the operator. When glucose is present and lactose is absent, a repressor protein binds tightly to the operator, preventing RNA polymerase from binding to the promoter. Thus the operon is transcriptionally inactive.

• When glucose is absent and lactose is present, the latter acts as an inducer by binding to the repressor. This changes the repressor's shape so that it no longer recognizes the operator. With the operator unoccupied, RNA polymerase binds to the promoter, and transcription occurs to produce the proteins for lactose utilization. **Review Figures 13.23, 13.24, 13.25**

• The *trp* operon is a repressible system in which the presence of the end product of a biochemical pathway, tryptophan, represses the syntheses of enzymes involved in its synthesis. Tryptophan acts as a corepressor, binding to an inactive repressor protein, giving it a strong affinity for the operator. Where the operator is occupied with repressor, transcription of the structural genes coding for the tryptophan biosynthesis pathway is blocked. **Review Figure 13.24**

• The efficiency of RNA polymerase can be increased by regulation of the level of cyclic AMP, which binds to a protein, CRP. The altered CRP then binds to a site near the promoter of a target gene, enhancing the binding of RNA polymerase and hence transcription. **Review Figure 13.25**

Self-Quiz

1. In bacterial conjugation
   a. each cell donates DNA to the other.
   b. a bacteriophage carries DNA between bacterial cells.
   c. one partner possesses a fertility plasmid.
   d. the two parent bacteria merge like sperm and egg.
   e. all the progeny are recombinant.

2. Which statement about the bacterial fertility factor is *not* true?
   a. It is a plasmid.
   b. It confers "maleness" on the cell in which it resides.
   c. It can be transferred to a female cell, making the female cell male.
   d. It has thin projections called F pili.
   e. It can become part of the bacterial chromosome.

3. Hfr mutants
   a. are female bacteria that are highly efficient recipients of genes.
   b. rarely transfer all the genes on the chromosome.
   c. keep their F plasmid separate from the chromosome at all times.
   d. are unable to conjugate with other bacteria.
   e. transfer genes in random order.

4. Lysogenic bacteria
   a. lack a prophage.
   b. are accompanied by free phages when growing in culture.
   c. lyse immediately.
   d. cannot by induced to enter a lytic cycle.
   e. are susceptible to further attack by the same strain of phage.

5. Which statement about transduction is *not* true?
   a. The viral DNA is an episome.
   b. Transduction is a useful tool for mapping a bacterial chromosome.
   c. In specialized transduction, the newly infected cell becomes lysogenic.
   d. Transduction can result in genetic recombination.
   e. To carry bacterial genes, the viral coat must contain viral DNA.

6. Plasmids
   a. are circular protein molecules.
   b. are required by bacteria.
   c. are tiny bacteria.
   d. may confer resistance to antibiotics.
   e. are a form of transposable element.

7. Which statement about transposable elements is *not* true?
   a. They can be copied to another DNA molecule.
   b. They can be copied to the same DNA molecule.
   c. They are typically 100 to 500 base pairs long.
   d. They may be part of a plasmid.
   e. They encode the enzyme transposase.

8. In an inducible operon
   a. an outside agent switches on enzyme synthesis.
   b. a corepressor unites with the repressor.
   c. an inducer affects the rate at which repressor is made.
   d. the regulatory gene lacks a promoter.
   e. the control mechanism is positive.

9. The promoter is
   a. the region that binds the repressor.
   b. the region that binds RNA polymerase.

c. the gene that codes for the repressor.

d. a structural gene.

e. an operon.

10. The CRP–cAMP system
    a. produces many catabolites.
    b. requires ribosomes.
    c. operates by an operator–repressor mechanism.
    d. is an example of positive control of transcription.
    e. relies on operators.

Applying Concepts

1. Viruses sometimes carry DNA from one cell to another by transduction. Sometimes a segment of bacterial DNA is incorporated into a phage protein coat without any phage DNA. These particles can infect a new host. Would the new host become lysogenic if the phage originally came from a lysogenic host? Why or why not?

2. Genetic markers A^+, B^+, and C^+ are in one strain of *E. coli* that is used for transduction into a second strain: $A^-B^-C^-$. After transduction, recombinants are selected for in the recipient bacteria. The frequencies of single recombinants, which are positive (+) for only one of A, B, or C, are all 1 percent. The frequencies of double transductants are as follows: $A^+B^+ = 10^{-6}$ percent; $A^+C^+ = 10^{-6}$ percent; $B^+C^+ = 10^{-3}$ percent. Explain these data.

3. You are provided with two strains of *E. coli*. One, an Hfr strain that is sensitive to streptomycin, carries the alleles A^+, B^+, and C^+. The other is an F⁻ strain that is resistant to streptomycin and carries the alleles A^-, B^-, and C^-. You mix the two cultures. After 20, 30, and 40 minutes you take samples of the mixed culture and swirl them vigorously in a blender. Next you add streptomycin to the swirled cultures. You examine surviving bacteria to determine their phenotypes. Some of the bacteria from the 20-minute sample are B^+; in the 30-minute sample there are both B^+ and C^+ bacteria; but A^+ bacteria are found only in the 40-minute sample. What can you say about the arrangement of the A, B, and C loci on the *E. coli* chromosome? Explain your answer fully.

4. You have isolated three strains of *E. coli*, which you name 1, 2, and 3. You attempt to cross these strains, and you find that recombinant progeny are obtained when 1 and 2 are mixed or when 2 and 3 are mixed, but not when 1 and 3 are mixed. By diluting a suspension of strain 2 and plating it out on solid medium, you isolate some separate clones. You find that almost all these clones can conjugate with strain 1 to produce recombinant offspring. One of the clones derived from strain 2, however, lacks the ability to conjugate with strain 1. Characterize strains 1, 2, and 3 and the nonconjugating clone of strain 2 in terms of the fertility (F) plasmid.

5. In the lactose (*lac*) operon of *E. coli*, repressor molecules are encoded by the regulatory gene. The repressor molecules are made in very small quantities and at a constant rate per cell. Would you surmise that the promoter for these repressor molecules is efficient or inefficient? Is synthesis of the repressor constitutive, or is it under environmental control?

6. A key characteristic of a repressible enzyme system is that the repressor molecule must react with a corepressor (typically, the end product of a pathway) before it can combine with the operator of an operon to shut the operon off. How is this different from an inducible enzyme system?

Readings

Griffiths, A. J. F., J. H. Miller, D. T. Suzuki, R. C. Lewontin and W. M. Gelbart. 1996. *An Introduction to Genetic Analysis*, 6th Edition. W. H. Freeman, New York. A revision of one of the best genetics textbooks, with excellent coverage of prokaryotic genetics.

Judson, H. F. 1996. *The Eighth Day of Creation: Makers of the Revolution in Biology*, Expanded Edition. CSHL Press, Plainview, NY. A constantly fascinating history of molecular biology, with much attention to the regulation of gene expression. For a lay audience.

Lodish, H., D. Baltimore, A. Berk, S. L. Zipursky, P. Matsudaira and J. Darnell. 1995. *Molecular Cell Biology*, 3rd Edition. Scientific American Books, New York. A comprehensive yet comprehensible summary of gene control.

Matthews, K. S. 1996. "The Whole Lactose Repressor." *Science*, vol. 271, pages 1245–1246. This nontechnical summary and the subsequent, technical article by Lewis (pages 1247–1254) illustrate the current sense of the structure and allosteric changes of the *lac* repressor.

Nomura, M. 1984. "The Control of Ribosome Synthesis." *Scientific American*, February. A discussion of how ribosomes are assembled and the roles of operons in regulating ribosome production in bacteria.

Ptashne, M. 1992. *A Genetic Switch,* 2nd Edition. Cell Press and Blackwell Scientific Publications, Cambridge, MA. An in-depth look at the life cycles of a single bacteriophage type, lambda, that reveals elegant control mechanisms.

Watson, J. D., M. Gilman, J. Wikowski and M. Zoller. 1992. *Recombinant DNA*, 2nd Edition. Scientific American Books, New York. Superbly written descriptions of prokaryotic and viral molecular genetics.

Chapter 14

The Eukaryotic Genome and Its Expression

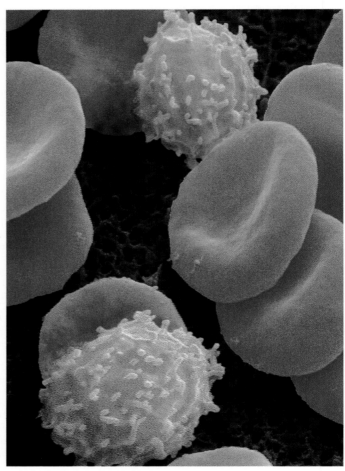

Two Cells, Two Different Protein Products
Although both cell types have the genes for both proteins, the red blood cells produce abundant hemoglobin, while the white blood cells synthesize proteins of the immune system.

When Tom was diagnosed with acute leukemia—cancer of the blood cells—his initial treatment included chemotherapy, in which combinations of powerful antimitotic drugs were administered to kill the dividing cancer cells that were rapidly spreading through his body. But the dosages his physicians prescribed were not up to the task, and the cells continued to spread. Higher dosages of these drugs would be lethal, as they would kill not only the cancer cells, but healthy and essential dividing cells—such as the stem cells in bone marrow that divide by the hundreds of millions to form blood cells. Without these stem cells, Tom's body would no longer produce red blood cells with their vital oxygen-carrying protein hemoglobin, nor would he be able to produce white blood cells, which make the proteins of the immune system that can eliminate infectious diseases as well as some tumors.

Tom's doctors tried a new approach. They extracted some of his bone marrow and removed the cancer cells from it, then stored the marrow in a refrigerator. Then they gave Tom extremely high doses of the chemotherapeutic drugs, which killed the cancer cells. Finally, the stored bone marrow was replaced into Tom's body. The stem cells began to divide, and after a few weeks they were differentiating into adequate populations of normal red and white blood cells. Tom's leukemia had disappeared.

The success of Tom's bone marrow transplant depended on many things, but the principle behind it is based on the specificity of gene expression during cell differentiation. What are the genetic mechanisms that ensure that healthy

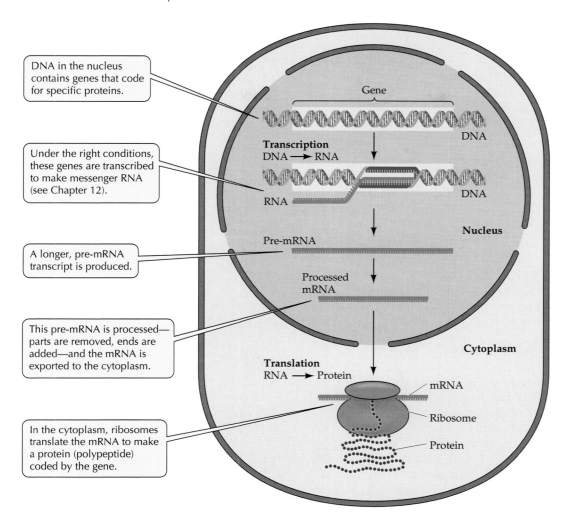

DNA in the nucleus contains genes that code for specific proteins.

Under the right conditions, these genes are transcribed to make messenger RNA (see Chapter 12).

A longer, pre-mRNA transcript is produced.

This pre-mRNA is processed—parts are removed, ends are added—and the mRNA is exported to the cytoplasm.

In the cytoplasm, ribosomes translate the mRNA to make a protein (polypeptide) coded by the gene.

14.1 Eukaryotic mRNA Is Processed in the Nucleus and Exported to the Cytoplasm Compare this "road map" to Figure 12.3.

red blood cells produce hemoglobin, and that white blood cells are able to create the vital antibody proteins of the immune system? What features of the DNA sequences of eukaryotic genes determine these mechanisms, and how do they differ from the genes that code for proteins in prokaryotes?

In this chapter, you will see that although the eukaryotic genetic material, DNA, is the same as that of prokaryotes, its *organization* in the chromosomes is often very different. Eukaryotic chromosomes contain vast amounts of repetitive sequences. Some of these sequences have vital roles, such as the genes for ribosomal and transfer RNAs, and others have unknown functions. Of special interest are the extreme ends of chromosomal DNA, the telomeres, which play a role in DNA replication and chromosome integrity.

The transcription and later tailoring of RNAs are more complicated processes in eukaryotes than in prokaryotes. A large pre-mRNA is transcribed from

each gene (Figure 14.1), and this molecule is extensively modified; even internal stretches of nucleotides are removed before the mRNA is exported from the nucleus to the cytoplasm.

Finally, we consider the fascinating molecular machinery that allows the precise regulation of gene expression needed for a eukaryote to develop and function. In contrast to the operons of prokaryotes, related eukaryotic genes are often scattered throughout the genome. Many elegant mechanisms control the selective transcription and translation of eukaryotic genes.

The Eukaryotic Genome

As biologists unraveled the intricacies of gene structure and expression in prokaryotes, they tried to generalize their findings by saying, "What's true for *E. coli* is also true for elephants." Although much of prokaryotic biochemistry does apply to eukaryotes, the old saying has its limitations (Table 14.1). For example, eukaryotes have much more DNA in each cell than prokaryotes have.

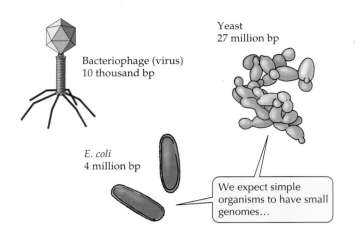

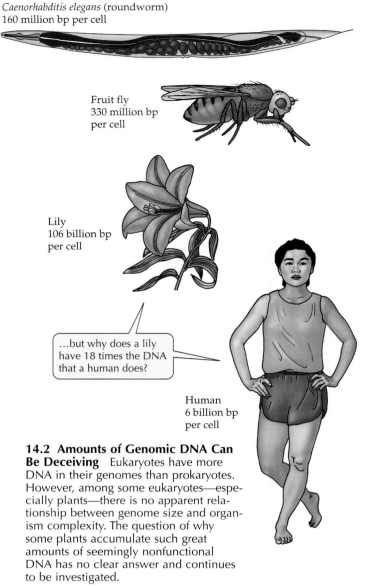

14.2 Amounts of Genomic DNA Can Be Deceiving Eukaryotes have more DNA in their genomes than prokaryotes. However, among some eukaryotes—especially plants—there is no apparent relationship between genome size and organism complexity. The question of why some plants accumulate such great amounts of seemingly nonfunctional DNA has no clear answer and continues to be investigated.

The eukaryotic genome is larger and more complex than the prokaryotic genome

The fact that the haploid DNA content (genome size) of eukaryotes is larger than that of prokaryotes might be expected, given that in multicellular organisms there are many cell types, many jobs to do, and many proteins—coded for by DNA—to do the jobs. A typical virus contains only enough DNA to code for a few proteins—about 10,000 base pairs (bp). The most thoroughly studied prokaryote, *E. coli*, has several thousand different proteins and sufficient DNA (about 4.7 million bp) to regulate the synthesis of those proteins. Humans have considerably more genes and complex controls; nearly 6 billion bp (2 meters of DNA) are crammed into each human cell. However, the idea of a more complex organism needing more DNA seems to break down with some plants. For example, the lily (which produces beautiful flowers each spring but produces fewer proteins than a human does) has 18 times more DNA than humans have (Figure 14.2).

Unlike prokaryotic DNA, most eukaryotic DNA does not code for proteins. Instead, interspersed

TABLE 14.1 A Comparison of Prokaryotic and Eukaryotic Genes and Genomes

CHARACTERISTIC	PROKARYOTES	EUKARYOTES
Genome size (base pairs)	$10^4–10^7$	$10^8–10^{11}$
Repeated sequences	Few	Many
Noncoding DNA within coding sequences	Rare	Common
Transcription and translation separated in cell	No	Yes
DNA segregated within a nucleus	No	Yes
DNA bound to proteins	Some	Extensive
Promoter	Yes	Yes
Enhancer/silencer	Rare	Common
Capping and tailing of mRNA	No	Yes
RNA splicing required	Rare	Common
Number of chromosomes in genome	One	Many

THE EUKARYOTIC GENOME AND ITS EXPRESSION

throughout the eukaryotic genome are various kinds of repeated DNA sequences. Even the coding regions of genes contain sequences that do not end up in mature mRNA. Some noncoding DNA maintains chromosomal integrity at the ends (telomeres), and some helps control gene expression. But the presence of much of this noncoding DNA remains an enigma.

In contrast to the single main DNA molecule of most prokaryotes, the eukaryotic genome is partitioned into several separate molecules of DNA, or chromosomes. In humans, each chromosome contains a double helix with 20 million to 100 million bp. This separation of the DNA into different volumes of the genomic encyclopedia requires that each chromosome have at a minimum three defining DNA sequences: recognition sequences for the DNA replication machinery, a *centromere region* that holds the replicated sequences together before mitosis, and a *telomeric sequence* at the ends of the chromosome.

In eukaryotes, the nuclear envelope separates DNA and its transcription (inside the nucleus) from the cytoplasmic site where mRNA is translated into protein. This separation allows for many points of control in the synthesis, processing, and export of mRNA to the cytoplasm. The organization of the nuclear eukaryotic genome is fundamentally about regulation: Great complexity requires a great deal of regulation, and this fact is evident in the many processes associated with the expression of the eukaryotic genome.

In addition, most eukaryotic DNA is not even fully exposed to the nuclear environment. Instead, it is extensively packaged by proteins into nucleosomes, fibers, and ultimately chromosomes (Figure 14.3). This extensive compaction is a means for restricting access of the RNA synthesis machinery to the DNA, as well as a way to segregate replicated DNAs during mitosis and meiosis.

Like the genes of prokaryotes that code for proteins, eukaryotic genes have noncoding flanking sequences that control their transcription. These include the *promoter* region, where RNA polymerase ultimately binds to begin transcription. Of equal importance in eukaryotes, but rare in prokaryotes, is a second set of controlling DNA sequences, the *enhancers* and *silencers*. Enhancers and silencers are often located quite distant from the promoter and act by binding proteins that then stimulate or inhibit transcription.

Within a protein-coding gene (that is, between the DNA bases coding for start and stop codons) are base sequences that interrupt the coding region and do not code for amino acids. These noncoding DNA sequences within protein-coding genes present a special problem: How do cells ensure that these noncoding regions do not end up in the mature mRNA that exits the nucleus?

The answer lies in an elaborate cutting and splicing mechanism within the nucleus that, after transcrip-

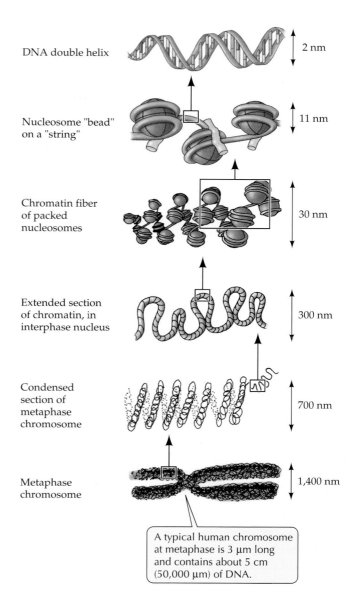

14.3 DNA is Packed into Chromosomes Proteins are essential in winding and folding DNA into a very compact structure.

tion, modifies an initial transcript called **pre-mRNA**. The noncoding sequences are cut out of the pre-mRNA and the coding regions are spliced together. Thus, in contrast to the "what is transcribed is what is translated" scheme of most prokaryotic genes, the mature mRNA at the eukaryotic ribosome is a modified and much smaller molecule than the one initially made in the nucleus.

Hybridization is a tool for genome and gene analysis

Investigations of eukaryotic genome and gene sequence organization (referred to as gene structure) have been greatly aided by **nucleic acid hybridization**.

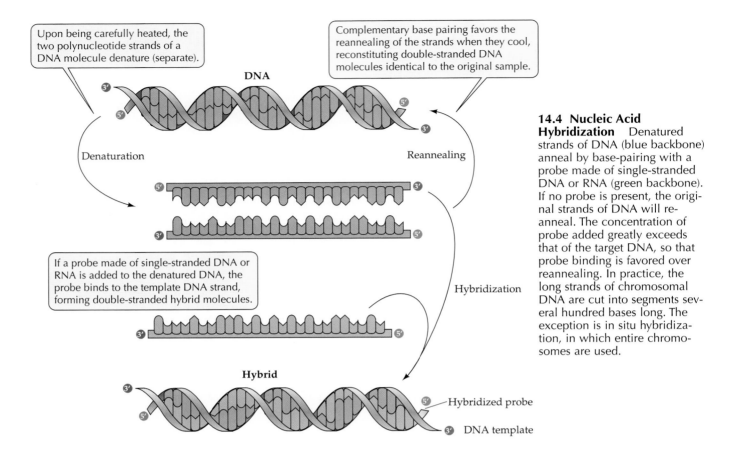

Upon being carefully heated, the two polynucleotide strands of a DNA molecule denature (separate).

Complementary base pairing favors the reannealing of the strands when they cool, reconstituting double-stranded DNA molecules identical to the original sample.

DNA

Denaturation

Reannealing

If a probe made of single-stranded DNA or RNA is added to the denatured DNA, the probe binds to the template DNA strand, forming double-stranded hybrid molecules.

Hybridization

Hybrid

Hybridized probe

DNA template

14.4 Nucleic Acid Hybridization Denatured strands of DNA (blue backbone) anneal by base-pairing with a probe made of single-stranded DNA or RNA (green backbone). If no probe is present, the original strands of DNA will reanneal. The concentration of probe added greatly exceeds that of the target DNA, so that probe binding is favored over reannealing. In practice, the long strands of chromosomal DNA are cut into segments several hundred bases long. The exception is in situ hybridization, in which entire chromosomes are used.

This technique depends on the association, through complementary base pairing, of single-stranded nucleic acids (Figure 14.4). It begins with separation of the strands of DNA. If DNA is carefully heated, the hydrogen bonds between base pairs break and the two strands of each double helix separate. This process is termed *DNA denaturation*. If this single-stranded mixture is allowed to cool, the two complementary strands eventually line up beside one another and re-form hydrogen bonds by the AT and GC base-pairing rules. This process, called *DNA reannealing*, has been very useful in research on eukaryotic genomes.

Suppose that instead of being allowed to reanneal, the single strands of denatured DNA are mixed with a high concentration of a second single-stranded nucleic acid, with a sequence complementary to one of the immobilized DNA strands. This second molecule is given time to find and anneal to its target sequence. The second molecule is called a **probe** because it is "probing" the target DNA, "looking" for its complementary sequence.

A probe may be DNA or RNA. If the probe is labeled with nucleotides that are radioactive or fluorescent, the formation of a base-paired hybrid between the probe and target can be identified by the experimenter. Hybridization has been invaluable in the ex-

amination of eukaryotic gene structures, as well as in mapping genes (see Chapters 16 and 17). It has also been essential to revealing several types of repetitive DNA in the eukaryotic genome.

Highly repetitive sequences contribute to chromosome structure and spindle attachment

If the DNA of a prokaryote is broken into fragments of about 1,000 bp and these fragments are denatured and reannealed in solution, each single-stranded region takes a long time to find its complementary partner. But when the same experiment is done with eukaryotic DNA, although some parts of the genome do reanneal slowly, others reanneal much faster.

Why do these rates differ in different sequences of the eukaryotic genome? The answer is that in eukaryotic DNA there are multiple copies of some, but not all, DNA sequences. If a particular single strand of DNA has, say, a few hundred complementary partners with which it can anneal, it will be able to find a partner much more rapidly than one for which only a single acceptable partner exists.

By measuring rates of reannealing, researchers discovered that there are three classes of eukaryotic DNA: single-copy sequences ("slow" reannealers), highly repetitive sequences ("fast" reannealers), and moderately repetitive sequences ("moderate" reannealers). The class that reanneals the slowest consists of single-copy sequences—genes that, like most prokaryotic genes, have only one copy in each genome. Single-copy sequences code for most enzymes and structural proteins in eukaryotes. Some noncoding single-copy sequences form long spacers between genes on a chromosome.

The class of DNA that reanneals the fastest consists of **highly repetitive DNA** sequences. These sequences typically are simple, 5 to 50 bp long, and repeated up to millions of times. Typically, there are several such blocks of repeated sequences in the genome. For example, in the guinea pig the sequence CCCTAA* is repeated at least 10,000 times at the centromere region of each chromosome. At the centromere, the highly repetitive DNA may contribute to the functional integrity of the chromosomes and the attachment of spindle fibers.

In most cases, however, the location of these simple sequences shows no apparent logic, and their role is unclear. What is clear is that they are usually not transcribed, since no RNAs that contain these tandemly repeated sequences are found. Humans have at least ten simple, highly repetitive sequence types.

Telomeres are repetitive sequences at the ends of chromosomes

There are several types of **moderately repetitive DNA** sequences in the eukaryotic genome. One type is important in maintaining the ends of chromosomes when DNA is replicated. Recall from Chapter 11 that replication proceeds differently on the two strands of a DNA molecule. Both new strands form in the 5'-to-3' direction, but one strand (the leading strand) grows continuously from one end to the other, while the other (the lagging strand) grows as a series of short Okazaki fragments (see Figure 11.18).

With the circular prokaryotic chromosome, as both DNA strands grow around the chromosome, production of a complete series of Okazaki fragments is not a problem. There is always some DNA at the 5' end of a primer, ready to replace the primer after it is removed. But things are more complicated in the eukaryotic chromosome. The leading strand can grow without interruption to the very end. How does the last Okazaki fragment for the end of the lagging strand form? Replication must begin with an RNA primer at the 5' end of the forming strand, but there is nothing beyond the primer in the 5' direction to replace the RNA. So the new chromosome formed after DNA replication lacks a bit of double-stranded DNA at each end. The ends of the chromosome have been clipped off.

In many eukaryotes, there are moderately repetitive sequences at the ends of chromosomes called **telomeres**; in humans the sequence is TTAGGG and it is repeated about 2,500 times (Figure 14.5a). The need for these repeats can be shown experimentally by putting fragments of DNA from a human into a yeast cell. The fragments will be stably maintained in their new home only if they have the telomeric repeats at their ends. Otherwise, they rapidly break down.

When human cells are removed from the body and put in a nutritious medium in the laboratory, they will grow and divide. But each chromosome can lose about 50 to 200 bp of telomeric DNA after each round of DNA replication and cell division. This shortening compromises the stability of the chromosomes, and after 20 to 30 cell divisions they are unable to take part in cell division. The same thing happens in the body, and explains in part why cells do not last the entire lifetime of the organism.

Yet constantly dividing cells, such as bone marrow cells and germ line cells, manage to maintain their moderately repetitive telomeric DNA. An enzyme, appropriately called telomerase, prevents the loss of this DNA by catalyzing the addition of any lost telomeric sequences (Figure 14.5b). Telomerase is made up not only of proteins but also of an RNA sequence that acts as the template for the telomeric sequence addition.

Considerable interest has been generated by the finding that telomerase is expressed in more than 90 percent of human cancers. Telomerase may be an important factor in the ability of cancer cells to divide continuously. Since most normal cells do not have this activity, telomerase is an attractive target for drugs designed to attack tumors specifically.

Some moderately repetitive sequences are transcribed

Some moderately repetitive DNA sequences code for tRNA and rRNA, which are used in protein synthesis. These RNAs are constantly being made, but even the maximum rate of transcription of single-copy sequences would be inadequate to supply the large amounts of these molecules needed by most cells; hence there are multiple copies of the DNA sequences coding for them. Since these moderately repetitive sequences are transcribed into RNA, they are properly termed "genes," and we can speak of rRNA genes and tRNA genes.

As an example, in mammals there are four sizes of rRNA in the ribosome—the 18S, 5.8S, 28S, and 5S rRNAs.[†] The 18S, 5.8S, and 28S rRNAs are transcribed

---

*When a DNA sequence such as CCCTAA is written, the complementary bases are assumed.
[†]The term "S" refers to the movement of a molecule in a centrifuge: In general, larger molecules have a higher S value.

14.5 Telomeres and Telomerase (a) The loss of repeat sequences from the telomere leads to cell death. (b) In cells that divide continuously, the enzyme telomerase prevents the loss of repeat sequences.

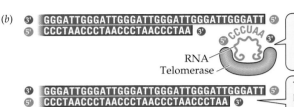

Human telomeres have about 2,500 repeats of this sequence.

Normally, DNA replicates from a primer in a 5'-to-3' direction. Because there is no primer at the extreme 5' end of a chromosome, there is a gap in replication, leading to shortening of the chromosome after each round of replication. Chromosome shortening leads in turn to cell death.

An RNA in telomerase acts as a template for DNA. This adds the telomeric sequence to the end of the chromosome.

The original length of the chromosomal DNA has been restored. Note the gap where the primer for DNA replication has been removed.

from a repeated unit of DNA that, as a single precursor, is twice the size of the three ultimate products (Figure 14.6). Several posttranscriptional steps cut this precursor into its final three rRNAs and discard the nonuseful, or "spacer," RNA. The DNA coding for these RNAs is moderately repetitive in humans: A total of 280 copies of the unit are located in clusters on five different chromosomes.

Other moderately repetitive sequences in mammals are not clustered, but instead are scattered throughout the genome. These DNAs usually are not transcribed and usually are short, about 300 bp long. In humans, half of these DNAs are of a single type, called the *Alu* family (because they have a sequence in them that is recognized by a nuclease enzyme, Alu I). There are 300,000 copies of the *Alu* family in the genome, and they may act as multiple origins for DNA replication.

Surprising those who long believed in genetic stability, some moderately repetitive DNA sequences move about the genome.

Transposable elements move about the genome

Most of the remaining scattered moderately repetitive DNA is not stably integrated into the genome. Instead, these DNA sequences can move from place to place in the genome and are called **transposable DNA**. There are four main types of transposable elements, or transposons, in eukaryotes (Table 14.2).

SINEs (short *in*terspersed *e*lements) are up to 300 bp long and are transcribed but not translated. LINEs (*long in*terspersed *e*lements) are up to 7,000 bp long, and some are transcribed and translated into proteins. Both of these elements are present in more than 100,000 copies and move about the genome in a distinctive way: They make an RNA copy, which acts as a template for the new DNA, which then inserts at a new location in the genome.

This mechanism is also employed by the third type of movable DNA, the retrotransposons, which are rare in mammals but are more common in yeasts and animals other than mammals. The genetic organization of viral retrotransposons resembles that of retroviruses such as HIV, but these segments lack the genes for protein coats and thus cannot produce viruses.

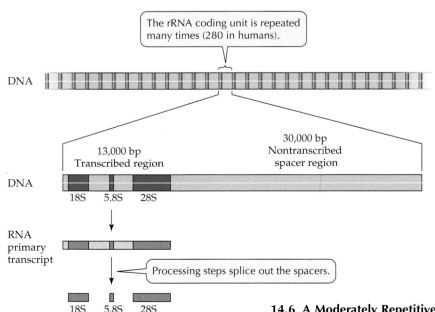

The rRNA coding unit is repeated many times (280 in humans).

DNA

13,000 bp Transcribed region

30,000 bp Nontranscribed spacer region

DNA

18S   5.8S   28S

RNA primary transcript

Processing steps splice out the spacers.

18S   5.8S   28S

Three rRNAs are the result.

14.6 A Moderately Repetitive Sequence Codes for rRNAs This rRNA gene, along with its nontranscribed spacer, is repeated 280 times in the human genome.

TABLE 14.2 Transposable Elements in Eukaryotes

TYPE OF ELEMENT	DESCRIPTION
DNA transposons	DNA segments that move or are copied at new sites
Retrotransposons	DNA segments that are copied by means of an RNA intermediate
Viral retrotransposons	Retrotransposons that possess long terminal repeats
Nonviral retrotransposons	Retrotransposons that lack long terminal repeats
LINEs	Long interspersed elements
SINEs	Short interspersed elements

The fourth type of moderately repetitive transposable element is the DNA transposon. Like its counterpart in prokaryotes, the eukaryotic DNA transposon does not use an RNA intermediate but actually moves to a new spot in the genome without replicating (Figure 14.7).

What role do these repeated sequences play in the cell? There are few answers. The best answer so far seems to be that transposons are cellular parasites that simply replicate themselves. But these replications can lead to the insertion of a transposon at a new location, and this event has important consequences. For example, insertion of a transposon into the coding region of a gene causes a mutation because of the addition of new base pairs.

If this process takes place in the germ line, a gamete with a new mutation results. If the process takes place in a somatic cell, cancer may result. If a transposon replicates not just itself but also an adjacent gene, the result may be a gene duplication. Clearly, transposition stirs the genetic pot in the eukaryotic genome and thus contributes to genetic variability.

In Chapter 4, we described the endosymbiosis theory of the origin of chloroplasts and mitochondria, which proposes that these organelles are the descendants of once free-living prokaryotes. Transposable elements may have played a role in this process. In living forms, although the organelles have some DNA, the nucleus has most of the genes that encode the organelle proteins. If the organelles were once independent, they must originally have had all of these genes. How did the genes move to the nucleus? The answer may lie in DNA transposition. Genes once in the organelle may have moved to the nucleus by well-known molecular events that still occur today. The current DNA in the organelles may be the remnants of more complete genomes.

The Structures of Protein-Coding Genes

Like their prokaryotic counterparts, protein-coding genes in eukaryotes are generally single-copy DNA sequences. But there are two distinctive characteristics of the eukaryotic genes that we will examine. First, they contain noncoding internal sequences, and second, they form gene families with structurally and functionally related cousins in the genome.

Protein-coding genes contain noncoding internal and flanking sequences

Eukaryotic proteins are usually encoded by single-copy genes. Preceding the beginning of the coding re-

14.7 Transposons and Transposition
At the end of each transposable element in DNA is an inverted repeat sequence that helps in the transposition process. If the transposon inserts within a gene, it disrupts the coding sequence and an abnormal protein can result.

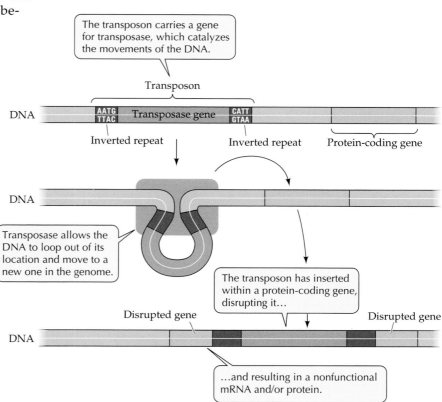

The transposon carries a gene for transposase, which catalyzes the movements of the DNA.

Transposon

DNA — AATG TTAC Transposase gene CATT GTAA

Inverted repeat   Inverted repeat   Protein-coding gene

DNA

Transposase allows the DNA to loop out of its location and move to a new one in the genome.

The transposon has inserted within a protein-coding gene, disrupting it...

DNA — Disrupted gene   Disrupted gene

...and resulting in a nonfunctional mRNA and/or protein.

gion is a **promoter**, where RNA polymerase begins the transcription process. Unlike the prokaryotic enzyme, the eukaryotic RNA polymerase does not itself recognize the promoter sequence, as we'll see later. At the other end of the gene, after the coding region, is a DNA sequence appropriately called the **terminator**, which RNA polymerase recognizes as the end point for transcription (Figure 14.8).

Both the promoter and terminator sequences are parts of the gene that do not become transcribed into RNA. Base sequences called **introns** (intervening sequences within the gene) also do not end up in mRNA. Although the introns are transcribed into RNA, they are *spliced* out within the nucleus. The locations of these noncoding base pairs can be determined by comparing the base sequences of a gene (DNA) with those of its final mRNA.

Remarkably, these extra base pairs are within the coding region of the gene. One or more introns are present in most eukaryotic genes (see Figure 14.8). These intron sequences appear in the primary tran-

script of RNA, the pre-mRNA within the nucleus, but by the time the mature mRNA exits the organelle, the introns have been removed (the pre-mRNA has been cut) and the dangling ends of the mRNA have been reconnected (spliced). The parts of the gene present in mRNA are called **exons**.

Although direct sequencing of the DNA that codes for an mRNA is the easiest way to map the locations of introns within a gene, nucleic acid hybridization is the method that originally revealed the existence of introns in protein-coding genes (Figure 14.9). Biologists denatured the DNA coding for one of the globin proteins that make up hemoglobin and added globin mRNA. From the resulting hybridization they expected to obtain a linear matchup of the mature mRNA to the globin-coding DNA.

They got their wish, in part: There were indeed stretches of RNA–DNA hybridization. But there were also some double-stranded, looped structures. These loops were the introns, stretches of DNA that did not have complementary bases on the mRNA. Later evidence showed that hybridization to the gene using the initial RNA transcript was complete and the introns were indeed transcribed. Somewhere on the path from transcript to mRNA the introns had been removed and the exons had been spliced together. We will examine this splicing process later in the chapter.

To summarize, most (but not all) vertebrate genes and many other eukaryotic genes, and even a few prokaryotic ones, contain introns. Introns interrupt but do not scramble the DNA sequence that codes for a polypeptide chain. The base sequence of the exons, taken in order, is exactly complementary to that of the mature mRNA product. The introns, therefore, separate a gene's protein-coding region into distinct parts. In some cases, these parts code for different functional regions, or domains, of the protein. For example, the

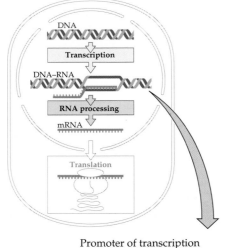

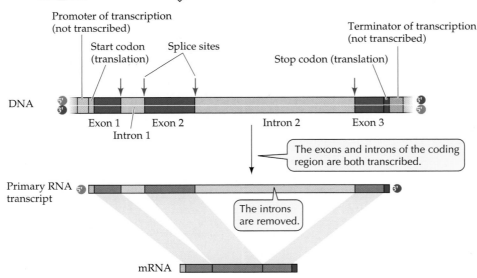

14.8 The Structure and Transcription of a Member of the β-Globin Gene Family
This entire gene is about 1,600 bp long. The coding region (blue) has 441 base pairs (triplets coding for 146 amino acids plus a triplet stop codon). Noncoding sequences of DNA—introns—are initially transcribed between codons 30 and 31 (130 bp long) and 104 and 105 (850 bp long) but are spliced out of the final gene product.

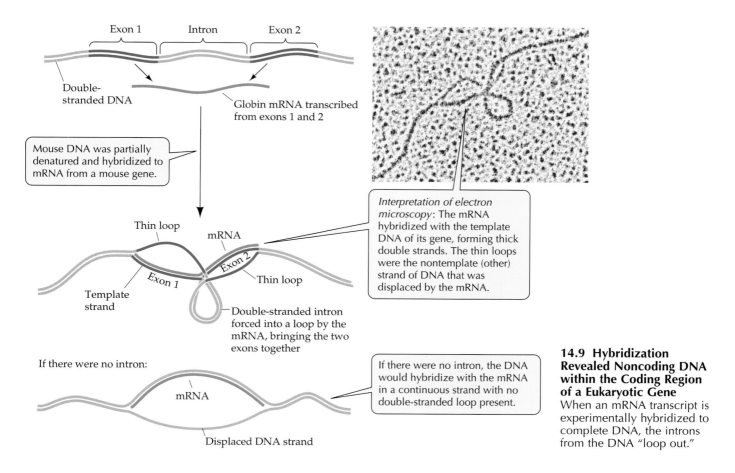

Exon 1   Intron   Exon 2

Double-stranded DNA

Globin mRNA transcribed from exons 1 and 2

Mouse DNA was partially denatured and hybridized to mRNA from a mouse gene.

Thin loop   mRNA

Exon 1   Exon 2

Template strand

Thin loop

Double-stranded intron forced into a loop by the mRNA, bringing the two exons together

Interpretation of electron microscopy: The mRNA hybridized with the template DNA of its gene, forming thick double strands. The thin loops were the nontemplate (other) strand of DNA that was displaced by the mRNA.

If there were no intron:

mRNA

Displaced DNA strand

If there were no intron, the DNA would hybridize with the mRNA in a continuous strand with no double-stranded loop present.

14.9 Hybridization Revealed Noncoding DNA within the Coding Region of a Eukaryotic Gene
When an mRNA transcript is experimentally hybridized to complete DNA, the introns from the DNA "loop out."

globin proteins that make up hemoglobin have two domains: one for binding to heme and another for binding to the other globin chains. These two domains are coded for by different exons in the globin gene.

Many eukaryotic genes are members of gene families

About half of all eukaryotic protein-coding genes are present in only one copy in the haploid genome. The rest have multiple copies. Often inexact, nonfunctional copies of a particular gene, called *pseudogenes*, are located nearby. The duplicates may have arisen by an abnormal event in chromosomal crossing over during meiosis or by retrotransposition. A set of duplicated or related genes is called a **gene family**. Some families, such as the β-globins that are part of hemoglobin, contain only a few members; other families, such as the immunoglobulins that make up antibodies, have hundreds of members.

Like the members of any family, the DNA sequences in a gene family are usually different from each other to a certain extent. As long as one member retains the original DNA sequence and thus codes for the proper protein, the other members can mutate slightly, extensively, or not at all. The availability of extra genes is important for "experiments" in evolu-

tion: If the mutated gene is useful, it will be selected for in succeeding generations. If the gene is a total loss (a pseudogene), the functional copy is still there to save the day.

A good example of gene families found in vertebrates is the gene family for the globins. As mentioned earlier, these are the proteins found in hemoglobin and also in myoglobin (an oxygen-binding protein present in muscle). The globin genes probably all arose from a single common ancestor gene long ago. In humans, there are three functional members of the alpha (α) globin family and five in the beta (β) globin family (Figure 14.10). Each hemoglobin contains the heme pigment held inside four globin polypeptides, two identical α-globins and two identical β-globins.

During human development, different members of the globin gene family are expressed at different times and in different tissues (Figure 14.11). This differential gene expression has great physiological significance. For example, the form of hemoglobin found in the fetus ($\alpha_2\gamma_2$) binds O_2 more tightly than adult hemoglobin ($\alpha_2\beta_2$) does. (Note that both γ and β are members of the β-globin family.) This specialized form of hemoglobin ensures that in the placenta, where maternal and fetal bloods come near each other, O_2 will be transferred from the mother to the developing child's

14.10 Gene Families   The human α- and β-globin gene families are located on different chromosomes. Each family is organized into a cluster of genes separated by "spacer" DNA. The nonfunctional pseudogenes are indicated by the Greek letter psi (ψ).

Spacer DNA is noncoding DNA between gene family members.

β-Globin gene cluster

ε   Gγ   Aγ   ψβ1   δ   β

α-Globin gene cluster

ζ2   ψζ1   ψα1   α2   α1

Pseudogenes

Pseudogenes are family members that do not code for functional mRNAs or proteins.

circulation. Just before birth, the synthesis of fetal hemoglobin in the liver stops, and the adult form takes over in bone marrow cells. The precise developmental regulation of transcription of this gene family is controlled at many levels, which we will discuss later.

In addition to genes that encode proteins, the globin family includes nonfunctional genes called **pseudogenes**, designated with the Greek letter psi (ψ). These pseudogenes are the "black sheep" of any gene family: They are experiments in evolution that went wrong.

The DNA sequence of a pseudogene may not differ vastly from that of other family members. It may just lack a promoter, for example, and thus cannot be transcribed. Or it may lack the recognition sites for the removal of introns and thus will be transcribed into pre-mRNA but not correctly processed into a useful mRNA. In some gene families, pseudogenes outnumber functional genes. However, since some members of the family are functional, there appears to be little selective pressure in evolution to eliminate pseudogenes.

RNA Processing

Unlike the situation in prokaryotic genes, the primary RNA transcript (pre-mRNA) of a eukaryotic gene is not the same as the mature mRNA. To produce the mRNA, the primary transcript is processed by the ad-

14.11 Differential Expression in the Globin Gene Families   During human development, different members of the globin gene family are expressed at different times and in different tissues.

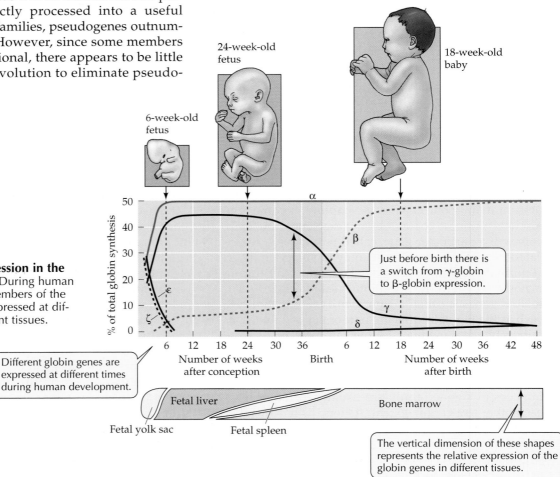

24-week-old fetus

18-week-old baby

6-week-old fetus

Just before birth there is a switch from γ-globin to β-globin expression.

Different globin genes are expressed at different times during human development.

% of total globin synthesis

Number of weeks after conception

Birth

Number of weeks after birth

Fetal yolk sac

Fetal liver

Fetal spleen

Bone marrow

The vertical dimension of these shapes represents the relative expression of the globin genes in different tissues.

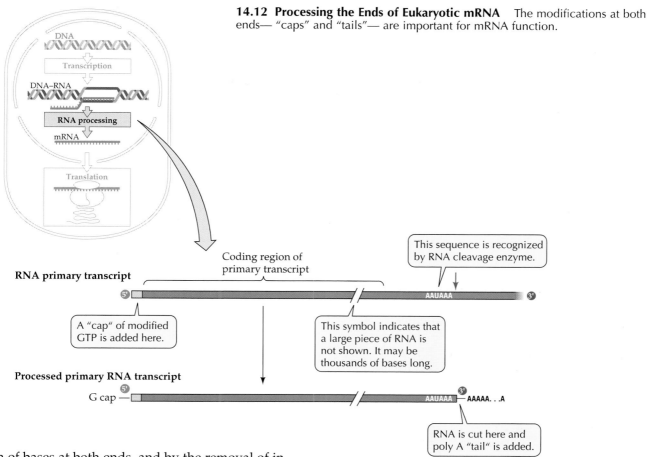

14.12 Processing the Ends of Eukaryotic mRNA   The modifications at both ends— "caps" and "tails"— are important for mRNA function.

dition of bases at both ends, and by the removal of introns and joining of exons.

The primary transcript of a protein-coding gene is modified at both ends

Two early steps in the processing of mRNA are the addition of a "cap" at the 5′ end and the addition of a poly A "tail" at the 3′ end (Figure 14.12). The cap is a chemically modified molecule of guanosine triphosphate (GTP). It apparently facilitates the binding of mRNA to the ribosome for translation and protects the mRNA from breaking down.

Following the last codon of most eukaryotic mRNAs, but not at the end of the pre-mRNA, is the sequence AAUAAA. This sequence acts as a signal for an enzyme to cut the pre-mRNA. Immediately after cleavage, another enzyme adds 100 to 300 residues of adenine (poly A) to the 3′ end of the mRNA. This "tail" may assist in export of the mRNA from the nucleus.

Splicing removes introns from the primary transcript

The next step in the processing of eukaryotic mRNA within the nucleus is the removal of the introns. If these RNA regions were not removed, a nonfunctional mRNA producing an improper amino acid sequence in the protein would result. The process called RNA

splicing removes introns and splices the exons together (see Figure 14.8).

After the primary transcript is made, it is quickly bound to several **small nuclear ribonucleoprotein particles** (**snRNPs**, commonly pronounced "snurps"), which begin the splicing process. How do these particles recognize the mRNA? The answer lies in the base sequences at the junctions between exons and introns and in the sequences of the RNAs in the snRNPs.

At the boundaries between introns and exons there are **consensus sequences**—short stretches of DNA that appear, with little variation, in many different genes. The RNA in one of the snRNPs (called U1) has a stretch of bases complementary to the consensus 5′ exon–intron boundary. Another snRNP (U2) binds near the 3′ intron–exon boundary. In both cases, the two RNAs—snRNP and pre-mRNA—bind by base pairing (Figure 14.13). Still other snRNPs bind sequences within the intron itself, forming a large RNA–protein complex called the **spliceosome**. The spliceosome uses ATP energy to cut the RNA, releases the introns, and joins the exons to produce mature mRNA (Figure 14.14).

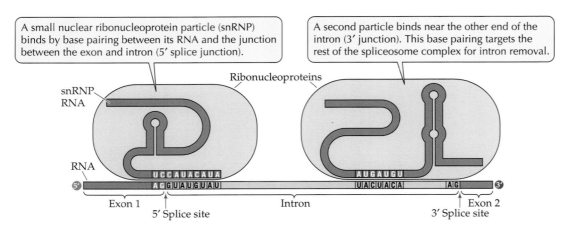

A small nuclear ribonucleoprotein particle (snRNP) binds by base pairing between its RNA and the junction between the exon and intron (5′ splice junction).

A second particle binds near the other end of the intron (3′ junction). This base pairing targets the rest of the spliceosome complex for intron removal.

Ribonucleoproteins

snRNP RNA

RNA

UCCAUACAUA   AUGAUGU

AGGUAUGUAU   UACUACA   AG

5′   3′

Exon 1   Intron   Exon 2

5′ Splice site   3′ Splice site

14.13 Recognition of Exon–Intron Junctions for Splicing
Two large RNA–protein complexes line up the splicing machinery.

Molecular studies of human diseases have been valuable tools in the investigation of consensus sequences and splicing machinery. Beta thalassemia is a human genetic disease inherited as an autosomal recessive trait in which people make an inadequate amount of the β-globin polypeptide that is part of hemoglobin. These people suffer from severe anemia because they have an inadequate supply of red blood cells. In some cases, the genetic mutation occurs at the consensus sequence at the boundary between an intron and an exon. Consequently, the mRNA cannot be spliced correctly, and nonfunctional globin mRNA is made.

This is an excellent example of the use of mutations in determining a cause-and-effect relationship in biology. In the logic of science, merely linking two phenomena (for example, consensus sequences and splicing) does not prove that one is necessary for the other. In an experiment, the scientist alters one phenomenon (for example, the base sequence at the consensus region) to see whether the other event (for example, splicing) occurs. In thalassemia, nature has done the experiment for us.

People who have certain connective-tissue diseases make antibodies against their own proteins (autoimmunity). In the rheumatic disease lupus, for example, antibodies are made against the U1 snRNP used in mRNA splicing. Addition of these antibodies to the spliceosome inhibits splicing. Although it is not clear whether the antibodies actually cause this inhibition in the patient, they have proved invaluable as a tool for extracting and studying the splicing complex.

After the processing events are completed in the nucleus, the mRNA exits the organelle, apparently through the nuclear pores (see Figure 4.8). A receptor at the nuclear pore recognizes the processed mRNA (or a protein bound to it). Unprocessed or incompletely processed primary transcripts remain in the nucleus.

Transcriptional Control

In a multicellular organism with specialized cells and tissues, every cell has every gene for that organism. For development to proceed normally and for each cell to acquire and maintain its proper function, certain specific proteins must be synthesized at just the right times and in just the right cells. Thus, the expression of eukaryotic genes is precisely regulated. The methods for this regulation are varied (Figure 14.15).

In some cases, gene regulation depends on changes in the DNA itself: Genes are rearranged on the chromosome or even selectively replicated to give more templates to transcribe. In other cases, changes in the proteins that bind to DNA can make genes more or less available for transcription. Both the transcription and the processing of pre-mRNA can be controlled. Transport of the mRNA into the cytoplasm and its stability in the new location can also be controlled. Once the mRNA is in the cytoplasm, its translation into protein can be regulated. Finally, once the protein itself is made, its structure can be modified, thereby affecting its activity.

In this section, we will describe several ways in which cells control the transcription of specific genes. First, we will examine how a high degree of selectivity in gene transcription can be achieved by specific activation and inhibition of proteins that bind to DNA. Then we will see how the overall structure and compactness of the protein–DNA complex that makes up chromatin within the nucleus often influences whether transcription can occur. In addition, the locations of genes can be changed to make them more accessible to the transcriptional apparatus or genes can be selectively replicated to make more templates for mRNA production.

14.14 The Spliceosome, an RNA Splicing Machine
After the snRNPs bind to initiate the process (see Figure 14.13), other proteins come to the complex and form the spliceosome.

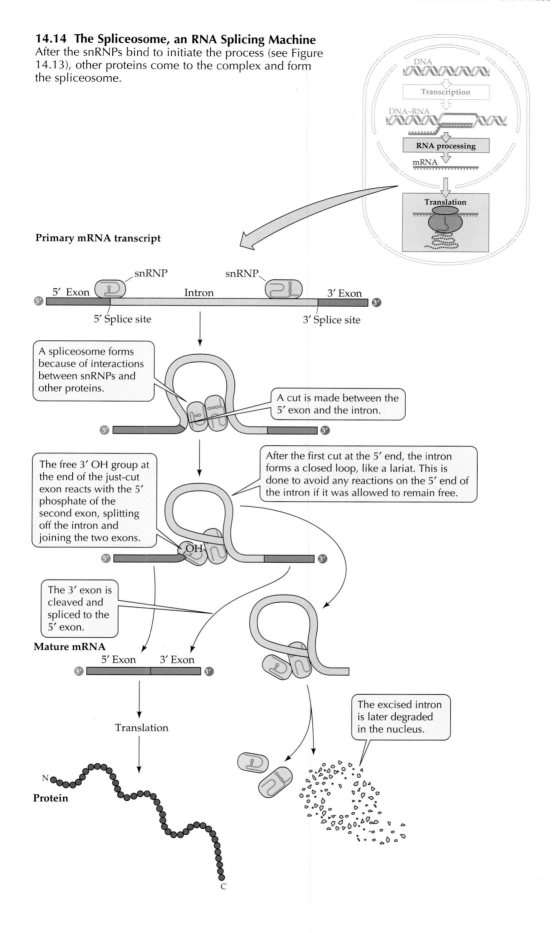

DNA

Transcription

DNA–RNA

RNA processing

mRNA

Translation

Primary mRNA transcript

snRNP   snRNP

5' Exon   Intron   3' Exon

5' Splice site   3' Splice site

A spliceosome forms because of interactions between snRNPs and other proteins.

A cut is made between the 5' exon and the intron.

After the first cut at the 5' end, the intron forms a closed loop, like a lariat. This is done to avoid any reactions on the 5' end of the intron if it was allowed to remain free.

The free 3' OH group at the end of the just-cut exon reacts with the 5' phosphate of the second exon, splitting off the intron and joining the two exons.

OH

The 3' exon is cleaved and spliced to the 5' exon.

Mature mRNA

5' Exon   3' Exon

Translation

The excised intron is later degraded in the nucleus.

N

Protein

C

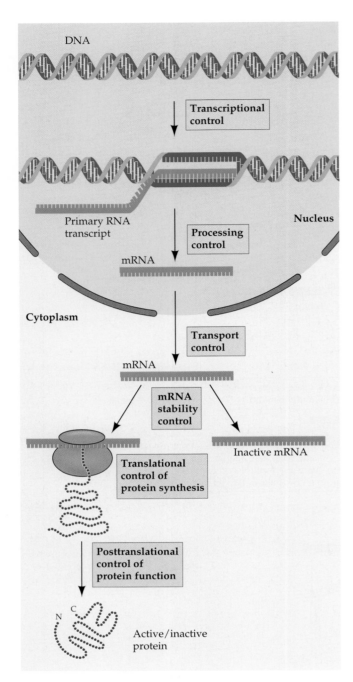

14.15 Potential Places for the Regulation of Gene Expression Genes are rearranged on the chromosome or even selectively replicated to give more templates to transcribe. In other cases, changes in the proteins that bind to DNA can make genes more or less available for transcription. And, of course, both transcription of pre-mRNA and its processing are controllable. Once the mRNA is in the cytoplasm, its translation into protein can be regulated. Finally, once the protein itself is made, modifications to its structure can occur which affect its activity.

Specific genes can be selectively transcribed

Brain and liver cells in a mouse have some proteins in common and some that are distinctive for each cell type. Yet both cells have the same DNA sequences and, therefore, the same genes. Are the differences in protein content due to differential transcription of genes, or is it that all the genes are transcribed in both cell types and a posttranscriptional mechanism (splicing, export of mRNA to the cytoplasm, translation of the mRNA, or protein longevity) is responsible for the differences in proteins?

The two alternatives—transcriptional or posttranscriptional control—can be distinguished by examination of the actual RNA sequences made within the nucleus of each cell type. Such analysis indicates that the major mechanism is differential gene transcription. Both brain and liver cells transcribe "housekeeping" genes, such as those for glycolysis enzymes and ribosomal RNAs. But liver cells transcribe some genes for liver-specific proteins, and brain cells transcribe some genes for brain-specific proteins. And neither cell type transcribes the special genes for proteins that are characteristic of muscle, blood, bone, and the other cell types in the body.

CONTRASTING EUKARYOTES AND PROKARYOTES. Unlike prokaryotes, in which related genes are transcribed as a unit in operons and under coordinate control, eukaryotes tend to have solitary genes. Thus, to regulate several genes at once requires common control elements in each gene. For example, steroid hormones such as estrogen stimulate the transcription of several different genes because each of these widely separated genes has, near its promoter, a specific sequence in common called the *hormone response element*.

In contrast to the single RNA polymerase in bacteria, eukaryotes have three different RNA polymerases. Each eukaryotic polymerase catalyzes the transcription of a specific type of gene. Only one (RNA polymerase II) transcribes protein-coding genes to mRNA. The other two transcribe the DNA that codes for rRNA (polymerase I) and for tRNA and small nuclear RNPs such as U1 (polymerase III).

The diversity of eukaryotic polymerases is reflected in the diversity of promoters, which tend to be much more variable than are prokaryotic promoters. In addition, most eukaryotic genes have enhancer and silencer elements (which we will discuss shortly) that can be quite distant from the protein-coding sequence and that control its rate of transcription. Whether a eukaryotic gene is transcribed depends on the sum total of the effects of all of these elements; thus there are many points of possible control.

Finally, the transcriptional apparatus is very different in eukaryotes than in prokaryotes, in which a sin-

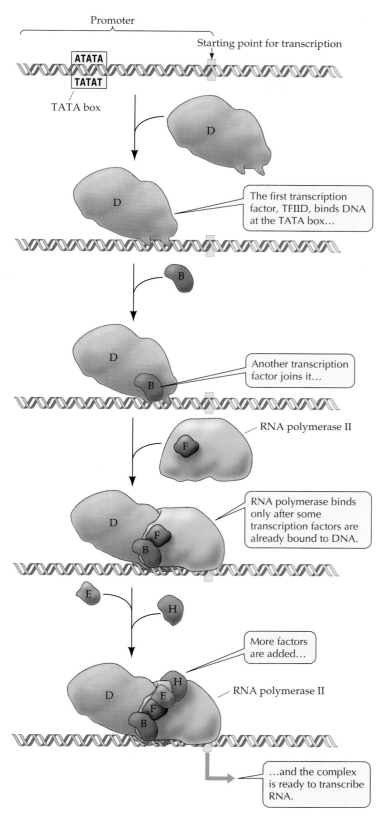

Promoter

ATATA
TATAT

TATA box

Starting point for transcription

The first transcription factor, TFIID, binds DNA at the TATA box…

Another transcription factor joins it…

RNA polymerase II

RNA polymerase binds only after some transcription factors are already bound to DNA.

More factors are added…

RNA polymerase II

…and the complex is ready to transcribe RNA.

14.16 The Initiation of Transcription on Eukaryotic DNA
Each transcription factor binds to specific regions on DNA. Except for TFIID, which also binds to the TATA box, each factor also has binding sites only for the other proteins.

gle peptide subunit can cause RNA polymerase to recognize the promoter. In eukaryotes, many proteins are involved at the initiation stage of transcription. We will confine our subsequent discussion to RNA polymerase II, which catalyzes the transcription of most protein-coding genes. The mechanisms for the other two polymerases are similar.

TRANSCRIPTION FACTORS. Recall from Chapter 13 that the *promoter* is a sequence of DNA near the 5' end of the the coding region where RNA polymerase begins transcription. A prokaryotic promoter has two essential regions: One, about 40 bp 5' to the initiation point of transcription, is the sequence recognized by RNA polymerase. The second, nearer to the initiation point, is rich in AT base pairs (it is called the TATA box) and is the site at which DNA begins to denature so that its templates can be exposed. In eukaryotes, there is a TATA box about 25 bp away from the initiation site for transcription, and one or two recognition sequences about 50 to 70 bp 5' from the TATA box.

Eukaryotic RNA polymerase II by itself cannot simply bind to the DNA at the promoter and initiate transcription. Rather, it binds and acts only after various regulatory proteins, called **transcription factors**, have assembled on the chromosome (Figure 14.16). First, the protein TFIID ("TF" stands for *t*ranscription *f*actor) binds to the TATA box. The binding event changes the shapes of both the protein and the DNA, presenting a new surface that attracts the binding of other transcription factors. RNA polymerase II does not bind until several other proteins have already bound to the complex.

Some sequences, such as the TATA box, are common to the promoters of many genes and are recognized by transcription factors found in all the cells of an organism. Other sequences in promoters may be specific to only a few genes and are recognized by transcription factors found only in certain tissues. These specific transcription factors play an important role in differentiation, the specialization of cells during development.

How do these transcription factors deal with the histone proteins present in nucleosomes? The nucleosomal proteins could block the binding of transcription factors and inhibit the initiation of transcription, but several mechanisms appear to be at work to prevent this inhibition. In some cases, the transcription factors simply bind to DNA just after it is replicated and before the histones have had a chance to bind. In other cases, proteins called nucleosome disruptors open up the nucleosome complex and allow the transcription factors to bind.

REGULATORS, ENHANCERS, AND SILENCERS. In addition to the initiation complex of transcription factors, two

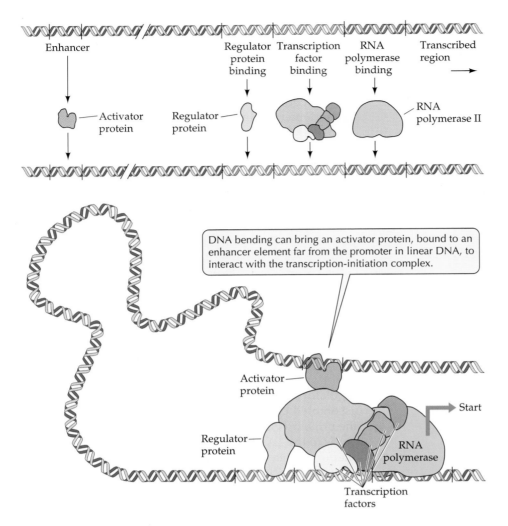

Enhancer

Regulator protein binding

Transcription factor binding

RNA polymerase binding

Transcribed region

Activator protein

Regulator protein

RNA polymerase II

DNA bending can bring an activator protein, bound to an enhancer element far from the promoter in linear DNA, to interact with the transcription-initiation complex.

Activator protein

Regulator protein

RNA polymerase

Start

Transcription factors

14.17 The Roles of Transcription Factors, Regulators, and Activators

The actions of many proteins determines whether and where RNA polymerase will transcribe DNA.

bination of all the factors is what determines the maximum rate of transcription. In the immature red blood cells of bone marrow, which make a large amount of the protein β-globin, the transcription of globin genes is stimulated by the binding of seven inducers and six activators. By contrast, in white blood cells in the same bone marrow, these regulatory proteins are not made and they do not bind to their sites adjacent to the β-globin genes; consequently these genes are hardly transcribed at all.

COORDINATING THE EXPRESSION OF GENES. In eukaryotes, how do cells coordinate the regulation of several genes whose transcription must be turned on at the same time? In prokaryotes, where related genes are linked together in an operon, coordination is clear: A single regulatory system can regulate several adjacent genes. But in eukaryotes, the several genes whose regulation requires coordination may be on different chromosomes.

In this case, regulation can be achieved if the various genes all have the same controlling sequences near them, which bind to the same activators and regulators. One of the many examples of this phenomenon is provided by the *stress proteins* (or heat shock proteins), which are made when eukaryotic cells are exposed to an elevated temperature. Under conditions of stress, various scattered genes become transcriptionally active to make the various stress proteins. Each of these genes has a specific regulatory sequence near its promoter called the heat shock element. Binding of a specific protein to this element causes the stimulation of RNA synthesis.

other regions of DNA bind proteins that activate RNA polymerase. The recently discovered **regulator** regions are clustered just upstream of the promoter. Various different proteins (in the β-globin gene, seven) may bind here. Their net effect is to bind to the adjacent transcription factor complex and activate it. Much farther away—up to 20,000 bp away—are the **enhancer** regions. Enhancer regions also bind *activator* proteins that strongly stimulate the transcription complex. How they can exert this influence from a distance is not clear. In one model, the DNA bends—it is known to do so—so that the activator is in contact with the RNA polymerase complex (Figure 14.17). Finally, there are negative regulatory regions on DNA called **silencers**, which have the reverse effect of enhancers. Silencers turn off transcription by binding to proteins appropriately called *repressors.*

How do these various proteins and DNA sequences—transcription factors, activators, repressors, regulators, enhancers, and silencers—regulate transcription? Apparently, all genes in most tissues can transcribe a small amount of RNA. But the right com-

THE BINDING OF PROTEINS TO DNA. A key to transcriptional control in eukaryotes is that transcription factors, activators, and repressors all bind to specific DNA sequences. How do they recognize and bind to DNA? There are four common structural themes for protein structures that bind to DNA. These themes are

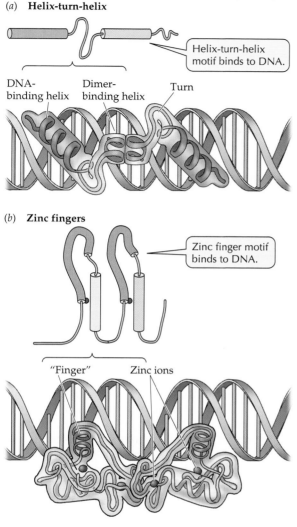

(a) **Helix-turn-helix**

Helix-turn-helix motif binds to DNA.

DNA-binding helix   Dimer-binding helix   Turn

(b) **Zinc fingers**

Zinc finger motif binds to DNA.

"Finger"   Zinc ions

14.18 DNA-Binding Proteins Most transcription factors and other DNA-binding proteins have one of these four structural regions (motifs) by which they bind to DNA.

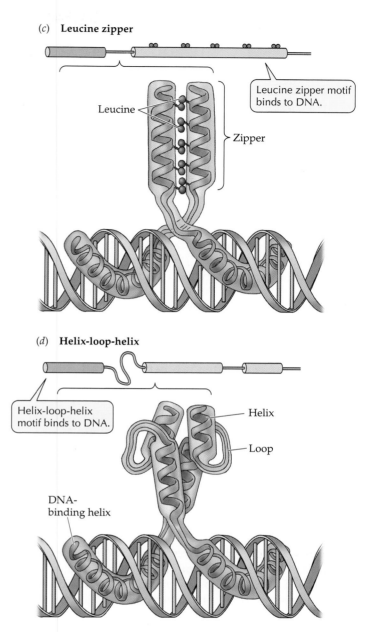

(c) **Leucine zipper**

Leucine zipper motif binds to DNA.

Leucine

Zipper

(d) **Helix-loop-helix**

Helix-loop-helix motif binds to DNA.

Helix

Loop

DNA-binding helix

called *motifs* and consist of combinations of structures and special components. The *helix-turn-helix* motif (Figure 14.18a) occurs in many transcription factors that stimulate specific genes during development. This motif appears in the proteins that activate genes involved in development of the embryo (homeobox proteins; see Chapter 15) and in the proteins that regulate development of the immune and central nervous systems.

The *zinc finger* motif (Figure 14.18b) occurs in transcription factors, most notably the steroid hormone receptors. Steroid hormones, such as estrogen, can enter cells though the lipid membrane because they are nonpolar. The hormones then bind to receptor proteins in

the cytoplasm, and the complex enters the nucleus. There, the complex binds to the hormone response elements near promoters to activate transcription.

The *leucine zipper* motif (Figure 14.18c) occurs in many DNA-binding proteins—for example, the inducer AP-1, which binds near promoters of genes involved in mammalian cell growth and division. Overactivity of AP-1 has been linked to several types of cancer.

The *helix-loop-helix* motif (Figure 14.18d) occurs in the activator proteins that bind to enhancers for the immunoglobulin genes that synthesize antibodies, as well as in the inducers involved in muscle protein synthesis.

Genes can be inactivated by chromatin structure

The packaging of DNA by nuclear proteins (see Figure 14.3) can make DNA physically inaccessible to RNA polymerase and the rest of the transcription apparatus, much like the binding of repressor to the operator in the prokaryotic *lac* operon prevents transcription. As mitosis or meiosis ends, chromosomes partly uncoil (see Chapter 9). Two kinds of chromatin can be distinguished by staining of the interphase nucleus: euchromatin and heterochromatin. *Euchromatin* is diffuse and stains lightly, and is transcribed into mRNA. *Heterochromatin* stains densely and is generally not transcribed; any genes that it contains are thus inactivated. Heterochromatin contains much of the highly repeated DNA.

How chromatin inactivation controls gene expression is best understood for the X and Y chromosomes of mammals. The normal female has two X chromosomes, the normal male an X and a Y. The Y chromosome has few genes that are also present on the X, and the Y is largely transcriptionally inactive in most cells. So there is a great difference between females and males in the "dosage" of X chromosome genes.

In other words, each female cell has two copies of the genes on the X chromosome and therefore has the potential of producing twice as much protein product of these genes as a male has. When the gene involved is on an autosome, the result is usually lethal and the embryo fails to develop. How do both sexes develop, when one of them obviously has an "extra" (or one too few, depending on your viewpoint) chromosome?

The answer was found in 1961 independently by Mary Lyon, Liane Russell, and Ernest Beutler. They suggested that one of the X chromosomes in each cell of an XX female is transcriptionally inactivated early in the life of the embryo and remains inactive, as do all the cells arising from it. In a given cell, the "choice" of which X in the pair of Xs to inactivate is usually random. Recall that one of the Xs in a female came from her father and one from her mother. Thus, in one embryonic cell the paternal X might be the one remaining active in mRNA synthesis, but in its neighboring cell, the maternal X might be active.

This suggestion is supported by genetic, biochemical, and cytological evidence. Interphase cells of XX females have a single, stainable nuclear body called a **Barr body** after its discoverer, Murray Barr (Figure 14.19). This clump of heterochromatin, which is not present in males, is the inactivated X chromosome.

The number of Barr bodies in each nucleus is equal to the number of X chromosomes minus one (the one represents the X chromosome that remains transcriptionally active). So a female with the normal two X chromosomes will have one Barr body, one with three X's has two, an XXXX female has three, and an XXY male has one. We may infer that interphase cells of each person, male or female, have a single *active* X chromosome, making the dosage of the expressed X chromosomes genes constant in both sexes.

The mechanism of *X inactivation* is chromosome condensation that makes the DNA sequences physically unavailable for the transcription machinery. This process involves chromosomal proteins. One method may be the addition of a methyl group to the 5' position of cytosine on DNA. Methylation seems to be most prevalent in transcriptionally inactive genes. For example, most of the DNA of the inactive X chromosome has many cytosines methylated, while few of them on the active X are methylated. Methylated DNA appears to bind certain specific chromosomal proteins, and these may be responsible for heterochromatin formation. However, this has not yet been proved.

The otherwise inactive X chromosome has one gene that is lightly methylated and *is* transcriptionally active. This gene is called *XIST* (for *X inactivation specific transcript*) and is heavily methylated and *not* transcribed from the other, "active" X chromosome in a female. The RNA transcribed from *XIST* does not leave the nucleus and is not an mRNA. Instead, it appears to bind to the X chromosome that transcribes it, and this binding somehow leads to a spreading of inactivation along the chromosome.

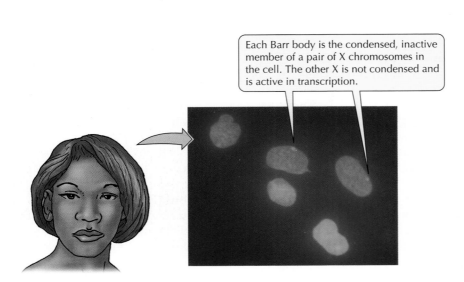

Each Barr body is the condensed, inactive member of a pair of X chromosomes in the cell. The other X is not condensed and is active in transcription.

14.19 Barr Bodies in the Nuclei of Female Cells The number of Barr bodies per nucleus is equal to the number of X chromosomes minus one. Thus males (XY) have no Barr body, whereas females (XX) have one.

A DNA sequence can move to a new location to activate transcription

In some instances gene expression is regulated by movement of a gene to a new location on the chromosome. One example is in the yeast *Saccharomyces cerevisiae*. The haploid single cells of this fungus exist in one of two mating types, *a* and α, which fuse to form a diploid zygote. Although all cells have an allele for each of these types, the allele that is expressed determines the mating type. In some yeasts, the mating type changes with almost every cell division cycle. How does it change so rapidly?

The yeast cell keeps the two different alleles (coding for type α and type *a*) at separate locations on the chromosome, away from a third site, the MAT locus, which is the site of expression. These two alleles are usually transcriptionally silent because a repressor protein binds to them. However, when a copy of the α or *a* allele inserts at the MAT region, the gene for proteins of the appropriate mating type is transcribed.

A change in mating type requires three steps: First, a new DNA copy of the nonexpressed allele is made (if the cell is now α, the new copy will be the *a* allele). Second, the current occupant of the MAT region (in this case, the α DNA) is removed by a nuclease. Third, the new allele (*a*) is inserted at the MAT region and transcribed. The *a* proteins are now made, and the mating type is changed.

DNA rearrangement is also important in producing the highly variable proteins that make up the human repertoire of antibodies. We will return to this subject in Chapter 18.

Selective gene amplification results in more templates for transcription

Another way for a cell to make more of a gene product than another cell does is to have more copies of the appropriate gene and to transcribe them all. The process of creating more copies of a specific gene in order to increase transcription is called **gene amplification**.

As described earlier, the genes that code for three of the four human ribosomal RNAs are linked together in a unit, and this unit is repeated several hundred times in the genome to provide more templates for rRNA synthesis (rRNA is the most abundant kind of RNA in the cell). In some instances, however, even this moderate repetition is not enough to satisfy the demands of the cell. For example, the mature eggs of frogs and fish have up to 1 trillion ribosomes. These ribosomes are used for the massive protein synthesis that follows fertilization. How can a single cell end up with so many ribosomes (and so much rRNA)? Adding to the problem is the fact that the cell that differentiates into the egg would take 50 years to make 1 trillion ribosomes if it transcribed its rRNA genes at peak efficiency.

The egg cell solves this problem by selectively amplifying its rRNA gene clusters until there are more than a million copies of a cluster that was originally present in fewer than 1,000 copies. In fact, this gene complex goes from being 0.2 percent of the total genome DNA to being 68 percent. These million copies transcribing at maximum rate (Figure 14.20) are just enough to make the necessary trillion ribosomes in a few days.

The mechanism for selective overreplication of a single gene is not clearly understood, but it is important. As Chapter 17 will show, in some cancers, a cancer-causing gene called an oncogene becomes amplified. Also, in some tumors treated with a drug that targets a

14.20 Transcription from Multiple Genes for rRNA
Elongating strands of rRNA transcripts form arrowhead-shaped regions, each centered on a strand of DNA that codes for rRNA.

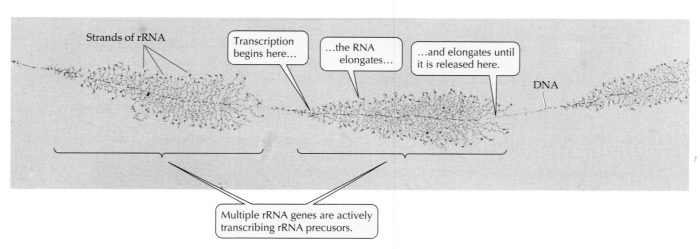

single protein, amplification of the gene for the target protein leads to an excess of that protein and the cell becomes resistant to the prescribed dose of the drug.

Posttranscriptional Control

There are many ways to control the appearance of mature mRNA even after a precursor has been transcribed. For example, the exons of the pre-mRNA can be recombined in different ways by alternate splicing. The result is that different proteins are synthesized, depending on which exons are used to make up the final mRNA. Or, the longevity of mRNA in the cytoplasm can be regulated. The longer an mRNA exists in the cytoplasm, the more of its coded protein can be made.

Different mRNAs can be made from the same gene by alternate splicing

Most primary transcripts have several introns (see Figure 14.8). The splicing mechanism recognizes the boundaries between exons and introns. What would happen if the β-globin pre-mRNA, which has two introns, was spliced from the start of the first intron to the end of the second? Not only the two introns but also the middle exon would be spliced out. An entirely new protein (certainly not a β-globin) would be made and the functions of normal β-globin would be lost.

Although alternate splicing is not common in normal RNA processing, it can be a deliberate mechanism to generate a family of different proteins from a single gene. For example, a single primary transcript for the structural protein tropomyosin can be alternately spliced to give five different mRNAs and five different proteins in five different tissues: skeletal muscle, smooth muscle, fibroblast, liver, and brain (Figure 14.21). The same mechanism is involved in generating the wide variety of antibodies in the immune system.

The stability of mRNA can be regulated

As the genetic material, DNA must remain stable, and there are elaborate repair mechanisms if it becomes damaged. RNA, on the other hand, has no such repair system. After it arrives in the cytoplasm, mRNA is subject to breakdown catalyzed by ribonucleases, which exist both in the cytoplasm and in lysosomes. But not all eukaryotic mRNAs have the same life span. The differences in how long mRNAs last make possible the control of protein synthesis by the differential stabilities of mRNAs.

Tubulin is a protein that polymerizes to form microtubules, a component of the cytoskeleton. When a large pool of free tubulin is available in the cytoplasm, there is no particular need for the cell to make more of it. Under these conditions some tubulin molecules bind to tubulin mRNA, and this binding makes the mRNA especially susceptible to breakdown, and less tubulin is made. Other examples illustrate the same mechanism—that the less time an mRNA stays in the cytoplasm, the less it can be translated into protein.

Translational and Posttranslational Control

Just as proteins can control the synthesis of mRNA by binding to DNA, they can also control the translation

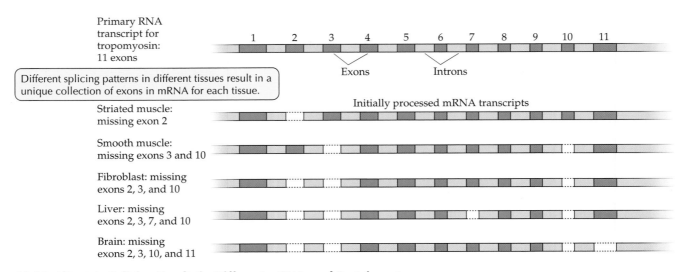

14.21 Alternate Splicing Results in Different mRNAs and Proteins In mammals, the protein tropomyosin is coded for by a gene that has 11 exons. Different tissues splice tropomyosin pre-mRNA differently, resulting in 5 different proteins. For example, in striated (skeletal) muscle, exon 2 (along with its flanking introns) is spliced out, resulting in a protein that is made up of amino acids coded for by exons 1 and 3–11.

of mRNA by binding to mRNA in the cytoplasm. This mode of control is especially important in long-lived mRNAs. A cell must not continue to make proteins that it does not need. For example, mammalian cells respond to certain stimuli by making cytokines, which stimulate specific cells to divide (see Chapter 9). If the mRNA for a cytokine is still in the cytoplasm and available for translation long after the cytokine is needed, it will be made and released inappropriately. This might cause a target cell population to divide inappropriately, forming a tumor.

Let's examine the role of translational repressors in protein synthesis and the ways in which the activity and lifetime of a protein can be regulated after it is made.

The translation of mRNA can be controlled

One way to control translation is by the capping mechanism on mRNA. As already noted, mRNA is capped at its 5′ end by a modified guanosine residue (see Figure 14.12). Messenger RNAs that have unmodified caps are not translated. For example, stored mRNA in the oocyte of the tobacco hornworm moth has the guanosine added to its end, but the G is not modified. Hence, this stored mRNA is not translated. However, after fertilization, the cap is modified, allowing the mRNA to be translated to produce proteins needed for early embryogenesis.

Free iron ions (Fe^{2+}) within a mammalian cell are bound by a storage protein, *ferritin*. When iron is in excess, ferritin synthesis rises dramatically. Yet the amount of ferritin mRNA remains constant. This increase in ferritin synthesis is due to an increased rate of mRNA translation. When the iron level in the cell is low, a *translational repressor* protein binds to ferritin mRNA and prevents its translation by blocking its attachment to the ribosome. When iron levels rise, the excess iron binds to the repressor and alters the three-dimensional structure of the translational repressor, causing it to detach from the mRNA, and translation proceeds.

Translational control also acts in the synthesis of hemoglobin. As we described earlier, hemoglobin consists of four polypeptide chains and a nonprotein pigment, heme. If heme synthesis does not equal globin synthesis, some polypeptide chains will stay free in the cell, waiting for their heme partner. One way that the balance is maintained is that excess heme in the cell stimulates the rate of translation of globin mRNA by removing a block of the initiation of translation at the ribosome.

Protein function and lifetime can be regulated after translation

We have considered how gene expression may be regulated by the control of transcription, of RNA process-

ing, and of translation. However, the story does not end here, because most proteins are modified after translation. Some of these changes are permanent, such as the addition of sugars (glycosylation) or the removal of a signal sequence after a protein has crossed a membrane (see Chapter 11). Other changes are reversible, such as the addition of phosphate groups (phosphorylation), which we will discuss further in later chapters.

An important way to regulate the action of a gene product (protein) in a cell is to regulate its lifetime in the cell. Proteins involved in cell division (e.g., the cyclins) are hydrolyzed at the right moment to time the sequence of events. Transcriptional inducers must be present only when they are needed, lest the affected genes be always "on." Proteins identified for breakdown are often linked to a small protein called *ubiquitin* (so called because it is ubiquitous, or widespread). Ubiquitin forms short chains on the targeted protein, and these chains attract a large complex of proteases that catalyze the breakdown of the protein.

Even single-celled eukaryotes such as yeasts have many of the complex mechanisms of gene regulation that we have described in this chapter. In multicellular organisms, from worms to wheat, these mechanisms must also coordinate the activities of different types of cells and tissues. Gene regulation is about the expression of genotype into phenotype. Nowhere is this more dramatic than in the unfolding of development from a fertilized egg to an adult organism. It is not surprising that many genes are expressed and then silenced during this process. We turn to these phenomena in the next chapter.

Summary of "The Eukaryotic Genome and Its Expression"

The Eukaryotic Genome

• Although eukaryotes have more DNA in their genomes than prokaryotes, in some cases there is no apparent relationship between genome size and organism complexity. **Review Figure 14.2**

• Unlike prokaryotic DNA, eukaryotic DNA in interphase cells is separated from the cytoplasm by being contained within a nucleus.

• Eukaryotic DNA is extensively packaged by proteins. **Review Figure 14.3**

• Nucleic acid hybridization is an important technique for analyzing eukaryotic genes. **Review Figure 14.4**

• Nucleic acid hybridization reveals that there are three classes of DNA. The fastest-reannealing DNA represents short sequences repeated up to millions of times in the genome. This highly repeated DNA is usually not tran-

scribed and may assist in spindle attachment to chromosomes during mitosis. DNA that reanneals more slowly consists of moderately repetitive sequences. The slowest-reannealing DNA consists of single-copy sequences.
• Some telomeric DNA may be lost during each DNA replication, eventually leading to chromosome instability and cell death. The enzyme telomerase catalyzes the restoration of the lost telomeric DNA. Most somatic cells lack telomerase and thus have limited life spans. Cancer cells are able to divide continuously because they express telomerase. **Review Figure 14.5**
• Some moderately repetitive DNA sequences, such as the sequences coding for rRNAs, are transcribed. **Review Figure 14.6**
• Some moderately repetitive DNA sequences are transposable, or able to move about the genome. **Review Figure 14.7**

The Structures of Protein-Coding Genes

• Eukaryotic protein-coding genes are usually present in only one copy per haploid genome.
• A typical protein-coding gene has noncoding internal sequences (introns) and flanking sequences that are involved in the machinery for transcription and translation. **Review Figures 14.8, 14.9**
• Some eukaryotic genes form families with related genes that have similar sequences and code for similar proteins. These related proteins may be made at different times and in different tissues. Some sequences in gene families are pseudogenes, which code for nonfunctional mRNAs or proteins. **Review Figures 14.10**
• Differential expression of different genes in the globin family ensures important physiological changes during human development. **Review Figure 14.11**

RNA Processing

• After transcription, the precursor to mature mRNA is covalently altered by the addition of a 5′ cap of a modified guanosine and a 3′ poly A tail. **Review Figure 14.12**
• The introns are removed from the mRNA precursor by the spliceosome, a complex of RNAs and proteins. **Review Figures 14.13, 14.14**

Transcriptional Control

• Eukaryotic gene expression can be controlled at the transcriptional, posttranscriptional, translational, and posttranslational levels. **Review Figure 14.15**
• The major method of control of eukaryotic gene expression is by selective transcription resulting from specific proteins binding to regulatory regions on DNA.
• A series of proteins must bind to the promoter before RNA polymerase can bind. Whether RNA polymerase will initiate transcription also depends on the binding of regulatory proteins and activator proteins (which are bound by enhancers and stimulate transcription), and repressor proteins (which are bound by silencers and inhibit transcription). **Review Figures 14.16, 14.17**
• Regulatory, enhancer, and silencer proteins bind to DNA by recognizing specific sequences. There are four domains, or regions, that are used in these DNA-binding proteins. **Review Figure 14.18**
• The clumping of chromatin by proteins bound to DNA can block transcription by making DNA physically unavailable to RNA polymerase. This clumping results in hetero-

chromatin, as in X inactivation in female mammals. Specific DNA modifications may be involved in this process. **Review Figure 14.19**
• In some cases, such as the change from one mating type to another in yeast, the movement of a gene to a new location on a chromosome may alter the ability of the gene to be transcribed.
• Some genes become selectively amplified, and the extra copies result in increased transcription. **Review Figure 14.20**

Posttranscriptional Control

• Because eukaryotic genes have several exons, alternate splicing can be used to produce different proteins. **Review Figure 14.21**
• The longevity of mRNA in the cytoplasm can be regulated by the binding of proteins.

Translational and Posttranslational Control

• Translational repressors can inhibit the translation of mRNA.
• The function and stability of a completed protein can be regulated by chemical modifications.

Self-Quiz

1. Eukaryotic protein-coding genes differ from their prokaryotic counterparts in that only eukaryotic genes
   a. are double-stranded.
   b. are present in only a single copy.
   c. contain introns.
   d. have a promoter.
   e. transcribe mRNA.

2. Which statement about nucleic acid hybridization is *not* true?
   a. Part of the process is complementary base pairing.
   b. A DNA strand can hybridize with another DNA strand.
   c. An RNA strand can hybridize with a DNA strand.
   d. A polypeptide can hybridize with a DNA strand.
   e. Double-stranded DNA is denatured at high temperatures.

3. Which statement about repetitive DNA is true?
   a. Highly repetitive DNA reanneals most slowly of the three classes of DNA.
   b. Highly repetitive DNA is usually transcribed rapidly.
   c. Much highly repetitive DNA is at the centromeres.
   d. Single-copy DNA is rare in eukaryotes.
   e. Transposable elements are single-copy genes.

4. Which of the following does *not* occur after mRNA is transcribed?
   a. binding of RNA polymerase II to the promoter
   b. capping of the 5′ end
   c. addition of a poly A tail to the 3′ end
   d. splicing out of the introns
   e. transport to the ribosome

5. Which statement about RNA splicing is *not* true?
   a. It removes introns.
   b. It is performed by small nuclear ribonucleoprotein particles (snRNPs).
   c. It always removes the same introns.
   d. It is directed by consensus sequences.
   e. It shortens the RNA molecule.

6. Telomeres are
   a. present in equal length in all cells.
   b. removed by telomerase.
   c. essential for the stability of chromosomes.
   d. located at the ends and middle of eukaryotic chromosomes.
   e. caused by errors in DNA replication.

7. Which statement about selective gene transcription in eukaryotes is *not* true?
   a. Different classes of RNA polymerase transcribe different parts of the genome.
   b. Transcription requires transcription factors.
   c. Genes are transcribed in groups called operons.
   d. Both positive and negative regulation occur.
   e. Many proteins bind at the promoter.

8. Heterochromatin
   a. contains more DNA than does euchromatin.
   b. is transcriptionally inactive.
   c. is responsible for all negative transcriptional control.
   d. clumps the X chromosome in human males.
   e. occurs only during mitosis.

9. Translational control
   a. is not observed in eukaryotes.
   b. is a slower form of regulation than transcriptional control.
   c. can be achieved by only one mechanism.
   d. requires that mRNA be uncapped.
   e. ensures that heme synthesis equals globin synthesis.

10. Which of the following are *not* used in transcriptional regulation in eukaryotes?
    a. enhancers
    b. silencers
    c. transcription factors
    d. RNA polymerase subunits
    e. promoters

Applying Concepts

1. In rats, a gene 1,440 bp long codes for an enzyme made up of 192 amino acid units. Discuss this apparent discrepancy. How long would the initial and final mRNA transcripts be?

2. The activity of the enzyme dihydrofolate reductase (DHFR) is high in some tumor cells. This activity makes the cells resistant to the anticancer drug methotrexate, which targets DHFR. Assuming that you had the complementary DNA for the gene that encodes DHFR, how would you show whether this increased activity was due to increased transcription of the single-copy DHFR gene or to amplification of the gene?

3. Describe the steps in the production of a mature, translatable mRNA from a eukaryotic gene that contains introns. Compare this to the situation in prokaryotes (see Chapter 13).

4. A protein-coding gene has three introns. How many different proteins can be made from alternate splicing of the pre-mRNA transcribed from this gene?

5. Most somatic cells in a mammal do not express telomerase. Yet the germ line cells that produce gametes by meiosis do express this enzyme. Explain.

Readings

Baeuerle, P. A. 1995. *Inducible Gene Expression.* Birkhauser, Boston, MA. A survey of the many ways that eukaryotic genes are regulated.

Cech, T. R. 1986. "RNA as an Enzyme." *Scientific American,* November. A description of the exciting discovery of RNA catalysts, some of which are involved in RNA splicing.

Elgin, S. (Ed.). 1995. *Chromatin Structure and Gene Expression.* IRL Press, Oxford. A description of how chromosomal proteins influence transcription.

Gesteland, R. F. and J. F. Atkins (Eds.). 1993. *The RNA World.* Cold Spring Harbor Laboratory Press, Cold Spring Harbor, NY. A collection of papers from experts on the occurrence and significance of RNA.

Grunstein, M. 1992. "Histones as Regulators of Genes." *Scientific American,* October. An account showing that histones not only organize DNA, but also regulate transcription.

Hershey, J. W. B., M. Matthews and N. Sonenberg (Eds.). 1996. *Translational Control.* Cold Spring Harbor Laboratory, Plainview, NY. An authoritative and extensive description of the control of translation, with an emphasis on eukaryotes.

Lewin, B. 1997. *Genes VI.* Oxford University Press, New York. This molecular biology text has excellent summaries of gene structure and transcription.

Rennie, J. 1993. "DNA's New Twists." *Scientific American,* March. A discussion of transposable elements and other features.

Rhodes, D. and A. Klug. 1993. "Zinc Fingers." *Scientific American,* February. A description of how zinc finger regions in proteins help them bind to DNA and regulate transcription.

Chapter 15

Development: Differential Gene Expression

New Organs from Small Changes
The fruit fly in the bottom picture has a genetic mutation that altered the pattern of its development and produced a complete and functional second set of wings.

*A*t one time, each of the *Drosophila melanogaster* in these photographs was a fertilized egg, a single cell. This cell contained the entire genome for the fly, with complete instructions for the production of a mature organism. In a matter of days, these instructions were carried out as a series of genes was expressed at just the right times and in just the right places. In addition, some cells produced growth factors and other signals, which interacted precisely with their target cells. Through a combination of cell division, cell movement, and cell differentiation (permanent changes in cell structure and function), the two eggs developed first into small, worm-like larvae and then into the handsome flies you see here.

These flies are obviously very different from each other. The normal one has a single pair of large wings. The other has two pairs of wings, both pairs properly constructed and fully functional. How did that happen? The fly with extra wings is a mutant—it has point mutations in just two genes. How can such a tiny genetic alteration produce such a dramatic phenotypic result? How does any multicellular organism develop its characteristic shape and functions? In this chapter we will consider how an organism's genome directs the development of an adult, with its many different cell types and complex body pattern, from a single cell.

Throughout this chapter we will describe how biologists have studied development in invertebrates, such as the fruit fly and nematode worm, as well as

330

"simpler" vertebrates such as the frog. Two major conclusions have come out of these studies: First, all types of somatic cells—all the cells except for gametes—in an organism retain all of the genes that were present in the fertilized egg. Cell differentiation does not result from a loss of DNA. Instead (and this is the second major theme of developmental biology) cellular actions during development and cell differentiation result from the differential expression of genes. During development, the various mechanisms described in Chapter 14, such as transcriptional and translational control, as well as intercellular signaling, work to produce a complex organism.

It is very appropriate to start (and end) this chapter with the fruit fly, for the mechanisms of development revealed by studying the fly turn out to be very similar to those of more complex eukaryotes, including humans.

The Processes of Development

Development is a process of progressive change during which an organism takes on the successive forms that characterize its life cycle (Figure 15.1). In its early stages of development a plant or animal is called an **embryo**. Sometimes the embryo is contained within a protective structure such as a seed coat, an eggshell, or a uterus. An embryo does not photosynthesize or actively feed; instead it obtains its food from its mother directly or indirectly (by way of nutrients stored in the egg, for example). Later stages, such as a human fetus, may precede the birth of the new, independent, organism. Development continues throughout an organism's life, ceasing only when the organism dies.

Growth (increase in size) is an important part of development, continuing throughout the individual's life in some species but reaching a more or less stable end point in others. A multicellular body cannot grow—indeed, it cannot arise from a single cell—without **cell division**. Repeated mitotic divisions generate the multicellular body. In plants, unless the daughter cells become longer (expand) after they form, the embryo does not grow; thus in plant development, cell expansion begins with the first cell divisions. In animal development, on the other hand, such **cell expansion** is often slow to begin: The animal embryo may consist of thousands of cells before it becomes larger than the fertilized egg.

Development has two major components in addition to cell division and cell expansion: differentiation and morphogenesis. The remainder of the chapter will focus on these processes.

Differentiation is the generation of cellular specificity; that is, differentiation defines the specific structure and function of a cell. Mitosis, as we have seen, produces daughter nuclei that are chromosomally and genetically identical to the nucleus that divides to pro-

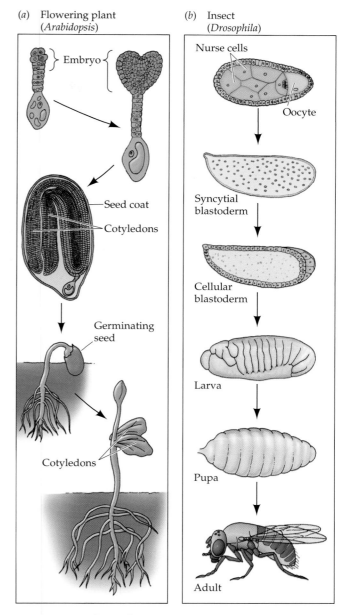

(a) Flowering plant (*Arabidopsis*)

Embryo

Seed coat

Cotyledons

Germinating seed

Cotyledons

(b) Insect (*Drosophila*)

Nurse cells

Oocyte

Syncytial blastoderm

Cellular blastoderm

Larva

Pupa

Adult

15.1 Stages of Development Stages from embryo to adult shown for a plant and an animal. Cell division and expansion, growth, cell differentiation, and the creation of the organs and tissues of the adult body are all part of the complex process.

duce them. However, the cells of an animal or plant body are not all identical in structure or function. The human body, with its approximately 100 trillion (10^{14}) cells, consists of about 200 functionally distinct cell types—for example, muscle cells, blood cells, and nerve cells. When the embryo consists of only a few cells, each cell has the potential to develop in many different ways. As development proceeds, the possibilities available to individual cells gradually narrow, until each cell's *fate* is fully determined and the cell has differentiated.

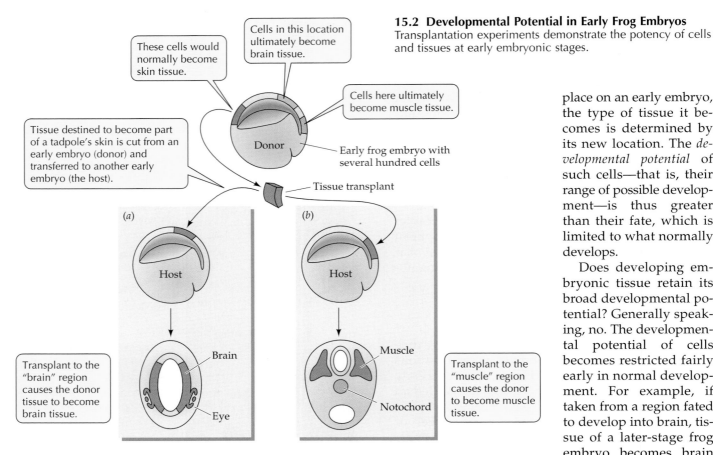

These cells would normally become skin tissue.

Cells in this location ultimately become brain tissue.

Cells here ultimately become muscle tissue.

Tissue destined to become part of a tadpole's skin is cut from an early embryo (donor) and transferred to another early embryo (the host).

Donor

Early frog embryo with several hundred cells

Tissue transplant

(a)

Host

(b)

Host

Transplant to the "brain" region causes the donor tissue to become brain tissue.

Brain

Eye

Muscle

Notochord

Transplant to the "muscle" region causes the donor to become muscle tissue.

15.2 Developmental Potential in Early Frog Embryos
Transplantation experiments demonstrate the potency of cells and tissues at early embryonic stages.

place on an early embryo, the type of tissue it becomes is determined by its new location. The *developmental potential* of such cells—that is, their range of possible development—is thus greater than their fate, which is limited to what normally develops.

Does developing embryonic tissue retain its broad developmental potential? Generally speaking, no. The developmental potential of cells becomes restricted fairly early in normal development. For example, if taken from a region fated to develop into brain, tissue of a later-stage frog embryo becomes brain tissue even if transplanted to parts of an early-stage embryo destined to become other structures. The tissue of the later-stage embryo is thus said to be *determined*: Its fate has been sealed, regardless of its surroundings. By contrast, the younger transplant tissue in Figure 15.2 has not yet become determined.

Determination precedes differentiation

Determination, the commitment of a cell to a particular fate, is a process influenced by the cellular environment and the cellular contents acting on the cell's genome. Determination is followed by differentiation, the actual changes in biochemistry, structure, and function that result in cells of different types. *Determination* is not something that is visible under the microscope—cells do not change appearance when they become determined. *Differentiation* often involves a change in appearance as well as function. Determination precedes differentiation. Determination is a commitment; the final realization of this commitment is differentiation.

The Role of Differential Gene Expression in Differentiation

Differentiated cells are recognizably different from each other, sometimes dramatically so. Certain cells in our

Whereas differentiation gives rise to cells of different kinds, **morphogenesis** (literally, "creation of form") gives rise to the shape of the multicellular body and its organs. Morphogenesis results from *pattern formation*, the organization of differentiated tissues into specific structures. (We will discuss pattern formation in detail later in the chapter.) The organized division and expansion of motionless cells are the major tools available for building the plant body form. In animals, cell movements are very important in morphogenesis, as is the programmed death of certain cells.

In plants and animals alike, differentiation and morphogenesis result ultimately from the regulated activities of genes and their products.

Cells have fewer and fewer options as development proceeds

The technique of marking specific cells of early embryos with stains and observing which cells of older embryos contain the stain enabled biologists to determine which adult structures develop from certain parts of the early embryo. For instance, the shaded area of the frog embryo shown in Figure 15.2 has the **fate** of becoming (that is, it is destined to become) part of the skin of the tadpole larva if left in place. However, if we cut out a piece from this region and transplant it to another

hair follicles continuously produce keratin, the protein of hair, nails, feathers, and porcupine quills. Blood cells do not produce keratin, nor do other kinds of cells in our bodies. In the hair follicle cells, the gene that encodes keratin is transcribed; in most other cells in our body that gene is not transcribed. During plant development, some cells activate a gene that encodes an enzyme that catalyzes the formation of suberin, the substance that gives cork its characteristic waxy feel. Activation of that gene is a key step in the differentiation of those cells. Generalizing from these observations, we may say that *differentiation results from differential gene expression*—that is, from the differential regulation of transcription, posttranscriptional events such as mRNA splicing, and translation (see Chapter 14).

Because the cells of a multicellular organism arise by mitotic divisions of a single-celled fertilized egg, or zygote, most of them are genetically identical. In the absence of mutation, all the cells in an organism have the same hereditary makeup; yet the adult organism is composed of many distinct types of cells. This apparent contradiction results from regulation of the expression of various parts of the genome.

Because the zygote has the ability to give rise to every type of cell in the adult body, we say it is **totipotent**. Its genetic "library" is complete, containing instructions for all the structures and functions that will arise throughout the life cycle. Later in the development of animals (and probably to a lesser extent in plants), the cellular descendants of the zygote lose their totipotency and become determined—that is, committed to form only certain parts of the embryo. Determined cells differentiate into specific types of cells such as neurons in the nervous system or muscle cells. When a cell achieves the fate for which it was determined, it is said to have differentiated.

The mechanisms of differentiation relate primarily to changes in the transcription and translation of genetic information.

Differentiation usually does not include a permanent change in the genome

An early explanation of the mechanisms of differentiation stated that the cell nucleus undergoes irreversible genetic changes during development. It was suggested that chromosomal material is lost, or that some of it is irreversibly inactivated.

Differentiation is irreversible in certain types of cells. Examples include the mammalian red blood cell, which loses its nucleus during development, and the tracheid, a water-conducting cell in vascular plants. Tracheid development culminates in the death of the cell, leaving only the pitted cell walls that formed while the cell was alive (see Chapter 31). In these two extreme cases, the irreversibility of differentiation can be explained by the absence of a nucleus.

Generalizing about mature cells that retain functional nuclei is more difficult. We tend to think of plant differentiation as reversible and of animal differentiation as irreversible, but this is not a hard-and-fast rule. A lobster can regenerate a missing claw, but a cat cannot regenerate a missing paw. Why is differentiation reversible in some cells but not in others? At some stage of development do changes within the nucleus permanently commit a cell to specialization?

At the Institute of Cancer Research in Philadelphia in the 1950s, Robert Briggs and Thomas J. King performed experiments to see whether all the genetic material of an organism is preserved in potentially active form, or if some of it is permanently inactivated or lost during normal development. To find out whether the nuclei of early frog embryos had lost the ability to do what the totipotent zygote nucleus could do, Briggs and King carried out a series of meticulous transplants. First they removed the nucleus from an unfertilized egg (thus forming what is called an *enucleated* egg). Then, with a very fine glass tube, they punctured a cell in an early embryo and drew up part of its contents, including a nucleus, which they then injected into the enucleated egg.

More than 80 percent of these nuclear transplant operations resulted in the formation, from the egg and its new nucleus, of a normal early embryo. Of these embryos, more than half developed into normal tadpoles and, ultimately, normal adult frogs. These experiments showed that no information is lost from the nuclei of cells as they pass through the early stages of embryonic development. On the other hand, Briggs and King found that when the donor nuclei were derived from older embryonic stages, fewer tadpoles developed (Figure 15.3).

The Briggs and King experiments demonstrated that every cell in the early frog embryo has all of the genes necessary to produce an adult frog. The same is probably true for humans, leading to a practical application in human genetic testing. An eight-celled human embryo can be isolated in the laboratory and a single cell removed to determine if a harmful allele is present in homozygous form. Each remaining cell, being totipotent, can be stimulated to divide and act as a zygote.

Briggs and King's work was carried further by John B. Gurdon at Oxford University, who performed similar transplants on a different species of frog, using nuclei from more advanced embryos and even swimming tadpoles. These nuclei, transplanted into an egg whose nucleus had been inactivated, could produce all the types of cells present in embryos, tadpoles, and, in some cases, adult frogs.

Gurdon's results confirmed the totipotency of the nucleus of a differentiated cell. Clearly, nuclei change in their activities during differentiation, but the changes need not be irreversible. The environment

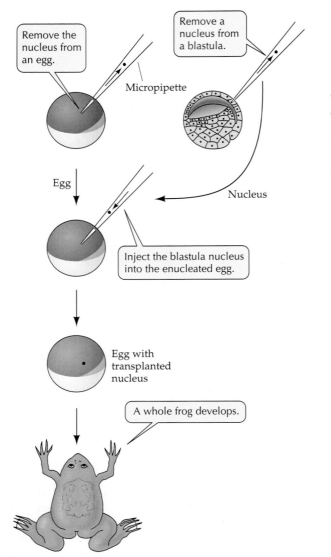

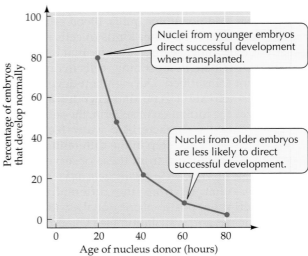

15.3 Nuclei Lose Differentiation Potential with Age
Nuclear transplant experiments on frogs by Briggs and King showed that the potential for cell differentiation declines with the age of the embryo.

around the nucleus—the cytoplasm—exerts a great influence over which nuclear genes are expressed. Thus a nucleus inside a cell in the tadpole's intestine is surrounded by "intestinal" signals, but it retains the ability to act as a fertilized egg nucleus if it is put into the cytoplasm of a fertilized egg.

A dramatic example of totipotency was reported in 1997, when Ian Wilmut and his colleagues at a biotechnology company in Scotland used a modification of Gurdon's nuclear transplant procedure to clone sheep (Figure 15.4). Previous attempts to produce mammals by this method had worked only if the donor nucleus was part of an entire cell and if the donor cell was from an early embryo. Using adult donor cells, as Gurdon did with frogs, resulted in chromosomal abnormalities and embryonic death. Apparently, when mammalian donor cells were in the G2 phase of the cell cycle and were fused with egg cytoplasm also in

G2, some extra DNA replication took place that created havoc with the cell cycle in the egg when it attempted to divide.

Wilmut took differentiated cells from a ewe's udder and starved them of nutrients for a week, thus halting the cells in G1. After one of these cells was fused with an enucleated egg from a different ewe (fusion of the donor cell and enucleated egg was achieved by electrical stimulation), mitotic inducers in the egg cytoplasm (see Chapter 9) were able to stimulate the donor nucleus to enter S phase and the rest of the cell cycle proceeded normally. After several cell divisions, the early embryo was transplanted into the womb of a surrogate mother. Of 272 successful attempts to fuse adult cells with enucleated eggs, one lamb, Dolly, survived to be born. Dolly was genetically identical to the ewe from whose udder the donor nucleus had been obtained.

The purpose of Wilmut's experiment was to clone sheep that have been genetically programmed to produce products such as pharmaceuticals in their milk (see Chapter 16). The cloning procedure could make multiple, identical copies of sheep that are reliable producers of a drug such as α-antitrypsin, which is used to treat people with cystic fibrosis. Not surprisingly, Wilmut's work touched off a storm of controversy, since Dolly was the first mammal to be born through cloning.

An example of nuclear totipotency gone awry occurs in a human tumor called a teratocarcinoma. Here, a differentiated cell "dedifferentiates" and divides, forming a tumor, as occurs in most cancers. But some

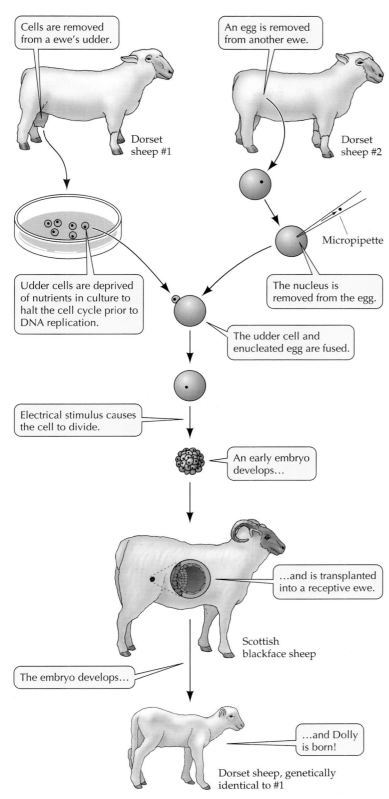

Cells are removed from a ewe's udder.

Dorset sheep #1

An egg is removed from another ewe.

Dorset sheep #2

Micropipette

Udder cells are deprived of nutrients in culture to halt the cell cycle prior to DNA replication.

The nucleus is removed from the egg.

The udder cell and enucleated egg are fused.

Electrical stimulus causes the cell to divide.

An early embryo develops…

…and is transplanted into a receptive ewe.

Scottish blackface sheep

The embryo develops…

…and Dolly is born!

Dorset sheep, genetically identical to #1

15.4 Cloning a Mammal Dolly, a cloned sheep resulting from this experiment, has the same genes as the ewe that donated the udder cells.

cells in the tumor redifferentiate to form specialized tissue arrangements. So the tumor can be a single ball of cells inside the abdomen, with some of them forming kidney tubes, others hair, and still others teeth! How this occurs is not clear.

For a clear demonstration that genes are not lost from differentiated cells, we consider next a phenomenon observed in developing insects.

Transdetermination shows that developing insects retain their genes

The overall pattern of development in butterflies, moths, and many other insects is probably familiar to you. From a fertilized egg there develops a creeping *larva* that feeds voraciously, growing through a series of molts, in which the external coat is shed to permit the body to grow. Some specialized cells, arranged in clusters called **imaginal discs**, remain undifferentiated throughout larval growth but later give rise to the tissues of the adult, such as wings, antennae, and legs. The larva eventually stops feeding and then surrounds itself with a cocoon and transforms into a *pupa*. In the pupa, tremendous changes take place. Some larval cells die, others are reprogrammed to make different products that are characteristic of the adult, and the imaginal discs differentiate into new adult structures (Figure 15.5). Such a major transformation between larva and adult is referred to as **complete metamorphosis**.

The fates of the imaginal discs of insects are determined long before metamorphosis. If transplanted from one larva to another, an imaginal disc still develops into the same type of organ (for example, a wing or an antenna) it would have if left undisturbed. In addition, that organ forms wherever the imaginal disc for it is placed in the host body.

If transplanted into an *adult* insect, an imaginal disc remains undifferentiated (because the hormonal signal for its development is lacking), but the cells of the disc continue to divide within the new host. Later these

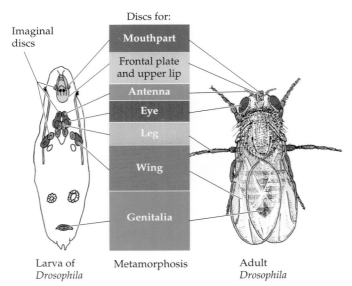

15.5 Metamorphosis in Fruit Flies In metamorphosis, most of the larval tissues die and are resorbed, providing building blocks for subsequent development. The remaining larval tissues are specialized imaginal discs, which differentiate to form the organs of the adult insect.

transplanted disc cells can be transplanted to other adult insects or to larvae. If returned to a larva, the disc cells almost always develop into the adult organ for which they were originally determined. Occasionally, however, an imaginal disc will **transdetermine** during a series of transplants; that is, it will develop into an organ other than that normally expected. Transdetermination shows that imaginal discs have not lost the genes that they normally do not express, since a disc *can*, under the right circumstances, express these genes and produce a different organ.

Many plant cells remain totipotent

A food storage cell in a carrot root faces a dark future. It is not destined to photosynthesize or to give rise to new carrot plants. However, if we isolate that cell from the root, maintain it in a suitable nutrient medium, and provide it with appropriate chemical cues, we can fool the cell into changing its behavior. In effect, the cell "thinks" it is a fertilized egg. It divides and gives rise to a typical carrot embryo and, eventually, a complete plant.

The ability to clone an entire carrot plant from a differentiated root cell indicates that the cell contains the entire carrot genetic library and that it can express the appropriate genes in the right sequence. Many cells from other plant species show similar behavior in the laboratory, and this ability to generate a whole plant from a single cell has been invaluable for genetically altering plants in biotechnology (see Chapter 16).

Genes are differentially expressed in cell differentiation

All of these experiments—nuclear transplants in frogs and sheep, imaginal disc manipulations in fruit flies, and plant cell cloning—point to the conclusion of genome constancy in all somatic cells of an organism. Molecular biology experiments have provided even more convincing evidence. For example, the gene for β-globin, one of the protein components of hemoglobin, is present and expressed in red blood cells as they form in the bone marrow of mammals. Is the same gene also present—but unexpressed—in nerve cells in the brain, which do not make the protein?

Nucleic acid hybridization (see Figure 14.4) can provide an answer. A probe for the β-globin gene can be applied to DNA from both immature red blood cells (recall that mature red blood cells lose their nuclei) and brain cells. In both cases, the probe finds its complement, showing that the β-globin gene is present in both types of cells.

On the other hand, if the probe is applied to cellular mRNAs rather than cellular DNA, it finds β-globin mRNA only in the red blood cells, and not in the brain cells. This result shows that the gene is expressed in only one of the two tissues. Many similar experiments have shown convincingly that differentiated cells lose none of the genes that were present in the fertilized egg.

What molecular program leads to this differential gene expression? One well-studied system is the conversion of undifferentiated muscle precursor cells, the myoblasts, into the large, multinucleated cells that make up skeletal muscle fibers (see Chapter 44). The key event that starts this conversion is the expression of *MyoD1* (*Myo*blast *D*etermining Gene 1). The protein product of this gene is a transcription factor (MyoD1) with a helix-loop-helix domain (see Figure 14.18) that not only binds to promoters of the muscle-determining genes to stimulate their transcription, but also acts at the gene promoter of *MyoD1* to keep its levels high in the cells and in the offspring resulting from cell divisions.

Strong evidence for the controlling role of MyoD1 comes from experiments in which *MyoD1* mRNA is injected into the precursors of other cell types. For example, if *MyoD1* mRNA is put into fat cell precursors, they become reprogrammed to become muscle cells. Genes such as *MyoD1*, which code for proteins that direct fundamental decisions in development, often by regulating genes on other chromosomes, are called **selector genes**. These genes usually code for transcription factors. We will describe other selector genes, such as the homeotic genes of *Drosophila*, later in this chapter.

The Role of Cytoplasmic Segregation in Cell Determination

What initially stimulates the *MyoD1* promoter to begin transcription is unclear, but a chemical signal clearly is involved. In general, two overall mechanisms have been found to cause such signals to stimulate their target cells. In one mechanism—*cytoplasmic segregation*—a factor within eggs, zygotes, or precursor cells is unequally distributed such that it ends up in some cells or regions of cells and not others. In the second mechanism—*induction*—a factor is actively produced and secreted to induce the target cells to differentiate.

First we will consider cytoplasmic segregation, beginning with its role in distinguishing one end of an organism from the other.

Eggs, zygotes, and embryos develop polarity

Polarity—the difference of one end from the other—is obvious in development. Our heads are distinct from our feet, and the distal ends of our arms (wrists and fingers) differ from the proximal ends (shoulders). An animal's polarity develops early, even in the egg itself. Yolk may be distributed asymmetrically in the egg and the embryo, and other chemical substances may be confined to specific parts of the cell or may be more concentrated at one pole than at the other.

In some animals, the original polar distribution of materials in the egg's cytoplasm changes as a result of fertilization. As division proceeds, the resulting cells contain unequal amounts of the materials that were not distributed uniformly in the zygote. As we learned from the work of Briggs and King, Gurdon, and Wilmut, cell nuclei do not always undergo irreversible changes during early development; thus we can explain some embryological events on the basis of the *cytoplasmic* differences in cells.

Even a structure as apparently simple as a sea urchin egg has polarity. A striking difference between cells can be demonstrated very early in embryonic development. The Swedish biologist Sven Hörstadius showed in the 1930s that the development of sea urchin embryos that have been divided in half at the eight-cell stage depends on how they are separated.

If the embryo is split into "left" and "right" halves, with each half containing cells from both the upper and the lower half, normal-shaped but dwarf larvae develop. If, however, the cut separates the upper four cells from the lower four, the result is different. The upper four cells develop into an abnormal early embryo with large cilia at one end that cannot form a larva. The lower four cells develop into a small, somewhat misshapen larva with an oversized gut (Figure 15.6). For fully normal development, factors from both the upper and lower parts of the embryo are neces-sary. In similar experiments on eggs, Hörstadius showed that material is distributed unequally between upper and lower parts already in the unfertilized egg.

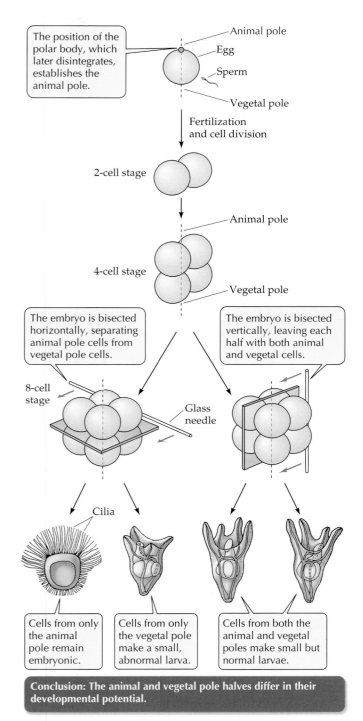

Cells from only the animal pole remain embryonic.

Cells from only the vegetal pole make a small, abnormal larva.

Cells from both the animal and vegetal poles make small but normal larvae.

Conclusion: The animal and vegetal pole halves differ in their developmental potential.

15.6 Early Asymmetry in the Embryo Experiments by Sven Hörstadius showed that the upper (animal pole) and lower (vegetal pole) halves of very early sea urchin embryos differ in their developmental potential, and that cells from both halves are necessary to produce a normal larva.

(a)

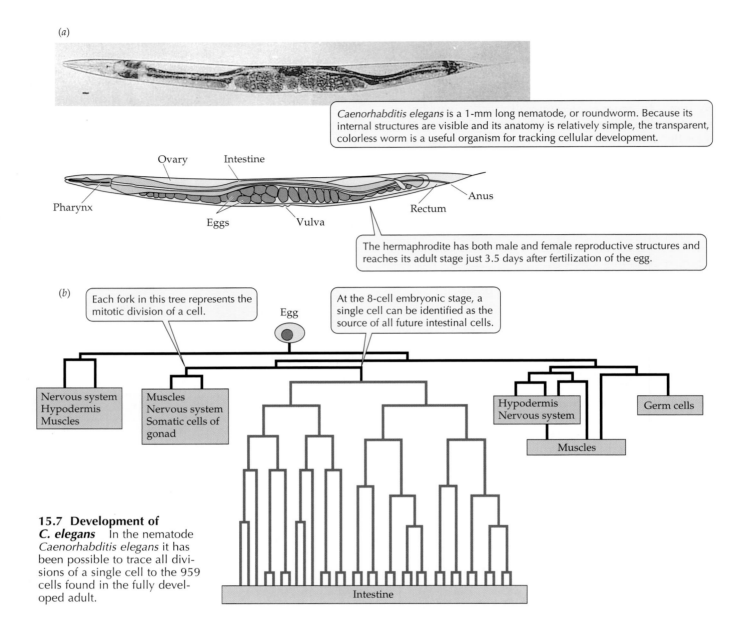

Caenorhabditis elegans is a 1-mm long nematode, or roundworm. Because its internal structures are visible and its anatomy is relatively simple, the transparent, colorless worm is a useful organism for tracking cellular development.

Ovary   Intestine

Pharynx

Eggs   Vulva

Rectum   Anus

The hermaphrodite has both male and female reproductive structures and reaches its adult stage just 3.5 days after fertilization of the egg.

(b)

Each fork in this tree represents the mitotic division of a cell.

Egg

At the 8-cell embryonic stage, a single cell can be identified as the source of all future intestinal cells.

Nervous system
Hypodermis
Muscles

Muscles
Nervous system
Somatic cells of gonad

Hypodermis
Nervous system

Muscles

Germ cells

Intestine

15.7 Development of *C. elegans*   In the nematode *Caenorhabditis elegans* it has been possible to trace all divisions of a single cell to the 959 cells found in the fully developed adult.

These and other experiments are among many that established that the unequal distribution of materials in the egg cytoplasm plays a role in directing embryonic development. Such materials are called **cytoplasmic determinants**.

Microfilaments distribute P granules in Caenorhabditis

The tiny nematode (roundworm) *Caenorhabditis elegans* (Figure 15.7) lives in the soil, where it feeds on bacteria. As we have seen in earlier chapters, bacteria can be grown in the laboratory on petri plates containing medium in agar. *C. elegans* roundworms are grown on such cultures of bacteria. The entire process of development from the egg to larva takes about 12 hours at 25°C and is easily observed using a low-magnification dis-

secting microscope, because the body covering is transparent. The facts that *C. elegans* is easy to culture, develops rapidly, and is easily observed have not surprisingly made this worm a favorite organism of developmental biologists. Indeed, its entire genome is being sequenced.

Because the development of *C. elegans* does not vary, it has been possible to construct a cellular "tree" that describes the origin of each of the 959 somatic cells of the adult form. One way to do this is to inject a marker dye into an embryonic cell: That cell and its descendants will all be labeled. But in some cases, the worm itself "marks" the cells that will differentiate along a certain pathway.

One pathway in which cells are marked is the *germ line*, which consists of cells that can form gametes. Particles called **P granules** appear to be cytoplasmic determinants for this line of differentiation. The positions of P granules in the zygote and embryo are determined by the action of microfilaments. Before the zygote divides, the P granules collect at the posterior end of the cell (thus the term P—for "pole"). Thus, all the granules appear in only one of the first two daughter cells (Figure 15.8). The P granules continue to be precisely distributed during the early cell divisions, ending up in only those cells—the germ cells—that will eventually give rise to eggs and sperm. It is uncertain whether P granules are cytoplasmic determinants themselves; they may simply be distributed together with something else that is the "real" cytoplasmic determinant. Their composition gives no obvious clue, since they are made up of many proteins and RNA.

Having seen some examples of determination by cytoplasmic segregation, next we will examine induction—the second general mechanism of determination—in which some tissues induce the determination of other tissues.

The Role of Embryonic Induction in Cell Determination

Experimental work has clearly established that the fates of particular tissues are determined by interactions with other specific tissues in the embryo. In developing animal embryos there are many such instances of induction, in which one tissue causes an adjacent tissue to develop in a different manner. We will describe two examples of such induction: one in the developing vertebrate eye, and the other in a developing reproductive structure in the nematode *C. elegans*.

Tissues direct the development of their neighbors by secreting inducers

The development of the lens in the vertebrate eye is a classic example of induction. In a frog embryo, the developing forebrain bulges out at both sides to form the *optic vesicles*, which expand until they come in contact with the cells at the surface of the head (Figure 15.9). The surface tissue in the region of contact with the optic vesicles thickens, forming a *lens placode*. The lens placode bends inward, folds over on itself, and ultimately detaches from the surface to produce a structure that will develop into the lens.

If the growing optic vesicle is cut away before it contacts the surface cells, no lens forms in the head region from which the optic vesicle has been removed. An impermeable barrier placed between the optic vesicle and the surface cells also prevents the lens from forming. These observations suggest that surface tissue begins to develop into a lens when it receives a signal—an **inducer**—from its contact with an optic vesicle.

Distribution of P granules

Distribution of nuclei

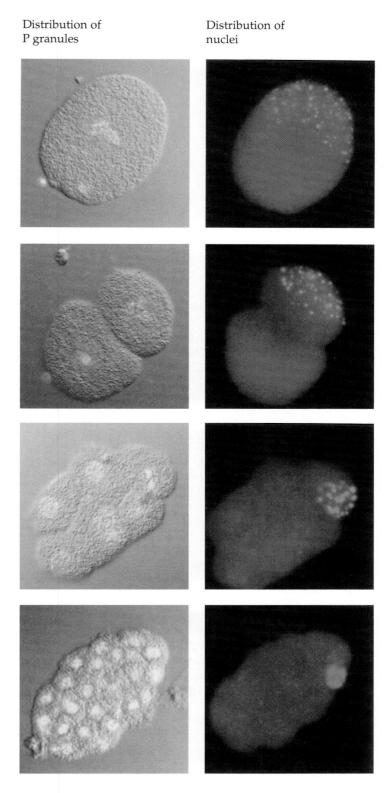

15.8 Distribution of P Granules in *C. elegans* These micrographs show a developing embryo of *Caenorhabditis elegans*. Whereas the P granules (bright spots) move to the posterior end of the embryo and are eventually confined to the single cell that gives rise to gametes, the nuclei (stained blue) of the same embryo are distributed evenly among its cells.

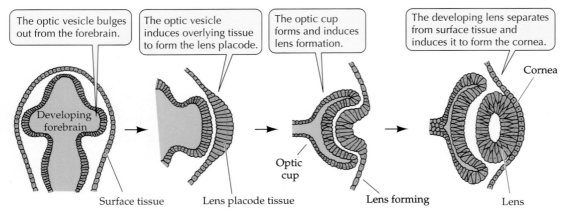

15.9 Inducers in the Vertebrate Eye The eye of a frog develops as inducers take their turns.

The interaction of tissues in eye development is a two-way street: There is a "dialogue" between the developing optic vesicle and the surface tissue. The developing lens determines the size of the optic cup that forms from the optic vesicle. If head surface tissue from a species of frog with small eyes is grafted over the optic vesicle of one with large eyes, both lens and optic cup are of intermediate size.

The developing lens also induces the surface tissue over it to develop into a cornea, a specialized layer that allows light to pass through and enter the eye. Thus a chain of inductive interactions participates in development of the parts required to make an eye. Induction triggers a sequence of gene expression in the responding cells. Tissues do not induce themselves; rather, different tissues interact and induce each other. We will return to embryonic induction in Chapter 40.

One of the most difficult problems in developmental biology has been identifying the chemical nature of the inducers. Often, only a few cells make a tiny amount of the substance. In some cases, specific diffusible proteins may be involved; the inducer that acts earliest in frog embryos appears to be a growth factor (see Chapter 40). In other cases, however, insoluble extracellular materials such as collagen and other proteins may be involved in induction.

Even single cells can induce changes in their neighbors

The tiny worm *Caenorhabditis elegans* is an excellent organism for studying the mechanisms of induction because, as we saw in Figure 15.7, the entire development of *C. elegans* takes only a few days and is clearly visible. The hermaphroditic form contains both male and female reproductive organs and lays eggs through a pore called the *vulva* on the ventral surface.

During development, a single cell, called the *anchor cell*, induces the vulva to form. If the anchor cell is de-

stroyed by laser surgery, no vulva forms. The anchor cell controls the fates of six cells on the animal's ventral surface. Each of these cells has three possible fates. By the manipulation of two genetic switches, a given cell becomes a primary vulval precursor, a secondary vulval precursor, or simply part of the worm's surface, an epidermal cell (Figure 15.10).

The anchor cell produces an inducer that diffuses out of the cell and interacts with adjacent cells. Cells that receive enough of the inducer become vulval precursors; cells slightly farther from the anchor cell become epidermis. The first switch, controlled by the inducer from the anchor cell, determines whether a cell takes the "track" toward becoming part of the vulva or the pathway toward becoming epidermis.

The cell closest to the anchor cell, having received the most inducer, becomes the primary vulval precursor and produces its own inducer, which acts on the two neighboring cells and directs them to become secondary vulval precursors. Thus the primary vulval precursor cell controls a second switch, determining whether a vulval precursor will take the primary track or the secondary track. The two inducers control the activation or inactivation of specific genes in the responding cells.

There is an important lesson to draw from this example. Much of development is controlled by switches that allow a cell to proceed down one of two alternative tracks. One challenge for the developmental biologist is to find these switches and determine how they work. In the case of vertebrates, some progress has been made.

As we will describe in Chapter 40, a key event early in embryology is the induction of differentiation of a layer of cells called the mesoderm. In this case, several proteins have been identified as having inducing properties. Three of these proteins are growth factors, which act by binding to receptors on recipient cells and causing a signal transduction cascade of events, ultimately leading to altered gene transcription. Another inducer is a transcription factor, the vertebrate version of a fruit fly protein called wingless. This

15.10 Two Gene "Switches" in *C. elegans*
Two secreted proteins act as the primary and secondary inducers. The gene activation patterns triggered by these switches determine cell fate.

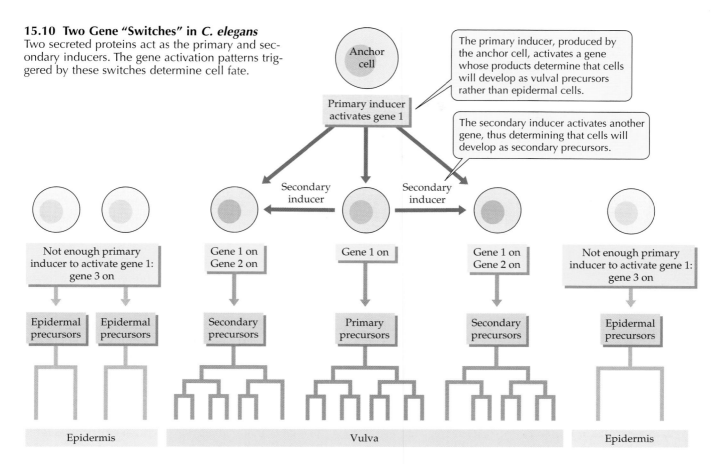

Anchor cell

The primary inducer, produced by the anchor cell, activates a gene whose products determine that cells will develop as vulval precursors rather than epidermal cells.

Primary inducer activates gene 1

The secondary inducer activates another gene, thus determining that cells will develop as secondary precursors.

Secondary inducer   Secondary inducer

Not enough primary inducer to activate gene 1: gene 3 on

Gene 1 on Gene 2 on

Gene 1 on

Gene 1 on Gene 2 on

Not enough primary inducer to activate gene 1: gene 3 on

Epidermal precursors | Epidermal precursors | Secondary precursors | Primary precursors | Secondary precursors | Epidermal precursors

Epidermis   Vulva   Epidermis

"vertebrate version" of a fruit fly protein is yet another example of the conservation of developmental mechanisms through evolutionary time.

The Role of Pattern Formation in Organ Development

A highly active area of current research in developmental biology is the study of **pattern formation**, the spatial organization of a tissue or organism. Pattern formation is inextricably linked to *morphogenesis*, the appearance of body form. The differentiation of cells is beginning to be understood in terms of molecular events, but how do molecular events contribute to the organization of multitudes of cells into specific body parts, such as a leaf, a flower, a shoulder blade, or a tear duct? There are several different processes, including apoptosis, cell adhesion, the establishment of chemical gradients, and cell movements.

Animal development results in part from the movement of cells, as we saw in the example of induction in the developing frog eye (see Figure 15.9). Plant development is restricted in this regard. The cell wall anchors a plant cell in place, preventing movement. Let's first see how a particular form can develop when cells cannot move around.

Directed cell division and cell expansion form the plant body pattern

Although a plant cell cannot move, it may be able to elongate preferentially in one direction. In addition, the direction in which a cell divides is often regulated genetically. Cytoskeletal elements play determining roles in both of these events. The strongest elements in the cell wall are cellulose microfibrils whose orientation is determined by microtubules of the cytoskeleton. The orientation of the cellulose microfibrils, in turn, determines the direction of cell elongation (Figure 15.11*a*). Other microtubules form a preprophase band that determines the plane of cell division (Figure 15.11*b*).

Some cells are programmed to die

As described in Chapter 9, *apoptosis* is programmed cell death, a series of events caused by the expression of certain genes (see Figure 9.22). Some of these "death genes" have been pinpointed, and there are related ones in organisms as diverse as worms and humans.

Apoptosis is vital to the normal development of all animals. For example, the nematode *C. elegans* produces precisely 1,030 somatic cells when it develops from a fertilized egg to an adult (see Figure 15.7). But 131 of these cells die. The sequential expression of two

(a) Expansion

Cellulose microfibrils encircle the cell in a specific orientation and constrain cell expansion.

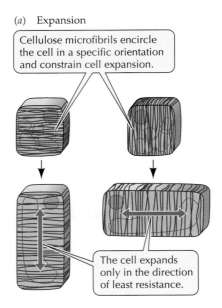

The cell expands only in the direction of least resistance.

(b) Division

The preprophase bands of cytoplasmic microtubules determine the orientation of cell division.

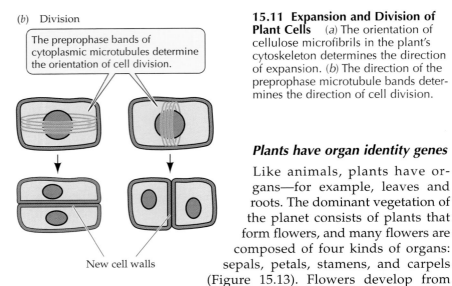

New cell walls

15.11 Expansion and Division of Plant Cells (a) The orientation of cellulose microfibrils in the plant's cytoskeleton determines the direction of expansion. (b) The direction of the preprophase microtubule bands determines the direction of cell division.

Plants have organ identity genes

Like animals, plants have organs—for example, leaves and roots. The dominant vegetation of the planet consists of plants that form flowers, and many flowers are composed of four kinds of organs: sepals, petals, stamens, and carpels (Figure 15.13). Flowers develop from groups of cells in the shape of domes at growing points on the stem. How does the dome give rise, in short order, to *whorls*—groups of organs that encircle a central axis—of four different organs? The answer involves the activities of a group of genes.

A group of **organ identity genes** work in concert to specify the successive whorls. We recognize the presence of organ identity genes because mutations in these genes lead to major alterations in flower structure. Table 15.1 describes the phenotypes of some of these mutations. Analyses of these mutant genes and their products are leading us to a preliminary understanding of how normal flowers develop. The developmental genetics of flowers is best understood for *Arabidopsis thaliana* (see Figure 15.1a) and snapdragons.*

*Not surprisingly, the gene products of organ identity genes are DNA-binding proteins. A single gene, *leafy*, appears to be important in the initiation of flower development in many species.

genes called *ced-4* and *ced-3* (for *cell death*) appears to control this process. In the nervous system, for example, there are 302 nerve cells that come from 405 precursors; thus 103 cells undergo apoptosis. If either *ced-3* or *ced-4* is nonfunctional, all 405 cells form neurons and organizational chaos results. A third gene, *ced-9*, acts as an inhibitor of apoptosis: that is, its protein blocks the function of the *ced-4* gene. So, where cell death is required, *ced-3* and *ced-4* are active and *ced-9* is inactive, and where cell death does not occur, the reverse is true.

Remarkably, a similar system of cell death genes acts in humans. During early development, human hands and feet look like tiny paddles—the fingers and toes are linked by webbing. Between days 41 and 56 of development, cells in the webbing die, freeing the individual fingers and toes (Figure 15.12). The gene (*caspase*) that stimulates this apoptosis is similar in DNA sequence to *ced-3*, and a second gene (*bcl-2*) that inhibits apoptosis is similar to *ced-9*. So humans and worms, two creatures separated by more than 600 million years of evolutionary time, have similar genes controlling programmed cell death.

Apoptosis is essential to pattern formation in animal development and plays other roles in your life. The lens of your eye consists of the specialized remains of cells that have undergone apoptosis. The dead cells that form the outermost layer of your skin and those from the uterine wall that are lost during menstruation have also undergone apoptosis. The white blood cells live only a few months in the circulation, then undergo apoptosis. In a form of cancer called follicular large-cell lymphoma, these white blood cells do not die but continue to divide. The reason for this malfunction is that a genetic mutation has caused the overexpression of *bcl-2*, the gene that inhibits cell death.

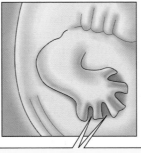

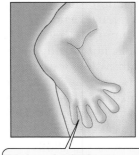

41 days after fertilization: Genes for programmed cell death are expressed only in the tissue between the digits.

56 days after fertilization: Apoptosis is complete. Cells of the digits have absorbed the remains of the dead cells.

15.12 Apoptosis Removes the Webbing between Fingers Early in the second month of human development, the webbing connecting the fingers is removed by apoptosis.

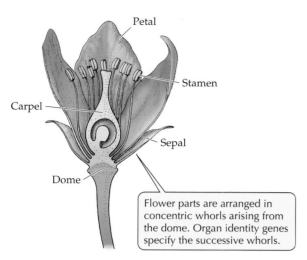

Petal
Stamen
Carpel
Sepal
Dome

Flower parts are arranged in concentric whorls arising from the dome. Organ identity genes specify the successive whorls.

15.13 The Organs of a Flower The sepal is the outermost whorl, enclosing the petals, then the stamen, and finally the carpel. Mutations in organ identity genes alter the flower's structure.

TABLE 15.1 Organ Identity Mutations in Flower Development

GENOTYPE	PHENOTYPE			
	WHORL 1	WHORL 2	WHORL 3	WHORL 4
Wild type	Sepals	Petals	Stamens	Carpels
apetala 2 mutant	Carpels	Stamens	Stamens	Carpels
apetala 3 mutant	Sepals	Sepals	Carpels	Carpels
agamous mutant	Sepals	Petals	Petals	Sepals

In addition to being fascinating to biologists, these organ identity genes have caught the attention of agricultural scientists. The foods that constitute much of the human diet are seeds and fruits, such as the grains rice, wheat, and corn. Seeds and fruits form from the female reproductive organs on the flower. Thus, genetically manipulating the number of these organs on a particular plant could increase the amount of grain a crop could produce. More carpels mean more seeds—that is, a larger crop.

Plants and animals use positional information

Certain cells in both plants and animals appear to "know" where they are with respect to the body as a whole; this spatial sense is called **positional information**. In plants, the pattern of development of two major types of conducting tissue—one for water and minerals and the other for the products of photosynthesis—suggested long ago that distance from the body surface may play a role in their formation.

Cells destined to become water conductors are farther from the body surface than are those destined to become photosynthate conductors. Thus those destined to become water conductors are exposed to lower concentrations of O_2 and higher concentrations of CO_2, and these differences may help determine which genes are expressed in which parts of the stem and root. Recently it has been suggested that the cells on the surface of the stem secrete a protein or other signal that is more concentrated close to the surface than deeper in the stem. Other signals may diffuse from the stem tip and root tip, establishing positional information along the plant's axis. These ideas are still speculative.

Morphogens provide positional information in the developing limbs of animals

More concrete evidence is available concerning positional information in animal embryos than in plants. In the 1960s and 1970s, the English developmental biologist Lewis Wolpert developed a theory of positional information, based on gradients of morphogens in developing limb buds of chick embryos. A **morphogen** is a substance, produced in one place, that diffuses and produces a concentration gradient, with the result that different cells are exposed to different concentrations of the morphogen and thus develop along different lines.

A morphogen concentration gradient results in the development of a chick wing from a wing bud, which is a bulge on the surface of a 3-day-old embryo. Like any other three-dimensional object, a wing can be described in terms of three perpendicular axes. The *anteroposterior axis* of the wing is the axis that corresponds to the axis of the body running from the head to the tail. The *proximodistal axis* runs from the base of the limb to its tip, and the *dorsoventral axis* from the top of the wing to the undersurface. Each axis has a corresponding type of positional information. Here we will consider just the anteroposterior axis.

Pattern formation along the anteroposterior axis can be modified experimentally in ways that suggest that it is controlled at least in part by a particular part of the wing bud, called the *zone of polarizing activity*, or **ZPA**, that lies on the posterior margin of the bud. Different parts of the limb appear to develop normally at *specific distances from the ZPA*. This hypothesis is supported by the results of delicate surgery and grafting experiments such as those depicted in Figure 15.14, in which a ZPA from one bud is grafted onto another bud that still has its own ZPA.

The donor ZPA can be placed in different positions on different hosts. If the extra ZPA is grafted on the anterior margin, opposite from the host ZPA, the anteroposterior axis of the wing tip is duplicated, and two complete mirror images form (Figure 15.14*b*). If the extra ZPA is placed somewhat closer to the host ZPA, incomplete mirror images appear (Figure 15.14*c*): One digit is missing from each of the units, as if the

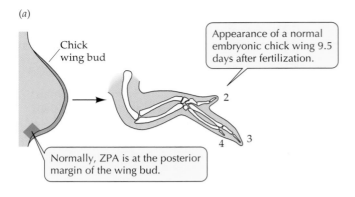

(a)

Chick wing bud

Appearance of a normal embryonic chick wing 9.5 days after fertilization.

2

4 3

Normally, ZPA is at the posterior margin of the wing bud.

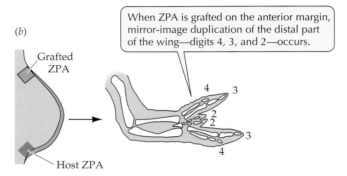

(b)

Grafted ZPA

When ZPA is grafted on the anterior margin, mirror-image duplication of the distal part of the wing—digits 4, 3, and 2—occurs.

4 3
2
2
3
4

Host ZPA

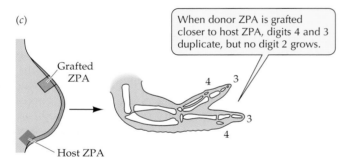

(c)

Grafted ZPA

When donor ZPA is grafted closer to host ZPA, digits 4 and 3 duplicate, but no digit 2 grows.

4 3
3
4

Host ZPA

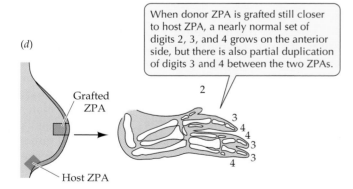

(d)

Grafted ZPA

When donor ZPA is grafted still closer to host ZPA, a nearly normal set of digits 2, 3, and 4 grows on the anterior side, but there is also partial duplication of digits 3 and 4 between the two ZPAs.

2
3
4
4
3
4 3

Host ZPA

Conclusion: Different parts of the limb develop at specific distances from the ZPA.

15.14 The ZPA Provides Positional Information In each experiment, the zone of polarizing activity from a donor wing bud was grafted onto a host wing bud that still had its own ZPA.

missing units are of a type that can form only if they are more than some minimum distance from a ZPA. In the third case (Figure 15.14d), with the two ZPAs close together, there is room for some duplication, but also enough room on the anterior side for a complete, nearly normal unit to form.

How might the ZPA produce these effects? The answers have been slow in coming and are still incomplete. However, the explanation includes at least two elements: a gene and a small molecule. The gene is called *sonic hedgehog* (for reasons not worth explaining here). This gene normally is expressed in the limb bud only where ZPA activity is greatest. If *sonic hedgehog* is inserted into other cells and caused to be expressed, those cells show ZPA activity when grafted into a developing limb bud. The sonic hedgehog protein appears to be a morphogen, secreted from the cells and forming a gradient in neighboring tissue.

The small molecule that plays a role in ZPA activity is retinoic acid, a derivative of vitamin A. Retinoic acid can replace the ZPA in transplant experiments. For example, if some retinoic acid inside a porous bead is placed in the anterior region, it induces a duplication of the digits. So which comes first—retinoic acid or sonic hedgehog? The answer appears to be the former. In the bead experiments, the vitamin derivative induces cells in the ZPA to produce sonic hedgehog, which is the true stimulus for cell differentiation.

Sonic hedgehog protein is also an important morphogen in mammals, and is especially active during the differentiation of parts of the nervous system. Mice that lack the gene for this protein die around the time of birth as a result of severe malformations of the brain. These abnormalities are very similar to a human disorder called holoprosencephaly, which is the cause of miscarriage pregnancies of one out of every 250 human fetuses. At least some cases of this devastating disease are caused by a mutation in the gene that codes for the human version of sonic hedgehog protein.

The Role of Differential Gene Expression in Establishing Body Segmentation

Another experimental subject that developmental biologists have used to study pattern formation is the *Drosophila* fruit fly. Insects (and many other animals) develop a highly modular body composed of different types of modules, called *segments*. Complex interactions of different sets of genes underlie the pattern formation of segmented bodies.

Unlike the body segments of segmented worms such as earthworms, the segments of the *Drosophila* body are different from one another. The *Drosophila* adult has a head (several fused segments), three different thoracic segments, eight abdominal segments, and a terminal segment at the posterior end. Thirteen seemingly identical segments in the *Drosophila* larva correspond to these specialized adult segments. Several types of genes are expressed sequentially in the embryo to define these segments. The first step in this process is to establish the polarity of the embryo.

Maternal effect genes determine polarity in Drosophila

In *Drosophila* eggs and larvae, polarity is based on the distribution of more than a dozen morphogens, of which some are mRNAs and some are proteins. These morphogens are products of specific **maternal effect genes** in the mother and are distributed to the eggs, often in a nonuniform manner. The cytoskeleton (especially microtubules) is essential to this process. Maternal effect genes produce effects on the embryo regardless of the genotype of the father. They determine the dorsoventral (back–belly) and anteroposterior (head–tail) axes of the embryo.

The fact that morphogens specify these axes has been established by the results of experiments in which cytoplasm was transferred from one egg to an-

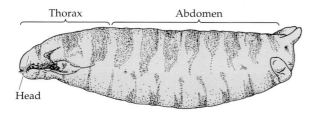

A normal larva produced by a wild-type female has normal body parts.

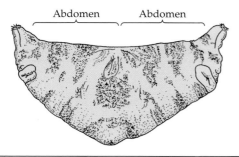

A larva produced by a female homozygous for a mutant allele of *bicoid*, one of the maternal effect genes, has no head or thorax. It consists of two hind ends.

15.15 Polarity Gone Wild The anteroposterior axis of *Drosophila* larvae arises from the interaction of several morphogens. If one of them is missing, the other predominates—in this case, the one that forms the abdomen.

other. Females homozygous for a particular mutation of the maternal effect gene *bicoid* produce larvae with no head and no thorax (Figure 15.15). However, if eggs of homozygous mutant *bicoid* females are inoculated at the anterior end with cytoplasm from the anterior region of a wild-type egg, the treated eggs develop into normal larvae—with heads developing from the part of the egg that receives the wild-type cytoplasm. On the other hand, removal of 5 percent or more of the cytoplasm from the anterior of a wild-type egg results in an abnormal larva that looks like a *bicoid* mutant larva.

Another maternal effect gene, *nanos*, plays a comparable role in the development of the posterior end of the larva. Eggs from homozygous mutant *nanos* females develop into larvae with missing abdominal segments. Injecting the *nanos* eggs with cytoplasm from the posterior region of wild-type eggs allows normal development. In wild-type larvae, the overall framework of anteroposterior and dorsoventral axes is laid down by the activity of the maternal effect genes. Their gene products are made by cells that surround and nurture the developing egg and are localized at certain specific regions of the egg as it forms.

After the axes of the embryo are determined, the next step in the segmentation process is to determine the larval segments.

Segmentation and homeotic genes act after the maternal effect genes

The number, boundaries, and polarity of the larval segments are determined by proteins encoded by the **segmentation genes**. The maternal effect genes set the segmentation genes in motion. Three classes of segmentation genes participate, one after the other, to regulate finer and finer details of the segmentation pattern (Figure 15.16).

First, **gap genes** organize large areas along the anteroposterior axis. Mutations in gap genes result in gaps in the body plan—the omission of several larval segments. Second, **pair rule genes** divide the embryo into units of two segments each. Mutations in pair rule genes result in embryos missing every second segment. Third, **segment polarity genes** determine the boundaries and anteroposterior organization of the segments. Mutations in segment polarity genes result in segments in which posterior structures are replaced by reversed (mirror-image) anterior structures.

Finally, after the basic pattern of segmentation has been established by the segmentation genes, differences between the segments are mediated by the activities of **homeotic genes**. These genes are expressed in different combinations along the length of the body and tell each segment what to become. Homeotic genes are analogous to the organ identity genes of plants, which are sometimes called homeotic-like genes.

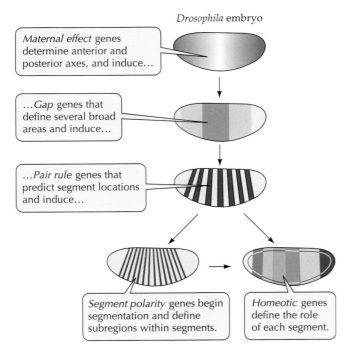

Drosophila embryo

Maternal effect genes determine anterior and posterior axes, and induce…

…*Gap* genes that define several broad areas and induce…

…*Pair rule* genes that predict segment locations and induce…

Segment polarity genes begin segmentation and define subregions within segments.

Homeotic genes define the role of each segment.

15.16 A Gene Cascade Controls Pattern Formation in the *Drosophila* Embryo Gap, pair rule, and segment polarity genes are collectively referred to as the segmentation genes. The shading shows the locations of the gene products, most of which are DNA-binding proteins.

To see how the maternal effect, segmentation, and homeotic genes interact, let's watch them "build" a *Drosophila* larva step by step, beginning with the unfertilized egg.

Drosophila *development results from a transcriptionally controlled cascade*

One of the most striking and important observations about development in *Drosophila*—and in other animals—is that it results from a sequence of changes, each change triggering the next. The sequence, or *cascade*, is controlled at transcription.

In general, unfertilized eggs are storehouses of mRNAs, which are made prior to fertilization to support protein synthesis during the early stages of embryo development. Indeed, early embryos do not carry out transcription. After several cell divisions, mRNA production resumes, forming the mRNAs needed for later development.

15.17 Bicoid and Nanos Protein Gradients Provide Positional Information Translation of mRNAs at the ends of the *Drosophila* larva leads to gradients of the morphogen products, which in turn control the expression of the gap genes.

Some of these prefabricated mRNAs in the egg provide positional information. Before the egg is fertilized, mRNA for the bicoid protein is localized at the end destined to become the anterior end of the animal. After the egg is fertilized and laid, and nuclear divisions begin, the *bicoid* mRNA is translated, forming bicoid protein that diffuses away from the anterior end, establishing a gradient of the protein (Figure 15.17). Another morphogen, the nanos protein, diffuses from the posterior end, forming a gradient in the other direction. Thus, each nucleus in the developing embryo is exposed to a different concentration of bicoid protein and to a different ratio of bicoid protein to nanos protein. What do these morphogens do?

The two morphogens regulate the expression of the gap genes, although in different ways. Bicoid protein affects transcription, while nanos affects translation. The high concentrations of bicoid protein in the anterior portion of the egg turn on a gap gene called *hunchback*, while simultaneously turning off another gap gene, *Krüppel*. Nanos at the posterior end reduces the translation of *hunchback*, so the difference in concentration of gap gene products at the two ends is established. The pattern of gap gene activity resulting from the actions of these morphogens is shown in Figure 15.18.

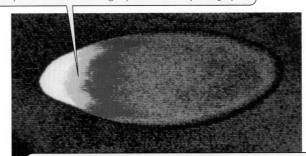

The concentration of bicoid protein is highest at the embryo's anterior end (bright yellow in this photograph).

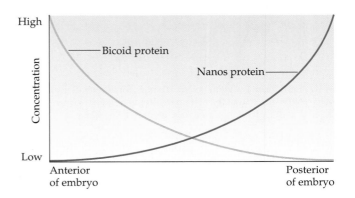

The color of the gradient moves from orange to red as bicoid concentration decreases into the dark blue posterior end.

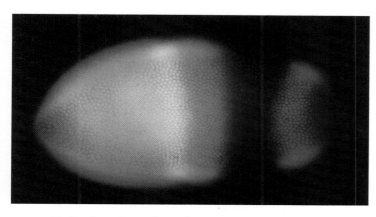

15.18 Gap Genes in Action Interactions of proteins encoded by gap genes define domains of the larval body in *Drosophila*. In this larva, hunchback (orange) and Krüppel (green) proteins overlap, forming a boundary (yellow) between two domains.

The proteins encoded by the gap genes are another set of transcription factors that control the expression of the pair rule genes. Many pair rule genes in turn encode transcription factors that control the expression of the segment polarity genes, giving rise to a complex, striped pattern (see Figure 15.16) that foreshadows the segmented body plan of *Drosophila*.

By this point, each nucleus of the embryo—it is still a single, multinucleated cell, or *syncytium*—is exposed to a distinctive set of transcription factors. The segmented body pattern of the larva is established even before any sign of segmentation is visible. When the segments do appear, they are not all identical, because the homeotic genes give the different segments their different structural and functional properties. Each homeotic gene is expressed over a characteristic portion of the embryo. The homeotic genes of *Drosophila* are arranged on the same chromosome—in the same order as the order in which they are expressed from the anterior to the posterior end of the larva.

Let's turn now to the homeotic genes and how their mutation can alter the course of development.

Homeotic mutations produce large-scale effects

Our present understanding of the genetics of pattern formation began with the discovery of dramatic mutations in *Drosophila* called homeotic mutations in which normal body parts are formed in inappropriate segments. Two bizarre examples are the *Antennapedia* mutant, in which legs grow in the place of antennae, and the *ophthalmoptera* mutant, in which wings grow in the place of eyes.

Homeotic genes fall into a few clusters on a chromosome. Of these, the best characterized is the **bithorax complex**. The eight or more genes of the bithorax complex control development of the abdomen and posterior thorax of the fly. The mutant fly shown at the beginning of this chapter resulted from mutations in the bithorax complex. Development of the head and anterior thorax is controlled by another homeotic gene cluster, the **Antennapedia complex**. The functions of the two complexes interact substantially.

If the entire bithorax complex is deleted, the larva produced, although highly abnormal, still has the normal number of segments. Thus the bithorax complex does not determine the number of segments. Rather, the segmentation genes determine the number and polarity of segments. During normal development, the homeotic genes act later than the segmentation genes and give each segment its distinctive character. For this reason, they are also called selector genes.

Homeobox-containing genes encode transcription factors

In the early 1980s Walter Gehring and his associates William McGinnis and Michael Levine, working in Switzerland, and, independently, Thomas Kaufman and Matthew Scott at Indiana University, undertook a study of the Antennapedia complex using the techniques of recombinant DNA technology.

These investigators set out to isolate and clone the *Antennapedia* (*Antp*) gene, a member of the Antennapedia complex, from *Drosophila*. As part of this study, they prepared a DNA that could hybridize to the *Antp* gene. To their surprise, the *Antp* DNA hybridized with both the *Antp* gene and a nearby segmentation gene (the *fushi tarazu* gene, abbreviated *ftz*, from the Japanese for "too few segments"). The *Antp* and *ftz* genes must have DNA sequences of close similarity, because part of each gene hybridizes with the same DNA.

Further hybridization studies demonstrated that this particular shared stretch of DNA is also found in the *bicoid* gene of the bithorax complex, in some other parts of the *Drosophila* genome, and in genes in other insect species. In fact, this important sequence of 180 base pairs of DNA, called the **homeobox**, has now been shown to be part of numerous genes of many animals and plants. The homeotic genes of animals contain the homeobox, but the organ identity genes of plants do not.

Mice and humans (the two best-studied mammals) have clusters of homeobox-containing genes. These genes are expressed in particular segments of the animal, just as in the fruit fly. Thirty-eight genes are divided into four clusters, each located on a different chromosome (Figure 15.19). As in *Drosophila*, these homeobox genes are arranged in the same order on each chromosome as they are expressed from anterior to posterior of the developing animal.

What is the function of the homeobox, which is present in almost all eukaryotes? The homeobox sequence

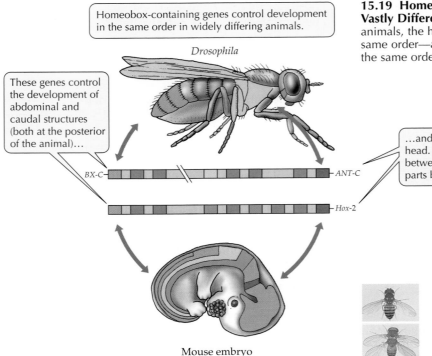

Homeobox-containing genes control development in the same order in widely differing animals.

Drosophila

These genes control the development of abdominal and caudal structures (both at the posterior of the animal)…

…and these control structures in the head. Homologous genes in between control the development of parts between the head and tail.

BX-C

ANT-C

Hox-2

Mouse embryo

15.19 Homeobox-Containing Genes Are Common to Vastly Different Organisms In both these very dissimilar animals, the homeobox-containing genes are lined up in the same order—and they produce effects that are lined up in the same order as the genes.

Summary of "Development: Differential Gene Expression"

The Processes of Development

• A multicellular organism develops through embryonic stages and eventually into an adult. Development continues until death. **Review Figure 15.1**
• Growth results from a combination of cell division and cell expansion.
• Differentiation produces specialized cell types. The overall form of the multicellular organism is the result of morphogenesis, resulting from pattern formation.
• In many organisms, the fates of the earliest embryonic cells have usually not yet been decided. These early embryonic cells may develop into different tissues if transplanted to other parts of an embryo. **Review Figure 15.2**
• As the embryo develops, its cells gradually become determined—committed to developing into particular parts of the embryo and into adult structures. Following determination, cells eventually differentiate into their final, often specialized, forms.

The Role of Differential Gene Expression in Differentiation

• The zygote is totipotent; it contains the entire genetic constitution of the organism and is capable of forming all adult tissues.
• Two lines of evidence show that differentiation does not involve permanent changes in the genome. First, nuclear transplant and cloning experiments show that the nucleus of a differentiated cell retains the ability to act like a zygote nucleus and stimulate the production of an entire organism. Second, molecular biological investigations have shown directly that all cells contain all genes for that organism, but that only certain genes are expressed in a given tissue. **Review Figures 15.3, 15.4**
• In insect metamorphosis, adult body parts arise from imaginal discs that normally have precise fates. Under certain circumstances imaginal discs may transdetermine and

codes for a region of 60 amino acids—the **homeodomain**—that is part of some proteins. Some of these proteins, the transcription factors, return to the nucleus and bind to DNA, regulating the transcription of other genes. A computerized search of published sequences of DNA from numerous species revealed a similarity between the homeobox and parts of certain regulatory genes in yeast—genes that produce proteins that also bind to specific DNA sequences. Some genes with homeoboxes are expressed only at certain times and in certain tissues as development proceeds, as would be expected if these proteins regulate development.

What are we to make of the presence of a homeobox in species as diverse as humans, fruit flies, frogs, nematodes, and tomatoes—and of its presence in several genes in the same organism? Its presence is consistent with the hypothesis that a single gene in an ancient organism was the evolutionary progenitor of what is now a widespread controlling system for development (see Chapter 23). One of the most astounding findings of recent developmental biology research has been this conservation of genes involved in developmental pathways.

As the DNA of more and more organisms is sequenced (see Chapter 17), more such similarities are emerging. Although there are certainly major differences in the end products of human and fruit fly development, the basic genetic mechanisms appear quite similar.

develop into quite different parts of the body, showing that determination does not entail the loss of genes. Instead, cells become differentiated by the differential expression of genes. **Review Figure 15.5.**

The Role of Cytoplasmic Segregation in Cell Determination

- Unequal distribution of cytoplasmic determinants in the egg, zygote, or embryo leads to patterns of cell determination in normal development. Experimentally altering this distribution can alter gene expression and produce abnormal or nonfunctional organisms. **Review Figure 15.6**
- The nematode *Caenorhabditis elegans* provides a striking example of cytoplasmic segregation. The hermaphroditic form of this worm consists of 959 cells that develop from the fertilized egg by a precise pattern of cell divisions and other events. The gametes arise from embryonic cells containing granules that are probably cytoplasmic determinants. **Review Figures 15.7, 15.8**

The Role of Embryonic Induction in Cell Determination

- Some embryonic animal tissues direct the development of their neighbors by secreting inducers.
- Induction is often reciprocal: One tissue induces a neighbor to change, and the neighbor, in turn, induces the first tissue to change, as in eye formation in vertebrate embryos. **Review Figure 15.9**
- Induction in *C. elegans* can be very precise, with individual cells producing specific effects in just two or three neighboring cells. **Review Figure 15.10**

The Role of Pattern Formation in Organ Development

- Pattern formation triggers the sequence of cell divisions, cell expansions, cell movements, and programmed cell deaths that constitutes organ development and morphogenesis.
- Plant cells do not move, but cytoskeletal elements direct cell division and expansion, which form the plant body pattern. **Review Figure 15.11**
- In plants and animals, programmed cell death (apoptosis) is important in development. Some genes whose protein products regulate apoptosis have been identified. **Review Figure 15.12**
- Plants have organ identity genes that interact to cause the formation of sepals, petals, stamens, and carpels. Mutation of these genes causes undifferentiated cells to form a different organ. **Review Figure 15.13 and Table 15.1**
- Both plants and animals use positional information as a basis for pattern formation.
- Positional information is well understood in chick limb formation. Gradients of morphogens are established in the embryo. Cells at different distances from the zone of polarizing activity are exposed to different concentrations of the morphogens and thus respond differently. Some morphogens are transcription factors. **Review Figure 15.14**

The Role of Differential Gene Expression in Establishing Body Segmentation

- The fruit fly *Drosophila melanogaster* has provided much information about the development of body segmentation; some of this information applies to mice and other mammals.
- The first genes to act in determining *Drosophila* segmentation are maternal effect genes, such as *bicoid* and *nanos*, which encode morphogens that form gradients in the egg. These morphogens act on segmentation genes to define the anteroposterior organization of the embryo. **Review Figures 15.15, 15.16, 15.17**
- Segmentation develops as the result of a transcriptionally controlled cascade, the product of one gene promoting or repressing the expression of another gene. There are three kinds of segmentation genes, each responsible for a different step in segmentation. Gap genes organize large areas along the anteroposterior axis, pair rule genes divide the axis into pairs of segments, and segment polarity genes see to it that each segment has an appropriate anteroposterior axis. **Review Figure 15.16**
- The bicoid and nanos proteins act as a transcription factor and translation regulator, respectively, to control the level of expression of gap genes. Gap genes encode transcription factors that regulate the expression of pair rule genes. The products of the pair rule genes are transcription factors that regulate the segment polarity genes. **Review Figures 15.16, 15.17, 15.18**
- Activation of the segmentation genes leads to the activation of the appropriate homeotic genes in different segments. The homeotic genes define the functional characteristics of the segments. **Review Figure 15.16**
- Mutations in homeotic genes often have bizarre effects, causing structures to form in inappropriate parts of the body. Homeotic genes contain the homeobox, which encodes an amino acid sequence that is part of many transcription factors. The homeobox is found in key genes of distantly related species; thus numerous regulatory mechanisms may trace back to a single evolutionary precursor. **Review Figure 15.19**

Self-Quiz

1. Which statement about determination is true?
   a. Differentiation precedes determination.
   b. All cells are determined after two cell divisions in most organisms.
   c. A determined cell will keep its determination no matter where it is placed in an embryo.
   d. A cell changes its appearance when it becomes determined.
   e. A differentiated cell has the same pattern of transcription as a determined cell.

2. The Briggs and King, Gurdon, and Wilmut experiments showed that
   a. all nuclei of an organism have the same genes.
   b. nuclei of embryonic cells can be totipotent.
   c. nuclei of differentiated cells have different genes than zygote nuclei have.
   d. differentiation is fully reversible in all cells of a frog.
   e. differentiation involves permanent changes in the genome.

3. If an imaginal disc for a wing is transplanted into an adult fruit fly, and then put into a larva, it will
   a. always form a wing.
   b. divide and remain undifferentiated.
   c. undergo programmed cell death.
   d. sometimes form an organ other than a wing.
   e. form an eye.

4. A major difference between early human and fruit fly embryology is that only in the latter
   a. does cytokinesis *not* occur.
   b. are polarity genes not expressed.
   c. is the fertilized egg totipotent.
   d. do nuclei become determined before differentiation.
   e. is *sonic hedgehog* expressed.

5. Which statement about cytoplasmic determinants in *Drosophila* is *not* true?
   a. They specify the dorsoventral and anteroposterior axes of the embryo.
   b. Their positions in the embryo are determined by microfilament action.
   c. They are products of specific genes in the mother insect.
   d. They often produce striking effects in larvae.
   e. They have been studied by the transfer of cytoplasm from egg to egg.

6. In fruit flies, the following genes are used to determine segment polarity: (k) gap genes; (l) homeotic genes; (m) maternal effect genes; (n) pair rule genes. In what order are these genes expressed during development?
   a. klmn
   b. lknm
   c. mknl
   d. nkml
   e. nmkl

7. Which statement about embryonic induction is *not* true?
   a. One tissue induces an adjacent tissue to develop in a certain way.
   b. It triggers a sequence of gene expression in target cells.
   c. It may be either instructive or permissive.
   d. A tissue may induce itself.
   e. The chemical identification of specific inducers has been difficult.

8. In the process of body segmentation in *Drosophila* larvae,
   a. the first steps are specified by homeotic genes.
   b. mutations in pair rule genes result in embryos missing every second segment.
   c. mutations in gap genes result in the insertion of extra segments.
   d. segment polarity genes determine the dorsoventral axes of segments.
   e. segmentation is the same as in earthworms.

9. Homeotic mutations
   a. are often so severe that they can be studied only in larvae.
   b. cause subtle changes in the forms of larvae or adults.
   c. occur only in prokaryotes.
   d. do not affect the animal's DNA.
   e. are confined to the zone of polarizing activity.

10. Which statement about the homeobox is *not* true?
    a. It is transcribed and translated.
    b. It is found only in animals.
    c. Some proteins containing the homeodomain bind to DNA.
    d. It is a stretch of DNA shared by many genes.
    e. Its activities often relate to body segmentation.

Applying Concepts

1. Molecular biologists can insert genes attached to high-level promoters into cells (see Chapter 16). What would happen if the following were inserted and overexpressed? Explain your answers.
   a. *ced-9* in embryonic nerve cell precursors in *C. elegans*
   b. *MyoD1* in undifferentiated myoblasts
   c. *sonic hedgehog* in a chick limb bud
   d. *nanos* at the anterior end of the *Drosophila* embryo

2. A powerful method to test for the function of a gene in development is to generate a "knockout" organism, in which the gene in question is inactivated. What do you think would happen in each of the following?
   a. knockout *C. elegans* for *ced-9*
   b. knockout *Drosophila* for *nanos*
   c. knockout *C. elegans* for P granules

3. Look at the chart of organ identity mutations in Table 15.1. What pattern do you perceive in the results of these mutations, and what might this pattern mean?

4. During development, the potential of a tissue becomes ever more limited, until, in the normal course of events, the potential is the same as the original prospective fate. On the basis of what you have learned in this chapter and in Chapter 14, discuss possible mechanisms for the progressive limitation of the potential.

5. How were biologists able to obtain such a complete accounting of all the cells in the roundworm *Caenorhabditis elegans*? Why can't we reason directly from studies of *C. elegans* to comparable problems in our own species?

Readings

Ameisen, J. C. 1996. "The Origin of Programmed Cell Death." *Science*, vol. 272, pages 1278–1279. Interesting speculation based on the observation of programmed cell death in four unicellular eukaryotes.

Duke, R. C., D. M. Ojcius and J. D.-E. Young. 1996. "Cell Suicide in Health and Disease." *Scientific American*, December. A discussion of the mechanisms and functions of apoptosis.

Gehring, W. J. 1985. "The Molecular Basis of Development." *Scientific American*, October. A clear account of homeotic mutations and the homeobox.

Gilbert, S. F. 1997. *Developmental Biology*, 5th Edition. Sinauer Associates, Sunderland, MA. An exceptionally well balanced treatment of developmental biology, covering both molecular and cellular concepts and embryology. Gives a feeling for the history of the discipline.

Holliday, R. 1989. "A Different Kind of Inheritance." *Scientific American*, June. A description of how patterns of gene activity are transmitted from one cell generation to the next.

Hynes, R. O. 1986. "Fibronectins." *Scientific American*, June. A look at proteins that guide migrating cells during development and their possible role in cancer.

Lawrence, P. A. 1992. *The Making of a Fly: The Genetics of Animal Design*. Blackwell Scientific, Oxford. A detailed, advanced treatment of the *Drosophila* material covered in this chapter.

Nüsslein-Volhard, C. 1996. "Gradients That Organize Embryo Development." *Scientific American*, August. Experiments conducted by the Nobel laureate author, her colleagues, and others to investigate morphogen gradients in *Drosophila* embryos.

Chapter 16

Recombinant DNA and Biotechnology

Making Medicine
This flask of genetically engineered mammalian cells is the starting point for the biotechnological production of a useful product, in this case a protein targeted to cancer cells.

*J*ohn is at home preparing dinner when one side of his face starts to twitch. He tries to call out to his wife, but his speech is slurred. Like several million people each year, John is having a stroke because a blood clot has blocked a major artery in his brain, depriving nerve cells of essential oxygen. Normally, the body's clot-dissolving mechanism would be activated slowly and blood flow might eventually resume. But if John's brain waits for this slow process, it will be deprived of oxygen for so long that essential nerve cells will die. Instead, his emergency room physicians inject directly onto the clot a substance that initiates the dissolving process, a protein called tissue plasminogen activator. The clot quickly dissolves, restoring blood flow to the brain. John leaves the hospital a day later, his face and speech entirely normal.

In a small village in Africa, a one-year-old boy, Kwame, is exposed to *Neisseria meningitidis.* Three years earlier, his sister, at the time also one year old, had been infected by this bacterium. She developed meningitis, with the bacteria infecting her central nervous system. Within a day, a stiff neck and high fever led to seizures; she was dead a week later. Fortunately, her brother is receiving an antimeningitis vaccine, but in a very unusual way. The bananas that Kwame eats have been genetically engineered to express the gene for the vaccine protein. The vaccine protein successfully protects him, and he does not develop the disease.

The gene for human tissue plasminogen activator has been inserted into bacteria, which are used as factories for the production of large quantities of this rare protein. Genes encoding vaccine proteins have been inserted into the banana genome, and the fruits are being tested for their effectiveness as oral vaccines in children with limited access to health care personnel. These are but two of many examples of the uses of **DNA technology**, the ability of humans to manipulate genes almost at will. DNA technology has revolutionized much of experimental biology. Its use has been important in most of the recent advances in our understanding of how eukaryotic genes are regulated and organized. It is also changing agriculture, medicine, and other areas of the chemical industry, as well as forensics and the battle against environmental pollution.

In this chapter, we consider the basic techniques of DNA manipulation and some of its applications. Although many of these techniques have been called revolutionary, most of them come from the knowledge of DNA transcription and translation that we described in earlier chapters. After a specific sequence of DNA is isolated from cells or made in the chemistry laboratory, it can be introduced into almost any prokaryotic or eukaryotic cell. Such genetic transformation, often across species lines, has provided an invaluable tool for studying molecular physiology.

We will also see how the new gene in a host cell can be coaxed into making its protein product, as is the case with tissue plasminogen activator and with the meningitis vaccine protein. Finally, a process called the polymerase chain reaction (PCR) allows DNA to replicate in a test tube—a capability that has wide applications.

Cleaving and Rejoining DNA

Scientists have realized that the chemical reactions used in living cells for one purpose may be applied in the laboratory for other, novel purposes. Recombinant DNA technology, the manipulation and combination of DNA molecules from different sources, is based on this realization, and on an understanding of the properties of certain enzymes and of DNA itself. In this section we will identify the numerous naturally occurring enzymes that cleave DNA, help it replicate, and repair it. Many of these enzymes have been isolated and purified, and are now used in the laboratory to manipulate and combine DNA. Then we will see how fragments of DNA can be separated and covalently linked to other fragments.

As we saw in previous chapters, the nucleic acid base-pairing rules underlie many fundamental processes of molecular biology. The mechanisms of DNA replication, transcription, and translation rely on complementary base pairing. Similarly, all the key techniques in recombinant DNA technology—sequencing, rejoining, amplifying, and locating DNA fragments—make use of the complementary base pairing of A with T (or U) and of G with C.

Restriction endonucleases cleave DNA at specific sequences

All organisms must have mechanisms to deal with their enemies. As we saw in Chapter 13, bacteria are attacked by viruses called bacteriophages that inject their genetic material into the host bacterial cell and disrupt its operations. Eventually the phage genes may be replicated by the enzyme systems of the host, produce new viruses, and kill the host. Some bacteria defend themselves against such invasions by producing enzymes called **restriction endonucleases** that catalyze the cleavage of double-stranded DNA molecules—such as those injected by phages—into smaller, noninfectious fragments (Figure 16.1). The bonds cut are between the 3' hydroxyl of one nucleotide and the 5' phosphate of the next one.

There are many such restriction enzymes, each of which cleaves DNA at a specific site defined by a sequence of bases called a **recognition site** or **restriction site**. The DNA of the host bacterial cell is not cleaved by its own restriction enzymes, because of specific modifying enzymes called methylases that add methyl (—CH$_3$) groups to certain bases at the restriction sites when host DNA is being replicated. The methylation of the host's bases makes the recognition sequence unrecognizable to the endonuclease, thus preventing cleavage of the host DNA. But the unmethylated phage DNA is efficiently cleaved.

A specific sequence of bases defines each recognition site. For example, the enzyme *Eco*RI (named after

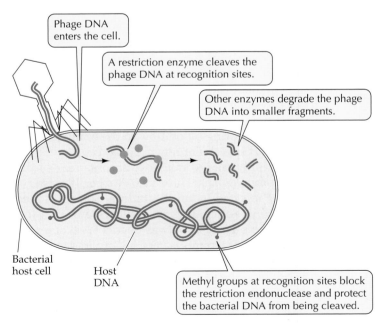

Phage DNA enters the cell.

A restriction enzyme cleaves the phage DNA at recognition sites.

Other enzymes degrade the phage DNA into smaller fragments.

Bacterial host cell

Host DNA

Methyl groups at recognition sites block the restriction endonuclease and protect the bacterial DNA from being cleaved.

16.1 Bacteria Fight Invading Viruses with Restriction Enzymes Bacteria produce restriction enzymes that cleave and degrade phage DNA. Other enzymes protect the bacteria's own DNA from being cleaved.

its source, a strain of the bacterium *E. coli*) cuts DNA only where it encounters the following paired sequence in the DNA double helix:

5′ ... GAATTC ... 3′
3′ ... CTTAAG ... 5′

Notice that this sequence reads the same in the 5′-to-3′ direction on both strands. It is palindromic, like the word "mom," in the sense that the double-stranded "word" is the same in both directions. *Eco*RI has two identical subunits that cleave each strand between the G and A.

This recognition sequence occurs on average about once in 4,000 base pairs in a typical genome—or about

once per four prokaryotic genes. So *Eco*RI can chop a large piece of DNA into smaller pieces containing, on average, just a few genes. For small genomes such as those of viruses that have only a few thousand base pairs, cleavage may result in a few fragments. For a huge eukaryotic chromosome with tens of millions of base pairs, the number of fragments is very large.

Of course, "on average" does not mean that the enzyme cuts all stretches of DNA at regular intervals. The *Eco*RI recognition sequence does not occur even once in the 40,000 base pairs of the genome of a phage called T7—a fact that is crucial to the survival of this virus, since its host is *E. coli*. Fortunately for the *E. coli* that makes *Eco*RI, the DNA of other phages does contain the recognition sequence. The ability to cleave the phage DNA at such sequences prevents the *E. coli* from being overrun by phages.

Hundreds of restriction enzymes have been purified from various microorganisms. In the test tube, different restriction enzymes that recognize different recognition sequences may cut the same sample of DNA. Thus, cutting a sample of DNA in many different, specific places is an easy task, and restriction enzymes can be used as "knives" for genetic "surgery."

Gel electrophoresis identifies the sizes of DNA fragments

After a sample of DNA has been digested with a restriction enzyme, the DNA is in fragments, each of which is bounded at its ends by the recognition sequence. As we noted, these fragments are not all the same size, and this property provides a way to separate the fragments from each other. Fragments are separated to determine the number and sizes of fragments produced, or to identify and purify an individual fragment.

The best way to separate DNA fragments is by **gel electrophoresis** (Figure 16.2). Because of its phosphate groups, DNA is negatively charged at

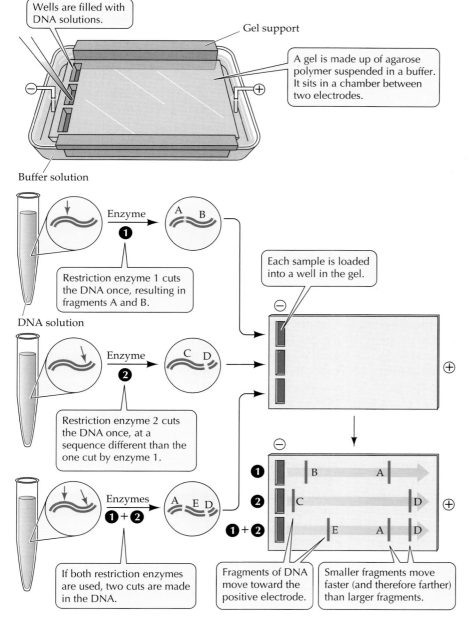

16.2 Separating Fragments of DNA by Gel Electrophoresis A mixture of DNA fragments is placed in a gel and an electric field is applied across the gel. The negatively charged DNA moves toward the positive end of the field, with smaller molecules moving faster than larger ones. When the electric field is shut off, the separate fragments can be analyzed.

(a)

(b)

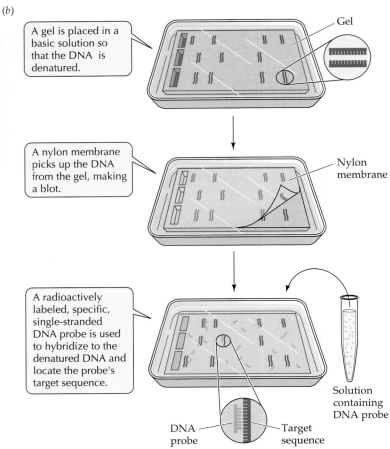

A gel is placed in a basic solution so that the DNA is denatured.

Gel

A nylon membrane picks up the DNA from the gel, making a blot.

Nylon membrane

A radioactively labeled, specific, single-stranded DNA probe is used to hybridize to the denatured DNA and locate the probe's target sequence.

Solution containing DNA probe

DNA probe

Target sequence

16.3 Analyzing DNA Fragments (*a*) Identifying the DNA fragments on an electrophoresis gel. (*b*) Technique for using a hybridization probe to identify a specific DNA fragment.

neutral pH. In gel electrophoresis, a mixture of fragments is placed in a porous gel and an electric field (with positive and negative ends) is applied across the gel. Because opposite charges attract, the DNA moves toward the positive end of the field. Since the porous gel acts as a sieve, the smaller molecules move faster than the larger ones. After a fixed time and while all fragments are still on the gel, the electric power is shut off and the separated fragments can be examined or removed individually.

Different DNA samples may be "run" on a gel side by side in different "lanes." DNA fragments of known molecular size (in base pairs) are often run in a lane on each gel to provide a size reference. We can visualize the separated DNA fragments by staining them with a dye that fluoresces under ultraviolet light. Or, we can identify a specific DNA sequence by denaturing the DNA in the gel, affixing the denatured DNA to a nylon membrane to make a "blot" of the gel, and hybridizing the fragments with a specific single-stranded DNA probe (Figure 16.3). The gel region containing a desired fragment can be removed when the gel is sliced, and then the pure DNA fragment can be removed from the gel.

Recombinant DNA can be made in the test tube

An important property of some restriction enzymes is that they make staggered cuts in DNA rather than cutting both strands at a single base pair. For example, *Eco*RI cuts DNA within the recognition sequence, as shown at the top of Figure 16.4. After the two cuts in the opposing strands are made, they are held together only by the hydrogen bonding between four base pairs. The hydrogen bonds of these few base pairs are too weak to persist at warm temperatures (above room temperature), so the two strands of DNA separate, or denature. As a result, there are single-stranded

"tails" at the location of each cut. These tails are called *sticky ends* because they have a specific base sequence that is capable of binding by base pairing with complementary sticky ends.

After a DNA molecule has been cut with a restriction enzyme, the complementary sticky ends can hydrogen-bond to one another. The original ends may rejoin, or an end may pair with another fragment. If more than one recognition site for a given restriction enzyme is present in a DNA sample, numerous fragments can be made, all with the same sequence at their sticky ends. Because all *Eco*RI ends are the same, fragments from one source, such as a human, can be joined to fragments from another, such as a bacterium, to create recombinant DNA. When the temperature is lowered, the fragments anneal (come together) at random, but these associations are unstable because they are held together by only four pairs of hydrogen bonds.

The associated sticky ends can be permanently united by a second enzyme, **DNA ligase**, which forms a covalent bond to "seal" each DNA strand. In the cell, this enzyme unites new fragments made during DNA

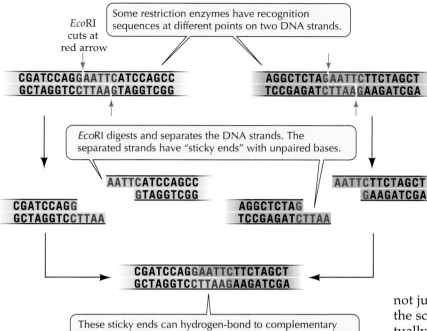

EcoRI cuts at red arrow

Some restriction enzymes have recognition sequences at different points on two DNA strands.

CGATCCAGGAATTCATCCAGCC
GCTAGGTCCTTAAGTAGGTCGG

AGGCTCTAGAATTCTTCTAGCT
TCCGAGATCTTAAGAAGATCGA

EcoRI digests and separates the DNA strands. The separated strands have "sticky ends" with unpaired bases.

AATTCATCCAGCC
GTAGGTCGG

AATTCTTCTAGCT
GAAGATCGA

CGATCCAGG
GCTAGGTCCTTAA

AGGCTCTAG
TCCGAGATCTTAA

CGATCCAGGAATTCTTCTAGCT
GCTAGGTCCTTAAGAAGATCGA

These sticky ends can hydrogen-bond to complementary sticky ends from other DNAs, and the resulting recombinant DNA can be sealed with DNA ligase.

16.4 Cutting and Splicing DNA Some restriction enzymes (*Eco*RI is shown here) make staggered cuts in DNA from two different sources (shown here in blue and green). At warm temperatures, the two DNA strands will separate (denature), leaving "sticky ends" that can recombine with complementary fragments when the temperature is lowered.

replication and mends breaks in DNA (see Chapter 11). In the laboratory, DNA ligase can be used to seal breaks in DNA generated by reannealing the fragments from restriction enzyme digestion.

With these two enzyme tools—restriction endonucleases and DNA ligases—scientists can cut and rejoin different DNA molecules to form recombinant DNA (see Figure 16.4). This simple concept has changed the directions of biological science. For example, a piece of nonbacterial DNA can be inserted into a plasmid, a small circular chromosome often present in bacteria, provided that both the plasmid and the nonbacterial DNAs have the same recognition sequence.

The nonbacterial DNA is cleaved at many places, producing fragments with sticky ends. The circular plasmid DNA with one restriction site is cut, transforming it into a linear molecule with sticky ends. The sticky ends of the nonbacterial DNA hydrogen-bond to the sticky ends of the plasmid DNA. If ligase is then added, its activity produces a circular plasmid containing the nonbacterial DNA.

Many restriction enzymes do not produce sticky ends. Instead, they cut both DNA strands at the same base pair within the recognition sequence, making

"blunt" ends. Chemical methods have been developed to ensure the ligation of a DNA fragment into blunt ends of a target DNA.

Cloning Genes

The goal of recombinant DNA work is to manipulate host cells to produce many copies of a particular gene, either for purposes of analysis or to produce its protein product in quantity. In this section, we will discuss host selection, the entry of DNA into cells, and genetic markers. The choice of host cell, prokaryotic or eukaryotic, is important. Once the host species is selected, the DNA of interest is mixed with the cells and, under specific conditions, can enter some of them.

Because all the host cells proliferate—not just the few that receive the recombinant DNA—the scientist must be able to determine which cells actually have the targeted DNA sequence. One common method of identifying recombinant DNA cells is to tag the inserted sequence with genetic markers whose phenotypes are easily observed. For example, the inserted sequence might carry a marker gene for resistance to an antibiotic; thus any cells that don't die when exposed to the antibiotic must contain the new DNA.

Genes can be cloned in prokaryotic or eukaryotic cells

The initial successes of recombinant DNA technology were achieved with bacteria as hosts. As noted in preceding chapters, bacterial cells are easily grown and manipulated in the laboratory. Much of their molecular biology is known, especially for certain bacteria, such as *E. coli*, and numerous genetic markers can be used to select for cells harboring the manipulated DNA. In some important ways, however, bacteria are not ideal organisms for studying and expressing eukaryotic genes. Bacteria lack the splicing complex to excise introns from the initial RNA transcript of eukaryotic genes.

Many eukaryotic proteins are extensively modified after translation by reactions such as glycosylation and phosphorylation (see Chapter 14). Often these modifications are essential for the activity of the protein. Unfortunately, prokaryotes usually lack the machinery to perform these eukaryotic modifications. Finally, in some instances the addition of a new gene and its expression in a eukaryote are the point of the experiment. That is, the aim is to produce a *transgenic organism*. In these cases, the host for the new DNA may be a mouse, a wheat plant, yeast, or a human, to name a few examples.

Yeasts, such as *Saccharomyces*, the baker's and brewer's yeasts, are common eukaryotic hosts for recombinant DNA studies. Advantages of using yeasts include rapid cell division (a life cycle completed in 2 to 8 hours), ease of growth in the laboratory, and a relatively small genome size (about 20 million base pairs)—several times larger than that of *E. coli*, yet 1/300 the size of the mammalian one.

Plant cells can be used as hosts, especially if the desired result is a transgenic plant. The property that makes plant cells good hosts is their *totipotency*—that is, the ability of a differentiated cell to act like a fertilized egg and produce an entire new organism. Isolated plant cells grown in culture can take up recombinant DNA, and by manipulation of the growth medium, these transgenic cells can be induced to form an entire new plant. This plant can then be reproduced naturally in the field and will carry and express the gene carried on the recombinant DNA.

Vectors can carry new DNA into host cells

In natural environments, DNA released from one bacterium can sometimes be taken up by another bacterium and genetically transform that bacterium (see Chapter 10), but this is not common. The challenge of inserting new DNA into a cell is not just entry, but the replication of the molecule in the host cell as it divides. DNA polymerase, the enzyme that catalyzes replication, does not bind to just any sequence of DNA and begin the replication. Rather, like any DNA-binding protein, it recognizes a specific sequence, the *origin of replication* (see Chapter 11).

There are two general ways in which the newly introduced DNA can be part of a *replicon*, or replication unit. First, it can insert into the host chromosome after entering the cell. Although this is often a random event, it is nevertheless a common method of integrating a new gene into the host cell. Alternatively, the new DNA can enter the host cell as part of a carrier DNA that already has the appropriate origin of replication.

This carrier DNA, targeted at the host cell, is called a **vector**. In addition to its ability to replicate independently in the host cell, a vector must have two other properties. First, it must have sequences that allow the new DNA to be added to it—a recognition sequence for a restriction enzyme. Thus the vector must be able to form recombinant DNA. Second, the vector should have a marker that will announce to the scientist its presence in the host cell. Typically, this marker is a gene that codes for a protein whose phenotype is easily detected, such as resistance to a drug. Another useful property for a vector is its ease of isolation and manipulation, which usually is reflected by its small size in comparison to host chromosomes.

PLASMIDS AS VECTORS. The properties of plasmids make them ideal vectors for genes in bacteria. Each plasmid is a naturally occurring bacterial chromosome, with an origin of DNA replication. The plasmid is small, usually 2,000 to 6,000 base pairs, as compared to the main *E. coli* chromosome, which has more than 4 million base pairs. Because it is so small, a plasmid not surprisingly has only single sites for various restriction enzymes (Figure 16.5a). The fact that there is only one restriction site for a given enzyme is essential because it allows for insertion of new DNA at only that location (see Figure 16.4).

Many plasmids contain genes for enzymes that confer antibiotic resistance. This characteristic provides the marker for a host cell carrying the plasmid. It is relatively easy to determine if a colony of bacteria is resistant to an antibiotic (see Chapter 13). A final useful property of plasmids is the capacity to replicate independently of the host chromosome, often many times more than the host. It is not uncommon for a

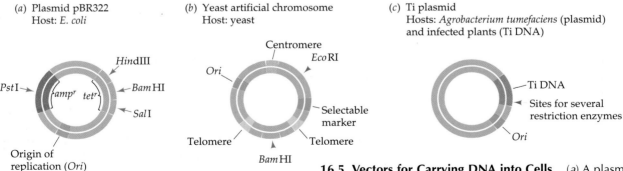

16.5 Vectors for Carrying DNA into Cells (a) A plasmid with genes for antibiotic resistance can be incorporated into an *E. coli* cell. (b) A DNA molecule synthesized in the laboratory becomes a chromosome that can carry its inserted DNA into yeasts. (c) The Ti plasmid, isolated from the bacterium *Agrobacterium tumefaciens*, is an important vector for inserting DNA into many types of plants.

(a) Plasmid pBR322 Host: *E. coli* — HindIII, PstI, BamHI, SalI, amp^r, tet^r, Origin of replication (*Ori*)

(b) Yeast artificial chromosome Host: yeast — Centromere, EcoRI, Ori, Selectable marker, Telomere, Telomere, BamHI

(c) Ti plasmid Hosts: *Agrobacterium tumefaciens* (plasmid) and infected plants (Ti DNA) — Ti DNA, Sites for several restriction enzymes, Ori

↓ Recognition site for restriction enzymes
amp^r: ampicillin resistance gene
tet^r: tetracycline resistance gene

bacterial cell with a single main chromosome to have hundreds of copies of a recombinant plasmid.

VIRUSES AS VECTORS.   Constraints to plasmid replication limit the size of new DNA that can be spliced into a plasmid to about 5,000 base pairs. Although a prokaryotic gene may be this small, 5,000 base pairs is much smaller than most eukaryotic genes, with their introns and extensive flanking sequences that are important in gene expression. So, a vector that accommodates larger DNA inserts is needed.

Both prokaryotic and eukaryotic viruses are often used as vectors for eukaryotic DNA. Bacteriophage lambda, which infects *E. coli*, has a DNA genome of more than 45,000 base pairs. If the genes that cause the host cell to die and lyse—about 20,000 base pairs—are eliminated, the virus can still infect a host cell and inject its DNA. The deleted 20,000 base pairs can be replaced with DNA from another organism, thereby creating recombinant viral DNA.

Because viruses infect cells naturally, they offer a great advantage as vectors over plasmids, which often require artificial means to coax them to enter cells. As we will see in Chapter 17, viruses are important vectors for delivering new genes to people in gene therapy.

ARTIFICIAL CHROMOSOMES AS VECTORS.   Bacterial plasmids are not good vectors for yeast hosts, because prokaryotic and eukaryotic DNA sequences use different origins of replication. Thus a recombinant bacterial plasmid will not replicate in yeast. To remedy this problem, scientists have created in the laboratory a "minimalist chromosome" called the **yeast artificial chromosome**, or **YAC** (Figure 16.5*b*).

This DNA molecule has not only the yeast origin of replication, but sequences for the yeast centromere and telomere as well, making it a true chromosome. With artificially synthesized single recognition sites for restriction enzymes and useful marker genes for yeast (nutritional requirements), YACs are only about 10,000 base pairs in size but can accommodate 50,000 to 1.5 million base pairs of inserted DNA.

PLASMID VECTORS FOR PLANTS.   An important vector for carrying new DNA into many types of plants is a plasmid that is found in *Agrobacterium tumefaciens*. This bacterium lives in the soil and causes a plant disease called crown gall, which is characterized by the presence of growths, or tumors, in the plant. *A. tumefaciens* contains a plasmid, called Ti (for *tumor-inducing*) (Figure 16.5*c*).

Part of the Ti plasmid is T DNA, a transposon that produces copies of itself in the chromosomes of infected plant cells. The T DNA has recognition sequences for restriction enzymes, and new DNA can be spliced into the T DNA region. When the T DNA is thus replaced, the plasmid no longer produces tumors, but the transposon, with the new DNA, is inserted into the host cell's chromosomes. The plant cell can then be grown in culture or induced to form a new, transgenic, plant.

There are many ways to insert recombinant DNA into host cells

Although some vectors, such as naturally infecting viruses, can enter host cells directly, most vectors require help to enter host cells. A major problem for DNA entry is that the exterior surface of the plasma membrane, with its phospholipid head groups, is negatively charged, and so is DNA. The resulting charge repulsion can be alleviated if the exterior of the cells and the DNA are neutralized with Ca^{2+} (calcium) salts. The salts reduce the charge effect, and the plasma membrane becomes permeable to DNA. In this way, almost any cell, prokaryotic or eukaryotic, can be transfected by taking up a DNA molecule from its environment. In plants and fungi, the cell walls must first be removed by hydrolysis with fungal enzymes; the resulting separated plant cells lacking cell walls are called protoplasts.

In addition to the "naked" DNA approach, DNA can be introduced into host cells by a variety of mechanical methods (Figure 16.6). In one, called *particle bombardment*, tiny high-velocity particles of tungsten or gold are coated with DNA and then shot into host cells. This "gene gun" approach must be undertaken with great care to prevent the cell contents from being damaged. In a second method, *electroporation*, cells are exposed to rapid pulses of high-voltage current. This treatment temporarily renders the plasma membrane permeable to DNA in the surrounding medium. A third mechanical way to insert DNA into cells is to inject it with a very fine pipette. This method is especially useful on large cells such as eggs. Finally, DNA can be coated in ways that allow it to pass through the plasma membrane. For example, it can be encased in liposomes, bubbles of lipid that fuse with the membranes of the host cell.

Genetic markers identify host cells that contain recombinant DNA

Following interaction with an appropriate vector, a population of host bacterial cells is heterogeneous, since only a small percentage of the cells have actually taken up the vector. Also, since the reaction of cutting the plasmid and inserting the new DNA to make recombinant DNA is far from perfect, only a few of the plasmids that have moved into the host cells actually contain the new DNA sequence. How can we select only the cells that contain the plasmid with the desired target DNA?

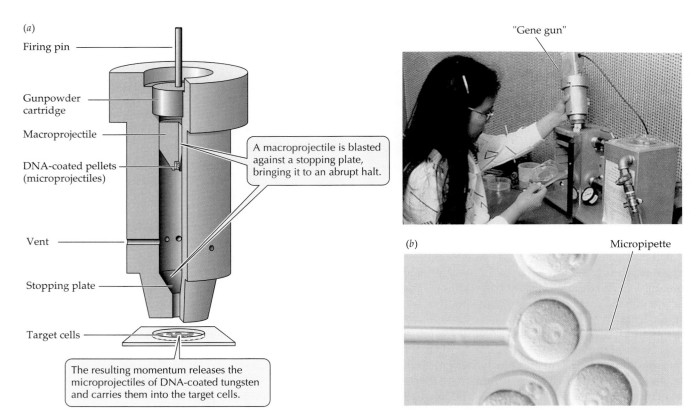

(a)

Firing pin

Gunpowder cartridge

Macroprojectile

DNA-coated pellets (microprojectiles)

A macroprojectile is blasted against a stopping plate, bringing it to an abrupt halt.

Vent

Stopping plate

Target cells

The resulting momentum releases the microprojectiles of DNA-coated tungsten and carries them into the target cells.

"Gene gun"

(b)

Micropipette

16.6 Methods of Mechanically Inserting DNA into Cells
(a) The "gene gun." A macroprojectile is blasted against a stopping plate, bringing it to an abrupt halt. The resulting momentum releases the microprojectiles of DNA-coated tungsten and carries them into the target cells. (b) A less equipment-intensive method is to inject DNA into a cell using a micropipette.

The experiment we are about to describe illustrates an elegant, commonly used approach to this problem. In this example, we use *E. coli* bacteria as hosts and the plasmid vector pBR322 (see Figure 16.5a), which carries the genes for resistance to the antibiotics ampicillin and tetracycline. When the plasmid is incubated with the restriction enzyme *Bam*HI, it encounters its recognition sequence, GGATCC, only once, at a site within the gene for tetracycline resistance.

If foreign DNA is inserted into this restriction site, the presence of these "extra" base pairs within the tetracycline resistance gene inactivates it. So plasmids containing the inserted DNA will carry an intact gene for ampicillin resistance but *not* an intact gene for tetracycline resistance. This is the key to selection of host bacteria that have the recombinant plasmid (Figure 16.7).

The cutting and splicing reaction results in three types of DNA, all of which can be taken up by the host bacteria, which normally are susceptible to killing by both ampicillin and tetracycline. The recombinant plas-

mid—the one we want—turns out to be the rarest type of DNA. Its uptake confers on host *E. coli* only resistance to ampicillin. More common are bacteria that take up pBR322 plasmid that has sealed back up on itself and retains intact genes for resistance to both antibiotics.

Still more common are bacteria that take up the targeted DNA sequence alone; since it is not part of a replicon, it does not survive as the bacteria divide. These host cells will remain susceptible to both antibiotics, as will the vast majority (more than 99.9 percent) of cells that take up no DNA. So the unique drug resistance phenotype of the cells with recombinant DNA marks them in a way that can be detected by simply altering the medium surrounding the cells with ampicillin and/or tetracycline.

In addition to genes for antibiotic resistance, several other marker genes are used to detect recombinant DNA in host cells. Scientists have created several artificial vectors in the laboratory that include sites for restriction enzymes within the *lac* operon (see Chapter 13). When this gene is inactivated by the insertion of foreign DNA, the vector no longer carries this operon's function into the host cell. Other "reporter genes" that have been used in vectors include the gene for luciferase, the enzyme that causes fireflies to glow in the dark when supplied with substrate. Green fluorescent protein, which does not require a substrate to glow, is now widely used (Figure 16.8).

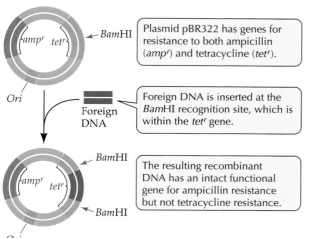

Plasmid pBR322 has genes for resistance to both ampicillin (*amp^r*) and tetracycline (*tet^r*).

Foreign DNA is inserted at the *Bam*HI recognition site, which is within the *tet^r* gene.

The resulting recombinant DNA has an intact functional gene for ampicillin resistance but not tetracycline resistance.

Detection of Recombinant DNA in *E. coli*

DNA TAKEN UP BY *AMP^S* AND *TET^S E. COLI*	PHENOTYPE FOR AMPICILLIN	PHENOTYPE FOR TETRACYCLINE
None	Sensitive	Sensitive
Foreign DNA only	Sensitive	Sensitive
pBR322 plasmid	Resistant	Resistant
pBR322 recombinant plasmid	Resistant	Sensitive

16.7 Marking Recombinant DNA by Inactivating a Gene
Scientists manipulate marker genes within plasmids so they will know which host cells have incorporated the recombinant genes. The bacteria in this experiment, which were "marked" by inactivating the gene for tetracycline resistance, could display any of the phenotype combinations indicated in the table; assuming we wish to select only those that have taken up the pBR322 *recombinant* plasmid, we can do so by altering antibiotic ingredients in the medium surrounding the cells.

After DNA uptake (or not), host cells are usually first grown on a solid medium for selection. If the concentration of cells dispersed on the solid medium is low, each cell will divide and grow into a distinct colony (see Chapter 13). The colonies that contain recombinant DNA can be picked off the medium and then grown in large amounts in liquid culture. The power of recombinant DNA technology to amplify a

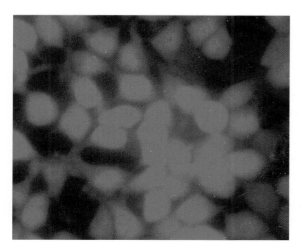

16.8 A Reporter Gene Announces the Presence of a Vector in Eukaryotic Cells These cells have taken up a vector that expresses a gene producing green fluorescent protein.

gene is indicated by the fact that a 1-liter culture of bacteria harboring the human β-globin gene in pBR322 plasmid has as many copies of the gene as the sum total of all the cells in a typical human being.

Sources of Genes for Cloning

There are three principal sources of the genes or DNA fragments used in recombinant DNA work. One source is random pieces of chromosomes inserted into vectors; these DNA–vector units are maintained as gene libraries. The second source is complementary DNA, obtained by reverse transcription from specific mRNAs. The third source is DNA synthesized by organic chemists in the laboratory. The specific fragments can be deliberately modified to create mutations or to revert a mutant sequence back to wild type. We will elaborate on these sources in the following sections.

Genomic libraries contain pieces of a genome

The 23 human chromosomes (or 24, in a male) are a library that contain all the genes of the species. Each chromosome, or "volume" in the library, contains on average 100 million base pairs of DNA. Such a huge molecule is not very useful for studying genome organization or for isolating a specific gene. Thus each chromosome must be broken into smaller pieces, and each piece is then analyzed. These smaller fragments still represent a library; however, the information is now in many more volumes than 23. Each DNA fragment can be inserted into a vector, which can then be taken up by a host cell (Figure 16.9). Each host cell colony, then, harbors a single fragment of human DNA.

Using plasmids, which are able to insert only a few thousand base pairs of foreign DNA, about a million separate fragments are required to make a library of the human genome. For phage lambda, which can

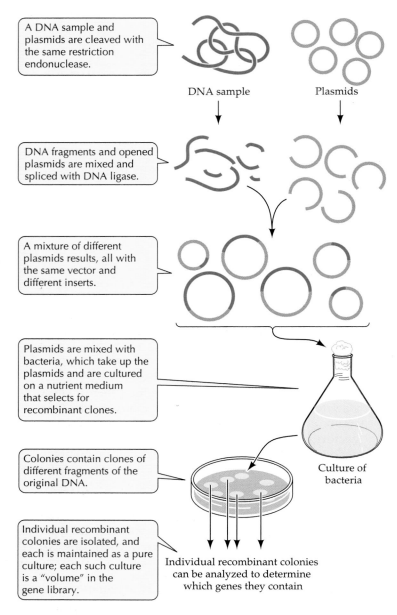

A DNA sample and plasmids are cleaved with the same restriction endonuclease.

DNA sample

Plasmids

DNA fragments and opened plasmids are mixed and spliced with DNA ligase.

A mixture of different plasmids results, all with the same vector and different inserts.

Plasmids are mixed with bacteria, which take up the plasmids and are cultured on a nutrient medium that selects for recombinant clones.

Culture of bacteria

Colonies contain clones of different fragments of the original DNA.

Individual recombinant colonies are isolated, and each is maintained as a pure culture; each such culture is a "volume" in the gene library.

Individual recombinant colonies can be analyzed to determine which genes they contain

16.9 Constructing a Gene Library Human chromosomes are broken up into fragments of DNA that are inserted into vectors (plasmids are shown here) and taken up by host bacterial cells, each of which harbors a single fragment of the human DNA. The information in these cell colonies constitutes a gene library.

carry four times as much DNA as a plasmid, the number of colonies for the library is reduced to about 250,000. Although this seems like a large number, a single growth plate can hold up to 80,000 phage colonies and is easily screened for the presence of a particular DNA sequence by denaturing the phage DNA and applying a particular probe for hybridization.

A DNA copy of mRNA can be made

A much smaller DNA library, that includes only genes transcribed in a particular tissue, can be made from complementary DNA, or cDNA (Figure 16.10). Recall that most eukaryotic mRNAs have a poly A tail—a string of adenosine residues at their 3′ end (see Chapter 14). The first step in cDNA production is to extract mRNA from a tissue and allow it to hybridize with a molecule (called oligo dT—the "d" indicates *deoxy*ribose) consisting of a string of T residues. After the hybrid forms, the oligo dT serves as a primer and the mRNA as a template for the enzyme reverse transcriptase, which synthesizes DNA from RNA templates. A cDNA strand complementary to mRNA is made; after it is removed from the template RNA, this DNA can be manipulated by cloning.

A collection of cDNAs from a particular tissue at a certain time in the lifetime of the organism is called a cDNA library. Because mRNAs are often present in small amounts and such libraries can be cloned, cDNA libraries have been invaluable in comparisons of gene expression in different tissues at different stages of development. For example, their use has shown that up to one-third of all the genes of an animal are expressed only during prenatal development. Complementary DNA is also a good starting point for the cloning of eukaryotic genes. It is especially useful for genes that are present in few copies and rarely expressed.

DNA can be synthesized chemically in the laboratory

When we know the amino acid sequence of a protein, we can obtain the DNA that codes for it by simply making it in the laboratory, using organic chemistry techniques. DNA synthesis has even been automated, and at many institutions a special service laboratory can make short to medium-length sequences overnight for any number of investigators.

How do we design a synthetic gene? Using the genetic code (see Figure 12.5) and the known amino acid sequence, we can figure out the appropriate base sequence for the gene. With this sequence as a starting point, we can add other sequences such as codons for translation initiation and termination, and flanking sequences for transcription initiation, termination, and regulation. Of course, these noncoding DNA sequences must be the ones actually recognized by the host cell if the synthetic gene is to be transcribed. It does no good to have a prokaryotic promoter sequence near a gene if that gene is to be inserted into a yeast for expression. Codon usage is also important: Many amino acids have more than one codon, and different organisms stress the use of different synonymous codons.

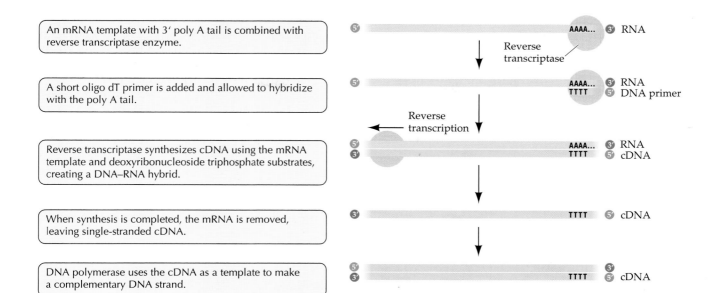

An mRNA template with 3′ poly A tail is combined with reverse transcriptase enzyme.

A short oligo dT primer is added and allowed to hybridize with the poly A tail.

Reverse transcriptase synthesizes cDNA using the mRNA template and deoxyribonucleoside triphosphate substrates, creating a DNA–RNA hybrid.

When synthesis is completed, the mRNA is removed, leaving single-stranded cDNA.

DNA polymerase uses the cDNA as a template to make a complementary DNA strand.

16.10 Synthesizing Complementary DNA Gene libraries that include only genes transcribed in a particular tissue at a particular time can be made from complementary DNA. cDNA synthesis is especially useful for identifying genes that are present only in a few copies, and is often a starting point for gene cloning.

DNA can be mutated in the laboratory

Mutations that occur in nature have been important in proving cause-and-effect relationships in biology. For example, in Chapter 14 we learned that some people with the disease beta thalassemia have a mutation at the splice site for intron removal and so cannot make proper β-globin mRNA. This example shows the importance of the splice site consensus sequence.

Recombinant DNA technology has allowed us to ask the "What if?" questions without having to look for mutations in nature. Because synthetic DNA can be made in any sequence desired, we can clone such DNA to create specific mutations and then see what happens when the mutant DNA expresses itself in the host cell. Additions, deletions, and base-pair substitutions are all possible on isolated or synthetic DNA.

These mutagenesis techniques have allowed scientists to bypass the search for naturally occurring mutant strains, leading to many cause-and-effect proofs. For example, it was proposed that the signal sequence at the beginning of a secreted protein is essential to its insertion across endoplasmic reticulum membranes. So, a gene coding for such a protein, with the codons for the signal sequence deleted, was made. Sure enough, when this gene was expressed in yeast cells, the protein did not cross the ER membrane.

Mutagenesis has also begun to be useful in the design of specific drugs. The advent of a new branch of biology called computational biology has led to so-phisticated studies of the three-dimensional shapes and chemical properties of enzymes, substrates, and their possible regulators. Attempts are being made to devise rules to predict the tertiary structure of a protein from its primary structure. For example, the three-dimensional design of a polypeptide that regulates an enzyme might be proposed. It could then be made and induced to mutate to test the relationship between structure and activity.

Some Additional Tools for DNA Manipulation

Biological methods are not the only ways that DNA can be managed in the laboratory. In this section, we examine two important enzyme-based techniques that are invaluable, and both can be automated. First, the nucleotide sequence of DNA can be determined, and second, a DNA sequence can be replicated many times by the polymerase chain reaction. (Not surprisingly, both of these techniques earned Nobel prizes for their discoverers.) Finally, we will describe the use of antisense RNA to block the translation of specific mRNAs.

The nucleotide sequence of DNA can be determined

Cloned DNA fragments can be isolated and their nucleotide sequences determined. This ability is important in distinguishing true protein-coding genes from any surrounding "spacer" DNA, in order to study gene structure and identify the base-pair changes in mutations. Although there are several methods for sequencing, a simple one invented by Frederick Sanger has been automated and is being used in the Human Genome Project, an undertaking in which the entire 3 billion bases of the human chromosome set are being sequenced.

The method for sequencing relies on the mechanism for DNA replication. The nucleoside triphosphates (NTPs) that are the normal substrates for DNA replication use the sugar 2-deoxyribose; thus they are referred to as dNTPs. If instead of deoxyribose, the sugar used in the NTPs is 2,3-dideoxyribose, the resulting nucleoside triphosphates (ddNTPs) will be picked up by DNA polymerase and added to the growing DNA chain.

However, because ddNTPs lack a hydroxyl group at the 3′ position, the next nucleotide cannot be added (Figure 16.11). (Recall that DNA replicates in a 5′-to-3′ direction.) Thus synthesis stops at the place where ddNTP is incorporated into the growing DNA strand.

In this technique, cloned DNA is denatured. Copies of the single-stranded DNA are combined with DNA polymerase (to synthesize the complementary strand),

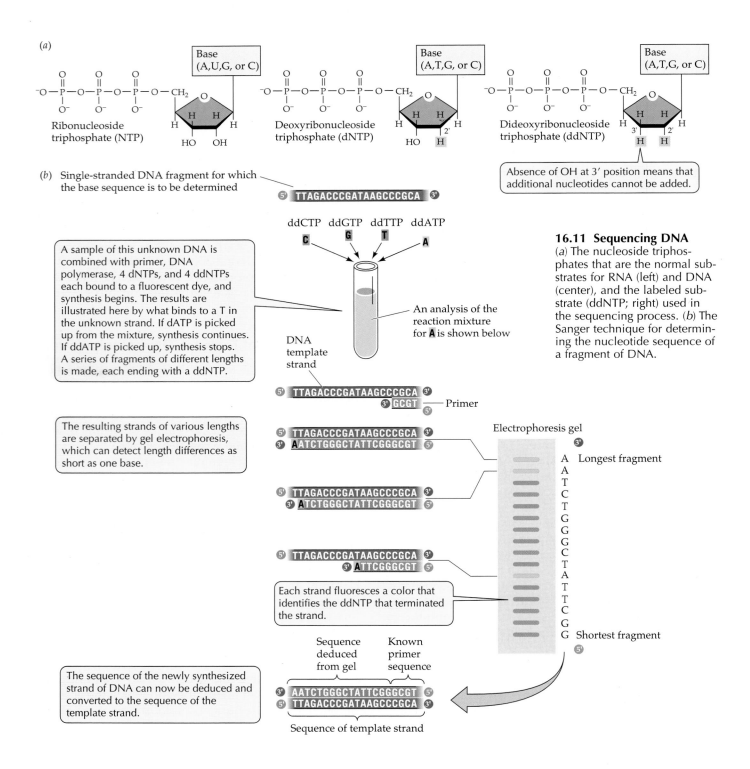

(a)

Ribonucleoside triphosphate (NTP)
Deoxyribonucleoside triphosphate (dNTP)
Dideoxyribonucleoside triphosphate (ddNTP)

Absence of OH at 3′ position means that additional nucleotides cannot be added.

(b) Single-stranded DNA fragment for which the base sequence is to be determined

5′ TTAGACCCGATAAGCCCGCA 3′

ddCTP ddGTP ddTTP ddATP

An analysis of the reaction mixture for **A** is shown below

A sample of this unknown DNA is combined with primer, DNA polymerase, 4 dNTPs, and 4 ddNTPs each bound to a fluorescent dye, and synthesis begins. The results are illustrated here by what binds to a T in the unknown strand. If dATP is picked up from the mixture, synthesis continues. If ddATP is picked up, synthesis stops. A series of fragments of different lengths is made, each ending with a ddNTP.

DNA template strand

5′ TTAGACCCGATAAGCCCGCA 3′
3′ GCGT 5′ — Primer

The resulting strands of various lengths are separated by gel electrophoresis, which can detect length differences as short as one base.

5′ TTAGACCCGATAAGCCCGCA 3′
3′ AATCTGGGCTATTCGGGCGT 5′

5′ TTAGACCCGATAAGCCCGCA 3′
3′ ATCTGGGCTATTCGGGCGT 5′

5′ TTAGACCCGATAAGCCCGCA 3′
3′ ATTCGGGCGT 5′

Each strand fluoresces a color that identifies the ddNTP that terminated the strand.

Electrophoresis gel

3′
A Longest fragment
A
T
C
T
G
G
G
C
T
A
T
T
C
G
G Shortest fragment
5′

16.11 Sequencing DNA
(a) The nucleoside triphosphates that are the normal substrates for RNA (left) and DNA (center), and the labeled substrate (ddNTP; right) used in the sequencing process. (b) The Sanger technique for determining the nucleotide sequence of a fragment of DNA.

Sequence deduced from gel
Known primer sequence

3′ AATCTGGGCTATTCGGGCGT 5′
5′ TTAGACCCGATAAGCCCGCA 3′

Sequence of template strand

The sequence of the newly synthesized strand of DNA can now be deduced and converted to the sequence of the template strand.

primers, the four dNTPs (dATP, dGTP, dCTP and dTTP), and small amounts of the four ddNTPs, each of them with a fluorescent "tag" that emits a different color of light. The reaction mixture soon contains a DNA mixture made up of the cloned DNA single strands (whose sequence will be determined) and shorter complementary strands. The latter, each ending with a ddNTP, are of varying lengths.

For example, each time a T is reached on the template strand, the growing complementary strand adds either dATP or ddATP. If dATP is added, the strand grows until a ddNTP stops it. If ddATP is added, chain growth terminates at that point. After DNA replication has been allowed to proceed for a while in a test tube, the numerous short fragments are denatured from the template and separated by gel electrophoresis (see Figure 16.11). During the electrophoresis run, the strands pass through a laser beam that excites the fluorescent tags. The light emitted is then detected, and the information—that is, which ddNTP is at the end of a strand of which length—is fed into a computer, which processes it and prints out the sequence.

Increasingly powerful analytical tools are being developed to analyze DNA sequences. In the computer, these sequences may be scanned for protein-coding regions, promoters, start and stop codons, and intron–exon boundaries. This type of analysis has been an essential step in the isolation of genes. Alternatively, sequences can be electronically "deposited" in a computerized data bank and compared to other known DNA sequences. Such comparisons have yielded many surprises, such as the presence of common protein-coding regions in regulatory genes for development in organisms ranging from fruit flies to humans (see Chapter 15).

The polymerase chain reaction makes multiple copies of DNA in the test tube

Cloning in a host organism is not the only way to obtain many copies of a particular DNA sequence for analysis. In fact, for DNA molecules whose sequences are at least partly known, cloning in cells has been largely replaced by the **polymerase chain reaction** (PCR) in test tubes. PCR is not very complicated: A short region of DNA, such as a gene, is copied many times in the test tube by DNA polymerase.

PCR is a cyclic process in which the following sequence of steps is repeated over and over again (Figure 16.12). Double-stranded DNA is denatured by heat into single strands, short primers for DNA replication are added to the 3' end of the single DNA template strands, and DNA polymerase catalyzes the production of complementary new strands. A single cycle, taking a few minutes, doubles the amount of DNA

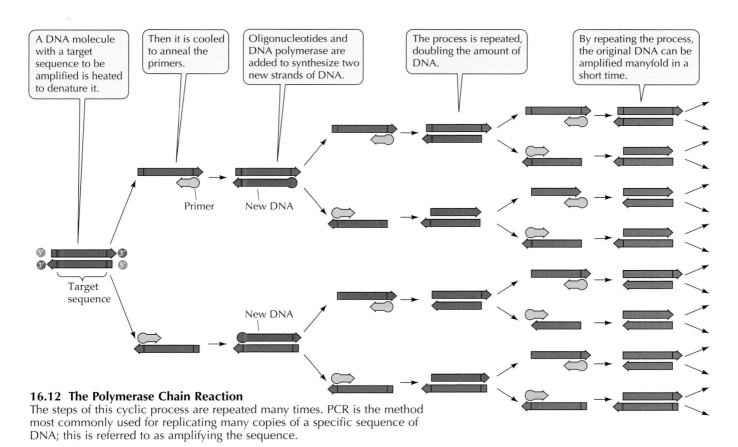

16.12 The Polymerase Chain Reaction
The steps of this cyclic process are repeated many times. PCR is the method most commonly used for replicating many copies of a specific sequence of DNA; this is referred to as amplifying the sequence.

and leaves the new strands in the double-stranded state. Repeating the cycle many times can theoretically lead to a geometric increase in the number of copies of the DNA sequence.

The PCR technique requires that the base sequences at the 3′ end of the target DNA be known so that a complementary primer, usually 15 to 20 bases long, can be made in the laboratory. Because of the uniqueness of DNA sequences, usually the two primers will bind to only one region of DNA in the organism's genome. For example, the two primers

$$5′–AGACTCAGAGAGAACCC–3′$$

and

$$3′–GGGGCACCAGAAACTTA–5′$$

will bind to a region of human DNA that surrounds the β-globin gene and no other region. Thus, in a mixture of fragments that contains the entire 3 billion base pairs of the human genome, these two primers can pinpoint the β-globin gene, which represents just 1 millionth of the human genome. This specificity in the face of the incredible diversity of target DNAs is a key to the power of PCR.

One potential problem with PCR involves its temperature requirements. To denature the DNA during each cycle, it must be heated to more than 90°C. Then it must be cooled to about 55°C to allow the primer to hydrogen-bond to the single strands of template DNA. Because the heating step destroys most DNA polymerases, the PCR method would not be feasible because new polymerase must be added during each cycle after denaturation—an expensive and laborious proposition.

This problem has been solved by nature: In the hot springs at Yellowstone National Park, as well as other locations, there live bacteria called, appropriately, *Thermus aquaticus*. The means by which these organisms survive temperatures up to 95°C was investigated by bacteriologist Thomas Brock and his colleagues. They discovered that *T. aquaticus* has an entire metabolic machinery that is heat-resistant, including DNA polymerase that does not denature at this high temperature.

Scientists pondering the problem of amplifying DNA by PCR read Brock's

basic research articles and got a clever idea: Why not use *T. aquaticus* DNA polymerase in the PCR reaction? It would not be denatured and thus would not have to be added during each cycle. The idea worked, and earned biochemist Kerry Mullis a Nobel prize. PCR has had enormous impact. Some of its most striking applications will be described later in this chapter and in Chapter 17.

Antisense RNA and ribozymes prevent specific gene expression

The base-pairing rules not only can be used to make genes; they can be employed to stop the translation of mRNA. As is often the case, this is an example of scientists imitating nature. In normal cells, a rare method of the control of gene expression is the production of an RNA molecule that is complementary to mRNA. This complementary molecule is called **antisense RNA** because it binds by base pairing to the "sense" bases that make up the coding sequence on mRNA. The formation of the double-stranded RNA hybrid prevents tRNA from binding to the mRNA, and the hybrid tends to be broken down in the cytoplasm. So, although the gene continues to be transcribed, this is an effective method of preventing translation—the synthesis of a protein from its mRNA.

In the laboratory, after determining the sequence of a gene and its mRNA, scientists can add antisense RNA to the cell to prevent translation of the mRNA (Figure 16.13). The antisense RNA can be added as itself—RNA can be taken up by cells in the same way that DNA is introduced—or it can be made in the cell by transcription from a DNA molecule introduced as a part of a vector. This technique is especially useful if a tissue-specific promoter is used to prime antisense transcription, so that its expression occurs only in a targeted tissue. In either case, translation of the targeted gene is prevented. An even more effective way to ensure that antisense RNA works is to couple the

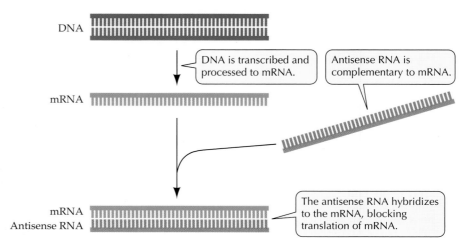

16.13 Using Antisense RNA to Block Translation of an mRNA Once a gene's sequence is determined in the laboratory, it may be desirable to prevent the synthesis of its protein. This can be done with antisense RNA that is complementary to the mRNA.

DNA

DNA is transcribed and processed to mRNA.

Antisense RNA is complementary to mRNA.

mRNA

mRNA
Antisense RNA

The antisense RNA hybridizes to the mRNA, blocking translation of mRNA.

antisense sequence to a special RNA sequence, a **ribozyme**, that catalyzes the cutting of its target RNA.

Antisense RNA (with or without a ribozyme) has been widely used to test cause-and-effect relationships. For example, it can be used to repress the synthesis of a specific protein—something that would otherwise be very difficult. Using antisense RNA, the synthesis of a protein essential for the growth of cancer cells has been blocked, and the cells have reverted to a normal state.

Biotechnology: Applications of DNA Manipulation

Biotechnology is the use of microbial, plant, and animal cells to produce materials useful to people. These products include foods, medicines, and chemicals. Some of them have been made biologically for a long time. For example, using yeasts to brew beer and wine dates back at least 8,000 years in human history, and using bacterial cultures to make cheese and yogurt is a technique many centuries old.

For a long time, people were not aware of the cellular bases of these biochemical transformations. About 100 years ago, thanks largely to Pasteur's work, it became clear that bacteria, yeasts, and other microbes could be harnessed as biological converters to make defined products. Alexander Fleming's discovery that the mold *Penicillium* makes the antibiotic penicillin led to the large-scale commercial culture of microbes to produce vast quantities of antibiotics, as well as other useful chemicals. Today, microbes are harnessed to make much of the industrial-grade alcohol, glycerol, butyric acid, and citric acid that are used by themselves or in manufacturing other products.

In the past, the list of such products was limited to those that were naturally made by the microbes. The many products that eukaryotic cells make, such as hormones and certain enzymes, had to be extracted from those complex organisms. Yields were low, and purification was difficult and costly. All this has changed with the advent of gene cloning. The insertion of almost any gene into bacteria or yeasts, along with methods to induce the genes to make their products, have turned these microbes into versatile factories for important products.

In this section we will describe how this is done and what results have been achieved in the production of pharmaceuticals. With increasing knowledge of the controls of gene activity, as well as new methods to insert genes into eukaryotic cells, it is possible to genetically transform complex organisms. Genetic transformation has been especially successful in agriculture, and a beginning at human gene therapy has been made (see Chapter 17). Finally, we describe some of the many uses of PCR.

Expression vectors turn cells into factories

If a eukaryotic gene is inserted into a plasmid such as pBR322 (see Figure 16.5a) and cloned in *E. coli*, little if any of the gene product of the eukaryotic gene will be made by the host cell. The reason is that the eukaryotic gene lacks the bacterial promoter for RNA polymerase binding, the terminator for transcription, and a special sequence on mRNA for ribosome binding. All of these are necessary for the eukaryotic gene to be expressed and its products synthesized in the bacterial cells.

Expression vectors have all the characteristics of typical vectors, as well as the extra sequences needed for the foreign gene to be expressed in the host cell. For bacteria, these additional sequences include the promoter, the transcription terminator, and the sequence for ribosome binding (Figure 16.14). For eukaryotes, also included might be the poly A addition site, transcription factor binding sites, and enhancers. Once these sequences are placed at the appropriate location on the vector, the gene will be expressed.

An expression vector can be refined in various ways. For example, a specific promoter can be used that responds to hormonal stimulation so that the foreign gene can be induced to transcribe its mRNA only when the scientist adds the hormone. Hormonal stimulation might also activate the promoter so that transcription and protein production occur at very high rates—a goal of obvious importance in the manufacture of an industrial product. Another refinement is to add a DNA sequence to the gene that leads either to the packaging of the expressed protein or to its secretion outside of the cell. Again, this is a great convenience for purification of the product.

Medically useful proteins can be made by DNA technology

Many medically useful products have been made by recombinant DNA technology (Table 16.1), and hundreds more are in various stages of development. We will focus on three of these products (the first of which was introduced at the beginning of this chapter) to illustrate exactly how scientists have developed pharmaceutically important products by recombinant DNA methods.

TISSUE PLASMINOGEN ACTIVATOR TO DISSOLVE BLOOD CLOTS. In most people, when a wound begins bleeding, a blood clot soon forms to stop the flow. Later, as the wound heals, the clot dissolves. How does the blood perform these conflicting functions at the right times? Mammalian blood has an enzyme called *plasmin* that catalyzes the dissolution of the clotting proteins. But plasmin is not always active; if it were, a blood clot would dissolve as soon as it formed! Instead, plasmin is "stored" in the blood in an inactive form called *plasminogen*. The conversion of plasminogen to plasmin is

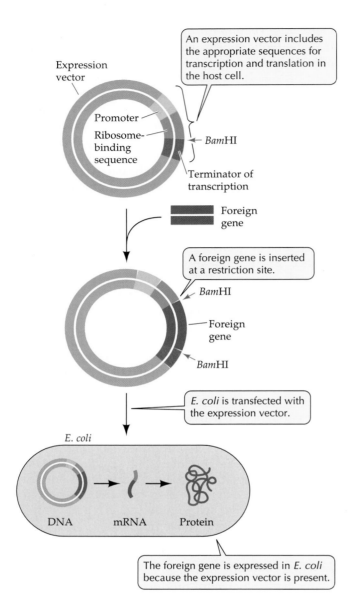

An expression vector includes the appropriate sequences for transcription and translation in the host cell.

Expression vector

Promoter

Ribosome-binding sequence

*Bam*HI

Terminator of transcription

Foreign gene

A foreign gene is inserted at a restriction site.

*Bam*HI

Foreign gene

*Bam*HI

E. coli is transfected with the expression vector.

E. coli

DNA → mRNA → Protein

The foreign gene is expressed in *E. coli* because the expression vector is present.

16.14 An Expression Vector Allows a Foreign Gene to Be Expressed An inserted eukaryotic gene may not be expressed in *E. coli* because the eukaryotic gene lacks the necessary bacterial sequences for promotion, termination, and ribosome binding. Expression vectors contain these additional sequences, enabling the eukaryotic protein to be synthesized in the prokaryotic cell.

activated by an enzyme appropriately called *tissue plasminogen activator* (TPA), which is produced by cells lining the blood vessels. Thus, the reaction is

Heart attacks and strokes are caused by blood clots

$$\text{plasminogen} \xrightarrow{\text{TPA}} \text{plasmin}$$
$$\text{(inactive)} \qquad\qquad \text{(active)}$$

that form in important blood vessels leading to the heart or the brain, respectively. During the 1970s, a bacterial enzyme, streptokinase, was found to stimulate the quick dissolution of clots in some patients with these afflictions. Treating people with this enzyme saved lives, but there were side effects. The drug was a protein foreign to the body, so people's immune systems reacted against it. Of more importance, the drug sometimes prevented clotting throughout the circulatory system, leading to an almost hemophilia-like condition in some people.

The discovery of TPA and its isolation from human tissues led to the hope of a human protein that would not provoke an immune reaction but would specifically bind at the clot, as TPA does. But the amounts of TPA available from human tissues were tiny, certainly not enough to inject at the site of a clot in a patient in the emergency room.

Recombinant DNA technology solved the problem. An antibody against human TPA was used to isolate

TABLE 16.1   Some Medically Useful Products of Biotechnology

PRODUCT	USE
Brain-derived neurotropic factor	Stimulates regrowth of brain tissue in patients with Lou Gehrig's disease
Colony-stimulating factor	Stimulates production of white blood cells in patients with AIDS
Erythropoietin	Prevents anemia in patients undergoing kidney dialysis
Factor VIII	Replaces clotting factor missing in patients with hemophilia A
Growth hormone	Replaces missing hormone in people of short stature
Insulin	Stimulates glucose uptake from blood in some people with diabetes
Platelet-derived growth factor	Stimulates wound healing
Tissue plasminogen activator	Dissolves blood clots after heart attacks and strokes
Vaccine proteins: hepatitis B, herpes, influenza, Lyme disease, meningitis, pertussis, etc.	Prevent and treat infectious diseases

the mRNA for TPA from cell extracts (Figure 16.15). The antibody precipitated not only finished TPA in the cytoplasm, but also TPA still being made on the ribosome but sufficiently folded for antibody recognition. Once TPA mRNA was available, it was used to make a cDNA copy, which was then inserted into an expression vector and introduced into *E. coli*. The transfected bacteria made the protein in quantity, and it soon became available commercially. This protein has had considerable success in dissolving blood clots in both heart attack and, especially, stroke patients.

ERYTHROPOIETIN TO REDUCE ANEMIA. Another widely used protein made through recombinant DNA methods is *erythropoietin* (EPO). The kidneys produce this hormone, which travels through the blood to bone marrow, where it stimulates the division of stem cells to produce red blood cells. People who have suffered kidney failure often require a procedure called kidney dialysis to remove toxins from the organ. However, because dialysis also removes EPO, these patients can become severely anemic (depleted of red blood cells).

As with TPA, the amounts of EPO that can be obtained from healthy people to give to these patients are extremely small, but once again, biotechnology has come to the rescue. The gene for EPO was isolated and cloned in an expression vector in bacteria. Large amounts of the protein are produced by the bacteria, and it is now given to tens of thousands of kidney dialysis patients, with great success at reducing anemia.

HUMAN INSULIN TO STIMULATE GLUCOSE UPTAKE. One of the first important medications made by recombinant DNA methods was human insulin. This hormone, normally made by the pancreas gland, stimulates cells to take up glucose from the blood. People who have certain forms of diabetes mellitus have a deficiency of pancreatic insulin. Injections of the hormone can compensate for this deficiency.

In the past, the injected insulin was obtained from the pancreases of cows and pigs, which caused two problems. First, animal insulin is laborious to purify; second, it is slightly different in its amino acid sequence from human insulin. Some diabetics' immune systems detect these differences and react against the foreign protein.

The ideal solution is to use human insulin, but until the advent of recombinant DNA technology, it was

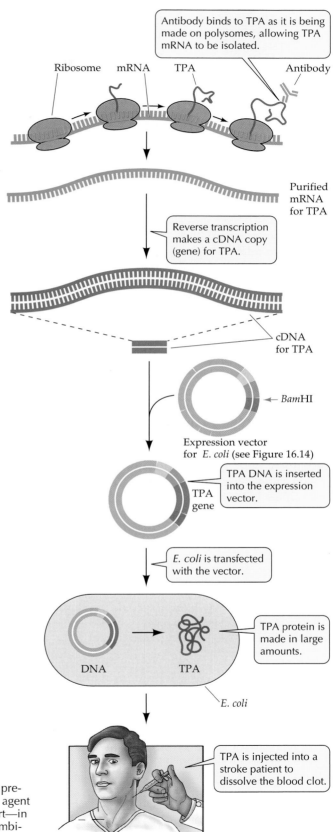

16.15 Tissue Plasminogen Activator: From Protein to Gene to Pharmaceutical TPA is a naturally occurring human protein that prevents blood from clotting. Its isolation and use as a pharmaceutical agent in treating patients suffering from blood clotting in the brain or heart—in other words, strokes and heart attacks—was made possible by recombinant DNA technology.

available only in minuscule amounts. Since insulin is made up of only 51 amino acids, scientists were able to synthesize a gene for this protein in the laboratory. This gene (there were actually two of them, one for each polypeptide chain of the protein) was inserted into *E. coli* via an expression vector, making feasible the use of human insulin in diabetics.

DNA manipulation is changing agriculture

The cultivation of plants and husbanding of animals that constitute agriculture give us the world's oldest examples of biotechnology, dating back more than 8,000 years in human history. Over the centuries, people have adapted crops and farm animals to their needs. Through cultivation and breeding (artificial selection), desirable characteristics such as ease of cooking the seeds or quality of the meat have been improved. In addition, people have selected crops with desirable growth characteristics, such as a reliable ripening season and resistance to diseases.

Until recently, the most common way to improve crop plants was to select varieties that existed in nature through mutational variation and that had the desired genotypes and phenotypes. The advent of genetics in the past century led to its application to plant and animal breeding. A crop plant with a desirable gene could be identified, and through deliberate crosses the single gene or, more usually, many genes could be introduced into a widely used variety of that crop.

Despite spectacular successes, such as the breeding of "supercrops" of wheat, rice, and corn, such deliberate crossing remains a hit-or-miss affair. Many desirable characteristics are complex in their genetics, and it is hard to predict accurately the results of a cross. Moreover, traditional crop plant breeding takes a long time: Many plants can reproduce only once or twice a year—a far cry from the rapid reproduction of bacteria or fruit flies.

Modern biotechnology based on recombinant DNA has two advantages over the traditional methods of breeding. First, the molecular approach allows a breeder to choose specific genes, making the process more precise and less likely to fail as a result of the incorporation of unforeseen genes. The ability to work with cells in the laboratory and then regenerate a whole plant by cloning makes the process much faster than the years needed for traditional breeding. The second advantage—and it is truly an amazing one—is that the molecular methods allow breeders to introduce any gene from any organism into a plant or animal species. This, along with deliberate mutagenesis, expands the range of possible new characteristics to an almost limitless horizon.

As with medicine, there are many examples of molecular biotechnology applied to agriculture (Table 16.2), ranging from improving the nutritional properties of crops, to using animals as gene product factories (transgenic goats are making human TPA in their milk), to using edible crops to make oral vaccines (as we saw at the beginning of the chapter). We will focus on a few examples to show the approaches used.

PLANTS THAT MAKE THEIR OWN INSECTICIDES. Humans are not the only species that consume crop plants. Plants are subject to infections by viruses, bacteria, and fungi, but probably the most important pests are insects. From the locusts of biblical (and modern) times to the cotton boll weevil, insects have continually plagued the crops people grow.

The development of insecticides has improved the situation somewhat, but insecticides have their problems. Most, such as the organophosphates, are relatively nonspecific, killing not only the pest in the field but beneficial insects in the ecosystem as well. Some even have toxic effects on other organisms, including people. What's more, insecticides are applied to the surface of the plants and tend to blow away to adjacent areas, where they may have unforeseen effects.

Some bacteria have solved their pest problem by producing proteins that kill insect larvae that eat the bacteria. For example, there are dozens of strains of *Bacillus thuringiensis*, each of which produces a protein toxic to the insect larvae that plague these bacteria. The toxicity of this protein is 80,000 times that of the usual commercial insecticides. When the hapless larva

TABLE 16.2 Agricultural Biotechnologies under Development

PROBLEM	TECHNOLOGY/GENES
Improving the environmental adaptations of plants	Drought tolerance; salt tolerance
Controlling crop pests	Herbicide tolerance; resistance to viruses, bacteria, fungi, and insects
Improving breeding	Male sterility for hybrid seeds
Improving nutritional traits	High-lysine seeds
Improving crops after harvest	Delay of fruit ripening; high-solids tomatoes; sweeter vegetables
Using plants as bioreactors	Plastics, oils, and drugs produced in plants

16.16 Some Transgenic Plants Make Their Own Insecticides The corn plant on the left has been eaten by earworms. The one on the right is healthy because it has been genetically transformed to make a natural insecticide, the toxin of *Bacillus thuringiensis*.

eats the bacteria, the toxin becomes activated, binding specifically to the insect gut to produce holes. The insect starves to death.

Dried formulations of *B. thuringiensis* have been sold for decades as a safe, biodegradable insecticide. But biodegradation is their limitation, because it means that the dried bacteria must be applied repeatedly during the growing season. A better approach would be to have the crop plants make the toxin themselves.

The toxin genes from different strains of *B. thuringiensis* have been isolated and cloned. They have been extensively modified by the addition of plant promoters and terminators, plant poly A signals, plant codon preferences, and plant controlling elements on DNA. Following introduction of the modified genes into cells in the laboratory using the Ti plasmid vector (see Figure 16.5c), transgenic plants have been grown and tested for insect resistance in the field. So far, transgenic tomato, corn, potato, and cotton crops have been successfully shown to have considerable resistance to their insect predators (Figure 16.16).

TOMATOES THAT RIPEN SLOWLY. Most people agree that for taste, tomatoes bought at the market are vastly inferior to the ones we can pick right off the vine. The problem is that tomatoes develop their full flavor while they ripen on the plant during a period of 6 weeks, so if they are picked early so that they are firm enough to be transported easily from farm to market, they ripen rapidly on the way and end up being bland. Could the ripening process be slowed down after picking?

Plant biochemists have identified an enzyme called *polygalacturonase* (PG) that is important in the ripening

process because it hydrolyzes components of the tomato cell walls. To specifically reduce PG synthesis, they used an antisense strategy (see Figure 16.13). They isolated the gene for PG from tomato plants and cloned it. They then used this DNA to make a gene that would code for an RNA complementary to PG mRNA. Then they attached the antisense gene to an active promoter in a Ti plasmid vector in the laboratory and inserted it into tomato protoplasts.

The transgenic tomatoes that resulted expressed the antisense RNA in their fruits as they developed, thus blocking PG production during the 6 weeks on the plant when ripening would normally occur. Fruit from these transgenic plants can be picked while immature, ships easily without ripening, and has a much longer and more flavorful shelf life than do conventional tomatoes.

CROPS THAT ARE RESISTANT TO HERBICIDES. Glyphosate (known by the trade name Roundup) is a widely used and effective weed killer, or herbicide. It works only on plants, by inhibiting an enzyme system in the chloroplast that is involved in the synthesis of amino acids. Glyphosate is truly a "miracle herbicide," killing 76 of the world's 78 worst weeds that grow in fields and rob crop plants of needed water and nutrients from the soil.

Unfortunately, it also kills crop plants, so great care must be taken with its use. In fact, it is best used to rid a field of weeds before the crop plant starts to grow. But as any gardener knows, when the crop begins to grow the weeds reappear. So it would be advantageous if the crop were not affected by the herbicide. Then, the herbicide could be applied at will to the field any time and would kill only the weeds.

Fortunately, some soil bacteria have mutated to develop an enzyme that breaks down glyphosate. Scientists have isolated the gene for this enzyme, cloned it, and added plant sequences for transcription, translation, and targeting to the chloroplast. The gene has been inserted into cotton and soybean plants and these transgenic crops are resistant to glyphosate, permitting the herbicide to be used more effectively.

DNA fingerprinting uses the polymerase chain reaction

"Everyone is unique." This old saying applies not only to human behavior but also to the human genome. Mutations, genetics, and recombination through sexual reproduction ensure that each member of a species (except identical twins) has a unique DNA sequence. An individual can be definitively characterized ("fingerprinted") by his or her DNA base sequence.

The ideal way to distinguish an individual from all the other people on Earth would be to describe his or her entire genomic DNA sequence. But since the human genome contains more than 3 billion nucleotides, this idea is clearly not practical. Instead, scientists have looked for genes that are highly polymorphic—that is, genes that have multiple alleles in the human population and are therefore different in different individuals. One easily analyzed genetic system consists of the short moderately repeated DNA sequences that occur side by side in the chromosomes.

These repeat patterns are inherited. For example, an individual might inherit a chromosome 15 with the short block repeated six times from her mother, and the same block repeated two times from her father. These repeats are easily detectable if they lie between two recognition sites for a restriction enzyme. If the DNA from this individual is isolated and cut with the restriction enzyme, she will have two different-sized fragments of DNA between the sites: one larger (the one from the mother) and the other smaller (the one from the father). These patterns are easily seen by use of gel electrophoresis (Figure 16.17). With several different repeated sequences (as many as eight are used, each with numerous alleles) an individual's unique pattern becomes apparent.

Typically, these methods require 1 µg of DNA, or the DNA content of about 100,000 human cells, but this amount is not always available. The power of PCR (see Figure 16.12) permits the amplification of the DNA from a single cell to produce in a few hours the necessary 1 µg for restriction and gel analysis.

In forensics, DNA fingerprints are used to help prove the innocence or guilt of a suspect. For example, in a rape case, DNA can be extracted from dried semen or hair left by the attacker. After PCR and restriction analyses, this extracted DNA can be compared to DNA from a suspect. So far, this method has been used to prove innocence (the DNA patterns are different) more often than guilt (the DNA patterns are the same). The reason is that it is easy to exclude someone on the basis of the tests, but two people could have the same patterns, since what is being tested is just a small sample of the genome, so proving that a suspect is guilty cannot rest on DNA fingerprinting alone.

In addition to this highly publicized use, there are many other applications of PCR-based DNA fingerprinting. In 1992, there were 52 California condors, looked after by the San Diego and Los Angeles zoos. Scientists made DNA fingerprints of all these birds so that the geneticists at the zoos can mate unrelated individuals that have many allelic differences in order to increase genetic variation in the offspring of this endangered species. A similar program is under way for the threatened Galapagos tortoises (see Chapter 55).

Plant scientists have found in nature or produced by breeding programs thousands of varieties of crops such as rice, wheat, corn, and grapes. Many of these varieties are kept in cold storage in "seed banks."

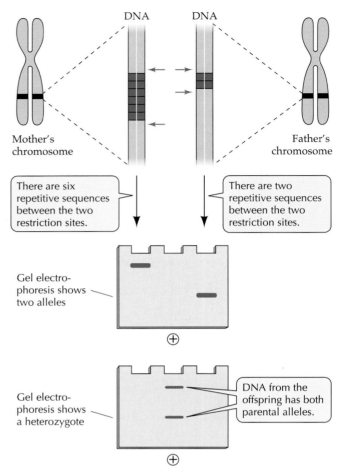

16.17 DNA Fingerprinting As many as eight different repeated gene sequences can be used to make a DNA fingerprint.

Samples of these plants are being DNA-fingerprinted to determine which varieties are genetically the same and which are the most diverse, as a guide to breeding programs.

A related use of PCR is in the diagnosis of infections. In this case, the test is whether the DNA of an infectious agent is present in blood or a tissue sample. Specifically, the question is whether a primer strand that matches the pathogen's DNA and is used for PCR will be amplified on DNA from the patient. Because so little of the target is needed and the primers can be made to bind only to a specific viral or bacterial genome, the PCR-based test is extremely sensitive. If an organism is present in small amounts, PCR testing will detect it.

Finally, the isolation and characterization of genes for various human diseases, such as sickle-cell anemia and cystic fibrosis, has made PCR-based genetic testing a reality. We will discuss this subject in depth in the next chapter.

Summary of "Recombinant DNA and Biotechnology"

Cleaving and Rejoining DNA

• Knowledge of DNA transcription, translation, and replication has been used to create recombinant DNA molecules, made up of sequences from different organisms.

• Restriction enzymes, which are made by microbes as a defense mechanism, bind to DNA at specific sequences and cut it. **Review Figure 16.1**

• DNA fragments generated from cleavage by restriction enzymes can be separated by size using gel electrophoresis. The sequences of these fragments can be further identifed by hybridization with a probe. **Review Figures 16.2, 16.3**

• Many restriction enzymes make staggered cuts in the two strands of DNA, creating "sticky ends," with unpaired bases. The sticky ends can be used to create recombinant DNA if DNA molecules from different species are cut with the same restriction enzyme and then mixed in the test tube. **Review Figure 16.4**

Cloning Genes

• Bacteria, yeasts, and cultured plant cells are commonly used as hosts for recombinant DNA experiments.

• Newly introduced DNA must be part of a replicon if it is to be propagated in host cells. One way to make sure that the introduced DNA is part of a replicon is to introduce it as part of a carrier DNA, or vector DNA, that has a replicon.

• There are specialized vectors for bacteria, yeasts, and plant cells. These vectors must contain a replicon, recognition sequences for restriction enzymes, and genetic markers to identify their presence in the host. **Review Figure 16.5**

• Naked DNA may be introduced by chemical or mechanical means. In this case, the DNA must integrate into the host DNA. **Review Figure 16.6**

• When vectors carrying recombinant DNA are incubated with host cells, nutritional, antibiotic resistance, or fluorescent markers can be used to identify which cells contain the vector. **Review Figures 16.7, 16.8**

Sources of Genes for Cloning

• The cutting of DNA by a restriction enzyme produces many fragments that can be individually and randomly put into a vector and inserted into a host to create a genomic library.

• A smaller library of cDNAs can be made from the mRNAs produced in a tissue or organism at a certain moment. In this case, mRNA is extracted and used to create DNA (cDNA) by reverse transcription. This cDNA is then used to make a library. **Review Figure 16.10**

• A third source of DNA for cloning is DNA made by chemists in the laboratory. Organic chemistry methods can be used to create specific, mutated DNA sequences.

Some Additional Tools for DNA Manipulation

• Cloned DNA fragments can be isolated so that their nucleotide sequences can be determined.

• Several hundred bases in DNA can be sequenced at a time. One method utilizes dideoxyribonucleotides, which cause the chain to terminate, to create fragments that are then separated by gel electrophoresis and analyzed by computer. **Review Figure 16.11**

• The polymerase chain reaction makes multiple copies of DNA in the test tube, using a heat-resistant DNA polymerase. If the sequences of the ends of DNA are known, primers can be made complementary to them. These primers can be extended in the test tube as part of the polymerase chain reaction. PCR can amplify a target DNA sequence manyfold in a few hours. **Review Figure 16.12**

• Preparation of an antisense RNA complementary to a specific mRNA can prevent translation when the antisense RNA hybridizes to the mRNA. A vector expressing antisense RNA can prevent translation continuously. **Review Figure 16.13**

Biotechnology: Applications of DNA Manipulation

• The ability to clone genes has made possible many new applications of biotechnology, such as the large-scale production of eukaryotic gene products.

• For a vector carrying a gene of interest to be expressed in a host cell, the gene must be adjacent to appropriate sequences for its transcription and translation in the host cell. **Review Figure 16.14**

• Recombinant DNA and expression vectors have been used to make medically useful proteins that would otherwise have been difficult to obtain in necessary quantities. **Review Figure 16.15 and Table 16.1**

• Because plant cells can be cloned to produce adult plants, the introduction of new genes into plants via vectors has been more rapid than with animal cells. The result is crop plants that carry new, useful genes. **Review Figure 16.16 and Table 16.2**

• Because the DNA of an individual or species is unique, the polymerase chain reaction can be used to identify an organism from a small sample of its cells—that is, to create a DNA fingerprint. **Review Figure 16.17**

Self-Quiz

1. Restriction endonucleases
   a. play no role in bacteria.
   b. cleave DNA at highly specific recognition sequences.
   c. are inserted into bacteria by bacteriophages.
   d. are made only by eukaryotic cells.
   e. add methyl groups to specific DNA sequences.

2. Sticky ends
   a. are double-stranded ends of DNA fragments.
   b. are identical for all restriction enzymes.
   c. are removed by restriction enzymes.
   d. are complementary to other sticky ends generated by the same restriction enzyme.
   e. are hundreds of bases long.

3. In gel electrophoresis,
   a. DNA fragments are separated on the basis of size.
   b. DNA does not have an electric charge.
   c. DNA fragments cannot be removed from the gel.
   d. the electric field separates positively charged DNA fragments from negatively charged DNA fragments.
   e. the DNA fragments are naturally fluorescent.

4. Possession of which feature is *not* desirable in a vector for gene cloning?
   a. An origin of DNA replication
   b. Genetic markers for the presence of the vector
   c. Multiple recognition sites for the restriction enzyme to be used
   d. One recognition site each for many different restriction enzymes
   e. Genes other than the target for cloning

5. DNA can be introduced into any cell by
   a. injection.
   b. being complexed with calcium salts.
   c. being placed along with the cell into a particle gun.
   d. gel electrophoresis.
   e. being heated to be denatured.

6. Complementary DNA (cDNA)
   a. is produced from ribonucleoside triphosphates.
   b. is produced by reverse transcription.
   c. is the "other strand" of single-stranded DNA.
   d. requires no template for its synthesis.
   e. cannot be placed into a vector, since it has the opposite base sequence of the vector DNA.

7. In a genomic library of frog DNA in *E. coli* bacteria,
   a. all bacterial cells have the same sequences of frog DNA.
   b. all bacterial cells have different sequences of DNA.
   c. each bacterial cell has a random fragment of frog DNA.
   d. each bacterial cell has many fragments of frog DNA.
   e. the frog DNA is transcribed into mRNA in the bacterial cells.

8. An expression vector requires all of the following, except
   a. genes for ribosomal RNA.
   b. a selectable genetic marker.
   c. a promoter of transcription.
   d. an origin of DNA replication.
   e. restriction enzyme recognition sites.

9. The polymerase chain reaction
   a. is a method for sequencing DNA.
   b. is used to transcribe specific genes.
   c. amplifies specific DNA sequences.
   d. does not require primers for DNA replication.
   e. uses a DNA polymerase that denatures at 55°C.

10. In DNA fingerprinting,
    a. a positive identification can be made.
    b. a gel blot is all that is required.
    c. multiple restriction digests generate unique fragments.
    d. the polymerase chain reaction amplifies finger DNA.
    e. the variability of repeated sequences between two restriction sites is evaluated.

Applying Concepts

1. Compare PCR and cloning as methods to amplify a gene. What are the requirements, benefits, and drawbacks of each method?

2. As specifically as you can, outline the steps you would take to (a) insert and express the gene for a new, nutritious seed protein in wheat; (b) insert and express a gene for a human enzyme in sheep's milk.

3. The *E. coli* plasmid pSCI carries genes for resistance to the antibiotics tetracycline and kanamycin. The *tet*^{r} gene has a single restriction site for the enzyme *Hin*dIII. Both the plasmid and the gene for corn glutein protein are cleaved with *Hin*dIII and incubated to create recombinant DNA. The reaction mixture is then incubated with *E. coli* that are sensitive to both antibiotics. What would be the characteristics, with respect to antibiotic sensitivity or resistance, of colonies of *E. coli* containing, in addition to its own genome: (a) no new DNA; (b) native pSCI DNA; (c) recombinant pSCI DNA; (d) corn DNA only? How would you detect these colonies?

4. Using a DNA synthesizer, you make a fragment of DNA with the proposed sequence 3'-ATTGTCCTCTGA-5'. Using the sequencing protocol described in Figure 16.11, how would you verify that this is the correct sequence? Describe the reactions and gel patterns that you would observe.

Readings

Chrispeels, M. J. and D. Sadava. 1994. *Plants, Genes and Agriculture.* Jones and Bartlett, Boston, MA. A comprehensive summary of the role of biotechnology in agriculture.

Gilbert, W. 1991. "Toward a Paradigm Shift in Biology." *Nature,* January 10, volume 349, page 99. A discussion of the impact of DNA technology on pure research in biology.

Lewin, B. 1997. *Genes VI.* Oxford University Press, Oxford. This text contains excellent outlines of gene libraries, cDNA cloning, and other specialized methods.

Lodish, H., D. Baltimore, A. Berk, S. L. Zipursky, P. Matsudaira and J. Darnell, 1995. *Molecular Cell Biology,* 3rd Edition. Scientific American Books, New York. Excellent chapters on recombinant DNA methods and the polymerase chain reaction.

Sambrook, J., E. Fritsch and T. Maniatis. 1989. *Molecular Cloning: A Laboratory Manual,* 2nd Edition. This is the book on the lab benches of most molecular biologists.

Chapter 17

Molecular Biology and Medicine

Dr. Asbjørn Følling
Trained as both a physician and a chemist, this Norwegian medical doctor (center) is shown at a 1962 ceremony where he was honored for his research establishing the genetic and biochemical nature of phenylketonuria. Følling's work led to effective treatment of the condition; today the idea of correcting the genetic defect that causes it is being contemplated.

The mother brought her two children to Dr. Asbjørn Følling in 1934 as a last resort. Ever since their births, she had watched the conditions of her 6-year-old daughter and 4-year old son deteriorate. Now both were severely mentally retarded. So far, all of the doctors who had examined the children had expressed sympathy but could do nothing. The mother had noticed a peculiar smell clinging to her children and she had heard that Dr. Følling was trained as both a chemist and a physician. Could he help? It turned out he couldn't, because their retardation was irreversible. But while examining these children, Dr. Følling made a major discovery.

As part of his examination, Dr. Følling tested the children's urine by adding a brown solution of ferric chloride to look for ketones, which are often excreted by diabetics. In normal people, this solution stays brown, but in diabetics it turns purple. To his surprise, the urine of these children turned the solution dark green. He had never seen this color before in the urine test, and it had not been described in any of his reference books. At first, he suspected the children were taking a medication that ended up in the urine and reacted with ferric chloride. So he asked the mother to refrain from giving her children any medications for a week and then to bring him two new urine samples. Once again, the samples tested green. Clearly, a substance unique to the bodies of these two children was responsible for the strange color.

Here's where Følling's chemistry training served him well. Using analytical chemistry, he purified the substance from the children's urine and tentatively identified it. Then, he used organic chemistry to make the substance in the lab and proved it to be identical to the one in the urine. This substance was phenylpyruvic acid. On the basis of the similarity between phenylpyruvic acid and the amino acid phenylalanine, Følling hypothesized that these children were unable to metabolize phenylalanine and that the excess was being converted to phenylpyruvic acid. He soon found other mentally retarded people who excreted this substance, and among the first ten were three pairs of siblings. Følling observed that the parents of these children were mentally normal and did not excrete phenylpyruvic acid. All of these observations fit the idea of an autosomally recessive inherited condition.

Dr. Følling had discovered the genetic disease *phenylketonuria*. But it was not the first such disease to be described in biochemical terms. In 1909, Dr. Archibald Garrod had found the cause of *alkaptonuria*—an inherited disorder in which the patients' urine turns black—to be on the same biochemical pathway as the cause of phenylketonuria. Garrod coined the term "inborn errors of metabolism" as a general description of diseases for which genetics and biochemistry are clearly linked.

Later the phenotypes of both phenylketonuria and alkaptonuria were identified as abnormalities in specific enzymes. Today the causes of hundreds of such single-gene, single-enzyme diseases are known. In some cases, these discoveries have led to the design of specific therapies and ways to screen for the abnormal proteins in people who do not overtly show the disease.

As we will see in this chapter, more precision in describing these abnormalities at the DNA level has come from molecular biology. Even cancer, it turns out, is caused in most cases by abnormalities in genes. The rise of "molecular medicine" is most dramatically shown by undertakings such as gene therapy and the Human Genome Project, which we will discuss at the end of the chapter.

Protein as Phenotype

The different types of a person's cells are distinguished by their unique proteins. For example, hair follicles make the hair protein keratin, and red blood cells contain hemoglobin. As we saw in Chapters 11 and 12, genetic mutations are often expressed phenotypically as proteins that differ from the normal, wild type. Many genetic diseases also have specific proteins that differ from the normal proteins.

In this first section of the chapter, we will identify and discuss the kinds of abnormal proteins that can result from inheritance of an abnormal allele or its origin by mutation. Then we will consider the role of the environment and patterns of inheritance resulting from autosomal recessives, autosomal dominants, X linkages, and chromosome abnormalities.

Many genetic diseases result from abnormal or missing proteins

Proteins have many roles in cells, and the genes that code for these proteins can be mutated to cause genetic diseases. Enzymes, carriers such as hemoglobin, receptors, transport proteins, and structural proteins have all been implicated in genetic diseases.

ENZYMES. Følling's patients with phenylketonuria (PKU)—the inability to metabolize phenylalanine and related compounds—had a dramatically unusual phenotype in that they were mentally retarded. In addition, they had a musty odor, resulting from phenylpyruvic acid, and they had lighter-colored hair and skin than their relatives. Although Følling made his discovery in 1934, not until 1957 was this complex phenotype traced back to its primary phenotype: a single abnormal protein.

As Følling had predicted, phenylalanine hydroxylase, the enzyme that catalyzes the conversion of dietary phenylalanine to tyrosine, was not active in these patients' livers (Figure 17.1). Lack of this conversion led to excess phenylalanine in the blood and explained the accumulation of phenylpyruvic acid, which is the product of an offshoot pathway activated because of the accumulation of phenylalanine.

Later, the sequences of the 451 amino acids that constitute phenylalanine hydroxylase in normal people were compared with those in individuals suffering from PKU, in many cases revealing only a single difference: Instead of the arginine at position 408 in the long polypeptide chain, many people with PKU have tryptophan. Once again, the ideas of

$$\text{one gene} \rightarrow \text{one enzyme}$$

and

$$\text{one mutation} \rightarrow \text{one amino acid change}$$

hold true in human diseases as they do in studies of so many other organisms.

How does the abnormality in PKU lead to its multitude of clinical symptoms? Since the pigment melanin is made from tyrosine, which is not synthesized but obtained mostly in the patients' diet, lighter skin and hair color seem reasonable to expect. But the mental retardation is much more difficult to explain. This symptom could result from the accumulation of the substrate (phenylalanine), due to the missing enzyme activity; alternatively, it could be due to a relative lack of the product (tyrosine), or to the accumulation of phenylpyruvic acid. While knowledge of liver biochemistry has led to specification of the enzyme abnormality, knowledge of brain biochemistry is more

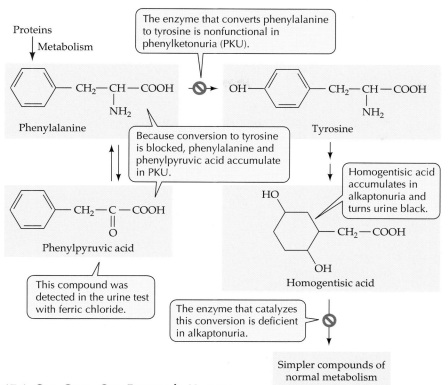

The enzyme that converts phenylalanine to tyrosine is nonfunctional in phenylketonuria (PKU).

Because conversion to tyrosine is blocked, phenylalanine and phenylpyruvic acid accumulate in PKU.

This compound was detected in the urine test with ferric chloride.

The enzyme that catalyzes this conversion is deficient in alkaptonuria.

Homogentisic acid accumulates in alkaptonuria and turns urine black.

Phenylalanine

Tyrosine

Phenylpyruvic acid

Homogentisic acid

Simpler compounds of normal metabolism

17.1 One Gene, One Enzyme in Humans
Phenylketonuria and alkaptonuria are caused by abnormalities in specific enzymes. Knowing the causes of such single-gene, single-enzyme metabolic diseases can aid in the development of screening tests and therapeutic regimens.

primitive. The exact cause of mental retardation in PKU remains elusive, but as we will see later in this chapter, it can be prevented.

Hundreds of human genetic diseases that result from enzyme abnormalities have been discovered, many of which lead to mental retardation and premature death (Table 17.1). Most of these diseases are rare; PKU, for example, shows up in one newborn infant out of every 12,000. But this is just the tip of the mutational iceberg. Undoubtedly, some mutations result in altered proteins that have no obvious clinical effects.

For example, there could be many amino acid changes in phenylalanine hydroxylase that do not affect its catalytic activity. Analysis of the same protein in different people often shows variations that have no functional significance. In fact, at least 30 percent of all proteins have detectable amino acid differences among individuals. If one protein variant exists in less than 99 percent of a population (that is, if the protein has another variant at least 2 percent of the time), the protein is said to be **polymorphic**. The key point is that polymorphism does not necessarily mean disease.

HEMOGLOBIN. The first human genetic disease for which an amino acid abnormality was tracked down as

the cause was not PKU. It was the blood disease *sickle-cell anemia*, which most often afflicts people whose ancestors came from the Tropics or from the Mediterranean. Among African-Americans, about 1 in 655 are homozygous for the sickle allele and have the disease. The abnormal allele produces an abnormal globin protein that leads to sickled red blood cells (see Figure 12.14). These cells tend to block narrow blood capillaries, especially when the O_2 concentration is low, and the result is tissue damage.

Human hemoglobin is a protein with quaternary structure, containing four polypeptide chains and the pigment heme. Hemoglobin has two α and two β chains. In sickle-cell anemia, one of the 146 amino acids in the β chain is abnormal: At position 6, the normal glutamic acid has been replaced by valine. This replacement changes the charge of the protein (glutamic acid is negatively charged and valine is neutral; see Table 3.1), and the protein can form long aggregates in the red blood cells. The result is anemia, a deficiency of red blood cells.

Because hemoglobin is easy to isolate and study, its variations in the human population have been extensively documented (Figure 17.2). Hundreds of single amino acid alterations in β-globin have been reported. Some of these polymorphisms are even at the same amino acid position. For example, at the same position

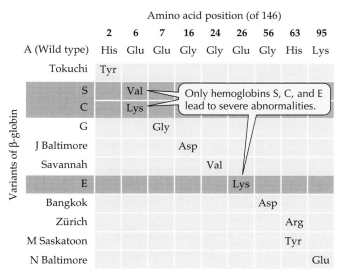

Amino acid position (of 146)

	2	6	7	16	24	26	56	63	95
A (Wild type)	His	Glu	Glu	Gly	Gly	Glu	Gly	His	Lys
Tokuchi	Tyr								
S		Val							
C		Lys							
G			Gly						
J Baltimore				Asp					
Savannah					Val				
E						Lys			
Bangkok							Asp		
Zürich								Arg	
M Saskatoon								Tyr	
N Baltimore									Glu

Only hemoglobins S, C, and E lead to severe abnormalities.

Variants of β-globin

17.2 Hemoglobin Polymorphism Only three of the many variants of hemoglobin lead to clinical abnormalities.

TABLE 17.1 Some Human Genetic Disorders

DISORDER	DESCRIPTION	NEWBORN FREQUENCY
AUTOSOMAL DOMINANT CONDITIONS		
Familial hypercholesterolemia	High cholesterol; early heart attacks	1 in 500
Huntington's disease	Neurological deterioration	1 in 2,500
Marfan syndrome	Tall, spindly; artery ruptures	1 in 20,000
AUTOSOMAL RECESSIVE CONDITIONS		
Sickle-cell anemia	Blocked capillaries; organ damage	1 in 655 (African-Americans)
Cystic fibrosis	Thick mucus; infections	1 in 2,500
Phenylketonuria	Mental retardation	1 in 12,000
X-LINKED CONDITIONS		
Duchenne's muscular dystrophy	Muscle weakness and deterioration	1 in 3,000 males
Hemophilia A	Lack of blood clotting	1 in 10,000 males
CHROMOSOME ABNORMALITIES		
Trisomy (extra) 21 (Down syndrome)	Mental retardation	1 in 600
XXY (Klinefelter syndrome)	Short stature; sterility	1 in 700 males
XO (Turner syndrome)	Sterility	1 in 1,500 females
XYY syndrome	Tall stature; acne	1 in 800 males
Fragile-X syndrome	Mental retardation	1 in 1,250 males; 1 in 2,000 females

that is mutated in sickle-cell anemia, the normal glutamic acid may be replaced by lysine in *hemoglobin C disease*. In this case, the anemia is usually not severe. Many hemoglobin variants, although obviously altering the primary phenotype, have no effect on the protein's function and thus no clinical phenotype. This is fortunate, since about 5 percent of all humans are carriers for one of these variants.

RECEPTORS AND TRANSPORT PROTEINS. Some of the most common human genetic diseases show their primary phenotype as altered membrane proteins. About one person in 500 is born with *familial hypercholesterolemia* (FH), in which levels of cholesterol in the blood are several times higher than normal. The excess cholesterol can accumulate on the inner walls of blood vessels, leading to complete blockage if a blood clot forms. If a blood clot forms in a major vessel serving the heart, the heart becomes starved of oxygen and a heart attack results. If a blood clot forms in the brain, the result is a stroke. People with FH often die of heart attacks before the age of 45, and in severe cases, before they are 20 years old.

Unlike PKU, which is characterized by the inability to convert phenylalanine to tyrosine, the problem in FH is not an inability to convert cholesterol to other products. People with FH have all the machinery to metabolize cholesterol, primarily in liver cells. The problem is that they are unable to transport the cholesterol into the cells that use it.

Cholesterol travels in the bloodstream in protein-containing particles called lipoproteins. One type of lipoprotein, *low-density lipoprotein*, carries cholesterol to the liver cells. After binding to a specific receptor on the membrane of the liver cell, the lipoprotein is taken up by the cell by endocytosis and then delivers its cholesterol "baggage" to the interior of the cell. People with FH lack a functional version of the receptor protein. Of the 840 amino acids that make up the receptor, often only one is abnormal in FH, but this is enough to change its structure so that it cannot bind to the lipoprotein and thus the lipoprotein (carrying the cholesterol) cannot enter the liver cell (Figure 17.3a).

In Caucasians, about one baby in 2,500 is born with *cystic fibrosis*. The clinical phenotype of this genetic disease is an unusually thick and dry mucus that lines organs such as the tubes that serve the respiratory system (see Chapter 45). This dryness prevents cilia on the surfaces of the epithelial cells from working efficiently to clear out bacteria and fungal spores that people take in with every breath. The results are recurrent and serious infections, as well as liver, pancreatic, and digestive failures. Patients often die in their 20s or 30s.

(a) Hypercholesterolemia

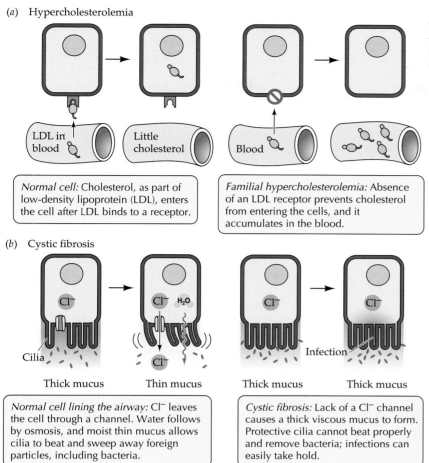

Normal cell: Cholesterol, as part of low-density lipoprotein (LDL), enters the cell after LDL binds to a receptor.

Familial hypercholesterolemia: Absence of an LDL receptor prevents cholesterol from entering the cells, and it accumulates in the blood.

(b) Cystic fibrosis

Normal cell lining the airway: Cl⁻ leaves the cell through a channel. Water follows by osmosis, and moist thin mucus allows cilia to beat and sweep away foreign particles, including bacteria.

Cystic fibrosis: Lack of a Cl⁻ channel causes a thick viscous mucus to form. Protective cilia cannot beat properly and remove bacteria; infections can easily take hold.

17.3 Genetic Diseases of Membrane Proteins
The left two panels illustrate normal cell function, while the right panels show the abnormalities caused by (a) hypercholesterolemia and (b) cystic fibrosis.

working copy of dystrophin, so their muscles also do not work.

A rarer genetic disease that is caused by a change in a structural protein is *osteogenesis imperfecta*, or brittle-bone disease, which occurs in one in 10,000 births. The culprit here is collagen, a protein that is essential in bone formation (see Chapter 44). Depending on the nature of the amino acid changes, the symptoms may be fairly minor—for example, skin, ligament, and tendon ailments. In more severe cases, bones are markedly malformed.

Coagulation proteins are involved in the clotting of blood at a wound. Inactive clotting proteins are always present in blood plasma (the noncellular portion of blood), but they become active only at a wound. People afflicted by the genetic disease *hemophilia* lack one of the coagulation proteins. Some people with this disease risk death from even minor cuts, since they cannot stop bleeding.

The reason for the thick mucus in patients with cystic fibrosis is a defective membrane protein, the *chloride transporter* (Figure 17.3b). In normal cells, this membrane channel opens after ATP hydrolysis to release Cl⁻ to the outside of the cell. The imbalance of ions (more are now on the outside of the cell) causes water to leave the cell by osmosis, resulting in a moist mucus. A single amino acid change in the transporter renders it nonfunctional, which leads to a dry mucus and the consequent clinical problems.

STRUCTURAL PROTEINS. About one boy in 3,000 is born with *Duchenne's muscular dystrophy*, in which the problem is not an enzyme or receptor but a protein involved in biological structure. People with this disease show progressively weaker muscles and are wheelchair-bound by their teenage years. Afflicted patients usually die in their 20s, when the muscles that serve their respiratory system fail. Normal people have a protein in their skeletal muscles called dystrophin that may bind the major muscle protein actin to the plasma membrane (see Chapter 44). Patients with Duchenne's muscular dystrophy do not have a

Most diseases are caused by both heredity and environment

Human diseases for which clinical phenotypes can be traced to a single altered protein may number in the thousands, and they are dramatic evidence of the one gene, one polypeptide idea. But taken together, these diseases have a frequency of about 1 percent in the total population.

Far more common are diseases that are *multifactorial*; that is, they are caused by many genes and proteins interacting with the environment. Although we tend to call humans either normal (wild type) or abnormal (mutant), the sum total of our genes is what determines which of us who eats a high-fat diet will die of a heart attack or which of us exposed to infectious bacteria actually come down with a disease. Estimates suggest that up to 60 percent of all people have diseases that are genetically influenced.

Human genetic diseases have several patterns of inheritance

PKU, sickle-cell anemia, and cystic fibrosis are inherited as **autosomal recessives**. This genetic term refers to the clinical phenotype: Typically, both parents are normal, heterozygous carriers of the abnormal allele. They have a 25 percent (one in four) chance of having an affected (homozygous recessive) boy or girl. Because of this low probability and the fact that many families in Western societies now have fewer than four children, it is unusual for more than one child in a family to have an autosomal recessive disease.

In the cells that produce the altered protein in people who have these recessive diseases—that is, liver cells for PKU, immature red blood cells for sickle-cell anemia, and epithelial cells for cystic fibrosis—only the nonfunctional, mutant, protein is made. In these homozygotes, a biochemical pathway or important cell function is disrupted, and disease results.

Not unexpectedly, heterozygotes, with one normal and one mutant allele, often have only 50 percent of the normal level of functional protein. For example, people who are heterozygous for the allele for PKU have half the number of active molecules of phenylalanine hydroxylase in their liver cells as individuals who carry two normal alleles for this enzyme. The cells compensate for this deficiency simply by having the active enzymes in the heterozygote do twice the work, and normal function is achieved.

Familial hypercholesterolemia and osteogenesis imperfecta are inherited as **autosomal dominants**. The presence of only one mutant allele is enough to produce the adverse clinical outcome, and direct transmission of the disease from parent to offspring is the rule. In people who are heterozygous for familial hypercholesterolemia, having half the number of functional receptors for low-density lipoprotein on the surface of liver cells is simply not enough to clear the cholesterol from the blood.

Osteogenesis imperfecta shows a *dominant negative* inheritance. Collagens are composed of three polypeptide chains, and when one of them is mutant, it prevents the other two from working properly, and the whole structure is disrupted. The expression of this abnormal collagen in the heterozygote is so harmful that it is often lethal.

The most prevalent types of hemophilia and Duchenne's muscular dystrophy are inherited as **X-linked** alleles. As we described in Chapter 10, a man always passes his X chromosome on to his daughters and not his sons. So if a father carries a mutant allele on the X chromosome, all of his daughters must inherit it, and none of his sons will inherit the mutant allele. On the other hand, a mother who is a carrier of the mutant allele can pass on the mutant X to both her sons and her daughters.

Both hemophilia and Duchenne's muscular dystrophy are inherited as recessives. Thus the son who inherits a mutant allele from his mother will have the disease, while a daughter will be a heterozygous carrier. Since until recently, few males with these diseases lived to reproduce, the most common pattern of inheritance has been from carrier mother to offspring, and these diseases are much more common in males than in females.

Chromosome abnormalities also cause human diseases. Such abnormalities include an excess or loss of one or two chromosomes (*aneuploidy*), loss of a piece of a chromosome (*deletion*), and transfer of a piece of one chromosome to another chromosome (*translocation*). About one newborn in 200 is born with a chromosome abnormality. While some of them are inherited, many are the result of meiotic problems such as nondisjunction (see Chapter 9).

Many zygotes that have abnormal chromosomes do not survive development and are spontaneously aborted. Of the 20 percent of pregnancies that are spontaneously aborted during the first 3 months of human development, an estimated half of them have chromosomal aberrations. For example, more than 90 percent of human zygotes that have only one X chromosome and no Y (Turner syndrome) do not live beyond the fourth prenatal month.

Why is the addition or deletion of chromosomal material so harmful to the developing human? This is the same question we ask in trying to explain X chromosome inactivation (see Chapter 14). The reasons are not yet clear. In translocations, the shuffling of chromosomal material to a new location causes a dramatic change in gene expression.

The most common form of inherited mental retardation is *fragile-X syndrome* (Figure 17.4). About one male in 1,500 and one female in 2,000 are affected. Near the tip of the abnormal X chromosome is a constriction that tends to break during preparation for microscopy, giving the name for this syndrome. Although the basic pattern of inheritance is that of an X-linked recessive trait, there are departures from this behavior. Not all people with the fragile-X abnormality are mentally retarded, as we will describe later.

Mutations and Human Diseases

The isolation and description of the precise nature of human mutations has proceeded rapidly since the development of molecular biology techniques (see Chapter 16). When the primary phenotypic expression was known, as in the case of abnormal hemoglobins, cloning the gene was straightforward, although time-consuming.

In other instances, such as Duchenne's muscular dystrophy, a chromosome deletion was associated with the disease, and this deletion pointed the way to the missing gene. In still other cases, such as cystic fibrosis, only a subtle molecular marker was available, leading investigators to the gene. In both of the latter

The constriction at the lower tip of this chromosome is the location of the fragile-X abnormality.

17.4 A Fragile-X Chromosome at Metaphase The chromosomal abnormality that causes the mental retardation symptomatic of fragile-X syndrome shows up physically as a constriction.

examples, the primary phenotype, the defective protein, was unknown when the gene was isolated; only then was the protein found.

In the discussions that follow, we will examine how mRNA, chromosome deletions, and DNA markers can be useful in identifying both mutant genes and abnormal proteins for genetic diseases such as muscular dystrophy and cystic fibrosis. We close this discussion by considering triplet repeats in several genetic diseases.

The logical way to isolate a gene is from its protein

As mentioned earlier, the primary phenotype for sickle-cell anemia was described in the 1950s as a single amino acid change in β-globin. On the basis of the clinical picture of sickled red blood cells, β-globin was certainly the right protein to examine. By the 1970s, β-globin mRNA could be isolated from immature red blood cells, which transcribe the globins as their major gene product. Then, a cDNA copy could be made and used to probe a human DNA library to isolate the entire β-globin gene (see Chapter 16), including introns and flanking sequences (Figure 17.5*a*). DNA sequencing could then be used to compare the normal gene with the gene from patients with sickle-cell anemia; sure enough, a single mutation was found.

Sometimes we can make an educated guess as to which protein is altered in a genetic disease. This was the case for *Marfan syndrome*. People with this dominantly inherited disorder often die suddenly when

17.5 Strategies for Isolating Human Genes (*a*) Once the seqence for the normal β-globin gene is established from the isolated protein, it can be compared to the gene sequence of patients with sickle-cell anemia. (*b*) When an abnormality is caused by a missing gene, as in Duchenne's muscular dystrophy, researchers compare the affected chromosome with a normal chromosome and isolate the DNA that is missing, then determine the protein for which this DNA codes.

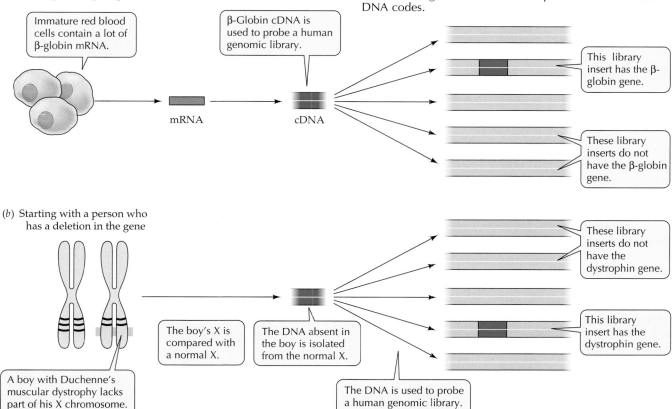

(*a*) Starting with a gene product

Immature red blood cells contain a lot of β-globin mRNA.

β-Globin cDNA is used to probe a human genomic library.

mRNA

cDNA

This library insert has the β-globin gene.

These library inserts do not have the β-globin gene.

(*b*) Starting with a person who has a deletion in the gene

A boy with Duchenne's muscular dystrophy lacks part of his X chromosome.

The boy's X is compared with a normal X.

The DNA absent in the boy is isolated from the normal X.

The DNA is used to probe a human genomic library.

These library inserts do not have the dystrophin gene.

This library insert has the dystrophin gene.

their heart beats vigorously, and, in trying to expand with the increased pressure from the blood, a major blood vessel ruptures. Until 1985, searching for a vessel wall protein in patients afflicted with Marfan syndrome that differed from the same protein in unaffected people proved fruitless. All the major isolated proteins seemed identical in the two groups of people. Then, electron microscopy revealed a new filamentous protein, called *fibrillin*, in the aorta wall. Fibrillin was purified, and in much the same way as β-globin, its gene was isolated and turned out to be mutated in people with Marfan syndrome.

Chromosome deletions can lead to gene and then protein isolation

The inheritance pattern of Duchenne's muscular dystrophy is consistent with an X-linked recessive trait. But until the late 1980s, neither the abnormal protein involved nor its gene had been described. This failure to describe either the protein or the gene was not from lack of effort: Almost every muscle protein that could be isolated at the time had been, and comparisons between affected people and normal muscles showed no differences.

Then several boys were found to have the disease because of a small deletion in their X chromosome. Comparison of the deleted and normal X chromosomes made possible isolation of the gene that was missing in the boys (Figure 17.5*b*). An important but not unexpected lesson that emerged from this research was that many different mutations can give rise to the Duchenne's muscular dystrophy phenotype. Some are deletions, and others are point mutations at intron splice sites.

DNA markers can point the way to important genes

In cases in which no candidate protein or visible chromosome deletion is available to help scientists in isolating a gene responsible for a disease, a technique called *positional cloning* has been invaluable. To understand this method, imagine an astronaut looking down from space, trying to find her son on a park bench on Chicago's North Shore. Unable to spot the boy with her naked eye, the astronaut picks out landmarks that will lead her to the park. She recognizes the shape of North America, then moves to Lake Michigan, the Sears Tower, and so on. Once she has zeroed in on the North Shore park, she can use advanced optical instruments to find her son.

The reference points for positional cloning are genetic markers on the DNA. These markers can be located within protein-coding regions, within introns, or in spacer DNA between genes. The only requirement is that they be polymorphic (recall that a sequence is polymorphic if one form is present less than 99 percent of the time—that is, if there is more than one form of the sequence).

As we described in Chapter 16, restriction enzymes cut DNA molecules at specific recognition sequences. On a particular human chromosome, a given restriction enzyme may make hundreds of cuts, producing many DNA fragments that can be probed on gel electrophoresis. For example, the enzyme *Eco*RI cuts DNA at

$$5' \ldots \text{GAATTC} \ldots 3'$$

Suppose this recognition site exists in a stretch of human chromosome 7. The enzyme will cut this stretch once and make two fragments of DNA. Now suppose in some people this sequence is mutated as follows:

$$5' \ldots \text{GAGTTC} \ldots 3'$$

This sequence will not be recognized by the restriction enzyme; thus it will remain intact and yield one larger fragment of DNA. The existence of such DNA differences is called a **restriction fragment length polymorphism**, or **RFLP** (Figure 17.6). An RFLP band pattern is inherited in a Mendelian fashion and can be followed through a pedigree. More than 1,000 such markers have been described for the human genome.

Genetic markers can be used as landmarks to find genes of interest if they too are polymorphic. The key to this method is the well-known observation from genetics that if two genes are located near each other on the same chromosome, they are passed on together from parent to offspring. The same holds true for any pair of genetic markers.

So, in order to narrow down the location of a gene on the human genome, a scientist must find a marker and gene that are always inherited together. The gene might be the one that causes cystic fibrosis, for example, and the marker might be the absence of the *Eco*RI site on the chromosome noted earlier. Family medical histories are taken and pedigrees are constructed. If the DNA marker and genetic disease are inherited together, they must be near each other on the chromosome. Unfortunately, "near each other" might be as much as a million base pairs away. This process of locating a marker and gene that are inherited together is thus similar to the astronaut focusing on Chicago: The landmarks lead to only an approximate location.

How can the gene be isolated? Many sophisticated methods are available. For example, the neighborhood around the RFLP can be screened for further RFLPs involving other restriction enzymes. With luck, one of them might be linked to the disease-causing gene, narrowing the search further. Then, DNA fragments from this region can be used to probe for sequences that are expressed and therefore encode protein. In the case of cystic fibrosis, a cDNA library from epithelial cells of sweat glands was used. Finally, the candidate gene is sequenced from normal people and from people who have the disease in question. If appropriate mutations

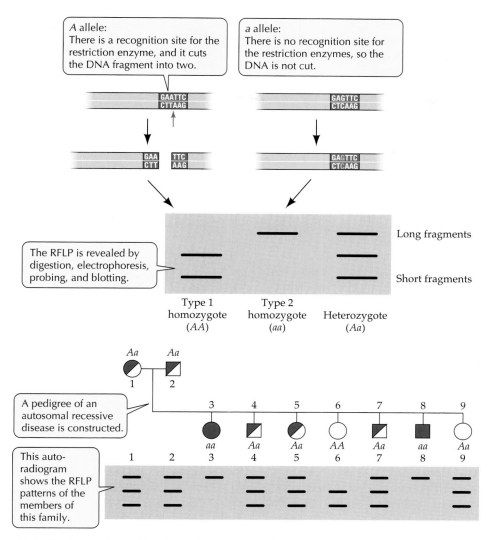

17.6 RFLP Mapping Restriction fragment length polymorphisms are differences in DNA sequences that serve as genetic markers. More than 1,000 such markers have been described for the human genome.

are found, the gene of interest has been isolated.

Since the isolation in 1989 of the gene responsible in mutant form for cystic fibrosis, positional cloning has been used to isolate dozens of important human genes, including those involved in fragile-X syndrome, Huntington's disease, and hereditary breast cancer.

Human gene mutations come in many sizes

The isolation of genes responsible for hereditary diseases has led to spectacular advances in the understanding of human biology. Before the genes, and then their proteins, were isolated for Duchenne's muscular dystrophy and for cystic fibrosis, proteins such as dystrophin and the chloride transporter had never been described. This identification of mutant genes has opened up new vistas in our understanding of how the human body works. Much more such information will come from the Human Genome Project, which we describe later in this chapter.

What mutations, then, give rise to human genetic diseases? Analyses of DNA have shown that the causes are both micro, at the level of point mutations, and macro, at the level of deletions, insertions, and a special case called expanding triplets. A single disease can have one of several micro and/or macro causes.

Phenylketonuria and sickle-cell anemia are caused by point mutations in the genes coding for the relevant proteins (Table 17.2). In addition, some of the variants of the β-globin gene cause disease, but others do not. Single base-pair mutations that alter a protein's function are usually important to the protein's three-dimensional structure—at the active site of an enzyme, for example.

Some mutations lead to not much of a protein at all. For example, some people with cystic fibrosis have a "nonsense" mutation such that a codon for an amino acid near the beginning of the

---

TABLE 17.2 Comparison of Two Genetic Diseases Caused by Point Mutations

VARIABLE	SICKLE-CELL ANEMIA	PHENYLKETONURIA
Protein in phenotype	β-globin	Phenylalanine hydroxylase
Length of chain	146 amino acids	451 amino acids
Normal protein	Glutamic acid at position 6	Arginine at position 408
Disease protein	Valine at position 6	Tryptophan at position 408
Length of gene	1512 base pairs	2448 base pairs
Normal allele	CGG at codon 6	GAA at codon 408
Disease allele	TGG at codon 6	GTA at codon 408

long protein chain has been changed to a stop codon, so protein translation stops at that point and a very short peptide is made. As we noted in Chapter 12, other point mutations affect RNA processing, leading to nonfunctional mRNA and a lack of protein synthesis.

DNA sequencing has revealed that mutations occur most often at certain base pairs in human DNA. These "hot spots" are often located where cytosine residues have been methylated to 5-methylcytosine (see Chapter 14). The explanation of this mutation phenomenon has to do with the natural instability of the bases in DNA.

Either spontaneously, or with chemical prodding, cytosine residues can lose their amino group and form uracil (Figure 17.7a). But the cell nucleus has a repair system that recognizes this uracil as being inappropriate for DNA: After all, uracil occurs only in RNA! So, the uracil is removed and cytosine replaced.

The fate of 5-methylcytosine that loses its amino group is rather different, since the result of that loss is thymine, a natural base for DNA. The uracil repair system ignores this thymine (Figure 17.7b). However, since the GC pair is now a mismatched GT pair, a different type of repair system comes in and tries to fix the mismatch. Half the time, the repair system matches a new C to the G, but the other half of the time, it matches a new A to the T, resulting in a mutation.

Larger mutations involve many base pairs of DNA. For example, earlier we described deletions in the X chromosome involving the dystrophin gene in Duchenne's muscular dystrophy. Some of these cover only part of the gene, leading to a partially complete protein and a mild form of the muscle disease. Others cover all of the gene and thus the protein is missing from muscle, resulting in the severe form of the disease. Still other deletions involve millions of base pairs, and cover not only the dystrophin gene but adjacent genes as well; the result may be several diseases simultaneously.

Insertions that cause disease can result from retrotransposition (see Chapter 14). For example, in some cases of hemophilia, a moderately repetitive LINE (long *in*terspersed *e*lement) sequence has been inserted within the protein-coding region of the clotting-factor gene. This insertion causes the gene to code for a larger and nonfunctional protein.

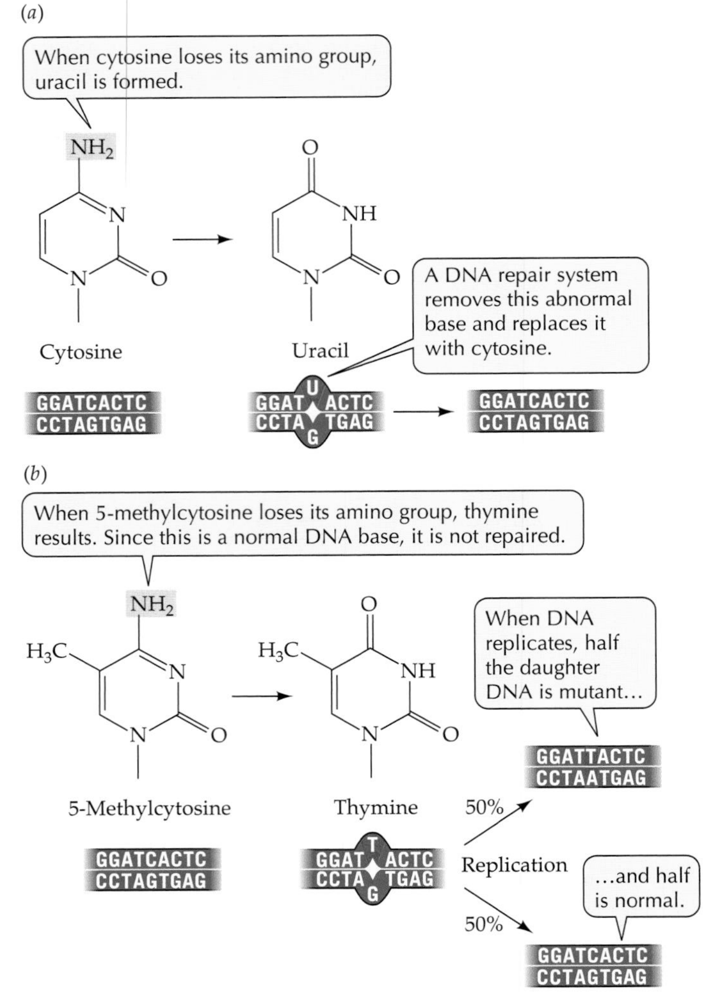

Triplet repeats show the fragility of some human genes

About one-fifth of all males that have a fragile-X chromosome are phenotypically normal, as are most of their daughters. But many of those daughters' sons are mentally retarded. In a family in which the fragile-X syndrome appears, later generations tend to show earlier onset and more severe symptoms of the

17.7 5-Methylcytosine in DNA Is a Hot Spot for Mutagenesis Cytosine can lose an amino group either spontaneously or because of exposure to certain chemical mutagens. The abnormality is usually repaired unless the cytosine residue has been methylated to 5-methylcytosine, in which case a mutation is likely to occur.

disease. It is almost as if the abnormal allele itself is changing—and getting worse. And that's exactly what's happening.

The gene responsible for fragile-X syndrome contains a repeated triplet of CGG at a certain point. In normal people, this triplet is repeated 6 to 54 times (average 29). In the alleles of mentally retarded people with fragile-X syndrome, the CGG region is repeated 200 to 1,300 times. In the "premutated" males who show no symptoms but who are likely to have affected offspring, the repeats are fewer—52 to 200 times. These repeats become more numerous as the daughters of these males pass the chromosome on to their children (Figure 17.8).

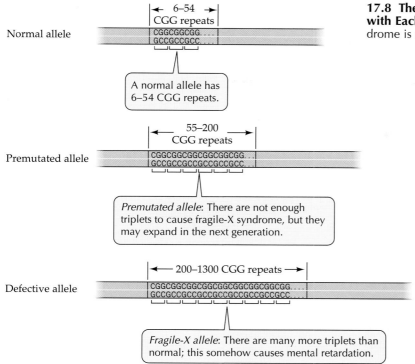

Normal allele

← 6–54 →
CGG repeats

CGGCGGCGG. . . .
GCCGCCGCC. . . .

A normal allele has
6–54 CGG repeats.

Premutated allele

55–200
CGG repeats

CGGCGGCGGCGGCGGCGG. . .
GCCGCCGCCGCCGCCGCC. . .

Premutated allele: There are not enough
triplets to cause fragile-X syndrome, but they
may expand in the next generation.

Defective allele

← 200–1300 CGG repeats →

CGGCGGCGGCGGCGGCGGCGGCGGCGG. . . .
GCCGCCGCCGCCGCCGCCGCCGCCGCC. . . .

Fragile-X allele: There are many more triplets than
normal; this somehow causes mental retardation.

**17.8 The CGG Repeat in the Fragile-X Gene Expands
with Each Generation** The genetic defect in fragile-X syndrome is caused by excessive repetitions of the CGG triplet.

posed to, a certain disease. It can be applied at many times and for many purposes: Prenatal testing can identify an embryo or fetus with a disease so that either a medical intervention can be applied or decisions about continuing the pregnancy can be considered. Newborn babies are tested so that proper medical intervention can be initiated. Asymptomatic people who have a relative with a genetic disease are tested to determine if they are carriers.

The goal of any screening is not just to provide information; it is to provide information that can be used to reduce an individual's burden resulting from the disease. In the broader arena of society, genetic screening poses ethical questions concerning our collective responsibility for people with genetic disorders.

Expanding triplet repeats have been found in other diseases, such as myotonic dystrophy (involving repeated CTG triplets) and Huntington's disease (in which CAG is repeated). Many non-disease-causing genes also appear to have these repeats, and because they are so common, they are assumed to play an important role. How they expand is not known. They may be found within the protein-coding region or outside of it.

Detecting Human Genetic Variations

Determination of the precise molecular phenotype and genotype for human genetic diseases has had three consequences. First, knowing what goes wrong is a way to find out what goes right: Mutants illuminate normal cell physiology. Second, knowing the cause of a disease at the biochemical level can lead to specific biochemical treatments and, potentially, cures. We return to some of these later in this chapter. Third, knowledge of molecular phenotypes and genotypes has led to ways of making precise diagnoses for many of these diseases, often before symptoms first appear, thus making medical intervention possible. DNA screening permits the identification of specific genotypes at any time of life and in any cell.

Genetic screening is the application of a test to identify people in a population who have, or are predis-

Screening for abnormal phenotypes makes use of protein expression

Screening for phenylketonuria in newborns is legally mandatory in many countries, including all of the United States and Canada. The reason for screening in the first days of life is that babies who are homozygous for this genetic disease are born with a normal phenotype, because excess phenylalanine in their blood before birth diffuses across the placenta to the mother's circulation. Since the mother is almost always heterozygous and therefore has adequate phenylalanine hydroxylase activity, the excess phenylalanine of the fetus is metabolized by its mother. Thus at birth the baby has not yet accumulated abnormal levels of phenylalanine.

After birth, however, the situation is different. The baby begins to consume protein-rich food (milk) and to break down some of its own proteins. In the process, phenylalanine enters the blood, and without the mother's enzyme to help, a baby with PKU, lacking its own phenylalanine hydroxylase activity, accumulates phenylalanine. The level of phenylalanine in its blood rises, and after a few days the level may be ten times higher than normal. Within days, the developing brain begins being damaged, and as Dr. Følling saw, untreated PKU patients are profoundly mentally retarded.

If PKU is detected early, it can be treated with a special diet (described later in this chapter). Thus detection is imperative. At first, physicians used Følling's ferric chloride test for phenylpyruvic acid in the urine. Unfortunately, babies with PKU do not start excreting large quantities of this substance until they are 4 to 8 weeks old, which can be too late to prevent brain damage. In 1963, Robert Guthrie described a simple screening test for PKU in newborns that today is used almost universally (Figure 17.9).

The Guthrie test uses a biological assay for blood phenylalanine. First, a drop of blood is collected from the baby's heel and placed on a piece of blotting paper. This paper is sent to the laboratory, where *Bacillus subtilis* bacteria are exposed to the dried blood spot. A small amount of a phenylalanine-like substance, 2-thienylalanine, is added to the growth medium. If there is no phenylalanine in the blood spot, the analog blocks bacterial growth.

A level of phenylalanine that is typical of normal individuals promotes a moderate amount of growth; more phenylalanine promotes more growth, and so on. Since the bacteria grow in a halo away from the dried spot, the diameter of the halo offers a simple estimate of the level of phenylalanine in the blood spot. This elegant application of bacterial physiology can be automated so that the screening laboratory can process many samples in a day.

If an infant tests positive for PKU in this screening, he or she must be retested using a more accurate chemical measurement for phenylalanine. If this test also shows a very high level in the blood, dietary intervention is begun. The whole process—the newborn genetic screen, confirmatory test, diagnosis, and initiation of treatment—are completed by the end of the second week of life. Since the screening test is inexpensive ($1) and since babies with PKU who receive early medical intervention are practically normal, the benefit is significant. Indeed, the benefit-to-cost ratio of newborn screening for PKU is one reason that legislators so readily make screening a legal requirement.

Even so, the screening test for PKU has come up against the concern with costs in contemporary medical care. Although the test itself is not costly, obtaining a blood sample may be. There is increasing financial pressure for mothers and their babies to leave the hospital less than a day after birth. Blood phenylalanine levels in some infants with PKU may rise slowly; they may not be significantly above normal levels during the first 2 days of life. So a blood sample taken before

the mother and child leave the hospital may not yet show a positive result on the screening test, even if the baby does have PKU. For this reason, it is recommended that the baby be tested a few days later.

Various other conditions are tested in newborns. Some of these conditions, such as galactosemia (an in-

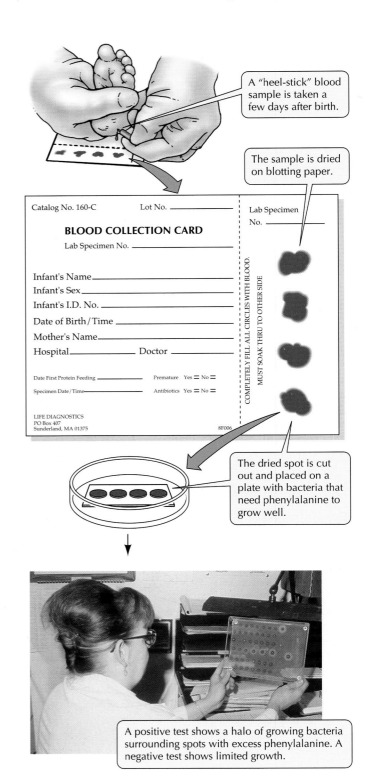

A "heel-stick" blood sample is taken a few days after birth.

The sample is dried on blotting paper.

Catalog No. 160-C   Lot No. _____   Lab Specimen No. _____

BLOOD COLLECTION CARD

Lab Specimen No. _____

Infant's Name _____
Infant's Sex _____
Infant's I.D. No. _____
Date of Birth/Time _____
Mother's Name _____
Hospital _____ Doctor _____

Date First Protein Feeding _____   Premature Yes ☰ No ☰

Specimen Date/Time _____   Antibiotics Yes ☰ No ☰

LIFE DIAGNOSTICS
PO Box 407
Sunderland, MA 01375   8F006

COMPLETELY FILL ALL CIRCLES WITH BLOOD.
MUST SOAK THRU TO OTHER SIDE

The dried spot is cut out and placed on a plate with bacteria that need phenylalanine to grow well.

A positive test shows a halo of growing bacteria surrounding spots with excess phenylalanine. A negative test shows limited growth.

17.9 Genetic Screening of Newborns for Phenylketonuria   A simple lab test devised by Robert Guthrie in 1963 is used today to screen newborns for phenylketonuria. Early detection means the symptoms of the condition can be prevented by following a therapeutic diet.

herited inability to break down milk sugar) have bacterial tests. Others, such as sickle-cell anemia, require differentiating between proteins with different charges (recall that normal hemoglobin is more negatively charged than is sickle hemoglobin). Still others use sophisticated immunoassays, such as the screening test for hypothyroidism. All of these tests provide valuable genetic information to the parents and clinical direction for the physicians to help the affected individuals.

There are several ways to screen for abnormal genes

Screening tests based on enzyme activity or protein structure, such as those for PKU and sickle-cell anemia, have limitations. They must be performed on tissues in which the relevant gene is expressed. For example, the blood level of phenylalanine is an indirect measure of phenylalanine hydroxylase activity in the liver, and hemoglobin electrophoresis shows the presence of sickle β-globin. But what if blood is difficult to test, as in a fetus? What about diseases that are expressed only in liver or brain and are not reflected in blood? What about proteins that are expressed under cellular controls, such that low activity might be the result of a simple dietary factor? Finally, since tissues in heterozygotes often compensate for having just one functional gene by raising the activity of the remaining active proteins to near normal levels, heterozygote testing is very difficult.

These problems are overcome by DNA testing, which is the most accurate way to test for an abnormal gene. With the description of the mutations in human genetic diseases (for example, see Table 17.2), it has become possible to directly examine any cell in the body at any time during the life span for gene mutations. However, these methods work best for diseases caused by one or a few mutations in the population. If there are dozens of possible gene mutations, simple tests such as the ones we will describe are much less informative.

The polymerase chain reaction (PCR) allows testing of the DNA of even one cell. For example, suppose a couple is heterozygous for the cystic fibrosis allele, have had a child with the disease, and want a normal child. If the woman is treated with the appropriate hormones, she can be induced to "superovulate," releasing several eggs. Following release, her eggs have completed the first meiotic division, with one large cell (the oocyte) and a small cell (the polar body) still attached to one another. In a heterozygote, one of these cells will have the normal allele and the other will have the allele for cystic fibrosis.

In polar body diagnosis, the polar body is removed and its DNA amplified by PCR. Then, a genetic test can be done to see which allele is present. If the polar body has the allele for cystic fibrosis, the oocyte must have the normal allele. The oocyte is then fertilized in the test tube with the husband's sperm, allowed to develop for a few days, and implanted in the mother's womb. The genetic diagnosis has been confirmed in all cases, and the result is normal children.

Polar body diagnosis is perhaps the extreme in DNA testing before birth. More typical are analyses of fetal cells after implantation in the womb. Fetal cells can be analyzed at about the tenth week of pregnancy (by chorionic villus sampling) or the thirteenth to seventeenth weeks (by amniocentesis). These two sampling methods are described in Chapter 39. In either case, only a few fetal cells are required for PCR and then genetic testing.

Newborns can also be screened for genes. The blood spots used for screening for PKU and other disorders contain enough of the baby's blood cells to permit extraction of the DNA, its amplification by PCR, and then genetic testing. Pilot studies are under way for testing for sickle-cell anemia and cystic fibrosis, and other genes will surely follow.

DNA testing is also now widely used to test adults for heterozygosity. For example, a sister or female cousin of a boy with Duchenne's muscular dystrophy may want to know if she is a carrier for the X chromosome that contains the dystrophin gene mutation. Similarly, the relatives of children with cystic fibrosis can determine their carrier status via DNA testing.

More problematic are genetic tests that show a predisposition to a disease. For example, as we will see later in this chapter, mutations in certain genes are associated with increased risk for certain types of cancer. If we can identify whether or not an individual carries such a mutation, who should have access to this information? DNA testing carries with it the potential for abuse. For example, given that there is a DNA marker for the Y chromosome, would it be ethical to test for the sex of an embryo or fetus and then choose to terminate the pregnancy on the basis of this information?

Of the numerous methods of genetic analysis, two are the most widespread. We will describe their use to detect the mutation in the DNA coding for β-globin, which results in sickle-cell anemia. The first method, *allele-specific cleavage*, employs differences between the normal and sickle alleles in the β-globin gene with respect to a restriction enzyme recognition site. Around position 6 in the normal gene is the sequence 5'...CCTGAGGAG...3'. This sequence is recognized by the restriction enzyme *Mst*II, which cleaves DNA at 5'...CCTNAGG...3', where "N" is any base. In the sickle mutation (see Table 17.2), the DNA sequence is changed to 5'...CCTGTGGAG...3'. The single base-pair alteration, while it results in the clinical phenotype of sickle-cell anemia, also makes this sequence unrecognizable by *Mst*II. So, when *Mst*II fails to make the cut in the mutant gene, gel electrophoresis detects

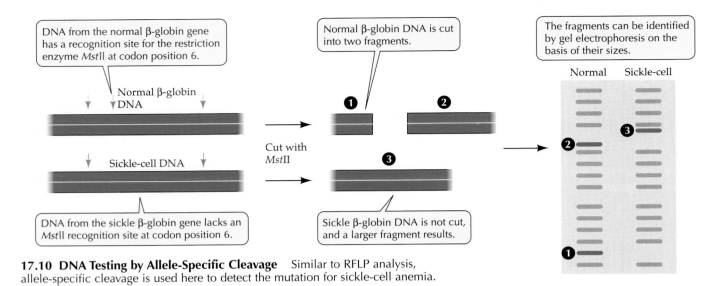

17.10 DNA Testing by Allele-Specific Cleavage Similar to RFLP analysis, allele-specific cleavage is used here to detect the mutation for sickle-cell anemia.

a larger DNA fragment (Figure 17.10). This analysis is similar to RFLP investigations (see Figure 17.6).

The second way to detect mutations is by *allele-specific oligonucleotide hybridization*. In this case, oligonucleotides are synthesized in the lab that will hybridize either to the denatured normal β-globin DNA sequence around position 6 or to the sickle mutant sequence. Usually, a probe of at least a dozen bases is needed to form a stable double helix with the target DNA, which may be fixed on a piece of filter paper. If the probe is labeled with radioactivity or with a colored or fluorescent substrate, hybridization is readily detected (Figure 17.11). This method is easier and faster than allele-specific cleavage.

Cancer: A Disease of Genetic Changes

Perhaps no malady affecting people in the developed world instills more fear than cancer. One in three Americans will have some form of cancer in their lifetime, and at present, one in four will die of it. With a million new cases and half a million deaths in the United States annually, cancer ranks second only to heart disease as a killer. Cancer was less common a century ago; then, as now in many poor regions of the world, people died of infectious diseases and did not live long enough to get cancer. Cancer is a disease of the later years of life; children are less frequently afflicted.

In 1970, the U.S. government declared "war on cancer." Although progress has been made, it is obvious from the statistics and from the people affected that the war is far from won. What has happened, however, is that the groundwork for winning the war has been laid during the past several decades. A tremendous amount of information on cancer cells—their growth and spread and their molecular changes—has been obtained by scientists and physicians mobilized to attack this challenging problem. Perhaps the most remarkable discovery is that cancer is a disease caused primarily by genetic changes.

In this discussion of cancer, we will examine the genetic changes in cancer cells and the distinctions between benign and malignant tumors. We will identify the viral causes of some cancers, and consider the roles of oncogenes and tumor suppressor genes. With these foundations, we will see why multiple genetic changes are often necessary to produce cancers.

Cancer cells differ from their normal cell counterparts

Cancer cells differ from the normal cells from which they originate in two major ways. First, a cancerous cell loses the control over cell division that exists in most tissues. Most cells in the body divide only if they are exposed to outside influences, such as growth factors and hormones. Cancer cells do not respond to these controls and instead divide more or less continuously, ultimately forming tumors (large masses of cells). By the time a physician can feel a tumor or see one on an X ray or CAT scan, it already has millions of cells.

Benign tumors resemble the tissue they came from, grow slowly, and remain localized where they develop. For example, a lipoma is a benign tumor of fat cells that may arise in the armpit and just stay there. Benign tumors are usually not a problem, but they must be removed if they impinge on an important organ, such as the brain.

Malignant tumors, on the other hand, are dedifferentiated and do not look like their parent tissue at all. A flat, specialized lung epithelial cell in a highly or-

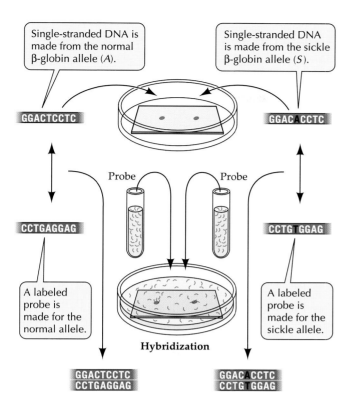

Single-stranded DNA is made from the normal β-globin allele (*A*).

GGACTCCTC

Single-stranded DNA is made from the sickle β-globin allele (*S*).

GGAC**A**CCTC

Probe

Probe

CCTGAGGAG

CCTG**T**GGAG

A labeled probe is made for the normal allele.

A labeled probe is made for the sickle allele.

Hybridization

GGACTCCTC
CCTGAGGAG

GGAC**A**CCTC
CCTG**T**GGAG

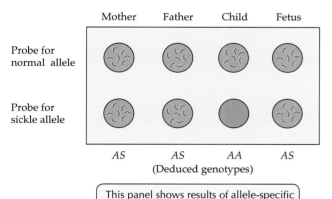

	Mother	Father	Child	Fetus
Probe for normal allele				
Probe for sickle allele				
	AS	*AS*	*AA*	*AS*

(Deduced genotypes)

This panel shows results of allele-specific hybridizations for a family.

17.11 DNA Testing by Allele-Specific Oligonucleotide Hybridization The hybridization process for this family reveals that three of them are carriers of the sickle-cell allele; the first child, however, inherited two normal alleles and is neither affected by the disease nor a carrier.

rounds them by actively secreting digestive enzymes to "eat" their way toward a blood vessel. Then some of the cancer cells enter either the bloodstream or the lymphatic system (Figure 17.12). Their journey through these vessels is perilous—imagine a small clump of cells going through the heart as it beats—and few cells survive, perhaps one in 10,000. When by chance cancer cells arrive at a suitable organ for further growth, they express cell surface proteins that allow them to bind to the new host tissue. This binding allows the tumor cells to grow away from the vessel, and they proceed to invade the host tissue. A malignant tumor also secretes chemical factors that cause blood vessels to grow to it and supply it with oxygen and nutrients.

Different forms of cancer affect different parts of the body. About 85 percent of all human tumors are *carcinomas*—cancers that arise in surface tissues such as the skin and linings (epithelial cells) of organs. Lung cancer, breast cancer, colon cancer, and liver cancer are all carcinomas. The first three account for more than half of all cancer deaths in Europe and North America; liver cancer is more common in Africa and Asia. *Sarcomas* are cancers of tissues such as bone, blood vessels, and muscle. *Leukemias* and *lymphomas* affect the cells that give rise to blood cells.

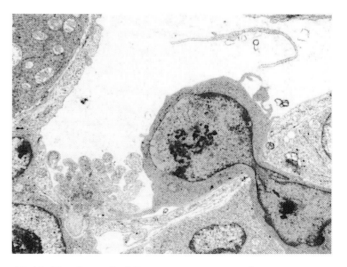

17.12 The Spread of Cancer A cancer cell squeezes into a small blood vessel through the vessel's wall. The cancer cell is then carried through the blood and, if it survives the journey, it may spread into other tissue.

dered arrangement in the wall of the lung may turn into an undistinguished, round lung cancer cell, clumps of which may break off. Malignant cells often have irregular structures, such as variable sizes and shapes of nuclei. Many malignant cells express the gene for telomerase (see Chapter 14) and thus do not shorten the ends of their chromosomes after each DNA replication.

The second, and most fearsome, characteristic of cancer cells that distinguishes them from normal cells is their ability to invade surrounding tissues and spread to other parts of the body. This spreading of cancer, called **metastasis**, occurs in several stages. First, the cancer cells extend into the tissue that sur-

Table 17.3 Human Cancers Known to Be Caused by Viruses

CANCER	ASSOCIATED VIRUS
Liver cancer	Hepatitis B virus
Lymphoma, nasopharyngeal cancer	Epstein–Barr virus
T cell leukemia	Human T cell leukemia virus
Anogenital cancers	Papilloma (wart) virus
Kaposi sarcoma	Herpes simplex virus

Some cancers are caused by viruses

Viruses cause many human diseases, and Peyton Rous's discovery in 1910 that a sarcoma in chickens is caused by a virus that is transmitted from one bird to another spawned an intensive search for cancer-causing viruses in humans. At least five types of human cancer are probably caused by viruses (Table 17.3).

Hepatitis B, a liver disease that affects people all over the world, is caused by a virus that contaminates blood or is carried from mother to child during birth. The viral infection can be long-lasting and may flare up numerous times. This virus is associated with liver cancer, especially in Asia and Africa, where millions of people are infected. But it does not act to cause cancer by itself: Some gene mutations also appear necessary in the tumor cells of Asians and Africans afflicted with the disease (although apparently not in Europeans and North Americans).

Similarly, Epstein–Barr virus, the agent that causes mononucleosis, is associated with certain types of lymphoma in Africa and Asia, but usually not in North America and Europe. Again, additional genetic events appear to occur in people who develop tumors. A rare form of leukemia is caused by HTLV-I, the human T cell leukemia virus. This virus is especially frequent in certain areas of Japan. Kaposi sarcoma, an otherwise rare tumor that is common in AIDS patients, is caused by a variant of herpes simplex virus. This virus is probably easily fought off by the immune systems of healthy people, but the crippled immune system of these patients cannot remove the virus, so it causes cancer.

An important group of virus-caused cancers in North Americans and Europeans consists of the various anogenital cancers caused by papillomaviruses. The genital and anal warts that these viruses cause often develop into tumors. These viruses seem to be able to act on their own, not needing mutations in the host tissue for tumors to arise. Sexual transmission of these papillomaviruses is unfortunately widespread.

Most cancers are caused by genetic mutations

Worldwide, no more than 15 percent of all cancers may be caused by viruses. What causes the other 85 percent? Because most cancers develop in older people, it is reasonable to assume that one must live long enough for a series of events to accumulate to finally produce the malignant cell. This assumption turns out to be correct, and the events are genetic mutations. But these are usually not the mutations found in genetic diseases that are present in germ line cells and passed on to offspring. Instead, the mutations in cancer cells are usually **somatic mutations**, which alter the genes of nonsex cells.

DNA can become damaged in many ways. Spontaneous mutations arise because of chemical changes in the nucleotides. For example, the conversion of cytosine to uracil, and of 5-methylcytosine to thymine (see Figure 17.7) occur with some frequency in DNA, as do errors in DNA replication. In addition, certain substances called *carcinogens* cause cancer. Familiar carcinogens include chemicals that are present in tobacco smoke and salted meats, ultraviolet light from the sun, and ionizing radiation from sources of radioactivity.

Less familiar but just as harmful are thousands of chemicals present naturally in the foods people eat. According to one estimate, these "natural" carcinogens account for well over 80 percent of the human exposure to agents that cause cancer. But the common theme in the natural and human-made carcinogens is that almost all of them damage DNA, usually by causing changes from one base to another (Figure 17.13). Dividing somatic cells, such as the cells in bone marrow that give rise to blood cells, are especially susceptible to genetic damage.

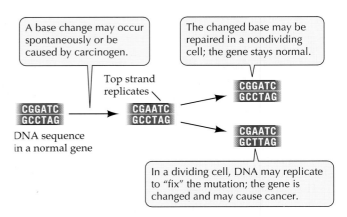

17.13 Dividing Cells Are Especially Susceptible to Genetic Damage and Cancer A base change is more likely to be repaired in a nondividing cell.

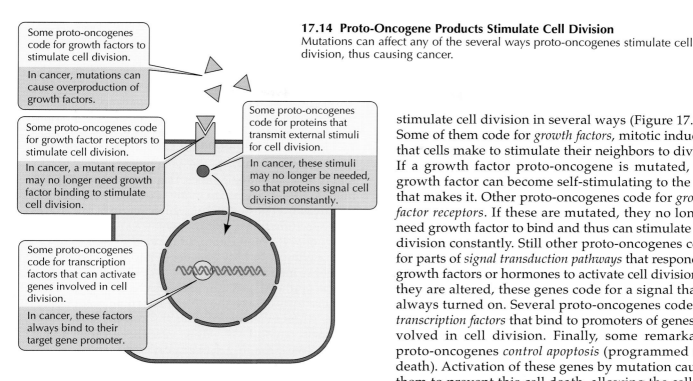

17.14 Proto-Oncogene Products Stimulate Cell Division
Mutations can affect any of the several ways proto-oncogenes stimulate cell division, thus causing cancer.

Some proto-oncogenes code for growth factors to stimulate cell division.

In cancer, mutations can cause overproduction of growth factors.

Some proto-oncogenes code for growth factor receptors to stimulate cell division.

In cancer, a mutant receptor may no longer need growth factor binding to stimulate cell division.

Some proto-oncogenes code for proteins that transmit external stimuli for cell division.

In cancer, these stimuli may no longer be needed, so that proteins signal cell division constantly.

Some proto-oncogenes code for transcription factors that can activate genes involved in cell division.

In cancer, these factors always bind to their target gene promoter.

stimulate cell division in several ways (Figure 17.14). Some of them code for *growth factors*, mitotic inducers that cells make to stimulate their neighbors to divide. If a growth factor proto-oncogene is mutated, the growth factor can become self-stimulating to the cell that makes it. Other proto-oncogenes code for *growth factor receptors*. If these are mutated, they no longer need growth factor to bind and thus can stimulate cell division constantly. Still other proto-oncogenes code for parts of *signal transduction pathways* that respond to growth factors or hormones to activate cell division. If they are altered, these genes code for a signal that is always turned on. Several proto-oncogenes code for *transcription factors* that bind to promoters of genes involved in cell division. Finally, some remarkable proto-oncogenes *control apoptosis* (programmed cell death). Activation of these genes by mutation causes them to prevent this cell death, allowing the cells to continue dividing.

The same types of mutations that we observed in human genetic diseases activate cellular proto-oncogenes. Some proto-oncogenes are activated by point mutations; others by chromosome changes such as translocations; still others by gene amplification. Whatever the mechanism, the result is the same: The "gas pedal" for cell division is pressed.

TUMOR SUPPRESSOR GENES. About 10 percent of all cancer is clearly inherited. Often the inherited form of the cancer is clinically similar to the noninherited kind that comes later in life, called the *sporadic form*. The major differences are that the inherited form strikes a person much earlier in life and usually shows up as multiple tumors.

In 1971, Alfred Knudson used these observations to predict that for a tumor to occur, a tumor suppressor gene must be inactivated. But unlike oncogenes, where one mutated form is all that is needed for activation, the full inactivation of a tumor suppressor requires that both alleles be turned off, which requires two mutational events. For people with sporadic cancer, it takes a long time for both genes in a single cell to mutate. But in inherited cancer, people are born with one mutant allele for the tumor suppressor and need just one more event for full inactivation of the "brakes" (Figure 17.15).

Various tumor suppressor genes have been isolated and confirm Knudson's "two-hit" hypothesis. Some of these tumor suppressor genes are involved in inher-

Two kinds of genes are changed in many cancers

The changes in the control of cell division that lie at the heart of cancer can be likened to the control of the speed of an automobile. To make a car move, two things must happen: The gas pedal must be pressed and the brake must be released. In the human genome, there are both **oncogenes**, which act as the gas pedal to stimulate cell division, and **tumor suppressor genes**, which put the brake on to inhibit it.

Normal, differentiated cells typically do not divide, so their oncogenes are turned off and their tumor suppressor genes are turned on. Somatic mutations cause these sets of genes to do the reverse: turn the oncogenes on and the tumor suppressors off.

ONCOGENES. The first hint that oncogenes (from the Greek *onco-*, "mass") were necessary for cells to become cancerous came with identification of the virally induced cancers. In many cases, these viruses infect their host cells and bring in a new gene, one that acts to stimulate cell division by being actively expressed in the viral genome. It soon became apparent that these oncogenes had counterparts in the genomes of the host cells, but some of these cellular genes, called *proto-oncogenes*, were not actively transcribed in differentiated, nondividing cells. So the search for genes that are damaged by carcinogens quickly zeroed in on the proto-oncogenes.

The cellular proto-oncogenes are attractive targets for carcinogenesis, since they have the capacity to

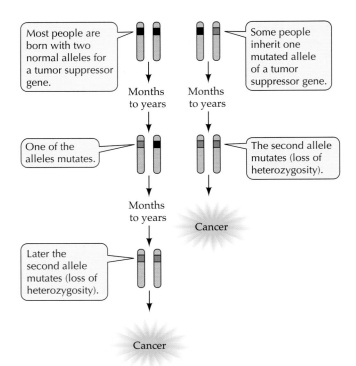

Most people are born with two normal alleles for a tumor suppressor gene.

Some people inherit one mutated allele of a tumor suppressor gene.

Months to years

Months to years

One of the alleles mutates.

The second allele mutates (loss of heterozygosity).

Months to years

Cancer

Later the second allele mutates (loss of heterozygosity).

Cancer

17.15 The "Two-Hit" Hypothesis for Cancer Although a single mutation can activate a proto-oncogene, *two* mutations are needed to inactivate a tumor suppressor gene. An inherited predisposition occurs in people born with one allele already mutated.

ited forms of rare childhood cancers such as retinoblastoma (a tumor of the eye) and Wilms' tumor of the kidney, as well as in inherited breast and prostate cancers. In all instances, both copies of the tumor suppressor gene must be inactivated for the tumor to develop.

People born with one inactivated copy are prone to early cancer and are much more likely to get the disease than are those born with two normal copies of the gene. A good example is breast cancer. The 9 percent of women who inherit one mutated copy of the gene *BRCA1* have a 60 percent chance of having breast cancer by age 50 and an 82 percent chance of developing it by age 70. The comparable figures for women who inherit two normal copies of the gene are 2 percent and 7 percent, respectively.

How do tumor suppressor genes act in the cell? Not surprisingly, they appear to control progress through the cell cycle. The protein encoded by *Rb*, a gene that was first described for its contribution to retinoblastoma, is active during G1. In the active form it binds some transcription factors that are necessary for progress to S phase and the rest of the cycle. In nondividing cells, *Rb* remains active, preventing cell division until the proper growth factor signals are present. When the Rb protein is inactive because of mutation, the cell cycle moves forward independent of growth factors.

Another widespread tumor suppressor gene, *p53*, has a protein product that also stops cells during G1. This gene is mutated in many types of cancers, including lung cancer and colon cancer.

The pathway from normal to cancerous is complex

The "gas pedal" and "brake" analogies for oncogenes and tumor suppressor genes, respectively, are elegant but simplified. There are many oncogenes and tumor suppressor genes, some of which act only in certain cells at certain times. The anogenital cancers caused by infection with human papillomaviruses offer a fascinating glimpse at how oncogenes and tumor suppressor genes interact.

The epithelial cells that are infected make the protein products of both the *Rb* and *p53* tumor suppressor genes, thereby ensuring that the cell cycle will not progress beyond G1. However, the viral genome makes two protein products, called E7 and E6 ("E" stands for the *e*arly time in the infection cycle that the proteins are made). Remarkably, the viral E7 protein binds to the Rb protein and inactivates it, and the E6 protein similarly binds to and inactivates the p53 protein. With both tumor suppressors inactive, cell division can proceed.

Because cancers progress to full malignancy slowly, it is possible to describe the oncogenes and tumor suppressor genes at each stage in great molecular detail. Such a description has been developed for colon cancer (Figure 17.16). At least three tumor suppressor genes and one oncogene must be mutated in sequence for a cell to become metastatic. Although the occurrence of all these events in a single cell might appear unlikely, remember that the colon has millions of cells, that the cells giving rise to epithelial cells are constantly dividing, and that these changes take place over many years of exposure to natural and human-made carcinogens and to spontaneous mutations.

The characterization of the molecular changes in tumor cells has opened up the possibility of genetic diagnoses and screening, as is done for genetic diseases. Many tumors are now commonly diagnosed in part by oligonucleotide-specific probes for oncogene and/or tumor suppressor gene alterations. For hereditary cancers, it is possible to detect early in life whether an individual has inherited a mutated tumor suppressor gene, making possible preventive action. For example, a person who inherits mutated copies of the tumor suppressor genes involved in colon cancer normally would have a high probability of developing this tumor by age 40. Surgical removal of the colon would prevent this metastatic tumor from arising.

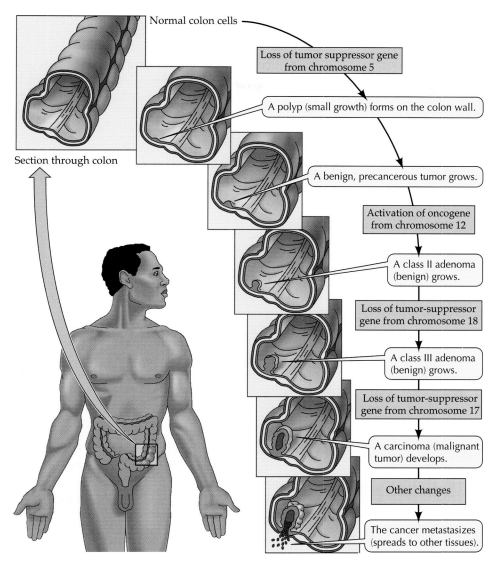

Normal colon cells

Section through colon

Loss of tumor suppressor gene from chromosome 5

A polyp (small growth) forms on the colon wall.

A benign, precancerous tumor grows.

Activation of oncogene from chromosome 12

A class II adenoma (benign) grows.

Loss of tumor-suppressor gene from chromosome 18

A class III adenoma (benign) grows.

Loss of tumor-suppressor gene from chromosome 17

A carcinoma (malignant tumor) develops.

Other changes

The cancer metastasizes (spreads to other tissues).

17.16 Multiple Mutations Transform a Normal Colon Epithelial Cell into a Cancerous Cell

Treating Genetic Diseases

Most treatments of genetic diseases try to alleviate the symptoms that affect the patient. But to effectively treat diseases caused by genes—whether they affect all cells, as in inherited disorders such as PKU, or only somatic cells, as in lung cancer—physicians must be able to diagnose the disease accurately, must know how the disease works at the biochemical level, and must be able to intervene early, before the disease ravages or kills the individual. As we have seen, basic research has provided the knowledge for accurate diagnostic tests, as well as a beginning at understanding pathogenesis (the cause of diseases) at the molecular level. Physicians are now applying this knowledge to treat genetic diseases. In this section, we will see that these treatments range

from specifically modifying the mutant phenotype, such as supplying a missing product of a defective enzyme, to supplying the normal version of a mutant gene.

One approach is to modify the phenotype

After a newborn baby is diagnosed with PKU, therapy is aimed at *restricting the substrate* of the deficient enzyme. In this case, the substrate is phenylalanine, which is not synthesized by the infant but comes mostly from its diet. So the infant is immediately put on a special diet that contains only enough of this amino acid for immediate use.

Lofenelac, a milk-based product that is low in phenylalanine, is fed to these infants just like formula. Later, certain fruits, vegetables, cereals, and noodles low in phenylalanine can be added to the diet. Meat, fish, eggs, dairy products, and bread, which contain high amounts of phenylalanine, must be avoided, especially during childhood, when brain development is most rapid. The artificial sweetener aspartame must also be avoided, because it is made of two amino acids, one of which is phenylalanine.

People with PKU are generally advised to stay on a low-phenylalanine diet for life. Although maintaining these dietary restrictions may be difficult, it is effective. Numerous follow-up studies since newborn screening was initiated have shown that people with PKU who stay on the diet are no different from the rest of the population in terms of mental ability. This is an impressive achievement in public health, given the extent of mental retardation in untreated patients.

Another way to modify a disease phenotype is by *metabolic inhibitors*. As we described earlier, people with familial hypercholesterolemia accumulate dangerous levels of cholesterol in their blood. Not only are these people unable to metabolize dietary cholesterol, but they also synthesize a lot of it. One effective treatment for people with this disease is the drug mevinolin, which blocks the patients' own cholesterol synthesis. Patients who receive this drug need only worry about cholesterol in their diet and not about what their cells are making.

Metabolic inhibitors also form the basis of cancer therapy with radiation and drugs. The strategy is to kill rapidly dividing cells, since rapid cell division is the hallmark of malignancy. Radiation, which can be focused on the tumor, damages DNA, breaks chromosomes, and generally kills cells by forming highly reactive chemicals called free radicals. If the tumor is localized, all of it may be killed. The same holds true of the most common way to treat cancers: surgical removal.

Unfortunately, by the time they are diagnosed, many tumors have already begun to metastasize. Treatment of metastatic cancers involves attempting to control them at many sites in the body. Many drugs have been designed to kill the dividing cells (Figure 17.17), but most of these drugs are given in the blood and thus also damage other, noncancerous, dividing cells in the body.

Given the broad scope of drug treatment, it is not surprising that people undergoing chemotherapy suffer side effects such as loss of hair (the skin epithelium constantly divides to replace the cells that die), digestive upsets (gut epithelial cells also divide constantly), and depletion of blood cells (bone marrow stem cells). The effective dose of these highly toxic drugs for treating the cancer is often just below the dose that would kill the patient, so they must be used with utmost care. Usually they can control the spread of cancer, but not cure it.

An obvious way to treat a disease phenotype in which a functional protein is missing is to *supply the missing protein*. This is the basis of treatment of hemophilia A, in which the missing blood clotting factor is supplied in pure form. The production of human clotting protein by DNA technology (see Chapter 16) is critically important here, because it allows a pure protein to be given instead of blood transfusions of crude blood products, which could be contaminated. Blood contamination was a problem in the early years of the AIDS epidemic, when testing for HIV in blood was not yet possible; as a result, many people with hemophilia developed AIDS because they were transfused with HIV-contaminated blood.

Unfortunately, the phenotypes of many diseases caused by abnormalities in genes are very complex. Simple interventions, such as some of those we have described, do not work for most such diseases. Indeed, a recent survey showed that current therapies for 351 diseases caused by single gene mutations improved the patients' life span by only 15 percent.

For the many polygenic disorders such as cancers, "magic bullet" specific therapies are unlikely. Knowledge of the precise genetic errors often far outstrips knowledge of clinical mechanisms. Is it possible to treat the root cause of these disorders at the DNA level?

Gene therapy offers the hope of specific therapy

Perhaps the most obvious thing to do when a cell lacks a functional allele is to provide one. Such **gene therapy** is under intensive investigation for diseases ranging from the rare inherited disorders caused by single mutations, to cancer, AIDS, and atherosclerosis. Like genetic transformation in other organisms (see Chapter 13), gene therapy in humans seeks to insert a new gene that will be expressed in the host. Thus, the new DNA is often attached to a promoter that will be active in the human cells.

Presenting the DNA for cellular uptake follows the methods used in biotechnology: Ca^{2+} complexes, liposomes, and viral vectors are used to enable uptake of the "good gene" into the human cells. Physicians who practice this "molecular medicine" are confronted by the challenges of genetic engineering: efficient uptake; effective vectors; appropriate expression and processing of mRNA and protein; and selection within the body for the cells that contain the recombinant DNA.

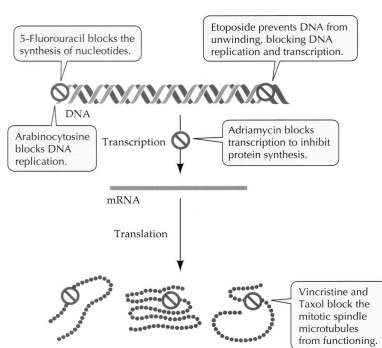

5-Fluorouracil blocks the synthesis of nucleotides.

Etoposide prevents DNA from unwinding, blocking DNA replication and transcription.

DNA

Arabinocytosine blocks DNA replication.

Transcription

Adriamycin blocks transcription to inhibit protein synthesis.

mRNA

Translation

Vincristine and Taxol block the mitotic spindle microtubules from functioning.

Cell division proteins

17.17 Drug Strategies for Killing Cancer Cells
These medications kill dividing cancer cells. Unfortunately, most of them also affect noncancerous dividing cells.

17.18 Gene Therapy New genes have been added to somatic cells taken from a patient's body, then returned to express normal alleles.

Which human cells should be the targets of gene therapy? The best idea for a genetic disease would be to replace the nonfunctional gene with a functional one in every cell of the body. But vectors to do this are simply not available, and delivery to every cell poses a formidable challenge. Until recently, the major attempts at gene therapy have been *ex vivo*. That is, the scientists and physicians have taken cells from the patient's body, added the new gene to the cells in the laboratory, and then returned the cells to the patient in the hope that the correct gene product would be made (Figure 17.18). A widely publicized example of this approach was the introduction of a functional gene for the enzyme adenosine deaminase into the white blood cells of a girl with a genetic deficiency of this enzyme. Unfortunately, these were mature white blood cells, and although they survived for a time in the girl and made the enzyme for some therapeutic benefit, they eventually died, as is the normal fate of such cells. It would be more effective to insert the functional gene into *stem cells*, the bone marrow cells that constantly divide. This is a major thrust of many current clinical experiments on gene therapy.

The other method of gene therapy is to insert the gene directly into cells in the body of the patient, using a vector or complex. This *in vivo* approach is being attempted for various types of cancer. For example, lung cancers are accessible if the DNA or vector is given as an aerosol through the respiratory system. Vectors expressing DNA for functional tumor suppressor genes that are mutated in the tumors, as well as vectors expressing antisense RNAs against oncogene mRNAs, have been successfully introduced in this way to patients with lung cancer, with some clinical improvement.

The Human Genome Project

In 1984 the U.S. government sponsored a conference to examine the problem of detecting DNA damage in people exposed to low levels of radiation, such as Japanese who had survived the atomic bombs 39 years earlier. The scientists attending this conference quickly realized that monitoring the human genome for changes would also be useful in evaluating environmental mutagens. But new, more efficient and sensitive technologies would be needed in order to be able to sequence the entire human genome. In 1986, Renato Dulbecco, who won the Nobel prize for his pioneering work on cancer

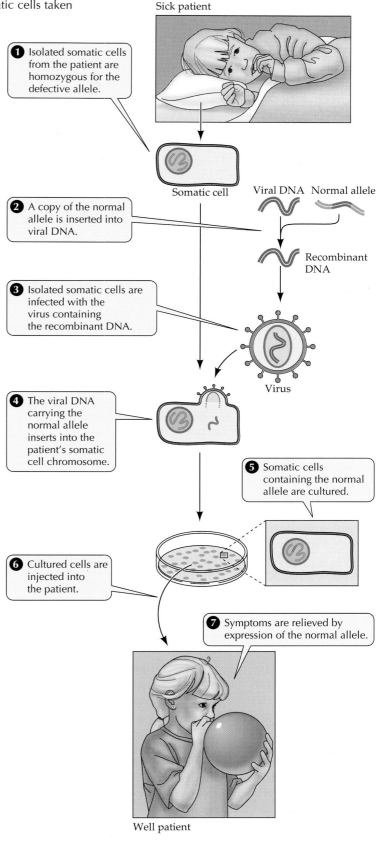

viruses, suggested that determining the entire sequence of human DNA could be a boon to cancer research. He proposed that the scientific community be mobilized to the task. By 1991, ambitious efforts were under way in the United States, England, France, Italy, Canada, and Japan. Progress has been rapid.

The major goal of the Human Genome Project (HGP) is to determine the nucleotide sequence of all the DNA of an individual. Of course, genetic polymorphisms ensure that every person is different. But some generalities will surely apply to us all. Knowing the sequences of the 50,000 to 100,000 human genes, most of which are not known today, could unlock the secrets of the polygenic disorders that plague humankind, as well as provide a wealth of basic information on who we are at the molecular level.

In this section, we will examine HGP approaches to its challenge and some of the benefits of knowing the entire sequence of the human genome. To reach this goal, the HGP has established three intermediate goals: (1) to develop ways of mapping the genome at increasingly fine levels of precision, (2) to test new techniques by determining the sequences of the genomes of some "simple" organisms, and (3) to study in advance some of the ethical issues that the knowledge will raise.

Mapping and sequencing entire genomes uses new molecular methods

Each human chromosome consists of a single double-stranded molecule of DNA, and because of their differing sizes (see Figure 9.14), the chromosomes can be separated from each other. So it is possible to carefully isolate the DNA of human chromosome 4, for example. The straightforward approach to sequencing the DNA of this chromosome would be to start at one end and simply do it all. Unfortunately, this approach does not work. The DNA of a molecule that is 50 million base pairs long cannot be sequenced in that form; only 500 base pairs at a time can be sequenced. (See Figure 16.11 for a review of DNA sequencing.)

In the HGP, then, chromosomal DNA is first cut into fragments, each 500 base pairs long, and then each fragment is sequenced. For the human genome, which has about 3 billion base pairs, there are about 6 million such fragments. Then the problem becomes one of putting these millions of pieces of the jigsaw puzzle together to form a long, linear molecule. This problem can be overcome by breaking up the DNA in several ways into "sub-jigsaws" that overlap, and using *mapped markers* to align the overlapping fragments. The data can be stored in powerful computers for analysis and ordering.

What are the mapped markers along the chromosome? The crudest of them are the **bands** that appear when chromosomes are stained with certain dyes (see

Figure 9.14). DNA in a chromosome is approximately evenly split among the 20 or so dark and light bands, so each band represents about 5 million base pairs.

Banding is useful to localize a gene or other DNA fragment by *fluorescence in situ hybridization* (FISH), in which a probe is hybridized to metaphase chromosomes whose DNA has been denatured (Figure 17.19). In interphase nuclei, the extended nature of the chromosomes can allow FISH to localize DNA sequences that are separated by only about 100,000 base pairs. By careful manipulation of separated chromosomes under the microscope, a single band can be dissected out and then further analyzed by fragmentation of its DNA.

In **physical mapping** of the chromosome, landmarks are ordered and the distances between them are determined (Figure 17.20). The result can be compared to a road map showing towns with the mileage separating them. The "towns" are DNA markers and the "mileage" is in base pairs. The simplest of the markers on DNA are the recognition sites for restriction enzymes. Since there are hundreds of these sequence–enzyme pairs, there are hundreds of ways to cut the DNA and then generate a restriction map. Physical mapping has been a very useful technique in generating maps for relatively small DNAs of thousands of base pairs, such as individual chromosome regions or the genomes of viruses.

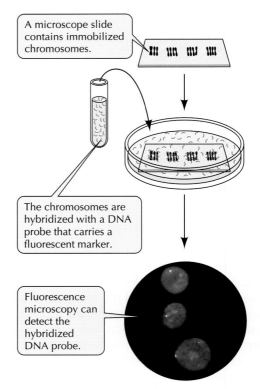

A microscope slide contains immobilized chromosomes.

The chromosomes are hybridized with a DNA probe that carries a fluorescent marker.

Fluorescence microscopy can detect the hybridized DNA probe.

17.19 Mapping a DNA Sequence by Fluorescence *in situ* Hybridization In the FISH technique, banding patterns are used to locate a gene using a fluorescent probe.

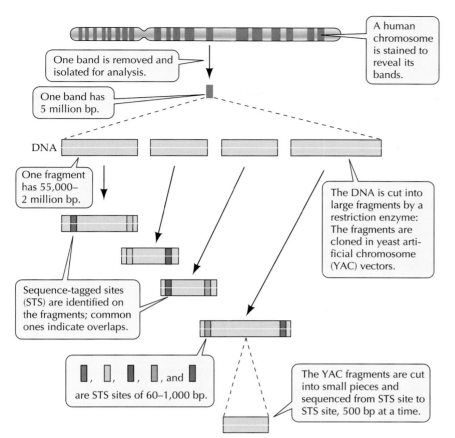

One band is removed and isolated for analysis.

A human chromosome is stained to reveal its bands.

One band has 5 million bp.

DNA

One fragment has 55,000–2 million bp.

The DNA is cut into large fragments by a restriction enzyme: The fragments are cloned in yeast artificial chromosome (YAC) vectors.

Sequence-tagged sites (STS) are identified on the fragments; common ones indicate overlaps.

are STS sites of 60–1,000 bp.

The YAC fragments are cut into small pieces and sequenced from STS site to STS site, 500 bp at a time.

17.20 Physical Mapping of Short Stretches of DNA Mapping a single band on a chromosome arranges the DNA fragments for sequencing.

base pairs long. These pieces are then sequenced.

In addition to physical maps, **genetic maps** of human chromosomes are being generated. Genetic maps also consist of signposts along the chromosome, but in this case they are DNA sequences whose locations are determined genetically. Many DNA sequences are polymorphic. If particular polymorphisms of two genes are always inherited together, they must be closely linked on the chromosome. Such linked DNA sequences might be a disease-causing gene and a particular RFLP. Short, side-by-side repeats of very simple sequences (such as CA) appear to occur in fixed numbers. For example, one sequence might be repeated five times (CACACACACA), another 36 times, and so on throughout the genome and are thus quite useful as genetic linkage markers.

Genetic maps are useful to the HGP in two ways. First, as we described earlier, linkage studies with markers have been very important in tracking down disease-causing genes by positional cloning. Second, the genetic and physical maps can complement each other—one providing new markers for the other.

Sequencing technologies for the final, 500-base-pair DNA fragments are being improved. The use of ultra-thin gels to speed the separation of DNA pieces, automation, and computer programs to process the huge amounts of information have reduced the cost and increased the efficiency of the sequencing process.

The genomes of several organisms have been sequenced

Although the "grunt work" of mapping and sequencing DNAs is not very exciting, the results for organisms whose complete genomes have been sequenced, as well as preliminary data from humans, have been spectacular. Complete genomic sequences have been determined for 141 viruses, 51 organelles, two bacteria, one archaean, and yeast, a single-celled eukaryote. We will describe some of the revelations from the sequences of a bacterium and from yeast.

The only host for the bacterium *Haemophilus influenzae* is humans, where it lives in the upper respiratory

Some restriction enzymes recognize 8 to 12 base pairs in DNA, not just the usual 4 to 6 base pairs. A DNA molecule with several million base pairs will have relatively few of these larger sites, and the enzyme thus will generate a small number of relatively large fragments. These large fragments can be put into yeast artificial chromosome (YAC) vectors (see Figure 16.5b) and made into a library. How are the books (fragments) in this library arranged?

A powerful approach to arranging the fragments involves the development of *sequence-tagged sites* (STSs)—unique stretches of DNA, 60 to 100 base pairs long, whose sequences are known. An STS can be obtained by simply taking a random piece of genomic DNA and seeing if it is unique to a location in the genome by FISH. If it is, the STS is now a marker for that particular chromosomal location. About 20,000 STSs have been mapped on human chromosomes, meaning that each is about 100,000 base pairs (or just a few genes) away from the next STS.

To arrange DNA fragments of a map, libraries made from different restriction enzyme cuts are compared. If two large fragments of DNA cut with different enzymes have the same STS, they must overlap. The individual fragments can be cut again—this time with more common restriction enzymes—into pieces 500

tract and can cause either ear infections or, more seriously, meningitis in children. Its 1,830,137 base pairs are in a single circular chromosome. In addition to its origin of DNA replication and genes coding for ribosomal and tRNAs, this bacterial chromosome has 1,743 regions containing amino acid codons along with the transcriptional (promoter) and translational (start and stop codons) information for protein synthesis. Only 1,007 (58 percent) of these have predicted amino acid sequences that correspond to known proteins. So there are probably many proteins from this bacterium to be discovered.

Of the genes and proteins with assigned roles, most confirm a century of biochemical description of bacterial enzymatic pathways. For example, there are genes for the entire pathways of glycolysis, fermentation, and electron transport. Some of the gene sequences for unknown proteins may code for membrane proteins, possibly involved in active transport. Another finding is that highly infective strains of this bacterium have genes coding for surface proteins that attach them to the respiratory tract, while noninfective strains lack these genes.

Not unexpectedly, the yeast genome is much more complex. It has 16 chromosomes and more than 12,068,000 base pairs. More than 600 scientists around the world collaborated in mapping and sequencing the yeast genome. When they began, they knew of about 1,000 yeast genes coding for RNA or protein. Now there are apparently 5,885! This means that there is a protein-coding gene every 2,000 base pairs in the genome, and 70 percent of the entire DNA is taken up by coding regions. The estimated 100,000 genes in the human genome, in contrast, take up less than 2 percent of all human DNA.

It is now possible to estimate what fractions of the yeast genome code for specific metabolic roles. Apparently 11 percent of the proteins are for general metabolism, 3 percent for energy production and storage, 3 percent for DNA replication and repair, and 13 percent for protein synthesis—there are more than 200 transcription factors! There are also many structural proteins, and, of course, many proteins whose functions are as yet unknown.

Comparisons with human genes indicate that many yeast genes have homologous sequences—similar but not identical—in humans. Since it is possible by molecular biology to selectively inactivate individual yeast genes, the resulting phenotypic changes will be relevant to the roles of these genes in humans.

Preliminary information reveals much about the human genome

Human DNA mapping has developed quickly, and large-scale sequencing is under way. Instead of taking the approach of the scientists studying yeast and bacteria—waiting for the complete genomic sequence before isolating protein-coding genes—molecular biologists have tried to get at the protein-coding genes first.

Since these regions are transcribed into mRNA, the approach has been to sequence cDNA. Actually, the entirety of each cDNA has not been sequenced from libraries of different tissues at different stages of life. Instead, only short yet unique regions, called *expressed-sequence tags*, have been developed to act as chromosome markers (STSs). These tags can be used to find the sequences of entire protein-coding genes.

Of the 30,000 human genes isolated and sequenced from this cDNA–STS approach, many have yeast homologs, so we can speculate as to their functions. About 40 percent of them appear to be genes for basic metabolism, about 22 percent are for protein synthesis, and 12 percent are involved in physiological signaling. Identification of many of these genes will no doubt lead to a better understanding of human biology and its pathologies. But these genes represent at most only half of the protein-coding regions of the human genome.

As the HGP progresses, more and more human genes are being identified, either by disease-related positional cloning or by cDNA analysis. Once a gene is sequenced from one person, it can be sequenced in others, enabling the discovery of common polymorphisms. Computerized databases of polymorphisms can then be matched to databases of diseases, to see if people with a certain gene variant have a certain disease. For example, an allele of the gene for apolipoprotein E is strongly correlated with increased risk for Alzheimer's disease.

The technologies developed for the HGP are already being applied to human diagnostics. Automation of PCR and DNA sequencing are making possible the diagnosis of many diseases caused by genes, from prenatal detection of cystic fibrosis to biopsy of developing tumors. "DNA-on-a-chip" technologies may allow the simultaneous amplification and detection of human variants by placing a drop of blood or other tissue fluid onto a computer chip–like structure made of oligonucleotides instead of silicon.

What will be the uses of genetic information?

After the primary genetic defect that causes cystic fibrosis was discovered, many people predicted a "tidal wave" of genetic testing for heterozygous carriers. Everyone would want the test, it was said—especially the relatives of patients with the genetic disease. But the tidal wave has not developed. To find out why, a team of psychologists, ethicists, and geneticists interviewed 20,000 people in the United States. What the researchers found surprised them: Most people are simply not very interested in their genetic makeup, unless they have a close relative with the disease and are involved in a pregnancy.

Some other people, however, might be very interested in genetic testing. People who test positive for genetic abnormalities, from hypercholesterolemia to cancer, might be denied employment or health insurance. The many linkages of genetic abnormalities to behavioral characteristics, ranging from manic depression to schizophrenia, has led to the potential for screening and then social manipulations of those at risk. Consequently, many legislative bodies are considering and passing bills that prohibit genetic discrimination. The HGP has set aside more than 5 percent of its budget for investigations into the ethical, legal, and social issues of its findings. Such an approach is unusual in scientific history, where the pattern has been to invent technology first and ask social questions later.

Although the HGP has the potential for social disruption, it also has enormous possible benefits. As the ultimate extension of molecular medicine, it will lead not only to the identification of the genes that are altered in disease, but also to targeted treatments, and possibly cures. It will also tell us a lot about ourselves. Comparing the DNA sequences of different people and other organisms will shed new light on how humans got to be where they are, genetically. And by focusing on polymorphisms—the differences between us—the HGP will help physicians understand the complex diseases that result from the interactions of many genes and the environment.

Summary of "Molecular Biology and Medicine"

Protein as Phenotype

• The idea of the one gene, one polypeptide relationship, developed with prokaryotic systems, also applies to human genetic diseases. In many human genetic diseases (for example, phenylketonuria), an enzyme is missing or inactive. **Review Figure 17.1 and Tables 17.1, 17.2**

• Point mutations in the genes that encode the protein components of hemoglobin either lead to clinical abnormalities such as sickle-cell anemia or are relatively benign. **Review Figure 17.2**

• Some diseases are caused by mutations that affect structural proteins; examples include Duchenne's muscular dystrophy, osteoporosis imperfecta, and hemophilia.

• The genes that code for receptors and membrane transport proteins can be mutated and cause diseases such as familial hypercholesterolemia and cystic fibrosis. **Review Figure 17.3**

• Few human diseases are caused by a single gene mutation. Most are caused by the interactions of many genes and proteins with the environment.

• Human genetic diseases show different patterns of inheritance. Phenylketonuria, sickle-cell anemia, and cystic fibrosis, for example, are inherited as autosomal recessives; familial hypercholesterolemia and osteoporosis imperfecta as autosomal dominants; hemophilia and Duchenne's muscular dystrophy as X-linked conditions; and fragile-X syndrome as a chromosome abnormality.

Mutations and Human Diseases

• Molecular biology techniques have made possible the isolation of many genes responsible for human genetic diseases. One method is to isolate the mRNA for the protein in question and then use the mRNA to isolate the gene from a genomic library. Another method is to compare DNA from a patient that lacks a piece of a chromosome with DNA from a person who does not show this deletion to isolate the missing DNA. **Review Figure 17.5**

• In positional cloning, DNA sequence landmarks are used as guides to point the way to a gene. These landmarks might be restriction fragment length polymorphisms, which are linked to a mutant gene. **Review Figure 17.6**

• Human mutations range from single point mutations to large deletions. Some of the most common mutations occur where the modifed base 5-methylcytosine is converted to thymine. **Review Figure 17.7 and Table 17.2**

• The fragile-X chromosome is inherited, and the allele's effects on mental retardation worsen with each generation. This effect is caused by a triplet repeat that tends to expand with each new generation. **Review Figure 17.8**

Detecting Human Genetic Variations

• Genetic screening detects human gene mutations. Some protein abnormalities can be detected by simple tests, such as detection of excess substrate or lack of product. **Review Figure 17.9**

• The advantage of testing DNA for mutations directly is that any cell can be tested at any time in the life of the organism.

• There are two methods of DNA testing: allele-specific cleavage and allele-specific oligonucleotide hybridization. **Review Figures 17.10, 17.11**

Cancer: A Disease of Genetic Changes

• Most cancers are caused by genetic changes.

• Tumors may be benign, which grow to a certain extent and then stop, or malignant, spreading through organs and to other places in the body.

• The most common cancers occur in dividing cells such as epithelia.

• At least five types of human cancers are caused by viruses, which account for about 15 percent of all cancers. **Review Table 17.3**

• Eighty-five percent of human cancers are caused not by viruses, but by genetic mutations of somatic cells. Normal cells contain proto-oncogenes, which when mutated can become oncogenes and cause cancer by stimulating cell division or preventing cell death. **Review Figures 17.13, 17.14**

• About 10 percent of all cancer is inherited as a result of the mutation of tumor suppressor genes, which normally act to slow down the cell cycle. For cancer to develop, both alleles of a tumor suppressor gene must be mutated. In inherited cancer, an individual inherits one mutant allele and then somatic mutation occurs in the second one. In sporadic can-

cer, two normal alleles are inherited, so two mutational events must occur in the same somatic cell. **Review Figure 17.15**
• Mutations must activate several oncogenes and inactivate several tumor suppressor genes for a cell to produce a malignant tumor. **Review Figure 17.16**

Treating Genetic Diseases

• Most genetic diseases are treated symptomatically. However, as more knowledge is accumulated, specific treatments are being devised.
• One treatment approach is to modify the phenotype—for example, by manipulating the diet or providing specific inhibitors to prevent the accumulation of a harmful substrate, or by supplying a missing protein. **Review Figure 17.17**
• In gene therapy, a mutant gene is replaced with a normal gene. Either the affected cells can be removed, the new gene added, and the cells returned to the body, or the new gene can be inserted via a vector directly into the patient. **Review Figure 17.18**

The Human Genome Project

• The aim of the Human Genome Project is to determine the entire DNA sequence of a human, which will require sequencing many 500-base-pair fragments and then putting the sequences together.
• The broadest genomic maps are chromosome bands. Individual sequences may be mapped on the bands by hybridization. **Review Figure 17.19**
• Various short DNA sequences have been mapped on chromosomes, physically or genetically, to act as guideposts for fragments of DNA. The various fragments may be ordered, cut with restriction enzymes, and sequenced. **Review Figure 17.20**
• The genome of the bacterium *Haemophilus* contains 1.8 million base pairs and has been sequenced. Many new proteins have been found encoded by this genome. The sequence of the yeast genome, which contains more than 12 million base pairs, reveals almost 6,000 protein-coding genes, of which only 1,000 are currently known. The various roles of the yeast genes can be inferred from their sequence.
• Sequencing has identified more than 30,000 human genes, some of which are responsible for diseases. As more genes relevant to human health are described, concerns about how such information is used are growing.

Self-Test

1. Phenylketonuria is an example of a genetic disease in which
   a. a single enzyme is not functional.
   b. inheritance is sex-linked.
   c. two parents without the disease cannot have a child with the disease.
   d. mental retardation always occurs, regardless of treatment.
   e. a transport protein does not work properly.

2. Mutations of the gene for β-globin
   a. are usually lethal.
   b. occur only at amino acid position 6.
   c. number in the hundreds.
   d. always result in sickling of red blood cells.
   e. can always be detected by gel electrophoresis.

3. Multifactorial diseases
   a. are less common than single-gene diseases.
   b. involve the interaction of many genes with the environment.
   c. affect less than 1 percent of humans.
   d. involve the interactions of several mRNAs.
   e. are exemified by sickle-cell anemia.

4. In fragile-X syndrome,
   a. females are affected more severely than males.
   b. a short sequence of DNA is repeated many times to create the fragile site.
   c. both the X and Y chromosomes tend to break when prepared for microscopy.
   d. all people who carry the gene that causes the syndrome are mentally retarded.
   e. the basic pattern of inheritance is autosomal dominant.

5. Which of the following is *not* a practical way to isolate a human gene that mutates to cause a disease?
   a. Use an antibody to isolate the mRNA for the protein involved, make a cDNA copy, and then use this cDNA to probe a genomic library.
   b. Compare the DNA of a person with the disease (who has a deleted chromosome) with the DNA from a person without the disease to see if the latter has extra sequences, which contain the target gene.
   c. Use DNA markers that are closely linked to the mutant gene and then isolate the gene by molecular methods.
   d. Isolate DNAs from people with and without the disease and insert each into cloning vectors to see what proteins are made.
   e. Sequence DNAs from people with and without the genetic mutation to determine the differences.

6. Mutational "hot spots" in human DNA
   a. always occur in genes that are transcribed.
   b. are common at cytosines that have been modified to 5-methylcytosine.
   c. involve long stretches on nucleotides.
   d. occur where there are triplet repeats, such as CTG.
   e. are very rare in genes that code for proteins.

7. Newborn genetic screening for PKU
   a. is very expensive.
   b. detects phenylketones in urine.
   c. has not led to the prevention of mental retardation resulting from this disorder.
   d. must be done during the first day of an infant's life.
   e. uses bacterial growth to detect excess phenylalanine in blood.

8. Genetic diagnosis by DNA testing
   a. detects only mutant and not normal alleles.
   b. can be done only on eggs or sperm.
   c. involves hybridization to rRNA.
   d. utilizes restriction enzymes and a polymorphic site.
   e. cannot be done with PCR.

9. Most human cancers
   a. are caused by viruses.
   b. are in blood cells or their precursors.
   c. involve mutations of somatic cells.
   d. spread through solid tissues rather than by the blood or lymphatic system.
   e. are inherited.

10. Current treatments for genetic diseases include all of the following *except*
    a. restricting a dietary substrate.
    b. replacing the mutated gene in all cells.
    c. alleviating the patient's symptoms.
    d. inhibiting the function of a harmful metabolite.
    e. supplying a protein that is missing.

Applying Concepts

1. Compare the roles of proto-oncogenes and tumor suppressor genes in normal cells. How do these genes and their functions change in tumor cells? Propose targets for cancer therapy involving these gene products.

2. In the past, it was common for people with phenylketonuria (PKU) who were placed on a low-phenylalanine diet after birth to be allowed to return to a normal diet during their teenage years. Although the levels of phenylalanine in their blood were high, their brains were thought to be beyond the stage of being harmed. If a woman with PKU becomes pregnant, however, a problem arises. Typically, the fetus is heterozygous but is unable at early stages of development to metabolize the high levels of phenylalanine that arrive from the mother's blood. Why is the fetus heterozygous? What do you think would happen to the fetus during this "maternal PKU" situation? What would be your advice to a woman with PKU who wants to have a child?

3. Cystic fibrosis is an autosomal recessive disease in which thick mucus is produced in the lungs and airway. The gene responsible for this disease codes for a protein composed of 1,480 amino acids. In most patients with cystic fibrosis, the protein has 1,479 amino acids: A phenylalanine is missing at position 508. A baby is born with cystic fibrosis. He has an older brother. How would you test the DNA of the brother to determine if he is a carrier for cystic fibrosis? How would you design a gene therapy protocol to "cure" the cells in the lung and airway?

4. The Human Genome Project aims to sequence an entire human genome. A related endeavor, the Human Genome *Diversity* Project, aims to sequence genomes from different groups of people from around the world by collecting hair and blood samples. What would be the value of such information? What concerns do you think are being raised by the people whose DNAs are being analyzed?

Readings

Anas, G. and S. Elias. 1992. "Gene Mapping: Using Law and Ethics as Guides." Oxford University Press, New York. A discussion of the social, legal, and ethical impacts of molecular genetic medicine.

"Cancer: Special Issue." 1996. *Scientific American*, September. A series of articles, covering topics ranging from causes to cures.

Mange, E. J. and A. P. Mange. 1994. *Basic Human Genetics*. Sinauer Associates, Sunderland, MA. A brief yet comprehensive text.

McKusick, V. A. 1994. *Mendelian Inheritance in Man: A Catalog of Human Genes and Genetic Disorders*, 11th Edition. Johns Hopkins University Press, Baltimore. The definitive work on human genetic variations.

"Molecular Medicine." 1994–1997. *New England Journal of Medicine*. An excellent series of articles on all aspects of molecular biology, including methods.

Pollock, R. 1995. "Signs of Life." Houghton-Mifflin, New York. A popular book on the molecular revolution in medicine, written by a distinguished scientist.

Scriver, C., A. Beaudet, W. Sly and D. Valle. 1996. *The Metabolic and Molecular Bases of Inherited Disease*. McGraw-Hill, New York. The definitive work on human biochemical genetics.

Chapter 18

Natural Defenses against Disease

Feel a Sneeze Coming On?
This artificially colored scanning electron micrograph shows highly magnified household dust. The unique shapes of these structures cause some of them to provoke reactions in the immune system.

Ouch! A splinter enters your finger. It smarts, and you know what's going to happen next. If you don't remove the splinter and cleanse the wound immediately, the tissue around the splinter will redden, heat up, swell, and start to throb. These four responses are the symptoms of inflammation, an effective mechanism our bodies use to limit the spread of infection by disease-producing organisms (pathogens) on the splinter. Our environment teems with pathogens such as viruses, bacteria, protists, and fungi. We are challenged by them daily. How do we survive?

You have probably been ill with the flu (influenza) at least a few times in your life. And you have almost certainly encountered the influenza virus since the last time you were actually ill with the flu. Here's just part of what happened in your recent encounter with the newest flu strain. In a lymph node, cells called macrophages phagocytosed some of the flu viruses and partly digested them. The macrophages displayed viral pieces on their plasma membranes, where specialized white blood cells called T cells recognized the pieces and were activated to divide and differentiate further. Many of the descendants of the activated T cells differentiated to attack your virus-infected cells, preventing viral reproduction and thus saving you from another case of the flu. Other descendants of the activated T cells persist in your body today as "memory cells" prepared to rapidly defend you the next time this variety of flu virus strikes.

These defensive events require the participation of many kinds of proteins. Some cellular proteins function as specific receptors, some as markers identify-

400

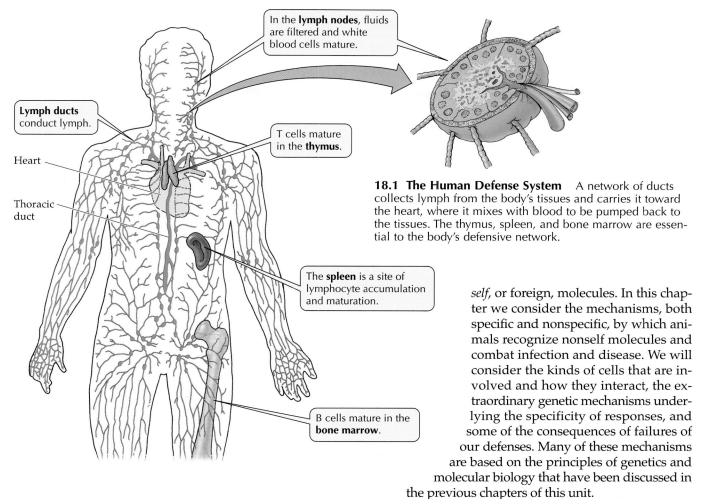

In the **lymph nodes**, fluids are filtered and white blood cells mature.

Lymph ducts conduct lymph.

T cells mature in the **thymus**.

Heart

Thoracic duct

The **spleen** is a site of lymphocyte accumulation and maturation.

B cells mature in the **bone marrow**.

18.1 The Human Defense System A network of ducts collects lymph from the body's tissues and carries it toward the heart, where it mixes with blood to be pumped back to the tissues. The thymus, spleen, and bone marrow are essential to the body's defensive network.

ing your cells, some as signals triggering events in the macrophages and T cells, and others as weapons leading to the breakdown of the infected cells.

Not all pathogens are greeted by the same set of defenses, but all do elicit a response—usually one that successfully wards off disease. Inflammation is one of several *nonspecific* defenses that we employ. Your response to the flu virus is an example of *specific* defenses, which can distinguish between one type of virus and another.

Our defensive responses typically involve interactions between different kinds of defender cells. They often result in the establishment of a memory of the invading pathogen that can last for years. Most of the responses are highly specific, tailored to meet very specific challenges. Each of us seems able to deal specifically with millions of different kinds of challenges. Genetic diversity underlies these specific defenses, as well as the differences between the individuals in a species—differences that lead, for example, to the rejection of tissues grafted from other individuals.

Animal defense systems are based on the distinction between *self*—the animal's own molecules—and *non-self*, or foreign, molecules. In this chapter we consider the mechanisms, both specific and nonspecific, by which animals recognize nonself molecules and combat infection and disease. We will consider the kinds of cells that are involved and how they interact, the extraordinary genetic mechanisms underlying the specificity of responses, and some of the consequences of failures of our defenses. Many of these mechanisms are based on the principles of genetics and molecular biology that have been discussed in the previous chapters of this unit.

Defensive Cells and Proteins

Components of our defense system (also called the **immune system**) are dispersed throughout the body and interact with all the other tissues and organs of the body. The thymus, bone marrow, spleen, and certain other lymphoid tissues are essential parts of our defense system (Figure 18.1), but central to their functioning are the blood and lymph.

Blood and lymph are fluid tissues that consist of water, dissolved solutes, and cells. About 60 percent of the **blood** volume is the yellowish solution called plasma. The remainder consists of red blood cells, white blood cells, and platelets (cell fragments essential to clotting) (Table 18.1). The red blood cells are normally confined to the closed circulatory system consisting of the heart, arteries, capillaries, and veins, but the other blood cells are also found in the lymph.

The **lymph** is a fluid that is derived from the blood and other tissues. It accumulates in spaces outside the blood vessels of the circulatory system and contains many of the components of blood, except red blood

TABLE 18.1   Cells and Cell Fragments in Blood and Lymph

TYPE OF CELL	FUNCTION
Red blood cells (erythrocytes)	Transport oxygen and carbon dioxide
White blood cells (leukocytes)	Defend against pathogens
Basophils	Release histamine; may promote the development of T cells
Phagocytes	Destroy nonself materials
Eosinophils	Kill antibody-coated parasites
Neutrophils	Phagocytize antibody-coated pathogens
Monocytes	Develop into macrophages
Macrophages	Engulf and digest microorganisms; activate T cells
Lymphocytes	Have many roles in the immune system
B cells	Differentiate to form antibody-producing cells and memory cells
T cells	Kill virus-infected cells; regulate activities of other white blood cells
Natural killer cells	Attack and lyse virus-infected or cancerous body cells
Platelets (cell fragments without nuclei)	Initiate blood clotting

cells. From the spaces around body cells, the lymph moves slowly into tiny lymph capillaries and then into larger vessels that join together, forming one large lymph duct that joins a major vein (the superior vena cava) near the heart.

By this system of vessels, the lymphatic fluid is eventually returned to the blood and the circulatory system. At many sites along the lymph vessels are small roundish structures called **lymph nodes** that contain a variety of white blood cells involved in nonspecific and specific defenses. Lymph nodes filter the blood and present foreign materials to the defensive cells.

In this first section of the chapter, we will introduce the variety of white blood cells, their defensive roles, and the secreted protein signals that coordinate their activities.

White blood cells play many defensive roles

One milliliter of blood typically contains about 5 billion red blood cells and 7 million larger white blood cells (Figure 18.2). White blood cells have nuclei and are colorless. They can receive signals that direct them to leave the circulatory system by squeezing through junctions between the cells that form the walls of blood capillaries. In response to invading pathogens, the number of white blood cells in the blood and lymph may rise sharply, providing medical professionals with a useful clue for detecting an infection.

The most abundant types of white blood cells are the **lymphocytes** (see Table 18.1). A healthy person has about a trillion lymphocytes, a number similar to the total number of cells in the brain. Two groups of lymphocytes, the **B cells** and **T cells**, together with spe-

cialized cells that arise from them, are the important cells of the immune system that detects specific stimuli and responds in a defensive way to them. B cells and T cells are found in the blood and lymphoid tissues (see Figure 18.1).

Both B and T cells originate from cells in the bone marrow. The precursors of T cells migrate via the blood

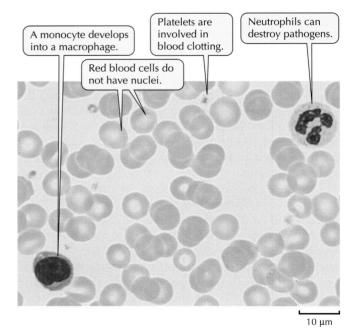

18.2 Blood Cells   Two white blood cells are shown among the many red blood cells in this micrograph. Table 18.1 defines the many types of white blood cells. Platelets are cell fragments.

to the thymus, where they become mature T cells. The B cells leave the bone marrow and circulate through the blood and lymph vessels, passing through the lymph nodes and spleen. B and T cells look the same under the light microscope, but they have quite different functions in immune responses, which are methods of reacting against nonself or altered-self substances. Discussion of these functions and their mechanisms will occupy most of this chapter.

In addition to lymphocytes are **phagocytes**, cells that engulf and digest nonself materials. The most important phagocytes are the neutrophils and the macrophages. **Macrophages** have the important additional function of presenting partly digested nonself materials to the T cells for inspection and response. Thus macrophages are essential to the functioning of specific defenses.

Immune system cells release proteins that bind pathogens or signal other cells

The various kinds of cells that defend our body work together like cast members in a drama, interacting with one another and with the cells of invading pathogens. These cell–cell interactions are accomplished by a variety of key proteins, including receptors, surface markers, signaling molecules, and toxins. They will be discussed later in the chapter, as they appear in the context of our story. However, let's take a brief look at them here, to help set the scene.

Lymphocytes, the prime agents in our specific defenses, receive protein signals from other cells, have unique receptors on their membranes, and influence the behavior of other cells by secreting special proteins—antibodies or signal molecules. B cells carry antibodies on their plasma membranes as receptors. **Antibodies** are proteins that bind specifically to substances identified by the immune system as nonself or altered self. Descendants of activated B cells secrete antibodies as weapons of defense. T cells also have surface receptors, called T cell receptors. Most cells of the human body display human leukocyte antigen (HLA) proteins on their plasma membranes. The HLA proteins are important "self"-identifying labels and play major parts in coordinating interactions among lymphocytes and macrophages.

T cells and macrophages communicate by secreting a variety of small, soluble signal proteins, called **cytokines**, that alter the behavior of their target cells. Different cytokines activate or inactivate B cells and macrophages, while others stimulate, inhibit, or kill other T cells. Certain cytokines limit tumor growth by killing tumor cells.

There are still other classes of proteins that participate in the body's nonspecific defenses, as we will see in the next section.

Nonspecific Defenses against Pathogens

Animals have defenses that stop pathogens from invading their bodies. Because these initial defenses give general protection against different pathogens, they are called **nonspecific defenses**, or the innate immune response. In humans, these nonspecific defenses include barriers and local conditions (such as the skin, the antibacterial enzyme lysozyme, mucus, and phagocytes) and cellular and chemical defenses (such as natural killer cells, defensive proteins such as interferons and the complement system, and the complex of responses collectively known as inflammation) (Table 18.2). Even bacteria and fungi that normally reside on body surfaces protect against invasion by pathogens.

Barriers and local agents can defend the body

Skin is a primary nonspecific defense against invasion. Bacteria and viruses rarely penetrate healthy, unbroken skin. But damaged skin or other surface tissue is another matter. The sensitivity of surface tissue accounts in part for the greatly increased risk of infection by pathogenic agents in a person who already has a disease that causes breaks in the skin.

The bacteria and fungi that normally live and reproduce in great numbers on our body surfaces without causing disease are referred to as **normal flora** (Figure 18.3). These natural occupants compete with pathogens for locations and nutrients.

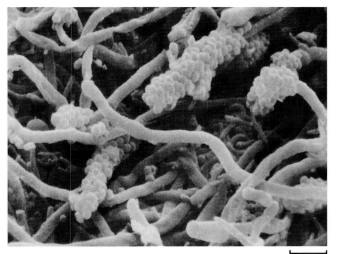

0.8 μm

18.3 Normal Flora Gone Rampant The human mouth harbors a wide variety of microorganisms, most of which cause no damage under normal conditions. When bacteria accumulate on the surfaces of teeth, the result is plaque, which contributes to tooth decay. This electron micrograph shows plaque on a tooth 3 days after the person stopped brushing.

TABLE 18.2 Nonspecific Defenses

DEFENSIVE AGENT	FUNCTION
SURFACE BARRIERS	
Skin	Prevents entry of pathogens and foreign substances
Acid secretions	Inhibit bacterial growth on skin
Mucous membranes	Prevent entry of pathogens
Mucus secretions	Trap bacteria and other pathogens in digestive and respiratory tracts
Nasal hairs	Filter bacteria in nasal passages
Cilia	Move mucus and trapped materials away from respiratory passages
Gastric juice	Concentrated HCl and proteases destroy pathogens in stomach
Acid vagina	Limits growth of fungi and bacteria in female reproductive tract
Tears, saliva	Lubricate and cleanse; contain lysozyme, which destroys bacteria
NONSPECIFIC CELLULAR, CHEMICAL, AND COORDINATED DEFENSES	
Normal flora	Compete with pathogens; may produce substances toxic to pathogens
Phagocytes (macrophages and neutrophils)	Engulf and destroy pathogens that enter body
Natural killer cells	Attack and lyse virus-infected or cancerous body cells
Antimicrobial proteins	
Interferons	Released by virus-infected cells to protect healthy tissue from viral infection; mobilize specific defenses
Complement proteins	Lyse microorganisms, enhance phagocytosis, and assist in inflammatory response
Inflammatory response (involves leakage of blood plasma and phagocytes from capillaries and some local heating)	Limits spread of pathogens to neighboring tissues, concentrates defenses, digests pathogens and dead tissue cells; released chemical mediators attract phagocytes and specific defense lymphocytes to site
Fever	Body-wide response inhibits microbial multiplication and speeds body repair processes
Coughing, sneezing	Expel pathogens from upper respiratory passages

Mucus-secreting tissues found in parts of the visual, respiratory, digestive, excretory, and reproductive systems have other defenses against pathogens. Secretions such as tears, nasal drips, and saliva possess an enzyme called **lysozyme** that attacks the cell walls of many bacteria. Mucus in our noses traps microorganisms in the air we breathe, and most of those that get past this filter end up trapped in mucus deeper in the respiratory tract. Mucus and trapped pathogens are removed by the beating of cilia in the respiratory passageway, which moves a sheet of mucus and the debris it contains up toward the nose and mouth. Another effective means of removing microorganisms from the respiratory tract is the sneeze.

Pathogens that reach the digestive tract (stomach, small intestine, and large intestine) are met by other defenses. The stomach is a deadly environment for many bacteria because of the hydrochloric acid and protein-digesting enzymes that are secreted into it. The intact lining of the small intestine cannot be penetrated by bacteria, and some pathogens are killed by bile salts secreted into this part of the tract. The large intestine harbors many bacteria, which multiply freely; however, these are usually removed quickly with the feces. Most of these bacteria in the large intestine are normal flora that provide benefits to the host. (The digestive system is described fully in Chapter 47.)

All of these barriers and secretions are *nonspecific* because they are the same for all invading pathogens. But there are more complicated nonspecific cellular chemical defenses.

Nonspecific defenses include cellular and chemical processes

Pathogens that manage to penetrate the surface cells of the body encounter additional nonspecific defenses. These defenses include the secretion of various defensive proteins (complement proteins and interferons) and cellular defenses involving phagocytosis.

COMPLEMENT PROTEINS. Vertebrate blood contains about 20 different antimicrobial proteins that make up the **complement system**. These proteins, in different combinations, provide three types of defenses. In each type,

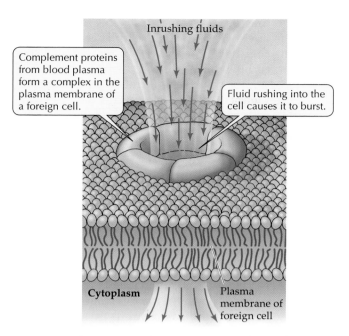

Complement proteins from blood plasma form a complex in the plasma membrane of a foreign cell.

Inrushing fluids

Fluid rushing into the cell causes it to burst.

Cytoplasm

Plasma membrane of foreign cell

18.4 Complement Proteins Destroy a Foreign Cell
Complement proteins attach to foreign cells such as bacteria and can form a porelike structure that allows fluids to pour in, until the foreign cells burst.

the complement proteins act in a characteristic sequence, or cascade, with each protein activating the next.

In one type of nonspecific defense, complement proteins help phagocytes destroy foreign microorganisms. The phagocytes can recognize foreign cells more easily after complement proteins attach to the foreign cells, and the phagocytes engulf complement-coated cells more readily than uncoated ones.

The second defensive activity of the complement system is to attract phagocytes to sites of infection. In concert with this activity, the complement system activates the inflammatory response (which we will describe shortly).

In the third and most impressive defense mounted by the complement system, complement proteins lyse (burst) foreign cells—bacteria, for example. Initial binding of antibodies to the surface of a foreign cell may also bring about the binding of complement proteins to the cell surface. What follows is a cascade of reactions, with different complement proteins acting on one an-

other in succession. The final product of the complement cascade is a doughnut-shaped structure in the membrane of the foreign cell that allows fluids to enter the cell rapidly, causing lysis (bursting) of the foreign cell (Figure 18.4).

INTERFERONS.   In the body, virus-infected cells produce small amounts of antimicrobial proteins called **interferons** that increase the resistance of neighboring cells to infection by the same *or other* viruses. Interferons have been found in many vertebrates and are one of the body's lines of nonspecific defense against viral infection.

Interferons differ from species to species, and each vertebrate species produces at least three different interferons. All interferons are glycoproteins (proteins with a carbohydrate component) consisting of about 160 amino acid units. By binding to receptors in the plasma membranes of cells, interferons inhibit the ability of the viruses to replicate. Interferons have been the subject of intensive research because of their possible applications in medicine—for example, the treatment of influenza and the common cold.

PHAGOCYTOSIS AND OTHER CELLULAR ASSAULTS.   Phagocytes provide an extremely important nonspecific defense against pathogens that penetrate the surface of the host. Some phagocytes travel freely in the circulatory system; others can move out of blood vessels and adhere to certain tissues. Pathogens become attached to the membrane of a phagocyte (Figure 18.5). Then the phagocyte ingests the pathogens by endocytosis. Once inside an endocytic vesicle in a phagocyte, pathogens are destroyed by enzymes when lysosomes fuse with the vesicle (see Figure 4.20b).

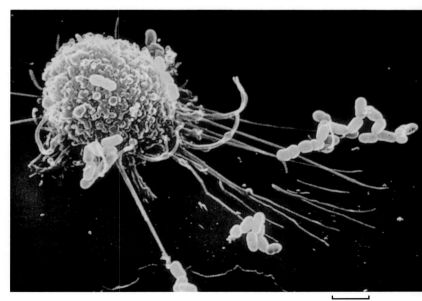

2 µm

18.5 A Phagocyte and Its Bacterial Prey   Some bacteria (appearing bright yellow in this artificially colored scanning electron micrograph) have become attached to the surface of a phagocyte in the human bloodstream. Many of these bacteria will be taken into the phagocyte and destroyed before they can multiply and damage the human host.

A single phagocyte can ingest 5 to 25 bacteria before it dies from the accumulation of toxic breakdown products. Even when phagocytes do not destroy all the invaders, they usually reduce the number of pathogens to the point where other defenses can finish the job. So important is the role of the phagocytes that if their functioning is impaired by disease, the animal usually soon dies of infection, not only because of decreased phagocytosis but also because these cells are important in the specific immune response.

There are two major types and one minor type of phagocyte (see Table 18.1). **Neutrophils** are by far the most abundant, but they are relatively short-lived. They recognize and attack pathogens in infected tissue. **Monocytes** mature into *macrophages*, which live longer than neutrophils and can consume large numbers of pathogens. Some macrophages roam through the body; others reside permanently in lymph nodes, the spleen, and certain other lymphoid tissues, "inspecting" the lymph fluid for pathogens. Finally, **eosinophils** are weakly phagocytic. Their primary function is to kill parasites, such as worms, that have been coated by antibodies.

One class of small white blood cells, known as **natural killer cells**, can initiate the lysis of some tumor cells and cells that are infected by a virus. The natural killer cells may seek out cancer cells that appear in the body. The targets of natural killer cells are the body's own cells that have gone awry. In recent years, we have learned how these cells can distinguish virus-infected and tumor cells from normal cells in the body. In addition to their roles in nonspecific defenses, natural killer cells are part of the specific defenses of the immune system.

Inflammation fights infections

Another important nonspecific defense is **inflammation**. The body employs this characteristic response in dealing with infections, and in any other process that causes tissue injury either on the surface of the body or internally. The damaged cells themselves cause the inflammation by releasing various substances, such as the chemical signal **histamine**. Cells adhering to the skin and linings of organs called **mast cells** contain histamine and release it when they are damaged, as do certain white blood cells called **basophils** (see Table 18.1).

You have experienced the symptoms of inflammation: redness and swelling, accompanied by heat and pain. The redness and heat result from histamine-induced dilation of blood vessels in the infected or injured area (Figure 18.6). Histamine also causes the capillaries (the smallest blood vessels) to become leaky, allowing some blood plasma and phagocytes to escape into the tissue, causing characteristic swelling. The pain results from increased pressure (from the

18.6 Interactions of Cells and Chemical Signals in Inflammation The histamine-induced swelling of the inflammation reaction is accompanied by redness, heat, and pain. The chemical signals associated with the reaction attract the phagocytes that are largely responsible for healing the wound.

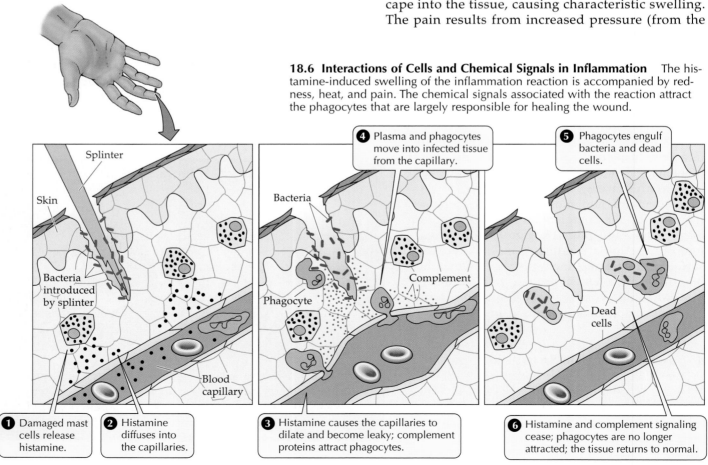

1 Damaged mast cells release histamine.

2 Histamine diffuses into the capillaries.

3 Histamine causes the capillaries to dilate and become leaky; complement proteins attract phagocytes.

4 Plasma and phagocytes move into infected tissue from the capillary.

5 Phagocytes engulf bacteria and dead cells.

6 Histamine and complement signaling cease; phagocytes are no longer attracted; the tissue returns to normal.

leakage) and from the action of some leaked enzymes. Some of the complement proteins and other chemical signals attract other phagocytes—neutrophils first, and then monocytes, which become macrophages. The neutrophils and macrophages are responsible for most of the healing associated with inflammation.

The heat may also play a healing role if it raises the temperature too high for pathogens to multiply effectively. In the aftermath of inflammation, pus may accumulate—it contains dead cells (spent neutrophils and the damaged body cells) and leaked fluid. A normal result of inflammation, pus is gradually cleaned up by macrophages.

Specific Defenses: The Immune Response

Nonspecific defenses are numerous and effective, but some invaders elude them and must be dealt with by defenses targeted against specific threats. The destruction of specific pathogens is an important function of an animal's immune system. In this section, we will identify and discuss the four characteristics and the two types of immune responses.

The immune system uses B and T lymphocytes to recognize and attack specific invaders, such as bacteria and viruses. It also detects normal cells that have been altered by viruses or mutation, and produces signals for their destruction. An animal with a defective immune system can die from infection by even "harmless" bacteria. Some microorganisms normally carried in or on a healthy animal's body are potentially pathogenic and will more readily cause disease if the body's immune system is defective. For example, certain immune deficiencies lead to recurrent, uncontrolled infections with bacteria as *Staphylococcus* or *Streptococcus*.

Four features characterize the immune response

The characteristic features of the immune response are specificity, the ability to respond to an enormous diversity of pathogens, the ability to distinguish self from nonself, and immunological memory.

SPECIFICITY. Pathogens and nonself molecules are diverse, yet their differences are often subtle, as between mutant strains of a single species of bacterium. **Antigens** are organisms or molecules that are recognized and/or interact with the immune system to initiate an immune response. The specific sites on antigens that the immune system recognizes are called **antigenic determinants** (Figure 18.7). Chemically, an antigenic determinant is a group of atoms that may be present on many different large molecules. A large antigen, such as a whole cell, may have many different antigenic determinants on its surface, each capable of being bound by a specific antibody or T cell. Even a single protein has multiple, different antigenic determinants. The host responds to the presence of an anti-

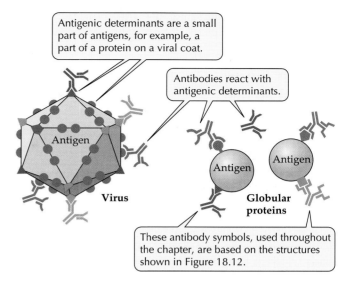

18.7 Each Antibody Matches an Antigenic Determinant
Each antigen has different antigenic determinants that are recognized by specific antibodies. An antibody recognizes and binds to its antigenic determinant to initiate defensive measures against the antigen.

gen by producing highly specific defenses—T cells or antibodies that correspond to the antigenic determinants on that antigen. Each T cell and each antibody is specific for a single antigenic determinant.

THE ENORMOUS SCOPE OF DIVERSITY. Challenges to the immune system are legion: viruses, bacteria, protists, and multicellular parasites. Each of these types of potential pathogens includes many species; each species includes many subtly differing genetic strains; each strain possesses multiple surface features. Estimates vary, but a reasonable guess is that our bodies can respond *specifically* to 10 million different antigenic determinants. On recognition of an antigenic determinant, the immune system responds by activating lymphocytes of the appropriate specificity.

DISTINGUISHING SELF FROM NONSELF. The human body contains tens of thousands of different proteins, each with specific antigenic determinants. Every cell in the body bears a tremendous number of antigenic determinants. A crucial attribute of an individual's immune system is that it recognizes the body's own antigenic determinants and does not attack them. Failure to make this distinction may lead to autoimmune disease—an attack on one's own body.

IMMUNOLOGICAL MEMORY. After responding to a particular type of pathogen once, the immune system "remembers" that pathogen and can usually respond more rapidly and powerfully to the same threat in the future. This memory usually saves us from repeats of childhood diseases such as chicken pox. Vaccination against disease works because the immune system remembers the antigenic determinants that are inoculated into the body.

There are two interactive immune responses

Foreign organisms and substances that invade the animal body and escape the nonspecific defenses come up against the specific defenses of the immune system. The specific immune system has two responses against invaders: the humoral immune response and the cellular immune response. The two responses operate in concert—simultaneously and cooperatively, sharing mechanisms.

In the **humoral immune response** (from the Latin *humor*, "fluid"), the highly specific antibodies react with antigenic determinants on foreign invaders in blood, lymph, and tissue fluids. An animal produces a vast diversity of antibodies that, among them, can react with almost any conceivable antigen in the bloodstream or lymph. Some antibodies travel free in the blood and lymph; others exist as integral membrane proteins on B cells. Antibodies are produced and secreted by specialized B cells that become **plasma cells**. Each antibody recognizes and is capable of binding to a particular antigenic determinant on one or more antigens that invade an animal's body (see Figure 18.7). On stimulation by antigen, a single plasma cell produces multiple soluble copies of antibody with the same specificity as the membrane antibody.

The **cellular immune response** is directed against an antigen that has become established within a cell of the host animal. It destroys virus-infected cells in the animal, as well as some tumor cells. Unlike the humoral response, the cellular immune response does not use antibodies. Instead, it is carried out by T cells within lymph nodes and, to a lesser extent, by T cells that roam through the bloodstream and tissue spaces. The T cells have *T cell receptors*—surface glycoproteins that recognize and bind to antigenic determinants while remaining part of the cell's plasma membrane.

Like antibodies, T cell receptors have specific molecular configurations that bind to specific antigenic determinants. There are two major types of T cells. *Helper T cells* (T_H cells) assist other lymphocytes of both the humoral and cellular immune responses. *Cytotoxic T cells* (T_C cells), or "killer" cells, cause certain other cells to lyse and die. Each of these types will be discussed in some detail later in this chapter. Once bound to a determinant, each type of T cell initiates characteristic activity. Because T cells recognize and mobilize attacks on foreign cells or altered self cells, they are also responsible for the rejection of certain types of organ or tissue transplants.

The specific immune responses act in concert not only with each other but also with the nonspecific defenses. Together, an animal's white blood cells (phagocytes and lymphocytes) defend it against disease.

Clonal selection accounts for the characteristic features of the immune response

Each person possesses an enormous number of different B cells and T cells, apparently capable of dealing with almost any antigen ever likely to be encountered. How does this diversity arise? And why don't our antibodies and T cells attack and destroy our own bodies?

An individual's immune system can mount an immune response against another person's proteins, yet it usually does not mount one against its own. The immune system can distinguish self (one's own antigens) from nonself (those foreign to the body). The versatility of immune responses and immunological memory can be explained satisfactorily by the **clonal selection theory**.

According to the theory, the individual animal contains an enormous variety of different B cells, and each type of B cell is able to produce *only one kind of antibody*. Antigenic determinants do not create an antibody structure and specificity. A surface receptor, already present on a B cell or T cell, binds to the antigen, and this binding signals the particular cell to divide and form a clone. In the case of B cells, the clone consists of plasma cells, all of which secrete a single kind of antibody (Figure 18.8). In the case of T cells, the clone makes signals that lead to various cell-killing mechanisms.

The clonal selection theory accounts nicely for the body's ability to respond rapidly to any of a vast number of different antigens. In the extreme case, even a single B cell might be sufficient for an immunological response by the body, provided that it encounters the antigen and then proliferates into a large clone rapidly enough to combat the invasion. Clonal selection accounts for the proliferation of both B and T cells.

The clonal selection theory also explains two other phenomena: recognition of self and immunological memory (the basis for natural and artificial immunity).

Immunological memory and immunity result from clonal selection

According to the clonal selection theory, an activated lymphocyte (B cell or T cell) produces two types of daughter cells. The ones that carry out the attack on the antigen are **effector cells**—either plasma cells that produce antibodies, or T cells that on binding antigenic determinants release messenger molecules called *interleukins*. The other products of an activated lymphocyte are called **memory cells**—long-lived cells that retain the ability to start dividing on short notice to produce more effector and more memory cells (see Figure 18.8). Effector cells live only a few days, but memory B and T cells may survive for decades.

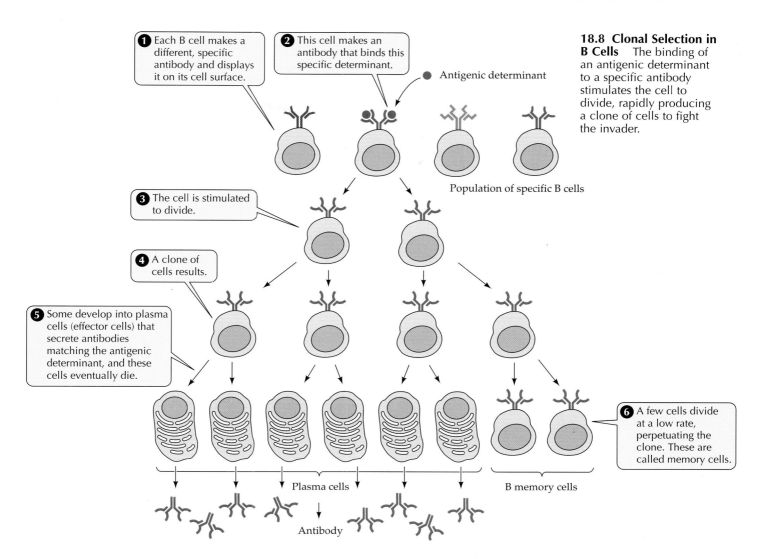

1 Each B cell makes a different, specific antibody and displays it on its cell surface.

2 This cell makes an antibody that binds this specific determinant.

Antigenic determinant

Population of specific B cells

18.8 Clonal Selection in B Cells The binding of an antigenic determinant to a specific antibody stimulates the cell to divide, rapidly producing a clone of cells to fight the invader.

3 The cell is stimulated to divide.

4 A clone of cells results.

5 Some develop into plasma cells (effector cells) that secrete antibodies matching the antigenic determinant, and these cells eventually die.

6 A few cells divide at a low rate, perpetuating the clone. These are called memory cells.

Plasma cells

B memory cells

Antibody

When the body first encounters a particular antigen, a *primary immune response* is activated and one or more types of lymphocytes produce clones of effector and memory cells. The effector cells destroy the invaders at hand and then die, but one or more clones of different memory cells have now been added to the immune system and provide **immunological memory**.

After the body's first immune response to a particular antigen, subsequent encounters with the same antigen will result in a greater and more rapid production of antigen-specific antibody or T cells. This response is called the *secondary immune response*. The first time a vertebrate animal is exposed to a particular antigen, there is a time lag (usually several days) before the number of antibody molecules and T cells slowly increases (Figure 18.9). But for years afterward—sometimes for life—the immune system "remembers" that particular antigen. The secondary immune response has a shorter lag phase, a greater rate of antibody production, and a larger production of total antibody (or T cells).

An early description of immunological memory came from the great historian of ancient Greece, Thucydides. Living in Athens, which was under siege from its rival, Sparta, in 430 B.C., Thucydides described the outbreak of a plague that killed about one-fourth of the Athenians, including their great leader, Pericles. However, some of the ill recovered, and, Thucydides noted, "the same man was never attacked twice, never at least fatally." Those who had been infected once and survived took care of the sick without fear. This pattern of recovery and immunity has been repeated throughout history, for many different diseases.

IMMUNITY: NATURAL AND ARTIFICIAL. Thanks to immunological memory, recovery from many diseases such as chicken pox provides a *natural immunity* to those diseases. However, we don't wait around to gain natural immunity to life-threatening diseases such as typhoid or tetanus. We minimize the risks by seeking *artificial immunity* by way of **immunization**.

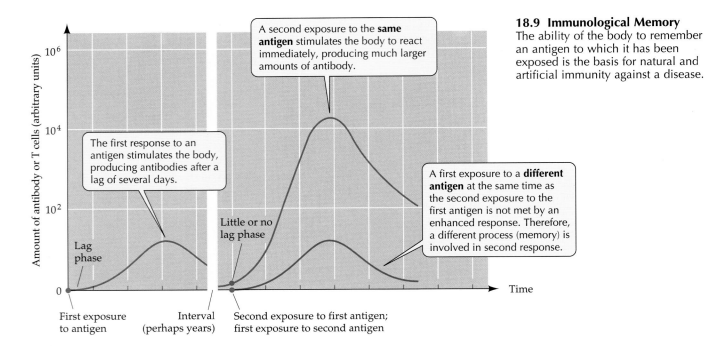

18.9 Immunological Memory
The ability of the body to remember an antigen to which it has been exposed is the basis for natural and artificial immunity against a disease.

The ability of the human body to remember a specific antigen explains why immunization has almost completely wiped out deadly diseases such as smallpox, diphtheria, and polio in medically sophisticated countries. In fact, smallpox has been eliminated worldwide from the spectrum of infectious diseases affecting humans, thanks to a concentrated international effort by the World Health Organization. As far as we know, the only remaining smallpox viruses on Earth are those kept in some laboratories.

Vaccination involves injecting a small amount of virus or bacteria or their proteins (usually treated to make them harmless) into the body. This injection initiates a primary immune response without making the person ill. Later, if the same or very similar disease organisms attack, T and B memory cells already exist. They recognize the antigen and quickly overwhelm the invaders with a massive production of lymphocytes and antibodies.

Animals distinguish self from nonself and tolerate their own antigens

Given the great array of different lymphocytes directed against particular antigens, why doesn't a healthy animal produce self-destructive immune responses? Self-tolerance seems to be based on two mechanisms: clonal deletion and clonal anergy.

Lymphocytes in the bone marrow that have not yet fully differentiated are not capable of attacking antigens. When "antiself" lymphocytes in this undifferentiated state encounter corresponding self antigens, the antiself lymphocytes are eliminated by programmed cell death (apoptosis); thus, no clones of antiself lymphocytes normally appear in the bloodstream. This phenomenon is referred to as **clonal deletion**; it eliminates about 90 percent of all the B cells made in the bone marrow.

If a T cell recognizes an antigenic complex on a body cell and yet does not form cytokines that result in clonal expansion, the cell is said to show **clonal anergy**. Anergy results apparently because the antigenic complex is not the only thing that signals the T cell response; a second, *costimulatory* signal on the surface of the body cell that presents the antigen also signals the T cell. This second signal, a protein called CD28, usually occurs on the cell that presents the antigen to the T cell. But when it is not present or is blocked in some way, anergy results and the clone does not develop.

The phenomenon of **immunological tolerance** was discovered through the observation that some *nonidentical* twin cattle with different blood types contained some of each other's red blood cells. Why did "foreign" blood cells not cause immune responses resulting in their elimination? The hypothesis suggested was that the blood cells had passed between the fetal animals in the womb before the differentiation of immune specificities was complete. Thus each calf regarded the other's red blood cells as self.

This hypothesis was confirmed when it was shown that injecting foreign antigen into an animal early in fetal development caused that animal henceforth to recognize that antigen as self. Two strains of mice were used, each so highly inbred that they were almost a clone. Cells from adult mice of one strain were injected into newborn mice of another strain. Other newborn mice of the second strain served as uninjected controls. Eight to ten weeks later, tolerance was examined in the treated and untreated mice of the second strain by grafting skin from the first strain or from a third strain onto them. The untreated mice rejected grafts from the first strain, but the treated mice accepted them (Figure 18.10). Grafts from the third strain were rejected. Thus it was concluded that immunological tolerance to an antigen can be induced by exposure to the antigen early in development.

Tolerance must be established repeatedly throughout the life of the animal because lymphocytes are produced constantly. Continued exposure to self antigen helps maintain tolerance. For unknown reasons, tolerance to self antigens may be lost. When this happens, the body produces antibodies or T cells against its own proteins, resulting in an *autoimmune disease* (for example, rheumatoid arthritis, psoriasis, or myasthenia gravis).

B Cells: The Humoral Immune Response

Every day, billions of B cells survive the test of clonal deletion and are released from the bone marrow to enter the circulation. The B cells are the basis for the humoral immune response. Since each B cell expresses on its surface an antibody that is specific for a foreign antigen, that antigen can bind to the B cell, causing it to expand to form a clone. In this section on B cells and humoral immunity, we will describe how plasma cells are generated from B cells, how the structure of antibody proteins relates to their functions, and how diverse antibody classes are.

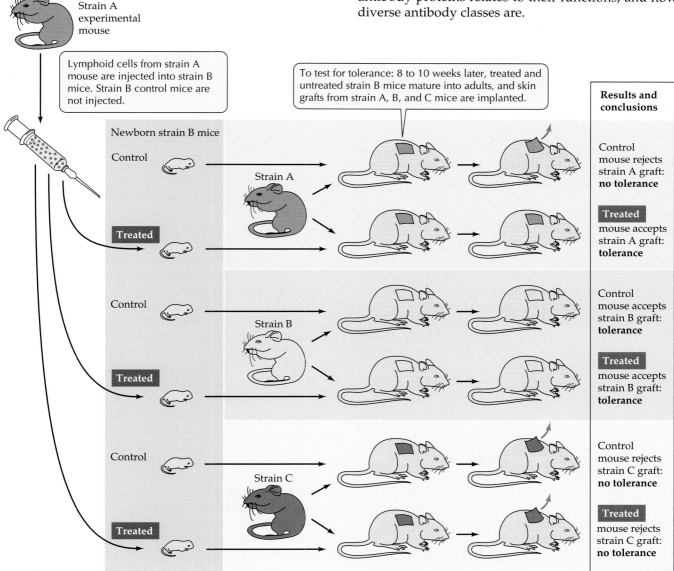

18.10 Making Nonself Seem Like Self   The ability of adult mice to recognize and reject grafts of foreign skin can be overcome by earlier exposure to nonself.

Conclusion: What is recognized as self and nonself depends partly on when it is first encountered.

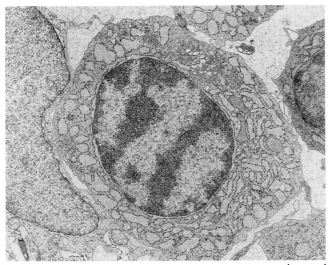

1.2 μm

18.11 A Plasma Cell The prominent nucleus (recognizable by the double membrane), the cytoplasm crowded with rough endoplasmic reticulum, and an extensive Golgi complex are all structural features of a cell actively synthesizing and exporting proteins.

Some B cells develop into plasma cells

The activation of a B cell begins with the binding of a particular antigen to the antibodies carried on the surface of the B cell. For plasma cells to develop and antibodies to be produced, a helper T cell (T_H) must also bind to the antigen. The cellular division and differentiation of the B cells is stimulated by the receipt of chemical signals from the antigen-responsive T cell. These events lead to the formation of **plasma cells** (the effector cells) and memory cells (see Figure 18.8).

As plasma cells develop, the number of ribosomes and the amount of endoplasmic reticulum in their cytoplasm increase greatly (Figure 18.11). These increases prepare the cells for synthesizing large amounts of antibodies for secretion. All the plasma cells arising from a given B cell produce antibodies of specificity identical to that of the surface receptors that originally bound antigen to the parent B cell. Thus specificity is maintained as cells proliferate.

Antibodies share a common structure but may be of different classes

Antibodies are proteins called **immunoglobulins**. There are several types of immunoglobulins, but all contain a tetramer consisting of four polypeptides. Two of these polypeptides are identical "light" chains, and two are identical "heavy" chains. Disulfide bonds (—S—S—) hold the chains together. Each chain consists of a constant region and a variable region (Figure 18.12). The **constant regions** of both light and heavy chains are similar in amino acid sequence from one immunoglobulin to another. The **variable regions** of

the heavy and light chains contribute directly to the binding site and are responsible for the diversity of antibody specificity.

The amino acid sequence of the variable region is unique in each of the millions of immune-specific immunoglobulins. Thus the variable regions of a light and a heavy chain together form a highly specific, three-dimensional structure. This part of a particular immunoglobulin molecule is what binds with a particular, unique antigenic determinant. The enormous range of antibody specificities is accomplished by a combination of genetic rearrangements and mutations, as we will see later in the chapter.

Although the variable regions are responsible for the *specificity* of an immunoglobulin, the constant regions are equally important, for the constant regions are what determine whether the antibody remains part of the cell's plasma membrane or is secreted into the bloodstream. The constant regions also determine the type of action to be taken in eliminating the antigen, as we will see in the next section. The two halves of an antibody, each consisting of one light and one heavy chain, are identical, so each of the two arms can combine with an identical antigen, leading sometimes to the formation of a large complex of antigen and antibody molecules (Figure 18.13).

IMMUNOGLOBULIN CLASSES AND FUNCTIONS. The five immunoglobulin classes are based on differences in the constant region of the heavy chain. One, called immunoglobulin M, or IgM, is always the first antibody product of a plasma cell. Most cells later switch to the production of other classes of immunoglobulins—all with equivalent specificity. The four other classes—IgA, IgD, IgE, and IgG—play different roles in the immune system (Table 18.3).

IgG molecules, which have the γ (gamma) heavy-chain constant region, make up about 85 percent of the total immunoglobulin content of the bloodstream. They consist of a single immunoglobulin unit (two identical heavy chains and two identical light chains) and are produced in greatest quantity during a second immune response (see Figure 18.9). IgG defends the body in several ways. For example, after some IgG molecules bind to antigens, they become attached by their heavy chains to macrophages. This attachment permits the macrophages to phagocytose the antigens (Figure 18.14). Another major function of IgG is to activate the complement system and enhance phagocytosis.

Most of the antibodies produced at the beginning of a primary immune response are IgM molecules. They differ from IgG in that they are composed of five immunoglobulin units (see Table 18.3). Because they have more binding sites, IgM molecules are more active than IgG molecules in activating the complement system and promoting the phagocytosis of antibody-coated cells.

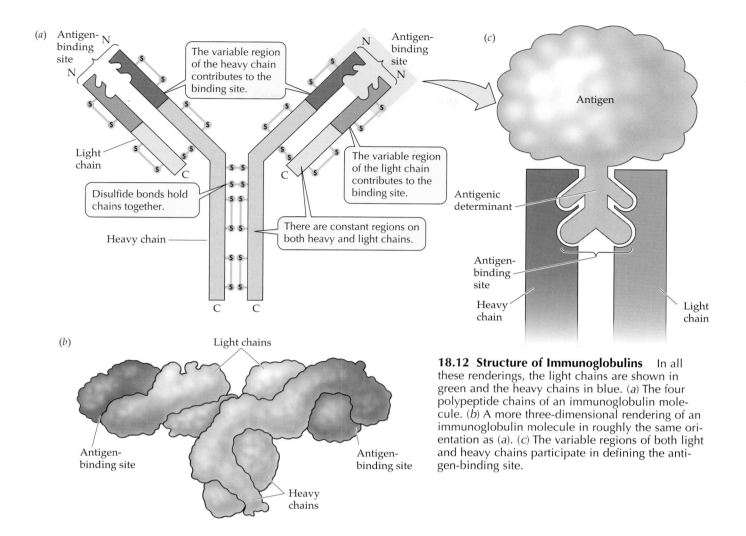

(a) Antigen-binding site

The variable region of the heavy chain contributes to the binding site.

Antigen-binding site

(c)

Antigen

Light chain

The variable region of the light chain contributes to the binding site.

Antigenic determinant

Disulfide bonds hold chains together.

There are constant regions on both heavy and light chains.

Heavy chain

Antigen-binding site

Heavy chain

Light chain

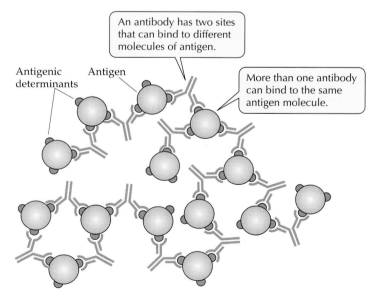

(b)

Light chains

Antigen-binding site

Antigen-binding site

Heavy chains

18.12 Structure of Immunoglobulins In all these renderings, the light chains are shown in green and the heavy chains in blue. (*a*) The four polypeptide chains of an immunoglobulin molecule. (*b*) A more three-dimensional rendering of an immunoglobulin molecule in roughly the same orientation as (*a*). (*c*) The variable regions of both light and heavy chains participate in defining the antigen-binding site.

Only small amounts of IgD antibody travel free in the bloodstream. The major role of IgD is to serve as membrane receptors on B cells.

IgE antibodies take part in inflammation and allergic reactions. IgE helps kill parasites such as the worms that cause the disease schistosomiasis, which affects some 200 million people in Asia, Africa, and South America (see Chapter 21). At inflammation sites, IgE may participate in bringing white blood cells, components of the complement system, and other factors into the inflamed region. For most people, the effect of IgE is most apparent in allergic reactions.

IgE molecules bind to antigenic determinants on the substances (allergens) that provoke the allergy, and they bind to receptor sites on the surfaces of mast cells. The mast cell–IgE–allergen complex stimulates the release of histamine and other compounds, leading in turn to inflammation, as we have already seen. Hives, hay fever, eczema, and asthma are all common allergic reactions.

Body secretions such as saliva, tears, milk, and gastric fluids all contain IgA. Soluble IgA molecules from

An antibody has two sites that can bind to different molecules of antigen.

Antigenic determinants

Antigen

More than one antibody can bind to the same antigen molecule.

18.13 Antibody–Antigen Complex When more than one antibody binds to the same antigen molecule, large antibody–antigen complexes facilitate neutralization or phagocytosis of the antigen.

TABLE 18.3 Antibody Classes

CLASS	GENERAL STRUCTURE	LOCATION	FUNCTION
IgG	Monomer	Free in plasma; about 80% of circulatory antibodies	Most abundant antibody in primary and secondary responses; crosses placenta and provides passive immunity to fetus
IgM	Pentamer	Surface of B cell; free in plasma	Antigen receptor on B cell membrane; first class of antibodies released by plasma cells during primary response
IgD	Monomer	Surface of B cell	Cell surface receptor of mature B cell; important in B cell activation
IgA	Monomer or polymer	Monomer found in plasma; polymer in saliva, tears, milk, and other body secretions	Protects mucosal surfaces; prevents attachment of pathogens to epithelial cells
IgE	Monomer	Secreted by plasma cells in skin and tissues lining gastrointestinal and respiratory tracts	Found on mast cells and basophils; when bound to antigens, triggers release of histamine from mast cell or basophil that contributes to inflammation and some allergic responses

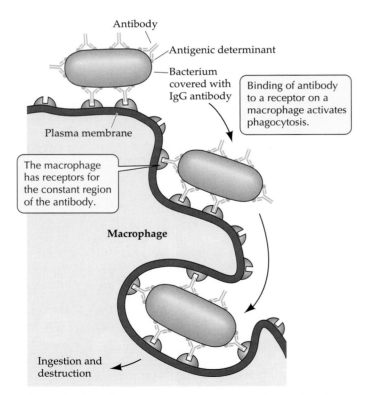

18.14 IgG Antibodies Promote Phagocytosis IgG antibodies cover a bacterium; receptors on the macrophage recognize the IgG molecules and phagocytose the cell they have coated.

the blood are taken up by epithelial cells, transported across the cells, and then secreted into the mucus that lines digestive, excretory, reproductive, and respiratory organs.

Because it can be a polymer, IgA is very effective at binding to complex antigens, such as bacteria and viruses. This binding prevents pathogens from attaching to the epithelial cells and then entering the tissues. The complexes of IgA and the pathogens are easily eliminated via the mucus. This is an important line of defense against bacteria such as *Salmonella*, which causes food poisoning, and *Neisseria gonorrhoeae*, which causes the sexually transmitted disease gonorrhea, as well as against the viruses that cause influenza and polio.

Hybridomas produce monoclonal antibodies

Because most antigens carry many different antigenic determinants, animals produce a complex mixture of antibodies when injected with a single antigen. It is very difficult to separate the antibodies with different specificities for chemical study. However, a single lymphocyte produces only a single species of antibody. Therefore, in principle, one might expect to be able to cause a single isolated lymphocyte to multiply in pure culture and give rise to a clone of cells, all dedicated to the production of the same antibody.

Unfortunately, cells that produce a single antibody cannot be cultured. On the other hand, cancerous tumors of plasma cells, called **myelomas**, do grow rapidly in culture. Each tumor arises from a single plasma cell. When the antibodies of different tumors are analyzed, each has a unique amino acid sequence. This discovery offered important evidence showing that each B cell clone produces a different antibody molecule.

Some myeloma cells in the laboratory have lost the ability to produce antibodies: These cells live long, but they do not secrete the protein. We now use both cell types—these myeloma cells and normal lymphocytes—to produce hybrid cells called **hybridomas**, which make specific normal antibodies in quantity and which, like the myeloma cells, proliferate rapidly and indefinitely in culture.

To make a clone of hybrid cells, an animal such as a mouse is first inoculated with an antigen to trigger specific lymphocyte proliferation. Later, the spleen is dissected out and B cells are collected from it (Figure 18.15). These B cells are mixed with myeloma cells from a single tumor in the presence of an agent that induces plasma membranes to fuse with one another when cells collide. Thus, some B cells fuse with myeloma cells, combining their contents and giving rise to hybridomas. The cell mixture is then cultured in a manner that destroys all nonhybrid cells.

The hybridomas are grown in a suitable medium so that each activated B cell forms a clone. Individual clones are tested, and the ones that produce the desired antibodies—specific for one antigenic determinant—are selected. These clones produce **monoclonal antibodies** (antibodies for a single antigenic determinant derived from a single clone of cells) in large quantities, either from a mass culture or after being transferred into an animal, where they can grow as a tumor. The hybridomas may be preserved and stored by freezing.

Monoclonal antibodies are ideal for the study of specific antibody chemistry, and they have been used to further our knowledge of cell membranes, as well as for specialized laboratory procedures such as tissue typing for grafts and transplants. Monoclonal antibodies have many practical applications. For example, they have been invaluable in the development of *immunoassays*, which use the great specificity of the antibodies to detect tiny amounts of molecules in tissues and fluids. Most human pregnancy tests use a monoclonal antibody to a hormone made by the developing embryo.

Radioactively tagged monoclonal antibodies are used to target antigens on the surface of cancer cells, enabling precise imaging of the tumor so that the physician can monitor the progress of therapy. The cancer cell–targeted antibody, when attached to a poison, can be used to kill the tumor cells. Monoclonals are also used for *passive immunization*—inoculation with specific antibody rather than with an antigen that causes the patient to develop his or her own antibody

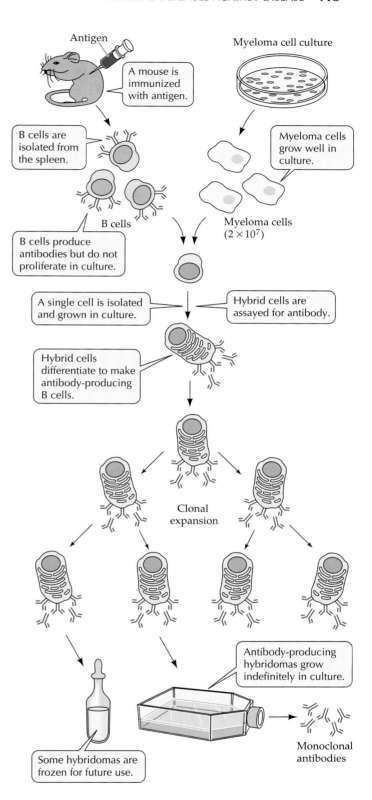

18.15 Creating Hybridomas for the Production of Monoclonal Antibodies Cancerous myeloma cells and normal lymphocytes are hybridized so that the proliferative properties of the myeloma cells can be merged with the antibody-producing lymphocytes. Clones of cells that produce desired monoclonal antibodies are selected.

(as most vaccines are designed to do). Passive immunization is the approach used to treat the early symptoms of rabies infection or a rattlesnake bite, cases in which the toxic nature of the infection is so serious that there is not enough time to allow the person's immune system to mount its own defense.

T Cells: The Cellular Immune Response

Thus far we have been concerned primarily with the humoral immune response, whose effector molecules are the antibodies secreted by plasma cells that develop from activated B cells. T cells are the effectors of the cellular immune response, which is directed against any factor, such as a virus or mutation, that changes a normal cell into an abnormal cell.

In this section, we will describe two types of T cells (helper T cells and cytotoxic T cells) and we will see that the binding of a T cell receptor to an antigenic determinant requires special proteins encoded by the HLA (human leukocyte antigen) genes.* These proteins underlie the immune system's tolerance for cells of its own body and are responsible for the rejection of foreign tissues by the body.

T cells of various types play several roles in immunity

Like B cells, T cells possess specific surface receptors. **T cell receptors** are not immunoglobulins, but glycoproteins with molecular weights about half that of an IgG. They are made up of two polypeptide chains, each encoded by a separate gene (Figure 18.16).

The genes that code for T cell receptors are similar to those for immunoglobulins, suggesting that both are derived from a single, evolutionarily more ancient group of genes. Like the immunoglobulins, T cell receptors include both variable and constant regions. Once formed, the receptors are bound to the plasma membrane of the T cell that produces them. In the sections that follow we discuss how T cell receptors bind antigens.

When T cells are activated by contact with a specific antigenic determinant, they develop and give rise to two types of effector cells. **Cytotoxic T cells**, or T_C **cells**, recognize virus-infected cells and kill them by causing them to lyse (Figure 18.17). **Helper T cells**, or T_H **cells**, assist both the cellular and humoral immune systems.

As mentioned already, a T_H cell of appropriate specificity must bind an antigen before a B cell can be activated by binding the antigen. The helper cell becomes the "conductor" of the "immunological orches-

*In mice, the analogous proteins are called the MHC (major histocompatibility complex).

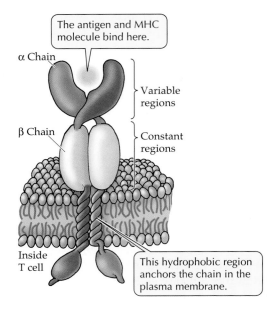

18.16 A T Cell Surface Receptor T cell receptors are glycoproteins, not immunoglobulins, although their structure is similar. The receptors are bound to the plasma membrane of the T cell that produces them.

tra," as it sends out chemical signals that not only result in its own clonal expansion but set in motion the actions of cytotoxic T cells as well as B cells.

Now that we are familiar with the major types of T cells, we can address the question of how T cells meet the antigenic determinants if the antigens themselves are inside host cells.

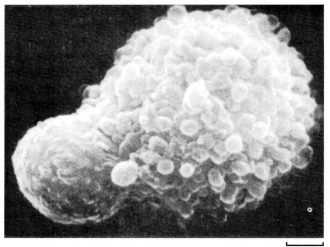

1.7 μm

18.17 A Cytotoxic T Cell in Action A cytotoxic T cell (smaller sphere) has come into contact with a virus-infected cell, causing the infected cell to lyse. The blisters on the infected cell's surface indicate that it is beginning to break up.

The major histocompatibility complex encodes many proteins

We have seen that a body's immune defenses recognize its own cells as self—proteins on our own cell surfaces are tolerated by our immune systems. There are several types of mammalian cell surface proteins, but we will focus on one very important group, the products of a cluster of genes in mice called the **major histocompatibility complex**, or **MHC**. The MHC gene products are plasma membrane glycoproteins, proteins with attached carbohydrate groups. The human counterparts of MHC are the human leukocyte antigens (HLA)

There are three classes of MHC proteins. **Class I MHC molecules** are present on the surface of every nucleated cell in the animal. These proteins function in antiviral T cell immunity. **Class II MHC molecules** are found only on the surfaces of B cells, T cells, and macrophages. This class of MHC proteins is responsible primarily for the interaction of T_H cells, macrophages, and B cells during antibody responses. Classes I and II bind with antigens or antigenic fragments bearing antigenic determinants inside the cell, are exported with antigen to the cell surface, and are then presented along with antigen to T cells. **Class III MHC molecules** include some of the proteins of the complement system that interact with antigen–antibody complexes and result in the lysis of foreign cells (see Figure 18.4).

The HLA complex in humans is highly polymorphic. It consists of more than 50 genes, and most of the genes have many alleles—more than 70 in some cases. Because of the number of HLA genes and the number of their alleles, different individuals are very likely to have different HLA genotypes and phenotypes—and that difference is what leads to the rejection of organ transplant between humans.

Similarities in base sequences between MHC and HLA genes, and the genes coding for antibodies and T cell receptors suggest that all three may have descended from the same ancestral genes and are part of a "superfamily." Major aspects of the immune systems seem to be woven together by a common evolutionary thread.

T cells and MHC molecules contribute to the humoral immune response

When a macrophage phagocytoses an antigen, it breaks the antigen into fragments, each with one or more antigenic determinants. This process is called **antigen processing**. Within the macrophage, class II MHC molecules bind processed antigen in the endoplasmic reticulum. The resulting complex moves by way of the Golgi and its vesicles to the plasma membrane of the cell, where it is displayed and is available for interaction with T_H cells (Figure 18.18). Because the MHC molecules *present* antigen, the macrophages are thus referred to as **antigen-presenting cells**.

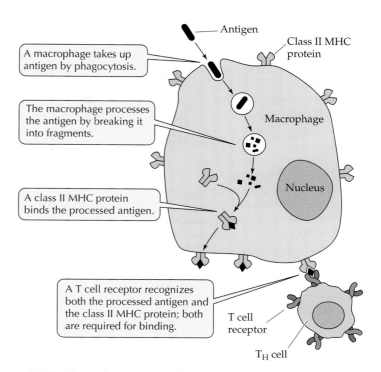

18.18 Macrophages Are Antigen-Presenting Cells
Processed antigen is displayed by MHC protein on the surface of a macrophage. Receptors on the helper T cell can then interact with the processed antigen/MHC protein complex.

A T_H cell can bind to processed antigen only if two criteria are met: (1) The T cell receptors correspond to the displayed antigenic determinant, and (2) the processed antigen is carried by an MHC molecule. The T cell receptor binds both the antigenic determinant and the MHC molecule.

Among the many T cell surface proteins, two—the CD4 and CD8 proteins—participate in the binding of some T cells and processed antigen. CD4, present on all T_H cells, binds to class II MHC molecules; CD8, present on all T_C cells, binds to class I MHC molecules. In each case, the effect is to enhance the binding of T cells to antigen-presenting cells.

When a macrophage phagocytoses a foreign particle, it becomes activated and produces cytokines. Similarly, when a T cell binds to the macrophage, the T cell releases cytokines, such as interleukin-1. These cytokines activate the T_H cell to produce a clone of differentiated cells capable of interacting with B cells. The steps to this point constitute the **activation phase** of the response, and they occur in lymphatic tissue. Next comes the **effector phase**, in which B cells are activated (Figure 18.19*a*).

B cells are also antigen-presenting cells. B cells take up by endocytosis antigen bound to their receptors, process it, and display it on class II MHC molecules. An activated T_H cell binds only if it recognizes both the dis-

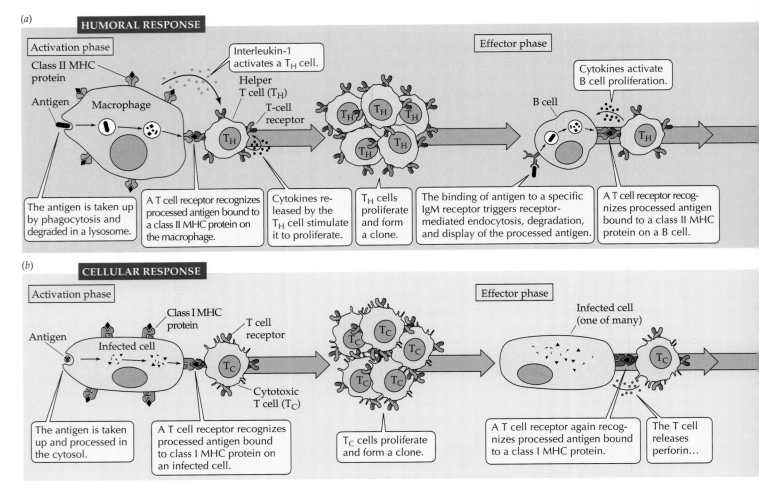

(a)

HUMORAL RESPONSE

Activation phase

Effector phase

Class II MHC protein

Interleukin-1 activates a T$_H$ cell.

Antigen

Macrophage

Helper T cell (T$_H$)

T-cell receptor

Cytokines activate B cell proliferation.

B cell

The antigen is taken up by phagocytosis and degraded in a lysosome.

A T cell receptor recognizes processed antigen bound to a class II MHC protein on the macrophage.

Cytokines released by the T$_H$ cell stimulate it to proliferate.

T$_H$ cells proliferate and form a clone.

The binding of antigen to a specific IgM receptor triggers receptor-mediated endocytosis, degradation, and display of the processed antigen.

A T cell receptor recognizes processed antigen bound to a class II MHC protein on a B cell.

(b)

CELLULAR RESPONSE

Activation phase

Effector phase

Class I MHC protein

T cell receptor

Infected cell (one of many)

Antigen

Infected cell

Cytotoxic T cell (T$_C$)

The antigen is taken up and processed in the cytosol.

A T cell receptor recognizes processed antigen bound to class I MHC protein on an infected cell.

T$_C$ cells proliferate and form a clone.

A T cell receptor again recognizes processed antigen bound to a class I MHC protein.

The T cell releases perforin…

18.19 Phases of the Humoral and Cellular Immune Responses Both types of immune response have an activation phase and an effector phase.

played antigenic determinant and the MHC molecule. Once again the bound T$_H$ cell releases interleukins that cause the B cell to produce a clone of plasma cells. Finally, the plasma cells secrete antibody, completing the effector phase of the humoral immune response.

T cells and MHC molecules contribute to the cellular immune response

Class I MHC molecules play a role in the cellular immune response that is similar to the role played by class II MHC molecules in the humoral immune response. In this case the virus-infected or mutated cells themselves are the antigen-presenting cells. Cellular DNA that has been altered by viral infection or mutation leads to the production of "foreign" proteins or peptide fragments, which combine with MHC class I molecules. As in the macrophage, the complex is displayed on the cell surface and presented to T$_C$ cells. When a T$_C$ cell binds the complex of processed antigen and class I MHC molecule, the cell becomes activated to proliferate and differentiate (Figure 18.19b).

In the effector phase of the cellular immune response, T cell receptors on the activated T$_C$ cells recognize processed viral antigen displayed by class I MHC molecules on the surface of other virus-infected cells. The T$_C$ cells then produce molecules that insert themselves into plasma membranes and result in cell lysis (see Figure 18.17). Because T cell receptors recognize self MHC molecules complexed with *nonself* antigens, they help rid the body of its own virus-infected cells. Because they also recognize MHC molecules complexed with *altered self* antigens (as a result of mutations), they help eliminate tumor cells, since most tumor cells have mutations (see Chapter 17).

Clearly, T$_H$ cells have a central role in both humoral and cellular immunity.

MHC molecules underlie the tolerance of self

MHC molecules play a key role in establishing tolerance to self, without which an animal would be destroyed by its own immune system. Throughout the animal's life, developing T cells are "tested" in the thy-

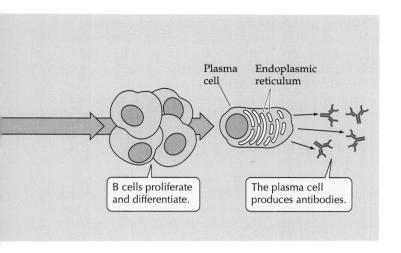

B cells proliferate and differentiate.

Plasma cell

Endoplasmic reticulum

The plasma cell produces antibodies.

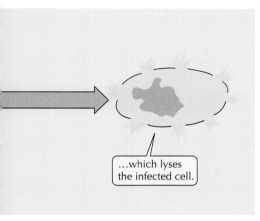

...which lyses the infected cell.

mus. One "question" the thymus "asks" is, Can this cell recognize the body's MHC proteins? A T cell unable to recognize self MHC proteins would be useless to the animal because it could not participate in any immune reactions. Such a cell fails the test and dies within about 3 days.

The second and more crucial question is, Does this cell bind to self MHC protein *and* to one of the body's own antigens? A T cell that satisfied both of these criteria would be harmful or lethal to the animal; it also fails the test and undergoes apoptosis (see Chapter 9).

T cells that survive this pair of tests mature into either T_C cells or T_H cells.

MHC molecules are responsible for transplant rejection

In humans, a major side effect of the HLA molecules (our version of the mouse's MHC) became important with the development of organ transplant surgery, sometimes with devastating results. Because the proteins produced by the HLA are specific to each individual, they act as antigens if transplanted into another individual. An organ or a piece of skin trans-

planted from one person to another is recognized as nonself and soon provokes an immune response; the tissue then is killed, or "rejected," by the cellular immune system. But if the transplant is performed immediately after birth or if it comes from a genetically identical person (an identical twin), the material is recognized as self and is not rejected.

Physicians can temporarily overcome the rejection problem by treating the patient with drugs, such as cyclosporin, that suppress the immune system. However, this approach compromises the ability of patients to defend themselves against bacteria and viruses. Cyclosporin and some other immunosuppressants interfere with communication between cells of the immune system. Specifically, they inhibit the production of interleukins.

The Genetic Basis of Antibody Diversity

A newborn mammal possesses a full set of genetic information for immunoglobulin synthesis. At each of the loci coding for the heavy and light chains, it has an allele from the mother and one from the father. Throughout the animal's life, each of its cells begins with the same full set. However, the genomes of B cells become modified during development in such a way that each cell eventually can produce one—and only one—specific type of antibody. Different B cells develop different antibody specificities. How can a single organism produce millions of different specific immunoglobulins with antibody specificities?

Research in recent years has answered this question. One suggestion was that we simply have millions of antibody genes. However, a simple calculation (the number of base pairs needed per antibody gene multiplied by millions) shows that our entire genome would thus be taken up by antibody genes. More than 30 years ago, an alternative hypothesis was proposed: that a relatively small number of genes recombine at the DNA level to produce many unique combinations, and it is this shuffling of the genetic deck that produces antibody diversity. Research since has proved this hypothesis.

In this section, we will describe the unusual DNA events that generate the enormous antibody diversity that normally characterizes each individual mammal. Then we will see how similar DNA events produce the five classes of antibodies by producing slightly different "constant regions" that have special properties.

Antibody diversity results from unusual genetic processes

In an unusual genetic process, functional immunoglobulin genes assemble from DNA segments that initially are spatially separate. Every cell in the body has hundreds of DNA segments potentially capable of participating in the synthesis of the variable regions, the parts of the antibody molecule that confer immunological specificity. However, during B cell development, these DNA segments are *rearranged*.

Pieces of the DNA are deleted, and DNA segments formerly distant from one another are joined together. In this fashion, a gene is assembled from randomly selected pieces of DNA. Each B cell precursor in the animal assembles its own unique set of immunoglobulin genes. This remarkable process generates many diverse antibodies from the same starting genome. The same type of process also accounts for the diversity of T cell receptors.

In both humans and mice, the DNA segments coding for immunoglobulin heavy chains are on one chromosome and those for light chains are on others. The variable region of the light chain is encoded by two families of DNA segments, and the variable region of the heavy chain is encoded by three families. (We discussed gene families in Chapter 14; see Figure 14.10.)

It is random which of the two variable light-chain families and which of the three variable heavy-chain families are used to make the final antibody. So the diversity afforded by several hundred DNA segments is multiplied by the combination of different families. Furthermore, since light and heavy chains are synthesized independently of one another, the combination of light and heavy chains introduces more diversity.

For example, in the human genome there are 100 different variable light-chain genes in the kappa family and another 100 genes in the lambda family. The heavy-chain family also has 100 members in its variable region. So the total number of possible variable regions is: (100 heavy × 100 light kappas) + (100 heavy × 100 light lambdas) = 20,000 different variable regions in immunoglobulins just by the random association of V (for *variable*) gene sequences.

But there are not just constant and variable regions in antibodies. Light chains have a third DNA region, which codes for amino acids between the variable and constant regions. This is the J (for *joining*) region, and in humans the kappa light-chain family has five members and the lambda family has six members. Thus the number of possible different kappa light chains is 5 × 100 = 500, and the number of lambda possibilities is 6 × 100 = 600. Similarly, there are six J family members in the genes for the variable region for the heavy chain, as well as 30 D (for *diversity*) genes. So the possible heavy chains are 6 × 30 × 100 = 18,000.

Now the possible antibodies that can be made through a random selection of one variable region from the heavy and light families is

18,000 heavy × 500 light kappas = 9 million

and

18,000 heavy × 600 light lambdas = 10.8 million

or a total of almost 20 million.

But there are more diversity mechanisms. When the DNA sequences for the V, J, and C (for *c*onstant) regions are rearranged so that they are next to one another, errors occur at the junctions such that the recombination event is not precise. This imprecise recombination can create new codons at the junction, with resulting amino acid changes. After the DNA fragments are cut out and before they are joined, an enzyme, **terminal transferase**, often adds some nucleotides to the free ends of the DNAs. These additional bases create insertion mutations.

Finally, there is a relatively high rate of mutation in immunoglobulin genes. Once again, this process creates many new alleles and antibody diversity. Adding these possibilities to the 20 million that can be made by random DNA rearrangements makes it not surprising that the immune system can mount a response to almost any natural or human-made substance.

How does a B cell produce a certain heavy chain?

To see how DNA rearrangement generates antibody diversity, let's consider how the heavy chain of IgM is produced. After B cells produce this antibody, it is inserted into the plasma membrane.

The gene families governing all heavy-chain synthesis are on chromosome 12 of mice and on chromosome 14 of humans. In mice, the gene families are arranged as shown in Figure 18.20, with a long stretch of DNA occupied by a family of 100 or more V segments. Humans have about 300 V segments. In a given B cell, only *one* of these segments is used to produce part of the variable region of the heavy chain; the remaining V segments are discarded or rendered inactive.

At some distance from the V segments is a family of ten or more D segments. Again, only one of these is used to produce part of the variable region of the heavy chain of a given B cell, as is only one J segment from the family of four such segments (in mice) lying yet farther along the chromosome. This combination of one each of V, D, and J segments forms a complete variable region for a functional gene.

Still farther along the chromosome, and separated from the suite of J segments, is a family of eight segments, one of which codes for the constant region of the heavy chain. Light chains are produced from similar families of DNA segments, but they lack D segments.

How does order emerge from this seeming chaos of DNA segments? Two important steps impose order: DNA rearrangements and RNA splicing. First, substantial chunks of DNA are deleted from the chromosome during rearrangement of the segments. As a result of these deletions, a particular D segment is rearranged to be directly beside a particular J segment, and then the DJ segment is rearranged to be adjacent to one of the V segments; thus a single "new" sequence, consisting of one V, one D, and one J segment, can now code for the variable region of the heavy chain. All the progeny of

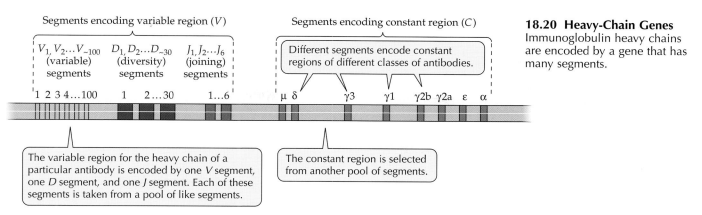

Segments encoding variable region (V)

Segments encoding constant region (C)

$V_1, V_2...V_{~100}$ (variable) segments $D_1, D_2...D_{~30}$ (diversity) segments $J_1, J_2...J_6$ (joining) segments

Different segments encode constant regions of different classes of antibodies.

1 2 3 4...100   1 2...30   1...6   μ δ   γ3   γ1   γ2b γ2a ε   α

The variable region for the heavy chain of a particular antibody is encoded by one V segment, one D segment, and one J segment. Each of these segments is taken from a pool of like segments.

The constant region is selected from another pool of segments.

18.20 Heavy-Chain Genes
Immunoglobulin heavy chains are encoded by a gene that has many segments.

this cell constitute a clone having the same sequence for the variable region (Figure 18.21*a*).

The second step in organizing the synthesis of an immunoglobulin chain follows transcription. Splicing of the RNA transcript (see Chapter 14) removes introns and any *J* segments lying between the selected *J* segment and the first constant-region segment (Figure 18.21*b*). The result is an mRNA that can be translated, directly yielding the heavy chain of the cell's specific antibody.

In summary, two distinct types of nucleic acid rearrangements contribute to the formation of an antibody. DNA rearrangements, before transcription, join the *V*, *D*, and *J* segments. RNA splicing, after transcription, joins the variable region (*VDJ*) to the constant region.

The constant region is involved in class switching

Early in its life, a B cell produces IgM molecules that are responsible for the specific recognition of a particular antigenic determinant. At this time, the constant region of the antibody's heavy chain is encoded by the

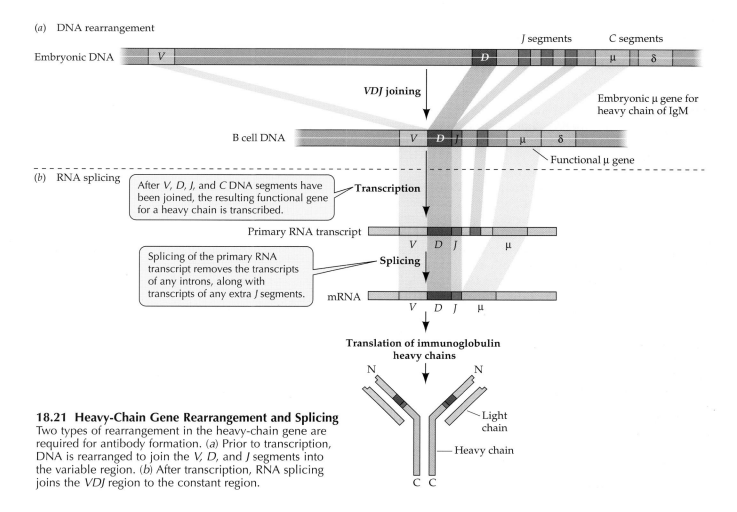

(*a*) DNA rearrangement

Embryonic DNA   V   D   *J* segments   *C* segments   μ   δ

VDJ joining

Embryonic μ gene for heavy chain of IgM

B cell DNA   V D *J*   μ   δ

Functional μ gene

(*b*) RNA splicing

After V, D, J, and C DNA segments have been joined, the resulting functional gene for a heavy chain is transcribed.

Transcription

Primary RNA transcript   V D J   μ

Splicing of the primary RNA transcript removes the transcripts of any introns, along with transcripts of any extra *J* segments.

Splicing

mRNA   V D J μ

Translation of immunoglobulin heavy chains

N   N

Light chain

Heavy chain

C C

18.21 Heavy-Chain Gene Rearrangement and Splicing
Two types of rearrangement in the heavy-chain gene are required for antibody formation. (*a*) Prior to transcription, DNA is rearranged to join the *V*, *D*, and *J* segments into the variable region. (*b*) After transcription, RNA splicing joins the *VDJ* region to the constant region.

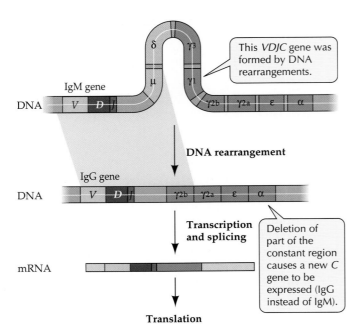

18.22 Class Switching The gene produced by joining *V*, *D*, *J*, and *C* segments (see Figure 18.21) may later be modified, causing a different *C* segment to be transcribed. This modification, known as class switching, is accomplished by deletion of part of the constant region. Shown here is class switching from an IgM gene to an IgG gene.

first constant-region segment, the μ segment (see Figure 18.20). During an immunological response—later in the life of the plasma cell—another deletion commonly occurs in the plasma cell's DNA, positioning the heavy-chain variable-region gene (consisting of the same *V*, *D*, and *J* segments) next to a constant-region segment farther down the original DNA, such as the γ, ε, or α constant region (Figure 18.22).

Such a DNA deletion, called **class switching**, results in the production of an antibody with a different *C* region and *function,* but the same *antigen specificity.* The new antibody has the same variable regions of the light and heavy chains but a different constant region of the heavy chain. This new antibody falls into one of the four other immunoglobulin classes (IgA, IgD, IgE, or IgG; see Table 18.3), depending on which of the constant-region segments is placed adjacent to the variable-region gene.

After switching classes, the plasma cell cannot go back to making the previous immunoglobulin class, because that part of the DNA has been deleted and lost. On the other hand, if additional constant-region segments are still present, the cell may switch classes again.

What triggers class switching, and what determines the class to which a given B cell will switch its antibody production? T_H cells influence these tasks, thus directing the course of an antibody response and de-

termining the nature of the attack on the antigen. These T cells induce class switching by sending cytokine signals (Figure 18.23). Interleukin-4 (IL-4) causes switching to IgG or IgE production; other cytokines cause other switches. Each cytokine produces its effect by binding to DNA at a specific point near a constant-region segment and changing the conformation of the chromosome, enabling an enzyme to bind and cause genetic rearrangement at that point.

The Evolution of Animal Defense Systems

The complex defense systems that we have been discussing are found in virtually all vertebrate animals, although there are some significant differences between most vertebrate immune systems and those of sharks, rays, and other evolutionarily ancient vertebrates (see Chapter 30). But did these defense systems first appear in the vertebrates? Not at all.

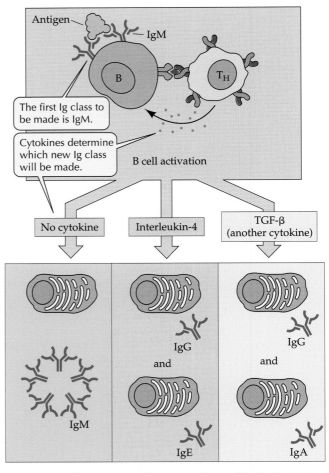

18.23 Cytokines Determine How the Antibody Switches Class A helper T cell initiates class switching in a B cell by secreting a cytokine. Different cytokines produce different switches.

The invertebrates, in all their diversity, have sturdy defense systems, and certain defense system elements are found even in unicellular protists. Many protists carry on phagocytosis, as do our own macrophages, and some protists use phagocytosis as a defense mechanism. Multicellular animals, both invertebrate and vertebrate, employ mobile phagocytic cells to patrol their bodies.

Like vertebrates, invertebrates (and probably some protists) distinguish between self and nonself. Making such distinctions enables invertebrates to reject tissue grafted from other individuals of the same species. Unlike vertebrates, however, invertebrates reject a second graft no more rapidly than a first graft—indicating that invertebrates lack immunological memory. This and other observations show that although immunological functions of invertebrates and vertebrates may be similar, their mechanisms often differ.

Invertebrates do not produce immunoglobulins, lymphocytes, or the complement system. However, they achieve similar protective goals by analogous methods, and the analogs are probably evolutionary precursors of the systems found in vertebrates. Many invertebrates make proteins very similar to interleukin-1, interleukin-2, and other vertebrate cytokines, and the proteins play regulatory roles similar to those in humans.

The evolutionary shift to new defense mechanisms probably occurred in the very earliest vertebrates, which have been extinct for 400 million years. The most ancient vertebrate groups still present today have clearly "vertebrate" immune systems. Studies of the immune systems of invertebrates and evolutionarily ancient vertebrates will help us understand our own immune systems better, and they will likely be helpful in understanding and dealing with human diseases.

Disorders of the Immune System

Immune deficiency diseases such as AIDS show us how much we depend on our immune system to protect us from pathogens. However, sometimes the immune system fails us in one way or another: It may overreact, as in an allergy; it may attack self antigens, as in an autoimmune disease; or it may function weakly or not at all, as in an immune deficiency disease. After a look at allergies and autoimmune conditions, we will examine the acquired immune deficiency that characterizes AIDS.

An inappropriately active immune system can cause problems

A common type of condition arises when the human immune system overreacts to a dose of antigen. Although the antigen itself may present no danger to the host, the immune response may produce inflammation and other symptoms that can cause serious illness or death. **Allergies** are the most familiar examples of such a problem. There are two types of allergic reaction. *Immediate hypersensitivity* occurs when an individual makes IgE that binds to a molecule or structure in a food, pollen, or venom of an insect. Mast cells and basophils bind the IgE on their surfaces, and when the antigen complexes with the IgE, these cells release amines such as histamine. The result is symptoms such as dilation of blood vessels, inflammation, and difficulty breathing. If untreated with antihistamines, a severe reaction can lead to death.

Delayed hypersensitivity does not begin until hours after exposure to the noxious antigen. In this case, the antigen is processed by antigen-presenting cells and a T cell response is initiated. This response can be so massive that the cytokines released cause macrophages to become activated and damage tissues. This is what happens when the bacteria that cause tuberculosis colonize the lung.

Sometimes prevention of the immune recognition of self fails, resulting in the appearance of one or more "forbidden clones" of B and T cells directed against self antigens. This failure does not always result in disease, but in some instances it can be disastrous. Among the **autoimmune diseases** of our species—diseases in which components of the body are attacked by its own immune system—are rheumatoid arthritis, ulcerative colitis, and myasthenia gravis (severe muscle weakness). Multiple sclerosis may result from a failure of tolerance induction in a type of T cell. The abnormal T cells mount an immune attack on the myelin sheath, an insulating material that surrounds many nerve cells (see Chapter 41). When the myelin sheath is damaged, the result can be loss of nerve function, including blindness and loss of motor control.

It now appears that most people have a few lymphocytes that react with self antigens, but this condition does not impair health. However, in autoimmune diseases these lymphocytes divide and become numerous, attacking the self cells and causing disease. In some cases, the underlying reason may be *molecular mimicry*, in which T cells that recognize a nonself antigen also recognize something on the self that has a similar structure.

For example, T cells that recognize an antigen on Coxsackie virus, a common infection, sometimes also react with part of an enzyme in the cells of the pancreas that make insulin. These T cells then attack and lead to the destruction of the pancreatic cells, leading to insulin-dependent diabetes, a disease that affects more than a million Americans.

AIDS is an immune deficiency disorder

There are various immune deficiency disorders, such as those in which T or B cells never form and others in which B cells lose the ability to give rise to plasma cells. In either case, the affected individual is unable to mount an immune response and thus lacks a major line of defense against microbial pathogens. The first human disorder to be treated in part with gene therapy was a hereditary immune deficiency caused by a mutation in the gene coding for the enzyme adenosine deaminase (see Figure 17.18).

Because of its essential roles in both antibody responses and cellular immune responses, the T_H cell is perhaps the most central of all the components of the immune system—a significant cell to lose to an immune deficiency disorder. This cell is the target of HIV (*h*uman *i*mmunodeficiency *v*irus), the virus that eventually results in AIDS (*a*cquired *i*mmune *d*eficiency *s*yndrome).

HIV is a retrovirus that infects mostly lymphoid tissues, where B cells mature, T cells reside, and the fluid that makes up lymph is filtered. Lymph is filtered on the highly folded cells called *follicular dendritic cells*. Soon after HIV infection, the virus spreads rapidly among T_H cells, since an immune response has not yet been mounted. But within days or weeks, when T cells recognize infected lymphocytes, an immune response is mounted and antibodies appear in the blood (Figure 18.24). These antibodies effectively block any further infections by circulating viruses.

By this time, several months after initial infection, the patient has a lot of circulating HIV complexed with antibodies that is gradually filtered out by the dendritic cells. But before they are filtered out, these antibody-complexed viruses can still infect T_H cells that come in contact with them. This secondary infection process is very slow and inefficient, which explains why it may take years after infection before an HIV carrier shows the clinical signs of severe immune deficiency.

The clinical signs include susceptibility to infections that cause otherwise rare cancers—lymphoma caused by Epstein–Barr virus and Kaposi sarcoma caused by a herpesvirus—as well as infections by *Pneumocystis carinii*, a fungus that causes severe pneumonia.

HIV is a retrovirus that infects immune system cells

As a retrovirus, HIV uses RNA as its genetic material and is capable of inserting a cDNA copy of its genome into the host cell's DNA. The structure of HIV is shown in Figure 13.2. A central core, with a protein coat (p24 core protein), contains two identical molecules of RNA, as well as the enzymes reverse transcriptase, integrase, and a protease. An envelope, derived from the plasma membrane of the cell in which the virus was formed, surrounds the core. The envelope is studded with envelope proteins (gp120 and gp41, where "gp" stands for *glycoprotein*) that enable the virus to infect its target cells.

HIV attaches to host cells at the membrane protein CD4, which is found primarily on T_H cells and macrophages. The initial infection may be in macrophages, and somewhat later, T_H cells are infected. CD4 acts as the receptor for the viral envelope protein gp120. The binding of gp120 to CD4 appears to require several additional cell surface molecules, and successful binding initiates a complex series of events (Figure 18.25; see also Figure 13.7). When HIV infects a cell, the viral core enters the cell and the core "uncoats" itself in the cytoplasm, releasing its contents (RNA and proteins).

Among the enzymes in the core is **reverse transcriptase**, which catalyzes the formation of a double-stranded DNA molecule encoding the same information as the viral RNA (see Chapters 13 and 14). The DNA transcript enters the nucleus of the host cell and is inserted into a chromosome, much as bacteriophage DNA may become incorporated into a bacterial chromosome as a prophage (see Chapter 12). Another HIV enzyme, called **integrase**, catalyzes this insertion. The DNA transcript of the HIV RNA thus becomes a permanent part of the T_H cell's DNA, replicating with it at each cell division.

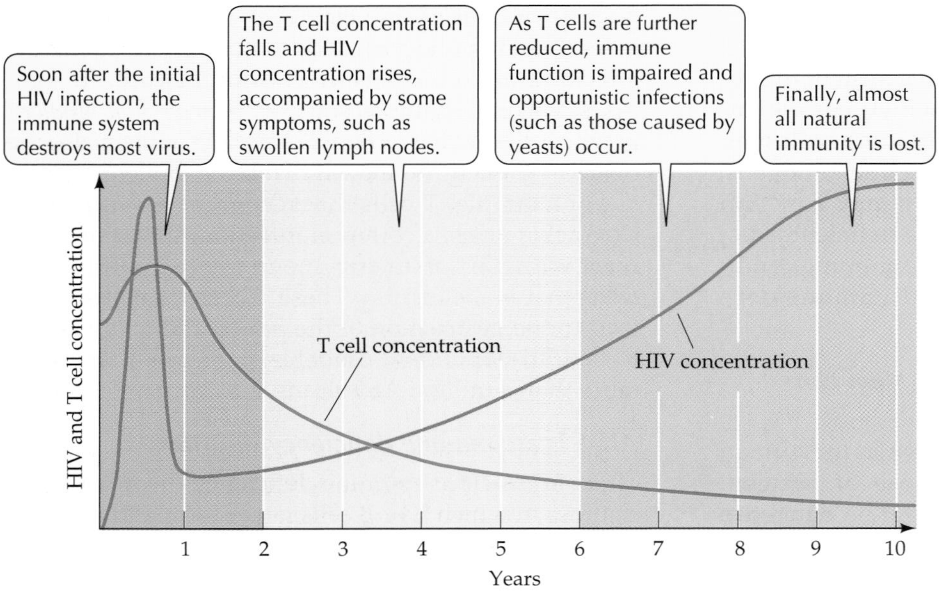

Soon after the initial HIV infection, the immune system destroys most virus.

The T cell concentration falls and HIV concentration rises, accompanied by some symptoms, such as swollen lymph nodes.

As T cells are further reduced, immune function is impaired and opportunistic infections (such as those caused by yeasts) occur.

Finally, almost all natural immunity is lost.

T cell concentration

HIV concentration

HIV and T cell concentration

Years

18.24 The Course of an HIV Infection HIV infection may be carried, unsuspected, for many years before the onset of symptoms. This long "dormant" period means the infection is often spread by people who are unaware they carry the virus.

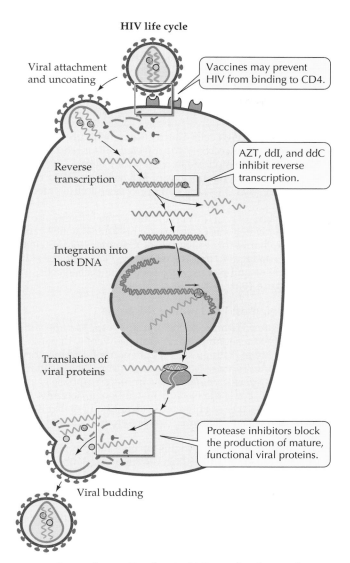

HIV life cycle

Viral attachment and uncoating

Vaccines may prevent HIV from binding to CD4.

Reverse transcription

AZT, ddI, and ddC inhibit reverse transcription.

Integration into host DNA

Translation of viral proteins

Protease inhibitors block the production of mature, functional viral proteins.

Viral budding

18.25 Strategies to Combat HIV Reproduction These widely used drugs block specific steps in the HIV life cycle.

A DNA transcript of the HIV RNA, once incorporated into the genome of a T_H cell, may remain there latent for days, months, or even a decade or more. The latent period ends if the HIV-infected T_H cell becomes activated, for example by infection or immunization. Then the viral DNA is transcribed, yielding many molecules of viral RNA. Some of these RNA transcripts are translated, forming the enzymes and structural proteins of a new generation of viruses. The **protease** encoded by HIV is needed to complete the formation of individual viral proteins from larger products of translation. Other viral RNA molecules are incorporated directly into the new viruses as their genetic material.

Formation of new viruses may be slow. The viruses bud from the infected cell, surrounding themselves with modified plasma membrane from the host.

Several viral genes control the rate of production of individual proteins and of whole viruses. HIV reverse transcriptase is a very fallible enzyme, making a mistake in one out of every 10,000 bases. Since the entire viral genome has about 10,000 bases, this means that nearly every HIV particle is genetically unique. For the host, this is good news—mutated viruses may not reproduce—and bad news—mutated viruses may escape antibodies made to specific structures.

HIV is a relatively fragile virus that needs the moist environment of body fluids in order to remain infective. HIV is transmitted with blood and certain other body fluids, including semen, vaginal fluid, and breast milk of infected persons. HIV can be transmitted by blood transfusions, by sexual activity (either homosexual or heterosexual), and from mother to fetus. HIV is not transmitted by mosquitoes or other insect vectors, or by casual contact. On a worldwide basis, the most common mode of transmission is heterosexual intercourse. However, in the United States the highest incidence of AIDS occurs among drug addicts (from the use of shared, contaminated needles) and male homosexuals.

At this writing, most of the HIV-infected people in the world live in Africa, but the virus is spreading fastest among people living in southern and Southeast Asia. By the year 2000, residents of developing countries will bear 90 percent of new HIV infections as the AIDS epidemic rages on throughout the world. Millions of people will have died before the AIDS epidemic ends. In the meantime, medical and biological research on the subject is proceeding intensely.

Are effective treatments for HIV infection on the horizon?

Prospects for a *cure* for AIDS appear dim at this time. What would an AIDS cure entail? It would include the detection and elimination of every HIV-infected cell in the body. Given the capacity of the viral genetic information to integrate into the host DNA, this task appears overwhelming. So what then?

Many investigators are seeking a vaccine to forestall HIV infection. One of the greatest problems is the exceptional genetic variation of HIV, largely because of unrepaired errors in the activity of reverse transcriptase. Dr. David Ho, a leading physician involved in AIDS treatment, estimates that in an infected person with replicating virus, more than 10 billion new viruses are made every day. It is not surprising that new strains are constantly appearing, and a promising new vaccine against one strain may afford no protection against others.

A second problem with a vaccine is the unusual life cycle of HIV. The rapid integration of the virus into T_H cells and the long life of infected cells make protective vaccines unlikely. One possibility, however, is treatment to diminish the impact of the virus on the body.

There is substantial agreement that we can learn ways to *control the replication* of HIV within the infected body and thus delay the onset of AIDS symptoms. There are many steps in the HIV life cycle. If we can block just *one* of them completely, we break the cycle and hold the infection in check. Potential therapeutic agents are being tested against the major steps shown in Figure 18.26. Of course, it is crucial to block steps that are unique to the virus, so that drug therapies do not harm the patient by blocking a step in the patient's own metabolism.

The first drugs in wide use that prolong the lives of HIV-infected people (AZT, ddI, and ddC) are all directed at the reverse transcription step. Protease inhibitors (saquinovir, indinovir, and ritinovir) can dramatically reduce HIV levels in the blood (recall that some HIV proteins are cut from larger precursor molecules by viral protease). HIV replication can be delayed best by a combination of these two approaches. Multiple-drug therapy using a protease inhibitor along with other drugs provides substantial relief, including reduction in HIV level and restored capability of the immune system—but the monetary cost of such multiple-drug therapy is high.

What can be done until biomedical science provides the tools to bring the worldwide AIDS epidemic to an end? Above all, people must recognize that they are in danger whenever they have sex with a partner whose total sexual history is not known. The danger rises as the number of sex partners rises, and the danger is much greater if partners participating in sexual intercourse are not protected by a latex condom. The danger that heterosexual intercourse will transmit HIV rises ten- to a hundredfold if either partner has another sexually transmitted disease.

Summary of "Natural Defenses against Disease"

• Our bodies defend against pathogens by both nonspecific and specific means.

Defensive Cells and Proteins

• Many of our defenses are implemented by cells and proteins carried in the bloodstream and in the lymphatic system. These cells and proteins also find their way into other parts of the immune system, including the thymus, spleen, and bone marrow. **Review Figure 18.1**
• White blood cells, including lymphocytes (B and T cells) and phagocytes such as neutrophils and macrophages, play many defensive roles. **Review Figure 18.2 and Table 18.1**
• Antibodies bind pathogens, HLA proteins help coordinate the recognition of pathogens and the activation of defensive cells, and cytokines produced by lymphocytes and macrophages alter the behavior of other cells.

Nonspecific Defenses against Pathogens

• An animal's nonspecific defenses include physical barriers, competing resident microorganisms, and hostile local environments that may be acidic or contain an antibacterial enzyme. **Review Table 18.2**
• Some nonspecific defenses are based on proteins. The complement system, composed of about 20 proteins, assembles itself in a cascade of reactions to cooperate with phagocytes or to lyse foreign cells directly. **Review Figure 18.4**
• Interferons produced by virus-infected cells inhibit the ability of viruses to replicate in neighboring cells.
• Macrophages and neutrophils phagocytose invading bacteria. Natural killer cells attack tumor cells and virus-infected body cells.
• Macrophages play an important role in the inflammation response. Activated mast cells release histamine, which causes blood capillaries to swell and leak. Complement proteins attract macrophages to the site, where they leak into the surrounding tissue and phagocytose bacteria and dead cells. **Review Figure 18.6**

Specific Defenses: The Immune Response

• Four features characterize the immune response: specificity, the ability to respond to an enormous diversity of pathogens, the ability to distinguish self from nonself, and memory.
• The immune response is directed against pathogens that evade the nonspecific defenses. The immune response is highly specific, being directed against antigenic determinants on the surfaces of antigens. Each antibody or T cell is directed against a particular antigenic determinant. **Review Figure 18.7**
• The immune response is highly diverse and can respond specifically to perhaps 10 million different antigenic determinants.
• The immune response distinguishes its own cells from foreign cells, attacking only the cells recognized as nonself.
• The immune system remembers; it can respond rapidly and effectively to a second exposure to an antigen.
• There are two interactive immune responses: the humoral immune response and the cellular immune reponse. The humoral immune response employs antibodies secreted by plasma cells to target pathogens in body fluids. The cellular immune response employs T cells to attack body cells that have been altered by viral infection or mutation or to target antigens that have invaded the body's cells.
• Clonal selection accounts for the rapidity, specificity, and diversity of the immune response. It also accounts for immunological memory, which is based on the production of both effector and memory cells as T cell and B cell clones expand. **Review Figure 18.8**
• Immunological memory plays roles in both natural immunity and artificial immunity based on vaccination. **Review Figure 18.9**
• Clonal selection also accounts for the immune system's recognition of self. Tolerance of self results from clonal deletion of antiself lymphocytes. **Review Figure 18.10**

B Cells: The Humoral Immune Response

• Activated B cells differentiate into memory cells and into plasma cells, which synthesize and secrete specific antibodies.
• Antibodies, or immunoglobulins, are of several classes. The basic immunoglobulin unit is a tetramer of four polypeptides: two identical light chains and two identical

heavy chains, each chain consisting of a constant and a variable region. **Review Figure 18.12**
- The variable regions of the light and heavy chains collaborate to form the antigen-binding sites of an antibody. The variable regions determine an antibody's specificity; the constant region determines the destination and function of the antibody. **Review Figure 18.13**
- The immunoglobulin classes are IgM, IgG, IgD, IgE, and IgA. IgM, formed first, is a membrane receptor on B cells, as is IgD. IgG is the most abundant antibody class and performs several defensive functions. IgE takes part in inflammation and allergies. IgA is present in various body secretions. **Review Figure 18.14 and Table 18.3**
- Monoclonal antibodies, produced by hybridomas, are useful in research and in medicine. A sample of a monoclonal antibody consists of identical immunoglobulin molecules directed against a single antigenic determinant. Hybridomas are produced by fusing B cells with myeloma cells (from cancerous tumors of plasma cells). **Review Figure 18.15**

T Cells: The Cellular Immune Response

- The cellular immune response is directed against altered or antigen-infected cells of the body. T_C cells attack virus-infected or tumor cells, causing them to lyse. T_H cells activate B cells (in the humoral immune response) and influence the development of other T cells and macrophages. **Review Figure 18.16**
- T cell receptors in the cellular immune response are analogous to immunoglobulins in the humoral immune response.
- The major histocompatibility complex (MHC) in mice encodes many membrane proteins. The human counterpart of the MHC is called human leukocyte antigen (HLA). In an immune response, MHC molecules in macrophages or lymphocytes bind processed antigen and present it to T cells.
- The humoral immune response requires collaboration of class II MHC molecules, the T cell surface protein CD4, and interleukin-1—all working together to activate T_H cells. The effector phase of the humoral immune response involves T cells, class II MHC molecules, B cells, and interleukins, resulting in the formation of active plasma cells. **Review Figures 18.18, 18.19**
- In the cellular immune response, class I MHC molecules, T_C cells, CD8, and interleukin-2 collaborate to activate T_C cells with the appropriate specificity. **Review Figures 18.17, 18.19**
- Developing T cells undergo two tests: They must be able to recognize self MHC molecules, but they must *not* bind to both self MHC and to any of the body's own antigens. T cells that fail either of these tests die.
- The rejection of organ transplants results from the great diversity of MHC molecules—so great that one individual's cells almost certainly differ antigenically (because of differing MHC molecules) from a given other individual's cells.

The Genetic Basis of Antibody Diversity

- Families of DNA segments underlie the incredible diversity of antibody and T cell receptor specificities.
- Antibody heavy-chain genes are constructed by unusual genetic processes from one each of numerous *V*, *D*, *J*, and *C* segments. The *V*, *D*, and *J* segments combine by DNA rearrangements, and transcription yields an RNA molecule that is spliced to form a translatable mRNA. **Review Figures 18.20, 18.21**
- Other gene families give rise to the light chains. There are millions of possible antibodies as a result of these DNA combinations.

- Imprecise DNA rearrangements, mutations, and random addition of bases to the ends of the DNAs before they are joined contribute even more diversity. **Review Figures 18.20, 18.21**
- A plasma cell produces IgM first, under the influence of cytokines released by T cells, but later it may switch to the production of other classes of antibodies. This class switching, resulting in antibodies with the same antigen specificity but a different function, is accomplished by DNA cutting and rejoining. **Review Figure 18.22, 18.23**

The Evolution of Animal Defense Systems

- Cellular defensive mechanisms such as phagocytosis are older than the animal kingdom, appearing even in some unicellular protists.
- Invertebrate animals reject nonself tissues but lack immunological memory. They possess cells and molecules analogous, but not identical, to lymphocytes, immunoglobulins, and cytokines.
- Even the most evolutionarily ancient groups among today's vertebrates have immune systems more similar to those of humans than to those of invertebrates.

Disorders of the Immune System

- Allergies result from an overreaction of the immune system to an antigen.
- Autoimmune diseases result from a failure in the immune recognition of self, with the appearance of antiself B and T cells that attack the body's own cells. Immune deficiency disorders result from failures of one or another part of the immune system.
- AIDS is an immune deficiency disorder arising from depletion of the body's T_H cells as a result of infection with the HIV retrovirus. Depletion of the T_H cells weakens and eventually destroys the immune system, leaving the host subject defenseless against "opportunistic" infections. **Review Figure 18.24**
- HIV inserts a copy of its genome into a chromosome of a macrophage or T_H cell, where it may lie dormant for years and many cell generations. When the viral genome is transcribed and translated, new viruses form slowly by budding from the infected cell—or rapidly, with lysis of the infected cell. The HIV reproductive cycle is complex. **Review Figures 13.7, 18.25**
- All steps in the reproductive cycle of HIV are under investigation as possible targets for drugs. Currently the most effective drugs are those directed against reverse transcriptase and protease.
- Some treatments may provide a dramatic reduction in HIV levels, but there is as yet no indication that we can prevent infection with HIV, as by vaccination. The only currently available strategy is for people to avoid behaviors that place them at risk.

Self-Quiz

1. Which statement about phagocytes is *not* true?
   a. Some travel in the circulatory system.
   b. They ingest microorganisms by endocytosis.
   c. A single phagocyte can ingest 5 to 25 bacteria before it dies.
   d. Although they are important, an animal can do perfectly well without them.
   e. Lysosomes play an important role in their function.

2. Which statement about immunoglobulins is true?
   a. They help antibodies do their job.
   b. They recognize and bind antigenic determinants.
   c. They encode some of the most important genes in an animal.
   d. They are the chief participants in nonspecific defense mechanisms.
   e. They are a specialized class of white blood cells.

3. Which statement about an antigenic determinant is *not* true?
   a. It is a specific chemical grouping.
   b. It may be part of many different molecules.
   c. It is the part of an antigen to which an antibody binds.
   d. It may be part of a cell.
   e. A single protein has only one on its surface.

4. T cell receptors
   a. are the primary receptors for the humoral immune system.
   b. are carbohydrates.
   c. cannot function unless the animal has previously encountered the antigen.
   d. are produced by plasma cells.
   e. are important in combating viral infections.

5. According to the clonal selection theory
   a. an antibody changes its shape according to the antigen it meets.
   b. an individual animal contains only one type of B cell.
   c. the animal contains many types of B cells, each producing one kind of antibody.
   d. each B cell produces many types of antibodies.
   e. many clones of antiself lymphocytes appear in the bloodstream.

6. Immunological tolerance
   a. depends on exposure to antigen.
   b. develops late in life and is usually life-threatening.
   c. disappears at birth.
   d. results from the activities of the complement system.
   e. results from DNA splicing.

7. The extraordinary diversity of antibodies results in part from
   a. the action of monoclonal antibodies.
   b. the splicing of protein molecules.
   c. the action of cytotoxic T cells.
   d. the rearrangement of gene segments.
   e. their remarkable nonspecificity.

8. Which of the following play(s) no role in the antibody response?
   a. Helper T cells
   b. Interleukins
   c. Macrophages
   d. Reverse transcriptase
   e. Products of class II MHC gene loci

9. The major histocompatibility complex
   a. codes for specific proteins found on the surface of cells.
   b. plays no role in T cell immunity.
   c. plays no role in antibody responses.
   d. plays no role in skin graft rejection.
   e. is encoded by a single locus with multiple alleles.

10. Which of the following plays no role in HIV reproduction?
    a. Integrase
    b. Reverse transcriptase
    c. gp120
    d. Interleukin-1
    e. Protease

Applying Concepts

1. Describe the part of an antibody molecule that interacts with an antigenic determinant. How is it similar to the active site of an enzyme? How does it differ from the active site of an enzyme?

2. Contrast immunoglobulins and T cell receptors with respect to structure and function.

3. Discuss the diversity of antibody specificities in an individual in relation to the diversity of enzymes. Does every cell in an animal contain genetic information for all the organism's enzymes? Does every cell contain genetic information for all the organism's immunoglobulins?

4. Discuss the roles of monoclonal antibodies in medicine and in biological research.

Readings

Ada, G. L. and G. Nossal. 1987. "The Clonal-Selection Theory." *Scientific American*, August. A fascinating historical account of the development of the central concept of immunology.

Beck, G. and G. S. Habicht. 1996. "Immunity and the Invertebrates." *Scientific American*, November. A description of the sophisticated immune systems of invertebrates and how they are similar to and different from the immune systems of vertebrates.

Feldman, M. and L. Eisenbach. 1988. "What Makes a Tumor Cell Metastatic?" *Scientific American*, November. A description of how MHC molecules on tumor cells enable the cells to evade the immune system.

Haynes, B. F. 1993. "Scientific and Social Issues of Human Immunodeficiency Virus Vaccine Development." *Science*, vol. 260, pages 1279–1286. A discussion of the enormous technical—as well as social and ethical—difficulties in conquering AIDS.

Janeway, C. A. and P. Travers. 1994. *Immunobiology: The Immune System in Health and Disease*. Garland, New York. An authoritative, concise text, with hundreds of clear diagrams.

Kuby, J. 1997. *Immunology*, 3rd Edition. W. H. Freeman, New York. A superbly written and illustrated text.

"Life, Death and the Immune System." 1993. *Scientific American*, September. A series of articles, each written by an expert, relating to the topics of this chapter.

Marrack, P. and J. Kappler. 1986. "The T Cell and Its Receptor." *Scientific American*, February. A detailed consideration of the key actors in the cellular immune system.

Tonegawa, S. 1985. "The Molecules of the Immune System." *Scientific American*, October. A beautifully illustrated account of the structures of antibodies and T cell receptors and of how they form.

von Boehmer, H. and P. Kisielow. 1991. "How the Immune System Learns about Self." *Scientific American*, October. A description of the experiments with transgenic mice that established the clonal deletion theory.

Part Three
Evolutionary Processes

Chapter 19

The History of Life on Earth

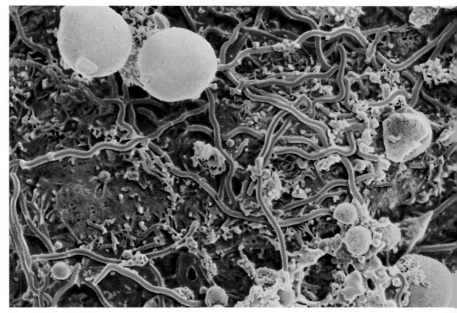

Borrelia bugdorferi

The Era of New Bacterial Diseases
The bacterium shown here causes Lyme disease. The disease has increased rapidly in North America since 1975.

*I*n 1967, the U.S. Surgeon General, William H. Stewart, announced that the time had come "to close the book on infectious diseases." Stewart and other world health officials were so confident of the power of their medical arsenal that they believed they were on the threshold of totally eradicating infectious diseases. However, the future did not conform to their prediction. In 1975, an era of new bacterial diseases began with legionnaires' disease, followed by Lyme disease, Brazilian purpuric fever, epithelioid angiomatosis, toxic shock syndrome, and others.

Since 1967, several hundred newly described bacteria have been associated with human disease, and dozens of other bacteria species that were thought to be harmless have been found to cause human disease. In addition, some historical diseases that had been nearly wiped out have resurged—some to epidemic proportions. Infectious diseases killed more than 16.5 million people in 1993, more than were killed by noninfectious diseases such as cancer and heart disease.

Health officials erred in their predictions because they did not understand the evolutionary capabilities of disease-causing organisms. Hippocrates described the symptoms of malaria, mumps, diphtheria, tuberculosis, and influenza 2,300 years ago, and human fossils show that our species has been plagued by infectious diseases throughout prerecorded history. Science, by making possible the rapid increase of Earth's human population, has helped create ideal conditions for the evolution of new pathogens.

Evolutionary changes are taking place all around us, and they have powerful implications for human welfare. Our attempts to control populations of undesirable species and increase populations of desirable species make us powerful agents of evolutionary change. In addition to producing the results we desire, we often cause undesirable outcomes, such as the evolution of resistance to medicines and pesticides and the evolution of greater virulence on the part of human pathogens. Medicine and agriculture can respond creatively to the evolutionary changes they are causing only if we understand why those changes happen.

What is biological evolution? **Biological evolution** is a change over time in the genetic composition of members of a population. Changes that take effect over a small number of generations constitute **microevolution**. Changes that take centuries, millennia, or longer to be completed are called **macroevolution**. The fossil record documents macroevolutionary changes among organisms. Many of these changes are dramatic. The goals of this part of the book are to document the history of life on Earth and to describe the processes of evolutionary change and the agents that cause them.

We begin with an overview of the history of life on Earth. Because the genetic material and basic cellular metabolic pathways are very similar or identical in all organisms, biologists believe that all living organisms have descended from a single ancestral lineage. How biologists determine the evolutionary histories of organisms, how they study the mechanisms of evolutionary change, and how the millions of species that live today (as well as those that became extinct) formed from a single common ancestor are addressed in subsequent chapters. Finally, in Chapter 24 we examine how life probably arose from nonliving matter several billion years ago.

To understand the long-term patterns of evolutionary change that we will document in this chapter, we must think in time frames spanning many millions of years and imagine events and conditions very different from those we now observe. The Earth of the distant past is, to us, a foreign planet inhabited by strange organisms. The continents were not where they are today, and climates were different. One of the remark-

TABLE 19.1   Earth's Geological History

EON	ERA	PERIOD	ONSET[a]	MAJOR PHYSICAL CHANGES ON EARTH
	Cenozoic	Quaternary	2 mya	Icehouse climate; repeated glaciations
		Tertiary	66 mya	Continents near current positions; climate cools
	Mesozoic	Cretaceous	138 mya	Northern continents attached; Gondwana begins to drift apart; meteorite strikes Yucatán Peninsula
		Jurassic	195 mya	Two large continents form: Laurasia (north) and Gondwana (south); climate warm
		Triassic	245 mya	Pangaea slowly begins to drift apart; hothouse climate
	Paleozoic	Permian	290 mya	Continents aggregate into Pangaea; large glaciers form; dry climates form in interior of Pangaea
		Carboniferous	345 mya	Climate cools; marked latitudinal climate gradients
		Devonian	400 mya	Continents collide at end of period; asteroid probably collides with Earth
		Silurian	440 mya	Sea levels rise; two large continents form; hothouse climate
		Ordovician	500 mya	Gondwana moves over South Pole; massive glaciation, sea level drops 50 m
		Cambrian	540 mya	O_2 levels approach current levels
Phanerozoic			600 mya	O_2 level at >5% of current level
Proterozoic			2.5 bya	O_2 level at >1% of current level
Archean			3.8 bya	O_2 first appears in atmosphere
Hadean			4.5 bya	

[a] In this chapter, figures that depict change over evolutionary time are shown with the most recent events at the top and the oldest at the bottom. mya, million years ago; bya, billion years ago.

able achievements of modern science has been the development of techniques that enable us to infer past conditions and to date them with some precision.

In this chapter we first examine how events in the distant past can be dated. Then we review the major changes in physical conditions on Earth during the past 4 billion years, look at how those changes affected life, and discuss the major patterns in the evolution of life.

Determining How Earth Has Changed

Geologists divide Earth's history into four **eons**: the Hadean eon, the Archean eon, the Proterozoic eon, and the Phanerozoic eon. The Phanerozoic eon is subdivided into **eras**, which are further subdivided into **periods** (Table 19.1). The boundaries between these divisions are based on major differences in the fossils contained in successive layers of rocks. The fossils of organisms, not the rocks themselves, provided the information geologists first used to order the sequence of events, because evolving life provided a record that

can be ordered through time. There is no directional evolution of rocks; a rock of a particular type can be formed at any time. However, scientists do have ways of determining the date a rock was formed.

Radioactivity provides a way to date rocks

Radioactivity can be used to date rocks precisely because in successive, equal periods of time, an equal fraction of the remaining radioactive material of any radioisotope decays, becoming the corresponding stable isotope. For example, in 14.3 days, one-half of any sample of phosphorus-32 (^{32}P), a radioactive isotope of phosphorus, decays. During the next 14.3 days, one-half of the remaining half decays, leaving one-fourth of the original sample of ^{32}P. After 42.9 days, three half-lives have passed, so one-eighth (that is, $\frac{1}{2} \times \frac{1}{2} \times \frac{1}{2}$) of the original radioactive material remains, and so forth.

Each radioisotope has a characteristic half-life. Tritium (^{3}H) has a half-life of 12.3 years, and carbon-14 (^{14}C) has a half-life of about 5,700 years. Some radioisotopes have much longer half-lives: The half-life of potassium-40 (^{40}K) is 1.3 billion years; that of uranium-238 (^{238}U) is about 10 billion years. Which isotope is used to estimate the ages of some ancient material depends on how old the material is thought to be. The decay of potassium-40 to argon-40 has been used to date most of the ancient events in the evolution of life that we will describe in this chapter.

To use radioisotopes to date past events, we must know the concentrations of the isotopes at the time of those events. In the case of carbon, we know the initial amounts of ^{14}C because the production of new ^{14}C in the upper atmosphere (by the reaction of neutrons with ^{14}N) just balances the natural radioactive decay of ^{14}C. Therefore a steady state exists.

The ratio of radioactive ^{14}C to nonradioactive ^{12}C in a living creature is always the same as that in the environment because carbon is constantly being exchanged between the environment and organisms. However, as soon as a tree or any other living thing dies, it ceases to equilibrate its carbon compounds with the rest of the world. Its decaying ^{14}C is not replenished from outside, and the ratio of ^{14}C to ^{12}C decreases. By measuring the fraction of ^{14}C in a carbon specimen, we can easily calculate how much time has elapsed since it died, but ^{14}C can be used to date events only within the last 30,000 years. Some radiocarbon dates of archaeological objects are shown in Figure 19.1.

Armed with information on the ages of mileposts in Earth's history, we can analyze what caused the changes in types of organisms preserved in the rocks that led geologists to identify boundaries between geological periods. The major physical events we will describe are listed in Table 19.1, along with the most important events in the history of life.

MAJOR EVENTS IN THE HISTORY OF LIFE
Humans evolve; large mammals become extinct
Radiation of birds, mammals, flowering plants, and insects
Dinosaurs continue to radiate; flowering plants and mammals diversify. **Mass Extinction** at end of period (≈76% of species disappear)
Diverse dinosaurs; first birds; two minor extinctions
Early dinosaurs; first mammals; marine invertebrates diversify. **Mass Extinction** at end of period (≈76% of species disappear)
Reptiles radiate; amphibians decline; **Mass Extinction** at end of period (≈96% of species disappear)
Extensive forests; first reptiles; insects radiate
Fishes diversify; first insects and amphibians. **Mass Extinction** at end of period (≈82% of species disappear)
Jawless fishes diversify; first bony fishes; plants and animals colonize land
Mass Extinction at end of period (≈85% of species disappear)
Most animal phyla present; diverse algae
Ediacaran fauna
Eukaryotes evolve; several animal phyla appear
Origin of life; prokaryotes flourish

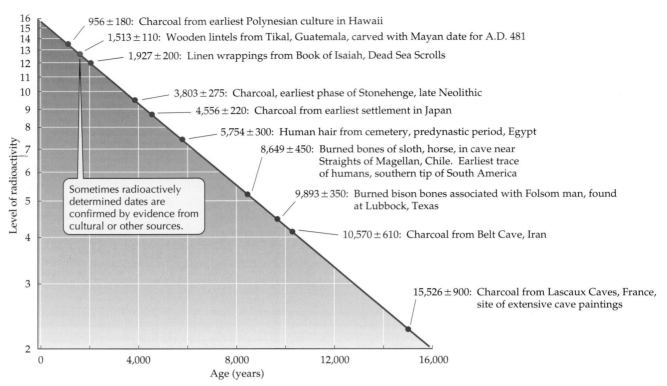

19.1 Radioactive Clocks Tell Time By measuring the fraction of radioactive ^{14}C in any remains that contain carbon, archeologists and biologists can determine the approximate age of specimens such as those described here.

Unidirectional Changes in Earth's Atmosphere

The atmosphere of early Earth was a reducing one; that is, it lacked free oxygen. Perhaps the most important environmental change since Earth cooled enough for water to condense on its surface is the largely unidirectional increase in atmospheric O_2 concentrations that began in the Phanerozoic eon. The oxygen concentration increased because certain sulfur bacteria evolved the ability to use water as the source of hydrogen during photosynthesis. The cyanobacteria that evolved from these sulfur bacteria became very abundant. They liberated enough O_2 to open the way for the evolution of oxidation reactions as the energy source for the synthesis of ATP.

An oxygenated atmosphere also made possible larger cells and more complicated organisms. Small, unicellular aquatic organisms can obtain enough O_2 by simple diffusion even when O_2 concentrations are very low. Larger unicellular organisms have lower surface area-to-volume ratios. In order to obtain enough O_2 by simple diffusion, they must live in an environment where concentrations of O_2 are higher than those that can support small prokaryotic cells. Bacteria can thrive on 1 percent of current atmospheric O_2

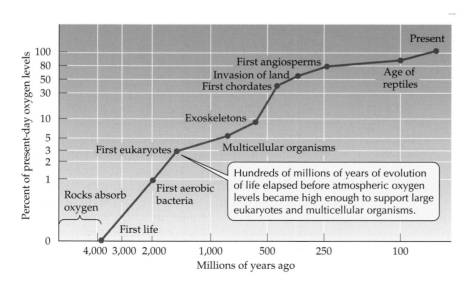

19.2 Large Cells Need More Oxygen Although aerobic prokaryotes flourish with less, larger eukaryotic cells with lower surface area-to-volume ratios require oxygen levels that are at least 2 to 3 percent of current atmospheric O_2 concentrations. (Both axes of the graph are on logarithmic scales.)

levels, but eukaryotic cells require oxygen levels that are at least 2 to 3 percent of current atmospheric concentrations.

About 1,500 million years ago (mya), O_2 concentrations became high enough for large eukaryotic cells to flourish and diversify (Figure 19.2). Further increases in atmospheric O_2 levels some 700 to 570 mya enabled multicellular organisms to evolve. The fact that it took millions of years for Earth to develop an oxygenated atmosphere probably explains why only unicellular prokaryotes lived on Earth for more than a billion years.

Processes of Major Change on Earth

Unlike the largely unidirectional change in O_2 concentrations in Earth's atmosphere, most major changes on Earth have been characterized by irregular oscillations in the planet's internal processes, such as the activity of volcanoes and the shifting and colliding of the continents. External events, such as collision with meteorites, have also left their mark, sometimes causing major disruptions in the history of life.

The continents have changed position

The maps and globes that adorn our walls, shelves, and books give an impression of a static Earth. It is easy for us to believe that the continents have always been where they are, but this conclusion would be quite incorrect. Earth's crust consists of solid plates approximately 40 km thick that float on a fluid mantle. The mantle fluid circulates because heat produced by radioactive decay sets up convection cells. The plates move because the seafloor spreads along ocean ridges where material from the mantle rises and pushes the plates aside. Where plates come together, either they move sideways or one plate moves under the other, creating mountain ranges. The movement of the plates and the continents they contain—a process known as **continental drift**—has had enormous effects on climate, sea levels, and the distribution of organisms.

At times, the drifting of the plates has brought the continents together; at other times the continents have drifted apart. The positions and sizes of the continents influence ocean circulation patterns and sea levels. Mass extinctions of species, particularly marine organisms, have usually accompanied major drops in sea level (Figure 19.3). Later in this chapter we will discuss how the positions and movements of the continents, vulcanism, and large meteorites influenced the major events in the evolution of life.

Earth's climate shifts between hothouse and icehouse conditions

Through much of its history, Earth's climate was considerably warmer than it is today, and temperatures

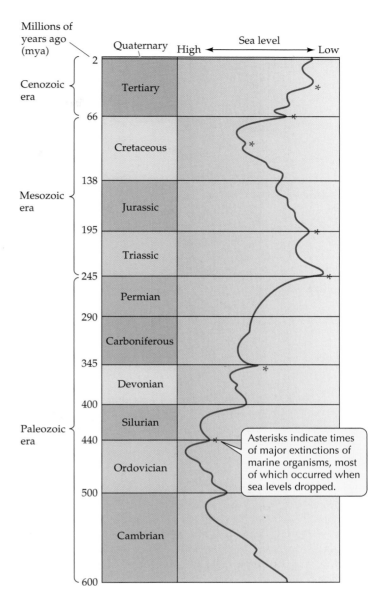

19.3 Sea Levels Have Changed Repeatedly As in Table 19.1, the most recent events are indicated at the top of the chart, the oldest at the bottom.

decreased more slowly toward the poles. At other times, however, Earth was colder than it is today. Large areas were covered with glaciers during the late Proterozoic, the Carboniferous, the Permian, and the Quaternary, but these cold periods were separated by long periods of milder climates (Figure 19.4). Because we live in one of the colder periods in the history of Earth, it is difficult for us to imagine the mild climates that were found at high latitudes during much of the history of life.

Usually climates change slowly, but major climatic shifts have taken place over periods as short as 5,000 to 10,000 years, primarily as a result of changes in

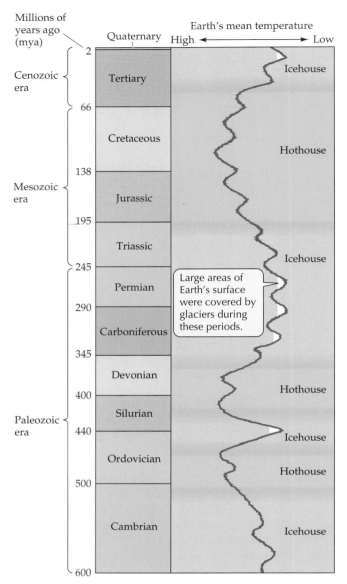

19.4 Hothouse and Icehouse Conditions Have Alternated   Throughout Earth's history, periods of cold climates and glaciations have been separated by long periods of milder climates.

Earth's orbit around the sun. A few shifts appear to have been even more rapid. For example, during one interglacial period, the Antarctic Ocean changed from being ice-covered to being nearly ice-free in less than 100 years. Such rapid changes are usually caused by sudden shifts in ocean currents. Climates have sometimes changed rapidly enough that extinctions caused by them appear "instantaneous" in the fossil record.

Volcanoes disrupt the evolution of life

On the morning of August 27, 1883, Krakatau, an island the size of Manhattan located in the Sunda Strait between Sumatra and Java, was devastated by a series of volcanic eruptions. Tidal waves caused by the eruption hit the shores of Java and Sumatra, demolishing towns and villages and killing 40,000 people. The 1883 explosion at Krakatau was not the largest in recorded history. The 1815 eruption of Tambora on the Indonesian island of Sumbawa was five times greater. An even larger eruption on Sumatra 75,000 years ago created the 65-km-long Lake Toba. Impressive though these eruptions were, their effects were either local or short-lived. They did not cause major changes in patterns of the evolution of life. But the even larger volcanic eruptions that occurred several times during the history of Earth, did have major consequences for life on Earth.

The collision of continents during the Permian period (260 to 250 mya) to form a single, gigantic land mass called Pangaea, caused massive volcanic eruptions. The ash the volcanoes ejected into Earth's atmosphere reduced the penetration of sunlight to Earth's surface, lowering temperatures and triggering the massive glaciation of that time. Massive volcanic eruptions also occurred as the continents drifted apart during the late Triassic period and at the end of the Cretaceous.

External events have triggered other changes on Earth

At least 30 meteorites between the sizes of baseballs and soccer balls hit Earth each year, but collisions with large meteorites are rare. A large meteorite, weighing 5,000 kg, fell in Norton County, Kansas, on February 18, 1948. A prehistoric gigantic meteorite formed Canyon Diablo in Arizona.

The hypothesis that the mass extinction of life at the end of the Cretaceous period, about 65 mya, might have been caused by the collision of Earth with a large meteorite was proposed in 1980 by Luis Alvarez and several of his colleages at the University of California, Berkeley. These scientists based their hypothesis on the finding of abnormally high concentrations of the element iridium in a thin layer separating rocks deposited during the Cretaceous from those of the Tertiary (Figure 19.5). Iridium is abundant in some meterorites but is exceedingly rare on Earth's surface.

To account for the estimated amount of iridium in this layer, Alvarez postulated that a meteorite 10 km in diameter collided with Earth at a speed of 72,000 km per hour. The force of such an impact would have ignited massive fires, created great tidal waves, and sent up an immense dust cloud that encircled Earth, blocking the sun and cooling the planet. The settling dust would have formed the iridium-rich layer.

This hypothesis generated a great deal of controversy, which continues today. Many paleontologists (scientists who study fossils) believe that the fossil record shows that the mass extinction took place over

19.5 Evidence of a Meteorite Collision Iridium is a metal common in some meteorites but rare on Earth. Its presence has been cited as evidence of a meteorite collision that may have been the cause of the mass extinction at the end of the Cretaceous.

A thin band rich in iridium marks the boundary between rocks deposited in the Cretaceous and Tertiary periods.

a period of millions of years, not instantaneously as the meteorite theory demands. Also, because some volcanoes emit substantial quantities of iridium, the massive vulcanism of the time could have generated the iridium layer.

This controversy has stimulated much activity. Some scientists have tried to locate the site of impact of the supposed meteorite. Others have worked to improve the precision with which events of that age could be dated. Still others have tried to determine more exactly the speed with which extinctions occurred at the Cretaceous–Tertiary boundary. Progress on all three fronts has tended to support the meteorite theory.

The theory was supported by the discovery of a circular crater 180 km in diameter buried beneath the north coast of the Yucatán Peninsula, Mexico, thought to have been formed by an impact 65 mya. Some new fossil evidence also suggests that there may have been a sudden extinction of organisms 65 mya, as required by the meteorite theory. Therefore, many scientists accept that the collision of Earth with a large meteorite contributed importantly to the mass extinctions at the boundary between the Cretaceous and Tertiary periods, but concensus has not yet been reached.

The Fossil Record of Life

Geological evidence is a major source of information about changes on Earth during the remote past. But the preserved remains of organisms that lived in the past, not the rocks themselves, are what have enabled geologists to order those events in time. What are these remains and what do they tell us about the influence of physical events on the evolution of life on Earth? After examining the conditions that preserve remains, we will consider the completeness of the fossil record, and how that record reveals patterns in life's history.

Much of what we know about the history of life is derived from **fossils**—the preserved remains of organisms or impressions of organisms in materials that formed rocks. An organism is most likely to be preserved if it dies or is deposited in an environment that lacks O_2. However, most organisms live in oxygenated environments and therefore decompose when they die. Thus many fossil assemblages are collections of organisms that were transported by wind or water to

sites that lacked O_2. Occasionally, however, organisms are preserved where they lived. In such cases—especially if the environment in question was a cool, anaerobic swamp, where conditions for preservation were excellent—we obtain a picture of communities of organisms that lived together.

How complete is the fossil record?

About 300,000 species of fossil organisms have been described, and the number is growing steadily. However, this number is only a tiny fraction of the species that have ever lived. We do not know how many species lived in the past, but we have ways of making reasonable estimates. Of the present-day **biota**—that is, the species in all kingdoms (bacteria, archaea, protists, plants, fungi, and animals)—approximately 1.5 million species have already been named. The actual number of living species is probably at least 10 million (and possibly as high as 50 million) because most species of insects, the richest animal group, have not yet been described. Thus the number of known fossil species is less than 2 percent of the probable total of living species.

Because life has existed on Earth for at least 3.5 billion years, and because species last, on average, less than 10 million years, Earth's biota must have turned over many times during geological history. The total number of species over evolutionary time greatly exceeds the number living today.

The sample of fossils, although small in relation to the total number of extinct species, is better for some groups than for others. The record is especially good for marine animals that have hard skeletons. Among the nine major animal groups that have hard-shelled members (see Chapters 29 and 30), approximately 200,000 species have been described from fossils, roughly twice the number of living marine species in

19.6 A Fossil Spider   Trapped in the sap of a tree in what is now Arkansas about 50 mya, this spider is exquisitely preserved in the amber that formed from the sap. The details of its external anatomy are clearly visible.

these same groups. Paleontologists lean heavily on these groups in their interpretations of the evolution of life in the past. Insects, although much rarer as fossils, are also relatively well represented in the fossil record (Figure 19.6).

The fossil record demonstrates several patterns

Despite its incompleteness, the fossil record reveals several patterns that are unlikely to be altered by future discoveries. First, great regularity exists: Organisms are not mixed together at random, but rather appear sequentially. Second, as we pass from ancient periods of geological time toward the present, fossils increasingly resemble species living today. The fossil record also tells us that extinction is the eventual fate of all species.

The fossil record contains many good series that demonstrate gradual change in lineages of organisms over time. A good example is the series of fossils showing the pathway by which whales evolved from hoofed terrestrial mammals, beginning about 50 mya. The intermediate fossils illustrate the major changes by which ancestors of whales became adapted for aquatic existence and lost their hind limbs (Figure 19.7). Interestingly, whales retain the genetic potential for developing legs; occasionally living whales have been found with small hind legs that extend outside their bodies. The claim made repeatedly by creationists that the fossil record does not contain examples of intermediates is false. Intermediates abound, and more and more of them are being discovered.

Nevertheless, the incompleteness of the fossil record can mislead our interpretations of what happened. Most described fossils come from a relatively small number of sites. Since any organism that was

evolving elsewhere would have been absent at these sites, many organisms may be entirely unrepresented. Moreover, when a species that has evolved in one place appears among the fossils of another site, it gives the false impression that it evolved very rapidly from one of the species that already lived there.

For example, horses evolved slowly over millions of years in North America. Many different lineages arose and died out (Figure 19.8). Ancestors of horses crossed the Bering land bridge into Asia several times,

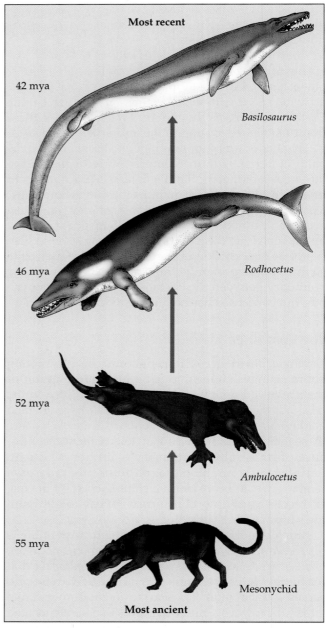

19.7 From Terrestrial to Aquatic Life   An artist's reconstruction, based on fossil skeletons, of four ancestors that represent different stages in the evolution of modern whales from an early terrestrial mammal.

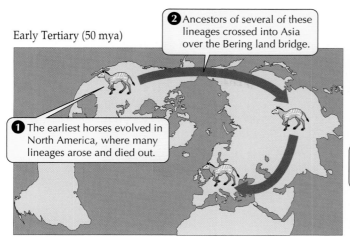

Early Tertiary (50 mya)

❷ Ancestors of several of these lineages crossed into Asia over the Bering land bridge.

❶ The earliest horses evolved in North America, where many lineages arose and died out.

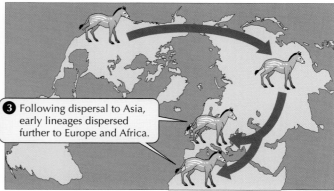

Mid Tertiary (20 mya)

❸ Following dispersal to Asia, early lineages dispersed further to Europe and Africa.

19.8 Horses Have a Complex Evolutionary History
Ancestors of horses crossed the Bering land bridge into Asia several times, the last one only a few million years ago (panel 4). If we lacked the earlier fossil evidence of horse evolution in North America, we might reach the false conclusion that horses evolved very rapidly somewhere in Asia.

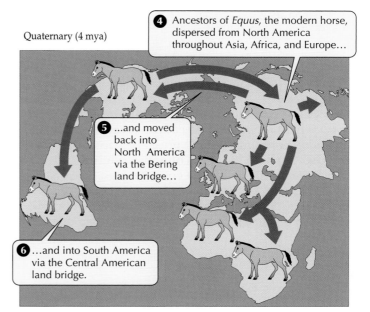

Quaternary (4 mya)

❹ Ancestors of *Equus*, the modern horse, dispersed from North America throughout Asia, Africa, and Europe…

❺ …and moved back into North America via the Bering land bridge…

❻ …and into South America via the Central American land bridge.

the last one only several million years ago. Evidence of each crossing appears suddenly in the Asian fossil record as a major new type of horse. If we lacked fossil evidence of horse evolution in North America, we might conclude that horses evolved very rapidly somewhere in Asia. On the other hand, an incomplete fossil record can also hide rapid changes.

Scientists have collected enough data to combine information of physical events during Earth's history with the course of evolution to compose pictures of what Earth and its inhabitants looked like at different times. We know in general where the continents were and how life changed, but many of the details are poorly known, especially for events in the more remote past.

Life in the Remote Past

The positions of the continents during the **Archean**, **Proterozoic**, and **Phanerozoic** eons are not well known. The major kingdoms of eukaryotic organisms evolved during the mid-Proterozoic eon. The fossil record shows that the volume of organisms increased dramatically in late Precambrian times, about 650 mya. The shallow Precambrian seas teemed with life. Protists and small multicellular animals fed on floating algae. Living plankton and plankton remains were devoured by animals that filtered food items from the water or ingested surface sediments and digested the organic remains in them. During the Phanerozoic, the continents were drifting apart.

The fossil record for Precambrian times is fragmentary. The best fossil assemblage of Precambrian ani-

mals, all soft-bodied invertebrates, is known as the Ediacaran fauna,* named after the Australian site where it was discovered (Figure 19.9). The Ediacaran animals are very different from any animals living today. Some of them may represent separate animal lineages that have no living descendants. In the discussions that follow in this section of the chapter, we will examine the changes that occurred in the Paleozoic, Mesozoic, and Cenozoic eras following the enormous proliferation of life in the Cambrian period.

Life exploded during the Cambrian period

During the **Cambrian period** (540 to 500 mya), O_2 levels in Earth's atmosphere approached their current concentration, and drifting of the plates resulted in several continental masses, the largest of which was

---

*The term *fauna* refers to all of the species of animals living in a particular area. The corresponding term for plants is *flora*.

Spriggina floundersi

Mawsonites

19.9 Ediacaran Animals Fossils of soft-bodied invertebrates excavated at Ediacara in southern Australia are 600 million years old and illustrate the diversity of life that evolved in Precambrian times.

Gondwana (Figure 19.10*a*). Thus, O_2 concentration was no longer a constraint on the evolution of large, multicellular organisms. All animal phyla that have species living today had already evolved, as revealed by the exceptionally well preserved fossils in the Burgess Shale in British Columbia (Figure 19.10*b*), which were deposited in an equatorial sea. The evolution of hard skeletons in representatives of so many phyla about 540 mya suggests that predation became very intense during this period when life "exploded."

The Paleozoic era was a time of major changes

THE ORDOVICIAN. During the **Ordovician period** (500 to 440 mya) the continents were located primarily in the Southern Hemisphere (Figure 19.11). No evidence exists of land or shallow marine environments in the Northern Hemisphere north of the Tropics during the Ordovician. Early during the period, the number of kinds of animals that filter small prey from the water, including brachiopods and mollusks, increased greatly. Floating graptolites, members of a now extinct phylum, were abundant. Ancestors of club mosses and horsetails colonized wet terrestrial environments, but they were still relatively small. At the end of the Ordovician, sea levels dropped about 50 m as massive glaciers formed over Gondwana (see Figure 19.3). Much of the continental shelf was exposed, and ocean temperatures dropped. About 85 percent of the species of marine animals became extinct, probably because of these major environmental changes.

19.10 Cambrian Continents and Animals (540–500 mya) (a) Positions of the continents during mid-Cambrian times. (b) The Burgess Shale has yielded fossils of Cambrian animals, some of which may not be members of any lineage that survived to the present.

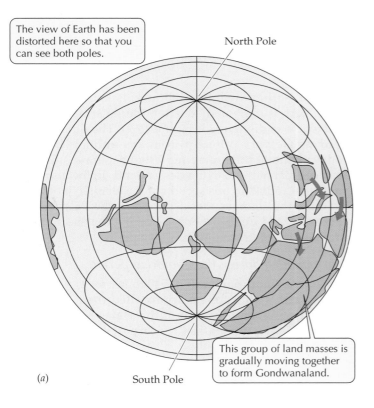

The view of Earth has been distorted here so that you can see both poles.

North Pole

This group of land masses is gradually moving together to form Gondwanaland.

(a) South Pole

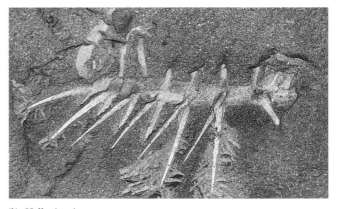

(b) Hallucigenia

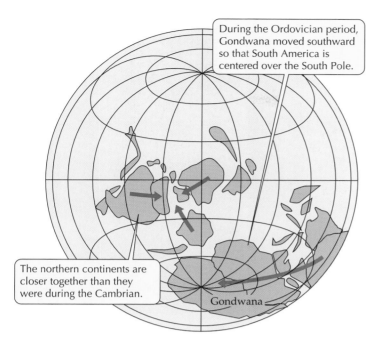

During the Ordovician period, Gondwana moved southward so that South America is centered over the South Pole.

The northern continents are closer together than they were during the Cambrian.

Gondwana

19.11 Ordovician Continents (500–440 mya)   The land mass of this period was still concentrated mainly in the Southern Hemisphere.

THE SILURIAN.   During the **Silurian period** (440 to 400 mya), the northern continents, which had been relatively close to one another during the Ordovician, coalesced, but the general positions of the continents did not change much. Marine life rebounded from the massive extinction at the end of the Ordovician, but no major new groups of marine organisms evolved. The tropical sea was uninterrupted by land barriers, and most marine genera were widely distributed. More significant changes happened on land where the first terrestrial arthropods—scorpions and millipedes—appeared.

THE DEVONIAN.   Rates of evolutionary change accelerated in many groups of organisms during the **Devonian period** (400 to 345 mya). Earth continued to be divided into northern and southern land masses, both of which were slowly moving northward (Figure 19.12*a*). There was a great evolutionary radiation of corals and shelled squidlike cephalopods (Figure 19.12*b*). (See Chapter 20 for a discussion of evolutionary radiation.) Fishes diversified as jawed forms replaced jawless ones, and the heavy armor that had characterized most earlier fishes gave way to the less rigid outer coverings of modern fishes.

19.12 Devonian Continents and Marine Communities (400–345 mya)   The reconstruction in (*b*) shows how a Devonian reef may have appeared.

Terrestrial communities also changed markedly during the Devonian period. Land plants became common, and some reached the size of trees. Most of them were club mosses and horsetails, along with some tree ferns. Toward the end of the period the first gymnosperms appeared. Distinct floras evolved on the two land masses. The first known fossils of centipedes, spiders, pseudoscorpions, mites, and insects date to this period, and the first fishlike amphibians began to occupy the land.

The end of the Devonian was marked by the extinction of about 80 percent of all marine species. Paleontologists disagree on the cause of this major extinction. Some believe that it was triggered by the col-

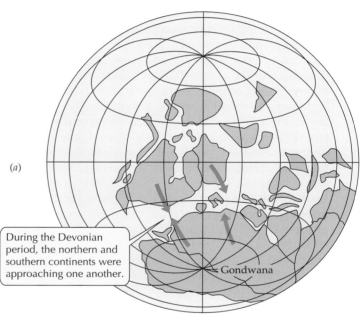

(*a*)

During the Devonian period, the northern and southern continents were approaching one another.

Gondwana

(*b*)

lision of the two continents, which destroyed much of the existing shallow, warm-water marine environment. This hypothesis is supported by the fact that extinction rates were much higher among tropical than among cold-water species. Other paleontologists believe that the extinction was caused by collision of a large asteroid with Earth.

THE CARBONIFEROUS.   Extensive forests grew on the tropical continents during the **Carboniferous period** (345 to 290 mya) (see Figure 27.12). The compressed remains of trees that grew in swampy forests where they fell into deep, anaerobic mud that preserved them from biological degradation are the coal we now mine for energy. Carboniferous beds are rich in fossils, many of which retain traces of the fine details of their structure (Figure 19.13). The diversity of terrestrial animals increased greatly in the Carboniferous period. Snails, scorpions, centipedes, and insects were present in great abundance and variety. Insects evolved wings, which enabled them to move readily among tall and structurally complex plants. Many Carboniferous plant fossils show evidence of damage by feeding insects. Amphibians became better adapted to terrestrial existence. Some of them were large animals more than 5 m long, quite unlike any surviving today. From one amphibian stock, the first reptiles evolved late in the period.

THE PERMIAN.   Deposits from the **Permian period** (290 to 245 mya) contain representatives of most modern groups of insects, including dragonflies with wingspreads that measured 2 feet, the largest insects that ever lived. By the end of the period, reptiles greatly outnumbered amphibians. These reptiles included a

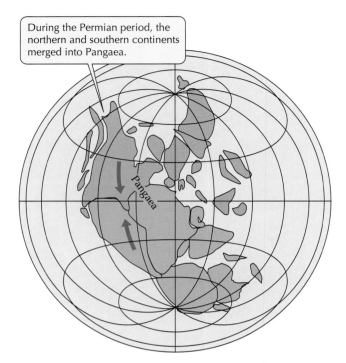

During the Permian period, the northern and southern continents merged into Pangaea.

Pangaea

19.14 Pangaea Formed in the Permian Period (290–250 mya)   The interior of this supercontinent experienced very harsh climates. The largest glaciers in Earth's history formed during this period.

variety of terrestrial forms, as well as species that were major predators in marine and freshwater environments. The lineage leading to mammals diverged from one reptilian lineage late in the period. In fresh waters, the Permian period was a time of extensive radiation of bony fishes.

At the end of the Permian period, about 90 percent of all species, both terrestrial and marine, became extinct. What caused this mass extinction? There was massive volcanic activity in Siberia at that time, but there was no collision of a large meteorite with Earth. The Permian extinctions may have happened slowly, perhaps over more than 10 million years, but the time frame is uncertain. The most probable cause was the coalescing of the continents into the supercontinent Pangaea (Figure 19.14). The interior of Pangaea, far removed from the oceans, experienced very harsh climates. The largest glaciers in Earth's history formed, causing sea levels to drop, drying out many of the shallow seas where most marine organisms lived (see Figure 19.3).

Geographic differentiation increased in the Mesozoic era

At the start of the **Mesozoic era,** the few surviving organisms found themselves in a relatively empty world. As Pangaea slowly separated into individual

Alethopteris serlii

19.13 A Carboniferous Fossil (345–290 mya)   A fossilized seed fern excavated near Washington, D.C. Like the fossil fuels we burn today, this fossil is a remnant of the massive forests of the Carboniferous period.

19.15 Mesozoic Dinosaurs Dinosaurs of the Mesozoic era continue to capture our imagination. This painting illustrates some of the large species from the Jurassic period (195–138 mya).

continents, the glaciers melted and the oceans rose and reflooded the continental shelves, forming huge, shallow inland seas. Life again diversified, but lineages different from those of the Permian period came to dominate Earth. The trees that dominated the great coal-forming forests were replaced by cycadeoids, cycads, ginkgos, and conifers.

Earth's biota, which until that time had been relatively homogeneous, became increasingly **provincialized**; that is, distinctive terrestrial floras and faunas evolved on each continent, and the biotas of shallow waters bordering the continents also diverged from one another. The provincialization that began during the Mesozoic continues to influence the geography of life today (see Chapter 54).

THE TRIASSIC. During the **Triassic period** (245 to 195 mya) many invertebrate lineages became more diverse, and many groups that previously had been restricted to living on surfaces of bottom sediments evolved burrowing forms. On land, gymnosperms and seed ferns became the dominant woody plants. A great radiation of reptiles began, which eventually gave rise to dinosaurs, crocodilians, and birds.

THE JURASSIC. The transition from the Triassic period to the **Jurassic period** (195 to 138 mya) was marked by a mass extinction that eliminated about 75 percent of species on Earth. Why they went extinct is not known, but meteor impact is suspected. The mass extinction was followed by another period of evolutionary diversification during the Jurassic. Bony fishes began the great radiation that culminated in their dominance of the seas. Frogs, salamanders, and lizards first appeared. Flying reptiles evolved, and dinosaur lineages evolved into bipedal predators and large quadrupedal herbivores (Figure 19.15). Several groups of mammals also evolved.

THE CRETACEOUS. By the early **Cretaceous period** (138 to 66 mya), the northern continents had completely separated from the southern ones and a continuous sea encircled the Tropics (Figure 19.16). Sea levels were high, and Earth existed in a hothouse state. Biological production was high both on land and in the oceans. Marine invertebrates increased in variety and number of species. On land, dinosaurs continued to diversify. The first snakes appeared during the Cretaceous, but their lineage did not radiate until much later.

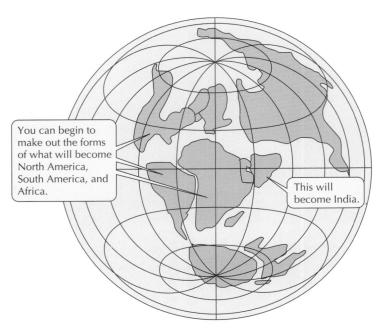

19.16 Continents during the Cretaceous Period (138–65 mya) Many groups of small mammals evolved during this period, and flowering plants began the radiation leading to their modern-day dominance on land.

By the end of the period, many groups of mammals had evolved, but these mammals were generally small. Early in the Cretaceous period, possibly somewhat earlier, flowering plants evolved from gymnosperm ancestors and began the radiation that led to their current dominance on land. And insects added angiosperms to the list of plants they ate.

Another mass extinction took place at the end of the Cretaceous period. On land, all vertebrates larger than about 25 kg in body weight apparently became extinct. In the seas, many planktonic organisms and bottom-dwelling invertebrates became extinct. As we discussed earlier, this extinction may have been caused by a large meteorite that collided with Earth off the Yucatán Peninsula.

The modern biota evolved during the Cenozoic era

By the early **Cenozoic era** (66 mya), the positions of the continents had begun to resemble those of today, but Australia was still attached to Antarctica, the Atlantic Ocean was much narrower, and the northern continents were

19.17 Continents during the early Tertiary Period (65 mya) The continents approached their modern positions by this time.

connected (Figure 19.17). The Cenozoic era was characterized by an extensive radiation of mammals, but other taxa were also undergoing important changes. Flowering plants diversified extensively and dominated world forests, except in cool regions.

THE TERTIARY. During the **Tertiary period** (66 to 2 mya), Australia began its northward drift, and by 20 mya it had nearly reached its current position. The map of the world during the Tertiary for the first time looks familiar to us.

In the middle of the Tertiary, when the climate became considerably drier and cooler, many lineages of flowering plants evolved herbaceous (nonwoody) forms; grasslands, with and without scattered trees, spread over much of Earth. By the beginning of the Cenozoic era, invertebrate faunas were already modern in most respects. It is among the vertebrates that evolutionary change during the Tertiary period was most rapid. Living groups of reptiles, such as snakes and lizards, underwent extensive radiations during this period, as did birds and mammals.

THE QUATERNARY. The present geological period, the **Quaternary period,** began with the Pleistocene epoch about 2 mya. The Pleistocene was a time of drastic cooling and climatic fluctuations. During four major and about 20 minor glacial episodes, Earth became much cooler and distributions of animals and plants shifted toward the equator. The last glaciers retreated from temperate latitudes less than 10,000 years ago.

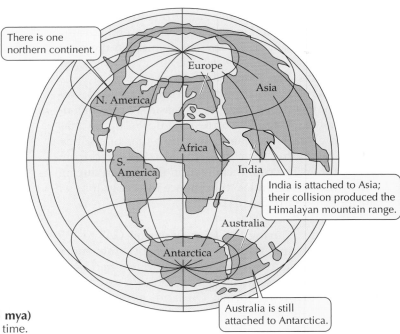

Organisms are still adjusting to these changes. Many high-latitude ecological communities have occupied their current locations no more than a few thousand years. Interestingly, most Pleistocene climatic fluctuations resulted in few extinctions. But many large birds and mammals became extinct in North and South America and in Australia when *Homo sapiens* arrived on those continents. Human hunting may have caused these extinctions, although existing evidence does not convince all paleontologists. How our species is affecting biological diversity today will be discussed in Chapter 55.

It took several million years for life to recover from mass extinctions

Fossil evidence shows that the number of species on Earth returned to the value it had before a mass extinction event within 1 to 7 million years. This recovery rate is rapid in geological time but slow relative to human life spans. Therefore, we cannot rely on evolution to replace the biological diversity we are currently losing as a result of human modification of Earth (see Chapter 55).

Although in the past the number of species has recovered within a few million years after each mass extinction, the kinds of organisms that have diversified have differed from those that dominated life on Earth before the extinctions. How have these different kinds of organisms arisen and achieved dominance?

Patterns of Evolutionary Change

Major new features, such as the feathers of birds or the legs of terrestrial vertebrates, that adapt organisms to a special way of life are called **evolutionary innovations**. How such novelties arise has been the subject of much debate from Darwin's time to the present. The variety of sizes and shapes among living organisms seems almost limitless, but when we take a closer look, the number of truly novel structures is remarkably small.

As fiction writers often do, we can imagine unusual vertebrates with wings sprouting from their shoulders, but in reality the wings of vertebrates are always modified front legs. Modern mammals are highly varied in their shapes, but all of their structures are modifications of structures found in ancestral mammals. As we saw earlier, transforming a terrestrial mammal into a whale did not require a drastic reorganization of the mammalian body plan (see Figure 19.7). Only a few evolutionary innovations, such as the notochord of chordates, do not appear to be modifications of a preexisting structure.

In the following discussions, we will see that minor developmental changes can provide major morphological changes. Then we will look at faunal patterns in time, in organism size, and in predator capacity.

Changes in growth rate alter shapes dramatically

A major reason why evolution typically proceeds via modifications of existing plans is that major changes in shape can result from minor changes in developmental processes. As first demonstrated by D'Arcy Thompson in his 1917 book *On Growth and Form*, dramatically different shapes of organisms can result simply from changes in rates of growth of different parts of the body (Figure 19.18). Changes in shape caused by differential growth rates are probably easy to evolve because they are often under very simple genetic control and because the sizes of all body parts are adjusted appropriately.

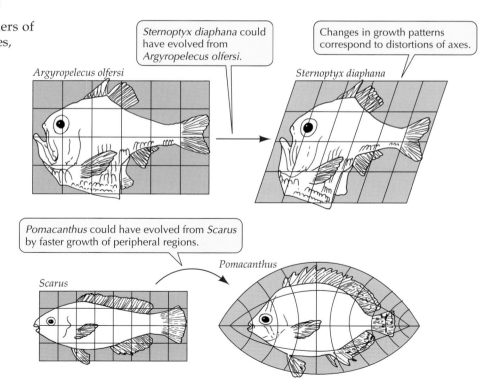

19.18 Changes in Growth Rate Alter Shapes Dramatically different shapes can result when different regions of the body grow at different rates.

Sternoptyx diaphana could have evolved from *Argyropelecus olfersi.*

Argyropelecus olfersi

Changes in growth patterns correspond to distortions of axes.

Sternoptyx diaphana

Pomacanthus could have evolved from *Scarus* by faster growth of peripheral regions.

Scarus

Pomacanthus

Three major faunas have dominated animal life on Earth

Only three events during the evolution of life have resulted in the evolution of major new faunas (Figure 19.19). The first one, known as the Cambrian explosion, took place about 500 mya. The second, about 60 million years later, resulted in the Paleozoic fauna. The great Permian extinctions 300 million years later were followed by the third event, the Triassic explosion, which led to our modern fauna.

During the Cambrian explosion all the major groups of present-day organisms appeared, along with a number of phyla that subsequently became extinct. The Paleozoic and Triassic explosions greatly increased the number of families, genera, and species,

but no new *phyla* of organisms evolved. These later explosions resulted in many new species of organisms, but all these species had modifications of the body plans already present when the great biological diversifications began.

Biologists have long puzzled over the striking differences between the Cambrian explosion and the two later explosions. The most commonly accepted theory is that because the Cambrian explosion took place in a world that contained few species of organisms, all of which were small, the ecological setting was favorable for the evolution of many new body plans and different ways of life. Many types of organisms were able to survive initially in this world, but as competition intensified and new types of predators evolved, many forms were unable to persist.

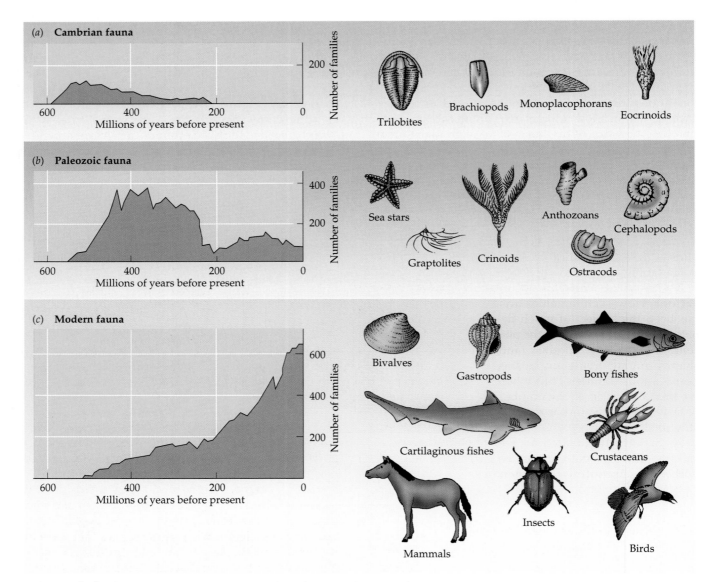

19.19 Evolutionary Faunas Representatives of the three "evolutionary faunas" are shown together with a graphic illustration of the richness of species in each fauna over time.

Although Earth was relatively poor in species at the time of the other two evolutionary explosions as well, the species that were already present included a wide array of body plans and ways of life. Therefore, new major innovations were less likely to evolve at these times than in the Cambrian period.

The size and complexity of organisms have increased

The earliest organisms were small prokaryotes. A modest increase in size and structural complexity accompanied the evolution of eukaryotes 2.5 billion years ago. Since then, the maximum sizes of organisms in many lineages have increased, irregularly to be sure. E. D. Cope noted in 1885 that sizes of vertebrates often increase with time within lineages, a trend that has since been verified in many other groups. The most striking exception to this trend is insects, which have remained relatively small throughout their evolutionary history, although some were much larger during the Carboniferous.

The overall increase in body size is the result of two opposing forces. Within a species, selection often favors larger size because larger individuals can dominate smaller ones. But larger species on average survive for less time than small species do, which is one reason why Earth is not populated primarily by large organisms.

Predators have become more efficient

During the Cretaceous period (138 to 66 mya) many species of crabs with powerful claws evolved, and carnivorous marine snails able to drill holes in shells began to fill the seas. Skates, rays, and bony fishes with powerful teeth capable of crushing mollusk shells also evolved, and large, powerful marine reptiles—the placodonts—fed heavily on clams. The increasing thickness and narrowing openings of snail shells during the Cretaceous is evidence that predation rates intensified (Figure 19.20). Other evidence of heavy predation pressure is the increase in the percentage of fossil shells that show signs of having been repaired following an attack that did not kill its owner.

Although shell thickness provided some protection from predators, predators were so effective that clams disappeared from the surfaces of most marine sediments. The survivors in those environments were species that burrowed into the substrate, where they were more difficult to capture.

The Speed of Evolutionary Change

Following each mass extinction, the diversity of life rebounded. How fast did evolution proceed during those times? Were rates too high to be accounted for by the speed at which current evolutionary changes are happening? Why did some lineages evolve rapidly

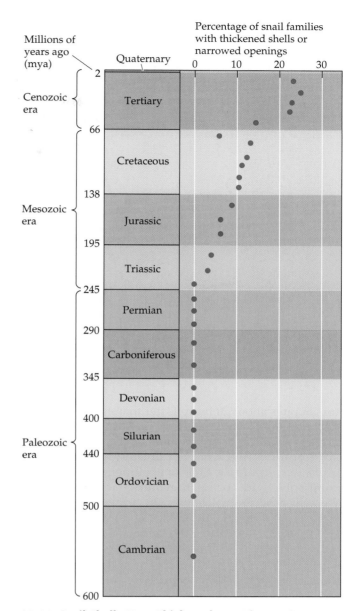

19.20 Snail Shells Have Thickened over Time The percentage of subfamilies of snails that have internally thickened or narrowed openings of their shells has increased with evolutionary time—evidence that predation on shelled animals intensified.

while others did not? Scientists have made enough progress in studying evolution to give at least tentative answers to these questions.

Fossil evidence can be used to estimate rates of change of size and shape in many lineages of organisms. One of the fastest known rates of evolution has been in the skeletons of house sparrows that were introduced into the United States from Europe in the nineteenth century, but that rate is many times slower than rates achieved by scientists performing evolutionary experiments in the laboratory. Nonetheless, house sparrows diverged from their European ances-

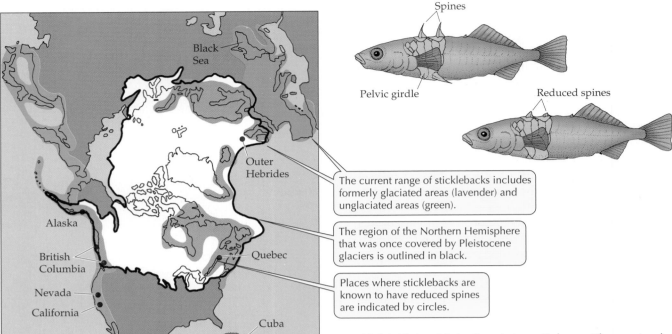

The current range of sticklebacks includes formerly glaciated areas (lavender) and unglaciated areas (green).

The region of the Northern Hemisphere that was once covered by Pleistocene glaciers is outlined in black.

Places where sticklebacks are known to have reduced spines are indicated by circles.

19.21 Natural Selection Acts on Spines   Three-spined stickleback populations with reduced spines are found primarily in young lakes that were covered by ice during the last glaciation. These lakes lack large predatory fishes but do have predatory insects.

tors at rates about 20 times faster than the highest rate of change that has been measured in characters among rapidly evolving fossil mammals. Rates of increase in sizes of dinosaurs were ten times slower than that. Thus, the most rapid rates of evolutionary change are compatible with the microevolutionary mechanisms we will discuss in Chapter 20. In considering the speed of evolution, we will look at variations in the rates of both evolution and extinctions.

Why are evolutionary rates uneven?

The fossil record shows that many species have experienced times of **stasis**, during which they change very little over long periods. For example, many marine lineages evolved very slowly during the Silurian period. The horseshoe crabs that lived 300 mya are almost identical in appearance to those living today (see Figure 29.34). The chambered nautiluses of the late Cretaceous are indistinguishable from living species.

"Living fossils" are found today in environments that have changed relatively little for millennia. These environments tend to be harsh in one way or another. The sandy coastlines where horseshoe crabs spawn have extremes in temperature and in salt concentration that are lethal to many organisms. Chambered nautiluses spend their days in deep, dark ocean waters, ascending to feed in food-rich surface waters only under the protective cover of darkness. Their intricate shells provide little protection against today's visually hunting fishes.

Periods of stasis may be broken by times during which changes, either in the physical or biological envi-

ronment, create conditions that favor new traits. How new conditions favor rapid evolutionary changes is illustrated by the sizes of spines on the three-spined stickleback (*Gasterosteus aculeatus*). This widespread marine fish has repeatedly invaded fresh water throughout its long evolutionary history (Figure 19.21). Sticklebacks are quite tiny (usually less than 10 cm long); all marine and most freshwater populations have well-developed pelvic girdles with prominent spines that make it difficult for other fishes to swallow them. However, large predatory insects can readily grasp the stickleback's spines. These insects prey selectively on stickleback individuals with the largest spines. When sticklebacks invade freshwater habitats where predatory fish are absent but predatory insects are present, the lineages rapidly evolve smaller spines. Populations with reduced spines are found primarily in young lakes that were covered by ice during the last glaciation. These lakes do not have large predatory fishes.

The extensive fossil record of sticklebacks shows that spine reduction evolved many times in different populations that invaded fresh water. In addition, molecular data reveal that each freshwater population is most closely related to an adjacent marine population, not to other freshwater populations. Therefore, spine reduction has evolved rapidly many times in different places in response to the same ecological situation: the absence of predatory fish.

Extinction rates vary over time

More than 99 percent of the species that have ever lived are extinct. However, we know relatively little about causes of extinctions, except for species that have become extinct in historical times. Species have become extinct throughout the history of life, but extinction rates have fluctuated dramatically. In some periods, some groups have had high extinction rates while others were proliferating. Paleontologists distinguish between normal or background extinction rates and rates during mass extinctions (see Table 19.1).

Each mass extinction changed the flora and fauna of the next period by selectively eliminating some types of organisms, thereby increasing the relative abundance of others. For example, among planktonic foraminifera, important marine protists (see Figure 26.7), the only survivors of mass extinction were relatively simple species with broad geographic ranges. Among the seashells of the Atlantic coastal plain of North America, species with a broad geographic range were less likely to become extinct during normal periods than were species with a small geographic range.

On the other hand, during the mass extinction of the late Cretaceous, *groups of closely related species* with large geographic ranges survived better than those with small ranges, even if the *individual species* had small ranges. Similar patterns are found in other molluscan groups elsewhere, suggesting that traits favoring long-term survival during normal times are often different from those that favor survival during times of mass extinctions.

At the end of the Cretaceous period, extinction rates on land were much higher among large vertebrates than among small ones. The same was true during the Pleistocene mass extinction, when extinction rates were high only among large mammals and large birds. In addition, as we have seen, during some mass extinctions marine organisms were heavily hit while terrestrial organisms survived well. Other extinctions affected organisms living in both environments. These differences are not surprising, given that major changes on land and in the oceans did not always coincide.

The Future of Evolution

The agents of evolution are operating today as they have been since life first appeared on Earth, but major changes are under way as a result of the dramatic increase of Earth's human population. The loss of large vertebrates that dominated human-caused extinctions until recently is being supplemented by increasing extinctions of small species, driven primarily by changes in Earth's vegetation. Deliberately or inadvertently, people are moving thousands of species around the globe, reversing the provincialization of Earth's biota that evolved during the Cenozoic era.

Humans have taken charge of the evolution of certain valuable species by means of artificial selection. Our ability to modify species has been enhanced by modern molecular methods that enable us to move genes at will among species, even among distantly related ones. In short, humans have become the dominant evolutionary agent on Earth today. How we handle our massive influence will powerfully affect the future of life on Earth.

Summary of "The History of Life on Earth"

- Biological evolution is change over time in the genetic composition of members of a population.

Determining How Earth Has Changed

- Microevolutionary changes take effect over a small number of generations, macroevolutionary changes over centuries, millennia, or longer.
- The relative ages of rock layers in Earth's crust were determined from their embedded fossils. The eons during which the rock layers were laid down are divided into the eras and periods of Earth's geological history. The boundaries between these units are based on differences between their fossil biotas. **Review Table 19.1**
- Radioisotopes supplied the key for assigning absolute ages to the boundaries between geological time units. **Review Figure 19.1**

Unidirectional Changes in Earth's Atmosphere

- The early atmosphere was a reducing atmosphere; it lacked free oxygen. Oxygen accumulated after prokaryotes evolved the ability to use water as their source of hydrogen ions in photosynthesis. Increasing concentrations of atmospheric oxygen made possible the evolution of eukaryotes and multicellular organisms. **Review Figure 19.2**

Processes of Major Change on Earth

- Physical conditions have changed dramatically and repeatedly during Earth's history. **Review Table 19.1**
- Throughout Earth's history continents have drifted about, sometimes separating from one another, at other times colliding. Collisions typically have led to periods of massive vulcanism, glaciations, and major shifts in sea levels and ocean currents. **Review Figures 19.3, 19.4, 19.10, 19.11, 19.12, 19.14, 19.16, 19.17**
- External events, such as collisions with meteorites, also have changed conditions on Earth. A meteorite may have caused the abrupt mass extinction at the end of the Cretaceous period. Most other mass extinctions were apparently caused by events originating on Earth.

The Fossil Record of Life

- Much of what we know about the history of life on Earth comes from the study of fossils.
- The fossil record, although incomplete, reveals broad patterns in the evolution of life. About 300,000 fossil species

have been described. The best record is that of hard-shelled animals fossilized in marine sediments.

• Fossils show that many evolutionary changes are gradual, but an incomplete record can falsely suggest or conceal times of rapid change. **Review Figures 19.7, 19.8**

Life in the Remote Past

• The fossil record for Precambrian times is fragmentary, but fossils from Australia show that many lineages that evolved then may not have left living descendants.

Patterns of Evolutionary Change

• Truly novel features of organisms have evolved infrequently. Most evolutionay changes are the result of modifications of already existing structures. Striking changes in form can be caused by simple genetic changes that alter the rates of growth of different body parts. **Review Figure 19.18**

• Extinctions of major groups opened evolutionary opportunities for other groups of organisms. **Review Figure 19.19**

• Throughout evolution, organisms have increased in size and complexity. Predation rates have also increased, resulting in the evolution of better armor among prey species. **Review Figure 19.20**

The Speed of Evolutionary Change

• After each mass extinction, the diversity of life rebounded within 7 million years, but the groups of organisms that dominated the new biotas differed markedly from those characteristic of earlier biotas.

• Rates of evolutionary change have been very uneven, but even the fastest rates are slow enough to have been caused by known evolutionary agents.

• Periods of rapid evolution have followed times of mass extinction. Rapid evolution also has been stimulated by the provincialization of biotas when continents drifted apart. **Review Figure 19.21**

The Future of Evolution

• The agents of evolution continue to operate today, but human intervention, whether deliberate or inadvertent, now plays an unprecedented role in the history of life.

Self-Quiz

1. The number of species of fossil organisms that has been described is about
   a. 50,000.
   b. 100,000.
   c. 200,000.
   d. 300,000.
   e. 500,000.

2. Radioactive carbon can be used to date the ages of fossil organisms because
   a. all organisms contain many carbon compounds.
   b. radioactive carbon has a regular rate of decay to nonradioactive carbon.
   c. the ratio of radioactive to nonradioactive carbon in living organisms is always the same as that in the atmosphere.
   d. the production of new radioactive carbon in the atmosphere just balances the natural radioactive decay of ^{14}C.
   e. all of the above

3. The total of all species of organisms living in a region is known as its
   a. biota.
   b. flora.
   c. fauna.
   d. flora and fauna.
   e. diversity.

4. The coal beds we now mine for energy are the remains of
   a. trees that grew in swamps during the Carboniferous period.
   b. trees that grew in swamps during the Devonian period.
   c. trees that grew in swamps during the Permian period.
   d. herbaceous plants that grew in swamps during the Carboniferous period.
   e. none of the above

5. The cause of mass extinctions at the end of the Ordivician was
   a. the collision of Earth with a large meteorite.
   b. massive vulcanism.
   c. massive glaciation in Gondwana.
   d. the uniting of all continents to form Pangaea.
   e. changes in Earth's orbit.

6. The cause of the mass extinction at the end of the Mesozoic era probably was
   a. continental drift.
   b. the collision of Earth with a large meteorite.
   c. changes in Earth's orbit.
   d. massive glaciation.
   e. changes in the salt concentration of the oceans.

7. The times during the history of life when many new evolutionary lineages appeared were the
   a. Precambrian, Cambrian, and Triassic.
   b. Precambrian, Cambrian, and Tertiary.
   c. Cambrian, Paleozoic, and Triassic.
   d. Cambrian, Triassic, and Devonian.
   e. Paleozoic, Triassic, and Tertiary.

8. Many scientists believe that the collision of Earth with a large meteorite was a major contributor to the mass extinction at the boundary between the Cretaceous and Tertiary periods, because
   a. there is an iridium-rich layer at the boundary of rocks from these two periods.
   b. a crater that may be the site of the collision has been found off the Yucatán Peninsula.
   c. the mass extinction at the end of the Cretaceous may have been very sudden.
   d. new methods have allowed scientists to date the iridium layer very precisely.
   e. all of the above

9. We know that organisms can evolve very rapidly, because
   a. the fossil record reveals periods of very rapid evolutionary change.
   b. theoretical models of evolutionary change show that rapid change can be produced by natural selection.
   c. rapid evolutionary changes have been produced under artificial selection.
   d. rapid evolutionary changes have been measured in natural populations of organisms during the past century.
   e. all of the above

10. At which of the following times was there *no* mass extinction?
    a. The end of the Cambrian period
    b. The end of the Devonian period
    c. The end of the Permian period
    d. The end of the Triassic period
    e. The end of the Silurian period

Applying Concepts

1. Some lineages of organisms have evolved to contain large numbers of species, whereas other lineages have produced only a few species. Is it meaningful to consider the former as more successful than the latter? What does the word "success" mean in evolution? How does your answer influence your thinking about *Homo sapiens*, the only surviving respresentative of the Hominidae—a family that never had many species in it?

2. If extinction rates in groups of organisms are relatively constant over long time spans, as they sometimes are, then recently evolved species should be no better adapted to their environments than older species are. Does this observation contradict the belief that natural selection adapts organisms to their environments?

3. Why is it useful to be able to date past events absolutely as well as relatively?

4. In a study of lungfish evolution, fossil lungfishes and modern specimens were assigned scores on the basis of the number of derived versus ancestral traits they possessed. On this scale, an organism that has only ancestral traits receives a score of 0. An organism that has only derived traits receives a score of 100. The rate of appearance of these traits in lungfishes is plotted in the two graphs that folllow. What does this pattern suggest about evolutionary radiation among lungfishes?

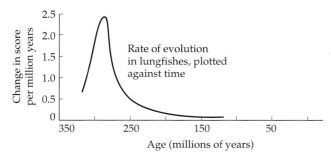

Rate of evolution in lungfishes, plotted against time

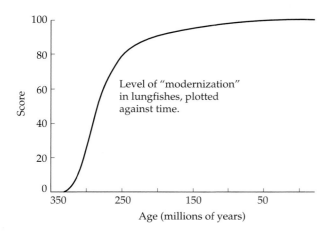

Level of "modernization" in lungfishes, plotted against time.

5. What factors favor increases in body size? How could *average* body size among species in a lineage decrease even if natural selection is favoring large body sizes in most of the species in the lineage?

Readings

Bonner, J. T. 1988. *The Evolution of Complexity by Means of Natural Selection.* Princeton University Press, Princeton, NJ. An excellent treatment of broad patterns in the evolutionary record and the developmental and physiological bases for them.

Burney, D. A. 1993. "Recent Animal Extinctions: Recipe for Disaster." *American Scientist*, vol. 81, pages 530–541. This account shows that late prehistoric extinctions were caused by factors, such as climate and vegetation change, human hunting, and the arrival of exotic animals, that are also serious today.

Dalziel, I. W. D. 1995. "Earth before Pangaea." *Scientific American*, January. An account of how geologists search for clues to the early wanderings of the continents.

Erwin, D. H. 1996. "The Mother of Mass Extinctions." *Scientific American*, July. A description of the mass extinction at the end of the Permian period.

Fenchel, T. and B. J. Finlay. 1994. "The Evolution of Life without Oxygen." *American Scientist*, vol. 82, pages 22–28. A discussion of clues about the evolution of early eukaryotes.

Futuyma, D. J. 1998. *Evolutionary Biology*, 3rd Edition. Sinauer Associates, Sunderland, MA. The most complete general treatment of evolution and its mechanisms.

Gates, D. M. 1993. *Climate Change and Its Biological Consequences*. Sinauer Associates, Sunderland, MA. An accessible book with extensive discussions of Earth's past climates and the methods scientists use to recreate climatic history.

Gordon, M. S. and E. C. Olson. 1995. *Invasions of the Land*. Columbia University Press, New York. A thorough account of the ways in which different lineages of organisms adapted to and colonized terrestrial environments.

Gould, S. J. 1989. *Wonderful Life: The Burgess Shale and the Nature of History*. W. W. Norton, New York. An engaging account of the remarkable fauna, containing representatives of phyla of animals that left no modern descendants, in the Burgess Shale. Explores the implications of this fauna for our view of the history of life.

Gould, S. J. 1994. "The Evolution of Life on Earth." *Scientific American*, October. A discussion of how catastrophes and random events have shaped the course of evolution.

Grimaldi, D. A. 1996. "Captured in Amber." *Scientific American*, April. This account shows how insects preserved in amber reveal aspects of genetic evolution.

Simpson, G. G. 1944. *Tempo and Mode in Evolution*. Columbia University Press, New York. A dated but classic book providing clear pictures of the long-term patterns of evolutionary change.

Stanley, S. M. 1987. *Extinction*. Scientific American Library, New York. A beautifully illustrated account of the fossil record and the phenomenon of extinction.

Ward, P. D. 1992. *On Methuselah's Trail: Living Fossils and the Great Extinctions*. W. H. Freeman, New York. A delightful personal account by a paleontologist who has studied both "living fossils" and true fossils embedded in rocks that formed at the times of the great extinctions.

Chapter 20

The Mechanisms of Evolution

Tuberculosis Is Again a Threat
In recent years there has been a significant increase in cases of tuberculosis, even in children under the age of 5. The bacterium responsible for the disease has evolved resistance to the medications humans use to combat it.

*B*y the 1980s, tuberculosis had almost disappeared from the United States. Doctors and public health officials, believing that they had won the war against the disease, turned their attention to other diseases. But while they looked away, the incidence of tuberculosis began to increase. It increased by nearly 20 percent between 1985 and 1992. The number of cases among children under 5 years old rose 30 percent between 1987 and 1990 alone.

Tuberculosis returned because populations of *Mycobacterium tuberculosis*, the bacterium that causes tuberculosis, evolved resistance to isoniazid, the principal drug used to combat the disease. The bacterium became resistant to the drug when it dropped from its chromosome a gene called *katG*, which codes for the production of two enzymes, catalase and peroxidase. Bacteria lacking this gene produce almost no catalase and peroxidase and thus are resistant to isoniazid. Proof that the lack of *katG* is responsible for the resistance came when investigators reinserted the missing gene into resistant bacteria and the bacteria subsequently were killed by isoniazid.

Tuberculosis is just one of many "nearly cured" diseases whose incidences are increasing. Most of these diseases are resurging because the microorganisms that cause them have developed resistance to drugs that formerly killed them. Many major medical and agricultural problems today are due to the rapid evolution of organisms that we have tried to control or eliminate. Why have so many disease-causing organisms evolved so rapidly? What can we learn by studying these examples of "evolution in action"? Several times in the previous chapter we referred to evolutionary agents, but we did not identify them. In this chapter we examine the processes that drive evolutionary changes and describe the agents that cause them. But first we need to define biological evolution.

Biological evolution is the change over time in the genetic composition of members of a population. Changes that occur over a small number of generations constitute **microevolution**. Changes that happen over centuries, millennia, and longer are called **macroevolution**. Genetic changes result when an evolutionary agent acts on genetic variability within the population. In the mid-1800s Charles Darwin identified natural selection as the key evolutionary agent. At about the same time, Gregor Mendel discovered the genetic basis of the variation on which evolutionary agents act (see Chapter 10). Mendel's and Darwin's insights form the basis for most current evolutionary hypotheses and the studies designed to test them.

In this chapter we will discuss genetically based variation and how it is measured. We will describe the agents of evolution and the short-term studies designed to investigate them. And we will show how genetic variation is maintained in populations over space and time. When you understand these processes, you will understand the mechanisms of evolution and why humans are powerful evolutionary agents.

Variation in Populations

Because biological evolution is a change over time in the genetic composition of members of a population, biologists studying evolution attempt to measure genetic variability and how it changes. The appropriate unit for defining and measuring genetic variation is a **population**: a group of organisms of the same species occupying a particular geographic region. Therefore, to understand evolutionary changes we need to engage in "population thinking."

The genetic constitution governing a heritable trait is called its **genotype** (see Chapter 10). A population evolves when individuals with different genotypes survive or reproduce at different rates. A single individual has only some of the alleles found in the population to which it belongs (Figure 20.1). The sum total of the alleles found in the population constitutes its **gene pool**.

The gene pool contains the variability on which agents of evolution act. Evolution may come about because heritable traits influence the ability of individuals to obtain mates or food or to avoid hazards. The agents of evolutionary change act on the physical expression of an organism's genotype, its **phenotype**. However, not all phenotypic variation is governed by genotypes. Some of the variation observed within populations is genetically determined, but some of it is not.

Some variation is environmentally induced

The shapes and sizes of many marine animals that live attached to a substratum (such as a rock) are affected by water temperature, concentration of nutrients (par-

ticularly calcium), competition with neighbors, and turbulence of the water. For example, limpets (*Acmaea*) growing high in the intertidal zone (the shallow water at the edge of land), where they experience heavy wave action, are more cone-shaped than are limpets of the same species growing in the subtidal zone, where they are protected from wave action. We know that this difference is not genetic, because individuals from high in the intertidal zone add new growth to their shells to produce a flatter, subtidal shape when they are transplanted to the subtidal zone (Figure 20.2a).

The cells of the leaves on a tree or shrub are normally genetically identical. Yet leaves on the same tree often differ in shape and size. Leaves closer to the top of an oak tree, for example, where they receive more wind and sunlight, may be more deeply lobed than leaves lower down on the same tree (Figure 20.2b). These within-plant variations, however, are not passed on to offspring. What *is* passed on is the ability to form various types of leaves from the same genotype in response to different environmental conditions.

Environmentally induced variation is often important in biology, but to investigate evolutionary questions, biologists must work with variation in traits that are heritable.

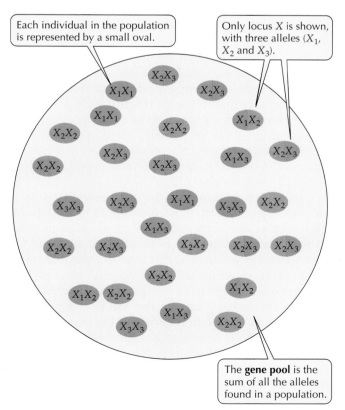

Each individual in the population is represented by a small oval.

Only locus X is shown, with three alleles (X_1, X_2 and X_3).

The **gene pool** is the sum of all the alleles found in a population.

20.1 A Gene Pool The allele frequencies in this drawing are 0.20 for X_1, 0.50 for X_2, and 0.30 for X_3.

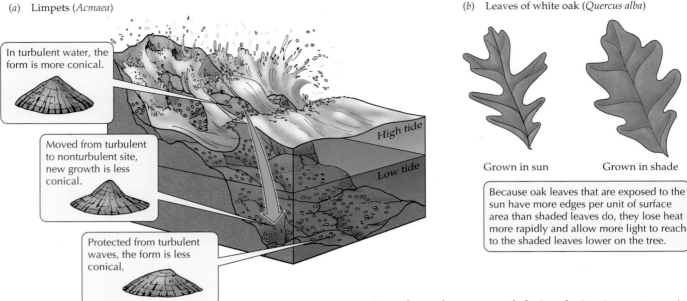

(a) Limpets (*Acmaea*)

In turbulent water, the form is more conical.

Moved from turbulent to nonturbulent site, new growth is less conical.

Protected from turbulent waves, the form is less conical.

High tide

Low tide

(b) Leaves of white oak (*Quercus alba*)

Grown in sun

Grown in shade

Because oak leaves that are exposed to the sun have more edges per unit of surface area than shaded leaves do, they lose heat more rapidly and allow more light to reach to the shaded leaves lower on the tree.

20.2 Environmentally Induced Variation Individuals that are genetically identical may vary morphologically if they live in different environments.

Much variation is genetically based

High levels of genetic variation characterize nearly all natural populations. This fact has been demonstrated over and over again for thousands of years by people attempting to breed desirable traits in plants and animals. For example, artificial selection for different traits in a common European wild mustard produced many important crop plants (Figure 20.3). Plant and animal breeders can achieve such results only if the original population has genetic variation for the traits of interest. The almost universal success of breeders indicates that genetic variation is common, but it does not tell us how much there is.

Laboratory experiments also demonstrate that considerable genetic variation is present in most populations. For example, investigators chose as parents for subsequent generations of a species of fruit fly (*Drosophila*) individuals with either high or low numbers of bristles on their bodies. After 35 generations, flies in both lineages had bristle numbers that fell well outside the range found in the original population (Figure 20.4). Notice that considerable variation that can be acted on by evolutionary agents still remains.

To understand evolution, we need to know more precisely how much genetic variation populations have, the sources of that genetic variation, and how genetic variation is maintained and expressed in populations in space and over time. We also need to know the agents that change the genetic variation in popula-

tions, how they act, and their relative importance in affecting the direction of evolutionary changes.

How do we measure genetic variation?

A locally interbreeding group within a geographic population is called a **deme** or a Mendelian population. Demes are often the subjects of evolutionary studies. In this section we will describe how genetic variation within demes is measured and how it is influenced by evolutionary agents. To measure the gene pool of a deme we would need to count every allele at every locus in every organism in that deme. By measuring all the individuals in a deme, we could determine the relative proportions, or frequencies, of all alleles in the deme.

This would be an impossible task, but biologists can *estimate* **allele frequencies** by measuring numbers of alleles in a *sample* of individuals from a deme. Measures of frequency range from 0 to 1; the sum of all allele frequencies at a locus is equal to 1. The percentages of different alleles at each locus and the percentages of different genotypes in a deme describe its **genetic structure**. An allele's frequency is calculated using the following formula:

$$\frac{\text{number of copies of the allele in the population}}{\text{sum of alleles in the population}}$$

If only two alleles (*A* and *a*) for a given locus are found among the members of a diploid population, they may combine to form three different genotypes: *AA*, *Aa*, and *aa*. Using the formula above, we can calculate the relative frequencies of alleles *A* and *a* in a population of *N* individuals as follows:

Let N_{AA} be the number of individuals that are homozygous for the *A* allele (*AA*);

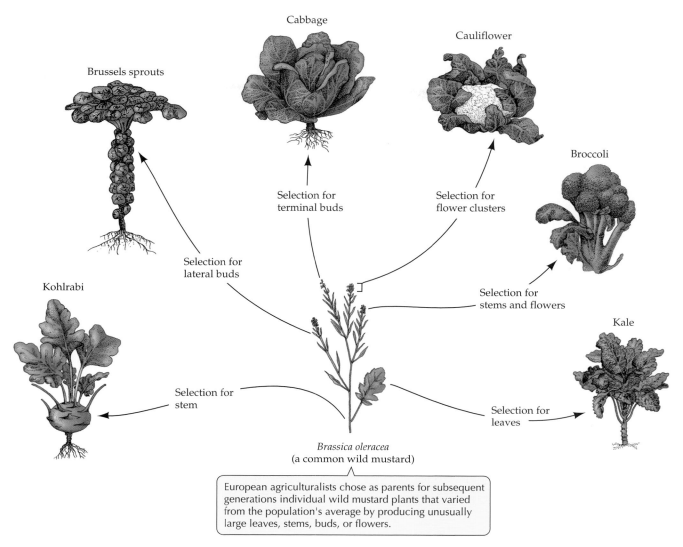

European agriculturalists chose as parents for subsequent generations individual wild mustard plants that varied from the population's average by producing unusually large leaves, stems, buds, or flowers.

20.3 Many Vegetables from One Species   All of these crop plants have been derived from one wild mustard species. They illustrate the vast amount of variation that can be present in a gene pool.

Let N_{Aa} be the number that are heterozygous (Aa);

and

Let N_{aa} be the number that are homozygous for the a allele (aa).

Note that $N_{AA} + N_{Aa} + N_{aa} = N$, the total number of individuals in the population, and that the total number of alleles present in the population is $2N$ because each individual is diploid. Each AA individual has two A alleles and each Aa individual has one A allele. Therefore, the total number of A alleles in the population is $2N_{AA} + N_{Aa}$, and the total number of a alleles in the population is $2N_{aa} + N_{Aa}$.

If p represents the frequency of A, and q represents the frequency of a, then

$$p = \frac{2N_{AA} + N_{Aa}}{2N}$$

and

$$q = \frac{2N_{aa} + N_{Aa}}{2N}$$

To see how this works, let's calculate allele frequencies in two populations, each consisting of 200 diploid individuals. Population 1 has mostly homozygotes (90 AA, 40 Aa, and 70 aa); population 2 has mostly heterozygotes (45 AA, 130 Aa, and 25 aa). In population 1, where $N_{AA} = 90$, $N_{Aa} = 40$, and $N_{aa} = 70$,

$$p_1 = \frac{2N_{AA} + N_{Aa}}{2N} = \frac{180 + 40}{400} = 0.55$$

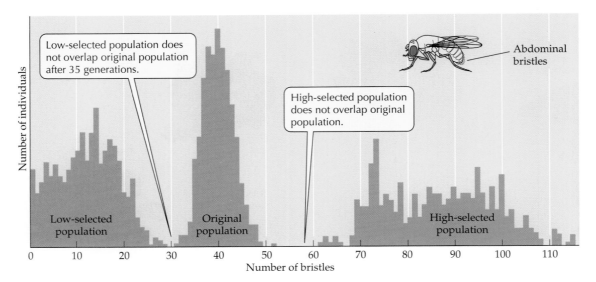

20.4 Artificial Selection   In laboratory experiments with *Drosophila*, changes in bristle number evolved rapidly when artificially selected for.

and

$$q_1 = \frac{2N_{aa} + N_{Aa}}{2N} = \frac{140 + 40}{400} = 0.45$$

In population 2, where $N_{AA} = 45$, $N_{Aa} = 130$, and $N_{aa} = 25$,

$$p_2 = \frac{2N_{AA} + N_{Aa}}{2N} = \frac{90 + 130}{400} = 0.55$$

and

$$q_2 = \frac{2N_{aa} + N_{Aa}}{2N} = \frac{50 + 130}{400} = 0.45$$

These calculations demonstrate two important points. First, notice that for each population, $p + q = 1$. If there is only one allele in a population, its frequency is 1. If an allele is missing from a population, its frequency is 0, and the locus in that population is represented by one or more other alleles. Because $p + q = 1$, $q = 1 - p$, which means that when there are two alleles at a locus in a population, we can calculate the frequency of one allele and then easily obtain the second frequency by subtraction.

The second thing to notice from these calculations is that the two populations—one consisting mostly of homozygotes and the other mostly of heterozygotes—have the same allele frequencies for *A* and *a*. Therefore, they have the same gene pool for this locus. However, because the alleles in the gene pool are distributed differently among genotypes, the populations have *different genetic structures*.

Although we began our calculations with *numbers* of genotypes, for many purposes genotypes, like alleles, are best thought of as frequencies. In population 1 of our example, the genotype frequencies, which we calculate as the number of individuals that have the genotype divided by the total number of individuals in the population, are 0.45 *AA*, 0.20 *Aa*, and 0.35 *aa*. What are the genotype frequencies of population 2?

Preserving Genetic Variability: The Hardy–Weinberg Rule

A population that is not changing genetically, that has the same allele and genotype frequencies from generation to generation, is said to be at **equilibrium**. The conditions that result in an equilibrium population were discovered independently by the British mathematician Godfrey H. Hardy and the German physician Wilhelm Weinberg in 1908. Hardy wrote his equations in response to a question posed to him by the Mendelian geneticist Reginald C. Punnett at the Cambridge University faculty club. Punnett was puzzled by the fact that although the allele for short fingers was dominant and the allele for normal-length fingers was recessive, most people in Britain had normal-length fingers. Hardy's equations explain why dominant alleles do not replace recessive alleles in populations. They also explain other features of the genetic structure of populations.

The Hardy–Weinberg rule, as these equations are now collectively called, consists of a set of assumptions and two major results. The assumptions are that the population is very large, that mating is random, and that no agents of evolution are acting on the population. The first major result of the rule is that if these conditions hold, the frequencies of alleles at a locus will remain constant from generation to generation. The second result is that after one generation of random mating, the genotype frequencies will remain in the proportions

$$p^{2(AA)} + 2pq^{(Aa)} + q^{2(aa)} = 1$$

To see why these results are true, consider an example in which the frequency of A alleles (p) is 0.6. Because we assume that individuals select mates without regard to genotype, gametes carrying A or a combine at random—that is, as predicted by the frequencies p and q. The probability that any given sperm or egg in this example will bear an A allele rather than an a allele is 0.6. In other words, 6 out of 10 random selections of a sperm or an egg will bear an A allele. Because $q = 1 - p$, the probability of an a allele is $1 - 0.6 = 0.4$.

To obtain the probability of two A-bearing gametes coming together at fertilization, we multiply the two independent probabilities of drawing them: $p \times p = p^2 = (0.6)^2 = 0.36$ (see the discussion of probability in Chapter 10). Therefore, 0.36, or 36 percent, of the offspring in the next generation will have the AA genotype. Similarly, the probability of bringing together two a-bearing gametes is $q \times q = q^2 = (0.4)^2 = 0.16$, so 16 percent of the next generation will have the aa genotype (Figure 20.5).

Figure 20.5 also shows that there are two ways of producing a heterozygote: an A sperm may combine with an a egg, the probability of which is $p \times q$; or an a sperm may combine with an A egg, the probability of which is also $p \times q$. Consequently, the overall probability of obtaining a heterozygote is $2pq$. What percentage of the next generation will be heterozygotes?

It is easy now to show that the allele frequencies p and q remain constant for each generation. Notice that the total of $p^2 + pq$ represents the total of the A alleles. The fraction that this frequency constitutes of all alleles is

$$\frac{p^2 + pq}{p^2 + 2pq + q^2} = \frac{p(p+q)}{(p+q)(p+q)} = \frac{p}{p+q} = \frac{p}{p+(1-p)} = p$$

Similarly, the frequency of a in the next generation will be

$$\frac{q^2 + pq}{p^2 + 2pq + q^2} = \frac{q(p+q)}{(p+q)(p+q)} = \frac{q}{p+q} = \frac{q}{(1-q)+q} = q$$

Thus the original allele frequencies are unchanged and the population is in Hardy–Weinberg equilibrium.

Why is the Hardy–Weinberg rule important?

The most important message of the Hardy–Weinberg rule is that allele frequencies remain the same from generation to generation unless a particular agent acts

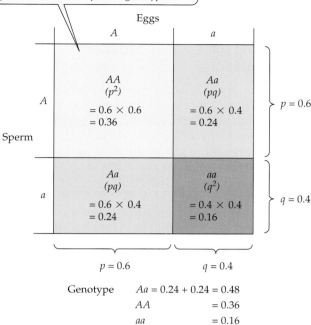

The areas within the squares are proportional to the expected frequencies of possible matings if mating is random with respect to genotype.

20.5 Calculating Hardy–Weinberg Genotype Frequencies
The probabilities of producing each genotype are calculated by assuming that mating is random. Because there are two ways of producing a heterozygote, the probability of this event occurring is the sum of the two Aa squares.

to change them. The rule also shows what distribution of genotypes to expect for a population at equilibrium at any value of p and q.

Given the stringency of the necessary conditions, you may already have recognized that populations in nature are rarely in Hardy–Weinberg equilibrium. Why, then, is the rule considered so important for the study of evolution? The answer is that without it, we cannot tell whether evolutionary agents are operating. If individuals in a population mate randomly and no other agents are operating to change allele frequencies, then the genotype frequencies we actually observe in the population will approximate those we calculate from the Hardy–Weinberg formula. However, if the frequencies of genotypes deviate significantly from the expected Hardy–Weinberg values, we have evidence that an agent of evolution is in action.

In practice it is not easy to measure allele frequencies precisely enough to determine if they are at Hardy–Weinberg proportions. Also, when we consider more than a single locus, the Hardy–Weinberg rule does not apply. Natural selection usually operates simultaneously on many traits or on traits governed jointly by more than one locus. Therefore to tell whether an evolutionary agent is influencing a population, biologists usually must measure the actual changes in allele frequencies in that population.

Changing the Genetic Structure of Populations

Evolutionary agents are forces that change the allele and genotype frequencies in a population. They cause deviations from the Hardy–Weinberg equilibrium. The known evolutionary agents are mutation, gene flow, genetic drift, nonrandom mating, and natural selection. We will discuss each in turn.

Mutations are random changes in genetic material

The origin of genetic variation is mutation (see Chapters 10 and 12). Most mutations are harmful or neutral to their bearers, but if the environment changes, previously neutral or harmful alleles may become advantageous. Mutation rates are very low for most loci that have been studied. Rates as high as one mutation in a thousand zygotes per generation are rare; one in a million is more typical. Nonetheless, these rates are sufficient to create considerable genetic variation over long time spans. In addition, mutations can restore to populations alleles that other evolutionary agents remove. Thus mutations both create and help maintain variation within populations.

Mutation rates are unusually high among some disease-causing microorganisms. This is especially true of viruses that are attacked by the immune systems of their hosts. For example, the human immunodeficiency virus type 1 (HIV-1; see Chapters 13 and 18), which is actively attacked by the human immune system, has such high mutation rates that it can overcome the host's defenses and evolve resistance faster than its host can respond. The viruses in a person with an advanced case of AIDS are genetically quite different from those that initially started the infection—they have evolved.

One condition for Hardy–Weinberg equilibrium is that there be no mutation. Although this condition is never strictly met, the rate at which mutations arise at single loci is usually so low that mutations result in only very small deviations from Hardy–Weinberg expectations. If large deviations are found, it is appropriate to dismiss mutation as the cause and to look for evidence of other evolutionary agents.

Migration of individuals followed by breeding produces gene flow

Because few populations are completely isolated from other populations of the same species, usually some migration between populations takes place. **Gene flow** happens when migration is *followed by breeding* in the new location. Gene flow ranges from extremely low to very high, depending on the number of migrating individuals and their genotypes.

Immigrants (individuals entering a population) may add new alleles to the pool of a population or may change the frequencies of alleles already present. Emigrants (individuals leaving a population) may completely remove alleles or may change the frequencies of alleles when they leave a population. For a population to be in Hardy–Weinberg equilibrium, there must be no net gene flow between it and other populations—in other words, the number of new alleles added by immigrants must equal the number of alleles removed by emigrants.

Genetic drift may cause large changes in small populations

In small populations, chance events can significantly alter allele frequencies. Such alteration is called **genetic drift**. Genetic drift is the reason that a population must be very large to be in Hardy–Weinberg equilibrium. If only a few individuals or a few gametes are drawn at random to form the next generation, the alleles they carry are not likely to be in the same proportions as alleles in the gene pool from which they were drawn. In very small populations, genetic drift may be strong enough to influence the direction of change of allele frequencies even when other evolutionary agents are pushing the frequencies in a different direction. Harmful alleles, for example, may increase because of genetic drift, and rare advantageous alleles may be lost. Two important causes of genetic drift are bottlenecks and founder effects.

BOTTLENECKS. Even organisms that normally have large populations may pass through occasional periods when only a small number of individuals survive. During these population **bottlenecks**, genetic variation can be lost by chance. This phenomenon is illustrated in Figure 20.6, which shows allele frequencies as proportions of red and yellow beans. Most of the beans that survive to germinate the next generation in this example are red, so the new population has a much higher frequency of red beans than the previous generation had.

Suppose we have performed a cross of $Aa \times Aa$ individuals of a species of *Drosophila* to produce an offspring population in which $p = q = 0.5$ and in which the genotype frequencies are 0.25 AA, 0.50 Aa, and 0.25 aa. If we randomly select four individuals from among the offspring to form the next generation, the allele frequencies in this small sample may differ markedly from $p = q = 0.5$. If, for example, we happen by chance to draw two AA homozygotes and two heterozygotes (Aa), the genotype frequencies in this "surviving population" are $p = 0.75$ and $q = 0.25$. If we replicate this sampling experiment 1,000 times, one of the two alleles will be missing entirely from about eight of the 1,000 "surviving populations."

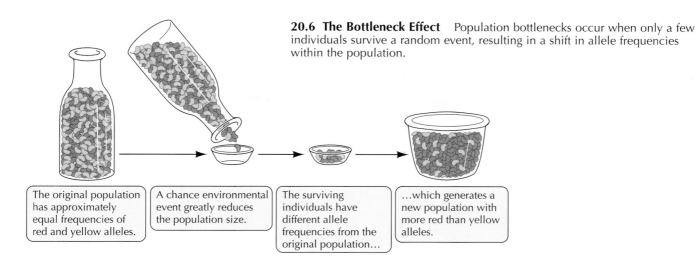

20.6 The Bottleneck Effect Population bottlenecks occur when only a few individuals survive a random event, resulting in a shift in allele frequencies within the population.

The original population has approximately equal frequencies of red and yellow alleles.

A chance environmental event greatly reduces the population size.

The surviving individuals have different allele frequencies from the original population…

…which generates a new population with more red than yellow alleles.

Populations in nature pass through bottlenecks for many different reasons. Predators may reduce populations of their prey to very small sizes. During the 1890s, hunting reduced the number of northern elephant seals to about 20 animals in a single population on the coast of Mexico. The actual breeding population may have been even smaller because only a few males mate with all the females and father the offspring in any generation (Figure 20.7).

By analyzing small samples of tissue collected from the current California population of northern elephant seals, which are descendants of the Mexican population that survived the bottleneck of the 1890s, scientists determined that these seals have less genetic variation than any other seal that has been studied. The investigators examined 24 proteins from each animal electrophoretically (see Chapter 16), looking for evidence of genetic variation among the seals at the loci encoding the proteins. They found no evidence of variation in any of the 24 proteins.

By contrast, the southern elephant seal, whose numbers were not severely reduced by hunting, has much more genetic variation. Currently, northern elephant seal populations are expanding rapidly, so their reduced genetic variability is not preventing high survival and reproductive rates. However, biologists worry that they may be vulnerable to a

disease outbreak or other sudden environmental change.

FOUNDER EFFECTS. When a species expands into new regions, populations may be started by a small number of pioneering individuals. These pioneers are not likely to have all the alleles found in their source population. Even if they do, the allele frequencies are likely to differ from those in the source population. The situation is equivalent to that for a large population reduced by a bottleneck, but rather than a small surviving population, there is a small founding population. This type of genetic drift is called a **founder effect**. Because many plant species reproduce sexually by self-fertilization, a new plant population may be started by a single seed—an extreme example of a founder effect.

20.7 A Species with Low Genetic Variation
A few males sire most of the offspring in this northern elephant seal breeding colony, and the size of the breeding population is smaller than the whole population. This nonrandom mating, together with a bottleneck that occurred when the seals were overhunted in the late nineteenth century, resulted in a population with very little genetic variation.

Scientists were given an opportunity to study the genetic composition of a founding population when *Drosophila subobscura*, a well-studied European species of fruit fly, was discovered near Puerto Montt, Chile, in 1978 and at Port Townsend, Washington, in 1982. In both South and North America, populations of the flies grew rapidly and expanded their ranges. Today in North America, *D. subobscura* ranges from British Columbia, Canada, to central California. In Chile it has spread across 23° of latitude, nearly as wide a range as the species has in Europe (Figure 20.8).

The *D. subobscura* founders probably reached Chile and the United States from Europe aboard the same ship, because both populations are genetically very similar. For example, the North and South American populations have only 20 chromosomal inversions (see Chapter 12), 19 of which are the same on the two continents, whereas 80 inversions are known from European populations. New World populations also have fewer enzyme alleles than Old World populations. Only alleles that have a frequency higher than 0.1 in European populations are present in the Americas. Thus, as expected from a small founding population, only a small part of the total genetic variability found in Europe reached the Americas. Geneticists estimate that at least ten, but no more than a hundred, flies initially arrived in the New World.

EFFECTIVE POPULATION SIZE.   Genetic drift can be important even in common species, many of which are divided into small, localized breeding populations. In addition, although there may be N individuals in a population, if some of them do not reproduce, the population is actually much smaller from a genetic point of view. For example, if only half of the individuals in a population of $N = 100$ breed, the population has an effective size, N_e, of 50, and N_e rather than N determines the rate of genetic drift. More generally, anything that causes some individuals to leave more offspring than others—such as unequal sex ratios, breeding systems in which only a few males in the population fertilize most of the females, and differential survival—reduces the effective population size.

Nonrandom mating increases the frequency of homozygotes

One Hardy–Weinberg assumption specifies that individuals do not choose mates on the basis of their genotypes; that is, mating must be random. In many cases, however, individuals with certain genotypes mate more often with individuals of either the same or different genotypes than would be expected on a random basis. Humans, for example, tend to choose mates that are similar in appearance to themselves. When such **assortative mating** takes place, in the next generation homozygous genotypes are overrepresented and heterozygous genotypes are underrepresented in comparison with Hardy–Weinberg expectations.

Self-fertilization (selfing), another form of nonrandom mating, is common in many groups of organisms, especially plants. Selfing tends to reduce the frequencies of heterozygous individuals in populations below Hardy–Weinberg expectations. Under assortative mating and self-fertilization, genotype frequencies change but allele frequencies remain the same. Nonrandom mating can alter allele frequencies, however, if some individuals are more successful in mating than others. We consider this situation, which is called **sexual selection**, in greater detail later in this chapter and in Chapter 50.

Natural selection leads to adaptation

Individuals vary in heritable traits that determine the success of their reproductive efforts. Not all individuals survive and reproduce equally well in a particular environment. Therefore, some individuals contribute more offspring to the next generation than do other individuals. As a result, allele frequencies in the population change.

The differential contribution of offspring resulting from variations in heritable traits was called **natural selection** by Charles Darwin. Natural selection is the only evolutionary agent that adapts organisms to their environments. Biologists investigate the action of natural selection by comparing genotype (or phenotype) frequencies between generations and attempting to

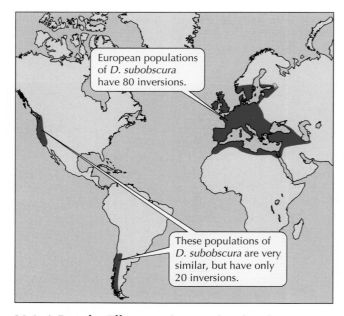

European populations of *D. subobscura* have 80 inversions.

These populations of *D. subobscura* are very similar, but have only 20 inversions.

20.8 A Founder Effect   Within two decades of arriving in the New World, populations of the fruit fly *Drosophila subobscura* have increased dramatically and spread widely in spite of having much-reduced genetic variation.

determine the reasons that some genotypes (or phenotypes) survived and reproduced better than others.

Natural selection produces variable results

Depending on which traits are favored in a population, natural selection can produce any one of several quite different results. Selection may (1) preserve the characteristics of a population by favoring average individuals, (2) change the characteristics of a population by favoring individuals that vary in one direction from the mean of the population, or (3) change the characteristics of a population by favoring individuals that vary in both directions from the mean of the population.

In the example in Figure 20.9, the variable trait is size, a trait likely to be controlled by many loci. If many genetic and environmental factors contribute to size—and there is no selection—then the distribution of sizes in a population should approximate the bell-shaped curve shown in the top row of the figure.

If both the smallest and the largest individuals contribute relatively fewer offspring to the next generation than those closer to the center do, **stabilizing selection** is operating (Figure 20.9a). Stabilizing selection

reduces variation but does not change the mean. Natural selection frequently acts in this way, countering increases in variation brought about by mutation or migration. We know from the fossil record that most populations evolve slowly most of the time. Rates of evolution are typically very slow because natural selection is usually stabilizing.

If individuals at one extreme of the size distribution—the larger ones, for example (as illustrated in Figure 20.9b)—contribute more offspring than other individuals do, then the mean size of individuals in the population will increase or decrease accordingly, and **directional selection** is operating. If directional selection operates over many generations, an evolutionary trend within the population results. Directional evolutionary trends often continue for many generations, but they may be reversed when the environment changes and different phenotypes are favored.

Individuals of plants growing on mine tailings may be able to tolerate high soil concentrations of heavy metals that are lethal to plants of the same species growing elsewhere. That ability resulted from directional selection. How tolerance might have evolved was demonstrated by an experimenter who planted seeds of

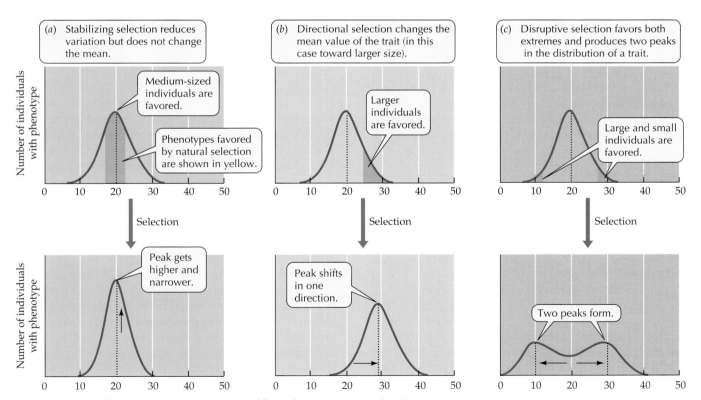

20.9 Natural Selection Operates on a Variable Trait   The curves plot the distributions of phenotypes (body size) in a population before selection (top) and after selection (bottom). Natural selection may change the shape and position of the original curves.

individuals from a nontolerant population in soil contaminated with copper. He harvested seeds from the few individuals that survived and reproduced, and planted them in contaminated soil. He repeated this procedure over several generations and found that the proportion of seeds that produced viable adults increased each generation. Tolerance to high concentrations of copper evolved rapidly because only tolerant individuals survived long enough to reproduce.

Disruptive selection is selection that simultaneously favors individuals at both extremes of the distribution (Figure 20.9*c*). This type of selection apparently is rare. When disruptive selection operates, individuals at the extremes contribute more offspring than those in the center, producing two peaks in the distribution of a trait. The strikingly bimodal (two-peaked) distribution of bill sizes in the black-bellied seed-cracker, *Pyrenestes ostrinus*, a West African finch (Figure 20.10), illustrates how disruptive selection can adapt populations in nature.

Seeds of two types of sedges (a marsh plant) are the most abundant food source for the finches during part of the year. Birds with large bills readily crack the hard seeds of the sedge *Scleria verrucosa*; birds with small bills crack *S. verrucosa* seeds only with difficulty, but small-billed birds feed more efficiently on the soft seeds of *S. goossensii* than do birds with larger bills.

Young finches whose bills deviate markedly from the two predominant bill sizes do not survive as well as finches whose bills are close to one of the two sizes represented by the distribution peaks. Because there are few abundant food sources in the environment and because the seeds of the two sedges do not overlap in hardness, birds with intermediate-sized bills are inefficient in utilizing either one of the principal food sources. Disruptive selection therefore maintains a bill size distribution with two peaks.

Sexual reproduction increases genetic variation

Recombination in sexually reproducing organisms multiplies the opportunities for natural selection to operate by increasing genetic variability. In asexually reproducing organisms, the daughter cells resulting from the mitotic division of a single cell normally contain identical genotypes at all loci.

In a population that reproduces asexually, there is no recombination of genetic material from different individuals, and every new individual is genetically identical to its parent unless there has been a mutation. However, when organisms exchange genetic material during sexual reproduction, the offspring differ from their parents because chromosomes assort randomly during meiosis and because fertilization brings together material from two different cells.

Sexual recombination generates an endless variety of genotypic combinations that increases the *evolutionary potential* of populations. Because it increases the

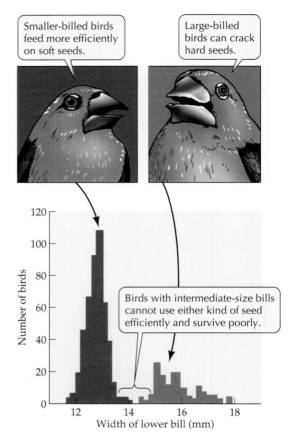

20.10 Natural Selection Alters Bill Sizes   The distribution of bill sizes in the black-bellied seedcracker of West Africa is an example of disruptive selection, which favors individuals with larger and smaller bill sizes over individuals with intermediate-size bills.

variation among the offspring produced by an individual, sexual recombination improves the chance that some of the offspring will be successful in the varying and often unpredictable environments they will encounter. As the Hardy–Weinberg rule shows, sexual recombination does not influence the frequency of different alleles. Rather, it generates new combinations of genetic material on which natural selection can act.

Adaptation is studied in various ways

Biologists use several different methods to study how natural selection adapts organisms to their environments. One way is to measure the consequences of altering the form of a particular feature of an organism. Another, known as the comparative method, is to predict patterns of adaptation among species. We provide an example of each method.

ALTERING TAIL LENGTH IN BARN SWALLOWS.   Both male and female barn swallows have long forked tails, but the tails of males are about 16 percent longer than the tails of females. Swallows molt their feathers every autumn, at which time they grow the new tails they

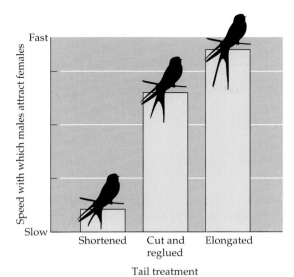

20.11 **Females Prefer Longer Tails**   This experiment, in which the tails of male barn swallows were artificially shortened or lengthened, showed that female barn swallows prefer long-tailed males.

will have the next spring. By cutting off the ends of the tails of some males and elongating the tails of others by gluing additional feathers to them, an investigator showed that females prefer to mate with males that have longer than average tails. (Control males were handled and their tails were cut and reglued but not changed in length.)

The males with long tails attracted females faster than the males with short tails (Figure 20.11). However, flying around with a long tail is energetically costly. Males whose tails were artifically elongated one year grew tails that were shorter than normal the next year, and they took longer to attract mates than males whose tails had not been elongated. In barn swallows,

then, sexual selection by females results in males having longer tails than females.

Why do females preferentially mate with males that have longer tails? They evolved this preference because by exercising it they produce healthier offspring. Their offspring are healthier because males with naturally longer tails are more vigorous than are males with shorter tails. Males that are resistant to the mites that often infest swallow nests become heavier and they grow longer tails than less resistant males do. Offspring of heavily parasitized males are more likely to become infested, too, even if they are raised in foster nests away from their fathers.

MALE PLUMAGE AND PARASITE RESISTANCE.   If resistance to parasites is heritable, females benefit by mating with resistant males. By choosing males on the basis of traits that signal their resistance, females may favor the evolution of a more striking expression of those traits. In many bird species, males with fewer parasites are healthier and thus more vividly colored than males who suffer with many parasites. Therefore, in species that typically are subject to heavy parasite infestation, females should pay particular attention to a male's bright plumage when choosing a mate; if, on the other hand, parasite infestation is not typically a problem for the species, females should pay little attention to the color of a male's feathers.

Many such evolutionary theories are tested by predicting and testing patterns among species. To test the prediction that colorful plumage would be more likely to occur in heavily parasitized than in lightly parasitized species, scientists compared the male plumages of many species. They did indeed find a positive correlation between brightly hued males and high normal levels of parasite infestation, giving the females of these species the opportunity to judge the health of prospective mates by their colors (Figure 20.12).

Piranga olivacea (scarlet tanager)

Passerina ciris (painted bunting)

Icterus galbula (northern oriole)

20.12 **Brightly Colored Males Are Healthy**   Sexual selection for healthy males has produced the striking plumages of male birds in (from left) the scarlet tanager, painted bunting, and northern oriole.

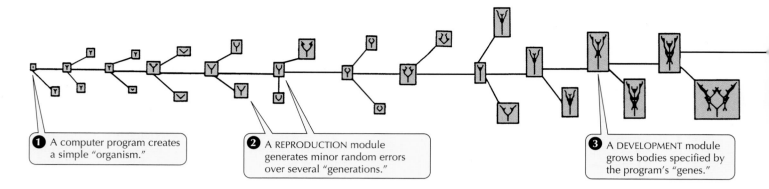

❶ A computer program creates a simple "organism."

❷ A REPRODUCTION module generates minor random errors over several "generations."

❸ A DEVELOPMENT module grows bodies specified by the program's "genes."

Natural selection acts nonrandomly

Many people have erroneously concluded that Darwinian evolution is a random process because mutation, which provides the raw material on which natural selection acts, is random. Nothing could be further from the truth. Although mutations appear randomly, natural selection increases the frequency of only those mutant alleles that improve survival and reproductive success. Thus, evolutionary changes are the result of nonrandom processes continuing over many generations.

In addition to being nonrandom, adaptation by natural selection is a cumulative process. Natural selection acts on modifications to already existing organisms. As Richard Dawkins pointed out in his book *The Blind Watchmaker*, cumulative evolutionary change is like a computer program in which in every generation a module called REPRODUCTION takes the genes handed to it by the previous generation and hands them to the next generation, but with minor random errors.

After being reproduced, the genes are handed to another module, called DEVELOPMENT, which grows the body specified by the genes. The offspring carrying mutant genes have bodies slightly different from those of their parents. The modified bodies are acted on by another module, called EVOLUTION, that favors some body forms over others. The favored ones become the parents of the next generation. By this process, surprising changes can accumulate over a small number of generations, even though each change is small.

Figure 20.13 shows the results that Dawkins obtained during 29 generations of the computer program he created. The program started with a simple dot. Subsequent "mutations" caused the dot to branch, the branches to curve, and then the branches to branch again. This artificial example shows how complex forms can arise by nonrandom selection of slight modifications of previous body plans.

Fitness is the relative reproductive contribution of genotypes

A central evolutionary concept is **fitness**. The fitness of a genotype or phenotype is its reproductive contribution to subsequent generations *relative* to the contribution of other genotypes or phenotypes. The word "relative" is critical; the absolute number of offspring produced by an individual does not influence allele frequencies. Changes in *absolute* numbers of offspring are responsible for increases and decreases in the *size* of a population, but the *relative* success among genotypes within a population is what leads to changes in allele frequencies—that is, to evolution.

An individual may influence its own fitness in two ways. First, it may produce its own offspring, contributing to its **individual fitness**. Second, it may help the survival of relatives that have the same alleles by descent from a common ancestor—a phenomenon known as **kin selection**. Together, individual fitness and kin selection determine the inclusive fitness of the individual.

Among species that either are solitary or reproduce in groups no larger than a pair and its offspring, individual fitness strongly dominates inclusive fitness. However, among highly social species, such as social insects, some birds, and many primates, kin selection also may be very important. We will return to this topic in Chapter 50. Now we will consider the role of individual selection in the distribution and maintenance of genetic variation in natural poplations.

Maintaining Genetic Variation

Genetic drift, stabilizing selection, and directional selection all tend to reduce genetic variation within populations. Nevertheless, most populations show considerable genetic variation. Why isn't the genetic variation of a species lost over time and space? To answer this question we will distinguish between heritable and nonheritable variation and examine how genetic variation is distributed among and within populations of a species. By combining these components we will be able to understand why genetic variation is maintained.

Not all genetic variation affects fitness

Genetic variation that does not affect the fitness of an organism is called **neutral variation**. For example, some amino acid substitutions caused by mutations

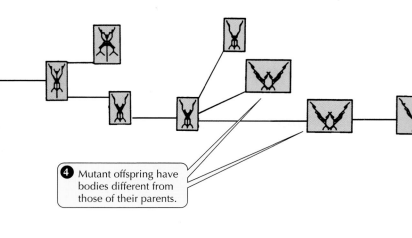

4 Mutant offspring have bodies different from those of their parents.

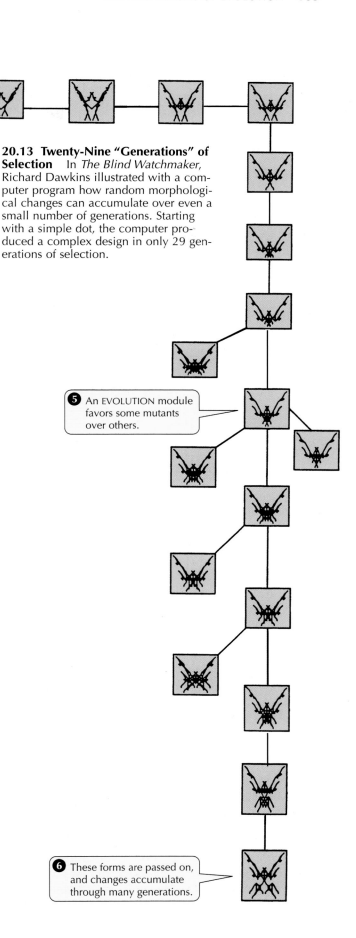

20.13 Twenty-Nine "Generations" of Selection In *The Blind Watchmaker*, Richard Dawkins illustrated with a computer program how random morphological changes can accumulate over even a small number of generations. Starting with a simple dot, the computer produced a complex design in only 29 generations of selection.

5 An EVOLUTION module favors some mutants over others.

6 These forms are passed on, and changes accumulate through many generations.

do not affect the functioning of the molecules of which they are a part. Such substitutions may be lost, or their frequencies may increase with time, untouched by natural selection. Therefore, much variation in neutral traits exists in most populations. Variation in the traits of organisms that we can observe with our unaided senses usually is not neutral, but much molecular variation apparently is neutral.

Modern molecular techniques enable us to measure neutral variation and provide the means by which to distinguish adaptive from neutral variation. How molecular techniques enable us to make those discriminations and how neutral variations can be used to estimate rates of evolution will be considered in Chapter 23.

Genetic variation is maintained spatially

Some of the genetic variation among populations of many species is correlated with the geographic distribution of the populations. For example, plant populations may vary geographically in the chemicals they synthesize to defend themselves against herbivores. Some individuals of the clover *Trifolium repens* produce the poisonous chemical cyanide. Poisonous individuals are less appealing to herbivores—particularly mice and slugs—than are nonpoisonous individuals. However, clover plants with cyanide are more likely to be killed by frost, because freezing damages cell membranes and releases the toxic cyanide into the plant's own tissues.

Thus in populations of *Trifolium repens*, the frequency of cyanide-producing individuals increases gradually from north to south and from east to west across Europe (Figure 20.14). Poisonous plants are abundant in clover populations only in areas where the winters are mild and the plants do not poison themselves. Cyanide-producing individuals are rare where winters are cold, even though herbivores graze them heavily in those areas because plants cannot use cyanide as a defense there.

Gradual geographic changes in phenotypes and genotypes, as illustrated by *Trifolium repens*, are called **clines**. Clines are widespread among most groups of organisms. In some regions, however, frequencies of

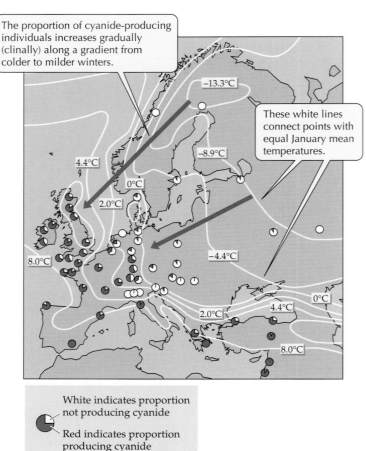

The proportion of cyanide-producing individuals increases gradually (clinally) along a gradient from colder to milder winters.

These white lines connect points with equal January mean temperatures.

−13.3°C

4.4°C

−8.9°C

0°C

2.0°C

8.0°C

−4.4°C

0°C

2.0°C

4.4°C

8.0°C

White indicates proportion not producing cyanide

Red indicates proportion producing cyanide

20.14 Geographic Variation in Poisonous Clovers The frequencies of cyanide-producing individuals in populations of white clover (*Trifolium repens*) are represented by the proportion of each circle that is red.

certain traits change abruptly, creating **step clines**. Color patterns in the rat snake, *Elaphe obsoleta*, are a good example (Figure 20.15). In this species the color differences are complex and striking. Single color patterns dominate extensive regions, and each change from one pattern to the next is abrupt. Because there are no obvious environmental changes in the regions where color patterns change, the causes of the abrupt changes are unknown.

Genetic variation exists within local populations

Natural selection often preserves variation by favoring different traits in different areas, as illustrated by clines and step clines. In such cases, much of the variation in a large population is preserved as differences between subpopulations. Natural selection also preserves variation as **polymorphisms**—genetic differences *within* a local population. For example, a polymorphism is maintained when the success of a genotype (or phenotype) depends on its frequency rel-

ative to other genotypes (or phenotypes), a process known as **frequency-dependent selection**.

A small fish that lives in Lake Tanganyika in east central Africa provides an example. The mouth of this scale-eating fish, *Perissodus microlepis*, opens either to the right or to the left as a result of an asymmetrical jaw joint (Figure 20.16). *P. microlepis* approaches its prey (another fish) from behind and dashes in to bite off several scales from its flank. "Right-mouthed" individuals always attack from the victim's left; "left-mouthed" individuals always attack from the victim's right. The distorted mouth enlarges the area of teeth in contact with the prey's flank, but only if the scale eater attacks from the appropriate side.

Prey fish are alert to approaching scale eaters, so attacks are more likely to be successful if the prey must watch both flanks. Guarding by the prey favors equal numbers of right-mouthed and left-mouthed *Perissodus*, because if one form were more common than the other, prey fish would pay more attention to potential attacks from the corresponding flank. Over an 11-year period in which the fish in Lake Tanganyika were studied, natural selection maintained the polymorphism, keeping the two forms of *P. microlepis* at about equal frequencies.

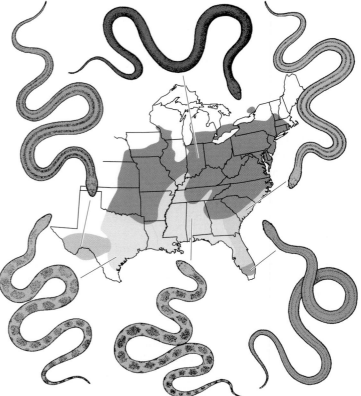

20.15 Step Clines in the Rat Snake There are no obvious environmental changes at the boundaries where the color patterns of these snakes change.

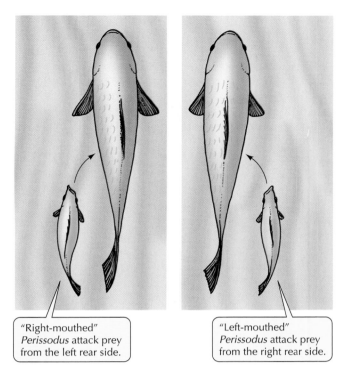

"Right-mouthed"
Perissodus attack prey
from the left rear side.

"Left-mouthed"
Perissodus attack prey
from the right rear side.

20.16 Balanced Polymorphism Natural selection maintains equal frequencies of left-mouthed and right-mouthed individuals of the scale-eating fish *Perissodus microlepis*.

Short-Term versus Long-Term Evolution

The short-term changes in populations—that is, changes in populations over years and decades—that we have been discussing in this chapter are microevolutionary changes. Studies of microevolution are an important part of evolutionary biology because short-term changes can be observed directly and can be manipulated experimentally.

Although studies of short-term changes reveal much about evolution, by themselves they cannot provide a complete explanation of the macroevolutionary changes we described in Chapter 19. We could measure the shapes of the legs of horses and correlate leg shape with the horses' running speed and their ability to escape from predators (microevolution). But such studies would not tell us why the small, four-toed, forest-dwelling horse *Hyracotherium* gave rise to large, plains-inhabiting descendants with a single toe (macroevolution).

Patterns of macroevolutionary changes can be strongly influenced by events that occur so infrequently or so slowly that they are unlikely to be observed during microevolutionary studies. The evolution of horses was influenced by long-term climatic changes that caused forests to shrink and grasslands to expand over large parts of Earth. In addition, the ways in which evolutionary agents act may change

with time; even among the descendants of a single ancestral species, different lineages may be evolving in different directions.

Therefore, we cannot interpret the past simply by extending the results of short-term experiments and observations backward in time. Additional types of evidence, such as the occurrence of rare and unusual events and trends in the fossil record, must be gathered if we wish to understand the course of evolution over billions of years.

Summary of "The Mechanisms of Evolution"

• Biological evolution is a change over time in the genetic composition of members of a population.
• A population evolves when individuals having different genotypes survive or reproduce at different rates.
• Biological evolution results from the actions of evolutionary agents over millions of years.

Variation in Populations

• For a population to evolve, its members must possess genetic variation, which is the raw material on which agents of evolution act. High levels of genetic variation characterize nearly all natural populations. **Review Figures 20.3, 20.4**
• Allele frequencies measure the amount of genetic variation in a population. Genotype frequencies show how a population's genetic variation is distributed among its members.
• Biologists estimate allele frequencies by measuring a sample of individuals from a population. The sum of all allele frequencies at a locus is equal to 1.
• Populations that have the same allele frequencies may nonetheless have different genotype frequencies.

Preserving Genetic Variability: The Hardy–Weinberg Rule

• A population that is not changing genetically is said to be in equilibrium. Hardy–Weinberg equilibrium is possible only if a population is very large, mating is random, and no evolutionary agents are acting on the population.
• In a population at Hardy–Weinberg equilibrium, allele frequencies remain the same from generation to generation. In addition, genotype frequencies will remain in the proportions $p^2(AA) + 2pq(Aa) + q^2(aa) = 1$. **Review Figure 20.5**
• Biologists can determine if an agent of evolution is acting on a population by comparing the genotype frequencies of that population with Hardy–Weinberg equilibrium frequencies.

Changing the Genetic Structure of Populations

• Changes in allele frequencies and genotype frequencies within populations are caused by the actions of several different evolutionary agents: mutation, gene flow, genetic drift, nonrandom mating, and natural selection.
• The origin of genetic variation is mutation. Most mutations are harmful or neutral to their bearers, but some are advantageous, particularly if the environment changes.

• The migration of individuals from one population to another, followed by breeding in the new location, produces gene flow. Immigrants may add new alleles to a population or may change the frequencies of alleles already present. Emigrants may remove alleles from a population when they leave.

• Genetic drift alters allele frequencies primarily in small populations. Organisms that normally have large populations may pass through occasional periods (bottlenecks) when only a small number of individuals survive. New populations established by a few founding immigrants also have gene frequencies that differ from those in the parent population. **Review Figures 20.6, 20.8**

• When individuals mate more often with individuals that have the same or different genotypes than would be expected on a random basis—that is, when mating is not random—frequencies of homozygous and heterozygous genotypes differ from Hardy–Weinberg expectations.

• Self-fertilization, an extreme form of nonrandom mating, reduces the frequencies of heterozygous individuals below Hardy–Weinberg expectations.

• Natural selection is the only agent of evolution that adapts populations to their environments. Natural selection may preserve allele frequencies or cause them to change with time.

• Stabilizing selection, directional selection, and disruptive selection change the distributions of phenotypes governed by more than one locus. **Review Figures 20.9, 20.10**

• Sexual recombination generates an endless variety of genotypic combinations that increases the evolutionary potential of populations, but it does not influence the frequencies of alleles. Rather it generates new combinations of genetic material on which natural selection can act.

• Biologists study adaptation by experimentally altering organisms or their environments and by comparing traits among species. **Review Figures 20.11, 20.12**

• Natural selection acts nonrandomly, and adaptation by natural selection is a cumulative process extending over many generations. Cumulative evolutionary change results from directional selection acting on variation in populations over many generations. **Review Figure 20.13**

• The fitness of a genotype or phenotype is its contribution to subsequent generations relative to the contributions of other genotypes or phenotypes. An individual may influence its fitness by producing offspring, which contributes to its individual fitness, and by helping the survival of relatives. Individual fitness and kin selection in combination determine the inclusive fitness of the individual.

Maintaining Genetic Variation

• Natural selection maintains genetic variation within a species when different traits are favored in different places and when the direction of selection changes over time in a given place. Most species may vary geographically; the variation can be gradual or abrupt. **Review Figures 20.14, 20.15**

• Genetic variation within a population may be maintained by frequency-dependent selection. **Review Figure 20.16**

Short-Term versus Long-Term Evolution

• Patterns of macroevolutionary change can be strongly influenced by events that occur so infrequently or so slowly that they are unlikely to be observed during microevolutionary studies. Additional types of evidence must be gathered to understand macroevolution.

Self-Quiz

1. The phenotype of an organism is
   a. the type specimen of its species in a museum.
   b. its genetic constitution, which governs its traits.
   c. the chronological expression of its genes.
   d. the physical expression of its genotype.
   e. the form it achieves as an adult.

2. The appropriate unit for defining and measuring genetic variation is
   a. the cell.
   b. the individual.
   c. the population.
   d. the community.
   e. the ecosystem.

3. Which statement about allele frequences is *not* true?
   a. The sum of any set of allele frequencies is always 1.
   b. If there are two alleles at a locus and we know the frequency of one of them, we can obtain the frequency of the other by subtraction.
   c. If an allele is missing from a population, its frequency is 0.
   d. If two populations have the same gene pool for a locus, they will have the same proportion of homozygotes at that locus.
   e. If there is only one allele at a locus, its frequency is 1.

4. In a population at Hardy–Weinberg equilibrium in which the frequency of A alleles (p) is 0.3, the expected frequency of Aa individuals is
   a. 0.21.
   b. 0.42.
   c. 0.63.
   d. 0.18.
   e. 0.36.

5. Natural selection that preserves existing allele frequencies is called
   a. unidirectional selection.
   b. bidirectional selection.
   c. prevalent selection.
   d. stabilizing selection.
   e. preserving selection.

6. Laboratory selection experiments with fruit flies have demonstrated that
   a. bristle number is not genetically controlled.
   b. bristle number is not genetically controlled, but changes in bristle number are caused by the environment in which the fly is raised.
   c. bristle number is genetically controlled, but there is little variation on which natural selection can act.
   d. bristle number is genetically controlled, but selection cannot result in flies having more bristles than any individual in the original population had.
   e. bristle number is genetically controlled, and selection can result in flies having more bristles than any individual in the original population had.

7. Disruptive selection maintains bill size variability in the West African seedcracker because
   a. bills of intermediate shapes are difficult to form.
   b. the two major food sources of the finches differ markedly in size and hardness.

c. males use their large bills in displays.

d. migrants introduce different bill sizes into the population each year.

e. older birds need larger bills than younger birds.

8. A population is said to be polymorphic for a locus if it has at least
   a. three different alleles at that locus.
   b. two different alleles at that locus.
   c. two genotypes for that locus.
   d. three genotypes for that locus.
   e. two genotypes for that locus, the rarest of which is more common than expected by mutation alone.

9. A cline is defined as
   a. the distribution of an organism across a slope.
   b. an abrupt change in frequencies of certain traits over time.
   c. a gradual change in frequencies of certain traits over time.
   d. an abrupt change in frequencies of certain traits over space.
   e. a gradual change in frequencies of certain traits over space.

10. Adaptation is studied by
    a. altering the form of an organism and observing the consequences.
    b. testing predictions by comparing traits in many species.
    c. developing theoretical models.
    d. selecting for traits in the laboratory.
    e. all of the above

Applying Concepts

1. During the past 50 years, more than 200 species of insects that attack crop plants have become highly resistant to DDT and other pesticides. Using your recently acquired knowledge of evolutionary processes, explain the rapid and widespread evolution of resistance. Propose ways of using pesticides that would slow down the rate of evolution of resistance. Now that DDT has been banned in the United States, what do you expect to happen to levels of resistance to DDT among insect populations? Justify your answer.

2. In nature, mating among individuals in a population is never truly random, and natural selection is seldom totally absent. Why, then, does it make sense to use the Hardy–Weinberg model, which is based on assumptions known generally to be false? Can you think of other models in science that are based on false assumptions? How are such models used?

3. An investigator is studying populations of house mice living in barns and sheds on a large farm. Each building has a population of between 25 and 50 mice. Populations in different buildings have strikingly different frequen-

cies of alleles determining coat color and tail length. By marking most individuals, the investigator determines that mice move only rarely between buildings. He interprets his study as providing evidence for random genetic drift. Could other agents of evolution plausibly account for the pattern? If so, how could they be distinguished?

4. As far as we know, natural selection cannot adapt organisms to future events. Yet many organisms exhibit responses in advance of natural events. For example, many mammals go into hibernation while it is still quite warm. Similarly, many birds leave the temperate zone for their southern wintering grounds long before winter has arrived. How can these "anticipatory" behaviors evolve?

5. Many people believe that species, like individual organisms, have life cycles. They believe that species are born by a process of speciation, grow and expand, and inevitably die out as a result of "species old age." Could any agent of evolution cause such a species life cycle? If not, how do you explain the high rates of extinction of species in nature?

Readings

Amábile-Cuevas, C. F., M. Cárdenas-García and M. Ludger. 1995. "Antibiotic Resistance." *American Scientist*, vol. 83, pages 320–329. An explanation of why bacteria are becoming increasingly resistant to antibiotics.

Dawkins, R. 1987. *The Blind Watchmaker*. W. W. Norton, New York. An engaging treatment showing how adaptation can happen in a world without design.

Dawkins, R. 1995. "God's Utility Function." *Scientific American*, November. An explanation of why patterns of seemingly intelligent design can be explained by basic evolutionary mechanisms.

Dennett, D. C. 1995. *Darwin's Dangerous Idea*. Simon and Schuster, New York. A compelling exposition of evolution and the meaning of life.

Endler, J. A. 1986. *Natural Selection in the Wild*. Princeton University Press, Princeton, NJ. A thorough review of the problems and successes in measuring natural selection in nature.

Futuyma, D. J. 1998. *Evolutionary Biology*, 3rd Edition. Sinauer Associates, Sunderland, MA. A comprehensive review of all aspects of evolutionary biology.

Hartl, D. L. and A. G. Clark. 1998. *Principles of Population Genetics*, 3rd Edition. Sinauer Associates, Sunderland, MA. An introduction to all aspects of modern population genetics.

Le Guenno, B. 1995. "Emerging Viruses." *Scientific American*, October. An explanation of the origins of the new viruses that increasingly cause outbreaks of disease.

Nowak, M. A. and A. J. McMichael. 1995. "How HIV Defeats the Immune System." *Scientific American*, August. A discussion of how proliferating, mutating viruses evolve faster than the immune system can change.

Williams, G. C. 1992. *Natural Selection: Domains, Levels, and Challenges*. Oxford University Press, New York. A thorough analysis of the mechanisms and meanings of natural selection as an evolutionary agent.

Chapter 21

Species and Their Formation

Trinidad Rain Forest
The mosquito that transmits malaria in Trinidad breeds in forests like this one.

During the 1940s officials in Trinidad launched an intensive campaign to control malaria. They spent much money on spraying and draining marshes, in the belief that malaria was transmitted by *Anopheles albimanus*, a swamp-breeding mosquito that is the principal vector of malaria in Latin America. The campaign failed because the principal vector of malaria in Trinidad was *Anopheles bellator*, a mosquito that breeds in water held within the leaves of bromeliads (relatives of pineapples) growing on the branches of palm trees.

Similarly, in Europe, malaria was believed to be transmitted by mosquitoes of a single species: *Anopheles maculipennis*. European efforts to control malaria sometimes succeeded and sometimes failed, because *A. maculipennis* is not a single species, but consists of at least 18 species that can be distinguished only by examination of their chromosomes. Some of the species breed in fresh water, others in brackish water. Some enter houses; others do not. Furthermore, which mosquito species is the vector of malaria changes regionally. Control efforts are successful only when directed against the species that actually transmits malaria in that area.

Millions of species inhabit Earth. Many of them are easy to identify, but some, such as the species in the *Anopheles maculipennis* complex, are not. All species, living and extinct, are believed to be descendants of a single ancestral species that lived several billion years ago. If speciation had been a rare event, the biological world would be very different than it is today. Speciation is an essential ingredient of evolutionary diversification, and species are the fundamental units of the biological classification systems we will discuss in Chapter 22.

But what are species and how did these millions of species form? How does one species become two? What factors stimulate such splitting? What conditions spur evolutionary radiations? Does speciation accelerate rates of evolutionary change? These and other related questions are the subject of this chapter.

What Are Species?

The word "Species" means, literally, "kinds;" but what do we mean by "kinds"? Someone who is knowledgeable about a group of organisms, such as orchids or lizards, usually can distinguish the different species of that group found in a particular area simply by examining them superficially. The patterns of morphological similarities that unite groups of organisms and separate them from other groups are familiar to all of us. The standard field guides to birds, mammals, insects, and flowers are possible only because most species are cohesive units that may change in appearance only gradually over large geographic distances. We can easily recognize red-winged blackbirds from New York and red-winged blackbirds from California as members of the same species (Figure 21.1).

But not all members of a species look that much alike. How do we decide whether similar but different individuals should be called different species or regarded as varieties within a species? The concept that has guided these decisions for a long time is genetic integration. If individuals within a population mate with one another but not with individuals of other populations, they constitute a reproductively isolated group within which genes recombine; that is, they are *independent evolutionary units*. These independent evolutionary groups are usually called species. Identifying

and naming species correctly is important because when different investigators report the results of studies of a particular species, we need to know that they are writing about the same species.

More than 200 years ago the Swedish biologist Carolus Linnaeus, who originated the system of naming organisms that we use today, described hundreds of species. Because he knew nothing about the mating patterns of the organisms he was naming, Linnaeus classified them on the basis of their appearance. Many species that were classified by their appearance, when nothing was known about their reproductive behavior, are actually independent evolutionary units. This is not surprising because the members of an evolutionary unit share genes inherited from common ancestors. These individuals look alike because they share many of the alleles that code for their structures.

The species definition that has been used by most biologists was proposed by Ernst Mayr in 1940. He stated, "Species are groups of actually or potentially interbreeding natural populations which are reproductively isolated from other such groups." The "groups" in this definition are collections of local populations. The words "actually or potentially" assert that, even if some members of a species are not in the same place and hence are unable to mate, they should not be placed in separate species if they would be likely to mate if they were together. The word "natural" is an important part of the definition because only *in nature* does the exchange of genes, which occurs within species, affect evolutionary processes; the interbreeding of two different species in captivity does not.

Gene exchange is the main reason why species are cohesive units. Individuals that mate with each other "recognize" one another as suitable mates. Many biologists study how individuals use visual, vocal, and chemical clues to recognize suitable mates and to avoid mating with individuals belonging to other species.

(a)

(b)

21.1 Redwings Are Redwings Everywhere   Both of these birds are obviously red-winged blackbirds, even though one (a) lives in Texas and the other (b) lives in Ontario.

Biologists attempting to reconstruct the evolutionary histories of organisms also try to identify independent evolutionary units. How biologists define species and identify evolutionary lineages will be the focus of Chapter 22.

How Do New Species Arise?

Not all evolutionary changes result in new species. Evolution creates two patterns across time and space: anagenesis and cladogenesis. **Anagenesis** is change in a single lineage through time. With sufficient time, the changes may be so great that the descendants are given another name, even though no "new" species has formed. Anagenetic changes are a common feature of the fossil record.

Cladogenesis (speciation) is the process by which one species splits into two species, which thereafter evolve as distinct lineages. Although Charles Darwin entitled his book *The Origin of Species,* he did not extensively discuss how a single species splits into two or more daughter species. Rather, he was concerned principally with demonstrating anagenesis, that species are altered by natural selection over time.

The critical process in the formation of two species from one ancestral species is the separation of the gene pool belonging to the ancestral species into two separate gene pools. Subsequently, within each isolated gene pool, allele and gene frequencies may change as a result of the action of evolutionary agents. If sufficient differences have accumulated during the period of isolation, the two populations may not exchange genes when they come together again.

Gene flow among populations may be interrupted in several ways, each of which characterizes a mode of speciation. The next three sections focus on these modes of speciation: sympatric speciation, allopatric speciation, and parapatric speciation.

Sympatric speciation occurs without physical separation

The subdividing of a gene pool even though members of the daughter species are not physically separated during the process is called **sympatric speciation** (*sym-*, "with"; *patris*, "country"). The most common means of sympatric speciation is **polyploidy**, a duplication of the number of chromosomes (see Chapter 9).

Polyploidy arises in two ways. One way is the accidental production during cell division of cells having four (tetraploid) rather than two (diploid) sets of chromosomes. This process produces an **autopolyploid** individual, one having more than two sets of chromosomes, all derived from a single species. This tetraploid individual cannot mate successfully with diploids, but if it self-fertilizes or mates with other tetraploids, a new evolutionary lineage may form.

A polyploid species can also be produced when individuals of two different species, whose chromosomes do not pair properly during meiosis, interbreed. The resulting individuals, called **allopolyploids**, are usually sterile, because the chromosomes from one species do not pair properly with those from the other species during meiosis, but they can reproduce asexually. After many generations, some of the individuals may become fertile as a result of further chromosome duplication.

Polyploidy can create a new species among plants much more easily than among animals because plants of many species can reproduce by self-fertilization as well as by crossing with a relatively unrelated individual. If the polyploidy arises in several offspring of a single parent, the siblings can fertilize one another. Speciation by polyploidy has been very important in the evolution of flowering plants. Botanists estimate that more than half of all species of flowering plants are polyploids. Most of these arose as a result of hybridization between two species, followed by self-fertilization.

The importance of allopolyploidy and the speed with which it can produce new species are illustrated by salsifies (*Tragopogon*), members of the sunflower family. Salsifies are weedy plants that thrive in disturbed areas around towns. People have inadvertently spread them around the world from their ancestral ranges in Eurasia. Three diploid species of salsify were introduced into North America early in this century: *T. porrifolius, T. pratensis,* and *T. dubius.* Two tetraploid hybrids—*T. mirus* and *T. miscellus*—between species of the original three were first reported in 1950. Both hybrids have spread since their discovery and today are more widespread than their diploid parents (Figure 21.2).

Studies of their cells have revealed that both types of hybrids have been formed more than once. Some populations of *T. miscellus,* for example, have the chloroplast genome of *T. pratensis,* whereas other populations have the chloroplast genome of *T. dubius.* Differences in the ribosomal genes of local populations of *T. miscellus* show that this allopolyploid has evolved independently at least three times. Scientists seldom know the dates and locations of species formation so well. The success of newly formed hybrid species of salsifies illustrates why so many species of flowering plants originated as polyploids.

Among animals, sympatric speciation by polyploidy is relatively rare, but a few cases are known. The tree frog *Hyla versicolor* is a tetraploid species with a broad range in eastern North America. *H. versicolor* arose recently as a result of at least three different hybridizations between individuals from eastern and western populations of its diploid relative *Hyla chrysocelis.* The eastern and western populations of *H. chrysocelis* were

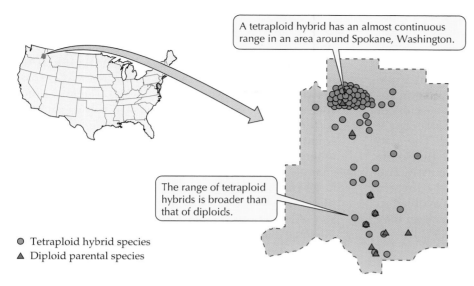

A tetraploid hybrid has an almost continuous range in an area around Spokane, Washington.

The range of tetraploid hybrids is broader than that of diploids.

● Tetraploid hybrid species
▲ Diploid parental species

21.2 Polyploids Can Outperform Their Parents   *Tragopogon* species (salsifies) are members of the sunflower family. The map shows the distribution of the diploid parent species and the tetraploid hybrid species of *Tragopogon* in eastern Washington and adjacent Idaho.

isolated from one another for about 4 million years, enough time for substantial genetic differences to accumulate. Today the ranges of the two populations abut in a north-south line west of the Mississippi River (Figure 21.3). Eastern and western individuals rarely hybridize in nature, probably because the calls of the males are so different that females respond only to the calls of males of their own type when they have a choice.

Among animals, sympatric speciation as a result of habitat selection is more common than speciation by polyploidy. A good example is speciation in the picture-winged fruit fly, *Rhagoletis pomenella*, in New York. Until the mid 1800s, these fruit flies courted, mated, and deposited their eggs only on hawthorn berries. The larvae learned the odor of hawthorn as they fed on the berries, and when they emerged from their pupae they used this food-based memory to locate other hawthorn plants.

About 150 years ago, large commercial apple orchards were planted in the Hudson River Valley. A few

21.3 Gray Tree Frogs Speciated by Intraspecific Hybridization   The tetraploid *Hyla versicolor* resulted from hybridization between eastern and western populations of the diploid *H. chrysocelis*.

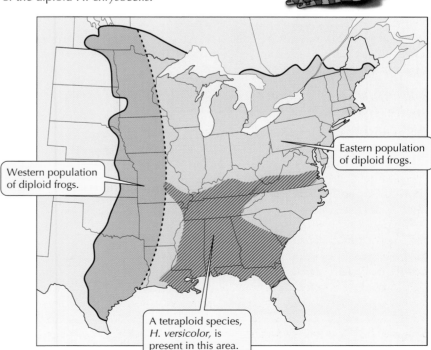

Western population of diploid frogs.

Eastern population of diploid frogs.

A tetraploid species, *H. versicolor,* is present in this area.

female *Rhagoletis* apparently laid their eggs on apples, perhaps by mistake. Their larvae did not grow as well as the larvae on hawthorn berries, but many did survive. These larvae had learned the odor of apples, so when they emerged as adults they sought out apple trees, where they mated with other flies reared on apples. Today there are two sympatric species of *Rhagoletis* in the Hudson River Valley. One feeds on hawthorn berries, the other on apples. The two species are reproductively isolated because they mate only with individuals raised on the same fruit, and because they emerge from their pupae at different times. In addition, apple-feeding flies have evolved so that they now grow more rapidly on apples than they originally did.

Allopatric speciation requires isolation by distance

Speciation that results when a population is divided by a barrier is known as **allopatric speciation** (*allo-*, "different"), or **geographic speciation** (Figure 21.4). Allopatric speciation is the dominant form of speciation among most groups of organisms. The range of a species may be divided by a physical barrier, such as water gaps for terrestrial organisms, dry land for aquatic organisms, and mountains. Barriers can form when continents drift, sea levels rise, or climates change. Populations separated in this way are often large initially. They evolve differences because the places in which they live are or become different.

Alternatively, allopatric speciation may result when some members of a population cross an existing barrier and found a new isolated population. Populations established in this way are typically small at first. They usually differ genetically from their parent populations because a small group of individuals has only an incomplete representation of the genes found in its parent population.

Allopatric speciation by this sampling effect is illustrated by the singing honeyeater, a common Australian bird that lives on the mainland and on coastal islands. The birds on Rottnest Island, 20 km off the coast of Western Australia, sing fewer song types than mainland birds do, and their songs have fewer syllables and notes. Evidently the island colonizers did not carry all of the alleles responsible for the full range and complexity of songs found in mainland individuals. Mainland singing honeyeaters do not respond to the songs of island birds and Rottnest birds do not respond to songs of mainland birds.

Dispersal across barriers often leads to species formation. For example, many of the hundreds of species of the fruit fly *Drosophila* in the Hawaiian Islands are restricted to a single island. They are almost

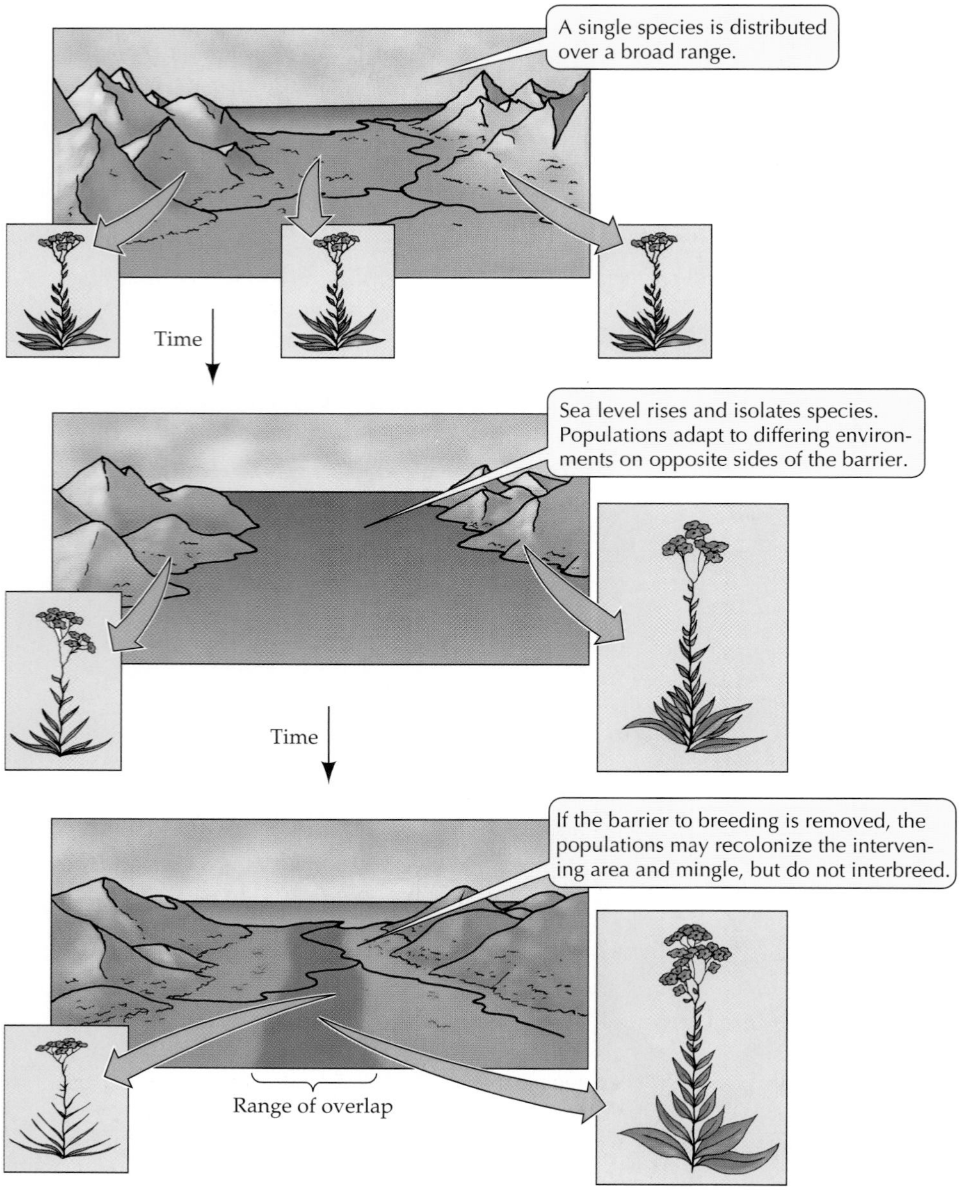

A single species is distributed over a broad range.

Sea level rises and isolates species. Populations adapt to differing environments on opposite sides of the barrier.

If the barrier to breeding is removed, the populations may recolonize the intervening area and mingle, but do not interbreed.

Range of overlap

21.4 Allopatric Speciation
Also known as geographic speciation, allopatric speciation may result when a population is divided by a physical barrier such as rising seas.

21.5 Founder Events Lead to Allopatric Speciation The extremely high level of speciation found among *Drosophila* in the Hawaiian Islands is almost certainly the result of founder events—new populations founded by individuals dispersing among the islands.

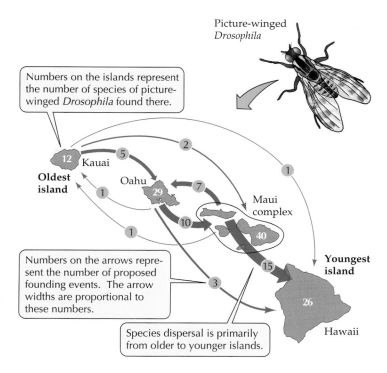

Picture-winged *Drosophila*

Numbers on the islands represent the number of species of picture-winged *Drosophila* found there.

Numbers on the arrows represent the number of proposed founding events. The arrow widths are proportional to these numbers.

Species dispersal is primarily from older to younger islands.

Kauai — **Oldest island** — 12

Oahu

Maui complex

Youngest island

Hawaii

certainly the result of new populations founded by individuals dispersing among the islands, because the closest relative of a species on one island is often a species on a neighboring island rather than a species on the same island. On the basis of studies of their chromosomes, speciation among the picture-winged *Drosophila* is believed to have been caused by at least 45 founder events (Figure 21.5).

If the environments on the two sides of the physical barrier differ, evolutionary agents may cause the populations to diverge further genetically. Differences that accumulate while a barrier is in place may become so large that the populations will fail to establish gene flow if the barrier later breaks down; that is, they will have evolved to be different species.

The finches of the Galapagos Islands, 1,000 km off the coast of Ecuador, demonstrate the importance of geographic isolation for speciation. Darwin's finches (as they are usually called, because Darwin was the first scientist to study them) arose on the Galapagos Islands by speciation from a single South American species that colonized the islands. Today there are 14 species of Galapagos finches, all of which differ strikingly from the probable mainland ancestor (Figure 21.6).

The islands of the Galapagos Archipelago are sufficiently isolated from one another that the finches seldom migrate between them. Also, environmental conditions differ among the islands. Some are relatively flat and arid; others have forested mountain slopes. Populations of finches on different islands have differentiated enough that when occasional immigrants arrive from other islands, they either do not breed with the residents, or if they do, the resulting offspring usually do not survive as well as those produced by pairs composed of island residents. The genetic distinctness of different populations is thus maintained.

A barrier's effectiveness at preventing gene flow depends on the size and mobility of species. What is a firm barrier for a terrestrial snail may be no barrier at all to a butterfly or a bird. Populations of wind-pollinated plants are totally isolated at the maximum distance pollen is blown by the wind, but individual plants are effectively isolated at much shorter distances. Among animal-pollinated plants, the width of the barrier is the distance that pollinators travel while carrying pollen (see Chapter 36). Even animals with great powers of dispersal are often reluctant to cross narrow strips of unsuitable habitat. For animals that cannot swim or fly, narrow water-filled gaps may be effective barriers.

Parapatric speciation separates adjacent populations

Sometimes reproductive isolation develops among adjacent members of a population in the absence of a geographic barrier. Known as **parapatric speciation** (*para*, "beside"), this type of speciation is much less common than allopatric or sympatric speciation because gene flow usually prevents differentiation between populations in contact. Occasionally, however, a species boundary forms where there is a marked change in environment. That is, allopatric speciation becomes parapatric speciation when the geographical boundary separating species becomes extremely small.

Unusually abrupt changes in soil are created by mining activities that leave rubble (tailings) with high concentrations of heavy metals that are detrimental to plant growth. For example, the soils developing on the tailings at the Goginian lead mine near Aberystwyth, Wales, are highly contaminated with lead, but where the tailings end, they suddenly give way to normal rich pastureland (Figure 21.7).

The pasture grass *Agrostis tenuis* is common on both types of soils, but there is a sharp gradient in lead tolerance among plants less than 20 m apart. Plants on the mine tailings grow well in lead concentrations that would be lethal to plants growing just a few meters away. Nearly complete reproductive isolation exists between plants on contaminated and normal soil because they flower at different times. These two populations have not yet been designated as separate species, but reproductive isolation between them has already evolved, demonstrating that gene

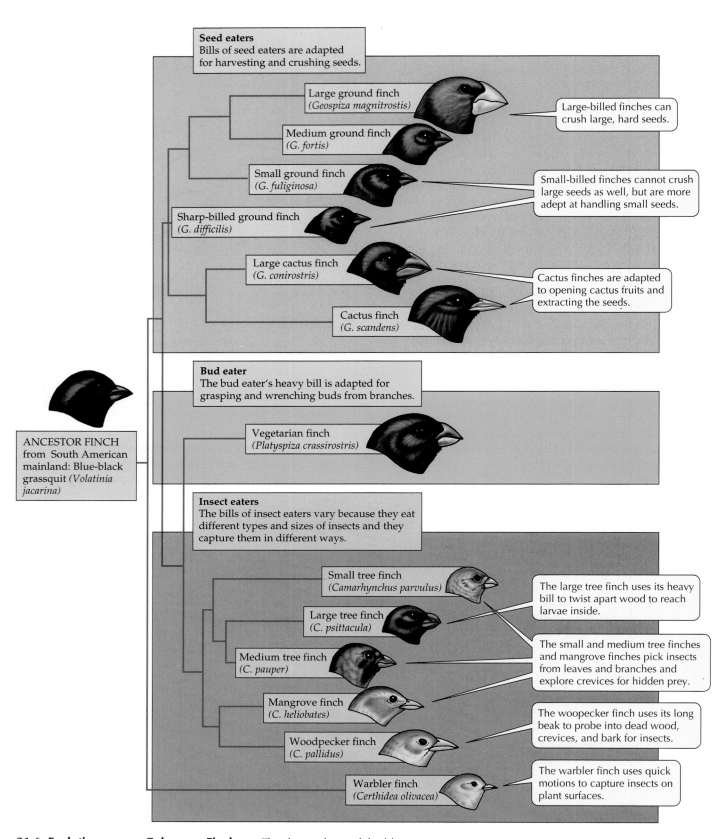

Seed eaters
Bills of seed eaters are adapted for harvesting and crushing seeds.

Large ground finch
(*Geospiza magnitrostis*)

Medium ground finch
(*G. fortis*)

Small ground finch
(*G. fuliginosa*)

Sharp-billed ground finch
(*G. difficilis*)

Large cactus finch
(*G. conirostris*)

Cactus finch
(*G. scandens*)

Large-billed finches can crush large, hard seeds.

Small-billed finches cannot crush large seeds as well, but are more adept at handling small seeds.

Cactus finches are adapted to opening cactus fruits and extracting the seeds.

Bud eater
The bud eater's heavy bill is adapted for grasping and wrenching buds from branches.

Vegetarian finch
(*Platyspiza crassirostris*)

ANCESTOR FINCH from South American mainland: Blue-black grassquit (*Volatinia jacarina*)

Insect eaters
The bills of insect eaters vary because they eat different types and sizes of insects and they capture them in different ways.

Small tree finch
(*Camarhynchus parvulus*)

Large tree finch
(*C. psittacula*)

Medium tree finch
(*C. pauper*)

Mangrove finch
(*C. heliobates*)

Woodpecker finch
(*C. pallidus*)

Warbler finch
(*Certhidea olivacea*)

The large tree finch uses its heavy bill to twist apart wood to reach larvae inside.

The small and medium tree finches and mangrove finches pick insects from leaves and branches and explore crevices for hidden prey.

The woopecker finch uses its long beak to probe into dead wood, crevices, and bark for insects.

The warbler finch uses quick motions to capture insects on plant surfaces.

21.6 Evolution among Galapagos Finches The descendants of the blue-black grassquits that colonized the Galapagos Islands several million years ago evolved into 14 species whose members are variously adapted to feed on seeds, buds, and insects. (The fourteenth species, not shown here, lives on Cocos Island, farther north in the Pacific Ocean.)

21.7 Parapatric Speciation *Agrostis tenuis* individuals growing on soil contaminated with heavy metals from this Welsh mine are reproductively isolated from nearby individuals growing on uncontaminated soil.

flow can slow or stop even in the absence of a distinct physical barrier.

Reproductive Isolating Mechanisms

Once a barrier to gene flow is established, by whatever means, the daughter populations may diverge genetically because of the action of evolutionary agents. Over many generations, differences that reduce the survivability of hybrid offspring may accumulate. In this way, reproductive isolation can evolve as an incidental by-product of other genetic changes in allopatric populations. For example, individuals in the two daughter populations may become so different that they are not recognized as suitable mates, as shown by the singing honeyeaters on Rottnest Island.

However, geographic isolation does not necessarily mean reproductive isolation. For example, American and European sycamores have been physically isolated from one another for at least 20 million years. Nevertheless, they are morphologically very similar (Figure 21.8), and they can form fertile hybrids. They lack traits that would prevent individuals of two different populations from producing fertile hybrids. In this section we examine the ways in which such traits—**reproductive isolating mechanisms**—arise. Then we explore what happens when reproductive isolation is incomplete.

Prezygotic barriers operate before mating

Individuals of different species may select different places in the environment in which to live. As a result, they may never come into contact during their respective mating seasons; that is, they are reproductively isolated by location (*spatial isolation*). Many organisms have mating periods that are as short as a few hours or days. If the mating periods of two species do not coincide, they will be reproductively isolated by time (*temporal isolation*). Differences in the sizes and shapes of reproductive organs may prevent the union of gametes

(a) *Plastanus occidentalis* (American sycamore)

(b) *Platanus hispanica* (European sycamore)

21.8 Geographically Separated, Morphologically Similar Although they have been separated on different continents for at least 20 million years, American and European sycamores are similar in appearance.

21.9 Prezygotic and Postzygotic Barriers to Gene Exchange Barriers to gene exchange exist both before and after mating.

Prezygotic barriers prevent fertilization from occurring.	**Postzygotic barriers** prevent normal development after mating and fertilization.
1. **Spatial isolation** (populations live in different habitats; do not meet during breeding season)	1. **Hybrid zygote abnormality** (hybrid zygotes fail to develop to sexual maturity)
2. **Temporal isolation** (populations reproduce at different seasons or different times of day; individuals do not meet when mating)	2. **Hybrid infertility** (hybrids fail to produce viable gametes)
3. **Behavioral isolation** (individuals do not find one another attractive as mates)	3. **Low viability of hybrid offspring** (hybrids have lower chances of surviving to reproductive age)
4. **Mechanical isolation** (structure of genitalia or flowers prevent successful copulation or pollen transfer)	
5. **Gametic isolation** (male and female gametes do not fuse)	

These barriers prevent mating from occurring.

This barrier prevents fertilization after mating.

from different species (*mechanical isolation*). Sperm of one species may not be attracted to the eggs of another species because the eggs do not release the appropriate attractive chemicals, or the sperm may be unable to penetrate the egg because it is chemically incompatible (*gametic isolation*). These mechanisms, all of which operate before mating, are called **prezygotic reproductive barriers** (Figure 21.9).

Postzygotic barriers operate after mating

If individuals of two different populations still recognize one another and mate, **postzygotic reproductive barriers** may prevent gene exchange (see Figure 21.9). Genetic differences that accumulated while the daughter populations were allopatric are likely to reduce the survivability of offspring produced by matings between individuals from the two daughter populations.

The offspring of parents from genetically dissimilar populations are known as **hybrids** (see Chapter 10). Hybrid zygotes may be abnormal (*hybrid zygote abnormality*), or the hybrids may mature normally but be infertile when they attempt to reproduce (*hybrid infertility*). For example, the offspring of matings between horses and donkeys—mules—are vigorous, but mules are sterile; they produce no descendants (Figure 21.10).

If hybrid offspring are less viable than offspring resulting from matings within populations of the daughter species, postzygotic barriers may be reinforced by the evolution of more effective prezygotic barriers. More effective prezygotic barriers should evolve because individuals engaging in hybrid matings leave fewer surviving offspring than those that mate only within their group. Reinforcement of prezygotic barriers has been demonstrated in a few laboratory populations, but evidence for it in nature has been slow to accumulate.

Sometimes reproductive isolation is incomplete

Sometimes contact is reestablished between formerly geographically isolated populations before many genetic differences have accumulated. Then the individuals may interbreed freely with members of the

21.10 Sturdy but Sterile Mules are widely used as pack animals because of their stamina. For that purpose, their infertility is unimportant.

21.11 Formation of a New Hybrid Zone Blue and snow geese are forming mixed pairs because they now winter together in Louisiana and Texas rice fields where they choose mates.

other population and produce hybrid offspring that are as successful as those resulting from matings within each population. If successful hybrids spread through both populations and reproduce with other individuals, the gene pools combine quickly, and no new species result from the period of isolation.

Alternatively, rather than thoroughly combining their gene pools, the two populations may interbreed only where they come into contact, resulting in a **hybrid zone**. To determine what happens when formerly separate populations come together, studies ideally begin when contact is first established. Blue and snow geese provide an opportunity to observe the formation of a hybrid zone (Figure 21.11). These geese breed in Arctic North America and spend the winter in the southern United States. Birds with white plumage (snow geese) dominate breeding populations in the West; birds with dark plumage (blue geese) dominate in the East. Historical evidence shows that the two color forms were almost completely separated geographically until the 1930s.

The recent hybrid zone has resulted from a change in the winter feeding ranges of the birds. Birds of both types now winter in large flocks in the rice-growing regions of inland Texas and Louisiana. The geese select mates while on the wintering grounds, and pairs migrate to nest on the breeding grounds from which the female came. Interbreeding is now common between the two forms, and a hybrid zone is developing in a small region of the Canadian Arctic. Biologists are monitoring the spread of this hybrid zone to determine whether isolating mechanisms are developing or whether the zone will continue to spread.

We can measure genetic differences among species

If two species hybridize, we know that they are very similar genetically, but the absence of interbreeding tells us nothing about how *dis*similar two species are. Not until modern molecular tools were developed could biologists measure genetic differences among species.

Molecular studies are now demonstrating that many sympatric species differ from one another very little genetically. For example, flies of different species of Hawaiian *Drosophila* share nearly all of their alleles. Most morphological differences *among* the species are based on variability already present *within* each of the species. All of the hundreds of species of this genus that have evolved in Hawaii during the past 40 million years, even those that have diverged morphologically, are relatively similar genetically (Figure 21.12). Other research confirms that the differences among species generally are similar in type to the differences within species.

21.12 Morphologically Different, Genetically Similar Although these fruit flies, a small sample of the hundreds of species found only on the Hawaiian Islands, are extremely variable in appearance, they are genetically almost identical.

Drosophila silvestris

Drosophila conspicua

Drosophila balioptera

Variation in Speciation Rates

Some lineages of organisms have many species; others have few. The hundreds of species of *Drosophila* found in the Hawaiian Islands have evolved within the last 40 million years. In contrast, there is only one species of horseshoe crab, even though its lineage has survived more than 200 million years. Why do rates of speciation vary so widely among lineages? In the sections that follow we will examine several factors that influence speciation rates: species diversity and range size, life history traits, environment, and generation times.

Species richness may favor speciation

The larger the number of species, the larger the number of opportunities for new species to form. This is particularly true of speciation by polyploidy because more species are available to hybridize with one another. It is also partly true for geographic speciation, because the number of ranges bisected by a given barrier should be positively correlated with the number of species living in the area.

However, the relationship between range size and speciation rate is not simple, because ranges of individual species tend to be smaller where there are many species. The larger the range of a species, the more likely a physical barrier is to subdivide it. Conversely, the smaller the range size, the less likely a particular randomly placed barrier will subdivide it. Also, species with large ranges are more likely than species with small ranges to establish isolated peripheral populations that survive long enough to form new species.

Random variation in rates of events that create barriers is an important cause of variable speciation rates. Where and when geographic barriers arise, and where and when genetic accidents that result in polyploid individuals happen, are unpredictable and variable. Nonetheless, traits of species may influence how often their ranges are divided by barriers.

Life history traits influence speciation rates

Individuals of species with poor dispersal abilities are unlikely to establish new populations by dispersing across barriers, and even narrow barriers are effective among species whose individuals are highly sedentary. Populations of land snails, which have speciated profusely on Pacific islands, are separated by barriers as narrow as a city street (Figure 21.13).

Animals with complex behavior are likely to speciate at a high rate because they make sophisticated discriminations among potential mating partners. They distinguish members of their own species from members of other species, and they make subtle discriminations among members of their own species on the basis of size, shape, appearance, and behavior.

These discriminations may be based on the quality of the genes of the potential partner, the quality of parental care likely to be given, or both. Such behavioral discrimination can greatly influence which individuals are most successful in producing offspring. Therefore, mate selection is probably a major cause of rapid evolution of reproductive isolation between species.

21.13 Mobility Affects Speciation Rate Even a narrow street presents a geographic barrier to land snails. Because populations are readily isolated, genetic differences accumulate rapidly.

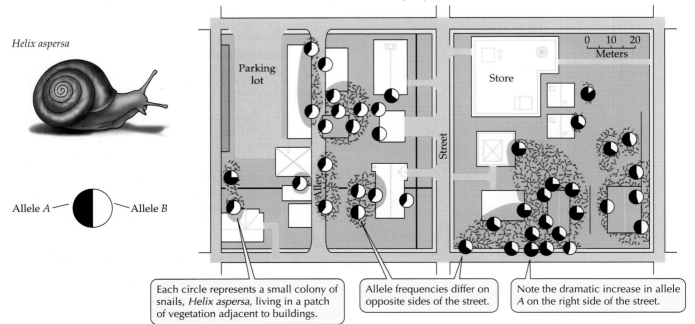

Helix aspersa

Allele *A* — Allele *B*

Each circle represents a small colony of snails, *Helix aspersa*, living in a patch of vegetation adjacent to buildings.

Allele frequencies differ on opposite sides of the street.

Note the dramatic increase in allele *A* on the right side of the street.

Parking lot

Store

0 10 20 Meters

Street

Alley

Heterogeneous environments favor high speciation rates

Speciation rates among different lineages of the large, hoofed mammals of Africa are correlated with diet: Grazers (which eat grass and other nonwoody plants) and browsers (which eat branches and leaves of woody plants) speciate faster than omnivores (which eat both plants and animals) and anteaters (Figure 21.14).

The grazers and browsers require large expanses of open grassland and woodland, respectively. In Africa, these resources disappeared from and reappeared in large areas during periods of climatic change, thus isolating populations and causing both high extinction rates and high rates of differentiation among populations between these isolated regions. Omnivores and anteaters, on the other hand, maintained more continuous populations during these climatic changes. Gene flow continued among their populations, and reproductive isolation was not established.

Short generation times enhance speciation

We have concentrated on factors that influence rates at which the ranges of species are subdivided by barriers. But the rate at which new species form also depends on how fast daughter populations diverge. The more rapidly they diverge, the sooner they are likely to evolve reproductive isolating mechanisms and the less likely they are to hybridize when they again become sympatric. Shorter generation times result in more generations per unit time and, as a result, generate the potential for more evolutionary changes per unit time.

Evolutionary Radiations: High Speciation and Low Extinction

As we learned in Chapter 19, the fossil record reveals that, at certain times in some lineages, speciation rates have been much higher than extinction rates. The result is an **evolutionary radiation** giving rise to a large number of daughter species. What conditions cause speciation rates to be much higher than extinction rates?

Evolutionary radiations are likely when a population colonizes an environment that has relatively few species. This condition typifies islands because many organisms disperse poorly across large water-filled gaps. Because islands lack many plant and animal groups found on the mainland, ecological opportunities exist that may stimulate rapid evolutionary changes. Water barriers also restrict gene flow among islands in an archipelago, so populations on different islands can evolve adaptations to their local environments. Together these two factors make it likely that speciation rates will exceed extinction rates.

Remarkable evolutionary radiations have occurred in the Hawaiian Archipelago, the most isolated islands in the world. The Hawaiian Islands lie 4,000 km from

21.14 Speciation Rates in Some African Hoofed Mammals Speciation rates of these animals correlate with their diets.

the nearest major land mass and 1,600 km from the nearest group of islands. The islands are arranged in a line of decreasing age—the youngest islands to the southeast, the oldest to the northwest (see Figure 21.5).

The native biota of the Hawaiian Islands includes 1,000 species of flowering plants, 10,000 species of insects, 1,000 land snails, and more than 100 birds. However, there were no amphibians, no terrestrial reptiles, and only one native mammal—a bat—until humans introduced additional species. The 10,000

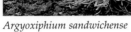
Argyoxiphium sandwichense

Wilkesia gymnoxiphium

Dubautia laxa

21.15 Rapid Evolution among Hawaiian Plants   Three closely related genera of the sunflower family are believed to have descended from a single ancestor, a tarweed that colonized Hawaii from the Pacific coast of North America. Their rapid evolution makes them appear more distantly related than they actually are.

known native species of insects on Hawaii are believed to have evolved from only about 400 immigrant species; only seven immigrant species are believed to account for all the native Hawaiian land birds.

More than 90 percent of all plant species on the Hawaiian Islands are **endemic**—that is, they are found nowhere else. Several groups of flowering plants have more diverse forms and life histories on the islands and live in a wider variety of habitats than do their close relatives on the mainland. An outstanding example is the group of sunflowers called silverswords and tarweeds (the genera *Argyroxiphium, Dubautia,* and *Wilkesia*). Chloroplast DNA data show that these species share a relatively recent common ancestor, which is believed to be a species of tarweed from the Pacific coast of North America.

Whereas all mainland tarweeds are small, upright, nonwoody plants (herbs), Hawaiian silversword species include prostrate and upright herbs, shrubs, trees, and vines (Figure 21.15). They occupy nearly all the habitats of the islands, from sea level to above timberline in the mountains. Despite their extraordinary diversification, however, the silverswords have differentiated very little genetically. In other words, the rate of morphological evolution has been much more rapid than the rate of evolution of chloroplast DNA in these plants.

The island silverswords are more diverse in size and shape than the mainland tarweeds because the

colonizers arrived on islands that had very few plant species. In particular, there were few trees and shrubs because such large-seeded plants only rarely disperse to oceanic islands; many island trees and shrubs have evolved from nonwoody ancestors. On the mainland, however, tarweeds have lived in ecological communities that contain tree and shrub lineages older than their own—that is, where opportunities to exploit the tree way of life were already preempted.

Evolutionary lineages may also radiate when they acquire a new adaptation that enables them to use the environment in new and varied ways. For example, ancestors of the 94 species of American blackbirds evolved powerful muscles for opening their bills. These muscles enable the birds to obtain food by opening their bills forcibly against objects they wish to move, exposing otherwise hidden prey (Figure 21.16). Such activity is called **gaping**. Birds lacking these powerful muscles can find prey only on exposed surfaces of objects. Blackbirds gape into wood, fruits, leaf clusters, and stems of nonwoody plants; under sticks, stones, and animal droppings; and into the soil. With this feeding method, they have come to occupy nearly all habitat types in North and South America, and they are among the most abundant birds throughout the region.

Speciation and Evolutionary Change

Does speciation stimulate evolutionary change? In 1972 Niles Eldredge and Stephen Jay Gould proposed that most evolutionary changes take place at the time

This yellow-headed blackbird is using gaping to turn over an object on the ground to expose prey.

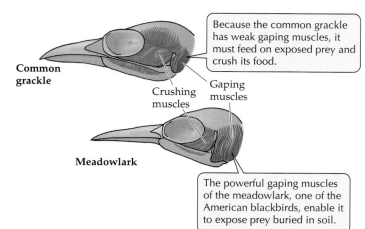

Because the common grackle has weak gaping muscles, it must feed on exposed prey and crush its food.

Common grackle

Crushing muscles

Gaping muscles

Meadowlark

The powerful gaping muscles of the meadowlark, one of the American blackbirds, enable it to expose prey buried in soil.

21.16 Blackbirds Expose Food by Gaping Gaping exposes food but blackbirds, as illustrated by grackles and meadowlarks, differ strikingly in the size and strength of their gaping muscles.

of speciation. They suggested that the isolation of small, peripheral populations was the most common event leading to rapid evolutionary changes. Their reasoning was that small founding populations lack some of the alleles found in the source populations and have different allele frequencies as a result of random genetic changes. These founding populations might change rapidly because they live in an environment that differs from the one from which they came. According to Eldredge and Gould, the speciation process stimulates evolutionary change, and once speciation has been completed, the better-integrated new genotypes resist change, leading to **stasis**—long periods of little change that are interrupted only by the next round of speciation. This pattern of evolution is called **punctuated equilibrium**.

The fossil record reveals examples of both stasis and long-term gradual evolution, but it does not tell us if periods of rapid change usually accompany times of speciation and are rare at other times. The fossil record is of limited help in testing this hypothesis because the ages of most fossils cannot be determined precisely enough. Molecular tools, on the other hand, enable evolutionary biologists to measure correlations between speciation and evolutionary change.

Lungless salamanders have been studied extensively at both molecular and morphological levels. The information from these studies allows us to compare the amount of genetic difference with the amount of morphological differences among these species to determine if speciation typically has been accompanied by morphological change. The results show that most speciation events are accompanied by almost no morphological changes. In these salamanders, speciation has proceeded at a much higher rate than morphological evolution. How often the rate of speciation exceeds the rate of morphological evolution is unknown.

The Significance of Speciation

The result of speciation processes operating over billions of years is a world in which life is organized into millions of species, each adapted to live in a particular place and to use environmental resources in a particular way. Earth would be very different if speciation had been a rare event in the history of life. How the millions of species are distributed over the surface of Earth and organized into ecological communities will be a major focus of Part Seven of this book, "Ecology and Biogeography," at which time we will also discuss how human activities are causing the extinction of many species and what we can do to reduce the rate of species loss.

Summary of "Species and Their Formation"

What Are Species?

• Species are independent evolutionary units. A generally accepted definition is that "species are groups of actually or potentially interbreeding natural populations which are reproductively isolated from other such groups."

How Do New Species Arise?

• Not all evolutionary changes result in new species.
• Evolution creates two patterns across time and space: anagenesis and cladogenesis. In anagenesis, a single lineage changes through time. In cladogenesis (speciation), one species splits into two separate species.
• Species may form quickly sympatrically by a multiplication of chromosome numbers because polyploid offspring are sterile in crosses with members of the parent species. Polyploidy has been a major factor in plant speciation but is rare among animals. **Review Figures 20.2, 20.3, 20.6**

• Allopatric (geographic) speciation is the most important means of speciation among animals and is common in other groups of organisms. **Review Figures 20.4, 20.5**
• Species may form parapatrically where marked environmental changes prevent gene flow among individual living in adjacent but different environments. **Review Figure 20.7**

Reproductive Isolating Mechanisms

• When previously allopatric species become sympatric, reproductive isolating mechanisms may prevent the exchange of genes.
• Barriers to gene exchange may operate before fertilization (prezygotic barriers) or after fertilization (postzygotic barriers). Hybrid zones may develop if barriers to gene exchange failed to develop during allopatry. **Review Figure 20.9**
• Hybrids may form if barriers break down before sufficient genetic differences have accumulated. Hybrids tell us that the two hybridizing species are very similar genetically, but species that do not hybridize may also differ from one another very little genetically.
• Genetic differences between species are similar in kind to those found within species, although differences between species usually are greater than differences within species.

Variation in Speciation Rates

• Rates of speciation differ greatly among lineages of organisms. Speciation rates are influenced by species diversity and range sizes, life history traits, environment, and generation times. **Review Figures 20.13, 20.14**

Evolutionary Radiations: High Speciation and Low Extinction

• Evolutionary radiations happen when speciation rates exceed extinction rates.
• High speciation rates often coincide with low extinction rates when species invade islands that have impoverished biotas or when a new way of exploiting the environment makes a different array of resources available to a species. **Review Figures 20.15, 20.16**

Speciation and Evolutionary Change

• Speciation may stimulate rapid evolutionary change, leading to a pattern known as punctuated equilibrium. Nonetheless, many speciation events are not accompanied by large evolutionary changes.

The Significance of Speciation

• As a result of speciation, Earth is populated with millions of species, each adapted to live in a particular place and to use environmental resources in a particular way.

Self-Quiz

1. A species is
   a. a group of actually interbreeding natural populations that is reproductively isolated from other such groups.
   b. a group of potentially interbreeding natural populations that is reproductively isolated from other such groups.
   c. a group of actually or potentially interbreeding natural populations that is reproductively isolated from other such groups.
   d. a group of actually or potentially interbreeding natural populations that is reproductively connected to other such groups.
   e. a group of actually interbreeding natural populations that is reproductively connected to other such groups.

2. Anagenesis is
   a. continuous change in a single lineage of organisms.
   b. the formation of two species by the splitting of one evolutionary lineage.
   c. the formation of a new species by the coming together of two evolutionary lineages.
   d. the reduction of two lineages by the extinction of one of them.
   e. the formation of new species by the reclassification of a group.

3. Allopatric speciation may happen when
   a. continents drift apart and separate previously connected lineages.
   b. a mountain range separates formerly connected populations.
   c. different environments on two sides of a barrier cause populations to diverge.
   d. the range of a species is separated by loss of intermediate habitat.
   e. all of the above

4. Finches speciated on the Galapagos Islands because
   a. the Galapagos Islands are a long way from the mainland.
   b. the Galapagos Islands are very arid.
   c. the Galapagos Islands are small.
   d. the islands in the Galapagos Archipelago are sufficiently isolated from one another that there is little migration among them.
   e. the islands in the Galapagos Archipelago are close enough to one another that there is considerable migration among them.

5. Which of the following is *not* a potential prezygotic isolating mechanism?
   a. Temporal segregation of breeding seasons
   b. Differences in chemicals that attract individuals
   c. Sterility of hybrids
   d. Spatial segregation of mating sites
   e. Inviability of sperm in female reproductive tracts

6. A common means of sympatric speciation is
   a. polyploidy.
   b. hybrid sterility.
   c. temporal segregation of breeding seasons.
   d. spatial segregation of mating sites.
   e. imposition of a geographic barrier.

7. Sympatric species are often very similar in appearance because
   a. appearances are often of little evolutionary significance.
   b. genetic changes accompanying speciation are often small.
   c. genetic changes accompanying speciation are usually large.
   d. speciation usually requires major reorganization of the genome.
   e. the traits that differ among species are not the same as the traits that differ among individuals within species.

8. Which statement about speciation is *not* true?
   a. It always takes thousands of years.
   b. It often takes thousands of years but may happen within a single generation.
   c. Among animals it usually requires a physical barrier.
   d. Among plants it often happens as a result of polyploidy.
   e. It has produced the millions of species living today.

9. Evolutionary radiations
   a. often happen on continents but rarely on island archipelagos.
   b. characterize birds and plants but not other taxonomic groups.
   c. have happened on all continents, as well as on islands.
   d. require major reorganizations of the genome.
   e. never happen in species-rich environments.

10. Speciation is often rapid within lineages in which species have complex behavior, because
   a. individuals of such species make very fine discriminations among potential mating partners.
   b. such species have short generation times.
   c. such species have high reproductive rates.
   d. such species have complex relationships with their environments.
   e. none of the above

Applying Concepts

1. Gene exchange between populations is prevented by geographic isolation, by behavioral responses before mating (for example, females rejecting courting males of other species), and by mechanisms that function after mating has occurred (for example, hybrid sterility). All of these are commonly called isolating mechanisms. In what ways are the three types very different? If you were to apply different names to them, which one would you call an isolating mechanism? Why?

2. The blue goose of North America has two distinct color forms: blue and white. As we have seen, matings between the two color types are common. On their breeding grounds in northern Canada, however, blue individuals pair with blue individuals much more frequently than would be expected by chance. Suppose that 75 percent of all mated pairs consisted of two individuals of the same color, what would you conclude about speciation processes in these geese? If 95 percent of pairs were of the same color? If 100 percent of pairs were of the same color?

3. Although many species of butterflies are divided into local populations among which there is little gene flow, these butterflies often show relatively little geographic variation. Describe the studies you would conduct to determine what maintains this morphological similarity.

4. Distinguish among the following: allopatric speciation, parapatric speciation, and sympatric speciation. For each of the three statements below, indicate which type of speciation is implied:
   a. This process in nature is most commonly a result of polyploidy.
   b. The present sizes of national parks and wildlife preserves may be too small to allow this type of speciation among organisms restricted to those areas.
   c. This process generally occurs in species that inhabit areas where sharp environmental contrasts exist.

5. Evolutionary radiations are common and easily studied on oceanic islands. In what types of mainland situations would you expect to find major evolutionary radiations? Why?

Readings

Endler, J. T. 1977. *Geographic Variation, Species, and Clines.* Princeton University Press, Princeton, NJ. A theoretical analysis of how sympatric and parapatric speciation might occur.

Futuyma, D. J. 1998. *Evolutionary Biology,* 3rd edition. Sinauer Associates, Sunderland, MA. Chapters 15 and 16 cover species and speciation in more detail.

Knowlton, N. 1994. "A Tale of Two Seas." *Natural History,* June. How allopatric speciation may have taken place when the Panamanian isthmus appeared, separating the Caribbean Sea from the Pacific Ocean.

Lambert, D. M. and H. G. Spencer. 1995. *Speciation and the Recognition Concept.* Johns Hopkins University Press, Baltimore. A collection of essays that examines the importance of mate recognition for processes of speciation.

Mayr, E. 1970. *Populations, Species, and Evolution.* Harvard University Press, Cambridge, MA. An abridged version of the most thorough work on speciation theory as applied to animals.

Otte, D. and J. A. Endler (Eds.). 1989. *Speciation and Its Consequences.* Sinauer Associates, Sunderland, MA. A comprehensive collection of essays on concepts, methods, and consequences of speciation. Includes general treatments and analyses of specific cases.

Ryan, M. J. 1990. "Signals, Species, and Sexual Selection." *American Scientist,* January/February. A study of the mate-recognition concept of species formation, using frog mating calls as the example.

Chapter 22

Constructing and Using Phylogenies

Asian Snails Can Transmit Schistosomiasis
Workers in the rice paddies of tropical Asia are at extreme risk of contracting schistosomiasis (known in some parts of the world as bilharzia). The disease is transmitted to humans via freshwater snails that thrive in the standing water of the paddies.

Schistosomiasis is a blood infection caused by a parasitic flatworm, *Schistosoma*. More than 200 million people in South America, Africa, China, Japan, and Southeast Asia have the disease. During part of its life cycle, *Schistosoma* inhabits a freshwater snail. People become infected when larval *Schistosoma* swim from a snail and penetrate their skin. The worm matures and lives in a person's abdominal blood vessels. The disease is progressively debilitating, causing a slow death.

For most of the twentieth century, only one species, *Schistosoma japonicum*, was known to infect humans, and it was thought to be transmitted by a single species of snail in the genus *Oncomelania*. Then in the 1970s researchers discovered that a different snail was transmitting *Schistosoma* to humans in the Mekong River in Laos. This discovery stimulated extensive field surveys and anatomical, genetic, and geographic research on the worms and snails in Southeast Asia.

Investigators found that *S. japonicum* was actually a cluster of at least six species. They also discovered that evolutionary relationships among snails determined which species could host *Schistosoma*. Evolutionary diversification from an ancestral stock of snails had produced a group of species of modern snails. Of these, only three retain the ability to host *Schistosoma*, and ten have a genetic trait that makes them unsuitable hosts for the disease.

This information is of great value in efforts to combat schistosomiasis. Most of the species of freshwater snails in Southeast Asia have not been described and named. By using information on evolutionary relationships among snails, scientists can quickly determine whether or not a newly discovered snail is a host for *Schistosoma*. Control efforts need to be directed toward only the snails that can transmit *Schistosoma* to humans, not to all freshwater snails in the region.

(a) *Campanula* sp.

(b) *Endymion nonscriptus*

(c) *Mertensia paniculata*

22.1 Many Different Plants Are Called Bluebells (a) These flowers from the plains of North Dakota are often called bluebells. (b) This English bluebell is a member of the lily family. (c) These Alaskan flowers are known as bluebells or chiming bells. None of these plants is closely related to the others.

How did investigators determine the evolutionary relationships among the snails that are hosts of *Schistosoma*? How could they determine that genes preventing snails from hosting *Schistosoma* arose three different times? How can evolutionary relationships be expressed in systems of classification that help guide further studies of organisms?

In this chapter, we present systematics, the science that provides answers to these questions. We consider the goals and methods of modern systematics, and we show how biologists determine evolutionary relationships among organisms—their **phylogeny**—and express those relationships in classification systems. Then we illustrate how knowledge of evolutionary relationships is used to solve other biological problems.

Classification systems are important. They improve our ability to explain relationships among things. Having classifications is especially important when biologists attempt to understand the evolutionary pathways that have produced the millions of species living today. Classification systems are also an aid to memory. It is impossible to remember the characteristics of many different things unless we can group them into categories based on shared characteristics.

Classification systems also provide unique names for organisms. If the names are changed, the systems provide means of tracing the changes. Common names, even if they exist (most organisms have no common names), are very unreliable and often confusing. For example, plants called bluebells are found in England, Scotland, Texas, and the Rocky Mountains—but none of the bluebells in any of those places is closely related evolutionarily to the bluebells in any of the other places (Figure 22.1).

Recognizing and interpreting similarities and differences among organisms is easier if the organisms are classified into groups that are ordered and ranked. Any group of organisms that is treated as a unit in a classification system is called a **taxon** (plural taxa). **Taxonomy** is the theory and practice of classifying organisms. **Systematics**, a larger field of which taxonomy is a part, is the scientific study of the diversity of organisms. Systematists study organisms to determine their evolutionary relationships and develop classification systems that reflect those relationships.

The Hierarchical Classification of Species

The biological classification system that is used today was developed by the great Swedish biologist Carolus Linnaeus in 1758. Linnaeus gave each species two names, one identifying the species itself and the other the genus to which the species belongs. This two-name system, referred to as **binomial nomenclature**,

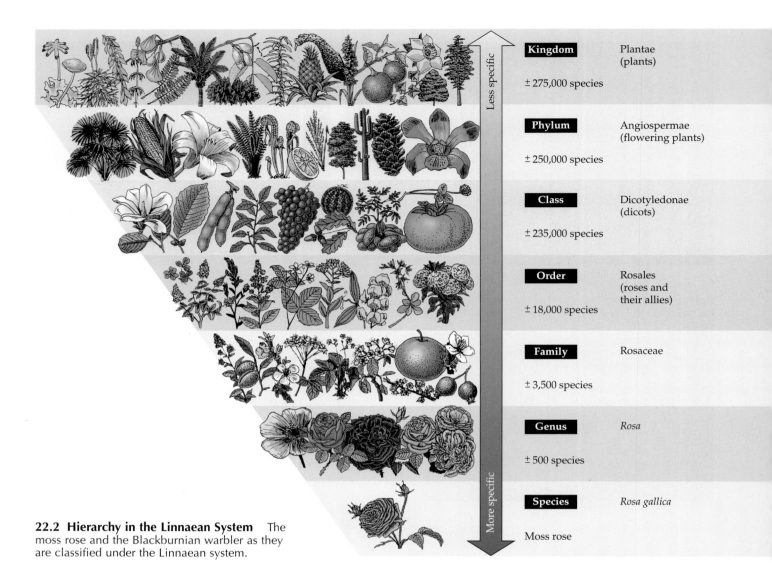

22.2 Hierarchy in the Linnaean System The moss rose and the Blackburnian warbler as they are classified under the Linnaean system.

Kingdom	Plantae (plants)	± 275,000 species
Phylum	Angiospermae (flowering plants)	± 250,000 species
Class	Dicotyledonae (dicots)	± 235,000 species
Order	Rosales (roses and their allies)	± 18,000 species
Family	Rosaceae	± 3,500 species
Genus	*Rosa*	± 500 species
Species	*Rosa gallica* Moss rose	

is universally employed in biology. Using this system, scientists throughout the world refer to the same organisms by the same names.

A **genus** (plural genera; adjectival form: generic) is a group of closely related species. In many cases the name of the taxonomist who first proposed the species name is added at the end. Thus, *Homo sapiens* Linnaeus is the name of the modern human species. *Homo* is the genus to which the species belongs, and *sapiens* identifies the species; Linnaeus proposed the species name *sapiens*. You can think of the generic name *Homo* as equivalent to your surname and the specific name *sapiens* as equivalent to your first name.

The generic name is always capitalized; the species name is not. Both names are always italicized, whereas common names are not. When referring to more than one species in a genus without naming each one, we use the abbreviation "spp." after the generic name (for

example, "*Drosophila* spp." means more than one species in the genus *Drosophila*). The abbreviation "sp." is used after a generic name if the identity of the species is uncertain. Rather than repeating a generic name when it is used several times in the same discussion, biologists often spell it out only once and abbreviate it to the initial letter thereafter (for example, *E. coli* is the abbreviated form of *Escherichia coli*).

In the Linnaean system, species are grouped into higher taxonomic categories. The category (taxon) above genus in the Linnaean system is **family**. The names of animal families end in the suffix "-idae." Thus Formicidae is the family that contains all ant species, and the family Hominidae contains humans, a few of our fossil relatives, and chimpanzees and gorillas. Family names are based on the name of a member genus. Formicidae is based on *Formica*, and Hominidae is based on *Homo*.

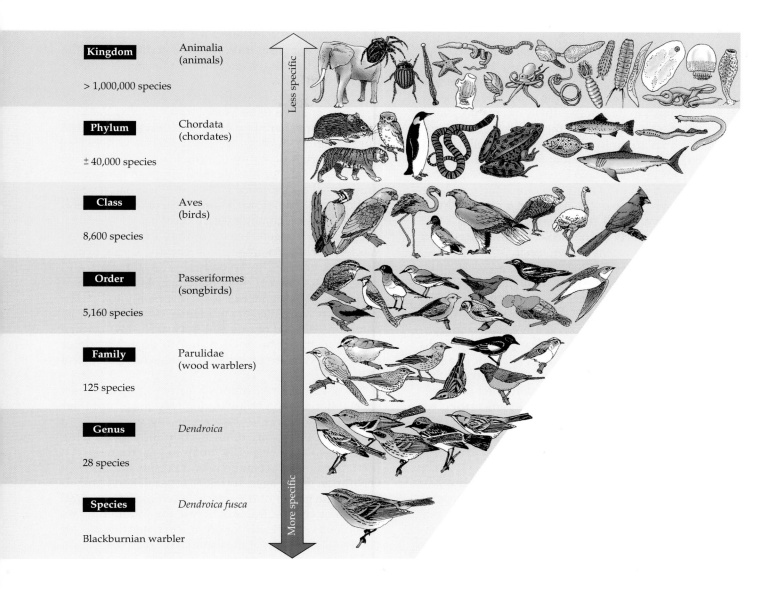

Kingdom	Animalia (animals)
> 1,000,000 species	
Phylum	Chordata (chordates)
± 40,000 species	
Class	Aves (birds)
8,600 species	
Order	Passeriformes (songbirds)
5,160 species	
Family	Parulidae (wood warblers)
125 species	
Genus	*Dendroica*
28 species	
Species	*Dendroica fusca*
Blackburnian warbler	

Less specific → More specific

Plant classification follows the same procedures except that the suffix "-aceae" is used with family names instead of "-idae." Thus Rosaceae is the family that includes the genus of roses (*Rosa*) and its close relatives. Unlike generic and species names, family names are not italicized, but they are capitalized.

Families, in turn, are grouped into **orders**, and orders into **classes**. Classes are grouped into **phyla** (singular phylum), and phyla into **kingdoms**. The hierarchical units of this classification system, as applied to the Blackburnian warbler (*Dendroica fusca*) and the moss rose (*Rosa gallica*) are shown in Figure 22.2. For their detailed studies of particular groups, taxonomists use additional categories, such as subfamilies and subspecies, to indicate finer degrees of relationships, but we need not worry about them here.

Systematists attempt to solve two distinct but related problems. The first is how to determine evolutionary re-

lationships among organisms—that is, how to construct phylogenetic trees. The second is how to express evolutionary relationships in a classification system. We turn first to the methods used to deduce evolutionary relationships. There must be a single "true" phylogeny for a particular group of organisms, but identifying it is very difficult. The key events happened in the distant past, and evidence of those events is incomplete and sometimes contradictory.

Inferring Phylogenies

To infer phylogenies, systematists use information provided by both fossils and living organisms. And the more information, the better. In the pages that follow, we discuss what fossils tell us about relationships among organisms. Then we describe the set of methods, known as cladistics, used to construct phylogenies of organisms.

Fossils are the key to the past

Fossils tell us where and when organisms lived and what they looked like (see Chapter 19). When available, this information is valuable, but sometimes few or no fossils have been found for a taxon whose phylogeny we wish to determine. Even if there are many fossils, the first individuals of the group that lived in a particular area are unlikely to become fossilized and even less likely to be discovered. Fossils have been described from relatively few areas, which means that most species lived in many places for which we have no record. At best we can say that a species lived where we have found its remains and that it lived there at least as early as the earliest known fossil and as late as the most recent known fossil.

The incompleteness of the fossil record means that the chance of finding a fossil of an organism that was the direct ancestor of another fossil we happen to find is very small. Indeed, we can probably never determine accurately the ancestors of any group of organisms we are studying. But we can infer that two groups of organisms had a common ancestor at a particular time in the past even if we cannot describe that ancestor precisely. Fortunately, to construct phylogenetic trees, we do not need to identify ancestors; we need only know when they existed and be able to recognize some of their features.

Cladistics is a powerful evolutionary tool

Cladistics is a powerful new method for determining the evolutionary histories of organisms and expressing those relationships in treelike diagrams. Cladistics gets its name from **clade**, which is the entire portion of a phylogeny that is descended from a **common ancestor** (a single ancestral species). An evolutionary tree constructed using cladistic methods is called a **cladogram**. A cladogram does not attempt to describe ancestors; instead it shows points at which lineages diverged from a common ancestral form.

The goal of cladistics is to construct phylogenies by analyzing evolutionary changes in the traits of organisms. A **phylogenetic tree** is a cladogram to which additional information, such as evidence of the dates of separation of lineages, has been added. Phylogenetic trees are rather like pedigrees of lineages of organisms, except that they are usually constructed with the ancestor at the bottom rather than at the top. The base of the "trunk" of the tree represents the point in the past when the lineage consisted of only the ancestor.

What evidence do cladistic systematists use to infer how traits changed during evolution? How do they use that evidence to draw branches on a cladogram? In order to identify how traits have changed during evolution, cladists must determine the original state of the trait and then determine how it has been modified. That is, they must distinguish between ancestral and derived traits. A trait shared with a common ancestor is called an **ancestral trait**. A trait that differs from the form of the trait in the ancestor of a lineage is called a **derived trait**.

Any two traits descended from a common ancestral structure are said to be **homologous**. *General homologous traits* are shared by most or all organisms in the lineage being studied. For example, all vertebrates have a vertebral column, which appears to have evolved only once. We infer this single evolution because fossil ancestors of vertebrates also had a vertebral column. Therefore, having vertebrae is a general homologous trait among vertebrates.

Special homologous traits are shared by only a few species. Such traits arose more recently during evolution than general homologous traits did. Rats and mice, but not dogs or other mammals, have long, continuously growing incisor teeth, for example. Continuously growing incisors evidently developed in the common ancestor of rats and mice after their lineage separated from the one leading to dogs and other mammals, because no other mammals have that kind of incisor.

Thus, having continuously growing incisors is a special homologous trait among rats, mice, and their relatives. However, if we were trying to construct a phylogeny of a group of mice, continuously growing incisors would be a *general* homologous trait. Special homologous traits can be used to order the times of separation of lineages. A trait that is a general homology in one group of organisms, such as the vertebral column of vertebrates, is of no use for determining relationships among those organisms, because all, or nearly all, members of the group have the trait.

The first step in constructing a cladogram is to select the group of organisms whose phylogeny is to be determined. The next step is to choose the traits that will be used in the analysis and to identify the possible forms of those traits. A trait may be present or absent, or it may exist in more than one form. The last and usually the most difficult step is to determine the ancestral and derived forms of the traits.

IDENTIFYING ANCESTRAL TRAITS. Distinguishing derived traits from ancestral traits is difficult because traits may **diverge** (become dissimilar), making ancestral states unrecognizable. For example, the leaves of plants have diverged to form many different structures. Several lines of evidence, especially details of their structure and development, indicate that protective spines, tendrils, and brightly colored structures that attract pollinators (Figure 22.3) are all modified leaves; they are homologs of one another even though they do not resemble one another closely.

Not all resemblances are products of a common ancestry. If a trait evolves more than once and thus is pos-

The orange spines of the barrel cactus and the colorful bracts of *Heliconia* are both modified leaves.

Ferrocactus acanthodes (barrel cactus)

Heliconia rostrata

22.3 Homologous Structures Derived from Leaves The leaves of plants have diverged during their evolution to form many different structures, some of which bear very little resemblance to each other.

by natural selection to perform similar functions. We call this process **convergent evolution**.

Bats, birds, and insects fly by flapping their wings, but the structure of insect wings is totally different from that of bat and bird wings (Figure 22.4). The skeletons of bat and bird wings are homologous, but the supports for insect wings are not homologous with the wing bones of bats and birds. Similarly, the structures that aid plants in climbing over other plants have evolved from several different structures, including stipules, leaflets, leaves, and inflorescences. Structures that perform similar functions but have resulted from convergent evolution (homoplasy) are said to be **analogous** to one another.

sessed by more than one species even though it was not found in their most recent common ancestor, it is said to exhibit **homoplasy**. Homoplasy can result when structures that were formerly very different come to resemble one another because they have been modified

The opening section of this chapter discussed the evolutionary diversification of the freshwater snails that act as hosts for the parasitic flatworm *Schistosoma* and transmit the parasite to humans. The ability of these snail species to host *Schistosoma* is a homologous, ancestral trait. However, Asian freshwater snails evolved resistance to the transmission of *Schistosoma* on three separate occasions; this resistance is an analogous trait (Figure 22.5).

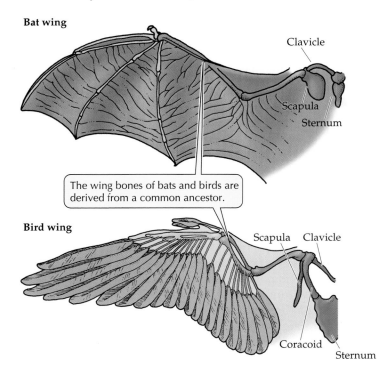

Bat wing

Clavicle
Scapula
Sternum

The wing bones of bats and birds are derived from a common ancestor.

Bird wing

Scapula   Clavicle
Coracoid
Sternum

22.4 Wing Structures May Be Homologous or Analogous The supporting structures of bat and bird wings are derived from a common tetrapod (four-limbed) ancestor and are thus homologous. Although they are also used for flight (and thus are analogous), insect wings evolved independently and their supports are not homologous with the wing bones of bats and birds.

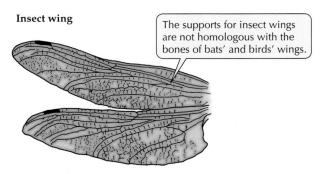

Insect wing

The supports for insect wings are not homologous with the bones of bats' and birds' wings.

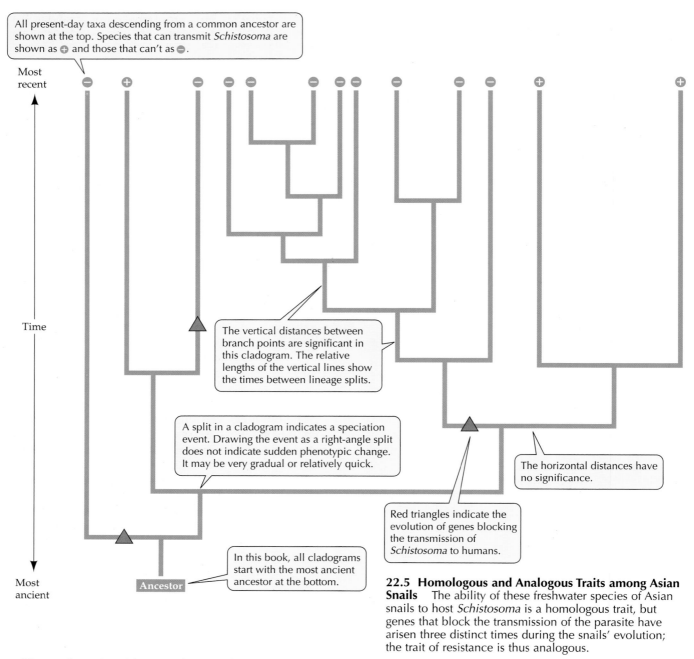

All present-day taxa descending from a common ancestor are shown at the top. Species that can transmit *Schistosoma* are shown as ⊕ and those that can't as ⊖.

Most recent

Time

The vertical distances between branch points are significant in this cladogram. The relative lengths of the vertical lines show the times between lineage splits.

A split in a cladogram indicates a speciation event. Drawing the event as a right-angle split does not indicate sudden phenotypic change. It may be very gradual or relatively quick.

The horizontal distances have no significance.

Red triangles indicate the evolution of genes blocking the transmission of *Schistosoma* to humans.

In this book, all cladograms start with the most ancient ancestor at the bottom.

Most ancient

Ancestor

22.5 Homologous and Analogous Traits among Asian Snails The ability of these freshwater species of Asian snails to host *Schistosoma* is a homologous trait, but genes that block the transmission of the parasite have arisen three distinct times during the snails' evolution; the trait of resistance is thus analogous.

Homoplasy is widespread in evolution. Its existence makes attempts to determine true phylogenies difficult because it is easy to assume that analogous traits are homologous. Taxonomists need a means by which to identify which forms of traits are ancestral and which are derived, and a means by which to distinguish homologies from analogies.

HENNIG'S METHOD FOR IDENTIFYING ANCESTRAL AND DERIVED TRAITS. A method for reconstructing phylogenies that serves both of these needs was developed by the German entomologist Willi Hennig in the 1950s. Hennig suggested that if two species possess the same trait, systematists should *provisionally* (that is, until proven otherwise) assume that the trait is homologous in the two species.

Hennig also proposed that general homology could be distinguished from special homology as follows. A general homologous trait is one that is found not only in one or more species of a group whose phylogeny is being reconstructed, but also appears *outside* this group in what is known as an outgroup. An **outgroup** is a taxon that is closely related to the group whose phylogeny is being reconstructed but that branched off from the lineage of the group below its base on the evolutionary tree. Traits found only *within* the group, on the other hand, are special homologous traits.

In Hennig's system, the members of the group whose phylogeny is being reconstructed are ordered according to the number of special derived homologous traits they share. Species that share a recent common ancestor should share many general homologous traits and a few special derived homologous traits. However, they should share very few homoplastic traits, because little time has been available for those traits to arise.

Using this system, as more and more traits are measured, the data are increasingly likely to support a single phylogenetic pattern, and biologists can more readily distinguish between homologies and homoplasies. A few of the traits originally assumed to be homologies may turn out to be homoplasies, but the best way to determine the true status of shared traits is to assume that they are homologous until additional evidence suggests they aren't.

To see how a cladogram is constructed, consider seven vertebrate animals—hagfish, perch, pigeon, chimpanzee, salamander, lizard, and mouse. We will assume initially that a given derived trait evolved only once during the evolution of these animals and that no derived traits were lost from any of the descendant groups. For simplicity, we have selected traits that are either present (+) or absent (–). As will become evident in the survey of animals in Chapter 30, hagfishes are believed to be more distantly related to the other vertebrates than other vertebrates are to each other. Therefore, we will choose hagfishes as the outgroup for our analysis. Derived traits are those that have been acquired by other members of the lineage since they separated from hagfishes. The taxa and the traits we will consider are shown in Table 22.1.

Cladistic methods infer the branching points in evolutionary trees by minimizing the number of evolutionary changes that need to be assumed—that is, by minimizing homoplasy. For example, the chimpanzee and mouse share two unique traits, fur and mammary glands. Because these traits are absent in both the outgroup and the other species whose relationships we are attempting to determine, we infer that fur and mammary glands are derived traits that evolved in a common ancestor of chimpanzees and mice after that lineage separated from the ones leading to the other lineages. In other words, we assume that fur and mammary glands evolved only once among the animals we are classifying.

The pigeon has one unique trait: feathers. Similarly, we assume that feathers evolved only once, after the lineage leading to birds separated from that leading to the mouse and chimpanzee. By the same reasoning, we assume that claws or nails evolved only once, after the lineage leading to salamanders separated from the lineage leading to those animals that have claws or nails. We make the same assumption for lungs and jaws, continuing to minimize the number of evolutionary events needed to produce the patterns of shared traits among these seven animals. We can see the pattern clearly by ordering the animals in Table 22.1 according to the number of derived traits they share.

Using this information, we can construct a cladogram. The taxon with no derived traits, the hagfish, is the outgroup, and we assume, following Hennig's rule, that the animals that share unique derived traits have a common ancestor not shared with animals lacking those traits. That is, we assume that feathers, which are found only in the pigeon, evolved after the lineage leading to birds separated from the lineage leading to mice and chimpanzees. Otherwise we would need to assume that the ancestors of mice and chimpanzees also had feathers but that the trait was subsequently lost—an unnecessary additional assumption.

A cladogram for these taxa, based on the traits we used and the assumption that each derived trait evolved only once, is shown in Figure 22.6. Notice that

TABLE 22.1   Some Vertebrates Ordered According to Unique Shared Derived Traits

TAXON	JAWS	LUNGS	CLAWS OR NAILS	FEATHERS	FUR	MAMMARY GLANDS
	DERIVED TRAIT[a]					
Hagfish (outgroup)	–	–	–	–	–	–
Perch	+	–	–	–	–	–
Salamander	+	+	–	–	–	–
Lizard	+	+	+	–	–	–
Pigeon	+	+	+	+	–	–
Mouse	+	+	+	–	+	+
Chimpanzee	+	+	+	–	+	+

[a] A plus sign indicates the trait is present, a minus sign that it is absent.

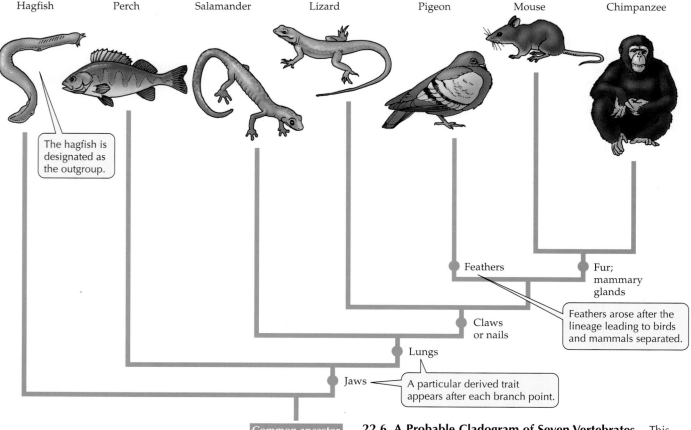

The hagfish is designated as the outgroup.

Feathers

Fur; mammary glands

Feathers arose after the lineage leading to birds and mammals separated.

Claws or nails

Lungs

Jaws — A particular derived trait appears after each branch point.

Common ancestor

22.6 A Probable Cladogram of Seven Vertebrates This cladogram can be constructed from the information given in Table 22.1. Except for the hagfish, all the groups display at least one derived trait.

the cladogram does not describe ancestors or date the splits between lineages. A cladogram shows only the temporal order of splits between lineages; the oldest splits are at the bottom, and the more recent ones are near the top. Notice also that the *x* axis has no scale. Horizontal distances between taxa do not correlate with degree of similarity or difference between them, and we show lineage separations with right angles only for graphical convenience. In other words, do not interpret the cladograms in this book to mean that changes occur suddenly when lineages split.

The cladogram of these seven vertebrates was easy to construct because the traits we chose fulfilled the assumptions that derived traits appeared only once in the lineage and were never lost once they appeared. However, if we had included a snake (a reptile like the lizard) in the group, the second assumption would have been violated, because the ancestors of snakes had limbs, which were subsequently lost, along with their claws.

We would need to examine additional traits to determine that the lineage leading to snakes separated from the one leading to lizards long after the lineage leading to lizards separated from the others. In fact,

the analysis of many traits shows that snakes evolved from burrowing lizards that lost their limbs during a long period of subterranean existence.

Outgroups have also been used to identify ancestral and derived traits in lineages of butterflies. Species in two families—the brush-footed butterflies (Nymphalidae) and the monarchs (Danaidae)—have four functional and two very small legs, whereas the swallowtails (Papilionidae), the sulfurs (Pieridae), and all other butterflies have six functional legs. Biologists assume that having six legs is ancestral because moths and all other orders of insects have six functional legs. The four-legged trait in monarchs and brush-footed butterflies is thus inferred to be a derived trait of butterflies descended from six-legged ancestors (Figure 22.7). If other special shared traits also united brush-footed butterflies and monarchs, we would conclude that these two groups of butterflies share a more recent common ancestor than either group shares with any other group of butterflies. If not, we would conclude that the four-legged condition arose twice during insect evolution.

Pieris protodice

22.7 Six Legs Is the Ancestral Number Having six functional legs is the ancestral trait among butterflies; some species, however, have four legs—a derived trait.

Danaus plexippus

Many traits must be analyzed to reconstruct a phylogeny, and systematists use various methods to combine information from the different traits. Each method is based on specific operating rules—provisional assumptions about how evolution proceeds. A simple method—the one we used in our vertebrate example—makes two assumptions: (1) that the evolution of traits is irreversible (that an ancestral trait can change into a derived one, but the reverse change does not happen), and (2) that each trait can change only once within a lineage. However, we know from fossil and other evidence that the states of traits *can* reverse and that traits *can* change more than once. Therefore, systematists must often relax the rules and allow a derived trait to be lost or to evolve more than once.

A cladogram that postulates fewer reversals and changes in traits is more likely to be accurate than one that requires more changes, because reversals and multiple origins of traits are relatively rare events. Therefore, systematists typically employ the principle of **parsimony** in reconstructing a phylogeny. That is, they arrange the organisms such that the number of changes in traits that must be postulated to account for the inferred lineage is minimized. Parsimony is used as an operating rule in many types of biological investigations.

Using parsimony is helpful not because evolutionary changes are necessarily parsimonious, but because it is generally wiser not to adopt complicated explanations when simpler ones are capable of explaining the known facts. More complicated explanations are accepted only when evidence requires them. Thus, cladograms are hypotheses about evolutionary relationships that are repeatedly tested as additional traits are measured and as new fossil evidence becomes available.

Constructing Phylogenies

As we have seen, many different traits must be measured if we wish to distinguish homologies from homoplasies. Because organisms differ in many ways, systematists use many traits to reconstruct phylogenies. Some traits are readily preserved in fossils; others, such as behavior and molecular structure, rarely survive fossilization processes. By using cladistic techniques, systematists can take into consideration behavioral and molecular traits, which are not preserved in fossils, as well as structural traits in both living and fossil organisms. There is only one correct phylogeny, so the more traits that are measured, the more inferred phylogenies should converge on one another and on the true phylogeny. In this section, we will show how structural, developmental, and molecular traits of organisms are used to infer phylogenies.

Structure and development show ancient connections

An important source of information for taxonomists is **gross morphology**—that is, sizes and shapes of body parts. These traits are useful because they are under genetic control and are relatively stable, but they do change over evolutionary time. Because living organisms have been studied for centuries, we have a wealth of morphological data and extensive museum and herbarium collections of organisms whose traits, including molecular ones, can be measured. Also, this is the type of information most readily available from fossils. Sophisticated methods are now available for measuring and analyzing morphology and for estimating the amount of morphological variation among individuals, populations, and species.

The early developmental stages of many organisms reveal similarities with other organisms that are lost by the time of adulthood. For example, the larvae of the marine creatures called sea squirts have a supporting rod in their backs—the notochord—that disappears as they develop into adults. Many other animals—all the animals called vertebrates—also have such a structure at some time during their development. This shared structure is one of the reasons for believing that sea squirts are more closely related to vertebrates than would be suspected by examination of adults only (Figure 22.8).

Molecular traits: getting close to the genes

Like the sizes and shapes of their body parts, the molecules of organisms are heritable characteristics that may diverge among lineages over evolutionary time; this *molecular evolution* will be discussed in detail in Chapter 23. Among the most important molecular traits of organisms for constructing phylogenies are the structures of their proteins and nucleic acids (DNA and RNA).

PROTEIN STRUCTURE. Relatively precise information about phylogenies can be obtained by comparison of the molecular structure of proteins. We estimate genetic differences between two taxa by obtaining homologous proteins from both and determining the number of amino acids that have changed since the lineages of the taxa diverged from a common ancestor. For example, determining the sequences of amino acids revealed a great deal about how natural selection influenced the evolution of cytochrome *c* (see Figure 23.1).

NUCLEIC ACID STRUCTURE. The base composition of DNA provides excellent evidence of evolutionary relationships among organisms. Cells of eukaryotes have genes in their nuclei and mitochondria. Plant cells also have genes in their chloroplasts. The chloroplast genome (cpDNA), which is used extensively in phylogenetic studies of plants, consists of a circular, double-stranded DNA molecule. This molecule is evolutionarily highly stable. All land plants have nearly the same complement of 100 chloroplast genes that code transfer RNAs and some ribosomal RNA subunits and proteins, particularly those involved in photosynthesis. Mitochondrial DNA (mtDNA), which is very similar to cpDNA, has been used extensively for evolutionary studies of animals (see Chapter 23 for details).

DNAs can be compared, even if the precise sequences of their bases are not known, by a process called **DNA hybridization**, in which the double-stranded DNAs of two different species are combined, then separated into single strands, or *denatured* (see Chapter 14) and allowed to reassociate. The degree of mismatching of the resulting hybrid DNA is related to its thermal stability in a very consistent way. DNA hybridization reveals that the DNAs of humans and chimpanzees are much more similar than would be expected

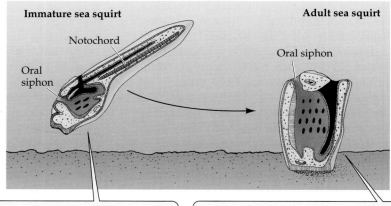

The free-swimming, immature form (larva) of the sea squirt and the vertebrate embryo (frog) both have a notochord for body support.

Both the adult form of the sea squirt and the adult frog lack notochords. In the adult frog, the vertebral column replaces the notochord as the support structure.

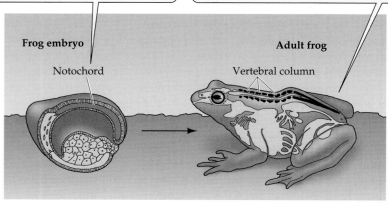

22.8 A Larva Reveals Evolutionary Relationships Sea squirt larvae, but not adults, have a well-developed notochord (blue) that reveals their relationship with vertebrates, all of which have one at some time during their life cycle.

given the considerable morphological differences between the two species (Table 22.2). This similarity indicates that humans and chimpanzees share a common ancestor that is more recent than previously thought.

A long-standing debate among biologists over whether the giant panda of China is more closely related to bears or to raccoons was resolved by DNA hybridization data, which clearly indicate that the giant panda is a bear (Figure 22.9). The non-bearlike features of the giant panda are recent adaptations to its specialized diet—bamboo.

Biological Classification and Evolutionary Relationships

Biological classification systems are designed to express relationships among organisms. The kind of relationships we wish to express determines which features we use to classify organisms. If, for instance, we were interested in a system that would help us decide what plants and animals were desirable as food, we might devise a classification based on tastiness, ease of capture, and the type of edible parts each organism

TABLE 22.2 Genetic Differences among Some Vertebrates as Estimated by DNA Hybridization

TAXA COMPARED	PERCENTAGE DIFFERENCE IN DNA SEQUENCES
Human/chimpanzee	1.6
Human/gibbon[a]	3.5
Human/rhesus monkey	5.5
Human/galago[b]	28.0
House mouse/Norway rat	20.0
Cow/sheep	7.5
Cow/pig	20.0

[a] The gibbon is the smallest member of the ape family, which also includes chimpanzees and gorillas.

[b] Galagos are small, nocturnal primates found in Africa. They are more closely related to the lemurs of Madagascar than they are to monkeys.

possessed. Early Hindu classifications of plants were designed according to these criteria. Biologists do not use such systems today, but those systems served the needs of the people who developed them.

Ailurus fulgens
(lesser panda)

Ailuropoda melanoleuca (giant panda)

22.9 The Giant Panda Is a Bear The DNA hybridization data used to construct the cladogram indicate that the giant panda of Asia is a bear. The lesser panda, also of Asia, was long thought to be closely related to the giant panda, but it is actually more closely related to raccoons and their allies of the Americas.

It is inappropriate to ask whether those classifications or any others, including contemporary ones, are right or wrong. Classification systems can be judged only in terms of their utility and consistency with their stated goals. To evaluate any classification system we must first ask the question, What relationships is it trying to express? Then, How well does it express those relationships? This section addresses these questions.

Early biological classifications were nonevolutionary

Many organisms were given species names and classified by Linnaeus and his followers before evolution became widely accepted as the central concept of biology. These workers described many features of organisms and grouped them according to which similarities seemed most important. They tried to develop "natural" systems of classification, but they had no basis for deciding what was natural or why some features of organisms were more important than others.

Because judging the "importance" of traits and identifying ancestors are difficult tasks, an approach to taxonomy that considered all traits to be of equal importance was developed in the 1950s. Numerical taxonomists, as practitioners of this approach came to be known, did not try to determine evolutionary relationships. Instead they measured as many traits as possible and used adding machines (at first) and computers to compute measures of differences between organisms. Today, many systematists use computer models developed by numerical taxonomists to assist them in developing cladograms and classifications based on evolutionary relationships.

Taxonomic systems should reflect evolutionary relationships

Most taxonomists today believe that classification systems should reflect the evolutionary relationships of organisms. However, they do not agree on the best criteria for doing so, even if they agree on the phylogenies. The reason for the lack of agreement is that lineages of organisms evolve at very different rates. Some "living fossils" have changed very little in the past 50 million years (Figure 22.10). Other lineages have undergone rapid evolutionary changes within the past few million years. Should a classification system be based only on the time since two lineages shared a common ancestor, or should it also reflect the rate of evolution of taxa after they separate from one another?

A clade contains all the descendants of a particular ancestor and no other organisms. For this reason, a clade is said to be **monophyletic**. A taxon consisting of members that do not share the same common ancestor is **polyphyletic**. A group that contains some but not all of the descendants of a particular ancestor is said to be **paraphyletic** (Figure 22.11). Taxonomists agree that polyphyletic groups are inappropriate as taxonomic

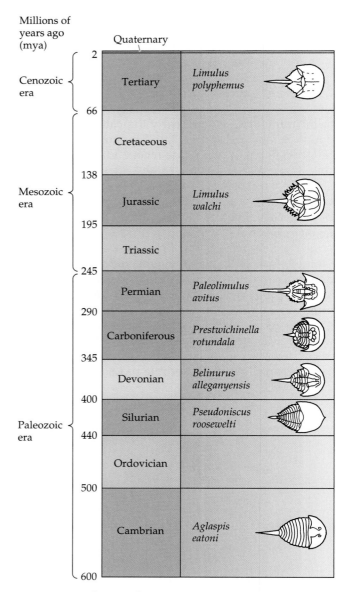

22.10 Horseshoe Crabs Are "Living Fossils" Horseshoe crabs that lived many millions of years ago are almost indistinguishable from modern horseshoe crabs. They have evolved very slowly.

units, but they disagree about the usefulness of paraphyletic groups.

Why they disagree can be illustrated by birds, crocodiles, and their relatives. We now know, primarily from fossil evidence, that birds and crocodilians (a group that includes crocodiles and alligators) share a more recent common ancestor than crocodilians share with snakes and lizards (Figure 22.12a). Until recently, crocodilians were grouped with snakes, lizards, and turtles as reptiles (class Reptilia). Birds were placed in a separate class, Aves (Figure 22.12b). This classification was used because crocodiles have evolved more slowly than birds since the two lineages separated. As

22.11 Mono-, Poly-, and Paraphyletic Taxa Polyphyletic groups are inappropriate as taxonomic units, but scientists disagree about the usefulness of paraphyletic groups.

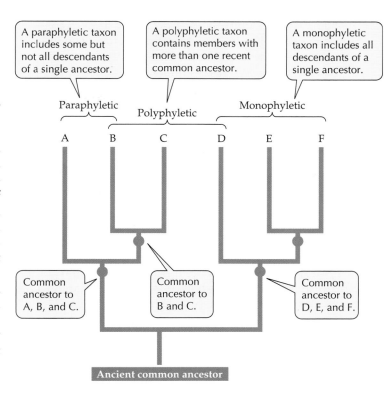

a result, crocodilians are more similar in many features to snakes, lizards, and turtles than they are to birds. They look like very large lizards.

The phylogeny shown in Figure 22.12*a* indicates that the class Reptilia is paraphyletic because the class does not include all the descendants of its common ancestor; that is, birds are not included in the class. If only monophyletic taxa were permitted, birds would be included with crocodilians and their ancestors in a single taxon separate from snakes, lizards, and turtles (Figure 22.12*c*). Retaining birds as a separate class (that is, recognizing the reptiles as a paraphyletic group) emphasizes that birds have undergone rapid evolution since they separated from reptiles and have developed major unique derived traits.

The classifications used today still contain many polyphyletic groups because many organisms have not

been studied enough to distinguish between homologies and convergent evolution. However, as soon as they detect convergent evolution, systematists change their classifications to eliminate polyphyletic taxa. Nevertheless, many systematists favor retaining paraphyletic taxa because doing so highlights groups that have undergone especially rapid evolutionary change.

Another reason for retaining paraphyletic taxa is to maintain stability in the taxonomic system. Strict avoidance of paraphyletic taxa would require reclassifying many higher-level taxa every time new phylogenetic information became available. Some systematists believe that the resulting chaos would seriously reduce the value of the classification system.

The Future of Systematics

The development of molecular methods and powerful computers has ushered in a new era of taxonomy. Many phylogenies are being reconstructed and classifications

(a) The evolutionary relationships among vertebrates

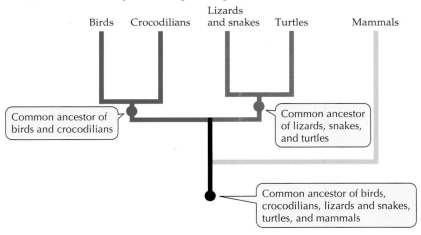

(b) The traditional classification of birds, reptiles, and mammals

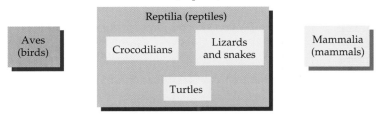

(c) A cladistic classification

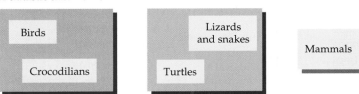

22.12 Phylogeny and Classification
A cladistic classification based on their evolutionary relationships includes the crocodilians in a subgroup with the birds.

are being revised. Information from many sources continues to be used in constructing phylogenies. The fossil record, which reveals when lineages diverged and began their independent evolutionary histories, is necessary to provide absolute dating for evolutionary events. Fossils provide important evidence that helps us distinguish ancestral from derived traits.

The range of data used in classification is likely to increase rather than decrease in the future because modern chemical, biochemical, and microscopic methods allow systematists to measure more traits of organisms than they could previously. Because systematics integrates activities from several different biological disciplines, a systematist needs to have a command of molecular techniques, natural history, and computer programming.

Typically, phylogenies are reconstructed as part of efforts to determine evolutionary relationships among organisms. Nevertheless, phylogenies can be used to answer many other types of biological questions. Indeed, it is difficult to think of any biological problem whose solution would not be assisted by having a reliable phylogeny of the organisms being studied.

Many biological statements are phylogenetic statements. Any statement claiming an association between a trait and a group of organisms is a claim about when during a lineage the trait first arose and about the fate of the trait since its first appearance. For example, the statement that DNA is the genetic material for all eukaryotes is a claim that organisms evolved DNA as their genetic material before eukaryotic cells appeared—that DNA is an ancestral and a general homologous trait among eukaryotes—and that DNA has been maintained as the genetic material in the subsequent evolution of all surviving eukaryote lineages.

The Three Domains and the Six Kingdoms

In Chapter 1 we introduced the three domains and six kingdoms into which organisms are generally classified. The domains are the most ancient divisions of the phylogenetic tree of life. They separated from one another so long ago that few fossils are available to help us determine when and in what order. The sequence of these ancient lineage splits that we use in this book is based on analyses of rRNAs and hundreds of amino acid sequences of dozens of enzymes from all the major groups of organisms.

These extensive molecular data suggest that the most ancient split in the phylogenetic tree of life, about 2,000 million years ago (mya), separated the ancestor of the domain Bacteria from the ancestor of all other life (Figure 22.13). The next split, about 1,800 mya, separated the ancestor of the domain Archaea from the ancestor of the domain Eukarya. Subsequent splits separated the ancestors of protists (1,230 mya),

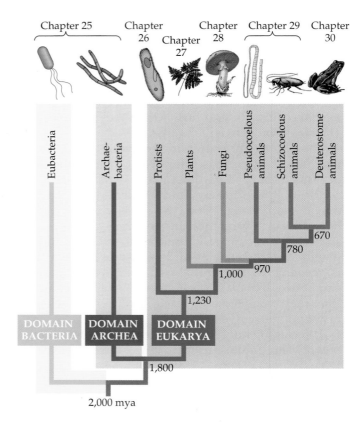

22.13 The Phylogeny of the Domains and Kingdoms
Approximate dates of lineage separations are shown at each node. mya, million years ago.

plants (1,000 mya), fungi (970 mya), and animals (780 mya). New information is likely to change the estimated dates of the splits, but the temporal ordering of the lineage separations is fairly certain.

Summary of "Constructing and Using Phylogenies"

• Classification systems improve our ability to explain relationships among things, aid our memory, and provide unique, universally used names for organisms.

The Hierarchical Classification of Species

• Biological nomenclature assigns to each organism a unique combination of a generic and a specific name. In the universally employed classification system, species are grouped into higher-level units called genera, families, orders, classes, phyla, and kingdoms. **Review Figure 22.2**

Inferring Phylogenies

• Systematists use data from fossils and the rich array of morphological and chemical data available from living organisms to determine evolutionary relationships.

- An ancestral trait is shared with a common ancestor. A derived trait differs from its form in the ancestors of a lineage.
- Homologous traits are descended from a common ancestor. Homoplastic traits evolved more than once. **Review Figures 22.3, 22.4**
- Structures that perform similar functions but have resulted from convergent evolution are said to be analogous to one another. **Review Figure 22.4**
- Cladistic methods were developed to help biologists distinguish between homologous and analogous traits. **Review Figure 22.5**
- To determine true evolutionary relationships, systematists must distinguish between ancestral and derived traits within a lineage, as well as between homologous and homoplastic traits. This task is often difficult because divergent evolution may make homologous traits appear dissimilar and convergent evolution may make nonhomologous traits appear similar. **Review Figures 22.3, 22.4, 22.5, 22.6, and Table 22.1**

Constructing Phylogenies

- Structures in early developmental stages sometimes show evolutionary relationships that are not evident in adults, and such early stages are often available in fossil material. **Review Figure 22.8**
- The structures of proteins and the base sequences of nucleic acids are important taxonomic data that can be obtained from living organisms.

Biological Classification and Evolutionary Relationships

- Taxonomists agree that taxa should share a common ancestor and that polyphyletic taxa are inappropriate taxonomic units. However, they disagree about the utility of paraphyletic taxa, which include some but not all of the descendants of a particular ancestor. **Review Figures 22.11, 22.12**
- Paraphyletic taxa may be retained to highlight the fact that members of some lineages evolved especially rapidly.

The Future of Systematics

- The development of molecular methods and powerful computers has ushered in a new era of systematics.
- Phylogenies are useful in solving many kinds of biological problems.

The Three Domains and the Six Kingdoms

- The domains are the most ancient divisions of the phylogenetic tree of life. They split more than 1,800 million years ago.
- Subsequent splits within the domain Eukarya separated the ancestors of protists (kingdom Protista), plants (kingdom Plantae), fungi (kingdom Fungi), and animals (kingdom Animalia). **Review Figure 22.13**

Self-Quiz

1. Which of the following is *not* a major role of a classification system?
   a. To aid memory
   b. To improve predictive powers
   c. To help explain relationships among things
   d. To provide relatively stable names for things
   e. To design identification keys

2. Any group of organisms treated as a unit in a classification system is
   a. a species.
   b. a genus.
   c. a taxon.
   d. a clade.
   e. a phylogen.

3. A genus is
   a. a group of closely related species.
   b. a group of genera.
   c. a group of similar genotypes.
   d. a taxonomic unit larger than a family.
   e. a taxonomic unit smaller than a species.

4. Outgroups are used in cladistic analyses to
   a. distinguish homoplasies from homologies.
   b. distinguish homoplasies from convergence.
   c. distinguish between general and special homologies.
   d. determine relationships between closely related taxa.
   e. distinguish between general and special homoplasies.

5. A clade contains
   a. all—and only—the descendants of a single ancestor.
   b. all the descendants of more than one ancestor.
   c. most but not all of the descendants of a single ancestor.
   d. members of two or more lineages.
   e. a few of the descendants of a single ancestor.

6. Which of the following is *not* a way of identifying ancestral traits?
   a. Determining which traits are found among fossil ancestors
   b. Using an outgroup in which the trait is also found
   c. Using more than one outgroup that has the trait
   d. Determining how many species in the lineage share the trait today
   e. Experimentally creating a known lineage

7. Traits that evolve very slowly are useful for determining relationships at the level of
   a. phyla.
   b. genera.
   c. orders.
   d. families.
   e. species.

8. Homologous traits are
   a. similar in function.
   b. similar in structure.
   c. similar in structure but derived from different ancestral structures.
   d. derived from a common ancestor whether or not they have the same function today.
   e. derived from different ancestral structures and have dissimilar structures.

9. The genes that are most extensively used to determine evolutionary relationships among plants are
   a. nuclear genes.
   b. chloroplast genes.
   c. mitochondrial genes.
   d. genes in flowers.
   e. genes in roots.

10. Which of the following is *not* a way in which phylogenies are used?
    a. To establish evolutionary relationships
    b. To determine how rapidly traits evolve
    c. To determine historical patterns of movement of organisms
    d. To help identify unknown organisms
    e. To infer evolutionary trends

Applying Concepts

1. The great blue heron (*Ardea herodias*) is found in most of North America. The very similar gray heron (*Ardea cinerea*) ranges over most of Europe and Asia. These two herons currently are treated as different species, but a colleague argues that they should be treated as a single species. What facts should you consider in evaluating your colleague's suggestion?

2. Why are systematists so concerned with identifying lineages descended from a single ancestor?

3. How are fossils used to identify ancestral and derived traits of organisms?

4. A student of the evolution of frogs has performed DNA hybridization experiments among about 25 percent of all frog species. As a result of these experiments, she proposes a new classification of frogs that differs strikingly from the traditionally accepted one. Should frog taxonomists immediately accept this new classification? Why or why not?

5. Linnaeus developed his system of classification before Darwin proposed his theory of evolution by natural selection, and most classifications of organisms were developed by people who were not evolutionists. Yet most of these classifications are still used today, with minor modifications, by evolutionists. Why?

Readings

Brooks, D. R. and D. H. McLennan. 1991. *Phylogeny, Ecology, and Behavior. A Research Program in Comparative Biology*. University of Chicago Press, Chicago. An excellent book that shows how rich insights can be derived by integrating phylogenetic analyses with studies of ecology and behavior. Includes a chapter describing the tools used to reconstruct phylogenies.

Eldredge, N. and J. Cracraft. 1980. *Phylogenetic Patterns and the Evolutionary Process*. Columbia University Press, New York. A good sampling of various perspectives on concepts and practices in systematics.

Harvey, P. H. and M. D. Pagel. 1991. *The Comparative Method in Evolutionary Biology*. Oxford University Press, Oxford. A good discussion of methods of reconstructing phylogenies and the use of comparative approaches in evolutionary studies.

Hillis, D. M., C. Moritz and B. K. Mable (Eds.). 1996. *Molecular Systematics*, 2nd Edition. Sinauer Associates, Sunderland, MA. A description of methods used currently in molecular systematics; designed to guide beginners through a molecular systematic study.

Mayr, E. 1982. *The Growth of Biological Thought*. Belknap Press, Cambridge, MA. A good treatment of the history of thinking about systematics, biological diversity, and evolution.

Ridley, M. 1986. *Evolution and Classification: The Reformation of Cladism*. Longman, London. A critical evaluation of current schools of thought in systematics that provides clear statements of the goals and methods of all approaches.

Wiley, E. O. 1981. *Phylogenetics: The Theory and Practice of Phylogenetic Systematics*. Wiley, New York. A clear overview of the phylogenetic approach to evolution and classification.

Chapter 23

Molecular Evolution

T here is a specialized secretion system in many types of pathogenic eubacteria that was first noticed in 1991. These bacteria, which cause diseases such as bubonic plague in humans and fire blight in fruit trees, had acquired a set of apparently identical genes coding for an elaborate protein delivery system. The system, which is activated by contact with host cells, injects a variety of damaging proteins directly into the host's cell, thereby avoiding most of the host's defenses. The system was first discovered by scientists studying a species of a gram-negative bacterium in the genus *Yersinia*, which causes bubonic plague and intestinal infections in humans. Because these proteins seemed to be associated with the outer membranes of the bacteria, they were called **Yops** (*Yersinia* outer proteins). (Yops are not actually outer membrane proteins, but the name has persisted.)

Other researchers soon found Yops in other species of bacteria only distantly related to *Yersinia*. How did so many species of bacteria acquire the same set of genes? Could they all have inherited them from a common ancestor in the distant past? Or did they acquire them by transfer from other bacteria? Why do molecular biologists believe that the latter explanation is the correct one? To answer this question, we need to understand patterns and processes in molecular evolution.

For much of its history, evolutionary biology has consisted of the study of the obvious morphological

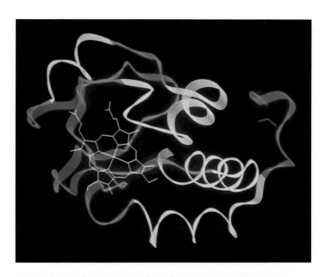

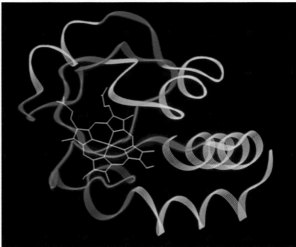

Protein Molecules Fold the Same Way in Different Species
Molecular sequencing techniques reveal that the three-dimensional structures of cytochrome *c* in rice (top) and tuna (bottom) are substantially similar. Such similarities indicate that all cytochrome *c* molecules have a common ancestor.

features of organisms. Much of what Charles Darwin documented during his 5-year voyage aboard the *Beagle* were detailed observations about morphological differences in the kinds of species found in different geographic areas. Darwin later synthesized these observations into descriptions of how species change over time.

His most famous example involved the finches of the Galapagos Islands. Darwin observed that insect-eating warblers and woodpeckers were absent from these isolated islands, but various species of finches, birds that usually eat seeds, had assumed the insect-eating habits of the missing species. In the process of adapting to their new diets, the finches had evolved beak morphologies that matched those of the missing species (see Figure 21.7).

Darwin was able to guess why many of these morphological changes had happened, but he could not determine the mechanisms by which those adaptations were expressed. Understanding of the mechanisms of morphological change had to await discoveries in biochemistry a century later. As you learned in Chapter 20, we now understand that genetic differences underlie the adaptive evolution of species. The goals of the study of *molecular evolution* are to determine the *patterns of evolutionary change* in the molecules of which organisms are composed and to determine the *processes that caused those changes*. If we understand the patterns and processes that underlie molecular evolution, we can use those insights to help us solve other biological problems.

To reveal patterns of molecular evolution, molecular biologists must determine the precise structure of molecules. Techniques for doing so were developed in the 1940s; the first complete sequence of a protein—insulin—was determined in 1952. Sequencing the insulins of different mammals revealed that amino acid substitutions were not randomly distributed across the molecules. Instead they were restricted to three positions on one chain of the molecule. In addition, the data revealed that most amino acid substitutions did not affect the biological activity of insulin. Insulin from one species was equally effective in other species.

As increasing numbers of proteins were sequenced, the same pattern was found: Most nucleotide substitutions were confined to particular regions of the molecules. This pattern could be explained by the assumption that most of the molecular changes being identified were substitutions that did not affect the functioning of the molecules; that is, they were neutral or nearly neutral substitutions.

The hypothesis of **neutral evolution** asserts that most of the variability in structures of molecules being measured by molecular biologists does not influence their functioning. In addition, the theory suggests that the rates at which neutral substitutions accumulate is not influenced by natural selection; it is determined simply by the mutation rate. As we will see, these discoveries and intepretations play a central role in the study of molecular evolution.

In this chapter, we briefly review how molecular biologists determine the structures of molecules. Then, by comparing the structures of molecules in living and fossil organisms, we show how biologists can infer both the patterns and the causal processes of molecular evolution. Finally, we will discuss how knowledge of the patterns of molecular evolution helps us solve other biological problems, including inferring phylogenetic relationships among organisms and determining how humans spread over Earth.

Determining the Structure of Molecules

As you learned in Chapter 11, biologists determined the structure of DNA by synthesizing evidence from several sources. The most important evidence came from X ray crystallography, which allows investigators to determine the positions of atoms in a crystalline substance by the diffraction pattern of X rays passed through the crystal. The fact that the total amount of purines in DNA equals the total amount of pyrimidines also provided important clues about the structure of DNA. By building three-dimensional models of possible molecular structures on the basis of this information, Francis Crick and James Watson proposed the structure of DNA that has been supported by all experimental evidence gathered since then.

The invention of the polymerase chain reaction (PCR) method (see Figure 16.12) to amplify DNA has allowed molecular biologists to determine the sequences of subunits of DNA from their fossilized remains, mummified tissues, or dried skins in museums, even though these objects contain only tiny amounts of DNA. DNA has been extracted and amplified from fossils up to 135 million years old (Table 23.1).

We infer past molecular structures by using the comparative method

Being able to measure directly the structure of molecules of fossil organisms is the most powerful way to determine patterns of molecular evolution. However, even when we cannot directly measure molecules from extinct organisms, we can nonetheless infer evolutionary patterns by comparing the molecules of living organisms. To make these comparisons, molecular evolutionists often use the cladistic methods described in Chapter 22. That is, they attempt to determine ancestral and derived states of molecules and develop cladograms based on shared derived traits.

A good example of the use of the comparative approach to determine patterns of molecular evolution is provided by studies of the enzyme cytochrome *c*. Cytochrome *c* is one component of the respiratory chain of mitochondria. Together with other proteins of the citric acid cycle and respiratory chain, cytochrome *c* is found in the cells of all eukaryotes.

TABLE 23.1 Biological Tissues from which Ancient DNA Has Been Extracted

TYPE OF MATERIAL	PROBABLE AGE IN YEARS
Human tissues	
Mummies	5,000
"Bog" bodies	7,500
Bones and teeth	10,000
Other animal material	
Feathers	130
Museum skins	140
Naturally preserved skins	13,000
Bones	>25,000
Amber specimens	135,000,000
Plant material	
Herbaria specimens	118
Charred seeds and cobs	4,500
Mummified seeds and embryos	44,600
Fossil magnolia leaf	20,000,000
Amber specimens	40,000,000

The amino acid sequences of cytochrome c are known for nearly 100 species of organisms, ranging from yeasts to humans (Figure 23.1). Among these cytochromes c are regions that accumulated changes relatively quickly; for example, positions 44, 89, and 100 differ among many of the organisms compared. There are also invariant positions, such as 14, 17, 18, and 80. This particular set of invariant residues is known to interact with the iron-containing heme group that is essential for enzyme functioning. Presumably, because any changes in these amino acids adversely affect the functioning of the heme group, they were removed by natural selection when they arose.

Molecular biologists now routinely use the comparative approach to identify regions of molecules that lack variation. They generally assume that changes in the amino acids in those positions would be likely to adversely affect the functioning of the molecule. In contrast, regions that change relatively quickly are believed to be functionally less significant. Consequently, amino acid substitutions in those regions are functionally neutral, and they are not influenced by natural selection.

The differences among the cytochromes c shown in Figure 23.1 are the result of accumulations of substitutions within the genomes of species over millions of years, but that is not the only means by which the genomes of organisms change. As you saw in Chapter 13, plasmids, episomes, and phage coats can transport genes from one bacterial cell to another. That is the mechanism by which Yops have been transferred among distantly related bacteria. Because Yops genes

are absent in the species most closely related to the ones that have them, the genes were not inherited from a common ancestor, but rather were transferred horizontally directly between distantly related species.

Another type of gene transport mechanism that functions within individual cells is provided by transposable elements, segments of chromosomal or plasmid DNA that can be inserted at other points in the same or other DNA molecules. Transposable elements may have played a major role in the development of the genome of eukaryotes. When the nuclear envelope of eukaryotes evolved, it separated the transcription of DNA from its translation. As you learned in Chapter 14, the exons of a eukaryotic gene code a single protein. Exons are separated from one another by noncoding introns. Genomes that lack introns, as found in prokaryotes, are probably the ancestral state of the genetic material. Introns are probably the remains of transposable elements that invaded the genome after the evolution of a nuclear membrane and subsequently lost the ability to transpose.

Globin diversity evolved via gene duplication

Molecular evolution also proceeds by gene duplication, which has been well studied in the globin family of genes. Globins were among the first proteins to be sequenced and their amino acid sequences compared. Several globins have been crystallized, such that details of their three-dimensional structure are known. This knowledge has greatly facilitated the study of the relationship between the rate at which amino acid sequences diverge and how details of the three-dimensional structures of proteins influence their functional properties.

Humans have three families of globin genes: the *myoglobin family*, whose single member is located on chromosome 22; the *α-globin family* on chromosome 16; and the *β-globin family* on chromosome 11 (Figure 23.2). Two types of proteins are produced by these three families: myoglobin and hemoglobin. Comparisons of amino acid sequences strongly suggest that rather than arising from different genes that have independently converged on some similar functions, the different forms of globins arose as gene duplications (Figure 23.3). After a duplication event, each copy could evolve along its own evolutionary path. How long the genes have been evolving separately can be inferred by comparison of their amino acid sequences. The greater the number of amino acid differences between two globins, the farther back in time was their most recent common ancestor.

The earliest organisms known to have both α- and β-globin genes lived about 500 million years ago (mya). Thus, the initial duplication event at which myoglobins diverged from all other globins probably happened at least that long ago (Figure 23.4). Assuming that the rate

23.1 Amino Acid Sequences of Cytochrome *c*

The protein sequence alignment for cytochrome *c* obtained from 33 species. Conservation of charge at a position is shown by the constancy of color in its column. Constancy of charge suggests that cytochrome *c* molecules in all these species fold in the same way and preserve the same three-dimensional structure; this structure is shown in the computer rendering below and at the beginning of this chapter.

of amino acid substitution has been relatively constant since then—about 100 substitutions per 500 million years—the two families, which differ at 77 sites, are estimated to have split about 450 mya.

Homologous gene families have similar effects on development

One way to detect genes that are homologous in distantly related organisms is to find identical or nearly identical families of genes that produce similar effects on developmental patterns. For example, some mutations in *Drosophila* cause appendages that are appro-

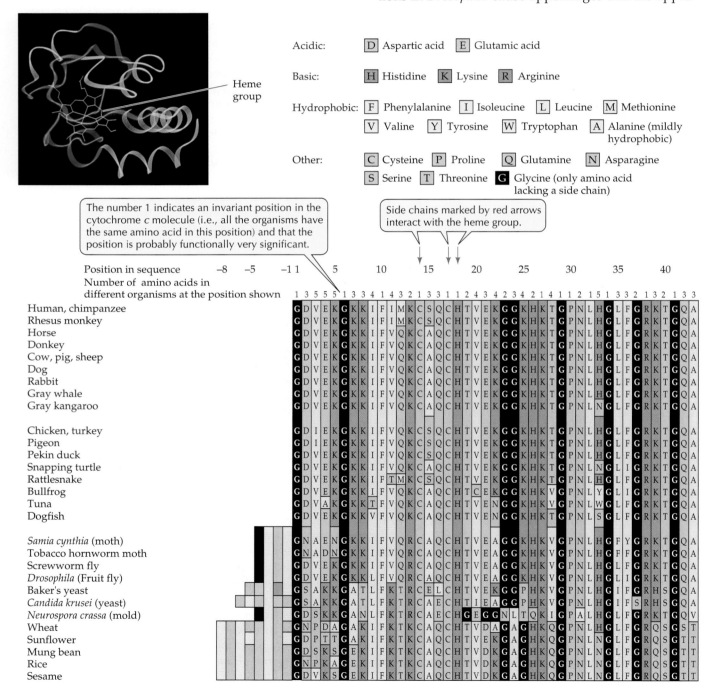

Heme group

Acidic: D Aspartic acid   E Glutamic acid

Basic: H Histidine   K Lysine   R Arginine

Hydrophobic: F Phenylalanine   I Isoleucine   L Leucine   M Methionine
V Valine   Y Tyrosine   W Tryptophan   A Alanine (mildly hydrophobic)

Other: C Cysteine   P Proline   Q Glutamine   N Asparagine
S Serine   T Threonine   G Glycine (only amino acid lacking a side chain)

The number 1 indicates an invariant position in the cytochrome *c* molecule (i.e., all the organisms have the same amino acid in this position) and that the position is probably functionally very significant.

Side chains marked by red arrows interact with the heme group.

23.2 Chromosomal Arrangement of the Human Globin Gene Families

The α-globin family of hemoglobin is found on chromosome 16, the β-globin family on chromosome 11, and the myoglobin gene on chromosome 22.

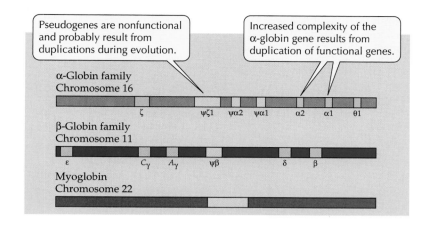

priate to one body segment to appear in another. Thus, leglike appendages may grow where there should be antennae, or a body segment may be duplicated (Figure 23.5). These unusual changes are caused by genes that occur in two tightly linked clusters that together constitute the **homeotic gene complex**. All homeotic genes contain a region, called the **homeobox**, which when transcribed specifies a sequence of 60 amino acids. Homeobox genes are active in specific body segments and, in the absence of mutations, are responsible for the appropriate development of those segments.

This pattern of developmental control is not unique to flies; more than 350 homeobox elements have been identified in fungi, plants, and animals. A similar set of genes, the *Hox* genes, has been found in house mice. The homeobox and *Hox* genes occur in the same order along the chromosome (see Figure 15.21). These struc-

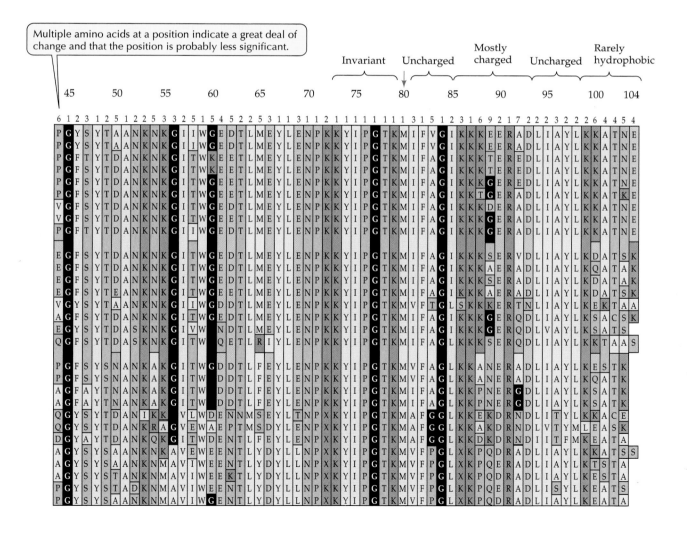

23.3 Amino Acid Sequences of Residues 1 through 39 of the Globin Superfamily A comparison of partial myoglobin and partial representative α and β family hemoglobin amino acid sequences. The single-letter abbreviations for the amino acids are as given in Figure 23.1. Identical residues are boxed.

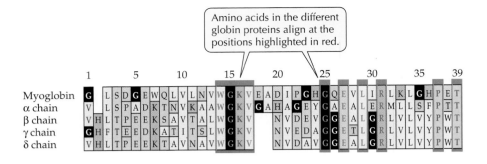

Amino acids in the different globin proteins align at the positions highlighted in red.

tural similarities strongly suggest that all homeobox genes have a common evolutionary origin, and that the mechanisms that determine the differentiation of the major morphological regions (body, head, trunk, and tail) may have arisen only once in animal evolution.

How Molecular Functions Change

Evolution as we know it would not have been possible if genes were unable to change their functional roles. As we have just seen with globin genes, gene duplication permits the genes, and the proteins they encode, to evolve different functions. Myoglobin, a monomer, has become the primary oxygen storage protein in muscle. Hemoglobin, a tetramer consisting of two α and two β chains, carries oxygen in blood. Myoglobin has evolved a specific affinity for oxygen that is much higher than that of hemoglobin.

In contrast to myoglobin, hemoglobin evolved to be much more refined and diversified in its role as the blood-oxygen carrier. Hemoglobin binds oxygen from the lungs or gills, where the partial pressure of oxygen is relatively high, transports it to regions of low oxygen partial pressure, and releases it in these areas. With its more complex, tetrameric structure, hemoglobin also is able to transfer hydrogen and carbon dioxide in the blood, bind four molecules of oxygen cooperatively, and respond to tissue acidity.

In humans, the α-hemoglobin family has four functional genes and three pseudogenes. The four functional genes diversified in function while the three pseudogenes lost all function. Thus, duplication events may result in increased genomic complexity (as seen with the alternate genes for α-hemoglobin) and produce nonfunctional DNA (pseudogenes). Pseudogenes may ultimately be removed via deletion.

Although point mutations can affect the function of a protein, duplication releases one copy of a gene from its original function. Duplication can result in the evolution of entirely novel functions, and, eventually, in the evolution of new gene families. The number of gene families in the human genome is not known, but much of the diversity within our genome arose as a result of ancient gene duplications. Ribosomal RNA and transfer RNA genes, lactate dehydrogenase genes, and pyruvate kinase genes are all examples of gene families that arose from gene duplication.

23.4 Globin Gene Tree The globin superfamily gene tree suggests that myoglobin diverged from modern hemoglobin precursors about 500 mya, at about the time of the origin of vertebrates.

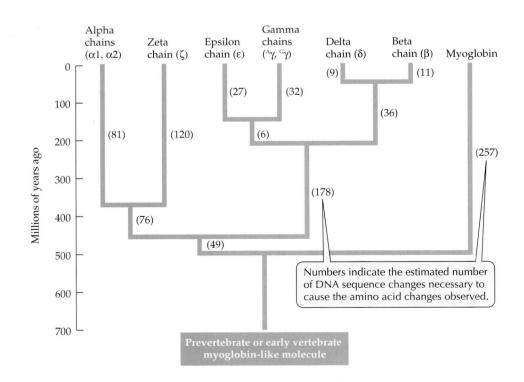

Normal *Drosophila*

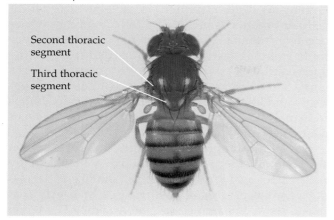

Second thoracic segment

Third thoracic segment

bithorax mutation

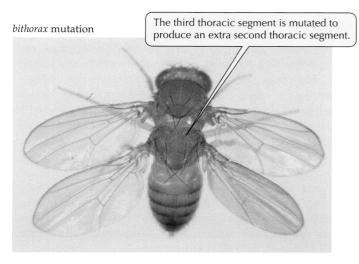

The third thoracic segment is mutated to produce an extra second thoracic segment.

23.5 The *bithorax* Mutation in *Drosophila* Mutations at one of the homeotic genes, *bithorax*, transform the third thoracic segment into a second copy of the second thoracic segment. The result is a fly with two sets of wings.

The globin genes illustrate that molecular functions may change after gene duplication, but how did these functional changes happen? We will explore this interesting component of molecular evolution by using lysozyme as an example.

Lysozyme evolved a novel function

Lysozyme is an enzyme found in almost all animals. It is produced in the tears, saliva, and milk of mammals and in the whites of bird eggs. Lysozyme digests the cell walls of bacteria, rupturing and killing them. As a result, lysozyme plays an important role as a first line of defense against invading bacteria. All animals digest bacteria, which is probably why all have lysozyme. However, in some animals lysozyme is used in the digestion of food.

Among mammals, a novel mode of digestion called *foregut fermentation* has evolved twice. The anterior part of the stomach (the foregut) becomes converted into a chamber for the bacterium-based digestion (fermentation) of ingested plant matter (see Chapter 47). The mammal uses bacteria to obtain nutrients from the otherwise indigestible cellulose of plant material. Foregut fermentation evolved independently in ruminants, such as cows, and certain leaf-eating monkeys, such as langurs.

We know these evolutionary events were independent because close relatives of langurs and ruminants do not ferment their food in their stomachs. In both foregut-fermenting lineages, lysozyme has been modified to play a new, nondefensive role in the stomach. Lysozyme ruptures some of the bacteria that normally live in the stomachs, releasing nutrients that the mammal absorbs through its stomach lining.

How many changes were incorporated in the lysozyme molecule to allow it to function amidst the harsh enzymes and low pH of the mammalian foregut? To answer this question, molecular evolutionists compared the amino acid sequences of lysozyme in foregut fermenters and several of their nonfermenting relatives. They then determined which amino acids differed and which were shared among the species (Table 23.2). Finally, they compared the patterns of these changes with the already known phylogenetic relationships among the species.

The most striking result is that amino acid changes have occurred about twice as rapidly in the lineage leading to langur lysozyme as in any other primate lineage. This increased rate of substitution shows that lysozyme went through a period of rapid adaptation in the stomach of langurs. The lysozymes of langurs and cows share five amino acid substitutions, all of which lie on the surface of the lysozyme molecule, well away from the active site. Several of the shared residues involve changes from arginine to lysine,

TABLE 23.2 Comparisons of Lysozyme Amino Acid Sequences of Different Species

SPECIES	LANGUR	BABOON	HUMAN	RAT	COW	HORSE
Langur*		14	18	38	32	65
Baboon	0		14	33	39	65
Human	0	1		37	41	64
Rat	0	1	0		55	64
Cow*	5	0	0	0		71
Horse	0	0	0	0	1	

Shown above the diagonal line is the number of amino acid sequence *differences* between the two species being compared; below the line are the number of sequences uniquely *shared* by the two species. Asterisks (*) indicate foregut-fermenting species.

which makes the proteins more resistant to attack by the pancreatic enzyme trypsin. By understanding the functional significance of amino acid substitutions, molecular evolutionists can explain the observed changes in amino acid sequences over time in terms of the changing function of the protein.

A large body of fossil, morphological, and physiological data shows that langurs and cows do not share a recent common ancestor. However, langur and ruminant lysozymes share many amino acid residues that neither animal shares with the lysozymes of their own closer relatives. In other words, the lysozymes have converged to a similar sequence despite having very distant ancestry. The amino acid residues they share give these lysozymes the ability to function in the unusual environment of the stomach. Langurs and ruminants both ferment their leafy food in their stomachs. If they are to profit from the metabolic activities of the bacteria that ferment the masticated food, they must have an enzyme capable of lysing bacteria that can function in that environment. Their nonfermenting relatives have no such need.

A group of birds uses lysozyme in digestion

An even more remarkable story emerges if we look at lysozyme in hoatzins, a group of Neotropical cuckoos that are the only known avian foregut fermenters. Hoatzins have an enlarged crop that contains resident bacteria and acts as a fermenting chamber, allowing hoatzins to survive on a diet of leaves. Lysozyme has also evolved to function in the crop of the hoatzin.

Many of the amino acid changes that occurred in the adaptation of hoatzin crop lysozyme are identical to the changes that evolved in ruminants and langurs. Thus, even though the three groups have evolved independently from one another for more than 300 million years, they have each evolved a similar molecule that enables them to recover increased amounts of nutrients from their fermenting bacteria in a highly acidic environment.

Genome Size: Surprising Variability

Organisms differ strikingly in genome size. Multicellular organisms have more DNA than do simpler organisms (Table 23.3). This is not surprising, since more complex instructions are needed for building and maintaining a large, complex organism than a small, simpler one. What is surprising is that lungfishes and lilies have about 40 times as much DNA as humans do. Clearly, a lungfish or a lily is not 40 times as complex as a human.

Some of the apparent difference disappears when we compare the percentage of DNA that actually codes functional RNA or proteins. The size of the *coding* genome of organisms increases in a way that makes sense. Eukaryotes have more coding DNA than prokaryotes have; vascular plants and invertebrate animals have more coding DNA than single-celled organisms have; invertebrates with wings, legs, and eyes have more coding DNA than a roundworm has; and vertebrates have more DNA than invertebrates have.

However, we still do not understand why the percentage of coding DNA varies so dramatically. Nor do we know how much genetic information is needed to program the development of a particular morphological structure.

Molecular Clocks

If we plot the time since the divergence of certain organisms, as determined by the fossil record, against the number of amino acids by which their cytochromes *c* differ, we find that differences in cytochrome *c* sequences have evolved at a relatively constant rate (Figure 23.6). Many proteins show this same constancy in the rate at which they have accumulated changes over time. If the structure of a molecule changes at a constant rate, we can use its **molecular clock** to determine dates of evolutionary events.

It would be convenient for biologists if the rates of change were the same for all molecules. Unfortunately,

---

TABLE 23.3   Genome Size and DNA Content of Some Organisms

ORGANISM	GENOME SIZE (BASE PAIRS × 10^9)	PERCENT OF DNA THAT CODES FOR SOMETHING	TOTAL CODING DNA
Bacterium (*Escherichia coli*)	0.0004	100	0.00004
Yeast (*Saccharomyces*)	0.0009	70	0.00063
Roundworm (*Caenorhabditis*)	0.09	25	0.023
Fruit fly (*Drosophila*)	0.18	33	0.059
Newt (*Triturus*)	19.0	1.5–4.5	0.29–0.855
Human (*Homo*)	3.5	9–27	0.32–0.945
Lungfish (*Protopterus*)	140.0	0.4–1.2	0.56–1.18
Mustard (*Arabidopsis*)	0.2	31	0.062
Lily (*Fritillaria*)	130.0	0.02	0.026

however, different molecular clocks tick at different rates. These differences arise because different molecules have different functional constraints on their evolution. The molecular clock for rRNA ticks very slowly; the cytochrome *c* clock ticks at a slightly more rapid rate; the lysozyme clock ticks at an even faster rate that is not constant for all lysozymes.

Despite these differences, the rates at which many molecular clocks tick appear to be relatively constant. The reason is that the vast majority of molecular changes involve nucleotide or amino acid substitutions that do not affect the functioning of the molecule and, hence, the fitness of the organism. These *neutral changes* are not influenced by natural selection, and they accumulate at a rate roughly equal to the mutation rate. When changes occur that do affect fitness, such as the variants of lysozyme that became adapted to life in the stomach, the rate of evolution is quite different from the neutral rate. However, if *adaptive changes* are few relative to all the changes that distinguish the molecules of any pair of species—that is, if most substitutions are neutral—the molecular clock can still be used to date lineage separations in the distant past.

To use molecular clocks to date events in the remote past, we need to know what proportion of molecular clocks tick at a constant rate. Fairly accurate molecular clocks are known for many proteins, but detailed investigations sometimes indicate that the rates of these clocks occasionally change. We have seen such changes among lysozymes. Fortunately, if we have a good un-

derstanding of the structure and function of a molecule, we can usually distinguish neutral from adaptive variation.

Even if the rate of ticking of a molecular clock changes slightly, the variations may not be great enough to seriously affect our estimates of the dates of lineage divergences. We can also check the constancy of molecular clocks by comparing the dates of lineage divergences calculated from molecular clocks with those estimated from well-dated fossil records. Typically these two estimates are close enough to one another to give us confidence in using molecular clocks to estimate dates of lineage separations among groups for which fossil data are rare or lacking.

Using Molecules to Infer Phylogenies

By comparing molecular structures from different species, we gain insights into how the molecules function and acquire a tool for inferring phylogenies. As we have seen, much evidence suggests that sequences of amino acids in proteins or base sequences in RNA and DNA that have changed very little during evolution probably have the same function in all species. Regions that have changed rapidly during evolution are likely to have fewer constraints on the functional

23.6 Cytochrome *c* Molecules Have Evolved at a Constant Rate   Many proteins accumulate amino acid substitutions at a constant rate over time. This constancy provides scientists with a molecular clock that can sometimes be used to date events in the remote past.

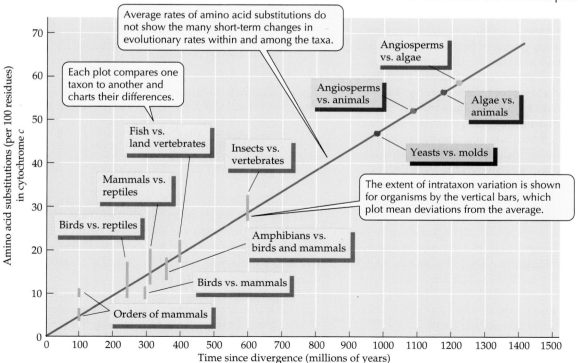

importance of specific sequences, or possibly have undergone major changes in function, as happened to lysozyme in foregut fermenters.

Because they differ in few amino acid or base sequences, molecules that have evolved slowly can be used to estimate relationships among organisms that diverged long ago. Molecules that have evolved rapidly are useful for studying organisms that share more recent common ancestors.

The first step in a molecular evolutionary analysis is to choose which molecule(s) to study. The best choice depends on the question at hand. If you are interested in determining the evolutionary relationships of all existing organisms, you must choose a molecule that all the organisms possess. Such a molecule might be a ribosomal RNA. Equally important is the fact that rRNAs experience very strong functional constraints (meaning that most changes in the RNA sequence prevent the ribosome from functioning properly). Thus, because rRNA evolves so slowly, comparisons of differences among the rRNAs of living organisms can be used to estimate when their lineages diverged, even if the split happened billions of years ago.

After the sequences of amino acids or bases have been determined, they must be compared—a task more difficult than might appear at a glance. A simple example illustrates. In Figure 23.7a two sequences (1 and 2) are being compared. The two sequences differ in number of residues and in identity. Our goal is to align these sequences so that we can compare homologous amino acids.

To do so, we first observe that, although the sequences appear quite different, they would become similar if we were to insert a space after the first amino acid in sequence 2 (after the leucine [leu] residue). In fact, these sequences then differ only by one amino acid at position 6 (serine or phenylalanine). A single insertion aligns the sequences in this case, but longer sequences and those that have diverged more extensively require more elaborate adjustments.

After we have aligned sequences, we can compare them in several ways. First, we can simply count the number of nucleotides or amino acids that differ between the sequences. Let's return to the previous example (see Figure 23.7a) and add additional sequences. We now have six amino acid sequences to align (Figure 23.7b). By adding up the number of similar and different amino acids, we can construct a **similarity matrix** (Figure 23.7c). The assumption is that the more recently two groups shared a common ancestor (for example, sequences 2 and 3 in Figure 23.7c), the more similar are their sequences. Having established similarities and differences between our sequences, we can then infer the evolutionary relationships among them.

A matrix of similarity (or difference) can be translated into a **molecular phylogeny**, or **gene tree**. A

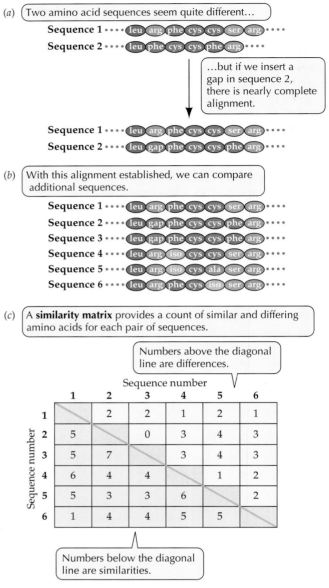

(a) Two amino acid sequences seem quite different...

Sequence 1 ···· leu arg phe cys cys ser arg ····
Sequence 2 ···· leu phe cys cys phe arg ····

...but if we insert a gap in sequence 2, there is nearly complete alignment.

Sequence 1 ···· leu arg phe cys cys ser arg ····
Sequence 2 ···· leu gap phe cys cys phe arg ····

(b) With this alignment established, we can compare additional sequences.

Sequence 1 ···· leu arg phe cys cys ser arg ····
Sequence 2 ···· leu gap phe cys cys phe arg ····
Sequence 3 ···· leu gap phe cys cys phe arg ····
Sequence 4 ···· leu arg iso cys cys ser arg ····
Sequence 5 ···· leu arg iso cys ala ser arg ····
Sequence 6 ···· leu arg phe cys iso ser arg ····

(c) A **similarity matrix** provides a count of similar and differing amino acids for each pair of sequences.

Numbers above the diagonal line are differences.

	Sequence number					
	1	2	3	4	5	6
1		2	2	1	2	1
2	5		0	3	4	3
3	5	7		3	4	3
4	6	4	4		1	2
5	5	3	3	6		2
6	1	4	4	5	5	

Numbers below the diagonal line are similarities.

23.7 Amino Acid Sequence Alignment The insertion of gaps allows us to align these sequences so that we can compare homologous amino acids. The similarity matrix in (c) sums similarities and differences; the larger the number of similarities, the more recent the presumed common ancestor of the species.

gene tree is a visual representation of the evolutionary relationship inferred for the genes being compared. Several methods are used to translate a similarity matrix into a gene tree. Each differs in the assumptions it makes about how DNA or protein sequences evolve. In our previous example (see Figure 23.7c), we could use the number of amino acids shared to infer the ancestry of the six sequences. In this case, the branches of our tree represent the number of events, including the deletion, that distinguish each of the six sequences.

Although often the only data available with which to estimate ancient lineage divisions or the phylogenies of prokaryotes, molecular data are also regularly used in combination with morphological and fossil data. Why do we use molecules when morphology is available? The answer is simple: The more characters that are used (morphological, molecular, fossil, and so on) to deduce a phylogeny, the less likely we are to be misled by loss of traits or evolutionary convergence of characters (homoplasy), as discussed in Chapter 22.

Similarly, if we use data from more than one type of molecule, we can better detect convergent evolution. For example, if we were to infer a phylogeny from only lysozyme sequences, we might falsely conclude that langurs, cows, and hoatzins are more closely related to one another than morphological and fossil data indicate. However, if we also compared the DNA sequences of the genes that encode lysozyme, rather than just amino acid sequences, we would find that most of the nucleotide substitutions that have accumulated since the divergence of langurs, cows, and birds were not influenced by strong selection.

Molecules provide the best evidence for ancient lineage splits

No fossils exist to document the most ancient splits in the lineages of life on Earth. To infer the times of these lineage separations we must use molecules that are found in all organisms and that evolve very slowly. A good candidate is 16S rRNA molecules, which are found in all organisms and evolve very slowly. They have been used to identify the three major branches, or domains, of life: the Bacteria, the Archaea, and the Eukarya (see Figure 22.14).

The greatest advantage of employing molecules to infer ancient lineages follows from the fact that we cannot use morphological characters for comparisons between microbes and more advanced organisms—or, in many cases, between microbes themselves—because they lack comparable structures. Thus, only molecules can be used for an across-the-board comparison of all living organisms.

The structure of DNA extracted from extinct organisms is being used to determine the evolutionary relationships between those organisms and their surviving relatives. For example, DNA was obtained from the bones and mummified soft tissues of moas—large, flightless birds (weighing up to 200 kg) that lived in New Zealand until humans arrived a thousand years ago and hunted them to extinction. The amplified DNA was compared with members of other groups of flightless birds, such as kiwis and rheas.

These comparisons suggest that although kiwis and moas both lived in New Zealand, they are not each other's closest relatives (Figure 23.8). The closest relatives of the moas are unknown, extinct flightless birds that also gave rise to flightless birds on Australia. Kiwis came to New Zealand more recently; their closest relatives, emus and cassowaries, live in their ancestral home, Australia and New Guinea.

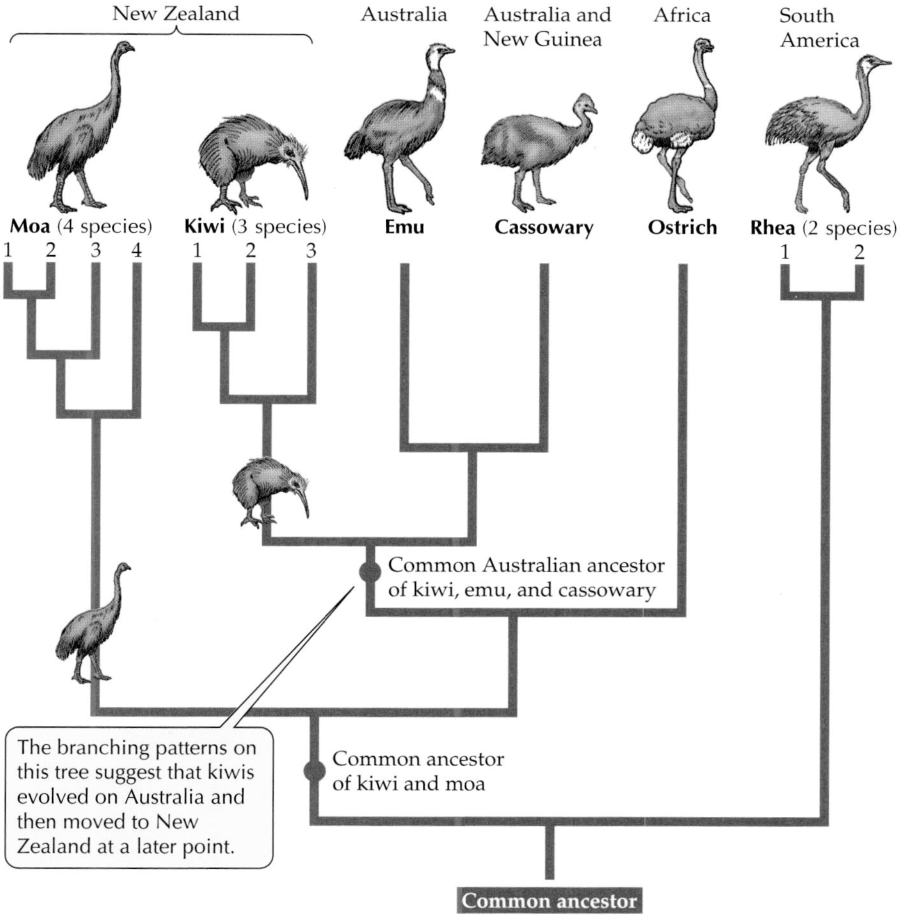

23.8 Flightless Bird Phylogeny
Molecular phylogenetics of flightless birds, living and extinct, illustrates that although moas and kiwis recently lived together in New Zealand, they are not each other's closest relatives.

Cloning extinct organisms is impossible

It is tempting to imagine that we could use molecular methods to reconstruct extinct organisms, as envisioned in Michael Crichton's novel *Jurassic Park*. However, we are unlikely to be able to do so. We are generally able to amplify or clone only small fragments of ancient DNA, because DNA deteriorates with time. Further, we have no idea how to piece together the millions of small fragments of DNA extracted from fossilized remains, nor do we have any idea how to regulate their proper expression during the development of an embryo. What we can accomplish by studying ancient DNA is to determine the place of extinct organisms in the phylogeny of life.

Molecules and Human Evolution

Molecular evolution has influenced our understanding of our own evolution. In particular, it suggests that the human species is of relatively recent origin. Fossil evidence suggests that the hominoid lineage leading to modern humans diverged from a chimpanzee-like lineage about 6 million years ago (mya) in Africa. Shortly after it arose in Africa, about 1.7 mya, *Homo erectus* then spread to other continents. Fossil remains have been found in Africa, Indonesia, China, the Middle East, and Europe. The transition from *Homo erectus* to *Homo sapiens* occurred about 400,000 years ago, but there is considerable controversy about the place of origin of modern humans. The "out of Africa" hypothesis suggests a single origin in Africa followed by several dispersals. The "multiple regions" hypothesis, in contrast, proposes parallel origins of *Homo sapiens* in different regions of Europe, Africa, and Asia (Figure 23.9).

The limited number of human fossils and their patchy distribution do not allow us to choose between these views. However, DNA sequences of several mitochondrial genes from individuals from more than 100 ethnically distinct modern human populations provide the needed information. Mitochondrial DNA (mtDNA) is useful for studying the recent evolution of closely related species and populations within species because it accumulates mutations rapidly and because it is maternally inherited. The Y chromosome serves the same role for following male lineages. The mtDNA sequences for ethnically distinct modern humans imply or suggest a common ancestry of all mtDNAs about 200,000 years ago. The date of the shared ancestry was calculated using the number of nucleotide differences among existing humans, and the rate of mtDNA sequence divergence was calibrated using mammals with better fossil records.

In contrast, the multiple-origins hypothesis requires at least 1 million years of divergence since the last common ancestor. Thus, mtDNA lends support to the "out of Africa" hypothesis, suggesting that all modern human populations share a recent mitochondrial ancestor. Of importance, the mtDNA data support the idea of a common female ancestral lineage that ultimately gave rise to all contemporary mtDNAs. In principle, the history of genes passed through the male lineage could be different. Studies of 26 nuclear genes and a family of repeated sequences (called the *Alu* family) also support a recent African ancestry.

Molecular phylogenetic techniques have also helped us resolve some outstanding archaeological debates. One example concerns the origin of the early human inhabitants of Easter Island, a remote island in

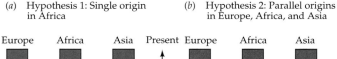

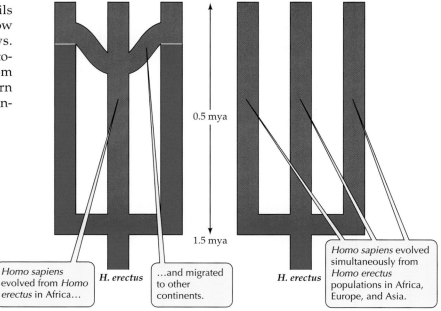

(a) Hypothesis 1: Single origin in Africa

(b) Hypothesis 2: Parallel origins in Europe, Africa, and Asia

Europe Africa Asia Present Europe Africa Asia

0.5 mya

1.5 mya

Homo sapiens evolved from *Homo erectus* in Africa…

H. erectus

…and migrated to other continents.

H. erectus

Homo sapiens evolved simultaneously from *Homo erectus* populations in Africa, Europe, and Asia.

23.9 Two Models for the Origin of Modern Humans *Homo erectus* (blue lineage) arose in Africa and spread to other continents. However, there is considerable controversy among scientists as to whether the transition to *Homo sapiens* (red lineage) took place only in Africa (*a*), or the evolution of modern humans occurred simultaneously on three continents (*b*).

the Pacific. Ancient DNA from human bones on Easter Island suggests that the inhabitants of this island were closely related to Polynesians from other Pacific islands, rather than to South American Indians, as was suggested by early Danish explorers.

Summary of "Molecular Evolution"

• The goals of the study of molecular evolution are to determine the patterns of evolutionary change in the molecules of which organisms are composed, to determine the processes that caused those changes, and to use those insights to help solve other biological problems. To achieve those goals, molecular evolutionists need to be able to determine the structures of molecules of living and fossil organisms.

Determining the Structure of Molecules

• The structure of DNA was deduced by combining data from X ray crystallography and base composition with three-dimensional models. The polymerase chain reaction method allows biologists to determine the sequence of DNA bases of organisms from their fossilized remains.
• Past molecular structures are inferred by comparison of molecules of existing organisms using cladistic techniques to identify ancestral and derived states.
• Changes evolve slowly in regions of molecules that are functionally significant, but more rapidly in regions where substitutions do not affect the functioning of the molecules. **Review Figure 23.1**
• Gene duplication, which frees one copy of a gene to evolve a novel function, has been responsible for much of the evolution of molecular diversity. **Review Figures 23.3, 23.4**
• Groups of genes that are aligned in the same order on chromosomes of distantly related species are likely to be homologs of one another. **Review Figures 23.5, 15.21**

How Molecular Functions Change

• Changes in the functions performed by molecules are stimulated by gene duplication, and by the traits that molecules must have to function in different situations, such as the acidic environment of the stomach.

Genome Size: Surprising Variability

• The genome sizes of organisms vary more than a hundredfold, but the amount of DNA that actually encodes protein varies much less. In general, eukaryotes have more coding DNA than do prokaryotes, vascular plants and invertebrate animals have more coding DNA than do single-celled organisms, and vertebrates have more coding DNA than do invertebrates. **Review Table 23.3**

Molecular Clocks

• Neutral molecular variation often accumulates at a constant rate determined by the mutation rate. Such a process is referred to as the ticking of a molecular clock.
• Molecular clocks tick more slowly for molecules, or parts of molecules, that experience strong constraints on their evolution than they do for molecules or parts of molecules influenced by strong directional selection. **Review Figure 23.6**
• We can assess the constancy of the ticking rates of molecular clocks by comparing the dates of lineage splits calculated assuming the operation of molecular clocks with those determined from accurately dated fossils. Typically these two estimates are reasonably similar, giving evolutionists confidence in using molecular clocks to date events for which there are no fossils.

Using Molecules to Infer Phylogenies

• Molecules are an important source of data that can be used to infer phylogenetic relationships among organisms. For ancient splits and phylogenies of prokaryotes, molecular data are the only source of information about phylogenetic relationships.
• The steps in a molecular evolutionary analysis are: choosing molecules to study; determining sequences of amino acids or bases; comparing the molecules; and constructing a gene tree. **Review Figure 23.7**
• Molecules that have evolved slowly are useful for determining ancient lineage splits. Molecules that evolve rapidly are useful for determining more recent lineage splits.

Molecules and Human Evolution

• Comparisons of mtDNA from more than 100 ethnically distinct modern human populations strongly suggest that all modern humans share a common African ancestor no more than 200,000 years old. **Review Figure 23.9**
• Molecules provide useful information about human evolution, helping to clarify the relationships between different peoples.

Self-Quiz

1. Questions about the *process* of molecular evolution address
   a. whether molecules evolve.
   b. how molecules change with time.
   c. how to reconstruct phylogenies of extinct organisms.
   d. the evolutionary relationships among molecules.
   e. the origin of organismal diversity.

2. Choosing the appropriate molecule for phylogenetic reconstruction does *not* require a consideration of
   a. the question at hand.
   b. the rate of evolution of the molecule.
   c. the phylogenetic distribution of the molecule.
   d. the function of the molecule.
   e. the completeness of the fossil record.

3. Ribosomal RNA sequences are useful for addressing the evolutionary relationships of highly divergent molecules because
   a. they evolve at a rapid rate.
   b. they have undergone convergent evolution in many lineages.
   c. they are molecules that all organisms have.
   d. they consist of mainly neutral characters.
   e. they are difficult to align.

4. Mitochondrial DNA sequences are useful in studying the recent evolution of closely related species because
   a. they accumulate mutations very rapidly.
   b. they are paternally inherited.
   c. they evolve only in a neutral fashion.

d. they are highly constrained in function.

e. they are easy to sequence.

5. Questions about the *pattern* of molecular evolution focus on
    a. the evolutionary relationships among molecules.
    b. the molecular clock.
    c. the rate of mutation for neutral characters.
    d. the importance of gene duplication in evolution.
    e. the neutral theory of evolution.

6. Molecules are used to reconstruct phylogenies, even if a fossil record is available, because
    a. the more characters the better.
    b. they are more accurate characters than are fossils.
    c. they undergo less homoplasy than do fossil characters.
    d. they are less subjective characters than are fossils.
    e. they give us the "right" phylogeny.

7. Neutral characters
    a. are not evolving under the influence of positive selection.
    b. have a neutral pH.
    c. are not useful in reconstructing phylogenies.
    d. are subject to strong functional constraints.
    e. are not likely to evolve.

8. *Jurassic Park* is unlikely ever to be a reality because
    a. we cannot obtain DNA from ancient organisms.
    b. we have no parks big enough for such large dinosaurs.
    c. although we can obtain DNA from dinosaurs, it is highly fragmented.
    d. the genetic code used by dinosaurs is different from all other organisms.
    e. we wouldn't know what to feed the dinosaurs.

9. The concept of a molecular clock implies
    a. that many proteins show a constancy in rate of change with time.
    b. that organisms evolve at a constant rate.
    c. that one can date evolutionary events with molecules alone.
    d. that all molecules change at the same rate in evolution.
    e. that we can predict how fast all genes will evolve.

10. The lysozyme story suggests that
    a. molecules cannot change function in evolution.
    b. that selection does not act at the molecular level.
    c. that molecules can help us understand the process of organismal evolution.
    d. that all organisms are capable of fermenting bacteria.
    e. that lysozyme has a very accurate molecular clock.

Applying Concepts

1. If you were interested in studying the molecular phylogeny of very closely related species of fruit flies, what kind of molecule(s) would you choose to examine? Why? If, on the other hand, you wanted to determine the phylogeny of all vertebrates, would you use the same molecule(s)? Why or why not?

2. How have our views about organismal evolution been affected by recent discoveries in molecular evolution?

3. Discuss the relative importance of molecular characters, versus morphological and fossil characters, in reconstructing the phylogeny of a group of organisms.

4. The existence of a "true" molecular clock is a contentious issue in molecular evolution. If it turns out that a molecular clock keeps very good time, of what use is it?

5. We are, by nature, interested in our own evolution. This chapter presented a brief introduction to the application of molecular methods to studying questions about human evolution. Make a short list of additional questions and develop a rough outline of the molecules and methods you might bring to bear in addressing these questions.

Readings

Avise, J. C. 1994. *Molecular Markers, Natural History and Evolution.* Chapman and Hall, New York. A thorough review of the use of molecules to infer evolutionary patterns.

De Robertis, E. M., O. Guillermo and C. V. E. Wright. 1990. "Homeobox Genes and the Vertebrate Body Plan." *Scientific American*, July. A well-illustrated review of current views regarding the evolution of the vertebrate body plan.

Gillespie, J. A. 1991. *The Causes of Molecular Evolution.* Oxford University Press, New York. A comprehensive and demanding review of the basic tenets of molecular evolution.

Kimura, M. 1983. *The Neutral Theory of Molecular Evolution.* Cambridge University Press, New York. An easy-to-read description of the process of neutral evolution by one of its original proponents.

Li, W.-H. 1997. *Molecular Evolution.* Sinauer Associates, Sunderland, MA. The best treatment of all aspects of molecular evolution.

Li, W.-H. and D. Graur. 1991. *Fundamentals of Molecular Evolution.* Sinauer Associates, Sunderland, MA. A brief, easy-to-read primer of the central topics in molecular evolution.

Olsen, G. J. and C. R. Woese. 1993. "Ribosomal RNA: A Key to Phylogeny." *Journal of the Federation of American Societies for Experimental Biology (FASEB)*, vol. 7, pages 113–123. A detailed and well-written summary of the use of rRNA in molecular phylogenetic reconstruction.

Chapter 24

The Origin of Life on Earth

In 1985, scientists discovered that the high-altitude ozone layer, which shields organisms from harmful ultraviolet radiation, had thinned greatly over Antarctica. Ozone (O_3) is produced by the action of sunlight on atmospheric oxygen (O_2) in tropical regions and is transported to high latitudes, where it is destroyed. Chlorine compounds, produced mainly by humans, appear to be the main cause of the unusually high rates of ozone destruction. Ozone is now seriously depleted at very high latitudes, where conditions favor its destruction.

Because ultraviolet radiation damages DNA, it increases the incidence of mutation and cancer. A 1 percent decrease in atmospheric ozone is estimated to result in a 6 percent increase in the incidence of skin cancer. Average ozone concentrations have decreased about 3 percent, enough to have caused a 20 percent increase in skin cancers at midlatitudes. Skin cancer, which is sometimes lethal, is a serious human health problem.

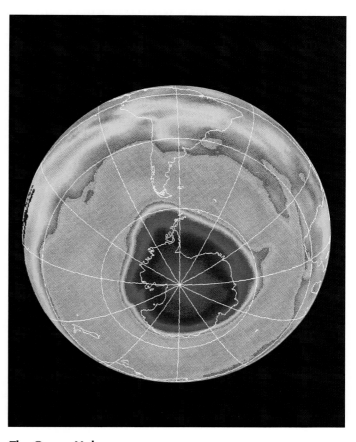

The Ozone Hole
A satellite image showing ozone depletion over Antarctica in 1990. The areas of most severe loss appear in violet and pink; orange and yellow indicate approximately normal concentrations of ozone. The continents are outlined in white.

The modification of Earth's atmosphere by human activity is just the latest of a series of atmospheric changes caused by organisms. The most dramatic of these changes was the generation of O_2 by photosynthetic prokaryotes 2.5 billion years ago. Organisms have also changed Earth's waters and soils. Conditions on Earth, in turn, have influenced the evolution of life. In this chapter we examine how conditions on Earth 3.8 billion years ago influenced where and how life evolved and how life in turn, during the early phases of its evolution, changed Earth.

The phenomenon of reproduction distinguishes living organisms from nonliving and was central to discussions at many points in Parts One and Two of this book. How could something that is self-replicating have existed before there were cells? What was the earliest self-replicating unit, and how did it reproduce? Remarkably, scientific evidence is good enough that we can suggest answers to these questions.

Earth's oldest rocks contain no fossils, showing that for at least several hundreds of thousands years after its formation, Earth had no life. Today we are confident that all organisms come from other organisms. The first life, however, must have come from nonliving matter. How did this happen? Under what conditions did life originate on Earth?

Is Life Forming from Nonlife Today?

Until the nineteenth century, people believed in **spontaneous generation**—the regular formation of living organisms from nonliving matter. Flies and maggots were believed to arise from rotting meat and barnyard manure, eels and fish from sea mud, and frogs and mice from moist soil. For more than 2,000 years, belief in spontaneous generation was accepted by most scholars, clergy, and plain folk, although attempts to disprove it had been made.

In 1862, the great French scientist Louis Pasteur performed a series of meticulous experiments showing that microorganisms come only from other microorganisms and that a genuinely sterile solution remains lifeless indefinitely unless contaminated by living creatures. His most elegant experiment relied on swan-necked flasks that were open to the air. Pasteur filled the flasks with nutrient medium, heated them to kill any microorganisms present, then cooled them slowly. No new growth appeared in these flasks (Figure 24.1). The shape of the necks kept any new organisms or minute dust particles from falling into the medium. However, in flasks without the narrow swan necks, microorganisms entered and grew rapidly.

As a result of Pasteur's experiments, similar experiments by other scientists, and other discoveries of nineteenth-century science, most people accepted that all life comes from existing life. These experiments suggested that life was not forming from nonliving matter under the current conditions on Earth, but they shed no light on when or how life did form on Earth 3.8 billion years ago.

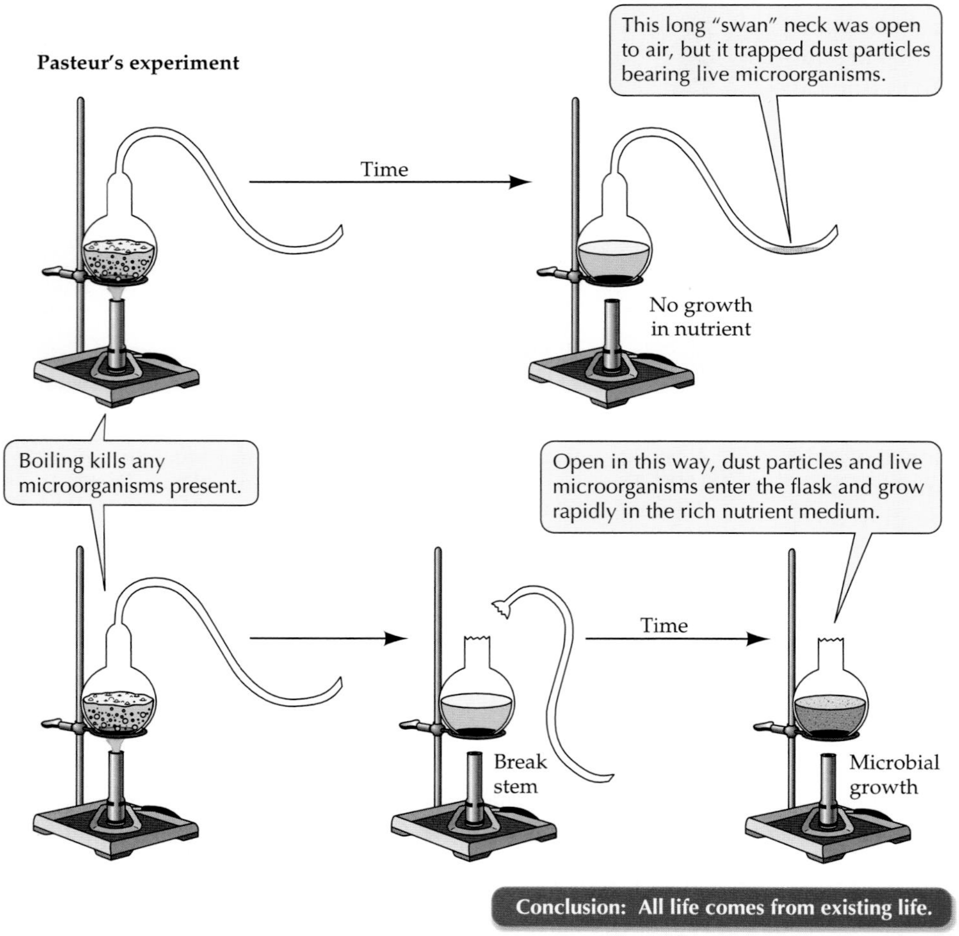

How Did Life Evolve from Nonlife?

To understand how life arose, we need to know the physical conditions that prevailed on Earth before there was life. And we need to know when life arose. The early conditions on Earth determined which types of chemical reactions could take place. If we know what types of reactions took place, we can suggest how those reactions could have led to the appearance of life.

For its first billion years, no life existed on Earth

How did the universe form? Scientists now believe that between 10 billion and 20 billion

Pasteur's experiment

This long "swan" neck was open to air, but it trapped dust particles bearing live microorganisms.

Time

No growth in nutrient

Boiling kills any microorganisms present.

Open in this way, dust particles and live microorganisms enter the flask and grow rapidly in the rich nutrient medium.

Break stem

Time

Microbial growth

Conclusion: All life comes from existing life.

24.1 Experiments Disproved the Spontaneous Generation of Life
Louis Pasteur's classic experiments showed that a genuinely sterile solution remains lifeless indefinitely; only "contamination" by living organisms could cause life to appear in the flasks.

years ago there was a mighty explosion. The matter of the universe, which had been highly concentrated, began to spread apart rapidly. Eventually clouds of gases collapsed on themselves through gravitational attraction, forming the galaxies—great clusters of hundreds of billions of stars.

Somewhat less than 5 billion years ago, toward the outer edge of our galaxy (the Milky Way), our solar system (the sun, Earth, and our sister planets) took form. Earth probably formed by gravitational attraction of rocks of various sizes. As Earth grew by this process over millions of years, the weight of the outer layers compressed the interior of the planet. The resulting pressures, combined with the energy from radioactive decay, heated the interior until it melted. Within this viscous liquid, the heavier elements settled to produce a fluid iron and nickel core with a radius of approximately 3,700 km that persists to this day. Around the core lies a mantle of dense silicate materials that is 3,000 km thick. Over the mantle is a lighter crust, more than 40 km thick under the continents but as little as 5 km thick in some places under the oceans.

Conditions on Earth 3.8 billion years ago differed from those of today

Before the evolution of life, Earth's mantle and crust released carbon dioxide, nitrogen, and other heavier gases. These gases were held by Earth's gravitational field and gradually formed a new atmosphere. Earth accumulated an atmosphere consisting mostly of methane (CH_4), carbon dioxide (CO_2), ammonia (NH_3), hydrogen (H_2), nitrogen (N_2), and water vapor (H_2O). Eventually Earth cooled enough that the water vapor escaping from inside the planet condensed to liquid water and formed the seas. Violent electrical storms battered Earth's surface.

No free oxygen (O_2) was present in this early atmosphere, because oxygen reacted with hydrogen to form water and with components of Earth's crust and atmosphere to form iron oxides, silicates, carbon dioxide, and carbon monoxide. For more than 2 billion years, all oxygen was bound up with other elements, and Earth existed as a re-

ducing environment. (See Chapter 7 for a discussion of oxidation and reduction.) As a setting for chemical reactions, then, early Earth differed fundamentally from present-day Earth, which has an oxidizing atmosphere with large quantities of O_2.

The sections that follow describe the sequence of events leading to the appearance of the main stages in the evolution of life (Figure 24.2).

Conditions favored the synthesis of compounds containing carbon, nitrogen, and hydrogen

What sort of chemical reactions could have occurred in Earth's early reducing environment? Could such reactions have been the first step toward the origin of life? To investigate these questions, in the 1950s Stanley Miller studied chemical reactions proceeding under conditions that resembled those believed to exist on Earth 4 billion years ago.

Within a closed system of glass tubes, he established a reducing atmosphere of hydrogen, ammonia,

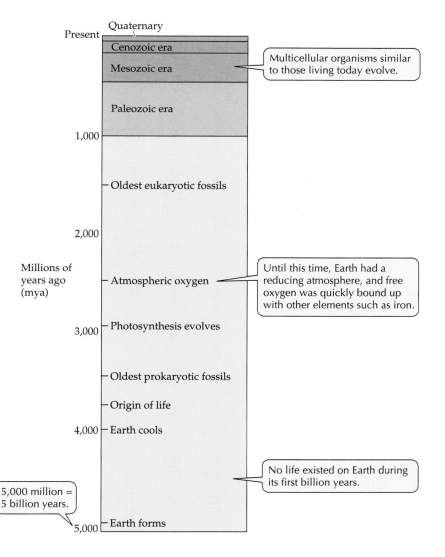

24.2 The Origin and Early Evolution of Life The sequence of events leading to the appearance of life and its earliest evolution, as described in this chapter and in Chapter 1.

methane gas, and water vapor. Through these gases, he passed a spark to simulate lightning (Figure 24.3). Within a few hours the system was found to contain numerous simple organic compounds, including hydrogen cyanide, formaldehyde, cyanogen, acetaldehyde, cyanoacetylene, and propionaldehyde. In water, these compounds dissolved and were rapidly converted into amino acids, simple acids, purines, and pyrimidines—the building blocks of life (Figure 24.4).

The same or similar compounds are produced under a variety of conditions, provided that free oxygen is absent—that is, if the environment is a reducing one. Thus, once Earth cooled enough for water to condense and form oceans, molecules of many kinds formed, and they probably accumulated until they reached relatively high concentrations. From such a prebiotic ("before life") molecular soup, life emerged from nonliving matter.

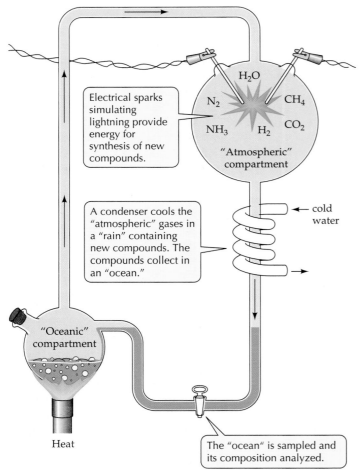

24.3 Synthesis of Molecules in an Experimental Atmosphere Stanley Miller used an apparatus similar to this one to determine which molecules could be produced in a reducing atmosphere such as existed on early Earth.

Labels in figure:
- Electrical sparks simulating lightning provide energy for synthesis of new compounds.
- H_2O
- N_2
- CH_4
- NH_3
- H_2
- CO_2
- "Atmospheric" compartment
- A condenser cools the "atmospheric" gases in a "rain" containing new compounds. The compounds collect in an "ocean."
- cold water
- "Oceanic" compartment
- Heat
- The "ocean" is sampled and its composition analyzed.

Polymerization provided macromolecules with diverse properties

The next stage in the sequence leading to life was the generation of large molecules by **polymerization** of small molecules. The important large molecules of which organisms are composed (polysaccharides, proteins, and nucleic acids) are polymers formed by the combination of subunits called monomers (see Chapter 3). The polymerization reactions that generate these large molecules belong to a class of reactions called condensations, or dehydrations. During a condensation reaction, water forms (see Figure 3.4). Large molecules are assembled through repeated condensations of monomers, and each condensation reaction requires energy. In the prebiotic soup, polymers that formed faster or were more stable would have come to predominate.

Some polymers can direct the synthesis of molecules identical to themselves. Which of the molecules on prebiotic Earth were most likely to reproduce themselves? Nucleic acids, the basis of today's genetic code, are clearly capable of self-copying, and the purine and pyrimidine constituents of nucleotides are formed under conditions similar to those believed to have prevailed on early Earth.

Sugars, another basic component of living cells, are generated by the polymerization of formaldehyde, another molecule formed in Miller's experiments. Solutions of formaldehyde spontaneously polymerize to form several different five- and six-carbon sugars, but these sugars are unstable in aqueous solutions and break down to alcohols and carboxylic acids. However, three- and four-carbon sugars are very stable and can accumulate for hundreds of years in aqueous solutions.

Over millions of years, these organic molecules would have accumulated in the oceans. They would have reached even higher concentrations in drying ponds. High concentrations of polymers, in turn, would have stimulated further polymerization and chemical reactions leading to the synthesis of sugars, which were important as components of nucleotides and for energy storage in early cells.

Phosphate-based polymerizations occurred in the prebiotic soup

Monomers formed under prebiotic conditions can polymerize by phosphorylation (the addition of a phosphate group). Phosphorylated monomers are stable enough to accumulate in solution but reactive enough to polymerize further. The first biologically active polymers of nucleic acid bases may have been compounds formed from three- and four-carbon sugars. They could have self-replicated by forming complementary double-stranded molecules, just as pentose-based nucleic acid polymers do today. At some

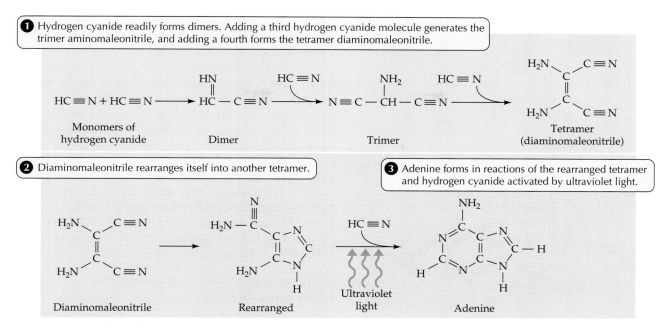

① Hydrogen cyanide readily forms dimers. Adding a third hydrogen cyanide molecule generates the trimer aminomaleonitrile, and adding a fourth forms the tetramer diaminomaleonitrile.

② Diaminomaleonitrile rearranges itself into another tetramer.

③ Adenine forms in reactions of the rearranged tetramer and hydrogen cyanide activated by ultraviolet light.

24.4 From Simple Carbon–Nitrogen Molecules to a Nucleic Acid Building Block In the oceans of prebiotic Earth, molecules of many kinds formed and accumulated. Like the chemicals in Miller's apparatus, these organic precursors were converted into the materials that become amino acids, simple acids, purines, and pyrimidines.

point, by as yet unidentified processes, polymers based on small sugars were replaced by those based on larger sugars, such as ribose, the sugar in the nucleotides of RNA.

RNA was probably the first biological catalyst

The enzymes that control the types and rates of reactions within organisms are proteins. As you learned in Chapter 12, proteins are synthesized by a process that begins with transcription of information from DNA to an RNA molecule that has a base sequence complementary to that of one strand of the DNA. This information is then translated into mRNA and is eventually used to synthesize a specific polypeptide from amino acids. Amino acids are brought to a ribosome by specific tRNA molecules and are attached sequentially to the growing polypeptide. But how could such a system have evolved if protein catalysts needed nucleic acids for their formation but nucleic acids needed proteins to catalyze their own replication? Which came first, if both are necessary?

The inability to solve this dilemma held up research on the origin of life for several decades. The discovery that provided a solution came in 1981 from researchers working with the unicellular protist *Tetrahymena thermophila*. These scientists were studying the excision of introns and the splicing together of exons. They found—entirely contrary to expectation—that excision

is catalyzed in the absence of enzymes. The intron itself—a 400-nucleotide sequence of RNA—carries out the excision and splicing. In addition, *Tetrahymena* ribosomes, which contain several molecules of RNA and a variety of proteins, have a catalytic RNA that operates in protein synthesis. RNAs that catalyze chemical reactions, called **ribozymes**, have now been found in many organisms (Figure 24.5).

The current system of macromolecular synthesis (DNA → RNA → protein) probably evolved gradually from much simpler processes. Biochemists believe that the first information-carrying molecules were short strands of RNA that replicated themselves without the help of enzymes. Evidence that RNAs can replicate themselves came first from experiments conducted by Manfred Eigen in the late 1970s. Eigen added RNA molecules to solutions containing monomers for making more RNA and found that sequences of five to ten nucleotides formed. If he added a simple inorganic molecule such as zinc, much longer sequences were copied.

Although these experiments showed that RNA could replicate itself, they did not demonstrate that RNA could catalyze the synthesis of other molecules as true enzymes do. Within two years, however, studies of a tRNA-processing enzyme that contains RNA and a protein (ribonuclease P) showed that the RNA alone can cut the pre-tRNA molecule at the correct spot, whereas the protein cannot.

Many scientists believe that the first genetic code was based on RNA that catalyzed both its own replication and other chemical reactions. In such conditions a high concentration of RNA would have been needed, so that it could participate in many different chemical

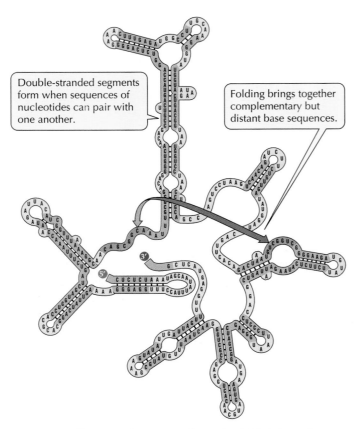

Double-stranded segments form when sequences of nucleotides can pair with one another.

Folding brings together complementary but distant base sequences.

24.5 A Ribozyme from a Protist The folded three-dimensional structure of this catalytic RNA, or ribozyme, enables it to catalyze chemical reactions during protein synthesis.

reactions. The accumulated products of RNA-catalyzed reactions could then participate in other reactions to form structures—for example, the production and accumulation of lipidlike molecules to form cell membranes, and the synthesis of proteins. However, after proteins evolved, they eventually took over most enzymatic functions because they are better catalysts than RNA and are capable of more diverse specificities.

In order to replicate, different RNAs would have competed with one another for monomers. Some RNA molecules would have been better at replicating in certain environments because their base sequences produced the most stable configurations (folded structures) under the particular conditions of temperature and salinity. With their higher rates of replication and greater stability, these molecules would have come to dominate the populations of RNA in the corresponding environments.

In the precellular world, RNA that contained both genetic information and catalytic capacity evolved. Molecules with these capacities have been produced in the laboratory by investigators who started with completely random-sequence RNA. These ribozymes

ligate (bind together) two RNA molecules in a reaction similar to that employed by the enzymes that synthesize RNA.

The investigators simulated the "evolution" of RNA molecules by selecting in test tubes for ribozymes with high ligation ability. By this method, they produced ribozymes with reaction rates 7 million times faster than the uncatalyzed reaction rate. These experiments suggest that ribozymes could have evolved rapidly when conditions on Earth became suitable for the formation of nucleic acids. But ribozymes in solution, even active ones, do not constitute life. Life required a barrier that permitted the homeostatic control of internal conditions; life could not appear until cells enclosed in membranes evolved. The prebiotic RNA world still lacked cells.

Membranes permitted a stable internal environment to form

The way in which ribozyme-based systems in solution might have evolved into cells was first suggested in the 1920s by the experiments of the Russian scientist Alexander Oparin, who spent much of his career studying complex solutions. Oparin observed that if he shook a mixture of a large protein and a polysaccharide, the drops that formed were divided into two "phases": an interior separated from but in contact with an exterior. These interiors, which were primarily protein and polysaccharide, with some water, were surrounded by an aqueous solution containing low concentrations of proteins and polysaccharides. These drops, known as **coacervates**, are quite stable and will form in solutions of many different types of polymers.

Coacervates have properties relevant to the origin of life. When added to a coacervate preparation, many substances are concentrated within the drops. Lipids coat the boundaries of drops with membranelike structures, which strengthen the drops and help contol the rates of passage of materials into and out of the drops. These properties simulate some characteristics of living cells. In addition, if enzyme molecules are added to them, coacervate drops exhibit a "metabolism": They can absorb substrates, catalyze reactions, and let the products diffuse back into the solution (Figure 24.6). Oparin even succeeded in making chlorophyll-containing coacervate drops that absorbed an oxidized dye from the solution, used light energy to reduce it, and released the reduced dye to the outside medium.

Coacervate drops are possible precursors to cells because they provide the physical structure by which internal conditions could be different from outside conditions. This structure permitted retention and concentration of some substances, new reactions, and release of products to the outside. Because drops in which chemical reactions were better controlled would have survived longer than drops with more poorly con-

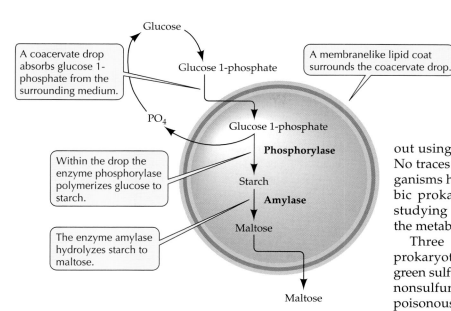

A coacervate drop absorbs glucose 1-phosphate from the surrounding medium.

A membranelike lipid coat surrounds the coacervate drop.

Glucose

Glucose 1-phosphate

PO₄

Glucose 1-phosphate

Phosphorylase

Within the drop the enzyme phosphorylase polymerizes glucose to starch.

Starch

Amylase

The enzyme amylase hydrolyzes starch to maltose.

Maltose

Maltose

24.6 "Metabolism" of a Coacervate Drop
The properties of coacervate drops are similar to some of the properites of living cells. Held intact by a membranelike lipid coating, chemical reactions take place inside the drops in the presence of enzymes.

trolled reactions, refinements of metabolic processes by the use of catalysts could have evolved within them.

In the environment, billions of droplets would have competed with one another to acquire substrate materials. Droplets that evolved a better capacity to replicate internal reactions would have grown larger. But when coacervates become large, they are less stable, and physical agitation breaks them apart into smaller droplets. During this early form of reproduction, they would have passed on at least rough copies of their molecules to daughter droplets. At some point, such droplets may have accumulated enough RNA to display the key property we associate with living cells—the use of energy to maintain homeostasis.

DNA evolved from an RNA template

If the first cells used RNA as their hereditary molecule, then RNA must have provided the template for the synthesis of DNA. In solution, DNA is less stable than RNA. Therefore, DNA probably did not evolve until RNA-based life became contained in membrane-enclosed cells where water concentrations were lower than in the surrounding environment. Because DNA is a more stable storage location for genetic information than is RNA, once the appropriate environments were available, DNA probably evolved rapidly, replacing RNA as the genetic code for most organisms. But by then, RNAs had assumed their current roles as intermediaries in the translation of genetic information into proteins.

Early Anaerobic Cells

Earth's atmosphere lacked the gas O_2 for about a billion years after life evolved. Therefore, the earliest organisms must have processed energy chemically without using O_2 as an electron acceptor (see Chapter 7). No traces of the metabolic pathways of those early organisms have been preserved, but some living anaerobic prokaryotes still employ similar pathways. By studying them, we can make reasoned guesses about the metabolism of ancient prokaryotes.

Three major types of anaerobic photosynthetic prokaryotes live today in sediments that lack oxygen: green sulfur bacteria, purple sulfur bacteria, and purple nonsulfur bacteria (see Chapter 25). In fact, oxygen is poisonous to them, as it must have been to organisms that evolved in a reducing atmosphere. These bacteria contain types of chlorophyll, called bacteriochlorophyll *a*, *c*, and *d*. Anaerobic photosynthetic bacteria also contain red and yellow carotenoids. These pigments absorb wavelengths of light that are not absorbed by chlorophyll and pass the absorbed energy along to chlorophyll for conversion into chemical energy.

The photosynthetic system of these bacteria is embedded in membrane complexes that also possess the electron transport chains and enzymes by which captured solar energy is converted to chemical energy and used to generate ATP (see Chapter 8). Photosynthetic prokaryotes similar in their metabolism to today's forms were so abundant about 3.4 billion years ago that their partly decomposed fossil remains formed extensive deposits of carbon, resembling the coal produced by vascular plant fossils 3 billion years later.

To reduce carbon dioxide (CO_2), a photosynthetic cell needs a source of electrons (hydrogen atoms). Many photosynthetic prokaryotes use light energy to generate ATP and NADPH + H⁺. The particular waste product that they liberate depends on the source of hydrogen atoms they use. The green and purple sulfur photosynthetic bacteria obtain their hydrogen atoms from hydrogen sulfide (H_2S) and generate sulfur as a waste product (Figure 24.7). The purple nonsulfur bacteria typically obtain hydrogen atoms from organic compounds such as ethanol, lactic acid, or pyruvic acid, or directly from hydrogen gas (H_2).

In some environments today, the H_2 is produced by other prokaryotes as the end product of their fermentations. Under the anaerobic conditions of early Earth, hydrogen sulfide and other compounds containing hydrogen were more abundant than they are now. Today O_2 directly oxidizes H_2S, H_2, pyruvic acid, and similar compounds into water, carbon dioxide, and sulfur oxides.

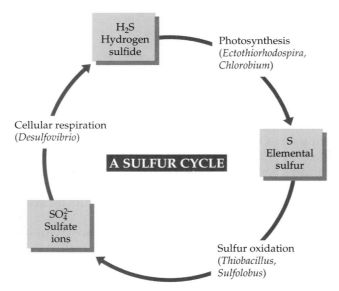

24.7 Oxidizing and Reducing Sulfur Several species of bacteria and archaea may function together to carry out sulfur cycles, one of which is shown here. Representative genera able to conduct the processes are given in parentheses. Many such cycles are found in nature.

Aerobic photosynthesis is the source of atmospheric O_2

The evolution of aerobic photosynthesis, slightly more than 1 billion years ago, changed the course of evolution and changed Earth. The key change was the acquisition of the ability to use water as the source of hydrogen: $2 H_2O \rightarrow 4 H + O_2$. The chemical splitting of H_2O produced O_2 as a waste product and made available hydrogen atoms for reducing CO_2.

This ability appeared first in certain sulfur bacteria that evolved into cyanobacteria (see Chapter 25). Remains of these bacteria are abundantly preserved in fossil **stromatolites**. Cyanobacteria are still forming stromatolites in a few very salty places on Earth (Figure 24.8).

The ability to split water was doubtless the cause of the extraordinary success of the cyanobacteria. The O_2 they liberated opened the way for the evolution of oxidation reactions as the energy source for the synthesis of ATP. Their success made possible the evolution of the full respiratory chain of reactions now carried out by all aerobic cells. The evolution of life irrevocably changed the nature of our planet. Life created the O_2 of our atmosphere, and it removed most of the carbon dioxide from the atmosphere by transferring it to ocean sediments.

Oxygen was poisonous to the anaerobic organisms that lived on Earth, but the prokaryotes that evolved a tolerance to it were able to live successfully in environments empty of other organisms. Also, aerobic (oxygenated) metabolism could proceed much more rapidly and efficiently than the anaerobic metabolism that had dominated life until then.

Eukaryotic Cells: Increasing Complexity

The generation of atmospheric O_2 by cyanobacteria probably set the stage for the evolution of eukaryotes. Small, prokaryotic cells can obtain enough oxygen by simple diffusion even when oxygen concentrations are very low. Larger cells, however, have a lower surface area–to–volume ratio and need higher environmental O_2 concentrations to meet their needs.

About 1.5 billion years ago oxygen concentrations became high enough for larger cells to flourish and diversify. Some of these cells grew big enough to consume smaller cells; they were the first predators. Most of the prey died after they were engulfed, but some of them survived and eventually evolved into the mitochondria and chloroplasts that now maintain mutually beneficial relationships with their host.

Chromosomes organize large amounts of DNA

All early organisms were haploid. They reproduced by dividing, and each division was accompanied by a duplication of their circular chromosomes. The single circular DNA molecule of prokaryotes fits into a small cell, and it can be duplicated rapidly. Over time, the amount of DNA in the cells of early organisms increased, as it proved advantageous to encode information for the enzymatic catalysis of new reactions and to pass this information to progeny cells.

But replicating a long DNA molecule would take many hours if it could be started only at one spot. Replication can proceed much more rapidly if genes are separated into chromosomes, each of which has multiple initiation sites for duplication. Division of the genome into chromosomes also makes it less likely that the DNA molecules will become tangled when they divide. For these reasons, biologists believe that chromosomes evolved soon after the appearance of eukaryotes, about 2.5 billion years ago.

Variable offspring are advantageous in variable environments

All early organisms reproduced by simple fission (see Chapter 9), and each division was accompanied by a duplication of their chromosomes. Occasionally and accidentally, the duplication of DNA was not accompanied by a cell division, and the result was a diploid cell. A diploid cell has certain advantages. It can repair more kinds of chromosome damage than a haploid cell because the undamaged duplicate copy can guide the repair. But even if repair was not possible, a duplicate copy of the genome offered assurance that the life of the cell could continue.

Today, chromosome breakage can be repaired only by diploid cells. Because chromosome damage is espe-

(a)

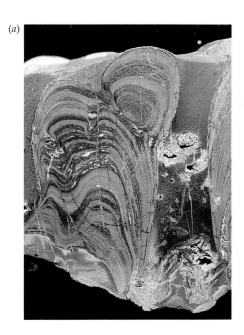

(b)

24.8 Stromatolites Survive in a Few Places (a) A section through a fossil stromatolite. (b) Living stromatolites forming off the coast of Shark Bay in Western Australia.

cially likely in stressful environments, diploidy is advantageous in those conditions. Many present-day unicellular organisms that live in a haploid state most of the time produce diploid cells when they are environmentally stressed. Diploid cells also have protection from point mutations, most of which alter only one copy of a gene. The unaltered copy on the homologous chromosome continues to function normally.

A diploid cell can also result if two haploid cells fuse. The chromosomes of two haploid cells are unlikely to have the same damages. Thus, a diploid cell formed by acquiring chromosomes from another cell would be able to repair all its damaged chromosomes. This ability may have provided the original advantage of sexual recombination, but we can only speculate. Whatever its original advantages, sex appeared early during the evolution of life, and nearly all living organisms engage in sexual recombination at least occasionally during their life cycles. Those that do not are believed to have evolved recently from sexual ancestors; they are destined to have short evolutionary lives.

Sex is maintained in multicellular organisms today by a benefit different from the ones that initially favored it. Whereas the original benefits accrued directly to the individuals accepting genetic material, sex is maintained because of its effects on offspring. Whereas in unicellular organisms genes are not exchanged during reproduction (cell division), in multicellular organisms genes are nearly always exchanged at the time of reproduction, which is the only time during the life cycle when a single-cell stage exists.

The offspring of sexual multicellular organisms are genetically highly variable. In an environment that varies in space and time, the genetically uniform offspring of an asexual organism are less likely to find suitable conditions than are the variable offspring of a sexual organism.

Matter, Time, and Inevitable Life

Scientists have been able to gather information that provides many insights into the origin of life on Earth. In laboratory experiments they have studied chemical reactions under conditions similar to those believed to have prevailed on early Earth. Under laboratory conditions, they have witnessed the "evolution" of ribozymes with catalytic activity from random-sequence RNA.

Taken together, this information suggests that the evolution of life as we know it was highly probable under the conditions that prevailed on Earth 3.8 billion years ago. The molecules on which life is based form readily under such conditions, as does the organization of those molecules into larger units. Much remains to be learned about the early evolution of the complex metabolism of living organisms, but techniques are available to help us.

If the origin of life was almost inevitable, why is new life not still being assembled from nonliving matter on today's Earth? The reason is that simple biological molecules released into today's environment are quickly consumed by existing life. They cannot accumulate to the densities that characterized the prebiotic "soup," even in anaerobic environments. In aerobic environments these molecules are quickly oxidized to other forms. They would not accumulate even if they were not consumed. Life was generated from nonlife on Earth, but it was an event of the remote past. Once life had evolved, it prevented the formation of other life from nonlife.

Summary of "The Origin of Life on Earth"

Is Life Forming from Nonlife Today?
• For its first billion years, no life existed on Earth. Life originated from nonliving matter 3.8 billion years ago, but experiments by Louis Pasteur and others convinced scientists that life does not come from nonlife on Earth today. **Review Figure 24.1**

How Did Life Evolve from Nonlife?
• Conditions on Earth at the time of the origin of life differed from those of today. No free oxygen was present in Earth's early atmosphere.
• The reducing atmosphere of early Earth was the setting for chemical reactions that produced a molecular soup. Under conditions that resemble Earth's early reducing atmosphere, small molecules essential to living systems form and polymerize. **Review Figure 24.3**
• Before life appeared, polymerization reactions generated the carbohydrates, lipids, amino acids, and nucleic acids of which organisms are composed. These molecules accumulated in the seas because the rate of their formation was greater than the rate at which they were destroyed. **Review Figure 24.4**
• The first genetic material may have been RNA that had both a catalytic function and an information transfer function. Some RNAs—called ribozymes—still have catalytic functions today. **Review Figure 24.5**
• The earliest cells may have been similar to the coacervates that can be produced in the laboratory by the mixing of proteins and polysaccharides. Once cells evolved, they outcompeted all acellular proto-life. **Review Figure 24.6**
• DNA probably evolved after RNA-based proto-life became surrounded by membranes that provided an environment in which DNA is stable.

Early Anaerobic Cells
• For more than 2 billion years, the only organisms were anaerobic prokaryotes. Many prokaryotes that live today in environments lacking O_2 use metabolic pathways similar to those of early prokaryotes.
• Cyanobacteria, which evolved the ability to split water into hydrogen atoms (for energy and reduction) and O_2, proliferated and created atmospheric O_2. The accumulation of free O_2 in Earth's atmosphere made possible the evolution of aerobic metabolism, eukaryotes, sexual recombination, and multicellularity.

Eukaryotic Cells: Increasing Complexity
• Division of the genome into chromosomes increased the rate of DNA replication and decreased the likelihood that DNA molecules would become tangled when they divide.
• All early cells were haploid. The first diploid cells appeared probably when some cells failed to divide after their DNA was duplicated. These "accidental" diploids had important advantages over haploid cells. Diploid cells were better at repairing damaged chromosomes and were better protected from harmful point mutations.
• Sex evolved because it was directly advantageous to the individuals that accepted genes from others, but sex is maintained today because the variable offspring that sex produces are more likely than the genetically uniform offspring of asexual organisms to survive in the variable environments in which they must live.

Matter, Time, and Inevitable Life
• Because most of the chemical reactions that gave rise to life proceed readily under the conditions that prevailed on early Earth, the evolution of life was probably nearly inevitable.
• New life is not being assembled from nonliving matter today because simple biological molecules that form in the current environment are quickly consumed by existing life.

Self-Quiz

1. The atmosphere of early Earth consisted largely of
   a. water vapor.
   b. hydrogen.
   c. carbon dioxide.
   d. helium.
   e. nitrogen.

2. Pasteur's experiments and similar ones that followed convinced most people that spontaneous generation of life did not happen, because
   a. Pasteur was extremely meticulous.
   b. Pasteur used very fine mesh screens to cover his flasks.
   c. Pasteur did not boil his flasks for a long time.
   d. Pasteur's swan-necked flasks ruled out the objection that spoiled air could have contaminated his experiments.
   e. by the time Pasteur performed his experiments, many people no longer believed in spontaneous generation.

3. To determine which molecules might have formed spontaneously on early Earth, Stanley Miller used an apparatus with an atmosphere containing
   a. oxygen, hydrogen, and nitrogen.
   b. oxygen, hydrogen, ammonia, and water vapor.
   c. oxygen, hydrogen, and methane.
   d. hydrogen, oxygen, and carbon dioxide.
   e. hydrogen, ammonia, methane, and water vapor.

4. Most biologists think that RNA was the first genetic material because
   a. amino acids were produced in Stanley Miller's apparatus.
   b. DNA is the universal genetic material of eukaryotes.
   c. the existence of ribozymes suggests that early cells could have used RNA to catalyze chemical reactions and transfer information.
   d. RNA is simpler than DNA.
   e. DNA is not stable in hydrophobic environments.

5. Biologists believe that the current DNA → RNA → protein system is the result of a long period of evolution because
   a. the transcription of DNA to mRNA and translation of mRNA into proteins consists of many steps.
   b. DNA replication is complicated but relatively error-free.
   c. the current system is very complex and precise.
   d. evidence indicates that RNA preceded DNA as the genetic material.
   e. all of the above

6. The question whose answer enabled research on the origin of catalysis to proceed rapidly was,
   a. How could complex life have evolved on such a young Earth?
   b. How could the precise duplication of DNA have evolved?
   c. How could catalysis have evolved, given that RNA needs proteins for its synthesis and proteins need RNA for their synthesis?
   d. How could eukaryotes evolve from prokaryotes?
   e. How did the first cells form?

7. The key process in the formation of nucleic acid bases is
   a. the polymerization of hydrogen cyanide.
   b. the polymerization of formaldehyde.
   c. the spontaneous formation of monomers.
   d. the spontaneous formation of proteins.
   e. the polymerization of proteins.

8. The metabolism of living prokaryotes provides important insights into the chemical processes used by early organisms because
   a. many prokaryotes live in environments similar to those in which life first evolved.
   b. prokaryotes are simpler to study and hence are better known than are eukaryotes.
   c. many prokaryotes are obligate aerobes.
   d. many prokaryotes use oxygen as their oxidizing agent.
   e. fermentation evolved before aerobic respiration.

9. The most important advantage of diploidy is that
   a. it gives a cell more genes.
   b. more genes better fill the larger volume of eukaryotic cells.
   c. two copies of a gene are better than one.
   d. the duplicate chromosomes increase the ability of cells to repair chromosomal damage.
   e. diploid cells can reproduce faster than haploid cells in constant environments.

10. Sex is maintained in multicellular organisms because
   a. without it life would not be worth living.
   b. the offspring of sexual organisms survive better in variable environments than do the offspring of sexual organisms.
   c. sexual individuals are better than asexual individuals at repairing damaged chromosomes.
   d. sexual individuals can produce more offspring than asexual individuals can.
   e. sexual individuals can reproduce faster than asexual individuals.

Applying Concepts

1. Why is determining the composition of Earth's early atmosphere a key step in inferring how life arose?

2. Why is the ability of ribozymes to catalyze both their own synthesis and the synthesis of proteins so important for understanding the origin of life?

3. On the one hand, scientists are confident that life no longer arises from nonliving matter. On the other hand, most biologists believe that life did arise on this planet, billions of years ago, from nonliving matter. How can scientists hold both of these beliefs?

4. Why do biologists believe that the evolution of life was highly probable on early Earth?

5. How might each of the following have been involved in the evolution of coacervate drops?
   a. coating of drop boundaries with lipids
   b. wave action in bodies of water
   c. catalysts within the drops

Readings

Cech, T. R. 1986. "RNA as an Enzyme." *Scientific American*, November. A well-illustrated discussion of the discovery of the catalytic abilities of RNA and their implications for the origin and early evolution of life.

de Duve, C. 1995. "The Beginning of Life on Earth." *American Scientist*, vol. 83, pages 428–437. An argument that life was an inevitable outcome of prebiotic physics and chemistry.

Dyson, F. J. 1985. *The Origins of Life*. Cambridge University Press, New York. A concise argument in favor of multiple origins of life.

Fenchel, T. and B. J. Finlay. 1994. "The Evolution of Life without Oxygen." *American Scientist*, vol. 82, pages 22–29. A discussion of recent clues about the evolution of early eukaryotic cells.

Loomis, W. F. 1988. *Four Billion Years*. Sinauer Associates, Sunderland, MA. A very readable book on the evolution of genes and organisms, concentrating on the first billion years.

Margulis, L. 1984. *Early Life*. Jones and Bartlett, Boston. An engaging account of the earliest organisms and how they evolved. Good treatment of the endosymbiosis theory of the origin of eukaryotes by its principal proponent.

Maynard Smith, J. and E. Szathmáry. 1995. *The Major Transitions in Evolution*. W. H. Freeman, San Francisco. The first book to describe and synthesize the evolution of the major changes in the way genetic information is organized and transmitted from generation to generation.

Orgel, L. 1994. "The Origin of Life on Earth." *Scientific American*, October. A discussion of how life evolved after self-replicating molecules appeared.

Schopf, J. W. and C. Klein (Eds.). 1992. *The Proterozoic Biosphere*. Cambridge University Press, New York. An excellent source of information on interactions between life and early Earth.

Taylor, S. R. and S. M. McLennan. 1996. "The Evolution of Continental Crust." *Scientific American*, June. Earth appears to be the only planet in our solar system that has sustained enough geologic activity to create large, stable land masses.

Weinberg, S. 1993. *The First Three Minutes*. Basic Books, New York. A stimulating but demanding account of the origin of Earth for those interested enough to make the investment.

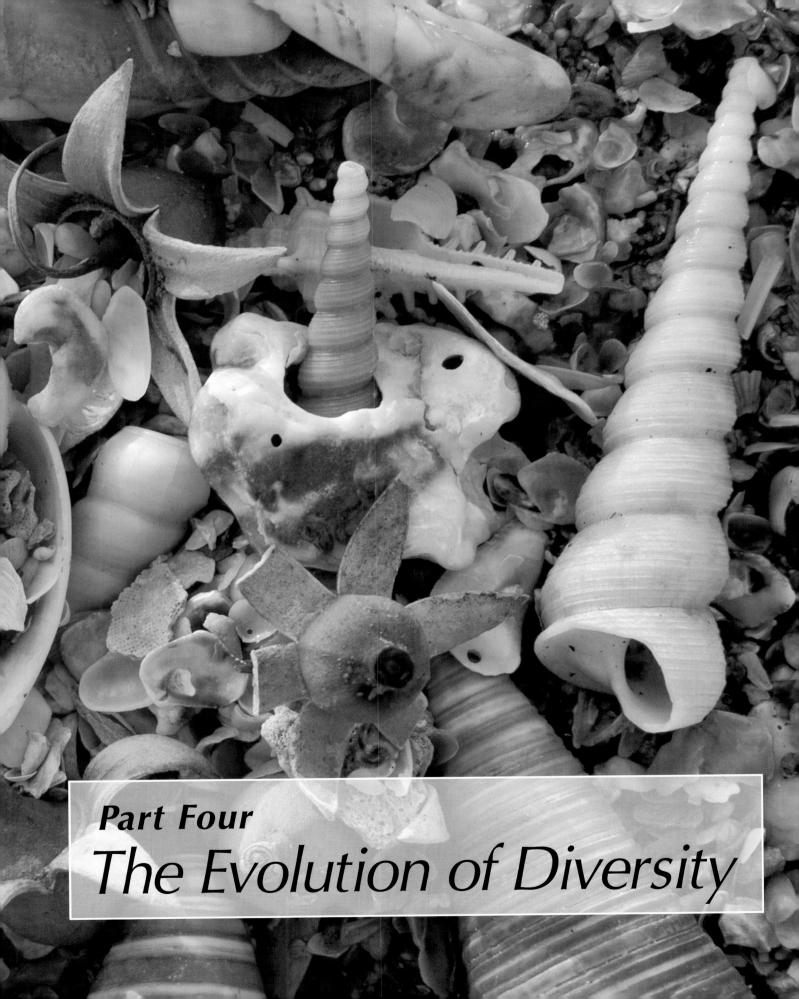

Part Four
The Evolution of Diversity

Chapter 25

Bacteria and Archaea: The Prokaryotic Domains

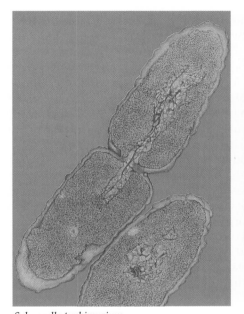

Salmonella typhimurium

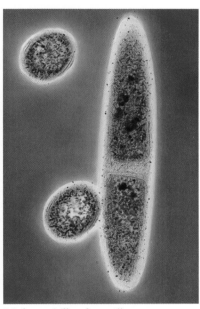

Methanospirillum hungatii

Very Different Prokaryotes
In each image, one of the cells has nearly finished dividing. On the left are bacteria; on the right are archaea, which are more closely related to you than they are to the bacteria.

How can two kinds of microscopic creatures look so much alike and yet be so vastly different? After examining a micrograph of either a bacterium or an archaean, you might say, "Looks like a bacterium to me." However, the cells shown here on the right are more closely related to you and me than they are to the cell on the left!

Both of these simple-looking organisms are prokaryotes, but they differ in numerous biochemical and genetic ways. Not until the 1970s did biologists discover how radically bacteria and archaea differ from each other. And only in 1996, with the sequencing of an archaean genome, did we realize just how extensively archaea differ from bacteria and from eukaryotes, whose genomes had been sequenced a little earlier.

Scientists acknowledge the antiquity of the lineage separations and the importance of their differences by recognizing three domains of living things: Bacteria, Archaea, and Eukarya. The domain Bacteria comprises one kingdom, the kingdom Eubacteria ("true bacteria"); the domain Archaea (from the Greek *archaios*, "ancient") comprises one kingdom, the kingdom Archaebacteria ("ancient bacteria"). The domain Eukarya consists of four kingdoms that include all other living things on Earth. Dividing the living world in this way, with two prokaryotic domains and a single domain for all the eukaryotes, fits with current trends to reflect evolutionary relationships in classification systems (see Chapter 22).

In the six chapters of Part Four, we celebrate and describe the diversity of the living world—the products of evolution. This chapter focuses on the prokaryotic kingdoms. Chapters 26 to 30 deal with the kingdoms Protista, Plantae, Fungi, and Animalia.

In this chapter, we will pay close attention to the ways in which the two domains of prokaryotic organisms resemble one another, and how they differ. Then we will survey the surprising diversity of organisms within each of these domains, relating the characteristics of different prokaryotic groups to their roles in the biosphere and in our lives.

Why Three Domains?

What does it mean to be *different*? You and the person nearest you look very different—certainly more different than the two cells shown at the beginning of this chapter look. But the two of you are members of the same species, and those two tiny organisms are classified in entirely separate domains. You (in the domain Eukarya) and those two prokaryotes (in the domains Bacteria and Archaea) have a lot in common. Members of all three domains conduct glycolysis and replicate their DNA semiconservatively. In all three, the DNA encodes polypeptides that are produced by transcription and translation, and all cells have plasma membranes and ribosomes in abundance.

As a member of the domain Eukarya, *you* have cells with nuclei, membrane-enclosed organelles, and a cytoskeleton—things that no prokaryote has. However, a glance at Table 25.1 will show you that there are also major differences, most of which cannot be seen even under the microscope, between the two prokaryotic

domains. In some ways the archaea are more like us; in other ways they are more like bacteria.

Comparisons of base sequences in ribosomal RNAs and of genomes themselves have led scientists to conclude that all three domains had a single common ancestor and that the present-day archaea share a more recent common ancestor with eukaryotes than they do with bacteria (Figure 25.1). In the same way, humans are more closely related to their sisters and brothers, with whom they share parents, than to their cousins, with whom they share grandparents.

Because of (1) the ancient time at which the three lineages diverged, (2) the major differences among these three kinds of organisms, and especially (3) the fact that the archaea are more closely related to the eukaryotes than are either of those groups to the bacteria, most biologists agree that it makes sense to treat the groups as domains—a higher taxonomic category than kingdom. To treat all the prokaryotes as a single kingdom within a five-kingdom classification of organisms would result in a kingdom that is paraphyletic. That is, the prokaryotic kingdom would not include all the descendants of their common ancestor (see Chapter 22, especially Figure 22.12, for a discussion of paraphyletic groups).

The Archaea, Bacteria, and Eukarya of today are all the products of billions of years of natural selection, and they are all highly adapted to present-day environments. None are "primitive." The common ancestor of the Archaea and the Eukarya lived 1.8 billion years ago by a conservative estimate, and the common

TABLE 25.1   The Three Domains of Life on Earth

CHARACTERISTIC	DOMAIN		
	BACTERIA	ARCHAEA	EUKARYA
Membrane-enclosed nucleus	Absent	Absent	*Present*
Membrane-enclosed organelles	Absent	Absent	*Present*
Peptidoglycan in cell wall	*Present*	Absent	Absent
Membrane lipids	Ester-linked	*Ether-linked*	Ester-linked
	Unbranched	*Branched*	Unbranched
Ribosomes[a]	70S	70S	*80S*
Initiator tRNA	*Formylmethionine*	Methionine	Methionine
Operons	Yes	Yes	*No*
Plasmids	Yes	Yes	*Rare*
RNA polymerases	One	Several	Three
Sensitive to chloramphenicol and streptomycin	*Yes*	No	No
Ribosomes sensitive to diphtheria toxin	*No*	Yes	Yes
Some are methanogens	No	*Yes*	No
Some fix nitrogen	Yes	Yes	*No*
Some conduct chlorophyll-based photosynthesis	Yes	*No*	Yes

[a] 70S ribosomes are smaller than 80S ribosomes.

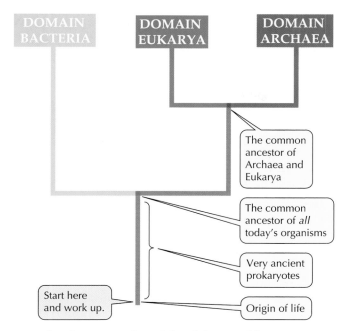

The common ancestor of Archaea and Eukarya

The common ancestor of *all* today's organisms

Very ancient prokaryotes

Start here and work up.

Origin of life

25.1 The Three Domains of the Living World

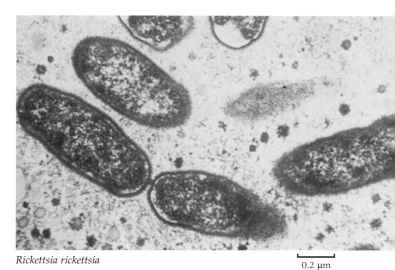

Rickettsia rickettsia

0.2 μm

25.2 Bacteria Living in Animal Cells These bacteria live within the mammalian body, where they can cause a lethal disease, Rocky Mountain spotted fever.

ancestor of the Archaea, the Eukarya, and the Bacteria lived more than 2 billion years ago.

The earliest prokaryotic fossils date back at least 3.5 billion years, and as we saw in Chapter 24, these ancient fossils indicate that there was considerable diversity among the prokaryotes even during the Archean eon. The prokaryotes reigned supreme on an otherwise sterile Earth for a very long time, adapting to new environments and to changes in existing environments.

General Biology of the Prokaryotes

There are many, many prokaryotes around us—everywhere. Although they are so small that we cannot see them with the naked eye, the prokaryotes are the most successful of all creatures on Earth if success is measured by numbers of individuals. The bacteria in one person's intestinal tract, for example, outnumber all the humans who have ever lived. Although small, prokaryotes play many critical roles in the biosphere, interacting in one way or another with every other living thing.

In this section on the general biology of the prokaryotes, we'll see that some perform key steps in the cycling of nitrogen, sulfur, and carbon. Other prokaryotes trap energy from the sun or from inorganic chemical sources, and some help animals digest their food. The members of the two prokaryotic domains outdo all other groups in metabolic diversity, and they occupy more—and more extreme—habitats than any other group.

They have spread to every conceivable habitat on the planet, from the coldest to the hottest, from the most acidic to the most alkaline, and to the saltiest. Some live where O_2 is abundant and others where there is no O_2 at all. They have established themselves at the bottom of the seas, in rocks more than 2 km into Earth's solid crust, and even inside other organisms, large and small (Figure 25.2). Their effects on our environment are diverse and profound.

Prokaryotes and their associations take a few characteristic forms

Three shapes are particularly common among the prokaryotes: spheres, rods, and curved or spiral forms (Figure 25.3). A spherical prokaryote is called a **coccus** (plural cocci). Cocci may live singly or may associate in two- or three-dimensional arrays as chains, plates, or blocks of cells. A rod-shaped prokaryote is called a **bacillus** (plural bacilli). Bacilli and spiral forms, the third main prokaryotic shape, may be single or may form chains.

Associations such as chains do not signify multicellularity, because each cell is fully viable and independent. Associations arise as cells adhere to one another after reproducing by fission, with each becoming two cells. Some bacteria associate in chains that become enclosed within delicate tubular sheaths. These associations are called filaments. All the cells of a filament divide simultaneously.

In addition to having distinguishing shapes, some prokaryotes are identified by other structural features. For example, some attach to their substrate by stalks that may be either an extension of the cell wall or a product secreted outside the cell.

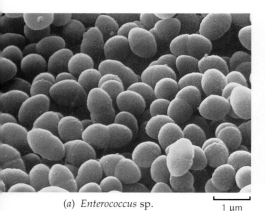

(a) *Enterococcus* sp.   1 μm

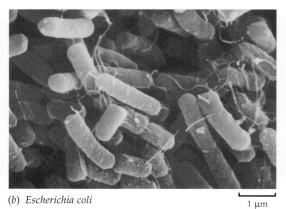

(b) *Escherichia coli*   1 μm

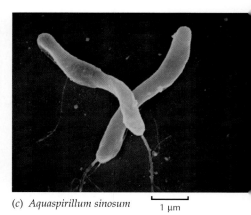

(c) *Aquaspirillum sinosum*   1 μm

25.3 Shapes of Prokaryotic Cells (a) These spherical cocci of an acid-producing bacterium grow in the mammalian gut. (b) Rod-shaped *E. coli* are the most thoroughly studied of all bacteria—indeed, of almost any organism. (c) A freshwater spiral bacteria species. The cells move by means of the tufts of flagella at each pole.

Prokaryotes lack nuclei, organelles, and a cytoskeleton

The architectures of prokaryotic and eukaryotic cells were compared in Chapter 4 (you may wish to review Figures 4.3, 4.4, 4.5, 4.7, and 4.8). The basic unit of archaea and bacteria is the prokaryotic cell, which contains a full complement of genetic and protein-synthesizing systems, including DNA, RNA, and all the enzymes needed to transcribe and translate the genetic information into protein (see Chapter 12). The prokaryotic cell also contains at least one system for generating the ATP it needs (see Chapter 7).

The prokaryotic cell differs from the eukaryotic cell in three important ways. First, the organization and replication of the genetic material differs. The DNA of the prokaryotic cell is not organized within a membrane-enclosed nucleus. DNA molecules in prokaryotes are circular; typically there is a single chromosome, but there are often plasmids as well (see Chapter 13). The elaborate mechanism of mitosis is missing; prokaryotic cells divide by their own elaborate method, fission, after replicating their DNA.

Second, prokaryotes have none of the membrane-enclosed cytoplasmic organelles that modern eukaryotes have—mitochondria, chloroplasts, Golgi apparatus, endoplasmic reticulum—but the cytoplasm of a prokaryotic cell may contain a variety of infoldings of the plasma membrane (see Figure 4.4) and photosynthetic membrane systems not found in eukaryotes. Membranous infoldings frequently associate with new cell walls during cell division, and in electron micrographs the DNA of a bacterial cell is often seen attached to such an infolding, called a *mesosome* (Figure 25.4). Third, prokaryotic cells lack a cytoskeleton.

Prokaryotes have distinctive modes of locomotion

Although many prokaryotes are not motile, others can move by one of several means. Spirochetes use a rolling motion made possible by internal fibrils (Figure 25.5a). Many cyanobacteria and the "gliding bacteria" use various poorly understood gliding mechanisms, including rolling. Some aquatic prokaryotes, including some cyanobacteria, can move slowly up and down in the water by adjusting the amount of gas in gas vesicles (Figure 25.5b). By far the most common type of locomotion in prokaryotes is that driven by flagella.

Prokaryotic flagella are whiplike filaments that extend singly or in tufts from one or both ends of the cell (see Figure 25.3c and 25.8a), or all around it (Figure 25.6). A prokaryotic flagellum consists of a single fibril made of the protein *flagellin*, projecting from the cell surface. In contrast, the flagellum of eukaryotes is enclosed by the plasma membrane and usually contains a circle of nine pairs of microtubules surrounding two

The large mesosome in this bacterium is continuous with the plasma membrane.

A large, fibrous mass of DNA attached to the mesosome fills most of the rest of the cell.

Plasma membrane

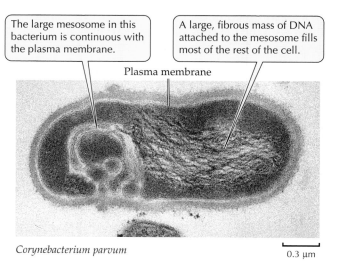

Corynebacterium parvum   0.3 μm

25.4 Some Prokaryotes Have Internal Membranes Unlike eukaryotic organelles, the infolding in this bacterial cell is not a separate, membrane-enclosed compartment.

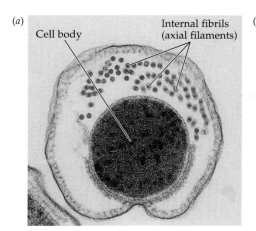

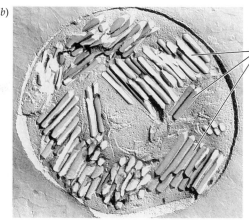

25.5 Structures Associated with Prokaryote Motility
(*a*) A spirochete from the gut of a termite, seen in cross section. (*b*) Gas vesicles in a cyanobacterium, visualized by the freeze-etch technique (see Chapter 5).

central microtubules, all made of the protein tubulin, along with other, associated proteins. The prokaryotic flagellum (see Figure 4.6) rotates about its base, rather than beating, as a eukaryotic flagellum or cilium does.

Prokaryotes have distinctive cell walls

Most prokaryotes have a thick and relatively stiff cell wall. This wall is quite different from the cell walls of plants and algae, which contain cellulose and other polysaccharides, and of fungi, which contain chitin. Almost all bacteria have cell walls containing peptidoglycan (a polymer of amino sugars). Archaean cell walls are of differing types, but most contain significant amounts of protein (often a glycoprotein). One group of archaea has pseudopeptidoglycan in its wall; as you have probably already guessed from the prefix

"pseudo-," pseudopeptidoglycan is similar to but distinct from the peptidoglycan of bacteria.

Peptidoglycan is a substance unique to bacteria; its absence from the walls of archaea indicates a key difference between the two prokaryotic domains, resulting from the separation of these two groups at the beginning of evolutionary history.

In 1884 Hans Christian Gram, a Danish physician, developed an uncomplicated staining process that has lasted into our high-technology era as the single most common tool in the identification of bacteria. The **Gram stain** separates most types of bacteria into two distinct groups: gram-positive and gram-negative, on the basis of their wall structure (Figure 25.7). A smear of cells on a microscope slide is soaked in a violet dye and treated with iodine; it is then washed with alcohol and counterstained with safranine (a red dye). **Gram-positive** bacteria retain the violet dye and appear blue to purple (Figure 25.7*a*). The alcohol washes the violet stain out of **gram-negative** cells; these cells then pick up the safranine counterstain and appear pink to red (Figure 25.7*b*). Gram-staining characteristics are a crucial consideration in classifying some kinds of bacteria and are important in determining the identity of bacteria in an unknown sample. Mycoplasmas, which lack walls, are not stained at all by the Gram stain.

The different staining reactions probably relate to differences in the physical structures of the cell walls of bacteria, but they correlate with the amount of peptidoglycan. The electron micrographs and associated sketches in Figure 25.7 show a thick layer of peptidoglycan outside the plasma membrane of gram-positive bacteria (see Figure 25.7*a*). The gram-negative cell wall usually has only one-fifth as much peptidoglycan, and outside the peptidoglycan layer the cell is surrounded by a second, outer membrane quite distinct in chemical makeup from the plasma membrane (see Figure 27.5*b*). The space between the inner (plasma) and outer membranes of gram-negative bacteria contains enzymes that are important in digesting some materi-

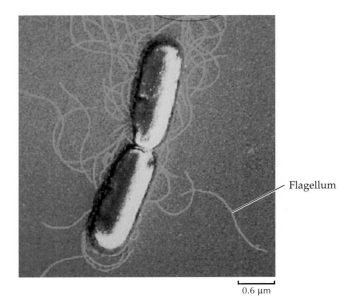

Flagellum

0.6 μm

25.6 Some Bacteria Use Flagella for Locomotion
Flagella surround the rod-shaped cells of this *Bacillus* species.

(a) *Bacillus subtilis*

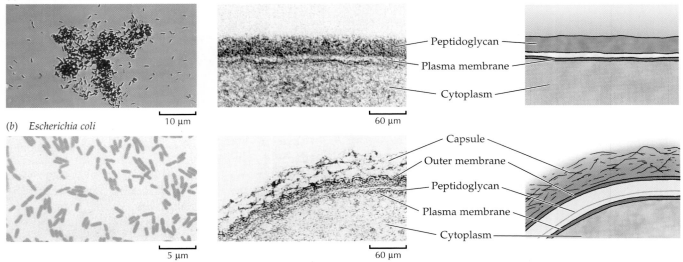

Peptidoglycan
Plasma membrane
Cytoplasm

(b) *Escherichia coli*

10 μm

60 μm

Capsule
Outer membrane
Peptidoglycan
Plasma membrane
Cytoplasm

5 μm

60 μm

25.7 The Gram Stain and the Cell Wall   When treated with a Gram stain, the cell wall components of different bacteria react in one of two ways. (a) Gram-positive bacteria retain the violet dye and appear deep blue or purple; the pink counterstain surrounds the cells in this micrograph. (b) Gram-negative bacteria internalize the counterstain and appear pink-to-red.

als, transporting others, and detecting chemical gradients in the environment.

The consequences of the different features of the cell walls are numerous and relate to the disease-causing characteristics of some prokaryotes. Indeed, the cell wall is a favorite target in medical combat against diseases that are caused by prokaryotes, because it has no counterpart in eukaryotic cells. Antibiotics and other agents that specifically attack peptidoglycan-containing cell walls tend to have little, if any, effect on the cells of humans and other eukaryotes.

Prokaryotes reproduce asexually, but genetic recombination does occur

Most prokaryotes reproduce by fission or by producing spores; both processes are asexual, or **vegetative**. Recall, however, that there are also processes—transformation, conjugation, and transduction—that allow the exchange of genetic information between some prokaryotes quite apart from either sex or reproduction.

Many prokaryotes multiply very rapidly. One of the fastest is the bacterium *Escherichia coli*, which under optimal conditions has a generation time of about 20 minutes. The shortest known prokaryote generation times are about 10 minutes; values of 1 to 3 hours are common; some extend to days. Bacteria living in rock deep in Earth's crust may suspend their growth for more than a century without dividing and then grow for a few days before suspending growth again.

Prokaryotes have exploited many metabolic possibilities

The long evolutionary history of bacteria and archaea, including their explorations of new environments, has led to extraordinary diversity of their metabolic "lifestyles"—their use or nonuse of O_2, their energy sources, the sources of their carbon atoms, and the materials they secrete.

ANAEROBIC VERSUS AEROBIC METABOLISM.   Some prokaryotes can live only by anaerobic metabolism, because they are poisoned by oxygen gas. These oxygen-sensitive fermenters are called **obligate anaerobes**. Fermentation is not the only anaerobic way to obtain energy, as we'll see shortly when we discuss the denitrifiers.

Other organisms can shift their metabolism between anaerobic and aerobic modes (see Chapter 7) and thus are called **facultative anaerobes**. Some facultative anaerobes cannot conduct cellular respiration but are not damaged by oxygen when it is present. Many types of prokaryotes are facultative anaerobes that alternate between anaerobic metabolism (such as fermentation) and cellular respiration as conditions dictate.

At the other extreme from the obligate anaerobes, some prokaryotes are **obligate aerobes**, unable to survive for extended periods in the *absence* of oxygen.

NUTRITIONAL CATEGORIES.   Biologists recognize four broad nutritional categories of organisms: photoautotrophs, photoheterotrophs, chemoautotrophs, and chemoheterotrophs. Prokaryotes are represented in all four groups (Table 25.2). **Photoautotrophs** are photosynthetic. They use light as their source of energy and carbon dioxide as their source of carbon. Like the photosynthetic eukaryotes, the cyanobacteria, one group of

TABLE 25.2 How Organisms Obtain Their Energy and Carbon

NUTRITIONAL CATEGORY	ENERGY SOURCE	CARBON SOURCE
Photoautotrophs (some Bacteria, some Eukarya)	Light	Carbon dioxide
Photoheterotrophs (some Bacteria)	Light	Organic compounds
Chemoautotrophs (some Bacteria)	Inorganic substances	Carbon dioxide
Chemoheterotrophs (found in all three domains)	Organic compounds	Organic compounds

photoautotrophs, photosynthesize with chlorophyll *a* as the key pigment and produce oxygen as a by-product of noncyclic photophosphorylation (see Chapter 8).

By contrast, the other photosynthetic bacteria use bacteriochlorophyll as their key photosynthetic pigment, and they do not release oxygen gas. Some of these photosynthesizers produce particles of pure sulfur instead because hydrogen sulfide (H_2S) rather than H_2O is the electron donor for photophosphorylation (Figure 25.8*a*). Bacteriochlorophyll absorbs light of longer wavelength than the chlorophyll used by all other photosynthesizing organisms does. As a result, bacteria using this pigment can grow in water beneath fairly dense layers of algae because light of the wavelengths they use is not appreciably absorbed by the algae (Figure 25.8*b*).

Photoheterotrophs use light as their source of energy but must obtain their carbon atoms from organic compounds made by other organisms. They use compounds such as carbohydrates, fatty acids, and alcohols as their organic "food." The purple nonsulfur bacteria, among others, are photoheterotrophs.

Chemoautotrophs obtain their energy by oxidizing inorganic substances, and they use some of that energy to fix carbon dioxide. Some chemoautotrophs use reactions identical to those of the photosynthetic carbon reduction cycle (see Chapter 8), but others use other pathways to fix carbon dioxide. The chemoautotrophs—some of them bacteria and others archaea—include the nitrifiers, which oxidize ammonia or nitrite ions to form nitrate ions that are taken up by plants, as well as other bacteria that oxidize hydrogen gas, hydrogen sulfide, sulfur, and other materials.

Scientists exploring the ocean bottom near the Galapagos Islands in 1977 discovered a spectacular example of chemoautotrophy. They found an entire ecosystem based on chemoautotrophic bacteria that are incorporated into a large community of crabs, mollusks, and giant worms (pogonophorans; see Figure 29)—all at a depth of 2,500 m, far below any hint of light from the sun but in the immediate neighborhood

25.8 Some Bacteria Photosynthesize (*a*) Cells of purple sulfur bacteria store granules of sulfur that they produce via anaerobic photosynthesis. (*b*) *Ulva*, a green alga, absorbs no light of wavelengths longer than 750 nm. Purple sulfur bacteria can conduct photosynthesis in the shade of the algae, using these longer wavelengths.

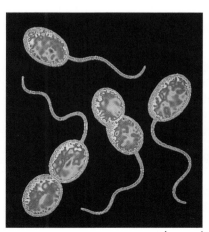

(*a*) *Thiocystis* sp.   0.6 μm   (*b*)

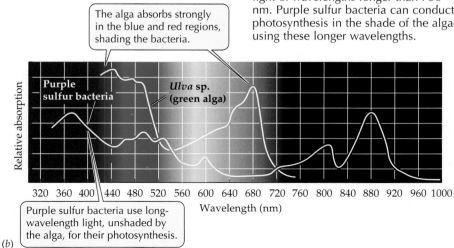

The alga absorbs strongly in the blue and red regions, shading the bacteria.

Purple sulfur bacteria

Ulva sp. (green alga)

Relative absorption

320 360 400 440 480 520 560 600 640 680 720 760 800 840 880 920 960 1000

Wavelength (nm)

Purple sulfur bacteria use long-wavelength light, unshaded by the alga, for their photosynthesis.

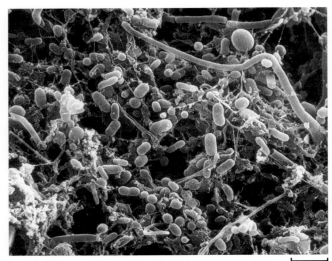

2 μm

25.9 Chemoautotrophs All of these archaea are chemoautotrophs that live near a hot-water vent along the Galapagos Rift in the eastern Pacific, where their fixing of carbon supports an entire community of organisms that thrives in total darkness.

of volcanic vents in the ocean floor (Figure 25.9). These bacteria obtain energy by oxidizing hydrogen sulfide and other substances released from the vents.

Finally, **chemoheterotrophs** typically obtain both energy and carbon atoms from one or more organic compounds. Most bacteria are chemoheterotrophs—as are all archaea, animals, and fungi, and many protists.

NITROGEN AND SULFUR METABOLISM. Some bacteria carry out respiratory electron transport without using oxygen as an electron acceptor. These forms use oxidized inorganic ions such as nitrate, nitrite, or sulfate as electron acceptors. Among these organisms are the **denitrifiers**, bacteria that return nitrogen to the atmosphere as nitrogen gas (N_2), completing the cycle of nitrogen in nature. These normally aerobic bacteria, mostly species of the genera *Bacillus* and *Pseudomonas*, use nitrate (NO_3^-) in place of oxygen if they are kept under anaerobic conditions:

$$2\ NO_3^- + 10\ e^- + 12\ H^+ \rightarrow N_2 + 6\ H_2O$$

Nitrogen fixers convert atmospheric nitrogen gas into chemical forms usable by the nitrogen fixers themselves and by other living things. For example,

$$N_2 + 6\ H \rightarrow 2\ NH_3\ (ammonia)$$

All organisms require nitrogen for their proteins, nucleic acids, and other important compounds. The vital process of nitrogen fixation is carried out by a wide variety of bacteria, including cyanobacteria, but by no other organisms. We'll discuss this process in detail in Chapter 34.

Ammonia is oxidized to nitrate by the process of **nitrification**. This process is carried out in the soil by bacteria called **nitrifiers**, mentioned earlier in our discussion of chemoautotrophs. Bacteria of two genera, *Nitrosomonas* and *Nitrosococcus*, convert ammonia to nitrite ions (NO_2^-), and *Nitrobacter* oxidizes nitrite to nitrate (NO_3^-). What do the nitrifiers get out of these reactions? These three genera are chemoautotrophs. Their chemosynthesis is powered by the energy released by oxidation of ammonia or nitrite.

For example, by passing the electrons from nitrite through an electron transport chain, *Nitrobacter* can make ATP, and using some of this ATP, it can also make NADH. With the ATP and NADH, the bacterium can convert CO_2 and H_2O to glucose and other foods. The nitrifiers base their entire biochemistry—their entire lives—on the oxidation of ammonia or NO_2^- ions. *Nitrobacter* can convert 6 molecules of CO_2 to 1 molecule of glucose for every 78 NO_2^- ions that they oxidize—not terribly efficient, but efficient enough to keep themselves living, growing, and reproducing.

What do plants get from the activities of nitrifiers? Under normal soil conditions, ammonium ions (NH_4^+) form readily from ammonia (NH_3) and bind to clay particles, which carry negative charges, in the soil. This binding makes NH_4^+ difficult for plants to take up. Nitrifiers have NH_4^+ transport proteins with a greater affinity than those of plants; they take up the NH_4^+ and oxidize it to NO_3^-, which, with its negative charge, doesn't bind to clay particles and hence is readily available to plants.

Numerous bacteria base their metabolism on the modification of sulfur-containing ions and compounds in their environment. As examples, we have already mentioned the photoautotrophic bacteria and chemoautotrophic archaea that use H_2S as an electron donor in place of H_2O. Such uses of nitrogen and sulfur have obvious environmental implications, as we'll see in the next section.

Prokaryotes in Their Environments

Prokaryotes live in and exploit all sorts of environments and are parts of many ecosystems. In the following pages, we'll examine prokaryotes in soils and water, and even in other living things, where they may exist in neutral, benevolent, or parasitic relationship with the host's tissues.

Prokaryotes are important players in soils and bodies of water

Animals depend on photosynthetic plants for their food, directly or indirectly. But plants depend on other organisms—prokaryotes—for their own nutrition. Without nitrogen fixers, no other life could exist, be-

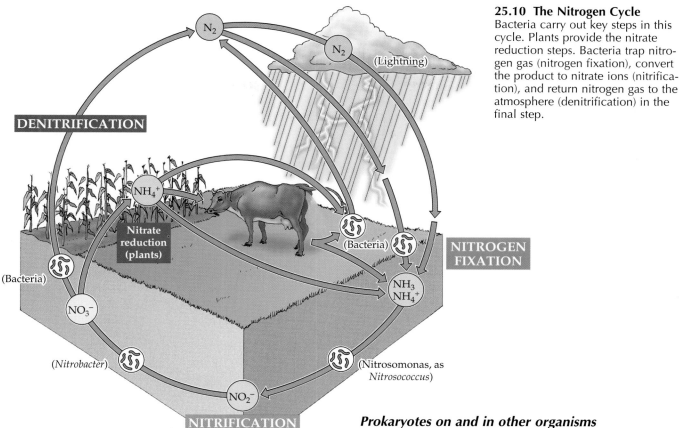

25.10 The Nitrogen Cycle
Bacteria carry out key steps in this cycle. Plants provide the nitrate reduction steps. Bacteria trap nitrogen gas (nitrogen fixation), convert the product to nitrate ions (nitrification), and return nitrogen gas to the atmosphere (denitrification) in the final step.

cause no other organisms can fix the nitrogen gas that surrounds us. Nitrifiers are also crucial to the biosphere, because plants often prefer nitrate ions (the end products of nitrification) over the products of nitrogen fixation for their own use. Nitrifiers produce the form of nitrogen most used by plants, which is the source of nitrogen compounds for animals and fungi.

If nitrogen fixers convert nitrogen gas into compounds that are used by the entire living world, why doesn't the atmosphere run out of N_2? Denitrifiers play a key role in keeping the nitrogen cycle going (Figure 25.10). Without denitrifiers, which convert nitrate ions back into nitrogen gas, all forms of nitrogen would leach from the soil and end up in lakes and oceans, making life on land impossible. Other prokaryotes crank a similar cycle for sulfur.

In the ancient past, the cyanobacteria had an equally dramatic impact: Their photosynthesis converted the global environment from anaerobic to aerobic. The result was the wholesale loss of species that couldn't tolerate the O_2 generated by the cyanobacteria, but this transformation made possible the evolution of cellular respiration and the subsequent explosion of eukaryotic life.

Prokaryotes on and in other organisms

Prokaryotes work together with eukaryotes in many ways. In fact, mitochondria and chloroplasts are descended from what were once free-living prokaryotes. Much later in evolutionary history, some plants learned to form associations with bacteria of the genus *Rhizobium* to form cooperative nitrogen-fixing nodules on their roots (see Chapter 34).

Many animals, including humans, harbor a variety of bacteria and archaea in their digestive tracts. Cows depend on prokaryotes in their complicated stomachs to perform important steps in digestion. Cows cannot produce cellulase, the enzyme needed to start the digestion of the cellulose that makes up the bulk of their food, but some of their stomach bacteria produce cellulase in sufficient quantity to process the cow's daily diet. We use some of the metabolic products—especially vitamins B_{12} and K—provided by the bacteria in our large intestine.

We are heavily populated, inside and out, by bacteria. Although very few of them are agents of disease, popular notions of bacteria as "germs" arouse our curiosity about those few, so we will briefly consider some bacterial pathogens.

A small minority of bacteria are pathogens

The late nineteenth century was a productive era in the history of medicine—a time during which bacteriologists, chemists, and physicians proved that many diseases are caused by microbial agents. During this time the German physician Robert Koch laid down a set of rules for testing the relationship between a disease and a microorganism.

According to Koch, the disease in question could be attributed to a particular microorganism if (1) the microorganism could always be found in diseased individuals; (2) the microorganism taken from the host could be grown in pure culture; (3) a sample of the culture produced the disease when injected into a new, healthy host; and (4) the newly infected host yielded a new, pure culture of microorganisms identical to that obtained in step 2. These rules—called **Koch's postulates**—were very important in a time when it was not widely accepted that microorganisms cause disease. Today, because of the availability of other diagnostic tools, medical science accepts less rigorous proofs in investigating new diseases.

Only a tiny percentage of all prokaryotes are pathogens (disease-producing organisms), and those that are are all bacteria. For an organism to be a successful pathogen, it must overcome several hurdles. It must arrive at the body surface of a potential host, enter the host, multiply inside the host, and, finally, prepare to infect the next host. Failure to overcome any of these hurdles ends the reproductive career of a pathogenic organism, and potential hosts have many defenses against pathogens. In Chapter 18 we considered the immune response and other modes of protection against diseases of microbial origin. But some bacteria *are* pathogenic, and they often succeed in infecting a host.

For the host, the consequences of an infection depend on several factors. One is the **invasiveness** of the pathogen—its ability to multiply within the body of the host. Another is its **toxigenicity**—its ability to produce chemical substances (toxins) harmful to the tissues of the host. *Corynebacterium diphtheriae*, the agent that causes diphtheria, has low invasiveness and multiplies only in the throat, but its toxigenicity is so great that the entire body is affected. By contrast, *Bacillus anthracis*, which causes anthrax (a disease primarily of cattle and sheep), has low toxigenicity but an invasiveness so great that the entire bloodstream ultimately teems with the bacteria.

There are two general types of bacterial toxins: exotoxins and endotoxins. **Endotoxins** are released when certain gram-negative bacteria lyse. They are lipopolysaccharides that form part of the outer membrane. Endotoxins are rarely fatal; they normally cause fever, vomiting, and diarrhea. Among the endotoxin producers are some species of *Salmonella* and *Escherichia*.

Exotoxins are proteins released by living, multiplying bacteria that may travel throughout the host's body. They are highly toxic—often fatal—but do not produce fevers. Many pathogenic bacteria produce exotoxins. Some examples of exotoxin-induced diseases are tetanus (from *Clostridium tetani*), botulism (from *Clostridium botulinum*), food poisoning (from *Bacillus cereus*), cholera (from *Vibrio cholerae*) and plague (from *Yersinia pestis*).

Remember that in spite of our frequent mention of human pathogens, only a small minority of the known prokaryotic species are pathogenic. Many more species play positive roles in our lives and in the biosphere. We make direct use of many bacteria and a few archaea in such diverse applications as cheese production, sewage treatment, and the industrial production of an amazing variety of antibiotics, vitamins, organic solvents, and other chemicals.

Prokaryote Phylogeny and Diversity

The prokaryotes comprise a diverse array of microscopic organisms. To explore this diversity, let's first consider how they are classified, and with what difficulty; then we'll look at some specific examples.

Nucleotide sequences of prokaryotes reveal evolutionary relationships

There are three primary motivations for classification schemes: to help identify unknown organisms, to reveal evolutionary relationships, and to provide universal names. Many scientists and medical technologists must be able to identify bacteria quickly and accurately. Lives depend on it.

Until recently taxonomists based their classification schemes for the prokaryotes on readily observable phenotypic characters such as color, motility, nutritional requirements, antibiotic sensitivity, and reaction to the Gram stain. Although such schemes have facilitated the identification of prokaryotes, they have failed miserably at giving insights into how these organisms evolved—a problem of great interest to microbiologists and to all students of evolution. The prokaryotes and protists (see Chapter 26) have long been major challenges to those who would classify them in a natural, evolution-based way. Only recently have systematists had the right tools for tackling this task.

Analyses of ribosomal RNAs now provide us with apparently reliable measures of evolutionary distance between taxonomic groups such as domains or genera. Ribosomal RNA (rRNA) is particularly useful for evolutionary studies of living organisms because (1) rRNA is evolutionarily ancient, (2) no living organism lacks rRNA, (3) rRNA plays the same role in translation in all organisms, and (4) rRNA itself has evolved slowly enough that sequence similarities between groups of organisms are easily found. (Most recently, biologists have been sequencing rRNA genes (DNA) rather than the rRNA itself, using the techniques described in Chapter 16.)

Comparisons of rRNAs from a great many sources showed that there are recognizable short base sequences characteristic of particular groups of organisms. These **signature sequences**, approximately 6 to 14 bases long, appear at the same approximate positions in rRNAs from related groups. For example, the signature sequence AAACUUAAAG occurs about 910 bases from one end of the RNA of the light subunit of ribosomes in 100 percent of the Archaea and Eukarya tested, but in *none* of the Bacteria tested. The signature sequence AAACUCAAA appears at the same position in all Bacteria but not in the Archaea or Eukarya. Several signature sequences distinguish each of the three domains. Similarly, each phylum of the kingdoms Archaebacteria and Eubacteria possesses a unique signature sequence.

These data sound promising, but as we will see, things aren't as easy as might seem possible. Although the evolutionary patterns revealed by signature sequences and other molecular tools are reliable, the groupings thus revealed are amazingly complex. A single phylum of bacteria or archaea may contain the most extraordinarily diverse species, and a species in one phylum may be phenotypically almost indistinguishable from one or many species in another phylum.

In the next section we'll explore some of the reasons for this spectacular diversity.

Mutations are the most important source of prokaryotic variation

Asexual reproduction, universal in prokaryotes, promotes genetic uniformity. Although prokaryotes can acquire different alleles by transformation, transduction, or conjugation (see Chapters 11 and 13), the most important source of genetic variation in populations of prokaryotes is mutation.

Mutations, especially recessive mutations, are slow to make their presence felt in populations of humans and other diploid organisms. In contrast, a mutation in a prokaryote, which is haploid, has immediate consequences for that organism, and if not lethal it will be transmitted to and expressed in the organism's daughter cells—and in their daughter cells, and so forth. A mutant allele spreads rapidly, if it is beneficial and favored by natural selection.

The rapid multiplication of many prokaryotes, coupled with mutation and selection, allows rapid changes in phenotype in a population. Important changes, such as loss of sensitivity to an antibiotic, can occur over broad geographic areas in just a few years. Think how many significant metabolic changes can occur over even modest time spans in relation to the history of life on Earth. When we introduce the largest phylum of bacteria, we will show how different groups within the phylum have easily and rapidly adopted and abandoned metabolic pathways under selective pressure from the environment.

The Bacteria

The great majority of prokaryotes are bacteria. They can at last be classified in a natural, phylogenetic way into 12 phyla, thanks to molecular tools such as rRNA sequencing that reveal evolutionary relationships. However, evolutionary relationships do not correspond closely with many of the important phenotypic traits of bacteria. The reason is that gene transfer has led to the abrupt appearance of certain traits (such as photosynthesis) multiple times during bacterial evolution, while other traits have been lost among the descendants of ancestors that had them. Therefore, the metabolic and ecological characteristics of bacteria, which are often the most important traits of the species, are shared among species that are only distantly related.

The rapid gain and loss of traits, although it makes classifying bacteria difficult, is in large part responsible for the success of these organisms. In the following discussion, rather than following evolutionary relationships, we will organize our analysis around some important functional traits of bacteria, such as their nutrition, locomotion, and ability to exploit habitats.

Metabolism in the Proteobacteria has evolved dramatically

By far the largest phylum of bacteria, in terms of number of species, is the phylum Proteobacteria, sometimes referred to as the Purple Bacteria. The classification problem begins right there—among the proteobacteria are many species of purple bacteria (gram-negative, bacteriochlorophyll-containing, sulfur-using photoautotrophs), but this phylum also includes a dramatically diverse group of bacteria that bear no resemblance to the purple bacteria in phenotype. It is this bacterial phylum to which the mitochondria of eukaryotes are most closely related.

No characteristic demonstrates the diversity of the proteobacteria more clearly than mode of nutrition (Figure 25.11). The common ancestor of all the proteobacteria was a photoautotroph. Early in evolution, two groups of proteobacteria lost their ability to photosynthesize and have been chemoheterotrophs ever since. The other three groups still have photoautotrophic members, but in *each* group, some evolutionary lines have abandoned photoautotrophy and taken up other modes of nutrition. There are chemoautotrophs and chemoheterotrophs in all of them. Why?

We can view each of the trends in Figure 25.11 as an evolutionary response to selective pressures encountered as these bacteria encountered new habitats that presented new challenges and opportunities. Much of the diversity of bacteria is metabolic, as illustrated here. But as we are about to see, there are other interesting differences.

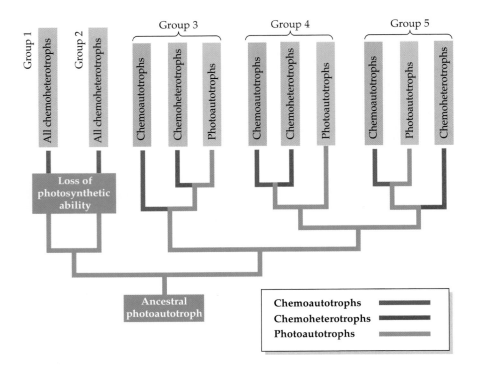

25.11 The Evolution of Metabolism in the Proteobacteria Different groups of proteobacteria adopted different patterns of nutrition as they dealt with new environments.

The chloroplasts of photosynthetic eukaryotes are more closely related to the cyanobacteria than to any other group of bacteria.

Cyanobacteria associate in colonies or live free as single cells. Depending on the species and on growth conditions, colonies of cyanobacteria range from flat sheets one cell thick to spherical balls of cells. Some filamentous colonies differentiate into three cell types: vegetative cells, spores, and heterocysts.

Heterocysts are cells specialized for nitrogen fixation; the enzyme nitrogenase gives them this ability. All the known cyanobacteria with heterocysts fix nitrogen. Heterocysts also have a role in reproduction: When filaments break apart to reproduce, the heterocyst may serve as a breaking point. Figure 25.13a and b shows heterocysts within filaments.

GLIDING BACTERIA. The gliding bacteria are filaments or rods whose movement resembles gliding. Members of the genus *Beggiatoa* (Figure 25.14a) are an example. *Beggiatoa* move up and down in the soil, lying deeper during the day. They are aerobes that need O_2 to

Regardless of classification, the bacteria are highly diverse

In addition to the large phylum Proteobacteria, the domain Bacteria (kingdom Eubacteria) includes eleven other phyla. But the discussion that follows considers nine groups of bacteria defined by *shared phenotypes* rather than strictly by phylogenetic relationships. In most cases, these groups do not correspond neatly to phyla.

THERMOPHILES. The three phyla branching out earliest during bacterial evolution are all heat lovers, as are the most ancient of the archaea. This observation supports the hypothesis that the first organisms were thermophiles that appeared in an environment much hotter than the ones that predominate today.

CYANOBACTERIA. The cyanobacteria (blue-green bacteria) are very independent nutritionally. They appeared in a time of intense competition among heterotrophic prokaryotes for resources, and they prospered on the basis of their autotrophy. Cyanobacteria photosynthesize by using chlorophyll *a* and liberating oxygen gas, and many species also fix nitrogen. They require only water, nitrogen gas, oxygen, a few mineral elements, light, and carbon dioxide.

Cyanobacteria carry out the type of photosynthesis otherwise characteristic only of eukaryotic photosynthesizers. They contain elaborate and highly organized internal membrane systems: the photosynthetic lamellae, or *thylakoids* (Figure 25.12; see also Figure 4.4a).

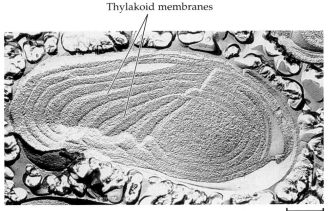

0.5 μm

25.12 Thylakoids in Cyanobacteria This cyanobacterial cell was prepared by freeze-etching (see Chapter 5) to emphasize the extensive system of internal membranes. These photosynthetic membranes are present through most of the cytoplasm.

Heterocyst   Resting spore

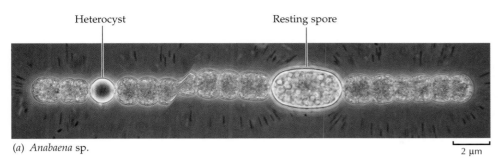

(a) *Anabaena* sp.   2 μm

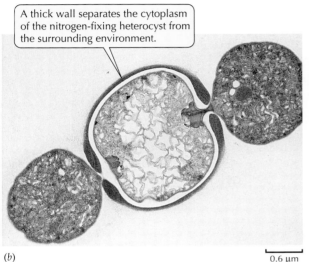

A thick wall separates the cytoplasm of the nitrogen-fixing heterocyst from the surrounding environment.

(b)   0.6 μm

(c)

25.13 Cyanobacteria (a) *Anabaena* is a genus of colonial, filamentous cyanobacteria; this colony displays both a heterocyst and a resting spore. (b) A thin neck attaches a heterocyst to each of two other cells in a colony. (c) Cyanobacteria appear in enormous numbers in some environments. This California pond has experienced eutrophication: Phosphorus and other nutrients generated by human activity have accumulated in the pond, feeding an immense green mat—commonly referred to as "pond scum"—made up of several species of unicellular cyanobacteria.

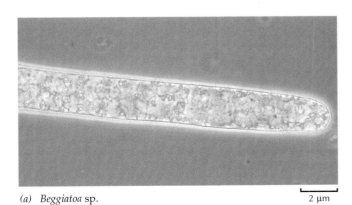

(a) *Beggiatoa* sp.   2 μm

25.14 Gliding Bacteria (a) This filamentous cell of a gliding bacterium was isolated from ocean mud. (b) Individual bacteria aggregated to make up the stalk and knobs of this fruiting body.

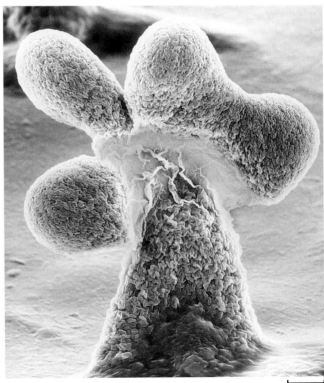

(b) *Stigmatella aurantiaca*   13 μm

metabolize, but they obtain their energy by oxidizing reduced sulfur-containing compounds such as H_2S. These sulfur compounds are produced deep in the soil, away from atmospheric O_2. Because O_2 itself oxidizes the sulfur compounds, *Beggiatoa* must grow in a narrow zone—a zone of compromise between being deep enough to have a supply of sulfur compounds but shallow enough to obtain enough O_2. Cyanobacteria also contribute to the O_2 supply, by photosynthesizing—which they do only during the day, allowing *Beggiatoa* to function deeper in the soil. At night, *Beggiatoa* must glide toward the surface to get enough O_2.

Some gliding bacteria form remarkable structures called fruiting bodies. A group of cells aggregates to make one of these structures (Figure 25.14*b*). The fruiting bodies of some species are simple globes more than 1 mm in diameter; those of other species are more complex, branched structures. Within the fruiting bodies of some species, single cells develop into thick-walled spores that can resist harsh environmental conditions; in other fruiting bodies, whole clusters of cells form a cyst that is resistant to drying. Both spores and cysts can germinate under favorable conditions to yield the next generation of typical gliding cells.

Beggiatoa and the gliding bacteria that form fruiting bodies belong to different groups of the phylum Proteobacteria; some other gliding bacteria belong to a different bacterial phylum. Gliding motion is not the signature of a single phylogenetic group.

SPIROCHETES. Spirochetes are characterized by unique structures called **axial filaments**, composed of flagella, running along the cell body between a thin, flexible cell wall and an outer envelope (see Figure 25.5*c*). In other respects the spirochete cell is similar to that of other gram-negative bacteria. The cell body is a long cylinder coiled into a spiral (Figure 25.15). The flagella constituting the axial filaments begin at either end of the cell and overlap in the middle. The axial filaments are responsible for the motility of these organisms,

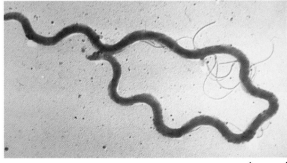

Treponema pallidum   0.8 μm

25.15 A Spirochete   This corkscrew-shaped spirochete causes syphilis in humans.

Salmonella typhosa   12 μm

25.16 Gram-Negative Rods   The cause of typhoid fever is this gram-negative rod. Recall from Figure 25.7 that the pink color is the "negative" response to the Gram stain.

and there are typical basal rings where the flagella are attached to the cell wall. Many spirochetes live in humans as parasites. Others live free in mud or water.

GRAM-NEGATIVE RODS.   Some gram-negative rods are aerobic, others are facultative anaerobes, and still others are obligate anaerobes. Some nitrogen-fixing genera such as *Rhizobium* (see Figure 34.10) are gram-negative rods, as are *Nitrobacter*, *Thiobacterium*, and other bacteria that use nitrogen or sulfur compounds instead of oxygen for respiration.

E. coli, the most studied organism, is a gram-negative rod. So, too, are many of the most famous human pathogens, such as *Yersinia pestis* (the cause of plague), *Shigella dysenteriae* (dysentery), *Vibrio cholerae* (cholera), and *Salmonella typhimurium* (a common agent of food poisoning in humans). A bacterium from this group is shown in Figure 25.16.

Certain gram-negative rods invade animal cells, where they survive and cause diseases. For example, *Yersinia pseudotuberculosis*, the agent of guinea pig plague, invades intestinal cells of guinea pigs and other mammals. Its ability to invade mammalian cells results from the possession of a single gene, called *inv*, that codes for a single large protein. Biologists using recombinant-DNA techniques have successfully transferred this gene from *Y. pseudotuberculosis* to *E. coli*, with the result that the recipient *E. coli* cells were able to invade mammalian cells, revealing the role of *inv*. Of course, these studies were carried out with great caution because of potential health hazards associated with such modified *E. coli* cells.

Most plant diseases are caused by fungi, and viruses cause others, but about 200 plant diseases are of bacterial origin. **Crown gall**, with its characteristic

25.17 Crown Gall This massive growth on the trunk of a white oak tree is crown gall, a plant disease caused by the gram-negative rod *Agrobacterium tumefaciens*.

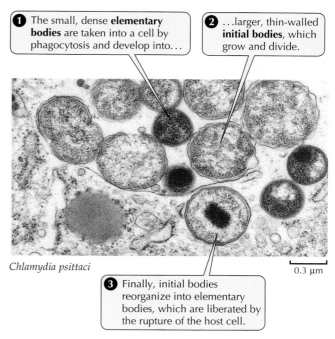

❶ The small, dense **elementary bodies** are taken into a cell by phagocytosis and develop into...

❷ ...larger, thin-walled **initial bodies**, which grow and divide.

Chlamydia psittaci

0.3 μm

❸ Finally, initial bodies reorganize into elementary bodies, which are liberated by the rupture of the host cell.

25.18 Chlamydias Change Form during Their Life Cycle Elementary bodies and initial bodies are the two major phases of the life cycle of a chlamydia.

tumors (Figure 25.17), is one of the most striking. The causal agent of crown gall is *Agrobacterium tumefaciens*, a gram-negative rod. *A. tumefaciens* harbors a plasmid containing the genes responsible for the crown gall disease. The plasmid is used in recombinant-DNA studies as a vehicle for inserting genes into new plant hosts, where they multiply along with the crown gall tumor cells (see Chapter 16).

CHLAMYDIAS. Chlamydias are among the smallest bacteria (0.2 to 1.5 μm in diameter). They can live only as parasites within the cells of other organisms. These tiny spheres are unique prokaryotes because of their

complex reproductive cycle, which involves two different types of cells (Figure 25.18). In humans, various strains of chlamydias cause eye infections (especially trachoma), sexually transmitted disease, and some forms of pneumonia.

GRAM-POSITIVE BACTERIA. Some gram-positive bacteria produce **endospores** (Figure 25.19)—heat-resistant resting structures. When nutrients become scarce, the bacterium produces an endospore. The bacterium

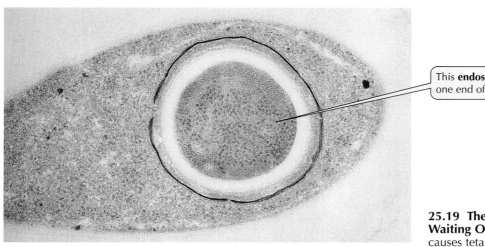

This **endospore** lies near one end of the cell.

Clostridium tetani

1.2 μm

25.19 The Endospore: A Structure for Waiting Out Bad Times The bacterium that causes tetanus produces endospores as resistant resting structures.

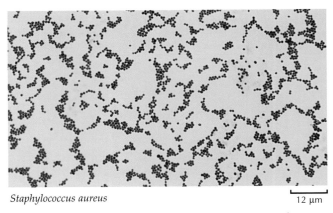

Staphylococcus aureus   12 μm

25.20 Gram-Positive Cocci "Grape clusters" are the usual arrangement of gram-positive staphylococci.

replicates its DNA and encapsulates one copy, along with some of its cytoplasm, in a tough cell wall heavily thickened with peptidoglycan and surrounded by a spore coat. The parent cell then breaks down, releasing the endospore. *This is not a reproductive process*; the endospore merely replaces the parent cell. The endospore can survive harsh environmental conditions, such as high or low temperatures or dryness, because it is dormant—its normal activity is suspended. Later, if it encounters favorable conditions, the endospore germinates; that is, it becomes metabolically active and divides, forming new cells like the parent. Some endospores apparently can germinate even after more than a thousand years of dormancy.

Members of this endospore-forming group include the many species of *Bacillus* and *Clostridium*. The toxins produced by *C. botulinum* are among the most poisonous ever discovered; the lethal dose for humans is about one-millionth of a gram (1 μg).

There are many gram-positive cocci. Cells of the genus *Staphylococcus* (Figure 25.20), called staphylococci, are abundant on the human body surface and are responsible for boils and many other skin problems. *S. aureus* is the best-known human pathogen; it is found in 20 to 40 percent of normal adults (and in 50 to 70 percent of hospitalized adults) and can cause respiratory, intestinal, and wound infections, in addition to skin diseases. Staphylococci produce toxins that are a major cause of food poisoning and that cause toxic shock syndrome.

ACTINOMYCETES. Actinomycetes develop an elaborately branched system of filaments (Figure 25.21). These bacteria closely resemble the filamentous bodies of fungi and, in fact, were once classified as fungi.

Some actinomycetes reproduce by forming chains of spores at the tips of the filaments. In the species that do not form spores, the branched, filamentous growth ceases and the structure breaks up into typical cocci or rods, which then reproduce by fission. The actino-

mycetes are part of the phylum called Gram-Positive Bacteria.

The actinomycetes include several medically important members: *Actinomyces israelii* causes infections in the oral cavity and elsewhere; *Mycobacterium tuberculosis* causes tuberculosis; and *Streptomyces* produces streptomycin, as well as hundreds of other antibiotics, including several dozen in general use. We derive most of our antibiotics from members of the actinomycetes. Why do these bacteria dedicate hundreds of genes to the production of antibiotics? We don't know for sure, but we do know that actinomycetes produce the antibiotics at the same time they produce their spores. Thus, the antibiotics may inhibit the growth of bacteria and fungi that would otherwise compete for nutrients with the germinating spores.

MYCOPLASMAS. Mycoplasmas lack cell walls, although some have a stiffening material outside their plasma membrane. Some of them are the smallest cellular creatures ever discovered—they are even smaller than chlamydias (Figure 25.22). The smallest mycoplasmas capable of growth have a diameter of about 0.2 μm, and they are small in another crucial sense: They have less than half as much DNA as do most other prokaryotes. It has been speculated that the amount of DNA in a mycoplasma, which is about the same as that in *Thermoplasma* (an archaean discussed in the next section), may be the minimum amount required to code for the essential properties of a cell.

The mycoplasmas are classified as members of the Gram-Positive Bacteria—even though the mycoplasmas lack peptidoglycan-containing cell walls. This apparent irregularity of classification resulted from the fact that the mycoplasmas are most closely allied with *Clostridium* and its relatives, all of which are Gram-Positive Bacteria. Remove the cell wall of a *Clostridium*

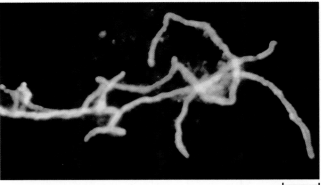

Actinomyces israelii   10 μm

25.21 Filaments of an Actinomycete Branching filaments are visualized with a fluorescent stain. This species is part of the normal flora in the human tonsils, mouth, intestinal tract, and lungs but will invade body tissues and cause severe abscesses when afforded the opportunity.

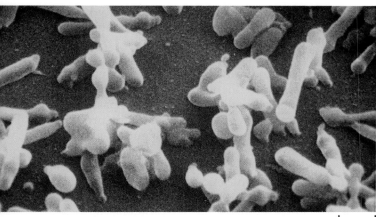

Mycoplasma gallisepticum　　　　　0.4 μm

25.22 The Tiniest Living Cells　Containing only about one-fifth as much DNA as *E. coli*, mycoplasmas are the smallest known bacteria.

and you have a mycoplasma—but the mycoplasmas never form a wall.

In the bodies of animals, mycoplasmas live on the surfaces of mucous membranes. Lacking cell walls, they cannot be killed with penicillin, which kills other bacteria by interfering with wall synthesis. Diseases caused by mycoplasmas include urinary tract infections and some forms of pneumonia; they must be treated with antibiotics other than those that act on wall synthesis.

The Archaea

The domain Archaea consists of a few prokaryotic genera that live in habitats notable for characteristics such as extreme salinity (salt content), low oxygen concentration, high temperature, or high or low pH. On the face of it, the archaea do not seem to belong together as a group. However, they do share certain characteristics.

The Archaea share some unique characteristics

Two characteristics shared by all archaea are a definitive lack of peptidoglycan in their walls and the possession of lipids of distinctive composition (see Table 25.1). The base sequences of their ribosomal RNAs confirm the close relationship. Their separation from the Bacteria and Eukarya was clarified when biologists sequenced the first archaean genome; it consisted of 1,738 genes, *more than half of which* were unlike any genes ever found in the other two domains.

The unusual lipids in the membranes of archaea deserve some discussion. They are found in all archaea, and in no bacteria or eukaryotes. Most membrane lipids of bacteria and eukaryotes contain long-chain fatty acids connected to glycerol by **ester linkages**:

$$O$$
$$-C-O-C-$$

(see also Figure 3.4). The fatty acids are straight and unbranched, and the lipids form a bilayer in the membranes of bacteria and eukaryotes.

The most distinctive feature of archaean membrane lipids is that they contain long-chain hydrocarbons connected to glycerol by **ether linkages**:

$$-C-O-C-$$

In addition, their long-chain hydrocarbons are branched. One class of these lipids contains glycerol at *both* ends of the hydrocarbons. This structure still fits in a biological membrane, as shown in Figure 25.23. In spite of the striking difference in membrane lipids, all three domains have membranes with similar overall structure, dimensions, and functions.

Archaea live in amazing places

Among the homes of archaea are extremely hot environments (sometimes also very acidic), extremely salty environments, and the guts of animals. The domain Archaea (thus the kingdom Archaebacteria) can be divided into four phyla: Hyperthermophiles, Methanogens, Extreme Halophiles, and one phylum consisting entirely of the genus *Thermoplasma*.

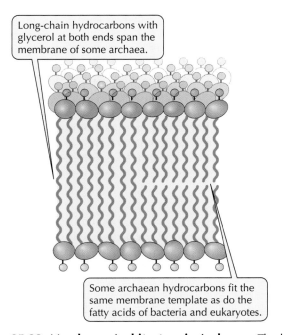

Long-chain hydrocarbons with glycerol at both ends span the membrane of some archaea.

Some archaean hydrocarbons fit the same membrane template as do the fatty acids of bacteria and eukaryotes.

25.23 Membrane Architecture in Archaea　The long-chain hydrocarbons of archaean membranes are branched and may contain glycerol at both ends. This structure still fits into a biological membrane, however; in fact, all three domains have similar membranes.

25.24 Some Would Call It Hell; These Archaea Call It Home Masses of heat- and acid-loving archaea form a salmon-pink mat inside a volcanic vent on the island of Kyushu, Japan. Sulfurous residue is visible at the edges of the archaean mat.

HYPERTHERMOPHILES. Some archaea are both thermophilic (heat-loving) and acidophilic (acid-loving). *Sulfolobus* is one hyperthermophilic genus. Organisms of this genus live in hot sulfur springs at temperatures of 70 to 75°C. They die of "cold" at 55°C (131°F). Hot sulfur springs are also extremely acidic. *Sulfolobus* grows best in the pH range from 2 to 3, but it readily tolerates pH values as low as 0.9. Acidophilic hyperthermophiles that have been tested maintain an internal pH near 7 (neutral) in spite of the acidity of their environment (see Chapter 2 for a discussion of pH). These and other hyperthermophiles thus thrive where very few other organisms can even survive (Figure 25.24).

METHANOGENS. Some species of prokaryotes, previously assigned to unrelated bacterial groups, share the property of producing methane (CH_4) by reducing carbon dioxide. All these methanogens are obligate anaerobes, and methane production is the key step in their energy metabolism. Comparison of rRNA base sequences revealed the close evolutionary relationship among all methanogens.

Methanogens release approximately 2 billion tons of methane gas into Earth's atmosphere each year, accounting for all the methane in our air, including that associated with mammalian flatulence (the passing of gas). Approximately a third of the methane production comes from methanogens in the guts of grazing herbivores such as cows.

One methanogen, *Methanopyrus*, lives on the ocean bottom near blazing volcanic vents. *Methanopyrus* can survive and grow at 110°C; it is the current record holder for temperature tolerance. It grows best at 98°C and not at all at temperatures below 84°C.

EXTREME HALOPHILES. The extreme halophiles live exclusively in very salty environments. Because they contain pink carotenoids, they can be seen easily under some circumstances (Figure 25.25). Halophiles grow in the Dead Sea and in brines of all types: Pickled fish may sometimes show reddish pink spots that are colonies of halophilic archaea. Photographs of salt flats taken from orbiting satellites show a distinct pink tinge resulting from the presence of vast numbers of *Halobacterium* and its halophilic relatives.

Few other organisms can live in the saltiest of the homes that the strict halophiles occupy; most would "dry" to death, losing too much water by osmosis to the hyperosmotic (more concentrated) environment.

25.25 Extreme Halophiles Commercial seawater evaporating ponds such as these in San Francisco Bay are attractive homes for salt-loving archaea.

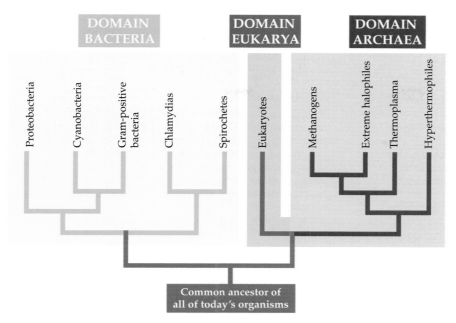

25.26 A Look Back at Two Domains and Forward to the Third   A brief summary classification of the domains Bacteria and Archaea; the many groups of Eukarya will be discussed in the next five chapters.

Strict halophiles have been found in lakes with pH values as high as 11.5—the most alkaline environment inhabited by living organisms, almost as alkaline as household ammonia.

Some of the extreme halophiles have a unique system for trapping light energy and using the energy to form ATP—without using any form of chlorophyll—but only when oxygen is in short supply. They use the pigment retinal, also found in the vertebrate eye, combined with a protein to form **bacteriorhodopsin**, which is incorporated into the plasma membrane. Retinal is purple. Clusters of bacteriorhodopsin molecules form purple patches that cover as much as half of the cell surface. When one of these molecules absorbs light, protons are pumped through the membrane out of the cell. ATP forms by a chemiosmotic mechanism of the sort described in Chapters 7 and 8.

THERMOPLASMA.   The fourth archaean phylum consists of a single genus, *Thermoplasma*. This prokaryote has no cell wall, it is thermophilic and acidophilic, its metabolism is aerobic, and it comes from coal deposits. It has the smallest genome among the archaea and perhaps the smallest genome (along with the mycoplasmas) of any free-living organisms—1,100 kilobase pairs.

The relationships of the three domains of the living world are summarized in Figure 25.26, which emphasizes the domains Bacteria and Archaea. For the rest of Part Four (Chapters 26 to 30) we'll consider the other domain: the Eukarya.

Summary of "Bacteria and Archaea: The Prokaryotic Domains"

Why Three Domains?

• Both the Archaea and the Bacteria are prokaryotic, but they differ from each other more radically than do the Archaea from the Eukarya, which constitute the rest of the living world.

• The evolutionary relationships of the three domains were first revealed by rRNA sequences. The common ancestor of all three domains lived more than 2 billion years ago, and the common ancestor of the Archaea and Eukarya at least 1.8 billion years ago. **Review Figure 25.1 and Table 25.1**

General Biology of the Prokaryotes

• The prokaryotes are the most numerous organisms on Earth, and they occupy an enormous variety of habitats.

• Most prokaryotes are cocci, bacilli, or spirals. Some link together to form associations, but none are truly multicellular. **Review Figure 25.3**

• Prokaryotes lack nuclei, membrane-enclosed organelles, and cytoskeletons. Their chromosomes are circular. They often contain plasmids. They reproduce asexually by fission.

• Many prokaryotes are motile by means of flagella, gas vesicles, or gliding mechanisms. Prokaryotic flagella rotate rather than beat.

• Prokaryotic cell walls differ from those of eukaryotes. Bacterial walls generally contain peptidoglycan. Differences in peptidoglycan content result in different reactions to the Gram stain. **Review Figure 25.7**

• Different prokaryotes have diverse metabolic pathways and nutritional modes. Prokaryotes are obligate anaerobes, facultative anaerobes, or obligate aerobes. The major nutritional types are photoautotrophs, photoheterotrophs, chemoautotrophs, and chemoheterotrophs. Some prokaryotes base their energy metabolism on nitrogen- or sulfur-containing ions. **Review Figure 25.8 and Table 25.2**

Prokaryotes in Their Environments

• Some prokaryotes play key roles in global nitrogen and sulfur cycles. Important players in the nitrogen cycle are the nitrogen fixers, nitrifiers, and denitrifiers. **Review Figure 25.10**
• Photosynthesis by cyanobacteria generated the O_2 that resulted in the aerobic environment that permitted the evolution of aerobic respiration, enabling the appearance of present-day eukaryotes.
• Many prokaryotes live in or on other organisms, with neutral, beneficial, or harmful effects.
• A small minority of bacteria are pathogens. Pathogens vary with respect to their invasiveness and toxigenicity. Some produce endotoxins, which are rarely fatal; others produce exotoxins, which tend to be highly toxic.

Prokaryote Phylogeny and Diversity

• Some classification schemes are designed to help us identify unknown organisms; others reflect evolutionary relationships. No scheme simultaneously achieves both of these goals for the prokaryotic domains.
• Phylogenetic ("natural") classification of prokaryotes is now based on rRNA sequences, with particular attention to signature sequences that are characteristic of individual groups.
• Evolution, powered by mutation and natural selection, can proceed rapidly in prokaryotes because they are haploid and can multiply rapidly.

The Bacteria

• There are far more bacteria than archaea. One phylogenetic classification of the domain Bacteria (kingdom Eubacteria) groups them into 12 phyla.
• Among bacteria, phenotypic characters often correlate poorly with phylogeny.
• All four nutritional types are observed in the Bacteria—and all four occur in the largest bacterial phylum, Proteobacteria. Metabolism in different groups of proteobacteria has evolved along different lines. **Review Figure 25.11**
• The most ancient bacteria, like the most ancient archaea, are thermophiles, suggesting that life originated in a hot environment.
• Cyanobacteria photosynthesize using the same pathways as plants use, in contrast with other bacteria. Many cyanobacteria fix nitrogen.
• Some gliding bacteria move up and down in the soil on a daily cycle. And some form fruiting bodies in response to unfavorable environmental conditions.
• Spirochetes move by means of axial filaments.
• There are many kinds of gram-negative rods, including *E. coli*; they differ dramatically from one another in metabolism and habitat.
• Chlamydias are tiny parasites that live within the cells of other organisms.
• Gram-positive bacteria are diverse; some of them produce endospores as resting structures that resist harsh conditions.
• Actinomycetes, some of which produce important antibiotics, grow as branching filaments.

• Mycoplasmas, the tiniest living things, lack conventional cell walls. They, like the archaean *Thermoplasma*, have very small genomes.
• Most of these phenotypically defined bacterial groups bear little relationship to phylogenetic groupings.

The Archaea

• Archaea have cell walls that differ from those of bacteria and eukaryotes. Their walls lack peptidoglycan, and their membrane lipids differ radically from those of bacteria and eukaryotes, containing branched long-chain hydrocarbons connected to glycerol by ether linkages. **Review Figure 25.23**
• The domain Archaea (kingdom Archaebacteria) is divided into four phyla: Hyperthermophiles, Methanogens, Extreme Halophiles, and *Thermoplasma*.
• Hyperthermophiles are heat-loving and often acid-loving archaea.
• Methanogens produce methane by reducing carbon dioxide. Some methanogens live in the guts of herbivorous animals; others occupy high-temperature environments on the ocean floor.
• Extreme halophiles are salt lovers that often lend a pinkish color to salty environments; some halophiles also grow in extremely alkaline environments.
• Archaea of the genus *Thermoplasma* lack cell walls, are thermophilic and acidophilic, and have a tiny genome (1,100 kilobase pairs).
• The three domains of life descended from a common ancestor. **Review Figure 25.26**

Self-Quiz

1. Most prokaryotes
   a. are agents of disease.
   b. lack ribosomes.
   c. evolved from the most ancient protists.
   d. lack a cell wall.
   e. are chemoheterotrophs.

2. The division of the living world into three domains
   a. is strictly arbitrary.
   b. was inspired by the morphological differences between archaea and bacteria.
   c. emphasizes the greater importance of eukaryotes.
   d. was proposed by the early microscopists.
   e. is strongly supported by data on rRNA sequences.

3. Which statement about the archaean genome is true?
   a. It is much more similar to the bacterial genome than to eukaryotic genomes.
   b. More than half of its genes are genes that are never observed in bacteria or eukaryotes.
   c. It is much smaller than the bacterial genome.
   d. It is housed in the nucleus.
   e. No archaean genome has yet been sequenced.

4. Which statement about nitrogen metabolism is *not* true?
   a. Certain prokaryotes reduce atmospheric N_2 to ammonia.
   b. Nitrifiers are soil bacteria.
   c. Denitrifiers are strict anaerobes.
   d. Nitrifiers obtain energy by oxidizing ammonia and nitrite.
   e. Without the denitrifiers, terrestrial organisms would lack a nitrogen supply.

5. All photosynthetic bacteria
   a. use chlorophyll *a* as their photosynthetic pigment.
   b. use bacteriochlorophyll as their photosynthetic pigment.
   c. release oxygen gas.
   d. produce particles of sulfur.
   e. are photoautotrophs.

6. Gram-negative bacteria
   a. appear blue to purple following Gram staining.
   b. are the most abundant of the bacterial groups.
   c. are all either rods or cocci.
   d. contain no peptidoglycan in their walls.
   e. are all photosynthetic.

7. Endospores
   a. are produced by viruses.
   b. are reproductive structures.
   c. are very delicate and easily killed.
   d. are resting structures.
   e. lack cell walls.

8. Actinomycetes
   a. are important producers of antibiotics.
   b. belong to the kingdom Fungi.
   c. are never pathogenic to humans.
   d. are gram-negative.
   e. are the smallest known bacteria.

9. Which statement about mycoplasmas is *not* true?
   a. They lack cell walls.
   b. They are the smallest known cellular organisms.
   c. They contain the same amount of DNA as do other prokaryotes.
   d. They cannot be killed with penicillin.
   e. Some are pathogens.

10. Archaea
    a. have cytoskeletons.
    b. have distinctive lipids in their plasma membranes.
    c. survive only at moderate temperatures and near neutrality.
    d. all produce methane.
    e. have substantial amounts of peptidoglycan in their cell walls.

Applying Concepts

1. Why do systematic biologists find rRNA sequence data more useful than data on metabolism or cell structure for classifying prokaryotes?

2. Differentiate among the members of the following sets of related terms:
   a. prokaryotic/eukaryotic
   b. obligate anaerobe/facultative anaerobe/obligate aerobe
   c. photoautotroph/photoheterotroph/chemoautotroph/chemoheterotroph
   d. gram-positive/gram-negative

3. For each type of organism listed below, give a single characteristic that may be used to differentiate it from the related organism(s) in parentheses.
   a. spirochetes (spiral bacteria)
   b. *Bacillus* (*Lactobacillus*)
   c. mycoplasmas (other bacteria)
   d. cyanobacteria (other photoautotrophic bacteria)

4. Until fairly recently, the cyanobacteria were called blue-green algae and were not grouped with the bacteria. Suggest several reasons for this (abandoned) tendency to separate the bacteria and cyanobacteria. Why are the cyanobacteria now grouped with the other bacteria?

5. Hyperthermophiles are of great interest to molecular biologists and biochemists. Why? What practical concerns might motivate that interest?

Readings

Balows, A., H. G. Trüper, M. Dworkin, W. Harder and K.-H. Schleifer (Eds.). 1992. *The Prokaryotes*, 2nd Edition. Four volumes. Springer-Verlag, New York. The ultimate reference on the prokaryotes, describing ecophysiology, isolation, identification, and applications.

Fischetti, V. A. 1991. "Streptococcal M Protein." *Scientific American*, June. A discussion of how rheumatic fever and strep throat bacteria evade the body's defenses.

Fredrickson, J. K. and T. C. Onstott. 1996. "Microbes Deep inside the Earth." *Scientific American*, October. An account of the discovery of bacteria and archaea by drilling. Probes the question, Are there microbes under the surface of Mars or one or more moons in the solar system?

Koch, A. L. 1990. "Growth and Form of the Bacterial Cell Wall." *American Scientist*, vol. 78, pages 327–341. An exploration into the question of how this cell wall, a single peptidoglycan molecule, allows the cell to grow but keeps it from bursting.

Losick, R. and D. Kaiser. 1997. "Why and How Bacteria Communicate." *Scientific American*, February. An examination of the important chemical means by which prokaryotes communicate with each other and with plants and animals.

Madigan, M. T. and B. L. Marrs. 1997. "Extremophiles." *Scientific American*, April. An interesting account of the archaea and their hardy enzymes.

Madigan, M. T., J. M. Martinko and J. Parker. 1997. *Brock Biology of Microorganisms*, 8th Edition. Prentice Hall, Upper Saddle River, NJ. An excellent general textbook, covering diversity, industrial microbiology, microbial ecology, and clinical and epidemiological topics.

McEvedy, C. 1988. "The Bubonic Plague." *Scientific American*, February. An account of how bubonic plague, which still exists, has shaped world history.

Shapiro, J. A. 1988. "Bacteria as Multicellular Organisms." *Scientific American*, June. A description of the behavior of highly regular bacterial colonies.

Woese, C. 1981. "Archaebacteria." *Scientific American*, June. An early treatment of the subject by its leading scholar.

Chapter 26

Protists and the Dawn of the Eukarya

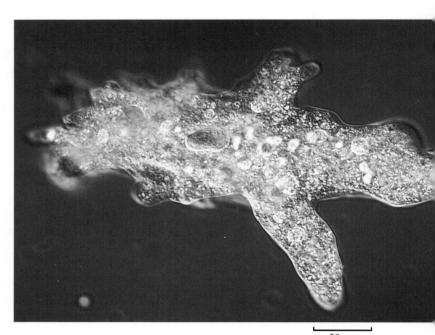

An Amoeba
Amoebas have a nucleus and several kinds of organelles; they are members of the domain Eukarya. Their flowing pseudopods are constantly changing shape as the amoeba moves and feeds.

50 µm

*A*fter their origin, prokaryotes had the living world to themselves for more than 1.5 billion years. Prokaryotes constitute two of the three domains of today's biosphere. As we saw in Chapter 25, the bacteria and the archaea differ sharply in several important ways—but neither looks much like the single-celled organism shown here. What strikes you the most about this amoeba? Probably the most obvious visible difference between it and the prokaryotes is that the amoeba has numerous compartments—membrane-enclosed organelles.

Amoebas are eukaryotic organisms: They have a nucleus enclosed by a nuclear envelope, they have several kinds of organelles, and they differ from members of the two prokaryotic domains in other important ways. They are members of the domain Eukarya. When eukaryotes appeared in the course of evolution, various members of the group experimented in many ways, resulting in a profusion of body forms. The evolution of the eukaryotes has produced great diversity, but also many cases of convergent evolution; for example, amoebalike organisms arose several times.

Look at the three eukaryotic organisms shown in Figure 26.1. Figures 26.1*a* and *b* show tiny unicellular organisms, both of which have a nucleus. The dinoflagellate is a swimming photosynthesizer. *Giardia* is a parasite that causes diarrhea and other symptoms in humans. Figure 26.1*c* shows a giant kelp—multicellular, photosynthetic, and very big, sometimes achieving lengths greater than that of a football field.

All three of these organisms belong to the Eukarya, but what kingdoms should we put them in? Is *Giardia* simply an unusually tiny animal? Are the photosynthetic organisms plants? Some biologists assign all three to a single

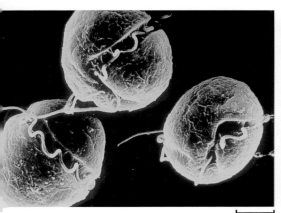

(a) *Gonyaulax* sp.

15 μm

26.1 Three Eukaryote Protists
(a) Dinoflagellates are photosynthetic unicellular algae. (b) *Giardia* is a unicellular parasite of humans. (c) Giant kelps are some of the world's longest organisms.

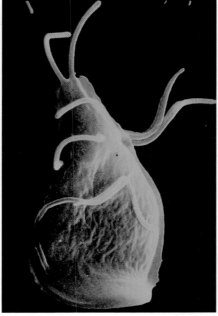

(b) *Giardia* sp.

4 μm

(c) *Macrocystis* sp.

kingdom, the Protista. Some others put them in three different kingdoms (Protista, Archezoa, and Chromista, respectively). Some consider the giant kelp (but not the microscopic dinoflagellate) a plant.

To keep the number of new terms to a minimum, and to simplify the presentation of the diversity in this fascinating group of "other eukaryotes," we will treat them as a single, paraphyletic kingdom, Protista. We hope this will be a convenient way of looking at these organisms. If you prefer to view them as representing more than one kingdom, it won't bother them or us.

The origin of the eukaryotic cell was one of the pivotal events in evolutionary history. In this chapter on the Protista we describe and celebrate the origin and early diversification of the eukaryotes and the complexity achieved in some single cells. Most of the organisms we call protists here are unicellular, but many are multicellular. This kingdom is a great evolutionary grab bag defined, for the purposes of this book, largely by exclusion: *the protists are all the eukaryotes that are not plants, fungi, or animals.* All protists are eukaryotic, and all evolved from prokaryotes.

We'll consider the origin of the eukaryotic cell, the shared characteristics of the protists, some of the diversity of protist body forms, and the relationships of certain protist groups to the other eukaryotic kingdoms.

Protista and the Other Eukaryotic Kingdoms

Part of the difficulty in placing certain eukaryotic organisms in the appropriate kingdom is a natural consequence of the fact that the other eukaryotic kingdoms have their evolutionary origin in the kingdom Protista. The other eukaryotic kingdoms—Plantae, Fungi, and Animalia—arose from protists in various ways.

Deciding just where to draw the lines between the Protista and the other eukaryotic kingdoms is difficult. Some protists, formerly classified as animals, are sometimes referred to as **protozoans**, although many biologists regard this term as inappropriate because it lumps together protist groups that are phylogenetically distant from one another. We will use the term "protozoan" for convenience in reference to certain protists, most of which ingest food by endocytosis. There are several kinds of photosynthetic protists that some biologists still refer to as **algae**. There are also some protists—the slime molds and water molds—that were once classified as fungi, and there are many others that look like nothing else on Earth.

In this book, we assign most unicellular and colonial eukaryotic organisms to the kingdom Protista. We base the separation of protists from fungi on the composition of the cell wall. Only fungi encase their absorptive (feeding) cells in a chitin-containing cell wall, and protists lack the dikaryotic phase that many fungal life cycles have. Development is what sets photosynthetic protists (algae) apart from plants as defined in this book: Whereas plants develop from embryos protected by tissues of the parent plant, algae develop from a single cell that has no such protection. The separation between protists and animals is relatively easy: An organism is an animal if it is a multicellular heterotroph with ingestive metabolism, passes through an embryonic stage called a blastula (see Chapter 40), and has an extracellular matrix containing collagen.

Tracing the natural phylogeny of the Eukarya presents some problems, and the phylogeny of protists is an area of intense, exciting research. The marvelous diversity of protist body forms and metabolic lifestyles seems reason enough for a fascination with these organisms, but questions about whether and how the multicellular eukaryotic kingdoms originated from the Protista stimulate further interest. Fortunately, the tools of molecular biology—rRNA sequencing in particular—make it possible to explore evolutionary relationships in greater detail and with greater confidence than previously (see Chapters 23 and 25).

The Origin of the Eukaryotic Cell

The eukaryotic cell differs in many ways from prokaryotic cells. How did it originate? Given the nature of evolutionary processes, the differences cannot all have arisen simultaneously. We think we can make some reasonable guesses about the sequence of events, bearing in mind that the global environment underwent an enormous change—from anaerobic to aerobic—during the course of these events. As you read this chapter, keep in mind that the steps we suggest are just that: guesses. This version of the story is one of a few under current consideration. We present it as a framework for thinking about this challenging problem, *not* as a set of facts.

The modern eukaryotic cell arose in several steps

The essential steps in the origin of the eukaryotic cell include

1. The origin of a flexible cell surface
2. The origin of a nuclear envelope
3. The appearance of digestive vesicles
4. The origin of a cytoskeleton
5. The endosymbiotic acquisition of certain organelles

WHAT A FLEXIBLE CELL SURFACE ALLOWS. Most present-day prokaryotic cells have firm cell walls. The first step toward the eukaryotic condition may have been the loss of the cell wall by an ancestral cell. This may not seem like an obvious first step, but consider the possibilities open to a flexible cell without a wall.

First, think of cell size. As a cell grows, its surface area-to-volume ratio decreases (see Chapter 4). Unless the surface is flexible and can fold inward and elaborate itself, creating more surface area for gas and nutrient exchange (Figure 26.2), the cell volume will reach an upper limit. With a surface flexible enough to allow infolding, the cell can exchange materials with its environment rapidly enough to sustain a larger volume and more rapid metabolism. Further, a flexible surface may pinch off bits of the environment, bringing them into the cell by endocytosis as compartments in which digestion may occur (Figure 26.3).

Also recall that the chromosome of a prokaryotic cell is attached to a site on its plasma membrane (see

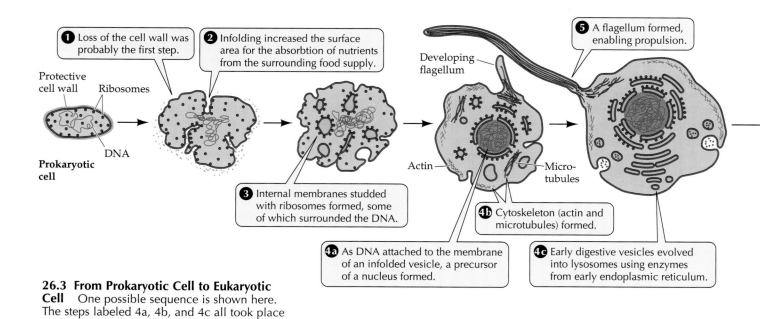

1 Loss of the cell wall was probably the first step.

2 Infolding increased the surface area for the absorbtion of nutrients from the surrounding food supply.

Protective cell wall — Ribosomes

Prokaryotic cell — DNA

3 Internal membranes studded with ribosomes formed, some of which surrounded the DNA.

4a As DNA attached to the membrane of an infolded vesicle, a precursor of a nucleus formed.

4b Cytoskeleton (actin and microtubules) formed.

4c Early digestive vesicles evolved into lysosomes using enzymes from early endoplasmic reticulum.

5 A flagellum formed, enabling propulsion.

Developing flagellum

Actin — Microtubules

26.3 From Prokaryotic Cell to Eukaryotic Cell One possible sequence is shown here. The steps labeled 4a, 4b, and 4c all took place at the same time.

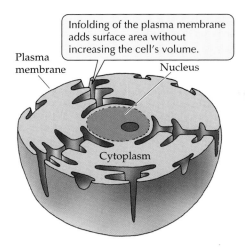

26.2 Membrane Infolding   The loss of the rigid prokaryotic cell wall meant the newly flexible cell could elaborate inward and create more surface area.

Figure 25.4). If that region of the plasma membrane were to fold into the cell, the first step would be taken toward the evolution of a nucleus, the key feature of the eukaryotic cell.

CHANGES IN CELL STRUCTURE AND FUNCTION.   Early steps in the evolution of the eukaryotic cell are likely to have included three advances: the formation of ribosome-studded internal membranes, some of which surrounded the DNA (steps 3 and 4a in Figure 26.3); the appearance of actin fibers and microtubules—a cytoskeleton to manage changes in shape and to move materials from one part of the now much larger cell to

other parts (step 4b); and the evolution of the early digestive vesicles into lysosomes (step 4c).

From this intermediate kind of cell, the next advance was probably to a truly eukaryotic cell that we could call a phagocyte—a motile cell that could prey on other cells by engulfing and digesting them. The first true eukaryote possessed a nuclear envelope and an associated endoplasmic reticulum and Golgi apparatus, and perhaps one or more flagella of the eukaryotic type. Notice how much of the progress to this point was made possible by the loss of the wall and the elaboration of what was originally the plasma membrane.

ENDOSYMBIOSIS AND ORGANELLES.   During the processes already outlined, the cyanobacteria were very busy, generating oxygen gas as a product of photosynthesis. The increasing O_2 levels in the atmosphere had disastrous consequences for most other living things, because most living things of the time (archaea and bacteria) were unable to tolerate the newly aerobic, oxidizing environment. But some prokaryotes managed to cope, and—fortunately for us—so did some of the ancient phagocytes.

According to one hypothesis, the key to the survival of early eukaryotes was the ingestion and incorporation of a prokaryote that became symbiotic within the phagocytes and evolved into the peroxisomes of today (step 6 in Figure 26.3). These organelles were able to disarm the toxic products of oxygen action, such as hydrogen peroxide. This association may have been the first important endosymbiosis in the evolution of the eukaryotic cell.

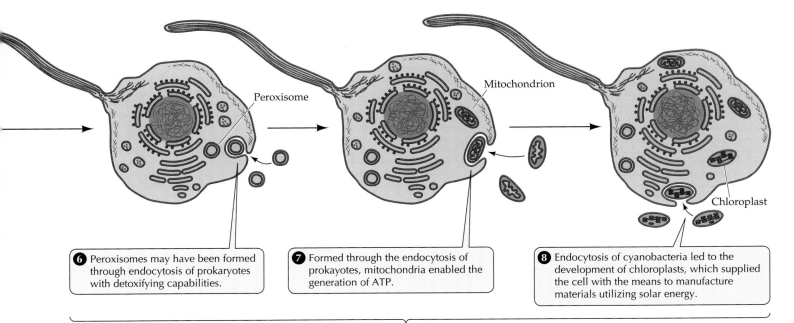

6 Peroxisomes may have been formed through endocytosis of prokaryotes with detoxifying capabilities.

7 Formed through the endocytosis of prokayotes, mitochondria enabled the generation of ATP.

8 Endocytosis of cyanobacteria led to the development of chloroplasts, which supplied the cell with the means to manufacture materials utilizing solar energy.

Endosymbiosis

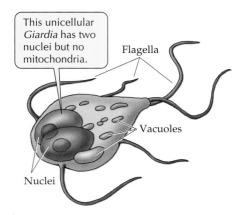

This unicellular *Giardia* has two nuclei but no mitochondria.

Flagella

Vacuoles

Nuclei

26.4 *Giardia*, a Protist without Mitochondria Present-day organisms such as this lend support to the hypothesis that the nucleus appeared before mitochondria in the evolution of eukaryotes.

Another key endosymbiotic event was the incorporation of a prokaryote, related to the proteobacteria, that was the precursor of mitochondria (step 7 in Figure 26.3). On completion of this step, the basic modern eukaryotic cell was complete. Some very important eukaryotes are the result of yet another endosymbiotic step, the incorporation of prokaryotes, related to today's cyanobacteria, that are now chloroplasts (step 8).

Reviewing the likely origin of eukaryotic cells, endosymbiosis played multiple roles, and elaboration of the plasma membrane resulted in various advances.

"Archezoans" resemble a proposed intermediate stage

The hypothesis that the eukaryotic nucleus evolved before the mitochondrion gains support from the existence of a polyphyletic group of protists sometimes called archezoans, consisting of a few organisms such as *Giardia*. *Giardia lamblia* is a familiar parasite that contaminates water supplies and causes the intestinal disease giardiasis (Figure 26.4). This tiny organism has no mitochondria, chloroplasts, or other membrane-enclosed organelles, but it contains a nucleus bounded by a nuclear envelope, and it has a cytoskeleton. We view *Giardia* as a modern descendant of a transitional stage in the evolution of the eukaryotes.

General Biology of the Protists

Most protists are aquatic. Some live in marine environments, others in fresh water, and still others in the body fluids of other organisms. The slime molds inhabit damp soil and the moist, decaying bark of rotting trees. Many other protists also live in soil water, some of them contributing to the global nitrogen cycle by preying on soil bacteria and recycling their nitrogen compounds to nitrates.

Protists are strikingly diverse in their metabolism, perhaps second only to the prokaryotes. Nutritionally, some are autotrophs, some absorptive heterotrophs, and others ingestive heterotrophs. Some switch with ease between the autotrophic and heterotrophic modes of nutrition.

Three protist phyla consist entirely of nonmotile organisms, but all the other phyla include cells that move by amoeboid motion, by ciliary action, or by means of flagella. Most unicellular protists are tiny, but the multicellular plantlike protists include the giant kelps (see Figure 26.1c), which live in the oceans and are among the longest organisms in existence.

Vesicles perform a variety of functions

Unicellular organisms tend to be of microscopic size. An important reason that cells are small is that they need enough membrane surface area in relation to their volume to support the exchange of materials required for them to live (see Figures 4.2 and 26.2). The size that unicellular protists can achieve is limited by their surface area–to–volume ratio. Many relatively large unicellular protists minimize this problem by having membrane-enclosed vesicles of various types that increase their effective surface area.

For example, many freshwater protists address their osmotic problems by using vesicles that contract to excrete excess water. Members of several of the protist phyla have such **contractile vacuoles**, which help them cope with their hypoosmotic environments.

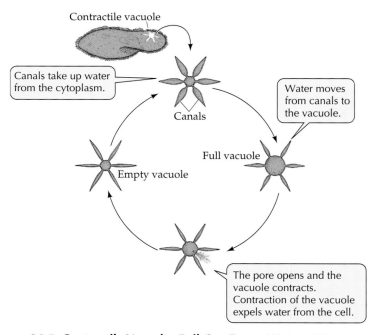

Contractile vacuole

Canals take up water from the cytoplasm.

Canals

Water moves from canals to the vacuole.

Full vacuole

Empty vacuole

The pore opens and the vacuole contracts. Contraction of the vacuole expels water from the cell.

26.5 Contractile Vacuoles Bail Out Excess Water Water constantly enters the cell by osmosis. A pore in the cell surface allows the contractile vacuole to expel the water it accumulates.

Because these organisms have a more negative water potential than their freshwater environment does, they constantly take in water by osmosis. Excess water collects in the contractile vacuole and is then pushed out (Figure 26.5).

A beautifully simple experiment confirms that bailing out water is the principal function of the contractile vacuole. First we observe cells under the microscope and note the rate at which the vacuoles are contracting—they look like little eyes winking. Then we place other cells of the same type in solutions of differing osmotic potential. The less negative the osmotic potential of the surrounding solution, the more hyperosmotic the cells are and the faster the water rushes into them, causing the contractile vacuoles to pump more rapidly. Conversely, the contractile vacuoles will stop pumping if the solute concentration of the medium is increased so that it is isosmotic with the cells.

A second important type of vesicle found in many protists is the **food vacuole**. Protists such as *Paramecium* engulf solid food, forming food vacuoles within which the food is digested (Figure 26.6). Smaller vesicles containing digested food pinch away from the food vesicle and enter the organism's cytoplasm. These tiny vesicles provide a large surface area across which the products of digestion may be absorbed by the rest of the cell.

The cell surfaces of protists are diverse

A few protists, such as some amoebas, are surrounded by only a plasma membrane, but most have stiffer surfaces that maintain the structural integrity of the cell. Many algae and other protists have cell walls, which are often complex in structure.

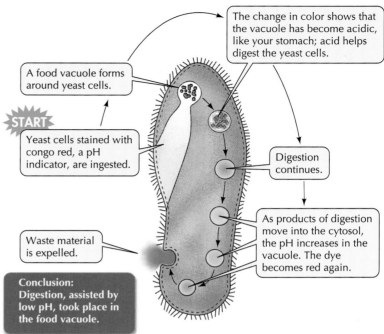

The change in color shows that the vacuole has become acidic, like your stomach; acid helps digest the yeast cells.

A food vacuole forms around yeast cells.

START

Yeast cells stained with congo red, a pH indicator, are ingested.

Digestion continues.

As products of digestion move into the cytosol, the pH increases in the vacuole. The dye becomes red again.

Waste material is expelled.

Conclusion: Digestion, assisted by low pH, took place in the food vacuole.

26.6 Food Vacuoles Handle Digestion and Excretion An experiment with *Paramecium* demonstrates the function of food vacuoles.

The protozoans, lacking cell walls, have a variety of ways of strengthening their surface. Some have internal "shells," which either the organism itself produces, as foraminiferans do, or is made of bits of sand and thickenings immediately beneath the plasma membrane, as in some amoebas (Figure 26.7).

26.7 Diversity in Protozoan Cell Surfaces (*a*) Foraminiferan shells are made of protein hardened with calcium carbonate. (*b*) This shelled amoeba constructed its shell by cementing sand grains together. (*c*) Spirals of protein make this *Paramecium*'s surface—known as its pellicle—flexible but resilient.

(*a*)

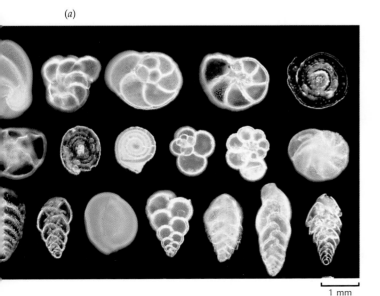

1 mm

(*b*) *Difflugia* sp.

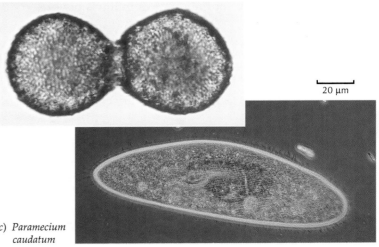

20 μm

(*c*) *Paramecium caudatum*

26.8 Protists within Protists Photosynthetic algae living as endosymbionts within these radiolarians provide food for the radiolarians, as well as part of the pigmentation seen through the glassy skeletons. Both the algae and the radiolarians are protists.

Many protists contain endosymbionts

In Chapter 4 we introduced the concept of *endosymbiosis* (organisms living together, one inside the other). Endosymbiosis is very common among the protists, and in some instances both the host and the endosymbiont are protists. Many radiolarians, for example, harbor photosynthetic protists as endosymbionts (Figure 26.8). As a result, these radiolarians appear greenish or yellowish, depending on the type of endosymbiont they contain.

This arrangement is beneficial to the radiolarian, for it can make use of the food produced by its photosynthesizing guest. The guest, in turn, may make use of metabolites made by the host, or it may simply receive physical protection. Alternatively, the guest may be a victim, exploited for its photosynthetic products while receiving no benefit itself.

Endosymbiosis is important in the lives of many protists. This and other phenomena have contributed to the great success of the kingdom Protista, a kingdom that flourished for hundreds of millions of years before the first multicellular species evolved. Another source of this extraordinary success of the protists is the remarkable diversity of their sexual and asexual reproductive strategies.

Both asexual and sexual reproduction occur in the kingdom Protista

Although most protists indulge in both asexual and sexual reproduction, some groups lack sexual reproduction. As we will see, some asexually reproducing protists do engage in genetic recombination, even though it does not relate directly to reproduction.

Asexual reproductive processes in the kingdom Protista include binary fission (simple splitting of the cell), multiple fission, budding (the outgrowth of a new cell from the surface of an old one), and the formation of spores. Sexual reproduction also takes various forms. In some protists, as in animals, the gametes are the only haploid cells. In many algae, by contrast, both diploid and haploid cells undergo mitosis, giving rise to an *alternation of generations* (which will be described later in the chapter; see Figure 26.26).

The diversity of form, habitat, metabolism, locomotion, reproduction, and life cycles of protists reflects the diversity of avenues pursued in the early evolution of eukaryotes. Many of these avenues led to great success, judging from the abundance and diversity of today's protists and other eukaryotes.

Protozoan Diversity

All protozoans are unicellular. Most ingest their food by endocytosis. Their diversity, which includes many of the most abundant or commonly observed protist types, is summarized in Table 26.1. In this and subsequent sections of this chapter, please be aware that our goal is to give you a clear sense of the *diversity* of protists, more than it is to reflect taxonomic groupings. Explorations of protist phylogeny are fascinating and rewarding, but they are not our primary concern here.

We will identify several groups of organisms as phyla, but other groupings, less natural, will not be

TABLE 26.1 Some Groups of Protozoans

COMMON NAME	PHYLUM	FORM	LOCOMOTION	EXAMPLES
Flagellates	(Several)	Unicellular, some colonial	One or more flagella	*Trypanosoma, Euglena,* Choanoflagellida
Amoebas	(Several)	Unicellular, no definite shape	Pseudopods	*Amoeba, Entamoeba, Chaos*
Actinopods	Actinopoda	Unicellular	Pseudopods	Radiolarians, heliozoans
Foraminiferans	Foraminifera	Unicellular	Pseudopods	Foraminiferans
Apicomplexans	Apicomplexa	Unicellular	None	*Plasmodium*
Ciliates	Ciliophora	Unicellular	Cilia	*Paramecium, Blepharisma, Vorticella*

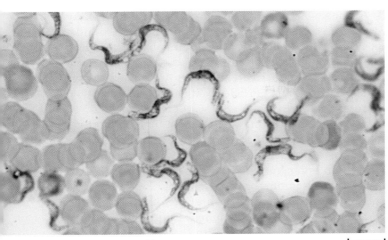

Trypanosoma gambiense   25 µm

26.9 A Parasitic Flagellate   Trypanosomes cause sleeping sickness in mammals. A flagellum runs along one edge of the cell as part of a structure called the undulating membrane. The other cells in the micrograph are mammalian red blood cells.

given the title of phylum. For example, certain body plans, such as those of amoebas and those of protists with flagella, arose again and again during evolution, in groups only distantly related to one another. We will emphasize diversity of body plans over assignment of species to phyla.

In this section on protozoan diversity, we'll describe euglenoids and other flagellates, amoebas, actinopods, foraminiferans, apicomplexans, and ciliates.

Many protists have flagella

Many protists—thousands of species—possess one or more flagella and hence are called flagellates. Flagellates do not constitute a single phylum. There are many flagellate groups, and they appear in different parts of the kingdom Protista. Flagellates reproduce vegetatively by binary fission (mitosis and cytokinesis)—the simplest and most direct way.

Some free-living flagellate species survive by preying on other protists. An impressive variety of other flagellates live as internal parasites on animals, including humans (Figure 26.9). Within the guts of certain wood-eating roaches and termites live an array of huge flagellates that possess some of the most bizarre and complicated body forms found anywhere among the protists.

One group of flagellates, the Choanoflagellida, is thought to comprise the closest relatives of the animals. Members of this group are especially closely related to the sponges, the most ancient of the surviving phyla of animals (see Chapter 29). Sponges are colonial, rather than truly multicellular. That is, sponges lack organized tissues, and their cells can be separated

and recombined. The Choanoflagellida bear a striking resemblance to the most characteristic type of cell found in the sponges (see Figure 29.5).

Some flagellates are human pathogens. Sleeping sickness, one of the most dreaded diseases of Africa, is caused by the parasitic flagellate *Trypanosoma* (see Figure 26.9). The vector (intermediate host) for sleeping sickness is an insect, the tsetse fly. Carrying its deadly cargo, the tsetse fly bites livestock, wild animals, and even humans, infecting all of them with *Trypanosoma*. *Trypanosoma* then multiplies in the mammalian bloodstream and produces toxic substances. When these parasites invade the nervous system, the neurological symptoms of sleeping sickness appear— and are followed by death. Another disease-causing flagellate is *Trichomonas vaginalis*, which causes a common but usually mild sexually transmitted disease.

Some euglenoids are photosynthetic

The 800 species of euglenoids* used to be claimed by the zoologists as animals and by the botanists as plants. They are unicellular flagellates, but many members of the group photosynthesize, as do the algae.

Figure 26.10 depicts a cell of the genus *Euglena*. Like most other members of the group, this common freshwater organism has a complex cell plan. It propels itself through the water with one of its two flagella, which may also serve as an anchor to hold the organism in place. The flagellum provides power by means of a wavy motion that spreads from base to tip. The second flagellum is often rudimentary. *Euglena* reproduces vegetatively by mitosis and cytokinesis.

---

* You may be wondering why we capitalized "Choanoflagellida" in the previous section but not "euglenoids" here. The official names of phyla and other taxonomic groups are capitalized; less formal names are not. Euglenoids belong to more than one phylum.

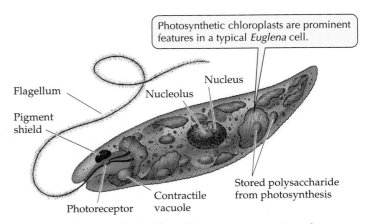

Photosynthetic chloroplasts are prominent features in a typical *Euglena* cell.

Flagellum

Pigment shield

Photoreceptor

Contractile vacuole

Nucleus

Nucleolus

Stored polysaccharide from photosynthesis

26.10 A Photosynthetic Flagellate   Several species of *Euglena* are among the best-known flagellates.

Euglena has very flexible nutritional requirements. In sunlight it is fully autotrophic, using its chloroplasts to synthesize organic compounds through photosynthesis. When kept in the dark, the organism loses its photosynthetic pigment and begins to feed exclusively on dead organic material floating in the water around it. Such a "bleached" cell of *Euglena* resynthesizes its photosynthetic pigment when it is returned to the light and hence becomes autotrophic again.

Euglena cells treated with certain antibiotics or mutagens lose their photosynthetic pigment completely; neither they nor their descendants are ever autotrophs again. Those descendants, however, function well as heterotrophs. We believe that a photosynthetic prokaryote became an endosymbiont inside an ancestor of the photosynthetic euglenoids, endowing its host with the ability to photosynthesize, in much the same way that chloroplasts became integral parts of the cells of green algae and plants.

Amoebas form pseudopods

Amoebas are protists that form **pseudopods**, extensions of their constantly changing body mass (see Figure 4.22*b* and the photo at the beginning of this chapter). As we have mentioned, the amoebas do not constitute a single, coherent group of organisms; rather, this body plan has appeared by convergent evolution in various groups.

Amoebas have often been portrayed in popular writing as simple blobs—the simplest form of "animal" life imaginable. Superficial examination of a typical amoeba shows how such an impression might have been obtained. An amoeba consists of a single cell. It feeds on small organisms and particles of organic matter by phagocytosis, engulfing them with its pseudopods. Particles of food are sealed off in food vacuoles within the cytoplasm of the amoeba. The material is then slowly digested and assimilated into the main body of the organism. Pseudopods are also the organs of locomotion. (The mechanism of amoeboid motion will be discussed in Chapter 44.)

Amoebas are specialized forms of protists. Many are adapted for life on the bottoms of lakes, ponds, and other bodies of water. Their creeping locomotion and their manner of engulfing food particles fit them for life close to a relatively rich supply of sedentary organisms or organic particles. Other amoebas are even more specialized.

All amoebas are animal-like, existing as predators, parasites, or scavengers. None are photosynthetic. Amoebas of the free-living genus *Naegleria*, some of which can enter humans and cause a fatal disease of the nervous system, have a two-stage life cycle, one stage having amoeboid cells and the other flagellated cells. Some amoebas are shelled, living in casings of sand grains glued together or in shells secreted by the organism itself (see Figure 26.7*b*).

Actinopods have thin, stiff pseudopods

The actinopods are recognizable by their thin, stiff pseudopods, which are reinforced by microtubules. The pseudopods play at least four roles: (1) They greatly increase the surface area of the cell for exchange of materials with the environment; (2) they help the cell float in its marine or freshwater environment; (3) they provide locomotion in heliozoans, a group of actinopods that roll along the substrate by shortening and elongating their pseudopods; and (4) they are the cell's feeding organs; the pseudopods trap smaller organisms, often taking up prey by endocytosis and transporting it to the main cell body.

Radiolarians, actinopods that are exclusively marine, are perhaps the most beautiful of all microorganisms (Figure 26.11*a*). Almost all radiolarian species secrete glassy endoskeletons (internal skeletons) from which needlelike pseudopods project. Part of the skeleton is a central capsule within the cytoplasm. The skeletons of the different species are as varied as

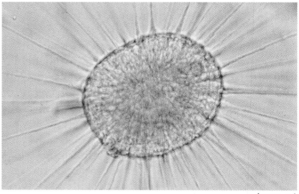

(*b*) *Actinosphaerium eichorni*   150 μm

26.11 Actinopods (*a*) A radiolarian displays its glassy skeleton of delicate intricacy. (*b*) A heliozoan with long pseudopods.

(*a*) Radiolarian (species not identified)

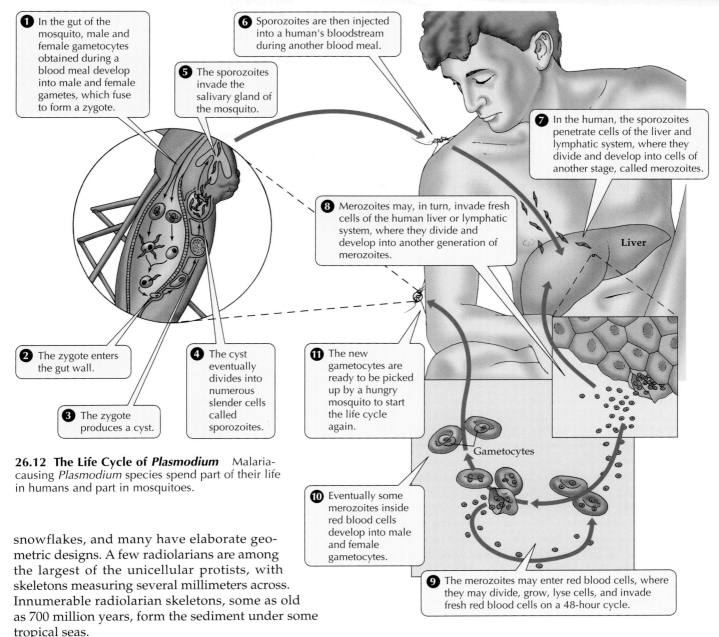

1 In the gut of the mosquito, male and female gametocytes obtained during a blood meal develop into male and female gametes, which fuse to form a zygote.

5 The sporozoites invade the salivary gland of the mosquito.

6 Sporozoites are then injected into a human's bloodstream during another blood meal.

7 In the human, the sporozoites penetrate cells of the liver and lymphatic system, where they divide and develop into cells of another stage, called merozoites.

8 Merozoites may, in turn, invade fresh cells of the human liver or lymphatic system, where they divide and develop into another generation of merozoites.

2 The zygote enters the gut wall.

4 The cyst eventually divides into numerous slender cells called sporozoites.

11 The new gametocytes are ready to be picked up by a hungry mosquito to start the life cycle again.

3 The zygote produces a cyst.

10 Eventually some merozoites inside red blood cells develop into male and female gametocytes.

9 The merozoites may enter red blood cells, where they may divide, grow, lyse cells, and invade fresh red blood cells on a 48-hour cycle.

Liver

Gametocytes

26.12 The Life Cycle of *Plasmodium* Malaria-causing *Plasmodium* species spend part of their life in humans and part in mosquitoes.

snowflakes, and many have elaborate geometric designs. A few radiolarians are among the largest of the unicellular protists, with skeletons measuring several millimeters across. Innumerable radiolarian skeletons, some as old as 700 million years, form the sediment under some tropical seas.

Heliozoans are actinopods that lack an endoskeleton (Figure 26.11*b*). Most heliozoans live in fresh water.

Apicomplexans are parasites with unusual spores

Exclusively parasitic protozoans, the apicomplexans are thus named because the apical end of their spore contains a mass of organelles. These organelles help the apicomplexan spore invade its host tissue. Unlike many other protists, apicomplexans lack contractile vacuoles. Because their rigid cell walls limit expansion, they do not take in excess water.

Apicomplexans generally have an indefinite body form like that of an amoeba. This body form has evolved over and over again in parasitic protists. The form has appeared, for example, even in parasitic dinoflagellates, a group of algae whose nonparasitic relatives have highly distinctive, regular body forms. Like many animal obligate parasites, apicomplexans

have elaborate life cycles featuring asexual and sexual reproduction by a series of very dissimilar life stages. Often these stages are associated with two types of host organisms.

The best-known apicomplexans are the malaria parasites of the genus *Plasmodium*, a highly specialized group of organisms that spend part of their life cycle within human red blood cells (Figure 26.12). Malaria continues to be a major problem in some tropical countries, although it has been almost eliminated from the United States; indeed, in terms of number of people infected, malaria is one of the world's most serious diseases.

Female mosquitoes of the genus *Anopheles* transmit *Plasmodium* to humans. *Plasmodium* enters the human circulatory system when an infected *Anopheles* mosquito penetrates the human skin in search of blood.

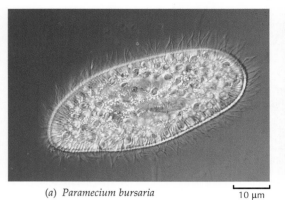

(a) Paramecium bursaria   ⊢─────┤ 10 µm

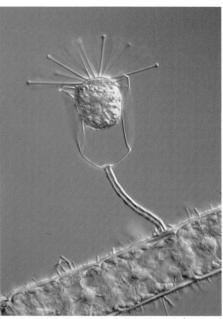

(c) Paracineta sp.   ⊢─────┤ 20 µm

26.13 Diversity in the Ciliates   (*a*) A free-swimming organism, this parame-cium belongs to the ciliate subgroup called holotrichs, which have many cilia of uniform length. (*b*) Members of the peritrich subgroup of Ciliophora have cilia on their mouthparts. (*c*) Tentacles replace cilia as suctorians (another subgroup) develop. (*d*) This ciliate "walks" on fused cilia called cirri that project from its body. Other cilia are fused into flat sheets that sweep in food particles; this individual has just fed on green algae.

(b) Epistylis sp.   ⊢─────┤ 60 µm

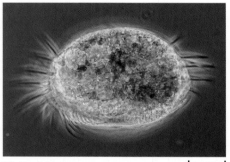

(d) Euplotes sp.   ⊢─────┤ 25 µm

The parasites find their way to cells in the liver and the lymphatic system, change their form, multiply, and reenter the bloodstream, attacking red blood cells. The attackers multiply in a red blood cell for approximately 2 days, producing as many as 36 new *Plasmodium* cells each. The victimized cell then bursts, releasing a new swarm of parasites to attack other red blood cells.

If another *Anopheles* bites the victim, some of the parasitic *Plasmodium* cells are taken into the mosquito along with the blood, thus infecting the mosquito. The infecting cells develop into gametes, which unite to form zygotes that lodge in the mosquito's gut, repro-duce, and move into the salivary glands, from which they can be passed to another human host.

Plasmodium is an intracellular parasite in the human host and an extracellular parasite in the mosquito. The *Plasmodium* life cycle that spreads malaria is best bro-ken by the removal of stagnant water, in which mos-quitoes breed. The use of insecticides to reduce the *Anopheles* population can be effective, but one must weigh possible ecological, economic, and health risks posed by the insecticides themselves.

Malaria kills more than a million people each year, and *Plasmodium* has proved to be a singularly difficult pathogen to attack. However, there is now new hope in the form of a genome-sequencing project that tar-gets *Plasmodium falciparum*. Scheduled to be completed by the year 2002, that project may provide the infor-mation needed to end this epidemic.

Foraminiferans have created vast limestone deposits

Foraminiferans are marine creatures—some floating as plankton and many others living at the bottom of the

sea—that secrete shells of calcium carbonate (see Figure 26.7*a*). Their long, threadlike, branched pseudo-pods reach out through numerous microscopic pores in the shell and interconnect to create a sticky net, which the foraminiferan uses to catch smaller plank-ton (free-floating microscopic organisms).

After foraminiferans reproduce—by mitosis and cy-tokinesis—the daughter cells abandon the parent shell and make new shells of their own. The discarded skeletons of ancient foraminiferans make up extensive limestone deposits in various parts of the world, form-ing a covering hundreds to thousands of meters deep over millions of square kilometers of ocean bottom. Foraminiferan skeletons also make up the sand of some beaches. A single gram of such sand may contain as many as 50,000 foraminiferan shells.

The shells of individual foraminiferan species have distinctive shapes and are easily preserved as fossils in marine sediments. Each geologic period has distinc-tive foraminiferan species. For this reason, and because they are so abundant, the remains of foraminiferans are especially valuable as indicators in the classification and dating of sedimentary rocks, as well as in oil prospecting.

Ciliates have two types of nuclei

Because members of the phylum Ciliophora characteristically have hairlike cilia, they have the common name ciliates. This protozoan group ranks with the flagellates in diversity and ecological importance (Figure 26.13). Almost all ciliates are heterotrophic (a few contain photosynthetic endosymbionts), and they are much more specialized in body form than are most flagellates and other protists. Ciliates are also characterized by the possession of two types of nuclei, a large **macronucleus** and, within the same cell, from 1 to as many as 80 **micronuclei**.

The micronuclei, which are typical eukaryotic nuclei, are essential for genetic recombination. The macronucleus contains many copies of the genetic information, packaged in units containing very few genes each; the macronuclear DNA is transcribed and translated to control the life of the cell. Although we do not know how this system of macro- and micronuclei came into being, we know something about the behavior of these nuclei, which we will discuss after describing the body plan of one important ciliate, *Paramecium*.

A CLOSER LOOK AT ONE CILIATE. *Paramecium*, a frequently studied ciliate genus, exemplifies the complex structure and behavior of ciliates (Figure 26.14*a*). The slipper-shaped cell is covered by an elaborate **pellicle**, a structure composed principally of an outer membrane and an inner layer of closely packed, membrane-enclosed sacs (called alveoli) that surround the bases of the cilia. Defensive organelles called trichocysts are also present in the pellicle. A microscopic explosion expels the trichocysts in a few milliseconds, and they emerge as sharp darts, driven forward at the tip of a long, expanding shaft (Figure 26.14*b*).

The cilia provide a form of locomotion that is generally more precise than that made possible by flagella or pseudopods. A paramecium can direct the beat of its cilia to propel itself either forward or backward in a spiraling manner (Figure 26.15). A paramecium can

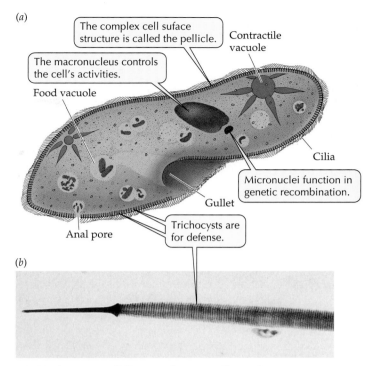

(a)

The complex cell suface structure is called the pellicle.

Contractile vacuole

The macronucleus controls the cell's activities.

Food vacuole

Cilia

Micronuclei function in genetic recombination.

Gullet

Anal pore

Trichocysts are for defense.

(b)

26.14 Anatomy of *Paramecium* (*a*) The major structures of a typical paramecium. (*b*) A trichocyst discharged from beneath the pellicle of a paramecium has a sharp point and a straight filament.

back off swiftly when it encounters a barrier or a negative stimulus. Some of these large ciliates hold the speed record for the kingdom Protista—faster than 2 mm/s. The coordination of ciliary beating is probably the result of a differential distribution of calcium and other ion channels near the two ends of the cell.

REPRODUCTION WITHOUT SEX, AND SEX WITHOUT REPRODUCTION. Paramecia reproduce by cell division. The micronuclei divide mitotically; the macronucleus simply pinches apart to yield two daughter macronuclei. Paramecia also have an elaborate sexual behavior called **conjugation** (Figure 26.16). Two paramecia line

26.15 "Swimming" with Cilia Beating its cilia in coordinated waves that progress from one end of the cell to the other, a paramecium can move in either direction with respect to the long axis of the cell; this one is moving from left to right. The cell also rotates in a spiral as it travels.

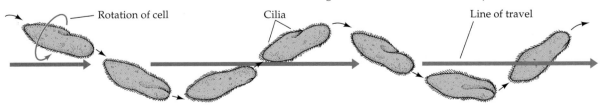

Rotation of cell

Cilia

Line of travel

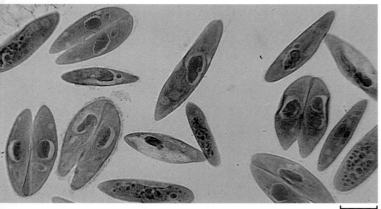

40 μm

26.16 Paramecia Achieve Genetic Recombination by Conjugating Conjugating *Paramecium* individuals exchange micronuclei, thereby permitting genetic recombination. After conjugation, the cells separate and continue their lives as two individuals.

Micronucleus

Macronucleus

Two paramecia conjugate; all but one micronucleus in each cell disintegrate. The remaining micronucleus undergoes meiosis.

Three of the four haploid micronuclei disintegrate; the remaining micronucleus undergoes mitosis.

Each conjugating paramecium donates a micronucleus to the other. The macronuclei disintegrate.

The (genetically) different micronuclei in each cell fuse.

The new diploid nuclei divide mitotically, eventually giving rise to a macronucleus and the appropriate number of micronuclei.

up tightly against each other and fuse in the oral region of the body. Nuclear material is extensively reorganized and exchanged during the next several hours. The reorganization includes both meiosis and the fusion of gametic nuclei from the two conjugating partners. The exchange of gametic nuclei is fully reciprocal—each of the two paramecia gives and receives an equal amount of DNA. Afterward the two organisms separate and go their own ways, each equipped with new combinations of alleles by the genetic recombination that occurred during conjugation.

Conjugation in *Paramecium* is a *sexual* process of genetic recombination, but it is not a *reproductive* process. The same two cells that begin the process are there at the end, and no new cells are created. As a rule, each clone of paramecia must periodically conjugate. Laborious experimentation has shown that if some species are not permitted to conjugate, the asexual clones can live through no more than approximately 350 cell divisions before they die out.

CYTOPLASMIC ORGANIZATION IN THE CILIATES. Most ciliates possess all the traits of *Paramecium*. Some, however, are notable for the exceptional degree of development of their individual organelle systems. Certain ciliates, for example, have the equivalent of legs. Fused cilia called cirri move in an independent, but coordinated, fashion and enable the organism to "walk" over surfaces (see Figure 26.13d). Cytoskeletal elements leading to individual cirri assist this locomotion. The coordination is lost if these structures are experimentally cut.

Many types of ciliates possess **myonemes**, muscle-like fibers within the cytoplasm. The contraction of myonemes causes a rapid retraction of the stalk in ciliates such as *Vorticella* (see Figure 26.13b) when the organism is disturbed. What may be the ultimate cytoplasmic organization is displayed by the highly specialized ciliates that live in the digestive tracts of cows and many other hoofed mammals. These ciliates possess not only myonemes and elaborately fused cilia, but also a cytoplasmic "skeleton" and a "cellular gut" complete with a "mouth," an "esophagus," and an "anus" (Figure 26.17).

When examining the intricate structures of these ciliates and many other protists, we must pause and remember that we are looking at only one cell. Structural complexity in multicellular organisms—fungi, animals, and plants—is based on the diversity

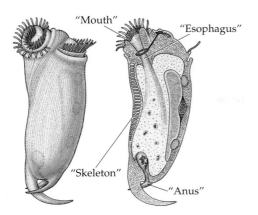

26.17 An Exceptional Ciliate Surface and cutaway views of *Diplodinium dentatum*, an amazingly complex single cell.

and coordination of cell types. These protists, on the other hand, owe their complexity to the diversity and coordination of organelles within a single cell.

Protists Once Considered To Be Fungi

Unlike fungi, the protists lack chitinous cell walls in their feeding phase and have no dikaryotic stage (in which cells contain two haploid nuclei each) in their life cycle (see Chapter 28). Table 26.2 summarizes some groups of protists that have been considered by some biologists to be fungi. In this section we'll describe four phyla: the Myxomycota, the Dictyostelida, the Acrasida, and the Oomycota.

The members of three of these groups seem so similar at first glance that they were once grouped in a single phylum. However, these so-called slime molds are actually so different that some biologists classify them in different *kingdoms.*

The three groups of slime molds share some characteristics. All are motile, all ingest particulate food by endocytosis, and all form spores on erect fruiting bodies. They undergo striking changes in organization during their life cycles, and one stage consists of isolated cells that engage in absorptive nutrition. Some

slime molds may attain areas of 1 meter or more in diameter while in their less-aggregated stage. Such a large slime mold may weigh more than 50 grams. Slime molds of all three types favor cool, moist habitats, primarily in forests. They range from colorless to brilliantly yellow and orange.

Acellular slime molds form multinucleate masses

If the nucleus of an amoeba began rapid mitotic division, accompanied by a tremendous increase in cytoplasm and organelles, the resulting organism might resemble the vegetative phase of the *acellular* slime molds (phylum Myxomycota). During most of its life history, an acellular slime mold is a wall-less mass of cytoplasm with numerous diploid nuclei. This mass streams very slowly over its substrate in a remarkable network of strands called a **plasmodium** (Figure 26.18*a*). A plasmodium of a myxomycete is an example of a **coenocyte**, a body in which many nuclei are enclosed in a single plasma membrane. The outer cytoplasm of the plasmodium (closest to the environment) is normally less fluid than the interior cytoplasm and thus provides some structural rigidity.

Myxomycetes such as *Physarum* (a popular research subject) provide a dramatic example of **cytoplasmic streaming**. The outer cytoplasmic region becomes more fluid in places, and cytoplasm rushes into those areas, stretching the plasmodium. This streaming somehow reverses its direction every few minutes as cytoplasm rushes into a new area and drains away from an older one, moving the plasmodium over its substrate in search of food.

The plasmodium engulfs food particles—predominantly bacteria, yeasts, spores of fungi, and other small organisms, as well as decaying animal and plant remains. Sometimes an entire wave of plasmodium moves across the substrate, leaving strands behind. Actin filaments and a contractile protein called myxomyosin interact to produce the streaming movement.

An acellular slime mold can grow almost indefinitely in its plasmodial stage, as long as the food supply is adequate and other conditions, such as moisture and pH, are favorable. However, one of two things can

TABLE 26.2 Classification of Protists with Absorptive Nutrition

PHYLUM	COMMON NAME	FORM	LOCOMOTION	EXAMPLES
Myxomycota	Acellular slime molds	Single cells and coenocytes	Amoeboid	*Physarum*
Dictyostelida	Dictyostelid cellular slime molds	Single cells and aggregates	Amoeboid	*Dictyostelium*
Acrasida	Acrasid cellular slime molds	Single cells and aggregates	Amoeboid	Acrasids
Oomycota	Water molds and downy mildews	Coenocytic mycelium	None	*Saprolegnia, Achlya, Phytophthora*

(a) *Physarum polycephalum*

(b) *Physarum* sp.

26.18 Acellular Slime Mold (a) Plasmodia of yellow slime mold cover a rock in Nova Scotia. (b) The fruiting structures—sporangiophore (yellow) and sporangia (black)—of *Physarum*.

happen if conditions become unfavorable. The plasmodium can form a resistant structure, an irregular mass of hardened cell-like components called a **sclerotium**, which rapidly becomes a plasmodium again when favorable conditions are restored; or the plasmodium can transform itself into spore-bearing fruiting structures (Figure 26.18b). Rising from heaped masses of plasmodium, these stalked or branched fruiting structures—called **sporangiophores**—derive their rigidity from the thickening of the walls of their component cells.

The nuclei of the plasmodium are diploid, and they divide by meiosis during the development of the sporangiophore. One or more knobs, called **sporangia**, develop on the end of the stalk. Within a sporangium, haploid nuclei become surrounded by walls and form spores. Eventually, as the sporangiophore dries, it sheds its spores.

The spores germinate into wall-less, flagellated, haploid cells called **swarm cells**, which either can divide mitotically to produce more haploid swarm cells or can function as gametes. Swarm cells can live on their own and can become walled and resistant cysts when conditions are unfavorable. When conditions improve again, the cysts release flagellated swarm cells. Two swarm cells can fuse to form a diploid zygote, which divides by mitosis (but without a wall forming between the nuclei) and thus forms a new, coenocytic plasmodium.

Cells retain their identity in the dictyostelid slime molds

The phylum Dictyostelida consists of *cellular* slime molds. Whereas the plasmodium is the basic vegeta-

tive (feeding) unit of acellular slime molds, an amoeboid cell is the vegetative unit of the dictyostelid slime molds. Large numbers of cells called **myxamoebas**, which have single haploid nuclei, engulf bacteria and other food particles by endocytosis and reproduce by mitosis and fission. This simple developmental stage, consisting of swarms of independent, isolated cells, can persist indefinitely (as long as food and moisture are available).

When conditions become unfavorable, however, dictyostelids aggregate and form fruiting structures, as do their acellular counterparts. The apparently independent myxamoebas aggregate into a mass called a **pseudoplasmodium** (Figure 26.19a). Unlike the true plasmodium of the acellular slime molds (see Figure 26.18a), this structure is not simply a giant sheet of cytoplasm with many nuclei; the individual myxamoebas retain their plasma membranes and, therefore, their identity.

The chemical signal that causes the myxamoebas of dictyostelid slime molds to aggregate into a pseudoplasmodium is 3',5'-cyclic adenosine monophosphate (cAMP), a compound that plays many important roles in chemical signaling in animals (see Chapter 38). A pseudoplasmodium may migrate over its substrate for several hours before becoming motionless and reorganizing to construct a delicate, stalked fruiting structure (Figure 26.19b). Cells at the top of the fruiting structure develop into thick-walled spores. The spores are released; later, under favorable conditions, they germinate, releasing myxamoebas.

The cycle from myxamoebas through a pseudoplasmodium and spores to new myxamoebas is asexual. There is also a sexual cycle, in which two myxamoebas (possibly of different mating types; see Chapter 28) fuse. The product of this fusion develops into a spherical structure that ultimately germinates, releasing new haploid myxamoebas.

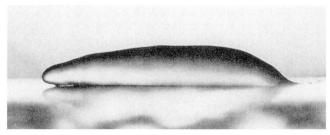

(a) *Dictyostelium discoideum*

26.19 A Cellular Slime Mold (a) A pseudoplasmodium migrates over its substrate. (b) Fruiting structures in various stages of development.

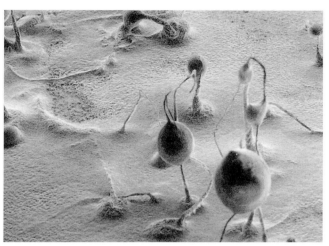

(b) *Dictyostelium discoideum*

Acrasids are also cellular slime molds

Members of the third group of slime molds, the phylum Acrasida, are also cellular slime molds, but they are only distantly related to the Dictyostelida and Myxomycota. Among their other differences from the dictyostelids, acrasids do not appear to use cAMP as an aggregation signal.

The oomycetes include water molds and their relatives

The three phyla of slime molds consist of motile organisms that feed by endocytosis. Members of the fourth phylum of protists that sometimes have been classified as fungi, the Oomycota, are filamentous and stationary, and they feed by absorption.

The phylum Oomycota consists in large part of the water molds and their funguslike terrestrial relatives, such as the downy mildews. If you have seen a whitish, cottony mold growing on dead fish or dead insects in water, it was probably a water mold of the common genus *Saprolegnia* (Figure 26.20). *Saprolegnia*

Saprolegnia sp.

26.20 Water Mold The filaments of a water mold radiate from the carcass of an insect.

itself is a common target of parasitism by the fungus *Rhizidiomyces*, described in Chapter 28.

The oomycetes are coenocytic: Their filaments have no cross-walls to separate the many nuclei into discrete cells. Their cytoplasm is continuous throughout the body of the mold, and there is no single structural unit with a single nucleus, except in certain reproductive stages. A distinguishing feature of the oomycetes is their flagellated reproductive cells. Oomycetes are diploid throughout most of their life cycle and have cellulose in their cell walls.

The water molds, such as *Saprolegnia*, are all aquatic and saprobic (feeding on dead organic matter). Some other ooymcetes are terrestrial. Although most terrestrial oomycetes are harmless or helpful decomposers of dead matter, a few are serious plant parasites that attack crops such as avocados, grapes, and potatoes. The mold *Phytophthora infestans*, for example, is the causal agent of late blight of potatoes, which brought about the great Irish potato famine of 1845 to 1847. *P. infestans* destroyed the entire Irish potato crop in a matter of days in 1846. Among the consequences of the famine were a million deaths from starvation and the emigration of about 2 million people, mostly to the United States.

Albugo, another oomycete, is a well-known parasitic genus that causes a mealy blight on sweet-potato leaves, morning glories, and numerous other plants. An obligate parasite, *Albugo* has never been grown on any medium other than its plant host.

The Algae

Algae (singular alga) are photosynthetic protists, carrying out probably 50 to 60 percent of all the photosynthesis on Earth (plants account for most of the rest). The overall contribution of cyanobacteria and other photosynthetic prokaryotes is smaller, although it is locally important in some aquatic ecosystems. Algae differ from plants in that the zygote of an alga is on its own; the parent gives the zygote no protection. A plant zygote, on the other hand, grows into a multicellular embryo that is protected by parental tissue.

Algae exhibit a remarkable range of growth forms. Some are unicellular; others are filaments composed either of distinct cells or of coenocytes (multinucleate structures that lack cross-walls). Still others—including the algae commonly known as seaweeds—are multicellular and intricately branched or arranged in leaflike extensions. The bodies of a few types of algae are even subdivided into tissues and organs. Certain algal phyla—for example, the phylum Chlorophyta (the green algae)—include representatives of almost all these growth forms.

Algal life cycles show extreme variation, but all algae except members of the phylum Rhodophyta (red algae) have forms with flagellated motile cells in at least one stage of their life cycle. Some algae—for example, the dinoflagellates in the phylum Pyrrophyta—are unicellular and motile throughout most of their existence.

Table 26.3 summarizes the classification of algae. Note that the algae do not constitute a natural group within the Protista—they are polyphyletic. For convenience, we describe them together here to emphasize the diversity of protists.

Some algae are unicellular flagellates

All members of the phylum Pyrrophyta are unicellular. A distinctive mixture of photosynthetic and accessory pigments gives their chloroplasts a golden-brown color. The dinoflagellates, the major group within this phylum, are of great ecological, evolutionary, and morphological interest. Dinoflagellates are probably second in importance only to the diatoms (see the next section) as primary photosynthetic producers of organic matter in the oceans.

Many dinoflagellates are endosymbionts, living within the cells of other organisms, including various invertebrates and even other marine protists. Dinoflagellates are particularly common endosymbionts in corals, to whose growth they contribute mightily by photosynthesis. Some dinoflagellates are nonphotosynthetic and live as parasites within other marine organisms.

Dinoflagellates are distinctive cells (see Figure 26.1a). They have two flagella, one in an equatorial groove around the cell, the other starting at the same point as the first and passing down a longitudinal groove before extending free into the surrounding medium. Most dinoflagellates are marine organisms. Some dinoflagellates reproduce in enormous numbers in warm and somewhat stagnant waters. The result can be a "red tide," so called because of the reddish color of the sea that results from pigments in the dinoflagellates (Figure 26.21). During a red tide, the concentration of dinoflagellates may reach 60 million cells per liter of ocean water. Certain red-tide species produce a potent nerve toxin that can kill tons of fish. The genus *Gonyaulax* produces a potent toxin that can accumulate in shellfish in amounts that, although not fatal to the shellfish, may kill a person who eats the shellfish.

Many dinoflagellates are bioluminescent. In complete darkness, cultures of these organisms emit a faint glow. If you suddenly stir or bubble air through a culture containing these dinoflagellates, the organisms each emit numerous bright flashes, producing a light that is perhaps a thousandfold brighter than the dim glow of an undisturbed culture. The flashing then rapidly subsides. A ship passing through a tropical ocean that contains a rich growth of these species produces a bow wave and a wake that glow eerily as billions of these dinoflagellates discharge their light systems.

TABLE 26.3   **Classification of Photosynthetic Protists**

PHYLUM	COMMON NAME	FORM	LOCOMOTION	EXAMPLES
Pyrrophyta	Dinoflagellates (and others)	Unicellular	Two flagella	*Gonyaulax, Ceratium, Noctiluca*
Chrysophyta	Diatoms	Usually unicellular	Usually none	*Diatoma, Fragilaria, Ochromonas*
Phaeophyta	Brown algae	Multicellular	Two flagella on reproductive cells	*Macrocystis, Fucus*
Rhodophyta	Red algae	Multicellular or unicellular	None	*Chondrus,* coralline algae
Chlorophyta	Green algae	Unicellular, colonial, or multicellular	Most have flagella at some stage	*Chlorella, Ulva, Acetabularia*

Architectural magnificence on a microscopic scale is the hallmark of the diatoms (Figure 26.22a). Despite their remarkable morphological diversity, however, all diatoms are symmetrical—either bilaterally (division along only one plane results in identical halves) or radially (division along any plane that passes through the center results in identical halves).

Many diatoms deposit silicon in their cell walls. The cell wall of some species is constructed in two pieces, with the wall of the top overlapping the wall of the bottom like the top and bottom of a petri plate (Figure 26.22b). The silicon-impregnated walls have intricate, unique patterns; in fact, the taxonomy of these marine or freshwater organisms is based entirely on their wall patterns.

Diatoms reproduce both sexually and asexually. Asexual reproduction is by cell division and is somewhat constrained by the stiff, silica-containing cell wall. Both the top and the bottom of the "petri plate" become tops of new "plates" without changing appreciably in size; as a result, the new cells made from former bottoms are smaller than the parent cells (Figure 26.23). If the process continued indefinitely, one cell line would simply vanish, but sexual reproduction largely solves this potential problem. Gametes form, shed their cell walls, and fuse. The resulting zygote then increases substantially in size before a new wall is laid down.

Diatoms are everywhere in the marine environment and are frequently present in great numbers, making them the leading photosynthetic producers in the oceans. Diatoms are also common in fresh water. Because the silicon-containing walls of dead diatom cells resist decomposition, certain sedimentary rocks are composed almost entirely of these silica-containing skeletons that sank to the seafloor. Diatomaceous earth, which is obtained from such rocks, has many industrial uses—from insulation and filtration to metal polishing. It has also been used as an "Earth-friendly" insecticide that clogs the tracheae (breathing structures) of insects.

26.21 A Red Tide of Dinoflagellates In astronomical numbers, the dinoflagellate *Gonyaulax tamarensis* causes a toxic red tide, as seen here off the coast of Baja California.

Diatoms in their glass "houses" account for most oceanic photosynthesis

Diatoms and their relatives constitute the phylum Chrysophyta. Some species are single-celled; others are filamentous. Many have sufficient carotenoids in their chloroplasts to give them a yellow or brownish color. All make chrysolaminarin (a carbohydrate) and oils as photosynthetic storage products.

(a)   30 μm

(b)   7 μm

26.22 Diatom Diversity (a) Diatoms exhibit a splendid variety of species-specific forms. (b) The dark and light areas of this scanning electron micrograph of a diatom emphasize the distinct two-piece construction of the cell wall.

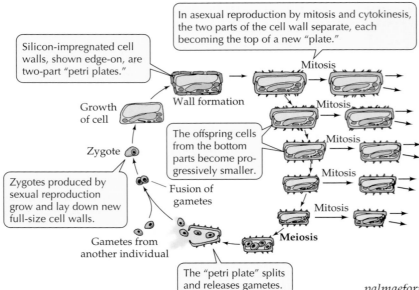

Silicon-impregnated cell walls, shown edge-on, are two-part "petri plates."

In asexual reproduction by mitosis and cytokinesis, the two parts of the cell wall separate, each becoming the top of a new "plate."

Growth of cell

Wall formation

Zygote

The offspring cells from the bottom parts become progressively smaller.

Zygotes produced by sexual reproduction grow and lay down new full-size cell walls.

Fusion of gametes

Gametes from another individual

The "petri plate" splits and releases gametes.

Mitosis

Meiosis

26.23 Diatom Reproduction Half of the cells created by asexual reproduction are always smaller than the parent cells; the sexual reproduction phase creates new parent cells with full-size cell walls.

may be up to 60 meters long (see Figure 26.1c). The brown algae are almost exclusively marine. Some float in the open ocean; the most famous example is the genus *Sargassum*, which forms dense mats of vegetation in the Sargasso Sea in the mid-Atlantic. Most brown algae, however, are attached to rocks near the shore. A few thrive only where they are regularly exposed to heavy surf; a notable example is the sea palm *Postelsia palmaeformis* of the Pacific coast (Figure 26.25a). All of the attached forms develop a specialized structure, called a **holdfast**, that literally glues them to the rocks (Figure 26.25b).

Some brown algae may differentiate extensively into stemlike stalks and leaflike blades, and some develop gas-filled cavities or bladders. For biochemical reasons that are only poorly understood, these gas cavities often contain as much as 5 percent carbon monoxide—a concentration high enough to kill a human.

In addition to organ differentiation, the larger brown algae also exhibit considerable tissue differentiation. Most of the giant kelps have photosynthetic filaments only in the outermost regions of the stalks and

Some brown algae are the largest protists

All members of the phylum Phaeophyta, commonly called brown algae, are multicellular, composed either of branched filaments or of leaflike growths called **thalli** (singular thallus) (Figure 26.24). The brown algae obtain their namesake color from the carotenoid fucoxanthin, which is abundant in the plastids. The combination of this yellow-orange pigment with the green of chlorophylls *a* and *c* yields a dirty brown color.

The Phaeophyta include the largest of the protists. Giant kelps, such as those of the genus *Macrocystis*,

(a) *Hormosira* sp.

26.24 Brown Algae (a) An intertidal brown alga growing in Australia. (b) A filamentous brown alga seen through a light microscope.

(b) *Ectocarpus* sp.

(b)

(a) *Postelsia palmaeformis*

26.25 Brown Algae in a Turbulent Environment   Algae growing in the intertidal zone on an exposed rocky shore take a tremendous pounding by the surf. (a) Sea palm growing along the California coast. (b) The tough, branched holdfast that anchors the sea palm.

metes fuse (syngamy), a diploid organism forms (Figure 26.26). The haploid organism, the diploid organism, or both may also reproduce asexually.

The two organisms (spore-producing and gamete-producing) differ genetically, in that one has haploid cells and the other has diploid cells, but they may or may not differ morphologically. In *heteromorphic* alternation of generations, the two organisms differ morphologically; in *isomorphic* alternation of generations, they do not, despite their genetic difference. We will see examples of both heteromorphic and isomorphic alternation of generations as we consider some representative algae.

Gametes are not generally produced directly by meiosis in plants or multicellular algae. Instead, specialized cells of the diploid organism, called **sporocytes**, divide meiotically to produce four spores. The spores may eventually germinate and divide mitotically to produce multicellular haploid organisms, the

blades. Within the photosynthetic region lie filaments of long cells that closely resemble the food-conducting tissue of plants (see Chapter 32). Called trumpet cells because they have flaring ends, these tubes rapidly conduct the products of photosynthesis through the body of the alga.

The cell walls of brown algae may contain as much as 25 percent alginic acid, a gummy polymer of sugar acids. Alginic acid cements cells and filaments together and provides good holdfast glue. It is used commercially as an emulsifier in ice cream, cosmetics, and other products.

Many algal life cycles feature an alternation of generations

In **alternation of generations**, a multicellular, diploid, spore-producing organism called the **sporophyte** gives rise to a multicellular, haploid, gamete-producing organism called the **gametophyte**. When two ga-

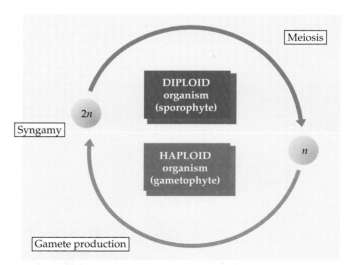

26.26 Alternation of Generations   A diploid generation that produces spores alternates with a haploid generation that produces gametes.

gametophyte generation, which produces gametes—by *mitosis* and cytokinesis.

Unlike spores, gametes can produce new organisms only by fusing with other gametes. The fusion of two gametes produces a diploid zygote, which then undergoes mitotic divisions to produce a diploid organism: the sporophyte generation. The sporocytes of the sporophyte generation undergo meiosis at some point and produce haploid spores, starting the cycle anew.

BROWN ALGAE HAVE ALTERNATING GAMETOPHYTE AND SPORO-PHYTE GENERATIONS. The brown algae exemplify the extraordinary diversity found among the algae. One genus of simple brown algae is *Ectocarpus* (see Figure 26.24*b*). Its branched filaments, a few centimeters long, commonly grow on shells and stones. The gametophyte and sporophyte phases of the alternation of generations of *Ectocarpus* can be distinguished only by chromosome number or reproductive products (zoospores or gametes). Thus the generations are isomorphic.

By contrast, some kelps of the genus *Laminaria* and some other brown algae show a more complex heteromorphic alternation of generations. The larger and more obvious generation of these species is the sporophyte. Meiosis in special fertile regions of the leaflike fronds produces haploid zoospores. These germinate to form a tiny, filamentous gametophyte that produces either eggs or sperm.

The genus *Fucus* carries reduction in gametophyte size still further: It has no multicellular haploid phase—only a multi*nucleate* haploid phase. The gametes themselves are formed directly by meiosis.

Red algae may have donated organelles to other protist phyla

Almost all Rhodophyta (red algae) are multicellular (Figure 26.27). The characteristic color of the red algae is a result of the pigment phycoerythrin, which is found in relatively large amounts in the chloroplasts of many species. In addition to phycoerythrin, red algae contain phycocyanin, carotenoids, and chlorophyll. The red algae include species that grow in the shallowest tide pools, as well as the algae found deepest in the ocean (as deep as 260 meters if nutrient conditions are right and the water is clear enough to permit the penetration of light). Very few red algae inhabit fresh water. Most grow attached to a substrate by a holdfast.

In a sense the red algae, along with several other groups of algae, are misnamed. They have the capacity to change the relative amounts of their various photosynthetic pigments depending on the light conditions where they are growing. Thus the leaflike *Chondrus crispus*, a common North Atlantic red alga, may appear bright green when it is growing at or near the surface of the water and deep red when growing at greater depths.

The pigmentation—the ratio of pigments present—depends to a remarkable degree on the intensity of the light that reaches the alga. In deep water, where the light is dimmest, the algae accumulate large amounts of phycoerythrin, an accessory photosynthetic pigment (see Figure 8.7). The algae in deeper water have as much chlorophyll as the green ones near the surface, but the accumulated phycoerythrin makes them look red.

In addition to being the only algae with phycoerythrin and phycocyanin among their pigments, the red

(*b*) *Polysiphonia* sp.

26.27 Red Algae (*a*) Dulse is a large, edible red alga growing on rocks in New Brunswick. (*b*) Both vegetative and reproductive structures of this alga can be seen under the light microscope.

(*a*) *Palmaria palmata*

algae have two other unique characteristics: They store the products of photosynthesis in the form of floridean starch, which is composed of very small, branched chains of approximately 15 glucose units. And they produce no motile, flagellated cells at any stage in their life cycle. The male gametes are naked and slightly amoeboid, and the female gametes are completely immobile.

Some red algal species enhance the formation of coral reefs. They share with the coral animals the biochemical machinery for depositing calcium carbonate both in and around their cell walls. After the death of the algal cells, the calcium carbonate persists, sometimes forming substantial rocky masses.

Some red algae also produce large amounts of mucilaginous polysaccharide substances, which contain the sugar galactose with a sulfate group attached. This material readily forms solid gels and is the source of agar, a substance widely used in the laboratory for making a solid aqueous medium on which tissue cultures and many microorganisms may be grown.

Some marine red algae are parasitic on other red algae. The hosts are photosynthetic, but the parasites are often colorless and nonphotosynthetic, deriving their nutrition from the host. The parasitic red alga *Choreocolax* inserts its nuclei into cells of its host red alga, *Polysiphonia*. This is the first example found of regular introduction of parasite nuclei into living host cells. Apparently some parasite genes are expressed in the host cytoplasm, diverting the host's metabolism.

Other red algae may at one time have been endosymbionts within the cells of certain other, nonphotosynthetic protists, eventually being reduced to chloroplasts. This seems to have been the evolutionary origin of the chloroplasts of a group of algae called cryptomonads. Endosymbiotic red algae may also be the ancestors of chloroplasts of the brown algae and the diatoms. Endosymbiosis has played several roles in the origins of different kinds of eukaryotic cells, as we have seen by now. Endosymbiotic cyanobacteria became chloroplasts, including those of the red algae, and red algal cells gave rise to the chloroplasts of certain other algae.

The green algae and plants form a monophyletic lineage

Many systematists place the green algae in the kingdom Plantae, and evidence from rRNA sequencing shows that the green algae and the plants form a monophyletic lineage. We will treat the green algae as the protist phylum Chlorophyta instead, as do some other systematists. Our primary motivation here is to group material for your convenience in learning; the topics in Part Five deal entirely with plants that have embryos and not at all with the green algae.

The kingdom Plantae evolved from one or more representatives of the phylum Chlorophyta. The green algae have uniform pigmentation and a characteristic storage product, starch. The Chlorophyta and the euglenoids are the only protist groups that contain the full complement of photosynthetic pigments that are also characteristic of the kingdom Plantae.

Chlorophyll *a* predominates, and a major pigment is chlorophyll *b*, which none of the other algae have. The carotenoids found in these groups, predominantly β-carotene and certain xanthophylls (carotenoids with one or more hydroxyl groups), are likewise those characteristic of plants. The principal photosynthetic storage product, like that of the plant kingdom, is long, straight or branched chains of glucose that together make up starch.

BODY SHAPE AND CELLULARITY IN THE CHLOROPHYTA. We find in the green algae an incredible variety in shape and construction of the algal body. *Chlorella* is an example of the simplest type: unicellular and flagellated. Surprisingly large and well-formed colonies of cells are found in such freshwater groups as the genus *Volvox* (Figure 26.28*a*). The cells are not differentiated into tissues and organs as in plants and animals, but the colonies show vividly how the preliminary step of this great evolutionary development might have been taken.

The intermediate stages between the one-celled state and the extreme colonial state of *Volvox* are preserved in loosely colonial forms, such as *Gonium* and *Pandorina*. By contrast, *Oedogonium* is multicellular and filamentous, and each of its cells has only one nucleus. *Cladophora* is multicellular, but each cell is multinucleate. *Bryopsis* is tubular and coenocytic, forming cross-walls only when reproductive structures form. *Acetabularia* is a single, giant uninucleate cell a few centimeters long and with remarkable morphology, becoming multinucleate only at the end of the reproductive stage. *Ulva lactuca* is a membranous sheet two cells thick; its unusual appearance justifies its common name: sea lettuce (Figure 26.28*b*). Finally, the remarkable unicellular desmids have elaborately sculptured cell walls (Figure 26.28*c*).

LIFE CYCLES IN THE CHLOROPHYTA. The life cycles of green algae are diverse. We will examine two algal life cycles in detail, beginning with that of the sea lettuce *Ulva lactuca* (Figure 26.29). The diploid sporophyte of this common seashore alga is a "leaf" a few centimeters in diameter. Specialized cells (sporocytes) differentiate and undergo meiosis and cytokinesis, producing motile haploid spores (zoospores). These swim away, each propelled by four flagella, and some eventually find a suitable place to settle. The spores then lose their flagella and begin to divide mitotically, producing a thin filament that develops into a broad sheet only two cells thick. The gametophyte thus produced looks just like the sporophyte.

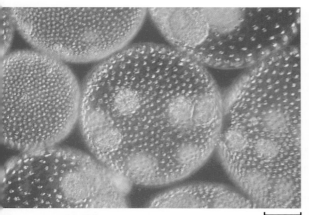

(a) *Volvox* sp.

18 µm

(b) *Ulva lactuca*

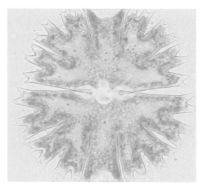

(c) *Micrasterias* sp.

26.28 Green Algae (a) Colonies showing the precise spacing of cells and containing numerous daughter colonies. (b) A stand of sea lettuce, submerged in a tidal pool. (c) A microscopic desmid. A narrow, nucleus-housing isthmus joins two elaborate semicells—halves of the unicellular organism. A single large, ornate chloroplast fills much of the volume of each semicell.

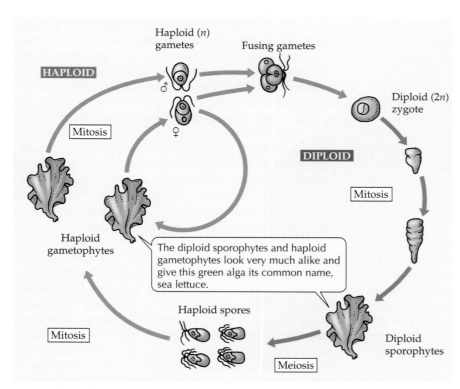

26.29 The Life Cycle of *Ulva lactuca*

The diploid sporophytes and haploid gametophytes look very much alike and give this green alga its common name, sea lettuce.

Each spore contains genetic information for just one mating type (see Chapter 28), and a given gametophyte can produce only male or female gametes—never both. The gametes arise mitotically within single cells (gametangia), rather than within a specialized multicellular structure, as in plants (see Chapter 27). Both types of gametes bear two flagella (in contrast to the four flagella of a haploid spore) and hence are motile.

In most species of *Ulva* the female and male gametes are indistinguishable structurally, making those species **isogamous**—having gametes of identical appearance. Some other algae and absorptive protists are also isogamous. Yet other algae, including some species of *Ulva*, are **anisogamous**—having female gametes that are distinctly larger than the male gametes. Female and male gametes come to-

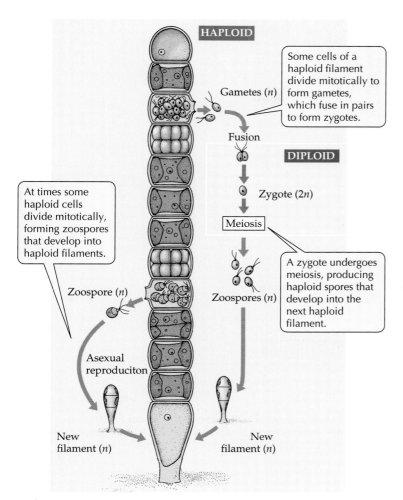

HAPLOID

Gametes (*n*)

Some cells of a haploid filament divide mitotically to form gametes, which fuse in pairs to form zygotes.

Fusion

DIPLOID

Zygote (2*n*)

Meiosis

At times some haploid cells divide mitotically, forming zoospores that develop into haploid filaments.

Zoospore (*n*)

A zygote undergoes meiosis, producing haploid spores that develop into the next haploid filament.

Zoospores (*n*)

Asexual reproduciton

New filament (*n*)

New filament (*n*)

26.30 A Haplontic Life Cycle   In the life cycle of *Ulothrix*, a filamentous, multicellular gametophyte generation alternates with a sporophyte generation consisting of a single cell.

meiosis to produce spores, which in turn produce a new haploid individual. In the entire haplontic life cycle, only one cell—the zygote—is diploid. The filamentous green algae of the genus *Ulothrix* are examples of haplontic algae.

Other algae have a **diplontic** life cycle like that of many animals. In the diplontic life cycle, meiosis of sporocytes produces gametes directly; the gametes fuse, and the resulting zygote divides mitotically to form a new multicellular sporophyte. In such organisms, every cell except the gametes is diploid. Between these two extremes are algae whose gametophyte and sporophyte generations are both multicellular, but that have one phase (usually the sporophyte) that is much larger and more prominent than the other.

EVOLUTIONARY TRENDS IN THE CHLOROPHYTA.   Among the green algae, some evolutionary trends have been identified. For example, some groups show a trend toward multicellularity. Another trend leads from the mass release of large numbers of tiny, seemingly undifferentiated gametes (isogamy) toward increased protection of a single, large egg (anisogamy). For example, the female gamete of *Oedogonium* is large and immobile, and the male gamete is free-swimming and flagellated—an extreme case of anisogamy known as oogamy.

Phylogeny of the Protists

Can we make sense of protist diversity? Molecular systematists still wrestle with the relationships among the groups we've considered here—some of which are themselves not monophyletic, natural groups. Just as in the kingdom Eubacteria, the systematics of which is complicated by the seemingly easy gain and loss by bacteria of metabolic pathways, it appears that chloroplasts have been gained and lost by protists.

Still, progress is being made, particularly but not exclusively in the application of molecular tools such as rRNA sequencing. Figure 26.31 summarizes the most convincing phylogeny currently available, focusing primarily on the protists. Evidence from molecular systematics supports each feature of the figure, but not all the evidence is consistent. We show only a few of the numerous protist groups, and some of the ones we omit present problems.

Biologists disagree about the number of kingdoms of living things. Figure 26.31 is not a lot of help in resolving the problem, and adding groups that we've omitted from the figure would not make it any easier. Biologists continue to pursue this vexing problem—how many kingdoms are there?—because we want to understand the full and true course of evolutionary history. Perhaps the kingdom concept will fade away and be replaced by an emphasis on the numerous protist lineages.

gether and unite, losing their flagella as the zygote forms and settles. After resting briefly, the zygote begins mitotic division, producing a multicellular sporophyte. Any gametes that fail to find partners can settle down on a favorable substrate, lose their flagella, undergo mitosis, and produce a new gametophyte directly; in other words, the gametes can also function as zoospores. Few algae other than *Ulva* have motile gametes that can also function as zoospores.

A life cycle such as that of *Ulva* is isomorphic: Sporophyte and gametophyte generations are identical in structure. By contrast, in many other algae, the generations are heteromorphic: Sporophyte and gametophyte generations differ in structure. In one variation of the heteromorphic cycle—the **haplontic** life cycle (Figure 26.30)—a multicellular haploid individual produces gametes that fuse to form a zygote. The zygote functions directly as a sporocyte, undergoing

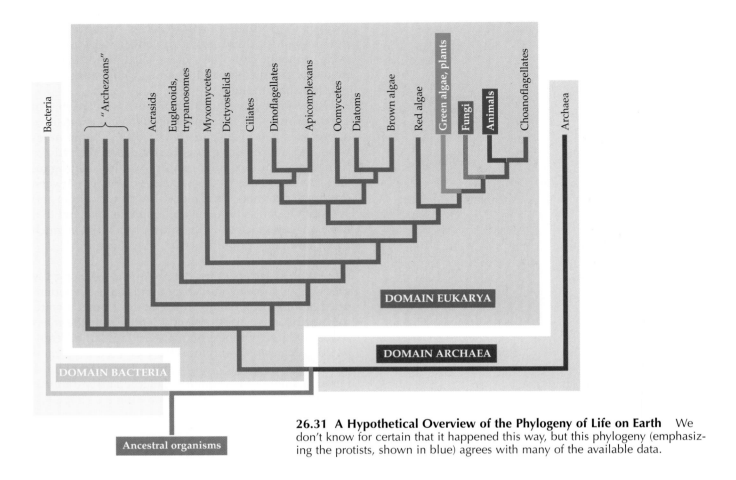

26.31 A Hypothetical Overview of the Phylogeny of Life on Earth We don't know for certain that it happened this way, but this phylogeny (emphasizing the protists, shown in blue) agrees with many of the available data.

Summary of "Protists and the Dawn of the Eukarya"

Protista and the Other Eukaryotic Kingdoms

• In this book we define the kingdom Protista as all eukaryotes that are not plants, fungi, or animals.
• The Protista arose from prokaryotic origins. The other eukaryotic kingdoms (Plantae, Fungi, Animalia) arose from protists.
• We distinguish protists from animals on the bases of how food enters the body (animals are ingestive heterotrophs), how development proceeds (animals have a blastula stage), and the presence (animals) or absence (protists) of an extracellular matrix.
• Protists that were once considered to be fungi differ from fungi in that they lack chitinous walls in their absorptive phase and their life cycles have no dikaryotic phase.
• Photosynthetic protists (algae) differ from plants in that they lack a protected embryo.

The Origin of the Eukaryotic Cell

• The modern eukaryotic cell arose in several steps. One possible scenario is as follows: An ancestral prokaryotic cell lost its cell wall, becoming more flexible. The plasma membrane folded into the cytoplasm, increasing the cell's surface area, thus allowing an increase in cell size and leading to the formation of vesicles that gained digestive and other capabilities. **Review Figure 26.2**
• In subsequent steps in the origin of the eukaryotic cell, infolded plasma membrane attached to the chromosome may have led to the formation of a nuclear envelope. Ribosomes became associated with internal membranes, some vesicles developed into lysosomes, and a primitive cytoskeleton evolved. **Review Figure 26.3**
• The first truly eukaryotic cell was much larger than the ancestral prokaryote, and it probably possessed one or more flagella of the eukaryotic type.
• The incorporation of prokaryotic cells as endosymbionts gave rise to eukaryotic organelles. Peroxisomes, protecting the host cell from an O_2-rich atmosphere, may have been the first of the organelles of endosymbiotic origin. Mitochondria evolved from once free-living proteobacteria. Then chloroplasts evolved from once free-living cyanobacteria. According to this hypothetical scenario, cells with nuclei appeared before the first cells with mitochondria. **Review Figure 26.3**
• Archezoans resemble a proposed intermediate evolutionary stage that lends support to the hypothesis that the eukaryotic nucleus evolved before the mitochondrion. **Review Figure 26.4**

General Biology of the Protists

• Most protists are aquatic; some live within other organisms. The great majority are unicellular and microscopic, but many are multicellular and a few are enormous.
• Protists vary widely in their nutrition, metabolism, and locomotion. Some protist cells contain contractile vacuoles, and some digest their food in food vacuoles. **Review Figures 26.5, 26.6**
• Protists have a variety of cell surfaces, some of them protective. **Review Figure 26.7**
• Many protists contain endosymbiotic prokaryotes. Some protists are endosymbionts in other cells, often other protists. Some endosymbiotic protists perform photosynthesis, to the advantage of their hosts.
• Most protists reproduce both asexually and sexually.

Protozoan Diversity

• As defined here, all protozoans are unicellular.
• Flagellates—protists with flagella—are the largest group of protozoans, but they are not a single, natural group. Some are parasitic, and some of these parasitic flagellates are important human pathogens. One flagellate group, the Choanoflagellida, is closely related to both the animal and fungal kingdoms. The animals and Choanoflagellida share an ancestor that no other organisms have. Some euglenoids, although flagellates, are photosynthetic. **Review Figure 26.10 and Table 26.1**
• Amoebas, which appear in many protist phyla, move by means of pseudopods. **Review Table 26.1**
• Actinopods have thin, stiff pseudopods that serve various functions, including food capture. **Review Table 26.1**
• Foraminiferans (phylum Foraminifera) also use pseudopods for feeding, and they secrete shells of calcium carbonate. **Review Table 26.1**
• Apicomplexans (phylum Apicomplexa) are parasites that have rigid cell walls and whose spores are adapted to the invasion of host tissue. The apicomplexan *Plasmodium*, which causes malaria, uses two alternate hosts (humans and *Anopheles* mosquitoes). **Review Figure 26.12 and Table 26.1**
• Ciliates (phylum Ciliophora) move rapidly by means of cilia and have two kinds of nuclei: a macronucleus and one or more micronuclei. The macronucleus controls the cell by way of transcription and translation. The micronuclei are responsible for genetic recombination, accomplished by conjugation that is sexual but not reproductive. Some ciliates have a remarkably complex internal structure. **Review Figures 26.14, 26.15, 26.16, 26.17**

Protists Once Considered To Be Fungi

• Acellular slime molds (phylum Myxomycota) and cellular slime molds (phyla Dictyostelida and Acrasida) are superficially very similar, moving as slimy masses and producing stalked fruiting structures. However, they differ at the cellular level. Acellular slime molds are coenocytes with diploid nuclei. Cellular slime molds consist of individual haploid cells that aggregate into masses consisting of distinct cells. **Review Table 26.2**
• Oomycetes (phylum Oomycota) include water molds, downy mildews, and some other protists. Like the acellular slime molds, the oomycetes are coenocytic. They are diploid for most of their life cycle. A few oomycetes are serious plant pathogens. **Review Table 26.2**

The Algae

• Algae are photosynthetic protists responsible for about half of the photosynthesis on Earth. Unlike plants, algae have no multicellular protection for their zygotes. As a group, the algae are polyphyletic.
• Dinoflagellates (phylum Pyrrophyta) are unicellular; they are major contributors to world photosynthesis. Many are endosymbionts; in that role they are important contributors to coral reef growth. Dinoflagellates are responsible for toxic "red tides." **Review Table 26.3**
• Diatoms (phylum Chrysophyta) are unicellular and have complex, two-part, glassy cell walls. They contribute extensively to world photosynthesis. **Review Figure 26.23 and Table 26.3**
• The brown algae (phylum Phaeophyta) and red algae (phylum Rhodophyta) are predominantly multicellular. The two differ in their pigmentation and in other ways. The brown algae include the largest of all protists, and some show considerable tissue differentiation. **Review Table 26.3**
• In many algae, both haploid and diploid cells undergo mitosis, leading to an alternation of generations. The diploid sporophyte generation forms spores by meiosis, and the spores develop into haploid organisms. This haploid gametophyte generation forms gametes by mitosis, and their fusion yields zygotes that develop into the next generation of sporophytes. **Review Figure 26.26**
• Red algae have a characteristic storage product (floridean starch) and differ from the other algal groups in lacking flagellated reproductive cells. Some red algae are parasitic; others gave rise to chloroplasts in some other algae.
• The green algae (phylum Chlorophyta) are often multicellular. They have the same chloroplast pigments and storage product as do plants, which descended from an ancestral green alga. Several life cycles are used by different green algae; among these are the isomorphic alternation of generations of *Ulva* and the haplontic cycle of *Ulothrix*. **Review Figures 26.29, 26.30**

Phylogeny of the Protists

• The evolutionary relationships among the various protist groups, and even their exact composition, remain controversial. However, some proposals for a natural phylogeny have been made, based in part on rRNA sequencing and other molecular techniques. Biologists have different views on the number of kingdoms into which the living world should be divided. **Review Figure 26.31**

Self-Quiz

1. Flagellates
   a. appear in several protist phyla.
   b. are all algae.
   c. all have pseudopods.
   d. are all colonial.
   e. are never pathogenic.

2. Which statement about amoebas is *not* true?
   a. They are specialized.
   b. They use amoeboid movement.
   c. They include both naked and shelled forms.
   d. They possess pseudopods.
   e. They appeared only once in evolutionary history.

3. The Apicomplexa
   a. possess flagella.
   b. possess chloroplasts.
   c. are all parasitic.
   d. are algae.
   e. include the trypanosomes that cause sleeping sickness.

4. The Ciliophora
   a. move by means of flagella.
   b. use amoeboid movement.
   c. include *Plasmodium*, the agent of malaria.
   d. possess both a macronucleus and micronuclei.
   e. are autotrophic.

5. The Myxomycota
   a. are also called the acellular slime molds.
   b. lack fruiting bodies.
   c. consist of large numbers of myxamoebas.
   d. consist at times of a mass called a pseudoplasmodium.
   e. possess flagella.

6. The Dictyostelida
   a. are also called the acellular slime molds.
   b. lack fruiting bodies.
   c. form a plasmodium that is a coenocyte.
   d. use cAMP as a "messenger" to signal aggregation.
   e. possess flagella.

7. Which statement about algae is *not* true?
   a. They differ from plants in lacking protected embryos.
   b. They are photosynthetic autotrophs.
   c. They have chitin in their cell walls.
   d. They include both unicellular and multicellular forms.
   e. Their life cycles show extreme variation.

8. Which statement about the Phaeophyta is *not* true?
   a. They are all multicellular.
   b. They use the same photosynthetic pigments as do plants.
   c. They are almost exclusively marine.
   d. A few are among the largest organisms on Earth.
   e. Some have extensive tissue differentiation.

9. The Rhodophyta
   a. are mostly unicellular.
   b. are mostly marine.
   c. owe their red color to a special form of chlorophyll.
   d. have flagella on their gametes.
   e. are all heterotrophic.

10. Which statement about the Chlorophyta is *not* true?
   a. They use the same photosynthetic pigments as do plants.
   b. Some are unicellular.
   c. Some are multicellular.
   d. All are microscopic in size.
   e. They display a great diversity of life cycles.

Applying Concepts

1. For each type of organism below, give a single characteristic that may be used to differentiate it from the other, related organism(s) in parentheses.
   a. foraminiferans (radiolarians)
   b. *Vorticella* (*Paramecium*)
   c. *Euglena* (*Volvox*)
   d. *Trypanosoma* (*Giardia*)
   e. amoeba (flagellate)
   f. *Physarum* (*Dictyostelium*)

2. For each of the following groups, give at least two characteristics used to distinguish the group from other groups. (*a*) Ciliophora; (*b*) Apicomplexa; (*c*) Phaeophyta; (*d*) Rhodophyta.

3. Identify one major role in the world that is played by the photosynthetic protists. Identify one major role in the world that is played by the absorptive protists.

4. Giant seaweed (mostly brown algae) have "floats" that aid in keeping their fronds suspended at or near the surface of the water. Why is it important that the fronds be suspended?

5. Justify the placement of *Euglena* among the flagellate protozoans. Why might it be considered part of the phylum Chlorophyta?

6. Why are algal pigments so much more diverse than those of plants?

Readings

Anderson, D. M. 1994. "Red Tides." *Scientific American*, August. A discussion of the origin and consequences of the dinoflagellate blooms.

Brusca, R. C. and G. J. Brusca. 1990. *Invertebrates*. Sinauer Associates, Sunderland, MA. Chapter 5, on the protozoans, also covers many algae—showing once again that different authorities follow different taxonomic schemes for the kingdom Protista.

de Duve, C. 1996. "The Birth of Complex Cells." *Scientific American*, April. One hypothesis, offered by a leading scholar of cell structure and function, concerning the evolution of the eukaryotic cell by loss, elaboration, and gain of parts.

Donelson, J. E. and M. J. Turner. 1985. "How the Trypanosome Changes Its Coat." *Scientific American*, February. A description of how trypanosomes evade the host's immune system by constantly switching on new genes that code for different surface antigens.

Farmer, J. N. 1980. *The Protozoa*. Mosby, St. Louis. A full treatment of the animal-like protists.

Fenchel, T. and B. J. Finlay. 1994. "The Evolution of Life without Oxygen." *American Scientist*, vol. 82, pages 22–29. Studies on eukaryote origins.

Godon, G. N. 1985. "Molecular Approaches to Malaria Vaccines." *Scientific American*, May. The life cycle and molecular biology of a problem pathogen.

Harrison, F. W. and J. O. Corliss (Eds.). 1991. *Microscopic Anatomy of Invertebrates*. Volume 1: *Protozoa*. Wiley-Liss, New York. Includes marvelous illustrations.

Jacobs, W. P. 1994. "Caulerpa." *Scientific American*, December. An exploration into the biology of the largest unicellular organism.

Kabnick, K. S. and D. A. Peattie. 1991. "*Giardia*: A Missing Link between Procaryotes and Eucaryotes." *American Scientist*, vol. 79, pages 34–43. A description of the intestinal parasite that appears to be an intermediate step in the origin of the eukaryotic cell.

Margulis, L. 1993. *Symbiosis in Cell Evolution*, 2nd Edition. W. H. Freeman, New York. It is interesting to review the theory as you study the protists, which may illustrate some of the stages proposed by Margulis.

Vidal, G. 1984. "The Oldest Eukaryotic Cells." *Scientific American*, February. A look at ancient marine protists.

Chapter 27

Plants: Pioneers of the Terrestrial Environment

What's missing?

Here they are!

*A*utotrophs are the vital base of every present-day food web—they feed the entire living world. Photosynthetic protists and bacteria feed the great bulk of oceanic life, and chemoautotrophic bacteria support thermal vent communities in the ocean deeps. But what about the terrestrial environment, the part of Earth that *we* inhabit? Terrestrial Earth is supported by the photosynthesis of plants. But at one time there were no plants.

What a thrill it would be to see the face of Earth at different stages of its evolution. Successive photographs from space would show the drifting of the continents. On Earth's surface we would see scenes scarcely imaginable to us today. For more than 4 billion years the terrestrial environment was basically a mass of rock. Its appearance changed relatively little until plants colonized the land and slowly spread over its surface.

Earth did not take on a green tint until less than half a billion years ago, long after the ancestors of today's plants had invaded the land sometime during the Paleozoic era (see Table 19.1). The earliest land plants were tiny, but their metabolic activities helped convert native rock to soil that could support the needs of their successors. Evolution led rapidly (in geologic terms) to larger and larger plants, and in the Carboniferous period (345 to 290 million years ago) great forests were widespread. Those forests would look unfamiliar to us, for few of their trees were like those we know today.

During the tens of millions of years since, these trees were replaced by others more familiar to us—pinelike conifers and their relatives and, in the last 100 million years, the broad-leaved forms of modern vegetation. Today Earth is a patchwork of widely differing environments, supporting differing communities of plants, animals, fungi, protists, archaea, and bacteria. What do we notice when we look at any of these environments? Consciously or not, we see its *plants*—or, in a very harsh environment, its relative *lack* of plants.

In this chapter we will see how the plant kingdom differs from the other kingdoms and how it conquered the land and then evolved. We will survey the diverse products of plant evolution, from evolutionarily ancient groups to the one that appeared most recently, the flowering plants.

The Plant Kingdom

Broadly defined, a plant is a *multicellular photosynthetic eukaryote*. The definition is not precise, because a few organisms that can be regarded only as plants are not photosynthetic. These parasitic species, such as Indian pipe (Figure 27.1), are clearly related to photosynthetic plants (they have leaves, roots, flowers, and so on) but possess adaptations that provide them with alternative modes of nutrition. The definition of plants casts an enormous net over a wide range of organisms that, while all multicellular and photosynthetic, differ in size, cellular organization, photosynthetic and associated pigments, and cell wall chemistry—and do not form a monophyletic group. At present we prefer to narrow the definition of a plant to exclude algae, which we discussed in Chapter 26.

According to this narrower definition, plants are multicellular, photosynthetic eukaryotes that have the following additional properties: *They develop from embryos protected by tissues of the parent plant.* (This characteristic is definitive; an older classification scheme used the term Embryophyta to refer to precisely those organisms that we assign here to the kingdom Plantae.) Their cells have walls that contain cellulose as the major strengthening polysaccharide. Their chloroplasts contain chlorophylls *a* and *b* and a limited array of specific carotenoids (see Chapter 8). Their major storage carbohydrate is starch. Their life cycles have important features in common, notably an alternation of haploid and diploid generations.

Life cycles of plants feature alternation of generations

A universal feature of the life cycles of plants is the alternation of generations (see Chapters 9 and 26). If we consider the plant life cycle to begin with a single cell, the diploid zygote, then the first phase of the cycle features the formation, by mitosis and cytokinesis, of a

Monotropa uniflora

27.1 Stretching the Definition Indian pipe is not photosynthetic, but it is a plant in all other respects.

multicellular embryo and eventually the mature diploid plant (see Figure 26.26).

This multicellular, diploid plant is the **sporophyte** ("sporophyte" means "spore plant"). Cells contained in **sporangia** (singular sporangium, "spore reservoir") on the sporophyte undergo meiosis to produce haploid, unicellular spores. By mitosis and cytokinesis a spore forms a haploid plant. This multicellular, haploid plant, the **gametophyte** ("gamete plant"), produces haploid gametes. The fusion of two gametes (syngamy, or fertilization) results in the formation of a diploid cell, the zygote, and the cycle begins again.

The sporophyte generation extends from the zygote through the adult, multicellular, diploid plant; the gametophyte generation extends from the spore through the adult, multicellular, haploid plant to the gamete. The transitions between the phases are accomplished by fertilization and meiosis.

The gametophyte and sporophyte of any plant look totally different, unlike the generations of some green algae, such as *Ulva*, which are indistinguishable to the eye (see Figure 26.29). In all plants and green algae, however, the sporophyte and gametophyte differ genetically: One has diploid cells, the other haploid cells. This alternation of diploid and haploid generations is found in the life cycles of all phyla of plants.

There are twelve surviving phyla of plants

The surviving members of the kingdom Plantae fall naturally into 12 phyla (Table 27.1). We will group some of these phyla to emphasize the evolutionary trends you'll be learning, but bear in mind that these groups are for convenience and that they are not natural groups in the sense that the phyla are.

All members of nine of the phyla possess well-developed "plumbing systems" that transport materials throughout the plant body. We call these nine phyla,

TABLE 27.1 Classification of Plants[a]

PHYLUM	COMMON NAME	CHARACTERISTICS
Nontracheophytes		
Hepatophyta	Liverworts	No filamentous stage
Anthocerophyta	Hornworts	Embedded archegonia
Bryophyta	Mosses	Filamentous stage
Tracheophytes		
Nonseed tracheophytes		
Lycophyta	Club mosses	Simple leaves in spirals
Sphenophyta	Horsetails	Simple leaves in whorls
Psilophyta	Whisk ferns	No true leaves
Pterophyta	Ferns	Complex leaves
Seed plants		
Gymnosperms		
Cycadophyta	Cycads	Fernlike leaves; swimming sperm; seeds in strobili
Ginkgophyta	Ginkgo	Deciduous; fan-shaped leaves
Gnetophyta	Gnetophytes	Vessels in vascular tissue
Coniferophyta	Conifers	Many have seeds in cones; needlelike leaves
Angiosperms		
Angiospermae	Flowering plants	Double fertilization; seeds in fruit

[a] No extinct groups are included in this classification.

collectively, the **tracheophytes**. The remaining three phyla (liverworts, hornworts, and mosses) were once considered classes of a single larger phylum, of which the most familiar examples are mosses. Now we use the term **nontracheophytes** to refer collectively to these three phyla. Although the older term "nonvascular plants" is a time-honored name, it is misleading in that some mosses, unlike liverworts and hornworts, have a limited amount of vascularlike tissue. The nontracheophytes are sometimes collectively called bryophytes, but we will reserve that term for the mosses.

Plants arose from green algae

The plant kingdom arose from the green algae (the protist phylum Chlorophyta). The characteristics of green algae that originally identified them as the most likely ancestors of the plants include the possession of photosynthetic and accessory pigments in their plastids that are similar to those of plants, the use of starch as their principal storage carbohydrate, and the presence of cellulose as the principal component of their cell walls.

In addition, some green algae (not all) show the same oogamous type of reproduction—which features a large, stationary egg—that is characteristic of sexual reproduction in plants. Like plants, a few green algae have a life cycle that includes both a multicellular gametophyte and a multicellular sporophyte generation. Furthermore, some green algae have bulky, three-dimensional bodies like plants, although most are unicellular, filamentous, or two-dimensional. And the chloroplasts of plants and many green algae have thylakoids that are arranged into grana. In addition, some green algae produce new cell walls after cell division using a mechanism similar to that in plants. No other phylum of algal protists shares so many traits with the Plantae. Molecular biological techniques such as rRNA sequencing have confirmed that plants evolved from green algal ancestors.

Much evidence indicates that the closest relatives of the plants are two groups of living green algae (stoneworts and *Coleochaete*-like algae), but we don't yet know which is the true sister group. The stoneworts are characterized by multicellular sex organs and complex body patterns, and they resemble the plants in terms of their rRNA and DNA sequences, peroxisome contents, mechanics of mitosis and cytokinesis, and chloroplast structure (Figure 27.2). Other strong evidence, from morphology-based cladistic analysis, suggests that the sister group of the plants is a group of green algae that includes the genus *Coleochaete*. Those algae have features, such as plas-

Chara sp. (stonewort)

27.2 The Closest Relatives of Plants
The plant kingdom evolved from an ancient organism that was also ancestral to this protist.

modesmata and embryolike structures, found in plants. The ancestral algae, whether more similar to stoneworts or to *Coleochaete*, lived at the margins of ponds or marshes, ringing them with a green mat. From the margins of ponds, plants moved onto the land.

Plants colonized the land

Plants or their immediate ancestors in the green mat pioneered and modified the terrestrial environment. That environment differs dramatically from the aquatic environment in several ways. The density of the plant body is much closer to that of water than of air, so in water aquatic plants are buoyant and are supported against gravity.

A multicellular plant on land must either have a support system against gravity or else sprawl unsupported on the ground. It also must have the means to transport water and minerals from the soil to its aerial (raised) parts or else live where water is abundant enough to bathe all of its parts regularly. It must be able to resist desiccation (drying). And it must use different mechanisms for dispersing its gametes and progeny than do its aquatic relatives. How did such organisms arise from aquatic ancestors to thrive in such a challenging environment?

Most present-day plants have vascular tissue

An ancestral organism gave rise to the first plants, which were, in turn, ancestral to all present-day plants. The first plants were truly nonvascular, lacking both water-conducting and food-conducting cells. Later, the first tracheophytes arose from a nontracheophyte ancestor (Figure 27.3).

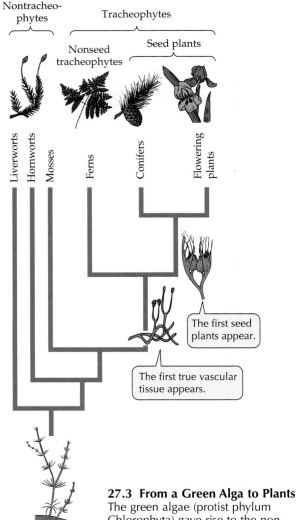

27.3 From a Green Alga to Plants
The green algae (protist phylum Chlorophyta) gave rise to the nontracheophytes, which in turn gave rise to the tracheophytes.

27.4 Some Nontracheophytes Form Mats Dense moss forms hummocks in a valley on New Zealand's South Island.

Liverworts, hornworts, and mosses never have been large plants. Except for some of the mosses, they have no water-transporting tissue, yet some are found in dry environments. Many grow in dense masses (Figure 27.4), through which water can move by capillary action. The "leafy" structure of nontracheophytes readily catches and holds water that splashes onto them. These plants are small enough that minerals can be distributed internally by diffusion. They lack the leaves, stems, and roots that characterize the tracheophytes, although they have structures analogous to each.

The tracheophytes include the ferns, conifers, and flowering plants. Tracheophytes differ from liverworts, hornworts, and mosses in crucial ways, one of which is the possession of a well-developed **vascular system** consisting of specialized tissues for the transport of materials from one part of the plant to another. One tissue, the **xylem**, conducts water and minerals from the soil to aerial parts of the plant; the other, the **phloem**, conducts the products of photosynthesis from sites of production or release to sites of utilization or storage (see Chapters 31 and 32).

The tracheophytes appear earlier than the nontracheophytes in the fossil record. The oldest tracheophyte fossils date back more than 410 million years, whereas the oldest nontracheophyte fossils are about 350 million years old, dating from a time when tracheophytes were already widely distributed. This does not mean that the tracheophytes are evolutionarily more ancient than the nontracheophytes; they are simply more likely to form fossils, given their structure and the chemical makeup of their cell walls.

Adaptations to life on land distinguish plants from algae

Most of the characteristics that distinguish plants from algae are evolutionary adaptations to life on land. Both nontracheophytes and tracheophytes have protective coverings that prevent drying, and both have means of taking up water from the soil (most tracheophytes have roots; nontracheophytes have rhizoids). Plants derive support against gravity from the internal fluid pressure (turgor pressure; see Chapter 32) of plant cells, from a woody stem in some tracheophytes, or from thickened cell walls in nontracheophytes. Unlike the algae, plants form embryos, young sporophytes contained within a protective structure.

We will examine the adaptations of the tracheophytes later in the chapter, but let's concentrate first on the nontracheophytes.

Nontracheophytes: Liverworts, Hornworts, and Mosses

Most liverworts, hornworts, and mosses are small and grow in dense mats in moist habitats (see Figure 27.4). The largest of these plants are only about 1 meter tall, but most are only a few centimeters tall or long. Why have larger nontracheophytes never evolved? The probable answer is that they lack an efficient system for conducting water and minerals from the soil to distant parts of the plant body.

Most nontracheophytes live on the soil or on other plants, but some grow on bare rock, dead and fallen tree trunks, and even on the buildings in which we live and work. These small plants have been here much longer than we have, and perhaps a quarter of a billion years longer than the tracheophytes have. Nontracheophytes are widely distributed over six continents and exist very locally on the coast of the seventh (Antarctica). They are very successful plants, well adapted to their environments. Most are terrestrial. Some live in wetlands. Although a few nontracheophytes live in fresh water, these aquatic forms are descended from terrestrial ones. There are no marine nontracheophytes. All nontracheophytes have a waxy covering that retards water loss, and their embryos are protected within layers of maternal tissue.

Nontracheophyte sporophytes are dependent on gametophytes

The life cycles of the nontracheophytes differ sharply from those of other plants in that the conspicuous green nontracheophyte that we recognize is the gametophyte, whereas the familiar forms of ferns and seed plants are sporophytes. The nontracheophyte gametophyte is photosynthetic and therefore nutritionally independent, whereas the sporophyte may or may not be photosynthetic but is *always* dependent on the gametophyte and remains permanently attached to it.

A sporophyte produces unicellular, haploid spores as products of meiosis. A spore germinates, giving rise to a multicellular, haploid gametophyte whose cells contain chloroplasts and are thus photosynthetic. Eventually, gametes form within specialized sex organs. The **archegonium** is a multicellular, flask-shaped female sex organ with a long neck and a swollen base (Figure 27.5a). The base contains a single egg. The **antheridium** is a male sex organ in which sperm, each bearing two flagella, are produced in large numbers (Figure 27.5b).

Once released, the sperm must swim or be splashed by raindrops to a nearby archegonium on the same or a neighboring plant. The sperm are aided in this task by chemical attractants released by the egg or the archegonium. Before sperm can enter the archegonium, certain cells in the neck of the archegonium must break down, leaving a water-filled canal through which the sperm swim to complete their journey. Note the dependence on liquid water for all these events.

On arrival at the egg, one of the sperm nuclei fuses with the egg nucleus to form the zygote. Mitotic divisions of the zygote produce a multicellular, diploid sporophyte embryo. The base of the archegonium grows to protect the embryo during its early growth. Eventually the developing sporophyte elongates sufficiently to break out of the archegonium, but it remains connected to the gametophyte by a "foot" that is embedded in the parent tissue and absorbs water and nutrients from it (see Figure 27.8). The sporophyte remains attached to the gametophyte throughout its life. The sporophyte produces a sporangium, or **capsule**, within which meiotic divisions produce spores and thus the next gametophyte generation.

The structure and pattern of elongation of the sporophyte differ among the three phyla of nontracheophytes—the liverworts (Hepatophyta), hornworts (Anthocerophyta), and mosses (Bryophyta). The evolutionary relationships of the three phyla and the tracheophytes can be seen in Figure 27.3.

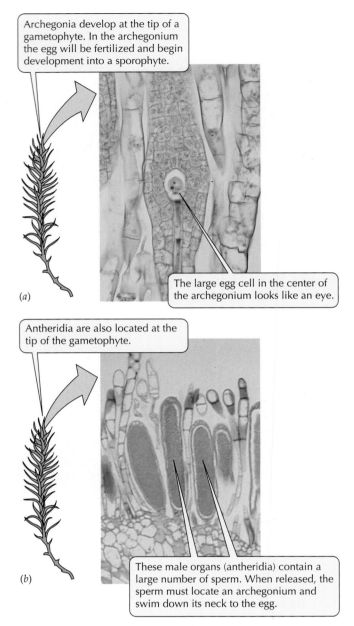

Archegonia develop at the tip of a gametophyte. In the archegonium the egg will be fertilized and begin development into a sporophyte.

(a)

The large egg cell in the center of the archegonium looks like an eye.

Antheridia are also located at the tip of the gametophyte.

(b)

These male organs (antheridia) contain a large number of sperm. When released, the sperm must locate an archegonium and swim down its neck to the egg.

27.5 Nontracheophyte Sex Organs Archegonia (a) and antheridia (b) of the moss *Mnium* (phylum Bryophyta).

Liverwort sporophytes have no specific growing zone

The gametophytes of some liverworts (phylum Hepatophyta) are "leafy" and prostrate. The simplest liverwort gametophytes, however, are flat plates of cells, a centimeter or so long, that produce antheridia or archegonia on their upper surfaces and rhizoids on the lower. Liverwort sporophytes are shorter than those of mosses and hornworts, rarely exceeding a few millimeters.

The sporophyte has a stalk that connects capsule and foot. The stalk elongates and thus raises the cap-

The finger-headed structures bear archegonia.

The disc-headed structures bear antheridia.

These cups are filled with gemmae—small, lens-shaped outgrowths of the body, each capable of developing into a new plant.

(a) *Marchantia* sp. (b) *Marchantia* sp. (c) *Lunularia* sp.

27.6 Liverwort Structures Members of the phylum Hepatophyta display various characteristic structures. (*a*) Gametophytes. (*b*) Antheridia and archegonia. (*c*) Gemmae.

sule above ground level, favoring dispersal of spores when they are released. The sporophyte elongates over the entire length of the stalk, which has no specific growing zone. The liverwort sporophyte has no stomata (pores allowing gas exchange between the atmosphere and the plant's interior).

The capsules of liverworts are simple: a globular capsule wall surrounding a mass of spores. In some species of liverworts spores are not released by the sporophyte until the surrounding capsule wall rots. In other liverworts, however, the spores are disseminated by structures called **elaters** located within the capsule. Elaters are long cells that have a helical thickening of the cell wall. As an elater loses water, the whole cell shrinks longitudinally to a fraction of its former length, thus compressing the helical thickening like a spring. When the stress becomes sufficient, the compressed "spring" snaps back to its resting position, throwing spores in all directions.

Among the most familiar liverworts are species of the genus *Marchantia* (Figure 27.6*a*). *Marchantia* is easily recognized by the characteristic structures on which its male and female gametophytes bear their antheridia and archegonia (Figure 27.6*b*). Like most liverworts, *Marchantia* also reproduces vegetatively by simple fragmentation of the gametophyte. *Marchantia* and some other liverworts and mosses also reproduce vegetatively by means of **gemmae** (singular gemma), lens-shaped clumps of cells loosely held in structures called gemma cups (Figure 27.6*c*).

Hornwort sporophytes grow at their basal end

The phylum Anthocerophyta, comprising the hornworts—so named because their sporophytes look like little horns—appear at first glance to be liverworts with very simple gametophytes (Figure 27.7*a*). These gametophytes consist of flat plates of cells, a few cells thick. However, hornworts—unlike liverworts—have stomata. And hornworts have two characteristics that distinguish them from both liverworts and mosses. First, the archegonia are embedded in the gametophytic tissue instead of being borne on stalks. Second, of all the nontracheophyte sporophytes, those of the hornworts come closest to being capable of indefinite growth (without a set limit).

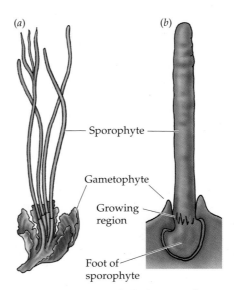

(a)

(b)

Sporophyte

Gametophyte

Growing region

Foot of sporophyte

27.7 Hornworts and Their Growth (*a*) The hornwort *Anthoceros*, drawn slightly larger than actual size. (*b*) The sporophyte of a hornwort grows from its basal end.

The stalk of either the liverwort or moss sporophyte stops growing as the capsule matures, so elongation of the sporophyte is strictly limited. In a hornwort such as *Anthoceros,* however, there is no stalk, but a basal region of the capsule remains capable of indefinite cell division, continuously producing new spore-bearing tissue above (Figure 27.7b). Sporophytes of some hornworts growing in mild and continuously moist conditions can become as tall as 20 cm, making them some of the tallest known nontracheophytes.

Whereas the photosynthetic cells in other plants have many small chloroplasts, each cell of a hornwort has only a single, large chloroplast. Hornworts have internal cavities filled with a mucilage, often populated by cyanobacteria that convert atmospheric nitrogen gas into a nutrient form usable by the host plant (see Chapter 34).

Moss sporophytes grow at their apical end

The most familiar nontracheophytes are the mosses (phylum Bryophyta). There are more species of mosses than of liverworts and hornworts combined, and these hardy little plants are found in almost every terrestrial environment. The mosses are sister to the tracheophytes (see Figure 27.3). The moss gametophyte that develops following spore germination is a branched, filamentous plant, or **protonema** (plural protonemata), that looks much like a filamentous green alga and is unique to this phylum. Some of the filaments contain chloroplasts and are photosynthetic; others are nonphotosynthetic and anchor the protonema to the substrate. The nonphotosynthetic filaments, called **rhizoids**, are the nontracheophyte counterpart of the root hairs (single-celled outgrowths) of tracheophytes, but they have no structural resemblance to true roots. After a period of growth, cells close to the tips of the photosynthetic filaments divide rapidly in three dimensions to form buds. The buds eventually differentiate a distinct apex and produce the familiar leafy moss plant with the "leaves" spirally arranged.

These leafy shoots produce antheridia or archegonia at their tips (Figure 27.8a). The antheridia release sperm that travel through liquid water to the archegonia, where they fertilize the eggs. Sporophyte development in most mosses follows a precise pattern, resulting ultimately in the formation of an absorptive foot, a stalk, and, at the tip, a swollen capsule. In contrast to hornworts, which grow from the base (see Figure 27.7b), the moss sporo-

(a)

Fertilization in mosses requires water so that sperm can swim to eggs.

Raindrop H₂O

Sperm (*n*)

Sperm (*n*)
Antheridium (*n*)

Egg (*n*)

Fertilization

Embryo (2*n*)

DIPLOID Sporophyte generation

Archegonium (*n*)

HAPLOID Gametophyte generation

Spores (*n*)

Sporophyte (2*n*)

Meiosis

Gametophyte (*n*)

Protonema with bud

Germinating spore

Gametophyte (*n*)

The "leafy" moss plant is the haploid gametophye.

The sporophyte is attached to and nutritionally dependent on the gametophyte.

(b)

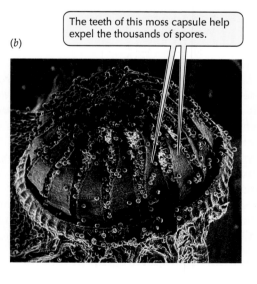

The teeth of this moss capsule help expel the thousands of spores.

27.8 The Moss Life Cycle

27.9 Peat Bogs The moss *Sphagnum* forms peat, which is used for fuel. This peat bog in Ireland is being harvested for commercial use.

phyte grows at its apical end, as do the tracheophytes. Cells at the tip of the stalk divide, supporting elongation of the structure and giving rise to the capsule. Unlike liverworts, moss sporophytes often possess stomata, allowing gas exchange with the atmosphere—the uptake of CO_2 for photosynthesis and the release of O_2.

The archegonial tissue grows rapidly as the stalk elongates, for a time keeping pace with the rapidly expanding sporophyte. Eventually, however, the archegonium is outgrown and is torn apart around its middle. The top portion of the archegonium frequently persists on the top of the rapidly elevating capsule as a little pointed cap, called the **calyptra**.

The top of the capsule is shed after meiosis and development of numerous mature spores within. Groups of cells just below the lid form a series of "teeth" surrounding the opening. Highly responsive to humidity, the teeth arch into the mass of spores when the atmosphere is dry and then out again, shoveling out spores, when the atmosphere becomes moist (Figure 27.8*b*). The spores are thus dispersed when the surrounding air is moist—that is, when conditions favor their subsequent germination.

Only a few mosses depart from this pattern of capsule development. A familiar exception is the genus *Sphagnum*, which occurs in tremendous quantities in northern bogs and tundra extending well into the Arctic (Figure 27.9) and whose global biomass probably exceeds that of all other mosses combined. Species

in this genus have a very simple capsule with an air chamber in it. Air pressure builds up in this chamber, eventually causing the capsule lid to pop open, dispersing the spores with an audible explosion.

Many mosses contain a type of cell, called a **hydroid**, that dies and leaves a tiny channel through which water may travel. The hydroid likely is a progenitor of the characteristic water-conducting cell of the tracheophytes, but it lacks lignin (a waterproofing substance) and the wall structure found in tracheophyte water-conducting cells. The possession of hydroids and of a limited system for transport of foods by some mosses shows that the old term "nonvascular plant" is somewhat misleading when applied to mosses.

Introducing the Tracheophytes

Although an extraordinarily large and diverse group, the tracheophytes can be said to have been launched by a single evolutionary event. Sometime during the Paleozoic era, probably well before the Silurian period, the sporophyte generation of a now long-extinct organism produced a new cell type, the **tracheid**. The tracheid is the principal water-conducting element of the xylem in all tracheophytes except the angiosperms (flowering plants); even in angiosperms the tracheid persists along with a more specialized and efficient system of vessels and fibers that evolved from tracheids.

The evolutionary appearance of a tissue composed of tracheids had two important consequences. First, it provided a pathway for long-distance transport of water and mineral nutrients from a source of supply to regions of need. Second, it provided something almost completely lacking—and unnecessary—in the largely aquatic algae: rigid structural support. Support is important in a terrestrial environment because land plants tend to grow upward as they compete for sunlight to power photosynthesis. Thus the tracheid set the stage for the complete and permanent invasion of land by plants.

The present-day evolutionary descendants of the early tracheophytes belong to nine distinct phyla (Figure 27.10). We can sort these phyla into two groups: those that produce seeds and those that do not. In the nonseed tracheophytes the haploid and diploid generations are independent at maturity. The sporophyte is the large and obvious plant that one normally notices in nature (in contrast to the nontracheophyte sporophyte, which is attached to and dependent on the gametophyte). Gametophytes of the nonseed tracheophytes are rarely more than 1 or 2 centimeters long and are short-lived, whereas the sporophyte of a tree fern, for example, may be 15 or 20 meters tall and may live for years.

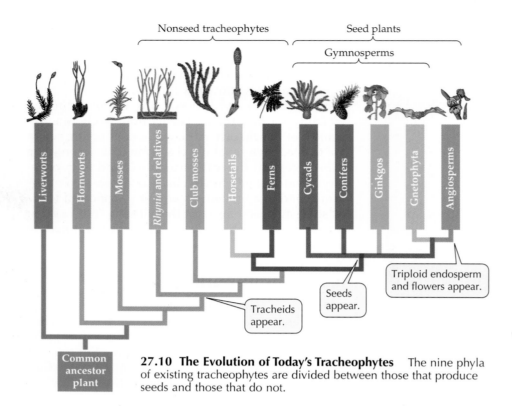

27.10 The Evolution of Today's Tracheophytes The nine phyla of existing tracheophytes are divided between those that produce seeds and those that do not.

The most prominent resting stage in the life cycle of a nonseed tracheophyte is the single-celled spore. This feature makes this life cycle similar to those of the fungi, the algae, and the nontracheophytes but not, as we will see, to that of the seed plants. Nonseed tracheophytes must have an aqueous environment for at least one stage of their life cycle because fertilization is accomplished by a motile, flagellated sperm.

We will discuss the life cycle of seed plants later in this chapter, but now we turn to a more detailed account of the evolution of the tracheophytes.

The tracheophytes have been evolving for almost half a billion years

The green algal ancestors of the plant kingdom successfully invaded the terrestrial environment between 400 and 500 million years ago. The evolution of a water-impermeable cuticle (waxy outer covering) and of protective layers for the gamete-bearing structures helped make the invasion successful, as did the initial absence of herbivores (plant-eating animals).

By the late Silurian period (see Table 19.1) tracheophytes were being preserved as fossils that we can study today. Several remarkable developments arose during the Devonian period, 400 to 345 million years ago. Three groups of nonseed tracheophytes that still exist made their first appearances during that period: the lycopods (club mosses), horsetails, and ferns. The proliferation of these plants made the terrestrial envi-

ronment ever more hospitable to animals; amphibians and insects arrived soon after the plants became established. Fossil remains about 360 million years old provide the first evidence of seed plants.

Trees of various kinds appeared in the Devonian, and came into their own in the Carboniferous period (345 to 290 million years ago). Mighty forests of lycopods up to 40 meters tall, horsetails, and tree ferns flourished in the tropical swamps of what would become North America and Europe (Figure 27.11). In the subsequent Permian period the continents came ponderously together to form a single, gigantic land mass, called Pangaea. The continental interior become warmer and drier, but late in the period glaciation was extensive. The 200-million-year reign of the lycopod–fern forests came to an end, to be replaced by forests of nonflowering seed plants (gymnosperms) that ruled throughout the Triassic and Jurassic periods. The gymnosperm forests changed with time as the gymnosperm groups evolved. Gymnosperm forests dominated during the era in which the continents drifted apart and dinosaurs strode the Earth.

The oldest evidence of angiosperms (flowering plants) dates to the Cretaceous period, about 120 million years ago. The angiosperms radiated almost explosively and, over a period of about 55 million years, became the dominant plant life of the planet.

The earliest tracheophytes lacked roots and leaves

The first tracheophytes belonged to the now-extinct phylum Rhyniophyta. The rhyniophytes appear to have been the only tracheophytes in the Silurian period of the Paleozoic era. The landscape at that time probably consisted of bare ground, with stands of rhyniophytes in low-lying moist areas. Early versions of the structural features of all the other tracheophyte phyla appeared in the plants of that time. These shared features strengthen the case for the origin of all tracheophytes from a common nontracheophyte ancestor.

In 1917 the British paleobotanists Robert Kidston and William H. Lang reported well-preserved fossils of tracheophytes embedded in Devonian rocks near Rhynie, Scotland. The preservation of these plants was remark-

Tree fern

Fern

Early seed plants

Seed ferns

Immature lycopods

Mature lycopods

Tree horsetails

Herbaceous horsetails

27.11 An Ancient Forest   A little more than 300 million years ago, a forest grew in a setting similar to tropical river delta habitats of today. Most of the plants depicted here were nonseed tracheophytes 10 to 20 m tall. Far in the distance, early seed plants—giants up to 40 m tall—towered over the forest.

able, considering that the rocks were more than 395 million years old. These fossil plants had a simple vascular system consisting of phloem and xylem (tracheids or hydroids). Flattened scales on the stems of some of the plants lacked vascular tissue and thus were not comparable with the true leaves of any other tracheophytes.

Lacking roots, the plants were apparently anchored in the soil by horizontal portions of stem (**rhizomes**) that bore rhizoids. These rhizomes also bore aerial branches. Sporangia—homologous with the nontracheophyte capsule—were found at the tips of the stems. Branching was dichotomous; that is, the shoot apex divided to produce two equivalent new branches, each pair diverging at approximately the same angle from the original stem (Figure 27.12). Scattered fragments of such plants had been found earlier, but never in such profusion or so well preserved as those discovered near Rhynie by Kidston and Lang.

The presence of xylem in these plants indicated that they, named *Rhynia* after the site of their discovery, were tracheophytes. But were they sporophytes or gametophytes? Close inspection of thin sections of fossil sporangia revealed that the spores were in groups of four. In almost all living nonseed tracheophytes (with no evidence to the contrary from fossil forms), the four

products of a meiotic division and cytokinesis remain attached to one another during their development into spores. The spores separate only when they are mature, and even after separation their walls reveal the exact geometry of how they were attached. Thus a group of four closely packed spores is found only immediately after meiosis, and a plant that produces

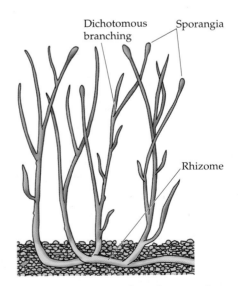

Dichotomous branching

Sporangia

Rhizome

27.12 The First Tracheophytes   This extinct plant in the genus *Rhynia* (phylum Rhyniophyta) lacked roots and leaves. The rhizome is a horizontal underground stem, not a root. The aerial shoots were less than 50 cm tall, and some were topped by sporangia.

such a group of four must be a diploid sporophyte. The gametophytes of the Rhyniophyta were branched. Depressions at the apices of the branches contained archegonia and antheridia.

Although apparently ancestral to the other tracheophyte phyla, the rhyniophytes themselves are long gone. None of their fossils appear anywhere after the Devonian period.

Early tracheophytes added new features

Within a few tens of millions of years, during the Devonian period, three new phyla—Lycophyta, Sphenophyta, and Pterophyta—of tracheophytes appeared on the scene, arising from rhyniophyte ancestors. These new groups featured specializations over the rhyniophytes, including one or more of the following: true roots, true leaves, and a differentiation between two types of spores.

THE ORIGIN OF ROOTS. *Rhynia* and its close relatives lacked true roots. They had only rhizoids, arising from a prostrate rhizome (see Figure 27.12), with which to gather water and minerals. How, then, did subsequent groups of tracheophytes come to have the complex roots we see today?

A French botanist, E. A. O. Lignier, proposed an attractive hypothesis in 1903 that is still widely accepted today. Lignier argued that the ancestors of the first tracheophytes branched dichotomously. This explanation accounts for the dichotomous branching observed in the rhyniophytes themselves. Lignier suggested that such a branch could bend and penetrate the soil, branching there (Figure 27.13). The underground portion could anchor the plant firmly, and even in this primitive condition it could absorb water and minerals. The subsequent discovery of fossil plants from the Devonian period, all having horizontal stems (rhizomes) with both underground and aerial branches, supported Lignier's hypothesis.

The underground branches, in an environment sharply different from that above the ground, were subjected to very different selection during the succeeding millions of years. Thus the two parts of the plant axis (the shoot and root systems) diverged in structure and came to have distinct internal and external anatomies (see Chapter 31). In spite of these differences, we believe that the root and shoot systems of tracheophytes are homologous—that that they were once part of the same organ.

THE ORIGIN OF TRUE LEAVES. Thus far, we have used the term "leaf" rather loosely. We spoke of "leafy" mosses; we also commented on the absence of "true leaves" in rhyniophytes. In the strictest sense, a **leaf** is a flattened photosynthetic structure emerging laterally from a main axis or stem and possessing true vascular tissue.

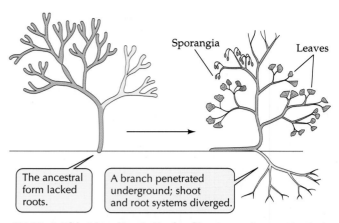

27.13 Is This How Roots Evolved? According to Lignier's hypothesis, branches from ancestral rootless plants could have penetrated the soil, where they gradually evolved into a root system.

This precise definition allows a closer look at true leaves in the tracheophytes, which shows us that there are two different types of leaves, very likely of different evolutionary origins.

The first type of leaf is usually small and only rarely has more than a single vascular strand, at least in plants alive today. Plants in the phylum Lycophyta

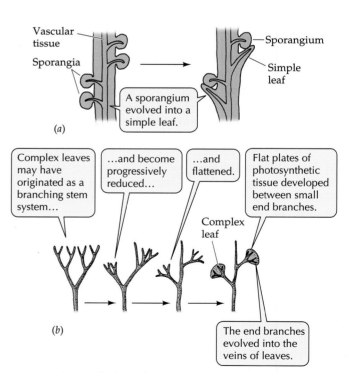

27.14 The Evolution of Leaves (a) Simple leaves are thought to have evolved from sterile sporangia. (b) The complex leaves of ferns and seed plants may have arisen as photosynthetic tissue developed between complex branching patterns.

(club mosses), of which only a few genera survive, have such *simple* leaves. The evolutionary origin of this type of leaf is thought to be sterile sporangia (Figure 27.14*a*). The principal characteristic of this type of leaf is that its vascular strand departs from the vascular system of the stem in such a way that the structure of the stem's vascular system is scarcely disturbed. This was true even in the fossil lycopod trees of the Carboniferous period, many of which had leaves several centimeters long.

The other type of leaf is encountered in ferns and seed plants. This larger, more *complex* type of leaf is thought to have arisen from the flattening of a dichotomously branching stem system, with the development of extensive photosynthetic tissue between the branch members (Figure 27.14*b*). The complex leaf may have evolved several times, in different phyla of tracheophytes.

HOMOSPORY AND HETEROSPORY.   In the most ancient of the present-day tracheophytes, both the gametophyte and the sporophyte are independent and usually photosynthetic. Spores produced by the sporophytes are of a single type, and they develop into a single type of gametophyte, bearing both female and male reproductive organs. Such plants, which bear a single type of spore, are said to be **homosporous** (Figure 27.15*a*). The sex organs on the gametophytes of homosporous plants are of two types. The female organ is a multicellular archegonium, typically containing a single egg. The male organ is an antheridium, containing many sperm.

A different system, with two distinct types of spores, evolved somewhat later. Plants of this type are said to be **heterosporous** (Figure 27.15*b*). One type of spore, the **megaspore**, develops into a larger, specifically female gametophyte (megagametophyte) that produces only eggs. The other type, the **microspore**, develops into a smaller, male gametophyte (microgametophyte) that produces only sperm. Megaspores are produced in small numbers in megasporangia on the sporophyte, and microspores in large numbers in microsporangia.

The most ancient tracheophytes were all homosporous. Heterospory evidently evolved several times, independently, in the early evolution of the tracheophytes descended from the rhyniophytes. The fact that heterospory evolved re-

peatedly suggests that it affords selective advantages. As we will see, subsequent evolution in the plant kingdom featured ever greater specialization of the heterosporous condition.

The Surviving Nonseed Tracheophytes

Ferns are now the most abundant and diverse phylum of nonseed tracheophytes, although the club mosses and horsetails were once dominant elements of Earth's

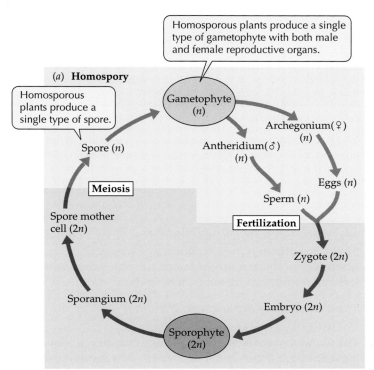

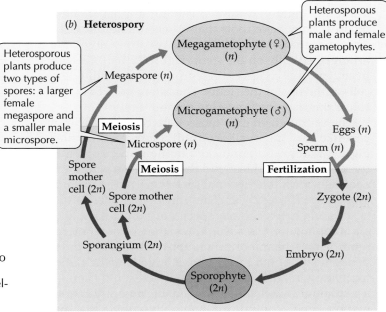

27.15 Homospory and Heterospory   (*a*) Homosporous plants bear a single type of spore. Each gametophyte has two types of sex organs, antheridia (male) and archegonia (female). (*b*) Heterospory, with two types of spores that develop into distinctly male and female gametophytes, evolved later.

(a) *Lycopodium flabellitorme*

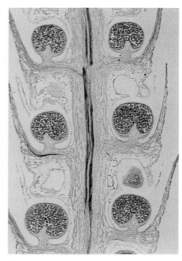

(b)

27.16 Club Mosses   (a) Conelike strobili are visible at the tips of this club moss. Club mosses have simple leaves arranged spirally on their stems. (b) Thin section through a strobilus of another club moss.

vegetation. A fourth phylum, the whisk ferns, contains only two genera. In this section we'll look at the characteristics of these four phyla and at some of the evolutionary advances that appeared in them.

The club mosses are sister to the other tracheophytes

The club mosses (lycopods, phylum Lycophyta) have ancient origins; the remaining tracheophytes share an ancestor that was not ancestral to the Lycophyta. There are relatively few surviving species of club mosses. They have roots that branch dichotomously. They bear only simple leaves, and the leaves are arranged spirally on the stem (Figure 27.16a). Growth in club mosses comes entirely from groups of dividing cells at the tips of the stems. Thus stem growth is apical, as it is in many flowering plants.

The sporangia in club mosses are contained within conelike structures called **strobili** (singular strobilus; Figure 27.16b) and are tucked in the upper angle between a specialized leaf and the stem. This placement contrasts with the terminal sporangia of the rhyniophytes (see Figure 27.12). There are both homosporous species and heterosporous species of club mosses. Like all other nonseed tracheophytes, they have a heteromorphic alternation of generations with a large, independent sporophyte and a small, independent gametophyte.

Although only a minor element of present-day vegetation, the Lycophyta are one of two phyla that ap-

pear to have been the dominant vegetation during the Carboniferous period. One abundant type of coal, Cannel coal, is formed almost entirely from fossilized spores of a tree lycopod named *Lepidodendron;* this finding is an indication of the abundance of this genus in the forests of that time. The other major element of the Carboniferous vegetation was the phylum Sphenophyta, the horsetails.

Horsetails grow at the bases of their segments

Like the club mosses, the horsetails (phylum Sphenophyta) are represented by only a few present-day species. They are sometimes called scouring rushes because silica deposits found in the cell walls made them once useful for cleaning. They have true roots that branch irregularly, as do the roots of all tracheophytes except the club mosses. Their sporangia curve back toward the stem on the ends of short stalks (sporangiophores) (Figure 27.17a). Horsetails have a large sporophyte and a small gametophyte, both independent.

The leaves of horsetails are simple and form distinct whorls (circles) (Figure 27.17b). Growth in horsetails originates to a large extent from discs of dividing cells just above each whorl of leaves, so each segment of the stem grows from its base. Such basal growth is uncommon in plants, but it is also found in the grasses, a major group of flowering plants.

Present-day whisk ferns resemble the most ancient tracheophytes

There once was some disagreement about whether rhyniophytes are entirely extinct. The confusion arose because of the existence today of two genera of rootless, spore-bearing plants, *Psilotum* and *Tmesipteris.*

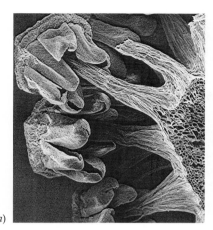

27.17 Horsetails (*a*) Sporangia and sporangiophores of the horsetail *Equisetum arvense*. (*b*) Vegetative and fertile shoots of the marsh horsetail. Leaves form in spaced whorls at nodes on the stems of the vegetative shoot on the right; the fertile shoot on the left is ready to disperse its spores.

(*b*) *Equisetum palustre*

the rhyniophytes, which apparently became extinct more than 300 million years ago, and *Psilotum* and *Tmesipteris*, which are modern plants. DNA sequence data finally settled the question in favor of a more modern origin. Most botanists now treat these two genera as their own phylum, the Psilophyta, or whisk ferns.

The whisk ferns are highly specialized plants that evolved fairly recently from anatomically more complex ancestors. Today *Psilotum* is widely distributed in the Tropics and Subtropics.

Ferns evolved large, complex leaves

The sporophytes of the ferns and seed plants have roots, stems, and leaves. Their leaves are typically large and have branching vascular strands. Some species have small leaves as a result of evolutionary reduction, but even the small leaves have more than one vascular strand. The sporangia of these plants are on the lower surfaces of leaves or leaflike structures or, more rarely, on their margins.

Psilotum nudum (Figure 27.18) has only minute scales instead of true leaves, but plants of the genus *Tmesipteris* have flattened photosynthetic organs with well-developed vascular tissue. Are these two genera the living relics of the rhyniophytes, or do they have more recent origins?

Psilotum and *Tmesipteris* once were thought to be evolutionarily ancient descendants of anatomically simple ancestors. That hypothesis was weakened by an enormous hole in the geologic record—between

The true ferns constitute the phylum Pterophyta, which first appeared during the Devonian period and today consists of about 12,000 species. Ferns are characterized by fronds (large leaves with complex vasculature) (Figure 27.19), by the absence of seeds, and by a requirement for water for the transport of the male gametes to the female gametes. Most ferns inhabit shaded, moist woodlands and swamps. Some, the tree ferns, reach heights of up to 20 m. Tree ferns lack the rigidity of woody plants, and thus do not grow in sites exposed directly to strong winds but rather in ravines or beneath trees in forests.

During its development, the fern frond unfurls from a tightly coiled "fiddlehead" (Figure 27.19*c*). Some fern leaves become climbing organs and may grow to be as much as 30 m long. The sporangia are found on the undersurfaces of the leaves, sometimes covering the whole undersurface and sometimes only at the edges; in some species the sporangia are clustered in groups called **sori** (singular sorus) (Figure 27.20).

Devonian fossil beds have yielded ferns with some characteristics that are rhyniophyte and some that resemble those of other tracheophyte phyla. For example, like a modern fern, the fossil genus *Protopteridium* had flattened branch systems with extensive photosynthetic tissue between the branches; but like a rhyniophyte, it bore terminal sporangia and lacked

Psilotum nudum

27.18 A Modern Whisk Fern Aerial branches of a whisk fern, a plant once considered by some to be a surviving rhyniophyte and by others to be a fern. It is now included in the phylum Psilophyta.

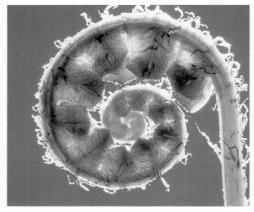

(a)

(b)

(c) *Marsilea mutica*

27.19 Fern Fronds Take Many Forms (a) Fern fronds blanket a forest floor in the Florida Everglades. (b) "Fiddleheads" (developing fronds) of a common forest fern; this structure will unfurl and expand to give rise to a complex adult frond such as those in (a). (c) The tiny fronds of a water fern.

true roots (Figure 27.21). During late Paleozoic times, the ferns underwent considerable evolutionary experimentation in the structure of their leaves and particularly in the arrangement of their vascular tissue.

The sporophyte generation dominates the fern life cycle

The undersides or edges of fern fronds carry sporangia in which cells undergo meiosis to form haploid spores (Figure 27.22). Once shed, spores often travel great distances and eventually germinate to form small, independent gametophytes. These gameto-

phytes produce antheridia and archegonia, although not necessarily at the same time or on the same gametophyte.

Sperm swim through water to archegonia, often on other gametophytes, where they unite with an egg. The resulting zygote develops into a new sporophyte embryo. The young sporophyte sprouts a root and can thus grow independently of the gametophyte. In the alternating generations of a fern, the gametophyte is small, delicate, and short-lived, but the sporophytes can be very large and can sometimes survive for hundreds of years.

Most ferns are homosporous. However, two groups of aquatic ferns, the Marsileales and Salviniales, have evolved heterospory. Megaspores and microspores of these plants (which germinate to produce female and male gametophytes, respectively) are produced in different sporangia, and the male spores are always

Dryopteris intermedia

27.20 Fern Sori Contain Sporangia Sori, each with many spore-producing sporangia, form on the underside of a frond of the midwestern fancy fern.

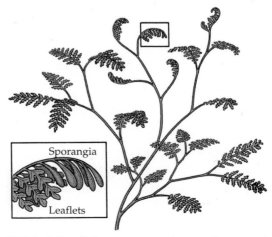

27.21 A Fossil Fern During their evolutionary history, ferns exhibited combinations of structures. This fossil fern (*Protopteridium*, from the Devonian period) lacked true roots and had branches with both leaflets and terminal sporangia.

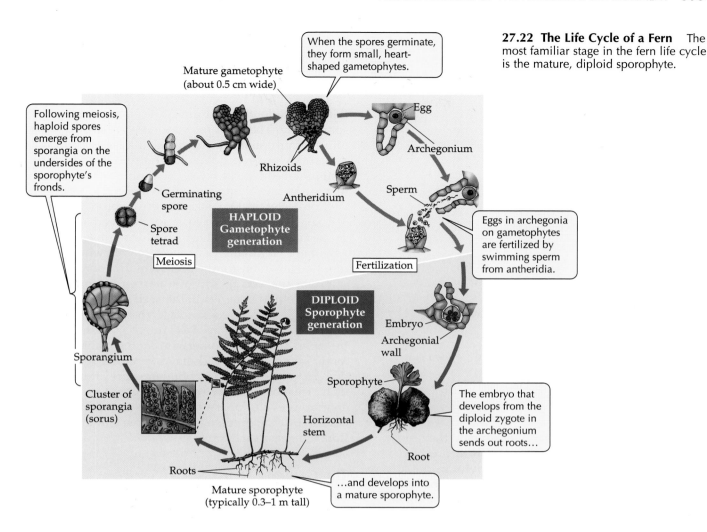

When the spores germinate, they form small, heart-shaped gametophytes.

27.22 The Life Cycle of a Fern The most familiar stage in the fern life cycle is the mature, diploid sporophyte.

Mature gametophyte (about 0.5 cm wide)

Following meiosis, haploid spores emerge from sporangia on the undersides of the sporophyte's fronds.

Egg

Archegonium

Rhizoids

Germinating spore

Antheridium

Sperm

Spore tetrad

HAPLOID Gametophyte generation

Meiosis

Fertilization

Eggs in archegonia on gametophytes are fertilized by swimming sperm from antheridia.

DIPLOID Sporophyte generation

Embryo

Archegonial wall

Sporangium

Sporophyte

Cluster of sporangia (sorus)

Horizontal stem

The embryo that develops from the diploid zygote in the archegonium sends out roots...

Root

Roots

Mature sporophyte (typically 0.3–1 m tall)

...and develops into a mature sporophyte.

much smaller and greater in number than the female spores.

A few genera of ferns produce a tuberous, fleshy gametophyte instead of the characteristic flattened, photosynthetic structure described earlier. These tuberous gametophytes depend on a mutualistic fungus for nutrition*; in some genera, even the sporophyte embryo must become associated with the fungus before extensive development can proceed. In Chapter 28 we will see that there are many important plant–fungus associations.

The Seeds of Success

The most recent group to appear in the evolution of the tracheophytes is the seed plants: the paraphyletic **gymnosperms** (such as pines and cycads) and the monophyletic **angiosperms** (flowering plants). In seed plants the gametophyte generation is reduced even

*In a mutualistic association, both partners—here, the gametophyte and the fungus—profit.

further than it is in ferns. The haploid gametophyte develops partly or entirely while attached to and nutritionally dependent on the diploid sporophyte. Among the seed plants, only the earliest types of gymnosperms and their few survivors (cycads, for example) had swimming sperm. All other seed plants have evolved other means of bringing gametes together. The culmination of this striking evolutionary trend in plants was independence from liquid water for the purposes of reproduction.

Seed plants are heterosporous, forming separate megasporangia and microsporangia—female and male sporangia, respectively—on modified leaves or leaflike structures that are grouped on short axes to form strobili (see Figure 27.16), such as the cones of conifers and the flowers of angiosperms.

As in other plants, spores of seed plants are produced by meiosis within the sporangia, but in this case they are not shed. Instead, the gametophytes develop within the sporangia and depend on them for food and water. In most species only one of the meiotic products in a megasporangium survives. The surviv-

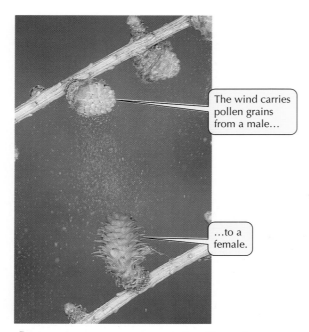

27.23 Gymnosperms Are Wind-Pollinated   Pollen grains are the male gametophytes of gymnosperms.

ing haploid nucleus divides mitotically, and the resulting cells (without walls) divide again to produce a small multicellular female gametophyte. In the angiosperms, female gametophytes do not normally include more than eight nuclei. The female gametophyte of a seed plant is retained within the megasporangium, where it matures, is fertilized, and undergoes the early development of the next sporophyte generation.

Within the microsporangium the meiotic products develop into microspores that divide one or a few times to form male gametophytes, the familiar **pollen grains** (Figure 27.23; see also Figure 36.2). Distributed by wind, by an insect, by a bird, or by a plant breeder, a pollen grain that reaches the appropriate surface of a sporophyte, near the female gametophyte, develops further. It produces a slender **pollen tube** that grows and digests its way through the sporophyte tissue toward the female gametophyte.

When the tip of the male gametophyte's pollen tube reaches the female gametophyte, one or two sperm cells are released from the tube and fertilization occurs. The resulting diploid zygote divides repeatedly, forming a young sporophyte that develops to an embryonic stage at which growth becomes temporarily suspended (often referred to as a dormant stage). The end product at this stage is a **seed**.

A seed may contain tissues from three generations. The seed coat develops from tissue of the diploid parent sporophyte. Within the seed coat is a layer of hap-

loid female gametophyte tissue from the next generation (this tissue is fairly extensive in most gymnosperm seeds, but in angiosperm seeds its place is taken by a tissue called endosperm, which we will discuss shortly). In the center of the seed package is the third generation, in the form of the embryo of the new sporophyte. The embryos of nonseed plants develop directly into sporophytes, which either survive or die, depending on environmental conditions; there is no resting stage in the life cycle. By contrast, the multicellular seed of a gymnosperm or an angiosperm is a well-protected resting stage for the embryo. Layers of cells enclose the embryo, and the seeds of some species may remain viable (capable of growth and development) for many years, germinating when conditions are favorable for the growth of the sporophyte.

When the young sporophyte begins to grow, it draws on food reserves in the seed. During the dormant stage the seed coat protects the embryo from excessive drying and may also protect against potential predators that would otherwise eat the embryo or the food reserves. Many seeds have structural adaptations that promote dispersal by wind or, more often, by animals. The possession of seeds is a major reason for the enormous evolutionary success of seed plants, which are the prominent elements of Earth's modern land flora in most areas.

The Gymnosperms: Naked Seeds

The gymnosperms are a paraphyletic group of seed plants that never form flowers. Although there are probably fewer than 750 species of living gymnosperms, these plants are second only to the angiosperms (flowering plants) in their dominance of the land masses.

There are four phyla of gymnosperms living today. The cycads (phylum Cycadophyta) are palmlike plants of the Tropics, growing as tall as 20 meters (Figure 27.24a). Ginkgos (phylum Ginkgophyta), which were common during the Mesozoic era, are represented today by a single genus and species, *Ginkgo biloba*, the maidenhair tree (Figure 27.24b). The phylum Gnetophyta consists of three very different genera that share a characteristic type of cell (the vessel element) in their xylem tissue that is found in no other group of plants except the angiosperms, which shared an ancestor with the Gnetophyta (see Figure 27.10). One member of this phylum is *Welwitschia* (Figure 27.24c), a long-lived desert plant with just two straplike leaves that sprawl on the sand and can become as long as 3 m. Far and away the most abundant of the gymnosperms are the conifers (phylum Coniferophyta), cone-bearing plants such as pines and redwoods (Figure 27.24d).

All living gymnosperms have active secondary growth (growth in the diameter of stems and roots; see Chapter 31), and all but the Gnetophyta have only tra-

(a) *Cycas revoluta*

(c) *Welwitschia mirabilis*

(b) *Ginkgo biloba*

27.24 Gymnosperms (a) This sago palm belongs to the cycads, the most ancient of the present-day gymnosperms. Many cycads have growth forms that resemble both ferns and palms. (b) Characteristic fruit and broad leaves of the maidenhair tree. (c) A gnetophyte growing in the Namib Desert of Africa. Two huge, straplike leaves grow throughout the life of the plant, breaking and splitting as they grow. (d) A dramatic conifer, this giant sequoia grows in Yosemite National Park, California.

(d) *Sequoiadendron giganteum*

cheids as water-conducting and support cells in the xylem. Despite this water transport and support system (which might seem suboptimal compared with that of angiosperms), the gymnosperms include some of the tallest trees known. The coastal redwoods of California are the tallest gymnosperms; the largest are well over 100 m tall. Xylem produced by gymnosperms is the principal resource of the lumber industry.

In this section on gymnosperms, we'll take a brief look at their fossil history and then examine the life cycle of the conifers in some detail.

We know the early gymnosperms only as fossils

The earliest fossil evidence of gymnosperms is found in Devonian rocks. The early gymnosperms combined characteristics of rhyniophytes and heterosporous ferns, but they had tracheids of the same type found in modern gymnosperms. They also differed from the plants around them by their extensively thickened stems.

By the Carboniferous period, several new lines of gymnosperms had evolved, including various seed ferns that possessed fernlike foliage but had characteristic gymnosperm seeds attached to the leaf margins.

The first true conifers appeared at approximately the same time. Either they were not dominant trees or they did not grow where conditions were right for fossilization, so we have few preserved examples. During the Permian period, however, the conifers and cycads came into their own. Gymnosperms dominated *all* forests until less than 80 million years ago, and they still dominate some present-day forests.

Conifers have cones but no motile cells

The great Douglas fir and cedar forests of the northwestern United States and the massive forests of pine, fir, and spruce that clothe the northern continental regions and upper slopes of mountain ranges rank among the great vegetation formations of the world (the "boreal forest"; see Chapter 54). All these trees belong to one phylum of gymnosperms, Coniferophyta—the conifers, or cone-bearers. Male and female spores are produced in separate male and female cones. Female cones are larger than male cones.

We will use the life cycle of a pine to illustrate reproduction in gymnosperms (Figure 27.25). The production of male gametophytes as pollen grains frees the plant once and for all from its dependence on external liquid water for fertilization. Instead of water, wind assists conifer pollen grains in their first stage of travel to the female gametophyte. The pollen tube provides the means for the last stage of travel by growing and digesting its way through maternal sporophytic tissue and eventually releasing a sperm cell near the egg.

The megasporangium, which will form the female gametophyte containing eggs within archegonia, is enclosed in a special layer of sporophytic tissue, called the **integument**, that will eventually develop into the seed coat. The integument, the megasporangium inside it, and the tissue attaching it to the maternal sporophyte form the **ovule**. The pollen grain travels through the small opening in the integument at the apex of the ovule, the **micropyle**.

Gymnosperms (meaning literally "naked-seeded") derive their name from the fact that their seeds are not protected by fruit tissue. Most conifer ovules (which on fertilization develop into seeds) are borne exposed on the upper surfaces of cone scales. The only protection from the environment is that the scales are tightly pressed against each other within the cone. Some pines have such tightly closed female cones that only fire suffices to split them open and release the seeds. One example is the lodgepole pine, which covers vast fire-prone areas in the Rocky Mountains and elsewhere.

About half of the conifer species have fruitlike, fleshy tissues associated with their seeds; examples are the "berries" of juniper and yew. Animals may eat these tissues and then disperse the seeds via the waste they deposit, often carrying them considerable distances—which may spread the conifer population. True fruits are one of the characteristics of the plant phylum that is dominant today: Angiospermae.

The Angiosperms: Flowering Plants

The phylum Angiospermae* consists of the flowering plants, also known as the angiosperms (literally "enclosed-seeded"). This highly diverse phylum includes about 275,000 species. In other chapters, when we mention plants in discussing processes such as long-distance transport in the xylem and phloem, or the chemical regulation of development, generally we are referring to the angiosperms.

The angiosperms represent the current extreme of an evolutionary trend that runs throughout the tracheophytes, in which the *sporophyte* generation becomes *larger* and *more independent* of the gametophyte, while the *gametophyte* generation becomes *smaller* and *more dependent* on the sporophyte. Angiosperms differ from other plants in several ways, although exceptions exist.

Double fertilization was long considered the single most reliable distinguishing characteristic of the angiosperms. *Two* male gametes, contained within a single microgametophyte (pollen grain), participate in fertilization events within the megagametophyte of an angiosperm. One sperm combines with the egg to produce a diploid zygote, the first cell of the sporophyte generation. The other sperm nucleus usually combines with two other haploid nuclei of the female gametophyte to form a triploid ($3n$) nucleus. This nucleus, in turn, divides to form a triploid tissue, the **endosperm**, that nourishes the embryonic sporophyte during its early development.

Double fertilization occurs in all present-day angiosperms and probably in all three existing genera of Gnetophyta: *Ephedra*, *Gnetum*, and *Welwitschia*. William Friedman, at the University of Georgia, confirmed in 1990 that *Ephedra* has double fertilization but that both fertilizations produce diploid products. Thus, we are left with the formation of an extensive triploid endosperm as the most definitively angiosperm trait. We are not sure when and how double fertilization evolved because there is no fossil evidence on this point.

A second consistent characteristic of angiosperms is the possession of specialized water-transporting cells called **vessel elements** in the xylem, but these are also found, in different form, in a few gymnosperms and

---

* The flowering plants have also been referred to as the phylum Anthophyta. However, that term is now typically used to denote the monophyletic group consisting of the Gnetophyta and the flowering plants.

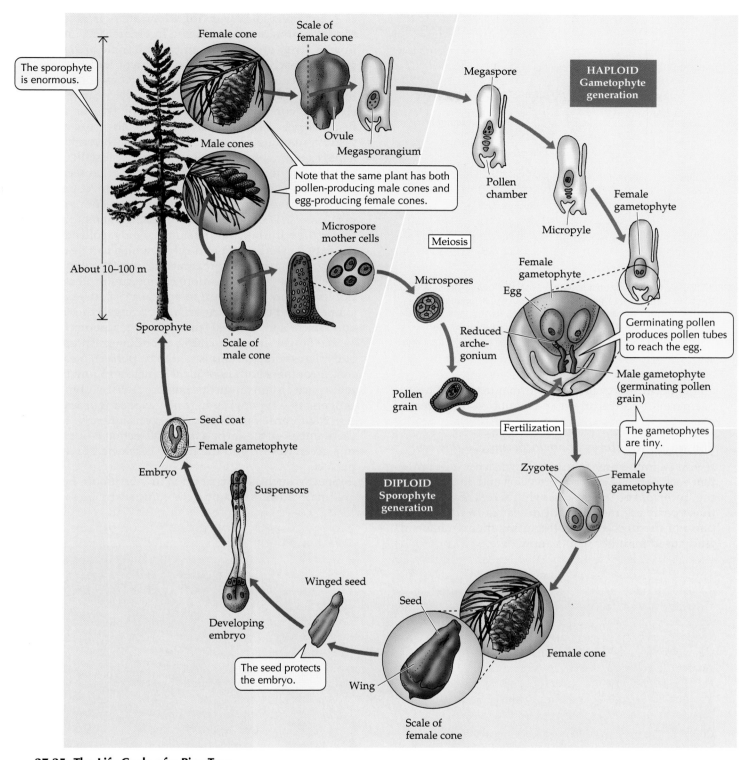

27.25 The Life Cycle of a Pine Tree

ferns, and even in a club moss and a horsetail. Another distinctive cell in angiosperm xylem is the **fiber**, which plays an important role in supporting the plant body. The name "angiosperm" refers to the diagnostic character that the seeds of these plants are enclosed in a modified leaflike structure called a *carpel*. Of course, the most evident diagnostic feature of angiosperms is that they have flowers.

In the following sections we'll examine the structure and function of flowers, evolutionary trends in flower structure, the functions of pollen and fruits, the

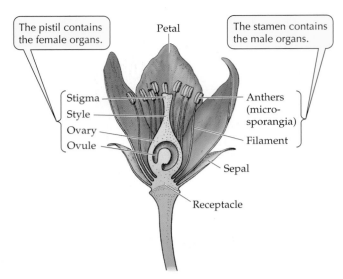

The pistil contains the female organs.

The stamen contains the male organs.

Petal

Stigma
Style
Ovary
Ovule

Anthers
(micro-
sporangia)

Filament

Sepal

Receptacle

27.26 A Generalized Flower Not all flowers possess all the structures shown here, but they must possess stamens, pistils, or both in order to play their role in sexual reproduction. Flowers that have both, such as this one does, are referred to as perfect.

angiosperm life cycle, the two major groups of angiosperms, and the origin and evolution of flowering plants.

The sexual structures of angiosperms are flowers

If you examine any familiar flower, you will notice that each part has one or more veins and, in the case of the outer parts, looks somewhat like a leaf. In fact, these parts are modified from leaves to function as parts of the flower. Thus the terms that we have already used for other plants apply here.

A generalized flower (for which there is no exact counterpart in nature) is shown in Figure 27.26. In the flower, structures bearing microsporangia are called **stamens**, each composed of a filament and two **anthers** that contain pollen-producing sporangia; structures bearing megasporangia are called **carpels**. A structure composed of one carpel or two or more fused carpels is called a **pistil**. The swollen base of the pistil, containing one or more ovules (each containing a megasporangium), is an **ovary**; the apical stalk of the pistil is a **style**; and the terminal surface that receives the pollen is called a **stigma**. In addition, a flower often has several specialized sterile (non-spore-bearing) leaves: The inner ones are called **petals** (collectively, the **corolla**), and the outer, **sepals** (collectively, the **calyx**). The corolla and calyx, which can be quite showy, often play roles in attracting animal pollinators to the flower. The calyx more commonly protects the flower in bud. From base to apex, the sepals, petals, stamens, and carpels are usually arranged in whorls and attached to a central stalk called the receptacle.

The flower in Figure 27.26 has both megasporangia and microsporangia and is said to be **perfect**, meaning that it contains both functional female and functional male parts. Many angiosperms produce two types of flowers, one type with only megasporangia and the other with only microsporangia; consequently, either the stamens or the carpels are nonfunctional or absent in a given flower, and the flower is referred to as **imperfect**.

Species such as corn or birch in which both female and male flowers occur on the same plant are said to

(a) *Daucus carota*

(b) *Echinacea purpurea*

(c) *Pennisetum setaceum*

27.27 Inflorescences (a) The inflorescence of Queen Anne's lace is an umbel. Each umbel bears flowers on stalks that arise from a common center. (b) Cornflowers are members of the aster family; their inflorescence is a head. In a head, each of the long, petal-like structures is a ray flower; the central portion of the head consists of dozens to hundreds of disc flowers. (c) Grasses such as this fountain grass have inflorescences called spikes.

(a)

(b)

27.28 Flower Form and Evolution (a) A magnolia flower shows the major features of early flowers: radial symmetry with the individual tepals, carpels, and stamens separate, numerous, and attached at their bases in a spiral arrangement. (b) Orchids have a bilaterally symmetrical structure that evolved much later than the form of the magnolia flower in (a). One of the three petals evolved into the complex lower "lip." Inside, the stamen and pistil are fused, and there is a single anther.

be **monoecious** (meaning "one-housed"—but, it must be added, one house with separate rooms). The sexes are completely separated in some other species of angiosperms, such as willows and date palms; in these species, a given plant produces either male or female flowers, but never both. Such species are said to be **dioecious** ("two-housed"). In other words, there are truly female plants and truly male plants.

In the generalized flower of Figure 27.26, we illustrated distinct petals and sepals arranged in distinct whorls. In nature, however, the petals and sepals sometimes are arranged in a continuous spiral and are indistinguishable. Such appendages are called **tepals**. In other cases petals, sepals, or tepals are completely absent.

Flowers may be single, or grouped together to form an **inflorescence**. Different families of flowering plants have their own, characteristic types of inflorescences, such as the umbels of the carrot family, the heads of the aster family, and the spikes of many grasses, among others (Figure 27.27).

There still are questions about evolutionary trends in flower structure

Botanists disagree about which type of flower is evolutionarily the most ancient. One of the earliest types has many tepals (or sepals and petals), carpels, and stamens, all spirally arranged (Figure 27.28a). Evolutionary change within the angiosperms included some striking modifications from this early condition: reduction in the number of each type of organ, differentiation of petals from sepals, stabilization of each type of organ to a fixed number, arrangement in whorls, and finally, change in symmetry from radial (as in a lily) to bilateral (as in a sweet pea or orchid), often accompanied by an extensive fusion of parts (Figure 27.28b). A great variety of corolla types have emerged in the course of evolution, as you will realize if you think of some of the flowers you recognize.

According to one theory, the first carpels to evolve were modified leaflike structures, folded but incompletely closed and thus differing from the scales of the gymnosperms and the true carpels of the angiosperms. In the groups of angiosperms that evolved later, the carpels fused and became progressively more buried in receptacle tissue (Figure 27.29a); in the flowers of the latest groups to evolve, the other flower parts are attached at the very top of the ovary rather than at the bottom. The stamens of the most-ancient flowers may have been leaflike (Figure 27.29b), little resembling those of the generalized flower in Figure 27.26. Botanists disagree as to whether carpels, stamens, and even petals evolved from leaves or from another structure.

Why do so many flowers have pistils with long styles and anthers with long filaments? Natural selection has favored length in both of these structures probably because length increases the likelihood of successful pollination. Long filaments may bring the anthers in contact with insect bodies, or they may put the anthers where they catch the wind better. Similar arguments apply to long styles.

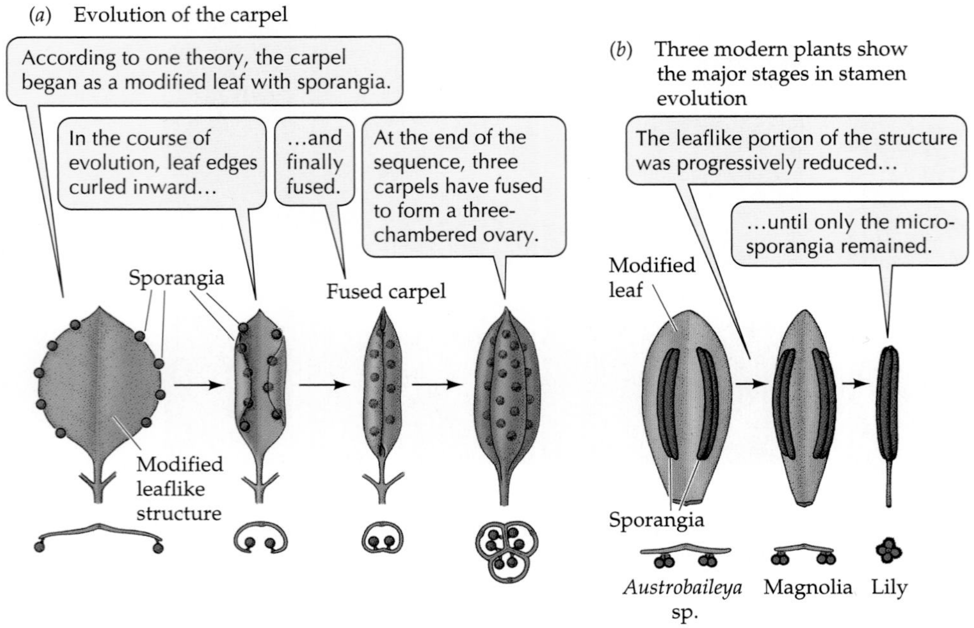

(a) Evolution of the carpel

> According to one theory, the carpel began as a modified leaf with sporangia.

> In the course of evolution, leaf edges curled inward...

> ...and finally fused.

> At the end of the sequence, three carpels have fused to form a three-chambered ovary.

(b) Three modern plants show the major stages in stamen evolution

> The leaflike portion of the structure was progressively reduced...

> ...until only the microsporangia remained.

Sporangia

Fused carpel

Modified leaflike structure

Modified leaf

Sporangia

Austrobaileya sp.   Magnolia   Lily

27.29 Evolution of Carpels and Stamens   Carpels and stamens may have evolved from leaflike structures. The small picture below each illustration is a cross section.

A long style may serve another purpose as well. If several pollen grains land on one stigma, a pollen tube will start growing from each grain toward the ovary. If there are more pollen grains than ovules, there is a "race" for the ovules. The race down the style can be viewed as "mate selection" by the plant holding that style. Pollen has played another major role in evolution, as we are about to see.

Pollen played a crucial role in the coevolution of angiosperms and animals

Whereas many gymnosperms are wind-pollinated, most angiosperms are animal-pollinated. Animals visit flowers to obtain nectar or pollen and in the process often carry pollen from one flower to another, or from one plant to another. Thus in its quest for food, the animal contributes to the genetic diversity of the plant population. Insects—especially bees—are among the most important pollinators; birds and some species of bats also play major roles.

For more than 100 million years, angiosperms and their animal pollinators have coevolved in the terrestrial environment. The animals have affected the evolution of the plants, and the plants have affected the evolution of the animals. Pollination by just one or a very few committed animal species provides a plant species with a reliable mechanism for transferring pollen from one to another of its members. Flower structure has become incredibly diverse under selection.

Some of the products of coevolution are highly spe-cific; for example, some yucca species are pollinated by one and only one species of moth. Most plant–pollinator interactions are much less specific; that is, many different animal species pollinate the same plant species, and the same animal species pollinate many plant species. However, even these less specific interactions have developed some specialization. Bird-pollinated flowers are often red and odorless; insect-pollinated flowers often have characteristic odors and may have conspicuous markings ("nectar guides") that are evident only in the ultraviolet region of the spectrum, where insects have better vision than in the red region.

We treat coevolution and other aspects of plant–animal interactions in more detail in Chapter 52. Here we'll consider what happens to the flower after fertilization.

Angiosperms produce fruits

The ovary of a flowering plant (together with its seeds) develops into a fruit after fertilization. Fruit production is thus another diagnostic character of angiosperms. A fruit may consist only of the mature ovary and its seeds, or it may include other parts of the flower or structures closely related to it.

A **simple fruit**, such as a cherry (Figure 27.30a), is one that develops from a single carpel or several united carpels. A raspberry is an example of an **aggregate fruit** (Figure 27.30b)—one that develops from several separate carpels of a single flower. Pineapples and figs are examples of **multiple fruits** (Figure 27.30c),

(a)

(b)

(c)

(d)

27.30 Fruits Come in Many Forms and Flavors (a) A simple fruit: sour cherry. (b) An aggregate fruit: raspberries. (c) A multiple fruit: pineapple. (d) An accessory fruit: pear.

formed from a cluster of flowers (an inflorescence). Fruits derived from parts in addition to the carpel and seeds are called **accessory fruits** (Figure 27.30d); examples are apples, pears, and bananas.

The development, ripening, and dispersal of fruits will be considered in Chapters 35 and 36.

The angiosperm life cycle features double fertilization

The life cycle of the angiosperms will be considered in detail in Chapter 36. The summarized life cycle in Figure 27.31 shows that the angiosperm life cycle has similarities with and differences from the conifer life cycle (see Figure 27.25). Like all other seed plants, angiosperms are heterosporous. The female gametophyte is even more reduced than in the gymnosperms. The ovules are contained within carpels, rather than being exposed on the surfaces of scales as in most gymnosperms. The male gametophytes are, again, pollen grains.

The ovule develops into a seed containing the products of the double fertilization that characterizes angiosperms. The triploid endosperm serves as storage tissue for starch or lipid reserves, storage proteins, and other reserve substances. The diploid zygote develops into an embryo consisting of an embryonic axis and one or two **cotyledons**. Also called seed leaves, cotyledons have different fates in different plants. In many, they serve as absorptive organs that take up and digest the endosperm. In others, they enlarge and become photosynthetic when the seed germinates. Often they play both of these roles.

There are two major groups of flowering plants

The angiosperms may be divided into two classes: Monocotyledones (the monocots; Figure 27.32) and Dicotyledones (the dicots; Figure 27.33). The monocots are so called because they have a single embryonic cotyledon; the dicots have two. And there are other major differences between the two classes (see Figure 31.5). Included are differences in leaf vein patterns, in the arrangement of vascular tissue in the stem and root, in the number of flower parts, and in the presence or absence of secondary growth (produced by a cambium; see Chapter 31). The cotyledons of some, but not all, dicots store the reserves originally present in the endosperm.

The monocots include grasses, cattails, lilies, orchids, and palm trees. The dicots include the vast number of familiar seed plants: most of the herbs, vines, trees, and shrubs. Among them are oaks, willows, violets, snapdragons, and sunflowers.

The origin and evolution of the flowering plants

How did the angiosperms arise? Modern cladistic analyses (see Chapter 22) have settled this once vexing question. It is widely agreed that the angiosperms and two groups of gymnosperms, the Gnetophyta and the long-extinct cycadeoids (so named because their leaves resembled cycad leaves), arose from a single common ancestor. A close relationship between the angiosperms and the Gnetophyta was long suspected, primarily on the grounds that some gnetophytes have vessel elements, which characterize the angiosperms. (Ironically, this was a poor argument, because vessel elements probably evolved independently in the two groups.)

In 1990 this theory was strengthened by the confirmation via the light microscope of double fertilization in *Ephedra*, a member of the Gnetophyta. The cycadeoids, which became extinct about the same time as the dinosaurs did, shared several important characteristics with the Gnetophyta and the angiosperms. The reproductive organ of one of the cycadeoids, although clearly a gymnosperm structure with naked seeds, was suggestively similar to the flower of *Magnolia*.

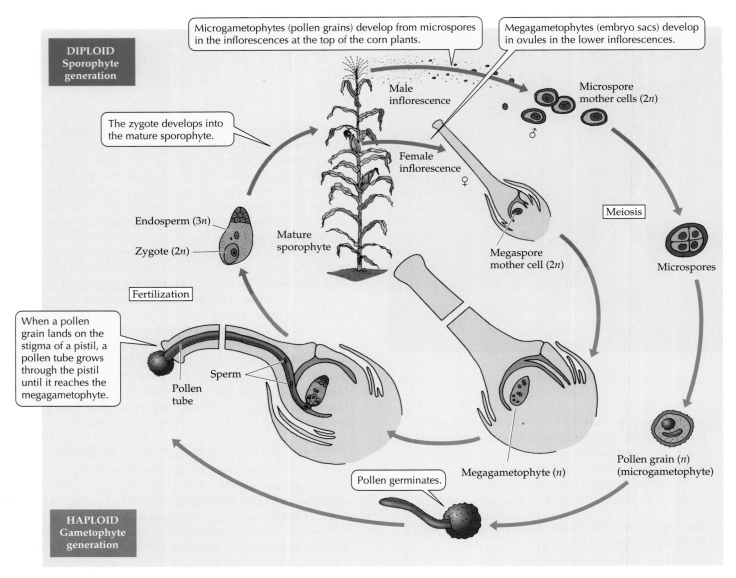

27.31 The Life Cycle of an Angiosperm *Zea mays*, popularly known as corn, is a thoroughly studied angiosperm with great economic value to humans.

(a)

(b)

(c)

27.32 Monocots (a) Palms are among the few monocot trees. Date palms are a major food source in some areas of the world. (b) Grasses such as this cultivated durham wheat and the fountain grass in Figure 27.27c are monocots. (c) Monocots include popular garden flowers such as these daylilies. Orchids (Figure 27.29b) are another highly prized monocot flower.

27.33 Dicots (a) The cactus family is a large group of dicots, with about 1,500 species in the Americas. This cactus bears scarlet flowers for a brief period of the year. (b) The flowering dogwood is a small dicot tree. (c) These climbing Cape Cod roses are members of the family Rosaceae, as are the familiar roses from your local florist.

(b)

(a)

(c)

27.34 Herbs and Tundra The landscape of this Alaskan tundra is dominated by herbs.

through good timing—growing, photosynthesizing, and reproducing early in the season, before the foliage develops fully on the trees above. When the leafy canopy of the forest becomes dense and blocks sunlight from reaching the forest floor, the shoots of these herbs die back, leaving underground organs (rhizomes or tubers) with stored food for the beginning of the following year's growth.

Herbs predominate in at least one extreme environment: tundra (Figure 27.34). Here the ability of some herbs to store photosynthate in underground organs stands them in particularly good stead, because they leave no aboveground parts to be damaged during the long cold season. The cold is so intense that shrubs and trees are unable to survive, but the dormant underground parts of herbs succeed.

Determining which angiosperms were the first flowering plants is a question of great controversy. Two candidates are the magnolia family (see Figure 27.28a) and another family, the Chloranthaceae, whose flowers are much simpler than those of the magnolias. Other candidates have also been suggested. One complication in the search for the earliest fossil angiosperms is that we are unlikely to be able to tell whether an ancient fossil plant practiced double fertilization. There continues to be disagreement as to whether the first angiosperms were trees or herbs. However that disagreement is being resolved, today the predominant ground cover consists of herbaceous angiosperms.

Herbaceous plants predominate in many situations

When a piece of ground becomes available for new colonization—perhaps as a result of fire, or the appearance of a new sandbar, or clearing by humans—the first colonizers are generally herbs. Their seeds may have been present in a dormant state in the soil, or they may have been brought in by wind or animals. The rapidly growing herbs produce seeds that germinate and contribute to the growth of the herb population.

Before the land is cleared, the foliage of shrubs and trees may shade the ground, making little light available for photosynthesis by small herbs. When the taller plants are removed, the herbs have their day in the sun. Later, as the land is modified by these herbaceous plants, other, larger forms appear (or reappear) and generally take over as succession proceeds (see Chapter 52).

Although not dominating the scene, some herbaceous angiosperms do well in forests. They succeed

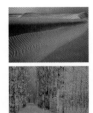

Summary of "Plants: Pioneers of the Terrestrial Environment"

The Plant Kingdom
• Plants are multicellular, photosynthetic eukaryotes that develop from embryos protected by parental tissue.
• Plant life cycles feature alternation of generations: gametophyte (haploid) and sporophyte (diploid).
• There are twelve surviving phyla of plants grouped into two main categories: tracheophytes and nontracheophytes. **Review Table 27.1**
• The plant kingdom arose from the green algae—specifically, a common ancestor of the stoneworts or of *Coleochaete*. Descendants of this ancestor colonized the land.
• Tracheophytes, characterized by possession of a vascular system, consisting of water- and mineral-conducting xylem and food-conducting phloem, evolved from nontracheophytes. **Review Figure 27.3**

Nontracheophytes: Liverworts, Hornworts, and Mosses
• The nontracheophytes include the liverworts (phylum Hepatophyta), hornworts (phylum Anthocerophyta), and mosses (phylum Bryophyta). **Review Table 27.1**
• Nontracheophytes either lack vascular tissues completely or, in the case of certain mosses, have only a rudimentary system of water- and food-conducting cells.
• The nontracheophyte sporophyte generation is smaller than the gametophyte generation and depends on the gametophyte for water and nutrition. **Review Figures 27.5, 27.8**
• Liverwort sporophytes have no specific growing zone. Hornwort sporophytes grow at their basal end, and moss sporophytes grow at their apical end. **Review Figure 27.7**
• The hydroids of mosses, through which water may travel, are probably ancestral to the water-conducting cells of the tracheophytes.

Introducing the Tracheophytes

• The tracheophytes have vascular tissue with tracheids and other specialized cells designed to conduct water, minerals, and foods.

• Present-day tracheophytes are grouped into nine phyla that form two major groups: nonseed tracheophytes and seed plants.

• In tracheophytes the sporophyte generation is larger than the gametophyte and independent of the gametophyte generation.

• The earliest tracheophytes, known to us only in fossil form, lacked roots and leaves. Roots may have evolved from branches that penetrated the ground. Simple leaves are thought to have evolved from sporangia, and complex leaves may have resulted from the flattening and reduction of a branching stem system. **Review Figures 27.12, 27.13, 27.14**

• Heterospory, the production of distinct female megaspores and male microspores, evolved on several occasions from homosporous ancestors. **Review Figure 27.15**

The Surviving Nonseed Tracheophytes

• Whisk ferns (phylum Psilophyta) look very much like the ancient (and extinct) first tracheophytes. Club mosses (phylum Lycophyta) have simple leaves arranged spirally. Horsetails (phylum Sphenophyta) have simple leaves in whorls. Leaves with more complex vasculature are characteristic of all other phyla of tracheophytes. **Review Table 27.1**

• Ferns (phylum Pterophyta) have complex leaves and a dominant sporophyte generation. **Review Figure 27.22 and Table 27.1**

The Seeds of Success

• The seed plants (gymnosperms and angiosperms) are heterosporous and have much reduced gametophytes.

• Most modern seed plants have no swimming gametes and are independent of liquid water for fertilization. The male gametophyte, the pollen grain, is dispersed by wind or by animals.

• The seed is a well protected resting stage that often contains food that supports the growth of the embryo.

The Gymnosperms: Naked Seeds

• The gymnosperms, once the dominant vegetation of all of Earth, still dominate forests in the northern parts of the Northern Hemisphere and at high elevations.

• The four surviving gymnosperm phyla are the Cycadophyta (the most ancient), Ginkgophyta (consisting of a single genus, the maidenhair tree), Gnetophyta (the sister group to the angiosperms), and Coniferophyta (the familiar cone-bearing trees). **Review Table 27.1**

• Modern gymnosperms all have abundant xylem and extensive secondary growth.

• Conifers have a life cycle in which naked seeds are produced on the scales of female cones. Male cones are smaller than female cones. Pollen is transferred from male to female cones by wind. **Review Figure 27.25**

The Angiosperms: Flowering Plants

• Angiosperms (phylum Angiospermae) are distinguished by the production of flowers and fruits. They demonstrate double fertilization resulting in a triploid nutritive tissue, the endosperm. Double fertilization is also characteristic of the Gnetophyta. **Review Figure 27.31**

• The vascular tissues of angiosperms contain cell types (such as vessel elements and fibers) rarely found elsewhere in the plant kingdom.

• Flowers are made up of various combinations of carpels, stamens, petals, and sepals. Perfect flowers have both carpels (female parts) and stamens (male parts). **Review Figure 27.26**

• Monoecious plant species have both female and male flowers on the same plant. Dioecious species have separate female and male plants.

• Carpels and stamens may have evolved from leaflike structures. **Review Figure 27.29**

• Angiosperms and the animals that pollinate them have coevolved.

• The two classes of flowering plants, Monocotyledones and Dicotyledones, differ in vein patterns in the leaves, in the number of cotyledons, and in the number of flower parts. Grasses are common monocots, and broad-leaved plants such as oaks and roses are common dicots.

Self-Quiz

1. Plants differ from algae in that only plants
   - *a.* are photosynthetic.
   - *b.* are multicellular.
   - *c.* possess chlorophyll.
   - *d.* have multicellular embryos protected by the parent.
   - *e.* are eukaryotic.

2. Which statement about the alternation of generations in plants is *not* true?
   - *a.* It is heteromorphic.
   - *b.* Meiosis occurs in sporangia.
   - *c.* Gametes are always produced by meiosis.
   - *d.* The zygote is the first cell of the sporophyte generation.
   - *e.* The gametophyte and sporophyte differ genetically.

3. Which statement is *not* evidence for the origin of plants from the green algae?
   - *a.* Some green algae have multicellular sporophytes and multicellular gametophytes.
   - *b.* Both plants and green algae have cellulose in their cell walls.
   - *c.* The two groups have the same photosynthetic and accessory pigments.
   - *d.* Both plants and green algae produce starch as their principal storage carbohydrate.
   - *e.* All green algae produce large, stationary eggs.

4. The nontracheophytes
   - *a.* lack a sporophyte generation.
   - *b.* grow in dense masses, allowing capillary movement of water.
   - *c.* possess xylem and phloem.
   - *d.* possess true leaves.
   - *e.* possess true roots.

5. The rhyniophytes
   - *a.* possessed vessel elements.
   - *b.* possessed true roots.
   - *c.* possessed sporangia at the tips of stems.
   - *d.* possessed leaves.
   - *e.* lacked branching stems.

6. Club mosses and horsetails
    a. have larger gametophytes than sporophytes.
    b. possess small leaves.
    c. are represented today primarily by trees.
    d. have never been a dominant part of the vegetation.
    e. produce only simple fruits.

7. Which statement about ferns is *not* true?
    a. The sporophyte is larger than the gametophyte.
    b. Most are heterosporous.
    c. The young sporophyte can grow independently of the gametophyte.
    d. The frond is a large leaf.
    e. The gametophytes produce archegonia and antheridia.

8. The gymnosperms
    a. dominate all land masses today.
    b. have never dominated land masses.
    c. have active secondary growth.
    d. all have vessel elements.
    e. lack sporangia.

9. Which statement about flowers is *not* true?
    a. Pollen is produced in the anthers.
    b. Pollen is received on the stigma.
    c. An inflorescence is a cluster of flowers.
    d. A species having female and male flowers on the same plant is dioecious.
    e. A flower with both mega- and microsporangia is said to be perfect.

10. Which statement about fruits is *not* true?
    a. They develop from ovaries.
    b. They may include other parts of the flower.
    c. A multiple fruit develops from several carpels of a single flower.
    d. They are produced only by angiosperms.
    e. A cherry is a simple fruit.

Applying Concepts

1. Mosses and ferns share a common trait that makes water droplets a necessity for sexual reproduction. What is this trait?

2. Ferns display a dominant sporophyte stage (with large fronds). Describe the major advance in anatomy that enables most ferns to grow much larger than mosses.

3. What features distinguish club mosses from horsetails? What features distinguish these groups from rhyniophytes and psilophytes? from ferns?

4. Suggest an explanation for the great success of the angiosperms in occupying terrestrial habitats.

5. Contrast simple leaves with complex leaves in terms of structure, evolutionary origin, and occurrence among plants.

6. In many locales, large gymnosperms predominate over large angiosperms. Under what conditions might gymnosperms have the advantage, and why?

Readings

Burnham, C. R. 1988. "The Restoration of the American Chestnut." *American Scientist*, vol. 76, pages 478–487. An account of the comeback of a species almost completely wiped out by disease. If you read the article on chestnut blight recommended in the next chapter, you may be encouraged to see that there are methods, based on Mendelian genetics, to rebuild the native population of this important tree species.

Crosson, P. R. and N. J. Rosenberg. 1989. "Strategies for Agriculture." *Scientific American*, September. A discussion of approaches to increasing yields for an expanding population, emphasizing that social and economic changes will also be required.

Friedman, W. E. 1990. "Double Fertilization in *Ephedra*, a Nonflowering Seed Plant: Its Bearing on the Origin of Angiosperms." *Science*, vol. 247, pages 951–954. A description of a breakthrough in understanding the origin of the flowering plants.

Graham, L. E. 1985. "The Origin of the Life Cycle of Land Plants." *American Scientist*, vol. 73, pages 178–186. A discussion of how plants made it to the terrestrial environment.

Heyler, D. and C. M. Poplin. 1988. "The Fossils of Montceau-les-Mines." *Scientific American*, September. A description of plants and animals of the Carboniferous period, discovered in a rich fossil lode in central France. Presents a detailed picture of life in a time long past.

Hinman, C. W. 1986. "Potential New Crops." *Scientific American*, July. A look at plants such as jojoba, buffalo gourd, and others as prospects for food and other materials.

Niklas, K. J. 1986. "Computer-Simulated Plant Evolution." *Scientific American*, March. An examination of some hypotheses concerning plant evolution. The hypotheses were modeled on computers, generating testable suggestions. The article presumes no knowledge of computers.

Niklas, K. J. 1987. "Aerodynamics of Wind Pollination." *Scientific American*, July. A discussion of the many adaptations of wind-pollinated plants that favor successful pollination.

Raven, P. H., R. F. Evert and S. Eichhorn. 1992. *Biology of Plants*, 5th Edition. Worth, New York. An excellent general botany textbook.

Chapter 28

Fungi: A Kingdom of Recyclers

Fairy-Tale Fungi
Forests full of elves might be one of the images evoked by members of the mushroom family of the kingdom Fungi. In fact fungi have many growth forms and play an indispensable role in the biosphere.

Surely you are familiar with mushrooms—in the market, on your plate, or in whimsical fairy-tale portraits, pleasant and picturesque. But what about the black spots that appear on your homemade bread, or the green fuzz that forms on your orange? These are molds, and they are neither pleasant nor picturesque. Molds and mushrooms are both fungi, and Earth would be a messy place without them. Because they are superbly adapted to absorptive nutrition, fungi are at work in forests, fields, and garbage dumps, breaking down the remains of dead organisms and even manufactured substances such as some plastics. For almost a billion years, the ability of fungi to decompose substances has been important for life on Earth, chiefly because by breaking down carbon compounds, fungi return carbon and other elements to the environment, where they can be used again by other organisms.

Already crucial for the continuation of life as we know it, the fungi may soon play enhanced roles in decomposing toxic environmental pollutants. This task is necessary because our species has been loading toxic substances into the environment for some time. Might the removal of these substances be a job for the fungi, naturally occurring or genetically engineered? Indeed it might, and biologists and chemists are exploring the ability of different fungi, as well as bacteria, to break down pesticides and other toxic materials into harmless substances. This overall process of degradation and decomposition of toxic materials is known as **bioremediation**.

28.1 Parasitic Fungi Attack Other Living Organisms (*a*) The gray masses on this ear of corn are the parasitic fungus *Ustilago maydis*, commonly called corn smut. (*b*) The tropical fungus whose fruiting body is growing out of the carcass of this ant has developed from a spore ingested by the ant. The spores of this fungus must be ingested by insects before they will germinate and develop. The growing fungus absorbs organic and inorganic nutrients from the ant's body, eventually killing it, after which the fruiting body produces a new crop of spores. (*c*) An amoeba (below) being parasitized by a fungus (above) of the genus *Amoebophilus* (which means "amoeba lover").

(*a*)

(*b*)

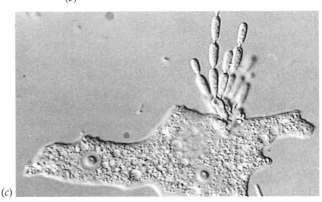

(*c*)

In this chapter we will examine the general biology of this very important group of organisms—the kingdom Fungi—which differs in interesting ways from the other kingdoms. We will also explore the diversity of body forms, reproductive structures, and life cycles of the four phyla of fungi, as well as the mutually beneficial associations of certain fungi with other organisms. As we begin our study, recall that the fungi and the animals are descended from a common ancestor—we are more closely related to molds and mushrooms than we are to the roses we admired in the last chapter (see Figure 26.31).

General Biology of the Fungi

We define the kingdom Fungi as encompassing *heterotrophic organisms* with *absorptive nutrition*. Fungi are **saprobes** (organisms that live on dead matter), **parasites** (organisms that absorb nutrients from living hosts; Figure 28.1), or **mutualists** (organisms that live in mutually beneficial symbiosis with other organisms). All fungi form spores, but only in one phylum (Chytridiomycota) do single-celled stages possess flagella. Fungi reproduce sexually in a variety of ways, ranging from fusion of filaments of different mating types to fusion of distinct gametes. The walls that enclose the filaments of all fungi contain at least some **chitin**, a polysaccharide that is also found in the skeletons of arthropods. Most fungi have complex forms.

These criteria enable us to distinguish between the fungi and some protists. The slime molds consist of two protist phyla (Dictyostelida and Acrasiomycota), whose members take up food by phagocytosis rather than by absorption, and a third protist phylum (Myxomycota), whose members have cells with flagella. The other funguslike protists (Oomycota) also have flagellated cells, and they have cellulose, rather than chitin, in their cell walls.

The kingdom Fungi consists of four phyla: Chytridiomycota, Zygomycota, Ascomycota, and Basidiomycota (Table 28.1). We distinguish the fungal phyla on the basis of the methods and structures they use for sexual reproduction and, to a lesser extent, on criteria such as the presence or absence of cross-walls separating their cell-like compartments.

Some fungi, called imperfect fungi, do not form sexual structures by which they might be easily identified as members of one of the four phyla. However, techniques of molecular taxonomy, such as DNA sequencing, now allow us to identify many imperfect fungi as asexual zygomycetes, ascomycetes, or basidiomycetes.

In the sections that follow we'll consider some aspects of the general biology of the fungi, including their body structure and its intimate relationship with the environment, their nutrition, and some special aspects of their sexual reproductive cycles, especially the occurrence of mating types, of a dikaryotic ($n + n$) nuclear condition, and of dual hosts.

The body of a fungus is composed of hyphae

The vegetative, feeding body of a fungus is called a **mycelium** (plural mycelia). It is composed of rapidly growing individual filaments called **hyphae** (singular

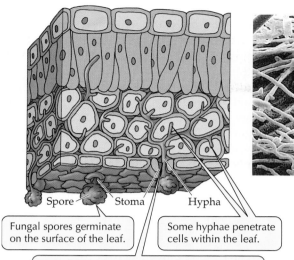

Spore   Stoma   Hypha

Fungal spores germinate on the surface of the leaf.

Some hyphae penetrate cells within the leaf.

Elongating hyphae pass through stomata into the interior of the leaf, elongate further, and branch.

28.2 A Fungus Attacks a Leaf The white structures in the micrograph are the hyphae of the fungus *Blumeria graminis*, which is growing on the dark surface of the leaf of a grass.

hypha), which are usually subdivided into cell-like compartments by *incomplete* partitions called **septa** (singular septum). In most hyphae, organelles (even nuclei) can move around, and there is no division into separate cells. Certain modified hyphae, the **rhizoids**, anchor saprobic fungi to their substrate (the dead organism or other matter upon which they feed). Parasitic fungi may have modified hyphae that take up nutrients from the host. The total hyphal growth of a mycelium (not the growth of an individual hypha) may exceed 1 km per day. The hyphae may be highly dispersed or may clump together in a cottony mass. Sometimes, when sexual spores are produced, the mycelium is organized into elaborate fruiting bodies such as mushrooms.

The way in which a parasitic fungus attacks a plant illustrates the roles of some fungal structures (Figure 28.2). The hyphae of a fungus invade a leaf through pores called stomata (see Chapter 8), through wounds, or, in some cases, by direct penetration of epidermal cells. Once inside, the hyphae form a mycelium. Some hyphae grow into living plant cells, absorbing the nutrients within the cells. Eventually fruiting bodies form, either within the plant body or on its surface.

Fungi are in intimate contact with their environment

The tubular hyphae of a multicellular fungus give it a unique relationship with its physical environment: The fungal mycelium has an enormous surface area–to–volume ratio compared with that of most large multicellular organisms. This large ratio of surface area to volume is a marvelous adaptation for absorptive nutrition, in which nutrients are absorbed across the hyphal surfaces. Throughout the mycelium, except in fruiting bodies, all the hyphae are very close to their environmental food source.

Another characteristic of some fungi is their tolerance for highly hyperosmotic environments (those with very negative osmotic potential; see Chapter 5). Many fungi are more resistant than are bacteria to damage in hyperosmotic surroundings. For example, jelly in the refrigerator will not become a growth medium for bacteria, because the jelly is too hyperosmotic to the bacteria, but it may eventually harbor mold colonies. You have probably seen the green mold *Penicillium* growing on oranges in the refrigerator. The refrigerator example illustrates another trait of many fungi: tolerance of temperature extremes. Many fungi

PHYLUM	COMMON NAME	FEATURES	EXAMPLES
Chytridiomycota	Chytrids	Aquatic; gametes have flagella	*Allomyces*
Zygomycota	Zygomycetes	No regularly occur ring hyphal cross-walls; usually no fleshy fruiting body	*Rhizopus*
Ascomycota	Ascomycetes	Ascus; perforated cross-walls	*Neurospora*, baker's yeast
Basidiomycota	Basidiomycetes	Basidium; perforated cross-walls	*Puccinia*, mushrooms

TABLE 28.1   Classification of Fungi

tolerate temperatures as low as 5 or 6°C below freezing, and some tolerate temperatures as high as 50°C or more.

Fungi are absorptive heterotrophs

All fungi are heterotrophs that obtain food by direct absorption from the immediate environment. The vast majority are saprobes, obtaining their energy, carbon, and nitrogen directly from dead organic matter by the action of enzymes. However, as we've learned already, some are parasites, and still others form mutualistic associations with other organisms.

Saprobic fungi, along with bacteria, are the major decomposers of the biosphere, contributing to decay and thus to the recycling of the elements used by living things. In the forest, for example, the invisible mycelia of fungi obtain nutrients from fallen trees, thus decomposing the trees. Fungi are the principal decomposers of cellulose and lignin, the main components of plant cell walls (most bacteria cannot break down plant cell wall material).

Because saprobic fungi are able to grow on artificial media, we can determine their exact nutritional requirements. Sugars are the favored source of carbon, and most fungi obtain nitrogen from proteins or the products of protein breakdown. Many fungi can use nitrate (NO_3^-) or ammonium (NH_4^+) ions as their sole source of nitrogen. No known fungus gets its nitrogen directly from nitrogen gas, as can some bacteria and plant–bacteria associations (see Chapter 34). Vitamins may play a role in fungal nutrition. Most fungi are unable to synthesize their own thiamin (vitamin B_1) or biotin (another B vitamin) and hence must absorb these vitamins from their environment. Other compounds that are vitamins for animals can be made by fungi. Like all other organisms, fungi also require some mineral elements.

Nutrition in the parasitic fungi is particularly interesting. **Facultative** parasites can grow parasitically or by themselves on defined artificial media. Biologists can work out the exact nutritional requirements by varying the composition of the growth medium. **Obligate** parasites cannot be grown on any available defined medium; they can grow only on their specific hosts, usually plants. Obligate parasites include various mildews. Because their growth is so limited, they must have unusual nutritional requirements. Biologists are thus very interested in learning more about them.

Some fungi have adaptations that enable them to function as active predators, trapping nearby microscopic protists or animals, from which they obtain nitrogen and energy. The most common approach is to secrete sticky substances from the hyphae so that passing organisms stick tightly to them. Fungal hyphae then quickly invade the prey, growing and branching within it, spreading through its body, absorbing nutrients, and eventually killing it.

A more dramatic adaptation for predation is the **constricting ring** formed by some species of *Arthrobotrys, Dactylaria,* and *Dactylella* (Figure 28.3). All of these fungi grow in soil; when nematodes (tiny roundworms) are present, these fungi form three-celled rings that have a diameter that just fits a nematode. A nematode crawling through one of these rings stimulates the ring, causing the cells of the ring to swell and trap the worm. Fungal hyphae quickly invade and digest the unlucky victim. Some species of wood-decaying mushrooms have the ability to supplement their nitrogen supply by feeding on nematodes.

Certain highly specific associations between fungi and other organisms have nutritional consequences for the fungal partner. **Lichens** are associations of a fungus with either a unicellular alga or a cyanobacterium; in this union the fungus draws nutrition from the photosynthetic bacterium or alga. **Mycorrhizae** (singular mycorrhiza) are associations between spe-

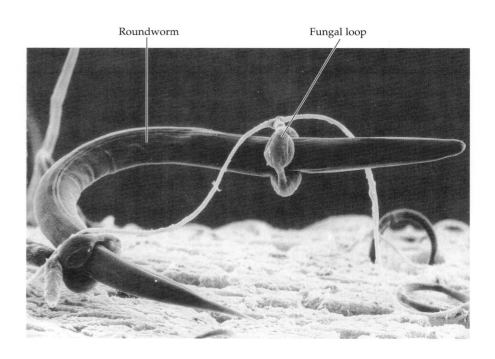

Roundworm   Fungal loop

28.3 Some Fungi Are Predators
A nematode (roundworm) is trapped in sticky loops of the soil-dwelling fungus *Arthrobotrys anchonia.*

Workers of the Costa Rican ant species *Atta cephalotes* add a cut piece of leaf to their fungal garden.

The fungus is the white material. In response to this care, the fungus grows and serves as food for the ants.

28.4 A Fungus Garden Central American leaf cutter ants cultivate fungi. These species of fungi are found only in association with the ant "farms."

cific fungi and the roots of plants. In such associations the fungus is fed by the plant but provides minerals (primarily phosphorus) to the plant root, so the plant's nutrition is also promoted. Seed germination in most orchid species depends on the presence of a specific mutualistic fungus, which itself derives nutrients from the seed and seedling of the orchid. We will discuss lichens and mycorrhizae more thoroughly later in this chapter.

Perhaps the most striking fungal associations are with insects. Some leaf-cutting ants "farm" fungi, feeding the fungi and later harvesting and eating them (Figure 28.4). The ants collect leaves and flower petals and chew them into small bits, on which they "plant" fungal mycelium. The ants even "weed" these gardens by removing other fungal species. The species of fungus cultivated by the ants are found nowhere other than in these tiny gardens. Scale insects live in association with the fungus *Septobasidium*. The fungus spreads over a colony of the insects, infecting some of them and thus parasitizing them—without killing them. As new insects hatch from the eggs within the colony, some of them become infected with the fungus. They take the fungus along as they establish new colonies. The fungus protects the colony against drying and against some predators but also draws its nutrition from the insects.

Most fungi reproduce asexually and sexually, and they have mating types

Both asexual and sexual reproduction are common in the fungi. Asexual reproduction takes several forms.

One form is the production of haploid spores within structures called sporangia. The second form is the production of naked spores at the tips of hyphae; these spores, called **conidia** (from the Greek *konis*, "dust"), are not produced in sporangia. The third form of asexual reproduction, performed by unicellular fungi, is cell division—either a relatively equal division or an asymmetric division in which a tiny bud is produced. The fourth form of asexual reproduction, seen in some hyphal fungi, is simple breakage of the mycelium.

Sexual reproduction in many fungi features an interesting twist. There is no distinction between female and male structures, or between female and male organisms. Rather, there is a genetically determined distinction between two or more **mating types**. Individuals of the same mating type cannot mate with one another, but they can mate with individuals of another mating type. This distinction prevents self-fertilization. Individuals of different mating types differ genetically from one another but are often visually and behaviorally indistinguishable.

The nuclei of most fungi are haploid, except in the zygotes formed in sexual reproduction, which are diploid. The zygotes undergo meiosis, producing haploid nuclei that become incorporated into spores. Haploid fungal spores, produced sexually or asexually, germinate, and their nuclei divide mitotically to produce hyphae. Fungal hyphae may have a nuclear configuration—a dikaryon stage, as described in the next section—other than the familiar haploid and diploid.

Many fungal life cycles include a dikaryon stage

Sexual reproduction begins in some fungi in an unusual way: The cytoplasms of two individuals of opposite mating type fuse long before their nuclei fuse, so *two genetically different haploid nuclei exist within the same hypha*. This hypha is called a **dikaryon** (having *two* nuclei). Because the two nuclei differ genetically, the hypha is also called a **heterokaryon** (having *different* nuclei).

Eventually specialized fruiting structures form, within which pairs of dissimilar nuclei—one from each parent—fuse, giving rise to zygotes long after the original "mating." The zygote nucleus—which may be the only diploid nucleus in the entire life cycle—undergoes meiosis, producing four haploid nuclei. The mitotic descendants of those nuclei become the nuclei of the next generation of hyphae.

The reproduction of such fungi displays several unusual features. First, there are no gamete cells, only gamete nuclei. Second, there is never any true diploid tissue, although for a long period during development the genes of both parents are present in the dikaryon and can be expressed. In effect, the cells are neither diploid (2*n*) nor haploid (*n*); rather they are dikaryotic

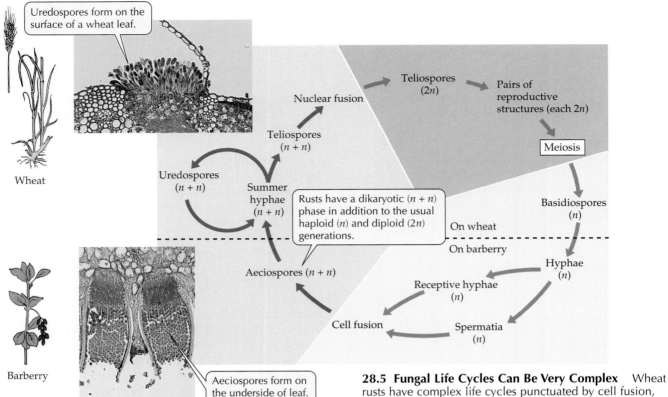

Uredospores form on the surface of a wheat leaf.

Aeciospores form on the underside of leaf.

Rusts have a dikaryotic (n + n) phase in addition to the usual haploid (n) and diploid (2n) generations.

Wheat

Barberry

28.5 Fungal Life Cycles Can Be Very Complex Wheat rusts have complex life cycles punctuated by cell fusion, nuclear fusion, and meiosis and divided between two host plants—wheat and barberry.

(n + n). A harmful recessive mutation in one nucleus may be compensated for by a normal allele on the same chromosome in the other nucleus. Dikaryosis is perhaps the most significant of the genetic peculiarities of the fungi.

Finally, although these organisms grow in moist places, the gamete nuclei are not motile and are not released into the environment; therefore, liquid water is not required for fertilization. (Some members of the plant kingdom require liquid water for the meeting of gametes, as do some aquatic animals.)

The life cycles of some parasitic fungi involve two hosts

Some parasitic fungi are very specific about what host organism they use as a source of nutrition, and some even use different hosts for different stages of their life cycle. (Remember that a host is an organism that harbors another organism—generally a parasite—and provides it with nutrition.) One of the most striking examples of a fungal life cycle that involves two different hosts is the complicated life cycle of *Puccinia graminis,* the agent of a major agricultural disease, black stem rust of wheat. In the epidemic of 1935, *P. graminis* was responsible for the loss of about one-fourth of the wheat crop in Canada and the United States.

Let's examine a year in the life cycle of *P. graminis* (Figure 28.5). During the summer, dikaryotic (n + n) hyphae of *P. graminis* proliferate in the stem and leaf tissues of wheat plants, drawing their nutrition from the wheat and damaging it severely. These dikaryotic hyphae produce extensive amounts of summer spores called **uredospores**. The one-celled, orange uredospores are scattered by the wind and infect other wheat plants, on which the summer hyphae then proliferate. Like the hyphae, the uredospores are dikaryotic.

Dark brown winter spores—**teliospores**—begin to appear on certain special hyphae in late summer. Each teliospore consists initially of two dikaryotic cells. The two haploid nuclei in each cell fuse to form a single, diploid (2n) nucleus. These are the only truly diploid cells in the entire life cycle. Both cells have thick walls and usually survive freezing. The teliospores remain dormant until spring, when they germinate. Each of the two cells develops into a reproductive structure, within which the nucleus divides meiotically to produce four haploid **basidiospores** (discussed later in this chapter, under the phylum Basidiomycota).

The basidiospores are of two different mating types: plus (+) and minus (–). They are carried by the wind, and if they land on a leaf of the common barberry plant, they germinate and produce haploid hyphae (+ or –, depending on the mating type of the germinating basidiospore). These hyphae invade the barberry leaf.

The hyphae form flask-shaped structures on the upper surfaces of the leaf, within which some of the hyphae pinch off tiny, colorless, haploid **spermatia** from their tips. The hyphae also form another type of structure, called an **aecial primordium**, near the lower surface of the barberry leaf. Thus the leaf contains "flasks" on the upper surface and aecial primordia near the lower one, and these are connected by the hyphae. Insects attracted to a sweetish liquid produced in the flasks carry spermatia from one flask to another, initiating the next step in the cycle.

A spermatium of one mating type fuses with a receptive hypha of the other type within a flask. The nucleus of the spermatium repeatedly divides mitotically, and the products move through the hyphae into immature aecial primordia, where they produce dikaryotic cells with two haploid nuclei, one of each mating type. These cells develop into **aeciospores**, each containing two unlike nuclei. The aeciospores are scattered by the wind, and some germinate on wheat plants, continuing the life cycle. When the dikaryotic aeciospores germinate, they produce the summer hyphae with two nuclei in each cell.

To summarize, in the stages from the aeciospore to the teliospore, *P. graminis* is dikaryotic; that is, the individual cells contain nuclei from both "parents." Teliospores, at first dikaryotic, become diploid. Basidiospores are produced by meiosis and cytokinesis, so they are haploid, each having a mating type of + or –.

The different types of spores produced during the life cycle of *P. graminis* play very different roles. Windborne uredospores are the primary agents for spreading the rust from wheat plant to wheat plant and to other fields. Resistant teliospores allow the rust to survive the harsh winter but contribute little to the spreading of the rust. Basidiospores spread the rust from wheat to barberry plants. Spermatia and receptive hyphae initiate the sexual cycle of the rust. Finally, aeciospores spread the rust from barberry to wheat plants.

The traditional means of combating *P. graminis* in wheat country was to remove barberry from the area, because without this obligate (necessary) alternate host, the fungus cannot complete its life cycle. However, this approach had limited success because some uredospores survived the winter and could infect a new generation of wheat. Modern control of the disease focuses on the breeding of resistant strains of wheat—a difficult task, given the rapid evolution of the rust. A new wheat variety carrying new resistance genes has to be released every year to keep up with the genetic changes in the rust.

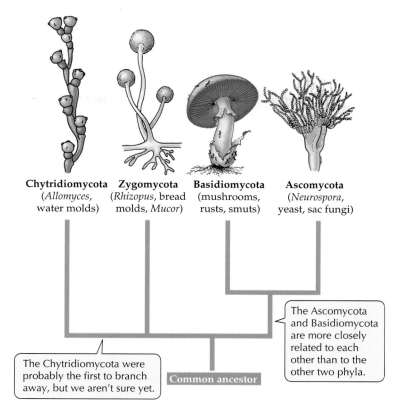

Chytridiomycota
(*Allomyces*, water molds)

Zygomycota
(*Rhizopus*, bread molds, *Mucor*)

Basidiomycota
(mushrooms, rusts, smuts)

Ascomycota
(*Neurospora*, yeast, sac fungi)

The Chytridiomycota were probably the first to branch away, but we aren't sure yet.

Common ancestor

The Ascomycota and Basidiomycota are more closely related to each other than to the other two phyla.

28.6 Phylogeny of the Fungi In addition to these four phyla, the imperfect fungi, or Deuteromycetes, is a "holding group" for fungal species whose status is yet to be determined.

Diversity in the Kingdom Fungi

Each phylum of the kingdom Fungi appears to be monophyletic, so our consideration of fungal diversity probably corresponds with a natural phylogeny (Figure 28.6). Because the imperfect fungi (deuteromycetes) almost certainly are polyphyletic, we will not give them phylum status. In this section on fungal diversity, we will consider the four phyla—Chytridiomycota, Zygomycota, Ascomycota, and Basidiomycota—and we'll discuss the status of the deuteromycetes.

Chytrids probably resemble the ancestral fungi

The most ancient fungal phylum is the Chytridiomycota—the chytrids, a group of aquatic microorganisms sometimes classified as protists. We place chytrids among the fungi because their cell walls consist primarily of chitin and because molecular evidence indicates that they are monophyletic with the other fungi.

Chytrids are either parasitic (on organisms such as algae, mosquito larvae, and nematodes) or saprobic, obtaining nutrients by breaking down dead organic

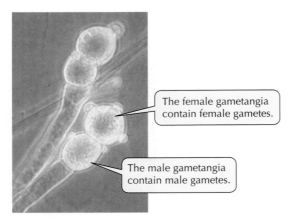

The female gametangia contain female gametes.

The male gametangia contain male gametes.

Allomyces sp.

28.7 Reproductive Structures of a Chytrid The haploid gametes produced in these gamete cases will fuse and form diploid zoospores.

matter. (Chytrids in the compound stomachs of cud-chewing animals may be an exception, living in a mutualistic association with their hosts.) Most chytrids live in freshwater habitats or in moist soil, but some are marine. Some chytrids are unicellular; others have mycelia made up of branching chains of cells. Chytrids reproduce both sexually and asexually.

Allomyces, a well-studied genus of chytrids, displays an alternation of haploid and diploid generations. A haploid **zoospore** (spore with flagella) comes to rest on dead plant or animal material in water and germinates to form a small haploid organism. This organism produces female and male **gametangia** (gamete cases) (Figure 28.7). The male gametangia are smaller than the female gametangia and possess a light orange pigment. Mitosis in the gametangia results in the formation of haploid gametes, each with a single nucleus.

Both female and male gametes have flagella. The motile female gamete produces a chemical attractant (pheromone; see Chapter 42) that attracts the swimming male gamete. The gametes fuse in pairs, and then their nuclei fuse to form a diploid zygote. Cell divisions give rise to a small diploid organism that produces numerous diploid flagellate zoospores by mitosis and cytokinesis. These diploid zoospores germinate to form more of the diploid organisms. Eventually, the diploid organism produces thick-walled resting sporangia that can survive unfavorable conditions such as dry weather or freezing. Nuclei in the resting sporangia eventually undergo meiosis, giving rise to haploid zoospores that are released into the water and begin the cycle anew.

Another chytrid, *Coelomomyces*, has a life cycle similar to that of *Allomyces*, but it is a parasite. *Coelomomyces* also requires two animal hosts: a mos-

quito larva and a copepod (a type of crustacean discussed in the next chapter).

Chytrids are the only fungi that have flagella. We speculate that the protist ancestor of the fungi possessed flagella, because the phylum Chytridiomycota was the first fungal group to diverge from the others (see Figure 28.6). As we learned in Chapter 26, the same protist ancestor gave rise to the protist phylum Choanoflagellida and to the animal kingdom. A key event when the chytrids branched off was the loss of flagella in the other branch of the fungal kingdom.

Zygomycetes reproduce sexually by fusion of two gametangia

Most zygomycetes (phylum Zygomycota) have hyphae without regularly occurring septa (cross-walls). They produce no motile cells, and only one diploid cell, the zygote, appears in the entire life cycle. Most zygomycetes form no fleshy fruiting body; rather, the hyphae spread in an apparently random fashion, with occasional stalked sporangia reaching up into the air (Figure 28.8). A very important group of zygomycetes serves as the fungal partners in the most common type of mycorrhizal association with plant roots (we discuss mycorrhizae in depth later in the chapter). Almost 900 species of zygomycetes have been described. A zygomycete that you may have seen at one time or another is *Rhizopus stolonifer,* the black bread mold. The mycelium of a zygomycete spreads over its substrate, growing forward by means of specialized hyphae. In vegetative reproduction of *Rhizopus,* many

Phycomyces sp.

28.8 A Zygomycete This small forest of filamentous structures is made up of sporangiophores. The stalks end in tiny, rounded sporangia.

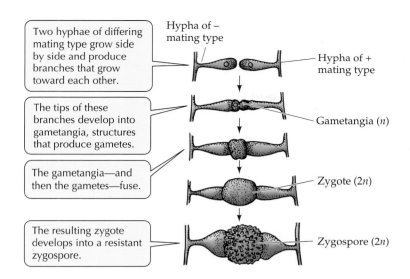

Two hyphae of differing mating type grow side by side and produce branches that grow toward each other.

The tips of these branches develop into gametangia, structures that produce gametes.

The gametangia—and then the gametes—fuse.

The resulting zygote develops into a resistant zygospore.

Hypha of – mating type

Hypha of + mating type

Gametangia (n)

Zygote (2n)

Zygospore (2n)

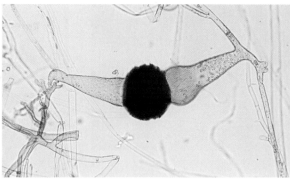

28.9 Conjugation in a Zygomycete The micrograph shows a zygospore produced by conjugation in the black bread mold.

Rhizopus stolonifer

stalked sporangiophores are produced, each bearing a single sporangium containing hundreds of minute spores. Other zygomycetes have sporangiophores with many sporangia. As in other filamentous fungi, the spore-forming structure is separated from the rest of the hypha by a wall.

Zygomycetes reproduce sexually when adjacent hyphae of two different mating types grow together and produce gametangia that fuse to form zygotes (Figure 28.9). Zygotes develop into thick-walled, highly resistant **zygospores** contained within the thickened walls of the old gametangia. Zygospores may remain dormant for months before their nuclei undergo meiosis.

A sporangium sprouts from the zygospore. The sporangium contains the products of meiosis: haploid nuclei that are incorporated into spores that germinate to form a new generation of haploid hyphae. What causes the hyphae to grow together and fuse? The two mating types release pheromones that direct this conjugation process. (Recall that a pheromone directs gamete attraction in *Allomyces* as well.)

The sexual reproductive structure of ascomycetes is a sac

The ascomycetes (phylum Ascomycota) are a large and diverse group of fungi distinguished by the production of sacs called **asci** (singular ascus) (Figure 28.10). The ascus is the characteristic sexual reproductive structure of the ascomycetes. Ascomycete hyphae are segmented by more or less regularly spaced septa. A pore in each septum permits extensive movement of cytoplasm and organelles (including the nuclei) from one "cell" to the next.

The approximately 30,000 known species of ascomycetes can be divided into two broad groups, depending on whether the asci are contained within a specialized fruiting structure. Species that have a fruit-

ing structure, the **ascocarp**, are collectively called euascomycetes ("true ascomycetes"); those without ascocarps are called hemiascomycetes ("half ascomycetes").

HEMIASCOMYCETES. In general, the hemiascomycetes are microscopic, and many species are unicellular. Perhaps the best known are the yeasts, especially baker's or brewer's yeast (*Saccharomyces cerevisiae*; Figure 28.11a). The yeasts are among the most important domesticated fungi. *S. cerevisiae* metabolizes glu-

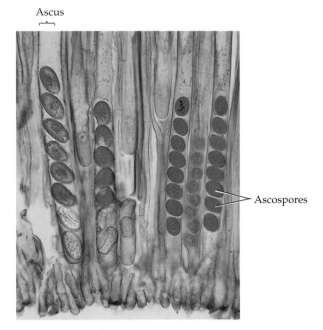

Ascus

Ascospores

28.10 Asci and Ascospores Each ascus contains eight ascospores—the products of meiosis followed by a single mitotic division and cytokinesis. The ascospores are stained red for this micrograph.

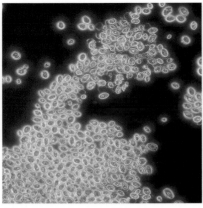

(a) *Saccharomyces cerevisiae*

(b) *Aleuria* sp.

(c) *Morchella esculenta*

28.11 Some Ascomycetes
(a) Some of these cells of baker's yeast are budding. These hemiascomycetes are facultative anaerobes—they can grow in either the presence or the absence of free oxygen. The brilliant red cups in (b) are cup fungi, as are the two yellow morels in (c). Morels, which have spongelike caps and a subtle flavor, are considered a gourmet delicacy.

cose obtained from its environment to ethanol and carbon dioxide. Carbon dioxide bubbles form in bread dough and give baked bread its light texture. Although baked away in bread making, the ethanol and carbon dioxide are both retained in beer. Other yeasts live on fruits such as figs and grapes and play an important role in the making of wine.

Yeasts reproduce asexually either by fission or, in the better-known genera, by **budding** (the outgrowth of a new cell from the surface of an old one). Sexual reproduction takes place only occasionally between two adjacent haploid cells of opposite mating type. (We discussed the genetics of yeast mating types in Chapter 14.) In some species, the resulting zygote buds to form a diploid cell population; in others, the zygote nucleus undergoes meiosis immediately. When diploid nuclei undergo meiosis, the entire cell becomes an ascus. Depending on whether the products of meiosis undergo mitosis, a yeast ascus usually has either eight or four **ascospores** (see Figure 28.10). The ascospores germinate to become haploid cells. Yeasts have no dikaryon stage.

Yeasts, especially *Saccharomyces cerevisiae*, are heavily used in molecular biological research. Just as *E. coli* is the most-studied prokaryote, *S. cerevisiae* is perhaps the most-studied eukaryote. Recall, for example, the creation and use of yeast artificial chromosomes, described in Chapter 16.

EUASCOMYCETES. Among the euascomycetes are several common molds, including *Neurospora*, the pink molds. *Neurospora* are found in a wide variety of habitats, and they are a serious contaminant in many laboratories. Recall that Beadle and Tatum used *Neurospora crassa* in their pioneering work on biochemical genetics (see Figure 12.1). Many euascomycetes are serious parasites on higher plants. Chestnut blight and Dutch elm disease are caused by euascomycetes. The powdery mildews are euascomycetes that infect cereal grains, lilacs, and roses, among many other plants. They can be a serious problem to grape growers, and a great deal of research has focused on ways to control these agricultural pests.

The euascomycetes also include the cup fungi (Figure 28.11b and c). In most of these organisms the fruiting structures are cup-shaped and can be as large as several centimeters across. The inner surfaces of the cups are covered with a mixture of both sterile filaments and asci, and they produce huge numbers of spores. Although these fleshy structures appear to be composed of distinct tissue layers, microscopic examination shows that their basic organization is still filamentous—a tightly woven mycelium. Such fruiting structures are formed only by the dikaryotic mycelium.

Two particularly delicious cup fungus fruiting structures are morels (Figure 28.11c) and truffles. Truffles grow in a mutualistic association with the roots of some species of oaks. Europeans traditionally used pigs to find truffles because some truffles secrete a substance that has an odor similar to a pig's sex attractant. Unfortunately, pigs also eat truffles, so dogs are now the usual truffle hunters.

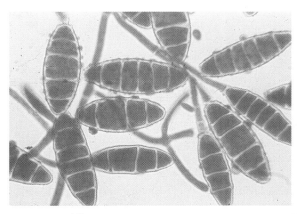

28.12 Conidia The large oval shapes are 4- to 6-celled macroconidia ("large conidia") at the tips of hyphae of an ascomycete. Each cell of a macroconidium can develop into a multicellular haploid individual, completing the asexual reproductive cycle.

The euascomycetes reproduce asexually by means of conidia that form at the tips of specialized hyphae (Figure 28.12). These small chains of conidia are produced by the millions and are sufficiently resistant to survive for weeks in nature. The conidia are what give molds their characteristic colors.

The sexual cycle of euascomycetes includes the formation of a dikaryon stage (Figure 28.13). Most euascomycetes form mating structures, some "female" and some "male." Nuclei from a male structure on one hypha enter a female mating structure on a hypha of compatible mating type. Dikaryotic ascogenous (ascus-forming) hyphae develop from the now dikaryotic female mating structure. The introduced nuclei divide simultaneously with the host nuclei. Eventually asci form at the tips of the ascogenous hyphae. Only with the formation of asci do the nuclei finally fuse. Both nuclear fusion and the subsequent meiosis of the resulting diploid nucleus take place within individual asci. The meiotic products are incorporated into ascospores that are ultimately shed by the ascus to begin the new haploid generation.

The sexual reproductive structure of basidiomycetes is a specialized cell bearing spores

About 25,000 species of basidiomycetes (phylum Basidiomycota) have been described. Basidiomycetes produce some of the most spectacular fruiting structures found anywhere among the fungi. These amazing fruiting structures are the puffballs (which may be more than half a meter in diameter), mushrooms of all kinds (more than 3,250 species, including the familiar *Agaricus campestris* you may enjoy on your pizza, as well as various poisonous mushrooms, such as members of the genus *Amanita*), and the giant bracket fungi often encountered on trees and fallen logs in a damp

forest (Figure 28.14). Bracket fungi do great damage to cut lumber and to stands of timber. Some basidiomycetes are among the most damaging plant pathogens, including wheat rust (*Puccinia graminis*; see Figure 28.5) and the smut fungi (see Figure 28.1*a*) that parasitize cereal grains and some noncereal crops. In sharp contrast, other basidiomycetes contribute to the well-being of plants as fungal partners in mycorrhizae (which we'll discuss shortly).

Basidiomycete hyphae characteristically have septa with small, distinctive pores. (Recall that hyphae of the zygomycetes lack regularly occurring septa.) The **basidium** (plural basidia), a swollen cell at the tip of a hypha, is the characteristic sexual reproductive structure of the basidiomycetes. It is the site of nuclear fusion and meiosis (Figure 28.15). Thus the basidium plays the same role in the basidiomycetes as the ascus does in the ascomycetes.

After the nuclei fuse in the basidium, the resulting diploid nucleus undergoes meiosis, and the four haploid nuclei migrate from the basidium into haploid **basidiospores**, which form on tiny stalks. These basidiospores typically are forcibly discharged from their basidia and then germinate, giving rise to haploid hyphae. As these hyphae grow, haploid hyphae of different mating types meet and fuse, forming dikaryotic hyphae, each cell of which contains two nuclei, one from each parent hypha. The dikaryotic mycelium grows and eventually produces fruiting structures. The dikaryotic phase may persist for years—some basidiomycetes live for decades or even centuries.

The elaborate fruiting structure of some fleshy basidiomycetes, such as the gill mushroom in Figure 28.15, is topped by a cap, or pileus, which has gills on its underside. Basidia develop in enormous numbers between the gills. The basidia discharge their spores into the air spaces between adjacent gills, and the spores sift down into air currents for dispersal and germination as new haploid mycelia.

The exact pattern of the gills and the spore color are criteria for distinguishing mushroom species. If a mature cap is placed gill side down on a piece of paper for a few hours in a quiet place, the ejected basidiospores settle from between the gills, leaving on the paper an elegant replica of the gill pattern and visible evidence of spore color.

Imperfect fungi lack a sexual stage

Mechanisms of sexual reproduction readily distinguish the zygomycetes, ascomycetes, and basidiomycetes. But many fungi, both saprobes and parasites, lack sexual stages entirely; presumably these stages have been lost during evolution. Classifying these fungi as belonging to any of the three major phyla was at one time difficult, but biologists now can classify most such fungi on the basis of DNA sequences.

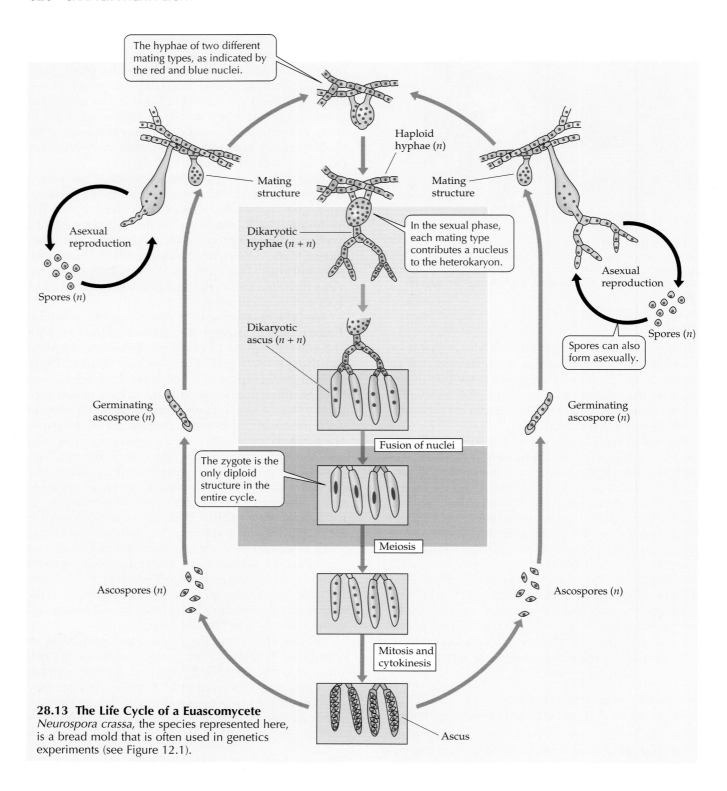

The hyphae of two different mating types, as indicated by the red and blue nuclei.

Haploid hyphae (*n*)

Mating structure

Mating structure

In the sexual phase, each mating type contributes a nucleus to the heterokaryon.

Asexual reproduction

Asexual reproduction

Spores (*n*)

Dikaryotic hyphae (*n* + *n*)

Spores (*n*)

Spores can also form asexually.

Dikaryotic ascus (*n* + *n*)

Germinating ascospore (*n*)

Germinating ascospore (*n*)

Fusion of nuclei

The zygote is the only diploid structure in the entire cycle.

Meiosis

Ascospores (*n*)

Ascospores (*n*)

Mitosis and cytokinesis

28.13 The Life Cycle of a Euascomycete
Neurospora crassa, the species represented here, is a bread mold that is often used in genetics experiments (see Figure 12.1).

Ascus

Fungi that lack sexual stages are found in the three phyla already discussed. Fungi that have not yet been placed in the existing phyla are grouped together in the "orphanage" known as the imperfect fungi, or deuteromycetes. Thus, the deuteromycete group is a holding area for species whose status is yet to be resolved. At present, about 25,000 species are classified as imperfect fungi.

If sexual structures are found on a fungus classified as a deuteromycete, the fungus is reassigned to the ap-

(a) *Lycoperdon perlatum*

(b) *Amanita muscaria*

(c) *Laetiporus sulphureus*

28.14 Basidiomycete Fruiting Structures   Although some persist only a few days or weeks, the fruiting structures of the basidiomycetes are probably the most familiar structures produced by fungi. (a) When raindrops hit them, these puffballs will release clouds of spores for dispersal. (b) A member of a highly poisonous mushroom genus, *Amanita*. (c) This bracket fungus is parasitizing a tree.

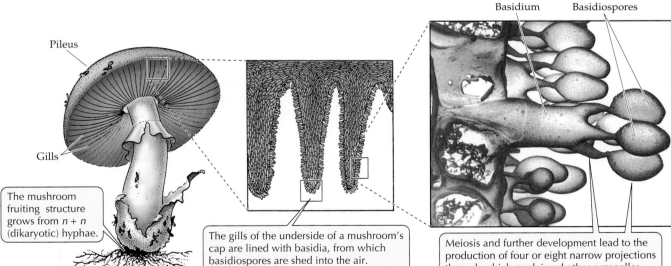

Basidium   Basidiospores

Pileus

Gills

The mushroom fruiting structure grows from $n + n$ (dikaryotic) hyphae.

The gills of the underside of a mushroom's cap are lined with basidia, from which basidiospores are shed into the air.

Meiosis and further development lead to the production of four or eight narrow projections through which nuclei and other organelles squeeze, forming basidiospores.

28.15 A Mushroom's Fruiting Structures   The basidium is the characteristic sexual reproductive structure of basidiomycetes and is the site of nuclear fusion. Basidiospores form on tiny stalks and are then forcibly dispersed to germinate into haploid hyphae, from which the familiar fruiting structure eventually grows.

propriate phylum. That happened, for example, with a fungus that produces plant growth hormones called gibberellins (see Chapter 35). Originally classified as the deuteromycete *Fusarium moniliforme,* this fungus was later found to produce asci, whereupon it was renamed *Gibberella fujikuroi* and transferred to the phylum Ascomycota.

Penicillium is a deuteromycete genus of green molds, of which some species produce the antibiotic penicillin, presumably for defense against competing bacteria. Two species, *P. camembertii* and *P. roquefortii,* are the organisms responsible for the characteristic flavors of Camembert and Roquefort cheeses, respectively.

Another deuteromycete genus of importance in some diets is *Aspergillus,* the genus of brown molds. *A. tamarii* acts on soybeans in the production of soy sauce, and *A. oryzae* is used in brewing the Japanese alcoholic beverage sake. Some species of *Aspergillus* that grow on nuts such as peanuts and pecans produce extremely carcinogenic (cancer-inducing) compounds called aflatoxins.

Fungal Associations

Earlier in this chapter we spoke briefly about mycorrhizae and lichens; in both, fungi live in intimate association with other organisms. Now that we have learned a bit about fungal diversity, it is appropriate to consider mycorrhizae and lichens in greater detail.

Mycorrhizae are essential to many plants

Many plants, including almost all tree species, depend on a mutually beneficial symbiotic association with fungi for an adequate supply of water and mineral elements. Unassisted, the root hairs of such plants do not absorb enough of these materials to sustain maximum growth. However, the roots become infected with fungi, forming the association called a *mycorrhiza* (Figure 28.16).

In ectomycorrhizae, the fungus wraps around the root; in endomycorrhizae, the infection is internal to the root, with no hyphae visible on the root surface. Infected roots characteristically branch extensively and become swollen and club-shaped. The hyphae of the fungi attached to the root increase the surface area for the absorption of water and minerals, and the mass of the mycorrhiza, like a sponge, holds water efficiently in the neighborhood of the root.

Most families of flowering plants contain some species that form mycorrhizae, as do liverworts, ferns, club mosses, and gymnosperms (see Chapter 27). Fossils of mycorrhizal structures more than 300 million years old have been found. Certain plants that live in nitrogen-poor habitats, such as cranberry bushes and orchids, invariably have mycorrhizae.

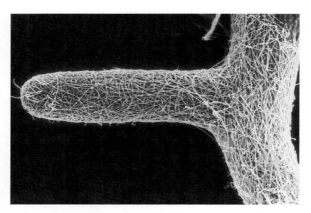

28.16 A Mycorrhizal Association   Hyphae of the fungus *Pisolithus tinctorius* cover this eucalyptus root, forming a mycorrhiza.

Orchid seeds will not germinate in nature unless they are already infected by the fungus that will form their mycorrhizae. Plants that lack chlorophyll always have mycorrhizae, which are often shared with the roots of green, photosynthetic plants.

The symbiotic fungus–plant association of a mycorrhiza is important to both partners. The fungus obtains important organic compounds, such as sugars and amino acids, from the plant. In return, the fungus greatly increases the absorption of water and minerals (especially phosphorus) by the plant. The fungus also often provides certain growth hormones (see Chapter 35), and it protects the plant against attack by microorganisms. Plants that have active mycorrhizae typically are a deeper green and may resist drought and temperature extremes better than plants of the same species that have little mycorrhizal development. Attempts to introduce some plant species to new areas have failed until a bit of soil from the native area (presumably containing the fungus necessary to establish mycorrhizae) was provided.

The partnership between plant and fungus results in a plant better adapted for life on land. It has been suggested that the evolution of this symbiotic association was the single most important step leading to the colonization of the terrestrial environment by living things. A hardy group of present-day colonizers is the lichens, which we consider next.

Lichens grow where no eukaryote has succeeded

A lichen is not a single organism, but is a meshwork of two radically different organisms: a fungus and a photosynthetic microorganism. Together the organisms constituting a lichen survive some of the harshest environments on Earth (Figure 28.17). In spite of this hardiness, lichens are very sensitive to air pollution because they are unable to excrete toxic substances that

28.17 Lichens in Frigid Environments The many types of lichens shown here are growing on an Alaskan tundra.

and photosynthetic partners separately and then reconstruct a lichen from the two.

There are about 13,500 "species" of lichens. They are found in all sorts of exposed habitats: tree bark, open soil, or bare rock. Reindeer "moss" (actually not a moss at all, but the lichen *Cladonia subtenuis*) covers vast areas in arctic, subarctic, and boreal regions, where it is an important part of the diets of reindeer and other large mammals. Lichens come in various forms and colors (Figure 28.18). Crustose (crustlike) lichens look like colored powder dusted over their substrate; foliose (leafy) and fruticose (shrubby) lichens may appear quite complex.

The most widely held interpretation of the lichen relationship is that it is a type of mutually beneficial symbiosis. Hyphae of the fungal mycelium are tightly pressed against the photosynthetic cells of the alga or cyanobacterium and sometimes even invade them. The bacterial or algal cells not only survive these indignities but continue their growth and photosynthesis. In fact, algal cells in a lichen "leak" photosynthetic products at a greater rate than do similar cells growing on their own. On the other hand, photosynthetic cells

they absorb. Hence they are not common in industrialized cities. Lichens are good biological indicators of air pollution because of their sensitivity.

The fungal components of most lichens are ascomycetes, but some are basidiomycetes or imperfect fungi (only one zygomycete serving as the fungal component of a lichen has been reported). The photosynthetic component may be either a cyanobacterium or a green alga. Relatively little experimental work has focused on lichens, perhaps because they grow so slowly—typically less than 1 cm per year. Thus only recently have workers been able to culture the fungal

(a)

(b)

(c)

28.18 Lichen Body Forms Lichens fall into three principal classes based on their form. (*a*) Crustose lichens such as the orange, white, and black species in this photograph often grow on otherwise bare rock, as shown here, or on tree bark. (*b*) A wet foliose lichen growing close to the ground. (*c*) A fruticose lichen grows among fallen leaves.

28.19 Lichen Anatomy   (a) Soredia of a fruticose lichen.
(b) Cross section showing layers of a lichen.

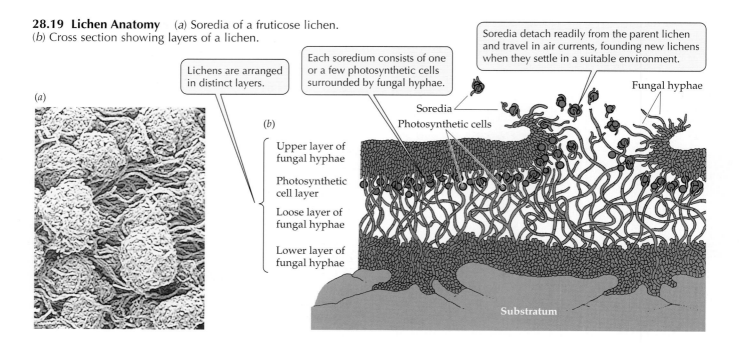

(a)

Lichens are arranged in distinct layers.

Each soredium consists of one or a few photosynthetic cells surrounded by fungal hyphae.

Soredia detach readily from the parent lichen and travel in air currents, founding new lichens when they settle in a suitable environment.

Fungal hyphae

Soredia

Photosynthetic cells

(b)

Upper layer of fungal hyphae

Photosynthetic cell layer

Loose layer of fungal hyphae

Lower layer of fungal hyphae

Substratum

from lichens grow more rapidly on their own than when combined with a fungus. On this ground, we may consider lichen fungi as parasitic on their photosynthetic partners.

Lichens can reproduce simply by fragmentation of the vegetative body, which is called the **thallus**, or by specialized structures called **soredia** (singular *soredium*). Soredia consist of one or a few photosynthetic cells surrounded by fungal hyphae (Figure 28.19a). The soredia become detached, move in air currents, and upon arriving at a favorable location, develop into a new lichen. If the fungal partner is an ascomycete or a basidiomycete, it may go through its sexual cycle, producing either ascospores or basidiospores. When these are discharged, however, they disperse alone, unaccompanied by the photosynthetic partner, and thus may not be capable of reestablishing the lichen association. Nevertheless, many lichens produce characteristic fruiting structures in which the asci or basidia are located.

Visible in a cross section of a typical foliose lichen are a tight upper region of fungal hyphae alone, a layer of cyanobacteria or algae, a looser hyphal layer, and finally hyphal rhizoids that attach the whole structure to its substrate (Figure 28.19b). The meshwork has properties that enable it to hold water fairly tenaciously. Some nutrients for the photosynthetic cells arrive through the fungal hyphae, the meshwork provides a suitably moist environment for the photosynthetic cells, and the fungi derive fixed carbon from the photosynthesis of the algal or cyanobacterial cells.

Lichens are often the first colonists on new areas of bare rock. They satisfy most of their needs from the air and from rainwater, augmented by minerals absorbed from their rocky substrate. A lichen begins to grow shortly after a rain, as it begins to dry. As it grows, the lichen acidifies its environment slightly, and this acid contributes to the slow breakdown of rocks, an early step in soil formation. After further drying, the lichen's photosynthesis ceases. The water content of the lichen may drop to less than 10 percent of its dry weight, at which point the lichen becomes highly insensitive to extremes of temperature. The flora of Antarctica features more than 100 times as many species of lichens as of plants.

Whether living on their own or in lichen associations, fungi have spread successfully over much of Earth since their origin from a protist ancestor. That ancestor also gave rise to the animal kingdom, the group we'll consider in the next two chapters.

Summary of "Fungi: A Kingdom of Recyclers"

• Fungi are the principal degraders of dead organic matter in the biosphere.

General Biology of the Fungi

• Fungi are heterotrophic eukaryotes with absorptive nutrition. They are saprobes, parasites, or mutualists. The body of a fungus is composed of chitinous-walled hyphae, often massed to form a mycelium. The filamentous hyphae give fungi a large surface area–to–volume ratio, enhancing their ability to absorb food.

• Fungi reproduce asexually by means of spores formed within sporangia, by conidia formed at the tips of hyphae, by budding, or by fragmentation. **Review Figures 28.5, 28.13**
• Fungi reproduce sexually when hyphae or motile cells of different mating types meet and fuse.
• The gametes and zoospores of chytrids are the only cells that have flagella in the kingdom Fungi. **Review Figure 28.9**
• In addition to the haploid and diploid states, many fungi demonstrate a third nuclear condition: the dikaryotic, or $n + n$, state. Some fungal life cycles involve two hosts. **Review Figures 28.5, 28.13**

Diversity in the Kingdom Fungi

• The kingdom Fungi consists of four phyla: Chytridiomycota, Zygomycota, Ascomycota, and Basidiomycota. These differ in their reproductive structures, spore formation, and less importantly, the cross-walls (if any) of their hyphae.
• Chytrids, with their flagellated zoospores and gametes, probably resemble the ancestral fungi.
• Zygomycetes reproduce sexually by fusion of gametangia. **Review Figure 28.9**
• The sexual reproductive structure of ascomycetes is an ascus containing ascospores. The ascomycetes are divided into two groups, euascomycetes and hemiascomycetes, on the basis of whether they have an ascocarp. **Review Figures 28.10, 28.12, 28.13**
• The sexual reproductive structure of basidiomycetes is a basidium, a swollen cell bearing basidiospores. **Review Figure 28.15**
• Imperfect fungi (deuteromycetes) lack sexual structures, but DNA sequencing helps identify the phyla to which they belong. **Review Figure 28.6 and Table 28.1**

Fungal Associations

• Mycorrhizae, associations of fungi with plant roots, enhance the ability of the roots to absorb water and nutrients.
• Lichens, mutualistic combinations of a fungus with a green alga or a cyanobacterium, are found in some of the most inhospitable environments on the planet. **Review Figure 28.19**

Self-Quiz

1. Which statement about fungi is *not* true?
   a. A hyphal fungus has a body called a mycelium.
   b. Hyphae are composed of individual mycelia.
   c. Many fungi tolerate highly hyperosmotic environments.
   d. Many fungi tolerate low temperatures.
   e. Most fungi are anchored to their substrate by rhizoids.

2. The absorptive nutrition of fungi is aided by
   a. dikaryon formation.
   b. spore formation.
   c. the fact that they are all parasites.
   d. their large surface area–to–volume ratio.
   e. their possession of chloroplasts.

3. Which statement about fungal nutrition is *not* true?
   a. Some fungi are active predators.
   b. Some fungi form mutualistic associations with other organisms.
   c. All fungi require mineral nutrients.
   d. Fungi can make some of the compounds that are vitamins for animals.
   e. Facultative parasites can grow only on their specific hosts.

4. Which statement about heterokaryosis is *not* true?
   a. The cytoplasm of two cells fuses before their nuclei fuse.
   b. The two haploid nuclei are genetically different.
   c. The two nuclei are of the same mating type.
   d. The heterokaryotic stage ends when the two nuclei fuse
   e. Not all fungi have a heterokaryotic stage.

5. Reproductive structures consisting of a photosynthetic cell surrounded by fungal hyphae are called
   a. ascospores.
   b. basidiospores.
   c. conidia.
   d. soredia.
   e. gametes.

6. The zygomycetes
   a. have hyphae without regularly occurring cross-walls.
   b. produce motile gametes.
   c. form fleshy fruiting bodies.
   d. are haploid throughout their life cycle.
   e. have structures homologous to those of the ascomycetes.

7. Which statement about ascomycetes is *not* true?
   a. They include the yeasts.
   b. They form reproductive structures called asci.
   c. Their hyphae are segmented by cross-walls.
   d. Many of their species have a dikaryotic stage.
   e. All have fruiting structures called ascocarps.

8. The basidiomycetes
   a. often produce fleshy fruiting structures.
   b. have hyphae without cross-walls.
   c. have no sexual stage.
   d. never produce large fruiting structures.
   e. form diploid basidiospores.

9. The deuteromycetes
   a. have distinctive sexual stages.
   b. are all parasitic.
   c. include some commercially important species.
   d. include the ascomycetes.
   e. are never components of lichens.

10. Which statement about lichens is *not* true?
    a. They can reproduce by fragmentation of their vegetative body.
    b. They are often the first colonists in a new area.
    c. They render their environment more basic (alkaline).
    d. They contribute to soil formation.
    e. They may contain less than 10 percent water by weight.

Applying Concepts

1. You are shown an object that looks superficially like a pale green mushroom. Describe at least three criteria (including anatomical and chemical traits) that would enable you to tell whether the object is a piece of a plant or a piece of a fungus.

2. Differentiate between the members of each of the following pairs of related terms.
   a. hypha/mycelium
   b. euascomycete/hemiascomycete
   c. ascus/basidium
   d. ectomycorrhiza/endomycorrhiza

3. For each type of organism listed below, give a single characteristic that may be used to differentiate it from the other, related organism(s) in parentheses.
   a. Zygomycota (Ascomycota)
   b. Basidiomycota (deuteromycetes)
   c. Ascomycota (Basidiomycota)
   d. baker's yeast (*Neurospora crassa*)

4. Many fungi are dikaryotic during part of their life cycle. Why are dikaryons described as $n + n$ instead of $2n$?

5. Review Figure 28.6. Why are the Ascomycota and Basidiomycota grouped together, apart from the other two phyla?

6. If all the fungi on Earth were suddenly to die, how would the surviving organisms be affected? Be thorough and specific in your answer.

Readings

Alexopoulos, C. J., C. W. Mims and M. Blackwell. 1996. *Introductory Mycology*, 4th Edition. John Wiley & Sons, New York. A new edition of a leading textbook on the biology of fungi.

Kendrick, B. 1992. *The Fifth Kingdom*, 2nd Edition. Focus Information Group, Newburyport, MA. An introduction to the fungi, with useful emphasis on taxonomy.

Kosikowski, E. V. 1985. "Cheese." *Scientific American*, May. A description of how fungi are involved in the production of many cheeses.

Moore-Landecker, E. 1996. *Fundamentals of Fungi*, 4th Edition. Prentice-Hall, Englewood Cliffs, NJ. Another introduction to the kingdom.

Newhouse, J. R. 1990. "Chestnut Blight." *Scientific American*, July. A description of a new biological method for controlling the fungus responsible for a disease that almost eliminated an important tree species.

Raven, P. H., R. F. Evert and S. Eichhorn. 1992. *Biology of Plants*, 5th Edition. Worth, New York. Includes an excellent discussion of the fungi.

Sternberg, S. 1994. "The Emerging Fungal Threat," *Science*, December 9, pages 1632–1634. Enumeration of the dangers posed to AIDS patients by fungi.

Chapter 29

Protostomate Animals: The Evolution of Bilateral Symmetry and Body Cavities

A Sea of Animals
The coral reef is formed by the skeletons of myriad animals, and a host of other animal species call the reef habitat home.

T ropical coral reefs abound with life. The corals themselves dominate the landscape, and their skeletons form its structures. Growing among the corals are many kinds of organisms. Some of them are algae; but many of them, although they may *look* like plants, are actually animals that do not at all resemble the terrestrial animals with which we are most familiar.

The members of the kingdom Animalia are among the most conspicuous living things in the world around us and, as members of this kingdom ourselves, we have a special interest in its other members. Humans tend to be most aware of the large animals that share our terrestrial environment, but these are only a few of Earth's myriad animal species. Even a brief excursion into a coral reef brings us into contact with individuals from many animal phyla—some of which we might not even recognize as animals.

The identification of Earth's animal species has barely begun. Millions of animal species live on Earth today, and describing them is an enormous task. Many are so inconspicuous they evade all but the most alert and trained eye. Many live in remote or harsh areas that are difficult for humans to visit. Some animals appear vastly different at different stages of their lives (the caterpillar and the butterfly are a familiar example) and without study and observation we might think they were two different species. Other animals that look very much alike might in fact belong to different species. Parasitic species, which live at least part of their life cycles inside other animals, are especially poorly known.

If many kinds of animals seem so unlike the animals with which we are familiar, how do we decide that they are indeed animals? We class them all as animals because they share both a common ancestry and certain traits not found in organisms belonging to the other kingdoms. In this chapter we will first discuss how biologists infer the evolutionary relationships among animals and some of the defining characteristics of the animal way of life. Then we'll describe the diversity of animals in one of the two great animal evolutionary lineages defined by their mode of embryonic development: the protostomes. Chapter 30 treats the other major animal lineage, the deuterostomes.

Descendants of a Common Ancestor

Biologists have long debated whether the animals arose once or several times from protist ancestors, but recent evidence strongly indicates that all animals are descendants of a single lineage. First, similarities in their 5S and 18S ribosomal RNAs suggest that all animals have a common ancestor. In addition, all animals share the same complex set of extracellular matrix (ECM) molecules, first discussed in Chapter 4.

These molecules, which make up the connective tissues of animals, form the basal laminae that underlie the sheets of the epithelial cells that all animals have. The ECM is the fabric that connects the cells of both embryos and adult animals (Figure 29.1). The molecular composition, structure, and functioning of the extracellular matrices of all animals are almost identical, and the ECM develops in similar ways in the embryos of all animals. The likelihood that such a complex system evolved more than once is remote.

Animals probably arose evolutionarily from ancestral colonial flagellated protists as a result of division of labor among their aggregated cells. Division of labor probably evolved because a mass of identical cells exchanges materials with its environment relatively slowly, and because some functions are performed better by specialized cells.

Within the ancestral colonies of cells—perhaps similar to those still existing in the chlorophyte *Volvox* or some colonial choanoflagellates (see Figure 26.28a)—some cells became specialized for movement, others became specialized for nutrition, and still others differentiated into gametes. Once the division of labor had begun, the units continued to differentiate, while improving their coordination with other working groups of cells. These coordinated groups of cells evolved into the larger and more complex organisms that we now call animals.

The development of a multicellular organism with differentiated cells requires the integration of several processes (see Chapter 15). First, genes must be regulated so that even though all cells contain all of the

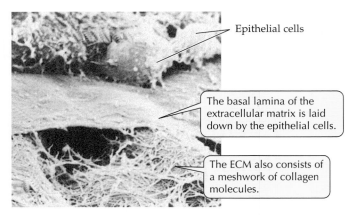

Epithelial cells

The basal lamina of the extracellular matrix is laid down by the epithelial cells.

The ECM also consists of a meshwork of collagen molecules.

29.1 An Extracellular Matrix is Found in All Animals Extracellular matrix molecules secreted by animal cells form connective tissue that holds the cells together. The similarity of this structure in all animals indicates a common ancestral origin for the kingdom.

animal's genes, particular genes are active only in particular cells at certain times during development. Second, the differentiation status of a cell must be transmitted during cell division. That is, muscle cells must give rise to muscle cells and epithelial cells must give rise to epithelial cells. Third, differentiated cells must become arranged in a specific and reliable spatial pattern. These processes require primarily changes in controls over developmental processes rather than new cellular structures or functions not present in unicellular organisms.

Food Procurement: The Key to the Animal Way of Life

Animals require a variety of complex organic molecules as sources of energy, and they obtain these molecules by active expenditure of energy. This energy is used to move the animals through their environment, to cause the environment and the food it contains to move to the animals, or to position the animals where food will pass by them. The foods animals eat include most other members of the animal kingdom, as well as members of all other kingdoms. Much of the diversity of animal sizes and shapes evolved as animals acquired the ability to capture and eat many different kinds of food.

The need to move in search for food has favored structures that provide animals with detailed information about their environment, and structures able to receive and coordinate this information. Consequently, most animals are behaviorally much more complex than plants. Because animals ingest chemically complex foods, they expend considerable energy to maintain a relatively constant internal composition while they take in foods that vary chemically.

A real appreciation of animal structure and functioning is best achieved through firsthand experience in the field and laboratory. The accounts in this chapter and the next serve as an orientation to the major groups of animals, their similarities and differences, and the evolutionary pathways that resulted in the current number and variety of animal species.

Clues to Evolutionary Relationships among Animals

Biologists try to classify animals in ways that reflect their evolutionary relationships. Determining the early evolutionary relationships of animals is difficult because fossils of most animal phyla appear simultaneously near the beginning of the Cambrian period, and because animal evolution is replete with examples of the convergence of traits. Therefore, biologists use a variety of traits in their attempts to infer animal phylogenies. Clues to these relationships are found in the fossil record, in the patterns of embryological development of animals, in the comparative morphology and physiology of living and fossil animals, and in their molecular biology (see Chapters 22 and 23).

Body plans are basic structural designs

The entire animal, its organ systems, and the integrated functioning of its parts are known as its **body plan**. Body plans are basic structural designs that reflect, and provide clues to, the evolutionary history of animal lineages. Consequently, we use them as one way to organize our treatment of animal groups. Animals in many, but not all, lineages evolved increasing body complexity. As you will see, animals with complex bodies can perform and control complex movements.

A fundamental aspect of an animal's body plan is its overall shape, described as its **symmetry**. A symmetrical animal can be divided along at least one plane into similar halves. Animals that have no plane of symmetry are said to be **asymmetrical** (Figure 29.2*a*). Many sponges are asymmetrical, but most animals have some kind of symmetry.

The simplest form is **spherical symmetry**, in which body parts radiate out from a central point. An infinite number of planes passing through the central point can divide a spherically symmetrical organism into similar halves. Spherical symmetry is widespread among protists (for example, radiolarians; see Figure 26.11*a*), but most animals possess other forms of symmetry.

An organism with **radial symmetry** has one main axis around which its body parts are arranged (Figure 29.2*b*). A perfectly radially symmetrical animal can be divided into similar halves by any plane that contains the main axis. Some simple sponges and a few other animals have such symmetry, but most radially symmetrical animals are modified such that only two

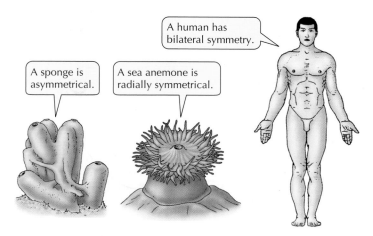

29.2 Body Symmetry Most animals above the level of sponges display some form of body symmetry. Any plane passing through the main axis of a radially symmetrical animal will divide it into equal halves, whereas there is only one plane that will divide a bilaterally symmetrical animal in half.

planes can divide them into similar halves. These animals are said to have **biradial symmetry**. Three animal phyla—Cnidaria, Ctenophora, and Echinodermata—are composed primarily of radially or biradially symmetrical animals. As we'll see, these animals move only slowly, if at all.

A **bilaterally symmetrical** animal can be divided into mirror images only by a single plane that passes through the midline of its body from the front (anterior) to the back (posterior) end (Figure 29.2*c*). The other (left–right) axis divides the body into two dissimilar sides; the side of a bilaterally symmetrical animal without a mouth is its **dorsal surface**; the side with a mouth is its **ventral surface**. Bilateral symmetry is a common characteristic of animals that move through their environments. Such symmetry is strongly correlated with the development of sense organs and central nervous tissues at the anterior end of the animal, a process known as **cephalization**. Cephalization may have been selected for among motile animals because the anterior end typically encounters new environments first.

Symmetry is an important way of comparing animal body plans and relating the plans to how the animals live. However, it is not very useful for determining evolutionary relationships among animals, because animals in closely related phyla with different lifestyles may have quite different symmetries.

Developmental patterns are often evolutionarily conservative

The early development of embryos has sometimes been **evolutionarily conservative**; that is, changes in early development have evolved very slowly in some lineages. For this reason, developmental patterns reveal a great deal about evolutionary relationships among some animals.

During development from a single-celled zygote to a multicellular adult, animals form layers of cells. These layers behave as units during early embryonic development and give rise to different tissues and organs in the adult animal. The embryos of **diploblastic** animals have only two cell layers: an outer *ectoderm* and an inner *endoderm.* The embryos of **triploblastic** animals have, in addition to ectoderm and endoderm, a third layer, the *mesoderm,* which lies between the ectoderm and the endoderm.

On the basis of differences in their early developmental patterns, combined with other shared, derived traits, zoologists divide animals other than sponges, cnidarians, and ctenophores into two major lineages. In the **protostomate lineage**, the pattern of early cleavage is spiral. The plane of division is oblique to the long axis of the egg, causing the cells to be arranged in a spiral pattern. Cleavage of the fertilized egg of protostomes is **determinate**; that is, if the egg is allowed to divide a few times and the cells are then separated, each cell develops into only a partial embryo (see Chapter 40).

In the **deuterostomate lineage**, the fertilized egg cleaves radially. Cells divide along a plane either parallel to or at right angles to the long axis of the fertilized egg. In addition, cleavage in deuterostomes typically is **indeterminate**; that is, cells separated after several cell divisions can still develop into complete embryos.

Finally, among deuterostomes, the mouth of the embryo originates some distance away from the embryonic structure called the *blastopore,* which becomes the anus. Among protostomes, the mouth arises from or near the blastopore. Table 29.1 summarizes the

TABLE 29.1 Developmental Differences between Protostomes and Deuterostomes.

PROTOSTOMES	DEUTEROSTOMES
Spiral cleavage	Radial cleavage
Determinate cleavage	Indeterminate cleavage
Blastopore becomes mouth	Blastopore becomes anus
Mesoderm derives from cells on lip of blastopore	Mesoderm derives from walls of developing gut
Mesoderm splits to form coelom	Mesoderm usually outpockets to form coelom

early embryological differences between protostomes and deuterostomes.

Fluid-filled spaces, called **body cavities**, lie between the cell layers of the bodies of many kinds of animals. These body cavities are of great functional significance to animals. The type of body cavity that animals have, in combination with other traits, is useful for comparing grades (levels of structural complexity) in animals. The type of body cavity an animal has strongly influences how it can move. Therefore, knowledge of a group's body cavities tells us a great deal about the way of life of its members. However, body cavities are not completely reliable traits for determining evolutionary relationships, because as animals adapt to different ways of life, their body cavities may be altered or even lost.

Some animals, such as flatworms, are called **acoelomates** because they lack an enclosed body cavity (Figure 29.3a). Their only internal cavity is the digestive cavity; the space between the gut and the body wall is filled with masses of cells that are collectively

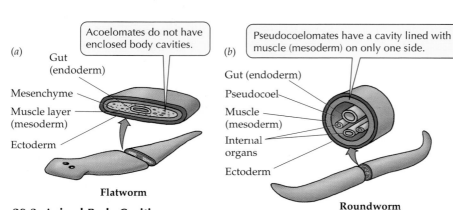

(a)
Acoelomates do not have enclosed body cavities.

Gut (endoderm)
Mesenchyme
Muscle layer (mesoderm)
Ectoderm

Flatworm

(b)
Pseudocoelomates have a cavity lined with muscle (mesoderm) on only one side.

Gut (endoderm)
Pseudocoel
Muscle (mesoderm)
Internal organs
Ectoderm

Roundworm

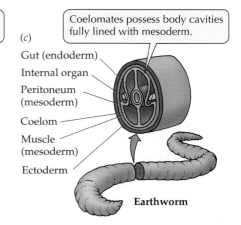

(c)
Coelomates possess body cavities fully lined with mesoderm.

Gut (endoderm)
Internal organ
Peritoneum (mesoderm)
Coelom
Muscle (mesoderm)
Ectoderm

Earthworm

29.3 Animal Body Cavities
The major types of animal body cavities differ in the types of cells that line them. Tissues derived from ectoderm are colored blue, those from mesoderm are pink, and those from endoderm are yellow; this color convention will be continued in Chapter 40, where these three cell layers will be discussed in detail.

called *mesenchyme.* Another group of animals, the **pseudocoelomates,** have a body cavity, called the **pseudocoel,** derived from the first cavity formed inside the proliferating ball of embryonic cells (Figure 29.3*b*). The pseudocoel provides a liquid-filled space in which many of the body organs are suspended, but control over body shape is crude because a pseudocoel has muscles on only one side.

Coelomate animals have a **coelom** (Figure 29.3*c*), a body cavity that develops within the embryonic mesoderm. It has muscles on both sides and is lined with a special structure derived from mesoderm, called the **peritoneum**. The internal organs of coelomates are slung in pouches of the peritoneum rather than being suspended within the cavity. Coelomate animals include peanut worms, echiuran worms, pogonophorans, annelids, mollusks, arthropods, echinoderms, and chordates. Among arthropods, the coelom evolved into a *hemocoel*—a cavity that is filled with blood.

The phylogeny of animals we adopt in this book is based on a combination of many developmental, structural, and molecular traits. Figure 29.4 shows the postulated order of splitting of the major lineages in animal evolution. Later in the chapter we will give details of phylogenies of protostomes and deuterostomes. New information continues to modify and refine our understanding of the details of phylogenetic relationships, and different types of traits often suggest different relationships. Nonetheless, some of the proposed lineages, such as the protostomes and deuterostomes, are supported by many types of data; they are unlikely to be altered by new information.

Early Animal Diets: Prokaryotes and Protists

When the first animals evolved, the primary food items available to them were algae floating in the water (called **phytoplankton**), extensive algal mats covering the shallow sea bottoms, protists, and prokaryotes.

The earliest animals were probably floating colonies of flagellated cells that fed on prokaryotes and protists. Some of these animals developed stinging tentacles housed in special cells. With such tentacles they could capture larger prey. Others evolved physical and behavioral adaptations for grazing in the algal mats. Both of these changes favored the evolution of larger animals that could move about.

Not surprisingly, the presence of these larger animals created opportunities for carnivores that fed on them and parasites that lived on or inside them. As defenses against predators, natural selection favored the evolution of shells and other protections, the ability to burrow, the use of safe refuges such as caves and crevices, and faster movement. These evolutionary themes of protection, defense, refuge, and movement continue today.

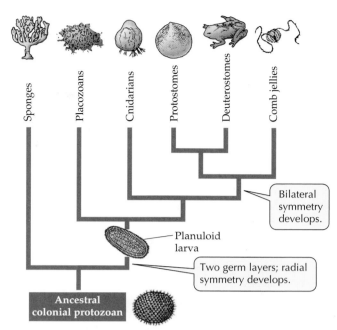

29.4 A Probable Phylogeny of Animals The evolutionary tree of animals we will use in this chapter and the next one postulates that animals are monophyletic. The two major animal lineages—protostomes and deuterostomes—are ancient divisions in animal evolution.

For the moment, however, let's examine how these selection pressures operated during the early evolution of animals, which began about 2.5 billion years ago. Remember that all life was confined to the seas before and during much of animal evolution.

Simple Aggregations: Sponges and Placozoans

Animals probably arose from protists whose cells remained together after division, forming a multicellular mass called a colony. It is difficult to distinguish between a protist colony and some simple multicellular animals that have little differentiation or coordination among their cells. The lineage leading to modern sponges separated from the lineage leading to all other animals very early in the evolution of animals (see Figure 29.4). Some living sponges are still very similar to the probable ancestral colonial protists.

Sponges are loosely organized

The sponges (phylum Porifera, Latin for "pore bearers") are *sessile*: They live attached to the substratum and do not move about. All sponges, even large ones, which may reach a meter in length, have a very simple body plan. The body of a sponge is a loose aggregation of cells built around a water canal system. A sponge has no mouth or digestive cavity, no muscles, and no nervous system. In fact, there are no organs in the usual sense of the word.

29.5 The Body Plan of a Simple Sponge   The flow of water through the sponge is shown by blue arrows.

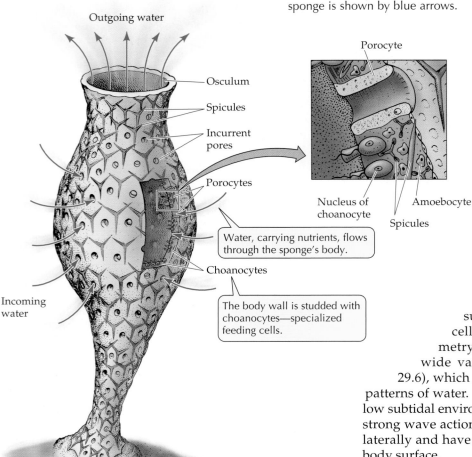

Outgoing water

Osculum

Spicules

Incurrent pores

Porocytes

Water, carrying nutrients, flows through the sponge's body.

Choanocytes

The body wall is studded with choanocytes—specialized feeding cells.

Incoming water

Porocyte

Nucleus of choanocyte

Amoebocyte

Spicules

through one or more larger openings (Figure 29.5).

Between the thin epidermis and the choanocytes lies a layer of cells, some of which are similar to amoebas and move about in the body. A supporting skeleton is also present, either in the form of simple or branching spines or as an elastic network of spongin fibers. The spicular skeletons of some sponges may be very complex, as is evident in the sponge skeletons you can find in a hardware store.

Unlike most other animals, sponges have no distinct body tissues, no cavity between the layers of cells, and no recognizable body symmetry. Nonetheless, sponges come in a wide variety of sizes and shapes (Figure 29.6), which are adapted to different movement patterns of water. Sponges living in intertidal or shallow subtidal environments, where they are subjected to strong wave action, hug the substratum. They spread laterally and have multiple pores scattered over their body surface.

Many sponges that live in calm waters are simple, with a single large opening on top of the body (see Figure 29.5). Water is taken in through pores on the sides of the body and expelled upward through the large opening on top. Sponges that live in flowing water do not need to exert much energy to move water through their bodies. Most of them are flattened and are oriented at right angles to the direction of current flow; they intercept water as it flows past them. Water is drawn out of the top pores of such tall, thin sponges just as air is drawn out of a tall chimney by the wind.

Sponges reproduce both sexually and asexually. In most species, a single individual produces both eggs and sperm. Water currents carry sperm from one individual to another. Asexual reproduction is by budding and other processes that produce fragments able to develop into new sponges. Most of the 10,000 species of sponges are marine animals, but about 50 species are found in fresh water.

Placozoans are the simplest multicellular animals

Molecular and morphological evidence suggests that the next evolutionary split separated placozoans from other animal lineages (see Figure 29.4). Placozoans

A sponge is so loosely organized that even if it is completely disassociated by being strained through a filter, its cells can reassociate into a new sponge. Nevertheless, sponges have at least 34 different cell types, most of which are specialized mobile cells. Thus sponges are more complex than their colonial flagellate ancestors.

Throughout most of its life, an individual sponge is attached to a substratum. It feeds by drawing water into itself and filtering out the small organisms and nutrient particles that flow past the walls of its inner cavity. Unique feeding cells called **choanocytes** line the inside of the water canals of sponges; these cells have a collar consisting of cytoplasmic extensions that surround a flagellum. The flagellated choanocytes set up the water currents. Water flows into the animal either by way of small pores that perforate special epidermal cells (in simple sponges) or through intercellular pores (in complex sponges). The water passes into small chambers within the body where food particles are captured by the choanocytes. Water then exits

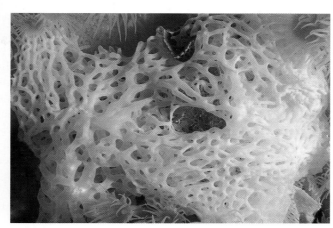

29.6 Sponges Differ in Size and Shape Two of the many different growth forms of sponges are illustrated.

Aplysina fistularia

Clathrina coriacea

(phylum Placozoa) are much smaller than sponges, less than 3 mm in diameter. Their bodies consist of no more than a few thousand cells and only four cell types, in contrast to the 34 cell types found in sponges. Placozoans lack any kind of symmetry, and they have no body cavity or distinct tissues or organs (Figure 29.7). Their body plan is a flat plate consisting of two layers of flagellated cells that enclose a fluid-filled area that contains fibrous cells.

For a long time biologists thought that placozoans were larvae of some type of sponge or cnidarian, but in the late 1960s placozoans were observed to achieve sexual maturity and produce gametes. Placozoans are widely distributed in shallow tropical ocean waters. They feed by endocytosis, taking prokaryotes or small protists into their own cells, or by secreting enzymes onto their protist prey, which then digest the prey out-

side the placozoan's body. The cells of the placozoan then absorb the prey's digested remains. Only two species have been described in this unusual phylum.

The Evolution of Cell Layers: The Cnidarians

Animals in all phyla other than Porifera and Placozoa have distinct cell layers and symmetrical bodies. Although the members of most animal phyla have three cell layers, the next lineage to split off from the main line of animal evolution resulted in a phylum of animals—the cnidarians—having only two cell layers (see Figure 29.4).

Cnidarians are simple but specialized carnivores

Cnidarians (phylum Cnidaria) appeared early in evolutionary history and radiated into many different species; they may have constituted more than half of the late Precambrian animal species. About 10,000 modern cnidarian species—jellyfish, sea anemones, corals, and hydrozoans—live today. All but a few species are marine.

The smallest cnidarians can hardly be seen without a microscope; the largest jellyfish is 2.5 meters wide and has tentacles 25 meters long. All cnidarians have only two cell layers and are radially symmetrical. A layer called the mesoglea between the two cell layers eventually develops from the ectoderm, but it never produces complex internal organs like those found in animals that have three true cell layers.

A key feature of cnidarians is their **cnidocytes**, specialized cells that contain stinging structures called **nematocysts** that can discharge toxins into their prey (Figure 29.8). Cnidocytes, which are borne on tentacles, allow cnidarians to capture large prey. The nematocysts paralyze and help hold prey; they are responsi-

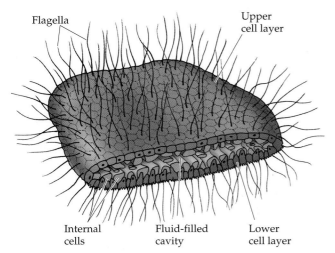

Flagella

Upper cell layer

Internal cells

Fluid-filled cavity

Lower cell layer

29.7 Placozoans Lack Symmetry This cross section illustrates the simple structure of a placozoan.

Portuguese man-of-war

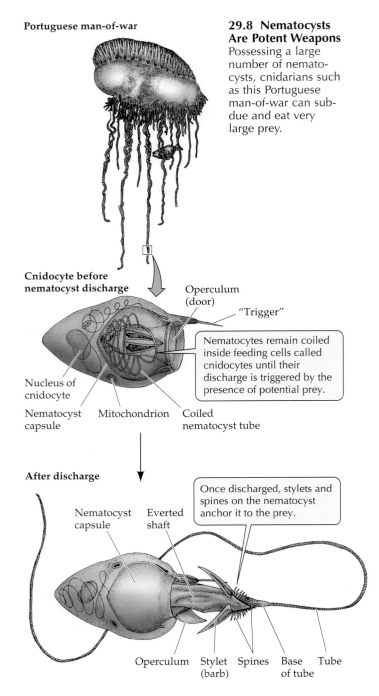

29.8 Nematocysts Are Potent Weapons
Possessing a large number of nematocysts, cnidarians such as this Portuguese man-of-war can subdue and eat very large prey.

Cnidocyte before nematocyst discharge

Operculum (door)

"Trigger"

Nematocytes remain coiled inside feeding cells called cnidocytes until their discharge is triggered by the presence of potential prey.

Nucleus of cnidocyte

Nematocyst capsule   Mitochondrion   Coiled nematocyst tube

After discharge

Once discharged, stylets and spines on the nematocyst anchor it to the prey.

Nematocyst capsule   Everted shaft

Operculum   Stylet (barb)   Spines   Base of tube   Tube

cnidarians. Cnidarians also have epithelial cells with muscle fibers whose contractions enable the animals to move, as well as nerve nets that integrate their body activities.

CNIDARIAN LIFE CYCLES. The generalized cnidarian life cycle has two distinct stages (Figure 29.9)—polyp and medusa—but one stage is lacking in many species. The **polyp** stage is a cylindrical stalk with tentacles surrounding a mouth that is at the opposite end from the site of attachment to the substratum. This stage is usually asexual, but individual polyps may reproduce by budding, thereby forming a colony.

The **medusa** (plural medusae) is a familiar, free-swimming, sexual stage shaped like a bell or an umbrella. It typically floats with its mouth and tentacles facing downward. Medusae produce eggs and sperm and release them into the water. When an egg is fertilized, it develops into a free-swimming, ciliated larva called a **planula** (plural planulae) that eventually settles to the bottom and transforms into a polyp.

Although the polyp and medusa stages appear very different, they share a similar body plan. A medusa is essentially a polyp without a stalk. Most of the outward differences between polyps and medusae are due to the development of the **mesoglea**, a middle body layer, composed of jellylike material, that is largely devoid of cells and has a very low metabolic rate. In polyps, the mesoglea is usually thin; in medusae it is very thick, constituting the bulk of the animal.

HYDROZOANS. Among members of the class Hydrozoa—the group containing the only freshwater cnidarians—life cycles are diverse. The polyp commonly dominates the life cycle, but some species have only medusae and others only polyps. A few species have solitary polyps, but most hydrozoans are colonial. A single planula eventually gives rise to a colony of many polyps, all interconnected and sharing a continuous gastrovascular cavity (Figure 29.10). Within such a colony, the polyps often differentiate. Some polyps have tentacles with many nematocysts; they capture prey for the colony. Others lack tentacles and are unable to feed, but are specialized for the production of medusae. Still others are fingerlike and defend the colony. However, all of these types are ultimately derived from a single, sexually produced planula.

The siphonophores, such as the Portugese man-of-war (see Figure 29.8), are free-floating hydrozoans in which medusae and polyps combine to form complex colonies. Individual units are modified for specific functions: to act as gas-filled floats, to move the colony through the water by jet propulsion, or to defend it. And of course there are also feeding and reproductive medusae.

ble for the sting that some jellyfish and other cnidarians can inflict on human swimmers. At the extreme, the tropical Pacific sea wasp (genus *Chironex*) can cause fatal injuries to humans.

A cnidarian transports its captured prey to its mouth by specialized structures called *tentacles*. The mouth is connected to a dead-end sac called the **gastrovascular cavity,** which functions in digestion, circulation, and gas exchange. The same opening, derived from the blastopore, serves as both mouth and anus in

29.9 The Life Cycle of a Cnidarian
Cnidarians typically have two body forms, one asexual and the other sexual.

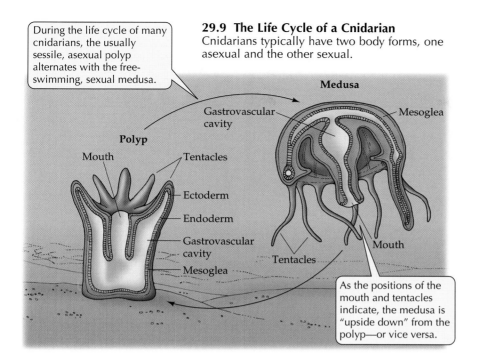

During the life cycle of many cnidarians, the usually sessile, asexual polyp alternates with the free-swimming, sexual medusa.

As the positions of the mouth and tentacles indicate, the medusa is "upside down" from the polyp—or vice versa.

Polyp — Mouth, Tentacles, Ectoderm, Endoderm, Gastrovascular cavity, Mesoglea

Medusa — Gastrovascular cavity, Mesoglea, Mouth, Tentacles

velops into a small, heavily ciliated planula that quickly settles on a substratum and changes into a small polyp. This polyp feeds and grows and may produce additional polyps by budding.

After a period of growth, the polyp begins to bud off small medusae by transverse division of its body column (Figure 29.11). These small medusae feed, grow, and transform themselves into adult medusae, which are commonly seen during summer in harbors and bays. Thus a polyp that grows from a single fertilized egg is capable of producing many genetically identical medusae that will eventually reproduce sexually.

ANTHOZOANS. The roughly 6,000 species of sea anemones and corals that constitute the class Anthozoa are all marine. Unlike other cnidarians, anthozoans entirely lack the medusa stage of the life cycle. The polyp produces eggs and sperm, and the fertilized egg develops into a planula that metamorphoses directly into another polyp. Many species can also reproduce asexually, by budding or fission.

Like all other cnidarians, anthozoans are carnivores that capture prey with nematocyst-studded tentacles. However, the digestive cavity of anthozoans is more complex than that of other cnidarians. It is partitioned

SCYPHOZOANS. The several hundred species of the class Scyphozoa are all marine. Some are as large as 2.5 meters in diameter. The mesoglea of their medusae is very thick and firm, giving rise to their common name, jellyfish. The medusa typically has the form of an inverted cup, and the tentacles with nematocysts extend downward from the margin of the cup.

Contraction of muscles ringing the margin expels water from the cup. When the muscles relax, the mesoglea expands the cup, which again fills with water. This muscular contraction cycle allows scyphozoan medusae to swim through the water. Food captured by the tentacles is passed to the mouth and distributed to one of four gastric pouches, where enzymes begin digesting the food.

The medusa, rather than the polyp, dominates the life cycle of scyphozoans. Gonads (sex organs) develop in tissues close to the gastric pouches. An individual medusa is male or female, releasing eggs or sperm into the open sea. The fertilized egg de-

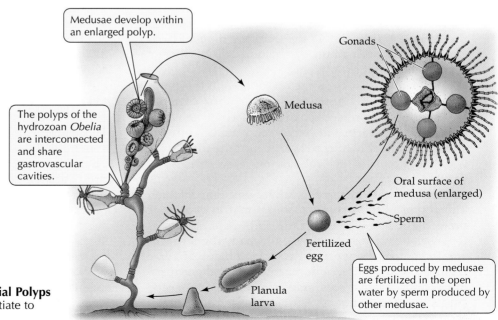

Medusae develop within an enlarged polyp.

The polyps of the hydrozoan *Obelia* are interconnected and share gastrovascular cavities.

Gonads, Medusa, Oral surface of medusa (enlarged), Sperm, Fertilized egg, Planula larva

Eggs produced by medusae are fertilized in the open water by sperm produced by other medusae.

29.10 Many Hydrozoans Have Colonial Polyps
The polyps within a colony may differentiate to perform specialized tasks.

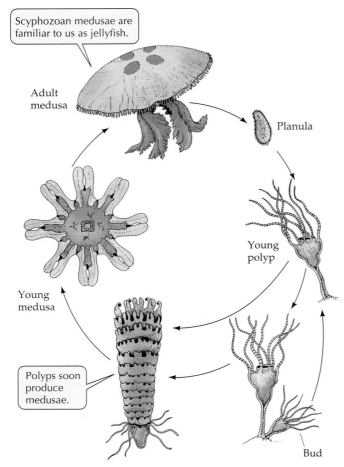

Scyphozoan medusae are familiar to us as jellyfish.

Adult medusa

Planula

Young polyp

Young medusa

Polyps soon produce medusae.

Bud

29.11 Medusae Dominate Scyphozoan Life Cycles
Scyphozoan medusae are the familiar jellyfish of coastal waters. The small polyps quickly produce medusae.

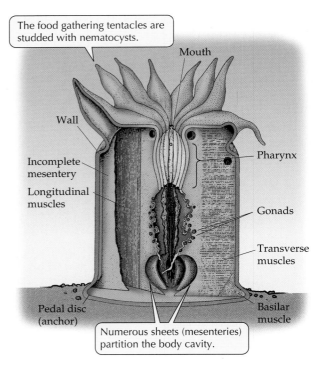

The food gathering tentacles are studded with nematocysts.

Mouth

Wall

Incomplete mesentery

Longitudinal muscles

Pharynx

Gonads

Transverse muscles

Pedal disc (anchor)

Basilar muscle

Numerous sheets (mesenteries) partition the body cavity.

29.12 Anthozoans Have Complex Polyps This sea anemone has the typical muscular structure of an anthozoan.

by numerous sheets, called *mesenteries*, that increase the surface area available for secreting digestive enzymes and absorbing nutrients (Figure 29.12). Gonads also develop on these mesenteries.

Sea anemones such as the one shown in the chapter-opening photograph are solitary anthozoans that lack specialized protective coverings. They are widespread in both warm and cold ocean waters. Many sea anemones are able to crawl slowly on the discs with which they attach to the substratum; a few species can swim.

Corals, by contrast, are usually sessile and colonial. The polyps of most corals secrete a matrix of organic molecules upon which calcium carbonate, the eventual skeleton of the colony, is deposited. The forms of coral skeletons are species-specific and highly diverse (Figure 29.13*a*). The common names of coral groups—horn corals, brain corals, staghorn corals, organ pipe corals, sea fans, and sea whips, among others—describe their appearance.

As a coral colony grows, old polyps die and leave their calcareous skeletons intact. The living members form a layer on top of a growing reef of skeletal remains. Reef-building corals are restricted to clear, warm waters. They are especially abundant in the Indo-Pacific region, where they form chains of islands and reefs. The Great Barrier Reef along the northeast coast of Australia is a system of coral formations more than 2,000 km long and as wide as 150 km. A continuous coral reef hundreds of kilometers long in the Red Sea has been calculated to contain more material than all the buildings in the major cities of North America combined.

Corals flourish in nutrient-poor, clear, tropical waters. For a long time scientists wondered how corals obtain enough nutrients to grow as rapidly as they do. The answer is that highly modified dinoflagellates live symbiotically within the cells of the corals. By their photosynthesis, the dinoflagellates provide carbohydrates to their hosts and help with calcium deposition. In turn, the dinoflagellates within the coral's tissues are protected from predators. This symbiotic relationship explains why reef-forming corals are restricted to surface waters, where light levels are high enough to allow photosynthesis (Figure 29.13*b*).

KEYS TO CNIDARIAN DIVERSITY. Cnidarians were the dominant marine organisms late in the Precambrian period, more than 600 million years ago, and they remain

(a)

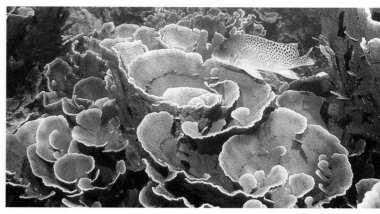

(b) *Montipora foliosa*

29.13 Corals (a) Many different species of corals grow together on this reef in Indonesia. (b) The ends of the branches of a cabbage coral spread outward, maximizing the amount of sunlight received by the photosynthetic dinoflagellates that live symbiotically inside their cells.

important components of marine ecological communities today (Figure 29.14). The cnidarian body plan combines a low metabolic rate with the ability to capture relatively large prey. The bulk of the body of medusae and many polyps is made up of the largely inert mesoglea. As a result, even a large cnidarian such as a sea anemone requires relatively little food and can fast for weeks or months.

Nematocysts allow cnidarians to subdue prey, such as fishes, that are more active and structurally complex than the cnidarians themselves. Nonetheless, many cnidarians eat microscopic prey, and ancestral cnidarians (as well as some living species) probably consumed bacteria and protists. Many corals and other cnidarians have symbiotic associations with photosynthetic dinoflagellates that enable them to grow where food is scarce.

These traits allow cnidarians to survive in environments where encounter rates with prey are much lower than required to sustain animals with higher

(a) *Gonothyraea loveni*

(b) *Pelagia panopyra*

29.14 Diversity among Cnidarians (a) The structure of the polyps of a North Atlantic coastal hydrozoan is visible here. (b) This purple jellyfish illustrates the complexity of some scyphozoan medusae. (c) The nematocyst-studded tentacles of the strawberry anemone are poised to capture large prey carried to the animal by water movement.

(c) *Corynactis californica*

metabolic needs. Cnidarians have persisted with few modifications for hundreds of millions of years.

The Evolution of Bilateral Symmetry

Although some cnidarians can move about slowly, they lack a front end that typically encounters new environments first. Bilateral symmetry probably arose first in simple organisms consisting of flattened masses of cells. These animals crawled over a substratum, feeding as they went, just as placozoans do today. Placozoans are no more complex than the earliest of these simple animals probably were, but the lineages that arose from these simple ancestors evolved into a great diversity of animals, many of which are highly complex structurally.

As inferred from traits of extant species, the first split in the lineage of bilaterally symmetrical animals separated the ctenophores (comb jellies) from the lineage leading to all other animals (see Figure 29.4).

Ctenophores capture prey with mucus and tentacles

Ctenophores, also known as comb jellies, constitute the phylum Ctenophora. Because the body plans of ctenophores and cnidarians are superficially similar, they were long considered to be closely related. Both have two cell layers separated by a thick, gelatinous mesoglea, and both have radial symmetry and feeding tentacles. Unlike cnidarians, however, ctenophores have a complete gut, with two anal pores through which wastes are voided.

Ctenophores have eight comblike rows of fused plates of cilia, called *ctenes*, and they move by beating these fused cilia rather than by muscular contractions. Ctenophoran tentacles are solid and lack nematocysts; instead, the tentacles are covered with sticky filaments to which prey adhere (Figure 29.15). After capturing its prey, the ctenophore retracts its tentacles to bring the food to its mouth. In some species, the entire surface of the body is coated with a sticky mucus that captures prey. All of the 100 known species of ctenophores are carnivorous marine animals.

Most ctenophores cannot capture large prey, but one group lacks tentacles and feeds on other ctenophores by ingesting them whole, using a muscular pharynx. The typical ctenophore feeding method is to dangle sticky tentacles in the water, where planktonic prey adheres to them. Like cnidarians, ctenophores have low metabolic rates because they are composed primarily of inert mesoglea. They are common in open seas, where prey are often scarce.

Ctenophore life cycles are simple. Gametes from gonads located on the walls of the gastrovascular cavity are liberated into the cavity and then discharged through the mouth or through pores. Fertilization takes place in the open seawater. In nearly all species the fertilized egg develops directly into a miniature ctenophore that gradually grows into an adult.

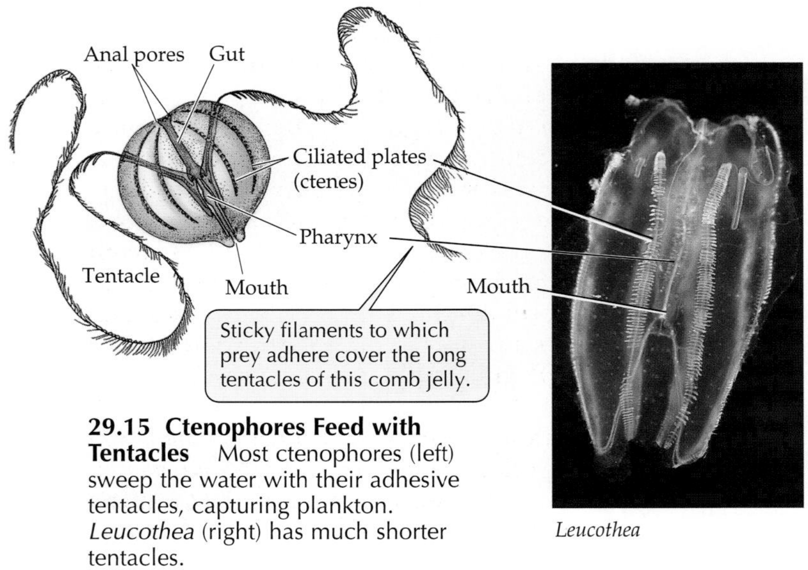

Protostomes and Deuterostomes: An Early Lineage Split

The next major split in the evolution of animal lineages separated the two groups that dominate today's biota—the protostomes and deuterostomes (see Figure 29.4). Because of the many differences in the structures and embryological development of protostomes and deuterostomes (see Table 29.1), evolutionary biologists agree that these are two monophyletic groups that have been evolving separately since the Cambrian period.

The major shared, derived traits that unite protostomes are a central nervous system consisting of an anterior brain that surrounds the entrance to the digestive tract; paired or fused longitudinal nerve cords, and a free-floating larva with a food-collecting system consisting of compound cilia on multiciliate cells. The major shared, derived traits that unite deuterostomes are a dorsal nervous system and larvae with a food-collecting system consisting of cells with single cilia.

Protostomes: A Diversity of Body Plans

Most of the world's living animal species are protostomes, among which are rotifers, roundworms, segmented worms, arthropods, and mollusks. The diversity of protostome body plans and lifestyles has posed many challenges to zoologists attempting to infer the evolutionary relationships among these animals. Developmental, structural, and molecular evidence often disagree, and the biology of some marine groups is poorly known.

Anal pores Gut
Ciliated plates (ctenes)
Pharynx
Tentacle
Mouth
Mouth

Sticky filaments to which prey adhere cover the long tentacles of this comb jelly.

Leucothea

29.15 Ctenophores Feed with Tentacles Most ctenophores (left) sweep the water with their adhesive tentacles, capturing plankton. *Leucothea* (right) has much shorter tentacles.

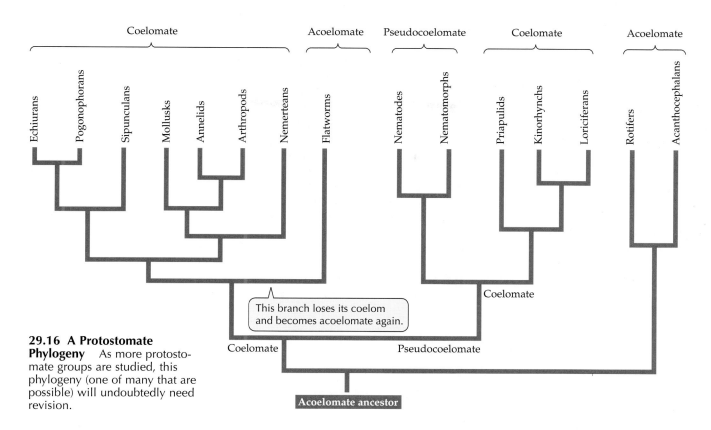

29.16 A Protostomate Phylogeny As more protostomate groups are studied, this phylogeny (one of many that are possible) will undoubtedly need revision.

The phylogeny we have adopted (Figure 29.16) is not accepted by all investigators, and changes are certain to be made in the near future. Our chosen phylogeny is based on the assumption that some animal groups with relatively simple body plans had more complex ancestors. In particular, we assume that possession of a pseudocoel arose independently several times during animal evolution in lineages that had coelomate ancestors.

Movement is a key feature of protostome evolution

Speed is often advantageous for both prey and the predators that pursue them. Fast-moving prey and predators evolved in the early Cambrian period (550 mya). Animals probably evolved fluid-filled body cavities because such cavities facilitate swift movement.

Hydrostatic skeletons, as fluid-filled cavities are called, facilitate movement because they are incompressible. When muscles around part of a fluid-filled body cavity contract, the fluid must move to another part of the cavity. If the body tissues around the cavity are flexible, fluids moving from one region cause the other region to expand. Fluids can thus move specific body parts or even the whole animal, provided that temporary attachments can be made to the substratum.

The types and numbers of body cavities provide a major key to the lives of these animals and the degree to which they can control and change their shapes and the complexity of the movements they can perform.

Rotifers are small but structurally complex

An abundant and widespread group of pseudocoelomate animals is the phylum Rotifera. Rotifers are bilateral, unsegmented animals that have three cell layers. Most rotifers are tiny (50 to 500 µm long)—smaller than some ciliate protists—but they have highly developed organs (Figure 29.17). A complete gut passes from an anterior mouth to a posterior anus, and the pseudocoel functions as a hydrostatic skeleton. Most rotifers are active, propelling themselves through the water by means of rapidly beating cilia rather than by muscular contraction. This type of movement is effective because rotifers are so small.

The most distinctive organs of rotifers are those used to collect and process food. A conspicuous, ciliated organ (the *corona*) surmounts the head of many species. Coordinated beating of cilia provides the force for locomotion and also sweeps particles of organic matter from the water into the mouth and down to a complicated structure (the *mastax*), which has teeth that grind the food. By contracting the muscles that surround the pseudocoel, a few rotifer species that prey on protists and small animals can protrude the mastax through the mouth and seize small objects with it.

Some rotifers are marine, but most of the 1,800 known living species live in fresh water. Members of a few species rest on the surface of mosses and lichens in a desiccated, inactive state until it rains. When rain

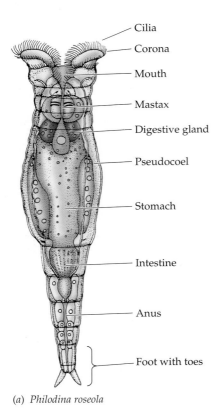

(a) *Philodina roseola*

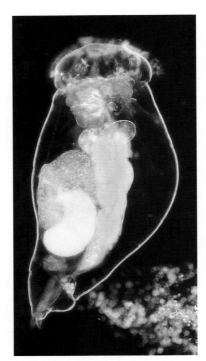

(b) *Epiphanes senta*

29.17 Rotifers (*a*) This rotifer reflects the general structure of many of the free-living species in this phylum. (*b*) The internal anatomy of a rotifer is clear in this micrograph.

Most of them are believed to be descendants of a lineage that has evolved independently since early animal evolution. This body form enables animals to move efficiently through muddy and sandy marine sediments.

Roundworms are simple but abundant

Roundworms (phylum Nematoda) have a thick, multilayered cuticle secreted by the underlying epidermis that gives their body its shape (Figure 29.19*a*). As a roundworm grows, it sheds and resecretes its cuticle four times. Because the cross-sectional shape of its body is round, a roundworm has a relatively small outer body surface for exchanging oxygen and other materials with the environment.

falls, they absorb water and swim about and feed in the films of water that temporarily cover the plants. Most rotifers live no longer than 1 or 2 weeks.

All spiny-headed worms are parasites

The spiny-headed worms, members of the phylum Acanthocephala (from the Greek *akantha*, "spine," and *kephale*, "head"), are obligate intestinal parasites of vertebrates, primarily freshwater fishes. Most species are less than 20 cm long, but a few extend nearly a meter. The anterior end has a hook-bearing structure by which the animal attaches to its host.

Acanthocephalans have no gut; they absorb their food directly through their body wall. Larvae develop in the body cavity of the female and are released with the host's feces. The larvae are eaten by an intermediate host (an insect or crustacean), which, in turn, is eaten by a fish or other vertebrate, thereby completing the life cycle (Figure 29.18). Many acanthocephalans change the behavior of their hosts in ways that enhance their transmission to the next host in their life cycle.

Wormlike Bodies: Movement through Sediments

Animals in 16 protostomate phyla are wormlike; that is, they are bilaterally symmetrical, legless, soft-bodied, and at least several times longer than they are wide.

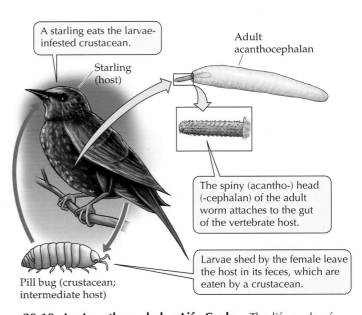

A starling eats the larvae-infested crustacean.

Starling (host)

Adult acanthocephalan

The spiny (acantho-) head (-cephalan) of the adult worm attaches to the gut of the vertebrate host.

Larvae shed by the female leave the host in its feces, which are eaten by a crustacean.

Pill bug (crustacean; intermediate host)

29.18 An Acanthocephalan Life Cycle The life cycle of a typical spiny-headed worm includes an arthropod and a vertebrate host. The parasites alter the behavior of the arthropods to increase their chances of being eaten by the vertebrate host.

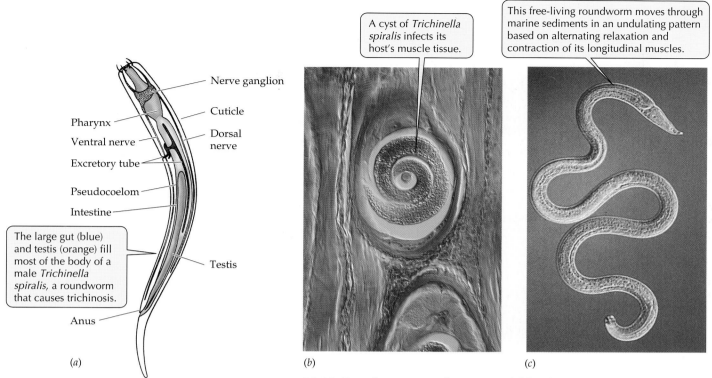

A cyst of *Trichinella spiralis* infects its host's muscle tissue.

This free-living roundworm moves through marine sediments in an undulating pattern based on alternating relaxation and contraction of its longitudinal muscles.

Nerve ganglion

Cuticle

Pharynx

Dorsal nerve

Ventral nerve

Excretory tube

Pseudocoelom

Intestine

The large gut (blue) and testis (orange) fill most of the body of a male *Trichinella spiralis*, a roundworm that causes trichinosis.

Testis

Anus

(a)

(b)

(c)

29.19 Roundworms   Both parasitic and free-living forms are illustrated.

Oxygen and nutrients are exchanged not only through the cuticle, but also through the intestine, which is only one cell layer thick. Materials are moved through the gut by rhythmic contraction of a highly muscular organ at the anterior end. The pseudocoelom of roundworms is small; the body organs fill most of the internal space.

Roundworms are one of the most abundant and universally distributed of all animal groups. We unintentionally eat and drink enormous numbers of roundworms in our lifetimes. For example, examination of a single rotting apple from the ground of an orchard revealed 90,000 roundworms contained in the flesh of the fruit. One square meter of mud off the coast of The Netherlands yielded 4,420,000 individuals. The topsoil of rich farmland has up to 3 billion nematodes per acre.

Countless numbers of roundworms live as scavengers in the upper layers of the soil, on the bottoms of lakes and streams, and as parasites in the bodies of most kinds of plants and animals. The largest known roundworm, which reaches a length of 9 m, is found in the placenta of female sperm whales. About 20,000 species have been described, but the actual number of living species may exceed 1 million.

The diets of roundworms are as varied as their habits. Many roundworms live parasitically within their hosts. Many are predators, preying on protists and other small animals (including other round-worms). The roundworms that are parasites of people, cats, dogs, cows, sheep, and economically important plants have been studied intensively in an effort to find ways of controlling them.

The structure of parasitic roundworms (see Figure 29.19a and b) does not differ much from that of free-living species (Figure 29.19c), but the life cycles of many parasitic species have special stages that facilitate the transfer of individuals among hosts. *Trichinella spiralis*, the parasite that causes the disease trichinosis, has a relatively simple life cycle. The larvae of *Trichinella* form cysts in the muscles of their mammalian hosts (see Figure 29.19b). If present in great numbers, these cysts cause severe pain or death.

Trichinella is transmitted to a mammal that eats the flesh of an infected individual. The activated larvae leave the cysts and attach to the intestinal wall, where they feed. Later, they bore through the host's intestinal wall and are carried in the bloodstream to muscles, where they form cysts. The alternate host is likely to be another mammal (usually a pig in the case of human infections). No special stage in the *Trichinella* life cycle lives in an alternate host. However, other roundworm life cycles are more complex, involving one or more alternate hosts.

Horsehair worms are parasites

The larvae of horsehair worms (phylum Nematomorpha) live inside terrestrial and aquatic insects and

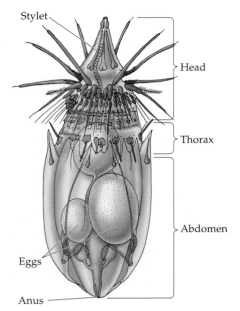

29.20 A Loriciferan
Even though they are abundant, the tiny (~0.3 μm) members of this phylum went unnoticed until 1983.

crabs. The gut is much reduced and is probably non-functional. Horsehair worms may feed only as larvae, absorbing nutrients from their hosts across their body wall, but many continue to grow after they have left their hosts, suggesting that adults may also absorb nutrients from their environment.

Several phyla of small worms live in ocean sediments

Members of several other small phyla of wormlike animals live in muddy and sandy ocean bottoms. Most are small and poorly known. Among these are members of a phylum, the Loricifera (Figure 29.20), that was first discovered in 1983. Loriciferans live at great depths and cling so tightly to ocean sediments that it is difficult to separate them from sand. As a result, even though abundant, these tiny worms escaped human detection for centuries.

Flatworms move by beating cilia

Flatworms (phylum Platyhelminthes) are bilaterally symmetrical animals whose internal organs, though simple, are more complex than those of cnidarians and ctenophores. Flatworms have no enclosed body cavity, they lack organs for transporting oxygen to internal tissues, and they have only simple organs for excreting metabolic wastes. This body plan dictates that each cell must be near a body surface in order to respire, a requirement aided by the flattened body form.

Flatworms have traditionally been thought to represent a protostomate lineage that was ancestral to nearly all other animal linages, but recent analyses suggest that many features of these worms are derived rather than ancestral traits. The digestive tract of a flatworm, if there is one, consists of a mouth opening into a dead-end sac. However, the sac is often highly branched, forming intricate patterns that increase the surface area available for absorption of nutrients. All living flatworms feed on animal tissues—some as carnivores and parasites, others as scavengers. Motile flatworms glide over surfaces, powered by broad layers of cilia. This form of movement is very slow but it is sufficient for small, scavenging animals.

The flatworms probably most similar to the ancestral forms are the turbellarians (class Turbellaria): small, free-living marine and freshwater animals (a

(a) *Pseudobiceros* sp.

29.21 Flatworms (a) Some flatworm species are free-living, like this marine flatworm of the South Pacific. (b) This flatworm lives parasitically in the gut of sea urchins. (c) The sheep liver fluke, also a parasite, is filled with highly branched large gonads.

As is typical of internal parasites, the flatworm's body is filled primarily with sex organs.

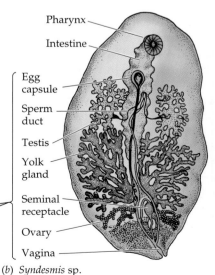

(b) *Syndesmis* sp.

(c) *Fasciola hepatica*

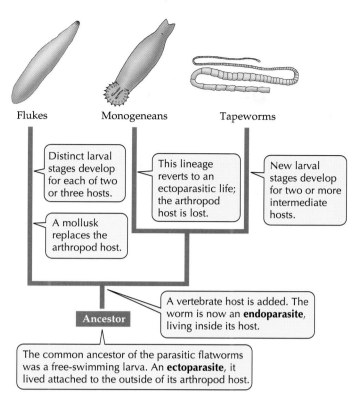

Flukes Monogeneans Tapeworms

> Distinct larval stages develop for each of two or three hosts.

> This lineage reverts to an ectoparasitic life; the arthropod host is lost.

> New larval stages develop for two or more intermediate hosts.

> A mollusk replaces the arthropod host.

Ancestor

> A vertebrate host is added. The worm is now an **endoparasite**, living inside its host.

> The common ancestor of the parasitic flatworms was a free-swimming larva. An **ectoparasite**, it lived attached to the outside of its arthropod host.

29.22 The Evolution of Life Cycles in Parasitic Flatworms Hosts have been added and subtracted during the evolution of parasitic flatworms.

few live in moist terrestrial habitats). Freshwater turbellarians of the genus *Dugesia*, better known as planarians, are the most familiar species of flatworms. At one end they have a head with chemoreceptor organs, two simple eyes, and a tiny brain composed of anterior thickenings of the longitudinal nerve cords.

Although the earliest flatworms were free-living (Figure 29.21*a*), the flatworm body plan readily adapted to a parasitic existence. A likely evolutionary transition was from feeding on dead organisms, to feeding on the body surfaces of dying hosts, to invading and consuming parts of living, healthy hosts. Most of the 25,000 species of living flatworms—the tapeworms (class Cestoda) and flukes (class Trematoda; Figure 29.21*b* and *c*)—are parasitic. These worms inhabit the bodies of many vertebrates and cause serious human diseases, such as trichinosis, filariasis, and elephantiasis. Monogeneans (class Monogenea) are external parasites of fishes and other aquatic vertebrates.

Cladistic analyses have been used to suggest the sequence in which lineages of parasitic flatworms added or subtracted hosts from their life cycle (Figure 29.22). For example, monogeneans, all of which are ectoparasites with simple life cycles, evolved from ancestors that had more than one host in their life cycle.

Parasites live in nutrient-rich environments in which food is delivered to them, but they face other challenges. To complete their life cycle, parasites must overcome the defenses of their host. Because they die when their host dies, they must disperse their offspring to new hosts. The eggs of some parasitic flatworms are voided with the host's feces. Later, these eggs are ingested directly by other host individuals. However, most parasitic species have complex life cycles involving two or more hosts and several larval stages (Figure 29.23).

The complex life cycles of internal parasites often evolve because intermediate hosts increase the probability that individuals of the primary host will be infected. An intermediate host may provide a good feeding environment for the parasite, allowing it to develop and reproduce within an organism that is likely to be eaten by the primary host. However, these very advantages provide scientists with opportunities to reduce infections by interrupting a transmission stage in the parasite's life cycle.

Coelomate Protostomes

Although the evolution of body cavities provided many new movement capabilities, control over body shape is crude if the cavity has muscles on only one side, as a pseudocoel does (see Figure 29.3*b*). A coelom, which is surrounded by muscles (see Figure 29.3*c*), allows better control over movement of the fluids it contains. Even with a coelom, however, control over movement is limited if an animal has only a single, large body cavity.

This limitation is overcome if a large cavity is separated into compartments or segments so that localized changes in shape are produced by circular and longitudinal muscles in individual segments. Thus the animal can change the shape of each segment independently of the others. Segmentation of the coelom evolved several different times among both protostomes and deuterostomes. Among the coelomate protostomes, segmentation is found in annelids, arthropods, and mollusks, but there are several lineages of unsegmented, coelomate worms. They are believed to be descendants of a lineage that long ago separated from the lineage in which body segmentation evolved (see Figure 29.16).

Ribbon worms are dorsoventrally flattened

Ribbon worms (phylum Nemertea) are dorsoventrally flattened and have nervous and excretory systems similar to those of flatworms, but unlike flatworms, they have a complete digestive tract with a mouth at one end and an anus at the other. Food items move in one direction through the digestive tract of a ribbon worm and are acted on by a series of digestive enzymes.

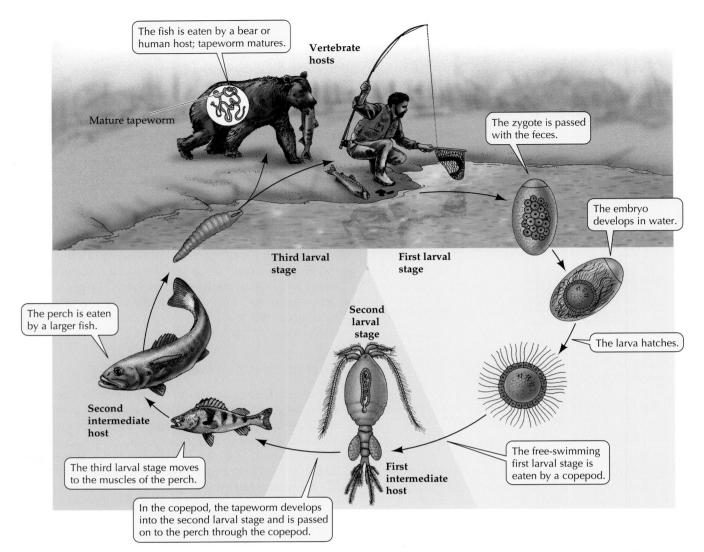

29.23 Returning to a Host by a Complex Route The broad fish tapeworm *Diphyllobothrium latum* must pass through the bodies of a copepod (a type of crustacean) and a fish before it can reinfect its primary host, a mammal. Such complex life cycles assist the flatworm's recolonization of hosts, but they also offer opportunities for humans to break the cycle with hygienic measures.

In the body of almost all ribbon worms is a fluid-filled cavity called the **rhynchocoel**, within which floats a hollow, muscular proboscis. The **proboscis** is the feeding organ; it may extend much of the length of the worm. Contraction of the muscles surrounding the rhynchocoel causes the proboscis to be ejected explosively through an anterior opening to obtain food (Figure 29.24). Thus the rhynchocoel transmits forces that move the proboscis rapidly, although it does not move the rest of the animal.

The proboscis of most ribbon worms is armed with a sharp hook (stylet) that pierces the prey. Paralysis-causing toxins produced by the proboscis are dis-charged into the wound made by the stylet. Some species lack a hook and capture prey by wrapping the muscular proboscis around it. The ribbon worm then withdraws its proboscis into the rhynchocoel by means of a retractor muscle and takes the prey into its mouth.

Small ribbon worms move by beating their cilia. Larger ones employ waves of contraction of body muscles to move on the surface of sediments or to burrow. Movement by both of these methods is slow. In addition to capturing prey, the rhynchocoel helps the ribbon worm to burrow. For burrowing, the worm pushes its proboscis into the substratum for attachment; when the proboscis contracts and fattens, the worm is pulled forward. Similar burrowing mechanisms are found in other phyla.

Nearly all of the approximately 900 species of ribbon worms are marine, but some are found in fresh water and a few live in moist tropical terrestrial environments. All ribbon worms are carnivores, feeding mostly on arthropods and small worms in several different phyla.

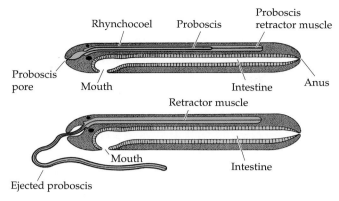

29.24 A Ribbon Worm The proboscis is the nemertean feeding organ. Floating in a cavity called the rhynchocoel, the proboscis can be moved rapidly. The worm, however, moves slowly.

Coelomate worms have varied body plans

Some phyla of coelomate worms are dismissed as "minor" groups because combined they have fewer than 600 species. However, because their body plans are quite different from those of other animals, they are worthy of mention.

The 250 species of peanut worms (phylum Sipuncula) are marine animals that burrow into sediments, live under rocks, or attach to the holdfasts of algae. They are common inhabitants of coral reefs, where they often burrow into calcareous coral skeletons. The body of a peanut worm is sausage-shaped and has a large, unsegmented coelom and feeding tentacles at the anterior end (Figure 29.25a). The coelom is filled with water and functions as a hydrostatic skeleton.

The echiuran worms (phylum Echiura) are similar in structure to peanut worms, and they often live in similar environments (Figure 29.25b). They are exclusively marine and most of them burrow in sand or mud. They gather their food with a proboscis that they extend over the surface of the substratum, where it gathers organic detritus.

Many early protostomes were small animals with thin body coverings. Gases and waste products were routinely exchanged across the body wall. Marine waters and sediments contain abundant food particles in the form of bacteria and dissolved organic matter that also can be taken in directly through the body wall.

Members of one lineage of protostomes, the phylum Pogonophora, evolved into burrowing forms with a crown of tentacles through which gases are exchanged, and they entirely lost their digestive tracts. Pogonophorans were not discovered until the twentieth century, when deep ocean exploration revealed them living many thousands of meters below the surface (Figure 29.25c). In these deep oceanic sediments, pogonophorans are abundant, reaching densities of many thousands per square meter. About 145 species have been described.

The coelom of a pogonophoran consists of an anterior compartment, into which the tentacles can be withdrawn, and a long, subdivided cavity that extends much of the length of its body. Experiments using radioactively labeled molecules have shown that pogonophorans take up dissolved organic matter at high rates from either the sediments in which they live or the surrounding water.

The largest and most remarkable pogonophorans, which grow to 2 m in length, live near deep-ocean hy-

(a) Peanut worm

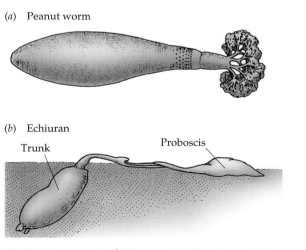

(b) Echiuran

(c)

The red bodies of these deep-sea-vent pogonophorans project from the long white tubes in which they live.

29.25 Unsegmented Worms (a) The sipunculids, or peanut worms, are burrowing marine animals. (b) Echiurans are also burrowing worms; they gather food by extending their proboscis. (c) Pogonophorans live in tubes from which they extend their large feeding tentacles.

drothermal vents—openings in the sea floor through which hot, sulfur-rich water pours. The tissues of these species are filled with prokaryotes that fix carbon using energy obtained from oxidation of hydrogen sulfide (H_2S). The pogonophorans either consume these prokaryotes directly or live on their metabolic by-products, while keeping the prokaryotes in optimal environments near the vents.

Hydrothermal vent ecosystems are not based on light as their energy source. They have attracted considerable interest because they may be the type of environment in which life on Earth originated (see Chapter 24).

Segmented Bodies: Improved Locomotion

A body cavity segmented into compartments allows an animal to alter the shape of its body in complex ways and to control its movements precisely. Fossils of segmented worms are known from the middle Cambrian; the earliest forms are thought to have been burrowing marine animals. Segmentation evolved several times among protostomes.

Annelids are the dominant segmented worms

The annelids (phylum Annelida) are a very diverse group of worms that have a segmented body cavity (Figure 29.26). The approximately 15,000 described annelid species live in marine, freshwater, and terrestrial environments. The bulging waves that undulate up and down the length of the body to move the worm are made possible by its segmentation. A separate nerve center controls each segment, but the centers are connected by nerve cords that coordinate their functioning. The coelom in each segment is isolated from those in other segments.

Most annelids lack a rigid, external protective covering. The thin body wall serves as a general surface for gas exchange in most species, but this thin, permeable body surface restricts annelids to moist environments; they lose body water rapidly in dry air.

POLYCHAETES. More than half of all annelid species are placed in the class Polychaeta. Nearly all polychaetes are marine animals. Most have one or more pairs of eyes and one or more pairs of tentacles at the anterior end of their body. The body wall in most segments extends laterally as a series of thin outgrowths, called **parapodia**, that have many blood vessels and function in gas exchange and locomotion. Stiff bristles protruding from each parapodium form temporary attachments to the substratum and prevent the animal from slipping backward when its muscles contract.

Many species live in burrows in soft sediments and capture prey from surrounding water with elaborate feathery tentacles (Figure 29.27a). The sexes are separate in most polychaetes. Typically, polychaetes release gametes into the water, where they become fertilized. A fertilized egg develops into a ciliated larva known as a **trochophore** (Figure 29.27b). As a trochophore develops, it forms body segments at its posterior end; eventually it metamorphoses into a small adult worm.

OLIGOCHAETES. More than 90 percent of the approximately 3,000 described species of oligochaetes (class Oligochaeta) live in freshwater or terrestrial habitats. Oligochaetes have no parapodia, eyes, or anterior tentacles, and they have relatively few bristles (setae). Earthworms—the most familiar oligochaetes—are scavengers and ingesters of soil, from which they extract food particles.

Unlike polychaetes, all oligochaetes are **hermaphroditic**: Each individual is both male and female. Sperm are exchanged simultaneously between two copulating individuals (Figure 29.27c). Eggs are laid in a cocoon outside the adult's body. The cocoon is shed, and

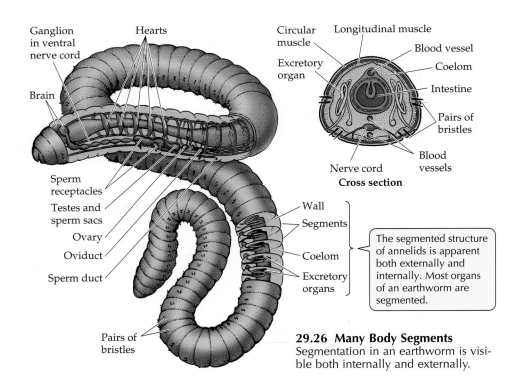

Ganglion in ventral nerve cord

Hearts

Brain

Sperm receptacles

Testes and sperm sacs

Ovary

Oviduct

Sperm duct

Pairs of bristles

Circular muscle

Longitudinal muscle

Blood vessel

Coelom

Intestine

Pairs of bristles

Excretory organ

Nerve cord

Blood vessels

Cross section

Wall

Segments

Coelom

Excretory organs

The segmented structure of annelids is apparent both externally and internally. Most organs of an earthworm are segmented.

29.26 Many Body Segments
Segmentation in an earthworm is visible both internally and externally.

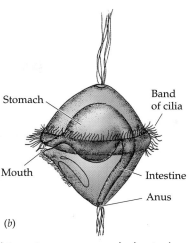

Stomach

Band
of cilia

Mouth

Intestine

Anus

During copulation, each
individual both donates
and receives sperm.

(*c*) *Lumbricus* sp.

(*a*) *Spirobranchus giganteus*

(*b*)

29.27 Annelid Diversity   (*a*) The Christmas tree worm, a polychaete, has striking feeding tentacles. (*b*) The trochophore is the polychaete larval form. (*c*) Individual earthworms are hermaphroditic (simultaneously male and female). (*d*) This freshwater leech has conspicuous anterior and posterior suckers.

(*d*) Leech (species not identified)

when development is complete, miniature worms emerge and begin independent life.

LEECHES.   Leeches (class Hirudinea) probably evolved from oligochaete ancestors. Most species live in freshwater or terrestrial habitats and, like oligochaetes, they lack parapodia and tentacles. Leeches are hermaphroditic; each individual serves as a sperm donor and a sperm recipient during copulation. The coelom of leeches is not divided into compartments, and the coelomic space is largely filled with mesenchyme tissue. Therefore, the movement of leeches differs radically from that of other annelids.

Groups of segments at each end of a leech are modified to form suckers, which serve as temporary anchors (Figure 29.27*d*). With its posterior anchor attached, the leech extends its body by contracting its circular muscles. The anterior sucker is then attached, the posterior one detached, and the leech shortens itself by contracting its longitudinal muscles.

Most leeches are external parasites of other animals, but some species also eat snails and other invertebrates. The mouth has three toothed jaws, with which the leech makes an incision in its host. It feeds on the host's blood. A leech feeds infrequently, but it can ingest so much blood in a single feeding that its body may enlarge several times. A substance secreted by the leech into the wound keeps the host's blood flowing.

For hundreds of years leeches were widely employed in medicine for bloodletting. Even today they are used to reduce fluid pressure in tissues damaged by, for example, a snake bite, and to eliminate pools of coagulated blood.

Mollusks lost segmentation but evolved shells

Mollusks (phylum Mollusca) underwent one of the most remarkable of animal evolutionary radiations. This dramatic radiation, which began with a segmented ancestor, was based on a body plan with three major structural components: the foot, the mantle, and the visceral mass, plus a rasping feeding structure known as the **radula** (plural radulae) at the anterior end.

Animals that appear very different, including snails, clams, and squids, are all built from these components, although individual components have been lost in some lineages (Figure 29.28). These three unique shared derived characteristics are the reason that zoologists place all 100,000 species of mollusks in one phylum.

The molluscan **foot** is a large, muscular structure that originally was the molluscan organ of locomotion, as well as the support for internal organs. In the lineage leading to squids and octopuses, the foot was modified to form arms and tentacles borne on a head with complicated sense organs. In other groups, such as clams, the foot was transformed into a burrowing organ. In some lineages the foot is greatly reduced.

The **mantle** is a fold of body wall that secretes the shell at its lip. The mantle partly encloses an external space called the mantle cavity. The internal organs lie in this cavity, covered by the outer layer of the **visceral mass**—the major internal organs. The gills of mollusks, which are used for gas exchange and feeding, lie

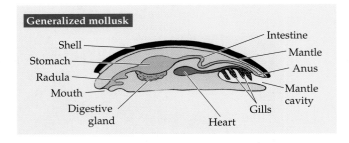

Generalized mollusk

Chitons

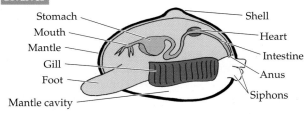

Gastropods

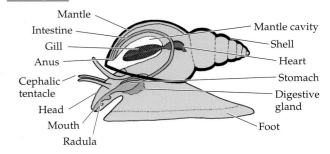

Bivalves

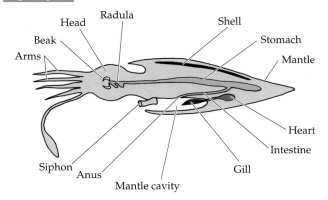

Cephalopods

29.28 Molluscan Body Plans The diverse modern mollusks all display variations of a general body plan that includes the foot, the mantle, and a visceral mass of internal organs.

in the mantle cavity. When the cilia on the gills beat, they create a flow of oxygenated water over the gills.

The coelom of mollusks is much reduced, but the open circulatory system has large fluid-filled cavities that are major components of the hydraulic skeleton. The radula was originally an organ for scraping algae from rocks, a function it retains in many living mollusks. However, in some mollusks the radula has been modified into a drill or a poison dart. In others, such as clams, it is absent.

Mollusks range in size from snails a mere 1 mm high to giant squids more than 18 meters long—the largest known protostomes. No fossils of the most ancient mollusks have yet been found, but they were probably segmented, as some mollusks are today.

MONOPLACOPHORANS. Unlike all other living mollusks, monoplacophorans (class Monoplacophora) have multiple gills, muscles, and excretory structures that are repeated along the length of the body. However, the shell is united into a single structure that covers the gills and other body organs. The gills are located in a large cavity under the shell, through which oxygen-bearing water circulates.

Monoplacophorans are believed to be the most ancient molluscan group. They were the most abundant mollusks during the Cambrian period, 550 million years ago. For a long time, it was thought that they had become extinct many millions of years ago, but in 1952, off the Pacific coast of Costa Rica, an oceanographic vessel dredged up ten specimens of an unusual little mollusk with a cap-shaped shell. Finding these animals, which were placed in the genus *Neopilina*, created a sensation because they turned out to be monoplacophorans.

CHITONS. Chitons (class Polyplacophora) have multiple gills and segmented shells, but other body parts are not segmented (Figure 29.29a). The chiton body is symmetrical, and its internal organs, particularly the digestive and nervous systems, are relatively simple. Chitons have trochophore larvae that are almost indistinguishable from those of annelids (see Figure 29.27b). Most chitons are marine herbivores that scrape algae from rocks with their sharp radulae. An adult chiton spends most of its life glued tightly to rock surfaces by its large, muscular, mucus-covered foot. A chiton can move slowly by means of rippling waves of muscular contractions in its foot.

(a) *Tonicella lineata*

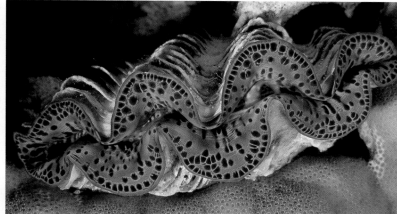

(b) *Tridacna* sp.

(c) *Tridachiella diomedea*

(d) *Helix aspera*

(e) *Octopus horridus*

(f) *Nautilus belavensis*

29.29 Mollusk Diversity (a) This chiton is common in the intertidal zone of the Pacific coast of North America. (b) Giant clams are common inhabitants of tropical coral reefs. (c) Terrestrial and marine slugs are gastropods that have lost their shells through evolution; this "Spanish dancer" nudibranch is very conspicuous. (d) These garden snails are shelled gastropods. (e) Octopuses are active predators. This one is a banded octopus found in the Indian Ocean. (f) The boundaries of the chambers are clearly visible on the outer surface of the shell of this nautilus.

BIVALVES AND GASTROPODS.   One lineage of early mollusks developed a two-part shell that extended over the sides of the body as well as the top, giving rise to the bivalves (class Bivalvia)—the familiar clams, oysters, scallops, mussels, and other important edible shellfish, together with many similar, less familiar forms (Figure 29.29*b*). The name "bivalve" derives from the structure of the shell of these animals, which has two major pieces connected by a flexible hinge. Bivalves are largely sedentary and have greatly reduced heads. The foot is compressed and, in many clams, is used for burrowing into mud and sand. Bivalves capture food from the water using their large gills, which are also the main sites of gas exchange.

Another lineage of early mollusks evolved into the gastropods (class Gastropoda), which includes the snails. Most gastropods are motile, using their large foot to move slowly across the substratum or to burrow through it. The shell and visceral mass of a gastropod larva undergo a 180° counterclockwise torsion relative to the foot during development. The result is that the digestive tract and nervous system are twisted so that the anus, the opening of the mantle, and the gills move to the front of the body, just behind and over the head. All gastropods undergo torsion, but in some the torsion is reversed later during development and they are not twisted as adults. The gills of gastropods are the primary sites of gas exchange. In some species, they are also feeding devices.

Gastropods are the most diverse and widely distributed of the molluscan classes (Figure 29.29*c,d*). Some species can crawl, including a rich variety of snails, whelks, limpets, slugs, abalones, and the often brilliantly ornamented nudibranchs. Still others—the sea butterflies and heteropods—have a modified foot that functions as a swimming organ with which they move through open ocean waters. The only mollusks that live in terrestrial environments—many snails and slugs—are gastropods. In terrestrial species the mantle cavity is modified into a highly vascularized lung.

CEPHALOPODS.   One lineage of mollusks evolved an exit tube for currents leaving the mantle cavity. At first, this tube probably simply improved the flow of water over the gills; subsequently it became modified to allow the early cephalopods (class Cephalopoda) to control the direction in which water leaves the shell cavity. By closing off shell chambers and then pumping out the water, the animals could also control their buoyancy. Together, these adaptations allow cephalopods to live in open water.

Cephalopods were the first large, shelled animals able to move vertically in the ocean. The modification of the mantle into a device for forcibly ejecting water from the cavity enabled them to move rapidly through the water. With this greatly enhanced mobility, some cephalopods, such as squids, became the major predators in open ocean waters (Figure 29.29*e*). They are still important marine predators today. As is typical of active predators, cephalopods have complicated sense organs, most notably eyes that are comparable to those of vertebrates in their ability to resolve images (see Chapter 42).

Cephalopods appeared near the beginning of the Cambrian period, about 600 million years ago, and by the Ordovician period a wide variety of types were present. Increases in size and reductions in external hard parts characterize the subsequent evolution of many lineages. The cephalopod foot is closely associated with the large, branched head that bears tentacles and a siphon. The large, muscular mantle is a solid external supporting structure. The gills hang within the mantle cavity. Cephalopods capture and subdue their prey with a sharp beak.

The earliest cephalopod shells were divided by partitions penetrated by tubes through which liquids could be removed. As fluid moves out of a chamber, gas diffuses into it, changing the buoyancy of the animal. Nautiloids (genus *Nautilus*) are the only cephalopods with external chambered shells that survive today (Figure 29.29*f*).

Arthropods: Segmented External Skeletons

Fluid-filled body cavities acted on by muscles surrounding them provide the skeletal support for most of the animals we have discussed so far. The body plans of these animals vary according to the size and extent of the cavities, how they are lined, and whether they are subdivided. Most of these animals have external coverings that are relatively thin and flexible and allow gas exchange (although some, such as roundworms, have tough cuticles).

In Precambrian times, the body covering in some wormlike lineages became thickened by the incorporation of layers of protein and a strong, flexible, waterproof polysaccharide called **chitin**. After this change, which was initially probably protective in function, the body covering acquired support and locomotor functions—it became an **exoskeleton**, a hard covering on the outside of the body.

To locomote, an animal with a rigid exoskeleton and no cilia needs appendages that are moved by muscles. Such appendages evolved several times in late Precambrian times, leading to the phyla collectively called **arthropods**. The appendages of these proto-arthropods were initially unjointed, but jointed appendages evolved in most arthropod lineages. The species with unjointed limbs that survive today show that effective locomotion is possible without joints.

The muscles of animals that have exoskeletons attach to the inside of the skeleton, and each segment

(a) Cross section through a segment of a generalized arthropod

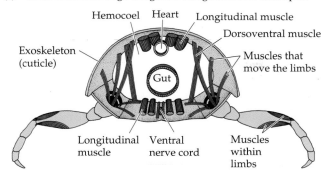

(b) Insect body plan

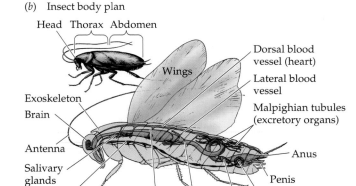

29.30 External and Jointed Skeletons (a) A cross section shows the arthropod exoskeleton with jointed appendages. (b) The body plan of this insect differs in many details from that of other arthropods, but the basic theme of a segmented body with modified appendages is general to most arthropod lineages.

has muscles that operate a particular segment and the appendages attached to it (Figure 29.30a). Arthropod appendages serve many functions, including walking and swimming, food capture and manipulation, copulation, and sensory perception. The division of labor among body regions afforded a versatility to each individual arthropod species not found in its less complex ancestors.

An exoskeleton affected its possessors in other ways. First, the coelom lost its function as a hydrostatic skeleton and became much reduced. The pseudocoel became a hemocoel, filled with blood that directly bathes the animal's organs and allows hydraulic extension of limbs. Second, because the exoskeleton slows gas exchange, arthropods evolved new means of taking up oxygen and releasing carbon dioxide: Many arthropods evolved gills for gas exchange; others evolved tubes, called *tracheae*, that carry oxygen deep into their bodies.

A rigid, nonliving exoskeleton prevents an animal from gradually increasing in size. Arthropods grow by **molting**: a periodic shedding of the exoskeleton followed by the rapid hardening of a new and larger exoskeleton formed from cells under the old one. Soon after the old exoskeleton is shed, the new one is pumped up, expands, and hardens. While this process is under way, movement is difficult or impossible and the animals are highly vulnerable to predators. Soft-shelled crabs, for example, are individuals captured just after they shed their old exoskeletons.

Arthropods diversified dramatically

The evolution of an exoskeleton had another profound influence on arthropod evolution. Encasement within armor does more than protect an animal from predators. It also provides support for walking on dry land, and, if it has special waterproofing, it keeps the animal from drying out quickly in air. Arthropods were, in short, excellent candidates to invade the land, and they did so—repeatedly.

Several lineages of arthropods colonized the land, but all other groups of arthropods are overshadowed in number and diversity by the insects. The great majority not only of arthropods, but of all animal species, are insects.

Ancient arthropod lineages are a mystery

The earliest known arthropods date from Precambrian times. The roots of the lineage are so ancient, and the rate of evolutionary diversification was so rapid, that the ancestors of arthropods are as yet unknown. The divisions among arthropod lineages are so ancient that it is reasonable to divide the species into four phyla: Trilobita, Chelicerata, Crustacea, and Uniramia, the latter three of which survive today. Because the sequence of splitting of the lineages is unknown, the phylogeny we have adopted shows the three lineages separating at the same time (Figure 29.31).

No trilobites survive

A once-dominant line of arthropods, the trilobites (phylum Trilobita) flourished in Cambrian and Ordovician seas, but were extinct by the close of the Paleozoic era, 245 million years ago. Trilobites were heavily armored, and their body segmentation and appendages followed a relatively simple, repetitive plan (Figure 29.32). Why trilobites declined in abundance and eventually became extinct is unknown.

Chelicerates invaded the land

The bodies of all chelicerates (phylum Chelicerata) are divided into two major regions, the anterior of which bears two pairs of appendages modified to form mouthparts and four pairs of walking legs. The 63,000 described species are usually placed in three classes,

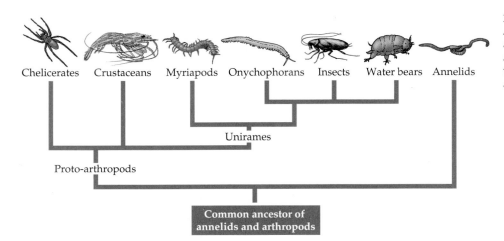

29.31 A Probable Phylogeny of Arthropods Lineage separations among surviving arthropods are so ancient that we divide them into three phyla: Chelicerata, Crustacea, and Uniramia.

only one of which contains many species.

The pycnogonids (class Pycnogonida), or sea spiders, are a small group of marine species that are seldom seen except by marine biologists (Figure 29.33a). The class Merostomata contains a single order, the Xiphosura, or horseshoe crabs. These marine animals, which have changed very little during their long fossil history, have a large horseshoe-shaped covering over most of the body. They are common in shallow waters along the eastern coasts of North America and Southeast Asia, where they scavenge and prey on bottom-dwelling invertebrates. Periodically they crawl into the intertidal zone to mate and lay eggs (Figure 29.33b).

The arachnids (class Arachnida) are abundant in terrestrial environments. Most arachnids have a simple life cycle in which miniature adults hatch from eggs and begin independent lives almost immediately. But some species have a more complex life cycle. Still others retain their eggs during development and give

(a) *Decolopoda* sp.

(b) *Limulus* sp.

Dalmanites limulurus

29.32 Trilobites The relatively simple, repetitive segments of the now-extinct trilobites are illustrated here by fossils of a species that lived during the Silurian period.

29.33 Minor Chelicerates (a) Although they are not spiders, it is easy to see why sea spiders were given their common name. (b) This spawning aggregation of horseshoe crabs was photographed on a sandy beach in Delaware.

(a) *Uroctonus mondax*

(c) Harvestman (species not identified)

(b) Salticid (jumping) spider (species not identified)

29.34 Arachnid Diversity   (a) Scorpions are nocturnal predators. (b) The evolution of webs is characteristic of spiders, which use them in many ways. This netcasting spider uses the web to envelop prey. (c) Harvestmen, often called daddy longlegs, are scavengers. (d) Ticks are blood-sucking, external parasites on vertebrates. This wood tick is piercing the skin of its human host.

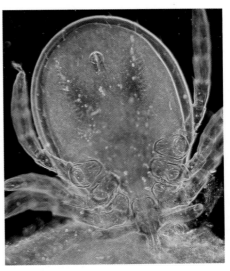

(d) *Ixodes ricinus*

birth to live young. The most diverse and ecologically important arachnids are the scorpions, harvestmen, spiders, mites, and ticks (Figure 29.34).

Spiders are important terrestrial predators. Some have excellent vision that enables them to chase and seize their prey. Others spin elaborate webs to snare prey. The webs of different groups of spiders are strikingly varied and enable spiders to position their snares in many different environments. Spiders also use protein threads to construct safety lines during climbing, and as homes, mating structures, protection for developing young, and dispersal. The threads are produced by modified abdominal appendages that are connected to internal glands that secrete the proteins of which the threads are constructed.

Crustaceans are diverse and abundant

Crustaceans (phylum Crustacea) are the dominant arthropods of the oceans. The individuals of one group alone, the copepods, are so numerous that they may be the most abundant of all animals. Most of the 40,000 species of crustaceans have a body that is divided into three regions: head, thorax, and abdomen. The segments of the head are fused together, and the head bears five pairs of appendages. Each of the multiple thoracic and abdominal segments usually bears one pair of appendages. In many species, a fold of the exoskeleton, the **carapace**, extends dorsally and laterally back from the head to cover and protect some of the other segments (Figure 29.35a).

The sexes are separate in nearly all crustaceans; males and females come together to copulate. The fertilized eggs of most crustacean species attach to the

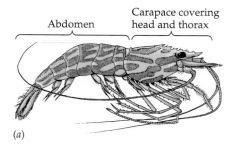

Abdomen | Carapace covering head and thorax

(a)

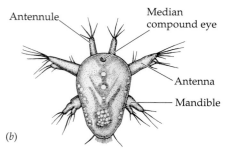

Antennule

Median compound eye

Antenna

Mandible

(b)

29.35 Crustacean Structure *(a)* The bodies of crustaceans are divided into three regions, each of which bears appendages. Two regions, the head and thorax, are covered by the protective carapace in many species. *(b)* A nauplius has one compound eye and three pairs of appendages.

(a) Decapod crustacean (species not identified)

(b) Ligia occidentalis

outside of the female's body, where they are held during their early development. At hatching, the young of some species are released as larvae; those of other species are released as juveniles similar in form to the adults. But some species release fertilized eggs into the water or attach them to an object in the environment. The typical crustacean larva, called a **nauplius**, has three pairs of appendages and one compound eye (Figure 29.35*b*).

The most familiar crustaceans are shrimp, lobsters, crayfish, and crabs (all decapods; Figure 29.36*a*); sow bugs (isopods; Figure 29.36*b*); and sand fleas (amphipods). Also included in the crustaceans are a wide variety of other small species, many of which superficially resemble shrimps. The abundant copepods (class Copepoda) already mentioned form one of these groups (Figure 29.36*c*). Barnacles (class Cirripedia) are unusual crustaceans that are sessile as adults (Figure 29.36*d*). With their calcareous shells, they superficially

(c) Eucheta sp.

(d) Lepas anatifera

29.36 Crustacean Diversity *(a)* This blue mountain crayfish is a terrestrial decapod crustacean from Australia. *(b)* This sow bug is an intertidal isopod found on the California coast. *(c)* A typical planktonic copepod. *(d)* Gooseneck barnacles attach to a substratum and feed by protruding and retracting a feeding appendage from their shells.

(a) *Scolopendra heros*

(b) Millipede (species not identified)

29.37 Myriapods (a) Centipedes have powerful jaws for capturing active prey. (b) Millipedes, which are scavengers and plant eaters, have smaller jaws and legs.

resemble mollusks, but as the zoologist Louis Agassiz remarked more than a century ago, a barnacle is "nothing more than a little shrimp-like animal, standing on its head in a limestone house and kicking food into its mouth."

Unirames are primarily terrestrial

The body of a unirame (phylum Uniramia) is divided into two or three regions. The anterior regions have few segments, but the posterior region, the abdomen, has many segments. Unirames are primarily terrestrial animals; most have elaborate systems of channels that bring oxygen to the cells of their internal organs.

MYRIAPODS. Centipedes, millipedes, and two other groups of animals (subphylum Myriapoda) have two body regions: a head and a trunk. Centipedes and millipedes have a well-formed head and a long, flexible, segmented trunk that bears many pairs of legs (Figure 29.37). Centipedes prey on insects and other small animals. Millipedes scavenge and eat plants. More than 3,000 species of centipedes and 10,000 species of millipedes have been described; many species remain unknown. Although most myriapods are less than a few centimeters long, some tropical species are ten times that size.

ONYCHOPHORANS. The 75 species of onychophorans (subphylum Onychophora) are similar to myriapods (Figure 29.38a). The soft bodies of onychophorans are covered by a thin, flexible cuticle that contains chitin; they use their body cavities as hydrostatic skeletons. Their soft, fleshy, unjointed, claw-bearing legs are formed by outgrowths of the body.

WATER BEARS. Like the onychophorans, water bears (subphylum Tardigrada) have fleshy, unjointed legs and use their fluid-filled body cavities as hydrostatic skeletons (Figure 29.38b). Unlike onychophorans, water bears are all extremely small, and they lack a circulatory system and gas exchange organs. The 550 extant species of water bears live in marine sands and gravels and on temporary water films on plants. When these films dry out, water bears also lose water and shrink to small, barrel-shaped objects that can survive at least a decade in a dehydrated state. Onychophorans and water bears are probably not ancestors of other arthropod lineages, but they may be similar in appearance to the early arthropods.

(a) *Peripatus* sp.

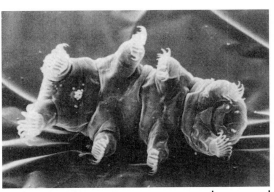

(b) *Echiniscoides sigismundi* 50 μm

29.38 Arthropods with Unjointed Legs
(a) Onychophorans have unjointed legs and use the body cavity as a hydrostatic skeleton. (b) The appendages and general anatomy of water bears superficially resemble those of onychophorans.

Insects are the dominant unirames

The 1.5 million species of insects (subphylum Insecta) that have been described are believed to be only a small fraction of the total number living on Earth today. They are the most numerous of all animal species, and the vast biological subdiscipline of entomology—the study of insects—attests to the importance of their interactions with humans.

Insects are found in nearly all terrestrial and freshwater habitats, and they utilize as food nearly all species of plants and many species of animals. Some are internal parasites of plants and animals; others suck blood externally or consume surface body tissues. Insects transmit many viral, bacterial, and protist diseases among plants and animals. Very few insect species live in the oceans. In freshwater environments, on the other hand, they are sometimes the dominant animals, burrowing through substrata, extracting suspended prey from the water, and actively pursuing other animals.

Insects have three basic body parts (head, thorax, abdomen), a single pair of antennae on the head, and three pairs of legs attached to the thorax (see Figure 29.30*b*). Insects exchange gases by means of air sacs and tubular channels called **tracheae** (singular trachea) that extend from external openings inward to tissues throughout the body. The adults of most flying insects have two pairs of stiff, membranous wings attached to the thorax—except for flies, which have only one pair, and beetles, in which the forewings form heavy, hardened wing covers.

Most insects have the full number of adult segments when they hatch from their eggs, but species differ strikingly in their state of maturity at hatching and in the processes by which they achieve adulthood. Wingless insects (class Apterygota), of modern insects probably the most similar to insect ancestors, have **simple development**. They hatch from the egg looking like small adults.

Development in the winged insects (class Pterygota) is more complex. The hatchlings are less similar in form to adults, and they undergo substantial changes at each molt in the process of becoming an adult. The immature stages of insects between molts are called **instars**. If changes between its instars are gradual, an insect is said to have **incomplete metamorphosis**. If dramatic changes occur between one instar and the next, an insect is said to have **complete metamorphosis** (see Figures 38.4).

Entomologists divide the winged insects into about 28 different orders; Figure 29.39 shows some of them. We can make sense out of this bewildering variety by recognizing three major types.

Members of one lineage cannot fold their wings back against the body. Although they are often excel-

29.39 Insect Diversity   (*a*) The firebrat is a typical member of the apterygote (wingless) order Thysanura. (*b*) Unlike most insects, this adult mayfly (order Ephemeroptera) cannot fold its wings over its back. Representatives of some of the largest insect orders are (*c*) a broad-winged katydid (Orthoptera), (*d*) harlequin bugs (Hemiptera), (*e*) a predaceous diving beetle (Coleoptera), (*f*) a Great Mormon butterfly (Lepidoptera), (*g*) a giant robberfly (Diptera), and (*h*) a red-tailed bumblebee (Hymenoptera). ▶

lent flyers, they require a great deal of open space in which to maneuver. The only surviving members of these insects are the orders Odonata (dragonflies and damselflies) and Ephemeroptera (mayflies). All members of these two orders have complete metamorphosis. Their aquatic larvae change into flying adults after they crawl out of the water. Dragonflies and damselflies are active predators as adults, but adult mayflies lack functional digestive tracts and live only long enough to mate and lay eggs.

A second major evolutionary lineage includes the orders Orthoptera (grasshoppers, crickets, roaches, mantids, and walking sticks), Isoptera (termites), Plecoptera (stone flies), Dermaptera (earwigs), Thysanoptera (thrips), Hemiptera (true bugs), and Homoptera (aphids, cicadas, and leafhoppers). These insects have incomplete metamorphosis and are able to fold their wings back. Hatchlings are sufficiently similar in form to adults to be recognizable. They acquire adult organ systems, such as wings and compound eyes, gradually through several juvenile instars.

Insects belonging to the third lineage also are able to fold their wings back, but like the first lineage, they undergo complete metamorphosis, with different life stages specialized for living in different environments and using different food sources. In many species the larvae are adapted for feeding and growing, and the adults are specialized for reproduction and dispersal. The adults of some species do not feed at all, living only long enough to mate, disperse, and lay eggs. In many species whose adults do feed, adults and larvae use different food resources.

The most familiar example of complete metamorphosis is a caterpillar that changes into a butterfly, but other insects, including beetles and flies, undergo similar transformations. Their wormlike larvae transform into the adult form during a specialized "inactive" phase, the **pupa**, in which many larval tissues are broken down and the adult form develops. About 85 percent of all winged insects have complete metamorphosis. Familiar examples are the orders Neuroptera (lacewings and their relatives), Coleoptera (beetles), Trichoptera (caddisflies), Lepidoptera (butterflies and moths), Diptera (flies), and Hymenoptera (sawflies, bees, wasps, and ants).

(*a*) *Thermobia domestica*

(*b*) Mayfly (species not identified)

(*c*) *Microcentrum rhombifolium*

(*d*) *Murgantia histrionica*

(*e*) *Dytiscus marginalis*

(*f*) *Papilio memnon*

(*g*) *Blepharotes asilidae*

(*h*) *Bombus lapidarius*

Because they can fold their wings over their backs, insects belonging to the latter two adaptive types are able to fly from one place to another and then, upon landing, tuck their wings out of the way and crawl into crevices and other tight places. Several orders, including the Phthiraptera (lice) and Siphonaptera (fleas) are parasitic. Although descended from flying ancestors, these insects have lost the ability to fly.

Why have the insects undergone such incredible evolutionary diversification? Insects may have originated from a centipedelike ancestor at least as far back as the Devonian period, more than 350 million years ago. With this early start, they were able to exploit the newly formed forests and other ancient forms of land vegetation. The terrestrial environments penetrated by insects were like a new planet, an ecological world with more complexity than the surrounding seas, but one with relatively few species of terrestrial animals.

By Carboniferous times, a great diversity of insect types already swarmed over the land, and winged insects—the first animals to fly—had appeared. Because they can fly, insects are able to move efficiently through complex vegetation, and they eat nearly all types of plant tissues.

Themes in Protostome Evolution

Most protostome evolution took place in the oceans. Because water is an incompressible liquid, early animals used body fluids as the basis for their support. When acted on by surrounding circular and longitudinal muscles, these fluid-filled spaces function as skeletons that enable large animals to move.

Subdivisions of the body cavity allow better control of movement and permit different parts of the body to be moved independently of one another. Thus some protostome lineages gradually evolved the ability to change their shape in complex ways and to move with great speed on and through sediments or in the water.

Predation may have been the major selective pressure for the development of hard, external body coverings. Such coverings evolved independently in many protostomate phyla. Originally protective in function, they became key elements in the development of new systems of locomotion. Locomotory abilities permitted prey to escape more readily from predators but also allowed predators to pursue their prey more effectively. Thus, the evolution of protostomes has been and continues to be a complex saga of arms races among predators and prey.

During much of protostome evolution, the only food in the water consisted of dissolved organic matter and very small organisms. Consequently, many different lineages of animals evolved feeding structures designed to extract small prey from water, as well as structures for moving water through or over their prey-collecting devices.

Because water flows readily, bringing food with it, sessile lifestyles also evolved repeatedly during protostome evolution. Most protostomate phyla today have at least some sessile members. A sessile animal gains access to local resources but forfeits access to more distant resources. A sessile animal is exposed to physical agents and predators from which it cannot escape by movement.

Sessile animals cannot come together to mate; instead most sessile animals rely on the fertilization of gametes that they have ejected into the water. Some species eject both eggs and sperm into the water; others retain their eggs within their bodies and extrude only their sperm, which are carried by the water to other individuals. Species whose adults are sessile often have motile larvae, many of which have complicated mechanisms for locating suitable sites on which to settle. Many colonial sessile protostomes are able to grow in the direction of better resources or into sites offering better protection.

For both animals and plants, a frequent consequence of a sessile existence is intense competition for space that provides access to light and other resources. Such competition is intense among plants in most terrestrial environments. In the sea, especially in shallow waters, animals also compete directly for space. They have evolved mechanisms for overgrowing one another and for engaging in toxic warfare where they come into contact.

Individual members of colonies, if they are directly connected, can share resources. The ability to share resources enables some individuals to specialize for particular functions, such as reproduction, defense, or feeding. The nonfeeding individuals can derive their nutrition from their feeding associates.

Although we have concentrated on the evolution of greater complexity in protostomate lineages, many lineages that remained simple survive today. Cnidarians are common in the oceans; roundworms are abundant in most aquatic and terrestrial environments. Parasites have retained simple body plans but have evolved complex life cycles.

All the phyla of protostomate animals had evolved by the Cambrian period, 600 million years ago, but extinction and diversification within those lineages continue. The characteristics of the existing protostomate phyla are summarized in Table 29.2. Many of the evolutionary trends demonstrated by protostomes also dominated the evolution of deuterostomes, the lineage that includes the chordates, the group to which humans belong. Hard external body coverings evolved and were later abandoned by many lineages. In the next chapter, after considering the evolution of deuterostomes, we will return to the major themes in animal evolution and describe patterns that characterize all groups of animals.

TABLE 29.2 General Characteristics of Protostomate Phyla

PHYLUM	SYMMETRY	BODY CAVITY	DIGESTIVE TRACT	CIRCULATORY SYSTEM
Rotifera	Bilateral	Pseudocoelom	Complete	None
Acanthocephala	Bilateral	Pseudocoelom	None	None
Nematoda	Bilateral	Pseudocoelom	Complete	None
Nematomorpha	Bilateral	Pseudocoelom	Greatly reduced	None
Loricifera	Bilateral	Pseudocoelom	Complete	None
Platyhelminthes	Bilateral	None	Dead-end sac	None
Nemertea	Bilateral	Coelom	Complete	Closed
Sipunculida	Bilateral	Coelom	Complete	None
Echiura	Bilateral	Coelom	Complete	None
Pogonophora	Bilateral	Coelom	None	None
Annelida	Bilateral	Coelom	Complete	Closed or open
Mollusca	Bilateral	Reduced coelom	Complete	Open except in cephalopods
Chelicerata	Bilateral	Hemocoel	Complete	Closed or open
Crustacea	Bilateral	Hemocoel	Complete	Closed or open
Uniramia	Bilateral	Hemocoel	Complete	Closed or open

Summary of "Protostomate Animals: The Evolution of Bilateral Symmetry and Body Cavities"

Descendants of a Common Ancestor

• All members of the kingdom Animalia are believed to have a common flagellated protist ancestor. **Review Figure 29.4**

• The specialization of cells by function made possible the complex, multicellular body plan of animals.

Food Procurement: The Key to the Animal Way of Life

• Animals obtain their food—complex organic molecules—by active expenditure of energy.

Clues to Evolutionary Relationships among Animals

• Most animals have either radial or bilateral symmetry. Radially symmetrical animals move slowly or not at all. Bilateral symmetry is strongly correlated with more rapid movement and the development of sense organs at the anterior end of the animal. **Review Figure 29.2**

• The two major animal lineages—protostomes and deuterostomes—are believed to have separated early in animal evolution, because they differ in several components of their early embryological development. **Review Figure 29.4 and Table 29.1**

The body cavity of an animal is strongly related to its ability to move. On the basis of body cavity, animals are classified as acoelomates, pseudocoelomates, or coelomates. **Review Figure 29.3 and Table 29.2**

Early Animal Diets: Prokaryotes and Protists

• The earliest animals probably ate prokaryotes and protists, but some evolved specialized cells with stinging tentacles that allowed them to capture larger prey.

Simple Aggregations: Sponges and Placozoans

• All sponges (phylum Porifera) are simple animals that lack cell layers and body symmetry, but they have many different cell types.

• Sponges feed via choanocytes, feeding cells that draw water through the sponge body and filter out small organisms and nutrient particles. **Review Figure 29.5**

• Sponges come in a wide variety of sizes and shapes that are adapted to different movement patterns of water.

• Placozoans (phylum Placozoa) are very small and simple and lack symmetry. **Review Figure 29.7**

The Evolution of Cell Layers: The Cnidarians

• Cnidarians (phylum Cnidaria) are radially symmetrical and have only two cell layers, but with their nematocyst-studded tentacles, they can take prey larger and more complex than themselves. **Review Figure 29.8**

• Most cnidarian life cycles have a sessile polyp and a free-swimming, sexual, medusa stage, but some species lack one of the stages. **Review Figures 29.9, 29.10, 29.11, 29.12**

• Cnidarians have persisted through evolution because of their low metabolic needs.

The Evolution of Bilateral Symmetry

• Bilateral symmetry probably arose first in simple animals consisting of flattened masses of cells.

• Ctenophores (phylum Ctenophora), descendants of the first split in the lineage of bilaterally symmetrical animals, are marine carnivores that have simple life cycles. **Review Figure 29.15**

Protostomes and Deuterostomes: An Early Lineage Split

• Protostomes and deuterostomes are monophyletic groups that have been evolving separately since the Cambrian period. Protostomes have a central nervous system, paired nerve cords, and larvae with compound cilia. Deuterostomes have a dorsal nervous system and larvae with single cilia. **Review Figure 29.4 and Table 29.1**

Protostomes: A Diversity of Body Plans

• Most of Earth's living animal species are protostomes. The diversity of protostomate body plans and lifestyles makes it a challenge to determine their evolutionary relationships. **Review Figure 29.16**
• Improved control of movement, which is determined by body cavity, is a key feature of protostome evolution.
• Although no larger than many ciliated protists, rotifers (phylum Rotifera) have highly developed internal organs. **Review Figure 29.17**
• Spiny-headed worms (phylum Acanthocephala) are parasites of vertebrates, primarily freshwater fishes. They have complex life cycles. **Review Figure 29.18**

Wormlike Bodies: Movement through Sediments

• Members of 16 protostomate phyla are wormlike.
• Roundworms (phylum Nematoda) are one of the most abundant and universally distributed of animal groups. Many are parasitic. **Review Figure 29.19**
• Flatworms (phylum Platyhelminthes) have no body cavity, lack organs for oxygen transport, have only one entrance to their gut, and move by beating their cilia. Many species are parasitic. **Review Figures 29.21, 29.23**
• Flatworms have added and subtracted hosts from their life cycle throughout evolution. **Review Figure 29.22**

Coelomate Protostomes

• A coelom, which is surrounded by muscles, allows better control over movement than a pseudocoel, which has muscles on only one side. **Review Figure 29.3**
• Ribbon worms (phylum Nemertea) have a complete digestive tract and capture prey with an ejectable proboscis. **Review Figure 29.24**
• Several small phyla (Sipuncula, Echiura, and Pogonophora) of unsegmented worms with simple coeloms live in marine sediments.

Segmented Bodies: Improved Locomotion

• Segmentation of the coelom, which evolved several times among protostomate lineages, allows precise control of movement. **Review Figure 29.26**
• Annelids (phylum Annelida) are a diverse group of segmented worms that live in marine, freshwater, and terrestrial environments.
• Mollusks (phylum Mollusca) evolved from segmented ancestors but subsequently became unsegmented. The molluscan body plan has three basic components: foot, mantle, and visceral mass. **Review Figure 29.28**
• The molluscan body plan has been modified to yield a diverse array of animals that superficially appear very different from one another.

Arthropods: Segmented External Skeletons

• The arthropod lineages evolved external skeletons and jointed appendages, with which they move. **Review Figure 29.30**
• The arthropod skeleton provides support for walking on dry land and, if waterproofed, keeps the animal from drying out quickly in air. Several lineages of arthropods colonized terrestrial environments.
• The arthropods contain more species than do any other animal groups, and most arthropods are insects.

• Arthropods are divided into three extant phyla: Chelicerata, Crustacea, and Uniramia. The roots of each of these lineages are very ancient and thus unknown. **Review Figure 29.31**
• Insects, of which there is staggering variety, live in nearly all terrestrial and freshwater environments. Some insects evolved wings, becoming the first animals to fly.

Themes in Protostome Evolution

• Most of protosome evolution took place in the oceans; early protostomes used their body fluids as hydrostatic skeletons.
• Protostomes in several lineages evolved improved abilities to change their shapes and control their movements.
• Protostomes evolved the ability to capture and eat a great variety of sizes and shapes of prey.
• Most protostomate phyla have some sessile members, many of which are colonial.

Self-Quiz

1. The body plan of an animal is
   a. its general structure.
   b. the functional interrelationship of its parts.
   c. its general form and the functional interrelationship of its parts.
   d. its general form and its evolutionary history.
   e. the functional interrelationship of its parts and its evolutionary history.

2. A bilaterally symmetrical animal can be divided into mirror images by
   a. any cut through the midline of its body.
   b. any cut from its anterior to its posterior end.
   c. any cut from its dorsal to its ventral surface.
   d. only a cut through the midline of its body from its anterior to its posterior end.
   e. only a cut through the midline of its body from its dorsal to its ventral surface.

3. Among protostomes, cleavage of the fertilized egg is
   a. delayed while the egg continues to mature.
   b. determinate; cells separated after a few divisions develop into only partial embryos.
   c. indeterminate; cells separated after a few divisions develop into complete embryos.
   d. triploblastic.
   e. diploblastic.

4. The sponge body plan is characterized by
   a. a mouth and digestive cavity but no muscles or nerves.
   b. muscles and nerves but no mouth or digestive cavity.
   c. a mouth, digestive cavity, and spicules.
   d. muscles and spicules but no digestive cavity or nerves.
   e. no mouth, digestive cavity, muscles, or nerves.

5. The phyla of diploblastic animals are
   a. Porifera and Cnidaria.
   b. Cnidaria and Ctenophora.
   c. Cnidaria and Platyhelminthes.
   d. Ctenophora and Platyhelminthes.
   e. Porifera and Ctenophora.

6. Cnidarians are abundant, perhaps because of their ability
   a. to live in both salt and fresh water.
   b. to move rapidly in the water column.
   c. to capture and consume large numbers of small prey.
   d. to capture large prey, and their low metabolic rate.
   e. ability to capture large prey and to move rapidly.

7. Many parasites evolved complex life cycles because
   a. they are too simple to disperse readily.
   b. they are poor at recognizing new hosts.
   c. they were driven to it by host defenses.
   d. having an intermediate host usually increases the probability of transfer to a new individual of the primary host.
   e. their ancestors had complex life cycles and they simply retained them.

8. Which of the following is *not* part of the molluscan body plan?
   a. Mantle
   b. Foot
   c. Radula
   d. Visceral mass
   e. Jointed skeleton

9. Insects that hatch from eggs into juveniles that resemble the adults are said to have
   a. instars.
   b. neopterous development.
   c. simple development.
   d. incomplete metamorphosis.
   e. complete metamorphosis.

10. Many lineages of protostomes evolved feeding structures designed to extract small prey from the water because
    a. during much of protostome evolution, the only food available was dissolved organic matter and very small organisms.
    b. during much of protostome evolution, small animals were more abundant than large animals.
    c. large animals were available as food but they were difficult to capture.
    d. to be successful in competition for space, protostomes had to feed on small prey.
    e. water flowed naturally over their feeding structures, so early protostomes did not have to work to get food.

Applying Concepts

1. Differentiate among the members of each of the following sets of related terms:
   a. radial symmetry/bilateral symmetry
   b. protostome/deuterostome
   c. indeterminate cleavage/determinate cleavage
   d. spiral cleavage/radial cleavage
   e. incomplete metamorphosis/complete metamorphosis
   f. coelomate/pseudocoelomate/acoelomate

2. For each of the types of organisms listed below, give a single trait that may be used to distinguish them from the organisms in parentheses:
   a. cnidarians (sponges)
   b. gastropods (all other mollusks)
   c. polychaetes (other annelids)
   d. ribbon worms (roundworms)

3. Segmentation has arisen various times during protostome evolution. What advantages does segmentation provide? Given these advantages, why do so many unsegmented animals survive?

4. Many animals extract food from the surrounding medium. What protostomate phyla contain animals that extract suspended food from the water column? What structures do these animals use to capture prey?

5. Which protostomate phyla have some species that form large colonies of attached individuals? What advantages does coloniality provide to animals?

6. Discuss the structures that different lineages of animals evolved that enable them to capture large prey.

7. A major factor influencing the evolution of animals was predation. What major animal features appear to have evolved in response to predation? How do they help their bearers avoid becoming a meal for another animal?

Readings

Barrington, E. J. W. 1979. *Invertebrate Structure and Function*, 2nd Edition. Halsted-Wiley, New York. Engaging coverage of the invertebrates, with an emphasis on morphology.

Brusca, R. C. and G. J. Brusca. 1990. *Invertebrates.* Sinauer Associates, Sunderland, MA. A thorough account of the invertebrates that provides detailed treatments of body plans and includes excellent discussions of phylogenies.

Chapman, R. F. 1982. *The Insects: Structure and Function*, 3rd Edition. Harvard University Press, Cambridge, MA. A good overview of the rich diversity of insect life.

Cloudsley-Thompson, J. L. 1976. *Insects and History.* St. Martin's Press, New York. A vivid and readable account of instances in which insects have affected human society; special emphasis is placed on insects as carriers of disease organisms.

Kozloff, E. N. 1990. *Invertebrates.* Saunders, Philadelphia. An excellent reference book on all invertebrate taxa.

Morris, S. C. and H. B. Whittington. 1979. "The Animals of the Burgess Shale." *Scientific American*, July. A brief overview of the rich collection of fossils from this important site.

Nielsen, C. 1995. *Animal Evolution: Interrelationships among the Living Phyla.* Oxford University Press, Oxford. A book that emphasizes evolutionary relationships among animal phyla and patterns of animal evolution.

Noble, E. R. and G. A. Noble. 1982. *Parasitology: The Biology of Animal Parasites*, 5th Edition. Lea & Febiger, Philadelphia. A general text covering the major parasitic protists and worms that infect animals, including people.

Ruppert, E. E. and R. D. Barnes. 1994. *The Invertebrates*, 6th edition,. Saunders, Philadelphia. A thorough general textbook with a strong functional approach to the invertebrates.

Chapter 30

Deuterostomate Animals: Evolution of Larger Brains and Complex Behaviors

Predators as Pets
All these animals belong to the same species, *Canis familiaris*, the domestic dog. The great variety in their appearance is the result of artificial selection by humans.

P redators are typically behaviorally more complex than their prey because capturing an animal with well-developed escape behavior is more difficult than finding suitable leaves to chew. As a result, predators make more interesting pets than herbivores do. Dogs were probably the first animals to be domesticated by people, and a remarkable variety of dog breeds has been produced by artificial selection. Some breeds were developed for purely utilitarian purposes, such as herding of sheep, assistance in hunting, transportation in snow, and protection of camps against predators. But many breeds have been developed simply for aesthetic or fanciful purposes, as a visit to any dog show will demonstrate.

Increasing behavioral complexity has been a common theme during evolution within many deuterostomate lineages, but early deuterostomes were structurally and behaviorally simple animals, as are many of their living descendants. The ancestral traits shared by all members of the deuterostomate lineage include indeterminate cleavage in the early embryo, formation of the mesoderm from outpocketing of the embryonic gut, a blastopore that becomes the anus, three body layers, and a well-developed coelom (see Table 29.1). Living deuterostomes vary in their type of body symmetry and circulatory systems (Table 30.1). No fossils of ancestral deuterostomes that lived before the lineage split into several lineages have been found.

There are fewer lineages and many fewer species of deuterostomes than of protostomes (Figure 30.1), but we have a special interest in deuterostomes because humans are members of that lineage. In this chapter, we first describe and

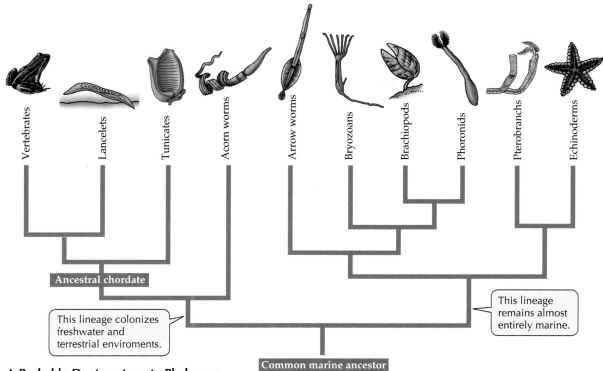

30.1 A Probable Deuterostomate Phylogeny
There are fewer lineages and many fewer species of deuterostomes than of protostomes.

discuss the major deuterostomate phyla. Then we direct special attention to the primate lineage that gave rise to our own species. Finally, we provide an overview of the major themes in the evolution of animals on Earth.

Three Body Parts: An Ancient Body Plan

The most ancient division of the deuterostomes resulted in two lineages that followed very different evolutionary pathways (see Figure 30.1). One lineage remained almost entirely marine and gave rise to the lophophorate animals (phoronids, brachiopods, bry-

ozoans, and pterobranchs), arrow worms, and echinoderms. The other lineage invaded freshwater and terrestrial environments, where it underwent a rich evolutionary radiation that gave rise to the chordates.

Early deuterostomes probably obtained food by filtering it from ocean waters, a trait they shared with many protostomes. Members of one deuterostomate lineage evolved a new structure—the **lophophore**—with which they filtered their food. The most conspicuous feature of these animals, the lophophore is a circular or U-shaped ridge around the mouth that bears one or two rows of ciliated, hollow tentacles (Figure 30.2). This large and complex structure is an organ for both food collection and gas exchange. It is held in position and moved by contraction of muscles surrounding the middle body part, the mesocoel. All adult lophophorate animals are sessile, and they use the tentacles and cilia of their lophophores to capture phytoplankton and zooplankton.

The body of a lophophorate is divided into three parts: the *prosome* (anterior), *mesosome* (middle), and *metasome* (posterior). In most species, each region has a separate coelomic compartment: the protocoel, mesocoel, and metacoel, respectively. Typically, these animals secrete a tough outer body covering. All lophophorates also have a U-shaped gut; the anus is located close to the mouth but outside the tentacles.

Four phyla of lophophorate animals survive today: Phoronida, Brachiopoda, Bryozoa, and Pterobranchia. Nearly all members of these phyla are marine; only a

TABLE 30.1	General Characteristics of Deuterostomate Animal Phyla[a]	
PHYLUM	SYMMETRY	CIRCULATORY SYSTEM
Lophophorate phyla	Bilateral	None in most
Chaetognatha	Bilateral	None
Echinodermata	Biradial	Open or none
Hemichordata	Bilateral	Closed
Chordata	Bilateral	Closed

[a]Members of all phyla have a coelom and a complete digestive tract.

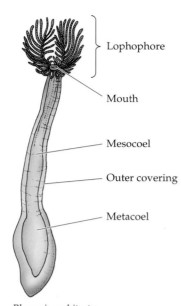

30.2 Lophophore Artistry
The lophophore dominates the anatomy of a phoronid.

Lophophore

Mouth

Mesocoel

Outer covering

Metacoel

Phoronis architecta

The lophophores of these phoronids are extended in the feeding position.

Phoronis sp.

30.3 Phoronids Phoronids live attached to the sediments of their marine habitat. They secrete a tube made mainly of chitin and live within it, extending their lophophores to feed.

few species live in fresh water. About 4,500 living species are known, but many times that number of species existed during the Paleozoic and Mesozoic eras.

Phoronids live in chitinous tubes

The 15 known species of phoronids (phylum Phoronida) are sedentary worms that live in muddy or sandy sediments or attached to a rocky substratum. Phoronids are found in waters ranging from intertidal zones to about 400 m deep. They range from 5 to 25 cm in length, and they secrete chitinous tubes in which they live.

The lophophore is the most conspicuous external feature of phoronids (Figure 30.3). Cilia drive water into the top of the lophophore. Water exits through the narrow spaces between the tentacles. Suspended food particles are caught and transported by ciliary action to the food groove and into the mouth.

Brachiopods superficially resemble bivalve mollusks

Brachiopods, or lampshells (phylum Brachiopoda), are solitary, marine, lophophorate animals that superficially resemble bivalve mollusks (Figure 30.4). Most brachiopods are between 4 and 6 cm long, but some are as large as 9 cm. Brachiopods have a mantle and a two-valved shell that can be pulled shut to protect the soft body. The shell differs from that of bivalves in that the two halves are dorsal and ventral rather than lateral. The two-armed lophophore of a brachiopod is located within the shell. The beating of cilia on the lophophore draws water into the slightly opened shell. Food is trapped in the lophophore and directed to a ridge along which it is transferred to the mouth.

Brachiopods are either attached to a solid substratum or are embedded in soft sediments. Most species are attached by means of a long, flexible stalk that holds the animal above the substratum. Gases are exchanged across nonspecialized body surfaces, especially the tentacles of the lophophore. Most brachiopods release their gametes into the water, where they are fertilized. The larvae, which resemble the adults, remain in the plankton for only a few days before they settle and metamorphose into adults.

Brachiopods reached their peak abundance and diversity in Paleozoic and Mesozoic times. More than 26,000 fossil species have been described. Only about 350 species survive, but they are common in some marine environments.

Lophophore

Laqueus sp.

30.4 Brachiopods You can see the lophophore of this North Pacific brachiopod between the valves of the shell.

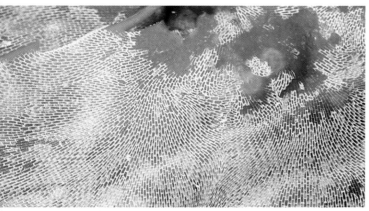

(a) *Membranipora membranacea*

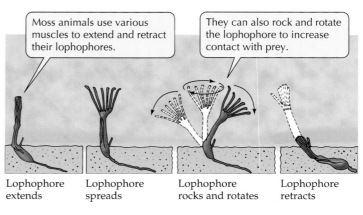

Moss animals use various muscles to extend and retract their lophophores.

They can also rock and rotate the lophophore to increase contact with prey.

Lophophore extends   Lophophore spreads   Lophophore rocks and rotates   Lophophore retracts

(b)

30.5 Moss Animals (a) Branching colonies of moss animals appear plantlike. Some members of the colony are specialized for certain functions. (b) Moss animals have greater control over their lophophores than members of other lophophorate species.

Bryozoans are colonial lophophorates

Moss animals (phylum Bryozoa) are colonial lophophorates with a "house" secreted by the body wall. A colony consists of many small individuals connected by strands of tissues along which materials can be moved. Bryozoans are called moss animals because their colonies appear plantlike (Figure 30.5a). Most moss animals are marine, but a few live in fresh water. They are the only lophophorates able to completely retract their lophophores, which they can also rock and rotate to increase contact with prey (Figure 30.5b).

A colony of moss animals is created by the asexual reproduction of its founding members. One colony may contain as many as 2 million individuals. In some species, individual colony members are specialized for feeding, reproduction, defense, or support. Moss animals reproduce sexually by releasing sperm into the water, where they are collected by other individuals. Eggs are fertilized internally, and developing embryos are brooded before they exit as larvae that seek suitable sites for attachment.

Pterobranchs capture prey with their tentacles

The fourth lophophorate group of animals consists of the pterobranchs (phylum Pterobranchia), which has only ten living species. They appear to have changed relatively little from the ancestors of their lineage. Pterobranchs are sedentary animals up to 12 mm long that live in tubes secreted by a proboscis, which is homologous to the prosome of phoronids and brachiopods. Some species are solitary; others form colonies of individuals joined together (Figure 30.6). Behind the proboscis is a collar with one to nine pairs of arms bearing long tentacles that capture prey and

permit gas exchange. The digestive tract is U-shaped, with the anus situated next to the tentacles. The proboscis encloses a coelomic cavity that has a pair of openings to the exterior, through which excretory wastes leave the body.

Arrow worms are active marine predators

Arrow worms (phylum Chaetognatha) also have three-part, streamlined bodies, but they do not have lophophores. Most of them swim in the open sea, but a few live on the seafloor. Their abundance as fossils indicates that they were already common more than 500 million years ago (mya). The 100 or so living species of arrow worms are small marine carnivores, all less than 12 cm long (Figure 30.7). They are so small that their gas exchange and excretion requirements are met by diffusion through the body surface.

Arrow worms lack a circulatory system. Wastes and nutrients are moved around the body in the coelomic fluid, which is propelled by cilia that line the coelom. The body plan is based on a coelom that is divided into head, trunk, and tail compartments. There is no

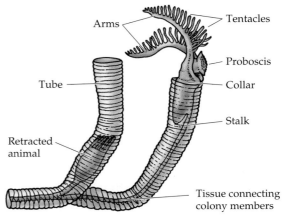

Arms — Tentacles
Tube — Proboscis
— Collar
— Stalk
Retracted animal
Tissue connecting colony members

30.6 Pterobranchs May Be Colonial or Solitary This drawing of *Rhabdopleura* depicts two members of a colony.

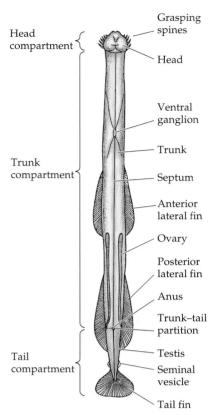

Head compartment

Trunk compartment

Tail compartment

Grasping spines

Head

Ventral ganglion

Trunk

Septum

Anterior lateral fin

Ovary

Posterior lateral fin

Anus

Trunk–tail partition

Testis

Seminal vesicle

Tail fin

30.7 Arrow Worms Arrow worms have a tripartite body plan: head, trunk, and tail. The fins and grasping spines are adaptations for a predatory life.

distinct larval stage. Miniature adults hatch directly from eggs that are released into the water.

Arrow worms are among the dominant predators of small organisms in the open oceans. They typically lie motionless in the water until movement of the water signals the approach of a prey item, which range from small protists to young fish as large as an arrow worm. At that time the arrow worm darts forward and grasps the prey with the stiff spines adjacent to its mouth. Arrow worms are stabilized in the water by means of one or two pairs of lateral fins and a caudal fin.

Arrow worms are not powerful enough to swim against strong currents, but several species undertake daily vertical migrations of up to several hundred meters, moving to deeper water during the day and back to the surface at night. (Many other small oceanic animals make such vertical treks, probably because surface waters are dangerous during the day, when visually hunting predators are active.)

Echinoderms: Complex Radial Symmetry

Members of one lineage whose ancestors had lophophores evolved calcified internal plates covered by thin layers of skin and some muscles. The calcified plates of the early ancestors became enlarged and thickened until they fused inside the entire body, giving rise to an internal skeleton. This skeleton gave its bearers protection against predators and enabled them to live above bottom sediments.

A second major change in this lineage was the evolution of a **water vascular system**, a series of seawater channels and spaces derived by the enlargement and extension of the mesocoel. The water vascular system is a network of hydraulic canals leading to extensions called **tube feet** that function in gas exchange, locomotion, and feeding (Figure 30.8*a*). Seawater enters the water vascular system through a perforated *sieve plate* from which a calcified canal leads to another canal that rings the esophagus. From this *ring canal*, other canals radiate, extending through the arms (in species that have arms) and connecting with the tube feet (Figure 30.8*a* and *b*). The development of these two structural innovations—calcified internal skeleton and water vascular system—resulted in one of the most striking evolutionary radiations, the one that gave rise to the echinoderms (phylum Echinodermata).

Echinoderms have an extensive fossil record. About 23 classes have been described, of which only six survive. About 7,000 species of echinoderms exist today, but an additional 13,000 species, probably only a small fraction of those that actually lived, have been described from their fossil remains. Nearly all living species have a bilaterally symmetrical, ciliated larva that feeds for a while as a planktonic organism before settling and transforming into a radially symmetrical adult (Figure 30.8*c*).

Living echinoderms are members of two lineages: subphyla Pelmatozoa and Eleutherozoa. The two groups differ in the form of their water vascular systems and in their number of arms.

PELMATOZOANS. Sea lilies and feather stars (class Crinoidea) are the only surviving pelmatozoans. Sea lilies were abundant 300 to 500 mya, but only about 80 species survive today. Sea lilies attach to a substratum by means of a flexible stalk consisting of a stack of calcareous discs. The main body of the animal is a cup-shaped structure that contains a tubular digestive system. Five to several hundred arms, usually in multiples of five, extend outward from the cup. Jointed calcareous plates that cover the arms enable them to bend. A ciliated groove runs down the center of each arm. On both sides of the groove are tube feet covered with mucus-secreting glands.

A sea lily feeds by orienting its arms in passing water currents. Food particles strike and stick to the tube feet, which transfer the particles to the groove, where the action of the cilia carries the food to the mouth. The tube feet of sea lilies are also used for gas exchange and elimination of nitrogenous wastes.

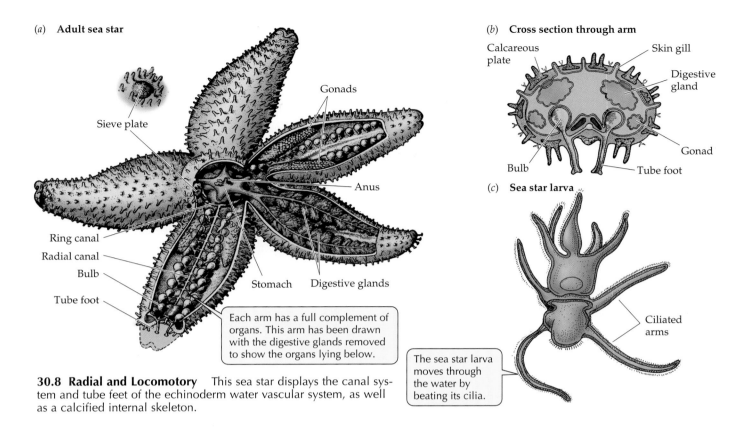

(a) **Adult sea star**

Sieve plate

Gonads

Anus

Ring canal

Radial canal

Bulb

Tube foot

Stomach   Digestive glands

Each arm has a full complement of organs. This arm has been drawn with the digestive glands removed to show the organs lying below.

(b) **Cross section through arm**

Calcareous plate

Skin gill

Digestive gland

Bulb

Gonad

Tube foot

(c) **Sea star larva**

Ciliated arms

The sea star larva moves through the water by beating its cilia.

30.8 Radial and Locomotory   This sea star displays the canal system and tube feet of the echinoderm water vascular system, as well as a calcified internal skeleton.

Feather stars are similar to sea lilies, but they have flexible appendages with which they grasp the substratum while they are feeding and resting (Figure 30.9a). Feather stars feed in much the same manner as sea lilies. Feather stars can walk on the tips of their arms or swim by rhythmically beating their arms. About 600 living species of feather stars have been described.

ELEUTHEROZOANS.   Most surviving echinoderms are members of the eleutherozoan lineage. The most familiar echinoderms are the sea stars (class Asteroidea; Figure 30.9b). Tube feet serve as organs of locomotion, and because their walls are thin, they are important sites for gas exchange. Each tube foot of a sea star is also a little adhesive organ consisting of an internal bulb connected by a muscular tube to an external sucker. The tube foot is moved by hydraulic expansion and contraction. When the bulb and the circular muscles of the tube contract, the tube foot elongates. The tube foot adheres to a surface by secreting a sticky substance around the sucker.

Many sea stars prey on polychaetes, gastropods, bivalves, and fishes. With hundreds of tube feet acting simultaneously, a sea star can exert an enormous and continuous force. It can grasp a clam in its arms, anchor the arms with tube feet, and, by steady contraction of the muscles in its arms, gradually exhaust the clam's muscles.

Sea stars that feed on bivalves are able to push their stomach out through their mouth and through the narrow space between the two halves of the clam's shell. The sea star's stomach then secretes enzymes into the soft parts of the bivalve, which digest the clam. Other sea star species feed on smaller prey or suspended particles and do not extrude their stomachs. Sea stars are important predators in many marine environments, such as coral reefs and rocky intertidal zones.

Brittle stars (class Ophiuroidea) are similar in structure to sea stars, but their flexible arms are composed of jointed hard plates (Figure 30.9c). Brittle stars can move by thrashing their arms, but burrowing forms and young individuals of most species also use their tube feet to move. Brittle stars generally have five arms, but each arm may branch several times. Most of the 2,000 species of brittle stars ingest particles from the surfaces of sediments and assimilate the organic material from them. They eject indigestible particles through their mouths, because, unlike most other echinoderms, brittle stars have only one opening to their digestive tract. Some brittle star species remove suspended food particles from the water; others capture small animals.

The remaining three classes of echinoderms lack arms. The sea daisies (class Concentricycloidea) were not discovered until 1986, and little is known about

30.9 A Diversity of Echinoderms (*a*) The flexible arms of these golden feather stars are clearly visible. (*b*) The cushion star is a typical five-armed sea star. Some species have many more arms. (*c*) This brittle star is resting on a sponge. (*d*) Purple sea urchins are important grazers of algae in the intertidal zone of the Pacific coast of North America. (*e*) This sea cucumber extends its feeding tentacles so food particles can adhere.

(*a*) Feather star (species not identified)

(*d*) *Strongylocentrotus purpuratus*

(*b*) *Pentaceraster cumingi*

(*e*) *Parastichopus californicus*

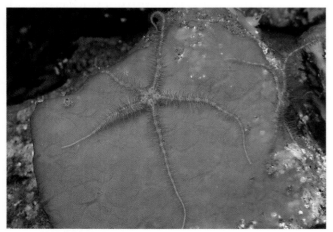

(*c*) *Ophiothrix spiculata*

them. They have tiny disc-shaped bodies with a ring of marginal spines but no arms. Sea daisies are the only echinoderms in which the water vascular system has two ring canals and in which the tube feet are arranged in a circle around the edge of the disc rather than along grooves radiating from the center. Sea daisies are found on rotting wood in deep ocean waters. They apparently feed on prokaryotes, which they digest and absorb either through a membrane that covers the oral surface or via a shallow, saclike stomach.

Sea urchins (class Echinoidea; Figure 30.9*d*) are armless, hemispherical animals that are covered with spines attached to the underlying skeleton via ball-

and-socket joints. Grooves run up the sides of the body toward the apex, instead of along the undersurfaces of arms as in sea stars. The spines of sea urchins come in varied sizes and shapes; a few produce highly toxic substances. Many sea urchins consume algae, scraping them from the rocks with a complex rasping structure. Others feed on small organic debris that they collect with their tube feet or spines.

Sea cucumbers (class Holothuroidea; Figure 30.9*e*) resemble stretched, flexible sea urchins that lack spines and have greatly reduced skeletal plates. Tube feet located on either side of five grooves along the body of the animal are used primarily for attaching to the substratum rather than for moving. Some species have tube feet only at the anterior end. The anterior tube feet are modified into large, feathery tentacles that can be protruded around the mouth. The tentacles are coated with a sticky substance to which prey or the surrounding substratum adhere. Periodically, a sea cucumber sticks its tentacles into its mouth, wipes them off, and then digests the adhered material.

Modified Lophophores: New Ways of Feeding

Evolution in the other major lineage of deuterostomes resulted in several different modifications of the lophophore and the coelomic cavity. These modifications, which include the enlargement of the proboscis and the development of pharyngeal gill slits, provided new ways of capturing and handling food. Some living representatives of this lineage, such as acorn worms, are wormlike animals that live buried in marine sands or muds, under rocks, or attached to algae. They are probably similar to the ancestors of this group of animals. Another lineage evolved the strikingly different body plan of the chordates.

Acorn worms capture prey with a proboscis

The acorn worms (phylum Hemichordata) have not changed much from the ancestral condition of animals in this lineage (Figure 30.10). They have a three-part body plan. The three regions of the body—proboscis, collar, and trunk—appear to be homologous to the prosome, mesosome, and metasome of lophophorates. In the hemichordate lineage, the lophophore was apparently lost and the proboscis grew larger and became a digging organ. The survivors of this lineage are the 70 species of acorn worms. These animals live in burrows in muddy and sandy sediments.

The enlarged proboscis is coated with a sticky mucus that traps prey items. The mucus and its attached prey are conveyed by ciliary action to the mouth. In the esophagus, the food-laden mucus is compacted into a ropelike mass that is moved through the digestive tract by ciliary action. Behind the mouth is a pharynx that opens to the outside through several

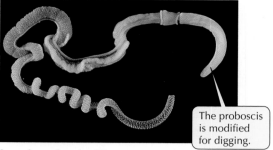

Saccoglossus kowalevskii

30.10 A Hemichordate Acorn worms have lost the lophophore. This one has been extracted from its burrow.

slits that allow water to exit. Highly vascularized tissue surrounding the slits serves as a gas exchange apparatus. An acorn worm breathes with the anterior end of its gut by pumping water into its mouth and out through its **pharyngeal slits**.

In chordates, the pharynx became a feeding device

The requirement for effective gas exchange—a large surface area—also serves well for capturing prey. In many lophophorates, the lophophore serves this dual function. The pharyngeal slits, which originally functioned as sites for gas exchange and eliminating water, as they do in modern acorn worms, were further enlarged in a sister lineage. This enlargement of the pharyngeal slits eventually led to remarkable evolutionary developments that gave rise to the chordates.

Members of the chordate lineage lost the lophophore and proboscis, replacing them with enlarged pharyngeal slits as a feeding device. The chordates (phylum Chordata) are bilaterally symmetrical animals that have pharyngeal slits at some stage in their development. The main shared ancestral features of their body plan are a dorsal, hollow nervous system; a ventral heart; and a tail that extends beyond the anus. All species have a dorsal supporting rod, the **notochord**, at some stage during their development (see Figure 30.12). In some species, the notochord is lost during metamorphosis to the adult stage. In other species, it is replaced by other skeletal structures that have the same function.

The tunicates (subphylum Urochordata) may be similar to the ancestors of all chordates. All 2,500 species of tunicates are marine animals, most of which are attached to a substratum as adults. The swimming, tadpolelike larvae reveal the close evolutionary relationships between tunicates and other chordates (see Figure 22.9).

In addition to its pharyngeal slits, a tunicate larva has a dorsal, hollow nerve cord and a notochord. Muscles attach to the notochord, which provides relatively rigid support. After a short time as a member of

30.11 Tunicates In a photograph these transparent "sea squirts" take on the blue of their environment. Pharyngeal baskets occupy most of the tunicate body cavity.

the plankton, the larva settles on the seafloor and transforms into a sessile adult that feeds by extracting plankton from the water with its pharynx, which is enlarged into a pharyngeal basket.

More than 90 percent of known species of tunicates are sea squirts (class Ascidiacea). Some sea squirts are solitary, but others produce colonies by asexual budding from a single founder. Individual sea squirts range in size from less than 1 mm to 60 cm, but colonies may measure several meters across. The baglike bodies of the adults are surrounded by a tough tunic, composed of protein and a complex polysaccharide, secreted by the epidermal cells. Much of the body is occupied by the large pharyngeal basket lined with cilia, whose beating moves water through the animal (Figure 30.11). The cilia also move the thin layer of mucus that lines the basket and to which the food particles adhere. Water enters the body through an anterior opening, passes through the pharyngeal basket into a chamber that is enclosed by the tunic, and out through another opening well removed from the site where the water entered.

The larvaceans (class Appendicularia) become reproductively mature and complete their life cycle as small planktonic organisms. They swim, filtering prey through screens made of mucopolysaccharides that collect small food particles. The larval form that functioned as a dispersal stage in their ancestors became a new lifestyle in these animals. There are only a few species of larvaceans, but they are widespread in the world's oceans.

The 25 species of lancelets (subphylum Cephalochordata) are small, fishlike animals that rarely exceed 5 cm in length. With their notochord, extending the entire length of the body throughout their lives, and their pharyngeal baskets, they resemble small fishes. Lancelets live partly buried in soft marine sediments worldwide, and they extract small prey from the water with their pharyngeal baskets (Figure 30.12). They may be living descendants of the ancestors of the vertebrates, another group of chordates.

The Origin of the Vertebrates: Sucking Mud

In one chordate lineage, which gave rise to the **vertebrates** (subphylum Vertebrata), the pharyngeal basket became enlarged. With its many exit openings, the enlarged basket was effective in extracting prey from mud, where many inedible particles are ingested along with the food. In the late Cambrian period, more than 500 mya, these early vertebrates evolved improved structures for extracting food from mud and sand and for swimming on the surface of the substratum.

A jointed, dorsal **vertebral column** replaced the notochord as the primary support in these chordates. Attached to the vertebral column, which gives the vertebrates their name, are two pairs of appendages. The faster locomotion made possible by these appendages favored the evolution of an anterior skull with a large brain (Figure 30.13). Vertebrates have a large coelom in which the body organs are slung, but the coelom serves neither as a hydrostatic skeleton nor as a gas exchange structure. Instead, a well-developed circulatory system, driven by contractions of a ventral heart, delivers oxygen to internal organs.

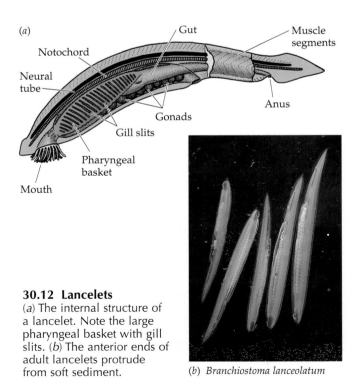

30.12 Lancelets
(a) The internal structure of a lancelet. Note the large pharyngeal basket with gill slits. (b) The anterior ends of adult lancelets protrude from soft sediment.

(a) Neural tube, Notochord, Gut, Muscle segments, Gonads, Anus, Gill slits, Pharyngeal basket, Mouth

(b) *Branchiostoma lanceolatum*

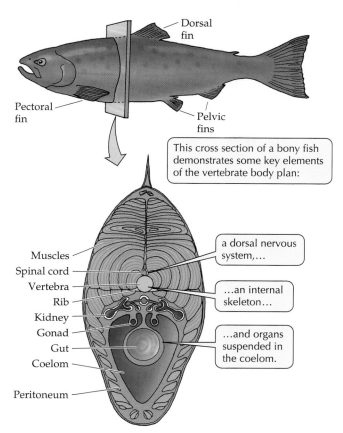

30.13 The Vertebrate Body Plan A bony fish is used here to illustrate the elements common to all vertebrates.

Dorsal fin
Pectoral fin
Pelvic fins

This cross section of a bony fish demonstrates some key elements of the vertebrate body plan:

Muscles
Spinal cord
Vertebra
Rib
Kidney
Gonad
Gut
Coelom
Peritoneum

a dorsal nervous system,…
…an internal skeleton…
…and organs suspended in the coelom.

The unique ability of living fishes to maintain nearly constant internal salt concentrations when environmental salinities change (see Chapter 37) leads many biologists to believe that vertebrates evolved in estuaries. Salinities change frequently in estuaries, and animals able to maintain relatively constant internal salt concentrations under those conditions would have been able to exploit environments from which other animals were excluded.

According to this view, filter-feeding ancestors of vertebrates evolved the ability to live in es-

tuaries and mouths of rivers, where they escaped from their predators and competitors, most of which could not handle varying salinities. The early vertebrates fed on microscopic organisms, and after growing for a while in estuaries, they could have returned to the oceans larger and better able to compete and defend themselves. They probably did not venture into rivers, because at that time rivers and terrestrial environments lacked multicellular plants or animals on which the early vertebrates could feed.

The first vertebrates lacked jaws. These jawless fishes (class Agnatha) probably swam over the bottom, sucking mud and extracting microscopic food from it as they moved. The lineage leading to modern hagfishes separated first from other groups and returned to live entirely in the sea (Figure 30.14). Hagfishes lack the osmoregulatory mechanisms found in all other fishes, suggesting that they have had a long period of life in the sea.

Other early jawless fishes probably continued to live primarily in estuaries. One group, called ostracoderms, meaning "shell-skinned," evolved a bony external armor. With their heavy armor, these small fishes could swim only slowly, but swimming above the substratum was easier than having to burrow through it, as all previous sediment feeders had done.

This new mobility enabled vertebrates to exploit their environments in new ways. One of those ways was to attach to dead, rotting flesh and use the pharynx to create a suction to pull fluids and partly decomposed tissues into the mouth. Hagfishes and lampreys, the only jawless fishes to survive beyond the Devonian period, feed in this way (Figure 30.15). These fishes have tough scaly skins instead of external armor. Hagfishes ingest the tissues of dead animals; most

30.14 A Probable Vertebrate Phylogeny This phylogeny incorporates the view that vertebrates evolved in estuaries, where their ability to handle varying salinities allowed them to exploit habitats not available to marine animals.

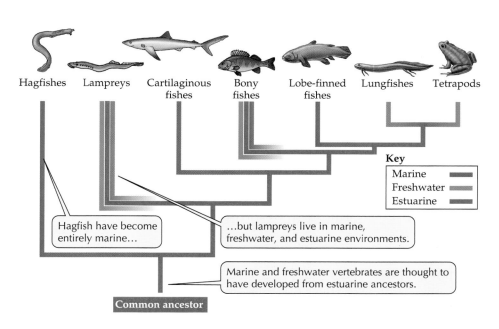

Hagfishes
Lampreys
Cartilaginous fishes
Bony fishes
Lobe-finned fishes
Lungfishes
Tetrapods

Key
Marine
Freshwater
Estuarine

Hagfish have become entirely marine…

…but lampreys live in marine, freshwater, and estuarine environments.

Marine and freshwater vertebrates are thought to have developed from estuarine ancestors.

Common ancestor

Petromyzon marinus

30.15 An Agnathan This sea lamprey has a large, jawless mouth, which it uses to suck blood and flesh from other fishes.

adult lampreys suck the blood of living fishes or eat the flesh of dying fishes. The round mouth is a sucking organ with which the animals attach to their prey and rasp at the flesh. Lampreys live in both fresh and salt water; many species move between the two environments, spawning in rivers and maturing in the sea.

Jaws were a key evolutionary novelty

During the Devonian period, about 400 mya, many new kinds of fishes evolved in the seas, estuaries, and fresh waters. Although most of these were jawless, members of one lineage evolved jaws from some of the skeletal arches that supported the gill region (Figure 30.16). A jaw allows a fish to grasp and subdue relatively large, living prey. Further development of the jaws and teeth among fishes led to the ability to chew both soft and hard body parts of prey. Chewing aided chemical digestion and improved nutrition obtained from prey. Although many intermediates must have existed between jawless fishes and the fully jawed ancestors of modern fishes, it is not difficult to imagine how each stage would have functioned better than those that preceded it.

The most important early jawed fishes were the heavily armored placoderms (class Placodermi). Some of these fishes evolved elaborate fins and sleek body forms that improved their ability to maneuver in open water. A few became huge and, together with squids, were probably the most important predators in the Devonian oceans. Despite their early abundance, however, most placoderms disappeared by the end of the Devonian period, 345 mya; none survived to the end of the Paleozoic era.

Cartilaginous and bony fishes evolved fins to control motion

Two other groups of fishes that survive today—the cartilaginous fishes and the bony fishes—became nu-

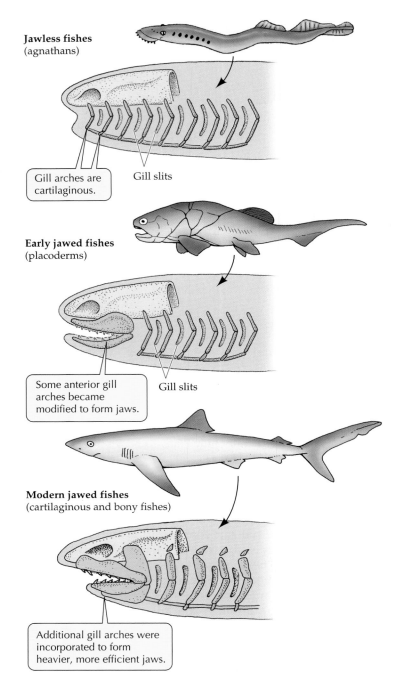

Jawless fishes (agnathans)

Gill arches are cartilaginous.

Gill slits

Early jawed fishes (placoderms)

Some anterior gill arches became modified to form jaws.

Gill slits

Modern jawed fishes (cartilaginous and bony fishes)

Additional gill arches were incorporated to form heavier, more efficient jaws.

30.16 Jaws from Gill Arches A probable scenario for the evolution of jaws from the anterior gill arches of fishes.

merically important during the Devonian period. The sharks, skates and rays, and chimaeras—all cartilaginous fishes (class Chondrichthyes)—have a skeleton composed entirely of a firm but pliable material called **cartilage** (Figure 30.17). Their skin is flexible and leathery, sometimes bearing bristly projections that give it the consistency of sandpaper. The loss of external armor, which increased mobility and ability to es-

(a) *Triaenodon obesus*

30.17 Cartilaginous Fishes (a) Most sharks, such as this Pacific white tip reef shark, are active predators living in open waters. (b) Skates and rays, represented here by a stingray, have their mouths on the ventral surface of their body and feed on the ocean bottom.

(b) *Trygon pastinaca*

cape from predators, was accompanied by the evolution of rapid swimming.

Control of swimming is provided by pairs of unjointed fins: a pair of pectoral fins just behind the gill slits and a pair of pelvic fins just in front of the anal region. A dorsal median fin stabilizes the fish as it moves (see Figure 30.13). Sharks move forward by means of their tail and pelvic fins. Skates and rays propel themselves by means of the undulating movements of their greatly enlarged pectoral fins.

Most sharks are predators, but some feed by straining plankton. The world's largest fish, the whale shark (*Rhincodon typhus*), which may grow to more than 15 m long and weigh more than 9,000 kg, feeds on plankton. Most skates and rays live on the ocean floor, where they feed on mollusks and other invertebrates buried in the sediments. Chimaeras feed on mollusks, whose shells they crack with their hard, flat teeth. Nearly all cartilaginous fishes live in the oceans.

Bony fishes evolved buoyancy

Fishes in another lineage that evolved in estuaries have internal skeletons of bone rather than cartilage, giving them their common name, bony fishes (class Osteichthyes) (see Figure 30.14). The pharyngeal slits in bony fishes open into a single chamber covered by a hard flap. Movement of the flap improves the flow of water over the gills and brings more oxygen in contact with the gas exchange surfaces.

Early bony fishes also evolved lunglike sacs that supplemented the gills in respiration. These features enabled these fishes to live where oxygen was periodically in short supply, as it often is in estuarine and freshwater environments. In most fishes, the lunglike sacs evolved into larger **swim bladders**, which serve as organs of buoyancy that help keep the fish suspended in water. By adjusting the amount of gas in its swim bladder, a fish can control the depth in the water column at which it is stable. Lungfishes are the only fishes that still use their lunglike sacs for respiration.

The external armor of bony fishes is greatly reduced, but most species are covered with flat, smooth, thin, lightweight scales that provide some protection. With their light skeletons and their swim bladders, some bony fishes recolonized the seas to become major components of marine ecological communities.

The more than 20,000 species of bony fishes have a remarkable diversity of sizes, shapes, and lifestyles (Figure 30.18). The smallest bony fish is a goby that is only 1 cm long as an adult. The largest are ocean sunfishes that weigh as much as 900 kg. Bony fishes exploit nearly all types of aquatic food sources. In the oceans they filter plankton from the water, rasp algae from rocks, eat corals and other colonial invertebrates, dig invertebrates from sediments, and prey on all other vertebrates except large whales and dolphins. In fresh water they eat plankton, devour insects of all aquatic orders, consume fruits that fall into the water in flooded forests, and prey on other aquatic vertebrates.

Many bony fishes live buried in soft sediments, where they capture passing prey or from which they emerge at night to feed in the water column above. Many are solitary, but in open water others form large aggregations called schools. Many fishes perform complicated behaviors by means of which they maintain schools, build nests, court and choose mates, and

(a) *Acipenser* sp.

(c) *Salmo gairdneri*

30.18 A Diversity of Bony Fishes (a) Sturgeon are survivors of an ancient lineage of bony fishes. (b) Coral reef fishes such as this Australian sweetlips are usually colorful and small. (c) These rainbow trout are members of a commercially important group, the salmonids. (d) This wolf eel lacks pelvic and dorsal fins but has well developed pectoral fins.

(b) Sweetlips (species not identified)

(d) *Anarrhichthys ocellatus*

care for their young. They are a group in which behavior has stimulated many evolutionary changes.

With their fins and swim bladders, fishes can readily control their position in open water, but their eggs tend to sink. Therefore, most fishes attach their eggs to a substratum. A few species, however, discharge their very small eggs directly into surface waters, where they are buoyant enough to complete their development before they sink very far. Most marine fishes move to food-rich shallow waters to lay their eggs, which is why coastal waters and estuaries are so important in the life cycles of many species of fishes. Some fishes, such as salmon, ascend rivers to spawn in freshwater streams and lakes. Conversely, some species, such as eels, that live most of their lives in fresh water, migrate to the sea to lay their eggs there.

Colonizing the Land

Although the evolution of lunglike sacs in early bony fishes was a response to the inadequacy of gill respiration in oxygen-poor waters, it also set the stage for the invasion of land by some of their descendants. Early bony fishes probably used their lungs to supplement their gills when oxygen levels in the water were low. This ability would also have allowed them to leave the water temporarily when pursued by predators unable to breathe air. But with their unjointed fins, bony fishes could only flop around on land as most fishes do today if placed out of water.

The evolution of joints in their fins enabled fishes to move over land to find new bodies of water when those in which they had been living dried up or became overpopulated. Descendants of these fishes began to use terrestrial food sources and become more fully adapted to life on land.

A group of bony fishes thought to be similar to the fishes that invaded the land are the lobe-finned fishes (subclass Crossopterygii). Lobe-fins flourished from the Devonian period (400 mya) until about 25 mya, when they were thought to have become extinct. However, in 1939 a lobe-fin was caught by a commercial fisherman off the east coast of Africa. Since that time, several dozen specimens of this extraordinary fish, *Latimeria chalumnae*, have been collected.

30.19 A Modern Lobe-Fin *Latimeria chalumnae* is the sole survivor of a lineage that was thought to be extinct.

Latimeria, a predator on other fishes, reaches a length of about 1.5 m and weighs up to 82 kg (Figure 30.19). The skeleton of *Latimeria* is composed mostly of cartilage rather than bone, and because it lives in very deep water, its swim bladder contains fat, which is less compressible than the gases found in the swim bladders of the early crossopterygians.

(a) *Dermophus mexicanus*

(b) *Scaphiophryne gottlebei*

(c) *Salamandra salamandra*

30.21 Amphibians (a) Burrowing caecilians superficially look more like worms than amphibians. (b) A rare frog species discovered in a national park on the island of Madagascar. (c) A European fire salamander.

Ancestral fish

Only a modest enlargement of the bones of a fish with fins like these…

Early amphibian

…is needed to convert them into walking legs.

30.20 Legs from Fins The evolutionary step from lobe-fin fish to amphibian is a fairly small one.

Amphibians were the first invaders of the land

During the Devonian period (400 to 345 mya), amphibians (class Amphibia) arose from an ancestor they shared with lungfishes (see Figure 30.14). These ancestral fishes had thin skins through which they exchanged gases, and stubby, jointed fins that evolved into walking legs (Figure 30.20). The design of these legs has remained largely unchanged throughout the evolution of terrestrial vertebrates.

Devonian predecessors of amphibians were the first *tetrapods*—animals with two pairs of limbs. They gradually evolved to be able to live on swampy land and, eventually, on drier land. They were probably able to crawl from one pond or stream to another by pulling themselves along on their finlike legs, as do some modern species of catfishes and other fishes. Living amphibians have relatively small lungs, and most species still exchange gases through their skin.

About 4,500 species of amphibians live on Earth today, many fewer than those known only from fossils. The living amphibians belong to three orders (Figure 30.21): the wormlike, tropical, burrowing caecilians (order Gymnophiona); frogs and toads (order Anura, "tail-less"); and salamanders (order Urodela or Caudata, "tailed"). Most species of frogs and toads live in tropical and warm temperate regions, although a few are found at very high latitudes. Salamanders are more diverse in temperate regions, but many species are found in cool, moist environments in the mountains of Central America.

Most amphibians live in water at some time in their lives. In the typical life cycle, an amphibian spends part or all of its adult life on land, usually in a moist habitat, but adults return to fresh water to lay their eggs (Figure 30.22). An amphibian egg must remain moist because it is surrounded by a delicate envelope through which it loses water readily if its surroundings are dry. The egg of most species gives rise to a larva that lives in water until it metamorphoses into a terrestrial adult.

There are interesting variations on this life cycle. Some amphibians are entirely aquatic, never leaving the water. Others are entirely terrestrial, laying their eggs in moist places on land. Many lungless salamanders live in rotting logs or moist soil and exchange gases entirely through their skin and mouth lining. Terrestrial species are confined to moist environments because most amphibians rapidly lose water through their skins when exposed to dry air, but some toads have tough skins that enable them to live for long periods of time in dry places.

Amphibians are the focus of much attention today because populations of many species are declining rapidly. For example, the golden toad has disappeared from the Monteverde Cloud Forest Reserve in Costa Rica, which was established primarily to protect this rare species. The reasons for the declines are not known, but biologists are monitoring amphibian populations closely to learn more about the causes of their difficulties and to determine the implications of amphibian declines for other organisms, including humans.

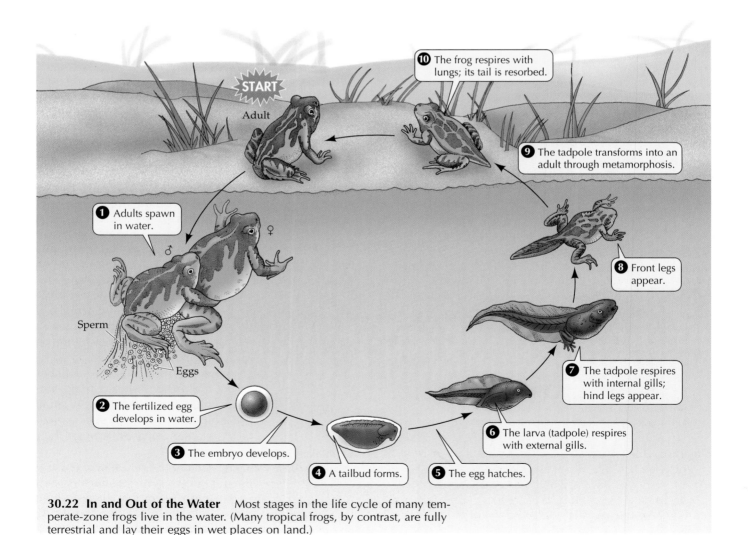

30.22 In and Out of the Water Most stages in the life cycle of many temperate-zone frogs live in the water. (Many tropical frogs, by contrast, are fully terrestrial and lay their eggs in wet places on land.)

Reptiles colonized dry environments

Two morphological changes allowed one lineage of vertebrates to control water loss and therefore to exploit the full range of terrestrial habitats. The first was an egg with a shell that is relatively impermeable to water and can be laid in dry places. The second was a combination of traits that included a tough skin impermeable to water and kidneys that could excrete concentrated urine.

The vertebrates that evolved both of these morphological changes are called *amniotes.* They were the first vertebrates to become common over much of the terrestrial surface of Earth. The amniote egg has a leathery or brittle calcium-impregnated shell that retards evaporation of the fluids inside but permits O_2 and CO_2 to pass. Within the shell and surrounding the embryo are membranes that protect the embryo from desiccation and assist its excretion and respiration. The egg also supplies the embryo with large quantities of food—yolk—that permit it to attain a relatively advanced state of development before it hatches and must feed itself (Figure 30.23).

An early amniote lineage, the reptiles (class Reptilia) arose from early tetrapods in the Carboniferous period, some 300 mya. About 6,000 species of reptiles live today. Most reptiles do not care for their eggs after laying them. In some species eggs do not develop shells, but are retained inside the female's body until they hatch. Still other species evolved placentas that nourish the developing embryos (see Chapter 39).

The skin of a reptile, which is covered with horny scales that greatly reduce loss of water from the body surface, is unavailable as an organ of gas exchange. In reptiles gases are exchanged almost entirely by the lungs, which are proportionally much larger in surface area than those of amphibians. A reptile forces air into and out of its lungs by bellowslike movements of its ribs. The reptilian heart is divided into chambers that separate oxygenated from unoxygenated blood. With this heart, reptiles can generate higher blood pressures than amphibians, and can sustain higher levels of muscular activity, although they tire much more rapidly than do birds or mammals.

MODERN REPTILES ARE MEMBERS OF THREE LINEAGES. Only three of the many reptilian lineages survive today. Turtles and tortoises (subclass Chelonia) have an armor of dorsal and ventral bony plates that form a shell into which the head and limbs can be withdrawn (Figure 30.24a). Most turtles live in lakes and ponds, but tortoises are terrestrial, and sea turtles spend their entire lives at sea except when they come ashore to lay eggs. Most turtles and tortoises are herbivores that eat a variety of aquatic and terrestrial plants, but some species are strongly carnivorous.

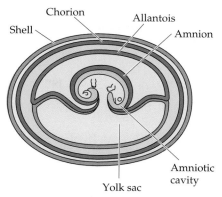

30.23 An Egg for Dry Places The evolution of the amniote egg, with its shell, extraembryonic membranes (amnion, chorion, and allantois), and embryo-nourishing yolk sac, was a major step in colonization of the terrestrial environment.

The only survivors of the reptilian subclass Sphenodontida are the two species of tuataras restricted to a few islands off the coast of New Zealand (Figure 30.24b). Tuataras superficially resemble lizards, but they differ in several internal anatomical features.

The third reptilian subclass, Squamata, includes the lizards, amphisbaenians—a group of legless, worm-like, burrowing animals with greatly reduced eyes—and snakes (Figure 30.24c and d). Most lizards are insectivores, but some are herbivores, and still others prey on other vertebrates. The largest lizards, growing as long as 3 m, are some species of monitors that live in the East Indies. Most lizards walk on four limbs, but some are limbless, as are all snakes.

Recent phylogenetic analyses show that snakes are monophyletic and that they evolved from burrowing lizards. All snakes are carnivores that can swallow objects much larger than their own diameter. Three groups of snakes evolved poison glands and inject venom into their prey with their teeth. The largest snakes—the pythons—are more than 10 m long.

Surviving Dinosaur Lineages

During the Mesozoic era (245 to 66 mya) one amniote lineage, the thecodonts (class Archosauria) split from other reptiles and underwent an extraordinary evolutionary diversification. One thecodont lineage gave rise to crocodilians (subclass Crocodylia)—crocodiles, caimans, gharials, and alligators—which are confined to tropical and warm temperate environments (Figure 30.25).

Crocodilians spend much of their time in water, but they build nests on land or on floating piles of vegetation. The eggs, which are warmed by heat generated by the decaying organic matter of the nest, typically are tended by the female until they hatch. All crocodil-

(a) *Chelonia mydas*

(b) *Sphenodon punctatus*

(c) *Chamaeleo* sp.

(d) *Tropidolaemus* sp.

30.24 Reptilian Diversity (a) The green sea turtle is widely distributed in tropical oceans. (b) The tuatara looks like a typical lizard, but it is one of only two survivors of a lineage that separated from lizards long ago. (c) The carpet chameleon from Madagascar has a long tail with which it can grasp branches and large eyes that move independently in their sockets. (d) The McGregor's viper is a tree snake found in the Philippines.

(a) *Alligator mississippiensis*

(b) *Crocodylus niloticus*

30.25 Crocodilians (a) Most crocodilians are tropical, but alligators live in warm temperate environments in Asia and, like this one, in the southeastern United States. (b) A Nile crocodile photographed in the Serengeti of Tanzania.

ians are carnivorous; they prey on vertebrates of all classes, including large mammals.

Another thecodont lineage led to the dinosaurs (subclass Dinosauria), the prevalent large terrestrial reptiles for millions of years and the largest terrestrial animals ever to inhabit Earth. Some of the largest dinosaurs weighed up to 100 tons, and many were agile and fast-moving (see Figure 19.15). The ability to move actively on land was not achieved easily. The first terrestrial vertebrates probably moved only very slowly, much more slowly than their aquatic relatives. The reason is that they apparently could not walk and breathe at the same time.

Not until the evolution of the lineages leading to the mammals, dinosaurs, and birds did the legs assume more vertical positions, which reduced the lateral forces on the body during locomotion. Special ventilatory muscles that enabled the lungs to be filled and emptied while the limbs moved also evolved. These muscles are visible in living birds and mammals. We can infer their existence in dinosaurs from the structure of the vertebral column and the capacity of many dinosaurs for bounding, bipedal (two-legged) locomotion.

The ability to breathe and run simultaneously, which we take for granted, was a major innovation in the evolution of terrestrial vertebrates. This capacity enabled its bearers to maintain steady, high levels of activity, which generated enough heat to result in high body temperatures.

Ancestral birds evolved feathers

During the Mesozoic era, about 175 mya, another thecodont lineage gave rise to the birds (subclass Aves). Fossils of the earliest birds have not yet been found. The oldest known avian fossil, *Archaeopteryx*, was covered with feathers that are almost identical to those of modern birds, and had well-developed wings and a long tail (Figure 30.26). It also had a wishbone, which in modern birds is an anchoring site for flight muscles. *Archaeopteryx* had typical perching bird claws, suggesting that it lived in trees and shrubs and used the clawed fingers on its forearms to assist it in clambering over branches. Thus, *Archaeopteryx* was already a highly evolved bird 150 mya.

During the Cretaceous period (138 to 66 mya), birds underwent an extensive evolutionary radiation. The dominant Cretaceous lineage was the "opposite birds," so named because the tarsal bones of their legs fused in the opposite direction from the way in which fusion happens in all modern birds. All lineages of opposite birds died out at the end of the Cretaceous, but paleontologists disagree over how many other avian lineages survived the mass extinction. Some believe, on the basis of the fossil record, that members of only one lineage, collectively known as the transitional

30.26 A Mesozoic Bird An artist's recreation of *Archaeopteryx* shows its modern feathers and arboreal habits.

shorebirds, survived (Figure 30.27*a*). Others believe, on the basis of molecular data, that at least some representatives of many Cretaceous avian lineages were not exterminated (Figure 30.27*b*).

The single most characteristic feature of birds is their feathers, which are highly modified scales. The flying surface of the wing is created by large quills that arise from the forearm and the reduced, stubby fingers. Other strong feathers sprout like a fan from the shortened tail and serve as stabilizers during flight. The contour feathers and down feathers, which arise from well-defined tracts, cover the body like a garment and provide insulation to control loss of body heat.

The light but strong bones of birds are hollow and have internal struts for strength. The sternum (breastbone) forms a large, vertical keel to which the breast muscles are attached. These muscles pull the wings downward during the main propulsive movement in flight. Flight is metabolically expensive, and a flying bird consumes energy at a very high rate.

Because birds have high metabolic rates, they generate large amounts of heat. They control the rate of heat loss using their feathers, which may be held close

30.27 The Evolution of Birds
(*a*) Some paleontologists believe that modern birds are descendants of a small group that survived the mass extinction at the end of the Cretaceous. (*b*) Others believe that many lineages of birds persisted during the transition from the Cretaceous to the Tertiary.

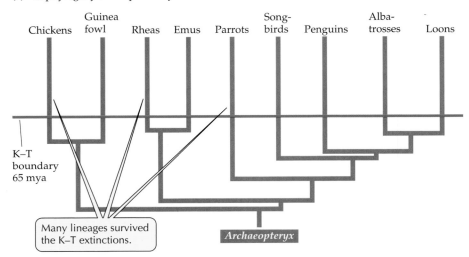

TWO COMPETING VIEWS

(*a*) A phylogeny based on the fossil record

Explosive radiation of all other living bird lineages took place.

Many lineages became extinct at the K–T boundary.

Opposite birds (many lineages)

Other lineages

Cretaceous–Tertiary (K–T) boundary 65 mya

Transitional shorebirds survived the K–T extinctions.

Archaeopteryx

(*b*) A phylogeny based primarily on molecular data

Chickens | Guinea fowl | Rheas | Emus | Parrots | Songbirds | Penguins | Albatrosses | Loons

K–T boundary 65 mya

Many lineages survived the K–T extinctions.

Archaeopteryx

to the body or elevated to alter the amount of air trapped as insulation. A bird's metabolic rate is so high that it consumes about eight times the amount of energy per day as a lizard of the same weight. The brain of a bird is relatively large in proportion to its body size, primarily because the cerebellum, the center of sight and muscular coordination, is enlarged. The beaks of modern birds lack teeth.

Most birds lay their eggs in a nest, where they are incubated by the body heat of an adult. Because birds have high body temperatures, the eggs of most species hatch in about 2 weeks. Nestlings of many species hatch at a relatively helpless stage and are fed for some time by their parents. Young of other species can feed themselves shortly after hatching. Adults of all species attend their offspring for some period of time, warning them of and protecting them from predators, guiding them to good foraging places, protecting them from bad weather, and feeding them.

As a group, birds eat almost all types of animal and plant material. A few aquatic species have bills modified for filtering small food particles from the water. In terrestrial environments, insects are the most important dietary item for birds. In addition, birds eat fruits and seeds, nectar and pollen, leaves and buds, carrion, and other vertebrates. Birds are major predators of flying insects during the day, and some species exploit that food source at night. By eating the fruits and seeds of vascular plants, birds serve as major agents of seed dispersal.

As adults, birds range in size from the 2-g bee hummingbird of the West Indies to 150-kg ostriches. Some flightless birds of Madagascar and New Zealand were even larger, but they were exterminated by early humans when the humans first reached those islands. Although there are about 8,600 species of living birds, more species than in any other vertebrate group except fishes, birds are less diverse structurally than are other vertebrates, probably because of the constraints imposed by flying (Figure 30.28).

The Tertiary: Mammals Diversify

Mammals (the chordate class Mammalia) appeared in the early part of the Mesozoic era (about 225 mya), branching from the now extinct, mammal-like reptiles known as therapsids. Small mammals coexisted with reptiles and dinosaurs for at least 150 million years, but when the large reptiles and dinosaurs disappeared during the mass extinction at the close of the Mesozoic era, mammals increased dramatically in number, diversity, and size.

(a) *Aptenodytes patagonicus*

(b) *Eclectus roratus*

(c) *Cardinalis cardinalis*

(d) *Cathartes aura*

30.28 A Diversity of Birds (a) Penguins such as these king penguins are adapted to the harsh Antarctic environment. They are expert swimmers, although they do not fly. (b) Parrots are a diverse group of birds, especially in the tropics of Asia and the Pacific islands. This eclectus parrot is native to Australia and New Guinea. (c) The perching birds are found worldwide. This female northern cardinal is a common example from eastern North America. (d) The turkey vulture, the most widespread New World vulture, glides on updrafts in search of carrion.

Mammals acquired simplified skeletons

Skeletal simplification accompanied the evolution of mammals from their therapsid ancestors. During mammalian evolution, most lower-jaw bones became incorporated into the middle ear, leaving a single bone in the lower jaw, and the number of bones in the skull decreased. The bulk of both the limbs and the bony girdles from which they are suspended was reduced, and the limbs became oriented beneath the body, as

they are in dinosaurs and birds, rather than poking out to the side and then down, as in reptiles. Fossils of later therapsids suggest that their legs also were positioned underneath the body. Thus the early mammals represent a continuation of changes that were already under way (Figure 30.29).

The skeletal features we have been discussing are readily fossilized, but the important soft parts of mammals are seldom preserved in fossils. Mammalian features such as mammary glands, sweat glands, hair, and a four-chambered heart may have evolved among the later therapsids, but the existing record does not tell us when this happened.

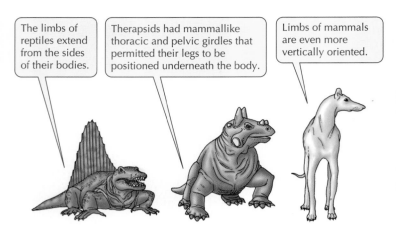

> The limbs of reptiles extend from the sides of their bodies.

> Therapsids had mammallike thoracic and pelvic girdles that permitted their legs to be positioned underneath the body.

> Limbs of mammals are even more vertically oriented.

30.29 Reptile to Mammal   The legs of therapsid reptiles gradually became positioned under the body; early mammals represent a continuation of this evolutionary trend.

Mammals are unique among animals in that they suckle their young with a nutritive fluid (milk) secreted by mammary glands. Mammalian eggs are fertilized within the female's body, and the embryos develop somewhat within the uterus before being born. In addition, mammals have a protective and insulating covering of hair, which is luxuriant in some species but has been largely lost in adult whales and dolphins in favor of thick layers of insulating fat (blubber).

Mammals have far fewer, but more highly differentiated, teeth than those of reptiles. Differences in the number, type, and arrangement of teeth in mammals reflect their varied diets. Mammals range in size from tiny shrews weighing only about 2 g to the blue whale, which measures up to 31 m long and weighs up to 160,000 kg—the largest animal ever to live on Earth.

Mammals have varied reproductive patterns

The approximately 4,000 species of living mammals are placed into three major groups: prototherians, marsupials, and eutherians. The subclass Prototheria contains a single order, the Monotremata, represented by two families and a total of three species, which are found only in Australia and New Guinea. These mammals, the duck-billed platypus and spiny anteaters, or echidnas, differ from other mammals in that they lay eggs and have some reptilelike anatomical features (Figure 30.30). Monotremes nurse their young on milk, but they have no nipples on their mammary glands; rather the milk simply oozes out and is lapped off the fur by the offspring.

The other two groups of mammals are members of the subclass Theria. Females of one group, the order Marsupialia, which contains about 240 species, have a ventral pouch in which they carry and feed their

Ornithorhyncus anatinus

30.30 A Monotreme   The Australian duck-billed platypus is one of three surviving species of monotremes.

young (Figure 30.31). Gestation (pregnancy) in marsupials is short; the young are born tiny but with well-developed forelimbs, with which they climb to the pouch. Once her offspring has left the uterus, a female marsupial may become sexually receptive again. She can then

(a) Macropus rufus   *(b) Caluromys phicander*

30.31 Marsupials   (a) Australia's kangaroos are perhaps the most familiar marsupials. This female red kangaroo is carrying her offspring in her distinctive pouch. (b) The wooly opossum is a South American marsupial from Guyana.

(a) *Citellus parryi*

(c) *Felis onca*

(b) *Eptesicus fuscus*

30.32 Eutherian Diversity (*a*) The Arctic ground squirrel is one of many species of small, diurnal rodents of western North America. (*b*) With their powers of echolocation (see Chapter 42), many bats can locate and capture prey even in complete darkness. This big brown bat is about to capture a large moth. (*c*) Cats ambush their prey and capture them after short chases. The jaguar, whose coloration camouflages it in the foliage, is the largest cat in the Americas. (*d*) Large hoofed mammals are important herbivores over much of Earth. This caribou bull is grazing by himself, although caribou are often seen in huge herds (see Figure 51.15*b*).

(d) *Rangifer tardanus*

carry a fertilized egg capable of initiating development and replacing the offspring in the pouch should something happen to it. The marsupial mode of reproduction functions well in varying arid environments, where adults must travel long distances to find food and where droughts may cause young offspring to die.

At one time marsupials were widely distributed on Earth, but today most species are restricted to the Australian region, with a modest representation in South America (Figure 30.31*a*). One species, the Virginia opossum, is widely distributed in the United States. Marsupials radiated into terrestrial herbivores, insectivores, and carnivores, but no species are marine or can fly. The largest living marsupial is the red kangaroo of Australia (Figure 30.31*b*), which weighs up to 90 kg, but much larger marsupials existed in Australia until quite recent times. These large marsupials were probably exterminated by humans soon after they reached Australia about 50,000 years ago.

Most living mammals are eutherians (sometimes called placentals, but this is not a good name because some marsupials also have placentas). Eutherians are more highly developed at birth than are marsupials, and no external pouch houses them after birth. The nearly 4,000 species of eutherians are placed into 16 major groups (Figure 30.32), the largest of which is the rodents, with about 1,700 species. The next-largest group, the bats, has about 850 species, followed by the insectivores (moles and shrews), with slightly more than 400 species.

Several lineages of terrestrial mammals subsequently colonized marine environments to become whales, dolphins, seals, and sea lions. Today the largest mammals are marine, but some terrestrial mammals, such as elephants and rhinoceroses, weigh more than several thousand kilograms.

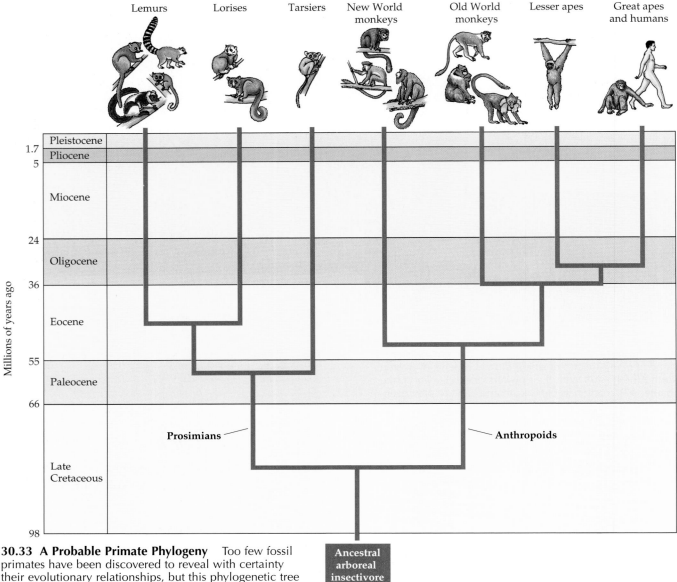

30.33 A Probable Primate Phylogeny   Too few fossil primates have been discovered to reveal with certainty their evolutionary relationships, but this phylogenetic tree is consistent with existing evidence.

Eutherians are extremely varied in form and ecology. They are, or were until recently, the most important grazers and browsers in most terrestrial ecosystems. Grazing and browsing by eutherian mammals has been an evolutionary force intense enough to select for the spines, tough leaves, and difficult-to-eat growth forms found on many plants.

Humans Evolve: Earth's Face Is Changed

The primate lineage, to which humans belong, has undergone extensive recent evolutionary radiation. Primates probably descended from small, arboreal (tree-inhabiting) insectivores sometime during the Cretaceous period. The major traits that distinguish primates from other mammals are all adaptations to arboreal life. They include dexterous hands with opposable thumbs that can grasp branches and manipulate food, nails rather than claws, eyes on the front of the face that provide good depth perception, and very small litters (usually one) of offspring that receive extended parental care.

The primate lineage split into two main branches—prosimians and anthropoids—early in its evolutionary history (Figure 30.33). The prosimians—lemurs, tarsiers, pottos, and lorises (Figure 30.34)—once lived on all continents, but today they are restricted to Africa, tropical Asia, and Madagascar. All mainland species are arboreal and nocturnal, but on Madagascar, the site of a remarkable prosimian radiation, there are also diurnal and terrestrial species. Until the recent arrival of *Homo sapiens*, there were no other primates on Madagascar.

(a) *Propithecus verreauxi*

(b) *Tarsius syrichta*

(c) *Nycticebus coucang*

30.34 Prosimians (a) The sifaka lemur is one of many lemur species of Madagascar, where it is part of a unique assemblage of plants and animals. (b) An inhabitant of the rainforests of the Philippines, the tarsier seems other-worldly to our eyes. (c) The slow loris lives in southeast Asia.

The anthropoids—monkeys, apes, and humans (see Figure 30.33)—evolved from an early primate stock about 55 mya in Africa or Asia. New World monkeys have been evolving separately from Old World monkeys long enough that they could have reached South America from Africa when those two continents were still close to one another. Perhaps because tropical America has been heavily forested for a long time, all New World monkeys are arboreal (Figure 30.35a). Many of them have long, prehensile tails with which

(a) *Leontopithecus rosalia*

30.35 Monkeys (a) Golden lion tamarins are endangered New World monkeys living in coastal Brazilian rainforests. (b) Many Old World species, such as these Barbary macaques, live and travel in social groups. Here two members of the group groom each other.

(b) *Macaca sylvanus*

(a) Hylobates lar

(b) Pan troglodytes

(c) Pongo pygmaeus

(d) Gorilla gorilla

30.36 Apes (*a*) The gibbons are the smallest of the apes. This common gibbon is found in Asia, from India to Borneo. (*b*) A mother chimpanzee carries her offspring on her back. Her "knuckle walk" position is characteristic of apes when they move on the ground. (*c*) Intelligent and endangered, orangutans face massive habitat destruction in their native Indonesia. (*d*) This mother also demonstrates the knuckle walk of lowland gorillas in Africa.

they can grasp branches. Many Old World primates are arboreal, but a few species are terrestrial, some of which, such as baboons and macaques, live and travel in large groups (Figure 30.35*b*). No Old World primates have prehensile tails.

About 35 mya the lineage leading to modern apes separated from other Old World primates. The first apes were arboreal, but some species came to live in drier habitats with scattered trees, where they obtained most of their food from the ground. Jawbones of apes that lived between 15 and 8 mya have been found in Africa, the Near East, and Asia. These extinct apes, genus *Ramapithecus*, have features suggesting that they were the beginning of the lineage leading to humans.

Like us, ramapithecines had short muzzles and small canines. Their chewing teeth were worn down flat, indicating that they chewed from side to side as we do, rather than up and down as chimpanzees and gorillas do. The four living genera of apes—gorillas (*Gorilla*), chimpanzees (*Pan*), orangutans (*Pongo*), and gibbons (*Hylobates*)—are restricted to tropical Africa and Asia (Figure 30.36).

Human ancestors descended to the ground

Some experts believe that *Ramapethicus* was the direct ancestor of both modern apes and humans; others believe that *Ramapithecus* is a member of only the hominid (human) lineage. But there is no disagreement that

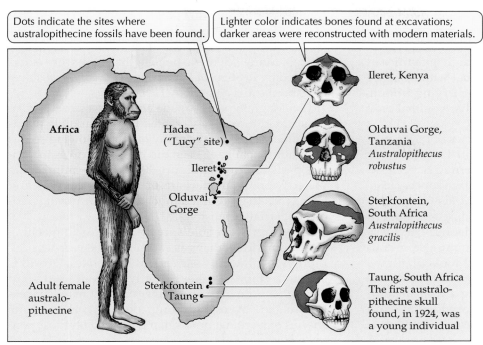

Dots indicate the sites where australopithecine fossils have been found.

Lighter color indicates bones found at excavations; darker areas were reconstructed with modern materials.

Africa

Hadar ("Lucy" site)

Ileret

Olduvai Gorge

Adult female australo-pithecine

Sterkfontein
Taung

Ileret, Kenya

Olduvai Gorge, Tanzania
Australopithecus robustus

Sterkfontein, South Africa
Australopithecus gracilis

Taung, South Africa
The first australo-pithecine skull found, in 1924, was a young individual

30.37 Australopithecine Fossils Few fossilized remains are complete, but skull shapes can be reconstructed accurately.

members of another lineage, the australopithecines, are direct human ancestors. The australopithecines had distinct morphological adaptations for **bipedalism**, locomotion in which the body is held erect and moved exclusively by movements of the hind legs. Bipedal locomotion frees the hands to manipulate objects and to carry them while walking. It also elevates the eyes, enabling the animal to see over tall vegetation to spot predators and prey. Both advantages were probably important for early australopithecines.

The first australopithecine skull was found in South Africa in 1924; since then, other fragments have been found at various sites in Africa (Figure 30.37). The oldest and most complete fossil skeleton of an australopithecine, approximately 3.5 million years old, was discovered in Ethiopia in 1974. That individual, a young female known to the world as Lucy, attracted a great deal of attention because her remains were so complete and well preserved. Lucy has been assigned to the species *Australopithecus afarensis*, the most likely ancestor of later hominids. All the evidence from different parts of her skeleton suggest that Lucy was only about 1 m tall and walked upright.

From *Australopithecus afarensis* ancestors, several species of australopithecines evolved. Several million years ago, two distinct types of australopithecines lived together over much of eastern Africa. The more robust type (about 40 kg) is represented by at least two species, both of which died out suddenly about 1.5

mya. The smaller (25–30 kg), more slender *A. africanus* is much rarer as a fossil, suggesting that it was less common than the other species.

Because they were less agile, members of the robust species probably stayed relatively close to trees, to which they retreated at night and when predators were near. Members of the smaller *A. africanus* were able to run faster and, because they were smaller, needed less food per day to survive. They probably lived in more open, drier savannas where food was less abundant than in the moister areas inhabited by the more robust species.

Humans arose from australopithecine ancestors

Many experts believe that a population of *Australopithecus africanus* or a similar species gave rise to the genus *Homo* about 2.5 mya. Early members of *Homo* lived contemporaneously with australopithecines for perhaps half a million years. Two major changes accompanied the evolution of *Homo* from *Australopithecus*: an increase in body size and a doubling of brain size.

The oldest fossil remains of members of the genus *Homo*, named *Homo habilis*, were discovered in the Olduvai Gorge, Tanzania, and are estimated to be 2 million years old. Other fossils of *Homo habilis* have been found in Kenya, Ethiopia, and South Africa, indicating that the species had a wide range in Africa. Tools used by these early hominids to obtain food were found with the fossils.

Homo habilis lived in relatively dry areas where, for much of the year, the main food reserves are subterranean roots, bulbs, and tubers. To exploit these food resources an animal must dig into hard, dry soils, something that cannot be done with an unaided hand. However, roots can be harvested in large quantities in a relatively short time by an individual with a simple digging tool. *Homo habilis* females carrying infants could have done so, freeing males to hunt animal prey to provide the proteins that roots lack.

The only other known extinct species of our genus, *Homo erectus*, evolved in Africa about 1.6 mya. Soon thereafter it had spread as far as eastern Asia. As it expanded its range and increased in abundance, *Homo erectus* may have exterminated *Homo habilis*.

Members of *Homo erectus* were as large as modern people, but their bones were somewhat heavier. *Homo erectus* used fire for cooking and for hunting large animals, and made characteristic stone tools that have been found in many parts of the Old World. These tools were probably used for digging, capturing animals, cleaning and cutting meat, scraping hides, and cutting wood. Although *Homo erectus* survived in Eurasia until about 250,000 years ago, it was replaced in tropical regions by our species, *Homo sapiens*, about half a million years ago.

Brains steadily become larger

The trends that accompanied the transition from *Australopithecus* to *Homo erectus* continued during the evolution of our own species. The earliest humans (*Homo sapiens*) had larger brains than did members of the earlier species of *Homo*, a change that was probably favored by an increasingly complex social life. The ability of group members to communicate with one another was valuable for cooperative hunting and gathering, for sharing information about the location and use of food sources, and for improving one's status in the complex social interactions that must have characterized those societies, just as they do ours today.

Several types of *Homo sapiens* existed during the mid-Pleistocene epoch, from about 1.5 million to about 300,000 years ago. All were skilled hunters of large mammals, but plants continued to be an important component of their diet. During this period two other distinctly human traits emerged: rituals and a concept of life after death. Deceased individuals were buried with tools and clothing in their graves, presumably for their existence in the next world.

One type of *Homo sapiens*, generally known as Neanderthal because it was first discovered in the Neander Valley in Germany, was widespread in Europe and Asia between about 75,000 and 30,000 years ago. Neanderthals were short, stocky, and powerfully built humans whose massive skulls housed brains somewhat larger than our own. They manufactured a variety of tools and hunted large mammals, which they probably ambushed and subdued in close

combat. For a short time, their range overlapped that of Cro-Magnon people, a more modern form of *Homo sapiens*, but then the Neanderthals abruptly disappeared. Many scientists believe that they were exterminated by the Cro-Magnons, just as *Homo habilis* may have been exterminated by *Homo erectus*.

Cro-Magnon people made and used a variety of sophisticated tools. They created the remarkable paintings of large mammals, many of them showing scenes of hunting, that have been discovered in caves in various parts of Europe (Figure 30.38a). The animals depicted were characteristic of the cold steppes and grasslands that occupied much of Europe during periods of glacial expansion.

Cro-Magnon people spread across Asia, reaching North America perhaps as early as 20,000 years ago, although the date of their arrival in the New World is still uncertain. As they rapidly spread southward through North and South America, they may have exterminated, by overhunting, populations of many species of large mammals that had lived on those continents.

Humans evolve language and culture

As our ancestors evolved larger brains, they also increased their behavioral capabilities, especially the capacity for language. Most nonhuman animal communication consists of a limited number of signals, which pertain mostly to immediate circumstances and are associated with changed emotional states induced by those circumstances. (The language of honeybees is unusual in that it contains a symbolic component referring to events distant in both space and time; see Chapter 49.)

Human language is far richer in its symbolic character than are any other animal vocalizations. Our words can refer to past and future times and to distant places. We are capable of learning thousands of words, many of them referring to abstract concepts. We can rearrange words to form sentences with complex meanings.

The expanded mental abilities of humans are largely reponsible for the development of **culture**, the process by which knowledge and traditions are passed along from one generation to another by teaching and observation. Culture can change rapidly because genetic changes are not necessary for a cultural trait to spread through a population. The primary disadvantage of culture is that its norms must be taught to each generation.

The tools and other implements associated with human fossils, as well as the cave paintings that early humans created, reveal cultural traditions. Cultural learning greatly facilitated the spread of domestic plants and animals and the resulting conversion from societies in which food was obtained by hunting and

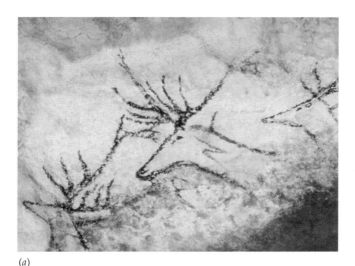

(a)

(b)

(c)

30.38 Hunting, Pastoralism, and Agriculture (a) Lascaux Cave in France provides striking examples of the artistry of Cro-Magnons. Cave walls often depict animals of the glacial steppes; these paintings may have served as part of rituals to increase success during hunts. (b) Pastoralism displaced hunting in many societies. This hut is the temporary dry-season quarters of a herder in Burkina Faso, Africa. (c) Agricultural development has totally transformed the landscape in these hills above Port-au-Prince, Haiti.

gathering to societies in which **pastoralism** (the raising of domestic animals to provide food and other products) and **agriculture** dominated (Figure 30.38*b* and *c*). The agricultural revolution, in turn, led to an increasingly sedentary life, the development of cities, greatly expanded food supplies, rapid growth of the human population, and the appearance of occupational specializations, such as artisans and shamans.

Agriculture developed in the Middle East approximately 11,000 years ago. From there it spread rapidly northwestward across Europe, finally reaching the British Isles about 4,000 years ago. The plants that these early agriculturalists domesticated were cereal grains such as wheat and barley; legumes—beans, lentils, and peas; and woody plant crops such as grapes and olives. Other plants, such as rye, cabbage, celery, and carrots, were domesticated as agriculturalists spread across Europe. Cattle, sheep, goats, horses, dogs, and cats were their most important domesticated animals.

Agriculture developed independently in eastern Asia, contributing to our modern diets soybeans, rice, citrus fruits, and mangoes, and pigs and chickens. There was some exchange, even in early times, among agricultural centers in the Old World, but when people crossed the cold and barren Bering land bridge into the New World, they apparently brought no domesticated plants with them. These people subsequently developed rich and varied agricultural systems based on corn, tomatoes, kidney and lima beans, peanuts, potatoes, chili peppers, and squashes.

The human population has had three major growth increases. The first, stimulated by toolmaking, lasted about a million years. During that time, human numbers increased to about 5 million. During the second surge, which followed the domestication of plants and animals and the invention of agriculture, the human population increased to about 500 million people within 8,000 years. We are currently in the middle of the third great surge, which was triggered by the Industrial Revolution (Figure 30.39).

The current human population is nearly 6 billion and is projected to increase to more than 11 billion within 50 years. The first two population surges were followed by periods of relative stability. Whether the current one will follow the same path, at what size the population might level off, and the environmental consequences of continued growth are questions that are fiercely debated (see Chapter 55).

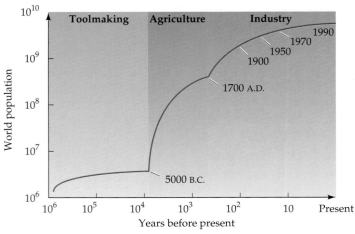

30.39 Human Population Surges The human population surged after the invention of tools, the domestication of plants and animals, and the Industrial Revolution. Note that the axes are scaled logarithmyically.

which the animal is stable (Figure 30.40b). Similar planktonic larval stages evolved in marine members of many protostomate and deuterostomate phyla; all of these fed on tiny planktonic organisms in the open water.

Both protostomes and deuterostomes colonized the land—the former via beaches, the latter via fresh water—but the consequences were very different. The jointed external skeletons of arthropods, although they provide excellent support and protection in air, are not suitable for large animals. In addition, an arthropod must shed its skin and become temporarily vulnerable in order to grow. But the internal, jointed skeletons of vertebrates permit growth to large size without temporary vulnerable stages. Consequently, although arthropods are abundant and diverse on land, they never evolved into large animals.

Terrestrial lineages of vertebrates recolonized aquatic environments several times. Suspension feed-

Deuterostomes and Protostomes: Shared Evolutionary Themes

Table 30.2 summarizes the traits of all animal phyla. Deuterostome evolution paralleled protostome evolution in several important ways. Both lineages exploited the abundant food supplies buried in soft marine substrata, attached to rocks, or suspended in water columns. Because of the ease with which water can be moved, many groups of both lineages developed elaborate structures for moving water and extracting prey from it (Figure 30.40a).

In both groups, a coelomic cavity evolved and subsequently became divided into compartments that allowed better control of body shape and movement. Both groups evolved locomotor abilities. Some members of both groups evolved mechanisms for controlling their buoyancy in water, using gas-filled internal spaces whose contents can be adjusted to control the depth at

30.40 Parallel Evolution Devices for filtering food from the water (a) and maintaining buoyancy in the water (b) evolved in both protostomate and deuterostomate lineages.

Protostomate animals

(a) Filtering device

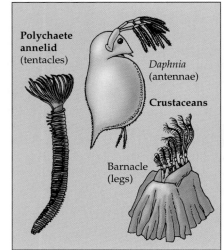

Polychaete annelid (tentacles)

Daphnia (antennae)

Crustaceans

Barnacle (legs)

Deuterostomate animals

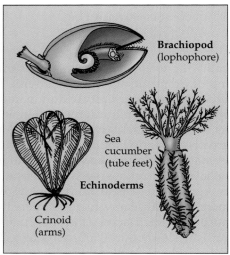

Brachiopod (lophophore)

Sea cucumber (tube feet)

Echinoderms

Crinoid (arms)

(b) Buoyancy control

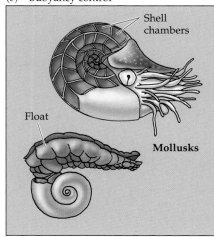

Shell chambers

Float

Mollusks

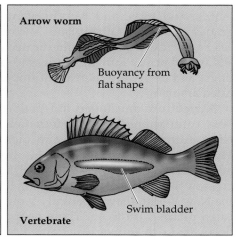

Arrow worm

Buoyancy from flat shape

Swim bladder

Vertebrate

TABLE 30.2   Summary of Living Members of the Kingdom Animalia[a]

PHYLUM	NUMBER OF LIVING SPECIES DESCRIBED	SUBGROUPS
Porifera: Sponges	10,000	
Cnidaria: Cnidarians	10,000	Hydrozoa: Hydrozoans Scyphozoa: Jellyfish Anthozoa: Corals, sea anemones
Ctenophora: Comb jellies	100	
Protostomes		
Rotifera: Rotifers	1,800	
Nematoda: Roundworms	20,000	
Platyhelminthes: Flatworms	25,000	Turbellaria: Free-living flatworms Trematoda: Flukes (all parasitic) Cestoda: Tapeworms (all parasitic) Monogenea: Ectoparasites of fish
Nemertea: Ribbon worms	900	
Pogonophora: Pogonophorans	145	
Annelida: Segmented worms	15,000	Polychaeta: Polychaetes (all marine) Oligochaeta: Earthworms, freshwater worms Hirudinea: Leeches
Mollusca: Mollusks	100,000	Monoplacophora: Monoplacophorans Polyplacophora: Chitons Bivalvia: Clams, oysters, mussels Gastropoda: Snails, slugs, limpets Cephalopoda: Squids, octopuses, nautiloids
Chelicerata: Chelicerates	63,000	Merostomata: Horseshoe crabs Arachnida: Scorpions, harvestmen, spiders, mites, ticks
Crustacea: Crustaceans	40,000	Crabs, shrimp, lobsters, barnacles, copepods
Uniramia: Unirames	1,500,000	Myriapoda: Centipedes, millipedes Onychophora: Onychophorans Tardigrada: Water bears Insecta: Insects
Deuterostomes		
Phoronida: Phoronids	15	
Brachiopoda: Lampshells	350	More than 26,000 fossil species described
Bryozoa: Moss animals	4,000	
Chaetognatha: Arrow worms	100	
Echinodermata: Echinoderms	7,000	Crinoidea: Sea lilies, feather stars
Pterobranchia: Pterobranchs		Asteroidea: Sea stars Ophiuroidea: Brittle stars Concentricycloidea: Sea daisies Echinoidea: Sea urchins Holothuroidea: Sea cucumbers
Hemichordata: Acorn worms	100	
Chordata: Chordates	40,000	Urochordata: Tunicates Cephalochordata: Lancelets Agnatha: Hagfishes, lampreys Chondrichthyes: Cartilaginous fishes Osteichthyes: Bony fishes Amphibia: Amphibians Reptilia: Reptiles Archosauria: Dinosaurs, crocodilians, birds Mammalia: Mammals

[a]A few small phyla are not included.

ing re-evolved in several of these lineages. The largest living mammals, the baleen whales (the toothless whales, including blue whales, humpback whales, and right whales), feed on relatively small prey that they extract from the water with large plates in their mouth.

Unlike the oceans, where the dominant photosynthetic plants are unicellular algae, most photosynthesis on land is carried out by vascular plants. Dominant plants in most terrestrial environments are large, complex organisms whose tissues are difficult to digest. Herbivores must ingest large quantities of fibers and defensive chemicals along with the energy-rich molecules they need. Because larger animals can exist on food of poorer quality than small animals can, a steady increase in body size is a common pattern in herbivore evolution.

This pattern is strikingly illustrated by the evolution of reptiles and the later evolution of mammalian herbivores. The evolution of large herbivores, in turn, favored the evolution of larger carnivores able to attack and overpower them. This evolutionary trend may have come to a temporary halt because of the invention of weapons by a moderately sized, omnivorous primate—*Homo sapiens*.

Summary of "Deuterostomate Animals: Evolution of Larger Brains and Complex Behaviors"

• There are fewer lineages and fewer species of deuterostomes than of protostomes, but as members of the lineage we have a special interest in its other members. **Review Figure 30.1**

Three Body Parts: An Ancient Body Plan

• Members of one deuterostomate lineage have bodies divided into three segments. They evolved a new structure—the lophophore—for extracting prey from the water. **Review Figure 30.2**
• Four lophophorate phyla survive today: Phoronida, Brachiopoda, Bryozoa, and Pterobranchia. Nearly all the members of these phyla are marine, but a few live in fresh water.
• Arrow worms, though divided into three body parts—head, trunk, and tail—lack a lophophore.

Echinoderms: Complex Radial Symmetry

• Echinoderms are deuterostomes that have a radially symmetrical body plan, a unique water vascular system, and a calcified internal skeleton. **Review Figure 30.8a**
• Nearly all living species of echinoderms have a bilaterally symmetrical, ciliated larva that feeds for a while as a planktonic organism. **Review Figure 30.8c**
• Of the 23 echinoderm lineages that have been described, only 6 groups survive. Some groups of echinoderms have arms, but others do not.

Modified Lophophores: New Ways of Feeding

• Evolution among acorn worms (phylum Hemichordata) and chordates (phylum Chordata) led to new ways of capturing and handling food.
• In the lineage leading to acorn worms, the lophophore was lost, but a proboscis grew larger and became a digging organ.
• Members of another lineage, the chordates (phylum Chordata), lost both the lophophore and the proboscis, replacing them with enlarged pharyngeal slits as a feeding device. They also evolved a dorsal supporting rod, the notochord.
• Tunicates (the chordate subphylum Urochordata) are sessile as adults and filter prey from seawater with large pharyngeal baskets. Only their larvae have notochords.
• Lancelets (the chordate subphylum Cephalochordata) are small, fishlike animals that live partly buried in soft marine sediments and extract small prey from the water with pharyngeal baskets.

The Origin of the Vertebrates: Sucking Mud

• Vertebrates (the chordate subphylum Vertebrata) evolved jointed internal skeletons that enabled them to swim rapidly. Early vertebrates lacked jaws and fed by filtering small animals from mud. **Review Figures 30.13, 30.14**
• Jaws, which evolved from anterior gill arches, enabled their possessors to grasp and chew their prey, expanding food sources and nutrition. Jawed fishes rapidly became dominant animals in both marine and fresh waters. **Review Figure 30.16**
• Cartilaginous and bony fishes evolved two pairs of unjointed fins, with which they control their swimming movements and stabilize themselves in the water.
• Bony fishes evolved lunglike sacs that help keep them suspended in open water. Bony fishes come in a wide variety of sizes and shapes, and many species have complex social systems.

Colonizing the Land

• One lineage of fishes evolved jointed fins that enabled them to move more effectively over land to find new bodies of water. One genus of these lobe-finned fishes—*Latimeria*—survives today.
• One group of lobe-fins gave rise to the amphibians, the first terrestrial vertebrates. **Review Figure 30.20**
• About 4,500 species of amphibians live today. They belong to three orders: caecilians, frogs and toads, and salamanders.
• Most amphibians live in water at some time in their lives, and their eggs must remain moist. **Review Figure 30.22**
• Amniotes, the common ancestors of reptiles and mammals, evolved eggs with shells impermeable to water and thus became the first vertebrates to be independent of water for breeding. **Review Figure 30.23**
• Modern reptiles are members of three lineages: turtles and tortoises, tuataras, and snakes and lizards.

Surviving Dinosaur Lineages

• One amniote lineage, the thecodonts, gave rise to the crocodilians.
• Another group of thecodonts led to dinosaurs, the dominant terrestrial reptiles for millions of years. The nearly 9,000 species of birds are the descendants of a third thecodont lineage. Birds are characterized by their feathers, high metabolic rates, and parental care.

The Tertiary: Mammals Diversify

- Mammals evolved during the Mesozoic era from now extinct therapsids. **Review Figure 30.29**
- The eggs of mammals are fertilized within the bodies of females, and mammalian embryos develop for some time before being born. Mammals are unique in that they suckle their young with milk secreted by mammary glands.
- The few species of monotremes lay eggs, but all other mammals give birth to live young.
- Marsupials give birth to tiny young that are housed and nursed in a pouch. Most marsupial species live in Australia.
- Eutherian mammals, which give birth to relatively well developed offspring, are placed in 16 major groups.

Humans Evolve: Earth's Face Is Changed

- The primates, to which humans belong, split into two major lineages, one leading to the prosimians (lemurs, lorises, and tarsiers), the other leading to the anthropoids (monkeys, apes, and humans). **Review Figure 30.33**
- Humans evolved in Africa from terrestrial, bipedal, australopithecine ancestors. **Review Figure 30.37**
- Early humans evolved large brains, language, and culture. They manufactured and used tools, developed rituals, and domesticated plants and animals. In combination, these traits enabled humans to increase greatly in number and to transform the face of Earth.
- The human population has increased greatly three times. We are currently in the middle of the third population surge. When and how it will end is hotly debated. **Review Figure 30.39**

Deuterostomes and Protostomes: Shared Evolutionary Themes

- Devices for extracting prey from water, for controlling buoyancy, and for moving rapidly evolved many times during the evolution of both protostomes and deuterostomes. Members of both lineages colonized terrestrial environments. **Review Figure 30.40**

Self-Quiz

1. Which of the following are deuterostomate phyla with a three-part body plan?
   a. Rotifera, Phoronida, Bryozoa, and Brachiopoda
   b. Phoronida, Bryozoa, Brachiopoda, and Hemichordata
   c. Phoronida, Bryozoa, Hemichordata, and Chordata
   d. Echinodermata, Bryozoa, Brachiopoda, and Chordata
   e. Phoronida, Bryozoa, Hemichordata, and Echinodermata

2. The structure used by brachiopods to capture food is a
   a. pharyngeal gill basket.
   b. proboscis.
   c. lophophore.
   d. mucous net.
   e. radula.

3. The water vascular system of echinoderms is a
   a. series of seawater channels derived by enlargement and extension of a coelomic cavity.
   b. series of seawater channels derived by enlargement and extension of the pharyngeal cavity.
   c. series of channels derived by enlargement and extension of a coelomic cavity, filled with coelomic fluid.
   d. series of channels derived by enlargement and extension of a coelomic cavity and filled with fresh water.
   e. series of channels that can be filled to different levels with water, enabling the animal to control its buoyancy.

4. The pharyngeal gill slits of chordates originally functioned as sites for
   a. uptake of oxygen only.
   b. release of carbon dioxide only.
   c. both uptake of oxygen and release of carbon dioxide.
   d. removal of small prey from the water.
   e. forcible expulsion of water to move the animal.

5. The key to the vertebrate body plan is
   a. a pharyngeal gill basket.
   b. a vertebral column to which internal organs are attached.
   c. a vertebral column to which two pairs of appendages are attached.
   d. a vertebral column to which a pharyngeal gill basket is attached.
   e. a pharyngeal gill basket and two pairs of appendages.

6. Which of the following fishes do *not* have a cartilaginous skeleton?
   a. Chimaeras
   b. Lungfishes
   c. Sharks
   d. Skates
   e. Rays

7. In most fishes, lunglike sacs evolved into
   a. pharyngeal gill slits.
   b. true lungs.
   c. coelomic cavities.
   d. swim bladders.
   e. none of the above

8. Most amphibians return to water to lay their eggs because
   a. water is isotonic to egg fluids.
   b. adults must be in water while they guard their eggs.
   c. there are fewer predators in water than on land.
   d. amphibians need water to produce their eggs.
   e. amphibian eggs quickly lose water and desiccate if their surroundings are dry.

9. The horny scales that cover the skin of reptiles prevent them from
   a. using their skin as an organ of gas exchange.
   b. sustaining high levels of metabolic activity.
   c. laying their eggs in water.
   d. flying.
   e. crawling into small spaces.

10. Which statement about bird feathers is *not* true?
    a. They are highly modified reptilian scales.
    b. They provide insulation for the body.
    c. They arise from well-defined tracts.
    d. They help birds fly.
    e. They are important sites of gas exchange.

11. Monotremes differ from other mammals in that they
    a. do not produce milk.
    b. lack body hairs.
    c. lay eggs.
    d. live in Australia.
    e. have a pouch in which the young are raised.

12. Bipedalism is believed to have evolved in the human lineage because
   a. bipedal locomotion is more efficient than quadrupedal locomotion.
   b. bipedal locomotion is more efficient than quadrupedal locomotion, and it frees the hands to manipulate objects.
   c. bipedal locomotion is less efficient than quadrupedal locomotion, but it frees the hands to manipulate objects.
   d. bipedal locomotion is less efficient than quadrupedal locomotion, but bipedal animals can run faster.
   e. bipedal locomotion is less efficient than quadrupedal locomotion, but natural selection does not act to improve efficiency.

Applying Concepts

1. In what animal phyla has the ability to fly evolved? How do structures used for flying differ among the animals in these phyla?

2. Extracting suspended food from the water column is a common mode of foraging among animals. Which groups contain species that extract prey from the air? Why is this mode of obtaining food so much less common than extracting prey from water?

3. Compare the buoyancy systems of cephalopods and fishes.

4. Why does possession of an external skeleton limit the size of a terrestrial animal more than possession of an internal skeleton does?

5. Large size both confers benefits and poses certain risks. What are these risks and benefits?

Readings

Alexander, R. M. 1975. *The Chordates.* Cambridge University Press, New York. A comprehensive and readable account of the biology of members of the phylum Chordata.

Bakker, R. T. 1975. "Dinosaur Renaissance," *Scientific American*, April. A discussion of the relationships between birds and dinosaurs, presenting evidence that dinosaurs were warm-blooded.

Bond, C. E. 1979. *Biology of Fishes.* Saunders, Philadelphia. A leading text for courses on ichthyology.

Carroll, R. C. 1987. *Vertebrate Paleontology and Evolution.* W. H. Freeman, San Francisco. A thorough account of the fascinating evolutionary history of the vertebrates.

Colbert, E. H. 1980. *Evolution of the Vertebrates: A History of the Backboned Animals through Time*, 3rd Edition. Wiley-Interscience, New York. A thoughtful discussion of the origins and evolutionary radiations of the vertebrate groups.

Diamond, J. 1992. *The Third Chimpanzee.* Harper Collins, New York. An engaging account of human evolution and how we acquired the unique traits that characterize our species.

Gill, F. B. 1990. *Ornithology.* W. H. Freeman, San Francisco. A technically accurate and readable introduction to bird biology for students at any level.

Langston, W., Jr. 1981. "Pterosaurs," *Scientific American*, February. An account of the largest animals ever to fly, including notes on evolutionary relationships among birds and reptiles.

Pough, F. H., J. B. Heiser, and W. N. McFarland. 1989. *Vertebrate Life.* Macmillan, New York. An excellent treatment of the evolution and ecology of the vertebrates.

Vaughan, T. A. 1978. *Mammalogy*, 2nd Edition. Saunders, Philadelphia. The leading textbook on mammals. Offers good coverage of both the orders of mammals and general aspects of mammalian biology.

Willson, M. F. 1984. *Vertebrate Natural History.* Saunders, Philadelphia. A thorough treatment of all aspects of the lives of vertebrates.

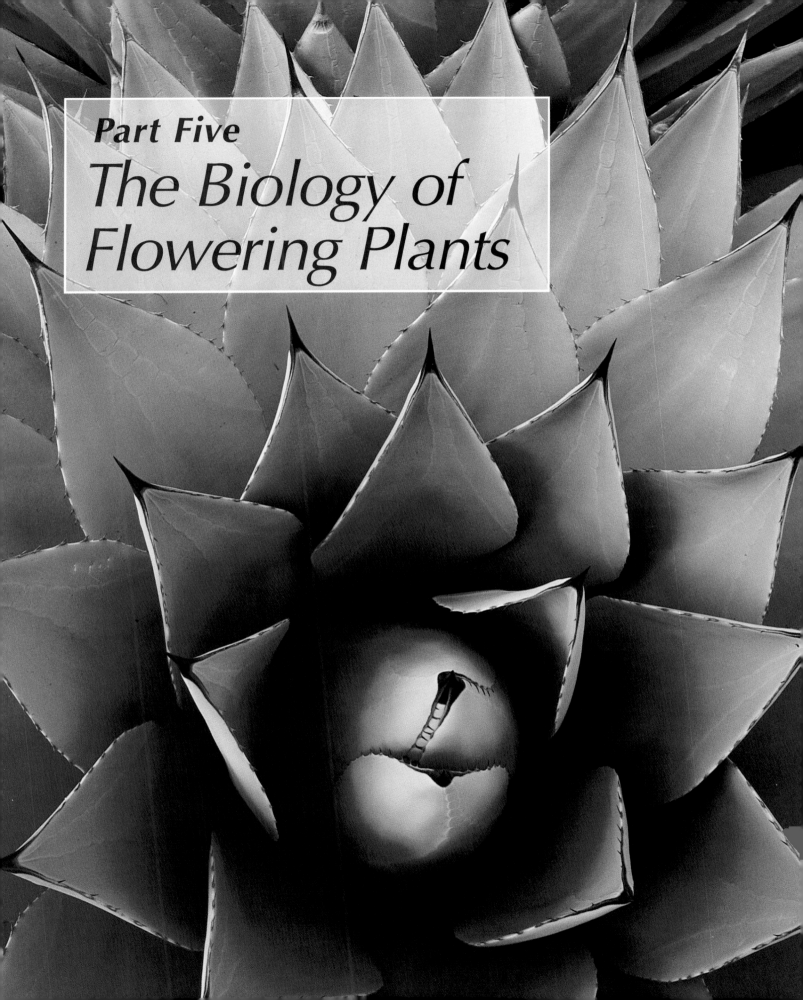

Part Five
The Biology of Flowering Plants

Chapter 31

The Flowering Plant Body

A Forest in Autumn
In the forested temperate latitudes, some deciduous trees can turn spectacular colors as winter approaches. Members of the maple family are particularly colorful.

*L*ong ago a young couple wandered in a forest far from town. Overcome by a romantic impulse, they wrote their initials on a scrap of paper, put it in a locket, and secured the locket to a young maple tree with a fine gold chain. At the time, the tree was only 4 meters tall; they wrapped the chain tightly around the trunk in the angle just above a slender branch a meter above the ground. The chain was long enough that the locket just touched the ground. By chance, *you* are the first person to pass the tree since that day. In the meantime the tree has grown to a height of 20 meters. Where do you find the locket? And where is the chain?

We won't answer these questions just yet. Instead, we'll let them introduce the main concerns of this chapter: the structure and patterns of growth of flowering plants, the angiosperms. Part Five as a whole (Chapters 31 through 36) deals with plant structure and function. There are many aspects of plant function to consider.

Plants—even the tallest trees—transport water from the soil to their tops, and they transport the products of photosynthesis from the leaves to the roots and other parts. Plants interact with the living and nonliving environment. They defend themselves against bacteria, fungi, animals, and other plants. Some plants can cope with hostile environments such as deserts, salt marshes, or sites polluted by mining and other human activities.

Although they have simpler nutritional needs than animals do, plants must nevertheless obtain nutrients—not only the raw materials of photosynthesis (carbon dioxide and water), but also mineral elements such as potassium and calcium. Plants respond to environmental cues as they develop. They produce

697

chemical signals that cause structural and functional changes appropriate to the environmental cues. Among the changes are ones that lead to reproduction and to the enhancement of photosynthesis.

Because we can understand plant function only in terms of the underlying structure, this chapter focuses on the structure of the flowering plant body. We'll consider contrasting body types in the flowering plants and then examine structure at the level of the organs, cells, tissues, and tissue systems. Then we'll see how meristems—organized groups of dividing cells—serve the growth of the plant body, both in length and, in woody plants, in width. The chapter concludes with a consideration of how flowering plants remain erect in a terrestrial environment, without water to buoy them.

Flowering Plants: Definition and Examples

Recall that flowering plants (phylum Angiospermae) are plants that have a vascular system and are characterized by double fertilization, by triploid endosperm, and by seeds enclosed in modified leaves; their xylem contains vessel elements and fibers. You might want to review the sections entitled "Most Present-Day Plants Have Vascular Tissue" and "The Angiosperms: Flowering Plants" in Chapter 27 before reading the rest of Chapter 31.

Flowering plants consist of a few important organs (roots, stems, leaves), the life-supporting functions of which can be understood in terms of their large-scale structure, as well as the microscopic structure of their component cells. These cells are grouped into tissues, and the tissues are grouped into organs. In this chapter we will present some anatomical features common to many flowering plants. As always in biology, it is important to remember that there are differences between organisms of the same species as well as between species.

Let's begin by looking at four important or familiar species: coconut palm, red maple, rice, and soybean. We have selected these four to illustrate both the similarities and the differences among plants.

Four examples show the similarities and differences among plants

COCONUT PALM.   In some cultures the coconut palm (*Cocos nucifera*; Figure 31.1) is called the tree of life because every aboveground part of the plant has value to humans. People use the stem—the trunk—of this tropical coastal lowland tree for lumber. They dry the sap from its trunk to use as a sugar, or they ferment it to drink. They use the leaves to thatch their homes and to make hats and baskets. They eat the apical bud at the top of the trunk in salads.

The coconut fruit also serves many purposes (see Figure 31.1*b*). The hard shell can be used as a container

(a)

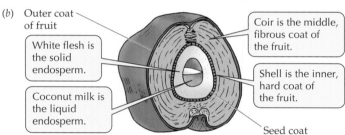

(b)

Outer coat of fruit

White flesh is the solid endosperm.

Coconut milk is the liquid endosperm.

Coir is the middle, fibrous coat of the fruit.

Shell is the inner, hard coat of the fruit.

Seed coat

31.1 Coconut Palm   (*a*) Coconut palms along the shore of the Indian Ocean in Sri Lanka. (*b*) A cross section of the coconut palm's fruit.

or burnt as fuel; the fibrous middle layer, or coir, of the fruit wall can be made into mats and rope. The seed of the coconut palm has both a liquid endosperm (coconut milk) and a solid endosperm (coconut meat). Because the refreshing and delicious milk contains no bacteria or other pathogens, it is a particularly important drink wherever the water is not fit for drinking.

Millions of people get most of their protein from coconut meat. Much coconut meat is dried and marketed as copra, from which coconut oil is pressed. Coconut oil is the most widely used vegetable oil in the world; it is used in the manufacture of a range of products from hydraulic brake fluid to synthetic rubber and, although nutritionally poor, as food. Ground copra serves as fertilizer and as food for livestock.

The trunk of a coconut palm differs in three basic ways from the trunks of many other familiar trees. The most striking difference is that it bears no branches, and all the leaves are borne in a cluster at the top of the trunk. Second, the coconut trunk tapers little from the base of the tree to the top—even the youngest part of the trunk is as thick as the base. Third, a cross section of the trunk reveals no annual rings.

Each coconut palm tree has separate male and female flowers; both are small and inconspicuous. The male flowers have six stamens. The leaves of the coconut palm are large and made up of numerous long, narrow leaflets, each having veins running parallel to one another.

RED MAPLE. One of the most familiar native trees in the eastern United States is the red, or scarlet, maple (*Acer rubrum*). Unlike the coconut palm, the red maple does not provide us with a great variety of useful products, but it enriches us by its beauty. Not only is it abundant in forests, but we admire it in parks and as a street tree growing as much as 10 to 30 m tall. We use its wood as lumber, although the sugar maple is a more important commercial source of maple wood; see the photo at the beginning of this chapter.

Microscopic examination of a very young maple stem reveals vascular bundles of water- and food-conducting cells, arranged in a cylindrical pattern (Figure 31.2). Like the coconut palm, the mature maple tree has a thick, massive trunk, but unlike the palm, the maple trunk is much thicker at the base than higher up the tree. A cross section of the maple trunk shows that the wood is made up of many annual rings. The roots, too, are woody.

The red maple leaf—the symbol of Canada—consists of a single blade with three to five lobes, and with veins that radiate from a single focal point. These leaves are among the brilliant contributors to the fall colors of eastern forests. The small, scarlet flowers

(a)

(b) Vascular bundles   (c) Red maple leaf   (d) Fruit

31.2 Red Maple   (a) The branches of a red maple tree reaching upward. (b) Vascular bundles in the stem of a young maple, diagrammed in cross section. (c) A leaf of red maple. (d) The characteristic winged fruit of the maple family.

have four sepals, four petals, eight stamens, and one pistil. The distinctive, winged fruit of the maple family contains two seeds.

RICE. More than half of the world's human population derives the bulk of its food energy from the seeds of a single plant: rice (*Oryza sativa*; Figure 31.3). Rice is particularly important in the diets of people in East and

(a)

(b)

31.3 Rice   (a) Terraced rice paddies in Bali, Indonesia. (b) A rice plant.

Southeast Asia, where it has been cultivated for nearly 5,000 years. People use rice straw in many ways, such as thatching for roofs, food and bedding for livestock, and clothing. Rice hulls also have many uses, ranging from fertilizer to fuel.

Rice is a fast-growing plant, yielding more than one crop per year. Some rice is fed to livestock, but most is eaten by humans. Processing by milling removes all parts of the seed except the endosperm. When milled for human consumption, rice is an incomplete food because milling removes the bran that contains B vitamins. Even unmilled rice is a poor source of protein; thus, it should be eaten with other, supplementary foods such as soybeans or fish. Most rice varieties are grown submerged in water for the bulk of the growing season.

The rice plant looks much like other cereal grain plants, such as wheat, barley, or oat. The leaves are long, narrow, flat, and more than half a meter long, with veins running parallel to one another along the length of the leaf. Rice stems do not thicken and become woody as do the stems of trees and shrubs. Rice flowers have six stamens and one ovary. The vascular bundles in the rice stem appear scattered, rather than in a ring as in the red maple stem.

SOYBEAN. Soybeans (*Glycine max*; Figure 31.4) were first grown in China thousands of years ago, but today the United States is the largest single producer. Soybeans are featured in many foods and sauces. They also yield a commercially important oil, used in the manufacture of adhesives, paints, inks, and plastics. After oil has been squeezed from the seeds, the residue may be fed to livestock or made into soy flour. Soybean stems may be used for straw.

The soybean plant stands from less than a meter to more than 2 m in height. Soybean leaves have three

31.4 Soybean This soybean plant is in bloom.

lobes, with veins radiating in a netlike pattern. The vascular bundles of the young soybean stem, like those of the red maple, are arranged in a cylindrical pattern. The flowers are small, either white or blue, and consist of five sepals, five petals, ten stamens, and one pistil. The fruit of a soybean, like that of other members of the pea family, is called a legume—think of a pea pod. Soybean plants tend to be drought-resistant because they have richly branching root systems, which often extend more than 1.5 m below the soil surface.

Flowering plants are divided into two classes: monocots and dicots

Comparison of the features of these four plants—coconut palm, red maple, rice, and soybean—suggests at least two ways in which they may be classified. We may divide them into trees (coconut palm, red maple) and herbs (rice, soybean) on the basis of their growth plans, or into monocots (coconut palm, rice) and dicots (red maple, soybean) on the basis of one clearly distinguishing character (possession of one or two seed leaves—cotyledons—in their embryo) and several important anatomical characters (Figure 31.5).

Monocots are generally narrow-leaved flowering plants such as grasses (including rice), lilies, orchids, and palms. **Dicots** are broad-leaved flowering plants such as soybeans, roses, sunflowers, and maples. As we learned in Chapter 27, the monocots and dicots are members of the two classes, Monocotyledones and Dicotyledones, respectively, that make up the phylum Angiospermae—flowering plants.

We'll consider other plant parts in addition to cotyledons, but first let's consider some general organizing concepts to remember as you read this chapter. They concern the partial independence of plant parts, the decentralization of control in the plant body, and the localized nature of plant growth.

Organs of the Plant Body

As the plant body grows, it may lose parts, and it forms new parts that may grow at different rates. Each branch of a plant may be thought of as a unit that is in many ways independent of the other branches. A branch of a plant does not bear the same relationship to the remainder of the body as an arm does to the remainder of the human body. Each branch lives out its own history, and branches grow independently, exploring different parts of the surrounding environment. Branches may respond differently to gravity, some growing more or less vertically and others horizontally.

Leaves are units of another sort, produced in fresh batches to take over the constant function of feeding the plant. Often the shapes of leaves on a plant differ

Cotyledons	Veins in leaves	Flower parts	Arrangement of primary vascular bundles in stem

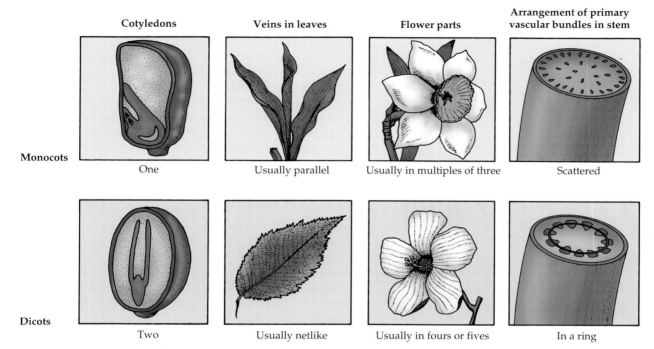

Monocots

One	Usually parallel	Usually in multiples of three	Scattered

Dicots

Two	Usually netlike	Usually in fours or fives	In a ring

31.5 Monocots versus Dicots The possession of one or two cotyledons clearly distinguishes the two types of angiosperms; several other anatomical characteristics also differ between the two.

depending on their differing local environments. Leaves are usually ephemeral, lasting weeks to a few years. Branches and stems are longer-lived, lasting from years to centuries.

Root systems are also branching structures, and branch roots are semi-independent units. As roots penetrate and explore the soil environment, there is considerable turnover; that is, many roots die and are replaced by new ones.

The partial independence of plant parts results in a decentralization of control systems. Branches experiencing different local environments send differing reports to the rest of the plant. If a plant has more than one stem, each stem may receive a different report as to the availability of water and minerals because each is served by different roots. But in spite of this decentralization, the plant functions as a coordinated, coherent unit.

All parts of the animal body grow as the individual develops from embryo to adult, but this growth is *determinate*; that is, growth of the individual and its parts ceases when the adult state is reached. Most plant growth, by contrast, is *indeterminate* and generated from specific regions of active cell division and cell expansion. The regions of cell division are called **meristems**, which are groups of cells that retain the ability to produce new organs indefinitely.

Meristems at the tips of the root and stem give rise to the plant body by producing the cells that subsequently expand and differentiate to form all plant organs. All plant organs arise from cell divisions in the meristems, followed by cell expansion and differentiation. Because meristems can continue to produce new organs, the plant body is much more plastic than the animal body, whose organs are laid down once, during embryonic development. As you read this chapter, notice the emphasis on the activities of meristems.

Now let's return to the tree you ran across at the beginning of this chapter. Where was the young lovers' locket when you found it? They had tied it 1 m above the ground, and the tree had grown from an initial height of 4 m to its current height of 20 m. The real question is, where did this growth take place? And the answer is that the tree grew at its meristems—not its middle. Thus the locket was still on the ground.

Plants have three principal organs

The body of most tracheophytes (as plants that have a vascular system are known) is divided into three principal organs: the **leaves**, the **stem**, and the **roots** (Figure 31.6). Taken together, a stem, its leaves, and any flowers (which contain leaflike parts) are called a shoot. The **shoot system** of a plant consists of all stems and all leaves. The **root system** consists of primary and branch roots.

Broadly speaking, the leaves are the chief organs of photosynthesis. The stem holds and displays the leaves to the sun, maximizing the photosynthetic yield, and provides transport connections between the

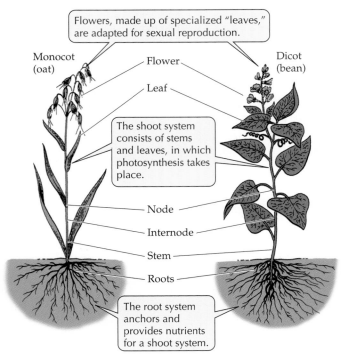

Flowers, made up of specialized "leaves," are adapted for sexual reproduction.

Monocot (oat)

Flower

Leaf

Dicot (bean)

The shoot system consists of stems and leaves, in which photosynthesis takes place.

Node

Internode

Stem

Roots

The root system anchors and provides nutrients for a shoot system.

31.6 Principal Plant Parts The basic plant body plan is illustrated for monocots and dicots.

(a)

(b)

roots and leaves. The points where leaves attach to the stem are called **nodes**, and the stem regions between nodes are **internodes** (see Figure 31.6). Roots anchor the plant in place, and their extreme branching and fine form adapt them to absorb water and mineral nutrients from the soil.

Each of the principal organs can best be understood in terms of its function and its structure. By structure we mean both its overall form and microscopic anatomy—the component cells and tissues, as well as their arrangement. Let's consider the major organs—roots, stems, and leaves—in order.

Roots anchor the plant and take up water and minerals

Water and minerals usually enter the plant through the root system, of which there are two principal types. Many dicots have a **taproot system**: a single, large, deep-growing root accompanied by less prominent secondary roots (Figure 31.7a). The taproot itself often functions as a food storage organ, as in carrots and radishes.

By contrast, monocots and some dicots have a **fibrous root system**, which is composed of numerous thin roots roughly equal in diameter (Figure 31.7b–e). Fibrous root systems often have a tremendous surface area for the absorption of water and minerals. A fibrous root system holds soil very well, giving grasses with such systems a protective role on steep hillsides where runoff from rain could cause erosion.

A tissue composed of rapidly dividing cells is located at the tip of the root proper, just behind the root cap. This tissue is the **root apical meristem**, which produces all the cells that contribute to growth in the length of the root. Some of the daughter cells from the root apical meristem contribute to the **root cap**, which

31.7 Root Systems The taproot system of a dandelion (a) contrasts with the fibrous root system of a violet (b). Fibrous root systems are diverse (c–e), their forms adapted to the different environments in which they grow.

(c)

(d)

(e)

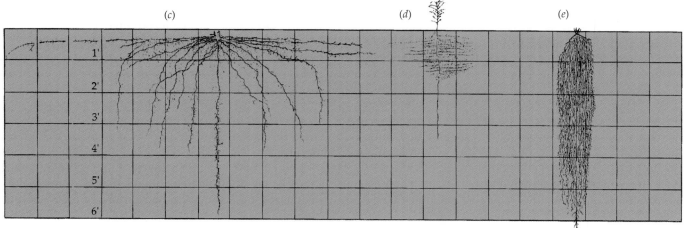

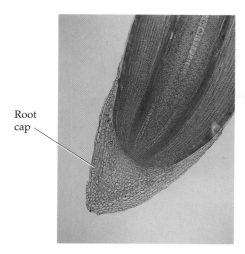

Root cap

31.8 The Root Cap Protects the Root Tip As the root grows through the soil, the root cap wears off and is replaced by cells from the root apical meristem.

protects the delicate growing region of the root as it pushes through the soil (Figure 31.8). Cells of the root cap are often damaged or scraped away and must therefore be replaced constantly. The root cap is also the structure that detects the pull of gravity and thus controls the downward growth of roots. Most root cells—those that are produced at the other end of the meristem—elongate. Following elongation, these cells differentiate, giving rise to the various tissues of the mature root.

If the romantic couple at the beginning of this chapter had decided to dig a shallow hole and tie their

locket to the taproot, perhaps half a meter below ground, would growth of the root have pulled the locket down into the soil? No, because the root elongates near its apical meristem, not along its entire length.

Some plants have roots that arise from points along the stem where roots would not usually occur, or even from the leaves. Known as **adventitious roots**, they also form (in many species) when a piece of shoot is cut from the plant and placed in water or soil. Adventitious rooting enables the cutting to establish itself in the soil. Some plants—corn, banyan trees, and some palms, for example—use adventitious roots as props to help support the shoot.

Stems bear the buds that represent the plant's future

Unlike most roots, a stem may be green and capable of photosynthesis. A stem bears leaves at its nodes, and where each leaf meets the stem there is a **lateral bud** (Figure 31.9a), which develops into a branch if it becomes active. A branch is also a stem. The branching patterns of plants are highly variable, depending on the species, environmental conditions, and a gardener's pruning activities.

Some stems are highly modified. The potato **tuber**, for example, is a portion of the stem rather than a root. Its "eyes" (Figure 31.9b) contain lateral buds; thus a sprouting potato is just a branching stem. The runners of strawberry plants and Bermuda grass are horizontal stems from which roots grow at frequent intervals (Figure 31.9c). If the links between the rooted portions are broken, independent plants can develop from each side of the break. This is a form of asexual reproduction, which we will discuss in Chapter 36.

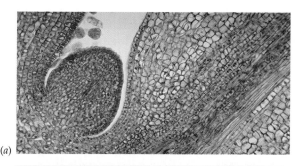

(a)

31.9 A Selection of Stems (a) Microscopic view of a lateral bud developing at the junction between leaf and stem of a lilac; the bud contains vascular tissue and may develop into a branch. (b) A potato is a modified stem called a tuber; the sprouts that grow from its eyes are branches. (c) Runners of beach strawberry are horizontal stems; such a stem produces roots at intervals, providing a local water supply and allowing rooted portions to live independently if the runner is cut.

(b)

(c)

31.10 Types of Shoot Systems You can probably find most of these stem types among the weeds on your campus. The type on the right is characteristic of aspen, May apple, water hyacinth, and some other plants.

Stems bear buds—embryonic shoots—of various types. We have already mentioned lateral buds, which give rise to branches. At the tip of each stem or branch is an **apical bud** containing a **shoot apical meristem**, which produces the cells for the growth and development of the stem. The shoot apical meristem also produces **leaf primordia**, lateral outgrowths that expand to become leaves in the apical bud (see the upper micrograph in Figure 31.16). At times that vary depending on the species, buds form that develop into flowers.

Shoot systems have various forms, in which branches take on different relationships to the plant as a whole (Figure 31.10). Some shoots branch underground, and their branches emerge from the soil looking like separate plants.

Leaves feed the plant—and the rest of us

In most flowering plants the leaves are responsible for most of the photosynthesis, producing food for the plant and releasing oxygen gas. Leaves also carry out metabolic reactions that make nitrogen available to the plant for the synthesis of proteins and nucleic acids (see Chapter 34). Leaves are important food storage organs in some species; in other species—the succulents—the leaves store water. The thorns of cacti are modified leaves. Certain leaves of poinsettias, dogwoods, and some other plants are brightly colored and help attract pollinating animals to the often less striking flowers. Many plants, such as peas and squash, have tendrils—modified leaves that support the plant by wrapping around other plants.

A less obvious but often crucial function of leaves is to shade neighboring plants. Like all other organisms, plants compete; if a plant can reduce the photosynthetic capability of its neighbors by intercepting sunlight, it can obtain a greater share of the available water and mineral nutrients. Finally, as we will see in Chapter 35, the "timer" by which some plants measure the length of the night is located in the leaves.

Leaves are marvelously adapted to serve as light-gathering, photosynthetic organs. Typically, the **blade** of a leaf is flat, and during the daytime it is held by its stalk, or **petiole**, at an angle almost perpendicular to the rays of the sun. Some leaves track the sun, moving so that they constantly face it. If leaves were thicker, the outer layers of cells would absorb so much of the light that the interior layers of cells would be too dark and thus would be unable to photosynthesize.

The different leaves of a single plant may have quite different shapes. The form of a leaf results from a combination of genetic, environmental, and developmental influences. Most species, however, bear leaves of a particular broadly defined type. For example, a leaf may be **simple**, consisting of a single blade, or **compound**, in which blades, or leaflets, are arranged along an axis or radiate from a central point. In a simple leaf, or in a leaflet of a compound leaf, the veins may be parallel to one another or in a netlike arrangement.

The general development of a specific leaf pattern is programmed in the individual's genes and is expressed by differential growth of the leaf veins and of the tissue between the veins. As a result, plant taxonomists have often found leaf forms (outline, margins, tips, bases, and patterns of arrangement) to be reliable characters for classification and identification. At least some of the forms in Figure 31.11 probably look familiar to you.

Leaves and other plant organs are composed of cells, tissues, and tissue systems. Next we'll examine these three levels of organization.

Levels of Organization in the Plant Body

Newly formed *cells* expand to their final size, and as they do so they differentiate—that is, become structurally or chemically specialized for particular functions. A *tissue* is an organized group of cells, working together as a functional unit. Some tissues are composed of a single type of cell; other tissues are composed of several cell types.

Plant tissues are organized into three *tissue systems* that extend throughout the body of the plant, from organ to organ. These three systems are the vascular

Shapes

Margins

Apices and bases

Arrangements on stem

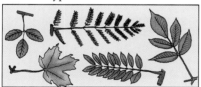

Parts and types

31.11 The Diversity of Leaves Simple leaves are those with a single blade. Compound leaves consist of leaflets arranged along a central axis.

tissue system (xylem and phloem), which conducts materials from one part of the body to another; the dermal tissue system, which protects the body surface; and the ground tissue system, which plays many roles, including producing and storing food materials.

To understand the structures and functions of the tissue systems, we must know the nature of their cellular building blocks. Some cells are alive when functional; others function only after their living parts have died and disintegrated. Some cells develop chemical capabilities not demonstrated by other cells. Several cell types differ dramatically in the structure of their cell walls. We will first consider the types of cells that make up the plant body and then see how aggregations of cells form functioning tissues and tissue systems.

Plant Cells

Living plant cells have all the essential organelles common to eukaryotes (see Chapter 4). In addition, they have some structures and organelles not shared by cells of the other kingdoms: Some plant cells contain chloroplasts and peroxisomes; some contain glyoxysomes. Many plant cells contain large central vacuoles. Every plant cell is surrounded by a cellulose-containing cell wall.

Cell walls may be complex in structure

The division of a plant cell is completed when cell walls form, separating the daughter cells. The first barrier to form between daughter cells is the **middle lamella** (Figure 31.12a). After this layer forms, the newly separated cells secrete structural materials, including cellulose. Each daughter cell, as it expands to its final size, secretes more cellulose and other polysaccharides to complete formation of the **primary wall** (Figure 31.12b).

Once cell expansion stops, a plant cell may deposit more polysaccharides and other materials—such as *lignin*, characteristic of wood, or *suberin*, a waxy substance characteristic of cork—in one or more layers in-

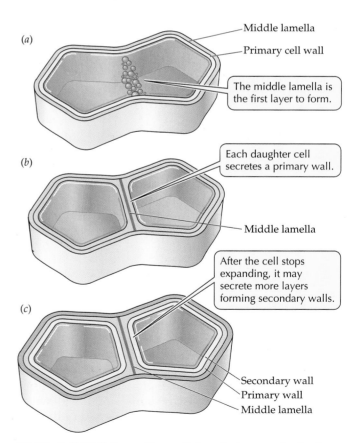

(a)
Middle lamella
Primary cell wall
The middle lamella is the first layer to form.

(b)
Each daughter cell secretes a primary wall.
Middle lamella

(c)
After the cell stops expanding, it may secrete more layers forming secondary walls.
Secondary wall
Primary wall
Middle lamella

31.12 Cell Wall Formation Cell walls form as the final step in plant cell division.

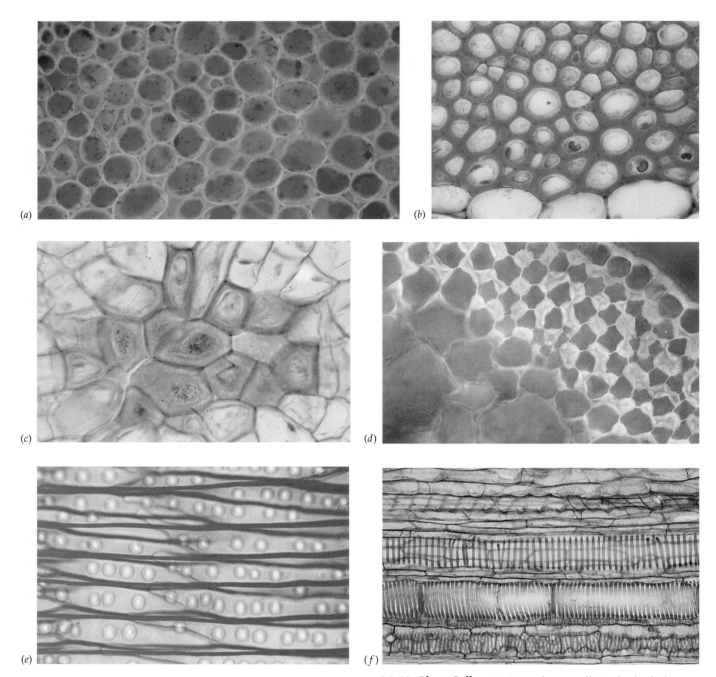

ternal to the primary wall. These layers collectively form the **secondary wall** (Figure 31.12c), which often serves supporting or waterproofing roles.

Although the cell wall lies outside the plasma membrane of the cell, it is not a chemically inactive region. Chemical reactions in the wall play important roles in cell expansion and defense responses. Cell walls may thicken or be sculpted or perforated as part of the differentiation into various cell types. Except where the secondary wall is waterproofed, the structure is porous to water and to most small molecules.

31.13 Plant Cells (a) Parenchyma cells in the leaf of a primrose plant; note the uniform cell walls. (b) Sclerenchyma: Fibers in a sunflower plant (*Helianthus*). The thickened walls are stained red. (c) Sclerenchyma: Thick-walled sclereids; these extremely thick secondary cell walls are laid down in layers. They provide support and a hard texture to structures such as nuts and seeds. (d) Collenchyma cells make up the five outer cell layers of this spinach leaf vein. They are recognizable because their cell walls are very thick at the corners of the cells and thin elsewhere. (e) Tracheids appear deep red in this micrograph of pinewood; note the complexity of the cell walls. (f) Vessel elements in the stem of a squash. The secondary walls are stained red; note the different patterns of thickening, including rings and spirals. Which cells in this figure function when they are alive and which only when they are dead?

However, water and dissolved materials can move directly from cell to cell, without passing into the cell wall space, by way of structures called **pit pairs**.

A pit is a thinning in the primary wall of a cell at a place where the secondary wall either is absent or is separated from the primary wall by a space. Where there is a pit in the wall of one cell, there is usually a corresponding pit in the wall of the adjacent cell; together they are a pit pair. Strands of cytoplasm called **plasmodesmata** pass through pit pairs and the middle lamella between them, allowing substances to move freely from cell to cell without having to cross a plasma membrane.

Parenchyma cells are alive when they perform their several functions

The most numerous cells in the young plant body are the **parenchyma** cells (Figure 31.13*a*). Some tissues consist primarily or solely of parenchyma cells, but parenchyma cells also play important roles in more complex tissues. Parenchyma cells are alive when they perform their functions in the plant. They usually have thin walls, consisting only of a primary wall and the shared middle lamella. Many parenchyma cells have shapes similar to those of soap bubbles crowded into a limited space—figures with 14 faces. They are usually not elongated or otherwise asymmetrical. Most have large central vacuoles.

Many parenchyma cells store various substances, such as starch or lipids. In the cytoplasm of these cells, starch is often stored in specialized plastids called leucoplasts (see Chapter 4). Lipids may be stored as oil droplets, also in the cytoplasm. Other parenchyma cells appear to serve as "packing material" and play a vital role in supporting the stem.

Leaves have a particularly important type of parenchyma cell that is specialized for photosynthesis and is equipped with abundant chloroplasts. Some other, nonphotosynthetic parenchyma cells retain the capacity to divide and hence may give rise to new meristems, as when a wound results in cell proliferation.

Sclerenchyma cells provide rigid support after they die

Sclerenchyma cells function when dead. A heavily thickened secondary wall performs their major function: support. There are two types of sclerenchyma cells: elongated fibers and variously shaped sclereids. **Fibers**, often organized into bundles, provide a relatively rigid support both in wood and in other parts of the plant (Figure 31.13*b*). The bark of trees owes much of its mechanical strength to long fibers. **Sclereids** may pack together very densely, as in a nut's shell or other types of seed coats (Figure 31.13*c*). Isolated clumps of sclereids, called stone cells, in pears and some other fruits give them their characteristic gritty texture.

Collenchyma cells provide flexible support while alive

Another type of supporting cell, the **collenchyma** cell, remains alive even after laying down thick cell walls (Figure 31.13*d*). Collenchyma cells are generally elongated. In these cells the primary wall thickens and no secondary wall forms. Collenchyma provides support to petioles, nonwoody shoots, and growing organs. Tissue made of collenchyma cells, although resistant to bending, is more flexible than sclerenchyma tissue; stems and leaf petioles strengthened by collenchyma can sway in the wind without snapping as they might if they were strengthened by sclerenchyma. The familiar "strings" in celery consist primarily of collenchyma.

The xylem contains water-conducting cells

The xylem of tracheophytes contains cells called **tracheary elements**, which undergo programmed cell death before they assume their ultimate function of transporting water and dissolved minerals. The tracheary elements of gymnosperms and angiosperms differ significantly. The tracheary elements of gymnosperms are **tracheids**—spindle-shaped cells interconnected by numerous pits in their cell walls (Figure 31.13*e*). Because the cell contents—nucleus and cytoplasm—disintegrate upon death, a group of dead tracheids forms a continuous hollow network through which water can readily be drawn.

Flowering plants evolved a water-conducting system made up of vessels. The individual cells, called **vessel elements**, also die before they become functional. Vessel elements are generally larger in diameter than tracheids; they are laid down end-to-end and lose all or part of their end walls, so that each vessel is a continuous hollow tube consisting of many vessel elements and providing a clear pipeline for water conduction (Figure 31.13*f*). In the course of angiosperm evolution, vessel elements have become shorter, and their end walls have become less and less obliquely oriented and less obstructed (Figure 31.14). The xylem of many angiosperms also includes tracheids.

The phloem contains food-conducting cells

The phloem, in contrast to xylem, consists primarily of living cells. In flowering plants the characteristic cell of the phloem is the **sieve tube member** (Figure 31.15*a*). Like vessel elements, these cells meet end-to-end and form long sieve tubes, which transport foods from their sources to tissues that consume or store them. In plants with mature leaves, for example, excess products of photosynthesis move from leaves to root tissues. As sieve tube members mature, enzymes expand small holes in the end walls, connecting the contents of neighboring cells. The result is that the end

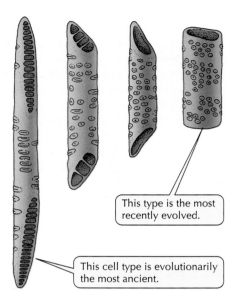

This type is the most recently evolved.

This cell type is evolutionarily the most ancient.

31.14 Evolution of the Conducting Cells of Vascular Systems The xylem of angiosperms, containing vessels that conduct water and minerals, has changed over time.

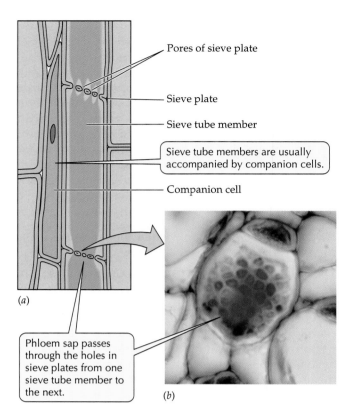

Pores of sieve plate

Sieve plate

Sieve tube member

Sieve tube members are usually accompanied by companion cells.

Companion cell

(a)

Phloem sap passes through the holes in sieve plates from one sieve tube member to the next.

(b)

31.15 Sieve Tubes Individual sieve tube members form long tubes that transport food. Sieve plates form at the ends of each sieve tube member.

walls look like sieves and are called **sieve plates** (Figure 31.15*b*).

As the holes in the sieve plates expand, the membrane around the central vacuole disappears, allowing some of the cytosol and the vacuole's contents to mingle and form a single fluid. This mixture is forced from cell to cell along the sieve tube, carrying its dissolved sugars and other important materials with it. The nucleus and some of the other organelles in the sieve tube member also break down and thus do not clog the holes of the sieve. A "fixed," stationary layer of cytoplasm remains, however, lining the cell wall and confining the remaining organelles.

In some flowering plants, the sieve tube members have adjacent **companion cells** (see Figure 31.15*a*), produced along with the sieve tube member when a parent cell divides. Companion cells retain all their organelles and may, through the activities of their nuclei, regulate the performance of the sieve tube members.

All these kinds of plant cells play important roles. Next let's see how they are organized into tissues and tissue systems.

Plant Tissues and Tissue Systems

Parenchyma cells make up parenchyma tissue, a *simple tissue*—that is, one composed of only one type of cell. Sclerenchyma and collenchyma are other simple tissues, composed, respectively, of sclerenchyma and collenchyma cells. Cells of various types also combine to form *complex tissues*. Xylem and phloem are complex tissues, composed of more than one type of cell.

All xylem contains parenchyma cells, which store food. The xylem of angiosperms contains vessel elements, as well as thick-walled sclerenchyma fibers that provide considerable mechanical strength to the xylem. In most gymnosperms, tracheids serve both in water conduction and in support because vessels and fibers are absent. In addition, old xylem that is no longer active in transport becomes compacted at the center of the tree trunk and continues to contribute support for the tree. As a result of its cellular complexity, xylem can perform a variety of functions, including transport, support, and storage. The phloem of angiosperms includes sieve tube members, companion cells, fibers, sclereids, and parenchyma cells.

The **vascular tissue system**, which includes the xylem and phloem, is the conductive, or "plumbing," system of the plant. All living cells require a source of energy and chemical building blocks. As already mentioned, the phloem transports food from the sites of production—called *sources*—commonly the leaves, to sites of utilization or storage—called *sinks*—elsewhere in the plant. The xylem distributes water and mineral ions taken up by the roots to the stem and leaves.

The **dermal tissue system** is the outer covering of the plant. All parts of the young plant body are covered by an **epidermis**, either a single layer of cells or several layers. The epidermis contains epidermal cells, a type of parenchyma, and may include specialized cell types such as the guard cells that form stomata (pores). The shoot epidermis secretes a layer of wax, the **cuticle**, that helps retard water loss. The protective covering of the stems and roots of older woody plants is the **periderm**, which is composed of cork and other tissues that will be discussed later in this chapter.

The **ground tissue system** makes up the rest of a plant and consists primarily of parenchyma tissue, often supplemented by collenchyma or sclerenchyma tissue. The ground tissues function primarily in storage, support, and photosynthesis.

In the discussions that follow, we'll examine how the tissue systems are organized in the different organs of a flowering plant. We will discover how this organization develops as the plant grows.

Growth: Meristems and Their Products

At the tip of each shoot or branch is a shoot apical meristem, and at the tip of each root is a root apical meristem. Growth from the apical meristems is called **primary growth**. These meristems give rise to the entire body of many plants (Figure 31.16).

Some plants develop what we commonly refer to as wood and bark. These complex tissues are derived

from other meristems: the vascular cambium and the cork cambium. The **vascular cambium** is a cylindrical tissue consisting primarily of vertically elongated cells that divide frequently. Toward the inside of the stem or root the dividing cells form new xylem, and toward the outside they form new phloem.

As trees grow in diameter, the outermost layers of the stem are sloughed off. Without the activity of **cork cambium**, which in a tree forms continuously in the bark, this sloughing off would expose the tree to potential damage, including excessive water loss or invasion by microorganisms. Cork cambium produces new cells, primarily in the outward direction. The walls of these cells become impregnated with the waxy substance suberin, thus augmenting the dermal tissue system. Growth in the diameter of stems and roots, produced by the vascular and cork cambia, is called **secondary growth**. It is the source of wood and bark.

Secondary growth of the maple tree visited by the young couple at the beginning of this chapter eventually broke the thin golden chain that the couple had tied around the trunk. Primary growth, taking place only at the tips of organs, did not move the chain upward, but secondary growth occurred along the entire length of the shoot, increased the tree's girth, and stretched the chain until it broke.

In some plants, meristems may remain active for years—even centuries. Such plants grow in size, or at least in diameter, throughout their life. This pattern of continuous growth, referred to earlier in this chapter, is known as indeterminate growth. Determinate growth, which stops at some point, is characteristic of most animals, as well as some plant parts, such as leaves, flowers, and fruits.

The life cycles of plants fall into three categories: annual, biennial, and perennial. **Annuals**, such as many food crops, live less than a year. **Biennials**—carrots and cabbage, for example—grow for all or part of one year

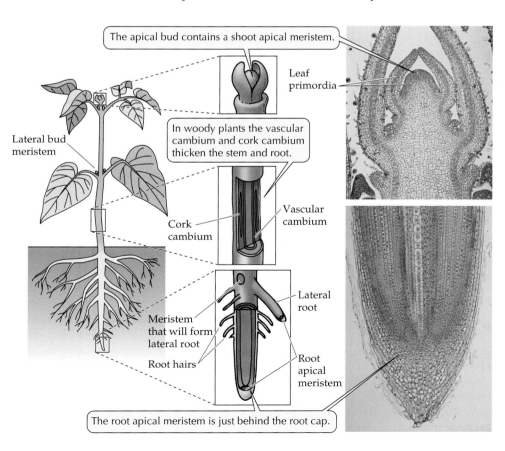

The apical bud contains a shoot apical meristem.

Leaf primordia

In woody plants the vascular cambium and cork cambium thicken the stem and root.

Lateral bud meristem

Vascular cambium

Cork cambium

Lateral root

Meristem that will form lateral root

Root hairs

Root apical meristem

The root apical meristem is just behind the root cap.

31.16 Meristems and the Plant Body The root apical meristem, shoot apical meristem, and lateral bud meristems give rise to the plant body.

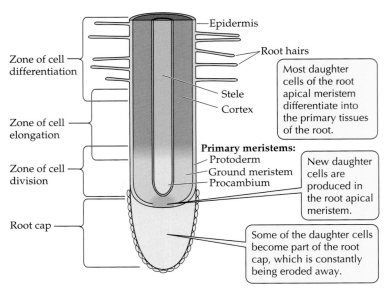

Most daughter cells of the root apical meristem differentiate into the primary tissues of the root.

New daughter cells are produced in the root apical meristem.

Some of the daughter cells become part of the root cap, which is constantly being eroded away.

31.17 Tissues and Regions of the Root Tip Extensive cell division creates the complex root structure.

and live on into a second year, during which they flower, set seed, and die. **Perennials**, such as oak trees, live for a few to many years.

In the sections that follow, we'll examine how the various meristems give rise to the plant body.

The root apical meristem gives rise to three primary meristems

Cell divisions in the root apical meristem produce both the protective root cap and the other primary tissues of the growing root. When a meristem cell divides, the products initially take up no more volume between them than did the dividing cell. One of the products of each cell division develops into another meristem cell the size of its parent, while the other product develops differently.

The products above the apical meristem—away from the root cap—constitute three cylindrical *primary meristems* that give rise to the three tissue systems of the root. The innermost primary meristem, the **procambium**, gives rise to the vascular tissue system; the **ground meristem** gives rise to the ground tissue system; and the outermost meristem, the **protoderm**, gives rise to the dermal tissue system (Figure 31.17).

The apical and primary meristems constitute the *zone of cell division*, the source of all the cells of the root's primary tissues. Just above this zone, where the cells are somewhat older, is the *zone of cell elongation*, in which cells are elongating and thus causing the root to reach farther into the soil. Where the cells are older yet, and where they are differentiating—taking on specialized forms and functions—is the *zone of cell differentiation*. These three zones grade imperceptibly

into one another; there is extensive cell division even as far up as the zone of cell differentiation, and some cells differentiate even in the zone of cell division.

Products of the primary meristems become the tissues of the young root

The protoderm gives rise to the outer layer of root cells, the epidermis, which is adapted for protection and for the absorption of mineral ions and water (Figure 31.18). Epidermal cells are flattened, and many of them produce amazingly long, delicate **root hairs** that vastly increase the surface area of the root (Figure 31.18*b*). It has been estimated that a mature rye plant has a total root surface of more than 1,500 km² (900 square miles), all contained within about 6 liters of soil. Root hairs grow out among the soil particles, probing nooks and crannies and taking up water and minerals.

Internal to the root's epidermis is a region of ground tissue that is many cells thick, called the **cortex** (see Figure 31.17). The cells of the cortex are relatively unspecialized and often function in food storage. In many plants, but especially in trees, epidermal and sometimes cortical cells form an association with a fungus. This association, called a **mycorrhiza**, increases the absorption of minerals and water by the plant (see Chapter 28). Some plant species have poorly developed root hairs or no root hairs. These plants cannot survive unless they develop mycorrhizae that help in mineral absorption.

Proceeding inward, we come to the **endodermis** of the root, a single layer of cells that is the innermost cell layer of the cortex. Endodermal cells differ markedly in structure from the rest of the cortical cells; parts of their walls contain the waxy substance suberin that forms a waterproof seal separating the inner core of vascular tissues from the cortex and the outer world. The endodermal cells control the access of water and dissolved substances to the inner, vascular tissues (see Figure 32.6). Elsewhere in the root, water can pass freely through cell walls and between cells.

Going past the endodermis, we enter the vascular cambium, or **stele**, produced by the procambium (see Figure 31.17). The stele consists of three tissues: pericycle, xylem, and phloem (Figure 31.19). The **pericycle** consists of one or more layers of relatively undifferentiated cells. It is the tissue within which branch roots arise (see Figure 31.18*c*); the pericycle also provides a few of the dividing cells that enable the root to grow in diameter.

At the very center of the root of a dicot lies the xylem—seen in cross section as a star with a variable number of points. Between the points of the xylem star are bundles of phloem. In a monocot root, a region of

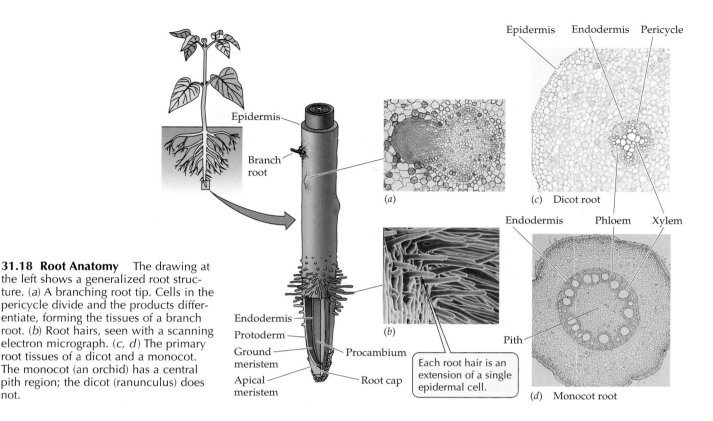

31.18 Root Anatomy The drawing at the left shows a generalized root structure. (*a*) A branching root tip. Cells in the pericycle divide and the products differentiate, forming the tissues of a branch root. (*b*) Root hairs, seen with a scanning electron micrograph. (*c, d*) The primary root tissues of a dicot and a monocot. The monocot (an orchid) has a central pith region; the dicot (ranunculus) does not.

Epidermis

Branch root

Epidermis

Endodermis — Protoderm — Ground meristem — Apical meristem

Procambium

Root cap

Each root hair is an extension of a single epidermal cell.

(*a*)

(*b*)

Epidermis   Endodermis   Pericycle

(*c*)   Dicot root

Endodermis   Phloem   Xylem

Pith

(*d*)   Monocot root

parenchyma cells, the **pith**, lies internal to the xylem. It is useful to try picturing these structures in three dimensions, as in Figure 31.19, rather than attempting to understand their functions solely on the basis of two-dimensional cross sections.

Three primary meristems give rise to the tissues of the stem

The shoot apical meristem, like the root apical meristem, forms three primary meristems: the procambium, ground meristem, and protoderm, which in turn give rise to the three tissue systems. Leaves arise from leaf primordia, bulges that form as cells divide on the sides of shoot apical meristems (see Figure 31.16). The growing stem has no cap analogous to the root cap, but the leaf primordia can act as a protective covering.

Dicot stems grow in length in a region of elongation directly below the shoot apical meristem. Grasses and some other monocots, however, elongate at the bases of internodes and leaves, where some meristem tissue remains. Lawn and range grasses can grow back after being mowed or grazed because they grow from basal meristems close to the soil surface. (In contrast, mown clover grows back by means of shoots that form from previously dormant lateral buds low on the plant.)

The plumbing of angiosperm stems differs from that of the roots. In roots, the vascular tissue lies in the middle, with the xylem at or near the very center. The vascular tissue of a young stem, however, is divided

into discrete **vascular bundles**, which in dicots generally form a cylinder (Figure 31.20*a*) but in monocots are seemingly scattered throughout a cross section of the stem (Figure 31.20*b*). Each vascular bundle contains both xylem and phloem.

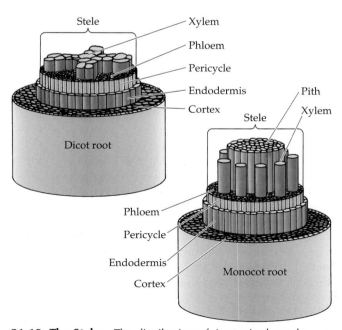

Stele   Xylem — Phloem — Pericycle — Endodermis — Cortex

Dicot root

Pith   Xylem   Stele

Phloem — Pericycle — Endodermis — Cortex

Monocot root

31.19 The Stele The distribution of tissues in the stele—the region internal to the endodermis—differs in the roots of dicots and monocots.

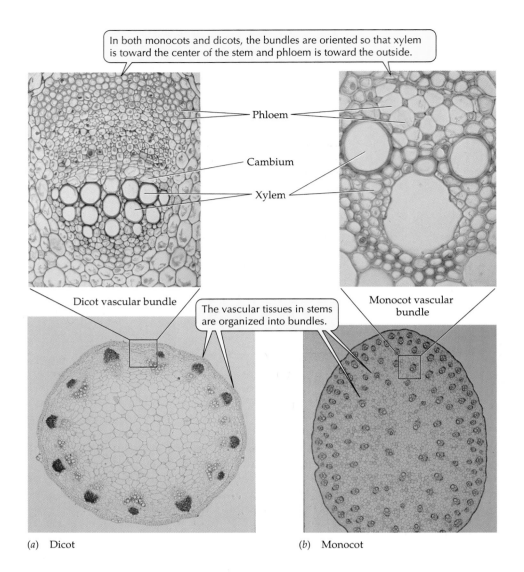

In both monocots and dicots, the bundles are oriented so that xylem is toward the center of the stem and phloem is toward the outside.

Phloem

Cambium

Xylem

Dicot vascular bundle

The vascular tissues in stems are organized into bundles.

Monocot vascular bundle

(a) Dicot

(b) Monocot

31.20 Vascular Bundles in Stems (a) In dicots the vascular bundles are arranged in a cylinder, with pith in the center and cortex outside the ring. (b) This scattered arrangement of bundles is typical of monocot stems.

The transitional region between root and stem is called the crown of a plant. Its vascular structure is complex because of the rearrangement of xylem and phloem leading from root to shoot.

The stem contains other important tissues in addition to the vascular tissues. Internal to the vascular bundles of dicots is a storage tissue, pith, and to the outside lies a similar storage tissue, cortex. The cortex may contain strengthening collenchyma cells with thickened walls. The pith, the cortex, and the regions between the vascular bundles in dicots—called pith rays—constitute the ground tissue system of the stem. The outermost cell layer of the young stem is the epidermis, the primary function of which is to minimize the loss of water from the cells within.

Many stems and roots show secondary growth

Some stems and roots show little or no growth in diameter (secondary growth), remaining slender, but many others, all of them dicots, thicken considerably.

This thickening is of great importance and interest because it gives rise to wood and bark, and it makes the support of large trees possible. Secondary growth results from the activity of two meristem tissues, vascular cambium and cork cambium (see Figure 31.16). Vascular cambium consists of cells that divide to produce new (secondary) xylem and phloem cells, while cork cambium produces mainly waxy-walled cork cells.

Initially, the vascular cambium is a single layer of cells between the primary xylem and the primary phloem. The root or stem increases in diameter when the cells of the vascular cambium divide, producing secondary xylem cells toward the inside of the root or stem and secondary phloem cells toward the outside (Figure 31.21a). In the stem of woody plants, cells of the pith rays between the vascular bundles also divide, forming a continuous cylinder of vascular cambium running the length of the stem. This cylinder in turn gives rise to complete cylinders of secondary xylem—wood—and secondary phloem—bark (Figure 31.21b).

As the vascular cambium produces secondary xylem and phloem, its principal products are vessel elements and supportive fibers in the xylem, and sieve tube members, companion cells, and fibers in the phloem. Not all xylem and phloem cells are adapted for transport or support; some store materials in the stem or root.

Living cells such as these storage cells must be connected to the sieve tubes of the phloem, or they will starve to death. The connections are provided by **vascular rays**, which are composed of cells derived from the vascular cambium. The rays, laid down progressively as the cambium divides, are rows of living

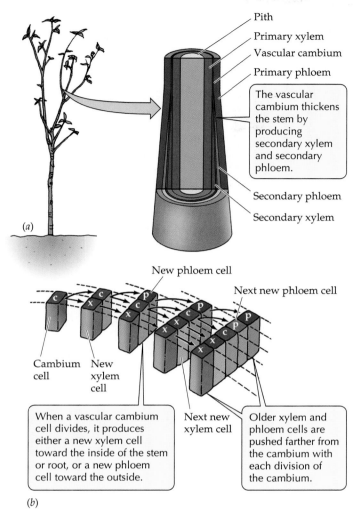

Pith
Primary xylem
Vascular cambium
Primary phloem

The vascular cambium thickens the stem by producing secondary xylem and secondary phloem.

Secondary phloem
Secondary xylem

(a)

New phloem cell
Next new phloem cell

Cambium cell
New xylem cell

When a vascular cambium cell divides, it produces either a new xylem cell toward the inside of the stem or root, or a new phloem cell toward the outside.

Next new xylem cell

Older xylem and phloem cells are pushed farther from the cambium with each division of the cambium.

(b)

31.21 Vascular Cambium Thickens Stems and Roots
Stems and roots grow thicker because a thin layer of cells, the vascular cambium, remains meristematic—capable of dividing.

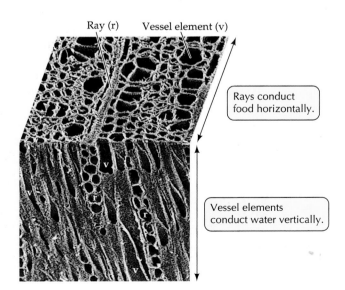

Ray (r)   Vessel element (v)

Rays conduct food horizontally.

Vessel elements conduct water vertically.

31.22 Vascular Rays and Vessel Elements Wood of the tulip poplar, showing that the orientation of xylem vessels is perpendicular to that of vascular rays. Rays transport food horizontally from the phloem to storage cells; xylem vessels conduct water vertically.

parenchyma cells that run perpendicular to the xylem vessels and phloem sieve tubes (Figure 31.22). As the root or stem continues to increase in diameter, new vascular rays are initiated so that this storage and transport tissue continues to meet the needs of the bark and of the living cells in the xylem. The cambium itself increases in circumference with the growth of the root or stem; if it did not, the cambium would split. The vascular cambium grows by the division of some of its cells in a plane at right angles to the plane that gives rise to secondary xylem and phloem. The products of each of these divisions lie within the vascular cambium itself.

Many dicots have a vascular cambium and a cork cambium and thus undergo secondary growth. In the rare cases in which monocots form thickened stems—palm trees, for example—they do so without using vascular cambium or cork cambium. Let's look at the wood and bark produced by woody dicots.

WOOD. Most trees in temperate-zone forests have annual rings (Figure 31.23), which result from seasonal environmental conditions during the year's growth. In spring, when water is relatively plentiful, the tracheids or vessel elements produced by the vascular cambium tend to be large in diameter and thin-walled. As water becomes less available during the summer, narrower cells with thicker walls are produced; this summer wood is darker and perhaps more dense. Thus each year is usually recorded in a tree trunk by a clearly visible annual ring consisting of one light and one dark layer. Trees in the wet tropics do not lay down such obvious, regular rings.

The difference between old and new regions also contributes to the appearance of wood. As a tree grows in diameter, the xylem toward the center becomes clogged with resins and ceases to conduct water and minerals; this is heartwood and appears darker. The portion of the xylem that is actively conducting all water and minerals in the tree is called sapwood and is lighter and more porous than heartwood. The knots that we find attractive in knotty pine but regard as a defect in structural timbers are branches: As a trunk grows, a branch extending out of it becomes buried in new wood and appears as a knot when the trunk is cut lengthwise.

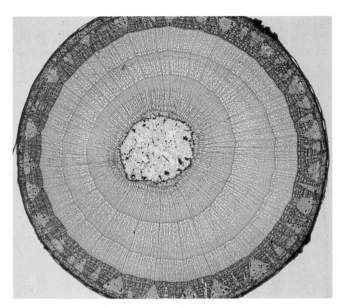

31.23 Annual Rings Rings of xylem vessels are the most noticeable feature of this cross section from a 3-year-old basswood stem.

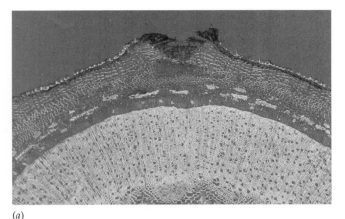

(a)

(b)

31.24 Lenticels Allow Gas Exchange through the Periderm (a) The region of periderm that appears broken open is a lenticel in a year-old elder twig; note the spongy tissue that constitutes the lenticel. (b) The rough areas on the trunk of this Chinese plum tree are lenticels. Most tree species have lenticels much smaller than these.

Let's recall the young couple with the locket one more time. Where was the chain when you found it? They had wrapped it tightly around the tree, a maple—a woody dicot. During all those years the trunk of the tree grew thicker, and the chain broke under the strain. It was lying on the ground by the locket, which itself had been pushed outward by the expanding trunk.

BARK. As secondary growth continues, something within the stem has to give. The vascular tissue expands, stretching and breaking the epidermis and cortex, which are ultimately lost like the locket's chain. Tissue derived from the phloem then becomes the outermost part of the stem. Woody roots behave similarly.

Since the epidermis is specialized in part for the retention of water, how does the plant cope if this tissue is shed? Before layers of epidermal cells are broken away, cells lying near the surface begin to divide and produce layers of cork, a tissue composed of cells with thickened, waterproof walls. The dividing cells, derived from the phloem, form a cork cambium. Sometimes cells are also produced to the inside by the cork cambium; these cells constitute what is known as the phelloderm.

Cork is waterproofed by suberin. The cork soon becomes the outermost tissue of the stem or root. Cork, cork cambium, and phelloderm—if present—make up the periderm (bark) of the secondary body. As the vascular cambium continues to produce secondary vascular tissue, the corky layers are in turn lost, but a simi-

lar process of cell division in the underlying phloem gives rise to new corky layers.

While bark forms, there is still a need for gas exchange with the environment. Carbon dioxide must be released and oxygen must be taken up for cellular respiration. **Lenticels** are spongy regions in the cork that allow such gas exchange (Figure 31.24). And just as stems and roots require gas exchange for cellular respiration, leaves require efficient gas exchange for photosynthesis. Let's see how the structure of a leaf relates to its function.

Leaf Anatomy to Support Photosynthesis

We can think of roots and stems as important supporting actors that sustain the activities of the real stars, the leaves—the organs of photosynthesis. Leaf anatomy is beautifully adapted to carry out photosynthesis. Figure 31.25a shows a typical dicot leaf in cross section.

(a)

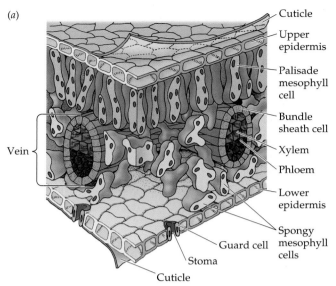

Cuticle
Upper epidermis
Palisade mesophyll cell
Bundle sheath cell
Xylem
Phloem
Lower epidermis
Spongy mesophyll cells
Guard cell
Stoma
Cuticle
Vein

(b)

(c)

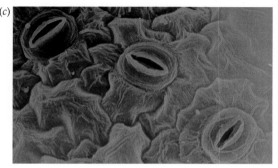

31.25 The Dicot Leaf (a) Cross section of a dicot leaf. (b) The network of fine veins in this red maple leaf (*Acer rubrum*) carries water to the mesophyll cells and carries photosynthetic products away from them. (c) The lower epidermis of a dicot leaf, stained. The small, heavily stained, paired cells are guard cells; the gaps between them are stomata, through which carbon dioxide enters the leaf.

Generally dicot leaves have two zones of photosynthesizing tissues referred to as **mesophyll**, which means "middle of the leaf." The upper layer or layers of mesophyll consist of roughly cylindrical cells. This zone is referred to as *palisade mesophyll*. The lower layer or layers consist of irregularly shaped cells called *spongy mesophyll*. Within the mesophyll is a great deal of air space through which carbon dioxide can diffuse to surround all photosynthesizing cells.

Vascular tissue branches extensively in the leaf, forming a network of **veins** (Figure 31.25b). These veins extend to within a few cell diameters of all the cells of the leaf, ensuring that the mesophyll cells are well supplied with water. The products of photosynthesis are loaded into the phloem of the veins for export to the rest of the plant.

Covering the entire leaf is a layer of nonphotosynthetic cells constituting the epidermis. To retard water loss, the epidermal cells and their overlying waxy cuticle must be highly impermeable, but this impermeability poses a problem: While keeping water in the leaf, the epidermis also keeps out carbon dioxide, the other raw material of photosynthesis.

The problem of balancing water retention and carbon dioxide availability is solved by an elegant regulatory system that will be discussed in more detail in Chapter 32. This system is based on pairs of **guard cells**, modified epidermal cells that change shape, thereby opening or closing pores called **stomata** (singular stoma) between the guard cells (Figure 31.25c;

see also Figure 32.10). When the stomata are open, carbon dioxide can enter, but some water can also be lost.

In Chapter 8 we described C_4 plants, which can fix carbon dioxide efficiently even when the carbon dioxide supply in the leaf decreases to a level at which the photosynthesis of C_3 plants is inefficient. One adaptation that helps C_4 plants do this is a modified leaf anatomy, as shown in Figure 8.25. Notice that the photosynthetic cells in the C_4 leaf are grouped around the veins in concentric layers: an outer mesophyll layer and an inner bundle sheath. These layers each contain different types of chloroplasts, leading to the biochemical division of labor described in Chapter 8.

To operate efficiently, leaves need to be held aloft for access to sunlight and carbon dioxide. Let's see how this is done.

Support in a Terrestrial Environment

Water buoys aquatic plants, but terrestrial plants must either sprawl on the ground or somehow be supported against gravity. There are two principal types of support for terrestrial plants, one based on the osmotic properties of cells and the other based on tissues stiffened by specialized cell walls. One type of support comes from the *turgor pressure* of the cells in the plant body. Turgor pressure (analogous to air pressure in a tire) develops as cell walls resist the osmotic entry of water. A small plant can maintain an erect posture if its cells are turgid, but it collapses—wilts—if the turgor pressure falls too low. Think about the difference between a wilted plant and a turgid one (see Figure

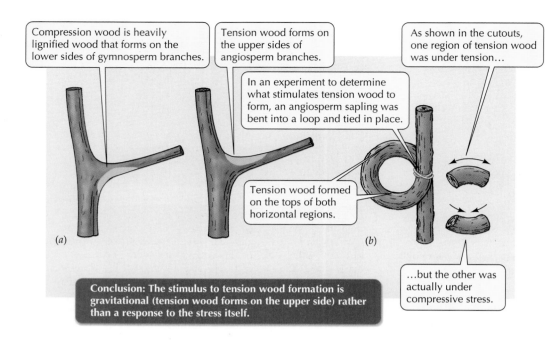

Compression wood is heavily lignified wood that forms on the lower sides of gymnosperm branches.

Tension wood forms on the upper sides of angiosperm branches.

As shown in the cutouts, one region of tension wood was under tension…

In an experiment to determine what stimulates tension wood to form, an angiosperm sapling was bent into a loop and tied in place.

Tension wood formed on the tops of both horizontal regions.

…but the other was actually under compressive stress.

Conclusion: The stimulus to tension wood formation is gravitational (tension wood forms on the upper side) rather than a response to the stress itself.

31.26 Reaction Wood
Reaction wood reduces the tendency of branches to sag.

(a)   (b)

32.2); the distinction dramatically illustrates the role of turgor pressure in supporting the body.

Support by the turgor pressure is often augmented by the second type of support, the presence of strengthening tissues such as collenchyma and sclerenchyma. Collenchyma is more flexible than sclerenchyma, which provides a more rigid, stronger support.

The most important support found in many plants is wood—a mass of secondary xylem. Wood is such a strong yet lightweight material that we have used it in buildings, furniture, and other structures for millennia. All dicot trees are supported by their woody stems, but not all wood is the same. Let's consider some of the special adaptations of secondary xylem.

Reaction wood helps support growing branches

As the branches of growing trees grow longer and heavier, why don't they simply sag to the ground? This problem is averted by means of a gravity-induced asymmetry in wood structure: Specialized **reaction wood**, so called because it forms as reaction to a gravitational stimulus, keeps the limb straight. Reaction wood differs structurally from normal wood; angiosperms and gymnosperms have different kinds of reaction wood, and in different places.

In gymnosperms, **compression wood** forms on the lower side of a branch. It is laid down under compressive stress and then expands, thus tending to push the branch upward (Figure 31.26*a*). Compression wood contains thicker and shorter tracheids, with more lignin and less cellulose in their walls than the tracheids of normal wood have. By contrast, the reaction wood of angiosperms, called **tension wood**, forms on the upper side of the branch. It is laid down under tension and it shrinks, thus tending to pull the branch up-

ward, or at least to resist downward bending (Figure 31.26*a*). In tension wood the fibers have thicker walls containing less lignin and more cellulose, and there are fewer and smaller vessels than in normal wood.

That a gravitational stimulus, not the sagging of the branch, determines the type of reaction wood that forms is illustrated by an experiment performed by the Australian plant physiologist A. B. Wardrop (Figure 31.26*b*). Wardrop bent the trunk of a young angiosperm sapling into a circle and allowed reaction wood to develop. Reaction wood formed on the upper side of both the top of the loop, where the branch was under tension, and the bottom of the loop, where the branch was under compression. Thus, the trunk was responding to a difference between up and down, not to the pattern of stress.

Trees grown indoors tend to be much more spindly than their outdoor counterparts, apparently because indoor trees are not subjected to buffeting by wind. They develop a firmer trunk if they are simply shaken from time to time. The change in wood deposition caused by such treatments may be akin to reaction wood formation.

Summary of *"The Flowering Plant Body"*

Flowering Plants: Definition and Examples

• Flowering plants (phylum Angiospermae) are plants with a vascular system, a triploid endosperm, and double fertilization. Their seeds are enclosed in modified leaves, and their xylem contains vessel elements and fibers.

- The Monocotyledones (monocots) and Dicotyledones (dicots) are the two classes of flowering plants.
- Monocots typically have a single cotyledon in their seeds, narrow leaves with parallel veins, flower parts in threes or multiples of three, and stems with scattered vascular bundles. **Review Figures 31.3, 31.5**
- Dicots typically have two cotyledons, broad leaves with netlike veins, flower parts in fours or fives, and vascular bundles in a ring. **Review Figures 31.2, 31.5**

Organs of the Plant Body

- The vegetative organs of flowering plants are roots, forming a root system, and stems and leaves, forming a shoot system. **Review Figure 31.6**
- Roots anchor the plant and take up water and minerals. A root apical meristem produces the cells that form the root and the root cap. **Review Figures 31.7, 31.8**
- Stems bear buds. Apical buds contain shoot apical meristems that give rise to flowers, leaves, and the growth of the stem. Lateral buds form branches. **Review Figures 31.9, 31.10**
- Leaves are responsible for most photosynthesis, for which their flat blades, oriented perpendicular to the sun's rays, are specially adapted. **Review Figure 31.11**

Levels of Organization in the Plant Body

- Cells are organized into tissues, which form three tissue systems in plants: the vascular tissue system, the dermal tissue system, and the ground tissue system.

Plant Cells

- The walls of plant cells have structures that often correspond with special functions of the cell. The wall includes the middle lamella common to two neighboring cells, and a primary wall. Some cells produce a thick secondary wall. Adjacent cells are connected by plasmodesmata that extend through both cell walls. **Review Figure 31.12**
- Parenchyma cells perform their functions, primarily storage, while living. Certain parenchyma cells carry out photosynthesis. **Review Figure 31.13a**
- Sclerenchyma cells provide strength and function when dead. **Review Figure 31.13b and c**
- Collenchyma cells provide flexible support. **Review Figure 31.13d**
- Tracheids and vessel elements are xylem cells that conduct water and minerals after the cells die. **Review Figures 31.13e and f, 31.14**
- Sieve tube members are food-conducting cells in the phloem. Their activities are often controlled by companion cells. **Review Figure 31.15**

Plant Tissues and Tissue Systems

- Plant cells are organized into tissues. Three tissue systems extend throughout the plant body. The vascular tissue system, consisting of xylem and phloem, conducts water, minerals, and the products of photosynthesis. The dermal tissue system protects the body surface. The ground tissue system produces and stores food materials and performs other functions.

Growth: Meristems and Their Products

- Apical meristems at the tips of stems, branches, and roots produce the primary growth of those organs. **Review Figure 31.16**

- In some plants, the products of primary growth constitute the entire plant body. Many other plants show secondary growth. The vascular cambium and cork cambium are responsible for secondary growth. **Review Figure 31.16**
- Plants generally show indeterminate growth because their meristems function throughout the life of the plant. Leaves, flowers, and fruits show determinate growth: They reach a certain size and stop growing.
- The young root has an apical meristem that gives rise to three primary meristems. The protoderm produces the dermal tissue system, the ground meristem produces the ground tissue system, and the procambium produces the vascular tissue system. Roots have three overlapping zones: the zone of cell division, the zone of cell elongation, and the zone of cell differentiation. **Review Figure 31.17**
- The dermal tissue system consists of the epidermis, part of which has the root hairs that are responsible for absorbing water and minerals. The ground tissue system of a young root is the cortex, whose innermost cell layer, the endodermis, controls access to the stele. The stele, consisting of the pericycle, xylem, and phloem, is the root's vascular tissue system. Branch roots arise in the pericycle. Monocot roots have a central pith region. **Review Figures 31.17, 31.18, 31.19**
- The shoot apical meristem also gives rise to three primary meristems, and their roles are similar to their counterparts in the root. Leaf primordia on the flanks of the apical meristem develop into leaves. The vascular tissue in young stems is divided into vascular bundles, each containing both xylem and phloem. Pith occupies the center of the dicot stem. Cortex lies to the outside of the vascular bundles in dicots, with pith rays lying between the vascular bundles. **Review Figure 31.20**
- Many stems and roots show secondary growth: Vascular and cork cambia give rise to secondary xylem and secondary phloem. As secondary growth continues, the products are wood and bark. **Review Figure 31.21**
- The vascular cambium must increase in girth to avoid splitting as the woody center of the stem grows and lays down annual rings served by vascular rays. **Review Figures 31.22, 31.23**
- Bark consists of cork, cork cambium, and phelloderm, all pierced at intervals by lenticels that allow gas exchange. **Review Figure 31.24**

Leaf Anatomy to Support Photosynthesis

- The photosynthetic tissue of a leaf is mesophyll. Veins bring water and minerals to the mesophyll and carry the products of photosynthesis to other parts of the body. A waxy cuticle prevents water loss but is impermeable to the carbon dioxide needed for photosynthesis. Guard cells control the opening of stomata, holes in the leaf that allow CO_2 to enter but also allow some water loss. **Review Figure 31.25**

Support in a Terrestrial Environment

- Small plants are held erect primarily by the turgor pressure of their cells. Collenchyma and sclerenchyma also contribute to the support of some plants. Wood supports shrubs and trees.
- Reaction wood provides support for branches, counteracting the tendency of the branches to sag. **Review Figure 31.26**

Self-Quiz

1. Which of the following is *not* a difference between monocots and dicots?
   a. Dicots more frequently have broad leaves.
   b. Monocots commonly have flower parts in multiples of three.
   c. Monocot stems do not generally undergo secondary thickening.
   d. The vascular bundles of monocots are commonly arranged as a cylinder.
   e. Dicot embryos commonly have two cotyledons.

2. Roots
   a. always form a fibrous root system that holds the soil.
   b. possess a root cap at their tip.
   c. form branches from lateral buds.
   d. are commonly photosynthetic.
   e. do not show secondary growth.

3. The plant cell wall
   a. lies immediately inside the plasma membrane.
   b. is an impenetrable barrier between cells.
   c. is always waterproofed with either lignin or suberin.
   d. always consists of a primary wall and a secondary wall, separated by a middle lamella.
   e. contains cellulose and other polysaccharides.

4. Which statement about parenchyma cells is *not* true?
   a. They are alive when they perform their functions.
   b. They typically lack a secondary wall.
   c. They often function as storage depots.
   d. They are the most numerous cells in the primary plant body.
   e. They are found only in stems and roots.

5. Tracheids and vessel elements
   a. die before they become functional.
   b. are important constituents of all plants.
   c. have walls consisting of middle lamella and primary wall.
   d. are always accompanied by companion cells.
   e. are found only in the secondary plant body.

6. Which statement about sieve tube members is *not* true?
   a. Their end walls are called sieve plates.
   b. They die before they become functional.
   c. They link end-to-end, forming sieve tubes.
   d. They form the system for translocation of foods.
   e. They lose the membrane that surrounds their central vacuole.

7. The pericycle
   a. separates the stele from the cortex.
   b. is the tissue within which branch roots arise.
   c. consists of highly differentiated cells.
   d. forms a star-shaped structure at the very center of the root.
   e. is waterproofed by Casparian strips.

8. Secondary growth of stems and roots
   a. is brought about by the apical meristems.
   b. is common in both monocots and dicots.
   c. is brought about by vascular and cork cambia.
   d. produces only xylem and phloem.
   e. is brought about by vascular rays.

9. Periderm
   a. contains lenticels that allow for gas exchange.
   b. is produced during primary growth.
   c. is permanent; it lasts as long as the plant does.
   d. is the innermost part of the plant.
   e. contains vascular bundles.

10. Which statement about leaf anatomy is *not* true?
    a. Stomata are controlled by paired guard cells.
    b. The cuticle is secreted by the epidermis.
    c. The veins contain xylem and phloem.
    d. The cells of the mesophyll are packed together, minimizing air space.
    e. C_3 and C_4 plants differ in leaf anatomy.

Applying Concepts

1. When a young oak was 5 m tall, a thoughtless person carved his initials in its trunk at a height of 1.5 m above the ground. Today that tree is 10 m tall. How high above the ground are those initials? Explain your answer in terms of the manner of plant growth.

2. Consider a newly formed sieve tube member in the secondary phloem of an oak tree. What kind of cell divided to produce the sieve tube member? What kind of cell divided to produce that parent cell? Keep tracing back until you arrive at a cell in the apical meristem.

3. Distinguish between sclerenchyma cells and collenchyma cells in terms of structure and function.

4. Distinguish between primary and secondary growth. Do all angiosperms undergo secondary growth? Explain.

5. What anatomical features make it possible for a plant to retain water as it grows? Describe the tissues and how and when they form.

Readings

Esau, K. 1977. *Anatomy of Seed Plants,* 2nd Edition. John Wiley, New York. A comprehensive treatment; particularly good on secondary growth.

Feldman, L. J. 1988. "The Habits of Roots." *BioScience,* vol. 38, pages 612–618. A consideration of many aspects of the biology of roots, including structure, competition, associations with soil microorganisms, and others.

Galston, A. W. 1994. *Life Processes of Plants.* W. H. Freeman, New York. A brief introduction to plant physiology.

Lee, D. W. 1997. "Iridescent Blue Plants." *American Scientist,* vol. 85, pages 56–63. An exploration of color as an aspect of plant structure, raising interesting questions about the roles of natural filters produced by some plants.

Moffat, A. S. 1996. "Higher Yielding Perennials Point the Way to New Crops." *Science,* vol. 274, pages 1469–1470. A brief note that draws attention to trade-offs and the energetic costs of producing different plant organs, a topic important in agricultural research.

Moore, R., W. D. Clark and K. R. Stern. 1995. *Botany.* W. C. Brown, Dubuque, IA. A comprehensive look at the biology of plants.

Niklas, K. J. 1989. "The Cellular Mechanics of Plants." *American Scientist,* vol. 77, pages 344–349. A fine article, subtitled "How Plants Stand Up," detailing how cell walls and other aspects of stem architecture enable terrestrial plants to stand erect.

Raven, P. H., R. F. Evert and S. Eichhorn. 1992. *Biology of Plants,* 5th Edition. Worth, New York. An excellent general botany textbook.

Wilson, B. F. and R. R. Archer. 1979. "Tree Design: Some Biological Solutions to Mechanical Problems." *BioScience,* vol. 29, pages 293–298. An examination of aspects of the process by which trees constantly redesign themselves as they grow, such as reaction wood formation, from an engineering viewpoint.

Chapter 32

Transport in Plants

*I*n the early 1960s, Per Scholander, a researcher at the Scripps Institution of Oceanography in California, wanted to know what was going on in the top of a tall tree. How could he collect his samples? He hired a sharpshooter, who aimed a high-powered rifle and fired high up into the mighty Douglas fir. From almost 80 meters up the tree, a twig fluttered to the ground, and Scholander quickly inserted it into a device called a pressure bomb and took a measurement. Scholander was making a key discovery leading to our understanding of how water and minerals reach the tops of tall trees.

The water and minerals in a plant's xylem must be transported to the entire shoot system, all the way to the highest leaves and apical buds. Food produced in all the leaves, including the highest, must be translocated to all the living nonphotosynthetic parts of the plant. Before we consider the mechanisms underlying these processes, we should know what needs to be explained: How much sap is transported? And how high must it go?

A Long Way to the Top
Water and minerals must defy gravity and climb over 80 meters to nourish the top branches of this Douglas fir.

In answer to the first question, consider the following example: A single maple tree 15 meters tall was estimated to have some 177,000 leaves, with a total leaf surface area of 675 square meters. During a summer day, that tree lost 220 liters of water *per hour* to the atmosphere by evaporation from the leaves. To prevent wilting, xylem transport in that tree needed to provide 220 liters of water to the leaves every hour.

The second question, How high must the xylem sap be transported?, can be rephrased: How tall are the tallest trees? The tallest gymnosperms, the coast redwoods—*Sequoia sempervirens*—exceed 110 meters in height, as do the tallest angiosperms, the Australian *Eucalyptus regnans*. Any successful explanation of transport in the xylem must account for transport to these great heights.

In this chapter we consider the uptake and transport of water and minerals, the control of evaporative water loss through the stomata, and the translocation of substances in the phloem.

Uptake and Transport of Water and Minerals

Terrestrial plants obtain both water and mineral nutrients from the soil, usually by way of their roots. You already know that leaves are loaded with chloroplasts and that water is one of the ingredients for food production by photosynthesis. Water is also essential for transporting solutes, for cooling the plant, and for developing the internal pressure that supports the plant body. How do leaves high in a tree obtain water from the soil? What are the mechanisms by which water and mineral ions enter the plant body through the dermal tissue of the root, pass through the ground tissue, enter the stele, and ascend as sap in the xylem?

Because neither water nor minerals can move through the plant into the xylem without crossing at least one plasma membrane, in this chapter we will first consider osmosis, the uptake of mineral ions, and the pathway by which water and minerals move through the root to gain entry to the xylem.

Water moves through a differentially permeable membrane by osmosis

Osmosis, the movement of a solvent such as water through a membrane in accordance with the laws of diffusion, was discussed in Chapter 5. Recall that the **osmotic potential** of a solution results from the presence of dissolved solutes. The greater the solute concentration, the more negative the osmotic potential and hence the greater the tendency of water to move into that solution from another solution of lower solute concentration. The two solutions must be separated by a differentially permeable membrane (permeable to water but relatively impermeable to the solute). Solutions—or cells—with identical osmotic potentials are *isosmotic*; if two solutions differ in osmotic potential, the one with the less negative osmotic potential is *hypoosmotic* to the other. Recall, too, that osmosis is a passive process—ATP is not required.

Unlike animal cells, plant cells are surrounded by a relatively rigid cell wall. After a certain amount of water enters a plant cell, the entry of more water is resisted by an opposing **turgor pressure**, owing to the rigidity of the wall. As more and more water enters, the turgor pressure becomes greater and greater. This hydraulic pressure is analogous to the air pressure in an automobile tire; it is a real pressure that can be measured with a pressure gauge. Cells with walls do not burst when placed in pure water, because of the rigidity of the walls— water enters by osmosis until the turgor pressure exactly balances the osmotic potential. At this point, the cell is quite turgid; that is, it has a high turgor pressure.

The overall tendency of a solution to take up water from pure water is called the **water potential**, represented as ψ, the Greek letter psi (Figure 32.1). The water potential is simply the sum of the (negative) osmotic potential (ψ_o) and the (posi-

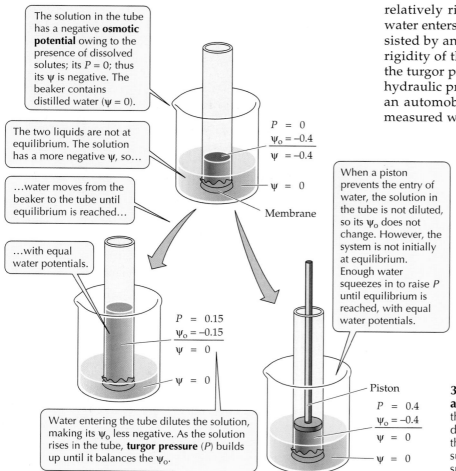

The solution in the tube has a negative **osmotic potential** owing to the presence of dissolved solutes; its $P = 0$; thus its ψ is negative. The beaker contains distilled water ($\psi = 0$).

The two liquids are not at equilibrium. The solution has a more negative ψ, so...

...water moves from the beaker to the tube until equilibrium is reached...

...with equal water potentials.

$P = 0$
$\psi_o = -0.4$
$\psi = -0.4$

$\psi = 0$

Membrane

$P = 0.15$
$\psi_o = -0.15$
$\psi = 0$

$\psi = 0$

Water entering the tube dilutes the solution, making its ψ_o less negative. As the solution rises in the tube, **turgor pressure** (*P*) builds up until it balances the ψ_o.

When a piston prevents the entry of water, the solution in the tube is not diluted, so its ψ_o does not change. However, the system is not initially at equilibrium. Enough water squeezes in to raise *P* until equilibrium is reached, with equal water potentials.

Piston

$P = 0.4$
$\psi_o = -0.4$
$\psi = 0$

$\psi = 0$

32.1 Water Potential, Osmotic Potential, and Turgor Pressure Water potential (ψ) is the tendency of a solution to take up water from distilled water. The water potential is the sum of the osmotic potential (ψ_o) and the turgor pressure (*P*). For pure water under no applied pressure, all three parameters are equal to zero.

This coleus plant remains turgid as long as the turgor pressure of its cells is high.

When cells lose too much water, their turgor pressure drops and the plant wilts.

32.2 Turgor in Plants   Without the turgor pressure provided by water, plants wilt.

tive) turgor pressure (P): $\psi = \psi_o + P$. For pure water under no applied pressure, all three of these parameters are defined as equal to zero. We measure osmotic potential, turgor pressure, and water potential in **megapascals** (MPa), a unit of pressure.

In all cases in which water moves between two cells, or between a cell and its environment, or between two solutions separated by a membrane, the following rule of osmosis applies: *Water always moves toward the region of more negative water potential* (see Figure 32.1).

Osmotic phenomena are of great importance. The structure of many plants is maintained by the turgor pressure of their cells; if the pressure is lost, a plant

wilts (Figure 32.2). The movement of water within a plant follows a gradient of water potential, and as we will see, the flow of phloem sap through the sieve tubes is driven by a gradient in turgor pressure.

Mineral uptake requires transport proteins

Mineral nutrient ions are taken up across plasma membranes with the help of proteins. (You may wish to review the sections "Passive Processes of Membrane Transport" and "Active Processes of Membrane Transport" in Chapter 5.) Some of these proteins are carriers for the facilitated diffusion of particular ions.

Facilitated diffusion does not require ATP. However, the concentrations of some ions in the soil solution are lower than those required inside the plant. Thus the plant must take up these ions against a concentration gradient. Such active transport is an energy-requiring process, depending on cellular respiration for ATP. Active transport, too, requires specific carrier proteins, which use ATP either directly or indirectly.

Unlike animals, plants do not have a sodium–potassium pump for active transport. Rather, plants have a **proton pump** that uses energy obtained from ATP to push protons out of the cell against a proton concentration gradient (Figure 32.3*a*). Because protons (H^+) are positively charged, their accumulation on one side of a membrane has two results. First, the region outside the membrane becomes positively charged with respect to the region inside. Second, a proton concentration difference develops.

32.3 The Proton Pump and Its Effects   The buildup of hydrogen ions transported across the cell membrane by the proton pump (*a*) triggers the movement of both cations (*b*) and anions (*c*).

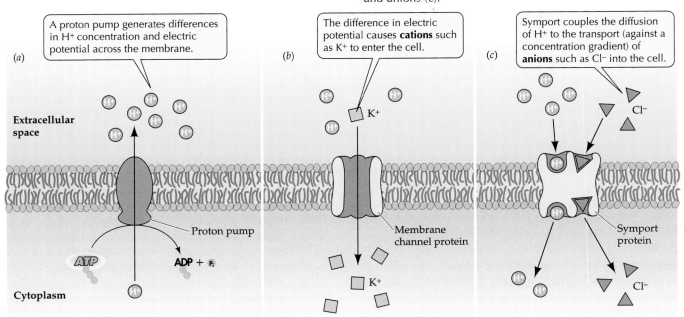

(*a*) A proton pump generates differences in H^+ concentration and electric potential across the membrane.

Extracellular space

Proton pump

ATP   ADP + P$_i$

Cytoplasm

(*b*) The difference in electric potential causes **cations** such as K^+ to enter the cell.

K^+

Membrane channel protein

K^+

(*c*) Symport couples the diffusion of H^+ to the transport (against a concentration gradient) of **anions** such as Cl^- into the cell.

Cl^-

Symport protein

Cl^-

Each of these results has consequences for the movement of other ions. Because of the charge difference across the membrane, the movement of positively charged ions such as K+ into the cell through their membrane channels is enhanced because these positive ions are moving into a region of negative charge (Figure 32.3b). In addition, the proton concentration difference can be harnessed to drive secondary active transport in which negatively charged ions such as Cl– are moved into the cell against a concentration gradient by a symport protein that couples their movement with that of H+ (Figure 32.3c). In sum, there is a vigorous traffic of ions across plant membranes.

Ion transport changes the electric potential across the plasma membrane. Biologists measure the changes with microelectrodes, just as they measure similar changes in neurons (nerve cells) and other animal cells (see Chapter 41). The proton pump and coordinated activities of other membrane transport proteins cause the interior of a plant cell to be strongly negative with respect to the exterior. Most plant cells have a membrane potential of at least –120 millivolts, and they control this difference in potential. This difference, which is much greater than that observed in animal cells, enables a plant cell to carry out most of its other transport processes. How can we study the traffic of ions?

A technique called **patch clamping** allows us to monitor the flow of ions through just a few carrier proteins—or even just one—at a time. First we remove the cell wall, by digesting it with enzymes, to expose the plasma membrane. Then we immobilize the naked cell by pulling it partway into a very fine glass micropipette (Figure 32.4a). Next, we press a still finer glass micropipette against the exposed plasma membrane and apply a slight suction so that a tiny patch of the membrane is effectively isolated from the rest of the surface (Figure 32.4b).

Once a tight seal has been made, we can proceed in one of various ways. In one approach, by pulling very carefully, we can tear the patch away from the rest of the membrane and study the flow of ions through the carrier or carriers contained in just that patch (Figure 32.4c). Because ions are electrically charged, we can measure their movement through the patch by recording the tiny electric current that flows. We can also experiment by altering the contents of the solutions on the two sides of the isolated patch.

We will give an example of results from patch clamping when we discuss stomata later in this chapter.

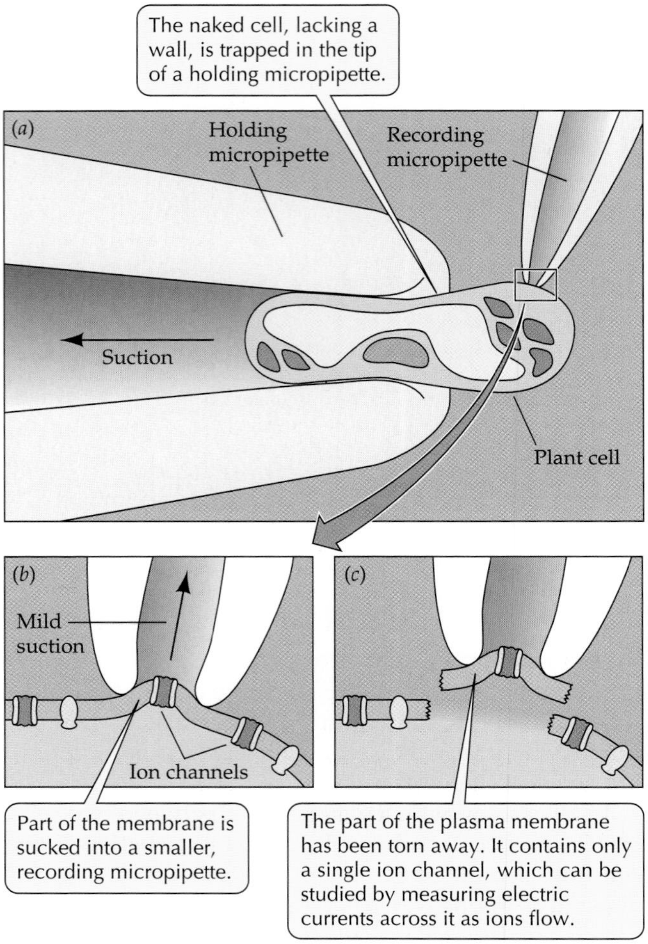

Water and ions pass to the xylem by way of the apoplast and symplast

Water moves along a gradient of water potential, toward ever more negative regions. Water moves into the stele of the root because the water potential is more negative in the stele than in the cortex. The cortex, in

32.4 Patch Clamping This laboratory technique allows us to monitor the flow of ions through one or a few carrier proteins in a "patch" of membrane.

turn, has a more negative water potential than does the soil solution. Minerals enter and move in plants in various ways. Where water is flowing, dissolved minerals are carried along in the mass flow. Where water is moving more slowly, minerals diffuse. At certain points, where plasma membranes are being crossed, some minerals are sped along by active transport.

Water and minerals from the soil may pass through the dermal and ground tissues to the stele via two pathways: the apoplast and the symplast. Plant cells are surrounded by cell walls that lie outside the plasma membrane, and intercellular spaces (spaces between cells) are common in many tissues. The walls and intercellular spaces together constitute the **apoplast** (from the Greek for "away from living material"). The apoplast is a continuous meshwork through which water and dissolved substances can flow or diffuse without ever having to cross a membrane (Figure 32.5). Movement of materials through the apoplast is thus unregulated.

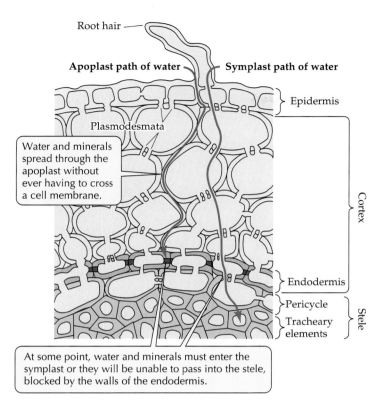

Root hair

Apoplast path of water — **Symplast path of water**

Plasmodesmata

Water and minerals spread through the apoplast without ever having to cross a cell membrane.

At some point, water and minerals must enter the symplast or they will be unable to pass into the stele, blocked by the walls of the endodermis.

Epidermis

Cortex

Endodermis

Pericycle

Tracheary elements

Stele

32.5 Apoplast and Symplast The plant cell walls and intercellular spaces constitute the apoplast. The symplast is the living cells, which are connected by plasmodesmata.

The remainder of the plant body is the **symplast** (from the Greek for "together with living material")— that is, the plant body enclosed by membranes, the continuous meshwork consisting of the living cells, connected by plasmodesmata (see Figure 32.5). The selectively permeable plasma membranes of the cells control access to the symplast, so movement of water and dissolved substances into the symplast is tightly regulated.

Water and minerals can pass from the soil solution through the apoplast to the inner border of the cortex. As you may recall from Chapter 31, to enter the stele, water and minerals must pass through the endodermis (the inner cell layer of the cortex; see Figure 31.17). What distinguishes the endodermis from the rest of the ground tissues is the presence of **Casparian strips**. These waxy, suberin-containing structures line the endodermal cells at their tops, bottoms, and sides, acting as a gasket that prevents water and ions from moving between the cells (Figure 32.6).

The Casparian strips of the endodermis thus completely separate the apoplast of the cortex from the apoplast of the stele. The Casparian strips do not obstruct the outer or inner faces of the endodermal cells. Accordingly, water and ions can enter the stele only by way of the symplast—that is, by entering and passing

through the cytoplasm of the endodermal cells. Thus transport proteins in membranes between the apoplast and symplast determine which minerals pass, and at what rates. This is one of several ways in which plants control their chemical composition and ensure an appropriate balance of their constituents. This balance is essential to plant life.

Once they have passed the endodermal barrier, water and minerals leave the symplast. Parenchyma cells in the pericycle or xylem help minerals move back into the apoplast. Some of these parenchyma cells, called **transfer cells**, are structurally modified for transporting mineral ions from their cytoplasm (part of the symplast) into their cell walls (part of the apoplast). The wall that receives the transported ions has many knobby growths extending into the transfer cell, increasing the surface area of the plasma membrane, the number of transport proteins, and thus the rate of transport.

Transfer cells also have many mitochondria that produce the ATP needed to power the active transport of mineral ions. As mineral ions move into the solution in the walls, the water potential of the wall solution becomes more negative; thus water moves out of the cells into the wall solution by osmosis. Active transport of ions moves the ions directly, and water follows passively. The end result is that water and minerals end up in the xylem, where they constitute the sap. Next we'll explore how sap rises once it is formed.

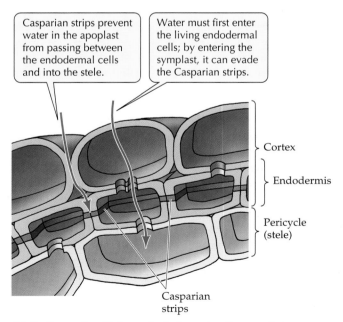

Casparian strips prevent water in the apoplast from passing between the endodermal cells and into the stele.

Water must first enter the living endodermal cells; by entering the symplast, it can evade the Casparian strips.

Cortex

Endodermis

Pericycle (stele)

Casparian strips

32.6 Casparian Strips Endodermal cells contain waxy Casparian strips that prevent water and ions from moving between the cells.

Transport of Water and Minerals in the Xylem

So far in this chapter we've described the movement of water and minerals into the root and their entry into the root xylem. Now we will consider how sap moves throughout the remainder of the plant. Let's first consider some early ideas about the ascent of sap and then turn to our current understanding of the mechanism. We'll describe some experimental evidence against an early model, as well as some evidence in support of the current model—and we'll find out what Per Scholander's sharpshooter was up to in the story that opened this chapter.

Experiments ruled out some early models of transport in the xylem

Some of the earliest models to explain the rise of sap in the xylem were based on a hypothetical pumping action by living cells in the stem, pushing the sap upward. However, experiments conducted and published in 1893 by the German botanist Eduard Strasburger definitively ruled out such models.

Strasburger worked with trees about 20 m tall. He sawed them through at their bases and plunged the cut ends into buckets containing solutions of poisons such as picric acid. The solutions rose through the trunks, as was readily evident from the progressive death of the bark higher and higher up. When the solutions reached the leaves, the leaves died, too, and the solutions then stopped being transported (the liquid levels in the buckets stopped dropping at that time).

This simple experiment established three important points: (1) Living, "pumping" cells are not responsible for the upward movement of the solutions, because the solutions themselves killed all living cells with which they came in contact. (2) The leaves play a crucial role in causing the transport. As long as they were alive, the solutions continued to be transported upward; when the leaves died, transport ceased. (3) Transport in these experiments, which covered distances of 20 m and more, was not caused by the roots, for the trunks had been completely separated from the roots.

Another early suggestion was a model based on capillary action, the rising of watery solutions in very thin tubes or in woven materials like paper. At first glance this theory seems reasonable. The diameters of vessel elements and tracheids are tiny, and the narrower the tube, the higher water will rise by capillary action. However, the diameters of tracheids are small enough to support a capillary column of only about 40 cm, shorter than many shrubs, let alone a giant eucalyptus towering over us to a height greater than the length of a football field.

32.7 Guttation Root pressure forces water through openings in the tips of this strawberry leaf.

Root pressure doesn't account for the bulk of xylem transport

After the capillary model was questioned, some plant physiologists turned (in spite of Strasburger's results) to a model based on root pressure—a pressure exerted by the root tissues that would force liquid up the xylem. The basis for root pressure is a higher solute concentration, and accordingly a more negative water potential, in the xylem sap than in the soil solution. This negative potential draws water into the stele; once there, the water has nowhere to go but up.

There is good evidence for root pressure—for example, the phenomenon of **guttation**, in which liquid water is forced out through openings in the leaves (Figure 32.7). Guttation occurs only under conditions of high atmospheric humidity and plentiful water in the soil. Root pressure is also the source of the sap that oozes from the cut stumps of some plants, such as *Coleus,* when their tops are removed. However, root pressure cannot account for the ascent of sap in trees.

Root pressures seldom exceed one or two times atmospheric pressure, and they are weaker at times when transport in the xylem is most rapid. If root pressure were driving sap up the xylem, we would observe a positive turgor pressure in the xylem at all times. In fact, as we are about to see, the xylem sap is under a tension—a negative turgor pressure—when it is ascending. Furthermore, as Strasburger had already shown, materials can be transported upward in the xylem even when the roots have been removed.

The evaporation–cohesion–tension mechanism does account for xylem transport

To understand how sap rises in the xylem, even to the tops of the tallest trees, we must begin by looking at the final step in the process of water movement from soil to root to leaf and out to the atmosphere. At the end of the line, water evaporates from the moist walls of mesophyll cells, diffuses through the air spaces of the leaf, and finally leaves as water vapor through the open stomata.

The evaporation of water from mesophyll cells makes their water potential more negative (as water is lost, tension develops in the remaining water adhering to the cell walls) so more water enters from the nearest tiny vein. Removing water from the xylem of the veins establishes a tension, or pull, on the entire column of water contained within the xylem, so the column is drawn upward all the way from the roots.

The ability of water to be pulled upward through a tiny tube results from the remarkable *cohesiveness* of water—the tendency of water molecules to cohere to one another through hydrogen bonding (see Chapter 2). The narrower the tube, the greater the tension the water column can withstand without breaking. The in-

tegrity of the column is also maintained by the adhesion of water to the cell walls. As the water column in the xylem is pulled upward, more water enters the xylem in the root from surrounding cells.

In summary, the key elements of water transport in the xylem are *evaporation* from the moist cells in the leaves and a resulting *tension* in the remainder of the xylem's water owing to its *cohesion,* which pulls up more water to replace water that has been lost (Figure 32.8). All this requires no work on the part of the plant. At each step between soil and atmosphere, water moves passively to a region with a more strongly negative water potential. Dry air has the most negative water potential, and the soil solution has the least negative water potential; xylem sap has a water potential more negative than that of cells in the root but less negative than that of mesophyll cells in the leaf.

Mineral ions contained in the xylem sap rise passively with the solution as it ascends from root to leaf. In this way the nutritional needs of the shoot are met. Some of the mineral elements brought to the leaves are subsequently redistributed to other parts of the plant by way of the phloem, but the initial delivery from the roots is through the xylem.

The evaporative loss of water from the shoot is called **transpiration**. In addition to promoting the transport of minerals, transpiration contributes to temperature regulation. As water evaporates from mesophyll cells, heat is taken up from the cells, and the leaf temperature drops. This cooling effect is important in enabling plants to live in hot environments.

A pressure bomb measures tension in the xylem sap

The evaporation–cohesion–tension model can be true only if the column of solution in the xylem is under tension. The most elegant demonstrations of this tension, and of tension's

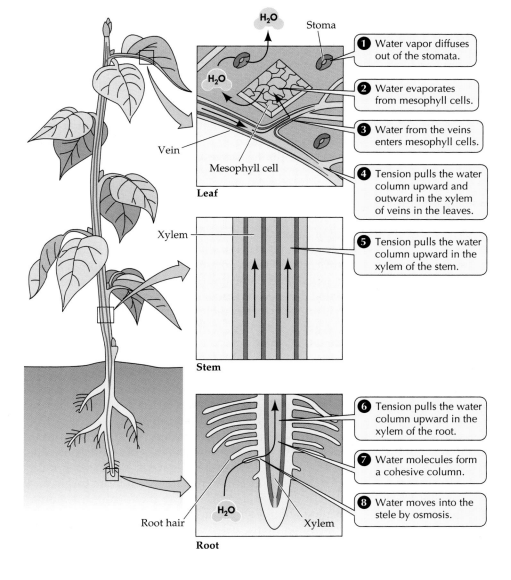

1. Water vapor diffuses out of the stomata.
2. Water evaporates from mesophyll cells.
3. Water from the veins enters mesophyll cells.
4. Tension pulls the water column upward and outward in the xylem of veins in the leaves.
5. Tension pulls the water column upward in the xylem of the stem.
6. Tension pulls the water column upward in the xylem of the root.
7. Water molecules form a cohesive column.
8. Water moves into the stele by osmosis.

32.8 Water Transport in Plants
Evaporation, cohesion, and tension account for the movement of water from the soil to the atmosphere.

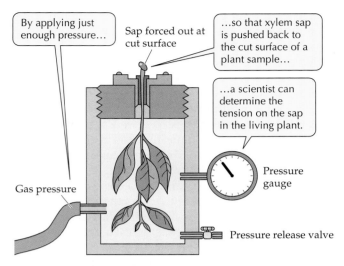

By applying just enough pressure...

Sap forced out at cut surface

...so that xylem sap is pushed back to the cut surface of a plant sample...

...a scientist can determine the tension on the sap in the living plant.

Pressure gauge

Gas pressure

Pressure release valve

32.9 A Pressure Bomb As measured by this laboratory device, the amount of tension on the sap of different types of plants reveals information about their functioning.

adequacy to account for the ascent of sap in tall trees, were performed by Per Scholander. He measured tension in stems with a device called a pressure bomb, as described at the start of this chapter.

The principle of the pressure bomb is as follows: Consider a stem in which the xylem sap is under tension. If the stem is cut, the sap pulls away from the cut, into the stem. Now the tissue is placed in a device—the bomb—in which the pressure may be raised. The cut surface remains outside the bomb. As pressure is applied, the xylem sap is forced back to the cut surface. When the sap first becomes visible again at the cut surface, the pressure in the bomb is recorded. This pressure is the reciprocal of the tension that was originally present in the xylem (Figure 32.9).

Scholander used the pressure bomb to study dozens of plant species, from diverse habitats, growing under a variety of conditions. In all cases in which the xylem sap was ascending, it was found to be under tension. The tension disappeared in some of the plants at night. In developing vines, the xylem sap was under no tension until leaves formed. Once leaves appeared, transport in the xylem began, and tensions were recorded.

Suppose you want to measure tensions in the xylem at various heights in a large tree, to confirm that the tensions are sufficient to account for the rate at which sap is moving up the trunk. How would you obtain stem samples for measurement? Scholander used surveying instruments to determine the heights of particular twigs and then had a sharpshooter shoot the twigs from the tree. As quickly as the twigs fell to the ground, Scholander inserted them in the pressure bomb and recorded their xylem tension.

Scholander had twigs shot from a tree at heights of 27 and 79 m at four different times of day. At each hour the difference in tensions was great enough to keep the xylem sap ascending, and that tension was established by transpiration from the leaves. However, although transpiration provides the impetus for transport of water and minerals in the xylem, it also results in the loss of tremendous quantities of water from the plant. How do plants control this loss?

Transpiration and the Stomata

The epidermis of leaves and stems minimizes transpirational water loss by secreting a waxy **cuticle**, which is impermeable to water. However, the cuticle is also impermeable to carbon dioxide, posing a problem: How can the leaf balance its need to retain water with its need to obtain carbon dioxide for photosynthesis?

An elegant compromise has evolved, in the form of stomata (singular stoma). A **stoma** is a gap in the epidermis; its opening and closing are controlled by a pair of specialized epidermal cells called **guard cells** (Figure 32.10a). When the stomata are open, carbon dioxide can enter the leaf by diffusion, but water vapor may also be lost in the same way. Closed stomata prevent water loss but also exclude carbon dioxide from the leaf.

Most plants compromise by opening the stomata only when the light intensity is sufficient to maintain a good rate of photosynthesis. At night, when darkness precludes photosynthesis, the stomata remain closed; no carbon dioxide is needed at this time, and water is conserved. Even during the day, the stomata close if water is being lost at too great a rate.

The stoma and guard cells in Figure 32.10a are typical of dicots. Monocots typically have specialized epidermal cells associated with their guard cells. The principle of operation, however, is the same for both monocot and dicot stomata. In what follows, we'll describe the mechanism of stomatal opening, the normal cycle of opening and closing, and the modified cycle used by some plants that live in dry or saline environments.

The guard cells control the size of the stomatal opening

When the stomata are about to open, potassium ions (K^+) are actively transported into the guard cells from the surrounding epidermis (Figure 32.10b). The accumulation of potassium ions makes the water potential of the guard cells more negative. Water enters the guard cells by osmosis, making them more turgid and stretching them in such a way that a gap, the stoma, appears between them. The pattern of stretching is controlled by the orientation of cellulose microfibrils in the walls of the guard cells.

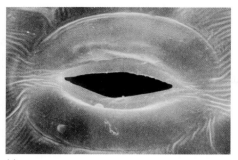

(a)

32.10 Stomata (a) A scanning electron micrograph of a gaping stoma between two sausage-shaped guard cells. (b) Potassium ion concentrations and water potential control the opening and closing of stomata. Negatively charged ions traveling with K^+ maintain electrical balance and contribute to the changes in osmotic potential that affect the guard cells.

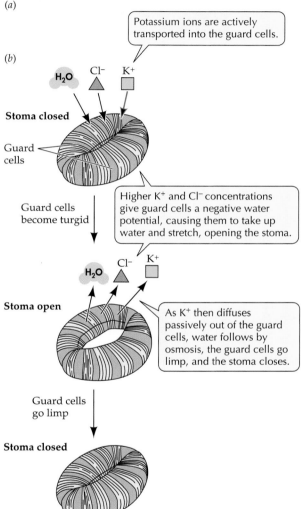

(b)

Potassium ions are actively transported into the guard cells.

H_2O Cl^- K^+

Stoma closed

Guard cells

Guard cells become turgid

Higher K^+ and Cl^- concentrations give guard cells a negative water potential, causing them to take up water and stretch, opening the stoma.

H_2O Cl^- K^+

Stoma open

As K^+ then diffuses passively out of the guard cells, water follows by osmosis, the guard cells go limp, and the stoma closes.

Guard cells go limp

Stoma closed

with more than one type of sensor system. For one thing, the level of carbon dioxide in the spaces inside the leaf is monitored; a low level favors opening of the stomata, thus allowing an increased carbon dioxide level and an enhanced rate of photosynthesis. On the other hand, certain cells monitor their own water potential. If they are lacking enough water—that is, if their water potential is too negative—they release a plant hormone called *abscisic acid*. According to one hypothesis, abscisic acid then acts on the guard cells, causing them to release potassium ions, thus closing the stomata and preventing further drying of the leaf. (Some scientists think that abscisic acid serves only to *keep* the stomata closed, rather than *causing* the closure; further experiments on the timing of abscisic acid production are needed to resolve this point.)

Light also controls the opening of the stomata, which makes sense in view of the fact that most plants conduct photosynthesis only in the light; we would expect stomata to be closed in the dark, thus preventing unnecessary water loss. Under certain conditions, brief exposures to blue light cause guard cells to acidify their environment—that is, to pump out protons. Patch-clamping experiments have revealed further details about the relationship between blue light and proton pumping (Figure 32.11). The proton pump, as described earlier in this chapter (see Figure 32.3), enables the guard cell to take up K^+ and Cl^- ions.

The stoma closes by the reverse process: Potassium ions diffuse passively out of the guard cells, water follows by osmosis, turgidity is lost, and the guard cells collapse together and seal off the stoma. Negatively charged chloride and organic ions also move along with the potassium ions, maintaining electrical balance and contributing to the change in osmotic potential of the guard cells.

What controls the movement of potassium into and out of guard cells? The control system is complex,

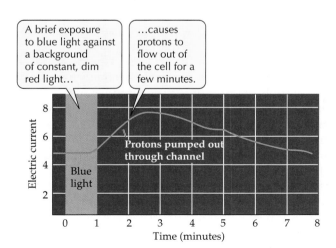

A brief exposure to blue light against a background of constant, dim red light…

…causes protons to flow out of the cell for a few minutes.

Protons pumped out through channel

Blue light

Electric current

Time (minutes)

32.11 Patch Clamping Reveals Light-Induced Proton Pumping A trace of the tiny electric current that results from the flow of protons across the plasma membrane of a guard cell.

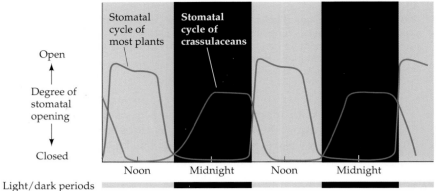

Sedum spathulifolium

32.12 Stomatal Cycles   Most plants open their stomata during the day. Plants of the family Crassulaceae, such as the stonecrop at right, have evolved means to reverse this stomatal cycle: Crassulacean stomata open during the night.

Crassulacean acid metabolism correlates with an inverted stomatal cycle

Most plants change their stomatal openings on a schedule shown by the blue curve of Figure 32.12: The stomata are typically open for much of the day and closed at night (they may, however, close during very hot days). But not all plants follow this pattern. Many **succulent plants**—fleshy plants, such as *Sedum* and *Kalanchoë*, that live in dry areas or near the ocean—are members of the flowering plant family Crassulaceae.

The crassulacean plants have some unusual biochemical and behavioral features. One that was particularly surprising to its discoverers is their "backward" stomatal cycle: Their stomata are open at night and closed by day (see the red curve in Figure 32.12). It was then discovered that crassulacean leaf tissues become acidic at night and more neutral in the daytime.

The mystery was resolved by additional discoveries. At night, while the stomata are open, carbon dioxide diffuses freely into the leaf and reacts in the mesophyll cells with phosphoenolpyruvic acid to produce organic acids such as malic acid and aspartic acid (see Chapter 8). These acids accumulate to high concentrations in the vacuoles of the mesophyll cells. At daybreak the stomata close, thus preventing water loss. Throughout the day, the organic acids are broken down to release the carbon dioxide once again—behind closed stomata. Because the carbon dioxide cannot diffuse out of the plant, it is available for photosynthesis. This set of chemical reactions, discussed in Chapter 8, is referred to as crassulacean acid metabolism, or **CAM**. CAM and the accompanying stomatal behavior were subsequently observed in species of many other plant families besides the Crassulaceae.

Notice that the formation of organic acids is absolutely essential to the functioning of the reversed stomatal pattern of CAM plants. Without the acid formation, carbon dioxide could still be admitted to the leaf at night and saved for daytime. However, it could build up in the intercellular spaces of the leaf only to the same level—0.035 percent of the atmosphere—as in the surrounding air. This amount would be used up by the Calvin–Benson cycle of photosynthesis in a matter of minutes in the daytime. Instead, during the night a CAM plant makes the carbon dioxide into organic acids as fast as it comes in, thus allowing more carbon dioxide to enter. Acid formation in effect fills the leaf with carbon dioxide.

CAM is well adapted to environments where water is scarce: A leaf with its stomata open only at night loses much less water, because the environment is cooler then, than does a leaf with its stomata open by day.

In both CAM and non-CAM plants, the carbon dioxide is converted to the products of photosynthesis. How does the plant deliver these products to the parts of the plant that do not perform photosynthesis? How are substances translocated in the phloem?

Translocation of Substances in the Phloem

The means by which substances in the phloem move from sources, such as leaves, to sinks, such as the root system, remains a topic of interest in plant physiology. Sugars, amino acids, some minerals, and a variety of other substances are translocated in the phloem. Any model to explain this translocation must account for a few important facts: (1) Translocation stops if the phloem tissue is killed by heating or other methods; thus the mechanism must be different from that of transport in the xylem. (2) Translocation often proceeds in both directions—up and down the stem or petiole—simultaneously. (3) Translocation is inhibited by compounds that inhibit cellular respiration and thus limit the ATP supply.

To answer some of the most pressing questions about translocation, plant physiologists needed to obtain samples of pure phloem sap from individual sieve tube members. This task was simplified when scientists recognized that a common garden pest, the aphid, feeds by drilling into a sieve tube. An aphid inserts its stylet, or feeding organ, into a stem until the stylet enters a sieve tube. Within the sieve tubes, the pressure is much greater than in the surrounding plant tissues, so phloem sap is forced up the stylet and into the aphid's digestive tract. So great is the pressure that sugary liquid is forced out the insect's anus (Figure 32.13). At times, ants collect this sugary discharge as food, and some species of ants actually "farm" colonies of aphids, moving them from place to place and protecting them from enemies.

Plant physiologists use the aphid somewhat differently. When liquid appears on the aphid's abdomen, indicating a connection with a sieve tube, the physiologist quickly freezes the aphid and cuts its body away from the stylet. Phloem sap continues to exude from the cut stylet, where it may be collected for analysis. Study of the sap gives accurate information about the chemical composition of the contents of a single sieve tube member over time. From that information one can infer things such as translocation rates. One can infer, too, that bidirectional transport may be explained in terms of neighboring sieve tubes conducting in opposite directions, with each sieve tube transporting all its contents in a single direction. Data obtained by this and other means led to the general adoption of the pressure flow model as an explanation for transport in the phloem.

The pressure flow model appears to account for phloem transport

Phloem sap flows, under pressure, through the sieve tubes. Two important steps in the flow are the active, ATP-requiring transport of sugars and other solutes *into* the sieve tubes in source areas where sugars are produced and the *removal* of the solutes by active transport where the sieve tubes enter sinks—sites where the sugars are needed.

According to the **pressure flow model**, the solute concentration at the source end of a sieve tube is higher than at the sink end, so water has a greater tendency to enter the sieve tube by osmosis at the source end. In turn, this entry of water causes a greater turgor pressure at the source end, so the entire fluid content of the sieve tube is, in effect, squeezed toward the sink end of the tube (Figure 32.14). This mechanism was first proposed more than half a century ago, but some of its features are still debated.

Other mechanisms have been proposed to account for translocation in sieve tubes. Some have been dis-

32.13 Phloem Sap Gets Around Aphids—the white organisms with "sculpted" abdomens—are drilling into a plant to obtain phloem sap. A drop of the sap has formed at the anus of one of the aphids, and an ant is about to collect it.

proved, and none of the rest have been supported by a weight of evidence comparable to that for the pressure flow model. The pressure flow model depends on two things: (1) The sieve plates must be unclogged, so that bulk flow from one sieve tube member to the next is possible, and (2) there must be an effective method for loading sucrose and other solutes into the phloem in source tissues and removing them in sink tissues. Are these conditions met?

Are the sieve plates clogged or free?

Early electron microscopic studies of phloem samples cut from plants produced results that seemed to contradict the pressure flow model. The pores in the sieve plates always appeared to be plugged with masses of a fibrous protein, suggesting that phloem sap could not flow freely. But what is the function of the fibrous protein?

One possibility is that this protein is usually distributed more or less at random throughout the sieve tube members until the sieve tube is damaged; then the sudden surge of sap toward the cut surface carries the protein into the pores, blocking the pores and effectively caulking the leak. That is, the protein does *not* block the pores unless the phloem is damaged. How might this be tested? How could we cut samples of phloem for microscopic observation without causing the sap to surge to the cut surface?

One way to prevent the surge of the sap is to freeze the tissue before cutting it. Another way is to let the tissue wilt so that there is no pressure in the phloem. These and related ideas were tested, and sure enough, when the tissue is cut in this way, the sieve plates are unclogged by protein. Thus, the first condition of the pressure flow model is met. Now, what about the need

A B

Differentially
permeable
membrane

Less concentrated   More concentrated
solution (sink)   solution (source)

32.14 The Pressure Flow Model This experimental demonstration of the pressure flow model describes how turgor pressure and water potential combine to drive sugars and other solutes from the source (B) to the sink (A). Phloem sap may flow through sieve tubes in this manner.

A B

H_2O H_2O

Water enters both
funnels by osmosis.

A B

H_2O

Internal pressure builds until water
can no longer enter funnel A.

A B

H_2O H_2O

Water enters B because of greater
solute concentration; water is
forced out at A; the solution flows
slowly from B to A.

In the plant, solutes are constantly added to
the phloem at the sources and removed at
the sinks, maintaining the flow.

the phloem from source regions and for unloading them into sink regions. One mechanism of phloem loading has been demonstrated in various plants. Sugars and other solutes to be transported pass from cell to cell through the symplast in the mesophyll. After these substances reach cells adjacent to the ends of leaf veins, they leave the mesophyll cells and enter the apoplast, sometimes with the help of transfer cells. Then specific sugars, amino acids, some mineral elements, and a few other compounds are actively transported into cells of the phloem, thus reentering the symplast (Figure 32.15).

Passage through the apoplast selects substances to be accumulated for translocation because substances

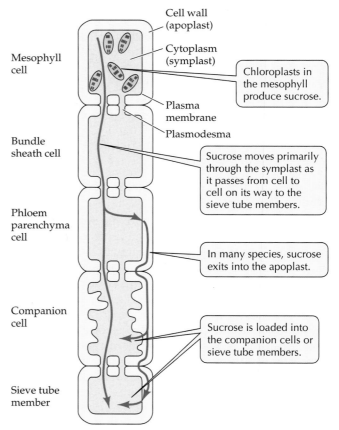

Mesophyll cell

Bundle sheath cell

Phloem parenchyma cell

Companion cell

Sieve tube member

Cell wall (apoplast)

Cytoplasm (symplast)

Chloroplasts in the mesophyll produce sucrose.

Plasma membrane

Plasmodesma

Sucrose moves primarily through the symplast as it passes from cell to cell on its way to the sieve tube members.

In many species, sucrose exits into the apoplast.

Sucrose is loaded into the companion cells or sieve tube members.

32.15 Solutes May Enter Sieve Tubes via the Apoplast Sugars pass from cell to cell through the symplast in the mesophyll. After these substances reach cells adjacent to the ends of leaf veins, they enter the apoplast, sometimes with the help of transfer cells. Specific compounds are actively transported into cells of the phloem, thus reentering the symplast.

for an effective method for loading and unloading solutes?

Neighboring cells load and unload the sieve tube members

If the pressure flow model is correct, there must be mechanisms for loading sugars and other solutes into

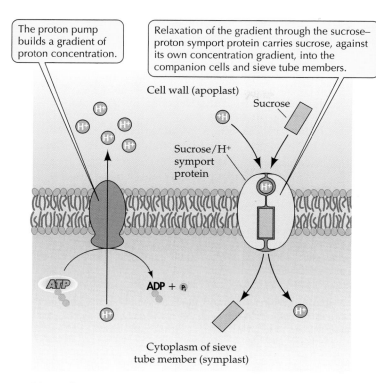

The proton pump builds a gradient of proton concentration.

Relaxation of the gradient through the sucrose–proton symport protein carries sucrose, against its own concentration gradient, into the companion cells and sieve tube members.

Cell wall (apoplast)

Sucrose

Sucrose/H$^+$ symport protein

ATP

ADP + P$_i$

Cytoplasm of sieve tube member (symplast)

32.16 Sucrose–Proton Symport The reentry of sucrose into the symplast is coupled with the entry of protons, which are pumped into the apoplast by the proton pump.

can enter the phloem only after passing through a differentially permeable membrane. In many plants, the cells thus loaded are companion cells (see Chapter 31), which then transfer the solutes to the adjacent sieve tube members. Loading of the phloem with solutes results in a very negative water potential in the sieve tubes; thus water enters by osmosis from the surrounding tissue and maintains a high turgor pressure within the sieve tubes. As Figure 32.15 shows, sucrose moves from the mesophyll to the sieve tube members entirely within the symplast in some species; that is, transfer of solutes from symplast to apoplast and back again is not a universal feature of phloem loading.

A form of secondary active transport (see Chapter 5) loads sucrose into the companion cells and sieve tube members. Sucrose is carried through the plasma membrane from apoplast to symplast by sucrose–proton symport—the entry of sucrose and of protons is strictly coupled. For this symport to work, the apoplast must have a high concentration of protons; the protons are supplied by a primary active transport system, the proton pump. The protons then "relax" back into the cell through the symport protein, bringing sucrose with them (Figure 32.16).

In sink regions, the transported solutes are actively transported out of the sieve tubes and into the surrounding tissues. This unloading serves two purposes:

It helps maintain the gradient of osmotic potential and hence of turgor pressure in the sieve tubes, and it promotes the buildup of sugars and starch to high concentrations in storage regions, such as developing fruits and seeds.

Some trees store starch in the xylem

The xylem sap, too, may occasionally contain sugars. In sugar maples and many other deciduous trees and shrubs of the temperate zones, excess photosynthate produced in late summer and early fall is stored as starch in living xylem cells of the trunk and twigs. Later, in early spring, the starch is digested into sugars that appear initially in the xylem sap, which may be collected and concentrated into syrup. The activities of plants may vary with the time of year, and patterns of transport and storage may change accordingly.

Summary of "Transport in Plants"

Uptake and Transport of Water and Minerals

- Plant roots take up water and minerals from the soil.
- Water moves through biological membranes by osmosis, always moving toward cells with a more negative water potential. The water potential of a cell or solution is the sum of the osmotic potential and the turgor pressure. All three parameters are expressed in megapascals. **Review Figure 32.1**
- Mineral uptake requires transport proteins. Some minerals enter the plant by facilitated diffusion, but others enter by active transport. A proton pump facilitates the active transport of many solutes across membranes in plants. **Review Figure 32.3**
- We can monitor the flow of ions through carrier proteins by patch clamping. **Review Figure 32.4**
- Water and minerals pass from the soil to the xylem by way of the apoplast and symplast. In the root, water and minerals may pass from the cortex into the stele only by way of the symplast; Casparian strips in the endodermis block water and solute movement in the apoplast. **Review Figures 32.5, 32.6**

Transport of Water and Minerals in the Xylem

- Early experiments established that sap does not move via the pumping action of living cells.
- Root pressure is responsible for guttation and for the oozing of sap from cut stumps, but it cannot account for the ascent of sap in trees.
- Xylem transport is the result of the combined effects of evaporation, water cohesion, and tension. Evaporation from moist-walled cells in the leaf lowers the water potential of those cells and thus pulls water—held together by its cohesiveness—up through the xylem from the root. Dissolved minerals go along for the ride. **Review Figure 32.8**

- Support for the evaporation–cohesion–tension model of xylem transport came from studies using a pressure bomb. **Review Figure 32.9**

Transpiration and the Stomata

- Evaporation of water cools the leaves, but a plant cannot afford to lose too much water. Transpirational water loss is minimized by the waxy cuticle of the leaves.
- Stomata allow a compromise between water retention and carbon dioxide uptake by leaves. The guard cells control the size of the stomatal opening. Guard cells take up potassium ions actively, causing water to follow osmotically, swelling the guard cells and opening the stomata. Potassium movement is affected by carbon dioxide level, water availability, and light. **Review Figures 32.10, 32.11**
- In most plants the stomata are open during the day and closed at night. CAM plants have an inverted stomatal cycle, enabling them to conserve water. **Review Figure 32.12**

Translocation of Substances in the Phloem

- Products of photosynthesis, and some minerals, are translocated through sieve tubes in the phloem—by way of living cells. Translocation proceeds in both directions in the stem, although in a single sieve tube it goes only one way, and it requires a supply of ATP.
- Translocation in the phloem proceeds in accordance with a pressure flow model: The difference in solute concentration between sources and sinks allows a difference in turgor pressure along the sieve tubes, resulting in bulk flow. **Review Figure 32.14**
- The pressure flow model succeeds because the sieve plates are unclogged, allowing bulk flow, and because neighboring cells load the sieve tube members in source regions and unload them in sink regions. **Review Figures 32.15, 32.16**

Self-Quiz

1. Osmosis
   a. requires ATP.
   b. results in the bursting of plant cells placed in pure water.
   c. can cause a cell to become turgid.
   d. is independent of solute concentrations.
   e. continues until the turgor pressure equals the water potential.

2. Water potential
   a. is the difference between the osmotic potential and the turgor pressure.
   b. is analogous to the air pressure in an automobile tire.
   c. is the movement of a solvent through a membrane.
   d. determines the direction of water movement between cells.
   e. is defined as 1.0 for pure water under no applied pressure.

3. Which statement about proton pumping across the plasma membranes of plants is *not* true?
   a. It requires ATP.
   b. The region inside the membrane becomes positively charged with respect to the region outside.
   c. It enhances the movement of K^+ ions into the cell.
   d. It pushes protons out of the cell against a proton concentration gradient.
   e. It can drive the secondary active transport of negatively charged ions.

4. Patch clamping
   a. is a mechanism that protects plants against insect pests.
   b. is a mechanism for sealing leaks in the plasma membrane.
   c. can be performed with the cell wall intact.
   d. can be used as a remedy for mineral deficiencies.
   e. can be used to monitor the flow of ions through channel proteins.

5. Which statement is *not* true?
   a. The symplast is a meshwork consisting of the (connected) living cells.
   b. Water can enter the stele without entering the symplast.
   c. The Casparian strips prevent water from moving between endodermal cells.
   d. The endodermis is a cell layer in the cortex.
   e. Water can move freely in the apoplast without entering cells.

6. Which of the following is *not* part of the evaporation–cohesion–tension model?
   a. Water evaporates from the walls of mesophyll cells.
   b. Removal of water from the xylem exerts a pull on the water column.
   c. Water is remarkably cohesive.
   d. The wider a tube, the greater the tension its water column can withstand.
   e. At each step, water moves to a region with a more strongly negative water potential.

7. Stomata
   a. control the opening of guard cells.
   b. release less water to the environment than do other parts of the epidermis.
   c. are usually most abundant on the upper epidermis of a leaf.
   d. are covered by a waxy cuticle.
   e. close when water is being lost at too great a rate.

8. Plants that perform crassulacean acid metabolism
   a. incorporate carbon dioxide into organic acids.
   b. have leaves that become more acidic during the day.
   c. close their stomata at night.
   d. are also called C_4 plants.
   e. must live in environments where water is plentiful.

9. Which statement about phloem transport is *not* true?
    a. It takes place in sieve tubes.
    b. It depends on mechanisms for loading solutes into the phloem in sources.
    c. It stops if the phloem is killed by heat.
    d. In sinks, solutes are actively transported into sieve tube members.
    e. A high turgor pressure is maintained in the sieve tubes.

10. The fibrous protein in sieve tube members
    a. clogs the sieve plates at all times.
    b. never clogs the sieve plates.
    c. serves no known function.
    d. may caulk leaks when a plant is damaged.
    e. provides the motive force for transport in the phloem.

Applying Concepts

1. Epidermal cells protect against excess water loss. How do they perform this function?

2. Phloem transports material from sources to sinks. What is meant by "source" and "sink"? Give examples of each.

3. Contrast the transport of organic substances through the phloem with the transport of water and minerals through the xylem. Mention mechanisms and overall direction.

4. Transpiration exerts a powerful pulling force on the water column in the xylem. When would you expect transpiration to proceed most rapidly? Why? Describe the source of the pulling force.

5. Plants that can perform crassulacean acid metabolism (CAM) are adapted to environments in which water is available in limited supply—they open their stomata only at night. Could a non-CAM plant, such as a pea plant, enjoy a similar advantage if it opened its stomata only at night? Explain.

Readings

Neher, E. and B. Sakmann. 1992. "The Patch Clamp Technique." *Scientific American,* March. Although this article deals with animal cells rather than plants, it gives a good account of the technique.

Raven, P. H., R. F. Evert and S. Eichhorn. 1992. *Biology of Plants,* 5th Edition. Worth, New York. A sound general botany textbook.

Salisbury, F. B. and C. W. Ross. 1992. *Plant Physiology,* 4th Edition. Wadsworth, Belmont, CA. An authoritative textbook with excellent chapters on transport and translocation.

Taiz, L. and E. Zeiger. 1998. *Plant Physiology,* 2nd Edition. Sinauer Associates, Sunderland, MA. Chapters 3, 4, 5, 6, and 10 give more advanced treatments of the topics presented in this chapter.

Chapter 33

Plant Responses to Environmental Challenges

A Pathogen's Prey
These tomato fruits, weakened by a calcium deficiency, were easy prey for a fungal pathogen.

P lants can get sick. The environment teems with plant pathogens. We know of more than 100 diseases that can kill a tomato plant, each of them caused by a pathogen—a living disease-producing agent. Tomato mosaic virus attacks the leaves, as do fungi such as powdery mildew, and some bacteria. The root knot nematode is a tiny roundworm whose target is obvious from its name. Some fungi attack the stem. The crown gall bacterium *Agrobacterium tumefaciens* attacks any part of the plant, producing a tumor that gives rise to further tumors, spreading like a cancer to other parts.

Why are tomatoes and the other plants not extinct by now, given the presence of so many pathogens? Like animals, plants have a variety of defense mechanisms. And like the defenses of our bodies, these mechanisms are not perfect, but they keep the plant world in competitive balance with its pathogens.

Pathogens aren't the only environmental challenges that plants face. Herbivores consume parts of plants and sometimes even entire plants. Some physical environments pose exceptional problems and thus drastically limit the kinds of plants that can live in them. The most challenging physical environments include ones that are very dry (deserts), ones at the other extreme (that are waterlogged and thus limit the availability of oxygen), environments that are dangerously salty, and ones that contain high concentrations of toxic substances such as heavy metals.

This chapter focuses on how plants meet the myriad challenges presented by their biological and physical environment. We'll begin by examining interactions between plants and pathogens and go on to consider interactions between plants and herbivores. Then we will discuss the adaptations of some types of plants to dry, waterlogged, salty, and toxic environments.

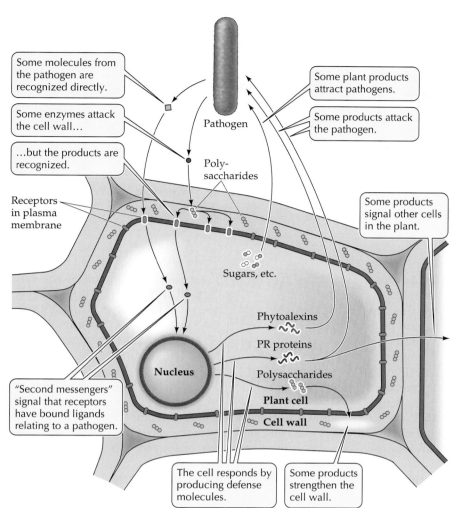

33.1 Signaling between Plants and Pathogens Chemical interactions between plants and pathogens are highly coevolved. Plants produce defensive molecules that work in many different ways.

pathogens pass these barriers, differences between the defense systems of plants and animals become apparent. Animals generally repair tissues that have been infected; they heal, through appropriate developmental pathways. Plants, on the other hand, do not make repairs. Instead, they develop in ways that seal off the damaged tissue so that the rest of the plant does not become infected. This approach is feasible because most plants have many "redundant" parts—for example, multiple branches, as compared with our two arms.

One of the first defensive responses is the rapid deposition of polysaccharides against the cell walls, forming a barrier to invasion by the pathogen. One effect of these polysaccharides is to block the plasmodesmata, limiting the ability of viral pathogens to move from cell to cell. The polysaccharides also serve as a base upon which lignin may be laid down, enhancing the barrier and rendering cells inhospitable, because the building blocks of lignin are toxic to some pathogens. Trees seal off damaged tissue by producing new wood that is different in orientation and chemical composition from the previously deposited wood. Sealing off damaged tissue is primarily a mechanical mechanism for limiting infection and preventing further damage.

Plant–Pathogen Interactions

Plants and pathogens evolve together, resulting in complex relationships. Pathogens have evolved mechanisms by which to attack plants, and plants have evolved defense mechanisms. Both sets of mechanisms are based on chemicals that provide information to the other partner—basically, the players "leak" information (Figure 33.1). The enzymes the pathogen employs to break down the plant's cell walls alert the plant to what the pathogen is up to. In turn, the plant's defenses alert the pathogen that it is under attack.

What determines the outcome of a battle between a plant and a pathogen? The key to success for the plant is to respond to the information about the pathogen quickly and massively. Plants use both mechanical and chemical defenses in this effort.

Some plant defenses are mechanical

The outer surfaces of plants are protected by tissues such as the epidermis or cork, and these tissues are generally covered by cutin, suberin, or waxes. If

Plants have potent chemical defenses against pathogens

When infected by certain fungi and bacteria, plants produce a variety of defense molecules, among which are small molecules called phytoalexins and larger proteins called pathogenesis-related (PR) proteins.

Phytoalexins are not present until the plant is infected; they are produced in the infected area and in immediately neighboring cells within hours of the onset of infection. Phytoalexins are toxic to many fungi and bacteria. Their antimicrobial activity is nonspecific: They can destroy many species of fungi and bacteria in addition to the one that originally triggered their production. Physical injuries, viral infections, and even certain chemical compounds can also induce the production of phytoalexins.

There are several types of **PR proteins**. Some are enzymes that may break down the cell walls of pathogens. These enzymes destroy some of the invading cells, and the breakdown products of the walls may be signals that trigger further defensive responses. Other PR proteins may serve as alarm signals to cells that have not yet encountered the pathogen. In general, PR proteins appear not to be rapid-response weapons; rather, they act more slowly, perhaps in cleanup operations.

The hypersensitive response combines defensive tactics

Plants that are resistant to fungal, bacterial, or viral diseases generally owe this resistance to what is known as the **hypersensitive response**. In this response cells around the site of microbial infection produce phytoalexins and other chemicals and then die, leaving a "dead spot," called a necrotic lesion, that contains what is left of the microbial invasion (Figure 33.2). The rest of the plant remains free of the infecting microbe. The hypersensitive response can have a long-lasting effect in some plants, imparting resistance to subsequent attacks by pathogens. One of the chemicals that may contribute to this long-term disease resistance may surprise you. Let's look at it now.

Aspirin—acetylsalicylic acid—is one of the best-selling drugs in the world and has been for many years. People commonly take aspirin to reduce fever and pain. In our bodies aspirin is hydrolyzed to produce the substance that actually causes the effects, salicylic acid. Since ancient times, people in Asia, Europe, and the Americas have used willow leaves and bark to relieve pain and fever. The active ingredient contained in willow (*Salix*) is salicylic acid, and it now appears that all plants contain at least some salicylic acid. This compound plays various roles in the plants themselves—notably a role in the hypersensitive response.

33.2 The Aftermath of a Hypersensitive Response The necrotic spots on this leaf are a response to the fungus that causes tomato leaf blight.

The acquired resistance that sometimes follows the hypersensitive response is accompanied by the synthesis of PR proteins. Treatment of plants with salicylic acid or with aspirin leads to the production of PR proteins and to a resistance to pathogens. Salicylic acid treatment provides substantial protection against tobacco mosaic virus (a well-studied plant pathogen) and some other viruses.

It has been proposed that salicylic acid also serves as a signal for disease resistance. In some cases, microbial infection in one part of a plant leads to the export of salicylic acid to other parts of the plant, where it causes the production of PR proteins before the infection can spread. The PR proteins would, according to this hypothesis, limit the extent of the infection.

Some plant genes match up with pathogen genes

Many plants use the hypersensitive response as a nonspecific defense against various pathogens. There are other defense mechanisms, some of which are highly specific. An important example is **gene-for-gene resistance**, in which the ability of a plant to defend itself against a specific strain of a pathogen depends on the plant's having a particular allele of a gene that matches a particular allele of a corresponding gene in the pathogen (Figure 33.3).

Plants have a large number of *R* genes (resistance genes), and many pathogens have sets of *Avr* genes (avirulence genes). Dominant *R* alleles favor resistance, and dominant *Avr* alleles render a pathogen less effective. If a particular plant has the dominant allele of an *R* gene corresponding to an *Avr* gene represented

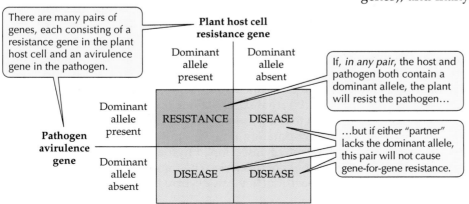

33.3 Gene-for-Gene Resistance A match in a single pair of genes promotes resistance even if all the other pairs are mismatches.

by the dominant allele in a pathogen strain, the plant will be resistant to that strain even if none of the other *R–Avr* pairs features matching dominant alleles. (This effect, one *R–Avr* pair overruling the others, is an example of epistasis, which was discussed in Chapter 10.)

The mechanism of gene-for-gene resistance is not completely understood. There are thousands of specific *R* genes in the plant kingdom, and their products have different functions. The *Avr* genes are simply genes, in pathogens, that cause the pathogen to produce a substance that elicits a defensive response in the plant (see Figure 33.1). Most gene-for-gene interactions trigger the hypersensitive response.

Not all biological threats to plants come from microorganisms and viruses that cause diseases. Many animals *eat* plants.

Plants and Herbivores: Benefits and Losses

Herbivores—animals that eat plants—depend on plants for energy and nutrients. Plants have defense mechanisms to protect themselves against herbivores, as we will see, but first let's consider some examples in which herbivores have a positive effect on the plants they eat.

Grazing increases the productivity of some plants

Consider the phenomenon of grazing, in which an animal predator eats part of its prey, such as the leaves of plants, without killing the prey organism, which has the potential to grow back (Figure 33.4). What are the consequences of grazing? Is it detrimental to the plants, or are they somehow adapted to their place in the food chain of nature? Certain plants and their predators evolved together, each acting as the agent of natural selection on the other. Because of this coevolution, grazing increases photosynthetic production in certain plant species.

The removal of some leaves from a plant typically increases the rate of photosynthesis of the remaining leaves. This phenomenon probably is the result of several factors. First, nitrogen obtained from the soil by the roots no longer needs to be divided among so many leaves. Second, the transport of sugars and other photosynthetic products from the leaves may be enhanced because the demand for those products in the sinks—such as roots—is undiminished, while the sources—leaves—have been decreased.

A third and particularly significant factor, especially in grasses, is an increase in the availability of light to the younger, more active leaves or leaf parts. The removal of older, dying—or even dead—leaves and leaf parts by a grazer decreases the shading of younger leaves, and unlike most other plants, which grow from their shoot and leaf tips, grasses grow from the base of the shoot and leaf.

A clear case of increased productivity resulting from grazing was reported in 1987 by Ken Paige of the University of Utah and Thomas Whitham of Northern Arizona University. Mule deer and elk graze many plants, including one called scarlet gilia. The grazing removes about 95 percent of the aboveground part of each scarlet gilia (Figure 33.5). However, each plant quickly regrows not one but four replacement stems.

33.4 Is Grazing Helpful or Harmful to Plants? Grazing mammals such as these North American elk exist in virtually all the world's biomes, and the foliage they feed on has evolved along with them.

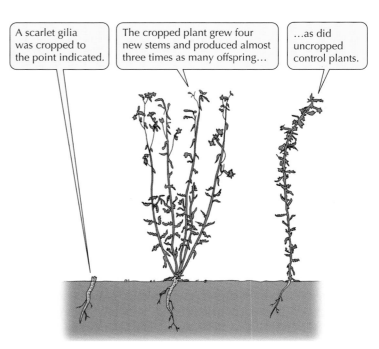

A scarlet gilia was cropped to the point indicated.

The cropped plant grew four new stems and produced almost three times as many offspring...

...as did uncropped control plants.

33.5 Overcompensation for Being Eaten Experiments reveal that some plants benefit from the effects of grazing.

The cropped (grazed) plants produce three times as many fruits by the end of the growing season as do ungrazed plants. Paige and Whitham cropped some scarlet gilia in the laboratory; these plants, too, produced more fruits than did uncropped plants. Not only does the productivity of cropped plants increase, but we may also conclude that grazing by herbivores increases the fitness of scarlet gilia, since the cropped plants pass their genes on to more surviving offspring than do uncropped plants.

Some grazed trees and shrubs continue to grow until much later in the season than ungrazed but otherwise similar plants do. This longer growing season results in part because the removal of apical buds by the grazers stimulates lateral buds to become active, thus producing a more heavily branched plant. In addition, leaves on ungrazed plants may die earlier in the growing season than leaves on grazed plants.

Some other plants also profit from moderate herbivory. In addition to the increase in its productivity, a plant benefits by attracting animals that spread its pollen or that eat its fruit and thus distribute its seeds through their feces. However, resisting attack by fungi and herbivorous animals and inhibiting the growth of neighboring plants are also to the advantage of a plant.

Some plants produce chemical defenses against herbivores

Rhus diversiloba is a gorgeous plant. Its flowers don't look like much, but its leaves are green and glistening in the spring and warm and red in late summer. It grows vigorously. It grows like a weed. It *is* a weed. It should be a great meal for many an animal, but it isn't. *Rhus diversiloba* is poison oak (Figure 33.6).

Poison oak and its close relative poison ivy are good examples of plants with a defense system that can't be seen. Their leaves are unmarked, no insect bites them, and no deer grazes on them. What keeps poison oak free of pests and predators? The plant produces a chemical substance that is highly toxic to animals and lethal to some. The substance causes most humans to break out in an extremely uncomfortable rash that lasts for days. You may have run afoul of one of these plants yourself. Poison oak and poison ivy are just two of thousands of examples of the evolution of chemical benefits to the plant kingdom.

Plants attract, resist, and inhibit other organisms often by producing special chemicals known as **secondary products**. (Primary products are substances, such as proteins, nucleic acids, carbohydrates, and lipids, that are produced and used by all living things.) Although different kinds of organisms share a biochemical heritage of primary products, they may also differ as radically in chemical content as in external appearance. Animals and fungi, for example, need various enzymes to digest their food, a need not

Rhus diversiloba

33.6 Watch Out! Beauty may be only cuticle-deep, as in this poison oak; its chemical defenses can sicken or even kill an animal that eats it.

shared by most plants. The plant kingdom is noteworthy for its profusion of secondary products that serve special functions. These compounds help plants compensate for being unable to move. Although a plant cannot flee its herbivorous enemies, it may be able to defend itself chemically.

The effects of defensive secondary products on animals are diverse. Some act on the nervous systems of herbivorous insects, mollusks, or mammals. Others mimic the natural hormones of animals, causing some insect larvae to fail to develop into adults. Still others damage the digestive tracts of herbivores. Some secondary products are toxic to fungal pests. We make commercial use of secondary plant products as fungicides, insecticides, and pharmaceuticals.

There are more than 10,000 secondary plant products, ranging in molecular weight from about 70 to more than 400,000; most, however, are of low molecular weight. Some are produced by only a single species, while others are characteristic of an entire genus or even family. Their roles are diverse, and in most cases unknown. While many secondary products have protective functions, as mentioned already, others are essential as attractants for pollinators and seed dispersers. Table 33.1 gives the major classes of secondary plant products and their roles. A few proteins and amino acids also protect plants against herbivores. In the next section we will look at a specific example of an insecticidal secondary product, canavanine.

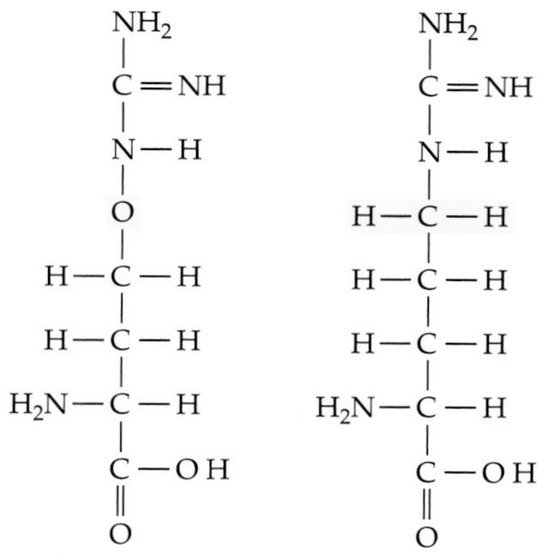

NH$_2$
|
C=NH
|
N—H
|
O
|
H—C—H
|
H—C—H
|
H$_2$N—C—H
|
C—OH
‖
O

Canavanine

NH$_2$
|
C=NH
|
N—H
|
H—C—H
|
H—C—H
|
H—C—H
|
H$_2$N—C—H
|
C—OH
‖
O

Arginine

33.7 A Toxic Secondary Product and Its Amino Acid Analog The chemical structures of canavanine and arginine are very similar; insect larvae may incorporate canavanine in place of the amino acid, with lethal results.

Some secondary products play multiple roles

Some plants produce *canavanine*, an amino acid that is not found in proteins but that is closely similar to the amino acid arginine, which is found in almost all proteins (Figure 33.7). Canavanine has recently been found to have two important roles in plants that produce it in significant quantity. The first role is as a nitrogen-storing compound in seeds. The second role is a defensive one and is based on the similarity of canavanine to arginine.

Many insect larvae that consume canavanine-containing plant tissue are poisoned: The canavanine is mistakenly incorporated into the insect's proteins in some of the places where the DNA has coded for arginine. The structure of canavanine is different enough from that of arginine that some of the proteins end up with modified tertiary structure and hence reduced biological activity. The defects in protein structure and function lead in turn to developmental abnormalities in the insect.

A few insect larvae are able to eat canavanine-containing plant tissue and still develop normally. How can this be? In these larvae, the enzyme that charges the tRNA specific for arginine discriminates accurately between arginine and canavanine. The canavanine they ingest is thus not incorporated into the proteins they form. The corresponding enzyme in susceptible larvae discriminates much less effectively between those two amino acids, so canavanine is frequently substituted for arginine.

In some cases, splicing in a gene may confer insect resistance

A group of scientists at the ARCO Plant Cell Research Institute and the University of Wisconsin recently studied the abilities of seeds of wild and domesticated common beans (*Phaseolus vulgaris*) to resist attack by two species of bean weevils. The investigation began with the observation that whereas some wild bean seeds are highly resistant to the weevils, no cultivated bean seeds show such resistance. The scientists discovered that all weevil-resistant bean seeds contain a specific seed protein, *arcelin*. This protein has never been found in cultivated bean seeds. Therefore, the scientists hypothesized that arcelin is responsible for the resistance of some seeds to predation by the weevils.

To rule out other differences between wild and cultivated beans as being responsible for the resistance, the scientists performed two series of experiments testing the relationship between resistance and the possession of arcelin. In one series, cultivated and wild bean plants were crossed. The progeny seeds of such crosses showed an absolute correlation between the presence of arcelin and resistance to weevils. In the other series of experiments, the scientists worked with artificial bean seeds made by removing seed coats of cultivated beans and grinding the remainder of the seeds into flour. Different concentrations of arcelin were added to different batches, and the flour was molded into artificial seeds. Bean weevils were then allowed to attack the artificial seeds. The more arcelin

TABLE 33.1 Secondary Plant Products

CLASS	SOME ROLES
Alkaloids	Affect herbivore nervous systems
Other nitrogen and sulfur compounds	Cause cancers, nerve damage, and pain in herbivores
Phenolics	Taste obnoxious to herbivores; affect herbivore nervous systems; act as fungicides
Quinones	Inhibit growth of competing plants
Terpenes	Act as fungicides and insecticides; attract pollinators
Steroids	Mimic animal hormones; prevent normal development of insect herbivores
Flavonoids	Attract pollinators and animals that disperse seeds

the artificial seeds contained, the more resistant they were to weevils.

Next the scientists prepared, cloned, and sequenced a cDNA (complementary DNA is discussed in Chapter 16) for arcelin so that they could compare the primary structure of arcelin with that of other, possibly related proteins that may confer insect resistance on seeds of beans and other legumes. The goal of this ongoing work is to introduce genes for arcelin or other resistance-conferring proteins into agriculturally important crops such as beans. In preliminary tests, arcelin in cooked beans was also shown not to be harmful to rats—a first step toward determining whether arcelin is safe in food for humans.

As we have seen, many plants produce their own toxic chemicals that protect them from herbivores and from pathogenic microbes. Why don't these secondary products kill the plants that produce them?

Why don't plants poison themselves?

Plants that produce toxic secondary products generally use one of the following measures to protect themselves: (1) keeping the toxic material isolated in a special compartment such as the central vacuole; (2) producing the toxic substance only after the plant's cells have already been damaged; or (3) using either modified enzymes or modified receptors that do not recognize the toxic substance. The first method is the most common.

Plants using the first method store their water-soluble poisons in vacuoles. If hydrophobic, the poison is stored in **laticifers** (tubes containing a white, rubbery latex) or dissolved in waxes on the epidermal surface. The compartmentalized storage keeps the toxic substance away from the mitochondria, chloroplasts, and other parts of the plant's own metabolic machinery.

Some plants store the precursors of toxic substances in one compartment, such as the epidermis, and store the enzymes that convert the precursors to the active poison in another compartment, such as the mesophyll. These plants use the second protective measure—that is, producing the toxic substance only after being damaged. When an herbivore chews part of the plant, cells rupture and the enzymes come in contact with the precursors, releasing the toxic product that repels the herbivore. The only part of the plant that is damaged by the toxic material is that which was already damaged by the herbivore. Plants that respond to attack by producing cyanide—a potent inhibitor of cellular respiration in all organisms that respire—are among those that use this protective measure.

The third protective measure is used by the canavanine-producing plants described earlier. These plants produce a tRNA-charging enzyme for arginine that does not bind canavanine. However, some herbivores can evade being poisoned by canavanine because their enzymes, like that of the canavanine-producing plants, do not use canavanine by mistake.

Not all plants use protective chemicals to defend themselves against herbivores or pathogens. Should we encourage such defenses in the plants that we cultivate? That is, should we be breeding crop plants that make their own pesticides?

Agricultural scientists must choose between protection and taste. A plant with sturdy chemical defenses may taste bad, make us sick, or even kill us. Not surprisingly, we have bred our food plants to minimize toxicity and obnoxious tastes—that is, to make the plants contain very little in the way of chemical defenses. As a result, we must take steps to save our relatively defenseless crops. A current goal of agricultural biotechnology is to develop crop plants that produce their own useful pesticides that are not harmful or offensive to us.

The plant doesn't always win, of course

Milkweeds such as *Asclepias syriaca* are latex-producing (laticiferous) plants. When damaged, a milkweed releases copious amounts of a toxic latex from its laticifers. Latex has long been suspected to deter insects from eating the plant, because laticiferous plants are not attacked by insects that feed on neighboring plants of other species. This observed behavior is consistent with, but does not prove, the hypothesis that the latex keeps the insects at bay. Stronger support for the hypothesis was obtained by two zoologists—David Dussourd, now of the University of Maryland, and Thomas Eisner of Cornell University—who studied field populations of *Labidomera clivicollis,* a beetle that is one of the few insects that feed on *A. syriaca.*

Dussourd and Eisner observed a remarkable prefeeding behavior: The beetles cut a few veins in the leaves before settling down to dine (Figure 33.8). In the undamaged plant the latex is under pressure, so cutting the veins, with their adjacent laticifers, causes massive leakage and depressurizes the system. By cutting a few veins, the beetles interrupt the latex supply to a downstream portion of the leaf. The beetles then move to the relatively latex-free portion and eat their fill.

Some other insects that do not feed on undamaged milkweeds will eat parts of leaves that have had the latex supply cut off. When presented simultaneously with leaves on undamaged plants and leaf parts that have had their laticifers cut, *L. clivicollis* and other insects that share this vein-cutting behavior select the relatively latex-free leaf parts.

Does this behavior of the beetles negate the adaptational value of latex protection? Not at all. There are still great numbers of potential insect pests that are effectively deterred by the latex. And evolution grinds on. Over time, milkweed plants producing higher con-

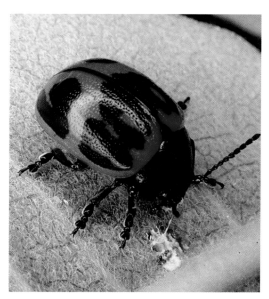

33.8 Disarming a Plant's Defenses   This beetle is inactivating a milkweed's defense system by cutting its laticifer supply lines.

centrations of toxins may be selected by virtue of their ability to kill beetles that cut the laticifers.

We have considered many examples of challenges presented to plants by the biological environment. Now we turn our attention to the physical environment, beginning with a look at plant adaptations to environments where water is scarce.

Dry Environments

Water for plants and other organisms is often in short supply in the terrestrial environment. Some terrestrial habitats, such as deserts, intensify this challenge, and many plants that inhabit particularly dry areas have one or more structural adaptations that allow them to conserve water. Plants adapted to dry environments are called **xerophytes**. In this section, we'll consider structural and chemical adaptations to dry environments, but first let's look at a behavioral adaptation.

Some plants evade drought

Some desert plants have no special *structural* adaptations for water conservation other than those found in almost all flowering plants. Instead they have an alternative *strategy*. These plants carry out their entire life cycle—from seed to seed—during a brief period in which the surrounding desert soil is sufficiently moist (Figure 33.9). Through the long dry periods that intervene, only the seeds remain alive, until enough moisture is present to trigger the next life cycle. These desert annuals simply evade the periods of drought. Plants that remain active during the dry periods must have special adaptations that enable them to survive.

Some leaves have special adaptations to dry environments

The secretion of a heavier layer of cuticle over the leaf epidermis to retard water loss is a common adaptation to dry environments. An even more common adaptation is a dense covering of epidermal hairs. Some species have stomata only in sunken cavities below the leaf surface, which reduces the drying effects of air currents; often these stomatal cavities contain hairs as well (Figure 33.10). Ice plants and their relatives have fleshy leaves in which water may be stored. Others, such as ocotillo, produce leaves only when water is abundant, shedding them as the soil dries out (Figure 33.11). Cacti and similar plants have spines rather than typical leaves, and photosynthesis is confined to the fleshy stems. The spines may reflect incident radiation, or they may dissipate heat. Corn and some related grasses have leaves that roll up during dry periods, thus reducing the leaf surface area through which water is lost. Some trees that grow in arid regions have

33.9 Desert Annuals Evade Drought   Seeds of desert plants often lie dormant for long periods awaiting conditions appropriate for germination. When they do germinate, they grow and reproduce rapidly before the short wet season passes. They cover the desert landscape with color for only a few weeks, since water is inadequate for their survival at other times.

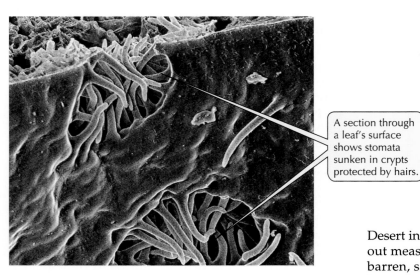

A section through a leaf's surface shows stomata sunken in crypts protected by hairs.

33.10 Stomatal Crypts   Stomata in some water-conserving leaves are in sunken pits called stomatal crypts. The hairs covering these crypts trap moist air. (A section of the leaf's interior can be seen at the top of the photo.)

leaves that hang vertically at all times, thus evading the midday sun. Characteristic examples are some eucalyptuses (Figure 33.12).

Xerophytic adaptations of leaves minimize water loss by the plant. However, such adaptations simulta-

neously minimize the uptake of carbon dioxide and thus limit photosynthesis. In consequence, most xerophytes grow slowly, but they utilize water more efficiently than do other plants; that is, they fix more grams of carbon by photosynthesis per gram of water lost to transpiration than other plants do.

There are other adaptations to a limited water supply

Roots may also be adapted to environments low in water. The Atacama Desert in northern Chile often goes several years without measurable rainfall. The landscape there is almost barren, save for a substantial number of surprisingly large mesquite trees (genus *Prosopis*). These trees obtain water by having taproots that grow to great depths, sufficient to reach underground water supplies (Figure 33.13)—as well as by condensation on their leaves. A more common adaptation of desert plants is to have a root system that grows during rainy seasons but dies back during dry periods. Cacti have shallow but extensive fibrous root systems that effectively trap water at the surface of the soil following even light rains.

During dry periods, the thorny, leafless stems of an ocotillo appear almost dead.

When water is on hand, leaves develop rapidly and provide the plant with photosynthetic products.

33.11 Opportune Leaf Production   The ocotillo is a xerophyte that lives in the lower deserts of the southwestern United States and northern Mexico.

33.12 Shade at Midday   Because eucalyptus leaves hang vertically, their flat surfaces are not presented directly to the midday sun. This adaptation minimizes heating as well as water loss.

Prosopis glandulosa

33.13 Mining Water with Deep Taproots Death Valley, California is not as arid as the Chilean Atacama, but the mesquite must reach far down into the sand dunes for its water supply.

Pneumatophores are root extensions that grow out of the water, under which the rest of the roots are submerged.

33.14 Coming Up for Air The roots of the mangroves in this tidal swamp obtain oxygen through pneumatophores.

Xerophytes and other plants that receive inadequate water may accumulate the amino acid proline to substantial concentrations in their vacuoles. As a consequence, the osmotic potential and water potential of the cells become more negative; thus these plants tend to extract more water from the soil than do plants that lack this adaptation.

As we have seen, there are many ways in which some plants eke out an existence in environments with very little water. What happens if there is too much water?

Where Water Is Plentiful and Oxygen Is Scarce

When soils become waterlogged, oxygen from the soil becomes less available. Most plants cannot tolerate this situation long. Some species, however, are adapted to life in a swampy habitat. Their roots grow slowly and hence do not penetrate deeply. With an oxygen level too low to support aerobic respiration, the roots carry on alcoholic fermentation (see Chapter 7), which provides ATP for the activities of the root system.

The root systems of some plants adapted to swampy environments have **pneumatophores**, extensions that grow out of the water and up into the air (Figure 33.14). Oxygen diffusing into pneumatophores aerates the submerged parts of the root system. Cypresses and some mangroves are examples of plants with pneumatophores.

Submerged or partly submerged aquatic plants often have large air spaces in the leaf parenchyma and in the petioles. Tissue with such air spaces is called **aerenchyma** (Figure 33.15). Aerenchyma stores oxygen

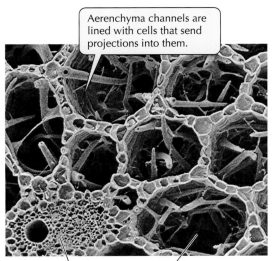

Aerenchyma channels are lined with cells that send projections into them.

Vascular bundle   Open channel

33.15 Aerenchyma Lets Oxygen Reach Submerged Tissues This scanning electron micrograph, a cross section of a petiole of the yellow water lily, shows a vascular bundle and aerenchyma. Because aerenchyma has far fewer cells than most petiole tissue, respiratory metabolism proceeds at a lower rate and the need for oxygen is much reduced.

produced by photosynthesis and permits its ready diffusion to places where it is needed for cellular respiration. Aerenchyma also imparts buoyancy.

Thus far we have considered water supply—either too little or too much—as a limiting factor in plant growth. Other substances also can make an environment inhospitable to plant growth. One of these is salt.

Saline Environments

Worldwide, no toxic substance restricts flowering plant growth more than salt (sodium chloride) does. Saline—salty—habitats support, at best, sparse vegetation. The **halophytes**, plants adapted to such a habitat, belong to a wide variety of flowering plant groups.

Saline environments themselves are diverse, ranging from hot, dry, salty deserts to moist, cool, salty marshes. Along the seashore are salty environments created by ocean spray. The ocean itself is a saline environment, as are river estuaries, where fresh and salt water meet and mingle. The salinization of agricultural land is an increasing world problem. Even where crops are irrigated with fresh water, low in salt, sodium ions in the water accumulate in the soil to ever greater concentrations. Biologists in Israel and elsewhere have had some success in breeding crops that can be watered with seawater or diluted seawater.

Saline environments pose an osmotic problem. Because of a high salt concentration, the environment has an unusually large negative water potential. To obtain water from such an environment, resident plants must have an even more negative water potential than that of plants in nonsaline environments; otherwise, the plants lose water and wilt. A second problem related to the saline environment is the potential toxicity of high concentrations of certain ions, notably sodium. Chloride ions may also be toxic at high concentrations.

How can halophytes cope with a highly saline environment?

Most halophytes accumulate salt; some have salt glands

Most halophytes share one adaptation: They accumulate sodium and, usually, chloride ions, and they transport these ions to the leaves. Nonhalophytes accumulate relatively little sodium, even when placed in a saline environment; of the sodium that is absorbed by their roots, very little is transported to the shoot. The increased salt concentration in halophytes makes their water potential more negative, so they can take up water from the saline environment. We still do not know how halophytes are able to tolerate such high internal sodium and chloride concentrations without being poisoned.

Some halophytes have other adaptations to life in saline environments. For example, some have salt

33.16 Secreting Salt This salty mangrove has used special glands to secrete salt, which now appears as crystals on the leaves.

glands in their leaves. These glands excrete salt, which collects on the leaf surface until it is removed by rain or wind (Figure 33.16). This adaptation, which reduces the danger of poisoning by accumulated salt, is found both in some desert plants, such as tamarisk, and in some mangroves growing in seawater in the Tropics.

Salt glands can play multiple roles, as in the arid-zone shrub *Atriplex halimus*. This shrub has glands that secrete salt into small bladders on the leaves where, by increasing the gradient in water potential, the salt helps the leaves obtain water from the roots. At the same time, by making the water potential of the leaves more negative, the salt reduces the transpirational loss of water to the atmosphere.

These adaptations are specific to halophytes. Several other adaptations are shared by halophytes and xerophytes.

Halophytes and xerophytes have some adaptations in common

Many halophytes accumulate the amino acid proline in their cell vacuoles. Unlike sodium, proline is relatively nontoxic. As in xerophytes, the accumulated proline makes the water potential more negative.

Succulence—the possession of fleshy, water-storing leaves—is an adaptation to dry environments. The same adaptation is common among halophytes, as might be expected, since saline environments also make water uptake difficult for plants. Succulence characterizes many halophytes that occupy salt marshes. There the salt concentration in the soil solution may change throughout the day; while the tide is out, for instance, evaporation increases the salt concentration. Succulence may offer a reserve of water for the plant during the period of maximum salinity;

when the salinity drops as the tide comes in, the leaf's store of water is replenished. Many succulents—both xerophytes and halophytes—use crassulacean acid metabolism (CAM) and have reversed stomatal cycles that enable them to conserve water by closing their stomata in the daytime (see Figure 32.12).

Other general adaptations to a saline environment are of the same sorts observed in xerophytes. These include high root-to-shoot ratios, sunken stomata, reduced leaf areas, and thick cuticles.

Some halophytes adjust to a range of salinities

The halophyte *Triglochin maritima* is unusual because it can adjust to a wide range of environmental salinities. Work in Toronto, Canada, established that *T. maritima* plants change in many ways when they are shifted to environments with lower or higher salt concentrations. Researchers watered some plants with seawater and some with diluted seawater. The plants watered with pure seawater produced much smaller leaves, with smaller cells, than those watered with diluted seawater, but they retained their leaves longer. The rate of leaf production is the same regardless of salinity, but when the environment changes, the pattern of leaf production changes accordingly.

These morphological differences in *T. maritima* leaves are accompanied by physiological changes. The leaf cells of the plants grown on undiluted seawater contained much higher concentrations of sodium and chloride ions, as well as of proline. The rates of photosynthesis in the leaves of the plants watered with undiluted seawater also were higher.

Salt is not the only toxic solute in soils. Some others are more toxic than salt when presented at the same concentration.

Habitats Laden with Heavy Metals

High concentrations of heavy-metal ions, such as copper, lead, nickel, and zinc, poison most plants, even though plants require some heavy metals at low concentrations. Some sites are naturally rich in heavy metals as a result of normal geological processes. Acid rain leads to the release of toxic aluminum ions in the soil. Other human activities, notably the mining of metallic ores, leave localized areas—known as tailings—with substantial concentrations of heavy metals and low concentrations of nutrients. Such sites are hostile to most plants, and seeds falling on them generally do not produce adult plants.

Mine tailings rich in heavy metals, however, generally are not completely barren (Figure 33.17). They may support healthy plant populations that differ genetically from populations of the same species on the surrounding normal soils. How can these plants survive? Within some species, a few individuals may

33.17 Life after Strip Mining Although high concentrations of metal kill most plants, grass is colonizing this eroded strip mine in North Park, Colorado.

have genotypes that allow them to survive in soils rich in heavy metals. Those individuals may grow poorly on such soils compared with their potential for growth on more normal soils, but they may nevertheless survive. Or, because most plants cannot survive in such habitats and hence the competition may be sharply reduced, the few plants growing in them may even thrive.

Initially, some plants were thought to tolerate heavy metals by excluding them: By not taking up the metal ions, the plant could avoid being poisoned. Measurements have shown, however, that tolerant plants growing on mine tailings do take up the heavy metals, accumulating them to concentrations that would kill most plants. Thus the tolerant plants must have a mechanism for dealing with the heavy metals they take up. Some tolerant plants may be found to be useful agents for bioremediation, to decrease the heavy metal content of some soils.

The British biologist D. Jowett made an interesting discovery about plants that tolerate heavy metals. In Wales and Scotland, bent grass (*Agrostis*) grows near many mines (see Figure 21.7). From mine to mine, the heavy metals in the soil differ. Jowett obtained samples of bent grass from several such sites and tested their ability to grow in various solutions, each containing only one of the heavy metals. In general, the plants tolerated a particular heavy metal—the one most abundant in their habitat—but were sensitive to other heavy metals. That is, they tolerated only one or two heavy metals, rather than heavy metals as a group.

Tolerant populations can evolve and colonize an area surprisingly rapidly. The bent grass population around a particular copper mine in Wales is resistant to copper and relatively abundant, even though the copper-rich soil dates from mining done late in the nineteenth century, only a century ago. If populations can evolve and cope with toxic soils, can they deal as well with soils in which nutrients are in short supply or in improper balance?

Serpentine Soils

Another unproductive soil type that is found in many parts of the world is derived from rock called **serpentine**. Calcium is in short supply in serpentine soils, as are some other essential plant nutrients. Magnesium is present in greater concentration than calcium, reducing plant growth. Chromium, nickel, and certain other heavy metals may be abundant. These factors make serpentine soils inhospitable to many plants. The vegetation on most serpentine soils differs dramatically from that on immediately adjacent nonserpentine soils, and the serpentine vegetation generally is more sparse and less diverse.

The shortage of calcium and the higher magnesium concentration probably are the principal challenges facing potential colonizers of a serpentine soil. Several plant species have physiological adaptations to meet the challenges. Different species exhibit striking differences in their response to calcium supply in the soil.

To test these differences, biologists divided some serpentine soil into several samples, adjusted the calcium level of each, and grew jewel flower (*Streptanthus glandulosus*), which grows on serpentine, and tomato—a crop plant intolerant of serpentine—on each sample. The growth of the tomato plants was sharply dependent on calcium concentration, while that of the jewel flower was remarkably insensitive to it (Figure 33.18). Serpentine plants such as jewel flower can absorb calcium efficiently even from soils highly deficient in that element. They may also be able either to exclude excess magnesium or to tolerate high internal magnesium concentrations. Such adaptations enable serpentine plants to grow in an environment hostile to most other plants.

Plants have effective mechanisms for coping with environmental challenges of many kinds. Their success is obvious—look around you.

Summary of "Plant Responses to Environmental Challenges"

Plant–Pathogen Interactions

• Plants and pathogens evolve together—the pathogens evolving attack mechanisms and the plants evolving defense mechanisms. **Review Figure 33.1**
• Some of the plant's defenses are mechanical—physical barriers to the entry of pathogens. Plants increase their mechanical defenses when attacked. Plants also have chemical defenses, including PR proteins and phytoalexins, that are produced in response to attack by a pathogen.
• In the hypersensitive response, cells produce phytoalexins and PR proteins and then die, trapping the pathogens in dead tissue. The hypersensitive response is often followed by an acquired resistance in which salicylic acid activates further synthesis of PR proteins and triggers responses in other parts of the plant.
• The hypersensitive response is nonspecific. A more specific response is gene-to-gene resistance. This mechanism is based on a matching up of appropriate alleles in a plant's resistance genes and a pathogen's avirulence genes. **Review Figure 33.3**

Plants and Herbivores: Benefits and Losses

• Grazing by herbivores increases the productivity of some plants. This increase results from a reduced competition among parts of the plant for nitrogen, a change in transport patterns, and especially a reduction in mutual shading of leaves. **Review Figure 33.5**
• Some plants produce secondary products that function as chemical defenses against herbivores. Some secondary products play multiple roles that may include toxicity to herbivores and functioning as storage compounds. **Review Table 33.1**
• To avoid poisoning themselves, plants may confine the toxic substances they produce to special compartments, or they may produce the substances only after cells have been damaged, or they may form enzymes and receptors that are not affected by the substances.

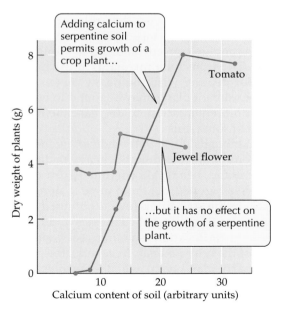

33.18 Responses to Added Calcium Experimenters added calcium to serpentine soil, allowing tomatoes to grow on it. The jewel flower *Streptanthus glandulosus*, which grows even on unsupplemented serpentine, was unaffected by the added mineral.

- In developing new pathogen-resistant strains for agriculture, scientists must also consider both taste and toxicity. Some insects, exhibiting coevolution with the plants on which they feed, have developed adaptations to overcome a plant's defenses.

Dry Environments

- Xerophytes are adapted to dry environments.
- Some xerophytic leaves have special adaptations to dry environments: a thickened cuticle, epidermal hairs, sunken stomata, fleshy leaves and stems, and spines. Other xerophytes display their leaves in ways that reduce the incident sunshine at midday. **Review Figures 33.10, 33.11**
- Other adaptations to dry environments include long taproots, root systems that die back each year, and proline formation to make the plant's water potential more negative.
- Desert annuals evade drought by living only long enough to take advantage of the brief period during which the soil has enough moisture to support them.

Where Water Is Plentiful and Oxygen Is Scarce

- The submerged roots of some plants form pneumatophores to promote oxygen uptake. The aerenchyma of submerged plant parts stores and permits the diffusion of oxygen. **Review Figures 33.14, 33.15**

Saline Environments

- Halophytes are adapted to environments that have high concentrations of sodium chloride. A saline environment can withhold water from plants as effectively as a dry environment.
- Most halophytes accumulate salt, and some have salt glands that excrete the salt to the leaf surface. Some halophytes can adjust to a range of salinities.
- Halophytes and xerophytes have some adaptations in common, including the accumulation of proline in vacuoles, succulence, crassulacean acid metabolism (CAM) with a reversed stomatal cycle, and specialized leaf adaptations.

Habitats Laden with Heavy Metals

- Copper, lead, nickel, and zinc are among the toxic heavy metals. Most species that grow where heavy metals contaminate the soil grow better on normal soils, but they are the dominant vegetation where heavy metals are abundant because of reduced competition from other species.
- Rather than excluding heavy metals, tolerant plants deal with them after taking them up. A given plant's tolerance is limited to only one or two heavy metals.

Serpentine Soils

- Serpentine soils are low in calcium and some other essential nutrient elements, and they are relatively high in magnesium. Serpentine soils often contain high levels of chromium, nickel, and some other heavy metals.
- Well-adapted serpentine-tolerant plants grow at rates independent of calcium concentration in the soil, and they deal successfully with high magnesium concentrations. **Review Figure 33.18**

Self-Quiz

1. Which of the following is *not* a common defense against bacteria, fungi, and viruses?
   a. New wood formation
   b. Phytoalexins
   c. A waxy covering
   d. The hypersensitive response
   e. Mycorrhizae

2. Plants sometimes protect themselves from their own toxic secondary products by
   a. producing special enzymes that destroy the toxic substances.
   b. storing precursors of the toxic substances in one compartment and the enzymes that convert precursors to toxic products in another compartment.
   c. storing the toxic substances in mitochondria or chloroplasts.
   d. distributing the toxic substances to all cells of the plant.
   e. performing crassulacean acid metabolism.

3. Herbivory
   a. is predation by plants on animals.
   b. always reduces plant growth.
   c. usually increases the rate of photosynthesis in the remaining leaves.
   d. reduces the rate of transport of photosynthetic products from the remaining leaves.
   e. always is lethal to the grazed plant.

4. Which statement about secondary plant products is *not* true?
   a. Some attract pollinators.
   b. Some are poisonous to herbivores.
   c. Most are proteins or nucleic acids.
   d. Most are stored in vacuoles.
   e. Some mimic the hormones of animals.

5. Which statement about latex is *not* true?
   a. It is sometimes contained in laticifers.
   b. It is typically white.
   c. It is often toxic to insects.
   d. It is a rubbery solid.
   e. Milkweeds produce it.

6. Which of the following is *not* an adaptation to dry environments?
   a. A less negative osmotic potential in the vacuoles
   b. Hairy leaves
   c. A heavier cuticle over the leaf epidermis
   d. Sunken stomata
   e. A root system that grows each rainy season and dies back when it is dry

7. Some plants adapted to swampy environments meet the oxygen needs of their roots by means of a specialized tissue called
   a. parenchyma.
   b. aerenchyma.
   c. collenchyma.
   d. sclerenchyma.
   e. chlorenchyma.

8. Halophytes
    a. all accumulate proline in their vacuoles.
    b. have osmotic potentials that are less negative than those of other plants.
    c. are often succulent.
    d. have low root-to-shoot ratios.
    e. rarely accumulate sodium.

9. Which of the following is *not* a commonly toxic heavy metal?
    a. Copper
    b. Lead
    c. Nickel
    d. Potassium
    e. Zinc

10. Plants that tolerate heavy metals commonly
    a. grow poorly where the soil contains heavy metals.
    b. do not take up the heavy metal ions.
    c. are tolerant to all heavy metals.
    d. are slow to colonize an area rich in heavy metals.
    e. weigh more than plants that are sensitive to heavy metals.

Applying Concepts

1. We mentioned the possibility of designing crop plants that produce their own pesticides. Now chemical companies are designing crop plants capable of detoxifying weed killers, so that crops grow after farmers have destroyed competing vegetation. Discuss the likely usefulness and possible drawbacks of such applications of recombinant-DNA technology.

2. How might plant adaptations affect the evolution of herbivores? How might adaptations of herbivores affect plant evolution?

3. The stomata of the common oleander (*Nerium oleander*) are sunk in crypts in its leaves. Whether or not you know what oleander is, you should be able to describe an important feature of its natural habitat. What is this feature?

4. Explain in detail why halophytes often use the same mechanisms for coping with their challenging environment as xerophytes do for coping with theirs.

5. In ancient times, people used less sophisticated methods for mining than we use today. Thus ancient mines often yield substantial profits to modern-day miners who find and work them. On the basis of material in this chapter, how might you try to locate an ancient mine site?

Readings

Barrett, S. C. H. 1987. "Mimicry in Plants." *Scientific American,* September. A discussion of how some plants use camouflage to avoid predation, and some weeds survive by mimicking crops so that humans will select them.

Dussourd, D. E. and T. Eisner. 1987. "Vein-cutting Behavior: Insect Counterploy to the Latex Defense of Plants." *Science,* vol. 237, pages 898–901. A clear, readable account of field observations and sound experimentation exploring the phenomenon discussed in this chapter.

Goulding, M. 1993. "Flooded Forests of the Amazon." *Scientific American,* March. A description of the adaptations of plants and other organisms in a seriously threatened environment.

Lewin, R. 1987. "On the Benefits of Being Eaten." *Science,* vol. 236, pages 519–520. A discussion of the work on scarlet gilia cited in this chapter; included are other examples of cropping.

Ronald, P. C. 1997. "Making Rice Disease-Resistant." *Scientific American,* November. A description of the genetic engineering of rice to protect the crop from bacterial infection.

Rosenthal, G. A. 1986. "The Chemical Defenses of Higher Plants." *Scientific American,* January. A look at the many chemicals that plants use to repel or poison herbivores or that retard the growth of herbivorous insects, and how some herbivores use these plant-derived compounds for their own defense.

Shigo, A. L. 1985. "Compartmentalization of Decay in Trees." *Scientific American,* April. A description of how trees defend themselves by sealing off the damage done to them.

Taiz, L. and E. Zeiger. 1998. *Plant Physiology*, 2nd Edition. Sinauer Associates, Sunderland, MA. A good general textbook. Chapters 13 and 25 focus on the topics of this chapter.

Chapter 34

Plant Nutrition

A Meat-Eating Plant
Dionaea, the Venus flytrap, has adapted to a nitrogen-poor environment by becoming carnivorous. It obtains this necessary mineral from the bodies of insects trapped inside the plant when its hinges snap shut.

*A*n insect has stepped on a trigger hair on the leaf of a Venus flytrap—a big mistake for the insect. The leaf secretes enzymes that will digest the trapped insect. Then the leaf will absorb the products of digestion, especially amino acids, and use them as a nutritional supplement. Why does the Venus flytrap go to this trouble? Few other plants are carnivorous; your petunia plant is not stalking you, after all. But the Venus flytrap (*Dionaea*) lives on soils in which nitrogen is less available than in many other soils. Its carnivorous adaptation affords *Dionaea* another way to obtain needed nitrogen.

Why do plants need nitrogen? The answer is simple if we recall the chemical structures of amino acids—and hence proteins—and nucleic acids. These vital components of all living things contain nitrogen, as do chlorophyll and many other important biochemical compounds. If a plant cannot get enough nitrogen, it cannot synthesize these compounds at a rate adequate to keep itself healthy.

In addition to nitrogen, plants need other nutrients from their environment. In this chapter we will explore basic strategy differences between plants and animals with respect to their nutrition. Then we'll see what nutrients plants require, and how they acquire them. Because most nutrients come from the soil, we will discuss the formation of soils and the impacts of plants on soils. As any farmer can tell you, nitrogen is the nutrient that most often limits plant growth, so we will devote a section specifically to nitrogen metabolism in plants. The chapter concludes with a section on plants that use means other than photosynthesis to supplement their nutrition.

Lifestyles and the Acquisition of Nutrients

Every living thing needs raw materials from its environment. These **nutrients** include the ingredients of macromolecules: carbon, hydrogen, oxygen, and nitrogen. Carbon and oxygen enter the living world through photosynthesis carried out by plants and by some bacteria and protists; these organisms obtain carbon and oxygen from atmospheric carbon dioxide. The principal source of hydrogen is water, usually taken up from the soil solution by plants. For hydrogen, too, photosynthesis is the gateway to the living world.

Nitrogen, which constitutes about four-fifths of the atmosphere, exists as the virtually inert gas N_2. A large amount of energy is required to break the triple covalent bond linking the two nitrogen atoms in a molecule of nitrogen gas and to convert N_2 to a reasonably reactive form from which amino acids and other nitrogen-containing organic compounds may be synthesized.

The movement of nitrogen into organisms begins with processing by some highly specialized bacteria living in the soil. The bacteria fix nitrogen (convert it into a form usable by other organisms) and oxidize the product, yielding materials that can be taken up readily by plants. The plants, in turn, provide organic nitrogen and carbon to animals, fungi, and many microorganisms.

The proteins of organisms contain sulfur, and their nucleic acids contain phosphorus. There is magnesium in chlorophyll, and iron in many important compounds, such as the cytochromes. In the three sections that follow, we'll examine alternative nutritional strategies used by organisms in different kingdoms. Within the soil, minerals dissolve in water, forming a solution that contacts the roots of plants. Plants take up most of these **mineral nutrients** from the soil solution in ionic form.

Autotrophs make their own food; heterotrophs steal theirs

The plant kingdom provides carbon, oxygen, hydrogen, nitrogen, and sulfur to most of the rest of the living world. Plants and some protists and bacteria are autotrophs; that is, they make their own organic food from simple *inorganic* nutrients—carbon dioxide, water, nitrate or ammonium ions containing nitrogen, and a few soluble minerals (Figure 34.1). Organisms that require at least one of their raw materials in the form of *organic* compounds (compounds that contain carbon and hydrogen) are called heterotrophs; herbivores depend directly and carnivores depend indirectly on autotrophs as their source of nutrition.

Most autotrophs are **photosynthetic**; that is, they use light as the source of energy for synthesizing or-

34.1 What Do Plants Need? Plants require only light plus carbon dioxide, water, nitrate or ammonium ions, and several essential minerals. These cucumber plants are growing on nothing more than a solution (dripping from above) that contains these ingredients. This technique is known as hydroponic culture.

ganic compounds from inorganic raw materials. Some autotrophs, however, are **chemosynthetic**, deriving their energy not from light but from reduced inorganic substances such as hydrogen sulfide (H_2S) in their environment. All chemosynthesizers are bacteria. The activities of chemosynthetic bacteria that fix nitrogen are vital to the nutrition of plants. But how does a plant get to the bacterial products—or to its nutrients in general?

How does a stationary organism find nutrients?

An organism that is *sessile* (stationary) must exploit energy that is somehow brought to it. Most sessile animals depend primarily on the movement of water to bring energy in the form of food to them, but a plant's supply of energy arrives at the speed of light from the sun. A plant's supply of essential materials, however, is strictly local, and the plant may deplete its local environment of water and minerals as it develops. How does a plant cope with such a problem?

One answer is to extend itself by growing into new resources—*growth is a plant's version of locomotion*. Root systems mine the soil. By growing, they reach new sources of minerals and water. Growth of leaves helps a plant secure light and carbon dioxide. A plant may compete with other plants for light by outgrowing them, both capturing more light for itself and preventing the growth of its neighbors by shading them.

As it grows, the plant, or even a single root, must deal with environmental diversity. Animal droppings give high local concentrations of nitrogen. A particle of calcium carbonate in the soil may make a tiny area alkaline, while dead organic matter may make a nearby area acidic. Does the root take up whatever materials it encounters?

Animals aren't very particular, but plants take up nutrients selectively

Animals ingest their meals, taking in unneeded and sometimes toxic materials along with needed nutrients; what an animal ingests is not determined by its needs. Part of what animals ingest must eventually be disposed of as waste products, such as urea. Plants, by contrast, do not urinate or produce wastes in other obvious ways. Instead, they control their uptake of most substances, matching the uptake rates to their biochemical needs. The major waste products released to the environment by plants are carbon dioxide or, during active photosynthesis, oxygen gas.

This is not to say that plants do not take up toxic substances from the soil. They do, at times, as discussed in Chapter 33. However, plants exert a more systematic control over what can enter their bodies than do animals. What important minerals are admitted, and what are their roles?

Mineral Nutrients Essential to Plants

Table 34.1 lists the mineral elements that are essential for plants. They all come from the soil solution and derive ultimately from rock. The criteria for calling something an **essential element** are the following: (1) The element must be necessary for normal growth and reproduction. (2) The element cannot be replaceable by another element. (3) The requirement must be direct—that is, not the result of an indirect effect, such as the need to relieve toxicity caused by another substance. In this section, we'll consider the symptoms of particular mineral deficiencies, the roles of some of the mineral nutrients, and the technique by which the essential elements were identified.

The essential minerals in Table 34.1 are divided into two categories: the macronutrients and the micronutrients. Plant tissues need **macronutrients** in concentrations of at least 1 mg (10^{-3} g) per gram of their dry matter, and they need **micronutrients** in concentrations of less than 100 µg (10^{-4} g) per gram of their dry matter. (Dry matter, or dry weight, is what remains after all the water has been removed from a tissue sample.) Both the macronutrients and the micronutrients are essential for the completion of a plant's life cycle.

TABLE 34.1   Elements Required by Plants

ELEMENT	SOURCE	ABSORBED FORM	MAJOR FUNCTIONS
Nonmineral elements			
Carbon (C)	Atmosphere	CO_2	In all organic molecules
Oxygen (O)	Atmosphere	CO_2	In most organic molecules
Hydrogen (H)	Soil	H_2O	In most organic molecules
Nitrogen (N)	Soil	NH_4^+ and NO_3^-	In proteins, nucleic acids, etc.
Mineral nutrients			
Macronutrients			
Phosphorus (P)	Soil	$H_2PO_4^-$	In nucleic acids, ATP, phospholipids, etc.
Potassium (K)	Soil	K^+	Enzyme activation; water balance; ion balance
Sulfur (S)	Soil	SO_4^{2-}	In proteins and coenzymes
Calcium (Ca)	Soil	Ca^{2+}	Affects the cytoskeleton, membranes, and many enzymes; second messenger
Magnesium (Mg)	Soil	Mg^{2+}	In chlorophyll; required by many enzymes; stabilizes ribosomes
Micronutrients			
Iron (Fe)	Soil	Fe^{3+}	In active site of many redox enzymes and electron carriers; chlorophyll synthesis
Chlorine (Cl)	Soil	Cl^-	Photosynthesis; ion balance
Manganese (Mn)	Soil	Mn^{2+}	Activation of many enzymes
Boron (B)	Soil	$H_2BO_3^-$, HBO_3^{2-}	Possibly carbohydrate transport (poorly understood)
Zinc (Zn)	Soil	Zn^{2+}	Enzyme activation; auxin synthesis
Copper (Cu)	Soil	Cu^{2+}	In active site of many redox enzymes and electron carriers
Molybdenum (Mo)	Soil	MoO_4^{3-}	Nitrogen fixation; nitrate reduction

Deficiency symptoms reveal inadequate nutrition

Before a plant that is deficient in an essential element dies, it usually displays characteristic **deficiency symptoms**. Table 34.2 describes the symptoms of some common mineral deficiencies. Such symptoms help horticulturists diagnose nutrient deficiencies in plants.

Nitrogen deficiency is the most common mineral deficiency of plants. The visible symptoms of nitrogen deficiency include uniform yellowing, or chlorosis, of older leaves because chlorophyll, which is responsible for the green color of leaves, contains nitrogen. Thus without nitrogen there is no chlorophyll, and without chlorophyll the leaves turn yellow.

Inadequate iron in the soil can also cause chlorosis, because, although it is not contained in the chlorophyll molecule, iron is required for chlorophyll synthesis. However, iron deficiency commonly causes chlorosis of the youngest leaves, with their veins remaining green, while nitrogen deficiency causes chlorosis of the oldest leaves. The reason for this difference is that nitrogen is readily translocated in the plant and can be redistributed from older tissues to younger tissues to favor their growth; iron, on the other hand, is contained in compounds that are not translocated and hence cannot be redistributed to the young tissues.

Several essential elements fulfill multiple roles

Several essential elements fulfill multiple roles—some structural, others catalytic. Magnesium, as we have mentioned, is a constituent of the chlorophyll molecule and hence is essential to photosynthesis. It is also required as a cofactor by numerous enzymes in cellular respiration and other metabolic pathways. Iron is a constituent of many molecules, including some proteins, that participate in oxidation–reduction reactions.

Phosphorus, usually in phosphate groups, is found in many compounds, particularly in nucleic acids and in pathways of energy metabolism such as photosynthesis and glycolysis. The transfer of phosphate groups is important in many energy-storing and energy-releasing reactions, notably those that use or produce ATP. Other roles of phosphate groups include the activation and inactivation of enzymes.

Plant tissues contain high concentrations of potassium, and this element plays a major role in maintaining the electrical neutrality of cells. Potassium ions (K^+) balance the negative charges of ionized carboxyl groups ($—COO^-$) of organic acids. Potassium also helps move water from cell to cell. There are no "pumps" for the active transport of water, yet water must be moved from place to place—for example, into and out of the guard cells surrounding stomata (see Chapter 32).

Plants and animals achieve water movement by actively transporting K^+ from one cell to another (see

TABLE 34.2	Some Mineral Deficiencies
DEFICIENCY	**SYMPTOMS**
Calcium	Growing points die back; young leaves are yellow and crinkly
Iron	Young leaves are white or yellow with green veins
Magnesium	Older leaves have yellow in stripes between veins
Manganese	Younger leaves are pale with stripes of dead patches
Nitrogen	Oldest leaves turn yellow and die prematurely; plant is stunted
Phosphorus	Plant is dark green with purple veins and is stunted
Potassium	Older leaves have dead edges
Sulfur	Young leaves are yellow to white with yellow veins
Zinc	Older leaves have many dead spots

Figure 32.3). Chloride ions (Cl^-) follow the K^+ passively, maintaining electrical balance. Movement of these ions changes the water potential of the cells, and water then moves passively to maintain osmotic balance. Both K^+ and Cl^- are also used to regulate the membrane potential (the electric charge difference across the plasma membrane), which is one of the factors determining mineral uptake or loss by plants.

Calcium plays many roles in plants. Its function in the processing of hormonal and environmental cues is the subject of great current interest (the analogous function of Ca^{2+} in animal cells is discussed in Chapter 38). Calcium also affects membranes and cytoskeleton activity, participates in spindle formation for mitosis and meiosis, and is a constituent of the middle lamella of cell walls.

All of these elements are essential to the life of all plants. Other elements may be essential to some plants—and perhaps to all, as we'll see in the next section.

The identification of essential elements

An element is considered essential if a plant does not grow, flower, or produce viable seed when deprived of that element. Plant physiologists identified most of the essential elements by the technique outlined in Figure 34.2. This technique is limited by the possibility that some elements thought to be absent from the solutions are present.

Some of the chemicals used in early experiments were so impure that they provided micronutrients that the first investigators thought they had excluded. Some minerals are required in such tiny amounts that there may be enough in a seed to feed the embryo and the resultant plant throughout its lifetime and leave

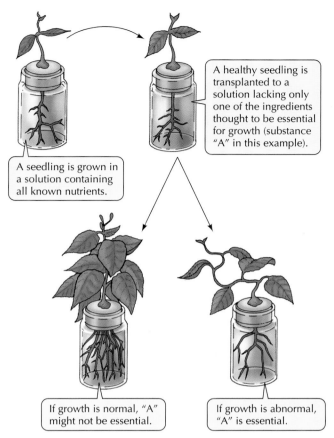

34.2 Identifying Essential Plant Nutrients The environment in this experiment must be rigorously controlled because some essential nutrients are needed in only tiny amounts that may be present as contaminants of other materials or on objects.

enough in the next seed to get the next generation well started. There was enough chloride on dust particles and water droplets in the air in Berkeley, California (where some of the work on essential elements was performed), for example, to provide the infinitesimal amounts needed to keep experimental plants growing. The essentiality of chlorine thus was not established until 1954, after special air filters had been installed in the laboratory. Simply touching a plant may give it a significant dose of chlorine in the form of chloride ions from sweat.

Only rarely are new essential elements reported now; either the list is nearly complete, or more likely, we will need more sophisticated techniques to add to it. The most recent addition to the list is nickel, which was shown in 1984 to be essential for legumes and is now known to be a cofactor for an enzyme that catalyzes a step in nitrogen metabolism.

Where does the plant find its essential minerals? And how does the plant get the environment to yield the minerals to it?

Soils and Plants

Soils are very important to plants, and plant interactions with the soil are complex. Plants obtain their mineral nutrients from the soil or the water in which they grow. Water for terrestrial plants also comes from the soil, as does the supply of oxygen for the roots. Soil also provides mechanical support for plants on land. Soil harbors bacteria that perform chemical reactions leading to products required for plant growth; on the other hand, soil may also contain organisms harmful to plants.

In the pages that follow, we'll examine the composition and structure of soils, their formation, their role in plant nutrition, their care and supplementation, and their modification by the plants that grow in them.

Soils are complex in structure

Soils have living and nonliving components. The living components include plant roots, as well as populations of bacteria and fungi (Figure 34.3). The nonliving portion of the soil includes rock fragments ranging in size from large boulders to particles 2 μm and less in diameter (Table 34.3). Soils also contain minerals, water, gases, and organic matter from animals, plants, fungi, and bacteria.

Soils change constantly because of both nonhuman natural causes—such as rain, temperature extremes, and the activities of plants and animals—and human activities, farming in particular. Soils from different parts of the world differ dramatically in their chemical composition and physical structure because the tem-

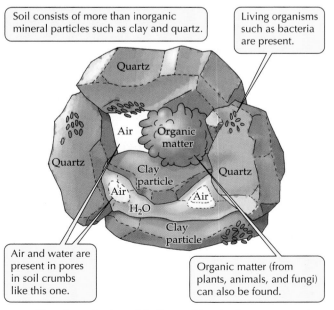

34.3 The Complexity of Soil Soil has both organic and inorganic components.

TABLE 34.3 Soil Particles

SOIL TYPE	PARTICLE SIZE (mm)
Coarse sand	0.2–2.0
Fine sand	0.02–0.2
Silt	0.002–0.02
Clay	<0.002

perature, water supply, and other factors during their formation differ from place to place.

The structure of any soil changes with depth, revealing a soil profile. Although soils differ greatly, almost all soils consist of two or more **horizons**—recognizable horizontal layers—lying on top of one another. Mineral nutrients tend to be *leached*—dissolved in rain or irrigation water and carried to deeper horizons. Other processes also move materials down—or up—in the soil.

Soil scientists recognize three major zones (A, B, and C) in the profile of a typical soil (Figure 34.4). The A horizon is the top zone, from which minerals have been depleted by leaching. Most of the organic matter in the soil is in the A horizon, as are most roots, earthworms, soil insects, nematodes, and soil protists. Successful agriculture depends on the presence of a suitable A horizon. The B horizon is the zone of infiltration and accumulation of materials leached from

A Topsoil

B Subsoil

C Weathering bedrock

34.4 A Soil's Profile The A, B, and C horizons can sometimes be seen in road cuts such as this one in Australia. The upper layers developed from the bedrock. The dark upper layer is home to most of the living organisms in the soil.

above, and the C horizon is the original parent material from which the soil is derived. Some deep-growing roots extend into the B horizon, but roots rarely enter the C horizon.

Soils form by weathering of rock

The type of soil in a given area depends on the rock from which it formed, the climate, the topography (features of the landscape), the organisms living there, and the length of time that soil-forming processes have been acting. Rocks are broken down in part by **mechanical weathering**, the physical breakdown—without any accompanying chemical changes—of materials by wetting, drying, and freezing. The most important parts of soil formation, however, include **chemical weathering**, the chemical alteration of at least some of the materials in the rocks.

The key process is the formation of clay. Both the physical and the chemical properties of soils depend on the amount and kind of clay particles they contain. Just grinding up rocks does not produce a clay that swells and shrinks and is chemically active. The rock must be chemically changed as well. The initial step in the chemical weathering of most soil minerals is hydrolysis, as illustrated for feldspar, a common soil mineral, in the following formula:

$$\langle Si, Al, O \rangle K^+ \ + \ H^+OH^- \longrightarrow \langle Si, Al, O \rangle H^+ \ + \ K^+OH^-$$

feldspar   water   hydrolyzed   potassium
                                   feldspar   hydroxide

Two examples illustrate the diversity of soil-forming processes. The first example is found in wet tropical regions. There, rainfall and temperatures are high and the soils are usually moist, and water moves rapidly downward through the soil. Silica and soluble nutrients are quickly leached, leaving insoluble iron and aluminum compounds in the A horizon. These are often oxidized, giving the soil bright reddish colors. This type of soil formation is known as **laterization.** The resulting soil is very poor in nutrients.

The second example is found in semiarid regions, where water from rainfall evaporates rapidly, and there is no net movement of water downward through the soil. Instead, water penetrates for a distance, stops, and then is drawn back up by the roots of plants and by evaporation from the soil surface. Under these conditions the soil remains rich in mineral nutrients, and a hard layer of calcium carbonate often forms in the B horizon. The B horizon may be so hard that it prevents deeper penetration of the soil by plant roots. These soils are very fertile, however, when supplied with additional water and nitrogen. Much of the success of irrigated agriculture depends on the high nutrient content of arid-zone soils.

Soils are the source of plant nutrition

The supply of minerals to plants depends on the presence of **clay particles**, which have a net negative charge, in the soil. Many of the minerals that are important for plant nutrition, such as potassium, magnesium, and calcium, exist in soil as positive ions chemically attached to clay particles. To become available to plants, the positive ions must be detached from the clay particles, a task accomplished by reactions with protons (hydrogen ions, H^+), which are released into the soil by roots or by the ionization of carbonic acid (H_2CO_3). (Carbonic acid is almost universally present in soils, because it forms whenever CO_2 from respiring roots or from the atmosphere dissolves in water, according to the reaction $CO_2 + H_2O \rightarrow H_2CO_3$.)

The clay particles get their net negative charge from negatively charged ions that are permanently attached to them. Positively charged ions in solution associate reversibly with these attached negative ions. Protons then trade places with ions such as potassium (K^+) and calcium (Ca^{2+}) on the clay particles, thus putting the nutrients back into the soil solution. This trading of places is called **ion exchange** (Figure 34.5). The fertility of a soil is determined primarily by its ability to provide nutrients such as potassium, magnesium, and calcium in this manner.

Clay particles effectively hold and exchange positively charged ions, but there is no comparable exchanger of negatively charged ions. As a result, important negative ions such as phosphate, nitrate, and sulfate—the primary sources of phosphorus, nitrogen, and sulfur, respectively—leach rapidly from soil, while positive ions tend to be retained in the A horizon.

Fertilizers and lime are used the world over

Agricultural soils often require fertilizers because irrigation leaches mineral nutrients from the soil and the harvesting of crops removes the nutrients that the plants took up from the soil during growth. Crop yields decrease if too much of any element is removed. Minerals may be replaced by organic fertilizers, such as rotted manure, or inorganic fertilizers of various types. The three elements most commonly added to agricultural soils are nitrogen (N), phosphorus (P), and potassium (K). The ratios of these elements vary among fertilizers, which are typically characterized by their N–P–K percentages. A 5–10–5 fertilizer, for example, contains 5 percent nitrogen, 10 percent phosphate, and 5 percent potassium. Sulfur, in the form of sulfate, also is often added.

Both organic and inorganic fertilizers can provide the necessary minerals. Organic fertilizers also contain materials that improve the physical properties of the soil, providing air pockets for gases, root growth, and drainage. Inorganic fertilizers, on the other hand, pro-

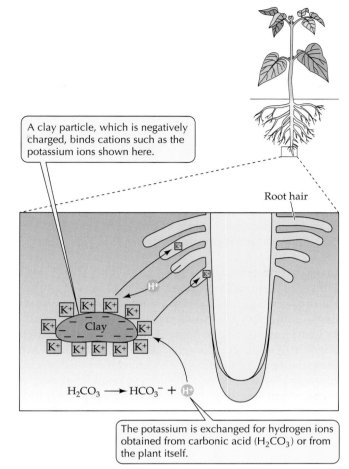

A clay particle, which is negatively charged, binds cations such as the potassium ions shown here.

Root hair

$H_2CO_3 \longrightarrow HCO_3^- + H^+$

The potassium is exchanged for hydrogen ions obtained from carbonic acid (H_2CO_3) or from the plant itself.

34.5 Ion Exchange Plants obtain mineral nutrients from the soil primarily in the form of ions; potassium is the example shown here.

vide an almost instantaneous supply of soil nutrients and can be formulated to meet the requirements of a particular soil and a particular crop.

The availability of nutrient ions, whether naturally present in the soil or added as fertilizer, is altered by changes in soil pH. Rainfall and the decomposition of organic substances in the soil lower its pH. Such acidification of the soil can be reversed by **liming**—the application of calcium-containing material (usually calcium carbonate or calcium hydroxide). This practice is older than agricultural history.

It is easy to guess how we learned the use of fertilizer; it didn't take much insight to notice improved plant growth around animal feces. Perhaps a similar observation about limestone, or chalk, or oyster shells—all sources of calcium—led to the practice of liming. Whatever its ancient source, the addition of Ca^{2+} allows H^+ to be released from soil particles by ion exchange and leached away, raising the soil pH (see Chapter 2 for a review of pH). Liming also increases the availability of Ca^{2+} to plants, which require it as a macronutrient. Sometimes a soil is not acidic enough; in this case, a farmer can add sulfate ions to decrease the pH. Iron and some other elements are more available at a slightly acidic pH.

Spraying leaves with a nutrient solution is another effective way to deliver some essential elements to growing plants. Plants take up more copper, iron, and manganese when these elements are applied as foliar (leaf) sprays rather than as soil fertilizer. Adjusting the concentrations of nutrient ions and the pH to optimize uptake and to minimize toxicity and controlling the time of spraying to avoid "burning" can yield excellent results. Foliar application of mineral nutrients is increasingly used in wheat production, but fertilizer is still delivered most commonly by way of the soil.

Plants affect soils

How soil forms in a particular place also depends on the types of plants growing there. Plant litter is a major source of carbon-rich materials that break down to form **humus**—dark-colored organic material, each particle of which is too small to be recognizable with the naked eye. Soils rich in exchangeable positive ions of mineral nutrients tend to support plants that extract large quantities of nutrients for incorporation into their tissues. When these tissues die and decompose, they produce a rich, alkaline humus called *mull*. Plants growing on nutrient-poor soils extract fewer nutrients and form tissues that yield a poor, acidic humus known as *mor*. The mor produced by conifers is particularly resistant to decay and may accumulate in thick layers on the surface of the soil.

Soils age. Young soils support rapidly increasing amounts of vegetation (biomass) that contribute materials for humus, which thus increases at the same rapid rate; the green and yellow lines in Figure 34.6 plot these increases for a hypothetical example. The increase in biomass comes at the expense of nutrients, which decline rapidly, as exemplified by the drop in calcium carbonate in Figure 34.6. The fraction of clay in the soil gradually increases, decreasing the availability of water for plant growth. The loss of mineral nutrients from the soil also contributes to the long-term decline in biomass.

Biologists were slow to recognize the importance of the long-term changes in soils because nearly all the soils of the northern temperate zone, where most soil scientists live and work, are very young, dating from the last glacial period, only a few thousand years ago. In large areas of the tropics and subtropics, however, especially away from areas of recent mountain formation, soils are ancient and have few remaining nutrients. The right-hand edge of Figure 34.6 suggests the makeup of such soils, which are unsuitable for agriculture unless heavily fertilized.

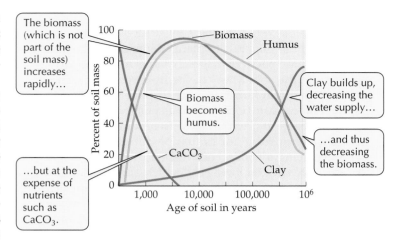

34.6 Aging of Soils   The rich humus and the biomass decrease over time, while the fraction of clay in the soil increases. Aging processes gradually remove exchangeable nutrient ions such as calcium carbonate from the soil, leaving an infertile residue.

Nitrogen is the essential element most often required as fertilizer. Its level in the soil can change dramatically, even during a single growing season. In the next few sections we will look at how nitrogen is made available and at how certain soil bacteria participate in this process.

Nitrogen Fixation

Plants cannot use nitrogen gas (N_2) directly as a nutrient. Making nitrogen from N_2 available to plants takes a great deal of energy because N_2 is a highly unreactive substance—the N≡N triple bond is extremely stable. A few species of bacteria can convert N_2 into a more useful form by a process called **nitrogen fixation**.

These prokaryotic organisms—nitrogen fixers—convert N_2 to ammonia (NH_3). Some of them must live in intimate association with specific eukaryotes before they develop functional nitrogen-fixing machinery. There are relatively few kinds of nitrogen fixers, and what few there are have a small biomass relative to the mass of other organisms on Earth. Without the nitrogen fixers, however, other organisms would not survive. This talented group of prokaryotes is just as essential in the biosphere as are the photosynthetic autotrophs.

Nitrogen fixers make all other life possible

Organisms fix approximately 90 million tons of atmospheric nitrogen per year. Tens of millions of tons are fixed industrially, mostly by a method called the Haber process, which requires a lot of energy. A smaller amount of nitrogen is fixed in the atmosphere by nonbiological means such as lightning, volcanic eruption, and forest fires; the products thus formed are brought to Earth by rainwater. By far the greatest share of total world nitrogen fixation, however, is that performed biologically by nitrogen-fixing organisms.

Nitrogen-fixing species are widely scattered in the kingdom Eubacteria. One group of microorganisms fixes nitrogen only in close association with the roots of certain seed plants; the best known of these bacteria belong to the genus *Rhizobium*. Some *Rhizobium* species live free in the soil, where they do not fix nitrogen. Other species of *Rhizobium* live in nodules on the roots of plants in the legume family, which includes peas, soybeans, alfalfa, and many tropical shrubs and trees (Figure 34.7).

These nodule-inhabiting *Rhizobium* species fix nitrogen. Some cyanobacteria fix nitrogen in association with fungi in lichens or with ferns, cycads, or nontracheophytes. Finally, the filamentous bacteria called actinomycetes fix nitrogen in association with root nodules on shrub species such as alder. How were the nitrogen-fixing roles of bacteria discovered?

The ancient Chinese, Greeks, and Romans—and probably members of other early civilizations—recognized that plants such as clover, alfalfa, and peas improve the soil in which they grow. Two German chemists, Hellriegel and Wilfarth , first showed in 1888 that the root nodules on these plants are caused by bacteria and are sites of nitrogen fixation. These particular plant-infecting bacteria all belong to the genus

34.8 Equipped to Colonize A retreating glacier in Alaska left this area of bare rock exposed. *Dryas drummondii*, the low-lying plant seen here in bloom, was one of the first plants to colonize the bare area. It and other rapid colonizers growing here share the characteristic of having nitrogen-fixing root nodules.

Rhizobium, and the various species of *Rhizobium* show a fairly high specificity for the species of legume they nodulate.

In the oceans, various photosynthetic bacteria, including cyanobacteria, fix nitrogen; in fresh water, cyanobacteria are the principal nitrogen fixers. On land, free-living soil bacteria make some contribution to nitrogen fixation, but bacteria in the root nodules of plants produce most of the fixed nitrogen. Unlike various free-living nitrogen fixers that fix what they need for their own uses and release the fixed nitrogen only when they die, bacteria in root nodules release up to 90 percent of the nitrogen they fix to the plant and excrete some amino acids into the soil, making nitrogen immediately available to other organisms. Some farmers alternate their crops, planting clover or alfalfa occasionally to increase the useful nitrogen content of the soil.

Root nodules permit some *non*leguminous plants to be pioneers, to occupy environments having few or no other plants (Figure 34.8; see also Figure 52.20). Shrubs such as alder thrive in mountainous areas, with their roots grasping chunks of rock in the talus (debris) slopes below the cliffs, which are poor in nitrogen. The western mountain lilac *Ceanothus* grows well in extremely gravelly soils that have little or no organic matter and hence no fixed nitrogen. Eastern sweet gale, *Myrica*, flourishes on almost pure sand; its growth depends on nitrogen from nodules. Pioneer plants, with their bacterial partners, make initially barren habitats available to other plants and to the animals that depend on them.

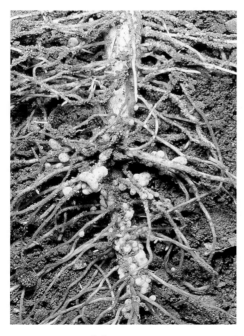

34.7 Root Nodules Large, round, tumorlike nodules are developing from a broad bean root system. These nodules house nitrogen-fixing bacteria.

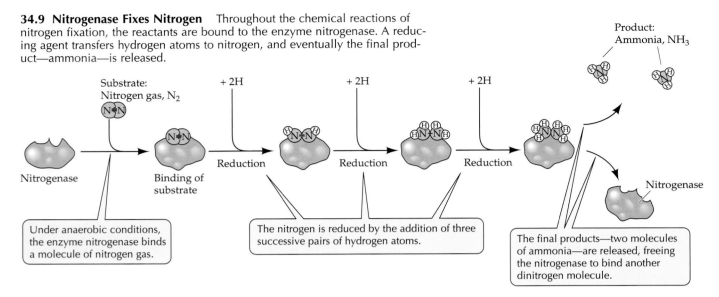

34.9 Nitrogenase Fixes Nitrogen   Throughout the chemical reactions of nitrogen fixation, the reactants are bound to the enzyme nitrogenase. A reducing agent transfers hydrogen atoms to nitrogen, and eventually the final product—ammonia—is released.

How does biological nitrogen fixation work? In the four sections that follow, we'll consider the role of the enzyme nitrogenase, the mutualistic collaboration of plant and bacterial cells in root nodules, the need to supplement biological nitrogen fixation in agriculture, and the contributions of plants and bacteria to the global nitrogen cycle.

Nitrogenase catalyzes nitrogen fixation

Nitrogen fixation proceeds by the addition of three pairs of hydrogen atoms:

$$N\equiv N \xrightarrow{\;2\,H\;} HN=NH \xrightarrow{\;2\,H\;} H_2N-NH_2 \xrightarrow{\;2\,H\;} 2\,NH_3$$

dinitrogen   ammonia

Throughout this series of reactions, the reactants are firmly bound to the surface of a single enzyme, called **nitrogenase** (Figure 34.9). The reactions require a strong reducing agent (see Chapter 7) to transfer hydrogen atoms to nitrogen and the intermediate products, as well as a great deal of energy, which is supplied by ATP. Depending on the species of nitrogen fixer, either respiration or photosynthesis may provide the necessary reducing agent and ATP.

Nitrogenase is so strongly inhibited by oxygen (O_2) that its discovery was delayed because investigators had not thought to seek it under anaerobic conditions, which are inconvenient to establish in the laboratory. Because nitrogenase cannot function in the presence of oxygen, it is not surprising that many nitrogen fixers are anaerobes. Free-living *Rhizobium* species respire aerobically, but they do not fix nitrogen under these conditions. Legumes respire aerobically, but their *Rhizobium*-containing, nitrogen-fixing root nodules maintain an anaerobic internal environment.

Aerobic nitrogen fixers must decrease their internal oxygen levels drastically in order for the process to work. One means of decreasing internal oxygen uses a special type of plant hemoglobin, which we will describe in the next section.

Some plants and bacteria work together to fix nitrogen

The legume nodule provides an excellent example of **symbiosis**, in which two different organisms live in physical contact and, in association, do things that neither organism can do separately. In the form of symbiosis called **mutualism**, both organisms benefit from the relationship. Neither free-living *Rhizobium* species nor uninfected legumes can fix nitrogen. Only when the two are closely associated in root nodules does the reaction take place.

The establishment of this symbiosis between *Rhizobium* and a legume requires a complex series of steps with active contributions by both the bacteria and the plant root. First the root releases chemical signals that attract the *Rhizobium* to the vicinity of the root. Then the bacteria produce substances that cause cell divisions in the root cortex, leading to the formation of a primary nodule meristem.

Next comes the infection of the plant by the bacteria. The bacteria attach to root hairs and then produce one or more growth substances that cause the cell walls of the root hairs to invaginate—fold inward (Figure 34.10). With help from the Golgi apparatus in the root's cells, the invagination proceeds inward through several cells as an infection thread; the bacteria in the thread continue to divide, although slowly.

At this stage the bacteria are still outside the plant in a sense, for the thread is lined with cellulose and other cell wall materials, and the bacteria remain exterior to the plasma membrane. The thread continues to grow into the cortex tissue of the root until it encounters cells that become the primary nodule meristem.

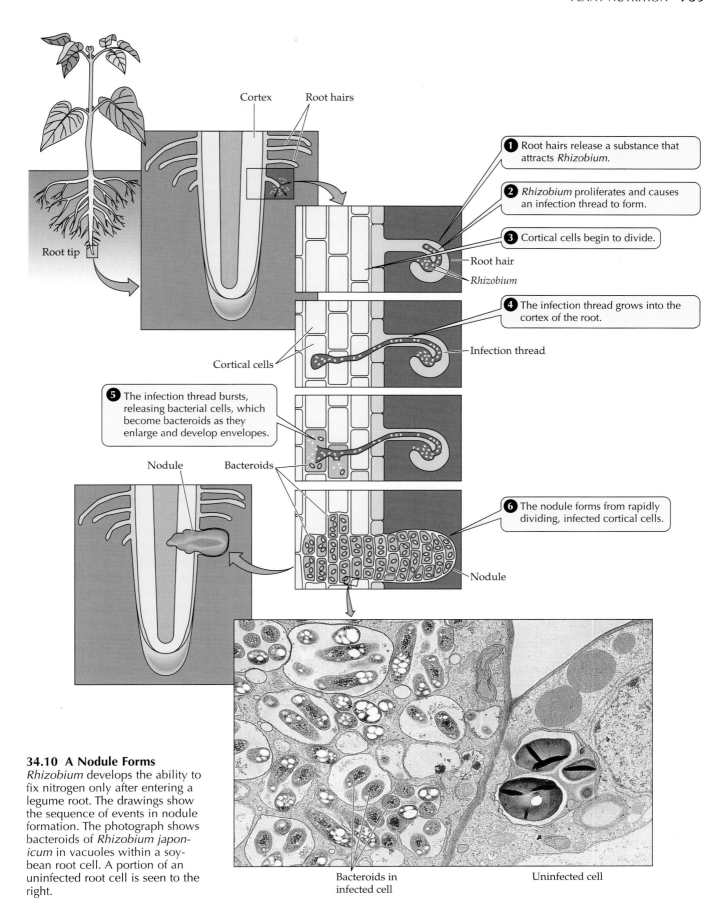

Cortex   Root hairs

Root tip

1 Root hairs release a substance that attracts *Rhizobium*.

2 *Rhizobium* proliferates and causes an infection thread to form.

3 Cortical cells begin to divide.

Root hair

Rhizobium

4 The infection thread grows into the cortex of the root.

Cortical cells

Infection thread

5 The infection thread bursts, releasing bacterial cells, which become bacteroids as they enlarge and develop envelopes.

Nodule   Bacteroids

6 The nodule forms from rapidly dividing, infected cortical cells.

Nodule

34.10 A Nodule Forms

Rhizobium develops the ability to fix nitrogen only after entering a legume root. The drawings show the sequence of events in nodule formation. The photograph shows bacteroids of *Rhizobium japonicum* in vacuoles within a soybean root cell. A portion of an uninfected root cell is seen to the right.

Bacteroids in infected cell

Uninfected cell

The meristem cells begin to divide rapidly, and the infection thread bursts, releasing the bacteria into the cytoplasm of these host cells.

The bacteria now undergo a remarkable transformation, increasing about tenfold in size, developing an outside membranous envelope, and forming an elaborately folded internal membrane. At this stage the infecting bacteria are called *bacteroids* (see Figure 34.10). The bacteroids are, in effect, organelles for nitrogen fixation; it has been suggested that, in the course of further evolution, such bacteroids eventually will give rise to permanent nitrogen-fixing organelles in some plant species.

The final step before the fixation of nitrogen can begin is that the plant produces leghemoglobin, which surrounds the bacteroids. Hemoglobin is an oxygen-carrying pigment of animals, not plants, but some plant nodules contain enough of its close relative leghemoglobin to be bright pink when viewed in cross section. The leghemoglobin traps oxygen, keeping it away from the oxygen-sensitive bacteroids and nitrogenase.

Biological nitrogen fixation isn't always sufficient

Bacterial nitrogen fixation does not suffice to support the needs of agriculture. Native Americans used to plant dead fish along with corn so that the decaying fish would release fixed nitrogen that the developing corn could use. Industrial nitrogen fixation is becoming ever more important to world agriculture because of the degradation of soils and the need to feed a rapidly expanding population. Research on biological nitrogen fixation is being vigorously pursued, with commercial applications very much in mind. An alternative to the current process for industrial nitrogen fixation is urgently needed because of the cost of energy and other economic factors associated with it. At present, the manufacture of nitrogen-containing fertilizer takes more energy than does any other aspect of crop production in the United States.

One line of investigation centers on recombinant DNA technology as a means of engineering new plants to produce nitrogenase. Workers in many industrial and academic laboratories are working on the insertion of bacterial genes coding for nitrogenase into the cells of angiosperms, particularly crop plants. However, to develop crops that can fix their own nitrogen will take more than just the insertion of genes for nitrogenase; there must also be provisions for excluding free oxygen and for obtaining strong reducing agents and an energy source.

Like industrial nitrogen fixation, biological nitrogen fixation is extremely expensive in terms of energy. Looking to nature for evidence of this energy toll, we find that legumes compete successfully with grasses only where there is a real shortage of nitrogen in the soil. Ultimately, the need for ATP represents a greater technical challenge than the insertion of nitrogenase genes. The stakes, however—especially the financial ones—are great, and a great amount of effort is being invested in research along these lines.

Plants and bacteria participate in the global nitrogen cycle

Fixed nitrogen released into the soil by nitrogen fixers is primarily in the form of ammonia (NH_3) and ammonium ions (NH_4^+). Although ammonia is toxic to plants, ammonium ions can be taken up safely at low concentrations. Most plants, however, grow better with nitrate (NO_3^-) than with ammonium ions as a source of nitrogen, because of the toxicity of ammonium ions at higher concentrations. The form of nitrogen taken up is affected by soil pH: Nitrate ions are taken up preferentially under more acidic conditions, ammonium ions under more basic ones.

But if ammonia and ammonium ions are the forms of nitrogen produced by nitrogen fixation, where do nitrate ions in the soil come from? Ammonia is oxidized to nitrate by the process of **nitrification**. This process is carried out in the soil by bacteria called nitrifiers, which we described in Chapter 25.

So far we have seen N_2 *reduced* to ammonia in nitrogen fixation and ammonia *oxidized* to nitrate in nitrification. We will now see that plants *reduce* the nitrate they have taken up all the way back to ammonia before using it further to manufacture amino acids (Figure 34.11). All the reactions of **nitrate reduction** are carried on by the plant's own enzymes. The later steps, from nitrite (NO^{2-}) to ammonia, take place in the chloroplasts; this conversion is not a part of photosynthesis. The final products of nitrate reduction are amino acids, from which the plant's proteins and all its other nitrogen-containing compounds form. Animals depend on plants to supply them with reduced nitrogenous compounds. Ultimately, bacteria called denitrifiers return nitrogen to the atmosphere as N_2. This process, described in Chapter 25, is called denitrification. In combination with leaching and the removal of crops, it keeps the level of available nitrogen in soils low.

Nitrogen metabolism, in bacteria and in plants, is complex. It is also of great importance. Nitrogen atoms constitute approximately 1 to 5 percent of the dry weight of a leaf, and nitrogen-containing compounds constitute 5 to 30 percent of the plant's total dry weight. The nitrogen content of animals is even higher, and all the nitrogen in the animal world arrives by way of the plant kingdom. As we are about to see, plants also play an important part in delivering sulfur to animals.

Sulfur Metabolism

All living things require sulfur, which is a constituent of two amino acids, cysteine and methionine, and

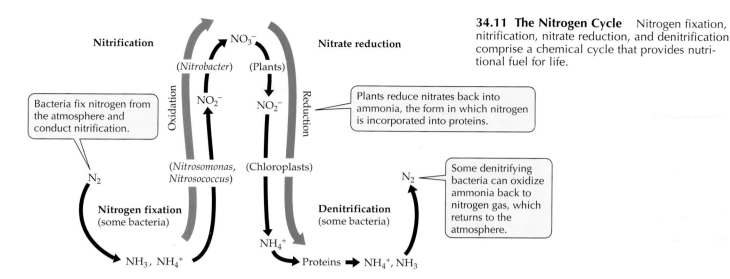

34.11 The Nitrogen Cycle Nitrogen fixation, nitrification, nitrate reduction, and denitrification comprise a chemical cycle that provides nutritional fuel for life.

Bacteria fix nitrogen from the atmosphere and conduct nitrification.

Plants reduce nitrates back into ammonia, the form in which nitrogen is incorporated into proteins.

Some denitrifying bacteria can oxidize ammonia back to nitrogen gas, which returns to the atmosphere.

hence of almost all proteins. Sulfur is also a component of other biologically crucial compounds, such as coenzyme A (see Chapter 7). Animals must obtain their cysteine and methionine from plants, but plants can start with sulfate ions (SO_4^{2-}) obtained from the soil or from a liquid environment.

Interestingly, all of the most abundant elements in plants are taken up from the environment in their most oxidized forms—sulfur as sulfate, carbon as carbon dioxide, nitrogen as nitrate, phosphorus as phosphate, and hydrogen as water. In plants, sulfate is reduced and incorporated into cysteine; from this amino acid all the other sulfur-containing compounds in the plant are made. These important processes—sulfate reduction and the utilization of cysteine—are closely analogous to the reduction of nitrate to ammonia and the subsequent utilization of ammonia by plants.

Thus far in this chapter we have considered the mineral nutrition of plants. As you already know, another crucial aspect of plant nutrition is photosynthesis—the principal source of energy and carbon for plants and for the biosphere as a whole. Not all plants, however, are photosynthetic autotrophs. A few, in the course of their evolution, have lost the ability to feed themselves by photosynthesis. How do these plants get their energy and carbon?

Heterotrophic and Carnivorous Seed Plants

A few plants are parasites that obtain their food directly from the living bodies of other plants. Perhaps the most familiar parasitic plants are the mistletoes and dodders (Figure 34.12). Mistletoes are green and carry on some photosynthesis, but they parasitize other plants for water and mineral nutrients and may derive photosynthetic products from them as well. Another parasitic plant, the Indian pipe, once was thought to obtain its food from dead organic matter; it

is now known to get its nutrients, with the help of fungi, from nearby actively photosynthesizing plants. Hence it too is a parasite.

Some other plants that don't live by photosynthesis alone are the 450 or so carnivorous species—those that augment their nitrogen and phosphorus supply by capturing and digesting flies and other insects (Figure 34.13; see also the photograph at the beginning of the chapter). The best-known plant carnivores are Venus flytraps (genus *Dionaea*), sundews (genus *Drosera*), and pitcher plants (genus *Sarracenia*).

34.12 A Parasitic Plant Orange-brown tendrils of dodder wrap around other plants. This parasitic plant obtains water, sugars, and other nutrients from its host through tiny, rootlike protuberances that penetrate the surface of the host.

34.13 A Carnivorous Sundew Sundews trap insects on their sticky hairs. Secreted enzymes digest the carcasses externally.

These plants are normally found in boggy regions where the soil is acidic. Most decay-causing organisms require a more neutral pH to break down the bodies of dead organisms, so relatively little available nitrogen is recycled into these acidic soils. Accordingly, the carnivorous plants have adaptations that allow them to augment their supply of nitrogen with animal proteins.

Sarracenia produces pitcher-shaped leaves that collect small amounts of rainwater. Insects are attracted into the pitchers either by bright colors or by scent and are prevented from getting out again by stiff, downward-pointing hairs. The insects eventually die and are digested by a combination of enzymes and bacteria in the water.

Sundews have leaves covered with hairs that secrete a clear, sticky liquid high in sugar. An insect touching one of these hairs becomes stuck, and more hairs curve over the insect and stick to it as well. The plant secretes enzymes to digest the insect and later absorbs the carbon- and nitrogen-containing products of digestion.

An insect entering the Venus flytrap springs a mechanical trap triggered electrically by three hairs in the center of a partly closed leaf lobe. The two halves of the leaves close, and spiny outgrowths at the margins of the leaves interlock to imprison the insect. Enzymes secreted by the plant digest the trapped insect.

None of the carnivorous plants must feed on insects; they grow adequately without insects, but in nature, they grow faster and are a darker green when insects are available to them. The extra supply of nitrogen is used to make more proteins, chlorophyll, and other nitrogen-containing compounds.

Summary of "Plant Nutrition"

Lifestyles and the Acquisition of Nutrients

• Plants are photosynthetic autotrophs that can produce all the compounds they need from carbon dioxide, water, a nitrogen source, and minerals. They obtain energy from sunlight, carbon dioxide from the atmosphere, and mineral salts from the soil.
• Plants explore new surroundings by growing rather than by locomotion.
• Unlike animals, plants take up nutrients selectively.

Mineral Nutrients Essential to Plants

• Plants require twelve essential mineral elements, all of which come from the soil. Several essential elements fulfill multiple roles. **Review Table 34.1**
• The five minerals required in substantial amount are called macronutrients; the seven required in much smaller amount are called micronutrients. **Review Table 34.1**
• Deficiency symptoms reveal what essential element a plant lacks. **Review Table 34.2**
• Biologists discovered each essential element by attempting to grow plants on artificial media lacking the test element. **Review Figure 34.2**

Soils and Plants

• Soils are complex in structure, with both living and nonliving components. They contain water, gases, and inorganic and organic substances and typically consist of two or three horizontal zones called horizons. **Review Figures 34.3, 34.4, and Table 34.3**
• Soils form by mechanical and chemical weathering of rock. They may degrade by processes such as laterization.
• Plants obtain mineral nutrients by ion exchange from the surface of clay particles. **Review Figure 34.5**
• In all parts of the world, farmers use fertilizer to make up for deficiencies in soil mineral content, and they apply lime (calcium carbonate or calcium hydroxide) to correct for excess soil acidity.
• Plants affect soils in various ways, helping them form, adding material such as humus, and damaging them by removing nutrients (especially in agriculture). Soils have complex histories. **Review Figure 34.6**

Nitrogen Fixation

• A few species of soil bacteria are responsible for almost all nitrogen fixation. Some nitrogen-fixing bacteria live free in the soil; others, such as some species of *Rhizobium*, live symbiotically within the roots of plants.
• In nitrogen fixation, nitrogen gas (N_2) is reduced to ammonia (NH_3) or ammonium ions (NH_4^+) in a reaction catalyzed by nitrogenase. **Review Figure 34.9**
• Nitrogenase requires anaerobic conditions; in root nodules, leghemoglobin helps lower the oxygen concentration. The formation of a nodule requires interaction between the plant and the invading *Rhizobium*. **Review Figure 34.10**

- Nitrogen-fixing bacteria reduce atmospheric N_2 to ammonia, but most plants preferentially take up nitrate ions under most conditions. Nitrifying bacteria oxidize the ammonia to nitrate. Plants take up nitrate and reduce it back to the ammonia level, a feat of which animals are incapable. Denitrifying bacteria return N_2 to the atmosphere, completing the biological nitrogen cycle. **Review Figure 34.11**

Sulfur Metabolism

- Plants take up sulfate ions and reduce them, forming the amino acids cysteine and methionine. Cysteine is the major precursor for other sulfur-containing compounds in plants and in animals, which must obtain their organic sulfur from plants.

Heterotrophic and Carnivorous Seed Plants

- A few plants are parasitic heterotrophs that feed on other plants.
- Carnivorous plant species are autotrophs that supplement their nitrogen supply by feeding on insects.

Self-Quiz

1. Which of the following is *not* an essential mineral element for plants?
   a. Potassium
   b. Magnesium
   c. Calcium
   d. Lead
   e. Phosphorus

2. Fertilizers
   a. are often characterized by their N–P–O percentages.
   b. are not required if crops are removed frequently enough.
   c. restore needed mineral nutrients to the soil.
   d. are needed to provide carbon, hydrogen, and oxygen to plants.
   e. are needed to destroy soil pests.

3. Which of the following is *not* an important step in soil formation?
   a. Removal of bacteria
   b. Mechanical weathering
   c. Chemical weathering
   d. Clay formation
   e. Hydrolysis of soil minerals

4. Laterization
   a. results in a very productive soil.
   b. often takes place in mine tailings.
   c. produces a soil rich in copper, lead, nickel, and zinc.
   d. produces a soil rich in chromium and poor in calcium.
   e. produces a soil rich in insoluble iron and aluminum compounds.

5. Nitrogen fixation
   a. is performed only by plants.
   b. is the oxidation of nitrogen gas.
   c. is catalyzed by the enzyme nitrogenase.
   d. is a single-step chemical reaction.
   e. is possible because N_2 is a highly reactive substance.

6. Nitrification
   a. is performed only by plants.
   b. is the reduction of ammonium ions to nitrate ions.
   c. is the reduction of nitrate ions to nitrogen gas.
   d. is catalyzed by the enzyme nitrogenase.
   e. is performed by certain bacteria in the soil.

7. Nitrate reduction
   a. is performed by plants.
   b. takes place in mitochondria.
   c. is catalyzed by the enzyme nitrogenase.
   d. includes the reduction of nitrite ions to nitrate ions.
   e. is known as the Haber process.

8. Which statement about sulfur is *not* true?
   a. All living things require it.
   b. It is a component of DNA and RNA.
   c. It is a constituent of two amino acids.
   d. Its metabolism is similar to the metabolism of nitrogen.
   e. Plants obtain it from the soil.

9. Which of the following is a parasite?
   a. Venus flytrap
   b. Pitcher plant
   c. Sundew
   d. Dodder
   e. Tobacco

10. All heterotrophic seed plants
    a. are parasites.
    b. are carnivores.
    c. are incapable of photosynthesis.
    d. derive their nutrition from animals.
    e. develop from multicellular embryos.

Applying Concepts

1. Methods for determining whether a particular element is essential have been known for more than a century. Since the methods are so well established, why was the essentiality of some elements discovered only recently?

2. If a Venus flytrap were deprived of soil sulfates and hence made unable to synthesize the amino acids cysteine and methionine, would it die from lack of protein?

3. Soils are dynamic systems. What changes might result when land is subjected to heavy irrigation for agriculture, after being relatively dry for many years? What changes in the soil might result when a virgin, deciduous forest is replaced by crops that are harvested each year?

4. We mentioned that important positively charged ions are held in the soil's A horizon by clay particles, but some equally important negatively charged ions are leached and carried down to the B horizon. Why doesn't this leaching cause an electrical imbalance in the soil? (*Hint*: Think of the ionization of water.)

Readings

Beardsley, T. 1991. "A Nitrogen Fix for Wheat." *Scientific American*, March. A discussion of how nonlegumes might be taught how to form root nodules.

Losick, R. and D. Kaiser. 1997. "Why and How Bacteria Communicate." *Scientific American*, February. This exploration into bacterial communication includes a nicely illustrated account of nitrogen fixation in legume nodules.

Power, J. F. and R. F. Follett. 1987. "Monoculture." *Scientific American*, March. A discussion of how the practice of growing the same crop year after year—an increasing tendency in agriculture—affects soils, and whether it is good or bad in general.

Raven, P. H., R. F. Evert and S. Eichhorn. 1992. *Biology of Plants*, 5th Edition. Worth, New York. An excellent general botany textbook.

Taiz, L. and E. Zeiger. 1998. *Plant Physiology*, 2nd Edition. Sinauer Associates, Sunderland, MA. An authoritative textbook with good chapters on mineral nutrition.

Chapter 35

Regulation of Plant Development

Reaching for the Light
In the dense tropical rainforest, this climbing plant uses its tendrils to gain access to the sun.

P lants are highly capable and adaptable organisms. They alter their pattern of development to deal with accidents and with environmental changes, and they seize opportunities that come their way. For example, some plants have tendrils—slender, coiling structures—that are sensitive to touch. If a tendril comes in contact with an object, it coils around the object, supporting the plant as it grows and gains improved access to light for photosynthesis, shading plants below it and seizing a competitive advantage.

If an herbivore (or a gardener) removes a bud from a plant, the stem produces new branches to replace the damaged one. If a seed germinates deep in the soil, the shoot grows rapidly until it reaches the surface and can "feed" on sunlight. As you have probably observed, shoots grow toward a light source, obtaining more light for photosynthesis. As you have probably not observed, roots grow away from a light source. If a plant is toppled by wind or by earth movement, the shoots grow upward and the roots downward. Another type of developmental response by plants to an environmental challenge is the hypersensitive response to pathogens (see Chapter 33).

In this chapter we will give a brief overview of the life of a plant and its developmental stages. We'll explore the internal and external regulators of plant development, and consider the multiple roles that each plays in normal development. We'll conclude with a brief look at how studies of mutant plants are shaping current research in plant development.

Four Major Regulators of Plant Development

The development of a plant—the progressive changes that take place throughout its life—is regulated in complex ways. Four major factors play roles in the regulation of plant development: the *environment, hormones,* the pigment *phytochrome,* and the plant's *genome.* These factors interact. Environmental cues, such as changes IBODN light or temperature, trigger some important developmental events, such as flowering and the onset and end of dormancy. Hormones and phytochrome mediate the effects of environmental cues.

Regulatory compounds that are active at very low concentrations are called **hormones**. They mediate many developmental phenomena, such as stem growth and autumn leaf fall. The plant hormones include *gibberellins, auxin, cytokinins, ethylene,* and *abscisic acid.* Unlike animals, which produce each hormone in a specific part of the body, plants produce hormones in most of their cells. Each plant hormone plays multiple regulatory roles, affecting several different aspects of development (Table 35.1). Interactions among the hormones are often complex.

Like the hormones, the pigment phytochrome regulates many processes. Unlike the hormones, which are small molecules, phytochrome is a protein. Light (an environmental cue) acts directly on phytochrome, which in turn regulates developmental processes such as the many changes accompanying the growth of a young plant out of the soil and into the light.

No matter what cues direct development, ultimately the plant's genome determines the limits within which the plant and its parts will develop. The genome is the master plan of the plant, but it is interpreted differently depending on the status of the environment. The genome encodes phytochrome and the enzymes that catalyze the formation of the hormones and mediate some of their actions; it is also the target for some hormone actions. For several decades hormones and phytochrome were the focus of most work on plant development, but recent advances in molecular genetics now allow us to focus on underlying processes.

From Seed to Death: An Overview of Plant Development

Let's review the life history of a flowering plant, from seed to death, focusing on how the developmental events are regulated. Keep in mind that as plants develop, the environment, hormones, and phytochrome affect three fundamental processes: cell division, cell expansion, and cell differentiation. Try to envision how the division, expansion, and differentiation of cells contribute to different developmental phenomena.

The seed germinates and forms a growing seedling

Consider the seed. All developmental activity may be suspended in the seed; in other words, it may be **dormant.** Typically, only 5 to 20 percent of its weight is water. Cells in dormant seeds do not divide, expand, or differentiate. Seed dormancy may be broken by one of several physical mechanisms—mechanical abrasion, fire, leaching of inhibitors by water—described later in this chapter.

As the seed **germinates** (begins to develop), it first imbibes (takes up) water. The growing embryo must then obtain building blocks—monomers—by digesting the polymeric reserve foods stored either in the cotyledons or in the endosperm. The embryos of some plant species secrete gibberellins that direct the mobilization of these reserves.

Early shoot development varies among the flowering plants. In some monocots, such as grasses, the shoot is initially protected by a leaf sheath, the **coleoptile** (Figure 35.1*a*). The developing shoot later grows out through the coleoptile. Dicots lack this protective structure. In most dicots, the hypocotyl (the part of the stem below the cotyledon) elongates and the cotyledons are carried to the surface, where they become the first important photosynthetic structures (Figure 35.1*b*). In other dicots, such as peas, the cotyledons remain below the soil surface, and tissue above them, the epicotyl, grows up through the soil (Figure 35.1*c*).

If the seed germinates underground, the new seedling must elongate rapidly and cope with life in the absence of light. Phytochrome controls this stage and ends it when the shoot reaches the light (Figure 35.2). The growth of the seedling, both in darkness and light, also involves the hormones auxin and gibberellin, and auxin is known to regulate tissue and organ formation but always in coordination with other hormones. Thus auxin affects cell differentiation. Other information regulating these phenomena comes from cytokinins.

Eventually the plant flowers and sets fruit

Flowering may be initiated when the plant reaches an appropriate age or size. Some plant species, however, flower at particular times of the year, meaning that the plant must sense the appropriate date. These plants are photoperiodic (see Chapter 36); they measure the length of the night (shorter in the summer, longer in the winter) with great precision. Although we don't know *how* it works, we do know a lot about photoperiodism. We know that leaves measure the length of darkness, that there is a biological clock—itself a mystery—and that light absorbed by phytochrome can affect the time-measuring process.

Once a leaf has determined, by measuring the night length, that it is time for the plant to flower, that information must be transported as a signal to the places

TABLE 35.1   Plant Hormones

HORMONE	SITE OF PRODUCTION	ACTIVITY				
		SEED DORMANCY	SEED GERMINATION	SEEDLING GROWTH	APICAL DOMINANCE	LEAF ABSCISSION
Gibberellins	Embryo, young leaves, root and shoot apices	Breaks	Promotes	Promotes cell division and expansion	—	—
Auxin	Embryo, young leaves, shoot apical meristem	—	—	Promotes cell expansion	Inhibits lateral buds	Inhibits
Cytokinins	Roots	—	—	Promotes cell division	Promotes lateral buds	Inhibits
Ethylene	Ripening fruit, senescing tissue, stem nodes	—	—	—	—	Promotes
Abscisic acid	Root cap, older leaves, stem	Imposes	Inhibits	—	—	—

where flowers will form. The means by which the signal is transmitted remains a mystery, but it seems likely that a "flowering hormone"—named florigen, but not yet isolated—travels from the leaf to the point of flower formation. If flowering information travels in another form, this form has not been identified.

After flowers form, hormones, including auxin and gibberellin, play further roles. Hormones and other substances control the growth of a pollen tube down the style of a pistil. Following fertilization, a fruit develops, controlled in several ways by gibberellin and auxin (Figure 35.3). The ripening of the fruit is also under chemical control, commonly by the gaseous hormone ethylene.

Dormancy, senescence, and death

Some perennials have buds that enter a state of winter dormancy during the cold season. (Perennial plants are those that continue to grow year after year.) Abscisic acid helps maintain such dormancy. Finally, the plant **senesces** (deteriorates because of aging) and dies. Death, which may be initiated by signals from the environment, follows senescent changes that are controlled by regulators such as ethylene (Figure 35.4).

By now, we have mentioned each of the five hormone classes. To keep them straight, learn a major, distinct role for each of them. Then add further, secondary roles for some of them as you read through this material. We'll consider in detail the elements of

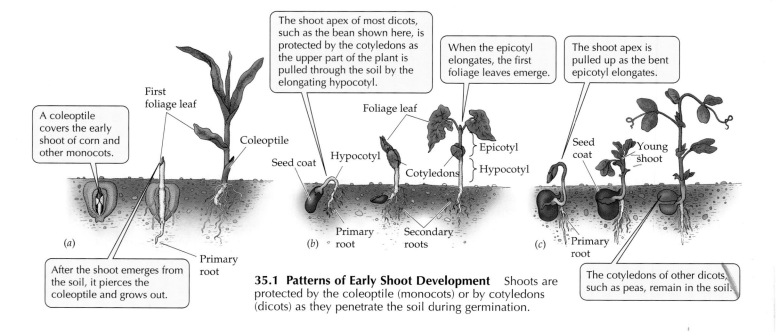

35.1 Patterns of Early Shoot Development Shoots are protected by the coleoptile (monocots) or by cotyledons (dicots) as they penetrate the soil during germination.

A coleoptile covers the early shoot of corn and other monocots.

After the shoot emerges from the soil, it pierces the coleoptile and grows out.

First foliage leaf

Coleoptile

Primary root

(a)

The shoot apex of most dicots, such as the bean shown here, is protected by the cotyledons as the upper part of the plant is pulled through the soil by the elongating hypocotyl.

Seed coat

Hypocotyl

Primary root

Secondary roots

(b)

When the epicotyl elongates, the first foliage leaves emerge.

Foliage leaf

Epicotyl

Hypocotyl

Cotyledons

The shoot apex is pulled up as the bent epicotyl elongates.

Seed coat

Young shoot

Primary root

(c)

The cotyledons of other dicots, such as peas, remain in the soil.

WINTER DORMANCY	FLOWERING	FRUIT DEVELOPMENT
Breaks	Stimulates in some plants	Promotes
—	—	Promotes
—	—	—
—	—	Promotes ripening
Imposes	—	—

this brief overview in this chapter and the next. Let's begin with regulation at the start of the life cycle—the seed and its germination.

Seed Dormancy and Germination

The seeds of some species are, in effect, instant plants, because all they need for germination is water. Many other species have seeds whose germination is regulated in more complex ways because they are initially dormant. Seed dormancy may last for weeks, months,

years, or even centuries. The mechanisms of dormancy are numerous and diverse, but three principal strategies dominate: exclusion of water or oxygen from the embryo, mechanical restraint of the embryo, and chemical inhibition of embryo development.

Some seeds exclude water or, sometimes, oxygen by having an impermeable seed coat. The dormancy of such seeds can be broken only if the seed coat is abraded as the seed tumbles across the ground or through creek beds, passes through the digestive tract of a bird, or by other means. Such modification of the seed coat is called **scarification.** The environment affects germination of these seeds by providing or withholding the means of scarification.

The seed coat may also impose dormancy by the simple mechanical restraint of the embryo; if the embryo cannot expand, the seed cannot germinate. In the laboratory we can promote germination of such a seed by simply cutting away part of the coat or by partly dissolving it with strong acid. In nature, soil microorganisms probably play a major role in softening seed coats of this type, and the action of digestive enzymes in the guts of birds or other animals is also important.

Another agent of scarification to release mechanical restraint is fire, which causes significantly increased germination in some natural habitats. Fire can also melt wax in seed coats, making water available to the embryo (Figure 35.5).

The action of chemical germination inhibitors is another mechanism of seed dormancy. As long as the concentration of inhibitor is high, the seed remains dormant. One means of reducing the level of inhibitor is by leaching—that is, prolonged exposure to water. Another is the scorching of the seeds by fire, which breaks down some inhibitors. Usually inhibitors of germination are already present in the dry seed, but in a few cases they are produced only after the seed has begun to take up water. The most common chemical inhibitor of seed germination is abscisic acid. In some seeds the level of abscisic acid or other inhibitors does not decline during germination; rather, the effect of the inhibitor is overcome by a gradual increase in the concentrations of growth promoters.

There are still other mechanisms for breaking dormancy. Some seeds, such as those of the tomato and lima bean, remain dormant until they have dried extensively. Temperature can be an environmental cue initiating germination.

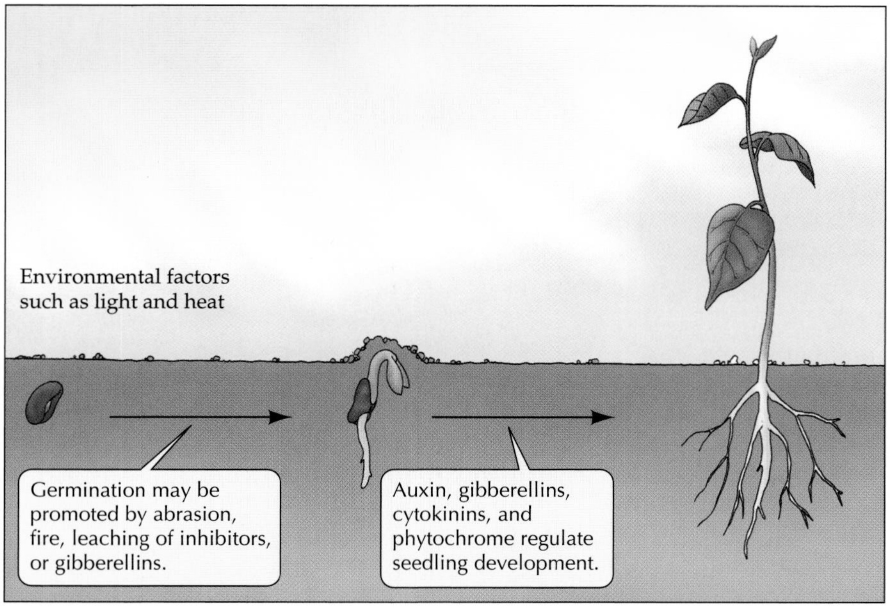

Environmental factors such as light and heat

Germination may be promoted by abrasion, fire, leaching of inhibitors, or gibberellins.

Auxin, gibberellins, cytokinins, and phytochrome regulate seedling development.

35.2 From Seed to Seedling   Environmental factors, hormones, and phytochrome regulate the first stages of plant growth.

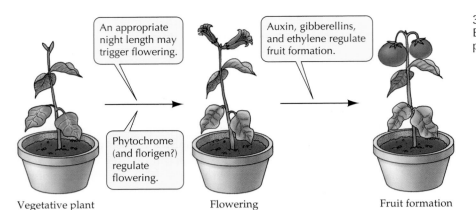

35.3 Flowering and Fruit Formation
Environmental factors, hormones, and phytochrome regulate plant reproduction.

Even a brief exposure to temperatures near freezing may stimulate germination, but more commonly a period of many days or weeks of low temperature is required to end dormancy. In agriculture and forestry, it is common practice to refrigerate seeds such as those of conifers to hasten germination. This refrigeration procedure is known as **stratification;** typically, it consists of a month or two at 5(C. The effects of cold treatment vary from species to species, but one result may be a gradual decrease in the content of germination inhibitors such as abscisic acid.

Although the means vary, environmental signals are important regulators of germination.

Seed dormancy affords adaptive advantages

What are the potential advantages of regulating the germination of seeds? For many species, dormancy results in germination at a favorable time. Seeds that require a long cold period for germination commonly germinate in the spring, when water is usually abundant—germination in the dry days of late summer could be risky. Some other seeds will not germinate

until a certain amount of time has passed, regardless of how they are treated. This period of **afterripening** prevents germination while the seed of a cereal grain, for example, is still attached to the parent plant, and it tends to favor dispersal of the seed.

Regulating germination may increase the likelihood of a seed's germinating in the right place. For example, some cypress trees grow in standing water, and their seeds germinate only if leached extensively by water (Figure 35.6). Many weed seeds must have their seed coats damaged before they will germinate, and other weed seeds will not germinate unless they have been

35.5 Fire and Seed Germination This fireweed germinated and flourished after a great fire along the Alaska Highway.

35.4 Senescence and Death Environmental factors and hormones regulate the final stages of plant growth.

35.6 Leaching of Germination Inhibitors   The seeds of water tupelo trees (*Nyssa aquatica*) germinate only after being leached by water, which increases the chances that they will germinate in a situation suitable for their growth.

exposed to light. Either type of weed germinates best in disturbed soils.

You may have noticed how a freshly cultivated patch of soil quickly teems with weeds that are free from competition with other plants. Seeds that must be scorched by fire in order to germinate also avoid competition; they germinate only when the area has been cleared by fire. Light-requiring seeds, which germinate only at or near the surface of the soil, are generally tiny seeds with few food reserves. The germination of some seeds is inhibited by light; these seeds germinate only when well buried. Light-inhibited seeds are usually large and well stocked with nutrients.

Seed dormancy helps annual species counter the effects of year-to-year variations in the environment. Some seeds remain dormant throughout an unfavorable year, and other seeds germinate at different times during the year. Seed dormancy can also contribute to the dispersal of a plant species. Seeds that remain dormant until they have passed through the guts of birds or other animals will likely be carried some distance before they are deposited. Seeds carried by birds in their digestive tracts can give rise to the first plants on newly formed volcanic islands, for example.

The seeds of some desert plants contain germination inhibitors that are leached by water. An amount of water sufficient to leach out the inhibitor is also enough to take the resulting plant through its entire life cycle.

Seed germination begins with the imbibition of water

The first step in seed germination is the uptake of water, called **imbibition**. Typically, the seed is dry before germination begins. Its water potential (see Chapter 32) is very negative, and water can be taken up readily if the seed coat allows it. The magnitude of the water potential is demonstrated by the force exerted by seeds expanding in water. Cocklebur seeds that are imbibing can exert a pressure of up to 1,000 atmospheres (about 100 megapascals) against a restraining force.

As a seed takes up water, it undergoes metabolic changes: Certain existing enzymes become activated, RNA and then enzymes are synthesized, the rate of cellular respiration increases, and other metabolic pathways become activated. Interestingly, in many seeds there is no DNA synthesis and no cell division during these early stages of seed germination. Emergent growth arises solely from the expansion of small, preformed cells. DNA is synthesized only after the **radicle**—the embryonic root—begins to grow and poke out beyond the seed coat (Figure 35.7).

The embryo must mobilize its reserves

Until the young plant (the **seedling**) becomes able to photosynthesize, it depends on built-in reserves from the seed to meet its needs for energy and materials. The reserves are stored in the endosperm or cotyledons. The principal reserve of energy and carbon in the seeds of some species is the carbohydrate starch. More species, however, store lipids—fats or oils—as reserves in their seeds. Typically, the endosperm of the seed holds amino acid reserves in the form of proteins, rather than as free amino acids.

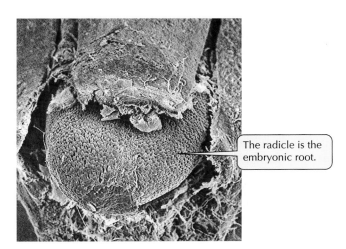

The radicle is the embryonic root.

35.7 The Radicle Emerges   The tip of this barley seed's radicle has just broken through its protective sheath. The appearance of the radicle is one of the first externally visible events in seed germination.

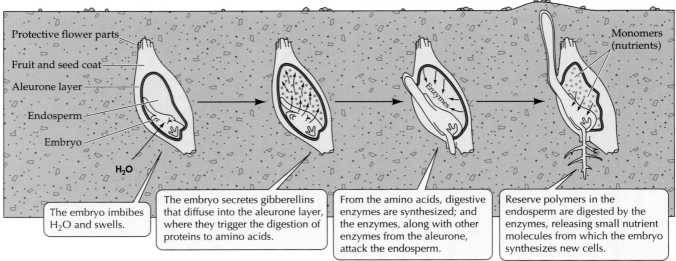

Protective flower parts
Fruit and seed coat
Aleurone layer
Endosperm
Embryo

H₂O

Monomers (nutrients)

Enzymes

The embryo imbibes H₂O and swells.

The embryo secretes gibberellins that diffuse into the aleurone layer, where they trigger the digestion of proteins to amino acids.

From the amino acids, digestive enzymes are synthesized; and the enzymes, along with other enzymes from the aleurone, attack the endosperm.

Reserve polymers in the endosperm are digested by the enzymes, releasing small nutrient molecules from which the embryo synthesizes new cells.

35.8 Embryos Mobilize Polymer Reserves Seed germination in cereal grasses consists of a cascade of processes. Gibberellin signals the conversion of reserve polymers into monomers that can be used by the developing embryo.

The giant molecules of starch, lipids, and proteins must be digested into monomers before they can enter the cells of the embryo to be used as building blocks and energy sources. Starch is a polymer of glucose, the starting point for glycolysis and cellular respiration. Starch is digested to glucose. The lipids can be digested to release glycerol and fatty acids, both of which yield energy through cellular respiration. Glycerol and fatty acids can also be converted to glucose, which permits fat-storing plants to make all the building blocks they need for growth.

Lipid-storing seeds can pack more energy in a smaller space than starch-storing seeds can because lipids contain more calories per unit of weight than does starch. Some species that store lipids in their seeds store starch in their roots or tubers, where space is not a concern. The growing embryo can also break down reserve proteins to obtain the amino acids it needs to assemble into its many new proteins. Plant species differ in their patterns of digestion of reserve polymers.

Germinating barley and other cereal seeds digest proteins and starch as follows (Figure 35.8). As the embryo becomes active, it secretes gibberellins. The gibberellins diffuse through the endosperm to a surrounding tissue called the **aleurone layer,** which lies inside the seed coat. The gibberellins trigger a crucial series of events in the aleurone layer. First, protein-containing bodies called aleurone grains break down, releasing amino acids. The aleurone layer then uses the amino acids in the assembly of digestive enzymes, including amylases (starch-degrading enzymes), proteases (protein-degrading enzymes), and ribonucleases (RNA-degrading enzymes). These enzymes, along with certain others already present in the aleurone layer, are next

secreted into the endosperm, where they catalyze the release of sugars and amino acid monomers from reserve polymers for use by the growing embryo.

We will now consider each of the major plant hormones, beginning with the gibberellins. Here is a study hint for the rest of this chapter. We have seen in a brief overview that information about hormones can be organized by considering the development of each part of the plant one after another. Now, to take a closer look at the hormones, we will organize the information by considering each hormone in turn. Try using the first method to outline the material presented in the sections that follow, showing in detail how the development of roots, stems, buds, and leaves is regulated. In general, this is an excellent way to study: Look for a different way to organize the material presented in lecture or in the book.

Gibberellins: A Large Family of Hormones

We just encountered the gibberellins in the mechanism by which the germinating seeds of barley and other cereals convert their reserve proteins and starch into soluble monomers (see Figure 35.8). Gibberellins produce a wide variety of other effects on plant development in addition to this triggering of digestive enzyme synthesis. In this section we'll discuss the discovery of the gibberellins, the startling number of different gibberellins, and the many effects of gibberellins.

Foolish seedlings led to the discovery of the gibberellins

The gibberellins are a large family of closely related compounds (Figure 35.9a and b), some found in plants and others in a pathogenic (disease-causing) fungus, where they were first discovered. The discovery of the gibberellins followed a crooked path, beginning with a book dictated in 1809 by a Japanese farmer named Konishi. Konishi described the symptoms of *ine bakanae-byo*, the "foolish seedling" disease of rice. Seedlings affected by the *bakanae* disease grow more rapidly than their healthy neighbors, but the rapid growth gives rise to spindly plants that die before producing seed. The disease has been of considerable economic importance in several parts of the world. In 1898 it was learned that the *bakanae* disease, including both its growth-promoting and toxic effects, is caused by a fungus now known as *Gibberella fujikuroi*.

In 1925 the Japanese biologist Eiichi Kurosawa studied how *G. fujikuroi* causes the spindly growth characteristic of the *bakanae* disease. He grew the fungus on a liquid medium and then separated the fungus from the medium by filtering. He heated the medium to kill any remaining fungus. The resulting heat-treated filtrate still stimulated the growth of uninfected rice seedlings. Kurosawa found no such effects using medium that had never contained the fungus (Figure 35.10). Thus, he established that *G. fujikuroi* produces a chemical substance that has growth-promoting properties—gibberellin. In the late 1930s it became clear that there was more than one gibberellin. It was also shown that the toxic effects of the fungus were caused by another, inhibitory substance.

Are the gibberellins simply exotic products of an obscure fungus, or do they play a more general role in the growth of plants? Bernard O. Phinney of the University of California, Los Angeles, answered this question in part in 1956, when he reported the spectacular growth-promoting effect of gibberellins on certain dwarf strains of corn. These plants were known to be genetic dwarfs; each phenotype was produced when a particular recessive allele was present in the homozygous condition (see Chapter 10 if you need to review these genetic terms). Gibberellin applied to nondwarf—tall—corn seedlings had almost no effect, but gibberellin applied to the dwarfs caused them to grow as tall as their normal relatives. (A comparable effect of gibberellin on dwarf mustard is shown in Figure 35.11.)

This result suggested to Phinney that (1) gibberellins are normal constituents of corn and perhaps of all plants, and (2) the dwarfs are short because they cannot produce their own gibberellin. According to these hypotheses, nondwarf plants manufacture enough gibberellins to promote their full growth. Phinney and other scientists tested extracts from numerous plant species to see if these extracts promoted growth in dwarf corn and they found that many such extracts did. These findings provided direct evidence that plants that are not genetic dwarfs contain gibberellin-like substances.

The roots, leaves, and flowers of a dwarf corn or dwarf tomato plant appear normal, but the stems are much shorter than their counterparts on other plants. What does this tell us? We know that all the cells of the dwarf plants have the same genome and all parts of the dwarf plants contain much less gibberellin than do the organs of other plants. From this knowledge we may infer that stem elongation *requires* gibberellin or the products of gibberellin action; on the other hand, gibberellin plays no comparable role in the development of roots, leaves, and flowers.

35.9 Plant Hormones The chemical structures of some representative compounds.

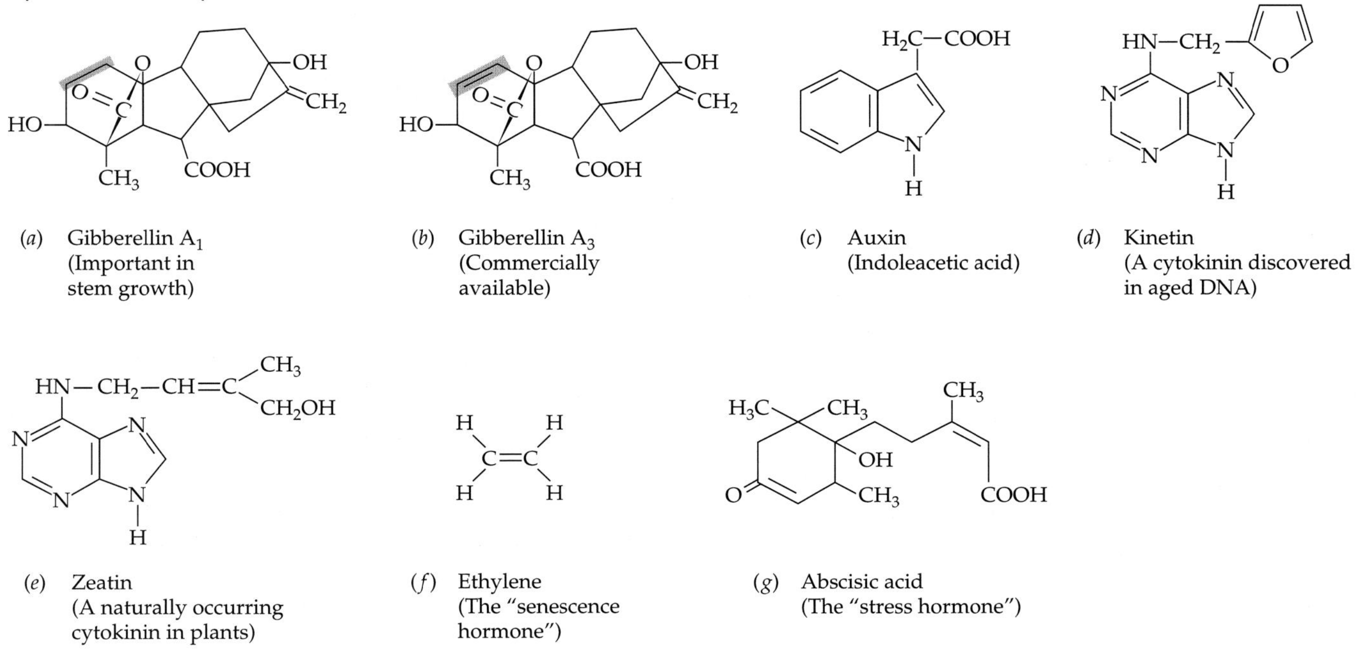

(*a*) Gibberellin A$_1$ (Important in stem growth)

(*b*) Gibberellin A$_3$ (Commercially available)

(*c*) Auxin (Indoleacetic acid)

(*d*) Kinetin (A cytokinin discovered in aged DNA)

(*e*) Zeatin (A naturally occurring cytokinin in plants)

(*f*) Ethylene (The "senescence hormone")

(*g*) Abscisic acid (The "stress hormone")

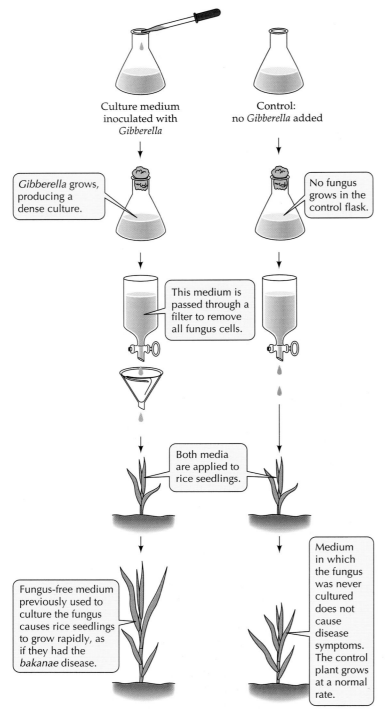

35.10 Kurosawa's Experiment Kurosawa demonstrated that *bakanae* disease is caused by substances produced by the fungus *Gibberella fujikuroi.*

Why are there so many gibberellins?

We do not yet know how many different gibberellins exist. Each year brings reports of more, and more than 80 are now characterized. Some gibberellins are produced in the root system, others in young leaves. For many years plant physiologists were puzzled by the existence of such a great number of different gibberellins. Recent work, however, has led to the conclusion that only one gibberellin, *gibberellin A₁*, actually controls stem elongation; the other gibberellins found in stems are simply intermediates in the production of gibberellin A₁.

As we will see in the next section, gibberellins affect processes other than stem elongation, but we do not yet know which gibberellin has any other particular effect.

The gibberellins have many effects

Gibberellins and other hormones regulate the growth of fruits. It has long been known that seedless varieties of grapes form smaller fruit than their seeded relatives. In one experiment, removal of seeds from very young seeded grapes prevented normal fruit growth, suggesting that the seeds are sources of a fruit growth regulator. It was then shown that spraying young seedless grapes with a gibberellin solution causes them to grow as large as seeded varieties. Subsequent biochemical studies showed that the developing seeds produce gibberellin, which diffuses out into the immature fruit tissue.

Some biennial plants respond dramatically to an increased level of gibberellin. Biennial plants grow vegetatively in their first year and flower and die in their second year. In their second year, in response either to

35.11 The Effect of Gibberellin on Dwarf Plant In this experiment, the effect of gibberellin is tested on two dwarf mustard plants (*Brassica rapa*).

the increasing length of days or to the winter cold period, the apical meristems of biennials produce elongated shoots that eventually bear flowers. This elongation is called **bolting**. When the plant senses the appropriate environmental cue—longer days or a sufficient winter chilling—it produces more gibberellins, raising the gibberellin concentration to a level that causes the shoot to bolt. Plants of some biennial species will bolt when sprayed with a gibberellin solution even if they have not experienced the environmental cue (Figure 35.12).

Gibberellins also cause fruit to grow from unfertilized flowers, promote seed germination in lettuce and some other species, and help bring spring buds out of winter dormancy. Hormones usually have multiple effects within the plant and often interact with one another in regulating developmental processes. In controlling stem elongation, for example, gibberellins interact with another hormone, auxin.

Auxin: A Hormone with Many Effects

If you pinch off the apical bud at the top of a bean plant, lateral buds that were once inactive become active. Similarly, pruning a shrub causes an increase in branching. If you cut off the blade of a leaf but leave its petiole attached to the plant, the petiole drops off sooner than it would if the leaf were intact. If a plant is kept indoors, its shoot system grows toward a bright window. What these diverse responses of shoot systems have in common is that they are mediated by a plant hormone called **auxin**—or *indoleacetic acid* (IAA) in chemical terms (see Figure 35.9c).

In the discussions that follow, we will look at the discovery of auxin, its transport within the plant, its role as mediator of light and gravitational effects on plant growth, its many effects on vegetative growth and its effects on fruit development.

Plant movements led to the discovery of auxin

The discovery of auxin and its numerous physiological activities traces back to work done in the 1880s by Charles Darwin and his son Francis, who were interested in plant movements. One type of movement they studied was **phototropism**, the growth of plant structures toward light (as in shoots) or away from it (as in roots). An obvious question they asked was, What part of the plant senses the light?

To answer this question, the Darwins worked with canary grass (*Phalaris canariensis*) seedlings grown in the dark. The dark-grown grass seedling has a coleoptile (leaf sheath) that covers the immature shoot. To find the light-receptive region of the coleoptile, the Darwins tried "blindfolding" it in various places and then illuminating it from one side (Figure 35.13). The coleoptile grew toward the light whenever its tip was

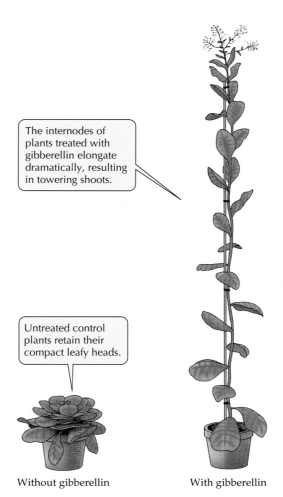

The internodes of plants treated with gibberellin elongate dramatically, resulting in towering shoots.

Untreated control plants retain their compact leafy heads.

Without gibberellin   With gibberellin

35.12 Bolting Spraying with gibberellin causes cabbage and some other plants to bolt.

exposed. If the top millimeter or more of the coleoptile was covered, however, there was no phototropic response. Thus the tip contains the photoreceptor. The bending, however, takes place in a growing region a few millimeters below the tip. Therefore, the Darwins reasoned, some type of message must travel within the coleoptile from the tip to the growing region.

Others later demonstrated that the message is a chemical substance; it cannot pass through a barrier impermeable to chemicals but does pass through certain nonliving materials, such as gelatin. The tip of the coleoptile produces a hormone that moves down the coleoptile to the growing region. If the tip is removed, the growth of the coleoptile is sharply inhibited; if the tip is then carefully replaced, growth resumes, even if the tip and base are separated by a thin layer of gelatin. The hormone moves down from the tip but does not move from one side of the coleoptile to the other. If the tip of an oat coleoptile is cut off and replaced so that it covers only one side of the cut end of the shoot, the coleoptile curves as the cells on the side

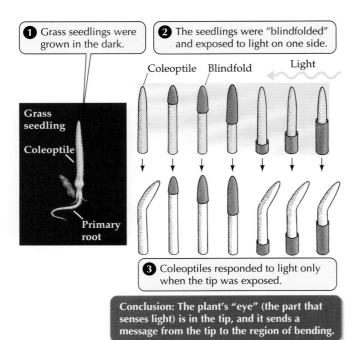

1 Grass seedlings were grown in the dark.

2 The seedlings were "blindfolded" and exposed to light on one side.

Coleoptile   Blindfold   Light

Grass seedling

Coleoptile

Primary root

3 Coleoptiles responded to light only when the tip was exposed.

Conclusion: The plant's "eye" (the part that senses light) is in the tip, and it sends a message from the tip to the region of bending.

35.13 The Darwins' Phototropism Experiment The top drawings show some of the ways in which seedlings grown in the dark were "blindfolded"; the lower drawings show what the Darwins observed in each case. Their observations led them to hypothesize the existence of a growth-promoting "messenger" substance.

below the replaced tip grow more rapidly than those on the other side.

With this information as a beginning, the Dutch botanist Frits W. Went succeeded where many had failed in isolating the hormone from oat coleoptiles. Went removed coleoptile tips and placed their cut surfaces on a block of gelatin, hoping the hormone would diffuse into the gelatin. Then he placed pieces of the gelatin block on decapitated coleoptiles—positioned to cover only one side, just as coleoptile tips had been placed in some of the earlier experiments (Figure 35.14). As they grew, the coleoptiles curved toward the side away from the gelatin. This curvature demonstrated that the hormone had indeed diffused into the gelatin block from the isolated coleoptile tips. The hormone had at last been isolated from the plant. It was later named auxin; still later, it was shown to be indoleacetic acid. Went's historic experiment was performed in 1926—the very year Kurosawa published his classic account of the isolation of a growth substance from the fungus *Gibberella fujikuroi*. We are still learning how these hormones—auxin and the gibberellins—work.

Auxin transport is often polar

After being isolated from the plant, auxin was studied in various ways. Early experiments showed that its

movement through certain tissues is strictly polar—that is, unidirectional along a line from apex to base. By inverting the setups in half of the experiments, scientists determined that the apex-to-base direction of auxin movement has nothing to do with gravity; the polarity of this movement is a totally biological matter. Many plant parts show at least partial polarity of auxin transport. For example, auxin moves in leaf petioles from the blade end toward the stem end.

Light and gravity affect the direction of plant growth

The *lateral* movement of auxin in the apex affects the direction of plant growth. When light strikes a coleoptile from one side, auxin at the tip moves toward the shaded side. The imbalance thus established is maintained down the coleoptile, so that in the growing region below there is more auxin on the shaded side, causing the unequal growth that results in a coleoptile bent toward the light. This bending is phototropism (Figure 35.15a).

Similarly, but even in the dark, auxin moves to the lower side of a shoot that has been tipped over, causing more rapid growth in the lower side and, hence, an upward bend of the shoot. This phenomenon is **gravitropism,** the growth of a plant part in a direction determined by gravity (Figure 35.15b). The upward gravitropic response of shoots is defined as negative; the gravitropism of roots, which bend downward, is positive.

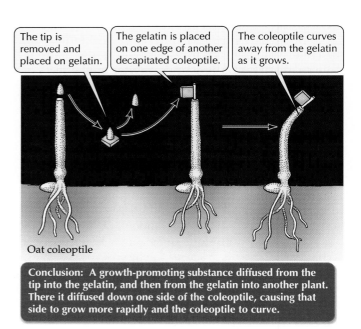

The tip is removed and placed on gelatin.

The gelatin is placed on one edge of another decapitated coleoptile.

The coleoptile curves away from the gelatin as it grows.

Oat coleoptile

Conclusion: A growth-promoting substance diffused from the tip to the gelatin, and then from the gelatin into another plant. There it diffused down one side of the coleoptile, causing that side to grow more rapidly and the coleoptile to curve.

35.14 Went's Experiment Went succeeded in isolating the growth-promoting chemical substance that the Darwins had hypothesized. This substance was named auxin; later chemical analysis revealed it to be indoleacetic acid.

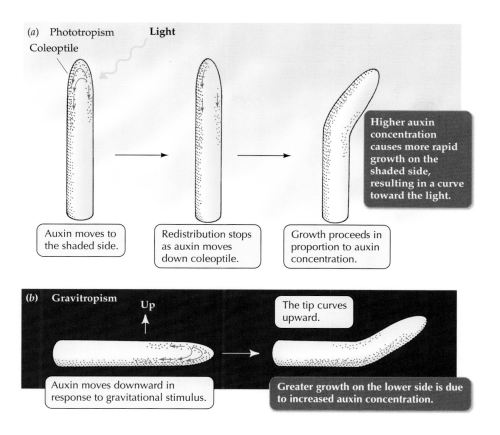

(a) Phototropism   **Light**
Coleoptile

Auxin moves to the shaded side.

Redistribution stops as auxin moves down coleoptile.

Growth proceeds in proportion to auxin concentration.

Higher auxin concentration causes more rapid growth on the shaded side, resulting in a curve toward the light.

(b) Gravitropism   **Up**

The tip curves upward.

Auxin moves downward in response to gravitational stimulus.

Greater growth on the lower side is due to increased auxin concentration.

35.15 Plants Respond to Light and Gravity Phototropism and gravitropism occur in response to auxin concentration.

Auxin maintains **apical dominance**, the tendency of some plants to grow a single main stem with minimal branching. This phenomenon can be shown by an experiment with young seedlings. If the plant remains intact, the stem elongates and the lateral buds remain inactive. Removal of the apical bud—the major site of auxin production—causes the lateral buds to grow out vigorously, but this growth is prevented if the cut surface of the stem is treated with an auxin solution (Figure 35.17). Apical buds of branches exert apical dominance: Their own lateral buds are inactive unless the apex is removed. In the two experi-

Auxin affects vegetative growth in several ways

Like the gibberellins, auxin has many roles in the plant. Cuttings from the shoots of some plants can produce roots and thus grow into entire new plants. For this to happen, certain undifferentiated cells in the *interior* of the shoot, originally destined to function only in food storage, must set off on an entirely new mission: to differentiate and become organized into the apical meristem of a new root.

These changes are very similar to those in the pericycle of a root when a branch root forms (see Chapter 31). Shoot cuttings of many species can be stimulated to grow profuse roots by dipping the cut surfaces into an auxin solution, suggesting that the plant's own auxin plays a role in the initiation of branch roots. Commercial preparations to enhance the rooting of plant cuttings typically contain mostly synthetic auxins.

The effect of auxin on the separation of old leaves from stems, called **abscission**, is quite different. Normally, if the blade of the leaf is excised, the petiole abscises more rapidly than if the leaf had remained intact (Figure 35.16*a*). If the cut surface is treated with an auxin solution, however, the petiole remains attached to the plant, often longer than an intact leaf would have (Figure 35.16*b*). The time of abscission of leaves in nature appears to be determined in part by a decrease in the movement of auxin, produced in the blade, through the petiole.

(a) Petiole   Blade

The leaf blade is removed.

No auxin is added.

The petiole abscises.

(b)

The leaf blade is removed.

Auxin in lanolin

Auxin is added.

The petiole remains on the plant.

35.16 Auxin and Leaf Abscission The leaf blade is a source of auxin throughout the growing season; without auxin, the petiole abscises.

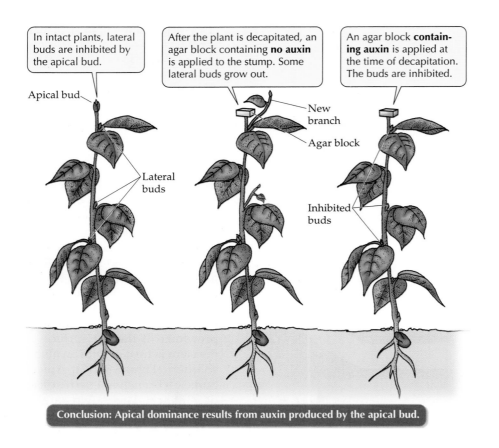

In intact plants, lateral buds are inhibited by the apical bud.

After the plant is decapitated, an agar block containing **no auxin** is applied to the stump. Some lateral buds grow out.

An agar block **containing auxin** is applied at the time of decapitation. The buds are inhibited.

Apical bud

New branch

Agar block

Lateral buds

Inhibited buds

Conclusion: Apical dominance results from auxin produced by the apical bud.

35.17 Auxin and Apical Dominance Auxin produced by the apical bud maintains apical dominance—the growth of a single main stem with minimal branching.

ments on leaves and stems that we have discussed, removal of a particular part of the plant produces an effect—(1) abscission or (2) loss of apical dominance—and the effect is prevented by treatment with auxin. These results are consistent with other data showing that the excised part of the leaf or stem is an auxin source and that auxin in the intact plant helps maintain apical dominance and delays the abscission of leaves.

Many synthetic auxins—chemical analogs of indoleacetic acid—have been produced and studied. One of them, 2,4-dichlorophenoxyacetic acid (2,4-D), has the striking property of being lethal to dicots at concentrations that are harmless to monocots. This property made 2,4-D a widely used **selective herbicide** that could be sprayed on a lawn or a cereal crop to kill the dicots, thus eliminating most of the weeds. However, because 2,4-D takes a long time to break down, it pollutes the environment, so scientists are seeking new approaches to selective weed killing.

Auxin controls the development of some fruits

Although fruit development normally depends on prior fertilization of the egg, in many species treatment of an unfertilized ovary with auxin or gibberellin causes **parthenocarpy**—fruit formation without fertilization of the egg. Parthenocarpic fruit form spontaneously in some plants, including dandelions, seedless

grapes, and the cultivated varieties of pineapple and banana.

The strawberry is an unusual "fruit." What we commonly call the fruit is actually a modified stem, or receptacle, and the tiny, dry "seeds," called achenes, are the true fruits. The achenes produce auxin, and the auxin induces the growth of the fleshy receptacle, as the French botanist Jean-Pierre Nitsch demonstrated (Figure 35.18). When Nitsch removed all the achenes within 3 weeks after pollination, the stem tissue did not develop into a "strawberry" (Figure 35.18b). If he pollinated only one to three of the many pistils and kept the others virgin, he observed localized receptacle growth in the area of pollination (Figure 35.18c). Nitsch could induce normal expansion of the receptacle even after removing all of the achenes, if he spread an auxin-containing paste over it (Figure 35.18d). These three results are consistent with the hypothesis that the achenes cause growth of the receptacle by producing auxin.

Plant hormones control other aspects of fruit physiology and of the development and senescence of flower parts. These activities illustrate again the great diversity of important roles that hormones play.

Master Reactions to Hormones

Each of the known plant hormones causes a variety of responses that are often seemingly unrelated, thus raising one of the major questions of plant physiology: Do all the effects of a particular hormone arise from a single mechanism—a common master chemical reaction? We know so little about how the plant hormones act at the molecular level that this question cannot yet be answered. For one hormone—auxin—we are beginning to gain some insight into its central mechanism or mechanisms, if indeed there is more than one molecular mechanism. To appreciate this mechanism, we must first briefly consider the architecture of the plant cell wall.

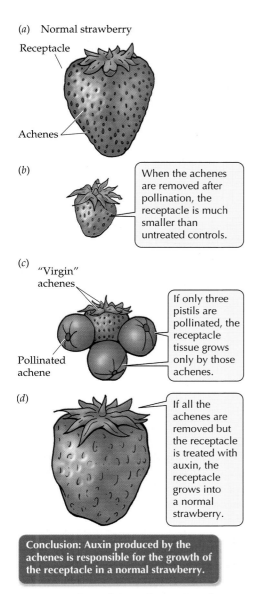

(a) Normal strawberry

Receptacle

Achenes

(b)

When the achenes are removed after pollination, the receptacle is much smaller than untreated controls.

(c) "Virgin" achenes

Pollinated achene

If only three pistils are pollinated, the receptacle tissue grows only by those achenes.

(d)

If all the achenes are removed but the receptacle is treated with auxin, the receptacle grows into a normal strawberry.

Conclusion: Auxin produced by the achenes is responsible for the growth of the receptacle in a normal strawberry.

35.18 Auxin and Strawberry Development   Auxin produced by the achenes is responsible for the normal growth of the strawberry "fruit."

Cell walls are a key to plant growth

The principal strengthening component of the plant cell wall is cellulose, a large polymer of glucose. In the wall, cellulose molecules tend to associate in parallel with one another, forming crystalline regions called micelles. Individual cellulose molecules may extend from one micelle across relatively noncrystalline regions to other micelles. Bundles of approximately 250 cellulose molecules, including many micelles, constitute microfibrils visible with an electron microscope (Figure 35.19*a*). What makes the cell wall rigid is a network of cellulose microfibrils connected by bridges of other, smaller polysaccharides (Figure 35.19*b*). As we have seen, the cellulose microfibrils are arranged in

such a way that they determine the direction of cell expansion (see Figure 15.11).

The growth of a plant cell is driven primarily by the uptake of water, which enters the cytoplasm of the cell and its vacuole (see Figure 4.22). As the vacuole expands, the cell grows rapidly, with the vacuole often making up more than 90 percent of the volume of a mature cell. As the vacuole expands, it presses the cytoplasm against the cell wall, and the wall resists this force.

For the cell to grow, its wall must loosen and be stretched. As the wall stretches, it should become thinner. However, because new polysaccharides are deposited throughout the wall and new cellulose microfibrils are deposited at the inner surface of the wall, the cell wall maintains its thickness. Thus the cellulose microfibrils in the outermost part of the wall are the oldest, and those in the innermost part the youngest.

The wall plays key roles in controlling the rate and direction of growth of a plant cell. How does the plant determine the behavior of its cell walls?

Auxin loosens the cell wall

Auxin can loosen cell walls—make them more stretchable—as the Dutch physiologist A. J. N. Heyn first demonstrated half a century ago. Heyn hung segments of oat coleoptiles on pins and hung weights on the ends of the segments, causing them to bend. Then he removed the weights, allowing the segments to bend back. Recovery was incomplete; that is, some of the bending was not reversible. Heyn called the reversible bending **elasticity** and the irreversible bending **plasticity** (Figure 35.20). Pretreating the coleoptile segments with auxin significantly increased their plasticity; it loosened the wall. This result suggested that auxin-induced cell expansion might result from just such a loosening effect.

It was later shown that auxin itself does not loosen cell walls upon contact. There is an intervening step; auxin may cause the release of a "wall-loosening factor" from the cytoplasm. Work in the 1970s in the United States and in Europe indicated that the wall-loosening factor was simply hydrogen ions (protons, H^+). Acidifying the growth medium (that is, adding H^+) causes segments of stems or coleoptiles to grow as rapidly as segments treated with auxin, and treating coleoptile segments with auxin causes acidification of the medium. Treatments that block acidification by auxin also block auxin-induced growth.

It was proposed that hydrogen ions, secreted into the cell wall as a result of auxin action, activate an enzyme or some enzymes in the wall. These enzymes might digest specific linkages connecting the cellulose microfibrils, or they might alter bonds in the matrix in which the microfibrils reside. The end result in either case would be a temporary loosening of the wall.

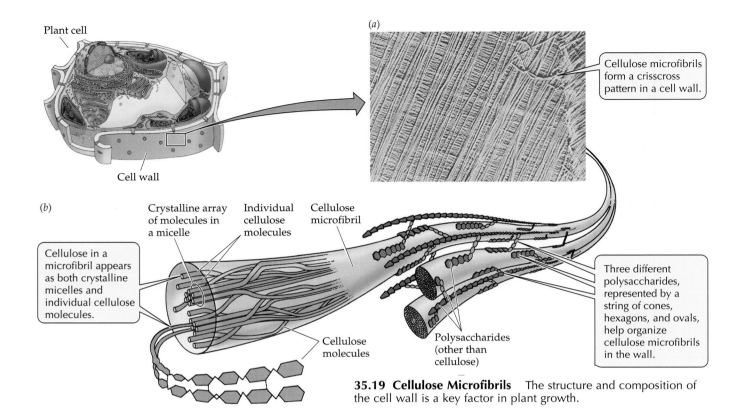

Plant cell

Cell wall

(a)

Cellulose microfibrils form a crisscross pattern in a cell wall.

(b)

Cellulose in a microfibril appears as both crystalline micelles and individual cellulose molecules.

Crystalline array of molecules in a micelle

Individual cellulose molecules

Cellulose microfibril

Cellulose molecules

Polysaccharides (other than cellulose)

Three different polysaccharides, represented by a string of cones, hexagons, and ovals, help organize cellulose microfibrils in the wall.

35.19 Cellulose Microfibrils The structure and composition of the cell wall is a key factor in plant growth.

In 1992, a group led by Daniel Cosgrove at Pennsylvania State University isolated two interesting proteins from cucumber cell walls. These proteins cause the extension of isolated cell walls of several species and appear to be the proteins activated by hydrogen ions. They seem not to digest linkages in polysaccharides such as cellulose, but rather to alter bonds in the matrix.

The cell wall is an important site for the major activities regulating plant development. Auxin is not the only agent that affects the properties of the cell wall. Gibberellins, too, can loosen the wall, which is not surprising in view of their growth-promoting activities. Other plant hormones also modify the cell wall.

Plants contain specific auxin receptor proteins

We will see in Chapter 38 that animal hormones begin to act only after binding with specific receptor proteins. Does the action of plant hormones also require their recognition by receptor proteins? Several proteins can bind various plant hormones. However, it has not been shown that all these proteins function in the living plant as receptors that mediate the effects of the regulators.

In 1989, Glenn Hicks, David Rayle, and Terri Lomax, at Oregon State University, provided the first solid evidence for a connection between such apparent receptor proteins and an auxin-related plant response. They were working with the *diageotropica* (*dgt*) mutation of tomato. Plants ho-

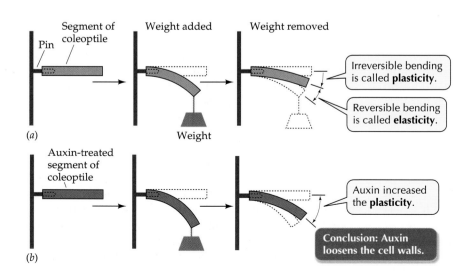

Pin

Segment of coleoptile

Weight added

Weight removed

Irreversible bending is called **plasticity**.

Reversible bending is called **elasticity**.

(a)

Weight

Auxin-treated segment of coleoptile

Auxin increased the **plasticity**.

Conclusion: Auxin loosens the cell walls.

(b)

35.20 Auxin Affects Cell Walls Auxin increases the plasticity, but not the elasticity, of cell walls.

mozygous for the *dgt* mutation fail to show normal gravitropism; instead of growing upright, they simply sprawl on the ground. They also show other symptoms indicating that they cannot respond to their own auxin. The Oregon State workers demonstrated that stems of the *dgt* homozygotes lack a pair of auxin receptor proteins that are present in normally gravitropic tomatoes and in many other plant species. It seems, then, that these proteins participate in auxin responses, and that the absence of these proteins in *dgt* homozygotes accounts for their aberrant growth.

Hormones evoke differentiation and organ formation

What substance, within a plant, signals the different types of cells and organs to form? Much of the research on these questions has been done with cultured tissues. One tissue that is easily grown in culture is pith—the spongy, innermost tissue of a stem. Pith tissue cultures proliferate rapidly but show no differentiation; all the cells are similar, and similarly unspecialized.

Cutting a notch in the cultured tissue and inserting a stem tip into the notch causes the pith cells below the inserted tip to differentiate. Some of the cells differentiate to form water-conducting cells of the sort found in xylem. Differentiation of pith cells also begins if, instead of a stem tip, a mixture of auxin and coconut milk (a rich source of plant hormones) is placed in the notch.

A similar effect can be observed in intact plants. If notches are cut in the stems of *Coleus blumei* plants, interrupting some of the strands of conducting tissues, the strands gradually regenerate from the upper side of the cut to the lower (recall that auxin moves from the tip to the base of a stem). If the leaves above the cut are removed, regeneration is slowed; if, however, the missing leaves are replaced with an auxin solution, conductive tissue regenerates. Auxin and other plant hormones signal the formation of specific cell types.

Experiments with cultured tissues have helped clarify which hormones control organ formation. Undifferentiated cultures of tobacco pith form roots when treated with an appropriate concentration of auxin. Another group of hormones—the cytokinins—causes buds and then shoots to form in such cultures. The pattern of organ formation depends on the ratio of auxin to cytokinin in the medium. A high proportion of auxin favors roots and a high proportion of cytokinins favors buds, but both processes are most active when both hormones are present.

Cytokinins: Hormones Active from Seed to Senescence

Besides stimulating bud formation, the cytokinins promote cell division in cultured tissues, an activity that led to their discovery. In addition, we'll see that cytokinins aid germination, inhibit stem elongation, stimulate lateral bud growth, and delay leaf senescence.

An old sample of DNA led to the discovery of the cytokinins

After studying cell division for many years, Folke Skoog, at the University of Wisconsin, and his associate Carlos Miller reasoned that, since a plant hormone (auxin) is what regulates cell *expansion,* a plant hormone must also regulate cell *division.* Because cell division requires DNA replication, Skoog and Miller suspected that the hypothetical hormone might regulate the metabolism of nucleic acids and that it might even be DNA itself.

In 1955 they and other members of Skoog's group were studying the effects of various compounds on the rate of cell division in cultures of carrot root tissue. They took an old bottle of herring sperm DNA (the only DNA on hand) off the shelf and added a bit to the medium in which some of the cultures were growing. They were gratified to observe that these cultures began to proliferate much more rapidly than the controls.

Determined to discover what component of the DNA preparation was the active material, they purchased some fresh herring sperm DNA, but they were disappointed to find that it was inactive. Deciding that the only difference between the two samples of DNA was age, they tried to "age" the new DNA by heating it—and it worked. The heated DNA preparation caused rapid cell division in the carrot cultures. Careful chemical work revealed that a single substance in the old or heated DNA preparations was the active material. Skoog's group named this substance kinetin (see Figure 35.9e) and suggested that it might be just one of a family of compounds, which are now called cytokinins.

For several years Skoog's group and other investigators tried in vain to find kinetin in plant tissues. What they did find were two closely related compounds called *zeatin* (see Figure 35.9f) and *isopentenyl adenine.* These two are naturally occurring cytokinins; kinetin, on the other hand, may be considered a synthetic cytokinin, because it has not been isolated from plant tissue.

Cytokinins have several effects

Cytokinins form primarily in the roots and move to other parts of the plant. Adding an appropriate combination of auxin and cytokinin to the medium yields rapid growth of plant tissues. Cytokinins can cause certain light-requiring seeds to germinate when the seeds are kept in constant darkness. Cytokinins usually inhibit the elongation of stems, but they cause lateral swelling of stems and roots; the fleshy roots of radishes are an extreme example. Cytokinins stimulate lateral buds to grow into branches; thus the balance between auxin and cytokinin levels controls the bushiness of a plant. Cytokinins increase the expansion of cut pieces of leaf tissue, so they may regulate normal leaf expansion.

Cytokinins also delay the senescence of leaves. If leaf blades are detached from a plant and floated on water or a nutrient solution, they quickly turn yellow and show other signs of senescence. If instead they are floated on a solution containing a cytokinin, they remain green and senesce much more slowly.

Cytokinins apparently regulate the redistribution of biologically active materials from one part of a plant to another. When one of a pair of leaves opposite each other on the stem of a bean plant is treated with a cytokinin, the treated leaf remains dark green and healthy. The untreated leaf opposite it, on the other hand, turns completely yellow and senesces rapidly as a result of its loss of nutrients to the treated leaf.

Ethylene: A Gaseous Hormone That Promotes Senescence

Whereas the cytokinins oppose or delay senescence, another plant hormone promotes it. This hormone is the gas ethylene, $H_2C=CH_2$, which is sometimes called the senescence hormone (see Figure 35.9f). Ethylene can be produced by all parts of the plant, and like all plant hormones, it has several effects.

Back when streets were lit by gas rather than by electricity, leaves on trees near street lamps abscised earlier than those on trees farther from the lamps. We now know that ethylene, a combustion product of the illuminating gas, is what caused the abscission. Auxin delays leaf abscission, but ethylene strongly promotes it; thus a balance of auxin and ethylene controls abscission (Figure 35.21). In this section we'll examine the effects of ethylene on senescence, fruit ripening, and stem elongation.

By controlling senescence, ethylene hastens the ripening of fruit

Another effect of ethylene that is related to senescence is the ripening of fruit. The old saying "one rotten apple spoils the barrel" is true. That rotten apple is a rich source of ethylene, which triggers the ripening and subsequent rotting of the others in the barrel. As the fruit ripens, it loses chlorophyll and its cell walls break down. Ethylene produced in the fruit tissue promotes both processes. Ethylene also causes an increase in its own production. Thus, once ripening begins, more and more ethylene forms, and because it is a gas, ethylene diffuses readily throughout the fruit and even to neighboring fruit on the same or other plants.

35.21 When a Leaf Is About to Fall The presence of auxin delays leaf abscission; the presence of ethylene promotes it.

Farmers in ancient times used to slash developing figs to hasten their ripening. Was this just superstition? No. We now know that wounding causes an increase in ethylene production by the fruit, and that the raised ethylene level promotes ripening. Today, commercial shippers and storers of fruit also hasten ripening by adding ethylene to the storage chambers. This use of ethylene is the single most important use of a plant hormone in agriculture and commerce.

Ripening can also be *delayed* by use of "scrubbers" and adsorbents to remove ethylene from the atmosphere in fruit storage chambers. Another way to slow ripening is to lower the atmospheric pressure in the chamber to accelerate the diffusion of ethylene from the fruits, lowering their internal ethylene concentration. Such low-pressure storage also results in a lowered O_2 concentration, which helps delay fruit ripening, but by a mechanism not yet known.

Abscission accompanies senescence

In some geographic regions, autumn is a time of striking change in the colors of the leaves of deciduous trees and shrubs. The display of colors is followed by the falling of the leaves, which have, in effect, passed through "old age" and died (Figure 35.22). Equally dramatic, in a different way, is the aging and death of entire plants, especially when they grow in great fields. Many crop plants grow vigorously throughout most of the year but then, after flowering and setting fruit, die together by the thousands. These examples show that senescence consists of irreversible, deteriorative changes controlled by internal factors.

As flowers senesce, their petals may abscise, to the detriment of the cut-flower industry. Florists or their suppliers often spray their flowers with dilute solutions of silver thiosulfate. Silver salts inhibit ethylene action, probably by interacting directly with the ethylene receptor, and thus delay senescence—enabling florists to keep their wares salable longer.

Is senescence simply an undesirable but unavoidable fact of life, or does it play a useful role? Leaf senescence and the subsequent abscission are of real importance for the survival of the plant. Leaves senesce and abscise at the end of the growing season, shortly before the onset of the severe conditions of winter. Many species of plants have delicate leaves that would be a liability during a typical winter in the temperate zone: The temperature would be too low for efficient photosynthesis (and the ground might be frozen, making water unavailable), yet water would still be lost from the stomata in the leaves. Damage from freezing would render the leaf unable to function normally during the next growing season in any case. Thus senescence is useful.

Before the leaves die and are shed, their proteins are hydrolyzed to yield amino acids, which are then exported from the leaves to the stems—an important form of resource conservation. Controlled leaf abscission thus costs the plant little and benefits it greatly. In other parts of the world, without cold seasons, plants shed their leaves during harsh dry periods and grow them during wet periods, which are more favorable for growth.

How are the senescence and death of the entire plant that follow flowering and seed setting in some species useful? This process appears to be an adaptation for producing more offspring by pumping energy (food) and nutrients into the seeds—the parent plant essentially starves itself to death.

Ethylene affects stems in several ways

Although associated primarily with senescence, ethylene is active at other stages of plant development as well. The stems of many dicot seedlings that have not yet seen light form an apical hook that protects the growing point while the stem grows through the soil (Figure 35.23). The apical hook is maintained through an asymmetric production of ethylene gas, which inhibits the elongation of cells on the inner surface of the hook. Once the seedling breaks through the soil surface, ethylene synthesis stops, the cells of the inner surface are no longer inhibited, the hook opens, and the leaves expand and are displayed to the sun for photosynthesis.

Ethylene inhibits stem elongation in general, promotes lateral swelling of stems (as do the cytokinins), and causes stems to lose their sensitivity to gravitropic stimulation.

35.22 Leaf Senescence Where leaves would be a liability under winter conditions, they senesce and die in autumn. Leaves are senescing in this forest in Vermont; only a skeleton of branches will remain to face the elements.

Several mutants contribute to studies of ethylene physiology

Plant biologists have isolated mutant plants that are defective in various aspects of ethylene synthesis and action. The tiny plant *Arabidopsis thaliana* has been particularly useful for mutant analysis (see Chapter 15). Among these mutants are some that do not respond to applied ethylene. Other mutants act as if they have been exposed to ethylene even though they haven't

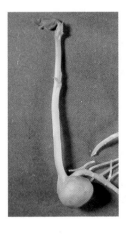

35.23 The Apical Hook of a Dicot Asymmetric production of ethylene is responsible for the apical hook of this young pea seedling, which was grown in the dark.

35.24 Ethylene Mutants of *Arabidopsis* Mutant plant types are important tools in the investigation of the actions of ethylene.

Wild-type		*ctr1*	*ein2*
Air	Ethylene	Air	Ethylene

Wild-type *Arabidopsis* seedlings respond to ethylene by developing a shorter root and stem, a thickened stem, and an exaggerated apical hook.

The *ctr1* mutant behaves as if there were ethylene present even when it is growing in air without added ethylene.

Grown in dark for 3 days

Even in the presence of ethylene, the *ein2* mutant grows as if it were in pure air; it is insensitive to ethylene.

Grown in light for 18 days (seen from above)

(Figure 35.24). The wild-type and mutant alleles have been cloned (see Chapter 16) and are important tools for investigations of ethylene action. Very recently, plant biologists have also cloned the gene that encodes an ethylene receptor protein and have begun to study the protein itself by comparing its amino acid sequence with those of other known enzymes.

Some *Arabidopsis* mutants cannot make auxin, and others cannot make gibberellin; such mutants are dwarfed. Other mutants overproduce a hormone and show phenotypes that are either taller than the wild type or are abnormal in their growth. Still other mutants cannot respond to a hormone.

Abscisic Acid: The Stress Hormone

Abscisic acid is another hormone that has multiple effects in the living plant (see Figure 35.9g). This compound inhibits stem elongation and is generally present in high concentrations in dormant buds and some dormant seeds. As we saw in Chapter 32, it also regulates gas and water vapor exchange between leaves and the atmosphere, through its effects on the guard cells of stomata in the leaf surface. Abscisic acid is sometimes referred to as the stress hormone of plants because it accumulates when plants are deprived of water and because of its possible role in maintaining winter dormancy of buds.

Abscisic acid appears to maintain winter dormancy in response to an environmental cue. In temperate zones the shoots of perennial plants do not grow constantly and in all seasons. Both the onset and termination of winter dormancy are precisely controlled, in response to environmental cues. At some time in the year the terminal buds of temperate-zone perennials become inactive, and growth ceases until the next spring. This dormancy minimizes damage to the plant during a harsh winter.

Buds on a typical deciduous tree of the northern temperate zone undergo various changes well in advance of winter. These changes include the formation of thickened, overlapping **bud scales** that are covered with wax, which helps waterproof the bud contents—the leaf primordia and the growing point of the stem

(Figure 35.25). The winter bud often contains an insulating material consisting of modified, cottony leaves.

Elsewhere in the plant are other changes. Leaves fall, and the scars produced where leaves were formerly attached to the stems are sealed with a corky material. Lateral growth of the trunk ceases, and the solute concentrations in the transport systems in-

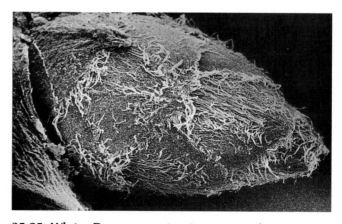

35.25 Winter Dormancy As winter approaches, many deciduous plants stop growing, cover their buds with scales, and shed their leaves. This scanning electron micrograph shows a winter bud from a maple tree. The bud is covered with scales, which are actually modified leaves.

crease, lowering the freezing point of the sap. These are several of the changes that constitute winter dormancy. In at least some species, some of these changes appear to be associated with an increased concentration of abscisic acid in the buds.

An environmental cue, the length of the night, determines the onset of winter dormancy. As summer wears on, the days become shorter (that is, the nights become longer). Leaves have a mechanism for measuring the length of the night, as we will see in the next chapter. This is a marvelous way to determine the season of the year. If a plant determined the season by the temperature, it could be fooled by a winter warm spell or by unseasonable cold weather in the summer. The length of the night, on the other hand, is determined by Earth's rotation around the sun and does not vary; plants use this accurate indicator to time several aspects of growth and development.

Length of night is one of several environmental cues detected by plants, or by individual parts such as leaves. Light—its presence or absence, its intensity, its color, and its duration—provides various cues. Temperature, too, provides important environmental cues, both by its value at any particular time and by the distribution of warmer and colder stretches over a period of time. The plant "reads" an environmental cue and then "interprets" it, often by stepping up or decreasing its production of hormones.

We'll discuss an example of a temperature cue in the next chapter. Now, however, let's see how phytochrome interprets light, its duration, and its wavelength distribution.

35.26 Sensitivity of Seeds to Light In each case the final exposure reversed the preceding exposure; seeds respond only to the wavelength of the final light exposure.

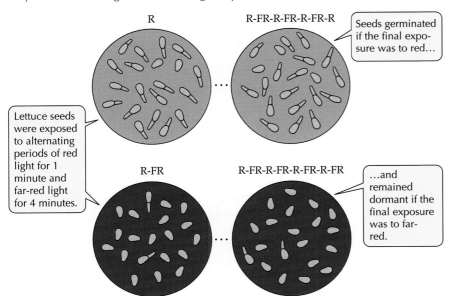

Light and Phytochrome

Light regulates many aspects of plant development. For example, some seeds will not germinate in darkness but do so readily after even a brief exposure to light. Studies have shown that blue and red light are highly effective in promoting germination, whereas green light is not. Of particular importance to plants is the fact that far-red light *reverses* the effect of a prior exposure to red light.

Far-red light is a very deep red, bordering on the limit of human vision and centered on a wavelength of 730 nm; red wavelengths are around 660 nm. If exposed to brief, alternating periods of red and far-red light in close succession, seeds respond only to the final exposure: If red, they germinate; if far red, they remain dormant (Figure 35.26). This reversibility of the effects of red and far-red light regulates many other aspects of plant development. In Chapter 36, for example, we will see how phytochrome participates in the photoperiodic timing mechanism of plants.

The basis for the red and far-red effects resides in a bluish pigment—a protein called **phytochrome**—that exists in the cytosol of plants in two interconvertible forms. Light drives the interconversion of the two forms of phytochrome, both in the test tube and in the living plant. The form that absorbs principally red light is called P_r. Upon absorption of a photon of red light, a molecule of P_r is converted into P_{fr}, the form that absorbs far-red light. P_{fr} has some important biological effects. As we have just seen, one of them is to initiate germination in certain seeds, such as lettuce. Proteins themselves do not absorb visible light; the absorption is accomplished by a small molecule, with alternating single and double bonds, that is bound to the polypeptide chain of phytochrome. The pigment is blue because it absorbs red and far-red light and reflects other light (Figure 35.27). Red light converts phytochrome into one form; far-red light converts the phytochrome back into the other form (Figure 35.28).

Phytochrome has many effects

Phytochrome helps regulate a seedling's early growth. The radicle, the embryonic root, is the first portion of the seedling to escape the seed coat. The shoot emerges later. Frequently seeds germinate below the soil surface, yielding a young plant that is **etiolated**—pale and spindly, with undeveloped leaves, as a result of being kept in darkness.

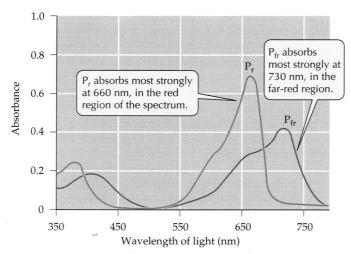

35.27 The Absorption Spectra of Phytochrome Phytochrome exists in two forms, each with a different absorption spectrum.

The seedling must reach the surface and begin photosynthesis before its food reserves are expended and it starves. Plants have evolved a variety of ways to cope with the problem of germinating underground. Etiolated flowering plants, for example, do not form chlorophyll. They turn green only when exposed to light, thereby conserving precious resources, for chlorophyll would be of no use in the dark. Only when light is available does it "pay" to expend metabolic energy on the production of chlorophyll. An etiolated shoot elongates rapidly to hasten its arrival at the soil surface, where photosynthesis quickly begins.

The shoot of etiolated dicot seedlings curves back into a hook at the apex during part of their development, in response to ethylene production (see Figure 35.23). The cotyledons and leaves of an etiolated seedling do not expand—an underground organ cannot photosynthesize—but the stem is greatly elongated.

All of these etiolation phenomena (lack of chlorophyll, rapid shoot elongation, production of an apical

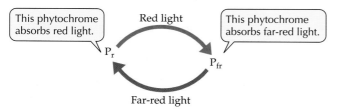

35.28 The Behavior of Phytochrome Each form of phytochrome is converted to the other by exposure to light of the appropriate color. This interconversion is the basis for the light-sensitivity of plant germination described in Figure 35.26.

hook, delayed leaf expansion) are adaptive, and they are regulated by the pigment phytochrome. In a seedling that has never been exposed to light, all the phytochrome is in the red-absorbing (P_r) form. Exposure to light converts P_r to P_{fr} (the far-red-absorbing form), and the P_{fr} initiates reversal of the etiolation phenomena: Chlorophyll synthesis begins, shoot elongation slows, the hook at the apex opens, and the leaves start to expand.

There are multiple forms of phytochrome

For years, plant biologists had difficulty accounting for some aspects of phytochrome action. Plant responses to long-term exposure to light, as opposed to the relatively brief exposures used in the early studies of phytochrome, were difficult to interpret. A solution to these problems may lie in the discovery of multiple forms of phytochrome. *Arabidopsis* has five genes that encode different phytochromes, and this diversity appears throughout the plant kingdom and in algae as well.

The several phytochromes may play differing roles in the various phytochrome-controlled responses. Some of them may even play off each other to fine-tune plant growth during the day. Consider, for example, the light spectrum available to a seedling that is growing in the shade of other plants in contrast to that available to a seedling in full sunlight. Because chlorophyll in the shading leaves absorbs the light first, the shaded seedling "sees" a spectrum relatively rich in far-red (and poor in red); the ratio of far-red to red is increased as much as 10- to 20-fold in the shade. The interplay among phytochromes with different actions and different absorption spectra leads to a "shade avoidance" response, featuring an increased rate of stem elongation.

Calcium, calmodulin, and cGMP may mediate phytochrome action

We do not yet know how the various forms of phytochrome produce their many effects, although it is evident that phytochrome acts through the plant's genome. The phytochromes are in solution in the cytosol, but they themselves do not move to the nucleus and activate genes. Instead, they appear to trigger sequences of events similar to those observed in the responses of animal cells to hormones (see Chapter 38). Although the supporting evidence is incomplete, we will outline some likely intermediate steps in the response to phytochrome.

Phytochrome appears to activate one or more G proteins. G proteins are membrane proteins that require guanosine triphosphate (GTP) to exert their effects (see Chapter 5). The phytochrome-activated G proteins, in turn, may do two things. One effect may be the conversion of GTP into a signaling substance called **cGMP** (cyclic guanosine monophosphate). The other effect may be to open channels that admit calcium ions to the cell, where Ca^{2+} binds to the protein **calmodulin**. Both cGMP and the calcium–calmodulin complex can trigger changes leading eventually to the activation of specific genes identified the genome, along with environmental cues, hormones, and phytochrome, as a major determinant of development.

Plant Research Today: Genes and Gene Products

Studies of plant genes represent the current state of technology in the understanding of plant development. As we have seen, the study of plant growth regulation had its beginnings in observations of developmental phenomena, such as the growth of plants toward a light source or the overgrowth of "foolish" rice seedlings. This stage was eventually followed by the isolation and chemical characterization of hormones responsible for the effects. When applied to plants, however, each hormone proved to have multiple effects on development. Researchers then undertook to learn the metabolism and biochemical effects of the hormones.

Now, we look for mutants that can't make the hormone, or that make too much of it, or that can't respond to it (Figure 35.29). We clone the genes and seek

35.29 Wild-Type and Mutant *Arabidopsis thaliana*
The large wild-type seedling is growing among tiny mutant seedlings that are unable to synthesize tryptophan. Addition of a small amount of that amino acid to the growth medium enabled the mutants to survive.

to determine the activities of the gene products. Some of the gene products are enzymes that catalyze processes such as hormone synthesis.

Other plant genes exert regulation at a higher level. Not surprisingly, we are discovering that plants and animals have various developmental elements in common (see Chapter 15). Plant genomes include information that specifies programmed cell death, in analogy to the apoptosis of animal cells. Mutations that change organ identities, as do the homeotic mutations studied in animals, are found in plants such as *Arabidopsis*, as we will see in the next chapter.

Summary of "Regulation of Plant Development"

Four Major Regulators of Plant Development
• The environment, hormones, the pigment phytochrome, and the plant's genome all play roles in the regulation of plant development. Important environmental features include light and temperature; the hormones include auxin, gibberellins, cytokinins, abscisic acid, and ethylene. **Review Figure 35.9 and Table 35.1**

From Seed To Death: An Overview of Plant Development
• Cell division, cell expansion, and cell differentiation all contribute to plant development.
• The dormant seed eventually germinates and forms a growing seedling. Phytochrome and hormones regulate seedling development, including growth. **Review Figures 35.1, 35.2**
• Eventually the plant flowers and forms fruit. Flowering in some plants is controlled by the length of day. Hormones, possibly including a flowering hormone, play roles in plant reproduction from flowering to fruit development. **Review Figure 35.3**
• Some plant buds demonstrate winter dormancy; plants finally senesce. Dormancy and senescence are triggered by environmental clues, mediated by phytochrome and hormones. **Review Figure 35.4**

Seed Dormancy and Germination
• Seed dormancy may be caused by exclusion of water or oxygen from the embryo, mechanical restraint of the embryo, or chemical inhibition of embryo development. In nature, seed dormancy is broken in various ways, including scarification, fire, leaching, and low temperatures.
• Seed dormancy offers adaptive advantages, such as an increased likelihood of germination in a place and at a time favorable for seedling growth.
• Seed germination begins with the imbibition of water. Then the embryo must mobilize its reserves to obtain building blocks and an energy supply. The embryos of cereal seeds secrete gibberellins that cause the aleurone layer to synthesize and secrete digestive enzymes that break down large molecules stored in the endosperm. **Review Figure 35.8**

Gibberellins: A Large Family of Hormones

• Studies of a fungal disease of rice led to the discovery and isolation of the gibberellins. There are dozens of gibberellins. One, gibberellin A_1, regulates stem growth. **Review Figure 35.10**

• Plants normally produce gibberellins. Mutant plants that cannot produce normal amounts of gibberellin are dwarfed: Their stems are shorter than wild-type stems. **Review Figure 35.11**

• In addition to regulating stem growth, gibberellins regulate the growth of some fruits, cause bolting in some biennial plants, induce parthenocarpic fruit development, promote seed germination, and help buds break winter dormancy. **Review Figure 35.12**

Auxin: A Hormone with Many Effects

• Studies of phototropism led to the discovery and isolation of auxin (indoleacetic acid). In grass seedlings the photoreceptor for phototropism is in the tip of the coleoptile, and auxin is a messenger from the photoreceptor to the growing region of the coleoptile. **Review Figures 35.13, 35.14**

• Auxin transport is often polar. Lateral movement of auxin establishes shoot and root responses to light and gravity: phototropism and gravitropism, respectively. **Review Figure 35.15**

• Auxin plays roles in root formation, leaf abscission, apical dominance, and parthenocarpic fruit development. Certain synthetic auxins are selective herbicides. **Review Figures 35.16, 35.17, 35.18**

Master Reactions to Hormones

• The arrangement of cellulose microfibrils in the wall limits the rate and direction of plant cell growth. Auxin binds receptor proteins and loosens the cell well, promoting cell expansion. Part of the auxin response results from the pumping of protons from the cytosol into the cell wall, where the lowered pH activates proteins. **Review Figures 35.19, 35.20**

• Auxin and other plant hormones signal cell differentiation and organ formation.

Cytokinins: Hormones Active from Seed to Senescence

• The first cytokinin, kinetin, was isolated from a deteriorating sample of DNA. Zeatin and isopentenyl adenine are naturally occurring cytokinins.

• First studied as promoters of plant cell division, cytokinins also promote seed germination in some species, inhibit stem elongation, promote lateral swelling of stems and roots, stimulate the growth of lateral buds, promote the expansion of leaf tissue, delay leaf senescence, and play a role in the redistribution of substances within the plant body.

Ethylene: A Gaseous Hormone That Promotes Senescence

• A balance between auxin and the gaseous hormone ethylene controls leaf abscission. **Review Figure 35.21**

• Ethylene promotes senescence and fruit ripening.

• Senescence of plant parts and of whole plants is not a random phenomenon but is programmed, as are all other aspects of plant development. Senescence and abscission are crucial to the survival of many plants.

• Ethylene causes the formation of a protective apical hook in dicot seedlings that have not seen light; in stems it inhibits elongation, promotes lateral swelling, and causes a loss of gravitropic sensitivity.

Abscisic Acid: The Stress Hormone

• Abscisic acid, sometimes referred to as the stress hormone, appears to maintain winter dormancy of terminal buds, and it may regulate other, correlated effects elsewhere in the plant. Through its effects on stomatal opening, abscisic acid also regulates gas and water exchange between leaves and the atmosphere. In addition, it inhibits stem elongation.

Light and Phytochrome

• Phytochrome is a bluish protein found in the cytosol. It exists in two forms, P_r and P_{fr}, that are interconvertible by light. P_r absorbs red light (with a maximum at 660 nm) and P_{fr} absorbs far-red light (730 nm). **Review Figures 35.26, 35.27, 35.28**

• Phytochrome has many effects, including the various manifestations of etiolation.

• There are multiple forms of phytochrome. These forms may play different roles in development, and they may interact to mediate the effects of light environments of differing spectral distribution.

• Calcium ions, calmodulin, and cGMP may play roles between the absorption of light by phytochrome in the cytosol and the activation of genes in the nucleus.

Plant Research Today: Genes and Gene Products

• Current research on plant hormones focuses on genes, using mutants as subjects.

Self-Quiz

1. Which of the following is *not* an advantage of seed dormancy?
   a. It makes the seed more likely to be digested by birds that disperse it.
   b. It counters the effects of year-to-year variations in the environment.
   c. It increases the likelihood that a seed will germinate in the right place.
   d. It favors dispersal of the seed.
   e. It may result in germination at a favorable time of year.

2. Which of the following does/do *not* participate in seed germination?
   a. Imbibition of water
   b. Metabolic changes
   c. Growth of the radicle
   d. Mobilization of food reserves
   e. Extensive mitotic divisions

3. To mobilize its food reserves, a germinating barley seed
   a. becomes dormant.
   b. undergoes senescence.
   c. secretes gibberellins into its endosperm.
   d. converts glycerol and fatty acids into lipids.
   e. embryo takes up proteins from the endosperm.

4. The gibberellins
   a. are responsible for phototropism and gravitropism.
   b. are gases at room temperature.
   c. are produced only by fungi.
   d. cause bolting in some biennial plants.
   e. inhibit the synthesis of digestive enzymes by barley seeds.

5. In coleoptile tissue, auxin
   a. is transported from base to tip.
   b. is transported from tip to base.
   c. can be transported toward either the tip or the base, depending on the orientation of the coleoptile with respect to gravity.
   d. is transported by simple diffusion, with no preferred direction.
   e. is not transported, because auxin is used where it is made.

6. Which process is *not* directly affected by auxin?
   a. Apical dominance
   b. Leaf abscission
   c. Synthesis of digestive enzymes by barley seeds
   d. Root initiation
   e. Parthenocarpic fruit development

7. Plant cell walls
   a. are strengthened primarily by proteins.
   b. often make up more than 90 percent of the total volume of an expanded cell.
   c. can be loosened by an increase in pH.
   d. become thinner and thinner as the cell grows longer and longer.
   e. are made more plastic by treatment with auxin.

8. Which statement about cytokinins is *not* true?
   a. They promote bud formation in tissue cultures.
   b. They delay the senescence of leaves.
   c. They usually promote the elongation of stems.
   d. They cause certain light-requiring seeds to germinate in the dark.
   e. They stimulate the development of branches from lateral buds.

9. Ethylene
   a. is antagonized by silver salts such as silver thiosulfate.
   b. is liquid at room temperature.
   c. delays the ripening of fruits.
   d. generally promotes stem elongation.
   e. inhibits the swelling of stems, in opposition to cytokinin effects.

10. Phytochrome
    a. is a nucleic acid.
    b. exists in two forms interconvertible by light.
    c. is a pigment that is colored red or far red.
    d. is sometimes called the stress hormone.
    e. is the photoreceptor for phototropism.

Applying Concepts

1. How may it be advantageous for some species to have seeds whose dormancy is broken by fire?

2. Cocklebur fruits contain two seeds each, and the two seeds are kept dormant by two different mechanisms. How may this use of two mechanisms of dormancy be advantageous to cockleburs?

3. Corn stunt virus causes a great reduction in the growth rate of infected corn plants, so the diseased plants take on a dwarfed form. Since their appearance is reminis-

cent of the genetically dwarfed corn studied by Phinney, you suspect that the virus may inhibit the synthesis of gibberellins by the corn plants. Describe two experiments you might conduct to test this hypothesis, only one of which should require chemical measurement.

4. Whereas relatively low concentrations of auxin promote the elongation of segments cut from young plant stems, higher concentrations generally inhibit growth, as shown in the figure. In some plants, the inhibitory effects of high auxin concentrations appear to be secondary: High auxin concentrations cause the synthesis of ethylene, which is what causes the growth inhibition. Silver thiosulfate inhibits ethylene action. How do you think the addition of silver thiosulfate to the solutions in which the stem segments grew would affect the appearance of the graph?

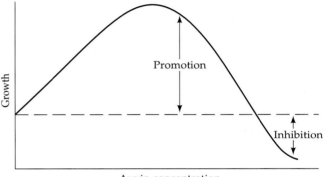

5. If you wanted to delay the ripening of the bananas you had just purchased at the grocery store, what effective steps might you take, given materials available in the laboratory?

Readings

Evans, M. L., R. Moore and K.-H. Hasenstein. 1986. "How Roots Respond to Gravity." *Scientific American*, December. A discussion of classical and modern experimentation on the mechanisms of gravitropism.

Mandoli, D. F. and W. R. Briggs. 1984. "Fiber Optics in Plants." *Scientific American*, August. A discussion of how plants guide light to regions of high phytochrome concentration, including an excellent description of the light environment in a wheat field.

Moses, P. B. and N.-H. Chua. 1988. "Light Switches for Plant Genes." *Scientific American*, April. An investigation into the question of how light absorption by phytochrome is transduced into developmental effects.

Raven, P. H., R. F. Evert and S. Eichhorn. 1992. *Biology of Plants*, 5th Edition. Worth, New York. A well-balanced general botany textbook.

Salisbury, F. B. and C. W. Ross. 1992. *Plant Physiology*, 4th Edition. Wadsworth, Belmont, CA. A sound textbook with excellent chapters on plant hormones and development.

Taiz, L. and E. Zeiger. 1998. *Plant Physiology*, 2nd Edition. Sinauer Associates, Sunderland, MA. Unit 3 of this authoritative textbook deals with plant hormones, phytochrome, and other aspects of development.

Chapter 36

Reproduction in Flowering Plants

A re all these flowers just for decoration? Why do plants expend the energy and resources needed to produce such structures? The answer is simple: Flowers are reproductive structures, and reproduction is one of the most important events in an organism's life.

Whatever its color, size, or shape, a flower is a sexual reproductive structure, containing either female or male organs or both. The female and male organs produce, respectively, eggs and sperm. Petals produce neither eggs nor sperm, but the showy petals of many plants attract animals that carry pollen from plant to plant, assisting in cross-fertilization. Thus the petals are part of a plant's way of uniting sperm and eggs.

Biologists still seek answers to some important questions about flowering and reproduction. We want to know how environmental clues lead to flower formation. The plants in the photo here are of several species, each having flowers with characteristic structures. The kinds of flower parts are very similar, but their numbers, shapes, colors, and arrangements are different. We want to know how flowers develop in general, and in the specific ways characteristic of different species.

What Are All the Flowers For?
Flowers exist not to give humans aesthetic pleasure, but to allow plants to reproduce.

In this chapter we deal with several aspects of plant reproduction, including some for which we are still in search of answers. We will contrast sexual and asexual reproduction and consider each in detail. We'll look at angiosperm gametophytes, pollination, double fertilization, embryonic development, and the roles of fruits in seed dispersal. The transition to the flowering state is a key event in plant development, and we'll see how the changing of seasons triggers this event in some plants—and speculate on the underlying mechanisms, including the possible existence of a flowering hormone. We conclude the chapter with an examination of asexual reproduction in nature and in agriculture.

Many Ways to Reproduce

Plants have many ways of reproducing themselves—and with humans helping, there are even more ways. Some flower parts are sex organs. It is thus no surprise that almost all flowering plants reproduce sexually. But many reproduce asexually as well; some even reproduce asexually most of the time. Which is the better way to reproduce?

Most of the answers to this question relate to genetic diversity or genetic recombination because sexual reproduction produces new genetic combinations. The details of sexual reproduction differ among different species of flowering plants. In our discussion of Mendel's work (see Chapter 10), we saw that some plants can reproduce sexually either by cross-pollinating or by self-pollinating. Self-pollination is possible because, as we explained in Chapter 27, in many species each individual has both male and female sex organs.

Both sexual and asexual reproduction are important in agriculture. Annual crops, including wheat, rice, millet, and corn—the great grain crops, all of which are grasses—as well as plants in other families, such as soybeans and safflower, are grown from seed—that is, from the products of sexual reproduction. Other crops begin asexually from grafts, or by other means.

Orange trees, which have been under cultivation for centuries, are grown from seed except for one type, the navel orange. This plant apparently arose only once in history. Early in the nineteenth century, on a plantation on the Brazilian coast, one seed gave rise to one tree that had aberrant flowers. Parts of the flowers aborted, and seedless fruits formed. Every navel orange in the world comes from a navel orange tree derived asexually from another, which came from another, and another, and so on back to that original Brazilian tree.

Navel oranges must be propagated asexually. Strawberries need not be, because they are capable of setting seed. Nonetheless, asexual propagation of strawberries is common because vast numbers of plants that are genetically identical to a particularly desirable plant can be produced from strawberry runners.

We will treat asexual reproduction in greater detail at the end of this chapter. We begin, however, by considering sexual reproduction.

Sexual Reproduction

Sexual reproduction provides genetic diversity through recombination. Meiosis and mating shuffle genes into new combinations, giving a population a variety of genotypes in each generation. This genetic diversity may serve well as the environment changes or as the population expands into new environments. The adaptability resulting from genetic diversity is the reason that sexual reproduction is often thought of as being "better" than asexual reproduction.

The flower is an angiosperm's device for sexual reproduction

A complete flower consists of four groups of modified leaves: the carpels, stamens, petals, and sepals (see Figures 36.1 and 15.13). To review briefly, the female organs, bearing megasporangia, are the carpels. Enclosed within each *carpel* are one or more ovules. Each *ovule* consists of the *megasporangium* it encloses, its protective layers of surrounding integuments, and its stalk. There is an opening in the integuments, the micropyle, through which the pollen tube grows on its way to the egg. A *pistil*, consisting of one or more fused carpels, bears a pollen-receptive *stigma* at the tip of its elongated style. The male organs are the stamens. Each *stamen* consists of a filament bearing an *anther* with four *microsporangia*. The parts of a flower are shown at the top of Figure 36.1.

In addition to these sexual parts, many flowers have petals and sepals arranged in whorls around the spore-bearing carpels and stamens. The petals, often colored, constitute the corolla. Below them the sepals, often green, constitute the calyx. All the parts of a flower are borne on a stem tip, the receptacle.

Flowering plants have microscopic gametophytes

Before reading this section, you may wish to review the section in Chapter 27 entitled "Life cycles of plants feature alternation of generations." The concept of alternation of generations is central to an understanding of plant reproduction.

In plants, the egg and sperm are produced not by the generation (the sporophyte) that produces the flower, but by a new, intermediate organism, the gametophyte. This is a unique strategy employed only by plants and some protists. The flower is more than just a place where the egg and sperm are eventually found—it is also the place where the alternate generation resides.

Gametophytes, the gamete-producing generation, develop from haploid spores in sporangia. Female gametophytes (megagametophytes), which are called **embryo sacs**, develop in megasporangia. Male gametophytes, the **pollen grains**, develop in microsporangia (Figure 36.1).

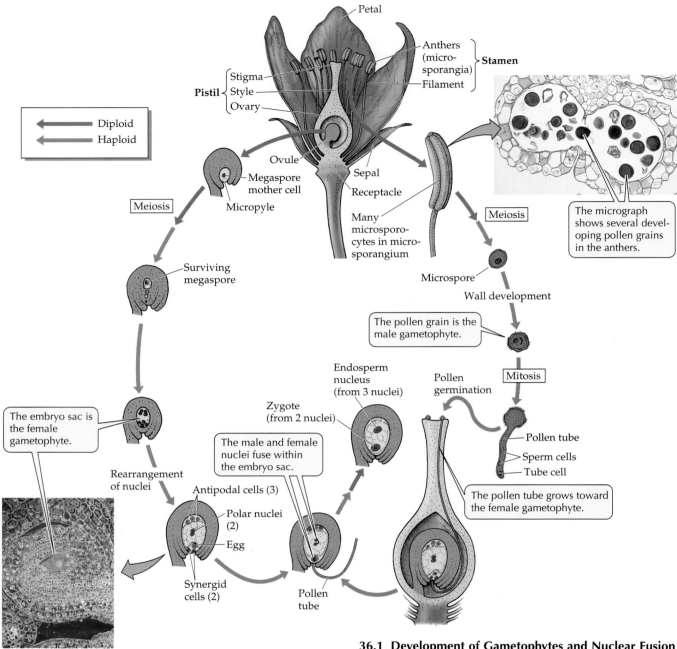

36.1 Development of Gametophytes and Nuclear Fusion
The embryo sac (left) is the female gametophyte; the pollen grain (right) is the male gametophyte. The male and female nuclei meet and fuse within the embryo sac.

The megasporocyte, a cell within the ovule's megasporangium, divides meiotically to produce four haploid megaspores. All but one of the megaspores then degenerate. The surviving megaspore undergoes mitotic divisions, usually producing eight nuclei, all initially contained within a single cell—three nuclei at one end, three at the other, and two in the middle. Subsequent cell wall formation leads to an elliptical, seven-celled megagametophyte with a total of eight nuclei. At the end of the megagametophyte nearest the micropyle are three tiny cells: the **egg** and two cells

called **synergids**. At the opposite end are three antipodal cells, and in the large central cell are two **polar nuclei**. Note that the embryo sac (the female gametophyte or megagametophyte) is the entire seven-cell, eight-nucleus structure. Follow the arrows down the left-hand side of Figure 36.1 to review the development of the embryo sac.

The male gametophyte, or pollen grain, consists of fewer cells than the female gametophyte has. As the

36.2 A Pollen Grain Sampler Pollen grains of dandelion, ragweed, and grass. Each species' pollen has a characteristic size, shape, and pattern of wall sculpturing.

36.3 Wind Pollination The numerous anthers on this flowering inflorescence of a grass all point away from the stalk and stand free of the plant, promoting dispersal of the pollen by wind.

right-hand side of Figure 36.1 illustrates, development of the pollen grain begins when the microsporocytes divide meiotically within the four microsporangia, which are fused into a two-lobed anther. Each resulting microspore normally undergoes one mitotic division within the spore wall before the anthers open and shed the pollen. Further development of the pollen grain, which we will describe shortly, is delayed until the pollen arrives at the stigma of a pistil. The transfer of pollen from the anther to the stigma is referred to as **pollination**.

Pollination enables fertilization in the absence of liquid water

Gymnosperms and angiosperms evolved independence from water as a medium for gamete travel and fertilization—a freedom not shared by other plant groups. The sperm nuclei of gymnosperms and angiosperms travel within a pollen grain (Figure 36.2), and the pollen tube provides a route to the ovary. But how do angiosperm pollen grains travel from an anther to a stigma? The mechanisms that have evolved for pollen transport are many and various. In one, pollination is accomplished before the flower bud opens, as in peas and their relatives—resulting in the self-fertilization described at the start of Chapter 10. In self-pollination, pollen is transferred by the direct contact of anther and stigma within the same flower.

Wind is the vehicle for pollen transport in many species, especially grasses (Figure 36.3). Wind-pollinated flowers have sticky or featherlike stigmas, and they produce pollen grains in great numbers. Some aquatic angiosperms are pollinated by water action, with water carrying pollen grains from plant to plant. Animals are important pollinators; the mutually beneficial aspects of such plant–animal associations are discussed in Chapter 52.

Angiosperms perform double fertilization

When a pollen grain lands on the stigma of a pistil, a pollen tube develops from the pollen grain and either grows downward on the inner surface of the style or digests its way down the spongy tissue of this female organ, growing millimeters or even centimeters in the process (Figure 36.4). The pollen tube follows a chemical gradient of calcium ions or other substances in the style until it reaches the micropyle.

Of the two cells in the pollen grain, one is the **tube cell** and the other, much smaller, is the **generative cell** (Figure 36.5). The pollen tube eventually grows through megasporangial tissue and reaches the female gametophyte. The generative cell meanwhile has undergone one mitotic division to produce two **sperm**

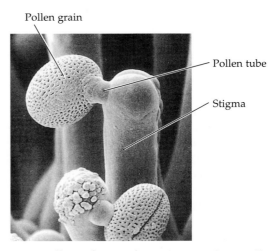

Pollen grain

Pollen tube

Stigma

36.4 Pollen Tubes Begin to Grow These pollen grains have landed on hairlike structures on the stigma of an *Arabidopsis* flower. Pollen tubes have started to form.

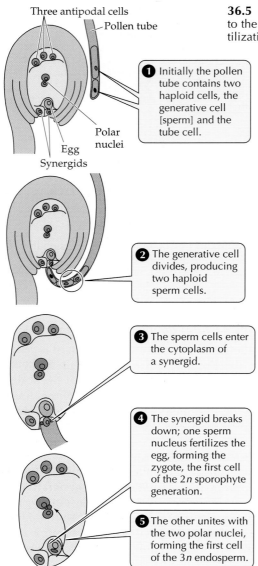

Three antipodal cells

Pollen tube

Egg

Synergids

Polar nuclei

❶ Initially the pollen tube contains two haploid cells, the generative cell [sperm] and the tube cell.

❷ The generative cell divides, producing two haploid sperm cells.

❸ The sperm cells enter the cytoplasm of a synergid.

❹ The synergid breaks down; one sperm nucleus fertilizes the egg, forming the zygote, the first cell of the 2n sporophyte generation.

❺ The other unites with the two polar nuclei, forming the first cell of the 3n endosperm.

36.5 Pollen Nuclei and Double Fertilization The sperm nuclei contribute to the formation of the diploid zygote and the triploid endosperm. Double fertilization is a characteristic feature of angiosperm reproduction.

trients are moved in from other parts of the plant, the endosperm begins accumulating starch, lipids, and proteins. The integuments develop into a double-layered seed coat, sometimes fleshy and sometimes heavily lignified and hard. The carpel ultimately becomes the wall of the fruit that encloses the seed.

Of all the characteristic traits of the angiosperms, only one trait, the formation of a triploid endosperm, is found *only* in angiosperms.

In vitro fertilization of plant eggs is not easily accomplished

Hundreds of children have been conceived by means of *in vitro* fertilization, the combination of egg and sperm in a petri plate rather than in the mother's body. Can we do the same thing with plant gametes, making test tube plants? *In vitro* fertilization would enable the study of some interesting questions about the reproductive development of plants. It sounds simple, but *in vitro* fertilization in plants did not succeed fully until 1993.

What made this achievement so difficult? In animals, sperm and eggs are set free (that is, they are not contained in other cells), and they can unite readily. In plants, the gametes are not free of other cells. The sperm are contained within the pollen tube, and biologists must use osmotic treatments to free the sperm from the pollen. The egg presents an even greater problem because it is contained in the embryo sac.

Treatment with enzymes that digest cell walls, followed by microdissection, can release the egg. However, the freed sperm and egg appear to be incapable of fusing spontaneously. What is wrong? One possibility is that fertilization depends on activities of other cells in the embryo sac. Another possibility is that the egg nucleus is damaged by the enzymes used to remove the egg cell wall.

A combination of techniques has now made *in vitro* fertilization in plants possible. One key technique is to treat both gametes with bursts of electricity. The second technique is to use "feeder cells," derived from normal embryos and separated from the treated gametes by a filter, to provide as yet unknown substances that support development of the new zygote. Under these conditions the isolated gametes fuse successfully, producing zygotes that develop normally and give rise to normal adult plants. Analysis of the products of the cultured feeder cells will shed light on the role of the synergids in normal, *in vivo* fertilization in plants.

cells, *both* of which enter the female gametophyte and are later released into the cytoplasm of one of the synergids.

From this synergid, which degenerates, each sperm enters a different cell. One sperm nucleus enters the egg cell and fuses with its nucleus, producing the diploid zygote. The other sperm nucleus enters the central cell of the embryo sac and fuses with the two polar nuclei to form a triploid (3n) nucleus. While the zygote nucleus begins division to form the new sporophyte embryo, the triploid nucleus undergoes rapid mitosis to form a specialized nutritive tissue, the **endosperm**. The female antipodal cells and synergids eventually degenerate.

Shortly after fertilization, highly coordinated growth and development of embryo, endosperm, integuments, and carpel ensues. As large amounts of nu-

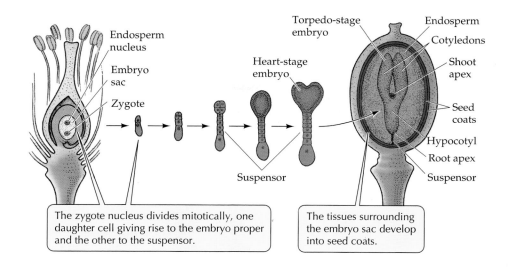

36.6 Early Development of a Dicot The suspensor and the embryo are distinguishable after just four mitotic divisions. The embryo develops through intermediate stages, including a characteristic heart-shaped form, to reach the torpedo stage.

The zygote nucleus divides mitotically, one daughter cell giving rise to the embryo proper and the other to the suspensor.

The tissues surrounding the embryo sac develop into seed coats.

Embryos develop within seeds

The first step in the normal formation of the embryo is a mitotic division of the zygote, the fertilized egg in the embryo sac, giving rise to two daughter cells. Even at this stage the two cells face different fates. An asymmetric (uneven) distribution of contents within the zygote causes one end to produce the embryo proper and the other end to produce an early supporting structure, the **suspensor**, that pushes the embryo into the endosperm (Figure 36.6).

With the asymmetric division, polarity has been established, as has the longitudinal axis of the new plant. A filamentous suspensor and a globular embryo are distinguishable after just four mitotic divisions. The suspensor soon ceases to elongate, and as development continues, the first organs take form within the embryo.

In dicots (monocots are somewhat different), the embryo takes on a characteristic heart-shaped form as the cotyledons start to grow. Further elongation of the cotyledons and of the main axis of the embryo gives rise to what is called the torpedo stage (see Figure 36.6), during which some of the internal tissues begin to differentiate. The elongated region below the cotyledons is the hypocotyl. At the top of the hypocotyl, between the cotyledons, is the shoot apex; at the other end is the root apex. Each of these apical regions contains an apical meristem whose dividing cells give rise to the organs of the mature plant.

In many species, the cotyledons absorb the food reserves from the surrounding endosperm and grow very large in relation to the rest of the embryo (Figure 36.7a). In others, the cotyledons remain thin (Figure 36.7b); they will draw on the reserves in the endosperm as needed when the seeds germinate. In either case, the endosperm is the maternal plant's contribution to the nutrition of the next generation.

In the late stages of embryonic development, the seed loses water—sometimes as much as 95 percent of its original water content. In its dried state, the embryo is incapable of further development. It remains in this dormant state until the conditions are right for germination. (Recall from Chapter 35 that a necessary first step in seed germination is the massive imbibition of water.)

Some fruits assist in seed dispersal

After fertilization, the ovary wall of a flowering plant—together with its seeds—develops into a fruit. A **fruit** may consist of only the mature ovary and its seeds, or it may include other parts of the flower or structures that are closely related to it. The major variations on this theme are illustrated in Figure 27.31.

Some fruits play a major role in reproduction because they help disperse seeds over substantial distances. Various trees, including ash, elm, maple, and tree of heaven, produce a dry, winged fruit called a samara (see Figure 31.2c). A samara spins like a helicopter blade and, while whirling downward, holds the fruit aloft long enough for it to be blown some dis-

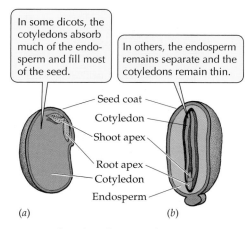

In some dicots, the cotyledons absorb much of the endosperm and fill most of the seed.

In others, the endosperm remains separate and the cotyledons remain thin.

Seed coat
Cotyledon
Shoot apex
Root apex
Cotyledon
Endosperm

(a)   (b)

36.7 Variety in Dicot Seeds In some dicots (a), the food reserves of the endosperm are absorbed by the cotyledons at the seed stage; in others (b), the reserves will be drawn on throughout the course of development.

tance from the parent tree. The dandelion fruit is also marvelously adapted for dispersal by wind. Water disperses some fruits; coconuts have been spread in this way from island to island in the Pacific (Figure 36.8*a*). Still other fruits travel by hitching rides with animals—either inside or outside them (Figure 36.8*b*). Burdocks, for example, have hooks that adhere to animal fur, and many other plants have prickled, barbed, hairy, or sticky fruits. Fleshy fruits such as berries provide food for mammals or birds; their seeds travel safely through the animal's digestive tract or are regurgitated, in either case being deposited some distance from the parent plant.

Although there are many mechanisms for seed and pollen dispersal, most seeds, and most pollen grains, end up close to their sources. Long-range dispersal is the exception, not the rule. In cases of extensive dispersal, paternal genes usually travel farther than maternal genes because, although both maternal and paternal genes travel with the seed, only paternal genes travel with the pollen grain.

We have traced the sexual life cycle from the flower to the fruit to the dispersal of seeds. We discussed seed germination and vegetative development of the seedling in Chapter 35. Now let's complete the sexual life cycle by considering the transition from the vegetative to the flowering state, and how this transition is regulated.

36.8 Dispersal of Fruits   (*a*) A coconut seed germinates where it washed ashore on a beach in the South Pacific. (*b*) Many sticky fruits, such as the burrs of the common burdock, hitch rides on animals, traveling far from their parent plants.

The Transition to the Flowering State

Flowering may terminate, repeatedly interrupt, or accompany vegetative growth. The transition to the flowering state often marks the end of vegetative growth for a plant. If we view a plant as something produced by a seed for the purpose of bearing more seeds, then the act of flowering is one of the supreme events in a plant's life.

Apical meristems can become inflorescence meristems

The first visible sign of the transition to the flowering state may be a change in one or more apical meristems in the shoot system. During vegetative growth an apical meristem continues to produce leaves, lateral buds, and **internodes**, the regions of stem between the **nodes** where leaves and buds form (Figure 36.9*a*). This growth is *indeterminate* (see Chapter 31), in contrast to the usually determinate growth of an animal to a standard size.

However, if a vegetative meristem becomes an **inflorescence meristem**, it generally produces several other structures: smaller leafy structures called **bracts**, as well as new meristems in the angles between the bracts and the internodes (Figure 36.9*b*). These new meristems may be inflorescence meristems or **floral meristems**, which give rise to the flowers themselves.

Each floral meristem typically produces four consecutive whorls of organs—the sepals, petals, stamens, and carpels—separated by very short internodes, keeping the flower compact (Figure 36.9*c*). In contrast to vegetative meristems and some inflorescence meristems, floral meristems are responsible for *determinate* growth—the limited growth of the flower.

How does the floral meristem give rise, in short order, to whorls of four different organs (sepals, petals, stamens, and carpels)? This problem is not unlike that of accounting for the construction of a fruit fly larva (see Chapter 15). Recent work underscores some of the similarities. Most strikingly, a group of **organ identity genes**—analogous to the homeotic genes of animals—work in concert to specify the successive whorls. These genes were discussed in Chapter 15 (see Figure 15.13).

Earlier work, which we will consider next, helped explain how the transition from the vegetative to the flowering state is initiated. Flowering is triggered in very different ways in different plant species. Environmental cues include seasonal changes in the lengths of days and nights, and seasonal changes in temperature.

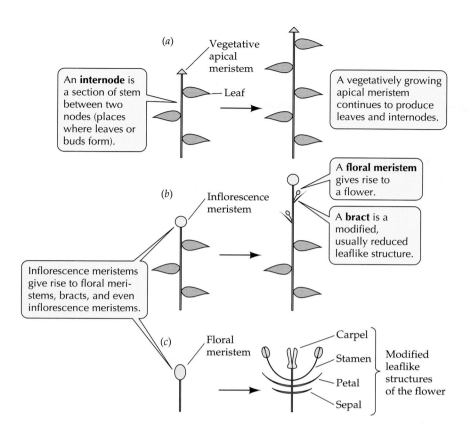

An **internode** is a section of stem between two nodes (places where leaves or buds form).

(a) Vegetative apical meristem

Leaf

A vegetatively growing apical meristem continues to produce leaves and internodes.

(b) Inflorescence meristem

Inflorescence meristems give rise to floral meristems, bracts, and even inflorescence meristems.

A **floral meristem** gives rise to a flower.

A **bract** is a modified, usually reduced leaflike structure.

(c) Floral meristem

Carpel

Stamen

Petal

Sepal

Modified leaflike structures of the flower

36.9 Flowering and the Apical Meristem
A vegetative apical meristem will grow without producing flowers. Once the transition to the flowering state is made, inflorescence meristems give rise to bracts and to floral meristems, which become the flowers.

of control by lengths of days and nights is called **photoperiodism**.

There are both short-day and long-day plants

Soybeans and Maryland Mammoth tobacco are **short-day plants** (SDPs), which flower only when the day is shorter than a critical maximum. Spinach and clover are examples of **long-day plants** (LDPs), which flower only when the day is longer than a critical minimum. Generally LDPs are triggered to flower in midsummer and SDPs in late summer, or sometimes in the spring.

It is a historical accident that we use the terms "short-day plant" and "long-day plant." The SDPs could as well have been called long-night plants and the LDPs short-night plants, because the natural day has a fixed length of 24 hours. Which do plants measure—the length of day or the length of night? After some further consideration of photoperiodic phenomena, we'll return to this question.

Some plants require more complex photoperiodic signals

Some plants require photoperiodic signals that are more complex than just "short-day" or "long-day" in order to flower. One group, the short-long-day plants, must first experience short days and then long ones. Accordingly, because they pass first through the short days of early spring and then through ever longer ones, they flower during the long days before midsummer. Another group, the long-short-day plants, cannot flower until the long days of summer have been followed by shorter ones, so they bloom only in the fall. Long-short-day plants will not bloom in the spring, in spite of its short days, nor will a short-long-day plant flower in late summer.

Other effects besides flowering are also under photoperiodic control. We have learned, for example, that short days trigger the onset of winter dormancy. (Animals, too, show a variety of photoperiodic behaviors; in aphids, for example, long days favor the development of sexually reproducing females, whereas females that reproduce asexually develop when days are short.)

The flowering of some angiosperms is not photoperiodic. In fact, there are more of these **day-neutral**

Photoperiodic Control of Flowering

In 1920, W. W. Garner and H. A. Allard of the U.S. Department of Agriculture studied the behavior of a newly discovered mutant tobacco plant. The mutant, named Maryland Mammoth, had large leaves and exceptional height. When the other plants in the field flowered, the Maryland Mammoth continued to grow. Garner and Allard took cuttings of the Maryland Mammoth into their greenhouse, and the plants that grew from the cuttings finally flowered in December. Garner and Allard also noticed that some soybean plants all flowered at about the same time, in late summer, even though they had been planted at different times in the spring.

Garner and Allard guessed that both of these observations had something to do with the seasons. They tested several likely seasonal variables, such as temperature, but the key variable proved to be the length of day. They moved plants between light and dark rooms at different times to vary the length of day artificially and were able to establish a direct link between flowering and day length.

The **critical day length** for Maryland Mammoth tobacco proved to be 14 hours (Figure 36.10). The plants did not flower if the light period was longer than 14 hours each day, but flowering commenced after the days became shorter than 14 hours. This phenomenon

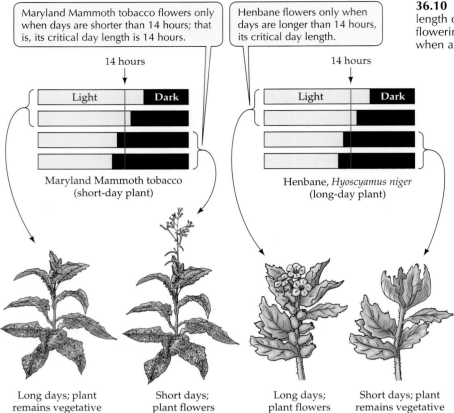

Maryland Mammoth tobacco flowers only when days are shorter than 14 hours; that is, its critical day length is 14 hours.

Henbane flowers only when days are longer than 14 hours, its critical day length.

14 hours

| Light | Dark |

Maryland Mammoth tobacco
(short-day plant)

14 hours

| Light | Dark |

Henbane, *Hyoscyamus niger*
(long-day plant)

Long days; plant remains vegetative

Short days; plant flowers

Long days; plant flowers

Short days; plant remains vegetative

36.10 Day Length and Flowering By varying the length of day, Garner and Allard showed that the flowering of Maryland Mammoth tobacco is initiated when a critical day length is reached.

In cocklebur, a single long night is enough of a photoperiodic stimulus to trigger full flowering some days later, even if the intervening nights are short ones. Most plants are less sensitive than the cocklebur, requiring from two to many nights of appropriate length to induce flowering. Plants of some species must experience an appropriate night length every night before they can flower. For these plants, a single shorter night, even one day before flowering would have commenced, inhibits flowering.

Hamner and Bonner showed that plants measure the length of the night using another method as well. They grew SDPs and LDPs on a variety of light regimes. In some regimes the dark period was interrupted by a brief exposure to light; in others, the light period was in-

plants than there are short-day and long-day plants. Some plants are photoperiodically sensitive only when young and become day-neutral as they grow older. Others require specific combinations of day length and other factors to flower.

Is it the length of day or of night that determines whether a plant will flower?

The terms "short-day plant" and "long-day plant" became entrenched before scientists learned that plants actually measure the length of *night*, or of darkness, rather than the length of day. This fact was demonstrated by Karl Hamner of the University of California at Los Angeles and James Bonner of the California Institute of Technology.

Working with cocklebur, an SDP, Hamner and Bonner ran a series of experiments in which (1) the light period was kept constant—either shorter or longer than the critical day length—and the dark period was varied, or (2) the dark period was kept constant and the light period was varied (Figure 36.11). The plants flowered under all treatments in which the dark period exceeded 9 hours, regardless of the length of the light period. Thus it is the length of the *night* that matters; for cocklebur, the critical night length is about 9 hours.

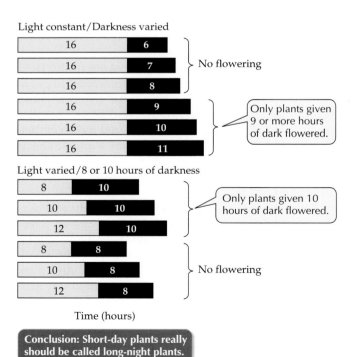

Light constant/Darkness varied

16	6
16	7
16	8

No flowering

16	9
16	10
16	11

Only plants given 9 or more hours of dark flowered.

Light varied/8 or 10 hours of darkness

8	10
10	10
12	10

Only plants given 10 hours of dark flowered.

8	8
10	8
12	8

No flowering

Time (hours)

Conclusion: Short-day plants really should be called long-night plants.

36.11 Night Length and Flowering The length of the darkness, not the light, triggers flowering.

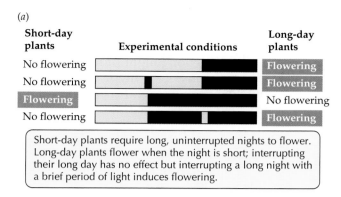

(a)

Short-day plants	Experimental conditions	Long-day plants
No flowering		Flowering
No flowering		Flowering
Flowering		No flowering
No flowering		Flowering

Short-day plants require long, uninterrupted nights to flower. Long-day plants flower when the night is short; interrupting their long day has no effect but interrupting a long night with a brief period of light induces flowering.

(b)

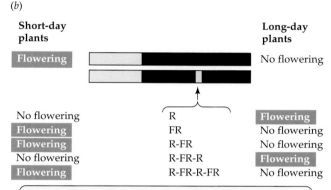

Short-day plants		Long-day plants
Flowering		No flowering

No flowering	R	Flowering
Flowering	FR	No flowering
Flowering	R-FR	No flowering
No flowering	R-FR-R	Flowering
Flowering	R-FR-R-FR	No flowering

When plants are exposed to red (R) and far-red (FR) light in alternation, the final treatment determines the effect of the light interruption, suggesting that phytochrome participates in photoperiodic responses.

36.12 The Effect of Interrupted Days and Nights
(a) Experiments suggest that plants are able to measure the length of a continuous dark period and use this information to trigger flowering. (b) Phytochrome (see Chapter 35) seems to be involved in the photoperiodic timing mechanism.

terrupted briefly by darkness. Interruptions of the light period by darkness had no effect on the flowering of either short-day or long-day plants. Even very brief interruptions of the dark period, however, completely nullified the effect of a long night (Figure 36.12a). An SDP flowered only if the long nights were uninterrupted. An LDP experiencing long nights flowered if these nights were broken by exposure to light. Thus a plant must have a timing mechanism that measures the length of a continuous dark period and uses the result to trigger flowering or to remain vegetative. Despite much study, the nature of the timing mechanism is still unknown.

Phytochrome seems to participate in the photoperiodic timing mechanism. In the interrupted-night experiments, the most effective wavelengths of light were in the red range (Figure 36.12b), and the effect of a red-light interruption of the night was fully reversed by a subsequent exposure to far-red light. It was once thought that the timing mechanism might simply be

the slow conversion of phytochrome during the night from the P_{fr} form—produced during the light hours—to the P_r form (see Chapter 35 for a discussion of phytochrome). But this suggestion is inconsistent with most of the experimental observations and must be wrong. Phytochrome must be only a photoreceptor. The timekeeping role is played by a biological clock.

Circadian rhythms are maintained by the biological clock

It is abundantly clear that organisms have a way of measuring time and that they are well adapted to the 24-hour day–night cycle of our planet. Some sort of biological clock resides within the cells of all eukaryotes, and the major outward manifestations of this clock are known as **circadian rhythms** (from the Latin *circa*, "about," and *dies*, "day").

Plants provide innumerable examples of approximately 24-hour cycles. The leaflets of a plant such as clover or the tropical tree *Albizia* normally hang down and fold at night and rise and expand during the day. Flowers of many plants show similar "sleep" movements, closing at night and opening during the day. They continue to open and close on an approximately 24-hour cycle even when the light and dark periods are experimentally modified (Figure 36.13).

Circadian rhythms of protists, animals, fungi, and plants share some important characteristics. First, the **period** is remarkably insensitive to temperature, although lowering the temperature may drastically reduce the **amplitude** of the fluctuation. (Figure 36.14 explains these terms.) Second, circadian rhythms are

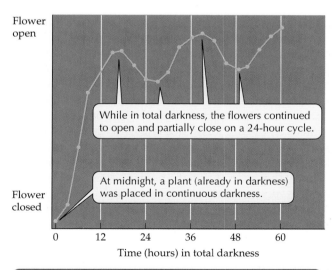

While in total darkness, the flowers continued to open and partially close on a 24-hour cycle.

At midnight, a plant (already in darkness) was placed in continuous darkness.

Conclusion: *Kalanchoe* flowers open and close in response to a biological clock even in the absence of external light/dark cues.

36.13 Sleep Movements of Flowers Many species of plants exhibit circadian (daily) rhythms.

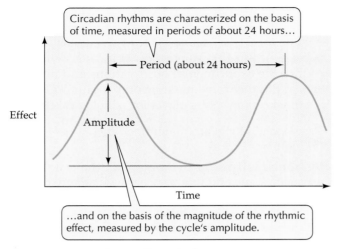

Circadian rhythms are characterized on the basis of time, measured in periods of about 24 hours...

...and on the basis of the magnitude of the rhythmic effect, measured by the cycle's amplitude.

36.14 Features of Circadian Rhythms   Plant rhythms share these features with the circadian rhythms of other organisms.

highly persistent; they continue even in an environment in which there is no alternation of light and dark.

Third, circadian rhythms can be **entrained**, within limits, by light–dark cycles that differ from 24 hours. That is, the period an organism expresses can be made to coincide with that of the light–dark regime. The period in nature is approximately 24 hours. If an *Albizia* tree, for example, were to be placed under artificial light on a day–night cycle totaling exactly 24 hours, the rhythm expressed would show a period of exactly 24 hours. If, however, an experimenter used a day–night cycle of, say, 22 hours, then with time the rhythm would change to a 22-hour period.

If light–dark cycles can entrain circadian rhythms, light alone should also be able to shift the rhythm—to cause phase shifts. If an organism is maintained under constant darkness, with its circadian rhythm expressed on the approximately 24-hour period, a brief exposure to light can make the next peak of activity appear either later or earlier than one would have predicted, depending on when the exposure is given. Moreover, the organism does not then return to its old schedule if kept in darkness. If the first peak is delayed by 6 hours, the subsequent peaks are all 6 hours late. Such phase shifts are permanent—until the organism receives more exposures to light.

Important questions about circadian rhythms remain to be answered. We do not know, for example, how light resets the biological clock. In fact, we still do not understand the biochemical or biophysical basis of the clock.

There is now ample evidence that the photoperiodic behavior of plants is based on the interaction of night length with the biological clock. But how the clock is coupled with flowering is unclear. Experi-

ments with the small lawn weed *Chenopodium rubrum* (commonly called red goosefoot) provided one sort of evidence. A short-day plant, *Chenopodium* will flower in response to a single long night. It shows at least some flowering response to a night as long as 96 hours. A single red flash during this extremely long night can either enhance or inhibit this response, depending on precisely when it is administered (Figure 36.15). Whether the red flash enhances or inhibits varies on an approximately daily basis (like the clock), but just how the underlying daily rhythm relates to flowering remains to be learned.

Is there a flowering hormone?

Is the timing device for flowering located in a particular part of an angiosperm, or are all parts able to sense the length of night? As in the Darwins' study of the light receptor for phototropism (see Figure 35.13), this question was resolved by "blindfolding" different parts of the plant.

It quickly became apparent that each leaf is capable of timing the night. If a short-day plant is kept under a regime of short nights and long days, but a leaf is covered so as to give it the needed long nights, the plant will flower (Figure 36.16a). This type of experiment works best if only one leaf is left on the plant. If one leaf is given a photoperiodic treatment conducive to flowering—an inductive treatment—other leaves kept under noninductive conditions will tend to inhibit flowering.

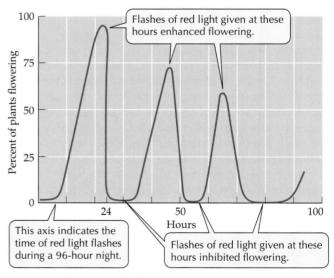

Flashes of red light given at these hours enhanced flowering.

This axis indicates the time of red light flashes during a 96-hour night.

Flashes of red light given at these hours inhibited flowering.

36.15 A Flowering Rhythm   The short-day plant *Chenopodium rubrum* flowers in response to a single 96-hour "night." If single flashes of red light are given during this 96-hour dark period, they either enhance or inhibit the flowering response. The peaks of enhancement are on about a 24-hour cycle initiated by the light-to-dark transition that started the night.

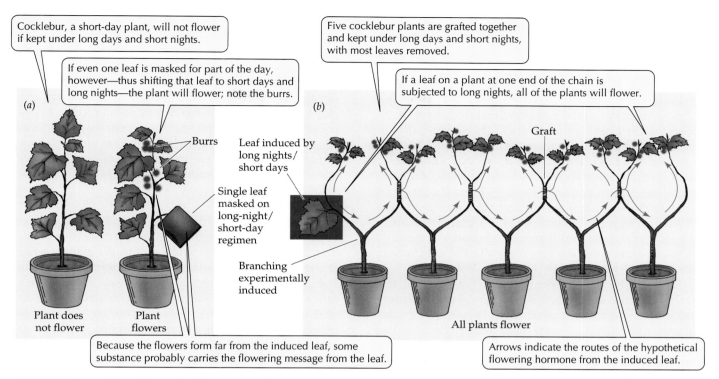

Cocklebur, a short-day plant, will not flower if kept under long days and short nights.

If even one leaf is masked for part of the day, however—thus shifting that leaf to short days and long nights—the plant will flower; note the burrs.

Five cocklebur plants are grafted together and kept under long days and short nights, with most leaves removed.

If a leaf on a plant at one end of the chain is subjected to long nights, all of the plants will flower.

(a)

Burrs

Leaf induced by long nights/ short days

Single leaf masked on long-night/ short-day regimen

Branching experimentally induced

Plant does not flower

Plant flowers

(b)

Graft

All plants flower

Because the flowers form far from the induced leaf, some substance probably carries the flowering message from the leaf.

Arrows indicate the routes of the hypothetical flowering hormone from the induced leaf.

36.16 Evidence for a Flowering Hormone It appears that if even a single leaf is induced to flower, the "message" will travel to the entire plant (and to other plants, if they are grafted to the flowering plant).

Although the leaves are what sense an inductive night period, the flowers form elsewhere on the plant. Thus a message must be sent from the leaf to the site of flower formation. Three lines of evidence suggest that this message is a chemical substance—a flowering hormone. First, if a photoperiodically induced leaf is removed from the plant shortly after the inductive night period, the plant does not flower. If, however, the induced leaf remains attached for several hours, the plant flowers. This result suggests that something—the hypothetical hormone—must be synthesized in the leaf in response to the inductive night, then move out of the leaf to induce flowering.

The second line of evidence for the existence of a flowering hormone comes from grafting experiments. If two cocklebur plants are grafted together, and if one plant is given inductive long nights and its graft partner is given noninductive short nights, both plants flower (Figure 36.16b).

Grafting experiments also provided the third line of evidence for a flowering hormone. Jan A. D. Zeevaart, a plant physiologist now at Michigan State University, exposed a single leaf of the SDP *Perilla* to a short-day/long-night regime, inducing the plant to flower. Then he detached this leaf and grafted it onto another, noninduced, *Perilla* plant—which responded by flow-

ering. The same leaf grafted onto successive hosts caused each of them to flower in turn. As long as 3 months after the leaf was exposed to the short-day/long-night regime, it could still cause plants to flower.

Does florigen exist?

Experiments such as Zeevaart's suggest that the photoperiodic induction of a leaf causes a more or less permanent change in it, inducing it to start and continue producing a flowering hormone that is transported to other parts of the plant, switching those target parts to the reproductive state. So reasonable is this idea that biologists have named the hormone **florigen**, even though after decades of active searching, the compound has not been isolated and characterized. The direct demonstration of florigen activity remains a cherished goal of plant physiologists. Gibberellins regulate the flowering of many species, especially of long-day plants, but these hormones do not have the properties of florigen.

As a final teaser, we will describe an experiment that suggests that the florigen of short-day plants is identical to that of long-day plants, even though SDPs produce it only under long nights and LDPs only under short nights. An SDP and an LDP were grafted together, and both flowered, as long as the photoperiodic conditions were inductive for one of the partners. Either the SDP or the LDP could be the one induced, but both would always flower. These results suggest that a flowering-inducing hormone, the elusive florigen, is being transferred from one plant to the other.

Vernalization and Flowering

In both wheat and rye, we distinguish two categories of flowering behavior. Spring wheat, for example, is sown in the spring and flowers in the same year. It is an annual plant. Winter wheat is biennial and must be sown in the fall; it flowers in its second summer. If winter wheat is not exposed to cold after its first year, it will not flower normally the next year. The implications of this finding were of great agricultural interest in Russia because winter wheat is a better producer than spring wheat, but it cannot be grown in parts of Russia because the winters there are too cold for its survival.

Several studies performed in Russia during the early 1900s demonstrated that if seeds of winter wheat are premoistened and prechilled, they develop and flower normally the same year when sown in the spring. Thus, high-yielding winter wheat could be grown even in previously hostile regions. This phenomenon—the induction of flowering by low temperatures—is called **vernalization**. Another example of vernalization is the cold requirement of some fruit trees. If the trees are planted too far south (in the Northern Hemisphere), they will not flower vigorously and therefore not produce much fruit.

Vernalization may require as many as 50 days of low temperature (in the range from about −2 to +12°C). Some plant species require both vernalization and long days to flower. There is a long wait from the cold days of winter to the long days of summer, but because the vernalized state easily lasts at least 200 days, these plants do flower when they experience the appropriate night length.

Asexual Reproduction

Although sexual reproduction takes up most of the space in this chapter's discussion of how plants reproduce, asexual reproduction is responsible for an important fraction of the new plant individuals appearing on Earth. This fact suggests another answer to the question asked at the beginning of this chapter: In some circumstances, asexual reproduction is better.

For example, think about genetic recombination. We have already noted that when a plant self-fertilizes, there are fewer opportunities for genetic recombination than there are with cross-fertilization. With self-fertilization, the only genetic variation that can be arranged into new combinations is that possessed by the single parental plant. A plant that is heterozygous for a locus can produce among its progeny both kinds of homozygotes for that locus plus the heterozygote,

but it cannot produce any progeny that carry alleles that it does not itself possess.

Asexual reproduction goes farther than self-fertilization: It eliminates genetic recombination altogether. When a plant reproduces asexually, it produces a clone of progeny with genotypes identical to its own. If a clone is highly adapted to its environment, the many copies of that genotype that may form by asexual reproduction may spread throughout that environment. This ability to exploit a particular environment is an advantage of asexual reproduction.

There are many forms of asexual reproduction

We call stems, leaves, and roots vegetative parts, distinguishing them from flowers, the reproductive parts. The modification of a vegetative part of a plant is what makes **vegetative reproduction** possible. The stem is the part that is modified in many cases. Strawberries and some grasses produce **stolons**, horizontal stems that form roots at intervals and establish potentially independent daughter plants (see Figure 31.9c). Other stolons are branches that sag to the ground and put out roots. The rapid multiplication of water hyacinths demonstrates the effectiveness of stolons for vegetative reproduction.

Some plants, such as potatoes, form **tubers**, the fleshy tips of underground stems. **Rhizomes** are underground stems that can give rise to new shoots. Bamboo is a striking example of a plant that reproduces vegetatively by means of rhizomes. A single bamboo plant can give rise to a stand—even a forest—of plants constituting a single, physically connected entity.

Whereas stolons and rhizomes are horizontal stems, bulbs and corms are short, vertical, underground stems. Lilies and onions form **bulbs** (Figure 36.17a), short stems with many fleshy, modified leaves. The leaves make up most of the bulb. Bulbs are thus large buds that store food and can later give rise to new plants. Crocuses, gladioli, and many other plants produce **corms**, underground stems that function very much as bulbs do. Corms are conical and consist primarily of stem tissue; they lack the fleshy modified leaves that are characteristic of bulbs.

Not all vegetative reproduction arises from modified stems. Leaves may also be the source of new plantlets, as in the succulent plants of the genus *Kalanchoe* (Figure 36.17b). Many kinds of angiosperms, ranging from grasses to trees such as aspens and poplars, form interconnected, genetically homogeneous populations by means of **root suckers**—horizontal roots. What appears to be a whole stand of aspen trees, for example, is a clone derived from a single tree by root suckers (see Figure 51.8).

Plants that reproduce vegetatively often grow in physically unstable environments such as eroding hillsides. Plants with stolons or rhizomes, such as beach

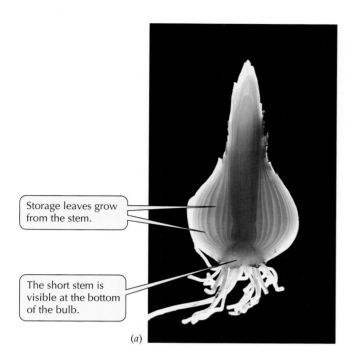

Storage leaves grow from the stem.

The short stem is visible at the bottom of the bulb.

(a)

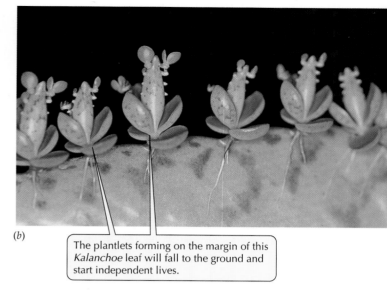

(b)

The plantlets forming on the margin of this *Kalanchoe* leaf will fall to the ground and start independent lives.

36.17 Vegetative Reproduction Two types are illustrated. (a) Bulbs are large buds that store food and can give rise to new plants. (b) New plantlets can form on leaves.

grasses, rushes, and sand verbena, are common pioneers on coastal sand dunes. Rapid vegetative reproduction enables the plants, once introduced, not only to multiply but also to survive burial by the shifting sand; in turn, the dunes are stabilized by the extensive network of rhizomes or stolons.

Dandelions and some other plants reproduce by **apomixis**, the asexual production of seeds. As you learned in Chapter 9, meiosis reduces the number of chromosomes in gametes, and fertilization restores the sporophytic number of chromosomes in the zygote. Some plants can skip over *both* meiosis and fertilization and still produce seeds. Apomixis produces seeds within the female gametophyte without the mingling and segregation of chromosomes and without the union of gametes. The ovule develops into a seed, and the ovary wall develops into a fruit. An apomictic embryo has the sporophytic number of chromosomes. The result of apomixis is a fruit with seeds that are genetically identical to the parent plant.

Interestingly, apomixis sometimes requires pollination. In some apomictic species a sperm nucleus must combine with the polar nuclei in order for the endosperm to form. In other apomictic species, the pollen provides the signals for embryo and endosperm formation, although neither sperm nucleus participates in fertilization. Pollination and fertilization are not the same thing!

Asexual reproduction plays a large role in agriculture

Farmers take advantage of some natural forms of vegetative reproduction. Farmers and scientists have also added new types of asexual reproduction by manipulating plants. One of the oldest methods of vegetative reproduction used in agriculture consists simply of making cuttings, or **slips**, of stems, inserting them in soil, and waiting for them to form roots and thus become autonomous plants. The slips are sometimes encouraged to root by treatment with a plant hormone, auxin, as described in Chapter 35.

Agriculturists reproduce many woody plants by **grafting**—attaching a piece of one plant to the stem or root of another plant. The part of the resulting plant that comes from the root-bearing "host" is called the **stock**; the part grafted on is called the **scion** (Figure 36.18). In order for a graft to succeed, the cambium of the scion must become associated with the cambium of the stock (see Chapter 31).

By cell division, cambia of scion and stock both form masses of wound tissue. If the two masses meet and fuse, the resulting continuous cambium can produce xylem and phloem, allowing transport of water and minerals to the scion and of photosynthate to the stock. Grafts are most often successful when the stock and scion belong to the same or closely related species. Grafting techniques are of great importance in agriculture; most fruit grown for market in the United States is produced on trees grown from grafts.

There are many reasons for grafting plants for fruit production. The most common is the desire to combine a hardy root system with a shoot system that produces the best-tasting fruit. This motive is illustrated by the story of the wine grape *Vitis vinifera*. In 1863, plant lice of the genus *Phylloxera* inflicted great damage in French vineyards. The roots of vines on more

(a) Cleft grafting (b) Whip grafting

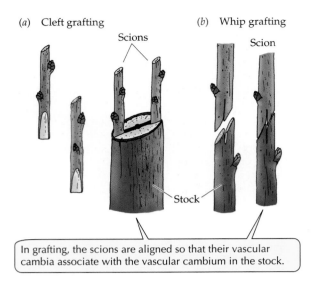

Scions

Scion

Stock

In grafting, the scions are aligned so that their vascular cambia associate with the vascular cambium in the stock.

36.18 Grafting Grafting—attaching a piece of a plant to the stem or root of another plant—is common in agriculture. The "host" stem or root is the stock; the grafted piece is the scion.

than 2.5 million acres were destroyed. The problem was overcome by the importation of great numbers of *V. vinifera* plants, which have *Phylloxera*-resistant root systems, from California. These plants were used as stocks to which French vines were grafted as scions. Thus the fine French grapes could be grown using roots resistant to the lice. (But the battle continues; in recent years, a new strain of *Phylloxera* has been damaging the grapevines in California.)

Scientists in universities and industrial laboratories have been developing new ways to produce valuable plant materials. For example, recombinant-DNA techniques can provide plants with capabilities they previously lacked (see Chapter 16). By causing cells of different sorts to fuse, one can obtain plants with exciting new combinations of properties. By cloning—making genetically identical copies of—small bits of tissue, one can obtain large numbers of equally desirable plants. But how can one efficiently take such small, delicate materials and get them to grow in the field?

Plants in nature solved this problem long ago. The product of sexual and apomictic reproduction in flowering plants is a compact package, protectively wrapped, containing an embryonic member of the next generation along with a supply of the nutrients it needs to begin its independent existence. This package is the seed. What was needed as a tool of plant biotechnology, and what has since been developed, is an artificial seed.

Artificial seeds contain a multicellular "somatic embryo." This is not a sexually produced embryo, but an embryolike product of mitotic divisions in tissue culture. Individual cells or small clusters of cells isolated

from the body of a suitable parent plant may develop in liquid culture into structures similar to normal embryos derived from zygotes. So that the somatic embryo does not dry out, and so that it may be stored and transported before planting, it is embedded in a water-soluble gel; then the combined embryo and gel are encapsulated in a protective plastic coat. The coat and gel dissolve away after the artificial seed is planted. Other materials may be added to the gel, among them suitable inorganic nutrients, fungicides, and pesticides.

Such scientifically designed artificial seeds will likely be used more and more often as the remaining problems are solved and methods are perfected for the mass production of these tiny packages.

Summary of "Reproduction In Flowering Plants"

Many Ways to Reproduce

• Almost all flowering plants reproduce sexually, and many also reproduce asexually.
• Both sexual and asexual reproduction are important in agriculture.

Sexual Reproduction

• Sexual reproduction promotes genetic diversity in a population.
• In the sexual cycle of angiosperms, a gametophyte (gamete-producing) generation alternates with a sporophyte (spore-producing) generation. The flower is an angiosperm's device for sexual reproduction.
• Flowering plants have microscopic gametophytes that develop in the flowers of the sporophytes. The (female) megagametophyte typically contains eight nuclei in a total of seven cells. The (male) microgametophyte is the pollen grain, which delivers two sperm nuclei to the megagametophyte by means of a long pollen tube. **Review Figure 36.1**
• Pollination enables fertilization in the absence of liquid water.
• Angiosperms perform double fertilization: One sperm nucleus fertilizes the egg, forming a zygote, and the other sperm nucleus unites with the two polar nuclei to form a triploid endosperm nucleus. **Review Figure 36.5**
• *In vitro* fertilization in plants is difficult because the gametes, unlike those of animals, are not free of other cells. The gametes require treatment both to be freed and to be able to fuse after being freed.
• The zygote develops into an embryo (with attached suspensor), which remains dormant in the seed until the conditions are right for germination. The endosperm is the nutritive reserve upon which the embryo depends at germination. **Review Figures 36.6, 36.7**
• Flowers develop into seed-containing fruits, which often play important roles in the dispersal of the species.

The Transition to the Flowering State

• For a vegetatively growing plant to flower, an apical meristem in the shoot system must change into an inflorescence meristem, which gives rise to bracts and meristems. The meristems thus produced may be floral meristems or additional inflorescence meristems. **Review Figure 36.9**
• Organ identity genes are expressed in floral meristems to give rise to sepals, petals, stamens, and carpels.

Photoperiodic Control of Flowering

• Some plants regulate their flowering by photoperiodism.
• Short-day plants flower when the days are shorter than a species-specific critical day length; long-day plants flower when the days are longer than the critical day length. **Review Figure 36.10**
• Some angiosperms have more complex photoperiodic requirements than short-day or long-day plants have, but most are day-neutral.
• The length of the night is what actually determines whether a photoperiodic plant will flower. **Review Figure 36.11**
• Interruption of the nightly dark period by a brief exposure to light undoes the effect of a long night. **Review Figure 36.12**
• The mechanism of photoperiodic control appears to include a biological clock and phytochrome. **Review Figures 36.13, 36.14, 36.15**
• Evidence suggests that there is a flowering hormone, florigen, but the substance has yet to be convincingly isolated from any plant. **Review Figure 36.16**

Vernalization and Flowering

• In some species, exposure to low temperatures—vernalization—is required for the plants to flower.

Asexual Reproduction

• Asexual reproduction allows rapid multiplication of organisms well suited to their environment. It often affords advantages in physically unstable environments, such as sand dunes.
• Stolons, tubers, rhizomes, bulbs, corms, and root suckers are means by which plants may reproduce vegetatively (asexually). Some species produce seeds asexually by apomixis.
• Agriculturists often plant vegetatively propagated slips or graft different plants together to increase yields or to take advantage of favorable properties of both stock and scion. **Review Figure 36.18**

Self-Quiz

1. Sexual reproduction in angiosperms
   a. is by way of apomixis.
   b. requires the presence of petals.
   c. can be accomplished by grafting.
   d. gives rise to genetically diverse offspring.
   e. cannot result from self-pollination.

2. The typical angiosperm female gametophyte
   a. is called a megaspore.
   b. has eight nuclei.
   c. has eight cells.
   d. is called a pollen grain.
   e. is carried to the male gametophyte by wind or animals.

3. Pollination in angiosperms
   a. never requires water.
   b. never occurs within a single flower.
   c. always requires help by animal pollinators.
   d. is also called fertilization.
   e. makes most angiosperms independent of water for reproduction.

4. Which statement about double fertilization is *not* true?
   a. It is found in most angiosperms.
   b. It is found in no plants other than angiosperms.
   c. One of its products is a triploid nucleus.
   d. One sperm nucleus fuses with the egg nucleus.
   e. One sperm nucleus fuses with two polar nuclei.

5. The suspensor
   a. gives rise to the embryo.
   b. is heart-shaped in dicots.
   c. separates the two cotyledons of dicots.
   d. ceases to elongate early in embryo development.
   e. is larger than the embryo.

6. Which statement about photoperiodism is *not* true?
   a. It is related to the biological clock.
   b. Phytochrome plays a role in the timing process.
   c. It is based on measurement of the length of the night.
   d. Most plant species are day-neutral.
   e. It is limited to the plant kingdom.

7. Although florigen has never been isolated, we think it exists because
   a. night length is measured in the leaves, but flowering occurs elsewhere.
   b. it is produced in the roots and transported to the shoot system.
   c. it is produced in the coleoptile tip and transported to the base.
   d. we think that gibberellin and florigen are the same compound.
   e. it may be activated by prolonged (more than a month) chilling.

8. Which statement about vernalization is *not* true?
   a. It may require more than a month of low temperature.
   b. The vernalized state generally lasts for about a week.
   c. Vernalization makes it possible to have two winter wheat crops each year.
   d. It is accomplished by subjecting moistened seeds to chilling.
   e. It was of interest to Russian scientists because of their native climate.

9. Which of the following does *not* participate in asexual reproduction?
   a. Stolon
   b. Rhizome
   c. Fertilization
   d. Tuber
   e. Apomixis

10. Apomixis includes
    a. sexual reproduction.
    b. meiosis.
    c. fertilization.
    d. a diploid embryo.
    e. no production of a seed.

Applying Concepts

1. For a crop plant that reproduces both sexually and asexually, which method of reproduction might the farmer prefer?

2. Thompson seedless grapes are produced by vines that are triploid. Think about the consequences of this chromosomal condition for meiosis in the flowers. Why are these grapes seedless? Describe the role played by the flower in fruit formation when no seeds are being formed. How do you suppose Thompson seedless grape plants are propagated?

3. Poinsettias are popular ornamental plants that typically bloom just before Christmas. Their flowering is photoperiodically controlled. Are they long-day or short-day plants? Explain.

4. You plan to induce the flowering of a crop of long-day plants in the field by using artificial light. Is it necessary to keep the lights on continuously from sundown until the point at which the critical day length is reached? Explain.

Readings

Barrett, S. C. H. 1987. "Mimicry in Plants." *Scientific American*, September. An exploration of how some plants mimic insects, encouraging the mimicked insects to act as pollinators.

Cleland, C. E. 1978. "The Flowering Enigma." *BioScience*, April. Florigen eluded physiologists when this article was published, and it still does.

Cox, P. A. 1993. "Water-Pollinated Plants." *Scientific American*, October. An unusual environment calls for different pollination strategies, and it provides unusual opportunities for research.

Goodman, B. 1993. "A 'Shotgun Wedding' Finally Produces Test-Tube Plants." *Science*, vol. 261, page 430. A brief account of the discovery of *in vitro* fertilization in plants.

Handel, S. N. and A. J. Beattie. 1990. "Seed Dispersal by Ants." *Scientific American*, August. A discussion of seed dispersal mechanisms, focusing especially on plants that produce seeds bearing specialized fat bodies that are eaten by ants after the ants carry the seeds to their nests.

Meyerowitz, E. M. 1994. "The Genetics of Flower Development." *Scientific American*, November. An exploration of what mutant strains are telling us about the formation of floral organs.

Niklas, K. J. 1987. "Aerodynamics of Wind Pollination." *Scientific American*, July. An investigation of the adaptations of wind-pollinated plants that favor successful pollination.

Pfennig, D. W. and P. W. Sherman. 1995. "Kin Recognition." *Scientific American*, June. Although this paper is concerned primarily with animals, it also touches on the ability of mountain delphiniums to distinguish between the pollen of relatives and nonrelatives and thus avoid breeding with close relatives.

Raven, P. H., R. F. Evert and S. Eichhorn. 1992. *Biology of Plants*, 5th Edition. Worth, New York. A well-balanced general botany textbook.

Salisbury, F. B. and C. W. Ross. 1992. *Plant Physiology*, 4th Edition. Wadsworth, Belmont, CA. An authoritative textbook with good chapters relating to reproductive development.

Schiebinger, L. 1996. "The Loves of the Plants." *Scientific American*, February. A fascinating discussion of Linnaeus's scheme for classifying plants on the basis of their reproductive parts, by a professor of history and women's studies.

Seymour, R. S. 1997. "Plants That Warm Themselves." *Scientific American*, March. Speculations on how heat generated by some plants may create a favorable environment for pollinators.

Taiz, L. and E. Zeiger. 1998. *Plant Physiology*, 2nd Edition. Sinauer Associates, Sunderland, MA. Chapter 24 deals with the control of flowering.

Part Six
The Biology of Animals

Chapter 37

Physiology, Homeostasis, and Temperature Regulation

A Cold Breeding Ground
In their frigid Antarctic homeland, emperor penguins migrate between the sea and breeding grounds such as this one. The distance may be more than 100 km.

*A*nimals live in amazing places, including the most extreme environments on our planet. Consider emperor penguins. All penguins are flightless, marine birds that spend most of their lives at sea but must come to land to breed. Emperors are the largest of the penguins. We don't know where emperors concentrate their summer activities at sea, but as winter seizes Antarctica, the sea freezes to form a huge ice shelf that surrounds the continent. The emperors that will breed gather together at the edge of the ice. They then walk more than 100 km across the Antarctic ice in the austral winter to breed in one of the coldest, most inhospitable places on Earth.

After the female lays an egg, she walks back to the sea to feed; the male incubates the egg, then protects and feeds the chick from his own fat reserves until the female returns with her body fat replenished. The female then takes over feeding the chick, and the male walks back to the sea, having fasted for more than 4 months. To understand how the natural history of emperor penguins is possible, we must study their anatomy and their physiology—form and function. The seemingly endless variations on form and function that we call adaptations are the essence of animal biology.

In Part Six (Chapters 37 through 49) you will learn about many fascinating animal adaptations—for example, adaptations that allow birds to fly at altitudes that humans cannot survive without technological assistance, seals to remain underwater for more than half an hour, insects to live without water in the hottest deserts, and bats to capture flying prey in total darkness. These adaptations are variations on basic themes. The basic themes are the physiological functions of cells, tissues, organs, and organ systems—themes that have much in common across the broad spectrum of animals from worms to humans.

Our goal will be to understand basic, common, underlying, physiological mechanisms, and then to explore their specialized adaptations that make unusual life histories possible. We'll take a four-step approach to each organ system. First, we'll consider the physical problem between animals and their environments that the organ system must solve. Second, we'll consider molecular, cellular, anatomical, and physiological mechanisms on which solutions to the problem are based. Third, we'll describe some of the evolutionary variation in adaptations based on those mechanisms. Finally, we'll explore how the resulting functions of organ systems are controlled and regulated.

In this chapter, we set the stage for our study of physiology by outlining the various organs and organ systems, explaining how their activities are integrated to maintain an internal environment that provides for the needs of all cells of the body, and discussing the general issue of control and regulation. We then move to a detailed consideration of one physiological function—temperature regulation. Keep in mind the four-step approach: How are living systems affected by temperature? What molecular, cellular, anatomical, and physiological mechanisms do animals have for dealing with thermal problems? How are these mechanisms used by different species? And how are thermal adaptations controlled and regulated?

The Needs of the One versus the Needs of the Many

Single-celled organisms meet all their needs by direct exchange with the environment. Even some simple, small multicellular animals meet all their needs by direct exchange with the environment as long as the design of their bodies enables each cell to be in contact with that environment. Such single-celled and simple multicellular animals are common in the sea.

Seawater contains nutrients, it has a suitable composition of salts, and it provides a stable physical environment. Each cell of a sponge or a jellyfish, for example, is in direct contact with the environment and functions independently. It receives its nutrients directly from and releases its wastes directly into seawater. But this lifestyle is quite limiting. No part of the animal's body can be more than two cell layers thick, every cell must be able to take care of all its needs, and the animal is limited to environments that will provide for all of those cellular needs.

The evolutionary steps that made possible complex multicellular animals that could occupy all sorts of extreme habitats created an **internal environment**. That internal enviroment consists of extracellular fluids that can differ in composition from the external environment. If every cell of the body can have its needs met through exchange with an internal environment, it does not have to be in contact with the external environment. And to the extent that the internal environment can be maintained independently of the external environment, the animal can occupy habitats that would not support the lives of cells if they were exposed to it directly.

Another advantage of an internal environment is that if its composition can be maintained at optimal levels, then each cell of the body does not have to be such a generalist to take care of its needs. A constant internal environment enables cells to become more specialized, and therefore more efficient at a particular task. Can you see the enormous evolutionary potential in this concept?

A relatively constant internal environment enables cells to become specialized in tasks that can contribute to maintenance of the internal environment. Thus, some cells can become organized into tissues specialized to maintain the ionic composition of the internal environment, others can become specialized to provide simple nutrients, and still others to maintain appropriate levels of oxygen and carbon dioxide. The evolutionary rationale of animal physiology is that an internal environment makes possible the specializations of tissues, organs, and organ systems that in turn maintain the optimality of the internal environment. Thus, through its effects on the internal environment, an organ provides for a specific need of all the cells of the animal over a wide range of external environments (Figure 37.1).

Staying the same while constantly changing: Homeostasis

What must constantly change to stay the same? The answer to this old riddle is a river. The internal environment of a body is similar because its components are constantly changing. Cells all over the body are taking nutrients from and giving wastes to the extracellular fluids that bathe them. The activities of the various organs must counteract these influences on the internal environment so that its composition does not deviate from an optimal range. The maintenance of a constant composition of the internal environment—constant in spite of continual and variable turnover of its components—is referred to as **homeostasis**.

Homeostasis of the internal environment is an essential feature of complex animals. It enables emperor penguins to function well in the Antarctic winter, and it enables other species to occupy other challenging environments. However, if an organ fails to function properly, homeostasis of the internal environment is compromised, and as a result cells become sick and die. The sick cells are not just those of the organ that

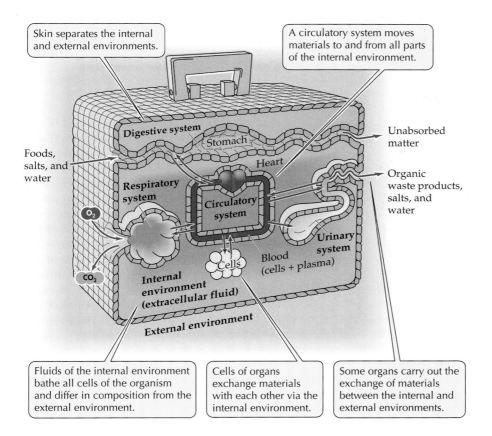

Skin separates the internal and external environments.

A circulatory system moves materials to and from all parts of the internal environment.

37.1 Maintaining Internal Stability While On the Go Organ systems maintain a constant internal environment that provides for the needs of all cells of the body, making it possible for animals to travel among different and often highly variable external environments.

Digestive system
Stomach
Unabsorbed matter

Foods, salts, and water

Heart

Respiratory system

Circulatory system

Organic waste products, salts, and water

O_2

Urinary system

CO_2

Blood (cells + plasma)

Cells

Internal environment (extracellular fluid)

External environment

Fluids of the internal environment bathe all cells of the organism and differ in composition from the external environment.

Cells of organs exchange materials with each other via the internal environment.

Some organs carry out the exchange of materials between the internal and external environments.

tems distinguished by function: information and control; protection, support, and movement; reproduction; nutrition; transport and defense; and excretion.

Organs are composed of tissues

Organs are made up of tissues. Tissues consist of cells that have similar structure and function. Biologists who study cells and tissues recognize many types of cells but group them into only four general types of tissue: epithelial, connective, muscle, and nervous (Figure 37.2).

Most **epithelial tissues** are sheets of tightly connected cells such as those that cover the body surface and those that line various hollow organs of the body, including the digestive tract and the lungs. Some epithelial cells have secretory functions—for example, those that secrete mucus, digestive enzymes, or sweat. Other epithelial cells have cilia and help substances move over surfaces or through tubes. Since epithelial cells create boundaries between the inside and the outside of the body and between body compartments, they frequently have absorptive and transport functions. An epithelium can be stratified, as is skin, which consists of many layers of cells, or it can be simple, as is the lining of the gut, which consists of a single layer of cells.

The skin and the lining of the gut are examples of epithelial tissues that receive much wear and tear. Accordingly, cells in these tissues have a high rate of cell division to replace older cells that die and are shed. Dandruff is discarded epithelial cells, and the well-known Pap smear test for cancer of the female reproductive tract is based on examination of shed epithelial cells. We'll encounter epithelial tissues in our discussions of linings and tubules of reproductive systems (Chapter 39), linings of gas exchange systems (Chapter 45), linings of digestive tracts (Chapter 47), and tubules of excretory systems (Chapter 48).

Connective tissues support and reinforce other tissues. Unlike epithelial tissues, which consist of

functions improperly, but the cells of all other organs as well. Loss of homeostasis is a serious problem that makes itself worse. To avoid loss of homeostasis, the activities of organs must be controlled and regulated in response to changes in both the external and the internal environments.

Control and regulation require information; hence the organ systems of information—the endocrine and nervous systems—must be included in our discussions of every physiological function. For that reason, we treat the endocrine and nervous systems early in this part of the book. Subsequent chapters deal with organ systems responsible for homeostasis of various aspects of the internal environment.

Organs and Organ Systems

The diversity of adaptations that enables animals to live in almost any environment presents us with a bewildering number of details that can make the study of physiology seem daunting. In this section, we begin with a road map of the organs, organ systems, and physiological functions of one species—*Homo sapiens*—but the road map applies to most other vertebrates and to some invertebrates as well.

We will examine the four basic types of tissues, then provide a brief overview of the vertebrate organ sys-

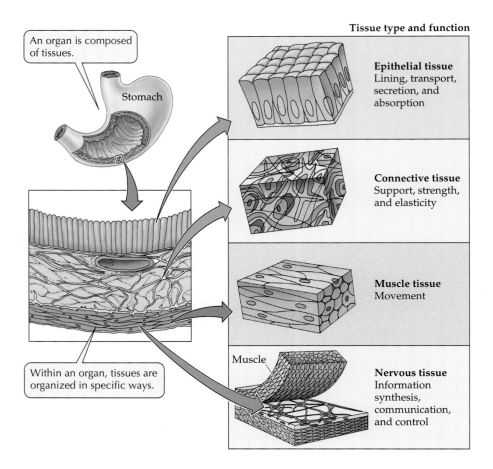

An organ is composed of tissues.

Stomach

Within an organ, tissues are organized in specific ways.

Tissue type and function

Epithelial tissue
Lining, transport, secretion, and absorption

Connective tissue
Support, strength, and elasticity

Muscle tissue
Movement

Muscle

Nervous tissue
Information synthesis, communication, and control

37.2 Four Types of Tissue All cells are classified into one of four tissue types. The cells of a given type have similar structure and function.

major types of connective tissue are **adipose tissue** (fat cells) and the cellular components of the blood (see Chapter 46).

Muscle tissue consists of cells that can contract and therefore can cause movements of organs, limbs, or most other parts of the body (see Chapter 44). There are three kinds of muscle tissue in our bodies. **Skeletal muscle** connects bones to bones and is mostly under conscious control. Skeletal muscles are responsible for our behavior. **Smooth muscle** is found in internal organs and is not under conscious control. Smooth muscles make food move through the gut (see Chapter 47) and constrict blood vessels (see Chapter 46). **Cardiac muscle** makes up the mass of the heart and pumps the blood (see Chapter 46).

Nervous tissue (see Chapters 41, 42, and 43) enables animals to deal with information. There are two basic cell types in nervous tissue. **Neurons**, which are extremely diverse, generate electrochemical signals. Some neurons respond to specific types of stimuli, such as light, sound, pressure, or certain molecules, by generating sudden voltage changes in their membranes. These nerve impulses can be conducted via long extensions of the cells to other parts of the body, where they are communicated to other neurons, to muscle cells, or to secretory cells by chemical signals that pass between cells. Neurons are involved in controlling the activities of most organ systems. The other type of cell in nervous tissue is the **glial cell**. Glial cells do not generate or conduct electric signals, but they provide a variety of supporting functions for neurons. There are more glial cells than neurons in our brains.

Organs are made up of more than one tissue type, and most organs include all tissue types. The gut—the group of organs that digest food and absorb nutrients—is a good example (see Figure 37.2). The surface of the gut that contacts food is lined internally with a single layer of epithelial cells. Some secrete mucus or enzymes; others mainly absorb nutrients. Beneath the gut lining is connective tissue. Within this connective tissue are embedded nerves, glands, and blood ves-

densely packed, tightly connected populations of cells, most connective tissues consist of a dispersed population of cells embedded in an extracellular matrix. The composition and properties of the matrix differ in different types of connective tissue.

Connective tissue underlying skin is flexible and contains fibers that give it both tensile strength and elasticity. Most of these fibers are various forms of the protein **collagen**, the most abundant protein in our bodies. In fact, collagen is the most abundant protein among vertebrates. One kind of collagen has high tensile strength and is abundant in tissues that connect muscle to bone and bones to each other. Another type of collagen is transparent and allows light to enter our eyes. Still another type, found in the smallest blood vessels, has such small spaces between its fibers that water and salts can pass into the intercellular fluids, while large protein molecules and blood cells stay inside the vessels. The protein **elastin** gives skin its elasticity. But, elastin has a very slow turnover, so it gradually degrades with age, resulting in a skin that is less resilient.

Bone is a dense connective tissue in which the extracellular matrix of collagen fibers has been hardened by mineral deposition (see Chapter 44). Two other

sels. Concentric layers of muscle tissue move food through the gut and mix it with the secretions of the epithelial cells. A network of neurons between the muscle layers controls the movements of the gut. Surrounding the gut is a layer of connective tissue called the serosa.

An individual organ may be part of an organ system. The major organ systems of the human body are outlined in Figures 37.3 through 37.8. We'll discuss each of these systems in the chapters that follow.

Nervous and endocrine systems process information and maintain control

The principal organ system that processes information and uses that information to control the physiology and behavior of the animal is the **nervous system** (Figure 37.3a, and Chapters 41, 42, and 43). It consists of the *brain* and *spinal cord* (the central nervous system) along with *peripheral nerves* that conduct electrochemical signals (*nerve impulses*) from *sensors* to the central nervous system and conduct signals from the central nervous system to effectors, which are either muscle tissue or secretory tissue. The sensors of the nervous system are diverse; they include eyes, ears, organs of taste and smell, organs of balance and orientation, and cells sensitive to temperature, touch, pressure, stretch, and pain.

The **endocrine system** (Figure 37.3b, and Chapter 38) also processes information and controls the functions of organs, but its messages are distributed mostly in the extracellular fluids, including blood, to the entire body as chemical signals called *hormones.* The principal organs of the endocrine system are *ductless glands* that secrete specific hormones into blood. For example, the pancreas secretes the hormone insulin into the blood, and this insulin stimulates most cells of the body to take up and use glucose. In addition to distinct endocrine glands, many other tissues such as the gut contain cells that secrete hormones.

There are strong interactions between the nervous system and the endocrine system. Cells in the brain produce hormones that control parts of the endocrine system. In turn, some cells in the brain respond to hormones produced by endocrine glands. In most cases, when a nerve impulse reaches the farthest extension of the nerve cell, the nerve cell communicates to its target through a chemical message called a *neurotransmitter.* In general, however, the chemical messages of nerve cells travel only very short distances to target cells, whereas the chemical messages of endocrine glands travel greater distances. You could think of the nervous system as being more like a telephone system in which the message (nerve impulse) goes to a specific

37.3 Organs of Information and Control

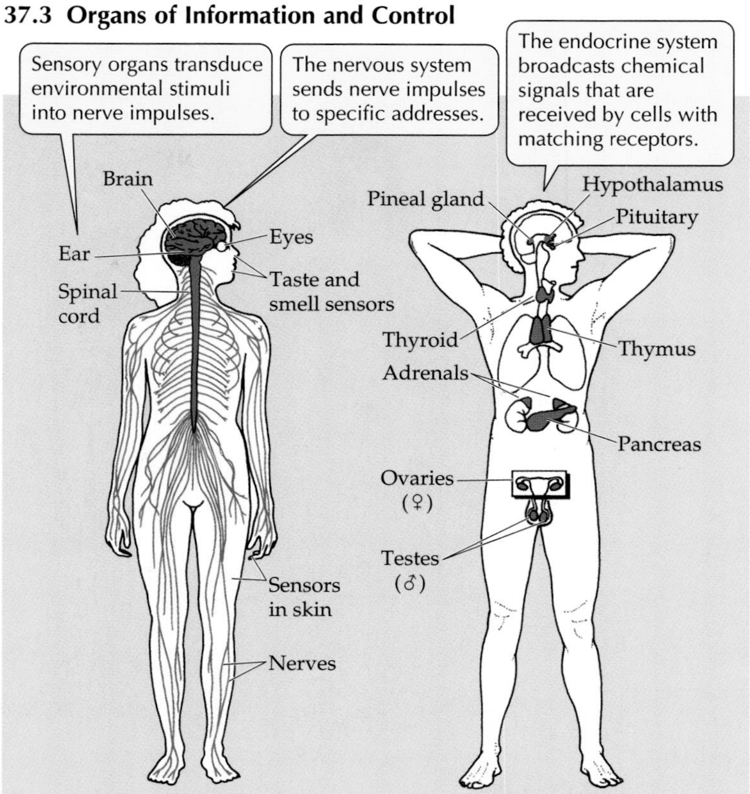

(a) **Nervous system** (b) **Endocrine system**

address, and the endocrine system as being more like radio communication in which the signal (hormone) is broadcast and can be picked up by anyone with a properly tuned receiver (a receptor in the case of target cells).

Skin, skeletal, and muscle systems provide protection, support, and movement

The largest organ of the body is the **skin**, along with its special elaborations, *hair* and *nails* (Figure 37.4a). The skin protects the body from organisms that cause disease, from the physical environment, and from excessive water loss. Because the skin contains nervous tissue that is sensitive to various stimuli, it is a major sense organ. The skin is also an effector: It can control the rate of heat exchange between the body and the environment, thereby helping to regulate body temperature.

The **skeletal system** supports and protects the body (Figure 37.4b, and Chapter 44). The skeleton is also an important effector: It forms the supports and the levers that muscles pull on to cause movement and behavior.

The **muscle system** (Figure 37.4c, and Chapter 44) includes the muscles that are under our conscious control and cause all voluntary movements (*skeletal mus-*

37.4 Organs of Protection, Support, and Movement

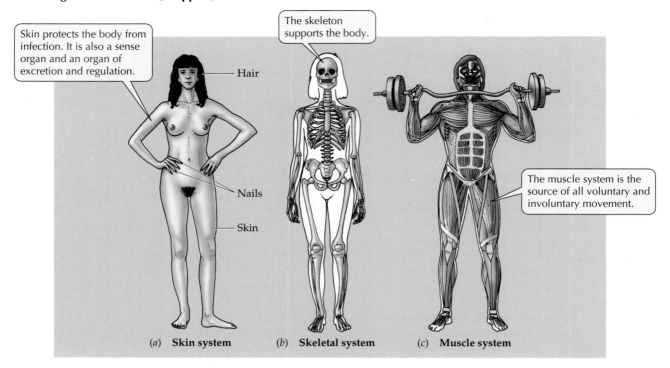

Skin protects the body from infection. It is also a sense organ and an organ of excretion and regulation.

The skeleton supports the body.

The muscle system is the source of all voluntary and involuntary movement.

Hair

Nails

Skin

(a) **Skin system**   *(b)* **Skeletal system**   *(c)* **Muscle system**

cles), the muscles of the internal organ systems that are not under our conscious control (*smooth muscles*), and the muscles that constitute the heart (*cardiac muscles*).

Reproductive systems produce offspring

The male and female **reproductive systems** consist of *gonads* (testes and ovaries, respectively), which produce sex cells (*gametes*), and other organs that deliver the sex cells to the site of fertilization (Figure 37.5, and Chapter 39). The female reproductive system includes the *uterus*, the organ that supports the development of the embryo. The female's *mammary glands* provide nutrients for the infant. In the gonads of both sexes are tissues that secrete hormones that play roles in sexual development, behavior, and reproduction.

Digestive and gas exchange systems supply nutrients to the body

The **digestive system** is largely a continuous tubular structure that extends from mouth to anus (Figure 37.6a, and Chapter 47). This tube, also called the *gut*, is divided into different segments that serve different functions in the processing and digestion of food and the absorption of nutrients. *Glands* associated with the gut deliver into it digestive enzymes and other molecules that break down complex food molecules. The lower gut resorbs water from wastes, stores wastes, and periodically eliminates them.

37.5 Organs of Reproduction

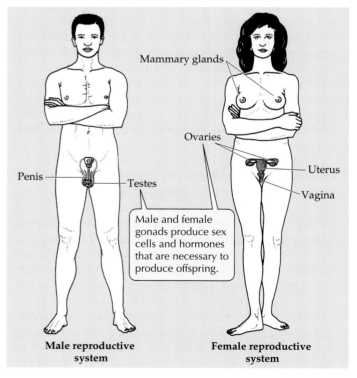

Mammary glands

Ovaries

Uterus

Penis

Testes

Vagina

Male and female gonads produce sex cells and hormones that are necessary to produce offspring.

Male reproductive system   **Female reproductive system**

37.6 Organs of Nutrition

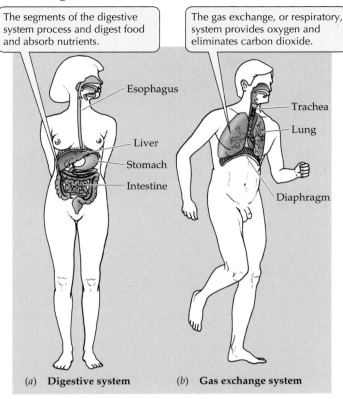

The segments of the digestive system process and digest food and absorb nutrients.

The gas exchange, or respiratory, system provides oxygen and eliminates carbon dioxide.

Esophagus

Trachea

Liver

Lung

Stomach

Intestine

Diaphragm

(a) **Digestive system**   (b) **Gas exchange system**

The **gas exchange system**, also called the respiratory system, provides oxygen, which is essential for cellular respiration (Figure 37.6b, and Chapter 45). Carbon dioxide, a waste product of cellular respiration, is eliminated by the gas exchange system. The gas exchange organs of humans are lungs, which are a system of progressively dividing airways leading to tiny but numerous sacs with membranous gas exchange surfaces that have a very large combined surface area. The diaphragm and other muscles that move air into and out of the lungs are another component of the gas exchange system.

Circulatory and lymphatic systems provide transport and defense

Oxygen must be transported from the lungs to the tissues of the body, and carbon dioxide must be transported from the tissues to the lungs. These gases are transported by the **circulatory system**, which includes a *pump* (the heart), a system of *blood vessels* (veins, arteries, and capillaries), and *blood* (Figure 37.7a, and Chapter 46).

The circulatory system also transports nutrients from the gut, delivers nitrogenous wastes to the excretory system, transports hormones, transports heat, and generates mechanical forces. Blood is made up of cellular components in a liquid medium called *plasma*.

The plasma is continuous with the extracellular fluids that make up the internal environment of the body.

The **lymphatic system** is another transport system consisting of a set of *vessels* that extend throughout the body, but unlike the circulatory system, it does not include a pump and its vessels do not form a complete circuit (Figure 37.7b). The lymphatic system picks up extracellular fluid and delivers it to the blood circulatory system.

The lymphatic and circulatory systems also encompass the body's major defense against foreign organisms and materials—the *immune system*. The cellular elements of the immune system are specialized blood cells produced in the bone marrow, thymus, spleen, and lymph nodes. As you already learned in Chapter 18, through various mechanisms the immune system recognizes, eliminates, and remembers foreign organisms and materials that invade the body.

The excretory system eliminates nitrogenous wastes and excess salts

Urine forms in the *kidneys*. Urine includes nitrogenous wastes from the metabolism of proteins and nucleic acids, as well as excess salts and other substances that the body excretes. The

37.7 Organs of Transport

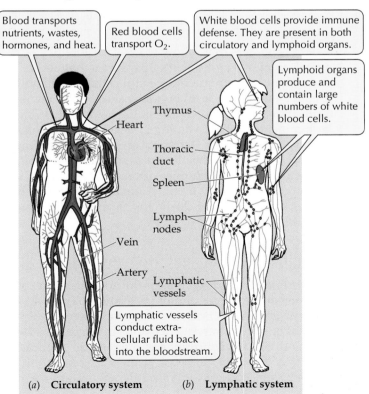

Blood transports nutrients, wastes, hormones, and heat.

Red blood cells transport O_2.

White blood cells provide immune defense. They are present in both circulatory and lymphoid organs.

Lymphoid organs produce and contain large numbers of white blood cells.

Thymus

Heart

Thoracic duct

Spleen

Lymph nodes

Vein

Artery

Lymphatic vessels

Lymphatic vessels conduct extracellular fluid back into the bloodstream.

(a) **Circulatory system**   (b) **Lymphatic system**

37.8 Organs of Excretion and of Water and Salt Balance

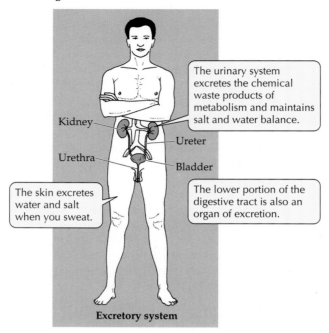

The urinary system excretes the chemical waste products of metabolism and maintains salt and water balance.

Kidney

Ureter

Urethra

Bladder

The skin excretes water and salt when you sweat.

The lower portion of the digestive tract is also an organ of excretion.

Excretory system

kidneys play crucial roles in maintaining the correct water content of the body and the correct salt composition of extracellular fluids. Urine passes to a *bladder* for storage until it is released to the exterior through the *urethra* (Figure 37.8, and Chapter 48). Remember that the skin also excretes water and salts as sweat, and the lower portions of the digestive tract excrete solid wastes.

In the chapters that follow we will explore physiological functions and the organ systems that accomplish them. In the remainder of this chapter we will discuss general principles of homeostasis, which have applications to all animal physiology. We'll illustrate these principles by considering in detail the regulation of body temperature. Temperature is an important physical parameter of the internal environment. It can be perturbed by the activities of cells and by changes in the outside environment. Animals have evolved various adaptations for dealing with changes in temperature.

Control, Regulation, and Homeostasis

Homeostasis depends on the control and regulation of functions of the organs and organ systems. These systems must be able to maintain a constant internal environment in spite of varying demands of the cells of the body and varying conditions in the outside environment.

The terms "control" and "regulation" might seem interchangeable, but their meanings differ. Control im-

plies the ability to *change* the rate of a reaction or process. Regulation—the more sophisticated and more specific physiological concept—refers to *maintaining* a variable within specific levels or limits. You control the speed of your car by using the accelerator and brake, but you regulate it by using the accelerator and brake to maintain a particular speed. In this section we will discuss the general properties of regulatory systems—set points and feedback—using temperature regulation as an example.

Set points and feedback information are required for regulation

Regulation requires, in addition to control mechanisms, the ability to obtain and use information. You can regulate the speed of a car only if you know the speed at which you are traveling and the speed you wish to maintain. The desired speed is a **set point**, and the reading on your speedometer is **feedback**. When the set point and the feedback are compared, any difference is an **error signal**. Error signals suggest corrective actions, which you make by using the accelerator or brake (Figure 37.9).

Understanding physiological regulation requires knowledge not only of the mechanisms of action of the molecules, cells, tissues, organs, and organ systems—the **controlled systems**—but also knowledge of how relevant information is obtained, processed, integrated, and converted into commands by the **regulatory systems**. A fundamental way to analyze a regulatory system is to identify its source of feedback.

Negative feedback is the most common type of feedback in regulatory systems. The word "negative" is use to indicate that this feedback consists of information used to reduce or reverse change. In our car analogy, the recognition that you are over the speed limit is negative feedback if it causes you to slow down. Conversely, the recognition that you are slowing down while going up a hill is negative feedback if it causes you to step harder on the accelerator.

Thermostats regulate temperature

To understand the features of a regulatory system, consider a thermostat—a relevant analogy, since we will be looking at the biological thermostats of vertebrate animals later in this chapter. The thermostat that is part of the heating–cooling system of a house is a regulatory system. It has upper and lower set points that you can adjust, and it receives feedback from a sensor. The circuitry of the thermostat converts any differences between the set points and the sensor information into signals that activate the controlled systems—the furnace and the air conditioner.

When room temperature rises above the upper set point, the thermostat activates the air conditioner, thus reducing room temperature below the set point. When

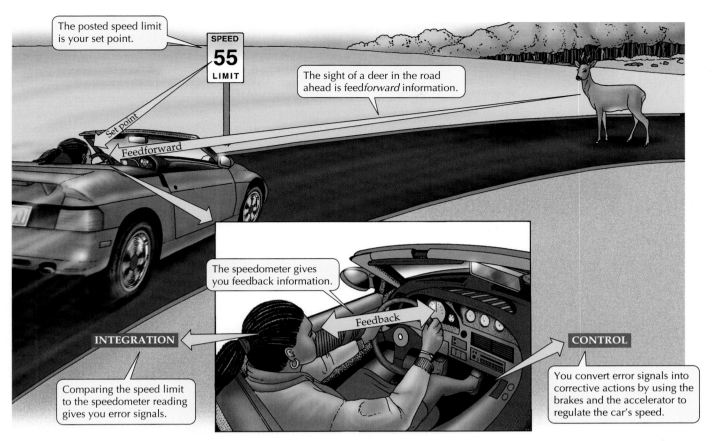

The posted speed limit is your set point.

SPEED 55 LIMIT

The sight of a deer in the road ahead is feed*forward* information.

Set point

Feedforward

The speedometer gives you feedback information.

Feedback

INTEGRATION

Comparing the speed limit to the speedometer reading gives you error signals.

CONTROL

You convert error signals into corrective actions by using the brakes and the accelerator to regulate the car's speed.

37.9 Control, Regulation, and Feedback A person driving a car obtains information and uses this information to control and regulate their actions and the car's performance.

thermostat activates the furnace, thus raising room temperature toward that set point. Hence the sensor of room temperature provides negative feedback (Figure 37.10).

Negative feedback makes good sense for physiological regulatory systems, so you may wonder if there is any such thing as positive feedback in physiology. Although not as common as negative feedback, it does exist. Rather than returning a system to a set point, **positive feedback** amplifies a response. One example is sexual behavior, in which a little stimulation can cause more behavior, which causes more stimulation, and so on. Positive feedback is not used by regulatory systems that maintain stability!

Feedforward information is another feature of regulatory systems. The function of feedforward is to change the set point. Seeing a deer ahead on the road when you are driving is an example of feedforward (see Figure 37.9); this information takes precedence over the posted speed limit, and you change your set point to a slower speed. If you want the temperature

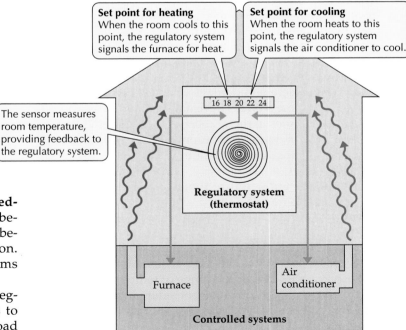

Set point for heating
When the room cools to this point, the regulatory system signals the furnace for heat.

Set point for cooling
When the room heats to this point, the regulatory system signals the air conditioner to cool.

The sensor measures room temperature, providing feedback to the regulatory system.

16 18 20 22 24

Regulatory system (thermostat)

Furnace

Air conditioner

Controlled systems

37.10 A Thermostat Regulates House Temperature
Changes in room temperature cause a sensor to move relative to set points on the thermostat and activate the furnace or the air conditioner.

point to a slower speed. If you want the temperature of your house to be lower at night than during the day, you can add a clock to the thermostat to provide feed-forward information about time of day.

These general considerations about control, regulation, and regulatory systems help organize our thinking about physiological systems, but the physiological systems can be far more complex than the thermostat and driving analogies. In some systems we do not even know the nature of the feedback. For most people, body weight is regulated, although it might not be at the level we prefer. Without consciously counting calories, the brain controls hunger so that food intake corresponds to energy expenditure. We do not understand what information the brain uses to achieve this remarkable feat of regulation, but there are some interesting hypotheses and active research on the problem.

One regulatory system that we understand well is the system that regulates body temperature. Let's begin our look at this regulatory system by asking two related questions: Why do organisms need to thermoregulate? and What is the effect of temperature on living systems?

Temperature and Life

Over the face of Earth, temperatures vary enormously—from the boiling hot springs of Yellowstone National Park to the frozen interior of Antarctica. Because heat always moves from a warmer to a cooler object, any change in the temperature of the environment causes a change in the temperature of an organism in that environment, unless the organism does something to regulate its temperature.

Living cells are restricted to a narrow range of temperatures. If cells cool to below 0°C, ice crystals can form within them and damage their structures, possibly fatally. Some animals have adaptations such as antifreeze molecules in their blood that helps them resist freezing; others have adaptations that enable them to survive freezing. Generally, however, cells must remain above 0°C to stay alive.

The upper temperature limit is around 45°C for most cells. Some specialized algae can grow in hot springs at 70°C, and some bacteria can live at near 100°C, but in general, proteins begin to denature as temperatures rise above 45°C. As proteins denature, they lose their functional properties. Most cellular functions are limited to the range of temperatures between 0 and 45°C, which are considered the thermal limits for life. In this section, we will examine temperature sensitivity within these limits.

Q_{10} is a measure of temperature sensitivity

Even within the range of 0 to 45°C, temperature changes can create problems for animals. Most physio-logical processes, like the biochemical reactions of which they are made, are temperature-sensitive, going faster at higher temperatures (see Figure 6.27). The temperature sensitivity of a reaction or process can be described in terms of Q_{10}, a quotient calculated by dividing the rate of a process or reaction at a certain temperature, R_T, by the rate of that process or reaction at a temperature 10°C lower, R_{T-10}:

$$Q_{10} = \frac{R_T}{R_{T-10}}$$

Q_{10} can be measured for a simple enzymatic reaction or for a complex physiological process, such as the rate of oxygen consumption. If a reaction or process is not temperature-sensitive, it has a Q_{10} of 1. Most biological Q_{10} values are between 2 and 3, which means that reaction rates double or triple as temperature increases by 10°C (Figure 37.11). For example, if a fish consumed 10 ml O_2/hour at 10°C and 25 ml O_2/hour at 25°C, the associated Q_{10} would be 2.5.

Changes in temperature can be particularly disruptive to an animal's functioning because all the component reactions in the animal do not have the same Q_{10}. Individual reactions with different Q_{10} values are linked together in complex networks that carry out physiological processes. Changes in temperature shift the rates of some reactions more than those of others, thus disrupting the balance and integration that the processes require. To maintain homeostasis, organisms

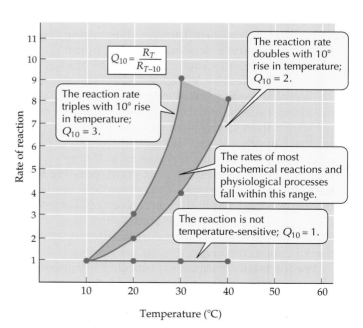

37.11 Q_{10} and Reaction Rate The larger the Q_{10}, the faster the reaction rate rises in response to an increase in temperature.

must be able to compensate for or prevent changes in temperature.

An animal's sensitivity to temperature can change

The body temperature of some animals is tightly coupled to environmental temperature. Think of a fish in a pond in a highly seasonal environment. As the temperature of the pond water changes from 4°C in midwinter to 24°C in midsummer, the body temperature of the fish does the same. If we bring such a fish into the laboratory in the summer and measure its **metabolic rate** (the sum total of the energy turnover of its cells) at different water temperatures, we might calculate a Q_{10} of 2 and plot our data as shown by the red line in Figure 37.12. We predict from our graph that in winter, when the temperature is 4°C, the fish's metabolic rate will be only one-fourth of what it was in the summer. We then return the fish to its pond.

When we bring the fish back to the laboratory in the winter and repeat the measurements, we find, as the blue line shows, that its metabolic rate at 4°C is not as low as we predicted; rather it is almost the same as it was at 24°C in the summer. If we repeat the measurement over a range of temperatures, we find that the fish's metabolic rate is always higher than the rate we predicted from the measurement we took at the same temperature in the summer fish. This difference is due to **acclimatization**, the process of physiological and biochemical change that an animal undergoes in response to seasonal changes in climate.

The reason for the difference between our prediction from the summer data and our measurements on the winter fish is that seasonal acclimatization in the fish has produced **metabolic compensation**. Metabolic compensation readjusts the biochemical machinery to counter the effects of temperature. What might account for such a change? Look again at Figure 6.26, which suggests a hypothesis: If the fish we are studying have duplicate enzymes that operate at different optimal temperatures, they may compensate metabolically by catalyzing reactions with one set of enzymes in summer and another set in winter. The end result of such readjustment is that metabolic functions are much less sensitive to seasonal changes in temperature than they are to shorter-term thermal fluctuations.

Maintaining Optimal Body Temperature

Animals can be classified by how they respond to the temperature of their environments. For example, we commonly describe animals as cold-blooded or warm-blooded. Taken at face value these terms could lead to some ridiculous errors. Desert lizards could be classified as warm-blooded during the day and cold-blooded at night, many insects might be classified as warm-blooded during flight and cold-blooded at rest,

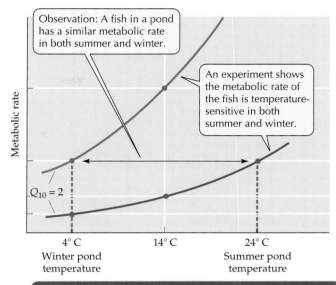

Conclusion: Between summer and winter a slow process of acclimatization compensates for seasonal temperature changes.

37.12 Metabolic Compensation   In its natural environment, a fish's metabolic rate readjusts, or acclimatizes, to compensate for seasonal changes in temperature.

and hibernating mammals would be cold-blooded during most of the winter and warm-blooded during the summer.

Biologists use a different set of terms. A **homeotherm** is an animal that regulates its body temperature at a constant level; a **poikilotherm** is an animal whose temperature changes. This system of classification says something about the biology of the animals, but it also presents problems. Should a fish in the deep ocean, where the temperature changes very little, be called a homeotherm? Should a hibernating mammal that allows its body temperature to drop to nearly the temperature of its environment be called a poikilotherm? The problem posed by the hibernator has been solved by creating a third category, the **heterotherm**: an animal that regulates its body temperature at a constant level *some* of the time. "Homeotherm," "poikilotherm," and "heterotherm" are useful descriptive terms.

Another set of terms classifies animals on the basis of the sources of heat that determine their body temperatures. **Ectotherms** depend largely on external sources of heat, such as solar radiation, to maintain their body temperature above the environmental temperature. **Endotherms** can regulate their body temperature by producing heat metabolically or by mobilizing active mechanisms of heat loss. Mammals and birds are endotherms; animals of all other species behave as ectotherms most of the time. In the discussions that follow, we will examine how animals regulate body temperature.

(a)

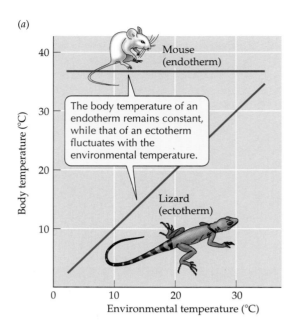

(b)

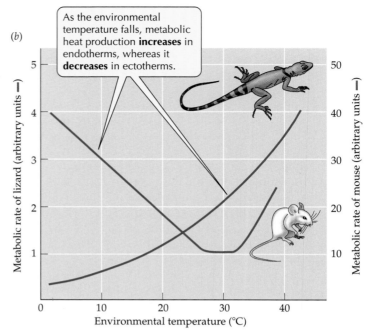

37.13 Ectotherms and Endotherms The body temperatures of a lizard and a mouse of the same body size respond differently to changes in environmental temperature.

Ectotherms and endotherms respond differently in metabolism chambers

In this discussion we'll use a small lizard to represent ectotherms and a mouse of the same body size to represent endotherms. We put each animal in a small metabolism chamber, and we measure body temperatures and metabolic rates as we change the temperature of the chamber from 0 to 35°C.

The results obtained from the two species are very different. The body temperature of the lizard always equilibrates with that of the chamber, whereas the body temperature of the mouse remains at 37°C (Figure 37.13a). The metabolic rate of the lizard increases with temperature (Figure 37.13b). Below about 27°C the metabolic rate of the mouse increases as chamber temperature decreases (notice that you must read the graph right to left to see this). The lizard apparently cannot regulate its body temperature or metabolism independently of environmental temperature. The mouse, however, regulates its body temperature by altering its rate of metabolic heat production.

Ectotherms use behavior to regulate body temperature

A logical next step would be to test in nature our laboratory conclusion that the lizard cannot regulate its body temperature. To do this we can release the lizard in its desert environment and measure its temperature as it goes about its normal behavior in an environment where temperature can change 40°C in a few hours.

Unlike the experiment in the metabolism chamber, the body temperature of the lizard in its real environ-

ment can be very different from ambient temperature. At night the temperature in the desert may drop close to freezing, but the temperature of the lizard remains stable at 16°C. This is not difficult to explain; the lizard spends the night in a burrow where soil temperature is a constant 16°C. Early in the morning, soon after sunrise, the lizard emerges from its burrow. The air temperature is still quite cool, but the body temperature of the lizard exceeds 30°C in less than 30 minutes.

The lizard achieves this rapid rise in temperature by basking on a rock with maximum exposure to the sun. As its dark skin absorbs solar radiation, its body temperature rises considerably above the surrounding air temperature. By altering its exposure to the sun, the lizard maintains its body temperature at around 35°C all morning as it seeks food, avoids predators, and interacts with potential mates or competitors.

By noon the air temperature near the surface of the desert has risen to 50°C, but the lizard's body temperature remains around 35°C. The lizard is now staying mostly in shade, frequently up in bushes where there is a cooling breeze. As afternoon progresses, the air cools, and the lizard again spends more of its time in the sun and on hot rocks to maintain its body temperature around 35°C. The lizard returns to its burrow just before sunset, and its body temperature rapidly drops to 16°C. Figure 37.14 shows the patterns of the lizard's behavior and the temperature changes over the course of a day.

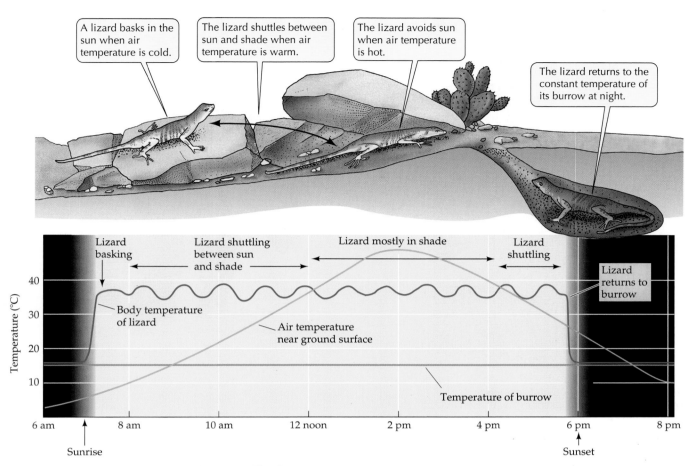

37.14 An Ectotherm Must Use Behavior to Regulate Its Body Temperature This lizard's body temperature is dependent on the ambient air temperature.

This field experiment shows that the lizard can regulate its body temperature quite well by behavioral mechanisms rather than by internal, metabolic mechanisms. The deficiency in our laboratory experiment was that the lizard in the chamber could not use its thermoregulatory behavior. If we give a lizard access to a thermal gradient in the laboratory, it is capable of regulating its body temperature by selecting the right place on the gradient. If only a hot place and a cold place are available, it will shuttle back and forth. It will maintain a different body temperature during the night than during the day, and if it is infected with pathogenic bacteria it will give itself a fever by selecting higher temperatures on the gradient.

The lesson to be learned from the discrepancies between the results of the laboratory and field experiments on the lizard is encapsulated by a quotation from the German embryologist Hans Spemann: "An experiment is like a conversation with an animal, but the animal must be permitted to answer in its own language."

Endotherms also use behavior to regulate body temperature

Behavioral thermoregulation is not the exclusive domain of ectotherms. It is also the first line of defense for endotherms. When the option is available, most animals select thermal microenvironments that are best for them. They may change their posture, orient to the sun, move between sun and shade, and move between still air and moving air, as demonstrated by the lizard in our field experiment. Examples of more complex thermoregulatory behavior are nest construction and social behavior such as huddling. In humans, the selection of clothing is important. Behavioral thermoregulation is widespread in the animal kingdom (Figure 37.15).

Both ectotherms and endotherms alter heat exchange by controlling blood flow to the skin

Just as behavioral thermoregulation is not the exclusive domain of ectotherms, physiological thermoregulation is not the exclusive domain of endotherms. Ectotherms exhibit various physiological thermoregulatory adaptations. Both ectotherms and endotherms can alter the rate of heat exchange between their bod-

(a)

(b)

37.15 Endotherms Can Use Behavior to Thermoregulate
Humans and other endotherms sometimes adjust their behavior to control their heat exchange with the environment. (*a*) Humans must put on many layers of insulating clothing to help their thermoregulatory mechanisms keep pace with the extreme cold of western Siberia. (*b*) When air temperatures on the African savanna soar, an elephant may use a cool shower to thermoregulate.

ies and their environments by controlling the flow of blood to the skin. For example, when a person's body temperature rises as a result of exercise, blood flow to the skin increases and the skin surface becomes quite warm. The extra heat brought from the body core to the skin by the blood is lost to the environment, thus tending to bring body temperature back to normal. By contrast, when a person is exposed to cold, the blood vessels supplying the skin constrict, decreasing blood flow and heat transport to the skin, thus reducing heat loss to the environment.

The control of blood flow to the skin is an important adaptation for ectotherms like the marine iguana of the Galapagos Islands. The Galapagos are volcanic islands on the equator, bathed by cold oceanic currents. Marine iguanas are reptiles that bask on black lava rocks near the ocean and swim in the sea, where they feed on submarine algae. When the iguanas cool to the temperature of the sea, they are slower and more vulnerable to predators, and probably incapable of efficient digestion. They therefore alternate between feeding in the sea and basking on the rocks. It is advantageous for iguanas to retain body heat as long as possible while swimming and to warm up as fast as

possible when basking. They adjust their cooling and heating rates by changing the flow of blood to the skin.

Blood vessels to the skin constrict when an iguana is in the ocean and dilate when it is basking. In addition, an iguana's heart rate is slower when it is swimming than when it is basking. Slowed heart rate and constricted vessels when swimming mean that less blood is being pumped through the skin, and therefore less heat is being transported from deep in the iguana's body to its skin to be lost to the water. Faster heart rate and dilated vessels when the iguana is basking increase the transport of heat from the skin to the rest of its body.

Of course, basking on black rocks under the equatorial sun can be too much of a good thing. When an iguana reaches an optimal body temperature, it lifts its body off the rocks and orients itself to minimize the surface area of its body that is directly exposed to solar radiation. Thus, the marine iguana uses both physiological and behavioral mechanisms to regulate its body temperature.

Some ectotherms produce heat

Even some ectotherms have adaptations that enable them to raise their body temperatures by producing heat. For example, the powerful flight muscles of many insects, such as dragonflies, moths, bees, and beetles, must reach a fairly high temperature (35 to 40°C) before the insects can fly, and they must maintain these high temperatures during flight, even at air temperatures around 0°C. Such insects use their flight muscles to produce the required heat. About 20 percent of the energy these muscles consume goes into useful work and 80 percent is lost as heat, which can raise body temperature (Figure 37.16). During flight the wing muscles contract alternately to move the wings up and down. During warm-up to flight temperature, however, the muscles contract simultaneously. Pulling against each other, they produce heat but do not cause the wings to beat. This behavior is similar to our shivering.

The heat-producing ability of insects can be quite remarkable. It enables moths to fly at night when air temperatures are low and solar basking is not possible. The heat-producing ability of a species of scarab beetle that lives in the mountains north of Los Angeles, California, has made it possible for these beetles to have an unusual mating behavior. The beetles spend most of their life cycle in the soil; the exception is mating, at which time females and males emerge from the soil and the males fly in search of females. What is unusual is that they engage in this behavior in winter, at night, during snowstorms. The drop in barometric pressure associated with a storm probably triggers their emergence from the soil. These beetles were long (and incorrectly) considered to be very rare because very few entomologists look for beetles in the mountains, in winter, at night, during snowstorms. Presumably the same is true for potential predators!

Honeybees regulate temperature as a group. They live in large colonies consisting mostly of female worker bees that maintain the hive and rear young that are hatched from eggs laid by the single queen bee in the colony. During winter, honeybee workers combine their individual heat-producing abilities to regulate the temperature of the brood. They cluster in the area of the hive where the brood is located and adjust their joint metabolic heat production and density of clustering so that the brood temperature remains remarkably constant, at about 34°C, even as outside air temperature drops below freezing.

Some reptiles use metabolic heat production to raise their body temperature above the air temperature. The female Indian python protects her eggs by coiling her body around them. If the air temperature falls, she contracts the muscles of her body wall to generate heat. Like the use of flight muscles by insects, this adaptation of the python is analogous to shivering in mammals. The python is able to maintain the temperature of her body—and therefore that of her eggs—considerably above air temperature.

Some fish elevate body temperature by conserving metabolic heat

It is particularly difficult for fish to raise body temperature, because blood pumped from their hearts comes into close contact with water flowing over the thin gill membranes before it travels through the body. Therefore, any heat transferred to the blood from active muscles is lost rapidly to the environment. It is thus surprising to find that some large, rapidly swimming fishes, such as bluefin tuna and great white sharks, can maintain temperature differences as great as 10 to 15°C between their body and the surrounding water (Figure 37.17). The heat comes from their powerful swimming muscles, and the ability of these "hot" fish to conserve that heat is due to remarkable arrangements of the blood vessels.

In the usual ("cold") fish circulatory system, oxygenated blood from the gills collects in a large, dorsal vessel, the aorta, which travels through the center of the fish, distributing blood to all organs and muscles (Figure 37.18a). "Hot" fish have a smaller central dorsal aorta. Most of their oxygenated blood is transported in large vessels just under the skin (Figure 37.18b). Hence the cold blood from the gills is kept close to the surface of the fish. Smaller vessels transporting this cold blood into the muscle mass run parallel to the vessels transporting warm blood from the muscle mass back toward the heart. Since the vessels carrying the cold blood into the muscle are in close contact with vessels carrying warm blood away, heat flows from the warm to the cold blood and is therefore trapped in the muscle mass.

Because heat is exchanged between blood vessels carrying blood in opposite directions, this adaptation is called a **countercurrent heat exchanger**. It keeps the heat within the muscle mass, enabling the fish to have an internal body temperature considerably above the

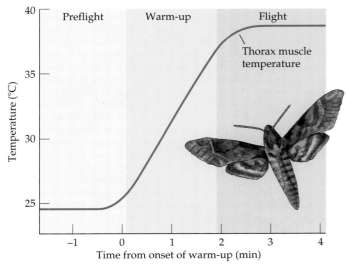

37.16 A Moth's Preflight Warm-Up Before takeoff, insects such as the sphinx moth contract the flight muscles in their thorax to generate heat and warm the muscles up to the temperature required for flight.

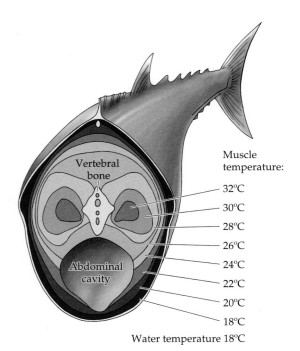

37.17 Cold Water, Warm Fish   Some actively swimming fish, such as the bluefin tuna, can conserve the heat produced by the muscles that power swimming; therefore, the fish's internal body temperature can be much higher than that of the surrounding water.

water temperature. Why is it advantageous for the fish to be warm? Each 10°C rise in muscle temperature increases the fish's sustainable power output almost threefold!

Thermoregulation in Endotherms

As we saw in Figure 37.13, endotherms can respond to changes in ambient temperature by changing their rate of heat production (which can be measured as metabolic rate). Within a narrow range of environmental temperatures called the **thermoneutral zone**, the metabolic rate of endotherms is low and independent of temperature. The metabolic rate of a resting animal at a temperature within the thermoneutral zone is called the **basal metabolic rate**. It is usually measured on animals that are quiet but awake and that are not using energy in the digestive processes or for reproduction. A resting animal consumes energy at the basal metabolic rate just to carry out all of its minimal metabolic functions other than thermoregulation.

37.18 Comparing Cold and Hot Fish   Most species are "cold" fish whose circulatory systems conduct oxygenated blood from the gills through a large dorsal aorta, bringing blood temperature into equilibrium with the water temperature. But the blood vessel anatomy of "hot" fish species such as the bluefin tuna allows for heat exchange between the cold arterial blood entering the muscles and the departing venous blood, which has been warmed by the action of the muscles.

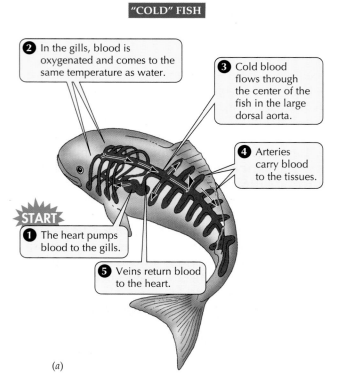

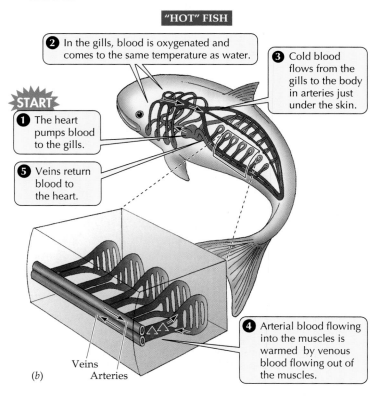

The basal metabolic rate of an endotherm is about six times greater than the metabolic rate of a similarly sized ectotherm at rest and at the same body temperature (see Figure 37.13*b*). A gram of mouse tissue consumes energy at a much higher rate than does a gram of lizard tissue when both tissues are at 37°C. This difference is due to a basic change in cell metabolism that accompanied the evolution of endotherms from their ectothermic ancestors.

The cells of endotherms are "leakier" to ions than are the cells of ectotherms. Thus endotherm cells must constantly expend considerable energy to pump potassium ions in and sodium ions out. The higher level of cell metabolism and therefore heat production by endotherms makes it easier for them to maintain a temperature difference between the body and the environment. In the sections that follow, we will discuss methods of regulating heat production and heat loss.

Endotherms actively increase heat production or heat loss

The thermoneutral zone is bounded by a lower critical temperature and an upper critical temperature (Figure 37.19). Below the lower critical temperature an endotherm's metabolic rate increases as environmental temperature declines, because the animal must produce more and more heat to maintain a constant body temperature as heat loss to the environment increases. As the environment becomes colder, eventually the animal reaches its maximum possible thermoregulatory heat production. If the environmental temperature falls still lower, the animal's body temperature will begin to drop. When the environmental temperature rises above the upper critical temperature, the animal pants or sweats. Since these active heat loss responses require an increased expenditure of energy, the metabolic rate rises above the basal level (see Figure 37.19).

Mammals use two mechanisms—shivering and nonshivering heat production—to create heat for thermoregulation. Birds use only shivering heat production. Shivering uses the contractile machinery of skeletal muscles to consume ATP without causing observable behavior. The muscles pull against each other so that little movement other than a tremor results. All the energy from the conversion of ATP to ADP in this process is released as heat.

Most nonshivering heat production occurs in specialized tissue called **brown fat**, which was discussed briefly at the beginning of Chapter 7. This tissue looks brown because of its abundant mitochondria and rich blood supply (Figure 37.20). In brown fat cells, a protein called **thermogenin** uncouples oxidative phosphorylation. Thus metabolic fuels are consumed to produce heat without producing ATP. Brown fat is especially abundant in newborn infants of many mam-

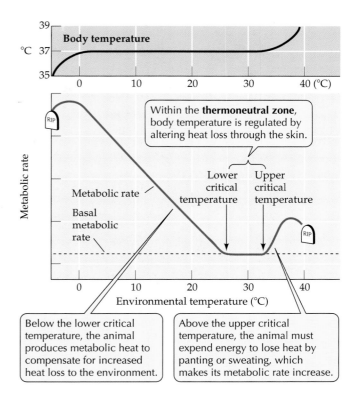

37.19 Environmental Temperature and Mammalian Metabolic Rates   Outside the thermoneutral zone, maintaining a constant body temperature requires mammals to expend energy.

malian species, in some adult mammals that are acclimatized to cold, and in mammals that hibernate.

Decreasing heat loss is important for life in the cold

The coldest habitats on Earth are in the Arctic, the Antarctic, and at the tops of high mountains. Many birds and mammals, but almost no reptiles or amphibians, live in very cold habitats. The ability to produce a substantial amount of heat metabolically has enabled endotherms to exploit formidable, frigid environments. But most tropical species of birds and mammals would not fare well in those environments. What adaptations besides endothermy characterize species that live in the cold?

The most important adaptations of endotherms to cold environments are those that reduce their heat loss to the environment. Since most heat is lost from the body surface, many cold-climate species have a smaller surface area than their warm-climate cousins, even when their body masses are the same. Rounder body shapes and shorter appendages reduce the surface area-to-volume ratios of some cold-climate species; compare, for example, the desert jackrabbit and the arctic hare (Figure 37.21).

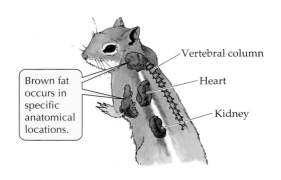

Brown fat occurs in specific anatomical locations.

Vertebral column

Heart

Kidney

37.20 Brown Fat In many mammals, specialized "brown fat" cells produce heat biochemically.

Capillary

Lipid droplet

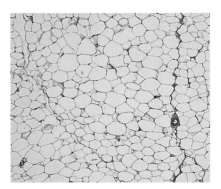

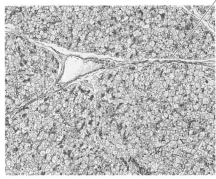

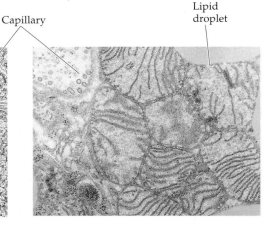

White fat viewed through a light microscope. Each cell is filled with a globule of lipid and has few organelles. The tissue has few blood vessels.

Brown fat viewed through a light microscope at the same magnification reveals cells with many intracellular structures and multiple droplets of lipid.

An electron micrograph of brown fat shows the tight packing of mitochondria in a brown fat cell.

Another means of decreasing heat loss is to increase thermal insulation. Arctic and alpine animals, and other animals adapted to cold, have much thicker layers of fur, feathers, or fat than do their warm-climate equivalents. The fur of an arctic fox or a northern sled dog provides such good thermal insulation that those animals don't even begin to shiver until the air temperature drops as low as –20 to –30°C. Fur and feathers are good insulators because they trap a layer of still, warm air close to the skin surface. If that air is displaced by water, insulation is drastically reduced. In many species, oil secretions spread through fur or feathers by grooming is critical for resisting wetting and maintaining a high level of insulation.

37.21 Adaptations to Hot and Cold Climates (*a*) The blacktail jackrabbit has a large surface area for its body mass, mainly due to its long extremities. Large ears serve as heat exchangers, passing heat from the rabbit's blood to the surrounding air. (*b*) The arctic hare has shorter extremities and therefore a smaller surface area for its body size. Its fur, longer and thicker than that of the jackrabbit, provides insulation.

(*a*)

(*b*)

Thermal insulation can be changed to control rate of heat loss

Humans change their thermal insulation by putting on or taking off clothes. How do other animals do it? We have already discussed one example, the marine iguana. By changing the blood flow to its skin, the marine iguana increases or decreases the exchange of heat between the environment and its body; in other words, the iguana changes its thermal insulation. Increasing or decreasing blood flow to the skin is an important thermoregulatory adaptation for endotherms as well. In a hot environment, your skin feels hot because of the high rate of blood flow through it, but when you are sitting in an overly air-conditioned theater, your hands, feet, and other body surfaces feel cold as blood flow to those areas decreases.

The wolf has an elegant mechanism for decreasing heat loss from its feet without the risk of freezing them. As long as the wolf's foot temperature is more than a few degrees above freezing, certain blood vessels in its foot are constricted and blood flow to the foot is minimal. As foot temperature approaches 0°C, these vessels open and allow more warm blood to flow through the foot, thus keeping it from freezing.

For highly insulated arctic animals and for many large mammals from all climates, getting rid of excess heat can be a serious problem, especially during exercise. Arctic species usually have a place on the body surface, such as the abdomen, that has only a thin layer of fur and can act as a window for heat loss. Large mammals, such as elephants, rhinoceroses, and water buffalo, have little or no fur and seek places where they can wallow in water when the air temperature is too high. Having water in contact with the skin greatly increases heat loss because water has a much greater capacity for absorbing heat than air does.

Evaporation of water is an effective way to lose heat

A gram of water absorbs about 580 calories when it evaporates. However, water is heavy, animals do not carry an excess supply of it, and hot environments tend to be arid places where water is a scarce resource. Therefore, evaporation of water by sweating or panting is usually a last resort for animals adapted to hot environments.

Sweating and panting are active processes that require the expenditure of metabolic energy. That's why the metabolic rate increases when the upper critical temperature is exceeded (see Figure 37.19). A sweating or panting animal is producing heat in the process of dissipating heat. This can be a losing battle. Animals can survive in environments that are below their lower critical temperature much better than in environments above their upper critical temperature.

The Vertebrate Thermostat

The thermoregulatory mechanisms and adaptations we have discussed are the controlled systems for the regulation of body temperature. These controlled systems must receive commands from a regulatory system that integrates information relevant to the regulation of body temperature. A convenient name for the regulatory system in this case is **thermostat**. All animals that thermoregulate, both vertebrate and invertebrate, must have regulatory systems, but here we will focus on the vertebrate thermostat.

Where is the vertebrate thermostat? The major integrative center is at the bottom of the brain in a structure called the **hypothalamus**. If you slide your tongue back as far as possible along the roof of your mouth, it will be just a few centimeters below your hypothalamus. The hypothalamus is a part of many regulatory systems, so we will refer to it many times in the chapters to come. If the hypothalamus of a mammal's brain is damaged, the animal loses its ability to regulate its body temperature, which then rises in warm environments and falls in cold ones.

In this section, we will discuss properties of vertebrate thermostats, as well as the adaptive significance of fevers in humans.

The vertebrate thermostat has set points and uses feedback

What information does the vertebrate thermostat use? In many species the temperature of the hypothalamus itself is the major source of feedback information to the thermostat. Cooling the hypothalamus causes fishes and reptiles to seek a warmer environment, and heating the hypothalamus causes them to seek a cooler environment. In mammals, cooling the hypothalamus can stimulate constriction of blood vessels to the skin and increase metabolic heat production. Because of the activation of these thermoregulatory responses, body temperature rises when the hypothalamus is cooled. Conversely, warming of the hypothalamus stimulates dilation of blood vessels to the skin and sweating or panting, and the overall body temperature falls when the hypothalamus is warmed (Figure 37.22).

The hypothalamus appears to generate a set point like a setting on a thermostat. When the temperature of the hypothalamus exceeds or drops below that set point, thermoregulatory responses (the controlled system) are activated to reverse the direction of temperature change. Hence, hypothalamic temperature is a negative feedback signal.

Heating and cooling the hypothalamus show that an animal has separate set points for activating different thermoregulatory responses. If the hypothalamus of a mammal is cooled, the vessels supplying blood to the skin constrict at a specific hypothalamic temperature. A slightly lower hypothalamic temperature initi-

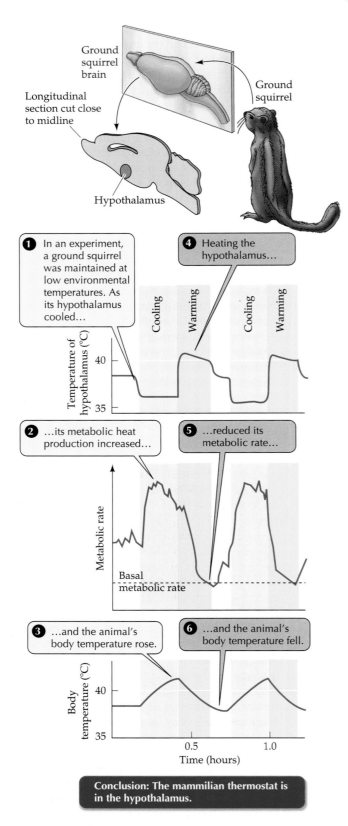

1 In an experiment, a ground squirrel was maintained at low environmental temperatures. As its hypothalamus cooled…

4 Heating the hypothalamus…

2 …its metabolic heat production increased…

5 …reduced its metabolic rate…

3 …and the animal's body temperature rose.

6 …and the animal's body temperature fell.

Cooling Warming Cooling Warming

Conclusion: The mammilian thermostat is in the hypothalamus.

37.22 The Hypothalamus Controls Metabolic Rate The brain's "thermostat," the hypothalamus, is also a part of many other vertebrate regulatory systems.

ates shivering. If the hypothalamic temperature is then raised, shivering ceases; then blood vessels to the skin dilate; and at still higher hypothalamic temperatures, panting starts.

We can describe the characteristics of hypothalamic control of each response. For example, if we measure metabolic heat production while heating and cooling the hypothalamus (see Figure 37.22), we can describe the results graphically (Figure 37.23). Within a certain range of hypothalamic temperatures, metabolic heat production remains low and constant, but cooling the hypothalamus below a certain level—a set point—stimulates increased metabolic heat production. The increase in heat production is proportional to how much the hypothalamus is cooled below the set point. This regulatory system is much more sophisticated than a simple on–off thermostat like the one in a house.

The vertebrate thermostat integrates other sources of information in addition to hypothalamic temperature. It integrates information about the temperature of the environment as registered by temperature sensors in the skin. Changes in skin temperature shift the hypothalamic set points for responses. As Figure 37.23 shows, in a warm environment you might have to cool the hypothalamus of a mammal to stimulate it to shiver, but in a cold environment you would have to

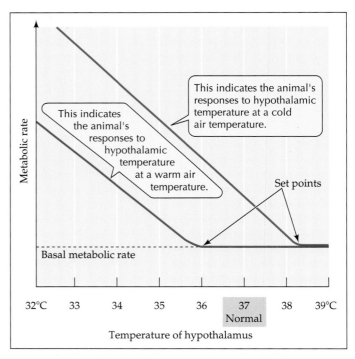

37.23 Adjustable Set Points Mammals have different set points for the metabolic heat production response to hypothalamic temperature at different environmental temperatures. Other factors, such as being asleep or awake, the time of day, or the presence of a fever, can also affect the set point.

warm the hypothalamus of the same animal to stop it from shivering. The set point for the metabolic heat production response is higher when the skin is cold and lower when the skin is warm.

Temperatue of the skin can be considered feedforward information that adjusts the hypothalamic set point. Many other factors also shift hypothalamic set points for responses. Set points are higher during wakefulness than during sleep, and they are higher during the active part of the daily cycle than during the inactive part, even if the animal is awake at both times.

Fevers help the body fight infections

You respond to many infectious illnesses by getting a fever and, in general, feeling crummy. Growing evidence suggests that getting fevers and feeling crummy are adaptive responses that help the body fight disease-causing organisms. A *fever* is a rise in body temperature in response to substances called **pyrogens** that are derived from bacteria or viruses that invade the body. Injections of the killed bacteria or even the purified cell walls of the killed bacteria can also cause fever. The presence of the pyrogen in the body causes a rise in the hypothalamic set point for the heat production response.

If you are the unlucky person developing a fever, you may feel unbearably cold (have the chills) even though your body temperature is normal. You shiver, put on more clothes, and turn up your room temperature or electric blanket. As a result, your body temperature rises until it matches the new set point. At the higher body temperature you no longer feel cold, and you may not feel hot, but someone touching your forehead will say that you are "burning up." If you take an aspirin, it lowers your set point to normal. Now you feel hot, take off clothes, and even sweat until your elevated body temperature returns to normal.

Why do we take aspirin for fevers and for feeling crummy? The foreign pyrogens entering the body are attacked by cells of the immune system called macrophages (see Chapter 18). One of the things the macrophages do when activated by pyrogens is to release chemicals called **interleukins** that sound the alarm to other cells of the immune system throughout the body and trigger responses that contribute to feeling crummy. The interleukins also raise the hypothalamic set point for metabolic heat production. Important intracellular signals triggered by interleukins are **prostaglandins**. Aspirin is a potent inhibitor of prostaglandin synthesis, thus explaining how this miracle drug reduces fever and makes us feel better. However, aspirin does not cure the disease. Is it a good idea to eliminate the symptoms without curing the disease?

Evidence suggests that moderate fevers help the body fight an infection. Some interesting studies were done on lizards that had access to a heat lamp. These animals regulated their body temperature at about 38°C by adjusting their position with respect to the lamp. After they were injected with disease-causing bacteria, they spent more time close to the lamp and raised their body temperature to 40°C and higher—they gave themselves a fever. To find out if the fever helped the lizards fight the bacteria, groups of animals were given equal innoculations of bacteria but were then placed in different incubators at 34, 36, 38, 40, and 42°C, respectively. All of the lizards at 34 and 36°C died, about 25 percent at 38°C survived, and about 75 percent at 40 and 42°C survived. Apparently fever helped the lizards fight the disease organisms.

Extreme fevers (for example, 40°C) can be dangerous to humans and must be reduced. Even more modest fevers can be dangerous to people who have weakened hearts or those who are seriously ill. A fetus can be endangered when a pregnant woman has a fever. Drugs that reduce fever may be important in such cases, but perhaps they should not be taken by most people at the first sign of aches or chills.

Animals can save energy by turning down the thermostat

In some adaptations the thermostat is turned down so that body temperature is regulated at a lower level, but not all decreases in body temperature are regulated. *Hypothermia* is the condition in which body temperature is below normal. It can result from a natural turning down of the thermostat, or from traumatic events such as starvation (lack of fuel), exposure, serious illness, or treatment by anesthesia. Because of Q_{10} effects, hypothermia slows metabolism, slows the heart, weakens muscle contractions (including those of the heart), decreases nerve conduction, and causes unconsciousness. This is not a happy state of affairs for most endotherms and can lead to death, but it can also be somewhat protective.

Some drowning victims—mostly small children drowned in cold water—who have been underwater for 10 or 15 minutes or more and have shown no pulse when pulled out of the water nevertheless have been revived by paramedics and have recovered to a remarkable extent. The rapid decrease in the body temperature of these victims slowed metabolism and thereby slowed the progress of cell damage caused by lack of oxygen. In this way, the hypothermia prevented irreversible brain damage even though it was a pathological condition induced by drowning. However, some animals induce the protective effects of hypothermia by turning down their thermostats when they anticipate unfavorable circumstances.

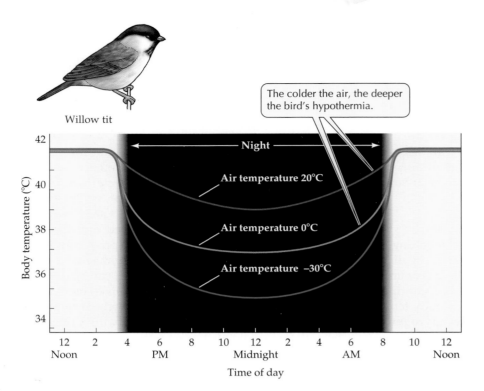

Willow tit

The colder the air, the deeper the bird's hypothermia.

Night

Air temperature 20°C

Air temperature 0°C

Air temperature –30°C

Body temperature (°C)

Time of day

37.24 Regulated Hypothermia Deepens with Cold and Dark The depth of the willow tit's hypothermia is set early in the night and must therefore be a result of information available at that time—not simply a consequence of running out of fuel reserves faster during colder nights. If nights are made shorter (which is possible in the laboratory), the birds maintain a higher body temperature at these same air temperatures.

TORPOR. Hypothermia conserves metabolic energy. Many species of birds and mammals use regulated hypothermia as a means of surviving periods of cold and food scarcity. Because of their extremely high surface area–to–volume ratios, very small endotherms such as hummingbirds and pocket mice may exhaust their metabolic reserves just getting through a single day without food if they are at normal body temperature. Animals of such species can extend the period over which they can survive without food by dropping body temperature during the portion of day they would normally be inactive. Hence, this adaptive hypothermia is called **daily torpor**. Body temperature can drop 10 to 20°C during daily torpor, resulting in an enormous saving of metabolic energy.

A small bird, the willow tit, provides an example of daily torpor that shows how well regulated this process can be. Willow tits live through the winter above the Arctic Circle. In spite of their good thermal insulation, these tiny birds must become hypothermic to survive the long, cold arctic nights. Each evening the bird lowers its metabolic rate to a level that it maintains all night, and its body temperature falls as a result of the decreased heat production.

How low the bird's metabolic rate drops is different on different nights (Figure 37.24). The decrease in its metabolism depends on air temperature, on season (and hence on length of night), and on the bird's fat reserves at roosting time. Every morning the bird has depleted its fat reserves and must immediately feed on seeds it has stored nearby. This is living on the razor's edge! The brain of this small animal integrates all the relevant information, resulting in just the right resetting of its thermostat to get it through the night.

HIBERNATION. Regulated hypothermia can last for days or even weeks, with drops to very low temperatures; this phenomenon is called **hibernation**. For the deep sleep of hibernation, the body's thermostat is turned down to an extremely low level to maximize energy conservation. Body temperature falls during hibernation because the hypothalamic set point drops, and arousal from hibernation results from a return of the hypothalamic set point to a normal level.

Many hibernators maintain body temperatures close to the freezing point during hibernation. The metabolic rate needed to sustain an animal in hibernation may be only one-fiftieth its basal metabolic rate, an enormous saving of metabolic energy. Many species of mammals hibernate, such as bats, bears, and ground squirrels; but only one species of bird, the poorwill, has been shown to hibernate.

Animals hibernate when temperatures are low and food is scarce. Individual bouts of hibernation may last for days (Figure 37.25). A bout terminates when the animal's body temperature spontaneously returns to normal. The animal may remain at its normal temperature for a few hours to a day, during which time it may eat. (Some hibernators store food in their well-insulated nests.) Then the animal enters another bout of hibernation.

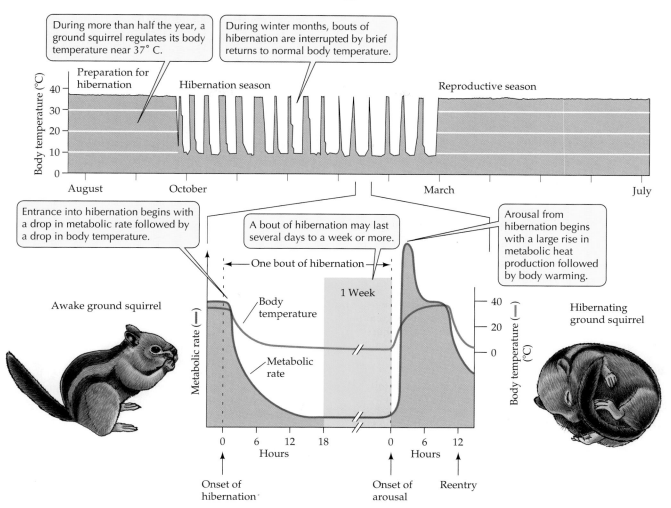

During more than half the year, a ground squirrel regulates its body temperature near 37° C.

During winter months, bouts of hibernation are interrupted by brief returns to normal body temperature.

Entrance into hibernation begins with a drop in metabolic rate followed by a drop in body temperature.

A bout of hibernation may last several days to a week or more.

Arousal from hibernation begins with a large rise in metabolic heat production followed by body warming.

Awake ground squirrel

Hibernating ground squirrel

37.25 A Ground Squirrel Enters Repeated Bouts of Hibernation during Winter   During late summer and fall the squirrel fattens in preparation for hibernation. It spends the hibernation season in a nest. At the beginning of each bout of hibernation, its metabolic rate and body temperature fall. Its body temperature may come into equilibrium with the temperature of its nest and stay at that level for many days. The bout is ended by a rise in metabolic heat production that returns body temperature to the normal level.

The hibernation season is controlled by an internal biological clock (or calendar), which continues to run with a periodicity of about a year even when animals are kept under constant conditions in the laboratory. This clock is called a **circannual rhythm** (from the Latin *circa*, "about," and *annus*, "year"). A typical circannual cycle for a hibernator such as a ground squirrel includes an active season, during which it cannot hibernate even if exposed to cold temperatures and deprived of food.

During the active season, usually spring through fall, the animals breed, raise their young, prepare their nests for winter, fatten their bodies, and store food.

During the hibernation season, as Figure 37.25 shows, animals hibernate in recurrent bouts. They progressively lose body weight even if food is available. Toward the end of the hibernation season the reproductive organs grow and become functional. The ability of hibernators to reduce the set point so dramatically probably evolved as an extension of the set point decrease that accompanies sleep even in nonhibernating species of mammals and birds.

Summary of "Physiology, Homeostasis, and Temperature Regulation"

The Needs of the One versus the Needs of the Many
• Single-celled organisms and some small, simple multicellular animals meet their needs by direct exchange between their cells and an aqueous environment, but larger, more complex animals must provide for the needs of their cells via an internal environment.

• The internal environment consists of the extracellular fluids. Organs and organ systems have specialized functions to maintain certain aspects of the composition of the internal environment constant. **Review Figure 37.1**
• Homeostasis is the maintenance of constancy in the internal environment. Homeostasis depends on the ability to control and regulate the functions of organs and organ systems.

Organs and Organ Systems

• Cells that have similar structure and function make up a tissue. Organs are made of tissues, of which there are four general types: epithelial, connective, muscle, and nervous. **Review Figure 37.2**
• Organs and organ systems can be grouped according to functions. Nervous and endocrine systems process information and control the functions of other tissues and organs. Skin, skeleton, and muscles provide protection, support, and the ability to move. Sex organs produce and nurture offspring. Digestive and gas exchange systems provide the internal environment with nutrients. Circulatory and lymphatic systems transport nutrients, wastes, hormones, heat, and blood cells. The excretory system eliminates nitrogenous wastes and maintains salt and water balance. **Review Figures 37.3, 37.4, 37.5, 37.6, 37.7, 37.8**

Control, Regulation, and Homeostasis

• Regulatory systems have set points and feedback information. Negative feedback corrects deviations from set point, positive feedback amplifies responses, and feedforward information anticipates future conditions. **Review Figure 37.9**
• The regulation of house temperature with a thermostat, furnace, and air conditioner is an example of a regulatory system. **Review Figure 37.10**

Temperature and Life

• Living systems require a range of temperatures between the freezing point of water and the temperature that denatures proteins.
• Most biological processes and reactions are temperature-sensitive. Q_{10} is a measure of temperature sensitivity. **Review Figure 37.11**
• Animals that cannot avoid seasonal changes in body temperature have biochemical adaptations that compensate for the changes. These adaptations enable animals to acclimatize to seasonal changes. **Review Figure 37.12**

Maintaining Optimal Body Temperature

• Homeotherms maintain a fairly constant body temperature most of the time; poikilotherms do not. Endotherms produce metabolic heat to elevate body temperature; ectotherms depend mostly on environmental sources of heat. **Review Figure 37.13**
• Ectotherms and endotherms can regulate body temperature through behavior and control heat exchange with the environment by altering blood flow to the skin. **Review Figure 37.14**
• Some ectotherms, such as bees, nocturnal moths, and beetles, can produce metabolic heat to raise their body temperature. **Review Figure 37.16**
• Some fish have vascular countercurrent heat exchangers to conserve heat produced by muscle metabolism. **Review Figures 37.17, 37.18**

Thermoregulation in Endotherms

• Endotherms have high basal metabolic rates. Over a range of environmental temperatures called the thermoneutral zone, the metabolic rate of resting endotherms remains at basal levels.
• When the environmental temperature falls below a lower critical temperature, endotherms elevate metabolic rate through shivering and nonshivering heat production. **Review Figures 37.19, 37.20**
• When the environmental temperature rises above an upper critical temperature, metabolic rate increases as a consequence of active evaporative water loss through sweating or panting—an effective but costly means of dissipating heat.
• Endotherms that live in cold climates have adaptations that minimize heat loss, including reduced surface area-to-volume ratio and increased insulation. **Review Figure 37.21**
• The primary adaptation of endotherms to different climates is their level of insulation.

The Vertebrate Thermostat

• The vertebrate thermostat is in the hypothalamus. It has set points for activating thermoregulatory responses. Hypothalamic temperature provides feedback information.
• Cooling the hypothalamus induces the constriction of blood vessels and increased metabolic heat production. Heating the hypothalamus induces the dilation of blood vessels and active evaporative water loss. Thermoregulatory behaviors are also induced by changes in hypothalamic temperature. **Review Figure 37.22**
• Changes in set point reflect the integration of information, such as environmental temperature and time of day, that is relevant to the regulation of body temperature. **Review Figure 37.23**
• Fever, which results from a rise in set point, helps the body fight infections.
• Adaptations in which set points are reduced to conserve energy include shallow daily torpor and deep hibernation. **Review Figures 37.24, 37.25**

Self-Quiz

1. If the Q_{10} of the metabolic rate of an animal is 2, then
   a. the animal is better acclimatized to a cold environment than if its Q_{10} is 3.
   b. the animal is an ectotherm.
   c. the animal consumes half as much oxygen per hour at 20°C as it does at 30°C.
   d. the animal's metabolic rate is not at basal levels.
   e. the animal produces twice as much heat at 20°C as it does at 30°C.

2. Which statement about brown fat is true?
   a. It produces heat without producing ATP.
   b. It insulates animals acclimatized to cold.
   c. It is a major source of heat production for birds.
   d. It is found only in hibernators.
   e. It provides fuel for muscle cells responsible for shivering.

3. What is the most important and most general difference between mammals and birds adapted to cold climates in comparison to species adapted to warm climates?
   a. Higher basal metabolic rates
   b. Higher Q_{10} values

  c. Brown fat
  d. Greater insulation
  e. Ability to hibernate

4. Which of the following would cause a *decrease* in the hypothalamic temperature set point for metabolic heat production?
  a. Entering a cold environment
  b. Taking an aspirin when you have a fever
  c. Arousing from hibernation
  d. Getting an infection that causes a fever
  e. Cooling the hypothalamus

5. Mammalian hibernation
  a. occurs when animals run out of metabolic fuel.
  b. is a regulated decrease in body temperature.
  c. is less common than hibernation in birds.
  d. can occur at any time of year.
  e. lasts for several months, during which body temperature remains close to environmental temperature.

6. Which of the following is an important difference between an ectotherm and an endotherm of similar body size?
  a. The ectotherm has higher Q_{10} values.
  b. Only the ectotherm uses behavioral thermoregulation.
  c. Only the endotherm can constrict and dilate the blood vessels to the skin to alter heat flow.
  d. Only the endotherm can have a fever.
  e. At a body temperature of 37°C, the ectotherm has a lower metabolic rate than the endotherm.

7. The function of the countercurrent heat exchanger in "hot" fish is
  a. to trap heat in the muscles.
  b. to produce heat.
  c. to heat the blood returning to the heart.
  d. to dissipate excess heat generated by powerful swimming muscles.
  e. to cool the skin.

8. What is the difference between a winter- and a summer-acclimatized fish that is termed metabolic compensation?
  a. The winter-acclimatized fish has a higher Q_{10}.
  b. The winter-acclimatized fish develops greater insulation.
  c. The winter-acclimatized fish hibernates.
  d. The summer-acclimatized fish has a countercurrent heat exchanger.
  e. The summer-acclimatized fish has a lower metabolic rate at any given water temperature.

9. Which of the following is an important characteristic of epithelial cells?
  a. They generate electric signals.
  b. They contract.
  c. They have an extensive extracellular matrix.
  d. They have secretory functions.
  e. They are found only on the surface of the body.

10. Negative feedback
  a. works in opposition to positive feedback to achieve homeostasis of a physiological variable.
  b. always turns off a process.
  c. reduces an error signal in a regulatory system.
  d. is responsible for metabolic compensation.
  e. is a feature of the thermoregulatory systems of endotherms but not of ectotherms.

Applying Concepts

1. Make a table that lists all the properties of the internal environment that you think are critical to keeping the cells of the body healthy. Next to each property list the organs or organ system responsible for maintaining it.

2. What are the major differences between ectotherms and endotherms? Compare and contrast their major thermoregulatory adaptations.

3. Why is an environment above the upper critical temperature of an endotherm more dangerous for that animal than is an environment below its lower critical temperature?

4. Why is it difficult for a fish to be endothermic? How do "hot" fish overcome these difficulties?

5. If the temperature of the hypothalamus of a mammal is the feedback information for its thermostat, why does the hypothalamic temperature scarcely change when that animal moves between environments hot enough and cold enough to stimulate the animal to pant and to shiver, respectively?

Readings

Crawshaw, L. I., B. P. Moffitt, D. E. Lemons and J. A. Downey. 1981. "The Evolutionary Development of Vertebrate Thermoregulation." *American Scientist*, vol. 69, pages 543–550. An exploration of the apparent common origin of the nervous-system mechanisms involved in the thermoregulation of all vertebrates.

Croxall, J. P. 1997. "Emperor Ecology in the Antarctic Winter." *Trends in Ecology and Evolution*, September. The remarkable story of how emperor penguins survive the coldest environment on Earth.

French, A. R. 1986. "The Patterns of Mammalian Hibernation." *American Scientist*, vol. 76, pages 569–575. A discussion of the variety of patterns of hibernation that have evolved to compensate for the difficulties that animals of different sizes have in metabolizing energy.

Heinrich, B. 1981. "The Regulation of Temperature in the Honeybee Swarm." *Scientific American*, June. A look at how honeybees thermoregulate when they leave their hive in a swarm.

Heinrich, B. 1994. "Some Like it Cold." *Natural History*, February. The remarkable abilities of some insects, such as moths, to produce and conserve heat, thus enabling them to exploit cold environments.

Madigan, M. T. and B. L. Marrs. 1997. "Extremophiles." *Scientific American*, April. A discussion of the adaptations of many microorganisms that enable them to live in extremely hot, cold, acidic, basic, or salty environments.

Schmidt-Nielsen, K. 1981. "Countercurrent Systems in Animals." *Scientific American*, May. A look at some countercurrent exchangers, which are the basis for a variety of physiological adaptations. Developed in special detail is the case of water conservation in the camel's nose.

Schmidt-Nielsen, K. 1997. *Animal Physiology: Adaptation and Environment*, 5th Edition. Cambridge University Press, New York. An excellent textbook on comparative animal physiology. The chapter "Temperature Regulation" expands on many of the topics presented in this chapter.

Seymour, R. S. 1997. "Plants That Warm Themselves." *Scientific American*, March. A look at endothermic plants.

Storey, K. B. and J. M. Storey. 1990. "Frozen and Alive." *Scientific American*, December. An explanation of the adaptations of some ectotherms (those that freeze solid in winter and survive) that cause ice to form between rather than within cells, thus protecting cell organelles from damage.

Chapter 38

Animal Hormones

Wimpy and Macho Males
In some species of cichlid fish, a dominant male displays bright colors that help attract females to his spawning pit. Most males are not dominant at any given time, but hormonal changes can transform the silver "wimp" into a "macho male."

A species of cichlid fish that lives in Lake Tanganyika in east central Africa gives biological meaning to the words "macho" and "wimpy." In shallow pools around the edge of the lake, big, brightly colored males stake out and vigorously defend territories against neighboring males. These "macho" males constantly patrol their territories and display their sexual adornments for the benefit of females who assemble in groups at the edge of the colony. The females are hard to see because they are inactive and protectively colored. When a female is ready to spawn and is impressed by a male's territory and display, she enters his territory and lays her eggs in a spawning pit that the male has prepared. The male then fertilizes the eggs.

At any one time, only about 10 percent of the males in the population are displaying and holding territories. All the other males are small and nondescript like the females, nonaggressive, and incapable of fertilizing eggs—in short, "wimpy." However, if a "macho" male is removed by a predator, wimpy males fight over the vacated territory. The winner rapidly turns into a macho male—brightly colored, big, aggressive, and able to attract females and fertilize eggs.

What accounts for this dramatic change? Soon after the wimpy male's victory, certain cells in its brain enlarge and secrete a chemical message that triggers cells in the pituitary gland, which is outside of the brain, to secrete chemical messages. Although secreted in tiny quantities, these molecules enter the blood circulation and are transported around the body. Wherever they encounter a cell that has appropriate receptors, the molecules bind to the receptors and initiate reactions in the cell.

The sum total of these cellular reactions to the chemical message converts the once wimp into a macho male. This is only one example of how chemical messages, or hormones, released in this case by a behavioral stimulus, can produce and coordinate developmental, physiological, and behavioral changes in an animal. There are many other examples, some of which we will explore in this chapter.

In Chapter 37 we discussed the importance of control and regulation in animal physiology. We learned that control and regulation require information. Information is transmitted in two forms in animals: as electric signals and as chemical signals. *Electric signals* are nerve impulses, a major focus of Chapters 41, 42, and 43. Nerve impulses are rapidly conducted over long extensions of nerve cells to specific targets. *Chemical signals* are hormones that are secreted by cells, diffuse locally in the extracellular fluid, and are distributed throughout the body by the circulation of blood.

The secretion, diffusion, and circulation of hormones is much slower than the transmission of nerve impulses. Therefore, hormonal information is not useful for controlling the rapid actions involved in cichlid fighting, but hormonal information is good for coordinating longer-term developmental processes such as the transition of wimpy male cichlids into macho males. Hormones also control many slower, longer-term physiological responses, such as the secretion of digestive enzymes by our guts as we digest a meal.

Hormones provide feedback information in many regulatory systems. For example, the male sex hormone produced in increasing quantities by the testes of the developing macho cichlid will circulate back to that fish's victory-stimulated brain cells and cause them to decrease their release of chemical message (negative feedback). Hormones are an important kind of information used by developmental and physiological systems.

In this chapter we will learn about the nature of hormones and their evolution. We will examine some of their roles in the control of invertebrate life cycles. But most of the chapter will be devoted to vertebrate hormones: their functions, control, and molecular mechanisms of action. We will pay particular attention to the extensive interactions between the systems of neural and hormonal information. In the process, we will discuss several cases of diseases involving hormonal dysfunction.

In Chapter 39 we will build on what we have learned about hormones by studying reproduction, since reproductive systems depend heavily on hormonal mechanisms for control and regulation.

Hormones and Their Actions

A **hormone** is a chemical message produced by certain cells of a multicellular organism and received by other cells. Cells that secrete hormones are called **endocrine cells** because they secrete their chemical messages directly into the internal (extracellular) environment. To receive the hormonal message, a **target cell** must have appropriate **receptors** to which the hormone can bind. The binding of a hormone to its receptor activates mechanisms within the target cell that eventually lead to a response, which may be developmental, physiological, or behavioral. In the case of the male cichlid, the release of hormones stimulates all three of these types of responses.

Hormonal signaling systems can be distinguished according to the distance over which their messages operate: Are their effects local, or are they distributed throughout the body? Some chemical messages even exert their effects on other organisms in the environment; these chemicals are called exocrine messages or *pheromones.* We will learn about pheromones in Chapter 49.

Some hormones act locally

Some chemical messages are released into the extracellular fluids in such tiny quantities, or they are so rapidly inactivated by degradative enzymes, or they are taken up so efficiently by local cells that they never enter the circulation and become distributed to distant target cells. In cases in which the receptors for the hormone are on the secreting cell itself, the hormone is called an **autocrine** message (Figure 38.1*a*). A common function of an autocrine hormone is to shut down its own production. Thus it serves as a negative feedback signal to prevent its concentration from rising too high in the extracellular environment.

When the locally acting hormone affects cells in its vicinity, it is called a **paracrine** message (see Figure 38.1*a*). An example of a paracrine message is **histamine**, one of the mediators of inflammation, a tissue response that can help protect the body from invasion by foreign organisms or materials. Histamine is released in damaged tissues by specialized cells called mast cells. When the skin is cut by a dirty object, the area around the cut becomes inflamed—red, hot, and swollen. Histamine causes this response by dilating the local blood vessels and making them more permeable, or leaky, allowing blood plasma, including protective blood proteins and white blood cells, to move into the damaged tissue.

Local responses to histamine are protective, but histamine responses that spread over large areas of the body can cause serious problems. In a person who is extremely allergic, or in a person who has a blood-

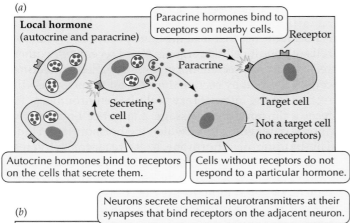

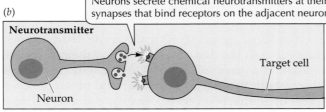

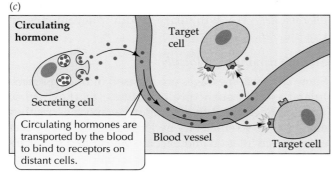

38.1 Chemical Signaling Systems   (*a*) A hormone that influences the cell that released it is called autocrine; paracrine hormones influence nearby cells. (*b*) Neurotransmitters (to be discussed in Chapter 40) can technically be classed as paracrine hormones. (*c*) The classic hormone is a secreted chemical message that is distributed throughout the body by the circulatory system.

borne infection, histamine and other mediators of inflammation may be released in such large amounts that they enter the blood and circulate around the body. The resulting expansion of blood vessels and leakage of fluid from the circulatory system can cause blood pressure to drop severely. Fluid may leak into the lungs, and the airway may become severely congested. The resulting failure of the circulatory and respiratory systems, termed **anaphylactic shock**, can be lethal.

Some highly sensitive people can go into anaphylactic shock from a single bee sting, injection of a particular antibiotic, or ingestion of a particular food. The counter measure for anaphylactic shock is injection of the hormone epinephrine (also called adrenaline), which circulates around the body and has the opposite effects of histamine. Epinephrine causes blood vessels to constrict and raises blood pressure by increasing the rate and strength of the heartbeat.

A major class of paracrine hormones consists of the various **growth factors** that stimulate the growth and differentiation of cells. Growth factors were first discovered in efforts to culture cells outside of the body. Given all sorts of nutrients and optimal conditions, cells did not grow well unless blood plasma or some tissue extract was added to the medium. The critical components were found to be specific molecules present in very small quantities that acquired the obvious name growth factors. At present, about 50 specific growth factors are known, along with a correspondingly complex group of receptors.

In the intact animal, growth factors are secreted by cells in response to damage, growth cues, or developmental cues. The secreted growth factors act in a variety of ways on cells in the vicinity. For example, nerve growth factor promotes the survival of nerve cells and stimulates growth and branching of the long extensions that are characteristic of nerve cells. Epidermal growth factor stimulates many kinds of cells to divide and is an inductive signal in developing embryos. Vascular endothelial growth factor stimulates the growth and branching of blood vessels; it is released in damaged tissues.

Technically, nerve cells can also be considered paracrine cells. As we will study in Chapter 41, a nerve cell communicates to another cell by a chemical message called a *neurotransmitter* that travels over a very small distance to the target cell (Figure 38.1*b*).

Most hormones are distributed in the blood

The classic hormone is a chemical message secreted by cells and distributed throughout the body by the circulatory system (Figure 38.1*c*). Wherever such a hormone encounters a cell with a receptor to which it can bind, it triggers a response. The nature of the response depends on the responding cell. The same hormone can cause different responses in different types of cells. For example, consider the hormone epinephrine. If a lion creeps up behind you and roars, your brain sends signals through your nervous system to epinephrine-containing cells, which immediately release epinephrine. The hormone diffuses into the blood and rapidly circulates around your body to activate your fight-or-flight response to emergency situations.

Epinephrine acts on the heart, blood vessels, liver, and fat cells. When it binds to its receptors in the heart, it causes the heart to beat faster and stronger. It activates receptors in the vessels that supply blood to your digestive tract, causing those vessels to constrict (you can digest lunch later!). Your heart is pumping more blood, and a greater percentage of that blood is going to the muscles needed for your escape. In the liver, epinephrine stimulates the breakdown of glycogen into glucose for a quick energy supply. In fatty tissue, epinephrine stimulates the breakdown of fats as another source of energy. These are some of the many actions triggered by this one hormone. They all contribute to increasing your chances of escaping the lion.

Not all cells in the body react in the same way to a hormone. Whether a cell responds to the surge of epinephrine depends on whether it has epinephrine receptors. How a cell responds to the surge of epinephrine depends on the cell's complement of receptors, second messengers, and enzymes—in other words, what type of cell it is.

Now that we have a logical classification of chemical messages, we must add the caveat that the same molecule can serve in more than one of the roles described. For example, several molecules first described as hormones in the digestive tract are now known to be neurotransmitters in the brain as well. Within the same animal and in different animals, a given molecule can play multiple roles.

Hormones don't evolve as rapidly as their functions

Chemical signaling between cells is extremely primitive; it exists in single-celled organisms. Recall the life cycle of slime molds as described in Chapter 26. These protists lead solitary lives and reproduce by mitosis and fission as long as conditions are good. But when food and moisture become scarce, they secrete a molecule called acrasin, which is really 3′,5′-cyclic adenosine monophosphate (known as cAMP). Individual slime mold cells move toward the highest concentration of acrasin and gradually aggregate into a multicellular, sluglike organism that can migrate to a new location before producing a reproductive structure that releases spores. Thus, in this protist, a chemical message passed between cells influences and coordinates their behavior, development, and physiology.

The molecule responsible for this very primitive form of chemical communication between slime mold cells—cAMP—is involved in many hormonal signaling systems in multicellular animals. As we will see later in this chapter, when many hormones bind with their receptors on the cell surface, they cause the production of cAMP within the cell. This "second messenger" then mediates a variety of responses within the cell via the phosphorylation of enzymes.

With the evolution of increased complexity of multicellular animals, more and more molecules acquired signaling functions. Yet, as animals evolved, those functions changed as the physiology of the animals changed. The evolution of hormonal systems has been efficient. Rather than adopt a new molecule for any new process that arises, available molecules have been co-opted for that purpose. As a result, we now find the same chemical substances used as hormones in widely divergent species, but they have completely different actions.

Many vertebrate hormones have molecular structures similar or identical to those of invertebrate hormones, but their functions are different. The hormone thyroxine, for example, is found in animal species ranging from mollusks and tunicates (sea squirts) to humans. Its function is unknown in invertebrates, but it is produced in an organ that is involved in feeding. In frogs, thyroxine is essential for the metamorphosis from tadpole to adult. In mammals, thyroxine elevates cellular metabolism.

Another example of evolutionary changes in hormone function is provided by the hormone prolactin, which stimulates milk production in female mammals after they give birth. In pigeons and doves, prolactin stimulates the production of crop milk for nourishment of young. Crop milk is really not a milk secretion at all, but a sloughing off of cells lining the upper digestive tract. In amphibians, prolactin causes the animals to prepare for reproduction by seeking water; and in fishes, such as salmon, that migrate between salt water and fresh water to breed, prolactin regulates the mechanisms that maintain osmotic balance with the changing environment.

In all of these cases prolactin is involved in reproductive processes, but as those processes have changed through evolution, so has the information signaled by the hormone. In summary, the molecules involved in chemical signaling have changed little through evolution, but their functions have changed dramatically.

Endocrine glands secrete hormones

Some endocrine cells are distributed singly within a tissue. For example, many hormones of the digestive tract are produced and secreted by isolated cells in the lining of the tract. As contents of the digestive tract come into contact with these cells, the cells release their hormones, which enter the blood and, like epinephrine, circulate throughout the body and activate cells that have appropriate receptors. Many hormones, however, are secreted by aggregations of endocrine cells that form secretory organs called **glands**.

Animals have two types of glands. Some, such as sweat glands and salivary glands, release secretions that are not hormones through ducts that lead outside the body. Sweat gland ducts open onto the surface of the skin, salivary gland ducts open into the mouth, and the duct from the pancreas carries digestive enzymes into the digestive tract. Such glands are called **exocrine glands** because they secrete their products to the outside of the body (the Greek *exo-* means "outside of").

Glands that secrete hormones do not have ducts; they are called **endocrine glands** because they secrete

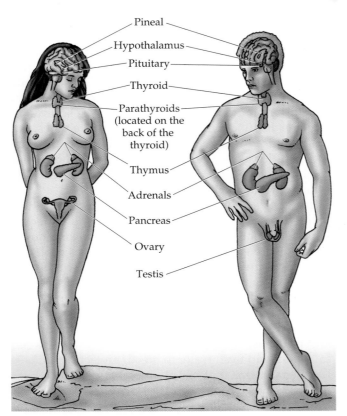

Pineal
Hypothalamus
Pituitary
Thyroid
Parathyroids
(located on the
back of the
thyroid)
Thymus
Adrenals
Pancreas
Ovary
Testis

38.2 The Endocrine Glands of Humans

their products into extracellular fluid, from which the hormones enter the blood, which is inside the body. Cells of most endocrine glands synthesize hormones and store them until they are stimulated to secrete their signals. Collectively, endocrine glands make up the endocrine system (Figure 38.2).

Hormonal Control of Molting and Development in Insects

The hormones of invertebrate animals have been well studied, and many of them have a plethora of functions. In this chapter we cannot do justice to the diversity of hormones in invertebrates, but we'll discuss two important aspects of the lives of many invertebrates that are controlled by hormonal mechanisms: molt and metamorphosis.

Hormones from the head control molting in insects

The British physiologist Sir Vincent Wigglesworth was a pioneer in the study of the hormonal control of growth and development in insects. He conducted experiments on the blood-sucking bug *Rhodnius*. Upon hatching from the egg, *Rhodnius* is nearly a miniature version of the adult, but it lacks some adult features. Because insects have rigid exoskeletons, their growth is episodic, punctuated with molts (shedding) of the

exoskeleton. Each growth stage between two molts is called an *instar*.

A blood meal triggers each episode of molting and growth in *Rhodnius*, which molts five times before developing into a complete adult. *Rhodnius* is a hardy experimental animal; it can live a long time even after it is decapitated. If decapitated about an hour after it has a blood meal, *Rhodnius* may live up to a year, but it does not molt. If decapitated a week after its blood meal, it will molt (Figure 38.3). These observations led to the hypothesis that something diffusing slowly from the region of the head controls the molt.

The proof that one or more diffusing substances cause the molt came from a clever experiment in which Wigglesworth decapitated two *Rhodnius*: one that had just had its blood meal and another that had had its blood meal 1 week earlier. The two decapitated bodies were connected with a short piece of glass tubing—and they both molted (Figure 36.3b). Thus one or more substances from the bug fed a week earlier crossed through the glass tube and stimulated molting in the other bug.

We now know that two hormones regulate molting. Cells in the brain produce a substance, called simply **brain hormone**, that is transported to a pair of neuroendocrine structures attached to the brain called the **corpora cardiaca** (singular corpus cardiacum). After appropriate stimulation—which for *Rhodnius* is a blood meal—the corpora cardiaca release brain hormone, which diffuses to an endocrine gland, the **prothoracic gland**. The prothoracic gland then produces and releases a hormone called **ecdysone**, which directly stimulates molting.

The control of molting by brain hormone and ecdysone is a general mechanism in insects. The nervous system receives various types of information that may be relevant in determining the optimal timing for growth and development. It makes sense, therefore, that the nervous system should control the endocrine glands that produce hormones that orchestrate all the physiological processes involved in development and molting. Later in this chapter we will see similar links between the nervous system and certain endocrine glands in vertebrates.

Juvenile hormone controls development in insects

The *Rhodnius* decapitation experiments yielded a curious result: Regardless of the instar used, the decapitated bug always molted directly into an adult form. Additional experiments by Wigglesworth demonstrated that a hormone other than those responsible for molting, and located in the rear of the head, determines whether, under natural conditions, a bug molts into another juvenile instar or into an adult.

Because the head of *Rhodnius* is long, it was possible to remove just the front part of the head, which

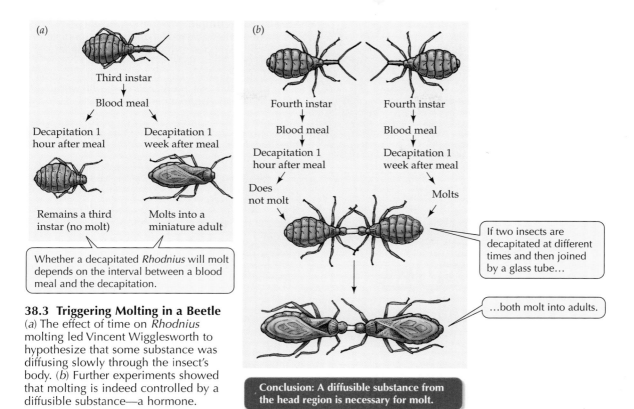

38.3 Triggering Molting in a Beetle
(a) The effect of time on *Rhodnius* molting led Vincent Wigglesworth to hypothesize that some substance was diffusing slowly through the insect's body. (b) Further experiments showed that molting is indeed controlled by a diffusible substance—a hormone.

contains the neuroendocrine cells that secrete and release brain hormone, while leaving intact the rear part, which contains two other endocrine structures called the **corpora allata** (singular corpus allatum). When fourth-instar bugs were partly decapitated, leaving the corpora allata intact, they molted into fifth instars and not into adults.

This experiment, indicating a role of the posterior part of the head in maintaining juvenile status, was followed up by more experiments with glass tubes. When an unfed, completely decapitated, fifth-instar bug was connected to a fourth-instar bug that had been fed but had only the front part of its head removed, both bugs molted into juvenile forms. A substance coming from the rear part of the head of the fourth-instar bug prevented the expected result that both bugs would molt into adult form.

We now know that the substance is **juvenile hormone** and that it comes from the corpora allata. As long as juvenile hormone is present, *Rhodnius* molts into another juvenile instar. The corpora allata normally stop producing juvenile hormone during the fifth instar. If juvenile hormone is absent, the bug molts into the adult form.

The control of development by juvenile hormone is more complex in most insects that, like butterflies, undergo **complete metamorphosis**. These animals undergo dramatic developmental differences between in-

stars. The fertilized egg hatches into a *larva*, which feeds and molts several times, becoming bigger and bigger. Then it enters an inactive stage called a *pupa*. It undergoes major body reorganization as a pupa, and finally emerges as an *adult. Rhodnius*, on the other hand, undergoes incomplete metamorphosis, a gradual progression from hatchling to adult that has no inactive stage.

An excellent example of complete metamorphosis is provided by the silkworm moth, *Hyalophora cecropia* (Figure 38.4). As long as juvenile hormone is present in high concentrations, larvae molt into larvae. When the level of juvenile hormone falls, larvae molt into pupae. Because no juvenile hormone is produced in pupae, they molt into adults.

In our perpetual war against insects, juvenile hormone is a new weapon. Synthetic forms of juvenile hormone can be distributed in the environment to prevent the development of juvenile insects into adults capable of reproduction.

The existence and function of insect hormones was experimentally demonstrated many years before the hormones were identified chemically. That is not surprising when you consider the tiny amounts of certain hormones that exist in an organism. In one of the earliest studies of ecdysone, biochemists produced only 250 mg of pure ecdysone (about one-fourth the weight of an apple seed) from 4 tons of silkworms!

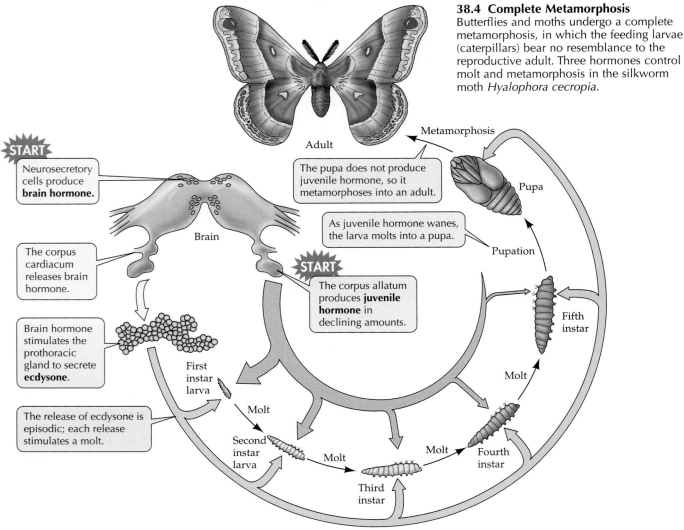

38.4 Complete Metamorphosis
Butterflies and moths undergo a complete metamorphosis, in which the feeding larvae (caterpillars) bear no resemblance to the reproductive adult. Three hormones control molt and metamorphosis in the silkworm moth *Hyalophora cecropia*.

Vertebrate Endocrine Systems

The list of chemical messages in the bodies of vertebrates is long and growing longer. To make the subject manageable, in this section we will focus mostly on the hormones of humans—how they function and how they are controlled. Table 38.1 presents an overview of the hormones of humans (most of which are found in all other mammals as well). Notice that the column listing the target tissues of these hormones includes every organ system of the body. We'll begin this survey with the pituitary gland because it produces several hormones, some of which control other endocrine tissues.

The pituitary develops from outpocketings of mouth and brain

The **pituitary gland** sits in a depression at the bottom of the skull just over the back of the roof of the mouth (Figure 38.5). It is attached to the part of the brain called the hypothalamus, which is involved in many homeostatic regulatory systems, including endocrine systems. The distinct anterior and posterior divisions of the pituitary have separate origins during development.

The **anterior pituitary** originates as an outpocketing of the embryonic mouth cavity, and the **posterior pituitary** originates as an outpocketing of the developing brain in the region that becomes the hypothalamus.

THE POSTERIOR PITUITARY. The posterior pituitary releases two neurohormones: vasopressin and oxytocin. These are small peptides synthesized in nerve cells in the hypothalamus. Vasopressin and oxytocin move down long extensions of these nerve cells that stretch down the pituitary stalk into the posterior pituitary, where the hormones are stored in the nerve endings (see Figure 38.5).

The posterior pituitary increases its release of **vasopressin** whenever blood pressure falls or the blood becomes too salty. The main action of vasopressin is to increase the amount of water conserved by the kid-

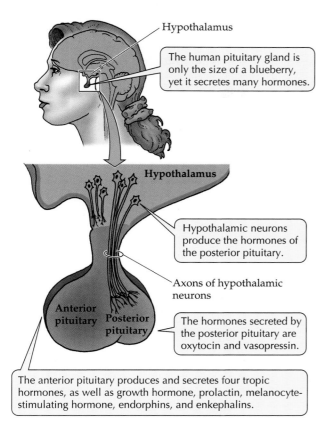

Hypothalamus

The human pituitary gland is only the size of a blueberry, yet it secretes many hormones.

Hypothalamus

Hypothalamic neurons produce the hormones of the posterior pituitary.

Axons of hypothalamic neurons

Anterior pituitary Posterior pituitary

The hormones secreted by the posterior pituitary are oxytocin and vasopressin.

The anterior pituitary produces and secretes four tropic hormones, as well as growth hormone, prolactin, melanocyte-stimulating hormone, endorphins, and enkephalins.

38.5 The Pituitary Is Two Glands in One The two hormones stored and released by the posterior pituitary are neurohormones produced in the hypothalamus. The anterior pituitary produces many hormones; the four tropic hormones (thyrotropin, adrenocorticotropin, luteinizing hormone, and follicle-stimulating hormone) act on other endocrine glands of the body.

neys; hence it is also called **antidiuretic hormone** or **ADH**. When vasopressin secretion is high, the kidneys resorb more water and produce only a small volume of highly concentrated urine. When vasopressin secretion is low, the kidneys produce a large volume of dilute urine. We will discuss the mechanism of vasopressin action in Chapter 48.

When a woman is about to give birth, her posterior pituitary releases **oxytocin**, which stimulates the contractions of the muscles that push the baby out of her body. Oxytocin also brings about the flow of milk from the mother's breasts. The baby's suckling stimulates nerve cells in the mother, causing the secretion of oxytocin. Even the sight and sounds of her baby can cause a nursing mother to secrete oxytocin and release milk from her breasts.

THE ANTERIOR PITUITARY. Four of the anterior pituitary hormones (*thyrotropin, adrenocorticotropin, luteinizing hormone,* and *follicle-stimulating hormone*) control the

activities of other endocrine glands and thus are called **tropic hormones** (see Figure 38.5). Each tropic hormone is produced by a different type of pituitary cell. We will say more about these hormones when we describe their target glands (thyroid, adrenal cortex, testes, and ovaries) later in this chapter and in the next. The other hormones produced by the anterior pituitary influence tissues that are not endocrine glands. These hormones are growth hormone, prolactin, melanocyte-stimulating hormone, endorphins, and enkephalins.

Growth hormone consists of about 200 amino acids and acts on a wide variety of tissues to promote growth directly and indirectly. One of its important direct effects is to stimulate cells to take up amino acids. Growth hormone promotes growth indirectly by stimulating the liver to produce growth-regulating chemical messages called **somatomedins**, which circulate in the blood and stimulate the growth of bone and cartilage.

Overproduction of growth hormone in children causes gigantism, and underproduction causes dwarfism (Figure 38.6). In adults, high levels of growth hormone cannot cause increased height, because the shafts and the growth plates of the long bones have fused. Rather, abnormally high levels of growth hormone in adults cause thickening of the hands, feet, jaw, nose, and ears—a condition known as acromegaly.

Beginning in the late 1950s, children diagnosed as having a serious deficiency of growth hormone, and therefore destined to become dwarfs, were treated with human growth hormone extracted from human pituitaries in cadavers. The treatment was successful in stimulating substantial growth, but it was extremely costly: A year's supply of human growth hormone for one individual required up to 50 pituitaries. In the mid-1980s, scientists using genetic-engineering technology isolated the gene for human growth hormone and introduced it into bacteria, which produced enough of the hormone to make it commercially available.

Preventing pituitary dwarfism is now feasible and affordable, but the availability of growth hormone raises new questions. Should every child at the lower end of the height charts be treated? Should a normal child whose parents want him or her to play basketball be given growth hormone? These types of questions are impossible to answer with scientific data alone. The controversy around growth hormone has become even more complex because of recent suggestions that growth hormone administered to older persons may reverse some of the effects of aging.

Earlier in the chapter we described the evolutionary diversity of the functions of **prolactin**, another hormone produced by the anterior pituitary. In human females the major function of prolactin is to stimulate

TABLE 38.1   Principal Hormones of Humans

SECRETING TISSUE OR GLAND	HORMONE	CHEMICAL NATURE	TARGET(S)	IMPORTANT PROPERTIES OR ACTIONS
Hypothalamus	Releasing and release-inhibiting hormones (see Table 38.2)	Peptides	Anterior pituitary	Control secretion of hormones of anterior pituitary
	Oxytocin, vasopressin	Peptides	(See Posterior pituitary)	Stored and released by posterior pituitary
Anterior pituitary: Tropic hormones	Thyrotropin	Glycoprotein	Thyroid gland	Stimulates synthesis and secretion of thyroxine
	Adrenocorticotropin	Polypeptide	Adrenal cortex	Stimulates release of hormones from adrenal cortex
	Luteinizing hormone	Glycoprotein	Gonads	Stimulates secretion of sex hormones from ovaries and testes
	Follicle-stimulating hormone	Glycoprotein	Gonads	Stimulates growth and maturation of eggs in females; stimulates sperm production in males
Anterior pituitary: Other hormones	Growth hormone	Protein	Bones, liver, muscles	Stimulates protein synthesis and growth
	Prolactin	Protein	Mammary glands	Stimulates milk production
	Melanocyte-stimulating hormone	Peptide	Melanocytes	Controls skin pigmentation
	Endorphins and enkephalins	Peptides	Spinal cord neurons	Decrease painful sensations
Posterior pituitary	Oxytocin	Peptide	Uterus, breasts	Induces birth by stimulating labor contractions; causes milk flow
	Vasopressin (antidiuretic hormone)	Peptide	Kidneys	Stimulates water reabsorption
Thyroid	Thyroxine	Iodinated amino acid derivative	Many tissues	Stimulates and maintains metabolism necessary for normal development and growth
	Calcitonin	Peptide	Bones	Stimulates bone formation; lowers blood calcium
Parathyroids	Parathormone	Protein	Bones	Absorbs bone; raises blood calcium
Thymus	Thymosins	Peptides	Immune system	Activate immune responses of T cells in the lymphatic system
Pancreas	Insulin	Protein	Muscles, liver, fat, other tissues	Stimulates uptake and metabolism of glucose; increases conversion of glucose to glycogen and fat
	Glucagon	Protein	Liver	Stimulates breakdown of glycogen and raises blood sugar
	Somatostatin	Peptide	Digestive tract; other cells of the pancreas	Inhibits insulin and glucagon release; decreases secretion, motility, and absorption in the digestive tract

the production and secretion of milk. In some mammals prolactin also functions as an important hormone during pregnancy. In human males prolactin plays a role along with other pituitary hormones in controlling the endocrine function of the testes.

Melanocyte-stimulating hormone is produced in very low amounts in the human anterior pituitary, and its functions in humans are not well understood. Melanocytes are cells that contain melanin, a black pigment. In many fishes, amphibians, and reptiles that can change their color, melanocyte-stimulating hormone changes the way melanin is distributed in the melanocytes, thereby darkening or lightening the tissue containing them.

Endorphins and **enkephalins** are referred to as the body's "natural opiates." In the brain these molecules act as neurotransmitters in pathways that control pain. The significance of their release from the anterior pituary is unknown. Interestingly, the production of endorphins and enkephalins in the pituitary is encoded

TABLE 38.1 Principal Hormones of Humans (continued)

SECRETING TISSUE OR GLAND	HORMONE	CHEMICAL NATURE	TARGET(S)	IMPORTANT PROPERTIES OR ACTIONS
Adrenal medulla	Epinephrine, norepinephrine	Modified amino acids	Heart, blood vessels, liver, fat cells	Stimulate fight-or-flight reactions: increase heart rate, redistribute blood to muscles, raise blood sugar
Adrenal cortex	Glucocorticoids (cortisol)	Steroids	Muscles, immune system, other tissues	Mediate response to stress; reduce metabolism of glucose, increase metabolism of proteins and fats; reduce inflammation and immune responses
	Mineralocorticoids (aldosterone)	Steroids	Kidneys	Stimulate excretion of potassium ions and reabsorption of sodium ions
Stomach lining	Gastrin	Peptide	Stomach	Promotes digestion of food by stimulating release of digestive juices; stimulates stomach movements that mix food and digestive juices
Lining of small intestine	Secretin	Peptide	Pancreas	Stimulate secretion of bicarbonate solution by ducts of pancreas
	Cholecystokinin	Peptide	Pancreas, liver, gallbladder	Stimulates secretion of digestive enzymes by pancreas and other digestive juices from liver; stimulates contractions of gallbladder and ducts
	Enterogastrone	Polypeptide	Stomach	Inhibits digestive activities in the stomach
Pineal	Melatonin	Modified amino acid	Hypothalamus	Involved in biological rhythms
Ovaries	Estrogens	Steroids	Breasts, uterus, other tissues	Stimulate development and maintenance of female characteristics and sexual behavior
	Progesterone	Steroid	Uterus	Sustains pregnancy; helps maintain secondary female sexual characteristics
Testes	Androgens	Steroids	Various tissues	Stimulate development and maintenance of male sexual behavior and secondary male sexual characteristics; stimulate sperm production
Most cells	Prostaglandins	Modified fatty acids	Various tissues	Have many diverse actions
Heart	Atrial natriuretic hormone	Peptide	Kidneys	Increases sodium ion excretion

by the same gene that codes at least two other pituitary hormones. The gene really codes for a large parent molecule called pro-opiomelanocortin. This large protein molecule is cleaved to produce several peptides, some of which have hormonal functions. Adrenocorticotropin, melanocyte-stimulating hormone, endorphins, and enkephalins all result from the cleavage of the protein pro-opiomelanocortin.

The anterior pituitary is controlled by hypothalamic neurohormones

Because it produces tropic hormones that control other endocrine glands, the anterior pituitary acquired the designation "master gland" in the early days of endocrinology. But the idea of the anterior pituitary as the master gland received quite a blow when it was discovered instead to be a "middleman" controlled by the hypothalamus.

The hypothalamus receives information about conditions in the body and in the external environment through the nervous system. If the connection between the hypothalamus and the pituitary is cut, pituitary hormones are no longer released in response to changes in the environment or in the body. If pituitary cells are maintained in culture, extracts of hypothalamic tissue stimulate some of those cells to release their hormones into the culture medium. Therefore, scientists hypothesized that secretions of the hypothalamic cells control the activities of anterior-pituitary cells.

Although hypothalamic neurons do not extend into the anterior pituitary as they do into the posterior pituitary, a special set of blood vessels called **portal blood vessels** connect the hypothalamus and the anterior pituitary (Figure 38.7). It was thus proposed that secretions from nerve endings in the hypothalamus enter the blood and are conducted down the portal

(a)

(b)

38.6 Effects of Abnormal Amounts of Growth Hormone
(a) Overproduction of growth hormone in childhood causes gigantism. This photo from 1939 shows a young man who is more than 8 feet tall standing next to his father, who is just under 6 feet tall. (b) Underproduction of growth hormone during childhood results in pituitary dwarfism. The man on the left is P. T. Barnum, the circus entrepreneur. The man on the right is Charles Straton, a dwarf, who appeared in Barnum's circus under the name General Tom Thumb.

vessels to the anterior pituitary, where they cause the release of anterior pituitary hormones.

In the 1960s two large teams of scientists, led by Roger Guillemin and Andrew Schally, initiated the search for the hypothalamic releasing neurohormones. Because the amounts of such hormones in any individual mammal would be tiny, "bucket biochemistry" was required. The scientists set up teams in slaughterhouses to collect massive numbers of hypothalami from pigs and sheep. The resulting tons of tissue was shipped to laboratories in refrigerated trucks.

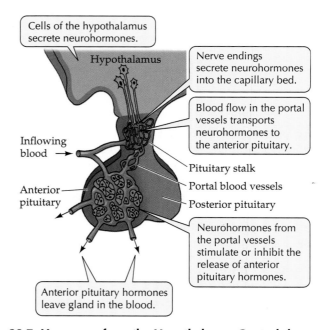

Cells of the hypothalamus secrete neurohormones.

Hypothalamus

Nerve endings secrete neurohormones into the capillary bed.

Blood flow in the portal vessels transports neurohormones to the anterior pituitary.

Inflowing blood →

Pituitary stalk

Portal blood vessels

Anterior pituitary

Posterior pituitary

Neurohormones from the portal vessels stimulate or inhibit the release of anterior pituitary hormones.

Anterior pituitary hormones leave gland in the blood.

38.7 Hormones from the Hypothalamus Control the Anterior Pituitary Neurohormones produced in tiny quanties by cells in the hypothalamus are transported to the anterior pituitary in a system of delicate portal blood vessels. These releasing and release-inhibiting hormones control the activities of anterior pituitary cells.

TABLE 38.2 Releasing and Release-Inhibiting Neurohormones of the Hypothalamus

NEUROHORMONE	ACTION
Thyrotropin-releasing hormone (TRH)	Stimulates thyrotropin release
Gonadotropin-releasing hormone	Stimulates release of follicle-stimulating hormone and luteinizing hormone
Prolactin release-inhibiting hormone	Inhibits prolactin release
Prolactin-releasing hormone	Stimulates prolactin release
Somatostatin (growth hormone release-inhibiting hormone)	Inhibits growth hormone release; interferes with thyrotropin release
Growth hormone–releasing hormone	Stimulates growth hormone release
Adrenocorticotropin-releasing hormone	Stimulates adrenocorticotropin release
Melanocyte-stimulating hormone release-inhibiting hormone	Inhibits release of melanocyte-stimulating hormone

One extraction effort began with the hypothalami from 270,000 sheep and yielded only 1 mg of purified **thyrotropin-releasing hormone**, or **TRH**, which was the first hypothalamic releasing (that is, release-stimulating) hormone isolated and characterized. Biochemical analysis of this pure sample revealed that TRH contains only three amino acids; it is a tripeptide. TRH causes certain anterior pituitary cells to release the tropic hormone thyrotropin, which in turn stimulates the activity of the thyroid gland.

Soon after discovering thyrotropin-releasing hormone, Guillemin and Schally's teams identified **gonadotropin-releasing hormone**, which stimulates certain anterior pituitary cells to release the tropic hormones that control the activity of the gonads, the ovaries and the testes. For these discoveries Guillemin and Schally received the 1972 Nobel prize in medicine. Many more hypothalamic neurohormones, including releasing hormones and release-inhibiting hormones, are now known (Table 38.2).

The thyroid hormone thyroxine controls cell energy metabolism

The **thyroid gland** consists of two lobes, one on either side of the trachea (windpipe), connected by a strip of thyroid tissue that wraps around the front of the trachea. The thyroid gland produces the hormones thyroxine and calcitonin.

Thyroxine is synthesized in thyroid cells from two molecules of diiodotyrosine, which is the amino acid tyrosine with two atoms of iodine chemically bonded to it. Thus, a thyroxine molecule has four atoms of iodine and is called T_4:

Thyroxine (T_4)

Thyroid cells can also combine a molecule of diiodotyrosine with a molecule of monoiodotyrosine to make triiodothyronine, or T_3:

Triiodothyronine (T_3)

The thyroid usually makes and releases about four times as much T_4 as T_3. T_3 is the more active hormone in cells of the body, but when T_4 is in circulation it can be converted to T_3 by an enzyme. Therefore, when you read about thyroxine, keep in mind that the actions discussed are primarily due to T_3.

Thyroxine in mammals plays many roles in regulating cell metabolism. It elevates the metabolic rates of most cells and tissues and promotes the use of carbohydrates rather than fats for fuel. Exposure to cold for several days leads to an increased release of thyroxine, an increased conversion of T_4 to T_3, and an increase in basal metabolic rate. Thyroxine is especially crucial during development and growth. It promotes amino acid uptake and protein synthesis by cells. Insufficient thyroxine in a human fetus or growing child greatly retards physical and mental growth, resulting in a condition known as cretinism.

The tropic hormone **thyrotropin**, which is secreted into the blood by the anterior pituitary, determines the activity of the thyroid gland. Thyrotropin activates the thyroid gland cells that produce thyroxine. TRH (thyrotropin-releasing hormone) produced in the hypothalamus and transported to the pituitary through the portal blood vessels activates thyrotropin-producing pituitary cells. The brain uses environmental information such as temperature or day length to determine whether to increase or decrease the secretion of TRH. There is a very important negative feedback loop in

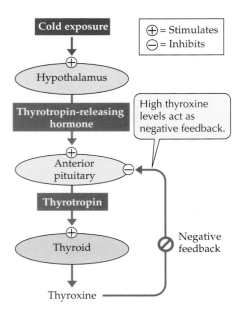

38.8 Regulation of Thyroid Function Exposure to cold stimulates the hypothalamus to produce thyrotropin-releasing hormone (TRH), which stimulates the anterior pituitary to release thyrotropin, which in turn stimulates the thyroid to release thyroxine. High thyroxine levels act as negative feedback to inhibit the release of thyrotropin from the anterior pituitary.

this sequence of steps: Circulating thyroxine inhibits the response of the pituitary cells to TRH. Less thyrotropin is released when thyroxine levels are high, and more thyrotropin is released when thyroxine levels are low (Figure 38.8).

Thyroid dysfunction causes goiter

A goiter is an enlarged thyroid gland causing a pronounced bulge on the front of the neck and extending around the sides of the neck. Goiter can be associated with either **hyperthyroidism** (very high levels of thyroxine) or **hypothyroidism** (very low levels of thyroxine). The control diagram in Figure 38.8 helps explain how two very different conditions can result in the same symptom. Hyperthyroid goiter results when the negative feedback mechanism fails to turn off the activity of thyroid cells even though blood levels of thyroxine are high.

The most common cause of hyperthyroidism is an autoimmune disease (see Chapter 18) in which an antibody to the thyrotropin receptor is produced. Such an antibody can bind to the receptor and activate the cells to produce and release thyroxine. The blood levels of thyrotropin may be quite low because of the negative feedback from thyroxine, but the thyroid remains maximally stimulated and it grows bigger. Another cause of hyperthyroidism is a pituitary tumor that produces thyrotropin or a hypothalamic tumor

that produces TRH. In either case the thyroid is maximally stimulated and grows. Hyperthyroid patients have high metabolic rates, are jumpy and nervous, usually feel hot, and may have a buildup of fat behind the eyeballs, causing their eyes to bulge.

Hypothyroid goiter results when there is not enough circulating thyroxine to turn off thyrotropin production. Its most common cause is a deficiency of dietary iodide, without which the thyroid gland cannot make thyroxine. Without thyroxine, thyrotropin levels remain high, so the thyroid continues to produce large amounts of nonfunctional thyroxine and becomes very large. The symptoms of hypothyroidism are low metabolism, intolerance of cold, and general physical and mental sluggishness.

Hypothyroid goiter used to be extremely common in mountainous areas and regions far from the oceans, where there is little iodide in the soil or water. Worldwide, goiter affects about 5 percent of the population. The addition of iodide to table salt has greatly reduced the incidence of the condition in industrialized nations, but goiter is still common in the less developed countries of the world.

The thyroid hormone calcitonin reduces blood calcium

Another hormone of the mammalian thyroid gland is **calcitonin**, although it is not produced by the same cells that produce thyroxine. Calcitonin helps regulate the levels of calcium circulating in the blood. Bone is a huge repository of calcium in the body and is continually being remodeled. Cells called **osteoclasts** break down bone and release calcium; **osteoblasts**, on the other hand, use circulating calcium to deposit new bone (Figure 38.9).

Calcitonin decreases the activity of osteoclasts and stimulates the activity of osteoblasts, thus shifting the balance from adding calcium ions to the blood to removing calcium ions from the blood. Calcitonin plays an important role in preventing bone loss in women during pregnancy. However, regulation of blood calcium levels is influenced more strongly by parathormone than it is by calcitonin.

Parathormone elevates blood calcium

The **parathyroid glands** are four tiny structures embedded on the surface of the thyroid gland. Their single hormone product is **parathyroid hormone**, or **parathormone**, a critical control element in the regulation of blood calcium levels. Growth and remodeling of bone require calcium; many cellular processes, such as nerve and muscle functions, are sensitive to changes in calcium concentration. Muscle contraction and nerve function are severely impaired if the blood calcium level rises or falls by as little as 30 percent of normal values.

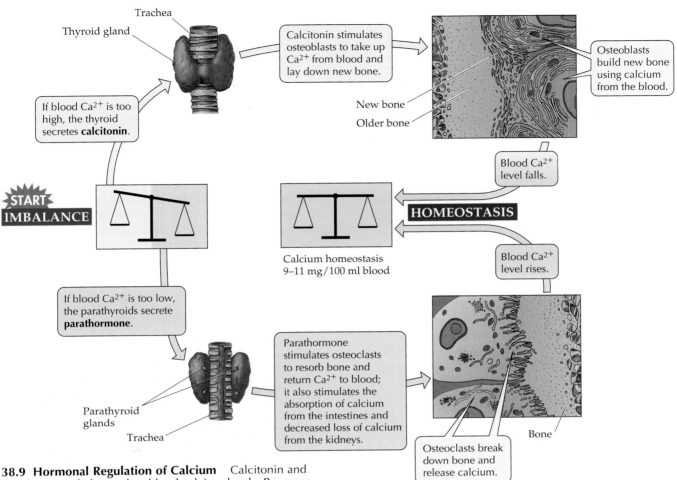

38.9 Hormonal Regulation of Calcium Calcitonin and parathormone help regulate blood calcium levels. Bone can be a source (site of production) of calcium or a sink (site of utilization or storage) for calcium.

A decrease in blood calcium triggers the release of parathormone, which in turn stimulates actions that add calcium to the blood. Parathormone stimulates osteoclasts to dissolve bone and release calcium (see Figure 38.9). It prevents this liberated calcium from being lost in the urine by promoting its reabsorption from the kidneys. In addition, parathormone plays an indirect role in stimulating the digestive tract to absorb calcium from food—by promoting the activation of vitamin D, which is essential for calcium uptake by the gut.

Thus, parathormone and calcitonin act antagonistically to regulate blood calcium levels: Parathormone elevates and calcitonin reduces. A similar antagonistic relationship is true of hormones of the pancreas, which regulate blood glucose levels.

Insulin and glucogon regulate blood glucose

Before the 1920s, diabetes mellitus was a fatal disease, characterized by weakness, lethargy, and body wasting. The disease was known to be connected with a gland located just below the stomach, the **pancreas**, and with abnormal glucose metabolism, but the link was not clear. Today we know that diabetes mellitus is caused by a lack of the hormone **insulin** or by a lack of responsiveness of target tissues to the hormone. In cases in which the hormone is lacking, insulin replacement therapy is an extremely successful treatment. At present, more than 1.5 million diabetics in the United States lead almost normal lives through the use of injected insulin.

We now know that insulin binds to a receptor on the cell membrane, and this insulin–receptor complex allows glucose to be taken into the cell. In a sense, insulin is the key that unlocks doors allowing glucose to enter most cells. In the absence of insulin or insulin receptors, glucose accumulates in the blood until it is lost in the urine. High levels of blood glucose cause water to move from cells into the blood by osmosis, and the kidneys increase urine output to excrete excess fluid volume from the blood. The name "diabetes" refers to the copious production of urine, and "mellitus" (from the Greek for "honey") reflects the fact that the urine of an untreated diabetic is sweet. Since the cells of the body cannot use blood glucose for fuel, they must burn fat and protein. As a result, the body of the untreated diabetic wastes away, and critical tissues and organs are damaged.

The change in outlook for diabetics came almost overnight in 1921, when medical doctor Frederick Banting and medical student Charles Best of the University of Toronto discovered that they could reduce the symptoms of diabetes by injecting an extract they had prepared from pancreatic tissue. The active component of the extract that Banting and Best prepared was a small protein hormone, insulin, consisting of 51 amino acids.

Insulin is produced in clusters of endocrine cells in the pancreas, called **islets of Langerhans** after a German medical student who discovered them. There are several types of cells in the islets: Beta (β) cells produce and secrete insulin; alpha (α) cells produce and secrete the hormone **glucagon**, which has effects opposite those of insulin; and delta (δ) cells produce the hormone **somatostatin**. The rest of the pancreas is an exocrine gland that produces enzymes and secretions that travel through ducts to the intestine, where they play roles in digestion.

After a meal, the concentration of glucose in the blood rises as glucose is absorbed from the gut. This increase in glucose concentration stimulates the pancreas to release insulin. Insulin stimulates cells to use glucose as fuel and to convert it into storage products such as glycogen and fat. When the gut contains no more food, the glucose concentration in the blood falls, and the pancreas stops releasing insulin. As a result, most cells of the body shift to using glycogen and fat rather than glucose for fuel.

If the concentration of glucose in the blood falls below normal, islet cells release glucagon, which stimulates the liver to convert glycogen back to glucose to resupply the blood. These effects and conversions will be discussed in greater detail in Chapter 47.

Somatostatin is a hormone of the brain and pancreas

Somatostatin is released from the pancreas in response to rapid rises of glucose and amino acids in the blood. This hormone has paracrine functions within the islets; it inhibits the release of both insulin and glucagon. Its actions outside the pancreas slow the digestive activities of the gut. Pancreatic somatostatin extends the period of time during which nutrients are absorbed from the gut and used by the cells of the body.

Somatostatin was first discovered as a hypothalamic neurohormone that inhibits the release of growth hormone and thyrotropin by the pituitary. It was therefore called growth hormone release-inhibiting hormone, but somatostatin is a more convenient name. Somatostatin has been highly conserved during the course of evolution and is even found in some single-celled organisms.

The adrenal is two glands in one

An adrenal gland sits above each kidney just below the middle of your back. Functionally and anatomically an adrenal gland consists of a gland within a gland (Figure 38.10). The core, called the **adrenal medulla**, produces the hormone **epinephrine** (also known as adrenaline) and, to a lesser degree, **norepinephrine** (or noradrenaline). Surrounding the medulla (as an apricot surrounds its pit) is the **adrenal cortex**, which produces other hormones. The medulla develops from nervous-system tissue and is under the control of the nervous system; the cortex is under hormonal control, largely by adrenocorticotropin from the anterior pituitary.

THE ADRENAL MEDULLA.   By producing epinephrine, the adrenal medulla arouses the body to action. As we saw earlier in the chapter, in stressful situations epinephrine increases heart rate, breathing rate, and blood pressure, and it diverts blood flow to active skeletal muscles and away from the gut. These fight-or-flight reactions can be stimulated by physically threatening events, such as encountering a mugger, or by events that are mentally stressful, such as giving a public

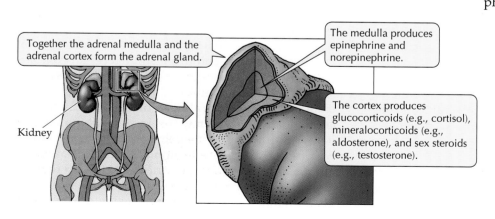

Together the adrenal medulla and the adrenal cortex form the adrenal gland.

Kidney

The medulla produces epinephrine and norepinephrine.

The cortex produces glucocorticoids (e.g., cortisol), mineralocorticoids (e.g., aldosterone), and sex steroids (e.g., testosterone).

38.10 The Adrenal Gland Has an Outer and an Inner Portion
An adrenal gland, consisting of an outer medulla and an inner cortex, sits on top of each kidney. The medulla and the cortex produce different hormones.

38.11 Steroid Hormones Are Built from Cholesterol

Different side groups on the sterol backbone give different properties to different steroid hormones. This simplified outline of steroid biosynthesis leaves out many intermediate steps. Sex steroids are produced in small amounts by the adrenal cortex and in much greater amounts by the gonads.

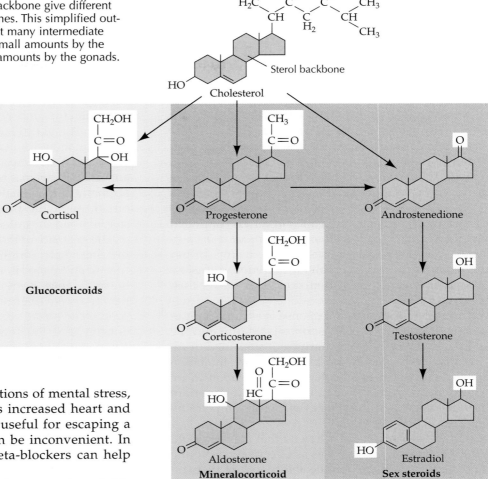

speech or taking a test. In situations of mental stress, many of the responses such as increased heart and breathing rates that would be useful for escaping a mugger are not useful and can be inconvenient. In extreme cases, drugs called beta-blockers can help reduce these responses.

Epinephrine and norepinephrine are water-soluble hormones that bind to receptors on the surfaces of cells. These receptors can be grouped into two general types, α-adrenergic and β-adrenergic receptors, which stimulate different actions within cells. Epinephrine acts equally on both types, but norepinephrine acts mostly on α-adrenergic receptors. Therefore, drugs that selectively block β-adrenergic receptors can reduce the fight-or-flight reactions to epinephrine without disrupting regulatory functions of norepinephrine.

THE ADRENAL CORTEX. The cells of the adrenal cortex use cholesterol to produce three classes of hormones: glucocorticoids, mineralocorticoids, and sex steroids (Figure 38.11). Collectively they are called the **corticosteroids**. The **glucocorticoids** influence blood glucose concentrations as well as other aspects of fat, protein, and carbohydrate metabolism. The **mineralocorticoids** influence the ionic balance of extracellular fluids. The **sex steroids** stimulate sexual development and reproductive activity.

Sex steroids are secreted in only negligible amounts by the adrenal cortex and will be discussed in the next section, on gonadal hormones. Many of the steroids found in the adrenal cortex are intermediates in biosynthetic pathways. Only two of the steroids produced and secreted by the adrenal cortex are significant physiologically. They are the mineralocorticoid called aldosterone and the glucocorticoid called cortisol.

Aldosterone stimulates the kidney to conserve sodium and to excrete potassium. If the adrenal glands are removed from an animal, the animal must add sodium to its diet to keep from becoming sodium depleted and dying. In one human case, a patient with adrenal insufficiency compensated by salting her food heavily and, in addition, eating a 60-pound block of salt in the course of a year. We'll discuss aldosterone in more detail in Chapter 48.

Cortisol is a critical hormone for mediating the body's response to stress. Imagine being on safari in Africa and encountering a lion that roars and leaps at you. Your immediate reaction, which we called the fight-or-flight reaction, is stimulated by your nervous system and by the release of epinephrine. Your heart is beating faster, you are breathing faster, and your running muscles are getting maximal supplies of oxygen and glucose. Let's assume your Land Rover is not nearby and the lion continues in pursuit as you try to escape in tall grass. You need to mobilize more energy reserves. Within about 5 minutes cortisol levels rise and help you sustain your escape.

Because you need a high level of blood glucose for your brain and muscles to function, cortisol stimulates the other cells of your body to decrease their use of glucose and start to metabolize fats and proteins for energy. This is not a time to feel sick, have allergic reactions, or heal wounds, so cortisol blocks immune-system reactions. This is why cortisol is useful for reducing inflammations and allergies.

Cortisol release is controlled by the pituitary hormone **adrenocorticotropin**, which in turn is controlled by the hypothalamic *adrenocorticotropin-releasing hormone*. Because the cortisol response to a stressor has this chain of steps, each involving secretion, diffusion, circulation, and cell activation, it is much slower than the epinephrine response.

Turning off the cortisol response is as important as turning it on. The stress response is adaptive in the short run, but it would be harmful if sustained or repeatedly activated. Elevated blood pressure, inhibition of immune responses, and elevated fat metabolism are not healthy conditions in the long run. It is important to turn these stress responses off when they are no longer needed.

A study of stress in rats showed that old rats could turn on their stress responses as effectively as young rats, but that they had lost the ability to turn them off as rapidly. As a result they suffered from the well-known consequences of stress—ulcers, cardiovascular problems, strokes, impaired immune-system function, and increased susceptibility to cancers and other diseases. Further research showed that turning off stress responses involves negative feedback actions of cortisol on cells in the brain, which in turn cause a decrease in the release of adrenocorticotropin-releasing hormone. Repeated activation of this negative feedback mechanism leads to gradual loss of cortisol-sensitive cells in the brain, and therefore a decreased ability to terminate stress responses.

The sex hormones are produced by the gonads

The testes of the male and the ovaries of the female produce hormones as well as gametes. Most of the gonadal hormones are steroids synthesized from cholesterol (see Figure 38.11). The male steroids are collectively called **androgens**, and the dominant one is **testosterone**. The female steroids are **estrogens** and **progesterone**. The dominant estrogen is **estradiol**.

The sex steroids have important developmental effects: They determine whether a fetus develops into a female or a male. (A fetus is the latter stage of an embryo; a human embryo is called a fetus from the eighth week of pregnancy to the moment of birth.) After birth the sex steroids control the maturation of the reproductive organs and the development and maintenance of secondary sexual characteristics, such as breasts and facial hair.

The sex steroids begin to exert effects in the human embryo in the seventh week of development. Until that time, the embryo can develop into either sex. In mammals and birds the ultimate instructions for sex determination reside in the genes. Individuals that receive two X chromosomes normally become females, and individuals that receive an X and a Y chromosome normally become males. These instructions are carried out through the production and action of the sex steroids, and the potential for error exists.

The presence of the Y chromosome normally causes the embryonic, undifferentiated gonads to begin producing androgens in the seventh week. In response to the androgens, the reproductive system develops into that of a male (Figure 38.12). If the androgens are not produced at that time, or the androgen receptors do not function, the female reproductive structures develop even if the fetus is genetically a male. In other words, androgens are required to trigger male development in humans. The opposite situation exists in birds: Male characteristics develop unless estrogens are present to trigger female development.

Occasionally the hormonal control of sexual development does not work perfectly, resulting in intersex individuals. The most extreme (but rare) case is a true **hermaphrodite**, who has both testes and ovaries. **Pseudohermaphrodites** have the gonads of one sex and the external sex organs of the other. For example, an XY fetus will develop testes, but if his tissues are insensitive to the androgens produced by those testes because of a receptor defect, they will remain within the abdomen, and the external sex organs and the secondary sexual characteristics of a female will develop.

Sex steroids have dramatic effects at the time of puberty—around the age of 12 to 13 years. Throughout childhood the production of sex steroids by the gonads is extremely low. As puberty approaches, the hypothalamus begins to produce and secrete *gonadotropin-releasing hormone* (GnRH), which causes the pituitary to produce two gonadotropins—**luteinizing hormone** (**LH**) and **follicle-stimulating hormone** (**FSH**).

In the female, the increasing levels of LH and FSH stimulate the ovaries to begin producing the female sex hormones. The increased circulating levels of these sex steroids initiate the development of the traits characteristic of a sexually mature woman: enlarged breasts, vagina, and uterus; broad hips; increased subcutaneous fat; pubic hair; and the initiation of the menstrual cycle.

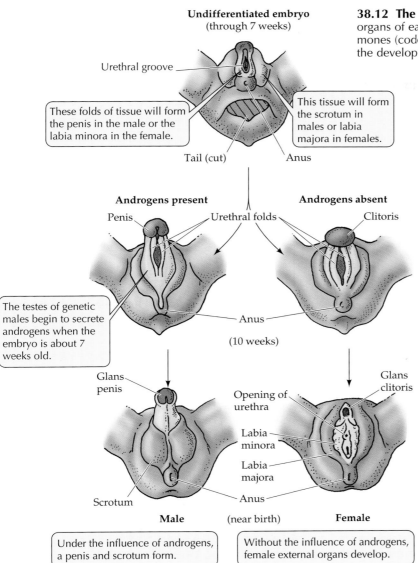

Undifferentiated embryo
(through 7 weeks)

Urethral groove

These folds of tissue will form the penis in the male or the labia minora in the female.

This tissue will form the scrotum in males or labia majora in females.

Tail (cut)   Anus

Androgens present   **Androgens absent**

Penis — Urethral folds   Clitoris

The testes of genetic males begin to secrete androgens when the embryo is about 7 weeks old.

Anus

(10 weeks)

Glans penis   Glans clitoris

Opening of urethra

Labia minora

Labia majora

Scrotum   Anus

Male   (near birth)   **Female**

Under the influence of androgens, a penis and scrotum form.

Without the influence of androgens, female external organs develop.

38.12 The Development of Human Sex Organs   The sex organs of early human embryos are similar. Androgen hormones (coded for by genes on the Y chromosome) promote the development of male sex organs.

thetic androgens—called *anabolic steroids.* However, anabolic steroids have serious negative side effects. In women, use of artificial androgens causes the breasts and uterus to shrink, the clitoris to enlarge, menstruation to become irregular, facial and body hair to grow, and the voice to deepen. In men, the testes shrink, hair loss increases, the breasts enlarge, and sterility can result.

You can understand the causes of some of these side effects by considering the negative feedback effects of sex steroids on the production of LH and FSH. Other effects are even more serious. Continued use of anabolic steroids greatly increases the risk of heart disease, certain cancers, kidney damage, and personality disorders such as depression, mania, psychoses, and extreme aggression. Most official athletic organizations, including the International Olympic Committee, ban the use of anabolic steroids.

The list of other hormones is long

We have discussed all the major endocrine glands and "classic" hormones in this chapter, but there are many hormones we have not mentioned. Examples include hormones produced in the digestive tract that help organize the way the gut processes food (see Table 38.1 and Chapter 47). Even the heart has endocrine functions. When blood pressure rises and causes the walls of the heart to stretch, certain cells in the walls of the heart release **atrial natriuretic hormone**. This hormone increases the excretion of sodium ions and water by the kidneys, thereby lowering blood volume and blood pressure. As we discuss the organ systems of the body in the chapters that follow, we will frequently mention hormones that their tissues produce or hormones that control their functions.

Mechanisms of Hormone Action

The hormones we have discussed are released in very small quantities, yet they can cause large responses in cells or tissues all over the body, and these responses can be quite specific in different cells. For example, we

In the male, an increasing level of LH stimulates groups of cells in the testes to synthesize androgens, which in turn initiate the profound physiological, anatomical, and psychological changes associated with adolescence. The voice deepens, hair begins to grow on the face and body, and the testes and penis grow. Androgens help skeletal muscles grow, especially when they are exercised regularly. The bulging biceps, triceps, pectorals, and deltoids of bodybuilders are extreme examples of the skeletal-muscle growth that occurs in every male past puberty. In preadolescent boys and in women, a rigorous program of weight lifting will not lead to massive muscle development, because of the absence of androgens.

Natural muscle development can be exaggerated by both men and women who want to increase their maximum strength in athletic competition if they take syn-

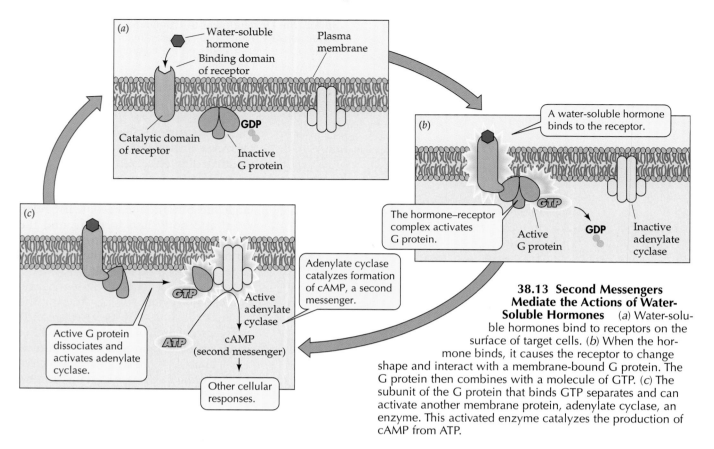

38.13 Second Messengers Mediate the Actions of Water-Soluble Hormones (*a*) Water-soluble hormones bind to receptors on the surface of target cells. (*b*) When the hormone binds, it causes the receptor to change shape and interact with a membrane-bound G protein. The G protein then combines with a molecule of GTP. (*c*) The subunit of the G protein that binds GTP separates and can activate another membrane protein, adenylate cyclase, an enzyme. This activated enzyme catalyzes the production of cAMP from ATP.

have discussed the many dramatic effects of testosterone, yet its concentration in the plasma of adult human males is only about 30 to 100 *nanograms* per ml. How can hormones in such tiny quantities have such widespread and selective actions?

In this section we will answer this question by considering the mechanisms whereby hormones induce their actions at the cellular and molecular levels. The first step is for the hormone to bind to a receptor. Only cells with appropriate receptors will respond to a hormone—one basis for the selective actions of hormones. The ability of receptors to initiate a cascade of responses that amplify the original signal helps explain why hormones are such powerful molecules.

Receptors can be found either on the surface of cells or within cells, giving us a basis for dividing all hormones into two groups. **Lipid-soluble hormones** can diffuse through plasma membranes and reach intracellular receptors; examples include the steroid hormones and thyroxine. **Water-soluble hormones**, such as epinephrine, and peptide or protein hormones, cannot readily pass through plasma membranes, and their receptors are on the cell surface.

Receptors for water-soluble hormones activate protein kinases

Water-soluble hormones bind with receptors on the surface of target cells. The receptors are glycoproteins that have a binding domain that projects beyond the outside of the cell membrane, a transmembrane domain, and a catalytic domain that extends into the cytoplasm of the cell (Figure 38.13*a*).

The catalytic domain initiates cell responses by directly or indirectly activating protein kinases. There are many different kinds of protein kinases, but they all catalyze the transfer of a phosphate group from ATP to a specific protein. In some cases the added phosphate activates the protein; in other cases phosphorylation inactivates the protein. Thus the binding of the hormone to its receptor can result in activation or inhibition of an enzyme.

When the catalytic domains of water-soluble hormone receptors activate protein kinases indirectly, they do so by way of **second messengers**. The hormone itself is the first messenger. When it binds to its receptor, the hormone stimulates a chain of reactions, producing small, diffusible molecules that are second messengers within the cell. The second messengers activate protein kinases. Cyclic adenosine monophosphate, or **cAMP**, is a well-studied second messenger that activates protein kinases in many different kinds of cells.

The actions of many hormones include cAMP as a second messenger. Glucagon and epinephrine stimulate glycogen breakdown in liver cells using cAMP as the second messenger. Pituitary tropic hormones such as thyrotropin, luteinizing hormone, and adrenocorticotropin stimulate their target cells to produce cAMP. Parathormone acts through cAMP as second messenger. How can the same second messenger induce completely different responses in different cells?

Different types of cells respond differently to cAMP because they have different enzymes that are phosphorylated by the protein kinases. Some of these enzymes are activated and some inactivated by phosphorylation. *The specificity of hormone action depends not only on which cells have receptors for a given hormone, but also on what response mechanisms the cells have.*

The sequence of events from binding of a hormone to its receptor to the ultimate response of the cell is called a **signal transduction pathway**. Second messengers are components of signal transduction pathways, but there are usually many more elements as well.

G proteins control the production of second messengers

A complex set of reactions takes place between binding of the hormone and production of the second messenger. When a receptor binds a hormone molecule, the receptor's shape changes. In its new form, the receptor can interact with a second membrane protein, enabling that protein to bind a molecule of guanosine triphosphate (GTP). The membrane proteins that bind GTP, called *G proteins*, are made of three subunits, only one of which binds GTP (Figure 38.13a and b).

The subunit of the G protein that binds GTP separates and moves to another membrane protein, the enzyme adenylate cyclase. The complex consisting of G protein subunit and GTP activates the adenylate cyclase. The active **adenylate cyclase** catalyzes the conversion of ATP to cAMP within the cell. Thus, the G protein is the link between the receptor and the production of the second messenger (Figure 38.13c).

What turns off the activation of adenylate cyclase by the G protein? The G protein subunit that binds the GTP eventually hydrolyzes the GTP to GDP (guanosine diphosphate). When that happens, the G protein dissociates from the adenylate cyclase, thus inactivating itself and helping to terminate hormone-induced responses. When the three G protein subunits recombine, they are again capable of being activated by the membrane receptor.

There are many kinds of G proteins. Some are inhibitory, inactivating adenylate cyclase and decreasing the production of second messenger. The presence of excitatory and inhibitory G protein–coupled receptors in the same cells enables those cells to be controlled by two or more hormones that act antagonistically.

Signal transduction pathways amplify responses to hormones

The action of epinephrine on liver cells is an example of how cAMP-dependent protein kinases work (Figure 38.14). Epinephrine binds to its receptor on the plasma membrane, thereby activating adenylate cyclase and causing cAMP to form. Cyclic AMP activates an enzyme called **protein kinase A** (PKA) that phosphorylates certain proteins. PKA influences glycogen metabolism by acting on two critical enzymes, glycogen synthase and glycogen phosphorylase, adding a phosphate group from ATP to each one.

Glycogen synthase catalyzes the joining of glucose molecules to synthesize the energy-storing molecule glycogen, but it is *inactivated* by the addition of the phosphate group. The other enzyme phosphorylated by PKA, phosphorylase kinase, is *activated* by the addition of the phosphate group. Phosphorylase kinase, itself a protein kinase, catalyzes phosphorylation and activation of the enzyme glycogen phosphorylase. Glycogen phosphorylase liberates glucose molecules

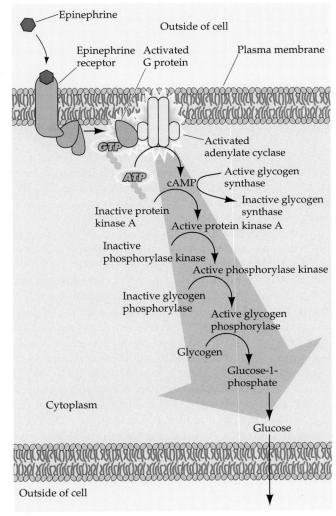

38.14 Liver Cells Respond to Epinephrine with a Cascade of Reactions The cAMP produced in liver cells by epinephrine triggers a cascade of reactions that stimulates the breakdown of glycogen to glucose and the inhibition of glycogen synthesis from glucose. The steps in the cascade amplify the response.

from glycogen. Thus cAMP, through its effects on two protein kinases, inhibits the storage of glucose as glycogen and promotes the release of glucose through glycogen breakdown. Both of these effects increase glucose levels in liver cells and hence in blood as well.

This cascade of regulatory steps amplifies enormously the effect of a single hormone molecule. A molecule of epinephrine binds to a single receptor molecule, but the activated receptor activates many molecules of the G protein. Each activated G protein activates one molecule of adenylate cyclase, but adenylate cyclase catalyzes production of cAMP as long as it remains activated by the G protein. It takes two cAMP molecules to activate one protein kinase molecule, but each protein kinase may activate many phosphorylase kinase molecules, each phosphorylase kinase may activate many glycogen phosphorylase molecules, and each glycogen phosphorylase molecule may catalyze the liberation of many molecules of glucose from glycogen. This response cascade enormously amplifies the original signal. A single molecule of epinephrine can cause the production of millions or billions of molecules of glucose.

Several factors turn off the signal transduction pathway

Unless there is a continuing supply of the hormone, it diffuses away from the receptor or is enzymatically degraded, allowing the receptor to revert to its inactive state, which does not bind G protein. As already mentioned, the G protein subunits that activate adenylate cyclase gradually dissociate from the adenylate cyclase as they hydrolyze their bound GTP to GDP. As a result, cAMP stops forming. The cAMP still present is quickly removed by the action of enzymes that convert it to an inactive product.

An active mechanism turns off the cascade when cAMP levels fall. Besides activating protein kinases, cAMP inhibits phosphoprotein phosphatases, which remove phosphate groups from proteins phosphorylated by protein kinases. Thus when cAMP levels fall, phosphoprotein phosphatase is no longer inhibited and is free to reverse the work done by protein kinases. For example, when liver cells are no longer stimulated by epinephrine or glucagon, their phosphoprotein phosphatase levels rise. As a result, glycogen phosphorylase is inactivated and glycogen synthase is activated. The cells switch from breaking down glycogen to building glycogen from glucose.

A cell can have multiple signal transduction pathways

In addition to cAMP, there are other second-messenger systems and, therefore, other signal transduction pathways in animal cells. Two second messengers, **inositol trisphosphate** (IP_3) and **diacylglycerol** (DAG), are produced from a membrane phospholipid that contains the alcohol inositol. This molecule is **phosphatidylinositol**; its two lipophilic fatty acid chains are embedded in the plasma membrane, and its inositol group, which is hydrophilic, projects into the cytoplasm.

More than 25 receptors on cell surfaces use this signal transduction pathway. For example, norepinephrine, GnRH, and vasopressin act through the IP_3–DAG system. The binding of hormones or neurotransmitters to those receptors activates a G protein that in turn activates a phosphatidylinositol-specific phospholipase C. The activated phospholipase C cleaves phosphatidylinositol (PTI) to form inositol trisphosphate and diacylglycerol:

$$PTI \xrightarrow{\text{phospholipase C}} IP_3 + DAG$$

IP_3 enters the cytoplasm, and DAG remains in the plasma membrane.

IP_3 and DAG have different modes of action that are synergistic (that is, they build on each other). DAG activates protein kinase C (PKC), which is also in the plasma membrane. Protein kinase C derives its name from the fact that it is Ca^{2+}-dependent—that's where IP_3 comes in. IP_3 diffuses through the cytoplasm and causes the release of intracellular calcium ions. DAG and increased Ca^{2+} work together to activate PKC. PKC then phosphorylates a wide variety of proteins, leading to the ultimate response of the cell (Figure 38.15). In addition, the increased Ca^{2+} alone can activate various cellular responses that are independent of PKC.

Different signal transduction pathways can act in a complementary way or in an antagonistic way to regulate the cell's responses. Earlier we discussed the cAMP pathway by which epinephrine and glucagon stimulate liver cells to break down stored glycogen to supply glucose to the blood. Liver cells also respond to norepinephrine. Even though norepinephrine is very similar to epinephrine, its receptor on liver cells activates the phosphatidylinositol signal transduction pathway rather than the cAMP pathway. In liver cells, however, PKA and PKC phosphorylate the same enzymes, resulting in inhibition of glycogen synthesis and stimulation of glycogen breakdown. In other situations the cAMP signal transduction pathway and the phosphatidylinositol pathway can act antagonistically in the regulation of a cell response.

Calcium ions control many intracellular responses

We have just seen how Ca^{2+} ions are involved in one signal transduction pathway by activating PKC (see Figure 38.16). Some other direct roles of Ca^{2+} ions are to control membrane ion channels and to stimulate secretory processes. As we'll see in Chapters 41 and 44, Ca^{2+} ions play crucial roles in the functions of both nerve cells and muscle cells.

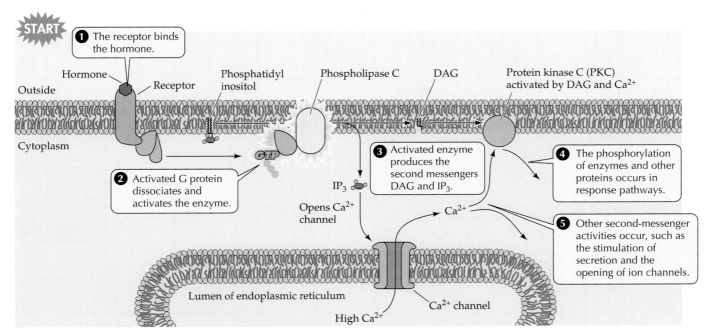

38.15 The IP₃ and DAG Second-Messenger System
A G protein activates the enzyme phospholipase C, which catalyzes the hydrolysis of a membrane phospholipid to form inositol trisphosphate (IP₃) and diacylglycerol (DAG). DAG stays in the membrane and IP₃ enters the cytoplasm. Both serve as second messengers.

Diagram labels:
- START
- **1** The receptor binds the hormone.
- Hormone
- Outside
- Receptor
- Phosphatidyl inositol
- Phospholipase C
- DAG
- Protein kinase C (PKC) activated by DAG and Ca²⁺
- Cytoplasm
- GTP
- **2** Activated G protein dissociates and activates the enzyme.
- IP₃
- **3** Activated enzyme produces the second messengers DAG and IP₃.
- Opens Ca²⁺ channel
- Ca²⁺
- **4** The phosphorylation of enzymes and other proteins occurs in response pathways.
- **5** Other second-messenger activities occur, such as the stimulation of secretion and the opening of ion channels.
- Lumen of endoplasmic reticulum
- High Ca²⁺
- Ca²⁺ channel

In some cases Ca²⁺ ions act via a calcium-binding protein. The most widely distributed calcium-binding protein is a small molecule called **calmodulin**. Calmodulin is activated by binding with Ca²⁺, and the active complex then triggers cell responses, including activation of more protein kinases, smooth-muscle contraction, microtubule assembly, protein synthesis, and various secretory events. Another calcium-binding protein, troponin, regulates a key reaction in the contraction of the skeletal muscles of vertebrates.

Cell surface receptors can trigger gene transcription

All the targets of the signal transduction systems we have discussed up to now have been in the cytoplasm or the plasma membrane of the cell, and not in the nucleus. However, cell surface receptors are involved in activating a broad range of gene expression responses. For example, some of the PKA activated by cAMP finds its way into the nucleus. There it activates a **cAMP response element–binding** protein, or **CREB**. Activated CREB can then bind to a cAMP response element (CRE) that stimulates the transcription of specific genes (Figure 38.16).

As an example, let's continue our discussion of the response of liver cells to epinephrine and glucagon. In addition to stimulating the breakdown of glycogen to supply glucose to the blood, these hormones stimulate the liver cells to synthesize glucose out of certain amino acids through a process called **gluconeogenesis**, which we'll discuss in Chapter 47. Gluconeogenesis requires a variety of special enzymes, and the transcription of the genes for these enzymes is stimulated by CREB in liver cells. In addition, intracellular signals triggered by PKC are translocated to the nucleus, where they cause gene transcription.

Growth factors are a large group of protein signaling molecules that bind to cell surface receptors and stimulate gene expression. The receptors for many growth factors belong to a family called **receptor tyrosine kinases**. These receptors span the plasma membrane and have regions projecting out of both sides. One serves as the receptor; the other acts as an enzyme that phosphorylates, hence activates, intracellular signaling molecules. This phosphorylation always occurs on tyrosine side chains, which explains the general name for the receptors. Many of the cell responses controlled by receptor tyrosine kinases are changes in gene expression.

Lipid-soluble hormones have intracellular receptors

Steroid hormones—such as progesterone, testosterone, and estradiol, and the hormones of the adrenal cortex—as well as thyroxine, generally do not react with receptors on the target cell surface. These hormones are all lipid-soluble, which, as you may recall from Chapter 5, means that they pass readily through lipid-rich plasma membranes. They act by stimulating the synthesis of new kinds of proteins through gene activation rather than by altering the activity of proteins already present in the target cells.

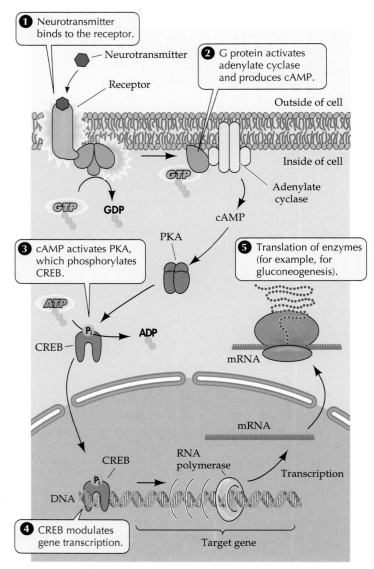

38.16 Cyclic AMP Can Trigger Gene Expression PKA activated by cAMP can enter the nucleus and activate a transcription factor, CREB. CREB binds to the cAMP response element (CRE) and enhances gene expression.

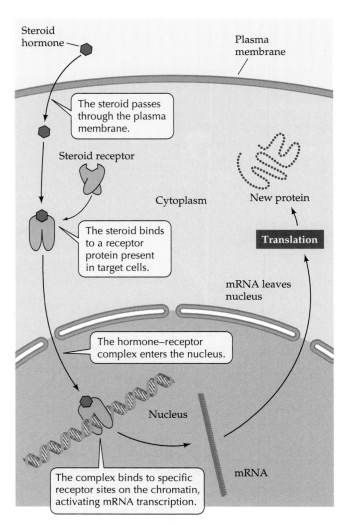

38.17 Lipid-Soluble Hormones Trigger Gene Expression The receptors of lipid-soluble hormones, such as steroids, are inside of cells. After the steroid binds to the receptor, the resulting hormone–receptor complex moves to the nucleus and acts as a transcription factor.

Once inside a cell, a lipid-soluble hormone binds to a receptor protein in the cytoplasm (Figure 38.17) or in the nucleus. The presence of a receptor protein is what distinguishes a responsive cell from a nonresponsive cell. A receptor protein is specific for a particular hormone, and it changes shape when it binds its hormone. The hormone–receptor complex moves to the nucleus and associates with a DNA sequence adjacent to the gene it regulates and either stimulates or in some cases inhibits transcription of the gene.

The actions of lipid-soluble hormones are slower and last longer than the actions of water-soluble hormones. When water-soluble hormones bind to their re-

ceptors on cell surfaces, they stimulate rapid changes, such as increased membrane permeability to ions or activation of an enzyme. In contrast, the actions induced by lipid-soluble hormones include transcription, translocation of the mRNA to the ribosomes, translation of the message, and frequently posttranslational modification of the protein product. This may take 30 minutes or more, and often that is just the primary response to the hormone. These protein products may in turn activate expression of a still different set of genes, which constitutes the secondary or longer-term response to the hormone.

Steroid hormones and the water-soluble tropic hormones that control them play major roles in reproductive systems, to which we turn in the next chapter.

Summary of "Animal Hormones"

Hormones and Their Actions

• Hormonal signaling systems involve endocrine cells that secrete chemical messages that bind to receptors on or in target cells.
• Most hormones diffuse through the extracellular fluids and are picked up by the blood, which distributes them throughout the body. Some hormones diffuse to targets near the site of secretion: Autocrine hormones influence receptors on the cell that secretes them; paracrine hormones reach nearby cells. **Review Figure 38.1**
• Hormones cause different responses from different target cells.
• The chemical structure of hormones has changed little through evolution, but their functions have changed dramatically.
• Hormones may be secreted by single cells or by cells organized into discrete glands.

Hormonal Control of Molting and Development in Insects

• Insects must molt their exoskeletons to grow. Progressive growth stages between molts are instars. Two diffusible substances, brain hormone and ecdysone, control molting. **Review Figure 38.3**
• Juvenile hormone, another diffusible substance, prevents maturation so that juvenile instars molt into bigger juvenile instars. When the level of juvenile hormone falls low enough, the juvenile molts into the adult form.
• Some insects, such as butterflies, go through complete metamorphosis. Juvenile butterflies are larvae that grow bigger with each molt. When juvenile hormone drops to a low level, the larval form becomes a pupa. Because no juvenile hormone is secreted during pupation, the pupa molts into an adult. **Review Figure 38.4**

Vertebrate Endocrine Systems

• Vertebrates have nine endocrine glands and many hormones. **Review Figure 38.2 and Table 38.1**
• The pituitary gland is divided into two parts. The anterior develops from embryonic mouth tissue; the posterior develops from the brain.
• The posterior pituitary secretes the neurohormones vasopressin and oxytocin. The anterior pituitary secretes tropic hormones (thyrotropin, adrenocorticotropin, and two gonadotropins), as well as growth hormone, prolactin, melanocyte-stimulating hormone, endorphins, and enkephalins. **Review Figure 38.5 and Table 38.1**
• The anterior pituitary is controlled by hypothalamic neurohormones produced by cells in the hypothalamus and transported through portal vessels to the anterior pituitary. **Review Figure 38.7 and Table 38.2**
• The thyroid gland is controlled by thyrotropin and secretes thyroxine, which controls cell metabolism. Goiter can be associated with too little or too much thyroxine. **Review Figure 38.8**

• The level of calcium in the blood is regulated by two hormones. Calcitonin, produced by the thyroid, lowers blood calcium. Parathormone, produced by the parathyroid glands, raises it. **Review Figure 38.9**
• The pancreas secretes three hormones: Insulin stimulates glucose uptake by cells and lowers blood glucose, glucagon raises blood glucose, and somatostatin slows the rate of nutrient absorption from the gut.
• Cortisol from the adrenal medulla decreases glucose utilization by most cells. The hormones of the adrenal cortex, epinephrine and norepinephrine, stimulate the liver to supply glucose to the blood, as well as other fight-or-flight reactions. **Review Figures 38.10, 38.11**
• Sex hormones (androgens in males, estrogens and progesterone in females) are produced by the gonads in response to tropic hormones. Sex hormones control sexual development, secondary sex characteristics, and reproductive functions. **Review Figure 38.12**

Mechanisms of Hormone Action

• Water-soluble hormones bind to cell surface receptors that activate protein kinases. The chain of events from receptor binding to intracellular signals to ultimate cell response is a signal transduction pathway.
• Some receptors activate G proteins in the plasma membrane, which in turn cause a second messenger to be produced. Cyclic AMP is a common second messenger that activates protein kinases. **Review Figure 38.13**
• Signal transduction pathways amplify responses to the binding of a hormone to a receptor through a cascade of reactions. **Review Figure 38.14**
• Cells can have multiple signal transduction pathways and different ultimate responses controlled by those pathways.
• Some cell surface receptors cause the splitting of phosphatidylinositol into inositol trisphosphate and diacylglycerol, both of which are second messengers that have different, synergistic modes of action. **Review Figure 38.15**
• Changes in intracellular calcium are an important part of many signal transduction pathways.
• Some cell surface receptors can trigger gene transcription. **Review Figure 38.16**
• Lipid-soluble hormones (steroid hormones and thyroxine) pass through membranes and bind to receptors in the cell cytoplasm or nucleus. The hormone–receptor complexes associate with DNA sequences that either stimulate or inhibit the transcription of adjacent genes. **Review Figure 38.17**

Self-Quiz

1. Which statement is true for *all* hormones?
    a. They are secreted by glands.
    b. They have receptors on cell surfaces.
    c. They may stimulate different responses in different cells.
    d. They target cells distant from their site of release.
    e. When the same hormone occurs in different species, it has the same action.

2. The hormone ecdysone
    a. is released from the posterior pituitary.
    b. stimulates molt and metamorphosis in insects.
    c. maintains an insect in larval stages unless brain hormone is present.
    d. stimulates the secretion of juvenile hormone from the prothoracic glands.
    e. keeps the insect exoskeleton flexible to permit growth.

3. The posterior pituitary
   a. produces oxytocin.
   b. is under the control of hypothalamic releasing neuro-hormones.
   c. secretes tropic hormones.
   d. secretes neurohormones.
   e. is under feedback control by thyroxine.

4. Growth hormone
   a. can cause adults to grow taller.
   b. stimulates protein synthesis.
   c. is released by the hypothalamus.
   d. can be obtained only from cadavers.
   e. is a steroid.

5. Both epinephrine and cortisol are secreted in response to stress. Which of the following statements is also true for *both* of these hormones?
   a. They act to increase blood glucose.
   b. Their receptors are on the surfaces of target cells.
   c. They are secreted by the adrenal cortex.
   d. Their secretion is stimulated by adrenocorticotropin.
   e. They are secreted into the blood within seconds of the onset of stress.

6. Before puberty
   a. the pituitary secretes luteinizing hormone and follicle-stimulating hormone, but the gonads are unresponsive.
   b. the hypothalamus does not secrete much gonadotropin-releasing hormone.
   c. males can stimulate massive muscle development through a vigorous training program.
   d. testosterone plays no role in development of the male sex organs.
   e. genetic females will develop male genitals unless estrogen is present.

7. Which statement about cyclic AMP is *not* true?
   a. It is broken down by adenylate cyclase.
   b. It is involved in the chain of events by which epinephrine stimulates liver cells to break down glycogen.
   c. It is a second messenger that mediates intracellular responses to many hormones.
   d. Many of its effects are mediated by protein kinases.
   e. Two molecules of cAMP are needed to activate a single protein kinase molecule.

8. Steroid hormones
   a. are all produced by the adrenal cortex.
   b. have only cell surface receptors.
   c. are lipophobic.
   d. act through altering the activity of proteins in the target cell.
   e. act through stimulating the production of new proteins in the target cell.

9. Which of the following is a likely cause of goiter?
   a. The thyroid gland is producing too much parathormone.
   b. Circulating levels of thyrotropin are too low.
   c. There is an inadequate supply of functional thyroxine.
   d. There is an oversupply of functional thyroxine.
   e. The diet contains too much iodine.

10. Parathormone
    a. stimulates osteoblasts to lay down new bone.
    b. reduces blood calcium levels.
    c. stimulates calcitonin release.
    d. is produced by the thyroid gland.
    e. is released when blood calcium levels fall.

Applying Concepts

1. Explain how both hyperthyroidism and hypothyroidism can cause goiter. Include the roles of the hypothalamus and the pituitary in your answers.

2. Explain the developmental abnormalities that can produce a genetic male with female secondary sexual characteristics. Describe the gonads of such an individual.

3. Various side effects of anabolic steroid use were mentioned in this chapter. Some of these effects are due to the direct action of the steroid, but others are due to the negative feedback action of the steroid. Discuss an example of each and explain possible mechanisms.

4. Explain what is meant by the statement that a cell's response to a hormone depends on what response mechanisms the cell has. Use examples.

5. How can cAMP working through a protein kinase activate one enzyme while inactivating another enzyme in the same cell?

Readings

Atkinson, M. A. and N. K. MacLaren. 1990. "What Causes Diabetes?" *Scientific American*, July. A discussion of how malfunctions of the immune system cause insulin-dependent diabetes, a major disease that involves a hormone deficiency.

Fernald, R. D. 1993. "Cichlids in Love." *The Sciences*, July/August. A fascinating study of "wimpy" and "macho" behavior among cichlid fish.

Hoberman, J. M. and C. E. Yesalis. 1995. "The History of Synthetic Testosterone." *Scientific American*, February. An examination of the use of testosterone during the twentieth century, including a look at its banning from sports and current research into its benefits and risks.

Lacy, P. 1995. "Treating Diabetes with Transplanted Cells." *Scientific American*. July. A discussion of a possible cure for diabetes: the transplantation of pancreatic cells.

Norris, D. O. 1997. *Vertebrate Endocrinology*, 3rd Edition. Academic Press, San Diego. Definitive comparative coverage of the hormonal systems of all vertebrate classes.

Snyder, S. H. 1985. "The Molecular Basis of Communication between Cells." *Scientific American*, October. An overview of the relationships between the nervous and endocrine systems, focusing on chemical messengers and their molecular biology.

Vander, A. J., J. H. Sherman and D. S. Luciano. 1994. *Human Physiology: The Mechanisms of Body Function*, 6th Edition. McGraw-Hill, New York. Chapter 10 deals specifically with hormonal regulation, but material on hormones appears throughout this fine text.

Chapter 39

Animal Reproduction

Piranga rubra (summer tanager)

Reproduction: An Essential Feature of Animal Life
Like humans, birds are sexually reproducing animals. The cycle of birth, sexual maturation, and reproduction provides a continuous line of genetic information from generation to generation.

We are sexually reproducing animals, and we think of eggs and sperm as specialized sex cells that we use to reproduce ourselves. But let's consider reproduction from an evolutionary point of view. The evolution of multicellular animals has produced an amazing diversity of adaptations to unite egg and sperm. An evolutionary view of reproduction, therefore, might be that the tissues, organs, bodies, and behaviors of multicellular animals are specialized adaptations that sex cells use to reproduce themselves. You might recognize this as the classic "what came first, the chicken or the egg?" question, but there are some interesting facts to consider.

Very early in the life of a new human embryo, cells arise that have no fixed location, but move around in the body until the primary sex organs form. After the ovaries and testes develop, these nomadic cells migrate to those organs and take up residence. Eventually these cells produce the eggs and sperm. The sex cells themselves are part of a lineage of cells called the germ line, which has been continuous, with gradual modification through mutation and recombination, throughout the evolution of sexually reproducing animal life. This view of the relation between the organism and the sex cells reveals the centrality of reproductive processes in the lives of animals.

In this chapter we will learn about the diversity of ways that animal life reproduces. *Asexual* mechanisms of reproduction are efficient means of producing large numbers of genetically identical individuals; *sexual* reproduction creates genetic diversity through recombination of parent genotypes. Sexual reproduction requires individuals of different sexes, or in some cases, male and female sex organs within the same individual. To reproduce, an organism must produce haploid sex cells, and those cells must unite to create a new diploid individual. These processes require specialized organs and behaviors.

After examining the reproductive systems of some invertebrates and nonmammalian vertebrates, we will focus in detail on the anatomy, function, and endocrine control of the human reproductive system. This information will allow us to understand technologies used to limit fertility and those used to treat infertility.

Asexual Reproduction

Sexual reproduction is a nearly universal trait among animals, but many species can reproduce asexually as well. Offspring produced asexually are genetically identical to one another and to their parents. Asexual reproduction is highly efficient because there is no mating; mating requires energy and involves risks. In addition, all individuals of an asexual population can convert resources into offspring, allowing the population to grow as rapidly as resources permit. However, asexual reproduction does not generate genotypic diversity.

An asexually reproducing population does not have a wide variety of genotypes on which natural selection can act as the environment changes. Nevertheless, a variety of animals, mostly invertebrates, reproduce asexually. They tend to be species that are sessile and therefore cannot search for mates, or species that live in sparse populations and rarely encounter potential mates. Furthermore, asexually reproducing species are more likely to be found in relatively constant environments in which the potential for rapid evolutionary change is not as important as in more variable environments.

In this section we will examine three common modes of asexual reproduction: budding, regeneration, and parthenogenesis.

Budding and regeneration produce new individuals by mitosis

A common mode of asexual reproduction in simple multicellular animals is **budding**, a process in which a new individual arises as an outgrowth of an older one. Some sponges form buds of undifferentiated cells on the outside of their bodies. These buds grow by mitotic cell division, and the cells differentiate before the buds break away from the parents and become independent sponges that are genetically identical to the parent. Budding is part of the life cycle of some cnidarians, such as *Hydra* (Figure 39.1*a*). The bud resembles the parent and may grow as large as the parent before it becomes independent.

The capacity of some cells in sponges and certain cnidarians to form buds that give rise to new, complete organisms is exceptional. Most animal cells lose this ability, known as *totipotency*, as they differentiate. It is possible that budding begins with cells that have not differentiated but have retained totipotency. However, there are numerous examples in which pieces of organisms give rise to new individuals. In these cases of **regeneration**, differentiated cells must dedifferentiate to play a role in the formation of a new individual.

In a classic experiment demonstrating regeneration, a sponge is pushed through a wire mesh, producing many little clusters of cells. Each cluster grows into a

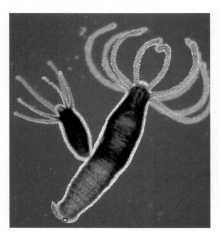

(a) *Hydra* sp.

(b) *Linickia multiflora*

(c) *Macrosiphum rosae*

39.1 Asexual Reproduction in Animals (a) Budding: A new individual forms as an outgrowth from an adult hydra. (b) Regeneration: A single amputated arm from a sea star develops into a new animal. (c) Parthenogenesis: These rose aphids are hatching from unfertilized eggs.

new, small but complete sponge. This ability of sponges to regenerate was used off the coast of Florida to restore the commercial bath sponge fishery, which was endangered by overfishing. Individual sponges were cut into small pieces, which were then seeded in appropriate habitats. A new population of sponges grew from the pieces. Echinoderms also have remarkable abilities to regenerate. If sea stars are cut into pieces, each piece that includes a portion of the central disc grows into a new animal (Figure 39.1b).

Regeneration frequently results when an animal is broken by an outside force. For example, a storm can cause a heavy surf that breaks colonial cnidarians such as corals. Pieces broken off the colony can regenerate into new colonies. In some species the breakage is a normal event initiated by the animal itself. Certain species of segmented worms (annelids) develop segments with rudimentary heads bearing sensory organs; then they break apart. Each fragmented segment forms a new worm.

Parthenogenesis is the development of unfertilized eggs

A common mode of asexual reproduction in arthropods is the development of offspring from unfertilized eggs. This phenomenon, called **parthenogenesis**, also occurs in some species of fish, amphibians, and reptiles. Most species that reproduce parthenogenetically also engage in sexual reproduction or sexual behavior. Aphids, for example, are parthenogenetic in the spring and summer, multiplying rapidly while conditions are favorable (Figure 39.1c). Some of the unfertilized eggs laid in spring and summer develop into male aphids, others into females. As conditions become less favorable, the aphids mate and the females lay fertilized eggs. These eggs do not hatch until the following spring, and they yield only females.

In some species, parthenogenesis is part of the mechanism that determines sex. For example, in ants and in most species of bees and wasps, females develop from unfertilized eggs and are haploid. Males develop from fertilized eggs and are diploid. Most females are sterile workers, but a select few become fertile queens. After a queen mates, she has a supply of sperm that she controls, enabling her to produce either fertilized or unfertilized eggs. Thus the queen determines when and how much of the colony resources are expended on males.

Parthenogenetic reproduction in some species requires a sex act, even though this act does not fertilize the egg. The eggs of parthenogenetically reproducing ticks and mites, for example, develop only after the animals have mated, but the eggs remain unfertilized. Some species of beetles have no males and can reproduce only parthenogenetically, yet their eggs require sperm to trigger development. These beetles mate with males of closely related, but different, species. Similarly, in some species of whiptail lizards there are no males, and females stimulate each other to activate the neuroendocrine mechanisms necessary for egg maturation.

Sexual Reproduction

One evolutionary result of sexual reproduction is that it produces genetic diversity. Contributing to this genetic diversity are processes that occur during the production of haploid sex cells and processes associated with the joining of those sex cells to produce new diploid individuals. Haploid sex cells are **gametes**, which are produced through **gametogenesis**. Genetic diversity is created during gametogenesis through recombination of genes by means of crossing over and the independent assortment of chromosomes in meiosis (see Chapter 10).

The joining of gametes to form new diploid individuals is **fertilization**. In most species fertilization is preceded by mating behaviors that bring potential parents and their gametes together. The genetic diversity in the gametes of a single individual and the genetic diversity between possible parents produce an enormous potential for genotypic variation between any two offspring of a sexually reproducing pair of individuals. This genotypic diversity is the raw material for natural selection; thus evolutionary change in sexually reproducing animals can be quite rapid.

In this section, we focus on the three fundamental phenomena of sexual reproduction in animals: gametogenesis, mating, and fertilization. Gametogenesis and fertilization are very similar across all groups of animals. Mating, however, shows incredible evolutionary diversity. Therefore, our discussion of gametogenesis will not refer to specific groups of animals or species, but will apply to the vast majority of sexually reproducing animals. On the other hand, our discussion of mating will focus on specific examples as representative of the fascinating diversity that exists. We will delay detailed discussion of fertilization until the next chapter, on animal development, since the union of egg and sperm is the first event in the development of a new individual.

Eggs and sperm form through gametogenesis

The tissues where haploid gametes are produced from germ cells through gametogenesis are the primary sex organs—**testes** (singular testis) in males and **ovaries** in females. The tiny gametes of males are **sperm**, which are mobile and move by beating their flagella. The much larger female gametes are **eggs**, or **ova** (singular ovum), and are nonmotile (Figure 39.2). In addition to primary sex organs, most animals (except sponges and cnidarians) have accessory sex organs, including

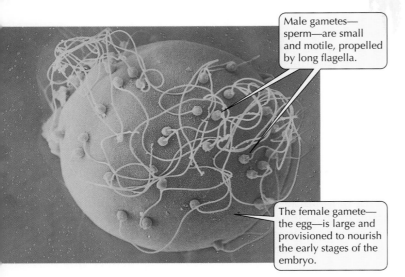

Male gametes—sperm—are small and motile, propelled by long flagella.

The female gamete—the egg—is large and provisioned to nourish the early stages of the embryo.

39.2 Gametes Differ in Size In this micrograph of mammalian fertilization, many sperm have attached to a single egg, although only one will enter.

ducts, glands, and structures that deliver and receive gametes. The primary and accessory sex organs of an animal constitute its reproductive system.

As we described in the introduction to this chapter, the gametes derive from a special lineage of cells called the germ line. Germ cells are not produced by the gonads; they come to reside in the gonads only after the gonads have formed in the embryo. Germ cells are diploid and proliferate by mitosis. Cells resulting from the mitotic proliferation of germ cells in the gonads of females are **oogonia** (singular oogonium); those in the gonads of males are **spermatogonia** (singular spermatogonium).

Meiosis, the next step in gametogenesis, reduces the chromosomes to the haploid number, and the haploid cells mature into sperm and ova. Because meiosis is central to the formation of both sperm and ova, you may want to review the discussion of meiosis in Chapter 9 before reading further. Although the steps of gametogenesis in males and females are very similar, there are also some significant differences, so we will examine gametogenesis in the two sexes in parallel (Figure 39.3).

SPERMATOGENESIS PRODUCES SPERM. Spermatogenesis begins when the diploid spermatogonia divide by mitosis to become **primary spermatocytes**. Primary spermatocytes undergo the first meiotic division to form secondary spermatocytes, which are haploid. (Recall that the first meiotic division halves the number of chromosomes.) In mammals these cells remain connected by cross-bridges of cytoplasm after each division. The second meiotic division produces four

haploid spermatids for each primary spermatocyte that entered meiosis (see Figure 39.3a).

The reason that mammalian spermatocytes remain in cytoplasmic contact throughout their development probably is the asymmetry of sex chromosomes in male mammals (in birds and many other vertebrates the females are the heterogametic sex). Half of the secondary spermatocytes receive an X chromosome, the other half a Y chromosome. The Y chromosome contains only a small number of genes in comparison to the X chromosome, and apparently some of the products of genes not included in the Y chromosome are essential for spermatocyte development. By remaining in cytoplasmic contact, all spermatocytes can share the gene products of the X chromosomes, even though only half of them have an X chromosome.

Spermatids differ from one another genetically because random orientation of the chromosomes at the first meiotic metaphase shuffles the parent genomes. A given spermatid contains some maternal chromosomes and some paternal chromosomes; the particular combination is a matter of chance. Crossing over during the first meiotic division also contributes to genetic differences among spermatids.

Just after being produced by meiosis, a spermatid bears little resemblance to a sperm. As it differentiates into a sperm, its nucleus becomes compact, its motile flagellum develops into a tail, and most of its cytoplasm is lost. As the head of the sperm forms, it is capped by an **acrosome**, which contains enzymes that will enable the sperm to digest its way into an egg. Between the head and tail of the mature sperm is a midpiece containing mitochondria that provide energy for locomotion (see Figure 39.10).

OOGENESIS PRODUCES EGGS. Oogenesis begins when the oogonia divide by mitosis. In species that produce and repeatedly release hundreds or thousands of eggs, oogonia remain capable of mitotic division throughout life. In animals that produce relatively few eggs throughout their life span (for example, humans), the proliferation of oogonia is limited to an early period during which all the egg precursor cells the individual will ever have form. This limited period of oogonia proliferation in oogenesis is a major contrast to spermatogenesis. Spermatogonia continue to divide mitotically throughout the life of the male. In humans, the period of oogonia proliferation ends before birth.

When the oogonia stop proliferating, the resulting egg precursor cells differentiate into **primary oocytes**, which immediately enter prophase of the first meiotic division. At this point their development arrests, and they may remain in this state for days, months, or years. In contrast, the development of male gametes does not arrest, but goes steadily to completion once the primary spermatocytes have been created. In the

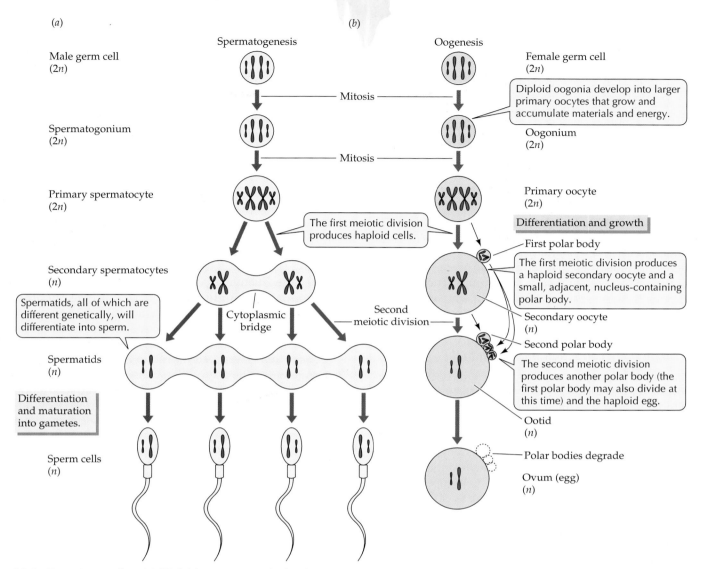

(a) Spermatogenesis

(b) Oogenesis

Male germ cell
(2n)

Female germ cell
(2n)

Mitosis

Diploid oogonia develop into larger primary oocytes that grow and accumulate materials and energy.

Spermatogonium
(2n)

Oogonium
(2n)

Mitosis

Primary spermatocyte
(2n)

Primary oocyte
(2n)

Differentiation and growth

The first meiotic division produces haploid cells.

First polar body

Secondary spermatocytes
(n)

The first meiotic division produces a haploid secondary oocyte and a small, adjacent, nucleus-containing polar body.

Spermatids, all of which are different genetically, will differentiate into sperm.

Cytoplasmic bridge

Second meiotic division

Secondary oocyte
(n)

Second polar body

Spermatids
(n)

The second meiotic division produces another polar body (the first polar body may also divide at this time) and the haploid egg.

Differentiation and maturation into gametes.

Ootid
(n)

Polar bodies degrade

Sperm cells
(n)

Ovum (egg)
(n)

39.3 Gametogenesis (a) Diploid spermatogonia develop into haploid spermatids. Spermatids, all of which are different genetically, differentiate into sperm. (b) Diploid oogonia develop into haploid secondary oocytes, which mature into ova.

human female some primary oocytes may remain in arrested prophase I for 50 years! During this arrest, or shortly before it ends, the primary oocytes enlarge in reponse to the production of ribosomes, RNA, cytoplasmic organelles, and energy stores. The primary oocyte acquires all of the energy, raw materials, and RNA that the egg will need to survive its first cell divisions after fertilization.

When a primary oocyte receives appropriate hormonal signals to resume meiosis, its nucleus completes the first meiotic division near the surface of the cell. Daughter cells of this division receive grossly unequal shares of the cytoplasm of the primary oocyte.

This is another major difference from spermatogenesis, in which cell divisions apportion cytoplasm equally. The daughter cell that receives almost all of the cytoplasm becomes the **secondary oocyte**, and the one that receives almost none forms the **first polar body** (see Figure 39.3b).

The second meiotic division of the large secondary oocyte is also accompanied by an asymmetric division of the cytoplasm. One daughter cell forms the large, haploid **ootid**, which eventually differentiates into an ovum, and the other forms the **second polar body**. Polar bodies eventually degenerate, so the end result of oogenesis is that each primary oocyte produces only one mature egg. However, that egg is a very large cell that is well provisioned for the rapid cell divisions that follow fertilization.

A second period of arrested egg development occurs after the first meitoic division forms the secondary oocyte. In this condition the egg may be expelled from the ovary during ovulation. In many species, including humans, the second meiotic division does not continue until the egg is fertilized by a sperm.

You may wonder which of the events of oogenesis is modified in animals that reproduce parthenogenetically. Parthenogenesis is a reproductive adaptation that has evolved in different species quite independently, so there are various patterns. In bees and ants oogenesis is the same as in sexually reproducing species, but haploid eggs can develop into haploid individuals, which are then female. In some other insects meiosis does not occur; instead oogenesis involves two mitotic divisions, so the resulting eggs are diploid and develop into diploid individuals. In still other insects and in some lizards, the oogonia double their chromosomes before meiosis (becoming tetraploid), so the resulting eggs are diploid. The most intriguing case is a species of fruit fly in which oogenesis is normal but one of the polar bodies serves as a "sperm" and fertilizes the haploid egg.

A single body can function as both male and female

Sexual reproduction requires both male and female haploid gametes. In most species, but not all, these gametes are produced by individuals that are either male or female. Species that have male and female members are called **dioecious** (from the Greek for "two houses"). By contrast, in some species a single individual may possess both female and male reproductive systems. Such species are called **monoecious** ("one house") or **hermaphroditic**.

Almost all invertebrate groups have hermaphroditic species. An earthworm is an example of a **simultaneous hermaphrodite**, meaning that it is both male and female at the same time. When two earthworms mate, both are fertilized and produce offspring (Figure 39.4). Some animals are **sequential hermaphrodites**, meaning that they function as a male or a female at different times in their lives.

Why has natural selection produced hermaphroditism? Some simultaneous hermaphrodites have a low probability of meeting a potential mate. An example is a parasitic tapeworm. Even though it may be large and cause lots of trouble, it may be the only tapeworm in your intestine. Tapeworms and some other simultaneous hermaphrodites can fertilize themselves, but most simultaneous hermaphrodites must mate with another individual.

Sequential hermaphroditism can confer several advantages. It can reduce the possibility of inbreeding among siblings by making them all the same sex at the same time and therefore incapable of mating with one another. In a species in which only a few males fertilize all females, sequential hermaphroditism can maxi-

Lumbricus sp.

39.4 Hermaphroditic Mating Each earthworm contains both male and female reproductive organs. They will fertilize each other, and both will produce offspring.

mize reproductive success by making it possible for an individual to reproduce as a female until the opportunity arises for it to function as a male.

An excellent example of a sequential hermaphrodite that reaps this benefit is the tropical Pacific fish *Labroides dimidiatus*. All individuals of this species are born female. The population consists of social groups consisting of three to six females controlled by one male. The male defends the group's territory from intruders. If the male dies or is removed, the largest, most dominant female in the group changes sex and becomes a functional male, assuming control of the group.

Anatomical and behavorial adaptations bring eggs and sperm together

Sexual reproduction requires that haploid gametes come into close proximity so that they can join into a single diploid **zygote** through fertilization. Because gametes are so small, uniting them is not a trivial problem. To solve this problem, many anatomical and behavioral adaptations have evolved to support a wide variety of mating strategies.

The simplest distinction in mating strategies is whether fertilization occurs externally or internally. Sexually reproducing animals may release their gametes into the environment, where the chance meeting of gametes results in fertilization, or male gametes may be inserted into the female's reproductive tract, where fertilization occurs.

EXTERNAL FERTILIZATION REQUIRES AN AQUATIC HABITAT. External fertilization is common among simple animals, especially those that are not very mobile. These animals produce huge numbers of gametes. A female oyster, for example, may produce 100 million eggs per year, and the number of sperm produced by a male oyster is astronomical. But numbers alone do not guarantee that gametes will meet. Timing is also important. Reproductive activities of males and females of a population must be synchronized. Seasonal breeders may use day length cues, changes in temperature, or changes in weather to time their production and release of gametes.

Sexual behavior can play an important role in bringing gametes together even when fertilization is external. Many species travel great distances to congregate with potential mates and release their gametes at the same time in a suitable environment. Salmon are an example (Figure 39.5). These fish hatch and develop through juvenile stages in fresh water. They then migrate to the ocean, where they live and grow for 3 to 5 years. When finally ready to breed (spawn), they migrate back to the stream where they hatched. When spawning, a female salmon expels her mass of eggs; then the male swims over the egg mass and releases a cloud of sperm. After spawning, adult salmon die.

INTERNAL FERTILIZATION ENABLES TERRESTRIAL LIFE. Delicate gametes released into air would dry out and die. Terrestrial animals avoid this problem by engaging in internal fertilization. Another great advantage of internal fertilization is protection for the early developmental stages of the organism. Animals have evolved an incredible diversity of sexual behaviors and accessory sex organs that facilitate internal fertilization. In general, a tubular structure, the **penis**, enables the male to deposit sperm in the female's accessory sex organ, the **vagina**, or, in some species, the **cloaca** (a cavity common to the digestive, urinary, and reproductive systems).

Copulation (the physical joining of male and female in sexual intercourse) is an act that permits sperm to move directly from the male's reproductive system into the female's reproductive system. Transfer of sperm in internal fertilization can also be indirect. Males of some species of mites and scorpions (among the arthropods) and salamanders (among the vertebrates) deposit **spermatophores**—containers filled with sperm—in the environment. When a female mite finds a spermatophore, she straddles it and opens a pair of plates in her abdomen so that the tip of the spermatophore enters her reproductive tract and allows the sperm to enter. Some female salamanders use the lips of their cloacae to scoop up the portion of the gelatinous spermatophore that contains sperm.

Male squid and spiders play a more active role in spermatophore transfer. The male spider secretes a drop containing sperm onto a bit of web; then, with a special structure on his foreleg, he picks up the sperm-containing web and inserts it through the female's genital opening. Male squid use one special tentacle to pick up a spermatophore and insert it into the female's genital opening. Thus two very different species have evolved similar behavior to ensure fertilization.

Oncorhynchas nerka

39.5 External Fertilization is Common in Aquatic Species
External fertilization requires an aqueous environment. Spawning female salmon expel a mass of eggs, which are fertilized externally by sperm expelled in a cloud by the males.

Most male insects copulate and transfer spermatophores to the female's vagina through a tubular penis. The **genitalia**—external sex organs—of insects often have species-specific shapes that match in a lock-and-key fashion. This mechanism ensures a tight, secure fit between the mating pair during the prolonged period of sperm transfer. The males of some insect species have an elaborate structure on their penis that enables them to remove sperm deposited previously by other males. Following this cleaning, a male transfers his own sperm into the tract.

The evolution of vertebrate reproductive systems parallels the move to land

The earliest vertebrates evolved in aquatic environments. The closest living relatives of those earliest vertebrates are fish. They remain exclusively aquatic animals, and most practice external fertilization. The most primitive of the fishes, the lampreys and hagfishes, broadcast their gametes into the environment as do many aquatic invertebrates. In most fishes, however, fertilization is more selective: Mating behaviors result in specific females coming into close proximity with specific males at the time of gamete release.

In some early vertebrates, especially sharks and rays, certain fins have evolved into structures that hold the male and female together for sperm transfer. The most elaborate of these structures, called **claspers**, have taken on the function of an intromittent organ. One clasper enters the cloaca of the female and directs sperm along a groove in its surface into the female reproductive tract. The evolution of internal fertilization in sharks and rays has made it possible for the females of some species to encase fertilized eggs in protective egg cases before depositing them in the environment.

39.6 Approximation: Bringing Sperm and Eggs Together
Fertilization in frogs is external, but amplexus—a behavior in which the male holds the female with his forelegs until she releases her egg mass—helps guarantee that sperm and egg will get together.

Smilisca baudinii

The first vertebrates to invade terrestrial environments were the amphibians. They had to deal with the problem that sperm cannot swim through air and that gametes are killed by desiccation in air. The solution that most amphibians still use is to reproduce in water. Some modern species of amphibians never leave water, but most of those that do return to water for reproduction. An exception is terrestrial salamanders, which use spermatophores to transfer sperm, as mentioned earlier. (Recall from Chapter 38 that the hormone prolactin in amphibians stimulates the animals to seek a wet habitat where the female can lay her egg mass and the male can fertilize the eggs.)

In another amphibian, the frog, this mating behavior is characterized by **amplexus**, a behavior in which the male grasps the female around the middle with his forelegs and holds on until she releases her egg mass (Figure 39.6). This behavior guarantees maximal probability that the male's sperm will fertilize the female's eggs. In the humid tropical rainforest, some frogs lay their eggs and fertilize them outside of water. The gelatinous mass that surrounds the eggs serves as the aqueous environment for sperm to reach the eggs, and it protects the eggs from drying. Frequently the mating pair create their gelatinous mass of fertilized eggs at the end of a branch over a stream, pond, or puddle so that as the eggs hatch, the tadpoles fall into water.

Reptiles were the first vertebrate group to solve the problem of reproduction in the terrestrial environment. Their solution, the shelled egg, is shared by birds (Figure 39.7*a*). The shell must be permeable to respiratory gases and reasonably impermeable to water. But the shelled egg created a new problem for fertilization. Sperm cannot penetrate the shell, so they have to reach the egg before the shell forms.

The solution to this problem was internal fertilization; hence the necessary accessory sex organs evolved. Male snakes and lizards have paired **hemipenes**, which can be filled with blood and thereby extruded

(*a*) *Elaphe guttata*

39.7 Reptiles and Birds: The Shelled Egg
(*a*) The shelled egg was a major evolutionary step enabling reproduction in the terrestrial environment. Here a corn snake has just laid its eggs. (*b*) The terrestrial environment offers no water to bring sperm and egg together; fertilization thus must take place internally, as with these fairy terns.

(*b*) *Gygis alba*

from the male's cloaca to form intromittent organs. Only one hemipene enters the female's cloaca at a time. It is usually rough or spiny at the end to achieve a secure hold while sperm are transferred down a groove on its surface. Retractor muscles pull the hemipene back into the male's body when mating is completed. Birds have erectile penises that channel sperm along a groove into the female cloaca (Figure 39.7*b*).

Mammals have retained internal fertilization (Figure 39.8) but have done away with the shelled egg by keeping the developing embryo in the female reproductive tract until it is capable of independent existence in the outside environment.

Reproductive systems can be classifed by where the embryo develops

Two patterns of care and nurture of the embryo have evolved in animals: oviparity and viviparity. **Oviparous** animals lay eggs in the environment, and their embryos develop outside the mother's body. **Viviparous** animals retain the embryo within the mother's body during its early developmental stages.

Oviparity is possible because eggs are stocked with abundant nutrients to supply the needs of the embryo until it is capable of actively taking up nutrients from the environment. As mentioned earlier, terrestrial animals that are oviparous—for example, reptiles, birds, and insects—must protect their eggs from desiccation. Their eggs have tough, waterproof membranes or shells, which can also be effective protection against predators. However, these egg coverings must be permeable to oxygen and carbon dioxide. Oviparous parents may practice various forms of parental behavior to protect the eggs, but until the eggs hatch, the embryos depend entirely on the nutrients stored in the egg at the time of fertilization. After hatching, the offspring may or may not receive continuing parental care. Among mammals, the only species that are oviparous are the monotremes: the spiny anteater and the duck-billed platypus (see Figure 30.30).

Most viviparous animals are mammals, and most mammals are viviparous. There are examples of viviparity in all other vertebrate groups except the crocodiles, turtles, and birds. Even some sharks retain fertilized eggs in their bodies and give birth to free-living offspring. But there is a big difference between viviparity in these species and viviparity in mammals. Mammals have a specialized portion of the female reproductive tract, the uterus, that holds and provides nutrients for the embryo. In contrast, nonmammalian viviparous animals simply retain the fertilized eggs in the mother's body until they hatch. The embryos receive their nutrition from the stores that were in their egg. For this reason, this reproductive system is called **ovoviviparity**.

Diceros bicornis

39.8 All Mammals Practice Internal Fertilization These black rhinoceroses are displaying the most prevalent mammalian mating behavior, in which the male mounts the female from the rear.

Even among mammals there are various degrees of uterine adaptation. In marsupials the uterus simply holds the embryo and has limited capability of supplying it with nutrients. Thus marsupials are born at a very early developmental stage, crawl into a pouch called a marsupium on the mother's belly, attach to a nipple of a mammary gland, and complete development outside of the mother's uterus (see Figure 30.31). Mammals other than monotremes and marsupials are called **eutherian mammals**, and they are characterized by the intimate association of blood supplies of mother and embryo in the walls of the uterus. We will discuss this special adaptation in the next section.

Reproductive Systems in Humans and Other Mammals

So far we have seen only a small sampling of the fascinating diversity of animal reproductive systems. We will now look at the mammalian reproductive system

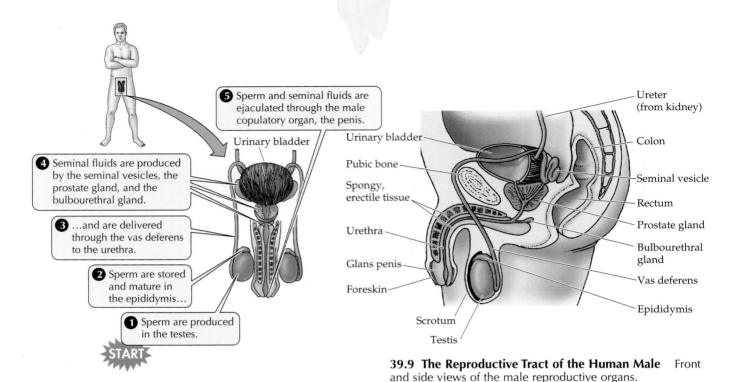

5 Sperm and seminal fluids are ejaculated through the male copulatory organ, the penis.

4 Seminal fluids are produced by the seminal vesicles, the prostate gland, and the bulbourethral gland.

3 ...and are delivered through the vas deferens to the urethra.

2 Sperm are stored and mature in the epididymis...

1 Sperm are produced in the testes.

START

Urinary bladder

Ureter (from kidney)

Urinary bladder

Colon

Pubic bone

Seminal vesicle

Spongy, erectile tissue

Rectum

Prostate gland

Urethra

Bulbourethral gland

Glans penis

Vas deferens

Foreskin

Epididymis

Scrotum

Testis

39.9 The Reproductive Tract of the Human Male Front and side views of the male reproductive organs.

in greater depth, using the human as our model. In this section we will discuss male and female sexual organs and their hormonal regulation, emphasizing ovarian and uterine cycles. We'll examine the human sexual response and technologies for both contraception and enhanced fertility. Finally, we'll close the section with a discussion of sexually transmitted diseases.

Male sex organs produce and deliver semen

The product of the male reproductive system is **semen**. Besides sperm, semen includes a complex mixture of fluids and molecules that support the sperm and facilitate fertilization. Sperm make up less than 5 percent of the volume of the semen.

Sperm are produced in the testes, the paired male gonads. In most mammals, the testes are outside the body cavity in a pouch of skin, the **scrotum**, a part of the male genitalia (Figure 39.9). The optimal temperature for spermatogenesis in most mammals is slightly lower than normal body temperature. The scrotum keeps the testes at this optimal temperature. Muscles in the scrotum contract in a cold environment, bringing the testes closer to the warmth of the body; in a hot environment they relax, suspending the testes farther from the body.

A testis consists of tightly coiled **seminiferous tubules** within which spermatogenesis takes place. The tubule walls are lined with spermatogonia. From the tubule wall going toward the center, germ cells are in successive stages of spermatogenesis (Figure 39.10).

Fully differentiated spermatids are shed into the lumen of the tubule.

These germ cells are intimately associated with **Sertoli cells**, which have multiple functions. Sertoli cells protect sperm by providing a barrier between sperm and any noxious substances that might be circulating in the blood. Sertoli cells also provide nutrients for developing sperm, and are involved in the hormonal control of spermatogenesis. Between the seminiferous tubules are clusters of **Leydig cells** that produce male sex hormones.

From the seminiferous tubules, sperm move into a storage structure called the **epididymis**, where they mature and become motile. The epididymis connects to the **urethra** by a tube called the **vas deferens** (plural vasa deferentia). The urethra comes from the bladder, runs through the penis, and opens to the outside of the body at the tip of the penis. The urethra is the common duct for the urinary and reproductive systems (see Figure 39.9).

The penis and the scrotum are the male genitalia. The shaft of the penis is covered with normal skin, but the tip, or **glans penis**, is covered with thinner, more sensitive skin that is especially responsive to sexual stimulation. A fold of skin called the foreskin covers the glans of the human penis. The practice of circumcision removes a portion of the foreskin. Although circumcision has cultural and religious origins rather than medical ones, evidence suggests that the risk of certain genital infections is lower in circumcised individuals.

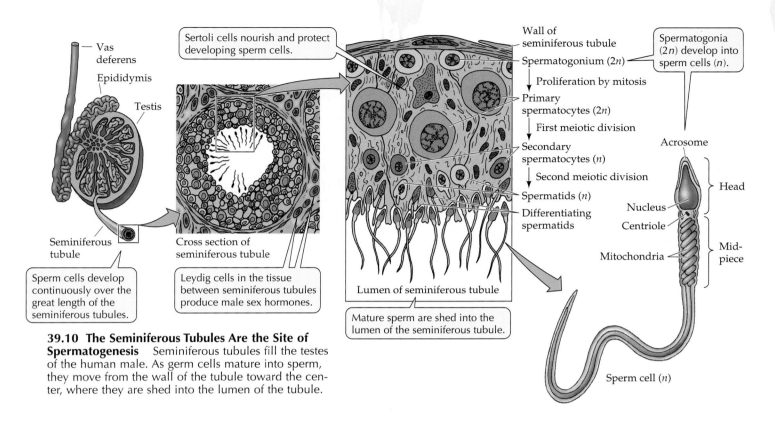

Vas deferens

Epididymis

Testis

Seminiferous tubule

Sertoli cells nourish and protect developing sperm cells.

Cross section of seminiferous tubule

Sperm cells develop continuously over the great length of the seminiferous tubules.

Leydig cells in the tissue between seminiferous tubules produce male sex hormones.

Wall of seminiferous tubule

Spermatogonium (2n)

Proliferation by mitosis

Primary spermatocytes (2n)

First meiotic division

Secondary spermatocytes (n)

Second meiotic division

Spermatids (n)

Differentiating spermatids

Lumen of seminiferous tubule

Mature sperm are shed into the lumen of the seminiferous tubule.

Spermatogonia (2n) develop into sperm cells (n).

Acrosome

Head

Nucleus

Centriole

Mitochondria

Mid-piece

Sperm cell (n)

39.10 The Seminiferous Tubules Are the Site of Spermatogenesis Seminiferous tubules fill the testes of the human male. As germ cells mature into sperm, they move from the wall of the tubule toward the center, where they are shed into the lumen of the tubule.

This lower risk is due probably to the fact that hygiene is easier to achieve when the foreskin is absent.

The penis becomes hard and erect during sexual arousal because blood fills shafts of spongy, erectile tissue along its length. The presence of this blood creates pressure that closes off the vessels that normally drain the penis. Thus, the penis becomes engorged with blood, facilitating insertion into the vagina. In some species of mammals, but not humans, the penis has a bone called the baculum or the *os penis*; however, even those species depend on erectile tissue for copulation.

The components of the semen other than sperm come from several accessory glands that contribute secretions to the urethra. About two-thirds of the volume of semen is seminal fluid, which comes from the **seminal vesicles**. Seminal fluid is thick because it contains mucus and protein. It also contains fructose, an energy source for the sperm, and modified fatty acids called prostaglandins that stimulate contractions in the female reproductive tract. These contractions are believed to help move the sperm up through the female reproductive tract.

One-fourth to one-third of the volume of semen is a thin, milky fluid that comes from the **prostate gland**. Prostate fluid makes the uterine environment more hospitable to sperm. The prostate also secretes a clotting enzyme that works on the protein in seminal fluid to convert semen into a gelatinous mass.

The prostate gland completely surrounds the urethra as it leaves the bladder. With age the prostate gland tends to enlarge, and thus creates problems for elderly men by blocking the urethra and making urination difficult. Prostate cancer is the second most common cancer in men. However, prostate cancer is relatively easy to diagnose, through a blood test that measures an enzyme produced by the prostate and can detect a rise in that enzyme resulting from cancerous growth of the gland. If detected early, prostate cancer is highly curable.

A relatively small volume of seminal fluid comes from the **bulbourethral glands**. This alkaline and mucoid secretion precedes the others and neutralizes acidity in the urethera and lubricates the tip of the penis.

The culmination of the male sex act propels semen through the vasa deferentia and the urethra in two steps, emission and ejaculation. During **emission**, rhythmic contractions of smooth muscles of the ducts containing sperm and of the accessory glands move sperm and the various secretions into the urethra at the base of the penis. **Ejaculation**, which follows emission, is caused by contractions of other muscles at the base of the penis surrounding the urethra. The rigidity of the erect penis allows these contractions to force the gelatinous mass of semen through the urethra and out of the body.

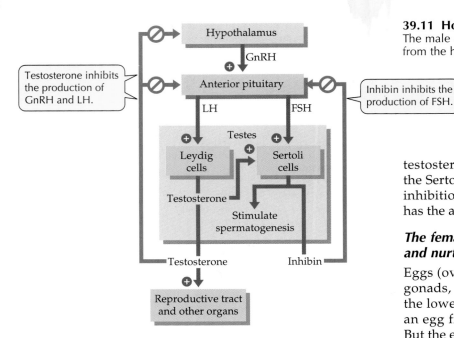

Testosterone inhibits the production of GnRH and LH.

Inhibin inhibits the production of FSH.

39.11 Hormones Control the Male Reproductive System
The male reproductive system is under hormonal control from the hypothalamus and the anterior pituitary.

testosterone. In addition, one hormone produced by the Sertoli cells seems to have as its only function the inhibition of LH and FSH production. This hormone has the appropriate name **inhibin**.

The female sex organs produce eggs, receive sperm, and nurture the embryo

Eggs (ova) are produced and released by the female gonads, the ovaries. Ovaries are paired structures in the lower part of the body cavity. Ovulation releases an egg from the ovary directly into the body cavity. But the egg can't go far. The ovary is enveloped by the fringes at the opening of the oviducts, which sweep the eggs into these tubes leading to the outside world.

Oviducts (also known as fallopian tubes) are paired, one for each ovary, and they are the site of fertilization if sperm are present in the female reproductive tract. Cilia lining the oviduct propel the egg slowly toward the **uterus**, or womb, which is a muscular, thick-walled cavity shaped like an upside-down pear. The embryo develops in the uterus, receiving its nutrition through the highly vascularized uterine wall. At the bottom of the uterus is an opening, the **cervix**, that leads into the **vagina**. Sperm are ejaculated into the vagina during copulation, and the baby passes through the vagina during birth. Figure 39.12 shows the female reproductive organs.

Two sets of skin folds surround the opening of the vagina and the opening of the urethra, through which urine passes. The inner, more delicate folds are the **labia minora** (singular labium minus); the outer, thicker folds are the **labia majora** (singular labium majus). At the anterior tip of the labia minora is the **clitoris**, a small bulb of erectile tissue that is the anatomical homolog of the penis. The clitoris is highly sensitive and plays an important role in sexual response. The labia minora and the clitoris consist of erectile tissue that becomes engorged with blood during sexual excitation.

The opening of an infant female's vagina is partly covered by a thin membrane, the *hymen*, which has no known function. Eventually the hymen becomes ruptured by vigorous physical activity or first sexual intercourse; it can make first intercourse difficult or painful for the female.

To fertilize an egg, sperm swim from the vagina up through the cervix, the uterus, and most of the

Male sexual function is controlled by hormones

Spermatogenesis and maintenance of male secondary sexual characteristics depend on **testosterone** produced by Leydig cells in the testes. In Chapter 38 we learned that increased production of testosterone at the time of puberty is due to an increased release of **gonadotropin-releasing hormone (GnRH)** by the hypothalamus, which in turn stimulates cells in the anterior pituitary to increase their secretion of **luteinizing hormone (LH)** and **follicle-stimulating hormone (FSH)**.

Leydig cells are stimulated by LH to produce testosterone. The rise in the level of testosterone in the prepubertal male causes the development of secondary sexual characteristics and the growth spurt, promotes increased muscle mass, and stimulates growth and maturation of the testes. If a male is castrated before puberty, he will not develop a deep voice, typical patterns of body hair, or a muscular build, and his external genitalia will remain childlike.

Continued production of testosterone after puberty is essential to maintain secondary sexual characteristics and to produce sperm. Testosterone production is controlled by the influence of LH on Leydig cells. Spermatogenesis is controlled by the influence of FSH and testosterone on Sertoli cells in the seminiferous tubules. In turn, the production of LH and FSH by the anterior pituitary is controlled by GnRH from the hypothalamus.

There are several negative feedback signals in the regulation of testes function (Figure 39.11). Hypothalamic production of GnRH and pituitary production of LH are inhibited by high levels of circulating

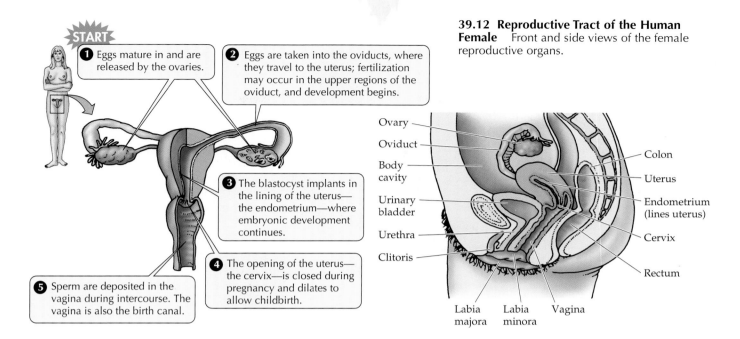

START

❶ Eggs mature in and are released by the ovaries.

❷ Eggs are taken into the oviducts, where they travel to the uterus; fertilization may occur in the upper regions of the oviduct, and development begins.

❸ The blastocyst implants in the lining of the uterus—the endometrium—where embryonic development continues.

❹ The opening of the uterus—the cervix—is closed during pregnancy and dilates to allow childbirth.

❺ Sperm are deposited in the vagina during intercourse. The vagina is also the birth canal.

39.12 Reproductive Tract of the Human Female Front and side views of the female reproductive organs.

Ovary
Oviduct
Body cavity
Urinary bladder
Urethra
Clitoris
Labia majora
Labia minora
Vagina
Colon
Uterus
Endometrium (lines uterus)
Cervix
Rectum

oviduct. The egg is fertilized in the upper region of the oviduct. The resulting zygote undergoes its first cell divisions, becoming a **blastocyst**, as it continues to move down the oviduct. When the blastocyst reaches the uterus, it attaches itself to the epithelial lining of the uterus, the **endometrium**. The endometrium and underlying cells of the uterine wall are a remarkable organ. Under hormonal control the uterine wall proliferates in anticipation of receiving a blastocyst.

Once attached to the endometrium, the blastocyst burrows into the uterine lining (a process called implantation) and interacts with these tissues to form a structure called the **placenta**. The placenta exchanges nutrients and waste products between the mother's blood and the baby's blood. If a blastocyst does not arrive in the uterus, the uterine lining regresses or is sloughed off. Thus, the female reproductive cycle consists of an *ovarian cycle* that produces eggs and hormones, and a *uterine cycle* that creates an appropriate environment for the embryo should fertilization occur.

The ovarian cycle produces a mature egg

At birth, a female has about a million primary oocytes in each ovary. By the time she reaches sexual maturity, she has only about 200,000 primary oocytes in each ovary; the rest have degenerated. During a woman's fertile years, her ovaries will go through about 450 ovarian cycles, and during each of these cycles one oocyte will fully mature and be released (Figure 39.13). At about 50 years old, she reaches **menopause**, the end of fertility, and may have only a few oocytes left in each ovary. Throughout a woman's life, oocytes are degenerating, and no new ones are produced.

Each primary oocyte in the ovary is surrounded by a layer of **follicle cells**. These cells, together with the eggs, constitute the functional unit of the ovary, the **follicle**. Between puberty and menopause, 6 to 12 follicles mature within the ovaries of a human female each month. In each of these follicles, the egg enlarges and the surrounding cells proliferate. After about a week one of these follicles is larger than the rest and continues to grow, while the others cease to develop and shrink. In the enlarged follicle, the follicle cells nurture the growing egg, supplying it with nutrients and even with macromolecules that it will use in early stages of development if it is fertilized.

After 2 weeks of growth, the follicle ruptures and releases an egg. Following **ovulation**, as this release is called, the follicle cells continue to proliferate and form a mass of endocrine tissue about the size of a marble. This structure, which remains in the ovary, is the **corpus luteum**. It functions as an endocrine gland, producing estrogen and progesterone for about 2 weeks. It then degenerates unless the egg meets a sperm and is fertilized. We will return to the corpus luteum and the hormonal aspects of the ovarian cycle later in this chapter.

The uterine cycle prepares an environment for the fertilized egg

The ovarian cycle in human females is about 28 days long, with ovulation occurring in the middle, and is paralleled by a cycle of buildup and breakdown of the endometrium, or uterine lining. About 5 days into the cycle, the uterine lining starts to grow and prepare to receive a blastocyst. The uterus attains its maximum

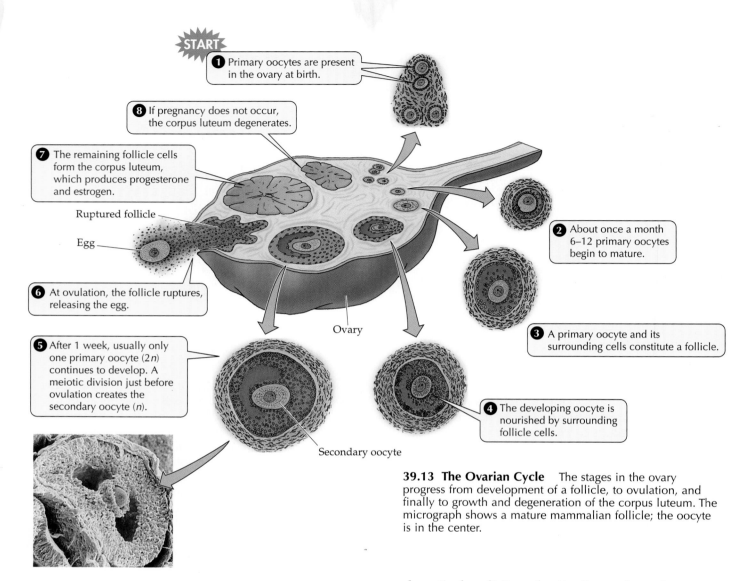

START

1 Primary oocytes are present in the ovary at birth.

8 If pregnancy does not occur, the corpus luteum degenerates.

7 The remaining follicle cells form the corpus luteum, which produces progesterone and estrogen.

Ruptured follicle

Egg

6 At ovulation, the follicle ruptures, releasing the egg.

5 After 1 week, usually only one primary oocyte (2n) continues to develop. A meiotic division just before ovulation creates the secondary oocyte (n).

Ovary

2 About once a month 6–12 primary oocytes begin to mature.

3 A primary oocyte and its surrounding cells constitute a follicle.

4 The developing oocyte is nourished by surrounding follicle cells.

Secondary oocyte

39.13 The Ovarian Cycle The stages in the ovary progress from development of a follicle, to ovulation, and finally to growth and degeneration of the corpus luteum. The micrograph shows a mature mammalian follicle; the oocyte is in the center.

state of preparedness about 5 days after ovulation and remains in that state for about another 9 days. If a blastocyst has not arrived by that time, the endometrium begins to break down and slough off—the process of **menstruation**, from *menses*, the Latin word for "months."

Some mammals have ovarian cycles shorter than 28 days; others have longer cycles. Rats and mice have ovarian cycles of about 4 days; many other mammalian species have only one ovarian cycle per year. The ovarian cycles of most mammals do not include a uterine cycle; instead, the uterine lining is reabsorbed. In these species the most obvious correlate of their ovarian cycles is a state of sexual receptivity called **estrus** at the time of ovulation. Therefore reproductive cycles in such species are frequently termed **estrous cycles**. When the female comes into estrus, or "heat,"

she actively solicits male attention and may be aggressive to other females. In many species of mammals, the female attracts males by releasing chemical signals, as well as through her sexual behavior. The human female is unusual among mammals in that she is potentially sexually receptive throughout her ovarian cycle and at all seasons of the year.

Hormones control and coordinate the ovarian and uterine cycles

The ovarian and uterine cycles of human females are coordinated and timed by the same hormones that initiate sexual maturation. Gonadotropins secreted by the anterior pituitary are the central elements of this control. Before puberty (that is, before about 11 years of age), secretion of gonadotropins is low, and ovaries are inactive. At puberty, the hypothalamus increases its release of gonadotropin-releasing hormone (GnRH), thus stimulating the anterior pituitary to secrete follicle-stimulating hormone (FSH) and luteinizing hormone (LH).

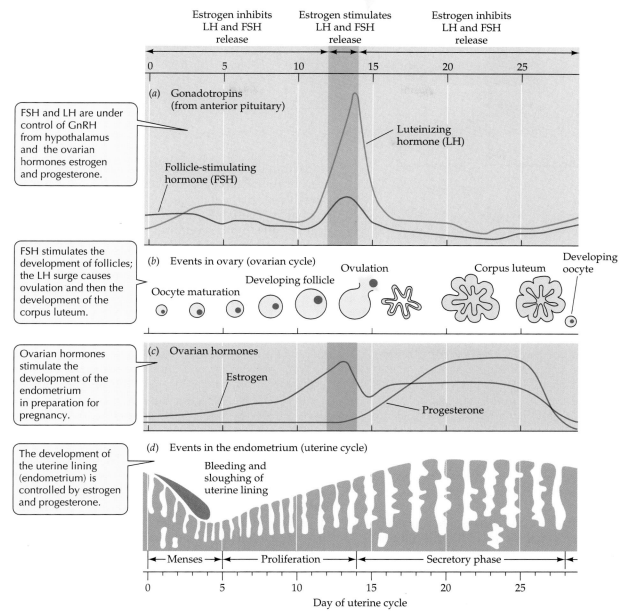

FSH and LH are under control of GnRH from hypothalamus and the ovarian hormones estrogen and progesterone.

FSH stimulates the development of follicles; the LH surge causes ovulation and then the development of the corpus luteum.

Ovarian hormones stimulate the development of the endometrium in preparation for pregnancy.

The development of the uterine lining (endometrium) is controlled by estrogen and progesterone.

39.14 The Uterine and Ovarian Cycles During a woman's uterine and ovarian cycles there are coordinated changes in (a) gonadotropins, (b) the ovary, (c) female sex hormones, and (d) the uterus. The cycle begins with the onset of menstruation; ovulation is at midcycle.

In response to FSH and LH, ovarian tissue grows and produces estrogen, and the follicles go through early stages of maturation (Figure 39.14). The rise in estrogen causes the development of secondary sexual characteristics, including maturation of the uterus. Between puberty and menopause (at which time ovarian cycles cease), interactions of gonadotropin-releasing hormone, gonadotropins, and sex steroids control the ovarian and uterine cycles.

Menstruation marks the beginning of the uterine and ovarian cycles (see Figure 39.14). A few days before menstruation begins, the anterior pituitary begins to increase its secretion of FSH and LH. In response, follicles begin to mature in the ovaries and follicle cells gradually increase production of estrogen. After about a week of growth, usually all but one of these follicles wither away. Occasionally more than one follicle continues to develop, making it possible for the woman to bear fraternal (nonidentical) twins. The one follicle that is still growing secretes increasing amounts of estrogen, stimulating the uterus to grow.

Estrogen exerts a negative feedback on gonadotropin release by the pituitary during the first 12 days of the ovarian cycle. Then, on about day 12, estrogen

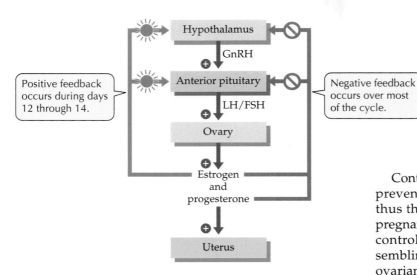

39.15 Hormonal Control of the Ovarian and Uterine Cycles With menstruation as the starting point, the ovarian and uterine cycles are under a complex series of positive and negative feedback controls involving several hormones.

exerts positive rather than negative feedback on the pituitary (Figure 39.15). As a result, there is a great surge of LH, and a lesser surge of FSH. The LH surge triggers the mature follicle to rupture and release the egg, and it stimulates follicle cells to develop into the corpus luteum and to secrete estrogen and progesterone.

Estrogen and especially progesterone secreted by the corpus luteum following ovulation are crucial to the continued development and maintenance of the uterine lining. In addition, these sex steroids send negative feedback to the pituitary, inhibiting gonadotropin release and thus preventing new follicles from beginning to mature.

If the egg is not fertilized, the corpus luteum degenerates on about day 26 of the cycle. Without the production of progesterone by the corpus luteum, the uterine lining sloughs off and menstruation occurs. The decrease in circulating steroids also relieves the negative feedback on the hypothalamus and pituitary, so GnRH, FSH, and LH all increase. The increase in these hormones induces the next round of follicles to develop, and the ovarian cycle begins again.

If the egg is fertilized, it divides numerous times, becoming a blastocyst as it travels down the oviduct. When the blastocyst arrives in the uterus and implants itself in the endometrium, a new hormone comes into play. A layer of cells covering the blastocyst begins to secrete **human chorionic gonadotropin (hCG)**. This gonadotropin, a molecular homolog of LH, keeps the corpus luteum functional. Because hCG is present in the blood only of pregnant women, the presence of this hormone is the basis for pregnancy testing. These tissues derived from the blastocyst also begin to produce estrogen and progesterone, eventually replacing the corpus luteum as the most important source of these sex steroid hormones.

Continued high levels of estrogen and progesterone prevent the pituitary from secreting gonadotropins; thus the ovarian cycle ceases for the duration of the pregnancy. The same mechanism is exploited by birth control pills, which contain synthetic hormones resembling estrogen and progesterone that prevent the ovarian cycle (but not the uterine cycle) by providing negative feedback to the hypothalamus and pituitary.

Human sexual responses consist of four phases

The sexual responses of both women and men consist of four phases: excitement, plateau, orgasm, and resolution (Figure 39.16). As sexual *excitement* begins in a woman, her heart rate and blood pressure rise, muscular tension increases, her breasts swell, and her nipples become erect. Her external genitals, including the sensitive clitoris, swell as they become filled with blood, and the walls of the vagina secrete lubricating fluid that facilitates copulation.

As a woman's sexual excitement increases, she enters the *plateau* phase. Her blood pressure and heart rate rise further, her breathing becomes rapid, and the clitoris begins to retract—the greater the excitement, the greater the retraction. The sensitivity that once focused in the clitoris spreads over the external genitals, and the clitoris itself becomes even more sensitive. *Orgasm* may last as long as a few minutes, and, unlike men, some women can experience several orgasms in rapid succession. During the *resolution* phase, blood drains from the genitals and body physiology returns to close to normal.

In the male, as in the female, the excitement phase is marked by an increase in blood pressure, heart rate, and muscle tension. The penis fills with blood and becomes hard and erect. In the plateau phase, breathing becomes rapid, the diameter of the glans increases, and a clear lubricating fluid oozes from the penis. Pressure and friction against the nerve endings in the glans and in the skin along the shaft of the penis eventually trigger orgasm. Massive spasms of the muscles in the genital area and contractions in the accessory reproductive organs result in ejaculation.

Within a few minutes after ejaculation, the penis shrinks to its normal size, and body physiology returns to resting conditions. The male sexual response includes a *refractory period* immediately after orgasm.

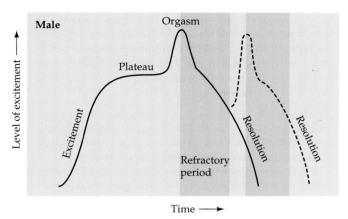

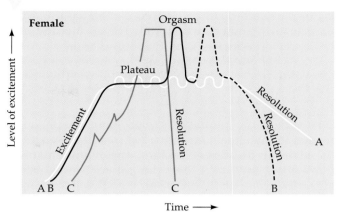

39.16 Human Sexual Responses The dashed lines show that both males and females may have repeated orgasms, but in males orgasms are separated by refractory periods during which sexual excitement cannot be maintained. Females have a diversity of response cycles, as shown by the three sets of lines labeled A, B, and C. The response cycle may be similar to that of the male (B). Alternatively, a female may experience sustained multiple orgasms (A) or may omit the plateau phase in a surge toward a very intense orgasm (C).

During this period, which may last 20 minutes or more, a man cannot achieve a full erection or another orgasm, regardless of the intensity of sexual stimulation.

A variety of technologies are used to control fertility

People use many methods to control the number of their children and the spacing between their children's births. The only absolutely sure method of preventing fertilization and pregnancy is **abstinence** from sexual activity. Since that approach is not acceptable to most people, they turn to a variety of methods to prevent pregnancy or conception, which therefore are called methods of **contraception**.

Some of these methods are used by the woman, others by the man. They vary from means of keeping sperm from meeting egg to means of preventing implantation of the blastocyst. In other words there are contraceptive methods targeted at just about every step of the reproductive process from gametogenesis to development of the embryo. Contraceptive methods vary enormously in their effectiveness and in their acceptability to those who use them. Here we review some of the most common contraception methods and their relative failure rates. Figure 39.17 gives an overview of birth control methods.

NATURAL APPROACHES. One "natural" contraceptive approach is to separate sperm and egg in time through the **rhythm method**. The couple avoids sex from day 10 to day 20 of the ovarian cycle, when the woman is most likely to be fertile. The cycle can be tracked by

METHOD	MODE OF ACTION	FAILURE RATE[a]
Rhythm method	Abstinence near time of ovulation	15–35
Coitus interruptus	Prevents sperm from reaching egg	10–40
Condom	Prevents sperm from entering vagina	3–20
Diaphragm/jelly; film	Prevents sperm from entering uterus; kills sperm	3–25
Vaginal jelly or foam	Kills sperm; blocks sperm movement	3–30
Douche	Supposedly flushes sperm from vagina	80
Birth control pills	Prevent ovulation	0–3
Vasectomy	Prevents release of sperm	0.0–0.15
Tubal ligation	Prevents egg from entering uterus	0.0–0.05
Intrauterine device (IUD)	Prevents implantation of fertilized egg	0.5–6
RU-486	Prevents development of fertilized egg	0–15
(Unprotected)	(No form of birth control)	(85)

[a] Number of pregnancies per 100 women per year

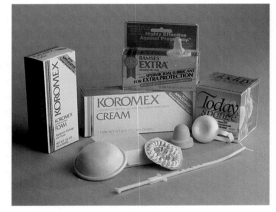

39.17 Methods of Contraception
Reproductive technologies are designed to block ovulation, fertilization, or gestation.

use of a calendar, supplemented by the basal body temperature method, which determines the day of ovulation on the basis of the observation that a woman's body temperature drops on the day of ovulation and rises sharply on the day afterward. Changes in the stickiness of cervical mucus also help determine the day of ovulation.

However, the problem is that some sperm deposited in the female reproductive tract may remain viable for up to 6 days. Similarly, the ovum remains viable for 2 to 3 days after ovulation. These facts, added to the variability in the timing of ovulation, result in an annual failure rate of between 15 and 35 percent for the rhythm method. In other words, 15 to 35 percent of women using the rhythm method for 1 year will become pregnant during that time.

Another natural approach is to try to separate the sperm and egg in space through **coitus interruptus**, withdrawal of the penis before ejaculation. The annual failure rate of this method can be as high as almost 40 percent.

BARRIER METHODS.   Two techniques for placing a physical barrier between egg and sperm have been used for more than a century. The **condom** is a sheath made of an impermeable material such as latex that can be fitted over the erect penis. A condom traps the ejaculate so that sperm do not enter the vagina. Condoms also help prevent the spread of sexually transmitted diseases such as AIDS, syphilis, and gonorrhea. In theory, the use of a condom can be highly effective, with failure rate near zero; in practice, the annual failure rate is about 15 percent, because of faulty technique.

The **diaphragm** is a dome-shaped piece of rubber with a firm rim that fits over the woman's cervix and thus blocks sperm from entering the uterus. Smaller than the diaphragm is the **cervical cap**, which fits snugly just over the tip of the cervix. Both the diaphragm and the cervical cap are treated first with contraceptive jelly or cream and then inserted through the vagina before intercourse. Annual failure rates are about the same as for condoms—about 15 percent.

Spermicidal foams, jellies, and creams can be used alone by being applied to the vagina with applicators. Used in this way, they have an annual failure rate of 25 percent or more. Recently spermicides have been made in the form of small, thin sheets of film that disolve when inserted in the vagina.

Douching (flushing the vagina with liquid after intercourse), in spite of popular belief, is almost entirely useless as a method of birth control. Remember that sperm can reach the upper regions of the oviducts within 10 minutes after ejaculation.

PREVENTING OVULATION.   The widely used **birth control pills** work by preventing ovulation. The mechanisms of action are based on the roles of estrogens and progesterone as negative feedback signals that work on both hypothalamus and pituitary to inhibit the release of luteinizing hormone and follicle-stimulating hormone. The most common pills contain low doses of synthetic estrogens and synthetic progesterones (progestins). By keeping the circulating levels of gonadotropins low, the birth control pill interferes with the maturation of follicles and ova. The ovarian cycle (but not the uterine cycle) is suspended.

The negative side effects of oral contraceptives have been the topic of extensive discussion. These side effects include increased risk of blood clot formation, heart attack, and stroke, but the once higher risk of these side effects was associated mostly with pills containing higher hormone concentrations than are used in current pills. For pills in use today, the risk of these side effects is low, except for women over 35 years old who smoke, for whom the risk is significantly greater. Risk of death from using the pill is less than that associated with a full-term pregnancy, and the pill is the most effective method of contraception other than sterilization. It has an annual failure rate of less than 1 percent.

Long-lasting injectable or implantable steroids are also used to block ovulation. DepoProvera® is an injectable progestin that blocks normal pituitary function for several months. Another device, called Norplant®, consists of thin, flexible tubes filled with progestin. Several of these tubes are inserted under the skin, where they continue to slowly release progestin for years.

The "mini-pill" is an oral birth control pill that contains very low doses of progestins. Although it may interfere with normal maturation and release of ova, its principal mode of action is to alter the environment of the female reproductive tract so that it is not hospitable to sperm. Cervical mucus normally becomes watery at the time of ovulation, but low levels of progestin keep the mucus thick and sticky so that it blocks the passage of sperm.

PREVENTING IMPLANTATION.   A highly effective method of contraception (with a failure rate varying from 1 percent to about 7 percent) is the intrauterine device, or **IUD**. The IUD is a small piece of plastic or copper that is inserted in the uterus. The IUD probably works by preventing implantation of the fertilized egg.

Another way of interfering with implantation is through the use of "morning-after pills," which deliver high doses of steroids, primarily estrogens. By acting in several ways on the oviduct and the uterine lining, this treatment prevents implantation. Morning-after pills can be effective for up to several days following intercourse.

A recent addition to birth control technology is a drug, **RU-486**, developed in France. RU-486 is not a

(a) Vasectomy

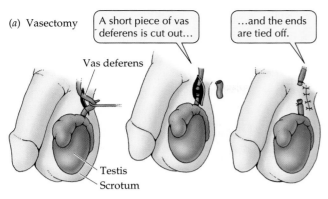

A short piece of vas deferens is cut out...

...and the ends are tied off.

Vas deferens

Testis

Scrotum

(b) Tubal ligation

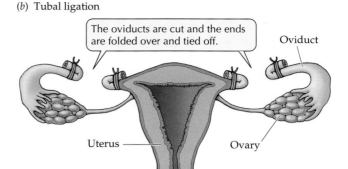

The oviducts are cut and the ends are folded over and tied off.

Oviduct

Uterus

Ovary

39.18 Sterilization Techniques (a) Vasectomy, the technique for male sterilization. (b) Tubal ligation is the sterilization procedure performed on human females.

contraceptive pill, but a *contragestational* pill. It is a progesterone-like molecule that blocks progesterone receptors. It therefore interferes with the normal role of progesterone produced by the corpus luteum, which is necessary for the maintenance of the uterine lining in early pregnancy. If RU-486 is administered as a morning-after pill, it prevents implantation. However, RU-486 can be effective even if taken at the time of the first missed menses after fertilization, after implantation has been initiated. After a few days of treatment with RU-486, the endometrium regresses and sloughs off along with the embryo, which is in very early stages of development.

STERILIZATION. The only certain methods of contraception are irreversible ones—sterilization of either the man or the woman. Male sterilization by vasectomy is a simple operation performed under a local anesthetic in a doctor's office. As shown in Figure 39.18*a*, each vas deferens is cut and the cut ends are tied. After this minor surgery, the ejaculate no longer contains sperm, because sperm cannot pass through the vas deferens after leaving the epididymis. Sperm production continues, but since the sperm cannot move out of the testes, they are destroyed by macrophages. Vasectomy does not affect a man's hormone levels or his sexual responses, and even the amount of semen he ejaculates is unchanged, because the sperm constitute so little of its volume.

In theory, vasectomies can be reversed by suturing the cut ends of the vas deferens back together, but the success rate is only around 50 percent. Even when the surgical repair is successful, the man may remain sterile because his immune system has begun to make antibodies that attack his own sperm. Because of the high probability that vasectomy will prove to be irreversible, before undergoing this procedure many men elect to store some of their sperm in a sperm bank,

where it is kept frozen and available to be used in artificial insemination.

In female sterilization, the aim is to make it impossible for the egg to travel to the uterus and to block sperm from reaching the egg. The most common method is **tubal ligation** ("tying the tubes"). A small piece is removed from each oviduct, and the ends of the oviduct are tied (Figure 39.18*b*). Alternatively, the oviducts may be burned (cauterized) to seal them off, a process called endoscopy. As in the male, these procedures do not alter reproductive hormones or sexual responses, and they can sometimes be reversed surgically.

ABORTION. Once a fertilized egg has successfully implanted itself in the uterus, the termination of a pregnancy is an abortion. A spontaneous abortion is the medical term for what is usually called a *miscarriage.* Miscariages are common early in a pregnancy; most of them occur because of an abnormality in the fetus or in the process of implantation. Abortions that are not spontaneous, but are due to intervention, are done either for therapeutic purposes or for birth control.

A therapeutic abortion may be necessary to protect the health of the mother, or it may be performed because the fetus has a severe defect. When performed in the first third of a pregnancy, a medical abortion carries less risk than a full-term pregnancy. The method is to dilate the cervix and then remove the fetus and the endometrium by physical means. After the first 12 weeks of pregnancy, the risk associated with a medical abortion rises substantially.

CONTROLLING MALE FERTILITY. You may ask why all the chemical approaches to controlling fertility apply to women and none have been devised to block male fertility. The control of male fertility is a difficult problem. First, since the production of sperm is not a cyclical event, it is not possible to block a particular step in the control of a cyclical process. The ovarian cycle is vulnerable to manipulation because certain events must happen at certain times for ovulation and implantation to occur. Second, the suppression of spermatogenesis must be total to be effective, since it takes only a single sperm to fertilize an egg, and normally millions are produced. Such suppression requires powerful chemical intervention, which has many unacceptable side effects.

Reproductive technologies help solve problems of infertility

There are many reasons why an apparently normal man and woman may not be able to have children. The man's rate of sperm production may be low, or his sperm may lack motility. The woman's oviducts may be blocked by scar tissue or by *endometriosis,* a proliferation of endometrial cells outside of the uterus. In some cases treatment with powerful chemicals to cure cancer damages the ability of the man's or the woman's gonads to produce gametes.

There are also reasons why a couple who wants children and can have children may voluntarily choose not to. If the man and the woman are both carriers for a recessive genetic disease, the probability that they will have an afflicted child is high, and they may not want to take that risk. Major recent developments in **assisted reproductive technologies** (ARTs) make it possible to overcome many of these kinds of barriers to childbearing.

The first successful ART was *in vitro* fertilization (IVF). In IVF, the woman is treated with hormones that stimulate many follicles in her ovaries to mature. Eggs are harvested from these follicles, and sperm are collected from the father. Eggs and sperm are combined in a culture medium (*in vitro*) where fertilization takes place. The resulting embryos in the blastocyst stage can then be injected into the mother's uterus or kept frozen for implantation later. The first "test tube baby" resulting from IVF was born in 1978. Since that time, thousands of babies have been produced by this ART. IVF is useful when the woman's oviducts are blocked, and its success rate is about 20 percent.

A technique called **gamete intrafallopian transfer** (GIFT) has been developed for cases in which only the entrance to the oviducts or the upper segment of the oviducts is blocked. This procedure injects harvested eggs and sperm directly into the upper regions of the oviduct, where fertilization produces a blastocyst, which enters the uterus via the normal route. GIFT has a success rate of about 30 percent.

A major cause of failure of IVF and GIFT is the lack of ability of the sperm to penetrate and fertilize eggs. In attempts to solve this problem, methods have been developed to inject a sperm cell directly into the cytoplasm of an egg. A harvested egg is held in place by

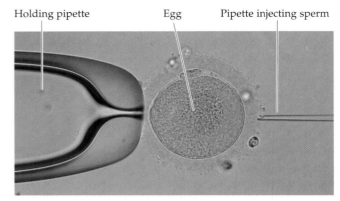

Holding pipette   Egg   Pipette injecting sperm

39.19 Assisted Reproductive Technologies Increase Fertility In this procedure, sperm are injected directly into a mature egg cell. The fertilized egg can then be placed back in the female reproductive tract.

suction applied to a polished glass pipette. A slender, sharp pipette is then used to penetrate the egg and inject a sperm (Figure 39.19). This ART was successful for the first time in 1992; now thousands of these procedures are performed in U.S. clinics each year, with a success rate of about 25 percent.

IVF, with or without introcytoplasmic injection of sperm, coupled with sensitive techniques of genetic analysis, can eliminate the risk that parents who are carriers for genetic diseases will produce afflicted children. With *in vitro* embryos, it is possible to take a cell from the 4- or 8-celled blastocyst without damaging the developmental potential of the embryo. The sampled cell can then be subjected to molecular techniques to determine if it carries the harmful gene. This information guarantees that only embryos that will not develop the genetic disease are implanted into the mother's uterus.

Sexual behavior transmits many disease organisms

Disease-causing organisms are parasites; they have very limited ability to lead a free existence not in association with a host organism. Hosts have finite life spans, which may be shortened considerably by the actions of the parasite. Therefore, the means of transmission from host to host is a major feature in the life cycles of disease-causing organisms. For example, viruses that cause colds and flu are transmitted in aerosols when infected individuals sneeze or cough.

Despite the success of cold and flu viruses, their means of transmission has a high probability of failure. Except in crowded conditions, there may not be another potential host in the immediate vicinity, and the longer the viruses are outside a host, the more they are exposed to unfavorable conditions, such as the ultraviolet radiation in sunlight.

One of the most intimate types of contact that host organisms can have is sexual interaction. It is not surprising then that many parasitic organisms have evolved to depend on sexual contact between their hosts as their means of transmission. These organisms are the causes of **sexually transmitted diseases** (commonly referred to as **STDs**), and they include viruses, bacteria, yeasts, and protozoans.

From Conception to Birth in Humans

Successful fertilization of a haploid egg by haploid sperm marks the beginning of the development of a new, genetically unique organism. In the next chapter we will examine fertilization in detail and see how different types of animals develop from a single-celled zygote to a fully formed individual. Here, however, it is important to recognize that in humans and most other mammals, reproduction involves an intimate partnership between the mother and her developing **embryo**, which is what the zygote becomes once cell division commences.

For the remainder of this chapter we will focus on the interactions between the mother and the embryo, beginning with fertilization, or conception, and continuing through the period of pregnancy, or **gestation**, until the birth of the baby.

The embryo implants itself in the lining of the uterus

The first cell divisions of a fertilized egg create an embryo that is a hollow ball of cells called a **blastocyst** (see Figure 40.7). It takes more than 3 days for the human blastocyst to travel down the oviduct to the uterus, where it lives free for the next 2 to 3 days. About the sixth day after fertilization, the blastocyst attaches to the lining of the uterus, thus beginning the process of implantation. The tissues of both the mother and the embryo

participate in implantation. The tissues of the embryo appear to invade the maternal endometrium, which responds by proliferating and growing more blood vessels (see Figure 40.15). For about the first 8 weeks of gestation, the embryo receives its nutrients directly from the endometrium in which it is embedded.

The placenta is an organ for nutrient exchange

As the embryo grows, its nutrient needs increase, and more efficient exchange between the mother and the embryo is required. This more efficient exchange of nutrients is accomplished by a specialized organ, the **placenta**, that arises out of the interaction between the endometrium and tissues of the embryo. Composed of tissues of both the mother and the embryo, the placenta lines a large area of the uterine wall. The embryo is connected to the placenta by means of its **umbilical cord**. Blood vessels from the embryo grow down the umbilical cord and branch in the placenta to form many fingerlike projections of tiny blood vessels (Figure 39.20).

The maternal blood flows into and out of the placenta in maternal blood vessels. Within the placenta, however, the maternal blood leaves the blood vessels and forms pools surrounding the fingerlike projections of the embryo's blood vessels. This is the site where the nutrients, dissolved gases, and wastes are exchanged between the blood systems of the mother and her offspring. The mother's blood and the embryo's blood remain separated in the placenta by very thin membranes of the embryo's blood vessels, but nutri-

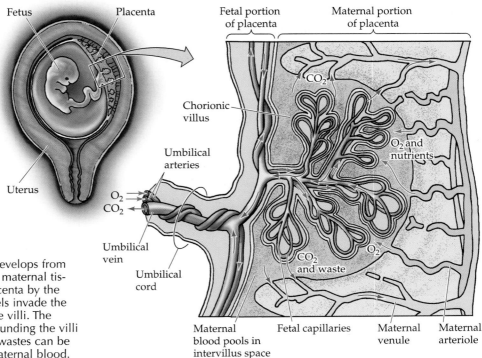

39.20 The Placenta The placenta develops from an interaction of embryonic tissue and maternal tissue. The embryo is attached to the placenta by the umbilical cord. Embryonic blood vessels invade the placental tissue to form many fingerlike villi. The maternal blood flows into spaces surrounding the villi so that nutrients, dissolved gases, and wastes can be exchanged between embryonic and maternal blood.

ents like oxygen and glucose can diffuse from mother's blood to embryo's blood, and wastes such as carbon dioxide can diffuse in the opposite direction.

The events of pregnancy can be divided into trimesters

The duration of pregnancy in mammals correlates positively with body size; in mice it is about 21 days, in cats and dogs about 60 days, in humans about 266 days, in horses about 330 days, and in elephants about 600 days. In discussing the events of human pregnancy, we divide it into three trimesters of about 3 months each.

THE FIRST TRIMESTER. The first trimester begins with fertilization, the formation of a ball of cells, or blastocyst, and the implantation of that blastocyst in the endometrial lining of the uterus. After implantation, tissues and organs of the embryo begin to differentiate (Figure 39.21). The human heart begins to beat in week 4, and limbs form by week 8. Most organs are present in at least primitive form by the end of the first trimester.

Because the first trimester is a time of rapid cell division and differentiation, it is the period during which the embryo is most sensitive to radiation, drugs, and chemicals that can cause birth defects. An embryo can be damaged before the mother even knows she is pregnant. By the end of the first trimester, the embryo ap-

pears to be a miniature version of the adult and is called a **fetus**.

Hormonal changes cause major and noticeable responses in the mother during the first trimester, even though the fetus at the end of that time is still so small that it would fit into a teaspoon. Soon after the blastocyst implants itself in the uterus, it begins to secrete human chorionic gonadotropin (hCG), the hormone that stimulates the corpus luteum to continue producing estrogen and progesterone. The high levels of these hormones prevent menstruation, which would abort the embryo, and exert negative feedback on the hypothalamus and the pituitary, inhibiting the release of follicle-stimulating hormone and luteinizing hormone and preventing a new round of ovulation. Side effects of these hormonal shifts are the well-known symptoms of pregnancy: morning sickness, mood swings, changes in the senses of taste and smell, and swelling of the breasts.

THE SECOND TRIMESTER. During the second trimester the fetus grows rapidly to a weight of about 600 g, and the mother's abdomen enlarges considerably. The limbs of the fetus elongate, and the fingers, toes, and facial features become well formed. Fetal movements are first felt by the mother early in the second trimester, and they become progressively stronger and more coordinated. By the end of the second trimester, the fetus may suck its thumb (Figure 39.22).

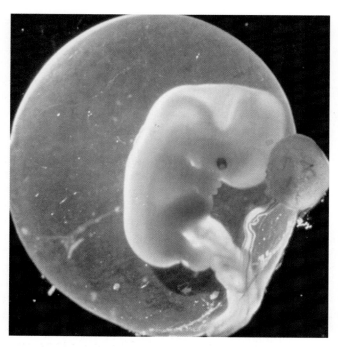

39.21 A Human Embryo The first trimester of pregnancy is a period of rapid cell division and differentiation; the organs and body structures of this 6-week-old embryo are forming rapidly.

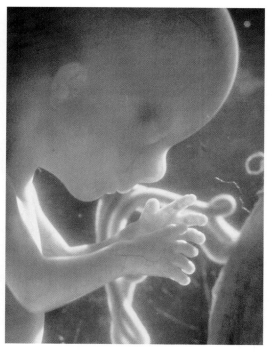

39.22 A Human Fetus At 4 months the fetus moves freely within its protective amniotic membrane. The fingers and toes are fully formed.

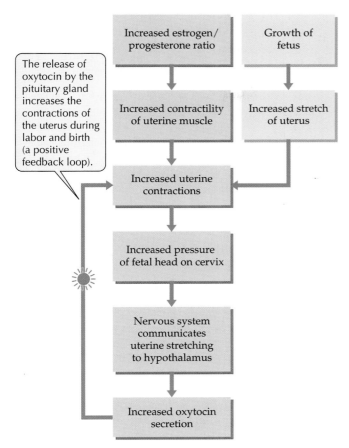

39.23 Increasing Oxytocin Production and Birth The release of oxytocin by the pituitary gland increases uterine contractions during labor and birth. Note the positive feedback loop.

The production of estrogen and progesterone by the placenta increases during the second trimester. As placental production of these hormones increases, the level of hCG and the activity of the corpus luteum decrease. The corpus luteum degenerates by the second trimester, but ovulation and menstruation are still inhibited by the high levels of steroids secreted by the placenta. Along with these hormonal changes, the unpleasant symptoms of early pregnancy usually disappear.

THE THIRD TRIMESTER. The fetus and the mother continue to grow rapidly during the third trimester. As the fetus approaches its full size, pressure on the mother's internal organs can cause indigestion, constipation, frequent urination, shortness of breath, and swelling of the legs and ankles. Throughout pregnancy the circulatory system of the fetus has been functioning, and as the third trimester approaches its end, other internal organs mature. The digestive system begins to function, the liver stores glycogen, the kidneys produce urine, and the brain undergoes cycles of sleep and waking.

Birth

Throughout pregnancy the uterus periodically undergoes slow, weak, rhythmic contractions called Braxton Hicks contractions. These contractions become gradually stronger during the third trimester and are sometimes called false labor contractions. True labor contractions usually mark the beginning of childbirth, or **parturition**. In some women, however, the first signs of labor are the discharge of the mucous plug that blocks the uterus during pregnancy ("a bloody show") or the rupture of the fluid-filled sack that surrounds the fetus ("water breaking").

LABOR. Many factors contribute to the onset of labor. Hormonal and mechanical stimuli increase the contractility of the uterus. Progesterone inhibits and estrogen stimulates contractions of uterine muscle. Toward the end of the third trimester the estrogen–progesterone ratio shifts in favor of estrogen. Oxytocin stimulates uterine contraction; its secretion by the pituitaries of both mother and fetus increases at the time of labor.

Mechanical stimuli for labor come from stretching of the uterus by the fully grown fetus and pressure of the fetal head on the cervix. These mechanical stimuli increase the pituitary release of oxytocin, which in turn increases activity of the uterine muscle, which causes even more pressure on the cervix. This positive feedback converts the weak, slow, rhythmic contractions of the uterus into stronger labor contractions (Figure 39.23).

In the early stage of labor, the contractions of the uterus are 15 to 20 minutes apart, and each lasts 45 to 60 seconds. During this time the contractions pull the cervix open until it is large enough to allow the baby to pass through. This stage of labor lasts an average of 12 to 15 hours in a first pregnancy and 8 hours or less in subsequent ones. Gradually the contractions become more frequent and more intense.

DELIVERY. In the second stage of labor, which begins when the cervix is fully dilated, the baby's head moves into the vagina and becomes visible from the outside (Figure 39.24). The usual head-down position of the baby at the time of delivery comes about when the fetus shifts its orientation during the seventh month. If the fetus fails to shift its position thus, at delivery a different part of the fetus enters the vagina first, and the birth is more difficult.

Passage of the fetus through the vagina is assisted by the woman's bearing down with her abdominal and other muscles to help push the baby along. Once the head and shoulders of the baby clear the cervix, the rest of its body eases out rapidly, but it is still connected to the placenta in the mother by the umbilical cord. This second stage of labor may take as little as a minute, or up to half an hour or more in a first pregnancy.

As soon as the baby clears the birth canal, it can start breathing and become independent of its mother's circulation. The umbilical cord may then be clamped and cut. The segment still attached to the baby dries up and sloughs off in a few days, leaving behind its distinctive signature, the belly button—more properly called the umbilicus. The detachment and expulsion of the placenta and fetal membranes take from a few minutes to an hour, and may be accompanied by uterine contractions.

In humans, the completion of delivery is the start of many years of nurturing and care for the young organism.

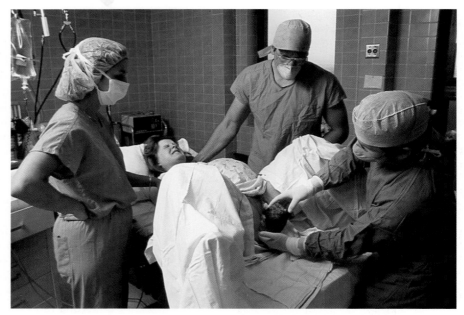

39.24 Delivery A new human being enters the world head-first.

Summary of "Animal Reproduction"

Asexual Reproduction

• Sexual reproduction is almost universal in animals, but some animals can reproduce asexually to produce offspring that are genetically identical to the parent and to each other. Advantages of asexual reproduction are avoidance of risks associated with mating, and efficient use of resources. A disadvantage of asexual reproduction is that no genetic diversity is produced.

• Means of asexual reproduction include budding, regeneration, and parthenogenesis. **Review Figure 39.1**

Sexual Reproduction

• Sexual reproduction consists of three basic steps: gametogenesis, mating, and fertilization. Gametogenesis and fertilization are similar in all animals, but mating includes a great variety of anatomical, physiological, and behavioral adaptations.

• In sexually reproducing species, genetic diversity is created by the recombination of genes during gametogenesis and by the independent assortment of chromosomes. Mating and fertilization also contribute to genetic diversity.

• Gametogenesis occurs in testes and ovaries. In spermatogenesis (the production of sperm) and oogenesis (the production of eggs), the primary germ cells proliferate mitotically, undergo meiosis, and mature into gametes. Spermatogonia continue to proliferate by mitosis throughout the life span of the male, and each spermatogonium can produce four haploid sperm through the two divisions of meiosis. **Review Figure 39.3**

• The females of some species can produce enormous numbers of diploid oogonia throughout life, but in other species, such as humans, the production of oogonia is limited to

before birth. These oogonia then mature sequentially throughout the reproductive years of the individual. To concentrate resources in the egg, each oogonium produces only one egg through meiosis. **Review Figure 39.3**

• Hermaphroditic species have both male and female reproductive systems in the same individual, either sequentially or simultaneously.

• Fertilization can occur externally in aquatic species that broadcast their gametes, or internally in species that copulate. An incredible diversity of anatomical and behavioral adaptations have evolved to bring about fertilization through copulation.

• Internal fertilization is necessary for nonaquatic species. The shelled egg is an important adaptation to the desiccating terrestrial environment, but it must be fertilized before the shell forms. All mammals except monotremes retain the embryo internally and thereby have done away with a shelled egg.

• Animals can be classified as oviparous or viviparous depending on whether the early stages of development occur inside or outside the mother's body.

Reproductive Systems in Humans and Other Mammals

• Male mammals produce and deliver semen into the female reproductive tract. Semen consists of sperm suspended in a fluid that nourishes them and facilitates fertilization.

• Sperm are produced in the seminiferous tubules of the testes, mature in the epididymis, and are delivered to the urethra through the vas deferens. Other components of semen are produced in the seminal vesicles, prostate gland, and bulbourethral glands. These secretions include fructose as a fuel supply, mucus and protein to create a gelatinous mass, alkaline substances to neutralize acidity in the uterus, and prostaglandins to stimulate motility in the uterus. All components of the semen join in the urethra at the base of the penis and are ejaculated through the erect penis by muscle contractions at the culmination of the sex act. **Review Figures 39.9, 39.10**

- Spermatogenesis depends on testosterone secreted by Leydig cells in the testes, which are under control of luteinizing hormone from the pituitary. Spermatogenesis is also controlled by follicle-stimulating hormone from the pituitary. Hypothalamic gonadotropin-releasing hormone controls pituitary secretion of LH and FSH. The production of these hormones by the hypothalamus and pituitary is controlled by negative feedback from testosterone and another testes hormone, inhibin. **Review Figure 39.11**
- Eggs (ova) mature in the ovaries and are released into the oviducts, which deliver the eggs to the uterus. Sperm deposited in the vagina during copulation swim up through the cervix into the uterus, some continuing up through the oviducts. **Review Figure 39.12**
- Fertilization occurs in the upper regions of the oviducts. The zygote becomes a blastocyst through repeated cell divisions as it passes down the oviduct. Upon arrival in the uterus, the blastocyst implants itself in the lining of the uterus, where a placenta forms and the embryo develops.
- The maturation and release of ova constitute an ovarian cycle under the control of the pituitary hormones FSH and LH. In humans this cycle takes 28 days. During the first half the ovum contained in a follicle matures under the influence of FSH. At midcycle a surge of FSH and LH causes ovulation. The remaining follicle cells form a corpus luteum, which secretes the hormones estrogen and progesterone. **Review Figures 39.13, 39.14**
- The uterus also has a cycle that prepares it for receipt of a blastocyst. During the first half of the uterine cycle estrogen stimulates the lining of the uterus to proliferate and become highly vascularized. As the uterine cycle continues, the lining is maintained by high levels of progesterone produced by the corpus luteum. If a blastocyst implants itself in the uterine wall, cells of the blastocyst produce human chorionic gonadotropin, which continues to stimulate the corpus luteum. If implantation does not occur, the secretory activity of the corpus luteum declines and the lining of the uterus deteriorates and sloughs off, which is the process of menstruation. **Review Figure 39.14**
- Both the ovarian and the uterine cycles are under control of hypothalamic and pituitary hormones, which in turn are under the feedback control of estrogen and progesterone. **Review Figure 39.15**
- Human sexual responses consist of four phases: excitement, plateau, orgasm, and resolution. In addition, males have a refractory period during which renewed excitement is not possible. **Review Figure 39.16**
- Methods to prevent pregnancy include abstention from copulation or the use of technologies that decrease the probability of fertilization. **Review Figure 39.17**
- Barrier methods of contraception, such as condoms, diaphragms, and spermacidal substances, block the passage of sperm in the female reproductive tract or weaken and kill them.
- Methods to prevent ovulation, such as birth control pills and other hormonal treatments, interfere with the ovarian cycle so that mature, fertile ova are not produced and released.
- Males and females can be sterilized by surgical blockage of the vasa deferentia (vasectomy) or oviducts (tubal ligation). **Review Figure 37.18**
- Methods to prevent implantation include intrauterine devices, excess doses of steroids, and a progesterone receptor blocker. After implantation, the termination of a pregnancy is called an abortion.

- New assisted reproductive technologies have been developed to increase fertility. ARTs include *in vitro* fertilization and gamete intrafallopian transfer. **Review Figure 39.19**
- Sexually transmitted diseases result from the transmission of disease-causing organisms through sexual behavior. Such organisms include yeasts, protozoans, bacteria, and viruses. Many STDs are curable if treated early, but they can have serious long-term consequences if not treated.

From Conception to Birth in Humans

- The first stage of the embryo, the blastocyst, implants itself in the endometrium of the uterus, from which the embryo receives nutrients for about the first 8 weeks of gestation.
- After the first 8 weeks of gestation, the embryo is nourished through the placenta, in which nutrients pass from the maternal to the fetal blood and wastes pass in the opposite direction. **Review Figure 39.20**
- Pregnancy is divided into three trimesters. In the first trimester, tissues and organs of the embryo differentiate rapidly; by the end of this trimester, the embryo has become a fetus. In the second trimester, the fetus grows rapidly and the mother's abdomen enlarges considerably. In the third trimester, the fetus continues its rapid growth to full size while its internal organs mature and begin functioning.
- Birth is initiated by uterine contractions stimulated by hormonal changes and by mechanical forces that stretch the uterus and apply pressure on the cervix. Toward the end of the third trimester, progesterone levels fall and estrogen levels rise, thus increasing the sensitivity of the uterine muscle and the cervix to being stretched. The response to these mechanical stimuli is the release of oxytocin by the posterior pituitary, which increases the contractions of the uterine muscle. Thus a positive feedback system is created that drives labor and results in delivery. **Review Figure 39.23**

Self-Quiz

1. Match each of the following modes of asexual reproduction with the statement or description that characterizes it. (Each letter may be used more than once, and more than one letter may apply to each statement.)
   a. Budding
   b. Regeneration
   c. Parthenogenesis
   (*i*) This form of asexual reproduction usually follows an animal's being broken by an external force, but it can also be initiated by the animal itself.
   (*ii*) Many freshwater sponges produce clusters of undifferentiated cells that eventually "escape" the parent and become free-living organisms genetically identical to the parent.
   (*iii*) Offspring develop from unfertilized eggs.
   (*iv*) This process requires totipotent cells.
   (*v*) Species that reproduce in this way may also engage in sexual reproduction.

2. A species in which the individual possesses both male and female reproductive systems is termed (choose all that apply)
   a. dioecious.
   b. parthenogenetic.
   c. hermaphroditic.
   d. diploid.
   e. monoecious.

3. The major advantage of internal fertilization is that
   a. it ensures paternity.
   b. it permits the fertilization of many gametes.
   c. it reduces the incidence of destructive competitive interactions between the members of a group.
   d. it results in the formation of a stable pair bond between mates.
   e. it gives the developing organism a greater degree of protection during the early phases of development.

4. Which statement about oocytes is true?
   a. At birth, the human female has produced all the oocytes she will ever produce.
   b. At the onset of puberty, ovarian follicles produce new ones in response to hormonal stimulation.
   c. At the onset of menopause, the human female stops producing them.
   d. They are produced by the human female throughout adolescence.
   e. Those produced by the female are stored in the seminiferous tubules.

5. Spermatogenesis and oogenesis differ in that
   a. spermatogenesis produces gametes with greater energy stores than those produced by oogenesis.
   b. spermatogenesis produces four equally functional diploid cells per meiotic event and oogenesis does not.
   c. oogenesis produces four equally functional haploid cells per meiotic event and spermatogenesis does not.
   d. spermatogenesis produces many gametes with meager energy reserves, whereas oogenesis produces relatively few, well-provisioned gametes.
   e. in humans, spermatogenesis begins before birth, whereas oogenesis does not start until the onset of puberty.

6. Semen contains all of the following except
   a. fructose.
   b. mucus.
   c. clotting enzymes.
   d. substances to reduce the pH of the uterine environment.
   e. substances to increase the motility of the uterine muscles.

7. During oogenesis in mammals, the second meiotic division occurs
   a. in the formation of the primary oocyte.
   b. in the formation of the secondary oocyte.
   c. before ovulation.
   d. after fertilization.
   e. after implantation.

8. One of the major differences between the sexual response cycles in human males and females is
   a. the increase in blood pressure in males.
   b. the increase in heart rate in females.
   c. the presence of a refractory period in females after orgasm.
   d. the presence of a refractory period in males after orgasm.
   e. the increase in muscle tension in males.

9. Which of the following is true of sexually transmitted diseases?
   a. They are always caused by viruses or bacteria.
   b. Using any form of contraception will prevent them.
   c. The organisms that cause them have evolved to depend on intimate physical contact as their means of transmission.

d. Their mode of transmission has a high probability of failure.
e. You cannot catch one from someone you love.

10. Contractions of muscles in the uterine wall and in the breasts are stimulated by
    a. progesterone.
    b. estrogen.
    c. prolactin.
    d. oxytocin.
    e. human chorionic gonadotropin.

Applying Concepts

1. Compare and contrast spermatogenesis and oogenesis in terms of their products and their timetables. Draw a diagram for each that shows the hormonal control mechanisms.

2. Explain how birth control pills prevent fertilization.

3. Ovarian and uterine events in the month following ovulation differ depending on whether fertilization occurs. Describe the differences and explain their hormonal controls.

4. Explain how positive feedback plays a role in birth.

Readings

Beaconsfield, P., G. Budwood and R. Beaconsfield. 1980. "The Placenta." *Scientific American*, August. A description of the implantation and development of the organ, including its many functions, that is the intermediary between fetus and mother.

Crews, D. 1994. *Scientific American*, January. A look at the range of mechanisms that determine whether an individual takes on masculine or feminine traits. Cross-species comparisons offer some surprising insights into the nature of sexuality.

Epel, D. 1977. "The Program of Fertilization." *Scientific American*, November. A discussion of the initial events in the interaction between sperm and egg.

Gilbert, S. F. 1997. *Developmental Biology*, 5th Edition. Sinauer Associates, Sunderland, MA. An excellent text on animal development that includes chapters on fertilization, as well as on the germ line and gametogenesis.

Johnson, M. H. and B. J. Everitt. 1995. *Essential Reproduction*, 4th Edition. Blackwell Scientific, Oxford. A concise and comprehensive technical account of the biology of gametogenesis, fertilization, and pregnancy.

Jones, R. E. 1996. *Human Reproductive Biology*, 2nd Edition. Academic Press, New York. An up-to-date treatment of the biological and biomedical aspects of human reproduction that includes extensive coverage of the technologies of fertility control, as well as discission of controversial topics of sex differences in the brain and the causes and patterns of gender identity and sexual behavior.

Katchadourian, H. A. 1989. *Fundamentals of Human Sexuality*, 5th Edition. Saunders, Philadelphia. An introductory text that covers the anatomy and physiology of sex and reproduction in the first two chapters and then discusses developmental, behavioral, and social aspects of sex.

Short, R. V. 1984. "Breast Feeding." *Scientific American*, April. A discussion of how trends toward bottle feeding and away from breast feeding, which has hormonally derived contraceptive effects, in developing countries may be causing population growth rates to rise.

Wassarman, P. M. 1988. "Fertilization in Mammals." *Scientific American*, December. An examination of the molecular and cellular events that surround the fusion of sperm and egg.

Chapter 40

Animal Development

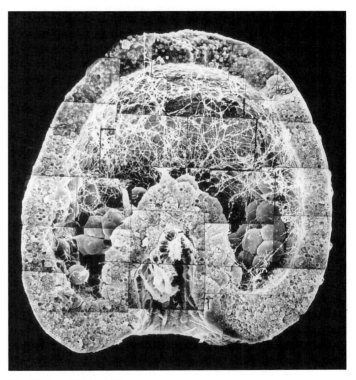

The Time of Your Life
For almost all animal species, gastrulation is an extremely important developmental stage in which a hollow ball of cells becomes an embryo with three distinct layers and well-defined body axes. This scanning electron micrograph of the gastrula of a sea urchin (*Lytechinus pictus*) was created from 40 separate images.

"It is not birth, marriage, or death, but gastrulation, which is truly the most important time of your life." This is the opinion of developmental biologist Lewis Wolpert. Events that you might consider to be the most important in your life—such as graduating from high school or college, getting married, getting your first job, having a child, or winning the Nobel prize—Wolpert finds trivial in comparison to gastrulation. What, you may ask, *is* gastrulation? It is an early stage of animal development that we will learn about in this chapter, but to appreciate its importance we need to consider how remarkable development is.

All animals begin life as a fertilized egg, a single cell called a *zygote*. The zygotes of sea urchins, frogs, alligators, eagles, blue whales, and humans all look very much alike, but through development they acquire the distinguishing features that make them very different animals. Through development, a single cell produces trillions of cells of enormous variety, which make up different tissues, organs, and organ systems.

We learned in Chapter 15 that differentiation of cell types involves differential gene expression, but what determines when and where cells differentiate? As cells differentiate, how do all the tissues end up in the right locations, how do all the organs have the right form and organization, and how do nerves and blood vessels make proper connection? In other words, how does the alligator acquire forelegs, the eagle wings, the blue whale flippers, and the human arms?

The developmental program depends heavily on positional information. That is, a cell possesses and receives certain information that will affect its development by virtue of its position in the embryo. Even in the earliest cell divisions, cells receive differential amounts of the contents of the cytoplasm of the fertilized egg. Some of these cytoplasmic contents are nutrients and some are informational molecules such as mRNA. Thus even very early in development, cells in different parts of the embryo have different developmental potential. They can differentiate into different tissues, but more importantly they may have different abilities to produce signals that will influence the developmental processes in neighboring cells.

Cell movements in the early embryo, such as those that occur during gastrulation, put cells in different associations with each other. New contacts are new opportunities for developmental signals to pass between cells, inducing the cells to follow certain paths of differentiation. As they differentiate, cells may become capable of producing new signals and making new movements that will put them in contact with still different populations of cells. This cascade of cell interactions results in progressive changes in gene expression and growth, which are the essence of development.

For all this to work correctly, cells must come into contact with each other at the right times and in the right places. The cell movements that occur during gastrulation set the stage for the positional information cascade that controls development. Any mistake during this early stage of development will be amplified throughout later stages and will be disastrous. That's why Professor Wolpert is so emphatic in his view that gastrulation is the most important time in your life.

In this chapter we will learn about the early stages of embryological development, during which the positional information cascade is creating the basic body form of the animal. We will begin with the events of fertilization, which can be taken as the starting point for the development of the organism. From there we will see how the zygote is converted by rapid cell divisions into a mass of similar cells. These cells then change position relative to each other during gastrulation, establishing new contacts between different groups of cells. These contacts initiate sequences of changes in cell growth, cell movements, and cell differentiation from which emerge the overall body plan, the various tissues, and the rudimentary organs of the adult.

To appreciate both the diversity and the similarity in the development of different animals, we will compare these early developmental steps in four favorite organisms of developmental biologists: sea urchins (invertebrates), and frogs, chicks, and humans (all vertebrates).

Fertilization: Changes in Sperm and Egg

Development of the embryo begins with **fertilization**—the union of sperm and egg, creating a diploid **zygote**. Fertilization is not a single event, but a com-

plex series of processes. As we learned in Chapter 39, fertilization begins with the meeting of sperm and egg, accomplished in most species by sexual behavior. The final distance between sperm and egg must be bridged by the motility of sperm because eggs are nonmotile.

The meeting of sperm and egg triggers several sequential events: The sperm is activated so that it is capable of gaining access to the plasma membrane of the egg, the membranes of the sperm and the egg fuse, and the egg is activated. Activation of the egg establishes blocks to entry by additional sperm, stimulates the final meiotic division of the egg nucleus, and initiates the first stages of development. The last event of fertilization is the fusion of the egg and sperm nuclei to create the diploid nucleus of the zygote. In the discussion that follows, we will look at each of these steps in turn.

Sperm activation guarantees species specificity

EXTERNAL FERTILIZATION. Animals such as sea urchins that use external fertilization must first solve two problems. The sperm must find an egg, and when sperm and egg do meet, something must guarantee that they are of the same species. These problems have been solved by a variety of species through different mechanisms of chemotaxis. The eggs release an attractant molecule, which for one species of sea urchin is a peptide consisting of 14 amino acids. This peptide influences only sperm of the same species, causing them to swim vigorously toward the site of maximum attractant concentration.

Further guarantees of species specificity for fertilization in many marine invertebrates are built into a process by which the sperm is activated by the jelly coat surrounding the egg. Recall from Chapter 39 that the sperm head is capped by a vesicle called the acrosome, which contains digestive enzymes. Exocytosis of these enzymes enables the sperm to create a passage through the egg's jelly coat and gain access to the proteinaceous **vitelline envelope** immediately surrounding the egg.

When a sperm contacts the jelly coat of an egg of the same species and the acrosomal membrane breaks down, an acrosomal process extends out of the head of the sperm. The **acrosomal process** is a long extension that forms from globular actin behind the acrosome, which polymerizes when the acrosomal membrane breaks down (Figure 40.1).

The acrosomal process extends through the jelly coat to make contact with the vitelline envelope. Herein lies another species recognition mechanism. The acrosomal process of the sea urchin is coated with membrane-bound protein molecules called **bindin**. Different species have different bindin molecules. The membrane of the vitelline envelope has species-spe-

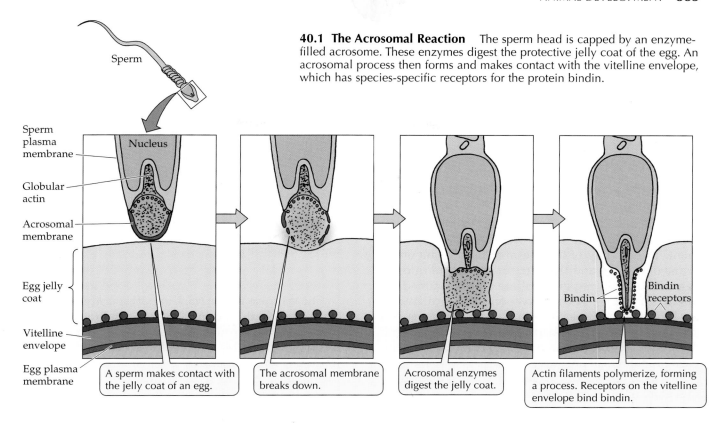

40.1 The Acrosomal Reaction The sperm head is capped by an enzyme-filled acrosome. These enzymes digest the protective jelly coat of the egg. An acrosomal process then forms and makes contact with the vitelline envelope, which has species-specific receptors for the protein bindin.

Sperm

Sperm plasma membrane

Nucleus

Globular actin

Acrosomal membrane

Egg jelly coat

Vitelline envelope

Egg plasma membrane

A sperm makes contact with the jelly coat of an egg.

The acrosomal membrane breaks down.

Acrosomal enzymes digest the jelly coat.

Actin filaments polymerize, forming a process. Receptors on the vitelline envelope bind bindin.

Bindin

Bindin receptors

cific bindin receptors. The reaction of acrosomal bindin with receptors of the vitelline envelope attaches the sperm to the egg and stimulates the egg membrane to form a depression, called the *fertilization cone*, that facilitates entry of the sperm into the egg.

In summary, the meeting and matching of sperm and egg in marine invertebrates is assisted by species-specific mechanisms of sperm attraction, sperm activation, and sperm binding to the egg.

INTERNAL FERTILIZATION. In animals such as mammals that practice internal fertilization, mating behaviors guarantee species specificity; thus egg–sperm recognition mechanisms are not so highly developed. However, mammalian sperm still face a formidable task after they are ejaculated into the female's reproductive tract. They must swim up from the vagina, through the uterus, and into the oviducts, where they might find an egg. The mammalian egg is surrounded by a glycoprotein envelope called the **zona pellucida** that is functionally similar to the vitelline envelope of sea urchin eggs. The egg and the zona pellucida are surrounded by another, thicker layer called the **cumulus**, which consists of follicle cells in a jellylike matrix (Figure 40.2).

When sperm are first deposited in the vagina, they are not capable of mounting an acrosomal reaction, but after being in the female reproductive tract for a time the sperm undergo **capacitation**. A capacitated sperm is capable of interacting with an egg and its barriers.

In mice, the first of those interactions is binding of the sperm by the zona pellucida. Specific glycoproteins in the zona pellucida bind to receptors on the acrosomal membrane. This binding attaches the sperm and induces release of the enzymes contained in the

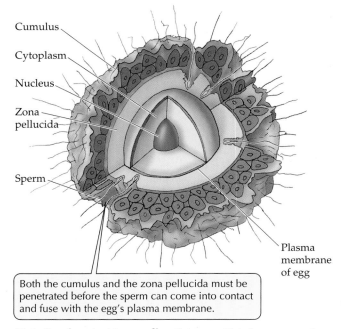

Cumulus

Cytoplasm

Nucleus

Zona pellucida

Sperm

Plasma membrane of egg

Both the cumulus and the zona pellucida must be penetrated before the sperm can come into contact and fuse with the egg's plasma membrane.

40.2 Barriers to Mammalian Sperm This human egg is protected by its cumulus and zona pellucida, both of which the sperm must digest in order to fertilize the egg.

acrosome. These enzymes digest a path through the zona pellucida so that the sperm can come into contact with the plasma membrane of the egg (see Figure 40.1). Upon contact with the plasma membrane, proteins, called fertilins, on the head of the sperm bind with receptors on the egg membrane, bringing about fusion of egg and sperm membrane and entry of the sperm nucleus, mitochondria, centriole, and flagellum into the egg cytoplasm.

Sperm entry activates the egg and triggers blocks to polyspermy

The unfertilized egg is metabolically sluggish, conserving its resources for the early stages of development. The binding of the sperm to the plasma membrane of the egg and the entry of the sperm into the egg activate the egg and initiate a programmed sequence of events. The first responses to fertilization are blocks to polyspermy—that is, mechanisms that prevent more than one sperm from entering the egg. If more than one sperm enters the egg, the resulting embryo probably will not survive.

Blocks to polyspermy have been studied intensively in sea urchins, which have large eggs that can be fertilized in a dish of seawater. Within a tenth of a second after a sperm enters a sea urchin egg, there is an influx of sodium ions. This increase in sodium concentration within the egg changes the electric potential across the egg's plasma membrane. The change prevents the entry of additional sperm and is called the *fast block* to polyspermy (Figure 40.3).

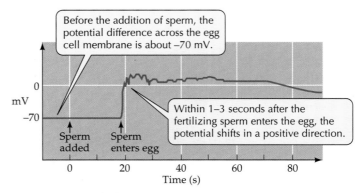

40.3 The Fast Block to Polyspermy The fast block to polyspermy is a change in the electric potential across the egg cell membrane.

There is also a *slow block* to polyspermy that takes 20 to 30 seconds and involves the vitelline envelope immediately surrounding the plasma membrane of the egg (Figure 40.4). The vitelline envelope is bonded to the plasma membrane, and just under the plasma membrane are *cortical vesicles* filled with enzymes. The sea urchin egg, like all other animal eggs, also contains calcium that is sequestered in organelles within the cell.

When a sperm enters, the egg releases calcium into its own cytoplasm. The increase in calcium causes the cortical vesicles to fuse with the plasma membrane

40.4 The Slow Block to Polyspermy Enzymes from the egg's cortical vesicles trigger the reactions of the slow block to polyspermy.

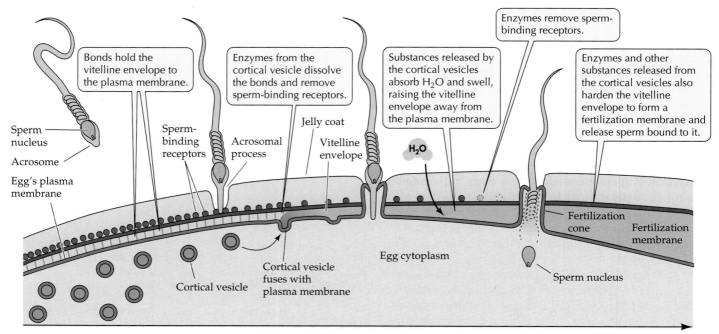

and release their enzymes outside the cell (through exocytosis). The enzymes break the bonds between the vitelline envelope and the plasma membrane. Water then flows by osmosis into the space between the vitelline envelope and the plasma membrane. This inflow of water raises the vitelline envelope to form what is called the *fertilization membrane*. The enzymes also remove unused sperm-binding receptors from the surface of the fertilization membrane and cause it to harden, preventing the passage of additional sperm.

After fertilization, the release of calcium ions activates the egg metabolically. The pH of the cytoplasm increases, oxygen consumption rises, and protein synthesis increases. However, the nuclei of the sperm and the egg do not fuse until some time after the sperm enters the egg's cytoplasm—about 1 hour in sea urchins and about 12 hours in mammals. The nucleus of the egg must complete its second meiotic division (see Figure 39.3) before egg and sperm nuclei unite.

The sperm and the egg make different contributions to the zygote

Most of the cytoplasm of the fertilized egg comes from the egg. This cytoplasm is well stocked with nutrients and a variety of molecules, such as mRNAs, that will be important in early development. Moreover, because most of the mitochondria in the zygote come from the egg, most of the mitochondria in the developing animal will be derived from its mother. In addition to its haploid nucleus, the sperm makes an important contribution, its centriole. The flagellum of the sperm disintegrates in the egg cytoplasm, but the centriole that gave rise to the microtubules of the flagellum remains. This centriole becomes the centrosome of the zygote that produces the mitotic spindles for subsequent cell divisions.

For a long time it was assumed that the one thing that sperm and egg contribute equally to the zygote is their haploid nuclei. However, we now know that even though they are equivalent in terms of genetic material, mammalian sperm and eggs are not equivalent in terms of their roles in development. In the laboratory, it is possible to construct zygotes in which both haploid nuclei come from the mother or both come from the father. In neither case does development progress normally. Apparently certain genes involved in development are active only if they come from a sperm and others are active only if they come from an egg.

Fertilization causes rearrangements of egg cytoplasm

As already mentioned, the egg cytoplasm is a storehouse of nutrients and molecules that play roles in early development. These cytoplasmic components are not homogeneously distributed and therefore will not be divided equally among all daughter cells when cell divisions begin. This unequal division of cytoplasmic components sets the stage for the unfolding program of positional information that orchestrates sequential steps of determination, differentiation, and morphogenesis (the development of form). The entry of the sperm into the egg can stimulate rearrangements of egg cytoplasm that will determine how the radial symmetry of the egg is transformed into the body axes of the embryo.

Upon fertilization, the rearrangements of egg cytoplasm in some species of frogs can be observed because of pigments in the egg cytoplasm. The nutrient molecules in an unfertilized frog egg are dense and therefore concentrated by gravity in the lower hemisphere of the egg, which is called the **vegetal hemisphere**. The haploid nucleus of the egg is located at the opposite end of the egg, the **animal hemisphere**.

The outermost cytoplasm of the animal hemisphere is heavily pigmented, and the underlying cytoplasm has more diffuse pigmentation. The vegetal hemisphere is not pigmented. Sperm always enter the egg in the animal hemisphere; afterward, the heavily pigmented cytoplasm rotates toward the site of sperm entry. This rotation of the outermost cytoplasm creates a band of diffusely pigmented cytoplasm on the side of the egg opposite the site of sperm entry. This band, called the **gray crescent**, will be the site of important developmental events (Figure 40.5).

The cytoplasmic rearrangements that create the gray crescent bring different regions of cytoplasm into contact on opposite sides of the egg. Therefore, bilateral symmetry is imposed on what was a radially symmetrical egg. There are now left–right and ante-

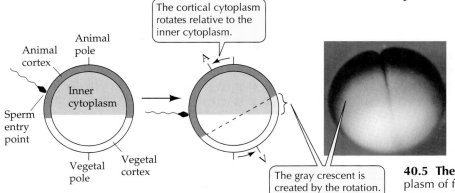

40.5 The Gray Crescent Rearrangements of cytoplasm of frog eggs after fertilization create the gray crescent.

rior–posterior axes, in addition to the up–down difference of the animal and vegetal hemispheres. In the frog, the site of sperm entry will become the head end, and the gray crescent will become the tail end of the embryo.

Cleavage: Repackaging the Cytoplasm

The first stage of development is a rapid series of cell divisions called **cleavage**. Because the cytoplasm of the zygote is not homogeneous, these first cell divisions differentially distribute nutrients and informational molecules such as mRNA into the cells of the early embryo. During this process, DNA synthesis and cell divisions proceed rapidly with no growth and little gene expression. Thus, the embryo becomes a ball of smaller and smaller cells. This ball or mass of cells is called a **blastula** when it forms a central cavity called a **blastocoel** (a process called blastulation). The individual cells are called **blastomeres**. Blastulation happens at different stages in different animals. For example, in sea urchins and frogs it occurs at about the 128-cell stage, in mammals it occurs at the 32-cell stage, but in birds it occurs much later, when there are about 60,000 cells.

In the discussions that follow, you will see that cleavage, and therefore the formation of the blastula, is influenced in different species by two major factors. First, when massive amounts of nutrients, or **yolk**, are stored in the egg, they will influence the patterns of cell divisions by impeding the formation of cleavage furrows. Second, proteins and mRNAs stored in the egg by the mother will guide the formation of mitotic spindles and the timing of cell divisions. We'll also consider unique characteristics of mammalian cleavage, and the construction of maps that identify the contribution of specific blastomeres to later tissues.

The amount of yolk influences cleavage

When the yolk content of the egg is sparse, there is little interference with cleavage furrow formation and daughter cells are of similar size; the sea urchin egg provides an example (Figure 40.6a). More yolk means more resistance to cleavage furrow formation; therefore, cell divisions progress more rapidly in the animal hemisphere than in the vegetal hemisphere. As a result, the cells derived from the vegetal hemisphere are fewer and larger; the frog egg provides an example (Figure 40.6b). In spite of this difference between sea urchin and frog eggs, in both cases the cleavage furrows completely divide the egg mass; thus these animals are said to have *complete cleavage*.

In contrast, when the egg contains a lot of yolk, such as in the chicken egg, the cleavage furrows cannot penetrate the yolk. As a result, cleavage is incomplete, and the embryo forms as a disc of cells, a **blas-**todisc, on top of the yolk mass (Figure 40.6c). This type of cleavage, called *incomplete cleavage,* and the resulting formation of a blastodisc is common in fish, reptiles, and birds.

Another type of incomplete cleavage occurs in insects such as the fruit fly, *Drosophila*. In the insect egg, the massive yolk is centrally located (Figure 40.6d). Early in development, the nucleus of the fertilized egg divides repeatedly, forming many daughter nuclei, but cytokinesis is not completed and plasma membranes do not form. The zygote remains a single cell with many nuclei. After seven or eight mitotic cycles, the nuclei migrate to the periphery of the egg and continue to divide. After several more mitotic cycles, the original plasma membrane of the egg begins to fold inward between the individual nuclei, thus partitioning them off into individual cells. At the end of this process the embryo consists of a single layer of cells surrounding a yolky core. This type of cleavage, seen in most arthropods, is called *superficial cleavage.*

The orientation of mitotic spindles influences the pattern of cleavage

Molecules in the egg cytoplasm also influence cleavage patterns, as is best illustrated in eggs that undergo complete cleavage. The positions of the mitotic spindles during cleavage are not random; they are determined by cytoplasmic factors. In turn, the orientation of the mitotic spindles determines the cleavage planes and, therefore, the arrangement of the daughter cells.

If the mitotic spindles of successive cell divisions form at right angles to each other, the cleavage pattern is *radial*, as in the sea urchin and the frog (see Figure 40.6a and b). In mollusks, however, the successive mitotic spindles are not at right angles, and the resulting pattern formed by cleavage has a twist and is called *spiral*. The coiling of snail shells is an expression of this form of cleavage. The orientations of successive mitotic spindles in mammals are at right angles, but the sequence is different from that in sea urchins, resulting in a pattern called *rotational cleavage.*

Cleavage in mammals is unique

Mammalian eggs are small and not easy to study, because they develop within the female reproductive tract. Therefore, knowledge of early stages of mammalian development lagged behind knowledge of all the other species we have been discussing. It turns out that mammalian development is quite different. Part of that difference has to do with the fact that the fertilized mammalian egg must produce both an embryo and elaborate extraembryonic structures that provide the interface between the embryo and the maternal uterus. Unique features of early mammalian development are that cell divisions are slower and genes are expressed during cleavage. As a result, proteins pro-

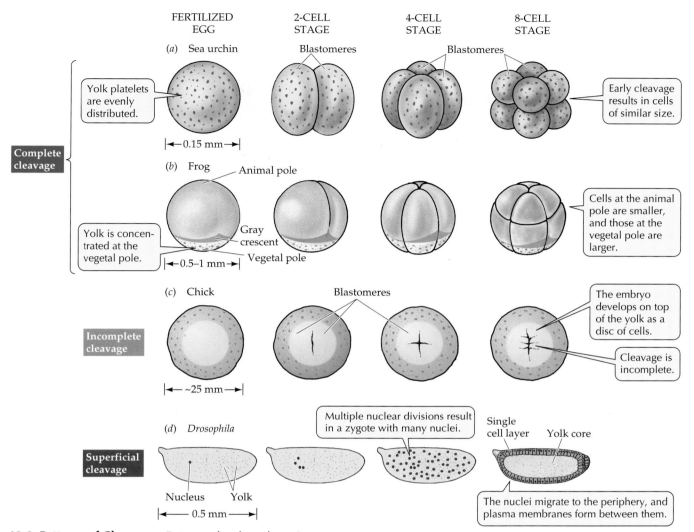

FERTILIZED EGG | 2-CELL STAGE | 4-CELL STAGE | 8-CELL STAGE

Complete cleavage

(a) Sea urchin

Blastomeres

Blastomeres

Yolk platelets are evenly distributed.

Early cleavage results in cells of similar size.

|←—0.15 mm—→|

(b) Frog — Animal pole

Gray crescent

Yolk is concentrated at the vegetal pole.

Vegetal pole

Cells at the animal pole are smaller, and those at the vegetal pole are larger.

|←—0.5–1 mm—→|

Incomplete cleavage

(c) Chick

Blastomeres

The embryo develops on top of the yolk as a disc of cells.

Cleavage is incomplete.

|←— ~25 mm —→|

Superficial cleavage

(d) *Drosophila*

Multiple nuclear divisions result in a zygote with many nuclei.

Single cell layer   Yolk core

Nucleus   Yolk

The nuclei migrate to the periphery, and plasma membranes form between them.

|←— 0.5 mm —→|

40.6 Patterns of Cleavage   Patterns of early embryonic development reflect differences in the way egg cytoplasm is organized.

duced from the genes of the mammalian embryo play a role in cleavage. In most species, cleavage is directed almost entirely by products of the maternal genome.

As in other animals that have complete cleavage, the early cell divisions in a mammalian embryo produce a loosely associated ball of cells. However, at about the 8-cell stage the behavior of the cells changes. They suddenly maximize their surface contact with each other, form tight junctions (see Figure 5.6a), and become a very compact mass of cells. At the transition from the 16-cell stage to the 32-cell stage, the innermost cells remain a compact mass that will become the embryo, while the outermost cells become an encompassing sac called the **trophoblast** that will become part of the placenta. The trophoblast cells secrete fluid, thus creating the blastocoel, with the innermost cell mass (the embryo) at one end (Figure 40.7). At this

stage the mammalian embryo is called a **blastocyst** to distinguish it from the blastula of nonmammals. All the cells of a blastula contribute to the developing embryo, but only the innermost cell mass of the blastocyst will become the embryo.

Until the blastocoel forms, the developing mammalian embryo is contained within the zona pellucida. The zona pellucida prevents the egg from implanting itself too soon—that is, in the upper regions of the oviduct, where it is fertilized, or in other parts of the oviduct as it undergoes cleavage while traveling to the uterus. The outermost cells of the resulting blastocyst—the trophoblast cells—will attach to the cells that line the uterus, initiating *implantation* of the embryo into the uterine wall.

If the egg implants itself in the oviduct, the result is an ectopic, or tubal, pregnancy—a very dangerous condition. In this situation, because there is not enough room for the embryo to develop in the oviduct, it will be aborted. An embryo growing in the

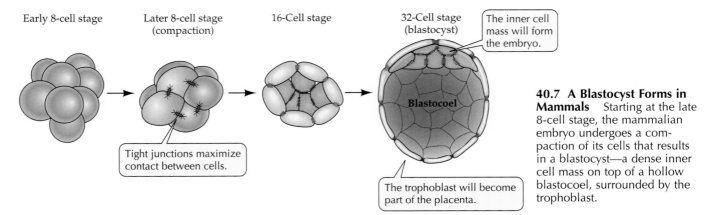

Early 8-cell stage

Later 8-cell stage (compaction)

16-Cell stage

32-Cell stage (blastocyst)

The inner cell mass will form the embryo.

Blastocoel

Tight junctions maximize contact between cells.

The trophoblast will become part of the placenta.

40.7 A Blastocyst Forms in Mammals Starting at the late 8-cell stage, the mammalian embryo undergoes a compaction of its cells that results in a blastocyst—a dense inner cell mass on top of a hollow blastocoel, surrounded by the trophoblast.

oviduct can cause the oviduct to rupture and hemorrhaging to occur. The zona pellucida prevents ectopic implantation. At about the time the blastocyst reaches the uterus, it "hatches" from the zona pellucida.

Fate maps show that specific blastomeres generate specific tissues and organs

Cleavage in all species results in a repackaging of the cytoplasm of the egg into a large number of small cells surrounding a central cavity. No cells have differentiated, and in most nonmammalian species none of the genome of the embryo has been expressed. Nevertheless, cells in different regions of the resulting blastula possess different complements of nutrients and informational molecules.

The blastocoel prevents the cells from different regions of the blastula from interacting, but that will soon change. During the next stage of development, the cells of the blastula will move around and come into new associations with each other, engage in inductive interactions, and begin to differentiate (see Chapter 15). The blastomere movements are so regular and well orchestrated that it is possible to label a specific cell of the blastula and identify what tissues and organs will form from its progeny (Figure 40.8). Such labeling experiments produce **fate maps**.

Separation of blastomeres can produce twins

Blastomeres do not differentiate, but they become committed to specific fates at different times in different species. In some species, such as roundworms and clams, blastomeres at the 8-cell stage are already committed. If one of these blastomeres is eliminated, a particular portion of the embryo will not form. This type of development has been called **mosaic development** because each blastomere seems to contribute a specific set of "tiles" to the final "mosaic," which is the adult animal. In contrast, other species, such as sea urchins, frogs, and vertebrates, have **regulated development**: The loss of some cells during cleavage will not affect the developing embryo; the remaining cells compensate for the loss.

If some cells can change their fate to compensate for the loss of other cells during cleavage and blastula formation, are these groups of cells capable of forming entire embryos? To a certain extent they are. During cleavage or early blastula formation, if the cells are physically separated into two groups, both groups can produce complete embryos (Figure 40.9a). Since these embryos come from the same zygote, they will be **monozygotic twins**—genetically identical.

There is an interesting variation on this process in humans and other mammals. Until the trophoblast forms—about the 16-cell stage—division of the blastocyst will result in twins that have separate placentas. After the trophoblast forms, separated trophoblast cells are not capable of developing into an embryo, but division of the inner cell mass will produce twins. In this case, however, the twins will have a shared placenta (Figure 40.9b).

Gastrulation: Producing the Body Plan

The blastula is a hollow ball of cells. Through gastrulation this ball of cells becomes an animal, made up of

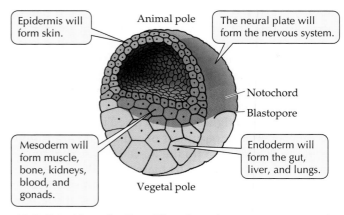

Epidermis will form skin.

Animal pole

The neural plate will form the nervous system.

Notochord

Blastopore

Mesoderm will form muscle, bone, kidneys, blood, and gonads.

Endoderm will form the gut, liver, and lungs.

Vegetal pole

40.8 Fate Map of a Frog Blastula The various tissues and organs of the animals will form from cells indicated in the same color.

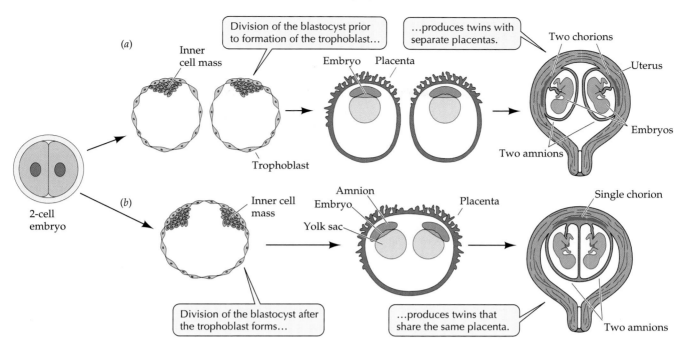

40.9 Twinning in Humans In humans, monozygotic (identical) twins result when groups of cells in the early stages of dividing become physically separated and both groups produce embryos. Monozygotic twins can form with separate placentas (*a*) or a shared placenta (*b*).

multiple tissue layers, that has head and tail ends and dorsal and ventral sides. Consider the general structure of an animal: skin is a tissue layer on the outside, the gut is a tissue layer on the inside, and between the two are muscles, bones, cartilage, and organs such as gonads, heart, blood vessels, and kidneys.

During gastrulation, blastomeres move as a sheet to the inside of the embryo, creating an inner germ layer, the **endoderm**, which will give rise to gut tissues. The cells remaining on the outside become the outer germ layer, the **ectoderm**, and give rise to skin and other outer tissues. Other cells migrate between these two layers to become the middle germ layer, or **mesoderm**, which will contribute tissues to many organs. Table 40.1 lists these embryonic germ layers and some of the tissues and organs they produce. **Gastrulation** is the process by which the germ layers form and take specific positions relative to each other. The resulting spatial relations between tissues make possible inductive tissue interactions that trigger processes of differentiation and organ formation.

In the discussion that follows we will consider similarities and differences of gastrulation in sea urchins, frogs, reptiles, birds, and mammals. In each, the site at which cells begin to move into the blastula plays an important role.

Invagination of the vegetal pole characterizes gastrulation in sea urchins

The sea urchin blastula is a simple, hollow ball of cells that is only one cell layer thick. The end of the blastula stage is marked by a dramatic slowing of the rate of mitosis, and the beginning of gastrulation is marked by a flattening of the vegetal hemisphere (Figure 40.10). Some cells of the vegetal pole bulge into the blastocoel, break free, and migrate into the cavity. These cells become **primary mesenchyme** cells—cells of the middle germ layer, the mesoderm. (The word

TABLE 40.1	Fates of Embryonic Germ Layers in Vertebrates[a]
GERM LAYER	**FATE**
Ectoderm	Brain and nervous system; lens of eye; inner ear; lining of mouth and of nasal canal; epidermis of skin; hair and nails; sweat glands, oil glands, milk secretory glands
Mesoderm	Skeletal system; bones, cartilage, notochord; gonads; muscle; outer coverings of internal organs; dermis of skin; circulatory system: heart, blood vessels, blood cells; kidneys
Endoderm	Inner linings of gut, respiratory tract (including lungs), liver, pancreas, thyroid, and urinary bladder

[a] The final structures are complex, containing cells from more than one germ layer. Interactions among tissues are usually important in determining the composition and structure of an organ.

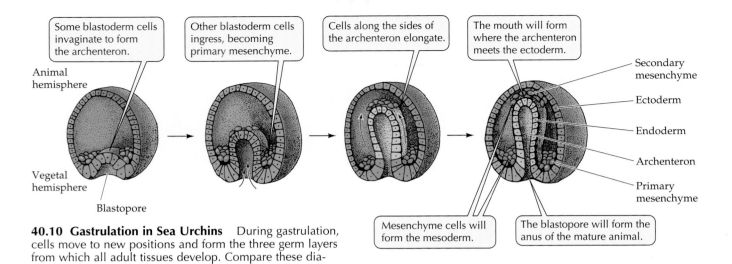

Some blastoderm cells invaginate to form the archenteron.

Other blastoderm cells ingress, becoming primary mesenchyme.

Cells along the sides of the archenteron elongate.

The mouth will form where the archenteron meets the ectoderm.

Animal hemisphere

Vegetal hemisphere

Blastopore

Secondary mesenchyme
Ectoderm
Endoderm
Archenteron
Primary mesenchyme

Mesenchyme cells will form the mesoderm.

The blastopore will form the anus of the mature animal.

40.10 Gastrulation in Sea Urchins During gastrulation, cells move to new positions and form the three germ layers from which all adult tissues develop. Compare these diagrams to the chapter-opening photo.

"mesenchyme" means "a loosely organized group of cells," in contrast to cells formed into a tightly packed sheet.)

The flattening at the vegetal pole becomes an *invagination*, as if someone were poking a finger into a hollow ball. The cells that invaginate become the endoderm and form the primitive gut, the **archenteron**. At the tip of the archenteron more cells break free, entering the blastocoel to form more mesoderm, the **secondary mesenchyme**. The archenteron continues to move forward, partly because of changes in the shapes of its cells and partly because of being pulled by mesenchyme cells that send out extensions, which attach to the overlying ectoderm and contract. Where the

archenteron makes contact with the ectoderm, the mouth will form, and the **blastopore**, an opening created by the invagination of the vegetal pole, will become the anus of the animal.

Where the mouth and the anus form is an important characteristic in establishing the evolutionary lineages of animals. In the lineage that includes crustaceans, mollusks, annelids, and roundworms, the mouth forms from the blastopore. Members of this lineage are called **protostomes** because the mouth forms first. In the other major lineage of animals, which includes echinoderms (such as sea urchins), chordates, and several other phyla, the mouth forms where the archenteron meets the ectoderm. Since the mouth forms later than the blastopore in the animals of this lineage, they are called the **deuterostomes**, which means literally

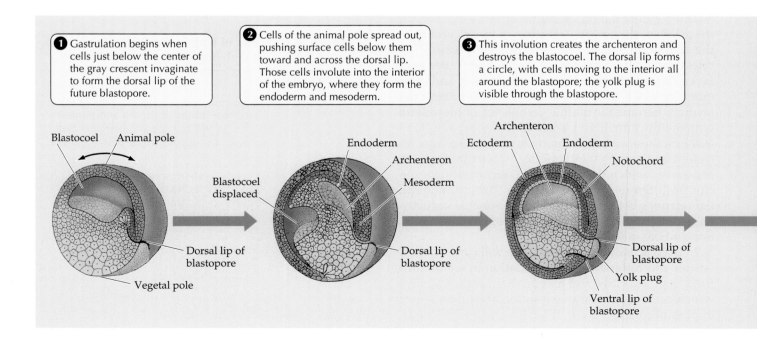

1 Gastrulation begins when cells just below the center of the gray crescent invaginate to form the dorsal lip of the future blastopore.

2 Cells of the animal pole spread out, pushing surface cells below them toward and across the dorsal lip. Those cells involute into the interior of the embryo, where they form the endoderm and mesoderm.

3 This involution creates the archenteron and destroys the blastocoel. The dorsal lip forms a circle, with cells moving to the interior all around the blastopore; the yolk plug is visible through the blastopore.

Blastocoel Animal pole

Blastocoel displaced

Endoderm
Archenteron
Mesoderm

Archenteron
Ectoderm Endoderm
Notochord

Dorsal lip of blastopore
Vegetal pole

Dorsal lip of blastopore

Dorsal lip of blastopore
Yolk plug
Ventral lip of blastopore

"mouth second." Other developmental differences between these two groups are listed in Table 29.1.

An important question associated with gastrulation is, What mechanisms control the various cell movements? The mechanisms of some movements are characteristics of individual cells. For example, vegetal cells migrate into the blastocoel to form the primary mesenchyme because they lose their attachments to neighboring cells. Once they bulge into the blastocoel, they move by extending long processes called filopodia along an extracellular matrix of proteins such as fibronectin that is laid down by the ectodermal cells lining the blastocoel.

The initial invagination of the vegetal pole involves hydration of a protein layer external to the cells of the vegetal pole. As water is added to this protein layer, it expands and bulges inward. Continued invagination and extension of the archenteron involves rearrangements and flattening of the endodermal cells through changes in their cytoskeleton. And as already mentioned, the secondary mesenchymal cells assist the invagination by contractions. Mechanisms of cell movements in development are a rich area of research.

Gastrulation in frogs begins at the gray crescent

Amphibian blastulas have considerable yolk and for the most part they are more than one cell layer thick; therefore, gastrulation is a little more complex in frogs than in sea urchins. Gastrulation begins when some cells in the region that had been the gray crescent change shape. The bodies of these cells bulge in toward the blastocoel while they remain attached to the outer surface by slender necks. Because of their shape, these cells are called *bottle cells*.

As the bottle cells move into the interior of the blastula, they appear to pull other cells of the surface of the blastula after them (Figure 40.11). This process creates a lip over which a sheet of cells moves into the interior. These invaginating cells are the prospective endoderm, and they form the archenteron. The initial site of invagination is called the **dorsal lip of the blastopore**, which, as we will see, plays a central role in vertebrate development.

As gastrulation proceeds, cells from all over the surface of the blastula move toward the site of invagination. This movement of cells toward the blastopore is called **epiboly**. The dorsal lip of the amphibian blastopore widens and eventually forms a complete circle surrounding a plug of yolk-rich cells. As prospective endodermal cells continue to move in through the blastopore, the archenteron grows and gradually displaces the blastocoel. In addition, some of the cells pouring over the lip of the blastopore move between the ectodermal and endodermal layers to form mesoderm (see Figure 40.11).

As gastrulation comes to an end, the blastopore narrows, and the yolk plug is completely internalized as part of the prospective endoderm. At this point the embryo consists of three germ layers: ectoderm on the outside, endoderm on the inside, and mesoderm in the middle. The embryo also has a dorsal–ventral and anterior–posterior organization. Most importantly, however, the fates of specific regions of endoderm, mesoderm, and ectoderm have become "determined" such that from here on in they will differentiate into specific tissue types. The discovery of determination, our next topic, is one of the most exciting stories in animal development.

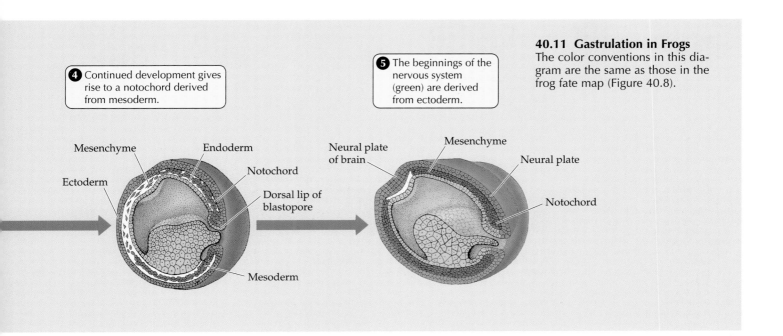

4 Continued development gives rise to a notochord derived from mesoderm.

5 The beginnings of the nervous system (green) are derived from ectoderm.

40.11 Gastrulation in Frogs
The color conventions in this diagram are the same as those in the frog fate map (Figure 40.8).

Mesenchyme
Endoderm
Ectoderm
Notochord
Dorsal lip of blastopore
Mesoderm

Neural plate of brain
Mesenchyme
Neural plate
Notochord

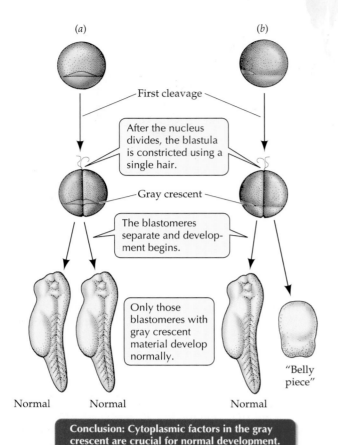

First cleavage

After the nucleus divides, the blastula is constricted using a single hair.

Gray crescent

The blastomeres separate and development begins.

Only those blastomeres with gray crescent material develop normally.

"Belly piece"

Normal   Normal   Normal

Conclusion: Cytoplasmic factors in the gray crescent are crucial for normal development.

40.12 Spemann's Experiment   Spemann confined nuclei to one side of a bisected salamander blastula, which did not prevent normal development as long as material from the gray crescent was present.

The dorsal lip of the blastopore can organize formation of the embryo

Early in the 1900s, German biologist Hans Spemann was studying the development of salamander eggs. He was interested in finding out if nuclei during cleavage and blastula formation remain capable of directing the development of a complete embryo. With great patience and dexterity, he used a loop of human baby hair to constrict fertilized eggs. When Spemann's loop bisected the gray crescent, the cells on both sides of the constriction developed into complete embryos (Figure 40.12). But when he tied his loop so that the gray crescent was on only one side of the constriction, only that side developed into a complete embryo. Spemann thus hypothesized that cytoplasmic factors in the region of the gray crescent are necessary for gas-

40.13 Tissue Transplants Reveal the Process of Determination   (a) Tissue transplant experiments by Spemann and Mangold revealed that cell fate becomes increasingly determined during gastrulation. (b) Transplanting the dorsal blastopore lip of an early gastrula resulted in a second initiation of gastrulation and the formation of a second embryo.

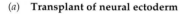

(a) **Transplant of neural ectoderm**

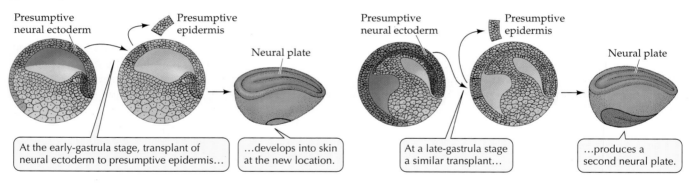

Presumptive neural ectoderm — Presumptive epidermis — Neural plate

At the early-gastrula stage, transplant of neural ectoderm to presumptive epidermis… …develops into skin at the new location.

Presumptive neural ectoderm — Presumptive epidermis — Neural plate

At a late-gastrula stage a similar transplant… …produces a second neural plate.

(b) **Transplant of blastopore lip**

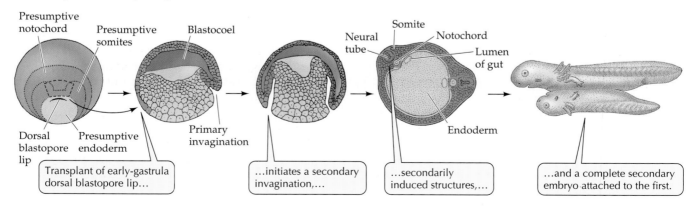

Presumptive notochord — Presumptive somites — Blastocoel — Neural tube — Somite — Notochord — Lumen of gut

Dorsal blastopore lip — Presumptive endoderm — Primary invagination — Endoderm

Transplant of early-gastrula dorsal blastopore lip… …initiates a secondary invagination,… …secondarily induced structures,… …and a complete secondary embryo attached to the first.

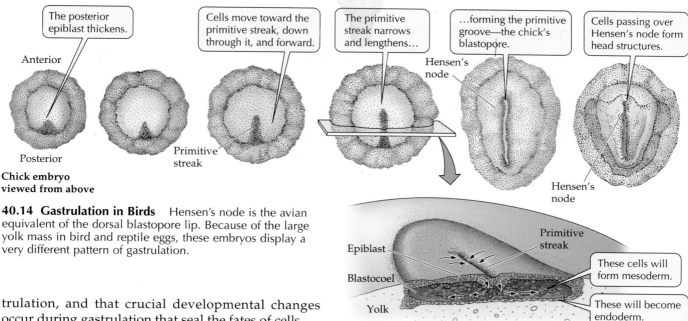

The posterior epiblast thickens.

Anterior

Posterior

Chick embryo viewed from above

Primitive streak

Cells move toward the primitive streak, down through it, and forward.

The primitive streak narrows and lengthens...

...forming the primitive groove—the chick's blastopore.

Cells passing over Hensen's node form head structures.

Hensen's node

Hensen's node

Epiblast

Blastocoel

Yolk

Primitive streak

These cells will form mesoderm.

These will become endoderm.

Cross section through chick embryo

40.14 Gastrulation in Birds Hensen's node is the avian equivalent of the dorsal blastopore lip. Because of the large yolk mass in bird and reptile eggs, these embryos display a very different pattern of gastrulation.

trulation, and that crucial developmental changes occur during gastrulation that seal the fates of cells.

To test his hypotheses, Spemann and his student Hilde Mangold conducted a series of delicate tissue transplantation experiments. They transplanted pieces of early gastrulas to various locations on other gastrulas. Guided by fate maps (see Figure 40.8), they were able to take a piece of a gastrula they knew would develop into skin and transplant it to a region that normally becomes nervous system, and vice versa (Figure 40.13a).

When they did these transplants in early gastrulas, the transplanted pieces always developed into tissues that were appropriate for the location where they were placed. Donor presumptive epidermis (that is, a piece of donor gastrula destined to become skin) developed into host nervous system, and donor presumptive neural ectoderm developed into host skin. Thus, the fates of the transplanted cells had not been determined before the transplantation.

In late gastrulas, however, the same experiment yielded opposite results. Donor presumptive epidermis produced patches of skin cells in the host nervous system, and donor presumptive neural ectoderm produced neurons in the host skin. Something had occurred during gastrulation to determine the fates of the embryonic cells. Thus, as Spemann hypothesized, the path of differentiation a cell would follow was *determined* during gastrulation, even though there was no evidence of differential gene expression.

Spemann and Mangold produced dramatic results by transplanting the dorsal lip of the blastopore (Figure 40.13b). When this small piece of tissue was transplanted into the presumptive belly area of another gastrula, it stimulated a second site of gastrulation. As a result of that second initiation of gastrulation, another whole embryo formed belly to belly with the original embryo. Thus, Spemann and Mangold

called the dorsal lip of the blastopore the **primary embryonic organizer**. Was this transplanted tissue, like an isolated blastomere, capable of forming an entire embryo? Clearly not, because many of the cells passing over the new dorsal lip were host cells, and these host cells that normally would have become belly of the original embryo were now involved in producing different parts of a new embryo. The dorsal lip of the blastopore seemed to have the ability to induce the cells that passed over it to develop in specific ways.

We now know that development involves a cascade of inductive tissue interactions. The classic experiments of Spemann and Mangold revealed some of the earliest and most dramatic inductive events in amphibian development. Understanding the molecular mechanisms of determination and induction remains an exciting area of research in developmental biology, as was discussed in Chapter 15.

Reptilian and avian gastrulation is an adaptation to yolky eggs

The eggs of reptiles and birds have a massive yolk content, and therefore blastulas in these species develop as discs of cells on top of the yolk. We will use the chicken egg to show how gastrulation occurs in a flat disc of cells rather than in a ball of cells as it does in sea urchins and amphibians.

The upper layer of the blastula is called the *epiblast*. All tissues of the chick embryo will derive from the epiblast. Gastrulation begins with a thickening of a posterior region of the epiblast (Figure 40.14). This thickening results from the movement of cells from the lateral regions of the posterior epiblast toward the midline and then forward along the midline. The result is a midline ridge called the **primitive streak**. As the primitive streak narrows and progresses forward, a crease develops along its length. This depression, the **primitive groove**, is the blastopore; through it, cells migrate into the blastocoel to become endoderm and mesoderm.

In the avian embryo, no archenteron forms, but the prospective endoderm and mesoderm migrate forward to form gut and other structures. At the extreme forward end of the primitive groove is a thickening called **Hensen's node**, which is the equivalent of the dorsal lip of the amphibian blastopore. Cells passing over Hensen's node become determined to differentiate into many tissues and structures that make up the head and the dorsal midline of the embryo.

Mammals have no yolk, but retain reptile-type gastrulation

Mammals and birds evolved from reptilian ancestors, so it is not surprising that they share patterns of early development even though mammalian eggs have no yolk. Earlier we described the development of the mammalian trophoblast and the inner cell mass, which is the equivalent of the avian epiblast. Keeping avian gastrulation in mind, think of the mammalian inner cell mass as sitting on top of an imaginary body of yolk (Figure 40.15).

As in avian development, cells separate from the inner cell mass to form a layer of cells surrounding a blastocoel. These cells, called the **hypoblast**, do not contribute to any part of the embryo, but they do participate in forming extraembryonic membranes, which we will discuss shortly. The remaining cells of the inner cell mass are now the epiblast, and they split off an upper layer of cells that form one of the extraembryonic membranes, the *amnion*.

The amnion eventually will grow to surround the developing embryo as a sac filled with amniotic fluid. The remaining epiblast contains all the cells that will form the embryo, and gastrulation occurs as in the avian epiblast. A primitive groove forms, epiblast cells migrate through the groove to become layers of endoderm and mesoderm, and Hensen's node is a site of inductive tissue interactions.

Neurulation: Initiating the Nervous System

Gastrulation produces an embryo with three germ layers that are positioned to influence each other through inductive tissue interactions. During the next phase of development many organs and organ systems are developing in parallel and in coordination with each other. One process of organogenesis is the formation of a neural tube that will give rise to the nervous system. The stage of development following gastrulation

Human embryo at 9 days

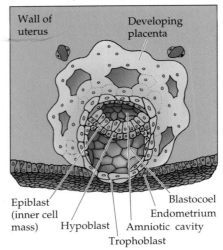

40.15 A Human Blastocyst Before Gastrulation The mammalian inner cell mass becomes the epiblast; it can be compared to the avian embryo by picturing it sitting on top of a mass of imaginary yolk.

takes its name from the formation of the neural tube, **neurulation**. We will examine this event in the amphibian embryo; it occurs in a very similar fashion in reptiles, birds, and mammals.

The stage is set by the dorsal lip of the blastopore

Some of the cells that pass over the dorsal lip and move anteriorly in the blastocoel are determined to become mesoderm. Some of that mesoderm closest to the midline is further determined to become **chordomesoderm**, which forms a rod of cartilage along the dorsal midline. This rod, called the **notochord**, gives structural support to the developing embryo. The notochord eventually will be replaced by the vertebral column, but after gastrulation it has an important role to play in the induction of the overlying ectoderm to begin to form the nervous system. This process involves the formation of an internal tube from an external sheet of cells.

The first signs of neurulation are flattening and thickening of the ectoderm overlying the notochord; this thickened area forms the **neural plate** (Figure 40.16). The edges of the neural plate that run in an anterior–posterior direction continue to thicken to form ridges or folds. Between the folds a groove forms and deepens as the folds roll over it to converge on the midline. The folds fuse, forming a cylinder, the **neural tube**. This neural tube will develop bulges at the anterior end that will develop into the major divisions of the brain. The rest of the tube will become the spinal cord, and outgrowths will form the peripheral nervous system. We will return to the development of the nervous system in Chapter 43.

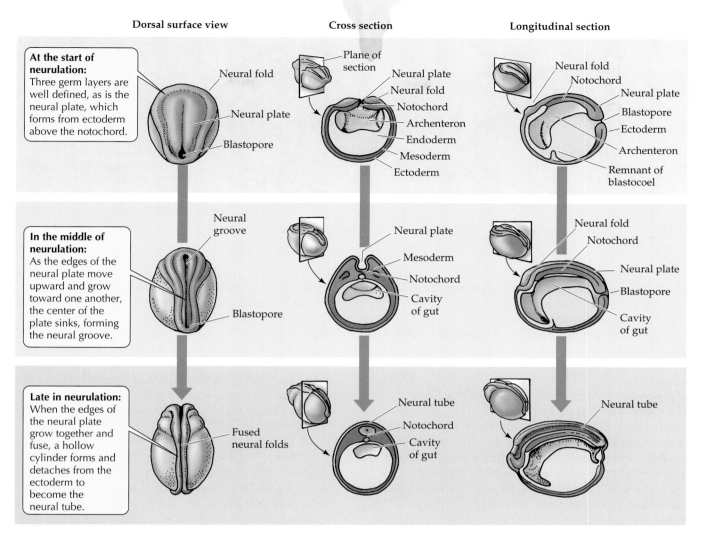

Dorsal surface view **Cross section** **Longitudinal section**

At the start of neurulation: Three germ layers are well defined, as is the neural plate, which forms from ectoderm above the notochord.

In the middle of neurulation: As the edges of the neural plate move upward and grow toward one another, the center of the plate sinks, forming the neural groove.

Late in neurulation: When the edges of the neural plate grow together and fuse, a hollow cylinder forms and detaches from the ectoderm to become the neural tube.

40.16 Neurulation in a Frog Continuing the sequence from Figure 40.11, these drawings outline the development of the frog's neural tube. The longitudinal views (far right) show the development of the notochord and its position relative to the neural tube.

Failure of the neural tube to develop normally can result in serious birth defects. If the neural tube fails to close in a posterior region, the result is a condition known as spina bifida. If it fails to close at the anterior end, an infant can develop without a forebrain—a condition called anencephaly. Such neural-tube defects occur in about 1 in 500 live births, but they can be detected during pregnancy.

Body segmentation develops during neurulation

As the neural tube is forming, along the sides of the notochord mesodermal tissues are congregating to form a repeating pattern of blocks of cells called **somites**. The vertebrate body plan consists of repeating segments that are modified during development. These segments are most evident as the repeating pat-

terns of vertebrae, ribs, nerves, and muscles along the anterior–posterior axis. The somites will produce cells that will become vertebrae and ribs, muscles of the trunk and limbs, and skin of the back. In addition, they will direct the outgrowth of nerves and the organization of blood vessels.

Extraembryonic Membranes

The embryos of reptiles, birds, and mammals are surrounded by several tissue membranes. These **extraembryonic membranes** are basic features of the amniotic egg, which was a major step in the evolution of reptiles from amphibian ancestors about 300 million years ago. The amniotic egg was the adaptation that freed terrestrial vertebrates from dependence on an aquatic environment for reproduction. Fish or amphibian eggs rapidly dry out and die if they are exposed to air, but the amniotic egg provides an aqueous environment within which the embryo can develop. Anything that

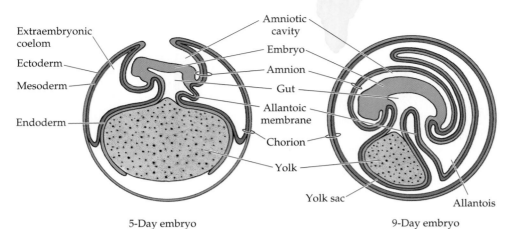

Extraembryonic coelom

Ectoderm

Mesoderm

Endoderm

Amniotic cavity

Embryo

Amnion

Gut

Allantoic membrane

Chorion

Yolk

Yolk sac

Allantois

5-Day embryo

9-Day embryo

40.17 The Extraembryonic Membranes of a Chick In birds, reptiles, and mammals, the embryo is enclosed within the amnion, allantois, and chorion. Fluids secreted by the amnion fill the amniotic cavity, providing an aqueous environment for the embryo. The chorion mediates gas exchange between the embryo and its environment, and the allantois stores the embryo's waste products.

reaches the embryo from the environment must pass through these membranes.

We will consider the embryonic origin of the extraembryonic membranes, their role in forming the placenta in mammals, and their role in technologies to detect genetic diseases in humans.

Four extraembryonic membranes form with contributions from all germ layers

We will use the chick to demonstrate how the extraembryonic membranes form from the germ layers created during gastrulation. The chick gastrula is a disc-shaped mass of cells that sits on top of the yolk. The invagination of epiblast cells through the primitive streak produces an innermost layer of cells, the endoderm, and an intermediate mass of cells, the mesoderm. The first extraembryonic membrane to form is the **yolk sac**.

The yolk sac develops from an extension of the hypoblast of the embryo, which has incorporated invaginating endodermal cells. This membrane extends to enclose the entire body of yolk in the egg (Figure 40.17). The yolk sac constricts at the top to create a tube that is continuous with the gut of the embryo. However, yolk does not pass through this tube. Yolk is digested by the endodermal cells of the yolk sac, and the nutrients are then transported to the embryo thorough blood vessels lining the outer surface of the yolk sac.

Just as the endoderm grows out from the embryo to form the yolk sac, the mesoderm and ectoderm extend beyond the limits of the disc of cells that is the embryo. These two layers of cells extend all along the inside of the eggshell both over the embryo and below the yolk sac. Where they meet over the embryo they fuse to form the **amnion**, thereby creating a cavity, the **amniotic cavity**, that encloses the embryo. The amnion protects the embryo from desiccation. Where the ex-

tending ectoderm and mesoderm fuse below the yolk, they form a continuous membrane under the eggshell. This membrane, called the **chorion**, controls gas exchange between the embryo and the outside world.

The fourth membrane to form, the allantoic membrane, is another outgrowth of the embryonic endoderm, and it forms the **allantois**, creating a sac for storage of metabolic wastes.

Extraembryonic membranes in mammals form the placenta

In mammals, the first extraembryonic membrane to form is the trophoblast. The trophoblast is already apparent by the fifth cell division (see Figure 40.7). When the blastocyst reaches the uterus and "hatches" out of its zona pellucida covering, the trophoblast cells interact directly with the lining of the uterus. Adhesion molecules expressed on the surfaces of these cells attach them to the uterine wall, and by excreting proteolytic enzymes the trophoblast burrows in, beginning the process of implantation. Eventually the entire trophoblast is within the wall of the uterus. The trophoblastic cells then send out numerous projections, or villi, to increase the surface area of contact with maternal blood.

Meanwhile, the hypoblast cells extend to form what in the bird is the yolk sac. But because there is no yolk in mammals, the yolk sac becomes a lining of the trophoblast, and it contributes to the trophoblast tissue by forming blood vessels. The yolk sac and trophoblast together make up the chorion. The interactions of the chorion with the uterine wall produce the **placenta**, the organ of nutrition for the embryo (Figure 40.18).

At the same time as the yolk sac is forming from the hypoblast, the epiblast splits off a layer of cells that expand to become the amnion enclosing the amniotic cavity. The extension of the amnion eventually encloses the entire embryo in a fluid-filled space. The

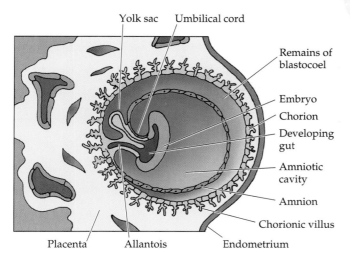

40.18 The Extraembryonic Membranes of Humans
Continuing the sequence of Figures 40.7 and 40.15, this drawing diagrams the membranes that have formed in a 40-day-old human embryo.

rupturing of the amnion and the loss of the amniotic fluids ("water breaking") are what frequently herald the onset of labor in humans.

An allantois also develops in mammals, but its importance depends on how well the nitrogenous wastes can be handled across the placenta. In humans the allantois is minor; in pigs it is important. In humans, allantoic tissues contribute to the formation of the umbilical cord by which the embryo is attached to the chorionic placenta. It is through the blood vessels of the umbilical cord that nutrients and O_2 from the mother reach the developing fetus and wastes, including CO_2 and urea, are removed.

The extraembryonic membranes provide means of detecting genetic diseases

Cells slough off of the embryo and float in the amniotic fluid that bathes it. Later in development, a small sample of the amniotic fluid may be withdrawn with a needle as the first step of a process called **amniocentesis**. Some of these cells can be cultured and used for biochemical and genetic analyses that can reveal the sex of the fetus, as well as genetic markers for diseases such as cystic fibrosis, Tay-Sachs disease, and Down syndrome (Figure 40.19).

Amniocentesis usually is not performed until after the fourteenth week of pregnancy, and the tests require 2 weeks. If abnormalities in the fetus are detected, termination of the pregnancy at that stage would put the woman's health at greater risk than would an earlier abortion. Therefore a newer technique, called **chorionic villus sampling**, has been developed. In this test a small sample of the tissue from

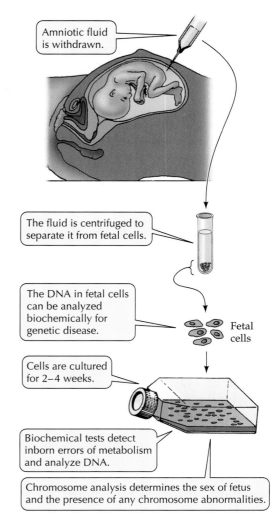

40.19 Amniocentesis   Embryonic cells in the amniotic fluid can be withdrawn and analyzed to reveal the sex of the fetus and the presence of any abnormal genetic markers. The test is usually performed late in the third or early in the fourth month of pregnancy.

the surface of the chorion is taken. This test can be done as early as the eighth week, and the results are available in several days.

Development Continues Throughout Life

We have compared some aspects of early development in some representative animals: sea urchins, insects, amphibians, birds, and mammals. With minor variations, very different groups of animals have common processes for fertilization, cleavage, gastrulation, and neurulation. However, more and more developmental differences appear among groups as development progresses and the animals get farther along the cascade

(a)

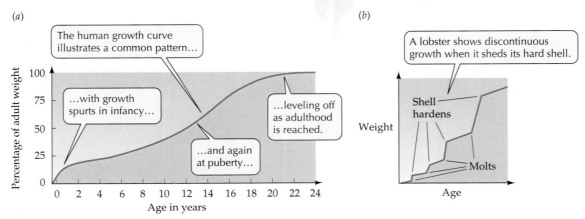

The human growth curve illustrates a common pattern...

...with growth spurts in infancy...

...leveling off as adulthood is reached.

...and again at puberty...

Percentage of adult weight

Age in years

(b)

A lobster shows discontinuous growth when it sheds its hard shell.

Weight

Shell hardens

Molts

Age

40.20 Growth Curves   (a) Many animals, from tiny shrews to humans to the massive blue whale, follow a similar curve in their growth pattern; it is illustrated here for humans. (b) A discontinuous growth pattern occurs when growth is limited to periods of molt. However, unlike humans, the lobster continues to grow over its entire lifespan.

of tissue interactions that mediate differentiation and organ formation. What is remarkable, however, is that the more we know about the molecular processes involved, the more we see common mechanisms underlying the developmental processes.

From our familiarity with our own species, we can appreciate that development does not end at birth. Nor does it end with the transition to adulthood. Developmental processes continue throughout the life of an animal. For example, one common feature of later development is growth—an irreversible increase in size. Growth results from cell multiplication and cell expansion. Whereas in early development cell multiplication and expansion frequently resulted in changes in shape and form, once an adult stage is reached, shape and form remain fairly constant even as size increases.

Growth rates and extents vary greatly. For example, consider growth in a human and a blue whale. Their zygotes are about equal in size, and their early developmental stages are almost identical. By birth, however, the human infant weighs only about 4 kg and the blue whale weighs 1,500 kg. By adulthood, the human may weigh 80 kg and the adult blue whale may weigh 100,000 kg. Between birth and weaning, the blue whale will grow about 4 to 5 cm in length and 100 kg in mass per day.

Add to this comparison a shrew, which starts as a similar-sized zygote but reaches only 10 g as an adult, and you can see the enormous variation in growth rates within one group of animals, the mammals, even though they all show similar patterns of growth (Figure 40.20*a*). In contrast to the mammalian pattern of growth, some animals grow throughout their lives.

The lobster is an example, but because of its exoskeleton, its growth is limited to times of molt (Figure 40.20*b*).

How can an organism grow without its organs changing in shape or structure? The answer is that many of the processes by which neighboring cells influence each other during development continue throughout life. Cells in embryos control the activities of their neighbors to create structure, and in adulthood they continue to control the activities of each other to maintain structure.

This is true even in adult tissues characterized by high rates of cell division. For example, the cells in the lining of the gut and the cells that make up our skin actively divide without changing the shape or structure of those organs. When controls on cell multiplication, shape, and expansion fail, growth becomes abnormal, which is the case in cancerous diseases. Cancer could be considered a failure of normal developmental mechanisms and controls.

Another developmental process that occurs throughout life is the determination and differentiation of tissues. For example, cells in our bone marrow remain capable of producing progeny that can differentiate into a wide variety of blood cell types. Also, in response to injury, cells can dedifferentiate and then respond to local conditions to regenerate structures that have been damaged or removed. The ability to regenerate structures through dedifferentiation is more limited in mammals than in some other organisms, such as amphibians, which can regenerate whole limbs. By learning more about the mechanisms of development we may discover means to enhance the recovery of structure and function in cases of injury.

Summary of "Animal Development"

Fertilization: Changes in Sperm and Egg

• Fertilization involves interactions between sperm and egg, including sperm activation, species-specific binding of sperm to the outer covering of the egg, the acrosomal reaction, digestion of the outer covering of the egg, fusion of sperm and egg plasma membranes, initiation of fast and slow reactions of the egg to prevent entry by additional sperm, and activation of the egg to complete meiosis and begin development. **Review Figures 40.1, 40.2, 40.3, 40.4**

• Sperm and egg contribute differentially to the zygote (the fertilized egg). The sperm contributes a haploid nucleus and its centriole, which becomes the centrosome that produces the mitotic spindles of the dividing zygote. The egg contributes a haploid nucleus, nutrients, and informational molecules that will control the early stages of development.

• Cytoplasmic contents of the egg are not distributed homogeneously, and they are rearranged after fertilization to set up the major axes of the future embryo. **Review Figure 40.5**

Cleavage: Repackaging the Cytoplasm

• Cleavage is a period of rapid cell division without cell growth and with almost no gene expression in most species. During cleavage the cytoplasm of the zygote is repackaged into smaller and smaller cells.

• The pattern of cleavage is influenced by the amount of yolk that impedes cleavage furrow formation and by the orientation of the mitotic spindles. The result of cleavage is a ball or mass of cells called a blastula (in nonmammals) or a blastocyst (in mammals). **Review Figure 40.6**

• Cleavage in mammals is unique in that cell divisions are much slower and genes are expressed early in the process. At an early stage, some cells lose their ability to form an embryo and will form only an extraembryonic membrane. **Review Figure 40.7**

• Fate maps, which identify what tissues and organs will form from particular cells, can be created for the cells of the blastula. **Review Figure 40.8**

• Some species have mosaic development, in which each cell is determined. Other species, including vertebrates, have regulated development, in which cells are not determined and can change developmental paths. In regulated cleavage, blastomeres separated at early stages can develop into complete embryos, which are then identical, or monozygotic, twins. **Review Figure 40.9**

Gastrulation: Producing the Body Plan

• Gastrulation involves massive cell movements that produce three primary germ layers and put cells from various regions of the late blastula into new associations with each other. **Review Table 40.1**

• The initial step of gastrulation is invagination of a portion of the blastula. The site of invagination becomes the blastopore. **Review Figure 40.10**

• Cells that invaginate become the endoderm and mesoderm; cells remaining on the outside become the ectoderm. New cell contacts result in induction of determination, leading to differentiation. **Review Figure 40.11**

• The dorsal lip of the blastopore is a critical site for the determination of tissues. **Review Figures 40.12, 40.13**

• Gastrulation in reptiles and birds is very different from that in sea urchins and frogs because the large amount of yolk in their eggs causes the blastula to be a flattened disc of cells. Mammals have the same pattern even though they have no yolk. **Review Figures 40.14, 40.15**

Neurulation: Initiating the Nervous System

• Neurulation marks the stage of development following gastrulation.

• The dorsal rod of mesoderm, the notochord, induces the overlying ectoderm to thicken, form parallel ridges, and fold in on itself to form a neural tube below the ectoderm. The nervous system develops from this neural tube. **Review Figure 40.16**

• The notochord participates in the segmental organization of tissues, or somites, along the body axis. Rudimentary organs and organ systems form during this stage.

Extraembryonic Membranes

• The embryos of reptiles, birds, and mammals are protected and nurtured by four extraembryonic membranes that form from the embryonic germ layers. In birds and reptiles the yolk sac surrounds the yolk and provides nutrients to the embryo; the chorion lines the eggshell and participates in gas exchange; the amnion surrounds the embryo and encloses it in an aqueous environment; the allantois stores metabolic wastes. **Review Figure 40.17**

• In mammals the chorion and the trophoblast cells interact with the maternal uterus to form a protective cavity for the embryo and, by forming a placenta, to provide for nutrient exchange. The amnion encloses the embryo in an aqueous environment. **Review Figure 40.18**

• Samples of amniotic fluid or pieces of chorion can be taken during pregnancy and analyzed for evidence of genetic disease. **Review Figure 40.19**

Development Continues Throughout Life

• Growth continues after birth. **Review Figure 40.20**

• Cells can influence their neighbors so that postnatal growth does not usually result in changes in shape or form. Failure of controls on cellular growth result in cancer.

• For repair of damage, cells can dedifferentiate and respond to local signals to regenerate lost tissues.

Self-Quiz

1. Fertilization involves all of the following except
   a. second meiotic division of the sperm nucleus.
   b. metabolic activation of the egg.
   c. breakdown of the acrosomal membrane.
   d. change in electric charge across the cell membrane of the egg.
   e. binding of the sperm cell to coatings surrounding the egg.

2. Which of the following does *not* occur during cleavage in frogs?
   a. High rate of mitosis
   b. Reduction in the size of cells
   c. Expression of genes critical for blastula formation
   d. Orientation of cleavage planes at right angles
   e. Unequal division of cytoplasmic determinants

3. How does cleavage in mammals differ from cleavage in frogs?
   a. Slower rate of cell division
   b. Formation of tight junctions
   c. Control involving the embryo's genome
   d. Early separation of cells that will not contribute to the embryo
   e. All of the above

4. Which statement about gastrulation is true?
   a. In frogs, it begins in the animal hemisphere.
   b. In sea urchins, it produces the notochord.
   c. In birds, it produces an epiblast and hypoblast.
   d. In mammals, the blastopore is a long groove in the epiblast.
   e. In sea urchins, it produces only two germ layers.

5. Which of the following was a conclusion from the experiments of Spemann and Mangold?
   a. Cytoplasmic determinants of development are homogeneously distributed in the amphibian zygote.
   b. In the late blastula, regions of cells are determined to form skin or nervous tissue.
   c. The dorsal lip of the blastopore can be isolated and will form a complete embryo.
   d. The dorsal lip of the blastopore can initiate gastrulation.
   e. The dorsal lip of the blastopore gives rise to the neural tube.

6. The acrosome of the sperm
   a. carries genetic information.
   b. provides energy for movement.
   c. carries the enzymes that facilitate fertilization.
   d. induces ovulation.
   e. prevents polyspermy.

7. Which of the following characterizes neurulation?
   a. Chordomesoderm forms a neural tube.
   b. Ectoderm forms a neural tube.
   c. A neural tube forms around the notochord.
   d. The neural tube induces somite formation.
   e. In chicks, the neural tube forms from the primitive groove.

8. Which statement about trophoblast cells is true?
   a. They are capable of producing monozygotic twins.
   b. They are derived from the hypoblast of the blastocyst.
   c. They are endodermal cells.
   d. They secrete proteolytic enzymes.
   e. They prevent the zona pellucida from attaching to the oviduct.

9. Which membrane is part of the embryonic contribution to placenta formation?
   a. Amnion
   b. Chorion
   c. Yolk sac
   d. Allantois
   e. Zona pellucida

10. Which of the following is a defining feature of deuterostomes?
    a. The blastopore becomes the mouth.
    b. Cleavage is incomplete.
    c. The mouth forms where the archenteron contacts ectoderm.
    d. Development is mosaic.
    e. A notochord is present.

Applying Concepts

1. Research some examples of maternal effect genes and explain how they control developmental processes.

2. Find out what a chimeric mouse is, and how it is produced.

3. Research and discuss some of the mechanisms that could explain how sheets of cells move during gastrulation.

4. Discuss how cells passing over the dorsal lip of the blastopore at different times could be determined to differentiate into different tissues.

5. Research and compare the development of the heart in fish, amphibians, reptiles, and mammals.

Readings

De Robertis, E. M., G. Oliver and C. V. E. Wright. 1990. "Homeobox Genes and the Vertebrate Body Plan." *Scientific American*, July. A good discussion of the development of anatomy and a presentation of how a highly conserved family of genes directs important aspects of development.

Epel, D. 1977. "The Program of Fertilization." *Scientific American*, November. A description of the initial events in the interaction between sperm and egg.

Gilbert, S. F. 1997. *Developmental Biology*, 5th Edition. Sinauer Associates, Sunderland, MA. An excellent advanced text covering all aspects of animal development.

Gilbert, S. F. and A. M. Raunio (Eds.). 1997. *Embryology: Constructing the Organism*. Sinauer Associates, Sunderland, MA. A textbook of comparative embryology that describes the processes of development across a wide range of animal phyla.

Horowitz, A. F. 1997. "Integrins and Health." *Scientific American*, May. An examination of research on the molecules reponsible for the important developmental mechanisms by which cells attach to and detach from each other and the extracellular matrix.

Nüsslein-Volhard, C. 1996. "Gradients That Organize Embryo Development." *Scientific American*, August. Presentation by a Nobel laureate of the exciting story of research on cytoplasmic determinants of development in fruit flies, which are an important model in current developmental biology.

Stern, C. and P. Ingham (Eds.). 1992. "Gastrulation." *Development*, Supplement. This special edition of the journal *Development* is a compendium of research papers. Included is a paper by Lewis Wolpert ("Gastrulation and the Evolution of Development"), whom we quoted at the beginning of this chapter.

Wassarman, P. M. 1988. "Fertilization in Mammals." *Scientific American*, December. An examination of the molecular and cellular events that surround the fusion of sperm and egg.

Chapter 41

Neurons and Nervous Systems

E very summer evening just after sunset, hundreds of thousands of Mexican free-tailed bats (*Tadarida brasiliensis*) swarm out of the entrance of the Bracken Cave in Texas for their nightly feeding trips. These bats roost during the day, hanging from the ceilings of passages deep within the cave. The daily commute to and from their roosts requires the bats to fly in huge numbers through narrow, winding passageways in total darkness. Once out of the cave, the bats pursue small moths. In the dark, they must be able to detect these tiny, erratic targets while avoid-

Blind as a Bat
These bats emit very loud sound pulses and then construct their "view" of the world from the echoes that come back.

ing obstacles such as leaves, branches, bushes, rocks, and telephone wires. How do they manage to avoid collisions and fill their stomachs?

The answer is that bats produce very high pitched and very loud sounds that bounce off of objects and create echoes. By listening to the echoes, the bats detect obstacles and tasty insects—a process called echolocation. Bats that are blinded fly quite well and catch insects, but bats with their ears plugged cannot avoid obstacles or catch prey in total darkness.

The bats emit about 20 to 80 echolocating calls per second. That rate requires very fine control of the muscles used in making vocalizations. These pulses of sound are exceedingly loud, but we cannot hear them because they are far above our range of hearing. The loudness of the pulse may be 120 decibels—louder than a jet engine at takeoff—but the echo that comes back from a tiny object only 5 to 10 m away is softer than a faint whisper. The bat's ability to hear these whispers between the blasts comes from tiny muscles in its ears that contract and relax 20 to 80 times per second in precise coordination with the vocal muscles.

Hearing the echoes is a remarkable accomplishment, but even more remarkable is the fact that the bat constructs its "view" of its complex environment from these echoes so that it can dart and dive and capture its flying prey while avoiding obstacles. The incredible echolocation feats of bats are possible because they have a *nervous system* that can extract and rapidly process information from their environment and convert that information to signals that control their muscles. The fundamental units of all nervous systems are cells called *neurons.*

Some neurons can convert environmental energy such as sound into a simple set of electric signals called *nerve impulses*, some can integrate nerve impulses, and others can send nerve impulses to muscles to make them contract. The basic story of any nervous system has three components: sensory input, integration, and output to muscles and other tissues. The story can be complex, as in the case of the bat, but the language is always simple—the nerve impulse.

A remarkable feature of nervous systems is that they can carry out many different tasks at the same time. The nervous system that is creating an auditory map of the bat's environment is also responsible for its flight; for controlling its heart, breathing, and body temperature, and other body functions; and for remembering its way back to the cave and to its roost within the cave.

This chapter is about neurons and nervous systems. We will learn how nerve impulses are generated, transmitted, and communicated from cell to cell—all tasks that are accomplished by the neurons. That knowledge will enable us to see how neurons are arranged in circuits to process information, and how even simple circuits can learn and remember.

Nervous Systems: Cells and Functions

Nervous systems are built of cells called **neurons**. In this section, we examine some general characteristics of neurons and their relationships with other cells. Throughout the animal kingdom, neurons are specialized to receive information, encode it, and transmit information to other cells. Together with their specialized supportive cells, neurons make up nervous systems, whose functions can be described in terms of circuits of interacting neurons.

Nervous systems process information

Various kinds of information from both inside and outside the animal are received and converted or transduced by **sensors** (see Chapter 42) into electric signals that can be transmitted and processed by neu-

41.1 Nervous Systems Vary in Complexity As we compare animals that have increasingly complex sensory and behavioral abilities, we find information processing increasingly centralized in ganglia (collections of nerve cells) or in a brain.

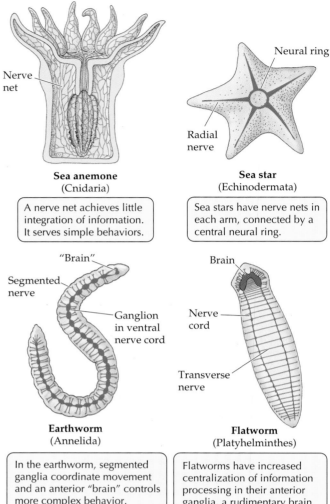

Sea anemone
(Cnidaria)

A nerve net achieves little integration of information. It serves simple behaviors.

Sea star
(Echinodermata)

Sea stars have nerve nets in each arm, connected by a central neural ring.

Earthworm
(Annelida)

In the earthworm, segmented ganglia coordinate movement and an anterior "brain" controls more complex behavior.

Flatworm
(Platyhelminthes)

Flatworms have increased centralization of information processing in their anterior ganglia, a rudimentary brain.

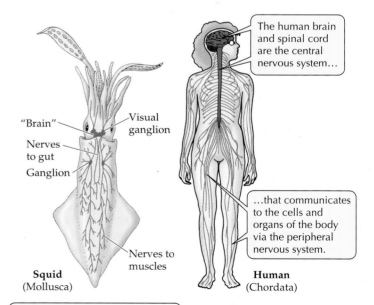

Squid
(Mollusca)

In squid, more complex behaviors are served by collections of neurons in specialized ganglia.

The human brain and spinal cord are the central nervous system…

…that communicates to the cells and organs of the body via the peripheral nervous system.

Human
(Chordata)

rons. To cause behavior or physiological responses, a nervous system communicates signals to **effectors**, such as muscles and glands (see Chapter 44).

A simple animal that remains fixed to its substrate, such as a cnidarian, can process information by a simple network of neurons that do little more than provide direct lines of communication from sensors to effectors (Figure 41.1). The cnidarian's nerve net merely detects food or danger and causes its tentacles to retract and its body to constrict.

More complex animals that move around the environment and hunt for food and mates need to process and integrate larger amounts of information. Even simple animals such as flatworms fit this description, and their increased need for information processing is met by clusters of neurons called **ganglia**. Ganglia serving different functions may be distributed around the body as in the earthworm or the squid, but frequently a pair of ganglia are larger and more central and are therefore given the designation of **brain**.

In vertebrates, most of the cells of the nervous system are found in the brain and the spinal cord, the site of most information processing, storage, and retrieval. Therefore, the brain and spinal cord are called the **central nervous system** (**CNS**). Information from sensors to the CNS and from the CNS to effectors is transmitted via neurons that extend or reside outside of the brain and the spinal cord; these neurons are called the **peripheral nervous system**.

Vertebrates have highly developed central nervous systems, but they vary greatly in behavioral complexity and physiological capabilities. Figure 41.2 shows the brains of similar-sized vertebrates drawn to the same scale. But even small nervous systems are remarkably complex. Consider the nervous systems of small spiders that have programmed within them the thousands of precise movements necessary to construct a beautiful web without prior experience.

One of the greatest challenges of biology is to understand how the human brain functions, but it would be a major breakthrough even to understand a much simpler nervous system. Much progress in neurobiology (the science that studies nervous systems) has come from research on the simpler nervous systems of invertebrates.

Neurons are the functional units of nervous systems

Even though nervous systems vary enormously in structure and function, neurons function almost identically in animals as different as squids and humans. The important property of neurons is that their plasma membranes can generate electric signals called **nerve impulses**. Their plasma membranes can also conduct nerve impulses from one location on a cell to the most distant reaches of that cell—a distance that can be more than a meter.

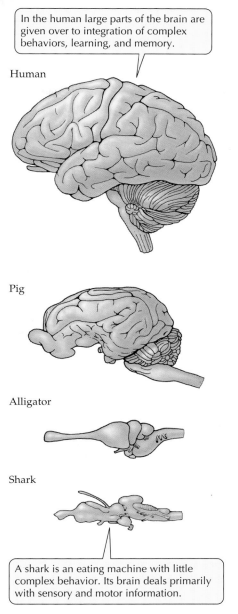

41.2 Brains Vary in Size and Complexity The brains of four vertebrate species—all of which may have a similar body mass—show immense differences.

Most neurons have four regions: a cell body, dendrites, an axon, and axon terminals (Figure 41.3a), but the variation in different types of neurons is considerable (Figure 41.3b). The **cell body** contains the nucleus and most of the cell's organelles. Many projections may sprout from the cell body. Most nerve cell projections are bushlike **dendrites** (from the Greek *dendron*, "tree"), which bring information from other neurons or sensory cells to the neuron's cell body. In most neurons, one projection is much longer than the others

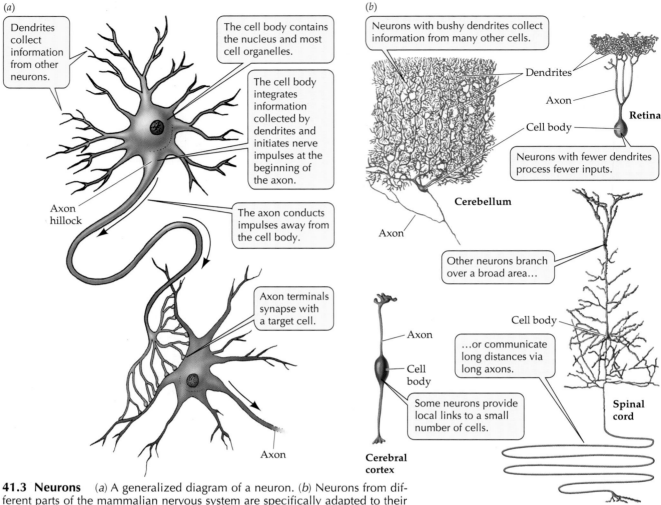

(a)

Dendrites collect information from other neurons.

The cell body contains the nucleus and most cell organelles.

The cell body integrates information collected by dendrites and initiates nerve impulses at the beginning of the axon.

Axon hillock

The axon conducts impulses away from the cell body.

Axon terminals synapse with a target cell.

Axon

(b)

Neurons with bushy dendrites collect information from many other cells.

Dendrites

Axon

Cell body

Retina

Neurons with fewer dendrites process fewer inputs.

Cerebellum

Axon

Other neurons branch over a broad area...

Cell body

...or communicate long distances via long axons.

Axon

Cell body

Some neurons provide local links to a small number of cells.

Cerebral cortex

Spinal cord

41.3 Neurons (a) A generalized diagram of a neuron. (b) Neurons from different parts of the mammalian nervous system are specifically adapted to their functions.

and is called the **axon**. Axons usually carry information away from the cell body. The length of the axon varies greatly in different types of neurons—as does the degree of branching of the dendrites.

The axons of some neurons are remarkably long. Axons are the "telephone lines" of the nervous system. Information received by the dendrites can influence the cell body to generate a nerve impulse that is then conducted along the axon to the cell that is its target. At the target cell, the axon divides, like the frayed end of a rope, into a spray of fine nerve endings. At the tip of each of these tiny nerve endings is a swelling called an **axon terminal** that comes very, very close to another neuron, a muscle cell, or a gland cell.

Where an axon terminal comes close to another cell, the membranes of both cells are modified to form a *synapse*. In most cases a space or cleft only about 25 nm wide separates the two membranes at the synapse. A nerve impulse arriving at an axon terminal that forms such a synapse causes molecules called *neuro-*

transmitters that are stored in the axon terminal to be released into the cleft. Neurotransmitters are released from the *presynaptic membrane*, and they diffuse across the cleft and bind to receptors on the *postsynaptic membrane*.

Most individual neurons make and receive thousands of synapses. A neuron integrates information (synaptic inputs) from many sources by producing nerve impulses that travel down its single axon to target cells. We will discuss synaptic transmissions in more detail later in the chapter.

Glial cells are also important components of nervous systems

Much of neurobiology focuses on the structure and function of neurons, but they are not the only type of cell in the nervous system. In fact, there are more **glial cells** than neurons in the human brain. Like neurons, glial cells come in many forms. Some glial cells physically support and orient the neurons and help the neu-

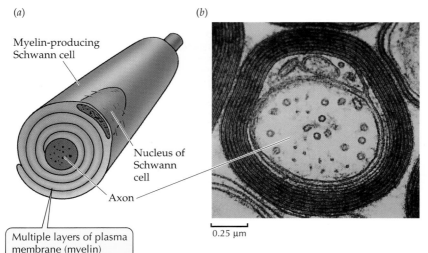

(a)

Myelin-producing
Schwann cell

Nucleus of
Schwann
cell

Axon

Multiple layers of plasma
membrane (myelin)
insulate the axon.

(b)

0.25 μm

41.4 Wrapping up an Axon
(a) Schwann cells wrap axons with layers
of myelin, a type of plasma membrane
that provides electrical insulation. (b) A
myelinated axon, seen in cross section
through an electron microscope.

rons make the right contacts during their embryonic development. Other glial cells insulate axons.

In the peripheral nervous system **Schwann cells**, one type of glial cell, wrap around the axons of neurons, thereby covering the axon with concentric layers of insulating plasma membrane (Figure 41.4). Glial cells called **oligodendrocytes** perform a similar function in the central nervous system. The covering produced by Schwann cells and oligodendrocytes, called **myelin**, gives many parts of the nervous system a glistening white appearance. Later in the chapter we will see how the electrical insulation that myelin provides for axons increases the speed with which they can conduct nerve impulses.

Glial cells are well known for the many housekeeping functions they perform. Some glial cells supply neurons with nutrients; others consume foreign particles and cell debris. Glial cells also help maintain the proper ionic environment around neurons. Although they have no axons and they do not generate or conduct nerve impulses, some glial cells communicate with one another electrically through a special type of contact called a **gap junction**, a connection that enables ions to flow between cells (see Chapter 5). Because we know very little about the functions of glial cells, they are an exciting area for future research.

Glial cells called **astrocytes** (because they look like stars) create the **blood–brain barrier**. Blood vessels throughout the body are very permeable to many chemicals, some of which are toxic. The brain is better protected from toxic substances than are most tissues of the body because of the blood–brain barrier. Astrocytes form this barrier by surrounding the smallest, most permeable blood vessels in the brain. Protection of the brain is crucial because, unlike other tissues of the body, it cannot recover from damage by generating new cells.

Shortly after birth, neurons in the mammalian brain cease cell division, at which time we have the greatest number of neurons we will ever have in our lives. Throughout the rest of life neurons are progressively lost as they die. Without the blood–brain barrier, the rate of neuron loss could be much greater. However, the barrier is not perfect. Because it consists mostly of plasma membranes, it is permeable to fat-soluble substances. Anesthetics and alcohol, both of which have well-known effects on the brain, are fat-soluble chemicals.

Neurons function in circuits

As we learn more about the properties of neurons, it is important to keep in mind that nervous systems depend on neurons working together in circuits. Toward the end of this chapter we will begin to explore the properties of simple neuronal circuits, and in Chapter 43 we will explore some of the more complex circuits in the human brain, which contains between 10^9 and 10^{11} neurons. Most of these neurons make 1,000 or more synapses with other neurons. Thus there may be as many as 10^{14} synapses in the human brain. Therein lies the incredible ability of the brain to process information.

This astronomical number of neurons and synapses is divided into thousands of distinct but interacting circuits that function in parallel and accomplish the many different tasks of the nervous system. But before we can understand how even one of these circuits works, we must understand the properties of individual neurons.

Neurons: Generating and Conducting Nerve Impulses

As in all other cells, the cytoplasm of a neuron has an excess of negative electric charges. The inside of a neuron is usually about 60 to 70 millivolts (mV) more negative than the outside of the cell. In this section, we will explore the consequences of the electrical properties of cells. After reviewing some basic electrical concepts, we will examine in detail the roles of ions, ion pumps, and ion channels in establishing and altering the electrical properties of neurons. These electrical changes generate nerve impulses whose conduction

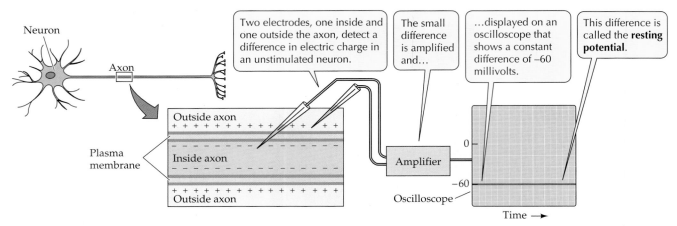

41.5 Measuring the Resting Potential   The difference in electric charge across the plasma membrane of a neuron can be measured using two electrodes, one inside and one outside the cell. In an unstimulated neurons, this difference is constant and is known as the resting potential.

and frequency establish the language by which the nervous system communicates information.

Simple electrical concepts underlie neuron functions

The difference in electric charge across the plasma membrane of a neuron is called its **membrane potential**. In an unstimulated neuron, it is identified as the **resting potential**. Electric potentials across membranes can be measured with electrodes. An electrode can be made from a glass pipette pulled to a very sharp tip and filled with a solution that conducts electric charge. Using such electrodes, we can record very tiny local electrical events that occur across plasma membranes. If a pair of electrodes are placed one on each side of the plasma membrane of a resting axon, they measure a voltage difference of about –60 mV, the resting potential (Figure 41.5).

The resting potential provides a means for neurons to be responsive to specific stimuli. A neuron is sensitive to any chemical or physical factor that causes a change in the resting potential across a portion of its plasma membrane. The most extreme change in membrane potential is the nerve impulse, which is a sudden and rapid reversal in the charge across a portion of the plasma membrane. For a brief moment, only 1 or 2 milliseconds, the inside of a part of the plasma membrane becomes more positive than the outside. A nerve impulse can move along a membrane from one part of a neuron to its farthest extensions. Nerve impulses are also called **action potentials**, a name that conveys the contrast with resting potential.

To understand how resting potentials are created, how they are perturbed, and how action potentials are

generated and conducted along plasma membranes, it is necessary to know a little about electricity, ions, and the special ion channel proteins in the plasma membranes of neurons. **Voltage** (potential) is the tendency for electrons to move between two points. Voltage is to the flow of electrons what pressure is to the flow of water. If the negative and the positive poles of a battery are connected by a copper wire, electrons flow from negative to positive. This flow of electrons is an electric **current**, and it can be used to do work, just as a current of water can be used to do work such as turning a mill wheel.

As you may recall from earlier chapters, electric charges cross cell membranes not as electrons but as charged ions. The major ions that carry electric charges across the membranes of neurons are sodium (Na^+), chloride (Cl^-), potassium (K^+), and calcium (Ca^{2+}). It is also important to remember that ions with opposite charges attract each other. With these basics of bioelectricity in mind, we can ask how the resting potential of the membrane is maintained, and how the flow of ions through membrane channels is turned on and off to generate nerve impulses—action potentials, as we will call them.

Ion pumps and channels generate resting and action potentials

The plasma membranes of neurons, like those of all other cells, are lipid bilayers that are impermeable to ions. However, these impermeable lipid bilayers contain at many points protein molecules that serve as ion channels and ion pumps. Pumps and channels are responsible for resting and action potentials.

Membrane **pumps** use energy to move ions or other molecules against their concentration gradients. The major pump in neuronal membranes is the sodium–potassium pump that we learned about in Chapter 5. The action of this pump expels Na^+ ions from inside the cell, exchanging them for K^+ ions from outside the

(a) Na⁺–K⁺ pump

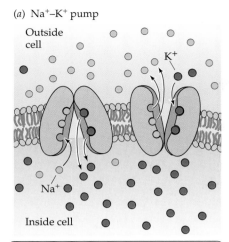

(b) Na⁺ and K⁺ channels

The Na⁺–K⁺ pump moves Na⁺ and K⁺ ions against their concentration gradients which would otherwise run down because…

…K⁺ and Na⁺ ions tend to leak down their concentration gradients through ion-specific channels.

41.6 Ion Pumps and Channels
(a) The Na⁺–K⁺ pump actively moves K⁺ ions to the inside of neurons and Na⁺ ions to the outside. (b) Ion channels allow K⁺ ions to leave cells and Na⁺ ions to enter cells. These Na⁺ channels are gated.

The resting potential results from the departure of potassium ions

Potassium channels are the most common type of open channel in the plasma membranes of neurons. These channels make neurons much more permeable to K⁺ than to any other ions, and as Figure 41.7 shows, this characteristic explains the resting potential. Because the plasma membrane of the neuron is permeable to K⁺, and because the sodium–potassium pump keeps the concentration of K⁺ inside the cell much higher than that outside the cell, K⁺ tends to diffuse out of the cell.

If a K⁺ ion leaves the cell, it leaves behind an unmatched negative charge. Most of the negative charges are on large molecules such as proteins in the cytoplasm, and those large molecules with their unbalanced negative charges cannot cross the plasma membrane. As a result, the tendency of the K⁺ ions to diffuse out of the cell through the open channels causes the inside of the cell to become negatively charged in comparison to the external medium. Thus

cell. The pump expels about three Na⁺ ions for every two K⁺ ions it brings in (Figure 41.6a; see also Figure 5.12). The sodium–potassium pump keeps the concentration of K⁺ inside the cell greater than that of the external medium, and the concentration of Na⁺ inside the cell less than that of the external medium. The concentration differences established by the pump mean that K⁺ would diffuse out of the cell and Na⁺ would diffuse into the cell if the ions could cross the lipid bilayer. By itself, the unequal distribution of K⁺ and Na⁺ ions on the two sides of the plasma membrane does not create the resting potential. To explain the resting potential, we must introduce ion channels.

Ion channels are pores formed by proteins in the lipid bilayer. These water-filled pores allow ions to pass through (Figure 41.6b). They are selective; that is, they may permit the passage of only one type of ion. Thus there are potassium channels, sodium channels, chloride channels, and calcium channels. Ions can move in either direction through a channel. Most ion channels of neurons behave as if they contain a "gate" that opens to allow ions to pass under some conditions, but closes under other conditions.

Voltage-gated channels open or close in response to a change in the voltage across the plasma membrane, and **chemically gated** channels open or close depending on the presence or absence of a specific chemical that binds to the channel protein or to a separate receptor that in turn alters the channel protein. Both voltage-gated and chemically gated channels play important roles in neuronal functions, as we will see, but first we will see how nongated channels, channels that are always open, are responsible for maintaining the resting potential in neurons and other cells.

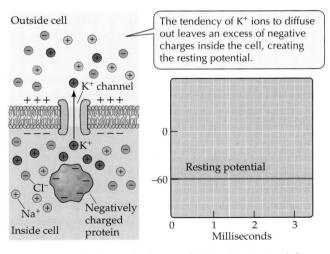

The tendency of K⁺ ions to diffuse out leaves an excess of negative charges inside the cell, creating the resting potential.

41.7 Open K⁺ Channels Create the Resting Potential
Open K⁺ channels allow potassium ions to diffuse out of the cell, leaving unbalanced negative charges behind (mostly on chloride ions and protein molecules).

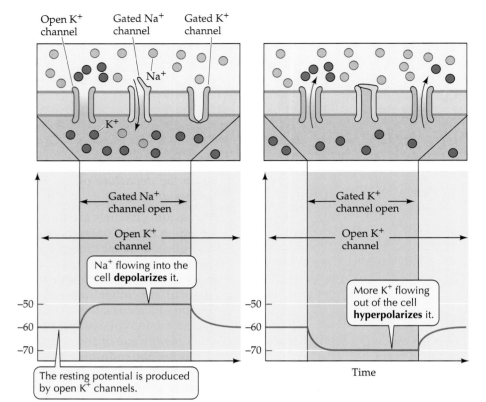

Open K⁺ channel   Gated Na⁺ channel   Gated K⁺ channel

Na⁺

K⁺

Gated Na⁺ channel open

Open K⁺ channel

Na⁺ flowing into the cell **depolarizes** it.

−50

−60

−70

The resting potential is produced by open K⁺ channels.

Gated K⁺ channel open

Open K⁺ channel

More K⁺ flowing out of the cell **hyperpolarizes** it.

−50

−60

−70

Time

41.8 Membranes Can Be Depolarized or Hyperpolarized

The resting potential created by the diffusion of K⁺ out of the cell (see Figure 41.7) can change. A shift to a less negative membrane potential, as occurs when a gated Na+ channel opens, is called depolarization. Hyperpolarization occurs when the membrane potential becomes more negative, as when more K⁺ enters the cell.

the plasma membrane is **polarized**—that is, regions of unequal electric charge are separated, resulting in a resting potential. But what happens when the resting potential is altered?

Changes in membrane permeability to ions alter membrane potential

If a stimulus perturbs the resting potential and the inside of the cell becomes *more* negative, the membrane is **hyperpolarized**. If the inside of the cell becomes *less* negative, the membrane is **depolarized**. Changes in the gated channels can cause a membrane to hyperpolarize or depolarize.

For example, what would happen if some sodium channels in the plasma membrane opened. Na⁺ ions would diffuse into the cell because of their higher concentration on the outside, and they would also be attracted into the cell by the excess negative charges on the protein molecules. As a result of the entry of Na⁺ ions, the plasma membrane would become depolarized in comparison to its resting condition. An opposite change in the resting potential would occur if gated K⁺ channels opened. The leak of positively charged ions from the inside of the cell would increase, and the plasma membrane would become hyperpolarized (Figure 41.8).

There are also gated Cl⁻ channels in the plasma membranes of neurons, and the concentration of Cl⁻ ions is greater in the extracellular fluid than in the in-

tracellular fluid. What would happen to membrane polarity if some of these Cl⁻ channels opened? The opening and closing of ion channels, which result in changes in the polarity of the plasma membrane, are the basic mechanisms by which neurons respond to electrical, chemical, or other stimuli, such as touch, sound, and light.

How does a neuron benefit from a change in its resting membrane potential at a particular location? Can that information be passed on to other parts of the cell? A local perturbation of membrane potential causes electric current to flow, and that flow of current spreads the change in membrane potential. However, the nerve cell is a poor conductor of electricity, so the change in membrane polarity diminishes and disappears before it travels very far from the site where the resting potential was perturbed.

The reason is that the axonal membrane is leaky to many charged ions; therefore, a potential difference does not move very far down an axon. It is like trying to send water through a leaky hose. Communication of a stimulus by the flow of electric current along axons is useful over only very short distances. As we will see, however, local flow of electric current is an important part of the mechanism that generates action potentials and moves action potentials along axons.

Sudden changes in ion channels generate action potentials

An action potential is a sudden and major change in membrane potential that lasts for only 1 or 2 milliseconds. This pulse of electric charge is conducted along the axon of a neuron at speeds of up to 100 m/s, which is equivalent to running the length of a football field in 1 s. We will see first how an action potential is generated, then how it moves down an axon.

If we place the tips of a pair of electrodes on either side of the plasma membrane of an axon in a resting state and measure the voltage difference, the reading is about –60 mV, as we saw in Figure 41.5. If these electrodes are exposed to an action potential traveling down the axon, they register a rapid change in membrane potential, from –60 mV to about +50 mV (Figure 41.9). The membrane potential rapidly returns to its resting level of –60 mV as the action potential passes.

Voltage-gated sodium channels in the plasma membrane of the axon are primarily responsible for the action potential. At a normal membrane resting potential these channels are mostly closed. They are called voltage-gated because when the plasma membrane is depolarized to a certain threshold level (less negative

than the normal resting potential), many of these Na^+ channels flip open briefly—for less than a millisecond. Because of the action of the sodium–potassium pump, the Na^+ concentration is much higher outside the axon than inside, so whenever the sodium channels open, Na^+ ions from the outside rush into the axon. The entering Na^+ ions make the inside of the plasma membrane electrically positive (see Figure 41.9).

41.9 The Course of an Action Potential Action potentials result from rapid changes in voltage-gated Na^+ and K^+ channels. Like the resting potential (see Figure 41.5), they can be measured using two electrodes.

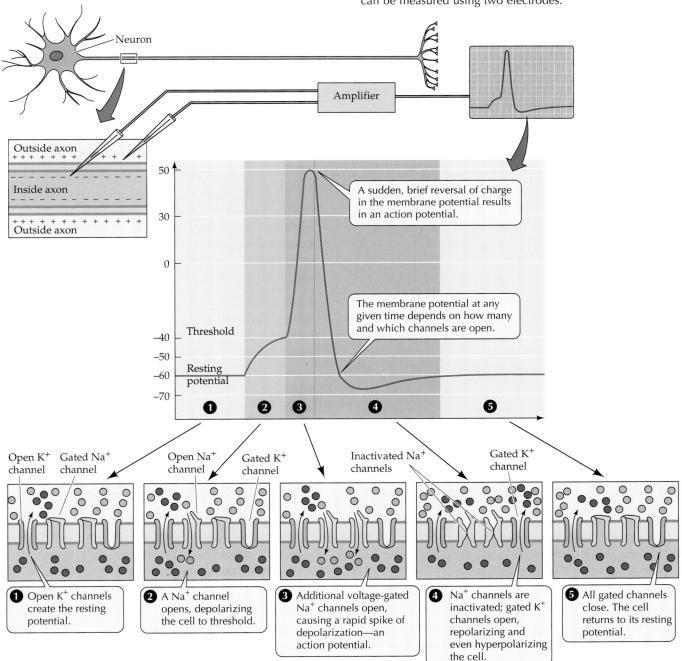

A sudden, brief reversal of charge in the membrane potential results in an action potential.

The membrane potential at any given time depends on how many and which channels are open.

Open K^+ channel   Gated Na^+ channel

Open Na^+ channel   Gated K^+ channel

Inactivated Na^+ channels

Gated K^+ channel

1 Open K^+ channels create the resting potential.

2 A Na^+ channel opens, depolarizing the cell to threshold.

3 Additional voltage-gated Na^+ channels open, causing a rapid spike of depolarization—an action potential.

4 Na^+ channels are inactivated; gated K^+ channels open, repolarizing and even hyperpolarizing the cell.

5 All gated channels close. The cell returns to its resting potential.

The opening of the Na⁺ channels causes the rise of the action potential—what neurobiologists call the spike. What causes the return to resting potential? The main reason for the drop is the closing of the sodium channels, which allows the plasma membrane to return to resting potential. Some axons also have voltage-gated potassium channels. These channels open more slowly than the sodium channels and they stay open longer; therefore, they help return the voltage across the membrane to its resting level by allowing K⁺ ions to carry excess positive charges out of the axon.

Another feature of the voltage-gated sodium channels is that once they open and close, they cannot be triggered again for several milliseconds. The properties of the voltage-gated sodium channels can be explained by the assumption that they have two voltage-sensitive gates, an activation gate and an inactivation gate (see Figure 41.9). Under resting conditions the activation gate is closed and the inactivation gate is open. Depolarization of the membrane to a threshold level causes both gates to change state, but the activation gate responds faster. As a result, the channel is open for the passage of Na⁺ ions for a brief time between the opening of the activation gate and the closing of the inactivation gate.

The inactivation gates remain closed for a few milliseconds before they spontaneously open again, thus explaining why the membrane has a **refractory period** (a period during which it cannot act) before it can fire another action potential. When the inactivation gates finally open, the activation gates are closed and the membrane is poised to respond once again to a depolarizing stimulus by firing another action potential.

The difference in concentration of Na⁺ ions across the plasma membrane of neurons is the "battery" that drives the action potential. How rapidly does the battery run down? It might seem that a substantial number of Na⁺ and K⁺ ions would have to cross the membrane for the membrane potential to change from –60 mV to +50 mV, and back to –60 mV again. In fact, only about one Na⁺ (or K⁺) ion in 10 million actually moves through the channels during the passage of an action potential.

Thus the effect of a single action potential on the concentration ratios of Na⁺ or K⁺ is very small. Even hundreds of action potentials barely change the concentration differences of Na⁺ and K⁺ on the two sides of the membrane. Consequently, it is not difficult for the sodium–potassium pump to keep the "battery" charged, even when the cell is generating many action potentials every second.

Action potentials are conducted down axons without reduction in the signal

Action potentials enable neurons to convey information over long distances with no loss of the signal. If

41.10 Action Potentials Travel along Axons When an action potential occurs in a piece of membrane, electric currents flow to adjacent areas of membrane and depolarize them. As voltage-gated channels in these areas reach threshold, they create an action potential. In this way, an action potential continuously regenerates itself along the axon.

we place two pairs of electrodes at two locations along an axon, we will record the same action potential as it travels down the axon. The height of the action potential does not change as it travels along the axon. The action potential is an all-or-nothing, self-regenerating event (Figure 41.10a).

How does an action potential move over long distances? We can start an action potential traveling down an axon by shocking the membrane with an electric current delivered through a stimulating electrode that depolarizes the membrane. Now we can observe the changes in membrane potential associated with the passage of that action potential past our recording electrodes. When one part of an axon fires an action potential, the adjacent regions of membrane also become slightly depolarized because of the spread of electric current (Figure 41.10b).

Positive ions flood into the axon at the site of the action potential. Once they are inside the axon, the positive ions spread to adjacent regions of the axon, making those regions less negative. If this depolarization of the adjacent region of plasma membrane brings it to the threshold level that causes the opening of voltage-gated sodium channels, an action potential is generated. Because an action potential always brings the adjacent area of membrane to threshold, the action potential is self-regenerating and propagates itself along the axon. The action potential propagates itself in only one direction; it cannot reverse itself, because the part of the membrane it came from is refractory.

The action potential does not travel along all axons at the same speed. Action potentials travel faster in large-diameter axons than in small-diameter axons. In invertebrates, the axon diameter determines the rate of conduction, and axons that transmit messages involved in escape behavior are very thick. The axons that enable squid to escape predators are almost 1 mm in diameter. These giant axons were the most important experimental material used in the classic studies in neurophysiology that produced basic discoveries about action potentials and their conduction.

Ion channels and their properties can now be studied directly

The large size of the squid axon made it possible for the British physiologists A. L. Hodgkin and A. F. Huxley to study the electrical properties of axonal membranes 50 years ago. They used fine electrodes to

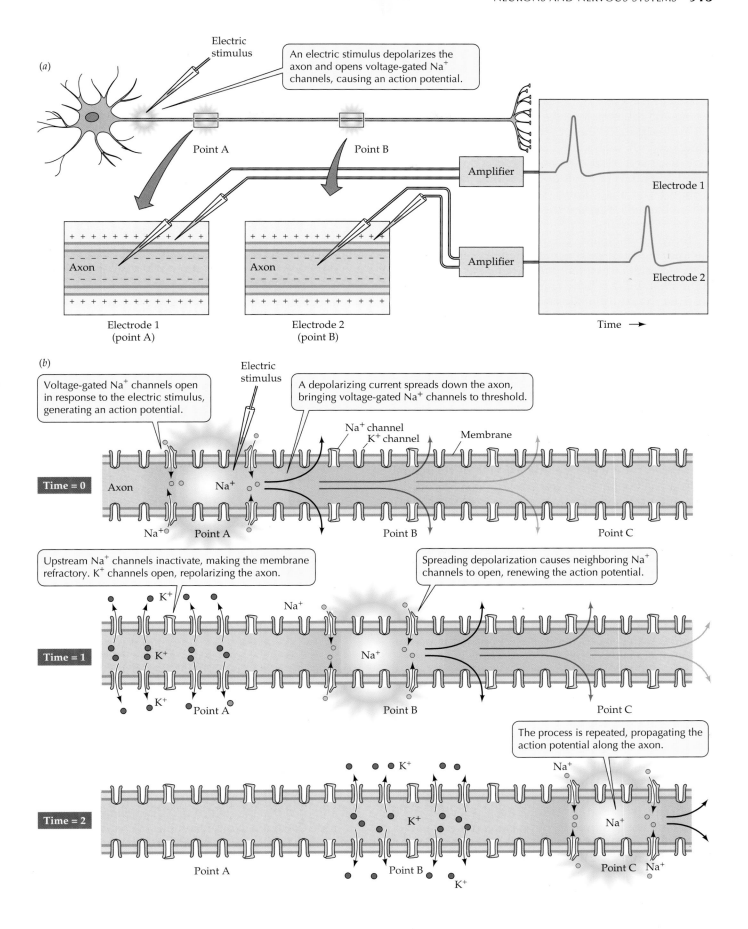

(a)

Electric stimulus

An electric stimulus depolarizes the axon and opens voltage-gated Na⁺ channels, causing an action potential.

Point A

Point B

Amplifier

Electrode 1

+ + + + + + + + + + +
Axon
– – – – – – – – – – –

+ + + + + + + + + + + +

Electrode 1
(point A)

+ + + + + + + + + + + +
Axon
– – – – – – – – – – –

+ + + + + + + + + + + + +

Electrode 2
(point B)

Amplifier

Electrode 2

Time ⟶

(b)

Voltage-gated Na⁺ channels open in response to the electric stimulus, generating an action potential.

Electric stimulus

A depolarizing current spreads down the axon, bringing voltage-gated Na⁺ channels to threshold.

Na⁺ channel
K⁺ channel
Membrane

Time = 0

Axon

Na⁺

Na⁺

Point A

Point B

Point C

Upstream Na⁺ channels inactivate, making the membrane refractory. K⁺ channels open, repolarizing the axon.

Spreading depolarization causes neighboring Na⁺ channels to open, renewing the action potential.

Time = 1

K⁺

Na⁺

K⁺

Na⁺

K⁺

Point A

Point B

Point C

The process is repeated, propagating the action potential along the axon.

Na⁺

Time = 2

K⁺

K⁺

Na⁺

Point A

Point B

K⁺

Point C

Na⁺

measure voltage differences across the plasma membrane and to pass electric current into the axon to change its resting potential. They also changed the concentrations of Na⁺ and K⁺ ions both inside and outside of the axon and measured the changes in the resting and action potentials.

On the basis of their many careful experiments, Hodgkin and Huxley developed the whole story we have discussed so far explaining resting potentials and action potentials in terms of Na⁺ and K⁺ ion channels in the plasma membrane. They had no methods to study channels directly; they had to postulate them and their properties.

Now neurobiologists can record currents caused by the openings and closings of single ion channels directly. A technique called **patch clamping**, developed in the 1980s by Bert Sakmann and Erwin Neher, made possible the study of single ion channels in plasma membranes. In patch clamping, a very fine, polished glass pipette is placed in contact with the plasma membrane. Slight suction makes a seal between the pipette and the patch of membrane under the tip of the pipette. Movements of ions, and therefore electric charges, through channels in the patch of membrane can be recorded through the pipette (Figure 41.11). In patch clamping, the solution filling the pipette determines the ion concentrations on the outside of the patch, and if the patch is torn lose from the cell, the ion concentrations on the interior side of the membrane can also change. Frequently, a patch will contain only one or a few ion channels; thus the electrical recording from that patch can show individual channels opening and closing. Neher and Sakmann received the Nobel prize in 1991 for their invention (28 years after Hodgkin and Huxley received their Nobel prizes).

Action potentials can jump down axons

In nervous systems that are more complex than those of invertebrates, increasing the speed of action potentials by increasing the diameter of axons is not feasible, because of the sheer number of axons in more complex animals. For example, about a million axons extend from each of our eyes. Evolution has increased propagation velocity in vertebrate axons in a way that does not require large size.

When we discussed glial cells earlier in the chapter, we mentioned that glial cells of one type, called Schwann cells, send out projections that wrap around axons, covering them with concentric layers of plasma membrane (see Figure 41.4). These wrappings of myelin are not continuous, but have regularly spaced gaps called **nodes of Ranvier** where the axon is not covered (Figure 41.12).

The myelin wrap electrically insulates the axon; that is, charged ions cannot cross the parts of the

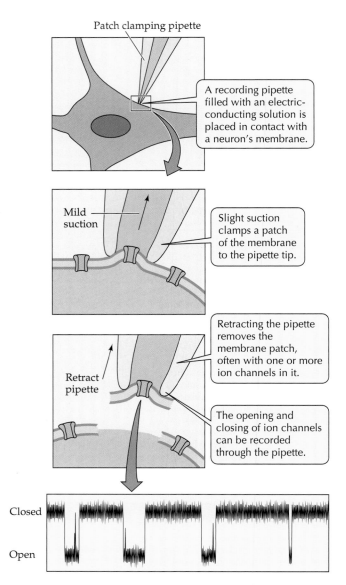

41.11 Patch Clamping A patch clamp electrode can record the opening and closing of a single ion channel.

plasma membrane that are wrapped in myelin. Thus, an axon can fire an action potential at a node of Ranvier, but that action potential cannot be conducted through the adjacent patch of axon covered with myelin. The positive charges that flow into the axon at the node, however, spread down the axon.

When this spread of current causes the plasma membrane at the next node of Ranvier to depolarize to threshold, an action potential is fired at that node. Action potentials therefore appear to jump from node to node down the axon. The speed of conduction is increased in these myelin-wrapped axons because electric current flows very fast compared to how long ion channels take to open and close. This form of impulse propagation is called **saltatory** (jumping) **conduction**

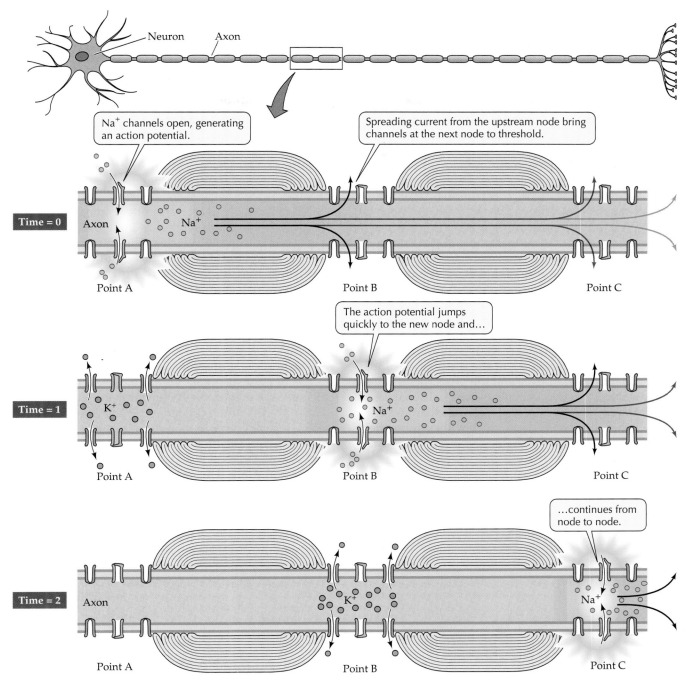

Na⁺ channels open, generating an action potential.

Spreading current from the upstream node bring channels at the next node to threshold.

Time = 0

Axon

Na⁺

Point A   Point B   Point C

The action potential jumps quickly to the new node and...

Time = 1

K⁺   Na⁺

Point A   Point B   Point C

...continues from node to node.

Time = 2

Axon   K⁺   Na⁺

Point A   Point B   Point C

41.12 Saltatory Action Potentials Action potentials appear to jump from node to node in myelinated axons.

and is much quicker than continuous propagation of action potentials down an unmyelinated axon.

You have probably experienced the difference in the velocity at which action potentials travel down myelinated and unmyelinated axons. If you touch a very hot or very cold object, you experience a sharp pain before you sense whether the object is hot or cold. With the sensing of the temperature also comes a burning pain different from the first sharp pain. Sensory axons that carry sharp-pain sensation are myelinated, but most axons carrying information about temperature, as well as axons carrying the sensation of burning, aching pain, are unmyelinated.

As a result, you know that something is wrong before you know what it is or how bad it is. The unmyelinated temperature and burning-pain axons conduct impulses at only 1 to 2 m/s, whereas the myelinated sharp-pain axons conduct impulses at velocities of up to 5 or 6 m/s. The largest myelinated axons in the human nervous system conduct impulses at velocities of up to 120 m/s.

Neurons, Synapses, and Communication

The most remarkable abilities of nervous systems stem from the interactions of neurons. These interactions process and integrate information. Our nervous systems can orchestrate complex behaviors, deal with complex concepts, and learn and remember because large numbers of neurons interact with one another. The mechanisms for these interactions lie in the synapses between cells.

Synapses are structurally specialized junctions where one cell influences another cell directly through the transfer of a chemical or an electrical message. We will focus first on chemical synapses, which include most of the synapses in vertebrates, using as our example the neuromuscular junction—that is, the synapse where a motor neuron stimulates a muscle to contract.

In our discussion of synapses, we'll examine the specializations and functions of the presynaptic and the postsynaptic membranes, and the properties of different kinds of synapses, including electrical synapses, which are more common in invertebrates than in vertebrates. At the end of this section we will examine the diversity of chemical signals used by chemical synapses.

The classic synapse connects a neuron and a muscle

A motor neuron that innervates a muscle has only one axon like other neurons, but that axon can branch into many axon terminals that form synaptic junctions with many individual muscle fibers. The synaptic junctions between neurons and muscle fibers are called **neuromuscular junctions**, and they are an excellent model for how fast, excitatory chemical synaptic transmissions work. At the very end of the axon terminal is an enlarged knob or buttonlike structure that contains many spherical vesicles filled with chemical messenger molecules called **neurotransmitters** (Figure 41.13). The neurotransmitter of all neurons that innervate vertebrate skeletal muscle is **acetylcholine**.

Where does the neurotransmitter come from? Some, like acetylcholine, are synthesized in the axon terminal and packaged in the vesicles. The enzymes required for this biosynthesis are produced in the cell body of the neuron and are transported down the axon to the nerve terminals. But there are many different kinds of neurotransmitters, and some, such as peptide neurotransmitters, are, like the synthetic enzymes, produced in the cell body and transported down the axon to the terminals.

The portion of the axon terminal plasma membrane that is in closest contact with the muscle is called the **presynaptic membrane**. Neurotransmitter is released by exocytosis when the membrane of a vesicle fuses with the presynaptic membrane. The **postsynaptic membrane** of the neuromuscular junction is a modified part of the muscle cell plasma membrane called a **motor end plate**. The space between the presynaptic membrane and the postsynaptic membrane is the **synaptic cleft**. On average, the cleft is about 20 to 40 nm wide in chemical synapses, and the neurotransmitter released into the cleft must diffuse across to the postsynaptic membrane (see Figure 41.13).

The modification that makes a patch of muscle cell plasma membrane a motor end plate is the presence of acetylcholine receptor molecules. The receptors function as chemically gated channels that allow both Na^+ and K^+ ions to pass through. Since the resting membrane is already fairly permeable to K^+ ions, the major change that occurs when these channels open is the movement of Na^+ ions into the cell. When a receptor binds acetylcholine, a channel opens and

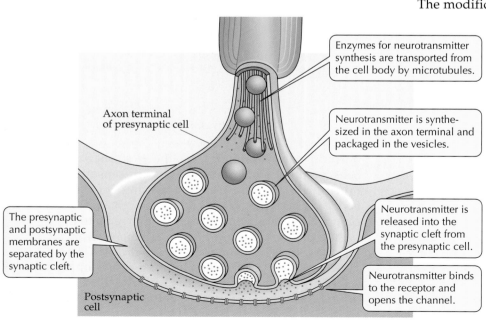

Enzymes for neurotransmitter synthesis are transported from the cell body by microtubules.

Axon terminal of presynaptic cell

Neurotransmitter is synthesized in the axon terminal and packaged in the vesicles.

The presynaptic and postsynaptic membranes are separated by the synaptic cleft.

Neurotransmitter is released into the synaptic cleft from the presynaptic cell.

Neurotransmitter binds to the receptor and opens the channel.

Postsynaptic cell

41.13 The Neuromuscular Junction Is a Chemical Synapse A motor neuron communicates chemically with muscle cells at the neuromuscular junction when neurotransmitter (red dots) crosses the synaptic cleft.

Na⁺ ions move into the cell, making the cell more positive inside (Figure 41.14).

The action of this neurotransmitter is limited by the presence of the enzyme **acetylcholinesterase**, which is found in and around the synapse. This powerful enzyme cleaves any acetylcholine molecules it encounters, including those that are bound to receptors. Thus, the activity of the neuromuscular junction is a balance between acteylcholine release by the presynaptic membrane and its destruction by acetylcholinesterase in the synaptic cleft (see Figure 41.14). The breakdown products of this neurotransmitter degradation are taken up by the presynaptic terminal and resynthesized into acetylcholine.

The arrival of the action potential causes the release of neurotransmitter

Neurotransmitter is released from the presynaptic membrane when an action potential arrives at the axon terminal. The plasma membrane of the axon terminal has a type of voltage-gated ion channel found nowhere else on the axon: the voltage-gated calcium channel. The action potential causes these calcium channels to open (Figure 41.15). Because Ca^{2+} ions are in greater concentration outside the cell than inside the cell, they rush in.

The increase in Ca^{2+} inside the cell causes the vesicles full of acetylcholine to fuse with the presynaptic membrane and spill their contents into the synaptic cleft. The acetylcholine molecules diffuse across the cleft and bind to the receptors on the motor end plate, causing the sodium channels to open briefly and depolarize the postsynaptic cell membrane.

The postsynaptic membrane integrates synaptic input

Postsynaptic membranes differ from presynaptic membranes in an important way. Motor end plates have very few voltage-gated sodium channels; therefore, they do not fire action potentials. This is true not only of motor end plates, but also of dendrites and of most regions of nerve cell bodies. The binding of neurotransmitter to receptors at the motor end plate and the resultant opening of chemically gated ion channels perturb the resting potential of the postsynaptic membrane. This local change in membrane potential spreads to neighboring regions of the plasma membrane of the postsynaptic cell.

Eventually, the spreading depolarization may reach an area of membrane that does contain voltage-gated channels. The entire plasma membrane of a skeletal muscle fiber, except for the motor end plates, has voltage-gated sodium channels. If the axon terminal of a presynaptic cell releases sufficient amounts of neurotransmitter to depolarize a motor end plate enough to bring the surrounding membrane to threshold, action potentials are fired. These action potentials are then conducted throughout the muscle fiber's system of membranes, causing the fiber to contract. (We'll learn about the coupling of muscle membrane action potentials and contraction of muscle fibers in Chapter 44.)

How much neurotransmitter is enough? Neither a single acetylcholine molecule nor the contents of an entire vesicle (about 10,000 acetylcholine molecules) are enough to bring the plasma membrane of a muscle cell to threshold. However, a single action potential in an axon terminal releases about 100 vesicles, which is enough to fire an action potential in the muscle fiber and cause it to twitch.

Synapses between neurons can be excitatory or inhibitory

In vertebrates, the synapses between motor neurons and skeletal

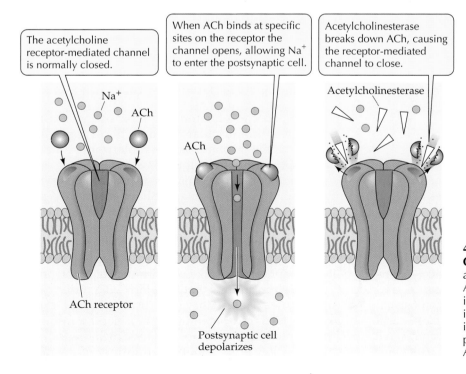

The acetylcholine receptor-mediated channel is normally closed.

Na⁺

ACh

ACh receptor

When ACh binds at specific sites on the receptor the channel opens, allowing Na⁺ to enter the postsynaptic cell.

ACh

Postsynaptic cell depolarizes

Acetylcholinesterase breaks down ACh, causing the receptor-mediated channel to close.

Acetylcholinesterase

41.14 The Actions of Acetylcholine Are Opposed by Acetylcholinesterase When a receptor on the postsynaptic cell binds ACh, a channel opens and Na⁺ ions move into the cell, making the cell more positive inside (depolarization). Acetylcholinesterase in the synapse destroys ACh; the breakdown products are then resynthesized into more ACh.

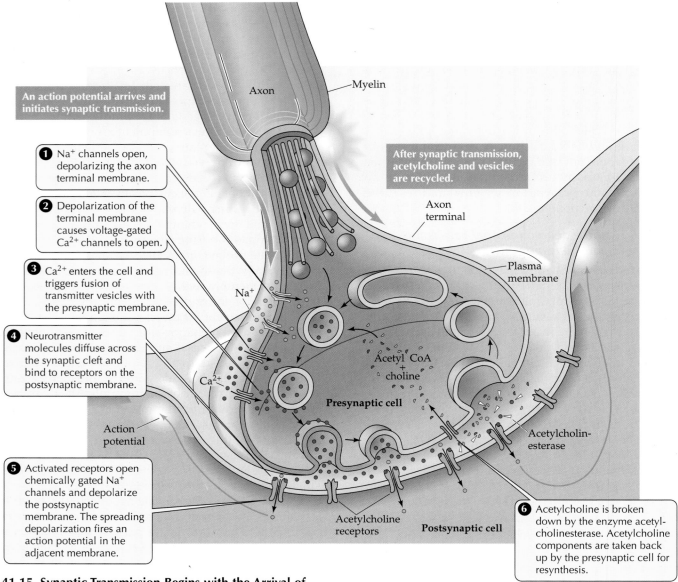

An action potential arrives and initiates synaptic transmission.

1 Na$^+$ channels open, depolarizing the axon terminal membrane.

2 Depolarization of the terminal membrane causes voltage-gated Ca^{2+} channels to open.

3 Ca^{2+} enters the cell and triggers fusion of transmitter vesicles with the presynaptic membrane.

4 Neurotransmitter molecules diffuse across the synaptic cleft and bind to receptors on the postsynaptic membrane.

5 Activated receptors open chemically gated Na$^+$ channels and depolarize the postsynaptic membrane. The spreading depolarization fires an action potential in the adjacent membrane.

After synaptic transmission, acetylcholine and vesicles are recycled.

Axon

Myelin

Axon terminal

Plasma membrane

Na$^+$

Acetyl CoA + choline

Ca^{2+}

Presynaptic cell

Action potential

Acetylcholinesterase

Acetylcholine receptors

Postsynaptic cell

6 Acetylcholine is broken down by the enzyme acetylcholinesterase. Acetylcholine components are taken back up by the presynaptic cell for resynthesis.

41.15 Synaptic Transmission Begins with the Arrival of an Action Potential The figure diagrams the sequence of events involved in transmission at a typical chemical synapse.

muscle are always excitatory; that is, motor end plates always respond to acetylcholine by depolarizing the postsynaptic membrane. However, synapses between neurons can be excitatory or inhibitory.

Recall that a neuron may have many dendrites. Axon terminals from many neurons may make synapses with those dendrites and with the cell body. The axon terminals of different presynaptic neurons may store and release different neurotransmitters, and membranes of the dendrites and cell body of a postsynaptic neuron may have receptors to a variety of neurotransmitters. Thus a postsynaptic neuron can receive various chemical messages. If the postsynaptic neu-

ron's response to a neurotransmitter is depolarization, as at the neuromuscular junction, the synapse is **excitatory**; if the response is hyperpolarization, the synapse is **inhibitory** (see Figure 41.8).

How do inhibitory synapses work? The postsynaptic cells in inhibitory synapses have chemically gated potassium or chloride channels. When these channels are activated by a neurotransmitter, they hyperpolarize the postsynaptic membrane. Thus the release of neurotransmitter at an inhibitory synapse makes the postsynaptic cell *less* likely to fire an action potential.

Neurotransmitters that depolarize the postsynaptic membrane are excitatory; they bring about an **excitatory postsynaptic potential** (EPSP). Neurotransmitters that hyperpolarize the postsynaptic membrane are inhibitory; they bring about an **inhibitory postsynaptic**

potential (IPSP). However, whether a synapse is excitatory or inhibitory depends not on the neurotransmitter but on the postsynaptic receptors—on what kind of ion channels the postsynaptic cell has. The same neurotransmitter can be excitatory at some synapses and inhibitory at others.

The postsynaptic membrane sums excitatory and inhibitory input

Individual neurons "decide" whether or not to fire action potentials by summing excitatory and inhibitory postsynaptic potentials. This summation ability of neurons is the major mechanism by which the nervous system integrates information. Each neuron may receive 1,000 or more synaptic inputs, but it has only one output: an action potential in a single axon. All the information contained in the thousands of inputs a neuron receives is reduced to the rate at which that neuron generates action potentials in its axon.

For most neurons the critical area for "decision making" is the **axon hillock**, the region of the cell body at the base of the axon. The plasma membrane of the axon hillock is not insulated by glial cells and has many voltage-gated channels. Excitatory and inhibitory postsynaptic potentials from anywhere on the dendrites or the cell body spread to the axon hillock. If the resulting combined potential depolarizes this area of membrane to threshold, the axon fires an action potential. Because postsynaptic potentials decrease as they spread from the site of the synapse, all postsynaptic potentials do not have equal influences on the axon hillock. A synapse at the end of a dendrite has less influence than a synapse on the cell body near the axon hillock.

Excitatory and inhibitory postsynaptic potentials can be summed over space or over time. Spatial summation adds up the simultaneous influences of synapses from different sites on the postsynaptic cell (Figure 41.16a). Temporal summation adds up postsynaptic potentials generated at the same site in a rapid sequence (Figure 41.16b).

Synapses on presynaptic axon terminals modulate the release of neurotransmitter

All the neuron-to-neuron synapses that we have discussed up to now are between the axon terminals of a presynaptic cell and the cell body or dendrites of a postsynaptic cell. Synapses can also form between the axon terminals of one cell and the axon terminals of another cell. Such a synapse can modulate how much neurotransmitter the second cell releases in response to action potentials traveling down its axon. We refer to this mechanism of regulating synaptic strength as **presynaptic excitation** or **presynaptic inhibition**. We will see an example later in the chapter.

Neurotransmitters of slow synapses activate second messengers

Synapses that use chemically gated ion channels are fast; their actions take a few milliseconds. However, some chemically mediated synapses are slow; their actions take hundreds of milliseconds or even many minutes. Neurotransmitters at these slow synapses activate second-messenger systems, rather than directly controlling ion channels in the postsynaptic cell.

Presynaptic events are the same in fast and slow synapses, but when the neurotransmitter of a slow synapse binds to a receptor, it activates a G protein and initiates a

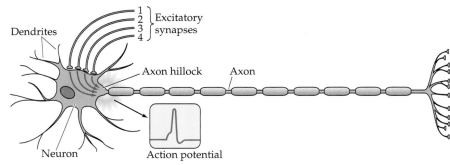

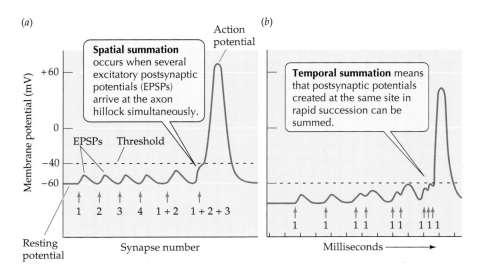

41.16 The Postsynaptic Membrane Sums Information Individual neurons sum excitatory and inhibitory postsynaptic potentials over space and time. When the sum exceeds a threshold amount, an action potential is generated.

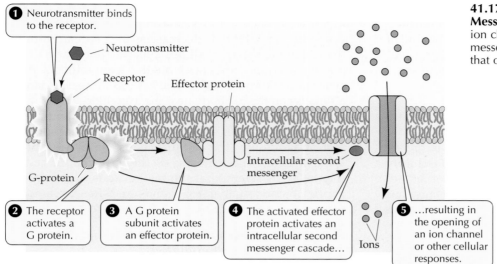

❶ Neurotransmitter binds to the receptor.

Neurotransmitter

Receptor

Effector protein

G-protein

❷ The receptor activates a G protein.

❸ A G protein subunit activates an effector protein.

Intracellular second messenger

❹ The activated effector protein activates an intracellular second messenger cascade...

Ions

❺ ...resulting in the opening of an ion channel or other cellular responses.

41.17 Slow Synapses Act Via Second Messengers Instead of acting directly on ion channels, slow synapses initiate second-messenger cascades. Their action is similar to that of many hormones.

second-messenger cascade (Figure 41.17). The mechanisms of slow synapses are therefore similar to the mechanisms of certain hormones that bind to receptors in the plasma membranes of their target cells (see Chapter 38). Slow synapses may open ion channels, influence membrane pumps, activate enzymes, and induce gene expression.

Electrical synapses are fast but do not integrate information well

Electrical synapses, or gap junctions, are completely different from chemical synapses because they directly couple neurons electrically. At gap junctions, the presynaptic and postsynaptic cell membranes are separated by a space of 2 to 3 nm, but the membrane proteins of the two neurons form connexons—molecular tunnels that bridge the two cells—through which ions and small molecules can readily pass (see Figure 5.6c). Electrical transmission across gap junctions is very fast and can proceed in either direction; that is, stimulation of either neuron can result in an action potential in the other.

Gap junctions are less common in the complex nervous systems of vertebrates than they are in the simple nervous systems of invertebrates, for several very important reasons. First, electrical continuity between neurons does not allow temporal summation of synaptic inputs, which is one way that complex nervous systems integrate information. Second, an effective electrical synapse requires a large area of contact between the presynaptic and postsynaptic cells. This condition rules out the possibility of thousands of synaptic inputs to a single neuron—which is the norm in complex nervous systems. Third, electrical synapses cannot be inhibitory; and fourth there is little plasticity (modifiability) in electrical synapses. Thus, electrical synapses are good for rapid communication but not good for

processes of integration and learning.

Neurotransmitter actions depend on their receptors

More than 25 neurotransmitters are recognized, and more will surely be discovered. Table 41.1 gives some examples. No transmitter is as thoroughly understood as acetylcholine, the neurotransmitter at all synapses between motor neurons and skeletal muscles—the synapses that control voluntary movement. As we will learn in Chapter 43, the vertebrate nervous system also controls many involuntary, physiological functions of the body, by controlling the contraction of smooth muscle.

Neuromuscular junctions with smooth muscle use both acetylcholine and norepinephrine as neurotransmitters. Many smooth muscles respond to both of these neurotransmitters, but in opposite ways. For example, smooth muscle in your intestine is depolarized by acetylcholine and hyperpolarized by norepinephrine. Acetylcholine and norepinephrine also play roles in synapses between neurons in the central nervous system, but they constitute only a small percentage of the neurotransmitter content of the brain.

The workhorse neurotransmitters of the brain are simple amino acids. Glutamic acid and aspartic acid are excitatory, whereas glycine and gamma-aminobutyric acid (GABA) are inhibitory. Another important group of neurotransmitters in the central nervous system consists of the monoamines, which are derivatives of amino acids. They include dopamine and norepinephrine (derivatives of tyrosine) and serotonin (a derivative of tryptophan). Peptides also function as neurotransmitters. A very exciting recent discovery revealed that two gases, carbon monoxide and nitric oxide, are used by neurons as intercellular messengers.

A neurotransmitter may have several different types of receptors in different tissues and may induce different actions. For example, acetylcholine has two well-known receptor types, called **muscarinic** receptors and **nicotinic** receptors because of other compounds that also bind to them. Nicotine, the active ingredient in tobacco, binds to acetylcholine receptors in the skeletal muscles, but not to those in heart muscle.

TABLE 41.1 Some Well-Known Neurotransmitters

NEUROTRANSMITTER	ACTIONS	COMMENTS
Acetylcholine	The neurotransmitter of vertebrate motor neurons and of some neural pathways in the brain.	Broken down in the synapse by acetyl cholinesterase; blockers of this enzyme are powerful poisons.
Monoamines		
Norepinephrine	Used in certain neural pathways in the brain, but also found in the peripheral nervous system, where it causes gut muscles to relax and the heart to beat faster.	Related to epinephrine and acts at some of the same receptors.
Dopamine	A neurotransmitter of the central nervous system.	Involved in schizophrenia. Loss of dopamine neurons is the cause of Parkinson's disease.
Histamine	A minor neurotransmitter in the brain.	Thought to be involved in maintaining wakefulness.
Serotonin	A neurotransmitter of the central nervous system that is involved in many systems, including pain control, sleep/wake control, and mood.	Certain medications that elevate mood and counter anxiety act by increasing serotonin levels.
Amino acids		
Glutamate	The most common excitatory neurotransmitter in the central nervous system.	Some people have reactions to the food additive monosodium glutamate because it can affect the nervous system.
Glycine Gamma-aminobutyric acid (GABA)	Common inhibitory neurotransmitters.	Drugs called benzodiazepines, used to reduce anxiety and produce sedation, mimic the actions of GABA.
Peptides Endorphins Enkephalins Substance P	Used by certain sensory nerves, especially in pain pathways.	Receptors are activated by narcotic drugs: opium, morphine, heroin, codeine.

Muscarine, a compound originally isolated from the toxic mushroom *Amanita muscaria* (but not found in high quantities in that mushroom, and not responsible for its toxic effects), binds to the acetylcholine receptors in heart muscle and to many in the peripheral nervous system, but not to those in skeletal muscle.

Both types of acetylcholine receptors are found in the central nervous system, where nicotinic receptors tend to be excitatory and muscarinic tend to be inhibitory. These receptors are the reason that smoking tobacco has behavioral and physiological effects and is addictive and why numerous cultures around the world have used *Amanita* mushrooms as hallucinogenic drugs.

The drug **curare**, extracted from the bark of a South American plant and used by native peoples to make poisoned darts and arrows, binds to nicotinic receptors but does not activate them. Therefore, skeletal muscles in an animal poisoned by curare cannot respond to motor neuron activation. The animal goes into flaccid (relaxed) paralysis and dies because its respiratory muscles stop contracting. Curare is used medically to treat severe muscle spasms and to prevent muscle contractions that would interfere with surgery.

Another compound, **atropine**, which is extracted from the plant *Atropa belladonna,* binds to muscarinic receptors and prevents acetylcholine from activating them. Atropine is used medically to increase heart rate, decrease secretions of digestive juices, and decrease spasms of the gut. Most people have encountered atropine; it is what the eye doctor uses to dilate the pupils for eye examinations. In the past atropine was used cosmetically to make the eyes look big and dark—hence the plant's species name, *belladonna,* meaning "beautiful lady." Of course, these beautiful ladies could not see very well.

The ability of compounds extracted from plants and animals to bind to certain neurotransmitter receptors is the basis for neuropharmacology, the study and development of drugs that influence the nervous system. Natural products are still an important source of drugs, but today many drugs are designed and synthesized by chemists. For example, a major group of drugs called benzodiazepines, which are used as tranquilizers, muscle relaxants, and sleeping pills, are synthetic molecules that act on GABA receptors, open Cl⁻ channels, hyperpolarize cells, and inhibit neural activity.

To turn off responses, synapses must be cleared of neurotransmitter

Turning off the action of neurotransmitters is as important as turning it on. If released neurotransmitter molecules simply remained in the synaptic cleft, the postsynaptic membrane would become saturated with neurotransmitter, and its receptors would be constantly bound. As a result, the postsynaptic neuron would remain hyperpolarized or depolarized and would be unresponsive to short-term changes in the presynaptic neuron. The more discrete each separate neural signal is, the more information can be processed in a given time. Thus, neurotransmitter must be cleared from the synaptic cleft shortly after it is released by the axon terminal.

Neurotransmitter action may be terminated in several ways. First, enzymes may destroy the neurotransmitter. As we already discussed, acetylcholine is rapidly destroyed by the enzyme acetylcholinesterase, which is present in the synaptic cleft in close association with the acetylcholine receptors on the postsynaptic membrane. Some of the most deadly nerve gases that were developed for chemical warfare work by inhibiting acetylcholinesterase. As a result, acetylcholine lingers in the synaptic clefts, causing the victim to die of spastic muscle paralysis. Some agricultural insecticides, such as malathion, also inhibit acetylcholinesterase and can poison farm workers if used without safety precautions. Second, neurotransmitter may simply diffuse away from the cleft. Third, neurotransmitter may be taken up via active transport by nearby cell membranes. The mode of action of Prozac, a commonly prescribed drug for mood enhancement, is to *slow* the reuptake of a neurotransmitter, serotonin, thus enhancing its activity at the synapse.

Neurons in Circuits

Because neurons can interact in the complex ways we have just discussed, networks of neurons can process and integrate information. Next we will examine some simple neuronal networks and their electrical circuits that explain certain behaviors.

The simplest neural circuit contains two neurons

Many invertebrates have a fairly simple nervous system in terms of the number of cells. Of more importance, these nervous systems consist of identifiable neurons; that is, all individuals of the same species have exactly the same neurons in the same places making the same connections. Thus, in these animals it is possible to develop a circuit diagram of the species'

nervous system. The challenge then is to explain the behavior of the animal in terms of this blueprint.

A model species that has received a lot of attention by researchers, especially Dr. Eric Kandel and his colleagues at Columbia University, is a marine mollusk called a sea slug or a sea hare (*Aplysia californica*). Even though it is a mollusk, *Aplysia* has no shell (so its nervous system is accessible). In addition, *Aplysia* can be raised in the laboratory, and it has several very simple behaviors that researchers can study in terms of the individual identifiable neurons in its nervous system.

One simple behavior of *Aplysia* is a protective reaction when its siphon—a tasty morsel for predators—is touched (Figure 41.18*a*). When *Aplysia* is undisturbed, its siphon is extended to take in water to ventilate its gill membranes. But when the siphon is touched, the animal withdraws it. Kandel and his fellow investigators found that this defensive behavior can be explained by a circuit of only two neurons. A sensory neuron responds to the touch by firing action potentials down its axon, which synapses onto a motor neuron that causes the muscle that withdraws the siphon to contract.

Even this simple circuit is capable of learning. Your nervous system learns to ignore information that is not of immediate importance. This is called **habituation**. *Aplysia* can learn that the repeated light touches on its siphon are not dangerous and do not require a withdrawal response. The animal habituates to the stimulus. By studying the characteristics of the one synapse in this gill withdrawal circuit, investigators found that the observed habituation was the result of a decrease in the amount of neurotransmitter released by the axon terminals of the presynaptic cells. This reduced synaptic activity lasted a considerable time and was not due simply to fatigue or depletion of neurotransmitter.

Another simple form of learning is **sensitization**—an *increase* in the responsiveness to a stimulus (Figure 41.18*b*). The investigators could also sensitize the siphon withdrawal response of *Aplysia*, but this involved adding two more neurons to the circuit. They applied a mild electrical stimulus to the tail of the animal before gently touching its siphon. Now the animal responded to the touch of its siphon with a much more vigorous withdrawal.

The electric shock excited a sensory neuron in the tail that formed a synapse with another neuron, which we'll call an **interneuron** because it is neither a sensory nor a motor neuron. The interneuron formed synapses with the axon terminals of the sensory cell from the siphon. Activity in this interneuron contributed *presynaptic excitation* and caused the axon terminal of the sensory neuron to release more transmitter in response to each action potential coming down its axon.

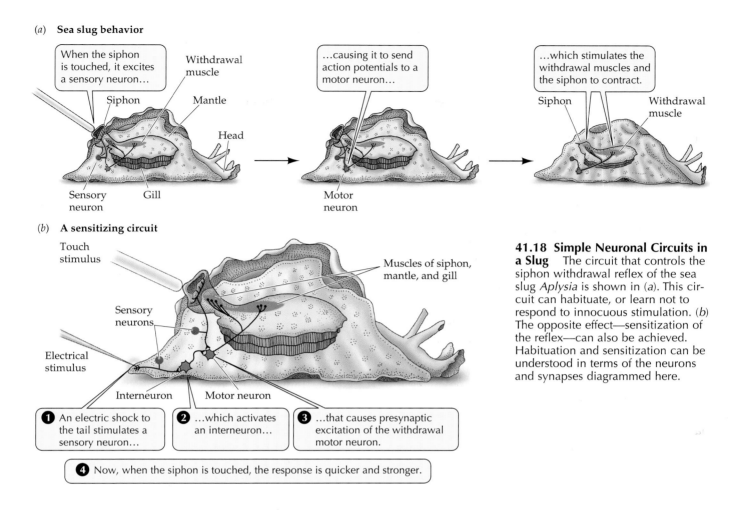

(a) **Sea slug behavior**

When the siphon is touched, it excites a sensory neuron…

Withdrawal muscle

Siphon

Mantle

Head

Sensory neuron

Gill

…causing it to send action potentials to a motor neuron…

Motor neuron

…which stimulates the withdrawal muscles and the siphon to contract.

Siphon

Withdrawal muscle

(b) **A sensitizing circuit**

Touch stimulus

Sensory neurons

Electrical stimulus

Interneuron

Motor neuron

Muscles of siphon, mantle, and gill

1 An electric shock to the tail stimulates a sensory neuron…

2 …which activates an interneuron…

3 …that causes presynaptic excitation of the withdrawal motor neuron.

4 Now, when the siphon is touched, the response is quicker and stronger.

41.18 Simple Neuronal Circuits in a Slug The circuit that controls the siphon withdrawal reflex of the sea slug *Aplysia* is shown in (a). This circuit can habituate, or learn not to respond to innocuous stimulation. (b) The opposite effect—sensitization of the reflex—can also be achieved. Habituation and sensitization can be understood in terms of the neurons and synapses diagrammed here.

Monosynaptic reflexes in humans are circuits composed of two neurons

We'll study the structure and function of the spinal cord in Chapter 43, but here we can note that much information is processed through neural circuits within the spinal cord, some of which are fairly simple. The simplest example of a spinal neural circuit controls a behavior called the **monosynaptic reflex**. This type of reflex depends on a neural circuit made up of a sensory neuron that enters the spinal cord through its upper (dorsal) root, and a motor neuron that leaves the spinal cord through its lower (ventral) root (Figure 41.19). Within the spinal cord there is just one synapse (hence the term "monosynaptic") between the sensory and the motor neurons in this circuit. It is common in a visit to a physician to have your reflexes tested by being struck below the kneecap with a rubber mallet. The response is a rapid, involuntary extension of the leg—the knee jerk reflex. Similar reflexes can be elicited by sharp blows to tendons of the wrist, elbow, ankle, and other joints.

The mallet blow on the tendon causes a quick stretch of the muscle attached to that tendon. Within the muscle are modified muscle fibers wrapped in connective tissue. These **muscle spindles** are stretch sensors that activate sensory neurons when they are stretched. The number of action potentials per second carried by the sensory neuron signals the degree of stretch.

The cell body of the sensory neuron from the muscle spindle is in a ganglion on the dorsal root of the spinal cord, and the axon of this neuron extends all the way into the core of the spinal cord. This core area is called **gray matter** because it contains lots of cell bodies of neurons and looks gray. Surrounding the core of the spinal cord is **white matter**, which is made up of bundles of axons running up and down the cord. The white matter looks white because most of these axons are covered with myelin.

In the lower region of the gray matter, the sensory fiber branches and forms synapses with motor neurons for the same muscle from which the sensory neuron originated. Each motor neuron sends impulses along its axon, which exits the spinal cord through the ventral root. The axon of the motor neuron synapses on the stretched muscle, causing it to contract. The

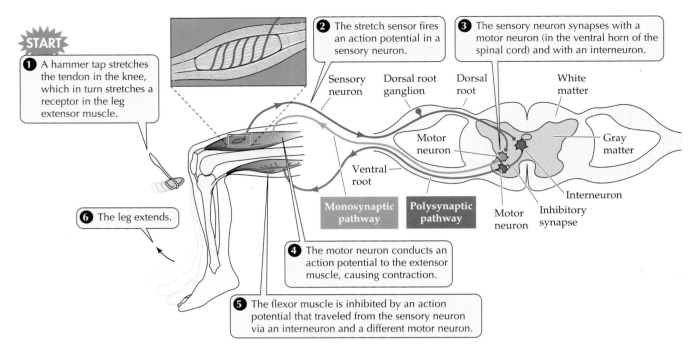

1 A hammer tap stretches the tendon in the knee, which in turn stretches a receptor in the leg extensor muscle.

2 The stretch sensor fires an action potential in a sensory neuron.

3 The sensory neuron synapses with a motor neuron (in the ventral horn of the spinal cord) and with an interneuron.

6 The leg extends.

4 The motor neuron conducts an action potential to the extensor muscle, causing contraction.

5 The flexor muscle is inhibited by an action potential that traveled from the sensory neuron via an interneuron and a different motor neuron.

Sensory neuron — Dorsal root ganglion — Dorsal root — White matter — Motor neuron — Gray matter — Ventral root — Interneuron — Monosynaptic pathway — Polysynaptic pathway — Motor neuron — Inhibitory synapse

41.19 Monosynaptic and Polysynaptic Spinal Reflexes
The knee jerk reflex is a *monosynaptic* reflex loop (shown in orange). Because muscles work in antagonistic pairs, however, one must relax while the other contracts. Therefore, the knee jerk reflex is accompanied by a parallel reflex that involves spinal interneurons and therefore more than one synapse; that is, it is *polysynaptic* (shown in red).

function of this reflex loop is to adjust the contraction in the muscle to changing loads. An increased load on the limb stretches the muscle, and the stretch reflex returns the limb to the desired position by increasing the strength of contraction of the muscle.

Even though the knee jerk reflex is involuntary, you are aware of being struck on the knee. This awareness means that the information also travels to your brain. In addition to forming the monosynaptic reflex circuit, the sensory neuron has branches that form synapses with interneurons in the gray matter of the spinal cord. These interneurons send axons up the white-matter tracts to the brain. Motor commands from the brain descend the spinal cord in other white-matter tracts to form synapses with the same motor neurons that are involved in the reflex circuit. Thus the same muscle can be controlled both by involuntary reflexes and by conscious commands.

Most spinal reflexes involve polysynaptic circuits

Interneurons in the spinal cord are involved in most spinal circuits. For example, a limb moves in opposite directions because muscles work in pairs. One muscle of a pair is an extensor and the other is a flexor. At the same time a flexor contracts, the extensor must relax,

or the limb cannot move. A polysynaptic reflex is added to the monosynaptic circuit that controls the stretch reflex to cause relaxation of the opposing muscle (see Figure 41.19).

Much more complex polysynaptic reflexes are responsible for coordinated escape movements. If you step on a tack, many muscles in your foot and leg work together to produce a coordinated withdrawal response of that limb, while muscles on the opposite side of the body cause extension of your other leg to support your body weight and maintain your balance. The large number of muscle contractions and relaxations required for this sequence of movements are initially orchestrated by interneurons in the spinal cord working together in a complex circuit.

Summary of "Neurons and Nervous Systems"

Nervous Systems: Cells and Functions

• Nervous systems integrate, process, and transmit information received from sensors, and they communicate commands to effectors.
• In vertebrates, the brain and spinal cord form the central nervous system, which communicates with other tissues of the body via the peripheral nervous system.
• Nervous systems of different species vary, but all are composed of cells called neurons, which receive information mostly via their dendrites and transmit information over their axons. Neurons function in circuits. **Review Figures 41.1, 41.2, 41.3**

- The information that neurons process is in the form of electrical events in their plasma membranes. Where cells meet, information is transmitted between cells mostly by the release of chemical signals called neurotransmitters.
- Glial cells physically support neurons and perform many housekeeping functions. Glial cells do not carry electric signals, but they are important components of the nervous system. Schwann cells and oligodendrocytes produce myelin, which insulates neurons. Astrocytes create the blood–brain barrier. **Review Figure 41.4**

Neurons: Generating and Conducting Nerve Impulses

- Neurons have an electric charge difference across their plasma membranes. This resting potential is created by (1) ion pumps that concentrate K^+ ions on the insides of neurons and Na^+ ions on the outsides of neurons, and (2) ion channels that allow K^+ ions to leak out, leaving behind unbalanced negative charges on chloride ions and protein molecules. **Review Figures 41.5, 41.6, 41.7**
- When the plasma membrane changes in its permeability to ions via the opening or closing of ion channels, its resting potential changes. This is the mechanism by which neurons respond to stimuli. **Review Figure 41.8**
- Rapid reversals in charge across portions of the plasma membrane resulting from the opening and closing of voltage-gated Na^+ and K^+ channels produce nerve impulses, or action potentials. These channels respond when the potential across the plasma membrane depolarizes to a threshold level. **Review Figure 41.9**
- Action potentials are conducted down axons because of the local spread of current, which depolarizes adjacent regions of membrane and brings them to threshold for the opening of voltage-gated channels. **Review Figure 41.10**
- Patch clamping lets us study single ion channels. **Review Figure 41.11**
- In myelinated axons, the action potentials appear to jump between patches of plasma membrane that are not covered by myelin (nodes of Ranvier). **Review Figure 41.12**

Neurons, Synapses, and Communication

- Neurons communicate with each other and with other cells at specialized junctions, synapses, where the plasma membranes of two cells come close together but still have a space between them. **Review Figure 41.13**
- When an action potential reaches the termination of the presynaptic cell, it causes the release of neurotransmitters, chemical signals that diffuse across the synaptic cleft and bind to receptors on the postsynaptic cell. **Review Figures 41.14, 41.15**
- The classical synapse is the junction between a motor neuron and a muscle fiber where the neurotransmitter is acetylcholine, which causes a depolarization of the postsynaptic membrane when it binds to its receptor. **Review Figure 41.13**
- Synapses between neurons can be either excitatory or inhibitory. Synapses can also form on presynaptic membranes and thereby influence the release of transmitter by the presynaptic cell.
- Postsynaptic neurons integrate information by summing its synaptic input in both space and time. **Review Figure 41.16**
- Some neurotransmitter receptors are ion channels, but others influence the postsynaptic cell through various signal transduction pathways and are called slow synapses. **Review Figure 41.17**

- Electrical synapses pass electric signals between cells without the use of neurotransmitters.
- There are many different neurotransmitters. Some of the most important ones include acetylcholine and various amino acids and monoamines. **Review Table 41.1**
- In chemical synapses, the transmitter must be cleared rapidly from the synapse to increase the speed of information processing.

Neurons in Circuits

- The simplest neural circuit has only two neurons. Examples are the circuit that controls the siphon withdrawal reflex in the sea slug and the circuit that produces the knee jerk reflex in humans.
- Even a simple neural circuit can show forms of learning called habituation and sensitization. **Review Figure 41.18**
- Monosynaptic reflexes such as the knee jerk reflex are produced by circuits composed of only two neurons. However, most neural circuits involve interneurons and are therefore polysynaptic. **Review Figure 41.19**

Self-Quiz

1. In the nervous system, the most abundant cell type is the
   a. motor neuron.
   b. sensory neuron.
   c. preganglionic parasympathetic neuron.
   d. glial cell.
   e. preganglionic sympathetic neuron.

2. Within the nerve cell, information moves from
   a. dendrite to cell body to axon.
   b. axon to cell body to dendrite.
   c. cell body to axon to dendrite.
   d. axon to dendrite to cell body.
   e. dendrite to axon to cell body.

3. The resting potential of a neuron is due mostly to
   a. local current spread.
   b. open Na^+ channels.
   c. synaptic summation.
   d. open K^+ channels.
   e. unequal pumping of K^+ and Na^+ ions across the membrane.

4. Which statement about synaptic transmission is *not* true?
   a. The synapses between neurons and muscles use acetylcholine as their neurotransmitter.
   b. A single vesicle of neurotransmitter cannot cause a muscle fiber to contract.
   c. The release of neurotransmitter at the neuromuscular junction causes the motor end plate to fire action potentials.
   d. In vertebrates, the synapses between motor neurons and muscle fibers are always excitatory.
   e. Inhibitory synapses cause the resting potential of the postsynaptic membrane to become more negative.

5. Which statement accurately describes an action potential?
   a. Its magnitude increases along the axon.
   b. Its magnitude decreases along the axon.
   c. All action potentials in a single neuron are of the same magnitude.
   d. During an action potential the transmembrane potential of a neuron remains constant.
   e. It permanently shifts a neuron's transmembrane potential away from its resting value.

6. A neuron that has just fired an action potential cannot be immediately restimulated to fire a second action potential. The short interval of time during which restimulation is not possible is called
   a. hyperpolarization.
   b. the resting potential.
   c. depolarization.
   d. repolarization.
   e. the refractory period.

7. The rate of propagation of an action potential depends on
   a. whether or not the axon is myelinated.
   b. the axon's diameter.
   c. whether or not the axon is insulated by glial cells.
   d. the cross-sectional area of the axon.
   e. all of the above

8. The binding of neurotransmitter to the postsynaptic receptors in an inhibitory synapse results in
   a. depolarization of the transmembrane potential.
   b. generation of an action potential.
   c. hyperpolarization of the transmembrane potential.
   d. increased permeability of the membrane to sodium ions.
   e. increased permeability of the membrane to calcium ions.

9. Whether a synapse is excitatory or inhibitory depends on
   a. the type of neurotransmitter.
   b. the presynaptic terminal.
   c. the size of the synapse.
   d. the nature of the postsynaptic neurotransmitter receptors.
   e. the concentration of neurotransmitter in the synaptic space.

10. In the knee jerk reflex
    a. spinal interneurons inhibit the motor neuron of the antagonistic muscle.
    b. the cell body of the muscle stretch receptor is in the dorsal horn of the spinal cord.
    c. the cell body of the motor neuron is in the dorsal horn of the spinal cord.
    d. action potentials in the sensory neuron release inhibitory neurotransmitter onto the motor neurons.
    e. the sensory neuron forms a monosynaptic loop with the motor neuron to the antagonistic muscle.

Applying Concepts

1. Describe the electrochemical and structural elements involved in the establishment and maintenance of the resting potential of a neuron.

2. The language of the nervous system consists of one word, "action potential." How can this single message convey a diversity of information, how can that information be quantitative, and how can it be integrated?

3. Describe two properties of axonal plasma membranes that limit the duration of the action potential. How does one of these mechanisms cause the action potential to be propagated in one direction only?

4. Discuss at least four mechanisms that could be involved in learning at the synaptic level.

5. What is the difference between fast and slow chemical synapses?

Readings

Dowling, John E. 1992. *Neurons and Networks: An Introduction to Neuroscience.* The Belknap Press of Harvard University Press, Cambridge, MA. A very readable introduction to cellular and systems neurobiology.

Kandel, E. R., J. H. Schwartz and T. M. Jessell. 1991. *Principles of Neural Science*, 3rd Edition. Elsevier, New York. A very thorough, advanced text in neurobiology.

Llinás, R. R. 1988. *The Biology of the Brain: From Neurons to Networks*, and Llinás, R. R. 1990. *The Workings of the Brain: Development, Memory, and Perception.* W. H. Freeman, New York. These two volumes are a rich collection of articles on aspects of neuroscience that have been published in *Scientific American* since 1976. They provide a broad yet selectively in-depth survey of modern neurobiology.

Nicholls, J. G., A. R. Martin and B. G. Wallace. 1992. *From Neuron to Brain*, 3rd Edition. Sinauer Associates, Sunderland, MA. An advanced text that is especially thorough in describing electrophysiology.

Purves, D., G. J. Augustine, D. Fitzpatrick, L. C. Katz, A.-S. LaMantia and J. O. McNamara. 1997. *Neuroscience.* Sinauer Associates, Sunderland, MA. Despite the formidable number of authors, this textbook of human neurobiology is delightfully readable and lucid. At just over 550 pages, it is one of the most concise of the thorough treatments of the subject.

Shepherd, G. M. 1994. *Neurobiology*, 3rd Edition. Oxford University Press, New York. A comprehensive advanced text that covers the full range of neurobiology from molecular mechanisms to human behavior.

Thompson, R. F. 1993. *The Brain: An Introduction to Neuroscience.* W. H. Freeman, New York. A well-written and easy-to-understand introductory text on neuroscience, covering topics from membrane events to the neural basis of behavior.

Chapter 42

Sensory Systems

Sniffing It Out
Dogs have an exceptionally keen sense of smell, which they rely on more than their vision.

Wcannot discuss anything for very long without using words that refer to our senses. Our senses are the window through which we view the world, and the world is what our senses tell us it is. Different species look through different sensory windows, so their views of the world differ. Dogs do not see color well, but they have keener senses of hearing and smell than humans do. Bees can see patterns on flowers that reflect ultraviolet light; we cannot. In environments that are totally dark to us, some snakes can "see" the infrared radiation emitted by bodies warmer than the environment. Porpoises, like the bats discussed at the beginning of Chapter 41, use reflected sound to avoid obstacles and catch their prey. How the environment "looks" to any animal depends on what information that animal receives from its sensory systems.

In this chapter, we look at some of the general properties of sensory cells and their conversions of stimuli to information, and we will examine in detail chemosensors, mechanosensors, and photosensors. We will show how sensory systems gather and filter stimuli, transduce stimuli into action potentials, and transmit these action potentials to the central nervous system for integration.

Sensors, Sensory Organs, and Transduction

In this first section, we will examine some of the general properties of animal sensors and their sensitivity to certain kinds or amounts of energy. We will ask, How can sensors change stimulus energy into action potentials? We will discover that the answers reveal a general pattern involving membranes and ion channels. Then we will ask, How are these neural messages interpreted by the central nervous system? We'll see that neural connections are important in de-

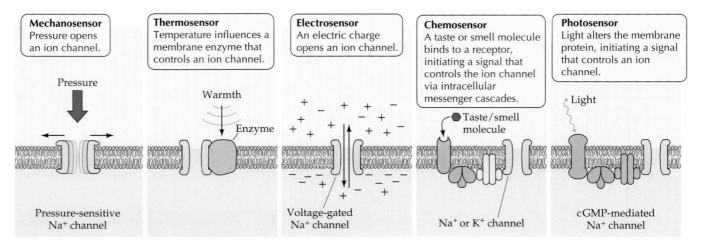

| Mechanosensor
Pressure opens
an ion channel. | Thermosensor
Temperature influences a
membrane enzyme that
controls an ion channel. | Electrosensor
An electric charge
opens an ion channel. | Chemosensor
A taste or smell molecule
binds to a receptor,
initiating a signal that
controls the ion channel
via intracellular
messenger cascades. | Photosensor
Light alters the membrane
protein, initiating a signal
that controls an ion
channel. |

Pressure-sensitive
Na^+ channel

Voltage-gated
Na^+ channel

Na^+ or K^+ channel

cGMP-mediated
Na^+ channel

42.1 Sensory Cell Membrane Proteins Respond to Stimuli Sensory stimuli modify receptor proteins in the membranes of sensors, which in turn modify ion channels. The receptors in mechanosensors, thermosensors, and electrosensors are themselves ion channels. In chemosensors and photosensors, activated receptor proteins initiate biochemical cascades that eventually open or close ion channels.

termining whether action potentials signal sound, odor, light, color, or touch.

Sensors respond to stimuli with changes in membrane potential

Sensory cells transduce (convert) physical or chemical stimuli into signals that are transmitted to other parts of the nervous system for processing and interpretation. Most sensory cells are modified neurons, but some are other types of cells closely associated with neurons. Sensors are specialized for specific types of stimuli—chemical, mechanical, and light.

In general, the sensor possesses a membrane protein that detects the stimulus and responds by altering the flow of ions across the cell membrane (Figure 42.1). The resulting change in membrane potential causes the sensor either to fire action potentials itself or to secrete neurotransmitter onto an associated cell that fires action potentials. Ultimately the stimulus is transduced into the universal message of the nervous system: the action potential (see Chapter 41). The intensity of the stimulus is coded as the frequency of action potentials.

Sensation depends on which neurons in the CNS receive nerve impulses from sensors

If the messages derived from all sensors are the same, how can we perceive different sensations? Sensations such as temperature, itch, pressure, pain, light, smell, and sound differ because the messages from sensors arrive at different places in the central nervous system (CNS). Action potentials arriving in the visual cortex are interpreted as light, in the auditory cortex as sound, in the olfactory bulb as smell, and so forth.

A small patch of skin on your arm has sensors that increase their firing rates when the skin is warmed and others that increase activity when the skin is cooled. Other types of sensors in the same patch of skin respond to touch, movement of hairs, irritants such as mosquito bites, and pain from cuts or burns. As we learned in Chapter 41, all these sensors in your arm transmit their messages through axons that enter the CNS through the dorsal horn of the spinal cord. The synapses made by those axons in the CNS and the subsequent pathways of transmission determine whether the stimulation of the patch of skin on your arm is perceived as warmth, cold, pain, touch, itch, or tickle.

The specificity of sensory circuits is dramatically illustrated in persons who have had a limb or part of a limb amputated. Although the sensors from that region are gone, the axons that communicated information from those sensors to the CNS may remain. If those axons are stimulated, the person feels specific sensations as if they were coming from the limb that is no longer there—a phantom limb.

The messages from some sensors communicate information about internal conditions in the body, but we may not be consciously aware of that information. The brain receives continuous information about things like body temperature, blood sugar, arterial pressure, muscle tensions, and positions of limbs. All this information is important for maintenance of homeostasis, but thankfully we don't have to think about it. If we did, there would be no time to think about anything else. Sensors produce information that the nervous system can use, but we are not always conscious of that.

Sensory organs are specialized for specific stimuli

Some sensors are assembled with other types of cells into sensory organs such as eyes, ears, and noses that

42.2 Sensory Transduction Is a Series of Steps The sensory cell itself may generate action potentials as a result of stimulation, or it may just vary its release of neurotransmitter in response to changes in its membrane potential, and the neurotransmitter may induce another cell to generate action potentials.

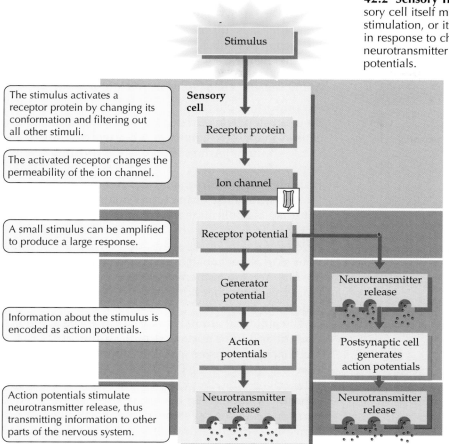

The stimulus activates a receptor protein by changing its conformation and filtering out all other stimuli.

The activated receptor changes the permeability of the ion channel.

A small stimulus can be amplified to produce a large response.

Information about the stimulus is encoded as action potentials.

Action potentials stimulate neurotransmitter release, thus transmitting information to other parts of the nervous system.

Stimulus

Sensory cell

Receptor protein

Ion channel

Receptor potential

Generator potential

Action potentials

Neurotransmitter release

Neurotransmitter release

Postsynaptic cell generates action potentials

Neurotransmitter release

Sensory transduction involves changes in membrane potentials

In this chapter we will examine several sensor types and the sensory organs with which they are associated. In each case we can ask the general question, How does the sensory cell transduce stimulus energy into action potentials? The details differ for different sensors, but those details all fit into a general pattern. We have already seen the first steps of sensory transduction in Figure 42.1.

A receptor protein in the plasma membrane of the sensory cell is activated by a specific stimulus. The activated receptor protein opens or closes specific ion channels in the membrane by one of several mechanisms. The receptor protein may be part of the ion channel, and by changing its conformation, it may open or close the channel directly. Alternatively, the activated receptor protein may set in motion intracellular events that eventually affect the ion channel. Figure 42.2 reviews these first steps of sensory transduction and outlines the subsequent steps.

The opening or closing of ion channels in response to a stimulus changes the membrane potential of the sensor, which is called the **receptor potential**. Such changes in membrane potential can spread by ionic currents over short distances, but to travel long distances in the nervous system, receptor potentials must be converted into action potentials. The intracellular events involved in the conversion of the original stimulus-induced alteration of the ion channels to the generation of action potentials can amplify the signal. In other words, the energy in the output of the sensor can be greater than the energy in the stimulus.

Receptor potentials produce action potentials in two ways: by generating action potentials within the sensors or by causing the release of neurotransmitter that induces an associated neuron to generate action potentials (see Figure 42.2). In the first case, the sensor has a region of plasma membrane with voltage-gated Na^+ channels. A receptor potential that spreads to this

enhance the ability of the sensors to collect, filter, and amplify stimuli. For example, a photosensor detects electromagnetic radiation of only a particular range of wavelength, filtering out radiation of other wavelengths. This filtering is the basis for color vision, and the specificity of photosensors explains why some insects can see ultraviolet light and some snakes can see infrared radiation.

In some simple organisms photosensors sense only the presence of light, but in more complex animals, photosensors are combined with other cell types into eyes. We'll learn how eyes collect light and focus it onto sheets of photosensors so that patterns of light can be detected.

Similarly, we'll see that the sense of hearing depends on mechanosensors, but the accessory structures that constitute the ear make it possible to amplify low levels of sound and filter it so that it also conveys directional information. Some sensory organs can reduce the level of stimulus energy that reaches sensors. For example, the pupillary reflex of vertebrate eyes varies the amount of light falling on the photosensors, and tiny muscles in ears can dampen the energy from loud sounds before it reaches the sensitive mechanosensors.

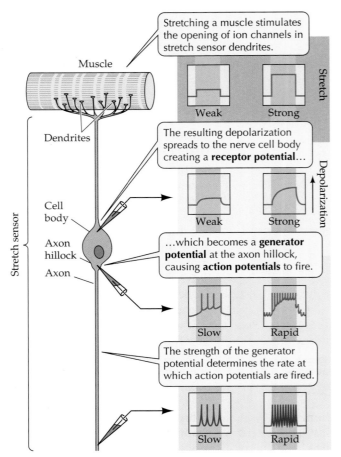

Stretching a muscle stimulates the opening of ion channels in stretch sensor dendrites.

Muscle

Stretch

Dendrites

The resulting depolarization spreads to the nerve cell body creating a **receptor potential**...

Depolarization

Cell body

Weak   Strong

Axon hillock

...which becomes a **generator potential** at the axon hillock, causing **action potentials** to fire.

Axon

Stretch sensor

Slow   Rapid

The strength of the generator potential determines the rate at which action potentials are fired.

Slow   Rapid

42.3 Stimulating a Sensor Produces a Generator Potential The stretch sensor of a crayfish produces a generator potential when the muscle is stretched. At the axon hillock the receptor potential becomes a generator potential, firing action potentials that travel down the axon.

region is called a **generator potential** because it generates action potentials by causing the voltage-gated Na^+ channels to open.

A good example of generator potentials is found in the stretch sensors of crayfish (Figure 42.3). By placing an electrode in the cell body of the crayfish stretch sensor, we can record the changes in the receptor potential that result from stretching the muscle to which the dendrites of the cell are attached. These changes in receptor potential become a generator potential, at the base of the sensor's axon, where there are voltage-gated Na^+ channels. Action potentials generated here travel down the axon to the CNS. The rate that action potentials are fired from the axon depends on the magnitude of the generator potential, and that, in turn, depends on how much the muscle is stretched.

In sensors that do not fire action potentials, the spreading receptor potential reaches a presynaptic patch of plasma membrane and induces the release of neurotransmitter. The neurotransmitter can then activate chemically gated ion channels on a postsynaptic

membrane and cause the postsynaptic cell to fire action potentials. Whether or not the sensor cell fires action potentials, ultimately the stimulus is transduced into action potentials, and the intensity of the stimulus is coded by the frequency of action potentials.

Many sensors adapt to repeated stimulation

An important characteristic of many sensors is that they can stop being excited by a stimulus that initially caused them to be active. In other words, they adapt to the stimulus. **Adaptation** enables an animal to ignore background or unchanging conditions while remaining sensitive to changes or to new information. (Note that this use of the term "adaptation" is different from its application in an evolutionary context.)

When you dress, you feel each item of clothing touch your skin, but the sensation of clothes touching your skin is not constantly on your mind throughout the day. You are immediately aware, however, when a seam rips, your shoe comes untied, or someone lightly touches your back.

The ability of animals to discriminate between important and unimportant stimuli is due partly to the fact that some sensors adapt; it is also due to information processing by the CNS. Some sensors adapt very little or very slowly; examples are pain receptors and sensors for balance. In the rest of this chapter we will learn how sensory systems gather and filter stimuli, transduce specific stimuli into action potentials, and transmit action potentials to the CNS. Sensors are also very important components of autonomic regulatory mechanisms.

Chemosensors: Responding to Specific Molecules

Animals receive information about chemical stimuli through **chemosensors**. Chemosensors are responsible for smell, taste, and the monitoring of aspects of the internal environment such as the level of carbon dioxide in the blood. Chemosensitivity is universal among animals.

A colony of corals responds to a small amount of meat extract in the seawater around it by extending bodies and tentacles and searching for food. A solution of a single kind of amino acid can stimulate this response. However, if an extract from an injured individual of the colony is released into the water, the colony members retract their tentacles and bodies to avoid danger. Humans have similar reactions to chemical stimuli. When we smell freshly baked bread, we salivate and feel hungry, but we gag and retch when we smell diamines from rotting meat.

Information from chemosensors can cause powerful behavioral and physiological responses. In this section we will examine olfaction and gustation, which we know as the senses of smell and taste.

The female moth releases the attractant from a gland at the tip of her abdomen.

A male moth detects this chemical attractant in the air passing over his antennae, which are covered with chemosensitive hairs.

42.4 Some Scents Travel Great Distances Mating in silk-worms of the genus *Bombyx* is coordinated by a chemical attractant, a pheromone.

Arthropods provide good systems for studying chemosensation

Arthropods use chemical signals to attract mates. These signals, called **pheromones**, demonstrate the sensitivity of chemosensory systems. The female silk-worm moth releases a pheromone called bombykol from a gland at the tip of her abdomen. The male silk-worm moth has sensors for this molecule on his antennae (Figure 42.4). Each feathery antenna carries about 10,000 bombykol-sensitive hairs, and each hair has a dendrite of a sensory cell at its core. A single molecule of bombykol is sufficient to activate a dendrite and

generate action potentials in the antennal nerve that transmits the signal to the CNS.

When approximately 200 hairs per second are activated, the male flies upwind in search of the female. Because of the male's high degree of sensitivity, the sexual message of a female moth is likely to reach any male that happens to be within a downwind area stretching over several kilometers. Because the rate of firing in the male's sensory nerves is proportional to bombykol concentrations in the air, the male can follow a concentration gradient and home in on the emitting female.

Many arthropods have chemosensory hairs, each containing one or more specific types of sensors. For example, crabs and flies have chemosensory hairs on their feet; they taste potential food by stepping in it. These hairs have sensors for sugars, amino acids, salts, and other molecules (Figure 42.5). After a fly tastes a drop of sugar water by stepping in it, its proboscis (a

42.5 Tasting with the Feet Flies such as *Drosophila melanogaster* can identify a potential food source by stepping in it.

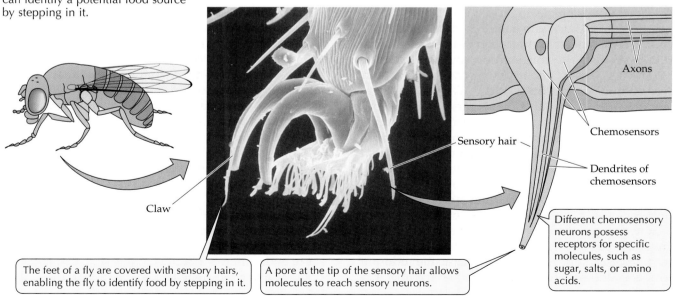

Claw

Sensory hair

Axons

Chemosensors

Dendrites of chemosensors

Different chemosensory neurons possess receptors for specific molecules, such as sugar, salts, or amino acids.

The feet of a fly are covered with sensory hairs, enabling the fly to identify food by stepping in it.

A pore at the tip of the sensory hair allows molecules to reach sensory neurons.

42.6 Olfactory Sensors Communicate Directly with the Brain The sensors of the human olfactory system are embedded in tissues lining the nasal cavity and send their axons to the olfactory bulb of the brain.

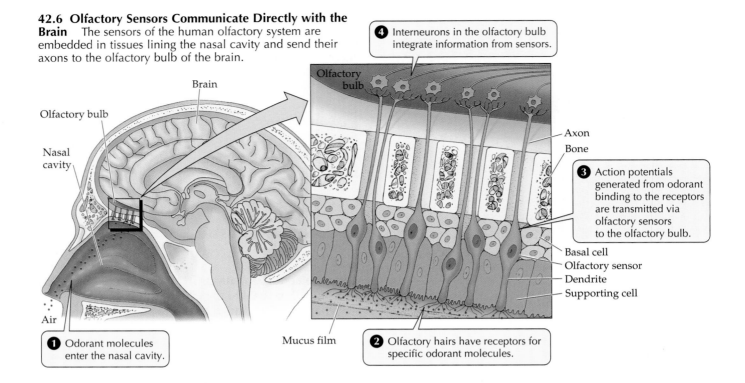

④ Interneurons in the olfactory bulb integrate information from sensors.

③ Action potentials generated from odorant binding to the receptors are transmitted via olfactory sensors to the olfactory bulb.

Brain

Olfactory bulb

Nasal cavity

Air

Olfactory bulb

Axon
Bone

Basal cell
Olfactory sensor
Dendrite
Supporting cell

Mucus film

① Odorant molecules enter the nasal cavity.

② Olfactory hairs have receptors for specific odorant molecules.

tubular feeding structure) extends to feed. Potential food items stimulate extension of the proboscis; other substances do not.

Olfaction is the sense of smell

In vertebrates, the olfactory sensors are neurons embedded in a layer of epithelial cells at the top of the nasal cavity (Figure 42.6). The axons of these neurons project to the olfactory bulb of the brain, and their dendrites end in olfactory hairs that project through to the surface of the nasal epithelium. A protective layer of mucus covers the epithelium. Molecules must diffuse through this mucus to reach the receptors on the olfactory hairs. When you have a cold or an attack of hay fever, the amount of mucus increases and the epithelium swells. With this in mind, study Figure 42.6 and you will easily understand why you lose your sense of smell at those times.

A dog has up to 40 million nerve endings per square centimeter of nasal epithelium, many more than humans do. Humans have a sensitive olfactory system, but we are unusual among mammals in that we depend more on vision than on olfaction (we tend to join bird-watching societies more often than mammal-smelling societies). Whales and porpoises have no olfactory sensors and hence no sense of smell.

How does an olfactory sensor transduce the structure of a molecule into action potentials? A molecule that triggers an olfactory sensor is called an odorant molecule. Odorants bind to receptors on the olfactory

hairs of the sensors. Olfactory receptors are specific for particular odorant molecules, and they work in the same way a lock-and-key mechanism does.

If a "key" (an odorant molecule) fits the "lock" (the receptor), then a G protein is activated, which in turn activates an enzyme (adenylate cyclase, for example) that causes an increase of a second messenger in the cytoplasm of the sensor. The second messenger binds with sodium channel proteins in the sensor's plasma membrane and opens the channels, causing an influx of Na^+. The sensor thus depolarizes to threshold and fires action potentials (Figure 42.7).

The olfactory world has an enormous number of "keys"—molecules that produce distinct smells. Are there a correspondingly large number of "locks"—receptor proteins? Indeed there are. Researchers have recently discovered a huge family of genes that code for olfactory receptor proteins.

How does the sensor signal the intensity of a smell? It responds in a graded fashion to the concentration of odorant molecules: The more odorant molecules that bind to receptors, the more action potentials that are generated and the greater the intensity of the perceived smell.

Gustation is the sense of taste

The sense of taste in humans and other vertebrates depends on clusters of sensors called taste buds. The taste buds of terrestrial vertebrates are confined to the mouth cavity, but some fishes have taste buds in their

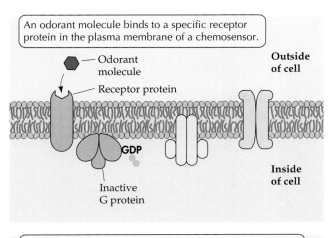

An odorant molecule binds to a specific receptor protein in the plasma membrane of a chemosensor.

Odorant molecule

Receptor protein

Outside of cell

Inside of cell

GDP

Inactive G protein

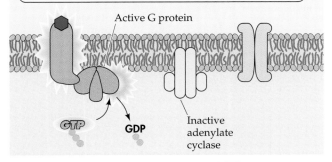

The receptor–odorant complex activates a G protein, which combines with a molecule of GTP, displacing GDP.

Active G protein

GTP

GDP

Inactive adenylate cyclase

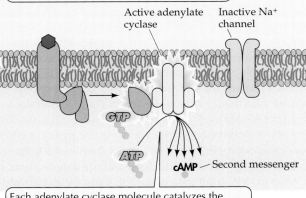

The G protein subunit dissociates and activates adenylate cyclase, which produces cAMP.

Active adenylate cyclase

Inactive Na+ channel

GTP

ATP

cAMP — Second messenger

Each adenylate cyclase molecule catalyzes the production of many molecules of cAMP, amplifying the response to a single odorant molecule.

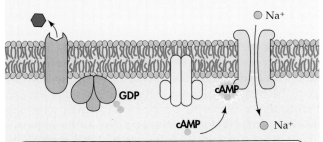

Na+

GDP

cAMP

cAMP

Na+

cAMP (the second messenger) binds to a sodium channel, opens it, and Na+ enters the cell, creating a generator potential. The receptor protein returns to the unbound state.

42.7 Olfactory Receptor Proteins Start a Cascade

In order for us to perceive a smell, odorant molecules must bind to receptors on the olfactory hairs of sensors. Odorant molecules initiate the cAMP cascade by activating G proteins.

skin that enhance the fish's ability to sense its environment. Some fishes living in murky water are very sensitive to small amounts of amino acids in the water around them and can find food without the use of vision. The duck-billed platypus, a monotreme mammal (see Chapter 30), has similar talents as a result of taste buds on the sensitive skin of its bill. What is a taste bud and how does it work?

A **taste bud** is a cluster of many taste sensors. A human tongue has approximately 10,000 taste buds. The taste buds are embedded in the epithelium of the tongue, and many are found on the raised papillae of the tongue. Look at your tongue in a mirror—the papillae make it look fuzzy. Each papilla has many taste buds. The outer surface of a taste bud has a pore that exposes the tips of the taste sensors (Figure 42.8). Microvilli (tiny hairlike projections) increase the surface area of the sensors where their tips converge at the taste pore. Taste sensors, unlike olfactory sensors, are not neurons. At their bases, taste sensors form synapses with dendrites of sensory neurons.

Gustation begins at receptors in the membranes of the microvilli. As with olfactory transduction, receptors on the sensors bind molecules, and the binding causes changes in the membrane polarity of the sensors. Because the taste sensors are not neurons, however, they do not fire action potentials. Instead, they release signaling molecules onto the dendrites of the sensory neurons.

The sensory neurons respond by firing action potentials that are conducted to the CNS. The tongue does a lot of hard work, so its epithelium is shed and replaced at a rapid rate. Individual taste buds last only a few days before they are replaced, but the sensory neurons associated with them live on, always forming new synapses as new taste buds form.

You have probably heard that humans can perceive only four tastes: sweet, salty, sour, and bitter. In actuality, taste buds can distinguish among a variety of sweet-tasting molecules and a variety of bitter-tasting molecules. The full complexity of the chemosensitivity that enables us to enjoy the subtle flavors of food comes from the combined activation of gustatory and olfactory sensors; this is the reason you lose some of your sense of "taste" when you have a cold.

Why does a snake continually sample the air by darting its forked tongue in and out? The forks of the snake's tongue fit into cavities in the roof of its mouth that are richly endowed with olfactory sensors. The

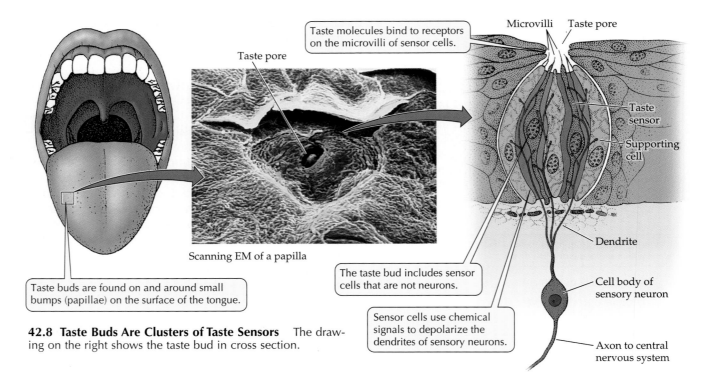

Taste molecules bind to receptors on the microvilli of sensor cells.

Taste pore

Microvilli Taste pore

Taste sensor

Supporting cell

Scanning EM of a papilla

Taste buds are found on and around small bumps (papillae) on the surface of the tongue.

The taste bud includes sensor cells that are not neurons.

Sensor cells use chemical signals to depolarize the dendrites of sensory neurons.

Dendrite

Cell body of sensory neuron

Axon to central nervous system

42.8 Taste Buds Are Clusters of Taste Sensors The drawing on the right shows the taste bud in cross section.

tongue samples the air and presents the sample directly to olfactory sensors. Thus the snake is really using its tongue to smell its environment, not to taste it. Why doesn't the snake simply use the flow of air to and from its lungs as we do to smell the environment? In reptiles, air flows to and from the lungs slowly, but the tongue can dart in and out many times in a second. It is a quick source of olfactory information.

Mechanosensors: Stimuli That Distort Membranes

Mechanosensors are specialized cells that are sensitive to mechanical forces. In this section we will examine a variety of mechanosensors. In the skin, different kinds of mechanosensors are responsible for the perception of touch, pressure, and tickle. Stretch sensors in muscles, tendons, and joints give information about the position of the parts of the body in space and the forces acting on them. Stretch sensors in the walls of blood vessels signal blood pressure. "Hair" cells with extensions that are sensitive to being bent are incorporated into mechanisms for hearing and also into mechanisms for signaling the body's position with respect to gravity.

Physical distortion of a mechanosensor's plasma membrane causes ion channels to open and alters the resting potential of the cell, which in turn leads to the generation of action potentials as shown for a stretch sensor in Figure 42.3. The rates of action potentials in the sensory nerves tell the CNS the strengths of the stimuli exciting the mechanosensor.

Many different sensors respond to touch and pressure

Objects touching our skin generate varied sensations because our skin is packed with diverse mechanosensors (Figure 42.9). The outer layers of skin, especially hairless skin such as lips and fingertips, contain whorls of nerve endings enclosed in connective-tissue capsules. These very sensitive mechanosensors are called **Meissner's corpuscles**, and they respond to objects that touch the skin even lightly. However, Meissner's corpuscles adapt very rapidly. That is why you roll a small object between your fingers, rather than holding it still—to discern its shape and texture. As you roll it, you continue to stimulate sensors anew.

Also in the outer regions of the skin are **expanded-tip tactile sensors** of various kinds. They differ from Meissner's corpuscles in that they adapt only partly and slowly. They are useful for providing steady-state information about objects that continue to touch the skin.

The density of the tactile sensors varies across the surface of the body. A two-point discrimination test demonstrates this fact. If you lightly touch someone's back with two toothpicks, you can determine how far apart the two stimuli have to be before the person can distinguish whether he or she was touched with one or two points. The same test applied to the person's lips or fingertips reveals a finer spatial discrimination; that is, the person can identify as separate two stimuli that are closer together.

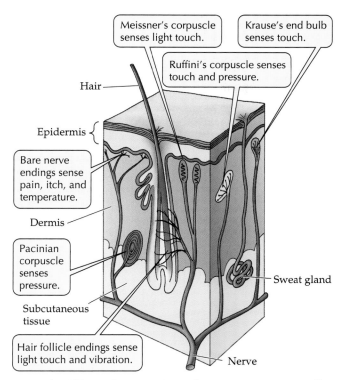

Meissner's corpuscle senses light touch.

Krause's end bulb senses touch.

Ruffini's corpuscle senses touch and pressure.

Hair

Epidermis

Bare nerve endings sense pain, itch, and temperature.

Dermis

Pacinian corpuscle senses pressure.

Subcutaneous tissue

Hair follicle endings sense light touch and vibration.

Sweat gland

Nerve

42.9 The Skin Feels Many Sensations Even a very small patch of skin contains a diversity of sensors that send information to the brain.

Deep in the skin, extensions of neurons wrap around hair follicles. When the hairs are displaced, those neurons are stimulated. Also deep within the skin is another type of mechanosensor, the **Pacinian corpuscle**. Pacinian corpuscles look like onions because they are made up of concentric layers of connective tissue that encapsulate an extension of a sensory neuron.

Pacinian corpuscles respond especially well to vibrations applied to the skin, but they adapt rapidly to steady pressure. The connective-tissue capsule is important in the adaptation of these sensors. An initial pressure distorts the corpuscle and the membrane of the neuron at its core, but the layers of the capsule rapidly rearrange to redistribute the force, thus eliminating the distortion of the membrane of the neuron.

Stretch sensors are found in muscles, tendons, and ligaments

An animal receives information from **stretch sensors** about the position of its limbs and the stresses on its muscles and joints. These mechanosensors are activated by being stretched. They continuously feed information to the CNS, and that information is essential for the coordination of movements. We encountered one important type of stretch sensor, the **muscle spindle**, when we discussed the monosynaptic reflex loop in Chapter 41.

Muscle spindles are embedded in connective tissue within skeletal muscles. They consist of modified muscle fibers that are innervated in the center with extensions of sensory neurons. Whenever the muscle stretches, muscle spindles are also stretched, and the neurons transmit action potentials to the CNS. Earlier in this chapter (see Figure 42.3), we learned how crayfish stretch sensors transduce physical force into action potentials.

Another stretch sensor is found in tendons and ligaments. It is called the **Golgi tendon organ**. Its role is to provide information about the force generated by a contracting muscle. When a contraction becomes too forceful, the information from the Golgi tendon organ feeds into the spinal cord, inhibits the motor neuron, and causes the contracting muscle to relax, thus protecting the muscle from tearing.

Hair cells provide information for orientation and equilibrium

Hair cells are mechanosensors that are not neurons. From one surface they have projections called **stereocilia** that look like a set of organ pipes. When these stereocilia are bent, they alter receptor proteins in the hair cell's plasma membrane. When the stereocilia of some hair cells are bent in one direction, the receptor potential becomes more negative, and when they are bent in the opposite direction, it becomes more positive. When the receptor potential becomes more positive, the hair cell releases neurotransmitter to the sensory neurons associated with them, and the sensory neurons send action potentials to the brain.

Hair cells are found in the **lateral line** sensory system of fishes. The lateral line consists of a canal just under the surface of the skin that runs down each side of the fish. The canal has numerous openings to the external environment. Many structures called cupulae project into the lateral line canal. Each cupula contains hair cells whose stereocilia are embedded in gelatinous (jellylike) material. Movements of water in the lateral line canal move the cupulae and stimulate the hair cells (Figure 42.10). Thus the lateral line provides information about movements of the fish through the water, as well as about the moving objects, such as predators or prey, that cause pressure waves in the surrounding water.

Many invertebrates have equilibrium organs called **statocysts** that use hair cells to signal the position of the animal with respect to gravity. In the case of the lobster, the statocyst is a chamber lined with hollow, nonliving hairs made of chitin. Each hair receives the dendrite of a sensory neuron. In the center of the statocyst is a dense **statolith** consisting of grains of sand Figure 42.11).

Because of gravity, the statolith stimulates the sensory hairs that are lowest, as determined by the posi-

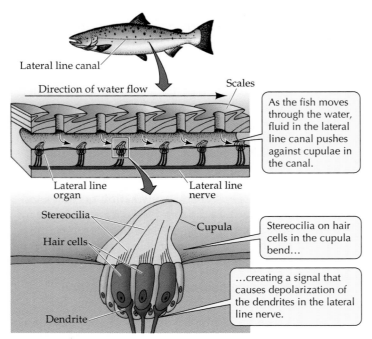

As the fish moves through the water, fluid in the lateral line canal pushes against cupulae in the canal.

Stereocilia on hair cells in the cupula bend…

…creating a signal that causes depolarization of the dendrites in the lateral line nerve.

42.10 The Lateral Line System Contains Mechanosensors Hair cells in the lateral line organs of a fish detect movement of the water around the animal, giving the fish information about its own movements and the movements of objects nearby.

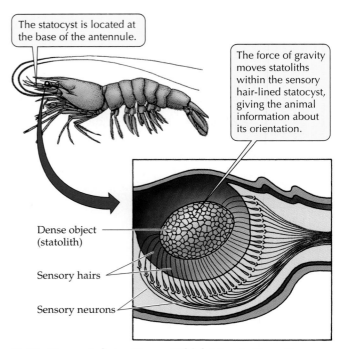

The statocyst is located at the base of the antennule.

The force of gravity moves statoliths within the sensory hair-lined statocyst, giving the animal information about its orientation.

42.11 How a Lobster Knows Which Way Is Up The statocyst is a sense organ found in many invertebrates.

tion of the animal. Replacing the statoliths of a lobster with iron filings and holding a magnet over the animal causes it to swim upside down. When a magnet is held to the lobster's side, it swims on its side. This experiment proved the role of the statoliths.

Vertebrates also have equilibrium organs. The mammalian inner ear, for example, has two organs of equilibrium that also use hair cells to detect the position of the body with respect to gravity. In the next section we will examine the structure of the ear. For the moment it is enough to know that the inner ear contains three **semicircular canals** at right angles to one another (Figure 42.12).

Each semicircular canal has a swelling called an ampulla, which contains a group of hair cells with their stereocilia embedded in a gelatinous cupula. The canals are filled with fluid. As an animal's head changes position, the fluid in its semicircular canals moves, puts pressure on the cupulae, and bends the stereocilia of the hair cells.

The second equilibrium organ is found in the vestibule. This organ, the **vestibular apparatus**, has two chambers that perform a function similar to that of the statocysts of invertebrates. Hair cells line the floors of the chambers; their stereocilia are embedded in a layer of gelatinous material. On top of this layer are **otoliths** (literally, "ear stones"), which are granules

of calcium carbonate. As the head moves, gravity pulls on the dense otoliths, which bend the stereocilia of the hair cells.

Auditory systems use hair cells to sense sound waves

The stimuli that animals perceive as sound are pressure waves. Auditory systems use mechanosensors to transduce pressure waves into action potentials. These systems include special structures to gather sound, direct it to the sensors, and amplify it.

Human hearing provides a good example of these aspects of auditory systems. The organs of hearing are the ears. The two prominent structures on the sides of our heads usually thought of as ears are the **ear pinnae**. The pinna of an ear collects sound waves and directs them into the auditory canal, which leads to the actual hearing apparatus in the middle ear and the inner ear. If you have ever watched a rabbit or a horse change the orientation of its ear pinnae to focus on a particular sound, then you have witnessed the role of ear pinnae in hearing.

The human ear is diagrammed at progressively higher levels of magnification in Figure 42.13. The eardrum, or **tympanic membrane**, covers the end of the auditory canal. The tympanic membrane vibrates in response to pressure waves traveling down the auditory canal. The chamber of the middle ear, an air-filled cavity, lies on the other side of the tympanic membrane.

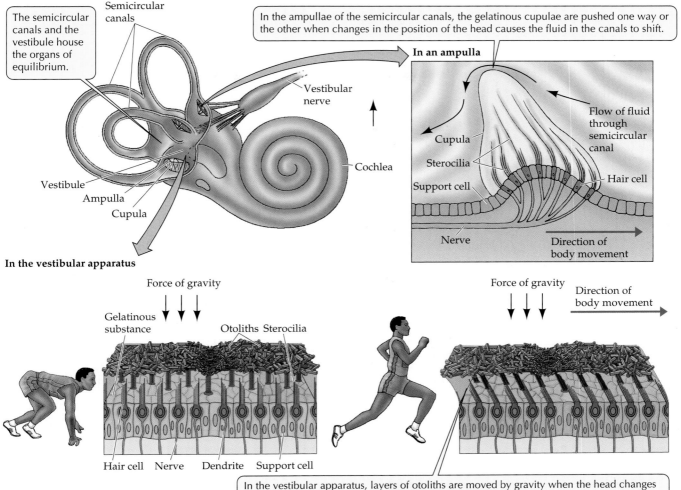

The semicircular canals and the vestibule house the organs of equilibrium.

Semicircular canals

In the ampullae of the semicircular canals, the gelatinous cupulae are pushed one way or the other when changes in the position of the head causes the fluid in the canals to shift.

In an ampulla

Vestibular nerve

Cochlea

Vestibule

Ampulla

Cupula

In the vestibular apparatus

Flow of fluid through semicircular canal

Cupula

Sterocilia

Support cell

Hair cell

Nerve

Direction of body movement

Force of gravity

Gelatinous substance

Otoliths Sterocilia

Force of gravity

Direction of body movement

Hair cell   Nerve   Dendrite   Support cell

In the vestibular apparatus, layers of otoliths are moved by gravity when the head changes position, or they are moved by their inertia if the head accelerates or decelarates.

42.12 Organs in the Inner Ear of Mammals Provide the Sense of Equilibrium   The bony inner ear has three parts: the snail-shaped cochlea, the semicircular canals, and the vestibule. The cochlea is for hearing; the semicircular canals and the vestibule are for equilibrium.

The middle ear is open to the throat at the back of the mouth through the **eustachian tube**. Because air flows through the eustachian tube, pressure equilibrates between the middle ear and the outside world. When you have a cold or hay fever, the tube becomes blocked by mucus or by tissue swelling, and you have difficulty "clearing your ears," or equilibrating the pressure in the middle ear with the outside air pressure. Then, the flexible tympanic membrane bulges in or out, dampening your hearing and sometimes causing earaches.

The middle ear contains three delicate bones called the **ear ossicles**, individually named the **malleus** (hammer), **incus** (anvil), and **stapes** (stirrup). The ossicles transmit the vibrations of the tympanic membrane to the fluid-filled inner ear, where they will be trans-

duced into action potentials. The leverlike action of the ossicles amplifies the vibrations about 20-fold. The malleus is attached to the center of the tympanic membrane, and at the other end of the chain of ossicles the stapes is attached to a smaller membrane called the **oval window**, which covers an opening into the inner ear.

The incus serves as a pivot point. When the tympanic membrane moves in, the lever action of the ossicles pushes the stapes, and the oval window bulges into the inner ear. When the tympanic membrane moves out, the stapes and the oval window are also pulled out. In this way, pressure waves in the auditory canal are converted into pressure waves in the fluid-filled inner ear.

Pressure waves are transduced into action potentials in the inner ear. The inner ear is a long, narrow, coiled chamber called the **cochlea** (from Latin and Greek words for "snail" or "shell"). A cross section of this chamber reveals that it is composed of three paral-

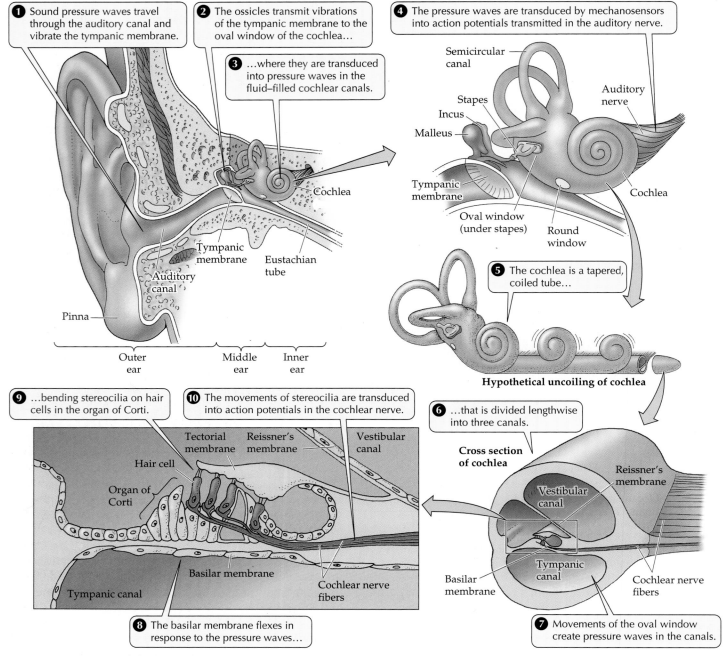

1 Sound pressure waves travel through the auditory canal and vibrate the tympanic membrane.

2 The ossicles transmit vibrations of the tympanic membrane to the oval window of the cochlea...

3 ...where they are transduced into pressure waves in the fluid-filled cochlear canals.

4 The pressure waves are transduced by mechanosensors into action potentials transmitted in the auditory nerve.

Semicircular canal

Stapes
Incus
Malleus

Auditory nerve

Tympanic membrane

Oval window (under stapes)
Round window

Cochlea

Cochlea
Tympanic membrane
Auditory canal
Eustachian tube
Pinna

Outer ear
Middle ear
Inner ear

5 The cochlea is a tapered, coiled tube...

Hypothetical uncoiling of cochlea

9 ...bending stereocilia on hair cells in the organ of Corti.

10 The movements of stereocilia are transduced into action potentials in the cochlear nerve.

6 ...that is divided lengthwise into three canals.

Tectorial membrane
Reissner's membrane
Vestibular canal

Hair cell

Organ of Corti

Basilar membrane

Tympanic canal

Cochlear nerve fibers

Cross section of cochlea

Vestibular canal

Reissner's membrane

Basilar membrane

Tympanic canal

Cochlear nerve fibers

8 The basilar membrane flexes in response to the pressure waves...

7 Movements of the oval window create pressure waves in the canals.

42.13 Structures of the Human Ear

lel canals separated by two membranes: **Reissner's membrane** and the **basilar membrane** (see Figure 42.13).

Sitting on the basilar membrane is the **organ of Corti**, the apparatus that transduces pressure waves into action potentials in the auditory nerve, which conveys information from the ear to the brain. The organ of Corti contains hair cells whose stereocilia are in contact with an overhanging, rigid shelf called the **tectorial membrane**. Whenever the basilar membrane

flexes, the tectorial membrane bends the hair cell stereocilia. As a consequence, the hair cells depolarize or hyperpolarize, altering the rate of action potentials transmitted to the brain by their associated sensory neurons.

What causes the basilar membrane to flex, and how does this mechanism distinguish sounds of different frequencies? In Figure 42.14, the cochlea is shown uncoiled to make it easier to understand its structure and function. To simplify matters, we have left out Reissner's membrane, thus combining the upper and

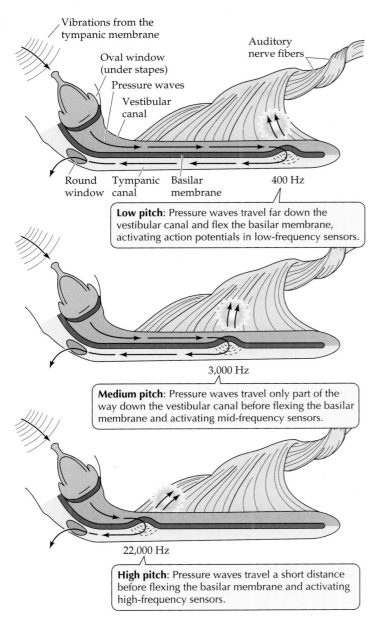

Vibrations from the tympanic membrane

Oval window (under stapes)

Pressure waves

Vestibular canal

Auditory nerve fibers

Round window

Tympanic canal

Basilar membrane

400 Hz

Low pitch: Pressure waves travel far down the vestibular canal and flex the basilar membrane, activating action potentials in low-frequency sensors.

3,000 Hz

Medium pitch: Pressure waves travel only part of the way down the vestibular canal before flexing the basilar membrane and activating mid-frequency sensors.

22,000 Hz

High pitch: Pressure waves travel a short distance before flexing the basilar membrane and activating high-frequency sensors.

42.14 Sensing Sound Pressure Waves in the Inner Ear
For simplicity, this diagram illustrates the cochlea as uncoiled. Pressure waves of different frequencies flex the basilar membrane at different locations. Information about sound frequency is specified by which hair cells—the sensors—are activated.

the middle canals into one upper canal. The purpose of Reissner's membrane is to contain a specific aqueous environment for the organ of Corti separate from the aqueous environment in the rest of the cochlea. This is important for the nutrition of the sensitive organ of Corti, but it does not play a role in the transduction of sound waves.

The simplified diagram of the cochlea shown in Figure 42.14 reveals two additional features that are

important for its function. First, the upper and lower chambers separated by the basilar membrane are joined at the distal end of the cochlea, making one continuous canal that folds back on itself. Second, just as the oval window is a flexible membrane at the beginning of the cochlea, the **round window** is a flexible membrane at the end of the long cochlear canal.

Air is highly compressible, but fluids are not. Therefore, a sound pressure wave can travel through air without much displacement of the air, but a sound pressure wave in fluid causes displacement of the fluid. Imagine holding the spring of screen door slightly stretched between your two hands. Someone could grab the spring in the center and move it back and forth without moving its ends—it is compressible. Now imagine holding a broomstick in the same way. If someone grabs it by the middle and moves it back and forth, its ends will move too—the broomstick is incompressible.

How does this comparison of springs and broomsticks relate to the inner ear? When the stapes pushes the oval window in, the fluid in the upper canal of the cochlea is displaced. Think about what happens if the oval window moves in very slowly. The cochlear fluid displacement travels down the upper canal, around the bend, and back through the lower chamber. At the end of the lower canal the displacement is absorbed by the outward bulging of the round window.

Now what happens if the oval window vibrates in and out rapidly? The waves of fluid displacement do not have enough time to travel all the way to the end of the upper canal and back through the lower canal. Instead, they take a shortcut by crossing the basilar membrane, causing it to flex. The more rapid the vibration, the closer to the oval and round windows the wave of displacement will flex the basilar membrane. Thus, different pitches of sound flex the basilar membrane at different locations and activate different sets of hair cells (see Figure 42.14).

The ability of the basilar membrane to respond to vibrations of different frequencies is enhanced by its structure. Near the oval and round windows, at the proximal end, the basilar membrane is narrow and stiff, but it gradually becomes wider and more flexible toward the opposite (distal) end. So it is easier for the proximal basilar membrane to resonate with high frequencies and for the distal basilar membrane to resonate with lower frequencies. A complex sound made up of many frequencies distorts the basilar membrane at many places simultaneously and activates a unique subset of hair cells. Action potentials generated by the mechanosensors at different places along the organ of Corti travel to the brain stem along the auditory nerve.

Deafness, the loss of the sense of hearing, has two general causes. **Conduction deafness** is caused by the loss of function of the tympanic membrane and the ossicles of the middle ear. Repeated infections of the middle ear can cause scarring of the tympanic membrane and stiffening of the connections between the ossicles. The consequence is less efficient conduction of sound waves from the tympanic membrane to the oval window. With increasing age, the ossicles progressively stiffen, resulting in a gradual loss of the ability to hear high-frequency sounds. **Nerve deafness** is caused by damage to the inner ear or the auditory pathways. A common cause of nerve deafness is damage to the hair cells of the delicate organ of Corti by exposure to loud sounds such as jet engines, pneumatic drills, or highly amplified rock music. This damage is cumulative and permanent.

Photosensors and Visual Systems: Responding to Light

Sensitivity to light—photosensitivity—confers on the simplest animals the ability to orient to the sun and sky and gives more complex animals instantaneous and extremely detailed information about objects in the environment. It is not surprising that simple and complex animals can sense and respond to light. What is remarkable is that across the entire range of animal species, evolution has conserved the same basis for photosensitivity: the molecule *rhodopsin.*

In this section we will learn how rhodopsin responds when stimulated by light energy and how that response is transduced into neural signals. We will also examine the structures of eyes, the organs that gather and focus light energy onto photosensitive cells. In the Chapter 43 we will learn how the brain uses action potentials from the retina to create our mental image of the visual world.

Rhodopsin is responsible for photosensitvity

Photosensitivity depends on the ability of a molecule to absorb photons of light and to respond by changing its conformation. The molecule that performs this task in the eyes of all animals is **rhodopsin** (Figure 42.15). Rhodopsin consists of a protein, **opsin**, which by itself does not absorb light, and a light-absorbing group, **11-*cis* retinal**. The light-absorbing group is cradled in the center of the opsin, and the entire rhodopsin molecule sits in the plasma membrane of a photosensor cell.

When the 11-*cis* retinal absorbs a photon of light energy, its shape changes into a different isomer of retinal—all-*trans* retinal. This conformational change puts a strain on the bonds between retinal and opsin, and the two components break apart. The dissociation of retinal and opsin—referred to as bleaching—causes the molecule to lose its photosensitivity. When the retinal spontaneously returns to its 11-*cis* isomer and recombines with opsin, it once again becomes photosensitive rhodopsin.

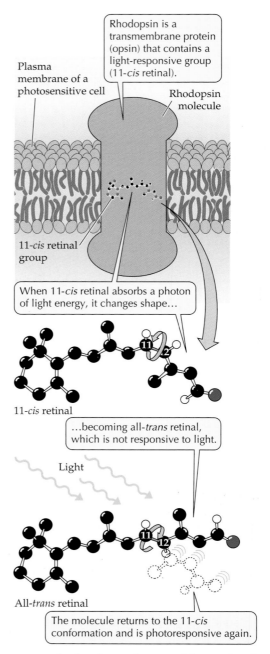

Rhodopsin is a transmembrane protein (opsin) that contains a light-responsive group (11-*cis* retinal).

Plasma membrane of a photosensitive cell

Rhodopsin molecule

11-*cis* retinal group

When 11-*cis* retinal absorbs a photon of light energy, it changes shape…

11-*cis* retinal

…becoming all-*trans* retinal, which is not responsive to light.

Light

All-*trans* retinal

The molecule returns to the 11-*cis* conformation and is photoresponsive again.

42.15 Rhodopsin: A Photosensitive Molecule
Rhodopsin changes its conformation when it absorbs light.

How does the light-induced conformational change of rhodopsin transduce light into a cellular response? As retinal converts from the 11-*cis* to the all-*trans* forms, its interactions with opsin pass through several unstable intermediate stages. One stage is known as photoexcited rhodopsin because it triggers a cascade of reactions that changes ion flows, producing the alteration of membrane potential that is the photosensor's response to light. Let's explore these events of transduction in more detail.

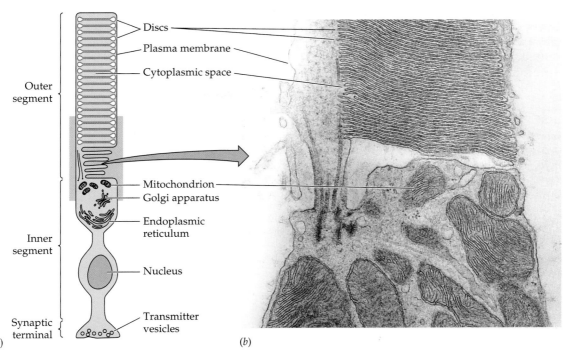

Outer segment

Inner segment

Synaptic terminal

Discs
Plasma membrane
Cytoplasmic space

Mitochondrion
Golgi apparatus
Endoplasmic reticulum

Nucleus

Transmitter vesicles

(a)

(b)

42.16 The Rod Cell: A Vertebrate Photosensor (a) The rod cell of the vertebrate retina is a neuron modified for photosensitivity. The membranes of a rod cell's discs are densely packed with rhodopsin. (b) A transmission electron micrograph of a section through a photosensor.

The rhodopsin molecule sits in the membrane of a photosensor. How does this molecule communicate to the cell that it has absorbed a photon? And how does the sensor then communicate to the nervous system that its rhodopsin molecules are receiving light? To answer these questions, we must see how photosensors respond to light. A good example of a photosensor cell is a vertebrate **rod cell**, which is a modified neuron

(Figure 42.16). At the back of the eye is the **retina**, which consists of several layers of neurons. One of these layers is photosensor cells, and rod cells are one type of photosensor. The retina transduces the visual world into the language of the nervous system.

Each rod cell in the retina has an inner segment and an outer segment. The inner segment contains the usual organelles of a cell and has a synaptic terminal at its base, where the cell communicates with other neurons. The outer segment is highly specialized and contains a stack of discs. These discs form by the invagination (folding inward) and pinching off of the plasma membrane of the outer segment. The membranous discs are densely packed with rhodopsin; their function is to capture photons of light passing through the rod cell.

To see how the rod cell responds to light, we can penetrate a single rod cell with an electrode. With this electrode we can record the receptor potential of the rod cell in the dark and in the light (Figure 42.17). From what we have learned about other types of sensors, we might expect stimulation of the photosensor with light to make its receptor potential less

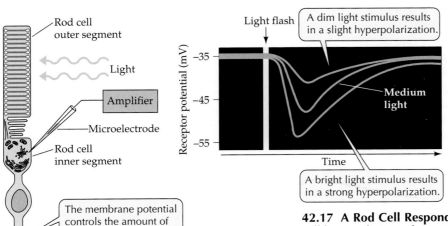

Rod cell outer segment

Light

Amplifier

Microelectrode

Rod cell inner segment

The membrane potential controls the amount of transmitter released.

Light flash

A dim light stimulus results in a slight hyperpolarization.

Receptor potential (mV)

−35

−45

−55

Medium light

Time

A bright light stimulus results in a strong hyperpolarization.

42.17 A Rod Cell Responds to Light The receptor potential of a rod cell hyperpolarizes—becomes more negative—in response to a flash of light.

negative, but photosensors are unusual and the opposite is true. When the rod cell is kept in the dark, it already has a very high resting potential in comparison with other neurons. In fact, the plasma membrane of the rod cell is fairly permeable to Na⁺ ions, so these positive charges are continually entering the cell.

When a light is flashed on the dark-adapted rod cell, its receptor potential becomes more negative—it hyperpolarizes. The rod cell itself *does not* generate action potentials. However, the rod cell changes its rate of neurotransmitter release as its membrane polarity changes. Later in this section we will learn how other cells in the visual pathway respond to neurotransmitter released from the photosensors so that information is communicated to the brain in the form of action potentials.

How does the absorption of light by rhodopsin hyperpolarize the rod cell? When rhodopsin is excited by light, it initiates a cascade of events. The photoexcited rhodopsin combines with and activates another protein, a G protein called **transducin**. Activated transducin in turn activates a phosphodiesterase. Active phosphodiesterase converts cyclic GMP (cGMP) to 5′-GMP. Before being converted to 5′-GMP, that cGMP was holding open the sodium channels and keeping the cell depolarized. As cGMP is converted to 5′-GMP, the sodium channels close and the cell hyperpolarizes. This may seem like a roundabout way of doing business, but its significance is its enormous amplification ability.

Each molecule of photoexcited rhodopsin can activate about 500 transducin molecules, thus activating about 500 phosphodiesterase molecules. The catalytic prowess of a molecule of phosphodiesterase is great; it can hydrolyze more than 4,000 molecules of cGMP per second. The bottom line is that a single photon of light can cause the closing of more than a million sodium channels and thereby change the rod cell's receptor potential (Figure 42.18). Now let's see how photosensors work in animals.

Invertebrates have a variety of visual systems

Flatworms, simple multicellular animals, obtain directional information about light from photosensitive cells organized into eye cups (Figure 42.19). The eye cups are bilateral structures, and each is partly shielded from light by a layer of pigmented cells lining the cup. Because the openings of the eye cups face in opposite directions, the photosensors on the two sides of the animal are unequally stimulated unless the animal is facing directly toward or away from a light source. Using directional information about light

42.18 Light Absorption Closes Na⁺ Channels The absorption of light by rhodopsin initiates a cascade resulting in the hyperpolarization of the rod cell.

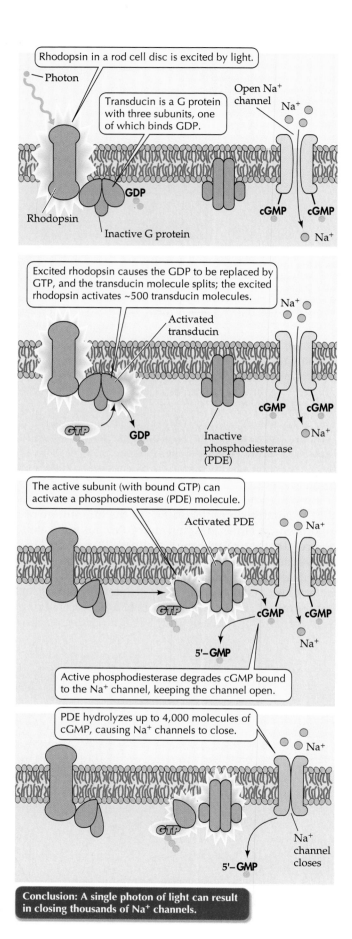

Rhodopsin in a rod cell disc is excited by light.

Photon

Transducin is a G protein with three subunits, one of which binds GDP.

Open Na⁺ channel

Na⁺

GDP

Rhodopsin

Inactive G protein

cGMP   cGMP

Na⁺

Excited rhodopsin causes the GDP to be replaced by GTP, and the transducin molecule splits; the excited rhodopsin activates ~500 transducin molecules.

Na⁺

Activated transducin

GTP   GDP

Inactive phosphodiesterase (PDE)

cGMP   cGMP

Na⁺

The active subunit (with bound GTP) can activate a phosphodiesterase (PDE) molecule.

Activated PDE

Na⁺

GTP

cGMP   cGMP

5′–GMP

Na⁺

Active phosphodiesterase degrades cGMP bound to the Na⁺ channel, keeping the channel open.

PDE hydrolyzes up to 4,000 molecules of cGMP, causing Na⁺ channels to close.

Na⁺

GTP

Na⁺ channel closes

5′–GMP

Conclusion: A single photon of light can result in closing thousands of Na⁺ channels.

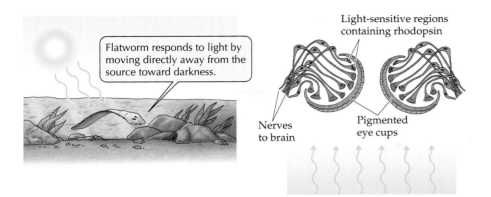

Flatworm responds to light by moving directly away from the source toward darkness.

Light-sensitive regions containing rhodopsin

Nerves to brain

Pigmented eye cups

42.19 A Simple Photosensory System Although flatworms do not "see" as we understand vision, the eye cups of this flatworm enable it to move away from a light source to an area where it may be less visible to predators.

sources, the flatworm generally moves away from light (and hence to better protection).

Arthropods (crustaceans, spiders, and insects) have evolved **compound eyes** that provide them with information about patterns or images in the environment. Each compound eye consists of many optical units called **ommatidia** (singular ommatidium) (Figure 42.20). The number of ommatidia in a compound eye varies from only a few in some ants, to 800 in fruit flies, to 10,000 in some dragonflies.

Each ommatidium has a lens structure that directs light onto photosensors called *retinula cells.* Flies have seven elongated retinula cells in each ommatidium. The inner borders of the retinula cells are covered with microvilli that contain rhodopsin and thus trap light. Since the microvilli of the different retinula cells overlap, they appear to form a central rod, called a *rhabdom,* down the center of the ommatidium.

Axons from the retinula cells communicate with the nervous system. Since each ommatidium of a compound eye is directed at a slightly different part of the visual world, only a crude, or perhaps a broken-up,

image of the visual field can be communicated from the compound eye to the CNS.

Image-forming eyes evolved independently in vertebrates and cephalopods

Both vertebrates and cephalopod mollusks have evolved eyes with exceptional abilities to form images of the visual world. These eyes operate like cameras, and considering that they evolved independently of

42.20 Ommatidia: The Functional Units of Insect Eyes The rhodopsin-containing retinula cells are the photosensors in ommatidia.

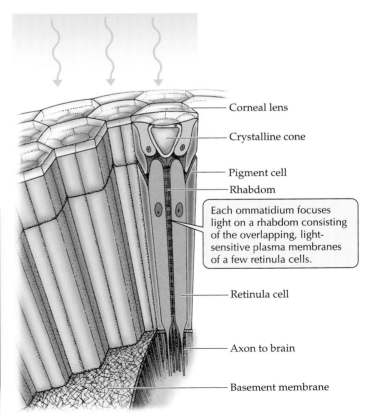

Corneal lens
Crystalline cone
Pigment cell
Rhabdom

Each ommatidium focuses light on a rhabdom consisting of the overlapping, light-sensitive plasma membranes of a few retinula cells.

Retinula cell
Axon to brain
Basement membrane

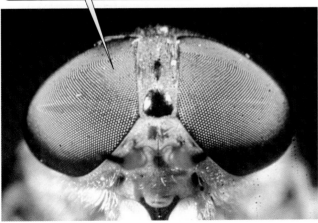

The compound eyes of a horsefly each contain hundreds of ommatidia.

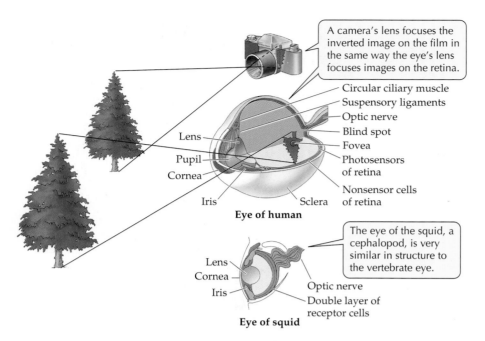

A camera's lens focuses the inverted image on the film in the same way the eye's lens focuses images on the retina.

Circular ciliary muscle
Suspensory ligaments
Optic nerve
Blind spot
Fovea
Photosensors of retina
Nonsensor cells of retina

Lens
Pupil
Cornea
Iris
Sclera

Eye of human

The eye of the squid, a cephalopod, is very similar in structure to the vertebrate eye.

Lens
Cornea
Iris
Optic nerve
Double layer of receptor cells

Eye of squid

42.21 Eyes Like Cameras The lenses of cephalopod and vertebrate eyes focus images on layers of photosensors, just as a camera's lens focuses images on film.

The lens is contained in a connective-tissue sheath that tends to keep it in a spherical shape, but it is also suspended by suspensory ligaments that pull it into a flatter shape. Circular muscles called the *ciliary muscles* counteract the pull of the suspensory ligaments and permit the lens to round up. With the ciliary muscles at rest, the flatter lens has the correct optical properties to focus distant images on the retina, but not close images. Contracting the ciliary muscles rounds up the lens, changing its light-bending properties to bring close images into focus (Figure 42.22). As we age, our lenses become less elastic, and we lose the ability to focus on objects close at hand without the help of corrective lenses. Prolonged concentration on small, close objects (such as the type on these pages) tires and strains the eyes by overworking the ciliary muscles.

each other, their high degree of similarity is remarkable (Figure 42.21).

The vertebrate eye is a spherical, fluid-filled structure bounded by a tough connective tissue layer called the *sclera*. At the front of the eyeball, the sclera forms the transparent **cornea** through which light enters the eye. Just inside the cornea is the pigmented **iris**, which gives the eye its color. The important function of the iris is to control the amount of light that reaches the photosensors at the back of the eyeball, just as the diaphragm of the camera controls the amount of light reaching the film. The central opening of the iris is the **pupil**. The iris is under neural control. In bright light the iris constricts and the pupil is very small, but as light levels fall, the iris relaxes and the pupil enlarges.

Behind the iris is the crystalline lens, which helps focus images on the photosensitive layer, the retina, at the back of the eye. The cornea and the fluids of the eye chambers also help focus light on the retina, but the lens is responsible for the ability to *accommodate*—to focus on objects at various locations in the near visual field. To focus a camera on objects close at hand, you must adjust the distance between the lens and the film. Fishes, amphibians, and reptiles accommodate in a similar manner, moving the lenses of their eyes closer to or farther from their retinas. Mammals and birds use a different method; they alter the shape of the lens.

The vertebrate retina processes information

During development, neural tissue grows out from the brain to form the retina. In addition to a layer of photosensors, the retina includes layers of cells that process the visual information from the photosensors and transmit it to the brain in the form of action potentials in the optic nerves. A curious feature of the anatomy of the retina is that the outer segments of the

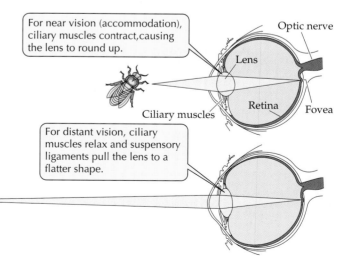

For near vision (accommodation), ciliary muscles contract, causing the lens to round up.

Optic nerve
Lens
Ciliary muscles
Retina
Fovea

For distant vision, ciliary muscles relax and suspensory ligaments pull the lens to a flatter shape.

42.22 Focusing Mammals and birds focus their eyes by changing the shape of the lens.

photosensors are all the way at the back of the retina, so light must pass through all the layers of retinal cells before reaching the place where photons are captured by rhodopsin. We will examine in detail how the cells of the retina process information, but first let's describe some general features of retinal organization.

THE CELLULAR STRUCTURE OF THE RETINA. The density of photosensors is not the same across the entire retina. Light coming from the center of the field of vision falls on an area of the retina called the **fovea**, where the density of sensors is the highest. The human fovea has about 160,000 sensors per square millimeter. A hawk has about 1,000,000 sensors per square millimeter of fovea, making its vision about eight times sharper than ours. In addition, the hawk has two foveas in each eye. One fovea receives light from straight ahead and the other receives light from below. Thus, while the hawk is flying, it sees both its projected flight path and the ground below, where it might detect a mouse scurrying in the grass.

The fovea of a horse is a long, vertical patch of retina. The horse's lens is not good at accommodation, but it focuses distant objects that are straight ahead on one part of this long fovea and close objects that are below the head on another part of the fovea. When horses are startled by an object close at hand, they pull their heads back and rear up to bring the object into focus on the close-vision part of the fovea.

Where blood vessels and the bundle of axons going to the brain pass through the back of the eye, there is a blind spot on the retina. You are normally not aware of your blind spot, but you can find it. Stare straight ahead, holding a pencil in your outstretched hand so that the eraser is in the center of your field of vision. While continuing to stare straight ahead, slowly move the pencil to the side until the eraser disappears. When this happens, the light from the eraser is focused directly on your blind spot.

Until now we have referred to only one kind of photosensor, the rod cell. But there are two major kinds of photosensors, both named for their shapes—rod cells and **cone cells** (Figure 42.23). A human retina has about 3 million cones and about 100 million rods. Rod cells are more sensitive to light but do not contribute to color vision. Cones are responsible for color vision but are less sensitive to light. Cones are also responsible for our sharpest vision. Even though there are many more rods than cones in human retinas, our foveas contain mostly cones.

Because cones have low sensitivity to light, they are of no use at night. At night our vision is not very sharp, and we see mostly in shades of gray. You may have trouble seeing a small object such as a keyhole at night when you are looking straight at it—that is, when its image is falling on your fovea. If you look a little to the

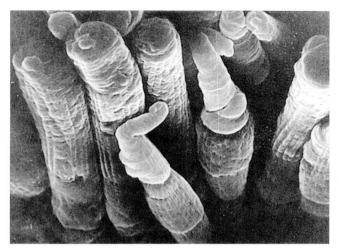

42.23 Rods and Cones This scanning electron micrograph of photosensors in the retina of a mud puppy (an amphibian) shows cylindrical rods and tapered cones.

side, so that the image falls on a rod-rich area of retina, you can see the object better. Astronomers looking for faint objects in the sky learned this trick a long time ago. Animals that are nocturnal (such as flying squirrels) may have only rods in their retina and have no color vision. By contrast, some animals that are active only during the day (such as chipmunks and ground squirrels) have only cones in their retinas.

How do cone cells enable us to see color? There are at least three kinds of cone cells, each possessing slightly different types of opsin molecules. Because different cone cells have different opsin molecules, they differ in the wavelengths of light they absorb best. Although the retinal group is the light absorber (see Figure 42.15), its molecular interactions with opsin tune its spectral sensitivity.

Some opsins cause retinal to absorb most efficiently in the blue region, some in the green, and some in the red (Figure 42.24). Intermediate wavelengths of light excite these classes of cones in different proportions. The genes that encode the different opsins of humans have been cloned: One codes for blue-sensitive opsin, one for red-sensitive opsin, and several for green-sensitive opsin.

The human retina is organized into five layers of neurons that receive visual information and process it before sending it to the brain (Figure 42.25). As mentioned earlier, the layer of photosensors is all the way at the back of the retina, so light must traverse all the other layers before reaching the rods and cones. The disc-containing ends of the rods and cones are partly buried in a layer of pigmented epithelium that absorbs photons not captured by rhodopsin and prevents any backscattering of light that might decrease visual sharpness.

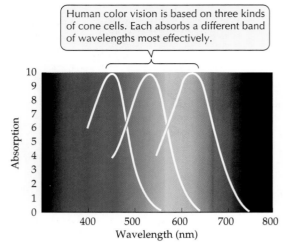

Human color vision is based on three kinds of cone cells. Each absorbs a different band of wavelengths most effectively.

42.24 Absorption Spectra of Cone Cells The three kinds of cone cells contain slightly different opsin molecules.

Nocturnal animals such as cats have a highly reflective layer, called the tapetum, behind the photosensors. Photons not captured on their first pass through the photosensors are reflected back, thus increasing visual sensitivity (but not sharpness) in low-light conditions. The reflective tapetum is what makes cats' eyes appear to glow in the dark.

INFORMATION FLOW IN THE RETINA. A first step in investigating how the human retina processes visual information is to study how its five layers of neurons are interconnected and how they influence one another. The neurons at the back of the retina are the photosensors. As we know, the photosensors hyperpolarize in response to light, and therefore they do not generate action potentials. The cells at the front of the retina are ganglion cells. Ganglion cells fire action potentials, and their axons

form the optic nerves that travel to the brain. The photosensors and ganglion cells are connected by bipolar cells. The changes in membrane potential of rods and cones in response to light alter the rate at which the rods and cones release neurotransmitter at their synapses with the bipolar cells.

Like rods and cones, bipolar cells do not fire action potentials. In response to neurotransmitter from the photosensors, the membrane potentials of bipolar cells change, altering the rate at which they release neurotransmitter onto ganglion cells. The ganglion cells generate action potentials, and the rate of neurotransmitter release from the bipolar cells determines the rate at which ganglion cells fire action potentials. Thus the direct flow of information in the retina is from photosensor to bipolar cell to the ganglion cell. Ganglion cells send the information to the brain (see Figure 42.25).

What are the functions of the other two layers—the horizontal cells and the amacrine cells? They communicate laterally across the retina. Horizontal cells connect neighboring pairs of photosensors and bipolar cells. Thus the communication between a photosensor and its bipolar cell can be influenced by the amount of light absorbed by neighboring photosensors. This lateral flow of information sharpens the perception of contrast between light and dark patterns falling on the retina. Amacrine cells connect neighboring pairs of bipolar cells and ganglion cells. One role of amacrine cells is to adjust the sensitivity of the eyes according to the overall level of light falling on the retina. When light levels change, amacrine cell connections to the ganglion cells help the ganglion cells remain sensitive

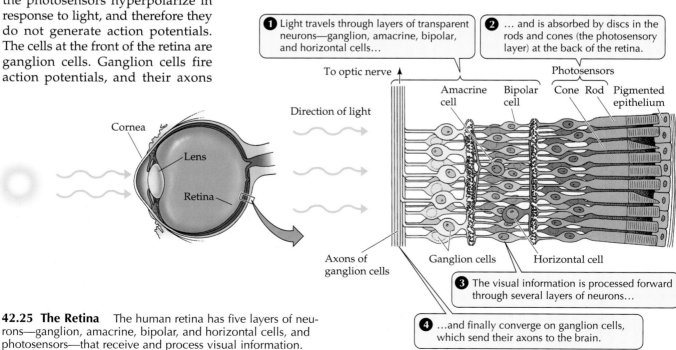

1 Light travels through layers of transparent neurons—ganglion, amacrine, bipolar, and horizontal cells...

2 ... and is absorbed by discs in the rods and cones (the photosensory layer) at the back of the retina.

3 The visual information is processed forward through several layers of neurons...

4 ...and finally converge on ganglion cells, which send their axons to the brain.

42.25 The Retina The human retina has five layers of neurons—ganglion, amacrine, bipolar, and horizontal cells, and photosensors—that receive and process visual information.

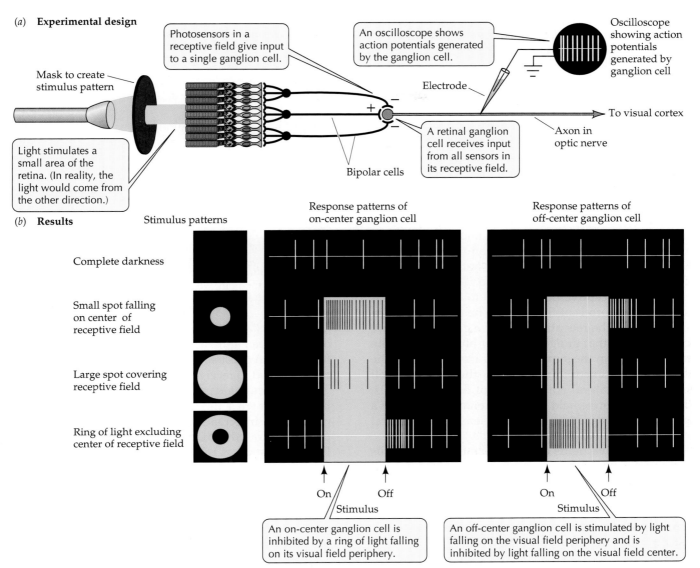

(a) **Experimental design**

Mask to create stimulus pattern

Photosensors in a receptive field give input to a single ganglion cell.

An oscilloscope shows action potentials generated by the ganglion cell.

Oscilloscope showing action potentials generated by ganglion cell

Electrode

Light stimulates a small area of the retina. (In reality, the light would come from the other direction.)

A retinal ganglion cell receives input from all sensors in its receptive field.

To visual cortex

Axon in optic nerve

Bipolar cells

(b) **Results**

Stimulus patterns

Response patterns of on-center ganglion cell

Response patterns of off-center ganglion cell

Complete darkness

Small spot falling on center of receptive field

Large spot covering receptive field

Ring of light excluding center of receptive field

On   Off
Stimulus

On   Off
Stimulus

An on-center ganglion cell is inhibited by a ring of light falling on its visual field periphery.

An off-center ganglion cell is stimulated by light falling on the visual field periphery and is inhibited by light falling on the visual field center.

42.26 What Does the Eye Tell the Brain? When the retina is stimulated with dots and rings of light, individual ganglion cells show different responses. Each ganglion cell responds to light shining on a small, circular area of retina in its receptive field. Some ganglion cells are stimulated and some are inhibited by a spot of light falling on the receptive field centers. An "on-center" ganglion cell is inhibited by a ring of light falling on the peripheral area of its receptive field. The opposite is true of "off-center" ganglion cells.

to temporal changes in stimulation. Thus even with large changes in background illumination, the eyes are sensitive to smaller, more rapid changes in the pattern of light falling on the retina.

Knowing the paths of information flow through the retina still doesn't tell us how that information is processed. What does the eye tell the brain in response to a pattern of light falling on the retina? One aspect of information processing in the retina is convergence of information. There are more than 100 million photosensors in each retina, but only about 1 million gan-

glion cells sending messages to the brain. How is the information from all those photosensors integrated to form the messages sent to the brain by the ganglion cells?

This question was addressed in elegant, classic experiments in which scientists used electrodes to record the activity of single ganglion cells in living animals while their retinas were stimulated with spots of light. These studies revealed that each ganglion cell has a well-defined **receptive field** that consists of a specific group of photosensors. Stimulating these photosensors with light activates the ganglion cell (Figure 42.26). Information from many photosensors is integrated in this way to produce a single message.

The receptive fields of ganglion cells are all circular. However, the way a spot of light influences the activity of the ganglion cell depends on where in the receptive field it falls. The receptive field of each ganglion cell is divided into two concentric areas, called the *cen-*

ter and the *surround.* There are two kinds of receptive fields, *on-center* and *off-center.* Stimulating the center of an on-center receptive field excites the ganglion cell, and stimulation of the surround inhibits it. Stimulation of the center of an off-center receptive field inhibits the ganglion cell, and stimulating the surround excites it. Center effects are always stronger than surround effects.

The response of a ganglion cell to stimulation of the center of its receptive field depends on how much of the surround area is also stimulated. A small dot of light directly on the center has the maximal effect, a bar of light hitting the center and pieces of the surround has less of an effect, and a large, uniform patch of light falling equally on center and surround has very little effect. Ganglion cells communicate to the brain information about contrasts between light and dark that fall on their receptive fields.

How are receptive fields related to the connections among the neurons of the retina? The photosensors in the center of the receptive field of a ganglion cell are connected to that ganglion cell by bipolar cells. The photosensors in the surround send information to the center photosensors and thus to the ganglion cell through the lateral connections of horizontal cells. Thus the receptive field of a ganglion cell consists of a pattern of synapses among photosensors, horizontal cells, bipolar cells, and ganglion cells.

The receptive fields of neighboring ganglion cells can overlap greatly; a given photosensor can be connected to several ganglion cells. The eye sends the brain simple messages about the pattern of light intensities falling on small, circular patches of retina. In Chapter 44 we will see how the brain reassembles that information into our view of the world.

Sensory Worlds Beyond Our Experience

Humans make use of only a subset of the information available to us in the environment. Other animals have sensory systems that enable them to use different subsets and different types of information.

Some species can see infrared and ultraviolet light

When discussing vision, we use the term "visible spectrum," but what we really mean is light visible to humans. Our visible spectrum is a very narrow region of the continuous range of electromagnetic radiation in the environment (see Figure 8.5). For example, we cannot see ultraviolet radiation, but many insects can.

One of the seven photosensors in each ommatidium of a fruit fly is sensitive to ultraviolet light. The visual sensitivity of many pollinating insects includes the ultraviolet part of the spectrum. Some flowers have patterns that are invisible to us but that show up if we photograph them with film that is sensitive to ultravi-

The "hole" just below and in front of the eye is a pit organ that senses infrared radiation.

42.27 Stalking in the Dark Pit vipers, such as this Pope's tree viper, are snakes that can locate prey in total darkness. They sense infrared radiation via the pit organ just in front of the eye; thus the body warmth from potential prey allows them to orient and strike with precision.

olet light (see Figure 52.23). Those patterns provide information to prospective pollinators, but humans are not equipped to receive that information.

At the other end of the spectrum is infrared radiation, which we sense as heat. Other animals extract much more information from infrared radiation—especially infrared radiation emitted by potential prey. Pit vipers such as rattlesnakes have pit organs, one just in front of each eye that can sense infrared radiation (Figure 42.27). In total darkness these snakes can locate a prey item such as a mouse, orient to it, and strike it with great accuracy on the basis of the directional information that comes to them from the warmth of the mouse's body.

Echolocation is sensing the world through reflected sound

Some species emit intense sounds and create images of their environments from the echoes of those sounds. Bats, porpoises, dolphins, and, to a lesser extent, whales, are accomplished echolocators. Some species of bats have elaborate modifications of their noses to direct the sounds they emit, as well as impressive ear pinnae to collect the returning echoes. The sounds they emit as pulses (about 20 to 80 per second) are above our range of hearing, but they are extremely loud in contrast to the resulting faint echoes bouncing off small insects (see the opening discussion in Chapter 41).

An echolocating bat is similar to a construction worker who is trying to overhear a whispered conversation on a street corner while using a pneumatic drill. To avoid deafening themselves, bats use muscles in their middle ears to dampen their sensitivity while they emit sounds, then relax them quickly enough to hear the echoes. The ability of bats to use echolocation

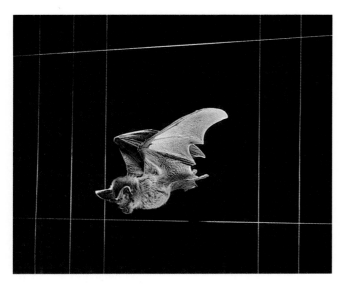

42.28 Navigating in the Dark Using echolocation, this bat avoids an obstacle course of thin wires while flying in complete darkness.

to "see" their environment is so good that in a totally dark room strung with fine wires, bats can capture tiny flying insects while navigating around the wires (Figure 42.28).

Magnetic sensors are built-in compasses

As astonishing as this may seem, the magnetic lines of force of Earth provide directional information for some animals. From experiments with homing pigeons, biologists have accumulated evidence of this phenomenon. Normally, homing pigeons released from locations distant from their homes can orient and find their way home, even if they are fitted with translucent contact lenses that eliminate all directional information from the sun. However, pigeons that are released on cloudy days with tiny magnets glued to their heads become disoriented and cannot navigate. The magnetic sense is poorly understood at present, but magnetosensory neurons have been discovered.

Summary of "Sensory Systems"

Sensors, Sensory Organs, and Transduction

• Sensors transduce information about an animal's external and internal environments into action potentials that are transmitted to the CNS. **Review Figures 42.1, 42.2**
• The interpretation of action potentials as light, sound, odor, touch, etc. depends on where in the CNS the action potentials are transmitted.
• Sensors have membrane proteins that alter ion channels and therefore membrane potential in response to specific stimuli. This change in membrane potential is called a receptor potential. **Review Figures 42.1, 42.3**
• Receptor potentials can directly influence the release of signaling molecules from the sensor cell, or they can spread to regions of the sensor cell plasma membrane that have voltage-gated channels that generate action potentials. **Review Figure 42.3**
• Adaptation enables the nervous system to remain responsive to important new stimuli.

Chemosensors: Responding to Specific Molecules

• Smell, taste, and sensing of pheromones are examples of chemosensitivity. Chemosensors have receptors that can bind to specific molecules that come into contact with the sensor cell membrane. **Review Figures 42.5, 42.6, 42.8**
• The binding of odorant molecule to receptor protein causes the production of a second messenger in the sensor cell. The second messenger alters ion channels and creates a receptor potential. **Review Figure 42.7**
• Chemosensors in the mouth cavities of vertebrates are responsible for gustation, the sense of taste. **Review Figure 42.8**

Mechanosensors: Stimuli That Distort Membranes

• In the skin there are a diversity of mechanosensors that respond to touch and pressure. The density of mechanosensors in any skin area determines the sensitivity of that area to touch and pressure. **Review Figure 42.9**
• Mechanosensors respond to distortions of their plasma membrane produced by mechanical forces. **Review Figures 42.9, 42.10, 42.11**
• Stretch sensors in muscles, tendons, and ligaments inform the CNS of the positions of and the loads on parts of the body. **Review Figure 42.3**
• Hair cells are mechanosensors that are not neurons. The bending of their stereocilia alters membrane proteins and therefore receptor potentials. Hair cells are found in organs of equilibrium and orientation such as the lateral line system of fishes, the statocysts of invertebrates, and the semicircular canals and vestibular apparatus of mammals. **Review Figures 42.10, 42.11, 42.12**
• Hair cells are responsible for mammalian auditory sensitivity. Ear pinnae collect and direct sound waves to the tympanic membrane, which vibrates in response to sound waves. The movements of the tympanic membrane are amplified through a chain of three ear bones that conduct the movement to the membrane-covered oval window that leads into the fluid-filled inner ear, or cochlea. Movements of the oval window create pressure waves in cochlear fluid, and depending on the frequencies of the pressure waves, the basilar membrane running down the center of the cochlea is distorted at specific locations. These distortions cause bending of hair cells in the organ of Corti, which rests on the basilar membrane. Changes in hair cell receptor potentials create action potentials in the auditory nerve, which conducts the information to the CNS. **Review Figures 42.13, 42.14**

Photosensors and Visual Systems: Responding to Light

• Photosensitivity in animals depends on the capture of photons of light by rhodopsin, a visual pigment that consists of a protein opsin and a light-absorbing group called retinal. Rhodopsin is found in the membranes of neurons specialized for photosensing. **Review Figure 42.15**
• Absorption of light by retinal induces a conformational change that results in the separation of retinal from opsin,

which in turn initiates a cascade of intracellular events leading to a change in receptor potential of the sensor neuron. **Review Figure 42.18**

• Vertebrate photosensors are rod cells, responsible for dim light and black-and-white vision, and cone cells, responsible for color vision by virtue of their spectral sensitivities. When excited by light, rods and cones hyperpolarize and release less neurotransmitter onto the neurons with which they form synapses. Rods and cones do not fire action potentials. **Review Figures 42.16, 42.17, 42.24**

• Vision results when eyes focus patterns of light onto layers of photosensors. Eyes vary from the simple eye cups of flatworms, which enable the animal to sense directionality of a light source, to the compound eyes of arthropods, which enable the animal to detect shapes and patterns in the visual field, to the lensed eyes of cephalopod mollusks, which focus detailed images of the visual field onto dense arrays of photosensors that transduce the visual image into neural signals that are transmitted to the CNS. **Review Figures 42.19, 42.20, 42.21, 42.22**

• The vertebrate retina is a dense array of neurons lining the back of the eyeball. It consists of five layers of cells. The outermost layer consists of the rods and cones; the innermost layer consists of the ganglion cells, which send their axons in the optic nerve to the brain. Between the photosensors and the ganglion cells are neurons that process the information from the photosensors so that each ganglion cell is stimulated by light falling on a small circular patch of retina called a receptive field. **Review Figures 42.25, 42.26**

• The area of the retina that receives light from the center of the visual field, the fovea, has the greatest density of photosensors. In humans the fovea contains almost exclusively cone cells, which are responsible for color vision but are not very sensitive in dim light.

• Receptive fields have a center and a surround, which have opposing effects on the ganglion cell. If the center is excitatory, the surround is inhibitory and vice versa. **Review Figure 42.26**

Sensory Worlds Beyond Our Experience

• Many animals have sensory worlds that we do not share. Bats echolocate, insects see ultraviolet radiation, pit vipers "see" in the infrared, fish sense electric fields, and homing pigeons can orient to Earth's magnetic fields.

Self-Quiz

1. Which statement about sensory systems is *not* true?
   a. Sensory transduction in vertebrate sensory systems involves the conversion (direct or indirect) of a physical or chemical stimulus into action potentials.
   b. In general, a stimulus causes a change in the flow of ions across the plasma membrane of a sensor.
   c. The term "adaptation" is given to the process by which a sensory system becomes insensitive to a continuing source of stimulation.
   d. The more intense a stimulus, the greater the magnitude of each action potential fired by a receptor.
   e. Sensory adaptation plays a role in the ability of organisms to discriminate between important and unimportant information.

2. The female silkworm moth releases a chemical called bombykol from a gland at the tip of her abdomen. Bombykol is
   a. a sex hormone.
   b. detected by the male only when present in large quantities.
   c. not species-specific.
   d. detected by hairs on the antennae of male silkworm moths.
   e. a chemical basic to the taste process in arthropods.

3. Which statement about olfaction is *not* true?
   a. Dogs are unusual among mammals in that they depend more on olfaction than on vision as their dominant sensory modality.
   b. Olfactory stimuli are recognized by the interaction between the stimulus and a specific macromolecule on olfactory hairs.
   c. The more odorant molecules binding to receptors, the more action potentials generated.
   d. The greater the number of action potentials generated by an olfactory receptor, the greater the intensity of the perceived smell.
   e. The perception of different smells results from the activation of different combinations of olfactory receptors.

4. The touch receptors that are located very close to the skin surface
   a. are relatively insensitive to light touch.
   b. adapt very quickly to stimuli.
   c. are uniformly distributed throughout the surface of the body.
   d. are called Pacinian corpuscles.
   e. adapt slowly and only partially to stimuli.

5. The membrane that gives us the ability to discriminate different pitches of sound is the
   a. round window.
   b. oval window.
   c. tympanic membrane.
   d. tectorial membrane.
   e. basilar membrane.

6. Which statement is *not* true?
   a. The transmembrane potential of a rod cell becomes more negative when the rod cell is exposed to light after a period of darkness.
   b. A photosensor releases the most neurotransmitter (per unit time) when in total darkness.
   c. Whereas in vision the intensity of a stimulus is encoded by the degree of hyperpolarization of photosensors, in hearing the intensity of a stimulus is encoded by changes in firing rates of sensory cells.
   d. Stiffening of the ossicles in the middle ear can lead to deafness.
   e. The interaction among hammer (malleus), anvil (incus), and stirrup (stapes) conducts sound waves across the fluid-filled middle ear.

7. In primates, the region of the retina where the central part of the visual field falls is called the
   a. central ganglion cell.
   b. fovea.
   c. optic nerve.
   d. cornea.
   e. pupil.

8. The region of the vertebrate eye where the optic nerve passes out of the retina is called the
   a. fovea.
   b. iris.
   c. blind spot.
   d. pupil.
   e. optic chiasm.

9. Which statement about the cones in a human eye is *not* true?
   a. They are responsible for high-acuity vision.
   b. They encode color vision.
   c. They are more sensitive to light than rods are.
   d. They are fewer in number than rods.
   e. They exist in high numbers at the fovea.

10. The color in vision results from the
    a. ability of each cone to absorb all wavelengths of light equally.
    b. lens of the eye acting like a prism and separating the different wavelengths by light.
    c. different absorption of wavelengths of light by different classes of rods.
    d. three different isomers of opsin in different classes of cone cells.
    e. absorption of different wavelengths of light by amacrine and horizontal cells.

Applying Concepts

1. Using your knowledge of the structure and function of the human auditory system, describe how the brain is able to distinguish sounds of different frequencies.

2. Describe the molecular mechanisms whereby a single photon is able to cause hyperpolarization of a rod cell.

3. Compare and contrast the functioning of olfactory receptors and photoreceptors. What is the basis whereby each system discriminates between an apple and an orange?

4. Describe and contrast two sensory systems that enable animals to "see" in the dark. What problems or limitations are inherent in these systems in comparison with vision?

5. Describe what is meant by a receptor potential and how it encodes intensity of stimulus. Use a specific sensor as your example.

Readings

Hubel, D. H. 1988. *Eye, Brain, and Vision.* Scientific American Library Series No. 22. W. H. Freeman, New York. A comprehensive and beautifully illustrated book about the neurophysiology and neuroanatomy of vision. While very readable for the nonexpert, it presents the depth and breadth of knowledge and experience of someone who has been a major contributor to this area of research.

Hudspeth, A. J. 1983. "The Hair Cells of the Inner Ear." *Scientific American*, January. A discussion of the cells in the inner ear that transduce mechanical forces of pressure waves into action potentials transmitted to the brain and how they do it.

Knudsen, E. I. 1981. "The Hearing of the Barn Owl." *Scientific American*, 1981. An explanation of the neurophysiological basis for the extreme accuracy of the barn owl's auditory system in locating prey in complete darkness.

Newman, E. A. and P. H. Hartline. 1982. "The Infrared 'Vision' of Snakes." *Scientific American*, March. An examination of snakes that are able to use infrared radiation emitted by objects in their environment to construct a sensory world.

Purves, D., G. J. Augustine, D. Fitzpatrick, L. C. Katz, A.-S. LaMantia and J. O. McNamara. 1997. *Neuroscience.* Sinauer Associates, Sunderland, MA. Despite the formidable number of authors, this textbook of human neurobiology is delightfully readable and lucid. At just over 550 pages, it is one of the most concise of the thorough treatments of the subject.

Stryer, L. 1987. "The Molecules of Visual Excitation." *Scientific American*, July. A description of the molecular chain of events that transduce photons of light falling on the retina into action potentials that are transmitted to the brain.

Suga, N. 1990. "Biosonar and Neural Computation in Bats." *Scientific American*, June. An exploration into the complex neural processing of information that enables bats to construct a view of their environment using their auditory systems.

Chapter 43

The Mammalian Nervous System: Structure and Higher Functions

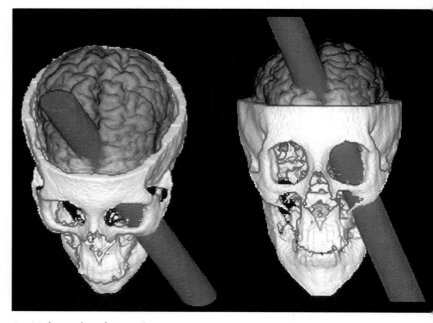

An Unintentional Experiment
In a nineteenth-century railroad construction accident, an explosion blew a tamping rod through the brain of Phineas P. Gage. Unbelievably, Gage survived, and the changes observed in his behavior stimulated early theories about complex brain functions.

Phineas Gage was an industrious, responsible, considerate young man. He was 25 years old in 1848 and was a railroad construction foreman. He had the respect of his men, and he looked out for them to the extent that he took on himself the most dangerous tasks associated with blasting the rocks in the path of the railroad.

Late one afternoon the last hole had been drilled for the day. Gage poured blasting powder into the hole and began tamping it with a meter-long, 6 kg iron rod. The iron rod hit the side of the hole and struck a spark, which ignited the powder. The explosion shot the tamping iron out of the hole like a bullet. It struck Phineas below his left eye, penetrated his skull, and the 3 cm–wide rod passed through the part of his brain behind his forehead and out the top of his head. Was this the end of Phineas Gage? Yes and no.

Gage regained consciousness within minutes and could speak. He was taken to a hotel where a physician dressed his wounds, but the doctor could do little else. Infections were a problem, but Gage's senses and memory were intact. In 3 weeks he left his bed, but he did not return to his work at the railroad. The recovered Phineas Gage was quarrelsome, bad tempered, lazy, and irresponsible. He was impatient and obstinate, and he used profane language, which he had never done before.

The body of Phineas Gage survived the accident, but he was an entirely different person. He spent the rest of his days as a drifter, earning money by telling his story, exhibiting his scars and his tamping iron. If you are in Cambridge, Massachusetts, you can pay him a visit. His skull, death mask, and tamping iron are on display in the Museum of the Medical College of Harvard University.

The sad story of Phineas Gage reveals that the essence of our individuality and personality resides in our brain. You probably could imagine remaining yourself after replacing any organ of your body except your brain. What is this miraculous organ and what does it do?

The human brain weighs about 1.5 kg, is mostly water, and has the consistency and color of vanilla custard. Yet the complexity of this small mass of tissue exceeds that of any other known matter. The work of the brain is to process and store information and to control the physiology and behavior of the body. The brain is constantly receiving, integrating, and interpreting information from all the senses. To respond to that information, it generates commands to the muscles and organs of the body.

The brain senses the need to act, decides on the appropriate action, orchestrates it, initiates and coordinates it, monitors it, and remembers it. Every second of its life, the brain is processing thousands of bits of information. The unit of function of the brain is the neuron. The human brain consists of about 10 billion neurons, which account for its ability to handle vast amounts of information.

In the previous two chapters we learned about the cellular properties of neurons and neurons in simple circuits: how neurons generate and conduct action potentials, how neurons communicate with each other at synapses, and how information is integrated at synapses. In this chapter we will take on the challenge of understanding functions of the human nervous system in terms of these cellular mechanisms.

We'll begin by learning about the general structure and the distribution of functions in the nervous system, with an emphasis on the central nervous system. Then we'll explore some examples of information processing that are fairly well understood in terms of neuronal circuitry. Finally, we'll discuss some of the complex, higher functions of the nervous system—including learning, memory, language, and consciousness—that cannot yet be explained in terms of circuits.

The Nervous System: Structure, Function, and Information Flow

The human nervous system is more than the brain. The brain and spinal cord are called the **central nervous system** (**CNS**), and information is brought to and from the CNS by means of an enormous network of nerves that make up the **peripheral nervous system** (Figure 43.1*a*). The peripheral nervous system connects to the CNS via spinal nerves and cranial nerves.

It is important to distinguish between the axons of single neurons and the spinal and cranial nerves. A **nerve** is a bundle of axons (Figure 43.1*b*), and it carries information about many things simultaneously. Some axons in a nerve may be carrying information to the CNS while other axons in the same nerve are carrying information from the CNS to the organs of the body. The peripheral nervous system reaches to every tissue of the body.

A *conceptual diagram of the nervous system traces information flow*

The nervous system is a very complex information-processing system that handles many tasks simultaneously. It will help to organize our thinking about the nervous system by beginning with a conceptual diagram of information flow (Figure 43.2). We can then put anatomical reality and functional details into this general conceptual scheme.

Information traveling through the peripheral nervous system to-

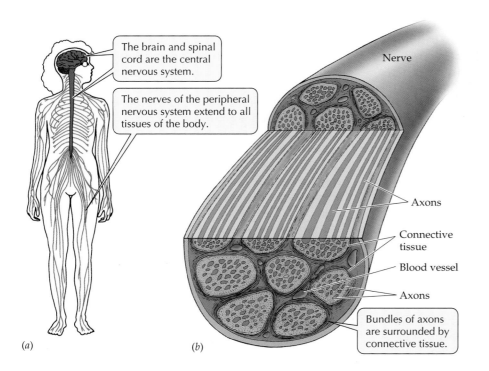

The brain and spinal cord are the central nervous system.

The nerves of the peripheral nervous system extend to all tissues of the body.

Nerve

Axons

Connective tissue

Blood vessel

Axons

Bundles of axons are surrounded by connective tissue.

(a) *(b)*

43.1 Anatomy of the Human Nervous System (*a*) Information is communicated between the central nervous system and the other tissues of the body through the peripheral nervous system. (*b*) A nerve contains the axons of many neurons. Some neurons conduct information to and others from the central nervous system.

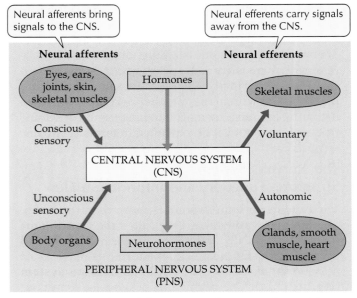

Neural afferents bring signals to the CNS.

Neural efferents carry signals away from the CNS.

Neural afferents | **Neural efferents**

Eyes, ears, joints, skin, skeletal muscles

Hormones

Skeletal muscles

Conscious sensory

Voluntary

CENTRAL NERVOUS SYSTEM (CNS)

Unconscious sensory

Autonomic

Body organs

Neurohormones

Glands, smooth muscle, heart muscle

PERIPHERAL NERVOUS SYSTEM (PNS)

43.2 Organization of the Nervous System The peripheral nervous system carries information both to and from the CNS. The CNS also receives hormonal inputs and produces hormonal outputs.

ward the CNS is **afferent** information. We are consciously aware of much of this information (for example, vision, hearing, temperature, touch, taste, pain, balance, and the position of limbs), but we are not consciously aware of other afferent information that is important for physiological regulation (for example, blood pressure, deep body temperature, and blood oxygen supply).

The **efferent** side of the peripheral nervous system carries information from the CNS to the muscles and glands of the body. Efferent pathways can be divided into a **voluntary** division that executes our conscious movements and an **involuntary**, or **autonomic**, division that controls physiological functions.

In addition to neural information, the CNS receives chemical information in the form of hormones circulating in the blood. Neurohormones released by neurons into the extracellular fluids of the brain can influence other neurons in the brain or can enter the circulation and leave the brain. In Chapter 38 we learned of the important role of neurohormones in the control of the anterior pituitary, as well as the fact that other neurohormones are released from the posterior pituitary into the circulation.

Now we can begin to translate our conceptual scheme of information flow into an anatomical view of the nervous system. It can be difficult to learn the relationships between the different structures of the adult nervous system, but the task is much easier if we begin with the development of the nervous system from a simple tubular structure in the embryo.

The human CNS develops from the embryonic neural tube

Early in the development of all vertebrate embryos, a hollow tube of neural tissue forms (see Chapter 40). The tube runs the length of the embryo on its dorsal side. At the head end of the embryo, this neural tube forms three swellings that become the basic divisions of the brain: the **hindbrain**, the **midbrain**, and the **forebrain**. The rest of the neural tube becomes the spinal cord (Figure 43.3). The cranial and spinal nerves, which are the peripheral nervous system, sprout from the neural tube and grow throughout the embryo.

Each of the three regions in the embryonic brain develops into several structures in the adult brain. From the hindbrain come the **medulla**, the **pons**, and the **cerebellum**. The medulla is continuous with the spinal cord. The pons is in front of the medulla, and the cerebellum is a dorsal outgrowth of the pons. The medulla and pons contain distinct groups of neurons (called *nuclei*) that are involved in the control of physiological functions such as breathing and circulation or basic motor patterns such as swallowing and vomiting. All neural information traveling between the spinal cord and higher brain areas must pass through the pons and the medulla.

The cerebellum orchestrates and refines behavior patterns. The cerebellum is like the conductor of an orchestra; it receives "copies" of the commands going to the muscles from higher brain areas, and it receives information coming up the spinal cord from the joints and muscles. Thus it can compare the motor "score" with the actual behavior of the muscles and refine the motor commands.

From the embryonic midbrain come structures that process aspects of visual and auditory information. In addition, all information between higher brain areas and the spinal cord must pass through the midbrain.

The embryonic forebrain develops a central region called the **diencephalon** and surrounding structures called the **telencephalon**. The telencephalon consists of two **cerebral hemispheres** (also referred to as the **cerebrum**). In humans the telencephalon is by far the largest part of the brain, and it plays major roles in sensory perception, learning, memory, and conscious behavior. The diencephalon is the core of the forebrain and consists of an upper structure called the **thalamus** and a lower structure called the **hypothalamus**. The thalamus is the final relay station for sensory information going to the telencephalon, and the hypothalamus is responsible for the regulation of many physiological functions and biological drives (see Chapters 37 and 38).

The important thing to remember from this brief consideration of the development of the nervous system is that it has a linear organization. Understanding the relationships among the many structures of the complex adult brain is a little easier if you keep this linear organization of the neural axis in mind. Communication between the spinal cord and the telen-

cephalon travels through the medulla, pons, midbrain, and diencephalon. These structures are referred to collectively as the **brain stem**. In general, more primitive and autonomic functions are localized farther down this neural axis, and more complex and evolutionarily advanced functions are found higher on the axis.

As we go up the vertebrate phylogenetic scale from fish to mammals, the telencephalon increases in size, complexity, and importance. The forebrain dominates the nervous systems of mammals, and damage to this region results in severe impairment or even coma. In contrast, a shark with its telencephalon removed can swim almost normally.

Functional Subsystems of the Nervous System

When we talk about development of the nervous system, we describe it in terms of anatomically distinct structures. However, the nervous system is always engaged in many tasks at the same time. This is referred to as parallel processing of information. Any one task usually involves many different anatomical structures. Understanding the nervous system is made simpler if we recognize functional subsystems, such as the spinal cord, reticular system, limbic system, and cerebrum. Any one anatomical structure may play roles in several functional subsystems.

The spinal cord receives and processes information from the body

The spinal cord conducts the bidirectional information flow between the brain and the organs of the body. It also integrates a great deal of the information coming from the peripheral nervous system and responds to that information by issuing motor commands. The conversion of afferent to efferent information in the spinal cord without participation of the brain is called a spinal reflex. We described the simplest spinal reflex circuits in Chapter 41, but will review and extend that information here.

A cross section of the spinal cord reveals a central area of gray matter in the shape of a butterfly that is surrounded by an area of white matter (Figure 43.4). The gray matter contains the cell bodies of the spinal neurons, and the white matter contains the axons that conduct information up and down the spinal cord. Spinal nerves leave the spinal cord at regular intervals on each side. Each spinal nerve has two roots connecting to the gray area—one connecting with the **dorsal horn**, the other with the **ventral horn**. Each spinal nerve carries both afferent information from the sense organs and efferent information to the muscles and glands of the body. The afferent nerves enter the spinal cord through the dorsal roots, and the efferent nerves leave the spinal cord through the ventral roots.

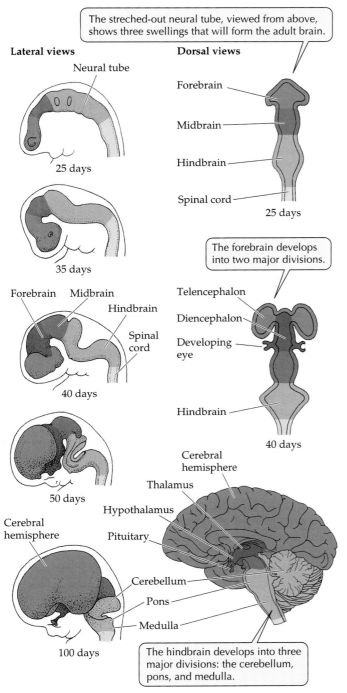

43.3 The Human Nervous System Develops Three swellings at the anterior end of the hollow neural tube in the early embryo develop into the parts of the adult brain. The final view is an adult brain section cut in half through the midline.

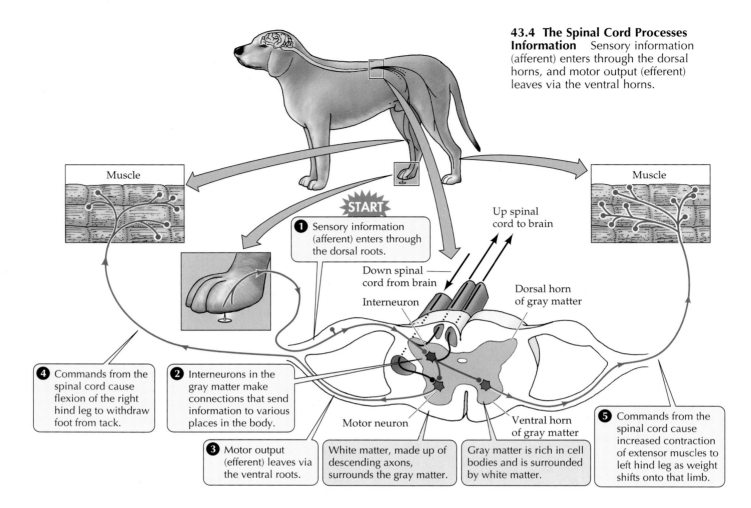

43.4 The Spinal Cord Processes Information Sensory information (afferent) enters through the dorsal horns, and motor output (efferent) leaves via the ventral horns.

Muscle

Muscle

START

Up spinal cord to brain

1 Sensory information (afferent) enters through the dorsal roots.

Down spinal cord from brain

Interneuron

Dorsal horn of gray matter

4 Commands from the spinal cord cause flexion of the right hind leg to withdraw foot from tack.

2 Interneurons in the gray matter make connections that send information to various places in the body.

Motor neuron

Ventral horn of gray matter

5 Commands from the spinal cord cause increased contraction of extensor muscles to left hind leg as weight shifts onto that limb.

3 Motor output (efferent) leaves via the ventral roots.

White matter, made up of descending axons, surrounds the gray matter.

Gray matter is rich in cell bodies and is surrounded by white matter.

Information entering the dorsal horn can be transmitted to neurons that will carry it to the brain, or it can be relayed to efferent neurons or to other neurons, called **interneurons**, that reside entirely in the gray matter of the spinal cord. Interneurons make connections with efferent neurons in the ventral horns, and they communicate with other spinal neurons up and down the spinal axis to amplify the response and to generate more complex motor patterns.

The spinal cord processes and integrates a lot of information. For example, when you step on a tack, spinal circuits coordinate the rapid pulling-back movements that are carried out by many muscles on both sides of your body. Spinal circuits can also generate repetitive motor patterns such as those of walking.

The reticular system alerts the forebrain

The **reticular system** of the brain stem is a highly complex network of neuronal fibers that includes many discrete groups of neurons. Such a group of neurons that is anatomically or neurochemically distinct is called a **nucleus**.

The reticular system is distributed through the core of the medulla, pons, and midbrain (Figure 43.5).

Afferent information coming up the neural axis passes through the reticular system, where many connections are made to various neurons involved in controlling many functions of the body. For example, information from joints and muscles is directed to nuclei in the pons and cerebellum that are involved in balance and coordination, and information from pain receptors is directed to nuclei in the brain stem that control sensitivity to pain. In addition, this information continues up the neural axis to the forebrain, where it results in conscious sensations that can be localized to the specific sites in the body where the information originated.

The information routed through the reticular system also influences the level of arousal of the nervous system. Nuclei in the reticular system are involved in the control of sleep and wakefulness. High levels of activity in the reticular system influence these nuclei to maintain the brain in a waking condition; low levels of activity enable sleep. Because of the alerting function of the reticular core of the brain stem, it has been called the reticular activating system.

If the brain stem of a person is damaged at midbrain or higher levels, and thus the alerting action of

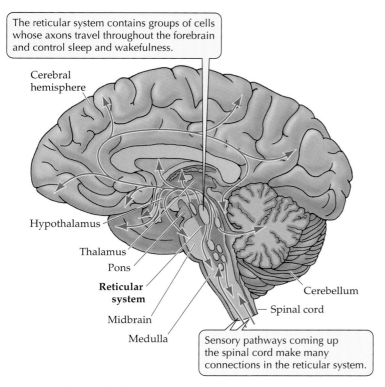

The reticular system contains groups of cells whose axons travel throughout the forebrain and control sleep and wakefulness.

Cerebral hemisphere

Hypothalamus

Thalamus

Pons

Reticular system

Midbrain

Medulla

Cerebellum

Spinal cord

Sensory pathways coming up the spinal cord make many connections in the reticular system.

43.5 The Reticular System Neuronal activity within the reticular network controls levels of alertness and sleep.

the reticular system cannot reach the forebrain, the person loses the ability to be in a conscious, waking state and becomes comatose. Damage in the neural axis below the reticular system does not interfere with the ascending alerting actions of the reticular system and leaves the person with normal patterns of sleep and wakefulness, even though the damage causes lack of sensation (paresthesia) and loss of motor function (paralysis).

The limbic system supports basic functions of the forebrain

The telencephalon of fishes, amphibians, and reptiles consists of only a few structures surrounding the diencephalon. In birds and mammals, these primitive forebrain structures are completely covered by the evolutionarily more recent elaborations of the telencephalon called the **neocortex**. The primitive parts of the forebrain still have important functions in birds and mammals, and are referred to as the **limbic system** (Figure 43.6).

The limbic system is responsible for basic physiological drives, instincts, and emotions. Within the limbic system are areas that when stimulated with small electric currents can cause intense sensations of pleasure, pain, or rage. If a rat is given the opportunity to stimulate its own pleasure centers by pressing a switch, it will ignore food, water, and even sex, push-

ing the switch until it is exhausted. Pleasure and pain centers in the limbic system are believed to play roles in learning and in physiological drives.

One part of the limbic system, the **hippocampus**, is necessary in humans for the transfer of short-term memory to long-term memory. If you are told a new telephone number, you may be able to hold it in short-term memory for a few minutes, but within half an hour it is forgotten unless you make a real effort to remember it. The phenomenon of remembering something for more than a few minutes requires the transfer from short-term to long-term memory.

Regions of the cerebrum interact for consciousness and control of behavior

The cerebral hemispheres are the dominant structures in the mammalian brain. In humans they are so large that they cover all other parts of the brain except the cerebellum (Figure 43.7). A sheet of gray matter (tissue rich in neuronal cell bodies) called the **cerebral cortex** covers each cerebral hemisphere. The cortex is about 4 mm thick and covers a total surface area over both hemispheres of 1 m^2.

The cerebral cortex is convoluted or folded into ridges called **gyri** and valleys called **sulci** so that it fits into the skull. Under the cerebral cortex is white matter, made up of the axons that connect the cell bodies in the cortex with each other and with other areas of the brain.

Regions of the cerebral hemispheres have specific functions. Some of those functions are easily defined,

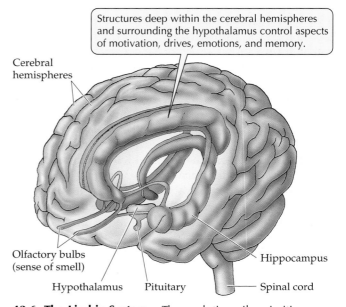

Structures deep within the cerebral hemispheres and surrounding the hypothalamus control aspects of motivation, drives, emotions, and memory.

Cerebral hemispheres

Olfactory bulbs (sense of smell)

Hypothalamus

Pituitary

Hippocampus

Spinal cord

43.6 The Limbic System The evolutionarily primitive parts of the forebrain are referred to as the limbic system (shown in blue). Pleasure and pain centers in the limbic system are believed to play roles in learning and in physiological drives.

(a)

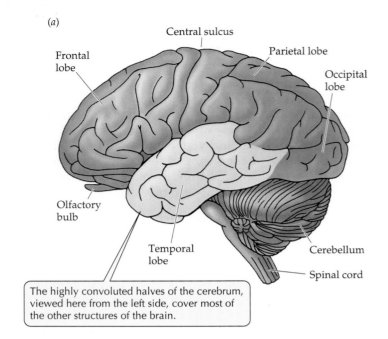

Central sulcus

Frontal lobe

Parietal lobe

Occipital lobe

Olfactory bulb

Temporal lobe

Cerebellum

Spinal cord

The highly convoluted halves of the cerebrum, viewed here from the left side, cover most of the other structures of the brain.

(b)

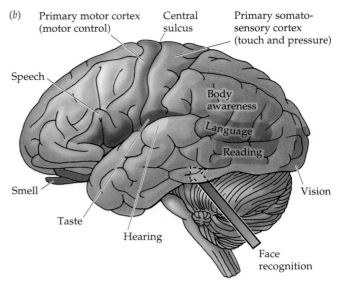

Primary motor cortex (motor control)

Central sulcus

Primary somato-sensory cortex (touch and pressure)

Speech

Body awareness

Language

Reading

Smell

Vision

Taste

Hearing

Face recognition

43.7 The Human Cerebrum (a) Each cerebral hemisphere is divided into four lobes. (b) Different functions are localized in particular areas of the cerebral lobes.

such as receiving and processing sensory information, but most of the cortex is involved in higher-order information processing that is less easy to define. These areas are given the general name of **association cortex**, but their functions differ regionally.

To understand the cerebral cortex, it helps to have an anatomical road map. As viewed from the left side, a left cerebral hemisphere looks like a boxing glove for the right hand with the fingers pointing forward, the thumb pointing out, and the wrist at the rear (see Figure 43.7a). The "thumb" area is the **temporal lobe**, the fingers the **frontal lobe**, the back of the hand the **parietal lobe**, and the wrist the **occipital lobe**. A mirror image of this arrangement characterizes the right cerebral hemisphere. Let's look at each lobe of the cerebrum separately.

THE TEMPORAL LOBE. The upper region of the temporal lobe receives and processes auditory information. The association areas of the temporal lobe are involved in the recognition, identification, and naming of objects. Damage to the temporal lobe results in disorders called **agnosias** in which the individual is aware of a stimulus but cannot identify it. Some of these deficits can be quite specific.

For example, damage to one area of the temporal lobe results in the inability to recognize faces. Even old aquaintances cannot be identified by facial features, but may be identified by other attributes such as voice, body features, and characteristic style of walking. Using monkeys, it has been possible to record from neurons in this region that respond selectively to faces

in general. These neurons do not respond to other stimuli in the visual field, and their responsiveness decreases if some of the features of the face are missing or appear in inappropriate locations (Figure 43.8). Other association areas of the temporal lobes are involved in deficits in understanding spoken language even though speaking, reading, and writing abilities may be intact.

THE OCCIPITAL LOBE. The occipital lobes receive and process visual information. We'll learn more about the details of that process later in this chapter. The association areas of the occipital cortex are essential for making sense out of the visual world and translating visual experience into language. Some deficits resulting from damage to association areas of the occipital cortex are quite specific. For example, in one case a woman with limited damage was unable to see motion. Her vision was intact, but she could see a waterfall only as a still image, and a car approaching only as a series of scenes of a stationary object at different distances.

THE PARIETAL LOBE. The frontal and parietal lobes are separated by a deep valley called the **central sulcus**. The strip of parietal-lobe cortex just behind the central sulcus is the **primary somatosensory cortex** (see Figure 43.7b). This area receives information through the thalamus about touch and pressure sensations.

The whole body surface is represented in this strip of cortex—the head at the bottom and the legs at the top (Figure 43.9). Areas of the body that have lots of

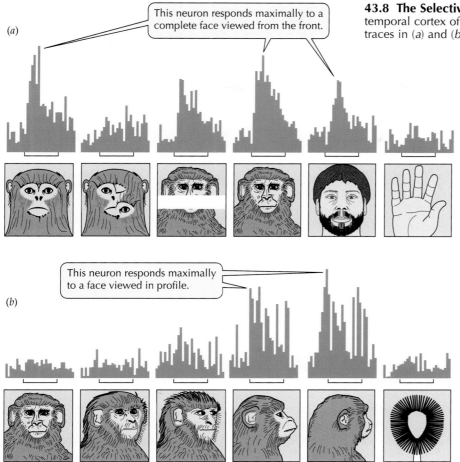

43.8 The Selective Response of Neurons Neurons in the temporal cortex of monkeys respond selectively to faces. The traces in (a) and (b) represent two different neurons.

THE FRONTAL LOBE. A strip of the frontal lobe cortex just in front of the central sulcus is called the **primary motor cortex** (see Figure 43.7b). The cells in this region have axons that extend to muscles in specific parts of the body. As with the primary somatosensory cortex, parts of the body map onto the primary motor cortex—the head region on the lower side, and the lower part of the body at the top; areas with fine motor control have the greatest representation (see Figure 43.9). If a very small region of the frontal lobe is electrically stimulated, the response is the twitch of a muscle, but not a coordinated, complex behavior.

The association functions of the frontal cortex are diverse and are best described as having to do with planning. The story at the beginning of this chapter about Phineas Gage is an example of deficits of the association areas of the frontal cortex. People with such deficits have drastic alterations of personality because they cannot create an accurate view of themselves in the context of the world around them, and they cannot plan for future events.

sensory neurons and are capable of making fine distinctions in touch (such as the lips and the fingers) have disproportionately large representation. If a very small area of the somatosensory cortex is stimulated electrically, the subject reports feeling specific sensations, such as touch, from a very localized part of the body.

A major association function of the parietal cortex is attending to complex stimuli. Damage to the right parietal lobe causes a condition called **contralateral neglect syndrome**, in which the individual tends to ignore stimuli from the left side of the body or the left visual field. Such individuals have difficulty performing complex tasks such as dressing the left side of the body, and an afflicted man may not be able to shave the left side of his face. When asked to copy simple drawings, a person who exhibits this syndrome can do well with the right side of the drawing, but not the left (Figure 43.10). The parietal cortex is not symmetrical with respect to its role in attention. Damage to the left parietal cortex does not cause neglect of the right side of the body. We will see similar asymmetries in cortical function when we discuss language.

As mentioned earlier, the size of the telencephalon relative to the rest of the brain increases substantially as we go from fish to amphibian, to reptile, to birds and mammals. Even when we consider only mammals, the cerebral cortex increases in size and complexity when we compare animals such as rodents, whose behavioral repertoires are relatively simple, with animals such as primates that have much more complex behavior.

The most dramatic increase in the size of the cerebral cortex took place during the last several million years of human evolution. The incredible intellectual capacities of *Homo sapiens* are associated with enlargement of the cerebral cortex. Humans do not have the largest brains in the animal kingdom; elephants, whales, and porpoises have larger brains in terms of mass. If we compare brain size to body size, however, humans and dolphins top the list. Humans have the

43.9 The Body Is Represented in the Primary Somatosensory Cortex and the Primary Motor Cortex Cross sections through the primary somatosensory and primary motor cortexes can be represented as maps of the human body. Body parts are shown in relation to the brain area devoted to them.

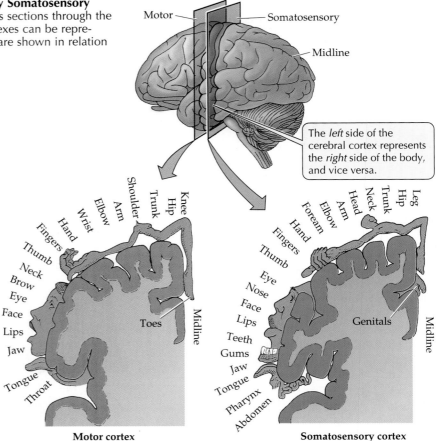

The *left* side of the cerebral cortex represents the *right* side of the body, and vice versa.

Motor cortex

Somatosensory cortex

largest ratio of brain size to body size, and they have the most highly developed cerebral cortex. Another feature of the cerebral cortex that reflects increasing behavioral and intellectual capabilities is the ratio of association cortex to primary sensory and motor cortices. Humans have the largest relative amount of association cortex.

Information Processing by Neuronal Circuits

In Chapter 41 we learned how neurons interact to process information. A goal of neurobiology is to understand complex functions of the nervous system in terms of the properties of neurons and the synapses between them. We will use two examples to show how the functions of subsystems can be understood in terms of neuronal circuits.

The first example, the autonomic nervous system, consists of efferent pathways; the second, the visual system, consists of afferent and integrative pathways. Techniques that have allowed neurobiologists to trace neuronal connections, chemically characterize synapses, and record the activities of single cells and groups of cells have advanced our understanding of how certain subsystems of the nervous system work.

The autonomic nervous system controls organs and organ systems

The autonomic nervous system is divided into two divisions: **sympathetic** and **parasympathetic**. These two divisions work in opposition to each other in their effects on most organs, one causing an increase in activity and the other causing a decrease. The most commonly known functions of the autonomic nervous system are those of the sympathetic division called the "fight-or-flight" mechanisms that increase heart rate, blood pressure, and cardiac output and prepare the body for emergencies. In contrast, the parasympathetic division slows the heart and lowers blood pressure.

It is tempting to think of the sympathetic division as the one that speeds things up and the parasympathetic division as the one that slows things down, but

that is not always a correct distinction. The sympathetic division slows the digestive system and the parasympathetic division accelerates it. The two divisions of the autonomic nervous system are easily distinguished from each other by their anatomy, their neurotransmitters, and their actions (Figure 43.11).

Both divisions of the autonomic nervous system are efferent pathways of the CNS. Each autonomic efferent begins with a neuron that has its cell body in the brain stem or spinal cord and uses acetylcholine as its neurotransmitter. These cells are called **preganglionic neu-**

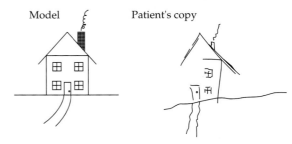

Model   Patient's copy

43.10 Contralateral Neglect Syndrome A person with damage to the right parietal association cortex will neglect the left side of a drawing when asked to copy a model.

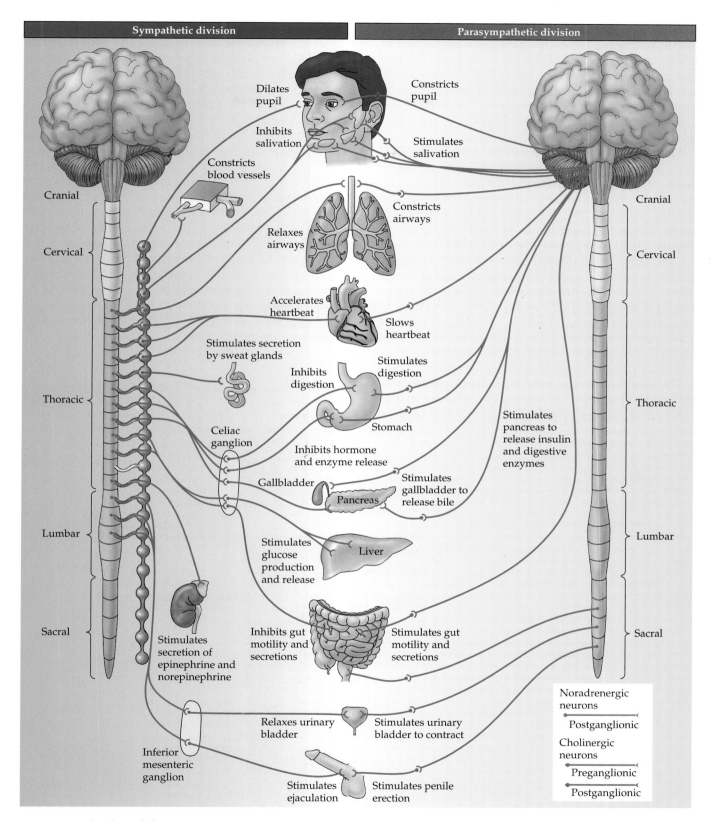

Sympathetic division

Parasympathetic division

Cranial

Cervical

Thoracic

Lumbar

Sacral

Cranial

Cervical

Thoracic

Lumbar

Sacral

Dilates pupil

Constricts pupil

Inhibits salivation

Stimulates salivation

Constricts blood vessels

Constricts airways

Relaxes airways

Accelerates heartbeat

Slows heartbeat

Stimulates secretion by sweat glands

Inhibits digestion

Stimulates digestion

Stomach

Celiac ganglion

Inhibits hormone and enzyme release

Gallbladder

Pancreas

Stimulates gallbladder to release bile

Stimulates pancreas to release insulin and digestive enzymes

Stimulates glucose production and release

Liver

Stimulates secretion of epinephrine and norepinephrine

Inhibits gut motility and secretions

Stimulates gut motility and secretions

Relaxes urinary bladder

Stimulates urinary bladder to contract

Inferior mesenteric ganglion

Stimulates ejaculation

Stimulates penile erection

Noradrenergic neurons
  Postganglionic
Cholinergic neurons
  Preganglionic
  Postganglionic

43.11 Organization of the Autonomic Nervous System
The autonomic nervous system is divided into the sympathetic and parasympathetic divisions, which work in opposition to each other in their effects on most organs (one causing an increase and the other a decrease in activity).

rons because the second neuron in each autonomic output pathway resides in a ganglion (a collection of neuron cell bodies) that is outside of the CNS. The second neuron in an output pathway of the autonomic nervous system is a **postganglionic neuron** because its axon extends out from the ganglion. The axons of the postganglionic neurons end on the cells of the target organs.

The postganglionic neurons of the sympathetic division use norepinephrine as their neurotransmitter; those of the parasympathetic division use acetylcholine. In organs that receive both sympathetic and parasympathetic input, the target cells respond in opposite ways to norepinephrine and acetylcholine (Figure 43.12). A region of the heart called the pacemaker, which generates the heartbeat, is an example. Norepinephrine increases the firing rate of pacemaker cells and causes the heart to beat faster. Acetylcholine decreases the firing rate of cardiac pacemaker cells and causes the heart to beat slower. In the digestive tract norepinephrine causes muscle cells to hyperpolarize, which slows digestion; acetylcholine depolarizes muscle cells in the gut, which accelerates digestion.

The sympathetic and parasympathetic divisions of the autonomic nervous system can also be distinguished by anatomy (see Figure 43.11). The pregan-

glionic neurons of the parasympathetic division come from the brain stem and the last segment of the spinal cord. The preganglionic neurons of the sympathetic division come from the upper regions of the spinal cord below the neck—the thoracic and lumbar regions. The ganglia of the sympathetic nervous system are mostly lined up in two chains, one on either side of the spinal cord. The parasympathetic ganglia are close to, sometimes sitting on, the target organs.

The autonomic nervous system is an important link between the CNS and many physiological functions of the body. It is crucial to the maintenance of homeostasis through the control of diverse organs and tissues. Yet in spite of its complexity, work by neurobiologists and physiologists over many decades has made it possible to understand its functions in terms of neuronal properties and circuits. For example, in Chapter 46 we will see how information from pressure sensors in the blood vessels is transmitted to the CNS, where it produces signals in autonomic pathways that control the rate at which the heart beats.

43.12 Autonomic Postganglionic Neurotransmitters Elicit Different Responses Many tissues that are innervated by both sympathetic and parasympathetic divisions of the autonomic nervous system respond differently to their postganglionic neurotransmitters.

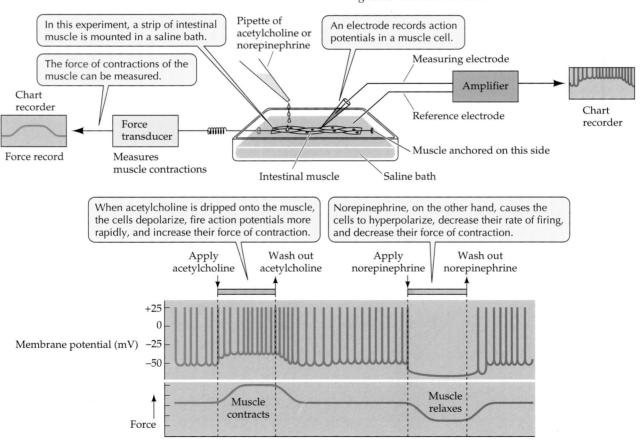

Neurons and circuits in the occipital cortex integrate visual information

In Chapter 42 we learned that the information conveyed to the brain in the optic nerve consists of action potentials that are stimulated by light falling on small circular areas of the retina called receptive fields. A receptive field contains many photosensors connected together in a circuit in such a way that the signals they produce are integrated and transmitted to the brain by a single retinal ganglion cell. The axon of each ganglion cell travels to the brain in the optic nerve. The question is, How does the brain construct the visual world from information about circular patches of light falling on the retina?

Information from the retina is transmitted through the optic nerves to a relay station in the thalamus, and then to the brain's visual-processing area, in the occipital cortex at the back of the cerebral hemispheres (see Figure 43.7b). David Hubel and Torsten Wiesel of Harvard University recorded the activities of single cells in the visual-processing areas of the brains of living animals while they stimulated the animals' retinas with spots and bars of light. They found that cells in the visual cortex, like ganglion cells, have receptive fields—specific areas of the retina that when stimulated by light influence the rate at which the cortical cells fire action potentials.

Cells in the visual cortex have receptive fields that differ from the simple circular receptive fields of retinal ganglion cells. Cortical cells called **simple cells** are maximally stimulated by bars of light that have specific orientations. So, simple cells probably receive input from several ganglion cells whose circular receptive fields are lined up in a row.

Complex cells are also maximally stimulated by a bar of light with a particular orientation, but the bar may fall anywhere on a large area of retina described as that cell's receptive field. Complex cells seem to receive input from several simple cells that share a certain stimu-lus orientation but that have receptive fields in different places on the retina (Figure 43.13). Some complex cells respond best when the bar of light moves in a particular direction.

The concept that emerges from these experiments is that the brain assembles a mental image of the visual world by analyzing edges of patterns of light falling on the retina. This analysis is conducted in a massively parallel fashion. Each retina sends 1 million axons to the brain, but there are at least 200 million neurons in the visual cortex. Each bit of information from a retinal ganglion cell is received by hundreds of cortical cells, each responsive to a different combination of orientation, position, and even movement of contrasting lines in the pattern of light falling on the retina.

Cortical cells receive input from both eyes

How do we see objects in three dimensions? The quick answer is that our two eyes see overlapping, yet slightly different, fields; that is, we have *binocular vision*. Turn a typical conical flowerpot upside down and look down at it so that the bottom of the pot is exactly in the center of your overall field of vision. You see the bottom of the pot, and you see equal amounts of the sides and rim of the pot as concentric circles around the bottom. Now if you close your left eye, you see more of the right side and right rim of the pot. With your right eye closed you see more of the left side and left rim of the pot.

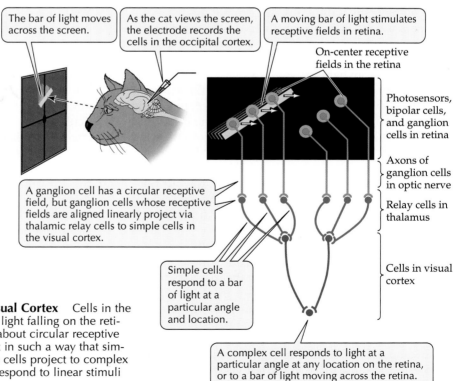

The bar of light moves across the screen.

As the cat views the screen, the electrode records the cells in the occipital cortex.

A moving bar of light stimulates receptive fields in retina.

On-center receptive fields in the retina

Photosensors, bipolar cells, and ganglion cells in retina

Axons of ganglion cells in optic nerve

Relay cells in thalamus

Cells in visual cortex

A ganglion cell has a circular receptive field, but ganglion cells whose receptive fields are aligned linearly project via thalamic relay cells to simple cells in the visual cortex.

Simple cells respond to a bar of light at a particular angle and location.

A complex cell responds to light at a particular angle at any location on the retina, or to a bar of light moving across the retina.

43.13 Receptive Fields of Cells in the Visual Cortex   Cells in the visual cortex respond to specific patterns of light falling on the retina. Ganglion cells that project information about circular receptive fields converge on simple cells in the cortex in such a way that simple cells have linear receptive fields. Simple cells project to complex cells in such a way that complex cells can respond to linear stimuli falling on different areas of the retina.

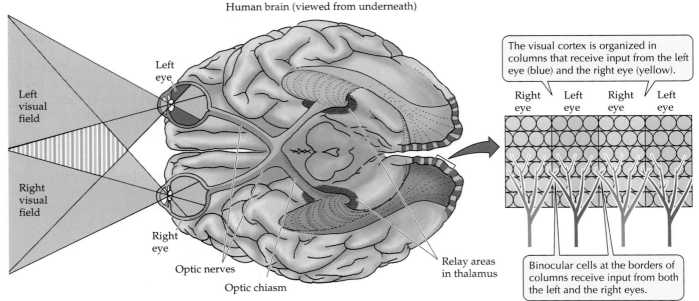

Human brain (viewed from underneath)

The visual cortex is organized in columns that receive input from the left eye (blue) and the right eye (yellow).

Binocular cells at the borders of columns receive input from both the left and the right eyes.

43.14 The Anatomy of Binocular Vision   Each eye transmits information to both sides of the brain; however, the right brain processes all information from the left visual field, and the left brain processes all information from the right visual field. The visual cortex sorts visual field information according to whether it comes from the right eye or the left eye.

The discrepancies in the information coming from your two eyes are interpreted by the brain to provide information about the depth and the three-dimensional shape of the flowerpot. If you are blind in one eye, you have great difficulty discriminating distances. Animals whose eyes are on the sides of their heads can have nonoverlapping fields of vision and, as a result, poor depth vision, but they can see predators creeping up from behind.

The story of how the brain integrates information from two eyes begins with the paths of the optic nerves. If you look at the underside of the brain, the optic nerves from the two eyes appear to join together just under the hypothalamus and then separate again. The place where they join is called the **optic chiasm**. Axons from the half of each retina closest to your nose cross in the optic chiasm and go to the opposite side of your brain. The axons from the other half of each retina go to the same side of the brain.

The result of this division of axons in the optic chiasm is that all visual information from your left visual field (everything left of straight ahead) goes to the right side of your brain, as shown in red in Figure 43.14. All visual information from your right visual field goes to the left side of your brain, as indicated in green in the figure. Both eyes transmit information about a specific spot in your right visual field, for example, to the same place in the left visual cortex. How are the two sources of information integrated?

Cells in the visual cortex are organized in columns. These columns alternate: left eye, right eye, left eye, right eye, and so on. Cells closest to the border between two columns receive input from both eyes and are therefore **binocular cells**. Binocular cells interpret distance by measuring the disparity between where the same stimulus falls on the two retinas.

What is disparity? Hold your finger out in front of you and look at it, closing one eye and then the other. Your finger appears to jump back and forth because its image falls on a different position on each retina. Repeat the exercise with an object at a distance. It doesn't appear to jump back and forth as much because there is less disparity in the positions of the image on the two retinas. Certain binocular cells respond optimally to a stimulus falling on both retinas with a particular disparity. Which set of binocular cells is stimulated depends on how far away the stimulus is.

When we look at something, we can detect its shape, color, depth, and movement. Where does all this information come together? Is there a single cell that fires only when a red sports car drives by? Probably not. A specific visual experience comes from simultaneous activity in a large collection of cells. In addition, most visual experiences are enhanced by information from the other senses and from memory as well. This realization helps explain why about 75 percent of the cerebral cortex is association cortex.

Understanding Higher Brain Functions in Cellular Terms

Very few functions of the nervous system have been worked out to the point of identifying the underlying

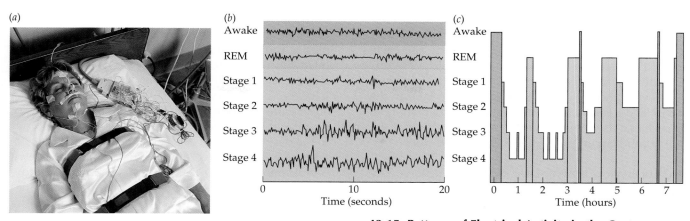

43.15 Patterns of Electrical Activity in the Cortex Characterize Stages of Sleep
(a) Electrical activity of the cerebral cortex is detected by electrodes placed on the scalp and recorded on moving chart paper by a polygraph. (b) The resulting record is the electroencephalogram (EEG). (c) During a night we cycle through the different stages of sleep.

neuronal circuits. The processes responsible for higher brain functions of dreaming, learning, memory, language use, and consciousness discussed in the remaining pages are undeniably complex. Nevertheless, neurobiologists using a wide range of techniques are making considerable progress in understanding some of the mechanisms involved in such higher functions of the nervous system. The following discussion presents several complex aspects of brain and behavior that present challenges. Neurobiologists want to learn how neurons interacting in circuits produce these phenomena.

Sleeping and dreaming involve electrical modulations in the cerebrum

A dominant feature of our behavior is the daily cycle of being asleep and awake. All birds and mammals, and probably all other vertebrates, sleep. We spend one-third of our lives sleeping, yet we do not know why or how. However, we do know that we need to sleep.

Loss of sleep impairs alertness and performance. Most people in our society—certainly most college students—are chronically sleep-deprived. Large numbers of accidents and serious mistakes that endanger lives can be attributed to impaired alertness due to sleep loss. Yet insomnia (difficulty in falling or staying asleep) is one of the most common medical complaints. To discover ways of dealing with these problems, it is important to learn more about the neural control of sleep.

THE ELECTROENCEPHALOGRAM. The classic tool of sleep researchers is the **electroencephalogram** (EEG). To record an EEG, the researcher or the clinician places electrodes on the scalp and uses electronic devices to record electric-potential differences between any two electrodes. These electric-potential differences reflect the electrical activity of the neurons in the brain regions under the electrodes, primarily regions of the cerebral cortex. These electric signals are amplified and used to deflect pens that write on a moving chart

(Figure 43.15a and b). The electrical activity of one or more skeletal muscles is recorded similarly on the chart, and these lines are called an **electromyogram** (EMG).

EEG and EMG patterns reveal the transition from being awake to being asleep; they also reveal that there are different states of sleep. In mammals other than humans, two major sleep states are easily distinguished. They are called **slow-wave sleep** and **rapid-eye-movement sleep** (**REM sleep**). In humans we characterize sleep as REM sleep and **non-REM sleep**.

Human non-REM sleep is divided into four stages, and only the two deepest stages are considered true slow-wave sleep. When a person falls asleep at night, the sleep state entered is non-REM sleep and it progresses from stage 1 to stage 4. Stages 3 and 4 are deep, restorative, slow-wave sleep. After this first episode of non-REM sleep follows an episode of REM sleep. Throughout the night, we have four or five cycles of non-REM/REM sleep (Figure 43.15c). About 80 percent of our sleep is non-REM sleep and 20 percent REM sleep.

We have vivid dreams and nightmares during REM sleep, which gets its name from the jerky movements of the eyeballs that occur during this state. The most remarkable feature of REM sleep is that inhibitory commands from the brain almost completely paralyze the skeletal muscles. Occasional muscle twitches break through the paralysis, as can be seen in a dog that appears to be trying to run in its sleep. If you look closely at a sleeping dog when its legs and paws are twitching, you will be able to see the rapid eye movements as well. Probably the function of muscle paralysis during REM sleep is to prevent the acting out of dreams. Sleepwalking, however, occurs during non-REM sleep.

The EEG characterization of sleep suggests many questions. Why do we have such very different states of sleep? Why does non-REM sleep precede and cycle with REM sleep? Why do we dream in REM sleep? Why do we have deeper non-REM sleep (stages 3 and 4) earlier in the night and more REM sleep later in the night? The answers to these questions are beginning to be revealed as we improve our understanding of the cellular and circuit properties that underlie sleep.

CELLULAR CHANGES DURING SLEEP. There are striking neurophysiological differences between non-REM and REM sleep. Non-REM sleep is characterized by a decrease in the responsiveness of neurons in the thalamus and cerebral cortex. Remember that neurons have a resting membrane potential that is negative (the inside of the cell is negative relative to the outside), and they have a threshold potential for firing action potentials.

Usually the resting potential is below the threshold potential, so the neuron is not firing. When synaptic input causes the cell membrane to become less negative (a depolarizing input), the cell can reach the threshold potential and fire action potentials. When we are awake, several nuclei in the brain stem are continuously active. The axons from these nuclei project widely to the thalamus and the cortex, and the neurotransmitters released by the nerve endings of these axons are generally depolarizing. Therefore, these broadly distributed neurotransmitters keep the neurons of the thalamus and cortex close to threshold and sensitive to inputs.

When we fall asleep, the neuronal activity in these brain stem nuclei decreases, and their nerve endings in the thalamus and cerebral cortex release less transmitter. With the withdrawal of these depolarizing neurotransmitters, the resting potentials of the cells of the thalamus and cerebral cortex become more negative (hyperpolarized) and, therefore, less sensitive to input. Their processing of information is inhibited, and consciousness is lost.

Another interesting event happens as a result of this hyperpolarization. The cells begin to fire in a bursting mode. The synchronization of these bursts over broad areas of cerebral cortex results in the EEG slow waves that characterize non-REM sleep. Studies of neurons of the thalamus and the cortex using intracellular recording techniques have shown that the hyperpolarization during non-REM sleep is due to increased opening of K^+ channels, and the bursting is due to Ca^{2+} channels that rapidly deactivate and require hyperpolarization to be reactivated.

In addition to the several brain stem nuclei that bring on sleep by decreasing their activity and causing general hyperpolarization, it has recently been discovered that another substance that is broadly distributed in the extracellular fluid of the brain has a strong hyperpolarizing influence. This substance is adenosine—part of the molecule ATP, which supplies energy for most cellular processes. When cells cannot maintain an adequate supply of ATP, they release adenosine. It has therefore been suggested that increased release of adenosine in brain signals reflects the depletion of brain energy reserves during waking and contributes to the depth of non-REM sleep because of its hyperpolarizing influence. A corollary of this hypothesis is that one function of non-REM sleep could be to restore brain energy reserves. There are many other molecules in the brain that can influence sleep. Understanding their role in sleep control will provide new information on the possible functions of sleep.

At the transition from non-REM to REM sleep, dramatic changes occur. Some of the brain stem nuclei that were inactive during non-REM sleep again increase their activity, causing a general depolarization of neurons of the cortex. Thus the bursting of these neurons ceases, the slow waves in the EEG disappear, and the EEG resembles that of the awake brain. Because the resting potentials of the neurons return to near threshold levels, the cortex can process information and dreaming occurs.

During REM sleep, however, the brain inhibits both afferent (sensory) and efferent (motor) pathways; therefore the activity in the cortex is unconstrained by its usual sources of information. One example of the effect of this loss of motor output and sensory input is the frequently reported dream in which a person is trying to run but is not able to move. We do not know the function of REM sleep, but since all mammals have approximately the same percentage of total sleep time that is REM sleep, it is probably a rather basic, cellular function.

Some learning and memory can be localized to specific brain areas

Learning is the modification of behavior by experience. *Memory* is the ability of the nervous system to retain what is learned and what is experienced. Even very simple animals can learn and remember, but these two abilities are most highly developed in humans. Language, culture, artistic creativity, and scientific progress are made possible by these abilities.

Consider the amount of information associated with learning a language. The capacity of human memory and the rate at which items can be retrieved are remarkable features of the nervous system. A major challenge in neurobiology is to understand these phenomena in terms of the cells and molecules that make up the brain. Such knowledge could help us find ways to prevent the tragic loss of learning abilities and memory that occur in the common condition of the elderly called **Alzheimer's disease**.

LEARNING.   In Chapter 41 we learned how two simple forms of learning, habituation and sensitization, have been studied in terms of cellular and circuit mechanisms in the marine snail *Aplysia*. Similar synaptic mechanisms likely are involved in habituation and sensitization in other animals, including humans. However, learning that leads to long-term memory and modification of behavior must involve long-lasting synaptic changes.

Such synaptic changes that lasted weeks were discovered in identifiable circuits of the mammalian hippocampus. If one of these circuits was given high-frequency electrical stimulation for a short time, it was subsequently much more sensitive to stimulation. This increased sensitivity lasted a long time and was called **long-term potentiation** (**LTP**). We now know a great deal about its molecular basis. Briefly, the high level of stimulation results in an increased entry of Ca^{2+} ions into the postsynaptic cell. This increase in intracellular Ca^{2+} activates enzymes that cause modifications in the postsynaptic cell that enhance its responsiveness.

In contrast, continuous, repetitive, low-level stimulation of the hippocampal circuit reduces its responsiveness, a phenomenon that has been called **long-term depression** (**LTD**). LTP and LTD have been demonstrated in circuits other than hippocampal circuits, and they may be fundamental cellular or molecular mechanisms involved in learning and memory.

A form of learning that is widespread among animal species is **associative learning**, in which two unrelated stimuli become linked to the same response. The simplest example of associative learning is the **conditioned reflex** discovered by the Russian physiologist Ivan Pavlov. Pavlov was studying the control of digestive functions in dogs and observed that a dog salivates at the sight or smell of food—a simple autonomic reflex. He discovered that if he rang a bell just before the food was presented to the dog, after a few trials the dog would salivate at the sound of the bell, even if no food followed. The salivation reflex was conditioned to be associated with the sound of a bell, which normally is unrelated to feeding.

This simple form of learning, the conditioned reflex, has been studied extensively in efforts to understand its underlying neural mechanisms (Figure 43.16). In a series of studies led by Richard Thompson, the eye-blink reflex of rabbits to puffs of air directed at their eyes was conditioned to be associated with a tone stimulus. After conditioning, the rabbit blinks when presented with just the tone. A small and specific area

43.16 The Conditioned Eye Blink Reflex Depends on a Cerebellar Circuit   Rabbits whose cerebellar nuclei were destroyed surgically did not form the conditioned eye-blink response.

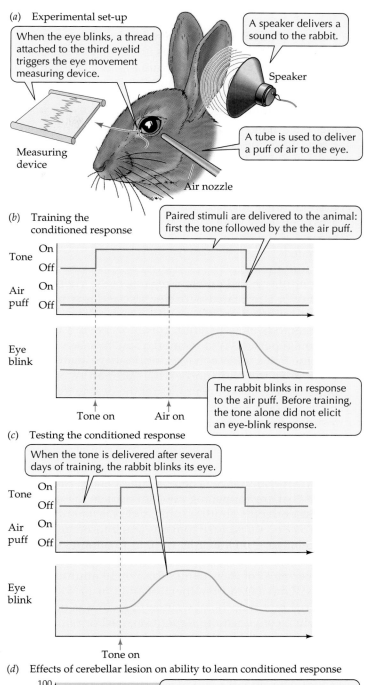

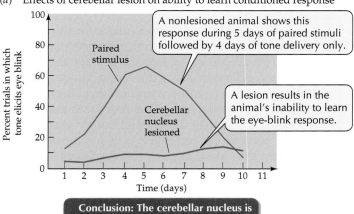

of the cerebellum was discovered to be necessary for this conditioned reflex. Thus, as in the *Aplysia* studies, it was possible to localize learning in an identifiable set of synapses in the mammalian brain.

MEMORY. Attempts to treat human neurological diseases have led to discoveries of the localization of areas of the brain involved in the formation and recall of memories. Epilepsy is a disorder characterized by uncontrollable increases in neural activity in specific parts of the brain. The resulting "epileptic fits" can endanger the afflicted individual. In the past, serious cases of epilepsy were sometimes treated by destroying the part of the brain from which the surge of activity originated.

To find the right area, the surgery was done under local anesthesia and different regions of the brain were electrically stimulated with fine electrodes while the patient reported on the resulting sensations. When some regions of association cortex were stimulated, patients reported vivid memories. Such observations were the first evidence that memories have anatomical locations in the brain and exist as properties of neurons and networks of neurons. Yet the destruction of a small area does not completely erase a memory, so it is postulated that memory is a function distributed over many brain regions and that a memory may be stimulated via many different routes.

You can recognize several forms of memory from your own experience. There is *immediate memory* for events that are happening now. Immediate memory is almost perfectly photographic, but it lasts only seconds. *Short-term memory* contains less information, but it lasts longer—on the order of 10 to 15 minutes. If you are introduced to a group of new people, you may remember most of their names for 5 or 10 minutes, but you will have forgotten them in an hour or so if you have not repeated them, written them down, or used them in a conversation. Repetition, use, or reinforcement by something that gets your attention (such as the title President or Queen) facilitates the transfer of short-term memory to *long-term memory*, which can last for days, months, or years.

Knowledge about neural mechanisms for the transfer of short-term memory to long-term memory has come from observations of patients who have lost parts of the limbic system, notably the hippocampus. A famous case is that of a patient identified as H.M., whose hippocampus on both sides of the brain was removed in an effort to control severe epilepsy. Since that surgery, H.M. has not been able to transfer information to long-term memory. If someone is introduced to him, has a conversation with him, and then leaves the room, when that person returns he or she is unknown to H.M.—it is as if the previous conversation had never taken place. H.M. retains his memories of events that

happened before his surgery, but he remembers post-surgery events for only 10 or 15 minutes.

Memory of people, places, events, and things is called *declarative* memory because you can consciously recall and describe it. Another type of memory, called *procedural*, cannot be consciously recalled and described; it is the memory of how to perform a motor task. When you learn to ride a bicycle, ski, or use a computer keyboard, you form procedural memories. Although H.M. is incapable of learning that involves declarative memory, he is capable of learning that involves procedural memory. When taught a motor task day after day, he cannot recall the lessons of the previous day, yet his performance steadily improves. Thus, procedural learning and memory must involve mechanisms different from those used in declarative learning and memory.

Language abilities are localized mostly in the left cerebral hemisphere

No aspect of brain function is as integrally related to consciousness and intellect as is language. Therefore, studies of the brain mechanisms that underlie the acquisition and use of language are extremely interesting to neuroscientists. A curious fact of language abilities is that they are located mostly in one cerebral hemisphere—which in 97 percent of all people is the left hemisphere. This phenomenon is referred to as the **lateralization** of language functions.

Some of the most fascinating research on this subject was conducted by Roger Sperry and his colleagues at the California Institute of Technology; Sperry received the Nobel prize in medicine for this work. The two cerebral hemispheres are connected by a white-matter tract called the **corpus callosum**. In one severe form of epilepsy, bursts of action potentials travel from hemisphere to hemisphere across the corpus callosum. Cutting the tract eliminates the problem, and patients function normally following the surgery. But experiments revealed interesting deficits in the language abilities of these "split-brain" patients. With the connections between the two hemispheres cut, the knowledge or experience of the right hemisphere could no longer be expressed in language, nor could language be used to communicate with the right hemisphere.

A curious feature of our nervous systems is that the left side of the body is served (in both sensory and motor aspects) mostly by the right side of the brain, and the right side of the body is served mostly by the left side of the brain. Thus, sensory input from the right hand goes to the left cerebral hemisphere, and sensory input from the left hand goes to the right cerebral hemisphere.

If a split-brain patient is blindfolded and a familiar tool is placed in his or her right hand, the patient can identify the tool and describe its use. If the tool is

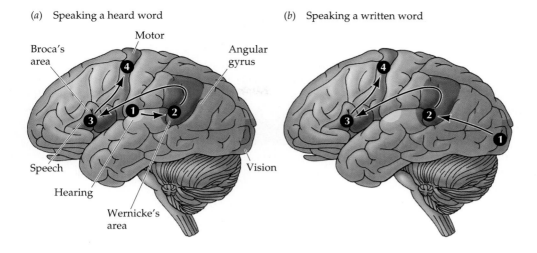

(a) Speaking a heard word

Broca's area

Motor

Speech

Hearing

Wernicke's area

(b) Speaking a written word

Angular gyrus

Vision

43.17 Language Areas of the Cortex Different regions of the left cerebral cortex participate in the processes of (a) repeating a word that is heard versus (b) repeating a written word.

placed in the left hand, however, the patient can use the tool correctly but cannot name it or describe its use. In split-brain individuals, the right hemisphere has lost access to the language abilities that reside predominantly in the left hemisphere.

The brain mechanisms of language in the left hemisphere have been the focus of much research. Again, the experimental subjects are persons who have suffered damage to a region of the left hemisphere and are left with one of many forms of **aphasias**, deficits in the abilities to use or understand words. Language areas of the left hemisphere are shown in Figure 43.17.

Broca's area, located in the frontal lobe just in front of the motor cortex, is essential for speech. Damage to Broca's area results in halting, slow, poorly articulated speech or even complete loss of speech, but the patient can still read and understand language. In the temporal lobe, close its border with the occipital lobe, *Wernicke's area* 'is more involved with sensory than with motor aspects of language. Damage to Wernicke's area can cause a person to lose the ability to speak sensibly while retaining the abilities to form the sounds of normal speech and to imitate its cadence. Moreover, such a patient cannot understand spoken or written language. Near Wernicke's area is the *angular gyrus*, which is believed to be essential for integrating spoken and written language.

Normal language ability depends on the flow of information among various areas of the left cerebral cortex. Input from spoken language travels from the primary auditory cortex to Wernicke's area (see Figure 43.17). Input from reading language travels from the primary visual cortex to the angular gyrus to Wernicke's area. Commands to speak are formulated in Wernicke's area and travel to Broca's area and from there to the primary motor cortex. Damage to any one of those areas or the pathways between them can result in aphasia. Using modern methods of functional brain imaging, it is possible to see the metabolic activ-

ity in different brain areas when the brain is using language (Figure 43.18).

Young children can recover remarkably from even severe damage to the left cerebral hemisphere because lateralization of language abilities to the left hemisphere has not fully developed, and the right hemisphere can take over all of the language-related functions. One such patient in Sperry's split-brain studies produced provocative results with respect to the relationship between language and intellect.

This patient had language functions in both hemispheres. As a result of left-brain damage in childhood, his right hemisphere had developed language functions, and over time his left hemisphere recovered. Then, for treatment of his epilepsy, his corpus callosum was cut. Afterward, it was possible to communicate with this individual through either his left or his right cerebral hemisphere. Each had a separate personality with its individual likes and dislikes. Each responded differently to evaluating events and projecting plans into the future. It was as if there were two persons housed in one brain.

What is consciousness?

This chapter has only scratched the surface of knowledge about the organization and functions of the human brain, but it may give you some idea of the incredible challenge that neurobiologists face in trying to understand their own brains. Progress is being aided by powerful new technologies, such as patch-clamp recording, functional imaging, and neurochemical and molecular methods. However, even this armamentarium of research tools may not enable an answer to the question "What is consciousness?"

If you look at a black dog, you are conscious of the fact that it is a dog, it is black, and it is a Labrador retriever, and you may remember that its name is Hunky-Dory, it is 4 years old, and it belongs to your friend Caterina, and so on. From what you have

43.18 Imaging Techniques Reveal Active Parts of the Brain Positron emission tomography (PET) scanning reveals brain regions that are activated by different aspects of language use.

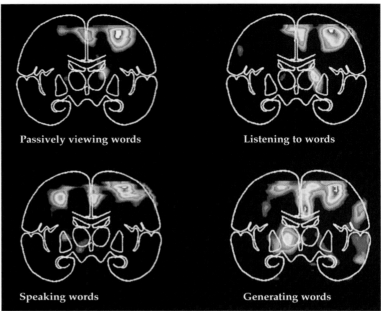

Passively viewing words

Listening to words

Speaking words

Generating words

learned in this chapter, imagine the large set of neurons that would be active during this experience: neurons in the visual system, neurons in language areas, and neurons in different regions of association cortex. But is being conscious of the black dog simply a result of the fact that all of these neurons are firing at the same time? Simultaneously, the brain is processing many other sensory inputs, but we are not necessarily conscious of those inputs. What makes us conscious of the black dog and associated memories and not conscious of other information the brain is processing at the same time?

If we could describe all the neurons and all the synapses involved in the conscious experience of seeing and naming a black dog and then build a computer with devices that modeled all these neurons and connections, would that computer be conscious? It has been said that the question of consciousness resolves into two types of problems: "easy" and "hard." The easy problems deal with all the cellular and systems neurophysiology of information processing that is involved in conscious experience. The implication of "easy" is that we seem to have the tools to solve these kinds of problems, as complex as they may be. The hard problems are explaining how properties of cells and circuits result in consciousness, and we seem to lack the proper tools or concepts even to begin to solve these problems.

Summary of "The Mammalian Nervous System: Structure and Higher Functions"

The Nervous System: Structure, Function, and Information Flow

- The brain and spinal cord are the central nervous system, and the cranial and spinal nerves are the peripheral nervous system. A nerve is a bundle of many axons carrying information to and from the central nervous system. **Review Figure 43.1**
- The nervous system can be modeled conceptually in terms of information flow and whether or not we are conscious of that information. **Review Figure 43.2**
- The vertebrate nervous system develops as a dorsal hollow neural tube, and the brain forms from three swellings at the anterior end of this neural tube, which become the hindbrain, the midbrain, and the forebrain, respectively. **Review Figure 43.3**

- The forebrain develops into the cerebral hemispheres and the underlying thalamus and hypothalamus. The midbrain and hindbrain develop into the brain stem. More primitive and autonomic functions are localized in the brain stem, and conscious experience depends on the cerebrum.

Functional Subsystems of the Nervous System

- The nervous system is composed of many subsystems that function simultaneously. Some important subsystems are the spinal cord, the reticular and limbic systems, and the cerebrum.
- The spinal cord communicates information between the brain and the body. It also processes and integrates much information resulting in spinal reflexes, which can be quite complex. **Review Figure 43.4**
- The reticular system of the brain stem is a complex network of fibers that directs incoming information to appropriate brain stem nuclei that control autonomic functions, as well as transmitting the information to the forebrain to result in conscious sensation. The reticular system controls the level of arousal of the nervous system. **Review Figure 43.5**
- The limbic system is an evolutionarily primitive part of the forebrain that is involved in emotions, physiological drives, instincts, and memory. **Review Figure 43.6**
- The convoluted cerebral hemispheres are the dominant structures of the human brain. The cerebral surfaces consist of a layer of neurons called the cerebral cortex.
- Most of the cerebral hemispheres are involved in higher-order information processing, and these areas are generally called association cortex.
- The cerebral hemispheres can be divided into temporal, frontal, parietal, and occipital lobes. Many motor functions are localized in parts of the frontal lobe, information from many sensors around the body projects to a region of the parietal lobe, visual information projects to the occipital lobe, and auditory information projects to a region of the temporal lobe. **Review Figures 43.7, 43.8, 43.9, 43.10**

Information Processing by Neuronal Circuits

• Functions of the nervous system are beginning to be understood in terms of the properties of cells organized in neuronal circuits.

• The autonomic nervous system consists of efferent pathways that control the organs and organ systems of the body. It is composed of sympathetic and parasympathetic divisions that normally work in opposition to each other. These divisions are characterized by their anatomy, neurotransmitters, and effects on target tissues. **Review Figures 43.11, 43.12**

• Neuronal circuits in the occipital cortex integrate visual information. Receptive-field responses of retinal ganglion cells are communicated to the brain in the optic nerves. This information is projected to the visual cortex in such a way as to create receptive fields for cortical cells.

• A simple cell is stimulated by a bar of light with a specific orientation falling at a specific location on the retina. A complex cell is maximally stimulated by such a stimulus moving across the retina. The visual cortex seems to assemble a mental image of the visual world by analyzing edges of patterns of light. **Review Figure 43.13**

• Binocular vision results from circuits that communicate information from both eyes to single cells (binocular cells) in the visual cortex. Light from an object falls at different locations on the retinas of the two eyes depending on how far away it is. Thus, which binocular cells respond maximally to a stimulus indicate how far away it is. Cells in the visual association cortex respond to many aspects of visual experience. **Review Figure 43.14**

Understanding Higher Brain Functions in Cellular Terms

• We have a daily cycle of sleep and wakefulness. Sleep can be divided into distinct states of slow-wave sleep and rapid-eye-movement (REM) sleep. Human non-REM sleep is divided into four stages of increasing depth. **Review Figure 43.15**

• During wakefulness and REM sleep, the brain stem influence on the forebrain is to keep the neurons sensitive to input, but during non-REM sleep the withdrawal of these influences makes the neurons of the forebrain much less sensitive to input. Activity of the cortex during REM sleep results in dreams, but inhibition of sensory and motor pathways disconnects the dream process from sensory input and motor output.

• Some learning and memory processes have been localized to specific brain areas. Repeated activation of identified circuits in brain regions such as the hippocampus have revealed long-lasting changes in synaptic properties referred to as long-term potentiation and long-term depression, which may be involved in learning and memory.

• Language abilities are localized mostly in the left cerebral hemisphere, a phenomenon known as lateralization. Different areas of the left hemisphere—including Broca's area, Wernicke's area, and the angular gyrus—are responsible for different aspects of language. **Review Figures 43.17, 43.18**

Self-Quiz

1. Which of the following describes the route of sensory information from the foot to the brain?
   a. ventral horn, spinal cord, medulla, cerebellum, midbrain, thalamus, parietal cortex
   b. dorsal horn, spinal cord, medulla, pons, midbrain, hypothalamus, frontal cortex
   c. dorsal horn, spinal cord, medulla, pons, midbrain, thalamus, parietal cortex
   d. ventral horn, spinal cord, pons, cerebellum, midbrain, thalamus, parietal cortex
   e. dorsal horn, spinal cord, medulla, pons, midbrain, thalamus, frontal cortex

2. Which statement about the reticular system is *not* true?
   a. Increased activity in the reticular system induces sleep.
   b. It is located in the brain stem.
   c. Lesions of the reticular system in the midbrain can result in coma.
   d. Information from the spinal cord is routed to different brain stem nuclei and to the forebrain in the reticular system.
   e. There are groups of neurons called nuclei in the reticular system.

3. Which statement is *not* true?
   a. Sensory afferents carry information of which we are consciously aware.
   b. Visceral afferents carry information about physiological functions of which we are not consciously aware.
   c. The voluntary motor division of the efferent side of the peripheral nervous system executes conscious movements.
   d. The cranial nerves and spinal nerves are part of the peripheral nervous system.
   e. Afferent and efferent axons never travel in the same nerve.

4. Which statement is *not* true?
   a. In the spinal cord, the white matter contains the axons conducting information up and down the spinal cord.
   b. The limbic system is involved in basic physiological drives, instincts, and emotions.
   c. The limbic system consists of primitive forebrain structures.
   d. The vast majority of the nerve cell bodies in the human nervous system are contained within the limbic system.
   e. In humans, a part of the limbic system is necessary for the transfer of short-term memory to long-term memory.

5. Which of the following describes the largest amount of the human cerebral cortex?
   a. the frontal lobes
   b. the primary somatosensory cortex
   c. the temporal cortex
   d. association cortex
   e. the occipital cortex

6. Which statement about the autonomic nervous system is true?
   a. The sympathetic division is afferent, and the parasympathetic division is efferent.
   b. The transmitter norepinephrine is always excitatory and acetylcholine is always inhibitory.
   c. Each pathway in the autonomic nervous system includes two neurons, and the neurotransmitter of the first neuron is acetylcholine.
   d. The cell bodies of many sympathetic preganglionic neurons are in the brain stem.
   e. The cell bodies of most parasympathetic postganglionic neurons are in or near the thoracic and lumbar spinal cord.

7. Which statement about cells in the visual cortex is *not* true?
   a. Many cortical cells receive inputs directly from single retinal ganglion cells.
   b. Many cortical cells respond most strongly to bars of light falling at a specific location on the retina.
   c. Some cortical cells respond most strongly to bars of light falling anywhere over large areas of the retina.
   d. Some cortical cells receive inputs from both eyes.
   e. Some cortical cells respond most strongly to an object when it is a certain distance from the eyes.

8. Which of the following characterizes non-REM sleep?
   a. dreaming
   b. hyperpolarization of cortical neurons
   c. inhibition of skeletal muscles
   d. rapid and jerky eye movements
   e. it makes up about 20 percent of total sleep time

9. Which conclusion was supported by experiments on split-brain patients?
   a. Language abilities are localized mostly in the left cerebral hemisphere.
   b. Language abilities require both Wernicke's area and Broca's area.
   c. The ability to speak depends on Broca's area.
   d. The ability to read depends on Wernicke's area.
   e. The left hand is served by the left cerebral hemisphere.

10. The part of the brain that differs the most in complexity between mammals and amphibians is
    a. the midbrain.
    b. the forebrain.
    c. the cerebellum.
    d. the limbic system.
    e. the hippocampus.

Applying Concepts

1. Compare and contrast the two divisions of the autonomic nervous system. Emphasize distinctions with respect to their anatomical organization, neurotransmitters used, and general effects on the functions of specific organ systems.

2. Outline the development of the vertebrate nervous system. Where on the neural axis are the more evolutionarily primitive and advanced functions localized?

3. Compare the sequential activation of brain areas involved in speaking a written word versus repeating a statement that is heard.

4. Diagram the patterns of neuronal connections that could explain a cortical cell that responds optimally to a waving U.S. flag.

5. Explain why inhibition of skeletal muscles is a characteristic of REM sleep but not of non-REM sleep.

Readings

Chalmers, D. J. 1995. "The Puzzle of Conscious Experience." *Scientific American*, December This article is a thought-provoking discussion of what does it mean to be conscious.

Damasio, H., T. G. Grabowski, R. Frank, A. M. Galaburda and A. R. Damasio. 1994. "The Return of Phineas Gage: Clues About the Brain from the Skull of a Famous Patient." *Science* , vol. 264, p. 1102. A modern reevaluation of the case history presented in the introduction to this chapter.

Llinás, R. R. 1988. *The Biology of the Brain: From Neurons to Networks*. 1990. *The Workings of the Brain: Development, Memory, and Perception*. W. H. Freeman, New York. These two volumes are a rich collection of articles on various aspects of neuroscience that have been published in *Scientific American* over the past 15 years. They provide a wonderfully broad yet selectively in-depth survey of modern neurobiology.

Purves, D., G. J. Augustine, D. Fitzpatrick, L. C. Katz, A.-S. LaMantia and J. O. McNamara. 1997. *Neuroscience*. Sinauer Associates, Sunderland, MA. This text offers a highly readable medical prespective on the subjects covered in this chapter.

Rosenzweig, M. R., A. L. Leiman and S. M. Breedlove. 1996. *Biological Psychology*. Sinauer Associates, Sunderland, MA. This text is thorough in its coverage of higher functions of vertebrate and especially human nervous systems.

Thompson, R. F. 1993. *The Brain: A Neuroscience Primer*. W. H. Freeman, New York. An easily understood introductory text on neuroscience, covering topics from membrane events to the neural basis of behavior. Treatments of higher brain functions amplify several of the topics introduced in this chapter.

Youdin, M. B. H. and P. Riederer. 1997. "Understanding Parkinson's Disease." *Scientific American*, January. One higher function of the nervous system that was not discussed in detail in this chapter is the control of movement. Parkinson's disease is a disorder of motor control. This article is an excellent example of how neurobiological research at many levels is contributing to understanding a complex function of the nervous system in both health and disease.

Chapter 44

Effectors

A Kangaroo's Jump Seems Effortless
Recoil from its stretched tendons helps to power the kangaroo's jump. A jumping kangaroo can increase its speed without the input of more metabolic energy.

*I*nformation from sensors and integrative abilities of a central nervous system are not of much value to an animal unless it can respond. An obvious response is movement, and a fascinating array of adaptations have evolved that enable animals to move. Consider jumping. When bending down to smell a flower, you see a spider. Startled by the spider, neural signals from your brain routed through spinal circuits activate muscles in your legs to contract, causing your legs to extend, and you jump. At the same time, the spider jumps in the opposite direction. Did its nervous system also cause muscles to contract and extend its jumping legs? No; the spider doesn't have such muscles. Instead, extracellular fluids were squeezed into its hollow jumping legs, and the increased pressure in the legs caused them to extend and the spider to jump.

While this drama was playing out, a flea sensed your body heat and jumped 20 cm onto your leg. The flea's jump involves still a different mechanism, one that works like a slingshot. The flea is so small and its initial acceleration so great that no muscle can contract fast enough to cause such a movement. Instead, at the base of its jumping legs is an elastic material that is compressed by slow muscles while the flea is resting. When a trigger mechanism is released, the elastic material recoils like a slingshot and propels the flea a distance perhaps 200 times its body length.

Another remarkable jumper that combines the mechanisms used by humans with those used by the flea is a kangaroo. As you run faster and faster, the number of strides you take and the energy you expend per minute increase rapidly. Neither effect is true for the kangaroo. When running at speeds from about 5 to 25 km/h, the kangaroo takes the same number of strides per minute and its metabolic rate does not increase. How can this be?

As we will learn in this chapter, muscles are attached to bones by tendons. Like the material at the base of the flea's legs, tendons can be elastic. The kangaroo's tendons stretch when it lands, and their recoil helps power the next jump. To move faster, the kangaroo increases the length of its stride, thereby increasing the stretch on its tendons each time it lands and the magnitude of the recoil at the initiation of each jump.

Jumping is just one adaptation an animal can use to respond to information received by its sensors. Adaptations that animals use to respond to information that is sensed, integrated, and transmitted by its neural and endocrine systems are called **effectors**. By this definition, effectors include the internal organs and organ systems that the animal uses to control its internal environment; these effectors are the subjects of subsequent chapters in this unit on animals.

In this chapter, we will focus on the mechanisms of creating mechanical forces and using those forces to change shape and to move—the basis for most animal behavior. A fish swims, an earthworm crawls, a mosquito flies, and a kangaroo jumps because cells can move. We will focus first on the molecular basis for movement, then on how movements of molecules move cilia, flagella, and muscle filaments. We will then examine different muscle types and how neurons enable muscles to contract. Since muscles with no rigid supports to pull against would be nothing but quivering masses of tissue, we will also study the skeletal systems that enable animals to use the contractile forces of muscles for specific tasks.

Cilia, Flagella, and Cell Movement

Two subcellular structures, microtubules and microfilaments (see Figure 4.21) generate cell movement by sliding long protein molecules past one another.

Microtubules create the small movements of cilia and flagella. Microfilaments reach their highest level of organization in muscle cells, which generate large-scale movements.

Cilia are tiny, hairlike appendages of cells

Certain animal cells are covered by dense patches of cilia. Each cilium is tiny, about 0.25 μm in diameter. Animals of most species use ciliated cells to move liquids and particles over cell surfaces. Many invertebrates use ciliated cells to obtain food and oxygen. Cilia circulate a current of water across the gill surfaces of some mollusks, for example. Oxygen diffuses across the gill membranes, and food, consisting of tiny organisms and detritus, is filtered from the water by the netlike gills and ingested (Figure 44.1). Cilia around the mouths of rotifers sweep microorganisms and detritus directly into the gut.

The airways of many animals are lined with and cleaned by ciliated cells (Figure 44.2). In humans, the cilia continuously sweep a layer of mucus from deep down in the lungs, up through the windpipe, and into the throat. The mucus carries particles of dirt and dead cells. We can then either swallow or spit out the mucus, and with it the trapped detritus. Ciliated cells lining the female reproductive tract create currents that sweep eggs from the ovaries into the oviducts and all the way down to the uterus (see Chapter 39).

A cilium pushes with the same basic motion as a swimmer's arms during the breaststroke (Figure 44.3*a*). During the **power stroke**, the cilium projects

44.1 Cilia Create Water Currents in a Clam's Siphons
(*a*) In burrowing mollusks such as clams, cilia lining the siphons maintain a unidirectional flow of water: in one siphon, over the gills, and out the other siphon. The gills extract oxygen and food from this flow of water. (*b*) The siphons of several clams protrude in this photograph; the clam bodies are buried in the sand.

(*a*) A drop of dye demonstrates the unidirectional flow of water through a clam's siphons.

Dye

Incurrent siphon — Excurrent siphon

(*b*)

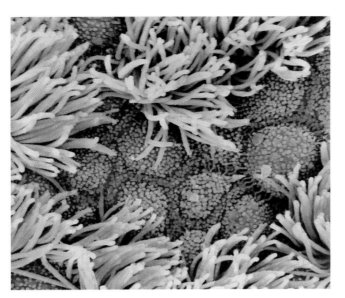

44.2 Cilia Line Respiratory Passages   A scanning electron micrograph of the epithelium of the windpipe.

stiffly outward and moves backward, propelling the cell forward (or the medium backward). During the **recovery stroke**, the cilium folds as it returns to its original position. The power stroke is fast, the recovery stroke slow. As you know from moving your arm or leg in water, there is less resistance the slower you move. The resistance of the medium to the recovery stroke thus is slight compared with its resistance to the power stroke. Fluids exposed to the beating of cilia are

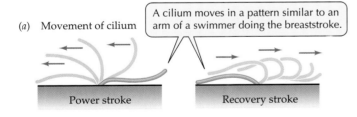

(a)   Movement of cilium

A cilium moves in a pattern similar to an arm of a swimmer doing the breaststroke.

Power stroke   Recovery stroke

(b)   Movement of flagellum

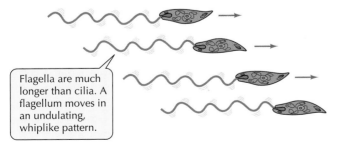

Flagella are much longer than cilia. A flagellum moves in an undulating, whiplike pattern.

44.3 Cilia and Flagella Move Differently   Cilia (a) have a "swimmer's stroke," whereas flagella (b) move with a whiplike effect.

propelled in the direction of the power stroke. Cilia typically beat in coordinated waves. At any particular moment, some cilia of a cell are moving through the power stroke and others are recovering.

Flagella are like long cilia

The **flagella** of eukaryotes are identical to cilia except that they are longer and occur singly or in groups of only a few on any one cell. Flagellated cells maintain a flow of water through the bodies of sponges, bringing in food and oxygen and removing carbon dioxide and wastes. Flagella power the movement of the sperm of most species. Because of their greater length, flagella have a whiplike stroke pattern rather than the "swimming" stroke pattern of cilia (Figure 44.3b).

Cilia and flagella are moved by microtubules

The central structure of a cilium or a flagellum, called the **axoneme**, contains a ring of nine pairs of microtubules. In the center of the ring may be an additional pair of microtubules, a single microtubule, or no microtubule (see Figure 4.25). Microtubules are hollow tubes formed from polymerization of the globular polypeptide **tubulin**. Other proteins in the axoneme form spokes, side arms, and cross-links (Figure 44.4a). Side arms composed of the protein **dynein** generate force. Dynein is an enzyme that catalyzes the hydrolysis of ATP and uses the released energy to change its orientation, thereby generating mechanical force.

When the dynein arms on one microtubule pair contact a neighboring microtubule pair and bind to it, ATP is broken down, and the resulting conformational changes in the dynein molecules cause the arms to point toward the base of the axoneme. This action pushes the microtubule pair ahead in relation to its neighbor. The dynein arms then detach from the neighboring pair and reorient to their starting horizontal position. As the cycle is repeated, adjacent microtubule pairs try to "row" past each other, with the dynein side arms acting as "oars" (Figure 44.4b).

Because the microtubules are anchored at the bottom, the axoneme bends, instead of elongating, as the microtubule pairs slide past one another (Figure 44.4c). In ways not fully understood, the central microtubule, together with the other proteins that bind the axoneme, control the dynein action so that not all the microtubule pairs are "rowing" at the same time.

Scientists have investigated the motile mechanisms of the axoneme by selectively removing its proteins. Axonemes severed from cells continue to flex in a normal pattern if exposed to calcium ions (Ca^{2+}) and ATP, demonstrating that the motile mechanism is part of the axoneme itself. However, gently treating the axoneme with enzymes that hydrolyze proteins disrupts the spokes and cross-links, leaving only the microtubules and dynein arms intact.

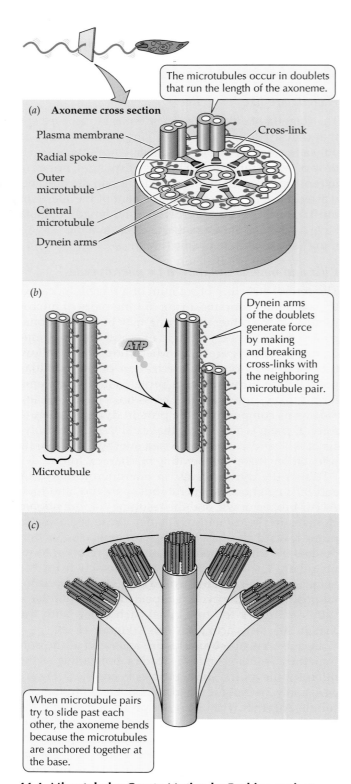

(a) **Axoneme cross section**

The microtubules occur in doublets that run the length of the axoneme.

Plasma membrane

Radial spoke

Outer microtubule

Central microtubule

Dynein arms

Cross-link

(b)

ATP

Microtubule

Dynein arms of the doublets generate force by making and breaking cross-links with the neighboring microtubule pair.

(c)

When microtubule pairs try to slide past each other, the axoneme bends because the microtubules are anchored together at the base.

44.4 Microtubules Create Motion by Pushing against Each Other Cilia and flagella move because of the actions of microtubules in the axoneme.

If the isolated microtubules are then exposed to Ca^{2+} and ATP, they row past one another and the whole structure elongates manyfold, demonstrating that the forces that move microtubule pairs along one another are the basis for the bending of the intact axoneme. When dynein (and only dynein) is removed from isolated axonemes, the microtubules lose their ATPase activity and their motility. Restoring purified dynein to the axonemes restores ATPase activity and motility.

Microtubules are intracellular effectors

Microtubules play important roles in cell movements. As components of the cytoskeleton, microtubules also contribute to the cell's shape. Cells can change shape and move by polymerizing and depolymerizing the tubulin in their microtubules. During mitosis, the spindle that moves chromosomes to the mitotic poles at anaphase forms by the polymerization of tubulin.

Another example of microtubule involvement in cell movement is the growth of the axons of neurons in the developing nervous system. Neurons find and make their appropriate connections by sending out long extensions that search for the correct contact cells. If polymerization of tubulin is chemically inhibited, the neurons do not extend. Microtubules are important intracellular effectors for changing cell shape, moving organelles, and enabling cells to respond to the environment.

Microfilaments change cell shape and cause cell movements

The dominant microfilament in cells is the protein **actin**. Bundles of cross-linked actin strands form important structural components of cells. For example, the microvilli (tiny projections) that increase the absorptive surface area of the cells lining the gut are stiffened by actin microfilaments (Figure 44.5a), as are the stereocilia of the sensory hair cells mentioned in Chapter 42 (Figure 44.5b). Like microtubules, actin microfilaments can change the shape of a cell simply by polymerizing and depolymerizing. The projections sent out by phagocytic cells (see Figure 18.5) are an example of this process.

Together with the protein **myosin**, actin microfilaments generate the contractile forces responsible for many aspects of cell locomotion and changes in cell shape. For example, the contractile ring that divides a cell undergoing mitosis into two daughter cells is composed of actin microfilaments in association with myosin. The mechanisms that many cells employ to engulf materials (phagocytosis and pinocytosis; see Chapter 5) also rely on actin microfilaments and myosin. Nets of actin and myosin beneath the cell membrane change a cell's shape as it moves to surround the particle being ingested.

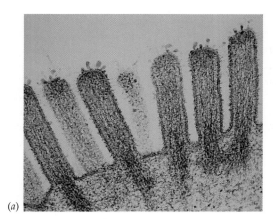

44.5 Cell Projections Supported by Microfilaments (*a*) The cells lining the gut have numerous fingerlike projections called microvilli. (*b*) Some mechanosensors, such as these hair cells in the organ of Corti, have stereocilia. Both microvilli and hair cells are stiffened by microfilaments.

(*b*)

The movement of certain cells in multicellular animals results from the activity of actin microfilaments and myosin. During development, many cells migrate by such **amoeboid movement**; throughout an animal's life, phagocytic cells circulate in the blood, squeeze through the walls of the blood vessels, and wander through the tissues by amoeboid movement. The mechanisms of amoeboid movement have been studied extensively in the protist for which this type of movement was named—the amoeba, which lives in freshwater streams and ponds.

Amoebas move by extending lobe-shaped projections called *pseudopods* and then seemingly squeezing themselves into the pseudopods (see the chapter-opening photograph in Chapter 26). The cytoplasm in the core of the amoeba is relatively liquid and is called *plasmasol*, but just beneath the plasma membrane the cytoplasm is much thicker and is called *plasmagel*.

To form a pseudopod, the thick plasmagel in a certain area of the cell thins, allowing a bulge to form. Just under the cell surface, in the plasmagel, is a network of actin microfilaments that interacts with myosin to squeeze plasmasol into the bulge, thus forming a pseudopod. As the network continues to contract, cytoplasm streams in the direction of the pseudopod. Eventually the cytoplasm at the leading edge of the pseudopod converts to gel and the pseudopod stops forming. Thus the basis for amoeboid motion is the ability of the cytoplasm to cycle through sol and gel states and the ability of the microfilament network under the cell membrane to contract and cause the cytoplasmic streaming that pushes out a pseudopod.

Muscle Contraction: Actin and Myosin Interactions

Muscle contraction is the most important effector mechanism that animals have for responding to their environments. All behavioral and most physiological responses depend on muscle cells. Muscle cells are specialized for contraction and have high densities of actin microfilaments and myosin. Such cells are found throughout the animal kingdom.

Muscle cells account for the thrashing movements of nematodes, the expansion–contraction movements of earthworms, the pulsating movements of jellyfish, and the limb movements of arthropods and vertebrates. Muscle cells are found in the walls of blood vessels, guts, bladders, and hearts. Wherever whole tissues contract in animals, muscle cells are responsible. In all cases, the molecular mechanism of contraction is the same, but many specializations fit muscle cells to the wide variety of functions they serve.

We begin our study of muscle by looking at the three types of muscle cells found in vertebrates: smooth muscle, skeletal muscle, and cardiac (heart) muscle. After examining the mechanism for skeletal muscle contraction, we describe the calcium-based system by which contractions are triggered and show how graded muscle responses are accomplished in whole muscles. Not all skeletal muscle cells are identical in the force they can generate. We will explore the basis for slow-twitch and fast-twitch muscle fibers and see how their relative proportions greatly influence the operation of whole muscles and the behavior of organisms.

Smooth muscle causes slow contractions of many internal organs

Smooth muscle provides the contractile forces for most of our internal organs, which are under the control of the autonomic nervous system (see Chapter 43).

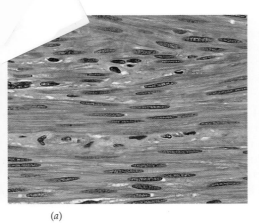
(a)

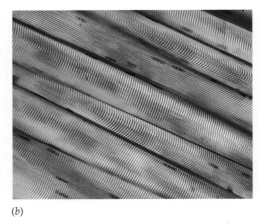
(b)

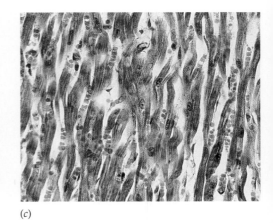
(c)

44.6 The Three Types of Vertebrate Muscle Tissue   (a) Smooth muscle cells are usually arranged in sheets such as those that make up the walls of the stomach and the intestine. The dark structures in this sheet of muscle cells are nuclei. (b) Skeletal, or striated, muscle appears to be striped, or banded, because of its highly regular arrangement of contractile filaments. (c) Although cardiac muscle is striated (striped), its fibers branch and create a meshwork that resists tearing or breaking.

Smooth muscle moves food through the digestive tract, controls the flow of blood through blood vessels, and empties the urinary bladder. Structurally, smooth muscle cells are the simplest muscle cells. They are usually long and spindle-shaped, and each cell has a single nucleus. Because the filaments of actin and myosin in smooth muscle are not as regularly arranged as those in the other muscle types, the contractile machinery is not obvious when the cells are viewed under the light microscope (Figure 44.6a).

If we study smooth muscle from a particular organ, such as the walls of the digestive tract, we find it has interesting properties. The cells are arranged in sheets, and individual cells in the sheets are in electrical contact with one another through gap junctions. As a result, an action potential generated in the membrane of one smooth muscle cell can spread to all the cells in the sheet of tissue.

Another interesting property of a smooth muscle cell is that the resting potential of its membrane is sensitive to being stretched. If the wall of the digestive tract is stretched in one location (such as by receiving a mouthful of food), the membranes of the stretched cells depolarize, reach threshold, and fire action potentials that cause the cells to contract. Thus smooth muscle contracts after being stretched, and the harder it is stretched, the stronger the contraction. Later in this chapter we will discuss how membrane depolarization triggers contraction.

Skeletal muscle causes behavior

Skeletal muscle carries out all voluntary movements, such as running or playing a piano, and generates the movements of breathing. Skeletal muscle is also called **striated muscle** because the highly regular arrangement of its actin microfilaments and myosin gives it a striped appearance (Figure 44.6b). Skeletal muscle cells, or **muscle fibers**, are large. Unlike smooth muscle and cardiac muscle cells, each of which has a single nucleus, skeletal muscle fibers have many nuclei because they develop through the fusion of many individual cells. A muscle such as your biceps (which bends your arm) is composed of many muscle fibers bundled together by connective tissue.

What is the relation between a muscle fiber and the actin and myosin filaments responsible for contraction? Each muscle fiber is composed of **myofibrils**—bundles of contractile filaments made up of actin and myosin (Figure 44.7). Within each myofibril are thin filaments, which are actin microfilaments, and thick filaments, which are composed of myosin. If we cut across the myofibril at certain locations, we see only thick filaments, in other locations only thin filaments, but in most regions of the myofibril, each thick myosin filament is surrounded by six thin actin filaments.

A longitudinal view of a myofibril reveals the striated appearance of skeletal muscle. The band pattern of the myofibril is due to repeating units called **sarcomeres**, which are the units of contraction (see Figure 44.7). Each sarcomere is made of overlapping filaments of actin and myosin. As the muscle contracts, the sarcomeres shorten, and the appearance of the band pattern changes.

The observation that the widths of the bands in the sarcomeres change when a muscle contracts led two British biologists, Hugh Huxley and Andrew Huxley,

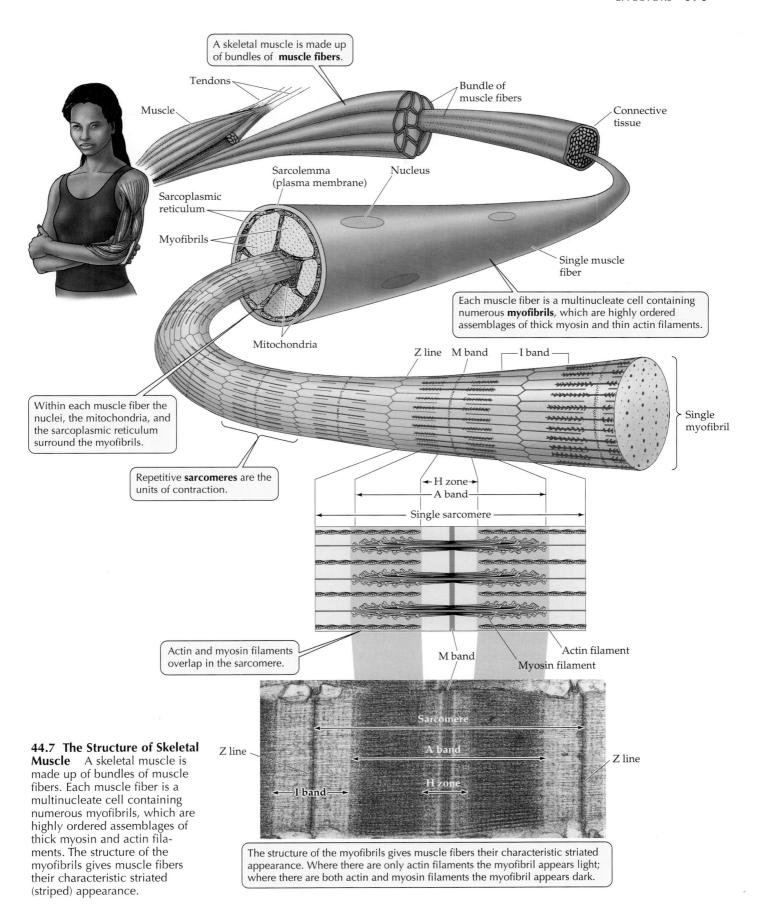

A skeletal muscle is made up of bundles of **muscle fibers**.

Tendons

Muscle

Bundle of muscle fibers

Connective tissue

Sarcolemma (plasma membrane)

Nucleus

Sarcoplasmic reticulum

Myofibrils

Single muscle fiber

Each muscle fiber is a multinucleate cell containing numerous **myofibrils**, which are highly ordered assemblages of thick myosin and thin actin filaments.

Mitochondria

Within each muscle fiber the nuclei, the mitochondria, and the sarcoplasmic reticulum surround the myofibrils.

Z line   M band   I band

Single myofibril

Repetitive **sarcomeres** are the units of contraction.

H zone

A band

Single sarcomere

Actin and myosin filaments overlap in the sarcomere.

M band

Actin filament

Myosin filament

44.7 The Structure of Skeletal Muscle A skeletal muscle is made up of bundles of muscle fibers. Each muscle fiber is a multinucleate cell containing numerous myofibrils, which are highly ordered assemblages of thick myosin and actin filaments. The structure of the myofibrils gives muscle fibers their characteristic striated (striped) appearance.

Z line

Sarcomere

A band

Z line

I band

H zone

The structure of the myofibrils gives muscle fibers their characteristic striated appearance. Where there are only actin filaments the myofibril appears light; where there are both actin and myosin filaments the myofibril appears dark.

to propose a molecular mechanism of muscle contraction. Let's look at the band pattern of the myofibril in detail (see the micrograph in Figure 44.7). Each sarcomere is bounded by *Z lines* that anchor the thin actin filaments. Centered in the sarcomere is the *A band*, which contains all the myosin filaments. The *H zone* and the *I band*, which appear light, are regions where actin and myosin filaments do not overlap in the relaxed muscle. The dark stripe within the H zone is called the *M band*; it contains proteins that help hold the myosin filaments in their regular hexagonal arrangement.

When the muscle contracts, the sarcomere shortens. The H zone and the I band become much narrower, and the Z lines move toward the A band as if the actin filaments were sliding into the region occupied by the myosin filaments. This observation led Huxley and Huxley to propose the **sliding-filament theory** of muscle contraction: Actin and myosin filaments slide past each other as the muscle contracts.

To understand what makes the filaments slide, we must examine the structure of actin and of myosin (Figure 44.8). The myosin molecule consists of two long polypeptide chains coiled together, each ending in a globular head. The myosin filament is made up of many myosin molecules arranged in parallel, with their heads projecting laterally from one or the other end of the filament. The actin filament consists of a helical arrangement of two chains of monomers like two strands of pearls twisted together. Twisting around the actin chains are two strands of another protein, **tropomyosin**.

The myosin heads have sites that can bind to actin and thereby form bridges between the myosin and the actin filaments. The myosin heads also have ATPase activity; they bind and hydrolyze ATP. The energy released changes the orientation of the myosin head.

Together, these details explain the cycle of events that cause the actin and myosin filaments to slide past one another and shorten the sarcomere. A myosin head binds to an actin filament. Upon binding, the head changes its orientation with respect to the myosin filament, thus exerting a force that causes the actin and myosin filaments to slide about 5 to 10 nm relative to each other. Next, the myosin head binds a molecule of ATP, which causes it to release the actin. When the ATP is hydrolyzed, the energy released causes the myosin head to return to its original conformation, in which it can bind again to actin. ATP hydrolysis is like the cocking of the hammer of a pistol, and binding with actin pulls the trigger.

We have been discussing the cycle of contraction in terms of a single myosin head. Don't forget that each myosin filament has many myosin heads at both ends and is surrounded by six actin filaments; thus the contraction of the sarcomere involves a great many cycles of interaction between actin and myosin molecules. That is why when a single myosin head breaks its contact with actin, the actin filaments do not slip backward.

An interesting aspect of this contractile mechanism is that ATP is needed to break the actin–myosin bonds but not to form them. Thus muscles require ATP to stop contracting. This fact explains why muscles stiffen soon after animals die, a condition known as *rigor mortis*. Death stops the replenishment of the ATP stores of muscle cells, so the myosin–actin bridges cannot be broken, and the muscles stiffen. Eventually the proteins begin to lose their integrity and the muscles soften. These events have regular time courses that differ somewhat for different regions of the body; therefore, an examination of the stiffness of the mus-

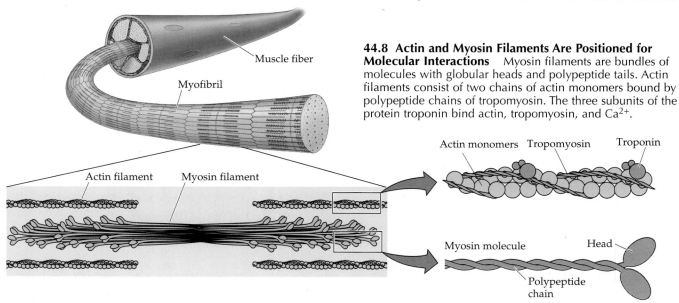

44.8 Actin and Myosin Filaments Are Positioned for Molecular Interactions Myosin filaments are bundles of molecules with globular heads and polypeptide tails. Actin filaments consist of two chains of actin monomers bound by polypeptide chains of tropomyosin. The three subunits of the protein troponin bind actin, tropomyosin, and Ca^{2+}.

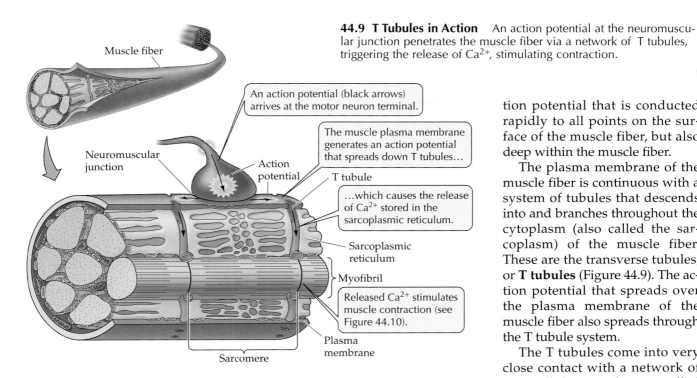

44.9 T Tubules in Action An action potential at the neuromuscular junction penetrates the muscle fiber via a network of T tubules, triggering the release of Ca^{2+}, stimulating contraction.

Muscle fiber

Neuromuscular junction

An action potential (black arrows) arrives at the motor neuron terminal.

Action potential

The muscle plasma membrane generates an action potential that spreads down T tubules…

T tubule

…which causes the release of Ca^{2+} stored in the sarcoplasmic reticulum.

Sarcoplasmic reticulum

Myofibril

Released Ca^{2+} stimulates muscle contraction (see Figure 44.10).

Plasma membrane

Sarcomere

cles of a corpse sometimes can help a coroner estimate the time of death.

Actin–myosin interactions are controlled by Ca^{2+}

Muscle contractions are initiated by nerve action potentials arriving at the neuromuscular junction. Motor neurons are generally highly branched and can innervate up to 100 muscle fibers each. All the fibers innervated by a single motor neuron are a **motor unit** and contract simultaneously in response to the action potentials fired by that motor neuron. To understand the fine control the nervous system has over the sliding of actin and myosin filaments, we must examine the membrane system of the muscle fiber and some additional protein components of the actin filaments.

Like neurons, vertebrate skeletal muscle fibers are excitable cells: When they are depolarized to a threshold that opens their voltage-gated sodium channels, their plasma membranes generate action potentials, just as the membranes of axons do. The action potential that spreads across the plasma membrane of the muscle fiber is generated at the neuromuscular junction—the synapse between the motor neuron and the muscle fiber.

As we saw in Chapter 41, neurotransmitter from the motor neuron binds to receptors in the postsynaptic membrane, causing ion channels to open. The depolarization of the postsynaptic membrane spreads to the surrounding plasma membrane of the muscle fiber, which contains voltage-gated ion channels. When threshold is reached, the plasma membrane fires an ac-

tion potential that is conducted rapidly to all points on the surface of the muscle fiber, but also deep within the muscle fiber.

The plasma membrane of the muscle fiber is continuous with a system of tubules that descends into and branches throughout the cytoplasm (also called the sarcoplasm) of the muscle fiber. These are the transverse tubules, or **T tubules** (Figure 44.9). The action potential that spreads over the plasma membrane of the muscle fiber also spreads through the T tubule system.

The T tubules come into very close contact with a network of intracellular membranes called the **sarcoplasmic reticulum** that surrounds every myofibril. Calcium pumps in the membranes of the sarcoplasmic reticulum cause this membrane-enclosed compartment of the fiber to take up Ca^{2+} ions from the cytoplasm (**sarcoplasm**) of the muscle fiber. The result is a high concentration of Ca^{2+} in the sarcoplasmic reticulum and a low concentration of Ca^{2+} in the sarcoplasm surrounding the filaments.

When an action potential spreads through the T tubule system, it causes Ca^{2+} channels in the sarcoplasmic reticulum to open, resulting in the diffusion of Ca^{2+} ions out of the sarcoplasmic reticulum and into the sarcoplasm surrounding the microfilaments. The Ca^{2+} stimulates the interaction of actin and myosin and the sliding of the filaments. How does this work?

An actin filament is a helical arrangement of two strands of actin monomers, and lying in the grooves between the two actin strands is the two-stranded protein tropomyosin (see Figure 44.8). At regular intervals the filament also includes another globular protein, **troponin**. The troponin molecule has three subunits; one binds actin, one binds tropomyosin, and one binds Ca^{2+}. When Ca^{2+} is sequestered in the sarcoplasmic reticulum, the tropomyosin strands block the sites where myosin heads can bind to the actin. When the T tubule system depolarizes, Ca^{2+} is released into the sarcoplasm, where it binds to the troponin, changing the shape of the troponin molecule.

Because the troponin is also bound to the tropomyosin, this conformational change of the troponin twists the tropomyosin enough to expose the actin–myosin binding sites. Thus the cycle of making and breaking actin–myosin bridges is initiated; the filaments are pulled past one another, and the muscle fiber contracts. When the T tubule system repolarizes, the calcium pumps remove the Ca²⁺ ions from the sarcoplasm, causing the tropomyosin to return to the position in which it blocks the binding of the myosin heads to the actin strands, and the muscle fiber returns to its resting condition. Figure 44.10 summarizes this cycle.

Calmodulin mediates Ca²⁺ control of contraction in smooth muscle

Smooth muscle cells do not have the troponin–tropomyosin mechanism for controlling contraction, but Ca²⁺ still plays a critical role. The Ca²⁺ influx into the cytoplasm of the smooth muscle cell can be stimulated by action potentials, by hormones, and by stretch. The Ca²⁺ that enters the cell combines with a protein called **calmodulin**, which acts as a second messenger. The calmodulin–Ca²⁺ complex activates an enzyme called myosin kinase, which can phosphorylate myosin heads.

When the myosin heads in smooth muscle are phosphorylated, they can undergo cycles of binding and release with actin, causing contraction. As Ca²⁺ is removed from the cytoplasm, it dissociates from calmodulin, and the levels of active myosin kinase fall. In addition, another enzyme, myosin phosphatase, dephosphorylates the myosin and helps stop the actin–myosin interactions.

44.10 The Result of Action Potentials at the Neuromuscular Junction The propagation of an action potential through the T tubules results in the release of Ca²⁺. When Ca²⁺ binds to troponin, it exposes cross-bridge binding sites. As long as cross-bridge sites and ATP are available, the cycle of actin and myosin interactions continues and the filaments slide.

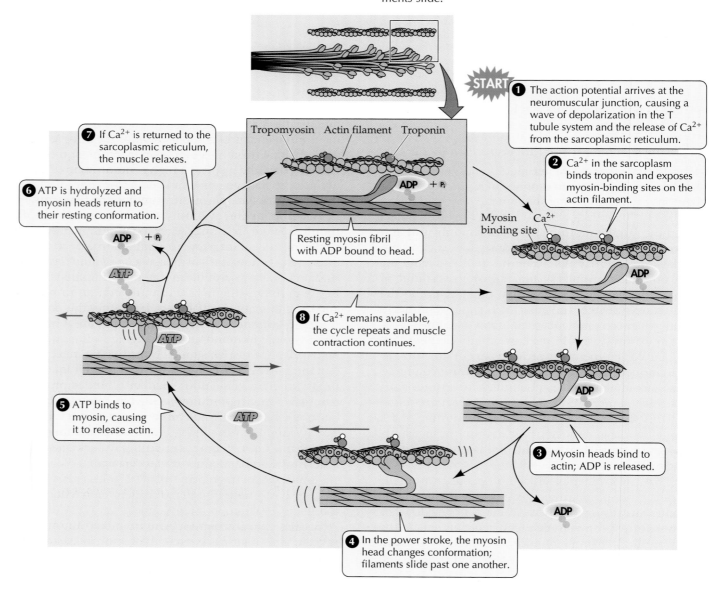

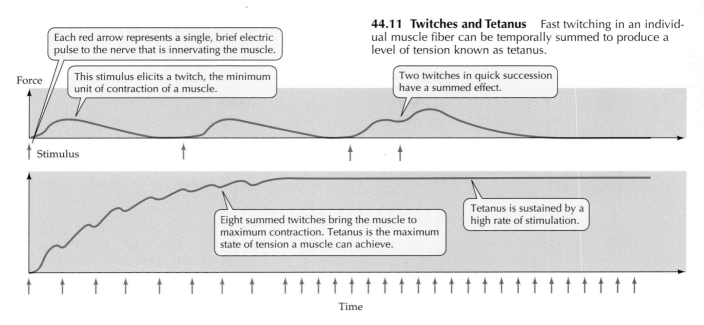

44.11 Twitches and Tetanus Fast twitching in an individual muscle fiber can be temporally summed to produce a level of tension known as tetanus.

(Figure labels within the image:)
- Each red arrow represents a single, brief electric pulse to the nerve that is innervating the muscle.
- This stimulus elicits a twitch, the minimum unit of contraction of a muscle.
- Two twitches in quick succession have a summed effect.
- Eight summed twitches bring the muscle to maximum contraction. Tetanus is the maximum state of tension a muscle can achieve.
- Tetanus is sustained by a high rate of stimulation.
- Force
- Stimulus
- Time

Single muscle twitches are summed into graded contractions

In vertebrate skeletal muscle, the arrival of an action potential at the neuromuscular junction causes an action potential in the muscle fiber. The spread of the action potential through the membrane system of the muscle fiber causes a minimum unit of contraction, called a **twitch**. A twitch can be measured in terms of the tension, or force, it generates.

If action potentials in the muscle fiber are adequately separated in time, each twitch is a discrete, all-or-none phenomenon. If action potentials are fired more rapidly, however, new twitches are triggered before the filaments have had a chance to return to their resting condition. As a result, the twitches sum, and the tension generated by the fiber increases and becomes more continuous (Figure 44.11). Thus the individual muscle fiber can show a graded response to increased levels of stimulation by its motor neuron.

At high levels of stimulation, the calcium pumps in the sarcoplasmic reticulum can no longer remove Ca^{2+} ions from the sarcoplasm between action potentials, and the contractile machinery generates maximum tension—a condition known as **tetanus**. (Do not confuse this condition with the disease tetanus, which is caused by a bacterial toxin and is characterized by spastic contractions of skeletal muscles.)

How long a muscle fiber can maintain a tetanic contraction depends on its supply of ATP. Eventually the fiber will fatigue. It may seem paradoxical that the lack of ATP causes fatigue, since the action of ATP is to break the actin–myosin bonds. But remember that the energy released from the hydrolysis of ATP "recocks" the myosin heads, allowing them to cycle through another power stroke. The situation is like rowing a boat upstream: You cannot maintain your position relative to the stream bank by just holding the oars out against the current; you have to keep rowing.

The ability of a whole muscle to generate different levels of tension depends also on how many muscle fibers in that muscle are activated. Whether a muscle contraction is strong or weak depends both on how many motor neurons to that muscle are firing and on the rate at which those neurons are firing. These two factors can be thought of as spatial and temporal summation, respectively.

Both types of summation increase the strength of contraction of the muscle as a whole. Faster twitching of individual fibers causes temporal summation (see Figure 44.11), and an increase in the number of motor units involved in the contraction causes spatial summation. (Remember that a motor unit consists of all the muscle fibers innervated by a single neuron, and that a single muscle consists of many motor units.)

Many muscles of the body maintain a low level of tension called **tonus** even when the body is at rest. For example, the muscles of our neck, trunk, and limbs that maintain our posture against the pull of gravity are always working, even when we are standing or sitting still. Muscle tonus comes from the activity of a small but changing number of motor units in a muscle; at any one time some of the muscle's fibers are contracting and others are relaxed. Tonus is constantly being readjusted by the nervous system.

Muscle fiber types determine aerobic capacity and strength

Not all skeletal muscle fibers are alike in how they twitch, and a single muscle may contain more than one type of fiber. The two major types of skeletal muscle fibers are slow-twitch fibers and fast-twitch fibers

(a) Cross section of a skeletal muscle

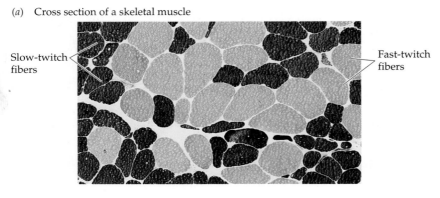

Slow-twitch fibers

Fast-twitch fibers

(b)

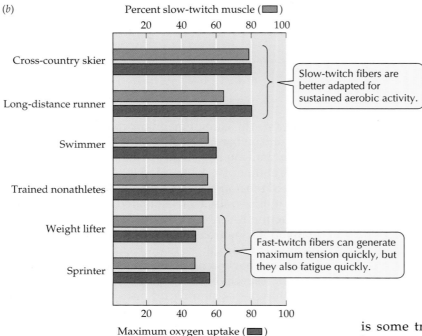

Percent slow-twitch muscle (▨)

Slow-twitch fibers are better adapted for sustained aerobic activity.

Fast-twitch fibers can generate maximum tension quickly, but they also fatigue quickly.

Maximum oxygen uptake (▨)
(ml/min/kg)

44.12 Two Types of Muscle Fibers Are Specialized for Fast or for Sustained Contraction (a) Skeletal muscle consists of fast- and slow-twitch fibers. (b) World-class athletes in different sports have different distributions of fiber types. For comparison, we include here average values for amateur athletes in good condition.

fewer mitochondria, little or no myoglobin, and fewer blood vessels. The white meat of domestic chickens is composed of fast-twitch fibers. Fast-twitch fibers can develop maximum tension more rapidly than slow-twitch fibers can, and that maximum is greater, but fast-twitch fibers fatigue rapidly. The myosin of fast-twitch fibers has high ATPase activity, so they can put the energy of ATP to work very rapidly, but these fibers cannot replenish it quickly enough to sustain contraction for a long time. Fast-twitch fibers are especially good for short-term work that requires maximum strength. Champion weight lifters and sprinters have leg and arm muscles with high proportions of fast-twitch fibers (see Figure 44.12b).

What determines the proportion of fast- and slow-twitch fibers in your skeletal muscles? The most important factor is your genetic heritage, so there is some truth to the statement that champions are born, not made. To a certain extent, however, you can alter the properties of your muscle fibers through training. With aerobic training, the oxidative capacity of fast-twitch fibers can improve substantially. But a person born with a high proportion of fast-twitch fibers will never become a champion marathon runner, and a person born with a high proportion of slow-twitch fibers will never become a champion sprinter.

Cardiac muscle causes the heart to beat

Cardiac muscle looks different from smooth muscle when viewed under the microscope (see Figure 44.6c). The cells appear striped, or striated, because of the regular arrangement of bundles of actin and myosin microfilaments in these cells. Actin and myosin are arranged in a similar way in skeletal muscle, which we'll learn more about in the next section.

The unique feature of cardiac cells is that they branch. The branches of adjoining cells are interdigitated into a meshwork that gives cardiac muscle an ability to resist tearing. As a result, the heart walls can withstand high pressures during the pumping of

(Figure 44.12a). **Slow-twitch fibers** are also called red muscle because they have lots of the oxygen-binding molecule myoglobin, they have lots of mitochondria, and they are well supplied with blood vessels. A single twitch of a slow-twitch fiber produces low tension.

The maximum tension a slow-twitch fiber can produce is low and develops slowly, but these fibers are highly resistant to fatigue. Because slow-twitch fibers have substantial reserves of glycogen and fat, their abundant mitochondria can maintain a steady, prolonged production of ATP if oxygen is available. Muscles with high proportions of slow-twitch fibers are good for long-term, aerobic work (that is, work that requires lots of oxygen). Champion long-distance runners, cross-country skiers, swimmers, and bicyclists have leg and arm muscles consisting mostly of slow-twitch fibers (Figure 44.12b).

Fast-twitch fibers are also called white muscle because, in comparison to slow-twitch fibers, they have

blood without the danger of developing leaks. Also adding to the strength of cardiac muscle are **intercalated discs** that provide strong mechanical adhesions between adjacent cells.

As is true of smooth muscle, the individual cells in a sheet of cardiac muscle are in electrical continuity with each other. Gap junctions present in the intercalated discs present low resistance to ions or electric currents. Therefore, a depolarization initiated at one point in the heart spreads rapidly through the mass of cardiac muscle.

An interesting feature of vertebrate cardiac muscle is that certain muscle cells are specialized for pacemaking function; they initiate the rhythmic contractions of the heart. We'll learn about the molecular basis for this pacemaking function in Chapter 46. Because of these specialized pacemaking cells, the heartbeat is **myogenic**—generated by the heart itself. The autonomic nervous system modifies the rate of the pacemaker cells but is not essential for their continued rhythmic function. A heart removed from an animal continues to beat with no input from the nervous system. The myogenic nature of the heartbeat is a major factor in making heart transplants possible, because the implanted heart does not depend on neural connections to beat.

Skeletal Systems Provide Rigid Supports

Muscles can only contract and relax. Without something rigid to pull against, muscles can do little more than lie in a formless mass that twitches and changes shape. Skeletal systems provide rigid supports against which muscles can pull, thereby creating directed movements. In this section, we'll examine the three types of skeletal systems found in animals: hydrostatic skeletons, exoskeletons, and endoskeletons.

A hydrostatic skeleton consists of fluid in a muscular cavity

The simplest type of skeleton is the **hydrostatic skeleton** of cnidarians, annelids, and many other soft-bodied invertebrates. It consists of a volume of incompressible fluid (water) enclosed in a body cavity surrounded by muscle. When muscles oriented in a certain direction contract, the fluid-filled body cavity bulges out in the opposite direction.

The sea anemone has a hydrostatic skeleton (see Figure 29.12); its body cavity is filled with seawater. To extend its body and its tentacles, the anemone closes its mouth and constricts muscle fibers that are arranged in circles around its body. Contraction of these circular muscles puts pressure on the liquid in the body cavity, and that pressure forces the body and tentacles to extend. If alarmed, the anemone retracts its tentacles and body by contracting muscle fibers that are arranged

longitudinally in the body wall and in the long dimension of the tentacles.

The hydrostatic skeletons of some animals have become adapted for locomotion. An annelid such as the earthworm uses its hydrostatic skeleton to crawl (see Figure 29.27). The earthworm's body cavity is divided into many separate segments. The body wall has two muscle layers: one in which the muscle fibers are arranged in circles around the body cavity, and another in which the muscle fibers run lengthwise (Figure 44.13a). A closed compartment in each segment of the worm is filled with fluid. If the circular muscles in a segment contract, the compartment in that segment narrows and elongates. If the lengthwise (longitudinal) muscles of a segment contract, the compartment shortens and bulges outward.

Alternating contractions of the earthworm's circular and longitudinal muscles create waves of narrowing and widening, lengthening and shortening, that travel down the body. The bulging, short segments serve as anchors as the narrowing, expanding segments project forward, and longitudinal contractions pull other segments forward. Bristles help the widest parts of the body to hold firm against the substrate. The alternating waves of contraction and extension along its body allow the earthworm to make fairly rapid progress through or over the soil (Figure 44.13b).

Another type of locomotion made possible by adaptation of the hydrostatic skeleton is the jet propulsion used by squid and octopuses. Muscles surrounding a water-filled cavity in these cephalopods contract, putting the water under pressure and expelling it from the animal's body. As the water shoots out under pressure, the animal is propelled in the opposite direction.

Exoskeletons are rigid outer structures

An **exoskeleton** is a hardened outer surface to which internal muscles can be attached. Contractions of these muscles cause jointed segments of the exoskeleton to move relative to each other. The simplest example of an exoskeleton is the shell of a mollusk, which generally consists of just one or two pieces. Some marine bivalves and snails have shells composed of protein strengthened by crystals of calcium carbonate (a rock-hard material). These shells can be massive, affording significant protection against predators. The shells of land snails generally lack the hard mineral component and are much lighter.

Molluscan shells can grow as the animal grows, and growth rings are usually apparent on the shells. The soft parts of the molluscan body have a hydrostatic skeleton. The hydrostatic skeleton is used in locomotion; the exoskeleton mainly provides protection. (Some scallops, however, swim by opening their shells and snapping them shut—another version of jet propulsion.)

(a)

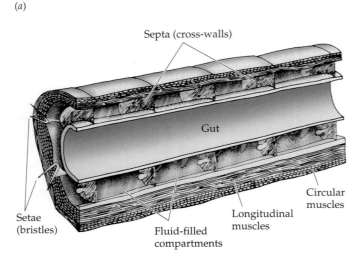

Septa (cross-walls)

Gut

Circular muscles

Longitudinal muscles

Setae (bristles)

Fluid-filled compartments

(b)

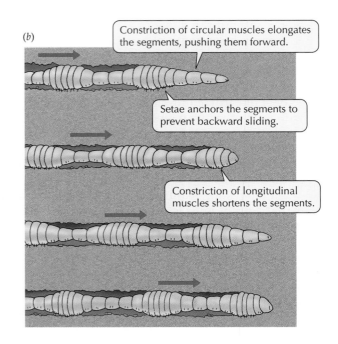

Constriction of circular muscles elongates the segments, pushing them forward.

Setae anchors the segments to prevent backward sliding.

Constriction of longitudinal muscles shortens the segments.

44.13 A Hydrostatic Skeleton and Locomotion *(a)* An earthworm's hydrostatic skeleton consists of fluid-filled compartments separated by septa. *(b)* Alternating waves of elongation and contraction move the earthworm through the soil.

The most complex exoskeletons are found among the arthropods. Plates of exoskeleton, or **cuticle**, cover all the outer surfaces of the arthropod's body and all its appendages. The plates are secreted by a layer of cells just below the exoskeleton. A continuous, layered, waxy coating covers the entire body. The skeleton contains stiffening materials everywhere except at the joints, where flexibility must be retained.

The layers of cuticle include an outer, thin, waxy epicuticle that protects the bodies of terrestrial arthropods from drying out, and a thicker, inner endocuticle that forms most of the structure. The endocuticle is a tough, pliable material found only in arthropods. It consists of a complex of protein and **chitin**, a nitrogen-containing polysaccharide. In marine crustaceans the endocuticle is further toughened by insoluble calcium salts. The thickness of the exoskeleton varies and it can form a protective armor. Muscles attached to the inner surfaces of the arthropod exoskeleton move its parts around the joints (Figure 44.14).

An exoskeleton protects all the soft tissues of the animal, but it is itself subject to damage such as abrasion and crushing. The greatest drawback of the arthropod exoskeleton is that it cannot grow. Therefore, if the animal is to become larger, it must **molt**, shedding its exoskeleton and forming a new, larger one. During this process the animal is vulnerable because the new exoskeleton takes time to harden. The animal's body is temporarily unprotected, and without the firm exoskeleton against which its muscles can exert maximum tension, it is unable to move rapidly. Soft-shelled crabs, a gourmet delicacy, are crabs caught when they are molting.

Vertebrate endoskeletons are attached to muscles

The **endoskeleton** of vertebrates is an internal scaffolding to which the muscles attach and against which they pull. It is composed of rodlike, platelike, and tubelike bones, which are connected to each other at a variety of joints that allow a wide range of movements.

The human skeleton consists of 206 bones (some of which are shown in Figure 44.15) and is divided into

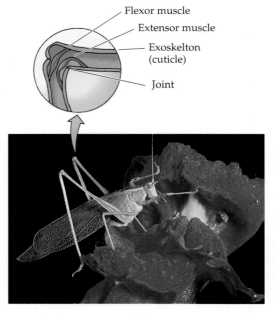

Flexor muscle
Extensor muscle
Exoskelton (cuticle)
Joint

44.14 An Insect's Exoskeleton Muscles attached to the exoskeleton move parts around flexible joints.

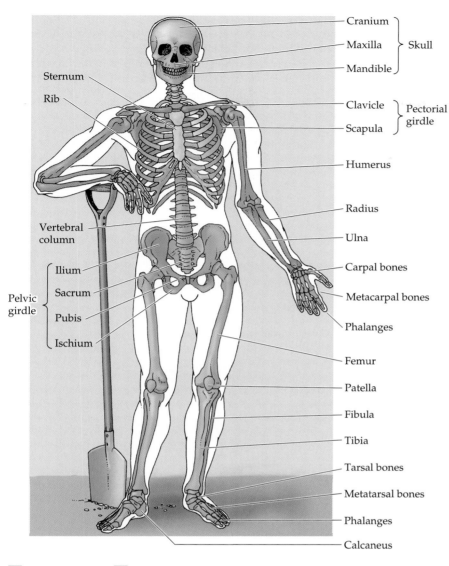

Cranium
Maxilla   } Skull
Mandible

Sternum
Rib
Clavicle   } Pectoral girdle
Scapula

Humerus

Radius

Ulna

Carpal bones

Metacarpal bones

Phalanges

Femur

Patella

Fibula

Tibia

Tarsal bones

Metatarsal bones

Phalanges

Calcaneus

Vertebral column

Ilium
Sacrum
Pubis
Ischium

Pelvic girdle

☐ Axial skeleton   ☐ Apendicular skeleton

44.15 The Human Endoskeleton
Cartilage and bone make up the internal skeleton of a human being.

Cartilage is found in parts of the endoskeleton where both stiffness and resiliency are required, such as on the surfaces of joints, where bones move against each other. Cartilage is also the supportive tissue in stiff but flexible structures such as the larynx (voice box), the nose, and the ears. Sharks and rays are called cartilaginous fishes (see Figure 27.14) because their skeletons are composed entirely of cartilage. In all other vertebrates, cartilage is the principal component of the embryonic skeleton, but during development it is gradually replaced by bone.

Bone consists mostly of extracellular matrix material that contains collagen fibers as well as crystals of insoluble calcium phosphate, which give bone its rigidity and hardness. The skeleton serves as a reservoir of calcium for the rest of the body and is in dynamic equilibrium with soluble calcium in the extracellular fluids of the body. This equilibrium is under hormonal control by calcitonin and parathyroid hormone (see Figure 38.9). If too much calcium is taken from the skeleton, the bones are seriously weakened.

The living cells of bone—called osteoblasts, osteocytes, and osteoclasts—are responsible for the dynamic remodeling of bone structure that is constantly under way. **Osteoblasts** lay down new matrix on bone surfaces. These cells gradually become surrounded by matrix and eventually become enclosed within the bone, at which point they cease laying down matrix but continue to exist within small lacunae (cavities) in the bone. In this state they are called **osteocytes**. In spite of the vast amounts of matrix between them, osteocytes remain in contact with one another through long cellular extensions that run through tiny channels in the bone. Communication between osteocytes is believed to be important in controlling the activities of the cells that are laying down new bone or eroding it away.

The cells that erode or reabsorb bone are the **osteoclasts**. They are derived from the same cell lineage that produces the white blood cells. Osteoclasts burrow into bone, forming cavities and tunnels. Osteo-

an *axial skeleton,* which includes the skull, vertebral column, and ribs, and an *appendicular skeleton,* which includes the pectoral girdle, the pelvic girdle, and the bones of the arms, legs, hands, and feet. Endoskeletons do not provide the protection that exoskeletons do, but their advantage is that bones continue to grow. Because bones are inside the body, the body can enlarge without shedding its skeleton.

The endoskeleton consists of two kinds of connective tissue: cartilage and bone. Connective tissue cells produce large amounts of extracellular matrix material. The matrix material produced by cartilage cells is a rubbery mixture of proteins and polysaccharides. The principal protein in the matrix is collagen. Collagen fibers run in all directions through the gel-like matrix and give it the well-known strength and resiliency of "gristle."

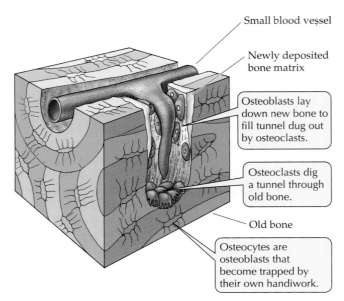

44.16 Renovating Bone Bones are constantly being remodeled by osteoblasts, which lay down bone, and osteoclasts, which dissolve bone.

blasts follow osteoclasts, depositing new bone (Figure 44.16). Thus the interplay of osteoblasts and osteoclasts constantly replaces and remodels the bones.

How the activities of these cells are coordinated is not understood, but stress placed on bones provides information used in the process. A remarkable finding in studies of astronauts who had spent long periods in zero gravity was that their bones had decalcified. Conversely, certain bones of athletes can become considerably thicker than they were before training or than the same bones in nonathletes. Both thickening and thinning of bones are experienced by someone who has a leg in a cast for a long time. The bones of the uninjured leg carry the person's weight and thicken, while the bones of the inactive leg in the cast thin. The jawbones of people who lose their teeth experience less compressional force during chewing and become considerably remodeled.

Bones develop from connective tissues

Bones are divided into two types on the basis of how they develop. *Membranous bone* forms on a scaffolding of connective-tissue membrane; *cartilage bone* forms first as cartilaginous structures and is gradually hardened (ossified) to become bone. The outer bones of the skull are membranous bones; the bones of the limbs are cartilage bones.

Cartilage bones can grow throughout ossification. For example, the long bones of the legs and arms ossify first at the centers and later at each end (Figure 44.17). Growth can continue until these areas of ossification join. The membranous bones forming the skull

cap grow until their edges meet. The soft spot on the top of a baby's head is the point at which the skull bones have not yet joined.

The composition of bone may be **compact** (solid and hard) or **cancellous** (having numerous internal cavities that make it appear spongy, even though it is

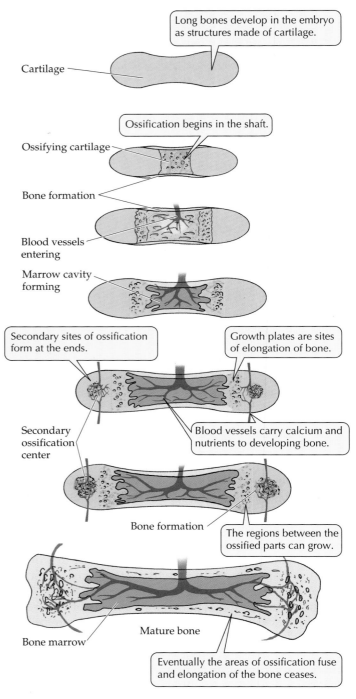

44.17 The Growth of Long Bones In the long bones of human limbs, ossification occurs first at the centers and later at each end.

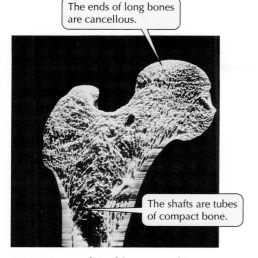

The ends of long bones are cancellous.

The shafts are tubes of compact bone.

44.18 Internal Architecture of Bone Bone may have cancellous (spongy) and compact (hard) regions, as can be seen in this longitudinal section of a long bone.

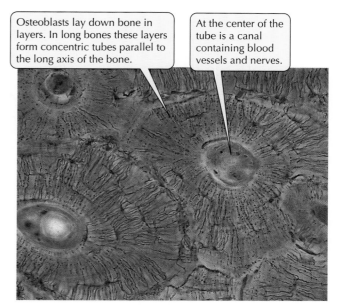

Osteoblasts lay down bone in layers. In long bones these layers form concentric tubes parallel to the long axis of the bone.

At the center of the tube is a canal containing blood vessels and nerves.

44.19 Haversian Systems in Bone This micrograph of a cross section of bone is colored because it was illuminated with polarized light.

rigid). The architecture of a specific bone depends on its position and function, but most bones have both compact and cancellous regions. The shafts of the long bones of the limbs, for example, are cylinders of compact bone surrounding central cavities that contain the bone marrow, where the cellular elements of the blood are made. The ends of long bones are cancellous (Figure 44.18).

Cancellous bone is lightweight because of its numerous cavities, but it is also strong because its internal meshwork constitutes a support system. It can withstand considerable forces of compression. The rigid, tubelike shaft of compact bone can withstand compression and bending forces. Architects and nature alike use hollow tubes as lightweight structural elements. In a solid rod that is subjected to a bending force, one side of the rod is compressed while the other side is stretched, and both help to resist the force. Because the center of a solid rod contributes very little to its ability to resist bending, hollowing out a rod reduces its weight but not its strength.

Most of the compact bone in mammals is called Haversian bone because it is composed of structural units called **Haversian systems** (Figure 44.19). Each system is a set of thin, concentric bony cylinders, between which are the osteocytes in their lacunae. Through the center of each Haversian system runs a narrow canal containing blood vessels (see Figure 44.16). The osteocytes in one Haversian system connect only with osteocytes in the same system; no channels cross the boundaries (called glue lines) between systems. An important feature of Haversian bone is its resistance to fracturing. If a crack forms in one Haversian system, it tends to stop at the nearest glue line.

Bones and joints can work as a lever

Muscles and bones work together around **joints**, where two or more bones come together. Since muscles can only contract and relax, they create movement around joints by working in antagonistic pairs: When one contracts, the other relaxes. With respect to a particular joint, such as the knee, we can refer to the muscle that flexes the joint as the **flexor** and the muscle that extends the joint as the **extensor** (Figure 44.20). The bones that meet at the joint are held in place by **ligaments**, which are flexible bands of connective tissue. Other straps of connective tissue, **tendons**, attach the muscles to the bones. In many kinds of joints, only the tendon spans the joint, sometimes moving over the surfaces of the bone like a rope over a pulley. The tendon of the quadriceps muscle traveling over the knee joint is what is tapped to elicit the knee jerk reflex.

The human skeleton has a wide variety of joints with different ranges of movement (Figure 44.21). The knee joint is a simple *hinge* that has almost no rotational movement and can flex in one direction only. At the shoulders and hips are *ball-and-socket joints* that allow movement in almost any direction. A *pivotal joint* between the two bones of the forearm where they meet at the elbow allows the smaller bone, the radius, to rotate when the wrist is twisted from side to side. Several kinds of joints permit some rotation, but not in all directions as do the ball-and-socket joints.

Bones around joints and the muscles that work with these bones can be thought of as levers. A lever has a *power arm* and a *load arm* that work around a *fulcrum*

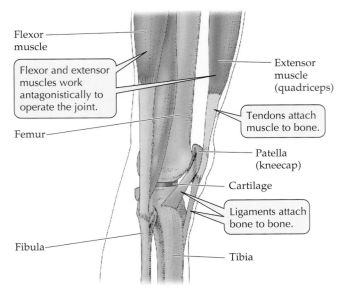

44.20 Joints, Ligaments, and Tendons A side view of the knee shows the interactions of muscle, bone, cartilage, ligaments, and tendons at this crucial and vulnerable human joint.

(pivot). The ratio of these two arms determines whether a particular lever can exert a lot of force over a short distance or is better at translating force into big or fast movements.

Compare the jaw joint and the knee joint (Figure 44.22). The power arm of the jaw is long relative to the load arm, allowing the jaw to apply great pressures over a small distance, such as when you crack a nut with your teeth. The power arm of the lower leg, on the other hand, is short relative to the load arm, so you can run fast, jump high, and deliver swift kicks, but you cannot apply nearly the pressure with a leg that you can with your jaws.

Other Effectors in the Animal Kingdom

Muscles in animals serve almost universally as effectors. Other effectors are more specialized and are not shared by many animal species. Some specialized effectors are used for defense, some for communication, some for capture of prey or avoidance of predators. A discussion of all the effectors animals use would take an entire book, so in this section we mention only a few to give a sampling of their evolutionary diversity.

44.21 Types of Joints The designs of joints are similar to mechanical counterparts and enable a variety of movements.

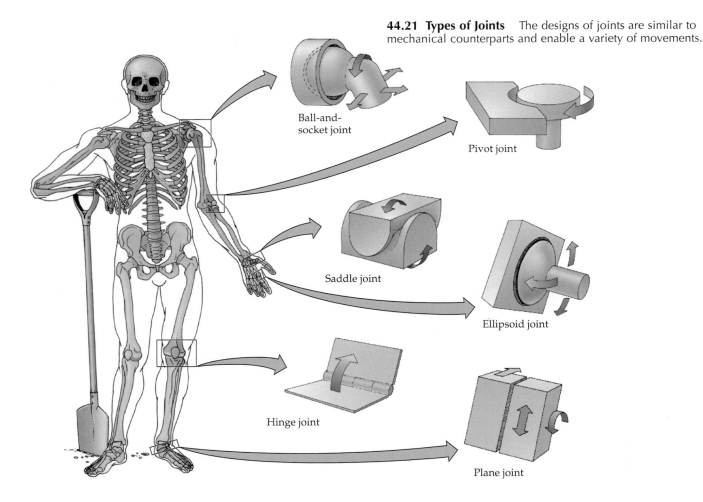

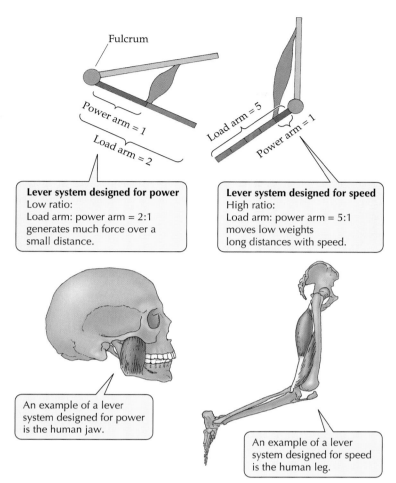

Fulcrum

Power arm = 1
Load arm = 2
Load arm = 5
Power arm = 1

Lever system designed for power
Low ratio:
Load arm: power arm = 2:1
generates much force over a
small distance.

Lever system designed for speed
High ratio:
Load arm: power arm = 5:1
moves low weights
long distances with speed.

An example of a lever
system designed for power
is the human jaw.

An example of a lever
system designed for speed
is the human leg.

44.22 Bones and Joints Work Like Systems of Levers
The lever system can be designed for power or speed.

Nematocysts capture prey and repel enemies

Some animals possess highly specialized organs that are fired like miniature missiles to capture prey and repel enemies. **Nematocysts** are elaborate cellular structures produced only by hydras, jellyfish, and other cnidarians. They are concentrated in huge numbers on the outer surface of the tentacles of the animal. Each nematocyst is made up of a slender thread coiled tightly within a capsule, which is armed with a spine-like trigger projecting to the outside (see Figure 29.8).

When potential prey brushes the trigger, the nematocyst fires, turning the thread inside out and exposing little spines along its base. The thread either entangles or penetrates the body of the victim, and a poison may be simultaneously released around the point of contact. Once the prey is subdued, it is pulled into the mouth of the cnidarian and swallowed. A jellyfish called the Portuguese man-of-war has tentacles that can be several meters long. These animals can capture, subdue, and devour full-grown mackerel, and the poi-

son of their nematocysts is so potent that it can kill a human who becomes tangled in the tentacles.

Chromatophores enable animals to change color

A change in body color is an effector response that some animals use to camouflage themselves in a particular environment or to communicate with other animals. **Chromatophores** are pigment-containing cells in the skin that can change the color and pattern of the animal. Chromatophores are under nervous or hormonal control, or both; in most cases they can effect a change within minutes or even seconds.

In squid, sole, and flounder, all of which spend much time on the seafloor, and in the famous chameleons (a group of African lizards; see Figure 30.24c) and a few other animals, chromatophores enable the animal to blend in with the background on which it is resting and thus be more likely to escape discovery by predators (Figure 44.23a). In other kinds of fishes and lizards, a color change sends a signal to potential mates and territorial rivals of the same species.

There are three principal types of chromatophore cells. The most common type has fixed cell boundaries, within which pigmented granules may be moved about by microfilaments. When the pigment is concentrated in the center of each chromatophore, the animal is pale; the animal turns darker when the pigment is dispersed throughout the cell (Figure 44.23b). Some other chromatophores are capable of amoeboid movement. They can mold themselves into shapes with a minimal surface area, leaving the tissue relatively pale, or they can flatten out to make the tissue appear darker.

Cephalopods have chromatophores that can change shape as a result of the action of muscle fibers radiating outward from the cell. When the muscles are relaxed, the chromatophores are small and compact and the animal is pale. To darken the animal, the muscles contract and spread the chromatophores over more of the surface. Chromatophores with different pigments enable animals to assume different hues or to become mottled to match the background more precisely.

Glands can be effectors

Glands are effector organs that produce and release chemicals. Some glands produce chemicals that are responsible for communication among animal cells (see Chapter 38). Other glands produce chemicals that are used defensively or to capture prey.

Certain snakes, frogs, salamanders, spiders, mollusks, and fish have poison glands. Many of the poisons produced by these glands have proven to be of practical use to humans. The poison dendrotoxin, which certain tribes of the Amazonian jungles use on the tips of their arrows, comes from the skin of a frog

(a)

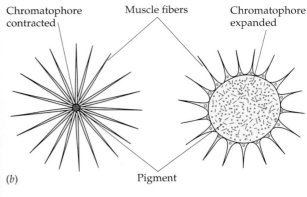

(b)

Chromatophore contracted   Muscle fibers   Chromatophore expanded

Pigment

44.23 Chromatophores Help Animals Camouflage Themselves (a) This octopus has adapted its chromatophores so that it is almost indistinguishable from its background of coral, sponges, and other animals. The octopus is the largest element in the photograph; its eyes are partly closed and its head fills much of the lower left quarter of the scene. (b) The chromatophores of the octopus are highly elastic.

(see Figure 30.21b) and blocks certain potassium channels. The snake venom bungarotoxin inactivates the neuromuscular acetylcholine receptors. The puffer fish poison tetrodotoxin blocks voltage-gated sodium channels. A poison from a mollusk, conotoxin, blocks calcium channels. There are many such examples, and they are useful research tools for neurobiologists.

Not all defensive secretions are poisonous. A well-known example is the odoriferous chemical mercaptan, which is sprayed by skunks. Human olfaction is more sensitive to mercaptan than to any other compound. The bombardier beetle (see Figure 52.11a) is another spectacular example of an animal that releases an irritating defensive secretion.

The glands that produce and release pheromones, chemicals used in communication among animals—which we mentioned when we discussed chemical sensors in Chapter 42 and which we will discuss further in Chapter 49—are another category of effectors. Still other effector glands produce secretions necessary to facilitate physiological functions. Examples are salivary gland secretions that aid digestion and sweat gland secretions that are an effective means of heat loss.

Electric organs can be shocking

Various fishes can generate electricity, including the electric eel, the knife fish, the torpedo (a type of ray), and the electric catfish. The electric fields they generate are used for sensing the environment, for commu-nication, and for stunning potential predators or prey. The electric organs of these animals evolved from muscle, and they produce electric potentials in the same general way as nerves and muscles do.

Electric organs consist of very large, disc-shaped cells arranged in long rows like stacks of coins. When the cells discharge simultaneously, the electric organ can generate far more current than can nerve or muscle. Electric eels, for example, can produce up to 600 volts with an output of approximately 100 watts—enough to light a row of lightbulbs or to temporarily stun a person.

Summary of "Effectors"

• Effectors enable animals to respond to stimulation from their internal and external environments. Most effector mechanisms generate mechanical forces and use those forces to change the shapes of or to move cells or whole animals.

Cilia, Flagella, and Cell Movement
• Cellular movement comes from two structures, microtubules and microfilaments, both of which consist of long protein molecules that can slide past each other.
• The movements of cilia and flagella depend on microtubules. **Review Figures 44.1, 44.3, 44.4**

Muscle Contraction: Actin and Myosin Interactions
• The three types of vertebrate muscle are smooth, cardiac, and skeletal (striated). **Review Figure 44.6**
• Smooth muscle provides contractile forces for internal organs. The cells in sheets of smooth muscle are electrically coupled through gap junctions, so action potentials that cause contraction can spread rapidly throughout the sheet. Because smooth muscle cell membranes can be depolarized by stretching, smooth muscle can respond to being stretched by contracting.

- The walls of the heart consist of sheets of branching cardiac muscle cells that are resistant to tearing. The cells are electrically coupled through gap junctions, so action potentials spread rapidly throughout sheets of cardiac muscle and cause coordinated contraction of the cells. Some cardiac cells depolarize spontaneously and serve as pacemaker cells responsible for generating the heartbeat.
- Skeletal, or striated, muscle consists of bundles of muscle fibers. Each muscle fiber is a huge cell containing multiple nuclei and numerous myofibrils, which are bundles of the microfilaments actin and myosin. The regular, overlapping arrangement of the actin and myosin microfilaments into sarcomeres gives skeletal muscle its striated appearance. During contraction, the actin and myosin microfilaments slide in a telescoping fashion. **Review Figure 44.7.**
- The molecular mechanism of muscle contraction involves binding between the globular heads of myosin molecules and actin. Upon binding, the myosin head changes conformation, causing the two microfilaments to move relative to each other. Release of the myosin heads from actin and their return to their original conformation requires ATP. **Review Figure 44.8**
- The plasma membrane of the muscle fiber is continuous with a system of T (transverse) tubules that extend deep into the cytoplasm of the cell—the sarcoplasm. **Review Figure 44.9**
- Under resting conditions, strands of the protein tropomyosin, which are wrapped around the actin microfilaments, block the myosin-binding sites, so the actin and myosin microfilaments cannot interact.
- At intervals along the actin microfilament are troponin molecules that are bound to the actin and to the tropomyosin and have an additional binding site for calcium ions. When an action potential spreads across the membrane and through the T tubules, it causes Ca^{2+} ions to be released from the sarcoplasmic reticulum. The Ca^{2+} ions bind to the troponin, change its conformation, which in turn pulls the tropomyosin strands away from the myosin-binding sites. Cycles of binding and release occur, and the muscle fiber contracts until the Ca^{2+} is returned to the sarcoplasmic reticulum. **Review Figure 44.10**
- In striated muscle, a single action potential causes a minimum unit of contraction called a twitch. Twitches occurring in rapid succession can be summed, thus increasing the strength of contraction. **Review Figure 44.11**
- Slow-twitch muscle fibers are adapted for extended, aerobic work; fast-twitch fibers are adapted for generating maximum forces for short periods of time. The ratio of slow-twitch fibers to fast-twitch fibers in the muscles of an individual are genetically determined. **Review Figure 44.12**

Skeletal Systems Provide Rigid Supports

- Skeletal systems provide rigid supports against which muscles can pull.
- Hydrostatic skeletons are fluid-filled cavities that can be squeezed by muscles. **Review Figure 44.13**
- Exoskeletons are hardened outer surfaces to which internal muscles are attached. **Review Figure 44.14**
- Endoskeletons are an internal, articulated system of rodlike, platelike, and tubelike rigid supports consisting of bone and cartilage to which muscles attach. **Review Figure 44.15**

- Bone is continually being remodeled by osteoblasts, which lay down new bone, and osteoclasts, which absorb bone. **Review Figure 44.16**
- Bones develop from connective tissue membranes or from cartilage through ossification, and can grow until centers of ossification meet. **Review Figure 44.17**
- Bone can be solid and hard (compact bone), or it can contain numerous internal spaces (cancellous bone). **Review Figures 44.18, 44.19**
- Tendons connect muscles to bones; ligaments connect bones to each other and also help direct the forces generated by muscles by holding tendons in place. **Review Figure 44.20**
- Muscles and bones work together around joints as systems of levers. **Review Figures 44.21, 44.22**

Other Effectors in the Animal Kingdom

- Effector organs other than muscles include nematocysts, chromatophores, glands, and structures that produce sound, light, and electric pulses. **Review Figure 44.23**

Self-Quiz

1. The movement of cilia and flagella is due to the
   - *a.* polymerization and depolymerization of tubulin.
   - *b.* making and breaking of cross-bridges between actin and myosin.
   - *c.* contractions of microtubules.
   - *d.* changes in conformations of dynein molecules.
   - *e.* use of ATP by spokes of the axoneme to contract.

2. Smooth muscle differs from both cardiac and skeletal muscle in that
   - *a.* it can act as a pacemaker for rhythmic contractions.
   - *b.* contractions of smooth muscle are not due to interactions between neighboring microfilaments.
   - *c.* neighboring cells can be in electrical continuity through gap junctions.
   - *d.* neighboring cells are tightly coupled by intercalated discs.
   - *e.* the membranes of smooth muscle cells are depolarized by stretching.

3. Fast-twitch fibers differ from slow-twitch fibers in that
   - *a.* they are more common in the leg muscles of champion sprinters.
   - *b.* they have more mitochondria.
   - *c.* they fatigue less rapidly.
   - *d.* their abundance is more a product of genetics than of training.
   - *e.* they are more common in the leg muscles of champion cross-country skiers.

4. The role of Ca^{2+} in the control of muscle contraction is
   - *a.* to cause depolarization of the T tubule system.
   - *b.* to change the conformation of troponin, thus exposing myosin-binding sites.
   - *c.* to change the conformation of myosin heads, thus causing microfilaments to slide past each other.
   - *d.* to bind to tropomyosin and break actin–myosin cross-bridges.
   - *e.* to block the ATP-binding site on myosin heads, enabling muscle to relax.

5. Which statement about muscle contractions is *not* true?
   a. A single action potential at the neuromuscular junction is sufficient to cause a muscle to twitch.
   b. Once maximum muscle tension is achieved, no ATP is required to maintain that level of tension.
   c. An action potential in the muscle cell activates contraction by releasing Ca^{2+} into the sarcoplasm.
   d. Summation of twitches leads to a graded increase in the tension that can be generated by a single muscle fiber.
   e. The tension generated by a muscle can be varied by controlling how many of its motor units are active.

6. Which statement about the structure of skeletal muscle is true?
   a. The bright bands of the sarcomere are the regions where actin and myosin filaments overlap.
   b. When a muscle contracts, the A bands of the sarcomere (dark regions) lengthen.
   c. The myosin filaments are anchored in the Z lines.
   d. When a muscle contracts, the H bands of the sarcomere (light regions) shorten.
   e. The sarcoplasm of the muscle cell is contained within the sarcoplasmic reticulum.

7. The long bones of our arms and legs are strong and can resist both compressional and bending forces because
   a. they are solid rods of compact bone.
   b. their extracellular matrix contains crystals of calcium carbonate.
   c. their extracellular matrix consists mostly of collagen and polysaccharides.
   d. they have a very high density of osteoclasts.
   e. they consist of lightweight cancellous bone with an internal meshwork of supporting elements.

8. If we compare the jaw joint with the knee joint as lever systems,
   a. the jaw joint can apply greater compressional forces.
   b. their ratios of power arm to load arm are about the same.
   c. the knee joint has greater rotational abilities.
   d. the knee joint has a greater ratio of power arm to load arm.
   e. only the jaw is a hinged joint.

9. Which statement about skeletons is true?
   a. They can consist only of cartilage.
   b. Hydrostatic skeletons can be used only for amoeboid locomotion.
   c. An advantage of exoskeletons is that they can continue to grow throughout the life of the animal.
   d. External skeletons must remain flexible, so they never include calcium carbonate crystals as do bones.
   e. Internal skeletons consist of four different types of bone: compact, cancellous, dermal, and cartilage.

10. Chemicals used by neurophysiologists to block voltage-gated sodium channels have come from
    a. chromatophores.
    b. nematocysts.
    c. electric eels.
    d. luciferase.
    e. poison glands of fish.

Applying Concepts

1. Describe in outline form all the events that occur between the arrival of an action potential at a motor nerve terminal and the contraction of a muscle fiber.

2. How do we know that the basis for the movement of cilia resides in the dynein components of the axoneme?

3. Wombats are powerful digging animals, and kangaroos are powerful jumping animals. How do you think the structures of their legs would compare in terms of their designs as lever systems?

4. Maria and Margaret are identical twin sisters. Their mother was an Olympic marathon runner and their father was on the varsity rowing team in college. Maria has become a serious cross-country skier and Margaret has joined the track team as a sprinter. Which one do you think will have the greatest chance of becoming a champion in her sport, and why?

5. If an adolescent breaks a leg bone close to the ankle joint, after the break heals that leg may not grow as long as the other one. Explain why, including an explanation of why the leg grows at all.

Readings

Cameron, J. N. 1985. "Molting in the Blue Crab." *Scientific American*, May. An explanation of how an arthropod deals with its exoskeleton in order to grow.

Caplan, A. I. 1984. "Cartilage." *Scientific American*, October. An exploration into the surprisingly diverse group of roles played by cartilage, a component of the vertebrate skeletal system, in the developing and mature animal.

Carafoli, E. and J. T. Penniston. 1985. "The Calcium Signal." *Scientific American*, November. A discussion of the roles of calcium as a second messenger and in muscle contraction.

Cohen, C. 1975. "The Protein Switch of Muscle Contraction." *Scientific American*, November. An examination of the interaction between muscle proteins and calcium ions.

Gans, C. 1974. *Biomechanics: An Approach to Vertebrate Biology*. University of Michigan Press, Ann Arbor. A small classic on the architecture of animals and how their structure is adapted to their environment and lifestyle.

Hadley, N. F. 1986. "The Arthropod Cuticle." *Scientific American*, July. A description of the properties of the exoskeleton of the most abundant and diversified phylum of animals.

Lazarides, E. and J. P. Revel. 1979. "The Molecular Basis of Cell Movement." *Scientific American*, May. An exploration of microtubules and microfilaments in action.

Randall, D., W. Burggren and K. French. 1997. *Animal Physiology: Mechanisms and Adaptations*, 4th Edition. W.H. Freeman, New York. An outstanding textbook of animal physiology, with chapters on muscle action and on cell motility.

Schmidt-Nielsen, K. 1997. *Animal Physiology: Adaptation and Environment*, 5th Edition. Cambridge University Press, New York. Chapter 11 gives a marvelous treatment of skeletons, muscles, and other effectors.

Vander, A. J., J. H. Sherman and D. S. Luciano. 1994. *Human Physiology: The Mechanisms of Body Function*, 6th Edition. McGraw-Hill, New York. Chapter 11 deals with the structure and function of human muscle.

Chapter 45

Gas Exchange in Animals

Holding the Breath of Life
Who can swim the farthest under water? We may hold our breath for a couple of minutes while we find out, but the urge to breathe becomes overpowering. If deprived of oxygen-rich air, we would die quickly.

*T*he wail that heralds the birth of an infant and brings joy to its parents also initiates the infant's breathing, a process that continues every minute of a person's life. Whether awake or asleep, exercising or resting, talking or eating, we must breathe. About 15 times per minute, 900 times per hour, 22,000 times per day, 8 million times per year.

For brief moments, you can hold your breath, but the urge to breathe mounts rapidly and soon becomes overwhelming. Anything that prevents breathing, such as drowning or choking, rapidly leads to death. And in spite of the sophisticated technology of medical intensive care that enables a machine to take over the breathing functions of an otherwise terminal patient, we still think of "the last breath" as the moment of death. Breathing is the most visible evidence of life.

The purpose of breathing is to facilitate gas exchange—the exchange of O_2 and CO_2 between the environment and the body. The cells of the body need oxygen because they require a constant supply of energy in the form of ATP to carry out their functions. ATP is produced through the oxidation of nutrient molecules (see Chapter 7). As long as oxygen is available, tissues can use their nutrients efficiently to produce an adequate supply of ATP.

Some tissues and some species have adaptations that enable them to function without oxygen, anaerobically, for considerable amounts of time. Your muscle tissue can produce ATP by glycolysis for many minutes without oxygen, a goldfish can depend on glycolysis for days without oxygen, and some turtles can live at low temperatures for months without oxygen. In all of these cases, however, oxygen debt accumulates and must be paid back eventually if life is to continue.

For us humans, our brain cells limit how long we can go without oxygen. Our brains have little capacity to function in the absence of oxygen, and without oxygen we lose consciousness in only a few minutes. The oxidation of nutrients to produce ATP also produces waste products. The major waste produced is carbon dioxide, which must be carried away from the cells to prevent their environment from becoming too acidic. Breathing facilitates the loss of CO_2 from the body to the environment and the uptake of O_2 from the environment to the body.

It is misleading that we use the word "respiration" to refer both to breathing and to the cellular utilization of oxygen and production of carbon dioxide. Breathing is not even the same as the exchange of oxygen and carbon dioxide between the body tissues and the external medium. These exchanges occur over highly specialized, delicate membranous surfaces, and only by diffusion—the random movement of molecules. Breathing only brings the outside medium, air or water, to and from the gas exchange surfaces where diffusion takes place.

As we will see, some animals can accomplish adequate gas exchange without breathing. These are usually small, not very active, aquatic animals. Bigger, more active, nonaquatic animals have evolved hand in hand with a diversity of physiological and anatomical adaptations that maximize respiratory gas exchange.

The importance of these adaptations for gas exchange is made clear by a comparison of the metabolic demands of some different species. Compare, for example, the oxygen requirement of a frog sitting at room temperature (about 0.01 liters of O_2 per hour) with that of a human running a marathon (about 150 liters of O_2 per hour). Then continue the comparison by considering that the same runner can be outdistanced and outlasted by some species of fish breathing water and by birds flying at very high altitudes where there is little O_2.

In this chapter we will learn about the physical limits on respiratory gas exchange. Then we'll examine a variety of respiratory gas exchange systems and adaptations that have enabled animals to achieve high levels of metabolic performance in spite of those physical limits.

Respiratory Gas Exchange

Diffusion is the only means by which respiratory gases are exchanged between the internal body fluids of an animal and the outside medium, air or water. No active transport mechanisms move respiratory gases across membranes. This is true for all gas exchange systems, despite their diverse structures.

To help you understand the adaptations of gas exchange systems that we will cover in this chapter, review the discussion of diffusion in Chapter 5 (pages 105–109). Because diffusion is strictly a physical phenomenon, it is limited by physical factors such as whether gases are exchanged from air or water and the temperature at which such exchanges take place.

Air is a better respiratory medium than water

Animals that breathe water must have more efficient gas exchange systems than animals that breathe air. O_2 can be exchanged more easily in air than in water for several reasons. First, the oxygen content of air is much higher than the oxygen content of an equal volume of water. The maximum O_2 content of a rapidly flowing stream splashing over rocks and tumbling over waterfalls is less than 10 ml of O_2 per liter of water. The O_2 content of fresh air is about 200 ml of O_2 per liter of air. Second, O_2 diffuses much more slowly in water than in air. In a still pond, the O_2 content of the water can be zero only a few millimeters below the surface. Third, when an animal breathes, it performs work to move air or water over its gas exchange surfaces. More energy is required to move water than to move air because water is denser and more viscous than air.

The slow diffusion of O_2 molecules in water imposes a gas exchange constraint on air-breathing animals as well as on water-breathing animals. Eukaryotic cells respire in their mitochondria, which are in the cytoplasm—an aqueous medium. Cells are bathed with extracellular fluid—also an aqueous medium. The slow rate of O_2 diffusion in water limits the efficiency of O_2 distribution from gas exchange membranes to the sites of cellular respiration in both air-breathing and water-breathing animals.

In water, a supply of O_2 that is adequate to support the metabolism of a typical animal cell can be obtained only if the diffusion path is shorter than about 1 mm. Therefore, if an animal lacks an internal system for transporting gases, no cell may be more than about 1 mm from the outside world. This is a severe size limit, but one way to accommodate it and still grow bigger is to have a flat, leaflike body plan, which is common among simple invertebrates (Figure 45.1a). Another way is to have a very thin body built around a central cavity through which water circulates (Figure 45.1b). Otherwise, specialized structures are required to provide an increased surface area for diffusion, an internal circulatory system is needed to carry gases to and from these exchange structures, and the outer surfaces of these exchange structures must be continuously bathed with fresh air or water (Figure 45.1c).

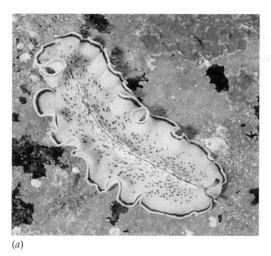

(a)

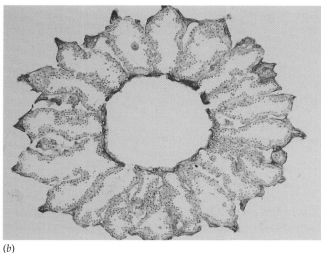

(b)

(c)

45.1 Keeping in Touch with the Medium (a) No cell in the leaflike body of this marine flatworm is more than a millimeter away from seawater. (b) The same is true of sponges; they have body walls perforated by many channels lined with flagellated cells. These channels communicate with the outside world and with a central cavity. The flagella maintain currents of water through the channels, through the central cavity, and out of the animal. Every cell in the sponge is very close to the respiratory medium. (c) The gills of this newt project like a feathery fringe and provide a large surface area for gas exchange. Blood circulating through the gills comes into close contact with the respiratory medium.

Rising temperature creates respiratory problems for aquatic animals

Temperature is a crucial factor influencing gas exchange in animals that breathe water. All water breathers are ectotherms. The body temperatures of aquatic ectotherms are tied to the temperature of the water around them. As the water temperature rises, so does their body temperature, causing an exponential increase in energy expenditure and oxygen demand, as described by a Q_{10} relationship (see Chapter 37).

But warm water holds less oxygen than cold water does (just think of what happens when you open a warm bottle of soda). Aquatic ectotherms are in a double bind: As the temperature of their environment goes up, their need for O_2 increases but the availability of O_2 in their environment decreases (Figure 45.2). In addition, if the animal performs work to move water across its gas exchange surfaces (as fish do, for example), the energy the animal must expend to breathe increases as water temperature rises. Thus as water temperature goes up, the water breather must extract more and more O_2 from its environment, and a lower

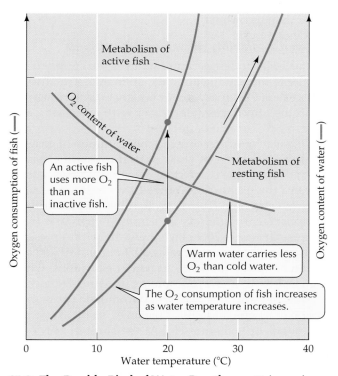

Metabolism of active fish

O_2 content of water

Oxygen consumption of fish (—)

Oxygen content of water (—)

An active fish uses more O_2 than an inactive fish.

Metabolism of resting fish

Warm water carries less O_2 than cold water.

The O_2 consumption of fish increases as water temperature increases.

Water temperature (°C)

45.2 The Double Bind of Water Breathers Fish need more oxygen in warmer water—but warm water carries *less* oxygen than cold.

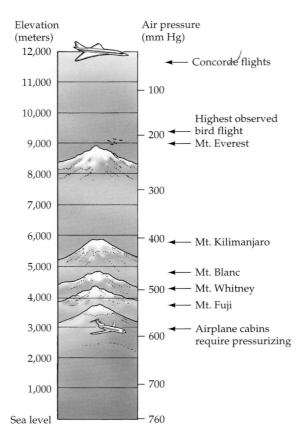

Elevation (meters) / Air pressure (mm Hg)

- 12,000 — Concorde flights
- 11,000 — 100
- 10,000
- 9,000 — 200 — Highest observed bird flight / Mt. Everest
- 8,000 — 300
- 7,000
- 6,000 — 400 — Mt. Kilimanjaro
- 5,000 — Mt. Blanc
- 4,000 — 500 — Mt. Whitney / Mt. Fuji
- 3,000 — 600 — Airplane cabins require pressurizing
- 2,000
- 1,000 — 700
- Sea level — 760

45.3 Scaling Heights The oxygen content of the atmosphere decreases with altitude. Therefore, airplane cabins must be pressurized and mountain climbers must carry pressurized containers of oxygen. Birds, however, have been observed flying over the highest peaks.

percentage of that O_2 is available to support activities other than breathing.

Oxygen availability decreases with altitude

Just as a rise in temperature reduces the supply of O_2 available for aquatic animals, an increase in altitude reduces the O_2 supply for air breathers. The amount of O_2 in the atmosphere decreases with increasing altitude.

One way of expressing the concentration of gases in air and in water is by their **partial pressures**. At sea level, the pressure exerted by the atmosphere is equivalent to the pressure produced by a column of mercury 760 mm high. Therefore, **barometric pressure** (atmospheric pressure) is 760 mm of mercury (Hg). Because dry air is 20.9 percent O_2, the partial pressure of oxygen (P_{O_2}) at sea level is 20.9 percent of 760 mm Hg, or about 159 mm Hg. At higher elevations, where there is less air above, barometric pressure declines.

At an altitude of 5,300 m, barometric pressure is only half as much as it is at sea level, so the P_{O_2} at that altitude is only 80 mm Hg. At the summit of Mount Everest (8,848 m) the P_{O_2} is only about 50 mm Hg, roughly one-third what it is at sea level. Remember that the diffusion of O_2 into the body depends on O_2 concentration differences between the air and the body fluids, so the drastically reduced O_2 concentra-

tion in the air at a high altitude constrains O_2 uptake. This low O_2 concentration is the reason that mountain climbers who venture to the heights of Mt. Everest breathe O_2 from pressurized bottles that they carry with them (Figure 45.3).

Carbon dioxide is lost by diffusion

Respiratory gas exchange is a two-way street. CO_2 diffuses out of the body as O_2 diffuses in. Given the same concentration gradient, CO_2 and O_2 molecules diffuse at about the same rate, whether in air or in water. However, the concentration gradients for the diffusion of O_2 and CO_2 across gas exchange membranes are generally not the same.

The concentration of CO_2 in the atmosphere is so low, and its solubility in the aquatic environment is so high, that the diffusion of CO_2 from an animal is usually not a problem. However, transporting CO_2 from where it is produced in the body to where it diffuses into the environment can be a limiting factor in gas exchange and hence metabolism.

Adaptations for Gas Exchange

In this section we present a diversity of animal adaptations that maximize rates of gas exchange. After considering the physics of diffusion, we'll examine the roles of surface area and fluid movement and apply these to understanding gas exchange systems in insects, fish, birds, and mammals. We'll end this section with a detailed description of the human lung and its associated structures.

Fick's law applies to all systems of gas exchange

Animals have evolved a great diversity of adaptations to maximize their rates of gas exchange. But all these adaptations work through influencing a few physical parameters that are described by a simple equation called **Fick's law of diffusion**:

$$Q = DA \frac{C_1 - C_2}{L}$$

Fick's law describes the rate, Q, at which a substance diffuses between two locations. D is the diffusion coefficient, which is a characteristic of a substance diffusing in a particular medium at a particular temperature. For example, perfume has a higher D than motor oil, and substances diffuse faster in air than in water. A is the cross-sectional area over which the substance is diffusing. C_1 and C_2 are the concentrations of the substance at two locations, and L is the distance between those locations. Therefore, $(C_1 - C_2)/L$ is a concentration gradient.

Animals can maximize D for respiratory gases by using air rather than water for the gas exchange medium whenever possible. All other adaptations for maximizing respiratory gas exchange must influence the surface area of exchange or the concentration gradient across that surface area.

Respiratory organs have large surface areas

Many anatomical adaptations maximize specialized body surface areas over which gases can diffuse.

External gills are highly branched and folded elaborations of the body surface that provide a large surface area for gas exchange (Figure 45.4a). They consist of thin, delicate membranes that minimize the length of the path traversed by diffusing molecules of O_2 and CO_2 (see Figure 45.1c). Because external gills are vulnerable to damage and are tempting morsels for carnivorous organisms, it is not surprising that protective body cavities for gills have evolved. Many mollusks, arthropods, and fishes have **internal gills** in such cavities.

Like water breathers, air-breathing animals have adapted by increasing their surface area for gas exchange, but their structures are quite different from gills (Figure 45.4b). First, gas exchange surfaces in air breathers must be in moist internal cavities to protect them from drying. Second, surface elaborations such as gills work only in water; without water for support, they collapse and stick together like the pages of a wet magazine and thus lose effective surface area. That's why a fish suffocates in air in spite of the much higher O_2 concentration.

The gas exchange structures, or **lungs**, of most air-breathing vertebrates are highly divided, elastic air sacs, which we describe in greater detail later in the chapter. The gas exchange structures of insects are highly branched systems of air-filled tubes that branch through all the tissues of the insect's body. In both cases the surface areas for gas exchange are greatly enhanced by their division into many small units.

Ventilation and perfusion maximize concentration gradients

Fick's law of diffusion points to other possible adaptations besides increasing surface area that can increase respiratory gas exchange. Animals can maximize the concentration gradients for the respiratory gases across the gas exchange membranes in several ways.

First, gill and lung membranes can be very thin so that the path length for diffusion (L) is small. Second, the environmental side of the exchange surfaces can be exposed to fresh respiratory medium (air or water) with the highest possible O_2 concentration and the lowest possible CO_2 concentration. Third, the internal sides of the exchange surfaces can be

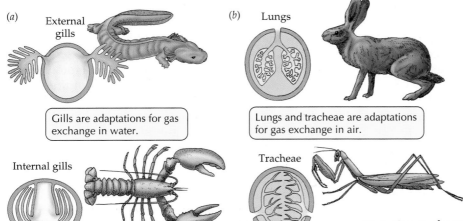

(a) External gills

Gills are adaptations for gas exchange in water.

Internal gills

(b) Lungs

Lungs and tracheae are adaptations for gas exchange in air.

Tracheae

45.4 Gas Exchange Organs   Increased surface area for the diffusion of respiratory gases is a common feature of animals.

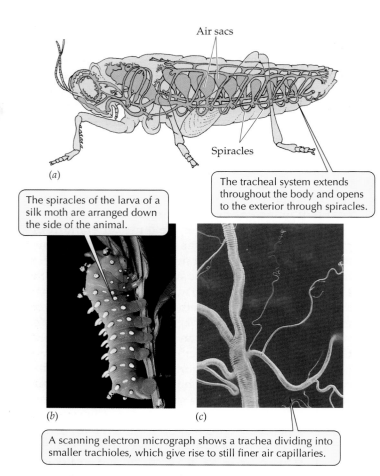

Air sacs

(a)

Spiracles

The spiracles of the larva of a silk moth are arranged down the side of the animal.

The tracheal system extends throughout the body and opens to the exterior through spiracles.

(b)   (c)

A scanning electron micrograph shows a trachea dividing into smaller trachioles, which give rise to still finer air capillaries.

45.5 The Tracheal Gas Exchange System of Insects
Respiratory gases diffuse through a system of air tubes (tracheae) that open to the external environment through holes called spiracles.

maintained at the lowest possible O_2 concentration and highest possible CO_2 concentration. Mechanisms that move substances over both sides of the gas exchange surfaces are important for maximizing concentration gradients.

External gills are exposed to fresh respiratory medium as they wave around in the environment. But gas exchange surfaces that are enclosed in body cavities must be **ventilated**; that is, the animal must move fresh respiratory medium over internal gills or lungs. Breathing consists of the movements that ventilate gills or lungs. **Perfusion** is the movement of blood across the internal side of the gas exchange membranes. Blood carries O_2 away as it diffuses across from the environmental side, and it brings CO_2 to the exchange surfaces so that it can diffuse in the opposite direction.

An animal's **gas exchange system** is made up of its gas exchange surfaces and the mechanisms it uses to ventilate and perfuse those surfaces. The following sections describe four gas exchange systems. First we

will look at the unique gas exchange system of insects. Then we'll describe two remarkably efficient systems: fish gills and bird lungs. Finally, we'll discuss mammalian lungs, which in comparison to fish gills and bird lungs are a relatively inefficient gas exchange system.

INSECT TRACHEAE. Respiratory gases diffuse through air most of the way to and from every cell of an insect's body. This diffusion is achieved through a system of air tubes, or **tracheae**, that open to the outside environment through holes called **spiracles** in the sides of the abdomen (Figure 45.5a and b). The tracheae branch into even finer tubes, or tracheoles, until they end in tiny **air capillaries** (Figure 45.5c). In the insect's flight muscle and other highly active tissue, no mitochondrion is more than a few micrometers away from an air capillary.

Because the diffusion rate of oxygen is about 300,000 times higher in air than in water, air capillaries enable insects to supply oxygen to their cells at high rates. Many insects metabolize at extremely high rates, but their relatively simple gas exchange system is well able to provide them with the oxygen they need. However, the rate of diffusion in insect tracheae and air capillaries is limited by the small diameter and by the length of these dead-end airways, so insects are relatively small animals.

Some species of bugs that dive and stay under water for long periods make use of an interesting variation on diffusion. These bugs carry with them a bubble of air. A small bubble may not seem like a very large reservoir of oxygen, yet these bugs can stay under water almost indefinitely with their small air tanks. The secret has to do with the partial pressure of O_2 in the bubble.

When the bug dives, the air bubble contains about 80 percent nitrogen and 20 percent O_2. As the insect consumes the O_2 in its bubble, the bubble shrinks a little. The bubble doesn't disappear, because it consists mostly of nitrogen, which the insect doesn't consume. When the partial pressure of O_2 in the bubble falls below the partial pressure of O_2 in the surrounding water, O_2 diffuses from the water into the bubble. For these small bugs, the rate of O_2 diffusion into the bubble is enough to meet the O_2 demand of the animal while it is under water.

FISH GILLS. The internal gills of fishes are marvelously adapted for gas exchange. They offer a large surface area for the exchange of gases between blood and water. They are supported by five or six bony **gill arches** on either side of the fish between the mouth cavity and the protective **opercular flaps**. Water flows into and then out the fish's mouth from under the opercular flaps.

Each gill arch is lined with hundreds of leaf-shaped gill filaments arranged in two columns. These columns of gill filaments point toward the opercular opening, which is the direction of water flow. The tips of the gill filaments of adjacent arches interlock. The upper and lower flat surfaces of each gill filament have rows of evenly spaced folds, or **lamellae**, which greatly increase the gill surface area. The enormous surface area of the lamellae is the site of gas exchange (Figure 45.6). The interlocking network of gill filaments and lamellae directs the flow so that practically all water that passes across the gills comes into close contact with the gas exchange surfaces.

The flow of blood perfusing the inner surfaces of the lamellae is unidirectional because of the arrangement of the **afferent blood vessels**, which bring blood to the gills, and the **efferent blood vessels**, which take blood away from the gills (see Figure 45.6). Blood flows through the lamellae in the opposite direction of the flow of water over the lamellae. Such **countercurrent flow** makes gas exchange much more efficient than parallel flow, as explained in the figure. Countercurrent exchange is an important principle in several different physiological systems. It applies to the transfer of heat (see the discussion of heat exchangers in "hot" fish; Figure 37.18), as well as to substances other than oxygen. In Chapter 48 we will see how countercurrent exchange in a structure called the loop of Henle allows the formation of a steep salt concentration gradient within the vertebrate kidney.

The very delicate structure of the lamellae minimizes the path length for diffusion of gases between blood and water. Blood travels in blood vessels through the gill arches and the gill filaments, but in the lamellae blood flows between the two surfaces of the lamellae as a sheet not much more than one red blood cell thick. The surfaces of the lamellae consist of highly flattened epithelial cells with almost no cytoplasm, so the water and the red blood cells are separated by little more than 1 or 2 μm.

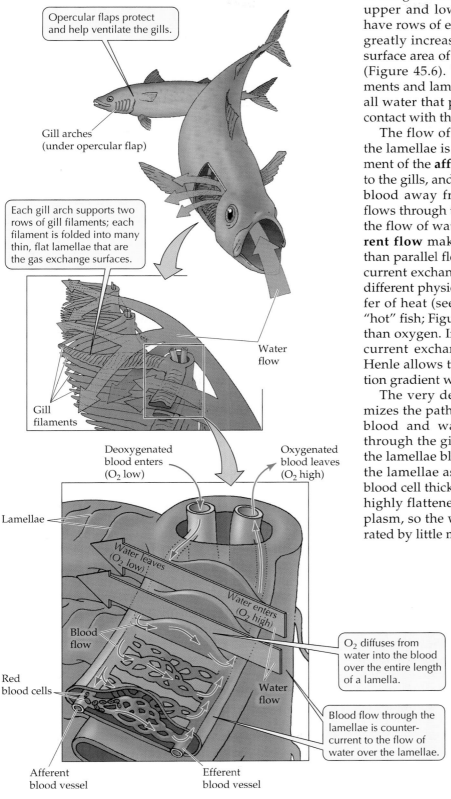

Opercular flaps protect and help ventilate the gills.

Gill arches (under opercular flap)

Each gill arch supports two rows of gill filaments; each filament is folded into many thin, flat lamellae that are the gas exchange surfaces.

Water flow

Gill filaments

Deoxygenated blood enters (O₂ low)

Oxygenated blood leaves (O₂ high)

Lamellae

Water leaves (O₂ low)

Water enters (O₂ high)

Blood flow

O₂ diffuses from water into the blood over the entire length of a lamella.

Red blood cells

Water flow

Blood flow through the lamellae is countercurrent to the flow of water over the lamellae.

Afferent blood vessel

Efferent blood vessel

45.6 Countercurrent Exchange in Gills
Blood flows through the lamellae of fish gills in the opposite direction (left to right, in this depiction) of the flow of water (right to left) over the lamellae. Oxygen begins to diffuse into the blood at the left; even though O₂ levels in the water are low at this point, they are still higher than in the deoxygenated blood. In countercurrent flow the water *always* has more O₂ than the blood, so O₂ diffuses into the blood for the entire length of the lamellae. If the blood and water flow were parallel rather than countercurrent, at some point their O₂ levels would equalize and oxygen would stop diffusing from the water into the blood.

Besides enlarging the surface area and shortening the diffusion path, what more can be done to maximize the rate of diffusion? The concentration difference of O_2 between water and blood can be maximized. Fish accomplish this task by ventilating the external surface and perfusing the internal surface of the lamellae. A constant flow of water moving over the gills maximizes the O_2 concentration on the external surfaces. On the internal side, the circulation of blood minimizes the concentration of O_2 by sweeping the O_2 away as rapidly as it diffuses across.

Most fishes ventilate the external surfaces of their gills by means of a two-pump mechanism that maintains a unidirectional and constant flow of water over the gills. The closing and contracting of the mouth cavity acts as a *positive-pressure pump*, pushing water over the gills. The opening and closing of the opercular flaps acts as a *negative-pressure pump*, or suction pump, pulling water over the gills. Because these pumps are slightly out of phase, they maintain an almost continuous flow of water across the gills.

In summary, fish can extract an adequate supply of O_2 from meager environmental sources by maximizing the surface area for diffusion, minimizing the path length for diffusion, and maximizing oxygen extraction efficiency by constant, unidirectional, countercurrent flow of blood and water over the opposite sides of the gas exchange surfaces.

UNIDIRECTIONAL AIRFLOW IN BIRD LUNGS.   Birds can sustain extremely high levels of activity for much longer periods than mammals can—and at very high altitudes, where mammals cannot even survive because the oxygen content of the air is so low. Yet the lungs of a bird are smaller than those of a mammal of a similar size.

A unique feature of birds is that in addition to lungs they have **air sacs** at several locations in their bodies. The air sacs connect with one another, with the lungs, and with air spaces (another unique feature of birds) in some of the bones of the bird (Figure 45.7). Even though the air sacs receive inhaled air as the lungs do, the air sacs are not gas exchange surfaces. If a sample of air or pure oxygen is tied off in an air sac, its composition does not change rapidly, as it would if O_2 were diffusing into the blood and CO_2 were diffusing into the air sac.

As in other air-breathing vertebrates, air enters and leaves a bird's gas exchange system through a **trachea** (commonly known as the windpipe), which divides into smaller airways called **bronchi** (singular bronchus). But the anatomy of the bird lung is unique. In air-breathing vertebrates other than birds, the bronchi generate trees of branching airways that become finer and finer until they dead-end in clusters of microscopic, membrane-enclosed air spaces, where gases are exchanged. In bird lungs, however, there are no

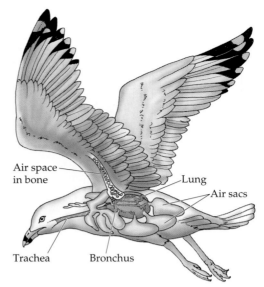

45.7 The Respiratory System of a Bird   The air sacs and air spaces in the bones are unique to bird anatomy.

dead ends, so air can flow completely through the lungs.

The bronchi distribute air to the air sacs and to the lungs. In the lungs the bronchi divide into tubelike **parabronchi** that conduct air unidirectionally through the lungs (Figure 45.8). Air capillaries run between the parabronchi. Thus although air flows unidirectionally through the lungs in the parabronchi, the air can cross between parabronchi through the air capillaries. The air capillaries are the gas exchange surfaces. Since the air capillaries are so small and so numerous, they provide an enormous surface area for gas exchange.

Another unusual feature of bird lungs is that in comparison to mammalian lungs they expand and contract relatively little during a breathing cycle. To make things even more puzzling, bird lungs contract during inhalation and expand during exhalation! The puzzle of how birds breathe was solved by an experiment that placed small oxygen sensors at different locations in the air sacs and airways of birds. When the bird was exposed to pure oxygen for just a single breath, that breath was labeled by its high oxygen concentration. The progress of that breath through the bird's gas exchange system could be followed by its effect on the oxygen sensors. This experiment showed that a single breath remains in the bird's gas exchange system for two cycles of inhalation and exhalation. Here's how it works.

First, the air sacs are divided into an anterior group and a posterior group (Figure 45.9). When the bird takes a breath (labeled with pure oxygen), that breath flows through the trachea primarily to the posterior air sacs. At the same time, the air that is in the lungs (which are now contracting) is flowing into the anterior air sacs. When the bird exhales, the labeled air in

45.8 Air Flows through Bird Lungs Constantly and Unidirectionally The gas exchange surfaces of birds are air capillaries branching off the parabronchi that run through the lungs; these structures are shown in the scanning electron micrograph and in the drawing below.

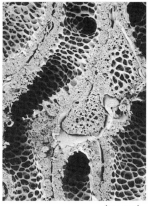

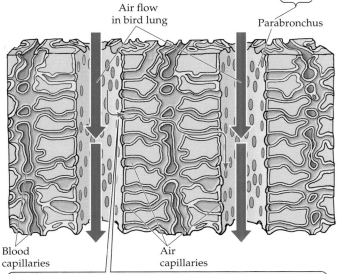

Air flow in bird lung

Parabronchus

Blood capillaries

Air capillaries

Air capillaries carry air from a parabronchus, over blood capillaries where O_2 is absorbed, and out through other parabronchi.

ternal side of the exchange surfaces. In birds, however, the blood flow appears to be crosscurrent (at right angles) rather than countercurrent to the airflow.

It is now clear how birds can fly over Mount Everest. A bird is able to supply its gas exchange surfaces with a continuous flow of fresh air that has an oxygen concentration close to that of the ambient air. Even though the P_{O_2} of the ambient air is only slightly above the P_{O_2} of the blood, O_2 can diffuse from air to blood.

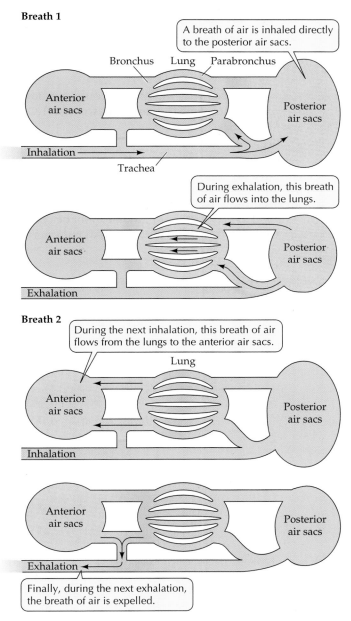

Breath 1

A breath of air is inhaled directly to the posterior air sacs.

Bronchus Lung Parabronchus

Anterior air sacs

Posterior air sacs

Inhalation

Trachea

During exhalation, this breath of air flows into the lungs.

Anterior air sacs

Posterior air sacs

Exhalation

Breath 2

During the next inhalation, this breath of air flows from the lungs to the anterior air sacs.

Lung

Anterior air sacs

Posterior air sacs

Inhalation

Anterior air sacs

Posterior air sacs

Exhalation

Finally, during the next exhalation, the breath of air is expelled.

45.9 The Path of Airflow through Bird Lungs The fresh air a bird takes in with one breath (blue) will travel through the lungs in one direction, from the posterior to the anterior air sacs. Two cycles of inhalation and exhalation are required for the air to travel through the bird's respiratory tract.

the posterior air sacs flows into the lungs (which are now expanding), and the air the bird is exhaling is coming mostly from the anterior air sacs.

During the next inhalation, the labeled air in the lungs flows into the anterior air sacs while the posterior air sacs are filling once again with the fresh air. On the second exhalation after the bird took the labeled breath, the labeled air passes from the anterior air sacs, through the trachea, to the outside. Thus, the air sacs work as bellows maintaining a continuous and unidirectional flow of fresh air through the parabronchi and the air capillaries.

The advantages of this unique gas exchange system are similar to those of fish gills. Because air from the outside flows unidirectionally and practically continuously over the gas exchange surfaces, the concentration of O_2 on the environmental side of those surfaces is maximized. Furthermore, the unidirectional flow of air through the system makes possible a pattern of blood flow to minimize the O_2 concentration on the in-

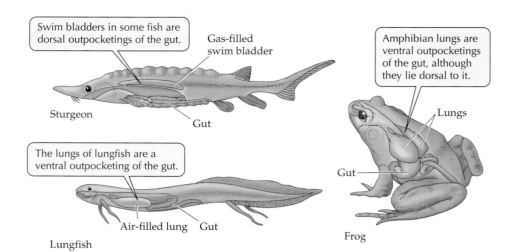

45.10 Lungs Evolved from the Digestive Tract Outpocketings of the digestive tract evolved into swim bladders in some fish. In lungfish and terrestrial vertebrates such as amphibians, gut outpocketings evolved into lungs.

Swim bladders in some fish are dorsal outpocketings of the gut.

Gas-filled swim bladder

Sturgeon

Gut

The lungs of lungfish are a ventral outpocketing of the gut.

Air-filled lung   Gut

Lungfish

Amphibian lungs are ventral outpocketings of the gut, although they lie dorsal to it.

Lungs

Gut

Frog

Next we will see why mammals would not be able to fly over Mount Everest—even if they could fly.

TIDAL BREATHING IN MAMMALS. Vertebrate lungs have their origins in outpocketings of the digestive tract (Figure 45.10). At the beginning of their evolution, lungs were dead-end sacs, and they remain so today in all air-breathing vertebrates except birds. Because lungs are dead-end sacs, ventilation cannot be constant and unidirectional, but must be tidal: Air comes in and flows out by the same route. A spirometer shows how we use our lung capacity in breathing (Figure 45.11). When we are at rest, the amount of air that our normal breathing cycle moves per breath is called the **tidal volume** (about 500 ml for an average human adult).

We can breathe much more deeply and inhale more air than our resting tidal volume; the additional vol-

45.11 Measuring Lung Ventilation with a Spirometer Breathing from a closed reservoir of air and measuring the changes in the volume of that reservoir demonstrate the characteristics of mammalian tidal breathing.

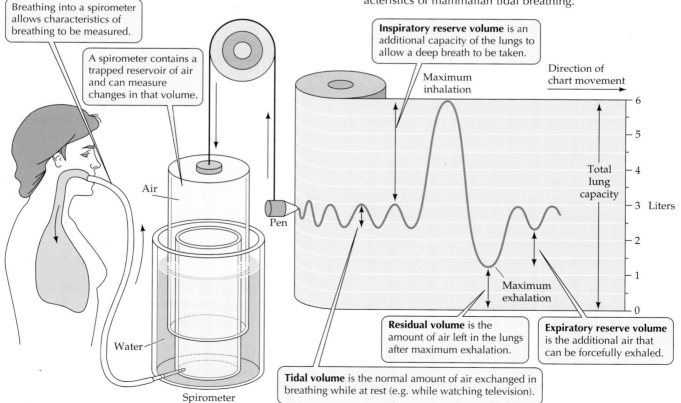

Breathing into a spirometer allows characteristics of breathing to be measured.

A spirometer contains a trapped reservoir of air and can measure changes in that volume.

Air

Pen

Water

Spirometer

Inspiratory reserve volume is an additional capacity of the lungs to allow a deep breath to be taken.

Maximum inhalation

Direction of chart movement

Total lung capacity

Liters

Maximum exhalation

Residual volume is the amount of air left in the lungs after maximum exhalation.

Expiratory reserve volume is the additional air that can be forcefully exhaled.

Tidal volume is the normal amount of air exchanged in breathing while at rest (e.g. while watching television).

ume of air we can take in above normal tidal volume is our **inspiratory reserve volume**. Conversely, we can forcefully exhale more air than we normally do during a resting exhalation. This additional amount of air that can be forced out of the lungs is the **expiratory reserve volume**. But even after the most extreme exhalation possible, some air remains in the lungs. The lungs and airways cannot be collapsed completely; they always contain a **residual volume**. The **total lung capacity** is the sum of the residual volume, expiratory reserve volume, tidal volume, and inspiratory reserve volume.

Tidal breathing severely limits the concentration difference available to drive the diffusion of O_2 from air into the blood. Fresh air is not moving into the lungs during half of the respiratory cycle; therefore, the average O_2 concentration of air in the lungs is considerably less than it is in the air outside the lungs. The incoming air also mixes with the stale air that was not expelled by the previous exhalation. The lung volume that is not ventilated with fresh air is called **dead space**. This dead space consists of the residual volume and, depending on the depth of breathing, some or all of the expiratory reserve volume.

The scale in Figure 45.11 tells us that a tidal volume of 500 ml of fresh air mixes with up to 2,000 ml of stale moist air before reaching the gas exchange surfaces. When the P_{O_2} in the ambient air is 150 mm Hg, the P_{O_2} of the air that reaches the gas exchange surfaces is only about 100 mm Hg. By contrast, the P_{O_2} in the water that bathes the lamellae of the fish gills or in the air that flows through the air capillaries of the bird lung is the same as the P_{O_2} in the outside water or air.

In addition to reducing the concentration difference, tidal breathing reduces the efficiency of gas exchange in another way. It does not allow countercurrent gas exchange between air and blood. Because the air enters and leaves the gas exchange structures by the same route, there is no anatomical way that blood can flow parallel to and countercurrent with the airflow.

Mammalian Lungs and Gas Exchange

Mammalian lungs possess some interesting and important design features that maximize the rate of gas exchange: an enormous surface area and a very short path length for diffusion. Mammalian lungs serve the respiratory needs of mammals well, considering the ecologies and lifestyles of these animals.

Air enters the lungs through the oral cavity or nasal passage, which join in the pharynx (Figure 45.12). The pharynx gives rise both to the esophagus, through which food reaches the stomach, and to the airways. At the beginning of the airways is the **larynx**, or voice box, which houses the vocal cords. The larynx is the "Adam's apple" that you can see and feel on the front of your neck. The larynx opens into the major airway, the *trachea*, which is about the diameter of a garden hose. The thin walls of the trachea are prevented from collapsing by rings of cartilage that support them as the air pressure changes during the breathing cycle. If you run your fingers down the front of your neck just below your larynx, you can feel a couple of these rings of cartilage.

The trachea branches into two slightly smaller *bronchi*, one leading to each lung. The bronchi branch repeatedly to generate a treelike structure of progressively smaller airways extending to all regions of the lungs. Structurally, each of these bronchi is a smaller version of the trachea; they all have supporting cartilage rings. As the branching of the bronchial tree continues to produce still smaller airways, the cartilage supports eventually disappear, marking the transition to **bronchioles**. The branching continues until the bronchioles are smaller than the diameter of a pencil lead, at which point the tiny, thin-walled air spaces called **alveoli** begin.

Alveoli resemble clusters of grapes on a system of stems (see Figure 45.12). The "stems" are the bronchioles, which continue to branch about six more times. Finally, terminal bronchioles end in alveoli. The alveoli are the sites of gas exchange. Because the airways only conduct the air to and from the alveoli, their volume is physiological dead space. If you trace an airway from the primary bronchus leaving the trachea down to the terminal bronchiole, you pass about 23 branching points. Thus there are 2^{23} terminal bronchioles—a very large number. The number of alveoli is even larger, about 300 million in human lungs. Even though each alveolus is very small, the combined surface area for diffusion of respiratory gases is about 70 m^2—the size of a badminton court.

Each alveolus consists of very thin cells. Between and surrounding the alveoli are networks of the smallest of blood vessels, the **capillaries**, also made up of exceedingly thin cells. Where blood vessel meets alveolus, very little space separates them (see Figure 45.12), so the length of the diffusion path between the air and the blood is only 2 μm. Even the diameter of a red blood cell is much greater—about 7 μm.

Respiratory tract secretions aid breathing

Mammalian lungs have two other adaptive features, although these features do not directly influence gas exchange properties. They are the production of mucus and the production of surfactant. (A surfactant is any substance that reduces the surface tension of a liquid.)

What is surface tension and why do we need to consider it as we study lung function? Surface tension arises from the attractive (cohesive) forces between the molecules of a liquid. At the surface of the liquid,

45.12 The Human Respiratory System The diagram traces the hierarchy of human respiratory structures, from the lungs down to the minuscule alveoli.

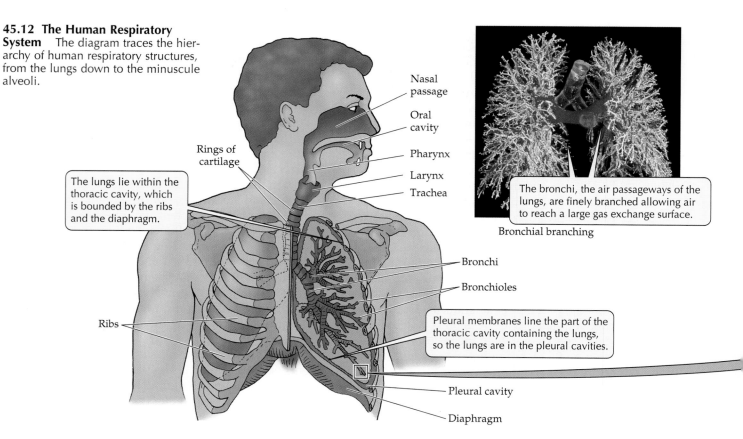

Nasal passage

Oral cavity

Pharynx

Larynx

Trachea

Rings of cartilage

The lungs lie within the thoracic cavity, which is bounded by the ribs and the diaphragm.

The bronchi, the air passageways of the lungs, are finely branched allowing air to reach a large gas exchange surface.

Bronchial branching

Bronchi

Bronchioles

Pleural membranes line the part of the thoracic cavity containing the lungs, so the lungs are in the pleural cavities.

Ribs

Pleural cavity

Diaphragm

these cohesive forces are unbalanced and give the surface the properties of an elastic membrane. Surface tension explains why certain insects called water striders can walk on the surface of water (see Figure 2.20) and why a carefully placed razor blade can "float." A surfactant interferes with the cohesive forces that create surface tension. Detergent is a surfactant; when added to water it causes the floating razor blade to sink and can make walking on water difficult for the water strider.

The thin, aqueous layer that lines the alveoli has surface tension, which can make inflation of the lungs very difficult. Surface tension in the alveoli normally is reduced by surfactant molecules produced by certain cells in the alveoli. If a baby is born more than a month prematurely, its alveoli may not be producing surfactant. Such a baby has great difficulty breathing because an enormous inhalation effort is required to stretch the alveoli against the surface tension. This condition, called **respiratory distress syndrome**, may cause a baby to die from exhaustion and suffocation. The common treatment is to put the baby on a respirator to assist its breathing and to give the baby hormones to speed its lung development. A new approach, applying surfactant to the lungs via an aerosol, is very promising.

Many cells lining the airways produce a sticky mucus that captures bits of dirt and microorganisms as they are inhaled. However, the mucus must be continually cleared from the airways; the beating of cilia lining the airways accomplishes this task (see Figure 44.2). The cilia move the mucus with its trapped debris up toward the pharynx, where it is swallowed. This phenomenon, called the *mucus escalator*, can be adversely affected by inhaled pollutants. Smoking one cigarette can immobilize the cilia of the airways for hours. Hacking, or smoker's cough, results from the need to clear the obstructing mucus from the airways when the mucus escalator is out of order.

Lungs are ventilated by pressure changes in the thoracic cavity

As Figure 45.12 shows, human lungs are suspended in the **thoracic cavity**, which is bounded on the top by the shoulder girdle, on the sides by the rib cage, and on the bottom by a domed sheet of muscle, the **diaphragm**. The thoracic cavity is lined on the inside by the **pleural membranes**, which divide it into right and left **pleural cavities**. Because the pleural cavities are closed spaces, any effort to increase their volume creates negative pressure—suction—inside them.

Negative pressure within the pleural cavities causes the lungs to expand as air flows into them from the outside. This is the mechanism of inhalation. The diaphragm contracts to begin an inhalation. This contraction pulls the diaphragm down, increasing the vol-

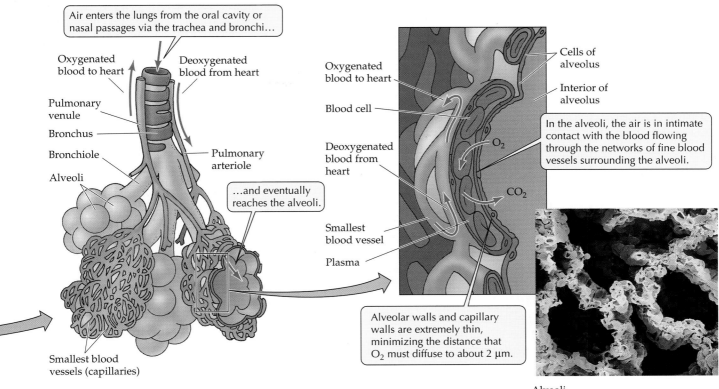

Air enters the lungs from the oral cavity or nasal passages via the trachea and bronchi...

Oxygenated blood to heart

Deoxygenated blood from heart

Pulmonary venule

Bronchus

Bronchiole

Alveoli

Pulmonary arteriole

...and eventually reaches the alveoli.

Smallest blood vessels (capillaries)

Oxygenated blood to heart

Blood cell

Deoxygenated blood from heart

Smallest blood vessel

Plasma

Cells of alveolus

Interior of alveolus

In the alveoli, the air is in intimate contact with the blood flowing through the networks of fine blood vessels surrounding the alveoli.

O_2

CO_2

Alveolar walls and capillary walls are extremely thin, minimizing the distance that O_2 must diffuse to about 2 μm.

Alveoli

ume of the thoracic and pleural cavities (Figure 45.13). As pressure in the pleural cavities becomes more negative, air enters the lungs. Exhaling begins when the contraction of the diaphragm ceases, the diaphragm relaxes and moves up, and the elastic recoil of the lungs pushes air out through the airways.

The diaphragm is not the only muscle that increases the volume of the thoracic cavity. Between the ribs are **intercostal muscles**. One set of these intercostal muscles expands the thoracic cavity by lifting the ribs up and outward. When heavy demands are placed on the gas exchange system, such as during strenuous exercise, these intercostal muscles and the diaphragm contract together and increase the volume of air inhaled.

Inhalation is always an active process, with muscles contracting; exhalation is usually a passive process,

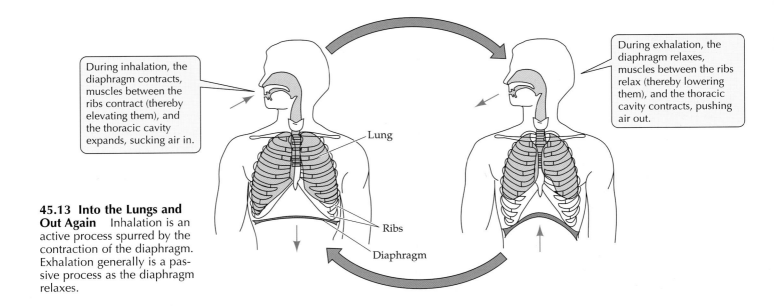

During inhalation, the diaphragm contracts, muscles between the ribs contract (thereby elevating them), and the thoracic cavity expands, sucking air in.

During exhalation, the diaphragm relaxes, muscles between the ribs relax (thereby lowering them), and the thoracic cavity contracts, pushing air out.

Lung

Ribs

Diaphragm

45.13 Into the Lungs and Out Again Inhalation is an active process spurred by the contraction of the diaphragm. Exhalation generally is a passive process as the diaphragm relaxes.

with muscles relaxing. However, another set of intercostal muscles can be called into play for forceful exhalations. Place your hands on your ribs and on your abdomen while breathing first shallowly and then deeply. Feel which muscles are active during inhalation and which are active during exhalation.

When the diaphragm is at rest between breaths, the pressure in the pleural cavities is still slightly negative. This slight suction keeps the alveoli partly inflated. If the thoracic wall is punctured, by a knife wound for example, air leaks into the pleural cavity, and the pressure from this air causes the lung to collapse. If the hole in the thoracic wall is not sealed, the breathing movements of diaphragm and intercostal muscles pull air into the pleural cavity rather than into the lung, and ventilation of the alveoli in that lung ceases.

Blood Transport of Respiratory Gases

The circulatory system is the subject of the next chapter, but since two of the substances it transports are the respiratory gases (O_2 and CO_2), we must mention aspects of it here. The circulatory system uses a pump (the heart) and a network of blood vessels to transport blood and the substances it carries around the body. As O_2 diffuses across the gas exchange surfaces into the vessels, the circulating blood sweeps it away. This internal perfusion of the gas exchange surfaces minimizes the concentration of O_2 on the internal side and promotes the diffusion of O_2 across the surface at the highest possible rate. The blood then delivers this O_2 to the cells and tissues of the body.

The liquid part of the blood, the **blood plasma**, carries some O_2 in solution, but the ability of the blood to pick up and transport O_2 would be quite limited if plasma were the only means available. Blood plasma carries about 0.3 ml of oxygen per 100 ml. To support the O_2 needs of a person at *rest*, the heart would have to pump about 5,000 liters of blood plasma *per hour* (enough to fill the gas tanks of about 100 cars). Fortunately, the blood also contains **red blood cells**, which are red because they are loaded with the oxygen-binding pigment hemoglobin. Hemoglobin increases the capacity of the blood to transport oxygen by about 60-fold.

In this section, we will examine the involvement of O_2 with the proteins hemoglobin and myoglobin and the transport of CO_2 in the liquid portion of the blood.

Hemoglobin combines reversibly with oxygen

Red blood cells contain enormous numbers of hemoglobin molecules. Hemoglobin is a protein consisting of four polypeptide subunits (see Figure 3.19). Each of these polypeptides surrounds a heme group—an iron-containing ring structure that can reversibly bind a molecule of O_2. As O_2 diffuses into the red blood cells, it binds to hemoglobin.

Once O_2 is bound, it cannot diffuse back across the cell membrane. By mopping up O_2 molecules as they enter the red blood cells, hemoglobin maximizes the concentration difference driving the diffusion of O_2 into the red blood cells. In addition, hemoglobin enables the red blood cells to carry a large amount of O_2 for use by the tissues of the body.

The ability of hemoglobin to pick up or release O_2 depends on the concentration or partial pressure of O_2 in its environment. When the P_{O_2} of the blood plasma is high, as it usually is in the lung capillaries, each molecule of hemoglobin can carry its maximum load of four molecules of O_2. As the blood circulates through capillary beds elsewhere in the body, it encounters lower P_{O_2} values. At these lower P_{O_2} values the hemoglobin releases some of the O_2 it is carrying.

The lower the P_{O_2} of the environment, the more O_2 hemoglobin releases to diffuse out of the red blood cells and into the tissues. However, the relation between P_{O_2} and the amount of O_2 bound to hemoglobin is not linear. This relationship is described by a sigmoid (S-shaped) curve (Figure 45.14), and it is an important property of hemoglobin.

Remember that the hemoglobin molecule consists of four subunits, each of which can bind one molecule of O_2. At very low P_{O_2} values, one subunit will bind an O_2 molecule. As a result, the shape of this subunit changes, causing an alteration in the quaternary structure of the whole hemoglobin molecule (see Chapter 3). This structural change makes it easier for the other subunits to bind a molecule of O_2; that is, their O_2 affinity is increased. Only small increases in the P_{O_2} cause most hemoglobin molecules to pick up a second and then a third molecule of O_2.

The influence of the binding of O_2 by one subunit on the binding affinity of the other subunits is called **positive cooperativity**, because binding of the first molecule makes binding of the second easier, and so forth. After binding of the third molecule of O_2, a large increase in P_{O_2} is required for the fourth subunits of all the hemoglobin molecules to be loaded because there are fewer and fewer available binding sites as more and more of the hemoglobin molecules become fully saturated.

The significance of the interactions of the hemoglobin subunits that result in the sigmoid shape of the hemoglobin–oxygen binding curve in Figure 45.14 is best appreciated by considering the dynamics of unloading the O_2 in the tissues. The normal P_{O_2} in the alveoli of the lungs is about 100 mm Hg, and at this P_{O_2} the hemoglobin is 100 percent saturated (each molecule of hemoglobin carries four molecules of O_2). The P_{O_2} in mixed venous blood is usually about 40 mm Hg, at which level the average hemoglobin molecule is binding only three molecules of O_2. Thus the hemoglobin returning to the heart from the body is still about 75 percent saturated. That means that most hemoglobin

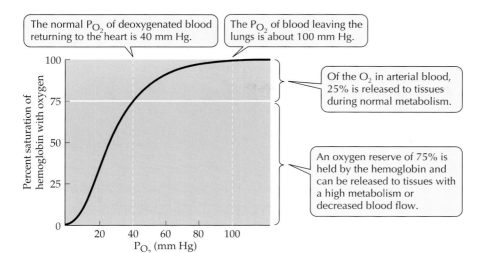

The normal P_{O_2} of deoxygenated blood returning to the heart is 40 mm Hg.

The P_{O_2} of blood leaving the lungs is about 100 mm Hg.

Of the O_2 in arterial blood, 25% is released to tissues during normal metabolism.

An oxygen reserve of 75% is held by the hemoglobin and can be released to tissues with a high metabolism or decreased blood flow.

45.14 The Binding of Oxygen to Hemoglobin Depends on Oxygen Concentration in the Plasma
Hemoglobin in blood leaving the lungs is 100 percent saturated (four molecules of O_2 on each hemoglobin). Most hemoglobin molecules will drop only one of their four O_2 molecules as they circulate through the body and are still 75 percent saturated when the blood returns to the lungs. The steep portion of this oxygen-binding curve comes into play when tissue P_{O_2} falls below the normal 40 mm Hg, at which point the hemoglobin will "unload" its oxygen reserves.

molecules drop only one of their four O_2 molecules as they circulate through the body.

This system may seem very inefficient for delivery of oxygen to the tissues, but it is actually extremely adaptive. When a tissue becomes starved of oxygen and its local P_{O_2} falls below 40 mm Hg, the hemoglobin flowing through that tissue will drop much more of its oxygen load with only small additional decreases in P_{O_2}. The steep portion of the sigmoid hemoglobin–oxygen binding curve comes into play when tissue P_{O_2} falls below the normal 40 mm Hg. Thus the cooperative oxygen-binding property of hemoglobin is very effective in making O_2 available to the tissues precisely when and where it is most needed.

Myoglobin holds an oxygen reserve

Muscle cells have their own oxygen-binding molecule, **myoglobin**. Myoglobin consists of just one polypeptide chain associated with an iron-containing ring structure that can bind one molecule of oxygen. Myo-

globin has a higher affinity for O_2 than hemoglobin does (see Figure 45.15), so it picks up and holds oxygen at P_{O_2} values at which hemoglobin is releasing its bound O_2.

Myoglobin provides a reserve of oxygen for the muscle cells for times when metabolic demands are high and blood flow is interrupted. Interruption of blood flow in muscles is common because contracting muscles constrict blood vessels. When hemoglobin has no more O_2 to give up, and tissue P_{O_2} falls even lower, myoglobin releases its bound O_2. Diving mammals such as seals that can remain active under water for many minutes have high concentrations of myoglobin in their muscles.

Muscles called on for extended periods of work frequently have more myoglobin than muscles that are used for short, intermittent periods. This is one of the

45.15 Oxygen-Binding Adaptations Llamas are used as pack animals because they are so well adapted to high altitudes, where the partial pressure of oxygen is low. Evolution has adapted the oxygen-binding properties of different hemoglobins and of myoglobin.

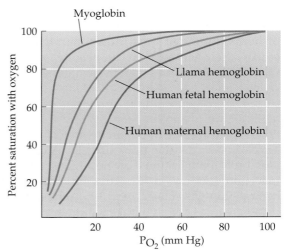

reasons for the difference in appearance of the white and dark meat of chickens and turkeys. These birds are not long-distance fliers, and their flight muscles (the white meat) have little myoglobin (but, since they use their legs for locomotion, their legs have higher amounts of myoglobin). On the other hand, ducks and geese come from distinguished lineages of long-distance fliers. Their flight muscles have much myoglobin, as well as more mitochondria and more blood vessels, and thus appear dark.

The affinity of hemoglobin for oxygen is variable

Various factors influence the oxygen-binding properties of hemoglobin, and thereby influence oxygen delivery to tissues. In this section we examine three of these factors: the chemical composition of hemoglobin, pH, and the presence of diphosphoglyceric acid.

HEMOGLOBIN COMPOSITION.   The chemical composition of the polypeptide chains that form the hemoglobin molecule varies. The normal hemoglobin of adult humans has two each of two kinds of polypeptide chains—two *alpha* chains and two *beta* chains—and the oxygen-binding characteristics shown in Figure 45.15.

Before birth, the fetus has a different form of hemoglobin, consisting of two *alpha* chains and two *gamma* chains. The chemical composition of fetal hemoglobin enables fetal blood to pick up O_2 from maternal blood when both are at the same P_{O_2}. Fetal hemoglobin thus has an oxygen-binding curve that plots to the left of the adult curve (Figure 45.16). This difference between maternal and fetal hemoglobin facilitates the transfer of O_2 from the mother's blood to the blood of the fetus in the placenta.

Llamas and vicuñas are mammals native to high altitudes in the Andes Mountains of South America (see Figure 45.15). The hemoglobins of these animals, like those of the human fetus, must pick up O_2 in an environment that has a low P_{O_2}. In the animals' natural habitat, more than 5,000 m above sea level, the P_{O_2} is below 85 mm Hg, and the P_{O_2} in their lungs is about 50 mm Hg. The hemoglobins of llamas and vicuñas have oxygen-binding curves much to the left of the curves of hemoglobins of most other mammals—in other words, their hemoglobin can become saturated with O_2 at lower P_{O_2} values than those of other animals can.

pH.   The oxygen-binding properties of normal adult hemoglobin are influenced by physiological conditions. The influence of pH on the function of hemoglobin is known as the **Bohr effect**. As the pH of the blood plasma falls, the oxygen-binding curve shifts to the right (Figure 45.16). This shift means that the hemoglobin will release more O_2 to the tissues.

Where does hemoglobin encounter a decreased pH as it circulates through the body? In tissues with very

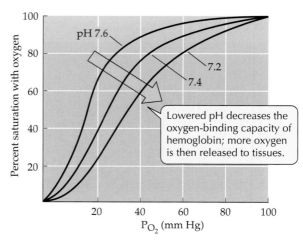

45.16 The Oxygen-Binding Properties of Hemoglobin Can Change   Changes in pH affect the oxygen-binding capacity of hemoglobin.

high metabolic rates the pH is reduced by the release of acidic metabolites such as lactic acid, fatty acids, and CO_2 (which combines with water to form carbonic acid). Because of the Bohr effect, hemoglobin releases more of its bound oxygen in these tissues—another way that O_2 is supplied where and when it is most needed.

DIPHOSPHOGLYCERIC ACID.   **Diphosphoglyceric acid** is a normal intermediate metabolite that plays an important role in regulating hemoglobin function. The mature mammalian red blood cell is a simple cell. It is little more than a sac of hemoglobin, but it has a very high content of diphosphoglyceric acid. The concentration of diphosphoglyceric acid in red blood cells increases in response to exercise and during acclimation to high altitude. Diphosphoglyceric acid reversibly combines with deoxygenated hemoglobin and changes the shape of the hemoglobin such that it has a lower affinity for O_2. The result is that at any P_{O_2}, hemoglobin releases more of its bound O_2 than it otherwise would. In other words, diphosphoglyceric acid shifts the oxygen-binding curve of mammalian hemoglobin to the right.

Llamas and humans employ opposite adjustments of hemoglobin function as adaptations for life at high altitudes. The llama's hemoglobin has a left-shifted oxygen-binding curve, which means that it can become 100 percent saturated with O_2 at the low P_{O_2} values at high altitudes. As a consequence, the llama's tissues must operate at a lower P_{O_2}. By contrast, human hemoglobin acquires, through acclimation, a right-shifted oxygen-binding curve. The result is that human hemoglobin never becomes fully saturated with O_2 at high altitudes, but more of the O_2 carried by that hemoglobin is released to the tissues.

Carbon dioxide is transported as bicarbonate ions in the blood

Delivering O_2 to the tissues is only half of the respiratory function of the blood. The blood also must take metabolic waste products away from the tissues. Because we are concerned with respiratory gases in this chapter, the metabolic waste product we will consider here is carbon dioxide. CO_2 is highly soluble and readily diffuses through cell membranes, moving from its site of production in a cell into the blood, where its concentration is lower. However, very little CO_2 is transported by the blood in this dissolved form. Most CO_2 produced by tissues is transported to the lungs in the form of the **bicarbonate ion**, HCO_3^-. How and where CO_2 becomes HCO_3^-, is transported, and then is converted back to CO_2 is an interesting story.

When CO_2 dissolves in water, some of it slowly reacts with the water molecules to form carbonic acid (H_2CO_3), some of which then dissociates into a proton (H^+) and a bicarbonate ion (HCO_3^-). This sequence of events is expressed as follows:

$$CO_2 + H_2O \rightleftharpoons H_2CO_3 \rightleftharpoons H^+ + HCO_3^-$$

In the blood plasma, the reaction between CO_2 and H_2O proceeds too slowly to have much effect. But it is a different story in the red blood cells, where the enzyme **carbonic anhydrase** speeds up the conversion of CO_2 to H_2CO_3. The newly formed carbonic acid dissociates, and the resulting bicarbonate ion diffuses back out into the plasma (Figure 45.17). This action of carbonic anhydrase in the red blood cells creates a sink for CO_2, thus facilitating the diffusion of CO_2 from tissue cells to plasma to red blood cells. Most CO_2 is transported by the blood as bicarbonate ions in the plasma. Some CO_2 is also carried in chemical combination with deoxygenated hemoglobin as *carboxyhemoglobin*.

In the lungs, the reactions involving CO_2 and bicarbonate are reversed. CO_2 diffuses from the blood plasma into the air in the alveoli and is exhaled. Exhalation creates a concentration difference, so CO_2 leaves the red blood cells and enters the plasma. The loss of CO_2 from the red blood cells shifts the equilibrium between CO_2 and bicarbonate. As the HCO_3^- in the red blood cells is converted back to CO_2, more HCO_3^- moves into the red blood cells from the plasma. Remember that an enzyme like carbonic anhydrase only speeds up a reversible reaction; it does not determine its direction. Direction is determined by concentrations of reactants and products (see Chapter 6).

Regulating Breathing

We must breathe every minute of our lives, but we don't worry about our need to breathe or even think

45.17 Carbon Dioxide Is Transported as Bicarbonate Ions CO_2, a waste product of cellular metabolism, is transported from the tissues to the lungs via the red blood cells. During this trip the CO_2 undergoes several enzyme-catalyzed chemical transformations and is finally exhaled.

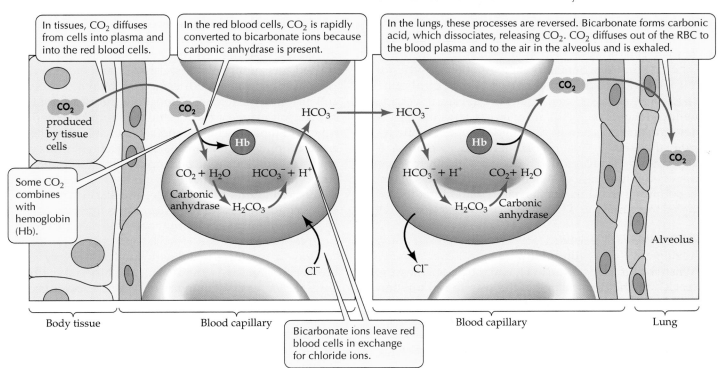

about it very often. Breathing is an autonomic function of the nervous system. The breathing pattern easily adjusts itself around other activities, such as speech and eating. Most impressively, our breathing rates change to match the metabolic demands of our bodies. In this section we will examine how the regular respiratory cycle is generated, and how it is controlled so that we get the oxygen we need and eliminate the carbon dioxide we produce as our level of activity changes.

The ventilatory rhythm is generated in the brain stem

The autonomic nervous system maintains a ventilatory rhythm and modifies its depth and frequency to meet the demands of the body for O_2 supply and CO_2 elimination. The ventilatory rhythm ceases if the spinal cord is severed in the neck region, showing that the rhythm is generated in the brain. If the brain stem is cut just above the medulla, the segment of the brain stem just above the spinal cord, a crude ventilatory rhythm remains (Figure 45.18).

Groups of neurons within the medulla increase their firing rates just before an inhalation begins. As more and more of these neurons fire, and fire faster and faster, the inhalation muscles contract. Suddenly the neurons stop firing, the inhalation muscles relax, and exhalation begins. Exhalation is usually a passive process that depends on the elastic recoil of the lung tissues. When respiratory demand is high, however, as during strenuous exercise, exhalation neurons in the medulla increase their firing rates and accelerate the ventilatory rhythm by adding an active component to the exhalation phase of the cycle. Brain areas above the medulla modify the ventilatory rhythm. The rhythm is modified to accommodate speech, ingestion of food, coughing, and emotional states.

As respiratory demands increase, the activities of the inhalation neurons also increase, thus contributing to greater depth of inhalation. However, an override reflex prevents the ventilatory muscles from overdistending and damaging the lung tissue. This reflex is named the *Hering–Breuer reflex* (after the two physiologists who discovered it). It begins with stretch sensors in the lung tissue. When stretched, these sensors send impulses via the vagus nerve that inhibit the inhalation neurons.

Coordinating ventilation to metabolic needs requires feedback information

When the partial pressure of O_2 and the partial pressure of CO_2 in the blood change, the respiratory rhythm changes to return these values to normal levels. An early experimental approach to understanding gas exchange in humans addressed the reasonable expectation that the blood P_{O_2} or P_{CO_2}, or both, should provide feedback to the respiratory centers in the brain.

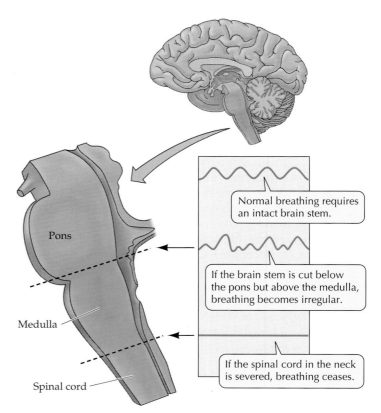

45.18 The Brain Stem Generates and Controls Breathing Rhythm   Severing the brain stem at different levels reveals that the basic breathing rhythm is generated in the medulla and modified by neurons in or above the pons.

Dramatic and disastrous insight regarding this expectation was provided by three French physiologists in 1875. They wanted to investigate the physiological effects of breathing low concentrations of O_2. Sophisticated gas pumps and pressure chambers did not exist in 1875, so the three decided to go up in a balloon to very high altitudes and observe the effects of the rarefied atmosphere on one another. They noted no ill effects in their notebooks and continued to throw out ballast, going higher than 8,000 m. Then all three became unconscious. The balloon finally descended on its own, and one of the physiologists regained consciousness to find his two colleagues dead. This infamous flight of the balloon *Zenith* is tragic proof that the human body is not very good at sensing its own need for O_2.

However, humans and other mammals are very sensitive to increases in the P_{CO_2} of the blood, whether caused by energy demands or by the composition of the air breathed. If you rebreathe a small volume of air, thereby increasing the P_{O_2} of that reservoir, your breathing becomes deeper and more rapid, and you become anxious and agitated. You react the same way even if pure O_2 is continually released into the reser-

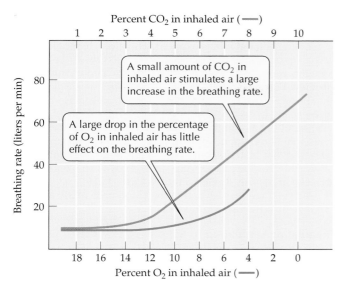

A small amount of CO_2 in inhaled air stimulates a large increase in the breathing rate.

A large drop in the percentage of O_2 in inhaled air has little effect on the breathing rate.

45.19 Carbon Dioxide Affects Breathing Rate Breathing is more sensitive to increased carbon dioxide than to decreased oxygen.

voir to keep its P_{O_2} constant. Typical ventilatory responses to changes in blood P_{O_2} and P_{CO_2} are shown in Figure 45.19.

It makes sense that CO_2 rather than O_2 is the dominant feedback stimulus for ventilation. As we have seen, animals have evolved gas exchange systems and hemoglobin properties that work to keep the blood that is leaving the gas exchange surfaces fully saturated with O_2 over a broad range of alveolar P_{CO_2} values and metabolic rates. Normal fluctuations in metabolism and ventilation have very little effect on the maximum amount of O_2 carried by the blood. By contrast, small changes in metabolism and alveolar P_{CO_2} do influence the concentration of CO_2 in the blood. Changes in blood P_{CO_2} are a much finer index of energy demands and respiratory performance than is the O_2 content of the blood.

Where are gas concentrations in the blood sensed? The major site of CO_2 sensitivity is an area on the ventral surface of the medulla, not far from the groups of neurons that generate the ventilatory rhythm. Sensitivity to the O_2 concentration of the blood resides in nodes of tissue on the large blood vessels leaving the heart: the aorta and the carotid arteries (Figure 45.20). These carotid and aortic bodies receive enormous supplies of blood, and they contain chemosensory nerve endings. If blood supply to these structures decreases, or if the P_{O_2} of the blood falls, the chemosensors send impulses to the respiratory centers. Although we are not very sensitive to changes in blood P_{O_2}, the carotid and aortic bodies can stimulate increases in ventilation during exposure to very high altitude or when blood volume or blood pressure is very low.

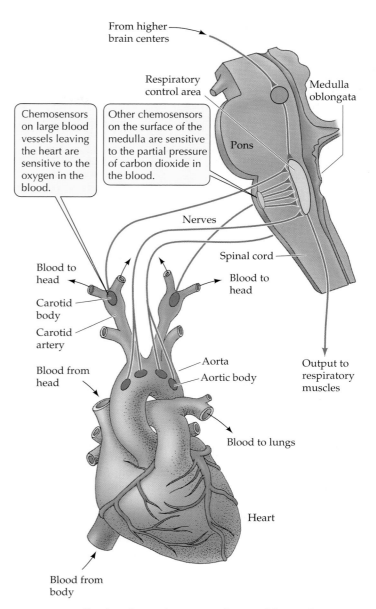

Chemosensors on large blood vessels leaving the heart are sensitive to the oxygen in the blood.

Other chemosensors on the surface of the medulla are sensitive to the partial pressure of carbon dioxide in the blood.

45.20 Feedback Information Controls Breathing The body uses feedback information from chemosensors in the heart and the brain to match breathing rate to metabolic demand.

Summary of "Gas Exchange in Animals"

Respiratory Gas Exchange

• Most cells require a constant supply of O_2 and continuous removal of CO_2. These respiratory gases are exchanged between the body fluids of an animal and the outside world by diffusion only.

• In aquatic animals, gas exchange is limited by the low diffusion rate and concentration of oxygen in water, the fact that oxygen concentration in water decreases but animals' oxygen needs increase as temperature rises, and the fact that considerable work is required to move water over gas exchange surfaces. **Review Figure 45.2**

Adaptations for Gas Exchange

• The evolution of larger animals with high metabolic rates has required the evolution of adaptations to maximize the rates of diffusion of respiratory gases between animals and their environments. These adaptations include increased surface areas for gas exchange, maximizing the concentration gradients across those exchange surfaces by decreasing their thickness, ventilating the outer surface with respiratory medium, and perfusing the inner surface with blood. **Review Figure 45.4**

• Since gases diffuse more rapidly in air than in water, it is important to minimize the path length for diffusion in an aqueous medium. This principle is most evident in insects, which distribute oxygen throughout their bodies in a system of air capillaries. **Review Figure 45.5**

• Fish have maximized their rates of gas exchange by having large surface areas for exchange that are ventilated continuously and unidirectionally with fresh water. Countercurrent blood flow helps increase the efficiency of gas exchange. **Review Figure 45.6**

• The gas exchange system of birds includes air sacs that communicate with the lungs but are not used for gas exchange. Air flows undirectionally through bird lungs in parabronchi. Gases are exchanged in air capillaries that run between parabronchi. **Review Figures 45.7, 45.8**

• Each breath of air remains in the bird gas exchange system for two breathing cycles. The air sacs work as bellows to supply the air capillaries with a continuous flow of fresh air at the highest possible oxygen concentration. **Review Figure 45.9**

• Breathing in mammals is tidal and therefore less efficient than gas exchange in fish or birds. Even though the volume of air exchanged with each breath can vary considerably, the inhaled air is always mixed with the stale air that fills the nonventilated spaces of the gas exchange system, and the gas exchange surfaces experience the maximum concentration of oxygen in the inhaled air for only a brief portion of each breathing cycle. **Review Figure 45.11**

Mammalian Lungs and Gas Exchange

• The gas exchange surface area of the millions of alveoli that make up mammalian lungs is enormous, and the path length between the air and perfusing blood is very short. **Review Figure 45.12**

• Surface tension in the alveoli would make their inflation difficult if the lungs did not produce surfactant.

• Lungs are inflated when contractions of the diaphragm and the intercostal muscles create a negative pressure in the thoracic cavity and air is inhaled. Relaxation of the diaphram and some intercostals and contraction of other intercostals increases pressure in the thoracic cavity and causes exhalation. **Review Figure 45.13**

Blood Transport of Respiratory Gases

• Oxygen is carried in the blood in reversible combination with hemoglobin. Each molecule of hemoglobin can carry a maximum of four molecules of oxygen.

• Because of positive cooperativity, the affinity of hemoglobin for oxygen depends on the partial pressure of oxygen to which the hemoglobin is exposed. Therefore, hemoglobin gives up oxygen in metabolically active tissues and picks up oxygen as it flows through respiratory exchange structures. **Review Figure 45.14**

• The affinity of hemoglobin for oxygen is decreased by the presence of hydrogen ions or diphosphoglyceric acid. Fetal hemoglobin has a higher affinity for oxygen than does maternal hemoglobin, allowing fetal blood to pick up oxygen from the maternal blood in the placenta. Myoglobin has a very high affinity for oxygen and serves as an oxygen reserve in muscle. **Review Figures 45.15, 45.16**

• Carbon dioxide is carried in the blood principally as bicarbonate ions. **Review Figure 45.17**

Regulating Ventilation to Supply O_2

• The breathing rhythm is an autonomic function generated by neurons in the medulla of the brain stem and modulated by higher brain centers. **Review Figure 45.18**

• The most important feedback stimulus for breathing is the level of CO_2 in the blood. **Review Figure 45.19**

• The breathing rhythm is sensitive to feedback from chemosensors on the ventral surface of the brain stem and in the carotid and aortic bodies on the large vessels leaving the heart. **Review Figure 45.20**

Self-Quiz

1. Which statement is *not* true?
   a. Respiratory gases are exchanged by diffusion only.
   b. Oxygen has a lower rate of diffusion in water than in air.
   c. The oxygen content of water falls as the temperature of water rises, all other things being equal.
   d. The amount of oxygen in the atmosphere decreases with increasing altitude.
   e. Birds have evolved active transport mechanisms to augment their respiratory gas exchange.

2. Which statement about the gas exchange system of birds is *not* true?
   a. Respiratory gases are not exchanged in the air sacs.
   b. It can achieve more complete exchange of O_2 from air to blood than the human gas exchange system can.
   c. Air passes through birds' lungs in only one direction.
   d. The gas exchange surfaces in bird lungs are the alveoli.
   e. A breath of air remains in the system for two breathing cycles.

3. In fish
   a. blood flows across the gas exchange surfaces in a direction opposite to the flow of water.
   b. gases are exchanged across the gill filaments.
   c. ventilation of the gills is tidal.
   d. less work is needed to ventilate gills in warm water than in cold water.
   e. the path length for diffusion of respiratory gases is determined by the length of the gill filaments.

4. In the human gas exchange system
   a. the lungs and airways are completely collapsed after a forceful exhalation.
   b. the average O_2 concentration of air inside the lungs is always lower than that in the air outside the lungs.

c. the P_{O_2} of the blood leaving the lungs is greater than the P_{O_2} of the exhaled air.

d. the amount of air that is moved per breath during normal, at-rest breathing is termed the total lung capacity.

e. oxygen and carbon dioxide are actively transported across the alveolar and capillary membranes.

5. Which statement about the human gas exchange system is *not* true?

a. During inhalation a negative pressure exists in the space between the lung and the thoracic wall.

b. Smoking one cigarette can immobilize the cilia lining the airways for hours.

c. The respiratory control center in the medulla responds more strongly to changes in arterial O_2 concentration than to changes in arterial CO_2 concentrations.

d. Without surfactant, the work of breathing is greatly increased.

e. The diaphragm contracts during inhalation and relaxes during exhalation.

6. The hemoglobin of a human fetus

a. is the same as that of an adult.

b. has a higher affinity for O_2 than adult hemoglobin has.

c. has only two protein subunits instead of four.

d. is supplied by the mother's red blood cells.

e. has a lower affinity for O_2 than adult hemoglobin has.

7. The amount of oxygen carried by hemoglobin depends on the partial pressure of oxygen in the blood. Hemoglobin in active muscles

a. becomes saturated with oxygen.

b. takes up only a small amount of oxygen.

c. readily unloads oxygen.

d. tends to decrease the partial pressure of oxygen in the muscle tissues.

e. is denatured.

8. Most carbon dioxide is carried in the blood

a. in the cytoplasm of red blood cells.

b. dissolved in the plasma.

c. in the plasma as bicarbonate ions.

d. bound to plasma proteins.

e. in red blood cells bound to hemoglobin.

9. Myoglobin

a. binds O_2 at P_{O_2} values at which hemoglobin is releasing its bound O_2.

b. has a lower affinity for O_2 than hemoglobin does.

c. consists of four polypeptide chains, just as hemoglobin does.

d. provides an immediate source of O_2 for muscle cells at the onset of activity.

e. can bind four O_2 molecules at once.

10. Carbon dioxide, a product of cellular respiration, is carried in the bloodstream. When the level of CO_2 in the blood becomes greater than the set operating range,

a. the rate of respiration decreases.

b. the pH of the blood rises.

c. the respiratory centers become dormant.

d. the rate of respiration increases.

e. the blood becomes more alkaline.

Applying Concepts

1. Compare and contrast the gas exchange systems of birds, fish, and humans. Why can birds and fish outperform mammals in environments where the concentration of O_2 is low?

2. What does the following chemical equation represent, and how does it relate to gas exchange in human lungs and in active tissues?

$$CO_2 + H_2O \rightleftharpoons H_2CO_3 \rightleftharpoons H^+ + HCO_3^-$$

3. Describe and contrast the adaptations of llamas and humans for gas exchange at high altitudes.

4. Describe, in terms of the structure of the human respiratory system, how air is brought into and then expelled from the lungs.

5. Workers A and B must inspect two large gas storage tanks. Tank 1 contains 100 percent N_2 (nitrogen gas); tank 2 contains 100 percent CO_2. The tanks are *not* flushed out before inspection. Worker A goes into tank 1, becomes unconscious, and dies. Worker B goes into tank 2, feels strangely short of breath, and leaves the tank feeling somewhat dizzy. Explain why worker A died and worker B lived.

Readings

Feder, M. E. and W. W. Burggren. 1985. "Skin Breathing in Vertebrates." *Scientific American*, November. A discussion of the adaptations of many vertebrates for gas exchange through the skin.

Houston, C. S. 1992. "Mountain Sickness." *Scientific American*, October. Humans have lived at high altitudes for centuries, even though breathing becomes difficult and lethal complications are common.

Moon, R. E. R. D. Vann and P. B. Bennett. 1995 "The Physiology of Decompression Illness." *Scientific American*, August. Breathing air at high pressure when diving in ocean depths forces nitrogen into solution in the plasma. If pressure is reduced too fast, the dissolved nitrogen can come out of solution, forming bubbles that cause serious problems and can even be fatal.

Perutz, M. F. 1978. "Hemoglobin Structure and Respiration." *Scientific American*, December. An authoritative article on hemoglobin structure and function by the man who received the Nobel prize for his work on the subject.

Randall, D., W. Burggren and K. French. 1997. *Animal Physiology: Mechanisms and Adaptations*, 4th Edition. W. H. Freeman, New York. An outstanding textbook of animal physiology, with a good chapter on gas exchange.

Schmidt-Nielsen, K. 1997. *Animal Physiology: Adaptation and Environment*, 5th Edition. Cambridge University Press, New York. An outstanding textbook that emphasizes the comparative approach.

Vander, A. J., J. H. Sherman and D. S. Luciano. 1994. *Human Physiology: The Mechanisms of Body Function*, 6th Edition. McGraw-Hill, New York. Chapter 15 deals with human respiration.

Welsh, M. J. and A. E. Smith. 1995. "Cystic Fibrosis." *Scientific American*, December. A discussion of current research and knowledge about a genetic disease characterized by abnormal lung secretions.

Chapter 46

Circulatory Systems

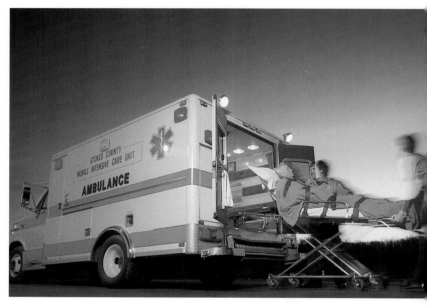

Emergency!
Cardiovascular failure is often the reason for a call to emergency paramedical services. Fast response is essential to the patient's survival.

S weating, severe chest pain, fainting, 911, flashing lights, paramedics, emergency room, intensive care—heart attack! Most of us will experience this sequence of events directly or indirectly. More than one-third of the deaths in the United States are due to heart failure, and many such deaths are those of a sudden nature that we call heart attacks. Why is it so serious when the heart fails? The answer is that all organs of the body depend on the heart.

The needs of all the cells in the body of an animal are served by the internal fluid environment (see Chapter 37). Cells take up nutrients from the fluid that bathes them, and they release their waste products into that fluid. Thus, the activities of cells change the internal fluid environment in ways that are not healthy for the cells themselves. The functions of the organ systems are to return different aspects of the internal environment to optimal levels.

The lungs take in oxygen and eliminate carbon dioxide, the digestive tract takes in nutrients, the liver controls nutrient levels and eliminates toxic compounds, and the kidneys control salt concentrations and eliminate toxic wastes. The cells of each organ depend on the activities of all the other organs, and only through a system of transport—circulation—can the activities of each organ influence the internal environment of the entire body. So when the heart stops, transport stops, the internal environment deteriorates, and cells become sick and die.

In this chapter we describe a variety of circulatory systems ranging from none at all to open systems to closed systems. *Closed circulatory systems* include a heart and a closed system of blood vessels. We will discuss the evolution, structure, and function of vertebrate hearts, which vary in complexity from the two-chambered heart of fish to the four-chambered heart of mammals. Taking the human circulatory system for in-depth study, we will see how the heart is really two pumps in one. The right heart pumps blood through the lung circuit, and the left heart pumps blood through the body circuit.

We will describe the characteristics of the major components of the vascular system: the arteries, capillaries, and veins. We will explain how materials are exchanged between the blood and the intercellular fluids across the walls of the capillaries. The third component of a circulatory system, after the heart and the vessels, is the blood. We will describe the features of this fluid tissue. We will end the chapter with a discussion of the hormonal and neural control and regulation of the human circulatory system.

Circulatory Systems: Vessels, Blood, and a Pump

A heart is not all there is to a transport system; it is just the pump. In addition, the system must have a vehicle (blood) to transport materials and a series of conduits (blood vessels) through which the materials can be pumped around the body. Heart, blood, and vessels together constitute a circulatory system, also known as a cardiovascular system (from the Greek *kardia*, "heart," and the Latin *vasculum*, "small vessel"). Only small, simple, aquatic animals do not have circulatory systems to transport substances to and away from the cells of their bodies.

In this section, we consider diverse circulatory systems of different groups of animals, emphasizing the nonmammalian vertebrates: fish, amphibians, reptiles, and birds.

Some simple aquatic animals do not have circulatory systems

A circulatory system is unnecessary if all the cells of an organism are close enough to the external environment that nutrients, respiratory gases, and wastes can diffuse directly between the cells and the outside environment. Small aquatic invertebrates have various structures and shapes that permit direct exchange by diffusion. The hydra, a cnidarian, is a good example (see Figure 29.10). This aquatic animal is cylindrical and only two cell layers thick. Each of the hydra's cells contacts water that is either surrounding the animal or circulating through its **gastrovascular cavity**, a dead-end sac that serves both for digestion ("gastro-") and for transport ("vascular") (Figure 46.1*a*).

The cells of some other invertebrates are served by diffusion from highly branched gastrovascular systems. In addition, flattened body shapes minimize the diffusion path length—the distance that molecules have to diffuse between cells and the external environment (Figure 46.1*b*). However, a central gastrovascular system cannot serve the needs of larger animals with

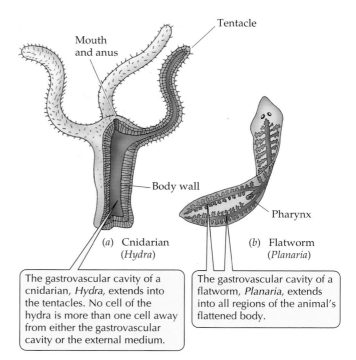

The gastrovascular cavity of a cnidarian, *Hydra*, extends into the tentacles. No cell of the hydra is more than one cell away from either the gastrovascular cavity or the external medium.

The gastrovascular cavity of a flatworm, *Planaria*, extends into all regions of the animal's flattened body.

46.1 Gastrovascular Cavities Gastrovascular cavities in animals without circulatory systems serve the metabolic needs of the innermost cells of the body.

many layers of cells. Transport in such animals requires a circulatory system. All terrestrial animals require circulatory systems because none of their cells are bathed by an external medium; therefore, all their cells must be served by intercellular fluids.

Open circulatory systems move intercellular fluid

In the simplest circulatory systems the interstitial fluid is simply squeezed through intercellular spaces as the animal moves. In these **open circulatory systems** there is no distinction between intercellular fluid and blood. Usually a muscular pump, or heart, assists the distribution of the fluid through the tissues. The contractions of the heart propel the intercellular fluid through vessels leading to different regions of the body, but the fluid leaves those vessels to trickle through the tissues and eventually to return to the heart. In the arthropod shown in Figure 46.2*a*, the fluid returns to the heart through valved holes called **ostia** when the heart relaxes. In the mollusk in Figure 46.2*b*, open vessels aid in the return of intercellular fluid to the heart.

Closed circulatory systems circulate blood through tissues

In a **closed circulatory system** some components of the blood never leave the vessels. Blood circulates through the vascular system, pumped by one or more muscular hearts. The system keeps the circulating blood separate from the intercellular fluid.

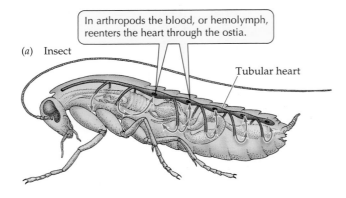

In arthropods the blood, or hemolymph, reenters the heart through the ostia.

(a) Insect

Tubular heart

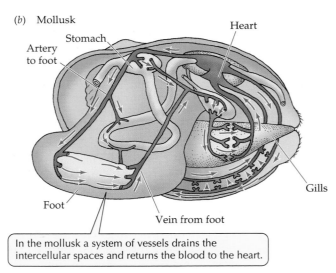

(b) Mollusk

Heart

Stomach

Artery to foot

Gills

Foot

Vein from foot

In the mollusk a system of vessels drains the intercellular spaces and returns the blood to the heart.

46.2 Open Circulatory Systems In both arthropods and mollusks, blood is pumped by a tubular heart and directed to regions of the body through vessels that open into interstitial spaces.

Closed circulatory systems have several advantages over open systems. First, closed systems can deliver oxygen and nutrients to the tissues and carry away metabolic wastes more rapidly than open systems can. Second, closed systems can direct blood to specific tissues. Third, cellular elements and large molecules that function within the vascular system can be kept within it; examples are red blood cells and large molecules that help in the distribution of hormones and nutrients.

Overall, closed circulatory systems can support higher levels of metabolic activity, especially in larger animals. How then do highly active insect species achieve high levels of metabolic output with their open circulatory systems? The key to answering this question is to remember something you learned in Chapter 45: Insects do not depend on their circulatory system for respiratory gas exchange (see Figure 45.5).

All vertebrates have closed circulatory systems and chambered hearts. Chambered hearts have valves that prevent the backflow of blood when the heart contracts. From fishes to amphibians to reptiles to birds and mammals, the complexity and number of chambers of the heart increase. An important consequence of this increased complexity is the gradual separation of the circulation into two circuits, one to the lungs and one to the rest of the body.

In fishes, blood is pumped from the heart to the gills and then to the tissues of the body and back to the heart. In birds and mammals, blood is pumped from the heart to the lungs and back to the heart in the **pulmonary circuit**, and from the heart to the rest of the body and back to the heart in the **systemic circuit**.

A simple example of a closed circulatory system is that of the common earthworm, an annelid (see Figure 29.27). One large blood vessel on the ventral side of the earthworm carries blood from its anterior end to its posterior end. In each segment of the worm, smaller vessels branch off and transport the blood to even smaller vessels in the tissues of that segment. Here, respiratory gases, nutrients, and metabolic wastes diffuse between the blood and the intercellular fluids. The blood then flows into larger vessels that lead into a single large vessel on the dorsal side of the worm. The dorsal vessel carries the blood from the posterior to the anterior end. Five pairs of vessels connect the large dorsal and ventral vessels in the anterior end, thus completing the circuit (Figure 46.3). The dorsal vessel and the five pairs of connecting vessels serve as hearts for the earthworm; their contractions keep the blood circulating. The direction of circulation is determined by one-way valves in the dorsal vessel and in the five pairs of connecting vessels.

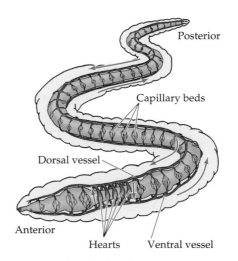

Posterior

Capillary beds

Dorsal vessel

Anterior

Hearts   Ventral vessel

46.3 A Closed Circulatory System In a closed circulatory system, blood is confined to the blood vessels, kept separate from the interstitial fluid, and pumped by one or more muscular hearts. The earthworm, with large dorsal and ventral blood vessels and a branching network of smaller vessels, exemplifies this type of system.

We will see how the separation of the circulation into two circuits improves the efficiency and capacity of the circulatory system.

A closed vascular system includes **arteries** that carry blood away from the heart, and **veins** that carry blood back to the heart. **Arterioles** are small arteries, and **venules** are small veins. **Capillaries** are very small, thin-walled vessels that connect arterioles and venules. Materials are exchanged between the blood and the intercellular fluid only across capillary walls.

Fishes have two-chambered hearts

The fish heart has a less muscular chamber, called the **atrium**, that receives blood from the body and pumps it into a more muscular chamber, the **ventricle**. The ventricle pumps the blood to the gills, where gases are exchanged. Blood leaving the gills collects in a large dorsal artery, the **aorta**, which distributes blood to smaller arteries and arterioles leading to all the organs and tissues of the body. In the tissues, blood flows through capillary beds, collects in venules and veins, and eventually returns to the heart (Figure 46.4a).

Most of the pressure imparted to the blood by contraction of the ventricle is dissipated by the high resistance of the many tiny, narrow spaces through which blood flows in the gills. As a result, blood entering the aorta of the fish is under low pressure, limiting the ability of the fish circulatory system to supply the tissues with oxygen and nutrients. This limitation on arterial blood pressure does not seem to hamper the performance of many rapidly swimming species, such as tuna and marlin.

An important evolutionary step is reflected in the circulatory systems of African lungfish. These fish are periodically exposed to water with low oxygen content or to situations in which their aquatic environment dries up. Their adaptation for dealing with these conditions is an outpocketing of the gut that serves as a lung (see Figure 45.11b). This lung contains many thin-walled blood vessels, so deoxygenated blood flowing through those vessels can pick up oxygen from air gulped into the lung.

What blood vessels serve this new organ? The last pair of gill arteries is modified to carry blood to the lung, and a new vessel carries oxygenated blood from the lung back to the heart. In addition, two other gill arches have lost their gills, and their blood vessels deliver blood from the heart directly to the dorsal aorta (Figure 46.4b). Because a few gill arches retain gills, the African lungfish can breathe either air or water.

In the evolution of vertebrate circulatory systems, the lungfish reveals the transition step leading to separate pulmonary and systemic circuits. Adaptations of the lungfish heart are also evolutionarily important. Unlike the hearts of other fish, the lungfish heart has a partly divided atrium; the left side receives oxygenated blood from the lungs, and the right side receives deoxygenated blood from the other tissues. The two bloodstreams stay mostly separate as they flow through the ventricle and the large vessel leading to the gill arches, so the oxygenated blood goes to the gill arteries leading to the dorsal aorta, and the deoxygenated blood goes to the arches with functional gills and to the lung.

Amphibians have three-chambered hearts

Pulmonary and systemic circulation are partly separated in adult amphibians such as frogs and toads. A single ventricle pumps blood to the lungs, where it picks up oxygen and dumps carbon dioxide. The ventricle also pumps blood to the rest of the body, where it picks up carbon dioxide and dumps oxygen. Separate atria receive the oxygenated blood from the lungs and the deoxygenated blood from the body (Figure 46.4c).

Because both of these atria deliver blood to the same ventricle, the oxygenated and deoxygenated blood could mix, in which case the blood going to the tissues would not carry a full load of oxygen. Mixing is limited, however, because anatomical features of the ventricle tend to direct the flow of deoxygenated blood from the right atrium to the pulmonary circuit and the flow of oxygenated blood from the left atrium to the aorta. The advantage of this partial separation of pulmonary and systemic circulation is that the high resistance of the capillary beds of the gas exchange organ no longer lies between the heart and the tissues. Therefore, the amphibian heart delivers blood to the aorta, and hence to the body, at a higher pressure than the fish heart does, which pumps the blood through the gills first.

Reptiles and crocodilians have exquisite control of pulmonary and systemic circulation

Turtles, snakes, and lizards are commonly said to have three-chambered hearts, while crocodilians (crocodiles and alligators) have four-chambered hearts. But to say that turtles, snakes, and lizards have three-chambered hearts is an oversimplification. The hearts of these animals have two separate atria and a ventricle that is partly divided in a complex way that decreases the mixing of oxygenated and deoxygenated blood (Figure 46.4d). When you explore in detail the complexity of reptilian hearts, you are more likely to conclude on a functional basis that they have five chambers because of the subdivisions of the ventricles. However, we can understand the important adaptations of reptilian hearts without going into that level of detail.

The most important and unusual feature of reptilian and crocodilian hearts is their ability to alter the distribution of blood going to the lungs and to the rest of the body. Think about the behavior, ecology, and

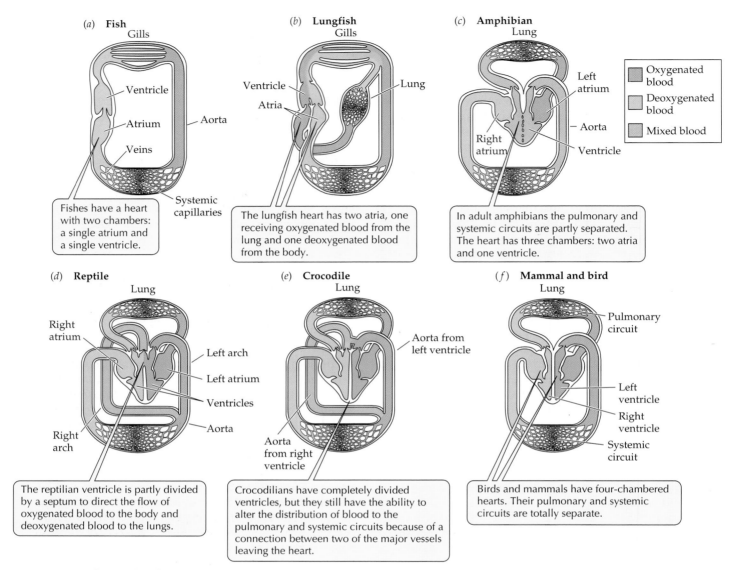

(a) **Fish**

Gills

Ventricle

Atrium

Aorta

Veins

Systemic capillaries

Fishes have a heart with two chambers: a single atrium and a single ventricle.

(b) **Lungfish**

Gills

Ventricle

Atria

Lung

The lungfish heart has two atria, one receiving oxygenated blood from the lung and one deoxygenated blood from the body.

(c) **Amphibian**

Lung

Left atrium

Right atrium

Aorta

Ventricle

☐ Oxygenated blood
☐ Deoxygenated blood
☐ Mixed blood

In adult amphibians the pulmonary and systemic circuits are partly separated. The heart has three chambers: two atria and one ventricle.

(d) **Reptile**

Lung

Right atrium

Left arch

Left atrium

Ventricles

Aorta

Right arch

The reptilian ventricle is partly divided by a septum to direct the flow of oxygenated blood to the body and deoxygenated blood to the lungs.

(e) **Crocodile**

Lung

Aorta from left ventricle

Aorta from right ventricle

Crocodilians have completely divided ventricles, but they still have the ability to alter the distribution of blood to the pulmonary and systemic circuits because of a connection between two of the major vessels leaving the heart.

(f) **Mammal and bird**

Lung

Pulmonary circuit

Left ventricle

Right ventricle

Systemic circuit

Birds and mammals have four-chambered hearts. Their pulmonary and systemic circuits are totally separate.

46.4 Vertebrate Circulatory Systems All vertebrates have closed circulatory systems. Evolutionary progression has led to an increasing separation of blood flow to the gas exchange organs (pulmonary circuit) and blood flow to the rest of the body (systemic circuit).

physiology of these animals. Despite the common image of the turtles as being slow and plodding, reptiles and crocodilians can be highly active, fast, powerful animals. Also, they can be inactive for long periods of time during which they have metabolic rates much lower than those of birds and mammals. The enormous range of metabolic demands in these animals means that they do not have to breathe continuously. Some species are also accomplished divers and spend long periods of time underwater without breathing.

To understand the wonderful adaptations of the reptilian and crocodilian hearts, you have to realize that there is no benefit from sending blood to the lungs when the animals are not breathing. The hearts of these animals enable them to circulate blood through their lungs and then the rest of their bodies when they are breathing, but when they are not breathing, they can bypass the lung circuit and pump all blood around the body. How do they accomplish this switching?

Reptiles and crocodilians have two aortas instead of one. The simplified representation of this anatomy in Figure 46.4d shows that one of these aortas can receive blood from either the right side or the left side of the ventricle. When the animal is breathing air, two factors cause blood from the right side of the ventricle to go preferentially into the lung circuit rather than into the systemic circuit. First, the resistance in the pulmonary circuit is low; second, there is a slight asynchrony in the timing of ventricular contraction, so the blood in the right side of the ventricle tends to be ejected slightly before the blood in the left side of the ventricle. Thus, as the ventricle contracts, the deoxygenated blood in the right side of the ventricle moves first into the lung circuit.

When the oxygenated blood in the left side of the ventricle starts to move, it encounters resistance in the lung circuit, which is already filled with the deoxygenated blood. Therefore the blood from the left side tends to flow into the two aortas. Blood flow is rerouted when the snake, lizard, or turtle stops breathing because of constriction of vessels in the lung circuit. As resistance in the lung circuit increases, the blood from the right side of the ventricle tends to be directed into one of the aortas. As a result, blood from both sides of the ventricle flows through the aortas to the systemic circuit.

The ability of snakes, lizards, and turtles to redirect blood flow from the lung circuit to the systemic circuit depends on the incomplete division of their ventricles. Crocodilians have true four-chambered hearts with completely divided ventricles. Yet, the crocodilians have not lost this ability to shunt blood from the lung circuit when they are not breathing. The crocodilians have one aorta originating in the left ventricle and one aorta originating in the right ventricle. However, a short channel connects these two aortas just after they leave the heart (Figure 46.4e).

Because the crocodile's ventricles are separate, they can generate different pressures when they contract. When the animal is breathing, the pressure in the left ventricle and its aorta is higher than the pressure in the right ventricle. This higher pressure is communicated through the connecting channel to the other aorta, and this back pressure prevents right-ventricle blood from entering that aorta. As a result, both aortas carry blood from the left ventricle, and the blood from the right ventricle flows to the lung circuit. When the animal is not breathing, constriction of vessels in the lung circuit increases the resistance in that circuit. As a result, pressure builds up in the right ventricle to a level that opens the valve to its aorta. Under these conditions, blood from both ventricles flows through the two aortas and little blood flows in the lung circuit.

You can now appreciate the fact that reptilian and crocodilian hearts are not primitive. Rather, these hearts and their major vessels are highly adapted to operate efficiently over a wide range of metabolic demands. Perhaps their hearts are one reason why these groups of animals have been so successful over such a long period of geological time.

Birds and mammals have fully separated pulmonary and systemic circuits

The four-chambered hearts of birds and mammals enable full separation of their pulmonary and systemic circuits (Figure 46.4f). This design has several advantages. First, since the oxygenated and deoxygenated bloodstreams cannot mix, the systemic circuit is always receiving arterial blood with the highest oxygen content. Second, respiratory gas exchange is maximized because the blood with the lowest oxygen content and highest CO_2 content is sent to the lungs. Third, because the systemic and the pulmonary circuits are completely separate, they can operate at different pressures.

Why is separation of systemic and pulmonary circuits important? Mammalian and bird tissues have high nutrient demands and thus a very high density of the smallest vessels, the capillaries. Many small vessels present lots of resistance to the flow of blood. Therefore, higher pressure is required to maintain adequate blood flow in the systemic circuit. The pulmonary circuit does not have such a large number of capillaries and such a high resistance, so it doesn't require such high pressure.

The Human Heart: Two Pumps in One

Like all other mammalian hearts, the human heart has four chambers: two atria and two ventricles (Figure 46.5). The atrium and ventricle on the right side of your body are called the right atrium and right ventricle. They can be thought of as the right heart. The atrium and ventricle on the left side of your body are called the left atrium and left ventricle. They can be thought of as the left heart.

Each atrium pumps blood into its respective ventricle, and the ventricles pump blood into arteries. The right ventricle pumps blood through the pulmonary circuit, and the left ventricle pumps blood through the systemic circuit. Valves between the atria and ventricles, the **atrioventricular valves**, prevent backflow of blood into the atria when the ventricles contract. The **pulmonary valves** and the **aortic valves** positioned between the ventricles and the arteries prevent the backflow of blood into the ventricles.

Here we focus on the flow of blood through the heart and through the body. Then we'll examine the unique electrical properties of cardiac muscle and how the heart's electrical activity can be monitored by the EKG (electrocardiogram).

Blood flows from right heart to lungs to left heart to body

Let's follow the circulation of the blood through the heart. The right atrium receives blood from the **superior vena cava** and the **inferior vena cava** (see Figure 46.5), large veins that collect blood from the upper and lower body, respectively. The veins of the heart itself also drain into the right atrium. From the right atrium, the blood flows into the right ventricle. Most of the filling of the ventricle is due to passive flow while the heart is relaxed between beats. Just at the end of this

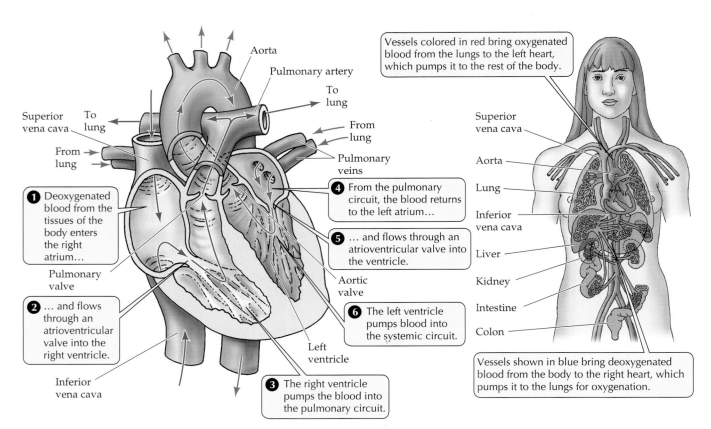

Aorta

Pulmonary artery

To lung

Superior vena cava

To lung

From lung

From lung

Pulmonary veins

1 Deoxygenated blood from the tissues of the body enters the right atrium…

Pulmonary valve

4 From the pulmonary circuit, the blood returns to the left atrium…

5 … and flows through an atrioventricular valve into the ventricle.

2 … and flows through an atrioventricular valve into the right ventricle.

Aortic valve

6 The left ventricle pumps blood into the systemic circuit.

Left ventricle

Inferior vena cava

3 The right ventricle pumps the blood into the pulmonary circuit.

Vessels colored in red bring oxygenated blood from the lungs to the left heart, which pumps it to the rest of the body.

Superior vena cava

Aorta

Lung

Inferior vena cava

Liver

Kidney

Intestine

Colon

Vessels shown in blue bring deoxygenated blood from the body to the right heart, which pumps it to the lungs for oxygenation.

46.5 The Human Heart and Circulation The atrioventricular valves prevent blood from flowing back into the atria when the ventricles contract. Pulmonary and aortic valves prevent blood from flowing back into ventricles from the arteries when the ventricles relax.

period of ventricular filling, the atrium contracts and adds a little more blood to the ventricular volume.

The right ventricle pumps blood into the **pulmonary artery**, which transports the blood to the lungs. The **pulmonary veins** return the oxygenated blood from the lungs to the left atrium, from which the blood enters the left ventricle. As with the right side of the heart, most left ventricular filling is passive and ventricular volume is topped off by contraction of the atrium just at the end of the period of filling (see Figure 46.6). The walls of the left ventricle are powerful muscles that contract around the blood with a wringing motion starting from the bottom. When pressure in the left ventricle is high enough to push open the aortic valve, the blood rushes into the aorta to begin its circulation throughout the body and eventually back to the right atrium. Note in Figure 46.5 how much more massive the left ventricle is than the right ventricle. Because there are many more arterioles and capillaries in the systemic circuit than in the pulmonary circuit, the resistance is higher in the systemic circuit and the left ventricle must squeeze with greater force than the right, even though both are pumping the same volume of blood.

The pumping of the heart—contraction of the two atria followed by contraction of the two ventricles, and then relaxation—is the **cardiac cycle**. Contraction of the ventricles is called **systole**, and relaxation of the ventricles **diastole** (Figure 46.6). Just at the end of diastole, the atria contract and top off the volume of blood in the ventricles. The sounds of the cardiac cycle, the "lub-dub" heard through a stethoscope placed on the chest, are created by the slamming shut of the heart valves. As the ventricles begin to contract, the atrioventricular valves close ("lub"), and when the ventricles begin to relax, the pressure in the aorta and pulmonary artery causes the aortic and pulmonary valves to bang shut ("dub"). Defective valves produce the sounds of heart murmurs. For example, if an atrioventricular valve is defective, blood will flow back into the atria with a "whoosh" sound following the "lub."

The cardiac cycle can be felt in the pulsation of arteries such as the one that supplies blood to your hand. You can feel your pulse by placing two fingers from one hand lightly over the wrist of the other hand just below the thumb. During systole, blood surges through the arteries of your arm and hand and you can feel that as a pulsing of the artery in your wrist.

You can measure blood pressure changes associated with the cardiac cycle in the large artery in your arm by using an inflatable pressure cuff called a sphygmomanometer and a stethoscope (Figure 46.7). When the inflation pressure of the cuff exceeds maximum (sys-

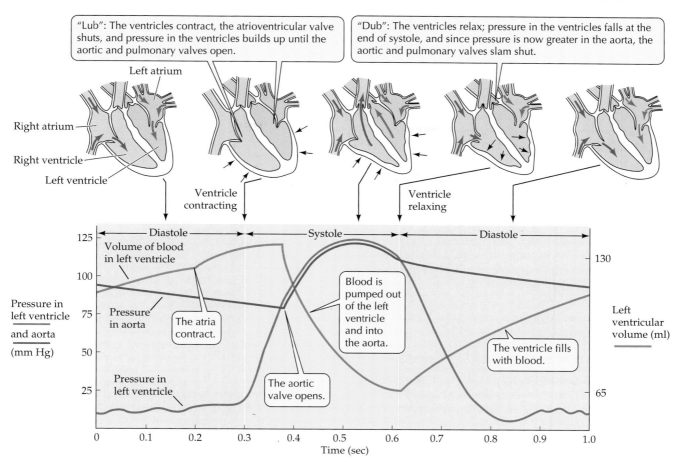

"Lub": The ventricles contract, the atrioventricular valve shuts, and pressure in the ventricles builds up until the aortic and pulmonary valves open.

"Dub": The ventricles relax; pressure in the ventricles falls at the end of systole, and since pressure is now greater in the aorta, the aortic and pulmonary valves slam shut.

Left atrium

Right atrium

Right ventricle

Left ventricle

Ventricle contracting

Ventricle relaxing

Pressure in left ventricle and aorta (mm Hg)

Volume of blood in left ventricle

Pressure in aorta

The atria contract.

Blood is pumped out of the left ventricle and into the aorta.

The ventricle fills with blood.

Pressure in left ventricle

The aortic valve opens.

Left ventricular volume (ml)

Diastole — Systole — Diastole

Time (sec)

46.6 The Cardiac Cycle The rhythmic contraction (systole) and relaxation (diastole) of the atria and ventricles is called the cardiac cycle.

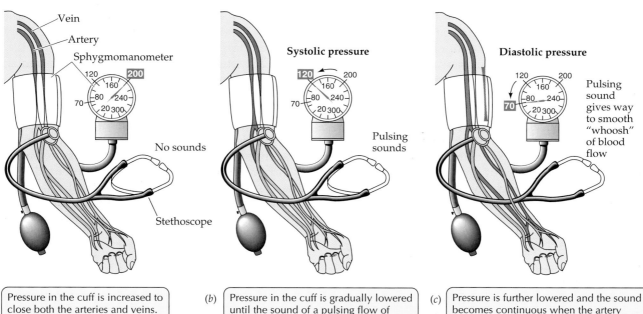

Vein

Artery

Sphygmomanometer

Systolic pressure

Diastolic pressure

No sounds

Stethoscope

Pulsing sounds

Pulsing sound gives way to smooth "whoosh" of blood flow

(a) Pressure in the cuff is increased to close both the arteries and veins. No sound is audible.

(b) Pressure in the cuff is gradually lowered until the sound of a pulsing flow of blood through the constriction in the artery during systole is heard. At this time, pressure in the cuff is just below the peak systolic pressure in the artery.

(c) Pressure is further lowered and the sound becomes continuous when the artery remains open for an entire heart cycle. The cuff is just below the diastolic pressure in the artery at this time.

46.7 Measuring Blood Pressure Blood pressure in a major artery of the arm can be measured with an inflatable pressure cuff called a sphygmomanometer.

Blood pressure in this person is 120/70.

tolic) blood pressure in the artery, blood flow in the artery stops. As the pressure in the cuff is gradually released, a point is reached when blood pressure at the peak of systole is greater than the pressure in the cuff. At this point, a little blood squirts through the closed artery and the artery slams shut, producing a sound that can be heard through a stethoscope applied to the arm. The pressure at which these slamming sounds are first heard is the systolic blood pressure.

As the cuff pressure is reduced even more, the slamming sounds gradually disappear to be replaced by a "whoosh" sound as the blood flow in the artery becomes more continuous. The pressure at which slamming sounds are no longer heard is the diastolic blood pressure. In a conventional blood pressure reading, the systolic value is placed over the diastolic value. Normal values for a young adult might be 120 mm of mercury (Hg) during systole and 80 mm Hg during diastole, or 120/80.

The heartbeat originates in the cardiac muscle

Cardiac muscle has unique properties that allow it to function as an effective pump. One such property is that the cardiac-muscle cells are in electrical continuity with one another. Special junctions called gap junctions enable action potentials to spread rapidly from cell to cell. (Recall what we learned about gap junctions in Chapters 5 and 41.) Because a spreading action potential stimulates contraction, large groups of cardiac-muscle cells contract in unison; this coordinated contraction is important for pumping blood.

Another important property of cardiac-muscle cells is that some of them have the ability to initiate action potentials without stimulation from the nervous system. These cells stimulate neighboring cells to contract, thereby acting as pacemakers. The important characteristic of a pacemaker cell is that its resting membrane potential gradually becomes less negative until it reaches the threshold voltage for initiating an action potential (Figure 46.8). These action potentials look different from the neuronal action potentials you saw in Chapter 41 because the depolarization is due primarily to the opening of voltage-gated calcium channels rather than voltage-gated sodium channels.

As with neuronal action potentials, however, cardiac muscle repolarizes in part as a response to the opening of potassium channels. These potassium channels in cardiac pacemaker cells are unique. After an action potential, they are open, causing the membrane potential to fall to its most negative level. Then they gradually close, and as they do so the membrane potential becomes less negative—it slowly depolarizes. When it reaches threshold for the voltage-gated calcium channels, another action potential occurs.

The nervous system controls the heartbeat (speeds it up or slows it down) by influencing the rate at

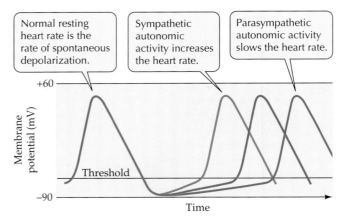

46.8 The Autonomic Nervous System Controls Heart Rate The resting potentials of the plasma membranes of pacemaker cells spontaneously depolarize to threshold and fire action potentials. Signals from the two divisions of the autonomic nervous system raise and lower the heart rate, respectively.

which pacemaker cells gradually depolarize between action potentials. Acetylcholine released by parasympathetic nerve endings onto the pacemaker cells slows the rate at which the potassium channels close and thereby slows the heart rate. Norepinephrine released by sympathetic nerve endings onto the pacemaker cells increases the rate at which the potassium channels close and thereby speeds the heart rate (see Figure 46.8).

Under normal circumstances the pacemaker activity of the heart originates from modified cardiac muscle cells located at the junction of the superior vena cava and right atrium, in the **sinoatrial node** (Figure 46.9). An action potential spreads from the sinoatrial node across the atrial walls, causing the two atria to contract in unison. However, there are no gap junctions between the atria and the ventricles.

The action potential initiated in the atria passes to the ventricles through another node of modified cardiac-muscle cells, the **atrioventricular node**. This atrioventricular node passes the action potential along to the ventricles via modified muscle fibers called the **bundle of His**. The bundle of His divides into right and left bundle branches. The branches connect with **Purkinje fibers** that, in turn, branch throughout the ventricular muscle.

The timing of the spread of the action potential from atria to ventricles is important. The atrioventricular node imposes a short delay in the spread of the action potential from atria to ventricles. Then the action potential spreads very rapidly throughout the ventricles, causing them to contract. Thus the atria contract before the ventricles do, so the blood passes progressively from the atria to the ventricles to the arteries.

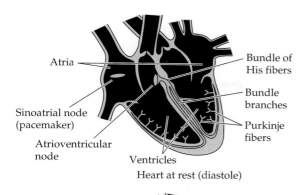

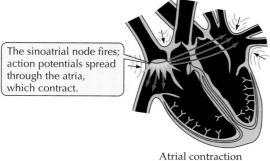

The sinoatrial node fires; action potentials spread through the atria, which contract.

Atrial contraction

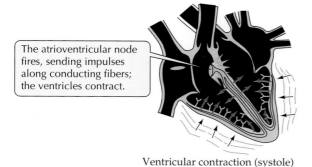

The atrioventricular node fires, sending impulses along conducting fibers; the ventricles contract.

Ventricular contraction (systole)

46.9 The Heartbeat Pacemaker cells in the sinoatrial node initiate action potentials that spread through the walls of the atria, causing them to contract.

The EKG records the electrical activity of the heart

During the cardiac cycle, electrical events in the cardiac muscle can be recorded by electrodes placed on the surface of the body. The recording is called an *electrocardiogram*, or *EKG* ("EKG" because the Greek word for heart is *kardia*, but "ECG" is also used). The EKG is an important tool for diagnosing heart problems.

The action potentials that sweep through the muscles of the atria and the ventricles before they contract are such massive, localized electrical events that they cause electric currents to flow outward from the heart to all parts of the body. Electrodes placed on the surface of the body at different locations—usually on the wrists and ankles—detect those electric currents at different times. The appearance of the EKG depends on the exact placement of the electrodes used for the recording. Electrodes placed on the right wrist and left ankle produced the normal EKG shown in Figure 46.10a. The waves of the EKG are designated P, Q, R, S, and T, each letter representing a particular event in the cardiac muscle, as indicated on Figure 46.10a.

The EKG is used by cardiologists (heart specialists) to diagnose heart problems. Figure 46.10b shows some abnormal EKGs that result from different heart problems. For patients who have had heart attacks it is possible to determine which region of the heart has been damaged by placing EKG electrodes at different locations on the chest (Figure 46.10c). Comparing EKGs from the different electrodes tells the cardiologist which region of the heart is behaving abnormally.

The Vascular System: Arteries, Capillaries, and Veins

The blood circulates around the body in a system of blood vessels: arteries, capillaries, and veins. Arteries receive blood from the heart; accordingly, they have properties that enable them to withstand high pressure. The arteries are important in controlling the distribution of blood to different organs and in controlling central blood pressure. Veins have characteristics that enable them to return blood to the heart at low pressure and to serve as a blood reservoir. The properties of capillaries make them the site of all exchanges between the blood and the internal environment. It is important to understand how the structure of the different vessels supports their functions. In addition to arteries, capillaries, and veins, this section considers the lymphatics and the relation of intercellular fluid to lymph and blood.

Arteries and arterioles have abundant elastic and muscle fibers

Blood pressure is highest in the vessels that carry blood away from the heart—the arteries and arterioles—and their structure reflects this fact. The walls of the large arteries have many elastic fibers that enable them to withstand high pressures (Figure 46.11). These elastic fibers have another important function as well: During systole they are stretched and thereby store some of the energy imparted to the blood by the heart. During diastole they return this energy by squeezing the blood and pushing it forward. As a result, even though the flow of blood through the arterial system pulsates, it is smoother than it would be through a system of rigid pipes.

Abundant smooth-muscle cells in the arteries and arterioles contract and expand the diameter of those vessels. These changes in the diameter alter the resistance of the vessels, which controls the flow of blood. By decreasing the resistance of these vessels, neural and hormonal mechanisms control the distribution of

(a)   A normal EKG

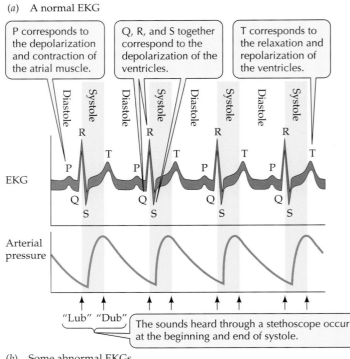

P corresponds to the depolarization and contraction of the atrial muscle.

Q, R, and S together correspond to the depolarization of the ventricles.

T corresponds to the relaxation and repolarization of the ventricles.

EKG

Arterial pressure

"Lub" "Dub"   The sounds heard through a stethoscope occur at the beginning and end of systole.

(b)   Some abnormal EKGs

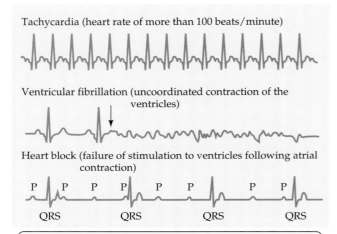

Tachycardia (heart rate of more than 100 beats/minute)

Ventricular fibrillation (uncoordinated contraction of the ventricles)

Heart block (failure of stimulation to ventricles following atrial contraction)

Besides detecting rhythmic irregularities in the heartbeat (arrhythmias), EKGs can detect damage to the heart muscle (infarctions) or decreased blood supply to the heart muscle (ischemias) by changes in the size and shape of the EKG curves.

(c)

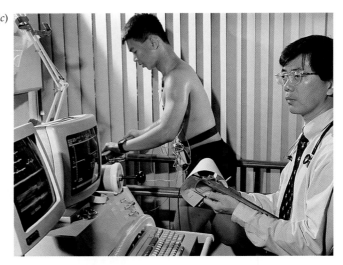

46.10 The Electrocardiogram   An EKG monitors heart function.

blood to the different tissues of the body; they also control central blood pressure. The arteries and arterioles are referred to as the **resistance vessels** because their resistance varies.

Materials are exchanged between blood and intercellular fluids in the capillaries

Beds of capillaries connect arterioles to venules. No cell of the body is more than a couple of cell diameters away from a capillary. The needs of cells are served by the exchange of materials between blood and intercel-

lular fluid. This exchange takes place across the capillary walls. It is possible because capillaries have thin, permeable walls and because blood flows through them slowly under very low pressure (see Figure 46.11).

To anyone who has played with a garden hose, it may seem strange that big arteries have high pressure and fast flow and that when the blood flows into the small capillaries the pressure and flow decrease. When you restrict the diameter of the garden hose by placing your thumb over the opening, the pressure in the hose increases, which in turn increases the velocity of the water spraying out of the hose.

This puzzle is solved by one more piece of information. Arterioles branch into so many capillaries that the total cross-sectional area of capillaries is much greater than that of any other class of vessels. Even though each capillary is so small that the red blood cells pass through in single file (Figure 46.12), each arteriole gives rise to such a large number of capillaries that together they have a much greater capacity for blood than do the arterioles. An analogy is a fast-flowing river dividing up into many small rivulets flowing across a broad delta. Each rivulet may be small and its flow sluggish, yet all together they accommodate all of the water poured into the delta by the larger, faster-flowing river.

Materials are exchanged in capillary beds by filtration, osmosis, and diffusion

The walls of capillaries are permeable to water and small molecules, but not to large molecules such as proteins. Blood pressure therefore tends to squeeze water and small molecules out of the capillaries and into the surrounding intercellular spaces. This process

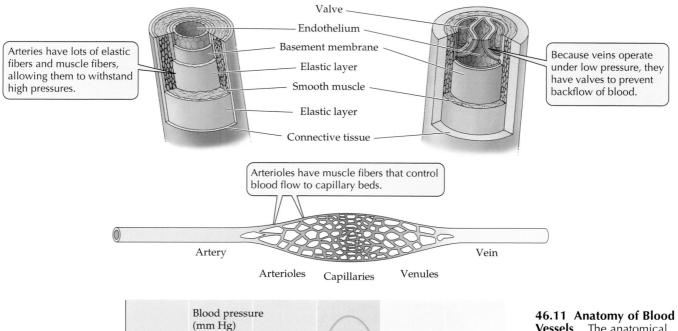

Valve

Endothelium

Basement membrane

Elastic layer

Smooth muscle

Elastic layer

Connective tissue

Arteries have lots of elastic fibers and muscle fibers, allowing them to withstand high pressures.

Because veins operate under low pressure, they have valves to prevent backflow of blood.

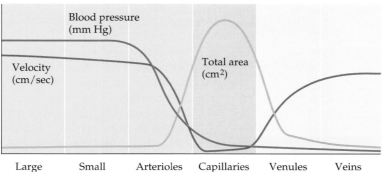

Arterioles have muscle fibers that control blood flow to capillary beds.

Artery

Vein

Arterioles   Capillaries   Venules

Blood pressure (mm Hg)

Velocity (cm/sec)

Total area (cm²)

Large arteries | Small arteries | Arterioles | Capillaries | Venules | Veins

46.11 Anatomy of Blood Vessels The anatomical characteristics of blood vessels match their functions. The total cross-sectional area of capillaries is larger than for any other class of vessels, and they are more permeable, thus suiting them for their function of exchange of nutrients and wastes with the extracellular fluids.

is *filtration.* The large molecules that cannot cross the capillary wall create an osmotic potential (also called osmotic pressure) that tends to draw water back into the capillary.

Blood pressure is highest on the arterial side of a capillary bed and steadily decreases as the blood flows to the venous side. Therefore, more water is squeezed out of the capillaries on the arterial side of the bed. The osmotic potential pulling water back into the capillary gradually becomes the dominant force as the blood flows toward the venous side of the bed. The interactions of the two opposing forces—blood pressure versus osmotic potential—determine the net flow of water (Figure 46.13).

The balance between blood pressure and osmotic potential changes if the blood pressure in the arterioles and the permeability of the capillary walls change. An example of such a change is associated with the inflammation that accompanies injuries to the skin or allergic reactions. The inflamed area becomes hot and red because blood flow to the area increases. The inflamed tissue also swells. The major mediator of these events is a chemical called **histamine** that is released

mainly by certain white blood cells flowing through the damaged tissue.

Histamine makes blood vessels expand, thus increasing blood flow to the area and increasing pressure in the capillaries. Because histamine also in-

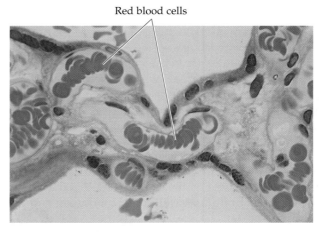

Red blood cells

46.12 A Narrow Lane Red blood cells pass through capillaries slowly, and in single file.

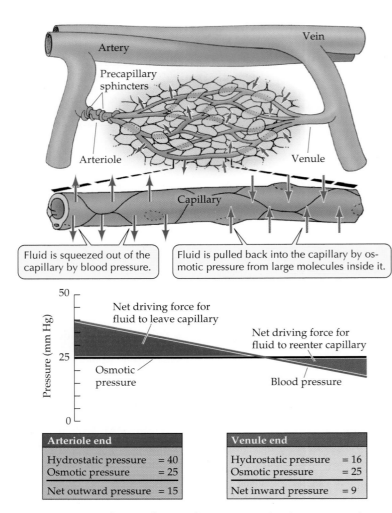

Fluid is squeezed out of the capillary by blood pressure.

Fluid is pulled back into the capillary by osmotic pressure from large molecules inside it.

Arteriole end	
Hydrostatic pressure	= 40
Osmotic pressure	= 25
Net outward pressure	= 15

Venule end	
Hydrostatic pressure	= 16
Osmotic pressure	= 25
Net inward pressure	= 9

46.13 A Balance of Opposite Forces Blood pressure and osmotic attraction control the exchange of fluids between blood vessels and intercellular space.

creases the permeability of the capillaries and venules, more water leaves the capillaries and venules, and the tissue swells because of the accumulation of intercellular fluids, a condition known as **edema**. The use of drugs called **antihistamines** can alleviate inflammation and allergic reactions.

The loss of water from the capillaries increases if the osmotic potential of the blood decreases, as is seen in the disease kwashiorkor (see Figure 47.20). This disease is caused by severe protein starvation. When the body has no amino acids available for the synthesis of essential proteins, it begins to break down its own blood proteins. Thus fewer molecules are available in the blood to maintain the osmotic potential that pulls water back into the capillaries. The result is that intercellular fluids build up, swelling the abdomen and the extremities.

Whether specific small molecules cross a capillary wall depends on the architecture of the capillary, the

type of substance, and the concentration difference between the blood and the intercellular fluid. Capillary walls are membranous, and lipid-soluble substances and many small solute molecules can pass through them from the area of higher concentration to that of lower concentration (see Chapter 5).

Consider what happens in the capillary beds of skeletal-muscle tissue. Because the concentration of oxygen is high in the blood coming from the arteriole but very low in active skeletal-muscle tissue, oxygen readily moves from the blood into the muscle. At the same time, carbon dioxide rapidly moves into the blood because its concentration is high in the working muscle but low in the blood. The concentrations of these gases in the blood thus change rapidly as the blood travels through the capillary beds.

Small molecules in the blood generally can pass through the capillary walls, but the capillaries in different tissues are differentially selective to the sizes of molecules that can pass through them. In all capillaries, O_2, CO_2, glucose, lactate, and small ions such as Na^+ and Cl^- can cross. In the capillaries of the brain, not much else can cross unless it is a lipid-soluble substance such as alcohol; this high selectivity of brain capillaries is known as the *blood–brain barrier* (see Chapter 41).

In other tissues the capillaries are much less selective. Such capillaries are found in the digestive tract, where nutrients are absorbed, and in the kidneys, where wastes are filtered. Some capillaries have large gaps that permit the movement of even larger substances. These capillaries are found in the bone marrow, spleen, and liver. Substances move across many capillary walls by endocytosis.

Lymphatics return intercellular fluid to the blood

The intercellular fluid that accumulates outside the capillaries contains water and small molecules but no red blood cells, and less protein than is in blood. A separate system of vessels—the **lymphatic system**—returns the intercellular fluid to the blood.

After entering the lymphatic vessels, the intercellular fluid is called **lymph**. Fine lymphatic capillaries merge progressively into larger and larger vessels and end in a major vessel—the **thoracic duct**—which empties into the superior vena cava returning blood to the heart (see Figure 18.1). Lymphatic vessels have one-way valves that keep the lymph flowing toward the thoracic duct. The propelling force moving the lymph is pressure on the lymphatic vessels from the contractions of nearby skeletal muscles.

Mammals and birds have lymph nodes along the major lymphatic vessels. Lymph nodes are an important component of the defensive machinery of the body (see Chapter 18). They are a major site of lymphocyte production and of the phagocytic action that removes microorganisms and other foreign materials from the circulation. The lymph nodes also act as mechanical filters. Particles become trapped there and are digested by the phagocytes that are abundant in the nodes.

Lymph nodes swell during infection. Some of them, particularly those on the side of the neck or in the armpit, become noticeable when they swell. The nodes also trap metastasized cancer cells—that is, those that have broken free of the original tumor. Because such cells may start additional tumors, surgeons often remove the neighboring lymph nodes when they excise a malignant tumor.

Blood flows back to the heart through veins

The pressure of the blood flowing from capillaries to venules is extremely low, and insufficient to propel blood back to the heart. If the veins are above the level of the heart, gravity helps blood flow, but below the level of the heart, blood must be moved against the pull of gravity. Blood tends to accumulate in veins, and the walls of veins are more expandable than the walls of arteries. As much as 80 percent of the total blood volume may be in the veins at any one time. Because of their high capacity to store blood, veins are called **capacitance vessels**.

Blood must be returned from the veins to the heart so that circulation can continue. If too much blood remains in the veins, then too little blood returns to the heart, and thus too little blood is pumped to the brain; a person may faint as a result. Fainting is self-correcting: A fainting person falls, thereby changing from the position in which gravity caused blood to accumulate in the lower body. But means other than fainting also move blood from the tissues back to the heart.

The most important of the forces that propel venous and lymphatic return from the regions of the body below the heart is the milking action caused by skeletal-muscle contraction around the vessels. As muscles contract, the vessels are squeezed and the blood moves through them. Blood flow might be temporarily obstructed during a prolonged muscle contraction, but with relaxation of the muscles the blood is free to move again. Within the veins are valves that prevent the backflow of blood. Thus whenever a vein is squeezed, blood is propelled forward because the valves prevent it from flowing backward. In this way blood is gradually pushed toward the heart (Figure 46.14). As we already noted, the lymphatic vessels have similar valves.

Gravity causes edema, as well as blood accumulation in veins. The back pressure that builds up in the capillaries when blood accumulates in the veins shifts the balance between blood pressure and osmotic potential so that there is a net movement of fluid into the interstitial spaces. This is why you have trouble putting your shoes back on after you sit for a long time

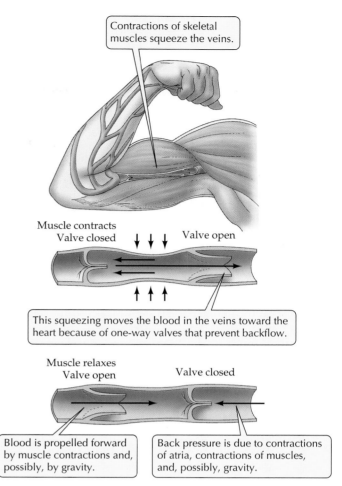

46.14 One-Way Flow The veins have valves that prevent blood from flowing backward.

with your shoes off, such as on an airline flight. In persons with very expandable veins, the veins may become so stretched that the valves can no longer prevent backflow. This condition produces varicose (swollen) veins. Draining these veins is highly desirable and can be aided by wearing support hose and periodically elevating the legs above the level of the heart.

During exercise, the milking action of muscles speeds blood toward the heart to be pumped to the lungs and then to the respiring tissues. As an animal runs, its legs act as auxiliary vascular pumps, returning blood to the heart from the veins of the lower body. As a greater volume of blood is returned to the heart, the heart contracts more forcefully, and its pumping action becomes more effective. This strengthening of the heartbeat is due to a property of cardiac-muscle cells referred to as the **Frank–Starling law**: If the cells are stretched, as they are when the volume of returning blood increases, they contract more forcefully. This principle holds (within a certain range) whenever venous return increases, by any mechanism.

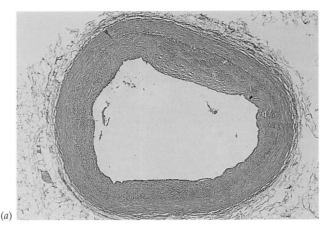

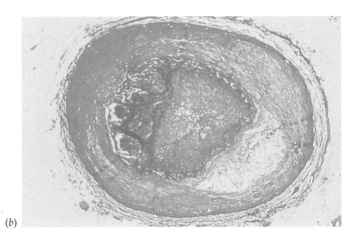

(a)

(b)

46.15 Atherosclerotic Plaque   (a) A healthy, clear artery. (b) An atherosclerotic artery, clogged with plaque and a thrombus.

The actions of breathing also help return venous blood to the heart. The ventilatory muscles create suction that pulls air into the lungs (see Chapter 45), and this suction also pulls blood and lymph toward the chest, increasing venous return to the right atrium.

Some smooth muscle in the walls of veins moves venous blood back to the heart by constricting the veins and moving the blood forward. These muscles are rare in most veins and are totally absent from lymphatic vessels in humans. They do not play a major role in venous return. However, in the largest veins closest to the heart, contraction of smooth muscle at the onset of exercise can suddenly increase venous return and stimulate the heart in accord with the Frank–Starling law, thus increasing cardiac output.

Will you die of cardiovascular disease?

Cardiovascular disease is by far the largest single killer in the United States and Europe; it is responsible for about half of all deaths each year. The immediate cause of most of these deaths is heart attack or stroke, but those events are the end result of a disease called **atherosclerosis** (hardening of the arteries) that begins many years before symptoms are detected. Hence atherosclerosis is called the silent killer. What is atherosclerosis, and how can it be prevented?

Healthy arteries have a smooth internal lining of endothelial cells (Figure 46.15a). This lining can be damaged by chronic high blood pressure, smoking, a high-fat diet, or microorganisms. Fatty deposits called **plaque** begin to form at sites of endothelial damage. First the endothelial cells at the damaged site swell and proliferate; then they are joined by smooth-muscle cells migrating from below. Lipids, especially cholesterol, are deposited in these cells, so the plaque becomes fatty. Fibrous connective tissue invades the plaque and, along with deposits of calcium, makes the artery wall

less elastic; this process is what gives us the terms "atherosclerosis" and "hardening of the arteries."

The growing plaque narrows the artery and causes turbulence in the blood flowing over it. Blood platelets (discussed later in this chapter) stick to the plaque and initiate the formation of a blood clot, called a **thrombus**, that further blocks the artery (Figure 46.15b). The blood supply to the heart itself flows through the *coronary arteries*. These arteries are highly susceptible to atherosclerosis; as they narrow, blood flow to the heart muscles decreases. Chest pains and shortness of breath during mild exertion are symptoms of this condition.

A person with atherosclerosis is at high risk of forming a thrombus in a coronary artery. This condition, called **coronary thrombosis**, can totally block the vessel, causing a heart attack, or **coronary infarction**. A piece of a thrombus breaking loose, called an **embolus**, is likely to travel to and become lodged in a vessel of smaller diameter, blocking its flow (a condition referred to as an **embolism**). Arteries already narrowed by plaque formation are likely places for an embolus to lodge. An embolism in an artery in the brain causes the cells fed by that artery to die. This is called a **stroke**. The specific damage resulting from a stroke, such as memory loss, speech impairment, or paralysis, depends on the location of the blocked artery.

The most important solution to cardiovascular disease is prevention, not treatment. The risk factors for developing atherosclerosis are high-fat and high-cholesterol diet, smoking, a sedentary lifestyle, hypertension (high blood pressure), obesity, certain medical conditions such as diabetes, and genetic predisposition. Changes in diet and behavior can prevent and reverse atherosclerosis and help fend off the silent killer.

Blood: A Fluid Tissue

We have considered the circulation of the blood in detail. Now let's examine the blood itself. Blood is a tissue; it has cellular elements suspended in an aqueous

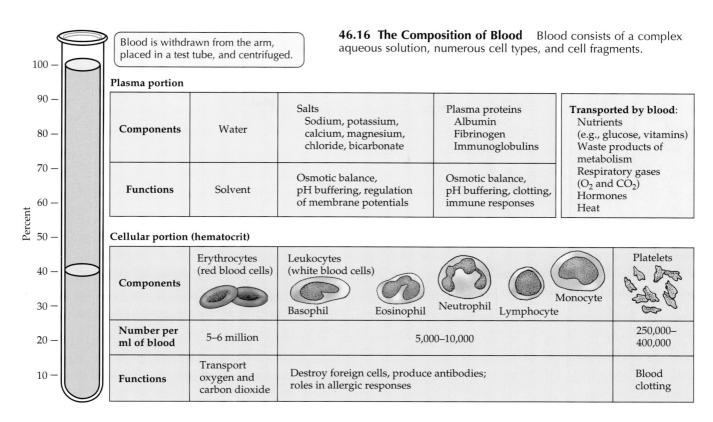

46.16 The Composition of Blood Blood consists of a complex aqueous solution, numerous cell types, and cell fragments.

medium of specific, yet complex, composition. The cells of the blood can be separated from the aqueous medium, called **plasma**, by centrifugation (Figure 46.16).

If we take a 100 ml sample of blood and spin it in a centrifuge, all the cells move to the bottom of the tube, leaving the straw-colored, clear plasma on top. The **packed-cell volume**, or **hematocrit**, is the percentage of the blood volume made up by cells. Normal hematocrit is about 38 percent for women and 46 percent for men, but the values can vary considerably. They are usually higher, for example, in people who live and do heavy work at high altitudes.

In this section, we will consider the three classes of cellular elements in blood: the red blood cells (erythrocytes); the white blood cells (leukocytes); and the platelets, which are pinched-off fragments of cells. Finally, we will take a closer look at the content of plasma.

Red blood cells transport respiratory gases

Most of the cells in the blood are **erythrocytes**, or red blood cells. The function of red blood cells is to transport the respiratory gases (see Chapter 45). There are about 5 million red blood cells per milliliter of blood but only 5,000 to 10,000 white blood cells in the same volume. Red blood cells form from special cells in the bone marrow called **stem cells**, particularly in the ribs, breastbone, pelvis, and vertebrae (Figure 46.17a). Red blood cell production is controlled by a hormone, **ery-**

thropoietin, which is released by cells in the kidney in response to insufficient oxygen (Figure 46.17b).

Erythropoietin stimulates stem cells to produce red blood cells. Under normal conditions your bone marrow produces about 2 million red blood cells every second. The developing red blood cells divide many times while still in the bone marrow, and during this time they are producing hemoglobin. When the hemoglobin content of a red blood cell approaches about 30 percent, its nucleus, endoplasmic reticulum, Golgi apparatus, and mitochondria begin to break down. This process is almost complete when the new red blood cell squeezes through pores in the endothelial walls of blood vessels and enters the circulation.

Each red blood cell circulates for about 120 days and then breaks down. The iron from its hemoglobin molecules is recycled to the bone marrow. Mature red blood cells are biconcave, flexible discs packed with hemoglobin. Their shape gives them a large surface area for gas exchange, and their flexibility enables them to squeeze through the capillaries (see Figures 45.18 and 46.12).

White blood cells protect the body

Leukocytes, or white blood cells, defend the body against foreign objects infection. Leukocytes squeeze through capillary walls and spend a great deal of time outside the vascular system. They move about by amoeboid motion and travel to sites of infection and

(a)

Vascular space

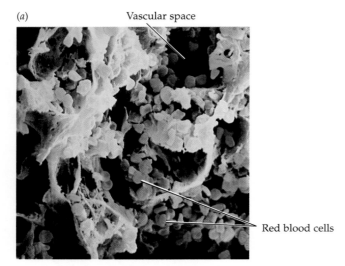

Red blood cells

(b)

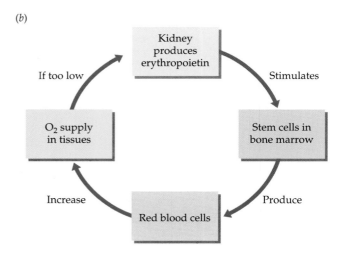

46.17 Red Blood Cells Form in the Bone Marrow
(a) This artificially colored electron micrograph shows red blood cells forming in the bone marrow. As new red blood cells mature, they squeeze through the endothelium lining the vessels and enter the blood. (b) Erythropoietin stimulates stem cells in the bone marrow to produce red blood cells, increasing oxygen supply in tissues.

cell damage by following cues from chemicals released by dead or sick cells.

Leukocytes arise from the same bone marrow stem cells that produce erythrocytes; hence these stem cells are called *totipotent*. Totipotent stem cells produce two populations of stem cells that are more differentiated (they have more limited potentials). One lineage is only capable of producing leukocytes known as lymphocytes. This lineage leaves the bone marrow and takes up residence in lymphoid tissue (such as the lymph nodes, thymus, spleen, and tonsils), where they produce the lymphocytes that in turn produce immunoglobulins (see Chapter 18).

The second lineage, called *pluripotent* stem cells, remains in the bone marrow, producing erythrocytes along with several types of leukocytes that protect the body by ingesting and destroying foreign objects and organisms by phagocytosis. Pluripotent stem cells also produce megakaryotcytes, which are large cells that produce cell fragments called **platelets**. Platelets circulate in the blood and participate in clotting reactions.

Platelets are essential for blood clotting

A platelet is just a tiny fragment of a cell, but it is packed with enzymes and chemicals necessary for its function of sealing leaks in the blood vessels—that is, clotting the blood. In a damaged vessel, collagen fibers are exposed. When a platelet encounters collagen fibers, it is activated. It swells, becomes irregularly shaped and sticky, and releases chemicals that activate other platelets and initiate the clotting of blood. The

sticky platelets form a plug at the damaged site, and the subsequent clotting forms a stronger patch on the vessel.

The clotting of blood requires many steps and many **clotting factors**. The absence of any one of these factors can cause excessive bleeding and thus can be lethal. Because the liver produces most of the clotting factors, liver diseases such as hepatitis and cirrhosis can result in excessive bleeding. The sex-linked trait hemophilia (see Chapter 10) is an example of a genetic inability to produce one of the clotting factors.

Blood clotting factors participate in a cascade of steps that activate other substances circulating in the blood. The cascade begins with cell damage and platelet activation and continues with the conversion of an inactive circulating enzyme, **prothrombin**, to its active form, **thrombin**. Thrombin causes circulating protein molecules called **fibrinogen** to polymerize and form **fibrin** threads. The fibrin threads form the meshwork that clots the blood cells, seals the vessel, and provides a scaffold for the formation of scar tissue (Figure 46.18).

Plasma is a complex solution

Plasma contains gases, ions, nutrient molecules, proteins, and other molecules, such as nonprotein hormones. Most of the ions are Na^+ and Cl^- (hence the salty taste of blood), but many other ions are also present. Nutrient molecules in plasma include glucose, amino acids, lipids, cholesterol, and lactic acid. The circulating proteins have many functions. We have just noted proteins that function in blood clotting; others of interest include albumin, which is largely responsible for the osmotic potential in capillaries that prevents a massive loss of water from plasma to intercellular spaces; antibodies (the immunoglobulins); hormones; and various carrier molecules, such as **transferrin**, which carries iron from the gut to where it is stored or used.

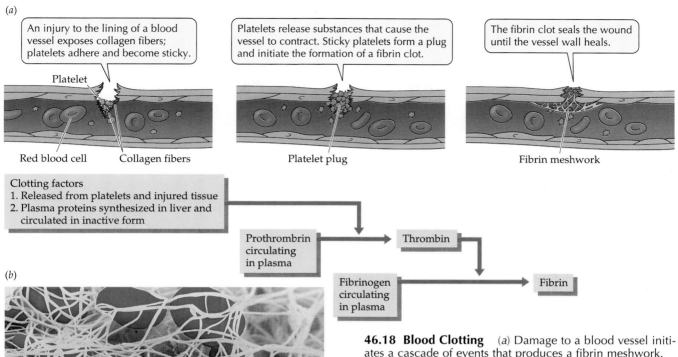

(a)

An injury to the lining of a blood vessel exposes collagen fibers; platelets adhere and become sticky.

Platelets release substances that cause the vessel to contract. Sticky platelets form a plug and initiate the formation of a fibrin clot.

The fibrin clot seals the wound until the vessel wall heals.

Platelet

Red blood cell   Collagen fibers

Platelet plug

Fibrin meshwork

Clotting factors
1. Released from platelets and injured tissue
2. Plasma proteins synthesized in liver and circulated in inactive form

Prothrombin circulating in plasma

Thrombin

Fibrinogen circulating in plasma

Fibrin

(b)

46.18 Blood Clotting   (a) Damage to a blood vessel initiates a cascade of events that produces a fibrin meshwork. (b) As the meshwork forms, red blood cells clot as they are enmeshed in the fibrin threads, as this scanning electron micrograph shows.

Plasma is very similar to intercellular fluid in composition, and most of its components move readily between these two fluid compartments of the body. The main difference between the two fluids is the higher concentration of proteins in the plasma.

Control and Regulation of Circulation

The circulatory system is controlled and regulated by neural, hormonal, local, and systemic mechanisms. Every tissue requires an adequate supply of blood that is saturated with O_2, that carries essential nutrients, and that is relatively free of waste products. The nervous system cannot monitor and control every capillary bed in the body. Instead, each bed regulates its own blood flow through **autoregulatory mechanisms** that cause the arterioles supplying the bed to constrict or dilate.

The autoregulatory actions of every capillary bed in every tissue influence the pressure and composition of the arterial blood leaving the heart. For example, if

many arterioles suddenly dilate, allowing blood to flow through many more capillary beds, arterial blood pressure falls. If all the newly filled capillary beds contribute metabolic waste products to the blood at one time, the concentration of wastes in the blood returning to the heart increases. Thus events in all capillary beds throughout the body produce combined effects on arterial blood pressure and composition. The nervous and endocrine systems respond to changes in arterial blood pressure and composition by changing breathing, heart rate, and blood distribution to match the metabolic needs of the body.

Autoregulation matches local flow to local need

The autoregulatory mechanisms that adjust the flow of blood to a tissue are part of the tissue itself, but they can be influenced by the nervous system and certain hormones. The amount of blood that flows through a capillary bed is controlled by the degree of contraction of the smooth muscles of the arteries and arterioles feeding that bed: As the muscles contract, they constrict the vessels, thereby decreasing the flow.

The flow of blood in a typical capillary bed is diagrammed in Figure 46.19. Blood flows into the bed from an arteriole. Smooth-muscle "cuffs," or precapillary sphincters, on the arteriole can completely shut off the supply of blood to the capillary bed. When the precapillary sphincters are relaxed and the arteriole is

46.19 Local Control of Blood Flow Low O_2 concentrations and/or high levels of metabolic byproducts cause the smooth muscle to relax, thus increasing the supply of blood to the capillary bed.

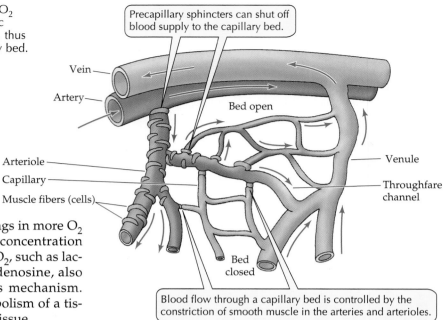

Precapillary sphincters can shut off blood supply to the capillary bed.

Vein

Artery

Bed open

Arteriole

Capillary

Muscle fibers (cells)

Venule

Throughfare channel

Bed closed

Blood flow through a capillary bed is controlled by the constriction of smooth muscle in the arteries and arterioles.

open, the arterial blood pressure pushes blood into the capillaries.

Autoregulation depends on the sensitivity of the smooth muscle to the composition of its chemical environment. Low O_2 concentrations and high CO_2 concentrations cause the smooth muscle to relax, thus increasing the supply of blood, which brings in more O_2 and carries away CO_2. Increases in the concentration of products of metabolism other than CO_2, such as lactate, hydrogen ions, potassium, and adenosine, also promote increased blood flow by this mechanism. Hence, activities that increase the metabolism of a tissue also increase the blood flow to that tissue.

Arterial pressure is controlled and regulated by hormonal and nervous mechanisms

The same smooth muscles of arteries and arterioles that respond to autoregulatory stimuli also respond to signals from the endocrine and central nervous systems. Most arteries and arterioles are innervated by the autonomic nervous system, particularly the sympathetic division. Most sympathetic neurons release norepinephrine, which causes the smooth-muscle cells to contract, thus constricting the vessels and increasing their resistance to blood flow. An exception is found in skeletal muscle, where specialized sympathetic neurons release acetylcholine and cause the smooth muscles of the arterioles to relax and the vessels to dilate, causing more blood to flow to the muscle.

Hormones also can cause arterioles to constrict. Epinephrine has actions similar to those of norepinephrine; it is released from the adrenal medulla during massive sympathetic activation—the fight-or-flight response. Angiotensin, produced when blood pressure to the kidneys falls, causes arterioles to constrict. Vasopressin, released by the posterior pituitary, has similar effects (Figure 46.20). These hormones influ-

46.20 Hormonal Control of Blood Pressure through Vascular Resistance A drop in arterial pressure reduces blood flow to tissues, resulting in local accumulation of metabolic wastes. This change in the extracellular environment stimulates autoregulatory opening of the arteries that would lead to a further decrease in central blood pressure if this were not prevented by the negative-feedback mechanisms shown in this diagram, which work through promoting the constriction of arteries in less essential tissues.

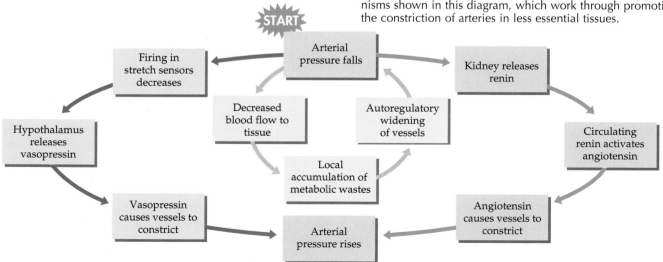

START

Firing in stretch sensors decreases

Arterial pressure falls

Kidney releases renin

Decreased blood flow to tissue

Autoregulatory widening of vessels

Hypothalamus releases vasopressin

Circulating renin activates angiotensin

Local accumulation of metabolic wastes

Vasopressin causes vessels to constrict

Arterial pressure rises

Angiotensin causes vessels to constrict

ence arterioles located for the most part in peripheral tissues (extremities) or in tissues whose functions need not be maintained continuously (such as the gut). By reducing blood flow in those arterioles, the hormones increase the central blood pressure and blood flow to essential organs such as the heart, brain, and kidneys.

The autonomic nervous system activity that controls heart rate and constriction of blood vessels originates in cardiovascular centers in the medulla of the brain stem. Many inputs converge on this central integrative network and influence the commands it issues via parasympathetic and sympathetic fibers (Figure 46.21). Of special importance is information about changes in blood pressure from stretch sensors in the

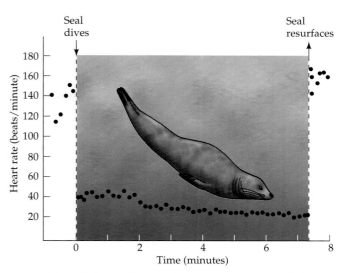

46.22 The Diving Reflex When a marine mammal dives, its heart rate slows and the arteries to most of its organs constrict, so almost all blood flow and available oxygen goes to the animal's heart and brain. These adaptations enable some seals to remain under water for up to an hour.

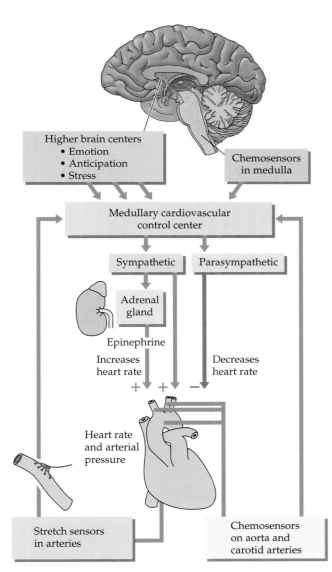

46.21 Regulating Blood Pressure The autonomic nervous system controls heart rate in response to information about blood pressure and blood composition that is integrated by regulatory centers in the medulla.

walls of the great arteries leading to the brain—the aorta and the carotid arteries.

Increased activity in the stretch sensors indicates rising blood pressure and inhibits sympathetic nervous system output. As a result, the heart slows and arterioles in peripheral tissues dilate. If pressure in the great arteries falls, the activity of the stretch sensors decreases, stimulating sympathetic output. Increased sympathetic output causes the heart to beat faster and the arterioles in peripheral tissues to constrict. When arterial pressure falls, the change in stretch sensor activity also causes the hypothalamus to release vasopressin, which helps increase blood pressure by stimulating peripheral arterioles to constrict.

Other information that causes the medullary regulatory system to increase heart rate and blood pressure comes from the carotid and aortic bodies (see Figure 45.21). These nodules of modified smooth-muscle tissue are chemosensors that respond to inadequate O_2 supply. If arterial blood flow slows or the O_2 content of the arterial blood falls drastically, these sensors are activated and send signals to the regulatory center.

The regulatory center also receives input from other brain areas. Emotions and the anticipation of intense activity, such as at the start of a race, can cause the center to increase heart rate and blood pressure. A reflex that slows the heart is the so-called diving reflex, which is highly developed in marine mammals (Figure 46.22). Humans also have a diving reflex that causes the heart to beat more slowly when the face is immersed in water.

A question we can ask about any physiological system is, What is being regulated and how? In the respiratory system (see Chapter 45), it is primarily the CO_2 concentration of the blood, and to a lesser extent the O_2 concentration, that is regulated by changes in the depth and frequency of breathing. Regulation in the circulatory system is more complex. The blood flow to individual tissues is regulated by local, autoregulatory mechanisms that cause dilation of local arterioles and precapillary sphincters when the tissue needs more oxygen or has accumulated wastes. As more blood flows into such tissues, the central blood pressure falls and the composition of the blood returning to the heart reflects the exchanges that occur in those tissues. Changes in central blood pressure and composition are sensed, and both endocrine and central nervous system responses are activated in order to return blood pressure and composition to normal. Thus circulatory functions are matched to the regional and overall needs of the body.

Summary of "Circulatory Systems"

Circulatory Systems: Vessels, Blood, and a Pump

- The metabolic needs of the cells of very small animals are met by direct exchange of materials with the external medium. The metabolic needs of cells of larger animals are met by a circulatory system that transports nutrients, respiratory gases, and metabolic wastes. **Review Figure 46.1**
- In open circulatory systems the blood or hemolymph leaves vessels and percolates through tissues. **Review Figure 46.2**
- In closed circulatory systems the blood is contained in a system of vessels. **Review Figure 46.3**
- The circulatory systems of vertebrates consist of a heart and a closed system of vessels. Arteries and arterioles carry blood from the heart; capillaries are the site of exchange between blood and intercellular fluids; venules and veins carry blood back to the heart. **Review Figure 46.9**
- The vertebrate heart evolved from two chambers in fishes to three in amphibians and reptiles, and four in crocodilians, mammals, and birds. This evolutionary progression has led to an increasing separation of blood flow to the gas exchange organs and blood flow to the rest of the body. **Review Figure 46.4**

The Human Heart: Two Pumps in One

- In mammals, blood circulates through two circuits: the pulmonary circuit and the systemic circuit.
- The human heart has four chambers. Valves in the heart prevent the backflow of blood. **Review Figure 46.5**
- The cardiac cycle has two phases: systole, in which the ventricles contract; and diastole, in which the ventricles relax. The heart sounds ("lub-dub") are the closure of heart valves. **Review Figure 46.6**
- The measurement of blood pressure using a sphygmomanometer and a stethoscope is based on detecting the pressures necessary to compress an artery so that blood does not flow through it at all, or flows through it intermittently, during systole. **Review Figure 46.7**
- The autonomic nervous system controls heart rate. Sympathetic activity increases it; parasympathetic activity decreases it. These actions are due to the effects of norepinephrine and acetylcholine on the rate of depolarization of the membranes of pacemaker cells. **Review Figure 46.8**
- A pacemaker (the sinoatrial node) controls the cardiac cycle by initiating a wave of depolarization in the atria, which is conducted to the ventricles through the atrioventricular node. **Review Figure 46.9**

The Vascular System: Arteries, Capillaries, and Veins

- Arteries and arterioles, which carry blood away from the heart, have many elastic fibers that enable them to withstand high pressures. In addition, abundant smooth-muscle cells allow these vessels to contract and expand, thereby altering their resistance and thus blood flow. **Review Figure 46.11**
- Capillary beds are the site of fluid exchange between the blood and the interstitial fluids.
- The exchange of fluids between the blood and interstitial fluids is determined by the balance between blood pressure and osmotic potential in the capillaries. The ability of a specific molecule to cross a capillary wall depends on the architecture of the capillary, the type of substance, and the concentration difference between the blood and the intercellular fluid. **Review Figure 46.13**
- A separate system of vessels, the lymphatic system, returns the intercellular fluid to the blood.
- Veins have a high capacity for storing blood. Aided by gravity, by contractions of skeletal muscle, and by the actions of breathing, they carry blood back to the heart. **Review Figure 46.14**
- Cardiovascular disease is responsible for about half of all deaths in the United States and Europe. Atherosclerosis and thrombus formation can lead to potentially fatal conditions such as coronary thrombosis, coronary infarction, embolism, and stroke. Diet and behavior are the keys to good cardiovascular health. **Review Figure 46.15**

Blood: A Fluid Tissue

- Blood can be divided into a plasma portion (water, salts, and proteins) and a cellular portion (red blood cells, white blood cells, and platelets), all of which are produced from totipotent cells in the bone marrow. **Review Figure 46.16**
- Red blood cells are non-nucleated, hemoglobin-containing cells that transport respiratory gases. **Review Figure 46.17**
- White blood cells defend the body against foreign substances by phagocytosis and by mechanisms of cell-mediated immune reactions.
- Platelets, along with circulating proteins, are involved in clotting responses. **Review Figure 46.18**
- Plasma is a complex solution that contains gases, ions, nutrient molecules, proteins, and other molecules.

Control and Regulation of Circulation

- Blood flow through capillary beds is controlled by local conditions, hormones, and the autonomic nervous system. **Review Figure 46.19**
- Heart rate is controlled by the autonomic nervous system, which responds to information about blood pressure and blood composition that is integrated by regulatory centers in the brain. **Review Figures 46.21, 46.22**

Self-Quiz

1. An open circulatory system is characterized by
   a. the absence of a heart.
   b. the absence of blood vessels.
   c. blood with a composition different from that of intercellular fluid.
   d. the absence of capillaries.
   e. a higher-pressure circuit through gills than to other organs.

2. Which statement about vertebrate circulatory systems is *not* true?
   a. In fish, oxygenated blood from the gills returns to the heart through the left atrium.
   b. In mammals, deoxygenated blood leaves the heart through the pulmonary artery.
   c. In amphibians, deoxygenated blood enters the heart through the right atrium.
   d. In reptiles, the blood in the pulmonary artery has a lower oxygen content than the blood in the aorta.
   e. In birds, the pressure in the aorta is higher than the pressure in the pulmonary artery.

3. Which statement about the human heart is true?
   a. The walls of the right ventricle are thicker than the walls of the left ventricle.
   b. Blood flowing through atrioventricular valves is always deoxygenated blood.
   c. The second heart sound is due to the closing of the aortic valve.
   d. Blood returns to the heart from the lungs in the vena cava.
   e. During systole the aortic valve is open and the pulmonary valve is closed.

4. Pacemaker actions of cardiac muscle
   a. are due to opposing actions of norepinephrine and acetylcholine.
   b. are localized in the bundle of His.
   c. depend on the gap junctions between cells that make up the atria and those that make up the ventricles.
   d. are due to spontaneous depolarization of the plasma membranes of some cardiac-muscle cells.
   e. result from hyperpolarization of cells in the sinoatrial node.

5. Blood flow through capillaries is slow because
   a. lots of blood volume is lost from the capillaries.
   b. the pressure in venules is high.
   c. the total cross-sectional area of capillaries is larger than that of arterioles.
   d. the osmotic pressure in capillaries is very high.
   e. red blood cells are bigger than capillaries and must squeeze through.

6. How are lymphatic vessels like veins?
   a. Both have nodes where they join together into larger common vessels.
   b. Both carry blood under low pressure.
   c. Both are capacitance vessels.
   d. Both have valves.
   e. Both carry fluids rich in plasma proteins.

7. The production of red blood cells
   a. ceases if the hematocrit falls below normal.
   b. is stimulated by erythropoietin.
   c. is about equal to the production of white blood cells.
   d. is inhibited by prothrombin.
   e. occurs in bone marrow before birth and in lymph nodes after birth.

8. Which of the following does *not* increase blood flow through a capillary bed?
   a. High concentration of CO_2
   b. High concentration of lactate and hydrogen ions
   c. Histamine
   d. Vasopressin
   e. Increase in arterial pressure

9. Blood clotting
   a. is impaired in patients with hemophilia because they don't produce platelets.
   b. is initiated when platelets release fibrinogen.
   c. involves a cascade of factors produced in the liver.
   d. is initiated by leukocytes forming a meshwork.
   e. requires conversion of angiotensinogen to angiotensin.

10. Autoregulation of blood flow to a tissue is due to
    a. sympathetic innervation.
    b. the release of vasopressin by the hypothalamus.
    c. increased activity of baroreceptors.
    d. chemosensors in carotid and aortic bodies.
    e. the effect of local environment on arterioles.

Applying Concepts

1. How is cardiac output increased at the beginning of a race? Include the Frank–Starling law in your answer.

2. The final stages of alcoholism include loss of liver function and accumulation of fluids in extremities and the abdominal cavity. Explain how these two consequences of alcoholism are related.

3. A sudden and massive loss of blood results in a decrease in blood pressure. Describe several mechanisms that help return blood pressure to normal.

4. You can describe the cycle of events in a ventricle of the heart by a graph that plots the pressure in the ventricle on the y axis and the volume of blood in the ventricle on the x axis. What would such a graph look like? Where would the heart sounds be on this graph? How would the graph differ for the left and the right ventricles?

5. Why doesn't diastolic blood pressure fall to zero between heartbeats? Why does systolic blood pressure increase with (a) sympathetic activity, (b) increased venous return, and (c) age?

Readings

Golde, D. W. and J. C. Gasson. 1988. "Hormones That Stimulate the Growth of Blood Cells." *Scientific American*, July. A discussion of how hemopoietins, now made by recombinant-DNA methods, promise to transform the practice of medicine.

Randall, D., W. Burggren and K. French. 1997. *Animal Physiology: Mechanisms and Adaptations*, 4th Edition. W. H. Freeman, New York. An outstanding textbook of animal physiology, with excellent coverage of the circulatory system.

Robinson, T. F., S. M. Factor and E. H. Sonnenblick. 1986. "The Heart as a Suction Pump." *Scientific American*, July. A new proposal concerning the filling of the heart, along with information on cardiac muscle and the connective tissues of the heart.

Vander, A. J., J. H. Sherman and D. S. Luciano. 1994. *Human Physiology: The Mechanisms of Body Function*, 6th Edition. McGraw-Hill, New York. Chapter 14 deals with circulation.

Zapol, W. M. 1987. "Diving Adaptations of the Weddell Seal." *Scientific American*, June. An exploration by breath-holding master divers into the diving reflex.

Chapter 47

Animal Nutrition

Eating on the Run
Sometimes good nutrition is the last thing on our minds.

How many different chemical compounds did you eat for lunch? In the course of a day while you eat, drink, and breathe, you are exposed to hundreds, perhaps thousands, of compounds from the environment. Some of these compounds—carbohydrates, proteins, and fats—are the body's sources of energy and the building blocks for the body's cells and tissues. Compounds such as vitamins, minerals, and water are essential for efficient functioning of the body's cells and tissues. Compounds such as coloring or spices may have little or no biological importance, but they may make your lunch look and taste good. Other compounds may be toxic.

Food, air, and water contain complex mixtures of chemicals, and often you get the good and the bad in the same serving. Many of our food plants, for example, contain complex chemicals that can be harmful to us in large doses. These compounds evolved in plants to discourage hungry animals. In return, animals have developed metabolic capabilities to break down a vast array of toxic plant products. In fact, some animals, such as monarch butterflies, have become capable of using the toxins in their food (milkweed plants in this case) to make themselves distasteful or toxic to their predators. Humans exploit the noxious and toxic compounds found in many plants and animals by using them to develop medicines and other useful chemicals.

In this chapter we will first consider the nutrients that organisms require. We will see that nutrients are needed for energy, for molecular building blocks, and for specific biochemical functions.

What animals eat and how they eat are sometimes among their most distinguishing characteristics, so we will examine some of the diverse means by which animals procure nutrients. Acquiring and ingesting food, however, do not

(a)

(b)

(c)

(d)

47.1 The Consumers Heterotrophs have evolved an amazing range of adaptations for exploiting sources of energy. (a) The manatee is an herbivore whose source of food is aquatic vegetation in tropical and subtropical rivers and lagoons. (b) The long bill and hovering flight pattern of this hummingbird, a fluid feeder, enables it to harvest the tiny amounts of nectar in individual flowers. (c) The red file clam of the South Pacific islands obtains food from the ocean water it constantly filters through its system. (d) The carnivorous African lion is a fearsome predator; the young join their parents in feasting on a fresh kill.

provide nutrients to the cells of the body until the food is digested and absorbed, so we will also study those processes. Finally, we will learn how the body regulates its traffic in molecules used for metabolic fuel.

By taking in nutrients as food, animals also take in compounds that are not needed for nutrition and can be toxic. We will briefly consider how animals deal with toxic compounds and how human activities that contribute new and highly dangerous toxic compounds to the environment are affecting human health and other organisms in the environment.

Nutrient Requirements

Animals must eat to stay alive; they are heterotrophs. Unlike **autotrophs** (most plants, some bacteria, and some protists), which trap solar energy through photosynthesis and use that energy to synthesize all of their structures from inorganic materials, **heterotrophs** must derive structural molecules and energy from their food.

Heterotrophs have evolved an enormous diversity of adaptations to exploit, directly or indirectly, the resources made available through the actions of autotrophs (Figure 47.1). We will first examine the use of nutrients for energy, then we'll look at the use of nutrients for carbon skeletons that animals cannot synthesize but need to build larger organic molecules, and finally we'll learn about nutrients that serve specific biochemical functions. Included in the latter category are mineral nutrients, such as iron and calcium, that animals require to build functional and structural molecules, and complex organic molecules called vitamins that are needed in small quantities as cofactors for enzymes and for other purposes.

Nutrient use for energy can be measured in calories

In Chapters 6 and 7 we learned that energy in the chemical bonds of food molecules is transferred to the high-energy phosphate bonds of ATP. ATP then provides energy for active transport, biosynthesis of molecules, degradation of molecules, muscle contraction, and other work. Because they are never 100 percent efficient, these energy conversions produce heat as a by-product. Even the useful energy conversions eventually are reduced to heat, as molecules that were synthesized are broken down and energy of movement is dissipated by friction.

In time, all the energy that is transferred to ATP from the chemical bonds of food molecules is released to the environment as heat. It is convenient, therefore, to talk about the energy requirements of animals and the energy content of food in terms of a measure of heat energy: the **calorie**. A calorie is the amount of heat necessary to raise the temperature of 1 g of water 1°C. Since this value is a tiny amount of energy in comparison to the energy requirements of many animals, physiologists commonly use the **kilocalorie** (kcal) as a unit (1 kcal = 1,000 calories).

Nutritionists also use the kilocalorie as a standard unit of energy, but they traditionally refer to it as the **Calorie**, which is always capitalized to distinguish it from the single calorie. A person on a daily diet of 1,000 Calories may consume up to 1,000 kcal/day. The confusion among these terms is unfortunate, but we live with it. (Physiologists are gradually abandoning the calorie as an energy unit as they switch to the International System of Units. In this system the basic unit of energy is the joule; 1 calorie = 4.184 joules.)

The metabolic rate of an animal (see Chapter 37) is a measure of the overall energy needs that must be met by an animal's ingestion and digestion of food. The components of food that provide energy are fats, carbohydrates, and proteins. Fat yields 9.5 kcal/g when it is metabolically oxidized, carbohydrate 4.2 kcal/g, and protein about 4.1 kcal/g. The basal metabolic rate of a human is about 1,300 to 1,500 kcal/day for an adult female and 1,600 to 1,800 kcal/day for an adult male. Physical activity adds to this basal energy requirement. Some equivalences of food, energy, and exercise are shown in Figure 47.2.

Sources of energy can be stored

Although the cells of the body use energy continuously, most animals do not eat continuously. Humans generally eat several meals a day, a lion may eat once in several days, a boa constrictor may eat once a month, and hibernating animals may go 5 to 6 months without eating. Therefore, animals must store fuel molecules that can be released as needed between meals.

47.2 Food Energy and How Fast We Burn It The energy in kilocalories for several common food items is shown on the left. The graph indicates about how long it would take a person with a basal metabolic rate of about 1,800 kcal/day to utilize the equivalent amount of energy while involved in various activities.

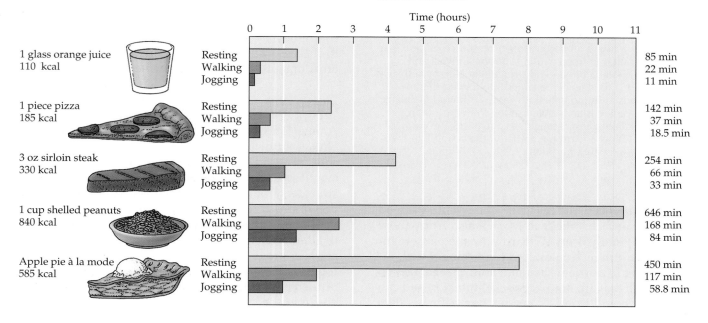

		Time (hours)	
1 glass orange juice 110 kcal	Resting Walking Jogging		85 min 22 min 11 min
1 piece pizza 185 kcal	Resting Walking Jogging		142 min 37 min 18.5 min
3 oz sirloin steak 330 kcal	Resting Walking Jogging		254 min 66 min 33 min
1 cup shelled peanuts 840 kcal	Resting Walking Jogging		646 min 168 min 84 min
Apple pie à la mode 585 kcal	Resting Walking Jogging		450 min 117 min 58.8 min

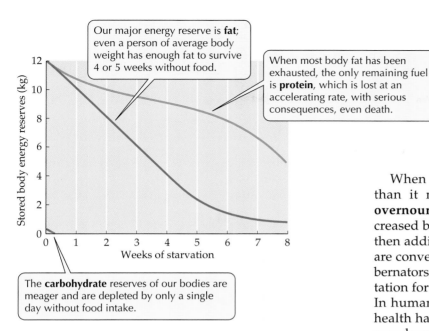

Our major energy reserve is **fat**; even a person of average body weight has enough fat to survive 4 or 5 weeks without food.

When most body fat has been exhausted, the only remaining fuel is **protein**, which is lost at an accelerating rate, with serious consequences, even death.

The **carbohydrate** reserves of our bodies are meager and are depleted by only a single day without food intake.

47.3 Depletion of Body Energy Reserves
Body fat is a person's major defense against starvation.

When an animal consistently takes in more food than it needs to meet its energy demands, it is **overnourished**. The excess nutrients are stored as increased body mass. First, glycogen reserves build up; then additional dietary carbohydrate, fat, and protein are converted to body fat. In some species, such as hibernators, seasonal overnutrition is an important adaptation for surviving periods when food is unavailable. In humans, however, overnutrition can be a serious health hazard, increasing the risk of high blood pressure, heart attack, diabetes, and other disorders.

A common clay building brick weighs about 5 pounds, so a person who is 50 pounds overweight is constantly carrying around the equivalent of ten bricks. That alone is quite a strain on the heart, but in addition, each extra pound of body tissue includes miles of additional blood vessels through which the heart must pump blood. Obesity is a health hazard, but so are poorly planned fad or crash diets that can lead to malnutrition (discussed in the next section). People spend billions of dollars every year on schemes to lose weight, even though all they need to do is follow a simple rule: Take in fewer calories than your body burns, but maintain a balanced diet.

Carbohydrates are stored in liver and muscle cells as glycogen (see Chapter 3), but the total glycogen store is usually not more than the equivalent of a day's energy requirements. Fat is the most important form of stored energy in the bodies of animals. Not only does fat have the highest energy content per gram, but it can be stored with little associated water, making it more compact. If migrating birds had to store energy as glycogen rather than fat to fuel long flights, they would be too heavy to fly! Protein is not used to store energy, although body protein can be metabolized as an energy source as a last resort.

If an animal takes in too little food to meet its needs for metabolic energy, it is **undernourished** and must make up the shortfall by metabolizing some of the molecules of its own body. Consumption of self for fuel begins with the storage compounds glycogen and fat. Protein loss is minimized for as long as possible, but eventually the body has to use its own proteins for fuel.

The breakdown of body proteins impairs body functions and eventually leads to death. Blood proteins are among the first to go, resulting in loss of fluid to the intercellular spaces (edema; see Chapter 46). Muscles atrophy (waste away) and eventually even brain protein is lost. Figure 47.3 shows the course of starvation. Undernourishment is rampant among people in underdeveloped and war-torn nations, and a billion people—one-fifth of the world's population—are undernourished. (Ironically, one cause of life-threatening undernourishment in Western developed nations is a self-imposed starvation called **anorexia nervosa** that results from a psychological aversion to body fat.)

Food provides carbon skeletons for biosynthesis

Every animal requires certain basic organic molecules (carbon skeletons) that it cannot synthesize for itself but must have in order to build the complex organic molecules needed for life. An example of a required carbon skeleton is the acetyl group (Figure 47.4). Animals cannot make acetyl groups from carbon, oxygen, and hydrogen molecules; they obtain acetyl groups by metabolizing carbohydrates, fats, or proteins.

From these acquired acetyl groups, animals create a wealth of other necessary compounds, including fatty acids, steroid hormones, electron carriers for cellular respiration, certain amino acids, and, indirectly, legions of other compounds. The three major classes of nutrients—carbohydrates, fats, and proteins—provide both the energy and the carbon skeletons for biosynthesis.

Because the acetyl group can be derived from the metabolism of almost any food, it is unlikely ever to be in short supply for an animal with an adequate food supply. However, other carbon skeletons are de-

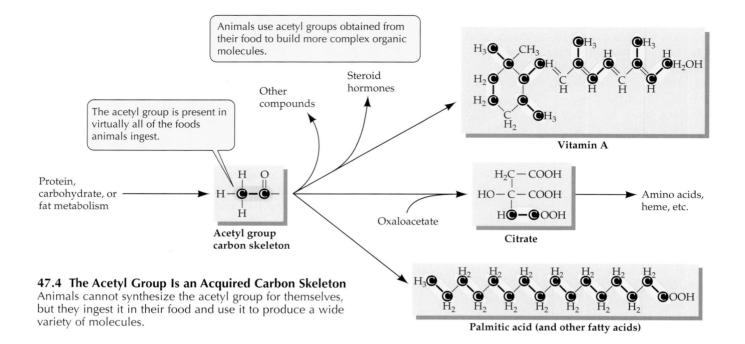

Animals use acetyl groups obtained from their food to build more complex organic molecules.

The acetyl group is present in virtually all of the foods animals ingest.

Other compounds

Steroid hormones

Vitamin A

Protein, carbohydrate, or fat metabolism

Acetyl group carbon skeleton

Oxaloacetate

Citrate

Amino acids, heme, etc.

47.4 The Acetyl Group Is an Acquired Carbon Skeleton
Animals cannot synthesize the acetyl group for themselves, but they ingest it in their food and use it to produce a wide variety of molecules.

Palmitic acid (and other fatty acids)

rived from more limited sources, and an animal can suffer a deficiency of these materials even if its caloric intake is adequate. This state of deficiency is called **malnutrition**.

Amino acids, the building blocks of protein, are a good example of such substances. Humans obtain amino acids by digesting protein from food and absorbing the resulting amino acids. The body then synthesizes its own protein molecules, as specified by its DNA, from these dietary amino acids. Another source of amino acids is the breakdown of existing body proteins, which are in constant turnover as the tissues of the body undergo normal remodeling and renewal.

Animals can synthesize some of their own amino acids by taking carbon skeletons synthesized from acetyl or other groups and transferring to them amino groups ($-NH_2$) derived from other amino acids. But most animals cannot synthesize all the amino acids they need. Each species has certain **essential amino acids** that must be obtained from food. Different species have different essential amino acids, and in general, herbivores have fewer essential amino acids than carnivores have.

If an animal does not take in one of its essential amino acids, its protein synthesis is impaired. Think of protein synthesis as writing a story on a typewriter. If the typewriter is missing a key, the story either comes to a stop or has an error in it wherever the letter represented by that key is needed. In protein synthesis, the story usually comes to a stop and a functional protein is not produced.

Humans require eight essential amino acids in their diet: isoleucine, leucine, lysine, methionine, phenylalanine, threonine, tryptophan, and valine (see Table 3.1).

All eight are available in milk, eggs, or meat; no plant food, however, contains all eight. A strict vegetarian thus runs a risk of protein malnutrition. An appropriate dietary *mixture* of plant foods, however, supplies all eight essential amino acids (Figure 47.5). Wheat, corn, rice, and other grains are deficient in lysine and isoleucine but are well stocked with most of the other amino acids. Beans, lentils, and other legumes have lots of lysine but are low in methionine and tryptophan. Eating only grains or only beans would lead to a serious deficiency of one or more amino acids. If grains and beans are eaten together, however, the diet includes all the essential amino acids.

In general, grains are complemented by legumes or by milk products; legumes are complemented by

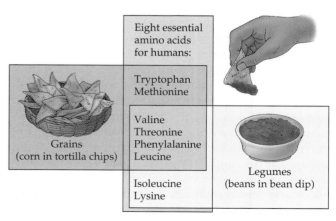

Eight essential amino acids for humans:

Tryptophan
Methionine

Valine
Threonine
Phenylalanine
Leucine

Isoleucine
Lysine

Grains (corn in tortilla chips)

Legumes (beans in bean dip)

47.5 A Strategy for Vegetarians By combining cereal grains and legumes, a vegetarian can obtain all eight essential amino acids.

grains and by seeds and nuts. Long before the chemical basis for this complementarity was understood, societies with little access to meat learned appropriate dietary practices through trial and error. Many Central and South American peoples traditionally ate beans with corn, and the native peoples of North America complemented their beans with squash. Remember that we do not retain great stores of free amino acids in our bodies, yet we synthesize proteins continuously. It makes little nutritional sense to eat grains one day and beans the next; they must be eaten together for proper amino acid balance. Excess amino acids are burned for fuel, converted to fat, or excreted as waste.

Why are dietary proteins completely digested to their constituent amino acids before being used by the body? Wouldn't it be more energy-efficient to reuse some dietary proteins directly? There are several reasons why ingested proteins are not used "as is." First, macromolecules such as proteins are not readily taken up through plasma membranes, but their constituent monomers (such as amino acids) are readily transported. Second, protein structure and function (see Chapter 3) are highly species-specific. A protein that functions optimally in one species might not function well in another species. Third, foreign proteins entering the body directly from the gut would be recognized as invaders and would be attacked by the immune system. Most animals avoid these problems by digesting food proteins extracellularly and then absorbing the amino acids into the body. The new proteins formed from these amino acids are recognized as "self" by the immune system.

From acetyl units obtained from food we can synthesize almost all the lipids required by the body, but we must have a dietary source of **essential fatty acids**—notably linoleic acid. Essential fatty acids are necessary components of membrane phospholipids, and a deficiency can lead to problems such as infertility and impaired lactation.

Animals need mineral elements in different amounts

Table 47.1 lists the principal mineral elements required by animals. Certain species require additional elements. Elements required in large amounts are known as **macronutrients**; elements required in only tiny amounts are called **micronutrients**. Some essential elements are required in such minute amounts that deficiencies are never observed, but these elements are nevertheless essential.

Animals need calcium and phosphorus in great quantity. Calcium phosphate is the principal structural material in bones and teeth. Muscle contraction, nerve function, and many other intracellular functions in animals require calcium. Phosphorus is an integral component of nucleic acids. We learned in Chapter 7 about the role of phosphate groups in biological energy transfers. Sulfur is part of the structure of two amino acids and is therefore found in almost all proteins. Other essential compounds also contain sulfur.

Iron is the oxygen-binding atom in both hemoglobin and myoglobin, the oxygen-carrying proteins in vertebrate blood and muscle. In addition, iron undergoes redox reactions in some of the electron-carrying proteins of cellular respiration. Some mineral nutrients—among them magnesium, manganese, zinc, and cobalt—act as cofactors for enzymes.

Potassium, sodium, and chloride ions are particularly important in the osmotic balance of tissues and in the electrical properties of membranes, including resting potentials and action potentials. Animals require large amounts of both sodium and chloride ions. Because plants contain few of these ions, herbivores may travel considerable distances to natural salt licks. Ranchers and game wardens frequently supply salt licks for animals that do not have access to natural sources.

Specific requirements for individual elements are different in different species. In vertebrates, copper is essential in trace amounts for certain enzymes to function properly; for example, hemoglobin synthesis requires copper, even though copper is not part of the hemoglobin molecule. In many invertebrate species, however, copper is part of the respiratory pigment hemocyanin, and those animals require more copper than vertebrates do.

An animal must obtain vitamins from its food

Another group of essential nutrients consists of the **vitamins**. Like essential amino acids and fatty acids, vitamins are organic compounds that an animal cannot make for itself but that are required for its normal growth and metabolism. Most vitamins function as coenzymes or parts of coenzymes and are required in very small amounts compared with essential amino acids and fatty acids that have structural roles.

The list of required vitamins varies from species to species. For example, ascorbic acid (vitamin C) is not a vitamin for most mammals, because they can make it themselves. However, primates (including humans) do not have this ability, so ascorbic acid for them is a vitamin. If we do not get vitamin C in our diet, we develop the disease known as scurvy, which over the centuries was a serious and frequently fatal problem for sailors on long voyages until it was discovered by a British physician that the disease could be prevented by having sailors eat citrus fruit. Ever since, British sailors have been called "limeys." There are 13 such compounds that humans cannot synthesize in sufficient quantities. They are divided into two groups: water-soluble vitamins and fat-soluble vitamins (Table 47.2).

TABLE 47.1 Mineral Elements Required by Animals

ELEMENT	SOURCE IN HUMAN DIET	MAJOR FUNCTIONS
Macronutrients		
Calcium (Ca)	Dairy foods, eggs, green leafy vegetables, whole grains, legumes, nuts	Found in bones and teeth; blood clotting; nerve and muscle action; enzyme activation
Chlorine (Cl)	Table salt (NaCl), meat, eggs, vegetables, dairy foods	Water balance; digestion (as HCl); principal negative ion in fluid around cells
Magnesium (Mg)	Green vegetables, meat, whole grains, nuts, milk, legumes	Required by many enzymes; found in bones and teeth
Phosphorus (P)	Dairy foods, eggs, meat, whole grains, legumes, nuts	Found in nucleic acids, ATP, and phospholipids; bone formation; buffers; metabolism of sugars
Potassium (K)	Meat, whole grains, fruits, vegetables	Nerve and muscle action; protein synthesis; principal positive ion in cells
Sodium (Na)	Table salt, dairy foods, meat, eggs, vegetables	Nerve and muscle action; water balance; principal positive ion in fluid around cells
Sulfur (S)	Meat, eggs, dairy foods, nuts, legumes	Found in proteins and coenzymes; detoxification of harmful substances
Micronutrients		
Chromium (Cr)	Meat, dairy foods, whole grains, dried beans, peanuts, brewer's yeast	Glucose metabolism
Cobalt (Co)	Meat, tap water	Found in vitamin B_{12}; formation of red blood cells
Copper (Cu)	Liver, meat, fish, shellfish, legumes, whole grains, nuts	Found in active site of many redox enzymes and electron carriers; production of hemoglobin; bone formation
Fluorine (F)	Most water supplies	Resistance to tooth decay
Iodine (I)	Fish, shellfish, iodized salt	Found in thyroid hormones
Iron (Fe)	Liver, meat, green vegetables, eggs, whole grains, legumes, nuts	Found in active sites of many redox enzymes and electron carriers, hemoglobin, and myoglobin
Manganese (Mn)	Organ meats, whole grains, legumes, nuts, tea, coffee	Activates many enzymes
Molybdenum (Mo)	Organ meats, dairy foods, whole grains, green vegetables, legumes	Found in some enzymes
Selenium (Se)	Meat, seafood, whole grains, eggs, chicken, milk, garlic	Fat metabolism
Zinc (Zn)	Liver, fish, shellfish, and many other foods	Found in some enzymes and some transcription factors; insulin physiology

Water-soluble vitamins (the B complex and vitamin C) play roles in both vertebrates and invertebrates. The B vitamins are coenzymes or parts of coenzymes. The B vitamin niacin, for example, we have encountered already under another name, nicotinamide (see Chapter 7). It is the portion of NAD (nicotinamide adenine dinucleotide) and NADP that undergoes oxidation and reduction in the respiratory chain and in other key redox systems in all living things.

Riboflavin (vitamin B_2) similarly is the site of oxidation and reduction in the respiratory chain intermediates FAD (flavin adenine dinucleotide) and FMN (flavin mononucleotide). Vitamin C (ascorbic acid) has many functions, among them an essential role in the formation of the structural protein collagen. Collagen is a fibrous protein that is a major constituent of bone, cartilage, tendons, ligaments, and skin. The water-soluble compounds that are vitamins for humans are essential to all animals. However, some species can make some of these compounds in sufficient quantity that they do not require the compounds in their diet.

Fat-soluble vitamins—A, D, E, and K—have diverse functions. Vitamin A (retinol) is a precursor of retinal, the visual pigment in our eyes. Vitamin D (calciferol) regulates the absorption and metabolism of calcium. Although vitamin D may be obtained in the diet, it can also be produced in human skin by the action of ultraviolet wavelengths of sunlight on certain lipids already present in the body. Thus vitamin D is a vitamin only for individuals with inadequate exposure to the sun, such as people who live in cold climates where clothing usually covers most of the body and where the sun may not shine for long periods of time.

TABLE 47.2 Vitamins in the Human Diet

VITAMIN	SOURCE	FUNCTION	DEFICIENCY SYMPTOMS
Water-soluble			
B₁, thiamin	Liver, legumes, whole grains, yeast	Coenzyme in cellular respiration	Beriberi, loss of appetite, fatigue
B₂, riboflavin	Dairy foods, organ meats, eggs, green leafy vegetables	Coenzyme in cellular respiration (in FAD and FMN)	Lesions in corners of mouth, eye irritation, skin disorders
Niacin (nicotinamide, nicotinic acid)	Meat, fowl, liver, yeast	Coenzyme in cellular metabolism (in NAD and NADP)	Pellagra, skin disorders, diarrhea, mental disorders
B₆, pyridoxine	Liver, whole grains, dairy foods	Coenzyme in amino acid metabolism	Anemia, slow growth, skin problems, convulsions
Pantothenic acid	Liver, eggs, yeast	Found in acetyl CoA	Adrenal problems, reproductive problems
Biotin	Liver, yeast, bacteria in gut	Found in coenzymes	Skin problems, loss of hair
B₁₂, cobalamin	Liver, meat, dairy foods, eggs	Coenzyme in formation of nucleic acids and proteins, and in red blood cell formation	Pernicious anemia
Folic acid	Vegetables, eggs, liver, whole grains	Coenzyme in formation of heme and nucleotides	Anemia
C, ascorbic acid	Citrus fruits, tomatoes, potatoes	Aids formation of connective tissues; prevents oxidation of cellular constituents	Scurvy, slow healing, poor bone growth
Fat-soluble			
A, retinol	Fruits, vegetables, liver, dairy foods	Found in visual pigments	Night blindness, damage to mucous membranes
D, calciferol	Fortified milk, fish oils, sunshine	Absorption of calcium and phosphorus	Rickets
E, tocopherol	Meat, dairy foods, whole grains	Muscle maintenance, prevents oxidation of cellular components	Anemia
K, menadione	Intestinal bacteria, liver	Blood clotting	Blood-clotting problems (in newborns)

The need for vitamin D may have been an important factor in the evolution of skin color. Human races that are adapted to equatorial and low latitudes have dark skin pigmentation as a protection against the damaging effects of ultraviolet radiation. These peoples generally have extensive skin areas exposed to the sun on a regular basis, so their skin synthesizes adequate amounts of vitamin D. In general, races that became adapted to higher latitudes lost dark skin pigmentation. Presumably, lighter skin facilitates vitamin D production in the relatively small areas of skin exposed to sunlight during the short days of winter. An exception to the correlation between latitude and skin pigmentation is the Inuit peoples of the Arctic. These dark-skinned people obtain plenty of vitamin D from the large amounts of fish oils in their diets; for them, exposure to sunlight is not a factor in obtaining this vitamin.

Vitamin E is poorly understood. Its principal function may be to protect unsaturated fatty acids in cellular membranes from oxidation. Vitamin K functions in blood clotting following an injury and hence plays a crucial role in protection of the body. The fat-soluble vitamins generally are required by vertebrates but not by invertebrates.

When water-soluble vitamins are ingested in excess of bodily needs, they are simply eliminated in the urine. (This is the fate of much of the vitamin C that people take in excessive doses.) The fat-soluble vitamins, however, accumulate in body fat and may build up to toxic levels if taken in excess.

Adaptations for Feeding

The ways in which an animal acquires its nutrients, and its adaptations for acquiring nutrients, are frequently its most distinguishing characteristics. We can even group heterotrophic organisms by how they acquire their nutrition. **Saprotrophs** (also called saprobes or decomposers) are mostly microbes that absorb nutrients from dead organic matter. **Detritivores**, such as earthworms and crabs, actively feed on dead organic material. Animals that feed on living organisms are **predators**. **Herbivores** are predators that prey on plants; **carnivores** prey on animals; and **omnivores** prey on both. **Filter feeders**, such as clams and blue whales, prey on small organisms by filtering them out of the environment. Finally, we are only too familiar with **fluid feeders**, which include mosquitoes, aphids, and leeches, as well as birds that feed on plant nectar.

The adaptations that enable a species to exploit a particular source of nutrition are frequently physiological and biochemical. For example, the Australian koala eats nothing but leaves of eucalyptus trees. Eucalyptus leaves are tough, low in nutrient content, and loaded with pungent, toxic compounds that evolved to protect the trees from predators. Yet the koala's gut can digest and detoxify the leaves and absorb all the nutrients the animal needs from this formidably specialized diet. The attributes of the koala's gut, however, are less obvious to us than are the anatomical and behavioral features that animals use to acquire and ingest their food.

We will move now to a discussion of the mechanisms used by herbivores to extract nutrition from the environment, followed by a discussion of carnivore behavior. Our attention from herein in focuses mainly on the vertebrates.

The food of herbivores is often low in energy and hard to digest

Cows or sheep graze in grassy meadows; caterpillars munch steadily on leaves. Some herbivores have striking adaptations for feeding, such as the trunk (a flexible, gripping nose) of the elephant or the long neck of the giraffe.

Vegetation is frequently coarse and difficult to break down physically, but animals must process large amounts of it, since its energy content is low. Many types of grinding, rasping, cutting, and shredding mouthparts have evolved in invertebrates for ingesting plant material, and in herbivorous vertebrates teeth have been shaped by selection to process coarse plant matter, as we'll see shortly.

Carnivores must detect, seize, and kill prey

The predatory behaviors of many carnivores are legendary. One need only call to mind the hunting skills of hawks, wolves, or any member of the cat family. Carnivores have evolved stealth, speed, power, large jaws, sharp teeth, and strong gripping appendages. A cheetah, for instance, first stalks its prey stealthily from downwind, aided by its natural camouflage. When close enough, it dashes after the prey at speeds as fast as 110 km/hour. It then brings the prey down with its sharp, powerful claws and teeth. Carnivores also have evolved remarkable means of detecting prey. Bats use echolocation, pit vipers sense infrared radiation from the warm bodies of their prey, and certain

47.6 Inside-Out Digestion This sea star is eating two mollusks. While it holds them with its arms, tissue from its everted stomach digests them.

fishes detect electric fields created in the water by their prey (see Chapter 42).

Adaptations for killing and ingesting prey are diverse and highly specialized. These adaptations can be especially important when the prey species are capable of inflicting damage on the predators. Many species of snakes take relatively large prey that are well equipped with sharp teeth and claws. A snake may strike with poisonous fangs and immobilize its prey before ingesting it. A boa or python immobilizes and kills its prey by squeezing it with coils of its powerful body. To swallow large prey, a snake's lower jaw disengages from its joint with the skull. The tentacles of jellyfish, corals, squid, and octopuses, the long, sticky tongues of frogs and chameleons, and the webs of spiders are other examples of fascinating adaptations for capturing and immobilizing prey.

Because some prey items are impossible to ingest, predators sometimes digest their prey externally. Sea stars evert their stomachs (turn them inside out) and digest their molluscan prey while they are still in their shells (Figure 47.6). Spiders usually prey on insects with indigestible exoskeletons. The spider can inject its prey with digestive enzymes and then suck out the liquefied contents, leaving behind the empty exoskeletons frequently seen in old spider webs.

Vertebrate species have distinctive teeth

Teeth are adapted for the acquisition and initial processing of specific types of foods, and because they are one of the hardest structures of the body, they remain in the environment a long time after an animal dies. Paleontologists use teeth to identify animals that lived in the distant past and to understand their behavior.

All mammalian teeth have a general structure consisting of three layers (Figure 47.7a). An extremely hard material called **enamel**, composed principally of calcium phosphate, covers the crown of the tooth.

Both the crown and the root contain a layer of a bony material called **dentine**, within which is a **pulp cavity** containing blood vessels, nerves, and the cells that produce the dentine.

The shapes and organization of mammalian teeth, however, can be very different, since they are adapted to specific diets (Figure 47.7b). In general, incisors are used for cutting, chopping, or gnawing; canines are used for stabbing, ripping, and shredding; and molars and premolars (the cheek teeth) are used for shearing, crushing, and grinding. The highly varied diet of humans is reflected by our multipurpose set of teeth, as is common among omnivores.

Digestion

Most animals digest food extracellularly. Animals take food into a body cavity that is continuous with the outside environment and then secrete digestive enzymes into that cavity. The enzymes act on the food, reducing it to nutrient molecules that can be absorbed by the cells lining the cavity. Only after they are absorbed by the cells are the nutrients within the body of the animal.

The simplest digestive systems are gastrovascular cavities that connect to the outside world through a single opening. Examples are found in cnidarians and flatworms. After a cnidarian captures a prey with its stinging nematocysts (see Figure 29.8), its tentacles cram the prey into the gastrovascular cavity (see Figure 46.1a), where enzymes partly digest it. Extracellular digestion in cnidarians is supplemented by intracellular digestion: Cells lining the gut take in some small food particles by endocytosis. The branching gastrovascular cavity of a flatworm opens to the outside through a muscular pharynx that projects out of the body during feeding.

Some multicellular animals have no digestive system. Many of these animals are internal parasites such as tapeworms. They live in an environment so rich in already digested nutrient molecules that they just absorb them directly into their cells.

Tubular guts have an opening at each end

The guts of all animal groups other than sponges, cnidarians, and flatworms are tubular. A mouth takes in food; molecules are digested and absorbed throughout the length of the gut; and solid digestive wastes, or feces, are excreted through an anus. Different regions in the tubular gut are specialized for particular functions (Figure 47.8). Keep in mind as we discuss these regions that all locations within the tubular gut are really outside the body of the animal. Only by

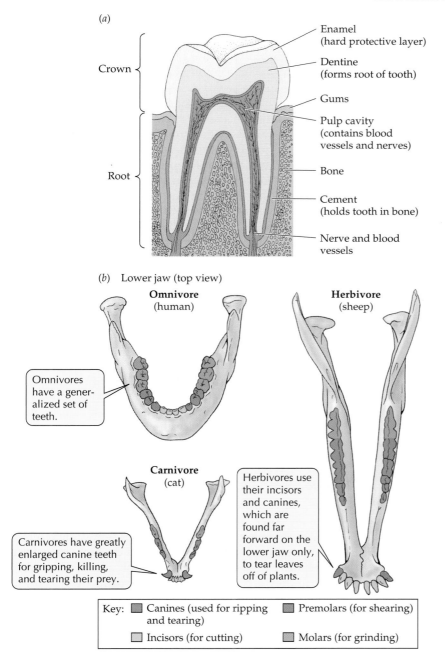

(a)

Crown

Root

Enamel (hard protective layer)

Dentine (forms root of tooth)

Gums

Pulp cavity (contains blood vessels and nerves)

Bone

Cement (holds tooth in bone)

Nerve and blood vessels

(b)   Lower jaw (top view)

Omnivore (human)

Herbivore (sheep)

Omnivores have a generalized set of teeth.

Carnivore (cat)

Herbivores use their incisors and canines, which are found far forward on the lower jaw only, to tear leaves off of plants.

Carnivores have greatly enlarged canine teeth for gripping, killing, and tearing their prey.

Key: ■ Canines (used for ripping and tearing)   ■ Premolars (for shearing)

■ Incisors (for cutting)   ■ Molars (for grinding)

47.7 Mammalian Teeth   (a) A mammalian tooth has three layers: enamel, dentine, and pulp cavity. (b) The teeth of different mammal species are specialized for different diets.

Nematode

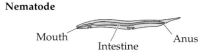

Earthworm

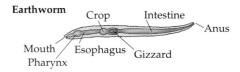

Snail

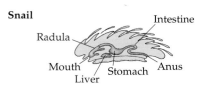

Cockroach

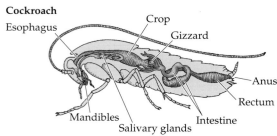

Rabbit

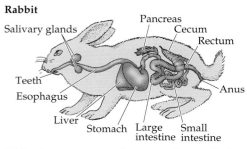

47.8 Compartments for Digestion and Absorption Most animal groups have tubular guts that begin with a mouth that takes in food and ends in an anus that excretes wastes. Between these two structures are specialized regions for digestion and nutrient absorption; these vary from species to species.

crossing the membranes lining the gut does a nutrient molecule enter the body.

At the anterior end of the gut are the mouth (the opening itself) and **buccal cavity** (mouth cavity). Food may be broken up by teeth (in some vertebrates), by the radula in snails, or by mandibles (in insects), or somewhat farther along the gut by structures such as the gizzards of birds and earthworms, where muscular contractions of the gut grind the food together with small stones. Some animals simply ingest large chunks of food, with little or no fragmentation.

Stomachs and **crops** are storage chambers, enabling animals to ingest relatively large amounts of food and digest it at leisure. Food may or may not be digested in

such a storage chamber, depending on the species. Food delivered into the next section of gut, the **midgut** or **intestine**, is well minced and well mixed. Most materials are digested and absorbed here. Specialized glands such as the pancreas in mammals secrete digestive enzymes into the intestine, and the gut wall itself secretes other digestive enzymes. The **hindgut** recovers water and ions and stores feces so that they can be released to the environment at an appropriate time or place. A muscular **rectum** near the anus assists in the expulsion of undigested wastes, the process of **defecation**.

Within the hindgut of many species are colonies of bacteria that live in cooperation (symbiosis) with their hosts. The bacteria obtain their own nutrition from the food passing through the host's gut while contributing to the digestive processes of the host. Members of the leech genus *Hirudo* produce no enzymes that can digest the proteins in the blood they suck from vertebrates; however, a colony of gut bacteria produces the enzymes necessary to break down those proteins into amino acids, which are subsequently used by both the leeches and the bacteria.

Many animals obtain vitamins from the bacteria in their hindgut. Herbivores such as rabbits, cattle, termites, and cockroaches depend on microorganisms in their guts for the digestion of cellulose. In some, specialized regions of the gut may even serve as microbial fermenters. An example is the cecum of the rabbit (see Figure 47.8).

In many animals, the parts of the gut that absorb nutrients have evolved extensive surface areas for absorption. The earthworm has a long, dorsal infolding of its intestine, called the typhlosole (Figure 47.9a), that provides extra absorptive surface area. The shark's intestine has a spiral valve, forcing food to take a longer path and thus encounter more absorptive surface (Figure 47.9b). In many vertebrates the wall of the gut is richly folded, with the individual folds bearing legions of tiny fingerlike projections called **villi** (Figure 47.9c). The villi in turn have microscopic projections, called microvilli, on the cells that line their surfaces. The microvilli present an enormous surface area for the absorption of nutrients.

Digestive enzymes break down complex food molecules

Protein, carbohydrate, and fat macromolecules are broken down into their simplest monomeric units by digestive enzymes. All of these enzymes cleave the chemical bonds of macromolecules through a reaction that adds a water molecule at the site of cleavage; hence they are generally called **hydrolytic enzymes**. Examples of hydrolytic cleavage are the breaking of the bonds between adjacent amino acids of a protein or peptide and the breaking of the bonds between adjacent glucose units in starch.

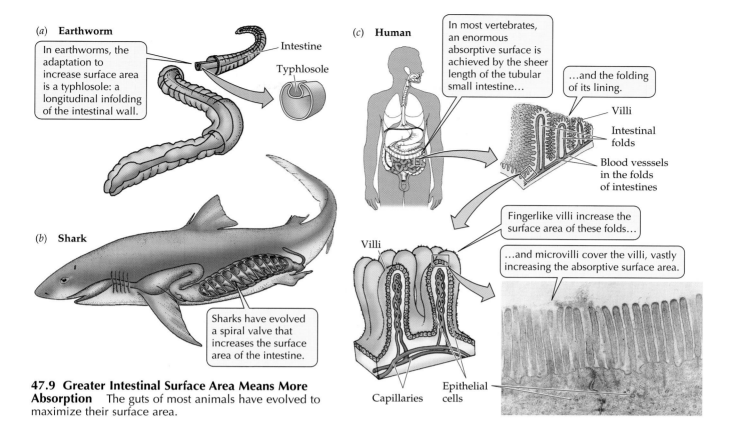

47.9 Greater Intestinal Surface Area Means More Absorption The guts of most animals have evolved to maximize their surface area.

Digestive enzymes are classified according to the substances they hydrolyze: carbohydrases hydrolyze carbohydrates; proteases, proteins; peptidases, peptides; lipases, fats; and nucleases, the nucleic acids. The prefixes "exo-" ("outside") and "endo-" ("within") indicate where the enzyme cleaves the molecule. Thus an endoprotease hydrolyzes a protein at an internal site along the polypeptide chain, and an exoprotease snips away amino acids at the ends of the molecule.

How can an organism produce enzymes to digest biological macromolecules without digesting itself? The answer is that the digesting, as you already know, is usually done *outside* the animal. The gut is simply a tunnel through the animal; food in the gut is outside the body and hence can receive treatment (such as high acidity or potent enzymes) that would be intolerable within a cell or a tissue. (In the cases in which digestion is intracellular, such as in cnidarians, the hydrolytic reactions are localized within food vacuoles.) Most digestive enzymes are produced in an inactive form, known as a **zymogen**. Thus they do not act on the cells that produce them. When secreted into the gut, the zymogens are activated, sometimes by exposure to a different pH, but more often by the action of another enzyme.

The gut itself is not digested by activated enzymes, because it is protected by a covering of mucus, a slimy material secreted by special cells in the lining of the gut. The mucus also lubricates the gut and protects it from abrasion. If mucus production is inadequate, digestive enzymes or stomach acid can act on the gut, producing sores called **ulcers**. Insects rely on a different trick to prevent digestion of the gut lining. Within the gut they secrete a thin tube of chitin, a modified polysaccharide (see Figure 3.14c) that is also found in the insect's protective exoskeleton. The chitin tube encloses the food and enzymes and protects the gut from abrasion and self-digestion.

Structure and Function of the Vertebrate Gut

The separate compartments that have specific functions in the digestive tract of vertebrates are all part of a continuous tube that runs from mouth to anus. The specific functions must be coordinated so that they occur in proper sequence and at appropriate rates. Let's take a tour of the vertebrate gut to see how structure and function work together to move food through the gut and bring about its sequential digestion and the absorption of nutrients.

Similar tissue layers are found in all regions of the vertebrate gut

The cellular architecture of the tube that forms the vertebrate gut follows a common plan throughout. Four major layers of different cell types form the wall of the tube (Figure 47.10). These layers differ somewhat from compartment to compartment, but they are always present.

Starting in the cavity, or **lumen**, of the gut, the first tissue layer is the **mucosa**. Nutrients are absorbed across the membranes of the mucosal cells; in some regions of the gut, those membranes have many folds that increase their surface area (see Figure 47.9c). Mucosal cells also have secretory functions. Some secrete mucus that lubricates the food and protects the walls of the gut. Others secrete digestive enzymes, and still others in the stomach secrete hydrochloric acid (HCl).

At the base of the mucosa are some smooth-muscle cells, and just outside the mucosa is the second layer of cells, the **submucosa**. Here we find the blood and lymph vessels that carry absorbed nutrients to the rest of the body. The submucosa also contains a network of nerves; these neurons are both sensory (responsible for stomachaches) and regulatory (controlling the various secretory functions of the gut).

External to the submucosa are two layers of smooth-muscle cells responsible for the movements of the gut. Because the cells of the innermost layer are oriented *around* the gut, this layer is called **circular muscle**. It constricts the lumen. The cells of the outermost layer of the tube's wall, the **longitudinal muscle**, are arranged along the length of the gut. When this layer contracts, the gut shortens. Between these two

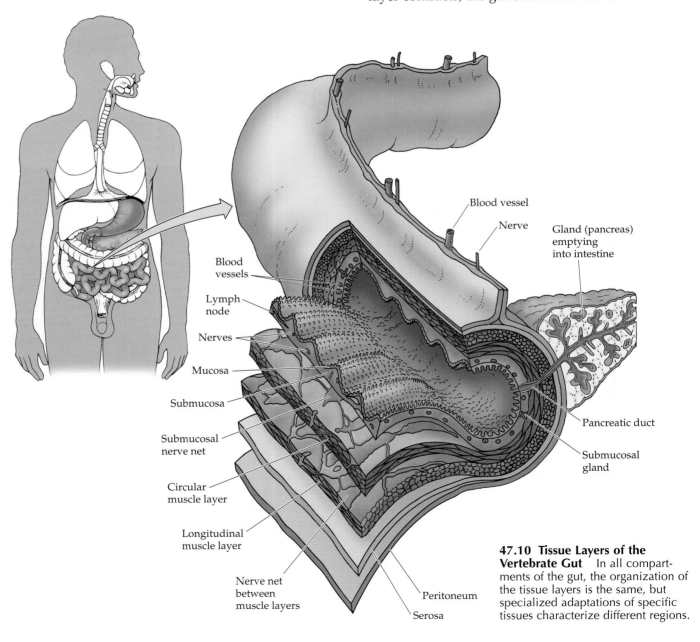

47.10 Tissue Layers of the Vertebrate Gut In all compartments of the gut, the organization of the tissue layers is the same, but specialized adaptations of specific tissues characterize different regions.

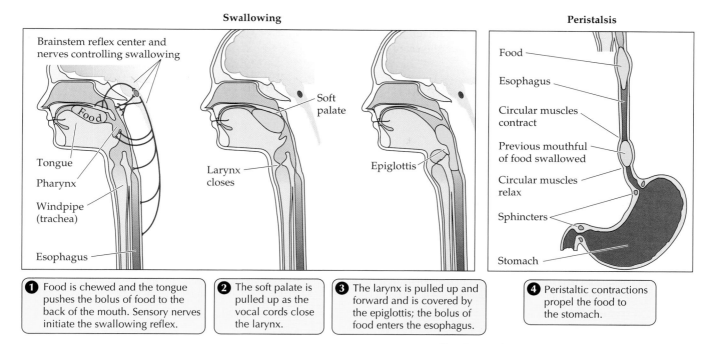

Swallowing

Brainstem reflex center and nerves controlling swallowing

Food

Tongue

Pharynx

Windpipe (trachea)

Esophagus

Soft palate

Larynx closes

Epiglottis

Peristalsis

Food

Esophagus

Circular muscles contract

Previous mouthful of food swallowed

Circular muscles relax

Sphincters

Stomach

❶ Food is chewed and the tongue pushes the bolus of food to the back of the mouth. Sensory nerves initiate the swallowing reflex.

❷ The soft palate is pulled up as the vocal cords close the larynx.

❸ The larynx is pulled up and forward and is covered by the epiglottis; the bolus of food enters the esophagus.

❹ Peristaltic contractions propel the food to the stomach.

47.11 Swallowing and Peristalsis Food pushed to the back of the mouth triggers the swallowing reflexes. Once food enters the esophagus, peristalsis propels it to the stomach.

layers of muscle is a network of nerves that controls the movements of the gut, coordinating the different regions with one another.

Surrounding the gut is a fibrous coat called the **serosa**. Like other abdominal organs, the gut is also covered and supported by a tissue, the **peritoneum**.

Peristalsis moves food through the gut

Food entering the mouth of most vertebrates is chewed and mixed with the secretions of salivary glands. The muscular tongue then pushes the mouthful, or bolus, of food toward the back of the mouth cavity. By making contact with the soft tissue at the back of the mouth, the food initiates a complex series of neural reflex actions known as swallowing. Stand in front of a mirror and gently touch this tissue at the back of your mouth with the eraser of your pencil or with a cotton swab. You may gag slightly, but you will also experience an uncontrollable urge to swallow. Swallowing involves many muscles doing a variety of jobs that propel the food through the **pharynx** and into the **esophagus** without allowing any of it to enter the windpipe (trachea) or nasal passages (Figure 47.11).

Once the food is in the esophagus, peristalsis takes over and pushes the food toward the stomach. **Peristalsis** is a wave of smooth-muscle contraction that moves progressively down the gut from the pharynx toward the anus. The smooth muscle of the gut contracts in response to being stretched. Swallowing a bolus of food stretches the upper end of the esophagus, and this stretching initiates a wave of contraction that slowly pushes the contents of the gut toward the anus.

Peristalsis can occasionally run in the opposite direction. When your stomach is very full, pressure on your abdomen or a sudden movement can push some stomach contents into the lower end of your esophagus. This action can initiate peristaltic movements that bring the acidic, partly digested food into your mouth. When you vomit, contractions of the abdominal muscles explosively force stomach contents out through the esophagus. Before you vomit, waves of reverse peristalsis can even bring the contents of the upper regions of the intestine back into the stomach to be expelled.

The movement of food from the stomach into the esophagus is normally prevented by a thick ring of circular smooth muscle at the junction of the esophagus and the stomach. This ring of muscle, called a **sphincter**, is normally constricted. Waves of peristalsis cause it to relax enough to let food pass through. Sphincter muscles are found elsewhere in the digestive tract as well. The pyloric sphincter governs the passage of stomach contents into the intestine. Another important sphincter surrounds the anus.

Digestion begins in the mouth and the stomach

In addition to physically disrupting food, the mouth initiates the digestion of carbohydrates through the action of the enzyme amylase, which is secreted with the saliva and mixed with the food as it is chewed. Amylase is a carbohydrase; it hydrolyzes the bonds between the six-carbon sugar units that make up the long-chain starch molecules. The action of amylase is what makes a piece of bread or cracker taste sweet if you chew it long enough.

Most vertebrates can rapidly consume a large volume of food, but digesting that food is a long, slow process. The stomach stores the food devoured during a meal and continues breaking it down physically. The secretions of the stomach kill microorganisms that are taken in with the food and begin digesting proteins. The major enzyme produced by the stomach is an endopeptidase called **pepsin**. Pepsin is secreted as the zymogen **pepsinogen** by cells in the **gastric pits**— deep folds in the stomach lining (Figure 47.12). Other cells in the gastric pits produce hydrochloric acid, and still others near the openings of the gastric pits and throughout the stomach mucosa secrete mucus.

Hydrochloric acid (HCl) maintains the stomach fluid (the gastric juice) at a pH between 1 and 3. This low pH activates the conversion of pepsinogen to pepsin. The conversion is amplified as the newly formed pepsin activates other pepsinogen molecules, a process called **autocatalysis**. Hydrochloric acid also provides the right pH for the enzymatic action of pepsin. In addition, the low pH helps dissolve the intercellular substances holding the ingested tissues together. Breakdown of the ingested tissues exposes more food surface area to the action of digestive enzymes. Mucus secreted by the stomach mucosa coats and protects the walls of the stomach from being eroded and digested by the HCl and pepsin.

Contractions of the muscles in the walls of the stomach churn its contents, thoroughly mixing them with the stomach secretions. The acidic, fluid mixture of digestive juices and partly digested food in the stomach is called **chyme**. A few substances can be absorbed from the chyme across the stomach wall, including alcohol (hence its rapid effects), aspirin, and caffeine, but even these substances are absorbed in rather small quantities in the stomach.

Peristaltic contractions of the stomach walls push the chyme toward the bottom end of the stomach. These waves of peristalsis cause the pyloric sphincter to relax briefly so that little squirts of the chyme can enter the first region of the intestine, where digestion of carbohydrates and proteins continues, digestion of fats begins, and absorption of nutrients begins. The human stomach empties itself gradually over a period of approximately 4 hours. This slow passage of food enables the intestine to work on a little material at a time and prolongs the digestive and absorptive processes throughout much of the time between meals.

The small intestine is the major site of digestion

Although the **small intestine** takes its name from its diameter, it is a very large organ and the site of the

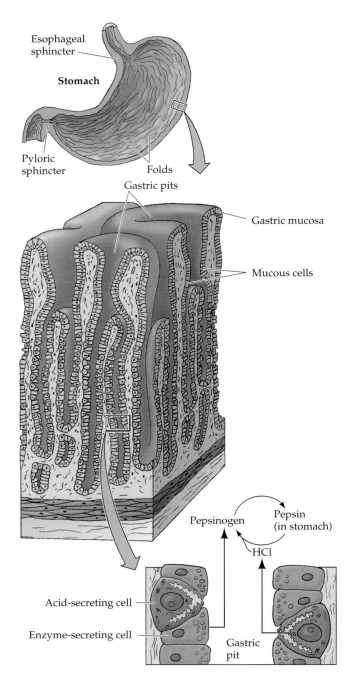

47.12 The Stomach The human stomach stores and breaks down ingested food. Cells in the gastric pits secrete digestive acids and enzymes. Both the pits and gastric mucosa secrete mucus that aids in digestion.

major events of digestion and absorption. The small intestine of an adult human is more than 6 m long; its coils fill much of the lower abdominal cavity (Figure 47.13). Because of its length, and because of the folds, villi, and microvilli of its lining (see Figure 47.9c), the inner surface area of the small intestine is enormous: about 550 m^2, or roughly the size of a tennis court.

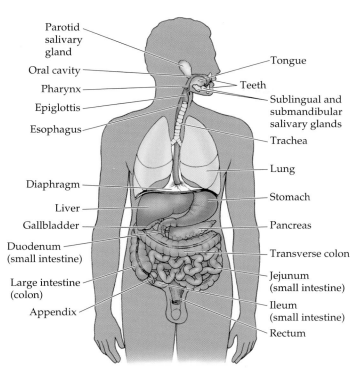

47.13 The Human Gut Compartments within this long tube specialize in digesting food, absorbing nutrients, and storing and expelling wastes.

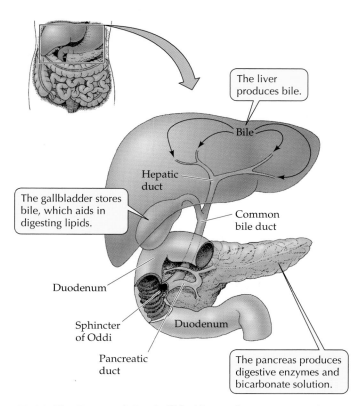

The liver produces bile.

Bile

The gallbladder stores bile, which aids in digesting lipids.

Hepatic duct

Common bile duct

Duodenum

Sphincter of Oddi

Duodenum

Pancreatic duct

The pancreas produces digestive enzymes and bicarbonate solution.

47.14 The Ducts of the Gallbladder and Pancreas Bile produced in the liver leaves the liver via the hepatic duct. Branching off this duct is the gallbladder, which stores bile. Below the gallbladder, the hepatic duct is called the common bile duct and is joined by the pancreatic duct before entering the duodenum.

Across this surface the small intestine absorbs all the nutrient molecules derived from food. The small intestine has three sections. The initial section—the **duodenum**—is the site of most digestion; the **jejunum** and the **ileum** carry out 90 percent of the absorption of nutrients.

Digestion requires many specialized enzymes, as well as several other secretions. Two accessory organs that are not part of the digestive tract—the liver and the pancreas—provide many of these enzymes and secretions. The liver synthesizes a substance called **bile** from cholesterol. Bile emulsifies fats just as soap emulsifies grease on your clothes or hands. Bile secreted from the liver flows through the **hepatic duct**.

A side branch of the hepatic duct delivers bile to the gallbladder, where it is stored until it is needed to assist in fat digestion (Figure 47.14). Below the branching point to the gallbladder, the hepatic duct is called the **common bile duct**. When undigested fats enter the duodenum, the gallbladder releases bile, which flows down the common bile duct and enters the duodenum.

To understand the role of bile in fat digestion, think of the oil in salad dressing; it is not soluble in water (it is hydrophobic), and it tends to aggregate together in large globules. The enzymes that digest fat, the **lipases**, are water-soluble and must do their work in an aqueous medium, but the fats are not water-soluble. Therefore, the interface between the aqueous digestive juices and large globules of fat would be very small if it weren't for bile.

Bile stabilizes tiny droplets of fat so that they cannot aggregate into large globules. One end of each bile molecule is soluble in fat (lipophilic, or hydrophobic); the other end is soluble in water (hydrophilic, or lipophobic). Bile molecules bury their lipophilic ends in fat droplets, leaving their lipophobic ends sticking out. As a result, they prevent the fat droplets from sticking together. These very small fat particles are called **micelles**, and their small size maximizes the surface area exposed to lipase action (see Figure 47.16).

The **pancreas** is a large gland that lies just beneath the stomach (see Figure 47.13). It has both endocrine (secreting to the inside of the body) and exocrine (secreting to the outside of the body) functions. Here we will consider the exocrine products, which it delivers to the gut through the pancreatic duct. The pancreatic duct joins the common bile duct before entering the duodenum (see Figure 47.14).

TABLE 47.3   Sources and Functions of the Major Digestive Enzymes of Humans

ENZYME	SOURCE	ACTION	SITE OF ACTION
Salivary amylase	Salivary glands	Starch → Maltose	Mouth
Pepsin	Stomach	Proteins → Peptides; autocatalysis	Stomach
Pancreatic amylase	Pancreas	Starch → Maltose	Small intestine
Lipase	Pancreas	Fats → Fatty acids and glycerol	Small intestine
Nuclease	Pancreas	Nucleic acids → Nucleotides	Small intestine
Trypsin	Pancreas	Proteins → Peptides; activation of zymogens	Small intestine
Chymotrypsin	Pancreas	Proteins → Peptides	Small intestine
Carboxypeptidase	Pancreas	Peptides → Peptides and amino acids	Small intestine
Aminopeptidase	Small intestine	Peptides → Peptides and amino acids	Small intestine
Dipeptidase	Small intestine	Dipeptides → Amino acids	Small intestine
Enterokinase	Small intestine	Trypsinogen → Trypsin	Small intestine
Nuclease	Small intestine	Nucleic acids → Nucleotides	Small intestine
Maltase	Small intestine	Maltose → Glucose	Small intestine
Lactase	Small intestine	Lactose → Galactose and glucose	Small intestine
Sucrase	Small intestine	Sucrose → Fructose and glucose	Small intestine

The pancreas produces a host of digestive enzymes (Table 47.3). As in the stomach, these enzymes are released as zymogens; otherwise they would digest the pancreas and its ducts before they ever reached the duodenum. Once in the duodenum, one of these inactive enzymes, **trypsinogen**, is activated by **enterokinase**, which is produced by cells lining the duodenum (Figure 47.15).

Active **trypsin** can cleave other trypsinogen molecules to release even more active trypsin (another example of autocatalysis). Similarly, trypsin acts on the other zymogens secreted by the pancreas and releases their active enzymes. The mixture of zymogens produced by the pancreas can be very dangerous if the pancreatic duct is blocked or if the pancreas is injured by an infection or a severe blow to the abdomen. A few trypsinogen molecules spontaneously converting to trypsin can initiate a chain reaction of enzyme activation that digests the pancreas in a very short period of time, destroying both its endocrine and exocrine functions.

The pancreas produces, in addition to digestive enzymes, a secretion rich in bicarbonate ions (HCO_3^-). Bicarbonate ions neutralize the pH of the chyme that enters the duodenum from the stomach. This neutralization is essential because intestinal enzymes function best at a neutral or slightly alkaline pH.

Nutrients are absorbed in the small intestine

Only the smallest products of digestion pass through the mucosa of the small intestine and into the blood and lymphatic vessels that lie in the submucosa. The final digestion of proteins and carbohydrates that produces these absorbable products takes place among the microvilli. The mucosal cells with microvilli produce dipeptidase, which cleaves larger peptides into tripeptides, dipeptides, and individual amino acids that the cells can absorb. These cells also produce the enzymes maltase, lactase, and sucrase, which cleave the common disaccharides into their constituent, absorbable monosaccharides—glucose, galactose, and fructose.

Disaccharides are not absorbed. Many humans stop producing the enzyme lactase around the age of 4 years and thereafter have difficulty digesting lactose, which is the sugar in milk. Lactose is a disaccharide and cannot be absorbed without being cleaved into its

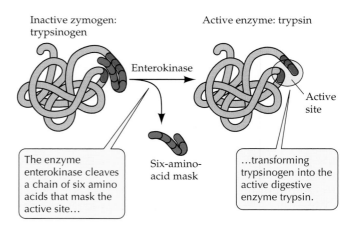

47.15 Zymogen Activation Powerful digestive enzymes often exist as inactive zymogens until their catalytic activity is required. For example, when the zymogen trypsinogen, which is secreted by the pancreas, reaches the small intestine, the enzyme enterokinase transforms it into the active digestive enzyme trypsin.

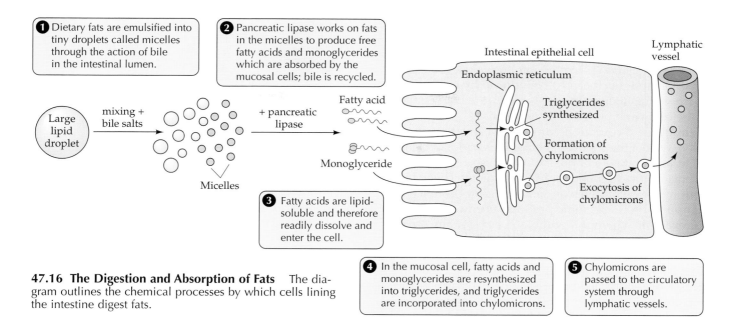

① Dietary fats are emulsified into tiny droplets called micelles through the action of bile in the intestinal lumen.

② Pancreatic lipase works on fats in the micelles to produce free fatty acids and monoglycerides which are absorbed by the mucosal cells; bile is recycled.

③ Fatty acids are lipid-soluble and therefore readily dissolve and enter the cell.

④ In the mucosal cell, fatty acids and monoglycerides are resynthesized into triglycerides, and triglycerides are incorporated into chylomicrons.

⑤ Chylomicrons are passed to the circulatory system through lymphatic vessels.

Large lipid droplet

mixing + bile salts

+ pancreatic lipase

Fatty acid

Monoglyceride

Micelles

Intestinal epithelial cell

Lymphatic vessel

Endoplasmic reticulum

Triglycerides synthesized

Formation of chylomicrons

Exocytosis of chylomicrons

47.16 The Digestion and Absorption of Fats The diagram outlines the chemical processes by which cells lining the intestine digest fats.

constituent units, glucose and galactose. If a substantial amount of lactose is unabsorbed and passes into the large intestine, its metabolism by bacteria in the large intestine causes abdominal cramps, gas, and diarrhea.

The mechanisms by which the cells lining the intestine absorb nutrient molecules and inorganic ions are diverse and not completely understood. Many inorganic ions are actively transported into the cells. For example, active transporters exist for sodium, calcium, and iron. Transporters also exist for certain classes of amino acids and for glucose and galactose, but curiously their activity is much reduced if active sodium transport is blocked.

Sodium diffuses from the gut contents into the mucosal cells and is actively transported from the mucosal cells into the submucosa. To diffuse into a mucosal cell, a sodium ion binds to a carrier molecule in the mucosal cell membrane, which also binds a nutrient molecule such as glucose or an amino acid. The diffusion of the sodium ion, driven by a concentration difference, therefore drives the absorption of the nutrient molecule. This mechanism is called *sodium cotransport*.

The absorption of the products of fat digestion is much simpler (Figure 47.16). Lipases break down fats into fatty acids and monoglycerides, which are lipid-soluble and are thus able to dissolve in the membranes of the microvilli and diffuse into the mucosal cells. Once in the cells, the fatty acids and monoglycerides are resynthesized into triglycerides, combined with cholesterol and phospholipids, and coated with protein to form water-soluble **chylomicrons**, which are really little particles of fat. The chylomicrons pass into the lymphatic vessels in the submucosa and into the bloodstream through the thoracic duct. After a meal

rich in fats, the chylomicrons can be so abundant in the blood that they give it a milky appearance.

The bile that emulsifies the fats is not absorbed along with the monoglycerides and the fatty acids, but is recycled back and forth between the gut contents and the microvilli. Finally, in the ileum, bile is actively reabsorbed and returned to the liver via the bloodstream. Bile is synthesized in the liver from cholesterol. Cholesterol is synthesized in the liver and also comes from our diet.

As we learned in Chapter 46, high cholesterol levels contribute to arterial plaque formation and therefore to cardiovascular disease. The body has no way of breaking down excess cholesterol, so high dietary intake or high levels of synthesis create problems. One major way that cholesterol leaves the body is through the elimination of unreabsorbed bile in the feces. The rationale for including certain kinds of fiber in our diet is that the fiber binds the bile, decreases its reabsorption in the ileum, and thus helps to lower body cholesterol.

Water and ions are absorbed in the large intestine

Peristalsis gradually pushes the contents of the small intestine into the large intestine, or **colon**. The rate of peristalsis is controlled so that food passes through the small intestine slowly enough for digestion and absorption to be complete, but quickly enough to ensure an adequate supply of nutrients for the body. Most of the nutrients have been removed from material that enters the colon, but the material contains a lot of water and inorganic ions.

The colon reabsorbs water and ions, producing semisolid feces from the slurry of indigestible materials it receives from the small intestine. Reabsorption of too much water can cause constipation. The opposite condition, diarrhea, results if too little water is reabsorbed or if water is secreted into the colon. (Both constipation and diarrhea can be induced by toxins from microorganisms.) Feces are stored in the last segment of the colon and periodically excreted.

Immense populations of bacteria live within the colon. One of the resident species is *Escherichia coli*, the bacterium that is so popular among researchers in biochemistry, genetics, and molecular biology. This inhabitant of the colon lives on matter indigestible to humans and produces some products useful to the host. For example, vitamin K and biotin are synthesized by *E. coli* and absorbed across the wall of the colon. Many species of mammals maximize the nutritional benefits from such bacterial activity by reingesting their own feces, a behavior called **coprophagy**. Excessive or prolonged intake of antibiotics can lead to vitamin deficiency because the antibiotics kill the normal intestinal bacteria at the same time they are killing the disease-causing organisms for which they are intended.

The intestinal bacteria produce gases such as methane and hydrogen sulfide as by-products of their largely anaerobic metabolism. Humans expel gas after eating beans because the beans are rich in carbohydrates that bacteria—but not humans—can break down.

The large intestine of humans has a small, fingerlike pouch called the **appendix**, which is best known for the trouble it causes when it becomes infected. The human appendix plays no essential role in digestion, but it does contribute to immune-system function. It can be surgically removed without serious consequences.

The part of the gut that forms the appendix in humans forms the much larger cecum in herbivores (see Figure 47.8), where it functions in cellulose digestion. As our primate ancestors evolved to exploit diets less rich in indigestible cellulose, the cecum no longer served an essential function and gradually became **vestigial** (reduced to a trace). Other examples of vestigial structures are the nonfunctional eyes of cave fish and the dewclaws of dogs and cats.

Herbivores have special adaptations to digest cellulose

Cellulose is the principal organic compound in the diets of herbivores. However, most herbivores cannot produce **cellulases**, the enzymes that hydrolyze cellulose. Exceptions include silverfish (well known for eating books and stored papers), earthworms, and shipworms. Other herbivores, from termites to cattle, rely on microorganisms living in their digestive tracts to digest cellulose for them. These microorganisms inhabit various parts of the gut, where they may be pres-

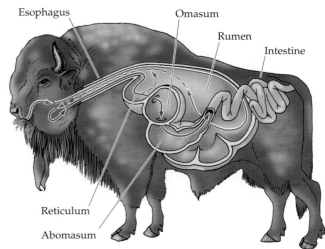

47.17 The Ruminant Stomach Specialized stomach compartments—the rumen, reticulum, omasum, and abomasum—enable ruminants to digest and subsist on protein-poor plant material.

ent in the billions. Most are bacteria, but some are fungi or protists.

The digestive tracts of **ruminants** (cud chewers) such as cattle, goats, and sheep are specialized to maximize benefits from microorganisms. In place of the usual mammalian stomach, ruminants have a large, four-chambered organ (Figure 47.17). The first and largest of these chambers is the rumen; the second is the reticulum. Both are packed with anaerobic microorganisms that break down cellulose. These two chambers serve as fermentation vats for the digestion of cellulose. The ruminant periodically regurgitates the contents of the rumen (the cud) into the mouth for rechewing. When the more thoroughly ground-up vegetable fibers are swallowed again, they present more surface area to the microorganisms for their digestive actions.

The microorganisms in the rumen and reticulum metabolize cellulose and other nutrients to simple fatty acids. Ruminants produce and swallow large quantitites of alkaline saliva to buffer this acid production. Although fatty acids are the major nutrients the host derives from its microorganisms, the microorganisms themselves provide an important source of protein. A cow can derive more than 100 g of protein per day from digestion of its own microorganisms. The plant materials ingested by a ruminant are a poor source of protein, but they contain inorganic nitrogen that the microorganisms use to synthesize their own amino acids.

The by-products of the fermentation of cellulose are carbon dioxide and methane, which the animal belches. A single cow can produce 400 liters of methane a day. Methane is the second most abundant "greenhouse gas," whose concentration in the atmosphere is increasing, and domesticated ruminants are second only to industry as a source of methane emitted into the atmosphere.

The food leaving the rumen carries with it enormous numbers of the cellulose-fermenting microorganisms. This mixture passes through the omasum, where it is concentrated by water reabsorption. It then enters the true stomach, the abomasum, which secretes hydrochloric acid and proteases. The microorganisms are killed by the acid, digested by the proteases, and passed on to the small intestine for further digestion and absorption. The rate of multiplication of microorganisms in the rumen is great enough to offset their loss, so a well-balanced, mutually beneficial relationship is maintained.

Mammalian herbivores other than ruminants have microbial farms and cellulose fermentation vats in a branch off the large intestine called the cecum. Rabbits and hares are good examples. Since the cecum empties into the large intestine, the absorption of the nutrients produced by the microorganisms is inefficient and incomplete. Therefore, these animals practice coprophagy. They frequently produce two kinds of feces, one of pure waste (which they discard), and one consisting mostly of cecal material, which they ingest directly from the anus. As this cecal material passes through the stomach and small intestine, the nutrients it contains are digested and absorbed.

Control and Regulation of Digestion

The vertebrate gut could be described as an assembly line in reverse—a disassembly line. As with a standard assembly line, control and coordination of sequential processes is critical. Both neural and hormonal controls thus govern gut functions.

Neural reflexes coordinate functions in different regions of the gut

Almost everyone has experienced salivation stimulated by the sight or smell of food. That response is a neural reflex, as is the act of swallowing following tactile stimulation at the back of the mouth. Many neural reflexes coordinate activities in different regions of the digestive tract so that it works in a properly timed, assembly line manner. For example, loading the stomach with food stimulates increased activity in the colon, which can lead to a bowel movement. This phenomenon is the **gastrocolic reflex**.

The digestive tract is unusual in that it has an intrinsic (that is, its own) nervous system. So in addition to neural reflexes involving the central nervous system, such as salivation and swallowing, neural messages can travel from one region of the digestive tract to another without being processed by the central nervous system.

Hormones control many digestive functions

Several hormones control the activities of the digestive tract and its accessory organs (Figure 47.18). The first hormone ever discovered came from the duodenum; it was called **secretin** because it causes the pancreas to secrete digestive juices. We now know that secretin is one of several hormones that control pancreatic secretion; specifically, it stimulates the pancreas to secrete a solution rich in bicarbonate ions.

In response to fats and proteins in the chyme, the mucosa of the small intestine also secretes **cholecystokinin**, a hormone that stimulates the gallbladder to release bile and the pancreas to release digestive enzymes. Cholecystokinin and secretin also slow the movements of the stomach, which slows the delivery of chyme into the small intestine.

The stomach secretes a hormone called **gastrin** into the blood. Cells in the lower region of the stomach release gastrin when they are stimulated by the presence of food. Gastrin circulates in the blood until it reaches

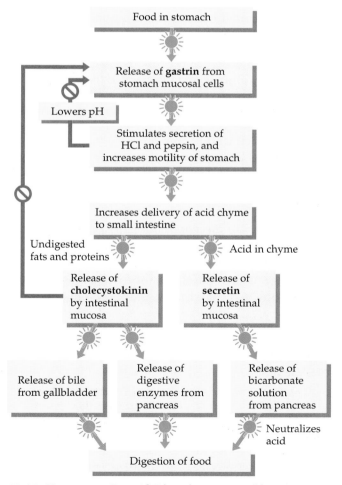

47.18 Hormones Control Digestion Several hormones are involved in feedback loops that control the sequential processing of food in the digestive tract.

cells in the upper areas of the stomach wall, where it stimulates the secretions and movements of the stomach. Gastrin release is inhibited when the stomach contents become too acidic—another example of negative feedback.

Control and Regulation of Fuel Metabolism

Most animals do not eat continuously. At times, food is in the gut and nutrients are being absorbed and are readily available to supply energy and molecular building blocks. When nutrients are not being absorbed, however, the continuous processes of energy metabolism and biosynthesis must run off of internal reserves. For this reason nutrient traffic must be controlled so that reserves accumulate during absorption and so that those reserves are used appropriately when the gut is empty.

The liver directs the traffic of fuel molecules

The liver directs the traffic of nutrient molecules used in energy metabolism (fuel molecules). When nutrients are abundant in the circulatory system, the liver can store them in the forms of glycogen (animal starch) and fat. The liver also synthesizes plasma proteins from circulating amino acids. When the availability of fuel molecules in the bloodstream declines, the liver delivers glucose and fats back to the blood.

The liver has an enormous capacity to interconvert fuel molecules. Liver cells can convert monosaccharides into either glycogen or fat, and vice versa. Certain amino acids and some other molecules, such as pyruvate and lactate, can be converted into glucose—a process called **gluconeogenesis**. The liver is also the major controller of fat metabolism, through its production of lipoproteins.

Lipoproteins are the good, the bad, and the ugly

In the intestine, bile solves the problem of processing hydrophobic fats in an aqueous medium. The transportation of fats in the circulatory system presents the same problem, but in this case lipoproteins offer the solution.

A lipoprotein is a particle made up of a core of fat and cholesterol and a covering of protein that makes it hydrophilic. The largest lipoprotein particles are the chylomicrons produced by the cells that line the intestine to transport dietary fat and cholesterol into the circulation (see Figure 47.16). As lipoproteins circulate through the liver and the fat tissues around the body, receptors on the capillary walls recognize the protein coat, and lipases begin to hydrolyze the fats, which are then absorbed into fat or liver cells. Thus the protein coat of the lipoprotein makes it water-soluble and serves as an "address" that targets the lipoprotein to a specific tissue.

Other lipoproteins originate in the liver and are classified according to their density. Because fat has a low density (it floats in water), the more fat a lipoprotein contains, the lower its density. Very-low-density lipoprotein (VLDL) produced by the liver contains mostly triglyceride fats that are being transported to fat cells in tissues around the body. Low-density lipoproteins (LDL) contain mostly cholesterol, which they transport to tissues around the body for use in biosynthesis and to be deposited. High-density lipoproteins (HDL) serve as acceptors of cholesterol and are believed to remove cholesterol from tissues and return it to the liver, where it can be used to synthesize bile.

Because of their differing functions in cholesterol regulation, LDL is sometimes called "bad cholesterol" and HDL "good cholesterol," but those designations are somewhat controversial. However, we do know that a high ratio of LDL to HDL in a person's blood is a risk factor for atherosclerotic heart disease. Cigarette smoking lowers HDL levels, and regular exercise increases HDL levels.

Fuel metabolism is controlled by hormones

The **absorptive period** refers to the time that food is in the gut and nutrients are being absorbed and circulated in the blood. During this time the liver converts glucose to glycogen and fat, the body fat tissues convert glucose and fatty acids to stored fat, and the cells of the body preferentially use glucose for metabolic fuel. When food is no longer in the gut—the **postabsorptive period**—these processes reverse. The liver breaks down glycogen to supply glucose to the blood, the liver and the fat tissues supply fatty acids to the blood, and most of the cells of the body preferentially use fatty acids for metabolic fuel.

The major exceptions to this rule are the cells of the nervous system, which require a constant supply of glucose for their energetic needs. Although the nervous system can use other fuels to a limited extent, its overall dependence on glucose is the reason it is so important for other cells of the body to shift to fat metabolism during the postabsorptive period. This shift preserves the available glucose and glycogen stores for the nervous system for as long as possible.

What directs the traffic in fuel molecules? Two hormones produced and released by the pancreas, **insulin** and **glucagon**, are largely responsible for controlling the metabolic directions that fuel molecules take (Figure 47.19). The most important of these hormones is insulin, which is produced in response to high blood glucose levels.

The pancreas releases insulin into the circulatory system when blood glucose rises above the normal postabsorptive level. Insulin facilitates the entry of glucose into most cells of the body. Thus when insulin

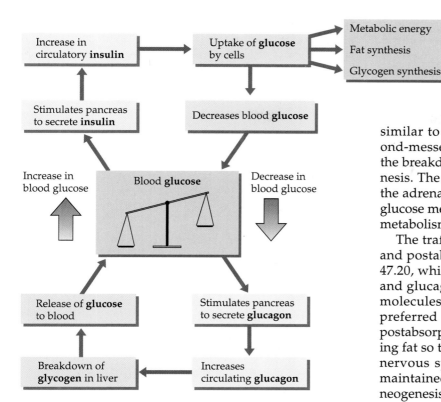

47.19 Regulating Glucose in the Blood
Insulin and glucagon maintain the homeostasis of blood glucose.

similar to that of glucagon. Through the cAMP second-messenger cascade (see Chapter 38), it increases the breakdown of glycogen and increases gluconeogenesis. The stress of low blood glucose also stimulates the adrenal cortex to secrete cortisol. Cortisol inhibits glucose metabolism in many cells while promoting the metabolism of fats and proteins.

The traffic of fuel molecules during the absorptive and postabsorptive periods is summarized in Figure 47.20, which indicates the steps controlled by insulin and glucagon. During the absorptive period, all fuel molecules move toward storage, and glucose is the preferred energy source for all cells. During the postabsorptive period, most cells switch to metabolizing fat so that blood glucose reserves are saved for the nervous system. The level of circulating glucose is maintained through glycogen breakdown and gluconeogenesis.

Nutrient Deficiency Diseases

Chronic shortage of a nutrient produces a characteristic deficiency disease. If the deficiency is not remedied, death may follow. An example is kwashiorkor, which results from protein deficiency. Kwashiorkor causes swelling of the extremities, distension of the abdomen (Figure 47.21), breakdown of the immune system, degeneration of the liver, mental retardation, and other problems.

Shortage of any of the vitamins results in specific deficiency symptoms (see Table 47.2). We mentioned the disease scurvy, which results from lack of vitamin C. Another deficiency disease, **beriberi**, was critically involved in the discovery of vitamins. *Beriberi* means "extreme weakness." This disease is found where unbalanced diets are common.

Beriberi was particularly prevalent in Asia in the nineteenth century, when it became standard practice to mill rice to a high, white polish and discard the hulls that are present in brown rice. It was discovered that chickens and pigeons showed beriberi-like symptoms when fed only polished rice. In 1912 Casimir Funk showed that pigeons with beriberi could be cured of their symptoms if they were fed the hulls that had been discarded to make the rice more "appealing."

Funk suggested the radical idea that beriberi and some other diseases are dietary in origin and result from deficiencies in specific substances. (At the time, all diseases were thought to be caused by microorgan-

is present, most cells burn glucose as their metabolic fuel, fat cells use glucose to make fat, and liver cells convert glucose to glycogen and fat. As soon as blood glucose falls back to postabsorptive levels, insulin release diminishes rapidly, and the entry of glucose into cells other than those of the nervous system is inhibited.

Without a supply of glucose, cells switch to using glycogen and fat as their metabolic fuels. In the absence of insulin, the liver and fat cells stop synthesizing glycogen and fat and begin breaking them down. As a result, the liver supplies glucose to the blood rather than taking it from the blood, and both the liver and the fat tissues supply fatty acids to the blood.

The pancreas releases glucagon when the blood glucose concentration falls below the normal postabsorptive level. Glucagon has the opposite effect of insulin; it stimulates liver cells to break down glycogen and carry out gluconeogenesis, thus releasing glucose into the blood. However, the major hormonal control of fuel metabolism during the postabsorptive period is the *lack of insulin*; glucagon plays a secondary role.

Two other hormones that help preserve blood glucose levels during the postabsorptive period are *epinephrine* and the glucocorticoid *cortisol*. Low blood glucose is a stress that triggers glucose-sensitive cells in the hypothalamus to signal the adrenal medulla, through the sympathetic nervous system, to secrete epinephrine. Epinephrine has an effect on liver cells

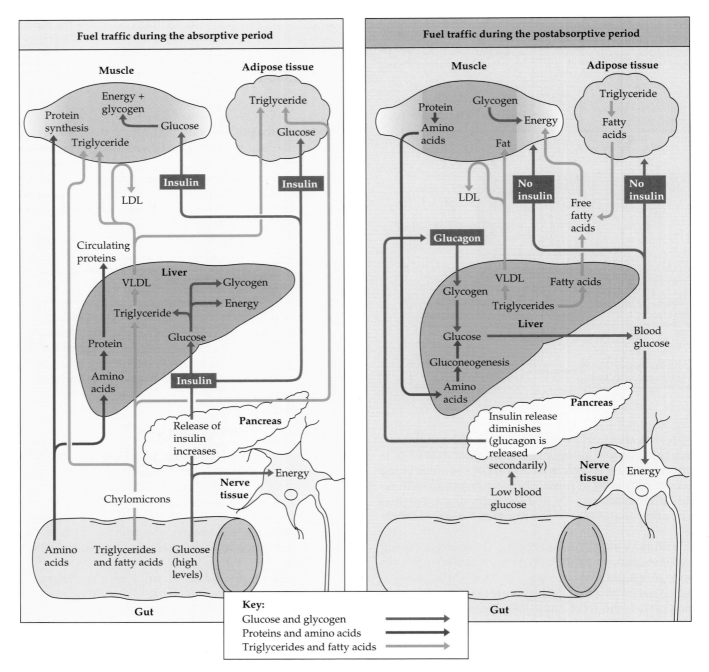

47.20 Fuel Molecule Traffic during the Absorptive and Postabsorptive Periods   Insulin promotes glucose uptake by liver, muscle, and fat cells during the absorptive period. During the postabsorptive period the lack of insulin blocks glucose uptake by these same tissues and promotes fat and glycogen breakdown to supply metabolic fuel.

isms.) Funk coined the term "vitamines" because he mistakenly thought that all these substances were amines vital for life. In 1926, thiamin (vitamin B_1)—the substance lost in the rice milling process—was the first vitamin to be isolated in pure form.

Pellagra, another vitamin B deficiency disease, results from a lack of niacin. A common and severe disease in many impoverished regions, pellagra also occurs in conjunction with chronic alcoholism. Its symptoms include diarrhea, itching skin, abdominal pain, and other problems.

Vitamin D deficiency decreases the absorption and use of calcium, leading to softening of the bones and a distortion of the skeleton. This deficiency disease is known as **rickets**.

Vitamin B_{12} (cobalamin) is produced by microorganisms that live in our intestines and use the cobalt in our diet. Cobalamin is present in all foods of animal origin.

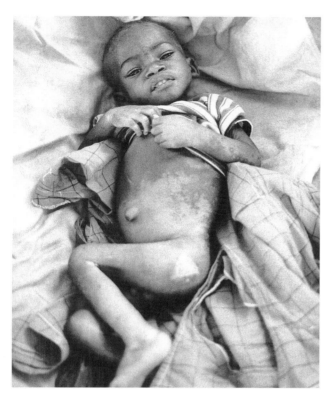

47.21 Kwashiorkor, "The Rejected One" Swollen abdomen, face, hands, and feet due to edema (fluid retention), as well as spindly limbs, are hallmarks of serious protein starvation. It can occur in children when their mother no longer breast-feeds and alternate protein sources are not available. The limbs are spindly because the body breaks down muscle tissue, as well as blood proteins, to obtain needed amino acids.

Since plants neither use nor produce vitamin B_{12}, a strictly vegetarian diet (not supplemented by vitamin pills) can lead to **pernicious anemia**, the B_{12} deficiency disease.

Inadequate mineral nutrition can also lead to deficiency diseases. Iodine, for example, is a constituent of the hormone thyroxine, which is produced in the thyroid gland. If insufficient iodine is obtained in the diet, the thyroid gland grows larger in an attempt to compensate for the inadequate production of thyroxine. The swelling of the neck that results is called a **goiter**.*

Toxic Compounds in Food

As mentioned early in the chapter, potential food can contain toxic compounds naturally. For example, some

*Such goiters are common in mountain areas such as the Andes of South America because of low iodine levels in the soil and hence in the crops grown there. Goiters were once common in Switzerland and in the Great Lakes area of the United States, but the problem was largely solved by the addition of small amounts of iodine to table salt or to drinking water.

mushrooms contain poisons and hallucinogens, some mollusks, fish, and amphibians contain neurotoxins, some plants contain compounds that stimulate or depress the heart, and of course tobacco contains nicotine, poppies contain opium, and marijuana contains tetrahydrocannabinol (Figure 47.22).

Ingesting natural materials can be dangerous. And human activities that add millions of tons of toxic compounds to our environment every year make the problem worse. Many of these compounds enter the air we breathe, the water we drink, and the food we eat. A whole new field called **environmental toxicology** has grown up around the problems of poisons in the environment.

Papaver somniferum

Arothron (Fugu) parda

47.22 Toxins Occur Naturally Many animals and plants contain dangerous, potentially deadly toxins. Opium poppies are the source of morphine and heroin. The puffer fish is a delicacy in Japanese sushi restaurants; restaurants that serve it have special licenses affirming their chef's ability to prepare the fish so that its deadly neurotoxins are not part of the meal.

Some toxins are retained and concentrated in organisms

The physical and chemical properties of a compound are important for its retention within a biological system. If a compound can dissolve in water (that is, it is **hydrophilic**), it may be quickly metabolized (thus detoxified) because it is easily accessible to the wide variety of enzymes that can break down complex molecules in food.

In addition to being broken down or metabolized, many hydrophilic compounds can be filtered out of the blood by the kidney, and therefore they do not accumulate in the body. The kidney reabsorbs biologically important sugars, amino acids, and minerals into the body's circulation; everything else is excreted into the urine. (That is why urine tests are used to detect illegal drug use by athletes and other individuals.) However, some potentially dangerous hydrophilic compounds are retained by the body because they slip into a natural biological slot even though they cannot perform the function of the molecule that normally occupies that slot. Such molecules can accumulate in the body and disrupt normal functions. An example is lead, which can replace iron in blood and calcium in bone.

Compounds that dissolve in oils but separate from water are termed **lipophilic** or hydrophobic. They are usually metabolized more slowly than hydrophilic compounds, and they are often stored in the body for a long time because they dissolve in the lipids in membranes and adipose tissues. Because they are retained in lipids, lipophilic compounds can accumulate in the body and reach very high concentrations. Even compounds that are beneficial or neutral to the body at low concentrations, such as fat-soluble vitamins, can become toxic at high levels.

Toxins can become concentrated in food chains

Lipophilic compounds can also **bioaccumulate**; that is, the compounds can become more and more concentrated as we move up a food chain.* For example, a lipophilic pesticide can wash off of a farmer's field into a local pond or stream. At this concentration the pesticide may not make the water dangerous to drink. However, algae (the autotrophs at the base of the food chain) and small organisms like copepods will pick it up, and the pesticide becomes a little more concentrated within these organism than it was in the water. In the next link of the food chain, a small invertebrate may eat the algae and copepods, transferring all of the pesticide contained within this food to its own body. The invertebrate now contains levels of pesticides

*See Chapter 53 for a discussion of the ecology of food chains and food webs.

47.23 DDT Affects Bird Eggs Before its use was banned, the bioaccumulation of DDT in birds resulted in severe thinning of the eggshells in many species, with resulting population declines. This brown pelican egg cracked open long before the embryo inside was ready to hatch.

greater than the level found in the water. Small fish eat the invertebrates, and these fish are eaten by larger fish. Eventually a bear or a human or a bird of prey—one of the predators at the top of the food chain—eats the large fish and its accumulated toxins.

Thus the pesticide load is passed along the links of the food chain, from individual to individual, growing increasingly concentrated in the tissues of each higher consumer. By the time a top predator feeds fish from this pond or stream to its young, the pesticide may be concentrated thousands or millions of times above the level found in the water.

Many species at the top of the food chain show high levels of pesticides and other synthetic chemicals in their tissues. These bioaccumulated toxins may be responsible for high rates of cancer and infertility in some wildlife populations. A well-documented case was the effect of the pesticide DDT on bird populations. The bioaccumulated DDT caused extreme thinning of the eggshells and therefore high mortality of embryos (Figure 47.23). As a result, many species of predatory birds, such as ospreys, eagles, and falcons, became endangered.

Since the banning of DDT use, these species have been recovering. Long-lived species like eagles or bears are particularly at risk for heavy **body burdens** and health effects from such a pesticide because they have many years to accumulate it. Scientists and policy makers are now setting standards for and researching alternatives to such pesticides to decrease the amount of human-synthesized compounds bioaccumulating within natural systems. Since humans are at the top of the food chain, eating at all levels, we must consider our own exposure to compounds that bioaccumulate.

The body cannot metabolize many synthetic toxins

How does the way that the body handles synthetic chemicals differ from how it handles natural chemicals? In most cases, the detoxification systems that metabolize natural chemicals can also metabolize synthetic chemicals, breaking them apart and eliminating them through the urine. Enzymes called **cytochrome P450s** are responsible for much of this detoxification.

P450s are less specific in their abilities to bind substrates than are most enzymes. Thus, each P450 can catalyze reactions with a wide range of compounds, and there are many P450s. Phase I P450s make small chemical changes to the substrate, such as adding an —OH group or an —SO$_3$ group, which prepares the substrate for a second reaction. Phase II P450s use the —OH or —SO$_3$ group to attach a hydrophilic group onto the substrate, which facilitates the elimination of that substrate from the body. Few natural compounds can escape the P450s, even when the body encounters them for the first time.

Some synthetic chemicals, however, fall outside the range of structures that P450s and other enzymes can metabolize. When such chemicals are lipophilic, they bioaccumulate, and any biological effect they have is greatly magnified. If a synthetic chemical that cannot be metabolized is structurally similar to a hormone, that synthetic chemical may activate the hormonal signaling pathway within target cells. Whereas the natural hormone signal can be turned off, the synthetic hormone cannot be turned off and control of function is lost.

One example of a class of synthetic chemicals that appear to mimic hormones in animals is the polychlorinated biphenyls, **PCBs**. PCBs were produced extensively for use as an insulating fluid in electrical transformers from the 1930s until recently. PCBs are chemically stable, lipophilic, and now found throughout the environment. They have been shown to bioaccumulate, reaching dangerously high levels in fish from contaminated waters like the Great Lakes.

The biological response to exposure to PCBs in the diet varies among species; in rhesus monkeys PCBs altered reproductive cycles, reduced weight gain in infants, depressed immune-system responsiveness, and increased the incidence of death in developing embryos. In communities around the Great Lakes, studies have indicated cognitive impairment in children of mothers with a high body burden of PCBs, probably from eating fish caught in the Great Lakes. In studies of animals and of humans that had been exposed accidentally to high levels of PCBs, even short-term effects of PCB exposure, such as hair loss and acne, were slow to reverse, lasting from several months to a year.

The PCB and the DDT cases are now clear, but it is usually difficult to make a causal connection between a toxin in the environment and specific health effects in a population. Environmental toxicologists must be able to study large populations, use powerful statistics, and do controlled laboratory studies to obtain evidence that will support policy changes to stop and reverse effects of human-made environmental toxins.

Summary of "Animal Nutrition"

Nutrient Requirements

• Autotrophic organisms can use the energy of sunlight to synthesize structural molecules from inorganic materials and to run biochemical processes. Heterotrophic animals must derive energy and structural building blocks from food and therefore ultimately from autotrophs.
• Carbohydrates, fats, and proteins in food supply animals with metabolic energy and carbon skeletons for biosynthesis.
• A measure of the energy content of food is the calorie. Excess caloric intake is stored as glycogen and fat. Overnutrition in humans can be a serious health hazard. An animal with insufficient caloric intake is undernourished and must metabolize its own carbohydrate, fat, and finally its own protein for energy.
• In addition to supplying energy, for many animals food provides essential carbon skeletons that they cannot synthesize, as well as mineral nutrients and vitamins. **Review Figure 47.4**
• Humans require eight essential amino acids in their diet. All are available in milk, eggs, or meat, but not in all vegetables. Thus, vegetarians must eat a mix of foods. **Review Figure 47.5**
• Malnutrition results when any essential nutrient is lacking from the diet.
• Different animals need mineral elements in different amounts. Macronutrients, such as calcium, phosphorus, sodium chloride, and iron, are needed in large amounts. Micronutrients, such as copper, magnesium, and zinc, are needed in trace amounts. **Review Table 47.1**
• Vitamins are organic molecules that must be obtained in food. The B vitamins and vitamin C are water-soluble; vitamins A, D, E, and K are fat-soluble. They have diverse functions. **Review Table 47.2**

Adaptations for Feeding

• Animals can be characterized by how they acquire nutrition: Saprotrophs and detritivores depend on dead organic matter, filter feeders strain the environment for small food items, herbivores eat plants, and carnivores eat animals.
• Behavioral and anatomical adaptations reflect feeding types. In vertebrates, teeth have evolved to match diet. **Review Figure 47.7**

Digestion

• In digestion, complex food molecules are broken down into units that can be absorbed and utilized by cells. In most animals, digestion is extracellular and external to the body, taking place in a tubular gut.
• Regions of guts are specialized for different digestive functions, such as physical or enzymatic breakdown or absorption. Absorptive areas are characterized by a large surface area. **Review Figures 47.8, 47.9**

• Hydrolytic enzymes break down proteins, carbohydrates, and fats into their monomeric units. To prevent the organism itself from being digested, these enzymes are released first as inactive zymogens, which become activated when secreted into the gut.

Structure and Function of the Vertebrate Gut

• The cells and tissues of the vertebrate gut are organized in the same way throughout its length. The innermost tissue layer, the mucosa, is the secretory and digestive surface. Just outside the mucosa is the submucosa, which contains secretory cells and glands, blood vessels, nerves, and lymph vessels. External to the submucosa are two smooth-muscle layers (circular and longitudinal) responsible for the movement of food through the gut. Between the two muscle layers is a nerve network that controls the movements of the gut. **Review Figure 47.10**

• When food is pushed to the back of the mouth, a swallowing reflex is initiated that pushes food into the esophagus. From that point on, the circular muscle progressively contracts behind the bolus of food while it relaxes anterior to the bolus of food, and the food is thus pushed along the gut in these waves of peristalsis. Sphincters block the gut at certain locations, but they relax as a wave of peristalsis approaches. **Review Figure 47.11**

• Enzymatic digestion begins in the mouth, where amylase is secreted with the saliva. Protein digestion begins in the stomach with pepsin and HCl secreted by the stomach mucosa. The mucosa also secretes mucus, which prevents the digestive enzymes from eating the tissues of the gut. **Review Figure 47.12**

• In the first region of the small intestine, the duodenum, pancreatic enzymes do most of the digestion of the food. Bile from the liver and gallbladder assist in the digestion of fats by emulsifying them. Bicarbonate ions from the ducts of the pancreas neutralize the pH of the chyme entering the small intestine from the stomach to produce an environment conducive to the actions of pancreatic enzymes. **Review Figures 47.14, 47.15**

• The site of the final enzymatic cleavage of peptides and disaccharides is the surface of the cells of the intestinal mucosa. Amino acids, monosaccharides, and many inorganic ions are absorbed by the microvilli of the mucosal cells. In many cases specific transport molecules exist in the membranes of these cells to facilitate transport of nutrients into the cells. Sodium cotransport is a common mechanism for actively absorbing molecules and ions.

• Fats are absorbed mostly as monoglycerides and fatty acids, which are the product of lipase action on triglycerides in food. These products of fat digestion dissolve in membranes of mucosal cells and are then resynthesized into triglycerides within the cells. The triglycerides are combined with cholesterol and coated with protein to form chylomicrons, which pass out of the mucosal cells and into lymphatic vessels in the submucosa. **Review Figure 47.16**

• Water and ions are absorbed in the large intestine so that waste matter is consolidated into feces that are periodically excreted.

• Especially in herbivores such as rabbits and ruminants, some compartments of the gut may have large populations of microorganisms that aid in digesting molecules that otherwise would be indigestible to the host. **Review Figure 47.17**

Control and Regulation of Digestion

• The processes of digestion are coordinated and controlled by neural and hormonal mechanisms. Whereas salivation and swallowing are strictly neural reflexes, actions of the stomach and small intestine are largely controlled by the hormones gastrin, secretin, and cholecystokinin. **Review Figure 47.18**

Control and Regulation of Fuel Metabolism

• The liver interconverts fuel molecules and plays a central role in directing their traffic. When food is being absorbed from the gut, the liver takes up and stores fats and carbohydrates (converting monosaccharides to glycogen or fat). The liver also takes up amino acids and uses them to produce plasma proteins.

• Fats and cholesterol are shipped out of the liver as low-density lipoproteins. High-density lipoproteins act as acceptors of cholesterol and are believed to bring fats and cholesterol back to the liver.

• Fuel metabolism during the absorptive period is controlled largely by the hormone insulin, which promotes glucose uptake and utilization by most cells of the body, as well as fat synthesis in adipose tissue. During the postabsorptive period, the lack of insulin blocks the uptake and utilization of glucose by most cells of the body except neurons. If blood glucose levels fall, the hormone glucagon is secreted, stimulating the liver to break down glycogen to release glucose to the blood. **Review Figures 47.19, 47.20**

Nutrient Deficiency Diseases

• Lack of essential nutrients causes deficiency diseases such as scurvy (lack of vitamin C), kwashiorkor (lack of protein), and anemia (lack of iron or vitamin B_{12}). **Review Table 47.2**

Toxic Compounds in Food

• Even natural foods can contain toxic compounds in addition to nutrients—a problem worsened by human activities such as the use of pesticides and the release of pollutants into the environment.

• An organism can accumulate toxic compounds in its body, especially if those compounds are lipid-soluble or take the structural place of a natural molecule.

• Toxins such as PCBs and DDT that accumulate in the body of a prey are transferred and further concentrated in the bodies of predators. This bioaccumulation produces high concentrations of toxins in animals high up the food chain.

Self-Quiz

1. Most of the metabolic energy that a bird requires for a long-distance migratory flight is stored as
   a. glycogen.
   b. fat.
   c. protein.
   d. carbohydrate.
   e. ATP.

2. Which statement about essential amino acids is true?
   a. They are not found in vegetarian diets.
   b. They are stored by the body for the times when they are needed.
   c. Without them, one is undernourished.
   d. All animals require the same ones.
   e. Humans can acquire all of theirs by eating milk, eggs, and meat.

3. Which statement about vitamins is true?
   a. They are essential inorganic nutrients.
   b. They are required in larger amounts than are essential amino acids.
   c. Many serve as coenzymes.
   d. Vitamin D can be acquired only by eating meat or dairy products.
   e. When vitamin C is eaten in large quantities, the excess is stored in fat for later use.

4. The digestive enzymes of the small intestine
   a. do not function best at a low pH.
   b. are produced and released in response to circulating secretin.
   c. are produced and released under neural control.
   d. are all secreted by the pancreas.
   e. are all activated by an acidic environment.

5. Which statement about nutrient absorption across the gut epithelium is true?
   a. Carbohydrates are absorbed as disaccharides.
   b. Fats are absorbed as fatty acids and monoglycerides.
   c. Amino acids move across only by diffusion.
   d. Bile salts transport fats across.
   e. Most nutrients are absorbed in the duodenum.

6. Chylomicrons are like the tiny particles of dietary fat in the lumen of the small intestine in that
   a. both are coated with bile salts.
   b. both are lipid-soluble.
   c. both travel in lacteals.
   d. both contain triglyceride.
   e. both are coated with lipoproteins.

7. Microbial fermentation in the gut of a cow
   a. produces fatty acids as a major nutrient for the cow.
   b. occurs in specialized regions of the small intestine.
   c. occurs in the cecum, from which food is regurgitated, chewed again, and swallowed into the true stomach.
   d. produces methane as a major nutrient.
   e. is possible because the stomach wall does not secrete hydrochloric acid.

8. Which of the following is stimulated by cholecystokinin?
   a. Stomach motility
   b. Release of bile
   c. Secretion of hydrochloric acid
   d. Secretion of bicarbonate ions
   e. Secretion of mucus

9. During the absorptive period
   a. breakdown of glycogen supplies glucose to the blood.
   b. glucagon secretion is high.
   c. the number of circulating lipoproteins is low.
   d. glucose is the major metabolic fuel.
   e. the synthesis of fats and glycogen in muscle is inhibited.

10. During the postabsorptive period
    a. glucose is the major metabolic fuel.
    b. glucagon stimulates the liver to produce glycogen.
    c. insulin facilitates the uptake of glucose by brain cells.
    d. fatty acids constitute the major metabolic fuel.
    e. liver functions slow down because of low insulin levels.

Applying Concepts

1. From what you have learned about nutrition in this chapter, discuss some of the problems with crash or fad diets. What should one take into account when considering or planning a diet aimed at weight reduction?

2. The digestive tract must move food slowly enough to enable digestion and absorption but fast enough to supply the animal's energetic needs. Describe controls that speed up and slow down the activities of the digestive tract.

3. Describe the role of the liver in the homeostasis of blood glucose. What are the controlling factors?

4. Why is obstruction of the common bile duct so serious? Consider in your answer the multiple functions of the pancreas and the way in which digestive enzymes are processed.

5. Trace the history of a fatty acid molecule from being on a piece of buttered toast to being in a plaque on a coronary artery. What possible forms and structures could it have passed through in the body? Describe a direct and an indirect route it could have taken.

Readings

Atkinson, M. A. and N. K. MacLaren. 1990. "What Causes Diabetes?" *Scientific American*, July. A look at how the body's own immune system causes this serious disease that destroys the ability to regulate fuel metabolism.

Brown, M. S. and J. L. Goldstein. 1984. "How LDL Receptors Influence Cholesterol and Atherosclerosis." *Scientific American*, November. A detailed discussion of what does and does not happen to the cholesterol and fatty acids we consume.

Davis, D. L. and H. L. Bradlow. 1995. "Can Environmental Estrogens Cause Breast Cancer?" *Scientific American*, October. An exploration into one example of the kind of problem that confronts enviromental toxicologists: Can estrogen-like compounds in pesticides be causing disease in humans?

Lienhard, G. E., J. W. Slot, D. E. James and M. M. Mueckler. 1992. "How Cells Absorb Glucose." *Scientific American*, January. A discussion of research on the glucose transporter and how it is regulated by insulin.

Moog, F. 1981. "The Lining of the Small Intestine." *Scientific American*, November. A description of how nutrients are absorbed.

Sanderson, S. L. and R. Wassersug. 1990. "Suspension-Feeding Vertebrates." *Scientific American*, March. A look at how some animals exploit an abundant food resource by filtering small particles out of the water.

Scrimshaw, N. S. 1991. "Iron Deficiency." *Scientific American*, October. A discussion of one of the most common dietary deficiencies in humans and its consequences.

Uvnas-Moberg, K. 1989. "The Gastrointestinal Tract in Growth and Reproduction." *Scientific American*, July. A concise treatment of hormonal controls of digestive-tract function and how they change to accommodate the special needs of pregnancy and lactation.

Weindruch, R. 1996. "Caloric Restriction and Aging." *Scientific American*, January. A discussion of whether humans can live longer on a well-balanced, low-calorie diet, as has been shown for many other animals.

Chapter 48

Salt and Water Balance and Nitrogen Excretion

Water Is Essential for Life
During the dry season on the African savanna, animals must congregate around the scarce sources of water.

*B*lood, sweat, and tears taste salty, like seawater. Blood is largely made up of extracellular fluid; sweat and tears are derived from extracellular fluid. As we learned in Chapter 37, the extracellular fluid provides the proper physical environment, as well as nutrients and waste removal, for all cells of multicellular animals. A very important feature of extracellular fluid is its salt and water composition.

Life evolved in the seas, and seawater is the extracellular environment for cells of the simplest marine animals. In more complex marine animals, extracellular fluids are separate from but have a composition similar to that of seawater. Animals that left the marine environment had to carry with them their own internal sea in the form of extracellular fluids with appropriate salt and water composition.

The availability of salts and water in a particular environment is a critical characteristic that determines which organisms can live there, in what numbers, and when. Aquatic animals that have invaded fresh water must acquire and conserve salts for their extracellular environment. If their extracellular fluids were fresh water, their cells would swell and burst. Conversely, animals that invaded terrestrial environments have to conserve and replenish water to sustain their extracellular fluids. Otherwise, their cells would shrivel and die.

Many species living in habitats that are seasonally dry migrate long distances to find water. During the dry season on the African plains, predators and prey alike congregate at scarce water holes. However, the availability of salts and water in the external environment is only part of the story. Animals must regulate the specific ionic composition of their extracellular fluids, and they must be able to eliminate waste products that accumulate in their extracellular fluids as a result of the metabolism of compounds that contain nitrogen.

In this chapter we will learn about the salt and water balance problems and excretory problems that animals encounter in different habitats, and about different adaptations for the excretion of the wastes produced by nitrogen metabolism. Using invertebrate examples, we will explore the basic mechanisms used in salt and water balance and nitrogen excretion in all animals: the filtration of the extracellular fluids, the active transport of solutes, and the osmotic movement of water.

The basic structure that accomplishes all of these tasks in vertebrates is the nephron. We will see how the nephron evolved from a structure to excrete water to the basic structural unit of the mammalian kidney, where its primary function is to conserve water by producing a concentrated urine. The chapter will end with discussions of the mechanisms that control and regulate salt and water balance in mammals.

Extracellular Fluids and Water Balance

The overarching theme of Chapters 37 through 48 is the concept that different organs and organ systems contribute to homeostasis of the body by controlling a particular feature of the extracellular fluids. The extracellular fluids in turn serve all the needs of the body's cells. In addition to supplying cells with oxygen and nutrients and carrying away waste products, these extracellular fluids determine the water balance of the cells.

To understand what is meant by **water balance**, recall that cell plasma membranes are permeable to water and that the movement of water across membranes depends on differences in osmotic potential. (You may find it useful to review the discussion of osmosis in Chapter 5.) If the osmotic potential of the extracellular fluid is less negative (that is, the fluid contains fewer solutes) than that of the intracellular fluids, water moves into the cells, causing them to swell and possibly burst. If the osmotic potential of the extracellular fluid is more negative (the fluid contains more solutes) than that of the intracellular fluids, the cells lose water and shrink. The osmotic potential of the extracellular environment determines both the volume and the osmotic potential of the intracellular environment.

Excretory organs regulate osmotic potential

Excretory systems consist of the organs that help regulate the osmotic potential and the volume of the extracellular fluids. In addition, excretory systems regulate the composition of the extracellular fluids by excreting molecules that are in excess (such as NaCl when we eat lots of salty popcorn) and conserving those that are valuable or in short supply (such as glucose and amino acids). In terrestrial organisms, excre-

tory systems also eliminate the toxic waste products of nitrogen metabolism.

The duties of the excretory system of a particular species depend on the environment in which that species lives. In this chapter we will examine excretory systems that maintain salt and water balance and eliminate nitrogen in marine, freshwater, and terrestrial habitats. In spite of the evolutionary diversity of the anatomical and physiological details, all these systems obey a common rule and employ common mechanisms. The common rule is that *there is no active transport of water*; water must be moved either by pressure or by a difference in osmotic potential (see Chapter 5).

Excretory systems use filtration, secretion, and reabsorption

The common mechanisms used by excretory systems are filtration of the extracellular fluids and processing of the filtrate through the active transport of solute molecules into and out of the filtrate. Most excretory organs consist of systems of tubules that receive a filtrate of the extracellular fluids. As this filtrate flows through the tubules, its composition changes, resulting in the fluid waste product **urine**, which is excreted. The extracellular fluid enters the excretory tubules by **filtration**, and its composition is changed by processes of active **secretion** and **reabsorption** of specific solute molecules by the cells of the tubules. These three mechanisms—filtration, secretion, and reabsorption—are used both in systems that excrete water and conserve salts and in systems that do the opposite, conserve water and excrete salts.

Distinguishing Environments and Animals in Terms of Salt and Water

We think of marine and freshwater environments as being distinctly different, one salty and the other not. In reality, aqueous environments grade continuously from fresh to extremely salty. Consider a place where a river enters the sea through a bay or a marsh. Aqueous environments within that bay or marsh range in salinity (salt content) from the fresh water of the river to the open sea. Evaporating tide pools can become much saltier than the sea. Animals live in all these environments.

Most marine invertebrates adjust to a wide range of environmental salinities by allowing their body fluids to have the same osmotic potential as the environment. Such animals are called **osmoconformers**. But there are limits to osmoconformity. No animal could have the same osmotic potential as fresh water and survive; nor could animals survive with internal salt concentrations as high as those that may be reached in an evaporating tide pool. Such concentrations cause proteins to denature.

(a) *Artemia* sp.

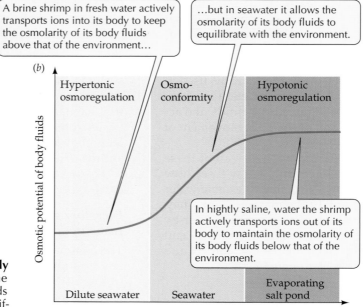

A brine shrimp in fresh water actively transports ions into its body to keep the osmolarity of its body fluids above that of the environment…

…but in seawater it allows the osmolarity of its body fluids to equilibrate with the environment.

In hightly saline, water the shrimp actively transports ions out of its body to maintain the osmolarity of its body fluids below that of the environment.

48.1 Environments Can Vary Greatly in Salt Concentration (a) Brine shrimp live in evaporating salt ponds and are exposed to a range of very different salinities. (b) Animals like the brine shrimp that live at the extremes of environmental salinities display osmoregulatory abilities. They become hypertonic osmoregulators in very dilute water, or hypotonic osmoregulators in very saline water.

Animals that maintain an osmotic potential of their internal fluids that is different from that of their environment are called **osmoregulators**. Even animals that osmoconform over a wide range of osmotic potentials must osmoregulate at the extremes of environmental salinity. To osmoregulate in fresh water, animals must excrete the water that invades their bodies by osmosis, but while doing so they must conserve solutes; hence they produce large amounts of dilute urine. To osmoregulate in salt water, animals must conserve water and excrete salts; thus they tend to produce small amounts of urine.

The brine shrimp *Artemia* (Figure 48.1a) is adapted to live in environments of almost any salinity. *Artemia* are found in huge numbers in the most saline environments possible, such as Great Salt Lake in Utah or coastal evaporation ponds where salt is obtained for commercial purposes (see Figure 25.25). The salinity of such water reaches 300 g/l (normal seawater contains about (35 g/l). At these high environmental salinities, *Artemia* is an exceptionally effective **hypotonic osmoregulator**, keeping the osmotic potential of its body fluids well below that of the water in which it is living. Very few organisms can survive in the crystallizing brine in which *Artemia* thrives. The main mechanism this small crustacean uses for osmoregulation is the active transport of NaCl across its gill membranes.

Artemia cannot survive long in fresh water, but they can live in very dilute seawater, in which they maintain the osmotic potential of their body fluids above

the osmotic potential of the environment. Under these conditions, *Artemia* is a **hypertonic osmoregulator**; that is, it regulates the concentration (the osmolarity) of its body fluids so that they are more concentrated than (hypertonic to) the environment (Figure 48.1b).

Osmoconformers can be **ionic conformers**, allowing the ionic composition, as well as the osmolarity, of their body fluids to match that of the environment. Most osmoconformers, however, are **ionic regulators** to some degree: They employ active transport mechanisms to maintain specific ions in their body fluids at concentrations different from those in the environment.

The terrestrial environment presents entirely different problems for salt and water balance. Because the terrestrial environment is extremely desiccating (drying), most terrestrial animals must conserve water. Exceptions are animals such as muskrats and beavers that live in water.

Terrestrial animals obtain their salts from their food. Because plants generally have low concentrations of sodium, most herbivores must be very effective in conserving sodium ions. As we mentioned in Chapter 47, some terrestrial herbivores travel long distances to naturally occurring salt licks to supplement their dietary intake of sodium. By contrast, birds that feed on marine animals must excrete the large excess of sodium they ingest with their food. Their **nasal salt glands** excrete a concentrated solution of sodium chloride via a duct that empties into the nasal cavity. Birds, such as penguins and seagulls, that have nasal salt glands can be seen frequently sneezing or shaking their heads to get rid of the very salty droplets that form (Figure 48.2).

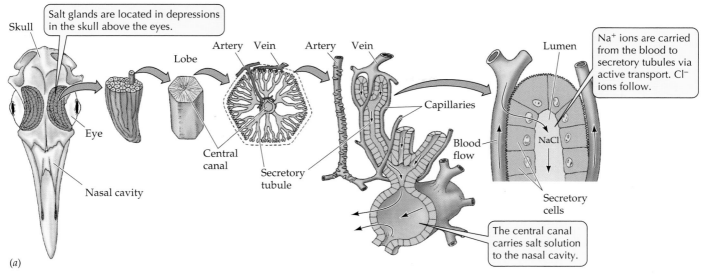

Salt glands are located in depressions in the skull above the eyes.

Skull

Eye

Nasal cavity

Lobe

Artery Vein

Central canal

Secretory tubule

Artery Vein

Capillaries

Blood flow

Lumen

Na+ ions are carried from the blood to secretory tubules via active transport. Cl⁻ ions follow.

NaCl

Secretory cells

The central canal carries salt solution to the nasal cavity.

(a)

48.2 Nasal Glands Excrete Excess Salt (a) Marine birds have nasal salt glands adapted to excrete the excess salt from the seawater they consume with their food. (b) This giant petrel has returned from a feeding trip at sea and is excreting salt through its nasal salt gland. Note the drop of excreted salt at the tip of the bird's beak.

Salty secretion runs out of nasal cavity. The secretions that have dripped onto the ground have left a crust of salt.

(b) Macronectus gigantus

Excreting Nitrogen as Ammonia, Uric Acid, or Urea

The end products of the metabolism of carbohydrates and fats are water (H_2O) and carbon dioxide (CO_2); they present no problems for excretion. Proteins and nucleic acids, however, contain nitrogen in addition to carbon, hydrogen, and oxygen. The metabolism of proteins and nucleic acids produces nitrogenous waste in addition to H_2O and CO_2. Most of that waste is ammonia (NH_3). Ammonia is highly toxic, and it must be excreted continuously to prevent accumulation or it must be detoxified by conversion into other molecules for excretion. Those molecules are principally **urea** and **uric acid** (Figure 48.3).

Ammonia excretion is relatively simple for aquatic animals. Ammonia diffuses and is highly soluble in water. Animals that breathe water continuously lose ammonia from their blood to the environment by diffusion across their gill membranes. Animals that excrete nitrogen as ammonia are called **ammonotelic**; they include aquatic invertebrates and bony fishes. Crocodiles and amphibian tadpoles are also ammonotelic.

Ammonia is a more dangerous metabolite for terrestrial animals that have limited access to water. In mammals, ammonia is lethal when it reaches only 5 mg/100 ml of blood. Therefore, terrestrial (and some aquatic) animals convert ammonia into either urea or uric acid.

Ureotelic animals, such as mammals, amphibians, and cartilaginous fishes (sharks and rays), excrete urea as their principal nitrogenous waste product.

Urea is quite soluble in water, but excretion of urea solutions at low concentrations could result in a large loss of water that many terrestrial animals can ill afford. Later in the chapter we will see that mammals have evolved excretory systems that produce urine, which contains urea and is hypertonic to the body, thereby conserving water while excreting the urea. The cartilaginous fishes are another story. These marine species maintain their body fluids hypertonic to the marine environment by retaining high concentrations of urea. Because water moves into their bodies by osmosis and must be excreted, water conservation is not a problem for them.

Terrestrial animals that conserve water by excreting nitrogenous wastes as uric acid are called **uricotelic**. Included among uricotelic animals are insects, reptiles, birds, and some amphibians. Uric acid is very insoluble in water and is excreted as a semisolid (for exam-

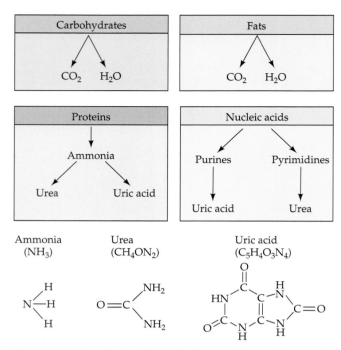

Ammonia
(NH₃)

Urea
(CH₄ON₂)

Uric acid
(C₅H₄O₃N₄)

48.3 Waste Products of Metabolism Whereas the metabolism of carbohydrates and fats yields only water and carbon dioxide, the metabolism of proteins and nucleic acids produces the nitrogenous wastes ammonia, uric acid, and urea. The metabolism of nucleic acids begins with their breakdown into their constituent bases—pyrimidines and purines (see Chapter 3).

ple, the whitish material in bird droppings). Therefore, a uricotelic animal loses very little water as it disposes of its nitrogenous waste.

Although we can classify animals, on the basis of their major nitrogenous waste product, as ammonotelic, ureotelic, or uricotelic, most species produce more than one nitrogenous waste. Humans are ureotelic, yet we also excrete uric acid and ammonia, as anyone who has changed diapers knows. (Actually, most of the ammonia in diapers is produced by the bacterial breakdown of urea. The bacterium that performs this reaction was first isolated by a microbiologist from a diaper of his child.) The uric acid in human urine comes largely from the metabolism of nucleic acids and of caffeine. In the condition known as gout, uric acid levels in the body fluids increase, and uric acid precipitates in the joints and elsewhere, causing swelling and pain.

In some species, different developmental forms live in quite different habitats and have different forms of nitrogen excretion. Tadpoles of frogs and toads, for example, excrete ammonia across their gill membranes, but when they develop into adult frogs or toads they generally excrete urea. Some adult frogs and toads that live in arid habitats excrete uric acid. These examples show the considerable evolutionary flexibility in how nitrogenous wastes are excreted.

The Diverse Excretory Systems of Invertebrates

Most marine invertebrates are osmoconformers, so they have few adaptations for salt and water balance other than the active transport of specific ions. To excrete nitrogen, they can passively lose ammonia by diffusion to the seawater. Freshwater and terrestrial invertebrates, however, display a variety of fascinating adaptations for maintaining salt and water balance and excreting nitrogen. Although diverse, all these adaptations are based on the same basic principles: filtration of body fluids and active secretion and reabsorption of specific ions. We will examine three such adaptations: protonephridia, metanephridia, and Malpighian tubules.

Protonephridia excrete water and conserve salts

Many flatworms, such as *Planaria*, live in fresh water and excrete water through an elaborate network of tubules running throughout their bodies. The tubules end in **flame cells**, so called because each one has a tuft of cilia beating inside the tubule, giving the appearance of a flickering flame (Figure 48.4). Flame cells and tubules together are called **protonephridia** (singular protonephridium; from the Greek *proto*, "before," and *nephros*, "kidney"). The beating of the cilia causes fluid in the tubules to flow toward the excretory pore of the animal. As it leaves the planarian, this fluid is hypotonic to the animal's internal body fluids.

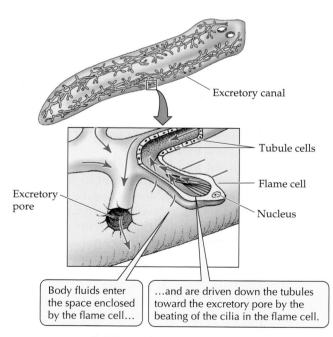

Body fluids enter the space enclosed by the flame cell… …and are driven down the tubules toward the excretory pore by the beating of the cilia in the flame cell.

48.4 Protonephridia in Flatworms The protonephridia of the flatworm *Planaria* consist of tubules ending in flame cells. The tubule cells modify the composition of the fluid passing through them.

48.5 Metanephridia in Earthworms
The metanephridia of annelids are arranged segmentally. The cross section (left) shows a pair of metanephridia. Longitudinal sections (right) show only one metanephridium of the two in each segment.

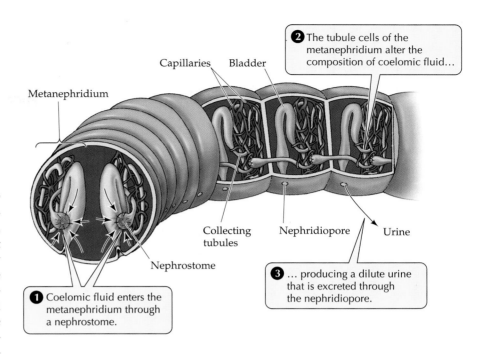

Capillaries Bladder

Metanephridium

2 The tubule cells of the metanephridium alter the composition of coelomic fluid...

Collecting tubules Nephridiopore Urine

Nephrostome

1 Coelomic fluid enters the metanephridium through a nephrostome.

3 ... producing a dilute urine that is excreted through the nephridiopore.

Metanephridia process coelomic fluid

Filtration of body fluids and tubular processing of urine are highly developed processes in annelid worms, such as the earthworm. Recall that annelids have a fluid-filled body cavity called a coelom (see Figure 29.27) and a closed circulatory system through which blood is pumped under pressure (see Figure 46.3). The pressure causes the blood to be filtered across the thin, permeable capillary walls into the coelom. This process is called filtration because the cells and large protein molecules of the blood stay behind in the capillaries while water and small molecules leave the capillaries and enter the coelom. Where does this coelomic fluid go?

Each segment of the earthworm contains a pair of **metanephridia** (singular metanephridium; from the Greek *meta*, "akin to," and *nephros*, "kidney"). Each metanephridium begins in one segment as a ciliated, funnel-like opening in the coelom called a **nephrostome**, which leads into a tubule in the next segment. The tubule ends in a pore called the nephridiopore, which opens to the outside of the animal (Figure 48.5). Coelomic fluid enters the metanephridia through the nephrostomes. As the coelomic fluid passes through the tubules, the cells of the tubules actively reabsorb certain molecules from it and actively secrete other molecules into it. What leaves the animal through the nephridiopores is a hypotonic urine containing nitrogenous wastes, among other solutes.

In the metanephridium we see all the basic processes used in the excretory systems of vertebrates that will be discussed later in this chapter: filtration of the body fluids, and tubular processing of the filtrate by active secretion and reabsorption of ions and molecules.

Malpighian tubules are the excretory organs of insects

Insects have remarkable systems for excreting nitrogenous wastes with very little loss of water. These animals can live in the driest habitats on Earth. The insect excretory system consists of blind tubules (from 2 to more than 100) attached to the gut between the midgut and hindgut and projecting into the fluid-filled coelom (Figure 48.6). The cells of these **Malpighian tubules** actively transport uric acid, potassium ions, and sodium ions from the coelomic fluid into the tubules. As solutes are secreted into the tubules, water follows because of the difference in osmotic potential. The walls of the Malpighian tubules have muscle fibers whose contractions help move the contents of the tubules toward the hindgut.

In the hindgut the tubular fluid continues to change in composition. The contents of the hindgut are more acidic than the tubular fluids; as a result, the uric acid becomes less soluble and precipitates out of solution as it approaches and enters the rectum. The cells of the walls of the hindgut and rectum actively transport sodium and potassium ions from the gut contents back into the coelom. Because the uric acid molecules have precipitated out of solution, water is free to follow the reabsorbed salts back into the coelom through osmosis.

Remaining in the rectum are crystals of uric acid mixed with undigested food, and this dry matter is all that the insect eliminates. The Malpighian tubule system is a highly effective mechanism for excreting nitrogenous wastes and some salts without giving up any significant fraction of the animal's precious water supply.

The Nephron: The Basic Unit of Vertebrate Excretion

The major vertebrate organ for salt and water balance and nitrogen excretion is the **kidney**. The functional unit of the kidney is the **nephron**. Each human kidney

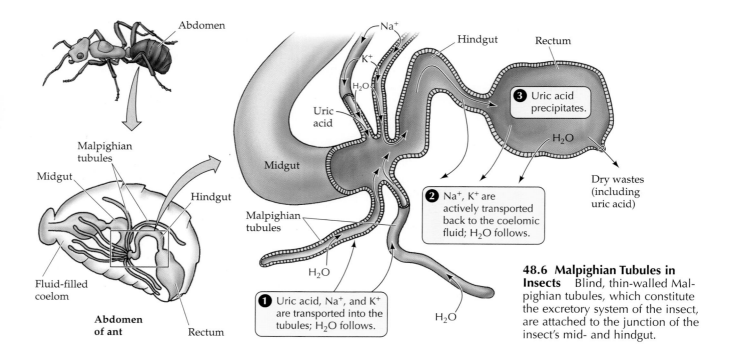

48.6 Malpighian Tubules in Insects Blind, thin-walled Malpighian tubules, which constitute the excretory system of the insect, are attached to the junction of the insect's mid- and hindgut.

Labels within figure 48.6:
- Abdomen
- Malpighian tubules
- Midgut
- Hindgut
- Fluid-filled coelom
- **Abdomen of ant**
- Rectum
- Na^+
- K^+
- H_2O
- Uric acid
- Midgut
- Hindgut
- Rectum
- ❸ Uric acid precipitates.
- H_2O
- Dry wastes (including uric acid)
- Malpighian tubules
- H_2O
- ❷ Na^+, K^+ are actively transported back to the coelomic fluid; H_2O follows.
- ❶ Uric acid, Na^+, and K^+ are transported into the tubules; H_2O follows.
- H_2O

has about 1 million nephrons. To understand how the kidney works, you must understand the structure and function of the nephron.

A remarkable fact about the kidneys of vertebrates is that in different species the same basic organ serves different needs. Whereas the kidneys of freshwater fishes excrete water, the kidneys of most mammals conserve water. To understand how the kidney can fulfill opposite functions in different animals, we need to examine structures of the nephron, their interactions with their immediate environment, and their functions.

Blood is filtered in the glomerulus

Each nephron has a vascular and a tubular component (Figure 48.7). The vascular component is unusual in that it consists of two capillary beds that lie between the arteriole that supplies it and the venule that drains it. The first capillary bed is a dense knot of

48.7 The Vertebrate Nephron The end of the renal tubule system envelops the glomerulus so that the filtrate from the glomerular capillaries enters the tubules.

Labels within figure 48.7:
- The nephron is a system of tubules closely associated with a system of blood vessels.
- Site of filtration (glomerulus)
- Site of tubular secretion and absorption
- Urine processing
- An efferent arteriole carries blood from the glomerulus.
- Podocytes
- Bowman's capsule receives H_2O and small molecules filtered from glomerular capillaries.
- The glomerulus, a knot of capillaries, is the site of blood filtration.
- Renal tubule cells alter composition of urine.
- Bowman's capsule
- An afferent arteriole supplies blood to the glomerulus.
- Peritubular capillaries carry away reabsorbed substances and bring materials to the tubules that will be secreted into the urine.
- The renal venule drains the peritubular capillaries.
- Blood out
- Urine
- The processed filtrate (urine) of the individual nephrons enters collecting ducts and is delivered to a common duct leaving the kidney.

very permeable vessels called the **glomerulus** (Figure 48.8*a*). Blood enters the glomerulus through an **afferent arteriole** and exits through an **efferent arteriole**. The efferent arteriole gives rise to the second set of capillaries, the **peritubular capillaries**, which surround the tubular component of the nephron (see Figure 48.7).

The tubular component of the nephron begins with **Bowman's capsule**, which encloses the glomerulus. The glomerulus appears to be pushed into Bowman's capsule much like a fist pushed into an inflated balloon. Together, the glomerulus and its surrounding Bowman's capsule are called the **renal corpuscle**. The cells of the capsule that come into direct contact with the glomerular capillaries are called **podocytes** (see Figure 48.7). These highly specialized cells have numerous armlike extensions, each with hundreds of fine, fingerlike projections. The podocytes wrap around the capillaries so that their fingerlike projections cover the capillaries completely (Figure 48.8*b* and *c*).

The glomerulus filters the blood to produce a tubular fluid that lacks cells and large molecules. The cells of the capillaries and the podocytes of Bowman's capsule participate in filtration. The walls of the capillaries have pores that allow water and small molecules to leave the capillary but that are too small to permit red blood cells and very large protein molecules to pass. Even smaller than the pores in the capillaries are the narrow slits between the fingerlike projections of the podocytes. The result of these anatomical adaptations is that water and small molecules pass from the capillary blood and enter the tubule of the nephron (Figure 48.8*d*), but red blood cells and proteins remain in the capillaries.

The force that drives filtration in the glomerulus is the pressure of the arterial blood. As in every other capillary bed, the pressure of the blood entering the permeable capillary causes the filtration of water and small molecules until the osmotic potential created by the large molecules remaining in the blood is sufficient to counter the outward flow of water. The glomerular filtration rate is high because glomerular capillary blood pressure is unusually high, and the capillaries of the glomerulus, along with their covering of podocytes, are much more permeable than other capillary beds in the body.

The tubules of the nephrons convert glomerular filtrate to urine

The composition of the fluid that is filtered into the tubules of the nephron, called **renal tubules** (see Figure 48.7), is similar to that of the blood plasma. This filtrate contains glucose, amino acids, ions, and nitrogenous wastes in the same concentrations as in the blood plasma, but it lacks the plasma proteins. As this fluid passes down the tubule, its composition

changes as the cells of the tubule actively reabsorb certain molecules from the tubular fluid and secrete other molecules into it. When the tubular fluid leaves the kidney as urine, its composition is very different from that of the original filtrate.

The function of the renal tubules is to control the composition of the urine by actively secreting and reabsorbing specific molecules. The peritubular capillaries serve the needs of the renal tubules by bringing to them the molecules to be secreted into the tubules and carrying away the molecules that are reabsorbed from the tubules back into the blood.

Both marine and terrestrial vertebrates must conserve water

If the vertebrate nephron evolved as a structure to excrete water while conserving salts and essential small molecules, how have vertebrates adapted to environments where water must be conserved and salts excreted? The answer to this question differs for each vertebrate group. Even among marine fishes, the adaptations of bony fishes are different from those of cartilaginous fishes.

MARINE BONY FISHES. Marine bony fishes cannot produce urine more concentrated than their body fluids, but they osmoregulate their body fluids to only one-fourth to one-third the osmotic potential of seawater. They prevent excessive loss of water by producing very little urine. Their urine production is low because their kidneys have fewer glomeruli than do the kidneys of freshwater fishes. In some species of marine bony fishes, the kidneys have no glomeruli. Even though the glomeruli are reduced or absent, renal tubules with closed ends are retained for active excretion of ions and certain molecules.

Marine bony fishes meet their water needs by drinking seawater, but this practice results in a large salt load. The fish handle salt loads by not absorbing divalent ions (such as Mg^{2+} or SO_4^{2-}) from their guts and by actively excreting monovalent ions (Na^+ and Cl^-) from gill membranes and from the renal tubules. Nitrogenous wastes are lost as ammonia from the gill membranes.

CARTILAGINOUS FISHES. Cartilaginous fishes are osmoconformers but not ion conformers. Unlike marine bony fishes, cartilaginous fishes convert nitrogenous waste to urea and retain large amounts of that urea in their body fluids so that their body fluids have the same osmotic concentration as seawater. In some cases these fluids are even slightly hypertonic to seawater, causing water to move into the body of the fish by osmosis. These species have adapted to a concentration of urea in the body fluids that would be fatal to other vertebrates.

(a)

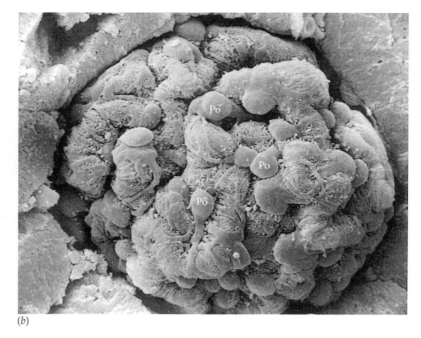

(b)

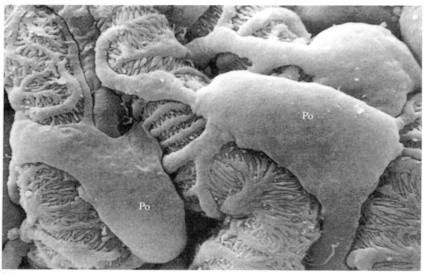

(c)

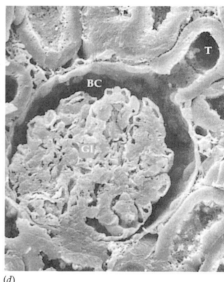

(d)

48.8 An SEM Tour of the Nephron   These scanning electron micrographs show the anatomical bases for kidney function. (*a*) When the blood vessels are filled with latex and all tissue etched away, we are left with a cast of the blood vessels in the kidney showing the knots of capillaries that are the glomeruli (Gl). Each glomerulus has an afferent and an efferent arteriole (Ar). Peritubular capillaries (Pt) are looser networks surrounding the tubules of the nephron. (*b*) In a live organism, the capillaries of the glomeruli are tightly wrapped by specialized tubule cells called podocytes (Po) derived from the cells of the inner wall of Bowman's capsule. Here and in parts (*c*) and (*d*) we are looking at the glomerulus from inside Bowman's capsule. (*c*) Each podocyte has hundreds of tiny, fingerlike projections that create filtration slits between them. Anything passing from the glomerular capillaries into the tubule of the nephron must pass through these slits. (*d*) Bowman's capsule (BC) surrounds the glomerulus (Gl), collects the filtrate, and funnels it into the tubule (T) of the nephron.

Sharks and rays have the problem of excreting the large amounts of salts that they take in with their food. They have several sites of active secretion of NaCl, but the major one is a salt-secreting **rectal gland**.

AMPHIBIANS.   Most amphibians live in or near fresh water and are limited to humid habitats when they venture from the water. Like freshwater fishes, typical amphibian species produce large amounts of dilute urine and conserve salts. Some amphibians, however, have adapted to habitats that require water conservation, and their adaptations are diverse.

Amphibians from very dry terrestrial environments have an important adaptation to prevent water loss: a

48.9 Waxy Frogs *Phyllomedusa* is a tree frog that lives in a seasonally dry and hot habitat. It reduces its evaporative water loss by secreting waxes and fats from skin glands and spreading these secretions all over its body.

reduction in the water permeability of their skin. Some secrete a waxy substance that they spread over the skin to waterproof it (Figure 48.9). Several species of frogs that live in arid regions of Australia have remarkable adaptations. These animals burrow deep in the ground and estivate during long dry periods. **Estivation** is a state of very low metabolic activity. When it rains, these frogs come out of estivation, feed, and reproduce. But their most interesting adaptation is that they have enormous urinary bladders. Before entering estivation, they fill their bladders with very dilute urine, which may make up one-third of their body weight. This dilute urine serves as a water reservoir that they use gradually during the long periods of estivation. Australian aboriginal peoples dig up estivating frogs as an emergency source of drinking water.

REPTILES. Reptiles occupy habitats ranging from aquatic to extremely hot and dry. Three major adaptations have freed the reptiles from maintaining the close association with water that is necessary for amphibians. First, reptiles do not need fresh water to reproduce, because they employ internal fertilization and lay eggs with shells that retard evaporative water loss. Second, they have scaly, dry skins that are much less permeable to water than is the skin of most amphibians. Third, they excrete nitrogenous wastes as uric acid solids, therefore losing little water in the process.

BIRDS. Birds have the same adaptations for water conservation as reptiles have: internal fertilization, shelled eggs, skin that retards water loss, and uric acid as the nitrogenous waste product. In addition, some birds can produce a urine that is hypertonic to their body fluids. This last ability is most developed in mammals.

The Mammalian Kidney

The ability of mammals and birds to produce urine that is hypertonic to their body fluids represents a major step in kidney evolution. In these species we see for the first time the kidney playing the major role in water conservation. A structure that originally evolved to excrete water has been converted to a structure to do the opposite, conserve water. To understand how this evolutionary switch occurred, we must examine the structure and function of the whole kidney.

The concentrating ability of the kidney depends on its anatomy

We will focus on humans as an example of the mammalian excretory system. Humans have two kidneys at the rear of the abdominal cavity (Figure 48.10). Each kidney releases the urine it produces into a tube (the **ureter**) that leads to the **urinary bladder**, where the urine is stored until it is excreted through the urethra. The **urethra** is a short tube that opens to the outside at the end of the penis in males or just anterior to the vagina in females.

Two sphincter muscles surrounding the base of the urethra control the timing of urination. One of these sphincters is a smooth muscle and is controlled by the autonomic nervous system. When the bladder is full, a spinal reflex relaxes this sphincter. This reflex is the only control of urination in infants, but the reflex gradually comes under the influence of higher centers in the nervous system as a child grows older. The other sphincter is a skeletal muscle and is controlled by the voluntary, or conscious, nervous system. When the bladder is *very* full, only serious concentration prevents urination.

ANATOMICAL REGIONS OF THE KIDNEY. The kidney is shaped like a kidney bean; when cut down its long axis and split open as a bean splits open, its important anatomical features are revealed (see Figure 48.10). The ureter and the **renal artery** and **renal vein** enter the kidney on its concave (punched-in) side. The ureter divides into several branches, the ends of which envelop projections of kidney tissue called **renal pyramids**. The renal pyramids make up the internal core, or **medulla**, of the kidney. The medulla is capped by a distinctly different tissue called the **cortex**. The renal artery and vein give rise to many arterioles and venules in the region between the cortex and the medulla.

FUNCTIONAL REGIONS OF THE RENAL TUBULES. The secret of the ability of the mammalian kidney to produce concentrated urine is in the relationship between the parts of the nephron (the functional unit of the kidney) and the anatomy of the kidney. Each human kidney con-

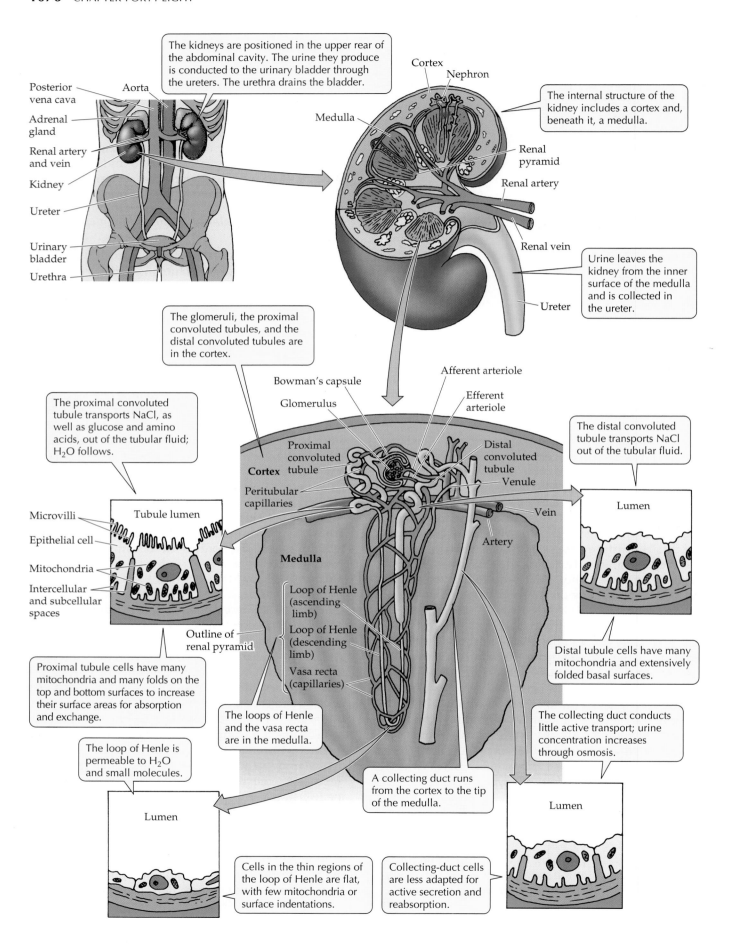

The kidneys are positioned in the upper rear of the abdominal cavity. The urine they produce is conducted to the urinary bladder through the ureters. The urethra drains the bladder.

Posterior vena cava
Aorta
Adrenal gland
Renal artery and vein
Kidney
Ureter
Urinary bladder
Urethra

Cortex
Nephron
Medulla

The internal structure of the kidney includes a cortex and, beneath it, a medulla.

Renal pyramid
Renal artery
Renal vein
Ureter

Urine leaves the kidney from the inner surface of the medulla and is collected in the ureter.

The glomeruli, the proximal convoluted tubules, and the distal convoluted tubules are in the cortex.

The proximal convoluted tubule transports NaCl, as well as glucose and amino acids, out of the tubular fluid; H_2O follows.

Bowman's capsule
Glomerulus
Afferent arteriole
Efferent arteriole

Proximal convoluted tubule
Distal convoluted tubule
Cortex
Venule

The distal convoluted tubule transports NaCl out of the tubular fluid.

Peritubular capillaries
Vein
Artery
Lumen

Microvilli
Epithelial cell
Mitochondria
Intercellular and subcellular spaces

Tubule lumen

Medulla

Proximal tubule cells have many mitochondria and many folds on the top and bottom surfaces to increase their surface areas for absorption and exchange.

Loop of Henle (ascending limb)
Loop of Henle (descending limb)
Vasa recta (capillaries)

Outline of renal pyramid

Distal tubule cells have many mitochondria and extensively folded basal surfaces.

The loops of Henle and the vasa recta are in the medulla.

The collecting duct conducts little active transport; urine concentration increases through osmosis.

The loop of Henle is permeable to H_2O and small molecules.

A collecting duct runs from the cortex to the tip of the medulla.

Lumen

Lumen

Cells in the thin regions of the loop of Henle are flat, with few mitochondria or surface indentations.

Collecting-duct cells are less adapted for active secretion and reabsorption.

tains about 1 million nephrons, and their organization within the kidney is very regular. All of the glomeruli are located in the cortex. The initial segment of the tubule of a nephron is called the **proximal convoluted tubule**—"proximal" because it is closest to the glomerulus and "convoluted" because it is twisted (see Figure 48.10). All the proximal convoluted tubules are also located in the cortex.

At a certain point the proximal tubule takes a dive directly down into the medulla. The portion of the tubule that is in the medulla is called the **loop of Henle**. It is called a loop because it runs straight down into the medulla, makes a hairpin turn, and comes straight back to the cortex. The ascending limb of the loop of Henle becomes the **distal convoluted tubule** when it reaches the cortex. The distal convoluted tubules of many nephrons join a common **collecting duct** in the cortex. The collecting ducts then run in parallel with the loops of Henle down through the medulla and empty into the ureter at the tips of the renal pyramids.

BLOOD FLOW IN THE KIDNEY. The organization of the blood vessels of the kidney closely parallels the organization of the nephrons (see Figure 48.10). Arterioles branch from the renal arteries and radiate into the cortex. An *afferent* arteriole carries blood to each glomerulus. Draining each glomerulus is an *efferent* arteriole that gives rise to the peritubular capillaries, which surround mostly the cortical portions of the tubules. A few peritubular capillaries run into the medulla in parallel with the loops of Henle and the collecting ducts. These capillaries are the **vasa recta**. All the peritubular capillaries from a nephron join back together into a venule that joins with venules from other nephrons and eventually leads to the renal vein, which takes blood from the kidney.

Remember that anything coming into the kidney comes through the renal artery, and everything that comes into the kidney must leave either through the renal vein or the ureter (there is also some drainage of lymph from the kidney, but it is minor). Only a small, selective percentage of everything that is filtered leaves the kidney in the urine. To understand kidney function, you must understand how most of the substances and water filtered from the blood in the glomerulus return to the extracellular fluids and to the venous blood draining the kidney.

◀ **48.10 The Human Excretory System** The structures of the cells of the renal tubules reflect the functions of the different tubule segments. Intercellular spaces and indentations at the basal end of the cells increase the area of cell contact with interstitial fluids.

The volume of glomerular filtration is greater than the volume of urine

Most of the water and solutes filtered in the glomerulus are reabsorbed and do not appear in the urine. We reach this conclusion by comparing the rate of filtration by the glomeruli with the rate of urine production. The kidneys receive about 20 percent of the blood pumped into arteries by the heart. The cardiac output of a human at rest is about 5 liters per minute, so the kidneys receive at least 1 liter of blood per minute, or more than 1,400 liters of blood per day.

How much of this huge volume is filtered in the glomeruli? The answer is about 12 percent. This is still a large number—180 liters per day! Since we normally only urinate 2 to 3 liters per day, about 98 to 99 percent of the fluid volume that is filtered in the glomerulus is being reabsorbed into the blood. Where and how is this enormous fluid volume reabsorbed from the renal tubules back into the blood?

Most of the glomerular filtrate is reabsorbed from the renal tubules

The proximal convoluted tubule is responsible for a major part of the reabsorption of the water and solutes in the glomerular filtrate. The cells of this section of the renal tubule are cuboidal, and their surfaces facing into the tubule have thousands of **microvilli**, which increase their surface area for reabsorption. These cells have lots of mitochondria—an indication that they are biochemically active. They transport NaCl and other solutes, such as glucose and amino acids, out of the tubular fluid. Almost all glucose molecules and amino acid molecules that are filtered from the blood are actively reabsorbed by the cells of the proximal convoluted tubules and transported into the extracellular fluids.

This movement of solutes makes the extracellular fluid hypertonic to the tubular fluid, and water flows from the tubular fluid in response to this difference. The water and solutes that are moved out of the tubular fluid by this process are taken up by the peritubular capillaries (see Figure 48.7) and thereby are returned to the venous blood leaving the kidney.

Despite the large volume of water and solutes reabsorbed by the proximal convoluted tubule, the overall concentration, or osmotic potential, of the fluid that enters the loop of Henle is not different from that of the blood plasma, even though the compositions of the two fluids are quite different. Next we have to consider how the kidney produces urine that is hypertonic to the blood plasma.

The loop of Henle creates a concentration gradient in the medulla

Humans can produce urine that is four times more concentrated than their plasma. Some mammals that

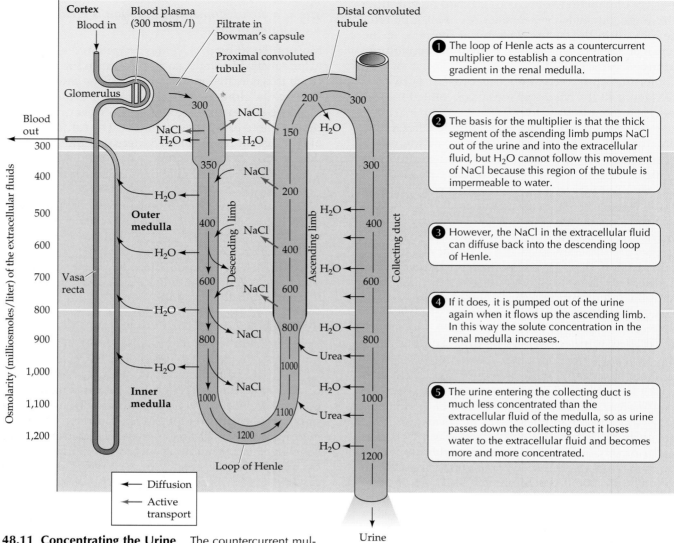

48.11 Concentrating the Urine The countercurrent multiplier system enables the loops of Henle to produce urine that is far more concentrated than mammalian blood plasma.

live in very dry deserts, such as kangaroo rats, are able to conserve water so well that their urine may be 12 to 15 times more concentrated than their blood plasma. How does the structure of the mammalian kidney enable it to produce urine that is hypertonic to the blood plasma?

This remarkable ability is due to the loops of Henle, which function as a **countercurrent multiplier system**. "Countercurrent" refers to the direction of urine flow in the descending versus the ascending limbs of the loop (see Chapter 46), and "multiplier" refers to the ability of this system to create a concentration gradient in the renal medulla. The loops of Henle *do not* concentrate the tubular fluid; rather they increase the osmotic concentration of the surrounding extracellular environment in the renal medulla. Let's see how they do it.

The cells of the descending limb of the loop of Henle, and the initial cells of the ascending limb, are unspecialized. These cells are flat, with no microvilli and few mitochondria (see Figure 48.10). The part of the tubule made up of these cells is permeable to water and small molecules. Partway up the ascending limb, however, the cells become specialized for transport again. They are cuboidal, with lots of mitochondria, and have some microvilli on their surfaces facing into the tubule. The portion of the ascending limb made up of these cells is impermeable to water, but the cells actively transport NaCl out of the tubular fluid. As a result, the urine becomes more dilute as it flows toward the distal convoluted tubule (Figure 48.11). Where does the NaCl that is transported out of the ascending limb go? It enters the extracellular fluid in the renal medulla, from which it can diffuse back into the descending limb of the loop of Henle. The NaCl that enters the descending limb flows up the ascending limb, where it is transported out of the urine once again. As a result of this process, NaCl accumulates in the extracellular fluid of the renal medulla, with the highest concentration near the tips of the renal pyramids.

How does this action of the loop of Henle concentrate the urine? Figure 48.11 shows that the urine is less concentrated when it leaves the loop than when it entered. The distal convoluted tubules continue to actively secrete substances to be excreted and reabsorb substances to be conserved. The urine becomes concentrated in the collecting duct.

The collecting duct runs from the cortex, where it receives filtrate from the distal tubules, down through the medulla, to the tip of the renal pyramids, where it discharges into the ureter. Over this distance the duct is surrounded by increasingly concentrated extracellular fluid. The collecting duct is permeable to water, but not to ions, so the osmotic potential of the extracellular environment draws water from the fluid in the duct and leaves behind an increasingly concentrated solution.

The urine that leaves the collecting duct at the tip of a renal pyramid can be almost as concentrated as the highest extracellular concentration established by the countercurrent multiplier system. The water reabsorbed from the collecting duct exits the renal medulla via the capillaries in the medulla (the vasa recta). The vas recta are highly permeable to salts and water.

The way the mammalian kidney works can be summarized as follows: The glomeruli filter large volumes of blood plasma. The proximal convoluted tubules reabsorb most of this volume, along with valuable molecules such as glucose and amino acids. The loops of Henle create a concentration gradient in the extracellular fluids in the medulla of the kidney. As the urine flows in the collecting ducts through this concentration gradient, water is reabsorbed, thus creating a urine that is hypertonic to the blood plasma.

Treating Kidney Failure

The function of the kidneys is to regulate the volume, the osmolarity, and the chemical composition of the extracellular fluids, including the blood. If these extracellular fluids do not have the right composition to meet the needs of the cells of the body, the cells cannot survive. Sudden and complete loss of kidney function is called acute renal failure. It results in the retention of salts and water (and hence high blood pressure), as well as the retention of urea (uremic poisoning) and metabolic acids (acidosis).

A person who suffers acute renal failure will die in 1 to 2 weeks if not treated. A drastic but effective treatment is kidney transplant, but it is usually necessary to maintain the patient in a healthy condition for a considerable period of time while waiting and preparing for the transplant. For this purpose, the older treatment option of artificial kidneys is used.

Artificial kidneys cleanse blood by dialysis

It is possible to compensate for renal failure and even surgical removal of the kidneys by using artificial kidneys. In an artificial kidney, or dialysis unit, the blood of the patient and a dialyzing fluid come into very close contact, separated only by a semipermeable membrane. This membrane allows small molecules to diffuse from the patient's blood into the dialysis fluid.

Because molecules and ions diffuse from an area of high concentration to an area of lower concentration, the composition of the dialysis fluid is crucial. The concentrations of the molecules or ions we want to conserve, such as glucose or sodium, must be the same in the dialysis fluid as in the plasma. The concentrations of molecules and ions that we want to clear from the plasma, such as urea and sulfate, must be zero in the dialysis fluid. The total osmotic potential of the dialysis fluid must equal the osmotic potential of the plasma.

Figure 48.12 shows how a dialysis machine works. Arterial blood flows between semipermeable membranes, which are surrounded with dialysis fluid at body temperature. The "cleansed" blood is returned to the body through a vein and the used dialysis fluid is discarded. At any one time, only about 500 ml of blood is in the dialysis unit, and the unit processes several hundred milliliters of blood per minute. A patient with no kidney function must be on the dialysis machine for 4 to 6 hours three times a week.

Control and Regulation of Kidney Functions

Control and regulatory mechanisms keep the kidneys functioning regardless of what is happening elsewhere in the body, and other mechanisms match kidney function to other needs of the body, such as maintenance of blood composition and control of blood pressure. Although we will discuss these various mechanisms separately, keep in mind that they are always working together.

The kidney acts to maintain its glomerular filtration rate

If the kidneys stop filtering the blood, they cannot accomplish any of their functions. Therefore, certain mechanisms keep the blood filtering through the glomeruli at a constant high rate, regardless of what is happening elsewhere in the body. Because these adaptations of the kidney are designed to maintain its own functions, these mechanisms are *autoregulatory*. The glomerular filtration rate (GFR) depends on an adequate blood supply to the kidneys at an adequate blood pressure. The autoregulatory mechanisms compensate

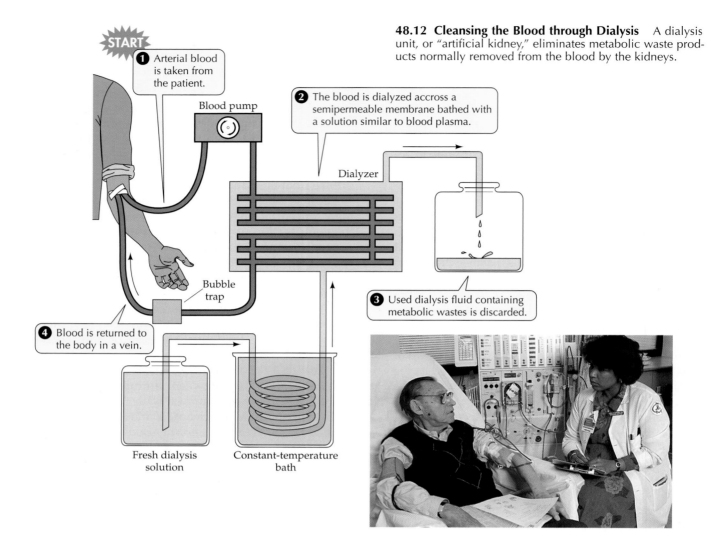

START

❶ Arterial blood is taken from the patient.

Blood pump

❷ The blood is dialyzed accross a semipermeable membrane bathed with a solution similar to blood plasma.

Dialyzer

48.12 Cleansing the Blood through Dialysis   A dialysis unit, or "artificial kidney," eliminates metabolic waste products normally removed from the blood by the kidneys.

Bubble trap

❸ Used dialysis fluid containing metabolic wastes is discarded.

❹ Blood is returned to the body in a vein.

Fresh dialysis solution

Constant-temperature bath

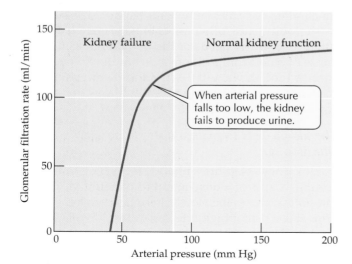

48.13 The Glomerular Filtration Rate   Even though glomerular filtration is driven by arterial pressure, autoregulatory mechanisms in the kidney make the glomerular filtration rate (GFR) independent of arterial pressure over a wide range.

Kidney failure | Normal kidney function

When arterial pressure falls too low, the kidney fails to produce urine.

Glomerular filtration rate (ml/min)

Arterial pressure (mm Hg)

for decreases in cardiac output or decreases in blood pressure so that the GFR remains high (Figure 48.13).

One autoregulatory mechanism is the dilation (expansion) of the afferent renal arterioles when blood pressure falls. This dilation decreases the resistance in the arterioles and helps maintain blood pressure in the glomerular capillaries. If that response does not keep the GFR from falling, then the kidney releases an enzyme, **renin**, into the blood. Renin acts on a circulating protein to begin converting this protein into an active hormone called **angiotensin**.

Angiotensin has several effects that help restore the GFR to normal. First, angiotensin causes the efferent renal arteriole to constrict, which elevates blood pressure in the glomerular capillaries. Second, angiotensin causes peripheral blood vessels all over the body to constrict—an action that elevates central blood pressure. Third, angiotensin stimulates the adrenal cortex to release the hormone **aldosterone**.

Aldosterone stimulates sodium reabsorption by the kidney, thereby making the reabsorption of water more effective. Enhanced water reabsorption helps maintain

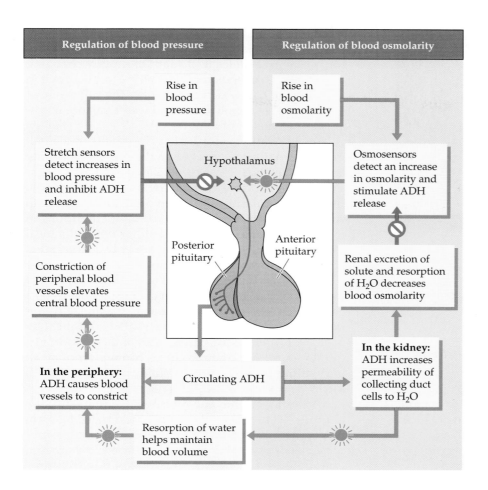

Regulation of blood pressure	Regulation of blood osmolarity

Rise in blood pressure

Rise in blood osmolarity

Stretch sensors detect increases in blood pressure and inhibit ADH release

Osmosensors detect an increase in osmolarity and stimulate ADH release

Hypothalamus

Posterior pituitary

Anterior pituitary

Constriction of peripheral blood vessels elevates central blood pressure

Renal excretion of solute and resorption of H_2O decreases blood osmolarity

In the periphery: ADH causes blood vessels to constrict

Circulating ADH

In the kidney: ADH increases permeability of collecting duct cells to H_2O

Resorption of water helps maintain blood volume

48.14 Antidiuretic Hormone Promotes Water Reabsorption
ADH controls the concentration of urine by increasing the permeability of the collecting duct to water. ADH is produced by neurons in the hypothalamus and released from their nerve endings in the posterior pituitary. The release of ADH is stimulated by hypothalamic osmosensors and inhibited by stretch sensors in the great arteries.

blood volume and therefore central blood pressure. Finally, angiotensin acts on structures in the brain to stimulate thirst. Increased water intake in response to thirst increases blood volume and blood pressure.

The kidneys respond to changes in blood volume

When you lose blood, your blood pressure tends to fall. Besides activating the kidney autoregulatory mechanisms described in the previous section, a drop in blood pressure decreases the activity of stretch sensors in the walls of the large arteries such as the aorta and the carotids. These stretch sensors provide information to cells in the hypothalamus that produce **antidiuretic hormone** (**ADH**, also called vasopressin) and send it down their axons to the posterior pituitary gland (see Chapter 38). As stretch sensor activity decreases, the production and release of this hormone increases (Figure 48.14).

ADH acts on the collecting ducts of the kidney to increase their permeability to water. When the circulating level of ADH is high, the collecting ducts are very permeable to water, more water is reabsorbed from the urine, and only small quantities of concentrated urine are produced, thus conserving blood volume and blood pressure. Without ADH, water cannot be reab-

sorbed from the collecting ducts, and lots of very dilute urine is produced.

The molecular mechanism by which ADH controls the permeability of the collecting duct has recently been revealed. The early 1990s yielded the discovery of a family of genes coding for membrane proteins that form water channels called **aquaporins**. Aquaporins are found in many tissues that are permeable to water—for example, capillary endothelium, red blood cells, and the proximal convoluted tubules of the kidney. Differences in water permeability can be related to the presence or absence of aquaporins. For example, aquaporins are expressed in the descending limb of the loop of Henle but not in the ascending limb.

One particular aquaporin (designated aquaporin-2) is found in collecting-duct cells and is controlled by ADH on both a long-term and a short-term basis. In the long term, ADH levels influence the expression of the gene for aquaporin-2; in the short term, ADH controls the insertion of the protein into the cell membranes.

Mutations of the gene that codes for aquaporin-2 can cause an insensitivity to ADH and therefore a condition called nephrogenic diabetes insipidus. "Nephrogenic" means that the defect is in the kidney and not in the synthesis or release of ADH, "diabetes" means a high rate of urine production, and "insipidus" means that the urine is tasteless. This condition contrasts with diabetes mellitus, which is caused by a lack of insulin and produces sweet-tasting urine (*mellitus* is Greek for "honey").

ADH helps regulate blood osmolarity. Sensors in the hypothalamus monitor the osmotic potential of the blood. If blood osmolarity increases, these **osmosensors** stimulate increased release of ADH to enhance water reabsorption from the kidneys. The osmosensors also stimulate thirst. The resulting increased water intake dilutes the blood as it expands blood volume.

48.15 The Ability to Concentrate   The ability of the mammalian kidney to concentrate urine depends on the lengths of its loops of Henle relative to the overall size of the kidney. Some desert rodents have single renal pyramids so long that they protrude out of the kidney and into the ureter.

The mammalian kidney can respond to extreme conditions

The ability of the mammalian kidney to produce a concentrated urine has made it possible for mammals to inhabit some of the most arid habitats on Earth. Some of these animals, such as the desert gerbil (*Meriones* sp.), have such extremely long loops of Henle that their renal pyramid (each of their kidneys has only one in contrast to ours) extends far out of the concave surface of the kidney (Figure 48.15). These animals are so effective in conserving water that they can survive on the water released by the metabolism of their dry food; they do not need to drink!

The concentrating ability of the mammalian kidney, coupled with the remarkable flexibility of its regulatory systems, enables it to adapt to rapidly changing conditions. This regulatory flexibility is pronounced in the vampire bat, which can display the full range of extremes of mammalian salt and water balance mechanisms in a matter of minutes.

The Australian vampire bat (*Desmodus*) is a small mammal that feeds at night on the blood of sleeping large mammals, such as cattle. Blood is a liquid, high-protein diet. To process this diet, the renal system of the vampire bat must shift from drought conditions to flood conditions and back to drought conditions in minutes. At sunset, when the bat has not had a meal for many hours, it is producing a highly concentrated urine at a low rate to conserve its precious body water (Figure 48.16). If it is successful in finding prey, the bat must process as much blood in as short a time as possible, before the victim wakes up. To maximize its nutrient intake, the bat concentrates its blood meal by rapidly excreting its water content. Accordingly, within minutes the bat produces copious amounts of very dilute urine. The warm fluid running down the victim's neck and waking it is not blood!

As soon as the bat finishes feeding—usually abruptly—it begins to digest the concentrated blood in its gut. Because the concentrated blood is mostly protein, a large amount of nitrogenous waste is produced and must be excreted as urea in solution. But now water is in short supply. The bat must limit its water loss because a long time may pass before the next meal. Consequently, the bat's kidneys produce small amounts of extremely concentrated urine. This urine can be more than 20 times more concentrated than the bat's plasma. Humans, in comparison, can produce a urine only about 4 times more concentrated than their plasma. In this way the remarkable regulatory abilities of the vampire bat kidney enable the animal to process its unusual diet.

48.16 Water Balance in the Vampire Bat Vampire bats can regulate kidney function to go from maximum water excretion to maximum water conservation in short time. The graph shows the changes in a vampire bat's urine concentration and urine flow rate before and after its meal of blood.

[Graph: Urine concentration (milliosm/l) on left y-axis (1,000–6,000) and Urine flow rate (ml/kg/min) on right y-axis (1–2), versus Time (hours) 0–3. Callout: "To maximize food intake, the bat rapidly excretes water as dilute urine." Callout: "To excrete nitrogenous waste with minimal loss of water, the bat concentrates its urine." Dashed line labeled "Plasma concentration." Arrow marking "Start to feed."]

Summary of "Salt and Water Balance and Nitrogen Excretion"

Extracellular Fluids and Water Balance

• The problems of salt and water balance and nitrogen excretion that animals face depend on their environments, but in all animal excretory systems there is no active transport of water.
• All adaptations for maintaining salt and water balance and for excreting nitrogen waste products employ the same basic principles: filtration of body fluids, and active secretion and reabsorption of specific ions.

Distinguishing Environments and Animals in Terms of Salt and Water

• Marine animals can be osmoconformers or osmoregulators. Freshwater animals must be osmoregulators and continually excrete water and conserve salts. All animals are ionic regulators to some degree. **Review Figure 48.1**
• On land, water conservation is essential, and diet determines whether salts must be conserved or excreted. **Review Figure 48.2**

Excreting Nitrogen as Ammonia, Uric Acid, or Urea

• Aquatic animals can eliminate nitrogenous wastes such as ammonia by diffusion across their gill membranes. Terrestrial animals detoxify ammonia by converting it to urea or uric acid for excretion. **Review Figure 48.3**
• Depending on the form in which they excrete their nitrogenous waste products, animals are classified as ammonotelic, ureotelic, or uricotelic.

The Diverse Excretory Systems of Invertebrates

• The protonephridia of flatworms consist of flame cells and excretory tubules. Extracellular fluids are filtered into the tubules, which process the filtrate to produce a dilute urine. **Review Figure 48.4**
• Annelid worms have a closed circulatory system, and blood pressure causes filtration of the blood across capillary walls. The filtrate enters the coelomic cavity, where it is taken up by open-ended tubules called metanephridia. As the filtrate passes through the tubules to the outside, its composition is changed by active transport mechanisms. **Review Figure 48.5**
• The Malpighian tubules of insects do not open into the body cavity; they receive ions and nitrogenous wastes by transport across the tubule cells. Water follows by osmosis. Ions and water are reabsorbed from the rectum, so the insect excretes semisolid wastes. **Review Figure 48.6**

The Nephron: The Basic Unit of Vertebrate Excretion

• The nephron, the functional unit of the vertebrate kidney, consists of a glomerulus in which the blood is filtered across the walls of a knot of capillaries, and a set of renal tubules that process the filtrate into urine to be excreted. **Review Figures 48.7, 48.8**
• The adaptations of marine fishes and terrestrial amphibians and reptiles to prevent water loss are diverse. Bony fishes have few glomeruli and produce little urine; they excrete salt from special glands and lose nitrogenous wastes as ammonia across their gills. Cartilaginous fishes retain urea so that the osmotic concentration of their body fluids is above that of seawater. Amphibians remain close to water or have waxy secretions on their skin to prevent water loss. Reptiles have water-impermeable skin, lay shelled eggs that resist desiccation, and excrete nitrogenous waste as uric acid, which requires little loss of water. Birds share the adaptations of reptiles; in addition, they can produce hypertonic urine—an ability shared only by the mammals.
• Only birds and mammals can produce urine more concentrated than their body fluids.

The Mammalian Kidney

• The concentrating ability of the mammalian kidney depends on its anatomy. Glomeruli and sections of the renal tubules called convoluted tubules are in the cortex of the kidney. Certain molecules, salts, and water are reabsorbed in bulk and other molecules are actively secreted in the convoluted tubules without the urine becoming more concentrated. Straight sections of renal tubules called loops of Henle and collecting ducts are arranged in parallel in the medulla of the kidney. **Review Figure 48.10**
• The loops of Henle create a concentration gradient in the extracellular fluids of the renal medulla by a countercurrent multiplier system. Urine flowing down the collecting ducts to the ureter is concentrated by the osmotic loss of water caused by the concentration of the surrounding extracellular environment. **Figure 48.11**

Treating Kidney Failure

• Without treatment, kidney loss is fatal. Artificial kidneys cleanse the blood by dialysis. **Review Figure 48.12**

Control and Regulation of Kidney Functions

• Kidney function in mammals is controlled by autoregulatory mechanisms for maintaining a constant high glomerular filtration rate even if blood pressure varies. An important autoregulatory mechanism is the release of renin by the kidney when blood pressure falls; renin activates angiotensin, which causes the constriction of peripheral blood vessels, causes the release of aldosterone (which enhances water reabsorption), and stimulates thirst. **Review Figure 48.13**
• Kidney function in mammals is also controlled by mechanisms responsive to blood pressure and blood composition. Changes in blood pressure and blood osmolarity influence the release of antidiuretic hormone, which controls the permeability of the collecting duct to water and therefore the amount of water that is reabsorbed from the urine. ADH stimulates the expression of proteins called aquaporins that serve as water channels in the membranes of collecting-duct cells. **Review Figure 48.14**
• The mammalian kidney is able to respond to extreme conditions by either diluting or concentrating its urine. **Review Figure 48.16**

Self-Quiz

1. Which statement about osmoregulators is true?
    a. Most marine invertebrates are osmoregulators.
    b. All freshwater invertebrates are hypertonic osmoregulators.
    c. Cartilaginous fishes are hypotonic osmoregulators.
    d. Bony marine fishes are hypertonic osmoregulators.
    e. Mammals are hypotonic osmoregulators.

2. The excretion of nitrogenous wastes
   a. by humans can be in the form of urea and uric acid.
   b. by mammals is never in the form of uric acid.
   c. by marine fishes is in the form of urea.
   d. does not contribute to the osmotic potential of the urine.
   e. requires more water if the waste product is the rather insoluble uric acid.

3. How are earthworm metanephridia like mammalian nephrons?
   a. Both process coelomic fluid.
   b. Both take in fluid through a ciliated opening.
   c. Both produce hypertonic urine.
   d. Both employ tubular secretion and reabsorption to control urine composition.
   e. Both deliver urine to a urinary bladder.

4. What is the role of renal podocytes?
   a. They control the glomerular filtration rate by changing the resistance of renal arterioles.
   b. They reabsorb most of the glucose that is filtered from the plasma.
   c. They prevent red blood cells and large molecules from entering the renal tubules.
   d. They provide a large surface area for tubular secretion and reabsorption.
   e. They release renin when the glomerular filtration rate falls.

5. Which of the following are *not* found in a renal pyramid?
   a. Collecting ducts
   b. Vasa recta
   c. Peritubular capillaries
   d. Convoluted tubules
   e. Loops of Henle

6. Which part of the nephron is responsible for most of the difference in mammals between the glomerular filtration rate and the urine production rate?
   a. The glomerulus
   b. The proximal convoluted tubule
   c. The loop of Henle
   d. The distal convoluted tubule
   e. The collecting duct

7. For mammals of the same size, what feature of their excretory systems would give them the greatest ability to produce a hypertonic urine?
   a. Higher glomerular filtration rate
   b. Longer convoluted tubules
   c. Increased number of nephrons
   d. More-permeable collecting ducts
   e. Longer loops of Henle

8. Which of the following would *not* be a response stimulated by a large drop in blood pressure?
   a. Constriction of afferent renal arterioles
   b. Increased release of renin
   c. Increased release of antidiuretic hormone
   d. Increased thirst
   e. Constriction of efferent renal arterioles

9. Which statement about angiotensin is true?
   a. It is secreted by the kidney when the glomerular filtration rate falls.
   b. It is released by the posterior pituitary when blood pressure falls.
   c. It stimulates thirst.
   d. It increases permeability of the collecting ducts to water.
   e. It decreases glomerular filtration rate when blood pressure rises.

10. Birds that feed on marine animals ingest a lot of salt, but they excrete most of it by means of
    a. Malpighian tubules.
    b. rectal salt glands.
    c. gill membranes.
    d. hypertonic urine.
    e. nasal salt glands.

Applying Concepts

1. What do marine fishes, reptiles, mammals, and insects have in common with respect to water balance? Compare their physiological adaptations for dealing with their common problem.

2. What are the relative advantages and disadvantages of ammonia, urea, and uric acid as nitrogenous waste products of animals?

3. Explain how the kidney is able to maintain a constant glomerular filtration rate over a wide range of arterial blood pressures. On the basis of what you learned about the regulation of circulatory function in Chapter 46, how can a decrease in glomerular filtration rate cause an increase in cardiac output?

4. Inulin is a molecule that is filtered in the glomerulus, but it is not secreted or reabsorbed by the renal tubules. If you injected inulin into a subject and after a brief time measured the concentration of inulin in the blood and in the urine of the subject, how could you determine the subject's glomerular filtration rate? Assume that the rate of urine production is 1 ml per minute.

5. Explain the roles of the loop of Henle and the collecting duct in producing a hypertonic urine in mammals. How is this mechanism controlled in response to changes in osmolarity of the blood and in blood pressure?

Readings

Cantin, M. and J. Genest. 1986. "The Heart as an Endocrine Gland." *Scientific American*, February. A discussion of the hormone secretion by the heart that helps control salt and water balance.

Heatwole, H. 1978. "Adaptations of Marine Snakes." *American Scientist*, vol. 66, pages 594–604. An exploration into the variety of adaptations, including means of maintaining salt and water balance, that allow several groups of snakes to exploit the marine environment.

McClanahan, L. L., R. Ruibal and V. H. Shoemaker. 1994. "Frogs and Toads in Deserts." *Scientific American*, March. A report of recent research on amphibians adapted to arid environments.

Schmidt-Nielsen, K. 1997. *Animal Physiology: Adaptation and Environment*, 5th Edition. Cambridge University Press, New York. An excellent textbook, emphasizing the comparative approach.

Stricker, E. M. and J. G. Verbalis. 1988. "Hormones and Behavior: The Biology of Thirst and Sodium Appetite." *American Scientist*, vol. 76, page 261. A discussion of the control of water and salt intake as an important part of osmoregulation.

Vander, A. J., J. H. Sherman and D. S. Luciano. 1994. *Human Physiology: The Mechanisms of Body Function*, 6th Edition. McGraw-Hill, New York. Chapter 16 deals with the regulation of water and salt balance.

Chapter 49

Animal Behavior

Charlotte's Web
The web of an orb-weaving spider, spun between stalks of Queen Anne's lace, is encrusted with morning dew.

Spider webs are objects of beauty and marvels of engineering. The construction of a classic web used to capture prey requires complex behavior. For example, a garden spider, as immortalized in E. B. White's book *Charlotte's Web*, spins a new web every day in the early morning hours before dawn. From an initial attachment point, she strings a horizontal thread. From the middle of that thread she drops a vertical thread to a lower attachment point. Pulling the thread taut creates a Y, the center of which will be the hub of the finished web.

The spider adds a few more radial supports and a few surrounding "framing" threads. Then she fills in all the radial spokes according to a set of rules. Finally, she lays down a spiral of sticky threads with regular spacing and attachment points to the radial spokes. This remarkable feat of construction takes only half an hour, but it requires thousands of specific movements performed in just the right sequence. Where is the blueprint for Charlotte's web? How does the garden spider acquire the construction skills needed to build it?

The blueprint is coded in genes and built into the spider's nervous system as a motor "program," or score. Learning plays no role in the expression of that complex blueprint. Newly hatched spiders disperse to new locations and usually spin their first webs without ever having experienced a web built by an adult of their species. Nevertheless, they build perfect webs the first time; each of the thousands of movements happens in just the right sequence. It is remarkable that the genetic code and the simple nervous system of a spider can contain and express behavior as complex as the spinning of an orb web.

Web spinning by spiders is an animal *behavior*—an act or set of acts performed with respect to another animal or the environment. Behavior falls into three general classes: acts to acquire food, acts to avoid environmental threats, and acts to reproduce. In studying any behavior we can ask *what*, *how*, and *why*.

What questions focus on the details of behavior, including the circumstances that influence when an animal acts in a certain way. Some *how* questions refer to the underlying neural, hormonal, and anatomical mechanisms that we have been studying in Part Six. Other *how* questions refer to the means by which an animal acquires a behavior. Some behaviors, such as web spinning, are genetically determined; others are learned. Many behaviors involve complex interactions of inheritance and learning.

From the large field of study that is animal behavior, this chapter presents some interesting examples of approaches to answering *what* and *how* questions. (*Why* questions, which have to do with the evolution of behavior, will be the major focus of the next chapter.) After a discussion of genetically determined behavior and behavior that results from a combination of genetics and learning, such as birdsong, we will briefly explore how hormones influence the development and expression of behavior. Next we will turn to animal communication. Studies of animal communication reveal the constraints that environment places on behavior. We will continue with a look at the timing of behaviors, or biological rhythms, and a discussion of how animals find their way through unfamiliar territory. Throughout the chapter, use what you read to raise your own questions about human behavior, which we will examine briefly at the end.

Genetically Determined Behavior

A behavior that is genetically determined rather than learned is called a **fixed action pattern**. Such behavior is highly *stereotypic*; that is, it is performed in the same way every time. It is also *species-specific*; there is very little variation in the way different individuals of the same species perform the behavior. In even closely related species, however, the behavior is expressed differently. For example, different species of spiders spin webs of different designs.

Fixed action patterns require no learning or prior experience for their expression, and they are generally not modified by learning. Another spider example illustrates this point. Spiders spin other structures in addition to webs for capturing prey. Most spiders lay their eggs in a cocoon that they form by spinning a base plate, building up the walls (inside of which they

lay the eggs), and spinning a lid to close the cocoon. Although this behavior requires thousands of individual movements, it is performed in exactly the same way every time and is not modified by experience.

If the spider is moved to a new location after she finishes the base plate, she will continue to spin the sides of the cocoon, lay her eggs (which then fall out the bottom), and spin the lid. If she is placed on her previously completed base plate the next time she is ready to begin a cocoon, she will spin a new base plate over the old one as if it were not there. If she is nutritionally deprived and runs out of silk in the middle of spinning a cocoon, she will continue spinning, completing all the thousands of movements in a pantomime of cocoon building. Once started, the cocoon-building motor score runs from beginning to end, and it can be started only at the beginning.

Fixed action patterns are good material for studying the mechanisms of animal behavior. In this chapter we will examine the genetics of fixed action patterns and the sequence of events whereby gene expression eventually results in a behavior. First, however, we must demonstrate that a given behavior *is* genetically determined. One powerful way of proving genetic determination is to deprive the animal of any opportunity to learn the behavior in question.

Deprivation experiments can reveal whether a behavior is genetically determined

In a **deprivation experiment**, an animal is reared so that it is deprived of all experience relevant to the behavior under study. For example, in one deprivation experiment a tree squirrel was reared in isolation, on a liquid diet, and in a cage without soil or other particulate matter. When the young squirrel was given a nut, it put the nut in its mouth and ran around the cage. Eventually it oriented toward a corner of the cage and made stereotypic digging movements, placed the nut in the corner, went through the motions of refilling the imaginary hole, and ended by tamping the nonexistent soil with its nose. The squirrel had never handled a food object and had never experienced soil, yet the fixed action patterns involved in burying its nut were fully expressed.

Deprivation experiments occur naturally. Many species, especially insects living in seasonal environments, have a life cycle that lasts only one year and the generations do not overlap: The adults lay eggs and die before the eggs hatch or the young mature into adults. Since learning from adults of the parent generation is impossible in such species, the complex behavior necessary for survival and reproductive success must be genetically programmed.

Web spinning by spiders is an example of complex behavior in species that may have no opportunity to learn from other members of their species. The courtship behavior of the triangular web spider is a similar example. The male spider must approach a female in her web. If he simply blundered into the web, he would give the same signals as a prey item caught in the web and the female would probably kill him and eat him. To avoid having his reproductive effort cut short, the male is genetically programmed to approach an anchor strand of the web and pluck it in just the right way to send a courtship message to the female.

If the message is correct and the female is receptive, he can enter the web and mate with the female rather than serve merely as her dinner. In some species, the female eats the male anyway after they mate, but at least then he has achieved reproductive success before becoming nutriment for his own offspring, and he is supplying energy for eggs fertilized only by him.

Fixed action patterns are triggered by simple stimuli

A behavior that is not expressed during a deprivation experiment may nonetheless be genetically programmed. The right conditions simply may not have been available to stimulate the behavior during the experiment. The squirrel described in the previous section, for example, had to be given a nut for its digging and burying behaviors to be triggered. Specific stimuli are usually required to elicit the expression of fixed action patterns. These stimuli are called sign stimuli or **releasers**. Two **ethologists** (scientists who study animal behavior), Konrad Lorenz and Niko Tinbergen, conducted classic studies of releasers and fixed action patterns. Their work provided insights into the properties of releasers.

Releasers are usually a simple subset of all the sensory information available to an animal. Adult male European robins have red feathers on their breasts. During the breeding season, the sight of an adult male robin stimulates another male robin to sing, perform aggressive displays, and attack the intruder if he does not heed the warnings. An immature male robin, whose feathers are all brown, does not elicit aggressive behavior. A tuft of red feathers on a stick, however, will elicit an attack (Figure 49.1). A patch of red in certain locations is a sufficient releaser for male aggressive behavior in robins.

Just as the motor score of the fixed action pattern is genetically programmed, so is the information that enables the animal to recognize the releaser for that fixed action pattern. Evolving a genetic mechanism to respond to a simple stimulus is more feasible than evolving a mechanism to recognize a complex set of stimuli. The simplicity of most releasers has resulted in some curious discoveries.

Tinbergen and A. C. Perdeck carefully examined the releasers and fixed action patterns involved in the interactions between herring gulls and their chicks dur-

A mounted immature male European robin with no red feathers does not stimulate aggression from a territorial adult male…

…but this formless clump of red feathers elicits strong aggressive attacks from the territorial adult.

49.1 Triggering Aggressive Behavior The stimulus that triggers a behavior is not always the one we might expect.

ing feeding. The adult gull has a red dot at the end of its bill (Figure 49.2). When the gull returns to its nest, the chicks peck at the red dot, thereby stimulating the adult to regurgitate food for the chicks to eat.

Tinbergen and Perdeck set out to determine the essential characteristics of the parent gull that release food solicitation behavior in the chicks. They made paper cutout models of gull heads and bills, varying the colors and the shapes. Then they rated each model according to how many pecks it received from chicks. The surprising results were that the shape or color of the head made no difference. In fact, a head was not even necessary; the chicks responded just as well to models of bills alone. The color of the bill and the dot also were not

49.2 The Dot Marks the Spot This gull is incubating eggs. When the young hatch they will peck at the red spot on the parent's bill, stimulating the parent to regurgitate food.

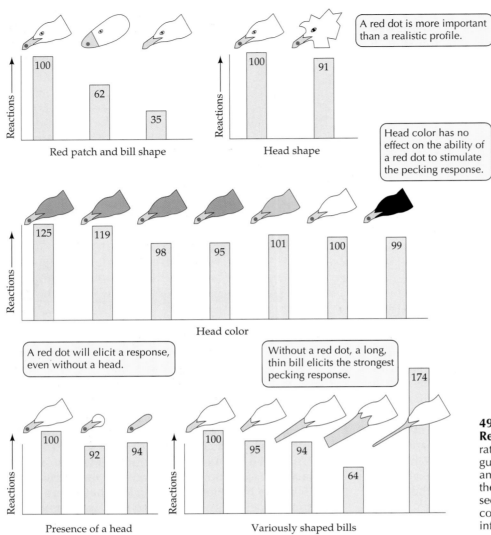

49.3 Releasing the Food Solicitation Response A series of experiments rated the pecking response of herring gull chicks to variations in the head and bill of the parent. The presence of the red dot and the shape of the bill seem to be releasers; the shape and color of the head have little or no influence on the response.

critical as long as there was a contrast between the two. Surprisingly, the most effective releaser for chick pecking was a long, thin object with a dark tip that bore no resemblance to an adult herring gull (Figure 49.3).

The simplicity of the properties of releasers makes possible the existence of a **supernormal releaser**, one that is more effective than the natural condition in eliciting a fixed action pattern. In the case of one wading bird, the oystercatcher, the sight of its clutch of eggs releases incubation behavior. However, given the choice between its own clutch of two eggs and a clutch of three artificial eggs, an oystercatcher will sit on the larger clutch of artificial eggs. And given the choice of its own clutch of two eggs or one very large artificial egg, it will try to incubate the large egg, even if it can hardly straddle it. Since these choices would not occur in nature, counterselection has not prevented the evolution of such maladaptive behavior.

Supernormal releasers, which are created by scientists, are a curiosity, but natural selection has produced its own dramatic results by favoring the exaggeration of releasers. Many of the elaborate behavior patterns and physical attributes used by species in courtship displays have arisen through natural selection favoring releasers that are more effective. Male bowerbirds, for example, even use colorful objects collected from the environment to enhance their courtship displays.

Motivation determines whether a fixed action pattern will be triggered

Another reason that a fixed action pattern may not be expressed in a deprivation experiment is that the animal is not in the appropriate developmental or physiological state. Juvenile animals do not show courtship behavior even if the appropriate releasers are present. An adult animal may not engage in aggressive display or courtship display when it is not in reproductive condition. The same animals that may be highly aggressive during the reproductive season may ignore one another at other times of the year.

49.4 Some Things an Animal Can't Afford to Learn by Trial and Error The sound of a striking rattlesnake triggers an automatic escape jump in a kangaroo rat; the rat likely would not survive if it had to learn this behavior.

The internal conditions of an animal determine its motivational state, and the motivational state determines which fixed action pattern is expressed at any particular time. The total behavior of an animal is not simply a random sequence of fixed action patterns that are triggered by whatever releasers it happens to encounter. Depending on its motivational state, an animal may search for the appropriate releaser and ignore many others. This search behavior may depend heavily on previous experience; hence, it is evidence of learning. For example, a hungry predator is likely to return to a site where it has encountered prey in the past.

Fixed action patterns and learning interact to produce some behaviors

The ability to learn and to modify behavior as a result of experience is often highly adaptive. Most human behavior is the result of learning. Why then are so many behavior patterns in so many species genetically determined and not modifiable? We have already considered one answer to this question. If role models and opportunities to learn are not available, there is no alternative to programmed behavior.

Fixed action patterns also can be adaptive when mistakes are costly or dangerous. Mating with a member of the wrong species is a costly mistake; thus the function of much courtship behavior is to guarantee correct species recognition. In an environment in which incorrect as well as correct models exist, learning the wrong pattern of courtship behavior is possible. Fixed action patterns that govern mating behavior can prevent such mistakes.

Behavior patterns such as avoidance of a predator or capture of dangerous prey allow no room for mistakes. If the behavior is not performed promptly and accurately the first time, there may not be a second chance. For example, rattlesnakes prey on kangaroo rats. A kangaroo rat that has never encountered a rattlesnake can avoid the snake in total darkness because the *sound* of the snake moving through the air to strike releases the powerful escape jump of the kangaroo rat (Figure 49.4). A kangaroo rat does not have time to learn what a rattlesnake sounds like when it is striking.

Let's look at another example. Although king snakes eat rattlesnakes, they are not immune to rattlesnake venom. When a king snake first encounters a rattlesnake, it strikes in such a way that its jaws clamp shut the mouth of the rattlesnake to prevent being bitten by its prey; then it begins the long process of swallowing. A king snake that grabs a rattlesnake at any other place on its body will not have the opportunity to learn by trial and error. Thus, *genetically programmed behavior is highly adaptive for species that have little opportunity to learn, for species that might learn the wrong behavior, and in situations where mistakes are costly or dangerous.*

Many behavior patterns are intricate interactions of genetically programmed elements and elements modified by experience. One example that has been the subject of elegant experiments is bird song. Adult male passerine birds use a species-specific song in territorial display and courtship. A few species, such as song sparrows, express their species-specific song even during deprivation experiments, but most do not.

For most species, such as the white-crowned sparrow, learning is an essential step in the acquisition of song. If the eggs of white-crowned sparrows are hatched in an incubator and the young male birds are reared in isolation from the song of their species, their adult songs will be an unusual assemblage of sounds (Figure 49.5a and b). This species cannot express its species-specific song without hearing that song as a nestling. But even though the white-crowned sparrow must hear the song of its own species during its nestling period in order to sing as an adult, it does not sing as a juvenile. It apparently uses auditory input as a nestling to form a template in its nervous system. It then matches the template through trial and error when it reaches sexual maturity the following spring.

If a bird that has heard its correct song as a juvenile is deafened before it begins to express its song, it will never develop its species-specific song (Figure 49.5c). The bird must be able to hear itself to match the template stored in its nervous system. If it is deafened *after* it expresses its correct song, it will continue to sing like a normal bird. Two periods of learning are essential: the first in the nestling stage, the second at the onset of sexual maturity.

Despite the crucial role of learning in a bird's ability to sing its species-specific song, studies of what birds can learn and when they can learn reveal strong genetic limits to the modifiability of their behavior through experience. The white-crowned sparrow, for example, must hear its species' song within a narrow

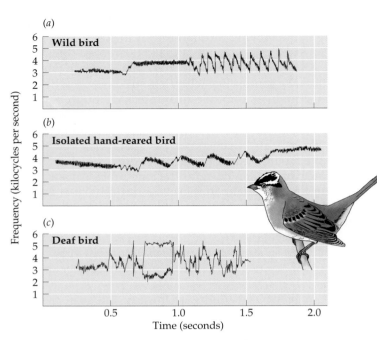

49.5 Song Learning in White-Crowned Sparrows These sonograms visually record sound frequencies and plot them over time. (a) The species-specific song of a male in its natural state. (b) The song of a male bird hatched in an incubator and reared in isolation; in this deprivation experiment, the bird did not learn the species-specific song. (c) In a variation on the deprivation experiment, a young bird that heard the correct song but was deafened before maturity cannot reproduce the song.

critical period during its development. Once this critical period has passed, the bird cannot learn to sing its species-specific song, regardless of how many role models it experiences.

What a bird can learn during its critical period is also severely limited, as revealed by experiments on hand-reared chaffinches that were played various tape recordings of birdsong during their critical periods. If exposed to the songs of other species, the chaffinches did not learn them. They also did not learn a chaffinch song played backward or with the elements scrambled. Even if they heard a chaffinch song played in pure tones, they did not form a template. However, if they heard a normal chaffinch song along with all these other sounds, they developed templates and learned to sing the proper song the following spring. Thus the chaffinch is genetically programmed to recognize the appropriate song to learn and when to learn it.

What advantage are genetic limits on what a bird can learn and when? A bird's acoustic environment can be quite complex. Many species of songbirds may be singing in the same area. The critical period limits learning to the period of most intimate contact between the young bird and its parents to ensure that the father's song is the one experienced most intensely.

Further limits on nestling song sensitivity help guarantee that the template it forms is not contaminated with other sounds it hears.

The learning of a song template by a nestling bird is an example of **imprinting**. We mentioned earlier that releasers generally are simple subsets of available information because there are limits to what can be programmed genetically. Imprinting makes it possible to encode complex information in the nervous system rather than in the genes. Offspring can imprint on their parents and parents on their offspring to ensure individual recognition, even in a crowded situation such as a colony or a herd. If a mother goat does not nuzzle and lick her newborn within 5 to 10 minutes after birth, she will not recognize it as her own later. In this case imprinting depends on olfactory cues, and the critical period is determined by the high levels of the hormone oxytocin circulating at the time of birth.

In the next section we will examine how hormones in general influence interactions between genetically determined behavior and learning.

Hormones and Behavior

All behavior depends on the nervous system for initiation, coordination, and execution. In addition, as we have discussed, fixed action patterns are built into the nervous system as a motor score. Yet that motor score is expressed only under certain conditions. The endocrine system, through its controlling influences on development and physiological state of the animal, plays a large role in determining when a particular motor pattern can be and is performed. In the previous section we saw that learning can play a role in the acquisiton of a species-specific behavior, but there may be narrow developmental windows or critical periods during which certain defined learning can occur.

Hormones control the complex interaction between genetically determined behavior and learning during specific stages of development. We have already seen one example of hormones controlling behavior: High levels of oxytocin in the female goat's blood at the time she gives birth determine a window of time during which she can imprint on her infant. In this section we will study two more complex cases in which hormones control the development, learning, and expression of species-specific behavior.

Sex steroids determine the development and expression of sexual behavior in rats

Differences in the behavior of males and females of a species point to clear examples of genetically determined behavior. Such differences in sexual behavior have been shown to be the result of actions of the sex steroids on the developing brain and on the mature brain. Like most other animals, rats behave sexually in accord with fixed action patterns. When a female rat is

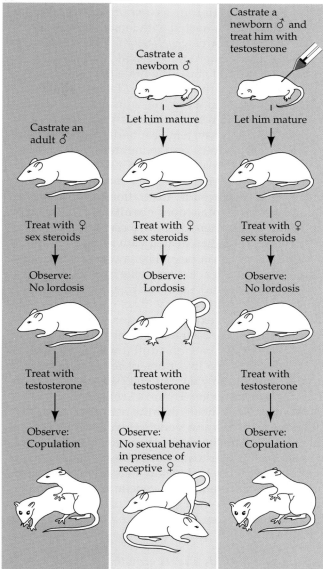

(a) **Treatment of female rats**

Spay an
adult ♀
|
Treat with ♀
sex steroids
↓
Observe:
Lordosis
↓
Treat with
testosterone
↓
Observe:
No sexual
behavior

Spay a
newborn ♀
|
Let her mature
↓
Treat with ♀
sex steroids
↓
Observe:
Lordosis
↓
Treat with
testosterone
↓
·Observe:
No sexual
behavior

Spay a newborn
♀ and treat
her with
testosterone
|
Let her mature
↓
Treat with ♀
sex steroids
↓
Observe:
No lordosis
↓
Treat with
testosterone
↓
Observe:
♂ sexual
behavior

(b) **Treatment of male rats**

Castrate an
adult ♂
|
Treat with ♀
sex steroids
↓
Observe:
No lordosis
↓
Treat with
testosterone
↓
Observe:
Copulation

Castrate a
newborn ♂
|
Let him mature
↓
Treat with ♀
sex steroids
↓
Observe:
Lordosis
↓
Treat with
testosterone
↓
Observe:
No sexual behavior
in presence of
receptive ♀

Castrate a
newborn ♂ and
treat him with
testosterone
|
Let him mature
↓
Treat with ♀
sex steroids
↓
Observe:
No lordosis
↓
Treat with
testosterone
↓
Observe:
Copulation

49.6 Hormonal Control of Sexual Behavior Sex steroids
control both the development and the expression of sexual
behavior in rats. The presence of testosterone in newborn rats
of both sexes whose reproductive organs (ovaries or testes)
have been removed establishes male behavior patterns, and
its absence establishes female patterns. Injections of sex
steroids in gonadectomized adult rats stimulate expression of
the sexual behavior pattern that developed in response to
genotype and early exposure to steroids.

Conclusion: In rats, the presence of testosterone establishes male behavior patterns, and its absence establishes female patterns.

in estrus (receptive to males), she responds to a tactile
stimulus of her hindquarters with a stereotypic posture
called **lordosis**. She lowers her front legs, extends her
hind legs, arches her back, and deflects her tail to one
side.

When a male rat encounters a female in estrus , he
copulates with her, performing the following sequence
of behaviors: He mounts her from the rear, clasps her
hindquarters, inserts his penis into her vagina, and
thrusts. The roles of genotype and sex hormones in the
development of the fixed action patterns of lordosis
and copulation have been investigated through manipulation of the exposure of the developing rat brain
to sex steroids.

If a female rat has her ovaries removed (that is, she
is spayed), either as a newborn or as an adult, she will
not show lordosis unless she is injected with female
sex steroids. The hormones are necessary for the expression of the female sexual behavior. If the same
adult female is then injected with testosterone, she
shows no sexual behavior (first two panels of Figure
49.6a). There is a surprising variation on this experi-

ment. If a female rat has her ovaries removed and is injected with testosterone as a *newborn*, she will not show lordosis when treated with female sex hormones as an adult. But if this *genetically female* adult rat is treated with testosterone, she will mount other females in estrus and show the male fixed action patterns associated with copulation. Testosterone masculinizes the developing nervous system of the ovariectomized newborn. When she reaches adulthood, her nervous system responds to male rather than to female steroids and generates male fixed action patterns (third panel of Figure 49.6a).

Similar experiments on genetic males do not yield entirely reciprocal results. Castration (removal of the testes) of an adult male does not alter its response to treatment with sex steroids. Such a castrated male does not show lordosis when treated with female hormones, and it does show male sexual behavior when injected with testosterone. If a *newborn* male is castrated, however, it *will* show lordosis when injected with female hormones as an adult, but it will *not* show male sexual behavior when injected with testosterone as an adult. If the newborn male is castrated *and* injected with testosterone, when it becomes an adult it will not show lordosis in response to injections of female hormones. However, it will show normal male sexual behavior in response to testosterone (Figure 49.6b).

These results indicate that the nervous systems of both genetic males and genetic females develop female fixed action patterns if testosterone is not present at an early stage. Testosterone in the newborn causes the nervous system to develop male fixed action patterns whether the animal is a genetic male or a genetic female. In all cases, the expression of the sexual fixed action patterns in the adult requires a certain level of appropriate sex steroids.

Testosterone in birds affects growth of the brain regions responsible for song

Learning is essential for the acquisition of birdsong. Both male and female nestlings hear their species-specific song, but only the males of most songbirds sing as adults. Males use song to claim territory, compete with other males, and declare dominance. They also use song to attract females, suggesting that the females know the song of their species even if they do not sing. Do sex steroids control the learning and expression of song in male and female songbirds?

After leaving the nest where they experienced their father's singing, young songbirds from temperate and arctic habitats may migrate and associate with other species in mixed flocks, but they do not sing and they do not hear their species-specific song again until the following spring. As that spring approaches and the days become longer, the young male's testes begin to grow and mature. As his testosterone level rises, he begins to try to sing.

Even if he is isolated from all other males of his species, his song will gradually improve until it is a proper rendition of his species-specific song. At that point the song is **crystallized**—the bird expresses it in similar form every spring thereafter. The juvenile bird's brain learns the pattern of the song by hearing the father. During the subsequent spring, under the influence of testosterone, the bird learns to express that song—a behavior that then becomes rigidly fixed in its nervous system.

Why don't the females sing? Can't they learn the patterns of their species-specific song? Don't they have the muscular or nervous system capabilities necessary to sing? Or do they simply lack the hormonal stimulus for developing the behavior? To answer these questions, investigators injected female songbirds with testosterone in the spring. In response to these injections, the females developed their species-specific song and sang just as the males do. Apparently females learn the song pattern of their species when they are nestlings and have the capability to express it, but they normally lack the hormonal stimulation.

What does testosterone do to the brain of the songbird? A remarkable discovery revealed that testosterone causes the parts of the brain necessary for learning and expressing song to grow larger (Figure 49.7). Each spring, certain regions of the males' brains grow. The individual cells increase in size and grow longer extensions, and—most surprisingly—the *numbers* of brain cells in those regions of the bird brain increase. Before these discoveries, newborn vertebrates were thought to have their full complement of brain cells, which they would lose progressively throughout life without replacing them.

Research on the neurobiology of birdsong has revealed that hormones can control behavior by influencing brain structure as well as brain functioning on both a developmental and a seasonal basis. As we'll see in the next section, genes also play a role in controlling behavior.

The Genetics of Behavior

To say that behavior is genetically determined does not mean that specific genes code for specific behavior. Genes code for proteins, and there are many complex steps between the expression of a gene as a protein product and the expression of a behavior. All genes are separated from their phenotypic expression by many intermediate steps, but the complexity is especially great when the phenotypic trait is a behavior. A specific protein may affect behavior by playing a critical role in the development of patterns in the nervous system or in the functioning of the nervous or endocrine system; such influence is indirect and difficult to discover.

In no case are all the steps between a gene and a behavior known. Nevertheless, the approaches of genet-

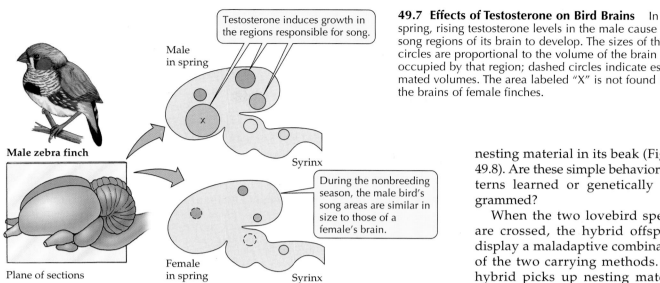

49.7 Effects of Testosterone on Bird Brains In spring, rising testosterone levels in the male cause the song regions of its brain to develop. The sizes of the circles are proportional to the volume of the brain occupied by that region; dashed circles indicate estimated volumes. The area labeled "X" is not found in the brains of female finches.

Male zebra finch

Testosterone induces growth in the regions responsible for song.

Male in spring

Syrinx

During the nonbreeding season, the male bird's song areas are similar in size to those of a female's brain.

Female in spring

Syrinx

Plane of sections

ics clearly show that behavior has genetic components and bring us closer to understanding the underlying mechanisms of inheritance of fixed action patterns. In this section we will look at three genetic approaches to learning how genes affect behavior: hybridization, artificial selection and crossing of the selected strains, and molecular analysis of genes and gene products.

Hybridization experiments show whether a behavior is genetically determined

The material for genetic analysis is variability, which is most pronounced between species. Closely related species frequently show large differences in fixed action patterns, and if such species can be hybridized, the offspring reveal interesting disruptions of their behavior. A classic case is nest building in lovebirds of the genus *Agapornis*. One species carries nesting material tucked under its tail feathers. Another species carries

nesting material in its beak (Figure 49.8). Are these simple behavior patterns learned or genetically programmed?

When the two lovebird species are crossed, the hybrid offspring display a maladaptive combination of the two carrying methods. The hybrid picks up nesting material and tucks it into its tail feathers, but does not release the material immediately. The hybrid inevitably pulls the nesting material out of its tail feathers and drops it. With years of experience, the hybrids learn to carry material in their beak, but they make an intention movement toward their tail feathers when they pick up nesting material. This study indicates that the ways in which birds of the two species carry nesting materials are genetically determined.

Konrad Lorenz conducted hybridization experiments on ducks to investigate the genetic determinants of their elaborate courtship. Dabbling ducks such as mallards, teals, pintails, and gadwalls are closely related and can interbreed, but because of the specificity of their courtship displays, in nature they rarely do so. Each male duck performs a carefully choreographed water ballet (Figure 49.9), and the female probably will not accept his advances unless the entire display is successfully completed.

The displays of dabbling duck species consist of about 20 components altogether. The display of each species includes a subset of these components put to-

(a) *Agapornis roseicollis*

(b) *Agapornis fischeri*

49.8 Nest-Building Behavior Is in the Genes (a) Peach-faced lovebirds carry nest-building materials tucked in their back feathers. (b) Fischer's lovebirds carry the objects in their bills. Hybrid offspring of the two species display a confused combination of the two behaviors, indicating that the behaviors are genetically programmed.

gether in a certain sequence. When Lorenz crossbred the species, he found that the hybrids expressed some components of the display of each parent in new combinations. Of particular interest, the hybrids sometimes showed display components that were not in the repertoire of either parent, but were characteristic of the displays of other species.

These hybridization studies clearly demonstrated that the motor patterns of the courtship displays are genetically programmed. Females were not interested in males showing the hybrid displays, thus demonstrating the adaptive significance of the species-specific fixed action patterns.

Artificial selection and crossbreeding reveal the genetic complexity of behaviors

Domesticated animals provide abundant evidence that artificial selection of mating pairs on the basis of their behavior can result in strains with distinct behavioral as well as anatomical characteristics. Among dogs, consider retrievers, pointers, and shepherds. Each has a particular behavioral tendency that can be honed to a fine degree by training, whereas other strains cannot be trained in this way. However, dogs and other large animals are not the best subjects for genetic studies. Most controlled selection experiments in behavioral genetics have been done on more convenient laboratory animals with short life cycle times and high numbers of offspring.

A favorite subject for such studies has been the fruit fly, genus *Drosophila*. Artificial selection has been successful in shaping a variety of behavior patterns in fruit flies, especially aspects of their courtship and mating behavior. Crossing of selected strains reveals that behavioral differences produced by artificial selection are usually due to multiple genes that probably influence the behavior indirectly by altering general properties of the nervous system.

Few behavioral genetic studies reveal simple Mendelian segregation of behavioral traits. One exception is nest-cleaning behavior in honeybees. Nest cleaning counteracts a bacterium that infects and kills the larvae of honeybees. One strain of honeybee that is resistant to this disease practices hygienic behavior: When a larva dies, workers uncap its brood cell and remove the carcass from the hive. Another strain of honeybee does not show hygienic behavior and therefore is more susceptible to the spread of the disease.

When these two strains were crossed, the results indicated that the hygienic behavior was controlled by two recessive genes. All members of the F_1 generation were nonhygienic, indicating that the behavior is controlled by recessive genes. Backcrossing the F_1 with the pure hygienic strain produced the typical 3:1 ratio ex-

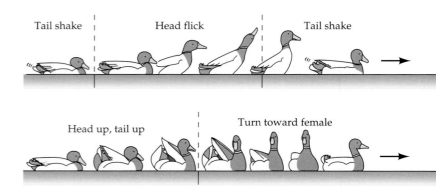

49.9 Courtship Ballet of the Mallard The courtship display of the male mallard duck contains about ten elements. Closely related duck species may display some of the same ten elements, but they have other elements not displayed by mallards. The elements of the courtship display and their sequence are species-specific and prevent hybridization.

pected for a two-gene trait. (To review the principles of Mendelian genetics, see Chapter 10.) The behavior of the nonhygienic individuals was very interesting. One-third of them showed no hygienic behavior at all; one-third uncapped the cells of dead larvae but did not remove them; and one-third did not uncap cells but did remove carcasses if the cells were open (Figure 49.10).

Even though these results appear to indicate a gene for uncapping and a gene for removal, these behavior patterns are complex. They involve sensory mechanisms, orientation movements, and motor patterns, each of which depends on multiple properties of many cells. The genetic deficits of nonhygienic bees could influence very small, specific, yet critical properties of some cells. If a single critical property, such as a crucial synapse or a particular sensory receptor, were lacking, the whole behavior would not be expressed. The responsible gene, then, is not a specific gene that codes for the entire behavior.

Molecular genetics reveals the roles of specific genes in determining behavior

The powerful techniques of molecular genetics enable investigation of the functions of specific genes that influence behavior. For example, in the marine mollusk *Aplysia*, egg laying consists of a sequence of fixed action patterns. The animal expels the eggs in a long string by contracting muscles of the reproductive duct. Then the animal stops whatever it is doing (usually eating or crawling) and takes the egg string in its mouth. With a series of stereotypic head movements, it pulls the egg string from the duct and coils it into a mass glued together by secretions from its mouth. Finally, with a strong head movement, it affixes the entire mass of eggs to a solid substratum.

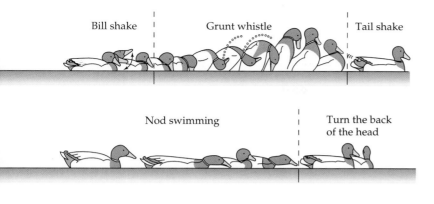

Bill shake Grunt whistle Tail shake

Nod swimming Turn the back of the head

Aplysia has a very simple nervous system, and specific cells in its nervous system were found to produce a peptide that can elicit certain aspects of egg laying. This peptide is called egg-laying hormone. The amino acid sequence of egg-laying hormone was determined, and then molecular genetic techniques were used to find the gene that codes for it.

The surprising discovery was that the gene codes for a precursor molecule that has almost 300 amino acids, whereas egg-laying hormone has only 36 amino acids. The precursor molecule also contains other peptides that function as neural signals controlling aspects of egg-laying behavior. One gene could thus code for a set of neural signals necessary to elicit the coordinated motor scores involved in egg-laying behavior. This example is about as close as we can get to making connections between a specific gene and a specific behavior, but the connections depend on the existence of a highly organized nervous system, which is a product of many genes.

Having discussed methods of investigating animal behavior, let's focus now on specific types of behavior. As investigations into animal communication illustrate, a diversity of issues can arise in studies of even a specific type of behavior.

Communication

Communication is behavior that influences the actions of other individuals. It consists of **displays** or **signals** that can be perceived by other individuals. Displays and signals are behaviors, anatomical features, or physiological responses that convey information to another indi-

vidual. The information they convey may be secondary or even incidental to their original function. If the transfer of information benefits the animal generating the signal or display, selection can shape the display to enhance its information content.

The displays or signals that an animal can generate depend on its physiology and anatomy. An animal's ability to perceive displays or signals depends on its sensory physiology and on the environment through which the display or signal must be transmitted. We learned in Chapter 42 that sensory physiology includes chemosensation, tactile sensation, audition, vision, and electrosensation. These are the forms of animal communication. Studies of communication can be complex because they must take into account the sender, the receiver, and the environment.

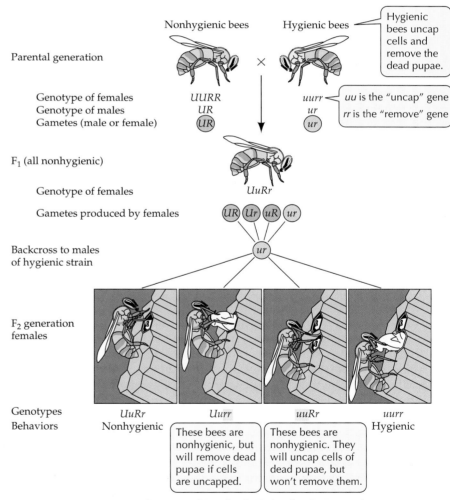

49.10 Genes and Hygienic Behavior in Honeybees
Some honeybee strains make a practice of removing the carcasses of dead larvae from their nests. This hygienic behavior seems to have two components—uncapping the larval cell (*u*), and removing the carcass (*r*)—each of which is under the control of a recessive gene.

(a)

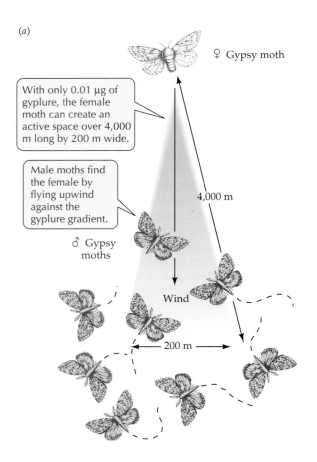

♀ Gypsy moth

With only 0.01 µg of gyplure, the female moth can create an active space over 4,000 m long by 200 m wide.

Male moths find the female by flying upwind against the gyplure gradient.

4,000 m

♂ Gypsy moths

Wind

200 m

(b)

Acinonyx jubatus

49.11 Many Animals Communicate with Pheromones
(a) A female gypsy moth secretes the pheromone gyplure, which can attract males thousands of meters downwind when it binds to sensors on their antennas. (b) To mark his territory, this male cheetah is spraying pheromones from a scent gland in his hindquarters onto a tree. Other cheetahs passing the spot will know that the area is "claimed," and they will know something about who claimed it.

In the discussion that follows, we will explore in turn each of the five forms of animal communication: chemical, visual, auditory, tactile, and electrical. We will conclude by delving briefly into the origins of communication signals.

Chemical signals are durable but inflexible

Chemosensation, which is based on the binding of signaling molecules at receptor sites on sensors that initiates cellular responses, is probably the oldest form of animal communication. Intracellular signaling involves molecules and sensors, and hormonal integration preceded neural integration in animal evolution. It is a very short evolutionary jump from internal coordination by signal molecules and sensors to the release of molecules into the environment for detection by other individuals. Even protists use such means of communication.

Slime molds spend most of their lives as individual amoeboid cells moving through soil and leaf litter, as long as food and moisture are adequate (see Chapter 26). When the environment becomes less favorable, the individuals aggregate, form a fruiting body, and release spores. The chemical signal that coordinates this aggregation is cAMP (cyclic adenosine monophosphate), a well-known intracellular signaling molecule (see Chapter 38). Stressed cells release cAMP into the environment, and other individuals sensing the cAMP move in the direction of higher concentrations.

Molecules used for chemical communication between individual animals are called **pheromones**. Because of the diversity of their molecular structures, pheromones can communicate very specific messages that contain a great deal of information. For example, when a female gypsy moth is ready to be inseminated, she releases a pheromone called gyplure. Male gypsy moths downwind by as much as thousands of meters are informed by these molecules that a female of their species is sexually receptive. By orienting to the wind direction and following the concentration gradient of the molecules, they can find her (Figure 49.11a). Territory marking is another example in which detailed information is conveyed by chemical communication (Figure 49.11b). The receiver of a message from a male cheetah, for example, detects information about the animal that is leaving the message: species, reproductive status, height of the message (indicating the size of the animal that left it), and strength of the scent (indicating the amount of elapsed time).

An important feature of pheromones is that once they are released, they remain in the environment for a long time. A pheromonal message can act as a territorial marker long after the animal claiming the territory has left. By contrast, the signals of a vocal or visual territorial display of a songbird disappear as soon as the bird stops singing and displaying. The durability of pheromonal signals enables them to be used to mark trails, as ants do, or to indicate directionality, as in the case of the gypsy moth sex attractant.

The chemical nature and the size of the pheromonal molecule determine its diffusion coefficient. The greater its diffusion coefficient, the more rapidly a pheromone diffuses and the farther the message will reach, but the sooner it will disappear. Trail-marking and territory-marking pheromones tend to be relatively large molecules with low diffusion coefficients; sex attractants tend to be small molecules with high diffusion coefficients. A disadvantage of pheromonal communication is that the message cannot be changed rapidly. A discussion based on smells would surely be less effective than one using speech or sign language.

Visual signals are rapid and versatile but limited by directionality

Many species use visual communication. Visual signals are easy to produce, come in an endless variety, can be changed very rapidly, and clearly indicate the position of the signaler. However, the extreme directionality of visual signals means that they are not the best means of getting the attention of a receiver. The sensors of the receiver must be focused on the signaler or the message will be missed. Most animals are sensitive to light and can therefore receive visual signals, but sharpness of vision limits the detail of the information that can be transmitted. Birds are highly visual and have evolved a vast diversity of patterns of colored feathers and body appendages that can be incorporated into complex displays used in communication (Figure 49.12).

Because visual communication requires light, it is not useful for many species at night or in environments that lack light, such as caves and the ocean depths. Some species have surmounted this constraint on visual communication by evolving their own light-emitting mechanisms. Fireflies use a enzymatic mechanism to create flashes of light. By emitting flashes in species-specific patterns, fireflies can advertise for mates at night by sending visual signals.

Firefly communication raises an interesting issue. Any system of communication is vulnerable to exploitation by illicit senders or receivers. Predators are common illicit receivers. When an animal emits a message in any form, it tends to signal its own position, and the information can be used by predators. Predators of fireflies can locate them in the dark by the flashes of light they emit. Some species of fireflies are themselves predators of other species of fireflies and have evolved the deceitful behavior of emitting signal patterns that mimic the mating messages of other species. When a prospective suitor homes in on the copied signal, it is eaten.

49.12 Selection for Effective Visual Communication Male peafowl—peacocks—have evolved brilliant tail plumage, which they display to the females (peahens) during courtship. These elaborate tail feathers are folded when the bird is not displaying.

Auditory signals communicate well over a distance

Humans are very familiar with communicating by sound. In Chapter 42 we discussed the physical properties of sound and the sensory structures that transduce sound pressure waves into neural signals. Compared with visual communication, auditory communication has several obvious advantages and disadvantages.

Sound can be used at night and in dark environments. Sound can go around objects that would interfere with visual signals. Sound is better than visual signals at getting the attention of a receiver, because the sensors do not have to be focused on the signaler for the message to be received. Like visual signals, sound can provide directional information, as long as the receiver has at least two sensors spaced somewhat apart. Differences in the intensity, in the time of arrival, and in the phase of the pressure wave of a sound reaching the two sensors can provide information about the direction of the sound source. By maximizing or minimizing these features of the sounds they emit, animals can make their location easier or more difficult to determine. Pure tones with sudden onsets and offsets are easier for the receiver to localize than are complex sounds with gradual onsets and offsets.

Sound is good for communicating over long distances. Even though the intensity of sound decreases with distance from the source, loud sounds can be used to communicate over distances much greater than those possible with visual signals. An extreme example is the communication of whales. Some whales, such as the humpback, have very complex songs. When these sounds are produced at a certain depth (around 1,000 m), the sound waves are channeled between the thermocline (a sudden temperature change in the water column) and much deeper waters, so they can be heard hundreds of kilometers away. In this way, humpback whales use auditory communication to locate each other over vast areas of ocean.

Auditory signals cannot convey complex information as rapidly as visual signals can, as the well-known expression "A picture is worth a thousand words" implies. When individuals are in visual contact, an enormous amount of information is exchanged instantaneously (for example, species, sex, individual identity, maturity, level of motivation, dominance, vigor, alliances with other individuals, and so on). Coding that amount of information, with all of its subtleties, as auditory signals would take considerable time, thus increasing the possibility that the communicators could be located by predators.

Remarkably few species produce sound or communicate by sound. The animal world is relatively silent. Most invertebrates do not produce sound; cicadas and crickets are marvelous exceptions. Most fish, amphibians, and reptiles also produce no sound.

Tactile signals can communicate complex messages

Communication by touch is extremely common, although not always obvious. Animals in close contact use tactile interactions extensively, especially under conditions that do not favor visual communication. When social insects such as ants, termites, or bees meet, they contact each other with their antennas and front legs. Some of these contacts may exchange chemical signals as well as tactile ones. In Chapter 42 we discussed how the lateral-line organs of fish are useful in sensing vibrations in the environment. Such vibrations can be used for intraspecific communication. At the beginning of this chapter we discussed how male spiders communicate with females by creating vibrations in their webs.

One of the most remarkable and best-studied uses of tactile communication is the dance of honeybee that is used to convey information about distance and direction to a food source. Dancing bees make sounds and carry odors on their bodies, but they convey a great deal of information by dance movement. The dances are monitored by other bees, who follow and touch the dancer to interpret the message.

When a foraging bee finds food, she returns to the hive and communicates her discovery by dancing in the dark on the vertical surface of the honeycomb. If the food is less than 80 to 100 m from the hive, she performs a **round dance**, running rapidly in a circle and reversing her direction after each circumference. The odor on her body indicates the flower to be looked for, but the dance contains no information about the direction to go—only that it is within 100 m of the hive.

If the food source is farther than 80 to 100 m, the bee performs a **waggle dance**, which conveys information about both the distance and the direction of the food source. In the waggle dance, the bee repeatedly traces out a figure eight pattern as she runs on the vertical surface. She alternates half circles to the left and right with vigorous wagging of her abdomen in the short, straight crossover between turns. The angle of the straight line indicates the direction of the food source relative to the direction of the sun (Figure 49.13). The speed of the dancing indicates the distance to the food source: The farther away it is, the slower the waggle run.

The dance of the honeybee is unusual because it is based on an arbitrary convention—that straight up represents the direction of the sun. Arbitrary, symbolic conventions like this have been developed to an extreme degree in human language.

Electric signals can communicate messages

Some fish use electrosensors to gather information about their murky environment. Other fish generate electric fields by emitting series of electric pulses. Although these trains of electric pulses can be used for sensing objects in the immediate surroundings of the fish, they can also be used for communication. An electrode connected to an amplifier and a speaker can be used to "listen" to the signal generated by each fish in a tank holding numerous electric fish. The amplifier reveals that each fish emits a pulse at a different frequency, and the frequency each fish uses relates to its status in the population.

Glass knife fish (*Eigenmannia*) males emit lower frequencies than females. The most dominant male has the lowest frequency, and the most dominant female has the highest frequency. When a new individual is introduced into the population, the other individuals adjust their frequencies so that they do not overlap, and the signal of the new individual indicates its position in the population. In their natural environment—the murky waters of tropical rainforests—these fish can tell the identity, sex, and social position of another member of the population by its electric signals. When two individuals interact directly, they interrupt their constant frequency emissions and modulate them to produce more complex, "chirplike" signals that perhaps communicate even more information.

Communication signals evolve by means of natural selection on behavior

Included among the constraints on the evolution of communication signals are the anatomical and physiological characteristics of a species that are available to be shaped by natural selection for the purpose of conveying information to other individuals. Charles Darwin was the first person to give serious attention to this problem of the origin and evolution of commu-

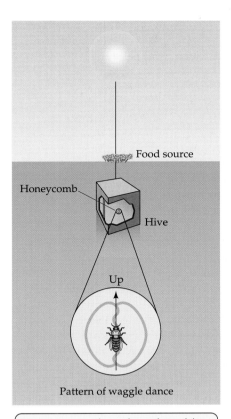

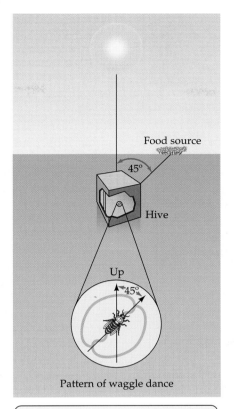

By running straight up the surface of the honeycomb in a dark hive, a honeybee tells her hive mates that there is a food source in the direction of the sun. The intensity of the waggle indicates exactly how far away the food source is. If the food source were in the opposite direction from the sun, she would orient her waggle runs straight down.

When the bee orients her waggle dance at an angle from the vertical, the other bees know that the same angle separates the direction of the food source from the direction of the sun.

49.13 The Waggle Dance of the Honeybee   This dance is performed by a bee to guide hivemates to a food source at least 80 m away.

blackbirds, for example, the red shoulder patches are most effectively displayed when the wings are slightly raised in the threat posture (see Figure 46.15*b*). Darwin pointed out the threat posture of the domestic cat, which appears to result from conflicting motivations. The forelegs push back while the hind legs push forward, resulting in the hunched back that makes the cat look bigger and more threatening (Figure 49.14).

Autonomic responses (see Chapter 41) are another possible source of material on which selection can operate to produce displays. The erection of fur or feathers under the control of the sympathetic nervous system is one obvious example. Urination and defecation are autonomic responses that have been used extensively as signals.

The mating display of the male frigate bird is a picturesque example of autonomic response that is used as a signal. Frigate birds are large and black and nest on oceanic islands

nication signals. In 1872 he published the results of his detailed studies in a book entitled *The Expression of the Emotions in Man and Animals*. In this perceptive book Darwin identified several important means by which communication signals originate.

One source of raw material for the evolution of signals is **intention movements**: movements that precede a particular behavior. For example, a bird ready to take off flexes its legs, sleeks its feathers, and raises its wings slightly at the shoulders. A bird in a threatening situation, such as facing a neighbor that is challenging its territorial boundary, will experience a conflict between motivation to flee and motivation to attack. Thus its intention movements are exaggerated and mixed. Raising its wings at the shoulders might make it appear bigger to its adversary, thus augmenting its attack intention movements. The combination of behavior patterns will convey information to the adversary about the degree of motivation of the defender.

Selection can work to enhance a threat display that originated in intention movements. In red-winged

49.14 Threat Display of a Cat   This drawing by Charles Darwin shows the arched-back posture of a cat when it is experiencing conflicting motivations to attack and to flee.

Fregata aquila

49.15 Frigate Bird Display   A male frigate bird displays his red throat pouch for the female next to him. This courtship display may have evolved from the throat-fluttering behavior that birds use in thermoregulation.

near the equator. The male builds a nest and then sits on it, trying to attract females flying overhead by spreading his wings, inflating a huge, bright red pouch in his throat, and shaking his head and pouch while vocalizing. This throat fluttering is a common thermoregulatory response of birds that involves rapid movements of air in and out of pouches of skin on the front of the neck. The bizarre and dramatic courtship display of the male frigate bird probably originated through exaggeration of autonomic thermoregulatory behaviors (Figure 49.15).

Displacement behavior is a third class of behavior patterns that can be shaped by natural selection into communication displays. In a tense situation that involves highly conflicting motivations, such as attack and escape, an animal sometimes does something completely irrelevant. It might groom itself, feed, or attack an object in the vicinity. If such behavior enhances the display of the animal, it can be favored by selection and incorporated into the display.

For example, male three-spined stickleback fish defend small territories when they build a nest to attract potential mates. At the boundary of his territory, the male stickleback can experience equally strong motivations to attack and to flee from a neighboring male. During boundary disputes, the males engage in head-down threat displays that resemble nest-building postures, but no nest-building takes place. This threat display probably evolved from nest-building displacement activities resulting from the approach–avoidance conflicts of boundary disputes.

This discussion of the origins of animal communication has focused on examples of how natural selection shapes the behavior of a species. In Chapter 50 we will study in more detail the evolution and functions of displays in the context of social behavior.

The Timing of Behavior: Biological Rhythms

An important aspect of behavior is its temporal organization. The neurophysiology of most behavior is poorly understood, but the study of biological rhythms has led to major discoveries about brain mechanisms. In the discussion that follows, we will examine two types of biological rhythms: circadian rhythms and circannual rhythms.

Circadian rhythms control the daily cycle of animal behaviors

Our planet turns on its axis once every 24 hours, creating a cycle of environmental conditions that has existed throughout the evolution of life. Many organisms thus evolved rhythmicity. In Chapter 36 we encountered rhythmicity in plants. Daily cycles are a characteristic of almost all organisms. What is surprising, however, is that daily rhythmicity does not depend on the 24-hour cycle of light and dark.

If animals are kept under absolutely constant environmental conditions, such as constant dark and constant temperature, with food and water available all the time, they still demonstrate daily cycles of activity: sleeping, eating, drinking, and just about anything else that can be measured. This persistence of the cycle in the absence of changes between light and dark suggests that the animal has an endogenous (internal) clock. Without time cues from the environment, however, these daily cycles are not exactly 24 hours long. They are therefore called **circadian rhythms** (from the Latin *circa*, "about," and *dies*, "day").

To discuss biological rhythms, we must introduce some terminology. A rhythm can be thought of as a series of cycles, and the length of one of those cycles is the **period** of the rhythm. Any point on the cycle is a **phase** of that cycle. Hence when two rhythms completely match, they are in phase, and if a rhythm is shifted (as in the resetting of a clock), it is phase-advanced or phase-delayed. Since the period of a circadian rhythm is not exactly 24 hours, it must be phase-advanced or phase-delayed every day to remain in phase with the daily cycle of the environment.

ENTRAINMENT.   The process of resetting the circadian rhythm by environmental cues is called **entrainment**. An animal kept in constant conditions will not be entrained to the 24-hour cycle of the environment, and its circadian clock will run according to its natural periodicity—it will be free-running. If its period is less than 24 hours, the animal will begin its activity a little earlier each day (middle panel of Figure 49.16).

Circannual rhythms control the yearly cycle of some animals

In addition to turning on its axis every 24 hours, our planet revolves around the sun once every 365 days. Because Earth is tilted on its axis, its revolution around the sun results in seasonal changes in day length at all locations except the equator. These changes in day length secondarily create seasonal changes in temperature, humidity, weather, and other variables. Because the behavior of animals must adapt to these seasonal changes, animals must be able to anticipate seasons and adjust their behavior accordingly. For example, most animals should not come into reproductive condition and mate just before winter, because if they did, their offspring would be born during a time of little food and harsh weather conditions.

For many species, the change in day length is an excellent and absolutely reliable indicator of seasonal changes to come. If photoperiod has a direct effect on the physiology and behavior of a species, that species is said to be photoperiodic. Deer are photoperiodic; male deer held in captivity and subjected to two cycles of change in day length in one year will grow and drop their antlers twice during that year. Many species of birds are also photoperiodic.

For some animals, change in day length is not a reliable cue. Hibernators, for example, spend long months in dark burrows underground but have to be physiologically prepared to breed almost as soon as they emerge in the spring. The timing of their breeding is important because their young must have time to grow and fatten before the next winter. Other examples of animals that receive little or ambiguous information from changes in day length are resident tropical species, migratory species that overwinter near the equator, or migratory species (mostly birds) that cross the equator.

A bird overwintering in the tropics cannot use change in photoperiod as a cue to time its migration north to the breeding grounds. A bird that crosses the equator must fly south as day length decreases at one time of year but fly north when day length decreases at another time of year. If change in day length is not a reliable cue for seasonal behavior, what else could be used as a calendar?

Hibernators and equatorial migrants have endogenous annual rhythms, called **circannual rhythms**. Their nervous systems have a built-in calendar. Just as circadian rhythms are not exactly 24 hours long, circannual rhythms are not exactly 365 days long. The circannual rhythm of an animal under constant conditions may be 360 days, or 345 days. But rarely is it longer than 365 days, because being late for an annual event such as breeding would be a very costly mistake.

Finding Their Way

Within a local environment, finding your way is not a problem. You remember landmarks and orient yourself with respect to those reference points. **Orientation** is a very common animal behavior. It means simply that the animal organizes its activity spatially with respect to reference points such as objects in the environment, a predator or prey, a mate or offspring, a nest or food source, or even a signal such as the call or display of another individual. But what if the destination is a considerable distance away? How does the animal orient to it and find its way?

Animals navigate over long distances and through unfamiliar territory by piloting, homing, and migrating. In this section we'll explore each of these navigational modes. Then we'll examine the underlying mechanisms of navigation.

Piloting animals orient themselves by means of landmarks

In most cases an animal finds its way using simple means: It knows and remembers the structure of its environment. It uses landmarks to find its nest, a safe hiding place, or a food source. Orienting by means of landmarks is called **piloting**. Even an animal that migrates long distances can pilot, as long as it does not depend on specific landmarks.

For example, the gray whales that spend the summer feeding in the Gulf of Alaska and the Bering Sea and migrate south in the winter to breed in lagoons on the Pacific coast of Baja California can find their way by following two simple rules (Figure 49.18): Keep the land to the left in the fall and to the right in the spring. By following the west coast of North America, they can travel from summer to winter areas and back again by piloting. Coastlines, mountain chains, rivers, water currents, and wind patterns serve as piloting cues for many species. But some remarkable cases of long-distance orientation and movement cannot be explained on the basis of piloting by landmarks.

Homing animals can return repeatedly to a specific location

The ability of an animal to return to a nest site, burrow, or any other specific location is **homing**. In most cases homing is merely piloting in a known environment, but some animals are capable of much more sophisticated feats of navigation. People who breed and race homing pigeons take the pigeons from their home loft and release them at a remote site where they have never been before. The first pigeon that reaches its home wins.

Data on departure directions, known flying speed, and distance traveled show that the pigeons fly fairly directly from the point of release to home. They do not

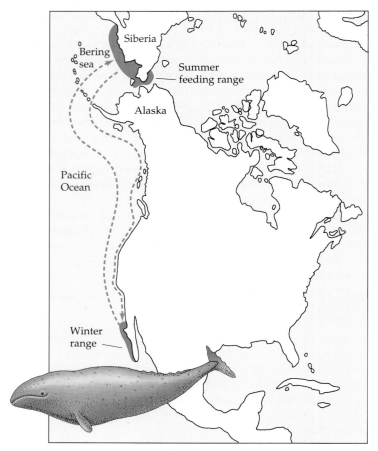

49.18 Piloting Gray whales migrate south in winter, from the Bering Sea to the coast of Baja California, by following a landmark: the west coast of North America.

randomly search until they encounter familiar territory. Scientists have used homing pigeons to investigate the mechanisms of how animals navigate. In one series of experiments the pigeons were fitted with frosted contact lenses. They could see no details other than degree of light and dark. These pigeons still homed and fluttered down to the ground in the vicinity of their loft. They were able to navigate without visual images of the landscape.

Marine birds provide many dramatic examples of homing over great distances in an environment where landmarks are rare. In daily feeding trips, many marine birds fly over hundreds of miles of featureless ocean and then return directly to a nest site on a tiny island. Albatrosses display remarkable feats of homing. When a young albatross leaves its nest on an oceanic island, it flies widely over the southern oceans for 8 or 9 years before it reaches reproductive maturity. At that time it flies back to the island where it was raised to select a mate and build a nest (Figure 49.19). After the first mating season the pair separate, and each bird resumes its solitary wanderings over the

oceans. The next year they return to the same nest site at the same time, reestablish their pair bond, and breed. Thereafter they return to the nest to breed every other year, spending many months in between at sea. These long-distance, synchronous homing trips are amazing feats of navigation and timing.

Migrating animals travel great distances with remarkable accuracy

Ever since humans inhabited temperate and subpolar latitudes, they must have been aware that whole populations of animals, especially birds, disappear and reappear seasonally—that is, they **migrate**. Not until the early nineteenth century, however, were patterns of migration established by the marking of individual birds with identification bands around their legs. Being able to identify individual birds in a population made it possible to demonstrate that the same birds and their offspring returned to the same breeding grounds year after year, and that these same birds were found during the nonbreeding season at distant locations hundreds or even thousands of kilometers from the breeding grounds.

How do migrants find their way over such great distances? A reasonable hypothesis is that young birds on their first migration follow experienced birds and learn the landmarks by which they will pilot in subsequent years. However, adult birds of many species leave the breeding grounds before the young have finished fattening and are ready to begin their first migration. These naive birds must be able to navigate accurately on their own, and with little room for mistakes.

Some species of small songbirds breed in the high latitudes of North America, fly to the coast for fattening, and then fly over the North Atlantic Ocean on a direct route to South America (Figure 49.20). They can-

Diomedea melanophris

49.19 Coming Home A pair of black-browed albatrosses engage in courtship display over their partly completed mud nest. Many albatrosses return to the site of their own birth to find a mate, and will return to the same site year after year.

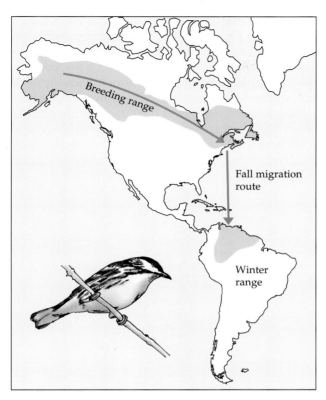

49.20 Songbirds Migrate over the Atlantic   The blackpoll warbler is one of many species that breeds over the northern United States and Canada and winters in South America. Its fall migration route takes it first to the northeast coast of North America, where it feeds in preparation for the nonstop overwater flight to South America.

not land on water, and their fuel reserves are limited by their small size. Considering distance, flight speed, and metabolic rate, they must be extremely efficient and accurate in navigating to their landfall on the coast of South America.

Navigation is based on internal and environmental cues

Homing and migrating animals find their way by several mechanisms of navigation. Although piloting is a type of navigation, it cannot explain the abilities of many species to take direct routes to their destinations through areas they have never experienced. Humans use two systems of navigation that differ in complexity: distance-and-direction navigation and bicoordinate navigation.

Required for *distance-and-direction navigation* is knowledge of the direction to reach the destination and of how far away that destination is. With a compass to determine direction and a means of measuring distance, humans can navigate. *Bicoordinate navigation*, also known as true navigation, requires knowledge of the latitude and longitude (the map coordinates) of

position and destination. From that information a route can be plotted to the destination. Do animals have these sophisticated abilities to navigate?

DISTANCE-AND-DIRECTION NAVIGATION.   Researchers conducted an experiment with European starlings to determine their method of navigation. These short-distance migrants travel between breeding grounds in the Netherlands and northern Germany and wintering grounds to the southwest, in southern England and western France (Figure 49.21). Birds were captured at the beginning of their fall migration and transported to Switzerland. When naive juvenile birds were released, they flew in the normal southwest direction and landed up in southwestern France and Spain. Experienced adult birds, on the other hand, were not as disrupted by the geographical displacement and traveled northwest, returning to their normal wintering grounds. Researchers concluded that the naive juvenile starlings used distance and direction for navigation.

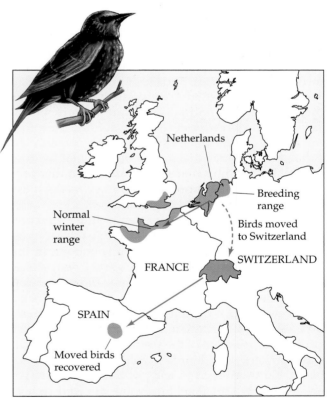

49.21 Navigation by Distance and Direction   European starlings normally make a short winter migration in a southwesterly direction, from the Netherlands to coastal France and southern England (red arrow). Experimental populations of juvenile starlings that were moved to a site in Switzerland did not fly northwest to their traditional grounds, but followed the same southwesterly route (blue arrow), which took them to Spain.

49.22 Raring to Go   A captive bird ready to migrate shows migratory restlessness in a circular cage. The cage is lined with a paper funnel, and on the floor is an ink pad. The bird's feet mark the orientation of its activity.

How do animals determine distance and direction? In many instances, distance is not a problem as long as the animal recognizes its destination. Homing animals recognize landmarks and can pilot once they reach familiar areas. Some evidence suggests that biological rhythms play a role in determining migration distances for some species. Birds kept in captivity display increased and oriented activity at the time of year when they would normally migrate (Figure 49.22). Such **migratory restlessness** has a definite duration, which corresponds to the usual duration of migration of the species. Since distance is determined by how long an animal moves in a given direction, the programming of the duration of migratory restlessness can set the distance for its migration.

Two obvious candidates for determining direction are the sun and the stars. During the day the sun is an excellent compass, as long as time is known. In the Northern Hemisphere the sun rises in the east, sets in the west, and points south at noon. Animals can tell the time of day by means of their circadian clocks. Furthermore, clock-shifting experiments demonstrate that animals use circadian clocks to determine direction from the position of the sun.

Researchers placed birds in a circular cage that enabled them to see the sun and sky but no other visual cues (Figure 49.23a). Food bins were arranged around the sides of the cage, and the birds were trained to expect food in the bin in one particular direction—south, for example (Figure 49.23b). After training, no matter when they were fed, and even with the cage rotated between feedings, the birds always went to the bin at the southern end of the cage for food, even if that bin contained no food. Next the birds were placed in a

room with a controlled light cycle, and their circadian rhythms were phase-shifted. For example, in the controlled light room the lights were turned on at midnight. After about 2 weeks the circadian clocks of the birds were phase-advanced by 6 hours. Then the birds were returned to the circular cage under natural light conditions, with sunrise at 6:00 A.M. Because of the shift in their circadian rhythms, their endogenous clocks were indicating noon at the time the sun came up.

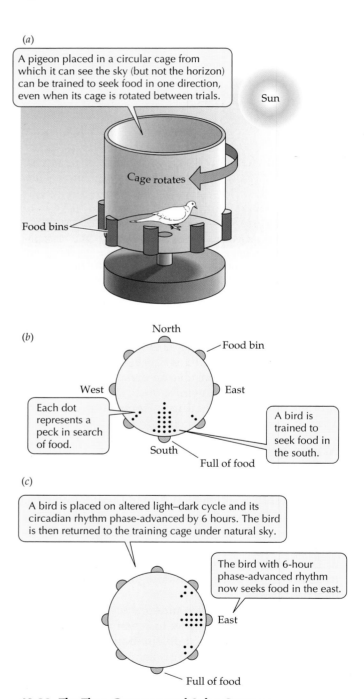

(a)

A pigeon placed in a circular cage from which it can see the sky (but not the horizon) can be trained to seek food in one direction, even when its cage is rotated between trials.

Sun

Cage rotates

Food bins

(b)

North

Food bin

West

East

Each dot represents a peck in search of food.

A bird is trained to seek food in the south.

South

Full of food

(c)

A bird is placed on altered light–dark cycle and its circadian rhythm phase-advanced by 6 hours. The bird is then returned to the training cage under natural sky.

The bird with 6-hour phase-advanced rhythm now seeks food in the east.

East

Full of food

49.23 The Time-Compensated Solar Compass

If food was always in the south, and it was sunup, they should have oriented 90 degrees to the right of the direction of the sun. But since their circadian clocks were telling them it was noon, they looked for food in the direction of the sun—the east bin (Figure 49.23c). The 6-hour phase shift in their circadian clocks resulted in a 90-degree error in their orientation. These types of experiments on many species have shown that animals can orient by means of a *time-compensated solar compass.*

Many animals are normally active at night; in addition, many day-active species of birds migrate at night and thus cannot use the sun to determine direction. The stars offer two sources of information about direction: moving constellations and a fixed point. The positions of constellations change because Earth is rotating. With a star map and a clock, direction can be determined from any constellation. But one point that does not change position during the night is the point directly over the axis on which Earth turns. In the Northern Hemisphere, a star called Polaris or the North Star lies in that position and always indicates north.

Stephen Emlen at Cornell University investigated whether birds use these sources of directional information from the stars. Emlen raised young birds in a planetarium, where star patterns are projected on the ceiling of large, domed room. The star patterns in the planetarium could be slowly rotated to simulate the rotation of Earth. When the star patterns were not rotated, birds caught in the wild could orient well in the planetarium, but birds raised in the planetarium under a nonmoving sky could not. If the star patterns in the planetarium were rotated each night as the young birds matured, they were able to orient in the planetarium, showing that birds can learn to use star patterns for orientation if the sky rotates (Figure 49.24).

However, these experiments provided no evidence that the birds use their circadian clocks to derive directional information from the star patterns. Experienced birds were not confused by a still sky or a sky that rotated faster than normal. The birds were orienting to the fixed point in the sky, the North Star. Young birds raised under a sky that rotated around a different star imprinted on that star and oriented to it as if it were the North Star. These studies showed that birds raised in the Northern Hemisphere learn a star map that they can use for orientation at night by imprinting on the fixed point in the sky.

Animals cannot use sun and star compasses when the sky is overcast, yet they still home and migrate under such conditions. Do other sources provide information they can use for orientation? There appears to be considerable redundancy in animals' abilities to sense direction. Pigeons are able to home as well on overcast days as on clear days, but this abil-

49.24 The Ability to Alter Star Patterns in a Planetarium Can Reveal Whether Birds Migrate by the Stars This scientist has placed birds in orientation cages (see Figure 49.22) in a planetarium. By changing the positions or movements of the stars projected on the planetarium ceiling, he can investigate what information the birds use to orient their migratory restlessness.

ity is severely impaired if small magnets are attached to their heads.

These experiments and subsequent ones with more sophisticated ways of disrupting the magnetic field around the bird have demonstrated a magnetic sense. Cells have been found that contain small particles of the magnetic mineral magnetite, but the neurophysiology of the magnetic sense is largely unknown. Another cue is the plane of polarization of light, which can give directional information even under heavy cloud cover. Very low frequencies of sound can give information about coastlines and mountain chains. Weather patterns can also provide considerable directional information.

BICOORDINATE NAVIGATION. Much less is known about the mechanisms of bicoordinate navigation than about distance-and-direction navigation, but some animals have the ability both to sense their geographical positions and to know where they should go. In other words, they have a map sense. Distance and direction capabilities are of little help without a map. Information about longitude and latitude is available from natural cues, but the evidence that animals can or do use these sources of information is meager.

Longitude can be determined by position of the sun and time of day: If the sun comes up earlier than expected, the animal must be east of home, and if the sun comes up later than expected, then it is west of home. Time and sun position give information about latitude as well. At a given time of day in the Northern

Hemisphere, a sun position higher in the sky than expected indicates that an animal is south of home, and if it is lower in the sky than expected the animal is north of home. Other information about longitude and latitude can come from sensing Earth's magnetic lines of force and from the positions of the stars.

To pinpoint home using these sources of information, an animal would have to have extremely precise and accurate sensory capabilities that have yet to be demonstrated. Perhaps, however, pinpointing home is not required of an animal's bicoordinate navigational abilities. If an animal gets anywhere near home, piloting can take over, and piloting can use a variety of long-distance cues—such as coastlines, smells, and low-frequency sounds—that have greater ranges than specific visual landmarks have.

Human Behavior

As we saw early in this chapter, the behavior of an animal is a mixture of components that are genetically programmed and components that can be molded by learning. However, even some aspects of learned behavior patterns—such as what can be learned and when it can be learned—may be genetically determined. Thus natural selection shapes not only the physiology and morphology of a species, but also its behavior. In some situations natural selection favors fixed action patterns; in others, learned behavior. In many cases the optimal adaptation is a mixture of fixed and learned behavioral components. Given these considerations, how would we characterize human behavior?

An important characteristic of human behavior is the extent to which it can be modified by experience. The transmission of learned behavior from generation to generation—culture—is the hallmark of humans. Nevertheless, the structure and many functions of our brain are coded in our genome, including drives, limits to and propensities for learning, and even some motor patterns. Biological drives such as hunger, thirst, sexual desire, and sleepiness are inherent to our nervous systems. Is it reasonable, therefore, to expect that emotions such as anger, aggression, fear, love, hate, and jealousy are solely the consequences of learning?

Our sensory systems enable us to use certain subsets of information from the environment; similarly, the structure of our nervous system makes it more or less possible to process certain types of information. Consider, for example, how basic and simple it is for an infant to learn spoken language, yet how many years that same child must struggle to master reading and writing. Verbal communication is deeply rooted in our evolutionary past, whereas reading and writing are relatively recent products of human culture.

Finally, evidence suggests that some motor patterns are programmed into our nervous systems. Studies of diverse human cultures from around the world reveal basic similarities of facial expressions and body language in human populations that have had little or no contact with one another. Infants born blind smile, frown, and show other facial expressions at appropriate times, even though they have never observed such expressions in others.

Acknowledging that our behavior has been shaped through evolution in no way detracts from the value we place on the learning abilities of humans. Human behavior is genetically determined in terms more of broad outline than of fine detail.

Summary of "Animal Behavior"

- There are three classes of behavior: acts to acquire food, acts to avoid environmental threats, and acts to reproduce.

Genetically Determined Behavior

- Behaviors that are genetically determined, called fixed action patterns, are stereotypic and species-specific. They neither require nor are modified by learning.
- Experiments that deprive an animal of all opportunities to learn a behavior can reveal whether a behavior is genetically determined.
- Fixed action patterns are triggered by simple stimuli called releasers. A supernormal releaser is more effective than the natural condition in eliciting a fixed action pattern. **Review Figures 49.1, 49.3**
- Releasers stimulate fixed action patterns only if the animal is in the correct motivational state, which is determined by its developmental and physiological state.
- Genetically programmed behavior is highly adaptive for species that have little opportunity to learn, for species that might learn the wrong behavior, and in situations where mistakes are costly or dangerous. **Review Figure 49.4**
- Genetic limitations on learned behavior can help an animal learn the correct behavior by focusing it on the correct stimuli at the correct times. **Review Figure 49.5**

Hormones and Behavior

- Hormones control the complex interaction between genetically determined behavior and learning.
- In rats, sex steroids present during development control what sexual behavior patterns develop, and sex steroids in the adult control the expression of those patterns. **Review Figure 49.6**
- In birds, testosterone determines a bird's ability to sing by causing the brain regions responsible for song to grow. **Review Figure 49.7**

The Genetics of Behavior

- There are many complex steps between the expression of a gene as a protein product and the expression of a behavior. Genetic experiments help reveal how genes affect behavior.

- Hybridization experiments reveal whether a behavior is genetically determined.
- Artificial selection and crossbreeding, which can produce individuals with particular behavioral traits, show that behaviors are controlled by the interactions of multiple genes. **Review Figure 49.10**
- Molecular-genetics techniques can reveal the functions of specific genes that influence behavior.

Communication

- Communication consists of chemical, visual, auditory, tactile, or electrical displays, or signals, that can be perceived by other individuals.
- At the simplest level, animals communicate by emitting pheromones into the environment and by sensing the pheromones of other animals. Pheromonal messages last a long time, but they cannot be changed quickly. **Review Figure 49.11**
- Visual communication is easy, versatile, and rapid, but it is limited by its directionality, by the visual acuity of the receiver, and by environmental conditions such as darkness. Many animals communicate via visual signals.
- Auditory signals can be used at night, can go around objects that would interfere with visual communication, can easily get the receiver's attention, can provide directional information, and can travel great distances. Compared to visual communication, however, auditory communication is slow. Few animals communicate via auditory signals.
- Tactile communication is very common. Tactile signals can communicate complex messages, as the dance of the honeybee demonstrates. **Review Figure 49.13**
- In addition to being used for sensing objects in the immediate surroundings, the electric signals generated by some animals, such as certain fish species, can be used for communication.
- The evolution of communication signals is constrained by the anatomical and physiological characteristics of a species that are available to be shaped by natural selection for the purpose of conveying information to other individuals.
- Sources of raw material on which natural selection can act to evolve communication signals include intention movements, autonomic responses, and displacement behavior.

The Timing of Behavior: Biological Rhythms

- Animal behaviors are expressed in daily cycles called circadian rhythms. To remain in phase with the 24-hour daily cycle of the environment, a circadian rhythm must be phase-shifted every day.
- Circadian rhythms can be entrained to periods that differ from the normal free-running period by environmental cues such as the onset of light and dark. **Review Figure 49.16**
- In mammals, the clock that controls the circadian rhythm is located in the suprachiasmatic nuclei of the brain. In other animals, different structures function as the circadian clock. **Review Figure 49.17**
- Circannual rhythms ensure that animals, such as hibernators and equatorial migrants, that cannot rely on changes in day length as seasonal cues perform the appropriate behaviors at the appropriate times of year.

Finding Their Way

- Methods by which animals find their way over long distances and through unfamiliar territory include piloting, homing, migration, and navigation.

- Piloting animals find their way by orienting to specific or general landmarks. **Review Figure 49.18**
- Homing animals find their way through unfamiliar territory to specific locations.
- Moving with the seasons, migrating animals travel great distances with remarkable accuracy. **Review Figure 49.20**
- Animals navigate by two different methods: distance-and-direction navigation and bicoordinate navigation.
- Animals that navigate by distance and direction may determine distance in part by recognizing landmarks in the vicinity of their destination and in part on the basis of biological rhythms, as evidenced by the migratory restlessness of migratory birds in captivity. Sources of directional information include environmental cues such as the sun and the stars. **Review Figures 49.21, 49.22, 49.23**
- Bicoordinate navigation requires a sense of longitude and latitude.

Human Behavior

- Like all other animal behavior, human behavior consists of genetically determined and learned components. What sets humans apart from other animals is the extent to which humans can modify their behavior on the basis of experience.

Self Quiz

1. The building of a web by a spider is an example of
   a. a fixed action pattern.
   b. a releaser.
   c. an open behavior program.
   d. imprinting.
   e. a learned behavior.

2. If you do not see courtship behavior in a deprivation experiment investigating the causes of sexual behavior, you can conclude that
   a. the animal is not sexually mature.
   b. the animal has low sexual drive.
   c. it is the wrong time of year.
   d. the appropriate releaser is not present.
   e. none of the above

3. Which statement about releasers is true?
   a. The appropriate releaser always triggers a fixed action pattern.
   b. They are simple subsets of sensory cues available to the animal.
   c. They are learned through imprinting.
   d. They trigger learned behavior patterns.
   e. An animal responds to a releaser only when it is sexually mature.

4. Which statement about the genetics of behavior is true?
   a. Approximately 20 genes control the courtship displays of male dabbling ducks.
   b. One gene can code for several neural signals involved in controlling a behavior.
   c. Genes for retrieving, pointing, and herding have been described in dogs.
   d. A single gene causes lovebirds to carry nesting material tucked in their tail feathers.
   e. Hygienic behavior in bees has been shown to be controlled by two dominant genes.

5. A display or signal is a behavior that
   a. has evolved to influence the behavior of other individuals.
   b. stimulates one or more types of sensors.
   c. stimulates the endocrine or reproductive systems of other individuals.
   d. began as an intention movement.
   e. began as a displacement behavior.

6. If the sun were to come up earlier than expected on the basis of a circadian rhythm,
   a. it could cause symptoms of jet lag.
   b. it could phase-advance the circadian rhythm.
   c. the animal could be east of home.
   d. it could entrain the circadian rhythm.
   e. all of the above

7. To be able to pilot, an animal must
   a. have a time-compensated solar compass.
   b. orient to a fixed point in the night sky.
   c. know the distance between two points.
   d. know landmarks.
   e. know its longitude and latitude.

8. Birds that migrate at night
   a. inherit a star map.
   b. determine direction by knowing the time and the position in the sky of a star constellation.
   c. orient to the fixed point in the sky.
   d. imprint on one or more key constellations.
   e. determine distance, but not direction, from the stars.

9. The most likely explanation for the observation that humans from entirely different societies smile when they greet a friend is that
   a. they share a common culture.
   b. they have imprinted on smiling faces when they were infants.
   c. they have learned that smiling does not stimulate aggression.
   d. smiling is a fixed action pattern.
   e. smiling is a learned behavior.

10. If (1) a bird is trained to seek food on the western side of a cage open to the sky, (2) the bird's circadian rhythm is then phase-delayed by 6 hours, and (3) after phase-shifting the bird is returned to the open cage at noon real time, it will seek food in the
    a. north.
    b. south.
    c. east.
    d. west.

Applying Concepts

1. Critique this statement: The hygienic behavior of bees is controlled by two genes, as demonstrated by hybridization and backcrossing experiments.

2. Since photoperiod (day length) can provide information about season (time of year), why do some birds have circannual rhythms?

3. If you raised a songbird in a deprivation experiment and it did not sing the song of its species the following fall, what hypotheses could you formulate about this result and how could you test them?

4. Male dogs lift a hind leg when they urinate; female dogs squat. If a male puppy receives an injection of estrogen when it is a newborn, it will never lift its leg to urinate for the rest of its life; it will always squat. How might this result be explained?

5. Pick an animal (other than a human) that you think would have mostly learned behavior and another animal that you think would have mostly genetically determined behavior. What differences in their biological characteristics could account for the differences in their behavioral repertoires?

Readings

Alcock, J. 1998. *Animal Behavior*, 6th Edition. Sinauer Associates, Sunderland, MA. A balanced textbook, recommended to readers searching for a good next step into the subject.

Emlen, S. 1975. "The Stellar-Orientation System of a Migratory Bird." *Scientific American*, August. A description of planetarium experiments studying the mechanisms by which birds orient to the stars.

Gould, J. L. 1981. *Ethology: The Mechanisms and Evolution of Behavior.* W. W. Norton, New York. A detailed treatment of behavior from a physiological point of view.

Gould, J. L. and P. Marler. 1987. "Learning by Instinct." *Scientific American*, January. A discussion of the interactions of learning and instinct, focusing on bees and on birdsong.

Gwinner, P. 1986. "Internal Rhythms in Bird Migration." *Scientific American*, April. A look at circannual rhythms, which play critical roles in long-distance migration.

Kirchner, W. H. and W. F. Towne. 1994. "The Sensory Basis of the Honeybee's Dance Language." *Scientific American*, June. An examination of new experiments that test what components of the honeybee's dance communicate information.

Lorenz, K. 1958. "The Evolution of Behavior." *Scientific American*, December. An essay on the evolution of releasers.

Moore-Ede, M. C., F. M. Sulzman and C. A. Fuller. 1982. *The Clocks That Time Us.* Harvard University Press, Cambridge, MA. A well-written book on circadian rhythmicity that covers most aspects of the subject.

Scheller, R. H. and R. Axel. 1984. "How Genes Control an Innate Behavior." *Scientific American*, March. A look at how a single gene codes for numerous neural signals.

Sherman, P. W. and J. Alcock. (Eds.). 1998. *Exploring Animal Behavior*, 2nd Edition. Sinauer Associates, Sundderland, MA. An anthology of articles on behavior that have appeared in *American Scientist* over the last 20 years.

Tinbergen, N. 1952. "The Curious Behavior of the Stickleback." *Scientific American*, December. A classic study of releasers and fixed action patterns.

Tinbergen, N. 1960. *The Herring Gull's World.* Doubleday, Garden City, NJ. A delightful account of the behavior of one species from the pen of one of the founders of modern ethology.

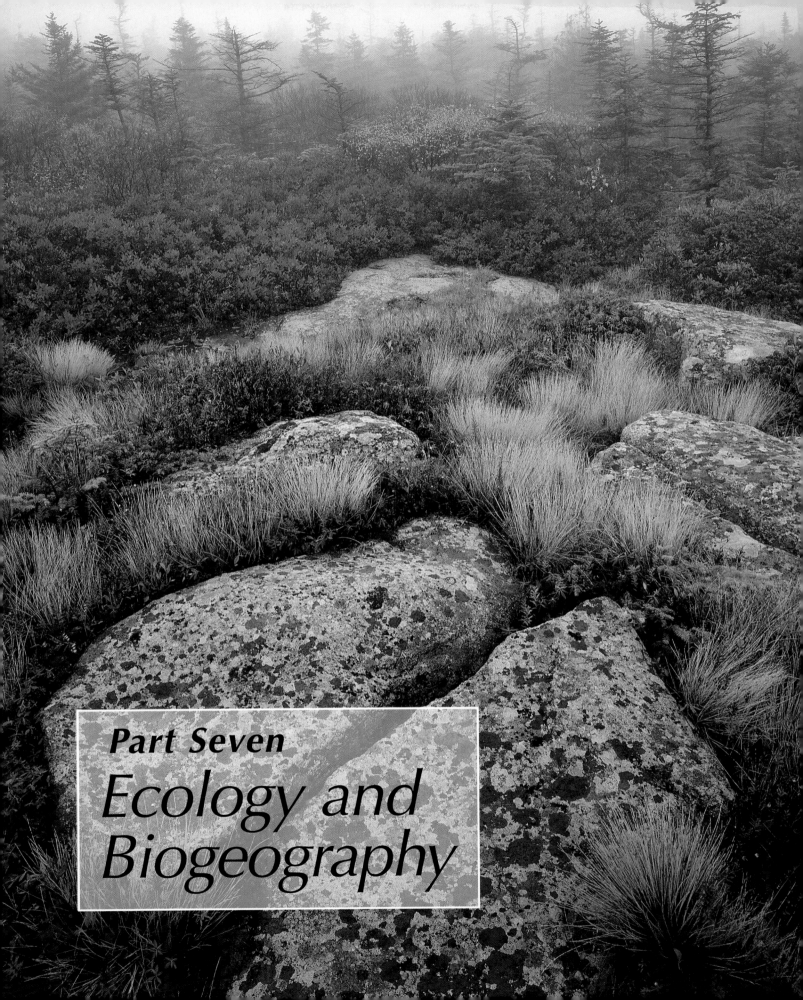

Part Seven
Ecology and Biogeography

Chapter 50

Behavioral Ecology

*I*ndividuals of all species interact in various ways with individuals of their own and other
species and with the physical environment. The task of ecology is
to understand the nature and consequences of these interactions.
Ecologists study patterns of distribution and abundance of organisms to determine how patterns
are established and maintained,
and how they change over short
and long time periods. From its
roots in descriptive natural history, ecology has developed into a
complex field of inquiry dealing

Choices Are the Foundation of Ecology
Choices made locally may influence distant events, and vice versa. Our
choices in a supermarket may influence interactions among species that
live far from where we shop. Do we prefer locally grown food or food
imported from remote places? Do we buy fresh or processed foods? Do we
choose foods produced by environmentally friendly methods?

with levels of organization ranging from relationships of individual organisms
to their physical and biological environments to the structure of communities
and ecosystems.

In this first chapter on ecology, we will discuss how animals behave, how
they make the decisions that influence their survival and reproductive success,
how ecologists study these decisions, and what they have learned from their
studies. Animals decide where to carry out their activities and how to select the
resources they need—food, water, shelter, nest sites. Animals respond to predators and competitors, and decide how to interact with other members of their
own species. (The use of the words "decision" and "decide" here does not
imply that the choices animals make are conscious, but rather that these choices
influence the survival and reproductive success of the animals and thus are
molded by natural selection.) Individual choices are the foundation of much of
ecology. Changes in densities and distributions of populations are the cumulative results of the decisions of myriad individuals.

The term *environment*, as used by ecologists, includes physical and chemical factors such as water, nutrients, light, temperature, and wind. It also includes all other organisms that influence the lives of individuals. Because species are adapted for life in many different environments, their interactions with their physical and biological environments are also varied. An environmental factor that exerts a strong influence on individuals of one species may have no influence on individuals of another species.

Interactions between organisms and their environments are two-way processes. Organisms both influence and are influenced by their environments. Indeed, managing environmental changes caused by our own species is one of the major problems of the modern world. For this reason, ecologists are often asked to help analyze causes of environmental problems and to assist in finding solutions for them. However, it is important not to confuse the science of ecology with the term "ecology" as it is often used in popular writing, to describe nature as some kind of superorganism.

Choosing Environments

Where an individual chooses to live, and how it uses that environment, strongly influence its survival and reproductive success. We will consider how animals choose places that provide adequate food and shelter, and how they select their food.

Features of an environment may indicate its suitability

Selecting a place to live is one of the most important decisions an individual makes. The environment in which an organism normally lives is called its **habitat**. Habitat selection requires a sequence of decisions, each one of which limits the options available for successive choices. Once a habitat is chosen, an animal must seek its food, resting places, nest sites, and escape routes within that habitat.

The cues organisms use to select suitable habitats are as varied as the organisms themselves, but all habitat selection cues have a common feature: they are good predictors of general conditions suitable for future survival and reproduction. For example, a young red abalone, a kind of gastropod mollusk, begins its life as an egg that is fertilized in the open ocean. About 14 hours after fertilization the egg hatches, but the motile larva emerges with enough yolk to continue developing for another 7 days without eating. At the end of 7 days the larva stops developing, swims to the seafloor, chooses a place in which to settle, and metamorphoses.

Red abalone larvae settle only on coralline algae, upon which they feed (Figure 50.1). They recognize coralline algae by a chemical signal—a water-soluble

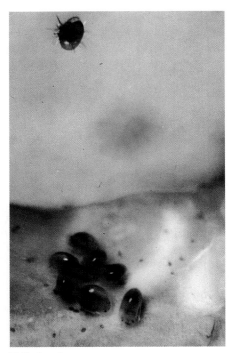

Haliotis rufescens

50.1 Chemistry Provides the Cue Red abalone larvae (dark ovals) settle only where they recognize the chemical composition of their food source, coralline algae.

peptide containing about 10 amino acids—that all these algae produce. In the laboratory, abalone larvae will settle on any surface on which this molecule has been placed, but in nature *only* coralline algae produce it. By using this simple cue, the larvae always settle on a surface that is suitable for their future development. Because red abalone larvae grow to adulthood by feeding on a single patch of coralline algae, their habitat is their food. However, most animals make many choices about where and how to seek and select food after they have settled in a habitat.

How do animals choose what to eat?

After choosing a habitat, individuals use the local resources, such as shelters from the physical environment, nest sites, and food. Because food is so important, we consider it here in some detail. When an animal is looking for food, how much time should it spend searching before moving to another site? When many different types of prey are available, which ones should a predator take, and which ones should it ignore? **Foraging theories** were developed to help answer these questions. To construct a theory to predict how a foraging animal should behave, a scientist first specifies the objective of the behavior and then attempts to determine the behavioral choices that would best achieve that objective. This approach is known as

optimality modeling, and its underlying assumption is that natural selection has molded the behavior of animals so that they solve problems by making the best choices available to them. There are many foraging theories, because a forager may be attempting to maximize the rate at which it obtains energy, to get enough vitamins or minerals, or to minimize its risk while foraging.

As an example, consider how a predator should choose among available prey in order to maximize the amount of energy it obtains. This is a reasonable objective because the more rapidly a predator captures food, the more time it will have for other activities, such as reproduction. Therefore, a more efficient predator should produce more offspring than a less efficient one, and animals should evolve to regularly make prey choice decisions that maximize their rate of energy intake.

To build a model of a predator that chooses prey in a way that maximizes its energy intake rate, we characterize each type of prey by the amount of time it takes the predator to pursue, capture, and consume one of them, and by the amount of energy an individual prey contains. We then rank the prey according to the amount of energy the predator gets relative to the amount of time the predator spends capturing and handling the prey. The most valuable prey type is the one that yields the most energy per unit of time invested.

With this information, we can build a model to calculate the rate at which a predator would obtain energy given a particular prey selection strategy. We can then compare alternative foraging strategies and determine the one that yields the highest rate of energy intake. The interesting result of such calculations is that, if the most valuable prey type is abundant enough, a predator gains the most energy per unit of time spent foraging by taking only the most valuable prey type and ignoring all others. The reason is that while the predator is capturing and consuming a less valuable prey individual, it could have found one of the most valuable ones. However, as the abundance of the most valuable prey type decreases, an energy-maximizing predator adds less valuable prey items to its diet in order of the energy per unit of time that those prey yield. Whether or not a certain prey type is included in the predator's diet does not depend on the abundance of that prey type. It depends only on the abundances of more valuable prey types.

Ecologists tested this theoretical model using bluegill sunfish (Figure 50.2a). They performed laboratory experiments with bluegills to measure the energy content of different prey types (water fleas of different sizes), the time needed to capture and eat different prey types, the energy spent searching for and capturing prey, and actual encounter rates with prey under different prey densities. Using these measurements,

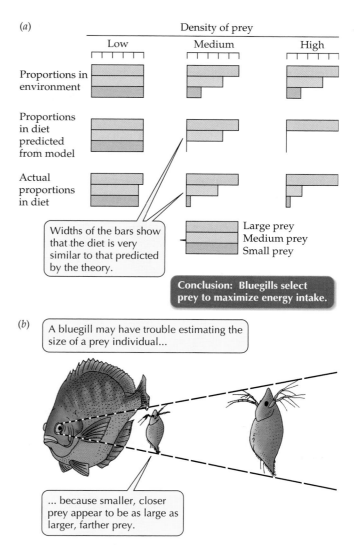

Conclusion: Bluegills select prey to maximize energy intake.

(b) A bluegill may have trouble estimating the size of a prey individual...

... because smaller, closer prey appear to be as large as larger, farther prey.

50.2 Energy Maximizers In an experiment, the prey choices of bluegill sunfish were very similar to those predicted by an energy-maximizing model.

the investigators predicted the proportions of large, medium, and small water fleas that bluegills would capture in environments stocked with different densities and proportions of water fleas of those sizes. They predicted that in an environment stocked with low densities of all three sizes of prey, the fish would take every water flea that they encountered, but that in an environment with abundant large water fleas, the bluegills would ignore smaller water fleas.

To test their predictions, the investigators put the bluegills in three different environments and observed the proportions of the water fleas of different sizes they actually captured (Figure 50.2b). The proportions of large, medium, and small water fleas taken by the fish were very close to those predicted by the model. Such tests of foraging theories using many different kinds of animals have provided ecologists with a set of rules showing how animals find and choose their prey. They have also provided estimates of the costs and benefits of foraging behavior.

Costs and Benefits of Behaviors

As shown by the example of prey selection behavior, ecologists who use optimality modeling to understand the evolution of behavior often analyze their observations in terms of costs and benefits. A cost–benefit analysis is based on the principle that an animal has only a limited amount of time, energy, and materials to devote to its activities. Animals generally do not perform behaviors whose total costs are greater than the sum of their benefits—the improvements in survival and reproductive success that the animal achieves by performing the behavior.

Of course, animals do not consciously calculate costs and benefits, but over many generations, natural selection molds behavior in accordance with costs and benefits. A cost–benefit approach provides a framework within which behavioral ecologists can design experiments and make observations that enable them to understand why behavior patterns evolve as they do. Even when costs and benefits cannot be measured directly, experiments can reveal which ones are important.

To understand what we mean by costs and benefits, consider the reproductive behavior of male elephant seals. A male elephant seal defends a small area of beach by threatening other males and fighting with challengers (Figure 50.3). Adult male elephant seals, which are several times larger than females, have large

Mirounga angustirostris

50.3 Loud Threats These mature male elephant seals are engaged in a mock battle for territory. This time, no blood will be drawn.

canine teeth with which they fight and thick skins that serve as shields. Their odd elephantlike snouts are resonating chambers that help amplify the roars that accompany their threat displays. Females congregate on a small number of beaches during the breeding season, where they give birth to their pups and mate again. Males that control good pupping beaches achieve most of the matings. Less than 10 percent of male elephant seals ever mate with a female. Of those that do, some mate with more than 100 females during their lifetimes. By contrast, nearly all females that survive to adulthood breed, but a female rarely weans as many as 10 pups during her lifetime.

The **energetic cost** of a behavior is the difference between the energy the animal would have expended had it rested and the energy expended in performing the behavior. A male elephant seal that fights with rivals expends more energy than a resting male. He therefore exhausts his fat reserves faster than if he had not attempted to defend a parcel of beach.

The **risk cost** of a behavior is the increased chance of being injured or killed as a result of performing it, compared with resting. A displaying male elephant seal is more likely to be injured by a rival than a male that avoids fights. The **opportunity cost** of a behavior is the sum of the benefits the animal forfeits by not being able to perform other behaviors during the same time interval. A male elephant seal cannot search for food while he defends a section of beach, and the longer he does so, the more time he needs to regain his energy reserves once he returns to the sea. The **benefit** a male elephant seal receives from incurring all those costs is access to reproductive females. A male elephant seal that does not defend a parcel of beach cannot mate with females and sire offspring.

Investigators used a cost–benefit approach to determine why female red-winged blackbirds regularly sing while they are on their nests. Most birds are silent while on their nests, presumably to reduce the chance that predators will find the nest. A behavioral ecologist guessed that these birds gained a benefit from singing that was greater than the risk of nest predation. To test this hypothesis, he placed redwing nests containing artificial eggs in breeding areas and placed loudspeakers near them. He broadcast recordings of female songs from some of the loudspeakers; the speakers at the control nests were silent. Ten of 15 nests near a noisy loudspeaker were destroyed by predators within 2 days, but none of the control nests were taken by predators. This experiment confirmed that predators find nests more readily if females are singing near them—a measure of the cost of singing.

To measure the benefits of singing, the ecologist compared the behavior of females that successfully raised their young with the behavior of those whose nests were lost to predators. He found that successful females sang significantly more often at their nests than unsuccessful females did. He then placed a model of a crow, an important nest predator in the area, near his mock nests. Male redwings scolded and attacked the crow more vigorously at nests located near loudspeakers that broadcast female songs than at silent control nests. Thus, both the costs and benefits of singing at the nest relate to predators: singing attracts both predators and males that defend the nest against them. The fact that most females sing regularly at their nests suggests that, on average, the benefits of singing more than compensates for the costs. Mock nests were often destroyed because males typically paid little attention to them.

Mating Tactics and Roles

Individuals choose their associates, how to interact with them, and when to leave them. The most important choice of associates an animal makes is mate selection. Mating behavior involves only a small set of choices. The most basic mating decision is choosing a partner of the correct species. Once the correct species has been determined, additional decisions can be based on the qualities of a potential mate, on the resources—food, nest sites, escape places—it controls, or on a combination of the two. Among those species in which individuals do not control any resources, traits of the partner are the only criteria for mate selection. Here we will discuss how individuals choose their mating partners and show why males and females approach courtship so differently.

Abundant small sperm and scarce large eggs drive mating behavior

The reproductive behavior of males and females is often very different. Males usually initiate courtship, and they often fight for opportunities to mate with females. Females seldom fight over males, and they often reject courting males. Why do males and females approach courtship and copulation so differently?

The answer lies in the costs of producing sperm and eggs. Because sperm are small and cheap to produce, one male produces enough to sire a very large number of offspring—usually many more than the number of eggs a female can produce or the number of young she can nourish. Therefore, males of most species can increase their reproductive success by mating with many females. Eggs, on the other hand, are typically much larger than sperm and are expensive to produce. Consequently, a female is unlikely to increase her reproductive output very much by increasing the number of males with which she mates. The reproductive success of a female depends primarily upon the resources her mate controls, the amount of assistance he provides in the care of her offspring, and the quality of

the genes she receives from him. By their choices among males, females cause the evolution of exaggerated traits that signal male quality.

Sexual selection leads to exaggerated traits

Traits may evolve among individuals of one sex as a result of **sexual selection**, the spread of traits that confer advantages to their bearers during courtship or when they compete for mates or resources. Successful competitors gain exclusive access to mates that are attracted to the resources they control. Traits that improve success in courtship evolve as a result of mating preferences by individuals of the opposite sex.

Sexual selection is responsible for the evolution of the remarkable tails of African long-tailed widowbirds, which are longer than their heads and bodies combined. To examine the role of the tail in sexual selection, an ecologist shortened the tails of some males by cutting them, and lengthened the tails of others by gluing on additional feathers. Both short-tailed and long-tailed males successfully defended their display sites, indicating that the long tail does not confer an advantage in male–male competition. However, males with artificially elongated tails attracted about four times more females than males with shortened tails (Figure 50.4). Given these experimental results, why don't male widowbirds have even longer tails than they do? A likely answer is that there are costs to producing and maintaining long tails that were not measured during these short-term experiments.

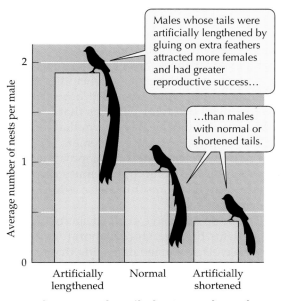

50.4 The Longer the Tail, the Better the Male Males with shortened tails defended their display sites successfully but attracted few females. Experiments showed how sexual selection favored long tails.

50.5 A Male Wins His Mate The male hanging fly on the left has just presented a moth to his mate, thus demonstrating his foraging skills. She feeds on the moth while they copulate.

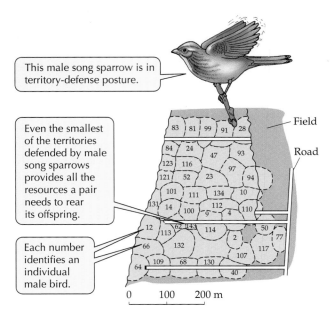

This male song sparrow is in territory-defense posture.

Even the smallest of the territories defended by male song sparrows provides all the resources a pair needs to rear its offspring.

Each number identifies an individual male bird.

Field

Road

0   100   200 m

50.6 Some Territories Provide Everything Male song sparrows defend territories that contain food, nesting sites, and protective cover.

Males attract mates in varied ways

Males employ a variety of tactics to induce females to copulate with them. If a male controls no resources, he uses courtship behavior that signals in some way that he is in good health, that he is a good provider of parental care, or that he has a good genotype. For example, males of some species of hanging flies court females by offering them dead insects. A female hanging fly will mate with a male only if he provides her with a morsel of food. The bigger the food item, the longer she copulates with him, and the more of her eggs he fertilizes (Figure 50.5). A female gains from this behavior because she obtains a better supply of energy for egg production. Also, because males fight for possession of dead insects, a male with a large insect is likely to be a good fighter and in good health.

Whether a male fertilizes the eggs of a female with whom he has copulated depends on when they copulate and whether she copulates with other males. Males have evolved behavior patterns that increase the probability that it will be *their* sperm that fertilize the eggs. The simplest method is to remain with the female and prevent other males from copulating with her, but this method has high opportunity costs because a male cannot do anything else while he is guarding a female.

Males of many species have evolved sperm competition mechanisms that are more elaborate but take less time. A male black-winged damselfly that grabs a receptive female uses his penis to scrub out sperm deposited by other males in her sperm storage chamber. The copulating male removes between 90 and 100 percent of competing sperm before he inserts his own sperm into the chamber. Males of many insects deposit a plug that effectively seals the opening to the female's genital chamber and prevents other sperm from entering.

Males of some species defend territories that contain food, nesting sites, or other resources. Some territories are all-purpose; they provide mating sites, nesting sites, and the food necessary to rear offspring (Figure 50.6). Other territories include a large breeding and nesting area, but do not supply all of the food necessary to rear young. Male red-winged blackbirds defend this type of territory in emergent vegetation in marshes (Figure 50.7). These territories provide nesting sites over water that are protected from some terrestrial mammalian predators, but both males and females get much of their food from upland areas near the marshes. The territories of many aquatic birds, such as gannets, penguins, and cormorants, are very small areas that individuals can defend while sitting on their nests (Figure 50.8).

Males of some species gather at display grounds, called **leks**, that may be used for many years. Within a lek each male defends a small display territory. Females come to the lek to choose a mate, but then leave the area and raise their young with no help from the males. Males battle intensely for possession of central display territories, and females usually mate with males holding central territories (Figure 50.9).

Females are the choosier sex

Females can improve their reproductive success if they can assess the genetic quality and health of potential mates, the quantity of parental care they may provide,

This display, which is accompanied by a song, warns other males to stay away from the territory but attracts females searching for breeding sites.

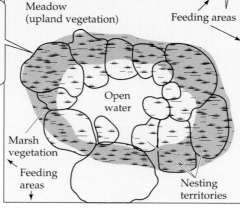

Feeding areas extend outside the territory boundaries.

Each outlined area is the territory of a single male red-winged blackbird.

Meadow (upland vegetation)

Feeding areas

Open water

Marsh vegetation

Feeding areas

Nesting territories

50.7 A Territory with Nesting Sites and Some Food
Red-winged blackbird nesting territories are located in emergent marsh vegetation, which provides some protection from predators. The territories are not large enough to provide all the food the birds need to raise their offspring.

Sula bassana

50.8 Some Territories Are Small   The spacing of individual breeding territories of cape gannets is determined by how far an incubating bird can reach to peck its neighbors without getting off its eggs.

had to assume that mated individuals were the parents of the offspring they raised. Recently, by using these new methods to compare the genetic material of offspring with that of their supposed parents and other individuals, ecologists have found that nestling

and the quality of the resources they control. But how can females make such assessments when all males attempt to signal that they are good in all three of these traits? The answer is that by paying particular attention to those signals at which males cannot cheat, females have favored the evolution of "reliable" signals. Possession of a large dead insect signals good fighting ability in a male hanging fly. A male elephant seal that controls a section of good pupping beach is certainly a high-quality mate.

Social and genetic partners may differ

Behavioral ecologists have known for many years that animals always copulate with their mates—the individuals with which they have established pair bonds—but that they sometimes also copulate with other individuals. However, until the recent development of DNA fingerprinting methods, investigators

50.9 Lek Territories Are Display Grounds   Fallow deer gather at leks, within which each antlered male establishes a small display ground to which he attempts to attract females. Dominant males command the territories closest to the center of the lek, and males may fight with each other over a specific display ground.

birds are nearly always the offspring of the female attending the nest—that is, females rarely lay eggs in other females' nests. However, the nestlings often have different fathers. For example, 34 percent of nestlings in a population of red-winged blackbirds in Washington State were fathered by a male other than the owner of the territory in which the nest was located. All these other fathers were males holding nearby territories; fertile females went to those territories and solicited copulations from the males.

Females that copulated with more than one male raised more offspring than females that remained faithful to their mates. Their reproductive success improved because neighboring males that had copulated with a female were more likely to defend her nest against predators than males that had not copulated with her. These males also let females with whom they had copulated look for food on their territories. Also, there were fewer infertile eggs in nests with multiple fathers than in nests with single fathers. Males try to prevent their mates from copulating with other males, but they must leave them unguarded at times, both to feed and to seek copulations with other females. Because all fathers hold territories, on average males gain and lose equally from extra-pair copulations.

Social Behavior: Costs and Benefits

When animals interact with individuals of the same species, we refer to their behavior as **social**. Social behavior evolves when individuals that cooperate with others of the same species have, on average, higher rates of survival and reproduction than those achieved by solitary individuals. Associations for reproduction may consist of little more than a coming together of eggs and sperm, but individuals of many species associate for longer times to provide care for their offspring. Associating with conspecific individuals may also improve survival for reasons unrelated to reproduction, such as reduced risk of predation.

We will describe only a few animal social systems, but these examples demonstrate two important concepts. First, social systems are best understood not by asking why they benefit the species as a whole, but by asking how the individuals that join together benefit by the association. Second, social systems are dynamic; individuals constantly communicate with one another and adjust their relationships.

Group living confers benefits

Living in groups may confer many types of benefits. It may improve hunting success or expand the range of prey that can be captured. For example, by hunting together, social carnivores improve their efficiency in bringing down prey (Figure 50.10). Such cooperative hunting was a key component of the evolution of

50.10 Group Hunting Improves Foraging Efficiency By hunting together lionesses can kill larger animals than a single female could subdue.

human sociality. By hunting in groups, our ancestors were able to kill large mammals they could not have subdued as individual hunters. These social humans could also defend their prey and themselves from other carnivores (Figure 50.11).

Many small birds form tight flocks when they are attacked by a hawk (Figure 50.12a). Clumping deters the predator because it risks injury if it hits one of the prey with its wing while dashing into a compact group. Also, predators may have greater difficulty in approaching groups of prey individuals without being detected. When a trained goshawk was released near wood pigeons in England, the hawk was most successful when it attacked solitary pigeons. Success in capturing a pigeon decreased as the number of pigeons in the flock increased (Figure 50.12b). To be successful, a goshawk must get close to foraging pigeons without being spotted so that it can launch a surprise attack. The larger the flock of pigeons, the sooner some individual in the flock spotted the hawk.

Group living imposes costs

Living in a group typically imposes costs as well as benefits because individuals in groups may compete for food, interfere with one another's foraging, injure one another's offspring, inhibit one another's reproduction, or transmit diseases to their associates. The effects of group living on the survival and reproductive success of an individual also depend on its age, sex, size, and physical condition. Individuals may be larger or smaller than the average for their age and sex. Variation in skills, competitive abilities, and at-

50.11 Early People Defended Their Prey Having killed their prey, a band of *Homo habilis* drives rival predators— spotted hyenas and sabertooth cats—from a fallen dinothere (an extinct relative of modern elephants).

50.12 Flocking Provides Defense against Hawks
Birds in flocks are less likely to be captured by a predator because (a) a falcon risks injury if it dashes into a compact flock and (b) the more birds in a flock, the sooner they spot the predator.

Undisturbed common starlings fly in loose flocks.

Starlings react to a peregrine falcon by flying closer together.

Peregrine falcon

(a)

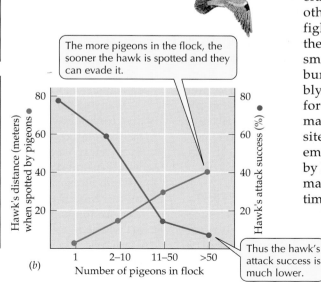

The more pigeons in the flock, the sooner the hawk is spotted and they can evade it.

Hawk's distance (meters) when spotted by pigeons ●

Hawk's attack success (%) ●

Number of pigeons in flock

1   2–10   11–50   >50

Thus the hawk's attack success is much lower.

(b)

tractiveness to potential mates is often associated with these size differences.

The relative sizes of individuals may determine their success when performing different types of behavior. For example, the largest males of *Centris pallida*, a solitary bee of the American Southwest, are three times the size of the smallest males (Figure 50.13). These size differences, which depend on the amount of food a male receives as a larva, are fixed for life. Large males search for females about to emerge from their buried pupae. When they detect a female, they attempt to dig her up and copulate with her, but the digging takes several minutes and often attracts other males. These new arrivals fight with the original male, and the largest male usually wins. If a small male searched and dug for buried females, he would probably just serve as a female-finder for larger males. Instead, small males patrol potential pupation sites and wait for females that emerge without being discovered by large males. Intermediate-sized males sometimes dig and sometimes patrol.

Centris pallida

50.13 Size Can Affect Behavior The large male bee on the right is digging a female out of the soil, while the small male on the left waits for a female to emerge.

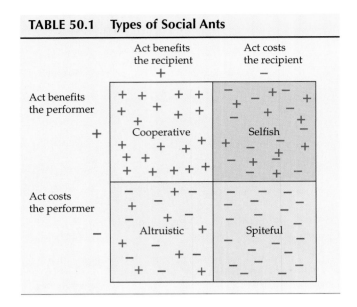

TABLE 50.1 **Types of Social Ants**

	Act benefits the recipient +	Act costs the recipient −
Act benefits the performer +	Cooperative	Selfish
Act costs the performer −	Altruistic	Spiteful

An almost universal cost associated with group living is higher exposure to diseases and parasites. Long before the causes of diseases were known, people knew that association with sick persons increased their chances of getting sick. Quarantine has been used to combat the spread of illness for as long as we have written records. The diseases of wild animals are not well known, but most of those that have been studied are spread by close contact.

Categories of Social Acts

Individuals living together perform many types of acts that are not performed by solitary animals. These acts can be grouped into four categories according to their effects on the interacting individuals (Table 50.1). An **altruistic act** benefits another individual at a cost to the performer. A **selfish act** benefits the performer but inflicts a cost on some other individual. A **cooperative act** benefits both the performer and the recipient. A **spiteful act** inflicts costs on both. These terms are purely descriptive; they do not imply conscious motivation or awareness on the part of the animal. If a genetic basis for a cooperative or selfish act exists, and if performing it increases the fitness of the performer, then the genes governing the act will increase in frequency in the lineage. In other words, cooperative or selfish behavior can evolve. We will now examine how altruistic behavior can evolve, both among close relatives and among unrelated individuals.

Altruism can evolve by means of natural selection

How could an act that *lowers* the performer's chances of survival evolve into a behavior pattern? One explanation for altruism lies in genetic relatedness: altruistic behaviors evolve most easily when performers and recipients are genetically related. Genetic relatedness, r,

is the probability that an allele in one individual is a copy that is identical, by descent, to an allele in another individual. To calculate r, we construct a diagram showing the individuals and their common ancestors, linked across generations by arrows. Because meiosis takes place at each generational link, the probability that a copy of any allele will be passed on is 0.5. For k generational links, the probability is (0.5^k). To calculate r, we sum this value for all possible pathways between the two individuals. Some examples are diagrammed in Figure 50.14.

Genetic relatedness is important because, as we learned in Chapter 21, an individual may influence its fitness in two different ways. First, it may produce its own offspring, contributing to its individual fitness. Second, it may help relatives that bear the same alleles it does because they are descended from a common ancestor. This process is called kin selection. Together, individual fitness and kin selection determine the inclusive fitness of an individual. Occasional altruistic acts may eventually evolve into altruistic behavior patterns if the benefits of increasing the reproductive success of related individuals exceed the costs of decreasing the altruist's own reproductive success.

Many social groups consist of some individuals that are close relatives and others that are unrelated or distantly related. Individuals of some species recognize their relatives and adjust their behavior accordingly. White-fronted bee-eaters (Figure 50.15) are colonial African birds in which most breeding pairs are assisted by nonbreeding adults that help incubate eggs and feed nestlings. Both males and females help; individuals whose nests fail may help at other nests later in the same breeding season. Nearly all of these helpers assist close relatives. When helpers have a choice of two nests at which to help, about 95 percent of the time

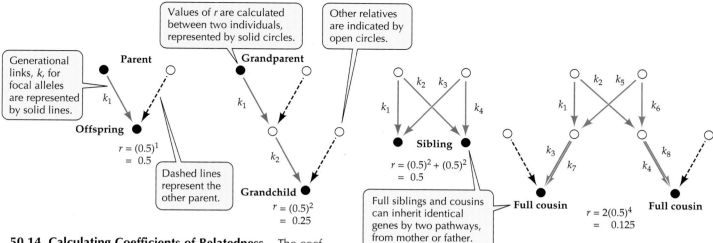

50.14 Calculating Coefficients of Relatedness The coefficient of relatedness, *r*, is the probability that an allele in one individual is identical, by descent, to an allele in another individual. To calculate *r*, we construct a diagram showing the individuals and their common ancestors, linked across generations by arrows.

they choose the nest with the young most closely related to them. Helping behavior among white-fronted bee-eaters is altruistic because individuals that help are not more successful when they become breeders than those that do not help, nor do they appear to gain any other advantage from helping. Extending help to non-relatives does not improve fitness, which is why white-fronted bee-eaters rarely help nonrelatives.

Species whose social groups include sterile individuals are said to be **eusocial**. This extreme form of sociality has evolved in termites and some Hymenoptera (the ant, bee, and wasp order) in which worker females defend the group against predators or bring food to the colony, but do not reproduce. Some species have specialized soldiers with large defensive weapons (Figure 50.16). Workers are at risk of being killed while defending the colony.

How could eusociality evolve? The British evolutionist W. D. Hamilton first suggested that eusociality evolved because worker female hymenopterans benefit by helping to raise their sisters, to which they are more similar genetically ($r = 0.75$) than they would be to any offspring they produced on their own ($r = 0.50$). Sterile hymenopteran workers are more closely related to their sisters than they would be to their offspring

50.15 White-Fronted Bee-Eaters Are Altruists These African birds are perched near entrances to their nests. Tags (visible on one bird) enable investigators to identify individual birds; studies show bee-eaters often help raise young that are not their own, but that are closely related to them.

50.16 Sterile Individuals Are Extreme Altruists Eusocial insect species contain classes of sterile individuals. These individuals defend and provide food for the colony, but do not reproduce. These soldier army ants from Panama have evolved to contain powerful weaponry in the form of large jaws.

because members of the order Hymenoptera have an unusual sex determination system in which males are haploid, but females are diploid. A fertilized egg hatches into a female; an unfertilized egg hatches into a male. If a female mates with only one male, all the sperm she receives are identical because the haploid males have only one set of chromosomes, all of which are transmitted to each sperm cell. Therefore, a female's daughters share all of these genes. They also share, on average, half of the genes they receive from their mother. As a result, on average they share 75 percent of their alleles, rather than 50 percent as they would if both parents were diploid. Sisters are therefore genetically more similar to one another ($r = 0.75$) than a mother is to her daughters and sons ($r = 0.50$).

If Hamilton's hypothesis is correct, there should be a "genetic conflict" between workers and their queen mother. The queen, who is equally related ($r = 0.50$) to her sons and her daughters, would maximize her fitness by producing equal numbers of sons and daughters. Her daughters, however, which are more closely related to their sisters ($r = 0.75$) than they are to their brothers ($r = 0.25$), would maximize their inclusive fitness with an investment ratio of 75:25 (3:1) in favor of sisters.

The queen can control the sex of the eggs she lays, but the workers that care for and feed the larvae could skew the sex ratio of surviving offspring by giving more food to their sisters than to their brothers. In fact, among species in which the queen normally mates only once, workers invest much more in caring for females than in caring for males. If the founding queen is removed, one of her daughters typically becomes the new queen. She produces offspring that are nieces and nephews of the workers produced by the first queen ($r = 0.375$). As predicted by Hamilton's hypothesis, these manipulated colonies produce more males than colonies with their original queen.

However, Hamilton's hypothesis cannot account for many cases of eusociality. It cannot explain eusociality among the many hymenopteran species in which queens mate with many males. Nor can it explain eusociality in species, such as termites and naked mole-rats, in which both sexes are diploid. Naked mole-rats, the only eusocial mammals, live in underground colonies containing 70 to 80 individuals whose tunnel systems are maintained by sterile workers. Breeding is restricted to a single queen and several kings that live in a nest chamber in the center of the colony. Other females and males are sterile.

The inability of Hamilton's hypothesis to explain many aspects of eusocial behavior suggests that other environmental factors may also favor helping. One clue is provided by the fact that nearly all eusocial animals construct elaborate nests or burrow systems within which their offspring are reared (Figure 50.17).

50.17 Termite Mounds Are Large and Complex These immense Australian termite mounds are very costly to construct and maintain. Elaborate nests or burrows are a common characteristic of nearly all eusocial animals.

Forming a new colony and constructing these nests or tunnels is costly. Founding individuals are at high risk of being captured by predators, and most founding events fail. Thus, high predation rates, which favor cooperation among founding individuals, may facilitate the evolution of eusociality. Colonies founded by multiple queens grow more rapidly, and are more likely to be successful, than colonies founded by solitary queens.

Unrelated individuals may behave altruistically toward one another

It is easy to understand how cooperative behavior can evolve among related individuals. It is more difficult to understand the evolution of warnings of danger, sharing of food, and grooming among unrelated individuals of the same species or between members of different species. How can we explain the evolution of such cooperative behavior? The proposed answer is a model called **reciprocal altruism**. According to this model, reciprocal altruism evolves if a helper is in turn the recipient of beneficial acts by the individuals it has helped. If there is a genetic basis for the acts, natural selection may increase the frequency of alleles governing the cooperative behavior.

The Evolution of Animal Societies

The decisions animals make about where to settle, with whom to mate, what resources to invest in a reproductive effort, and when to terminate investment all help determine the type of social system that animals have. Today's social systems are the result of long periods of evolution, but there are few records of past social systems because behavior leaves few traces in the fossil record. Possible routes of the evolution of social systems must therefore be inferred primarily from current patterns of social organization. Fortunately, many stages of complexity exist among species, and the simpler systems provide clues about the stages through which the more complex ones may have passed. First we will examine the origins of all animal societies, which lie in the association of parents with their offspring. Then we will discuss how environmental factors affect social behavior. Finally, we will consider the influence of ancient evolutionary history on current social behaviors.

Parents of many species care for their offspring

Individuals of many species invest time and energy in caring for offspring. Parental care increases the chances of an offspring's survival, but it may reduce the ability of the parent to produce additional offspring. Parental care may also lower the chances of survival of the parent itself, because the parent could have used the time and energy to engage in other activities that would improve its own survival or reproductive success.

Males and females often differ strikingly in the kinds and amounts of parental investment they can and do make. Birds, mammals, and fishes illustrate these differences and why they exist. Only female mammals have functional mammary glands; males cannot produce milk. Males of most mammal species contribute nothing to offspring nutrition. Birds, on the other hand, do not produce milk. Among birds, all aspects of reproduction except production of eggs and sperm can be performed readily by both males and females. Not surprisingly, both males and females feed offspring in about 90 percent of bird species.

Sex roles among fishes differ from those of birds and mammals because most fish species do not feed their young. Parental care consists primarily of guarding eggs and young from predators (Figure 50.18). In many fish species, males are the primary guardians. A male can guard a clutch of eggs while attracting additional females to lay eggs in his nest. A female, on the other hand, can produce another clutch of eggs sooner if she resumes foraging immediately after mating than if she spends time guarding her eggs.

The most widespread form of social system is the family, an association of one or more adults and their

This dark area is a large brood of tiny young fish.

50.18 Cichlid Fish Guard Their Young Both parents vigorously defend their offspring for as long as 48 days.

dependent offspring. If parental care lasts a long time, or if the breeding season is longer than the time it takes for the young to mature, the adults may still be caring for younger offspring when older offspring reach the age at which they could help their parents. Many communal breeding systems probably evolved by this route. Florida scrub jays live all year on territories, each of which contains a breeding pair and up to six helpers (Figure 50.19). About three-fourths of the helpers are offspring from the previous breeding season that remain with their parents.

Most mammals evolved sociality via the extended family route. In simple mammalian social systems, solitary females or male–female pairs care for their young. As the period of parental care increases, older offspring are still present when the next generation is born, and they often help rear their younger siblings. In most social mammal species, female offspring remain in the group in which they were born, but males tend to leave, or are driven out, and must seek other social groups. Therefore, among mammals, most helpers are females.

The environment influences the evolution of animal societies

The type of social organization a species evolves is strongly related to the environment in which it lives. Among the African weaverbirds, species that live in forests eat insects, feed alone, and build well-hidden nests. Most of these species are monogamous, and males and females look alike. In marked contrast, weaverbirds that live in tree-studded grasslands called savannas eat primarily seeds, feed in large flocks, and nest in colonies, usually in isolated *Acacia* trees, where their nests are large and conspicuous (Figure 50.20). In most of these species, males have several mates—that

50.19 Cooperation among Florida Scrub Jays Florida scrub jay helpers, most of which are offspring from the previous breeding season, are helping to feed nestlings and defend the nest against predators such as the approaching snake. By doing so, they are improving their inclusive fitness.

50.20 Savanna Weaverbirds Nest Colonially Many African weaverbirds nest in colonies in isolated trees. Although these nests are highly conspicuous, it is difficult for most avian, mammalian, and reptilian predators to get to them at the tips of small branches.

is, they are polygynous—and are brightly colored, but females are not.

These striking differences probably evolved because nesting sites and food in forests are common and widely dispersed. Solitary pairs can use these resources more efficiently than animals in groups can. In savannas, good nesting trees are scarce and highly clumped, and nests are easy for predators to find. Males compete for these limited nest sites; the males that hold the best sites attract the most females. Males spend their time attempting to attract additional mates rather than helping to rear the offspring they already have, which explains the evolution of brighter plumage among males.

Among the herbivorous hoofed mammals of Africa, social organization and feeding ecology are correlated with the size of the animal (Table 50.2). Smaller animals have higher metabolic demands per unit of body weight than do larger ones (see Chapter 37). Therefore, smaller hoofed mammals are very selective in what they eat. They feed preferentially on high-protein foods such as buds, young leaves, and fruits. These foods are dispersed throughout forests, which also provide cover in which to hide from predators. Hiding is a tactic that is effective for solitary animals. The largest hoofed mammal species are able to eat lower-quality food, but they must process great quantities of it each day. They feed in grasslands with high standing crops of herbaceous vegetation, follow the rains to areas where grass growth is best, and live in large herds (Figure 50.21).

Living in herds makes it possible for males to compete among themselves for control of females, and for

TABLE 50.2   Social Organization and Ecology in African Hoofed Animals

SPECIES	BODY WEIGHT (KG)	FEEDING ECOLOGY	GROUP SIZE	SOCIAL ORGANIZATION
Dikdik, duiker	3–60	Selective browsing and grazing	1 or 2	Pair
Reedbuck, gerenuk	20–60	Selective browsing and grazing	2–12	Male with small harem
Impala, gazelles	20–250	Grazing, browsing	2–100	Territory-defending male with harem
Wildebeest, hartebeest	90–270	Grazing	Up to thousands	Herd in which males defend females
Eland, buffalo	300–600	Unselective grazing	Up to thousands	Herd with male dominance hierarchy

Connochaetes sp.

50.21 Living in Large Herds East African wilde-beest live in large herds that move from place to place to obtain the fresh green grass upon which they feed. Their major predators—lions—live on permanent terri-tories and often have little to eat when the wildebeest herds are far away.

dominant males to defend medium to large harems. In these open environments, hiding is impossible, but it is also difficult for predators to approach the herds un-detected. Species of intermediate sizes have feeding ecologies and social systems intermediate between those of smaller and larger species.

Among primates, nocturnal forest-dwelling insect eaters, such as lorises, some lemurs, and the owl mon-key, live in pairs and are usually solitary foragers. Many diurnal species take insects and other animal food when they are available, but most of them eat fruits, seeds, and leaves. In Africa and Asia, group sizes are smallest among arboreal forest-dwelling species, whatever their diets, and largest among the ground-dwelling savanna species, such as baboons (Figure 50.22). In troops with more than one male,

Papio sp.

50.22 Baboons in Groups Baboons forage in open savan-nas and travel in large groups. If a predator approaches, the formidable males cooperate in defending the group.

(a) Concentrated lek
(*Pipra serena*)

Several males dance, moving randomly in a small area.

(b) Dispersed lek
(*Masius chrysopterus*)

He lands…

…hops…

…and hops back.

(c) Cooperative lek
(*Chiroxiphia caudata*)

Two males cooperate in a "round dance."

50.23 Manakin Males Display in Leks The three manakin species illustrated here display in different types of leks.

strong dominance hierarchies exist among the males, and one or two of the males do most of the copulating. Females may also have dominance relationships, and young females often assume the status of their mothers when they mature.

Social systems also reflect evolutionary history

Until now we have looked only at the current conditions, such as asymmetry in reproductive investment between the sexes and environmental conditions, that influence a species' social behavior. Until the development of new methods of determining the evolutionary relationships of species to one another, no other approach was possible. However, if a reliable phylogeny exists for an animal group, we can use it to tell whether differences in social behaviors among species in that lineage are determined solely by adaptation to current conditions or whether they also reflect longer-term history.

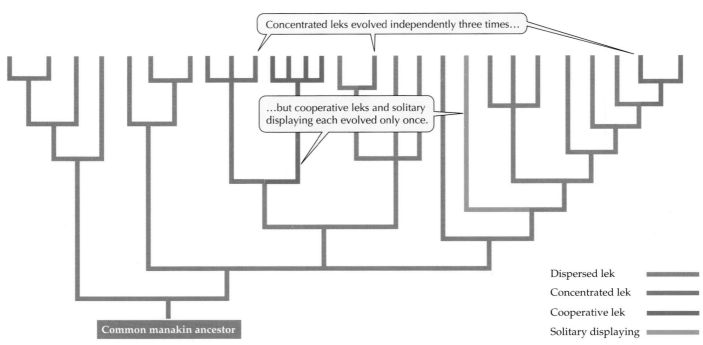

Concentrated leks evolved independently three times…

…but cooperative leks and solitary displaying each evolved only once.

Common manakin ancestor

Dispersed lek
Concentrated lek
Cooperative lek
Solitary displaying

50.24 A Phylogeny of Manakin Lekking Behaviors This tree shows the evolutionary relationships of the manakins and the types of leks they use. Concentrated leks evolved several times.

This approach has been applied to the social systems of manakins, a group of about 40 species of small fruit-eating birds that live in tropical American forests. Male manakins form leks where females choose and copulate with males. Females then carry out all other reproductive activities unassisted by the males. Manakin leks differ in how close the displaying males are to one another and in whether the males coordinate and cooperate with one another as they display (Figure 50.23). The phylogeny of manakins indicates that lek-breeding behavior evolved in the common ancestor of all manakins, and that lekking has been lost only once during manakin social evolution. By contrast, leks in which the males display close to one another have evolved independently three times among manakins (Figure 50.24). Leks at which males coordinate their displays with one another have evolved independently five times. Cooperative leks and solitary displaying each evolved once.

Which manakin species uses which type of lek is not correlated with the type of environment in which it lives, but species that share recent common ancestors have similar lek types. This pattern indicates that once a particular type of lek evolved in a lineage, it was maintained among the descendants even if they came to live in different habitats. As more complete phylogenies become available, behavioral ecologists will be able to assess the roles of phylogenetic history and adaptation to current environments in the evolution of animal societies.

Summary of "Behavioral Ecology"

• Ecologists study the nature and consequences of interactions among organisms and their environments at all scales, from local to global.
• Behavioral ecology is the study of how animals decide where to carry out different activities, how to select the resources they need, how to respond to predators and competitors, and how to interact with other member of their own species.

Choosing Environments
• Selecting a habitat in which to live is one of the most important decisions an individual makes.
• The cues animals use to select habitats are good predictors of conditions suitable for future survival and reproduction. **Review Figure 50.1**
• Foraging theories were developed to understand how animals select prey items from the array present in the environment. **Review Figure 50.2**

Costs and Benefits of Behaviors
• Cost–benefit analyses of behavior are based on the principle that animals have only limited amounts of time and energy to devote to their activities.

Mating Tactics and Roles
• Individuals choose their associates, how to interact with them, and when to leave them. The most widespread choice of associates is the choice of a mate.
• Because males produce enough sperm to fertilize many more eggs than a single female can produce, males typically increase their reproductive success by mating with many females. The reproductive success of females, on the other hand, is typically limited by the cost of producing eggs. As a result, males usually initiate courtship and often fight for opportunities to mate with females. Females seldom fight over males and often reject courting males.
• Sexual selection often leads to exaggerated traits such as long tails. **Review Figure 50.4**
• During courtship, males signal their desirability as mating partners and may perform behaviors that increase the probability that their sperm will fertilize eggs. **Review Figure 50.5**
• By paying particular attention to those signals at which males cannot cheat, females have favored the evolution of "reliable" signals.
• Males of some species defend territories that contain food, nesting sites, or other resources. **Review Figures 50.6, 50.7, 50.8**

Social Behavior: Costs and Benefits
• Benefits of social living include better opportunities to capture prey and to avoid predators. **Review Figures 50.10, 50.11, 50.12**
• Costs of social living include competition for food, interference, and transmission of diseases.

Categories of Social Acts
• The acts performed by individuals living together can be grouped into four descriptive categories: altruistic, selfish, cooperative, and spiteful. **Review Table 50.1**
• Altruism among closely related individuals can evolve by means of natural selection because individuals that help close relatives can improve their inclusive fitness by kin selection. **Review Figure 50.14**
• Eusocial systems with sterile individuals have evolved among termites, hymenopterans (ants, bees, and wasps), and in one mammal, the naked mole-rat. **Review Figure 50.16**

The Evolution of Animal Societies
• The origin of most animal societies is the family, an association of one or more adults and their dependent offspring. **Review Figure 50.19**
• The type of social organization a species evolves is strongly related to the environment in which it lives. **Review Table 50.2 and Figures 50.20, 50.21, 50.22**
• Evolutionary history also influences a species' social organization. **Review Figures 50.23, 50.24**

Self-Quiz

1. Which of the following is *not* a component of the cost of performing a behavioral act?
   a. Its energetic cost
   b. The risk of being injured
   c. Its opportunity cost
   d. The risk of being attacked by a predator
   e. Its information cost

2. An almost universal cost associated with group living is
   a. increased risk of predation.
   b. interference with foraging.
   c. higher exposure to diseases and parasites.
   d. poorer access to mates.
   e. poorer access to sleeping sites.

3. An act is said to be altruistic if it
   a. confers a benefit on the performer by inflicting some cost on some other individual.
   b. confers a benefit both on the performer and on some other individual.
   c. inflicts a cost both on the performer and on some other individual.
   d. confers a benefit on another individual at some cost to the performer.
   e. imposes a cost on the performer without benefiting any other individual.

4. Which of the following statements about male and female roles in social systems is *not* correct?
   a. Females invest more in gamete production, but they may invest more or less than males in care of offspring.
   b. Biparental care is prevalent among birds.
   c. Males of most mammal species help feed offspring.
   d. Males with a high probability of parentage invest more in parental care than males that are less certainly related to the offspring of their mates.
   e. Among fishes, if there is unequal parental care by individuals of the two sexes, it is nearly always the male that does more.

5. Male and female mating tactics usually differ because
   a. males are typically larger than females.
   b. males do not contribute as many genes to their offspring as females do.
   c. males, but not females, usually can increase their fitness by mating with more than one partner.
   d. males can control copulations to their advantage.
   e. males and females occupy different positions when they copulate.

6. Choice of mating partner may be based on
   a. the inherent qualities of a potential mate.
   b. the resources held by a potential mate.
   c. both the inherent qualities of a potential mate and the resources it holds.
   d. the success of individuals of the opposite sex in courtship.
   e. all of the above

7. A lek is
   a. a territory held by a single male.
   b. a territory held by two or more males.
   c. a display ground at which a single male displays.
   d. a display ground at which two or more males display.
   e. a territory held by two or more females.

8. Among social birds, there are usually more male than female helpers in those species with helpers because
   a. males are better helpers than females.
   b. males typically receive greater benefits from helping.
   c. males survive better than females.
   d. mothers do not allow their daughters to help.
   e. males often need to wait for an unoccupied territory elsewhere.

9. Small African hoofed mammals are usually solitary because
   a. they feed on scattered high-quality foods in forested environments.
   b. the low quality of their food does not permit them to assemble in groups.
   c. they are too small to defend themselves against predators.
   d. they are too small to follow the rains to areas where grass growth is best.
   e. they are usually driven from their natal groups.

10. A phylogenetic analysis of lekking behavior in manakins shows that the type of lek a manakin species uses
    a. is correlated with the type of environment in which it lives.
    b. depends on how well males can coordinate their dancing.
    c. is similar among species that share a recent common ancestor.
    d. is unrelated to the phylogeny of manakins.
    e. evolves rapidly among closely related species.

Applying Concepts

1. Most hawks are solitary hunters. Swallows often hunt in groups. What are some plausible explanations for this difference? How could you test your ideas?

2. Because costs and benefits of behaviors can seldom be measured directly, behavioral ecologists often use indirect measures. What are the strengths and weaknesses of some of these indirect measures?

3. Among birds, males of polygamous species that display in leks are usually much larger and more brightly colored than females, whereas among species that form monogamous pairs, males are usually similar in size to females, whether or not they are more brightly colored. What hypotheses can be advanced to explain this difference?

4. Polyandry is a mating system in which one female has a "harem" of several males. Why is polyandry much rarer among both birds and mammals than polygyny, the situation in which one male mates with several females?

5. Many animals defend space, but the sizes of the territories they defend and the resources these areas provide vary enormously. Why don't all animals defend the same type and size of territory?

6. When frogs mate, a male clasps a gravid female behind her front legs and stays with her until she lays her eggs, at which time he fertilizes them. In most species of frogs, the male remains clasped to the female for a short time, usually no longer than a few hours. However, in some species, pairs may remain together for up to several weeks. In view of the fact that a male cannot court or mate with any other female while clasping one, and that

a female lays only a single clutch of eggs, why is it advantageous for males to behave this way? What can you guess about the breeding ecology of frogs that remain clasped for long periods? Why should females permit males to clasp them for so long? (Females do not struggle!)

7. Among vertebrates, helpers are individuals capable of reproducing, and most of them later breed on their own. Among eusocial insects, sterile castes have evolved repeatedly. What differences between vertebrates and insects might explain the failure of sterile castes to evolve in the former?

Readings

Alcock, J. 1998. *Animal Behavior*, 6th Edition. Sinauer Associates, Sunderland, MA. An excellent text of animal behavior from an evolutionary perspective. Contains additional material on all the topics discussed in this chapter.

Borgia, G. 1995. "Why Do Bowerbirds Build Bowers?" *American Scientist*, vol. 83, pages 542–547. Argues that courtship areas provide choosy females with easy avenues of escape.

Buss, D. M. 1994. "The Strategies of Human Mating." *American Scientist*, vol. 82, pages 238–248. Shows that people worldwide are attracted to the same qualities in the opposite sex.

De Waal, F. B. 1995. "Bonobo Sex and Society." *Scientific American*, March. Describes how bonobos, the most humanlike of the apes, live in peaceful societies in which females dominate the hierarchy and casual sex soothes all conflicts.

Emlen, S. T., P. H. Wrege and N. J. Demong. 1995. "Making Decisions in the Family: An Evolutionary Perspective." *American Scientist*, vol. 83, pages 148–157. Describes the complex social interactions in a family of white-fronted bee-eaters.

Gordon, D. M. 1995. "The Development of Organization in an Ant Colony." *American Scientist*, vol. 83, pages 50–57. Shows how simple, local decisions can generate a complex society.

Heinrich, B. and J. Marzluff. 1995. "Why Ravens Share." *American Scientist*, vol. 83, pages 428–437. Shows how by feeding in groups, ravens eat regularly even when food is scarce.

Krebs, J. R. and N. B. Davies. 1993. *An Introduction to Behavioral Ecology*, 3rd Edition. Blackwell Scientific Publications, Oxford. An excellent account of the methods and results of modern studies of behavioral ecology.

Narins, P. M. 1995. "Frog Communication." *Scientific American*, August. Discusses why loudly croaking frogs are not a chorus, but rather an assembly of males, each of which employs acoustic adaptations to make himself heard above the din.

Pfennig, D. W. and P. W. Sherman. 1995. "Kin Recognition." *Scientific American*, June. Describes how many animals can recognize close kin by assessing genetic similarities, by olfaction, or by their having grown up together.

Sherman, P. W. and J. Alcock. (Eds.). 1998. *Exploring Animal Behavior*, 2nd Edition. Sinauer Associatres, Sunderland, MA. An anthology of articles on behavior that have appeared in *American Scientist* over the last 20 years.

Thornhill, R. and J. Alcock. 1983. *The Evolution of Insect Mating Systems*. Harvard University Press, Cambridge, MA. A comprehensive account of the amazing variety of insect mating systems.

Wilson, E. O. 1975. *Sociobiology: The New Synthesis*. Harvard University Press, Cambridge, MA. A classic review of all aspects of social evolution and animal societies.

Chapter 51

Population Ecology

Loxodonta africana

Stages of Life
Female elephants groom young calves in Etosha National Park, Namibia. African elephants produce comparatively few offspring over their lifetime, and the females spend a great deal of energy caring for each one. Humans use a similar reproductive strategy.

Elephants reproduce at a slower rate than most animals. A female African elephant does not reach reproductive age until she is about 10 years old, after which she produces a single calf at a time. Because she nurses each calf for about 2 years and the gestation period of elephants is 22 months, a female produces only one calf every 4 years. Thus, although she may live for many decades, a female African elephant produces only 10 to 12 calves during her lifetime.

Female elephants live in small herds, each of which is led by an old, experienced female. Female calves remain in the herd with their mothers, but young males are expelled and live most of their lives in small bachelor herds. Males join family herds only when females are in breeding condition and they take no part in raising the offspring they sire.

The reproductive traits of all organisms have been molded by natural selection acting over many generations. In each lineage, those traits that maximized reproductive success were favored, but natural selection has not produced a single, dominant pattern of reproduction. Some organisms, like elephants and humans, usually give birth to a single offspring in each reproductive episode; others spawn thousands or millions of eggs in one bout. Some organisms begin to reproduce within days or weeks of being born; others live for many years before reproducing. The traits that an individual expresses during its life constitute its **life history**. Life history traits influence how many individuals of a species are found in an area and how dramatically their densities vary.

Why are life history traits so variable? What causes a species to be common or rare? Why do the sizes of populations fluctuate yearly, seasonally, or not at all? Why is a species common in some parts of its geographic range and rare in others? What determines the limits of the ranges of species? Answering such questions is the major task of population ecology.

A **population** consists of all the individuals of a species within a given area. The boundaries of the area are usually determined by the goals of a scientific study. To understand population dynamics, population ecologists count individuals in different places and try to determine the factors that influence birth, death, immigration, and emigration rates. In this chapter we will discuss how and why the sizes of populations of species vary over space and time, and show how this ecological knowledge is used to predict and manage the growth of populations.

Varied Life Histories

The complete life history of an organism consists of its birth, growth to maturity, reproduction, and death. During its life an individual organism ingests nutrients or food, grows, interacts with other individuals of the same and other species, reproduces, and usually moves or is moved so that it does not die exactly where it was born. Life histories describe how an organism divides its efforts among these activities.

All life histories are based on a certain set of traits: size at birth; how fast the individuals grow; how long they live; how many times they reproduce; the rate and pattern of growth and development; the ages at which they die; how much they move around; the number, size, and sex composition of their offspring; and the number and timing of reproductive events in an organism's life.

Many organisms, such as the red-spotted newt (*Notophthalmus viridescens*), a colorful salamander that breeds in small ponds throughout eastern North America, have complex life histories. The olive-colored, red-spotted adult newts feed and lay their eggs in ponds. The eggs hatch into larvae that grow and develop in the ponds for several months. Eventually the larvae lose their gills and change into *efts*, immature newts that leave the water to live on land for 4 to 9 years (Figure 51.1). The bright red efts are highly toxic to their natural predators.

Most efts return to the ponds in which they were born to reproduce, but some travel long distances and colonize new ponds. Newly created farm ponds in rural areas and wildlife management ponds in national forests are populated quickly by red-spotted newts, showing that they are good colonizers of new habitats. Thus, during its life, a red-spotted newt goes through three distinct stages—larva, eft, and adult—and lives in two different habitats. We will now describe in more general terms the life history stages illustrated by red-spotted newts.

Notophthalmus viridescens

51.1 A Red Eft The toxic red eft is the terrestrial dispersal stage in the life cycle of the red-spotted newt.

Life histories include stages for growth, change, and dispersal

For at least part of their lives, all organisms grow by gathering and assimilating energy and nutrients. Some organisms, such as red-spotted newts, gather energy and nutrients throughout their lives, even after they reach adult size and stop growing. Energy gathered after growth stops maintains organisms and supports reproduction. In many species, however, energy gathering is confined to a particular stage. Most moths, for example, feed only when they are larvae. The adults lack mouthparts and digestive tracts, live on energy gathered by the larvae, and survive only long enough to disperse, mate, and lay eggs. Having different stages specialized for different activities increases the efficiency of performing a particular activity.

Individuals of many species also change form during their lives. Human babies are unmistakably human, but newborns of many species differ dramatically from adults. Some of the most striking changes are found among insects such as beetles, flies, moths, butterflies, and bees, which undergo radical metamorphoses from their larval to their adult forms (see Chapter 29). These metamorphoses take place during the pupal stage, during which larval tissues and organs are broken down, and adult tissues and organs are constructed from the larval material. Many plants have resting stages, such as spores and seeds, that have low metabolic rates and are highly resistant to changes in the physical environment. Growth typically does not take place in these stages, but dispersal is common.

At some time in their lives, all organisms disperse. Some, such as plants and sessile animals, disperse as small eggs, larvae, spores, or seeds. Others, such as insects and birds, disperse primarily as adults. Still others may disperse during several different stages. Individuals of some species can change their location many times during their lives in response to environmental changes. Others must remain in the first place they settle.

Life histories embody trade-offs and constraints

Life history traits are molded by natural selection, but they also involve trade-offs. Changes that improve fitness by means of one life history trait often reduce fitness by means of another. For example, a higher number of births is often traded off for better survival of offspring already produced. What are the major trade-offs in life history traits?

A universal trade-off exists between number and size of offspring. Every newborn individual begins to grow with energy and nutrients from its maternal parent, but how much energy and nutrients individuals receive from their mothers varies greatly. Orchid seeds receive very little, grass seeds slightly more. Coconuts, birds, and placental mammals receive large amounts of maternal energy. The larger the amount of energy provided to each offspring, the larger it can grow before it must gather its own energy, but the fewer offspring a parent can produce for a given amount of energy—a major trade-off.

A trade-off also exists between number of offspring produced and the amount of care parents provide to their offspring after birth. Individuals of many species are completely independent of their parents during their growth periods, obtaining all their own energy and providing for their own protection. However, in some animal species, parents provide additional care and protection that may extend, as it does in many birds and mammals, until the offspring have reached adult size. The more parental care the parents provide, the fewer offspring they can produce for the same investment in reproduction.

Different species produce their offspring at different times, and produce different numbers of offspring in a given batch (known as a *clutch* or *litter* in animals or a *seed crop* in vascular plants). Some organisms reproduce only once and then die. A bacterium that forms two daughter cells may be considered to have died when it divided. Some plants (called *annuals*) invest so much of the energy they gain during their single growing season in seed production that they do not survive long after reproducing. Most insects and spiders also live for less than a year and die soon after reproducing.

Some longer-lived organisms also reproduce once in their lifetimes and die very soon afterward. Pacific salmon (genus *Oncorhynchus*) hatch in fresh water, spend a number of years at sea, return to fresh water, spawn, and die. Most agaves (century plants) of the American Southwest likewise store up energy for many years before producing a large flowering stalk, forming many seeds, and dying (Figure 51.2). Yucca plants, which grow in the same environments, appear

51.2 Big Bang Reproduction This century plant has mobilized the energy stored during its long life to produce a large flowering stalk with hundreds of flowers, literally reproducing itself to death.

similar, but they invest less in each reproduction and live to reproduce many times. If two individuals have the same amount of energy to invest in reproduction, and one reproduces only once while the other reproduces several times, the former can produce more offspring in a single episode than the latter because it reserves no energy for its own future survival.

A major trade-off exists between reproduction and survival. Studies of many species of plants, shrimp, snails, fishes, birds, and mammals show that engaging in reproduction reduces adult survival rates. Female red deer that produce fawns and nurse them die at higher rates than females that do not produce fawns. The more an adult invests in reproduction, the shorter its lifetime.

A conflict between reproduction and growth is a major life history trade-off. Members of many species do not begin to reproduce until they have reached full size, but others, such as most plants, mollusks, fishes, and reptiles, start to reproduce while they are still relatively small and continue to reproduce as they grow. Reproduction usually reduces growth because these two processes compete for the limited amount of energy an individual has at its disposal. Beech trees in

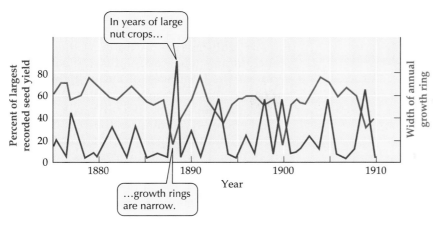

51.3 Reproduction Slows Growth Rates in Beech Trees The width of annual growth rings (plotted in red) reveals the growth rates of beech trees in different years. The trees grow slowly in years when they produce large crops of beechnuts.

Germany, for example, grew more slowly during years when they produced large seed crops than they did during years when their seed crops were small (Figure 51.3).

Offspring are like "money in the bank"

The potential contribution of an individual's offspring to future generations depends upon when they are produced. A useful analogy compares the production of offspring to earning interest on money deposited in a bank. It pays to deposit money in the bank as soon as possible so that it can begin earning interest. Offspring produced early in an adult's life likewise "yield interest" quickly—that is, they begin to reproduce sooner than offspring produced later. However, if juvenile survival is very poor and reproduction greatly reduces the life span of the parent, natural selection may favor delaying reproduction until the parent is older. For example, most birds begin to reproduce when they are 1 year old, but gulls, penguins, and albatrosses do not breed until they are 3 to 9 years old. Although individuals in these three groups reach full size within a year, they are not efficient foragers at this age and cannot gather enough energy to reproduce without jeopardizing their survival and future reproduction.

Reproductive value is the average number of offspring that remain to be born to individuals of a particular age. A newborn individual does not have the highest reproductive value, even though it has its full reproductive potential ahead of it, because many newborn individuals die before they have a chance to reproduce. Therefore we must discount the number of offspring an individual could produce if it survived by the chance that it will die before reaching reproductive age, or during reproduction. When we make the appropriate calculations, we find that the reproductive value of an individual steadily increases until it begins to reproduce, at which time it can no longer die before starting to produce offspring. After maturity, repro-

ductive value usually declines; in most species, it reaches zero when the individual has finished reproducing. However, individuals can still have positive reproductive value after they have stopped reproducing if they continue to assist the survival of their offspring and grandoffspring.

Because reproductive value declines with age in a mature individual, the power of natural selection acting on alleles that first produce their phenotypic effects at older ages grows increasingly weaker. Once reproductive value has dropped to zero, natural selection cannot influence phenotypic traits, even those that are highly detrimental to the individual's survival. As a result, increasing numbers of harmful alleles are expressed as individuals age, causing increased mortality rates, especially after reproduction has ceased. In this manner, **senescence**—an increased probability of dying per unit of time—has evolved.

Senescence poses serious social problems for people in modern industrial societies. As a result of improved hygiene and nutrition, most people in these societies are now spared the serious childhood infections that cause death rates to be high in nonindustrial societies. Most people live to the age when the so-called genetic diseases of old age begin to afflict them. Cancer and heart disease, the main killers in industrialized societies, are much more difficult to deal with than the contagious diseases that formerly caused most deaths. For this reason, despite the expenditure of enormous resources to extend life, the average age at death in the United States has changed very little during the past 30 years. As one source of mortality is eliminated, another takes its place (Figure 51.4). Life history theory suggests that this problem is likely to continue indefinitely.

Population Structure: Patterns in Space

At any given moment, an individual organism occupies only one spot and is of one particular age. The members of a population, however, are distributed over space and differ in age and size. These features determine **population structure**. Ecologists study population structure at different spatial scales ranging from local subpopulations to entire species. Spatial

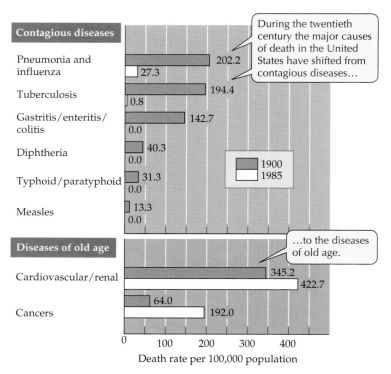

During the twentieth century the major causes of death in the United States have shifted from contagious diseases...

Contagious diseases	
Pneumonia and influenza	202.2 / 27.3
Tuberculosis	194.4 / 0.8
Gastritis/enteritis/ colitis	142.7 / 0.0
Diphtheria	40.3 / 0.0
Typhoid/paratyphoid	31.3 / 0.0
Measles	13.3 / 0.0

■ 1900
□ 1985

Diseases of old age	
Cardiovascular/renal	345.2 / 422.7
Cancers	64.0 / 192.0

...to the diseases of old age.

0   100   200   300   400
Death rate per 100,000 population

51.4 Causes of Human Death Today in the United States most people die of diseases of old age.

distributions of organisms influence the stability of populations and affect interactions among species. Geneticists and evolutionary biologists also study population structure, but they are interested primarily in distributions of genotypes and their degree of isolation from one another because that component of population structure influences how populations evolve (see Chapter 21).

Population density is an important feature

The number of individuals of a species per unit of area (or volume) is its **population density**. Ecologists are interested in population densities because dense populations often exert strong influences on their members and on populations of other species. Other scientists—those who work in agriculture, conservation, or medicine, for example—wish to manage species to raise their densities (crop plants, aesthetically attractive species, threatened or endangered species) or reduce their densities (agricultural pests, disease organisms). To manipulate population densities, we must know what factors make populations grow and shrink and how these factors work.

Because organisms and their environments differ, densities are measured in more than one way. Ecologists usually measure the densities of organisms in terrestrial environments as the number of individuals per unit of area, but number per unit of volume is

generally a more useful measure for organisms living in water. For species whose members differ markedly in size, as do most plants and some animals (such as mollusks, fishes, and reptiles), the total mass of individuals—their **biomass**—may be a more useful measure of density than the number of individuals.

Sometimes individuals can be counted directly without missing any of them or counting any of them twice, but this process is usually impossible or too laborious. Ecologists commonly estimate population densities by sampling a population in a representative area and extending their findings to a larger area. Estimates of population size can also be made by marking and recapturing individuals. For example, if we capture and mark 100 individuals in a population, we can take another sample later and determine what percentage of individuals captured in that sample were already marked. If 10 percent of the captured individuals were already marked, we would conclude that the population contained about 1,000 individuals.

Spacing patterns tell us much about populations

Ecologists studying population structure look at the way the individuals in a population are spaced. Spacing patterns often reveal why individuals settled and survived where they did. Individuals of a population may be tightly clumped together, evenly spaced,

51.5 Competitive Spacing In the sand dunes of this Australian desert, plant spacing is regulated by the availability of water. Each plant removes so much water from the surrounding soil that young plants are unable to establish themselves in the bare areas.

or randomly scattered. Distributions can become clumped when young individuals settle close to their birthplaces, when suitable habitat patches are "islands" separated by unsuitable areas, or by chance. The spacing of many plants is a result of competition for light, water, and soil nutrients (Figure 51.5), or because they rub against one another when moved by wind or water currents. Among animals, defense of space is the most common cause of even distributions (see Figure 50.8). Random distributions may result when many factors interact to influence where individuals settle and survive.

Age distributions reflect past events

Populations are composed of individuals ranging from newborns to postreproductive adults. The proportions of individuals in each age group in a population make up its **age distribution**. The density and spacing of individuals are spatial attributes of a population; age distribution is a *temporal* (time-oriented) attribute. The timing and rates of births and deaths determine age distributions. If birth rates and death rates are both high, a population is dominated by young individuals. If birth rates and death rates are low, a relatively even distribution of individuals of different ages results. The age distribution of a population reveals much about its recent history of births and deaths.

The timing of births and deaths may influence age distributions for many years in populations of long-lived species. The human population of the United States is a good example. Between 1947 and 1964, the United States experienced what is called the post–World War II baby boom. During these years, average family size grew from 2.5 to 3.8 children; an unprecedented 4.3 million babies were born in 1957. Birth rates declined during the 1960s, but Americans born during the baby boom will constitute the dominant age class into the twenty-first century (Figure 51.6a). "Baby boomers" became parents in the 1980s, producing another bulge in the age distribution—a baby boom echo—but they had, on average, fewer children than their parents, so the bulge is not as large. Fish populations may also be dominated by individuals of one age group. In Lake Erie, 1944 was such an excellent year for reproduction and survival of whitefish that individuals of that age group dominated whitefish catches in the lake for several years (Figure 51.6b).

Population Dynamics: Changes over Time

At any moment in time, a population has a particular structure determined by the distribution of its mem-

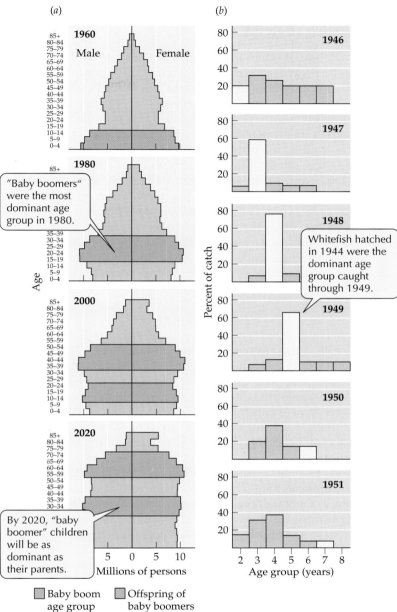

51.6 Age Distribution Can Be Influenced by the Timing of Births Individuals born during years of high birth rates may dominate age distributions for many years in populations of long-lived individuals; two examples are (a) humans in the United States and (b) whitefish in Lake Erie.

bers in space and their ages. However, as we have just seen, population structure is not static. Changes in population structure influence whether the population will increase or decrease; that is, they determine the *dynamics* of the population. We will now examine how ecologists measure changes in birth and death rates and use that information to understand changes in population densities.

Births, deaths, immigration, and emigration drive population dynamics

Knowledge of when individuals are born and when they die provides a surprising amount of information about a population. Births, deaths, immigration, and emigration are **demographic events**—events that determine the numbers of individuals in a population. Ecologists measure the *rates* at which these events take place—that is, the number of such events per unit of time. These rates are influenced both by environmental factors and by the life history traits of the species. The number of individuals in a population at any given time is equal to the number present at some time in the past, plus the number born between then and now, minus the number that died, plus the number that immigrated, minus the number that emigrated. That is, the number of individuals at a given time, N_1, is given by the equation

$$N_1 = N_0 + B - D + I - E$$

where N_1 is the number of individuals at time 1; N_0 is the number of individuals at time 0; B is the number of individuals born, D the number that died, I the number that immigrated, and E the number that emigrated between time 0 and time 1. If we measure these rates over many time intervals, we can determine why a population's density changes.

Life tables summarize patterns of births and deaths

Life tables can help us visualize patterns of births and deaths in a population. We can construct a life table by determining for a group of individuals born at the same time—a **cohort**—the number still alive at specific times and the number of offspring they produced during each time interval. An example is shown in Table 51.1.

As you can see from the table, members of a cohort of the grass *Poa annua* began producing seeds some time after they were 3 months old and continued to produce seeds for the rest of their lives. By the end of 2 years, all members of the cohort were dead. Note that the life table includes both numbers observed and rates calculated from those numbers. For example, we can calculate the probability of dying during the 6–9 month age interval by dividing the number of individuals that died by the number alive at the beginning of the interval: $211/527 = 0.4$. The data in Table 51.1 show that the number of seeds produced per individual peaks at 6 months of age, and that individuals produce very few seeds after they are a year old.

Ecologists often use graphs to highlight the most important changes in birth and death rates in populations. Graphs of survivorship—the mirror image of death rate—in relation to age show when individuals survive well and when they do not. To interpret survivorship data, ecologists have found it useful to compare real data with several hypothetical curves that illustrate a range of possible survivorship patterns. A useful type of graph plots the proportion of individuals of a cohort that are alive at different times during their total potential life span (Figure 51.7a). At one extreme, nearly all individuals survive for their entire potential life span and die almost simultaneously (hypothetical curve I). At the other extreme, the survivorship of young individuals is very low, but survivorship is high for the remainder of the life span (hypothetical curve III). An intermediate possibility is that survivorship is the same throughout the life span (hypothetical curve II).

Survivorship data from real populations often resemble one of these hypothetical curves. For example, survivorship of *Poa annua* seedlings is very high for the first 6 months, but then, as in hypothetical curve I, it declines significantly in older individuals (Figure

TABLE 51.1   Life Table for a Cohort of 843 Individuals of a Short-Lived Grass[a]

AGE INTERVAL (PERIOD BETWEEN TWO EXACT AGES, IN MONTHS)	NUMBER ALIVE AT BEGINNING OF AGE INTERVAL	PROPORTION ALIVE AT BEGINNING OF AGE INTERVAL	NUMBER DYING DURING AGE INTERVAL	DEATH RATE FOR AGE INTERVAL	NUMBER OF SEEDS PRODUCED DURING AGE INTERVAL
0–3	843	1.000	121	0.144	0
3–6	722	0.857	195	0.270	300
6–9	527	0.625	211	0.400	620
9–12	316	0.375	172	0.544	430
12–15	144	0.171	95	0.625	210
15–18	54	0.064	39	0.722	60
18–21	15	0.018	12	0.800	30
21–24	3	0.004	3	1.000	10
24	0	0.000	0	0.000	0

[a]*Poa annua*

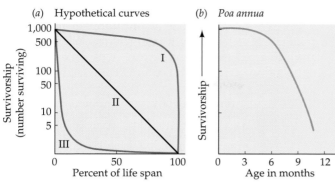

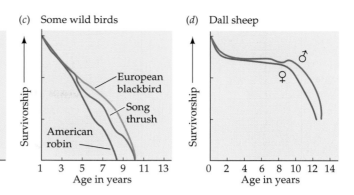

51.7 Survivorship Curves Survivorship curves show the proportion of individuals of a cohort still alive at different times over the life span. (*a*) The range of possible survivorship patterns. (*b*) *Poa annua*, a type of grass. (*c*) Three species of thrushes. (*d*) Male and female dall sheep.

51.7*b*). Many wild birds have survivorship curves similar to hypothetical curve II; the probability of their surviving is about the same over most of the life span once individuals are a few months old (Figure 51.7*c*).

A more common survivorship pattern, especially among organisms that produce large numbers of offspring, each of which receives little energy and no subsequent parental care, is one with low survivorship of young individuals, followed by high survivorship during the middle part of the life span, and then low survivorship toward the end of the life span. Dall sheep, although they do not have a high birth rate, have such a survivorship curve (Figure 51.7*d*), one that combines the first part of the type III curve with the middle and late parts of the type I curve. Survivorship curves help us understand how birth and death rates change over time, but the data ecologists use to estimate the curves are incomplete. None of them include deaths of zygotes prior to birth or seeds prior to germination.

Males and females of a single species may have different survivorship curves. For example, survivorship of adult female red deer is nearly constant once they reach reproductive age (2 to 3 years old), although females that reproduce do not survive as well as nonreproductive females. By contrast, the death rates of males rise sharply when they reach their breeding age of 7 to 8 years (Table 51.2) because males engage in heavy combat with each other during the breeding season. Many males are injured during these fights.

Individuals are not the most important units in modular organisms

For the kinds of organisms we have considered so far, called **unitary organisms**, individuals are easy to distinguish, and most adult members of a population are similar to one another in size and shape. For example,

all populations of mollusks, echinoderms, insects, and vertebrates consist of unitary organisms. But not all organisms are unitary.

Modular organisms are organisms whose bodies consist of repeated units. The fertilized egg of a modular organism develops into a unit of construction—a module—which then produces additional modules much like itself. Many plants are modular, and there are many important groups of modular protists, fungi, and animals (sponges, corals, moss animals, colonial tunicates). Modular organisms may grow to large sizes, and it is often difficult to distinguish a modular organism from a cluster of genetically separate individuals (Figure 51.8).

The effect of modular organisms on their environment often depends primarily on the number and size of the modules. An aspen clone, which is a single genetic individual, affects a much larger area than a single tree does. The modules of a single organism may differ markedly in size and age. Therefore, students of modular organisms are often concerned primarily with the number, size, and shape of modules rather than with the number of genetically distinct individuals. But all population ecologists are interested in how

TABLE 51.2 Death Rates of Male and Female Red Deer on Rhum Island, Scotland

	DEATH RATE[a]	
AGE	FEMALES	MALES
0–6 months	12.4	12.4
6–12 months	15.0	20.8
Yearlings	7.4	13.0
2 years	1.1	1.8
3–4 years	3.6	1.7
5–6 years	3.8	2.2
7–8 years	2.3	6.1
9–10 years	2.8	16.3
11–12 years	8.7	37.0

[a] Percentage of those alive at beginning of year

51.8 A Single Modular Organism May Look Like a Population Each clump of these quaking aspens is a single genetic individual that has spread by underground roots and has sent up many stems, each of which appears to be a separate tree.

population densities change over time, the topic to which we now turn.

Exponential Population Growth

If a single bacterium selected at random from the surface of this book, and all its descendants, were able to grow and reproduce in an unlimited environment, an explosive population growth would result. In a month this bacterial colony would weigh more than the visible universe and would be expanding outward at the speed of light. Similarly, a single pair of Atlantic cod and their descendants, reproducing without hindrance, would fill the Atlantic Ocean in 6 years.

All populations have the potential for explosive growth because as the number of individuals in the population increases, the number of new individuals added per unit of time accelerates, even if the rate of growth per individual—called the per capita growth rate—remains constant. This form of explosive increase is called **exponential growth**. If for the moment

we ignore immigration and emigration and assume that births and deaths occur continuously and at constant rates, such a growth pattern forms a continuous curve (Figure 51.9a) that is expressed mathematically in the following way:

Rate of increase in number of individuals

$$= \left(\begin{array}{l} \text{Average per capita birth rate} \\ - \text{ Average per capita death rate} \end{array} \right)$$

$\times$ Number of individuals

or, more concisely,

$$\frac{\Delta N}{\Delta t} = (b - d)N$$

where $\Delta N / \Delta t$ is the rate of change in size of the population (ΔN = change in number of individuals; Δt = change in time). The difference between the average per capita birth rate (b) and the average per capita death rate (d) is called r. In these equations, b includes both births and immigration and d includes both deaths and emigration. When conditions are optimal for the population, r has its highest value, called r_{max}, the **intrinsic rate of increase**; r_{max} has a characteristic value for each species. Therefore, the rate of growth of a population under optimal conditions is $\Delta N / \Delta t = r_{max} N$.

However, as we have already seen, for many organisms, births and deaths are not continuous processes. Many insects and annual plants reproduce only once each year and then die. Many other species reproduce only during a short period of the year—the breeding season—but deaths occur continuously during the year. These populations have discrete generations, and they are modeled with a different equation:

$$N_{t+1} = \lambda N_1$$

(a) Exponential (unrestricted) growth

Theoretically, a population in an imaginary environment with unlimited resources could grow like this.

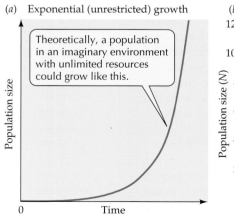

Population size

0   Time

51.9. Exponential Growth (a) A theoretical exponential growth curve. (b) The growth of a pheasant population introduced to Protection Island, Washington State.

(b) Pheasant population on protected island

The population grew exponentially for a few years after it was introduced to a favorable environment.

In populations with discrete generations, growth is jagged.

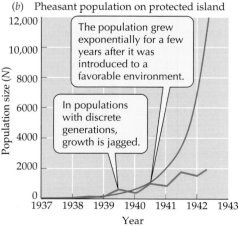

Population size (N)

1937 1938 1939 1940 1941 1942 1943
Year

where λ, the **finite rate of increase**, is the ratio of the population size during the next time period to the population size during the current time period. The growth curve of a population with a single breeding season each year resembles a saw blade, with a sharp vertical increase during the breeding season followed by a gradual decline due to deaths during the rest of the year (see Figure 51.9b).

Populations may experience exponential growth when they are introduced to or colonize previously unoccupied but favorable environments. For example, in 1937, eight pheasants were introduced onto Protection Island off the coast of Washington State. The island, which had abundant food and no mammalian predators, was too far from the mainland for pheasants to fly to and from it. By 1942, the population had increased to nearly 2,000 birds. Because pheasants lay eggs only once each year, in spring, the growth curve is jagged. The rate of population growth initially was close to the theoretical maximum (Figure 51.9b). However, by 1942, the population was no longer growing exponentially, as it did during the first few years. In the next section we explore why growth rates in all populations eventually slow down, often very soon.

Population Growth in Limited Environments

No real population can maintain exponential growth for very long because environmental limitations cause birth rates to drop and death rates to rise. In fact, over long time periods, the size of most populations fluctuates around a relatively constant number. The simplest way to picture the limits imposed by the environment is to assume that an environment can support no more than a certain number of individuals of any particular species. This number, called the environmental **carrying capacity**, is determined by the availability of resources—food, nest sites, shelter—as well as by disease, predators, and perhaps social interactions.

Population growth is influenced by the carrying capacity

The limitations of the carrying capacity mean that, rather than being exponential, population growth follows a curve that flattens out as the population approaches the carrying capacity, so that the curve has an S-shape (Figure 51.10). This is what happened to the pheasant population on Protection Island.

The S-shaped growth pattern, which is characteristic of many populations growing in environments with limited resources, can be represented mathematically by adding to the equation for exponential growth a term, $(K - N)/K$, that slows the population's growth as it approaches the carrying capacity. The simplest such equation is that for **logistic growth**, in which

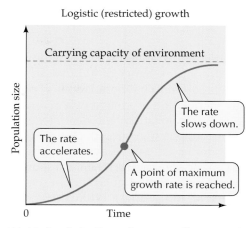

Logistic (restricted) growth

51.10 Logistic Growth   Typically, a population in an environment with limited resources stops growing exponentially when it reaches the environmental carrying capacity.

each individual added to the population depresses population growth equally:

$$\frac{\Delta N}{\Delta t} = r\left(\frac{K - N}{K}\right)N$$

where K is the carrying capacity and the other symbols are the same as in the equation for exponential growth. The biological assumption in this equation is that each additional individual makes things slightly worse for the others because it competes with them for available resources, or for other reasons. Population growth stops when $N = K$ because then $(K - N) = 0$, so $(K - N)/K = 0$, and thus $\Delta N/\Delta t = 0$.

The logistic growth equation contains some important simplifications that are not true for most populations. Its most critical assumptions are that (1) each individual exerts its effects immediately at birth; (2) all individuals produce equal effects; and (3) births and deaths are continuous. However, in nature, organisms grow during their lives, and their effects on others normally increase with age, so there may be a delay between the birth of an individual and the time at which it begins to affect the other members of its population. A seedling tree exerts a much smaller effect on its neighbors than a large adult tree does, and it does not begin to reproduce until it reaches a relatively large size.

In addition, the logistic equation models a population in a single, uniform habitat patch. Individuals that live elsewhere may immigrate into the local population and some individuals may leave it, but we have modeled only the density of individuals in the local population. Next we consider the dynamics of an assembly of local populations, which is more complex than the growth of a single population.

Many species are divided into discrete subpopulations

Populations that are divided into discrete subpopulations, among which some exchange of individuals occurs, are called **metapopulations** (populations of populations). Each subpopulation of a metapopulation has a probability of birth (colonization) and death (extinction). Within each subpopulation, growth occurs in the ways we have just discussed, but because subpopulations are typically much smaller than the metapopulation as a whole, local disturbances and random fluctuations in numbers of individuals often cause the extinction of a subpopulation. However, if individuals move frequently between subpopulations, immigrants may prevent declining subpopulations from becoming extinct. This process is known as the **rescue effect**.

The bay checkerspot butterfly (*Euphydryas editha bayensis*) provides a good illustration of metapopulation dynamics. The larvae of this butterfly feed on only a few species of annual plants that are restricted to outcrops of a particular kind of rock on hills south of San Francisco (Figure 51.11). The bay checkerspot has been studied for many years by Stanford University biologists. During drought years, most host plant individuals die early in spring, before the butterfly larvae have completed their development. At least three butterfly subpopulations became extinct during a severe drought in 1975–1977. The largest patch of suitable habitat, Morgan Hill, typically supported thousands of butterflies. It probably served as a source of individuals that dispersed to and colonized small patches where the butterflies had become extinct.

During the severe California drought of the 1990s, many subpopulations of the bay checkerspot became extinct. During the spring of 1996, no butterflies were found in any of the subpopulations. The metapopulation had become extinct. Drought was a major factor causing the extinction, but suburban expansion that destroyed or diminished the sizes of some of the patches probably also contributed to it.

The logistic growth equation is one way of expressing the fact that the births, deaths, and movements of organisms are influenced by the densities of individuals of the same and other species. The logistic growth equation incorporates only the influence of other individuals of the same species, but even this simple inclusion provides the foundation for considering the regulation of the density and distributions of individuals of a species, the topic to which we now turn.

Population Regulation

In a population that is growing logistically, growth slows down as density increases because the members increasingly affect one another adversely. As a result, a population above the environmental carrying capacity is more likely to decrease in density than one that is below the carrying capacity. In this section we discuss how populations may be regulated by interactions between their density and the carrying capacity of their environment, by disturbances, and by movements of individuals.

How does population density influence birth and death rates?

If the density of a population is determined primarily by changes in per capita birth or death rates in response to density, it is said to be regulated by **density-dependent** factors. Death or birth rates may be density-dependent for several reasons. First, as a species increases in abundance, it may deplete its food supply, reducing the amount of food that each individual gets. Poor nutrition may increase death rates and lower birth rates. Second, predators may be attracted to regions where densities of their prey have increased. If predators are able to capture a larger proportion of the prey than they did when the prey were scarce, the per capita death rate of the prey rises. Third, diseases, which may increase death rates, spread more easily in dense populations than in sparse populations.

If the per capita birth and death rates in a population are unrelated to its density, population regulation is said to be **density-independent**. A very cold spell in winter, for example, may kill a large proportion of the individuals in a population regardless of its density. However, even density-*independent* environmental fac-

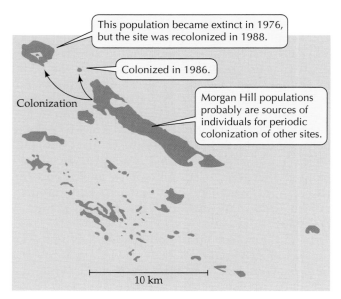

51.11 A Metapopulation The bay checkerspot butterfly metapopulation was divided into a number of subpopulations confined to habitats that contain the food plant of its larvae. (The entire metapopulation has since become extinct.)

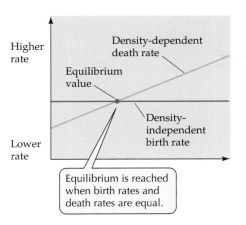
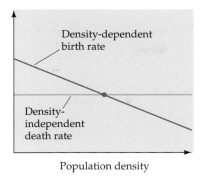
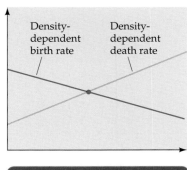

Higher rate

Equilibrium value

Density-dependent death rate

Density-independent birth rate

Lower rate

Equilibrium is reached when birth rates and death rates are equal.

Density-dependent birth rate

Density-independent death rate

Population density

Density-dependent birth rate

Density-dependent death rate

Conclusion: If birth rate or death rate, or both, are density-dependent, a population's size tends to fluctuate around an equilibrium value.

51.12 Density-Dependent Factors Regulate Population Size Densities of all populations fluctuate, but fluctuations are diluted by density-dependent birth and death rates.

tors may result indirectly in density-*dependent* mortality. The cold weather, for example, may not kill organisms directly, but may increase the amount of food individuals need to eat each day. Individuals pushed by population density into poorer foraging areas may be more likely to die than those in better foraging areas. Or the death rate may be related to the quality of sleeping places. If population density is high, a larger proportion of individuals may be forced to sleep in places that expose them to the cold.

Various combinations of density-dependent and density-independent events can regulate the density of a population. The hypothetical graphs in Figure 51.12 show how birth and death rates can change in relation to population density. When birth and death rates are equal—the point at which the lines cross—the population neither grows nor shrinks. If birth or death rates or both are density-dependent, the population responds to increases or decreases in its density by returning toward an equilibrium density. If neither rate is density-dependent, there is no equilibrium. The abundance of a species is determined by the combined effects of all the factors and processes, density-dependent and density-independent, impinging upon its populations.

Ecologists often infer the operation of density-dependent processes by observing correlations between population density and birth and death rates. Stronger evidence is provided by experiments in which densities are manipulated. An experimental test of density-dependent population regulation was performed on a gall-forming fly (*Eurosta solidaginis*) that infests goldenrod (*Solidago altissima*), a common herbaceous plant found in fields in eastern North America. *Eurosta* is widespread, but it usually exists at moderately low and relatively constant densities. Adult flies emerge in early June from galls in which they have overwin-

tered. A female fly lays her eggs in a bud on a growing goldenrod stem. After it hatches, the larva bores into the stem and develops inside a gall, where it feeds on the plant's tissues. The gall forms rapidly, reaching full size in early July (Figure 51.13). In September, the fully developed larva pupates within the gall after excavating an exit tunnel through which it will emerge.

Eurosta larvae and pupae are attacked by a suite of predators that includes a parasitoid wasp, a beetle, and birds. To determine whether predation rates on *Eurosta* act in a density-dependent manner, an ecologist collected all the galls from a number of plots, kept them in an unheated room during the winter, and released the flies and their wasp predators into the fields

51.13 A Gall-Forming Fly This goldenrod was stimulated to form multiple galls when a female *Eurosta* laid her eggs on the stem. A larval fly will hatch from each gall.

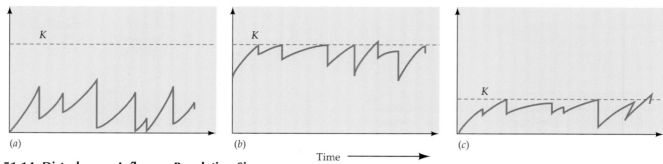

Time

51.14 Disturbances Influence Population Size
(a) Dynamics of a population dominated by phases of population growth after repeated disturbances. The population density is well below the environmental carrying capacity (*K*) most of the time. (b) Dynamics of a population dominated by limitations on environmental carrying capacity. The population density is close to *K* most of the time. (c) Same as in (b), but with a much lower carrying capacity.

at the appropriate time the following spring. On some plots, more flies were released than had been present the previous year; on others, fewer flies were released—in other words, population densities were augmented or reduced. At the end of the following winter all galls were collected, and survival rates of *Eurosta* were determined by dissecting the galls. Death rates during the year were strongly density-dependent; that is, they were highest in the experimental plots where the greatest density of flies had been released. Thus, populations of *Eurosta* in the study area appear to be maintained at relatively constant levels by density-dependent predation.

Disturbances affect population densities

Populations are repeatedly exposed to **disturbances**—short-term events that disrupt populations by changing their environment and, hence, its carrying capacity. Common physical disturbances are fires, hurricanes, ice storms, floods, landslides, and lava flows. Biological disturbances include tree falls, disease epidemics, and the burrowing and trampling activities of animals. Disturbances differ in their spatial distribution, frequency, predictability, and severity.

A disturbance may lower the environmental carrying capacity for only a short time. When it recovers, populations may grow rapidly. In general, smaller organisms are more strongly affected by environmental disturbances than larger ones are. Most insect populations in the temperate zone are constantly recovering from disturbances. In contrast, many bird species appear to be close to the environmental carrying capacity much of the time. Insect populations often behave according the pattern shown in Figure 51.14*a*, whereas birds generally follow a pattern that is similar to those in Figures 51.14*b* and *c*.

The responses of organisms to disturbances depend upon the frequency and severity of those disturbances. Organisms respond behaviorally and physiologically to disturbances that occur regularly and repeatedly during their lifetimes. Animals seek shelter in storms and go into hibernation in winter. Trees drop their leaves in winter and change physiologically so that they can tolerate frosts. However, if such a disturbance is unusually severe, the tolerances of individuals may be exceeded, and many may die. An organism is more likely to have adaptations for tolerating a particular kind of disturbance if the disturbance is frequent relative to the organism's life span.

Populations of organisms can also influence the frequency of some disturbances. Immediately after a fire, there is not enough combustible organic matter to carry another fire. However, as vegetation grows back, dead wood, branches, and leaves accumulate, gradually increasing the supply of fuel to support another fire. Thus the frequency of fires may be proportional to the rate at which fuel accumulates through the growth of plant populations, or the rate at which herbivores consume plant materials that would otherwise accumulate. Similarly, as many trees age, their roots become weakened by fungal infections. Old, large trees are thus susceptible to being toppled by high winds. Therefore, the likelihood of a major blowdown increases with forest age.

Organisms often cope with habitat changes by dispersing

A common response of animals to disturbances or other environmental changes is dispersal. If habitat quality declines greatly, individuals may be able to improve their survival and reproduction by going elsewhere. If regularly repeated seasons cause changes in a habitat, organisms may adjust their life cycles in equally regular ways, appearing to anticipate the changes.

One of the most spectacular responses to seasonal changes in habitat quality is **migration,** the regular seasonal movement of animals from one place to another. This behavior is most widespread among birds, but some insects, such as monarch butterflies, and

(a) *Danaus* sp.

(b) *Rangifer tarandus*

51.15 Animals Migrate to Remain in Suitable Environments   (a) Most of North America's monarch butterflies migrate to central Mexico. They spend the winter in cool mountain valleys, where they can reduce their energy expenditure by keeping their metabolic rates low. (b) These caribou in the American Arctic are migrating from the open tundra to their winter feeding grounds at the edge of the boreal forest, where food, some of it in the form of lichens on the branches of trees, is more readily available when the ground is covered with snow.

some mammals also migrate (Figure 51.15). The primary function of migration is to keep the animals in good foraging areas at all times of the year. In arctic regions, caribou migrate each year between winter and summer ranges (see Figure 50.21). Most insectivorous birds leave high latitudes in autumn for more favorable wintering grounds at low latitudes.

The limits of species ranges are dynamic

Environmental conditions, births, deaths, and dispersal determine the density and distribution of a species. Within the geographic ranges of most species, population densities are higher toward the center of the range and decline toward the periphery, becoming zero at the limit of the range of the species. The scissor-tailed flycatcher (*Tyrannus forficatus*), a bird that winters in Central America and breeds in the southern Great Plains of the United States, illustrates this pattern of population density (Figure 51.16). The density pattern of scissor-tailed flycatchers is simple because the region where they breed is relatively flat. Species

that live in more mountainous regions typically have many centers of high population density.

Some species achieve their highest population densities at the margins of their ranges. This situation often arises when range boundaries are at the edges of land masses, where there is an abrupt transition from suitable terrestrial habitats to unsuitable marine habitats. Competition between ecologically similar species can also result in abrupt changes in population densities at the edges of ranges, as we will see in Chapter 52.

Humans Manage Populations

For many centuries, people have tried to decrease populations of species they consider undesirable and

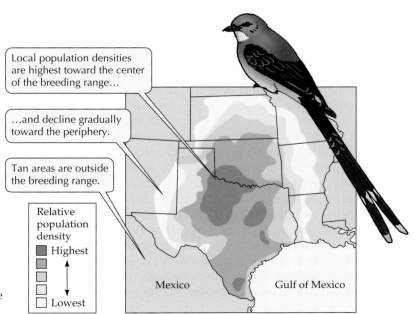

Local population densities are highest toward the center of the breeding range...

...and decline gradually toward the periphery.

Tan areas are outside the breeding range.

Relative population density

☐ Highest
☐
☐
☐
☐ Lowest

Mexico   Gulf of Mexico

51.16 Population Densities Are Greater toward the Center of a Range   The population density pattern of the scissor-tailed flycatcher is simple because its breeding range is relatively flat.

increase populations of desirable species. Strategies for controlling and managing populations of organisms are based on our understanding of how populations grow and are regulated. A general principle of population dynamics is that the total number of births and the growth rates of individuals tend to be highest when a population is well below its carrying capacity. Therefore, if we wish to maximize the number of individuals that can be harvested, we should manage a population so that it is far enough below carrying capacity to have high birth and growth rates. Hunting seasons for birds and mammals are determined with this objective in mind.

Life history traits determine how heavily a population can be exploited

Populations of some organisms can sustain their growth despite a high rate of harvest. Such populations (many species of fish, for example) reproduce at high rates, with each female laying thousands or millions of eggs. Another characteristic of these high-yielding populations is that individual growth is density-dependent. If prereproductive individuals are harvested at a high rate, the remaining individuals may grow faster. Many fish populations can be harvested heavily for many years because only a modest number of females must survive to reproductive age to produce the eggs needed to maintain the population.

Fish can, of course, be overharvested. Many populations have been greatly reduced because so many individuals were harvested that too few reproductive adults survived to maintain the population. The Georges Bank off the coast of New England—a source of cod, halibut, and other prime food fishes—has been exploited so heavily that many fish stocks have been reduced to levels insufficient to support a commercial fishery.

The whaling industry has also engaged in excessive harvests. The blue whale, Earth's largest animal, was the first whale species to be hunted nearly to extinction. The industry then turned to smaller species of whales that were still numerous enough to support commercially viable whaling operations (Figure 51.17). Management of whale populations is difficult for two reasons. First, whales reproduce at very low rates—they have long prereproductive periods, produce only one offspring at a time, and have long intervals between births. Thus many whales are needed to produce even a small number of offspring. Second, because whales are distributed widely throughout Earth's oceans, they are an international resource whose conservation and wise management depends upon cooperative action by *all* whaling nations. The recovery of whale populations will require observance by all nations of a moratorium on commercial harvesting.

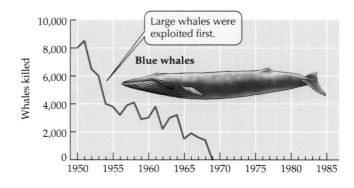

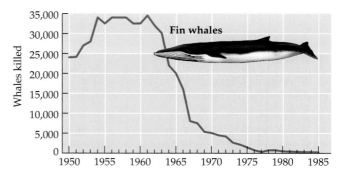

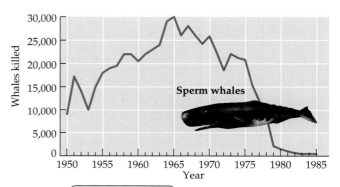

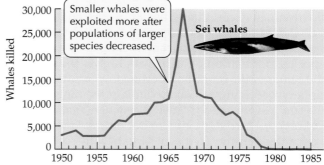

51.17 Overexploitation of Whales The graphs show the number of whales of four species killed each year from 1969 to 1985.

Life history information is used to control populations

The same principles apply if we wish to reduce the size of populations of undesirable species and keep

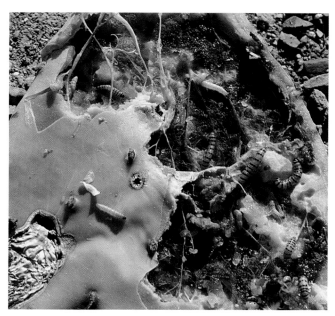

51.18 Biological Control of a Pest *Cactoblastis* caterpillars consume an *Opuntia* cactus, a pest in Australia.

them at low densities. At densities well below carrying capacity, populations have high birth rates, and can therefore withstand higher death rates than they could closer to carrying capacity. Killing part of a density-dependently regulated population only brings it back to the point at which it experiences the most rapid rate of growth. A far more effective approach to reducing the population of a species is to remove its resources, thereby lowering the carrying capacity of its environ-

ment. We can rid our dumps and cities of rats more easily by making garbage unavailable (reducing the carrying capacity of the rats' environment) than by poisoning the rats.

Similarly, if we wish to preserve a rare species, the most important step usually is to provide it with suitable habitat. If habitat is available, the species will usually reproduce at rates sufficient to maintain its population. If the habitat is insufficient, preserving the species usually requires expensive and continuing intervention, such as providing extra food.

Humans often introduce predators and parasites to regulate populations of undesirable species. For example, the cactus *Opuntia,* introduced into Australia from South America, spread rapidly and became a common pest species over vast expanses of valuable sheep-grazing land. It was controlled by introducing a moth species (*Cactoblastis cactorum*) whose larvae hatch in and completely destroy patches of *Opuntia* found by the egg-laying females (Figure 51.18). However, new patches of cactus arise in other places from seeds dispersed by birds. These new patches flourish until they are found and destroyed by *Cactoblastis*. Today, over a large region, the numbers of both *Opuntia* and *Cactoblastis* are fairly constant and low, but in the local areas that make up the whole, there are vigorous oscillations resulting from the extermination of first the plant and then the herbivore.

Can we manage our human population?

Managing our own population has become a matter of great concern because the size of the human population is responsible for most environmental problems,

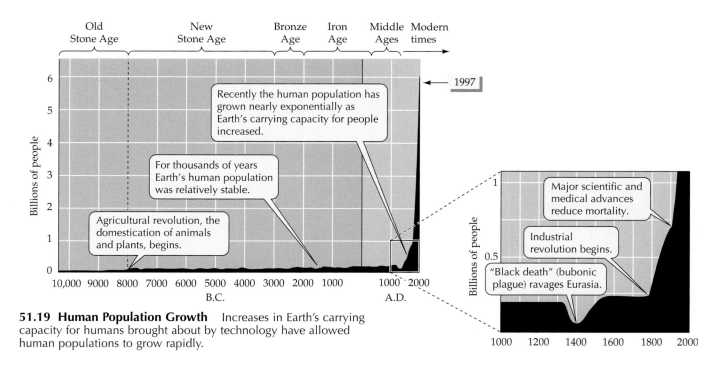

51.19 Human Population Growth Increases in Earth's carrying capacity for humans brought about by technology have allowed human populations to grow rapidly.

from pollution to extinctions of other species. For thousands of years, Earth's carrying capacity for human populations was set at a low level by food and water supplies and disease—that is, by the technology available to garner resources and combat diseases. Domestication of plants and animals and cultivation of the land enabled our ancestors to increase resources at their disposal dramatically. These developments stimulated rapid population growth up to the next carrying capacity limit, which was determined by the agricultural productivity possible with only human- and animal-powered tools. Agricultural machines and artificial fertilizers, made possible by the tapping of fossil fuels, greatly increased agricultural productivity, further raising Earth's carrying capacity for humans. The development of modern medicine reduced the effectiveness of disease as a limiting factor on human populations, raising the global carrying capacity still further (Figure 51.19). Medicine and better hygiene have allowed people to live in large numbers in areas where diseases formerly kept numbers very low.

What is Earth's present carrying capacity for people? Today's carrying capacity is set by Earth's ability to absorb the by-products of our enormous consumption of fossil fuel energy and by whether we are willing to cause the extinction of millions of other species to accommodate our increasing use of environmental resources. We will explore some of the consequences of high human population densities for the survival of other species in Chapter 55.

Summary of "Population Ecology"

Varied Life Histories

• The life history of a species consists of growth, dispersal, and reproductive stages. Many organisms have complex life histories.
• A population consists of all the individuals of a species within a given area.
• Trade-offs inevitably exist between number and size of offspring, between number of offspring and parental care, between survival and reproduction, and between growth and reproduction. **Review Figures 51.2, 51.3**
• Reproductive value is the average number of offspring that remain to be born to individuals of a particular age. Reproductive value rises to a peak when individuals first begin to reproduce and then declines to zero after reproduction ceases.

Population Structure: Patterns in Space

• The number of individuals of a species per unit of area (or volume) is its population density. Dense populations often exert strong influences on populations of other species.

• Its age distribution reveals much about the recent history of births and deaths in a population. The timing of births and deaths may influence age distributions for many years. **Review Figure 51.6**

Population Dynamics: Changes over Time

• Births, deaths, immigration, and emigration drive changes in population density and distribution.
• Life tables help us visualize patterns of births and deaths in a population. **Review Table 51.1**
• Graphs of survivorship in relation to age show when individuals survive well and when they do not. **Review Figure 51.7**

Exponential Population Growth

• All populations have the potential to grow exponentially when they colonize suitable environments. **Review Figure 51.9**

Population Growth in Limited Environments

• No population can maintain exponential growth for very long because environmental limitations cause birth rates to drop and death rates to rise.
• The number of individuals of a particular species that an environment can support—called the carrying capacity—is determined by the availability of resources and by disease and predators.
• A population in a constant but limited environment at first grows rapidly, but growth rates decrease as the carrying capacity is approached. **Review Figure 51.10**
• Metapopulation dynamics are determined by "births" (colonizations) and "deaths" (extinctions) of local subpopulations. Immigrants may prevent declining subpopulations from becoming extinct, a process known as the rescue effect. **Review Figure 51.11**

Population Regulation

• Regulation of a population by changes in per capita birth or death rates in response to density is said to be density-dependent.
• If per capita birth and death rates are unrelated to a population's density, population regulation is said to be density-independent.
• The abundance of a species is determined by the combined effects of all density-dependent and density-independent factors affecting it. **Review Figure 51.12**
• Populations of many organisms are below carrying capacity most of the time because they are recovering from disturbances. **Review Figure 51.14**
• Population densities of most species are high near the center of the species' range and decline toward the periphery. **Review Figure 51.16**

Humans Manage Populations

• Humans use the principles of population dynamics to control and manage populations of desirable and undesirable species. Nevertheless, many populations have been overexploited. **Review Figures 51.17, 51.18**
• Earth's carrying capacity for humans has been increased several times by technological developments. Whether the current human population exceeds Earth's carrying capacity is hotly debated. **Review Figure 51.19**

Self-Quiz

1. The number of individuals of a species per unit of area is known as its
   a. population size.
   b. population density.
   c. population structure.
   d. subpopulation.
   e. biomass.

2. The age distribution of a population is determined by
   a. the timing of births.
   b. the timing of deaths.
   c. the timing of both births and deaths.
   d. the rate at which the population is growing.
   e. all of the above

3. Which of the following is *not* a component of the life history of all organisms?
   a. Growth
   b. Dispersal
   c. Reproduction
   d. Reorganization
   e. Energy gathering

4. Which of the following is *not* a demographic event?
   a. Growth
   b. Birth
   c. Death
   d. Immigration
   e. Emigration

5. A group of individuals born at the same time is known as a
   a. deme.
   b. subpopulation.
   c. Mendelian population.
   d. cohort.
   e. taxon.

6. A population grows at a rate closest to its intrinsic rate of increase when
   a. its birth rates are the highest.
   b. its death rates are the lowest.
   c. environmental conditions are optimal.
   d. it is close to the environmental carrying capacity.
   e. it is well below the environmental carrying capacity.

7. Some organisms reproduce only once in their lifetimes because they
   a. invest so much in reproduction that they have insufficient reserves for survival.
   b. produce so many offspring at one time that they do not need to survive longer.
   c. don't have enough eggs to reproduce again.
   d. don't have enough sperm to reproduce again.
   e. have stopped growing.

8. Which of the following is *not* true of reproductive value?
   a. Reproductive value is the average number of offspring that remain to be born to an individual of a particular age.
   b. The reproductive value of an individual increases until it begins to reproduce.
   c. Reproductive value reaches its maximum when an individual completes reproduction.
   d. Reproductive value usually reaches its maximum when an individual begins to reproduce.
   e. Reproductive value usually declines during the reproductive life of an individual.

9. Density-dependent population regulation results when
   a. only birth rates change in response to density.
   b. only death rates change in response to density.
   c. diseases spread in populations at all densities.
   d. both birth and death rates change in response to density.
   e. population densities fluctuate very little.

10. The best way to reduce the population of an undesirable species in the long term is to
   a. reduce the carrying capacity of the environment for the species.
   b. selectively kill reproducing adults.
   c. selectively kill prereproductive individuals.
   d. attempt to kill individuals of all ages.
   e. sterilize individuals.

Applying Concepts

1. Huntington disease is a severe disorder of the human nervous system that generally results in death. It is caused by a dominant allele that does not usually express itself phenotypically until its bearer is 35 to 40 years old. How fast is the gene causing Huntington disease likely to be eliminated from the human population? How would your answer change if the gene expressed itself when its bearer was 20 years old? 10 years old?

2. Many people have improperly formed wisdom teeth and must spend considerable sums of money to have them removed. Assuming, as is probably the case, that the presence or absence of wisdom teeth and their mode of development are partly under genetic control, will we gradually lose our wisdom teeth by evolutionary processes?

3. Some organisms, such as oysters and elm trees, produce vast quantities of offspring, nearly all of which die before they reach adulthood. What fraction of such deaths are likely to be selective—that is, dependent on the genotypes of the individuals dying? If in fact most such deaths are nonselective, what does that imply for the rates of evolution of oysters and elms?

4. Most organisms whose populations we wish to manage for higher densities are long-lived and have low reproductive rates, whereas most organisms whose populations we attempt to reduce are short-lived but have high reproductive rates. What is the significance of this difference for management strategies and effectiveness of management practices?

5. In the mid-nineteenth century, the human population of Ireland was largely dependent upon a single food crop, the potato. When a disease caused the potato crop to fail, the Irish population declined drastically for three reasons: (1) a large percentage of the population emigrated to the United States and other countries; (2) the average age of a woman at marriage increased from about 20 to about 30 years; and (3) many families starved to death rather than accept food from Britain.

None of these social changes was planned at the national level, yet all contributed to adjusting population size to the new carrying capacity. Discuss the ecological strategies involved, using examples from other species. What would you have done had you been in charge of the national population policy for Ireland?

6. From a purely ecological standpoint, can the problem of world hunger ever be overcome by improved agriculture alone? What other components must a hunger-control policy include?

Readings

Begon, M., J. L. Harper and C. R. Townsend. 1996. *Ecology: Individuals, Populations and Communities*, 3rd Edition. Blackwell Scientific Publications, Oxford. A basic text for all aspects of contemporary ecology.

Begon, M. and M. Mortimer. 1996. *Population Ecology: A Unified Study of Animals and Plants*, 3rd Edition. Blackwell Scientific Publications, Oxford. This introduction to population dynamics stresses the differences between plants and animals.

Charnov, E. L. 1993. *Life History Invariants: Some Explorations of Symmetry in Evolutionary Ecology*. Oxford University Press, Oxford. An excellent but technical treatment of phylogenetic constraints and trade-offs in life histories.

Gotelli, N. J. 1995. *A Primer of Ecology*. Sinauer Associates, Sunderland, MA. A unique introduction to models of population growth, population structure, and metapopulation dynamics. Has problems to work at the end of each chapter.

Harper, J. L. 1977. *The Population Biology of Plants*. Academic Press, New York. The most complete review of the literature on plant populations; an excellent advanced reference for most topics in plant population biology.

Myers, J. H. 1993. "Population Outbreaks in Forest Lepidoptera." *American Scientist*, vol. 81, pages 240–251. Illustrates the role of viruses in causing the remarkable population cycles of tent caterpillars and other forest insects.

Ricklefs, R. E. 1997. *Ecology*, 4th Edition. W. H. Freeman and Company, New York. This text covers both the dynamic and the evolutionary aspects of ecology.

Stearns, S. C. 1992. *The Evolution of Life Histories*. Oxford University Press, Oxford. Thorough coverage of all aspects of life history evolution, including a discussion of the major analytic tools used in life history analyses.

Chapter 52

Community Ecology

Snow Triggers Community Interactions
Weather conditions in the mountains above the Four Corners desert affected the vegetation, which affected the insect and deer mouse populations, which resulted in the spread of hantaviruses carried by the mice. When predators controlled the deer mouse population, incidence of hantavirus-related respiratory illness among humans also declined.

During May and June, 1993, many people in the Four Corners region of Arizona, Colorado, New Mexico, and Utah contracted a rare respiratory disease. Scientists at the Center for Disease Control and Prevention (CDC) tested the blood of people stricken with the disease and found antibodies to hantaviruses in their tissues. Hantaviruses infest between 4,000 and 200,000 people each year in Europe and Asia; between 4,000 and 20,000 of them die. Because rodents are vectors of hantaviruses in the Old World, CDC scientists contacted ecologists in the Four Corners region to find out whether anything unusual had happened to rodent populations in the region. The ecologists, who had been studying rodents in the area for many years, knew that unusual events had indeed occurred.

Heavy snow and rain in the mountains during the previous winter and spring had resulted in an abundant crop of piñon pine cones and unusually large insect populations. Deer mice feasted on the abundant piñon nuts and grasshoppers; their population increased tenfold between May 1992 and May 1993. CDC scientists found that about 30 percent of the deer mice in the region were infected by hantaviruses, suggesting that they were the primary vectors of the disease. Fortunately, the population of deer mice declined rapidly during the summer of 1993 from 30 per hectare in June to 4 per hectare in August, probably because of heavy predation by foxes, owls, and snakes.

The rise and fall of populations of deer mice in the Four Corners region was the result of interactions between the mice, their food supply, and their predators. The populations of these organisms also interact with populations of other plants and animals, fungi, protists, bacteria, and archaebacteria. All of the organisms that live in a particular area constitute an **ecological community**. Each of the species interacts in unique ways with other species in its community. Some of these interactions are strong and important; others are weak and affect the functioning of the community very little. The study of such interactions, and how they determine which and how many species live in a place, is the focus of community ecology. In this chapter we consider the niches of organisms and the types of ecological interactions—predator–prey, competition, mutualism, and commensalism—and show how they determine the structure and functioning of ecological communities.

Ecological Interactions

Deer mice, piñon pines, hantaviruses, and the other organisms in the Four Corners region, and in all other ecological communities, interact with one another in four major ways (Table 52.1). One organism, by its activities, may benefit itself while harming the other, as when individuals of one species eat individuals of the other. The eater is called a predator or parasite and the eaten is its prey or host. These interactions are known as **predator–prey** or **host–parasite interactions** (a +/− interaction). Alternatively, two organisms may mutually harm one another. This type of interaction is common when two organisms use the same resources and those resources are insufficient to supply their combined needs. Such organisms are called competitors, and their interactions constitute **competition** (a −/− interaction). If both participants benefit from an interaction, we call them mutualists, and their +/+ interaction is a **mutualism**. If one participant benefits but the other is unaffected, the interaction is a **commensalism** (a +/0 interaction). If one participant is harmed but the other is unaffected, the interaction is an **amensalism** (a 0/− interaction).

Figs and fig wasps illustrate the complexity of ecological interactions

The categories of species interactions are not clear-cut, both because the strengths of interactions vary and because many cases do not fit the categories neatly. That

52.1 Fig Wasps Have Their Own Predators A female parasitoid wasp bores through a fig with her long ovipositor to lay her eggs on the fig wasp larvae inside.

nature is more complex than this simple classification system is illustrated by interactions between figs and fig wasps. Most species of fig trees can reproduce only with the help of certain wasps. Fig wasps of most species visit only one fig species, and most fig species are visited by only one wasp species. A fig tree begins reproduction by producing a large number of closed flower clusters, within which many female flowers form. A female fig wasp bearing both her own fertilized eggs and fig pollen enters a cluster through a small hole, which soon seals. She pollinates receptive female flowers, lays her eggs in the ovaries of some of the flowers, and dies. Each wasp larva develops within—and eats—one seed. Not all of the larvae develop; some may be parasitized by another wasp species. A parasitic female wasp may puncture the fig with her long ovipositor and lay her eggs on the fig wasp larvae (Figure 52.1).

TABLE 52.1 Types of Ecological Interactions

		EFFECTS ON ORGANISM 2		
		BENEFIT	**HARM**	**NO EFFECT**
EFFECTS ON ORGANISM 1	**BENEFIT**	Mutualism	Predation or parasitism	Commensalism
	HARM	Predation or parasitism	Competition	Amensalism
	NO EFFECT	Commensalism	Amensalism	—

As the fig develops, pollen-bearing male flowers mature. At this point, the young male fig wasps chew their way out of the seeds in which they developed and crawl around inside the fig searching for seeds housing female wasps. The males chew open these seeds and mate with the females. The fertilized females emerge from their seeds and collect pollen from male flowers. The females leave the fig through a hole cut in its wall by the males, fly to another fig tree, and begin the reproductive cycle again. The fig finishes ripening, becoming a soft, sweet fig that may eventually be eaten by a bird or mammal.

Fig wasps depend completely on figs to complete their life cycles. To produce offspring, a female fig wasp must carry pollen to a receptive cluster and pollinate its flowers. Otherwise the ovaries of the flowers will not develop into seeds that can be eaten by her offspring. In some fig species, only one female wasp normally enters a cluster. If she is not carrying pollen, none of her offspring survive. A fig tree pays a price to get its flowers pollinated: Wasp larvae consume many potential fig seeds. From the perspective of figs, fig wasps are both mutualists and predators, and their interaction with figs does not fit cleanly into the scheme shown in Table 52.1.

Although there are many other exceptions to the simple scheme shown in Table 52.1, most interactions fit the categories well enough for us to use them as a guide for exploring interactions among species in this chapter.

Interactions between resources and consumers influence community dynamics

Many interactions between organisms within communities center on resources and their consumers. A **resource** is any substance directly used by an organism that can potentially lead to the growth of its population and whose availability is reduced when it is used. We usually think first of resources that can be consumed by being eaten, but space, including hiding places and nest sites, becomes unavailable if it is occupied. Factors such as temperature, humidity, salinity, and pH, even though they may strongly affect population growth, are not resources because they are not used up or monopolized. Some resources, such as nest sites, are not altered by being used and immediately become available for occupancy when the user leaves. Other resources must regenerate before they are again available to consumers.

Niches:
The Conditions under which Species Persist

The set of environmental conditions under which a species can persist defines its **ecological niche**. If there were no competitors, predators, or disease organisms in its environment, a species would be able to persist under a broader array of conditions than it can in the presence of other species that negatively affect it. On the other hand, the presence of beneficial species may increase the range of environments in which a species can persist.

Both physical and biotic factors determine a species' niche

An experiment performed on two species of barnacles, *Balanus balanoides* and *Chthamalus stellatus*, demonstrated the importance of both physical and biotic factors in determining where the two species actually live. These barnacles live in the intertidal zone of rocky North Atlantic shores. Adult *Chthamalus* generally live higher in the intertidal zone than do adult *Balanus*, but young *Chthamalus* settle in large numbers in the *Balanus* zone. In the absence of *Balanus*, young *Chthamalus* survive and grow well in the *Balanus* zone, but if *Balanus* are present, the *Chthamalus* are eliminated by being smothered, crushed, or undercut by the larger, more rapidly growing *Balanus*. Young *Balanus* settle in the *Chthamalus* zone, but they grow poorly because they lose water rapidly when exposed to air; therefore *Chthamalus* can compete successfully with them there. However, *Balanus* would persist slightly higher in the intertidal zone in the absence of *Chthamalus*. The result is intertidal zonation, with *Chthamalus* growing above *Balanus* (Figure 52.2).

The actual distribution of *Chthamalus* is more restricted than its potential distribution because of competition with *Balanus* in the lower intertidal zone, and the distribution of *Balanus* is more restricted than its potential distribution because of the combined effects of desiccation and competition with *Chthamalus* in the higher intertidal zone. Experiments have shown that the vertical ranges of adults of both barnacles are greater if the other species is removed. Figure 52.2 shows the differences between the potential and actual distributions of the barnacles for only one dimension: height in the intertidal zone. Other important factors that influence barnacle distributions include the type of substrate, the amount of wave action, and water temperature.

Limiting resources determine the outcomes of interactions

Population ecologists typically direct their attention toward differences among species in their requirements for **limiting resources**—resources whose supply is less than the demand made upon them by organisms. Even though a species needs a resource, that resource may not be significant for understanding the species' population dynamics if it is not limiting. For example, most terrestrial animals have a strict but similar requirement for a certain minimum level of oxy-

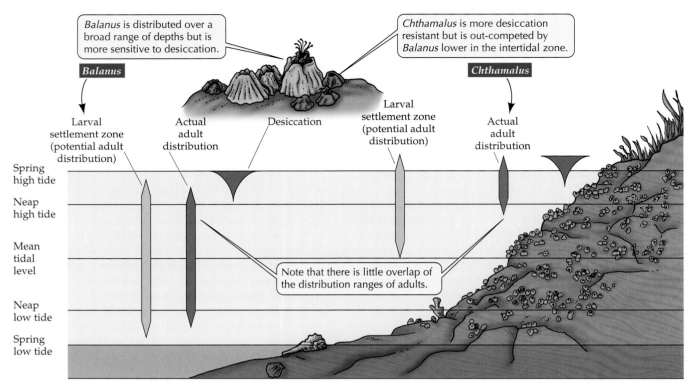

52.2 Potential and Actual Distributions of Two Barnacle Species
Because of interspecific competition, the zone each species occupies is different from the zone it could potentially occupy in the other's absence. (The width of the red and gold bars is proportional to the density of the populations.)

gen. However, studying the use of oxygen reveals very little about the structure of terrestrial communities because the supply of oxygen is nearly always above that minimum level. The primary resources that influence distributions and abundances of terrestrial species are those that are depletable and regenerate slowly, such as food. In some freshwater aquatic environments, however, organisms regularly deplete dissolved oxygen. Aquatic ecologists, unlike terrestrial ecologists, pay careful attention to oxygen levels.

Limiting resources differ among environments, but some resources, such as food supplies, are often limiting. Because of the importance of resources in the lives of all species, we first examine competition among organisms for scarce resources, and then consider predation.

Competition: Seeking Scarce Resources

If two or more organisms use the same resources, and those resources are insufficient to meet their demands, the individuals are competitors, whether they are members of the same or a different species. **Intraspecific competition,** competition among individuals of the same species, may result in reduced growth and

reproductive rates for some individuals, may exclude some individuals from better habitats, and may cause the deaths of others. **Interspecific competition,** competition among individuals of different species, affects individuals in the same way, but in addition, an entire species may be excluded from habitats where it cannot compete successfully. In extreme cases, a competitor may cause the extinction of another species. In this section we will show how ecologists study competition in the laboratory and in the field. Then we will discuss how competition influences the composition of ecological communities.

Competition can be studied in the laboratory

We can model competition by adding to the equation for logistic growth (see Chapter 51) a term that incorporates the effects of members of a competing species on the growth of the focal species:

$$\Delta N_1 / \Delta t = r_1 N_1 (K_1 - N_1 - \alpha N_2 / K_1)$$

where N_1 is the number of individuals in species 1, N_2 is the number of individuals in species 2, K_1 is the environmental carrying capacity for species 1, and α is a competition coefficient that measures the *per individual* effect of species 2 on the growth rate of the population of species 1.

We can write a similar equation for the growth of species 2 in which β is the competition coefficient that measures the *per individual* effect of species 1 on the growth of the population of species 2. α and β are usually, but not always, less than 1; that is, the addition of an individual of another species has a smaller competitive effect than the addition of an individual of the same species.

To determine the effects of competition, ecologists can perform laboratory experiments. The first such experiments were performed by the Russian ecologist G. F. Gause and reported in his influential book, *The Struggle for Existence,* published in 1934. Gause began by conducting experiments with protozoans in containers within which the environment was homogeneous. In every experiment, the population of one species grew faster, monopolizing the food resource until the other died out, an outcome called **competitive exclusion**. The results of one of Gause's experiments with two species of *Paramecium,* in which *P. aurelia* was the winner, are shown in Figure 52.3. In simple laboratory environments such as these, competitive exclusion is common.

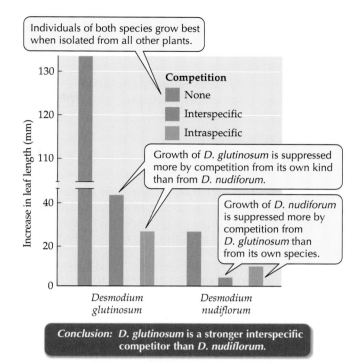

52.4 Interspecific Competition Is Strong among Plants
Plants compete with members of their own species and with members of other species for light, water, and nutrients, all of which can be manipulated in experiments.

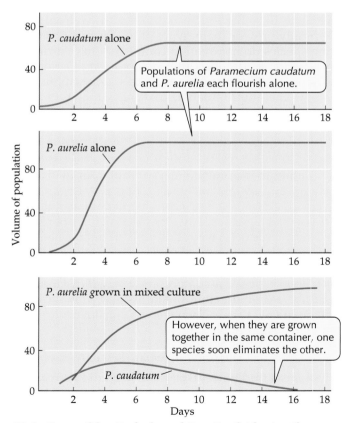

52.3 Competitive Exclusion of One Protist by Another
Competitive exclusion is common in simple laboratory environments.

In later experiments, Gause was able to prevent competitive exclusion by providing a heterogeneous environment containing some places where one species did better and other places where its competitor did better. For example, *Paramecium bursaria* and *P. aurelia* exclude one another in homogeneous environments, but they form stable mixtures if there is deoxygenated water at the bottom of their container. *P. bursaria,* which has mutualistic algae within its cell, can feed in deoxygenated water, whereas *P. aurelia* cannot.

Plants are good subjects for competition experiments because they compete for light, water, and nutrients, all of which can easily be manipulated. To measure the intensity of intraspecific and interspecific competition, investigators used two species of *Desmodium, D. glutinosum* and *D. nudiflorum,* both small herbaceous members of the pea family. To measure intraspecific competition, they planted small individuals of each species 10 cm from a large individual of the same species. To measure interspecific competition, they planted small individuals 10 cm from a large individual of the other species. Control individuals were planted at least 3 m from any other *Desmodium* plant.

Not surprisingly, individuals of both species grew best in the absence of competition (Figure 52.4). Growth of *D. nudiflorum* was depressed more by inter-

TABLE 52.2 Experiments Show How Ants and Rodents Interact with Their Food Supply

	RODENTS REMOVED	ANTS REMOVED	RODENTS AND ANTS REMOVED	CONTROL PLOTS
Number of ant colonies	543	0	0	318
Number of rodents	0	144	0	122
Density of seeds relative to control plots	1.0	1.0	5.5	1.0

specific competition than by intraspecific competition, whereas *D. glutinosum* was depressed more by inter-specific competition than by intraspecific competition. Clearly *D. glutinosum* was the stronger competitor of the two.

Competition can be studied experimentally in nature

Although it is more difficult to do so, competition experiments can also be performed in nature. Experiments in nature are important because laboratory experiments show only the effects of competition in artificial environments. Field experiments are needed to reveal whether competition is influencing abundances and distributions of species in nature.

To determine whether the many coexisting species of seed-eating ants and rodents that live together in the Sonoran Desert of Arizona compete with one another, ecologists removed ants from some sites, rodents from other sites, and both ants and rodents from a third set of sites. If they removed either ants or rodents, population densities of the other group increased relative to densities in control plots (Table 52.2). These experimental results demonstrate that competition for food links the ants and the rodents. Further observations showed that ants and rodents both greatly reduce seed densities.

To determine whether different species of rodents also compete with one another, ecologists erected rodent-proof fences around 50 m by 50 m desert plots. The fences around the experimental plots had holes through which small rodents could pass, but too small to allow the passage of large kangaroo rats. The holes in the fences surrounding the control plots were large enough for all rodents to pass through. Within 2 years of the removal of kangaroo rats from the experimental plots, densities of small seed-eating rodents increased more than twofold, and the plots without kangaroo rats supported more rodent species than the control plots. Thus kangaroo rats reduce populations of some rodent species and eliminate others from places where they live. Kangaroo rats compete with other seed-eating rodents by reducing their food supply and by aggressively defending space.

Species with similar requirements often coexist in nature

Laboratory experiments show that competitors cannot coexist in simple, homogeneous environments. Yet many species with similar ecological requirements live together in most natural communities, as seed-eating ants and rodents do in the Sonoran Desert. How can so many similar species live together in natural environments? Part of the answer is that nature differs from simple laboratory environments in several important ways.

First, natural environments are variable in space and time. Therefore, competing species often eliminate one another from some parts of the environment, but not from others. This was the outcome of the competition among intertidal barnacles that we discussed early in this chapter. Second, natural environments typically provide many types of food, so that competing species do not overlap completely in their use of resources. In addition, other factors, such as predators, disease, and bad weather, may keep populations well below the environmental carrying capacity so that they rarely compete.

Predator–Prey Interactions

By eating them, predators may reduce populations of their prey. Local populations of some prey are replaced very quickly after being eaten by predators; others increase much more slowly. On rocky marine shorelines, each wave brings a new and barely undiminished supply of planktonic food to suspension-feeding animals such as barnacles. In this section we will describe the different types of predators, and we will discuss why numbers of interacting predators and prey typically fluctuate over time. Then we will consider the evolutionary results of predator–prey interactions.

Predators are classified by what they eat

Biologists find it useful to classify predators by what they eat. Because many predators eat organisms of many species, such dietary classifications are crude, but they are useful for describing interactions between predators and their prey.

Herbivores eat the tissues of plants. Plant tissues are abundant, but many of them offer poor nutrition for other organisms. Wood is primarily cellulose, itself very difficult to break down, impregnated with lignins, which are even more difficult to digest. Because wood is so hard to digest, few organisms attack branches and trunks unless the plant is already weak or dead. Plants also produce roots, leaves, flowers, nectar, pollen, fruits, and seeds, which differ in chemistry, size, structure, and pattern of production. Organisms specialized for eating different plant tissues are correspondingly diverse.

Carnivores are organisms that eat other animals. They are generally moderately larger than their prey, and they pursue, capture, and eat their prey one by one. Predators eat many individual prey items during their lives. For example, a small bird in a temperate forest during winter must eat several average-sized insects or seeds every minute during the day to maintain itself.

Suspension feeders eat prey much smaller than themselves that are suspended in the water or air. As we saw in Chapters 29 and 30, suspension feeders are found in many animal phyla (Brachiopoda, Mollusca, Annelida, the arthropod phyla, Phoronida, Ectoprocta, Echinodermata, Chordata). Every suspension feeder has a filtering apparatus whose structure determines the upper and lower size limits of prey that can be captured. Prey that are too small pass through the mesh of the structure; prey that are too large bounce off it. Most suspension feeders depend on the movement of the surrounding medium through their filtering apparatus, which can be accomplished by movement of either the medium or the animal.

The relative sizes of predators and prey influence their interactions

The relative sizes of predators and prey strongly influence their interactions because they determine how a predator captures and handles its prey. If the predator is much larger than its prey, prey are handled in bulk. The world's largest predators, baleen whales, feed on very small prey that they filter from the water. Predators that are only moderately larger than their prey usually pursue, capture, and eat their prey one at a time. Typical **parasites**, which are much smaller than their hosts, often maintain large populations on or in their hosts, only sometimes killing them (Figure 52.5). The parasitic larvae of many wasps and flies, called **parasitoids**, which are as small as or even smaller than their prey, can complete their development in a single host, eventually killing it.

A single prey individual may harbor hundreds or thousands of parasites without being killed by them. In addition, hosts have defenses against parasites, and

52.5 Most Parasites Are Smaller Than their Hosts   This Caribbean soldierfish is host to the parasitic isopod attached to its head between its eyes. The fish has no way to remove the isopod, which feeds on its body tissues.

if a host is in good condition, it may be able to kill them. If a host is already weakened by stresses imposed by the physical environment or shortage of food, the parasites are more likely to succeed.

Pine trees, for example, defend themselves against parasites by exuding the sticky pitch for which pines are famous. Bark beetles attack pine trees by tunneling into the trunks of trees, but they find it difficult to chew through pitch. Weakened trees exude less pitch than healthy trees do, and are easier for the beetles to attack. The beetles lay eggs in their tunnels, and their larvae excavate more tunnels as they eat the nutritive layers just below the outer bark. Each colonizing beetle releases a powerful pheromone that attracts other individuals to the same tree. If enough beetles are attracted, their tunneling larvae may eventually kill the tree (Figure 52.6).

Numbers of predators and prey fluctuate in ecological time

When a typical predator captures and eats a prey individual, it reduces the size of the prey population by one, but the effects of predators on prey population dynamics cannot be determined simply by counting the number of prey eaten. We also need to know how prey densities influence the ease with which prey are captured and how rapidly they reproduce. To understand the direct and indirect interactions between predators and their prey, it is useful to consider the process of predation from the perspective of an individual predator.

Consider a predator that eats only one kind of prey. An individual predator can find enough to eat if the rate at which it encounters prey is above a certain threshold value. Below that threshold, it will lose weight and eventually starve. Nevertheless, the predator may continue to eat prey while slowly losing weight, driving the prey population down further. Eventually the number of predators may be reduced by starvation or emigration, which may allow the prey population to increase its numbers. This increase of prey may, in turn, permit the predator population to increase. Because of this pattern, the recovery of a predator population often lags behind the recovery of its prey population. Thus predator–prey interactions often change the population densities of both species, producing oscillations.

Population density changes among small mammals and their predators living at high latitudes are the best-known examples of predator–prey oscillations. Populations of Arctic lemmings and their chief predators—snowy owls, jaegers, and Arctic foxes—oscillate with a 3- to 4-year periodicity. Populations of Canadian lynx and their principal prey—snowshoe hares—oscillate on a 9- to 11-year cycle (Figure 52.7).

For many years, hare–lynx oscillations were believed to be driven only by interactions between hares and lynxes. Recently, ecologists performed experiments to find out whether any part of the lynx–hare oscillation could be explained by fluctuations in the hares' food supply in addition to predation by lynxes.

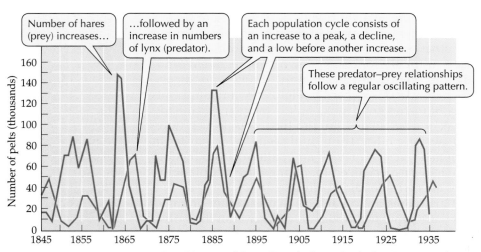

52.7 Hare and Lynx Populations Cycle in Nature The 9- to 11-year population cycles of the snowshoe hare and its major predator, the lynx, in Canada were revealed by the number of pelts sold by fur trappers to the Hudson Bay Company.

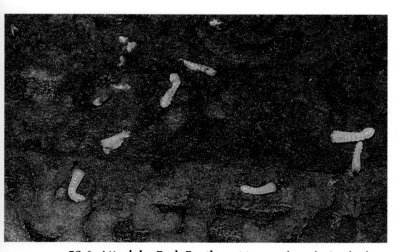

52.6 Attack by Bark Beetles Masses of egg-laying bark beetles have attacked this fallen Douglas fir tree. When the eggs hatched, the tunnels under the bark (which has been removed) were created by developing larvae burrowing through the tissues, eating as they went.

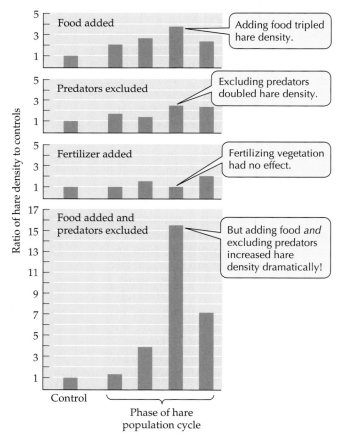

52.8 Prey Population Cycles May Have Multiple Causes Experiments showed that both food supply and predators (but not food quality) affect the population densities of snowshoe hares.

They selected nine 1 km² blocks of undisturbed coniferous forest in Yukon Territory, Canada. In two of the blocks, the hares were given supplemental food year-round. An electric fence with a grid large enough to allow hares, but not their mammalian predators, to pass through was erected around two other blocks. In one of these blocks, extra food was provided. In two other blocks, nitrogen-potassium-phosphorus fertilizer was added to increase plant growth. Three other blocks served as unmanipulated controls.

These experiments produced striking results. Excluding predators doubled, and adding food tripled, hare densities during the peak and decline phases of a cycle. Predator exclusion combined with food addition increased hare density 11-fold, but adding fertilizer had no effect on hare population density (Figure 52.8). Thus, the cycle is driven both by predation and by interactions between hares and their food supply.

The vegetation in the plots where these experiments on snowshoe hares were performed was relatively homogeneous. In nonuniform environments, predators may eliminate their prey in some places, but not in others. In ponds on islands in Lake Superior, chorus frogs (*Pseudacris triseriata*) are found in only some of the habitats that are suitable for them. Three major predators—larvae of a salamander, nymphs of a large dragonfly, and dytiscid beetles—eat chorus frog tadpoles. An ecologist noticed that the tadpoles were common in ponds with beetles, but were rare in ponds with salamander larvae and dragonfly nymphs. In laboratory experiments, he established that the salamander larvae could eat only small tadpoles, but that dragonfly nymphs could eat tadpoles of all sizes. Therefore, he hypothesized that dragonfly nymphs were responsible for eliminating chorus frogs from many ponds.

52.9 Nymphs Eliminate Tadpoles
The speed with which dragonfly nymphs can eliminate tadpoles of the chorus frog is illustrated by the results of two experiments, one in which dragonfly nymphs were added to pools with tadpoles, and the other in which tadpoles were added to pools with dragonfly nymphs.

To test this hypothesis, he selected two large ponds that contained dragonfly nymphs but no tadpoles, and two ponds that contained tadpoles but no nymphs. So that all the tadpoles were handled equally, he removed the tadpoles from the two ponds that lacked dragonfly nymphs and then reintroduced them at the same density. He introduced dragonfly nymphs into one of the ponds at typical densities. He also removed nearly all dragonfly nymphs from one of the ponds that had them and then introduced tadpoles to both of those ponds. The dramatic results of the experiment supported his hypothesis (Figure 52.9). Tadpoles were eliminated from ponds with dragonfly nymphs, but survived well in ponds from which dragonfly nymphs were absent, or nearly so. This experiment shows that a particular predator may eliminate its prey in certain environments; however, it does not tell us why drag-

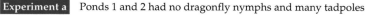

Experiment a   Ponds 1 and 2 had no dragonfly nymphs and many tadpoles

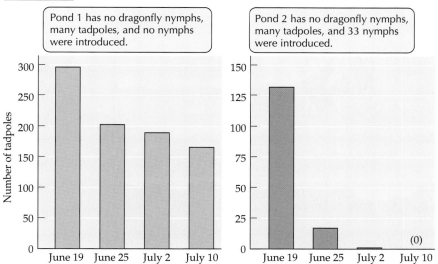

Pond 1 has no dragonfly nymphs, many tadpoles, and no nymphs were introduced.

Pond 2 has no dragonfly nymphs, many tadpoles, and 33 nymphs were introduced.

Experiment b   Ponds 3 and 4 had many dragonfly nymphs and no tadpoles

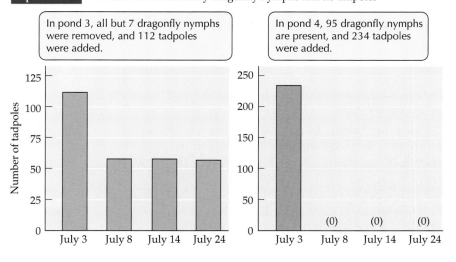

In pond 3, all but 7 dragonfly nymphs were removed, and 112 tadpoles were added.

In pond 4, 95 dragonfly nymphs are present, and 234 tadpoles were added.

(a) *Brachinus* sp.

(b) *Pterois volitans*

52.10 Defenses of Animal Prey (a) A bombardier beetle ejects a noxious spray at the temperature of boiling water in the direction of a predator. The spray is ejected in high-speed pulses more than 20 times in succession. (b) The Indo-Pacific lionfish is among the most toxic of all reef fishes. Glands at the base of its spines can inject poison into an attacker; its bright markings are thought to warn potential predators of this capability.

onfly nymphs were naturally absent from some of the ponds.

Predator–prey interactions change over evolutionary time

Because predators do not capture prey individuals randomly, they are agents of evolution as well as agents of mortality. As a consequence, prey have evolved a rich variety of adaptations that make them more difficult to capture, subdue, and eat. Among the evolutionary adaptations of prey are toxic hairs and bristles, tough spines, noxious chemicals and the means for ejecting them (Figure 52.10), camouflage,

and mimicry of inedible objects or of larger or more dangerous organisms. Predators, in turn, evolve to be more effective at overcoming these prey defenses.

MIMICRY. Among the best-studied adaptations to predation is **mimicry**: taking on the appearance of some inedible or unpalatable item. In **Batesian mimicry** a palatable species mimics a noxious or harmful species. Examples are the mimicry of ants by spiders and of bees and wasps by many different insects (Figure 52.11). Batesian mimicry works because a predator that captures an individual of an unpalatable species learns to avoid any prey of similar appearance. However, if a predator captures a palatable mimic, it is rewarded, and it learns to associate palatability with prey of that appearance. As a result, individuals of unpalatable species are attacked more often than they would be if there were no mimics. Because unpalatable individuals that differ from their mimics more than the average are less likely to be attacked by predators that have eaten a mimic, directional selection causes unpalatable species to evolve away from their mimics. Batesian mimicry systems are stable only if a mimic evolves toward a palatable species faster than the palatable species evolves away from it, which usually requires that the mimic be less common than the unpalatable species.

Another type of mimicry is **Müllerian mimicry**, the convergence over evolutionary time in the appearance of two or more unpalatable species. All species in a Müllerian mimicry system, including the predators, benefit when inexperienced predators eat individuals of any of the species because the predators learn rapidly that all similar species are unpalatable. Some of the most spectacular tropical butterflies are mem-

52.11 A Batesian Mimic Falsely Advertises Danger By mimicking a wasp, this otenucid moth is protected from predators.

<table>
<tr><td>■ Highly unpalatable</td></tr>
<tr><td>■ Moderately unpalatable</td></tr>
<tr><td>■ Highly palatable (Batesian mimics)</td></tr>
<tr><td>□ Palatability not yet tested with birds</td></tr>
<tr><td>* Müllerian mimics of butterflies in the same group</td></tr>
</table>

52.12 Müllerian and Batesian Mimics By converging in appearance, the unpalatable Müllerian mimics among these different species of Costa Rican butterflies and moths reinforce each other in deterring predators; the palatable Batesian mimics benefit because predators learn to associate these color patterns with unpalatability.

bers of Müllerian mimicry systems (Figure 52.12), as are many kinds of bees and wasps.

PARASITE VIRULENCE. Some parasites kill their hosts quickly; others live in or on their hosts without harming them very much. Why do parasites differ so much in virulence? To answer this question, we must consider the ability of a parasite to overcome the host's defenses, its rate of population growth within a host, and the length of time an infected host survives.

The faster parasites kill their hosts, the shorter the time during which they can be transferred to new hosts. Therefore, a parasite that does not kill its host, or kills it very slowly, is more likely to be transferred to another host than a parasite that kills its host quickly. On the other hand, a parasite that multiplies slowly within its host may be outcompeted by faster-multiplying individuals of its own species, with the result that more of the faster-multiplying individuals are transmitted to another host. However, if hosts are typically colonized by only one or a few genetically similar parasites, slow-growing genotypes may continue to dominate the parasite population. If parasites can continue to be transmitted to another host after the death of their host, as happens with many waterborne

diseases, rapidly growing parasite genotypes that may kill their hosts are often favored. This is why many waterborne human diseases, such as cholera, hepatitis, and dysentery, are so deadly.

PLANT DEFENSES. The leaves of many plants are defended physically against herbivores by being tough or having hairs or spines. Most leaves also contain chemicals called **secondary compounds** that have negative effects on herbivores. Defensive secondary compounds of one type—acute toxins—interfere with herbivore metabolism. Some of these toxins, such as nicotine, interfere with the transmission of nerve impulses to muscles. Other toxins are hallucinogens that cause individuals that ingest them to have a seriously distorted view of their environment. As a result, hallucinating herbivores are likely to ignore real environmental dangers. Some toxins imitate insect hormones and prevent insects from completing metamorphosis. Still other toxins are unusual amino acids that become incorporated into herbivore proteins and interfere with their functioning.

Defensive chemicals of the second type make leaves difficult to digest, reducing their suitability as food for herbivores. The most common of these substances are tannins, which are present in the leaves of some herbaceous and most woody species. As most leaves age, their tannin concentrations increase, and the leaves also become tougher. Tannins may be present in such large quantities that waters draining from areas dominated by tanniferous plants are tea-colored. The most famous of such "blackwater rivers" is the Río Negro in Brazil (Figure 52.13).

Some plants respond to being eaten by increasing the concentrations of defensive chemicals in their leaves. Mountain birches in northern Finland are eaten by caterpillars of the moth *Oporinia autumnata*. When caterpillars attack a birch, the tree responds by increas-

52.13 The Black River The dark, tannin-laden waters of the Río Negro (top left) show up against the waters of the Amazon as the two rivers join.

52.14 Damage Comes from Above Shrubs and herbaceous plants are often destroyed by dead branches falling from tall trees.

ing the concentrations of defensive chemicals in its leaves, sometimes within a few days of the initial attack. The larvae of *Oporinia*, of another moth, and of two sawflies—all of which feed on birches—developed more slowly when they were fed on birch leaves that grew close to leaves heavily eaten by caterpillars during the *same* growing season. Similarly, caterpillars to which ecologists fed leaves from birch trees that had been severely damaged the previous year grew more slowly than caterpillars fed leaves from birches only lightly damaged the previous year. Slower growth lowers larval survival and reproduction, thereby reducing the damage to the trees from the next generation of caterpillars.

Beneficial Interspecific Interactions

During predator–prey and competitive interactions, one or both participants in the interaction are harmed. Amensalism causes harm to one of the partners without affecting the other (–/0). In the other two types of interspecific interactions—commensalism (+/0) and mutualism (+/+)—neither partner is harmed, and one or both may benefit. We will examine these interactions in the sections that follow.

In amensalism and commensalism one participant is unaffected

An individual may harm another organism without benefiting itself (a 0/– interaction). Mammals, for example, create bare spaces around waterholes. They benefit by drinking water, but not by trampling the plants they kill. Leaves and branches falling from trees damage smaller plants beneath them. The trees drop old structures regardless of whether or not they damage other plants. Such interactions—amensalisms—are widespread and important. Herbs, shrubs, and small trees in tropical forests often are damaged more by falling objects than by herbivores (Figure 52.14).

Commensalism, a +/0 interaction, benefits one partner but has no effect on the other. An example is the relationship between cattle egrets and grazing mammals. Cattle egrets are found throughout the tropics and subtropics. They typically forage on the ground around cattle or other large mammals, concentrating their attention near the heads and feet of the mammals, where they catch insects flushed by their hooves and mouths (Figure 52.15). Cattle egrets foraging close to grazing mammals capture more food for less effort than egrets foraging away from grazing mammals. The benefit to the egrets is clear; the mammals neither gain nor lose.

52.15 Commensalism Is a +/0 Interaction Cattle egrets, such as these individuals foraging around elephants in East Africa, catch more insects with less work than do egrets foraging away from the larger beasts. The elephants are neither harmed nor helped by the egrets.

Mutualisms benefit both participants

Mutualisms—interactions that benefit both participants (+/+ interactions)—are important among virtually all groups of organisms. Mutualistic interactions exist between plants and microorganisms, protists and fungi, plants and insects, and among plants. Animals also have mutualistic interactions with protists and with one another. The evolution of eukaryotic organisms is believed to be the result of mutualistic interactions between previously free-living prokaryotes and the cells they originally infected (see Chapters 4 and 24).

Some of the most complex and ecologically important mutualisms are between members of different kingdoms. Nitrogen-fixing bacteria of the genus *Rhizobium* receive protection and nutrients from their host plant and provide their host with nitrogen (see Chapter 25). Lichens are compound organisms consisting of highly modified fungi that harbor either cyanobacteria or green algae among their hyphae (see Chapter 26). The fungi absorb water and nutrients from the environment and provide these as well as a supporting structure for the microorganisms, which conduct photosynthesis. This mutualistic combination is especially successful at occupying inhospitable habitats such as rock surfaces, tree bark, and bare, hard ground.

Animals have important mutualistic interactions with protists. Corals and some tunicates gain most of their energy from photosynthetic protists that live within their tissues. In exchange, they provide the pro-

tists with nutrients from the small animals they capture. Termites have nitrogen-fixing protists in their guts that help them digest cellulose in the wood they eat. Young termites must acquire their protists by eating the feces of other termites; if prevented from doing so, they soon die. The protists are provided with a suitable environment in which to live and an abundant supply of cellulose.

ANIMAL–ANIMAL MUTUALISMS. Many species of ants have mutualistic relationships with aphids. Ants "milk" these small plant-sucking insects by stroking them with their forelegs and antennae. The aphids respond by secreting droplets of partly digested plant sap that has passed through their guts. In return, the ants protect the aphids from predatory wasps, beetles, and other natural enemies. The aphids lose nothing, because plant sap is high in sugar but low in amino acids, with the result that aphids ingest more sugar than they need.

Some coral reef fishes and shrimps obtain their energy by eating parasites from the scales and gills of larger fish (Figure 52.16). It presumably benefits the hosts to have their parasites removed, but this benefit has never been quantified. These mutualisms are particularly interesting because the cleaners are suitable prey that may actually enter the mouths of dangerous predators. The predators usually refrain from attacking the cleaners, but such restraint could not have been present when the interactions first began to evolve. Such cleaning probably began with the cleaners removing parasites from less dangerous locations.

PLANT–ANIMAL MUTUALISMS. Terrestrial plants have many mutualistic interactions with animals (Table

52.16 An Animal–Animal Mutualism This prawn (*Lysmata grabhami*) is removing parasites from a moray eel.

TABLE 52.3 Mutualistic Relationships of Plants with Animals

BENEFIT TO PLANTS	BENEFIT TO ANIMALS	SOME EXAMPLES
Animals disperse pollen	Animals feed on pollen or nectar	Most plants with brightly colored flowers
Animals disperse seeds	Animals feed on fleshy rewards surrounding or attached to seeds	Conifers such as junipers and yews
Animals disperse both pollen and seeds	Animals feed on both floral and fruit rewards	Most tropical trees, shrubs at all latitudes

52.3). A complex mutualism between trees and ants that live in Central America illustrates the benefits of such interactions. Trees of the species *Acacia cornigera* have large, hollow thorns in which ants of the genus *Pseudomyrmex* construct their nests and raise their young (Figure 52.17). These ants live only on acacias. The ants feed on nectar produced at the bases of the leaf petioles and on special nutritive bodies on the leaves. The ants attack and drive off leaf-eating insects, eat the eggs and larvae of herbivorous insects, and even bite and sting browsing mammals. They also cut back the tips of other plants, particularly vines that grow over their host tree. The ants get room and board; the plants get protection against both predators and competitors.

Many angiosperms depend on animals to move their pollen and seeds (Table 52.3). The plants benefit by hav-ing their pollen carried to other plants and by receiving pollen to fertilize their ovules. The animals benefit by obtaining food in the form of nectar and pollen. Plants provide animals with attractive rewards—nutrient-rich nectar. Movement to another flower of the same species is encouraged by the limited amount of nectar on any one plant, and by the existence of similar rewards on other individuals of the same species. As a result, the foraging animals transfer pollen to the stigmas of plants belonging to the same species. But there is a price: The energy and materials the plant spends to produce nectar and other rewards for animals cannot be used for growth or seed production.

Animals are induced to move seeds by the presence of nutritive rewards attached to or surrounding them. Many seeds are surrounded by fleshy fruits that are eaten by animals, which either regurgitate or defecate

(a)

(b)

(c)

52.17 A Plant–Animal Mutualism Some acacia trees have large, swollen, hollow thorns (a) that house ants. The ants patrol the trees, attacking herbivorous insects and cutting away vines and branches of neighboring plants that would otherwise smother the acacia. In an experiment, small acacia trees were cut down, and ants were allowed to recolonize some trees as they regrew, but not others. Those trees with ant colonies (b) grew back quickly, but those without ants (c) were heavily attacked by other insects and regained their leaves very slowly.

(a) *Bombycilla garrulus*

(b) *Formica* sp.

The ant-attracting elaiosome is the white tissue wrapped around the black seed.

52.18 Fruits Attract Different Frugivores (a) Bright red fruits are attractive to many birds, such as this waxwing. (b) A *Formica* ant removes a ripe seed from a pod of a golden snake plant in the Colorado Rockies.

the seeds some time later away from the plant (Figure 52.18a). A nutritive body called an elaiosome is attached to many seeds that are dispersed by bats or ants. Ants carry the elaiosome and the seed back to their nests, but do not eat the seed, and eventually discard it on a refuse pile (Figure 52.18b). Although ants carry seeds only short distances, seeds in their underground nests are protected from fires and predators.

Interactions between plants and their pollinators and seed dispersers are clearly mutualistic, but they are not purely mutualistic. As we saw earlier, fig wasps are both pollinators and seed predators. Many seed dispersers are also seed predators that destroy some of the seeds they remove from plants. Some organisms that collect rewards are not mutualists at all. Many animals visit flowers without transferring any pollen, sometimes cutting holes in them to get to the nectar-producing regions at the base of the flower. On the other hand, some plants exploit pollinators. The flowers of certain orchids, for example, mimic female insects, enticing male insects to copulate with them (Figure 52.19). The male insects neither sire any offspring nor obtain any reward, but they transfer pollen between flowers, benefiting the orchid.

Temporal Changes in Communities

The species that live together in ecological communities constantly change as environmental shifts alter interspecific interactions. The plants that first colonize a site after a disturbance, for example, differ from those that colonize the site later. The process by which the species composition of a community changes over time is called **ecological succession**. Patterns and causes of ecological succession are varied, but the

early colonists always alter the conditions under which later-arriving species grow.

Succession may begin at sites that have never been modified by organisms. The retreat of a glacier in Glacier Bay, Alaska, over the last 200 years exposed unoccupied sites that were colonized by plants. The retreating glacier left a series of moraines—gravel deposits formed where the glacial front was stationary for a number of years. No scientist was present to measure changes over the 200-year period, but ecologists have inferred the temporal pattern of succession

52.19 Some Orchids Mimic Female Insects The flowers of some orchids so closely resemble female wasps that males are fooled into attempting to copulate with the flowers, as this male wasp is trying to do.

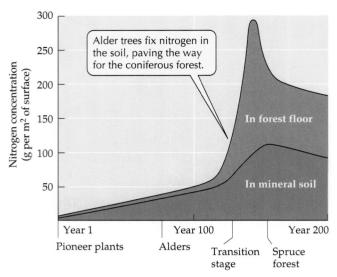

Alder trees fix nitrogen in the soil, paving the way for the coniferous forest.

In forest floor

In mineral soil

Year 1 — Pioneer plants
Year 100 — Alders
Transition stage
Year 200 — Spruce forest

52.20 Soil Properties Change during Vegetation Succession As the plant community occupying an Alaskan glacial moraine changed from pioneering plants to a spruce forest, nitrogen accumulated in both the forest floor and the mineral soil.

by measuring plant communities on moraines of different ages. The youngest moraines, close to the current glacial front, are populated with small organisms such as bacteria, fungi, and algae. Slightly older moraines have lichens, mosses, and a few species of shallow-rooted herbs. Successively older moraines have shrubby willows, alders, and eventually conifers. By comparing moraines of different ages, the ecologists deduced the pattern of plant succession and of changes in soil nitrogen content at Glacier Bay, shown in Figure 52.20. Succession was caused in part by changes in the soil brought about by the plants themselves. Alder trees have nitrogen-fixing fungi in nodules on their roots. Because nitrogen is virtually absent from glacial moraines, nitrogen fixation by alders improved the soil for the growth of conifers. Conifers then outcompeted and displaced the alders.

Succession that takes place when all or part of the dead body of some plant or animal is decomposed is called **degradative succession**. The succession of fungal species that decompose pine needles in litter beneath Scots pines (*Pinus sylvestris*) is shown in Figure 52.21. New litter is continuously deposited under pines, so that the surface layer of litter is young and deeper layers are progressively older. Degradative succession begins when the first group of organisms starts consuming the needles as soon as they fall. Each group of organisms degrades certain compounds, converting them to other compounds that are attacked by the next successional group. This process continues over about 7 years, by which time the last group of organisms—basidiomycetes—has decomposed the remaining cellulose and lignin. By then, the remains are no longer recognizable as pine needles.

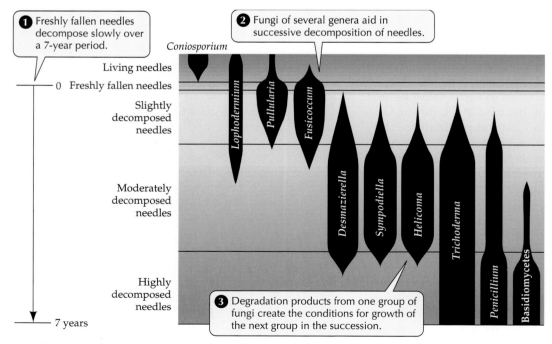

❶ Freshly fallen needles decompose slowly over a 7-year period.

❷ Fungi of several genera aid in successive decomposition of needles.

Coniosporium

Living needles

0 Freshly fallen needles

Slightly decomposed needles

Moderately decomposed needles

Highly decomposed needles

7 years

Lophodermium

Pullularia

Fusicoccum

Desmazierella

Sympodiella

Helicoma

Trichoderma

Penicillium

Basidiomycetes

❸ Degradation products from one group of fungi create the conditions for growth of the next group in the succession.

52.21 Degradative Succession on Pine Needles As indicated by the widths of the black bars, the abundances of ten types of fungi in pine litter change with time and with the depth of the layer.

(a) *Yucca brevifolia*

(b) *Tegeticula yuccasella*

52.22 Yucca–Yucca Moth Coevolution (a) The Joshua tree is pollinated only by (b) the yucca moth shown here on a flower.

Coevolution of Interacting Species

Over time, the evolution of many species traits has been influenced by interactions with other species. Species that have mutually influenced one another's evolution are said to have **coevolved**. Often coevolution is diffuse and general, but sometimes it is species-specific. The fig and fig wasp species discussed above have intimately coevolved—neither can reproduce without the other. So have yucca plants and the moths of the genus *Tegeticula* that pollinate them.

Female yucca moths lay their eggs only in the ovules of yucca flowers. A female *Tegeticula* lays no more than five eggs in one flower. After she has laid her eggs, she scrapes pollen from the flower's anthers, rolls it into a small ball, flies to another yucca plant, and places the pollen ball on the stigma of the flower before laying another batch of eggs. When the eggs hatch, the larvae burrow into the ovary and feed upon the developing seeds. *Yucca* has no other pollinators, *Tegeticula* larvae eat no other food, and each yucca species has a specific moth species associated with it (Figure 52.22).

One feature of the coevolved relationship between *Tegeticula* and *Yucca* is quite surprising: Why do female moths lay so few eggs per flower? Wouldn't a female moth that laid more than five eggs in a single flower produce more surviving offspring than moths that lay only the usual number? The evolutionary reason for their restraint is that *Yucca* plants selectively abort flowers in which more eggs are laid. As a result, fewer moth offspring are produced in flowers in which more than the normal number of eggs are laid. Thus, the mutualism is stabilized at a level that represents an "evolutionary compromise" between the fitness of both the moths and the yuccas.

Although species-specific coevolution is relatively rare, diffuse coevolution is widespread. In **diffuse coevolution**, species traits are influenced by interactions with a wide variety of predators, parasites, prey, and mutualists. Most flowers are pollinated by a number of animal species, and most pollinators visit many species of flowers. Most flowers adapted for bird pollination are red, a color that attracts most species of birds. Many flowers adapted for insect pollination have contrasting colors that form guidelines leading to the entrances to the flowers. These lines, which are conspicuous when viewed under ultraviolet light (Figure 52.23), are visible to bees and butterflies, but not to birds. Bat-pollinated flowers open at night and have wide openings into which a bat's head can enter, making the floral rewards accessible to many species of bats.

The traits of the fleshy fruits that surround many seeds are also the result of diffuse coevolution; very few fruits are adapted for dispersal by only a few species of animals. Most bird-dispersed fruits are red or some combination of red and another color. Fruits dispersed by nonflying mammals, many of which lack color vision, are typically purple and are not highly visible to birds. Bat-dispersed fruits are typically green and have a fruity odor when ripe. They are inconspicuous during the day but are easy for bats to detect at night. Many bird-dispersed fruits taste bitter to mammals, and vice versa.

(a) (b)

52.23 Bee Vision and Flower Colors Demonstrate Diffuse Coevolution
(a) Under normal sunlight, these black-eyed susans appear familiar to us.
(b) Bees, however, can see the patterns that appear in this ultraviolet photograph, in which part of the outer ray of petals blends with the central flowers in the heads to make a larger visual target.

Most traits of flowers and fruits are the result of diffuse coevolution because most flower visitors and fruit dispersers must use many different plant species to survive throughout the year. Most plant species produce flowers and fruits for only a few weeks or months each year. The animals must travel to where flowers and fruits are available and must feed on whichever plant species are flowering or bearing fruit.

Keystone Species: Major Influences on Community Composition

Organisms influence the communities in which they live by altering microclimate, soil structure and chemistry, and water movement. These alterations change the suitability of the physical environment for other organisms. As we have just seen, organisms also change the amount and distribution of resources, and they consume one another. Species whose influences on ecological communities are greater than would be expected on the basis of their abundance are called **keystone species.** Keystone species may influence the species richness of communities, the flow of energy and materials in ecosystems, or both. In this section we will focus on how keystone species affect the numbers and kinds of species that live in a community. In the next chapter, we will discuss the influence of keystone species on ecosystem processes.

Plants provide most of terrestrial ecosystem structure

In terrestrial communities, plants form most of the structural environment, are the major modifiers of the physical environment, and are the pathway through which energy and nutrients enter communities. Any-

one who has walked into the shade of a tree on a hot, sunny day knows that climate near the ground is strongly influenced by plants. Temperatures fluctuate less between day and night under trees than they do in the open, and light levels are much lower there. Leaves of trees intercept and evaporate much of the rain that falls on them so that less reaches the ground than in open areas. However, rain that does reach the ground evaporates more slowly inside a forest than in the open because temperatures are lower, humidities are higher, and there is less wind.

Keystone animals may change vegetation structure

Animals that are able to change vegetation structure can be powerful keystone species. Beavers, for example, cut trees and build dams; moose accelerate the successional change from deciduous trees to conifers by selectively browsing on deciduous trees. Ecologists determined how moose alter plant succession by building fences around four 100-m^2 plots of land on Isle Royale in Lake Superior to keep moose out. They also laid out, but did not fence, control plots outside the exclosures.

Moose preferentially feed on deciduous trees, such as mountain ash, mountain maple, aspen, and birch, that colonize disturbed sites. They rarely eat white spruce and balsam fir, species that replace deciduous trees during succession, because the foliage of these conifers has high concentrations of indigestible resins and low concentrations of nitrogen. In the control plots, deciduous species were so heavily eaten by moose that spruce and fir were the only plants that grew above the height at which moose feed. Inside the enclosures, on the other hand, deciduous trees remained abundant, and the succession to conifers was slower.

Predators may be keystone species

A predator, by consuming a prey that would dominate its environment if it were not eaten, may create openings for species that would otherwise be competitively excluded from the community. The sea star *Pisaster ochraceous*, an abundant predator in rocky intertidal communities on the Pacific coast of North America, functions as a keystone species in this way. In the absence of sea star predation, its preferred prey, the mussel *Mytilus californianus*, pushes out other competitors in a broad belt of the intertidal zone. By consuming mussels, *Pisaster* creates bare spaces that are taken over by a variety of other species (Figure 52.24).

The influence of *Pisaster* on community composition was demonstrated by experimentally removing them from selected parts of the intertidal zone repeatedly over a five-year period. The removals resulted in two major changes. First, the lower edge of the mussel bed extended farther down into the intertidal zone, showing that sea stars are able to eliminate the mussels completely where they are covered with water most of the time. Second, and more dramatically, 28 species of animals and algae disappeared from the removal zone, until only *Mytilus*, the competitive dominant, occupied the entire substratum. By altering competitive relationships, predation by *Pisaster*, in combination with physical factors such as desiccation and wave action, determines which species live in these rocky intertidal communities.

Some microorganisms are keystone species

Despite their small size, certain microorganisms strongly influence community structure. Wood is broken down primarily by microorganisms, especially

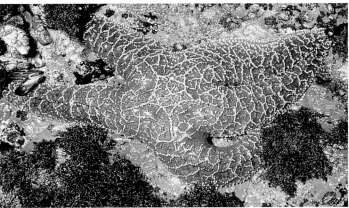

Pisaster ochraceous

52.24 A Sea Star Prevents Its Prey from Dominating Its Community This sea star is resting on rocks from which it has harvested all the mussels. Many other organisms, including the algae visible in the photograph, will now be able to colonize the site.

52.25 Nonregenerating Clear-Cuts in Oregon Because soil microorganisms were eliminated by burning and herbicides, no conifers are growing in these 15- to 20-year-old clear-cuts, even though each clearing has been planted with seedlings four times since the tree cover was cut.

fungi. Nitrogen fixation, the only source of biological nitrogen, is carried out only by prokaryotic microorganisms. Ignoring the importance of relationships between plants and microorganisms has sometimes led to the failure of reforestation attempts. For example, a 15-hectare plot in the Klamath Mountains of southern Oregon, clear-cut in 1968, has been replanted four times. All the plantings were failures, even though forests in this area regenerate readily after wildfires.

The reason for the failures is that the site was both burned and treated with herbicides to open it up for better growth of conifer seedlings. This treatment killed the early successional deciduous trees and shrubs that normally support soil organisms and ameliorate temperatures and moisture. When those plants were eliminated, most soil microorganisms, including those that form mutualistic associations with conifer roots, died, and conifers then were unable to grow. Many such nonregenerating clear-cuts dot high mountains in this region (Figure 52.25).

Indirect Effects of Interactions among Species

In the experiments we have described above, one member of a community was removed or excluded, and investigators measured the resulting changes. Such single-species removal experiments can demonstrate the direct effects of species on one another, but to detect their indirect effects, observations and manipulations of several species are needed.

52.26 Direct and Indirect Interactions Control Populations
Several species, including mice, gypsy moths, and oak trees, interact to influence one another's population densities.

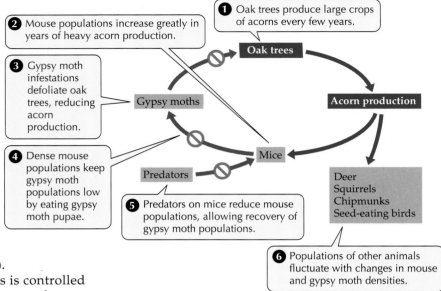

❶ Oak trees produce large crops of acorns every few years.

❷ Mouse populations increase greatly in years of heavy acorn production.

❸ Gypsy moth infestations defoliate oak trees, reducing acorn production.

❹ Dense mouse populations keep gypsy moth populations low by eating gypsy moth pupae.

❺ Predators on mice reduce mouse populations, allowing recovery of gypsy moth populations.

❻ Populations of other animals fluctuate with changes in mouse and gypsy moth densities.

Oak trees

Acorn production

Gypsy moths

Mice

Predators

Deer
Squirrels
Chipmunks
Seed-eating birds

Ecologists have assembled a variety of data to help them understand relationships between oak trees and the animals that eat their leaves and acorns. Most damage to leaves in the oak forests of eastern North America is caused by gypsy moths; most acorns are eaten by mice, chipmunks, deer, squirrels, and birds (Figure 52.26). The abundance of mice and chipmunks is controlled largely by acorn abundance, which varies greatly over the years. Deer consume large quantities of acorns during years of high acorn production, but during poor acorn years they shift to other foods.

Gypsy moths eat oak leaves, and during years when their populations are very large, they may defoliate large expanses of forest. Outbreaks of gypsy moths occur once every 6 to 10 years. Gypsy moth populations collapse after they defoliate a forest because most larvae die of starvation. In the year following defoliation, oak trees are full of leaves, but gypsy moth populations remain low for many years. Why is this?

To determine whether mice could prevent gypsy moth populations from recovering after a crash, ecologists measured predation rates by mice on gypsy moth pupae by attaching freeze-dried pupae to small squares of burlap. They placed the burlap panels on oak tree trunks at sites where gypsy moths typically pupate. During a year of moderately dense mouse populations, all of the introduced pupae were eaten within 8 days. During a year of low mouse density, half of the pupae survived more than 18 days, which is several days longer than it takes for gypsy moth caterpillars to complete metamorphosis and emerge from their pupae. Thus, in years when acorns are abundant, mice keep gypsy moth populations at low densities by eating most of their pupae. In so doing they allow the oak trees to recover from defoliation and accumulate enough energy reserves to produce another large crop of acorns.

Mouse populations typically drop precipitously about 1.5 years after a year of high acorn production, after which gypsy moth populations rebound and again defoliate the trees. If few mice were present, gypsy moths might rebound so quickly that oaks could never produce large crops of acorns. If the investigators had studied only the interactions between mice and acorns or between gypsy moths and oak trees, they would not have discovered the important influences of other species on the interactions.

Summary of "Community Ecology"

Ecological Interactions

• Species interact with one another in four major ways. **Review Table 52.1**

Niches: The Conditions under which Species Persist

• A species' niche is the range of environmental conditions under which it can persist.
• Interactions among species often restrict the range of a species to only part of its potential distribution. **Review Figure 52.2**

Competition: Seeking Scarce Resources

• If organisms use the same resources and those resources are in short supply, the individuals are competitors. Competition may be either intraspecific or interspecific.
• Competition is easily studied in the laboratory, where competitive exclusion is common in homogeneous environments, but not in heterogeneous ones. **Review Figure 52.3**
• Species that use similar resources commonly coexist in nature because nature is spatially and temporally complex, many resources are typically available, and other factors often keep populations below carrying capacity so that they do not compete strongly.

Predator–Prey Interactions

• Predators are classified as herbivores, carnivores, or suspension feeders depending upon what they eat. Herbivores consume large quantities of plant parts, which have low nutritional value. Carnivores eat other animals. Suspension feeders extract large numbers of small prey from water.

• Relative sizes of predators and prey influence their interactions. Parasites are typically much smaller than their prey and may live in or on their hosts without killing them.

• Because of time lags in the responses of both predators and prey, interactions dominated by one predator and one prey typically oscillate. **Review Figure 52.7**

• Experimental manipulation of predators in nature reveals that they are often important in determining both numbers and distributions of their prey. Predators may prevent prey from living in some environments that are otherwise suitable for them. **Review Figures 52.8, 52.9**

• Predators act as evolutionary agents against which prey evolve adaptations, such as toxic hairs and bristles, tough spines, noxious chemicals, and mimicry of inedible objects or dangerous organisms. **Review Figures 52.10, 52.11, 52.12**

Beneficial Interspecific Interactions

• Commensal interactions, in which one partner benefits while the other is unaffected, are common in nature. **Review Figure 52.15**

• Mutualistic interactions, in which both participants benefit, are also common in nature. Mutualistic interactions occur between members of different kingdoms (between plants and prokaryotes, between fungi and algae, and between animals and protists). Animals have mutualistic interactions with other animals and with plants (pollination, seed dispersal). **Review Figures 52.16. 52.17, 52.18**

Temporal Changes in Communities

• Ecological succession involves changes in the species composition of a community over time. Early colonists alter the conditions under which later-arriving species grow.

• Succession may begin at sites that have never been modified by organisms. **Review Figure 52.20**

• Succession may take place when all or part of the dead body of some organism is decomposed. **Review Figure 52.21**

Coevolution of Interacting Species

• Some mutualistic relationships, such as those between figs and fig wasps and yuccas and yucca moths, are tightly coevolved, but diffuse coevolution between many species is much more common. **Review Figures 52.22, 52.23**

Keystone Species: Major Influences on Community Composition

• Keystone species have influences on ecological communities that are greater than would be expected from their abundances.

• Plants, mammals that change vegetation structure, predators on dominant competitors, and microorganisms may function as keystone species. **Review Figures 52.24, 52.25**

Indirect Effects of Interactions among Species

• Indirect effects of species interactions affect many species populations. For example, mice prevent gypsy moth populations from recovering quickly after they defoliate oak trees, thereby allowing the trees to recover. **Review Figure 52.26**

Self-Quiz

1. Two organisms that use the same resources when those resources are in short supply are said to be
   a. predators.
   b. competitors.
   c. mutualists.
   d. commensalists.
   e. amensalists.

2. Which of the following is *not* a resource?
   a. Food
   b. Space
   c. Hiding places
   d. Nest sites
   e. Temperature

3. A species' potential distribution is the range of conditions under which it could survive if
   a. there were no predators or competitors.
   b. there were no predators or other negative influences.
   c. there were no competitors, predators, or disease-causing organisms.
   d. environmental conditions were ideal.
   e. the environment were fundamentally different.

4. An animal that is much smaller than its prey and which attacks it from the inside is called a
   a. predator.
   b. parasite.
   c. commensalist.
   d. competitor.
   e. parasitoid.

5. Which of the following factors tends to stabilize populations of predators and their prey?
   a. A high birth rate of the prey
   b. A high birth rate of the predator
   c. The ability of predators to further reduce prey when they are scarce
   d. The ability of predators to search widely for prey
   e. Environmental heterogeneity

6. The convergence over evolutionary time in the appearance of two or more unpalatable species is called
   a. cladism.
   b. mutual adaptation.
   c. Müllerian mimicry.
   d. Batesian mimicry.
   e. convergent mimicry.

7. Plants are good subjects for experiments to study competition because
   a. plants don't move around.
   b. the resources for which plants compete are easily measured.
   c. the resources for which plants compete are easily manipulated.
   d. plants often compete both in nature and in the laboratory.
   e. all of the above

8. Damage caused to shrubs by branches falling from overhead trees is an example of
   a. interference competition.
   b. partial predation.
   c. amensalism.
   d. commensalism.
   e. diffuse coevolution.

9. Ecological succession is
   a. the changes in species over time.
   b. the gradual process by which the species composition of a community changes.
   c. the changes in a forest as the trees grow larger.
   d. the process by which a species becomes abundant.
   e. the buildup of soil nutrients.

10. Keystone species
    a. influence the structure of the communities in which they live more than expected on the basis of their abundance.
    b. strongly influence the species composition of communities.
    c. may speed up the rate of vegetation succession.
    d. may be herbivores or carnivores.
    e. all of the above

Applying Concepts

1. A general rule commonly accepted by ecologists states that a "jack of all trades is master of none." Yet, most ecological communities are mixtures of jacks and masters—that is, generalists and specialists. Under what conditions would you expect jacks to be more successful? Masters? Why?

2. What features of predator–prey interactions tend to generate instabilities that lead to fluctuations in the densities of both species? Given that instabilities are expected, what keeps populations of either predator or prey from fluctuating to extinction?

3. Parasites usually have generation times much shorter than those of their hosts. Consequently, they should be able to evolve faster. What prevents them from evolving so fast that they completely overcome the resistances of their hosts and exterminate them?

4. Wind does not direct pollen toward conspecific stigmas. Given this inefficiency, why are there so many wind-pollinated plants? If seeds that land close to the parent plant survive less well than those that are carried farther away, why do so many plants produce seeds lacking dispersal devices?

5. On the eastern side of the Sierra Nevada in California, four species of chipmunks occupy adjacent habitats from which they exclude one another by direct aggressive interference. In the San Jacinto Mountains of southern California, three other chipmunk species similarly occupy adjacent habitats, but no interspecific aggression is observed. Each species simply remains in its own habitat. Which of these two assemblages do you think is the older one? Why?

6. Some direct interactions between two species benefit only one of those species. Give examples of such "one-way" benefits in each of the following cases:
   a. between two species of plants (give one example of energetic and another example of physical support)
   b. between a plant and an eater of its leaves
   c. between a predator and its prey

7. Wood is an abundant food source that has been available for millions of years. Why have so few animals evolved to be able to eat wood?

8. In the text, we showed how large mammals modify the communities in which they live. Give some examples of ways in which smaller animals modify their communities.

Readings

Barbour, M. G., H. J. Burk and W. D. Pitts. 1980. *Terrestrial Plant Ecology.* Benjamin Cummings, Menlo Park, CA. A general text on the interactions between plants and their surroundings, including several good chapters on communities and vegetation types.

Begon, M., J. L. Harper and C. R. Townsend. 1996. *Ecology: Individuals, Populations, and Communities,* 3rd Edition. Blackwell Scientific Publications, Oxford. A comprehensive treatment of all aspects of ecology.

Cox, P. A. and M. J. Balick. 1994. "The Ethnobotanical Approach to Drug Discovery." *Scientific American,* June. Discusses how biologists, by analyzing plants already used as drugs by indigenous cultures, can find new pharmaceutical compounds more rapidly than by randomly screening plants.

Fleming, T. H. 1993. "Plant-Visiting Bats." *American Scientist,* vol. 81, pages 460–467. Shows how the availability of fruits and flowers has resulted in evolutionary divergence in the behavior, ecology, and morphology of bats.

Fritz, R. S. and E. L. Simms (Eds.). 1992. *Plant Resistance to Herbivores and Pathogens.* University of Chicago Press, Chicago. Essays dealing with the ecology, evolution, and genetics of plant resistance mechanisms and how they work.

Futuyma, D. J. and M. Slatkin (Eds.). 1983. *Coevolution.* Sinauer Associates, Sunderland, MA. A collection of essays that summarize current knowledge of coevolutionary relationships among all types of living organisms.

Gotelli, N. J. 1995. *A Primer of Ecology.* Sinauer Associates, Sunderland, MA. An excellent introduction to population modeling.

Harborne, J. B. 1988. *Introduction to Ecological Biochemistry,* 3rd Edition. Academic Press, New York. An excellent presentation of the types of chemicals produced by living organisms and their roles in ecological interactions.

Ricklefs, R. E. 1997. *The Economy of Nature,* 4th Edition. W. H. Freeman, New York. A readable text covering all aspects of ecology.

Ricklefs, R. E. and D. Schluter (Eds). 1993. *Species Diversity in Ecological Communities: Historical and Geographical Perspectives.* University of Chicago Press, Chicago. A comprehensive multi-authored volume that explores species interactions at many different spatial and temporal scales.

Thompson, J. N. 1982. *Interaction and Coevolution.* Wiley, New York. A review of patterns in and conditions favoring the evolution of close interactions among species.

Thompson, J. N. 1994. *The Coevolutionary Process.* University of Chicago Press, Chicago. A thorough review of the processes by which coevolutionary relationships evolve.

Wickler, W. 1968. *Mimicry in Plants and Animals.* Wiedenfeld and Nicholson, London. A general and easily followed presentation of the evolution of close resemblances among organisms that are not closely related to one another.

Chapter 53

Ecosystems

Dams: Some Surprising Aftereffects
Damming rivers and lakes in the expectation of creating benefits for the human population often results in unexpected and unpredictable detrimental effects to the ecosystem.

*I*n 1976, the outlet of Southern Indian Lake in northern Manitoba, Canada, was dammed, raising the lake level 3 meters. Engineers then diverted the Churchill River so that rather than flowing into the lake, it flowed southward across a drainage divide and through a series of hydroelectric generating stations. Before the dam was built, ecologists studied the lake in detail to assess the likely consequences of raising its level and greatly reducing the flow of river water into it. They predicted that fewer nutrients would enter the lake, but that the reduction would be compensated by nutrients derived from increased soil erosion along the elevated shoreline. Based on their predictions, they saw no reason to believe that the Southern Indian Lake whitefish fishery, the most important commercial fishery in northern Manitoba, would be seriously adversely affected.

The ecologists' predictions of the future nutrient status of the lake and amounts of algal photosynthesis were correct. However, to everyone's surprise, the whitefish fishery collapsed. The greatly increased soil erosion on the new shoreline released large quantities of mercury into the lake. Mercury concentrations in fish in Southern Indian Lake now exceed Canadian safety standards and will probably remain above standard for many years. From 1977 to 1982, Manitoba Hydro, the builder of the dam, subsidized the commercial fishermen, and in 1982, it provided a one-time cash settlement of $2.5 million Canadian dollars for future losses to the fishermen.

Unexpected surprises commonly follow not only the damming of rivers and lakes, but any attempts to alter ecological systems. Surprises happen because the behavior of those systems is the result of interactions among many different processes, most of which are only incompletely understood. Ecologists now recognize that mercury pollution often results from the raising of lake levels, but they did not know that in the 1970s. As humans continue to alter Earth's ecological systems, new surprises confront us each year.

The organisms living in a particular area, such as Southern Indian Lake, together with the physical environment with which they interact, constitute an **ecosystem**. Ecosystems can be recognized and studied at many different spatial scales, ranging from local units, such as lakes, to the entire globe. At the global scale, Earth is a single ecosystem.

The dynamics of ecosystems are the result of the activities of myriad individual organisms, which are influenced by processes in the physical environment. Some of the processes are altered by organisms in turn, and some are not. Individuals of the many different species that interact in all ecosystems do so by capturing energy and materials, transforming and retaining them, and transferring them to other organisms that eat them.

The goal of ecosystem ecology, which we will discuss in this chapter, is to understand the factors that control the flow of energy and the cycling of materials through ecosystems. Ecologists use this knowledge to understand how and why ecosystems respond as they do to human-caused disturbances, and how society can best use the services that ecosystems provide for the benefit of humanity.

To set the stage for our study of energy flow and nutrient cycling in ecosystems, we will first describe climates on Earth and how they influence ecosystem functioning.

Climates on Earth

The sun drives the global circulation patterns of air and ocean waters and is the source of energy for photosynthesis and ecosystem energetics. The warming and cooling of moving masses of air and water explain much of Earth's climatic patterns. Climates, in turn, exert a powerful influence on the distributions, abundances, and evolution of species.

Climates vary greatly from place to place on Earth, primarily because different places receive different amounts of solar energy and because the monthly amount of incident solar energy is nearly constant at the equator, but varies dramatically at high latitudes. In this section we will examine how differences in solar energy input determine atmospheric and oceanic circulation.

Solar energy inputs drive global climates

Every place on Earth receives the same total number of hours of sunlight each year—an average of 12 hours per day—but not the same amount of *heat*. The rate at which heat arrives per unit of ground area depends primarily on the angle of sunlight. If the sun is low in the sky, a given amount of solar energy is spread over a larger area (and is thus less intense) than if the sun is directly overhead. In addition, when the sun is low in the sky, sunlight must pass through more atmosphere, with the result that more of its energy is absorbed and reflected before it reaches the ground. At higher latitudes (closer to the poles), there is more variation in both day length and the angle of arriving solar energy over the course of a year than at latitudes closer to the equator.

On average, the mean annual air temperature decreases about 0.4°C for every degree of latitude (about 110 kilometers) at sea level. Air temperature also decreases with elevation. The effect of elevation on temperature is due to the properties of gases. As a parcel of air rises, it expands, its pressure drops, and energy is expended in pushing molecules apart. With that loss of energy, the temperature of the air drops. When the parcel of air descends, it is compressed, its pressure rises, the same amount of energy is recovered, and its temperature increases.

When wind patterns bring air into contact with a mountain range, the air rises to pass over the mountains, cooling as it does so. Because cool air cannot hold as much moisture as warm air, clouds frequently form, and moisture is released as rain or snow. For this reason, the windward side of a mountain range generally receives more rainfall than the leeward side. On the leeward side, air descends and warms, and the dry air picks up moisture. Places on the leeward side of a mountain range that receive little rainfall for this reason are said to be in a **rain shadow** (Figure 53.1).

Global atmospheric circulation influences climates

Earth's climates are strongly influenced by global air circulation patterns. Air rises not only when it crosses mountains, but also when it is heated by the sun. Warm air rises in the tropics, which receive the greatest solar energy input. This air is replaced by air that flows toward the equator from the north and south. That air, in turn, is replaced by air from aloft that descends after having traveled away from the equator at great heights. At roughly 30° north and south latitudes, air that cooled and lost its moisture while rising at the equator descends and warms. Many of Earth's deserts, such as the Sahara and the Australian deserts, are located at these latitudes.

At about 60° north and south latitudes, air rises again, and cold, dense air descends at the poles, where there is little input of solar energy. The black arrows around the edge of Figure 53.2 show these vertical patterns, which are one component of Earth's winds.

The spinning of Earth on its axis influences surface winds because Earth's velocity is rapid at the equator,

53.1 A Rain Shadow Average annual rainfall tends to be lower on the leeward side of a mountain range than on the windward side.

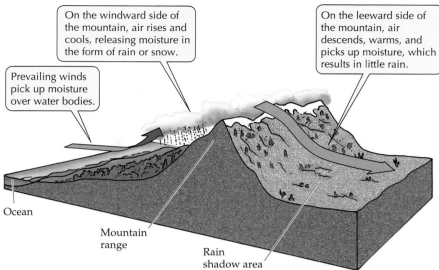

Prevailing winds pick up moisture over water bodies.

On the windward side of the mountain, air rises and cools, releasing moisture in the form of rain or snow.

On the leeward side of the mountain, air descends, warms, and picks up moisture, which results in little rain.

Ocean

Mountain range

Rain shadow area

but relatively slow close to the poles. An air mass at a particular latitude has the same velocity as Earth has at that latitude. As an air mass moves toward the equator, it confronts a faster and faster spin, and it slows down relative to Earth beneath it. As an air mass moves poleward, it confronts a slower and slower spin, and it speeds up relative to Earth beneath it. Therefore, air masses moving latitudinally are deflected to the right in the Northern Hemisphere and to the left in the Southern Hemisphere. Winds blowing toward the equator from the north and south veer to become the northeast and southeast trade winds, respectively. Winds blowing away from the equator also veer and become the westerlies that prevail at mid-latitudes. These surface winds are shown by the blue arrows in Figure 53.2.

Because Earth's axis is tilted, the amount of solar energy that reaches a given region varies seasonally as Earth orbits the sun. The amount of solar energy input is at its maximum at the time of year when the sun is closest to being overhead at noon. The **intertropical convergence zone**—the location of greatest solar energy input and the site where trade winds converge and air rises—thus shifts with the season. It shifts to

the north during the northern summer (southern winter) and to the south during the southern summer (northern winter).

However, the intertropical convergence zone lags behind the overhead passage of the sun by a bit more than a month because it takes that long to heat the surface mass of Earth. Seasonal changes in climate close to the equator are associated with the movement of the intertropical convergence zone because whenever an area is within the zone, air rises and heavy rains fall. When the zone is to the north or south of a tropical region, the prevailing winds are trade winds, which seldom yield rain unless forced to rise over mountains.

Global oceanic circulation is driven by the wind

The global pattern of wind circulation drives the circulation of ocean water. Ocean water generally moves in the direction of the prevailing winds (Figure 53.3). Winds blowing toward the equator from the northeast and southeast cause water to converge at the equator and move westward until it is blocked by a continental land mass. At that point

Jet stream

Cold deserts

60° N

Westerlies

Forests

30° N

Hot deserts

Northeast trades

Rising air

Forests

0° Equator

Descending air

30° S

Hot deserts

Southeast trades

Westerlies

Forests

Jet stream

60° S

Cold deserts

53.2 Circulation of Earth's Atmosphere If we could stand outside Earth and observe the movement of the air, we would see vertical movements like those indicated by the black arrows and surface winds like those shown by the blue arrows.

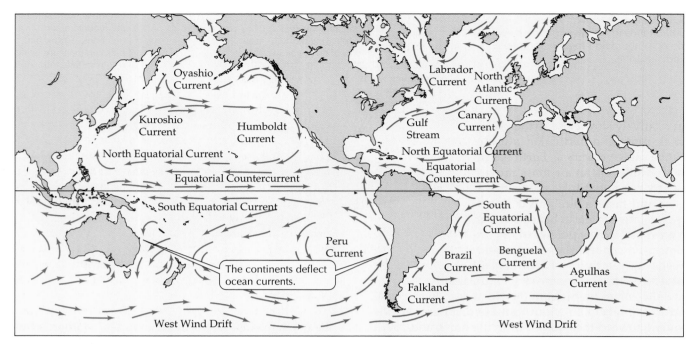

53.3 Global Oceanic Circulation To see that ocean currents are driven primarily by the wind, compare the surface currents shown here with the prevailing surface winds shown in Figure 53.2. Deep ocean currents differ strikingly from the surface ones shown here.

the water splits, some of it moving north and some of it moving south along continental shores. This poleward movement of ocean water is a major mechanism of heat transfer to high latitudes. As it moves toward the poles, the water veers right in the Northern Hemisphere and left in the Southern Hemisphere. Thus water turns eastward until it is blocked by another continent and is deflected laterally along its shores. In both hemispheres, water flows toward the equator along the west sides of continents, continuing to veer right or left until it meets at the equator and flows westward again.

The oceans play an important role in world climates, both because their waters move long distances and because water has a high specific heat. The **specific heat** of a substance is the amount of energy required to raise the temperature of 1 gram of the substance 1°C. For water, this value is 1 cal/g at 15°C. Similarly, 1 gram of water that cools 1°C gives off 1 cal/g. Air and land surfaces have a much lower specific heat. Consequently, in comparison with continents, oceans warm up more slowly in summer because it takes more heat to raise their temperature, and cool off more slowly in winter because more heat must be released to cool them.

At high latitudes, the temperatures of the interiors of large continents fluctuate greatly with the seasons,

becoming very cold in winter and hot in summer, a pattern called a **continental climate**. The coasts of continents, particularly those on west sides at middle latitudes, where the prevailing winds blow from ocean to land, have **maritime climates**, with smaller differences between winter and summer temperatures. Seasonal temperatures change the most on the largest land mass, Asia, where strong winter high pressure (descending air) over Siberia causes winds to blow from the continent toward the coasts. In summer, however, strong low pressure (rising air) over Siberia draws great quantities of moist air over the land from the Indian Ocean, producing the great summer monsoons (rainy seasons) characteristic of southern Asia.

The amount and annual pattern of energy input into different ecosystems determines the rates at which those ecosystems function and the kinds of organisms that live there. Next we will discuss how climates influence the amount of energy that flows through ecosystems.

Energy Flow through Ecosystems

Organisms depend on inputs of energy (in the form of sunlight or high-energy molecules), water, and minerals for metabolism and growth. Except for a few limited ecosystems (caves, deep-sea thermal systems) in which solar energy is not the main energy source, almost all energy utilized by organisms comes (or once came) from the sun. Even the fossil fuels—coal, oil, and natural gas—upon which the economy of modern civilization is based are reserves of captured solar energy locked up in the remains of organisms that lived millions of years ago.

Only about 5 percent of the solar energy that arrives on Earth is captured by photosynthesis. The remaining energy is either radiated back into the atmosphere as heat (especially in places where Earth's surface is bare because there is too little water to support plant growth) or consumed by the evaporation of water from plants. The energy that *is* captured powers the "metabolism" of ecosystems. How that energy is captured by green plants and subsequently passes through a series of organisms is the topic of the next section.

Photosynthesis drives energy flow in ecosystems

Energy flow in most ecosystems originates with photosynthesis (see Chapter 8). The major factors influencing the rate of photosynthesis are the amount of solar radiation, the availability of water, the abundance of mineral nutrients and carbon dioxide, and temperature. The total amount of energy that plants assimilate by photosynthesis is called **gross primary production**; the production that remains after subtracting the energy that plants use for maintenance and for building tissues is called **net primary production.** The rate at which plants assimilate energy is called **primary pro-**

ductivity. The distribution of primary production reflects the distribution of temperature and moisture on Earth (Figure 53.4).

Water availability and temperature are major determinants of gross primary productivity. To obtain minerals from the soil and to photosynthesize, plants must open their stomata, and when their stomata are open, they lose water. The rate of photosynthesis depends on temperature because most chemical reactions proceed faster as temperature increases, roughly doubling with every rise of 10°C. Temperature and moisture interact to determine primary productivity because the evaporative power of air, which affects transpiration and thus the flow of water within plants, is less at low temperatures than at high temperatures.

In many areas on Earth, the annual gross primary production is determined by available soil moisture and soil fertility. Shortage of water limits primary production during much of the year in most arid regions. Production in aquatic systems is limited by light, which decreases rapidly with depth; by nutrients, which sink and must be replaced by upwelling of water; and by temperature.

Close to the equator, temperatures are high throughout the year, and the water supply is adequate much of the time. In these climates, highly productive forests thrive. In lower- and mid-latitude deserts,

53.4 Primary Production in Different Ecosystems The (a) geographic extent, (b) annual production per unit of area, and (c) percentage of Earth's net primary production provided by different ecosystems.

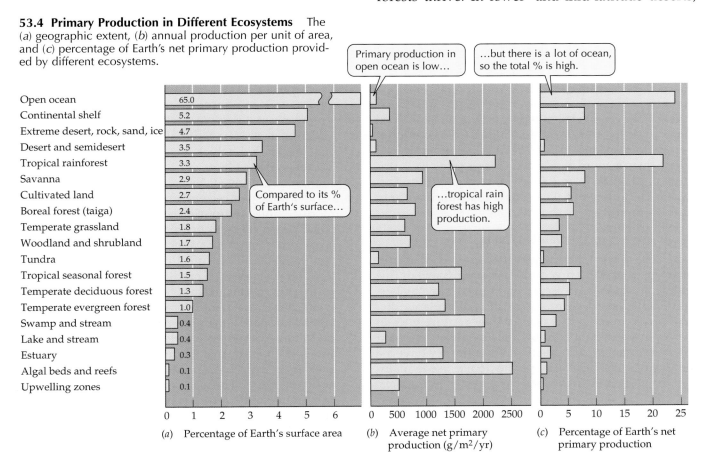

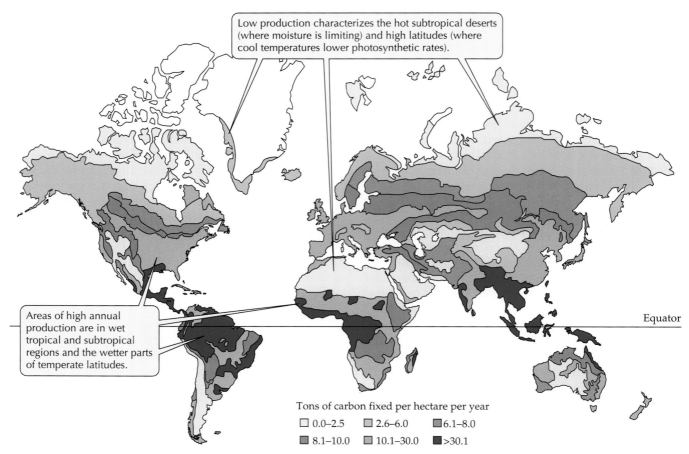

Low production characterizes the hot subtropical deserts (where moisture is limiting) and high latitudes (where cool temperatures lower photosynthetic rates).

Areas of high annual production are in wet tropical and subtropical regions and the wetter parts of temperate latitudes.

Equator

Tons of carbon fixed per hectare per year

▢ 0.0–2.5 ◻ 2.6–6.0 ◼ 6.1–8.0
◼ 8.1–10.0 ▨ 10.1–30.0 ■ >30.1

53.5 Net Primary Production of Terrestrial Ecosystems
Variations in temperature and water availability over Earth's land surface affect the productivity of its ecosystems.

where plant growth is limited by lack of moisture, primary production is low, and plants of low stature dominate the landscape. At still higher latitudes, where there is more moisture and trees grow well, primary production is limited by low temperatures during much of the year. The global distribution of net primary production is shown in Figure 53.5.

Plants use energy to maintain themselves

Plants use most of the energy they capture to maintain themselves and to grow and reproduce. Some of this energy produces new tissues that can be eaten by herbivores or used by organisms after the plants die. Because so much of the energy they capture goes to power their own metabolism, plants always contain much less energy than the total amount they have assimilated; only the energy plants do not use to maintain themselves is available to be harvested by animals.

Energy flows when organisms eat one another

Because energy flows through ecosystems when organisms eat one another, biologists find it useful to

group organisms according to their source of energy. The organisms that obtain their energy from a common source constitute a **trophic level** (Table 53.1). Organisms at a particular trophic level occupy a position in an ecosystem that is determined by the number of steps through which energy passes to reach them. Photosynthetic plants get their energy directly from sunlight. Collectively, they constitute the trophic level called photosynthesizers or **primary producers**. They produce the energy-rich organic molecules upon which nearly all other organisms feed.

All other organisms are called **consumers** because they consume, either directly or indirectly, the energy-rich organic molecules produced by photosynthetic organisms. Organisms that eat plants constitute the trophic level called herbivores. Organisms that eat herbivores are called primary carnivores. Those that eat primary carnivores are called secondary carnivores, and so on. Organisms that eat the dead bodies of organisms or their waste products are called detritivores or decomposers. The many organisms that obtain their food from more than one trophic level are called omnivores.

A sequence of linkages in which a plant is eaten by an herbivore, which is in turn eaten by a primary carni-

TABLE 53.1 The Major Trophic Levels

TROPHIC LEVEL	SOURCE OF ENERGY	EXAMPLES
Photosynthesizers (primary producers)	Solar energy	Green plants, photosynthetic bacteria, and protists
Herbivores	Tissues of primary producers	Termites, grasshoppers, water fleas, anchovies, deer, geese
Primary carnivores	Herbivores	Spiders, warblers, wolves, copepods
Secondary carnivores	Primary carnivores	Tuna, falcons, killer whales
Omnivores	Several trophic levels	Humans, opossums, crabs, robins
Detritivores	Dead bodies and waste products of other organisms	Fungi, many bacteria, vultures, earthworms

vore, and so on, is called a **food chain**. Food chains are usually interconnected to make a **food web**. Food webs result from the fact that most species in a community eat and are eaten by more than one other species.

The arrows in representations of food webs show who eats whom. A simplified food web, not including detritivores, for Gatun Lake, Panama, is shown in Figure 53.6. A food web is a useful summary of predator–prey interactions within a community. A complete food web, showing the position of every species in an ecosystem, would be confusingly complex because most biological communities contain so many species. Therefore, similar species, especially those at lower trophic levels, are usually lumped together, as they are in the diagram of the Gatun Lake food web.

Much energy is lost between trophic levels

Only a small portion of the energy captured at one trophic level is available to organisms at the next higher level because the energy that organisms use to

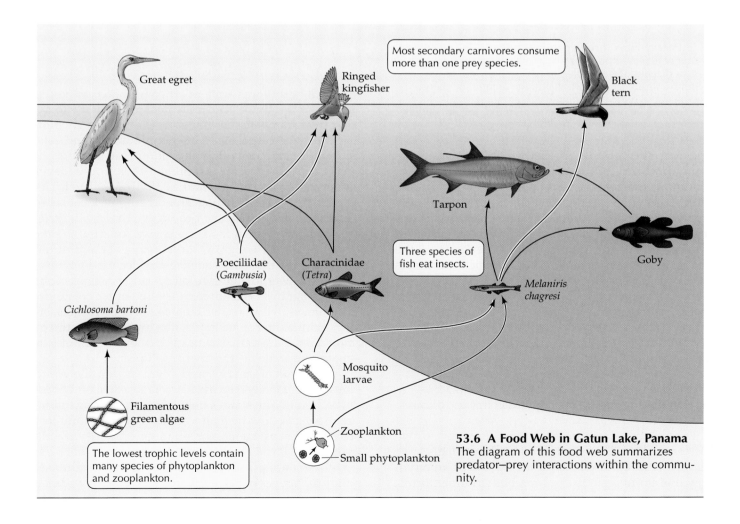

53.6 A Food Web in Gatun Lake, Panama
The diagram of this food web summarizes predator–prey interactions within the community.

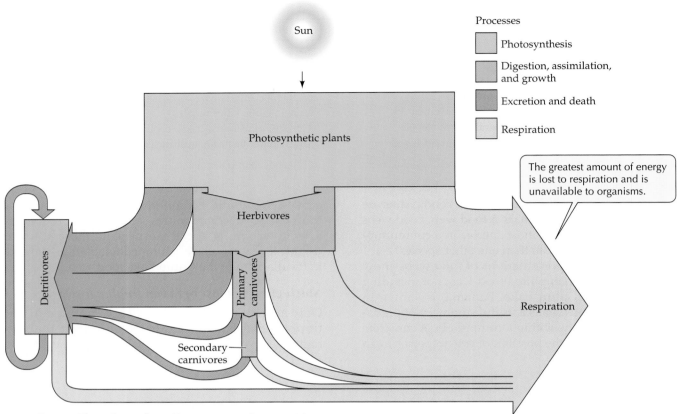

53.7 Energy Flow through an Ecosystem The quantities of energy flowing through an ecosystem can be visualized using a diagram like this one, in which the width of the channels is roughly proportional to the amount of energy flowing through them. Trophic levels are indicated by blocks of green or blue; energy channels are in red or gold, and the direction of energy flow is shown by arrows.

maintain themselves is dissipated as heat, a form of energy that cannot be used by other organisms. The energy content of an organism's net production—its growth plus reproduction—is available to organisms at the next trophic level (Figure 53.7). The efficiency of energy transfer through food webs depends on the fraction of net production at one trophic level that is consumed by organisms at the next level, and how those organisms divide the ingested energy between production and maintenance (respiration). We can calculate the efficiency of energy transfer of a species or group of species as

$$E = P/(P + R)$$

where E is efficiency, P is net production, and R is respiration.

The E values for different animal taxa reveal two patterns (Table 53.2). First, birds and mammals have very low efficiencies because they expend so much energy maintaining constant high body temperatures.

Second, herbivores are less efficient than carnivores because plant tissues generally take more energy to digest than animal tissues do.

Even when efficiencies of energy transfer and consumption rates are both high, seldom is as much as 20 percent of the energy assimilated by a trophic level converted to production at the next trophic level. The amount of energy reaching a higher trophic level is determined by net primary production and by the efficiencies with which food energy is converted to biomass (the total weight of organisms) at the trophic levels below it. To show how energy decreases in moving from lower to higher trophic levels, ecologists construct diagrams called **pyramids of energy**. A **pyramid of biomass**, which shows the mass of organisms existing at different trophic levels, illustrates the amount of biomass that is available at a given moment in time for organisms at the next trophic level (Figure 53.8).

Pyramids of energy and biomass for the same ecosystem usually have similar shapes, but sometimes they do not. The shapes depend on the dominant organisms and how they allocate their energies. In most terrestrial ecosystems, the dominant photosynthetic plants are large and store energy for long periods, much of it in difficult-to-digest forms (cellulose, lignin). However, terrestrial ecosystems may differ

53.8 Pyramids of Biomass and Energy Ecosystems can be compared in terms of the amount of material present in organisms at different trophic levels (left), and in terms of energy flow (right).

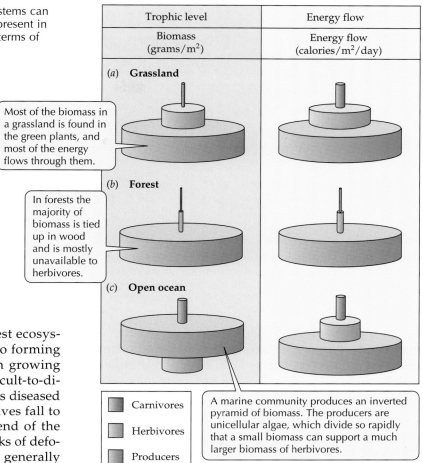

Most of the biomass in a grassland is found in the green plants, and most of the energy flows through them.

In forests the majority of biomass is tied up in wood and is mostly unavailable to herbivores.

A marine community produces an inverted pyramid of biomass. The producers are unicellular algae, which divide so rapidly that a small biomass can support a much larger biomass of herbivores.

strikingly in patterns of energy flow depending on the life forms of the dominant plants. In grassland ecosystems, because plants produce few hard-to-digest woody tissues, animals are able to consume most of the annual production of plant tissues each year. In grasslands, mammals—wild or domestic—may consume 30 to 40 percent of the annual aboveground net primary production. Insects may consume an additional 5 to 15 percent. Soil organisms, primarily nematodes, may consume 6 to 40 percent of the belowground biomass (Figure 53.8a).

By contrast, the dominant plants in forest ecosystems allocate a great deal of their energy to forming wood, which accumulates at high rates in growing forests. Wood, which is constructed of difficult-to-digest material, is rarely eaten unless a plant is diseased or otherwise weakened. In most forests leaves fall to the ground relatively undamaged at the end of the growing season. Although there are outbreaks of defoliating insects in forests, browsing rates are generally so low that forest ecologists often ignore losses to herbivores when calculating forest production (Figure 53.8b).

TABLE 53.2 Average Production Efficiencies (E) for Various Groups of Animals

GROUP	PRODUCTION EFFICIENCY (%) P/(P + R)
Insectivores (mammals)	0.9
Birds	1.3
Small mammals	1.5
Large mammals	3.1
Fishes and social insects	10.0
Invertebrates other than insects	
Herbivores	21
Carnivores	28
Detritivores	36
Nonsocial insects	
Herbivores	39
Carnivores	56
Detritivores	47

In most aquatic communities, on the other hand, the dominant photosynthesizers are bacteria and protists. They have such high rates of cell division that a small biomass of photosynthesizers can feed a much larger biomass of herbivores, which grow and reproduce much more slowly. This pattern can produce an inverted pyramid of biomass, even though the pyramid of energy for the same ecosystem has the typical shape (Figure 53.8c).

Much of the energy ingested by organisms is converted to biomass that is eventually consumed by detritivores (see Figure 53.7). Detritivores immobilize some nutrients, but they transform the remains and waste products of organisms (detritus) into carbon dioxide, water, and free mineral nutrients that can be taken up by plants again. If there were no detritivores, most nutrients would eventually be tied up in dead bodies, where they would be unavailable to plants. Therefore, continued ecosystem productivity depends on rapid decomposition of detritus.

Under the warm, wet conditions found in tropical forests, detritus is decomposed within a few weeks or months, and no litter accumulates on the soil surface.

53.9 Agriculture Requires Energy Inputs (*a*) In traditional agriculture, people supply most of the energy, as in this Bengali rice paddy. (*b*) Modern agriculture is based on high rates of consumption of fossil fuels and use of toxic chemicals.

Rates of decomposition are slower under colder and drier conditions (see Figure 52.22). At high altitudes and latitudes, decomposition of leaf litter may take decades; decomposition of tree trunks may take more than a century.

Agricultural Manipulation of Ecosystem Productivity

Humans exploit ecosystems by replacing species of low economic value with species of high value. By means of agriculture, we help some species compete with others and manipulate ecosystems so as to increase the yield of products useful to humans. Agriculture has several intricately intertwined components: We eliminate competition between crops and unwanted plants by cultivating and by applying herbicides; we reduce competing herbivores and disease-causing organisms, usually by applying toxic chemicals; we augment photosynthesis by fertilizing and irrigating; and we develop special high-yielding strains of plants that respond to additional fertilizer by increasing their growth rates. All these components must work together, because "miracle" strains of crops do not actually yield more than other strains unless they are provided with fertilizers and protected from competitors, herbivores, and pathogens. Agriculture also depends on energy from outside the system for cultivation and harvesting. In modern agriculture, this energy comes from fossil fuels (Figure 53.9).

Although human manipulations of agricultural systems have spectacularly increased food production per hectare, they have also created problems. Herbicides and insecticides have polluted lakes, rivers, and groundwater in most industrialized countries. Many agricultural pests have evolved resistances to pesticides. The usual response has been to increase the use of pesticides, creating even more severe pollution problems. Recently, however, agriculturists have developed new, less toxic methods of pest control in agricultural systems.

New methods of pest control, collectively known as **integrated pest management** (IPM), are growing in use. IPM combines chemical approaches to pest control with cultural practices—such as crop rotation, mixed plantings of crop plants, and mechanical tillage of the soil—and biological methods—such as development of pest-resistant strains of crops, use of natural predators and parasites, and use of chemical attractants. The reduced use of toxic chemicals avoids most pollution problems and reduces the chance that pests will evolve resistance to pesticides.

Cycles of Materials in Ecosystems

As we have just seen, energy is not fully recycled in ecosystems because at each transformation, much of it is dissipated as heat, a form that cannot be used by organisms to power their metabolism. Chemical elements, on the other hand, are not lost when they are transferred among organisms; they cycle through organisms and the physical environment. Carbon, nitrogen, phosphorus, calcium, sodium, sulfur, hydrogen, and oxygen, together with smaller amounts of other chemical elements, are the primary materials of which organisms are constructed. The quantities of these elements that are available to organisms are strongly influenced by how organisms get them, how long they hold onto them, and what they do with them while they have them.

To understand the cycling of elements, it is convenient to divide the global ecosystem into four com-

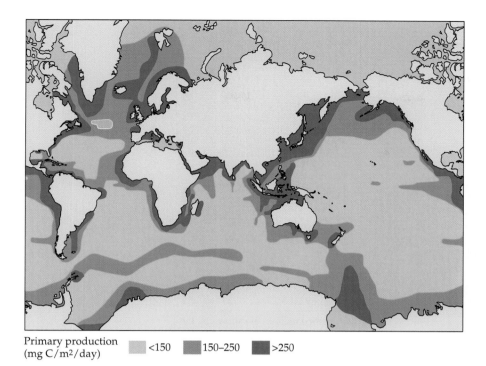

53.10 Primary Production Is High in Zones of Upwelling Primary production in the oceans is highest near continents in areas where surface waters, driven by the prevailing winds, move offshore and are replaced by cool, nutrient-rich water upwelling from below.

Primary production (mg C/m²/day)   <150   150–250   >250

are very low over most of the oceans. Oxygen is usually present at all depths, because even slow mixing suffices to replenish the oxygen consumed by the respiration and decomposition of the few organisms that live in the nutrient-poor water. Most of the elements that enter the oceans settle to the bottom and remain there until bottom sediments are elevated above sea level by movements of Earth's crust, but this process may take many millions of years.

partments: oceans, fresh waters, atmosphere, and land. The physical environments in each compartment and the types of organisms living there are different, and therefore the amounts of elements found in the different compartments, what happens to those elements, and the rates at which they enter and leave the compartments differ strikingly. After we have described these compartments, we will consider them together to illustrate how elements cycle through the global ecosystem.

Oceans receive materials from the land and atmosphere

Oceans receive materials from land as runoff from rivers. On time scales of hundreds to thousands of years, oceans are the ultimate repository of most materials produced by human activity, even though the immediate receivers are often other compartments of the global ecosystem. Because of their huge size, and because they exchange materials with the atmosphere only at their surface, oceans respond very slowly to outside disturbances.

Except on continental shelves, ocean waters mix very slowly and are strongly stratified. Elements that enter the oceans from other compartments gradually sink to the seafloor, unless they are brought back to the surface by the cool bottom water that rises—**up-wells**—near the coasts of continents (Figure 53.10). Waters in these zones of upwelling are rich in nutrients, and most of the world's great fisheries are concentrated there. Concentrations of mineral nutrients

Lakes and rivers contain only a small fraction of Earth's water

Lakes and rivers contain much less water than oceans do, and because these bodies of water are relatively small, most mineral nutrients entering them are not buried in bottom sediments for long periods of time. Some mineral nutrients enter fresh waters in rainfall, but most are released by the weathering of rocks and are carried to lakes and rivers via groundwater (the water that resides in the soil and in rocks) or by surface flow.

After entering rivers, mineral nutrients are usually carried rapidly to lakes or to the oceans. In lakes they are taken up by organisms and incorporated into their cells. These organisms eventually die and sink to the bottom, where decomposition of their tissues uses up the oxygen. Surface waters of lakes thus quickly become depleted of nutrients, while deeper waters become depleted of oxygen. However, this stratification process is countered by vertical movements of water—**turnover**—that bring nutrients to the surface and oxygen to deeper water. Wind is an important mixing agent in shallow lakes, but in deeper lakes it usually mixes only surface waters.

In temperate regions, lake waters turn over because water is most dense at 4°C; above and below that temperature it expands (Figure 53.11). In spring at mid-latitudes, the sun warms the surface layer of a lake. The depth of the warm layer gradually increases as spring and summer progress. However, there is still a well-defined zone—the **thermocline**—where the tempera-

53.11 Annual Temperature and Oxygen Cycles in a Temperate Lake   These vertical temperature profiles are typical of temperate-zone lakes that freeze in winter.

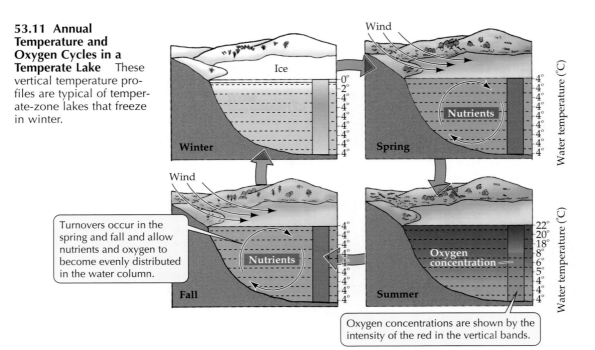

Turnovers occur in the spring and fall and allow nutrients and oxygen to become evenly distributed in the water column.

Oxygen concentrations are shown by the intensity of the red in the vertical bands.

ture drops abruptly to about 4°C. Only if the lake is shallow enough to warm to the bottom does the temperature of the deepest water rise above 4°C.

In autumn, as the surface of the lake cools, the cooler surface water, which is denser than the warmer water below it, sinks, and is replaced by warmer water from below. This process continues until the entire water column has reached 4°C. At this point, the density of the water is uniform throughout the lake, and even modest winds readily mix the entire water column. As colder weather then cools the surface water below 4°C, it becomes less dense than the 4°C water below it and floats at the top. Another turnover occurs in spring, when the surface layers above the thermocline warm to 4°C and the water column, again being of uniform density throughout, is easily mixed by wind.

Deep tropical and subtropical lakes may be permanently stratified because they never become cool enough to have uniformly dense water. Their bottom waters lack oxygen because decomposition quickly depletes any oxygen that reaches them. However, many tropical lakes are overturned at least periodically by strong winds so that their deeper waters are occasionally oxygenated. Arctic lakes turn over only once each year.

The atmosphere regulates temperatures close to Earth's surface

The atmosphere is a thin sphere of gases surrounding Earth. About 80 percent of the mass of the atmosphere lies in its lowest layer, the troposphere, which extends upward from Earth's surface about 17 km in the trop-

ics and subtropics, but only about 10 km at higher latitudes (Figure 53.12). Most global air circulation takes place within the troposphere, and virtually all atmospheric water vapor is located there.

Above the troposphere, the stratosphere, which extends upward to about 50 kilometers above Earth's surface, contains about 99 percent of the remaining atmospheric mass, but it is extremely dry. Materials enter the stratosphere from the troposphere near the equator, where air rises to high altitudes. These materials tend to remain in the stratosphere for a relatively long time because stratospheric air circulation is horizontal. Ozone (O^3) in the stratosphere absorbs most shorter wavelengths of biologically damaging ultraviolet radiation, which is why the development of the ozone hole, which we discussed in Chapter 19, is of great concern.

The atmosphere is 78.08 percent nitrogen as N_2, 20.95 percent oxygen, 0.93 percent argon, and 0.03 percent carbon dioxide. It also contains traces of hydrogen gas, neon, helium, krypton, xenon, ozone, and methane. The atmosphere contains Earth's biggest pool of nitrogen and large supplies of oxygen. Although carbon dioxide constitutes a very small fraction of the atmosphere, it is the source of the carbon used by terrestrial photosynthetic organisms. Concentrations of atmospheric water vapor are highly variable in space and time. The atmosphere is a transport medium for many gases, as well as for airborne particles containing carbon, nitrogen, sulfur, phosphorus, and other nutrient elements.

The atmosphere plays a decisive role in regulating temperatures at and close to Earth's surface. Without

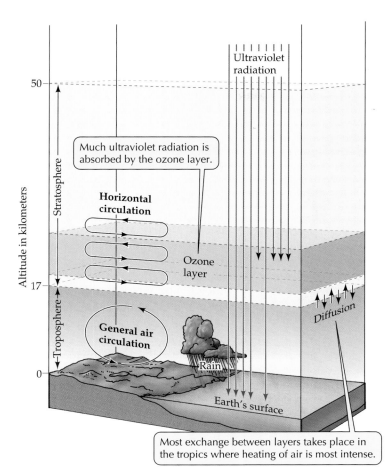

Ultraviolet
radiation

Much ultraviolet radiation is
absorbed by the ozone layer.

**Horizontal
circulation**

Ozone
layer

**General air
circulation**

Rain

Diffusion

Earth's surface

Most exchange between layers takes place in
the tropics where heating of air is most intense.

53.12 Earth's Atmosphere   The two lowest atmospheric
layers, the troposphere and the stratosphere, differ in their
circulation patterns, the amount of moisture they contain,
and the amount of ultraviolet radiation they receive.

an atmosphere, the average surface temperature of
Earth would be about –18°C rather than its actual
+17°C. Earth remains at this warm temperature because the atmosphere is relatively transparent to visible light, but traps a large part of the outgoing infrared
radiation (heat), the main radiation emitted by a cool
body like Earth. Water vapor, carbon dioxide, and
ozone are especially important trappers of infrared radiation. That is why, as we will see below, increased
concentrations of atmospheric carbon dioxide may
lead to important climatic changes.

Land covers about one-fourth of Earth's surface

About one-fourth of Earth's surface, most of it in the
Northern Hemisphere, is currently above sea level.
Most of this land is covered by a layer of soil of varying depths, weathered from parent rocks beneath it or
carried to its present location by agents of erosion.
Unlike their behavior in air and water, elements on
land move slowly, and they usually move only short
distances. For this reason, we will emphasize local

rather than global ecosystems in our discussion of the
land compartment.

The land is connected to the atmospheric compartment by terrestrial organisms that take chemical elements from and release them to the air. Chemical elements in soils are carried in solution into the
groundwater and eventually into rivers and oceans,
where they are lost to organisms until an episode of
uplifting raises marine sediments and a new cycle of
erosion and weathering begins.

As you may recall from Chapter 34, the type of soil
that forms in an area depends on the underlying rock,
as well as on climate, topography, the organisms living
there, and the length of time that soil-forming
processes have been acting. As a soil weathers, its clay
particles slowly decompose chemically. After hundreds of thousands of years, most nutrients needed by
plants have weathered out and have been carried to
the oceans. Therefore, very old soils are much less fertile than young soils are. Figure 34.6 shows a general
picture of these changes over a period of 1 million
years. Even though the global supply of nutrients is
constant, regional and local deficiencies strongly affect
ecosystem processes on land.

Biogeochemical Cycles

The chemical elements organisms need in large quantities—carbon, hydrogen, oxygen, nitrogen, phosphorus, and sulfur—cycle through organisms to the environment and back again. The pattern of movement of
a chemical element through organisms and reservoirs
in the physical environment is called its **biogeochemical cycle**. The carbon and nitrogen atoms of which life
is composed today are the same atoms that made up
dinosaurs, insects, and trees in the Mesozoic era. Some
chemical elements circulate continually, but large
quantities of other elements are temporarily lost from
circulation through deposition in deep-sea sediments.

Each chemical element used by organisms has a distinctive biogeochemical cycle whose properties depend
on the physical and chemical nature of the element and
how organisms use it. All chemical elements cycle
quickly through organisms because no individual,
even of the longest-lived species, lives very long in geologic terms. Chemical elements, such as carbon and
nitrogen, that exist in the atmosphere as a gas cycle
faster than nongaseous elements. After discussing the
movements of water, we'll discuss the cycling of the
most abundant chemical elements in organisms.

Water cycles through the oceans, atmosphere, and land

The cycling of water through the oceans, atmosphere,
and land is known as the **hydrological cycle**. Although water is a compound, not an element, we dis-

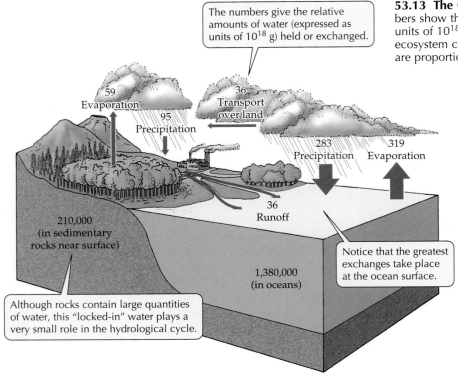

The numbers give the relative amounts of water (expressed as units of 10^{18} g) held or exchanged.

53.13 The Global Hydrological Cycle The numbers show the relative amounts of water (expressed as units of 10^{18} g) held in or exchanged annually by ecosystem compartments. The widths of the arrows are proportional to the size of the fluxes.

59 Evaporation

36 Transport over land

95 Precipitation

283 Precipitation

319 Evaporation

36 Runoff

210,000 (in sedimentary rocks near surface)

1,380,000 (in oceans)

Notice that the greatest exchanges take place at the ocean surface.

Although rocks contain large quantities of water, this "locked-in" water plays a very small role in the hydrological cycle.

Organisms profoundly influence the carbon cycle

Organisms are triumphs of carbon chemistry; to survive, they must have access to carbon atoms. Nearly all the carbon in organisms comes from carbon dioxide (CO_2) in the atmosphere or dissolved carbonate ions (HCO_3^-) in water. In the cells of some bacteria, algae, and leaves of plants, carbon is incorporated into organic molecules by photosynthesis. All organisms in other kingdoms get their carbon by consuming other organisms or their remains.

cuss its cycle here together with those of individual elements because of its importance to life. The hydrological cycle is driven by the evaporation of water, most of it from ocean surfaces. Some water returns to the oceans as precipitation, but much less falls back on the oceans than is evaporated from them. The remaining evaporated water is carried by winds over the land, where it falls as rain or snow.

Water also evaporates from soils, from freshwater lakes and rivers, and from the leaves of plants (transpiration), but the total amount evaporated is less than the amount that falls as precipitation. The excess water eventually returns to the oceans via rivers, coastal runoff, and subterranean flows (Figure 53.13).

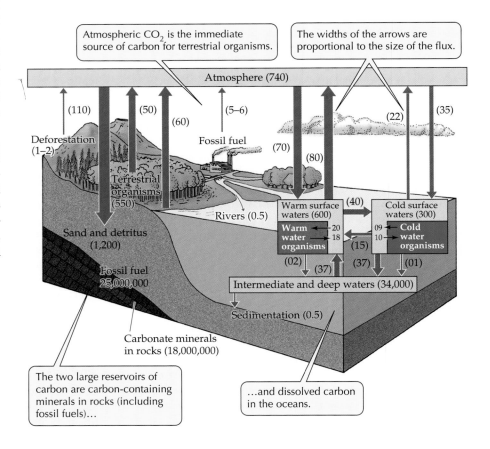

Atmospheric CO_2 is the immediate source of carbon for terrestrial organisms.

The widths of the arrows are proportional to the size of the flux.

Atmosphere (740)

(110)   (50)   (60)   (5–6)   (22)   (35)

Deforestation (1–2)

Fossil fuel

(70)   (80)

Terrestrial organisms (550)

Rivers (0.5)

Warm surface waters (600)   (40)   Cold surface waters (300)

Warm water organisms   20   18   09   10   Cold water organisms

(15)

Sand and detritus (1,200)

(02)   (37)   (37)   (01)

Intermediate and deep waters (34,000)

Fossil fuel 25,000,000

Sedimentation (0.5)

53.14 The Global Carbon Cycle The numbers show the quantities of carbon (expressed as units of 10^{15} g) in organisms and in various carbon reservoirs and the amounts that move annually between the various ecosystem compartments. The widths of the arrows are proportional to the size of the fluxes.

Carbonate minerals in rocks (18,000,000)

The two large reservoirs of carbon are carbon-containing minerals in rocks (including fossil fuels)...

...and dissolved carbon in the oceans.

Although atmospheric carbon dioxide is the immediate source of carbon for terrestrial organisms, only a small part of Earth's carbon is in the atmosphere. Most of it exists as nongaseous, dissolved carbon in oceans and carbon-containing minerals in rocks (Figure 53.14). Sedimentary rocks hold most of the carbon that is in rocks. Movement of carbon between rocks and other reservoirs of carbon is very slow.

Although marine organisms contain very little carbon, they have a profound influence on the distribution of carbon in the oceans. They convert soluble carbonate ions from seawater into insoluble ocean sediments by depositing carbon in their shells and skeletons, which eventually sink to the bottom.

Biological processes move carbon between the atmospheric and terrestrial compartments, removing it from the atmosphere during photosynthesis and returning it to the atmosphere during respiration. Growing plants at middle and high latitudes in the Northern Hemisphere incorporate so much carbon into their bodies during the summer that they reduce the concentration of atmospheric carbon dioxide from about 350 parts per million in winter to 335 parts per million in midsummer. This carbon is released back into the atmosphere by decomposition in autumn.

At times in the remote past, large quantities of carbon were removed from the global carbon cycle when organisms died in large numbers in environments without oxygen. In such environments, detritivores do not reduce organic carbon to carbon dioxide. Instead, organic molecules accumulate and eventually are transformed into oil, natural gas, coal, or peat. Humans have discovered and used these deposits, known as **fossil fuels**, at ever-increasing rates during the past 150 years. As a result, carbon dioxide, the final product of burning these fuels, is being released into the atmosphere faster than it is being transferred to the oceans and incorporated into terrestrial biomass (see Figure 53.18).

Elemental nitrogen can be used by few organisms

Nitrogen is an essential component of many organic molecules, such as nucleic acids and proteins. Although nitrogen (N_2) makes up 78 percent of the atmosphere, it cannot be used by most organisms in its gaseous form. It can be converted into biologically useful forms only by a few species of bacteria and cyanobacteria (see Chapter 23). Therefore, despite its abundance, usable nitrogen is often in short supply in ecosystems. This is why nearly all commercial fertilizers contain biologically useful compounds of nitrogen.

Just as organisms other than nitrogen fixers cannot take up nitrogen gas directly from the atmosphere, they do not respire nitrogen back to the atmosphere. Instead, organic molecules containing nitrogen are converted to inorganic molecules in several stages by different organisms. Most of the resulting nitrogen-containing compounds, such as nitrates or ammonia, are again taken up by plants. This movement of nitrogen among organisms accounts for about 95 percent of all nitrogen fluxes on Earth (Figure 53.15).

The phosphorus cycle has no gaseous phase

The phosphorus cycle differs from the other cycles discussed in this section in that it lacks a gaseous phase. Some phosphorus is transported on dust particles, but in general the atmosphere plays a very minor role in the phosphorus cycle. Phosphorus exists mostly as phosphate (PO_4^{3-}) or similar compounds.

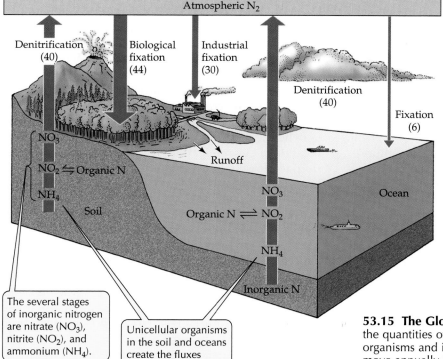

The several stages of inorganic nitrogen are nitrate (NO_3), nitrite (NO_2), and ammonium (NH_4).

Unicellular organisms in the soil and oceans create the fluxes (expressed as 10^9 kg of nitrogen per year).

53.15 The Global Nitrogen Cycle The numbers show the quantities of nitrogen (expressed as units of 10^9 kg) in organisms and in various reservoirs and the amounts that move annually between the various ecosystem compartments. The widths of the arrows are proportional to the size of the fluxes.

TABLE 53.3 Reservoirs, Fluxes, and Residence Times of Phosphorus in the Global Ecosystem

RESERVOIR	AMOUNT IN RESERVOIR (10^6 METRIC TONS)	FLUX (10^6 METRIC TONS/YEAR)	RESIDENCE TIME (YEARS)
Atmosphere	0.0028	4.5	0.0006 (53 hours)
Land biota	3,000	63.5	47.2
Land	2,000,000	88–100	2,000
Shallow ocean	2,710	1,058	2.56
Ocean biota	138	1,040	0.1327 (48 days)
Deep ocean	87,100	60	1,45
Sediments	4×10^9	214	1.87×10^8
Total ocean ecosystem	89,810	1.9	47,270

Most phosphate deposits are of marine origin. In nature, phosphorus becomes available through the slow weathering and dissolution of rocks and minerals. Organisms need phosphorus as a component of the energy-rich molecules involved in cellular metabolism. Phosphorus is often a limiting nutrient in soils and lakes. That is why phosphate is a component of most fertilizers and why adding phosphate to lakes causes marked increases in their biological productivity.

Table 53.3 shows how much more rapidly phosphorus is cycled through marine organisms than through terrestrial organisms. Phosphorus also moves readily between the surface and the bottom of the oceans. On average, a phosphorus atom cycles about 50 times between deep waters and surface waters before it is deposited in ocean sediments. Each time a phosphorus atom reaches surface waters, it cycles between the marine biota and the dissolved phosphate in the water about 25 times before it returns to deep waters. As a result, the average phosphorus atom is incorporated into marine organisms about 1,250 times during its stay in the ocean!

Organisms drive the sulfur cycle

Sulfur is biologically important because it is a component of proteins. Emissions of sulfur dioxide (SO_2) and hydrogen sulfide (H_2S) from volcanoes and fumaroles (vents for hot gases) are the only significant natural nonbiological fluxes of sulfur. These emissions release, on average, between 10 and 20 percent of the total natural flux of gaseous sulfur to the atmosphere, but they vary greatly in time and space. Large eruptions spread great quantities of sulfur over broad areas, but they are rare events.

Volatile sulfur compounds are also emitted by both terrestrial and marine organisms. Certain marine algae produce large amounts of dimethyl sulfide (CH_3SCH_3), which accounts for about half of the biotic component of the sulfur cycle; the other half is produced by terrestrial organisms. On land, the breakdown of organic sulfur compounds during fermentation is the most important mechanism of sulfur release (Figure 53.16).

Sulfur is apparently always abundant enough to meet the needs of living organisms. It also plays an important role in global climate. Even if air is moist, clouds do not form readily unless there are nuclei around which water can condense. Dimethyl sulfide is the major source of such nuclei. Therefore, increases or decreases in sulfur emissions can change cloud cover and hence climate.

Human Alterations of Biogeochemical Cycles

Human activity has greatly modified the quantities of elements being cycled as well as how and where they enter and exit ecosystems. These changes can increase metabolic rates if they increase the availability of nutrients, or decrease them if levels of elements become high enough to be toxic to organisms, or if high levels cause other detrimental environmental changes.

We will now consider several examples of consequences resulting from human modifications of biogeochemical cycles. These consequences range from local (lake eutrophication) to regional (acid precipitation) to global (climate change).

Lake eutrophication is a local effect

The most striking and best-studied example of a local effect of altered biogeochemical cycles is **eutrophication**—the addition of nutrients, especially phosphorus, to fresh water. Humans, who tend to live around water, dump large quantities of nutrients directly and indirectly into lakes and rivers. Most of these nutrients come from domestic and industrial sewage, but many come from leaching of fertilizers and pesticides from agricultural lands draining into rivers and lakes. Some nutrients also arrive naturally in precipitation, but human activities have greatly increased the quantities of nutrients that enter fresh water.

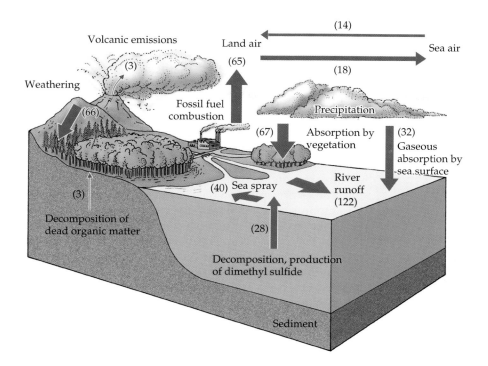

(14)

Volcanic emissions   Land air   Sea air

(3)   (65)   (18)

Weathering   Precipitation

(66)   Fossil fuel
               combustion

                              (67)   Absorption by   (32)
                                        vegetation   Gaseous
                                                             absorption by
                                              River   sea surface
(3)   (40) Sea spray   runoff
                                           (122)
Decomposition of
dead organic matter   (28)

               Decomposition, production
               of dimethyl sulfide

                      Sediment

53.16 The Global Sulfur Cycle The numbers show the quantities of sulfur (expressed as units of 10^9 kg) in organisms and in various reservoirs and the amounts that move annually between the various ecosystem compartments. The widths of the arrows are proportional to the size of the fluxes. The transfer rates shown include both natural and human-caused fluxes. The total fluxes are now more than twice what they were a century ago, primarily because of fossil fuel combustion.

In fresh water, photosynthesis is most often limited by supplies of phosphorus. In eutrophic (enriched) lakes, the extra phosphorus provided by fertilizers and detergents allows algae and bacteria to multiply, forming blooms that turn the water green. The decomposition of dead cells produced by this increased biological activity uses up all the oxygen in the lake, and anaerobic organisms come to dominate the sediments. These anaerobic organisms do not break down organic compounds all the way to carbon dioxide, and many of the end products of their activities have unpleasant odors.

Lake Erie is a eutrophic lake, but two hundred years ago it had only moderate levels of photosynthesis and clear, oxygenated water. Today more than 15 million people live in the Lake Erie basin. Nearby cities pour more than 250 billion liters of domestic and industrial wastes into the lake annually. The entire basin is intensely farmed and heavily fertilized.

In the early part of the twentieth century, nutrients in the lake increased greatly, and algae proliferated. At the water filtration plant in Cleveland, algae increased from 81 per milliliter in 1929 to 2,423 per milliliter in 1962. Algal blooms and populations of bacteria also increased. The numbers of *Escherichia coli* increased enough to cause the closing of many of the lake's beaches as health hazards.

As oxygen levels dropped in deeper lake waters, many native species that thrive only in oxygenated water declined. Before 1900, the dominant fishes in Lake Erie were lake herring, blue pike, carp, yellow perch, sauger, whitefish, and walleye. Lake trout were common in deeper waters. By 1925, herring had become too scarce to support the herring fishery. After

1945, blue pike, sauger, and white-fish became very scarce and lake trout disappeared. Currently the lake's fishery depends upon yellow perch, smelt, sheepshead, white bass, carp, catfish, and walleye, most of which are less valuable commercially than the species that declined.

Since 1972, the United States and Canada have invested more than 8 billion dollars to improve municipal waste facilities and reduce discharges of phosphorus into Lake Erie. As a result, the amount of phosphate added to Lake Erie has decreased more than 80 percent from the maximum level, and phosphorus concentrations in the lake have declined substantially. Since 1985, estimated inputs of phosphorus to Lake Erie have been reduced to about the target goal of 11,000 metric tons per year. The deeper waters of Lake Erie still become poor in oxygen during the summer months, but the rate of oxygen depletion is declining. Algal blooms have decreased, as have populations of small fishes that feed on algae.

The rate at which a eutrophic lake recovers depends on the replacement rate of its waters. Because water flows slowly through Lake Erie, it will take many years for the lake to recover from the heavy pollutant loads it has received. By contrast, the molecules of water in Lake Washington, a smaller lake adjacent to the city of Seattle, are replaced every 3 years. When sewage was diverted away from Lake Washington, the lake returned to its former condition within a decade.

Acid precipitation is a regional effect

An important regional effect of human alteration of two major biogeochemical cycles is **acid precipitation**—rain or snow whose pH is lowered by the presence of sulfuric acid (H_2SO_4) and nitric acid (HNO_3), derived in large part from the burning of fossil fuels. These acids enter the atmosphere and may travel hundreds of kilometers before they settle to Earth in precipitation or as dry particles.

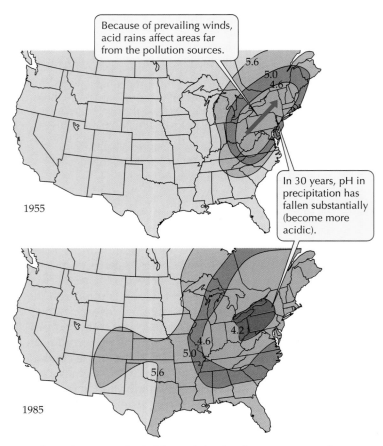

Because of prevailing winds, acid rains affect areas far from the pollution sources.

In 30 years, pH in precipitation has fallen substantially (become more acidic).

53.17 Increases in Acid Precipitation in Eastern North America The numbers represent the annual average pH of precipitation. Oxides of nitrogen and sulfur—the principal contributors to acid precipitation—travel far enough from their sources that the effects of many sources blend together to produce the pattern shown here.

Acid precipitation is now a phenomenon of all major industrial countries and is particularly widespread in eastern North America (Figure 53.17). The source of acid precipitation in New England is primarily the Ohio Valley; that of Scandinavia is primarily the industrial areas of England and Germany. The normal pH of precipitation in New England is about 5.6, but precipitation there now averages about pH 4.1, and there are occasional storms with a precipitation pH as low as 3.0. Precipitation with a pH of about 3.5 or lower causes direct damage to the leaves of plants and reduces photosynthetic rates. In central Europe, acid precipitation has contributed to moderate to severe damage to 15 to 20 percent of the growing stock of harvestable forest.

Ecologists in Canada studied the effects of acid precipitation on small lakes by adding enough sulfuric acid to two lakes to reduce their pH from about 6.6 to 5.2. In both lakes, nitrifying bacteria failed to adapt to these moderately acidic conditions, with the result that the nitrogen cycle was blocked and ammonium accu-

mulated in the water. When the ecologists stopped adding acid to one of the lakes, its pH increased to 5.4, and nitrification resumed after a lag of about 1 year. These experiments show that lakes are very sensitive to acidification, but can recover quickly when pH returns to normal values.

Acid precipitation also changes soil chemistry. Since 1963, scientists have monitored the chemistry of both precipitation and stream water at the Hubbard Brook Experimental Forest in New Hampshire. The pH of stream water leaving the forest has changed relatively little as precipitation has become more acid, but large amounts of calcium and magnesium have flowed out of the watershed. So much calcium has been lost from the soil that calcium supply has been limiting forest production at Hubbard Brook since 1987, and annual accumulation of biomass in the forest is now much lower than it was during the previous decades.

When pollution originates in one area but causes problems in another, as acid precipitation does, solving the problem is politically difficult. Acid precipitation is caused by the generation of the energy upon which modern societies depend. Oxides of nitrogen and sulfur can be removed from smokestack gases, but costs rise sharply as the percentage removed rises above 90 percent. The number of sources emitting these oxides is now so great that almost complete removal will be necessary to correct the problem, even if no new sources are added.

Alterations of the carbon cycle produce global effects

The biogeochemical cycle most seriously disturbed globally by human activity is the carbon cycle. Climatologists have measured atmospheric concentrations of carbon dioxide on top of Mauna Loa in Hawaii since 1958. Their measurements reveal a slow but steady increase in carbon dioxide concentrations (Figure 53.18). Based on a variety of calculations, atmospheric scientists believe that 150 years ago, before the Industrial Revolution, the concentration of atmospheric carbon dioxide was probably about 265 parts per million. Today it is 350 parts per million.

This increase has been caused primarily by combustion of fossil fuels and secondarily by the burning of forests. If current trends in both these activities continue, atmospheric carbon dioxide is expected to reach 580 parts per million by the middle of the twenty-first century. This carbon dioxide will eventually be transferred to the oceans and deposited in sediments as calcium carbonate ($CaCO_3$), but the rate of transfer is much slower than the rate at which humans are introducing carbon dioxide into the atmosphere.

Enough carbon is being released by the burning of fossil fuels to alter the heat balance of Earth, even though the absolute quantity is small relative to other

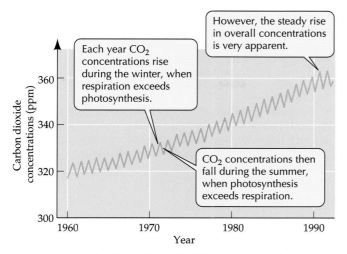

53.18 Atmospheric Carbon Dioxide Concentrations Are Increasing Carbon dioxide concentrations, expressed as parts per million by volume of dry air, are recorded on top of Mauna Loa, Hawaii.

components of the carbon cycle. The buildup of atmospheric carbon dioxide will warm Earth during the twenty-first century. Like the glass in a greenhouse, carbon dioxide is transparent to sunlight but opaque to radiated heat. Carbon dioxide thus permits sunlight to strike and warm Earth, but it traps some of the heat that Earth radiates back toward space.

The concentration of carbon dioxide in air trapped in the Antarctic and Greenland ice caps during the last Ice Age—between 15,000 and 30,000 years ago—was as low as 200 parts per million; during a warm interval 5,000 years ago it may have been slightly higher than it is today. This long-term record, which shows that Earth was warmer when CO_2 levels were higher than they are today, is a major reason why scientists expect Earth to warm as atmospheric CO_2 levels continue to increase.

In 1988, the United Nations Environment Programme and the World Meteorological Organization founded the Intergovernmental Panel on Climate Change (IPCC). The panel's mandate is to assess scientific and technical information about climate change. In its 1995 report, the work of 2,500 contributing scientists from more than 60 countries, the panel concluded that the average surface temperature over Earth has risen about 0.6°C since the late 1800s. More importantly, all of the 10 warmest years on record since 1860 have occurred during the last 15 years, despite the 3-year global cooling effect caused by the eruption of Mount Pinatubo in the Philippines in 1991, which ejected large amounts of dust into the atmosphere. The associated warming of ocean waters has resulted in a 3- to 6-centimeter rise in global sea level during the last 100 years.

Global climate models are used to predict the likely consequences of further increases in concentrations of atmospheric carbon dioxide. Predictions from these models are imprecise because nature is much more complicated than any model. Nonetheless, if the current concentration of atmospheric carbon dioxide doubles, the mean temperature of Earth is expected to increase 3 to 5°C, with greater increases at higher latitudes. A carbon dioxide doubling would also shift climatic patterns latitudinally, would probably cause droughts in the central regions of continents, and would increase precipitation in coastal areas. Global warming may result in the melting of the Greenland and Antarctic ice caps and will warm the oceans, which will thus expand, raising sea levels and flooding coastal cities and agricultural lands.

Because carbon dioxide is carried by air movements to places thousands of kilometers away from where it enters the atmosphere, the problem is global. Although it is very difficult for societies to decide how much they should invest today to avert potential future climatic problems, some nations have committed themselves to efforts at reducing their emissions of carbon dioxide. Such reductions will not be easy because carbon dioxide is the inevitable end product of fossil fuel combustion, and modern societies are powered by fossil fuels.

The amount of carbon dioxide emitted by a power plant burning fossil fuels cannot be reduced by cleansing of the gases leaving the smokestack. However, because so much energy is currently wasted, many steps can be taken to increase energy use efficiency so that we get more valuable services for the same amount of fuel burned. In addition, we can substitute energy sources that do not contain carbon (solar, wind, geothermal, nuclear) for fossil fuels.

Chemosynthesis-Powered Ecosystems

Most ecosystems are powered by sunlight falling directly on them, but some depend upon sunlight that falls elsewhere. For example, marine ecosystems below the level where enough light penetrates to permit photosynthesis depend on biomass produced in the well-lit zone above them. The productivity of most deep-sea ecosystems is low because only small amounts of detritus descend through the water column to reach them.

Some deep-sea ecosystems are totally independent of sunlight. The most striking are those around hot springs associated with seafloor spreading zones. The energy base of these ecosystems is chemoautotrophy by sulfur-oxidizing bacteria, which obtain energy by oxidizing hydrogen sulfide emitted from the vents. Most of the other organisms in these ecosystems, such as pogonophores, live directly or indirectly on these sulfur-oxidizing bacteria (see Figure 24.7).

Ecologists recently discovered a cave in southern Romania whose ecosystem is powered by bacteria that fix inorganic carbon by using hydrogen sulfide as an energy source. Chemoautotrophic production by these bacteria is the food base for 48 species of cave-adapted terrestrial and aquatic invertebrates. Air pockets in the submerged portions of the cave contain mats of bacteria and fungi that float on the water surface and grow on the limestone walls of the cave.

Summary of "Ecosystems"

• The organisms living in a particular area, together with the physical environment with which they interact, constitute an ecosystem. At a global scale, Earth is a single ecosystem.

Climates on Earth
• Biological processes on Earth are driven primarily by solar radiation.
• Climates determine the amount of heat, moisture, and sunlight available to living organisms in different places on Earth.
• Rising air expands and cools, releasing moisture. Descending air warms and dries and takes up moisture, creating rain shadows. **Review Figure 53.1**
• Global air circulation is driven by solar radiation and the spinning of Earth on its axis. **Review Figure 53.2**
• Oceanic currents are driven primarily by prevailing winds. **Review Figure 53.3**

Energy Flow through Ecosystems
• The capture of solar radiation by photosynthesis powers ecosystem productivity.
• The annual production of an area is determined primarily by temperature and moisture. **Review Figures 53.4, 53.5**
• Energy flows through ecosystems as organisms capture and store energy and transfer it to other organisms when they are eaten. Organisms are grouped into trophic levels according to the number of steps through which energy passes to get to them. **Review Table 53.1**
• Who eats whom in a ecosystem can be diagrammed as a food web. **Review Figure 53.6**
• The amount of energy flowing through an ecosystem depends on primary production and on the efficiency of transfer of energy from one trophic level to another. **Review Figures 53.7, 53.8**

Agricultural Manipulation of Ecosystem Productivity
• Humans manipulate ecosystem productivity by increasing rates of photosynthesis and reducing losses of crops to pests. In modern agriculture, the energy required to do this is provided by fossil fuels. **Review Figure 53.9**

Cycles of Materials in Ecosystems
• The main compartments of the global ecosystem are the oceans, fresh waters, land, and atmosphere, among which materials are constantly being exchanged.
• Primary production in oceans is highest adjacent to continents, where nutrient-rich waters rise to the surface. **Review Figure 53.10**
• Temperate-zone lakes turn over twice each year as water cools and warms. **Review Figure 53.11**
• The two lowest layers of Earth's atmosphere differ from each other in their circulation patterns, the amount of moisture they contain, and the amount of ultraviolet radiation they receive. **Review Figure 53.12**

Biogeochemical Cycles
• The elements organisms need in large quantities cycle though organisms to the environment and back again.
• The cycle of water—the hydrological cycle—is driven by evaporation of water, most of it from ocean surfaces. **Review Figure 53.13**
• Atmospheric carbon dioxide is the immediate source of carbon for terrestrial organisms, but only a small part of Earth's carbon is in the atmosphere. **Review Figure 53.14**
• Although nitrogen makes up 78 percent of Earth's atmosphere, nitrogen can be converted into biologically useful forms only by a few species of bacteria and cyanobacteria. **Review Figure 53.15**
• The phosphorus cycle differs from the cycles of carbon and nitrogen in that it lacks a gaseous phase. **Review Table 53.3**

Human Alterations of Biogeochemical Cycles
• The most striking example of a local effect of altered biogeochemical cycles is lake eutrophication.
• Acid precipitation is an important regional consequence of human modifications of the nitrogen and sulfur cycles. **Review Figure 53.17**
• Earth's climate is being changed as a result of increasing concentrations of carbon dioxide in the atmosphere. **Review Figure 53.18**

Chemosynthesis-Powered Ecosystems
• A few deep-sea and cave ecosystems are powered by chemosynthesis rather than photosynthesis.

Self-Quiz
1. Which of the following is true about the amount of sunlight and heat arriving on Earth?
   a. Every place on Earth receives the same annual number of hours of sunlight and the same amount of heat.
   b. Every place on Earth receives the same annual number of hours of sunlight, but not the same amount of heat.
   c. Every place on Earth receives the same annual amount of heat, but not the same number of hours of sunlight.
   d. Both the annual amount of sunlight and the amount of heat received vary over the surface of Earth.
   e. None of the above

2. When an area is within the intertropical convergence zone,
   a. the northeast trade winds blow steadily.
   b. the southeast trade winds blow steadily.
   c. air is descending and it seldom rains.
   d. air is rising and heavy rains fall frequently.
   e. westerly winds blow steadily.

3. Zones of marine upwelling are important because
   a. they help scientists measure the chemistry of deep ocean water.
   b. they bring to the surface organisms that are difficult to observe elsewhere.
   c. ships can sail faster in these zones.
   d. they increase marine productivity by bringing nutrients back to surface ocean waters.
   e. they bring oxygenated water to the surface.

4. Which of the following is *not* true of the troposphere?
   a. It contains nearly all atmospheric water vapor.
   b. Materials enter it primarily at the intertropical convergence zone.
   c. It is about 17 km deep in the tropics.
   d. Most global atmospheric circulation takes place there.
   e. It contains about 80 percent of the mass of the atmosphere.

5. Carbon dioxide is called a greenhouse gas because
   a. it is used in greenhouses to increase plant growth.
   b. it is transparent to heat radiation but opaque to sunlight.
   c. it is transparent to sunlight but opaque to heat radiation.
   d. it is transparent to both sunlight and heat radiation.
   e. it is opaque to both sunlight and heat radiation.

6. The phosphorus cycle differs from those of carbon and nitrogen in that
   a. it lacks a gaseous phase.
   b. it lacks a liquid phase.
   c. only phosphorus is cycled through marine organisms.
   d. living organisms do not need phosphorus.
   e. The phosphorus cycle does not differ importantly from the carbon and nitrogen cycles.

7. Acid precipitation results from human modifications of
   a. the carbon and nitrogen cycles.
   b. the carbon and sulfur cycles.
   c. the carbon and phosphorus cycles.
   d. the nitrogen and sulfur cycles.
   e. the nitrogen and phosphorus cycles.

8. The total amount of energy that plants assimilate by photosynthesis is called
   a. gross primary production.
   b. net primary production.
   c. biomass.
   d. a pyramid of energy.
   e. eutrophication.

9. The amount of energy reaching an upper trophic level is determined by
   a. net primary production.
   b. net primary production and the efficiencies with which food energy is converted to biomass.
   c. gross primary production.
   d. gross primary production and the efficiencies with which food energy is converted to biomass.
   e. gross primary production and net primary production.

10. Which of the following is *not* a component of integrated pest management?
    a. Use of cultural strategies such as crop rotation and mixed plantings
    b. Use of pest-resistant strains of crops
    c. Use of predators and parasites of crop pests
    d. Use of chemical attractants
    e. Use of chemical pesticides whenever pests are discovered

Applying Concepts

1. How would you expect temperature and oxygen profiles to appear in a broad, shallow tropical lake? In a very deep tropical lake? Why?

2. The waters of Lake Washington, adjacent to the city of Seattle, rapidly returned to their preindustrial condition when sewage was diverted from the lake to Puget Sound, an arm of the Pacific Ocean. Would all lakes being polluted with sewage clean themselves up as rapidly as Lake Washington if pollutant input were stopped? What characteristics of a lake are most important to its rate of recovery following removal of pollutant inputs? What is the diverted sewage likely to do to Puget Sound?

3. Tropical forests currently are being cut at a very rapid rate. Does this necessarily mean that deforestation is a major source of input of carbon dioxide to the atmosphere? If not, why not?

4. The two drawings below represent pyramids of biomass for (*a*) an old field in Georgia and (*b*) the English Channel. Explain the significance of the inversion of the second pyramid compared with the first.

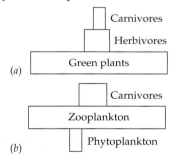

5. The amount of energy flowing through a food chain declines more or less rapidly depending upon the nature of the organisms in the chain. Which of the following simplified food chains is likely to be more efficient? Why? What criteria of efficiency are you using?
   a. phytoplankton → zooplankton → herring
   b. shrubs → deer → wolf

6. A government official authorizes the construction of a large power plant in a former wilderness area. Its smokestacks discharge great quantities of waste resulting from the combustion of coal. List and describe all *likely* ecological results at local, regional, and global levels. Now suppose the wastes were thoroughly scrubbed from the stack gases. Which of the ecological results you have just outlined would still happen?

Readings

Anderson, D. M. 1994. "Red Tides." *Scientific American*, August. Shows that the frequency of red tides, which can release potent toxins into the oceans, is increasing because pollution provides rich nutrients for the red tide organisms.

Bazzaz, F. A. and E. D. Fajer. 1992. "Plant Life in a CO_2-Rich World." *Scientific American*, January. Discusses the fact that even without considerations of global warming, increasing atmospheric levels of carbon dioxide may greatly alter the structure and functioning of ecosystems. These changes will not necessarily benefit plants.

Berner, R. A. and A. C. Lasaga. 1989. "Modeling the Geochemical Carbon Cycle." *Scientific American*, March. Discusses how natural geochemical processes that result in the slow buildup of atmospheric carbon dioxide may have caused past geologic intervals of global warming through the greenhouse effect.

Broeker, W. S. 1995. "Chaotic Climate." *Scientific American*, November. Describes the geologic record that shows that Earth's weather patterns have sometimes changed dramatically in a decade or less.

Butcher, S. S., R. J. Charlson, G. H. Orians and G. V. Wolfe (Eds). 1992. *Global Biochemical Cycles*. Academic Press, New York. A comprehensive review of all global biogeochemical cycles and the methods used to study them.

Gates, D. M. 1993. *Climate Change and Its Biological Consequences*. Sinauer Associates, Sunderland, MA. An introduction to Earth's climates, past and present, and the effect climate change has on organisms and ecosystems. An accessible explanation of the various computer models that predict massive climate warming in the twenty-first century.

Jordan, C. F. 1985. *Nutrient Cycling in Tropical Forest Ecosystems*. Wiley, New York. A useful summary of nutrient cycling patterns in tropical forests and how they differ from those of temperate forests.

Kusler, J. A., W. J. Mitsch and J. S. Larson. 1994. "Wetlands." *Scientific American*, January. Discusses how wetlands purify water, reduce floods, and serve as incubators for aquatic life. Assesses what happens when the protection of an essential natural resource must be weighed against society's demand for real estate, fossil fuels, and agricultural land.

Makarewicz, J. C. and P. Bertram. 1991. "Evidence for the Restoration of the Lake Erie Ecosystem." *BioScience*, vol. 41, pages 216–223. Describes changes in water quality, oxygen levels, and population dynamics of microorganisms, plants, and animals in Lake Erie.

Mohnen, V. A. 1988. "The Challenge of Acid Rain." *Scientific American*, August. Argues that acid rain's effects on soil and water leave no doubt that we need to control its causes, and shows how new advances in technology offer attractive solutions.

Nicol, S. and W. de la Mare. 1993. "Ecosystem Management and the Antarctic Krill." *American Scientist*, vol. 81, pages 36–47. Discusses how krill's unusual biology and ecological significance are driving an effort to apply systems theory to managing an ecosystem.

Rambler, M. B., L. Margulis and R. Fester (Eds.). 1989. *Global Ecology: Towards a Science of the Biosphere*. Academic Press, Boston. A collection of essays covering a wide variety of the interactions between organisms, global biogeochemical cycles, and global climate.

Rietzler, K. and I. C. Feller 1996. "Caribbean Mangrove Swamps." *Scientific American*, March. Describes the remarkable ecosystems that exist in shallow water on muddy tropical ocean shores.

Robison B. H. 1995. "Light in the Ocean's Midwater." *Scientific American*, July. Describes a rich and remarkable community of organisms below the photic zone that is illuminated only by the radiance of its residents.

Schlesinger, W. 1997. *Biogeochemistry*, 2nd Edition. Oxford University Press, Oxford. A thorough review of all aspects of biogeochemistry.

Schneider, S. H. 1989. "The Changing Climate." *Scientific American*, September. Argues that global warming should be unmistakable within a decade or two and shows how prompt emission cuts could slow the buildup of heat-trapping gases.

Whittaker, R. H. 1975. *Communities and Ecosystems*, 2nd Edition. Macmillan, New York. A good, short textbook with excellent coverage of theoretical topics; an excellent follow-up to the materials in this chapter.

Chapter 54

Biogeography

Phascolarctos cinerus

Strange Organisms in New Places
Animals like the koala were unfamiliar to the Europeans who colonized Australia.

When the first Europeans arrived in Australia, they saw plants and animals that differed in perplexing ways from the ones they had known at home, such as flowers pollinated by brush-tongued parrots and mammals that hopped around on their hind legs, carrying their young in pouches. On the other hand, the first Europeans to visit North America felt more at home because the plants and animals of North America were similar to those of Europe.

During their worldwide travels, European explorers found many vegetation types—tropical forests, mangrove forests, and deserts with tall cacti—that were unfamiliar to them, but they also found many areas where the vegetation was similar to what they knew back home, even though they seldom recognized any familiar species. The study of the diversity of organisms over space began when those eighteenth-century travelers who first noted intercontinental differences in biotas attempted to understand those differences.

Biogeography is the science that attempts to explain patterns in the variation of individuals, populations, species, and communities along geographic gradients. In this chapter, we will show how biogeographers determine how events in the remote past influenced distributional patterns today, and how biological differences among species in different geographic regions influence the species composition of ecological communities in different regions of Earth.

1195

Why Are Species Found Where They Are?

Superficially, explaining species' distributions seems to be a simple matter because the question of why a species is or is not found in a certain location has only a few possible answers. If a species occupies a particular area, either it evolved there, or it evolved elsewhere and dispersed to the area. If a species is *not* found in a particular area, either it evolved elsewhere and never dispersed to the area, or it was once present in the area but no longer lives there. Unfortunately, determining which of these answers is correct turns out to be difficult.

To explain the distributions of organisms, biogeographers must draw upon and interpret a broad array of knowledge. Finding the answers to the questions listed above requires information about the evolutionary histories of species, which comes from fossils and from knowledge of phylogenetic relationships (see Chapter 22). In addition, it requires information about Earth's changes—continental drift, glacial advances and retreats, sea level changes, and mountain building—during the time when the organisms were evolving. Such geologic information can tell us whether organisms evolved where they are currently found or dispersed and colonized new areas from a distant area of origin. In this section we show how the acceptance of continental drift and the use of cladistic methods of analysis changed the science of biogeography.

Ancient events influence current distributions

Early biogeographers believed in a relatively constant Earth that was too young to account for the diversity and distribution of life by any means except divine creation. They assembled much valuable information, but their interpretations were constrained by their belief that organisms had been created in their current forms in their current ranges. Linnaeus, for example, believed that all organisms had been created in one place—which he called Paradise—from which they later dispersed. Indeed, because most people believed that the continents were fixed in their positions, the only way to account for current distributions was to invoke massive dispersal.

The notion that the continents might have moved was not seriously considered until 1912. Alfred Wegener, the German meteorologist who argued that the continents had drifted, based his ideas on the shapes of continents (the outlines of Africa and South America seem to fit together like pieces of a puzzle), the alignment of mountains chains and rock strata, coal beds, glacial deposits, and biogeography (the distributions of plants and animals in Africa and South America were hard to explain if one assumed that the continents had always been where they are now).

When Wegener proposed his ideas, few scientists took them seriously, primarily because there were no known mechanisms for continental movement and because no convincing geologic evidence of such movements existed. As we learned in Chapter 22, geologic evidence and plausible mechanisms were eventually discovered, and the broad pattern of continental movement is now clear.

About 280 million years ago, the continents were united to form a single land mass, Pangaea. By the early Mesozoic era (about 245 million years ago), when the continents were still very close to one another, many groups of nonmarine organisms, including insects, freshwater fishes, and frogs, had already evolved. Some organisms that live on widely separated continents today were probably present on those land masses when they were all part of Pangaea.

By 100 mya, Pangaea had separated into northern (Laurasia) and southern (Gondwanaland) land masses, and the southern continents were drifting away from each other (see Figure 19.16). Eventually, continental drift, which continues today, brought India into contact with southern Asia, Australia closer to Southeast Asia, and South America into contact with North America (see Figure 19.17). Continental drift has thus influenced biogeographic patterns throughout the history of life on Earth.

Modern biogeographic methods

As the great age of Earth and the fact of evolution began to be understood, two groups of investigators developed new methods for generating testable hypotheses about geographic distributions. One group consisted of ecologists who studied how current distributions were influenced by interactions among species and by interactions between species and their physical environments. Because local habitats always contain fewer species than the number living in the surrounding region, these ecologists believed that interactions of the types discussed in Chapter 52 could explain many aspects of the distributions of organisms.

The other group consisted of historical biogeographers who used new techniques to investigate ancient influences on distributions. An important technique they developed was the transformation of taxonomic cladograms into **area cladograms** by substituting the species' geographic distributions for their names. The distribution patterns identified by area cladograms may suggest routes of dispersal or point to the splitting of biota due to the appearance of major barriers (Figure 54.1). Comparisons of area cladograms of many evolutionary lineages can demonstrate similarities and differences in their distribution patterns. Similarities suggest common responses to physical events, such as continental drift, mountain building, and sea level changes. Differences suggest that organisms in different lineages responded in different ways,

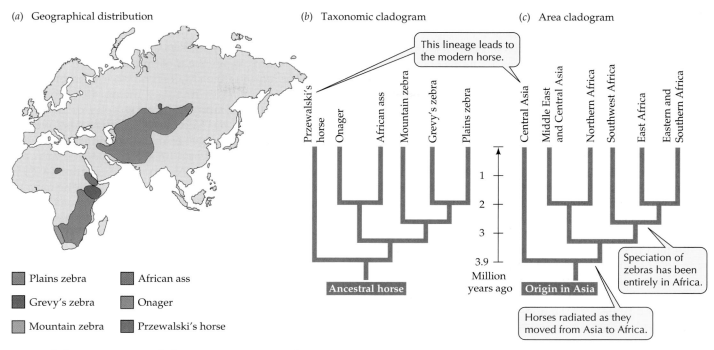

(a) Geographical distribution (b) Taxonomic cladogram (c) Area cladogram

This lineage leads to the modern horse.

Przewalski's horse
Onager
African ass
Mountain zebra
Grevy's zebra
Plains zebra

Central Asia
Middle East and Central Asia
Northern Africa
Southwest Africa
East Africa
Eastern and Southern Africa

1
2
3
3.9
Million years ago

Ancestral horse

Origin in Asia

Speciation of zebras has been entirely in Africa.

Horses radiated as they moved from Asia to Africa.

- Plains zebra
- Grevy's zebra
- Mountain zebra
- African ass
- Onager
- Przewalski's horse

54.1 Cladogram to Area Cladogram Conversion of taxonomic cladograms to area cladograms can aid in the understanding of distribution patterns.

or at different times, to past events, or that they have dispersed in unique ways.

These ecological and evolutionary approaches characterize the two main subdisciplines of biogeography: historical and ecological biogeography. *Historical biogeography* concerns itself primarily with the evolutionary histories of lineages of organisms: Where and when did they originate? How did they spread? What does their present-day distribution tell us about their past histories? *Ecological biogeography* concentrates on the current interactions of organisms with the physical environment and with one another. Ecological biogeographers seek to understand how ecological relationships influence where species and higher taxa are found today.

The names of these two subdisciplines are somewhat misleading. All biogeography is historical. Ecological biogeographers concentrate on recent history, current interactions, and changes within the past few thousand years. They also study patterns of distribution within local areas and regions. Historical biogeographers concentrate on longer time periods and larger spatial scales. Because time and space are continuous variables, these two subdisciplines blend together. The integration of knowledge from both of these subdisciplines is essential to a full understanding of the geographic distributions of organisms. We will first discuss the methods and findings of historical biogeographers.

Historical Biogeography

Historical biogeographers attempt to determine the influence of past events on today's patterns of distribution. We can never know past events with complete certainty, but by using a variety of types of evidence, historical biogeographers can develop, test, and adopt interpretations in which they have a high degree of confidence. As we have just seen, historical biogeographers often base their interpretations on phylogenies, which show the evolutionary relationships among organisms in a lineage. Phylogenies are most useful to biogeographers if the approximate times of evolutionary and geographic separations of lineages can be estimated.

Biogeographers use several approaches to infer the approximate times of separation of taxa within a lineage. First, if a "molecular clock" has been ticking, the degree of difference in the molecules of species is strongly correlated with the length of time their lineages have been separated (see Chapter 23). Second, fossils can show a biogeographer how long a taxon has been present in an area and whether its members formerly lived in areas where they are no longer found. The fossil record is helpful, but it is always incomplete. The first and last members of a taxon to live in an area are extremely unlikely to have become fossils that are discovered and described.

A third valuable source of information is the distributions of living species. Much more complete and extensive information can be gathered on such distributions than will ever be available from fossils. Much can be learned by examining the distribution patterns of *many different groups* of living organisms. Similarities in their distributions provide clues about past events that affected many of them.

Biogeographers wish to explain general patterns. To do so, they must compare the distributions of many different species, they must know the species' evolutionary relationships, and they must know the timing of geologic events that might have affected many species.

Vicariance and dispersal can both explain distributions

A species may be found in an area either because it evolved there or because it dispersed to the area. The appearance of a barrier that splits a species' previously continuous distribution is called a **vicariant event**, and the species is said to have a **vicariant distribution**. If, on the other hand, a barrier existed prior to the expansion of a species' range, and members of the species subsequently crossed it and established a new population, the species is said to have a **dispersal distribution**.

By studying a single lineage, a biogeographer may discover evidence suggesting that the distributions of ancestral species were influenced by some vicariant event, such as changes in sea level, mountain building, or continental movement. If that inference is correct, other lineages should also have been influenced by the same event—that is, they should have similar distribution patterns. Differences in distribution patterns among lineages indicate either that the lineages responded differently to the same vicariant events, or that the lineages separated at different times. By analyzing such similarities and differences, biogeographers can discover how vicariant events and dispersal may have influenced today's distribution patterns.

Species, genera, and families found in only one region are said to be **endemic** to that location. As far as we know, all species are endemic to Earth. Some species are endemic to one continent. Others are restricted to very small areas, such as tiny islands or single mountaintops. Because a species may disperse widely and then die out where it originated, biogeographers cannot assume that a species now endemic to a region actually evolved there. Endemic taxa can be very old ones that are in the process of becoming extinct, or very young taxa that have recently evolved in a restricted area.

The longer an area has been isolated from other areas by a vicariant event, such as continental drift, the more endemic taxa it is likely to have, because there has been more time for evolutionary divergence to take place. Australia, which has been separated the longest from the other continents (about 65 million years) has the most distinct biota. South America has the next most distinct biota, having been isolated from other continents for nearly 60 million years. North America and Eurasia, which were joined together for much of Earth's history, have very similar biotas. That is why the early European travelers felt more at home in North America than in Australia.

Biogeographers use parsimony to explain distributions

When several hypotheses can explain a pattern, scientists typically prefer the most parsimonious one—the one that requires the smallest number of unobserved events to account for it. To see the application of the **principle of parsimony** to biogeography, consider the distribution of the New Zealand flightless weevil *Lyperobius huttoni*, a species that is found in the mountains of South Island and on sea cliffs at the extreme southwestern corner of North Island (Figure 54.2). If you knew only its current distribution and the current positions of the two islands, you might guess, even though this weevil cannot fly, that *L. huttoni* had somehow managed to cross Cook Strait, the 25-kilometer body of water that separates the two islands.

However, more than 60 other animal and plant species, including other species of flightless insects, live on both sides of Cook Strait. It is unlikely that all of these species made the same ocean crossing. In fact, that assumption is unnecessary to explain the distribution patterns. Geologic evidence indicates that the present-day southwestern tip of North Island was formerly united with South Island. Therefore, none of the 60 species need have made a water crossing. A single vicariant event, the separation of the northern tip of South Island from the remainder of the island by the newly formed Cook Strait, could have split all of the distributions.

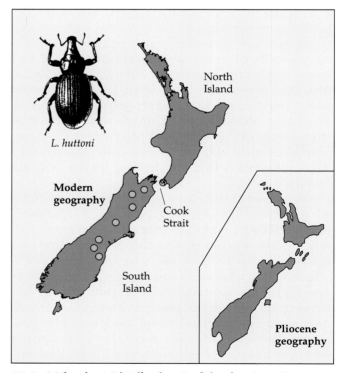

54.2 A Vicariant Distribution Explained The yellow circles indicate the current distribution of the weevil *Lyperobius huttoni*. Compare the present New Zealand geography with that in the Pliocene.

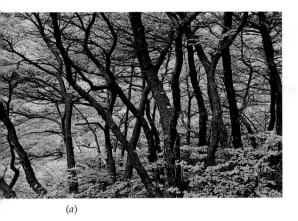

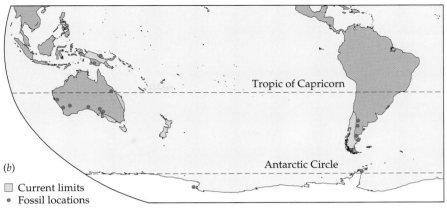

(b)

☐ Current limits
● Fossil locations

54.3 Southern Beeches Were Carried by Drifting Continents (a) *Nothofagus*, the southern beech tree, dominates the wet forests of southern Argentina. (b) Current and fossil distributions of *Nothofagus*.

Although organisms *do* cross major oceanic and terrestrial barriers, biogeographers often apply the rule of parsimony when interpreting distribution patterns, just as evolutionists do when reconstructing phylogenies (see Chapter 22). Although one vicariant event could have separated many formerly continuous distributions, some of the species found on both South and North Island, such as birds and flying insects, may indeed have dispersed across Cook Strait. However, in the absence of evidence indicating that they did, most biogeographers

favor the more parsimonious interpretation. The principle of parsimony is a useful operating rule, but evolutionary history has not always been parsimonious. Other evidence may strongly support a less parsimonious explanation of current distributions of organisms.

Biogeographic histories are reconstructed from various kinds of data

Biogeographers use distribution maps, phylogenies, and knowledge of paleoclimates and paleogeography to reconstruct the biogeographic histories of taxa. This kind of information suggests, for example, that the current distributions of southern beeches (genus *Nothofagus*: family Nothofagaceae; Figure 54.3a) were influenced by continental drift many millions of years ago. Today southern beeches are found in South America, Australia, Tasmania, New Zealand, New Guinea, New Britain, and New Caledonia (Figure 54.3b).

An extensive fossil pollen record shows that southern beeches were distributed across Gondwanaland when South America, Antarctica, Australia, and the associated islands were still united. But they apparently arrived there after Africa and India had drifted away from the other southern continents, because no fossil pollen of *Nothofagus* has been found on those two land masses. *Nothofagus* subsequently became extinct in Antarctica when it became too cold to support trees.

Another taxon of woody plants whose current distribution was influenced by the breakup of Gondwanaland is the proteads (family Proteaceae; Figure 54.4). Unlike

(a)

Leucospermum conocarpodendron   *Banksia integrifolia*

South American species have their closest relatives in Australia.

(b) Current range

54.4 Protead Distributions Reveal a Gondwanaland Ancestry (a) Proteads from South Africa (left) and Australia (right). (b) Current distribution of the family Proteaceae.

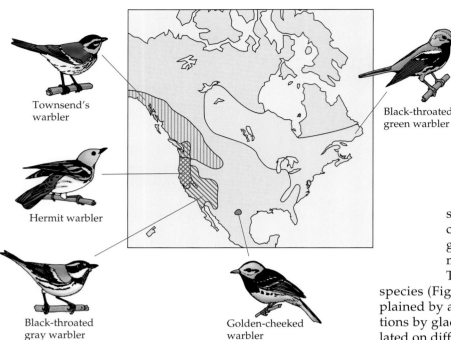

54.5 Why Are There Five Species?
The present-day breeding ranges of five closely related species of warblers suggested that glaciers may have isolated the western and eastern populations of a widespread ancestral species.

Nothofagus, proteads are found in Africa, but the African species are highly specialized members of an endemic subfamily. The South American species are members of a different subfamily, whose closest relatives are found in the Australian region. Thus, the phylogeny and distribution of the proteads suggests that they had a broad distribution in Gondwanaland earlier than *Nothofagus* did, before that large land mass began to separate.

Similar combinations of geologic and phylogenetic information can be used to investigate distribution patterns that have been strongly influenced by more recent events. For example, biogeographers have investigated the possibility that recent advances and retreats of glaciers in North America initiated speciation events among songbirds. During times of glacial advance, populations of many forest-dwelling birds were divided into two segments, one in the western mountains and the other in the Appalachian mountains, both far south of the ice. One possibility is that populations isolated by glaciation evolved enough differences during their separation that they failed to interbreed when they again became sympatric as the glaciers retreated.

A biogeographer postulated that several such advances and retreats of glaciers could explain the speciation pattern and current distributions of species in the black-throated green warbler complex (Figure 54.5). No fossils of these warblers exist, but a cladogram, based on their mitochondrial DNA, shows that the black-throated gray warbler lineage was the earliest to separate from the remainder of the lineage. This split preceded a glacial advance, which could have separated the black-throated green warbler from the ancestor of the hermit and Townsend's warblers. Hermit and Townsend's warblers, however, are sister species (Figure 54.6). Their separation cannot be explained by a splitting of eastern and western populations by glacial advances. They probably became isolated on different western mountain ranges by climate changes associated with glacial advances.

Earth can be divided into major terrestrial biogeographic regions

Although the drifting continents carried many kinds of organisms with them, the continents have been isolated from one another long enough to have evolved

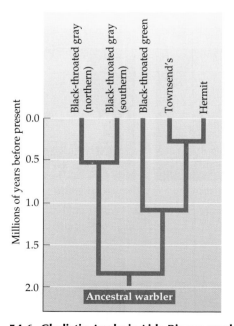

54.6 Cladistic Analysis Aids Biogeographic Reconstruction This cladogram, constructed from mitochondrial DNA data, shows that some lineage splits in the warblers shown in Figure 54.5 preceded the Pleistocene glacial advances.

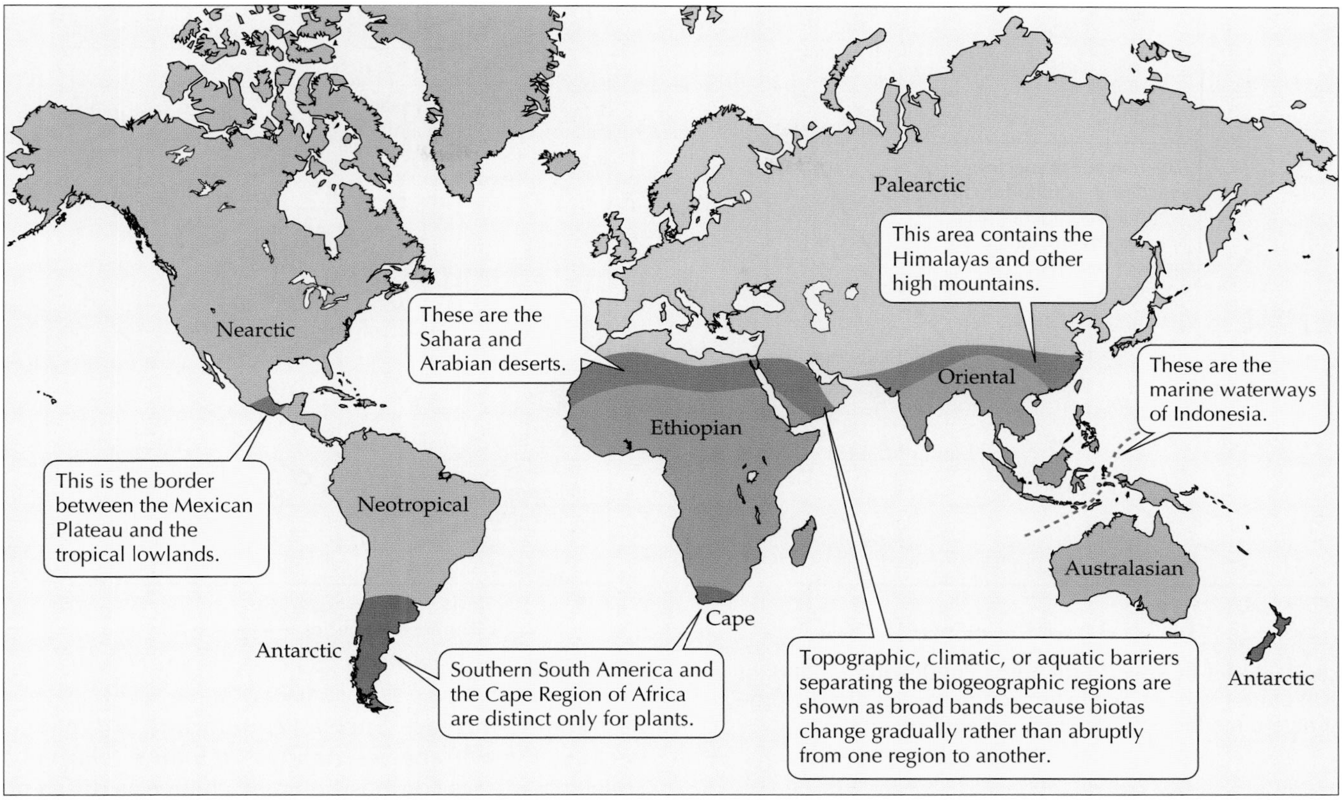
Palearctic
This area contains the Himalayas and other high mountains.
These are the Sahara and Arabian deserts.
Nearctic
Oriental
These are the marine waterways of Indonesia.
Ethiopian
This is the border between the Mexican Plateau and the tropical lowlands.
Neotropical
Australasian
Cape
Antarctic
Southern South America and the Cape Region of Africa are distinct only for plants.
Topographic, climatic, or aquatic barriers separating the biogeographic regions are shown as broad bands because biotas change gradually rather than abruptly from one region to another.
Antarctic

54.7 Major Biogeographic Regions The biotas of Earth's major biogeographic regions differ strikingly from one another, even though changes may be gradual.

distinct biotas. The differences among continental biotas were first recognized more than two centuries ago, and later formed the basis for dividing Earth into major biogeographic regions. Biogeo-graphers drew the boundaries of these regions where species compositions change dramatically over short distances (Figure 54.7).

All biogeographers agree on the boundaries of many of these regions, but plant biogeographers recognize two regions not used by zoogeographers: southern South America and the Cape Region of South Africa. The floras of these two regions are very distinct from those of adjacent areas on those continents, but the faunas of southern South America and the Cape Region are very similar to those of the remainder of those continents.

Except for the Australian region, the biogeographic regions are no longer separated from each other by water, as they were in the past (see Figure 19.16). The biological distinctness of these biogeographic regions is maintained today in part by mountain and desert barriers to dispersal and in part by major changes in climates over short distances.

Ecological Biogeography

Ecological biogeographers use the wealth of available information on current distributions of organisms to test theories that explain the numbers of species in different communities, the ways in which species disperse, and the effects of different types of barriers. They can also use experiments to test hypotheses, something that historical biogeographers cannot do.

First we will discuss a model that attempts to account for the number of species living in an area—its **species richness**—and then we will look at experiments conducted to test this model.

The species richness of an area is determined by rates of immigration and extinction

Over periods of a few thousand years (during which speciation is unlikely), the species richness of an area is influenced by the immigration of new species and the extinction of species already present. It is easiest to visualize the effects of these two processes if we consider, as did Robert MacArthur and Edward Wilson, an oceanic island that initially has no species.

Imagine a newly formed oceanic island that receives colonists from a mainland area. The list of species on the mainland that might possibly invade the island is called the *species pool*. The first colonists to arrive on the island are all "new" species because no

(a)

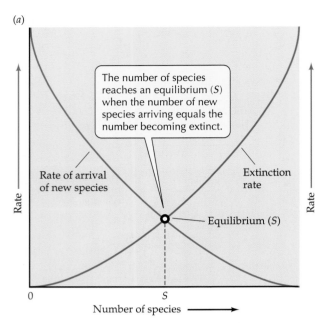

(b)

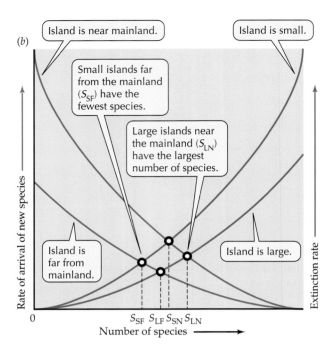

54.8 A Model of Species Richness Equilibrium Rates of arrival of new species and extinction of species already present determine the equilibrium number of species.

species live there initially. As the number of species on the island increases, a larger fraction of colonists will be members of species already present, so even if the same number of species arrive as before, the rate of arrival of *new* species decreases, until it reaches zero when the island has all the species in the species pool.

Now consider extinction rates. First there will be only a few species on the island, and their populations may grow large. As more species arrive and their numbers increase, the resources of the island will be divided among more species. We therefore expect the average population size of each species to become smaller as the number of species increases. The smaller a population, the more likely it is to become extinct. In addition, the number of species that can become extinct increases as species accumulate on the island. New arrivals to the island may include pathogens and predators that increase the probability of extinction of other species, further increasing the number of species becoming extinct per unit of time.

Because the rate of arrival of new species decreases and the extinction rate increases as the number of species increases, eventually the number of species should reach an equilibrium at which the rates of arrival and extinction are equal (Figure 54.8a). If there are more species than the equilibrium number, extinctions should exceed arrivals, and species richness should decline toward the equilibrium number. If there are fewer species than the equilibrium number, arrivals should exceed extinctions, and species rich-

ness should increase. Such an equilibrium is dynamic because even if species richness remains relatively constant, species composition may change as different species replace those that become extinct.

This equilibrium model does not predict which species will arrive and which will become extinct. It predicts only the equilibrium number of species if arrival and extinction rates are known and are constant. If either rate fluctuates, the equilibrium point shifts up and down.

If extinction and immigration rates are relatively constant, the equilibrium model can be used to predict how species richness should differ among islands of different sizes and different distances from the mainland. We expect extinction rates to be higher on small islands than on large islands because species' populations would, on average, be smaller there. Similarly, we expect fewer immigrants to reach islands more distant from the mainland. Figure 54.8b gives relative species richness equilibria for islands of different sizes and distances from the mainland. As you can see, the equilibrium number of species should be highest for islands that are relatively large and relatively close to the mainland.

The species richness equilibrium model has been tested

The species richness equilibrium model predicts that the equilibrium number of species should increase with island size and decrease with distance from the mainland. Bird species on islands in the Pacific Ocean exhibit these patterns (Figure 54.9). New Guinea supplies the mainland species pool for most of these small

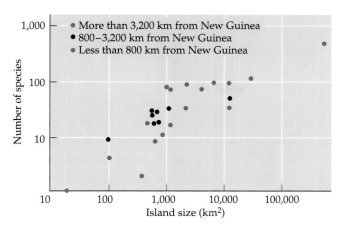

54.9 Small, Distant Islands Have Fewer Bird Species
The dots show the numbers of land and freshwater bird species on islands of different sizes in the Moluccas, Melanesia, Micronesia, and Polynesia. These islands have been divided into three groups according to their distance from the mainland, New Guinea.

islands. The species richness patterns of plants, insects, lizards, and mammals on Pacific islands are similar to those of birds.

The predictions of the equilibrium model are based on assumptions about rates of colonization and extinction. Major disturbances, which serve as "natural experiments," sometimes permit immigration and extinction rates to be estimated directly. The eruption of Krakatau in August 1883, described in Chapter 19, destroyed all life on the island's surface. After the lava cooled, Krakatau was colonized rapidly by plants and animals from Sumatra to the east and Java to the west. By 1933 Krakatau was again covered with a tropical evergreen forest, and 271 species of plants and 27 species of resident land birds were found there. During the 1920s, a forest canopy was developing, and there

were high rates of immigration of both birds and plants to Krakatau (Table 54.1). Birds probably brought the seeds of many plants because, between 1908 and 1934, both the percentage (from 20 to 25) and the absolute number (from 21 to 54) of plant species with bird-dispersed seeds increased.

Today the numbers of species of plants and birds are not increasing as fast as they did during the 1920s, and they may be approaching equilibrium on Krakatau. Future biological censuses of Krakatau will continue to measure the arrival and extinction rates of species and will show if and when different taxa reach an equilibrium number of species.

A manipulative experiment testing the equilibrium model of species richness was carried out in the Florida Keys, a region dotted with thousands of small islands consisting entirely of red mangrove trees rooted in shallow water. Six tiny islands were fumigated with methyl bromide, which destroyed all arthropods living on them (Figure 54.10). Methyl bromide decomposes rapidly and does not inhibit recolonization. The design of this experiment permitted the investigators to measure arrival and extinction rates directly. They found that rates of recolonization of the islands by arthropods were very high. Within a year the fumigated islands had about their original number of species, a result supporting the equilibrium model.

Habitat islands influence species richness

Our discussion of the equilibrium species richness model has used as examples oceanic islands, where water provides barriers to dispersal. Mainland areas are full of **habitat islands**, which are patches of habitat separated from other similar patches by different types of habitat. For many species living in habitat islands, such as ponds or montane forests surrounded by deserts, the intervening areas may be as unsuitable to live in as if they were covered with water.

54.10 Testing the Equilibrium Species Richness Model Scaffolding is erected by scientists to enclose a small mangrove island in the Florida Keys. Methyl bromide introduced into the enclosure killed all arthropods inside it. When the enclosure was removed, arthropods quickly recolonized the island.

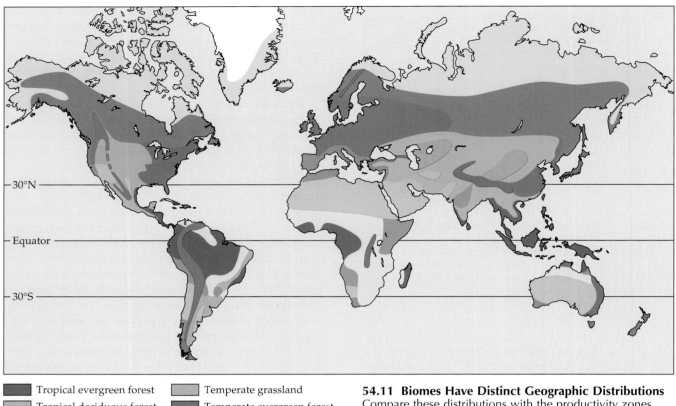

Tropical evergreen forest
Tropical deciduous forest
Tropical thorn forest
Savanna
Hot desert
Chaparral
Cold desert

Temperate grassland
Temperate evergreen forest
Temperate deciduous forest
Boreal forest
Tundra
Alpine
Polar ice cap

54.11 Biomes Have Distinct Geographic Distributions
Compare these distributions with the productivity zones shown in Figure 53.5.

logenetically. Although biomes are named for and identified by their characteristic vegetation, sometimes supplemented by their location or climate, each biome contains many species of organisms in other kingdoms adapted to its physical environment and the physical structure provided by the plants.

Biomes are identified by their distinctive climates and dominant plants

Because climate plays a key role in determining which types of plants live in a given environment, the distri-

The equilibrium species richness model can be applied to habitat islands if we recognize that some intervening areas, though unsuitable for permanent occupancy, may nonetheless permit a brief stopover. Therefore, arrival rates are higher for most habitat islands than they are for similarly sized oceanic islands. We will discuss the importance of habitat islands to the conservation of species richness in the next chapter.

Terrestrial Biomes

Ecologists apply the name **biomes** to major ecosystem types that differ from one another in the structure of their dominant vegetation. The vegetation of a biome appears similar wherever that biome is found, but the plant species in these communities, despite their similarities, may not be closely related phy-

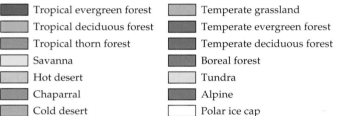

TABLE 54.1 **Number of Species of Resident Land Birds on Krakatau**

PERIOD	NUMBER OF SPECIES	EXTINCTIONS	COLONIZATIONS
1908	13		
1908–1919		2	17
1919–1921	28		
1921–1933		3	4
1933–1934	29		
1934–1951		3	7
1951	33		
1952–1984		4	7
1984–1996	36		

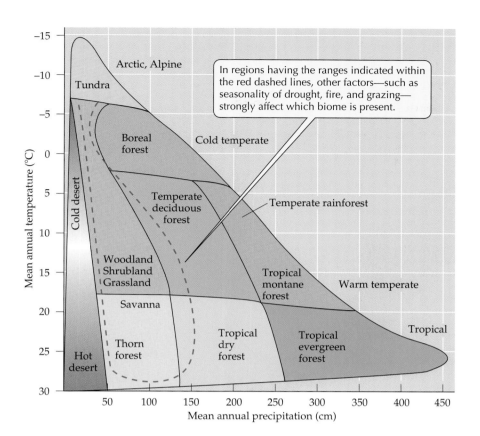

In regions having the ranges indicated within the red dashed lines, other factors—such as seasonality of drought, fire, and grazing—strongly affect which biome is present.

54.12 Most Biomes Have a Distinct Temperature and Precipitation Range Temperature and precipitation play key roles in determining which types of plants live in a given environment. The Mediterranean biome is not included because its distribution depends on the *seasonal* distribution of rainfall, a climate feature not shown here.

things, communities of streams, lakes, marshes, salt flats, dry slopes with shallow soils, moist valleys with deep soils, farmlands, pastures, and cities. Biomes usually blend into and intermingle with one another; sharp boundaries between them are rare. Nonetheless, by recognizing major biomes, we draw attention to the ecosystems that would predominate if natural processes had not been disturbed.

bution of biomes on Earth is strongly influenced by annual patterns of temperature and rainfall. The geographic distribution of biomes is shown in Figure 54.11; their distribution in relation to mean annual temperature and mean annual precipitation is shown in Figure 54.12. Taken together, these two figures show how each biome fits into global climate types.

In some biomes, such as temperate deciduous forest, precipitation is relatively constant throughout the year, but temperature varies strikingly between summer and winter. In other biomes, both temperature and precipitation change seasonally. In certain biomes, such as tropical rainforest, temperatures are nearly constant but rainfall varies seasonally. In the tropics, where seasonal temperature fluctuations are small, annual cycles are dominated by wet and dry seasons. In general, the length of time that a region is close to the intertropical convergence zone, and hence receives rainfall, increases toward the equator. The intertropical convergence zone shifts latitudinally in a seasonally predictable way (see Figure 53.2), resulting in a characteristic latitudinal pattern of distribution of rainy and dry seasons in tropical and subtropical regions (Figure 54.13).

Although terrestrial biomes are identified on the basis of their dominant plant communities, many other communities are found within each biome. For example, the deciduous forest biome contains, among other

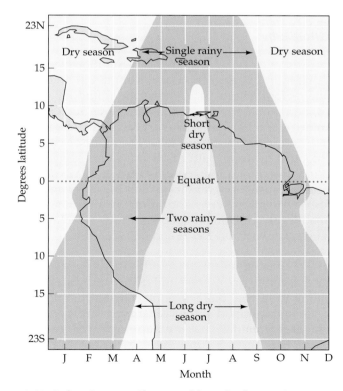

54.13 Rainy Seasons Change with Latitude In the tropics and subtropics, which months are rainy and which are dry is highly predictable based on the region's latitude.

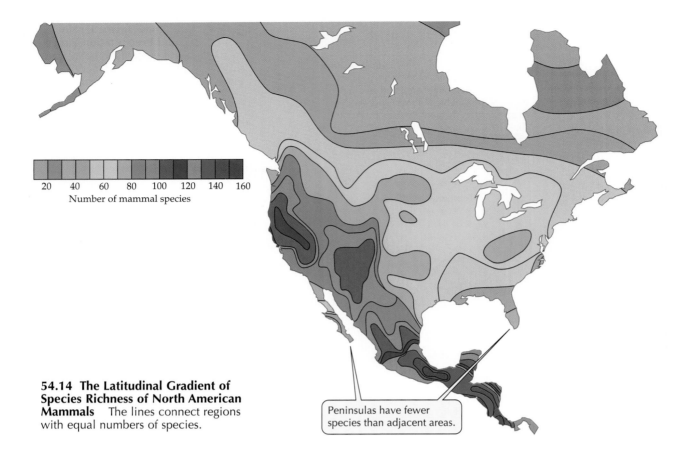

54.14 The Latitudinal Gradient of Species Richness of North American Mammals The lines connect regions with equal numbers of species.

Number of mammal species
20 40 60 80 100 120 140 160

Peninsulas have fewer species than adjacent areas.

Species richness varies latitudinally

A nearly universal pattern in the distribution of species is that more species live in tropical than in high-latitude regions. Figure 54.14 shows the latitudinal gradient in mammal species richness in North and Central America. Similar patterns exist for birds, frogs, and trees and for many marine taxa. The figure also shows two other general patterns. First, more species are found in mountainous regions than in relatively flat areas because topographically complex regions have more vegetation types and climates within a small area. Second, mammal species richness declines on peninsulas, such as Florida and Baja California.

Pictures capture the essence of terrestrial biomes

It is easiest to grasp the similarities and differences among terrestrial biomes by means of a combination of photographs and graphs of temperature, precipitation, and biological activity, supplemented by a few words that describe the species richness and other attributes of those biomes. We use this method in the following pages to describe the major terrestrial biomes of the world (Figures 54.15–54.24).

Each biome is illustrated by two photographs that show either the biome at different times of year or rep-resentatives of the biome in different parts of Earth. Below the photos are graphs that plot seasonal patterns of temperature and precipitation at a typical site in the biome, and graphs showing how active different kinds of organisms are during the year. Levels of biological activity, shown by the width of horizontal bars, change either because resident organisms become more active (produce leaves, come out of hibernation, hatch, or reproduce) or because organisms migrate into and out of the biome at different times of the year. A small box describes the kinds of plants that dominate vegetation in the biome and patterns of species richness there. These descriptions are very general and do not capture the variability that exists within each biome.

Tundra is found at high latitudes and in high mountains

The tundra biome is found in the Arctic and high in mountains at all latitudes. Because the climate is too cold for trees to grow, tundra vegetation is dominated by low-growing perennial plants (Figure 54.15).

In the Arctic, permanently frozen soil—**permafrost**—underlies the tundra's vegetation. The top few centimeters of soil thaw during the short (but often warm) summers, when the sun shines 24 hours a day. Even though there is little precipitation, lowland

54.15 Tundra Arctic tundra: Denali National Park, Alaska (top). Tropical alpine tundra: Teleki Valley, Mt. Kenya (below).

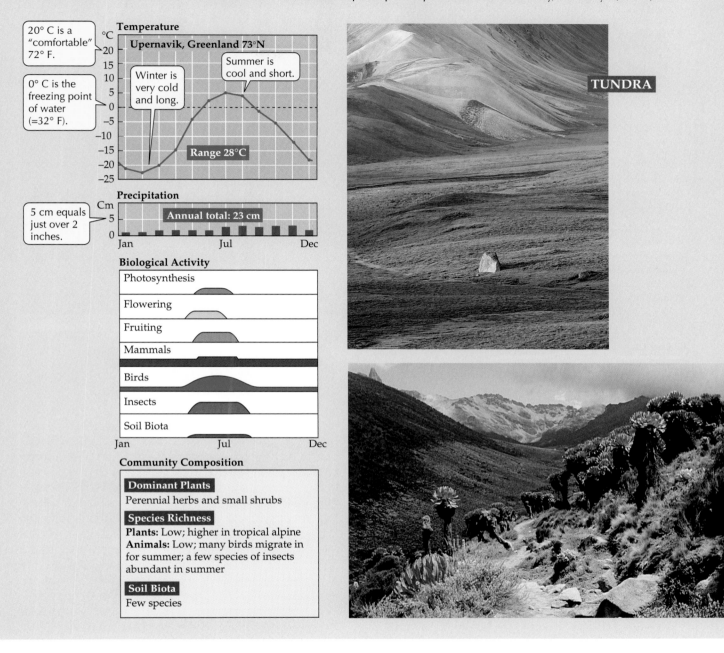

20° C is a "comfortable" 72° F.

0° C is the freezing point of water (=32° F).

Temperature

Upernavik, Greenland 73°N

Winter is very cold and long.

Summer is cool and short.

Range 28°C

Precipitation

5 cm equals just over 2 inches.

Annual total: 23 cm

Jan   Jul   Dec

Biological Activity

Photosynthesis

Flowering

Fruiting

Mammals

Birds

Insects

Soil Biota

Jan   Jul   Dec

Community Composition

Dominant Plants
Perennial herbs and small shrubs

Species Richness
Plants: Low; higher in tropical alpine
Animals: Low; many birds migrate in for summer; a few species of insects abundant in summer

Soil Biota
Few species

TUNDRA

Arctic tundra is very wet because water cannot drain down through the frozen soil. Upland, or montane, tundra in the Arctic is better drained than lowland tundra, and south-facing slopes may not be underlain by permafrost. Plants grow actively in the shallow, water-logged soils for a few months each year. Most Arctic tundra animals either migrate into the area for the summer and go elsewhere for the winter, or they are dormant for most of the year.

Tropical alpine tundra has days and nights of equal length (12 hours) all year. At these high altitudes the temperature is never very high, and it drops below freezing on most clear nights. Plants photosynthesize (although slowly) all year, and most animals are year-round residents.

Boreal forests are dominated by evergreen trees
Moving from the poles toward the equator, or to a lower elevation on temperate-zone mountains, we find the boreal forest (Figure 54.16). Over most of the boreal forest biome winters are long and very cold and summers are short (although often warm). The short-

54.16 Boreal Forest Northern spruce forest, Altay Mountains, Xinjiang, China (top). Bryophytes and lichens on southern evergreens, Tasmania, southern Australia (below).

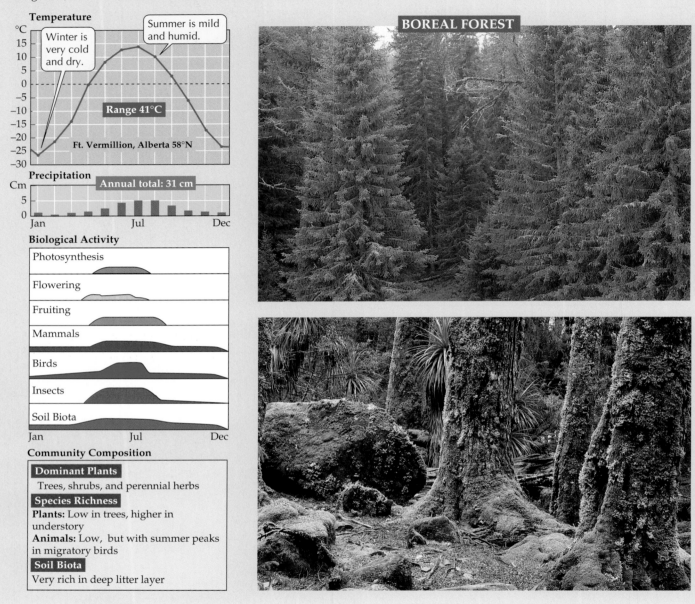

Temperature

Winter is very cold and dry.

Summer is mild and humid.

Range 41°C

Ft. Vermillion, Alberta 58°N

Precipitation

Annual total: 31 cm

Jan Jul Dec

Biological Activity

Photosynthesis

Flowering

Fruiting

Mammals

Birds

Insects

Soil Biota

Jan Jul Dec

Community Composition

Dominant Plants
Trees, shrubs, and perennial herbs

Species Richness
Plants: Low in trees, higher in understory
Animals: Low, but with summer peaks in migratory birds

Soil Biota
Very rich in deep litter layer

BOREAL FOREST

ness of the summers favors trees with evergreen leaves that live for several years. These trees are ready to photosynthesize as soon as temperatures become favorable in the spring.

The boreal forests of the Northern Hemisphere are dominated by coniferous evergreen gymnosperms, while in the Southern Hemisphere the dominant trees are southern beeches (*Nothofagus*), which are angiosperms with small leaves. Some species of *Nothofagus* are evergreen, others are deciduous. Coniferous evergreen forests also grow along the west coasts of continents at middle to high latitudes where winters are mild but very wet and summers are cool and dry;

these forests are home to Earth's tallest trees, and they support the highest standing biomasses of wood of all ecological communities.

Boreal forests have only a few tree species, nearly all of which are wind-pollinated and have wind-dispersed seeds. The dominant animals—such as insects, moose, and hares—eat leaves. The seeds in the cones of conifers also support a fauna of rodents and birds.

Temperate deciduous forests change with the seasons

The temperate deciduous forest biome (Figure 54.17) is found in eastern North America, eastern Asia, and in

54.17 Temperate Deciduous Forest   A Rhode Island forest in summer (left) and winter (right).

TEMPERATE DECIDUOUS FOREST

Temperature

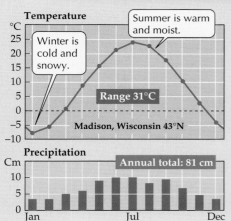

°C

Winter is cold and snowy.

Summer is warm and moist.

Range 31°C

Madison, Wisconsin 43°N

Precipitation

Cm

Annual total: 81 cm

Jan   Jul   Dec

Biological Activity

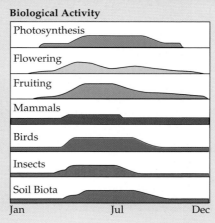

Photosynthesis

Flowering

Fruiting

Mammals

Birds

Insects

Soil Biota

Jan   Jul   Dec

Community Composition

| **Dominant Plants** |
| Trees and shrubs |

Species Richness

Plants: Many tree species in Southeast USA and East Asia, rich shrub layer
Animals: Rich; many migrant birds, richest amphibian communities on Earth, rich summer insect fauna

Soil Biota

Rich

parts of western Europe. These regions have striking seasonal cycles of activity. Temperatures fluctuate dramatically between summer and winter and ample precipitation falls throughout the year.

Changes in the leaves are the most conspicuous sign of the seasons. Deciduous trees lose their leaves during the cold winters and produce leaves that photosynthesize rapidly during the warm, moist summers. There are many more tree species here than in boreal forests, and many deciduous trees have animal-dispersed pollen and fruits. Most deciduous trees and shrubs produce their fruits in autumn, at the end of the growing season.

Annual primary production in these forests may equal that of many tropical forests. Many birds migrate into this biome in summer, when insects are abundant. The temperate forests richest in species are in the southern Appalachian Mountains of the United States and in eastern China and Japan. Each of these is an area that was not disturbed by the glaciations of the Pleistocene (see Chapter 19).

Grasslands are ubiquitous

The temperate grassland biome is found in many parts of the world, all of which are relatively dry much of the year (Figure 54.18). Grasslands often experience

54.18 Temperate Grassland Nebraska prairie in spring (top). The Veldt, Natal, South Africa (below).

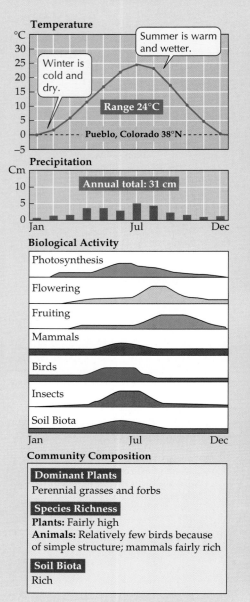

Temperature

°C
Summer is warm and wetter.

Winter is cold and dry.

Range 24°C

Pueblo, Colorado 38°N

Precipitation

Cm

Annual total: 31 cm

Jan   Jul   Dec

Biological Activity

Photosynthesis

Flowering

Fruiting

Mammals

Birds

Insects

Soil Biota

Jan   Jul   Dec

Community Composition

Dominant Plants
Perennial grasses and forbs

Species Richness
Plants: Fairly high
Animals: Relatively few birds because of simple structure; mammals fairly rich

Soil Biota
Rich

TEMPERATE GRASSLANDS

what is known as a continental climate, with hot summers and cold winters. Such regions as the pampas of Argentina, the veldt of South Africa, and the Great Plains of the United States are all part of this biome, much of which has been converted by humans for agricultural purposes.

Natural grasslands are structurally simple, but they are rich in species of perennial grasses, sedges, and forbs (broad-leaved herbaceous species). Many forbs have showy flowers, and grasslands are often riots of color when forbs are in bloom. Grasses are uniquely adapted to survive disturbances because they store much of their energy underground and quickly resprout after they are burned or heavily grazed. As we saw in Chapter 53, grasslands typically support large populations of grazing mammals, and fires are common. In some grasslands most of the precipitation falls in winter, while in others the majority of moisture falls in summer.

Cold deserts are high and dry

Deserts are characterized by low rainfall. The cold desert biome is found in dry regions of the middle to high latitudes, especially in the interiors of large conti-

54.19 Cold Desert Sagebrush steppe near Mono Lake, California (top). Los Glacieres National Park, Argentina (below).

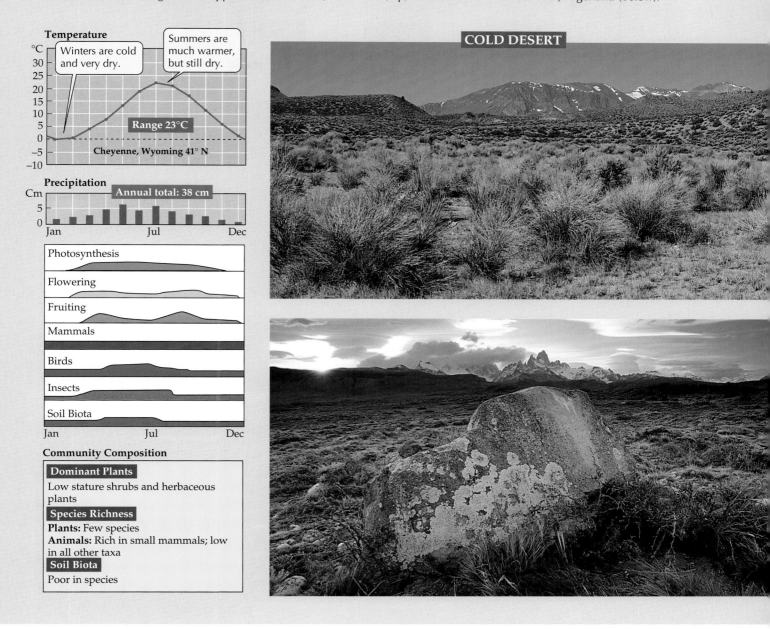

COLD DESERT

Temperature

Winters are cold and very dry.

Summers are much warmer, but still dry.

Range 23°C

Cheyenne, Wyoming 41° N

Precipitation

Annual total: 38 cm

Photosynthesis

Flowering

Fruiting

Mammals

Birds

Insects

Soil Biota

Community Composition

Dominant Plants
Low stature shrubs and herbaceous plants

Species Richness
Plants: Few species
Animals: Rich in small mammals; low in all other taxa

Soil Biota
Poor in species

nents. Cold deserts also are found at fairly high latitudes in the rain shadows of mountain ranges (see Figure 53.1). Seasonal changes in temperature are great, and most of the small amount of rain that falls does so in the winter months.

Cold deserts are dominated by a few species of low-growing shrubs (Figure 54.19). The surface layers of the soil are recharged with moisture in winter, and plant growth is concentrated in spring. By early summer cold deserts are usually barren; so little rain falls that plants cannot conduct much photosynthesis.

Cold deserts are relatively poor in species in most

taxonomic groups, although the plants of this biome tend to produce a great many seeds, supporting a rich fauna of seed-eating birds, ants, and rodents.

Hot deserts form at 30° latitude

The hot desert biome is found in two belts, centered at 30° north and 30° south latitudes, respectively. These are the regions where air from aloft in the atmosphere descends, warms, and picks up moisture (see Figure 53.2). Hot deserts receive most of their rainfall in summer, when the intertropical convergence zone described in Chapter 53 moves toward the

54.20 Hot Desert Anza Borrego Desert, California (left). Rainbow Valley in the desert of central Australia (right).

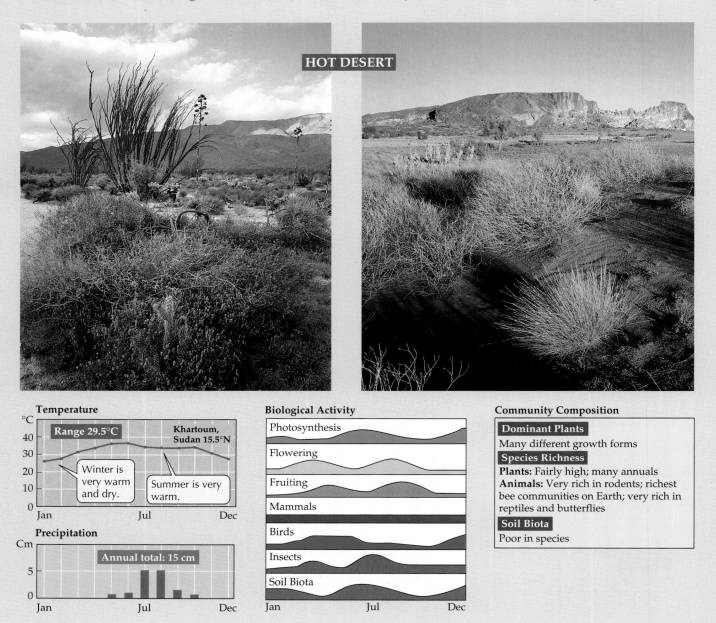

HOT DESERT

Temperature

°C

Range 29.5°C | Khartoum, Sudan 15.5°N

40
30
20
10
0

Jan | Jul | Dec

Winter is very warm and dry.

Summer is very warm.

Precipitation

Cm

Annual total: 15 cm

5

0

Jan | Jul | Dec

Biological Activity

Photosynthesis
Flowering
Fruiting
Mammals
Birds
Insects
Soil Biota

Jan | Jul | Dec

Community Composition

Dominant Plants
Many different growth forms

Species Richness
Plants: Fairly high; many annuals
Animals: Very rich in rodents; richest bee communities on Earth; very rich in reptiles and butterflies

Soil Biota
Poor in species

poles. However, they also receive winter rains from storms that form over the mid-latitude oceans. The driest regions, where summer and winter rains rarely penetrate, are in the center of Australia and the middle of the Sahara Desert of Africa.

Except in these driest regions, hot deserts have a richer and structurally more diverse vegetation than cold deserts do (Figure 54.20). Succulent plants such as cacti, which store large quantities of water in their expandable stems and that photosynthesize primarily with their stems rather than their leaves, are conspicu-

ous in many hot deserts. An abundance of annual plants springs suddenly into profusion when rain falls.

Desert plants produce great quantities of seeds. Pollination and dispersal of fruits by animals are typical. Population densities of rodents and ants are often remarkably high, and lizards and snakes typically are common.

The chaparral climate is dry and pleasant

The chaparral biome is found on the west sides of continents at moderate latitudes, where cool ocean

54.21 Chaparral Fynbos vegetation, Cape of Good Hope, South Africa (top). Mendocino, California (below).

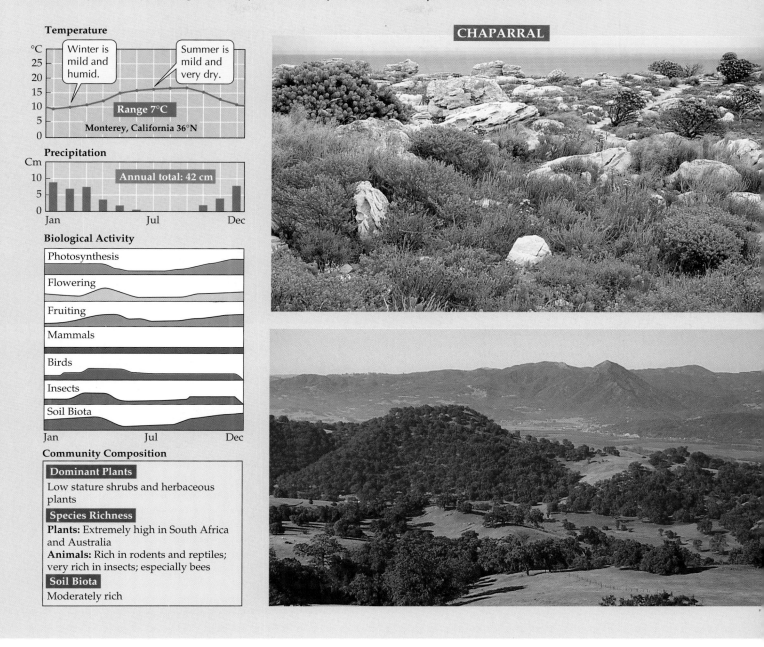

CHAPARRAL

Temperature

°C
Winter is mild and humid.
Summer is mild and very dry.

Range 7°C

Monterey, California 36°N

Precipitation

Cm

Annual total: 42 cm

Jan   Jul   Dec

Biological Activity

Photosynthesis

Flowering

Fruiting

Mammals

Birds

Insects

Soil Biota

Jan   Jul   Dec

Community Composition

Dominant Plants
Low stature shrubs and herbaceous plants

Species Richness
Plants: Extremely high in South Africa and Australia
Animals: Rich in rodents and reptiles; very rich in insects; especially bees

Soil Biota
Moderately rich

waters flow offshore. Winters in this biome are cool and wet and summers are hot and dry. Such climates are found in the Mediterranean region of Europe (hence the term "Mediterranean climate," which is sometimes applied to this biome), coastal California, central Chile, extreme southern Africa, and southwestern Australia.

Chaparral is dominated by low-growing shrubs that have tough, evergreen leaves (Figure 54.21). The shrubs carry out most of their growth and photosynthesis in early spring, which is when insects are active

and birds breed. Annual plants are abundant, providing seeds that store well during the hot, dry summers. This biome thus supports large populations of small rodents, most of which store seeds in underground burrows. The vegetation of the chaparral is naturally adapted to experience periodic fires.

Many shrubs of the Northern Hemisphere chaparral produce bird-dispersed fruits that ripen in the late fall or early spring, which is when large numbers of migrant birds arive from the north. One such fruit, the olive, has played a very important role in human his-

54.22 Thorn Forest and Savanna A thorn forest in Madagascar (top), and a grove of *Acacia* trees in Tanzania (below).

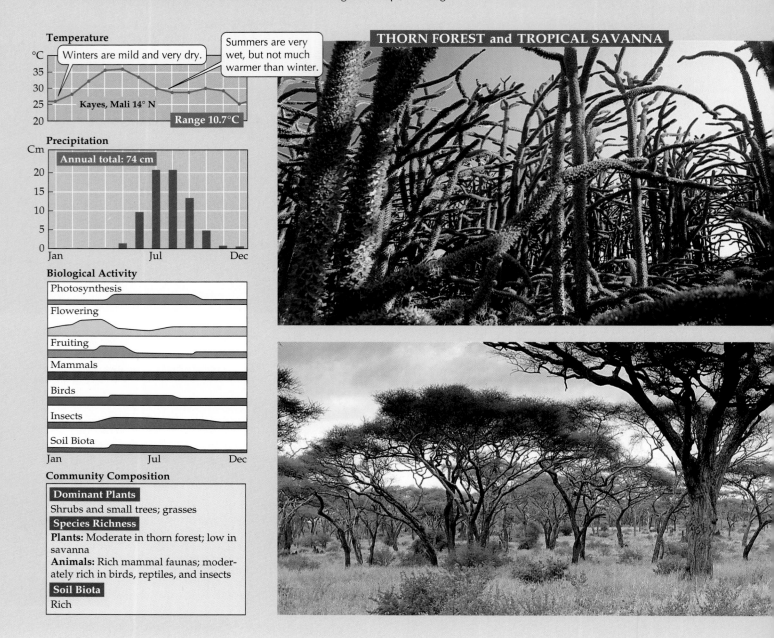

THORN FOREST and TROPICAL SAVANNA

Temperature

Winters are mild and very dry.

Summers are very wet, but not much warmer than winter.

Kayes, Mali 14° N

Range 10.7°C

Precipitation

Annual total: 74 cm

Biological Activity

Photosynthesis

Flowering

Fruiting

Mammals

Birds

Insects

Soil Biota

Community Composition

Dominant Plants
Shrubs and small trees; grasses

Species Richness
Plants: Moderate in thorn forest; low in savanna
Animals: Rich mammal faunas; moderately rich in birds, reptiles, and insects

Soil Biota
Rich

---

tory, providing a rich food source for people at a period of otherwise low food availability.

Thorn forests and savannas have similar climates

Thorn forests are found on the equatorial sides of hot deserts. The climate is semiarid; little or no rain falls during winter, but rainfall may be heavy during the summer wet season. Thorn forests contain many plants similar to those found in hot deserts. The dominant plants are small, spiny shrubs and trees (Figure 54.22). Members of the genus *Acacia* (see Figure 52.18) are common in thorn forests all over the world

The dry tropical and subtropical regions of Africa, South America, and Australia have extensive areas of savannas—expanses of perennial grasses and grasslike plants punctuated by scattered trees. The largest savannas are found in central and eastern Africa, where the biome supports huge numbers of grazing and browsing mammals, which in turn serve as prey for many large carnivores. Grazers and browsers maintain the savan-

54.23 Tropical Deciduous Forest Palo Verde National Park, Costa Rica, shown in the the rainy (top) and the dry season (below).

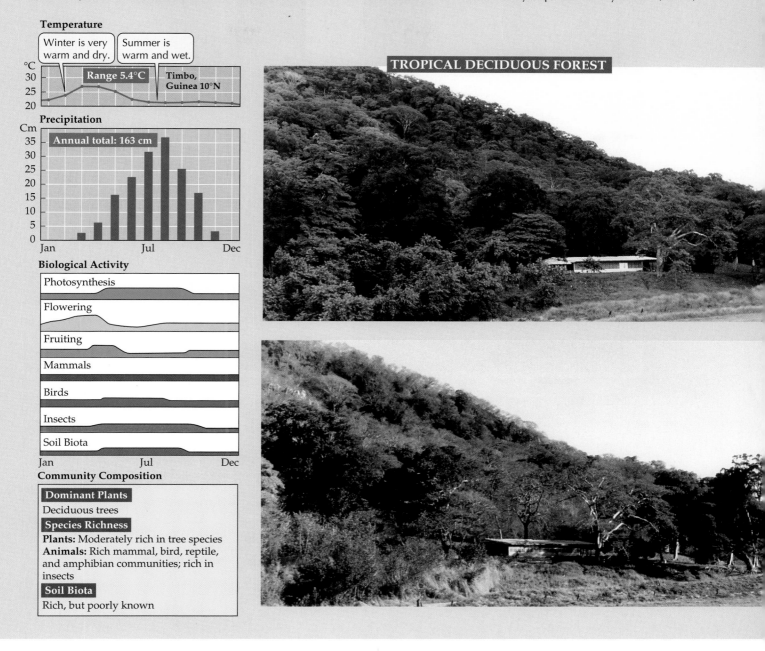

Temperature

Winter is very warm and dry.

Summer is warm and wet.

°C
30
25
20

Range 5.4°C

Timbo, Guinea 10°N

Precipitation

Cm
35
30
25
20
15
10
5
0

Annual total: 163 cm

Jan Jul Dec

Biological Activity

Photosynthesis

Flowering

Fruiting

Mammals

Birds

Insects

Soil Biota

Jan Jul Dec

Community Composition

Dominant Plants
Deciduous trees

Species Richness
Plants: Moderately rich in tree species
Animals: Rich mammal, bird, reptile, and amphibian communities; rich in insects

Soil Biota
Rich, but poorly known

TROPICAL DECIDUOUS FOREST

nas; if savanna vegetation is not grazed, browsed, or burned, it reverts to dense thorn forest.

Tropical deciduous forests occur in hot lowlands

As the length of the rainy season increases toward the equator, thorn forests are replaced by tropical deciduous forests. These forests have taller trees and fewer succulent plants than the thorn forests, and they are much richer in species (Figure 54.23). The long dry season is very hot and often windy. Most of the trees, except for those growing along rivers, lose their leaves during the dry season; many of them flower while they are in this leafless state. The biome is very rich in species of both plants and animals.

Because the soils of the tropical deciduous forest biome are less leached of nutrients than are the soils of wetter areas, they are some of the best soils in the tropics for agriculture. As a result, most tropical deciduous forests have been cleared for grazing cattle and growing crops.

54.24 Tropical Evergreen Forest Canopies of montane wet forest, Bwindi, Uganda (left) and lowland wet forest, Madagascar (right).

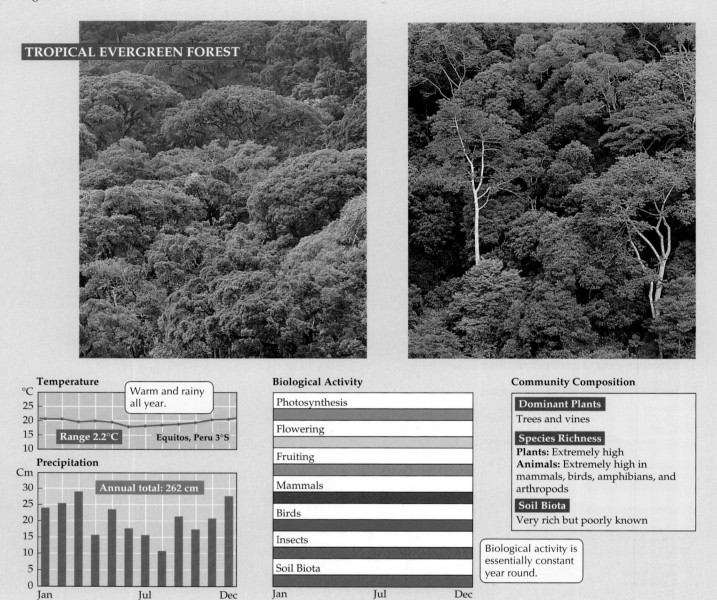

TROPICAL EVERGREEN FOREST

Temperature

Warm and rainy all year.

Range 2.2°C   Equitos, Peru 3°S

Precipitation

Annual total: 262 cm

Biological Activity

Photosynthesis
Flowering
Fruiting
Mammals
Birds
Insects
Soil Biota

Biological activity is essentially constant year round.

Community Composition

Dominant Plants
Trees and vines

Species Richness
Plants: Extremely high
Animals: Extremely high in mammals, birds, amphibians, and arthropods

Soil Biota
Very rich but poorly known

Tropical evergreen forests are species-rich

The tropical evergreen forest biome is found in equatorial regions where total rainfall exceeds 250 cm annually and the dry season lasts for no more than two or three months. It is the richest of all biomes in number of species of both plants and animals, with up to 500 species of trees per square kilometer. Many of these species are rare, and nearly all of them rely on animals to transport their pollen and disperse their fruits. Food webs in this community are extremely complex.

Along with their immense richness of species, tropical evergreen forests have the highest overall productivity of all ecological communities (Figure 54.24). However, most mineral nutrients are tied up in the vegetation; the soils are deeply weathered and usually cannot support agriculture without massive applications of fertilizers.

In the upland (montane) wet forests of the tropics, temperature decreases about 6° for each 1,000 m of elevation. Trees here are shorter than lowland tropical trees. Their leaves are smaller, and there are more epi-phytes—plants that derive their nutrients and moisture from air and water rather than soil. Epiphytes, which grow on other plants, thrive in the high forests where clouds form regularly, bathing the forest in moisture. Photosynthetic rates in the mountain forests are slower than in the lowlands because the temperature is lower and because the leaves are wet most of the time.

Human activities are destroying tropical evergreen forests at a very high rate. Many of the species living here, especially the invertebrates, have not yet been described or named by scientists. Many will pass into extinction without our knowledge.

Aquatic Biomes

Three-fourths of Earth's surface is covered by water. For organisms that cannot survive out of water, terrestrial habitats are barriers to dispersal. However, some aquatic species have flying adults that can disperse widely among water bodies. Others have windborne, desiccation-resistant spores and seeds, and still others are small enough to be transported by means such as mud on the feet of birds. Many freshwater taxa that are capable of dispersing across terrestrial barriers are distributed widely over several continents.

Freshwater biomes have little water but many species

Although a minute fraction of Earth's water is found in ponds, lakes, and streams, about 10 percent of all aquatic species live in freshwater habitats. Prominent among these are the more than 25,000 species of insects that have at least one aquatic stage in their life cycle. Most commonly, eggs and larvae are aquatic and the winged adults are the primary dispersers. Adults of some of these insects, such as dragonflies, are powerful flyers, but adults of mayflies and some other species are weak flyers, rapidly desiccate in air, and live no longer than a few days. As you would expect, oceanic islands have no or very few species of mayflies because of their inability to survive long enough to disperse across wide expanses of salt water.

Similarly, fishes unable to tolerate salt water can disperse only within the connected rivers and lakes of a river basin. Dispersal between river basins can occur if their headwaters become joined when erosion removes the barrier between them. This happened, for example, when large amounts of water released by melting glaciers at the end of the Ice Age connected the headwaters of the Yukon River to the basin of the Mackenzie River (Figure 54.25). Today these rivers share a number of freshwater fish species, even though they are no longer connected.

Most families of freshwater fishes that cannot tolerate salt water are restricted to a single continent. Those families with species distributed on both sides of major saltwater barriers are believed to be ancient lineages whose ancestors were distributed widely in Pangaea or Gondwanaland.

Boundaries between marine biomes are determined primarily by changes in water temperature

All oceans are connected, and ocean water moves in great circular patterns—clockwise in the Northern Hemisphere and counterclockwise in the Southern Hemisphere (see Figure 53.3). These movements disperse organisms with limited swimming abilities. Nevertheless, most marine organisms have restricted ranges, indicating that important environmental limits to their distributions exist in the oceans.

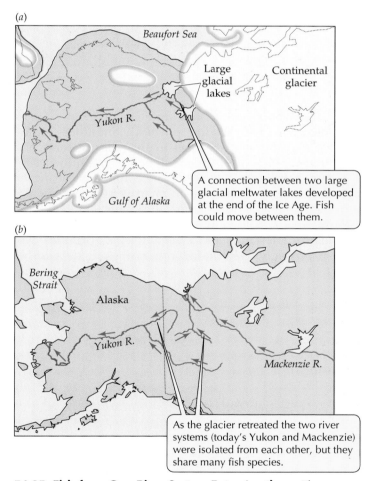

A connection between two large glacial meltwater lakes developed at the end of the Ice Age. Fish could move between them.

As the glacier retreated the two river systems (today's Yukon and Mackenzie) were isolated from each other, but they share many fish species.

54.25 Fish from One River System Enter Another The Yukon and Mackenzie Rivers of Canada share many freshwater fish species because they were connected by glacial melt at the end of the Ice Age. (*a*) The early stages of glacial retreat. Glaciers are shown in pale gray. (*b*) Today's drainage pattern.

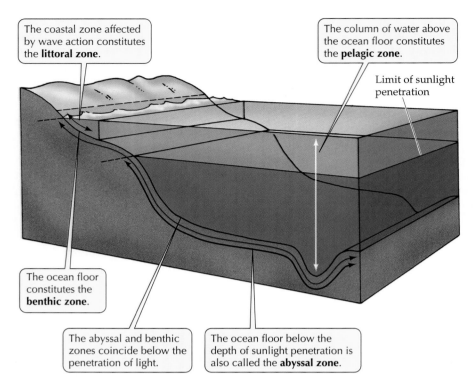

The coastal zone affected by wave action constitutes the **littoral zone**.

The column of water above the ocean floor constitutes the **pelagic zone**.

Limit of sunlight penetration

The ocean floor constitutes the **benthic zone**.

The abyssal and benthic zones coincide below the penetration of light.

The ocean floor below the depth of sunlight penetration is also called the **abyssal zone**.

54.26 Zones of the Ocean
Oceanic zones are shown schematically in relation to depth and sunlight penetration.

tween these zones, organisms from one zone survive poorly if they attempt to live in another.

Ocean temperatures are barriers to colonization because many marine organisms are well adapted to only relatively narrow temperature ranges. The main biogeographic divisions of the pelagic zone coincide with regions where the temperature of surface waters changes relatively abruptly as a result of horizontal and vertical ocean currents (Figure 54.27). These temperature changes, in combination with seasonal changes in daylight, determine the seasons of maximum primary production. Marine algae tend to be adapted to photosynthesize either in summer or in winter, but not during both seasons.

Because nutrients gradually sink to the ocean bottom, high concentrations of nutrients in the pelagic zone are restricted to areas where upwelling currents bring nutrient-rich deep waters to the surface. Most marine organisms that grow and reproduce well in nutrient-rich waters perform relatively poorly in nutrient-poor waters, and vice versa. Therefore, nutrient-rich waters typically have biotas that differ considerably from those of nutrient-poor waters in the same general region.

Deep ocean waters are barriers to the dispersal of marine organisms that live only in shallow water. The distances that the eggs and larvae of marine organisms can be carried by ocean currents are determined in large part by the time it takes for the larvae to metamorphose into sedentary adults. Relatively few species have eggs and larvae that can dis-

Horizontal and vertical gradients divide the oceans into zones with distinct physical conditions (Figure 54.26). Water temperatures, hydrostatic pressures, and food supplies all change with depth and influence biotic distributions. Food is scarce, for example, in the deep waters of the oceans, may arrive very infrequently, and cannot be hunted visually. Living successfully in different regions of the ocean requires different physiological tolerances and morphological attributes. Not surprisingly, even though organisms can disperse be-

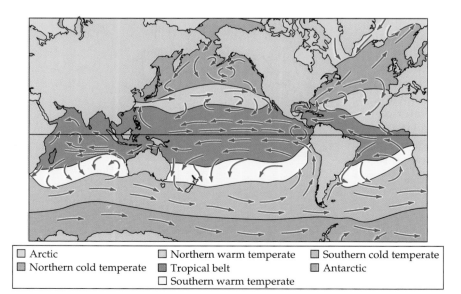

☐ Arctic
■ Northern cold temperate
☐ Northern warm temperate
■ Tropical belt
☐ Southern warm temperate
☐ Southern cold temperate
■ Antarctic

54.27 Pelagic Regions Are Determined by Ocean Currents The arrows represent ocean currents. Regions in which photosynthesis is maximized at different seasons are indicated by different colors.

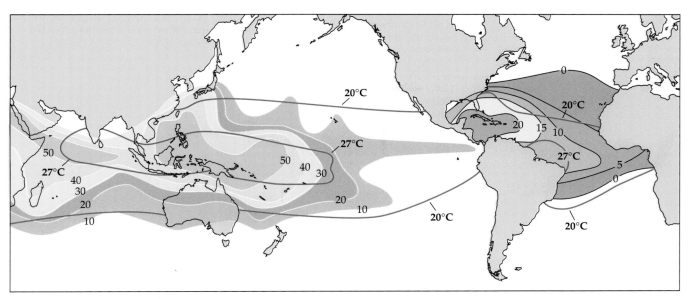

54.28 Generic Richness of Reef-Building Corals Declines with Distance from New Guinea The lines connect areas with equal numbers of genera. The 20° and 27° mean annual temperature isotherms are also shown.

perse across wide barriers of deep water. As a result, the richness of sedentary shallow-water species in the intertidal and subtidal zones of isolated islands in the Pacific Ocean decreases with distance from New Guinea (Figure 54.28).

Marine vicariant events influence species distributions

During the time of Pangaea, the seas were also united to form one world ocean, Panthalassa. Continental drift dramatically changed the sizes and shapes of the oceans, but ocean waters move so rapidly that ancient vicariant events may no longer influence distributions of marine organisms today. More recent marine vicariant events, however, have left their traces.

Relatively rapid changes in sea level are produced by major tectonic events and climate change. An important tectonic event was the elevation of the Panamanian Isthmus about 3 million years ago. This event separated the Pacific Ocean from the Caribbean Sea for the first time in more than 100 million years. Distinct marine biotas are now evolving on opposites sides of the isthmus, which also forms a barrier to the dispersal of Pacific species, such as sea snakes, that reached the west coast of the Americas after the barrier formed (Figure 54.29). If a sea level canal were constructed across the isthmus, poisonous sea snakes and other marine organisms would be able to disperse into the Caribbean. Currently the fresh waters of Gatun Lake form a barrier to the dispersal of marine organisms.

The total amount of water on Earth is roughly constant, so water that is tied up in continental glaciers is subtracted from the water that remains in the oceans. At numerous times during the Pleistocene, sea levels dropped more than 100 meters as continental glaciers formed and expanded, then rose when the glaciers re-

Pelamis platurus

54.29 A Block to Dispersal The presence of the Panamanian Isthmus means that poisonous sea snakes cannot enter the Caribbean Sea from the Pacific Ocean.

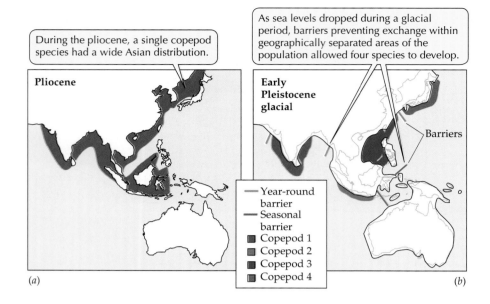

During the pliocene, a single copepod species had a wide Asian distribution.

As sea levels dropped during a glacial period, barriers preventing exchange within geographically separated areas of the population allowed four species to develop.

Pliocene

Early Pleistocene glacial

Barriers

— Year-round barrier
— Seasonal barrier
■ Copepod 1
■ Copepod 2
■ Copepod 3
■ Copepod 4

(a)   (b)

54.30 Sea Level Changes Permitted Speciation (*a*) Distribution of the ancestral species of the copepod genus *Labidocera*. (*b*) Barriers allowed the formation of four species.

ceded. The repeated formation and breakdown of seasonal and continental barriers as sea levels fluctuated in association with glacial cycles separated lineages of some marine organisms. These events resulted in repeated cycles of range expansion, contraction, and speciation in some taxa (Figure 54.30). As terrestrial animals we find it difficult to perceive marine barriers to dispersal, but they exist and strongly influence distributions of marine organisms.

Summary of "Biogeography"

Why Are Species Found Where They Are?

• If a species occupies an area, either it evolved there, or it evolved elsewhere and dispersed to the area. If a species is not found in a particular area, either it evolved elsewhere and never dispersed to the area, or it was once present in the area but no longer lives there.
• Continental drift has influenced the distributions of organisms throughout Earth's history.
• Biogeographers often analyze species distributions by converting taxonomic cladograms into area cladograms. **Review Figure 54.1**

Historical Biogeography

• Historical biogeographers attempt to determine the influence of past events on today's patterns of species distributions.
• Biogeographers use the principle of parsimony when they attempt to explain distribution patterns. **Review Figure 54.2**
• Vicariance and dispersal events have both influenced current distributions. **Review Figures 54.3, 54.4**
• Recent vicariant events, such as advances and retreats of glaciers, have affected current distributions of organisms.

Review Figures 54.5, 54.6
• Animal biogeographers divide Earth into six major biogeographic regions. Plant geographers recognize two additional regions. **Review Figure 54.7**

Ecological Biogeography

• Ecological biogeographers test theories that explain the numbers of species in different communities, how species disperse, and the effectiveness of barriers to movement.
• An equilibrium model of species richness, which predicts the number of species on islands, has been tested by examining patterns of distribution and by performing experiments. **Review Figures 54.8, 54.9, 54.10**

Terrestrial Biomes

• Terrestrial biomes are major ecosystem types that differ from one another in the structure of their dominant vegetation. These biomes are tundra, boreal forest, temperate deciduous forest, temperate grassland, cold desert, hot desert, chaparral, thorn forest and savanna, tropical deciduous forest, and tropical evergreen forest.
• The distribution of biomes on Earth is strongly influenced by annual patterns of temperature and rainfall. **Review Figures 54.11, 54.12, 54.13**
• The number of species in most lineages increases from polar to tropical regions. **Review Figure 54.14**
• Pictures and graphs capture the similarities and differences among Earth's terrestrial biomes. **Review Figures 54.15–54.24**

Aquatic Biomes

• No absolute barriers to the movement of marine organisms exist within the oceans, but most marine organisms have restricted ranges.
• Conditions in the oceans change dramatically with depth and sunlight penetration. **Review Figure 54.26**
• Boundaries between many pelagic regions are determined by ocean currents. **Review Figure 54.27**
• Species that live in shallow waters disperse with difficulty across wide deep-water barriers. **Review Figures 54.28**

• Changes in sea levels have resulted in speciation in some shallow-water lineages. **Review Figure 54.30**

Self-Quiz

1. Biogeography as a science began when
   a. eighteenth-century travelers first noted intercontinental differences in the distributions of organisms.
   b. Europeans went to the Middle East during the Crusades.
   c. cladistic methods were developed.
   d. the fact of continental drift was accepted.
   e. Charles Darwin proposed the theory of natural selection.

2. Historical and ecological biogeography differ in that
   a. only historical biogeography is concerned with history.
   b. historical biogeography is concerned with longer time periods and larger spatial scales.
   c. both are concerned with the same time scales, but historical biogeography deals with larger spatial scales.
   d. both are concerned with the same spatial scales, but historical biogeography deals with longer time scales.
   e. historical biogeography is not concerned with the current distributions of organisms.

3. Marine biogeographic regions are less distinct than terrestrial ones because
   a. the ocean biota is more poorly known than the terrestrial biota.
   b. there are currently fewer barriers to dispersal of marine organisms.
   c. most marine families and higher taxa evolved before the oceans were separated by continental drift.
   d. we know less about the distributions of marine organisms.
   e. oceanic circulation is faster than atmospheric circulation.

4. A parsimonious interpretation of a distribution pattern is one that
   a. requires the smallest number of undocumented vicariant events.
   b. requires the smallest number of undocumented dispersal events.
   c. requires the smallest total number of undocumented vicariant plus dispersal events.
   d. accords with the cladogram of a group.
   e. accounts for centers of endemism.

5. The only major biogeographic region that today is isolated by water from other regions is
   a. Greenland.
   b. Africa.
   c. South America.
   d. Australasia.
   e. North America.

6. Equilibrium species richness is reached in the MacArthur–Wilson model when
   a. immigration rates of new species and extinction rates of species are equal.
   b. immigration rates of all species and extinction rates of species are equal.
   c. the rate of vicariant events equals rates of dispersal.
   d. the rate of island formation equals the rate of island loss.
   e. No equilibrium number of species exists in that model.

7. Chaparral vegetation is dominated by
   a. deciduous trees.
   b. evergreen trees.
   c. deciduous shrubs.
   d. evergreen shrubs.
   e. grasses.

8. Which of the following is *not* true of tropical evergreen forests?
   a. They have large numbers of species of trees.
   b. Most plant species are animal-pollinated.
   c. Most plant species have animal-dispersed fruits.
   d. Biological energy flow is very high.
   e. Productivity depends on a rich supply of soil nutrients.

9. At all depths, the bottom of the ocean is known as the
   a. benthic zone.
   b. abyssal zone.
   c. pelagic zone.
   d. interoceanic convergence zone.
   e. subtidal zone.

10. Vicariant events that influence current distributions of marine organisms include
   a. the breakup of Gondwanaland.
   b. the breakup of Pangaea.
   c. sea level changes caused by the advance and retreat of Pleistocene glaciers.
   d. sea level changes caused by the advance and retreat of Permian glaciers.
   e. Ocean currents are too strong for past vicariant events to influence current distributions of marine organisms.

Applying Concepts

1. Horses evolved in North America but subsequently became extinct there. They survived to modern times only in Africa and Asia. In the absence of a fossil record we would probably infer that horses originated in the Old World. Today, the Hawaiian Islands have by far the greatest species richness of fruit flies (*Drosophila*). Would you conclude that the genus *Drosophila* originally evolved in Hawaii and spread to other regions? Under what circumstances do you think it is safe to conclude that a group of organisms evolved close to where the greatest number of species live today?

2. In nearly every ecological community, the number of species present is much smaller than the number potentially available to colonize it. Does this pattern constitute good evidence for species richness equilibrium? What do you consider the strongest evidence for species richness equilibrium?

3. A well-known legend states that Saint Patrick drove the snakes out of Ireland. Give some alternative explanations, based on sound biogeographic principles, for the absence of indigenous snakes in that country.

4. What are some significant present-day human problems whose solutions involve biogeographic considerations? What kinds of biogeographic knowledge are most important for each one?

5. Most of the world's flightless birds are either nocturnal and secretive (such as the kagu of New Caledonia) or large, swift, and well-armed (such as the ostrich of Africa). The exceptions are found primarily on islands,

and many of these island species have become extinct with the arrival of humans and their domestic animals. What special biogeographic conditions on islands might permit the survival of flightless birds? Why has human colonization so often resulted in the extinction of such birds? The power of flight has been lost secondarily in representatives of many groups of birds; what are some possible evolutionary advantages of flightlessness that might offset its obvious disadvantages?

Readings

Brown, J. H. and M. V. Lomolino. 1998. *Biogeography*, 2nd Edition. Sinauer Associates, Sunderland, MA. A comprehensive treatment of both ecological and historical biogeography.

Humphries, C. J. and L. R. Parenti. 1986. *Cladistic Biogeography.* Clarendon Press, Oxford. A concise treatment of the ways in which cladistic methods are used to determine the causes of current distributions of organisms.

MacArthur, R. H. and E. O. Wilson. 1967. *The Theory of Island Biogeography.* Princeton University Press, Princeton, NJ. The classic book that launched modern investigations of ecological biogeography.

Myers, A. A. and P. S. Giller. 1988. *Analytical Biogeography: An Integrated Approach to the Study of Animal and Plant Distributions.* Chapman & Hall, London. Contains chapters by different authors on many aspects of ecological and historical biogeography, including discussions of modern methods and their significance.

Terborgh, J. 1993. *Diversity and the Tropical Rain Forest.* W. H. Freeman, San Francisco. An exciting account of the richness of life in tropical evergreen forests.

Thornton, I. 1995. *Krakatau: The Destruction and Reassembly of an Island Ecosystem.* Harvard University Press, Cambridge, MA. The story of how scientists studied the recolonization of Krakatau.

Weiner, J. 1994. *The Beak of the Finch.* Alfred A. Knopf, New York. A stimulating account of field investigations of evolutionary changes in the beaks of Galapagos finches. The best book on modern evolutionary studies for the general reader.

Chapter 55

Conservation Biology

An Extinct Island Bird
This artist's reconstruction of a flightless Hawaiian goose shows one of the many bird species exterminated by the Polynesian settlers of the islands.

When Polynesian people settled in Hawaii about 2,000 years ago, they quickly exterminated, probably by overhunting, at least 39 species of endemic land birds, including 7 species of geese, 2 species of flightless ibises, a sea eagle, a small hawk, 7 flightless rails, 3 species of owls, 2 large crows, a honeyeater, and at least 15 finches.

No people lived in New Zealand until about 1,000 years ago, when the Polynesian ancestors of the Maori colonized the island. Hunting by the Maori caused the extinction of 13 species of flightless moas, some of which were larger than ostriches.

When humans arrived in North America over the Bering land bridge, about 20,000 years ago, they encountered a rich fauna of large mammals. Most of those species were exterminated within a few thousand years. A similar extermination of large animals followed the human colonization of Australia, about 40,000 years ago. At that time Australia had 13 genera of marsupials larger than 50 kg, a genus of gigantic lizards, and a genus of heavy flightless birds. All the species in 13 of those 15 genera had become extinct by 18,000 years ago.

The accelerating pace of human-caused extinctions of species, which raises serious concerns about the future of biological diversity on Earth, has led to the rapid development of **conservation biology**—the study of the diversity of life and how to preserve it. Conservation biology is an applied discipline that uses the best available science to preserve species and ecosystems. Conservation biologists study the causes of declines in species richness and develop methods by which genes, species, communities, and ecosystems can be preserved. The science of conservation biology draws heavily on concepts and knowledge from population genetics, evolution, ecology, biogeography, wildlife management, economics, and sociology. In turn, the needs of conservation are stimulating new research in those fields.

In this chapter we will see how biologists estimate rates of species extinctions. We will also explore why the accelerated extinction rates of other species are so important to humans, the causes of species endangerment, and how management plans that can reduce extinction rates and restore endangered species to a viable status are being developed.

Estimating Current Rates of Extinction

None of the human activities that are causing extinctions are new. It is just that many more of us are doing those things than ever before. We do not know how many species will become extinct during the next 100 years, in part because the number of extinctions will depend both on what we do and on unexpected events. Nevertheless, methods exist for estimating probable rates of extinction resulting from human actions. It is better to act using roughly correct estimates than to wait until precise estimates can be calculated. In this section we will discuss how conservation biologists estimate current rates of extinction and identify species at risk of extinction.

Species–area relationships are used to estimate extinction rates

Conservation biologists often use the well-established fact that the number of species present increases with the size of an area (see Figure 54.9) to estimate the number of species extinctions resulting from habitat destruction. The species–area relationship suggests that, on average, a 90 percent loss of habitat will result in the extirpation of half of the species living in that habitat. For example, if we assume that about half of existing terrestrial species live in tropical forests, and that about one-third of the remaining tropical forests will be logged during the next few decades, the species–area relationship suggests that about 1 million species will be extirpated during that period.

The rate at which tropical forests are being logged and converted to cropland and pastures is not precisely known, but it is currently very high (Figure 55.1). Moreover, even the lowest estimates of current extinction rates predict that at least 10 percent of Earth's species are likely to become extinct during the next two decades. Some estimates predict the extinction of 50 percent of Earth's species during the next 50 years.

Conservation biologists estimate risks of extinction

To estimate the risk of extinction of a particular species, conservation biologists develop models based on trends in its population size and changes in habitat availability to assess whether the species is at risk now or may become endangered in the future.

Although rarity itself is not always a cause for concern, species whose populations are shrinking rapidly usually are at risk. Species with only a few individuals confined to a small range are at risk of extinction because they can easily be eliminated by sudden local disturbances such as fires, unusual weather, disease, and predators.

Estimation of the risks faced by small populations is an important component of preservation analyses. The concept of a **minimum viable population** (MVP) is now widely used to estimate a population's risk of extinction. Its development was stimulated by the National Forest Management Act of 1976, which required the U.S. Forest Service to maintain viable popu-

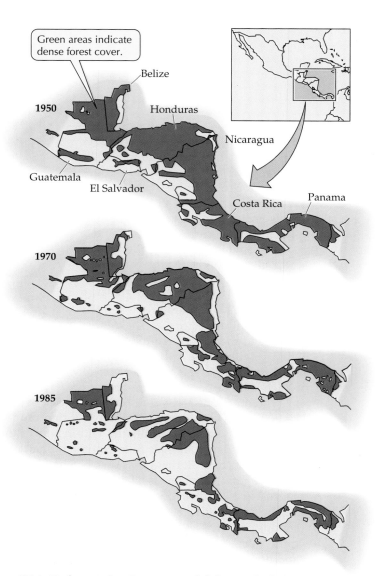

55.1 Deforestation Rates Are High in Tropical Forests
Central America provides an example of the high rate of destruction of tropical forests that has taken place in recent years. Less than half the forest that existed in 1950 remains, and much of that is in small patches.

lations of all native vertebrate species in each national forest. A minimum viable population is the estimated density or number of individuals necessary for the species to maintain or increase its numbers in a region. There is no sharp threshold above which a population is viable and below which it is not, but an MVP analysis can estimate a population's risk of extinction over decades and centuries, time frames that are appropriate for management plans.

A **population viability analysis** (PVA) is carried out to estimate how the size of a population influences its risk of becoming extinct within a specified time period—for example, 100 years. A PVA is based on knowledge of the interactions between the genetic variation, morphology, physiology, and behavior of a population and its environment, both physical and biological.

One component of a PVA is estimation of the extent and significance of *demographic stochasticity*—that is, the amount of random variation in birth and death rates. In a small population, extinction is likely when high death rates coincide with low birth rates. Estimates of the sizes of local populations at high risk of immediate extinction due to demographic stochasticity range from 10 individuals for microorganisms reproducing by fission to about 50 for sexually reproducing animals with lengthy prereproductive periods. Larger populations also may be at high risk because the same environmental conditions that cause low birth rates are likely to cause high death rates.

As we saw in Chapter 51, many species are distributed as metapopulations, collections of subpopulations each of which occupies a suitable patch in a landscape of otherwise unsuitable habitat (see Figure 51.11). The fraction of suitable habitat patches occupied by a metapopulation at any moment in time is determined by the rate at which local subpopulations become extinct and by the rate of colonization of empty patches by dispersing individuals. The prevention of extinction of a subpopulation by occasional immigrants from nearby patches is called the rescue effect.

The acorn woodpecker, a bird of the oak woodlands of western North America, exists as a metapopulation. The oak woodlands in New Mexico in which acorn woodpeckers live are found as small, isolated patches surrounded by open, more arid land. Conservation biologists performed a PVA for the acorn woodpecker by developing a computer simulation model to predict its survival and extinction. To run the model, they entered information on known birth and death rates of the woodpeckers.

The model predicted that, if birth and death rates remained constant at current levels, most of the acorn woodpecker subpopulations in isolated oak woodlands in New Mexico would become extinct within 20 years if no birds migrated between them. However, with only a small amount of migration, most simulated subpopulations survived more than 100 years (Figure 55.2). Many of these small populations of acorn woodpeckers are known to have survived for more than 70 years, suggesting that today birds occasionally fly between patches. However, if the patches of oak woodland were to become even more isolated, too few birds might disperse between patches to maintain the populations.

Another component of a PVA is analysis of **genetic stochasticity**, fluctuations in a population's level of ge-

55.2 A PVA Shows that Acorn Woodpeckers Occasionally Disperse between Woodland Patches
Acorn woodpeckers live in isolated oak woodlands in New Mexico. The graph shows population persistence as a function of the rate of dispersal between patches.

Acorn woodpecker

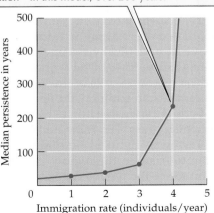

The immigration of as few as four individuals per year results in a dramatic increase in the persistence of the population—in this model, over 200 years.

Median persistence in years

Immigration rate (individuals/year)

netic variation. The level of inbreeding increases in small populations, leading to a lowering of fitness. Therefore, genetic information is important for planning the reintroduction of individuals from captivity or selecting individuals to reestablish extirpated populations.

Why Care about Species Extinctions?

The extinction of species is of great concern because, despite our increased ability to alter and restructure our surroundings, people depend on other species in many ways. We also derive enormous aesthetic pleasure from interacting with other organisms. Many people would consider a world with far fewer species a less desirable one in which to live.

More than half the medical prescriptions written in the United States contain a natural plant or animal product (Figure 55.3), yet the search for and exploitation of such products from the living world has barely begun. Many species may be eliminated by forest destruction before we find out if they might be sources of useful products.

Ecosystem processes produce many benefits to humanity, such as the generation and maintenance of fertile soils, prevention of soil erosion, detoxification and recycling of waste products, regulation of hydrological cycles and the composition of the atmosphere, control of agricultural pests, pollination, and the maintenance of the species richness upon which humanity depends for drugs, medicines, and aesthetic enjoyment. It is easy to list these benefits, but to justify the allocation of scarce public resources to maintain them, we need quantitative estimates of the value of ecosystem services.

A detailed study by economists, ecologists, and land managers in Western Cape Province, South Africa, has shown that an intensive program to eradicate invasive exotic plants in the mountains of the region is a highly cost effective way of maintaining a reliable regional supply of high-quality water. The native vegetation of the watersheds of the area is dominated by a species-rich community of shrubs known as fynbos (pronounced "fainbos") that can survive regular summer drought, nutrient-poor soils, and the fires that periodically sweep through the Cape mountains (Figure 55.4a). The fynbos-clad mountains provide about two-thirds of the Western Cape's water requirements. In addition, the species-rich endemic flora is widely harvested for cut and dried flowers and thatching grass. The combined value of these harvests in 1993 was about $19 million. Some of the income from tourism in the region comes from people who want to see the fynbos vegetation. About 400,000 people visit the Cape of Good Hope Nature Reserve each year, primarily to see plants.

Catharanthus roseus

55.3 Source of a Life-Saving Drug   A drug for combating leukemia was derived from the Madagascar rosy periwinkle.

During recent decades a number of exotic plants, introduced into South Africa to provide a source of fast-growing timber and as hedge plants, have invaded the shrub-clad mountains. Because they are taller and faster-growing than the native plants, the exotics increase the intensity and severity of fires. By transpiring larger quantities of water, they decrease stream flows to less than half the amount flowing from mountains covered with native plants (Figure 55.4b). Removing the exotic plants by felling and digging out invasive trees and shrubs and managing fire is estimated to cost between $140 and $830 per hectare, depending on the densities of invasive plants. Annual follow-up operations cost about $8 per hectare.

The costs of alternative methods to replace the water lost from watersheds taken over by alien plants are much higher. A sewage purification plant that would deliver the same volume of water as a well-managed catchment of 10,000 hectares would cost $135 million to build and $2.6 million per year to operate. Desalination of seawater would cost four times as much.

Thus, the available alternatives would deliver water at a cost between 1.8 and 6.7 times more than the cost of maintaining natural vegetation in the watershed. Modern industrial societies often favor technologically sophisticated methods of substituting for lost ecosystem services. The study of water resources in the Western Cape Region shows that simple but labor-intensive methods—cutting and burning—can be cheaper and, in addition, preserve other valuable services, such as tourism and harvested plant products.

Some ecosystem services, such as aesthetic benefits, cannot be replaced with technological inventions.

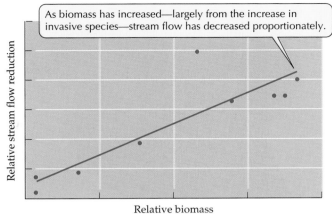

(b) Stream flow from fynbos catchments

> As biomass has increased—largely from the increase in invasive species—stream flow has decreased proportionately.

Relative stream flow reduction

Relative biomass

(a) **55.4 Water Flows from Fynbos Vegetation**   (a) The fynbos vegetation of South Africa. (b) Stream flow from fynbos watersheds in relation to plant biomass. (c) A computer simulations graphs water loss due to invasions of exotic trees.

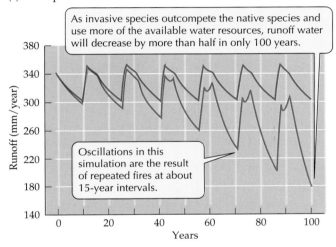

(c)   Computer simulation

> As invasive species outcompete the native species and use more of the available water resources, runoff water will decrease by more than half in only 100 years.

> Oscillations in this simulation are the result of repeated fires at about 15-year intervals.

Runoff (mm/year)

Years

Aesthetic benefits may contribute a great deal to a country's economy. One of the largest sources of foreign income in Kenya is nature tourism. The loss of a single species probably would not reduce the flow of tourists to Kenya, but if elephants, rhinoceroses, lions, leopards, and buffalo were all to disappear, few people would pay the high price of a Kenyan vacation (Figure 55.5). Populations of these species can be maintained only if large tracts of the ecosystems in which they live are preserved.

55.5 Large Mammals Support Ecotourism in Kenya
Without the wide array of large and impressive mammals that are present in Kenya's national parks, the country's flourishing ecotourism business would probably collapse.

Diceros bicornis

Panthera leo

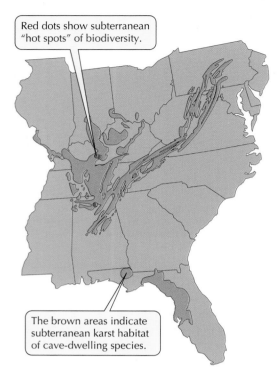

Red dots show subterranean "hot spots" of biodiversity.

The brown areas indicate subterranean karst habitat of cave-dwelling species.

55.6 Cave Animals Have Very Small Ranges   Cave animals tend to be rare because their habitat is rare. The map shows the distribution of cave habitat (subterranean karst) in the southeastern United States. This region has the highest biodiversity of cave animals in the Americas.

Determining Causes of Endangerment and Extinction

Species may be rare for any of several reasons. For example, they may live in a habitat that is rare. Cave-dwelling species typically have narrow ranges and small populations (Figure 55.6), as do many species that live only in desert lakes with high salt concentrations. Another reason is trophic—secondary carnivores are usually rare because so little energy is available to support their populations (see Chapter 53). Being rare may increase a species' chance of becoming extinct, but many rare species, especially those that have been rare for many years, are likely to persist for long time spans.

In this section we will examine the major causes of extinctions, which include overexploitation, the introduction of predators and diseases, the loss of mutualists, habitat destruction, and habitat fragmentation.

Overexploitation has driven many species to extinction

Until recently, humans caused extinctions primarily by overhunting. The passenger pigeon, the most abundant bird in North America in the early 1800s, became extinct by 1914, largely due to overhunting. Russian whalers exterminated the unusual Steller's sea cow of the North Pacific in the late 1800s, just 37 years after it was first described. The American bison was on the brink of extinction at the beginning of the twentieth century and might well be extinct today if its hunting had not been outlawed. Overexploitation continues today; for example, elephants and rhinoceroses are threatened in Africa because poachers kill them for their valuable tusks and horns.

Introduced pests, predators, and competitors have eliminated many species

Deliberately and accidentally, people move species from one continent to another. Pheasants and partridges were introduced into North America for hunting. European settlers took their crops and domesticated animals to Australia. They introduced other species, such as rabbits and foxes, for sport. Weed seeds were accidentally carried around the world in soil used as ballast in sailing ships and as contaminants in sacks of crop seeds. Despite quarantines, disease organisms spread rapidly, carried by infected plants, animals, and people.

A species that has evolved over time in a community with certain predators and competitors may be vulnerable to newly introduced predators and competitors. Introduced species have caused the extinctions of thousands of native species worldwide. Nearly half of the small to medium-sized marsupials and rodents of Australia have been exterminated during the last 100 years by a combination of competition with introduced rabbits and predation by introduced cats and foxes.

Black rats carried to remote oceanic islands on ships are especially destructive predators. On the Galapagos Islands, introduced pigs and rats exterminated several races of tortoises before Darwin explored the islands. Today on some islands they excavate all nests of the giant Galapagos tortoises and devour the eggs. Populations of tortoises on some islands are maintained today only because conservationists remove eggs and rear the young tortoises in captivity until they are large enough to defend themselves against pigs and rats (Figure 55.7).

Proliferation of pests with destructive consequences has quickly followed their introduction to new continents. Forest trees in eastern North America, for example, have been attacked by several European diseases. The chestnut blight, caused by a fungus originally from Europe, virtually eliminated the American chestnut, formerly a dominant tree in forests of the Appalachian Mountains. Some individuals still resprout, but the sprouts are soon infected by the fungus and killed. Nearly all American elms over large areas of the East and Midwest have been killed by Dutch elm disease, caused by the fungus *Ceratocystis ulmi*, which

55.7 Tortoises are Raised at Charles Darwin Station
Conservationists remove tortoise eggs from nests. When the eggs hatch, the young are reared in captivity until they are large enough to defend themselves.

55.8 Coevolved Mutualists As the iiwi, a Hawaiian honeycreeper, inserts its bill into the corolla tube to extract nectar from a lobelia flower, it deposits pollen from flowers it visited previously. Declining populations of the honeycreeper also threaten the plant, which is left without a pollinator.

reached North America via Europe in 1930. Ecologists suspect that intercontinental movement of disease organisms caused extinctions in the past, but evidence of disease outbreaks is not usually preserved in the fossil record.

Loss of mutualists threatens some species

Many plants have mutualistic relationships with pollinators, but most of these mutualisms are not highly species-specific (see Chapter 52). On islands, however, where ecological communities contain relatively few species, plant–pollinator interactions often evolve to be highly specific. For example, a single species of *Lobelia* colonized the Hawaiian Islands, where it eventually gave rise to 110 of the 350 species in the genus. A single colonizing species of songbird gave rise to at least 47 species of honeycreepers, some of which have long, slender, curved bills. These nectar-feeding birds were the only pollinators of many species of Hawaiian lobelias, whose long, curved tubular flowers match the shapes of their bills (Figure 55.8).

Today, half of the nectar-feeding birds of Hawaii are extinct, leaving many lobelias without pollinators. Many of these lobelias still survive, but populations of some species have been reduced to only a few individuals. A few species reproduce now only because biologists artificially pollinate them.

Habitat destruction and fragmentation are important causes of extinctions today

The 6 billion people that exist on Earth today are fed, clothed, and housed by agricultural and forestry industries that convert natural ecological communities containing many species into highly modified commu-

nities dominated by one or a few species of plants. Within these communities, humans discourage the presence of other species by applying chemicals that kill competing plants, bacteria, fungi, nematodes, insects and other arthropods, and vertebrates.

Although agricultural ecosystems have long harbored fewer species than the complex ecosystems they replaced, only recently have farmers planted large tracts of land in single crops. Traditional farmers planted many different crops together, maintaining some of the diversity that is key to the success of natural communities (Figure 55.9). Many species that can-

55.9. Many Species of Plants Grow in Traditional Gardens Traditional farmers plant many different crops together, as in this terrace garden in Honduras. The diversity maintained by these agricultural methods is vital to the continued well-being of the world's food crops.

TABLE 55.1   Co-option of Net Primary Production by Human Manipulations of Ecosystems

HABITAT CATEGORY	NET PRIMARY PRODUCTION CO-OPTED[a]
Cultivated land	15.0
Grazing land	11.6
Forest land	13.6
Human-occupied areas	0.4
Total	40.6
Total net primary production	132.1
Percent co-opted	30.7

[a] Values are in petagrams (one petagram = 10^{15} grams).

not survive in intensive modern agricultural systems thrive in traditional ones.

When ecosystems such as agricultural lands and plantation forests are managed so as to divert most of their primary production to certain species favored by people, we say that their production is **co-opted**. Agriculture and forestry today are so extensive that more than 30 percent of all terrestrial production is co-opted for human use (Table 55.1). All the other species on Earth have only two-thirds of the total global terrestrial production available for their use, and the fraction is steadily decreasing.

Because of increasing habitat modification and the co-option of primary production, habitat loss is certain to be the most important cause of species extinctions during the next century. The habitats required by some species are being completely destroyed. Other habitats, particularly old-growth forests, natural grasslands, and estuaries, are being reduced to widely separated patches that are too small to sustain populations of many of the species that live in them.

As habitats are progressively destroyed, the remaining patches become smaller and more isolated. Small habitat patches are qualitatively different from larger patches of the same habitat in ways that affect the survival of species. Small patches cannot maintain populations of species that require large areas, and they support only small populations of many of the species that can survive in them.

In addition, the fraction of a patch that is influenced by effects originating in adjacent habitats—**edge effects**—increases rapidly as patch size decreases (Figure 55.10). Close to the edges of forest patches, for example, winds are stronger, temperatures are higher, humidities are lower, and light levels are higher than they are farther inside the forest. Species from surrounding habitats often invade the edges of patches to compete with or prey upon the species living there.

More than 70 percent of the old-growth coniferous forests of the northwestern United States have been cut, mostly during the past 40 years. Second-growth forests in this area are being cut before they are 80 years old, so they do not acquire key characteristics of old-growth forests, such as trees of all ages (some of the conifers live more than 500 years), many standing dead trees, and large decomposing logs on the forest floor (Figure 55.11a). Among the species that require old growth to maintain viable populations are salamanders, which live inside rotting logs, the only microhabitat that retains moisture during the dry summers; the spotted owl, a species that hunts for rodents within mature forests; and the marbled murrelet, a seabird that nests on horizontal branches of large trees in coastal forests (Figure 55.11b).

As the fraction of the area remaining in old-growth forest is reduced, the home ranges of spotted owls must increase; hunting success, and hence reproductive success, thus decreases, and juvenile mortality during dispersal between patches increases. The spotted owl faces a high risk of being extirpated in most of Washington and Oregon within the next 50 years if clear-cutting of old-growth forests continues.

Usually we do not know which organisms lived in an area before its habitats became fragmented. To address this problem, a major research project near Manaus, Brazil, was launched before logging took place. The landowners agreed to preserve forest patches of certain sizes and locations, and censuses of those patches were conducted while the areas were still part of the continuous forest. Soon after the surrounding forest

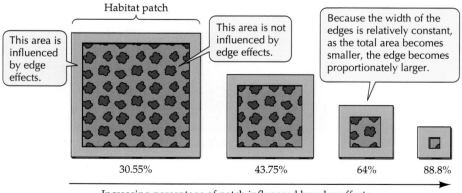

Habitat patch

This area is influenced by edge effects.

This area is not influenced by edge effects.

Because the width of the edges is relatively constant, as the total area becomes smaller, the edge becomes proportionately larger.

30.55%   43.75%   64%   88.8%

Increasing percentage of patch influenced by edge effects

55.10 Edge Effects   The smaller the patch, the greater the proportion of the patch that is influenced by detrimental edge effects.

(a)

(b) Strix occidentalis caurina

55.11 Old-Growth Forests and Owls (*a*) Old-growth coniferous forests of the Pacific Northwest have trees of all ages, and there are many large logs on the forest floor. (*b*) The northern spotted owl reproduces at high rates only in old-growth forests.

was cut, species began to disappear from the isolated patches (Figure 55.12). The first species to be eliminated were monkeys with large home ranges, such as the black spider monkey, the tufted capuchin, and the bearded saki (Figure 55.13*a*).

Birds that follow swarms of army ants to capture insects flushed by the ants also disappeared quickly from small patches (Figure 55.13*b*). A particular colony of army ants is a useful resource for the birds only

(a) Ateles sp.

55.12 Brazilian Forest Fragments Studied for Species Loss Isolated patches (foreground) lose species much more quickly than patches connected to the main forest do, even if the isolated patches, such as the one in the foreground, are larger than the connected ones (in background).

(b)

55.13 Extinction in Patches (*a*) The spider monkey was rapidly displaced by the logging industry in South America. (*b*) The white-plumed antbird, a species common in Brazilian forests, has become extinct in isolated forest plots, but survives in connected patches.

(a)

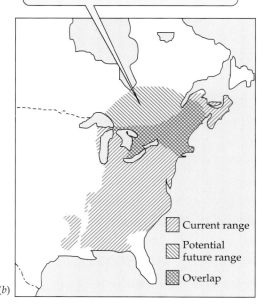

If the climate of eastern North America warms by as little as 4°C, about half the potential future range of beech trees will be beyond the northernmost extent of the current range.

Current range

Potential future range

Overlap

(b)

55.14 Beeches Are Threatened by Climate Warming
(a) Seedlings and saplings abound in this healthy beech forest. (b) If the climate of eastern North America warms by as little as 4°C, about half the potential future range of beech trees will be beyond the northernmost extent of their current range.

when it is raiding (about 27 days out of the 35-day period between colony moves). Therefore, the birds must have access to a number of colonies to be guaranteed that there is always at least one colony raiding. In small patches with few ant colonies, there are periods when none are raiding. During these days, ant-following birds have trouble finding enough to eat.

Climate change may cause species extinctions

Atmospheric scientists predict that, as a result of increasing concentrations of greenhouse gases (gases such as carbon dioxide and methane that are transparent to sunlight but opaque to heat radiated from Earth), average temperatures in North America will increase 2 to 5°C by the end of the twenty-first century. Conservation biologists are attempting to predict the effects of this warming trend on North American deciduous forest trees. If the climate warms by only 1°C, the average temperature formerly found at a particular location will shift 150 km north of that location. An organism that survives best at that average temperature will therefore need to shift its range 150 km north to remain in the same climate. Therefore, trees would need to shift their ranges as much as 500 to 800 km in a single century if the climate warmed 2 to 5°C.

Most deciduous forest trees grow for long periods before they begin to reproduce, and their seeds move only short distances. The American beech, like other large-seeded species, cannot shift its range rapidly. Beeches, whose seeds are dispersed primarily by jays,

advanced northward only about 20 km per century when glaciers retreated following the last glacial period. Beeches would thus have to migrate 40 times faster than they did in the past to keep up with the anticipated rate of climate change (Figure 55.14). Even though there might be areas of climate suitable for them, beeches probably could not reach them without human assistance. To maintain beech forests, we may need to intervene by moving seeds and by assisting seedling establishment.

In the past, global climate changed at a much slower rate than that predicted for the twenty-first century. Most organisms were able to shift their ranges rapidly enough to keep up with the changes. In addition, habitats were more continuous, and migration routes were not blocked by extensive areas of unsuitable human-modified habitats, as they are today. Organisms may therefore have much more difficulty dealing with climate changes during the twenty-first century than they did during glacial periods.

If Earth warms as predicted, climatic zones will not simply shift northward; new climates will develop and some existing climates will disappear. New climates are certain to develop at low elevations in the tropics. All models predict that the climate will warm less in tropical regions than at high latitudes, but a warming of even 2°C would result in climates near sea level that are hotter than those found anywhere in the humid tropics today. Adaptation to those climates may prove difficult for many tropical organisms.

Designing Recovery Plans

Once the causes of endangerment of species have been identified, appropriate remedies can be designed. If the cause of endangerment is overexploitation, harvesting can be controlled or banned. If habitat destruction is the cause, existing habitats can be protected and damaged habitats restored. If habitat isolation is the cause, individuals can be moved among subpopulations to increase dispersal rates. In this section we will describe how good analyses of population viability have been used to implement management actions designed to prevent species from becoming extinct by reducing mortality rates, reintroducing species to areas from which they have been extirpated, and raising endangered species in captivity.

Preserving habitats is always important

No population is viable without suitable habitat in which individuals can survive and reproduce. An excellent way to maintain viable populations is to set aside areas in which species and their habitats are protected. Protected areas may be established to preserve single species, but usually they are designed to main-

tain entire communities and ecosystems. Often single species of special economic, aesthetic, or ecological value serve as indicators of the viability of entire communities. The protection of single species that require large areas of suitable habitat—so-called "umbrella species" such as elephants, grizzly bears, and spotted owls—may promote the survival of most or all other species living in that ecosystem.

Demographic parameters may need to be manipulated

If the cause of a population decline is low reproductive success or unusually high mortality rates, managers may be able to intervene to change those rates. In the United States, losses of birds' nests to predators, such as crows, jays, native mammals, and domestic cats and dogs, are very high, especially in suburban areas. Many species of songbirds are declining in eastern North America because they cannot produce enough offspring to replace adult losses (Figure 55.15). Suburbanites need to recognize that their pets may be contributing to these declines.

Brown-headed cowbirds, which lay their eggs in other birds' nests, have increased dramatically in many parts of North America as a result of land clearing and associated high densities of domestic live-

55.15 Declining Songbird Populations in the Eastern United States These graphs show the results of long-term counts in specific forest patches in the Middle Atlantic states. Brood parasitism and other predation on songbird nests is making it difficult for some populations to replace themselves.

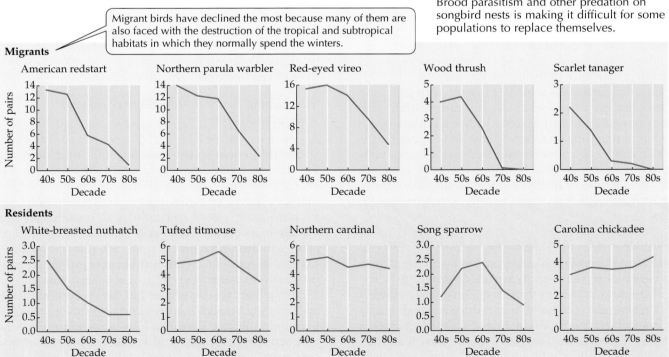

Migrant birds have declined the most because many of them are also faced with the destruction of the tropical and subtropical habitats in which they normally spend the winters.

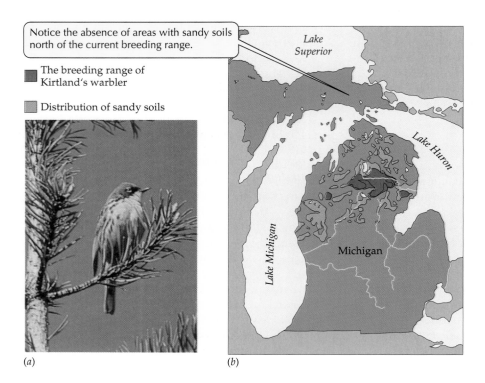

Notice the absence of areas with sandy soils north of the current breeding range.

■ The breeding range of Kirtland's warbler

□ Distribution of sandy soils

Lake Superior

Lake Huron

Lake Michigan

Michigan

(a)

(b)

55.16 The Kirtland's Warbler Is Threatened by Habitat Loss (*a*) A male Kirtland's warbler in a young jack pine. (*b*) The warbler's breeding range and the distribution of the sandy soils that support the stands of jack pines.

stock. Their avian hosts incubate the cowbird eggs and feed the cowbird nestlings in addition to, or instead of, their own offspring.

The Kirtland's warbler, an endangered species that nests only in young stands of jack pine on sandy soils in Michigan, is at risk from both heavy parasitism by cowbirds and loss of habitat (Figure 55.16). The current population of Kirtland's warbler is less than 1,000 individuals. Stands of jack pines depend on periodic fires for their persistence because jack pine cones remain closed on the branches. Only when heated do they open and release their seeds, which germinate in the ash on the floor of the burned forest.

Because Kirtland's warblers nest only in jack pine forests that are 8 to 18 years old, fire suppression measures have threatened to deprive the warblers of essential young forest habitat. To prevent further threat to the warblers, conservation biologists ignite controlled fires in jack pine forests to maintain a steady supply of forests of the right age. And they are removing brown-headed cowbirds to reduce nest parasitism rates.

Genotypes need to be matched to environments

Individuals are more likely to survive and reproduce successfully if their genotypes are appropriate to the environments in which they live. Thus, when conservation biologists prepared to introduce collared lizards to areas of the Ozark Mountains of southern Missouri from which the lizards had been extirpated, they thought carefully about where to get their colonists.

The open prairie vegetation required by collared lizards exists only in isolated, fire-prone glades on south-facing slopes with shallow soils, a habitat much reduced by agriculture and fire prevention (Figure 55.17). The Ozark populations of collared lizards, which have been isolated for about 2,000 lizard generations, are genetically distinct from those elsewhere in the range of the species. Therefore, the biologists decided to introduce lizards only from other glades in the Ozarks.

To obtain enough colonists, the founding lizards could not be taken from a single glade, because if fewer than ten mature lizards were introduced, the new population was unlikely to become established. And taking more than ten lizards from a single donor population would have threatened the future of the donor population. Therefore, lizards from at least five different glades were released together in a single new glade. Because individuals from different donor populations have distinct maternally inherited mitochondrial DNA markers, investigators will be able to determine which individuals have reproduced successfully. They can then use this information to select individuals to introduce to other glades.

Maintaining keystone species may prevent widespread extinctions

As we discussed in Chapter 52, keystone species exert strong influences on the structure and functioning of the ecological communities in which they live. We can characterize keystone species in terms of the strength of their effects on a community or ecosystem trait. One such measure, called **community importance**, is the change in a community or ecosystem trait per unit of change in the abundance of a species. A *trait* is a quantitative feature of a community or ecosystem, such as productivity, nutrient cycling, or species richness.

Mutualistic relationships that depend on keystone species may be vital for the survival of many species in a community. In Peruvian forests, for example, only a dozen species of figs and palms support an entire community of large fruit-eating birds and mammals

55.17 Both the Glades and the Lizards are Endangered (*a*) Open glades exist only as small patches on south-facing slopes in the Ozark Mountains. (*b*) The collared lizards of the Ozarks, which live only in these open glades, are genetically different from those elsewhere in the range of the species.

during the part of the year when fruits are least available. Loss of these few tree species would probably eliminate most of these birds and mammals, even if hundreds of other tree species remained. In turn, loss of the fruit-eaters might seriously impair the dispersal of the seeds of many other tree species. Thus, many mutualistic relationships are probably maintained by a few keystone tree species, which constitute only a small fraction of the 2,000 species of trees in the forest.

Because the extinction of keystone species could result in the extinction of many species in their communities, conservation biologists try to identify keystone species and take action to preserve them.

Captive propagation has a role in conservation

Species being threatened by overexploitation, loss of habitat, or environmental degradation through pollution can sometimes be maintained in captivity while the external threats to their existence are reduced or removed. Research on nutrition and the preparation of suitable diets, on the use of vaccinations and antibiotics, and on the control and enhancement of reproduction by both behavioral and technical means (such as artificial insemination and embryo transfers) supports captive propagation efforts.

Captive propagation is only a temporary measure that buys time. Existing zoos, aquariums, and botanical gardens do not have enough space to maintain adequate populations of more than a small fraction of Earth's rare and endangered species. Nonetheless, captive propagation can play an important role by maintaining species during critical periods and by providing a source of individuals for reintroduction into the wild. Captive propagation projects in zoos have also been very effective in raising public awareness of the rate at which species are threatened with extinction.

THE PEREGRINE FALCON. In 1942, about 350 pairs of peregrines bred in the United States east of the Mississippi River. This breeding population disappeared entirely by 1960, and no peregrines are known to have reproduced in the region during the next 20 years. The cause of their disappearance was the widespread use of organochlorine pesticides, such as DDT and dieldrin. These pesticides degrade very slowly in the environment and gradually accumulate in the prey of predators such as falcons. Their accumulation in the peregrines' bodies interfered with the deposition of calcium in eggshells. As a result, most of the falcons' eggs broke before they hatched. The successful reintroduction of the peregrine falcon to eastern North America depended on a strong captive propagation program.

Much of eastern North America became suitable habitat for peregrines again when the use of DDT was terminated by federal laws in the United States. Captive breeding of peregrines began at Cornell University in 1970, and by the end of 1986 more than 850 birds reared in captivity had been released in 13 eastern states, with spectacular success (Figure 55.18). In addition, some individuals adapted to live in urban environments, using ledges on buildings and other structures of cities as nesting sites (Figure 15.19). Peregrines probably would have recolonized the East by themselves, but they would have done so much more slowly without human assistance.

THE CALIFORNIA CONDOR. With its 9-foot wingspan, the California condor is North America's largest bird. Two hundred years ago, condors ranged from southern British Columbia to northern Mexico, but by the 1940s they were confined to a small region in the mountains and foothills north of Los Angeles. By 1978, the wild population was plunging toward extinction—only 25

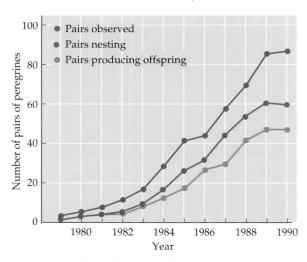

55.18 Peregrine Falcon Populations Have Been Reestablished Throughout the eastern United States, many pairs of peregrine falcons now attempt to reproduce, most of them successfully.

to 30 birds remained. In 1985 alone, 6 of the remaining 15 wild birds disappeared. To save the condor from extinction, biologists initiated a captive propagation program in 1983.

To maximize genetic variation in the captive population, all of the remaining wild birds were captured, the last one in April 1987. The first chick conceived in captivity hatched in 1988. By 1993, nine captive pairs were producing chicks, and the captive population had increased to more than 60 birds. The captive population was large enough to risk releasing 6 captive-bred birds in the mountains north of Los Angeles in 1992 (Figure 55.20).

55.19 A New Home in the City Peregrine falcons have responded well to captive propagation. Some individuals have adapted to urban life, nesting on the tall buildings of cities and feeding on pigeons.

Gymnogyps californianus

55.20 Back in the Wild Captive propagation of the California condor is providing the individuals that are being reintroduced into the species' former range.

The released condors are being provided with contaminant-free food in remote areas, and 3 of them were still alive in the spring of 1994. The released birds are using the same roosting sites, bathing pools, and mountain ridges that their predecessors did. More captive-reared birds also were released late in 1996 in northern Arizona. It is still too early to pronounce the program a success, but without captive propagation, the California condor would probably be extinct today.

CAPTIVE PROPAGATION HAS HIGH COSTS AND BENEFITS The California condor rehabilitation program costs about 1 million dollars a year. The Peregrine Fund at Cornell spent nearly 3 million dollars over the past 25 years; the expenses of other cooperating agencies add at least another half million to the total. These amounts may seem large, but they are small compared with the costs of other human activities, and compared with the cost of continued loss of species. The work needed to restore all of the world's threatened birds of prey probably could be accomplished with 5 million dollars per year, the approximate cost of one armored tank.

Establishing Priorities for Conservation Efforts

Many species and ecosystems are threatened, but the resources that can be allocated to preservation efforts are limited. How can those resources be allocated to achieve the most conservation benefits? Because many species can survive only in the ecological communities in which they evolved, preserving complete ecological communities and habitats is vital. Because only a small fraction of the landscape can be incorporated into parks and reserves, proper selection of those habitats is of great importance.

The primary function of parks, sanctuaries, and reserves is to maintain species and ecosystems relatively free of human disturbance. Parks are being established in many countries. Although their size and number will never be equal to the task of ecosystem and species preservation, more parks are urgently needed. Where should they be established?

Where should parks be established?

Two kinds of areas are high-priority sites for protection by parks and reserves—those that are home to unusually large numbers of species, and those that have many endemic species that live nowhere else. Because tropical ecosystems are generally richer in species than are ecosystems at higher latitudes (see Chapter 54), losses of tropical habitats threaten more species than do losses of comparable areas of temperate habitats.

The number of species that become extinct as a result of habitat destruction also depends on how many local species are found nowhere else. For example, nearly all the mammals and birds of Madagascar are found only on that island. Therefore, if the small fragments of tropical forests remaining on Madagascar are destroyed, the species dependent on them are certain to be exterminated in the wild (Figure 55.21).

Species with very small ranges are frequently found on islands, but some mainland regions also have many species found nowhere else. The Rift Valley lakes of Africa harbor more than 1,000 species of fish, most of which live in only one lake. The Atlantic coastal forests of southeastern Brazil are another center of endemism. Because only about 1 percent of the original extent of those forests remains, many species there have become extinct or are in danger of immediate extinction. Mountainous regions have many endemic species because temperature and rainfall change rapidly with elevation, creating many distinct habitats within a small area.

Centers of endemism are not the same for all groups of organisms. The Cape region at the southern tip of Africa, for example, has a flora of 8,500 species, 80 percent of which are endemic, but only 4 of the 187 species of birds found there are endemic.

Conservation biologists examine a region's conservation potential

To identify the places where conservation efforts are likely to provide the greatest benefits, conservation bi-

ologists compare countries with respect to estimates of the number of endemic species, the amount of land that is protected, and the amount of unprotected land that remains in a more or less natural state. Using these criteria, conservation biologists at the World Wildlife Fund divided the countries of the Indo-Pacific region into four categories (Figure 55.22). Category I countries have much land already protected and much land remaining in high-quality forests. Additional reserves can and should be established in these countries while the land is still available. Category III countries have little land already protected, but much land that could become reserves. They are high-priority countries for establishment of reserves that would transfer them to Category I. Countries in the other two categories have little land still available to be incorporated into protected reserves. Consequently, they are lower-priority targets for conservation efforts.

The Importance of Commercial Lands

In the majority of countries, most parks must be established in already settled areas because few pristine areas remain. The people living there cannot be evicted, nor is it appropriate, in most cases, to prevent hungry people from settling in or hunting in the parks. The high rates of human population growth in most tropical countries guarantee that pressures on parks from agricultural settlers will increase rather than decrease.

For these reasons, lands that are exploited for food, medicines, and fiber must play an important role in conservation. These lands are far more extensive than parks and reserves, and they include climates and ecosystems not represented in parks. Fortunately, many species can be preserved on lands that are being used in economically beneficial ways. Only a few species, such as predators on humans and domestic animals, or large, destructive herbivores, are incompatible with most human uses of land.

Some economic land uses are compatible with conservation

Forest reserves in which economically valuable products are harvested can support both species preservation and economic development. In Belize, about 75 percent of primary health care is provided by traditional healers using plant remedies. Persons known as *hierbateros* collect medicinal plants in the forests and sell them to the *curanderos* who actually provide the health care. A botanist and an economist determined that two 1-hectare plots in second-growth forest, one 30 years old, the other 50 years old, yielded 309 and 1,434 kg dry weight of medicines per year, which were sold at an average price of U.S.$2.80 per kilogram. Thus, these 1-hectare plots yielded gross revenues of

Andansonia grandieri (giant baobob tree)

55.21 Madagascar Abounds with Endemic Species The majority of plant and animal species found on the island of Madagascar off Africa's east coast are found nowhere else in the world.

Southern spiny forest

Indri indri (indri)

Furcifer revocosus (warty chameleon)

Enlemur fulvus fulvus (brown lemur)

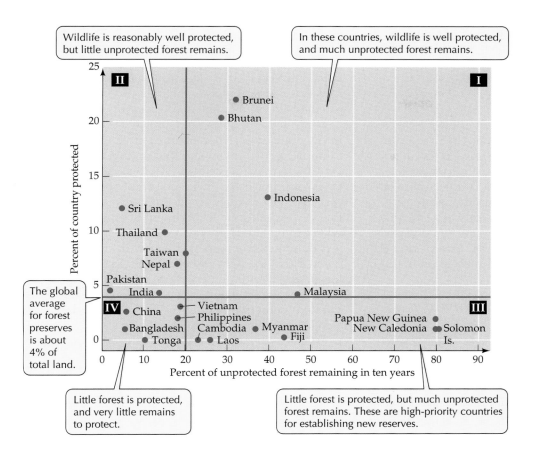

Wildlife is reasonably well protected, but little unprotected forest remains.

In these countries, wildlife is well protected, and much unprotected forest remains.

The global average for forest preserves is about 4% of total land.

Little forest is protected, and very little remains to protect.

Little forest is protected, but much unprotected forest remains. These are high-priority countries for establishing new reserves.

55.22 Identifying Conservation Priorities
In the World Wildlife Fund scheme, countries are grouped in four categories on the basis of the total area of their forest reserves and the area available to be incorporated into reserves.

areas managed for economically valuable products. Some of the reserves are the homes of indigenous people who continue to use the environment in their traditional ways.

The largest Costa Rican megareserve is La Amistad Biosphere Reserve, an area of more than 500,000 hectares that includes three national parks, a biological reserve, five Indian reservations, and two forest reserves. La Amistad contains the largest tract of highland vegetation in Central America, and it has considerable hydroelectric generating potential. All the native large predators still survive there. Managed properly, the reserve can provide drinking water, electricity, forest products, and nature tourism, protect indigenous cultures, and preserve

$865 and $4,017, which are greater than the incomes that would result from cultivating squash and corn on that land. Harvesting of medicinal plants can be compatible with agriculture if agricultural plots are allowed to regrow into forests for several decades before they are cleared again. These second-growth forests support many species that would otherwise not survive in the area.

Conservation requires large-scale planning

Knowing that large areas are required to preserve many species, conservation biologists promote the establishment of megareserves. A typical **megareserve** is a large area of land that has a central core of undisturbed habitat. Surrounding the core are buffer areas in which economic activity is permitted as long at it does not destroy the ecosystem. Appropriate activities may include sustainable harvesting of animal populations and plant products such as rubber, fruits, nuts, and wood. On the edges of a megareserve is a zone in which more intensive land uses, such as agriculture or plantation forestry, are permitted.

Costa Rica has pioneered the development of megareserves. It has consolidated its parks and reserves into eight megareserves that should maintain about 80 percent of the country's biodiversity (Figure 55.23). Each megareserve includes natural areas and

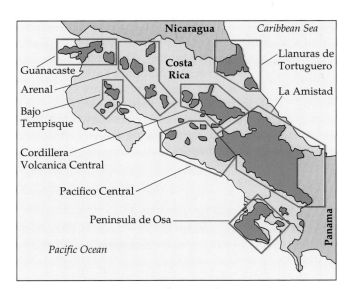

55.23 The Megareserves of Costa Rica The eight areas outlined in red are being managed for both biodiversity and economic activities. The green areas are the current reserves.

species. In combination, these benefits outweigh what could be gained by logging the steep slopes and converting them to low-productivity agricultural systems.

Restoring Degraded Ecosystems

Many areas that could be incorporated into reserves have been highly altered by human activities. Some of these areas can play their intended roles in biodiversity conservation only if they are restored to their original state. To accomplish this task, a subdiscipline of conservation biology, known as **restoration ecology**, is growing rapidly. Research on methods of restoring species and ecosystems is needed because many ecological communities will not recover, or will do so only very slowly, without creative intervention in the recovery process.

Tropical deciduous forests are being restored in Costa Rica

The world's largest restoration project is under way in Guanacaste National Park in northwestern Costa Rica. Its goal is to restore a large area of tropical deciduous forest, the most threatened ecosystem in Central America, from small fragments that remain in an area converted primarily to cattle pastures. One approach to restoration would be to exclude fire and domestic livestock from the park and let nature take its course. Grass patches of less than 120 hectares would be covered by woody vegetation within 20 years, but large expanses of pasture would require 50 to 200 years to regrow because tree seeds move into open pastures only very slowly.

Reforestation can be speeded up by manipulating the habitat, which is what Daniel Janzen, architect of the restoration project, is doing. To design his project, Janzen gathered and used basic ecological information about the abilities of different plant species to germinate and grow in open pastures. The single most important threat to Guanacaste National Park during the coming decades is fires, most of which are started by people. The introduced pasture grasses produce dense, highly flammable stands that spawn hot fires that penetrate far into surrounding forests. Domestic livestock keep these grasses under control by grazing, and they also disperse the seeds of native trees that are good at invading pastures. The restoration program is encouraging some grazing by domestic livestock in the park until plant succession has progressed to the point at which the grass no longer poses serious competition to the woody species and is no longer sufficiently dense to carry hot fires.

Restoring some habitats is difficult

Restoration of damaged and degraded habitats is an important activity, but ecologists still have limited ability to restore natural ecosystems. In the United States, the belief that existing ecosystems can be replaced has made it easy to get permits for development projects that destroy valuable habitats. Developers need only state that they will create substitutes for the areas they are destroying. Promising to restore a habitat, however, is much easier than doing it. Even the most experienced wetland ecologists, for example, are having great difficulty creating new wetlands that mimic those being destroyed.

An example is the "restored" wetland that was conceived as part of a compensation agreement that allowed the California Department of Transportation to widen Interstate Highway 5 near San Diego. The project damaged a marsh and jeopardized two endangered birds, the light-footed clapper rail and the least tern, and an endangered plant, the salt marsh bird's beak. Despite stringent, court-imposed standards and the involvement of wetland experts, the endangered birds were still not breeding in the "restored" marsh 9 years after it was created. The restoration ecologists have not given up, but the advice offered by a recent National Research Council committee on wetland restoration needs to be heeded: "Wetland restoration should not be used to mitigate avoidable destruction of other wetlands until it can be scientifically demonstrated that the replacement ecosystems are of equal or better functioning."

Markets and Conservation

Most species are common property resources—that is, they are "owned" by everybody. Because no individual or group of individuals typically has strong incentives to use common property resources sustainably, their preservation usually depends on central governments. Unfortunately, governments generally lack sufficient resources to do the job, and governmental actions sometimes are not well attuned to local situations. For these reasons, establishing property rights that allow owners to receive the economic benefits from managing biological resources can, under some conditions, assist conservation efforts. Such use of property rights and markets is most likely to be successful if the species being managed are relatively sedentary and if a reliable supply of the products derived from them is commercially important.

A good example of the use of property rights to preserve species is provided by butterfly farming in Papua New Guinea. Many species of butterflies in that country, and elsewhere in the tropics, are threatened by habitat destruction, environmental degradation, and overexploitation. By the mid-1960s, butterfly collecting and commercial harvesting, which provided butterflies for mounting in plastic and glass trays, tabletops, decorative screens, and even clear plastic

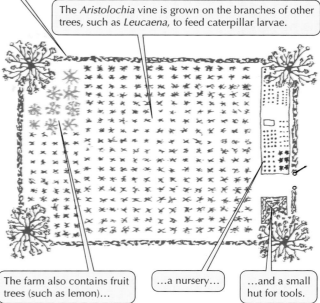

A typical butterfly farm is surrounded by a hedge of *Hibiscus*, *Ixora* and *Poinsettia* to keep pigs out and to provide nectar.

The *Aristolochia* vine is grown on the branches of other trees, such as *Leucaena*, to feed caterpillar larvae.

The farm also contains fruit trees (such as lemon)…

…a nursery…

…and a small hut for tools.

55.24 Butterfly Farm in Papua New Guinea Birdwing butterfly species are a valuable economic asset in Papua New Guinea.

toilet seats, had endangered some of the most striking and valuable birdwing butterflies.

In 1966, the government of Papua New Guinea prohibited collecting and trading in seven species of birdwing butterflies and established several butterfly reserves. In 1974, the government initiated a butterfly farming program designed to give it a monopoly on trade in its butterflies and to return all economic benefits to the country's people. By 1978, about 500 butterfly farms existed, operated by villagers who planted flowering hibiscus vines to attract adult butterflies and other plants on which the caterpillars feed (Figure 55.24). Local people tend the caterpillars and house the pupae so that the butterflies can be collected when they emerge. Many butterflies are still collected in the wild in Papua New Guinea, but the farms are a significant source of specimens, and they take pressure off the butterfly populations in the surrounding forest lands. Because farmed butterflies are generally in very good condition, they command high prices.

Investing to Preserve Biodiversity

Genes, species, and habitats are the sources of many economic benefits to people. However, most of these benefits will assist future generations, not the current one. We do not know how future generations will value these benefits or the biodiversity that generates them. We do not even know how the current generation values them. We also cannot predict which species will turn out to be sources of valuable foods, medicines, or drugs. Most species probably will not.

Between 1951 and 1981, the National Cancer Institute screened more than 100,000 extracts from 35,000 different species. To date, only one compound—taxol, derived from the Pacific yew tree—has received approval as a drug component. That drug is very valuable, but many species had to be searched to find it.

Although the chance that any given species will turn out to be the source of some valuable marketable product is small, biodiversity has another vital feature: extinction is forever. If we purposely or inadvertently exterminate a species, we have irreversibly destroyed a resource of unknown value. The irreversibility of ex-

tinction makes loss of biodiversity an urgent public issue.

How much should societies invest to preserve biodiversity? This question does not have an ecological answer. Economists and evolutionary biologists can contribute valuable information to the public debate, but the final decision is an ethical, spiritual, and political one that will depend on our beliefs about our responsibilities to the other organisms that share Earth with us.

The preservation of biological diversity and ecosystem services is one of the greatest challenges facing humankind. Many of the scientific tools needed for the task are already available, but appropriate use of these tools requires major changes in people's attitudes toward other species. If species are valued only because they are economically useful to us, increased losses of species are inevitable. Other uses of natural habitats are likely to be seen as more profitable, at least in the short run. Even though a wetland has great value, some people usually can make enough money in the short run by destroying it to be motivated to do so. Only when we value biological diversity and ecosystem functioning as the heritage of all humankind, a heritage to be passed on to our descendants as completely as possible, will we begin to reduce the current alarming rates of ecosystem destruction and species extinctions.

Summary of "Conservation Biology"

Estimating Current Rates of Extinction
• Estimates of current rates of extinction are based primarily on species–area relationships and rates of tropical deforestation. **Review Figure 55.1**

Why Care about Species Extinctions?
• Species provide the food, fiber, medicines, and aesthetic opportunities upon which human life depends.
• Ecosystems provide services that can be replaced only by expensive and continuing human effort. **Review Figure 55.4**

Determining Causes of Endangerment and Extinction
• Overexploitation, which historically resulted in most human-caused extinctions, is still an important cause of species extinctions today.
• Introduced species have caused many extinctions, particularly on islands. **Review Figure 55.7**
• Habitat destruction is the most important cause of species extinctions today. **Review Table 55.1**
• The fragmentation of habitat into patches that are too small to support populations is a major cause of extinctions.
• Edge effects that increase the detrimental effects of surrounding terrain on a habitat increase as habitat patch size decreases. **Review Figures 55.10, 55.12, 55.15**
• The predicted rapid warming of Earth's climate as a result of human activity could make survival difficult for many species that cannot shift their ranges quickly enough. **Review Figure 55.14**

Designing Recovery Plans
• The best way we know of to maintain populations is to set aside areas in which species and their habitats are protected. Human manipulation of these protected environments and the populations that inhabit them is sometimes needed.
• The identification and preservation of keystone species, whose loss could result in the extinction of many other species that depend on them, is a high priority of conservation biologists.
• Captive propagation can play a useful role in conservation. **Review Figures 55.18, 55.19**

Establishing Priorities for Conservation Efforts
• More parks and reserves are needed. High-priority areas for their establishment are regions of unusually high species richness and endemism and countries where substantial tracts of undisturbed habitats still remain. **Review Figure 55.22**

The Importance of Commercial Lands
• Megareserves combine areas under total protection with areas where commercial exploitation of the ecosystem is allowed as long as it does not damage ecosystem processes. **Review Figure 55.23**

Restoring Degraded Ecosystems
• Tropical deciduous forest, an endangered ecosystem in Central America, is being restored in Costa Rica. **Review Figure 55.24**

Markets and Conservation
• Properly employed, markets can help preserve biodiversity. **Review Figure 55.25**

Investing to Preserve Biodiversity
• Preserving biodiversity is not just a scientific issue. It raises serious moral and ethical concerns that define what it means to be a human being on Earth.

Self-Quiz

1. Which of the following is *not* currently a major cause of species extinctions?
   a. Habitat destruction
   b. Climate change
   c. Overexploitation
   d. Introduction of predators
   e. Introduction of diseases

2. When ecosystems are managed to favor strongly those species intended for human use, we say that their production is
   a. modified.
   b. diverted.
   c. co-opted.
   d. channeled.
   e. managed.

3. A minimum viable population is
   a. the estimated number of individuals necessary for the species to maintain genetic diversity.
   b. the estimated number of individuals necessary for the species to persist in all U.S. national forests.
   c. the estimated number of individuals necessary for the species to survive for several decades.
   d. the estimated number of individuals necessary for a species to maintain or increase its numbers in a region.
   e. the minimum density required for individuals to find mating partners.

4. Which of the following is *not* a component of a population vulnerability analysis?
   a. Spatial structure of a population
   b. Sex ratio within a population
   c. Amount of variation in birth and death rates
   d. Amount of heterozygosity and genetic variance
   e. Captive propagation of individuals

5. Which of the following is *not* an ecosystem service?
   a. Production of carbon dioxide
   b. Flood control
   c. Water purification
   d. Air purification
   e. Preservation of biological diversity

6. Conservation biologists are concerned about global warming because
   a. the rate of change in climate is projected to be faster than the rate at which many species can shift their ranges.

b. it is already too hot in the tropics.

c. climates have been so stable for thousands of years that many species lack the ability to tolerate variable temperatures.

d. climate change will be especially harmful to rare species.

e. none of the above

7. A species that is found only in a particular region is said to be

   a. an indicator species for that region.

   b. a restricted species.

   c. a vulnerable species.

   d. endemic to that region.

   e. demographically constrained.

8. A keystone species is one that

   a. preys heavily on a particular species.

   b. is especially vulnerable to extinction.

   c. is restricted to a small geographic area.

   d. experiences considerable demographic stochasticity.

   e. strongly influences the structure and functioning of its ecological community.

9. As a habitat patch gets smaller it

   a. cannot support populations of species that require large areas.

   b. supports only small populations of many species.

   c. is influenced to an increasing degree by edge effects.

   d. is invaded by species from surrounding habitats.

   e. all of the above.

10. Restoration ecology is an important discipline because

   a. many areas being incorporated into megareserves have been highly degraded.

   b. many areas being incorporated into megareserves are vulnerable to global climate change.

   c. many species suffer from demographic stochasticity.

   d. many species are genetically impoverished.

   e. fire is a threat to many reserves.

Applying Concepts

1. Most species driven to extinction by people in the past were large vertebrates. Do you expect this pattern to persist into the future? If not, why not?

2. Species endangered as a result of global climatic warming might be preserved if we move individuals from areas that are becoming unsuitable for them to areas likely to be better for them in the future. What are the major difficulties associated with such interventions? For what types of species would they work well? Poorly?

3. Conservation biologists have debated extensively which is better: many small reserves or a few large ones. What biological processes should be evaluated in making judgments about the size and location of reserves? To what extent should we be concerned with preserving the largest number of species rather than those species judged to be of unusual importance?

4. During World War I, French doctors adopted a "triage" system of dealing with wounded soldiers. The wounded were divided into three categories: those almost certain to die no matter what was done to help them, those likely to recover even if not assisted, and those whose prob-ability of survival would be greatly increased if they were given medical attention. The limited resources available to the doctors were directed primarily at the third category. What would be the implications of adopting a similar attitude toward species preservation?

5. Utilitarian arguments dominate discussions about the importance of preserving the biological richness of the planet. In your opinion, what role should moral arguments play?

Readings

Defenders of Wildlife. 1989. *Preserving Communities and Corridors.* Washington, D.C. A short collection of essays exploring the roles of corridors in preserving wildlife and how we can best implement important conservation legislation.

Flannery, T. 1994. *The Future Eaters.* Reed Books, Melbourne, Australia. A compelling examination of the consequences of unsustainable human activities. Focuses on the Australian region, but its message is applicable worldwide.

Gradwohl, J. and R. Greenberg. 1988. *Saving the Tropical Forests.* Island Press, Washington, D.C. A good account of the causes of tropical forest destruction and of successful projects throughout the world where local communities have averted forest destruction while reaping social and financial benefits.

Lawton, J. H. and R. M. May (Eds.). 1995. *Extinction Rates.* Oxford University Press, Oxford. Provides coverage of the quantitative and qualitative methods of estimating extinction rates and their ecological and evolutionary causes.

Meffe, G. K. and C. R. Carroll (Eds.). 1997. *Principles of Conservation Biology,* 2nd Edition. Sinauer Associates, Sunderland, MA. A multiauthored text that provides an extensive overview of conservation biology.

Perrings, C., K.-G. Mäler, C. Folke, C. S. Holling and B.-O. Jansson. 1995. *Biodiversity Loss: Economic and Ecological Issues.* Cambridge University Press, Cambridge. Reports findings of a research program that brought together economists and ecologists to consider the causes and consequences of biodiversity loss.

Primack, R. B. 1998. *Essentials of Conservation Biology,* 2nd Edition. Sinauer Associates, Sunderland, MA. An introductory text that combines theory with basic and applied research to explain the connections between conservation biology and other disciplines.

Repetto, R. 1990. "Deforestation in the Tropics." *Scientific American,* April. Argues that government policies that encourage exploitation (in particular, excessive logging and clearing for ranches and farms) are largely to blame for the accelerating destruction of tropical forests.

Tattersall, I. 1993. "Madagascar's Lemurs." *Scientific American,* January. Describes how diverse Madagascan habitats, and the endemic lemurs they house, are disappearing fast.

Terborgh, J. 1992. "Why American Songbirds Are Vanishing." *Scientific American,* May. Describes the sharp decline in the number of songbirds in eastern North America and tells why this trend will be difficult to reverse.

Western, D. and M. Pearl (Eds.). 1989. *Conservation for the Twenty-First Century.* Oxford University Press, New York. A set of essays by conservationists, governmental decision makers, and wildlife managers that identify gaps in knowledge and propose agendas for conservation action worldwide.

Wilson, E. O. 1992. *The Diversity of Life.* Belknap Press of Harvard University Press, Cambridge, MA. A readable book that outlines the processes that created the diversity of life, explains the threats to that diversity, and shows what we must do to preserve it.

Glossary

Abdomen (ab' duh mun) [L.: belly]   In arthropods, the posterior portion of the body; in mammals, the part of the body containing the intestines and most other internal organs, posterior to the thorax.

Abomasum (ab' oh may' sum)   The true stomach of ruminants (animals such as cattle, sheep, and goats).

Abscisic acid (ab sighs' ik) [L. *abscissio*: breaking off]   A plant growth substance having growth-inhibiting action. Causes stomata to close.

Abscission (ab sizh' un) [L. *abscissio*: breaking off]   The process by which leaves, petals, and fruits separate from a plant.

Absolute temperature scale   A temperature scale in which the degree is the same size as in the Celsius (centigrade) scale, and zero is the state of no molecular motion. Absolute zero is –273° on the Celsius scale.

Absorption   (1) Of light: complete retention, without reflection or transmission. (2) Of liquids: soaking up (taking in through pores or cracks).

Absorption spectrum   A graph of light absorption versus wavelength of light; shows how much light is absorbed at each wavelength.

Abyssal zone (uh biss' ul) [Gr. *abyssos*: bottomless]   That portion of the deep ocean where no light penetrates.

Accessory pigments   Pigments that absorb light and transfer energy to chlorophylls for photosynthesis.

Acclimatization   Changes in an organism that improve its ability to tolerate seasonal changes in its environment.

Acellular   Not composed of cells.

Acetylcholine   A neurotransmitter substance that carries information across vertebrate neuromuscular junctions and some other synapses. **Acetylcholinesterase** is an enzyme that breaks down acetylcholine.

Acetyl CoA (acetyl coenzyme A)   Compound that reacts with oxaloacetate to produce citrate at the beginning of the citric acid cycle; a key metabolic intermediate in the formation of many compounds.

Acid [L. *acidus*: sharp, sour]   A substance that can release a proton. (Contrast with base.)

Acid precipitation   Precipitation that has a lower pH than normal as a result of acid-forming precursors introduced into the atmosphere by human activities.

Acidic   Having a pH of less than 7.0 (a hydrogen ion concentration greater than 10^{-7} molar).

Acoelomate   Lacking a coelom.

Acquired Immune Deficiency Syndrome   See AIDS.

Acrosome (a' krow soam) [Gr. *akros*: highest or outermost + *soma*: body]   The structure at the forward tip of an animal sperm which is the first to fuse with the egg membrane and enter the egg cell.

ACTH (adrenocorticotropin)   A pituitary hormone that stimulates the adrenal cortex.

Actin [Gr. *aktis*: a ray]   One of the two major proteins of muscle; it makes up the thin filaments. Forms the microfilaments found in most eukaryotic cells.

Action potential   An impulse in a neuron taking the form of a wave of depolarization or hyperpolarization imposed on a polarized cell surface.

Action spectrum   A graph of biological activity versus wavelength of light. It compares the effectiveness of light of different wavelengths.

Activating enzymes (also called aminoacyl-tRNA synthetases)   These enzymes catalyze the addition of amino acids to their appropriate tRNAs.

Activation energy (E_a)   The energy barrier that blocks the tendency for a set of chemical substances to react. A reaction is speeded up if this energy barrier is surmounted by adding heat, or if the barrier is lowered by providing a different reaction pathway with the aid of a catalyst.

Active site   The region on the surface of an enzyme where the substrate binds, and where catalysis occurs.

Active transport   The transport of a substance across a biological membrane against a concentration gradient—that is, from a region of low concentration (of that substance) to a region of high concentration. Active transport requires the expenditure of energy and is a saturable process. (Contrast with facilitated diffusion, free diffusion; see primary active transport, secondary active transport.)

Adaptation (a dap tay' shun)   In evolutionary biology, a particular structure, physiological process, or behavior that makes an organism better able to survive and reproduce. Also, the evolutionary process that leads to the development or persistence of such a trait.

Adenine (a' den een)   A nitrogen-containing base found in nucleic acids, ATP, NAD, etc.

Adenosine triphosphate   See ATP.

Adenylate cyclase   Enzyme catalyzing the formation of cyclic AMP from ATP.

Adhesion molecules   See cell adhesion molecules.

Adrenal (a dree' nal) [L. *ad-*: toward + *renes*: kidneys]   An endocrine gland located near the kidneys of vertebrates, consisting of two glandular parts, the cortex and medulla.

Adrenaline   See epinephrine.

Adrenocorticotropin   See ACTH.

Adsorption   Binding of a gas or a solute to the surface of a solid.

Aerenchyma (air eng' kyma) [Gr. *aer*: air + *enchyma*: infusion]   Modified parenchyma tissue, with many air spaces, found in shoots of some aquatic plants. (See parenchyma.)

Aerobic (air oh' bic) [Gr. *aer*: air + *bios*: life]   In the presence of oxygen, or requiring oxygen.

Afferent (af' ur unt) [L. *ad*: to + *ferre*: to bear]   To or toward, as in a neuron that carries impulses to the central nervous system, or a blood vessel that carries blood to a structure. (Contrast with efferents.)

Age distribution   The proportion of individuals in a population belonging to each of the age categories into which the population has been divided. The number of divisions is arbitrary.

AIDS (Acquired immune deficiency syndrome)   Condition in which the body's helper T lymphocytes are destroyed, leaving the victim subject to opportunistic diseases. Caused by the HIV-I virus.

Air sacs   Structures in the avian respiratory system that facilitate unidirectional flow of air through the lungs.

Alcohol   An organic compound with one or more hydroxyl (–OH) groups.

Aldehyde (al' duh hide)   A compound with a –CHO functional group. Many sugars are aldehydes. (Contrast with ketone.)

Aldosterone (al dahs' ter own)   A steroid hormone produced in the adrenal cortex of mammals. Promotes secretion of potassium and reabsorption of sodium in the kidney.

Aleurone layer (al' yur own) [Gr. *aleuron*: wheat flour]   In grass seeds, a specialized cell layer just between the seed coat and the endosperm, synthesizing hydrolytic enzymes under the influence of gibberellin, and thus helping mobilize reserves for the developing embryo.

Alga (al' gah) (plural: algae) [L.: seaweed]   Any one of a wide diversity of protists belonging to the phyla Pyrrophyta, Chrysophyta, Phaeophyta, Rhodophyta, and Chlorophyta (and, formerly, Cyanophyta—"blue-green algae"). Most live in the water, where they are the dominant autotrophs; most are unicellular, but a minority are multicellular ("seaweeds" and similar protists).

Allele (a leel') [Gr. *allos*: other]   The alternate forms of a genetic character found at a given locus on a chromosome.

Allele frequency   The relative proportion of a particular allele in a specific population.

Allergy [Ger. *allergie*: altered reaction]   An overreaction to an antigen in amounts that do not affect most people; often involves IgE antibodies.

Allometric growth   A pattern of growth in which some parts of the body of an organism grow faster than others, resulting in a change in body proportions as the organism grows.

Allopatric (al' lo pat' rick) [Gr. *allos*: other + *patria*: fatherland]   Pertaining to populations that occur in different places.

Allopatric speciation   See geographical speciation.

Allopolyploid   A polyploid in which the chromosome sets are derived from more than one species.

Allostery (al' lo steer' y) [Gr. *allos*: other + *stereos*: structure]   Regulation of the activity of an enzyme by binding, at a site other than the catalytic active site, of an effector molecule that does not have the same structure as any of the enzyme's substrates.

Alpha helix   Type of protein secondary structure; a right-handed spiral.

Alternation of generations   The succession of haploid and diploid phases in a sexually reproducing organism. In most animals (male wasps and honey bees are notable exceptions), the haploid phase consists only of the gametes. In fungi, algae, and plants, however, the haploid phase may be the more prominent phase (as in fungi and mosses) or may be as prominent as the diploid phase (see the life cycle of *Ulva*, for example). In vascular plants, the diploid phase is more prominent.

Altruistic act   A behavior whose performance harms the actor but benefits other individuals.

Alveolus (al ve' o lus) (plural: alveoli) [L. *alveus*: cavity]   A small, baglike cavity, especially the blind sacs of the lung.

Amensalism (a men' sul ism)   Interaction in which one animal is harmed and the other is unaffected. (Contrast with commensalism, mutualism.)

Ames test   A test for mutagens (and possible carcinogens) based on the ability of a test compound to cause mutations in the bacterium, *Salmonella*.

Amine   An organic compound with an amino group (see Amino acid).

Amino acid   An organic compound of the general formula $H_2N–CHR–COOH$, where R can be one of 20 or more different side groups. An amino acid is so named because it has both a basic amine group, $–NH_2$, and an acidic carboxyl group, $–COOH$. Proteins are polymers of amino acids.

Ammonotelic (am moan' o teel' ic) [Gr. *telos*: end]   Describes an organism in which the final product of breakdown of nitrogen-containing compounds (primarily proteins) is ammonia. (Contrast with ureotelic, uricotelic.)

Amniocentesis   A medical procedure in which cells from the fetus are obtained from the amniotic fluid. The genetic material of the cells is then examined. (Contrast with chorionic villus sampling.)

Amniotic egg   The eggs of birds and reptiles, which can be incubated in air because the embryo is enclosed by a fluid-filled sac.

Amoeba (a mee' bah) [Gr. *amoibe*: change]   Any one of a large number of different kinds of unicellular protists belonging to the phylum Rhizopoda, characterized among other features by its ability to change shape frequently through the protrusion and retraction of cytoplasmic extensions called pseudopods.

Amoeboid (a mee' boid)   Like an amoeba; constantly changing shape by the protrusion and retraction of pseudopodia.

Amphi- [Gr.: both]   Prefix used to denote a character or kind of organism that occupies two or more states. For example, amphibian (an animal that lives both on the land and in the water).

Amphipathic (am' fi path' ic) [Gr. *amphi*: both + *pathos*: emotion]   Of a molecule, having both hydrophilic and hydrophobic regions.

amu (atomic mass unit, or dalton)   The basic unit of mass on an atomic scale, defined as one-twelfth the mass of a carbon-12 atom. There are 6.023×10^{23} amu in one gram. This number is known as Avogadro's number.

Amylase (am' ill ase)   Any of a group of enzymes that digest starch.

Anabolism (an ab' uh liz' em) [Gr. *ana*: up, throughout + *ballein*: to throw]   Synthetic reactions of metabolism, in which complex molecules are formed from simpler ones. (Contrast with catabolism.)

Anaerobic (an ur row' bic) [Gr. *an*: not + *aer*: air + *bios*: life]   Occurring without the use of molecular oxygen, O_2.

Anagenesis   Evolutionary change in a single lineage over time.

Analogy (a nal' o jee) [Gr. *analogia*: resembling]   A resemblance in function, and often appearance as well, between two structures which is due to convergence in evolution rather than to common ancestry. (Contrast with homology.)

Anaphase (an' a phase) [Gr. *ana*: indicating upward progress]   The stage in nuclear division at which the first separation of sister chromatids (or, in the first meiotic division, of paired homologues) occurs. Anaphase lasts from the moment of first separation to the time at which the moving chromosomes converge at the poles of the spindle.

Anaphylactic shock   A precipitous drop in blood pressure caused by loss of fluid from capillaries because of an increase in their permeability stimulated by an allergic reaction.

Ancestral trait   Trait shared by a group of organisms as a result of descent from a common ancestor.

Androgens (an' dro jens)   The male sex steroids.

Aneuploidy (an' you ploy dee)   A condition in which one or more chromosomes or pieces of chromosomes are either lacking or present in excess.

Angiosperm (an' jee oh spurm) [Gr. *angion*: vessel + *sperma*: seed]   One of the flowering plants; literally, one whose seed is carried in a "vessel," which is the fruit. (See fruit.)

Angiotensin (an' jee oh ten' sin)   A peptide hormone that raises blood pressure by causing peripheral vessels to constrict; maintains glomerular filtration by constricting efferent glomerular vessels; stimulates thirst; and stimulates the release of aldosterone.

Animal [L. *animus*: breath, soul]   A member of the kingdom Animalia. In general, a multicellular eukaryote that obtains its food by ingestion.

Animal hemisphere   The metabolically active upper portion of some animal eggs, zygotes, and embryos, which does *not* contain the dense nutrient yolk. The **animal pole** refers to the very top of the egg or embryo. (Contrast with vegetal hemisphere.)

Anion (an' eye one)   An ion with one or more negative charges. (Contrast with cation.)

Anisogamy (an' eye sog' a mee) [Gr. *aniso*: unequal + *gamos*: marriage]   The existence of two dissimilar gametes (egg and sperm).

Annual   Referring to a plant whose life cycle is completed in one growing season. (Contrast with biennial, perennial.)

Anorexia nervosa (an or ex' ee ah) [Gr. *an*: not + *orexis*: appetite]   Severe malnutrition and body wasting brought on by a psychological aversion to food.

Antennapedia complex A group of homeotic genes that control the development of the head and anterior thorax of the fruit fly, *Drosophila melanogaster*.

Anterior Toward the front.

Anterior pituitary The portion of the vertebrate pituitary gland that derives from gut epithelium and produces tropic hormones.

Anther (an' thur) [Gr. *anthos*: flower] A pollen-bearing portion of the stamen of a flower.

Antheridium (an' thur id' ee um) (plural: antheridia) [Gr. *antheros*: blooming] The multicellular structure that produces the sperm in bryophytes and ferns.

Antibody One of millions of blood proteins, produced by the immune system, that specifically recognizes a foreign substance and initiates its removal from the body.

Anticodon A "triplet" of three nucleotides in transfer RNA that is able to pair with a complementary triplet (a codon) in messenger RNA, thus aligning the transfer RNA on the proper place on the messenger. The codon (and, reciprocally, the anticodon) codes for a specific amino acid.

Antidiuretic hormone A hormone that controls water reabsorption in the mammalian kidney. Also called vasopressin.

Antigen (an' ti jun) Any substance that stimulates the production of an antibody or antibodies upon introduction into the body of a vertebrate.

Antigen processing The breakdown of antigenic proteins into smaller fragments, which are then presented on the cell surface, along with MHC proteins, to T cells.

Antigenic determinant A specific region of an antigen, which is recognized by and binds to a specific antibody.

Antiparallel Parallel but running in opposite directions. The two strands of DNA are antiparallel.

Antipodals (an tip' o dulls) [Gr. *anti*: against + *podus*: foot] Cells (usually three) of the mature embryo sac of a flowering plant, located at the end opposite the egg (and micropyle).

Antiport A membrane transport process that carries one substance in one direction and another in the opposite direction. (Contrast with symport.)

Antisense nucleic acid A single-stranded RNA or DNA complementary to and thus targeted against the mRNA transcribed from a harmful gene such as an oncogene.

Anus (a' nus) Opening through which digestive wastes are expelled, located at the posterior end of the gut.

Aorta (a or' tuh) [Gr. *aorte*: aorta] The main trunk of the arteries leading to the systemic (as opposed to the pulmonary) circulation.

Apex (a' pecks) The tip or highest point of a structure, as the apex of a growing stem or root.

Apical (a' pi kul) Pertaining to the apex, as the apical meristem, which is the actively growing tissue at the tip of a stem or root.

Apomixis (ap oh mix' is) [Gr. *apo*: away from + *mixis*: sexual intercourse] The asexual production of seeds.

Apoplast (ap' oh plast) In plants, the continuous meshwork of cell walls and extracellular spaces through which material can pass without crossing a plasma membrane. (Contrast with symplast.)

Apoptosis (ay' pu toh sis) A series of genetically programmed events leading to cell death.

Appendix A vestigial portion of the human gut at the junction of the ileum with the colon.

Apterous Lacking wings.

Assisted reproductive technologies (ART) Any of a number of technological approaches to improve human fertility. They include in vitro fertilization of eggs, injection of sperm and eggs into the oviduct, and injection of individual sperm into individual eggs.

Archaebacteria (ark' ee bacteria) [Gr. *archaios*: ancient] One of the two kingdoms of prokaryotes; the archaebacteria possess distinctive lipids and lack peptidoglycan. Most live in extreme environments. (Contrast with eubacteria.)

Archegonium (ar' ke go' nee um) [Gr. *archegonos*: first of a kind] The multicellular structure that produces eggs in bryophytes, ferns, and gymnosperms.

Archenteron (ark en' ter on) [Gr. *archos*: beginning + *enteron*: bowel] The earliest primordial animal digestive tract.

Area cladogram A cladogram in which the geographic ranges of the taxa are substituted for the names of the taxa.

Arteriosclerosis See atherosclerosis.

Artery A muscular blood vessel carrying oxygenated blood away from the heart to other parts of the body. (Contrast with vein.)

Artifact [L. *ars, artis*: art + *facere*: to make] Something made by human effort or intervention. In biology, something that was not present in the living cell or organism, but was unintentionally produced by an experimental procedure.

Ascospore (ass' ko spor) A fungus spore produced within an ascus.

Ascus (ass' cuss) [Gr. *askos*: bladder] In fungi belonging to the class Ascomycetes (sac fungi), the club-shaped sporangium within which spores are produced by meiosis.

Asexual Without sex.

Associative learning "Pavlovian" learning, in which an animal comes to associate a previously neutral stimulus (such as the ringing of a bell) with a particular reward or punishment.

Assortative mating A breeding system under which mates are selected on the basis of a particular trait or group of traits.

Assortment (genetic) The random separation during meiosis of nonhomologous chromosomes and of genes carried on nonhomologous chromosomes. For example, if genes *A* and *B* are borne on nonhomologous chromosomes, meiosis of diploid cells of genotype *AaBb* will produce haploid cells of the following types in equal numbers: *AB*, *Ab*, *aB*, and *ab*.

Asymmetric The state of lacking any plane of symmetry.

Asymmetric carbon atom In a molecule, a carbon atom to which four different atoms or groups are bound.

Atherosclerosis (ath' er oh sklair oh' sis) A disease of the lining of the arteries characterized by fatty, cholesterol-rich deposits in the walls of the arteries. When fibroblasts infiltrate these deposits and calcium precipitates in them, the disease become arteriosclerosis, or "hardening of the arteries."

Atmosphere The gaseous mass surrounding our planet. Also: a unit of pressure, equal to the normal pressure of air at sea level.

Atom [Gr. *atomos*: indivisible] The smallest unit of a chemical element. Consists of a nucleus and one or more electrons.

Atomic mass (also called atomic weight) The average mass of an atom of an element on the amu scale. (The average depends upon the relative amounts of different isotopes of an element on Earth.)

Atomic mass unit See amu.

Atomic number The number of protons in the nucleus of an atom, also equal to the number of electrons around the neutral atom. Determines the chemical properties of the atom.

ATP (adenosine triphosphate) A compound containing adenine, ribose, and three phosphate groups. When it is formed, useful energy is stored; when it is broken down (to ADP or AMP), energy is released to drive endergonic reactions. ATP is a universal energy storage compound.

ATP synthase An integral membrane protein that couples the transport of proteins with the formation of ATP.

Atrium (a' tree um) A body cavity, as in the hearts of vertebrates. The thin-walled chamber(s) entered by blood on its way to the ventricle(s). Also, the outer ear.

Autocatalysis An enzymatic reaction in which the inactive form of an enzyme is converted into its active form by the enzyme itself.

Autoimmune disease A disorder in which the immune system attacks the animal's own body.

Autonomic nervous system The system (which in vertebrates comprises sympathetic and parasympathetic subsystems) that controls such involuntary functions as those of guts and glands.

Autopolyploid A polyploid in which the sets of chromosomes are derived from the same species.

Autoradiography The detection of a radioactive substance in a cell or organism by putting it in contact with a photographic emulsion and allowing the material to "take its own picture." The emulsion is developed, and the location of the radioactivity in the cell is seen by the presence of silver grains in the emulsion.

Autoregulatory mechanism A feedback mechanism that enables a structure to regulate its own function.

Autosome Any chromosome (in a eukaryote) other than a sex chromosome.

Autotroph (au′ tow trow′ fik) [Gr. *autos*: self + *trophe*: food] An organism that is capable of living exclusively on inorganic materials, water, and some energy source such as sunlight or chemically reduced matter. (Contrast with heterotroph.)

Auxin (awk′ sin) [Gr. *auxein*: increase] In plants, a substance (indoleacetic acid) that regulates growth and various aspects of development.

Auxotroph (awks′ o trofe) [Gr. *auxanein*: to grow + *trophe*: food] A mutant form of an organism that requires a nutrient or nutrients not required by the wild type, or reference, form of the organism. (Contrast with prototroph.)

Avogadro's number The conversion factor between atomic mass units and grams. More usefully, the number of atoms in that quantity of an element which, expressed in grams, is numerically equal to the atomic weight in amu; 6.023×10^{23} atoms. (See mole.)

Axon [Gr.: axle] Fiber of a neuron which can carry action potentials. Carries impulses away from the cell body of the neuron; releases a neurotransmitter substance.

Axon hillock The junction between an axon and its cell body; where action potentials are generated.

Axon terminals The endings of an axon; they form synapses and release neurotransmitter.

Axoneme (ax′ oh neem) The complex of microtubules and their crossbridges that forms the motile apparatus of a cilium.

Bacillus (buh sil′ us) [L.: little rod] Any of various rod-shaped bacteria.

Bacteriophage (bak teer′ ee o fayj) [Gr. *bakterion*: little rod + *phagein*: to eat] One of a group of viruses that infect bacteria and ultimately cause their disintegration.

Bacterium (bak teer′ ee um) (plural: bacteria) [Gr. *bakterion*: little rod] A prokaryote. An organism with chromosomes not contained in nuclear envelopes.

Balanced polymorphism [Gr. *polymorphos*: having many forms] The maintenance of more than one form, or the maintenance at a given locus of more than one allele, at frequencies of greater than one percent in a population. Often results when heterozygotes are superior to both homozygotes.

Baroreceptor [Gr. *baros*: weight] A pressure-sensing cell or organ.

Barr body In mammals, an inactivated X chromosome.

Basal body Centriole found at the base of a eukaryotic flagellum or cilium.

Basal metabolic rate The minimum rate of energy turnover in an awake (but resting) bird or mammal that is not expending energy for thermoregulation.

Base A substance which can accept a proton (H^+). (Contrast with acid.) In nucleic acids, a nitrogen-containing base (purine or pyrimidine) is attached to each sugar in the backbone.

Base pairing See complementary base pairing.

Basic Having a pH greater than 7.0 (having a hydrogen ion concentration lower than 10^{-7} molar).

Basidium (bass id′ ee yum) In fungi of the class Basidiomycetes, the characteristic sporangium in which four spores are formed by meiosis and then borne externally before being shed.

Batesian mimicry Mimicry by a relatively harmless kind of organism of a more dangerous one, by which the mimic enjoys protection from predators that mistake it for the dangerous model. (Contrast with Müllerian mimicry.)

B cell A type of lymphocyte involved in the humoral immune response of vertebrates. Upon recognizing an antigenic determinant, a B cell develops into a plasma cell, which secretes an antibody. (Contrast with a T cell.)

Benefit An improvement in survival and reproductive success resulting from a behavior. (Contrast with cost.)

Benign (be nine′) A tumor that grows to a certain size and then stops, uaually with a fibrous capsule surrounding the mass of cells. Benign tumors do not spread (metastasize) to other organs.

Benthic zone [Gr. *benthos*: bottom of the sea] The bottom of the ocean. (Contrast with pelagic zone.)

Beta-pleated sheet Type of protein secondary structure; results from hydrogen bonding between polypeptide regions running antiparallel to each other.

Biennial Referring to a plant whose life cycle includes vegetative growth in the first year and flowering and senescence in the second year. (Contrast with annual, perennial.)

Bilateral symmetry The condition in which only the right and left sides of an organism, divided exactly down the back, are mirror images of each other. (Contrast with radial symmetry.)

Bile A secretion of the liver delivered to the small intestine via the common bile duct. In the intestine, bile emulsifies fats.

Binocular cells Neurons in the visual cortex that respond to input from both retinas; involved in depth perception.

Binomial (bye nome′ ee al) Consisting of two names; for example, the binomial nomenclature of biology which gives the name of the genus followed by the name of the species.

Bioaccumulation The ever-increasing concentration of a chemical compound in the tissues of an organism as it is passed up the food chain.

Biodiversity crisis The current high rate of loss of species, caused primarily by human activities.

Biogenesis [Gr. *bios*: life + *genesis*: source] The origin of living things from other living things.

Biogeochemical cycles Movement of elements through living organisms and the physical environment.

Biogeography The scientific study of the geographic distribution of organisms. Ecological biogeography is concerned with the habitats in which organisms live, historical biogeography with the complete geographic ranges of organisms and the historical circumstances that determine the ranges.

Biological species concept The view that a species is most usefully defined as a population or series of populations within which there is a significant amount of gene flow under natural conditions, but which is genetically isolated from other populations.

Bioluminescence The production of light by biochemical processes in an organism.

Biomass The total weight of all the living organisms, or some designated group of living organisms, in a given area.

Biome (bye′ ome) A major division of the ecological communities of Earth; characterized by distinctive vegetation.

Biota (bye oh′ tah) All of the organisms, including animals, plants, fungi, and microorganisms, found in a given area.

Biotechnology The use of cells to make medicines, foods and other products useful to humans.

Biradial symmetry Radial symmetry modified so that only two planes can divide the animal into similar halves.

Bithorax complex A group of homeotic genes that control the development of the abdomen and posterior thorax of *Drosophila melanogaster.*

Blastocoel (blass′ toe seal) [Br. *blastos*: sprout + *koilos*: hollow] The central, hollow cavity of a blastula.

Blastodisc (blass′ toe disk) A disk of cells forming on the surface of a large yolk mass, comparable to a blastula, but occurring in animals such as birds and reptiles, in which the massive yolk restricts cleavage to one side of the egg only.

Blastomere A cell produced by the division of a fertilized egg.

Blastopore The opening from the archenteron to the exterior of a gastrula.

Blastula (blass′ chu luh) [Gr. *blastos*: sprout] An early stage in animal embryology; in many species, a hollow sphere of cells surrounding a central cavity, the blastocoel. (Contrast with blastodisc.)

Blood–brain barrier A property of the blood vessels of the brain that prevents most chemicals from diffusing from the blood into the brain.

Bohr effect (boar) The reduction in affinity of hemoglobin for oxygen caused by acidic conditions, usually as a result of increased CO_2.

Bottleneck A combination of environmental conditions that causes a serious reduction in the size of the population.

Bowman's capsule An elaboration of kidney tubule cells that surrounds a knot of capillaries (the glomerulus). Blood is filtered across the walls of these capillaries and the filtrate is collected into Bowman's capsule.

Brain stem The portion of the vertebrate brain between the spinal cord and the forebrain.

Bronchus (plural: bronchi) The major airway(s) branching off the trachea into the vertebrate lung.

Brown fat Fat tissue in mammals that is specialized to produce heat. It has many mitochondria and capillaries, and a protein that uncouples oxidative phosphorylation. It is frequently found in newborns, in mammals acclimated to cold, and in hibernators.

Browser An animal that feeds on the tissues of woody plants.

Bryophyte (bri' uh fite') [Gr. *bruon*: moss + *phyton*: plant] Any nonvascular plant, including mosses, liverworts, and hornworts.

Bud primordium [L. *primordium*: the beginning] In plants, a small mass of potentially meristematic tissue found in the angle between the leaf stalk and the shoot apex. Will give rise to a lateral branch under appropriate conditions.

Budding Asexual reproduction in which a more or less complete new organism simply grows from the body of the parent organism and eventually detaches itself.

Buffering A process by which a system resists change—particularly in pH, in which case added acid or base is partially converted to another form.

Bulb In plants, an underground storage organ composed principally of enlarged and fleshy leaf bases.

Bundle sheath In C_4 plants, a layer of photosynthetic cells between the mesophyll and a vascular bundle of a leaf.

C_3 photosynthesis The form of photosynthesis in which 3-phosphoglycerate is the first stable product, and ribulose bisphosphate is the CO_2 receptor.

C_4 photosynthesis The form of photosynthesis in which oxaloacetate is the first stable product, and phosphoenolpyruvate is the CO_2 acceptor. C_4 plants also perform the reactions of C_3 photosynthesis.

Calcitonin A hormone produced by the thyroid gland; it lowers blood calcium and promotes bone formation. (Contrast with parathormone.)

Calmodulin (cal mod' joo lin) A calcium-binding protein found in all animal and plant cells; mediates many calcium-regulated processes.

calorie [L. *calor*: heat] The amount of heat required to raise the temperature of one gram of water by one degree Celsius (1°C) from 14.5°C to 15.5°C. In nutrition studies, "Calorie" (spelled with a capital C) refers to the kilocalorie (1 kcal = 1,000 cal), the amount of heat required to raise the temperature of one kilogram of water by 1°C.

Calvin–Benson cycle The stage of photosynthesis in which CO_2 reacts with RuBP to form 3PG, 3PG is reduced to a sugar, and RuBP is regenerated, while other products are released to the rest of the plant.

Calyptra (kuh lip' tra) [Gr. *kalyptra*: covering for the head] A hood or cap found partially covering the apex of the sporophyte capsule in many moss species, formed from the expanded wall and neck of the archegonium.

Calyx (kay' licks) [Gr. *kalyx*: cup] All of the sepals of a flower, collectively.

CAM See crassulacean acid metabolism.

Cambium (kam' bee um) [L. *cambiare*: to exchange] A meristem that gives rise to radial rows of cells in stem and root, increasing them in girth; commonly applied to the vascular cambium which produces wood and phloem, and the cork cambium, which produces bark.

cAMP (cyclic AMP) A compound, formed from ATP, that mediates the effects of numerous animal hormones. Also needed for the transcription of catabolite-repressible operons in bacteria. Used for communication by cellular slime molds.

Canopy The leaf-bearing part of a tree. Collectively the aggregate of the leaves and branches of the larger woody plants of an ecological community.

Capacitance vessels Refers to veins because of their variable capacity to hold blood.

Capillaries [L. *capillaris*: hair] Very small tubes, especially the smallest blood-carrying vessels of animals between the termination of the arteries and the beginnings of the veins.

Capping In eukaryote RNA processing, the addition of a modified G at the 5' end of the molecule.

Capsid The protein coat of a virus.

Capsule In bryophytes, the spore case. In some bacteria, a gelatinous layer exterior to the cell wall.

Carbohydrates Organic compounds with the general formula $C_nH_{2m}O_m$. Common examples are sugars, starch, and cellulose.

Carboxylic acid (kar box sill' ik) An organic acid containing the carboxyl group, –COOH, which dissociates to the carboxylate ion, –COO$^-$.

Carcinogen (car sin' oh jen) A substance that causes cancer.

Cardiac (kar' dee ak) [Gr. *kardia*: heart] Pertaining to the heart and its functions.

Carnivore [L. *carn*: flesh + *vovare*: to devour] An organism that feeds on animal tissue. (Contrast with detritivore, herbivore, omnivore.)

Carotenoid (ka rah' tuh noid) [L. *carota*: carrot] A yellow, orange, or red lipid pigment commonly found as an accessory pigment in photosynthesis; also found in fungi.

Carpel (kar' pel) [Gr. *karpos*: fruit] The organ of the flower that contains one or more ovules.

Carrier In facilitated diffusion, a membrane protein that binds a specific molecule and transports it through the membrane. In genetics, a person heterozygous for a recessive trait. In respiratory and photosynthetic electron transport, a participating substance such as NAD that exists in both oxidized and reduced forms.

Carrying capacity In ecology, the largest number of organisms of a particular species that can be maintained indefinitely in a given part of the environment.

Cartilage In vertebrates, a tough connective tissue found in joints, the outer ear, and elsewhere. Forms the entire skeleton in some animal groups.

Casparian strip A band of cell wall containing suberin and lignin, found in the endodermis. Restricts the movement of water across the endodermis.

Catabolism [Ge. *kata*: down + *ballein*: to throw] Degradational reactions of metabolism, in which complex molecules are broken down. (Contrast with anabolism.)

Catabolite repression The decreased synthesis of many enzymes that tend to provide glucose for a cell; caused by the presence of excellent carbon sources, particularly glucose.

Catalyst (cat' a list) [Gr. *kata-*, implying the breaking down of a compound] A chemical substance that accelerates a reaction without itself being consumed in the overall course of the reaction. Catalysts lower the activation energy of a reaction. Enzymes are biological catalysts.

Cation (cat' eye on) An ion with one or more positive charges. (Contrast with anion.)

Caudal [L. *cauda*: tail] Pertaining to the tail, or to the posterior part of the body.

cDNA See complementary DNA.

Cecum (see' cum) [L. *caecus*: blind] A blind branch off the large intestine. In many nonruminant mammals, the cecum contains a colony of microorganisms that contribute to the digestion of food.

Cell adhesion molecules Molecules on animal cell surfaces that affect the selective association of cells during development of the embryo.

Cell cycle The stages through which a cell passes between one division and the next. Includes all stages of interphase and mitosis.

Cell division The reproduction of a cell to produce two new cells. In eukaryotes, this process involves nuclear division (mitosis) and cytoplasmic division (cytokinesis).

Cell theory The theory, well established, that organisms consist of cells, and that all cells come from preexisting cells.

Cell wall A relatively rigid structure that encloses cells of plants, fungi, many protists, and most bacteria. The cell wall gives these cells their shape and limits their expansion in hypotonic media.

Cellular immune system That part of the immune system that is based on the activities of T cells. Directed against parasites, fungi, intracellular viruses, and foreign tissues (grafts). (Contrast with humoral immune system.)

Cellular respiration   See respiration.

Cellulose (sell' you lowss)   A straight-chain polymer of glucose molecules, used by plants as a structural supporting material. Cellulase is an enzyme that hydrolyzes cellulose.

Central dogma   The statement that information flows from DNA to RNA to polypeptide (in retroviruses, there is also information flow from RNA to cDNA).

Central nervous system   That part of the nervous system which is condensed and centrally located, e.g., the brain and spinal cord of vertebrates; the chain of cerebral, thoracic and abdominal ganglia of arthropods.

Centrifuge [L. *fugere*: to flee]   A device in which a sample can be spun around a central axis at high speed, creating a centrifugal force that mimics a very strong gravitational force. Used to separate mixtures of suspended materials.

Centriole (sen' tree ole)   A paired organelle that helps organize the microtubules in animal and protist cells during nuclear division.

Centromere (sen' tro meer) [Gr. *centron*: center + *meros*: part]   The region where sister chromatids join.

Centrosome (sen' tro soam)   The major microtubule organizing center of an animal cell.

Cephalization (sef' uh luh zay' shun) [Gr. *kephale*: head]   The evolutionary trend toward increasing concentration of brain and sensory organs at the anterior end of the animal.

Cerebellum (sair' uh bell' um) [L.: diminutive of *cerebrum*: brain]   The brain region that controls muscular coordination; located at the anterior end of the hindbrain.

Cerebral cortex   The thin layer of gray matter (neuronal cell bodies) that overlays the cerebrum.

Cerebrum (su ree' brum) [L.: brain]   The dorsal anterior portion of the forebrain, making up the largest part of the brain of mammals. In mammals, the chief coordination center of the nervous system; consists of two **cerebral hemispheres**.

Cervix (sir' vix) [L.: neck]   The opening of the uterus into the vagina.

cGMP (cyclic guanosine monophosphate)   An intracellular messenger that is part of signal transmission pathways involving G proteins. (See G protein.)

Channel   A membrane protein that forms an aqueous passageway though which specific solutes may pass by simple diffusion; some channels are gated: they open and close in response to binding of specific molecules.

Chaperone protein   A protein that assists a newly forming protein in adopting its appropriate tertiary structure.

Chemical bond   An attractive force stably linking two atoms.

Chemiosmotic mechanism   According to this model, ATP formation in mitochondria and chloroplasts results from a pumping of protons across a membrane (against a gradient of electrical charge and of pH), followed by the return of the protons through a protein channel with ATPase activity.

Chemoautotroph   An organism that uses carbon dioxide as a carbon source and obtains energy by oxidizing inorganic substances from its environment. (Contrast with chemoheterotroph, photoautotroph, photoheterotroph.)

Chemoheterotroph   An organism that must obtain both carbon and energy from organic substances. (Contrast with chemoautotroph, photoautotroph, photoheterotroph.)

Chemosensor   A cell or tissue that senses specific substances in its environment.

Chemosynthesis   Synthesis of food substances, using the oxidation of reduced materials from the environment as a source of energy.

Chiasma (kie az' muh) (plural: chiasmata) [Gr.: cross]   An X-shaped connection between paired homologous chromosomes in prophase I of meiosis. A chiasma is the visible manifestation of crossing over between homologous chromosomes.

Chitin (kye' tin) [Gr. *chiton*: tunic]   The characteristic tough but flexible organic component of the exoskeleton of arthropods, consisting of a complex, nitrogen-containing polysaccharide. Also found in cell walls of fungi.

Chlorophyll (klor' o fill) [Gr. *chloros*: green + *phyllon*: leaf]   Any of a few green pigments associated with chloroplasts or with certain bacterial membranes; responsible for trapping light energy for photosynthesis.

Chloroplast [Gr. *chloros*: green + *plast*: a particle]   An organelle bounded by a double membrane containing the enzymes and pigments that perform photosynthesis. Chloroplasts occur only in eukaryotes.

Choanocyte (cho' an oh cite)   The collared, flagellated feeding cells of sponges.

Cholecystokinin (ko' lee sis to kai nin)   A hormone produced and released by the lining of the duodenum when it is stimulated by undigested fats and proteins. It stimulates the gallbladder to release bile and slows stomach activity.

Chorion (kor' ee on) [Gr. *khorion*: afterbirth]   The outermost of the membranes protecting mammal, bird, and reptile embryos; in mammals it forms part of the placenta.

Chorionic villus sampling   A medical procedure that extracts a portion of the chorion from a pregnant woman to enable genetic and biochemical analysis of the embryo. (Contrast with amniocentesis.)

Chromatid (kro' ma tid)   Each of a pair of new sister chromosomes from the time at which the molecular duplication occurs until the time at which the centromeres separate at the anaphase of nuclear division.

Chromatin   The nucleic acid–protein complex found in eukaryotic chromosomes.

Chromatography   Any one of several techniques for the separation of chemical substances, based on differing relative tendencies of the substances to associate with a mobile phase or a stationary phase.

Chromatophore (krow mat' o for) [Gr. *chroma*: color + *phoreus*: carrier]   A pigment-bearing cell that expands or contracts to change the color of the organism.

Chromosomal aberration   Any large change in the structure of a chromosome, including duplication or loss of chromosomes or parts thereof, usually gross enough to be detected with the light microscope.

Chromosome (krome' o sowm) [Gr. *chroma*: color = *soma*: body]   In bacteria and viruses, the DNA molecule that contains most or all of the genetic information of the cell or virus. In eukaryotes, a structure composed of DNA and proteins that bears part of the genetic information of the cell.

Chromosome walking   A technique based on recognition of overlapping fragments; used as a step in DNA sequencing.

Chylomicron (ky low my' cron)   Particles of lipid coated with protein, produced in the gut from dietary fats and secreted into the extracellular fluids.

Chyme (kime) [Gr. *chymus*, juice]   Created in the stomach; a mixture of ingested food with the digestive juices secreted by the salivary glands and the stomach lining.

Ciliate (sil' ee ate)   A member of the protist phylum Ciliophora, unicellular organisms that propel themselves by means of cilia.

Cilium (sil' ee um) (plural: cilia) [L. *cilium*: eyelash]   Hairlike organelle used for locomotion by many unicellular organisms and for moving water and mucus by many multicellular organisms. Generally shorter than a flagellum.

Circadian rhythm (sir kade' ee an) [L. *circa*: approximately + *dies*: day]   A rhythm in behavior, growth, or some other activity that recurs about every 24 hours under constant conditions.

Circannual rhythm (sir can' you al) [L. *circa*: approximately + *annus*: year]   A rhythm of behavior, growth, or some other activity that recurs on a yearly basis.

Citric acid cycle   A set of chemical reactions in cellular respiration, in which acetyl CoA reacts with oxaloacetate to form citric acid, and oxaloacetate is regenerated. Acetyl CoA is oxidized to carbon dioxide, and hydrogen atoms are stored as NADH and FADH$_2$.

Clade (clayd) [Gr. *klados*: branch]   All of the organisms, both living and fossil, descended from a particular common ancestor.

Cladistic classification   A classification based entirely on the phylogenetic relationships among organisms.

Cladogenesis (clay doh jen' e sis) [Gr. *klados*: branch + *genesis*: source]   The formation of new species by the splitting of an evolutionary lineage.

Cladogram   The graphic representation of a clade.

Class   In taxonomy, the category below the phylum and above the order; a group of related, similar orders.

Class I MHC molecules   These cell surface proteins participate in the cellular immune response directed against virus-infected cells.

Class II MHC molecules   These cell surface proteins participate in the cell-cell interactions (of helper T cells, macrophages, and B cells) of the humoral immune response.

Class II MHC molecules   These proteins do not present processed antigen, as do classes I and II, but instead include some proteins of the complement system and certain cytokines.

Class switching   The process whereby a plasma cell changes the class of immunoglobulin that it synthesizes. This results from the deletion of part of the constant region of DNA, bringing in a new C segment. The variable region is the same as before, so that the new immunoglobulin has the same antigenic specificity.

Clathrin   A fibrous protein on the inner surfaces of animal cell membranes that strengthens coated vesicles and thus participates in receptor-mediated endocytosis.

Clay   A soil constituent comprising particles smaller than 2 micrometers in diameter.

Cleavages   First divisions of the fertilized egg of an animal.

Climax   In ecology, a community that terminates a succession and which tends to replace itself unless it is further disturbed or the physical environment changes.

Climograph (clime' o graf)   Graph relating temperature and precipitation with time of year.

Cline   A gradual change in the traits of a species over a geographical gradient.

Clitoris (klit' er us, klite' er us)   A structure in the human female reproductive system that is homologous with the male penis and is involved in sexual stimulation.

Cloaca (klo ay' kuh) [L. *cloaca*: sewer]   In some invertebrates, the posterior part of the gut; in many vertebrates, a cavity receiving material from the digestive, reproductive, and excretory systems.

Clonal anergy   When a naive T cell encounters a self-antigen, the T cell may bind to the antigen but does not receive signals from an antigen-presenting cell. Instead of being activated, the T cell dies (becomes anergic). In this way, we avoid reacting to our own tissue-specific antigens.

Clonal deletion   In immunology, the inactivation or destruction of lymphocyte clones that would produce immune reactions against the animal's own body.

Clonal selection   The mechanism by which exposure to antigen results in the activation of selected T- or B-cell clones, resulting in an immune response.

Clone [Gr. *klon*: twig, shoot]   Genetically identical cells or organisms produced from a common ancestor by asexual means.

Cnidocytes   The feeding cells of cnidarians, within which nematocysts are housed.

Coacervate (ko as' er vate) [L. *coacervare*: to heap up]   An aggregate of colloidal particles in suspension.

Coacervate drop   Drops formed when a mixture of large proteins and polysaccharides is shaken in water. The interiors of these drops, which are often very stable, contain most of the proteins and polysaccharides.

Coated vesicle   Vesicle, sometimes formed from a coated pit, with characteristic "bristly" surface; its membrane contains distinctive proteins, including clathrin.

Coccus (kock' us) [Gr. *kokkos*: berry, pit]   Any of various spherical or spheroidal bacteria.

Cochlea (kock' lee uh) [Gr. *kokhlos*: a land snail]   A spiral tube in the inner ear of vertebrates; it contains the sensory cells involved in hearing.

Codominance   A condition in which two alleles at a locus produce different phenotypic effects and both effects appear in heterozygotes.

Codon   A "triplet" of three nucleotides in messenger RNA that directs the placement of a particular amino acid into a polypeptide chain. (Contrast with anticodon.)

Coefficient of relatedness   The probability that an allele in one individual is an identical copy, by descent, of an allele in another individual.

Coelom (see' lum) [Gr. *koiloma*: cavity]   The body cavity of certain animals, which is lined with cells of mesodermal origin.

Coelomate   Having a coelom.

Coenocyte (seen' a sight) [Gr.: common cell]   A "cell" bounded by a single plasma membrane, but containing many nuclei.

Coenzyme   A nonprotein molecule that plays a role in catalysis by an enzyme. The coenzyme may be part of the enzyme molecule or free in solution. Some coenzymes are oxidizing or reducing agents, others play different roles.

Coevolution   Concurrent evolution of two or more species that are mutually affecting each other's evolution.

Cohort (co' hort) [L. *cohors*: company of soldiers]   A group of similar-age organisms, considered as it passes through time.

Coleoptile (koe' lee op' til) [Gr. *koleos*: sheath + *ptilon*: feather]   A pointed sheath covering the shoot of grass seedlings.

Collagen [Gr. *kolla*: glue]   A fibrous protein found extensively in bone and connective tissue.

Collecting duct   In vertebrates, a tubule that receives urine produced in the nephrons of the kidney and delivers that fluid to the ureter for excretion.

Collenchyma (cull eng' kyma) [Gr. *kolla*: glue + *enchyma*: infusion]   A type of plant cell, living at functional maturity, which lends flexible support by virtue of primary cell walls thickened at the corners. (Contrast with parenchyma, sclerenchyma.)

Colon [Gr. *kolon*: large intestine]   The large intestine.

Commensalism   The form of symbiosis in which one species benefits from the association, while the other is neither harmed nor benefited.

Common bile duct   A single duct that delivers bile from the gallbladder and secretions from the pancreas into the small intestine.

Communication   A signal from one organism (or cell) that alters the pattern of behavior in another organism (or cell) in an adaptive fashion.

Community   Any ecologically integrated group of species of microorganisms, plants, and animals inhabiting a given area.

Companion cell   Specialized cell found adjacent to a sieve tube member in some flowering plants.

Comparative analysis   An approach to studying evolution in which hypotheses are tested by measuring the distribution of states among a large number of species.

Compensation point   The light intensity at which the rates of photosynthesis and of cellular respiration are equal.

Competitive inhibitor   A substance, similar in structure to an enzyme's substrate, that binds the active site and thus inhibits a reaction.

Competition   In ecology, use of the same resource by two or more species, when the resource is present in insufficient supply for the combined needs of the species.

Competitive exclusion   A result of competition between species for a limiting resource in which one species completely eliminates the other.

Competitive inhibitor   A substance, similar in structure to an enzyme's substrate, that binds the active site and inhibits a reaction.

Complement system   A group of eleven proteins that play a role in some reactions of the immune system. The complement proteins are not immunoglobulins.

Complementary base pairing   The A–T (or A–U), T–A (or U–A), C–G and G–C pairing of bases in double-stranded DNA, in transcription, and between tRNA and mRNA.

Complementary DNA (cDNA)   DNA formed by reverse transcriptase acting with an RNA template; essential intermediate in the reproduction of retroviruses; used as a tool in recombinant DNA technology; lacks introns.

Complete metamorphosis   A change of state during the life cycle of an organism in which the body is almost completely rebuilt to produce an individual with a very different body form. Characteristic of insects such as butterflies, moths, beetles, ants, wasps, and flies.

Compound   (1) A substance made up of atoms of more than one element. (2) Made up of many units, as the compound eyes of arthropods (as opposed to the simple eyes of the same group of organisms).

Condensation reaction   A reaction in which two molecules become connected by a covalent bond and a molecule of water is released. ($AH + BOH \rightarrow AB + H_2O$.)

Cones (1) In the vertebrate retina: photoreceptors responsible for color vision. (2) In gymnosperms: reproductive structures consisting of many sporophylls packed relatively tightly.

Conidium (ko nid' ee um) [Gr. *konis*: dust] An asexual fungus spore borne singly or in chains either apically or laterally on a hypha.

Conifer (kahn' e fer) [Gr. *konos*: cone + *phero*: carry] One of the cone-bearing gymnosperms, mostly trees, such as pines and firs.

Conjugation (kahn' jew gay' shun) [L. *conjugare*: yoke together] The close approximation of two cells during which they exchange genetic material, as in *Paramecium* and other ciliates, or during which DNA passes from one to the other through a tube, as in bacteria.

Connective tissue An animal tissue that connects or surrounds other tissues; its cells are embedded in a collagen-containing matrix.

Connexon In a gap junction, a protein channel linking adjacent animal cells.

Consensus sequences Short stretches of DNA that appear, with little variation, in many different genes.

Constant region The constant region in an immunoglobulin is encoded by a single exon, and determines the function, but not the specificity, of the molecule. The constant region of the T cell receptor anchors the protein to the plasma membrane.

Constitutive enzyme An enzyme that is present in approximately constant amounts in a system, whether its substrates are present or absent. (Contrast with inducible enzyme.)

Consumer An organism that eats the tissues of some other organism.

Continental climate A pattern, typical of the interiors of large continents at high latitudes, in which bitterly cold winters alternate with hot summers. (Contrast with maritime climate.)

Continental drift The gradual drifting apart of the world's continents that has occurred over a period of billions of years.

Contractile vacuole An organelle, often found in protists, which pumps excess water out of the cell and keeps it from being "flooded" in hypotonic environments.

Convergent evolution The evolution of similar features independently in unrelated taxa from different ancestral structures.

Cooperative act Behavior in which two or more individuals interact to their mutual benefit. No conscious awareness by the actors of the effects of their behavior is implied.

Cooption The act of capturing something for a particular use. In ecology refers to the diversion of ecological production for human use. Such production is said to be coopted.

Copulation Reproductive behavior that results in a male depositing sperm in the reproductive tract of a female.

Corepressor A low molecular weight compound that unites with a protein (the repressor) to prevent transcription in a repressible operon.

Cork A waterproofing tissue in plants, with suberin-containing cell walls. Produced by a cork cambium.

Corm A conical, underground stem that gives rise to a new plant. (Contrast with bulb.)

Corolla (ko role' lah) [L.: diminutive of *corona*: wreath, crown] All of the petals of a flower, collectively.

Coronary (kor' oh nair ee) Referring to the blood vessels of the heart.

Corpus luteum (kor' pus loo' tee um) [L. *corpus*: body + *luteum*: yellow] A structure formed from a follicle after ovulation; it produces hormones important to the maintenance of pregnancy.

Cortex [L.: bark or rind] (1) In plants: the tissue between the epidermis and the vascular tissue of a stem or root. (2) In animals: the outer tissue of certain organs, such as the adrenal cortex and cerebral cortex.

Corticosteroids Steroid hormones produced and released by the cortex of the adrenal gland.

Cost See energetic cost, opportunity cost, risk cost.

Cotyledon (kot' ul lee' dun) [Gr. *kotyledon*: a hollow space] A "seed leaf." An embryonic organ which stores and digests reserve materials; may expand when seed germinates.

Countercurrent exchange An adaptation that promotes maximum exchange of heat or any diffusible substance between two fluids by the fluids flow in opposite directions through parallel tubes in close approximation to each other. An example is countercurrent heat exchange between arterioles and venules in the extremities of some animals.

Covalent bond A chemical bond that arises from the sharing of electrons between two atoms. Usually a strong bond.

Crassulacean acid metabolism (CAM) A metabolic pathway enabling the plants that possess it to store carbon dioxide at night and then perform photosynthesis during the day with stomata closed.

Crista (plural: cristae) A small, shelflike projection of the inner membrane of a mitochondrion; the site of oxidative phosphorylation.

Critical night length In the photoperiodic flowering response of short-day plants, the length of night above which flowering occurs and below which the plant remains vegetative. (The reverse applies in the case of long-day plants.)

Critical period The age during which some particular type of learning must take place or during which it occurs much more easily than at other times. Typical of song learning among birds.

Cross-pollination The pollination of one plant by pollen from another plant. (Contrast with self-pollination.)

Cross section (also called a transverse section) A section taken perpendicular to the longest axis of a structure.

Crossing over The mechanism by which linked markers undergo recombination. In general, the term refers to the reciprocal exchange of corresponding segments between two homologous chromatids. However, the reciprocity of crossing-over is problematical in prokaryotes and viruses; and even in eukaryotes, very closely linked markers often recombine by a nonreciprocal mechanism.

CRP The cAMP receptor protein that interacts with the promoter to enhance transcription; a lowered cAMP concentration results in catabolite repression.

Crustacean (crus tay' see an) A member of the phylum Crustacea, such as a crab, shrimp, or sowbug.

Cryptic appearance The resemblance of an animal to some part of its environment, which helps it to escape detection by predators.

Culture (1) A laboratory association of organisms under controlled conditions. (2) The collection of knowledge, tools, values, and rules that characterize a human society.

Cuticle A waxy layer on the outer surface of a plant or an insect, tending to retard water loss.

Cutin (cue' tin) [L. *cutis*: skin] A mixture of long, straight-chain hydrocarbons and waxes secreted by the plant epidermis, providing a water-impermeable coating on aerial plant parts.

Cyanobacteria (sigh an' o bacteria) [Gr. *kuanos*: the color blue] A division of photosynthetic bacteria, formerly referred to as blue-green algae; they lack sexual reproduction, and they use chlorophyll *a* in their photosynthesis.

Cyclic AMP See cAMP.

Cyclins Proteins that activate cyclin-dependent kinases, bringing about transitions in the cell cycle.

Cyclin-dependent kinase (cdk) A kinase is an enzyme that catalzyes the addition of phosphate groups from ATP to target molecules. Cdk's target proteins involved in transitions in the cell cycle and are active only when complexed to additional protein subunits, cyclins.

Cyst (sist) [Gr. *kystis*: pouch] (1) A resistant, thick-walled cell formed by some protists and other organisms. (2) An abnormal sac, containing a liquid or semisolid substance, produced in response to injury or illness.

Cytochromes (sy' toe chromes) [Gr. *kytos*: container + *chroma*: color] Iron-containing red proteins, components of the electron-transfer chains in photophosphorylation and respiration.

Cytokine A small protein that is made by one type of immune cell and stimulates a target cell which has a specific receptor for that cytokine.

Cytokinesis (sy' toe kine ee' sis) [Gr. *kytos*: container + *kinein*: to move] The division of the cytoplasm of a dividing cell. (Contrast with mitosis.)

Cytokinin (sy' toe kine' in) [Gr. *kytos*: container + *kinein*: to move]   A member of a class of plant growth substances playing roles in senescence, cell division, and other phenomena.

Cytoplasm   The contents of the cell, excluding the nucleus.

Cytoplasmic determinants   In animal development, gene products whose spatial distribution may determine such things as embryonic axes.

Cytosine (site' oh seen)   A nitrogen-containing base found in DNA and RNA.

Cytoskeleton   The network of microtubules and microfilaments that gives a eukaryotic cell its shape and its capacity to arrange its organelles and to move.

Cytosol   The fluid portion of the cytoplasm, excluding organelles and other solids.

Cytotoxic T cells   Cells of the cellular immune system that recognize and directly eliminate virus-infected cells. (Contrast with helper T cells, suppressor T cells.)

Dalton   See amu.

Deciduous (de sid' you us) [L. *decidere*: fall off]   Referring to a plant that sheds its leaves at certain seasons. (Contrast with evergreen.)

Degeneracy   The situation in which a single amino acid may be represented by any of two or more different codons in messenger RNA. Most of the amino acids can be represented by more than one codon.

Degradative succession   Ecological succession occuring on the dead remains of the bodies of plants and animals, as when leaves or animal bodies rot.

Dehydration   See condensation reaction.

Deletion (genetic)   A mutation resulting from the loss of a continuous segment of a gene or chromosome. Such mutations never revert to wild type. (Contrast with duplication, point mutation.)

Deme (deem) [Gr. *demos*: common people]   Any local population of individuals belonging to the same species that interbreed with one another.

Demographic processes   The events—such as births, deaths, immigration, and emigration—that determine the number of individuals in a population.

Demographic stochasticity   Random variations in the factors influencing the size, density, and distribution of a population.

Demography   The study of dynamical changes in the sizes, densities, and distributions of populations.

Denaturation   Loss of activity of an enzyme or nucleic acid molecule as a result of structural changes induced by heat or other means.

Dendrite [Gr. *dendron*: a tree]   A fiber of a neuron which often cannot carry action potentials. Usually much branched and relatively short compared with the axon, and commonly carries information to the cell body of the neuron.

Denitrification   Metabolic activity by which inorganic nitrogen-containing ions are reduced to form nitrogen gas and other products; carried on by certain soil bacteria.

Density dependence   Change in the severity of action of agents affecting birth and death rates within populations that are directly or inversely related to population density.

Density independence   The state where the severity of action of agents affecting birth and death rates within a population does not change with the density of the population.

Deoxyribonucleic acid   See DNA.

Depolarization   A change in the electric potential across a membrane from a condition in which the inside of the cell is more negative than the outside to a condition in which the inside is less negative, or even positive, with reference to the outside of the cell. (Contrast with hyperpolarization.)

Derived trait   A trait found among members of a lineage that was not present in the ancestors of that lineage.

Dermal tissue system   The outer covering of a plant, consisting of epidermis in the young plant and periderm in a plant with extensive secondary growth. (Contrast with ground tissue system and vascular tissue system.)

Desmosome (dez' mo sowm) [Gr. *desmos*: bond + *soma*: body]   An adhering junction between animal cells.

Determination   Process whereby an embryonic cell or group of cells becomes fixed into a predictable developmental pathway.

Detritivore (di try' ti vore) [L. *detritus*: worn away + *vorare*: to devour]   An organism that eats the dead remains of other organisms.

Deuterium   An isotope of hydrogen possessing one neutron in its nucleus. Deuterium oxide is called "heavy water."

Deuterostome   One of two major lines of evolution in animals, characterized by radial cleavage, enterocoelous development, and other traits.

Development   Progressive change, as in structure or metabolism; in most kinds of organisms, development continues throughout the life of the organism.

Dialysis (dye ahl' uh sis) [Gr. *dialyein*: separation]   The removal of ions or small molecules from a solution by their diffusion across a semipermeable membrane to a solvent where their concentration is lower.

Diaphragm (dye' uh fram) [Gr. *diaphrassein*, to barricade]   (1) A sheet of muscle that separates the thoracic and abdominal cavities in mammals; responsible for the action of breathing. (2) A method of birth control in which a sheet of rubber is fitted over the woman's cervix, blocking the entry of sperm.

Diastole (dye ahs' toll ee) [Gr.: dilation]   The portion of the cardiac cycle when the heart muscle relaxes. (Contrast with systole.)

Dicot (short for dicotyledon) [Gr. *dis*: two + *kotyledon*: a cup-shaped hollow]   Any member of the angiosperm class Dicotyledones, flowering plants in which the embryo produces two cotyledons prior to germination. Leaves of most dicots have major veins arranged in a branched or reticulate pattern.

Differentiation   Process whereby originally similar cells follow different developmental pathways. The actual expression of determination.

Diffuse coevolution   The situation in which the evolution of a lineage is influenced by its interactions with a number of species, most of which exert only a small influence on the evolution of the focal lineage.

Diffusion   Random movement of molecules or other particles, resulting in even distribution of the particles when no barriers are present.

Digestion   Enzyme-catalyzed process by which large, usually insoluble, molecules (foods) are hydrolyzed to form smaller molecules of soluble substances.

Dihybrid cross   A mating in which the parents differ with respect to the alleles of two loci of interest.

Dikaryon (di care' ee ahn) [Gr. *dis*: two + *karyon*: kernel]   A cell or organism carrying two genetically distinguishable nuclei. Common in fungi.

Dioecious (die eesh' us) [Gr.: two houses]   Organisms in which the two sexes are "housed" in two different individuals, so that eggs and sperm are not produced in the same individuals. Examples: humans, fruit flies, oak trees, date palms. (Contrast with monoecious.)

Diploblastic   Having two cell layers. (Contrast with triploblastic.)

Diploid (dip' loid) [Gr. *diploos*: double]   Having a chromosome complement consisting of two copies (homologues) of each chromosome. A diploid individual (or cell) usually arises as a result of the fusion of two gametes, each with just one copy of each chromosome. Thus, the two homologues in each chromosome pair in a diploid cell are of separate origin, one derived from the female parent and one from the male parent.

Diplontic life cycle   A life cycle in which every cell except the gametes is diploid.

Directional selection   Selection in which phenotypes at one extreme of the population distribution are favored. (Contrast with disruptive selection; stabilizing selection.)

Disaccharide   A carbohydrate made up of two monosaccharides (simple sugars).

Dispersal stage   Stage in its life history at which an organism moves from its birthplace to where it will live as an adult.

Displacement activity   Apparently irrelevant behavior performed by an animal under conflict situations, especially when tendencies to attack and escape are closely balanced.

Display A behavior that has evolved to influence the actions of other individuals.

Disruptive selection Selection in which phenotypes at both extremes of the population distribution are favored. (Contrast with directional selection; stabilizing selection.)

Distal Away from the point of attachment or other reference point. (Contrast with proximal.)

Disturbance A short-term event that disrupts populations, communities, or ecosystems by changing the environment.

Diverticulum (di ver tic' u lum) [L. *divertere*: turn away] A small cavity or tube that connects to a major cavity or tube.

Division A term used by some microbiologists and formerly by botanists, corresponding to the term phylum.

DNA (deoxyribonucleic acid) The fundamental hereditary material of all living organisms. In eukaryotes, stored primarily in the cell nucleus. A nucleic acid using deoxyribose rather than ribose.

DNA hybridization A process by which DNAs from two species are mixed and heated so that interspecific double helixes are formed.

DNA ligase Enzyme that unites Okazaki fragments of the lagging strand during DNA replication; also mends breaks in DNA strands. It connects pieces of a DNA strand and is used in recombinant DNA technology.

DNA methylation Addition of methyl groups to DNA; plays role in regulation of gene expression; protects a bacterium's DNA against its restriction endonucleases.

DNA polymerase Any of a group of enzymes that catalyze the formation of DNA strands from a DNA template.

Dominance In genetic terminology, the ability of one allelic form of a gene to determine the phenotype of a heterozygous individual, in which the homologous chromosome carries both it and a different allele. For example, if *A* and *a* are two allelic forms of a gene, *A* is said to be dominant to *a* if *AA* diploids and *Aa* diploids are phenotypically identical and are distinguishable from *aa* diploids. The *a* allele is said to be recessive.

Dominance hierarchy The set of relationships within a group of animals, usually established and maintained by aggression, in which one individual has precedence over all others in eating, mating, and other activities; a second individual has precedence over all but the highest-ranking individual, and so on down the line.

Dormancy A condition in which normal activity is suspended, as in some seeds and buds.

Dorsal [L. *dorsum*: back] Pertaining to the back or upper surface. (Contrast with ventral.)

Double fertilization Process virtually unique to angiosperms in which one sperm nucleus combines with the egg to produce a zygote, and the other sperm nucleus combines with the two polar nuclei to produce the first cell of the triploid endosperm.

Double helix Of DNA: molecular structure in which two complementary polynucleotide strands, antiparallel to each other, form a right-handed spiral.

Duodenum (doo' uh dee' num) The beginning portion of the vertebrate small intestine. (Contrast with ileum, jejunum.)

Duplication (genetic) A mutation resulting from the introduction into the genome of an extra copy of a segment of a gene or chromosome. (Contrast with deletion, point mutation.)

Dynein [Gr. *dunamis*: power] A protein that undergoes conformational changes and thus plays a part in the movement of eukaryotic flagella and cilia.

Ecdysone (eck die' sone) [Gr. *ek*: out of + *dyo*: to clothe] In insects, a hormone that induces molting.

Ecological biogeography The study of the distributions of organisms from an ecological perspective, usually concentrating on migration, dispersal, and species interactions.

Ecological community The species living together at a particular site.

Ecological niche (nitch) [L. *nidus*: nest] The functioning of a species in relation to other species and its physical environment.

Ecology [Gr. *oikos*: house + *logos*: discourse, study] The scientific study of the interaction of organisms with their environment, including both the physical environment and the other organisms that live in it.

Ecosystem (eek' oh sis tum) The organisms of a particular habitat, such as a pond or forest, together with the physical environment in which they live.

Ecto- (eck' toh) [Gr.: outer, outside] A prefix used to designate a structure on the outer surface of the body. For example, ectoderm. (Contrast with endo- and meso-.)

Ectoderm [Gr. *ektos*: outside + *derma*: skin] The outermost of the three embryonic tissue layers first delineated during gastrulation. Gives rise to the skin, sense organs, nervous system, etc.

Ectotherm [Gr. *ektos*: outside + *thermos*: heat] An animal unable to control its body temperature. (Contrast with endotherm.)

Edema (i dee' mah) [Gr. *oidema*: swelling] Tissue swelling caused by the accumulation of fluid.

Edge effect The changes in ecological processes in a community caused by physical and biological factors originating in an adjacent community.

Effector Any organ, cell, or organelle that moves the organism through the environment or else alters the environment to the organism's advantage. Examples include muscle, bone, and a wide variety of exocrine glands.

Effector cell A lymphocyte that performs a role in the immune system without further differentiation.

Effector phase In this phase of the immune response, effector T cells called cytotoxic T cells attack virus-infected cells, and effector helper T cells assist B cells to differentiate into plasma cells, which release antibodies.

Efferent [L. *ex*: out + *ferre*: to bear] Away from, as in neurons that conduct action potentials out from the central nervous system, or arterioles that conduct blood away from a structure. (Contrast with afferent.)

Egg In all sexually reproducing organisms, the female gamete; in birds, reptiles, and some other vertebrates, a structure witin which early embryonic development occurs.

Elasticity The property of returning quickly to a former state after a disturbance.

Electrocardiogram (EKG) A graphic recording of electrical potentials from the heart.

Electroencephalogram (EEG) A graphic recording of electrical potentials from the brain.

Electromyogram (EMG) A graphic recording of electrical potentials from muscle.

Electron (e lek' tron) [L. *electrum*: amber (associated with static electricity), from Gr. *slektor*: bright sun (color of amber)] One of the three most important fundamental particles of matter, with mass approximately 0.00055 amu and charge −1.

Electron microscope An instrument that uses an electron beam to form images of minute structures; the transmission electron microscope is useful for thinly-sliced material, and the scanning electron microscope gives surface views of cells and organisms.

Electrophoresis (e lek' tro fo ree' sis) [L. *electrum*: amber + Gr. *phorein*: to bear] A separation technique in which substances are separated from one another on the basis of their electric charges and molecular weights.

Electrotonic potential In neurons, a hyperpolarization or small depolarization of the membrane potential induced by the application of a small electric current. (Contrast with action potential, resting potential.)

Elemental substance A substance composed of only one type of atom.

Embolus (em' buh lus) [Gr. *embolos*: inserted object; stopper] A circulating blood clot. Blockage of a blood vessel by an embolus or by a bubble of gas is referred to as an **embolism**. (Contrast with thrombus.)

Embryo [Gr. *en-*: in + *bryein*: to grow] A young animal, or young plant sporophyte, while it is still contained within a protective structure such as a seed, egg, or uterus.

Embryo sac In angiosperms, the female gametophyte. Found within the ovule, it consists of eight or fewer cells, membrane bounded, but without cellulose walls between them.

Emergent property A property of a complex system that is not exhibited by its individual component parts.

Emigration The deliberate and usually oriented departure of an organism from the habitat in which it has been living.

3′ end (3-prime)   The end of a DNA or RNA strand that has a free hydroxyl group at the 3′-carbon of the sugar (deoxyribose or ribose).

5′ end (5-prime)   The end of a DNA or RNA strand that has a free phosphate group at the 5′-carbon of the sugar (deoxyribose or ribose).

Endemic (en dem′ ik) [Gr. *endemos*: dwelling in a place]   Confined to a particular region, thus often having a comparatively restricted distribution.

Endergonic reaction   One for which energy must be supplied. (Contrast with exergonic reaction.)

Endo- [Gr.: within, inside]   A prefix used to designate an innermost structure. For example, endoderm, endocrine. (Contrast with ecto-, meso-.)

Endocrine gland (en′ doh krin) [Gr. *endon*: inside + *krinein*: to separate]   Any gland, such as the adrenal or pituitary gland of vertebrates, that secretes certain substances, especially hormones, into the body through the blood.

Endocrinology   The study of hormones and their actions.

Endocytosis   A process by which liquids or solid particles are taken up by a cell through invagination of the plasma membrane. (Contrast with exocytosis.)

Endoderm [Gr. *endon*: within + *derma*: skin]   The innermost of the three embryonic tissue layers first delineated during gastrulation. Gives rise to the digestive and respiratory tracts and structures associated with them.

Endodermis [Gr. *endon*: within + *derma*: skin]   In plants, a specialized cell layer marking the inside of the cortex in roots and some stems. Frequently a barrier to free diffusion of solutes.

Endomembrane system   Endoplasmic reticulum plus Golgi apparatus plus, when present, lysosomes; thus, a system of membranes that exchange material with one another.

Endometrium (en do mee′ tree um) [Gr. *endon*: within + *metrios*: womb]   The epithelial cells lining the uterus of mammals.

Endoplasmic reticulum [Gr. *endon*: within + L. *plasma*: form; L. *reticulum*: little net]   A system of membrane-bounded tubes and flattened sacs, often continuous with the nuclear envelope, found in the cytoplasm of eukaryotes. Exists as rough ER, studded with ribosomes, and smooth ER, lacking ribosomes.

Endorphins   Naturally occurring, opiate-like substances in the mammalian brain.

Endoskeleton   A skeleton covered by other, soft body tissues. (Contrast with exoskeleton.)

Endosperm [Gr. *endon*: within + *sperma*: seed]   A specialized triploid seed tissue found only in angiosperms; contains stored food for the developing embryo.

Endosymbiosis [Gr. *endon*: within + *syn*: together + *bios*: life]   The living together of two species, with one living inside the body (or even the cells) of the other.

Endosymbiotic theory   Theory that the eukaryotic cell evolved from a prokaryote that contained other, endosymbiotic prokaryotes.

Endotherm [Gr. *endon*: within + *thermos*: hot]   An animal that can control its body temperature by the expenditure of its own metabolic energy. (Contrast with ectotherm.)

Energetic cost   The difference between the energy an animal would have expended had it rested, and that expended in performing a behavior.

Energy   The capacity to do work.

Enhancer   In eukaryotes, a DNA sequence, lying on either side of the gene it regulates, that stimulates a specific promoter.

Enterocoelous development   A pattern of development in which the coelum is formed by an outpocketing of the embryonic gut (enteron).

Enterokinase (ent uh row kine′ ase)   An enzyme secreted by the mucosa of the duodenum. It activates the zymogen trypsinogen to create the active digestive enzyme trypsin.

Entrainment   With respect to circadian rhythms, the process whereby the period is adjusted to match the 24-hour environmental cycle.

Entropy (en′ tro pee) [Gr. *en*: in + *tropein*: to change]   A measure of the degree of disorder in any system. A perfectly ordered system has zero entropy; increasing disorder is measured by positive entropy. Spontaneous reactions in a closed system are always accompanied by an increase in disorder and entropy.

Environment   An organism's surroundings, both living and nonliving; includes temperature, light intensity, and all other species that influence the focal organism.

Environmental toxicology   The study of the distribution and effects of toxic compounds in the environment.

Enzyme (en′ zime) [Gr. *en*: in + *zyme*: yeast]   A protein, on the surface of which are chemical groups so arranged as to make the enzyme a catalyst for a chemical reaction.

Epi- [Gr.: upon, over]   A prefix used to designate a structure located on top of another; for example: epidermis, epiphyte.

Epicotyl (epp′ i kot′ il) [Gr. *epi*: upon + *kotyle*: something hollow]   That part of a plant embryo or seedling that is above the cotyledons.

Epidermis [Gr. *epi*: upon + *derma*: skin]   In plants and animals, the outermost cell layers. (Only one cell layer thick in plants.)

Epididymis (epuh did′ uh mus) [Gr. *epi*: upon + *didymos*: testicle]   Coiled tubules in the testes that store sperm and conduct sperm from the seiminiferous tubules to the vas deferens.

Epinephrine (ep i nef′ rin) [Gr. *epi*: upon + *nephros*: a kidney]   The "fight or flight" hormone. Produced by the medulla of the adrenal gland, it also functions as a neurotransmitter. Also known as adrenaline.

Epiphyte (ep′ e fyte) [Gr. *epi*: upon + *phyton*: plant]   A specialized plant that grows on the surface of other plants but does not parasitize them.

Episome   A plasmid that may exist either free or integrated into a chromosome. (See plasmid.)

Epistasis   An interaction between genes, in which the presence of a particular allele of one gene determines whether another gene will be expressed.

Epithelium   In animals, a layer of cells covering or lining an external surface or a cavity.

Equilibrium   (1) In biochemistry, a state in which forward and reverse reactions are proceeding at counterbalancing rates, so there is no observable change in the concentrations of reactants and products. (2) In evolutionary genetics, a condition in which allele and genotype frequencies in a population are constant from generation to generation.

Error signal   In physiology, the difference between a set point and a feedback signal that results in a corrective response.

Erythrocyte (ur rith′ row sight) [Gr. *erythros*: red + *kytos*: hollow vessel]   A red blood cell.

Esophagus (i soff′ i gus) [Gr. *oisophagos*: gullet]   That part of the gut between the pharynx and the stomach.

Essential amino acid   An amino acid an animal cannot synthesize for itself and must obtain from its diet.

Essential element   An irreplaceable mineral element without which normal growth and reproduction cannot proceed.

Estivation (ess tuh vay′ shun) [L. *aestivalis*: summer]   A state of dormancy and hypometabolism that occurs during the summer; usually a means of surviving drought and/or intense heat. Contrast with hibernation.

Estrogen   Any of several steroid sex hormones, produced chiefly by the ovaries in mammals.

Estrous cycle   The cyclical changes in reproductive physiology and behavior in female mammals (other than some primates), culminating in estrus.

Estrus (es′ truss) [L. *oestrus*: frenzy]   The period of heat, or maximum sexual receptivity, in some female mammals. Ordinarily, the estrus is also the time of release of eggs in the female.

Ethology (ee thol′ o jee) [Gr. *ethos*: habit, custom + *logos*: discourse]   The study of whole patterns of animal behavior in natural environments, stressing the analysis of adaptation and evolution of the patterns.

Ethylene   One of the plant hormones, the gas $H_2C = 2CH_2$.

Etiolation   Plant growth in the absence of light.

Eubacteria (yew bacteria)   Kingdom including the great majority of bacteria, such as the gram negative bacteria, gram positive bacteria, mycoplasmas, etc. (Contrast with Archaebacteria.)

Euchromatin Chromatin that is diffuse and non-staining during interphase; may be transcribed. (Contrast with heterochromatin.)

Eukaryotes (yew car' ry otes) [Gr. *eu*: true + *karyon*: kernel or nucleus] Organisms whose cells contain their genetic material inside a nucleus. Includes all life other than the viruses, Archaebacteria, and Eubacteria.

Eusocial Term applied to insects, such as termites, ants, and many bees and wasps, in which individuals cooperate in the care of offspring, there are sterile castes, and generations overlap.

Eutrophication (yoo trofe' ik ay' shun) [Gr. *eu-*: well + *trephein*: to flourish] The addition of nutrient materials to a body of water, resulting in changes to species composition therein.

Evergreen A plant that retains its leaves through all seasons. (Contrast with deciduous.)

Evolution Any gradual change. Organic evolution, often referred to as evolution, is any genetic and resulting phenotypic change in organisms from generation to generation.

Evolutionary agent Any factor that influences the direction and rate of evolutionary changes.

Evolutionary biology The collective branches of biology that study evolutionary process and their products—the diversity and history of living things.

Evolutionarily conservative Traits of organisms that evolve very slowly.

Evolutionary innovations Major changes in body plans of organisms; these have been very rare during evolutionary history.

Evolutionary radiation The proliferation of species within a single evolutionary lineage.

Excision repair The removal and damaged DNA and its replacement by the appropriate nucleotides. Often, several bases on either side of the damaged base are removed by the action of an endonuclease. Then a DNA polymerase adds the correct bases according to the template still present on the other strand of DNA. DNA ligase catalyzes the sealing up of the repaired strand.

Excitatory postsynaptic potential (EPSP) A change in the resting potential of a postsynaptic membrane in a positive (depolarizing) direction. (Contrast with inhibitory postsynaptic potential.)

Excretion Release of metabolic wastes by an organism.

Exergonic reaction A reaction in which free energy is released. (Contrast with endergonic reaction.)

Exo- (eks' oh) Same as ecto-.

Exocrine gland (eks' oh krin) [Gr. *exo*: outside + *krinein*: to separate] Any gland, such as a salivary gland, that secretes to the outside of the body or into the gut.

Exocytosis A process by which a vesicle within a cell fuses with the plasma membrane and releases its contents to the outside. (Contrast with endocytosis.)

Exon A portion of a DNA molecule, in eukaryotes, that codes for part of a polypeptide. (Contrast with intron.)

Exoskeleton (eks' oh skel' e ton) A hard covering on the outside of the body to which muscles are attached. (Contrast with endoskeleton.)

Experiment A scientific method in which particular factors are manipulated while other factors are held constant so that the potential influences of the manipulated factors can be determined.

Exploitation competition Competition that occurs because resources are depleted. (Contrast with interference competition.)

Exponential growth Growth, especially in the number of organisms in a population, which is a simple function of the size of the growing entity: the larger the entity, the faster it grows. (Contrast with logistic growth.)

Expression vector A DNA vector, such as a plasmid, that carries a DNA sequence that includes the adjacent sequences for its expression into mRNA and protein in a host cell.

Expressivity The degree to which a genotype is expressed in the phenotype— may be affected by the environment.

Extensor A muscle that extends an appendage.

Extinction The termination of a lineage of organisms.

Extrinsic protein A membrane protein found only on the surface of the membrane. (Contrast with intrinsic protein.)

F-duction Transfer of genes from one bacterium to another, using the F-factor as a vehicle.

F-factor In some bacteria, the fertility factor; a plasmid conferring "maleness" on the cell that contains it.

F_1 generation The immediate progeny of a parental (P) mating; the first filial generation.

F_2 generation The immediate progeny of a mating between members of the F_1 generation.

Facilitated diffusion Passive movement through a membrane involving a specific carrier protein; does not proceed against a concentration gradient. (Contrast with active transport, free diffusion.)

Facultative Capable of occurring or not occurring, as in facultative aerobes. (Contrast with obligate.)

Family In taxonomy, the category below the order and above the genus; a group of related, similar genera.

Fat A triglyceride that is solid at room temperature. (Contrast with oil.)

Fatty acid A molecule with a long hydrocarbon tail and a carboxyl group at the other end. Found in many lipids.

Fauna (faw' nah) All of the animals found in a given area. (Contrast with flora.)

Feces [L. *faeces*: dregs] Waste excreted from the digestive system.

Feedback control Control of a particular step of a multistep process, induced by the presence or absence of a product of one of the later steps. A thermostat regulating the flow of heating oil to a furnace in a home is a negative feedback control device.

Fermentation (fur men tay' shun) [L. *fermentum*: yeast] The degradation of a substance such as glucose to smaller molecules with the extraction of energy, without the use of oxygen (i.e., anaerobically). Involves the glycolytic pathway.

Fertilization Union of gametes. Also known as syngamy.

Fertilization membrane A membrane surrounding an animal egg which becomes rapidly raised above the egg surface within seconds after fertilization, serving to prevent entry of a second sperm.

Fetus The latter stages of an embryo that is still contained in an egg or uterus; in humans, the unborn young from the eighth week of pregnancy to the moment of birth.

Fiber An elongated and tapering cell of vascular plants, usually with a thick cell wall. Serves a support function.

Fibrin A protein that polymerizes to form long threads that provide structure to a blood clot.

Filter feeder An organism that feeds upon much smaller organisms, that are suspended in water or air, by means of a straining device.

Filtration In the excretory physiology of some animals, the process by which the initial urine is formed; water and most solutes are transferred into the excretory tract, while proteins are retained in the blood or hemolymph.

First law of thermodynamics Energy can be neither created nor destroyed.

Fission Reproduction of a prokaryote by division of a cell into two comparable progeny cells.

Fitness The contribution of a genotype or phenotype to the composition of subsequent generations, relative to the contribution of other genotypes or phenotypes. (See inclusive fitness.)

Fixed action pattern A behavior that is genetically programmed.

Flagellin (fla jell' in) The protein from which prokaryotic (but not eukaryotic) flagella are constructed.

Flagellum (fla jell' um) (plural: flagella) [L. *flagellum*: whip] Long, whiplike appendage that propels cells. Prokaryotic flagella differ sharply from those found in eukaryotes.

Flexor A muscle that flexes an appendage.

Flora (flore' ah) All of the plants found in a given area. (Contrast with fauna.)

Florigen A plant hormone (not yet isolated) involved in the conversion of a vegetative shoot apex to a flower.

Flower The total reproductive structure of an angiosperm; its basic parts include the calyx, corolla, stamens, and carpels.

Fluorescence The emission of a photon of visible light by an excited atom or molecule.

Follicle [L. *folliculus*: little bag] In female mammals, an immature egg surrounded by nutritive cells.

Follicle-stimulating hormone A gonadotropic hormone produced by the anterior pituitary.

Food chain A portion of a food web, most commonly a simple sequence of prey species and the predators that consume them.

Food web The complete set of food links between species in a community; a diagram indicating which ones are the eaters and which are consumed.

Forb Any broad-leaved (dicotyledonous), herbaceous plant. Especially applied to such plants growing in grasslands.

Fossil Any recognizable structure originating from an organism, or any impression from such a structure, that has been preserved over geological time.

Founder effect Random changes in allele frequencies resulting from establishment of a population by a very small number of individuals.

Fovea [L. *fovea*; a small pit] The area, in the vertebrate retina, of most distinct vision.

Frame-shift mutation A mutation resulting from the addition or deletion of a single base pair in the DNA sequence of a gene. As a result of this, mRNA transcribed from such a gene is translated normally until the ribosome reaches the point at which the mutation has occurred. From that point on, codons are read out of proper register and the amino acid sequence bears no resemblance to the normal sequence. (Contrast with missense mutation, nonsense mutation.)

Free diffusion Diffusion directly across a membrane without the involvement of carrier molecules. Free diffusion is not saturable and cannot cause the net transport from a region of low concentration to a region of higher concentration. (Contrast with facilitated diffusion and active transport.)

Free energy That energy which is available for doing useful work, after allowance has been made for the increase or decrease of disorder. Designated by the symbol G (for Gibbs free energy), and defined by: $G = H - TS$, where H = heat, S = entropy, and T = absolute (Kelvin) temperature.

Frequency-dependent selection Selection that changes in intensity with the proportion of individuals having the trait.

Fruit In angiosperms, a ripened and mature ovary (or group of ovaries) containing the seeds. Sometimes applied to reproductive structures of other groups of plants, and includes any adjacent parts which may be fused with the reproductive structures.

Fruiting body A structure that bears spores.

Fundamental niche The range of condition under which an organism could survive if it were the only one in the environment. (Contrast with realized niche.)

Fungus (fung' gus) A member of the kingdom Fungi, a (usually) multicellular eukaryote with absorptive nutrition.

G_1 phase In the cell cycle, the gap between the end of mitosis and the onset of the S phase.

G_2 phase In the cell cycle, the gap between the S (synthesis) phase and the onset of mitosis.

G protein A membrane protein involved in signal transduction; characterized by binding guanyl nucleotides. The activation of certain receptors activates the G protein, which in turn activates adenylate cyclase. G protein activation involves binding a GTP molecule in place of a GDP molecule.

Gametangium (gam i tan' gee um) [Gr. *gamos*: marriage + *angeion*: vessel or reservoir] Any plant or fungal structure within which a gamete is formed.

Gamete (gam' eet) [Gr. *gamete*: wife, *gametes*: husband] The mature sexual reproductive cell: the egg or the sperm.

Gametocyte (ga meet' oh site) [Gr. *gamete*: wife, *gametes*: husband + *kytos*: cell] The cell that gives rise to sex cells, either the eggs or the sperm. (See oocyte and spermatocyte.)

Gametogenesis (ga meet' oh jen' e sis) [Gr. *gamete*: wife, *gametes*: husband + *genesis*: source] The specialized series of cellular divisions that leads to the production of sex cells (gametes). (Contrast with oogenesis and spermatogenesis.)

Gametophyte (ga meet' oh fyte) In plants with alternation of generations, the haploid phase that produces the gametes. (Contrast with sporophyte.)

Ganglion (gang' glee un) [Gr.: tumor] A group or concentration of neuron cell bodies.

Gap junction A 2.7-nanometer gap between plasma membranes of two animal cells, spanned by protein channels. Gap junctions allow chemical substances or electrical signals to pass from cell to cell.

Gas exchange In animals, the process of taking up oxygen from the environment and releasing carbon dioxide to the environment.

Gastrovascular cavity Serving for both digestion (gastro) and circulation (vascular); in particular, the central cavity of the body of jellyfish and other cnidarians.

Gastrula (gas' true luh) [Gr. *gaster*: stomach] An embryo forming the characteristic three cell layers (ectoderm, endoderm, and mesoderm) which will give rise to all of the major tissue systems of the adult animal.

Gastrulation Development of a blastula into a gastrula.

Gated channel A channel (membrane protein) that opens and closes in response to binding of specific molecules or to changes in membrane potential.

Gel electrophoresis (jel ul lec tro for' eesis) A semisolid matrix suspended in a salty buffer in which molecules can be separated on the basis of their size and change when current is passed through the gel.

Gene [Gr. *gen*: to produce] A unit of heredity. Used here as the unit of genetic function which carries the information for a single polypeptide.

Gene amplification Creation of multiple copies of a particular gene, allowing the production of large amounts of the RNA transcript (as in rRNA synthesis in oocytes).

Gene cloning Formation of a clone of bacteria or yeast cells containing a particular foreign gene.

Gene family A set of identical, or once-identical, genes, derived from a single parent gene; need not be on the same chromosomes; classic example is the globin family in vertebrates.

Gene flow The exchange of genes between different species (an extreme case referred to as hybridization) or between different populations of the same species caused by migration following breeding.

Gene pool All of the genes in a population.

Gene therapy Treatment of a genetic disease by providing patients with cells containing wild type alleles for the genes that are nonfunctional in their bodies.

Generalized transduction The transfer of any bacterial host gene in a virus particle to another bacterium.

Generative nucleus In a pollen tube, a haploid nucleus that undergoes mitosis to produce the two sperm nuclei that participate in double fertilization. (Contrast with tube nucleus.)

Generator potential A stimulus-induced change in membrane resting potential in the direction of threshold for generating action potentials.

Genet The genetic individual of a plant that is composed of a number of nearly identical but repeated units.

Genetic drift Changes in gene frequencies from generation to generation in a small population as a result of random processes.

Genetic stochasticity Variation in the frequencies of alleles and genotypes in a population over time.

Genetics The study of heredity.

Genetic structure The frequencies of alleles and genotypes in a population.

Genome (jee' nome) The genes in a complete haploid set of chromosomes.

Genotype (jean' oh type) [Gr. *gen*: to produce + *typos*: impression] An exact description of the genetic constitution of an individual, either with respect to a single trait or with respect to a larger set of traits. (Contrast with phenotype.)

Genus (jean' us) (plural: genera) [Gr. *genos*: stock, kind] A group of related, similar species.

Geographical (allopatric) speciation Formation of two species from one by the interposition of (or crossing of) a physical barrier. (Contrast with parapatric, sympatric speciation.)

Geotropism See gravitropism.

Germ cell A reproductive cell or gamete of a multicellular organism.

Germination The sprouting of a seed or spore.

Gestation (jes tay' shun) [L. *gestare*: to bear] The period during which the embryo of a mammal develops within the uterus. Also known as **pregnancy**.

Gibberellin (jib er el' lin) [L. *gibberella*: hunchback (refers to shape of a reproductive structure of a fungus that produces gibberellins)] One of a class of plant growth substances playing roles in stem elongation, seed germination, flowering of certain plants, etc. Named for the fungus *Gibberella*.

Gill An organ for gas exchange in aquatic organisms.

Gill arch A skeletal structure that supports gill filaments and the blood vessels that supply them.

Gizzard (giz' erd) [L. *gigeria*: cooked chicken parts] A very muscular port of the stomach of birds that grinds up food, sometimes with the aid of fragments of stone.

Gland An organ or group of cells that produces and secretes one or more substances.

Glans penis Sexually sensitive tissue at the tip of the penis.

Glia (glee' uh) [Gr.: glue] Cells, found only in the nervous system, which do not conduct action potentials.

Glomerulus (glo mare' yew lus) [L. *glomus*: ball] Sites in the kidney where blood filtration takes place. Each glomerulus consists of a knot of capillaries served by afferent and efferent arterioles.

Glucocorticoids Steroid hormones produced by the adrenal cortex. Secreted in response to ACTH, they inhibit glucose uptake by many tissues in addition to mediating other stress responses.

Glucagon A hormone produced and released by cells in the islets of Langerhans of the pancreas. It stimulates the breakdown of glycogen in liver cells.

Gluconeogenesis The biochemical synthesis of glucose from other substances, such as amino acids, lactate, and glycerol.

Glucose (glue' kose) [Gr. *gleukos*: sweet wine mash for fermentation] The most common sugar, one of several monosaccharides with the formula $C_6H_{12}O_6$.

Glycerol (gliss' er ole) A three-carbon alcohol with three hydroxyl groups, the linking component of phospholipids and triglycerides.

Glycogen (gly' ko jen) A branched-chain polymer of glucose, similar to starch (which is less branched and may be of lower molecular weight). Exists mostly in liver and muscle; the principal storage carbohydrate of most animals and fungi.

Glycolysis (gly kol' li sis) [from glucose + Gr. *lysis*: loosening] The enzymatic breakdown of glucose to pyruvic acid. One of the oldest energy-yielding machanisms in living organisms.

Glycosidic linkage The connection in an oligosaccharide or polysaccharide chain, formed by removal of water during the linking of monosaccharides.by root pressure.

Glyoxysome (gly ox' ee soam) An organelle found in plants, in which stored lipids are converted to carbohydrates.

Golgi apparatus (goal' jee) A system of concentrically folded membranes found in the cytoplasm of eukaryotic cells. Plays a role in the production and release of secretory materials such as the digestive enzymes manufactured in the pancreas. First described by Camillo Golgi (1844–1926).

Gonad (go' nad) [Gr. *gone*: seed, that which produces seed] An organ that produces sex cells in animals: either an ovary (female gonad) or testis (male gonad).

Gonadotropin A hormone that stimulates the gonads.

Grade The level of complexity found in an animal's body plan.

Gram stain A differential stain useful in characterizing bacteria.

Granum Within a chloroplast, a stack of thylakoids.

Gravitropism A directed plant growth response to gravity.

Grazer An animal that eats the vegetative tissues of herbaceous plants.

Green gland An excretory organ of crustaceans.

Gross morphology The sizes and shapes of the major body parts of a plant or animal.

Gross primary production The total energy captured by plants growing in a particular area.

Ground meristem That part of an apical meristem that gives rise to the ground tissue system of the primary plant body.

Ground tissue system Those parts of the plant body not included in the dermal or vascular tissue systems. Ground tissues function in storage, photosynthesis, and support.

Groundwater Water present deep in soils and rocks; may be stationary or flow slowly eventually to discharge into lakes, rivers, or oceans.

Group transfer The exchange of atoms between molecules.

Growth Irreversible increase in volume (probably the most accurate definition, but at best a dangerous oversimplification).

Growth factors A group of proteins that circulate in the blood and trigger the normal growth of cells. Each growth factor acts only on certain target cells.

Growth stage That stage in the life history of an organism in which it grows to its adult size.

Guanine (gwan'een) A nitrogen-containing base found in DNA, RNA, and GTP.

Guard cells In plants, paired epidermal cells which surround and control the opening of a stoma (pore).

Gut An animal's digestive tract.

Guttation The extrusion of liquid water through openings in leaves, caused by root pressure.

Gymnosperm (jim' no sperm) [Gr. *gymnos*: naked + *sperma*: seed] A plant, such as a pine or other conifer, whose seeds do not develop within an ovary (hence, the seeds are "naked").

Gyrus (plural: gyri) The raised or ridged portion of the convoluted surface of the brain. (Contrast to sulcus.)

Habit The form or pattern of growth characteristic of an organism.

Habitat The environment in which an organism lives.

Habituation (ha bich' oo ay shun) The simplest form of learning, in which an animal presented with a stimulus without reward or punishment eventually ceases to respond.

Hair cell A type of mechanosensor in animals.

Half-life The time required for half of a sample of a radioactive isotope to decay to its stable, nonradioactive form.

Halophyte (hal' oh fyte) [Gr. *halos*: salt + *phyton*: plant] A plant that grows in a saline (salty) environment.

Haploid (hap' loid) [Gr. *haploeides*: single] Having a chromosome complement consisting of just one copy of each chromosome. This is the normal "ploidy" of gametes or of asexual spores produced by meiosis or of organisms (such as the gametophyte generation of plants) that grow from such spores without fertilization.

Haplontic life cycle A life cycle in which the zygote is the only diploid cell.

Hardy–Weinberg rule The rule that the basic processes of Mendelian heredity (meiosis and recombination) do not alter either the frequencies of genes or their diploid combinations. The Law also states how the percentages of diploid combinations can be predicted from a knowledge of the proportions of alleles in the population.

Haustorium (haw stor' ee um) [L. *haustus*: draw up] A specialized hypha or other structure by which fungi and some parasitic plants draw food from a host plant.

Haversian systems Units of organization in compact bone that reflect the action of intercommunicating osteoblasts.

Helicase (heel' uh case) An enzyme that unwinds the DNA double helix for DNA relplication.

Helper T cells T cells that participate in the activation of B cells and of other T cells; targets of the HIV-I virus, the agent of AIDS. (Contrast with cytotoxic T cells, suppressor T cells.)

Hematocrit (heme at o krit) [Gr. *haima*: blood + *krites*: judge]   The proportion of 100 cc of blood that consists of red blood cells.

Hemizygous (hem' ee zie' gus) [Gr. *hemi*: half + *zygotos*: joined]   In a diploid organism, having only one allele for a given trait, typically the case for X-linked genes in male mammals and Z-linked genes in female birds. (Contrast with homozygous, heterozygous.)

Hemoglobin (hee' mo glow' bin) [Gr. *haima*: blood + L. *globus*: globe]   The colored protein of vertebrate blood (and blood of some invertebrates) which transports oxygen.

Hepatic (heh pat' ik) [Gr. *hepar*: liver] Pertaining to the liver.

Hepatic duct   The duct that conveys bile from the liver to the gallbladder.

Herbicide (ur' bis ide)   A chemical substance that kills plants.

Herbivore [L. *herba*: plant + *vorare*: to devour]   An animal which eats the tissues of plants. (Contrast with carnivore, detritivore, omnivore.)

Heritable   Able to be inherited; in biology usually refers to genetically determined traits.

Hermaphroditism (her maf' row dite' ism) [Gr. *hermaphroditos*: a person with both male and female traits]   The coexistence of both female and male sex organs in the same organism.

Hertz (abbreviated as Hz)   Cycles per second.

Hetero- [Gr.: other, different]   A prefix used in biology to mean that two or more different conditions are involved; for example, heterotroph, heterozygous.

Heterochromatin   Chromatin that retains its coiling during interphase; generally not transcribed. (Contrast with euchromatin.)

Heterocyst   A large, thick-walled cell in the filaments of certain cyanobacteria; performs nitrogen fixation.

Heterogeneous nuclear RNA (hnRNA) The product of transcription of a eukaryotic gene, including transcripts of introns.

Heterokaryon (het' er oh care' ee ahn) [Gr. *heteros*: different + *karyon*: kernel]   A cell or organism carrying a mixture of genetically distinguishable nuclei. A heterokaryon is usually the result of the fusion of two cells without fusion of their nuclei.

Heteromorphic (het' er oh more' fik) [Gr. *heteros*: different + *morphe*: form]   having a different form or appearance, as two heteromorphic life stages of a plant. (Contrast with isomorphic.)

Heterosporous (het' er os' por us) Producing two types of spores, one of which gives rise to a female megaspore and the other to a male microspore. Heterosporous plants produce distinct female and male gametophytes. (Contrast with homosporous.)

Heterotherm   An animal that regulates its body temperature at a constant level at some times but not others, such as a hibernator.

Heterotroph (het' er oh trof) [Gr. *heteros*: different + *trophe*: food]   An organism that requires preformed organic molecules as food. (Contrast with autotroph.)

Heterozygous (het' er oh zie' gus) [Gr. *heteros*: different + *zygotos*: joined]   Of a diploid organism having different alleles of a given gene on the pair of homologues carrying that gene. (Contrast with homozygous.)

Hfr (for "high frequency of recombination") Donor bacterium in which the F-factor has been integrated into the chromosome. This produces a bacterium that transfers its chromosomal markers at a very high frequency to recipient (F⁻) cells.

Hibernation [L. *hibernus*: winter]   The state of inactivity of some animals during winter; marked by a drop in body temperature and metabolic rate.

Highly repetitive DNA   Short DNA sequences present in millions of copies in the genome, next to each other (in tandem). In a In a reassociation experiment, denatured highly repetitive DNA reanneals very quickly.

Hippocampus   A part of the forebrain that takes part in long-term memory formation.

Histamine (hiss' tah meen)   A substance released within a damaged tissue by a type of white blood cell. Histamines are responsible for aspects of allergice reactions, including the increased vascular permeability that leads to edema (swelling).

Histology   The study of tissues.

Histone   Any one of a group of basic proteins forming the core of a nucleosome, the structural unit of a eukaryotic chromosome. (See nucleosome.)

Historical biogeography   The study of the distributions of organisms from a long-term, historical perspective.

hnRNA   See heterogeneous nuclear RNA.

Holdfast   In many large attached algae, specialized tissue attaching the plant to its substratum.

Homeobox   A segment of DNA, found in a few genes, perhaps regulating the expression of other genes and thus controlling large-scale developmental processes.

Homeostasis (home' ee o sta' sis) [Gr. *homos*: same + *stasis*: position]   The maintenance of a steady state, such as a constant temperature or a stable social structure, by means of physiological or behavioral feedback responses.

Homeotherm (home' ee o therm) [Gr. *homos*: same + *therme*: heat]   An animal which maintains a constant body temperature by virtue of its own heating and cooling mechanisms. (Contrast with heterotherm, poikilotherm.)

Homeotic genes (home' ee ott' ic)   Genes that determine what entire segments of an animal become.

Homeotic mutation   A drastic mutation causing the transformation of body parts in *Drosophila* metamorphosis. Examples include the *Antennapedia* and *ophthalmoptera* mutants.

Homolog (home' o log') [Gr. *homos*: same + *logos*: word]   One of a pair, or larger set, of chromosomes having the same overall genetic composition and sequence. In diploid organisms, each chromosome inherited from one parent is matched by an identical (except for mutational changes) chromosome—its homolog—from the other parent.

Homology (ho mol' o jee) [Gr. *homologi(a)*: agreement]   A similarity between two structures that is due to inheritance from a common ancestor. The structures are said to be homologous. (Contrast with analogy.)

Homoplasy (home' uh play zee) [Gr. *homos*: same + *plastikos*: to mold]   The presence in several species of a trait not present in their most common ancestor. Can result from convergent evolution, reverse evolution, or parallel evolution.

Homosporous   Producing a single type of spore that gives rise to a single type of gametophyte, bearing both female and male reproductive organs. (Contrast with heterosporous.)

Homozygous (home' o zie' gus) [Gr. *homos*: same + *zygotos*: joined]   Of a diploid organism having identical alleles of a given gene on both homologous chromosomes. An organism may be a "homozygote" with respect to one gene and, at the same time, a "heterozygote" with respect to another. (Contrast with heterozygous.)

Hormone (hore' mone) [Gr. *hormon*: excite, stimulate]   A substance produced in one part of a multicellular organism and transported to another part where it exerts its specific effect on the physiology or biochemistry of the target cells.

Host   An organism that harbors a parasite and provides it with nourishment.

Host–parasite interaction   The dynamic interaction between populations of a host and the parasites that attack it.

Humoral immune system   The part of the immune system mediated by B cells; it is mediated by circulating antibodies and is active against extracellular bacterial and viral infections.

Humus (hew' muss)   The partly decomposed remains of plants and animals on the surface of a soil. Its characteristics depend primarily upon climate and the species of plants growing on the site.

Hyaluronidase (hill yew ron' uh dase)   An enzyme that digests proteoglycans. Found in sperm cells, it helps digest the coatings surrounding an egg so the sperm can penetrate the egg cell membrane.

Hybrid (high' brid) [L. *hybrida*: mongrel] The offspring of genetically dissimilar parents. In molecular biology, a double helix formed of nucleic acids from different sources.

Hybridoma   A cell produced by the fusion of an antibody-producing cell with a myeloma cell; it produces monoclonal antibodies.

Hydrocarbon   A compound containing only carbon and hydrogen atoms.

Hydrogen bond A chemical bond which arises from the attraction between the slight positive charge on a hydrogen atom and a slight negative charge on a nearby fluorine, oxygen, or nitrogen atom. Weak bonds, but found in great quantities in proteins, nucleic acids, and other biological macromolecules.

Hydrological cycle The sum total of movement of water from the oceans to the atmosphere, to the soil, and back to the oceans. Some water is cycled many times within compartments of the system before completing one full circuit.

Hydrolyze (hi' dro lize) [Gr. *hydro*: water + *lysis*: cleavage] To break a chemical bond, as in a peptide linkage, with the insertion of the components of water, –H and –OH, at the cleaved ends of a chain. The digestion of proteins is a hydrolysis.

Hydrophilic [Gr. *hydro*: water + *philia*: love] Having an affinity for water. (Contrast with hydrophobic.)

Hydrophobic [Gr. *hydro*: water + *phobia*: fear] Molecules and amino acid side chains, which are mainly hydrocarbons (compounds of C and H with no charged groups or polar groups), have a lower energy when they are clustered together than when they are distributed through an aqueous solution. Because of their attraction for one another and their reluctance to mix with water they are called "hydrophobic." Oil is a hydrophobic substance; phenylalanine is a hydrophobic amino acid. (Contrast with hydrophilic.)

Hydrophobic interaction A weak attraction between highly nonpolar molecules or parts of molecules suspended in water.

Hydrostatic skeleton The incompressible internal liquids of some animals that transfer forces from one part of the body to another when acted upon by the surrounding muscles.

Hydroxyl group The —OH group, characteristic of alcohols.

Hyperosmotic Having a more negative osmotic potential, as a result of having a higher concentration of osmotically active particles. Said of one solution as compared with another. (Contrast with hypoosmotic, isosmotic.)

Hyperpolarization A change in the resting potential of a membrane so the inside of a cell becomes more electronegative. (Contrast with depolarization.)

Hypersensitive response A defensive response of plants to microbial infection; it results in a "dead spot."

Hypertension High blood pressure.

Hypha (high' fuh) (plural: hyphae) [Gr. *hyphe*: web] In the fungi, any single filament. May be multinucleate (zygomycetes, ascomycetes) or multicellular (basidiomycetes).

Hypocotyl That part of the embryonic or seedling plant shoot that is below the cotyledons.

Hypoosmotic Having a less negative osmotic potential, as a result of having a lower concentration of osmotically active particles. Said of one solution as compared with another. (Contrast with hyperosmotic, isosmotic.)

Hypothalamus The part of the brain lying below the thalamus; it coordinates water balance, reproduction, temperature regulation, and metabolism.

Hypothetico-deductive method A method of science in which hypotheses are erected, predictions are made from them, and experiments and observations are performed to test the predictions. The process may be repeated many times in the course of answering a question.

Icosahedron (eye kos a heed' ron) A 20-sided crystal. Some viruses have coat proteins which form a icosahedron.

Imaginal disc In insect larvae, groups of cells that develop into specific adult organs.

Imbibition [L. *imbibo*: to drink] The binding of a solvent to another molecule. Dry starch and protein will imbibe water.

Immune system A system in mammals that recognizes and eliminates or neutralizes either foreign substances or self substances that have been altered to appear foreign.

Immunization The deliberate introduction of antigen to bring about an immune response.

Immunoglobulins A class of proteins, with a characteristic structure, active as receptors and effectors in the immune system.

Immunological memory Certain clones of immune system cells made to respond to an antigen persist. This leads to a more rapid and massive response of the immune system to any subsequenct exposure to that antigen.

Immunological tolerance A mechanism by which an animal does not mount an immune response to the antigenic determinants of its own macromolecules.

Imprinting A rapid form of learning, in which an animal comes to make a particular response, which is maintained for life, to some object or other organism.

Inclusive fitness The sum of an individual's own fitness (the effect of producing its own offspring: the individual selection component) plus its influence on fitness in relatives other than direct descendants (the kin selection component).

Incomplete dominance Condition in which the heterozygous phenotype is intermediate between the two homozygous phenotypes.

Incomplete metamorphosis Insect development in which changes between instars are gradual.

Incus (in' kus) [L. *incus*: anvil] The middle of the three bones that conduct movements of the eardrum to the oval window of the inner ear. (See malleus, stapes.)

Individual fitness That component of inclusive fitness that results from an organism producing its own offspring. (Contrast with kin selection component.)

Indoleacetic acid See auxin.

Induced fit A change in the tertiary structures of some enzymes, caused by binding of substrate to the active site.

Inducer (1) In enzyme systems, a small molecule which, when added to a growth medium, causes a large increase in the level of some enzyme. (2) In embryology, a substance that causes a group of target cells to differentiate in a particular way.

Inducible enzyme An enzyme that is present in much larger amounts when a particular compound (the inducer) has been added to the system. (Contrast with constitutive enzyme.)

Inflammation A nonspecific defense against pathogens; characterized by redness, swelling, pain, and increased temperature.

Inflorescence A structure composed of several flowers.

Inhibitor A substance which binds to the surface of an enzyme and interferes with its action on its substrates.

Inhibitory postsynaptic potential A change in the resting potential of a postsynaptic membrane in the hyperpolarizing (negative) direction.

Initiation complex Combination of a ribosomal light subunit, an mRNA molecule, and the tRNA charged with the first amino acid coded for by the mRNA; formed at the onset of translation.

Initiation factors Proteins that assist in forming the translation initiation complex at the ribosome.

Inositol triphosphate (IP3) An intracellular second messenger derived from membrane phospholipids.

Insertion sequence A large piece of DNA that can give rise to copies at other loci; a type of transposable genetic element.

Instar (in' star) [L.: image, form] An immature stage of an insect between molts.

Instinct Behavior that is relatively highly steretoyped and self-differentiating, that develops in individuals unable to observe other individuals performing the behavior or to practice the behavior in the presence of the objects toward which it is usually directed.

Insulin (in' su lin) [L. *insula*: island] A hormone, synthesized in islet cells of the pancreas, that promotes the conversion of glucose to the storage material, glycogen.

Integrase An enzyme that integrates retroviral cDNA into the genome of the host cell.

Integrated pest management A method of control of pests in which natural predators and parasites are used in conjunction with sparing use of chemical methods to achieve control of a pest without causing serious adverse environmental side effects.

Integument [L. *integumentum*: covering] A protective surface structure. In gymnosperms and angiosperms, a layer of tissue around the ovule which will become the seed coat. Gymnosperm ovules have one integument, angiosperm ovules two.

Intention movement The preparatory motions that animals go through prior to a complete behavior response; for example, the crouch before flying, the snarl before biting, etc.

Intercalary meristem A meristematic region in plants which occurs not apically, but between two regions of mature tissue. Intercalary meristems occur in the nodes of grass stems, for example.

Intercostal muscles Muscles between the ribs that can augment breathing movements by elevating and suppressing the rib cage.

Interference competition Competition resulting from direct behavioral interactions between organisms. (Contrast with exploitation competition.)

Interferon A glycoprotein produced by virus-infected animal cells; increases the resistance of neighboring cells to the virus.

Interkinesis The phase between the first and second meiotic divisions.

Interleukins Regulatory proteins, produced by macrophages and lymphocytes, that act upon other lymphocytes and direct their development.

Intermediate filaments Fibrous proteins that stabilize cell structure and resist tension.

Internode Section between two nodes of a plant stem.

Interphase The period between successive nuclear divisions during which the chromosomes are diffuse and the nuclear envelope is intact. It is during this period that the cell is most active in transcribing and translating genetic information.

Interspecific competition Competition between members of two or more species.

Intertropical convergence zone The tropical region where the air rises most strongly; moves north and south with the passage of the sun overhead.

Intraspecific competition Competition among members of a single species.

Intrinsic protein A membrane protein that is embedded in the phospholipid bilayer of the membrane. (Contrast with extrinsic protein.)

Intrinsic rate of increase The rate at which a population can grow when its density is low and environmental conditions are highly favorable.

Intron A portion of a DNA molecule that, because of RNA splicing, is not involved in coding for part of a polypeptide molecule. (Contrast with exon.)

Invagination An infolding.

Inversion (genetic) A rare mutational event that leads to the reversal of the order of genes within a segment of a chromosome, as if that segment had been removed from the chromosome, turned 180°, and then reattached.

Invertebrate Any animal that is not a vertebrate, that is, whose nerve cord is not enclosed in a backbone of bony segments.

In vitro [L.: in glass] In a test tube, rather than in a living organism. (Contrast with in vivo.)

In vivo [L.: in the living state] In a living organism. Many processes that occur in vivo can be reproduced in vitro with the right selection of cellular components. (Contrast with in vitro.)

Ion (eye' on) [Gr.: wanderer] An atom or group of atoms with electrons added or removed, giving it a negative or positive electrical charge.

Ionic channel A membrane protein that can let ions pass across the membrane. The channel can be ion-selective, and it can be voltage-gated or ligand-gated.

Ionic bond A chemical bond which arises from the electrostatic attraction between positively and negatively charged ions. Usually a strong bond.

Iris (eye' ris) [Gr. iris: rainbow] The round, pigmented membrane that surrounds the pupil of the eye and adjusts its aperture to regulate the amount of light entering the eye.

Irruption A rapid increase in the density of a population. Often followed by massive emigration.

Islets of Langerhans Clusters of hormone-producing cells in the pancreas.

Isogamy (eye sog' ah mee) [Gr. isos: equal + gamos: marriage] A kind of sexual reproduction in which the gametes (or gametangia) are not distinguishable on the basis of size or morphology.

Isolating mechanism Geographical, physiological, ecological, or behavioral mechanisms that lead to a reduction in the frequency of hybrid matings.

Isomers Molecules consisting of the same numbers and kinds of atoms, but differing in the way in which the atoms are combined.

Isomorphic (eye' so more' fik) [Gr. isos: equal + morphe: form] having the same form or appearance, as two isomorphic life stages. (Contrast with heteromorphic.)

Isosmotic Having the same osmotic potential. Said of two solutions. (Contrast with hyperosmotic, hypoosmotic.)

Isotope (eye' so tope) [Gr. isos: equal + topos: place] Two isotopes of the same chemical element have the same number of protons in their nuclei, but differ in the number of neutrons.

Isozymes Chemically different enzymes that catalyze the same reaction.

Jejunum (jih jew' num) The middle division of the small intestine, where most absorption of nutrients occurs. (See duodenum, ileum.)

Joule (jool, or jowl) A unit of energy, equal to 0.24 calories.

Juvenile hormone In insects, a hormone maintaining larval growth and preventing maturation or pupation.

Karyotype The number, forms, and types of chromosomes in a cell.

Kelvin temperature scale See absolute temperature scale.

Keratin (ker' a tin) [Gr. keras: horn] A protein which contains sulfur and is part of such hard tissues as horn, nail, and the outermost cells of the skin.

Ketone (key' tone) A compound with a C=O group attached to two other groups, neither of which is an H atom. Many sugars are ketones. (Contrast with aldehyde.)

Keystone species A species that exerts a major influence on the composition and dynamics of the community in which it lives.

Kidneys A pair of excretory organs in vertebrates.

Kin selection component The component of inclusive fitness resulting from helping the survival of relatives containing the same alleles by descent from a common ancestor.

Kinase (kye' nase) An enzyme that transfers a phosphate group from ATP to another molecule. Protein kinases transfer phosphate from ATP to specific proteins, playing important roles in cell regulation.

Kinesis (ki nee' sis) [Gr.: movement] Orientation behavior in which the organism does not move in a particular direction with reference to a stimulus but instead simply moves at an increasing or decreasing rate until it ends up farther from the object or closer to it. (Contrast with taxis.)

Kinetochore (kin net' oh core) [Gr. kinetos: moving + khorein: to move] Specialized structure on a centromere to which microtubules attach.

Kingdom The highest taxonomic category in the Linnaean system.

Knockout mouse A genetically engineered mouse in which one or more functioning alleles have been replaced by defective alleles.

Lac operon A region of DNA in E. coli that contains a single promoter and operator controlling the expression of three adjacent genes involved in the utilization of the sugar, lactose.

Lactic acid The end product of fermentation in vertebrate muscle and some microorganisms.

Lagging strand In DNA replication, the daughter strand that is synthesized discontinuously.

Lamella Layer.

Larynx (lar' inks) A structure between the pharynx and the trachea that includes the vocal cords.

Larva (plural: larvae) [L.: ghost, early stage] An immature stage of any invertebrate animal that differs dramatically in appearance from the adult.

Lateral Pertaining to the side.

Lateral inhibition In visual information processing in the arthropod eye, the mutual inhibition of optic nerve cells; results in enhanced detection of edges.

Laterization (lat' ur iz ay shun) The formation of a nutrient-poor soil that is rich in nsoluble iron and aluminum compounds.

Law of independent assortment Alleles of different, unlinked genes assort independently of one another during gamete formation, Mendel's second law.

Law of segregation Alleles segregate from one another during gamete formation, Mendel's first law.

Leader sequence A sequence of amino acids at the N-terminal end of a newly synthesized protein, determining where the protein will be placed in the cell.

Leading strand In DNA replication, the daughter strand that is synthesized continuously.

Leaf axil The upper angle between a leaf and the stem, site of lateral buds which under appropriate circumstances become activated to form lateral branches.

Leaf primordium [L.: the beginning] A small mound of cells on the flank of a shoot apical meristem that will give rise to a leaf.

Lek A traditional courtship display ground, where males display to females.

Lenticel Spongy region in a plant's periderm, allowing gas exchange.

Leukocyte (loo' ko sight) [Gr. *leukos*: clear + *kutos*: hollow vessel] A white blood cell.

Leuteinizing hormone A peptide hormone produced by pituitary cells that stimulates follicle maturation in females.

Lichen (lie' kun) [Gr. *leikhen*: licker] An organism resulting from the symbiotic association of a true fungus and either a cyanobacterium or a unicellular alga.

Life cycle The entire span of the life of an organism from the moment of fertilization (or asexual generation) to the time it reproduces in turn.

Life history The stages an individual goes through during its life.

Life table A table showing, for a group of equal-aged individuals, the proportion still alive at different times in the future and the number of offspring they produce during each time interval.

Ligament A band of connective tissue linking two bones in a joint.

Ligand (lig' and) A molecule that binds to a receptor site of another molecule.

Lignin The principal noncarbohydrate component of wood, a polymer that binds together cellulose fibrils in some plant cell walls.

Limbic system A group of primitive vertebrate forebrain nuclei that form a network and are involved in emotions, drives, instinctive behaviors, learning, and memory.

Limiting resource The required resource whose supply most strongly influences the size of a population.

Linkage Association between genetic markers on the same chromosome such that they do not show random assortment and seldom recombine; the closer the markers, the lower the frequency of recombination.

Lipase (lip' ase; lye' pase) An enzyme that digests fats.

Lipids (lip' ids) [Gr. *lipos*: fat] Substances in a cell which are easily extracted by organic solvents; fats, oils, waxes, steroids, and other large organic molecules, including those which, with proteins, make up the cell membranes. (See phospholipids.)

Litter The partly decomposed remains of plants on the surface and in the upper layers of the soil.

Littoral zone The coastal zone from the upper limits of tidal action down to the depths where the water is thoroughly stirred by wave action.

Liver A large digestive gland. In vertebrates, it secretes bile and is involved in the formation of blood.

Lobes Regions of the human cerebral hemispheres; includes the temporal, frontal, parietal, and occipital lobes.

Locus In genetics, a specific location on a chromosome. May be considered to be synonymous with "gene."

Logistic growth Growth, especially in the size of an organism or in the number of organisms that constitute a population, which slows steadily as the entity approaches its maximum size. (Contrast with exponential growth.)

Loop of Henle (hen' lee) Long, hairpin loop of the mammalian renal tubule that runs from the cortex down into the medulla, and back to the cortex. Creates a concentration gradient in the interstitial fluids in the medulla.

Lophophore A U-shaped fold of the body wall with hollow, ciliated tentacles that encircles the mouth of animals in several different phyla. Used for filtering prey from the surrounding water.

Lordosis (lor doe' sis) [Gk. *lordosis*: curving forward] A posture assumed by females of some mammalian species (especially rodents) to signal sexual receptivity.

Lumen (loo' men) [L.: light] The cavity inside any tubular part of an organ, such as a piece of gut or a kidney tubule.

Lungs A pair of saclike chambers within the bodies of some animals, functioning in gas exchange.

Luteinizing hormone A gonadotropin produced by the anterior pituitary. It stimulates the gonads to produce sex hormones.

Lymph [L. *lympha*: water] A clear, watery fluid that is formed as a filtrate of blood; it contains white blood cells; it collects in a series of special vessels and is returned to the bloodstream.

Lymph nodes Specialized tissue regions that act as filters for cells, bacteria and foreign matter.

Lymphocyte A major class of white blood cells. Includes T cells, B cells, and other cell types important in the immune response.

Lysis (lie' sis) [Gr.: a loosening] Bursting of a cell.

Lysogenic The condition of a bacterium that carries the genome of a virus in a relatively stable form. (Contrast with lytic.)

Lysosome (lie' so soam) [Gr. *lysis*: a loosening + *soma*: body] A membrane-bounded inclusion found in eukaryotic cells (other than plants). Lysosomes contain a mixture of enzymes that can digest most of the macromolecules found in the rest of the cell.

Lysozyme (lie' so zyme) An enzyme in saliva, tears, and nasal secretions that attacks bacterial cell walls, as one of the body's nonspecific defense mechanisms.

Lytic Condition in which a bacterium lyses shortly after infection by a virus; the viral genome does not become stabilized within the bacterial cell. (Contrast with lysogenic.)

Macro- (mack' roh) [Gr. *makros*: large, long] A prefix commonly used to denote something large. (Contrast with micro-.)

Macroevolution Evolutionary changes occurring over long time spans and usually involving changes in many traits. (Contrast with microevolution.)

Macroevolutionary time The time required for macroveolutionary changes in a lineage.

Macromolecule A giant polymeric molecule. The macromolecules are proteins, polysaccharides, and nucleic acids.

Macronutrient A mineral element required by plant tissues in concentrations of at least 1 milligram per gram of their dry matter.

Macrophage (mac' roh faj) A type of white blood cell that endocytoses bacteria and other cells.

Major histocompatibility complex (MHC) A complex of linked genes, with multiple alleles, that control a number of immunological phenomena; it is important in graft rejection.

Malignant tumor A tumor whose cells can invade surrounding tissues and spread to other organs.

Malleus (mal' ee us) [L. *malleus*: hammer] The first of the three bones that conduct movements of the eardrum to the oval window of the inner ear. (See incus, stapes.)

Malpighian tubule (mal pee' gy un) A type of protonephridium found in insects.

Mammal [L. *mamma*: breast, teat] Any animal of the class Mammalia, characterized by the production of milk by the female mammary glands and the possession of hair for body covering.

Mantle A sheet of specialized tissues that covers most of the viscera of mollusks; provides protection to internal organs and secretes the shell.

Map unit In eukaryotic genetics, one map unit corresponds to a recombinant frequency of 0.01.

Mapping In genetics, determining the order of genes on a chromosome and the distances between them.

Marine [L. *mare*: sea, ocean] Pertaining to or living in the ocean. (Contrast with aquatic, terrestrial.)

Maritime climate Weather pattern typical of coasts of continents, particularly those on the western sides at mid latitudes, in which the difference between summer and winter is relatively small. (Contrast with continental climate.)

Marsupial (mar soo' pee al) A mammal belonging to the subclass Metatheria, such as opossums and kangaroos. Most have a pouch (marsupium) that contains the milk glands and serves as a receptacle for the young.

Mass extinctions Geological periods during which rates of extinction were much higher than during intervening times.

Mass number The sum of the number of protons and neutrons in an atom's nucleus.

Mast cells Typically found in connective tissue, mast cells can be provoked by antigens or inflammation to release histamine.

Maternal effect genes These genes code for morphogens that determine the polarity of the egg and larva in the fruit fly, *Drosophila melanogaster.*

Maternal inheritance (cytoplasmic inheritance) Inheritance in which the phenotype of the offspring depends on factors, such as mitochondria or chloroplasts, that are inherited from the female parent through the cytoplasm of the female gamete.

Mating type In some bacteria, fungi, and protists, sexual reproduction can occur only between partners of a different mating type. "Mating type" is not the same as "sex," since some species have as many as 8 mating types; mating may also be between hermaphroditic partners of opposite mating type, with both partners acting as both "male" and "female" in terms of donating and receiving genetic information.

Maturation The automatic development of a pattern of behavior, which becomes increasingly complex or precise as the animal matures. Unlike learning, the development does not require experience to occur.

Mechanosensor A cell that is sensitive to physical movement and generates action potentials in response.

Medulla (meh dull' luh) [L.: narrow] (1) The inner, core region of an organ, as in the adrenal medulla (adrenal gland) or the renal medulla (kidneys). (2) The portion of the brain stem that connects to the spinal cord.

Medusa (meh doo' suh) The tentacle-bearing, jellyfish-like, free-swimming sexual stage in the life cycle of a cnidarian.

Mega- [Gr. *megas*: large, great] A prefix often used to denote something large. (Contrast with micro-.)

Megareserve A large park or reserve; usually has associated buffer areas in which human use of the environment is restricted to activities that do not destroy the functioning of the ecosystem.

Megasporangium The special structure (sporangium) that produces the megaspores.

Megaspore [Gr. *megas*: large + *spora*:seed] In plants, a haploid spore that produces a female gametophyte. In many cases the megaspore is larger than the male-producing microspore.

Meiosis (my oh' sis) [Gr.: diminution] Division of a diploid nucleus to produce four haploid daughter cells. The process consists of two successive nuclear divisions with only one cycle of chromosome replication.

Membrane potential The difference in electrical charge between the inside and the outside of a cell, caused by a difference in the distribution of ions.

Memory cells Long-lived lymphocytes produced by exposure to antigen. They persist in the body and are able to mount a rapid response to subsequent exposures to the antigen.

Mendelian population A local population of individuals belonging to the same species and exchanging genes with one another.

Menopause The time in a human female's life when the ovarian and menstrual cycles cease.

Menstrual cycle The monthly sloughing off of the uterine lining if fertilization does not occur in the female. Occurs between puberty and menopause.

Meristem [Gr. *meristos*: divided] Plant tissue made up of actively dividing cells.

Mesenchyme (mez' en kyme) [Gr. *mesos*: middle + *enchyma*: infusion] Embryonic or unspecialized cells derived from the mesoderm.

Meso- (mez' oh) [Gr.: middle] A prefix often used to designate a structure located in the middle, or a stage that appears at some intermediate time. For example, mesoderm, Mesozoic.

Mesoderm [Gr. *mesos*: middle + *derma*: skin] The middle of the three embryonic tissue layers first delineated during gastrulation. Gives rise to skeleton, circulatory system, muscles, excretory system, and most of the reproductive system.

Mesoglea The jelly-like middle layer that constitutes the bulk of the bodies of the medusae of many cnidarians; not a true cell layer.

Mesophyll (mez' a fill) [Gr. *mesos*: middle + *phyllon*: leaf] Chloroplast-containing, photosynthetic cells in the interior of leaves.

Mesosome (mez' o soam') [Gr. *mesos*: middle + *soma*: body] A localized infolding of the plasma membrane of a bacterium.

Messenger RNA (mRNA) A transcript of one of the strands of DNA, it carries information (as a sequence of codons) for the synthesis of one or more proteins.

Meta- [Gr.: between, along with, beyond] A prefix used in biology to denote a change or a shift to a new form or level; for example, as used in metamorphosis.

Metabolic compensation Changes in biochemical properties of an organism that render it less sensitive to temperature changes.

Metabolic pathway A series of enzyme-catalyzed reactions so arranged that the product of one reaction is the substrate of the next.

Metabolism (meh tab' a lizm) [Gr. *metabole*: to change] The sum total of the chemical reactions that occur in an organism, or some subset of that total (as in "respiratory metabolism").

Metamorphosis (met' a mor' fo sis) [Gr. *meta*: between + *morphe*: form, shape] A radical change occurring between one developmental stage and another, as for example from a tadpole to a frog or an insect larva to the adult.

Metaphase (met' a phase) [Gr. *meta*: between] The stage in nuclear division at which the centromeres of the highly supercoiled chromosomes are all lying on a plane (the metaphase plane or plate) perpendicular to a line connecting the division poles.

Metapopulation A population divided into subpopulations, among which there are occasional exchanges of individuals.

Metastasis (meh tass' tuh sis) The spread of cancer cells from their original site to other parts of the body.

Methanogen Any member of a group of Archaebacteria that release methane as a metabolic product. This group is considered to be an extremely ancient one.

MHC See major histocompatibility complex.

Micelles (my sells') [L. *mica*: grain, crumb] The small particles of fat in the small intestine, resulting from the emulsification of dietary fat by bile.

Micro- (mike' roh) [Gr. *mikros*: small] A prefix often used to denote something small. (Contrast with macro-, mega-.)

Microbiology [Gr. *mikros*: small + *bios*: life + *logos*: discourse] The scientific study of microscopic organisms, particularly bacteria, unicellular algae, protists, and viruses.

Microevolution The small evolutionary changes typically occurring over short time spans; generally involving a small number of traits and minor genetic changes. (Contrast with macroevolution.)

Microevolutionary time The time required for microevolutionary changes within a lineage of organisms.

Microfilament Minute fibrous structure generally composed of actin found in the cytoplasm of eukaryotic cells. They play a role in the motion of cells.

Micromorphology The structure of the macromolecules of an organism.

Micronutrient A mineral element required by plant tissues in concentrations of less than 100 micrograms per gram of their dry matter.

Microorganism Any microscopic organism, such as a bacterium or one-celled alga.

Micropyle (mike' roh pile) [Gr. *mikros*: small + *pyle*: gate] Opening in the integument(s) of a seed plant ovule through which pollen grows to reach the female gametophyte within.

Microsporangium The special structure (sporangium) that produces the microspores.

Microspores [Gr. *mikros*: small + *spora*: seed] In plants, a haploid spore that produces a male gametophyte. In many cases the microspore is smaller than the female-producing megaspore.

Microtubules Minute tubular structures found in centrioles, spindle apparatus, cilia, flagella, and other places in the cytoplasm of eukaryotic cells. These tubules play roles in the motion and maintenance of shape of eukaryotic cells.

Microvilli (singular: microvillus) The projections of epithelial cells, such as the cells lining the small intestine, that increase their surface area.

Middle lamella A layer of derivative polysaccharides that separates plant cells; a common middle lamella lies outside the primary walls of the two cells.

Migration The regular, seasonal movements of animals between breeding and nonbreeding ranges.

Mimicry (mim' ik ree) The resemblance of one kind of organism to another, or to some inanimate object; serves the function of making the organism difficult to find, of discouraging potential enemies or of attracting potential prey. (See Batesian mimicry and Müllerian mimicry.)

Mineral An inorganic substance other than water.

Mineralocorticoid A hormone produced by the adrenal cortex that influences mineral ion balance; aldosterone.

Minimal medium A medium for the growth of bacteria, fungi, or tissue cultures, containing only those nutrients absolutely required for the growth of wild type cells.

Minimum viable population. The smallest number of individuals required for a population to persist in a region.

Mismatch repair When a single base in DNA is changed into a different base, or the wrong base inserted during DNA replication, there is a mismatch in base pairing with the base on the opposite strand. A repair system removes the incorrect base and inserts the proper one for pairing with the opposite strand.

Missense mutation A mutation that changes a codon for one amino acid to a codon for a different amino acid. (Contrast with frame-shift mutation, nonsense mutation.)

Mitochondrial matrix The fluid interior of the mitochondrion, enclosed by the inner mitochondrial membrane.

Mitochondrion (my' toe kon' dree un) (plural: mitochondria) [Gr. *mitos*: thread + *chondros*: cartilage, or grain] An organelle that occurs in eukaryotic cells and contains the enzymes of the ctric acid cycle, the respiratory chain, and oxidative phosphorylation. A mitochondrion is bounded by a double membrane.

Mitosis (my toe' sis) [Gr. *mitos*: thread] Nuclear division in eukaryotes leading to the formation of two daughter nuclei each with a chromosome complement identical to that of the original nucleus.

Mitotic center Cellular region that organizes the microtubules for mitosis. In animals a centrosome serves as the mitotic center.

Mobbing Gathering of calling animals around a predator; their calls and the confusion they create reduce the probability that the predator can hunt successfully in the area.

Moderately repetitive DNA DNA sequences that appear hundreds to thousands of times in the genome. They include the DNA sequences coding for rRNAs and tRNAs, as well as the DNA at telomeres.

Modular organism An organism which grows by producing additional units of body construction that are very similar to the units of which it is already composed.

Mole A quantity of a compound whose weight in grams is numerically equal to its molecular weight expressed in atomic mass units. Avogadro's number of molecules: 6.023×10^{23} molecules.

Molecular formula A representation that shows how many atoms of each element are present in a molecule.

Molecular weight The sum of the atomic weights of the atoms in a molecule.

Molecule A particle made up of two or more atoms joined by covalent bonds or ionic attractions.

Molting The process of shedding part or all of an outer covering, as the shedding of feathers by birds or of the entire exoskeleton by arthropods.

Monoecious (mo nee' shus) [Gr.: one house] Organisms in which both sexes are "housed" in a single individual, which produces both eggs and sperm. (In some plants, these are found in different flowers within the same plant.) Examples: corn, peas, earthworms, hydras. (Contrast with dioecious, perfect flower.)

Moneran (moh neer' un) A bacterium. This term was coined when both archaebacteria and eubacteria were considered to be members of a single kingdom, Monera.

Mono- [Gr. *monos*: one] Prefix denoting a single entity. (Contrast with poly.)

Monoclonal antibody Antibody produced in the laboratory from a clone of hybridoma cells, each of which produces the same specific antibody.

Monocot (short for monocotyledon) [Gr. *monos*: one + *kotyledon*: a cup-shaped hollow] Any member of the angiosperm class Monocotyledones, plants in which the embryo produces but a single cotyledon (seed leaf). Leaves of most monocots have their major veins arranged parallel to each other.

Monocytes White blood cells that produce macrophages.

Monohybrid cross A mating in which the parents differ with respect to the alleles of only one locus of interest.

Monomer A small molecule, two or more of which can be combined to form oligomers (consisting of a few monomers) or polymers (consisting of many monomers).

Monophyletic (mon' oh fih leht' ik) [Gk. *monos*: single + *phylon*: tribe] Being descended from a single ancestral stock.

Monosaccharide A simple sugar. Oligosaccharides and polysaccharides are made up of monosaccharides.

Monosynaptic reflex A neural reflex that begins in a sensory neuron and makes a single synapse before activating a motor neuron.

Morphogens Diffusible substances whose concentration gradients determine patterns of development in animals and plants.

Morphogenesis (more' fo jen' e sis) [Gr. *morphe*: form + *genesis*: origin] The development of form. Morphogenesis is the overall consequence of determination, differentiation, and growth.

Morphology (more fol' o jee) [Gr. *morphe*: form + *logos*: discourse] The scientific study of organic form, including both its development and function.

Mosaic development Pattern of animal embryonic development in which each blastomere contributes a specific part of the adult body. (Contrast with regulative development.)

Motor end plate The modified area on a muscle cell membrane where a synapse is formed with a motor neuron.

Motor neuron A neuron carrying information from the central nervous system to an effector such as a muscle fiber.

Motor unit A motor neuron and the set of muscle fibers it controls.

mRNA (See messenger RNA.)

Mucosa (mew koh' sah) An epithelial membrane containing cells that secrete mucus. The inner cell layers of the digestive and respiratory tracts.

Müllerian mimicry The resemblance of two or more unpleasant or dangerous kinds of organisms to each other.

Multicellular [L. *multus*: much + *cella*: chamber] Consisting of more than one cell, as for example a multicellular organism. (Contrast with unicellular.)

Muscle Contractile tissue containing actin and myosin organized into polymeric chains called microfilaments. In vertebrates, the tissues are either cardiac muscle, smooth muscle, or striated (skeletal) muscle.

Muscle fiber A single muscle cell. In the case of striated muscle, a syncitial, multinucleate cell.

Muscle spindle Modified muscle fibers encased in a connective sheat and functioning as stretch sensors.

Mutagen (mute' ah jen) [L. *mutare*: change + Gr. *genesis*: source] An agent, especially a chemical, that increases the mutation rate.

Mutation In the broad sense, any discontinuous change in the genetic constitution of an organism. In the narrow sense, the word usually refers to a "point mutation," a change along a very narrow portion of the nucleic acid sequence.

Mutation pressure Evolution (change in gene proportions) by different mutation rates alone.

Mutualism The type of symbiosis, such as that exhibited by fungi and algae or cyanobacteria in forming lichens, in which both species profit from the association.

Mycelium (my seel' ee yum) [Gr. *mykes*: fungus] In the fungi, a mass of hyphae.

Mycorrhiza (my' ka rye' za) [Gr. *mykes*: fungus + *rhiza*: root] An association of the root of a plant with the mycelium of a fungus.

Myelin (my' a lin) A material forming a sheath around some axons. It is formed by Schwann cells that wrap themselves about the axon. It serves to insulate the axon electrically and to increase the rate of transmission of a nervous impulse.

Myofibril (my' oh fy' bril) [Gr. *mys*: muscle + L. *fibrilla*: small fiber] A polymeric unit of actin or myosin in a muscle.

Myogenic (my oh jen' ik) [Gr. *mys*: muscle + *genesis*: source] Originating in muscle.

Myoglobin (my' oh globe' in) [Gr. *mys*: muscle + L. *globus*: sphere] An oxygen-binding molecule found in muscle. Consists of a heme unit and a single globiin chain, and carrys less oxygen than hemoglobin.

Myosin [Gr. *mys*: muscle] One of the two major proteins of muscle, it makes up the thick filaments. (See actin.)

NAD (nicotinamide adenine dinucleotide) A compound found in all living cells, existing in two interconvertible forms: the oxidizing agent NAD⁺ and the reducing agent NADH.

NADP (nicotinamide adenine dinucleotide phosphate) Like NAD, but possessing another phosphate group; plays similar roles but is used by different enzymes.

Natal group The group into which an individual was born.

Natural selection The differential contribution of offspring to the next generation by various genetic types belonging to the same population. The mechanism of evolution proposed by Charles Darwin.

Nauplius (no' plee us) [Gk. *nauplios*: shellfish] The typical larva of crustaceans. Has three pairs of appendages and a median compound eye.

Necrosis (nec roh' sis) Tissue damage resulting from cell death.

Negative control The situation in which a regulatory macromolecule (generally a repressor) functions to turn off transcription. In the absence of a regulatory macromolecule, the structural genes are turned on.

Negative feedback A pattern of regulation in which a change in a sensed variable results in a correction that opposes the change.

Nekton [Gr. *nekhein*: to swim] Animals, such as fish, that can swim against currents of water. (Contrast with plankton.)

Nematocyst (ne mat' o sist) [Gr. *nema*: thread + *kystis*: cell] An elaborate, threadlike structure produced by cells of jellyfish and other cnidarians, used chiefly to paralyze and capture prey.

Nephridium (nef rid' ee um) [Gr. *nephros*: kidney] An organ which is involved in excretion, and often in water balance, involving a tube that opens to the exterior at one end.

Nephron (nef' ron) [Gr. *nephros*: kidney] The basic component of the kidney, which is made up of numerous nephrons. Its form varies in detail, but it always has at one end a device for receiving a filtrate of blood, and then a tubule that absorbs selected parts of the filtrate back into the bloodstream.

Nephrostome (nef' ro stome) [Gr. *nephros*: kidney + *stoma*: opening] An opening in a nephridium through which body fluids can enter.

Nerve A structure consisting of many neuronal axons and connective tissue.

Net primary production Total photosynthesis minus respiration by plants.

Neural plate A thickened strip of ectoderm along the dorsal side of the early vertebrate embryo; gives rise to the central nervous system.

Neural tube An early stage in the development of the vertebrate nervous system consisting of a hollow tube created by two opposing folds of the dorsal ectoderm along the anterior–posterior body axis.

Neuromuscular junction The region where a motor neuron contacts a muscle fiber, creating a synapse.

Neuron (noor' on) [Gr. *neuron*: nerve, sinew] A cell derived from embryonic ectoderm and characterized by a membrane potential that can change in response to stimuli, generating action potentials. Action potentials are generated along an extension of the cell (the axon), which makes junctions (synapses) with other neurons, muscle cells, or gland cells.

Neurotransmitter A substance, produced in and released by one neuron, that diffuses across a synapse and excites or inhibits the postsynaptic neuron.

Neurula (nure' you la) [Gr. *neuron*: nerve] Embryonic stage during formation of the dorsal nerve cord by two ectodermal ridges.

Neutral alleles Alleles that differ so slightly that the proteins for which they code function identically.

Neutron (new' tron) [E.: neutral] One of the three main fundamental particles of matter, with mass approximately 1 amu and no electrical charge.

Nicotinamide adenine dinucleotide (See NAD.)

Nicotinamide adenine dinucleotide phosphate (See NADP.)

Nitrification The oxidation of ammonia to nitrite and nitrate ions, performed by certain soil bacteria.

Nitrogenase In nitrogen-fixing organisms, an enzyme complex that mediates the stepwise reduction of atmospheric N_2 to ammonia.

Nitrogen fixation Conversion of nitrogen gas to ammonia, which makes nitrogen available to living things. Carried out by certain prokaryotes, some of them free-living and others living within plant roots.

Node [L. *nodus*: knob, knot] In plants, a (sometimes enlarged) point on a stem where a leaf or bud is or was attached.

Node of Ranvier A gap in the myelin sheath covering an axons, where the axonal membrane can fire action potentials.

Noncompetitive inhibitor An inhibitor that binds the enzyme at a site other than the active site. (Contrast with competitive inhibitor.)

Nondisjunction Failure of sister chromatids to separate in meiosis II or mitosis, or failure of homologous chromosomes to separate in meiosis I. Results in aneuploidy.

Nonpolar molecule A molecule whose electric charge is evenly balanced from one end of the molecule to the other.

Nonsense (chain-terminating) mutation Mutations that change a codon for an amino acid to one of the codons (UAG, UAA, or UGA) that signal termination of translation. The resulting gene product is a shortened polypeptide that begins normally at the amino-terminal end and ends at the position of the altered codon. (Contrast with frame-shift mutation, missense mutation.)

Nonspecific defenses Immunologic responses directed against most or all pathogens, generally without reference to the pathogens' antigens. These defenses include the skin, normal flora, lysozyme, the acidic stomach, interferon, and the inflammatory response.

Nontracheophytes Those plants lacking well-developed vascular tissue; the liverworts, hornworts, and mosses. (Contrast with tracheophytes.)

Normal flora The bacteria and fungi that live on animal body surfaces without causing disease.

Norepinephrine A neurotransmitter found in the central nervous system and also at the postganglionic nerve endings of the sympathetic nervous system. Also called noradrenaline.

Notochord (no' tow kord) [Gr. *notos*: back + *chorde*: string] A flexible rod of gelatinous material serving as a support in the embryos of all chordates and in the adults of tunicates and lancelets.

Nuclear envelope The surface, consisting of two layers of membrane, that encloses the nucleus of eukaryotic cells.

Nucleic acid (new klay' ik) [E.: nucleus of a cell] A long-chain alternating polymer of deoxyribose or ribose and phosphate groups, with nitrogenous bases—adenine, thymine, uracil, guanine, or cytosine (A, T, U, G, or C)—as side chains. DNA and RNA are nucleic acids.

Nucleoid (new' klee oid)   The region that harbors the chromosomes of a prokaryotic cell. Unlike the eukaryotic nucleus, it is not bounded by a membrane.

Nucleolar organizer (new klee' o lar)   A region on a chromosome that is associated with the formation of a new nucleolus following nuclear division. The site of the genes that code for ribosomal RNA.

Nucleolus (new klee' oh lus) [from L. diminutive of *nux*: little kernel or little nut]   A small, generally spherical body found within the nucleus of eukaryotic cells. The site of synthesis of ribosomal RNA.

Nucleoplasm (new' klee o plazm)   The fluid material within the nuclear envelope of a cell, as opposed to the chromosomes, nucleoli, and other particulate constituents.

Nucleosome   A portion of a eukaryotic chromosome, consisting of part of the DNA molecule wrapped around a group of histone molecules, and held together by another type of histone molecule. The chromosome is made up of many nucleosomes.

Nucleotide   The basic chemical unit (monomer) in a nucleic acid. A nucleotide in RNA consists of one of four nitrogenous bases linked to ribose, which in turn is linked to phosphate. In DNA, deoxyribose is present instead of ribose.

Nucleus (new' klee us) [from L. diminutive of *nux*: kernel or nut]   (1) In chemistry, the dense central portion of an atom, made up of protons and neutrons, with a positive charge. Surrounded by a cloud of negatively charged electrons. (2) In cells, the centrally located chamber of eukaryotic cells that is bounded by a double membrane and contains the chromosomes. The information center of the cell.

Nutrient   A food substance; or, in the case of mineral nutrients, an inorganic element required for completion of the life cycle of an organism.

Obligate (ob' li gut)   Necessary, as in obligate anaerobe. (Contrast with facultative.)

Obligate anaerobe   An animal that can live only in oxygenated environments.

Oil   A triglyceride that is liquid at room temperature. (Contrast with fat.)

Okazaki fragments   Newly formed DNA strands making up the lagging strand in DNA replication. DNA ligase links the Okazaki fragments to give a continuous strand.

Olfactory   Having to do with the sense of smell.

Oligomer   A compound molecule of intermediate size, made up of two to a few monomers. (Contrast with monomer, polymer.)

Ommatidium [Gr. *omma*: an eye]   One of the units which, collected into groups of up to 20,000, make up the compound eye of arthropods.

Omnivore [L. *omnis*: all, everything + *vorare*: to devour]   An organism that eats both animal and plant material. (Contrast with carnivore, detritivore, herbivore.)

Oncogenic (ong' co jen' ik) [Gr. *onkos*: mass, tumor + *genes*: born]   Causing cancer.

Oocyte (oh' eh site) [Gr. *oon*: egg + *kytos*: cell]   The cell that gives rise to eggs in animals.

Oogenesis (oh' eh jen e sis) [Gr. *oon*: egg + *genesis*: source]   Female gametogenesis, leading to production of the egg.

Oogonium (oh' eh go' nee um)   In some algae and fungi, a cell in which an egg is produced.

Operator   The region of an operon that acts as the binding site for the repressor.

Operon   A genetic unit of transcription, typically consisting of several structural genes that are transcribed together; the operon contains at least two control regions: the promoter and the operator.

Opportunity cost   The sum of the benefits an animal forfeits by not being able to perform some other behavior during the time when it is performing a given behavior.

Opsin (op' sin) [Gr. *opsis*: sight]   The protein protion of the visual pigment rhodopsin. (See rhodopsin.)

Optic chiasm   Stucture on the lower surface of the vertebrate brain where the two optic nerves come together.

Optical isomers   Isomers that differ in the configuration of the four different groups attached to a single carbon atom; so named because solutions of the two isomers rotate the plane of polarized light in opposite directions. The two isomers are mirror images of one another.

Optimality models   Models developed to determine the structures or behaviors that best solve particular problems faced by organisms.

Order   In taxonomy, the category below the class and above the family; a group of related, similar families.

Organ   A body part, such as the heart, liver, brain, root, or leaf, composed of different tissues integrated to perform a distinct function for the body as a whole.

Organ identity genes   These plant genes specify the various parts of the flower.

Organ of Corti   Structure in the inner ear that transforms mechanical forces produced from pressure waves ("sound waves") into action potentials that are sensed as sound.

Organelles (or' gan els') [L.: little organ]   Organized structures that are found in or on cells. Examples: ribosomes, nuclei, mitochrondria, chloroplasts, cilia, and contractile vacuoles.

Organic   Pertaining to any aspect of living matter, e.g., to its evolution, structure, or chemistry. The term is also applied to any chemical compound that contains carbon.

Organism   Any living creature.

Organizer, embryonic   A region of an embryo which directs the development of nearby regions. In amphibian early gastrulas, the dorsal lip of the blastopore.

Origin of replication   A DNA sequence at which helicase unwinds the DNA double helix and DNA polymerase binds to initiate DNA replication.

Osmoregulation   Regulation of the chemical composition of the body fluids of an organism.

Osmosensor   A neuron that converts changes in the osmotic potential of interstial fluids into action potentials.

Osmosis (oz mo' sis) [Gr. *osmos*: to push]   The movement of water through a differentially permeable membrane from one region to another where the water potential is more negative. This is often a region in which the concentration of dissolved molecules or ions is higher, although the effect of dissolved substances may be offset by hydrostatic pressure in cells with semi-rigid walls.

Osmotic potential   A property of any solution, resulting from its solute content; it may be zero or have a negative value. A negative osmotic potential tends to cause water to move into the solution; it may be offset by a positive pressure potential in the solution or by a more negative water potential in a neighboring solution. (Contrast with turgor pressure.)

Ossicle (ah' sick ul) [L. *os*: bone]   The calcified construction unit of echinoderm skeletons.

Osteoblasts   Cells that lay down the protein matrix of bone.

Osteoclasts   Cells that dissolve bone.

Otolith (oh' tuh lith) [Gk.*otikos*: ear + *lithos*: stone]   Structures in the vertebrate vestibular apparatus that mechanically stimulate hair cells when the head moves or changes position.

Outgroup   A taxon that separated from another taxon, whose lineage is to be inferred, before the latter underwent evolutionary radiation.

Oval window   The flexible membrane which, when moved by the bones of the middle ear, produces pressure waves in the inner ear

Ovary (oh' var ee)   Any female organ, in plants or animals, that produces an egg.

Oviduct [L. *ovum*: egg + *ducere*: to lead]   In mammals, the tube serving to transport eggs to the uterus or to outside of the body.

Oviparous (oh vip' uh rus)   Reproduction in which eggs are released by the female and development is external to the mother's body. (Contrast with viviparous.)

Ovulation   The release of an egg from an ovary.

Ovule (oh' vule) [L. *ovulum*: little egg]   In plants, an organ that contains a gametophyte and, within the gametophyte, an egg; when it matures, an ovule becomes a seed.

Ovum (oh' vum) [L.: egg]   The egg, the female sex cell.

Oxidation (ox i day' shun)   Relative loss of electrons in a chemical reaction; either outright removal to form an ion, or the sharing of electrons with substances having a greater affinity for them, such as oxygen. Most oxidation, including biological ones, are associated with the liberation of energy. (Contrast with reduction.)

Oxidative phosphorylation ATP formation in the mitochondrion, associated with flow of electrons through the respiratory chain.

Oxidizing agent A substance that can accept electrons from another. The oxidizing agent becomes reduced; its partner becomes oxidized.

P generation Also called the parental generation. The individuals that mate in a genetic cross. Their immediate offspring are the F_1 generation.

Pacemaker That part of the heart which undergoes most rapid spontaneous contraction, thus setting the pace for the beat of the entire heart. In mammals, the sinoatrial (SA) node. Also, an artificial device, implanted in the heart, that initiates rhythmic contraction of the organ.

Pacinian corpuscle A sensory neuron surrounded by sheaths of connective tissue. Found in the deep layers of the skin, where it senses touch and vibration.

Pair rule genes Segmentation genes that divide the *Drosophila* larva into two segments each.

Paleontology (pale' ee on tol' oh jee) [Gr. *palaios*: ancient, old + *logos*: discourse] The scientific study of fossils and all aspects of extinct life.

Palisade parenchyma In leaves, one or several layers of tightly packed, columnar photosynthetic cells, frequently found just below the upper epidermis.

Pancreas (pan' cree us) A gland, located near the stomach of vertebrates, that secretes digestive enzymes into the small intestine and releases insulin into the bloodstream.

Pangaea (pan jee' uh) [Gk. *pan*: all, every] The single land mass formed when all the continents came together in the Permian period.

Parabronchi Passages in the lungs of birds through which air flows.

Paradigm A general framework within which some scientific discipline (or even the whole Earth) is viewed and within which questions are asked and hypotheses are developed. Scientific revolutions usually involve major paradigm changes.

Parapatric speciation Development of reproductive isolation among members of a continuous population in the absence of a geographical barrier. (Contrast with geographic, sympatric speciation.)

Paraphyletic taxon A taxon that includes some, but not all, of the descendants of a single ancestor.

Parasite An organism that attacks and consumes parts of an organism much larger than itself. Parasites sometimes, but not always, kill the host.

Parasitoid A parasite that is so large relative to its host that only one individual or at most a few individuals can live within a single host.

Parasympathetic nervous system A portion of the autonomic (involuntary) nervous system. Activity in the parasympathetic nervous system produces effects such as decreased blood pressure and decelerated heart beat. (Contrast with sympathetic nervous system.)

Parathormone Hormone secreted by the parathyroid glands. Stimulates osteoclast activity and raises blood calcium levels.

Parathyroids Four glands on the posterior surface of the thyroid that produce and release parathormone.

Parenchyma (pair eng' kyma) [Gr. *para*: beside + *enchyma*: infusion] A plant tissue composed of relatively unspecialized cells without secondary walls.

Parental investment Investment in one offspring or group of offspring that reduces the ability of the parent to assist other offspring.

Parsimony The principle of preferring the simplest among a set of plausible explanations of a phenomenon. Commonly employed in evolutionary and biogeographic studies.

Parthenocarpy Formation of fruit from a flower without fertilization.

Parthenogenesis (par' then oh jen' e sis) [Gr. *parthenos*: virgin + *genesis*: source] The production of an organism from an unfertilized egg.

Partial pressure The portion of the barometric pressure of a mixture of gases that is due to one component of that mixture. For example, the partial pressure of oxygen at sea level is 20.9% of barometric pressure.

Pasteur effect The sharp decrease in rate of glucose utilization when conditions become aerobic.

Pastoralism A nomadic form of human culture based on the tending of herds of domestic animals.

Patch clamping A technique for isolating a tiny patch of membrane to allow the study of ion movement through a particular channel.

Pathogen (path' o jen) [Gr. *pathos*: suffering + *gignomai*: causing] An organism that causes disease.

Pattern formation In animal embryonic development, the organization of differentiated tissues into specific structures such as wings.

Pedigree The pattern of transmission of a genetic trait in a family.

Pelagic zone (puh ladj' ik) [Gr. *pelagos*: the sea] The open waters of the ocean.

Penetrance Of a genotype, the proportion of individuals with that genotype who show the expected phenotype.

Penis (pee' nis) [L.: tail] The male organ inserted into the female during sexual intercourse.

PEP carboxylase The enzyme that combines carbon dioxide with PEP to form a 4-carbon dicarboxylic acid at the start of C_4

photosynthesis or of Crassulacean acid metabolism (CAM).

Pepsin [Gr. *pepsis*: digestion] An enzyme, in gastric juice, that digests protein.

Peptide linkage The connecting group in a protein chain, –CO–NH–, formed by removal of water during the linking of amino acids, –COOH to –NH₂. Also called an amide linkage.

Peptidoglycan The cell wall material of many prokaryotes, consisting of a single enormous molecule that surrounds the entire cell.

Perennial (per ren' ee al) [L. *per*: through + *annus*: a year] Referring to a plant that lives from year to year. (Contrast with annual, biennial.)

Perfect flower A flower with both stamens and carpels, therefore hermaphroditic.

Pericycle [Gr. *peri*: around + *kyklos*: ring or circle] In plant roots, tissue just within the endodermis, but outside of the root vascular tissue. Meristematic activity of pericycle cells produces lateral root primordia.

Periderm The outer tissue of the secondary plant body, consisting primarily of cork.

Period (1) A minor category in the geological time scale. (2) The duration of a cyclical event, such as a circadian rhythm.

Peripheral nervous system Neurons that transmit information to and from the central nervous system and whose cell bodies reside outside the brain or spinal cord.

Peristalsis (pair' i stall' sis) [Gr. *peri*: around + *stellein*: place] Wavelike muscular contractions proceeding along a tubular organ, propelling the contents along the tube.

Peritoneum The mesodermal lining of the coelom among coelomate animals.

Permease A protein in membranes that specifically transports a compound or family of compounds across the membrane.

Peroxisome An organelle that houses reactions in which toxic peroxides are formed. The peroxisome isolates these peroxides from the rest of the cell.

Petal In an angiosperm flower, a sterile modified leaf, nonphotosynthetic, frequently brightly colored, and often serving to attract pollinating insects.

Petiole (pet' ee ole) [L. *petiolus*: small foot] The stalk of a leaf.

pH The negative logarithm of the hydrogen ion concentration; a measure of the acidity of a solution. A solution with pH = 7 is said to be neutral; pH values higher than 7 characterize basic solutions, while acidic solutions have pH values less than 7.

Phage (fayj) Short for bacteriophage.

Phagocyte A white blood cell that ingests microorganisms by endocytosis.

Phagocytosis [Gr.: *phagein* to eat; cell-eating] A form of endocytosis, the uptake of a solid particle by forming a pocket of plasma membrane around the particle and pinching off the pocket to form an intracellular particle bounded by membrane. (Contrast with pinocytosis.)

Pharynx [Gr.: throat]   The part of the gut between the mouth and the esophagus.

Phenotype (fee' no type) [Gr. *phanein*: to show]   The observable properties of an individual as they have developed under the combined influences of the genetic constitution of the individual and the effects of environmental factors. (Contrast with genotype.)

Pheromone (feer' o mone) [Gr. *phero*: carry + *hormon*: excite, arouse]   A chemical substance used in communication between organisms of the same species.

Phloem (flo' um) [Gr. *phloos*: bark]   In vascular plants, the food-conducting tissue. It consists of sieve cells or sieve tubes, fibers, and other specialized cells.

Phosphate group   The functional group –OPO_3H_2; the transfer of energy from one compound to another is often accomplished by the transfer of a phosphate group.

Phosphodiester linkage   The connection in a nucleic acid strand, formed by linking two nucleotides.

3-Phosphoglycerate   The first product of photosynthesis, produced by the reaction of ribulose bisphosphate with carbon dioxide.

Phospholipids   Cellular materials that contain phosphorus and are soluble in organic solvents. An example is lecithin (phosphatidyl choline). Phospholipids are important constituents of cellular membranes. (See lipids.)

Phosphorylation   The addition of a phosphate group.

Photoautotroph   An organism that obtains energy from light and carbon from carbon dioxide. (Contrast with chemoautotroph, chemoheterotroph, photoheterotroph.)

Photoheterotroph   An organism that obtains energy from light but must obtain its carbon from organic compounds. (Contrast with chemoautotroph, chemoheterotroph, photoautotroph.)

Photon (foe' tohn) [Gr. *photos*: light]   A quantum of visible radiation; a "packet" of light energy.

Photoperiod (foe' tow peer' ee ud)   The duration of a period of light, such as the length of time in a 24-hour cycle in which daylight is present. The regulation of processes such as flowering by the changing length of day (or of night) is known as photoperiodism.

Photophosphorylation   Photosynthetic reactions in which light energy trapped by chlorophyll is used to produce ATP and, in noncyclic photophosphorylation, is used to reduce $NADP^+$ to NADPH.

Photorespiration   Light-driven uptake of oxygen and release of carbon dioxide, the carbon being derived from the early reactions of photosynthesis.

Photosensor   A cell that senses and responds to light energy. Also called a **photoreceptor**.

Photosynthesis (foe tow sin' the sis) [literally, "synthesis out of light"]   Metabolic processes, carried out by green plants, by which visible light is trapped and the energy used to synthesize compounds such as ATP and glucose.

Phototropism [Gr. *photos*: light + *trope*: a turning]   A directed plant growth response to light.

Phylogenetic tree   Graphic representation of lines of descent among organisms.

Phylogeny (fy loj' e nee) [Gr. *phylon*: tribe, race + *genesis*: source]   The evolutionary history of a particular group of organisms; also, the diagram of the "family tree" that shows genetic linkages between ancestors and descendants.

Phylum (plural: phyla) [Gr. *phylon*: tribe, stock]   In taxonomy, a high-level category just beneath kingdom and above the class; a group of related, similar classes.

Physiology (fiz' ee ol' o jee) [Gr. *physis*: natural form + *logos*: discourse, study]   The scientific study of the functions of living organisms and the individual organs, tissues, and cells of which they are composed.

Phytoalexins   Substances toxic to fungi, produced by plants in response to fungal infection.

Phytochrome (fy' tow krome) [Gr. *phyton*: plant + *chroma*: color]   A plant pigment regulating a large number of developmental and other phenomena in plants; can exist in two different forms, one of which is active and the other is not. Different wavelengths of light can drive it from one form to the other.

Phytoplankton (fy' tow plangk' ton) [Gr. *phyton*: plant + *planktos*: wandering]   The autotrophic portion of the plankton, consisting mostly of algae.

Pigment   A substance that absorbs visible light.

Pilus (pill' us) [Lat. *pilus*: hair]   A surface appendage by which some bacteria adhere to one another during conjugation.

Pinocytosis [Gr.: drinking cell]   A form of endocytosis; the uptake of liquids by engulfing a sample of the external medium into a pocket of the plasma membrane followed by pinching off the pocket to form an intracellular vesicle. (Contrast with phagocytosis and endocytosis.)

Pistil [L. *pistillum*: pestle]   The female structure of an angiosperm flower, within which the ovules are borne. May consist of a single carpel, or of several carpels fused into a single structure. Usually differentiated into ovary, style, and stigma.

Pith   In plants, relatively unspecialized tissue found within a cylinder of vascular tissue.

Pituitary   A small gland attached to the base of the brain in vertebrates. Its hormones control the activities of other glands. Also known as the hypophysis.

Placenta (pla sen' ta) [Gr. *plax*: flat surface]   The organ found in most mammals that provides for the nourishment of the fetus and elimination of the fetal waste products.

Placental (pla sen' tal)   Pertaining to mammals of the subclass Eutheria, a group characterized by the presence of a placenta; contains the majority of living species of mammals.

Plankton [Gr. *planktos*: wandering]   The free-floating organisms of the sea and fresh water that for the most part move passively with the water currents. Consisting mostly of microorganisms and small plants and animals. (Contrast with nekton.)

Plant   A member of the kingdom Plantae. Multicellular, gaining its nutrition by photosynthesis.

Planula (plan' yew la) [L. *planum*: something flat]   The free-swimming, ciliated larva of the cnidarians.

Plaque (plack) [Fr.: a metal plate or coin]   (1) A circular clearing in a turbid layer (lawn) of bacteria growing on the surface of a nutrient agar gel. Produced by successive rounds of infection initiated by a single bacteriophage. (2) An accumulation of prokaryotic organisms on tooth enamel. Acids produced by the metabolism of these microorganisms can cause tooth decay.

Plasma (plaz' muh) [Gr. *plassein*: to mold]   The liquid portion of blood, in which blood cells and other particulates are suspended.

Plasma cell   An antibody-secreting cell that developed from a B cell. The effector cell of the humoral immune system.

Plasma membrane   The membrane that surrounds the cell, regulating the entry and exit of molecules and ions. Every cell has a plasma membrane.

Plasmid   A DNA molecule distinct from the chromosome(s); that is, an extrachromosomal element. May replicate independently of the chromosome.

Plasmodesma (plural: plasmodesmata) [Gr. *plasma*: formed or molded + *desmos*: band]   A cytoplasmic strand connecting two adjacent plant cells.

Plasmodium   In the noncellular slime molds, a multinucleate mass of protoplasm surrounded by a membrane; characteristic of the vegetative feeding stage.

Plasmolysis (plaz mol' i sis)   Shrinking of the cytoplasm and plasma membrane away from the cell wall, resulting from the osmotic outflow of water. Occurs only in cells with rigid cell walls.

Plastid   Organelle in plants that serves for food manufacture (by photosynthesis) or food storage; bounded by a double membrane.

Platelet   A membrane-bounded body without a nucleus, arising as a fragment of a cell in the bone marrow of mammals. Important to blood-clotting action.

Pleiotropy (plee' a tro pee) [Gr. *pleion*: more]   The determination of more than one character by a single gene.

Pleural membrane [Gk. *pleuras*: rib, side]   The membrane lining the outside of the lungs and the walls of the thoracic cavity. Inflammation of these membranes is a condition known as *pleurisy*.

Podocytes   Cells of Bowman's capsule of the nephron that cover the capillaries of the glomerulus, forming filtration slits.

Poikilotherm (poy' kill o therm) [Gr. *poikilos*: varied + *therme*: heat]   An animal whose body temperature tends to vary with the surrounding environment. (Contrast with homeotherm, heterotherm.)

Point mutation   A mutation that results from a small, localized alteration in the chemical structure of a gene. Such mutations can give rise to wild-type revertants as a result of reverse mutation. In genetic crosses, a point mutation behaves as if it resided at a single point on the genetic map. (Contrast with deletion.)

Polar body   A nonfunctional nucleus produced by meiosis, accompanied by very little cytoplasm. The meiosis which produces the mammalian egg produces in addition three polar bodies.

Polar molecule   A molecule in which the electric charge is not distributed evenly in the covalent bonds.

Polar nucleus   One of two nuclei derived from each end of the angiosperm embryo sac, both of which become centrally located. They fuse with a male nucleus to form the primary triploid nucleus that will prduce the endosperm tissue of the angiosperm seed.

Polarity   In development, the difference between one end and the other. In chemistry, the property that makes a polar molecule.

Pollen [L.: fine powder, dust]   The fertilizing element of seed plants, containing the male gametophyte and the gamete, at the stage in which it is shed.

Pollination   Process of transferring pollen from the anther to the receptive surface (stigma) of the ovary in plants.

Poly- [Gr. *poly*: many]   A prefix denoting multiple entities.

Polygamy [Gr. *poly*: many + *gamos*: marriage]   A breeding system in which an individual acquires more than one mate. In polyandry, a female mates with more than one male, in polygyny, a male mates with more than one female.

Polygenes   Multiple loci whose alleles increase or decrease a continuously variable phenotypic trait.

Polymer   A large molecule made up of similar or identical subunits called monomers. (Contrast with monomer, oligomer.)

Polymerase chain reaction (PCR)   A technique for the rapid production of millions of copies of a particular stretch of DNA.

Polymerization reactions   Chemical reactions that generate polymers by means of condensation reactions.

Polymorphism (pol' lee mor' fiz um) [Gr. *poly*: many + *morphe*: form, shape]   (1) In genetics, the coexistence in the same population of two distinct hereditary types based on different alleles. (2) In social organisms such as colonial cnidarians and social insects, the coexistence of two or more functionally different castes within the same colony.

Polyp   The sessile, asexual stage in the life cycle of most cnidarians.

Polypeptide   A large molecule made up of many amino acids joined by peptide linkages. Large polypeptides are called proteins.

Polyphyletic group   A group containing taxa, not all of which share the most recent common ancestor.

Polyploid (pol' lee ploid)   A cell or an organism in which the number of complete sets of chromosomes is greater than two.

Polysaccharide   A macromolecule composed of many monosaccharides (simple sugars). Common examples are cellulose and starch.

Polysome   A complex consisting of a threadlike molecule of messenger RNA and several (or many) ribosomes. The ribosomes move along the mRNA, synthesizing polypeptide chains as they proceed.

Polytene (pol' lee teen) [Gr. *poly*: many + *taenia*: ribbon]   An adjective describing giant interphase chromosomes, such as those found in the salivary glands of fly larvae. The characteristic, reproducible pattern of bands and bulges seen on these chromosomes has provided a method for preparing detailed chromosome maps of several organisms.

Pons [L. *pons*: bridge]   Region of the brain stem anterior to the medulla.

Population   Any group of organisms coexisting at the same time and in the same place and capable of interbreeding with one another.

Population density   The number of individuals (or modules) of a population in a unit of area or volume.

Population dynamics   Changes in the distribution and abundance of individuals in a population.

Population structure   The proportions of individuals in a population belonging to different age classes (age structure). Also, the distribution of the population in space.

Population vulnerability analysis (PVA)   A determination of the risk of extinction of a population given its current size and distribution.

Portal vein   A vein connecting two capillary beds, as in the hepatic portal system.

Positive control   The situation in which a regulatory macromolecule is needed to turn transcription of structural genes on. In its absence, transcription will not occur.

Positive cooperativity   Occurs when a molecule can bind several ligands and each one that binds alters the conformation of the molecule so that it can bind the next ligand more easily. The binding of four molecules of O_2 by hemoglobin is an example of positive cooperativity.

Positive feedback   A regulatory system in which an error signal stimulates responses that increase the error.

Postabsorptive period   When there is no food in the gut and no nutrients are being absorbed.

Posterior   Toward or pertaining to the rear.

Postsynaptic cell   The cell whose membranes receive the neurotransmitter released at a synapse.

Postzygotic isolating mechanism   Any factor that reduces the viability of zygotes resulting from matings between individuals of different species.

Predator   An organism that kills and eats other organisms. Predation is usually thought of as involving the consumption of animals by animals, but it can also mean the eating of plants.

Presynaptic excitation/inhibition   Occurs when a neuron modifies activity at a synapse by releasing a neurotransmitter onto the presynaptic nerve terminal.

Prey [L. *praeda*: booty]   An organism consumed as an energy source.

Prezygotic isolating mechanism   A mechanism that reduces the probability that individuals of different species will mate.

Primary active transport   Form of active transport in which ATP is hydrolyzed, yielding the energy required to transport ions against their concentration gradients. (Contrast with secondary active transport.)

Primary growth   In plants, growth produced by the apical meristems. (Contrast with secondary growth.)

Primary producer   A photosynthetic or chemosynthetic organism that synthesizes complex organic molecules from simple inorganic ones.

Primary succession   Succession that begins in an areas initially devoid of life, such as on recently exposed glacial till or lava flows.

Primary structure   The specific sequence of amino acids in a protein.

Primary wall   Cellulose-rich cell wall layers laid down by a growing plant cell.

Primate (pry' mate)   A member of the order Primates, such as a lemur, monkey, ape, or human.

Primer   A short, single-stranded segment of DNA serving as the necessary starting material for the synthesis of a new DNA strand, which is synthesized from the 3' end of the primer.

Primitive streak   A line running axially along the blastodisc, the site of inward cell migration during formation of the three-layered embryo. Formed in the embryos of birds and fish.

Primordium [L. *primordium*: origin]   The most rudimentary stage of an organ or other part.

Pro- [L.: first, before, favoring]   A prefix often used in biology to denote a developmental stage that comes first or an evolutionary form that appeared earlier than another. For example, prokaryote, prophase.

Probe   A segment of single stranded nucleic acid used to identify DNA molecules containing the complementary sequence.

Procambium   Primary meristem that produces the vascular tissue.

Progesterone [L. *pro*: favoring + *gestare*: to bear]   A vertebrate female sex hormone that maintains pregnancy.

Prokaryotes (pro kar' ry otes) [L. *pro*: before + Gk. *karyon*: kernel, nucleus]   Organisms whose genetic material is not contained within a nucleus. The bacteria. Considered an earlier stage in the evolution of life than the eukaryotes.

Prometaphase The phase of nuclear division that begins with the disintegration of the nuclear envelope.

Promoter The region of an operon that acts as the initial binding site for RNA polymerase.

Proofreading The correction of an error in DNA replication just after an incorrectly paired base is added to the growing polynucleotide chain.

Prophage (pro' fayj) The noninfectious units that are linked with the chromosomes of the host bacteria and multiply with them but do not cause dissolution of the cell. Prophage can later enter into the lytic phase to complete the virus life cycle.

Prophase (pro' phase) The first stage of nuclear division, during which chromosomes condense from diffuse, threadlike material to discrete, compact bodies.

Proplastid [Gr. *pro*: before + *plastos*: molded] A plant cell organelle which under appropriate conditions will develop into a plastid, usually the photosynthetic chloroplast. If plants are kept in the dark, proplastids may become quite large and complex.

Prostaglandin Any one of a group of specialized lipids with hormone-like functions. It is not clear that they act at any considerable distance from the site of their production.

Prosthetic group Any nonprotein portion of an enzyme.

Protease (pro' tee ase) See proteolytic enzyme.

Protein (pro' teen) [Gr. *protos*: first] One of the most fundamental building substances of living organisms. A long-chain polymer of amino acids with twenty different common side chains. Occurs with its polymer chain extended in fibrous proteins, or coiled into a compact macromolecule in enzymes and other globular proteins.

Proteolytic enzyme An enzyme whose main catalytic function is the digestion of a protein or polypeptide chain. The digestive enzymes trypsin, pepsin, and carboxypeptidase are all proteolytic enzymes (proteases).

Protist A member of the kingdom Protista, which consists of those eukaryotes not included in the kingdoms Animalia, Fungi, or Plantae. Many protists are unicellular. The kingdom Protista includes protozoa, algae, and fungus-like protists.

Protoderm Primary meristem that gives rise to epidermis.

Proton (pro' ton) [Gr. *protos*: first] One of the three most fundamental particles of matter, with mass approximately 1 amu and an electrical charge of +1.

Proton motive force The proton gradient and electric charge difference produced by chemiosmotic proton pumping. It drives protons back across the membrane, with the concomitant formation of ATP.

Protonema (pro' tow nee' mah) [Gr. *protos*: first + *nema*: thread] The hairlike growth form that constitutes an early stage in the development of a moss gametophyte.

Proto-oncogenes The normal alleles of genes possessing oncogenes (cancer-causing genes) as mutant alleles. Proto-oncogenes encode growth factors and receptor proteins.

Protoplast A cell that would normally have a cell wall, from which the wall has been removed by enzymatic digestion or by special growth conditions.

Protostome One of two major lines of animal evolution, characterized by spiral, determinate cleavage of the egg, and by schizocoelous development. (Contrast with deuterostome.)

Prototroph (pro' tow trofe') [Gr. *protos*: first + *trophein*: to nourish] The nutritional wild type, or reference form, of an organism. Any deviant form that requires growth nutrients not required by the prototrophic form is said to be a nutritional mutant, or auxotroph.

Protozoa A group of single-celled organisms classified by some biologists as a single phylum; includes the flagellates, amoebas, and ciliates. This textbook follows most modern classifications in elevating the protozoans to a distinct kingdom (Protista) and each of their major subgroups to the rank of phylum.

Provincialized A biogeographic term referring to the separation, by environmental barriers, of the biota into units with distinct species compositions.

Provirus See prophage.

Proximal Near the point of attachment or other reference point. (Contrast with distal.)

Pseudocoelom A body cavity not surrounded by a peritoneum. Characteristic of nematodes and rotifers.

Pseudogene A DNA segment that is homologous to a functional gene but contains a nucleotide change that prevents its expression.

Pseudoplasmodium [Gr. *pseudes*: false + *plasma*: mold or form] In the cellular slime molds such as *Dictyostelium*, an aggregation of single amoeboid cells. Occurs prior to formation of a fruiting structure.

Pseudopod (soo' do pod) [Gr. *pseudes*: false + *podos*: foot] A temporary, soft extension of the cell body that is used in location, attachment to surfaces, or engulfing particles.

Pulmonary Pertaining to the lungs.

Punctuated equilibrium An evolutionary pattern in which periods of rapid change are separated by longer periods of little or no change.

Pupa (pew' pa) [L.: doll, puppet] In certain insects (the Holometabola), the encased developmental stage that intervenes between the larva and the adult.

Pupil The opening in the vertebrate eye through which light passes.

Purine (pure' een) A type of nitrogenous base. The purines adenine and guanine are found in nucleic acids.

Purkinje fibers Specialized heart muscle cells that conduct excitation throughout the ventricular muscle.

Pyramid of biomass Graphical representation of the total masses at different trophic levels in an ecosystem.

Pyramid of energy Graphical representation of the total energy contents at different trophic levels in an ecosystem.

Pyrimidine (peer im' a deen) A type of nitrogenous base. The pyrimidines cytosine, thymine, and uracil are found in nucleic acids.

Pyrogen A substance that causes fever.

Pyruvate A three-carbon acid; the end product of glycolysis and the raw material for the citric acid cycle.

Q_{10} A value that compares the rate of a biochemical process or reaction over a 10°C range of temperature. A process that is not temperature-sensitive has a Q_{10} of 1. Values of 2 or 3 mean the reaction speeds up as temperature increases.

Quantum (kwon' tum) [L. *quantus*: how great] An indivisible unit of energy.

Quaternary structure Of aggregating proteins, the arrangement of polypeptide subunits.

R factor (resistance factor) A plasmid that contains one or more genes that encode resistance to antibiotics.

Radial symmetry The condition in which two halves of a body are mirror images of each other regardless of the angle of the cut, providing the cut is made along the center line. Thus, a cylinder cut lengthwise down its center displays this form of symmetry. (Contrast with bilateral symmetry.)

Radioisotope A radioactive isotope of an element. Examples are carbon-14 (^{14}C) and hydrogen-3, or tritium (^3H).

Radiotherapy Treatment, as of cancer, with X- or gamma rays.

Rain shadow A region of low precipitation on the leeward side of a mountain range.

Ramet The repeated morphological units of sessile, modular organisms. (Contrast with genet.)

Random drift Evolution (change in gene proportions) by chance processes alone.

Rate constant Of a particular chemical reaction, a constant which, when multiplied by the concentration(s) of reactant(s), gives the rate of the reaction.

Reactant A chemical substance that enters into a chemical reaction with another substance.

Reaction, chemical A process in which atoms combine or change bonding partners.

Reaction wood Modified wood produced in branches in response to gravitational stimulation. Gymnosperms produce compression wood that tends to push the branch up; angiosperms produce tension wood that tends to pull the branch up.

Realized niche The actual niche occupied by an organism; it differs from the fundamental niche because of the presence of other species.

Receptacle [L. *receptaculum*: reservoir] In an angiosperm flower, the end of the stem to which all of the various flower parts are attached.

Receptive field Of a neuron, the area on the retina from which the activity of that neuron can be influenced.

Receptor-mediated endocytosis A form of endocytosis in which macromolecules in the environment bind specific receptor proteins in the plasma membrane and are brought into the cell interior in coated vesicles.

Receptor potential The change in the resting potential of a sensory cell when it is stimulated.

Recessive See dominance.

Reciprocal altruism The exchange of altruistic acts between two or more individuals. The acts may be separated considerably in time.

Reciprocal crosses A pair of crosses, in one of which a female of genotype A mates with a male of genotype B and in the other of which a female of genotype B mates with a male of genotype A.

Recognition site (also called a restriction site) A sequence of nucleotides in DNA to which a restriction enzyme binds and then cuts the DNA.

Recombinant An individual, meiotic product, or single chromosome in which genetic materials originally present in two individuals end up in the same haploid complement of genes. The reshuffling of genes can be either by independent segregation, or by crossing over between homologous chromosomes. For example, a human may pass on genes from both parents in a single haploid gamete.

Recombinant DNA technology The application of genetic tools (restriction endonucleases, plasmids, and transformation) to the production of specific proteins by biological "factories" such as bacteria.

Rectum The terminal portion of the gut, ending at the anus.

Redox reaction A chemical reaction in which one reactant becomes oxidized and the other becomes reduced.

Reducing agent A substance that can donate electrons to another substance. The reducing agent becomes oxidized, and its partner becomes reduced.

Reduction (re duk' shun) Gain of electrons; the reverse of oxidation. Most reductions lead to the storage of chemical energy, which can be released later by an oxidation reaction. Energy storage compounds such as sugars and fats are highly reduced compounds. (Contrast with oxidation.)

Reflex An automatic action, involving only a few neurons (in vertebrates, often in the spinal cord), in which a motor response swiftly follows a sensory stimulus.

Refractory period Of a neuron, the time interval after an action potential, during which another action potential cannot be elicited.

Regulative development A pattern of animal embryonic development in which the fates of the first blastomeres are not absolutely fixed. (Contrast with mosaic development.)

Regulatory gene A gene that contains the information for making a regulatory macromolecule, often a repressor protein.

Releaser A sensory stimulus that triggers a fixed action pattern.

Releasing hormone One of several hypothalamic hormones that stimulates the secretion of anterior pituitary hormone.

REM sleep A sleep state characterized by dreaming, skeletal muscle relaxation, and rapid eye movements.

Renal [L. *renes*: kidneys] Relating to the kidneys.

Replica plating A technique used in the selection of colonies of cells with a desired genotype.

Replication fork A point at which a DNA molecule is replicating. The fork forms by the unwinding of the parent molecule.

Repressible enzyme An enzyme whose synthesis can be decreased or prevented by the presence of a particular compound. A repressible opren often controls the synthesis of such an enzyme.

Repressor A protein coded by the regulatory gene. The repressor can bind to a specific operator and prevent transcription of the operon.

Reproductive isolating mechanism Any trait that prevents individuals from two different populations from producing fertile hybrids.

Reproductive isolation The condition in which a population is not exchanging genes with other populations of the same species.

Reproductive value The expected contribution of an individual of a particular age to the future growth of the population to which it belongs.

Rescue effect The avoidance of extinction by immigration of individuals from other populations.

Resolving power Of an optical device such as a microscope, the smallest distance between two lines that allows the lines to be seen as separate from one another.

Resource Something in the environment required by an organism for its maintenance and growth that is consumed in the process of being used.

Resource defense polygamy A breeding system in which individuals of one sex (usually males) defend resources that are attractive to individuals of the other sex (usually females); individuals holding better resources attract more mates.

Respiration (res pi ra' shun) [L. *spirare*: to breathe] (1) Cellular respiration; the oxidation of the end products of glycolysis with the storage of much energy in ATP. The oxidant in the respiration of eukaryotes is oxygen gas. Some bacteria can use nitrate or sulfate instead of O_2. (2) Breathing.

Respiratory chain The terminal reactions of cellular respiration, in which electrons are passed from NAD or FAD, through a series of intermediate carriers, to molecular oxygen, with the concomitant production of ATP.

Respiratory uncoupler A substance that allows protons to cross the inner mitochondrial membrane without the concomitant formation of ATP, thus uncoupling respiration from phosphorylation.

Resting potential The membrane potential of a living cell at rest. In cells at rest, the interior is negative to the exterior. (Contrast with action potential, electrotonic potential.)

Restoration ecology The science and practice of restoring damaged or degraded ecosystems.

Restriction endonuclease Any one of several enzymes, produced by bacteria, that break foreign DNA molecules at very specific sites. Some produce "sticky ends." Extensively used in recombinant DNA technology.

Restriction map A partial genetic map of a DNA molecule, showing the points at which particular restriction endonuclease recognition sites reside.

Reticular system A central region of the vertebrate brain stem that includes complex fiber tracts conveying neural signals between the forebrain and the spinal cord, with collateral fibers to a variety of nuclei that are involved in autonomic functions, including arousal from sleep.

Retina (rett' in uh) [L. *rete*: net] The light-sensitive layer of cells in the vertebrate or cephalopod eye.

Retinal The light-absorbing portion of visual pigment molecules. Derived from β-carotene.

Retrovirus An RNA virus that contains reverse transcriptase. Its RNA serves as a template for cDNA production, and the cDNA is integrated into a chromosome of the mammalian host cell.

Reverse transcriptase An enzyme that catalyzes the production of DNA (cDNA), using RNA as a template; essential to the reproduction of retroviruses.

Reversion (genetic) A mutational event that restores wild type phenotype to a mutant.

RFLP (Restriction fragment length polymorphism) Coexistence of two or more patterns of restriction fragments (patterns produced by restriction enzymes), as revealed by a probe. The polymorphism reflects a difference in DNA sequence on homologous chromosomes.

Rhizoids (rye' zoids) [Gr. *rhiza*: root] Hairlike extensions of cells in mosses, liverworts, and a few vascular plants that serve the same function as roots and root hairs in vascular plants. The term is also applied to branched, rootlike extensions of some fungi and algae.

Rhizome (rye' zome) [Gr. *rhizoma*: mass of roots] A special underground stem (as opposed to root) that runs horizontally beneath the ground.

Rhodopsin A photopigment used in the visual process of transducing photons of light into changes in the membrane potential of photosensory cells.

Ribonucleic acid See RNA.

Ribose (rye' bose) A sugar of chemical formula $C_5H_{10}O_5$, one of the building blocks of ribonucleic acids.

Ribosomal RNA (rRNA) Several species of RNA that are incorporated into the ribosome.

Ribosome A small organelle that is the site of protein synthesis.

Ribozyme An RNA molecule with catalytic activity.

Ribulose 1,5-bisphosphate (RuBP) The compound in chloroplasts which reacts with carbon dioxide in the first reaction of the Calvin-Benson cycle.

Risk cost The increased chance of being injured or killed as a result of performing a behavior, compared to resting.

RNA (ribonucleic acid) A nucleic acid using ribose. Various classes of RNA are involved in the transcription and translation of genetic information. RNA serves as the genetic storage material in some viruses.

RNA polymerase An enzyme that catalyzes the formation of RNA from a DNA template.

RNA splicing The last stage of RNA processing in eukaryotes, in which the transcripts of introns are excised through the action of small nuclear ribonucleoprotein particles (snRNP).

Rods Light-sensitive cells (photosensors) in the retina. (Contrast with cones.)

Root cap A thimble-shaped mass of cells, produced by the root apical meristem, that protects the meristem and that is the organ that perceives the gravitational stimulus in root gravitropism.

Root hair A specialized epidermal cell with a long, thin process that absorbs water and minerals from the soil solution.

Round dance The dance performed on the vertical surface of a honeycomb by a returning honeybee forager when she has discovered a food source less than 100 meters from the hive.

Round window A flexible membrane between the middle and inner ear that distributes pressure waves in the fluid of the inner ear.

rRNA See ribosomal RNA.

Rubisco (RuBP carboxylase) Enzyme that combines carbon dioxide with ribulose bisphosphate to produce 3-phosphoglycerate, the first product of C_3 photosynthesis. The most abundant protein on Earth.

Rumen (rew' mun) The first division of the ruminant stomach. It stores and initiates bacterial fermentation of food. Food is regurgitated from the rumen for further chewing.

Ruminant An herbivorous, cud-chewing mammal such as a cow, sheep, or deer, having a stomach consisting of four compartments.

S phase In the cell cycle, the stage of interphase during which DNA is replicated. (Contrast with G_1 phase, G_2 phase.)

Sap An aqueous solution of nutrients, minerals, and other substances that passes through the xylem of plants.

Saprobe [Gr. *sapros*:rotten + *bios*: life] An organism (usually a bacterium or fungus) that obtains its carbon and energy directly from dead organic matter.

Sarcomere (sark' o meer) [Gr. *sark*: flesh + *meros*: a part] The contractile unit of a skeletal muscle.

Saturated hydrocarbon A compound consisting only of carbon and hydrogen, with the hydrogen atoms connected by single bonds.

Schizocoelous development Formation of a coelom during embryological development by a splitting of mesodermal masses.

Schwann cell A glial cell that wraps around part of the axon of a peripheral neuron, creating a myelin sheath.

Sclereid A type of sclerenchyma cell, commonly found in nutshells, that is not elongated.

Sclerenchyma (skler eng' kyma) A plant tissue composed of cells with heavily thickened cell walls, dead at functional maturity. The principal types of sclerenchyma cells are fibers and sclereids.

Second messenger A signaling molecule that is created or actived inside the cell in response to activation of a receptor on the cell surface. The second messenger molecule then triggers the cell's response. An example is cyclic AMP.

Secondary active transport Form of active transport in which ions or molecules are transported against their concentration gradient using energy obtained by relaxation of a gradient of sodium ion concentration rather than directly from ATP. (Contrast with primary active transport.)

Secondary compound A compound synthesized by a plant that is not needed for basic cellular metabolism. Typically has an antiherbivore or antiparasite function.

Secondary growth In plants, growth produced by vascular and cork cambia, contributing to an increase in girth. (Contrast with primary growth.)

Secondary structure Of a protein, localized regularities of structure, such as the α helix and the β pleated sheet.

Secondary wall Wall layers laid down by a plant cell that has ceased growing; often impregnated with lignin or suberin.

Second law of thermodynamics States that in any real (irreversible) process, there is a decrease in free energy and an increase in entropy.

Second messenger A compound, such as cyclic AMP, that is released within a target cell after a hormone or other "first messenger" has bound to a surface receptor on a cell; the second messenger triggers further reactions within the cell.

Secretin (si kreet' in) A peptide hormone secreted by the upper region of the small intestine when acidic chyme is present. Stimulates the pancreatic duct to secrete bicarbonate ions.

Section A thin slice, usually for microscopy, as a tangential section or a transverse section.

Seed A fertilized, ripened ovule of a gymnosperm or angiosperm. Consists of the embryo, nutritive tissue, and a seed coat.

Seed crop The number of seeds produced by a plant during a particular bout of reproduction.

Seedling A young plant that has grown from a seed (rather than by grafting or by other means).

Segmentation genes In insect larvae, genes that determine the number and polarity of larval segments.

Segment polarity genes Genes that determine the boundaries and front-to-back organization of the segments in the *Drosophila* larva.

Segregation (genetic) The separation of alleles, or of homologous chromosomes, from one another during meiosis so that each of the haploid daughter nuclei produced by meiosis contains one or the other member of the pair found in the diploid mother cell, but never both.

Selective permeability A characteristic of a membrane, allowing certain substances to pass through while other substances are excluded.

Self-differentiating Behavior that develops without experience with the normal objects toward which it is usually directed and without any practice. (See also instinct.)

Selfish act A behavioral act that benefits its performer but harms the recipients.

Self-pollination The fertilization of a plant by its own pollen. (Contrast with cross-pollination.)

Semelparous organism An organism that reproduces only once in its lifetime. (Contrast with iteroparous.)

Semen (see' men) [L.: seed] The thick, whitish liquid produced by the male reproductive organ in mammals, containing the sperm.

Semicircular canals Part of the vestibular system of mammals.

Semiconservative replication The common way in which DNA is synthesized. Each of the two partner strands in a double helix acts as a template for a new partner strand. Hence, after replication, each double helix consists of one old and one new strand.

Seminiferous tubules The tubules within the testes within which sperm production occurs.

Senescence [L. *senescere*: to grow old] Aging; deteriorative changes with aging.

Sensor A sensory cell; a cell transduces a physical or chemical stimulus into a membrane potential change.

Sensory neuron A neuron leading from a sensory cell to the central nervous system. (Contrast with motor neuron.)

Sepal (see' pul) One of the outermost structures of the flower, usually protective in function and enclosing the rest of the flower in the bud stage.

Septum [L.: partition] A membrane or wall between two cavities.

Sertoli cells Cells in the seminiferous tubules that nuture the developing sperm.

Serum That part of the blood plasma that remains after clots have formed and been removed.

Sessile (sess' ul) [L. *sedere*: to sit] Permanently attached; not moving.

Sertoli cells Cells in the seminiferous tubules that nuture the developing sperm.

Set point In a regulatory system, the threshold sensitivity to the feedback stimulus.

Sex chromosome In organisms with a chromosomal mechanism of sex determination, one of the chromosomes involved in sex determination. One sex chromosome, the X chromosome, is present in two copies in one sex and only one copy in the other sex. The autosomes, as opposed to the sex chromosomes, are present in two copies in both sexes. In many organisms, there is a second sex chromosome, the Y chromosome, that is found in only one sex—the sex having only one copy of the X.

Sexduction See F-duction.

Sex linkage The pattern of inheritance characteristic of genes located on the sex chromosomes of organisms having a chromosomal mechanism for sex determination. The sex that is diploid with respect to sex chromosomes can assume three genotypes: homozygous wild type, homozygous mutant, or heterozygous carrier. The other sex, haploid for sex chromosomes, is either hemizygous wild type or hemizygous mutant.

Sexuality The ability, by any of a multitude of mechanisms, to bring together in one individual genes that were originally carried by two different individuals. The capacity for genetic recombination.

Sexual selection Selection by one sex of characteristics in individuals of the opposite sex. Also, the favoring of characteristics in one sex as a result of competition among individuals of that sex for mates.

Shoot The aerial part of a vascular plant, consisting of the leaves, stem(s), and flowers.

Sieve plate In sieve tubes, the highly specialized end walls in which are concentrated the clusters of pores through which the protoplasts of adjacent sieve tube members are interconnected.

Sieve tube A column of specialized cells found in the phloem, specialized to conduct organic matter from sources (such as photo-synthesizing leaves) to sinks (such as roots). Found principally in flowering plants.

Sieve tube member A single cell of a sieve tube, containing cytoplasm but relatively few organelles, with highly specialized perforated end walls leading to elements above and below.

Sign stimulus The single stimulus, or one out of a very few stimuli, by which an animal distinguishes key objects, such as an enemy, or a mate, or a place to nest, etc.

Signal sequence The sequence of a protein that directs the protein through a particular cellular membrane.

Signal transduction pathway The series of biochemical steps whereby a stimulus to a cell (such as a hormone or neurotransmitter binding to a receptor) is translated into a response of the cell.

Silencer A sequence of eukaryotic DNA that binds proteins that inhibit the transcription of an associated gene.

Silent mutations Genetic changes that do not lead to a phenotypic change. At the molecular level, these are DNA sequence changes that, because of the redundancy of the genetic code, result in the same amino acids in the resulting protein.

Similarity matrix A matrix to compare the structures of two molecules constructed by adding the number of their amino acids that are identical or different

Sinoatrial node (sigh' no ay' tree al) The pacemaker of the mammalian heart.

Sinus (sigh' nus) [L. *sinus*: a bend, hollow] A cavity in a bone, a tissue space, or an enlargement in a blood vessel.

Skeletal muscle See striated muscle.

Sliding filament theory A proposed mechanism of muscle contraction based on formation and breaking of crossbridges between actin and myosin filaments, causing them to slide together.

Small intestine The portion of the gut between the stomach and the colon, consisting of the duodenum, the jejunum, and the ileum.

Small nuclear ribonucleoprotein particle (snRNP) A complex of an enzyme and a small nuclear RNA molecule, functioning in RNA splicing.

Smooth muscle One of three types of muscle tissue. Usually consists of sheets of mononucleated cells innervated by the autonomic nervous system.

Society A group of individuals belonging to the same species and organized in a cooperative manner; in the broadest sense, includes parents and their offspring.

Sodium cotransport Carrier-mediated transport of molecules across membranes driven by sodium ions binding to the same carrier and moving down their concentration gradient.

Sodium–potassium pump The complex protein in plasma membranes that is responsible for primary active transport; it pumps sodium ions out of the cell and potassium ions into the cell, both against their concentration gradients.

Solute A substance that is dissolved in a liquid (solvent).

Solution A liquid (solvent) and its dissolved solutes.

Solvent A liquid that has dissolved or can dissolve one or more solutes.

Somatic [Gr. *soma*: body] Pertaining to the body, or body cells (rather than to germ cells).

Somite (so' might) One of the segments into which an embryo becomes divided longitudinally, leading to the eventual segmentation of the animal as illustrated by the spinal column, ribs, and associated muscles.

Southern blotting Transfer of DNA fragments from an electrophoretic gel to a sheet of paper or other absorbent material for analysis with a probe.

Spatial summation In the production or inhibition of action potentials in a postsynaptic neuron, the interaction of depolarizations and hyperpolarizations produced by several terminal boutons.

Spawning The direct release of sex cells into the water.

Specialized transduction In some types of bacteriophage (e.g., lambda), a prophage inserts at a specific location in the genome. When the prophage is induced to become lytic, it leaves the host chromosome and may take only the adjacent bacterial genes along with its phage DNA.

Speciation (spee' shee ay' shun) The process of splitting one population into two populations that are reproductively isolated from one another.

Species (spee' shees) [L.: kind] The basic lower unit of classification, consisting of a population or series of populations of closely related and similar organisms. The more narrowly defined "biological species" consists of individuals capable of interbreeding freely with each other but not with members of other species.

Species diversity A weighted representation of the species of organisms living in a region; large and common species are given greater weight than are small and rare ones. (Contrast with species richness.)

Species pool All the species potentially available to colonize a particular habitat.

Species richness The number of species of organisms living in a region. (Contrast with species diversity.)

Specific heat The amount of energy that must be absorbed by a gram of a substance to raise its temperature by one degree centigrade. By convention, water is assigned a specific heat of one.

Sperm [Gr. *sperma*: seed] A male reproductive cell.

Spermatocyte (spur mat' oh site) [Gr. *sperma*: seed + *kytos*: cell] The cell that gives rise to the sperm in animals.

Spermatogenesis (spur mat' oh jen' e sis) [Gr. *sperma*: seed + *genesis*: source]   Male gametogenesis, leading to the production of sperm.

Spermatogonia   Undifferentiated germ cells that give rise to primary spermatocytes and hence to sperm.

Sphincter (sfingk' ter) [Gr. *sphinkter*: that which binds tight]   A ring of muscle that can close an orifice, for example at the anus.

Spindle apparatus   An array of microtubules stretching from pole to pole of a dividing nucleus and playing a role in the movement of chromosomes at nuclear division. Named for its shape.

Spiracle (spy' rih kel) [L. *spirare*: to breathe]   An opening of the treacheal respiratory system of terrestrial arthropods.

Spiteful act   A behavioral act that harms both the actor and the recipient of the act.

Spliceosome   An RNA–protein complex that splices out introns from eukaryotic pre-mRNAs.

Splicing   The removal of introns and connecting of exons in eukaryotic pre-mRNAs.

Spongy parenchyma   In leaves, a layer of loosely packed photosynthetic cells with extensive intercellular spaces for gas diffusion. Frequently found between the palisade parenchyma and the lower epidermis.

Spontaneous generation   The idea that life is generated continually from nonliving matter. Usually distinguished from the current idea that life evolved from nonliving matter under primordial conditions at an early stage in the history of earth.

Spontaneous reaction   A chemical reaction which will proceed on its own, without any outside influence. A spontaneous reaction need not be rapid.

Sporangiophore [Gr. *phore*: to bear]   Any branch bearing one or more sporangia.

Sporangium (spor an' gee um) [Gr. *spora*: seed + *angeion*: vessel or reservoir]   In plants and fungi, any specialized stucture within which one or more spores are formed.

Spore [Gr. *spora*: seed]   Any asexual reproductive cell capable of developing into an adult plant without gametic fusion. Haploid spores develop into gametophytes, diploid spores into sporophytes. In prokaryotes, a resistant cell capable of surviving unfavorable periods.

Sporophyll (spor' o fill) [Gr. *spora*: seed + *phyllon*: leaf]   Any leaf or leaflike structure that bears sporangia; refers to carpels and stamens of angiosperms and to sporangium-bearing leaves on ferns, for example.

Sporophyte (spor' o fyte) [Gr. *spora*: seed + *phyton*: plant]   In plants with alternation of generations, the diploid phase that produces the spores. (Contrast with gametophyte.)

Stabilizing selection   Selection against the extreme phenotypes in a population, so that the intermediate types are favored. (Contrast with disruptive selection.)

Stamen (stay' men) [L.: thread]   A male (pollen-producing) unit of a flower, usually composed of an anther, which bears the pollen, and a filament, which is a stalk supporting the anther.

Starch [O.E. *stearc*: stiff]   An α-linked polymer of glucose; used by plants as a means of storing energy and carbon atoms.

Start codon   The mRNA triplet (AUG) that acts as signals for the beginning of translation at the ribosome. (Compare with stop codons. There are a few mnior exceptions to these codons.)

Stasis   Period during which little or no evolutionary change takes place within a lineage or groups of lineages.

Statocyst (stat' oh sist) [Gk. *statos*: stationary + *kystos*: pouch]   An organ of equilibrium in some invertebrates.

Statolith (stat' oh lith) [Gk. *statos*: stationary + *lithos*: stone]   A solid object that responds to gravity or movement and stimulates the mechanosensors of a statocyst.

Stele (steel) [Gr. *stele*: pillar]   The central cylinder of vascular tissue in a plant stem.

Stem cell   A cell capable of extensive proliferation, generating more stem cells and a large clone of differentiated progeny cells, as in the formation of red blood cells.

Step cline   A sudden change in one or more traits of a species along a geographical gradient.

Steroid   Any of numerous lipids based on a 17-carbon atom ring system.

Sticky ends   On a piece of two-stranded DNA, short, complementary, one-stranded regions produced by the action of a restriction endonuclease. Sticky ends allow the joining of segments of DNA from different sources.

Stigma [L.: mark, brand]   The part of the pistil at the apex of the style, which is receptive to pollen, and on which pollen germinates.

Stimulus   Something causing a response; something in the environment detected by a receptor.

Stolon   A horizontal stem that forms roots at intervals.

Stoma (plural: stomata) [Gr. *stoma*: mouth, opening]   Small opening in the plant epidermis that permits gas exchange; bounded by a pair of guard cells whose osmotic status regulates the size of the opening.

Stop codons   Triplets (UAG, UGA, UAA) in mRNA that act as signals for the end of translation at the ribosome. (See also start codon. There are a few mnior exceptions to these codons.)

Stratosphere   The part of the atmosphere above the troposphere; extends upward to approximately 50 kilometers above the surface of the earth; contains very little water.

Stratum (plural strata)   A layer or sedimentary rock laid down at a particular time in a past.

Striated muscle   Contractile tissue characterized by multinucleated cells containing highly ordered arrangements of actin and myosin microfilaments. Also known as **skeletal muscle**.

Strobilus (strobe' a lus) [Gr. *strobilos*: a cone]   The cone, or characteristic fruit, of the pine and other gymnosperms. Also, a cone-shaped mass of saprophylls found in club mosses.

Stroma   The fluid contents of an organelle, such as a chloroplast.

Stromatolite   A composite, flat-to-domed structure composed of successive mineral layers. Some are known to be produced by the action of bacteria in salt or fresh water, and some ancient ones are considered to be evidence for early life on the earth.

Structural formula   A representation of the positions of atoms and bonds in a molecule.

Structural gene   A gene that encodes the primary structure of a protein.

Style [Gr. *stylos*: pillar or column]   In flowering plants, a column of tissue extending from the tip of the ovary, and bearing the stigma or receptive surface for pollen at its apex.

Sub- [L.: under]   A prefix often used to designate a structure that lies beneath another or is less than another. For example, subcutaneous, subspecies.

Suberin   A waxy material serving as a waterproofing agent in cork and in the Casparian strips of the endodermis in plants.

Submucosa (sub mew koe' sah)   The tissue layer just under the epithelial lining of the lumen of the digestive tract. (Contrast with mucosa.)

Substrate (sub' strayte)   The molecule or molecules on which an enzyme exerts catalytic action.

Substrate level phosphorylation   ATP formation resulting from direct transfer of a phosphate group to ADP from an intermediate in glycolysis. (Contrast with oxidative phosphorylation.)

Succession   In ecology, the gradual, sequential series of changes in species composition of a community following a disturbance.

Sulcus (plural: sulci)   The valleys or creases between the raised portions of the convoluted surface of the brain. (Contrast to gyrus.)

Sulfhydryl group   The —SH group.

Summation   The ability of a neuron to fire action potentials in response to numerous subthreshold postsynaptic potentials arriving simultaneously at differentiated places on the cell, or arriving at the same site in rapid succession.

Suppressor T cells   T cells that inhibit the responses of B cells and other T cells to antigens. (Contrast with cytotoxic T cells, helper T cells.)

Surface-to-volume ratio   For any cell, organism, or geometrical solid, the ratio of surface area to volume; this is an important factor in setting an upper limit on the size a cell or organism can attain.

Surfactant   A substance that decreases the surface tension of a liquid. Lung surfactant, secreted by cells of the alveoli, is mostly phospholipid and decreases the amount of work necessary to inflate the lungs.

Survivorship curve   A plot of the logarithm of the fraction of individuals still alive, as a function of time.

Suspensor   In plants, a cell or group of cells derived from the zygote, but not actually part of the embryo proper, which in some seed plants pushes the young embryo deeper into nutritive gametophyte tissue or endosperm by its growth.

Swim bladder   An internal gas-filled organ that helps fishes maintain their position in the water column; later evolved into an organ for gas exchange in some lineages.

Symbiosis (sim' bee oh' sis) [Gr.: to live together]   The living together of two or more species in a prolonged and intimate ecological relationship. (See parasitism, commensalism, mutualism.)

Symmetry   In biology, the property that two halves of an object are mirror images of each other. (See bilateral symmetry and radial symmetry.)

Sympathetic nervous system   A division of the autonomic (involuntary) nervous system. Its activities include increasing blood pressure and acceleration of the heartbeat. The neurotransmitter at the sympathetic terminals is epinephrine or norepinephrine. (Contrast with parasympathetic nervous system.)

Sympatric (sim pat' rik) [Gr. *syn*: together + *patria*: homeland]   Referring to populations whose geographic regions overlap at least in part.

Sympatric speciation   Formation of new species even though members of the daughter species overlap in their distribution during the speciation process. (Contrast with geographic, parapatric speciation.)

Symplast   The continuous meshwork of the interiors of living cells in the plant body, resulting from the presence of plasmodesmata. (Contrast with apoplast.)

Symport   A membrane transport process that carries two substances in the same direction across the membrane. (Contrast with antiport.)

Synapse (sin' aps) [Gr. *syn*: together + *haptein*: to fasten]   The narrow gap between the terminal bouton of one neutron and the dendrite or cell body of another.

Synapsis (sin ap' sis)   The highly specific parallel alignment (pairing) of homologous chromosomes during the first division of meiosis.

Synaptic vesicle   A membrane-bounded vesicle, containing neurotransmitter, which is produced in and discharged by the presynaptic neuron.

Synergids (sin nur' jids)   Two cells found close to the egg cell in the angiosperm embryo sac; they disappear shortly after fertilization.

Syngamy (sing' guh mee) [Gr. *sun-*: together + *gamos*: marriage]   Union of gametes. Also known as fertilization.

Syrinx (sear' inks) [Gr.: pipe, cavity]   A specialized structure at the junction of the trachea and the primary bronchi leading to the lungs. The vocal organ of birds.

Systematics   The scientific study of the diversity of organisms.

Systemic circulation   The part of the circulatory system serving those parts of the body other than the lungs or gills.

Systole (sis' tuh lee) [Gr.: contraction]   Contraction of a chamber of the heart, driving blood forward in the circulatory system.

T cell   A type of lymphocyte, involved in the cellular immune response. The final stages of its development occur in the thymus gland. (Contrast with B cell; see also cytotoxic T cell, helper T cell, suppressor T cell.)

T cell receptor   A protein on the surface of a T cell that recognizes the antigenic determinant for which the cell is specific.

T tubules   A system of tubules that runs throughout the cytoplasm of muscle fibers, through which action potentials spread.

Target cell   A cell with the appropriate receptors to bind and respond to a particular hormone or other chemical mediator.

Taste bud   A structure in the epithelium of the tongue that includes a cluster of chemosensors innervated by sensory neurons.

TATA box   An eight-base-pair sequence, found about 25 base pairs before the starting point for transcription in many eukaryotic promoters, that binds a transcription factor and thus helps initiate transcription.

Taxis (tak' sis) [Gr. *taxis*: arrange, put in order]   The movement of an organism in a particular direction with reference to a stimulus. A taxis usually involves the employment of one sense and a movement directly toward or away from the stimulus, or else the maintenance of a constant angle to it. Thus a positive phototaxis is movement toward a light source, negative geotaxis is movement upward (away from gravity), and so on.

Taxon   A unit in a taxonomic system.

Taxonomy (taks on' oh me) [Gr. *taxis*: arrange, classify]   The science of classification of organisms.

Telomeres (tee' lo merz) [Gr. *telos*: end]   Repeated DNA sequences at the ends of eukaryotic chromosomes.

Telophase (tee' lo phase) [Gr. *telos*: end]   The final phase of mitosis or meiosis during which chromosomes became diffuse, nuclear envelopes reform, and nucleoli begin to reappear in the daughter nuclei.

Template   In biochemistry, a molecule or surface upon which another molecule is synthesized in complementary fashion, as in the replication of DNA. In the brain, a pattern that responds to a normal input but not to incorrect inputs.

Template strand   In a stretch of double-stranded DNA, the strand that is transcribed.

Temporal summation   In the production or inhibition of action potentials in a postsynaptic neuron, the interaction of depolarizations or hyperpolarizations produced by rapidly repeated stimulation of a single point.

Tendon   A collagen-containing band of tissue that connects a muscle with a bone.

Tepal   In an angiosperm flower, a sterile modified leaf. This term is used to refer to such flower parts when one is unable to distinguish between petals and sepals.

Terminal transferase   An enzyme that adds nucleotides to free ends of DNA, without reference to a template strand.

Terrestrial (ter res' tree al) [L. *terra*: earth]   Pertaining to the land.

Territory   A fixed area from which an animal or group of animals excludes other members of the same species by aggressive behavior or display.

Tertiary structure   In reference to a protein, the relative locations in three-dimensional space of all the atoms in the molecule. The overall shape of a protein. (Contrast with primary, secondary, and quaternary structures.)

Test cross   A cross of a dominant-phenotype individual (which may be either heterozygous or homozygous) with a homozygous-recessive individual.

Testis (tes' tis) (plural: testes) [L.: witness]   The male gonad; that is, the organ that produces the male sex cells.

Testosterone (tes toss' tuhr own)   A male sex steroid hormone.

Tetanus [Gr. *tetanos*: stretched]   (1) In physiology, a state of sustained, maximal muscular contraction caused by rapidly repeated stimulation. (2) In medicine, an often-fatal disease ("lockjaw") caused by the bacterium *Clostridium tetani*.

Thalamus   A region of the vertebrate forebrain; involved in integration of sensory input.

Thallus (thal' us) [Gr.: sprout]   Any algal body which is not differentiated into root, stem, and leaf.

Thermocline   In a body of water, the zone where the temperatures change abruptly to about 4°C.

Thermoneutral zone   The range of temperatures over which an endotherm does not have to expend extra energy to thermoregulate.

Thermosensor   A cell or structure that responds to changes in temperature.

Thoracic cavity   The portion of the mammalian body cavity bounded by the ribs, shoulders, and diaphragm. Contains the heart and the lungs.

Thorax   In an insect, the middle region of the body, between the head and abdomen. In mammals, the part of the body between the neck and the diaphragm.

Thrombin   An enzyme that converts fibrinogen to fibrin, thus triggering the formation of blood clots.

Thrombus (throm' bus) [Gk. *thrombos*: clot]   A blood clot that forms within a blood vessel and remains attached to the wall of the vessel. (Contrast with embolus.)

Thylakoid A flattened sac within a chloroplast. The membranes of the numerous thylakoids contain all of the chlorophyll in a plant, in addition to the electron carriers of photophosphorylation. Thylakoids stack to form grana.

Thymine A nitrogen-containing base found in DNA.

Thymus A ductless, glandular portion of the lymphoid system, involved in development of the immune system of vertebrates.

Thyroid [Gr. *thyreos*: door-shaped] A two-lobed gland in vertebrates. Produces the hormone thyroxin.

Thyrotropic hormone A hormone that is produced in the pituitary gland of amphibia such as frogs and transported in the bloodstream to the thyroid gland, inducing the thyroid gland to produce the thyroid hormone that regulates metamorphosis from tadpole to adult frog.

Tight junction A junction between epithelial cells, in which there is no gap whatever between the adjacent cells. Materials may get through a tight junction only by entering the epithelial cells themselves.

Tissue A group of similar cells organized into a functional unit and usually integrated with other tissues to form part of an organ such as a heart or leaf.

Tonus A low level of muscular tension that is maintained even when the body is at rest.

Tornaria (tor nare' e ah) [L. *tornus*: lathe] The free-swimming ciliated larva of certain echinoderms and hemichordates; its existence indicates the evolutionary relationship of these two groups.

Totipotency In a cell, the condition of possessing all the genetic information and other capacities necessary to form an entire individual.

Toxigenicity The ability of a bacterium to produce chemical substances injurious to the tissues of the host organism.

Trachea (tray' kee ah) [Gr. *trakhoia*: a small tube] A tube that carries air to the bronchi of the lungs of vertebrates, or to the cells of arthropods.

Tracheid (tray' kee id) A distinctive conducting and supporting cell found in the xylem of nearly all vascular plants, characterized by tapering ends and walls that are pitted but not perforated.

Tracheophytes [Gr. *trakhoia*: a small tube + *phyton*: plant] Those plants with xylem and phloem, including psilophytes, club mosses, horsetails, ferns, gymnosperms, and angiosperms. (Contrast with nontracheophytes.)

Trade winds The winds that blow toward the intertropical convergence zone from the northeast and southeast.

Trait One form of a character: Eye color is a character; brown eyes and blue eyes are traits.

Transcription The synthesis of RNA, using one strand of DNA as the template.

Transcription factors Proteins that assemble on a eukaryotic chromosome, allowing RNA polymerase II to perform transcription.

Transdetermination Alteration of the developmental fate of an imaginal disc in *Drosophila*.

Transduction (1) Transfer of genes from one bacterium to another, with a bacterial virus acting as the carrier of the genes. (2) In sensory cells, the transformation of a stimulus (e.g., light energy, sound pressure waves, chemical or electrical stimulants) into action potentials.

Transfection Uptake, incorporation, and expression of recombinant DNA.

Transfer cells A modified parenchyma cell that transports mineral ions from its cytoplasm into its cell wall, thus moving the ions from the symplast into the apoplast.

Transfer RNA (tRNA) A category of relatively small RNA molecules (about 75 nucleotides). Each kind of transfer RNA is able to accept a particular activated amino acid from its specific activating enzyme, after which the amino acid is added to a growing polypeptide chain.

Transformation Mechanism for transfer of genetic information in bacteria in which pure DNA extracted from bacteria of one genotype is taken in through the cell surface of bacteria of a different genotype and incorporated into the chromosome of the recipient cell. By extension, the term has come to be applied to phenomena in other organisms in which specific genetic alterations have been produced by treatment with purified DNA from genetically marked donors.

Transgenic Containing recombinant DNA incorporated into its genetic material.

Translation The synthesis of a protein (polypeptide). This occurs on ribosomes, using the information encoded in messenger RNA.

Translocation (1) In genetics, a rare mutational event that moves a portion of a chromosome to a new location, generally on a nonhomologous chromosome. (2) In vascular plants, movement of solutes in the phloem.

Transpiration [L. *spirare*: to breathe] The evaporation of water from plant leaves and stem, driven by heat from the sun, and providing the motive force to raise water (plus ions) from the roots.

Transposable element A segment of DNA that can move to, or give rise to copies at, another locus on the same or a different chromosome. May be a single insertion sequence or a more complex structure (transposon) consisting of two insertion sequences and one or more intervening genes.

Trichocyst (trick' o sist) [Gr. *trichos*: hair + *kystis*: cell] A threadlike organelle ejected from the surface of ciliates, used both as a weapon and as an anchoring device.

Triglyceride A simple lipid in which three fatty acids are combined with one molecule of glycerol.

Triplet See codon.

Triplet repeat Occurrence of repeated triplet of bases in a gene, often leading to genetic disease, as does excessive repetition of CGG in the gene responsible for fragile-X syndrome.

Triploblastic Having three cell layers. (Contrast with diploblastic.)

Trisomic Containing three, rather than two members of a chromosome pair.

tRNA See transfer RNA.

Trochophore (troke' o fore) [Gr. *trochos*: wheel + *phoreus*: bearer] The free-swimming larva of some annelids and mollusks, distinguished by a wheel-like band of cilia around the middle, and indicating an evolutionary relationship between these two groups.

Trophic level A group of organisms united by obtaining their energy from the same part of the food web of a biological community.

Tropic hormones Hormones of the anterior pituitary that control the secretion of hormones by other endocrine glands.

Tropism [Gr. *tropos*: to turn] In plants, growth toward or away from a stimulus such as light (phototropism) or gravity (gravitropism).

Tropomyosin (troe poe my' oh sin) A protein that, along with actin, constitutes the thin filaments of myofibrils. It controls the interactions of actin and myosin necessary for muscle contraction.

Troposphere The atmospheric zone reaching upward approximately 17 km in the tropics and subtropics but only to about 10 km at higher latitudes. The zone in which virtually all the water vapor in the atmosphere is located.

Trypsin A protein-digesting enzyme. Secreted by the pancreas in its inactive form (trypsinogen), it becomes active in the duodenum of the small intestine.

T-tubules A set of transverse tubes that penetrates skeletal muscle fibers and terminates in the sarcoplasmic reticulum. The T-system transmits impulses to the sacs, which then release Ca^{2+} to initiate muscle contraction.

Tube foot In echinoderms, a part of the water vascular system. It grasps the substratum, prey, or other solid objects.

Tube nucleus In a pollen tube, the haploid nucleus that does not participate in double fertilization. (Contrast with generative nucleus.)

Tuber [L.: swelling] A short, fleshy underground stem, usually much enlarged, and serving a storage function, as in the case of the potato.

Tubulin A protein that polymerizes to form microtubules.

Tumor A disorganized mass of cells, often growing out of control. Malignant tumors spread to other parts of the body.

Tumor suppressor genes Genes which, when homozygous mutant, result in cancer. Such genes code for protein products that inhibit cell proliferation.

Turgor pressure   The actual physical (hydrostatic) pressure within a cell. (Contrast with osmotic potential, water potential.)

Twitch   A single unit of muscle contraction.

Tympanic membrane [Gr. *tympanum*: drum] The eardrum.

Umbilical cord   Tissue made up of embryonic membranes and blood vessels that connects the embryo to the placenta in eutherian mammals.

Uncoupler   See respiratory uncoupler.

Understory   The aggregate of smaller plants growing beneath the canopy of dominant plants in a forest.

Unicellular (yoon' e sell' yer ler) [L. *unus*: one + *cella*: chamber]   Consisting of a single cell; as for example a unicellular organism. (Contrast with multicellular.)

Uniport   A membrane transport process that carries a single substance. (Contrast with antiport, symport.)

Unitary organism   An organism that consists of only one module.

Unsaturated hydrocarbon   A compound containing only carbon and hydrogen atoms. One or more pairs of carbon atoms are connected by double bonds.

Upwelling   The upward movement of nutrient-rich, cooler water from deeper layers of the ocean.

Urea   A compound serving as the main excreted form of nitrogen by many animals, including mammals.

Ureotelic   Describes an organism in which the final product of the breakdown of nitrogen-containing compounds (primarily proteins) is urea. (Contrast with ammonotelic, uricotelic.)

Ureter (your' uh tur) [Gr. *ouron*: urine]   A long duct leading from the vertebrate kidney to the urinary bladder or the cloaca.

Urethra (you ree' thra) [Gr. *ouron*: urine]   In most mammals, the canal through which urine is discharged from the bladder and which serves as the genital duct in males.

Uric acid   A compound that serves as the main excreted form of nitrogen in some animals, particularly those which must conserve water, such as birds, insects, and reptiles.

Uricotelic   Describes an organism in which the final product of the breakdown of nitrogen-containing compounds (primarily proteins) is uric acid. (Contrast with ammonotelic, ureotelic.)

Urinary bladder   A structure structure that receives urine from the kidneys via the ureter, stores it, and expels it periodically through the urethra.

Urine (you' rin) [Gk. *ouron*: urine]   In vertebrates, the fluid waste product containing the toxic nitrogenous by-products of protein and amino acid metabolism.

Uterus (yoo' ter us) [L.: womb]   The uterus or womb is a specialized portion of the female reproductive tract in certain mammals. It receives the fertilized egg and nurtures the embryo in its early development.

Vaccination   Injection of virus or bacteria or their proteins into the body, to induce immunization. The injected material is usually attenuated (weakened) before injection.

Vacuole (vac' yew ole) [Fr.: small vacuum]   A liquid-filled cavity in a cell, enclosed within a single membrane. Vacuoles play a wide variety of roles in cellular metabolism, some being digestive chambers, some storage chambers, some waste bins, and so forth.

Vagina (vuh jine' uh) [L.: sheath]   In female mammals, the passage leading from the external genital orifice to the uterus; receives the copulatory organ of the male in mating.

Van der Waals interaction   A weak attraction between atoms resulting from the interaction of the electrons of one atom with the nucleus of the other atom. This attraction is about one-fourth as strong as a hydrogen bond.

Variable regions   The part of an immunoglobulin molecule or T-cell receptor that includes the antigen-binding site.

Vascular (vas' kew lar)   Pertaining to organs and tissues that conduct fluid, such as blood vessels in animals and phloem and xylem in plants.

Vascular bundle   In vascular plants, a strand of vascular tissue, including conducting cells of xylem and phloem as well as thick-walled fibers.

Vascular ray   In vascular plants, radially oriented sheets of cells produced by the vascular cambium, carrying materials laterally between the wood and the phloem.

Vascular tissue system   The conductive system of the plant, consisting primarily of xylem and phloem. (Contrast with dermal tissue system, ground tissue system.)

Vasopressin   See antidiuretic hormone.

Vector   (1) An agent, such as an insect, that carries a pathogen affecting another species. (2) A plasmid or virus that carries an inserted piece of DNA into a bacterium for cloning purposes in recombinant DNA technology.

Vegetal hemisphere   The lower portion of some animal eggs, zygotes, and embryos, in which the dense nutrient yolk settles. The **vegetal pole** refers to the very bottom of the egg or embryo. (Contrast with animal hemisphere.)

Vegetative   Nonreproductive, or nonflowering, or asexual.

Vein [L. *vena*: channel]   A blood vessel that returns blood to the heart. (Contrast with artery.)

Vena cava [L.: hollow vein]   One of a pair of large veins that carry blood from the systemic circulatory system into the heart.

Ventral [L. *venter*: belly, womb]   Toward or pertaining to the belly or lower side. (Contrast with dorsal.)

Ventricle   A muscular heart chamber that pumps blood through the body.

Vernalization [L. *vernalis*: belonging to spring]   Events occurring during a required chilling period, leading eventually to flowering. Vernalization may require many weeks of below-freezing temperatures.

Vertebral column   The jointed, dorsal column that is the primary support structure of vertebrates.

Vertebrate   An animal whose nerve cord is enclosed in a backbone of bony segments, called vertebrae. The principal groups of vertebrate animals are the fishes, amphibians, reptiles, birds, and mammals.

Vessel [L. *vasculum*: a small vessel]   In botany, a tube-shaped portion of the xylem consisting of hollow cells (vessel elements) placed end to end and connected by perforations. Together with tracheids, vessel elements conduct water and minerals in the plant.

Vestibular apparatus (ves tib' yew lar) [L. *vestibulum*: an enclosed passage]   Structures associated with the vertebrate ear; these structures sense changes in position or momentum of the head, affecing balance and motor skills.

Vestigial (ves tij' ee al) [L. *vestigium*: footprint, track]   The remains of body structures that are no longer of adaptive value to the organism and therefore are not maintained by selection.

Vicariance (vye care' ee unce) [L. *vicus*: change]   The splitting of the range of a taxon by the imposition of some barrier to dispersal of its members. May lead to cladogenesis.

Vicariant distribution   A distribution resulting from the disruption of a formerly continuous range by a vicariant event.

Villus (vil' lus) (plural: villi) [L.: shaggy hair]   A hairlike projection from a membrane; for example, from many gut walls.

Virion (veer' e on)   The virus particle, the minimum unit capable of infecting a cell.

Viroid (vye' roid)   An infectious agent consisting of a single-stranded RNA molecule with no protein coat; produces diseases in plants.

Virus [L.: poison, slimy liquid]   Any of a group of ultramicroscopic infectious particles constructed of nucleic acid and protein (and, sometimes, lipid) that can reproduce only in living cells.

Visceral mass   The major internal organs of a mollusk.

Vitamin [L. *vita*: life]   Any one of several structurally unrelated organic compounds that an organism cannot synthesize itself, but nevertheless requires in small quantity for normal growth and metabolism.

Viviparous (vye vip' uh rus) [L. *vivus*: alive]   Reproduction in which fertilization of the egg and development of the embryo occur inside the mother's body. (Contrast with oviparous.)

Waggle dance   The running movement of a working honey bee on the hive, during which the worker traces out a repeated figure eight. The dance contains elements that transmit to other bees the location of the food.

Water potential   In osmosis, the tendency for a system (a cell or solution) to take up water from pure water, through a differentially permeable membrane. Water flows toward the system with a more negative water potential. (Contrast with osmotic potential, turgor pressure.)

Water vascular system   The array of canals and tubelike appendages that serves as the circulatory system, locomotory system, and food-capturing system of many echinoderms; is in direct connection with the surrounding sea water.

Wavelength   The distance between successive peaks of a wave train, such as electromagnetic radiation.

Wild type   Geneticists' term for standard or reference type. Deviants from this standard, even if the deviants are found in the wild, are said to be mutant.

Xanthophyll (zan' tho fill) [Gr. *xanthos*: yellowish-brown + *phyllon*: leaf]   A yellow or orange pigment commonly found as an accessory pigment in photosynthesis, but found elsewhere as well. An oxygen-containing carotenoid.

X chromosome   See sex chromosome.

X-linked (also called sex-linked)   A character that is coded for by a gene on the X chromosome.

Xerophyte (zee' row fyte) [Gr. *xerox*: dry + *phyton*: plant]   A plant adapted to an environment with a limited water supply.

Xylem (zy' lum) [Gr. *xylon*: wood]   In vascular plants, the woody tissue that conducts water and minerals; xylem consists, in various plants, of tracheids, vessel elements, fibers, and other highly specialized cells.

Y chromosome   See sex chromosome.

Yeast artificial chromosome   A laboratory-made DNA molecule containing sequences of yeast chromosomes (origin or replication, telomeres, centromere, and selectable markers) so that it can be used as a vector in yeast.

Yolk   The stored food material in animal eggs, usually rich in protein and lipid.

Z-DNA   A form of DNA in which the molecule spirals to the left rather than to the right.

Zooplankton (zoe' o plang ton) [Gr. *zoon*: animal + *planktos*: wandering]   The animal portion of the plankton.

Zoospore (zoe' o spore) [Gr. *zoon*: animal + *spora*: seed]   In algae and fungi, any swimming spore. May be diploid or haploid.

Zygospore   A highly resistant type of fungal spore produced by the zygomycetes (conjugating fungi).

Zygote (zye' gote) [Gr. *zygotos*: yoked]   The cell created by the union of two gametes, in which the gamete nuclei are also fused. The earliest stage of the diploid generation.

Zymogen   An inactive precursor of a digestive enzyme secreted into the lumen of the gut, where a protease cleaves it to form the active enzyme.

Answers to Self-Quizzes

Chapter 2

1. b	6. a
2. e	7. c
3. c	8. b
4. c	9. e
5. d	10. d

Chapter 3

1. e	6. a
2. d	7. c
3. c	8. e
4. d	9. a
5. b	10. d

Chapter 4

1. a	6. e
2. e	7. a
3. c	8. d
4. e	9. b
5. c	10. d

Chapter 5

1. e	6. b
2. d	7. c
3. a	8. b
4. d	9. e
5. c	10. c

Chapter 6

1. c	6. a
2. e	7. e
3. b	8. b
4. c	9. d
5. c	10. e

Chapter 7

1. a	6. d
2. d	7. a
3. c	8. e
4. e	9. c
5. c	10. e

Chapter 8

1. c	6. d
2. b	7. c
3. d	8. d
4. b	9. b
5. e	10. b

Chapter 9

1. e	6. a
2. c	7. e
3. b	8. d
4. d	9. b
5. c	10. a

Chapter 10*

1. d	6. d
2. a	7. b
3. e	8. a
4. d	9. b
5. d	10. c

Chapter 11

1. c	6. b
2. a	7. d
3. c	8. d
4. b	9. a
5. e	10. c

Chapter 12

1. c	6. d
2. d	7. b
3. c	8. d
4. d	9. d
5. b	10. a

Chapter 13

1. c	6. d
2. d	7. c
3. b	8. a
4. b	9. b
5. e	10. d

Chapter 14

1. c	6. c
2. d	7. c
3. c	8. b
4. a	9. e
5. c	10. d

Chapter 15

1. c	6. c
2. a	7. d
3. d	8. b
4. a	9. a
5. b	10. b

Chapter 16

1. b	6. b
2. d	7. c
3. a	8. a
4. c	9. c
5. b	10. e

Chapter 17

1. a	6. b
2. c	7. e
3. b	8. d
4. b	9. c
5. e	10. b

Chapter 18

1. d	6. a
2. b	7. d
3. e	8. d
4. e	9. a
5. c	10. d

Chapter 19

1. d	6. b
2. e	7. c
3. a	8. e
4. a	9. e
5. c	10. b

Chapter 20

1. d	6. e
2. c	7. b
3. d	8. e
4. b	9. d
5. d	10. e

Chapter 21

1. c	6. a
2. a	7. b
3. e	8. a
4. d	9. c
5. c	10. a

Chapter 22

1. e	6. d
2. c	7. a
3. a	8. d
4. c	9. b
5. a	10. d

Chapter 23

1. b	6. a
2. e	7. a
3. c	8. c
4. a	9. a
5. a	10. c

Chapter 24

1. e	6. c
2. d	7. a
3. e	8. a
4. c	9. d
5. e	10. b

Chapter 25

1. e	6. b
2. e	7. d
3. b	8. a
4. c	9. c
5. e	10. b

Chapter 26

1. a	6. d
2. e	7. c
3. c	8. b
4. d	9. b
5. a	10. d

Chapter 27

1. d	6. b
2. c	7. b
3. e	8. c
4. b	9. d
5. c	10. c

Chapter 28

1. b	6. a
2. d	7. e
3. e	8. a
4. c	9. c
5. d	10. c

Chapter 29

1. c	6. d
2. d	7. d
3. b	8. e
4. e	9. d
5. b	10. a

Chapter 30

1. b	7. d
2. c	8. e
3. a	9. a
4. d	10. e
5. c	11. c
6. b	12. c

Chapter 31

1. d	6. b
2. b	7. b
3. e	8. c
4. e	9. a
5. a	10. d

*Answers to the "Applying Concepts" questions in Chapter 10 appear at the end of this section.

Chapter 32
1. c	6. d
2. d	7. e
3. b	8. a
4. e	9. d
5. b	10. d

Chapter 33
1. e	6. a
2. b	7. b
3. c	8. c
4. c	9. d
5. d	10. a

Chapter 34
1. d	6. e
2. c	7. a
3. a	8. b
4. e	9. d
5. c	10. e

Chapter 35
1. a	6. c
2. e	7. e
3. c	8. c
4. d	9. a
5. b	10. b

Chapter 36
1. d	6. e
2. b	7. a
3. e	8. b
4. b	9. c
5. d	10. d

Chapter 37
1. c	6. e
2. a	7. a
3. d	8. e
4. b	9. d
5. b	10. c

Chapter 38
1. c	6. b
2. b	7. a
3. d	8. e
4. b	9. c
5. a	10. e

Chapter 39
1. (i) b	4. a
(ii) a	5. d
(iii) c	6. d
(iv) a,b,c	7. d
(v) a,b,c	8. d
2. c, e	9. c
3. e	10. d

Chapter 40
1. a	6. c
2. c	7. b
3. e	8. d
4. c	9. b
5. d	10. c

Chapter 41
1. d	6. e
2. a	7. e
3. d	8. c
4. c	9. d
5. c	10. a

Chapter 42
1. d	6. e
2. d	7. e
3. a	8. c
4. b	9. c
5. e	10. c

Chapter 43
1. c	6. c
2. a	7. a
3. e	8. b
4. d	9. a
5. d	10. b

Chapter 44
1. d	6. d
2. e	7. b
3. a	8. a
4. b	9. a
5. b	10. e

Chapter 45
1. e	6. b
2. d	7. c
3. a	8. c
4. b	9. a
5. c	10. d

Chapter 46
1. d	6. d
2. a	7. b
3. c	8. d
4. d	9. c
5. c	10. e

Chapter 47
1. b	6. d
2. e	7. a
3. c	8. b
4. a	9. d
5. b	10. d

Chapter 48
1. b	6. b
2. a	7. e
3. d	8. a
4. c	9. c
5. d	10. e

Chapter 49
1. a	6. c
2. e	7. d
3. b	8. c
4. b	9. d
5. a	10. a

Chapter 50
1. e	6. e
2. c	7. d
3. d	8. e
4. c	9. a
5. c	10. c

Chapter 51
1. b	6. e
2. e	7. a
3. d	8. c
4. a	9. d
5. d	10. a

Chapter 52
1. b	6. c
2. e	7. e
3. c	8. c
4. b	9. b
5. e	10. e

Chapter 53
1. b	6. a
2. d	7. e
3. d	8. a
4. b	9. d
5. c	10. e

Chapter 54
1. a	6. a
2. d	7. d
3. b	8. e
4. c	9. a
5. d	10. c

Chapter 55
1. b	6. a
2. c	7. d
3. c	8. e
4. e	9. e
5. a	10. a

Answers to "Applying Concepts" for Chapter 10, "Transmission Genetics: Mendel and Beyond"

1. Each of the eight boxes in the Punnet squares should contain the genotype Tt, regardless of which parent was tall and which dwarf.

2. Yellow parent = $s^Y s^b$; offspring 3 yellow (s^Y–): 1 black ($s^b s^b$). Black parent = $s^b s^b$; offspring all black ($s^b s^b$). Orange parent = $s^O s^b$; offspring 3 orange (s^O–): 1 black ($s^b s^b$). Both s^O and s^Y are dominant to s^b.

3. See Figure 10.5.

4. The trait is autosomal. Mother $dp\ dp$, father $Dp\ dp$. If the trait were sex-linked, all daughters would be wild-type and sons would be *dumpy*.

5. All females wild-type; all males spotted.

6. F_1 all wild-type, $PpSwsw$; F_2 9:3:3:1 in phenotypes. See Figure 10.8 for analogous genotypes.

7a. Ratio of phenotypes in F_2 is 3:1 (double dominant to double recessive).

7b. The F_1 are $Pby\ pB^Y$; they produce just two kinds of gametes (Pby and pBy). Combine them carefully and see the 1:2:1 phenotypic ratio fall out in the F_2.

7c. Pink-blistery.

7d. See Figures 9.15 and 9.17. Crossing over took place in the F_1 generation.

8. The genotypes are $PpSwsw$, $Ppswsw$, and $ppswsw$ in a ratio of 1:1:1:1.

9a. 1 black:2 blue:1 splashed white

9b. Always cross black with splashed white.

10a. $w^+ > w^e > w$

10b. Parents $w^e w$ and $w^+ Y$. Progeny $w+w^e$, $w+w$, $w^e Y$, and wY.

11. All will have normal vision because they inherit Dad's wild-type X chromosome, but half of them will be carriers.

12. Agouti parent $AaBb$. Albino offspring $aaBb$ and $aabb$; black offspring $Aabb$; agouti offspring $AaBb$.

13. Because the gene is carried on mitochondrial DNA, it is passed through the mother only. Thus if the woman does not have the disease but her husband does, their child will not be affected. On the other hand, if the woman has the disease but her husband does not, their child *will* have the disease.

son/Visuals Unlimited. 28.9: James W. Richardson/Visuals Unlimited. 28.10: John D. Cunningham/Visuals Unlimited. 28.11a: Jim Solliday/BPS. 28.11b: L. West, National Audubon Society/Photo Researchers, Inc. 28.11c: Ray Coleman/Photo Researchers, Inc. 28.12: Centers for Disease Control, Atlanta. 28.14a: Angelina Lax/Photo Researchers, Inc. 28.14b: Manfred Danegger/Photo Researchers, Inc. 28.14c: Stan Flegler/Visuals Unlimited. 28.15: Biophoto Associates. 28.16: R. L. Peterson/BPS. 28.17: Paul A. Zahl/Photo Researchers, Inc. 28.18a: Gregory G. Dimijian/Photo Researchers, Inc. 28.18b: George Herben Photo/Visuals Unlimited. 28.18c: David Sieren/Visuals Unlimited. 28.19a: J. N. A. Lott/BPS.

Chapter 29 *Opener*: Rod Salm/Planet Earth Pictures. 29.1: Courtesy of R. L. Trelsted. 29.6 *left*: Gillian Lythgoe/Planet Earth Pictures. 29.6 *right*: Christian Petron/Planet Earth Pictures. 29.13a: Georgette Douwma/Planet Earth Pictures. 29.13b: Rod Salm/Planet Earth Pictures. 29.14a: Robert Brons/BPS. 29.14b,c: Chuck Davis Photo. 29.15: David J. Wrobel/BPS. 29.17b, 29.19b: James Solliday/BPS. 29.19c: Jim Solliday/BPS. 29.21a: Fred McConnaughey/Photo Researchers, Inc. 29.21c: Robert Brons/BPS. 29.25c: R. R. Hessler, Scripps Institute of Oceanography. 29.27a: Georgette Douwma/Planet Earth Pictures. 29.27c: Roger K. Burnard/BPS. 29.27d: Roger & Linda Mitchell. 29.29a: Ken Lucas/Planet Earth Pictures. 29.29b: Pete Atkinson/Masterfile. 29.29c: Chuck Davis Photo. 29.29d: Richard Humbert/BPS. 29.29e: Alex Kerstitch/Planet Earth Pictures. 29.29f: David J. Wrobel/BPS. 29.32: J. N. A. Lott/BPS. 29.33a: Joel Simon. 29.33b: Barbara Miller/BPS. 29.34a: Peter J. Bryant/BPS. 29.34b: David Maitland/Masterfile. 29.34c: L. E. Gilbert/BPS. 29.34d: Robert Brons/BPS. 29.36a: Gary Bell/Masterfile. 29.36b: Peter J. Bryant/BPS. 29.36c: David Wrobel/BPS. 29.36d: Geoff du Feu/Planet Earth Pictures. 29.37a: Charles R. Wyttenbach/BPS. 29.37b: Roger K. Burnard/BPS. 29.38a: R. F. Ashley/Visuals Unlimited. 29.38b: From Kristensen and Hallas, 1980. 29.39a: Richard Humbert/BPS. 29.39b,c: Peter J. Bryant/BPS. 29.39d,g: David Maitland/Masterfile. 29.39e: Steve Nicholls/Planet Earth Pictures. 29.39f: Brian Kenney/Planet Earth Pictures. 29.39h: Steve Hopkin/Planet Earth Pictures.

Chapter 30 *Opener*: Tim Davis/Tony Stone Images. 30.3; 30.4: David Wrobel/BPS. 30.5a: Ken Lucas/Planet Earth Pictures. 30.9a: Doug Perrine/DRK PHOTO. 30.9b–e: David Wrobel/BPS. 30.10: C. R. Wyttenbach/BPS. 30.11: Gary Bell/Masterfile. 30.12: M. Laverack/Planet Earth Pictures. 30.15: H. W. Pratt/BPS. 30.17a: David Wrobel/BPS. 30.17b: Marty Snyderman/Masterfile. 30.18a,d: David Wrobel/BPS. 30.18b: Doug Perrine/Masterfile. 30.18c: Ken Lucas/Planet Earth Pictures. 30.19: Peter Scoones/Planet Earth Pictures. 30.21a: Ken Lucas/BPS. 30.21b: Nick Garbutt/Indri Images. 30.21c: Art Wolfe. 30.24a: David J.

Wrobel/BPS. 30.24b: Carl Gans/BPS. 30.24c,d: Brian Kenney/Masterfile. 30.25a: Gavriel Jecan/Art Wolfe Inc. 30.25b: Art Wolfe. 30.26: Courtesy of Carnegie Museum of Natural History, Pittsburgh. 30.28a: Peter Scoones/Masterfile. 30.28b: Larry Tackett/Masterfile. 30.28c,d: Adam Jones. 30.30: Fritz Prenzel/Animals Animals. 30.31a: Art Wolfe. 30.31b: Jany Sauvanet/Photo Researchers, Inc. 30.32a: Staffan Widstrand. 30.32b: Merlin D. Tuttle, Bat Conservation International. 30.32c: Carol Farneti/Masterfile. 30.32d: Theo Allofs. 30.34a,b,c: Art Wolfe. 30.35a: Steve Kaufman/DRK PHOTO. 30.35b: John Bracegirdle/Masterfile. 30.36a: Art Wolfe. 30.36b: Kennan Ward/DRK Photo. 30.36c: K. & K. Ammann/Masterfile. 30.36d: Brian Kenney/Planet Earth Pictures. 30.38a: The American Museum of Natural History. 30.38b,c: Terraphotographics/BPS.

Chapter 31 *Opener*: Rod Planck/Photo Researchers, Inc. 31.1a: Hubertus Kanus/Photo Researchers, Inc. 31.2a: Michael P. Gadomski/Bruce Coleman Inc. 31.3a: D. Waugh/Tony Stone Images. 31.4: Nigel Cattlin/Holt Studios International/Photo Researchers, Inc. 31.7a: John Kaprielian/Photo Researchers, Inc. 31.7b, 31.8: John D. Cunningham/Visuals Unlimited. 31.9a: J. Robert Waaland/BPS. 31.9b: Joyce Photographics/Photo Researchers, Inc. 31.9c: Renee Lynn/Photo Researchers, Inc. 31.13a,d: Phil Gates/BPS. 31.13b: Biophoto Associates/Photo Researchers, Inc. 31.13c: Jack M. Bostrack/Visuals Unlimited. 31.13e: John D. Cunningham/Visuals Unlimited. 31.13f; 31.15b; 31.16 *upper, lower*: J. Robert Waaland/BPS. 31.18a: Jim Solliday/BPS. 31.18b: Microfield Scientific LTD/Photo Researchers, Inc. 31.18c: Alfred Owczarzak/BPS. 31.18d: John D. Cunningham/Visuals Unlimited. 31.20a *upper*: J. Robert Waaland/BPS. 31.20a *lower*: Cabisco/Visuals Unlimited. 31.20b *upper*: J. Robert Waaland/BPS. 31.20b *lower*: Cabisco/Visuals Unlimited. 31.22: J. N. A. Lott/BPS. 31.23: Jim Solliday/BPS. 31.24a,b: Phil Gates/BPS. 31.25b: Jeff Lepore/Photo Researchers, Inc. 31.25c: C. G. Van Dyke/Visuals Unlimited.

Chapter 32 *Opener*: Peter K. Ziminski/Visuals Unlimited. 32.2: Biophoto Associates/Photo Researchers, Inc. 32.7: John D. Cunningham/Visuals Unlimited. 32.10a: David M. Phillips/Visuals Unlimited. 32.12 *right*: Derrick Ditchburn/Visuals Unlimited. 32.13: Thomas Eisner, Cornell Univ.

Chapter 33 *Opener* and 33.2: Nigel Cattlin/Holt Studios International/Photo Researchers, Inc. 33.4: Jess R. Lee/Photo Researchers, Inc. 33.6: Stephen Parker/Photo Researchers, Inc. 33.8: Thomas Eisner, Cornell Univ. 33.9: Jon Mark Stewart/BPS. 33.10: J. N. A. Lott/BPS. 33.11, 33.12: Richard Shiell. 33.13: Janine Pestel/Visuals Unlimited. 33.14: Barbara J. Miller/BPS. 33.15: J. N. A. Lott/BPS. 33.16: Robert & Linda Mitchell. 33.17: Budd Titlow/Visuals Unlimited.

Chapter 34 *Opener*: Cabisco/Visuals Unlimited. 34.1: The Photo Works/Photo Researchers, Inc. 34.4: Kathleen Blanchard/Visuals Unlimited. 34.7: Jeremy Burgess/Photo Researchers, Inc. 34.8: Barbara J. O'Donnell/BPS. 34.10: E. H. Newcomb and S. R. Tandon/BPS. 34.12: Richard C. Johnson/Visuals Unlimited. 34.13: Nuridsany et Pérennou/Photo Researchers, Inc.

Chapter 35 *Opener*: Mary Clay/Masterfile. 35.5: Tom J. Ulrich/Visuals Unlimited. 35.6: Barbara J. Miller/BPS. 35.7: J. N. A. Lott/BPS. 35.11: J. A. D. Zeevaart, Michigan State Univ. 35.19a: Biophoto Associates/Photo Researchers, Inc. 35.21: Cabisco/Visuals Unlimited. 35.22: Adam Jones. 35.23; 35.25: J. N. A. Lott/BPS. 35.29: R. Last, Cornell Univ. Courtesy of the Society for Plant Physiology.

Chapter 36 *Opener*: David Woodfall/Woodfall Wild Images. 36:2: Stan Flegler/Visuals Unlimited. 36.3: Nigel Cattlin, Holt Studios International/Photo Researchers, Inc. 36.4: Bowman, J. (ed.). 1994. Arabidopsis: *An Atlas of Morphology and Development.* Springer-Verlag, New York. Photos by S. Craig & A. Chaudhury, Plate 6.2. 36.8a: John Kaprielian/Photo Researchers, Inc. 36.8b: Renee Lynn/Photo Researchers, Inc. 36.17a: Nigel Cattlin, Holt Studios International/Photo Researchers, Inc. 36.17b: James W. Richardson/Visuals Unlimited.

Chapter 37 *Opener*: Daniel J. Cox/Natural Exposures. 37.15a: B. & C. Alexander/Photo Researchers, Inc. 37.15b: Timothy Ransom/BPS. 37.20 *left, center*: G. W. Willis/BPS. 37.20 *right*: Fran Thomas, Stanford Univ. 37.21a: Barbara Gerlach/Visuals Unlimited. 37.21b: Art Wolfe.

Chapter 38 *Opener*: R. D. Fernald, Stanford Univ. 38.6a: Associated Press Photo. 38.6b: The Bettmann Archive, Inc.

Chapter 39 *Opener*: Ron Austing/Photo Researchers, Inc. 39.1a: Biophoto Associates/Photo Researchers, Inc. 39.1b: Andrew J. Martinez/Photo Researchers, Inc. 39.1c: Peter J. Bryant/BPS. 39.2: David M. Phillips/Photo Researchers, Inc. 39.4: Paul W. Johnson/BPS. 39.5: Fletcher & Baylis/Photo Researchers, Inc. 39.6: James Beveridge/Visuals Unlimited. 39.7a: Jim Merli/Visuals Unlimited. 39.7b: Robert W. Hernandez. 39.8: Renee Lynn/Photo Researchers, Inc. 39.13 *inset*: P. Bagavandoss/Photo Researchers, Inc. 39.17: Lara Hartley/BPS. 39.19: CC Studio/Photo Researchers, Inc. 39.21: C. Eldeman/Photo Researchers, Inc. 39.22: Nestle/Photo Researchers, Inc. 39.24: S. I. U. School of Med./Photo Researchers, Inc.

Chapter 40 *Opener*: Courtesy of John Morrill, New College. 40.5 *inset*: Courtesy of Richard Elinson, Univ. of Toronto.

Chapter 41 *Opener*: Stephen Krasemann, Nature Conservancy/Photo Researchers, Inc. 41.4: C. Raines/Visuals Unlimited.

Index

INDEX

About the Book

Editor: Andrew D. Sinauer
Project Editor: Carol J. Wigg
Developmental Editor: James Funston
Copy Editor: Stephanie Hiebert
Production Manager: Christopher Small
Book Layout and Production: Janice Holabird and Jefferson Johnson
Art Editing and Illustration Program: J/B Woolsey Associates
Design: Jefferson Johnson and Christopher Small
Book Cover Design: MBDesign
Photo Research: Jane Potter
Color Separations: Vision Graphics, Inc.
Cover Manufacture: Henry N. Sawyer Company, Inc.
Book Manufacture: R. R. Donnelley & Sons Company